Inference and Learning from Data

Volume I

This extraordinary three-volume work, written in an engaging and rigorous style by a world authority in the field, provides an accessible, comprehensive introduction to the full spectrum of mathematical and statistical techniques underpinning contemporary methods in data-driven learning and inference.

This first volume, *Foundations*, introduces core topics in inference and learning, such as matrix theory, linear algebra, random variables, convex optimization, stochastic optimization, and decentralized methods, and prepares students for studying their practical application in later volumes.

A consistent structure and pedagogy is employed throughout this volume to reinforce student understanding, with over 600 end-of-chapter problems (including solutions for instructors), 180 solved examples, 100 figures, datasets, and downloadable Matlab code. Supported by sister volumes *Inference* and *Learning*, and unique in its scale and depth, this textbook sequence is ideal for early-career researchers and graduate students across many courses in signal processing, machine learning, statistical analysis, data science, and inference.

Ali H. Sayed is Professor and Dean of Engineering at École Polytechnique Fédérale de Lausanne (EPFL), Switzerland. He has also served as Distinguished Professor and Chairman of Electrical Engineering at the University of California, Los Angeles (UCLA), USA, and as President of the IEEE Signal Processing Society. He is a member of the US National Academy of Engineering (NAE) and The World Academy of Sciences (TWAS), and a recipient of several awards, including the 2022 IEEE Fourier Award and the 2020 IEEE Norbert Wiener Society Award. He is a Fellow of the IEEE, EURASIP, and AAAS.

Inference and Learning from Data

Volume I: Foundations

ALI H. SAYED
École Polytechnique Fédérale de Lausanne
University of California at Los Angeles

CAMBRIDGE
UNIVERSITY PRESS

CAMBRIDGE
UNIVERSITY PRESS

Shaftesbury Road, Cambridge CB2 8EA, United Kingdom

One Liberty Plaza, 20th Floor, New York, NY 10006, USA

477 Williamstown Road, Port Melbourne, VIC 3207, Australia

314–321, 3rd Floor, Plot 3, Splendor Forum, Jasola District Centre, New Delhi – 110025, India

103 Penang Road, #05–06/07, Visioncrest Commercial, Singapore 238467

Cambridge University Press is part of Cambridge University Press & Assessment, a department of the University of Cambridge.

We share the University's mission to contribute to society through the pursuit of education, learning and research at the highest international levels of excellence.

www.cambridge.org
Information on this title: www.cambridge.org/highereducation/isbn/9781009218122
DOI: 10.1017/9781009218146

First published 2023

Printed in the United Kingdom by Bell and Bain Ltd

A catalogue record for this publication is available from the British Library.

ISBN - 3 Volume Set 978-1-009-21810-8 Hardback
ISBN - Volume I 978-1-009-21812-2 Hardback
ISBN - Volume II 978-1-009-21826-9 Hardback
ISBN - Volume III 978-1-009-21828-3 Hardback

Additional resources for this publication at www.cambridge.org/sayed-vol1

In loving memory of my parents

Contents

Preface

Learning directly from data is critical to a host of disciplines in engineering and the physical, social, and life sciences. Modern society is literally driven by an interconnected web of data exchanges at rates unseen before, and it relies heavily on decisions inferred from patterns in data. There is nothing fundamentally wrong with this approach, except that the inference and learning methodologies need to be anchored on solid foundations, be fair and reliable in their conclusions, and be robust to unwarranted imperfections and malicious interference.

P.1 EMPHASIS ON FOUNDATIONS

Given the explosive interest in data-driven learning methods, it is not uncommon to encounter claims of superior designs in the literature that are substantiated mainly by sporadic simulations and the potential for "life-changing" applications rather than by an approach that is founded on the well-tested scientific principle to inquiry. For this reason, one of the main objectives of this text is to highlight, in a unified and formal manner, the firm mathematical and statistical pillars that underlie many popular data-driven learning and inference methods. This is a nontrivial task given the wide scope of techniques that exist, and which have often been motivated independently of each other. It is nevertheless important for practitioners and researchers alike to remain cognizant of the common foundational threads that run across these methods. It is also imperative that progress in the domain remains grounded on firm theory. As the aphorism often attributed to Lewin (1945) states, "*there is nothing more practical than a good theory.*" According to Bedeian (2016), this saying has an even older history.

Rigorous data analysis, and conclusions derived from experimentation and theory, have been driving science since time immemorial. As reported by Heath (1912), the Greek scientist Archimedes of Syracuse devised the now famous Archimedes' Principle about the volume displaced by an immersed object from observing how the level of water in a tub rose when he sat in it. In the account by Hall (1970), Gauss' formulation of the least-squares problem was driven by his desire to predict the future location of the planetoid Ceres from observations of its location over 41 prior days. There are numerous similar examples by notable scientists where experimentation led to hypotheses and from there to substantiated theories and well-founded design methodologies. Science is also full of progress in the reverse direction, where theories have been developed first to be validated only decades later through experimentation and data analysis. Einstein (1916) postulated the existence of gravitational waves over 100 years ago. It took until 2016 to detect them! Regardless of which direction one follows, experimentation to theory or the reverse, the match between solid theory and rigorous data analysis has enabled science and humanity to march confidently toward the immense progress that permeates our modern world today.

For similar reasons, data-driven learning and inference should be developed with strong theoretical guarantees. Otherwise, the confidence in their reliability can be shaken if there is over-reliance on "proof by simulation or experience." Whenever possible, we explain the underlying models and statistical theories for a large number of methods covered in this text. A good grasp of these theories will enable practitioners and researchers to devise variations with greater mastery. We weave through the foundations in a coherent and cohesive manner, and show how the various methods blend together techniques that may appear decoupled but are actually facets of the same common methodology. In this process, we discover that a good number of techniques are well-grounded and meet proven performance guarantees, while other methods are driven by ingenious insights but lack solid justifications and cannot be guaranteed to be "fail-proof."

Researchers on learning and inference methods are of course aware of the limitations of some of their approaches, so much so that we encounter today many studies, for example, on the topic of "explainable machine learning." The objective here is to understand why learning algorithms produce certain recommendations. While this is an important area of inquiry, it nevertheless highlights one interesting shift in paradigm. In the past, the emphasis would have been on designing inference methods that respond to the input data in certain desirable and controllable ways. Today, in many instances, the emphasis is to stick to the available algorithms (often, out of convenience) and try to understand or explain why they are responding in certain ways to the input!

Writing this text has been a rewarding journey that took me from the early days of statistical mathematical theory to the modern state of affairs in learning theory. One can only stand in awe at the wondrous ideas that have been introduced by notable researchers along this trajectory. At the same time, one observes with some concern an emerging trend in recent years where solid foundations receive less attention in lieu of "speed publishing" and over-reliance on "illustration by simulation." This is of course not the norm and most researchers in the field stay honest to the scientific approach to inquiry and design. After concluding this comprehensive text, I stand humbled at the realization of "*how little we know*!" There are countless questions that remain open, and even for many of the questions that have been answered, their answers rely on assumptions or (over)simplifications. It is understandable that the complexity of the problems we face today has increased manifold, and ingenious approximations become necessary to enable tractable solutions.

P.2 GLIMPSE OF HISTORY

Reading through the text, the alert reader will quickly realize that the core foundations of modern-day machine learning, data analytics, and inference methods date back for at least two centuries, with contributions arising from a range of fields including mathematics, statistics, optimization theory, information theory, signal processing, communications, control, and computer science. For the benefit of the reader, I reproduce here with permission from IEEE some historical remarks from the editorial I published in Sayed (2018). I explained there that these disciplines have generated a string of "big ideas" that are driving today multi-faceted efforts in the age of "big data" and machine learning. Generations of students in the statistical sciences and engineering have been trained in the art of modeling, problem solving, and optimization. Their algorithms power everything from cell phones, to spacecraft, robotic explorers, imaging devices, automated systems, computing machines, and also recommender systems. These students mastered the foundations of their fields and have been well prepared to contribute to the explosive growth of data analysis and machine learning solutions.

As the list below shows, many well-known engineering and statistical methods have actually been motivated by data-driven inquiries, even from times remote. The list is a tour of some older historical contributions, which is of course biased by my personal preferences and is not intended to be exhaustive. It is only meant to illustrate how concepts from statistics and the information sciences have always been at the center of promoting big ideas for data and machine learning. Readers will encounter these concepts in various chapters in the text. Readers will also encounter additional historical accounts in the concluding remarks of each chapter, and in particular comments on newer contributions and contributors.

Let me start with Gauss himself, who in 1795 at the young age of 18, was fitting lines and hyperplanes to astronomical data and invented the least-squares criterion for regression analysis – see the collection of his works in Gauss (1903). He even devised the recursive least-squares solution to address what was a "big" data problem for him at the time: He had to avoid tedious repeated calculations by hand as more observational data became available. What a wonderful big idea for a data-driven problem! Of course, Gauss had many other big ideas.

de Moivre (1730), Laplace (1812), and Lyapunov (1901) worked on the central limit theorem. The theorem deals with the limiting distribution of averages of "large" amounts of data. The result is also related to the law of "large" numbers, which even has the qualification "large" in its name. Again, big ideas motivated by "large" data problems.

Bayes (ca mid-1750s) and Laplace (1774) appear to have independently discovered the Bayes rule, which updates probabilities conditioned on observations – see the article by Bayes and Price (1763). The rule forms the backbone of much of statistical signal analysis, Bayes classifiers, Naïve classifiers, and Bayesian networks. Again, a big idea for data-driven inference.

Fourier (1822), whose tools are at the core of disciplines in the information sciences, developed the phenomenal Fourier representation for signals. It is meant to transform data from one domain to another to facilitate the extraction and visualization of information. A big transformative idea for data.

Forward to modern times. The fast Fourier transform (FFT) is another example of an algorithm driven by challenges posed by data size. Its modern version is due to Cooley and Tukey (1965). Their algorithm revolutionized the field of discrete-time signal processing, and FFT processors have become common components in many modern electronic devices. Even Gauss had a role to play here, having proposed an early version of the algorithm some 160 years before, again motivated by a data-driven problem while trying to fit astronomical data onto trigonometric polynomials. A big idea for a data-driven problem.

Closer to the core of statistical mathematical theory, both Kolmogorov (1939) and Wiener (1949) laid out the foundations of modern statistical signal analysis and optimal prediction methods. Their theories taught us how to extract information optimally from data, leading to further refinements by Wiener's student Levinson (1947) and more dramatically by Kalman (1960). The innovations approach by Kailath (1968) exploited to great effect the concept of orthogonalization of the data and recursive constructions. The Kalman filter is applied across many domains today, including in financial analysis from market data. Kalman's work was an outgrowth of the model-based approach to system theory advanced by Zadeh (1954). The concept of a recursive solution from streaming data was a novelty in Kalman's filter; the same concept is commonplace today in most online learning techniques. Again, big ideas for recursive inference from data.

Cauchy (1847) early on, and Robbins and Monro (1951) a century later, developed the powerful gradient-descent method for root finding, which is also recursive in nature. Their techniques have grown to motivate huge advances in stochastic approximation theory. Notable contributions that followed include the work by Rosenblatt (1957) on the perceptron algorithm for single-layer networks, and the impactful delta rule by Widrow and Hoff (1960), widely known as the LMS algorithm in the signal processing literature. Subsequent work on multilayer neural networks grew out of the desire to increase the approximation power of single-layer networks, culminating with the backpropagation method of Werbos (1974). Many of these techniques form the backbone of modern learning algorithms. Again, big ideas for recursive online learning.

Shannon (1948a, b) contributed fundamental insights to data representation, sampling, coding, and communications. His concepts of entropy and information measure helped quantify the amount of uncertainty in data and are used, among other areas, in the design of decision trees for classification purposes and in driving learning algorithms for neural networks. Nyquist (1928) contributed to the understanding of data representations as well. Big ideas for data sampling and data manipulation.

Bellman (1957a, b), a towering system-theorist, introduced dynamic programming and the notion of the curse of dimensionality, both of which are core underpinnings of many results in learning theory, reinforcement learning, and the theory of Markov decision processes. Viterbi's algorithm (1967) is one notable example of a dynamic programming solution, which has revolutionized communications and has also found applications in hidden Markov models widely used in speech recognition nowadays. Big ideas for conquering complex data problems by dividing them into simpler problems.

Kernel methods, building on foundational results by Mercer (1909) and Aronszajn (1950), have found widespread applications in learning theory since the mid-1960s with the introduction of the kernel perceptron algorithm. They have also been widely used in estimation theory by Parzen (1962), Kailath (1971), and others. Again, a big idea for learning from data.

Pearson and Fisher launched the modern field of mathematical statistical signal analysis with the introduction of methods such as principal component analysis (PCA) by Pearson (1901) and maximum likelihood and linear discriminant analysis by Fisher (1912, 1922, 1925). These methods are at the core of statistical signal processing. Pearson (1894, 1896) also had one of the earliest studies of fitting a mixture of Gaussian models to biological data. Mixture models have now become an important tool in modern learning algorithms. Big ideas for data-driven inference.

Markov (1913) introduced the formalism of Markov chains, which is widely used today as a powerful modeling tool in a variety of fields including word and speech recognition, handwriting recognition, natural language processing, spam filtering, gene analysis, and web search. Markov chains are also used in Google's PageRank algorithm. Markov's motivation was to study letter patterns in texts. He laboriously went through the first 20,000 letters of a classical Russian novel and counted pairs of vowels, consonants, vowels followed by a consonant, and consonants followed by a vowel. A "big" data problem for his time. Great ideas (and great patience) for data-driven inquiries.

And the list goes on, with many modern-day and ongoing contributions by statisticians, engineers, and computer scientists to network science, distributed processing, compressed sensing, randomized algorithms, optimization, multi-agent systems, intelligent systems, computational imaging, speech processing, forensics, computer visions, privacy and security, and so forth. We provide additional historical accounts about these contributions and contributors at the end of the chapters.

P.3 ORGANIZATION OF THE TEXT

The text is organized into three volumes, with a sizable number of problems and solved examples. The table of contents provides details on what is covered in each volume. Here we provide a condensed summary listing the three main themes:

1. (**Volume I: Foundations**). The first volume covers the *foundations* needed for a solid grasp of inference and learning methods. Many important topics are covered in this part, in a manner that prepares readers for the study of inference and learning methods in the second and third volumes. Topics include: matrix theory, linear algebra, random variables, Gaussian and exponential distributions, entropy and divergence, Lipschitz conditions, convexity, convex optimization, proximal operators, gradient-descent, mirror-descent, conjugate-gradient, subgradient methods, stochastic optimization, adaptive gradient methods, variance-reduced methods, distributed optimization, and nonconvex optimization. Interestingly enough, the following concepts occur time and again in all three volumes and the reader is well-advised to develop familiarity with them: convexity, sample mean and law of large numbers, Gaussianity, Bayes rule, entropy, Kullback–Leibler divergence, gradient-descent, least squares, regularization, and maximum-likelihood. The last three concepts are discussed in the initial chapters of the second volume.

2. (**Volume II: Inference**). The second volume covers inference methods. By "inference" we mean techniques that infer some unknown variable or quantity from observations. The difference we make between "inference" and "learning" in our treatment is that inference methods will target situations where some prior information is known about the underlying signal models or signal distributions (such as their joint probability density functions or generative models). The performance by many of these inference methods will be the ultimate goal that learning algorithms, studied in the third volume, will attempt to emulate. Topics covered here include: mean-square-error inference, Bayesian inference, maximum-likelihood estimation, expectation maximization, expectation propagation, Kalman filters, particle filters, posterior modeling and prediction, Markov chain Monte Carlo methods, sampling methods, variational inference, latent Dirichlet allocation, hidden Markov models, independent component analysis, Bayesian networks, inference over directed and undirected graphs, Markov decision processes, dynamic programming, and reinforcement learning.

3. (**Volume III: Learning**). The third volume covers learning methods. Here, again, we are interested in inferring some unknown variable or quantity from observations. The difference, however, is that the inference will now be solely data-driven, i.e., based on available data and not on any assumed knowledge about signal distributions or models. The designer is only given a collection of observations that arise from the underlying (unknown) distribution. New phenomena arise related to generalization power, overfitting, and underfitting depending on how representative the data is and how complex or simple the approximate models are. The target is to use the data to learn about the quantity of interest (its value or evolution). Topics covered here include: least-squares methods, regularization, nearest-neighbor rule, self-organizing maps, decision trees, naïve Bayes classifier, linear discrimi-

nant analysis, principal component analysis, dictionary learning, perceptron, support vector machines, bagging and boosting, kernel methods, Gaussian processes, generalization theory, feedforward neural networks, deep belief networks, convolutional networks, generative networks, recurrent networks, explainable learning, adversarial attacks, and meta learning.

Figure P.1 shows how various topics are grouped together in the text; the numbers in the boxes indicate the chapters where these subjects are covered. The figure can be read as follows. For example, instructors wishing to cover:

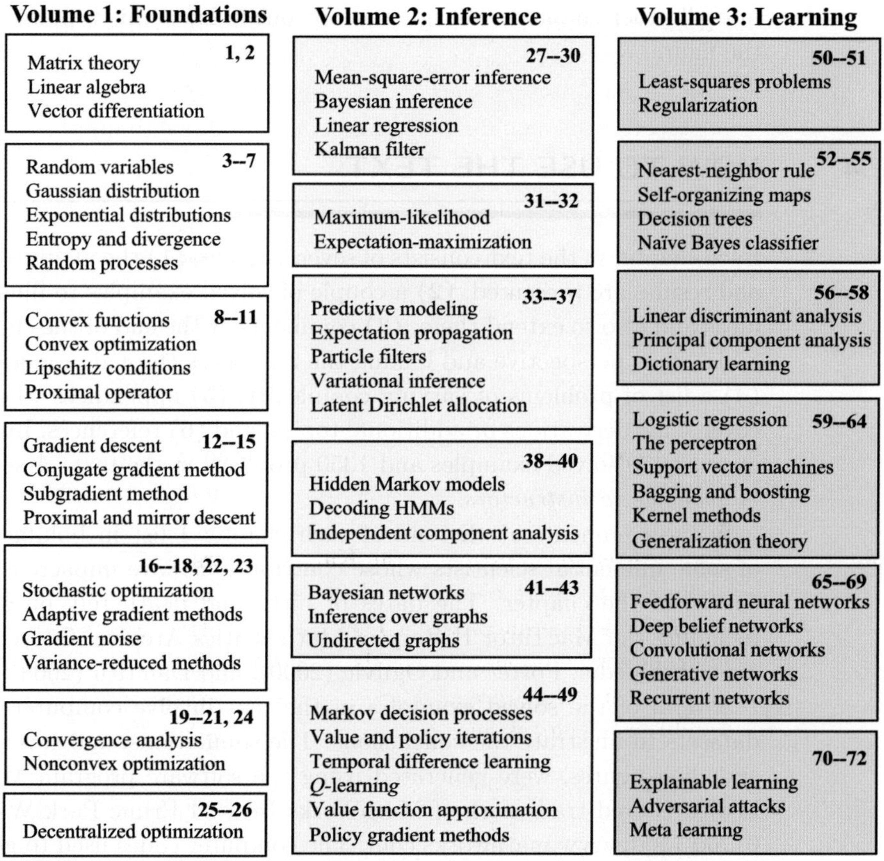

Figure P.1 Organization of the text.

(a) Background material on linear algebra and matrix theory: they can use Chapters 1 and 2.

(b) Background material on random variables and probability theory: they can select from Chapters 3 through 7.

(c) Background material on convex functions and convex optimization: they can use Chapters 8 through 11.

The three groupings **(a)–(c)** contain introductory core concepts that are needed for subsequent chapters. For instance, instructors wishing to cover gradient descent and iterative optimization techniques would then proceed to Chapters 12 through 15, while instructors wishing to cover stochastic optimization methods would use Chapters 16–24 and so forth. Figure P.2 provides a representation of the estimated dependencies among the chapters in the text. The chapters are color-coded depending on the volume they appear in. An arrow from Chapter a toward Chapter b implies that the material in the latter chapter benefits from the material in the earlier chapter. In principle, we should have added arrows from Chapter 1, which covers background material on matrix and linear algebra, into all other chapters. We ignored obvious links of this type to avoid crowding the figure.

P.4 HOW TO USE THE TEXT

Each chapter in the text consists of several blocks: **(1)** the main text where theory and results are presented, **(2)** a couple of solved examples to illustrate the main ideas and also to extend them, **(3)** comments at the end of the chapter providing a historical perspective and linking the references through a motivated timeline, **(4)** a list of problems of varying complexity, **(5)** appendices when necessary to cover some derivations or additional topics, and **(6)** references. In total, there are close to 470 solved examples and 1350 problems in the text. *A solutions manual is available to instructors.*

In the comments at the end of each chapter I list in boldface the life span of some influential scientists whose contributions have impacted the results discussed in the chapter. The dates of birth and death rely on several sources, including the MacTutor History of Mathematics Archive, Encyclopedia Britannica, Wikipedia, Porter and Ogilvie (2000), and Daintith (2008).

Several of the solved examples in the text involve computer simulations on datasets to illustrate the conclusions. The simulations, and several of the corresponding figures, were generated using the software program Matlab©, which is a registered trademark of MathWorks Inc., 24 Prime Park Way, Natick, MA 01760-1500, www.mathworks.com. The computer codes used to generate the figures are provided "as is" and without any guarantees. While these codes are useful for the instructional purposes of the book, they are not intended to be examples of full-blown or optimized designs; practitioners should use them at their own risk. We have made no attempts to optimize the codes, perfect them, or even check them for absolute accuracy. On the contrary, in order to keep the codes at a level that is easy to follow by students, we have often decided to sacrifice performance or even programming elegance in lieu of simplicity. Students can use the computer codes to run variations of the examples shown in the text.

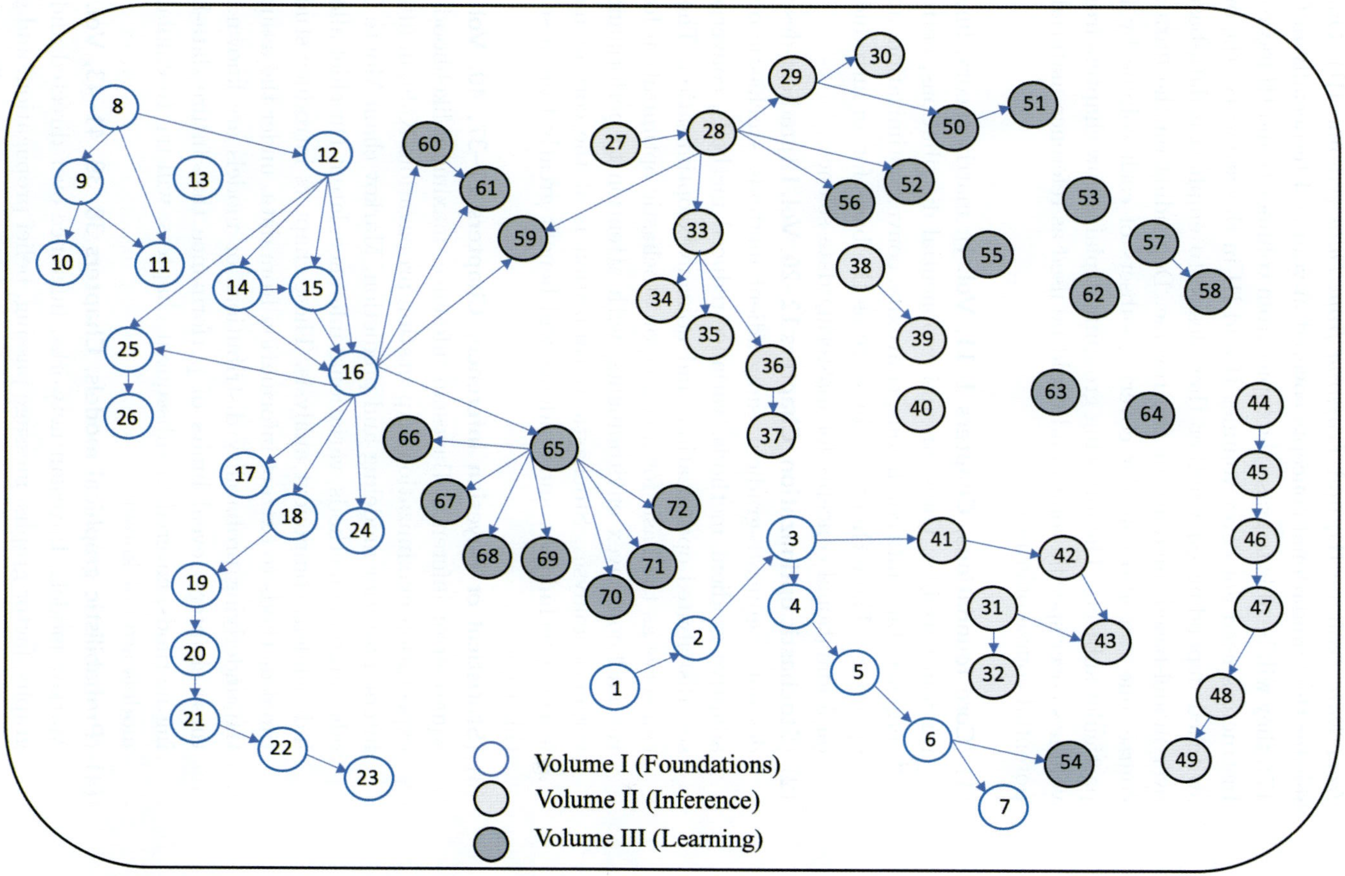

Figure P.2 A diagram showing the approximate dependencies among the chapters in the text. The color scheme identifies chapters from the same volume, with the numbers inside the circles referring to the chapter numbers.

In principle, each volume could serve as the basis for a master-level graduate course, such as courses on *Foundations of Data Science* (volume I), *Inference from Data* (volume II), and *Learning from Data* (volume III). Once students master the foundational concepts covered in volume I (especially in Chapters 1–17), they will be able to grasp the topics from volumes II and III more confidently. Instructors need not cover volumes II and III in this sequence; the order can be switched depending on whether they desire to emphasize data-based learning over model-based inference or the reverse. Depending on the duration of each course, one can also consider covering subsets of each volume by focusing on particular subjects. The following grouping explains how chapters from the three volumes cover specific topics and could be used as reference material for several potential course offerings:

(1) (**Core foundations, Chapters 1–11, Vol. I**): matrix theory, linear algebra, random variables, Gaussian and exponential distributions, entropy and divergence, Lipschitz conditions, convexity, convex optimization, and proximal operators. These chapters can serve as the basis for an introductory course on foundational concepts for mastering data science.

(2) (**Stochastic optimization, Chapters 12–26, Vol. I**): gradient-descent, mirror-descent, conjugate-gradient, subgradient methods, stochastic optimization, adaptive gradient methods, variance-reduced methods, convergence analyses, distributed optimization, and nonconvex optimization. These chapters can serve as the basis for a course on stochastic optimization for both convex and nonconvex environments, with attention to performance and convergence analyses. Stochastic optimization is at the core of most modern learning techniques, and students will benefit greatly from a solid grasp of this topic.

(3) (**Statistical or Bayesian inference, Chapters 27–37, 40, Vol. II**): mean-square-error inference, Bayesian inference, maximum-likelihood estimation, expectation maximization, expectation propagation, Kalman filters, particle filters, posterior modeling and prediction, Markov chain Monte Carlo methods, sampling methods, variational inference, latent Dirichlet allocation, and independent component analysis. These chapters introduce students to optimal methods to extract information from data, under the assumption that the underlying probability distributions or models are known. In a sense, these chapters reveal limits of performance that future data-based learning methods, covered in subsequent chapters, will try to emulate when the models are not known.

(4) (**Probabilistic graphical models, Chapters 38, 39, 41–43, Vol. II**): hidden Markov models, Bayesian networks, inference over directed and undirected graphs, factor graphs, message passing, belief propagation, and graph learning. These chapters can serve as the basis for a course on Bayesian inference over graphs. Several methods and techniques are discussed along with supporting examples and algorithms.

(5) (**Reinforcement learning, Chapters 44–49, Vol. II**): Markov decision processes, dynamic programming, value and policy iterations, temporal difference learning, Q-learning, value function approximation, and policy gradient methods. These chapters can serve as the basis for a course on reinforcement learning. They cover many relevant techniques, illustrated by means of examples, and include performance and convergence analyses.

(6) (**Data-driven and online learning, Chapters 50–64, Vol. III**): least-squares methods, regularization, nearest-neighbor rule, self-organizing maps, decision trees, naïve Bayes classifier, linear discriminant analysis, principal component analysis, dictionary learning, perceptron, support vector machines, bagging and boosting, kernel methods, Gaussian processes, and generalization theory. These chapters cover a variety of methods for learning directly from data, including various methods for online learning from sequential data. The chapters also cover performance guarantees from statistical learning theory.

(7) (**Neural networks, Chapters 65–72, Vol. III**): feedforward neural networks, deep belief networks, convolutional networks, generative networks, recurrent networks, explainable learning, adversarial attacks, and meta learning. These chapters cover various architectures for neural networks and the respective algorithms for training them. The chapters also cover topics related to explainability and adversarial behavior over these networks.

The above groupings assume that students have been introduced to background material on matrix theory, random variables, entropy, convexity, and gradient-descent methods. One can, however, rearrange the groupings by designing stand-alone courses where the background material is included along with the other relevant chapters. By doing so, it is possible to devise various course offerings, covering themes such as stochastic optimization, online or sequential learning, probabilistic graphical models, reinforcement learning, neural networks, Bayesian machine learning, kernel methods, decentralized optimization, and so forth. Figure P.3 shows several suggested selections of topics from across the text, and the respective chapters, which can be used to construct courses with particular emphasis. Other selections are of course possible, depending on individual preferences and on the intended breadth and depth for the courses.

P.5 SIMULATION DATASETS

In several examples in this work we run simulations that rely on publicly available datasets. The sources for these datasets are acknowledged in the appropriate locations in the text. Here we provide an aggregate summary for ease of reference:

(1) **Iris dataset**. This classical dataset contains information about the sepal length and width for three types of iris flowers: virginica, setosa, and

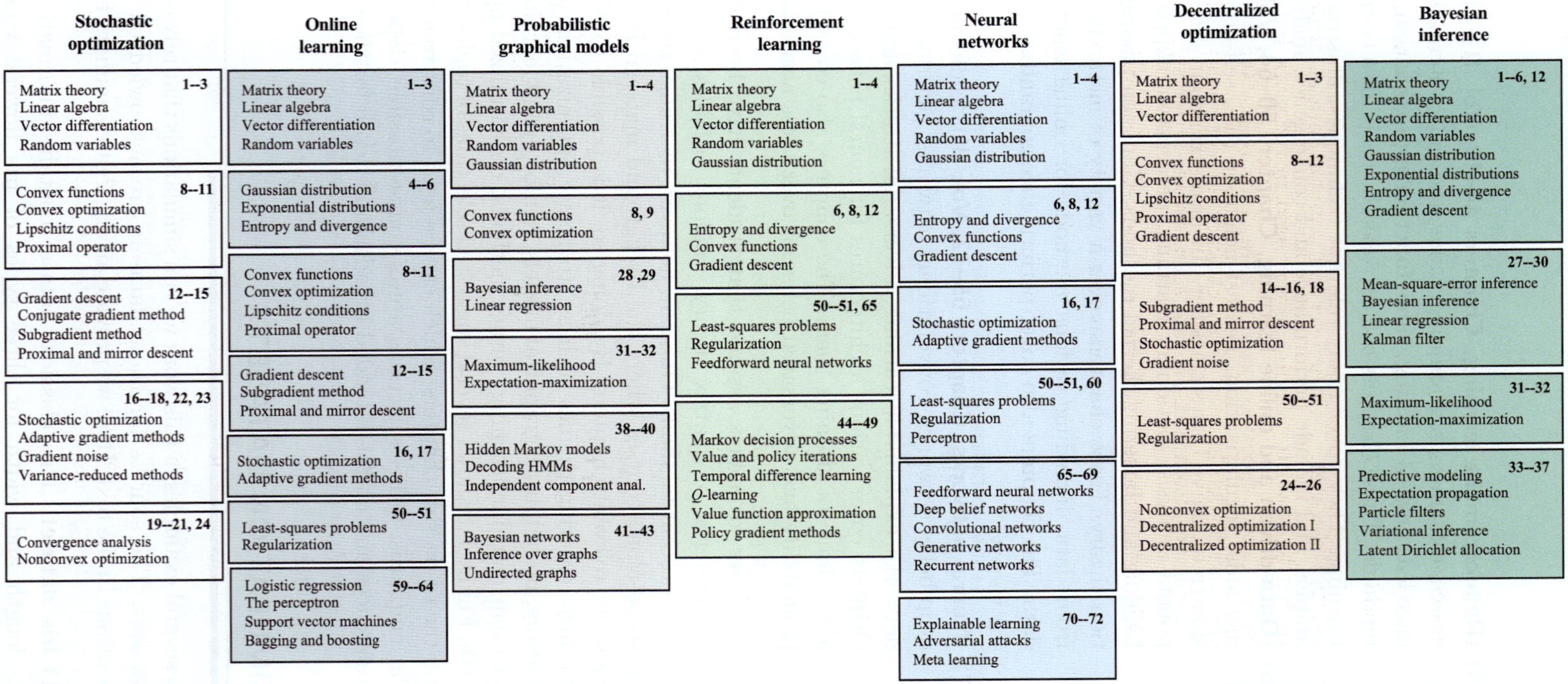

Figure P.3 Suggested selections of topics from across the text, which can be used to construct stand-alone courses with particular emphases. Other options are possible based on individual preferences.

versicolor. It was originally used by Fisher (1936) and is available at the UCI Machine Learning Repository at https://archive.ics.uci.edu/ml/datasets/iris. Actually, several of the datasets in our list are downloaded from this useful repository – see Dua and Graff (2019).

(2) **MNIST dataset**. This is a second popular dataset, which is useful for classifying handwritten digits. It was used in the work by LeCun *et al.* (1998) on document recognition. The data contains 60,000 labeled training examples and 10,000 labeled test examples for the digits 0 through 9. It can be downloaded from http://yann.lecun.com/exdb/mnist/.

(3) **CIFAR-10 dataset**. This dataset consists of color images that can belong to one of 10 classes: airplanes, automobiles, birds, cats, deer, dogs, frogs, horses, ships, and trucks. It is described by Krizhevsky (2009) and can be downloaded from www.cs.toronto.edu/∼kriz/cifar.html.

(4) **FBI crime dataset**. This dataset contains statistics showing the burglary rates per 100,000 inhabitants for the period 1997–2016. The source of the data is the US Criminal Justice Information Services Division at the link https://ucr.fbi.gov/crime-in-the-u.s/2016/crime-in-the-u.s.-2016/tables/table-1.

(5) **Sea level and global temperature changes dataset**. The sea level dataset measures the change in sea level relative to the start of year 1993. There are 952 data points consisting of fractional year values. The source of the data is the NASA Goddard Space Flight Center at https://climate.nasa. gov/vital-signs/sea-level/. For information on how the data was generated, the reader may consult Beckley *et al.* (2017) and the report GSFC (2017). The temperature dataset measures changes in the global surface temperature relative to the average over the period 1951–1980. There are 139 measurements between the years 1880 and 2018. The source of the data is the NASA Goddard Institute for Space Studies (GISS) at https://climate.nasa.gov/vital-signs/global-temperature/.

(6) **Breast cancer Wisconsin dataset**. This dataset consists of 569 samples, with each sample corresponding to a benign or malignant cancer classification. It can be downloaded from the UCI Machine Learning Repository at https://archive.ics.uci.edu/ml/datasets/Breast+Cancer+Wisconsin+(Diagnostic). For information on how the data was generated, the reader may consult Mangasarian, Street, and Wolberg (1995).

(7) **Heart-disease Cleveland dataset**. This dataset consists of 297 samples that belong to patients with and without heart disease. It is available on the UCI Machine Learning Repository and can be downloaded from https://archive.ics .uci.edu/ml/datasets/heart+Disease. The investigators responsible for the collection of the data are the four leading co-authors of the article by Detrano *et al.* (1989).

P.6 ACKNOWLEDGMENTS

A project of this magnitude is not possible without the support of a web of colleagues and students. I am indebted to all of them for their input at various stages of this project, either through feedback on earlier drafts or through conversations that deepened my understanding of the topics. I am grateful to several of my former and current Ph.D. students and post-doctoral associates in no specific order: Stefan Vlaski, Kun Yuan, Bicheng Ying, Zaid Towfic, Jianshu Chen, Xiaochuan Zhao, Sulaiman Alghunaim, Qiyue Zou, Zhi Quan, Federico Cattivelli, Lucas Cassano, Roula Nassif, Virginia Bordignon, Elsa Risk, Mert Kayaalp, Hawraa Salami, Mirette Sadek, Sylvia Dominguez, Sheng-Yuan Tu, Waleed Younis, Shang-Kee Tee, Chung-Kai Tu, Alireza Tarighat, Nima Khajehnouri, Vitor Nascimento, Ricardo Merched, Cassio Lopes, Nabil Yousef, Ananth Subramanian, Augusto Santos, and Mansour Aldajani. I am also indebted to former internship and visiting undergraduate and MS students Mateja Ilic, Chao Yutong, Yigit Efe Erginbas, Zhuoyoue Wang, and Edward Nguyen for their help with some of the simulations.

I also wish to acknowledge several colleagues with whom I have had fruitful interactions over the years on topics of relevance to this text, including coauthoring joint publications, and who have contributed directly or indirectly to my work: Thomas Kailath (Stanford University, USA), Vince Poor (Princeton University, USA), José Moura (Carnegie Mellon University, USA), Mos Kaveh (University of Minnesota, USA), Bernard Widrow (Stanford University, USA), Simon Haykin (McMaster University, Canada), Thomas Cover (Stanford University, USA, *in memoriam*), Gene Golub (Stanford University, USA, *in memoriam*), Sergios Theodoridis (University of Athens, Greece), Vincenzo Matta (University of Salerno, Italy), Abdelhak Zoubir (Technical University Darmstadt, Germany), Cedric Richard (Universite Côte d'Azur, France), John Treichler (Raytheon, USA), Tiberiu Constantinescu (University of Texas Dallas, USA, *in memoriam*), Shiv Chandrasekaran (University of California, Santa Barbara, USA), Ming Gu (University of California, Berkeley, USA), Babak Hassibi (Caltech, USA), Jeff Shamma (University of Illinois, Urbana Champaign, USA), P. P. Vaidyanathan (Caltech, USA), Hanoch Lev-Ari (Northeastern University, USA), Markus Rupp (Tech. Universität Wien, Austria), Alan Laub, Wotao Yin, Lieven Vandenberghe, Mihaela van der Schaar, and Vwani Roychowdhury (University of California, Los Angeles), Vitor Nascimento (University of São Paulo, Brazil), Jeronimo Arena Garcia (Universidad Carlos III, Spain), Tareq Al-Naffouri (King Abdullah University of Science and Technology, Saudi Arabia), Jie Chen (Northwestern Polytechnical University, China), Sergio Barbarossa (Universita di Roma, Italy), Paolo Di Lorenzo (Universita di Roma, Italy), Alle-Jan van der Veen (Delft University, the Netherlands), Paulo Diniz (Federal University of Rio de Janeiro, Brazil), Sulyman Kozat (Bilkent University, Turkey), Mohammed Dahleh (University of California, Santa Barbara,

USA, *in memoriam*), Alexandre Bertrand (Katholieke Universiteit Leuven, Belgium), Marc Moonen (Katholieke Universiteit Leuven, Belgium), Phillip Regalia (National Science Foundation, USA), Martin Vetterli, Michael Unser, Pascal Frossard, Pierre Vandergheynst, Rudiger Urbanke, Emre Telatar, and Volkan Cevher (EPFL, Switzerland), Helmut Bölcskei (ETHZ, Switzerland), Visa Koivunen (Aalto University, Finland), Isao Yamada (Tokyo Institute of Technology, Japan), Zhi-Quan Luo and Shuguang Cui (Chinese University of Hong Kong, Shenzhen, China), Soummya Kar (Carnegie Mellon University, USA), Waheed Bajwa (Rutgers University, USA), Usman Khan (Tufts University, USA), Michael Rabbat (Facebook, Canada), Petar Djuric (Stony Brook University, USA), Lina Karam (Lebanese American University, Lebanon), Danilo Mandic (Imperial College, United Kingdom), Jonathon Chambers (University of Leicester, United Kingdom), Rabab Ward (University of British Columbia, Canada), and Nikos Sidiropoulos (University of Virginia, USA).

I would like to acknowledge the support of my publisher, Elizabeth Horne, at Cambridge University Press during the production phase of this project. I would also like to express my gratitude to the publishers IEEE, Pearson Education, NOW, and Wiley for allowing me to adapt some excerpts and problems from my earlier works, namely, Sayed (*Fundamentals of Adaptive Filtering*, © 2003 Wiley), Sayed (*Adaptive Filters*, © 2008 Wiley), Sayed (*Adaptation, Learning, and Optimization over Networks*, © 2014 A. H. Sayed by NOW Publishers), Sayed ("Big ideas or big data," © 2018 IEEE), and Kailath, Sayed, and Hassibi (*Linear Estimation*, © 2000 Prentice Hall).

I initiated my work on this project in Westwood, Los Angeles, while working as a faculty member at the University of California, Los Angeles (UCLA), and concluded it in Lausanne, Switzerland, while working at the École Polytechnique Fédérale de Lausanne (EPFL). I am grateful to both institutions for their wonderful and supportive environments.

My wife Laila, and daughters Faten and Samiya, have always provided me with their utmost support and encouragement without which I would not have been able to devote my early mornings and good portions of my weekend days to the completion of this text. My beloved parents, now deceased, were overwhelming in their support of my education. For all the sacrifices they have endured during their lifetime, I dedicate this text to their loving memory, knowing very well that this tiny gift will never match their gift.

Ali H. Sayed
Lausanne, Switzerland
March 2021

References

Aronszajn, N. (1950), "Theory of reproducing kernels," *Trans. Amer. Math. Soc.*, vol. 68, no. 3, pp. 337–404.

Bayes, T. and R. Price (1763), "An essay towards solving a problem in the doctrine of chances," Bayes's article communicated by R. Price and published posthumously in *Philos. Trans. Roy. Soc. Lond.*, vol. 53, pp. 370–418.

Beckley, B. D., P. S. Callahan, D. W. Hancock, G. T. Mitchum, and R. D. Ray (2017), "On the cal-mode correction to TOPEX satellite altimetry and its effect on the global mean sea level time series," *J. Geophy. Res. Oceans*, vol. 122, no. 11, pp. 8371–8384.

Bedeian, A. G. (2016), "A note on the aphorism 'there is nothing as practical as a good theory'," *J. Manag. Hist.*, vol. 22, no. 2, pp. 236–242.

Bellman, R. E. (1957a), *Dynamic Programming*, Princeton University Press. Also published in 2003 by Dover Publications.

Bellman, R. E. (1957b), "A Markovian decision process," *Indiana Univ. Math. J.*, vol. 6, no. 4, pp. 679–684.

Cauchy, A.-L. (1847), "Methode générale pour la résolution des systems déquations simultanes," *Comptes Rendus Hebd. Séances Acad. Sci.*, vol. 25, pp. 536–538.

Cooley, J. W. and J. W. Tukey (1965), "An algorithm for the machine calculation of complex Fourier series" *Math. Comput.*, vol. 19, no. 90, pp. 297–301.

Daintith, J. (2008), editor, *Biographical Encyclopedia of Scientists*, 3rd ed., CRC Press.

de Moivre, A. (1730), *Miscellanea Analytica de Seriebus et Quadraturis*, J. Tonson and J. Watts, London.

Detrano, R., A. Janosi, W. Steinbrunn, M. Pfisterer, J. Schmid, S. Sandhu, K. Guppy, S. Lee, and V. Froelicher (1989), "International application of a new probability algorithm for the diagnosis of coronary artery disease," *Am. J. Cardiol.*, vol. 64, pp. 304–310.

Dua, D. and C. Graff (2019), *UCI Machine Learning Repository*, available at http://archive.ics.uci.edu/ml, School of Information and Computer Science, University of California, Irvine.

Einstein, A. (1916), "Näherungsweise Integration der Feldgleichungen der Gravitation," *Sitzungsberichte der Königlich Preussischen Akademie der Wissenschaften Berlin*, part 1, pp. 688–696.

Fisher, R. A. (1912), "On an absolute criterion for fitting frequency curves," *Messeg. Math.*, vol. 41, pp. 155–160.

Fisher, R. A. (1922), "On the mathematical foundations of theoretical statistics," *Philos. Trans. Roy. Soc. Lond. Ser. A.*, vol. 222, pp. 309–368.

Fisher, R. A. (1925), "Theory of statistical estimation," *Proc. Cambridge Philos. Soc.*, vol. 22, pp. 700–725.

Fisher, R. A. (1936), "The use of multiple measurements in taxonomic problems," *Ann. Eugenics*, vol. 7, no. 2, pp. 179–188.

Fourier, J. (1822), *Théorie Analytique de la Chaleur*, Firmin Didot Père et Fils. English translation by A. Freeman in 1878 reissued as *The Analytic Theory of Heat*, Dover Publications.

Gauss, C. F. (1903), *Carl Friedrich Gauss Werke,* Akademie der Wissenschaften.

GSFC (2017), "Global mean sea level trend from integrated multi-mission ocean altimeters TOPEX/Poseidon, Jason-1, OSTM/Jason-2," ver. 4.2 PO.DAAC, CA, USA. Dataset accessed 2019-03-18 at http://dx.doi.org/10.5067/GMSLM-TJ42.

Hall, T. (1970), *Carl Friedrich Gauss: A Biography*, MIT Press.

Heath, J. L. (1912), *The Works of Archimedes*, Dover Publications.

Kailath, T. (1968), "An innovations approach to least-squares estimation, part I: Linear filtering in additive white noise," *IEEE Trans. Aut. Control*, vol. 13, pp. 646–655.

Kailath, T. (1971), "RKHS approach to detection and estimation problems I: Deterministic signals in Gaussian noise," *IEEE Trans. Inf. Theory*, vol. 17, no. 5, pp. 530–549.

Kailath, T., A. H. Sayed, and B. Hassibi (2000), *Linear Estimation*, Prentice Hall.

Kalman, R. E. (1960), "A new approach to linear filtering and prediction problems," *Trans. ASME J. Basic Eng.*, vol. 82, pp. 34–45.
Kolmogorov, A. N. (1939), "Sur l'interpolation et extrapolation des suites stationnaires," *C. R. Acad. Sci.*, vol. 208, p. 2043.
Krizhevsky, A. (2009), *Learning Multiple Layers of Features from Tiny Images*, MS dissertation, Computer Science Department, University of Toronto, Canada.
Laplace, P. S. (1774), "Mémoire sur la probabilité des causes par les événements," *Mém. Acad. R. Sci. de MI (Savants étrangers)*, vol. 4, pp. 621–656. See also *Oeuvres Complètes de Laplace*, vol. 8, pp. 27–65 published by the L'Académie des Sciences, Paris, during the period 1878–1912. Translated by S. M. Sitgler, *Statistical Science*, vol. 1, no. 3, pp. 366–367.
Laplace, P. S. (1812), *Théorie Analytique des Probabilités*, Paris.
LeCun, Y., L. Bottou, Y. Bengio, and P. Haffner (1998), "Gradient-based learning applied to document recognition," *Proc. IEEE*, vol. 86, no. 11, pp. 2278–2324.
Levinson, N. (1947), "The Wiener RMS error criterion in filter design and prediction," *J. Math. Phys.*, vol. 25, pp. 261–278.
Lewin, K. (1945), "The research center for group dynamics at MIT," *Sociometry*, vol. 8, pp. 126–135. See also page 169 of Lewin, K. (1952), *Field Theory in Social Science: Selected Theoretical Papers by Kurt Lewin*, Tavistock.
Lyapunov, A. M. (1901), "Nouvelle forme du théoreme sur la limite de probabilité," *Mémoires de l'Académie de Saint-Petersbourg*, vol. 12, no. 8, pp. 1–24.
Mangasarian, O. L., W. N. Street, and W. H. Wolberg (1995), "Breast cancer diagnosis and prognosis via linear programming," *Op. Res.*, vol. 43, no. 4, pp. 570–577.
Markov, A. A. (1913), "An example of statistical investigation in the text of *Eugene Onyegin* illustrating coupling of texts in chains," *Proc. Acad. Sci. St. Petersburg*, vol. 7, no. 3, p. 153–162. English translation in *Science in Context*, vol. 19, no. 4, pp. 591–600, 2006.
Mercer, J. (1909), "Functions of positive and negative type and their connection with the theory of integral equations," *Philos. Trans. Roy. Soc. Lond. Ser. A*, vol. 209, pp. 415–446.
Nyquist, H. (1928), "Certain topics in telegraph transmission theory," *Trans. AIEE*, vol. 47, pp. 617–644. Reprinted as classic paper in *Proc. IEEE*, vol. 90, no. 2, pp. 280–305, 2002.
Parzen, E. (1962), "Extraction and detection problems and reproducing kernel Hilbert spaces," *J. Soc. Indus. Appl. Math. Ser. A: Control*, vol. 1, no. 1, pp. 35–62.
Pearson, K. (1894), "Contributions to the mathematical theory of evolution," *Philos. Trans. Roy. Soc. Lond.*, vol. 185, pp. 71–110.
Pearson, K. (1896), "Mathematical contributions to the theory of evolution. III. Regression, heredity and panmixia," *Philos. Trans. Roy. Soc. Lond.*, vol. 187, pp. 253–318.
Pearson, K. (1901), "On lines and planes of closest fit to systems of points in space," *Philos. Mag.*, vol. 2, no. 11, pp. 559–572.
Porter, R. and M. Ogilvie (2000), editors, *The Biographical Dictionary of Scientists*, 3rd ed., Oxford University Press.
Robbins, H. and S. Monro (1951), "A stochastic approximation method," *Ann. Math. Stat.*, vol. 22, pp. 400–407.
Rosenblatt, F. (1957), *The Perceptron: A Perceiving and Recognizing Automaton*, Technical Report 85-460-1, Project PARA, Cornell Aeronautical Lab.
Sayed, A. H. (2003), *Fundamentals of Adaptive Filtering*, Wiley.
Sayed, A. H. (2008), *Adaptive Filters*, Wiley.
Sayed, A. H. (2014a), *Adaptation, Learning, and Optimization over Networks,* Foundations and Trends in Machine Learning, NOW Publishers, vol. 7, no. 4–5, pp. 311–801.
Sayed, A. H. (2018), "Big ideas or big data?" *IEEE Sign. Process. Mag.*, vol. 35, no. 2, pp. 5–6.
Shannon, C. E. (1948a), "A mathematical theory of communication," *Bell Syst. Tech. J.*, vol. 27, pp. 379–423.
Shannon, C. E. (1948b), "A mathematical theory of communication," *Bell Syst. Tech. J.*, vol. 27, pp. 623–656.

Viterbi, A. J. (1967), "Error bounds for convolutional codes and an asymptotically optimal decoding algorithm," *IEEE Trans. Inf. Theory*, vol. 13, pp. 260–269.

Werbos, P. J. (1974), *Beyond Regression: New Tools for Prediction and Analysis in the Behavioral Sciences*, Ph.D. dissertation, Harvard University.

Widrow, B. and M. E. Hoff (1960), "Adaptive switching circuits," *IRE WESCON Conv. Rec.*, Institute of Radio Engineers, pt. 4, pp. 96–104.

Wiener, N. (1949), *Extrapolation, Interpolation and Smoothing of Stationary Time Series,* Technology Press and Wiley. Originally published in 1942 as a classified Nat. Defense Res. Council Report. Also published under the title *Time Series Analysis* by MIT Press.

Zadeh, L. A. (1954), "System theory," *Columbia Engr. Quart.*, vol. 8, pp. 16–19.

Notation

The following is a list of notational conventions used in the text:

(a) We use **boldface letters** to denote random variables and normal font letters to refer to their realizations (i.e., deterministic values), like $\boldsymbol{x}$ and x, respectively. In other words, we reserve the boldface notation for random quantities.

(b) We use **CAPITAL LETTERS** for matrices and small letters for *both* vectors and scalars, for example, X and x, respectively. In view of the first convention, $\boldsymbol{X}$ would denote a matrix with random entries, while X would denote a matrix realization (i.e., a matrix with deterministic entries). Likewise, $\boldsymbol{x}$ would denote a vector with random entries, while x would denote a vector realization (or a vector with deterministic entries). One notable *exception* to the capital letter notation is the use of such letters to refer to matrix dimensions or the number of data points. For example, we usually write M to denote the size of a feature vector and N to denote the number of data samples. These exceptions will be obvious from the context.

(c) Small **Greek letters** generally refer to scalar quantities, such as α and β, while **CAPITAL** Greek letters generally refer to matrices such as Σ and Γ.

(d) All vectors in our presentation are *column vectors* unless mentioned otherwise. Thus, if $h \in \mathbb{R}^M$ refers to a feature vector and $w \in \mathbb{R}^M$ refers to a classifier, then their inner product is $h^\mathsf{T} w$ where $\cdot^\mathsf{T}$ denotes the transposition symbol.

(e) If $P(w) : \mathbb{R}^M \rightarrow \mathbb{R}$ is some objective function, then its gradient relative to w^T is denoted by either $\nabla_{w^\mathsf{T}} P(w)$ or $\partial P(w)/\partial w$ and this notation refers to the *column* vector consisting of the partial derivatives of $P(w)$ relative to the individual entries of w:

$$\partial P(w)/\partial w \;=\; \nabla_{w^\mathsf{T}} P(w) = \begin{bmatrix} \partial P(w)/\partial w_1 \\ \partial P(w)/\partial w_2 \\ \vdots \\ \partial P(w)/\partial w_M \end{bmatrix} \qquad (M \times 1)$$

Symbols

We collect here, for ease of reference, a list of the main symbols used throughout the text.

$\mathbb{R}$	set of real numbers
$\mathbb{C}$	set of complex numbers
$\mathbb{Z}$	set of integer numbers
$\mathbb{R}^M$	set of $M \times 1$ real vectors
$.^\mathsf{T}$	matrix transposition
$.^*$	complex conjugation (transposition) for scalars (matrices)
$\boldsymbol{x}$	boldface letter denotes a random scalar or vector
$\boldsymbol{X}$	boldface capital letter denotes a random matrix
x	letter in normal font denotes a scalar or vector
X	capital letter in normal font denotes a matrix
$\mathbb{E}\,\boldsymbol{x}$	expected value of the random variable $\boldsymbol{x}$
$\mathbb{E}_{\boldsymbol{x}}\, g(\boldsymbol{x})$	expected value of $g(\boldsymbol{x})$ relative to pdf of $\boldsymbol{x}$
$\boldsymbol{x} \perp \boldsymbol{y}$	orthogonal random variables $\boldsymbol{x}$ and $\boldsymbol{y}$, i.e., $\mathbb{E}\,\boldsymbol{x}\boldsymbol{y}^\mathsf{T} = 0$
$x \perp y$	orthogonal vectors x and y, i.e., $x^\mathsf{T} y = 0$
$\boldsymbol{x} \perp\!\!\!\perp \boldsymbol{y} \mid \boldsymbol{z}$	$\boldsymbol{x}$ and $\boldsymbol{y}$ are conditionally independent given $\boldsymbol{z}$
$\|x\|^2$	$x^\mathsf{T} x$ for a real x; squared Euclidean norm of x
$\|x\|^2_W$	$x^\mathsf{T} W x$ for a real x and positive-definite matrix W
$\|x\|$ or $\|x\|_2$	$\sqrt{x^\mathsf{T} x}$ for a real column vector x; Euclidean norm of x
$\|x\|_1$	ℓ_1-norm of vector x; sum of its absolute entries
$\|x\|_\infty$	ℓ_∞-norm of vector x; maximum absolute entry
$\|x\|_\star$	dual norm of vector x
$\|A\|$ or $\|A\|_2$	maximum singular value of A; also the spectral norm of A
$\|A\|_\mathrm{F}$	Frobenius norm of A
$\|A\|_1$	ℓ_1-norm of matrix A or maximum absolute column sum
$\|A\|_\infty$	ℓ_∞-norm of matrix A or maximum absolute row sum
$\|A\|_\star$	dual norm of matrix A
$\mathrm{col}\{a, b\}$	column vector with a and b stacked on top of each other
$\mathrm{diag}\{a, b\}$	diagonal matrix with diagonal entries a and b
$\mathrm{diag}\{A\}$	column vector formed from the diagonal entries of A
$\mathrm{diag}\{a\}$	diagonal matrix with entries read from column a
$a \oplus b$	same as $\mathrm{diag}\{a, b\}$
$a = \mathrm{vec}\{A\}$	column vector a formed by stacking the columns of A
$A = \mathrm{vec}^{-1}\{a\}$	square matrix A recovered by unstacking its columns from a
$\mathrm{blkcol}\{a, b\}$	columns a and b stacked on top of each other
$\mathrm{blkdiag}\{A, B\}$	block diagonal matrix with blocks A and B on diagonal
$a \odot b$	Hadamard elementwise product of vectors a and b
$a \oslash b$	Hadamard elementwise division of vectors a and b
$A \otimes B$	Kronecker product of matrices A and B
$\mathcal{A} \otimes_b \mathcal{B}$	block Kronecker product of block matrices $\mathcal{A}$ and $\mathcal{B}$

$A^\dagger$	pseudo-inverse of A
$a \stackrel{\Delta}{=} b$	quantity a is defined as b
0	zero scalar, vector, or matrix
I_M	identity matrix of size $M \times M$
$P > 0$	positive-definite matrix P
$P \geq 0$	positive-semidefinite matrix P
$P^{1/2}$	square-root factor of $P \geq 0$, usually lower triangular
$A > B$	means that $A - B$ is positive-definite
$A \geq B$	means that $A - B$ is positive-semidefinite
$\det A$	determinant of matrix A
$\text{Tr}(A)$	trace of matrix A
$A = QR$	QR factorization of matrix A
$A = U\Sigma V^\mathsf{T}$	SVD factorization of matrix A
$\rho(A)$	spectral radius of matrix A
$\lambda(A)$	refers to a generic eigenvalue of A
$\sigma(A)$	refers to a generic singular value of A
$\mathcal{N}(A)$	nullspace of matrix A
$\mathcal{R}(A)$	range space or column span of matrix A
$\text{rank}(A)$	rank of matrix A
$\text{In}(A)$	inertia of matrix A
$\widehat{\boldsymbol{x}}$	estimator for $\boldsymbol{x}$
$\widetilde{\boldsymbol{x}}$	error in estimating $\boldsymbol{x}$
$\bar{x}$ or $\mathbb{E}\,\boldsymbol{x}$	mean of random variable $\boldsymbol{x}$
σ_x^2	variance of a scalar random variable $\boldsymbol{x}$
R_x	covariance matrix of a vector random variable $\boldsymbol{x}$
$\mathbb{P}(A)$	probability of event A
$\mathbb{P}(A\|B)$	probability of event A conditioned on knowledge of B
$f_{\boldsymbol{x}}(x)$	pdf of random variable $\boldsymbol{x}$
$f_{\boldsymbol{x}}(x;\theta)$	pdf of $\boldsymbol{x}$ parameterized by θ
$f_{\boldsymbol{x}\|\boldsymbol{y}}(x\|y)$	conditional pdf of random variable $\boldsymbol{x}$ given $\boldsymbol{y}$
$\mathcal{S}(\theta)$	score function, equal to $\nabla_{\theta^\mathsf{T}} f_{\boldsymbol{x}}(\boldsymbol{x};\theta)$
$F(\theta)$	Fisher information matrix: covariance matrix of $\mathcal{S}(\theta)$
$\mathcal{N}_{\boldsymbol{x}}(\bar{x}, R_x)$	Gaussian distribution over $\boldsymbol{x}$ with mean $\bar{x}$, covariance R_x
$\boldsymbol{x} \sim f_{\boldsymbol{x}}(x)$	random variable $\boldsymbol{x}$ distributed according to pdf $f_{\boldsymbol{x}}(x)$
$\mathcal{G}_{\boldsymbol{g}}(m(x), K(x,x'))$	Gaussian process $\boldsymbol{g}$; mean $m(x)$ and covariance $K(x,x')$
$K(x,x')$	kernel function with arguments x, x'
$H(\boldsymbol{x})$	entropy of random variable $\boldsymbol{x}$
$H(\boldsymbol{x}\|\boldsymbol{y})$	conditional entropy of random variable $\boldsymbol{x}$ given $\boldsymbol{y}$
$I(\boldsymbol{x};\boldsymbol{y})$	mutual information of random variables $\boldsymbol{x}$ and $\boldsymbol{y}$
$D_{\text{KL}}(p\\|q)$	KL divergence of pdf distributions $p_{\boldsymbol{x}}(x)$ and $q_{\boldsymbol{x}}(x)$
$D_\phi(a,b)$	Bregman divergence of points a, b relative to mirror $\phi(w)$

$S_{\boldsymbol{x}}(z)$	z-spectrum of stationary random process, $\boldsymbol{x}(n)$
$S_{\boldsymbol{x}}(e^{j\omega})$	power spectral density function
$\widehat{\boldsymbol{x}}_{\text{MSE}}$	mean-square-error estimator of $\boldsymbol{x}$
$\widehat{\boldsymbol{x}}_{\text{MAP}}$	maximum a-posteriori estimator of $\boldsymbol{x}$
$\widehat{\boldsymbol{x}}_{\text{MVUE}}$	minimum-variance unbiased estimator of $\boldsymbol{x}$
$\widehat{x}_{\text{ML}}$	maximum-likelihood estimate of x
$\ell(\theta)$	log-likelihood function; parameterized by θ
$\mathcal{P}_C(z)$	projection of point z onto convex set $\mathcal{C}$
$\text{dist}(x, \mathcal{C})$	distance from point x to convex set $\mathcal{C}$
$\text{prox}_{\mu h}(z)$	proximal projection of z relative to $h(w)$
$\mathcal{M}_{\mu h}(z)$	Moreau envelope of $\text{prox}_{\mu h}(z)$
$h^\star(x)$	conjugate function of $h(w)$
$\mathbb{T}_\alpha(x)$	soft-thresholding applied to x with threshold α
$\{x\}_+$	$\max\{0, x\}$
$P(w)$ or $\mathcal{P}(W)$	risk functions
$Q(w, \cdot)$ or $\mathcal{Q}(W, \cdot)$	loss functions
$\nabla_w P(w)$	row gradient vector of $P(w)$ relative to w
$\nabla_{w^\mathsf{T}} P(w)$	column gradient vector of $P(w)$ relative to w^T
$\partial P(w)/\partial w$	same as the column gradient $\nabla_{w^\mathsf{T}} P(w)$
$\nabla_w^2 P(w)$	Hessian matrix of $P(w)$ relative to w
$\partial_w P(w)$	subdifferential set of $P(\cdot)$ at location w
h_n	nth feature vector
$\gamma(n)$	nth target or label signal, when scalar
γ_n	nth target or label signal, when vector
$c(h_n)$	classifier applied to h_n
w_n	weight iterate at nth iteration of an algorithm
$w^\star$	minimizer of an empirical risk, $P(w)$
w^o	minimizer of a stochastic risk, $P(w)$
$\widetilde{w}_n$	weight error at iteration n
$R_{\text{emp}}(c)$	empirical risk of classifier $c(h)$
$R(c)$	actual risk of classifier $c(h)$
$q^\star(z)$	variational factor for estimating the posterior $f_{\boldsymbol{z}\|\boldsymbol{y}}(z\|y)$
$\mathbb{I}[a]$	indicator function: 1 when a is true; otherwise 0
$\mathbb{I}_{C,\infty}[x]$	indicator function: 0 when $x \in \mathcal{C}$; otherwise ∞
$\log a$	logarithm of a relative to base 10
$\ln a$	natural logarithm of a
$O(\mu)$	asymptotically bounded by a constant multiple of μ
$o(\mu)$	asymptotically bounded by a higher power of μ
$O(1/n)$	decays asymptotically at rate comparable to $1/n$
$o(1/n)$	decays asymptotically at rate faster than $1/n$
$\mathsf{mb}(\boldsymbol{x})$	Markov blanket of node $\boldsymbol{x}$ in a graph
$\mathsf{pa}(\boldsymbol{x})$	parents of node $\boldsymbol{x}$ in a graph
$\phi_C(\cdot)$	potential function associated with clique C

$\mathcal{N}_k$	neighborhood of node k in a graph
$\mathcal{M} = (\mathbb{S}, \mathbb{A}, \mathcal{P}, r)$	Markov decision process
$\mathbb{S}$	set of states for an MDP
$\mathbb{A}$	set of actions for an MDP
$\mathcal{P}$	transition probabilities for an MDP
r	reward function for an MDP
$\pi(a\|s)$	policy for selecting action a conditioned on state s
$\pi^\star(a\|s)$	optimal policy for an MDP
$v^\pi(s)$	state value function at state s
$v^\star(s)$	optimal state value function at state s
$q^\pi(s,a)$	state–action value function at state s and action a
$q^\star(s,a)$	optimal state–action value function at (s,a)
$\text{softmax}(z)$	softmax operation applied to entries of vector z
z_ℓ	pre-activation vector at layer ℓ of a neural network
y_ℓ	post-activation vector at layer ℓ of a neural network
W_ℓ	weighting matrix between layers ℓ and $\ell+1$
θ_ℓ	bias vector feeding into layer $\ell+1$
$w_{ij}^{(\ell)}$	weight from node i in layer ℓ to node j in layer $\ell+1$
$\theta_\ell(j)$	weight from bias source in layer ℓ to node j in layer $\ell+1$
δ_ℓ	sensitivity vector for layer ℓ

Abbreviations

ADF	assumed density filtering
ae	convergence almost everywhere
AIC	Akaike information criterion
BIBO	bounded-input bounded-output
BIC	Bayesian information criterion
BPTT	backpropagation through time
cdf	cumulative distribution function
CNN	convolutional neural network
CPT	conditional probability table
DAG	directed acyclic graph
EKF	extended Kalman filter
ELBO	evidence lower bound
ELU	exponential linear unit
EM	expectation maximization
EP	expectation propagation
ERC	exact recovery condition
ERM	empirical risk minimization
FDA	Fisher discriminant analysis
FGSM	fast gradient sign method
GAN	generative adversarial network
GLM	generalized linear model
GMM	Gaussian mixture model
GRU	gated recurrent unit

HALS	hierarchical alternating least-squares
HMM	hidden Markov model
ICA	independent component analysis
iid	independent and identically distributed
IRLS	iterative reweighted least-squares
ISTA	iterated soft-thresholding algorithm
JSMA	Jacobian saliency map approach
KKT	Karush–Kuhn–Tucker
KL	Kullback–Leibler divergence
LASSO	least absolute shrinkage and selection operator
LDA	latent Dirichlet allocation
LDA	linear discriminant analysis
LLMSE	linear least-mean-squares error
LMSE	least-mean-squares error
LOESS	locally estimated scatter-plot smoothing
LOWESS	locally weighted scatter-plot smoothing
LSTM	long short-term memory network
LWRP	layer-wise relevance propagation
MAB	multi-armed bandit
MAP	maximum a-posteriori
MCMC	Markov chain Monte Carlo
MDL	minimum description length
MDP	Markov decision process
ML	maximum likelihood
MMSE	minimum mean-square error
MP	matching pursuit
MRF	Markov random field
MSD	mean-square deviation
MSE	mean-square error
MVUE	minimum variance unbiased estimator
NMF	nonnegative matrix factorization
NN	nearest-neighbor rule
OCR	optical character recognition
OMP	orthogonal matching pursuit
PCA	principal component analysis
pdf	probability density function
PGM	probabilistic graphical model
pmf	probability mass function
POMDP	partially observable MDP
PPO	proximal policy optimization
RBF	radial basis function
RBM	restricted Boltzmann machine
ReLu	rectified linear unit
RIP	restricted isometry property

RKHS	reproducing kernel Hilbert space
RLS	recursive least-squares
RNN	recurrent neural network
RTRL	real time recurrent learning
SARSA	sequence of state, action, reward, state, action
SNR	signal-to-noise ratio
SOM	self-organizing map
SVD	singular value decomposition
SVM	support vector machine
TD	temporal difference
TRPO	trust region policy optimization
UCB	upper confidence bound
VAE	variational autoencoder
VC	Vapnik–Chervonenkis dimension
VI	variational inference
$\blacksquare$	end of theorem/lemma/proof/remark

1 Matrix Theory

We collect in this chapter useful background material on matrix theory and linear algebra. The emphasis is on results that are needed for future developments. Among other concepts, we review symmetric and nonnegative-definite matrices, range spaces and nullspaces, as well as several matrix decompositions, including the spectral decomposition, the triangular decomposition, the QR decomposition, and the singular value decomposition (SVD). We also discuss vector and matrix norms, Kronecker products, Schur complements, and the useful Rayleigh–Ritz characterization of the eigenvalues of symmetric matrices.

1.1 SYMMETRIC MATRICES

Symmetric and nonnegative-definite matrices play a prominent role in data analysis. We review some of their properties in this section. Thus, consider an arbitrary square matrix of size $N \times N$ with real entries, written as $A \in \mathbb{R}^{N\times N}$. The transpose of A is denoted by A^{T} and is obtained by transforming the rows of A into columns of A^{T}. For example,

$$A = \begin{bmatrix} 1 & -1 & 3 \\ -2 & 4 & 5 \\ 0 & 6 & 8 \end{bmatrix} \implies A^{\mathsf{T}} = \begin{bmatrix} 1 & -2 & 0 \\ -1 & 4 & 6 \\ 3 & 5 & 8 \end{bmatrix} \tag{1.1}$$

The matrix A is said to be symmetric if it happens to coincide with its matrix transpose, i.e., if it satisfies

$$A = A^{\mathsf{T}} \qquad \textbf{(symmetry)} \tag{1.2}$$

Real eigenvalues

A useful property of symmetric matrices is that they can only have *real* eigenvalues. To see this, let u represent a column eigenvector of A corresponding to some eigenvalue λ, i.e., u is nonzero and satisfies along with λ the relation:

$$Au = \lambda u \tag{1.3}$$

The eigenvector u may be complex-valued so that, in general, $u \in \mathbb{C}^N$. Let the symbol $*$ denote the operation of complex conjugate transposition, so that u^*

is the row vector that is obtained by transposing u and replacing its entries by their complex conjugate values, e.g.,

$$u \triangleq \begin{bmatrix} 1+j \\ 2 \\ -2+3j \end{bmatrix} \implies u^* = \begin{bmatrix} 1-j & 2 & -2-3j \end{bmatrix} \tag{1.4}$$

where $j \triangleq \sqrt{-1}$. The same complex conjugation operation can be applied to matrices as well so that, e.g.,

$$B = \begin{bmatrix} 1 & j & -2+j \\ 3-j & 1-2j & 0 \end{bmatrix} \implies B^* = \begin{bmatrix} 1 & 3+j \\ -j & 1+2j \\ -2-j & 0 \end{bmatrix} \tag{1.5}$$

Returning to (1.3) and multiplying from the left by the row vector u^* we get

$$u^*Au = \lambda u^*u \;=\; \lambda\|u\|^2 \tag{1.6}$$

where the notation $\|\cdot\|$ denotes the Euclidean norm of its vector argument. Note that the quantity u^*Au is a scalar. Moreover, it is real-valued because it coincides with its complex conjugate value:

$$(u^*Au)^* \;=\; u^*A^*(u^*)^* \;=\; u^*Au \tag{1.7}$$

where in the last step we used the fact that $A^* = A$ since A is real-valued and symmetric. Therefore, u^*Au is real and, from equality (1.6), we conclude that $\lambda\|u\|^2$ must also be real. But since $\|u\|^2$ is real and nonzero, we conclude that the eigenvalue λ must be real too.

One consequence of this conclusion is that we can always find *real-valued* eigenvectors for symmetric matrices. Indeed, if we express u in terms of its real and imaginary vector components, say, as

$$u = p + jq, \quad p, q \in \mathbb{R}^N \tag{1.8}$$

then, using (1.3) and the fact that λ is real, we conclude that it must hold:

$$Ap = \lambda p, \quad Aq = \lambda q \tag{1.9}$$

so that p and q are eigenvectors associated with λ.

Spectral theorem

A second important property of real symmetric matrices, one whose proof requires a more elaborate argument and is deferred to Appendix 1.A, is that such matrices always have a *full* set of orthonormal eigenvectors. That is, if $A \in \mathbb{R}^{N\times N}$ is symmetric, then there will exist a set of N orthonormal real eigenvectors $u_n \in \mathbb{R}^N$ satisfying

$$Au_n = \lambda_n u_n, \quad \|u_n\|^2 = 1, \quad u_n^{\mathsf{T}} u_m = 0 \;\text{ for } n \neq m \tag{1.10}$$

where all N eigenvalues $\{\lambda_n, n = 1, 2, \ldots, N\}$ are real, and all eigenvectors $\{u_n\}$ have unit norm and are orthogonal to each other. This result is known as the *spectral theorem*. For illustration purposes, assume A is 3×3. Then, the above statement asserts that there will exist three real orthonormal vectors $\{u_1, u_2, u_3\}$ and three real eigenvalues $\{\lambda_1, \lambda_2, \lambda_3\}$ such that

$$A \underbrace{\begin{bmatrix} u_1 & u_2 & u_3 \end{bmatrix}}_{\triangleq\, U} = \underbrace{\begin{bmatrix} u_1 & u_2 & u_3 \end{bmatrix}}_{U} \underbrace{\begin{bmatrix} \lambda_1 & & \\ & \lambda_2 & \\ & & \lambda_3 \end{bmatrix}}_{\triangleq\, \Lambda} \tag{1.11}$$

where we are introducing the matrices U and Λ for compactness of notation: U contains real eigenvectors for A and Λ is a diagonal matrix with the corresponding eigenvalues. Then, we can write (1.11) more compactly as

$$AU = U\Lambda \tag{1.12}$$

However, the fact that the columns of U are orthogonal to each other and have unit norms implies that U satisfies the important normalization property:

$$UU^{\mathsf{T}} = I_N \quad \text{and} \quad U^{\mathsf{T}}U = I_N \tag{1.13}$$

That is, the product of U with U^{T} (or U^{T} with U) results in the identity matrix of size $N \times N$ – see Prob. 1.1. We say that U is an *orthogonal* matrix. Using this property and multiplying the matrix equality (1.12) by U^{T} from the right we get

$$A\underbrace{UU^{\mathsf{T}}}_{=I} = U\Lambda U^{\mathsf{T}} \tag{1.14}$$

We therefore conclude that every real symmetric matrix A can be expressed in the following spectral (or eigen-) decomposition form:

$$\boxed{A = U\Lambda U^{\mathsf{T}}} \qquad \textbf{(eigen-decomposition)} \tag{1.15a}$$

where, for general dimensions, the $N \times N$ matrices Λ and U are constructed from the eigenvalues and orthonormal eigenvectors of A as follows:

$$\Lambda = \text{diag}\{\lambda_1, \lambda_2, \ldots, \lambda_N\} \tag{1.15b}$$

$$U = \begin{bmatrix} u_1 & u_2 & \ldots & u_N \end{bmatrix} \tag{1.15c}$$

Rayleigh–Ritz ratio

There is a useful characterization of the smallest and largest eigenvalues of real symmetric matrices, known as the Rayleigh–Ritz ratio. Specifically, if $A \in \mathbb{R}^{N\times N}$ is symmetric, then it holds that for all vectors $x \in \mathbb{R}^N$:

$$\boxed{\lambda_{\min}\|x\|^2 \leq x^{\mathsf{T}}Ax \leq \lambda_{\max}\|x\|^2} \tag{1.16}$$

as well as

$$\lambda_{\min} = \min_{x \neq 0} \left(\frac{x^\mathsf{T} A x}{x^\mathsf{T} x} \right) = \min_{\|x\|=1} x^\mathsf{T} A x \tag{1.17a}$$

$$\lambda_{\max} = \max_{x \neq 0} \left(\frac{x^\mathsf{T} A x}{x^\mathsf{T} x} \right) = \max_{\|x\|=1} x^\mathsf{T} A x \tag{1.17b}$$

where $\{\lambda_{\min}, \lambda_{\max}\}$ denote the smallest and largest eigenvalues of A. The ratio $x^\mathsf{T} A x / x^\mathsf{T} x$ is called the Rayleigh–Ritz ratio.

Proof of (1.16) and (1.17a)–(1.17b): Consider the eigen-decomposition (1.15a) and introduce the vector $y = U^\mathsf{T} x$ for any vector x. Then,

$$x^\mathsf{T} A x \;=\; x^\mathsf{T} U \Lambda U^\mathsf{T} x \;=\; y^\mathsf{T} \Lambda y \;=\; \sum_{n=1}^{N} \lambda_n y_n^2 \tag{1.18}$$

with the $\{y_n\}$ denoting the individual entries of y. Now since the squared terms $\{y_n^2\}$ are nonnegative and the $\{\lambda_n\}$ are real, we get

$$\lambda_{\min} \left(\sum_{n=1}^{N} y_n^2 \right) \leq \sum_{n=1}^{N} \lambda_n y_n^2 \leq \lambda_{\max} \left(\sum_{n=1}^{N} y_n^2 \right) \tag{1.19}$$

or, equivalently,

$$\lambda_{\min} \|y\|^2 \leq x^\mathsf{T} A x \leq \lambda_{\max} \|y\|^2 \tag{1.20}$$

Using the fact that U is orthogonal and, hence,

$$\|y\|^2 \;=\; y^\mathsf{T} y \;=\; x \underbrace{U U^\mathsf{T}}_{=I} x \;=\; \|x\|^2 \tag{1.21}$$

we conclude that (1.16) holds. The lower (upper) bound in (1.19) is achieved when x is chosen as the eigenvector $u_{\min}(u_{\max})$ corresponding to $\lambda_{\min}(\lambda_{\max})$.

■

Example 1.1 **(Quadratic curve)** Consider the two-dimensional function

$$g(r,s) = ar^2 + as^2 + 2brs, \quad r,s \in \mathbb{R} \tag{1.22}$$

We would like to determine the largest and smallest values that the function attains on the circle $r^2 + s^2 = 1$. One way to solve the problem is to recognize that $g(r,s)$ can be rewritten as

$$g(r,s) = \underbrace{\begin{bmatrix} r & s \end{bmatrix}}_{\triangleq\, x^\mathsf{T}} \underbrace{\begin{bmatrix} a & b \\ b & a \end{bmatrix}}_{\triangleq\, A} \underbrace{\begin{bmatrix} r \\ s \end{bmatrix}}_{=x} = x^\mathsf{T} A x \tag{1.23}$$

We therefore want to determine the extreme values of the quadratic form $x^\mathsf{T} A x$ under the constraint $\|x\| = 1$. According to (1.17a)–(1.17b), these values correspond to $\lambda_{\min}(A)$ and $\lambda_{\max}(A)$. It can be easily verified that the eigenvalues of A are given by $\lambda(A) = \{a-b, a+b\}$ and, hence,

$$\lambda_{\min}(A) = \min\{a-b, a+b\}, \qquad \lambda_{\max}(A) = \max\{a-b, a+b\} \tag{1.24}$$

1.2 POSITIVE-DEFINITE MATRICES

An $N \times N$ real symmetric matrix A is said to be nonnegative-definite (also called positive semi-definite) if it satisfies the property:

$$v^{\mathsf{T}} A v \geq 0, \qquad \text{for all column vectors } v \in \mathbb{R}^N \tag{1.25}$$

The matrix A is said to be positive-definite if $v^{\mathsf{T}} A v > 0$ for all $v \neq 0$. We denote a positive-definite matrix by writing $A > 0$ and a positive semi-definite matrix by writing $A \geq 0$.

Example 1.2 **(Diagonal matrices)** The notion of positive semi-definiteness is trivial for diagonal matrices. Consider the diagonal matrix

$$A = \text{diag}\{a_1, a_2, a_3\} \in \mathbb{R}^{3\times 3} \tag{1.26}$$

and let

$$v = \begin{bmatrix} v_1 \\ v_2 \\ v_3 \end{bmatrix} \in \mathbb{R}^3 \tag{1.27}$$

denote an arbitrary vector. Then, some simple algebra shows that

$$v^{\mathsf{T}} A v \;=\; a_1 v_1^2 + a_2 v_2^2 + a_3 v_3^2 \tag{1.28}$$

This expression will be nonnegative for *any* v if, and only if, the entries a_n are all nonnegative. This is because if any a_n is negative, say a_2, then we can select a vector v with an entry v_2 that is large enough to result in a negative term $a_2 v_2^2$ that exceeds the contribution of the other two terms in the sum $v^{\mathsf{T}} A v$. Therefore, for a diagonal matrix to be positive semi-definite, it is necessary and sufficient that its diagonal entries be nonnegative. Likewise, a diagonal matrix A is positive-definite if, and only if, its diagonal entries are positive. We cannot extrapolate and say that a general nondiagonal matrix A is positive semi-definite if all its entries are nonnegative; this conclusion is *not* true, as the next example shows.

Example 1.3 **(Nondiagonal matrices)** Consider the 2×2 matrix

$$A = \begin{bmatrix} 3 & -1 \\ -1 & 3 \end{bmatrix} \tag{1.29}$$

This matrix is positive-definite. Indeed, pick any nonzero column vector $v \in \mathbb{R}^2$. Then,

$$\begin{aligned} v^{\mathsf{T}} A v &= \begin{bmatrix} v_1 & v_2 \end{bmatrix} \begin{bmatrix} 3 & -1 \\ -1 & 3 \end{bmatrix} \begin{bmatrix} v_1 \\ v_2 \end{bmatrix} \\ &= 3v_1^2 + 3v_2^2 - 2v_1 v_2 \\ &= (v_1 - v_2)^2 \;+\; 2v_1^2 \;+\; 2v_2^2 \\ &> 0, \;\; \text{for any } v \neq 0 \end{aligned} \tag{1.30}$$

Among the several equivalent characterizations of positive-definite matrices, we note that an $N \times N$ real symmetric matrix A is positive-definite if, and only if, all its N eigenvalues are positive:

$$\boxed{A > 0 \iff \{\lambda_n > 0\}_{n=1}^{N}} \tag{1.31}$$

One proof relies on the use of the eigen-decomposition of A.

Proof of (1.31): We need to prove the statement in both directions. Assume initially that A is positive-definite and let us establish that all its eigenvalues are positive. Let $A = U\Lambda U^{\mathsf{T}}$ denote the spectral decomposition of A. Let also u_n denote the nth column of U corresponding to the eigenvalue λ_n, i.e., $Au_n = \lambda_n u_n$ with $\|u_n\|^2 = 1$. If we multiply this equality from the left by u_n^{T} we get

$$u_n^{\mathsf{T}} A u_n = \lambda_n \|u_n\|^2 = \lambda_n > 0 \tag{1.32}$$

where the last inequality follows from the fact that $u^{\mathsf{T}} A u > 0$ for any nonzero vector u since A is assumed to be positive-definite. Therefore, $A > 0$ implies $\lambda_n > 0$ for $n = 1, 2, \ldots, N$.

Conversely, assume all $\lambda_n > 0$ and let us show that $A > 0$. Multiply the equality $A = U\Lambda U^{\mathsf{T}}$ by any *nonzero* vector v and its transpose, from right and left, to get

$$v^{\mathsf{T}} A v = v^{\mathsf{T}} U \Lambda U^{\mathsf{T}} v \tag{1.33}$$

Now introduce the real diagonal matrix

$$D \triangleq \text{diag}\left\{\sqrt{\lambda_1},\ \sqrt{\lambda_2},\ \ldots,\ \sqrt{\lambda_n}\right\} \tag{1.34}$$

and the vector

$$s \triangleq DU^{\mathsf{T}} v \tag{1.35}$$

The vector s is nonzero. This can be seen as follows. Let $w = U^{\mathsf{T}} v$. Then, the vectors v and w have the same Euclidean norm since

$$\|w\|^2 = w^{\mathsf{T}} w = v^{\mathsf{T}} \underbrace{UU^{\mathsf{T}}}_{=I} v = v^{\mathsf{T}} v = \|v\|^2 \tag{1.36}$$

It follows that the vector w is nonzero since v is nonzero. Now since $s = Dw$ and all entries of D are nonzero, we conclude that $s \neq 0$. Returning to (1.33), we get

$$v^{\mathsf{T}} A v = \|s\|^2 > 0 \tag{1.37}$$

for any nonzero v, which establishes that $A > 0$.

■

In a similar vein, we can show that

$$\boxed{A \geq 0 \iff \{\lambda_n \geq 0\}_{n=1}^{N}} \tag{1.38}$$

Example 1.4 **(Positive-definite matrix)** Consider again the 2×2 matrix from Example 1.3:

$$A = \begin{bmatrix} 3 & -1 \\ -1 & 3 \end{bmatrix} \tag{1.39}$$

We established in that example from first principles that $A > 0$. Alternatively, we can determine the eigenvalues of A and verify that they are positive. The eigenvalues are the roots of the characteristic equation, $\det(\lambda I - A) = 0$, which leads to the quadratic equation $(\lambda - 3)^2 - 1 = 0$ so that $\lambda_1 = 4 > 0$ and $\lambda_2 = 2 > 0$.

A second useful property of positive-definite matrices is that they have positive determinants. To see this, recall first that for two square matrices A and B it holds that

$$\det(AB) = \det(A)\det(B) \tag{1.40}$$

That is, the determinant of the product is equal to the product of the determinants. Now starting with a positive-definite matrix A, and applying the above determinant formula to its eigen-decomposition (1.15a), we get

$$\det A = (\det U)\,(\det \Lambda)\,\big(\det U^{\mathsf{T}}\big) \tag{1.41}$$

But $UU^{\mathsf{T}} = I$ so that

$$(\det U)\,\big(\det U^{\mathsf{T}}\big) = 1 \tag{1.42}$$

and we conclude that

$$\det A = \det \Lambda = \prod_{n=1}^{N} \lambda_n \tag{1.43}$$

This result is actually general and holds for arbitrary square matrices A (the matrices do not need to be symmetric or positive-definite): the determinant of a matrix is always equal to the product of its eigenvalues (counting multiplicities) – see Prob. 1.2. Now, when the matrix A happens to be positive-definite, all its eigenvalues will be positive and, hence,

$$\boxed{A > 0 \implies \det A > 0} \tag{1.44}$$

Note that this statement goes in one direction only; the converse is not true.

1.3 RANGE SPACES AND NULLSPACES

Let A denote an $N \times M$ real matrix without any constraint on the relative sizes of N and M. When $N = M$, we say that A is a square matrix. Otherwise, when $N > M$, we say that A is a "tall" matrix and when $N < M$ we say that A is a "fat" matrix.

Definitions

The *column span* or the *range space* of A is defined as the set of all $N \times 1$ vectors q that can be generated by Ap, for all $M \times 1$ vectors p. We denote the column span of A by

$$\boxed{\mathcal{R}(A) \triangleq \left\{\text{set of all } q \in \mathbb{R}^N \text{ such that } q = Ap \text{ for some } p \in \mathbb{R}^M\right\}} \tag{1.45}$$

Likewise, the nullspace of A is the set of all $M \times 1$ vectors p that are annihilated by A, namely, that satisfy $Ap = 0$. We denote the nullspace of A by

$$\boxed{\mathcal{N}(A) \triangleq \left\{\text{set of all } p \in \mathbb{R}^M \text{ such that } Ap = 0\right\}} \tag{1.46}$$

The rank of a matrix A is defined as the number of linearly independent columns of A. It can be verified that, for any matrix A, the number of linearly independent columns is also equal to the number of linearly independent rows – see Prob. 1.5. It therefore holds that

$$\boxed{\text{rank}(A) \leq \min\{N, M\}} \tag{1.47}$$

That is, the rank of a matrix never exceeds its smallest dimension. A matrix is said to have *full rank* if

$$\text{rank}(A) = \min\{N, M\} \tag{1.48}$$

Otherwise, the matrix is said to be *rank deficient*.

If A is a square matrix (i.e., $N = M$), then rank deficiency is equivalent to a zero determinant, $\det A = 0$. Indeed, if A is rank deficient, then there exists a nonzero p such that $Ap = 0$. This means that $\lambda = 0$ is an eigenvalue of A so that its determinant must be zero.

Useful relations

One useful property that follows from the definition of range spaces and nullspaces is that any vector $z \in \mathbb{R}^N$ from the nullspace of A^T (not A) is orthogonal to any vector $q \in \mathbb{R}^N$ in the range space of A, i.e.,

$$\boxed{z \in \mathcal{N}(A^\mathsf{T}),\;\; q \in \mathcal{R}(A) \quad \Longrightarrow \quad z^\mathsf{T} q = 0} \tag{1.49}$$

Proof of (1.49): Indeed, $z \in \mathcal{N}(A^\mathsf{T})$ implies that $A^\mathsf{T} z = 0$ or, equivalently, $z^\mathsf{T} A = 0$. Now write $q = Ap$ for some p. Then, $z^\mathsf{T} q = z^\mathsf{T} Ap = 0$, as desired.

■

A second useful property is that the matrices $A^\mathsf{T} A$ and A^T have the *same* range space (i.e., they span the same space). Also, A and $A^\mathsf{T} A$ have the same nullspace, i.e.,

$$\boxed{\mathcal{R}(A^\mathsf{T}) = \mathcal{R}(A^\mathsf{T} A), \qquad \mathcal{N}(A) = \mathcal{N}(A^\mathsf{T} A)} \tag{1.50}$$

Proof of (1.50): Consider a vector $q \in \mathcal{R}(A^\mathsf{T}A)$, i.e., $q = A^\mathsf{T}Ap$ for some p. Define $r = Ap$, then $q = A^\mathsf{T}r$. This shows that $q \in \mathcal{R}(A^\mathsf{T})$ and we conclude that $\mathcal{R}(A^\mathsf{T}A) \subset \mathcal{R}(A^\mathsf{T})$. The proof of the converse statement requires more effort.

Consider a vector $q \in \mathcal{R}(A^\mathsf{T})$ and let us show by contradiction that $q \in \mathcal{R}(A^\mathsf{T}A)$. Assume, to the contrary, that q does not lie in $\mathcal{R}(A^\mathsf{T}A)$. This implies by (1.49) that there exists a vector z in the nullspace of $A^\mathsf{T}A$ that is not orthogonal to q, i.e., $A^\mathsf{T}Az = 0$ and $z^\mathsf{T}q \neq 0$. Now, if we multiply the equality $A^\mathsf{T}Az = 0$ by z^T from the left we obtain that $z^\mathsf{T}A^\mathsf{T}Az = 0$ or, equivalently, $\|Az\|^2 = 0$. Therefore, Az is necessarily the zero vector, $Az = 0$. But from $q \in \mathcal{R}(A^\mathsf{T})$ we have that $q = A^\mathsf{T}p$ for some p. Then, it must hold that $z^\mathsf{T}q = z^\mathsf{T}A^\mathsf{T}p = 0$, which contradicts $z^\mathsf{T}q \neq 0$. Therefore, we must have $q \in \mathcal{R}(A^\mathsf{T}A)$ and we conclude that $\mathcal{R}(A^\mathsf{T}) \subset \mathcal{R}(A^\mathsf{T}A)$.

The second assertion in (1.50) is more immediate. If $Ap = 0$ then $A^\mathsf{T}Ap = 0$ so that $\mathcal{N}(A) \subset \mathcal{N}(A^\mathsf{T}A)$. Conversely, if $A^\mathsf{T}Ap = 0$ then $p^\mathsf{T}A^\mathsf{T}Ap = \|Ap\|^2 = 0$ and we must have $Ap = 0$. That is, $\mathcal{N}(A^\mathsf{T}A) \subset \mathcal{N}(A)$. Combining both facts we conclude that $\mathcal{N}(A) = \mathcal{N}(A^\mathsf{T}A)$.

■

Normal equations

One immediate consequence of result (1.50) is that linear systems of equations of the following form:

$$\boxed{A^\mathsf{T}Ax = A^\mathsf{T}b} \qquad \textbf{(normal equations)} \tag{1.51}$$

are always consistent, i.e., they always have a solution x for *any* vector b. This is because $A^\mathsf{T}b$ belongs to $\mathcal{R}(A^\mathsf{T})$ and, therefore, also belongs to $\mathcal{R}(A^\mathsf{T}A)$. This type of linear system of equations will appear as normal equations when we study least-squares problems later in Chapter 50 – see (50.25); the reason for the designation "normal equations" will be explained there. We can say more about the solution of such equations. For example, when the coefficient matrix $A^\mathsf{T}A$, which is always square regardless of the column and row dimensions of the $N \times M$ matrix A, happens to be invertible, then the normal equations (1.51) will have a unique solution given by

$$x = (A^\mathsf{T}A)^{-1}A^\mathsf{T}b \tag{1.52}$$

We explain later in (1.58) that the matrix product $A^\mathsf{T}A$ will be invertible when the following two conditions hold: $N \geq M$ *and* A has full rank. In all other cases, the matrix product $A^\mathsf{T}A$ will be singular and will, therefore, have a nontrivial nullspace. Let p be any nonzero vector in the nullspace of $A^\mathsf{T}A$. We know from (1.50) that this vector also lies in the nullspace of A. Since we know that a solution x always exists for (1.51) then, by adding any such p to x, we obtain another solution. This is because:

$$\begin{aligned} A^\mathsf{T}A(x+p) &= A^\mathsf{T}Ax + A^\mathsf{T}Ap \\ &= A^\mathsf{T}Ax + 0 \\ &= A^\mathsf{T}Ax \\ &= A^\mathsf{T}b \end{aligned} \tag{1.53}$$

Knowing that there exist infinitely many vectors in $\mathcal{N}(A^{\mathsf{T}}A)$, e.g., any scaled multiple of p belongs to the same nullspace, we conclude that when $A^{\mathsf{T}}A$ is singular, there will exist infinitely many solutions to the normal equations (1.51). We therefore find that the normal equations (1.51) either have a unique solution (when $A^{\mathsf{T}}A$ is invertible) or infinitely many solutions (when $A^{\mathsf{T}}A$ is singular).

We can be more explicit about the latter case and verify that, when infinitely many solutions exist, they all differ by a vector in the nullspace of A. Indeed, assume $A^{\mathsf{T}}A$ is singular and let x_1 and x_2 denote two solutions to the normal equations (1.51). Then,

$$A^{\mathsf{T}}Ax_1 = A^{\mathsf{T}}b, \quad A^{\mathsf{T}}Ax_2 = A^{\mathsf{T}}b \tag{1.54}$$

Subtracting these two equalities we find that

$$A^{\mathsf{T}}A(x_1 - x_2) = 0 \tag{1.55}$$

which means that the difference $x_1 - x_2$ belongs to the nullspace of $A^{\mathsf{T}}A$ or, equivalently, to the nullspace of A in view of (1.50), namely,

$$x_1 - x_2 \in \mathcal{N}(A) \tag{1.56}$$

as claimed. We collect the results in the following statement for ease of reference.

Lemma 1.1. (Solution of normal equations) *Consider the normal system of equations $A^{\mathsf{T}}Ax = A^{\mathsf{T}}b$, where $A \in \mathbb{R}^{N\times M}$, $b \in \mathbb{R}^N$, and $x \in \mathbb{R}^M$. The following facts hold:*

(a) *A solution x always exists.*
(b) *The solution x is unique when $A^{\mathsf{T}}A$ is invertible (i.e., when $N \geq M$ and A has full rank). In this case, the solution is given by expression (1.52).*
(c) *There exist infinitely many solutions x when $A^{\mathsf{T}}A$ is singular.*
(d) *Under (c), any two solutions x_1 and x_2 will differ by a vector in $\mathcal{N}(A)$, i.e., (1.56) holds.*

The next result clarifies when the matrix product $A^{\mathsf{T}}A$ is invertible. Note in particular that the matrix $A^{\mathsf{T}}A$ is symmetric and nonnegative-definite; the latter property is because, for any nonzero x, it holds that

$$x^{\mathsf{T}}A^{\mathsf{T}}Ax = \|Ax\|^2 \geq 0 \tag{1.57}$$

Thus, let A be $N \times M$, with $N \geq M$ (i.e., A is a "tall" or square matrix). Then,

$$\boxed{A \text{ has full rank} \iff A^{\mathsf{T}}A \text{ is positive-definite}} \tag{1.58}$$

That is, every tall full rank matrix is such that the square matrix $A^{\mathsf{T}}A$ is invertible (actually, positive-definite).

Proof of (1.58): Assume first that A has full rank. This means that all columns of A are linearly independent, which in turn means that $Ax \neq 0$ for any nonzero x. Consequently, it holds that $\|Ax\|^2 > 0$, which is equivalent to $x^{\mathsf{T}}A^{\mathsf{T}}Ax > 0$ for any

$x \neq 0$. It follows that $A^{\mathsf{T}}A > 0$. Conversely, assume that $A^{\mathsf{T}}A > 0$. This means that $x^{\mathsf{T}}A^{\mathsf{T}}Ax > 0$ for any nonzero x, which is equivalent to $\|Ax\|^2 > 0$ and, hence, $Ax \neq 0$. It follows that the columns of A are linearly independent so that A has full column rank.

■

In fact, when A has full rank, not only $A^{\mathsf{T}}A$ is positive-definite, but also any product of the form $A^{\mathsf{T}}BA$ for any symmetric positive-definite matrix B. Specifically, if $B > 0$, then

$$\boxed{A:\; N \times M,\; N \geq M,\; \text{full rank} \quad \Longleftrightarrow \quad A^{\mathsf{T}}BA > 0} \tag{1.59}$$

Proof of (1.59): Assume first that A has full rank. This means that all columns of A are linearly independent, which in turn means that the vector $z = Ax \neq 0$ for any nonzero x. Now, since $B > 0$, it holds that $z^{\mathsf{T}}Bz > 0$ and, hence, $x^{\mathsf{T}}A^{\mathsf{T}}BAx > 0$ for any nonzero x. It follows that $A^{\mathsf{T}}BA > 0$. Conversely, assume that $A^{\mathsf{T}}BA > 0$. This means that $x^{\mathsf{T}}A^{\mathsf{T}}BAx > 0$ for any nonzero x, which allows us to conclude, by contradiction, that A must be full rank. Indeed, assume not. Then, there should exist a nonzero vector p such that $Ap = 0$, which implies that $p^{\mathsf{T}}A^{\mathsf{T}}BAp = 0$. This conclusion contradicts the fact that $x^{\mathsf{T}}A^{\mathsf{T}}BAx > 0$ for any nonzero x. Therefore, A has full rank, as desired.

■

1.4 SCHUR COMPLEMENTS

There is a useful block triangularization formula that can often be used to facilitate the computation of matrix inverses or to reduce matrices to convenient block diagonal forms.

In this section we assume inverses exist whenever needed. Thus, consider a real block matrix:

$$S = \begin{bmatrix} A & B \\ C & D \end{bmatrix} \tag{1.60}$$

The Schur complement of A in S is denoted by Δ_A and is defined as the quantity:

$$\boxed{\Delta_A \stackrel{\Delta}{=} D - CA^{-1}B} \tag{1.61}$$

Likewise, the Schur complement of D in S is denoted by Δ_D and is defined as

$$\boxed{\Delta_D \stackrel{\Delta}{=} A - BD^{-1}C} \tag{1.62}$$

Block triangular factorizations

In terms of these Schur complements, it is easy to verify by direct calculation that the block matrix S can be factored in either of the following two useful forms:

$$\begin{bmatrix} A & B \\ C & D \end{bmatrix} = \begin{bmatrix} I & 0 \\ CA^{-1} & I \end{bmatrix} \begin{bmatrix} A & 0 \\ 0 & \Delta_A \end{bmatrix} \begin{bmatrix} I & A^{-1}B \\ 0 & I \end{bmatrix} \tag{1.63a}$$

$$= \begin{bmatrix} I & BD^{-1} \\ 0 & I \end{bmatrix} \begin{bmatrix} \Delta_D & 0 \\ 0 & D \end{bmatrix} \begin{bmatrix} I & 0 \\ D^{-1}C & I \end{bmatrix} \tag{1.63b}$$

Two useful results that follow directly from these factorizations are the determinantal formulas:

$$\det \begin{bmatrix} A & B \\ C & D \end{bmatrix} = \det A \, \det(D - CA^{-1}B) \tag{1.64a}$$

$$= \det D \, \det(A - BD^{-1}C) \tag{1.64b}$$

Block inversion formulas

Moreover, by inverting both sides of (1.63b), we readily conclude that

$$\begin{bmatrix} A & B \\ C & D \end{bmatrix}^{-1} = \begin{bmatrix} I & -A^{-1}B \\ 0 & I \end{bmatrix} \begin{bmatrix} A^{-1} & 0 \\ 0 & \Delta_A^{-1} \end{bmatrix} \begin{bmatrix} I & 0 \\ -CA^{-1} & I \end{bmatrix} \tag{1.65a}$$

$$= \begin{bmatrix} I & 0 \\ -D^{-1}C & I \end{bmatrix} \begin{bmatrix} \Delta_D^{-1} & 0 \\ 0 & D^{-1} \end{bmatrix} \begin{bmatrix} I & -BD^{-1} \\ 0 & I \end{bmatrix} \tag{1.65b}$$

where we used the fact that for block triangular matrices it holds

$$\begin{bmatrix} I & 0 \\ X & I \end{bmatrix}^{-1} = \begin{bmatrix} I & 0 \\ -X & I \end{bmatrix}, \quad \begin{bmatrix} I & X \\ 0 & I \end{bmatrix}^{-1} = \begin{bmatrix} I & -X \\ 0 & I \end{bmatrix} \tag{1.66}$$

If we expand (1.65b) we can also write

$$\begin{bmatrix} A & B \\ C & D \end{bmatrix}^{-1} = \begin{bmatrix} A^{-1} + A^{-1}B\Delta_A^{-1}CA^{-1} & -A^{-1}B\Delta_A^{-1} \\ -\Delta_A^{-1}CA^{-1} & \Delta_A^{-1} \end{bmatrix} \tag{1.67a}$$

$$= \begin{bmatrix} \Delta_D^{-1} & -\Delta_D^{-1}BD^{-1} \\ -D^{-1}C\Delta_D^{-1} & D^{-1} + D^{-1}C\Delta_D^{-1}BD^{-1} \end{bmatrix} \tag{1.67b}$$

Matrix inertia and congruence

When the block matrix S is symmetric, its eigenvalues are real. We define the *inertia* of S as the triplet:

$$\text{In}\{S\} \triangleq \Big\{I_+, I_-, I_0\Big\} \tag{1.68}$$

in terms of the integers:

$$I_+(S) = \text{the number of positive eigenvalues of } S \tag{1.69a}$$

$$I_-(S) = \text{the number of negative eigenvalues of } S \tag{1.69b}$$

$$I_0(S) = \text{the number of zero eigenvalues of } S \tag{1.69c}$$

Now, given a symmetric matrix S and any invertible matrix Q, the matrices S and QSQ^T are said to be *congruent*. An important result regarding congruent matrices is that congruence preserves inertia, i.e., it holds that

$$\boxed{\text{In}\{S\} = \text{In}\left\{QSQ^\mathsf{T}\right\}} \qquad \textbf{(congruence)} \tag{1.70}$$

so that the matrices S and QSQ^T will have the same number of positive, negative, and zero eigenvalues for any invertible Q. This result is known as the Sylvester law of inertia.

Example 1.5 **(Inertia and Schur complements)** One immediate application of the above congruence result is the following characterization of the inertia of a matrix in terms of the inertia of its Schur complements. Thus, assume that S is symmetric with the block structure:

$$S = \begin{bmatrix} A & B \\ B^\mathsf{T} & D \end{bmatrix}, \qquad \text{where } A = A^\mathsf{T} \text{ and } D = D^\mathsf{T} \tag{1.71}$$

Consider the corresponding block factorizations (1.63a)–(1.63b):

$$\begin{bmatrix} A & B \\ B^\mathsf{T} & D \end{bmatrix} = \begin{bmatrix} I & 0 \\ B^\mathsf{T}A^{-1} & I \end{bmatrix} \begin{bmatrix} A & 0 \\ 0 & \Delta_A \end{bmatrix} \begin{bmatrix} I & A^{-1}B \\ 0 & I \end{bmatrix} \tag{1.72a}$$

$$= \begin{bmatrix} I & BD^{-1} \\ 0 & I \end{bmatrix} \begin{bmatrix} \Delta_D & 0 \\ 0 & D \end{bmatrix} \begin{bmatrix} I & 0 \\ D^{-1}B^\mathsf{T} & I \end{bmatrix} \tag{1.72b}$$

in terms of the Schur complements:

$$\Delta_A = D - B^\mathsf{T}A^{-1}B, \qquad \Delta_D = A - BD^{-1}B^\mathsf{T} \tag{1.73}$$

The factorizations (1.72a)–(1.72b) have the form of congruence relations so that we must have

$$\text{In}\{S\} = \text{In}\left\{\begin{bmatrix} A & 0 \\ 0 & \Delta_A \end{bmatrix}\right\} \quad \text{and} \quad \text{In}\{S\} = \text{In}\left\{\begin{bmatrix} \Delta_D & 0 \\ 0 & D \end{bmatrix}\right\} \tag{1.74}$$

When S is positive-definite, all its eigenvalues are positive. Then, from the above inertia equalities, it follows that the matrices $\{A, \Delta_A, \Delta_D, D\}$ can only have positive eigenvalues. In other words, it must hold that

$$\boxed{S > 0 \iff A > 0 \text{ and } \Delta_A > 0} \tag{1.75a}$$

Likewise,

$$\boxed{S > 0 \iff D > 0 \text{ and } \Delta_D > 0} \tag{1.75b}$$

Example 1.6 (**A completion-of-squares formula**) Consider a quadratic expression of the form

$$J(x) = x^{\mathsf{T}}Ax - 2b^{\mathsf{T}}x + \alpha, \quad A \in \mathbb{R}^{M\times M}, \quad b \in \mathbb{R}^{M\times 1}, \quad \alpha \in \mathbb{R} \tag{1.76}$$

where A is assumed invertible and symmetric. We write $J(x)$ as

$$J(x) = \begin{bmatrix} x^{\mathsf{T}} & 1 \end{bmatrix} \begin{bmatrix} A & -b \\ -b^{\mathsf{T}} & \alpha \end{bmatrix} \begin{bmatrix} x \\ 1 \end{bmatrix} \tag{1.77}$$

Usually, the block matrix

$$\begin{bmatrix} A & -b \\ -b^{\mathsf{T}} & \alpha \end{bmatrix} \tag{1.78}$$

is positive-definite, in which case A is positive-definite and the Schur complement relative to it, namely, $\alpha - b^{\mathsf{T}}A^{-1}b$, is also positive. Next, we introduce the triangular factorization

$$\begin{bmatrix} A & -b \\ -b^{\mathsf{T}} & \alpha \end{bmatrix} = \begin{bmatrix} I_M & 0 \\ -b^{\mathsf{T}}A^{-1} & 1 \end{bmatrix} \begin{bmatrix} A & 0 \\ 0 & \alpha - b^{\mathsf{T}}A^{-1}b \end{bmatrix} \begin{bmatrix} I_M & -A^{-1}b \\ 0 & 1 \end{bmatrix} \tag{1.79}$$

and substitute it into (1.77) to get

$$\boxed{J(x) = (x - \widehat{x})^{\mathsf{T}}A(x - \widehat{x}) + (\alpha - b^{\mathsf{T}}A^{-1}b)} \tag{1.80}$$

where $\widehat{x} = A^{-1}b$. Decomposition (1.80) is referred to as a "sum-of-squares" expression since it is the sum of two positive terms.

Matrix inversion formula

For matrices of compatible dimensions, and invertible A, it holds that

$$\boxed{(A + BCD)^{-1} = A^{-1} - A^{-1}B(C^{-1} + DA^{-1}B)^{-1}DA^{-1}} \tag{1.81}$$

This is a useful matrix identity that shows how the inverse of a matrix A is modified when it is perturbed by a product, BCD. The validity of the expression can be readily checked by multiplying both sides by $A + BCD$.

1.5 CHOLESKY FACTORIZATION

The Cholesky factorization of a positive-definite matrix is a useful computational tool and it can be motivated by means of the Schur decomposition results discussed above. Thus, consider an $M \times M$ symmetric positive-definite matrix A and partition it in the following manner:

$$A = \begin{bmatrix} \alpha & b^{\mathsf{T}} \\ b & D \end{bmatrix} \tag{1.82}$$

where α is its leading diagonal entry, b is an $(M-1)\times 1$ column vector, and D has dimensions $(M-1)\times(M-1)$. The positive-definiteness of A guarantees

$\alpha > 0$ and $D > 0$. Using (1.72a), let us consider the following block factorization for A:

$$A = \begin{bmatrix} 1 & 0 \\ b/\alpha & I_{M-1} \end{bmatrix} \begin{bmatrix} \alpha & 0 \\ 0 & \Delta_\alpha \end{bmatrix} \begin{bmatrix} 1 & b^\mathsf{T}/\alpha \\ 0 & I_{M-1} \end{bmatrix} \tag{1.83a}$$

$$\Delta_\alpha = D - bb^\mathsf{T}/\alpha \tag{1.83b}$$

We can rewrite the factorization more compactly in the form:

$$A = \mathcal{L}_0 \begin{bmatrix} d(0) & \\ & \Delta_0 \end{bmatrix} \mathcal{L}_0^\mathsf{T} \tag{1.84}$$

where $\mathcal{L}_0$ is the lower-triangular matrix

$$\mathcal{L}_0 \stackrel{\Delta}{=} \begin{bmatrix} 1 & 0 \\ b/\alpha & I_{M-1} \end{bmatrix} \tag{1.85}$$

and $d(0) = \alpha$, $\Delta_0 = \Delta_\alpha$. Observe that the first column of $\mathcal{L}_0$ is the first column of A normalized by the inverse of its leading diagonal entry. Moreover, the positive-definiteness of A guarantees $d(0) > 0$ and $\Delta_0 > 0$.

Expression (1.84) provides a factorization for A that consists of a lower-triangular matrix $\mathcal{L}_0$ followed by a *block-diagonal* matrix and an upper-triangular matrix. Now since Δ_0 is itself positive-definite, we can repeat the construction and introduce a similar factorization for it, which we denote by

$$\Delta_0 = L_1 \begin{bmatrix} d(1) & \\ & \Delta_1 \end{bmatrix} L_1^\mathsf{T} \tag{1.86}$$

for some lower-triangular matrix L_1 and where $d(1)$ is the leading diagonal entry of Δ_0. Moreover, Δ_1 is the Schur complement relative to $d(1)$ in Δ_0, and its dimensions are $(M-2) \times (M-2)$. In addition, the first column of L_1 coincides with the first column of Δ_0 normalized by the inverse of its leading diagonal entry. Also, the positive-definiteness of Δ_0 guarantees $d(1) > 0$ and $\Delta_1 > 0$. Substituting the above factorization for Δ_0 into the factorization for A we get

$$A = \mathcal{L}_0 \begin{bmatrix} 1 & \\ & L_1 \end{bmatrix} \begin{bmatrix} d(0) & & \\ & d(1) & \\ & & \Delta_1 \end{bmatrix} \begin{bmatrix} 1 & \\ & L_1^\mathsf{T} \end{bmatrix} \mathcal{L}_0^\mathsf{T} \tag{1.87}$$

But since the product of two lower-triangular matrices is also lower-triangular, we conclude that the product

$$\mathcal{L}_1 \stackrel{\Delta}{=} \mathcal{L}_0 \begin{bmatrix} 1 & \\ & L_1 \end{bmatrix}$$

is lower-triangular and we denote it by $\mathcal{L}_1$. Using this notation, we write instead

$$A = \mathcal{L}_1 \begin{bmatrix} d(0) & & \\ & d(1) & \\ & & \Delta_1 \end{bmatrix} \mathcal{L}_1^\mathsf{T} \tag{1.88}$$

Clearly, the first column of $\mathcal{L}_1$ is the first column of $\mathcal{L}_0$ and the second column of $\mathcal{L}_1$ is formed from the first column of L_1.

We can proceed to factor Δ_1, which would lead to an expression of the form

$$A = \mathcal{L}_2 \begin{bmatrix} d(0) & & & \\ & d(1) & & \\ & & d(2) & \\ & & & \Delta_2 \end{bmatrix} \mathcal{L}_2^{\mathsf{T}} \tag{1.89}$$

where $d(2) > 0$ is the $(0,0)$ entry of Δ_1 and $\Delta_2 > 0$ is the Schur complement of $d(2)$ in Δ_1. Continuing in this fashion we arrive after $(M-1)$ Schur complementation steps at a factorization for A of the form

$$A = \mathcal{L}_{M-1}\mathcal{D}\mathcal{L}_{M-1}^{\mathsf{T}} \tag{1.90}$$

where $\mathcal{L}_{M-1}$ is $M \times M$ lower-triangular and $\mathcal{D}$ is $M \times M$ diagonal with positive entries $\{d(m)\}$. The columns of $\mathcal{L}_{M-1}$ are the successive leading columns of the Schur complements $\{\Delta_m\}$, normalized by the inverses of their leading diagonal entries. The diagonal entries of $\mathcal{D}$ coincide with these leading entries.

If we define $\bar{L} \overset{\Delta}{=} \mathcal{L}_{n-1}\mathcal{D}^{1/2}$, where $\mathcal{D}^{1/2}$ is a diagonal matrix with the positive square-roots of the $\{d(m)\}$, we obtain

$$\boxed{A = \bar{L}\bar{L}^{\mathsf{T}}} \qquad \text{(\textbf{lower-upper triangular factorization})} \tag{1.91}$$

In summary, this constructive argument shows that every positive-definite matrix can be factored as the product of a lower-triangular matrix with positive diagonal entries by its transpose. This factorization is known as the *Cholesky* factorization of A. Had we instead partitioned A as

$$A = \begin{bmatrix} B & b \\ b^{\mathsf{T}} & \beta \end{bmatrix} \tag{1.92}$$

where $\beta > 0$ is now a scalar, and had we used the block factorization (1.72b), we would have arrived at a similar factorization for A albeit one of the form:

$$\boxed{A = \bar{U}\bar{U}^{\mathsf{T}}} \qquad \text{(\textbf{upper-lower triangular factorization})} \tag{1.93}$$

where $\bar{U}$ is an upper-triangular matrix with positive diagonal entries. The two triangular factorizations are illustrated in Fig. 1.1.

LEMMA 1.2. **(Cholesky factorization)** *Every positive-definite matrix A admits a unique factorization of either form $A = \bar{L}\bar{L}^{\mathsf{T}} = \bar{U}\bar{U}^{\mathsf{T}}$, where $\bar{L}$ ($\bar{U}$) is a lower (upper)-triangular matrix with positive entries along its diagonal.*

Proof: The existence of the factorizations was proved prior to the statement of the lemma. It remains to establish uniqueness. We show this for one of the factorizations. A similar argument applies to the other factorization. Thus, assume that

$$A = \bar{L}_1\bar{L}_1^{\mathsf{T}} = \bar{L}_2\bar{L}_2^{\mathsf{T}} \tag{1.94}$$

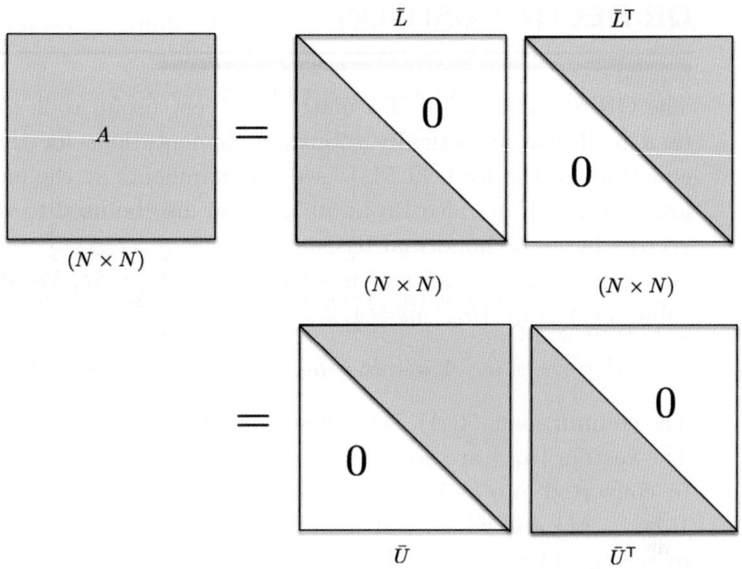

Figure 1.1 Two triangular factorizations for an $N \times N$ matrix A. The triangular factors $\{\bar{L}, \bar{U}\}$ are both $N \times N$ with positive entries on their main diagonal.

are two Cholesky factorizations for A. Then,

$$\bar{L}_2^{-1}\bar{L}_1 = \bar{L}_2^{\mathsf{T}}\bar{L}_1^{-\mathsf{T}} \tag{1.95}$$

where the compact notation $A^{-\mathsf{T}}$ stands for $[A^{\mathsf{T}}]^{-1}$. But since the inverse of a lower-triangular matrix is lower-triangular, and since the product of two lower-triangular matrices is also lower-triangular, we conclude that $\bar{L}_2^{-1}\bar{L}_1$ is lower-triangular. Likewise, the product $\bar{L}_2^{\mathsf{T}}\bar{L}_1^{-\mathsf{T}}$ is upper-triangular. Therefore, equality (1.95) will hold if, and only if, $\bar{L}_2^{-1}\bar{L}_1$ is diagonal, which means that

$$\bar{L}_1 = \bar{L}_2 D \tag{1.96}$$

for some diagonal matrix D. We want to show that D is the identity matrix. Indeed, it is easy to see from (1.94) that the $(0,0)$ entries of $\bar{L}_1$ and $\bar{L}_2$ must coincide so that the leading entry of D must be unity. This further implies from (1.96) that the first column of $\bar{L}_1$ should coincide with the first column of $\bar{L}_2$, so that using (1.94) again we conclude that the $(1,1)$ entries of $\bar{L}_1$ and $\bar{L}_2$ also coincide. Hence, the second entry of D is also unity. Proceeding in this fashion we conclude $D = I$.

■

Remark 1.1. (Triangular factorization) We also conclude from the discussion in this section that every positive-definite matrix A admits a unique factorization of either form $A = LDL^{\mathsf{T}} = UD_uU^{\mathsf{T}}$, where L (U) is a lower (upper)-triangular matrix with *unit* diagonal entries, and D and D_u are diagonal matrices with positive entries.

■

1.6 QR DECOMPOSITION

The QR decomposition of a matrix is a very useful tool; for example, it can be used to derive numerically robust implementations for the solution of normal equations of the form (1.51) – see the comments at the end of the chapter following (1.216) and also Prob. 50.5. It can also be used to replace a collection of vectors by an orthonormal basis.

Consider an $N \times M$ real matrix A with $N \geq M$. We denote the individual columns of A by $\{h_m,\ m = 1, 2, \ldots, M\}$:

$$A = \begin{bmatrix} h_1 & h_2 & \ldots & h_M \end{bmatrix}, \quad h_m \in \mathbb{R}^N \tag{1.97}$$

The column span, $\mathcal{R}(A)$, is the result of all linear combinations of these columns. The vectors $\{h_m\}$ are not generally orthogonal to each other. They can, however, be converted into an orthonormal set of vectors, which we denote by $\{q_m,\ m = 1, 2, \ldots, M\}$ and which will span the same $\mathcal{R}(A)$. This objective can be achieved by means of the *Gram–Schmidt procedure*. It is an iterative procedure that starts by setting

$$q_1 = h_1/\|h_1\|, \quad r_1 \stackrel{\Delta}{=} h_1 \tag{1.98}$$

and then repeats for $m = 2, \ldots, M$:

$$r_m = h_m - \sum_{j=1}^{m-1} \left(h_m^{\mathsf{T}} q_j\right) q_j \tag{1.99a}$$

$$q_m = r_m/\|r_m\| \tag{1.99b}$$

By iterating this construction, we end up expressing each column h_m as a linear combination of the vectors $\{q_1, q_2, \ldots, q_m\}$ as follows:

$$h_m = (h_m^{\mathsf{T}} q_1)q_1 + (h_m^{\mathsf{T}} q_2)q_2 + \ldots + (h_m^{\mathsf{T}} q_{m-1})q_{m-1} + \|r_m\|\, q_m \tag{1.100}$$

If we collect the coefficients of this linear combination, for all $m = 1, 2, \ldots, M$, into the columns of an $M \times M$ upper-triangular matrix R:

$$R \stackrel{\Delta}{=} \begin{bmatrix} \|r_1\| & h_2^{\mathsf{T}} q_1 & h_3^{\mathsf{T}} q_1 & \ldots & h_M^{\mathsf{T}} q_1 \\ & \|r_2\| & h_3^{\mathsf{T}} q_2 & \ldots & h_M^{\mathsf{T}} q_2 \\ & & \|r_3\| & & \vdots \\ & & & \ddots & h_M^{\mathsf{T}} q_{M-1} \\ & & & & \|r_M\| \end{bmatrix} \tag{1.101}$$

we conclude from (1.100) that

$$A = \widehat{Q} R \tag{1.102}$$

where $\widehat{Q}$ is the $N \times M$ matrix with orthonormal columns $\{q_m\}$, i.e.,

$$\widehat{Q} = \begin{bmatrix} q_1 & q_2 & \ldots & q_M \end{bmatrix} \tag{1.103}$$

with

$$q_m^{\mathsf{T}} q_s = \begin{cases} 0, & m \neq s \\ 1, & m = s \end{cases} \tag{1.104}$$

When A has full rank, i.e., when A has rank M, all the diagonal entries of R will be positive. The factorization $A = \widehat{Q}R$ is referred to as the *reduced* QR decomposition of A, and it simply amounts to the orthonormalization of the columns of A.

It is often more convenient to employ the *full* QR decomposition of $A \in \mathbb{R}^{N\times M}$, as opposed to its reduced decomposition. The full decomposition is obtained by appending $N - M$ orthonormal columns to $\widehat{Q}$ so that it becomes an orthogonal $N \times N$ (square) matrix Q. We also append rows of zeros to R so that (1.102) becomes (see Fig. 1.2):

$$\boxed{A = Q\begin{bmatrix} R \\ 0 \end{bmatrix}, \quad Q\,(N\times N),\ R\,(M\times M)} \tag{1.105}$$

where

$$Q = \begin{bmatrix} \widehat{Q} & q_{M+1} & \dots & q_N \end{bmatrix}, \quad Q^{\mathsf{T}}Q = I_N \tag{1.106}$$

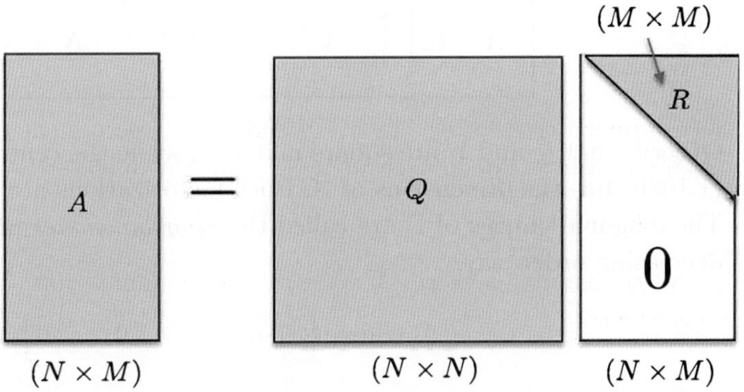

Figure 1.2 Full QR decomposition of an $N \times M$ matrix A, where Q is $N \times N$ orthogonal and R is $M \times M$ upper-triangular.

Example 1.7 (Cholesky and QR factorizations) Consider a full rank $N \times M$ matrix A with $N \geq M$ and introduce its QR decomposition (1.105). Then, $A^{\mathsf{T}}A$ is positive-definite and its Cholesky factorization is given by

$$A^{\mathsf{T}}A = R^{\mathsf{T}}R \tag{1.107}$$

1.7 SINGULAR VALUE DECOMPOSITION

The singular value decomposition (SVD) of a matrix is another powerful tool that is useful for both analytical and numerical purposes. It enables us to represent any matrix (square or not, invertible or not, symmetric or not) as the product of three matrices with special and desirable properties: two of the matrices are orthogonal and the third matrix is composed of a diagonal matrix and a zero block.

Definition

The SVD of a real matrix A states that if A is $N \times M$, then there exist an $N \times N$ orthogonal matrix U ($UU^{\mathsf{T}} = I_N$), an $M \times M$ orthogonal matrix V ($VV^{\mathsf{T}} = I_M$), and a diagonal matrix Σ with nonnegative entries such that:

(a) If $N \le M$, then Σ is $N \times N$ and

$$\boxed{A = U \left[\begin{array}{cc} \Sigma & 0 \end{array}\right] V^{\mathsf{T}}, \quad A \in \mathbb{R}^{N\times M}, \quad N \le M} \tag{1.108a}$$

(b) If $N \ge M$, then Σ is $M \times M$ and

$$\boxed{A = U \left[\begin{array}{c} \Sigma \\ 0 \end{array}\right] V^{\mathsf{T}}, \quad A \in \mathbb{R}^{N\times M}, \quad N \ge M} \tag{1.108b}$$

Observe that U and V are square matrices, while the central matrix in (1.108a)–(1.108b) has the dimensions of A. Both factorizations are illustrated in Fig. 1.3. The diagonal entries of Σ are called the *singular values* of A and are ordered in decreasing order, say,

$$\Sigma = \text{diag}\Big\{\sigma_1, \sigma_2, \ldots, \sigma_r, 0, \ldots, 0\Big\} \tag{1.109}$$

with

$$\sigma_1 \ge \sigma_2 \ge \ldots \ge \sigma_r > 0 \tag{1.110}$$

If Σ has r nonzero diagonal entries then A has rank r. The columns of U and V are called the left and right *singular vectors* of A, respectively. The ratio of the largest to smallest singular value of A is called the *condition number* of A and is denoted by

$$\kappa(A) \stackrel{\Delta}{=} \sigma_1/\sigma_r \tag{1.111}$$

One constructive proof for the SVD is given in Appendix 1.B.

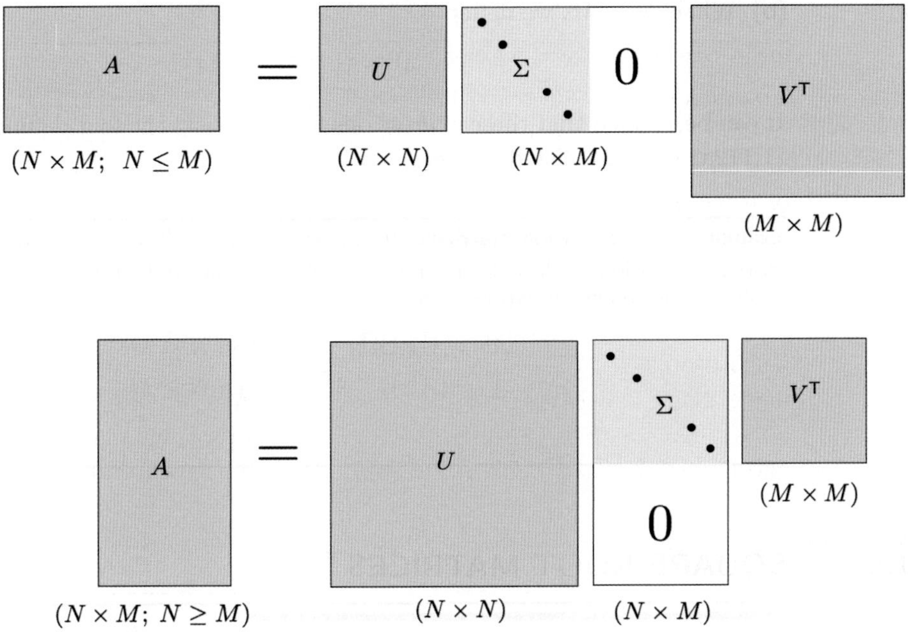

Figure 1.3 Singular value decompositions of an $N \times M$ matrix A for both cases when $N \geq M$ and $N \leq M$.

Pseudo inverses

The pseudo-inverse of a matrix is a generalization of the concept of inverses for square invertible matrices; it is defined for matrices that need not be invertible or even square.

Given an $N \times M$ matrix A of rank r, its pseudo-inverse is defined as the unique $M \times N$ matrix $A^\dagger$ that satisfies the following four requirements:

$$\textbf{(i)}\ AA^\dagger A = A \tag{1.112a}$$

$$\textbf{(ii)}\ A^\dagger A A^\dagger = A^\dagger \tag{1.112b}$$

$$\textbf{(iii)}\ (AA^\dagger)^{\mathsf{T}} = AA^\dagger \tag{1.112c}$$

$$\textbf{(iv)}\ (A^\dagger A)^{\mathsf{T}} = A^\dagger A \tag{1.112d}$$

The SVD of A can be used to determine its pseudo-inverse as follows. Introduce the matrix

$$\Sigma^\dagger = \text{diagonal}\left\{\sigma_1^{-1}, \sigma_2^{-1}, \ldots, \sigma_r^{-1}, 0, \ldots, 0\right\} \tag{1.113}$$

That is, we invert the nonzero entries of Σ and keep the zero entries unchanged.

(a) When $N \leq M$, we define

$$A^\dagger = V \begin{bmatrix} \Sigma^\dagger \\ 0 \end{bmatrix} U^{\mathsf{T}} \tag{1.114}$$

(b) When $N \geq M$, we define

$$A^\dagger = V \begin{bmatrix} \Sigma^\dagger & 0 \end{bmatrix} U^\mathsf{T} \tag{1.115}$$

It can be verified that these expressions for $A^\dagger$ satisfy the four defining properties (1.112a)–(1.112d) listed above.

Example 1.8 **(Full rank matrices)** It can also be verified, by replacing A by its SVD in the expressions below, that when $A \in \mathbb{R}^{N\times M}$ has *full rank*, its pseudo-inverse is given by the following expressions:

$$A^\dagger = A^\mathsf{T}(AA^\mathsf{T})^{-1}, \quad \text{when } N \leq M \tag{1.116a}$$
$$A^\dagger = (A^\mathsf{T}A)^{-1}A^\mathsf{T}, \quad \text{when } N \geq M \tag{1.116b}$$

1.8 SQUARE-ROOT MATRICES

One useful concept in matrix analysis is that of the *square-root matrix*. Although square-roots can be defined for nonnegative-definite matrices, it is sufficient for our purposes to focus on positive-definite matrices. Thus, consider an $N \times N$ positive-definite matrix A and introduce its eigen-decomposition

$$A = U\Lambda U^\mathsf{T} \tag{1.117}$$

where Λ is an $N \times N$ diagonal with positive entries and U is $N \times N$ orthogonal:

$$UU^\mathsf{T} = U^\mathsf{T}U = I_N \tag{1.118}$$

Let $\Lambda^{1/2}$ denote the diagonal matrix whose entries are the positive square-roots of the diagonal entries of Λ. Then, we can rewrite (1.117) as

$$A = \left(U\Lambda^{1/2}\right)\left(U\Lambda^{1/2}\right)^\mathsf{T} \tag{1.119}$$

which expresses A as the product of an $N \times N$ matrix and its transpose, namely,

$$A = XX^\mathsf{T}, \quad \text{with } X = U\Lambda^{1/2} \tag{1.120}$$

We say that X is *a* square-root for A.

Definition 1.1 **(Square-root factors)** *A square-root of an $N \times N$ positive-definite matrix A is any $N \times N$ matrix X satisfying $A = XX^\mathsf{T}$. The square-root is said to be symmetric if $X = X^\mathsf{T}$ in which case $A = X^2$.*

The construction prior to the definition exhibits one possible choice for X, namely, $X = U\Lambda^{1/2}$, in terms of the eigenvectors and eigenvalues of A. However, square-root factors are not unique. If we consider the above X and multiply it by any orthogonal matrix Θ, say, $\bar{X} = X\Theta$ where $\Theta\Theta^{\mathsf{T}} = I_N$, then $\bar{X}$ is also a square-root factor for A since

$$\bar{X}\bar{X}^{\mathsf{T}} = X\underbrace{\Theta\Theta^{\mathsf{T}}}_{=I}X^{\mathsf{T}} = XX^{\mathsf{T}} = A \tag{1.121}$$

In particular, for the same matrix $A = U\Lambda U^{\mathsf{T}}$, the matrix $X = U\Lambda^{1/2}U^{\mathsf{T}}$ is also a square-root for A. And this particular square-root factor happens to be symmetric. This argument shows that every symmetric positive-definite matrix A admits a *symmetric* square-root factor for which we can write $A = X^2$.

Notation

It is customary to use the notation $A^{1/2}$ to refer to a square-root of a matrix A and, therefore, we write

$$A = A^{1/2}\left(A^{1/2}\right)^{\mathsf{T}} \tag{1.122a}$$

It is also customary to employ the compact notation

$$A^{\mathsf{T}/2} \triangleq \left(A^{1/2}\right)^{\mathsf{T}}, \quad A^{-1/2} \triangleq \left(A^{1/2}\right)^{-1}, \quad A^{-\mathsf{T}/2} \triangleq \left(A^{1/2}\right)^{-\mathsf{T}} \tag{1.122b}$$

so that

$$\boxed{A = A^{1/2}A^{\mathsf{T}/2}, \qquad A^{-1} = A^{-\mathsf{T}/2}A^{-1/2}} \tag{1.122c}$$

Cholesky factor

One of the most widely used square-root factors of a positive-definite matrix is its Cholesky factor. Recall that we showed in Section 1.5 that every positive-definite matrix A admits a *unique* triangular factorization of the form $A = \bar{L}\bar{L}^{\mathsf{T}}$, where $\bar{L}$ is a lower-triangular matrix with positive entries on its diagonal. We could also consider the alternative factorization $A = \bar{U}\bar{U}^{\mathsf{T}}$ in terms of an upper-triangular matrix $\bar{U}$. Comparing these forms with the defining relation $A = XX^{\mathsf{T}}$, we conclude that $\bar{L}$ and $\bar{U}$ are valid choices for square-root factors of A. When one refers to the square-root of a matrix, it generally means its (lower- or upper-triangular) Cholesky factor. This choice has two advantages in relation to other square-root factors: it is triangular and is uniquely defined (i.e., there is no other triangular square-root factor with positive diagonal entries).

Example 1.9 **(Basis rotation)** The following is an important matrix result that is critical to the development of algorithms that compute and propagate square-root factors. Consider two $N \times M$ ($N \le M$) matrices A and B. Then $AA^{\mathsf{T}} = BB^{\mathsf{T}}$ if, and only if, there exists an $M \times M$ orthogonal matrix Θ such that $A = B\Theta$.

Proof: One direction is obvious. If $A = B\Theta$, for some orthogonal matrix Θ, then

$$AA^{\mathsf{T}} = (B\Theta)(B\Theta)^{\mathsf{T}} = B(\Theta\Theta^{\mathsf{T}})B^{\mathsf{T}} = BB^{\mathsf{T}} \tag{1.123}$$

One proof for the converse implication follows by using the SVDs of A and B:

$$A = U_A \begin{bmatrix} \Sigma_A & 0 \end{bmatrix} V_A^\mathsf{T}, \quad B = U_B \begin{bmatrix} \Sigma_B & 0 \end{bmatrix} V_B^\mathsf{T} \tag{1.124}$$

where U_A and U_B are $N \times N$ orthogonal matrices, V_A and V_B are $M \times M$ orthogonal matrices, and Σ_A and Σ_B are $N \times N$ diagonal matrices with nonnegative entries. The squares of the diagonal entries of Σ_A (Σ_B) are the eigenvalues of AA^T (BB^T). Moreover, U_A (U_B) are constructed from an orthonormal basis for the right eigenvectors of AA^T (BB^T). Hence, it follows from the identity $AA^\mathsf{T} = BB^\mathsf{T}$ that $\Sigma_A = \Sigma_B$ and $U_A = U_B$. Let $\Theta = V_B V_A^\mathsf{T}$. Then, it holds that $\Theta\Theta^\mathsf{T} = I$ and $B\Theta = A$.

■

1.9 KRONECKER PRODUCTS

Let $A = [a_{ij}]_{i,j=1}^N$ and $B = [b_{ij}]_{i,j=1}^M$ be $N \times N$ and $M \times M$ real-valued matrices, respectively, whose individual (i, j)th entries are denoted by a_{ij} and b_{ij}. Their Kronecker product is denoted by $\mathcal{K} = A \otimes B$ and is defined as the $NM \times NM$ matrix whose entries are given by:

$$\mathcal{K} \triangleq A \otimes B = \begin{bmatrix} a_{11}B & a_{12}B & \ldots & a_{1N}B \\ a_{21}B & a_{22}B & \ldots & a_{2N}B \\ \vdots & & \vdots & \\ a_{N1}B & a_{N2}B & \ldots & a_{NN}B \end{bmatrix} \tag{1.125}$$

In other words, each scalar entry a_{ij} of A is replaced by a block quantity that is equal to a scaled multiple of B, namely, $a_{ij}B$.

1.9.1 Properties

Let $\{\lambda_i(A), i = 1, \ldots, N\}$ and $\{\lambda_j(B), j = 1, \ldots, M\}$ denote the eigenvalues of A and B, respectively. Then, the eigenvalues of $A \otimes B$ will consist of all nm product combinations $\{\lambda_i(A)\lambda_j(B)\}$. A similar conclusion holds for the singular values of $A \otimes B$ in relation to the singular values of the individual matrices A and B, which we denote by $\{\sigma_i(A), \sigma_j(B)\}$. Table 1.1 lists several well-known properties of Kronecker products for matrices $\{A, B, C, D\}$ of compatible dimensions and column vectors $\{x, y\}$. The last four properties involve the trace and vec operations: the trace of a square matrix is the sum of its diagonal elements, while the vec operation transforms a matrix into a vector by stacking the columns of the matrix on top of each other.

Example 1.10 **(Derivation of select properties)** Property (2) in Table 1.1 follows by direct calculation from the definition of Kronecker products. Property (4) follows by using property (2) to note that

$$(A \otimes B)(A^{-1} \otimes B^{-1}) = I_N \otimes I_M = I_{NM} \tag{1.126}$$

Table 1.1 Properties for the Kronecker product (1.125).

	Relation	Property
1.	$(A+B)\otimes C = (A\otimes C) + (B\otimes C)$	distributive property
2.	$(A\otimes B)(C\otimes D) = (AC\otimes BD)$	multiplication property
3.	$(A\otimes B)^{\mathsf{T}} = A^{\mathsf{T}}\otimes B^{\mathsf{T}}$	transposition property
4.	$(A\otimes B)^{-1} = A^{-1}\otimes B^{-1}$	inversion property
5.	$(A\otimes B)^{\ell} = A^{\ell}\otimes B^{\ell}$, integer ℓ	exponentiation property
6.	$\{\lambda(A\otimes B)\} = \{\lambda_i(A)\lambda_j(B)\}_{i=1,j=1}^{N,M}$	eigenvalues
7.	$\{\sigma(A\otimes B)\} = \{\sigma_i(A)\sigma_j(B)\}_{i=1,j=1}^{N,M}$	singular values
8.	$\det(A\otimes B) = (\det A)^M(\det B)^N$	determinant property
9.	$\mathrm{Tr}(A\otimes B) = \mathrm{Tr}(A)\mathrm{Tr}(B)$	trace of Kronecher product
10.	$\mathrm{Tr}(AB) = \left(\mathrm{vec}(B^{\mathsf{T}})\right)^{\mathsf{T}}\mathrm{vec}(A)$	trace of matrix product
11.	$\mathrm{vec}(ACB) = (B^{\mathsf{T}}\otimes A)\mathrm{vec}(C)$	vectorization property
12.	$\mathrm{vec}(xy^{\mathsf{T}}) = y\otimes x$	vectorization of outer product

Property (6) follows from property (2) by choosing C as a right eigenvector for A and D as a right eigenvector for B, say, $C = q_i$ and $D = p_j$ where

$$Aq_i = \lambda_i(A)q_i, \quad Bp_j = \lambda_j(B)p_j \tag{1.127}$$

Then,

$$(A\otimes B)(q_i\otimes p_j) = \lambda_i(A)\lambda_j(B)(q_i\otimes p_j) \tag{1.128}$$

which shows that $(q_i\otimes p_j)$ is an eigenvector of $(A\otimes B)$ with eigenvalue $\lambda_i(A)\lambda_j(B)$. Property (9) follows from property (6) for square matrices since

$$\mathrm{Tr}(A) = \sum_{i=1}^{N}\lambda_i(A), \quad \mathrm{Tr}(B) = \sum_{j=1}^{M}\lambda_j(B) \tag{1.129}$$

and, therefore,

$$\begin{aligned}\mathrm{Tr}(A)\mathrm{Tr}(B) &= \left(\sum_{i=1}^{N}\lambda_i(A)\right)\left(\sum_{j=1}^{M}\lambda_j(B)\right)\\ &= \sum_{i=1}^{N}\sum_{j=1}^{M}\lambda_i(A)\lambda_j(B)\\ &= \mathrm{Tr}(A\otimes B)\end{aligned} \tag{1.130}$$

Property (8) also follows from property (6) since

$$\begin{aligned}\det(A\otimes B) &= \prod_{i=1}^{N}\prod_{j=1}^{M}\lambda_i(A)\lambda_j(B)\\ &= \left(\prod_{i=1}^{N}\lambda_i(A)\right)^M\left(\prod_{j=1}^{M}\lambda_j(B)\right)^N\\ &= (\det A)^M(\det B)^N\end{aligned} \tag{1.131}$$

Property (11) follows from the definition of Kronecker products and from noting that, for any two column vectors x and y, the vec representation of the rank one matrix xy^{T} is $y \otimes x$, i.e., $\text{vec}(xy^{\mathsf{T}}) = y \otimes x$, which is property (12). Finally, property (3) follows from the definition of Kronecker products.

Example 1.11 **(Discrete-time Lyapunov equations)** Consider $N \times N$ matrices X, A, and Q, where Q is symmetric and nonnegative-definite. The matrix X is said to satisfy a discrete-time Lyapunov equation, also called a Stein equation, if

$$\boxed{X - A^{\mathsf{T}} X A = Q} \tag{1.132}$$

Let $\lambda_k(A)$ denote any of the eigenvalues of A. We say that A is a *stable* matrix when all of its eigenvalues lie strictly inside the unit disc (i.e., their magnitudes are strictly less than 1). Using properties of the Kronecker product, it can be verified that the following important facts hold:

(a) The solution X of (1.132) is unique if, and only if, $\lambda_k(A)\lambda_\ell(A) \neq 1$ for all $k, \ell = 1, 2, \ldots, N$. In this case, the unique solution X is symmetric.

(b) When A is stable, the solution X is unique, symmetric, and nonnegative-definite. Moreover, it admits the series representation:

$$X = \sum_{n=0}^{\infty} (A^{\mathsf{T}})^n Q A^n \tag{1.133}$$

Proof: We call upon property (11) from Table 1.1 and apply the vec operation to both sides of (1.132) to get

$$(I - A^{\mathsf{T}} \otimes A^{\mathsf{T}})\text{vec}(X) = \text{vec}(Q) \tag{1.134}$$

This linear system of equations has a unique solution, $\text{vec}(X)$, if, and only if, the coefficient matrix, $I - A^{\mathsf{T}} \otimes A^{\mathsf{T}}$, is nonsingular. This latter condition requires $\lambda_k(A)\lambda_\ell(A) \neq 1$ for all $k, \ell = 1, 2, \ldots, N$. When A is stable, all of its eigenvalues lie strictly inside the unit disc and this uniqueness condition is automatically satisfied. If we transpose both sides of (1.132) we find that X^{T} satisfies the same Lyapunov equation as X and, hence, by uniqueness, we must have $X = X^{\mathsf{T}}$. Finally, let $F = A^{\mathsf{T}} \otimes A^{\mathsf{T}}$. When A is stable, the matrix F is also stable by property (6) from Table 1.1. In this case, the matrix inverse $(I - F)^{-1}$ admits the series expansion

$$(I - F)^{-1} = I + F + F^2 + F^3 + \ldots \tag{1.135}$$

so that using (1.134) we have

$$\begin{aligned}
\text{vec}(X) &= (I - F)^{-1}\text{vec}(Q) \\
&= \sum_{n=0}^{\infty} F^n \,\text{vec}(Q) \\
&= \sum_{n=0}^{\infty} \Big((A^{\mathsf{T}})^n \otimes (A^{\mathsf{T}})^n\Big) \,\text{vec}(Q) \\
&= \sum_{n=0}^{\infty} \text{vec}\,\Big((A^{\mathsf{T}})^n Q A^n\Big)
\end{aligned} \tag{1.136}$$

from which we deduce the series representation (1.133). The last equality in (1.136) follows from property (11) in Table 1.1.

Example 1.12 **(Continuous-time Lyapunov equations)** We extend the analysis of Example 1.11 to the following continuous-time Lyapunov equation (also called a Sylvester equation):

$$\boxed{XA^{\mathsf{T}} + AX + Q = 0} \tag{1.137}$$

where Q continues to be symmetric and nonnegative-definite. In the continuous-time case, a stable matrix A is one whose eigenvalues lie in the open left-half plane (i.e., they have strictly negative real parts). The following facts hold:

(a) The solution X of (1.137) is unique if, and only if, $\lambda_k(A) + \lambda_\ell(A) \neq 0$ for all $k, \ell = 1, 2, \ldots, N$. In this case, the unique solution X is symmetric.

(b) When A is stable (i.e., all its eigenvalues lie in the open left-half plane), the solution X is unique, symmetric, and nonnegative-definite. Moreover, it admits the integral representation

$$X = \int_0^\infty e^{At} Q e^{A^{\mathsf{T}} t} dt \tag{1.138}$$

where the notation e^{At} refers to the matrix exponential function evaluated at At. By definition, this function is equal to the following series representation:

$$e^{At} = I_N + \frac{1}{1!}At + \frac{1}{2!}A^2t^2 + \frac{1}{3!}A^3t^3 + \ldots \tag{1.139}$$

Proof: We use property (11) from Table 1.1 and apply the vec operation to both sides of (1.137) to get

$$\Big[(A^{\mathsf{T}} \otimes I) + (I \otimes A)\Big] \operatorname{vec}(X) = -\operatorname{vec}(Q) \tag{1.140}$$

This linear system of equations has a unique solution, $\operatorname{vec}(X)$, if, and only if, the coefficient matrix, $(A^{\mathsf{T}} \otimes I) + (I \otimes A)$, is nonsingular. This latter condition requires $\lambda_k(A) + \lambda_\ell(A) \neq 0$ for all $k, \ell = 1, 2, \ldots, N$. To see this, let $F = (A^{\mathsf{T}} \otimes I) + (I \otimes A)$ and let us verify that the eigenvalues of F are given by all linear combinations $\lambda_k(A) + \lambda_\ell(A)$. Consider the eigenvalue–eigenvector pairs $Ax_k = \lambda_k(A)x_k$ and $A^{\mathsf{T}} y_\ell = \lambda_\ell(A) y_\ell$. Then, using property (2) from Table 1.1 for Kronecker products we get

$$\begin{aligned} F(y_\ell \otimes x_k) &= \Big[(A^{\mathsf{T}} \otimes I) + (I \otimes A)\Big](y_\ell \otimes x_k) \\ &= (A^{\mathsf{T}} y_\ell \otimes x_k) + (y_\ell \otimes Ax_k) \\ &= \lambda_\ell(A)(y_\ell \otimes x_k) + \lambda_k(A)(y_\ell \otimes x_k) \\ &= (\lambda_k(A) + \lambda_\ell(A))(y_\ell \otimes x_k) \end{aligned} \tag{1.141}$$

so that the vector $(y_\ell \otimes x_k)$ is an eigenvector for F with eigenvalue $\lambda_k(A) + \lambda_\ell(A)$, as claimed. If we now transpose both sides of (1.137) we find that X^{T} satisfies the same Lyapunov equation as X and, hence, by uniqueness, we must have $X = X^{\mathsf{T}}$. Moreover, it follows from the integral representation (1.138) that X is nonnegative-definite since $Q \geq 0$ and $e^{A^{\mathsf{T}} t} = (e^{At})^{\mathsf{T}}$. To establish the integral representation we verify that it satisfies the Sylvester equation (1.137) so that, by uniqueness, the solution X should agree with it. Thus, let

$$Y \triangleq \int_0^\infty e^{At} Q e^{A^{\mathsf{T}} t} dt \tag{1.142}$$

and note that (refer to Prob. 1.20):

$$\begin{aligned} AY + YA^{\mathsf{T}} &= \int_0^\infty \left(Ae^{At}Qe^{A^{\mathsf{T}}t} + e^{At}Qe^{A^{\mathsf{T}}t}A^{\mathsf{T}} \right) dt \\ &= \int_0^\infty \frac{d}{dt}\left(e^{At}Qe^{A^{\mathsf{T}}t} \right) dt \\ &= e^{At}Qe^{A^{\mathsf{T}}t}\Big|_{t=0}^{\infty} \\ &= -Q \end{aligned} \tag{1.143}$$

so that $AY + YA^{\mathsf{T}} + Q = 0$ and Y satisfies the same Sylvester equation as X. By uniqueness, we conclude that the integral representation (1.138) holds.

1.9.2 Block Kronecker Products

Let $\mathcal{A}$ now denote a *block* matrix of size $NP \times NP$ with each block having size $P \times P$. We denote the (i,j)th block of $\mathcal{A}$ by the notation A_{ij}; it is a matrix of size $P \times P$. Likewise, let $\mathcal{B}$ denote a second *block* matrix of size $MP \times MP$ with each of its blocks having the same size $P \times P$. We denote the (i,j)th block of $\mathcal{B}$ by the notation B_{ij}; it is a matrix of size $P \times P$:

$$\mathcal{A} = \begin{bmatrix} A_{11} & A_{12} & \dots & A_{1N} \\ A_{21} & A_{22} & \dots & A_{2N} \\ \vdots & & & \vdots \\ A_{N1} & A_{N2} & \dots & A_{NN} \end{bmatrix} \tag{1.144a}$$

$$\mathcal{B} = \begin{bmatrix} B_{11} & B_{12} & \dots & B_{1M} \\ B_{21} & B_{22} & \dots & B_{2M} \\ \vdots & & & \vdots \\ B_{M1} & B_{M2} & \dots & B_{MM} \end{bmatrix} \tag{1.144b}$$

The *block* Kronecker product of these two matrices is denoted by $\mathcal{K} = \mathcal{A} \otimes_b \mathcal{B}$ and is defined as the following block matrix of dimensions $NMP^2 \times NMP^2$:

$$\mathcal{K} \stackrel{\Delta}{=} \mathcal{A} \otimes_b \mathcal{B} = \begin{bmatrix} K_{11} & K_{12} & \dots & K_{1N} \\ K_{21} & K_{22} & \dots & K_{2N} \\ \vdots & \vdots & \ddots & \vdots \\ K_{N1} & K_{N2} & \dots & K_{NN} \end{bmatrix} \tag{1.145}$$

where each block K_{ij} is $MP^2 \times MP^2$ and is constructed as follows:

$$K_{ij} = \begin{bmatrix} A_{ij} \otimes B_{11} & A_{ij} \otimes B_{12} & \dots & A_{ij} \otimes B_{1M} \\ A_{ij} \otimes B_{21} & A_{ij} \otimes B_{22} & \dots & A_{ij} \otimes B_{2M} \\ \vdots & \vdots & \ddots & \vdots \\ A_{ij} \otimes B_{M1} & A_{ij} \otimes B_{M2} & \dots & A_{ij} \otimes B_{MM} \end{bmatrix} \tag{1.146}$$

Table 1.2 lists some useful properties of block Kronecker products for matrices $\{\mathcal{A}, \mathcal{B}, \mathcal{C}, \mathcal{D}\}$ with blocks of size $P \times P$. The last three properties involve the block vectorization operation denoted by *bvec*: it vectorizes each block entry of the matrix and then stacks the resulting columns on top of each other, i.e.,

$$
\text{bvec}(\mathcal{A}) \triangleq \left[\begin{array}{c} \text{vec}(A_{11}) \\ \vdots \\ \text{vec}(A_{N1}) \\ \hline \text{vec}(A_{21}) \\ \vdots \\ \text{vec}(A_{N2}) \\ \hline \vdots \\ \hline \text{vec}(A_{1N}) \\ \vdots \\ \text{vec}(A_{NN}) \end{array}\right] \begin{array}{l} \left.\begin{array}{c} \\ \\ \\ \end{array}\right\} \text{first block column of } \mathcal{A} \\ \\ \\ \\ \\ \\ \left.\begin{array}{c} \\ \\ \\ \end{array}\right\} \text{last block column of } \mathcal{A} \end{array} \tag{1.147}
$$

Expression (1.148) illustrates one of the advantages of working with the bvec operation for block matrices. It compares the effect of the block vectorization operation to that of the regular vec operation. It is seen that bvec preserves the locality of the blocks from the original matrix: Entries arising from the same block appear together followed by entries from other blocks. In contrast, in the regular vec construction, entries from different blocks are mixed together.

$$
\left[\begin{array}{c} \circ \\ \circ \\ \square \\ \square \\ \hline \bullet \\ \bullet \\ \blacksquare \\ \blacksquare \\ \hline \triangle \\ \triangle \\ \star \\ \star \\ \hline \blacktriangle \\ \blacktriangle \\ \bigstar \\ \bigstar \end{array}\right] \overset{\text{vec}(\mathcal{A})}{\Longleftarrow} \underbrace{\left[\begin{array}{cc|cc} \circ & \bullet & \triangle & \blacktriangle \\ \circ & \bullet & \triangle & \blacktriangle \\ \hline \square & \blacksquare & \star & \bigstar \\ \square & \blacksquare & \star & \bigstar \end{array}\right]}_{=\mathcal{A}} \overset{\text{bvec}(\mathcal{A})}{\Longrightarrow} \left[\begin{array}{c} \circ \\ \circ \\ \bullet \\ \bullet \\ \hline \square \\ \square \\ \blacksquare \\ \blacksquare \\ \hline \triangle \\ \triangle \\ \blacktriangle \\ \blacktriangle \\ \hline \star \\ \star \\ \bigstar \\ \bigstar \end{array}\right] \tag{1.148}
$$

Table 1.2 Properties for the block Kronecker product (1.145).

	Relation	Property
1.	$(\mathcal{A}+\mathcal{B})\otimes_b \mathcal{C} = (\mathcal{A}\otimes_b \mathcal{C}) + (\mathcal{B}\otimes_b \mathcal{C})$	distributive property
2.	$(\mathcal{A}\otimes_b \mathcal{B})(\mathcal{C}\otimes_b \mathcal{D}) = (\mathcal{A}\mathcal{C}\otimes_b \mathcal{B}\mathcal{D})$	multiplication property
3.	$(\mathcal{A}\otimes_b \mathcal{B})^{\mathsf{T}} = \mathcal{A}^{\mathsf{T}}\otimes_b \mathcal{B}^{\mathsf{T}}$	transposition property
4.	$\{\lambda(\mathcal{A}\otimes_b \mathcal{B})\} = \{\lambda_i(\mathcal{A})\lambda_j(\mathcal{B})\}_{i=1,j=1}^{NP,MP}$	eigenvalues
5.	$\mathrm{Tr}(\mathcal{A}\mathcal{B}) = \left(\mathrm{bvec}(\mathcal{B}^{\mathsf{T}})\right)^{\mathsf{T}}\mathrm{bvec}(\mathcal{A})$	trace of matrix product
6.	$\mathrm{bvec}(\mathcal{A}\mathcal{C}\mathcal{B}) = (\mathcal{B}^{\mathsf{T}}\otimes_b \mathcal{A})\mathrm{bvec}(\mathcal{C})$	block vectorization property
7.	$\mathrm{bvec}(xy^{\mathsf{T}}) = y\otimes_b x$	vectorization of outer product

1.10 VECTOR AND MATRIX NORMS

We list in this section several useful vector and matrix norms, which will arise regularly in studies of inference and learning methods.

Definition of norms

If we let X denote an arbitrary real matrix or vector quantity, then a matrix or vector norm, denoted by $\|X\|$, is any function that satisfies the following properties, for any X and Y of compatible dimensions and scalar α:

$$\|X\| \geq 0 \tag{1.149a}$$
$$\|X\| = 0 \text{ if, and only if, } X = 0 \tag{1.149b}$$
$$\|\alpha X\| = |\alpha|\,\|X\| \tag{1.149c}$$
$$\|X+Y\| \leq \|X\| + \|X\| \quad (\textbf{triangle inequality}) \tag{1.149d}$$
$$\|XY\| \leq \|X\|\,\|Y\| \quad (\textbf{sub-multiplicative property}) \tag{1.149e}$$

For any column vector $x \in \mathbb{R}^N$ with individual entries $\{x_n\}$, any of the definitions listed in Table 1.3 constitutes a valid vector norm.

It is straightforward to verify the validity of the following inequalities relating these norms:

$$\|x\|_2 \leq \|x\|_1 \leq \sqrt{N}\,\|x\|_2 \tag{1.150a}$$
$$\|x\|_\infty \leq \|x\|_2 \leq \sqrt{N}\,\|x\|_\infty \tag{1.150b}$$
$$\|x\|_\infty \leq \|x\|_1 \leq N\,\|x\|_\infty \tag{1.150c}$$
$$\|x\|_\infty \leq \|x\|_2 \leq \|x\|_1 \tag{1.150d}$$
$$\|x\|_\infty \leq \|x\|_p \leq N^{1/p}\,\|x\|_\infty, \quad p \geq 1 \tag{1.150e}$$

There are similarly many useful matrix norms. For any matrix $A \in \mathbb{R}^{N\times M}$ with individual entries $\{a_{\ell k}\}$, any of the definitions listed in Table 1.4 constitutes a valid matrix norm. In particular, the 2-induced norm of A is a special case of the p-induced norm and reduces to the maximum singular value of A – see

Table 1.3 Useful vector norms, where the $\{x_n\}$ denote the entries of the vector x.

Vector norm	Name
$\|x\|_1 \triangleq \sum_{n=1}^{N} \|x_n\|$	(1-norm or ℓ_1-norm)
$\|x\|_\infty \triangleq \max_{1\leq n\leq N} \|x_n\|$	(∞-norm or ℓ_∞-norm)
$\|x\|_2 \triangleq \left(\sum_{n=1}^{N} \|x_n\|^2\right)^{1/2}$	(Euclidean or ℓ_2-norm, also written as $\|x\|$)
$\|x\|_p \triangleq \left(\sum_{n=1}^{N} \|x_n\|^p\right)^{1/p}$	(p-norm or ℓ_p-norm, for any real $p \geq 1$)

Example 1.13. A fundamental result in matrix theory is that all matrix norms in finite dimensional spaces are *equivalent*. Specifically, if $\|A\|_a$ and $\|A\|_b$ denote two generic matrix norms, then there exist positive constants c_ℓ and c_u that bound one norm by the other from above and from below, namely,

$$c_\ell \,\|A\|_b \;\leq\; \|A\|_a \;\leq\; c_u \,\|A\|_b \tag{1.151}$$

The values of $\{c_\ell, c_u\}$ are independent of the matrix entries but they may be dependent on the matrix dimensions. Vector norms are also equivalent to each other.

Table 1.4 Useful matrix norms, where the $\{a_{nm}\}$ denote the entries of A.

Matrix norm	Name
$\|A\|_1 \triangleq \max_{1\leq m\leq M} \left(\sum_{n=1}^{N} \|a_{nm}\|\right)$	(1-norm, or maximum absolute column sum)
$\|A\|_\infty \triangleq \max_{1\leq n\leq N} \left(\sum_{m=1}^{M} \|a_{nm}\|\right)$	(∞-norm, or maximum absolute row sum)
$\|A\|_{\mathrm{F}} \triangleq \sqrt{\mathrm{Tr}(A^\mathsf{T} A)}$	(Frobenius norm)
$\|A\|_p \triangleq \max_{x\neq 0} \frac{\|Ax\|_p}{\|x\|_p}$	(p-induced norm for any real $p \geq 1$)
$\|A\|_2 \triangleq \max_{x\neq 0} \frac{\|Ax\|}{\|x\|}$	(2-induced norm)

Example 1.13 (Spectral norm of a matrix) Assume $N \geq M$ and consider the $M \times M$ square matrix $A^\mathsf{T} A$. Using the Rayleigh–Ritz characterization (1.17b) for the maximum eigenvalue of a matrix we have that

$$\lambda_{\max}(A^{\mathsf{T}}A) = \max_{x \neq 0} \left(\frac{x^{\mathsf{T}}A^{\mathsf{T}}Ax}{x^{\mathsf{T}}x} \right) = \max_{x \neq 0} \left(\frac{\|Ax\|^2}{\|x\|^2} \right) \tag{1.152}$$

But we already know from the argument in Appendix 1.B that $\sigma_1^2 = \lambda_{\max}(A^{\mathsf{T}}A)$. We conclude that the largest singular value of A satisfies:

$$\sigma_1 = \max_{x \neq 0} \left(\frac{\|Ax\|}{\|x\|} \right) \tag{1.153}$$

This maximum value is achieved if we select $x = v_1$ (i.e., as the right singular vector corresponding to σ_1). Indeed, using $Av_1 = \sigma_1 u_1$ we get

$$\frac{\|Av_1\|^2}{\|v_1\|^2} = \frac{\|\sigma_1 u_1\|^2}{\|v_1\|^2} = \sigma_1^2 \tag{1.154}$$

since $\|u_1\| = \|v_1\| = 1$. We therefore find that the square of the maximum singular value, σ_1^2, measures the maximum energy gain from x to Ax. The same conclusion holds when $N \leq M$ since then $\sigma_1^2 = \lambda_{\max}(AA^{\mathsf{T}})$ and the argument can be repeated to conclude that

$$\sigma_1 = \max_{x \neq 0} \left(\frac{\|A^{\mathsf{T}}x\|}{\|x\|} \right) \tag{1.155}$$

The maximum singular value of a matrix is called its *spectral norm* or its 2-induced norm, also written as

$$\sigma_1 = \|A\|_2 = \|A^{\mathsf{T}}\|_2 = \max_{\|x\|=1} \|Ax\| = \max_{\|x\|=1} \|A^{\mathsf{T}}x\| \tag{1.156}$$

Dual norms

An important concept in matrix theory is that of the *dual norm*. Let $\|\cdot\|$ denote some vector norm in $\mathbb{R}^M$. The associated dual norm is denoted by $\|x\|_\star$ and defined as

$$\|x\|_\star \stackrel{\Delta}{=} \sup_{y} \left\{ x^{\mathsf{T}}y \mid \|y\| \leq 1 \right\} \tag{1.157}$$

In other words, we consider all vectors y that lie inside the ball $\|y\| \leq 1$ and examine their transformation by x through the inner product $x^{\mathsf{T}}y$. The largest value this transformation attains is taken as the dual norm of x. It is shown in Prob. 1.24 that $\|x\|_\star$ is a valid vector norm and that it can be expressed equivalently as

$$\|x\|_\star \stackrel{\Delta}{=} \sup_{y \neq 0} \left\{ \frac{x^{\mathsf{T}}y}{\|y\|} \right\} \tag{1.158}$$

where we scale the inner product by $\|y\|$ and compute the supremum over all nonzero vectors y. Using (1.158), we can readily write the following Cauchy–Schwarz inequality involving a norm and its dual:

$$\boxed{x^{\mathsf{T}}y \leq \|y\| \, \|x\|_\star} \tag{1.159}$$

This result will be useful in our analysis of inference methods. This is because we will often deal with norms other than the Euclidean norm, and we will need to bound inner products between vectors. The above inequality shows that the bound will involve the product of two norms: the regular norm and its dual.

For the sake of illustration, let us determine the dual norm of the Euclidean norm, $\|\cdot\|_2$. To do so, we need to assess:

$$\|x\|_\star = \sup_y \left\{ x^\mathsf{T} y \mid \|y\|_2 \leq 1 \right\} \tag{1.160}$$

We know from the classical Cauchy–Schwarz inequality for inner products in Euclidean space that $x^\mathsf{T} y \leq \|x\|_2 \, \|y\|_2$, with equality when the vectors are parallel to each other and pointing in the same direction, say, $y = \alpha x$ for some $\alpha > 0$. Given that the norm of y should be bounded by 1, we must set $\alpha \leq 1/\|x\|_2$. The inner product $x^\mathsf{T} y$ becomes $\alpha\|x\|_2^2$ and it is maximized when $\alpha = 1/\|x\|$, in which case we conclude that $\|x\|_\star = \|x\|_2$. In other words, the dual norm of the Euclidean norm is the Euclidean norm itself.

Several other dual norms are determined in Probs. 1.26–1.29 for both vector and matrix norms, where the definition of the dual norm for matrices is taken as:

$$\|A\|_\star \triangleq \sup_Y \left\{ \mathrm{Tr}(A^\mathsf{T} Y) \mid \|Y\| \leq 1 \right\} \tag{1.161}$$

We collect the results into Table 1.5. The last line introduces the nuclear norm of a matrix, which is equal to the sum of its singular values. The notation for the nuclear norm continues to use the star subscript. The results in the table show that we can interpret the $(\ell_1, \ell_2, \ell_p, \ell_\infty)$-norms of vectors in the equivalent forms:

$$\|x\|_2 = \sup_y \{ x^\mathsf{T} y \mid \|y\|_2 \leq 1 \} \tag{1.162a}$$

$$\|x\|_\infty = \sup_y \{ x^\mathsf{T} y \mid \|y\|_1 \leq 1 \} \tag{1.162b}$$

$$\|x\|_1 = \sup_y \{ x^\mathsf{T} y \mid \|y\|_\infty \leq 1 \} \tag{1.162c}$$

$$\|x\|_q = \sup_y \{ x^\mathsf{T} y \mid \|y\|_p \leq 1 \}, \quad p, q \geq 1, \quad 1/p + 1/q = 1 \tag{1.162d}$$

Likewise, the Frobenius and nuclear norms of a matrix A can be interpreted as corresponding to:

$$\|A\|_\mathrm{F} = \sup_Y \left\{ \mathrm{Tr}(A^\mathsf{T} Y) \mid \|Y\|_\mathrm{F} \leq 1 \right\} \tag{1.163a}$$

$$\|A\|_\star = \sup_Y \left\{ \mathrm{Tr}(A^\mathsf{T} Y) \mid \|Y\|_2 \leq 1 \right\} \tag{1.163b}$$

It is straightforward to verify the validity of the following inequalities relating several matrix norms for matrices A of size $N \times M$ and rank r:

Table 1.5 List of dual norms for vectors and matrices.

Original norm	Dual norm
$\|x\|_2$	$\|x\|_2$
$\|x\|_1$	$\|x\|_\infty$
$\|x\|_\infty$	$\|x\|_1$
$\|x\|_p$	$\|x\|_q,\ p,q \geq 1,\ 1/p + 1/q = 1$
$\|A\|_{\rm F}$	$\|A\|_{\rm F}$
$\|A\|_2$	$\|A\|_\star = \sum_{r=1}^{\text{rank}(A)} \sigma_r$ (nuclear norm)

$$M^{-1/2}\,\|A\|_\infty \leq \|A\|_2 \leq N^{1/2}\,\|A\|_\infty \tag{1.164a}$$
$$N^{-1/2}\,\|A\|_1 \leq \|A\|_2 \leq M^{1/2}\,\|A\|_1 \tag{1.164b}$$
$$\|A\|_\infty \leq M^{1/2}\,\|A\|_2 \leq M\,\|A\|_1 \tag{1.164c}$$
$$\|A\|_2 \leq \|A\|_{\rm F} \leq \sqrt{r}\,\|A\|_2 \tag{1.164d}$$
$$\|A\|_{\rm F} \leq \|A\|_\star \leq \sqrt{r}\,\|A\|_{\rm F} \tag{1.164e}$$
$$\|A\|_2^2 \leq \|A\|_1 \times \|A\|_\infty \tag{1.164f}$$

ρ-norm[1]

In this section and the next we introduce two specific norms; the discussion can be skipped on a first reading. Let B denote an $N \times N$ real matrix with eigenvalues $\{\lambda_n\}$. The spectral radius of B, denoted by $\rho(B)$, is defined as

$$\boxed{\rho(B) \overset{\Delta}{=} \max_{1\leq n\leq N} |\lambda_n|} \tag{1.165}$$

A fundamental result in matrix theory asserts that every matrix admits a so-called *canonical Jordan decomposition*, which is of the form

$$B = UJU^{-1} \tag{1.166}$$

for some invertible matrix U and where

$$J = \text{blkdiag}\{J_1, J_2, \ldots, J_R\} \tag{1.167}$$

is a block diagonal matrix, say with R blocks. When B happens to be symmetric, we already know from the spectral decomposition (1.15a) that J will be diagonal and U will be orthogonal so that $U^{-1} = U^{\mathsf{T}}$. More generally, each block J_r will have a Jordan structure with an eigenvalue λ_r on its diagonal entries, unit entries on the first sub-diagonal, and zeros everywhere else. For example, for a block of size 4×4:

[1] The two sections on the block-maximum and ρ-norms can be skipped on a first reading.

$$J_r = \begin{bmatrix} \lambda_r & & & \\ 1 & \lambda_r & & \\ & 1 & \lambda_r & \\ & & 1 & \lambda_r \end{bmatrix} \tag{1.168}$$

Let ϵ denote an arbitrary positive scalar that we are free to choose and define the $N \times N$ diagonal scaling matrix:

$$D \stackrel{\Delta}{=} \text{diag}\left\{\epsilon, \epsilon^2, \ldots, \epsilon^N\right\} \tag{1.169}$$

We can use the matrix U originating from B to define the following matrix norm, denoted by $\|\cdot\|_\rho$, for any matrix A of size $N \times N$:

$$\boxed{\|A\|_\rho \stackrel{\Delta}{=} \left\|DU^{-1}AUD^{-1}\right\|_1} \tag{1.170}$$

in terms of the 1-norm (i.e., maximum absolute column sum) of the matrix product on the right-hand side. It is not difficult to verify that the transformation (1.170) is a valid matrix norm, namely, that it satisfies the following properties, for any matrices A and C of compatible dimensions and for any scalar α:

$$\begin{array}{ll} \textbf{(a)} \ \|A\|_\rho \geq 0 \text{ with } \|A\|_\rho = 0 \text{ if, and only if, } A = 0 & \\ \textbf{(b)} \ \|\alpha A\|_\rho = |\alpha| \, \|A\|_\rho & \\ \textbf{(c)} \ \|A + C\|_\rho \leq \|A\|_\rho + \|C\|_\rho & \textbf{(triangular inequality)} \\ \textbf{(d)} \ \|AC\|_\rho \leq \|A\|_\rho \, \|C\|_\rho & \textbf{(sub-multiplicative property)} \end{array} \tag{1.171}$$

One important property of the ρ-norm defined by (1.170) is that when it is applied to the matrix B itself, it will hold that:

$$\rho(B) \leq \|B\|_\rho \leq \rho(B) + \epsilon \tag{1.172}$$

That is, the ρ-norm of B lies between two bounds defined by its spectral radius. It follows that if the matrix B happens to be stable to begin with, so that $\rho(B) < 1$, then we can always select ϵ small enough to ensure $\|B\|_\rho < 1$.

The matrix norm defined by (1.170) is an induced norm relative to the following vector norm:

$$\|x\|_\rho \stackrel{\Delta}{=} \|DU^{-1}x\|_1 \tag{1.173}$$

That is, for any matrix A, it holds that

$$\|A\|_\rho = \max_{x \neq 0} \left\{ \frac{\|Ax\|_\rho}{\|x\|_\rho} \right\} \tag{1.174}$$

Proof of (1.174): Indeed, using (1.173), we first note that for any vector $x \neq 0$:

$$\begin{aligned}\|Ax\|_\rho &= \|DU^{-1}Ax\|_1\\ &= \|DU^{-1}AUD^{-1}DU^{-1}x\|_1\\ &\le \|DU^{-1}AUD^{-1}\|_1\,\|DU^{-1}x\|_1\\ &= \|A\|_\rho\,\|x\|_\rho\end{aligned} \tag{1.175}$$

so that

$$\max_{x\neq 0}\left\{\frac{\|Ax\|_\rho}{\|x\|_\rho}\right\} \le \|A\|_\rho \tag{1.176}$$

To show that equality holds in (1.176), it is sufficient to exhibit one nonzero vector x_o that attains it. Let k_o denote the index of the column that attains the maximum absolute column sum in the matrix product $DU^{-1}AUD^{-1}$. Let e_{k_o} denote the column basis vector of size $N\times 1$ with one at location k_o and zeros elsewhere. Select

$$x_o \stackrel{\Delta}{=} UD^{-1}e_{k_o} \tag{1.177}$$

Then, it is straightforward to verify that

$$\|x_o\|_\rho \stackrel{\Delta}{=} \|DU^{-1}x_o\|_1 \stackrel{(1.177)}{=} \|e_{k_o}\|_1 = 1 \tag{1.178}$$

and

$$\begin{aligned}\|Ax_o\|_\rho &\stackrel{\Delta}{=} \|DU^{-1}Ax_o\|_1\\ &= \|DU^{-1}AUD^{-1}DU^{-1}x_o\|_1\\ &\stackrel{(1.177)}{=} \|DU^{-1}AUD^{-1}e_{k_o}\|_1\\ &\stackrel{(1.170)}{=} \|A\|_\rho\end{aligned} \tag{1.179}$$

so that, for this particular vector, we have

$$\frac{\|Ax_o\|_\rho}{\|x_o\|_\rho} = \|A\|_\rho \tag{1.180}$$

as desired.

■

Block maximum norm

Let

$$x \stackrel{\Delta}{=} \text{blkcol}\{x_1, x_2, \ldots, x_N\} \tag{1.181}$$

denote an $N\times 1$ *block* column vector whose individual entries $\{x_k\}$ are themselves vectors of size $M\times 1$ each. The block maximum norm of x is denoted by $\|x\|_{b,\infty}$ and is defined as

$$\|x\|_{b,\infty} \stackrel{\Delta}{=} \max_{1\le k\le N} \|x_k\| \tag{1.182}$$

That is, $\|x\|_{b,\infty}$ is equal to the largest Euclidean norm of its block components. This vector norm induces a block maximum matrix norm. Let $\mathcal{A}$ denote an arbitrary $N \times N$ block matrix with individual block entries of size $M \times M$ each. Then, the block maximum norm of $\mathcal{A}$ is defined as

$$\|\mathcal{A}\|_{b,\infty} \stackrel{\Delta}{=} \max_{x \neq 0} \left\{ \frac{\|\mathcal{A}x\|_{b,\infty}}{\|x\|_{b,\infty}} \right\} \tag{1.183}$$

The block maximum norm has several useful properties (see Prob. 1.23):

(a) Let $\mathcal{U} = \text{diag}\{U_1, U_2, \ldots, U_N\}$ denote an $N \times N$ block diagonal matrix with $M \times M$ orthogonal blocks $\{U_k\}$. Then, transformations by $\mathcal{U}$ do not modify the block maximum norm, i.e., it holds that $\|\mathcal{U}x\|_{b,\infty} = \|x\|_{b,\infty}$ and $\|\mathcal{U}\mathcal{A}\mathcal{U}^\mathsf{T}\|_{b,\infty} = \|\mathcal{A}\|_{b,\infty}$.

(b) Let $\mathcal{D} = \text{diag}\{D_1,\ D_2,\ \ldots, D_N\}$ denote an $N \times N$ block diagonal matrix with $M \times M$ symmetric blocks $\{D_k\}$. Then, $\rho(\mathcal{D}) = \|\mathcal{D}\|_{b,\infty}$.

(c) Let A be an $N \times N$ matrix and define $\mathcal{A} = A \otimes I_M$ whose blocks are therefore of size $M \times M$ each. If A is left-stochastic (i.e., the entries on each column of A add up to 1, as defined later by (1.193)), then $\|\mathcal{A}^\mathsf{T}\|_{b,\infty} = 1$.

(d) For any block diagonal matrix $\mathcal{D}$, and any left-stochastic matrices $\mathcal{A}_1 = A_1 \otimes I_M$ and $\mathcal{A}_2 = A_2 \otimes I_M$ constructed as in part (c), it holds that

$$\rho\left(\mathcal{A}_2^\mathsf{T}\,\mathcal{D}\,\mathcal{A}_1^\mathsf{T}\right) \;\le\; \|\mathcal{D}\|_{b,\infty} \tag{1.184}$$

(e) If the matrix $\mathcal{D}$ in part (d) has symmetric blocks, it holds that

$$\rho\left(\mathcal{A}_2^\mathsf{T}\,\mathcal{D}\,\mathcal{A}_1^\mathsf{T}\right) \;\le\; \rho(\mathcal{D}) \tag{1.185}$$

1.11 PERTURBATION BOUNDS ON EIGENVALUES[2]

We state below two useful results that bound matrix eigenvalues.

Weyl theorem

The first result, known as *Weyl theorem*, shows how the eigenvalues of a symmetric matrix are disturbed through additive perturbations to the entries of the matrix. Thus, let $\{A', A, \Delta A\}$ denote arbitrary $N \times N$ real symmetric matrices with ordered eigenvalues $\{\lambda_m(A'), \lambda_m(A), \lambda_m(\Delta A)\}$, i.e.,

$$\lambda_1(A) \;\ge\; \lambda_2(A) \;\ge\; \ldots \;\ge\; \lambda_N(A) \tag{1.186}$$

and similarly for the eigenvalues of $\{A', \Delta A\}$, with the subscripts 1 and N representing the largest and smallest eigenvalues, respectively. Weyl theorem states that if A is perturbed to

$$A' \;=\; A \;+\; \Delta A \tag{1.187}$$

[2] This section can be skipped on a first reading.

then the eigenvalues of the new matrix are bounded as follows:

$$\boxed{\lambda_n(A) + \lambda_N(\Delta A) \le \lambda_n(A') \le \lambda_n(A) + \lambda_1(\Delta A)} \tag{1.188}$$

for $1 \le n \le N$. In particular, it follows that the maximum eigenvalue is perturbed as follows:

$$\lambda_{\max}(A) + \lambda_{\min}(\Delta A) \le \lambda_{\max}(A') \le \lambda_{\max}(A) + \lambda_{\max}(\Delta A) \tag{1.189}$$

In the special case when $\Delta A \ge 0$, we conclude from (1.188) that $\lambda_n(A') \ge \lambda_n(A)$ for all $n = 1, 2, \ldots, N$.

Gershgorin theorem

The second result, known as *Gershgorin theorem*, specifies circular regions within which the eigenvalues of a matrix are located. Thus, consider an $N \times N$ real matrix A with scalar entries $\{a_{\ell k}\}$. With each diagonal entry $a_{\ell\ell}$ we associate a disc in the complex plane centered at $a_{\ell\ell}$ and with radius

$$r_\ell \stackrel{\Delta}{=} \sum_{k\neq\ell,k=1}^{N} |a_{\ell k}| \tag{1.190}$$

That is, r_ℓ is equal to the sum of the magnitudes of the nondiagonal entries on the same row as $a_{\ell\ell}$. We denote the disc by D_ℓ; it consists of all points that satisfy

$$D_\ell = \left\{ z \in \mathbb{C}^N \text{ such that } |z - a_{\ell\ell}| \le r_\ell \right\} \tag{1.191}$$

The theorem states that the spectrum of A (i.e., the set of all its eigenvalues, denoted by $\lambda(A)$) is contained in the union of all N Gershgorin discs:

$$\lambda(A) \subset \left\{ \bigcup_{\ell=1}^{N} D_\ell \right\} \tag{1.192}$$

A stronger statement of the Gershgorin theorem covers the situation in which some of the Gershgorin discs happen to be disjoint. Specifically, if the union of L of the discs is disjoint from the union of the remaining $N - L$ discs, then the theorem further asserts that L eigenvalues of A will lie in the first union of L discs and the remaining $N - L$ eigenvalues of A will lie in the second union of $N - L$ discs.

1.12 STOCHASTIC MATRICES

Consider $N \times N$ matrices A with nonnegative entries, $\{a_{\ell k} \ge 0\}$. The matrix $A = [a_{\ell k}]$ is said to be left-stochastic if it satisfies

$$A^\mathsf{T}\mathbb{1} = \mathbb{1} \qquad \textbf{(left-stochastic)} \tag{1.193}$$

where $\mathbb{1}$ denotes the column vector whose entries are all equal to 1. That is, the entries on each column of A should add up to 1. The matrix A is said to be right-stochastic if

$$A\mathbb{1} = \mathbb{1} \qquad \textbf{(right-stochastic)} \tag{1.194}$$

so that the entries on each row of A add up to 1. The matrix A is doubly stochastic if the entries on each of its columns and on each of its rows add up to 1 (i.e., if it is both left- and right-stochastic):

$$A\mathbb{1} = \mathbb{1}, \quad A^{\mathsf{T}}\mathbb{1} = \mathbb{1} \qquad \textbf{(doubly stochastic)} \tag{1.195}$$

Stochastic matrices arise frequently in the study of Markov chains, multi-agent networks, and signals over graphs. The following statement lists two properties of stochastic matrices:

(a) The spectral radius of A is equal to 1, $\rho(A) = 1$. It follows that all eigenvalues of A lie inside the unit disc, i.e., $|\lambda(A)| \leq 1$. The matrix A may have multiple eigenvalues with magnitude equal to 1 – see Prob. 1.49.

(b) Assume A is additionally a primitive matrix, i.e., there exists some finite integer power of A such that all its entries are strictly positive:

$$[A^{n_o}]_{\ell k} > 0, \quad \text{for some integer } n_o > 0 \tag{1.196}$$

and for all $1 \leq \ell, k \leq N$. Then, the matrix A will have a single eigenvalue at 1 (i.e., the eigenvalue at 1 will have multiplicity 1). All other eigenvalues of A will lie strictly inside the unit circle. Moreover, with proper sign scaling, all entries of the right-eigenvector of A corresponding to the single eigenvalue at 1 will be strictly positive, namely, if we let p denote this right-eigenvector with entries $\{p_k\}$ and normalize its entries to add up to 1, then

$$Ap = p, \quad \mathbb{1}^{\mathsf{T}}p = 1, \quad p_k > 0, \quad k = 1, 2, \ldots, N \tag{1.197}$$

We refer to p as the *Perron eigenvector* of A. All other eigenvectors of A associated with the other eigenvalues will have at least one negative or complex entry.

1.13 COMPLEX-VALUED MATRICES

Although the presentation in the earlier sections has focused exclusively on real-valued matrices, most of the concepts and results extend almost effortlessly to complex-valued matrices. For example, in relation to symmetry, if $A \in \mathbb{C}^{N\times N}$, then the matrix A will be said to be *Hermitian* if it satisfies

$$A = A^* \quad \textbf{(Hermitian symmetry)} \tag{1.198}$$

where the symbol $*$ denotes complex conjugate transposition – recall (1.5). This notion extends the definition of matrix symmetry (which requires $A = A^\mathsf{T}$) to the complex case. Hermitian matrices can again be shown to have only real eigenvalues and their spectral decomposition will now take the form:

$$A = U\Lambda U^* \tag{1.199}$$

where Λ continues to be a diagonal matrix with the eigenvalues of A, while U is now a *unitary* (as opposed to an orthogonal) matrix, namely, it satisfies

$$UU^* = U^*U = I_N \tag{1.200}$$

In other words, most of the results discussed so far extend rather immediately by replacing transposition by complex conjugation, such as replacing x^T by x^* and A^T by A^*. For example, while the squared Euclidean norm of a vector $x \in \mathbb{R}^N$ is given by $\|x\|^2 = x^\mathsf{T}x$, the same squared norm for a vector $x \in \mathbb{C}^N$ will be given by $\|x\|^2 = x^*x$. In this way, the Rayleigh–Ritz characterization of the smallest and largest eigenvalues of A will become

$$\lambda_{\min} = \min_{\|x\|=1} \left\{x^*Ax\right\}, \qquad \lambda_{\max} = \max_{\|x\|=1} \left\{x^*Ax\right\} \tag{1.201}$$

Likewise, positive-definite matrices will be ones that satisfy

$$v^*Av > 0, \;\; \text{for any } v \neq 0 \in \mathbb{C}^N \tag{1.202}$$

These matrices will continue to have positive eigenvalues and positive determinants. In addition, the range space and nullspace of A will be defined similarly to the real case:

$$\mathcal{R}(A) \triangleq \left\{q \in \mathbb{C}^N \mid \text{such that } q = Ap \text{ for some } p \in \mathbb{C}^M\right\} \tag{1.203a}$$

$$\mathcal{N}(A) \triangleq \left\{p \in \mathbb{C}^M \mid \text{such that } Ap = 0\right\} \tag{1.203b}$$

with the properties

$$z \in \mathcal{N}(A^*), \;\; q \in \mathcal{R}(A) \;\; \Longrightarrow \;\; z^*q = 0 \tag{1.204a}$$

$$\mathcal{R}(A^*) = \mathcal{R}(A^*A) \tag{1.204b}$$

$$\mathcal{N}(A) = \mathcal{N}(A^*A) \tag{1.204c}$$

Moreover, two square matrices A and B will be congruent if $A = QBQ^*$, for some nonsingular Q. Finally, the SVD of $A \in \mathbb{C}^{N\times M}$ will now take the following form with unitary matrices $U \in \mathbb{C}^{N\times N}$ and $V \in \mathbb{C}^{M\times M}$:

(a) If $N \leq M$, then Σ is $N \times N$ and

$$A = U \begin{bmatrix} \Sigma & 0 \end{bmatrix} V^* \tag{1.205}$$

(b) If $N \geq M$, then Σ is $M \times M$ and

$$A = U \begin{bmatrix} \Sigma \\ 0 \end{bmatrix} V^* \tag{1.206}$$

1.14 COMMENTARIES AND DISCUSSION

Linear algebra and matrix theory. The presentation in this chapter follows the overviews from Sayed (2003, 2008, 2014a). Throughout our treatment of inference and learning theories, the reader will be exposed to a variety of concepts from linear algebra and matrix theory in a motivated manner. In this way, after progressing sufficiently enough into our treatment, readers will be able to master many useful concepts. Several of these concepts are summarized in this chapter. If additional help is needed, some accessible references on matrix theory are the works by MacDuffee (1946), Gantmacher (1959), Bellman (1970), Horn and Johnson (1990), Golub and Van Loan (1996), Meyer (2001), Laub (2004), and Bernstein (2018). Accessible references on linear algebra are the books by Halmos (1974), Strang (1988, 2009), Gelfand (1989), Lay (1994), Lax (1997), Lay, Lay, and McDonald (2014), Hogben (2014), and Nicholson (2019).

Kronecker products. We introduced the Kronecker product in Section 1.9. This product provides a useful and compact representation for generating a block matrix structure from two separate matrices. We illustrated two useful applications of Kronecker products in Examples 1.11 and 1.12 in the context of Lyapunov equations. There are of course many other applications. The notion of Kronecker products was introduced by the German mathematician **Johann Zehfuss (1832–1901)** in the work by Zehfuss (1858) – see the historical accounts by Henderson, Pukelsheim, and Searle (1983) and Hackbusch (2012). In his article, Zehfuss (1858) introduced the determinantal formula

$$\det(A \otimes B) = (\det(A))^M \, (\det(B))^N \tag{1.207}$$

for square matrices A and B of sizes $N \times N$ and $M \times M$, respectively. This formula was later attributed erroneously by Hensel (1891) to the German mathematician **Leopold Kronecker (1823–1891)**, who discussed it in some of his lectures in the 1880s. The operation $\otimes$ became subsequently known as the Kronecker product instead of the more appropriate "Zehfuss product." Useful surveys on Kronecker products, their properties, and applications appear in Henderson and Searle (1981b), Regalia and Mitra (1989), and Van Loan (2000). Useful references on block Kronecker products are the works by Tracy and Singh (1972), Koning, Neudecker, and Wansbeek (1991), and Liu (1999). The block Kronecker product (1.145) is also known as the Tracy–Singh product.

Schur complements. According to the historical overview by Puntanen and Styan (2005), the designation "Schur complement" is due to Haynsworth (1968) in her study of the inertia of a block-partitioned matrix. If we consider a block symmetric matrix of the form

$$S = \begin{bmatrix} A & B \\ B^{\mathsf{T}} & D \end{bmatrix}, \quad A = A^{\mathsf{T}}, \; D = D^{\mathsf{T}} \tag{1.208}$$

with a nonsingular A, we recognize that the following block triangular factorization of S amounts to a congruence relation:

$$S = \begin{bmatrix} I & 0 \\ B^{\mathsf{T}} A^{-1} & I \end{bmatrix} \begin{bmatrix} A & 0 \\ 0 & \Delta_A \end{bmatrix} \begin{bmatrix} I & A^{-1}B \\ 0 & I \end{bmatrix} \tag{1.209}$$

where $\Delta_A = D - B^{\mathsf{T}} A^{-1} B$. It follows that

$$\text{In}(S) \; = \; \text{In}(A) \; + \; \text{In}(\Delta_A) \tag{1.210}$$

where the addition operation means that the individual components of the inertia measure, namely, I_+, I_-, I_0, are added together. This inertia additivity formula was derived by Haynsworth (1968) and the factor Δ_A was referred to as the "Schur complement" relative to A. The attribution to "Schur" is because the determinantal formula (1.211) involving Δ_A was given by the German mathematician **Issai Schur (1875–1941)** in a famous paper by Schur (1917), where he studied the characterization of functions that are analytic and contractive inside the unit disc – see the survey articles by Cottle (1974), Kailath (1986), and Kailath and Sayed (1995), and the books by Constantinescu (1996) and Zhang (2005). In Schur (1917), the following expression appeared, which is now easy to conclude from the block factorization expression (1.209):

$$\det(S) = \det(A)\,\det(\Delta_A) \tag{1.211}$$

Schur was a student of the German mathematician **Georg Frobenius (1849–1917)**. His determinantal formula extended a special case studied by Frobenius (1908), which corresponds to the situation in which D is a scalar δ, and B is a column vector b, i.e.,

$$\det\left(\begin{bmatrix} A & b \\ b^{\mathsf{T}} & \delta \end{bmatrix}\right) = \det(A)\,(\delta - b^{\mathsf{T}}A^{-1}b) \tag{1.212}$$

The fact that congruence preserves inertia, as illustrated in Example 1.5 in connection with Schur complements, was established by Sylvester (1852) and is known as the Sylvester law of inertia.

The block triangular factorization formulas (1.63a)–(1.63b), written in various equivalent forms, appear less directly in the work of Schur (1917) and more explicitly in the works by Banachiewicz (1937a, b), Aitken (1939), Hotelling (1943a, b), and Duncan (1944) – see the overviews by Henderson and Searle (1981a) and Puntanen, Styan, and Isotalo (2011). The matrix inversion formula (1.81) was apparently first given by Duncan (1944) and Guttman (1946), though it is often attributed to Sherman and Morrison (1949, 1950) and Woodbury (1950) and referred to as the Sherman–Morrison–Woodbury formula. A special case when $C = 1$, B is a column vector b, and D is a row vector d^{T} appeared in Bartlett (1951), namely,

$$(A + bd^{\mathsf{T}})^{-1} = A^{-1} - \frac{A^{-1}bd^{\mathsf{T}}A^{-1}}{1 + d^{\mathsf{T}}A^{-1}b} \tag{1.213}$$

This particular inversion formula was also introduced independently by Plackett (1950) in his study of recursive updates of least-squares problems, as we are going to discuss later in Section 50.3.2. Useful overviews on the origin of the matrix inversion formula appear in Householder (1953, 1957, 1964), Henderson and Searle (1981a), and Hager (1989). One of the first uses of the formula in the context of filtering theory appears to be Kailath (1960) and Ho (1963).

Spectral theorem. We established the spectral theorem for finite-dimensional matrices in Section 1.1, which states that every $N \times N$ symmetric matrix A has N real eigenvalues, $\{\lambda_n\}$, and a complete set of N orthonormal eigenvectors, $\{u_n\}$. This means that A can be expressed in either form:

$$A = U\Lambda U^{\mathsf{T}} = \sum_{n=1}^{N} \lambda_n u_n u_n^{\mathsf{T}} \tag{1.214}$$

where U is an orthogonal matrix whose columns are the $\{u_n\}$ and Λ is a diagonal matrix with real entries $\{\lambda_n\}$. For any vector $x \in \mathbb{R}^n$, we introduce the change of variables $y = U^{\mathsf{T}}x$ and let $\{y_n\}$ denote the individual entries of y. Then, one useful consequence of the spectral decomposition (1.214) is that every generic quadratic term of the form $x^{\mathsf{T}}Ax$ can be decomposed into the sum of N elementary quadratic terms, namely,

$$x^{\mathsf{T}}Ax = (x^{\mathsf{T}}U)\Lambda(U^{\mathsf{T}}x) = y^{\mathsf{T}}\Lambda y = \sum_{n=1}^{N} \lambda_n y_n^2 \tag{1.215}$$

The property that every symmetric matrix is diagonalizable holds more generally and is part of the spectral theory of self-adjoint linear operators in Hilbert space (which extends the study of symmetric finite-dimensional matrix mappings to infinite dimensional mappings defined over spaces endowed with inner products) – an accessible overview is given by Halmos (1963, 1974, 2013). It is sufficient for our purposes to focus on finite-dimensional matrices. The result in this case was first established by the French mathematician **Augustine Cauchy (1789–1857)** in the work by Cauchy (1829) – see the useful historical account by Hawkins (1975). It was later generalized to the operator setting by the Hungarian-American mathematician **John von Neumann (1903–1957)** in the work by von Neumann (1929, 1932) in his studies of linear operators in the context of quantum mechanics – see the overview in the volume edited by Bródy and Vámos (1995). The derivation of the spectral theorem in Appendix 1.A relies on the fundamental theorem of algebra, which guarantees that every polynomial of order N has N roots – see, e.g., the text by Fine and Rosenberger (1997) and Carrera (1992). The argument in the appendix is motivated by the presentations in Horn and Johnson (1990), Trefethen and Bau (1997), Calafiore and El Ghaoui (2014), and Nicholson (2019).

QR decomposition. The matrix decomposition (1.102) is a restatement of the Gram–Schmidt orthonormalization procedure, whereby the columns of a matrix A are replaced by an orthonormal basis for the linear subspace that is spanned by them. The Gram–Schmidt procedure is named after the Danish and German mathematicians **Jorgen Gram (1850–1916)** and **Erhard Schmidt (1876–1959)**, respectively. The latter published the procedure in the work by Schmidt (1907, 1908) and acknowledged that his algorithm is the same as one published earlier by Gram (1883). Nevertheless, a similar construction was already proposed over half a century before by the French mathematician **Pierre-Simon Laplace (1749–1827)** in the treatise by Laplace (1812). He orthogonalized the columns of a (tall) observation matrix $A \in \mathbb{R}^{N\times M}$ in order to solve a least-squares problem of the form:

$$\min_{w\in\mathbb{R}^M} \|y - Aw\|^2 \tag{1.216}$$

His solution method is a precursor of the QR method for solving least-squares problems. Specifically, assuming A has full rank, we introduce the full QR decomposition of A,

$$A = Q\begin{bmatrix} R \\ 0 \end{bmatrix} \tag{1.217}$$

where Q is $N \times N$ orthogonal and R is $M \times M$ upper-triangular with positive diagonal entries. We further let

$$\begin{bmatrix} z_1 \\ z_2 \end{bmatrix} \triangleq Q^{\mathsf{T}}y \tag{1.218}$$

where z_1 is $M \times 1$. It then follows that

$$\|y - Aw\|^2 = \|z_1 - Rw\|^2 + \|z_2\|^2 \tag{1.219}$$

so that the solution to (1.216) is obtained by solving the triangular system of equations $R\widehat{w} = z_1$, i.e.,

$$\widehat{w} = R^{-1}z_1 \tag{1.220}$$

According to Bjorck (1996), the earliest work linking the names of Gram and Schmidt to the orthogonalization procedure appears to be Wong (1935). Today, the QR decomposition is one of the main tools in modern numerical analysis. For further information and discussions, the reader may refer to the texts by Bjorck (1996), Golub and Van Loan (1996), and Trefethen and Bau (1997).

Singular value decomposition. The SVD is one of the most powerful matrix decompositions, valid for both rectangular and square matrices and useful for both analysis and numerical computations. It has a rich history with contributions from notable mathematicians. The article by Stewart (1993) provides an excellent overview of the historical evolution of the SVD and the main contributors to its development. The SVD was proposed independently by the Italian mathematician **Eugenio Beltrami (1835–1900)** and the French mathematician **Camille Jordan (1838–1922)** in the works by Beltrami (1873) and Jordan (1874a, b). According to Stewart (1993), they were both interested in bilinear forms of the form $f(x,y) = x^{\mathsf{T}} A y$, where A is a square real matrix of size $N \times N$ and x and y are column vectors. They were motivated to introduce and compute a decomposition for A in the form $A = U\Sigma V^{\mathsf{T}}$ in order to reduce the bilinear form to the canonical form

$$f(x,y) = \underbrace{x^{\mathsf{T}} U}_{\triangleq\, a^{\mathsf{T}}} \Sigma \underbrace{V^{\mathsf{T}} y}_{\triangleq\, b} = a^{\mathsf{T}} \Sigma b = \sum_{n=1}^{N} \sigma_n \, a_n b_n \tag{1.221}$$

in terms of the entries of $\{a, b\}$ and the diagonal entries of Σ. Beltrami (1873) focused on square and invertible matrices A while Jordan (1874a, b) considered degenerate situations with singularities as well. Beltrami (1873) exploits property (1.240) and uses it to relate the SVD factors U and V to the eigen-decompositions of the matrix products $A^{\mathsf{T}} A$ and AA^{T}. Unaware of the works by Beltrami and Jordan, the English mathematician **James Sylvester (1814–1897)** also introduced the SVD over a decade later in the works by Sylvester (1889a, b). The SVD was later extended to complex-valued matrices by Autonne (1913) and to rectangular matrices by Eckart and Young (1939). In the process, Eckart and Young (1936) rediscovered a *low-rank approximation theorem* established earlier by the same German mathematician **Erhard Schmidt (1876–1959)** of Gram–Schmidt fame in the work by Schmidt (1907), and which is nowadays known as the Eckart–Young theorem – see the account by Stewart (1993). The proof of the following statement is left to Prob. 1.56 – see also Van Huffel and Vandewalle (1987).

Eckart–Young theorem (Schmidt (1907), Eckart and Young (1936)): *Given an $N \times N$ real matrix A, consider the problem of seeking a low-rank approximation for A of rank no larger than $r < N$ by solving the problem:*

$$\widehat{A} \triangleq \underset{\{x_m, y_m\}}{\operatorname{argmin}} \left\{ \left\| A - \sum_{m=1}^{r} x_m y_m^{\mathsf{T}} \right\|_{\mathrm{F}}^{2} \right\} \tag{1.222}$$

in terms of the Frobenius norm of the difference between A and its approximation, and where $\{x_m, y_m\}$ are column vectors to be determined. If we introduce the SVD of A:

$$A = \sum_{n=1}^{N} \sigma_n u_n v_n^{\mathsf{T}} \tag{1.223}$$

and order the singular values $\{\sigma_n\}$ in decreasing order, i.e., $\sigma_1 \geq \sigma_2 \geq \cdots \geq \sigma_N$, then the solution to (1.222) is given by

$$\widehat{A} = \sum_{n=1}^{r} \sigma_n u_n v_n^{\mathsf{T}} \tag{1.224}$$

in terms of the singular vectors $\{u_n, v_n\}$ associated with the r largest singular values.

Today, the SVD is a widely adopted tool in scientific computing. One of the most widely used procedures for its evaluation is the algorithm proposed by Golub and Kahan (1965) and refined by Golub and Reinsch (1970). For additional discussion on the SVD and its properties, the reader may consult Horn and Johnson (1990) and Golub and Van Loan (1996). In Appendix 1.B we provide one constructive proof for the SVD motivated by arguments from Horn and Johnson (1990), Strang (2009), Calafiore and El Ghaoui (2014), Lay, Lay, and McDonald (2014), and Nicholson (2019).

Matrix norms. A useful reference for the induced matrix norm (1.170) and, more generally, for vector and matrix norms and their properties, is Horn and Johnson (1990) – see also Golub and Van Loan (1996) and Meyer (2001). References for the block maximum norm (1.182) and (1.183) are Bertsekas and Tsitsiklis (1997), Takahashi and Yamada (2008), Takahashi, Yamada, and Sayed (2010), and Sayed (2014c); the latter reference provides several additional properties in its Appendix D and shows how this norm is useful in the study of multi-agent systems where block vector structures arise naturally.

Rayleigh–Ritz ratio. We described in Section 1.1 the Rayleigh–Ritz characterization of the eigenvalues of symmetric matrices, which we already know are real-valued. For an $N \times N$ real symmetric matrix, the quantity $x^{\mathsf{T}} Ax/x^{\mathsf{T}} x$ is called the Rayleigh–Ritz ratio after Ritz (1908, 1909) and the Nobel Laureate in Physics **Lord Rayleigh (1842–1919)** in the works by Rayleigh (1877, 1878). Both authors developed methods for determining the natural frequencies of vibrating systems (such as strings or bars). They transformed the problem of finding the natural frequencies into equivalent problems involving the determination of the stationary points of the ratio of quadratic terms. It appears that the solution method by Ritz (1908, 1909) was more complete with performance guarantees and is more widely adopted. Nevertheless, both authors relied on the use of the ratio of quadratic terms, which justifies the designation Rayleigh–Ritz quotient or ratio. Accounts of the contributions of Rayleigh and Ritz are given by Courant (1943) and Leissa (2005), and by Lindsay (1945) in the introduction to the 1945 Dover editions of Rayleigh (1877, 1878).

Eigenvalue perturbations. We described in Section 1.11 two useful results that provide bounds on the eigenvalues of matrices. The first result is Weyl theorem, which shows how the eigenvalues of a symmetric matrix are disturbed through additive perturbations to the entries of the matrix. The second result is Gershgorin theorem (also known as Gershgorin circle theorem or Gershgorin disc theorem), which specifies circular regions within which the eigenvalues of a matrix are located. The original references for these results are Weyl (1909, 1912) and Gershgorin (1931). A useful overview of Weyl inequality and its ramifications, along with historical remarks, appear in Bhatia (2001) and Stewart (1993); the latter reference discusses the significance of Weyl (1912) in the development of the theory of the SVD. A second useful overview of eigenvalue problems and perturbation results from the twentieth century appears in Golub and van der Vost (2001). For extensions and generalized treatments of both theorems, the reader my refer to Feingold and Varga (1962), Wilkinson (1965), Horn and Johnson (1990), Stewart and Sun (1990), Brualdi and Mellendorf (1994), Golub and Van Loan (1996), Demmel (1997), Parlett (1998), and Varga (2004).

Stochastic matrices. These matrices are prevalent in the study of Markov chains involving a finite number of states and in the study of distributed learning over graphs (see Chapters 25 and 38). The matrices are used to represent the transition probabilities from one state to another in the Markov chain:

$$[A]_{nm} \;=\; \mathbb{P}\,(\text{transitioning from state } n \text{ to state } m) \tag{1.225}$$

Discussions on properties of stochastic matrices can be found in Minc (1988), Horn and Johnson (1990), Berman and Plemmons (1994), Meyer (2001), Seneta (2007), and Sayed (2014c, appendix C). The existence of the Perron vector defined by (1.197) is guaranteed by a famous result known as the *Perron–Frobenius theorem* due to Perron (1907) and Frobenius (1908, 1909, 1912). The theorem applies more generally to matrices with nonnegative entries (i.e., the columns or rows of A do not need to add up to 1). A useful survey appears in Pillai, Suel, and Cha (2005). To state the theorem, we first introduce the notions of irreducible and primitive matrices.

Let A denote an $N \times N$ matrix with nonnegative entries. We view each entry $a_{\ell k}$ as a weight from state ℓ to state k. The matrix A is said to be *irreducible* if, and only if, for every pair of indices (ℓ, k), there exists a finite integer $n_{\ell k} > 0$ such that

$$[A^{n_{\ell k}}]_{\ell k} > 0 \tag{1.226}$$

From the rules of matrix multiplication, the (ℓ, k)th entry of the $n_{\ell k}$th power of A is given by:

$$[A^{n_{\ell k}}]_{\ell k} = \sum_{m_1=1}^{N} \sum_{m_2=1}^{N} \cdots \sum_{m_{n_{\ell k}-1}=1}^{N} a_{\ell m_1} a_{m_1 m_2} \ldots a_{m_{n_{\ell k}-1} k} \tag{1.227}$$

Therefore, property (1.226) means that there should exist a sequence of integer indices, denoted by $(\ell, m_1, m_2, \ldots, m_{n_{\ell k}-1}, k)$, which forms a path from state ℓ to state k with $n_{\ell k}$ nonzero weights denoted by $\{a_{\ell m_1}, a_{m_1,m_2}, \ldots, a_{m_{n_{\ell k}-1},k}\}$ such that

$$\ell \overset{a_{\ell m_1}}{\longrightarrow} m_1 \overset{a_{m_1,m_2}}{\longrightarrow} m_2 \longrightarrow \ldots \longrightarrow m_{n_{\ell k}-1} \overset{a_{m_{n_{\ell k}-1},k}}{\longrightarrow} k \qquad [n_{\ell k} \text{ edges}] \tag{1.228}$$

We assume that $n_{\ell k}$ is the smallest integer that satisfies this property. Note that under irreducibility, the power $n_{\ell k}$ is allowed to be dependent on the indices (ℓ, k). Therefore, irreducibility ensures that we can always find a path with nonzero weights linking state ℓ to state k.

A primitive matrix A is an irreducible matrix where, in addition, at least one a_{k_o,k_o} is positive for some state k_o. That is, there exists at least one state with a self-loop. It can be shown that when this holds, then an integer $n_o > 0$ exists such that (see Meyer (2001), Seneta (2007), Sayed (2014a), Prob. 1.50, and Appendix 25.A):

$$[A^{n_o}]_{\ell k} > 0, \quad \text{uniformly for all } (\ell, k) \tag{1.229}$$

Observe that the value of n_o is now *independent* of (ℓ, k). The following statement lists some of the properties that are guaranteed by the Perron–Frobenius theorem.

Perron–Frobenius theorem (Perron (1907), Frobenius (1908, 1909, 1912)): *Let A denote a square irreducible matrix with nonnegative entries and spectral radius denoted by $\lambda = \rho(A)$. Then, the following properties hold:*

(1) *$\lambda > 0$ and λ is a simple eigenvalue of A (i.e., it has multiplicity 1).*
(2) *There exists a right-eigenvector, p, with all its entries positive, such that $Ap = \lambda p$.*
(3) *There exists a left-eigenvector, q, with all its entries positive, such that $A^\mathsf{T} q = \lambda q$.*
(4) *All other eigenvectors, associated with the other eigenvalues of A, do not share the properties of p and q, i.e., their entries are not all positive (they can have negative and/or complex entries).*
(5) *The number of eigenvalues of A whose absolute values match $\rho(A)$ is called the period of A; we denote it by the letter P. Then, all eigenvalues of A whose absolute value match $\rho(A)$ are of the form $e^{\frac{j2\pi k}{P}}\lambda$, for $k = 0, 1, \ldots, P-1$. The period P is also equal to the greatest common divisor of all integers n for which $[A^n]_{kk} > 0$.*
(6) *When A is primitive, there exists a single eigenvalue of A that matches $\rho(A)$.*

PROBLEMS

1.1 Consider a matrix $U \in \mathbb{R}^{N\times N}$ satisfying $UU^{\mathsf{T}} = I_N$. Show that $U^{\mathsf{T}}U = I_N$.
1.2 Consider a square matrix $A \in \mathbb{R}^{N\times N}$. As explained prior to (1.168), the matrix A admits the canonical Jordan decomposition $A = UJU^{-1}$, where $J = \text{blkdiag}\{J_1, \ldots, J_R\}$ is a block diagonal matrix, say, with R blocks. Each J_r has dimensions $N_r \times N_r$ where N_r represents the multiplicity of the eigenvalue λ_r. Show that $\det A = \prod_{r=1}^{R} (\lambda_r)^{N_r}$.
1.3 The trace of a square matrix is the sum of its diagonal entries. Use the canonical Jordan factorization of Prob. 1.2, and the property $\text{Tr}(XY) = \text{Tr}(YX)$ for any matrices $\{X, Y\}$ of compatible dimensions, to show that the trace of a matrix is also equal to the sum of its eigenvalues.
1.4 What are the eigenvalues of the 2×2 matrix:

$$A = \begin{bmatrix} \cos\theta & -\sin\theta \\ \sin\theta & \cos\theta \end{bmatrix}$$

for any angle $\theta \in [0, 2\pi]$? What are the eigenvectors of A for any $\theta \neq 0$?
1.5 Consider an arbitrary matrix $A \in \mathbb{R}^{N\times M}$. Show that its row rank is equal to its column rank. That is, show that the number of independent columns is equal to the number of independent rows (for any N and M).
1.6 Consider two square matrices $A, B \in \mathbb{R}^{N\times N}$. The matrices are said to be similar if there exists a nonsingular matrix T such that $A = TBT^{-1}$. Show that similarity transformations preserve eigenvalues, i.e., the eigenvalues of A and B coincide.
1.7 Consider an arbitrary matrix $A \in \mathbb{R}^{N\times M}$. Show that the nonzero eigenvalues of AA^{T} and $A^{\mathsf{T}}A$ coincide with each other.
1.8 Consider two symmetric and nonnegative-definite matrices A and B. Verify that $\lambda_{\min}(B)\text{Tr}(A) \leq \text{Tr}(AB) \leq \lambda_{\max}(B)\text{Tr}(A)$.
1.9 Consider two $N \times N$ matrices A and B with singular values $\{\sigma_{A,n}, \sigma_{B,n}\}$ ordered such that $\sigma_{A,1} \geq \sigma_{A,2} \geq \ldots \geq \sigma_{A,N}$ and $\sigma_{B,1} \geq \sigma_{B,2} \geq \ldots \geq \sigma_{B,N}$. Establish the following trace inequality due to von Neumann (1937): $|\text{Tr}(AB)| \leq \sum_{n=1}^{N} \sigma_{A,n}\sigma_{B,n}$.
1.10 Consider arbitrary column vectors $x, y \in \mathbb{R}^N$. Verify that

$$(I_N + xy^{\mathsf{T}})^{-1} = I_N - \frac{xy^{\mathsf{T}}}{1 + y^{\mathsf{T}}x}$$

1.11 Consider two $M \times M$ invertible matrices $\{R_a, R_b\}$ and two $M \times 1$ vectors $\{x_a, x_b\}$. In many inference problems, we will be faced with constructing a new matrix R_c and a new vector x_c using the following two relations (also known as fusion equations)

$$R_c^{-1} \triangleq R_a^{-1} + R_b^{-1}, \quad R_c^{-1}x_c \triangleq R_a^{-1}x_a + R_b^{-1}x_b$$

Show that these equations can be rewritten in any of the following equivalent forms:

$$\begin{aligned} R_c &= R_a - R_a(R_a + R_b)^{-1}R_a, & x_c &= x_a + R_cR_b^{-1}(x_b - x_a) \\ R_c &= R_b - R_b(R_a + R_b)^{-1}R_b, & x_c &= x_b + R_cR_a^{-1}(x_a - x_b) \end{aligned}$$

1.12 Verify that the vectors $\{q_m\}$ that result from the Gram–Schmidt construction (1.99a)–(1.99b) are orthonormal.
1.13 Consider an arbitrary matrix $A \in \mathbb{R}^{N\times M}$ of rank r and refer to its SVD representation (1.108a) or (1.108b). Introduce the partitioning $U = \begin{bmatrix} U_1 & U_2 \end{bmatrix}$ and $V = \begin{bmatrix} V_1 & V_2 \end{bmatrix}$ where U_1 is $N \times r$ and V_1 is $M \times r$. Show that the matrices $\{U_1, U_2, V_1, V_2\}$ provide orthonormal basis for the four fundamental spaces $\{\mathcal{R}(A), \mathcal{N}(A^{\mathsf{T}}), \mathcal{R}(A^{\mathsf{T}}), \mathcal{N}(A)\}$.

1.14 Assume A, B are symmetric positive-definite. Show that

$$\lambda_{\max}(B^{-1}A) \;=\; \max_{x\neq 0}\left\{\frac{x^{\mathsf T}Ax}{x^{\mathsf T}Bx}\right\}$$

and that the maximum is attained when x is an eigenvector of $B^{-1}A$ that is associated with its largest eigenvalue.

1.15 Consider matrices A, B, C, and D of compatible dimensions. Show that

$$\mathrm{Tr}(A^{\mathsf T}BCD^{\mathsf T}) \;=\; (\mathrm{vec}(A))^{\mathsf T}\,(D\otimes B)\mathrm{vec}(C)$$

1.16 Establish the singular value property (7) for Kronecker products from Table 1.1.

1.17 Establish the validity of property (10) from Table 1.1. Show that it can also be written in the equivalent form $\mathrm{Tr}(AB) = (\mathrm{vec}(B^*))^*\,\mathrm{vec}(A)$ in terms of the complex conjugation operation (instead of matrix transposition).

1.18 Verify that when $B\in\mathbb{C}^{N\times N}$ is Hermitian (i.e., $B = B^*$), it holds that $\mathrm{Tr}(AB) = (\mathrm{vec}(B))^*\,\mathrm{vec}(A)$.

1.19 Consider a Lyapunov recursion of the form $Z_{i+1} = AZ_iA^{\mathsf T} + B$, for $i\geq 0$ and where $Z_i\in\mathbb{R}^{N\times N}$ with square matrices $\{A, B\}$. Shown that when A is stable (i.e., when all its eigenvalues lie strictly inside the unit disc), the matrix Z_i will converge to the unique solution of the Lyapunov equation $Z = AZA^{\mathsf T} + B$.

1.20 Refer to the exponential function series (1.139). Show that $\frac{d}{dt}e^{At} = Ae^{At}$.

1.21 Show that the infinity and p-norms of a vector $x\in\mathbb{R}^M$ are related via $\|x\|_\infty = \lim_{p\to\infty}\|x\|_p$.

1.22 Partition a vector into sub-vectors as $x = \mathrm{blkcol}\{x_1, x_2, \ldots, x_K\}$ and define $\|x\|_{1,p} = \sum_{k=1}^{K}\|x_k\|_p$, which is equal to the sum of the ℓ_p-norms of the sub-vectors. Show that $\|x\|_{1,p}$ is a valid vector norm.

1.23 Establish properties (a)–(e) for the block maximum norm stated right after (1.183). *Remark.* More discussion on the properties of this norm appears in Appendix D of Sayed (2014c).

1.24 Let $\|\cdot\|$ denote some norm in $\mathbb{R}^M$. The associated dual norm is denoted by $\|x\|_\star$ and defined by (1.157). Show that $\|x\|_\star$ is a valid vector norm. Show that it can be expressed equivalently by (1.158).

1.25 Let $p, q\geq 1$ satisfy $1/p + 1/q = 1$. Show that the norms $\|x\|_q$ and $\|x\|_p$ are dual of each other.

1.26 Show that:

(a) The dual of the ℓ_1-norm is the ℓ_∞-norm.

(b) The dual of the ℓ_∞-norm is the ℓ_1-norm.

1.27 Show that the dual norm of a dual norm is the original norm, i.e., $\|x\|_{\star\star} = \|x\|$.

1.28 Show that the dual norm of the Frobenius norm is the Frobenius norm itself.

1.29 Show that the dual norm of the 2-induced norm of a matrix is the nuclear norm, which is defined as the sum of its singular values.

1.30 Refer to the matrix norms in Table 1.4. Does it hold that $\|A\| = \|A^{\mathsf T}\|$?

1.31 Show that, for any matrix norm, $|\mathrm{Tr}(A)|\leq c\|A\|$ for some constant $c\geq 0$.

1.32 Let R_u denote an $M\times M$ symmetric positive-definite matrix with eigenvalues $\{\lambda_m\}$. Show that $\mathrm{Tr}(R_u)\mathrm{Tr}(R_u^{-1})\geq M^2$ and $(\mathrm{Tr}(R_u))^2\leq M\mathrm{Tr}(R_u^2)$.

1.33 Consider two symmetric nonnegative-definite matrices, $A\geq 0$ and $B\geq 0$. Show that $\mathrm{Tr}(AB)\leq\mathrm{Tr}(A)\,\mathrm{Tr}(B)$.

1.34 For any symmetric matrices A and B satisfying $A\geq B\geq 0$, show that $\det A\geq \det B$. Here, the notation $A\geq B$ means that the difference $A-B$ is nonnegative-definite.

1.35 Consider a symmetric $N\times N$ nonnegative-definite matrix, $A\geq 0$, and a second arbitrary matrix B also of size $N\times N$. Show that $|\mathrm{Tr}(AB)|\leq\mathrm{Tr}(A)\,\|B\|$, in terms of the 2-induced norm (maximum singular value) of B.

1.36 Show that the spectral radius of a square symmetric matrix A agrees with its spectral norm (maximum singular value), i.e., $\rho(A) = \|A\|$.

1.37 For any matrix norm, show that the spectral radius of a square matrix A satisfies $\rho(A) \leq \|A\|$.

1.38 For any matrix norm and $\epsilon > 0$, show that $\|A^n\|^{1/n} \leq \rho(A) + \epsilon$ for n large enough.

1.39 For any matrix norm, show that the spectral radius of a square matrix A satisfies $\rho(A) = \lim_{n\to\infty} \|A^n\|^{1/n}$.

1.40 Let H denote a positive-definite symmetric matrix and let G denote a symmetric matrix of compatible dimensions. Show that $HG \geq 0$ if, and only if, $G \geq 0$, where the notation $A \geq 0$ means that all eigenvalues of matrix A are nonnegative.

1.41 Introduce the notation $\|x\|_\Sigma^2 = x^\mathsf{T}\Sigma x$, where Σ is a symmetric and nonnegative-definite matrix.

(a) Show that $\|x\|_\Sigma = \sqrt{x^\mathsf{T}\Sigma x}$ is a valid vector norm when Σ is positive-definite, i.e., verify that it satisfies all the properties of vector norms.

(b) When Σ is singular, which properties of vector norms are violated?

1.42 Show that the condition numbers of H and $H^\mathsf{T}H$ satisfy $\kappa(H^\mathsf{T}H) = \kappa^2(H)$.

1.43 Consider a linear system of equations of the form $Ax = b$ and let $\kappa(A)$ denote the condition number of A. Assume A is invertible so that the solution is given by $x = A^{-1}b$. Now assume that b is perturbed slightly to $b+\delta b$. The new solution becomes $x+\delta x$, where $\delta x = A^{-1}\delta b$. The relative change in b is $\beta = \|\delta b\|/\|b\|$. The relative change in the solution is $\alpha = \|\delta x\|/\|x\|$. When the matrix is ill-conditioned, the relative change in the solution can be much larger than the relative change in b. Indeed, verify that $\alpha/\beta \leq \kappa(A)$. Can you provide an example where the ratio achieves $\kappa(A)$?

1.44 Consider a matrix Y of the form $Y = I - \beta xx^\mathsf{T}$, for some real scalar β. For what condition on β is Y positive semi-definite? When this is the case, show that Y admits a symmetric square-root factor of the form $Y^{1/2} = I - \alpha xx^\mathsf{T}$ for some real scalar α.

1.45 Let $\Phi = \sum_{n=0}^{N} \lambda^{N-n} h_n h_n^\mathsf{T}$, where the h_n are $M \times 1$ vectors and $0 \ll \lambda \leq 1$. The matrix Φ can have full rank or it may be rank deficient. Assume its rank is $r \leq M$. Let $\Phi^{1/2}$ denote an $M \times r$ full rank square-root factor, i.e., $\Phi^{1/2}$ has rank r and satisfies $\Phi^{1/2}\Phi^{\mathsf{T}/2} = \Phi$. Show that h_n belongs to the column span of $\Phi^{1/2}$.

1.46 Refer to Example 1.9. Extend the result to the case $A^\mathsf{T}A = B^\mathsf{T}B$ where now $N \geq M$.

1.47 We provide another derivation for the basis rotation result from Example 1.9 by assuming that the matrices A and B have full rank. Introduce the QR decompositions

$$A^\mathsf{T} = Q_A \begin{bmatrix} R_A \\ 0 \end{bmatrix}, \qquad B^\mathsf{T} = Q_B \begin{bmatrix} R_B \\ 0 \end{bmatrix}$$

where Q_A and Q_B are $M \times M$ orthogonal matrices, and R_A and R_B are $N \times N$ upper-triangular matrices with positive diagonal entries (due to the full rank assumption on A and B).

(a) Show that $AA^\mathsf{T} = R_A^\mathsf{T} R_A = R_B^\mathsf{T} R_B$.

(b) Conclude, by uniqueness of the Cholesky factorization, that $R_A = R_B$. Verify further that $\Theta = Q_B Q_A^\mathsf{T}$ is orthogonal and maps B to A.

1.48 Consider a matrix $B \in \mathbb{R}^{N\times M}$ and let $\sigma_{\min}^2(B)$ denote its smallest nonzero singular value. Let $x \in \mathcal{R}(B)$ (i.e., x is any vector in the range space of B). Use the eigenvalue decomposition of $B^\mathsf{T}B$ to verify that $\|B^\mathsf{T}x\|^2 \geq \sigma_{\min}^2(B)\,\|x\|^2$.

1.49 Refer to Section 1.12 on stochastic matrices. Show that the spectral radius of a left- or right-stochastic matrix is equal to 1.

1.50 Let A denote an irreducible matrix with nonnegative entries. Show that if $a_{k_o,k_o} > 0$ for some k_o, then A is a primitive matrix.

1.51 Let A denote a matrix with positive entries. Show that A is primitive.

1.52 Show that to check whether an $N \times N$ left-stochastic matrix A is irreducible (primitive), we can replace all nonzero entries in A by 1s and verify instead whether the resulting matrix is irreducible (primitive).

1.53 Consider an $N\times N$ left-stochastic matrix A that is irreducible but not necessarily primitive. Let $B = 0.5(I + A)$. Is B left-stochastic? Show that the entries of B^{N-1} are all positive. Conclude that B is primitive.

1.54 Assume A is a left-stochastic primitive matrix of size $N \times N$.

(a) Show that A is power convergent and the limit converges to the rank-one product $\lim_{n\to\infty} A^n = p\mathbb{1}^\mathsf{T}$, where p is the Perron vector of A. Is the limit a primitive matrix?

(b) For any vector $b = \text{col}\{b_1, b_2, \ldots, b_N\}$, show that $\lim_{n\to\infty} A^n b = \alpha p$, where $\alpha = b_1 + b_2 + \ldots + b_N$.

(c) If A is irreducible but not necessarily primitive, does the limit of part (a) exist?

1.55 Consider an $N \times N$ left-stochastic matrix A. Let $n_o = N^2 - 2N + 2$. Show that A is primitive if, and only if, $[A^{n_o}]_{\ell k} > 0$ for all ℓ and k.

1.56 Establish the validity of the Eckart–Young approximation expression (1.224).

1.57 Consider an $M \times N$ matrix A (usually $N \geq M$ with more columns than rows). The *spark* of A is defined as the smallest number of linearly *dependent* columns in A, also written as:

$$\text{spark}(A) \stackrel{\Delta}{=} \min_{d\neq 0} \Big\{ \|d\|_0 \Big\}, \quad \text{subject to } Ad = 0$$

where $\|d\|_0$ denotes the number of nonzero arguments in the vector d. If A has full rank, then its spark is set to ∞. Now consider the linear system of equations $Ax^o = b$ and assume x^o is k-sparse in the sense that only k of its entries are nonzero. Show that x^o is the only solution to the following problem (i.e., the only k-sparse vector satisfying the linear equations):

$$x^o = \underset{x\in\mathbb{R}^M}{\text{argmin}} \Big\{ \|x\|_0 \Big\}, \quad \text{subject to } Ax = b$$

if, and only if, $\text{spark}(A) > 2k$. *Remark.* We describe in Appendix 58.A the *orthogonal matching pursuit* (OMP) algorithm for finding the sparse solution x^o.

1.58 Consider an $M \times N$ matrix A. The matrix is said to satisfy a *restricted isometry property* (RIP) with constant λ_k if for any k-sparse vector x it holds $(1 - \lambda_k)\|x\|_2 \leq \|Ax\|_2 \leq (1 + \lambda_k)\|x\|_2$. Now let $Ax_1 = b_1$ and $Ax_2 = b_2$. Assume A satisfies RIP for any sparse vectors of level $2k$. Show that for any k-sparse vectors $\{x_1, x_2\}$, the corresponding measurements $\{b_1, b_2\}$ will be sufficiently away from each other in the sense that $\|b_1 - b_2\|_2 \geq (1 - \lambda_{2k})\|x_1 - x_2\|_2$. *Remark.* For more discussion on sparsity and the RIP condition, the reader may refer to the text by Elad (2010) and the many references therein, as well as to the works by Candes and Tao (2006) and Candes, Romberg, and Tao (2006).

1.A PROOF OF SPECTRAL THEOREM

The arguments in this appendix and the next on the spectral theorem and the SVD are motivated by the presentations in Horn and Johnson (1990), Trefethen and Bau (1997), Strang (2009), Calafiore and El Ghaoui (2014), Lay, Lay, and McDonald (2014), and Nicholson (2019).

In this first appendix, we establish the validity of the eigen-decomposition (1.15a) for $N \times N$ real symmetric matrices A. We start by verifying that A has at least one real eigenvector. For this purpose, we first recall that an equivalent characterization of the eigenvalues of a matrix is that they are the roots of its characteristic polynomial, defined as

$$p(\lambda) \stackrel{\Delta}{=} \det(\lambda I_N - A) \tag{1.230}$$

in terms of the determinant of the matrix $\lambda I_N - A$. Note that $p(\lambda)$ is a polynomial of order N and, by the fundamental theorem of algebra, every such polynomial has N

roots. We already know that these roots are all real when A is symmetric. Therefore, there exists some real value λ_1 such that $p(\lambda_1) = 0$. The scalar λ_1 makes the matrix $\lambda_1 I_N - A$ singular since its determinant will be zero. In this way, the columns of the matrix $\lambda_1 I_N - A$ will be linearly dependent and there must exist some nonzero real vector u_1 such that

$$(\lambda_1 I_N - A)u_1 = 0 \tag{1.231}$$

This relation establishes the claim that there exists some nonzero real vector u_1 satisfying $Au_1 = \lambda_1 u_1$. We can always scale u_1 to satisfy $\|u_1\| = 1$.

Induction argument

One traditional approach to establish the spectral theorem is by induction. Assume the theorem holds for all symmetric matrices of dimensions up to $(N-1)\times(N-1)$ and let us prove that the statement also holds for the next dimension $N \times N$; it certainly holds when $N = 1$ (in which case A is a scalar). Thus, given an $N \times N$ real symmetric matrix A, we already know that it has at least one real eigenvector u_1 of unit-norm associated with a real eigenvalue λ_1. Since u_1 lies in N-dimensional space, we can choose $N-1$ real vectors $\{\bar{v}_2, \bar{v}_3, \ldots, \bar{v}_N\}$ such that the columns of the $N \times N$ matrix

$$\bar{V} = \left[\begin{array}{c|cccc} u_1 & \bar{v}_2 & \bar{v}_3 & \ldots & \bar{v}_N \end{array}\right] \tag{1.232}$$

are linearly independent. The columns of $\bar{V}$ constitute a basis for the N-dimensional space, $\mathbb{R}^N$. We apply the Gram–Schmidt orthogonalization procedure to the trailing columns of $\bar{V}$ and replace the $\{\bar{v}_n\}$ by a new set of vectors $\{v_n\}$ that have unit norm each, and such that the columns of the matrix V below are all orthogonal to each other:

$$V = \left[\begin{array}{c|cccc} u_1 & v_2 & v_3 & \ldots & v_N \end{array}\right] \stackrel{\Delta}{=} \left[\begin{array}{cc} u_1 & V_1 \end{array}\right] \tag{1.233}$$

Note that we kept u_1 unchanged; we are also denoting the trailing columns of V by V_1. Now, multiplying A by V^{T} from the left and by V from the right we get

$$V^{\mathsf{T}}AV = \begin{bmatrix} u_1^{\mathsf{T}}Au_1 & u_1^{\mathsf{T}}AV_1 \\ V_1^{\mathsf{T}}Au_1 & V_1^{\mathsf{T}}AV_1 \end{bmatrix} = \begin{bmatrix} \lambda_1\|u_1\|^2 & \lambda_1 u_1^{\mathsf{T}}V_1 \\ \lambda_1 V_1^{\mathsf{T}}u_1 & V_1^{\mathsf{T}}AV_1 \end{bmatrix} = \begin{bmatrix} \lambda_1 & 0 \\ 0 & V_1^{\mathsf{T}}AV_1 \end{bmatrix} \tag{1.234}$$

since $\|u_1\|^2 = 1$ and u_1 is orthogonal to the columns of V_1 (so that $V_1^{\mathsf{T}}u_1 = 0$). Note that we used the fact that A is symmetric in the above calculation to conclude that

$$u_1^{\mathsf{T}}A = u_1^{\mathsf{T}}A^{\mathsf{T}} = (Au_1)^{\mathsf{T}} = (\lambda_1 u_1)^{\mathsf{T}} = \lambda_1 u_1^{\mathsf{T}} \tag{1.235}$$

Thus, observe that the matrix product $V^{\mathsf{T}}AV$ turns out to be block diagonal with λ_1 as its $(1,1)$ leading entry and with the $(N-1)\times(N-1)$ real symmetric matrix, $V_1^{\mathsf{T}}AV_1$, as its trailing block. We know from the induction assumption that this smaller symmetric matrix admits a full set of orthonormal eigenvectors. That is, there exists an $(N-1)\times(N-1)$ real orthogonal matrix W_1 and a diagonal matrix Λ_1 with real entries such that

$$V_1^{\mathsf{T}}AV_1 = W_1\Lambda_1W_1^{\mathsf{T}}, \quad W_1^{\mathsf{T}}W_1 = I_{N-1} \tag{1.236}$$

or, equivalently,

$$W_1^{\mathsf{T}}V_1^{\mathsf{T}}AV_1W_1 = \Lambda_1 \tag{1.237}$$

Using this equality, we obtain from (1.234) that

$$\underbrace{\begin{bmatrix} 1 & \\ & W_1^{\mathsf{T}} \end{bmatrix} V^{\mathsf{T}}}_{=\,U^{\mathsf{T}}} A \underbrace{V \begin{bmatrix} 1 & \\ & W_1 \end{bmatrix}}_{\stackrel{\Delta}{=}\,U} = \underbrace{\begin{bmatrix} \lambda_1 & \\ & \Lambda_1 \end{bmatrix}}_{\stackrel{\Delta}{=}\,\Lambda} \tag{1.238}$$

with a diagonal matrix on the right-hand side. The matrix U can be verified to be orthogonal since

$$U^{\mathsf T}U = \begin{bmatrix} 1 & \\ & W_1^{\mathsf T} \end{bmatrix} \underbrace{V^{\mathsf T}V}_{=I} \begin{bmatrix} 1 & \\ & W_1 \end{bmatrix} = \begin{bmatrix} 1 & \\ & W_1^{\mathsf T}W_1 \end{bmatrix} = I_N \tag{1.239}$$

Therefore, we established that an orthogonal matrix U and a real diagonal matrix Λ exist such that $A = U\Lambda U^{\mathsf T}$, as claimed.

1.B CONSTRUCTIVE PROOF OF SVD

One proof of the SVD defined by (1.108a)–(1.108b) follows from the eigen-decomposition of symmetric nonnegative-definite matrices. The argument given here assumes $N \leq M$, but it can be adjusted to handle the case $N \geq M$. Thus, note that the product $AA^{\mathsf T}$ is a symmetric nonnegative-definite matrix of size $N \times N$. Consequently, from the spectral theorem, there exists an $N \times N$ orthogonal matrix U and an $N \times N$ diagonal matrix Σ^2, with nonnegative entries, such that

$$AA^{\mathsf T} = U\Sigma^2 U^{\mathsf T} \tag{1.240}$$

This representation corresponds to the eigen-decomposition of $AA^{\mathsf T}$. The diagonal entries of Σ^2 are the eigenvalues of $AA^{\mathsf T}$, which are nonnegative (and, hence, the notation Σ^2); the nonzero entries of Σ^2 also coincide with the nonzero eigenvalues of $A^{\mathsf T}A$ – see Prob. 1.7. The columns of U are the orthonormal eigenvectors of $AA^{\mathsf T}$. By proper reordering, we can arrange the diagonal entries of Σ^2, denoted by $\{\sigma_r^2\}$, in decreasing order so that Σ^2 can be put into the form

$$\Sigma^2 = \text{diagonal}\left\{\sigma_1^2, \sigma_2^2, \ldots, \sigma_r^2, 0, \ldots, 0\right\} \triangleq \begin{bmatrix} \Lambda^2 & \\ & 0_{N-r} \end{bmatrix} \tag{1.241}$$

where $r = \text{rank}(AA^{\mathsf T})$ and $\sigma_1^2 \geq \sigma_2^2 \geq \ldots \geq \sigma_r^2 > 0$. The $r \times r$ diagonal matrix Λ consists of the positive entries

$$\Lambda \triangleq \text{diagonal}\{\sigma_1, \sigma_2, \ldots, \sigma_r\} > 0 \tag{1.242}$$

We partition $U = \begin{bmatrix} U_1 & U_2 \end{bmatrix}$, where U_1 is $N \times r$, and conclude from the orthogonality of U (i.e., from $U^{\mathsf T}U = I$) that $U_1^{\mathsf T}U_1 = I_r$ and $U_1^{\mathsf T}U_2 = 0$. If we substitute into (1.240) we find that

$$AA^{\mathsf T} = U_1\Lambda^2 U_1^{\mathsf T} \tag{1.243a}$$

$$AA^{\mathsf T}U = U\Sigma^2 \implies AA^{\mathsf T}U_2 = 0 \overset{(1.50)}{\Longleftrightarrow} A^{\mathsf T}U_2 = 0 \tag{1.243b}$$

where the middle expression in the last line is indicating that the columns of U_2 belong to the nullspace of $AA^{\mathsf T}$. But since we know from property (1.50) that $\mathcal{N}(AA^{\mathsf T}) = \mathcal{N}(A^{\mathsf T})$, we conclude that the columns of U_2 also belong to the nullspace of $A^{\mathsf T}$. Next, we introduce the $M \times r$ matrix

$$V_1 \triangleq A^{\mathsf T}U_1\,\Lambda^{-1} \tag{1.244}$$

The columns of V_1 are orthonormal since

$$V_1^{\mathsf T}V_1 = \underbrace{\Lambda^{-1}U_1^{\mathsf T}A}_{=V_1^{\mathsf T}}\underbrace{A^{\mathsf T}U_1\Lambda^{-1}}_{V_1} = \Lambda^{-1}U_1^{\mathsf T}\underbrace{U_1\Lambda^2U_1^{\mathsf T}}_{=AA^{\mathsf T}}U_1\Lambda^{-1} = I_r \tag{1.245a}$$

Moreover, it also holds that

$$V_1^\mathsf{T} A^\mathsf{T} U_1 = \underbrace{\Lambda^{-1} U_1^\mathsf{T} A}_{=V_1^\mathsf{T}} A^\mathsf{T} U_1 = \Lambda^{-1} U_1^\mathsf{T} \underbrace{U_1 \Lambda^2 U_1^\mathsf{T}}_{=AA^\mathsf{T}} U_1 = \Lambda^{-1}\Lambda^2 = \Lambda \tag{1.245b}$$

and, similarly,

$$V_1^\mathsf{T} A^\mathsf{T} U_2 = 0_{r\times(N-r)}, \quad \text{since } A^\mathsf{T} U_2 = 0 \tag{1.245c}$$

Combining (1.245b)–(1.245c) gives

$$V_1^\mathsf{T} A^\mathsf{T} U = \begin{bmatrix} \Lambda & 0_{r\times(N-r)} \end{bmatrix} \tag{1.246}$$

Now, since V_1 has r orthonormal columns in M-dimensional space, we can add $M - r$ more columns to enlarge V_1 into an $M \times M$ orthogonal matrix V as follows:

$$V \triangleq \begin{bmatrix} V_1 & V_2 \end{bmatrix}, \quad V^\mathsf{T} V = I_M \tag{1.247}$$

which implies that the new columns in $V_2 \in \mathbb{R}^{M\times(M-r)}$ should satisfy $V_2^\mathsf{T} V_1 = 0$ and $V_2^\mathsf{T} V_2 = I$. It follows that

$$\begin{aligned} V_2^\mathsf{T} V_1 = 0 &\Longrightarrow V_2^\mathsf{T} \underbrace{A^\mathsf{T} U_1 \Lambda^{-1}}_{=V_1} = 0 \\ &\Longrightarrow V_2^\mathsf{T} A^\mathsf{T} U_1 = 0_{(M-r)\times r}, \quad \text{since } \Lambda > 0 \end{aligned} \tag{1.248}$$

and

$$V_2^\mathsf{T} A^\mathsf{T} U_2 = 0_{(M-r)\times(N-r)}, \quad \text{since } A^\mathsf{T} U_2 = 0 \tag{1.249}$$

Adding these conclusions into (1.246) we can write

$$\begin{bmatrix} V_1^\mathsf{T} \\ V_2^\mathsf{T} \end{bmatrix} A^\mathsf{T} U = \begin{bmatrix} \Lambda & 0_{r\times(N-r)} \\ 0_{(M-r)\times r} & 0_{(M-r)\times(N-r)} \end{bmatrix} \tag{1.250}$$

and we conclude that orthogonal matrices U and V exist such that

$$A = U \underbrace{\begin{bmatrix} \Lambda & 0_{r\times(M-r)} \\ 0_{(N-r)\times r} & 0_{(N-r)\times(M-r)} \end{bmatrix}}_{=\begin{bmatrix} \Sigma & 0_{N\times(M-N)} \end{bmatrix}} V^\mathsf{T} \tag{1.251}$$

as claimed by (1.108a). A similar argument establishes the SVD decomposition of A when $N > M$.

REFERENCES

Aitken, A. C. (1939), *Determinants and Matrices*, 9th ed., Interscience Publishers.

Autonne, L. (1913), "Sur les matrices hypohermitiennes et les unitairs," *Comptes Rendus de l'Academie Sciences*, vol. 156, pp. 858–860.

Banachiewicz, T. (1937a), "Sur l'inverse d'un cracovien et une solution générale d'un systeme d'équations linéaires," *Comptes Rendus Mensules des Séances de la Classe des Sciences Mathématique et Naturelles de l'Académie Polonaise des Sciences et des Lettres*, vol. 4, pp. 3–4.

Banachiewicz, T. (1937b), "Zur Berechungung der Determinanten, wie auch der Inversen, und zur darauf basierten Auflosung der Systeme lineare Gleichungen," *Acta Astronomica*, Sér. C, vol. 3, pp. 41–67.

Bartlett, M. S. (1951), "An inverse matrix adjustment arising in discriminant analysis," *Ann. Math. Statist.*, vol. 22, pp. 107–111.

Bellman, R. E. (1970), *Introduction to Matrix Analysis*, 2nd ed., McGraw Hill.

Beltrami, E. (1873), "Sulle funzioni bilineari," *Giornale di Matematiche ad Uso degli Studenti Delle Universita*, vol. 11, pp. 98–110.

Berman, A. and R. J. Plemmons (1994), *Nonnegative Matrices in the Mathematical Sciences*, SIAM.

Bernstein, D. S. (2018), *Scalar, Vector, and Matrix Mathematics: Theory, Facts, and Formulas*, revised ed., Princeton University Press.

Bertsekas, D. P. and J. N. Tsitsiklis (1997), *Parallel and Distributed Computation: Numerical Methods*, Athena Scientific.

Bhatia, R. (2001), "Linear algebra to quantum cohomology: The story of Alfred Horn's inequalities," *Am. Math. Monthly*, vol. 108, no. 4, pp. 289–318.

Bjorck, A. (1996), *Numerical Methods for Least Squares Problems,* SIAM.

Bródy, F. and T. Vámos (1995), *The Neumann Compendium*, vol. 1, World Scientific.

Brualdi, R. A. and S. Mellendorf (1994), "Regions in the complex plane containing the eigenvalues of a matrix," *Amer. Math. Monthly*, vol. 101, pp. 975–985.

Calafiore, G. C. and L. El Ghaoui (2014), *Optimization Models*, Cambridge University Press.

Candes, E. J., J. K. Romberg, and T. Tao (2006), "Robust uncertainty principles: Exact signal reconstruction from highly incomplete frequency information," *IEEE Trans. Inf. Theory*, vol. 52, no. 2, pp. 489–509.

Candes, E. J. and T. Tao (2006), "Near-optimal signal recovery from random projections: Universal encoding strategies," *IEEE Trans. Inf. Theory*, vol. 52, pp. 5406–5425.

Carrera, J. P. (1992), "The fundamental theorem of algebra before Carl Friedrich Gauss," *Publ. Mat.*, vol. 36, pp. 879–911.

Cauchy, A.-L. (1829), "Sur l'équation a l'aide de laquelle on determine les inégalités séculaires des mouvements des planetes," *Exer. de math.*, vol. 4, pp. 174–195.

Constantinescu, T. (1996), *Schur Parameters, Factorization, and Dilation Problems*, Birkhaüser.

Cottle, R. W. (1974), "Manifestations of the Schur complement," *Linear Algebra Appl.*, vol. 8, pp. 189–211.

Courant, R. (1943), "Variational methods for the solution of problems of equilibrium and vibrations," *Bull. Amer. Math. Soc.*, vol. 49, pp. 1–23.

Demmel, J. (1997), *Applied Numerical Linear Algebra*, SIAM.

Duncan, W. J. (1944), "Some of the solution of large sets of simultaneous linear equations (with an appendix on the reciprocation of partitioned matrices)," *Lond. Edinb. Dublin Philos. Mag. J. Sci.*, Seventh Series, vol. 35, pp. 660–670.

Eckart, C. and G. Young (1936), "The approximation of one matrix by another of lower rank," *Psychometrika*, vol. 1, no. 3, pp. 211–218.

Eckart, C. and G. Young (1939), "A principal axis transformation for non-Hermitian matrices," *Bull. Amer. Math. Soc.*, vol. 45, pp. 118–121.

Elad, M. (2010), *Sparse and Redundant Representations*, Springer.

Feingold, D. G. and R. S. Varga (1962), "Block diagonally dominant matrices and generalizations of the Gerschgorin circle theorem," *Pacific J. Math.*, vol. 12, pp. 1241–1250.

Fine, B. and G. Rosenberger (1997), *The Fundamental Theorem of Algebra*, Springer.

Frobenius, F. G. (1908), "Uber matrizen aus positiven elementen, 1" *Sitzungsber. Konigl. Preuss. Akad. Wiss.*, pp. 471–476.

Frobenius, G. (1909), "Uber matrizen aus positiven elementen, 2," *Sitzungsber. Konigl. Preuss. Akad. Wiss.*, pp. 514–518.

Frobenius, G. (1912), "Uber matrizen aus nicht negativen elementen," *Sitzungsber. Konigl. Preuss. Akad. Wiss.*, pp. 456–477.

Gantmacher, F. R. (1959), *The Theory of Matrices*, Chelsea Publishing Company.

Gelfand, I. M. (1989), *Lectures on Linear Algebra*, Dover Publications.

Gershgorin, S. (1931), "Über die abgrenzung der eigenwerte einer matrix," *Izv. Akad. Nauk. USSR Otd. Fiz.-Mat. Nauk*, vol. 7, pp. 749–754.

Golub, G. H. and W. Kahan (1965), "Calculating the singular values and pseudo-inverse of a matrix," *J. Soc. Indust. Appl. Math.: Ser. B, Numer. Anal.*, vol. 2, no. 2, pp. 205–224.

Golub, G. H. and C. Reinsch (1970), "Singular value decomposition and least squares solutions," *Numerische Mathematik*, vol. 14, no. 5, pp. 403–420.

Golub, G. H. and H. A. van der Vost (2001), "Eigenvalue computation in the 20th century," in *Numerical Analysis: Historical Developments in the 20th Century*, C. Brezinski and L. Wuytack, editors, pp. 209–238, North-Holland, Elsevier.

Golub, G. H. and C. F. Van Loan (1996), *Matrix Computations*, 3rd ed., John Hopkins University Press.

Gram, J. (1883), "Ueber die Entwickelung reeller Funtionen in Reihen mittelst der Methode der kleinsten Quadrate," *J. Reine Angew. Math.*, vol. 94, pp. 41–73.

Guttman, L. (1946), "Enlargement methods for computing the inverse matrix," *Ann. Math. Statist.*, vol. 17, pp. 336–343.

Hackbusch, W. (2012), *Tensor Spaces and Numerical Tensor Calculus*, Springer.

Hager, W. W. (1989), "Updating the inverse of a matrix," *SIAM Rev.*, vol. 31, no. 2, pp. 221–239.

Halmos, P. R. (1963), "What does the spectral theorem say?" *Am. Math. Monthly*, vol. 70, no. 3, pp. 241–247.

Halmos, P. R. (1974), *Finite-Dimensional Vector Spaces*, Springer.

Halmos, P. R. (2013), *Introduction to Hilbert Space and the Theory of Spectral Multiplicity*, Martino Fine Books.

Hawkins, T. (1975), "Cauchy and the spectral theory of matrices," *Historia Mathematica*, vol. 2, no. 1, pp. 1–29.

Haynsworth, E. V. (1968), "Determination of the inertia of a partitioned Hermitian matrix," *Linear Algebra Appl.*, vol. 1, pp. 73–81.

Henderson, H. V., F. Pukelsheim, and S. R. Searle (1983), "On the history of the Kronecker product," *Linear Multilinear Algebra*, vol. 14, pp. 113–120.

Henderson, H. V. and S. R. Searle (1981a), "On deriving the inverse of a sum of matrices," *SIAM Rev.*, vol. 23, pp. 53–60.

Henderson, H. V. and S. R. Searle (1981b), "The vec-permutation matrix, the vec operator, and Kronecker products: A review," *Linear Multilinear Algebra*, vol. 9, pp. 271–288.

Hensel, K. (1891), "Uber die Darstellung der Determinante eines Systems, welches aus zwei anderen componirt ist," *ACTA Mathematica*, vol. 14, pp. 317–319.

Ho, Y. C. (1963), "On the stochastic approximation method and optimal filter theory," *J. Math. Anal. Appl.*, vol. 6, pp. 152–154.

Hogben, L., (2014), editor, *Handbook of Linear Algebra*, 2nd ed., CRC Press.

Horn, R. A. and C. R. Johnson (1990), *Matrix Analysis*, Cambridge University Press.

Hotelling, H. (1943a), "Some new methods in matrix calculation," *Ann. Math. Statist.*, vol. 14, pp. 1–34.

Hotelling, H. (1943b), "Further points on matrix calculation and simultaneous equations," *Ann. Math. Statist.*, vol. 14, pp. 440–441.

Householder, A. S. (1953), *Principles of Numerical Analysis*, McGraw-Hill.

Householder, A. S. (1957), "A survey of closed methods for inverting matrices," *J. Soc. Ind. Appl. Math.*, vol. 5, pp. 155–169.

Householder, A. S. (1964), *The Theory of Matrices in Numerical Analysis*, Blaisdell.

Jordan, C. (1874a), "Mémoire sur les formes bilinéaires," *J. Math. Pures Appl.*, Deuxieme Série, vol. 19, pp. 35–54.

Jordan, C. (1874b), "Sur la réduction des formes bilinéaires," *Comptes Rendus de l'Academie Sciences*, pp. 614–617.

Kailath, T. (1960), "Estimating filters for linear time-invariant channels," *Quarterly Progress Report 58*, MIT Research Laboratory of Electronics, pp. 185–197.

Kailath, T. (1986), "A theorem of I. Schur and its impact on modern signal processing," in *Operator Theory: Advances and Applications*, vol. 18, pp. 9–30, Birkhauser.

Kailath, T. and A. H. Sayed (1995), "Displacement structure: Theory and applications," *SIAM Rev.*, vol. 37, no. 3, pp. 297–386.

Koning, R. H., H. Neudecker, and T. Wansbeek (1991), "Block Kronecker products and the vecb operator," *Linear Algebra Appl.*, vol. 149, pp. 165–184.

Laplace, P. S. (1812), *Théorie Analytique des Probabilités*, Paris.

Laub, A. J. (2004), *Matrix Analysis for Scientists and Engineers*, SIAM.

Lax, P. (1997), *Linear Algebra*, Wiley.

Lay, D. (1994), *Linear Algebra and Its Applications*, Addison-Wesley.

Lay, D., S. Lay, and J. McDonald (2014), *Linear Algebra and Its Applications*, 5th ed., Pearson.

Leissa, A. W. (2005), "The historical bases of the Rayleigh and Ritz methods," *J. Sound Vib.*, vol. 287, pp. 961–978.

Lindsay, R. B. (1945), "Historical introduction," in J. W. S. Rayleigh, *The Theory of Sound*, vol. 1, Dover Publications.

Liu, S. (1999), "Matrix results on the Khatri–Rao and Tracy–Singh products," *Linear Algebra Appl.*, vol. 289, no. 1–3, pp. 267–277.

MacDuffee, C. C. (1946), *The Theory of Matrices*, Chelsea Publishing Company.

Meyer, C. D. (2001), *Matrix Analysis and Applied Linear Algebra*, SIAM.

Minc, H. (1988), *Nonnegative Matrices*, Wiley.

Nicholson, W. K. (2019), *Linear Algebra with Applications*, open ed., revision A. Available online under the Creative Commons License on lyryx.com.

Parlett, B. N. (1998), *The Symmetric Eigenvalue Problem*, SIAM.

Perron, O. (1907), "Zur theorie der matrices," *Mathematische Annalen*, vol. 64, no. 2, pp. 248–263.

Pillai, S. U., T. Suel, and S. Cha (2005), "The Perron–Frobenius theorem: Some of its applications," *IEEE Signal Process. Mag.*, vol. 22, no. 2, pp. 62–75.

Plackett, R. L. (1950), "Some theorems in least-squares," *Biometrika*, vol. 37, no. 1–2, pp. 149–157.

Puntanen, S. and G. P. H. Styan (2005), "Historical introduction: Issai Schur and the early development of the Schur complement," in *The Schur Complement and Its Applications*, F. Zhang, editor, pp. 1–16, Springer.

Puntanen, S., G. P. H. Styan, and J. Isotalo (2011), "Block-diagonalization and the Schur complement,"' in *Matrix Tricks for Linear Statistical Models*, S. Puntanen, G. P. H. Styan, and J. Isotalo, editors, pp. 291–304, Springer.

Rayleigh, J. W. S. (1877), *The Theory of Sound*, vol. 1, The Macmillan Company. Reprinted in 1945 by Dover Publications.

Rayleigh, J. W. S. (1878), *The Theory of Sound*, vol. 2, The Macmillan Company. Reprinted in 1945 by Dover Publications.

Regalia, P. A. and S. Mitra (1989), "Kronecker products, unitary matrices, and signal processing applications," *SIAM Rev.*, vol. 31, pp. 586–613.

Ritz, W. (1908), "Uber eine neue methode zur Losung gewisser variationsprobleme der mathematischen," *Physik, Journal fur Reine und Angewandte Mathematik*, vol. 135, pp. 1–61.

Ritz, W. (1909), "Theorie der transversalschwingungen einer quadratische platte mit freien Randern," *Annalen der Physik*, vol. 28, pp. 737–786.

Sayed, A. H. (2003), *Fundamentals of Adaptive Filtering*, Wiley.

Sayed, A. H. (2008), *Adaptive Filters*, Wiley.

Sayed, A. H. (2014a), *Adaptation, Learning, and Optimization over Networks,* Foundations and Trends in Machine Learning, NOW Publishers, vol. 7, no. 4–5, pp. 311–801.

Sayed, A. H. (2014c), "Diffusion adaptation over networks," in *E-Reference Signal Processing*, R. Chellapa and S. Theodoridis, editors, vol. 3, pp. 323–454, Academic Press.

Schmidt, E. (1907), "Zur Theorie der linearen und nichtlinearen Integralgleichungen. I. Teil. Entwicklung willkurlichen Funktionen nach System vorgeschriebener," *Math. Ann.*, vol. 63, pp. 433–476.

Schmidt, E. (1908), "Uber die Auflosung linearen Gleichungen mit unendlich vielen Unbekanten," *Rend. Circ. Mat. Palermo. Ser.* 1, vol. 25, pp. 53–77.

Schur, I. (1917), "Über potenzreihen die inm inneren des einheitskreises beschränkt sind," *Journal für die Reine und Angewandte Mathematik*, vol. 147, pp. 205–232. [English translation in *Operator Theory: Advances and Applications*, vol. 18, pp. 31–88, edited by I. Gohberg, Birkhaüser, 1986.]

Seneta, E. (2007), *Non-negative Matrices and Markov Chains*, 2nd ed., Springer.

Sherman, J. and W. J. Morrison (1949), "Adjustment of an inverse matrix corresponding to changes in the elements of a given column or a given row of the original matrix," *Ann. Math. Statistics*, vol. 20, p. 621.

Sherman, J. and W. J. Morrison (1950), "Adjustment of an inverse matrix corresponding to a change in one element of a given matrix," *Ann. Math. Statist.*, vol. 21, pp. 124–127.

Stewart, G. W. (1993), "On the early history of the singular value decomposition," *SIAM Rev.*, vol. 35, pp. 551–566.

Stewart, G. W. and J.-G. Sun (1990), *Matrix Perturbation Theory*, Academic Press.

Strang, G. (1988), *Linear Algebra and Its Applications*, 3rd ed., Academic Press.

Strang, G. (2009), *Introduction to Linear Algebra*, 4th ed., Wellesley-Cambridge Press.

Sylvester, J. J. (1852), "A demonstration of the theorem that every homogeneous quadratic polynomial is reducible by real orthogonal substitutions to the form of a sum of positive and negative squares," *Philos. Mag.*, Ser. 4, vol. 4, no. 23, pp. 138–142.

Sylvester, J. J. (1889a), "A new proof that a general quadratic may be reduced to its canonical form (that is, a linear function of squares) by means of a real orthogonal substitution," *Messenger Math.*, vol. 19, pp. 1–5.

Sylvester, J. J. (1889b), "On the reduction of a bilinear quantic of the nth order to the form of a sum of n products by a double orthogonal substitution," *Messenger Math.*, vol. 19, pp. 42–46.

Takahashi, N. and I. Yamada (2008), "Parallel algorithms for variational inequalities over the Cartesian product of the intersections of the fixed point sets of nonexpansive mappings," *J. Approx. Theory*, vol. 153, no. 2, pp. 139–160.

Takahashi, N., I. Yamada, and A. H. Sayed (2010), "Diffusion least-mean-squares with adaptive combiners: Formulation and performance analysis," *IEEE Trans. Signal Process.*, vol. 58, no. 9, pp. 4795–4810.

Tracy, D. S. and R. P. Singh (1972), "A new matrix product and its applications in matrix differentiation," *Statistica Neerlandica*, vol. 26, no. 4, pp. 143–157.

Trefethen, L. N. and D. Bau (1997), *Numerical Linear Algebra*, SIAM.

Van Huffel, S. and J. Vandewalle (1987), *The Total Least Squares Problem: Computational Aspects and Analysis*, SIAM.

Van Loan, C. F. (2000), "The ubiquitous Kronecker product," *J. Comput. Applied Math.*, vol. 123, no. 1–2, pp. 85–100.

Varga, R. S. (2004), *Gersgorin and His Circles*, Springer.

von Neumann, J. (1929), "Zur algebra der funktionaloperatoren und theorie der normalen operatoren," *Math. Ann.*, vol. 102, pp. 370–427.

von Neumann, J. (1932), *The Mathematical Foundations of Quantum Mechanics,* translation published in 1996 of original text, Princeton University Press.

von Neumann, J. (1937), "Some matrix-inequalities and metrization of matrix-space" *Tomsk Univ. Rev.*, vol. 1, pp. 286–300. Reprinted in *Collected Works*, Pergamon Press, 1962, vol. IV, pp. 205–219.

Weyl, H. (1909), "Über beschrankte quadratiche formen, deren differenz vollsteig ist," *Rend. Circ. Mat. Palermo*, vol. 27, pp. 373–392.

Weyl, H. (1912), "Das asymptotische Verteilungsgesetz der Eigenwerte linearer partieller Differentialgleichungen," *Math. Ann.*, vol. 71, pp. 441–479.

Wilkinson, J. H. (1965), *The Algebraic Eigenvalue Problem*, Oxford University Press.

Wong, Y. K. (1935), "An application of orthogonalization process to the theory of least squares," *Ann. Math. Statist.*, vol. 6, pp. 53–75.

Woodbury, M. (1950), *Inverting Modified Matrices*, Mem. Rep. 42, Statistical Research Group, Princeton University.
Zehfuss, G. (1858), "Ueber eine gewisse Determinante," *Zeitschrift für Mathematik und Physik*, vol. 3, pp. 298–301.
Zhang, F. (2005), *The Schur Complement and Its Applications*, Springer.

2 Vector Differentiation

Gradient vectors and Hessian matrices play an important role in the development of iterative algorithms for inference and learning. In this chapter, we define the notions of first- and second-order differentiation for functions of vector arguments, and introduce the notation for future chapters.

2.1 GRADIENT VECTORS

We describe two closely related differentiation operations that we will be using regularly in our treatment of inference and learning problems.

Let $z \in \mathbb{R}^M$ denote a real-valued *column* vector with M entries denoted by

$$z = \begin{bmatrix} z_1 \\ z_2 \\ \vdots \\ z_M \end{bmatrix} \tag{2.1}$$

Let also $g(z) : \mathbb{R}^M \rightarrow \mathbb{R}$ denote a real-valued function of the vector argument, z. We denote the gradient vector of $g(z)$ with respect to z by the notation $\nabla_z\, g(z)$ and define it as the following *row* vector:

$$\nabla_z\, g(z) \;\stackrel{\Delta}{=}\; \begin{bmatrix} \dfrac{\partial g}{\partial z_1} & \dfrac{\partial g}{\partial z_2} & \cdots & \dfrac{\partial g}{\partial z_M} \end{bmatrix}, \quad \begin{cases} z \text{ is a } \textbf{column} \\ \nabla_z\, g(z) \text{ is a } \textbf{row} \end{cases} \tag{2.2}$$

Note that the gradient is defined in terms of the partial derivatives of $g(z)$ relative to the individual entries of z.

Jacobian

Expression (2.2) for $\nabla_z\, g(z)$ is related to the concept of a *Jacobian* matrix for *vector-valued* functions. Specifically, consider a second function $h(z) : \mathbb{R}^M \rightarrow \mathbb{R}^N$, which now maps z into a *vector*, assumed of dimension N and whose individual entries are denoted by

$$h(z) \;=\; \text{col}\Big\{h_1(z),\, h_2(z),\, h_3(z),\, \ldots,\, h_N(z)\Big\} \tag{2.3}$$

The Jacobian of $h(z)$ relative to z is defined as the matrix:

$$\nabla_z\, h(z) \;\stackrel{\Delta}{=}\; \begin{bmatrix} \partial h_1/\partial z_1 & \partial h_1/\partial z_2 & \partial h_1/\partial z_3 & \dots & \partial h_1/\partial z_M \\ \partial h_2/\partial z_1 & \partial h_2/\partial z_2 & \partial h_2/\partial z_3 & \dots & \partial h_2/\partial z_M \\ \partial h_3/\partial z_1 & \partial h_3/\partial z_2 & \partial h_3/\partial z_3 & \dots & \partial h_3/\partial z_M \\ \vdots & \vdots & \vdots & \vdots & \\ \partial h_N/\partial z_1 & \partial h_N/\partial z_2 & \partial h_N/\partial z_3 & \dots & \partial h_N/\partial z_M \end{bmatrix} \tag{2.4}$$

Thus, note that if $h(z)$ were scalar-valued, with $N = 1$, then its Jacobian will reduce to the first row in the above matrix, which agrees with the definition for the gradient vector in (2.2).

In the same token, we will denote the gradient vector of $g(z)$ relative to the *row* vector z^{T} by $\nabla_{z^{\mathsf{T}}}\, g(z)$ and define it as the *column* vector:

$$\nabla_{z^{\mathsf{T}}}\, g(z) \;\stackrel{\Delta}{=}\; \begin{bmatrix} \partial g/\partial z_1 \\ \partial g/\partial z_2 \\ \vdots \\ \partial g/\partial z_M \end{bmatrix}, \qquad \begin{cases} z^{\mathsf{T}} \text{ is a } \textbf{row} \\ \nabla_{z^{\mathsf{T}}}\, g(z) \text{ is a } \textbf{column} \end{cases} \tag{2.5}$$

It is clear that

$$\nabla_{z^{\mathsf{T}}}\, g(z) \;=\; \Big(\nabla_z\, g(z)\Big)^{\mathsf{T}} \tag{2.6}$$

Notation

Observe that we are defining the gradient of a function with respect to a *column* vector to be a *row* vector, and the gradient with respect to a *row* vector to be a *column* vector:

gradient vector relative to a column is a row (2.7a)

gradient vector relative to a row is a column (2.7b)

Some references may reverse these conventions, such as defining the gradient vector of $g(z)$ relative to the column z to be the column vector (2.5). There is no standard convention in the literature. To avoid any ambiguity, we make a distinction between differentiating relative to z and z^{T}. Specifically, we adopt the convention that the gradient relative to a column (row) is a row (column). The main motivation for doing so is because the results of differentiation that follow from this convention will be consistent with what we would expect from traditional differentiation rules from the calculus of single variables. This is illustrated in the next examples.

Observe further that the result of the gradient operation (2.5) relative to z^{T} is a vector that has the *same dimension* as z; we will also use the following alternative notation for this gradient vector when convenient:

$$\frac{\partial g(z)}{\partial z} \triangleq \begin{bmatrix} \partial g/\partial z_1 \\ \partial g/\partial z_2 \\ \vdots \\ \partial g/\partial z_M \end{bmatrix} \tag{2.8}$$

In this way, we end up with the following convention:

$$\begin{cases} \textbf{a)} \;\; z \text{ is an } M \times 1 \text{ column vector;} \\ \textbf{b)} \;\; \partial g(z)/\partial z \text{ and } \nabla_{z^\mathsf{T}}\, g(z) \text{ coincide and have the same dimensions as } z; \\ \textbf{c)} \;\; \nabla_z\, g(z) \text{ has the same dimensions as } z^\mathsf{T} \end{cases} \tag{2.9}$$

The notation $\partial g(z)/\partial z$ organizes the partial derivatives in column form, while the notation $\nabla_z\, g(z)$ organizes the *same* partial derivatives in row form. The two operations of differentiation and gradient evaluation are the transpose of each other. We will be using these forms interchangeably.

Example 2.1 **(Calculations for vector arguments)** We consider a couple of examples.

(1) Let $g(z) = a^\mathsf{T} z$, where $\{a, z\}$ are column vectors in $\mathbb{R}^M$ with entries $\{a_m, z_m\}$. Then,

$$\begin{aligned} \nabla_z\, g(z) &\triangleq \begin{bmatrix} \partial g(z)/\partial z_1 & \partial g(z)/\partial z_2 & \dots & \partial g(z)/\partial z_M \end{bmatrix} \\ &= \begin{bmatrix} a_1 & a_2 & \dots & a_M \end{bmatrix} \\ &= a^\mathsf{T} \end{aligned} \tag{2.10}$$

Note further that since $g(z)$ is real-valued, we can also write $g(z) = z^\mathsf{T} a$ and a similar calculation gives

$$\nabla_{z^\mathsf{T}}\, g(z) = a \tag{2.11}$$

(2) Let $g(z) = \|z\|^2 = z^\mathsf{T} z$, where $z \in \mathbb{R}^M$. Then,

$$\begin{aligned} \nabla_z\, g(z) &\triangleq \begin{bmatrix} \partial g(z)/\partial z_1 & \partial g(z)/\partial z_2 & \dots & \partial g(z)/\partial z_M \end{bmatrix} \\ &= \begin{bmatrix} 2z_1 & 2z_2 & \dots & 2z_M \end{bmatrix} \\ &= 2z^\mathsf{T} \end{aligned} \tag{2.12}$$

Likewise, we get $\nabla_{z^\mathsf{T}}\, g(z) = 2z$.

(3) Let $g(z) = z^\mathsf{T} C z$, where C is a *symmetric* matrix in $\mathbb{R}^{M\times M}$ that does not depend on z. If we denote the individual entries of C by C_{mn}, then we can write

$$g(z) = \sum_{m=1}^{M}\sum_{n=1}^{M} C_{mn} z_m z_n \tag{2.13}$$

so that

$$\begin{aligned}\frac{\partial g(z)}{\partial z_m} &= 2C_{mm}z_m + \sum_{n\neq m}^{M}(C_{mn}+C_{nm})z_n \\ &= 2\sum_{n=1}^{M} C_{nm}z_n \\ &= 2z^{\mathsf{T}}C_{:,m} \end{aligned} \tag{2.14}$$

where in the second equality we used the fact that C is symmetric and, hence, $C_{mn} = C_{nm}$, and in the last equality we introduced the notation $C_{:,m}$ to refer to the mth column of C. Collecting all partial derivatives $\{\partial g(z)/z_m\}$, for $m = 1, 2, \ldots, M$, into a row vector we conclude that

$$\nabla_z\, g(z) \;=\; 2z^{\mathsf{T}}C \tag{2.15}$$

(4) Let $g(z) = z^{\mathsf{T}}Cz$, where C is now an arbitrary (not necessarily symmetric) matrix in $\mathbb{R}^{M\times M}$. If we repeat the same argument as in part (3) we arrive at

$$\nabla_z\, g(z) \;=\; z^{\mathsf{T}}(C + C^{\mathsf{T}}) \tag{2.16}$$

(5) Let $g(z) = \kappa + a^{\mathsf{T}}z + z^{\mathsf{T}}b + z^{\mathsf{T}}Cz$, where κ is a scalar, $\{a, b\}$ are column vectors in $\mathbb{R}^M$, and C is a matrix in $\mathbb{R}^{M\times M}$. Then,

$$\nabla_z\, g(z) \;=\; a^{\mathsf{T}} + b^{\mathsf{T}} + z^{\mathsf{T}}(C + C^{\mathsf{T}}) \tag{2.17}$$

(6) Let $g(z) = Az$, where $A \in \mathbb{R}^{M\times M}$ does not depend on z. Then, the Jacobian matrix is given by $\nabla_z\, g(z) = A$.

2.2 HESSIAN MATRICES

Hessian matrices involve second-order partial derivatives. Consider again the real-valued function $g(z) : \mathbb{R}^M \to \mathbb{R}$. We continue to denote the individual entries of the column vector $z \in \mathbb{R}^M$ by $z = \text{col}\{z_1, z_2, \ldots, z_M\}$. The Hessian matrix of $g(z)$ is an $M \times M$ *symmetric* matrix function of z, denoted by $H(z)$, and whose (m, n)th entry is constructed as follows:

$$[H(z)]_{m,n} \stackrel{\Delta}{=} \frac{\partial^2 g(z)}{\partial z_m \partial z_n} = \frac{\partial}{\partial z_m}\left(\frac{\partial g(z)}{\partial z_n}\right) = \frac{\partial}{\partial z_n}\left(\frac{\partial g(z)}{\partial z_m}\right) \tag{2.18}$$

in terms of the partial derivatives of $g(z)$ with respect to the scalar arguments $\{z_m, z_n\}$. For example, for a two-dimensional argument z (i.e., $M = 2$), the four entries of the 2×2 Hessian matrix are:

$$H(z) \;=\; \begin{bmatrix} \dfrac{\partial^2 g(z)}{\partial z_1^2} & \dfrac{\partial^2 g(z)}{\partial z_1 \partial z_2} \\[2ex] \dfrac{\partial^2 g(z)}{\partial z_2 \partial z_1} & \dfrac{\partial^2 g(z)}{\partial z_2^2} \end{bmatrix} \tag{2.19}$$

It is straightforward to verify that $H(z)$ can also be obtained as the result of two successive gradient vector calculations with respect to z and $z^{\mathsf T}$ in the following manner (where the order of the differentiation does not matter):

$$H(z) \stackrel{\Delta}{=} \nabla_{z^{\mathsf T}}\Big(\nabla_z\, g(z)\Big) = \nabla_z\Big(\nabla_{z^{\mathsf T}}\, g(z)\Big) \tag{2.20}$$

For instance, using the first expression, the gradient operation $\nabla_z\, g(z)$ generates a $1 \times M$ (row) vector function and the subsequent differentiation with respect to $z^{\mathsf T}$ leads to the $M \times M$ Hessian matrix, $H(z)$. It is clear from (2.18) and (2.20) that the Hessian matrix is symmetric so that

$$H(z) = \Big(H(z)\Big)^{\mathsf T} \tag{2.21}$$

A useful property of Hessian matrices is that they help characterize the nature of stationary points for functions $g(z)$ that are twice differentiable. Specifically, if z^o is a stationary point of $g(z)$ (i.e., a point where $\nabla_z\, g(z) = 0$), then the following facts hold:

(a) z^o will correspond to a local minimum of $g(z)$ if $H(z^o) > 0$, i.e., if all eigenvalues of $H(z^o)$ are positive.

(b) z^o will correspond to a local maximum of $g(z)$ if $H(z^o) < 0$, i.e., if all eigenvalues of $H(z^o)$ are negative.

Example 2.2 (Quadratic cost functions) Consider $g(z) = \kappa + 2a^{\mathsf T}z + z^{\mathsf T}Cz$, where κ is a real scalar, a is a real column vector of dimensions $M \times 1$, and C is an $M \times M$ real *symmetric* matrix. We know from (2.17) that

$$\nabla_z\, g(z) = 2a^{\mathsf T} + 2z^{\mathsf T}C \tag{2.22}$$

Differentiating again gives:

$$H(z) \stackrel{\Delta}{=} \nabla_{z^{\mathsf T}}\left(\nabla_z\, g(z)\right) = \nabla_{z^{\mathsf T}}\Big(2a^{\mathsf T} + 2z^{\mathsf T}C\Big) = 2C \tag{2.23}$$

We find that for quadratic functions $g(z)$, the Hessian matrix is independent of z. Moreover, any stationary point z^o of $g(z)$ should satisfy

$$2a^{\mathsf T} + 2(z^o)^{\mathsf T}C = 0 \implies Cz^o = -a \tag{2.24}$$

The stationary point will be unique if C is invertible. And the unique z^o will be a global minimum of $g(z)$ if, and only if, $C > 0$.

2.3 MATRIX DIFFERENTIATION

We end the chapter with a list of useful matrix differentiation results, collected in Table 2.1. We leave the derivations to Probs. 2.10–2.14. The notation used in the table refers to the following definitions.

Table 2.1 Some useful matrix differentiation results; inverses are assumed to exist whenever necessary. The last column provides the problem numbers where these properties are established.

Property	Result	Problem
1.	$\partial(A(\alpha))^{-1}/\partial\alpha = -A^{-1}(\partial A/\partial\alpha)A^{-1}$	2.10
2.	$\partial \det A/\partial\alpha = \det(A)\,\mathrm{Tr}\{A^{-1}(\partial A/\partial\alpha)\}$	2.10
3.	$\partial \ln\lvert\det A\rvert/\partial\alpha = \mathrm{Tr}\Big\{A^{-1}(\partial A/\partial\alpha)\Big\}$	2.10
4.	$\partial A(\alpha)B(\alpha)/\partial\alpha = A(\partial B/\partial\alpha) + (\partial A/\partial\alpha)B$	2.10
5.	$\nabla_{X^\mathsf{T}}\,\mathrm{Tr}(AXB) = A^\mathsf{T}B^\mathsf{T}$	2.11
6.	$\nabla_{X^\mathsf{T}}\,\mathrm{Tr}(AX^{-1}B) = -(X^{-1}BAX^{-1})^\mathsf{T}$	2.11
7.	$\nabla_{X^\mathsf{T}}\,\mathrm{Tr}(X^{-1}A) = -(X^{-1}A^\mathsf{T}X^{-1})^\mathsf{T}$	2.11
8.	$\nabla_{X^\mathsf{T}}\,\mathrm{Tr}(X^\mathsf{T}AX) = (A + A^\mathsf{T})X$	2.11
9.	$\nabla_{X^\mathsf{T}}\,\mathrm{Tr}(X^\mathsf{T}X) = 2X$	2.11
10.	$\nabla_{X^\mathsf{T}}\,f(X) = -X^{-\mathsf{T}}\left(\nabla_{X^{-1}}f(X)\right)X^{-\mathsf{T}}$	2.11
11.	$\nabla_{X^\mathsf{T}}\,\det(X) = \det(X)X^{-\mathsf{T}}$	2.12
12.	$\nabla_{X^\mathsf{T}}\,\det(X^{-1}) = -\det(X^{-1})X^{-\mathsf{T}}$	2.12
13.	$\nabla_{X^\mathsf{T}}\,\det(AXB) = \det(AXB)(X^{-1})^\mathsf{T}$	2.12
14.	$\nabla_{X^\mathsf{T}}\,\det(X^\mathsf{T}AX) = 2\det(X^\mathsf{T}AX)(X^{-1})^\mathsf{T}$	2.12
15.	$\nabla_{X^\mathsf{T}}\,\ln\lvert\det(X)\rvert = (X^{-1})^\mathsf{T}$	2.12
16.	$\nabla_{X^\mathsf{T}}\,\lVert X\rVert_\mathrm{F}^2 = 2X$	2.13
17.	$\nabla_{X^\mathsf{T}}\,\mathrm{Tr}(X^p) = pX^{p-1},\; p \in \mathrm{I\!R}$	2.14

Let $X \in \mathrm{I\!R}^{M\times M}$ be a square matrix whose individual entries are functions of some real scalar α. We denote the individual entries of X by X_{mn} and define its derivative relative to α as the $M \times M$ matrix whose individual entries are the partial derivatives:

$$\frac{\partial X(\alpha)}{\partial\alpha} = \left[\frac{\partial X_{mn}}{\partial\alpha}\right]_{m,n}, \qquad (M\times M) \tag{2.25}$$

Likewise, let $f(X)$ denote some scalar real-valued function of a real-valued $M \times N$ matrix argument X. We employ two closely related operations to refer to differentiation operations applied to $f(X)$. The derivative of $f(\cdot)$ relative to X is defined as the $M \times N$ matrix whose individual entries are the partial derivatives of $f(X)$:

$$\frac{\partial f(X)}{\partial X} \overset{\Delta}{=} \left[\frac{\partial f(X)}{\partial X_{mn}}\right]_{m,n} = \nabla_{X^\mathsf{T}}\,f(X), \qquad (M\times N) \tag{2.26}$$

Observe that the result has the *same* dimensions as X. Likewise, we define the gradient matrix of $f(\cdot)$ relative to X as the $N \times M$ matrix

$$\nabla_X f(X) = \left(\left[\frac{\partial f(X)}{\partial X_{mn}} \right]_{m,n} \right)^{\mathsf{T}}, \qquad (N \times M) \tag{2.27}$$

The result has the same dimensions as the transpose matrix, X^{T}. This construction is consistent with our earlier convention for the differentiation and gradient operations for vector arguments. In particular, note that

$$\nabla_X f(X) \;=\; \left(\frac{\partial f(X)}{\partial X} \right)^{\mathsf{T}} \;=\; \left(\nabla_{X^{\mathsf{T}}} f(X) \right)^{\mathsf{T}} \tag{2.28}$$

2.4 COMMENTARIES AND DISCUSSION

Gradients and Hessians. Gradient vectors help identify the location of stationary points of a function and play an important role in the development of iterative algorithms for seeking these locations. Hessian matrices, on the other hand, help clarify the nature of a stationary point such as deciding whether it corresponds to a local minimum, a local maximum, or a saddle point. Hessian matrices are named after the German mathematician **Ludwig Hesse (1811–1874)**, who introduced them in his study of quadratic and cubic curves – see the work collection by Dyck *et al.* (1972). A useful listing of gradient vector calculations for functions of real arguments is given by Petersen and Pedersen (2012). Some of these results appear in Table 2.1. For more discussion on first- and second-order differentiation for functions of several variables, the reader may refer to Fleming (1987), Edwards (1995), Zorich (2004), Moskowitz and Paliogiannis (2011), Hubbard and Hubbard (2015), and Bernstein (2018).

PROBLEMS

2.1 Let $g(x,z) = x^{\mathsf{T}} C z$, where $x, z \in \mathbb{R}^M$ and C is a matrix. Verify that

$$\nabla_z\, g(x,z) = x^{\mathsf{T}} C, \qquad \nabla_x\, g(x,z) = z^{\mathsf{T}} C^{\mathsf{T}}$$

2.2 Let $g(z) = Ah(z)$, where $A \in \mathbb{R}^{M\times M}$ and both $g(z)$ and $h(z)$ are vector-valued functions. Show that the Jacobian matrices of $g(z)$ and $h(z)$ are related as follows:

$$\nabla_z\, g(z) \;=\; A\, \nabla_z\, h(z)$$

2.3 Let $g(z) = (h(z))^{\mathsf{T}} h(z) = \|h(z)\|^2$, where $g(z)$ is scalar-valued while $h(z)$ is vector-valued with $z \in \mathbb{R}^M$. Show that

$$\nabla_z\, g(z) \;=\; 2\,(h(z))^{\mathsf{T}}\, \nabla_z\, h(z)$$

in terms of the Jacobian matrix of $h(z)$.

2.4 Let $g(z) = \frac{1}{2}\,(h(z))^{\mathsf{T}} A^{-1} h(z)$, where $A > 0$ and $g(z)$ is scalar-valued while $h(z)$ is vector-valued with $z \in \mathbb{R}^M$. Show that

$$\nabla_z\, g(z) \;=\; (h(z))^{\mathsf{T}} A^{-1} \nabla_z\, h(z)$$

in terms of the Jacobian matrix of $h(z)$.

2.5 Let $g(z) = x^{\mathsf{T}}Cw$, where $x, w \in \mathbb{R}^P$ and both are functions of a vector $z \in \mathbb{R}^M$, i.e., $x = x(z)$ and $w = w(z)$, while C is a matrix that is independent of z. Establish the chain rule

$$\nabla_z\, g(z) = x^{\mathsf{T}} C\, \nabla_z\, w(z) \;+\; w^{\mathsf{T}} C^{\mathsf{T}}\, \nabla_z\, x(z)$$

in terms of the Jacobian matrices of $x(z)$ and $w(z)$ relative to z.

2.6 Let $g(z)$ be a real-valued differentiable function with $z \in \mathbb{R}^M$. Assume the entries of z are functions of a scalar parameter t, i.e., $z = \text{col}\{z_1(t), z_2(t), \ldots, z_M(t)\}$. Introduce the column vector $dz/dt = \text{col}\,\{dz_1/dt, dz_2/dt, \ldots, dz_M/dt\}$. Show that

$$\frac{dg}{dt} = \nabla_z\, g(z)\, \frac{dz}{dt}$$

2.7 Let $g(z)$ be a real-valued function with $z \in \mathbb{R}^M$. Let $f(t)$ be a real-valued function with $t \in \mathbb{R}$. Both functions are differentiable in their arguments. Show that

$$\nabla_z\, f(g(z)) = \left(\left. \frac{df(t)}{dt} \right|_{t=g(z)} \right) \times \nabla_z\, g(z)$$

2.8 Let $g(z)$ be a real-valued twice-differentiable function with $z \in \mathbb{R}^M$. Define $f(t) = g(z + t\Delta z)$ for $t \in [0, 1]$. Show from first principles that

$$\begin{aligned} \frac{df(t)}{dt} &= (\nabla_z\, g(z + t\Delta z))\, \Delta z \\ \frac{d^2 f(t)}{dt^2} &= (\Delta z)^{\mathsf{T}} \left(\nabla_z^2\, g(z + t\Delta z)\right) \Delta z \end{aligned}$$

2.9 Let $\|x\|_p$ denote the pth norm of vector $x \in \mathbb{R}^M$. Verify that

$$\nabla_{x^{\mathsf{T}}}\, \|x\|_p = \frac{x \odot |x|^{p-2}}{\|x\|_p^{p-1}}$$

where $\odot$ denotes the Hadamard (elementwise) product of two vectors, and the notation $|x|^{p-2}$ refers to a vector whose individual entries are the absolute values of the entries of x raised to the power $p-2$.

2.10 Let $A \in \mathbb{R}^{M\times M}$ be a matrix whose individual entries are functions of some real scalar α. We denote the individual entries of A by A_{mn} and define its derivative relative to α as the $M \times M$ matrix whose individual entries are given by the partial derivatives:

$$\left[\frac{\partial A}{\partial \alpha}\right]_{m,n} = \frac{\partial A_{mn}}{\partial \alpha}$$

Establish the following relations:
(a) $\partial A^{-1}/\partial\alpha = -A^{-1}(\partial A/\partial\alpha)A^{-1}$.
(b) $\partial AB/\partial\alpha = A(\partial B/\partial\alpha) + (\partial A/\partial\alpha)B$, for matrices A and B that depend on α.
(c) $\partial \det A/\partial\alpha = \det(A)\,\text{Tr}\Big\{A^{-1}(\partial A/\partial\alpha)\Big\}$.
(d) $\partial \ln|\det A|/\partial\alpha = \text{Tr}\Big\{A^{-1}(\partial A/\partial\alpha)\Big\}$.

2.11 Let $f(X)$ denote a scalar real-valued function of a real-valued $M \times N$ matrix argument X. Let X_{mn} denote the (m, n)th entry of X. The gradient of $f(\cdot)$ relative to X^{T} is defined as the $M \times N$ matrix whose individual entries are given by the partial derivatives:

$$\Big[\nabla_{X^{\mathsf{T}}}\, f(X)\Big]_{m,n} = \frac{\partial f(X)}{\partial X_{mn}}$$

Establish the following differentiation results (assume X is square and/or invertible when necessary):
(a) $\nabla_{X^{\mathsf{T}}}\, \text{Tr}(AXB) = A^{\mathsf{T}}B^{\mathsf{T}}$.

(b) $\nabla_{X^\mathsf{T}} \operatorname{Tr}(AX^{-1}B) = -(X^{-1}BAX^{-1})^\mathsf{T}$.
(c) $\nabla_{X^\mathsf{T}} \operatorname{Tr}(X^{-1}A) = -(X^{-1}A^\mathsf{T}X^{-1})^\mathsf{T}$.
(d) $\nabla_{X^\mathsf{T}} \operatorname{Tr}(X^\mathsf{T}AX) = (A + A^\mathsf{T})X$.
(e) $\nabla_{X^\mathsf{T}} \operatorname{Tr}(X^\mathsf{T}X) = 2X$.
(f) $\nabla_{X^\mathsf{T}} f(X) = -X^{-\mathsf{T}}\left(\nabla_{X^{-1}} f(X)\right) X^{-\mathsf{T}}$.

2.12 Consider the same setting of Prob. 2.11. Show that
(a) $\nabla_{X^\mathsf{T}} \det(X) = \det(X)X^{-\mathsf{T}}$.
(b) $\nabla_{X^\mathsf{T}} \det(X^{-1}) = -\det(X^{-1})X^{-\mathsf{T}}$.
(c) $\nabla_{X^\mathsf{T}} \det(AXB) = \det(AXB)(X^{-1})^\mathsf{T}$.
(d) $\nabla_{X^\mathsf{T}} \det(X^\mathsf{T}AX) = 2\det(X^\mathsf{T}AX)(X^{-1})^\mathsf{T}$, for square invertible X.
(e) $\nabla_{X^\mathsf{T}} \ln|\det(X)| = (X^{-1})^\mathsf{T}$.

2.13 Let $\|X\|_\mathrm{F}$ denote the Frobenius norm of matrix X, as was defined in Table 1.4. Show that $\nabla_{X^\mathsf{T}} \|X\|_\mathrm{F}^2 = 2X$.

2.14 Let $p \in \mathbb{R}$ and consider a positive-definite matrix, X. Show that $\nabla_{X^\mathsf{T}} \operatorname{Tr}(X^p) = pX^{p-1}$.

2.15 For the purposes of this problem, let the notation $\|X\|_1$ denote the sum of the absolute values of all entries of X. Find $\nabla_{X^\mathsf{T}} \|XX^\mathsf{T}\|_1$.

REFERENCES

Bernstein, D. S. (2018), *Scalar, Vector, and Matrix Mathematics: Theory, Facts, and Formulas*, revised ed., Princeton University Press.

Dyck, W., S. Gundelfinger, J. Luroth, and M. Noether (1972), *Ludwig Otto Hesse's Gesammelte Werke*, published originally in 1897 by the Bavarian Academy of Sciences and Humanities. Republished in 1972 by Chelsea Publishing Company.

Edwards, Jr., C. H. (1995), *Advanced Calculus of Several Variables*, Dover Publications.

Fleming, W. (1987), *Functions of Several Variables*, 2nd ed., Springer.

Hubbard, B. B. and J. Hubbard (2015), *Vector Calculus, Linear Algebra, and Differential Forms: A Unified Approach*, 5th ed., Matrix Editions.

Moskowitz, M. and F. Paliogiannis (2011), *Functions of Several Real Variables*, World Scientific.

Petersen, K. B. and M. S. Pedersen (2012), "The matrix cookbook," available at http://matrixcookbook.com.

Zorich, V. A. (2004), *Mathematical Analysis*, vols. I and II, Springer.

3 Random Variables

The material in future chapters will require familiarity with basic concepts from probability theory, random variables, and random processes. For the benefit of the reader, we review in this and the next chapters several concepts of general interest. The discussion is not meant to be comprehensive or exhaustive. Only concepts that are necessary for our treatment of inference and learning are reviewed. It is assumed that readers have had some prior exposure to random variables and probability theory.

3.1 PROBABILITY DENSITY FUNCTIONS

In loose terms, the designation "random variable" refers to a variable whose value cannot be predicted with certainty prior to observing it. This is because the variable may assume any of a collection of values in an experiment, and some of the values can be more likely to occur than other values. In other words, there is an element of *chance* associated with each possibility.

In our treatment we will often (but not exclusively) be interested in random variables whose observations assume *numerical* values. Obviously, in many situations of interest, the random variables need not be numerical but are categorical in nature. One example is when a ball is drawn from a bowl and is either blue with probability $1/4$ or red with probability $3/4$. In this case, the qualifications $\{\text{red}, \text{blue}\}$ refer to the two possible outcomes, which are not numerical. Nevertheless, it is common practice in scenarios like this to associate numerical values with each category, such as assigning the numerical value $+1$ to red and the numerical value -1 to blue. In this way, drawing a red ball amounts to observing the value $+1$ with probability $3/4$ and drawing a blue ball amounts to observing the value -1 with probability $1/4$.

We will use **boldface** symbols to refer to random variables and symbols in *normal* font to refer to their realizations or observations. For example, we let $\boldsymbol{x}$ denote the random variable that corresponds to the outcome of throwing a die. Each time the experiment is repeated, one of six possible outcomes can be observed, namely, $x \in \{1, 2, 3, 4, 5, 6\}$ – see Fig. 3.1. We cannot tell beforehand which value will occur (assuming a fair die). We say that the random variable, $\boldsymbol{x}$, represents the outcome of the experiment and each observation x is a realization for $\boldsymbol{x}$. In this example, the realization x can assume one of six possible integer values, which constitute the *sample space* for $\boldsymbol{x}$ and is denoted by the letter $\Omega = \{1, 2, 3, 4, 5, 6\}$. Our choice of notation $\{\boldsymbol{x}, x\}$ is meant to distinguish between a random variable and its realizations or observations.

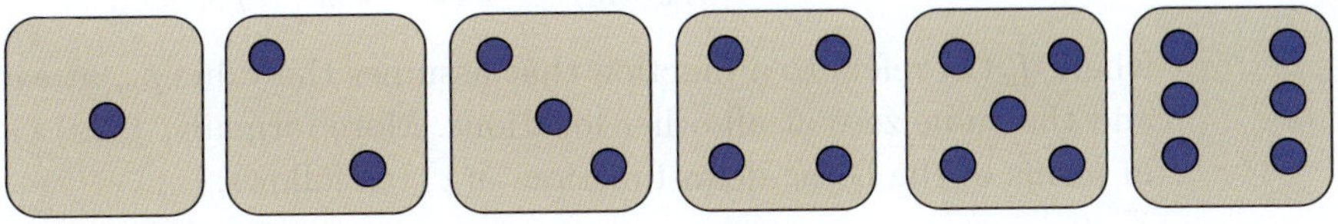

Figure 3.1 The sample space for a die consists of the outcomes $\Omega = \{1, 2, 3, 4, 5, 6\}$.

Discrete variables

A numerical random variable can be *discrete* or *continuous* depending on the range of values it assumes. The realizations of a discrete random variable can only assume values from a countable set of distinct possibilities. For example, the outcome of the throw of a die is a discrete random variable since the realizations can only assume one of six possible values from the set $\Omega = \{1, 2, 3, 4, 5, 6\}$. In contrast, the realizations of a continuous random variable can assume an infinite number of possible values, e.g., values within an interval on the real line.

When a random variable is discrete, we associate with each element of the sample space, Ω, a nonnegative number in the range $[0, 1]$. This number represents the probability of occurrence of that particular element of Ω. For example, assuming a fair die throw, the probability that the realization $x = 4$ is observed is equal to $1/6$. This is because all six possible outcomes are equally likely with the same probability of occurrence, which we denote by

$$p_m = 1/6, \quad m = 1, 2, 3, 4, 5, 6 \tag{3.1}$$

with one value p_m for each possible outcome m. Obviously, the sum of all six probabilities must add up to 1. We refer to the $\{p_m\}$ as representing the probability distribution or the *probability mass function* (pmf) that is associated with the die experiment. More generally, for a discrete random variable, $\boldsymbol{x}$, with M possible realizations, $\{x_m\}$, we associate with each outcome a probability value p_m for all $1 \leq m \leq M$. These probabilities need not be identical because some outcomes may be more likely to occur than others, but they must satisfy the following two conditions:

$$\boxed{0 \leq p_m \leq 1 \quad \text{and} \quad \sum_{m=1}^{M} p_m = 1} \tag{3.2}$$

with the number p_m corresponding to the probability of the mth event occurring, written as

$$p_m \triangleq \mathbb{P}(\boldsymbol{x} = x_m) \tag{3.3}$$

When convenient, we will also use the alternative function notation $f_{\boldsymbol{x}}(x_m)$ to refer to the probability of event x_m, namely,

$$f_{\boldsymbol{x}}(x_m) \triangleq \mathbb{P}(\boldsymbol{x} = x_m) = p_m \tag{3.4}$$

where $f_{\boldsymbol{x}}(x)$ refers to a function that assumes the value p_m at each location x_m, and the value zero at all other locations. More formally, $f_{\boldsymbol{x}}(x)$ can be expressed in terms of the Dirac delta function, $\delta(x)$, as follows:

$$f_{\boldsymbol{x}}(x) = \sum_{m=1}^{M} p_m \, \delta(x - x_m) \tag{3.5}$$

where the delta function is defined by the sifting property:

$$\int_{-\infty}^{\infty} g(x)\delta(x - x_m)dx \;=\; g(x_m) \tag{3.6}$$

for any function $g(x)$ that is well-defined at $x = x_m$. Representation (3.5) highlights the fact that the probability distribution of a discrete random variable, $\boldsymbol{x}$, is concentrated at a finite number of locations defined by the coordinates $\{x_m\}$.

Continuous variables

The function notation $f_{\boldsymbol{x}}(x)$ for the pmf of a discrete random variable is useful because, as explained next, it will allow us to adopt a common notation for both discrete and continuous random variables.

When the random variable $\boldsymbol{x}$ is continuous, the probability that $\boldsymbol{x}$ assumes any particular value x from its sample space is equal to zero. This is because there are now infinitely many possible realization values. For this reason, for continuous random variables, we are more interested in the probability of events involving a *range* of values rather than a specific value. To evaluate the probability of such events, we associate with the random variable $\boldsymbol{x}$ a *probability density function* (pdf), which we will denote by the same notation $f_{\boldsymbol{x}}(x)$. The pdf is a function of x and it is required to satisfy the following two conditions:

$$\boxed{f_{\boldsymbol{x}}(x) \geq 0 \quad \text{and} \quad \int_{-\infty}^{\infty} f_{\boldsymbol{x}}(x)dx = 1} \tag{3.7}$$

The pdf of $\boldsymbol{x}$ allows us to evaluate probabilities of events of the form

$$\mathbb{P}(a \leq \boldsymbol{x} \leq b) \tag{3.8}$$

which refer to the probability that $\boldsymbol{x}$ assumes values within the interval $[a, b]$. This probability is obtained through the integral calculation:

$$\mathbb{P}(a \leq \boldsymbol{x} \leq b) \;=\; \int_a^b f_{\boldsymbol{x}}(x)dx \tag{3.9}$$

We will use the terminology of "probability *mass functions*" for discrete random variables, and "probability *density functions*" for continuous random variables. Moreover, we will often use the same pdf notation, $f_{\boldsymbol{x}}(x)$, to refer to probability distributions in both cases.

Example 3.1 (Uniform random variable) A continuous random variable $\boldsymbol{x}$ is said to be uniformly distributed within the interval $[a, b]$ if its pdf is constant over this interval and zero elsewhere, namely,

$$f_{\boldsymbol{x}}(x) = \begin{cases} c, & a \le x \le b \\ 0, & \text{otherwise} \end{cases} \tag{3.10}$$

for some constant $c > 0$ and where $b > a$. The value of c can be determined from the normalization requirement

$$\int_{-\infty}^{\infty} f_{\boldsymbol{x}}(x)dx = 1 \tag{3.11}$$

so that we must have

$$\int_a^b c\,dx = 1 \implies c = \frac{1}{b-a} \tag{3.12}$$

We conclude that the pdf of a uniform random variable is given by

$$f_{\boldsymbol{x}}(x) = \begin{cases} \dfrac{1}{b-a}, & a \le x \le b \\ 0, & \text{otherwise} \end{cases} \tag{3.13}$$

3.2 MEAN AND VARIANCE

Consider a continuous real-valued random variable $\boldsymbol{x}$ and let $x \in \mathcal{X}$ denote the domain of its realizations (i.e., the range of values that can be assumed by x). For example, for the uniform variable described in Example 3.1, we have $\mathcal{X} = [a\ b]$.

Definitions

The mean of $\boldsymbol{x}$ is denoted by $\bar{x}$ or $\mathbb{E}\,\boldsymbol{x}$ and is defined as the calculation:

$$\mathbb{E}\,\boldsymbol{x} \;\stackrel{\Delta}{=}\; \bar{x} \;=\; \int_{x\in\mathcal{X}} x f_{\boldsymbol{x}}(x)dx \tag{3.14}$$

The mean of $\boldsymbol{x}$ is also called the expected value or the first-order moment of $\boldsymbol{x}$, and its computation can be interpreted as determining the center of mass of the pdf. This interpretation is illustrated by Example 3.3.

Likewise, the variance of a real-valued random variable $\boldsymbol{x}$ is denoted by σ_x^2 and defined by the following equivalent expressions:

$$\sigma_x^2 \;\stackrel{\Delta}{=}\; \mathbb{E}\,(\boldsymbol{x}-\bar{x})^2 \;=\; \int_{x\in\mathcal{X}} (x-\bar{x})^2 f_{\boldsymbol{x}}(x)dx \tag{3.15a}$$

$$= \mathbb{E}\,\boldsymbol{x}^2 \;-\; \bar{x}^2 \;=\; \left(\int_{x\in\mathcal{X}} x^2 f_{\boldsymbol{x}}(x)dx\right) \;-\; \bar{x}^2 \tag{3.15b}$$

Obviously, the variance is a nonnegative number,

$$\boxed{\sigma_x^2 \geq 0} \tag{3.16}$$

and its square-root, which we denote by σ_x, is referred to as the *standard deviation* of $\boldsymbol{x}$. When $\boldsymbol{x}$ has zero mean, it is seen from (3.15a) that its variance expression reduces to the second-order moment of $\boldsymbol{x}$, i.e.,

$$\sigma_x^2 = \mathbb{E}\,\boldsymbol{x}^2 \;=\; \int_{-\infty}^{\infty} x^2 f_{\boldsymbol{x}}(x)dx, \qquad \text{when } \mathbb{E}\,\boldsymbol{x} = 0 \tag{3.17}$$

Example 3.2 **(Mean and variance of a uniform random variable)** Let us reconsider the uniform pdf from Example 3.1. The mean of $\boldsymbol{x}$ is given by

$$\bar{x} = \int_a^b x\,\frac{1}{b-a}dx \;=\; \frac{1}{b-a}\frac{x^2}{2}\bigg|_a^b \;=\; \frac{1}{2}(a+b) \tag{3.18}$$

which is the midpoint of the interval $[a, b]$. The variance of $\boldsymbol{x}$ is given by

$$\begin{aligned} \sigma_x^2 &= \int_a^b x^2\,\frac{1}{b-a}dx \;-\; \left(\frac{a+b}{2}\right)^2 \\ &= \frac{1}{b-a}\frac{x^3}{3}\bigg|_a^b \;-\; \left(\frac{a+b}{2}\right)^2 \\ &= \frac{1}{12}(b-a)^2 \end{aligned} \tag{3.19}$$

Example 3.3 **(Center of mass)** Consider a rod of length ℓ and unit-mass lying horizontally along the x-axis. The left end of the rod is the origin of the horizontal axis, and the distribution of mass density across the rod is described by the function $f_{\boldsymbol{x}}(x)$ (measured in mass/unit length). Specifically, the mass content between locations x_1 and x_2 is given by the integral of $f_{\boldsymbol{x}}(x)$ over the interval $[x_1, x_2]$. The unit-mass assumption means that

$$\int_0^{\ell} f_{\boldsymbol{x}}(x)dx = 1 \tag{3.20}$$

If left unattended, the rod will swing around its left end. We would like to determine the x-coordinate of the center of mass of the rod where it can be stabilized. We denote this location by $\bar{x}$. The mass of the rod to the left of $\bar{x}$ exerts a torque that would make the rod rotate in an anti-clockwise direction, while the mass of the rod to the right of $\bar{x}$ exerts a torque that would make the rod rotate in a clockwise direction – see Fig. 3.2. An equilibrium is reached when these two torques are balanced against each other. Recall that torque is force multiplied by distance and the forces present are the cumulative weights of the respective parts of the rod to the left and right of $\bar{x}$. Therefore, the equilibrium condition amounts to:

$$\int_0^{\bar{x}} (\bar{x} - x)g f_{\boldsymbol{x}}(x)dx \;=\; \int_{\bar{x}}^{\ell} (x - \bar{x})g f_{\boldsymbol{x}}(x)dx \tag{3.21}$$

where g is the gravitational acceleration constant, approximately equal to $g = 9.8\,\text{m/s}^2$. Solving for $\bar{x}$ we find that

$$\bar{x} = \int_0^{\ell} x f_{\boldsymbol{x}}(x)dx \tag{3.22}$$

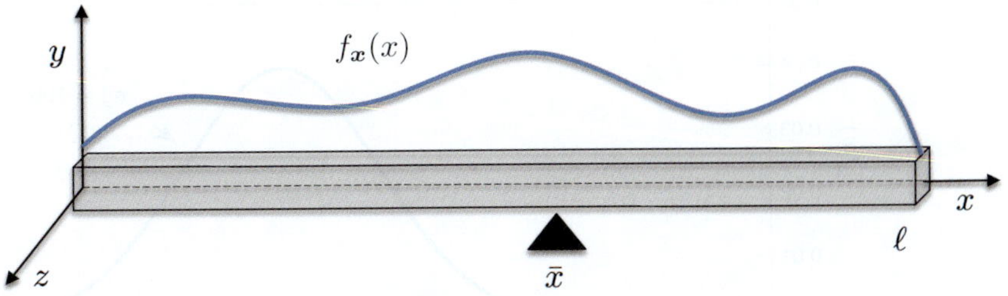

Figure 3.2 A rod of unit-mass and length ℓ is balanced horizontally at location $\bar{x}$.

Although the previous definitions for mean and variance assume a continuous random variable, they are also applicable to discrete random variables if we resort to the representation (3.5). Indeed, in this case, and assuming M possible outcomes $\{x_m\}$, each with a probability of occurrence p_m, the mean and variance relations (3.14) and (3.15b) simplify to the following expressions where integrals are replaced by sums:

$$\bar{x} = \sum_{m=1}^{M} p_m x_m \tag{3.23}$$

$$\sigma_x^2 = \left(\sum_{m=1}^{M} p_m x_m^2\right) - \bar{x}^2 \tag{3.24}$$

Measure of uncertainty

The variance of a random variable admits a useful interpretation as a measure of uncertainty. Intuitively, the variance σ_x^2 defines an interval on the real axis around the mean $\bar{x}$ where the values of the random variable $\boldsymbol{x}$ are most likely to occur:

(a) A small value of σ_x^2 indicates that $\boldsymbol{x}$ is more likely to assume values close to its mean, $\bar{x}$. In this case, we would have a reasonably good idea about what range of values are likely to be observed for $\boldsymbol{x}$ in experiments.

(b) A large value of σ_x^2 indicates that $\boldsymbol{x}$ can assume values over a wider interval around its mean. In this case, we are less certain about what values to expect for $\boldsymbol{x}$ in experiments.

For this reason, it is customary to regard the variance of a random variable as a measure of the *uncertainty* about the value it will assume in a given experiment. A small variance indicates that we are more certain about what values to expect for $\boldsymbol{x}$ (namely, values that are close to its mean), while a large variance indicates that we are less certain about what values to expect. These two situations are illustrated in Figs. 3.3 and 3.4 for two different pdfs.

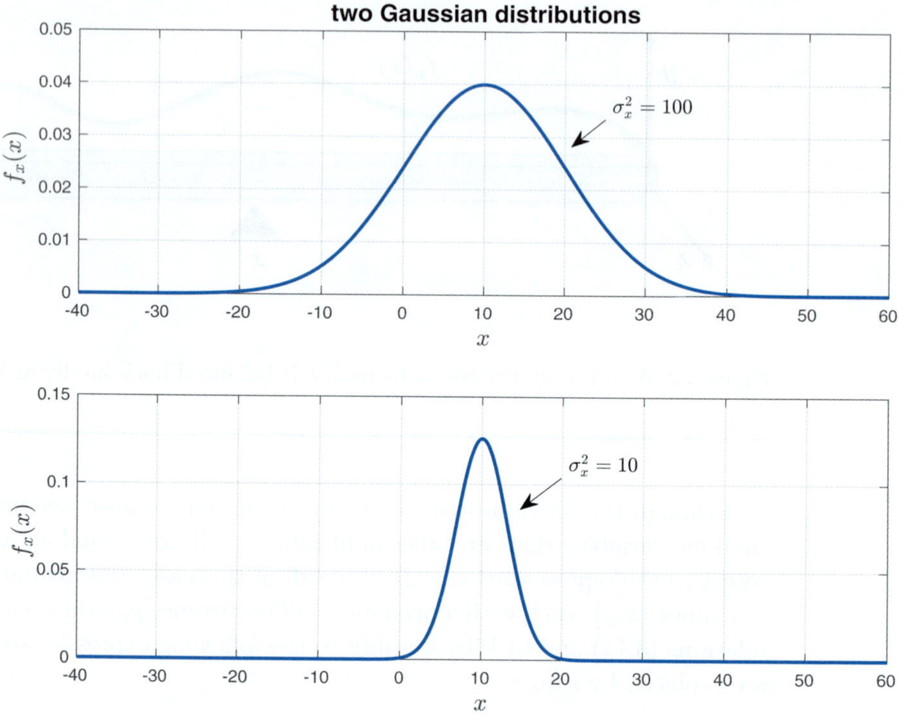

Figure 3.3 Probability density functions $f_{\boldsymbol{x}}(x)$ of a Gaussian random variable $\boldsymbol{x}$ with mean $\bar{x} = 10$, variance $\sigma_x^2 = 100$ in the top plot, and variance $\sigma_x^2 = 10$ in the bottom plot.

Figure 3.3 plots the pdf of a Gaussian-distributed random variable $\boldsymbol{x}$ for two different variances. In both cases, the mean of the random variable is fixed at $\bar{x} = 10$ while the variance is $\sigma_x^2 = 100$ in one case and $\sigma_x^2 = 10$ in the other. We explain in Chapter 4 that the pdf of a Gaussian random variable is defined in terms of $(\bar{x}, \sigma_x^2)$ by the following expression (see expression (4.4)):

$$\boxed{f_{\boldsymbol{x}}(x) = \frac{1}{\sqrt{2\pi}\,\sigma_x}\, e^{-(x-\bar{x})^2/2\sigma_x^2}, \quad x \in (-\infty, \infty)} \qquad \textbf{(Gaussian)} \qquad (3.25)$$

From Fig. 3.3 we observe that the smaller the variance of $\boldsymbol{x}$ is, the more concentrated its pdf is around its mean. Figure 3.4 provides similar plots for a second random variable $\boldsymbol{x}$ with a Rayleigh distribution, namely, with a pdf given by

$$\boxed{f_{\boldsymbol{x}}(x) = \frac{x}{\alpha^2}\, e^{-x^2/2\alpha^2}, \quad x \geq 0, \;\; \alpha > 0} \qquad \textbf{(Rayleigh)} \qquad (3.26)$$

where α is a positive parameter. The value of α determines the mean and variance of $\boldsymbol{x}$ according to the following expressions (see Prob. 3.15):

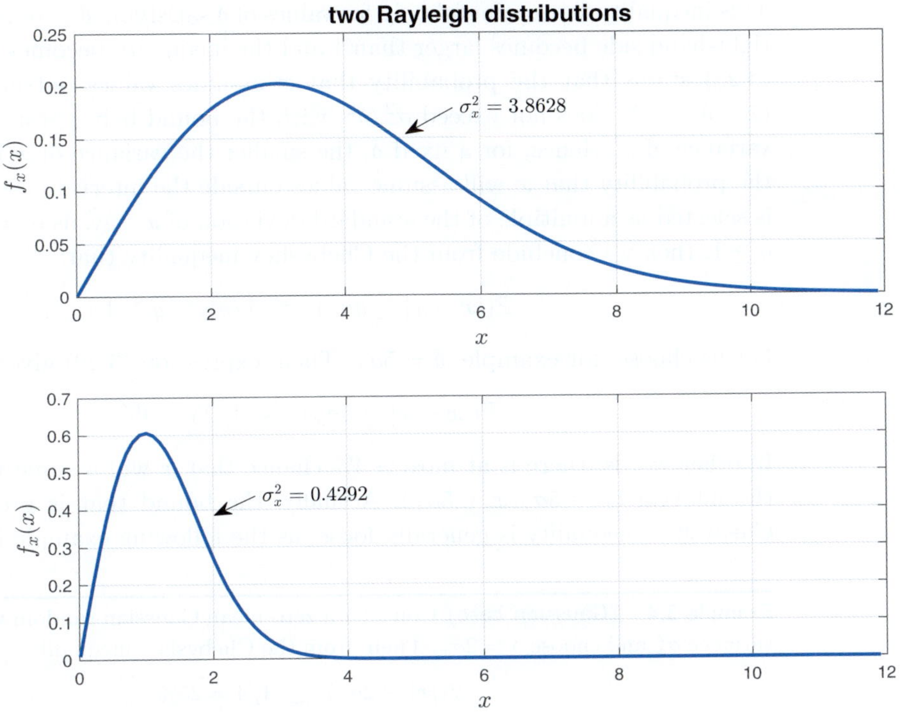

Figure 3.4 Probability density functions $f_{\boldsymbol{x}}(x)$ of a Rayleigh random variable $\boldsymbol{x}$ with mean $\bar{x} = 3.7599$ and variance $\sigma_x^2 = 3.8628$ in the top plot, and mean $\bar{x} = 1.2533$ and variance $\sigma_x^2 = 0.4292$ in the bottom plot.

$$\bar{x} = \alpha\sqrt{\frac{\pi}{2}}, \qquad \sigma_x^2 = \left(2 - \frac{\pi}{2}\right)\alpha^2 \tag{3.27}$$

so that, in contrast to the Gaussian case, the mean and variance of a Rayleigh-distributed random variable cannot be chosen independently of each other since their values are linked through α. In Fig. 3.4, the top plot corresponds to $\bar{x} = 3.7599$ and $\sigma_x^2 = 3.8628$, while the bottom plot corresponds to $\bar{x} = 1.2533$ and $\sigma_x^2 = 0.4292$.

Chebyshev inequality

The above remarks on the variance of a random variable can be qualified more formally by invoking a well-known result from probability theory known as *Chebyshev inequality* – see Probs. 3.17 and 3.18. The result states that for a random variable $\boldsymbol{x}$ with mean $\bar{x}$ and finite variance σ_x^2, and for any given scalar $\delta > 0$, it holds that

$$\boxed{\mathbb{P}(|\boldsymbol{x} - \bar{x}| \geq \delta) \;\leq\; \sigma_x^2/\delta^2} \tag{3.28}$$

This inequality is meaningful only for values of δ satisfying $\delta \geq \sigma_x$; otherwise, the right-hand side becomes larger than 1 and the inequality becomes trivial. Result (3.28) states that the probability that $\boldsymbol{x}$ assumes values outside the interval $(\bar{x} - \delta, \bar{x} + \delta)$ does not exceed σ_x^2/δ^2, with the bound being proportional to the variance of $\boldsymbol{x}$. Hence, for a fixed δ, the smaller the variance of $\boldsymbol{x}$ is, the smaller the probability that $\boldsymbol{x}$ will assume values outside the interval $(\bar{x} - \delta, \bar{x} + \delta)$. If δ is selected as a multiple of the standard deviation of $\boldsymbol{x}$, say, as $\delta = q\sigma_x$, for some $q \geq 1$, then we conclude from the Chebyshev inequality that

$$\mathbb{P}(|\boldsymbol{x} - \bar{x}| \geq q\sigma_x) \;\leq\; 1/q^2, \quad q \geq 1 \tag{3.29}$$

Let us choose, for example, $\delta = 5\sigma_x$. Then, expression (3.29) gives

$$\mathbb{P}(|\boldsymbol{x} - \bar{x}| \geq 5\sigma_x) \;\leq\; 1/25 = 4\% \tag{3.30}$$

In other words, there is at most a 4% chance that $\boldsymbol{x}$ will assume values outside the interval $(\bar{x} - 5\sigma_x, \bar{x} + 5\sigma_x)$. Actually, the bound that is provided by the Chebyshev inequality is generally loose, as the following example illustrates.

Example 3.4 **(Gaussian case)** Consider a zero-mean Gaussian random variable $\boldsymbol{x}$ with variance σ_x^2 and choose $\delta = 2\sigma_x$. Then, from the Chebyshev inequality (3.29) we obtain

$$\mathbb{P}(|\boldsymbol{x}| \geq 2\sigma_x) \;\leq\; 1/4 = 25\% \tag{3.31}$$

whereas direct evaluation of the probability using the Gaussian pdf (3.25) gives

$$\begin{aligned}
\mathbb{P}(|\boldsymbol{x}| \geq 2\sigma_x) &= \mathbb{P}(\boldsymbol{x} \geq 2\sigma_x) \;+\; \mathbb{P}(\boldsymbol{x} \leq -2\sigma_x) \\
&= \frac{1}{\sqrt{2\pi}\,\sigma_x}\left(\int_{2\sigma_x}^{\infty} e^{-x^2/2\sigma_x^2}\,dx \;+\; \int_{-\infty}^{-2\sigma_x} e^{-x^2/2\sigma_x^2}\,dx\right) \\
&= 1 - 2\left(\frac{1}{\sqrt{2\pi}\,\sigma_x}\int_0^{2\sigma_x} e^{-x^2/2\sigma_x^2}\,dx\right)
\end{aligned} \tag{3.32}$$

which can be evaluated numerically to yield:

$$\mathbb{P}(|\boldsymbol{x}| \geq 2\sigma_x) \;\approx\; 4.56\% \tag{3.33}$$

Example 3.5 **(Zero-variance random variables)** One useful consequence of the Chebyshev inequality (3.28) is that it allows us to interpret a zero-variance random variable as one that is equal to its mean in probability – see also Prob. 3.42. This is because when $\sigma_x^2 = 0$, we obtain from (3.28) that for *any* small $\delta > 0$:

$$\mathbb{P}(|\boldsymbol{x} - \bar{x}| \geq \delta) \;\leq\; 0 \tag{3.34}$$

But since the probability of any event is necessarily a nonnegative number, we conclude that

$$\mathbb{P}(|\boldsymbol{x} - \bar{x}| \geq \delta) = 0, \quad \text{for any } \delta > 0 \tag{3.35}$$

We say in this case that the equality $\boldsymbol{x} = \bar{x}$ holds in probability:

$$\boxed{\sigma_x^2 = 0 \;\Longrightarrow\; \boldsymbol{x} = \bar{x} \quad \text{in probability}} \tag{3.36}$$

For the benefit of the reader, we explain in Appendix 3.A various notions of convergence for random variables, including convergence in probability, convergence in distribution, almost-sure convergence, and mean-square convergence. For the current example, the convergence in probability result (3.36) is equivalent to statement (3.35).

3.3 DEPENDENT RANDOM VARIABLES

In inference problems, it is generally the case that information about one unobservable random variable is inferred from observations of another random variable. The observations of the second random variable will convey more or less information about the desired variable depending on how closely related (i.e., dependent) the two random variables are. Let us illustrate this concept using two examples.

Example 3.6 **(Independent random variables)** Assume a die is rolled twice. Let $\boldsymbol{x}$ denote the random variable that represents the outcome of the first roll and let $\boldsymbol{z}$ denote the random variable that represents the outcome of the second roll. These two random variables are independent of each other since the outcome of one experiment does not influence the outcome of the other. For example, if it is observed that $z = 5$, then this value does not tell us anything about what value $\boldsymbol{x}$ assumed in the first roll. Likewise, if $x = 4$, then this value does not tell us anything about what value $\boldsymbol{z}$ will assume in the second roll. That is, observations of one variable do not provide any information about the other variable.

Example 3.7 **(Two throws of a die)** Assume again that the die is rolled twice. Let $\boldsymbol{x}$ denote the random variable that represents the outcome of the first roll. Let $\boldsymbol{y}$ denote the random variable that represents the *sum* of the two rolls. Assume we only observe the outcome of $\boldsymbol{y}$ and are unaware of the outcome of $\boldsymbol{x}$. Obviously, the observation of $\boldsymbol{y}$ conveys some information about $\boldsymbol{x}$. For example, if the observation of $\boldsymbol{y}$ is $y = 10$, then x could not be $1, 2$, or 3 because in these cases the result of the second roll can never result in a sum that is equal to 10. We therefore say that the random variables $\boldsymbol{x}$ and $\boldsymbol{y}$ are dependent (the value assumed by one variable in a given experiment conveys some information about the potential value assumed by the other variable). When random variables are dependent in this way, it becomes possible to use observations of one variable to estimate the value of the other random variable. Obviously, the result of the estimation is generally imperfect and it is rarely the case that we can infer precisely the value of the unobserved variable. In most situations, we will be satisfied with close-enough guesses, where the measure of "closeness" will be formalized in some well-defined manner, e.g., by using the mean-square-error criterion or other criteria.

3.3.1 Bayes Rule

The dependency between two real-valued random variables $\{\boldsymbol{x}, \boldsymbol{y}\}$ is captured by their *joint* pdf, which is a two-dimensional function denoted by $f_{\boldsymbol{x},\boldsymbol{y}}(x, y)$. The joint pdf allows us to evaluate probabilities of events of the form:

$$\mathbb{P}(a \le \boldsymbol{x} \le b, c \le \boldsymbol{y} \le d) = \int_c^d \int_a^b f_{\boldsymbol{x},\boldsymbol{y}}(x, y) dx dy \tag{3.37}$$

namely, the probability that $\boldsymbol{x}$ and $\boldsymbol{y}$ assume values inside the intervals $[a,b]$ and $[c,d]$, respectively. We also introduce the *conditional* pdf of $\boldsymbol{x}$ given $\boldsymbol{y}$, which is denoted by $f_{\boldsymbol{x}|\boldsymbol{y}}(x|y)$; this function allows us to evaluate probabilities of events of the form:

$$\mathbb{P}(a \leq \boldsymbol{x} \leq b \mid \boldsymbol{y} = y) = \int_a^b f_{\boldsymbol{x}|\boldsymbol{y}}(x|y)dx \tag{3.38}$$

namely, the probability that $\boldsymbol{x}$ assumes values inside the interval $[a,b]$ given that $\boldsymbol{y}$ assumes the value y. It is a well-known result in probability theory that the joint and conditional pdfs of two random variables are related via the Bayes rule, which states that:

$$\begin{array}{l}(\boldsymbol{x} \textbf{ and } \boldsymbol{y} \textbf{ are continuous}) \\ f_{\boldsymbol{x},\boldsymbol{y}}(x,y) = f_{\boldsymbol{y}}(y)\, f_{\boldsymbol{x}|\boldsymbol{y}}(x|y) = f_{\boldsymbol{x}}(x)\, f_{\boldsymbol{y}|\boldsymbol{x}}(y|x)\end{array} \tag{3.39}$$

This relation expresses the joint pdf as the product of the individual and conditional pdfs of $\boldsymbol{x}$ and $\boldsymbol{y}$; it also implies that

$$f_{\boldsymbol{x}|\boldsymbol{y}}(x|y) = \frac{f_{\boldsymbol{x}}(x)\, f_{\boldsymbol{y}|\boldsymbol{x}}(y|x)}{f_{\boldsymbol{y}}(y)} \tag{3.40}$$

This relation will arise frequently in our study of inference problems, so much so that the different terms in this expression have their own terminology:

$$f_{\boldsymbol{y}}(y) \text{ is called the } \textbf{evidence} \text{ of } \boldsymbol{y} \tag{3.41a}$$
$$f_{\boldsymbol{y}|\boldsymbol{x}}(y|x) \text{ is called the } \textbf{likelihood} \text{ of } \boldsymbol{y} \tag{3.41b}$$
$$f_{\boldsymbol{x}}(x) \text{ is called the } \textbf{prior} \text{ of } \boldsymbol{x} \tag{3.41c}$$
$$f_{\boldsymbol{x}|\boldsymbol{y}}(x|y) \text{ is called the } \textbf{posterior} \text{ of } \boldsymbol{x} \tag{3.41d}$$

In inference problems we will deal frequently with the problem of observing realizations for some random variable and using them to infer the values for some other hidden or unobservable variable. Usually, the notation $\boldsymbol{y}$ plays the role of the observation and $f_{\boldsymbol{y}}(y)$ refers to its pdf, which is also called its *evidence*. The evidence provides information about the distribution of the observations. Likewise, the notation $\boldsymbol{x}$ plays the role of the hidden variable we are interested in estimating or learning about. Its distribution $f_{\boldsymbol{x}}(x)$ is called the *prior*: it represents the distribution of $\boldsymbol{x}$ prior to observing $\boldsymbol{y}$. In the same token, the conditional pdf $f_{\boldsymbol{x}|\boldsymbol{y}}(x|y)$ is called the *posterior* because it represents the distribution of $\boldsymbol{x}$ after observing $\boldsymbol{y}$. The second conditional pdf $f_{\boldsymbol{y}|\boldsymbol{x}}(y|x)$ is called the *likelihood* of $\boldsymbol{y}$ because it shows how likely the values of $\boldsymbol{y}$ are if $\boldsymbol{x}$ were known. The likelihood can also be interpreted as representing a model for the generation of the observation $\boldsymbol{y}$ from knowledge of $\boldsymbol{x}$.

Form (3.39) for the Bayes rule assumes that both variables $\boldsymbol{x}$ and $\boldsymbol{y}$ are continuous. There are variations of this rule when one or both of the random variables happen to be discrete, namely,

(x **continuous,** y **discrete**)

$$f_{\boldsymbol{x},\boldsymbol{y}}(x,y) = \mathbb{P}(\boldsymbol{y}=y)\, f_{\boldsymbol{x}|\boldsymbol{y}}(x|\boldsymbol{y}=y) = f_{\boldsymbol{x}}(x)\, \mathbb{P}(\boldsymbol{y}=y|\boldsymbol{x}=x) \tag{3.42a}$$

(x **discrete,** y **continuous**)

$$f_{\boldsymbol{x},\boldsymbol{y}}(x,y) = \mathbb{P}(\boldsymbol{x}=x)\, f_{\boldsymbol{y}|\boldsymbol{x}}(y|\boldsymbol{x}=x) = f_{\boldsymbol{y}}(y)\, \mathbb{P}(\boldsymbol{x}=x|\boldsymbol{y}=y) \tag{3.42b}$$

(x **and** y **discrete**)

$$\mathbb{P}(\boldsymbol{x}=x,\boldsymbol{y}=y) = \mathbb{P}(\boldsymbol{y}=y)\, \mathbb{P}(\boldsymbol{x}=x|\boldsymbol{y}=y) = \mathbb{P}(\boldsymbol{x}=x)\, \mathbb{P}(\boldsymbol{y}=y|\boldsymbol{x}=x) \tag{3.42c}$$

where under (3.42a) the notation $f_{\boldsymbol{x},\boldsymbol{y}}(x,y)$ for the joint pdf refers to a nonnegative function that allows us to recover the marginals of $\boldsymbol{x}$ and $\boldsymbol{y}$ as follows:

$$f_{\boldsymbol{x}}(x) = \sum_{y\in\mathcal{Y}} f_{\boldsymbol{x},\boldsymbol{y}}(x,y), \qquad \mathbb{P}(\boldsymbol{y}=y) = \int_{x\in\mathcal{X}} f_{\boldsymbol{x},\boldsymbol{y}}(x,y)dx \tag{3.43}$$

and similarly for the joint pdf in (3.42b) with the roles of $\boldsymbol{x}$ and $\boldsymbol{y}$ reversed. In the above, the sets $\mathcal{X}$ and $\mathcal{Y}$ refer to the domains of the variables $\boldsymbol{x}$ and $\boldsymbol{y}$.

Table 3.1 summarizes these variations. We will continue our discussion by considering form (3.39) for continuous random variables, but note that the conclusions can be easily extended to combinations of discrete and continuous variables.

Table 3.1 Different forms of the Bayes rule depending on the discrete or continuous nature of the random variables.

x	y	Bayes rule
continuous	continuous	$f_{\boldsymbol{x},\boldsymbol{y}}(x,y) = f_{\boldsymbol{y}}(y)\, f_{\boldsymbol{x}\|\boldsymbol{y}}(x\|y)$ $f_{\boldsymbol{x},\boldsymbol{y}}(x,y) = f_{\boldsymbol{x}}(x)\, f_{\boldsymbol{y}\|\boldsymbol{x}}(y\|x)$
discrete	discrete	$\mathbb{P}(\boldsymbol{x}=x,\boldsymbol{y}=y) = \mathbb{P}(\boldsymbol{y}=y)\, \mathbb{P}(\boldsymbol{x}=x\|\boldsymbol{y}=y)$ $\mathbb{P}(\boldsymbol{x}=x,\boldsymbol{y}=y) = \mathbb{P}(\boldsymbol{x}=x)\, \mathbb{P}(\boldsymbol{y}=y\|\boldsymbol{x}=x)$
discrete	continuous	$f_{\boldsymbol{x},\boldsymbol{y}}(x,y) = \mathbb{P}(\boldsymbol{x}=x)\, f_{\boldsymbol{y}\|\boldsymbol{x}}(y\|\boldsymbol{x}=x)$ $f_{\boldsymbol{x},\boldsymbol{y}}(x,y) = f_{\boldsymbol{y}}(y)\, \mathbb{P}(\boldsymbol{x}=x\|\boldsymbol{y}=y)$
continuous	discrete	$f_{\boldsymbol{x},\boldsymbol{y}}(x,y) = \mathbb{P}(\boldsymbol{y}=y)\, f_{\boldsymbol{x}\|\boldsymbol{y}}(x\|\boldsymbol{y}=y)$ $f_{\boldsymbol{x},\boldsymbol{y}}(x,y) = f_{\boldsymbol{x}}(x)\, \mathbb{P}(\boldsymbol{y}=y\|\boldsymbol{x}=x)$

Example 3.8 **(Observation under additive noise)** Assume $\boldsymbol{x}$ is a discrete random variable that assumes the values ± 1 with probability p for $+1$ and $1-p$ for -1. Assume further that $\boldsymbol{v}$ is a zero-mean Gaussian random variable with variance σ_v^2. In a given experiment, the user observes a scaled version of $\boldsymbol{x}$ in the presence of the noise variable $\boldsymbol{v}$. Specifically, the user observes the random variable $\boldsymbol{y} = \frac{1}{2}\boldsymbol{x} + \boldsymbol{v}$. Clearly, the random variables $\boldsymbol{x}$ and $\boldsymbol{y}$ are dependent since realizations of $\boldsymbol{x}$ alter the pdf of $\boldsymbol{y}$. For example,

if $x = +1$, then the random variable $\boldsymbol{y}$ will be Gaussian-distributed with mean $+\frac{1}{2}$ and variance σ_v^2, written as

$$f_{\boldsymbol{y}|\boldsymbol{x}}(y|x=+1) \;=\; \mathcal{N}_{\boldsymbol{y}}\left(1/2, \sigma_v^2\right) \tag{3.44}$$

where the notation $\mathcal{N}_{\boldsymbol{a}}(\bar{a}, \sigma_a^2)$ denotes a Gaussian random variable $\boldsymbol{a}$ with mean $\bar{a}$ and variance σ_a^2, namely, a random variable with pdf given by

$$\boxed{\mathcal{N}_{\boldsymbol{a}}(\bar{a}, \sigma_a^2) \;\equiv\; \frac{1}{\sqrt{2\pi}\,\sigma_a}\, e^{-(a-\bar{a})^2/2\sigma_a^2}} \qquad \textbf{(notation)} \tag{3.45}$$

On the other hand, if the realization for $\boldsymbol{x}$ happens to be $x = -1$, then the random variable $\boldsymbol{y}$ will be Gaussian-distributed with mean $-\frac{1}{2}$ and same variance σ_v^2:

$$f_{\boldsymbol{y}|\boldsymbol{x}}(y|x=-1) \;=\; \mathcal{N}_{\boldsymbol{y}}\left(-1/2, \sigma_v^2\right) \tag{3.46}$$

The overall pdf of $\boldsymbol{y}$ will then be given by:

$$f_{\boldsymbol{y}}(y) \;=\; p\,\mathcal{N}_{\boldsymbol{y}}\left(1/2, \sigma_v^2\right) \,+\, (1-p)\,\mathcal{N}_{\boldsymbol{y}}\left(-1/2, \sigma_v^2\right) \tag{3.47}$$

It is clear that $\boldsymbol{x}$ alters the pdf of $\boldsymbol{y}$ so that $\boldsymbol{x}$ and $\boldsymbol{y}$ are dependent random variables.

3.3.2 Marginal and Conditional Distributions

Given the *joint* pdf $f_{\boldsymbol{x},\boldsymbol{y}}(x, y)$ of two random variables $\boldsymbol{x}$ and $\boldsymbol{y}$, we can use this information to determine several other distributions related to the same random variables:

(a) We can determine the *marginal* pdfs corresponding to each of the variables separately, namely, the distributions $f_{\boldsymbol{x}}(x)$ and $f_{\boldsymbol{y}}(y)$. For continuous variables, these can be obtained by integrating the joint pdf over the relevant variables such as

$$f_{\boldsymbol{x}}(x) = \int_{y\in\mathcal{Y}} f_{\boldsymbol{x},\boldsymbol{y}}(x, y)dy, \qquad f_{\boldsymbol{y}}(y) = \int_{x\in\mathcal{X}} f_{\boldsymbol{x},\boldsymbol{y}}(x, y)dx \tag{3.48}$$

The first integral removes the contribution of $\boldsymbol{y}$ while the second integral removes the contribution of $\boldsymbol{x}$. If the variables $\boldsymbol{x}$ and $\boldsymbol{y}$ happen to be discrete and described by their joint pmf $\mathbb{P}(\boldsymbol{x}, \boldsymbol{y})$, we would determine the marginal pmfs by using sums rather than integrals:

$$\mathbb{P}(\boldsymbol{x} = x) = \sum_{y\in\mathcal{Y}} \mathbb{P}(\boldsymbol{x} = x, \boldsymbol{y} = y) \tag{3.49a}$$

$$\mathbb{P}(\boldsymbol{y} = y) = \sum_{x\in\mathcal{X}} \mathbb{P}(\boldsymbol{x} = x, \boldsymbol{y} = y) \tag{3.49b}$$

(b) We can also determine the *conditional* pdfs corresponding to each of the variables conditioned on the other variable, namely, the distributions $f_{\boldsymbol{x}|\boldsymbol{y}}(x|y)$ and $f_{\boldsymbol{y}|\boldsymbol{x}}(y|x)$. These can be obtained by appealing to the Bayes rule:

$$f_{\boldsymbol{x}|\boldsymbol{y}}(x|y) = f_{\boldsymbol{x},\boldsymbol{y}}(x,y)/f_{\boldsymbol{y}}(\boldsymbol{y}) \tag{3.50a}$$

$$f_{\boldsymbol{y}|\boldsymbol{x}}(y|x) = f_{\boldsymbol{x},\boldsymbol{y}}(x,y)/f_{\boldsymbol{x}}(\boldsymbol{x}) \tag{3.50b}$$

In other words, the joint pdf needs to be scaled by the marginal pdfs. For discrete random variables, we would use instead:

$$\mathbb{P}(\boldsymbol{x}=x|\boldsymbol{y}=y) = \frac{\mathbb{P}(\boldsymbol{x}=x,\boldsymbol{y}=y)}{\mathbb{P}(\boldsymbol{y}=y)} \tag{3.51a}$$

$$\mathbb{P}(\boldsymbol{y}=y|\boldsymbol{x}=x) = \frac{\mathbb{P}(\boldsymbol{x}=x,\boldsymbol{y}=y)}{\mathbb{P}(\boldsymbol{x}=x)} \tag{3.51b}$$

Example 3.9 **(Law of total probability)** Consider two random variables $\boldsymbol{x}$ and $\boldsymbol{y}$ with conditional pdf $f_{\boldsymbol{y}|\boldsymbol{x}}(y|x)$ and marginal pdf $f_{\boldsymbol{x}}(x)$. Using the Bayes rule we have that the joint pdf factorizes as

$$f_{\boldsymbol{x},\boldsymbol{y}}(x,y) = f_{\boldsymbol{x}}(x)\, f_{\boldsymbol{y}|\boldsymbol{x}}(y|x) \tag{3.52}$$

Marginalizing over $\boldsymbol{x}$ we arrive at the useful relation, also known as the *law of total probability*:

$$\boxed{f_{\boldsymbol{y}}(y) \;=\; \int_{x\in\mathcal{X}} f_{\boldsymbol{x}}(x) f_{\boldsymbol{y}|\boldsymbol{x}}(y|x)dx} \tag{3.53}$$

In other words, we can recover the marginal of $\boldsymbol{y}$ from knowledge of the marginal of $\boldsymbol{x}$ and the conditional of $\boldsymbol{y}$ given $\boldsymbol{x}$.

Example 3.10 **(Useful conditional relation)** Consider three discrete random variables $\{\boldsymbol{A},\boldsymbol{B},\boldsymbol{C}\}$ and note that

$$\mathbb{P}(\boldsymbol{A}|\boldsymbol{B},\boldsymbol{C}) = \frac{\mathbb{P}(\boldsymbol{A},\boldsymbol{B},\boldsymbol{C})}{\mathbb{P}(\boldsymbol{B},\boldsymbol{C})} = \frac{\cancel{\mathbb{P}(\boldsymbol{B})}\,\mathbb{P}(\boldsymbol{A}|\boldsymbol{B})\,\mathbb{P}(\boldsymbol{C}|\boldsymbol{A},\boldsymbol{B})}{\cancel{\mathbb{P}(\boldsymbol{B})}\,\mathbb{P}(\boldsymbol{C}|\boldsymbol{B})} \tag{3.54}$$

so that

$$\boxed{\mathbb{P}(\boldsymbol{A}|\boldsymbol{B},\boldsymbol{C}) = \frac{\mathbb{P}(\boldsymbol{A}|\boldsymbol{B})\,\mathbb{P}(\boldsymbol{C}|\boldsymbol{A},\boldsymbol{B})}{\mathbb{P}(\boldsymbol{C}|\boldsymbol{B})}} \tag{3.55}$$

Example 3.11 **(Finding marginal and conditional distributions)** Let us consider a situation involving two discrete random variables, denoted by $\boldsymbol{C}$ (cold) and $\boldsymbol{H}$ (headache). Each variable is binary and assumes the values $\{0,1\}$. For example, $\boldsymbol{C}=1$ and $\boldsymbol{H}=0$ means that the individual has a cold but does not have a headache. Likewise, the combination $\boldsymbol{C}=0$ and $\boldsymbol{H}=0$ means that the individual neither has a cold nor a headache. There are four combinations for the random variables and we describe their joint pmf in the following tabular form (the numbers in the table are for illustration purposes only and do not correspond to any actual measurements or have any medical significance):

C (cold)	H (headache)	$\mathbb{P}(C, H)$ (joint pmf)
0	0	0.60
0	1	0.10
1	0	0.10
1	1	0.20

Observe how all entries in the last column corresponding to the joint pmf add up to 1, as expected. Let us determine first the marginal pmf for the variable $\boldsymbol{C}$. For this purpose, we need to determine the *two* probabilities: $\mathbb{P}(\boldsymbol{C}=1)$ and $\mathbb{P}(\boldsymbol{C}=0)$. This is because the variable $\boldsymbol{C}$ can assume one of two values in $\{0,1\}$. For the first probability, we add over H when $\boldsymbol{C}=1$ to get

$$\begin{aligned}\mathbb{P}(\boldsymbol{C}=1) &= \sum_{H\in\{0,1\}} \mathbb{P}(\boldsymbol{C}=1, \boldsymbol{H}=H)\\ &= \mathbb{P}(\boldsymbol{C}=1,\boldsymbol{H}=1) + \mathbb{P}(\boldsymbol{C}=1,\boldsymbol{H}=0)\\ &= 0.20+0.10\\ &= 0.3 \end{aligned} \tag{3.56}$$

Observe that we simply added the probabilities in the last two rows of the table corresponding to $\boldsymbol{C}=1$. We repeat for $\mathbb{P}(\boldsymbol{C}=0)$ to find that

$$\begin{aligned}\mathbb{P}(\boldsymbol{C}=0) &= \sum_{H\in\{0,1\}} \mathbb{P}(\boldsymbol{C}=0, \boldsymbol{H}=H)\\ &= \mathbb{P}(\boldsymbol{C}=0,\boldsymbol{H}=1) + \mathbb{P}(\boldsymbol{C}=0,\boldsymbol{H}=0)\\ &= 0.10+0.60\\ &= 0.7 \end{aligned} \tag{3.57}$$

Here we added the probabilities in the first two rows of the table corresponding to $\boldsymbol{C}=0$. Note that the two probabilities for $\boldsymbol{C}$ add up to 1, as is expected for a valid pmf. In a similar manner we find that the pmf for the variable $\boldsymbol{H}$ is given by

$$\mathbb{P}(\boldsymbol{H}=1)=0.3, \quad \mathbb{P}(\boldsymbol{H}=0)=0.7 \tag{3.58}$$

Using the marginal pmfs, we can now determine conditional pmfs. For example, assume we observe that $\boldsymbol{H}=1$ (that is, the individual has a headache), and we would like to know the likelihood that the individual has a cold too. To do so, we appeal to the Bayes rule and write

$$\mathbb{P}(\boldsymbol{C}=1|\boldsymbol{H}=1) = \frac{\mathbb{P}(\boldsymbol{C}=1,\boldsymbol{H}=1)}{\mathbb{P}(\boldsymbol{H}=1)} = 0.2/0.3 = 2/3 \tag{3.59}$$

Accordingly, we also have

$$\mathbb{P}(\boldsymbol{C}=0|\boldsymbol{H}=1)=1/3 \tag{3.60}$$

since these two conditional probabilities need to add up to 1. In a similar manner, we find that

$$\mathbb{P}(\boldsymbol{C}=1|\boldsymbol{H}=0)=1/7, \quad \mathbb{P}(\boldsymbol{C}=0|\boldsymbol{H}=0)=6/7 \tag{3.61}$$

Observe that in order to specify the conditional pmf of $\boldsymbol{C}$ given $\boldsymbol{H}$ we need to determine four probability values since each of the variables $\boldsymbol{C}$ or $\boldsymbol{H}$ assumes two levels in $\{0,1\}$. We repeat similar calculations to compute the conditional pmf of $\boldsymbol{H}$ given $\boldsymbol{C}$ (with the roles of $\boldsymbol{C}$ and $\boldsymbol{H}$) reversed. We summarize these conditional calculations in tabular form:

C	H	$\mathbb{P}(C\|H)$ (conditional pmf)	$\mathbb{P}(H\|C)$ (conditional pmf)
0	0	6/7	6/7
0	1	1/3	1/7
1	0	1/7	1/3
1	1	2/3	2/3

Observe that the entries in the third column corresponding to the conditional pmf of $\boldsymbol{C}$ given $\boldsymbol{H}$ do *not* add up to 1. Likewise, for the last column corresponding to the conditional pmf of $\boldsymbol{H}$ given $\boldsymbol{C}$.

3.3.3 Dependent Random Variables

We say that two continuous random variables $\{\boldsymbol{x}, \boldsymbol{y}\}$ are *independent* of each other if, and only if:

$$f_{\boldsymbol{x}|\boldsymbol{y}}(x|y) = f_{\boldsymbol{x}}(x) \quad \text{and} \quad f_{\boldsymbol{y}|\boldsymbol{x}}(y|x) = f_{\boldsymbol{y}}(y) \tag{3.62}$$

In other words, the pdfs of $\boldsymbol{x}$ and $\boldsymbol{y}$ are not modified by conditioning on knowledge of $\boldsymbol{y}$ or $\boldsymbol{x}$. Otherwise, the random variables are *dependent*. It follows directly from the Bayes rule that dependency is equivalent to saying that the joint pdf factorizes as the product of the two marginal pdfs:

$$f_{\boldsymbol{x}}(\boldsymbol{x}, \boldsymbol{y}) = f_{\boldsymbol{x}}(x)\, f_{\boldsymbol{y}}(y) \tag{3.63}$$

We will employ the following notation to refer to the fact that the random variables $\boldsymbol{x}$ and $\boldsymbol{y}$ are independent:

$$\boxed{\boldsymbol{x} \perp\!\!\!\perp \boldsymbol{y}} \qquad \textbf{(independent random variables)} \tag{3.64}$$

When the variables are discrete, they will be independent of each other if, and only if,

$$\mathbb{P}(\boldsymbol{x}|\boldsymbol{y}) = \mathbb{P}(\boldsymbol{x}) \quad \text{and} \quad \mathbb{P}(\boldsymbol{y}|\boldsymbol{x}) = \mathbb{P}(\boldsymbol{y}) \tag{3.65}$$

or, equivalently,

$$\mathbb{P}(\boldsymbol{x}, \boldsymbol{y}) = \mathbb{P}(\boldsymbol{x})\, \mathbb{P}(\boldsymbol{y}) \tag{3.66}$$

The notation in the last two expressions needs some clarification. Consider the equality $\mathbb{P}(\boldsymbol{x}|\boldsymbol{y}) = \mathbb{P}(\boldsymbol{x})$, where the random variables are indicated in boldface and no specific values are listed for them. This compact relation is read as follows. Regardless of which value we observe for $\boldsymbol{y}$, and for any value of $\boldsymbol{x}$, the likelihood of observing that value for $\boldsymbol{x}$ given $\boldsymbol{y}$ will remain unchanged. We can write the relation more explicitly as follows:

$$\mathbb{P}(\boldsymbol{x} = x|\boldsymbol{y} = y) = \mathbb{P}(\boldsymbol{x} = x), \quad \text{for any } x \in \mathcal{X},\ y \in \mathcal{Y} \tag{3.67}$$

or, in terms of the joint and marginal pmfs:

$$\mathbb{P}(\boldsymbol{x} = x, \boldsymbol{y} = y) = \mathbb{P}(\boldsymbol{x} = x)\, \mathbb{P}(\boldsymbol{y} = y), \quad \text{for any } x \in \mathcal{X},\ y \in \mathcal{Y} \tag{3.68}$$

It is sufficient to find one combination of values (x, y) for which the equality does not hold to conclude that $\boldsymbol{x}$ and $\boldsymbol{y}$ are dependent.

Example 3.12 **(Checking for dependency)** Let us reconsider Example 3.11 involving the variables $\boldsymbol{C}$ (cold) and $\boldsymbol{H}$ (headache). We know from the calculations in the example that:

$$\mathbb{P}(\boldsymbol{C}=1)=0.3,\ \ \mathbb{P}(\boldsymbol{C}=1|\boldsymbol{H}=1)=2/3,\ \ \mathbb{P}(\boldsymbol{C}=1|\boldsymbol{H}=0)=1/7 \tag{3.69}$$
$$\mathbb{P}(\boldsymbol{C}=0)=0.7,\ \ \mathbb{P}(\boldsymbol{C}=0|\boldsymbol{H}=1)=1/3,\ \ \mathbb{P}(\boldsymbol{C}=0|\boldsymbol{H}=0)=6/7 \tag{3.70}$$

It is clear from these values that the random variables $\boldsymbol{C}$ and $\boldsymbol{H}$ are dependent; knowledge of one variable alters the likelihood of the other variable. For instance, knowing that the individual has a headache ($\boldsymbol{H}=1$), raises the likelihood of the individual having a cold from 0.3 to 2/3.

Example 3.13 **(Dependency is not causality)** The notion of statistical dependence is *bidirectional* or *symmetric*. When $\boldsymbol{x}$ and $\boldsymbol{y}$ are dependent, this means that $\boldsymbol{x}$ depends on $\boldsymbol{y}$ *and* $\boldsymbol{y}$ depends on $\boldsymbol{x}$ in the sense that their conditional pdfs (or pmfs) satisfy (3.62) or (3.65). This also means that observing one variable alters the likelihood about the other variable. Again, from Example 3.11, the variables $\boldsymbol{C}$ and $\boldsymbol{H}$ are dependent on each other. Observing whether an individual has a cold or not alters the likelihood of whether the same individual has a headache or not. Similarly, observing whether the individual has a headache or not alters the likelihood of that individual having a cold or not.

The notion of dependency between random variables is sometimes confused with the notion of *causality* or *causation*. Causality implies dependency but not the other way around. In other words, statistical dependence is a necessary but not sufficient condition for causality:

$$\textsf{\textbf{causality}} \Longrightarrow \textsf{\textbf{statistical dependence}} \tag{3.71}$$

One main reason why these two notions are not equivalent is because dependency is symmetric while causality is *asymmetric*. If a random variable $\boldsymbol{y}$ depends *causally* on another random variable $\boldsymbol{x}$, this means that $\boldsymbol{x}$ assuming certain values will contribute to (or have an effect on) $\boldsymbol{y}$ assuming certain values of its own. For example, if $\boldsymbol{x}$ is the random variable that indicates whether it is raining or not and $\boldsymbol{y}$ is the random variable that indicates whether the grass in the garden is wet or not, then having $\boldsymbol{x}$ assume the value $\boldsymbol{x}=1$ (raining) will cause $\boldsymbol{y}$ to assume the value $\boldsymbol{y}=1$ (wet grass):

$$\boldsymbol{x}=1 \text{ (raining)} \xrightarrow{\text{causes}} \boldsymbol{y}=1 \text{ (wet grass)} \tag{3.72}$$

Here, we say that raining ($\boldsymbol{x}$) influences the grass ($\boldsymbol{y}$) in a causal manner. As a result, the variables $\{\boldsymbol{x}, \boldsymbol{y}\}$ will be statistically dependent as well. This is because knowing that it has rained influences the likelihood of observing wet grass and, conversely, knowing the state of the grass influences the likelihood of having observed rain. However, while dependency is bidirectional, the same is not true for causality: observing wet grass does not cause the rain to fall. That is, the state of the grass variable ($\boldsymbol{y}$) does not have a cause effect on the state of the rain variable ($\boldsymbol{x}$).

One formal way to define causality is to introduce the do operator. Writing $\mathsf{do}(\boldsymbol{x}=1)$ means that we manipulate the value of the random variable $\boldsymbol{x}$ and set it to 1 (rather than observe it as having assumed the value 1). Using this abstraction, we say that a random variable $\boldsymbol{x}$ has a *cause* effect on another random variable $\boldsymbol{y}$ if, and only if, the following *two* conditional probability relations hold:

$$\mathbb{P}\Big(\boldsymbol{y} = y|\mathsf{do}(\boldsymbol{x} = x)\Big) \neq \mathbb{P}(\boldsymbol{y} = y) \tag{3.73a}$$

$$\mathbb{P}\Big(\boldsymbol{x} = x|\mathsf{do}(\boldsymbol{y} = y)\Big) = \mathbb{P}(\boldsymbol{x} = x) \tag{3.73b}$$

for any $(x, y) \in \mathcal{X} \times \mathcal{Y}$. The first relation says that having $\boldsymbol{x}$ assume particular values will alter the distribution of $\boldsymbol{y}$, while the second relation says that the reverse effect does not hold. These two conditions highlight the asymmetric nature of causality.

Example 3.14 **(Conditional independence)** We can extend the notion of independence to conditional distributions. Given three continuous random variables $\{\boldsymbol{x}, \boldsymbol{y}, \boldsymbol{z}\}$, we say that $\boldsymbol{x}$ and $\boldsymbol{y}$ are conditionally independent given $\boldsymbol{z}$, written as

$$\boldsymbol{x} \perp\!\!\!\perp \boldsymbol{y} \,|\, \boldsymbol{z} \tag{3.74}$$

if, and only if,

$$f_{\boldsymbol{x},\boldsymbol{y}|\boldsymbol{z}}(x, y|z) = f_{\boldsymbol{x}|\boldsymbol{z}}(x|z)\, f_{\boldsymbol{y}|\boldsymbol{z}}(y|z) \tag{3.75}$$

That is, the conditional pdf of $\{\boldsymbol{x}, \boldsymbol{y}\}$ given $\boldsymbol{z}$ decouples into the product of the individual conditional distributions of $\boldsymbol{x}$ and $\boldsymbol{y}$ given $\boldsymbol{z}$. For discrete random variables, the independence relation translates into requiring

$$\mathbb{P}(\boldsymbol{x}, \boldsymbol{y}|\boldsymbol{z}) = \mathbb{P}(\boldsymbol{x}|\boldsymbol{z})\, \mathbb{P}(\boldsymbol{y}|\boldsymbol{z}) \tag{3.76}$$

Conditional dependence will play a prominent role in the study of Bayesian networks and probabilistic graphical models later in our treatment. Let us illustrate the definition by means of a numerical example involving three binary random variables assuming the values $\{0, 1\}$:

$\boldsymbol{R}$ = indicates whether it is raining (1) or not (0) (3.77a)

$\boldsymbol{A}$ = indicates whether there has been a traffic accident (1) or not (0) (3.77b)

$\boldsymbol{L}$ = indicates whether the individual is late to work (1) or not (0) (3.77c)

For example, $\boldsymbol{R} = 1$, $\boldsymbol{A} = 0$, and $\boldsymbol{L} = 0$ means that it is raining, there has been no traffic accident on the road, and the individual is not late to work. Since each variable is binary, there are eight possible combinations. We describe the joint pmf in the following tabular form (the numbers in the table are for illustration purposes only):

$\boldsymbol{R}$ (rain)	$\boldsymbol{A}$ (accident)	$\boldsymbol{L}$ (late)	$\mathbb{P}(\boldsymbol{R}, \boldsymbol{A}, \boldsymbol{L})$ (joint pmf)
0	0	0	4/15
0	0	1	8/45
0	1	0	1/48
0	1	1	1/16
1	0	0	2/15
1	0	1	4/45
1	1	0	1/16
1	1	1	3/16

Assume we are able to observe whether there has been an accident on the road (i.e., we are able to know whether $\boldsymbol{A} = 1$ or $\boldsymbol{A} = 0$). Given this knowledge, we want to verify whether the random variables $\boldsymbol{R}$ and $\boldsymbol{L}$ are independent of each other. In particular, if these variables turn out to be dependent, then observing the individual arriving late to work would influence the likelihood of whether it has been raining or not.

To answer these questions, we need to determine the conditional pmfs $\mathbb{P}(\boldsymbol{R}, \boldsymbol{L}|\boldsymbol{A})$, $\mathbb{P}(\boldsymbol{R}|\boldsymbol{A})$, and $\mathbb{P}(\boldsymbol{L}|\boldsymbol{A})$, and then verify whether they satisfy the product relation:

$$\mathbb{P}(\boldsymbol{R}, \boldsymbol{L}|\boldsymbol{A}) \stackrel{?}{=} \mathbb{P}(\boldsymbol{R}|\boldsymbol{A})\,\mathbb{P}(\boldsymbol{L}|\boldsymbol{A}) \tag{3.78}$$

We start by computing the marginal pmf for the variable $\boldsymbol{A}$:

$$\begin{aligned}\mathbb{P}(\boldsymbol{A}=1) &= \sum_{R\in\{0,1\}} \sum_{L\in\{0,1\}} \mathbb{P}(\boldsymbol{R}=R, \boldsymbol{L}=L, \boldsymbol{A}=1) \\ &= 1/48 + 1/16 + 1/16 + 3/16 \\ &= 1/3 \end{aligned} \tag{3.79}$$

Observe that this probability is obtained by adding the entries in the last column of the table that correspond to the rows with $\boldsymbol{A}=1$. It follows that

$$\mathbb{P}(\boldsymbol{A}=0) = 2/3 \tag{3.80}$$

Next, we determine the joint pmfs $\mathbb{P}(\boldsymbol{R}, \boldsymbol{A}=1)$ and $\mathbb{P}(\boldsymbol{L}, \boldsymbol{A}=1)$:

$$\mathbb{P}(\boldsymbol{R}=1, \boldsymbol{A}=1) = \sum_{L\in\{0,1\}} \mathbb{P}(\boldsymbol{R}=1, \boldsymbol{L}=L, \boldsymbol{A}=1) = \frac{1}{16} + \frac{3}{16} = 1/4 \tag{3.81}$$

$$\mathbb{P}(\boldsymbol{R}=0, \boldsymbol{A}=1) = \sum_{L\in\{0,1\}} \mathbb{P}(\boldsymbol{R}=0, \boldsymbol{L}=L, \boldsymbol{A}=1) = \frac{1}{48} + \frac{1}{16} = 1/12 \tag{3.82}$$

and

$$\mathbb{P}(\boldsymbol{L}=1, \boldsymbol{A}=1) = \sum_{R\in\{0,1\}} \mathbb{P}(\boldsymbol{R}=R, \boldsymbol{L}=1, \boldsymbol{A}=1) = \frac{1}{16} + \frac{3}{16} = 1/4 \tag{3.83}$$

$$\mathbb{P}(\boldsymbol{L}=0, \boldsymbol{A}=1) = \sum_{R\in\{0,1\}} \mathbb{P}(\boldsymbol{R}=R, \boldsymbol{L}=0, \boldsymbol{A}=1) = \frac{1}{48} + \frac{1}{16} = 1/12 \tag{3.84}$$

We similarly determine the joint pmfs $\mathbb{P}(\boldsymbol{R}, \boldsymbol{A}=0)$ and $\mathbb{P}(\boldsymbol{L}, \boldsymbol{A}=0)$:

$$\mathbb{P}(\boldsymbol{R}=1, \boldsymbol{A}=0) = \sum_{L\in\{0,1\}} \mathbb{P}(\boldsymbol{R}=1, \boldsymbol{L}=L, \boldsymbol{A}=0) = \frac{2}{15} + \frac{4}{45} = 2/9 \tag{3.85}$$

$$\mathbb{P}(\boldsymbol{R}=0, \boldsymbol{A}=0) = \sum_{L\in\{0,1\}} \mathbb{P}(\boldsymbol{R}=0, \boldsymbol{L}=L, \boldsymbol{A}=0) = \frac{4}{15} + \frac{8}{45} = 4/9 \tag{3.86}$$

and

$$\mathbb{P}(\boldsymbol{L}=1, \boldsymbol{A}=0) = \sum_{R\in\{0,1\}} \mathbb{P}(\boldsymbol{R}=R, \boldsymbol{L}=1, \boldsymbol{A}=0) = \frac{8}{15} + \frac{4}{45} = 4/15 \tag{3.87}$$

$$\mathbb{P}(\boldsymbol{L}=0, \boldsymbol{A}=0) = \sum_{R\in\{0,1\}} \mathbb{P}(\boldsymbol{R}=R, \boldsymbol{L}=0, \boldsymbol{A}=0) = \frac{4}{15} + \frac{2}{45} = 2/5 \tag{3.88}$$

Next, appealing to the Bayes rule we get

$$\mathbb{P}(\boldsymbol{R}=1|\boldsymbol{A}=1)=\frac{\mathbb{P}(\boldsymbol{R}=1,\boldsymbol{A}=1)}{\mathbb{P}(\boldsymbol{A}=1)}=\frac{1/4}{1/3}=3/4 \tag{3.89a}$$

$$\mathbb{P}(\boldsymbol{R}=0|\boldsymbol{A}=1)=\frac{\mathbb{P}(\boldsymbol{R}=0,\boldsymbol{A}=1)}{\mathbb{P}(\boldsymbol{A}=1)}=\frac{1/12}{1/3}=1/4 \tag{3.89b}$$

$$\mathbb{P}(\boldsymbol{R}=1|\boldsymbol{A}=0)=\frac{\mathbb{P}(\boldsymbol{R}=1,\boldsymbol{A}=0)}{\mathbb{P}(\boldsymbol{A}=0)}=\frac{2/9}{2/3}=1/3 \tag{3.89c}$$

$$\mathbb{P}(\boldsymbol{R}=0|\boldsymbol{A}=0)=\frac{\mathbb{P}(\boldsymbol{R}=0,\boldsymbol{A}=0)}{\mathbb{P}(\boldsymbol{A}=0)}=\frac{4/9}{2/3}=2/3 \tag{3.89d}$$

and

$$\mathbb{P}(\boldsymbol{L}=1|\boldsymbol{A}=1)=\frac{\mathbb{P}(\boldsymbol{L}=1,\boldsymbol{A}=1)}{\mathbb{P}(\boldsymbol{A}=1)}=\frac{1/4}{1/3}=3/4 \tag{3.90a}$$

$$\mathbb{P}(\boldsymbol{L}=0|\boldsymbol{A}=1)=\frac{\mathbb{P}(\boldsymbol{L}=0,\boldsymbol{A}=1)}{\mathbb{P}(\boldsymbol{A}=1)}=\frac{1/12}{1/3}=1/4 \tag{3.90b}$$

$$\mathbb{P}(\boldsymbol{L}=1|\boldsymbol{A}=0)=\frac{\mathbb{P}(\boldsymbol{L}=1,\boldsymbol{A}=0)}{\mathbb{P}(\boldsymbol{A}=0)}=\frac{4/15}{2/3}=2/5 \tag{3.90c}$$

$$\mathbb{P}(\boldsymbol{L}=0|\boldsymbol{A}=0)=\frac{\mathbb{P}(\boldsymbol{L}=0,\boldsymbol{A}=0)}{\mathbb{P}(\boldsymbol{A}=0)}=\frac{2/5}{2/3}=3/5 \tag{3.90d}$$

We still need to compute the joint conditional pmf $\mathbb{P}(\boldsymbol{R},\boldsymbol{L}|\boldsymbol{A})$. Thus, note that

$$\mathbb{P}(\boldsymbol{R}=1,\boldsymbol{L}=1|\boldsymbol{A}=1)=\frac{\mathbb{P}(\boldsymbol{R}=1,\boldsymbol{L}=1,\boldsymbol{A}=1)}{\mathbb{P}(\boldsymbol{A}=1)}=(3/16)/(1/3)=9/16 \tag{3.91a}$$

$$\mathbb{P}(\boldsymbol{R}=1,\boldsymbol{L}=0|\boldsymbol{A}=1)=\frac{\mathbb{P}(\boldsymbol{R}=1,\boldsymbol{L}=0,\boldsymbol{A}=1)}{\mathbb{P}(\boldsymbol{A}=1)}=(1/16)/(1/3)=3/16 \tag{3.91b}$$

$$\mathbb{P}(\boldsymbol{R}=0,\boldsymbol{L}=1|\boldsymbol{A}=1)=\frac{\mathbb{P}(\boldsymbol{R}=0,\boldsymbol{L}=1,\boldsymbol{A}=1)}{\mathbb{P}(\boldsymbol{A}=1)}=(1/16)/(1/3)=3/16 \tag{3.91c}$$

$$\mathbb{P}(\boldsymbol{R}=0,\boldsymbol{L}=0|\boldsymbol{A}=1)=\frac{\mathbb{P}(\boldsymbol{R}=0,\boldsymbol{L}=0,\boldsymbol{A}=1)}{\mathbb{P}(\boldsymbol{A}=1)}=(1/48)/(1/3)=1/16 \tag{3.91d}$$

and

$$\mathbb{P}(\boldsymbol{R}=1,\boldsymbol{L}=1|\boldsymbol{A}=0) = \frac{\mathbb{P}(\boldsymbol{R}=1,\boldsymbol{L}=1,\boldsymbol{A}=0)}{\mathbb{P}(\boldsymbol{A}=0)} = (4/45)/(2/3) = 2/15 \tag{3.92a}$$

$$\mathbb{P}(\boldsymbol{R}=1,\boldsymbol{L}=0|\boldsymbol{A}=0) = \frac{\mathbb{P}(\boldsymbol{R}=1,\boldsymbol{L}=0,\boldsymbol{A}=0)}{\mathbb{P}(\boldsymbol{A}=0)} = (2/15)/(2/3) = 1/5 \tag{3.92b}$$

$$\mathbb{P}(\boldsymbol{R}=0,\boldsymbol{L}=1|\boldsymbol{A}=0) = \frac{\mathbb{P}(\boldsymbol{R}=0,\boldsymbol{L}=1,\boldsymbol{A}=0)}{\mathbb{P}(\boldsymbol{A}=0)} = (8/45)/(2/3) = 4/15 \tag{3.92c}$$

$$\mathbb{P}(\boldsymbol{R}=0,\boldsymbol{L}=0|\boldsymbol{A}=0) = \frac{\mathbb{P}(\boldsymbol{R}=0,\boldsymbol{L}=0,\boldsymbol{A}=0)}{\mathbb{P}(\boldsymbol{A}=0)} = (4/15)/(2/3) = 2/5 \tag{3.92d}$$

We collect the results in tabular form and conclude from comparing the entries in the fourth and last columns that the variables $\boldsymbol{R}$ and $\boldsymbol{L}$ are *independent* conditioned on $\boldsymbol{A}$.

$\boldsymbol{R}$	$\boldsymbol{A}$	$\boldsymbol{L}$	$\mathbb{P}(\boldsymbol{R},\boldsymbol{L}\|\boldsymbol{A})$	$\mathbb{P}(\boldsymbol{R}\|\boldsymbol{A})$	$\mathbb{P}(\boldsymbol{L}\|\boldsymbol{A})$	$\mathbb{P}(\boldsymbol{R}\|\boldsymbol{A})\times\mathbb{P}(\boldsymbol{L}\|\boldsymbol{A})$
0	0	0	2/5	2/3	3/5	2/5
0	0	1	4/15	2/3	2/5	4/15
0	1	0	1/16	1/4	1/4	1/16
0	1	1	3/16	1/4	3/4	3/16
1	0	0	1/5	1/3	3/5	1/5
1	0	1	2/15	1/3	2/5	2/15
1	1	0	3/16	3/4	1/4	3/16
1	1	1	9/16	3/4	3/4	9/16

Example 3.15 (**Other conditional independence relations**) We list additional properties for conditionally independent random variables; we focus on discrete random variables for illustration purposes, although the results are applicable to continuous random variables as well.

(a) First, consider two variables $\boldsymbol{x}$ and $\boldsymbol{y}$ that are independent given $\boldsymbol{z}$. It then holds:

$$\boxed{\boldsymbol{x} \perp\!\!\!\perp \boldsymbol{y}\,|\,\boldsymbol{z} \iff \mathbb{P}(\boldsymbol{x}|\boldsymbol{y},\boldsymbol{z}) = \mathbb{P}(\boldsymbol{x}|\boldsymbol{z})} \tag{3.93}$$

Proof: Indeed, note from the Bayes rule that

$$\begin{aligned} \mathbb{P}(\boldsymbol{x}|\boldsymbol{y},\boldsymbol{z}) &= \frac{\mathbb{P}(\boldsymbol{x},\boldsymbol{y},\boldsymbol{z})}{\mathbb{P}(\boldsymbol{y},\boldsymbol{z})} \\ &= \frac{\cancel{\mathbb{P}(\boldsymbol{z})}\,\mathbb{P}(\boldsymbol{x},\boldsymbol{y}|\boldsymbol{z})}{\cancel{\mathbb{P}(\boldsymbol{z})}\,\mathbb{P}(\boldsymbol{y}|\boldsymbol{z})} \\ &\overset{(3.76)}{=} \frac{\mathbb{P}(\boldsymbol{x}|\boldsymbol{z})\,\mathbb{P}(\boldsymbol{y}|\boldsymbol{z})}{\mathbb{P}(\boldsymbol{y}|\boldsymbol{z})}, \quad \text{since } \boldsymbol{x} \perp\!\!\!\perp \boldsymbol{y}\,|\,\boldsymbol{z} \\ &= \mathbb{P}(\boldsymbol{x}|\boldsymbol{z}) \end{aligned} \tag{3.94}$$

■

(b) Second, we consider the following result referred to as the *weak union* property for conditional independence:

$$\boxed{\boldsymbol{x} \perp\!\!\!\perp \{\boldsymbol{y}, \boldsymbol{z}\} \implies (\boldsymbol{x} \perp\!\!\!\perp \boldsymbol{y} \,|\boldsymbol{z}) \text{ and } (\boldsymbol{x} \perp\!\!\!\perp \boldsymbol{z} \,|\boldsymbol{y})} \tag{3.95}$$

That is, if $\boldsymbol{x}$ is independent of both $\boldsymbol{y}$ and $\boldsymbol{z}$, then $\boldsymbol{x}$ is conditionally independent of $\boldsymbol{y}$ given $\boldsymbol{z}$ and of $\boldsymbol{z}$ given $\boldsymbol{y}$.

Proof: Since $\boldsymbol{x}$ is independent of both $\boldsymbol{y}$ and $\boldsymbol{z}$, it holds that

$$\mathbb{P}(\boldsymbol{x}|\boldsymbol{y}, \boldsymbol{z}) = \mathbb{P}(\boldsymbol{x}), \quad \mathbb{P}(\boldsymbol{x}|\boldsymbol{y}) = \mathbb{P}(\boldsymbol{x}), \quad \mathbb{P}(\boldsymbol{x}|\boldsymbol{z}) = \mathbb{P}(\boldsymbol{x}) \tag{3.96}$$

Consequently, we have

$$\mathbb{P}(\boldsymbol{x}|\boldsymbol{y}, \boldsymbol{z}) = \mathbb{P}(\boldsymbol{x}|\boldsymbol{y}) \quad \text{and} \quad \mathbb{P}(\boldsymbol{x}|\boldsymbol{y}, \boldsymbol{z}) = \mathbb{P}(\boldsymbol{x}|\boldsymbol{z}) \tag{3.97}$$

which, in view of (3.93), allow us to conclude that $\boldsymbol{x}$ is independent of $\boldsymbol{z}$ given $\boldsymbol{y}$ and of $\boldsymbol{y}$ given $\boldsymbol{z}$.

■

(c) Third, we consider the following result referred to as the *contraction* property for conditional independence:

$$\boxed{(\boldsymbol{x} \perp\!\!\!\perp \boldsymbol{y} \,|\, \boldsymbol{z}) \text{ and } (\boldsymbol{x} \perp\!\!\!\perp \boldsymbol{z}) \implies (\boldsymbol{x} \perp\!\!\!\perp \{\boldsymbol{y}, \boldsymbol{z}\})} \tag{3.98}$$

That is, if $\boldsymbol{x}$ is independent of $\boldsymbol{z}$ and conditionally independent of $\boldsymbol{y}$ given $\boldsymbol{z}$, then $\boldsymbol{x}$ is independent of both $\boldsymbol{y}$ and $\boldsymbol{z}$.

Proof: From the assumed independence properties we have

$$\mathbb{P}(\boldsymbol{x}|\boldsymbol{y}, \boldsymbol{z}) = \mathbb{P}(\boldsymbol{x}|\boldsymbol{z}) = \mathbb{P}(\boldsymbol{x}) \tag{3.99}$$

from which we conclude that $\boldsymbol{x}$ is independent of both $\boldsymbol{y}$ and $\boldsymbol{z}$.

■

3.3.4 Conditional Mean

The conditional mean of a real-valued random variable $\boldsymbol{x}$ given observations of another real-valued random variable $\boldsymbol{y}$ is denoted by $\mathbb{E}\,(\boldsymbol{x}|\boldsymbol{y})$ and is defined as the calculation:

$$\mathbb{E}\,(\boldsymbol{x}|\boldsymbol{y}) \;=\; \int_{x \in \mathcal{X}} x f_{\boldsymbol{x}|\boldsymbol{y}}(x|y)dx \tag{3.100}$$

where both variables are assumed to be continuous in this representation. This computation amounts to determining the center of gravity of the conditional pdf of $\boldsymbol{x}$ given $\boldsymbol{y}$. When both variables are discrete, expression (3.100) is replaced by

$$\mathbb{E}\,(\boldsymbol{x}|\boldsymbol{y} = y) \;=\; \sum_{m=1}^{M} x_m \, \mathbb{P}(\boldsymbol{x} = x_m|\boldsymbol{y} = y) \tag{3.101}$$

where we are assuming that $\boldsymbol{x}$ admits M possible outcomes $\{x_m\}$ with probability p_m each. The next example considers a situation where one random variable is continuous and the other is discrete.

Example 3.16 **(Conditional mean computation)** Assume $\boldsymbol{y}$ is a random variable that is red with probability 1/3 and blue with probability 2/3:

$$\mathbb{P}(\boldsymbol{y} = \text{red}) = 1/3, \quad \mathbb{P}(\boldsymbol{y} = \text{blue}) = 2/3 \tag{3.102}$$

Likewise, assume $\boldsymbol{x}$ is a random variable that is Gaussian with mean 1 and variance 2 if $\boldsymbol{y}$ is red, and uniformly distributed between -1 and 1 if $\boldsymbol{y}$ is blue. Then, the conditional pdfs of $\boldsymbol{x}$ given observations of $\boldsymbol{y}$ are given by:

$$f_{\boldsymbol{x}|\boldsymbol{y}}(x|y = \text{red}) = \mathcal{N}_{\boldsymbol{x}}(1, 2), \quad f_{\boldsymbol{x}|\boldsymbol{y}}(x|y = \text{blue}) = \mathcal{U}[-1, 1] \tag{3.103}$$

where we are using the notation $\mathcal{U}[a, b]$ to refer to a uniform distribution in the interval $[a, b]$. It follows that the conditional means of $\boldsymbol{x}$ are:

$$\mathbb{E}\,(\boldsymbol{x}|\boldsymbol{y} = \text{red}) = 1, \quad \mathbb{E}\,(\boldsymbol{x}|\boldsymbol{y} = \text{blue}) = 0 \tag{3.104}$$

We can now employ these values, along with the pmf of the discrete variable $\boldsymbol{y}$, to compute the mean of $\boldsymbol{x}$. Thus, note that the mean of $\boldsymbol{x}$ is equal to 1 with probability 1/3 and to 0 with probability 2/3. It follows that:

$$\begin{aligned}\mathbb{E}\,\boldsymbol{x} &= \mathbb{E}\,(\boldsymbol{x}|\boldsymbol{y} = \text{red}) \times \mathbb{P}(\boldsymbol{y} = \text{red}) \;+\; \mathbb{E}\,(\boldsymbol{x}|\boldsymbol{y} = \text{blue}) \times \mathbb{P}(\boldsymbol{y} = \text{blue}) \\ &= 1 \times \frac{1}{3} \;+\; 0 \times \frac{2}{3} \;=\; \frac{1}{3}\end{aligned} \tag{3.105}$$

An alternative way to understand this result is to introduce the variable $\boldsymbol{z} = \mathbb{E}\,(\boldsymbol{x}|\boldsymbol{y})$. This is a discrete random variable with two values: $\boldsymbol{z} = 1$ (which happens when $\boldsymbol{y}$ is red with probability 1/3) and $\boldsymbol{z} = 0$ (which happens when $\boldsymbol{y}$ is blue with probability 2/3). That is,

$$\mathbb{P}(\boldsymbol{z} = 1) = 1/3, \quad \mathbb{P}(\boldsymbol{z} = 0) = 2/3 \tag{3.106}$$

Now, it is shown in Prob. 3.25 that, for any two random variables $\{\boldsymbol{x}, \boldsymbol{y}\}$, it holds that

$$\boxed{\mathbb{E}\left\{\mathbb{E}\,(\boldsymbol{x}|\boldsymbol{y})\right\} = \mathbb{E}\,\boldsymbol{x}} \tag{3.107}$$

where the outermost expectation is over the pdf of $\boldsymbol{y}$ while the innermost expectation is over the conditional pdf of $\boldsymbol{x}$ given $\boldsymbol{y}$. We can indicate these facts explicitly by adding subscripts and writing

$$\mathbb{E}_{\boldsymbol{y}}\left\{\mathbb{E}_{\boldsymbol{x}|\boldsymbol{y}}(\boldsymbol{x}|\boldsymbol{y})\right\} = \mathbb{E}\,\boldsymbol{x} \tag{3.108}$$

In this way, the desired mean of $\boldsymbol{x}$ is simply the mean of $\boldsymbol{z}$ itself. We conclude that

$$\begin{aligned}\mathbb{E}\,\boldsymbol{x} = \mathbb{E}\,\boldsymbol{z} &= 1 \times \mathbb{P}(\boldsymbol{z} = 1) \;+\; 0 \times \mathbb{P}(\boldsymbol{z} = 0) \\ &= 1 \times \frac{1}{3} \;+\; 0 \times \frac{2}{3} \;=\; 1/3\end{aligned} \tag{3.109}$$

Example 3.17 **(Another conditional mean computation)** Assume $\boldsymbol{x}$ is a binary random variable that assumes the values ± 1 with probability 1/2 each. Assume in addition that we observe a noisy realization of $\boldsymbol{x}$, say,

$$\boldsymbol{y} = \boldsymbol{x} + \boldsymbol{v} \tag{3.110}$$

where $\boldsymbol{v}$ is a zero-mean Gaussian random variable that is independent of $\boldsymbol{x}$ and has variance equal to 1, i.e., its pdf is given by

$$f_{\boldsymbol{v}}(v) = \frac{1}{\sqrt{2\pi}}\, e^{-v^2/2} \tag{3.111}$$

Let us evaluate the conditional mean of $\boldsymbol{x}$ given observations of $\boldsymbol{y}$. From definition (3.100), we find that we need to know the conditional pdf, $f_{\boldsymbol{x}|\boldsymbol{y}}(x|y)$, in order to evaluate the integral expression. For this purpose, we call upon future result (3.160), which states that the pdf of the sum of two independent random variables, namely, $\boldsymbol{y} = \boldsymbol{x} + \boldsymbol{v}$, is equal to the convolution of their individual pdfs, i.e.,

$$f_{\boldsymbol{y}}(y) = \int_{-\infty}^{\infty} f_{\boldsymbol{x}}(x) f_{\boldsymbol{v}}(y-x)dx \tag{3.112}$$

In this example we have

$$f_{\boldsymbol{x}}(x) = \frac{1}{2}\delta(x-1) + \frac{1}{2}\delta(x+1) \tag{3.113}$$

where $\delta(\cdot)$ is the Dirac delta function, so that $f_{\boldsymbol{y}}(y)$ is given by

$$f_{\boldsymbol{y}}(y) = \frac{1}{2} f_{\boldsymbol{v}}(y+1) + \frac{1}{2} f_{\boldsymbol{v}}(y-1) \tag{3.114}$$

Moreover, in view of the Bayes rule (3.42b), the joint pdf of $\{\boldsymbol{x}, \boldsymbol{y}\}$ is given by

$$\begin{aligned} f_{\boldsymbol{x},\boldsymbol{y}}(x,y) &= f_{\boldsymbol{x}}(x)\, f_{\boldsymbol{y}|\boldsymbol{x}}(y|x) \\ &= \left(\frac{1}{2}\delta(x-1) + \frac{1}{2}\delta(x+1)\right) f_{\boldsymbol{v}}(y-x) \\ &= \frac{1}{2} f_{\boldsymbol{v}}(y-1)\delta(x-1) + \frac{1}{2} f_{\boldsymbol{v}}(y+1)\delta(x+1) \end{aligned} \tag{3.115}$$

Using the Bayes rule again we get

$$\begin{aligned} f_{\boldsymbol{x}|\boldsymbol{y}}(x|y) &= \frac{f_{\boldsymbol{x},\boldsymbol{y}}(x,y)}{f_{\boldsymbol{y}}(y)} \\ &= \frac{f_{\boldsymbol{v}}(y-1)\delta(x-1)}{f_{\boldsymbol{v}}(y+1) + f_{\boldsymbol{v}}(y-1)} + \frac{f_{\boldsymbol{v}}(y+1)\delta(x+1)}{f_{\boldsymbol{v}}(y+1) + f_{\boldsymbol{v}}(y-1)} \end{aligned} \tag{3.116}$$

Substituting into expression (3.100) and integrating we obtain

$$\begin{aligned} \mathbb{E}\,(\boldsymbol{x}|\boldsymbol{y}) &= \frac{f_{\boldsymbol{v}}(y-1)}{f_{\boldsymbol{v}}(y+1) + f_{\boldsymbol{v}}(y-1)} - \frac{f_{\boldsymbol{v}}(y+1)}{f_{\boldsymbol{v}}(y+1) + f_{\boldsymbol{v}}(y-1)} \\ &= \frac{1}{\left(\frac{e^{-(y+1)^2/2}}{e^{-(y-1)^2/2}}\right) + 1} - \frac{1}{\left(\frac{e^{-(y-1)^2/2}}{e^{-(y+1)^2/2}}\right) + 1} \\ &= \frac{e^{y} - e^{-y}}{e^{y} + e^{-y}} \\ &\stackrel{\Delta}{=} \tanh(y) \end{aligned} \tag{3.117}$$

In other words, the conditional mean of $\boldsymbol{x}$ given observations of $\boldsymbol{y}$ is the hyperbolic tangent function, which is shown in Fig. 3.5.

3.3.5 Correlated and Orthogonal Variables

The covariance between two random variables, $\boldsymbol{x}$ and $\boldsymbol{y}$, is denoted by the symbol σ_{xy} and is defined by either of the following equivalent expressions:

$$\boxed{\sigma_{xy} \stackrel{\Delta}{=} \mathbb{E}\,(\boldsymbol{x} - \bar{x})(\boldsymbol{y} - \bar{y}) = \mathbb{E}\,\boldsymbol{x}\boldsymbol{y} - \bar{x}\bar{y}} \qquad \textbf{(covariance)} \tag{3.118}$$

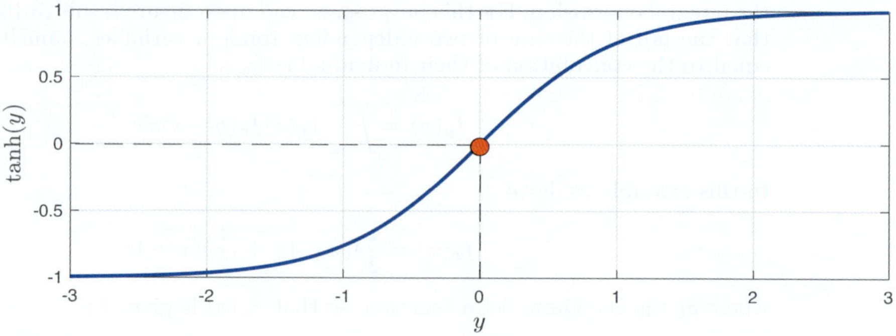

Figure 3.5 A plot of the hyperbolic tangent function, $\tanh(y) = \frac{e^y - e^{-y}}{e^y + e^{-y}}$.

We say that the random variables are uncorrelated if, and only if, their covariance is zero, i.e.,

$$\sigma_{xy} = 0 \tag{3.119}$$

which, in view of the defining relation (3.118), is also equivalent to requiring

$$\boxed{\mathbb{E}\,\boldsymbol{xy} \;=\; (\mathbb{E}\,\boldsymbol{x})\,(\mathbb{E}\,\boldsymbol{y})} \qquad \text{(}\textbf{uncorrelated random variables}\text{)} \tag{3.120}$$

so that the mean of the product is equal to the product of the means. On the other hand, we say that two random variables are *orthogonal* if, and only if,

$$\boxed{\mathbb{E}\,\boldsymbol{xy} = 0} \qquad \text{(}\textbf{orthogonal random variables}\text{)} \tag{3.121}$$

Observe that the means of $\boldsymbol{x}$ and $\boldsymbol{y}$ do not enter into this condition. It then follows that the concepts of orthogonality and uncorrelatedness coincide with each other if at least one of the random variables has zero mean.

When two random variables $\boldsymbol{x}$ and $\boldsymbol{y}$ are independent, it will also hold that

$$\mathbb{E}\,\boldsymbol{xy} = (\mathbb{E}\,\boldsymbol{x})\,(\mathbb{E}\,\boldsymbol{y}) \tag{3.122}$$

This is because

$$\begin{aligned}
\mathbb{E}\,\boldsymbol{xy} &= \int_{x\in\mathcal{X}}\int_{y\in\mathcal{Y}} xy f_{\boldsymbol{x},\boldsymbol{y}}(x,y)\,dxdy \\
&\overset{(3.62)}{=} \int_{x\in\mathcal{X}}\int_{y\in\mathcal{Y}} xy f_{\boldsymbol{x}}(x) f_{\boldsymbol{y}}(y)\,dxdy \\
&= \left(\int_{x\in\mathcal{X}} x f_{\boldsymbol{x}}(x)dx\right)\left(\int_{y\in\mathcal{Y}} y f_{\boldsymbol{y}}(y)dy\right) \\
&= (\mathbb{E}\,\boldsymbol{x})\,(\mathbb{E}\,\boldsymbol{y})
\end{aligned} \tag{3.123}$$

It follows that independent random variables are uncorrelated:

$$\textbf{independent random variables} \implies \textbf{uncorrelated random variables} \tag{3.124}$$

The converse statement is not true.

Example 3.18 **(Uncorrelatedness and dependency)** Let $\boldsymbol{\theta}$ be a random variable that is uniformly distributed over the interval $[0, 2\pi]$. Introduce the zero-mean random variables:

$$\boldsymbol{x} = \cos \boldsymbol{\theta} \quad \text{and} \quad \boldsymbol{y} = \sin \boldsymbol{\theta} \tag{3.125}$$

Then, it holds that $\boldsymbol{x}^2 + \boldsymbol{y}^2 = 1$ so that $\boldsymbol{x}$ and $\boldsymbol{y}$ are dependent. However,

$$\begin{aligned} \mathbb{E}\,\boldsymbol{xy} &= \mathbb{E}\,\cos \boldsymbol{\theta} \sin \boldsymbol{\theta} \\ &= \frac{1}{2}\mathbb{E}\,\sin 2\boldsymbol{\theta} \\ &= \frac{1}{2}\frac{1}{2\pi}\int_0^{2\pi} \sin 2\theta \, d\theta \\ &= 0 \end{aligned} \tag{3.126}$$

so that $\boldsymbol{x}$ and $\boldsymbol{y}$ are uncorrelated. Therefore, we have an example of two uncorrelated random variables that are dependent.

3.4 RANDOM VECTORS

It is common in applications to encounter *vector-valued* (as opposed to scalar) random variables, also known as random vectors. A random vector consists of a collection of scalar random variables grouped together either in column form or row form. For example, assume $\boldsymbol{x}_1$ and $\boldsymbol{x}_2$ are two scalar random variables. Then, the column vector

$$\boldsymbol{x} = \begin{bmatrix} \boldsymbol{x}_1 \\ \boldsymbol{x}_2 \end{bmatrix} \tag{3.127a}$$

is a vector-valued random variable in column form; its dimensions are 2×1. We sometimes use the compact notation

$$\boldsymbol{x} = \text{col}\Big\{\boldsymbol{x}_1, \boldsymbol{x}_2\Big\} \tag{3.127b}$$

to denote a column vector with entries $\boldsymbol{x}_1$ and $\boldsymbol{x}_2$ stacked on top of each other. Alternatively, we could have collected the entries $\{\boldsymbol{x}_1, \boldsymbol{x}_2\}$ into a row vector and obtained a row random vector instead, say,

$$\boldsymbol{x} = \begin{bmatrix} \boldsymbol{x}_1 & \boldsymbol{x}_2 \end{bmatrix} \tag{3.127c}$$

In this case, the dimensions of the random vector are 1×2. Working with either the column or row format is generally a matter of convenience.

We continue, for illustration purposes, with the 2×1 vector $\boldsymbol{x} = \text{col}\{\boldsymbol{x}_1, \boldsymbol{x}_2\}$. The mean of $\boldsymbol{x}$ is defined as the vector of individual means, namely,

$$\bar{x} = \mathbb{E}\,\boldsymbol{x} \stackrel{\Delta}{=} \begin{bmatrix} \bar{x}_1 \\ \bar{x}_2 \end{bmatrix} = \begin{bmatrix} \mathbb{E}\,\boldsymbol{x}_1 \\ \mathbb{E}\,\boldsymbol{x}_2 \end{bmatrix} \tag{3.128}$$

The definition extends trivially to vectors of larger dimensions so that the mean of a random vector is the vector of individual means.

With regards to the "variance" of a random vector, it will now be a matrix (and not a scalar) and will be referred to as the *covariance* matrix (rather than the variance). For the same 2×1 vector $\boldsymbol{x}$ as above, its covariance matrix is denoted by R_x and is defined as the following 2×2 matrix:

$$R_x = \begin{bmatrix} \sigma_{x_1}^2 & \sigma_{x_1,x_2} \\ \sigma_{x_2,x_1} & \sigma_{x_2}^2 \end{bmatrix} \tag{3.129}$$

in terms of the variances $\{\sigma_{x_1}^2, \sigma_{x_2}^2\}$ of the individual entries $\boldsymbol{x}_1$ and $\boldsymbol{x}_2$,

$$\sigma_{x_1}^2 = \mathbb{E}\,(\boldsymbol{x}_1 - \bar{x}_1)^2 \tag{3.130a}$$

$$\sigma_{x_2}^2 = \mathbb{E}\,(\boldsymbol{x}_2 - \bar{x}_2)^2 \tag{3.130b}$$

and the covariances between these individual entries:

$$\sigma_{x_1,x_2} \stackrel{\Delta}{=} \mathbb{E}\,(\boldsymbol{x}_1 - \bar{x}_1)(\boldsymbol{x}_2 - \bar{x}_2) \tag{3.130c}$$

$$\sigma_{x_2,x_1} \stackrel{\Delta}{=} \mathbb{E}\,(\boldsymbol{x}_2 - \bar{x}_2)(\boldsymbol{x}_1 - \bar{x}_1) = \sigma_{x_1,x_2} \tag{3.130d}$$

The matrix form (3.129) can be described in the form:

$$\boxed{R_x \stackrel{\Delta}{=} \mathbb{E}\,(\boldsymbol{x} - \bar{x})(\boldsymbol{x} - \bar{x})^\mathsf{T}} \quad \textbf{(when } \boldsymbol{x} \textbf{ is a column vector)} \tag{3.131}$$

Expression (3.131) is general and applies to random vectors $\boldsymbol{x}$ of higher dimension than 2. It is worth noting that if $\boldsymbol{x}$ were constructed instead as a *row* (rather than a column) vector, then the covariance matrix R_x in (3.131) would instead be defined as

$$R_x \stackrel{\Delta}{=} \mathbb{E}\,(\boldsymbol{x} - \bar{x})^\mathsf{T}(\boldsymbol{x} - \bar{x}) \quad \textbf{(when } \boldsymbol{x} \textbf{ is a row vector)} \tag{3.132}$$

with the transposed term coming first. This is because it is now the product $(\boldsymbol{x} - \bar{x})^\mathsf{T}(\boldsymbol{x} - \bar{x})$ that yields a matrix and leads to (3.129).

We can also extend the notion of correlations to random vectors. Thus, let $\boldsymbol{x}$ and $\boldsymbol{y}$ be two column random vectors. Their cross-covariance matrix is denoted by R_{xy} and is defined as

$$R_{xy} \stackrel{\Delta}{=} \mathbb{E}\,(\boldsymbol{x} - \bar{x})(\boldsymbol{y} - \bar{y})^\mathsf{T} \quad \textbf{(}\boldsymbol{x} \textbf{ and } \boldsymbol{y} \textbf{ are column vectors)} \tag{3.133a}$$

$$R_{xy} \stackrel{\Delta}{=} \mathbb{E}\,(\boldsymbol{x} - \bar{x})^\mathsf{T}(\boldsymbol{y} - \bar{y}) \quad \textbf{(}\boldsymbol{x} \textbf{ and } \boldsymbol{y} \textbf{ are row vectors)} \tag{3.133b}$$

Example 3.19 (Mean and covariance of a random vector) Let us reconsider the setting of Example 3.8 where $\boldsymbol{x}$ assumes the values ± 1 with probability p for $+1$ and $1-p$ for -1, and $\boldsymbol{v}$ is a zero-mean Gaussian random variable with variance σ_v^2. The variables $\boldsymbol{x}$ and $\boldsymbol{v}$ are further assumed to be uncorrelated and the measurement $\boldsymbol{y}$ is given by

$$\boldsymbol{y} = \frac{1}{2}\boldsymbol{x} + \boldsymbol{v} \tag{3.134}$$

Introduce the 2×1 column vector $\boldsymbol{z} = \text{col}\{\boldsymbol{x}, \boldsymbol{y}\}$ and let us evaluate its mean and 2×2 covariance matrix. To begin with, using (3.23), the mean of $\boldsymbol{x}$ is given by

$$\mathbb{E}\,\boldsymbol{x} = \bar{x} \;=\; p \times 1 + (1-p) \times (-1) \;= 2p - 1 \tag{3.135}$$

and the mean-square of $\boldsymbol{x}$ is given by

$$\mathbb{E}\,\boldsymbol{x}^2 = p \times (1)^2 + (1-p) \times (-1)^2 \;=\; 1 \tag{3.136}$$

so that the variance of $\boldsymbol{x}$ is

$$\sigma_x^2 = \mathbb{E}\,\boldsymbol{x}^2 \;-\; (\bar{x})^2 \;=\; 1 - (2p-1)^2 \;=\; 4p(1-p) \tag{3.137}$$

Now using the measurement relation (3.134), we find that

$$\mathbb{E}\,\boldsymbol{y} = \frac{1}{2}\mathbb{E}\,\boldsymbol{x} + \mathbb{E}\,\boldsymbol{v} \;=\; \frac{1}{2}(2p-1) + 0 \;=\; p - \frac{1}{2} \tag{3.138}$$

and we conclude that

$$\mathbb{E}\,\boldsymbol{z} \;=\; \bar{z} \;=\; \begin{bmatrix} \bar{x} \\ \bar{y} \end{bmatrix} \;=\; \begin{bmatrix} 2p-1 \\ p - \frac{1}{2} \end{bmatrix} \tag{3.139}$$

Moreover, using the assumed uncorrelatedness between $\boldsymbol{x}$ and $\boldsymbol{v}$ we further get:

$$\begin{aligned} \mathbb{E}\,\boldsymbol{y}^2 &= \mathbb{E}\left(\frac{1}{2}\boldsymbol{x} + \boldsymbol{v}\right)^2 \\ &= \frac{1}{4}\mathbb{E}\,\boldsymbol{x}^2 \;+\; \mathbb{E}\,\boldsymbol{v}^2 \;+\; \mathbb{E}\,\boldsymbol{x}\boldsymbol{v} \\ &= \frac{1}{4} \;+\; \sigma_v^2 + \mathbb{E}\,\boldsymbol{x}\,\mathbb{E}\,\boldsymbol{v} \\ &= \frac{1}{4} \;+\; \sigma_v^2 \;+\; 0, \quad \text{since } \mathbb{E}\,\boldsymbol{v} = 0 \\ &= \frac{1}{4} \;+\; \sigma_v^2 \end{aligned} \tag{3.140}$$

It follows that

$$\begin{aligned} \sigma_y^2 &= \mathbb{E}\,\boldsymbol{y}^2 - \bar{y}^2 \\ &= \frac{1}{4} \;+\; \sigma_v^2 - \left(p - \frac{1}{2}\right)^2 \\ &= p(1-p) \;+\; \sigma_v^2 \end{aligned} \tag{3.141}$$

The last expression is simply stating that

$$\sigma_y^2 = \frac{1}{4}\sigma_x^2 + \sigma_v^2 \tag{3.142}$$

when $\boldsymbol{x}$ and $\boldsymbol{v}$ are uncorrelated. Finally, the correlation between $\boldsymbol{x}$ and $\boldsymbol{y}$ is given by

$$\begin{aligned} \sigma_{xy} &= \mathbb{E}\,\boldsymbol{x}\boldsymbol{y} \;-\; \bar{x}\bar{y} \\ &= \mathbb{E}\,\boldsymbol{x}\left(\frac{1}{2}\boldsymbol{x} + \boldsymbol{v}\right) \;-\; (2p-1)\left(p - \frac{1}{2}\right) \\ &= \frac{1}{2}\mathbb{E}\,\boldsymbol{x}^2 \;+\; \mathbb{E}\,\boldsymbol{x}\boldsymbol{v} \;-\; (2p-1)\left(p - \frac{1}{2}\right) \\ &= \frac{1}{2} \;+\; 0 \;-\; (2p-1)\left(p - \frac{1}{2}\right) \\ &= 2p(1-p) \end{aligned} \tag{3.143}$$

We conclude that the covariance matrix of the vector $\boldsymbol{z} = \text{col}\{\boldsymbol{x}, \boldsymbol{y}\}$ is given by

$$R_z = \begin{bmatrix} \sigma_x^2 & \sigma_{xy} \\ \sigma_{xy} & \sigma_y^2 \end{bmatrix} = \begin{bmatrix} 4p(1-p) & 2p(1-p) \\ 2p(1-p) & p(1-p)+\sigma_v^2 \end{bmatrix} \tag{3.144}$$

3.5 PROPERTIES OF COVARIANCE MATRICES

Covariance matrices of random vectors satisfy two important properties: **(a)** they are symmetric and **(b)** they are nonnegative-definite. Both properties can be easily verified from the definition. Indeed, using the matrix property

$$(AB)^{\mathsf{T}} = B^{\mathsf{T}} A^{\mathsf{T}} \tag{3.145}$$

for any two matrices A and B of compatible dimensions, we find that for any column random vector $\boldsymbol{x}$,

$$\begin{aligned} R_x^{\mathsf{T}} &= \left(\mathbb{E}\,(\boldsymbol{x}-\bar{x})(\boldsymbol{x}-\bar{x})^{\mathsf{T}}\right)^{\mathsf{T}} \\ &= \mathbb{E}\left(\underbrace{(\boldsymbol{x}-\bar{x})}_{A}\underbrace{(\boldsymbol{x}-\bar{x})^{\mathsf{T}}}_{B}\right)^{\mathsf{T}} \\ &= \mathbb{E}\,(\boldsymbol{x}-\bar{x})(\boldsymbol{x}-\bar{x})^{\mathsf{T}}, \qquad \text{using (3.145)} \\ &= R_x \end{aligned} \tag{3.146}$$

so that

$$\boxed{R_x = R_x^{\mathsf{T}}} \tag{3.147}$$

and the covariance matrix is symmetric. It follows that covariance matrices can only have real eigenvalues.

Example 3.20 **(Eigenvalues of 2×2 covariance matrix)** Consider again Example 3.19 where we encountered the covariance matrix

$$R_z = \begin{bmatrix} 4p(1-p) & 2p(1-p) \\ 2p(1-p) & p(1-p)+\sigma_v^2 \end{bmatrix} \tag{3.148}$$

The result just established about the nature of the eigenvalues of a covariance matrix can be used to affirm that the eigenvalues of the above R_z will be real-valued no matter what the values of $p \in [0,1]$ and $\sigma_v^2 \geq 0$ are! Let us verify this fact from first principles by determining the eigenvalues of R_z. Recall that the eigenvalues $\{\lambda\}$ can be determined by solving the polynomial equation (also called the characteristic equation of R_z):

$$\det(\lambda I_2 - R_z) = 0 \tag{3.149}$$

in terms of the determinant of the difference $\lambda I_2 - R_z$, where I_2 is the 2×2 identity matrix. We denote the characteristic polynomial by the notation $q(\lambda)$:

$$q(\lambda) \stackrel{\Delta}{=} \det(\lambda I_2 - R_z) \tag{3.150}$$

In our example we have

$$\lambda I_2 - R_z \;=\; \begin{bmatrix} \lambda - 4p(1-p) & -2p(1-p) \\ -2p(1-p) & \lambda - p(1-p) - \sigma_v^2 \end{bmatrix} \tag{3.151}$$

so that

$$q(\lambda) = \lambda^2 \;-\; \lambda\left(5p(1-p) + \sigma_v^2\right) \;+\; 4p(1-p)\sigma_v^2 \tag{3.152}$$

which is a quadratic polynomial in λ. The discriminant of $q(\lambda)$ is given by

$$\begin{aligned} \Delta &= \left(5p(1-p) + \sigma_v^2\right)^2 \;-\; 16p(1-p)\sigma_v^2 \\ &= 25p^2(1-p)^2 \;+\; \sigma_v^4 \;-\; 6p(1-p)\sigma_v^2 \\ &= \left(5p(1-p) - \sigma_v^2\right)^2 \;+\; 4p^2(1-p)^2 \\ &\geq 0 \end{aligned} \tag{3.153}$$

so that $q(\lambda)$ has real roots. We conclude that the eigenvalues of R_z are real for any $p \in [0,1]$ and $\sigma_v^2 \geq 0$, as expected.

Covariance matrices are also nonnegative-definite, i.e.,

$$\boxed{R_x \geq 0} \tag{3.154}$$

To see this, let v denote an arbitrary column vector and introduce the scalar-valued random variable

$$\boldsymbol{y} = v^{\mathsf{T}}(\boldsymbol{x} - \bar{x}) \tag{3.155}$$

Then, the variable $\boldsymbol{y}$ has zero mean and its variance is given by

$$\begin{aligned} \sigma_y^2 &\stackrel{\Delta}{=} \mathbb{E}\,\boldsymbol{y}^2 \\ &= \mathbb{E}\,v^{\mathsf{T}}(\boldsymbol{x} - \bar{x})(\boldsymbol{x} - \bar{x})^{\mathsf{T}} v \\ &= v^{\mathsf{T}}\left(\mathbb{E}\,(\boldsymbol{x} - \bar{x})(\boldsymbol{x} - \bar{x})^{\mathsf{T}}\right) v \\ &= v^{\mathsf{T}} R_x v \end{aligned} \tag{3.156}$$

But the variance of any scalar-valued random variable is always nonnegative so that $\sigma_y^2 \geq 0$. It follows that $v^{\mathsf{T}} R_x v \geq 0$ for any v. This means that R_x is nonnegative-definite, as claimed. It then follows that the eigenvalues of covariance matrices are not only real but also nonnegative.

3.6 ILLUSTRATIVE APPLICATIONS

In this section we consider some applications of the concepts covered in the chapter in the context of convolution sums, characteristic functions, random walks, and statistical mechanics. The last two examples are meant to show how randomness is useful in modeling important physical phenomena, such as Brownian motion by particles suspended in fluids and the regulation of ion channels in cell membranes. We will encounter some of these concepts in future sections and use them to motivate useful inference and learning methods – see, e.g., the discussion in Section 66.2 on Boltzmann machines.

3.6.1 Convolution Sums

We examine first the pmf of the sum of two independent discrete random variables and show that this pmf can be obtained by convolving the individual pmfs.

Thus, let $\boldsymbol{a}$ and $\boldsymbol{b}$ be two independent *discrete* scalar-valued random variables: $\boldsymbol{a}$ assumes the integer values $n = 0, 1, \ldots, N_a$ with probabilities $\{p_a(n)\}$ each, while $\boldsymbol{b}$ assumes the integer values $n = 0, 1, \ldots, N_b$ with probabilities $\{p_b(n)\}$ each. The probabilities $\{p_a(n), p_b(n)\}$ are zero outside the respective intervals for n. Let

$$\boldsymbol{c} \stackrel{\Delta}{=} \boldsymbol{a} + \boldsymbol{b} \tag{3.157}$$

denote the sum variable. The possible realizations for $\boldsymbol{c}$ are the integer values that occur between $n = 0$ and $n = N_a + N_b$. Each possible realization n occurs with some probability $p_c(n)$ that we wish to determine. For each realization m for $\boldsymbol{a}$, the corresponding realization for $\boldsymbol{b}$ should be the value $n - m$. This means that the probability of the event $\boldsymbol{c} = n$ is given by the following expression:

$$\begin{aligned}
p_c(n) &\stackrel{\Delta}{=} \mathbb{P}(\boldsymbol{c} = n) \\
&= \sum_{m=-\infty}^{\infty} \mathbb{P}(\boldsymbol{a} = m, \boldsymbol{b} = n - m) \\
&= \sum_{m=-\infty}^{\infty} \mathbb{P}(\boldsymbol{a} = m)\, \mathbb{P}(\boldsymbol{b} = n - m) \quad \text{(by independence)} \\
&= \sum_{m=-\infty}^{\infty} p_a(m) p_b(n - m)
\end{aligned} \tag{3.158}$$

That is,

$$\boxed{p_c(n) \;=\; p_a(n) \star p_b(n)} \qquad \textbf{(convolution sum)} \tag{3.159}$$

where the equality in the third line is due to the assumed independence of the random variables $\boldsymbol{a}$ and $\boldsymbol{b}$, and the symbol $\star$ denotes the convolution operation. We therefore observe that the sequence of probabilities $\{p_c(n)\}$ is obtained by convolving the sequences $\{p_a(n), p_b(n)\}$.

Likewise, if $\boldsymbol{x}$ and $\boldsymbol{v}$ are two independent *continuous* random variables with pdfs $f_{\boldsymbol{x}}(x)$ and $f_{\boldsymbol{v}}(v)$, respectively, and $\boldsymbol{y} = \boldsymbol{x} + \boldsymbol{v}$, then the pdf of $\boldsymbol{y}$ is given by the convolution integral (see Prob. 3.29):

$$f_{\boldsymbol{y}}(y) = \int_{-\infty}^{\infty} f_{\boldsymbol{x}}(x) f_{\boldsymbol{v}}(y - x) dx \tag{3.160}$$

3.6.2 Characteristic Functions[1]

Our second example illustrates how the continuous-time Fourier transform is applicable to the study of random variables. Thus, consider a continuous random variable $\boldsymbol{x}$ with pdf $f_{\boldsymbol{x}}(x)$. The characteristic function of $\boldsymbol{x}$ is denoted by $\varphi_{\boldsymbol{x}}(t)$ and is defined as the mean value of the exponential variable $e^{jt\boldsymbol{x}}$, where t is a real-valued argument. That is,

$$\varphi_{\boldsymbol{x}}(t) \stackrel{\Delta}{=} \mathbb{E}\, e^{jt\boldsymbol{x}}, \quad t \in \mathbb{R} \tag{3.161}$$

or, more explicitly,

$$\varphi_{\boldsymbol{x}}(t) \stackrel{\Delta}{=} \int_{-\infty}^{\infty} f_{\boldsymbol{x}}(x) e^{jtx} dx \qquad \textbf{(characteristic function)} \tag{3.162}$$

Comparing this expression with the definition for the Fourier transform of a continuous-time signal $x(t)$, shown in the first line of Table 3.2, we see that the time variable t is replaced by x and the frequency variable Ω is replaced by $-t$.

Table 3.2 Analogy between the Fourier transform of a continuous-time signal $x(t)$ and the characteristic function of a random variable $\boldsymbol{x}$.

Signal	Transform	Name	Definition
$x(t)$	$X(j\Omega)$	Fourier transform	$X(j\Omega) = \int_{-\infty}^{\infty} x(t)e^{-j\Omega t} dt$
$f_{\boldsymbol{x}}(x)$	$\varphi_{\boldsymbol{x}}(t)$	characteristic transform	$\varphi_{\boldsymbol{x}}(t) = \int_{-\infty}^{\infty} f_{\boldsymbol{x}}(x)e^{jtx} dx$

A useful property of the characteristic function is that the value of its successive derivatives at $t = 0$ can be used to evaluate the moments of the random variable $\boldsymbol{x}$. Specifically, it holds that

$$\mathbb{E}\, \boldsymbol{x}^k = (-j)^k \left. \frac{d^k \varphi_{\boldsymbol{x}}(t)}{dt^k} \right|_{t=0}, \quad k = 1, 2, 3, \ldots \tag{3.163}$$

in terms of the kth-order derivative of $\varphi_{\boldsymbol{x}}(t)$ evaluated at $t = 0$, and where $j = \sqrt{-1}$. Moreover, in a manner similar to the inversion formula for Fourier transforms, we can recover the pdf, $f_{\boldsymbol{x}}(x)$, from knowledge of the characteristic function as follows (see Probs. 3.30–3.32):

$$f_{\boldsymbol{x}}(x) = \frac{1}{2\pi} \int_{-\infty}^{\infty} \varphi_{\boldsymbol{x}}(t) e^{-jtx} dt \tag{3.164}$$

3.6.3 Statistical Mechanics

Our next example deals with the Boltzmann distribution, which plays a critical role in statistical mechanics; it provides a powerful tool to model complex systems

[1] This section can be skipped on a first reading.

consisting of a large number of interacting components, including interactions at the molecular level. In statistical mechanics, complex systems are modeled as having a large number of microstates. Each microstate i occurs with probability p_i and is assigned an energy level E_i. The Boltzmann distribution states that the probability that a system is at state i is proportional to $e^{-\beta E_i}$, i.e.,

$$p_i \stackrel{\Delta}{=} \mathbb{P}(\text{complex system is at microstate } i) = \alpha e^{-\beta E_i} \tag{3.165}$$

where α describes the proportionality constant and

$$\beta = \frac{1}{k_B T} \tag{3.166}$$

which involves the temperature T measured in kelvin and the Boltzmann constant $k_B = 1.38 \times 10^{-23}$ J/K (measured in joules/kelvin).

Boltzmann distribution. If a complex system has N microstates, then it must hold that

$$\sum_{i=1}^{N} p_i = 1 \iff \alpha \left(\sum_{i=1}^{N} e^{-\beta E_i} \right) = 1 \tag{3.167}$$

which enables us to determine the value of α. It follows that the Boltzmann distribution (also called the Gibbs distribution) is given by

$$\boxed{\mathbb{P}(\text{complex system is at microstate } i) = \frac{e^{-\beta E_i}}{\sum_{k=1}^{N} e^{-\beta E_k}}} \tag{3.168}$$

Ion channels. Let us illustrate how these results can be used to model the behavior of ion channels, which regulate the flow of ions through biological cell membranes. A simple model for ion channels assumes that they can be either closed or open. A channel in its closed state has energy E_c and a channel in its open state has energy E_o – see Fig. 3.6. These energy levels can be measured through experimentation. Using the Boltzmann distribution (3.168), we can evaluate the probability that the channel will be open as follows:

$$\mathbb{P}(\text{channel open}) = \frac{e^{-\beta E_o}}{e^{-\beta E_o} + e^{-\beta E_c}} \tag{3.169}$$

Dividing the numerator and denominator by $e^{-\beta E_o}$, we can express the probability of an open or closed channel as

$$\mathbb{P}(\text{channel open}) = \frac{1}{1 + e^{-\beta(E_c - E_o)}} \tag{3.170a}$$

$$\mathbb{P}(\text{channel closed}) = \frac{1}{1 + e^{\beta(E_c - E_o)}} \tag{3.170b}$$

Thus, observe that the probabilities of either state depend only on the difference between the energies of the states and not on the individual values of their

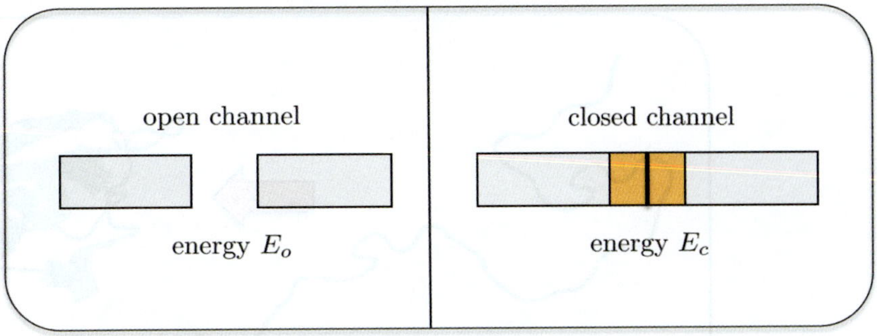

Figure 3.6 The energy of an open ion channel is assumed to be E_o and the energy of a closed ion channel is assumed to be E_c. The Boltzmann distribution allows us to evaluate the probability of encountering the channel in either state based on these energy values.

energies. Given the above probabilities, we can evaluate the average channel energy as

$$E_{\text{channel}} = \frac{E_o e^{-\beta E_o} + E_c e^{-\beta E_c}}{e^{-\beta E_o} + e^{-\beta E_c}} \tag{3.171}$$

Folding and unfolding of proteins. Similar arguments can be applied to the problem of protein folding, which refers to the process by which proteins fold into their three-dimensional structure – see Fig. 3.7. We assume the folded microstate has energy E_f and the unfolded microstate has energy E_u. We also assume that there is one folded microstate and L possible unfolded microstates. Assume we have a total of N proteins. Given this information, we would like to evaluate how many proteins on average we would encounter in their folded state among the N proteins.

The system has a total of $L+1$ microstates: one of them has energy E_f and L of them have energy E_u each. Using the Boltzmann distribution, we can evaluate the probability that the protein is folded as follows:

$$\mathbb{P}(\text{protein folded}) = \frac{e^{-\beta E_f}}{e^{-\beta E_f} + L e^{-\beta E_u}} \tag{3.172}$$

Dividing the numerator and the denominator by $e^{-\beta E_f}$, we can express the probability of a folded or unfolded protein as

$$\mathbb{P}(\text{protein folded}) = \frac{1}{1 + L e^{-\beta(E_u - E_f)}} \tag{3.173a}$$

$$\mathbb{P}(\text{protein unfolded}) = \frac{1}{1 + \frac{1}{L} e^{\beta(E_u - E_f)}} \tag{3.173b}$$

Observe again that the probabilities of either state depend only on the difference between the energies of the states and not on the individual values of their

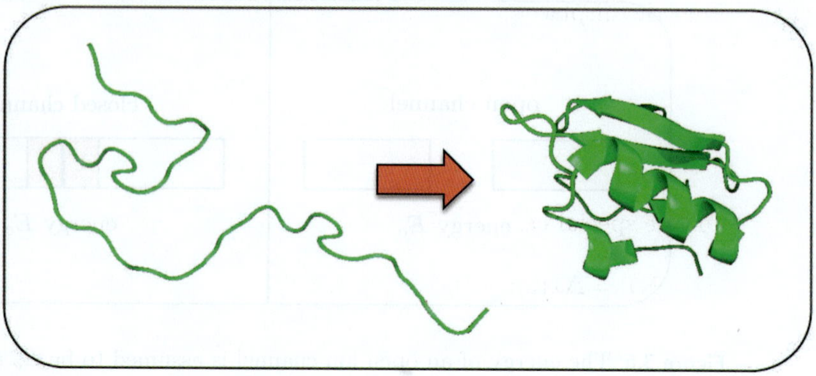

Figure 3.7 The figure shows a protein before (left) and after (right) folding into its three-dimensional structure. The source for the image is Wikimedia Commons, where the image is available in the public domain at https://commons.wikimedia.org/wiki/File:Protein_folding.png.

energies. It follows that the average number of proteins that will be encountered in the folded state from among a total of N proteins is:

$$N_f = N \times \mathbb{P}(\text{protein folded}) \;=\; \frac{N}{1 + L\,e^{-\beta(E_u - E_f)}} \tag{3.174}$$

3.6.4 Random Walks and Diffusion[2]

In this section we illustrate another useful application of randomness involving Brownian motion, which is used to model the random displacement of particles suspended in a fluid, such as a liquid or gas. In these environments, the random motion of particles results from collisions among molecules, leading to a random walk process where particles take successive random steps.

It is sufficient for our purposes to focus on one-dimensional (1D) random walks; these random walks are examples of Markov chains, to be discussed in greater detail in Chapter 38. Thus, consider a particle that is initially located at the origin of displacement. At each interval of time Δt, the particle takes one step randomly along the direction of a fixed line passing through the origin (say, along the direction of the horizontal axis). The particle may step a distance d_r to the right with probability p or a distance d_ℓ to the left with probability $1-p$, as illustrated in the top plot of Fig. 3.8. Let $\boldsymbol{r}$ denote the displacement of the particle after one step. Obviously, the quantity $\boldsymbol{r}$ is a random variable and we can evaluate its mean and variance. On average, the particle would be located at

$$\mathbb{E}\,\boldsymbol{r} = p\,d_r \;+\; (1-p)\,(-d_\ell) \;=\; (d_r + d_\ell)p - d_\ell \tag{3.175}$$

[2] This section can be skipped on a first reading.

and the displacement variance would be

$$\begin{aligned}\sigma_r^2 &= \mathbb{E}\,\boldsymbol{r}^2 - (\mathbb{E}\,\boldsymbol{r})^2 \\ &= pd_r^2 + (1-p)d_\ell^2 - ((d_r + d_\ell)p - d_\ell)^2 \\ &= (d_r + d_\ell)^2 p(1-p) \end{aligned} \tag{3.176}$$

In the special case when $d_r = d_\ell = \Delta x$, the above expressions simplify to

$$\mathbb{E}\,\boldsymbol{r} = \Delta x(2p-1), \quad \sigma_r^2 = 4\Delta x^2 p(1-p) \quad (\text{when } d_r = d_\ell = \Delta x) \tag{3.177}$$

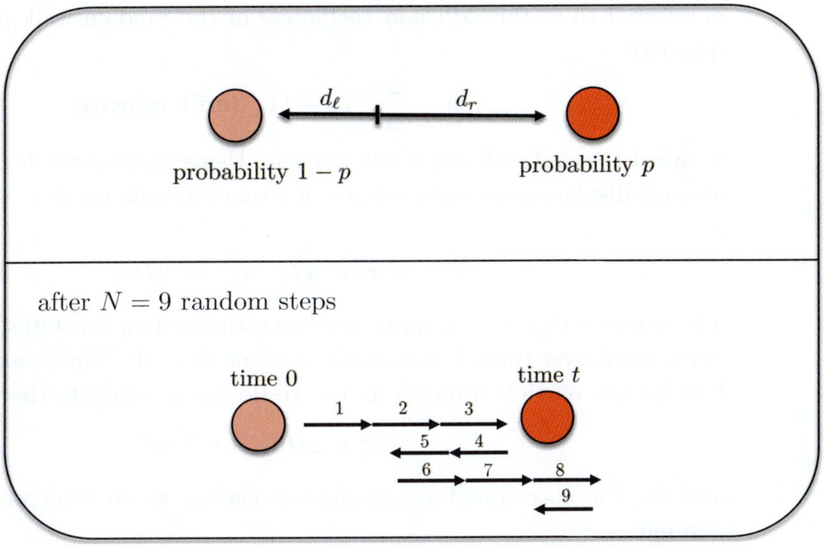

Figure 3.8 During each interval Δt, a particle takes random steps to the left (of size d_ℓ) or to the right (of size d_r) in 1-D space. After N such steps, and at time $t = N\Delta t$, the particle would be located at some random displacement that results from the aggregate effect of all individual steps.

After N successive independent random steps, and at time $t = N\Delta t$, the particle will be located at some displacement $\boldsymbol{x}$ along the same line. The quantity $\boldsymbol{x}$ is a random variable that is the result of summing N independent and identically distributed random variables $\{\boldsymbol{r}_n\}$ corresponding to the individual displacements over $n = 1, 2, \ldots, N$:

$$\boldsymbol{x} = \boldsymbol{r}_1 + \boldsymbol{r}_2 + \ldots + \boldsymbol{r}_N \tag{3.178}$$

Therefore, the mean and variance of $\boldsymbol{x}$ are given by

$$\mathbb{E}\,\boldsymbol{x} = N\,\mathbb{E}\,\boldsymbol{r} \;=\; N(d_r + d_\ell)p - Nd_\ell \tag{3.179}$$

$$\sigma_x^2 = N\,\sigma_r^2 \;=\; N(d_r + d_\ell)^2 p(1-p) \tag{3.180}$$

Diffusion coefficient and drift velocity. We focus henceforth on the case $d_r = d_\ell = \Delta x$, where the sizes of the steps to the left or to the right are identical. Replacing N by $t/\Delta t$, we find that

$$\mathbb{E}\boldsymbol{x} = \frac{\Delta x}{\Delta t}\,(2p-1)\,t \tag{3.181}$$

$$\sigma_x^2 = 2\,\frac{2\Delta x^2\,p(1-p)}{\Delta t}\,t \tag{3.182}$$

The quantity

$$D \overset{\Delta}{=} \frac{2\Delta x^2\,p(1-p)}{\Delta t} \qquad \textbf{(diffusion coefficient)} \tag{3.183}$$

is referred to as the *diffusion coefficient* of the random walk process, while the quantity

$$v = \frac{\Delta x}{\Delta t}\,(2p-1) \qquad \textbf{(drift velocity)} \tag{3.184}$$

is called the *drift velocity* of the particle. Using these variables, we find that the average displacement and variance of a random walk particle at time t are given by

$$\mathbb{E}\boldsymbol{x} \;=\; vt, \quad \sigma_x^2 \;=\; 2Dt \tag{3.185}$$

The variance expression can be used to estimate how far diffusing particles wander around over time. For example, assume $D = 10^{-6}\text{cm}^2/\text{s}$ and let us estimate how far the particle wanders in 5 s. To do so, we evaluate the variance:

$$\sigma_x^2 = 2Dt = 10^{-5}\text{cm}^2 \tag{3.186}$$

and use the corresponding standard deviation as an estimate for the distance covered:

$$\sigma_x = \sqrt{10^{-5}} \;\approx\; 0.0032 \text{ cm} \tag{3.187}$$

Central limit theorem. When the number of steps N is large (e.g., when t is large and Δt is small), then the variable $\boldsymbol{x}$ in (3.178) can be regarded as the sum of a large number of iid random variables. Accordingly, by the central limit theorem (see comments at the end of Chapter 4), the pdf of the displacement variable $\boldsymbol{x}$ will approach a Gaussian distribution with mean vt and variance $2Dt$, namely,

$$f_{\boldsymbol{x}}(x,t) \;=\; \frac{1}{\sqrt{4\pi Dt}}\exp\left\{-\frac{(x-vt)^2}{4Dt}\right\} \tag{3.188}$$

This distribution is a function of both x and time; it specifies the likelihood of the locations of the particle at any time t.

Einstein–Smoluchowski relation. Observe from (3.184) that the drift velocity of a diffusing particle is nonzero whenever $p \neq 1/2$. But what can cause a particle in a random walk motion to operate under $p \neq 1/2$ and give preference to one direction of motion over another? This preferential motion can be the result of an external force, such as gravity, which would result in a nonzero drift velocity. For example, assume a particle of mass m is diffusing down a fluid – see Fig. 3.9. Two forces act on the particle: the force of gravity, f, which acts downward, and a drag force, ζv, which opposes the motion of the particle and acts upward. The drag force is proportional to the velocity v through a frictional drag coefficient denoted by ζ.

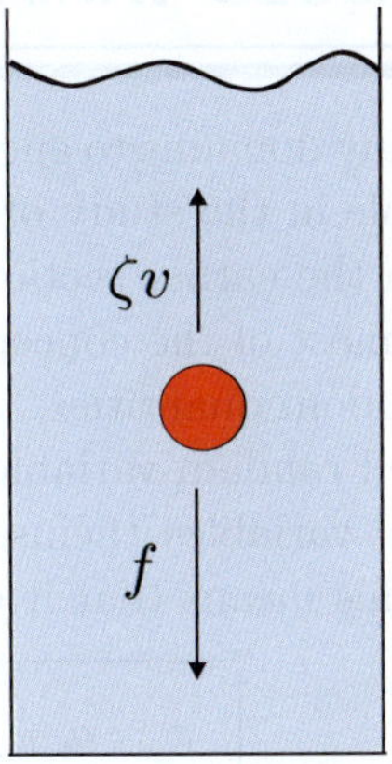

Figure 3.9 A particle of mass m diffuses down a fluid. Two forces act on it: the force of gravity, f, and the drag force, ζv.

Taking the downward direction to be the positive direction, a straightforward application of Newton's second law to the motion of the particle gives

$$m\frac{dv}{dt} = f - \zeta v \tag{3.189}$$

In steady-state, we must have $dv/dt = 0$ and, therefore, the particle attains the nonzero drift velocity $v = f/\zeta$. Substituting into (3.184), we can solve for p and deduce how the external force influences the probability value:

$$p = \frac{1}{2} + \frac{f}{2\zeta}\frac{\Delta x}{\Delta t} \tag{3.190}$$

It turns out that a fundamental relation exists between the diffusion coefficient, D, and the frictional coefficient, ζ, known as the Einstein–Smoluchowski relation:

$$\boxed{D\zeta = k_B T} \tag{3.191}$$

where T is the temperature in kelvin and k_B is the Boltzmann constant. In other words, the product $D\zeta$ is constant and independent of the size of the particle and the nature of the medium where diffusion is taking place. At room temperature ($T = 300$ K), the value of $k_B T$ is equal to

$$k_B T = 4.14 \times 10^{-14}\ \text{g}\,\text{cm}^2/\text{s}^2 \tag{3.192}$$

For a spherical particle of radius R diffusing in a medium with viscosity η ($\eta = 0.01$ g/cm s for water), the values of D and ζ are given by

$$\zeta = 6\pi\eta R, \qquad D = \frac{k_B T}{6\pi\eta R} \tag{3.193}$$

3.7 COMPLEX-VALUED VARIABLES[3]

It is common in many domains to encounter complex-valued random variables, as happens for example in the study of digital communications systems. Although the presentation in the earlier sections has focused on real-valued random variables and vectors, most of the concepts and results extend almost effortlessly to complex-valued random quantities.

A complex-valued random variable is one whose real and imaginary parts are *real*-valued random variables themselves. Specifically, if $\boldsymbol{x}$ is a scalar complex random variable, this means that it can be written in the form:

$$\boxed{\boldsymbol{x} = \boldsymbol{a} + j\boldsymbol{b}, \quad j \stackrel{\Delta}{=} \sqrt{-1}} \tag{3.194}$$

where $\boldsymbol{a}$ and $\boldsymbol{b}$ denote the real and imaginary parts of $\boldsymbol{x}$ and they are both real-valued random variables. Therefore, the pdf of $\boldsymbol{x}$ is completely characterized in terms of the joint pdf, $f_{\boldsymbol{a},\boldsymbol{b}}(a,b)$, of its real and imaginary parts. This means that we can regard (treat) a complex random variable as a function of two real random variables.

The mean of $\boldsymbol{x}$ is obtained as

$$\boxed{\bar{x} \stackrel{\Delta}{=} \mathbb{E}\boldsymbol{x} \stackrel{\Delta}{=} \mathbb{E}\boldsymbol{a} + j\mathbb{E}\boldsymbol{b} = \bar{a} + j\bar{b}} \tag{3.195}$$

in terms of the means of its real and imaginary parts. The variance of $\boldsymbol{x}$, on the other hand, continues to be denoted by σ_x^2 but is now defined by any of the following equivalent expressions:

$$\boxed{\sigma_x^2 \stackrel{\Delta}{=} \mathbb{E}\,(\boldsymbol{x} - \bar{x})(\boldsymbol{x} - \bar{x})^* = \mathbb{E}\,|\boldsymbol{x} - \bar{x}|^2 = \mathbb{E}\,|\boldsymbol{x}|^2 - |\bar{x}|^2} \tag{3.196}$$

where the symbol $*$ denotes complex conjugation. Comparing with the earlier definition (3.15a)–(3.15b) in the real case, we see that the definition in the complex case is different because of the use of the conjugation symbol (in the real case, the conjugate of $(\boldsymbol{x} - \bar{x})$ is itself and the above definitions reduce to (3.15a)–(3.15b)). The use of the conjugate term in (3.196) is necessary in order to guarantee that σ_x^2 will remain a nonnegative number. In particular, it is immediate to verify from (3.196) that

$$\sigma_x^2 = \sigma_a^2 + \sigma_b^2 \tag{3.197}$$

[3] This section can be skipped on a first reading.

in terms of the sum of the individual variances of $\boldsymbol{a}$ and $\boldsymbol{b}$.

Likewise, the covariance between two complex-valued random variables, $\boldsymbol{x}$ and $\boldsymbol{y}$, is now defined as

$$\boxed{\sigma_{xy} \stackrel{\Delta}{=} \mathbb{E}\,(\boldsymbol{x}-\bar{x})(\boldsymbol{y}-\bar{y})^*} \qquad \textbf{(covariance)} \tag{3.198}$$

with the conjugation symbol used in comparison with the real case in (3.118). We again say that the random variables are uncorrelated if, and only if, their covariance is zero, i.e.,

$$\sigma_{xy} = 0 \tag{3.199}$$

In view of the definition (3.198), this condition is equivalent to requiring

$$\boxed{\mathbb{E}\,\boldsymbol{x}\boldsymbol{y}^* = (\mathbb{E}\,\boldsymbol{x})\,(\mathbb{E}\,\boldsymbol{y})^*} \qquad \textbf{(uncorrelated random variables)} \tag{3.200}$$

On the other hand, we say that $\boldsymbol{x}$ and $\boldsymbol{y}$ are *orthogonal* if, and only if,

$$\boxed{\mathbb{E}\,\boldsymbol{x}\boldsymbol{y}^* = 0} \qquad \textbf{(orthogonal random variables)} \tag{3.201}$$

It can again be verified that the concepts of orthogonality and uncorrelatedness coincide if at least one of the random variables has zero mean.

Example 3.21 (QPSK constellation) Consider a signal $\boldsymbol{x}$ that is chosen uniformly from a quadrature-phase-shift-keying (QPSK) constellation, i.e., $\boldsymbol{x}$ assumes any of the four values:

$$x_m = \pm\frac{\sqrt{2}}{2} \pm j\frac{\sqrt{2}}{2} \tag{3.202}$$

with equal probability $p_m = 1/4$ (see Fig. 3.10). Clearly, $\boldsymbol{x}$ is a complex-valued random variable; its mean and variance are easily found to be $\bar{x} = 0$ and $\sigma_x^2 = 1$. Indeed, note first that

$$\begin{aligned}\bar{x} &= \frac{1}{4}\left(\frac{\sqrt{2}}{2}+\frac{\sqrt{2}}{2}-\frac{\sqrt{2}}{2}-\frac{\sqrt{2}}{2}\right)+j\frac{1}{4}\left(\frac{\sqrt{2}}{2}+\frac{\sqrt{2}}{2}-\frac{\sqrt{2}}{2}-\frac{\sqrt{2}}{2}\right)\\ &= 0\end{aligned} \tag{3.203}$$

while the variance of the real part of $\boldsymbol{x} = \boldsymbol{a} + j\boldsymbol{b}$ is given by:

$$\sigma_a^2 = \frac{1}{4}\left[\left(\frac{\sqrt{2}}{2}\right)^2+\left(\frac{\sqrt{2}}{2}\right)^2+\left(-\frac{\sqrt{2}}{2}\right)^2+\left(-\frac{\sqrt{2}}{2}\right)^2\right] = \frac{1}{2} \tag{3.204}$$

and, similarly, the variance of its imaginary part is $\sigma_b^2 = 1/2$. It follows that the variance of $\boldsymbol{x}$ is

$$\sigma_x^2 = \frac{1}{2}+\frac{1}{2} = 1 \tag{3.205}$$

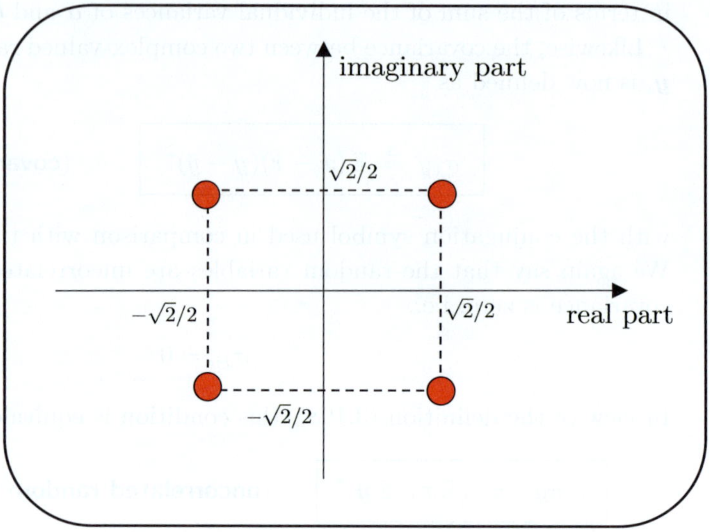

Figure 3.10 QPSK constellation with four equally probable complex symbols.

Alternatively, observe that $|\boldsymbol{x}| = 1$ for all four possibilities of $\boldsymbol{x}$, and each of these possibilities occurs with probability $1/4$. Therefore,

$$\begin{aligned} \sigma_x^2 &= \mathbb{E}\,|\boldsymbol{x}|^2 \;-\; |\bar{x}|^2 \\ &= \frac{1}{4}\,(1+1+1+1) - 0 \\ &= 1 \end{aligned} \tag{3.206}$$

When $\boldsymbol{x}$ is vector-valued, its mean consists of the vector of means and its covariance matrix is defined as

$$\boxed{R_x \;\stackrel{\Delta}{=}\; \mathbb{E}\,(\boldsymbol{x}-\bar{x})(\boldsymbol{x}-\bar{x})^*} \qquad \textbf{(when } x \textbf{ is a column vector)} \tag{3.207}$$

where the symbol $*$ now denotes complex-conjugate transposition (i.e., we transpose the vector and then replace each of its entries by the corresponding conjugate value). If $\boldsymbol{x}$ is instead a *row* random vector, then its covariance matrix is defined as

$$\boxed{R_x \;\stackrel{\Delta}{=}\; \mathbb{E}\,(\boldsymbol{x}-\bar{x})^*(\boldsymbol{x}-\bar{x})} \qquad \textbf{(when } x \textbf{ is a row vector)} \tag{3.208}$$

with the conjugated term coming first. This is because it is now the product $(\boldsymbol{x}-\bar{x})^*(\boldsymbol{x}-\bar{x})$ that yields a matrix.

3.8 COMMENTARIES AND DISCUSSION

Probability theory. The exposition in this chapter assumes some basic knowledge of probability theory, mainly with regards to the concepts of mean, variance, probability density function, and vector-random variables. Most of these ideas, including some of the examples, were introduced in the chapter from first principles following the overviews from Sayed (2003, 2008). If additional help is needed, some accessible references on probability theory and random variables are Kolmogorov (1960), Feller (1968, 1971), Billingsley (1986), Papoulis (1991), Picinbono (1993), Stark and Woods (1994), Durrett (1996), Gnedenko (1998), Chung (2000), Grimmett and Stirzaker (2001), Dudley (2002), Ash (2008), and Leon-Garcia (2008). For an insightful discussion on the notions of statistical dependence and causality, the reader may refer to Pearl (1995, 2000). In Section 3.6.2 we illustrated how Fourier analysis is useful in the study of randomness through the notion of the characteristic function – see, e.g., Bochner (1955), Lukacs (1970), Feller (1971), and Billingsley (1986). Some good references on Fourier analysis in mathematics and signal processing are Stein and Shakarchi (2003), Katznelson (2004), Oppenheim, Schafer, and Buck (2009), and Vetterli, Kovacevic, and Goyal (2014).

The modern formulation of probability theory is due to the Soviet mathematician **Andrey Kolmogorov (1903–1987)**, who put forward in Kolmogorov (1931, 1933) a collection of axioms that form the foundations for probabilistic modeling and reasoning – see the accounts by Kolmogorov (1960), Doob (1996), and Shafer and Vovk (2006). We illustrate these axioms for the case of discrete random variables.

To begin with, the finite or countable set of all possible outcomes in an experiment with discrete random results is called the *sample space*, and is denoted by the letter Ω. For example, in an experiment that involves rolling a die, the sample space is $\Omega = \{1, 2, 3, 4, 5, 6\}$. Any subset of the sample space is called an *event*, and is denoted by the letter E. For example, observing an even outcome in the roll of the die corresponds to observing an outcome from the event $E = \{2, 4, 6\}$. A probability measure, $\mathbb{P}(E)$, is assigned with every possible event. The three axioms of probability state that:

$$0 \leq \mathbb{P}(E) < \infty, \text{ for every } E \subset \Omega \tag{3.209a}$$

$$\mathbb{P}(\Omega) = 1 \tag{3.209b}$$

$$\mathbb{P}(E_1 \cup E_2) = \mathbb{P}(E_1) + \mathbb{P}(E_2), \text{ for mutually exclusive events} \tag{3.209c}$$

$$\mathbb{P}\left(\bigcup_{n=1}^{N} E_n\right) = \sum_{n=1}^{N} \mathbb{P}(E_n), \text{ for mutually exclusive events} \tag{3.209d}$$

where N can be countably infinite. The first axiom states that the probability of any event is a nonnegative real number that cannot be infinite. The second axiom means that the probability of at least one event from the sample space occurring is equal to 1; this statement assumes that the sample space captures all possible outcomes for the random experiment. The third equality is a special case of the last one for $N = 2$; these equalities constitute the third axiom.

Concentration inequalities. The Chebyshev inequality (3.28) is a useful result that reveals how realizations for random variables are more likely to concentrate around their means for distributions with small variances. We explain in Probs. 3.17 and 3.18 that the inequality is related to another result in probability theory known as the *Markov inequality*, namely, that for any scalar nonnegative real-valued random variable, $\boldsymbol{x}$, it holds that:

$$\mathbb{P}(\boldsymbol{x} \geq \alpha) \leq \mathbb{E}\,\boldsymbol{x}/\alpha, \quad \text{for any } \alpha > 0 \tag{3.210}$$

According to Knuth (1997), the Chebyshev inequality was originally developed by Bienaymé (1853) and later proved by the Russian mathematician **Pafnuty Chebyshev**

(1821–1894) in the work by Chebyshev (1867) and subsequently by his student Markov (1884) in his PhD dissertation – see also the accounts by Hardy, Littlewood, and Pólya (1934), Bernshtein (1945), Shiryayev (1984), Papoulis (1991), and Fischer (2011).

The Markov and Chebyshev bounds are examples of *concentration inequalities*, which help bound the deviation of a random variable (or combinations of random variables) away from certain values (typically their means). In Appendix 3.B we establish three famous results known as Azuma inequality, Hoeffding inequality, and McDiarmid inequality, which provide bounds on the probability of the sum of a collection of random variables deviating from their mean. The Hoeffding inequality is due to the Finnish statistician **Wassily Hoeffding (1914–1991)** and appeared in the work by Hoeffding (1963). Earlier related investigations appear in Chernoff (1952) and Okamoto (1958). The McDiarmid inequality extends the results of Hoeffding to more general functions that satisfy a bounded variations property. This extension was proven by McDiarmid (1989). Both inequalities play an important role in the analysis of learning algorithms and will be used, for example, in the derivation of generalization bounds in Chapter 64. For further details on concentration inequalities, readers may refer to Ledoux (2001), Boucheron, Lugosi, and Bousquet (2004), Chung and Lu (2006a, b), Massart (2007), Alon and Spencer (2008), Boucheron, Lugosi, and Massart (2013), Mohri, Rostamizadeh, and Talwalkar (2018), Vershynin (2018), and Wainwright (2019).

Bayes rule. Expression (3.39) is one manifestation of a fundamental result in probability theory known as the Bayes rule, which is applicable to both cases of discrete and continuous random variables. For instance, if the letters A, B, and C denote some discrete probability events, then the Bayes rule ensures that

$$\mathbb{P}(A, B) = \mathbb{P}(A|B)\mathbb{P}(B) = \mathbb{P}(B|A)\mathbb{P}(A) \tag{3.211}$$

in terms of the joint probability of events A and B, and their individual and conditional probabilities. In particular, it follows that

$$\mathbb{P}(A|B) = \frac{\mathbb{P}(B|A)\mathbb{P}(A)}{\mathbb{P}(B)} \tag{3.212}$$

which enables us to update the belief in event A following the observation of event B. In this way, the Bayes rule allows us to update prior probabilities into posterior (conditional) probabilities. A similar construction applies when one or both random variables happen to be continuous, in which case their distributions are described in terms of pdfs. In that situation, relation (3.211) would be replaced by (3.39), namely,

$$f_{\boldsymbol{x},\boldsymbol{y}}(x, y) \;=\; f_{\boldsymbol{x}|\boldsymbol{y}}(x|y)f_{\boldsymbol{y}}(y) \;=\; f_{\boldsymbol{y}|\boldsymbol{x}}(y|x)f_{\boldsymbol{x}}(x) \tag{3.213}$$

A special case of the Bayes rule (3.211) was first proposed by the English statistician **Thomas Bayes (1701–1761)** in his study of the problem of inferring the probability of success, p, based on observing S successes in N repeated Bernoulli trials. His work was published posthumously by the Welsh philosopher **Robert Price (1723–1791)** in the work by Bayes and Price (1763). Interestingly, the Bayes rule in its general form (3.211) appears to have been independently discovered by the French mathematician **Pierre-Simon Laplace (1749–1827)** and published about a decade later in the work by Laplace (1774). The article by Stigler (1983) suggests a different historical timeline and argues that the rule may have been discovered over a decade before Bayes by another English mathematician named Nicholas Saunderson (1682–1739). However, this interpretation is not universally accepted by statisticians and the controversy remains – see, e.g., Edwards (1986), Hald (1998), Dale (2003), and Feinberg (2003).

Random walks and Brownian motion. In Section. 3.6.4 we described one application of the central limit theorem by examining the diffusive behavior of particles. The example is motivated by the discussion in Berg (1993). The Einstein–Smoluchowski relation (3.191) is a fundamental result in physics relating the diffusion coefficient, D, of a particle and the frictional coefficient, ζ. It was discovered independently and almost simultaneously by the German-American physicist and Nobel Laureate **Albert Einstein (1879–1955)** in the work by Einstein (1905), and by Sutherland (1905) and Smoluchowski (1906) in their studies of Brownian motion.

Brownian motion refers to the random motion of particles suspended in a fluid, where the displacements of the particles result from collisions among molecules. This explanation was provided by Einstein (1905); it was subsequently used as one indirect proof for the existence of elementary particles such as atoms and molecules. The designation "Brownian motion" is after the Scottish botanist **Robert Brown (1773–1858)**, who observed under a microscope in 1827 the motion of tiny particles suspended in water – see the useful account by Pearle *et al.* (2010) and also Brown (1828, 1866). There have been earlier observations of "Brownian motion" and Brown (1828) lists in his paper the names of several researchers who have commented before on aspects of this behavior. For instance, the study by van der Pas (1971) notes that the Dutch biologist **Jan Ingenhousz (1730–1799)**, who is credited with discovering the process of photosynthesis, had also reported observing the motion of coal dust particles in a liquid almost four decades before Brown in 1784 – see Ingenhousz (1784) and the English translation that appears in van der Pas (1971). In this translation, Ingenhousz comments on how *"the entire liquid and consequently everything which is contained in it, is kept in continuous motion by the evaporation, and that this motion can give the impression that some of these corpuscules are living, even if they have not the slightest life in them."*

One useful way to describe Brownian motion is in terms of a random walk process, where a particle takes successive random steps. The designation "random walk" is due to the English statistician **Karl Pearson (1857–1936)**, who is credited along with **Ronald Fisher (1890–1962)** with establishing the modern field of mathematical statistics – see the exposition by Tankard (1984). We will comment on other contributions by Pearson later in this text, including his development of the method of principal component analysis (PCA) and the Neyman–Pearson technique for hypothesis testing. For further accounts on the theory of Brownian motion and random walks, the reader may refer to several texts, including those by Rogers and Williams (2000), Morters and Peres (2010), Lawler and Limic (2010), Bass (2011), and Gallager (2014).

Boltzmann distribution. We described the Boltzmann distribution in expression (3.168) and explained how it is useful in characterizing the probability distribution of the states of a complex system as a function of the energies of the state levels. This probability distribution is widely used in statistical mechanics, which is the field that deals with understanding how microscopic properties at the atomic level translate into physical properties at the macroscopic level. In particular, the Boltzmann distribution encodes the useful property that lower-energy states are more likely to occur than higher-energy states – see, e.g., the treatments in Gibbs (1902), Landau and Lifshitz (1980), Hill (1987), Tolman (2010), and Pathria and Beale (2011). The distribution is named after the Austrian physicist **Ludwig Boltzmann (1844–1906)**, who is regarded as one of the developers of the field of statistical physics/mechanics. He introduced it in the work by Boltzmann (1877, 1909) while developing a probabilistic view of the second law of thermodynamics. A useful historical overview of Boltzmann's work is given by Uffink (2014). Boltzmann's visionary contributions at the time, and his statistical analysis of the motion of atoms and the resulting macroscopic properties of matter, were harshly criticized by some fellow scientists who were unable to grasp his probabilistic reasoning. Unfortunately, he committed suicide in 1906. Since then, his theories and explanations have been validated by experimentation and the atomic theory of matter. We will encounter the Boltzmann distribution later in Section 66.2 when we study restricted Boltzmann machines in the context of deep learning networks. Applications of the Boltzmann distribution to molecular biology problems, such as the ion-channel states and protein folding examples discussed in the text, can be found in Onuchic,

Luthey-Schulten, and Wolynes (1997), Huang (2005), Santana, Larranaga, and Lozano (2008), Dubois, Gilles, and Rouzaire-Dubois (2009), and Phillips *et al.* (2012).

Law of large numbers. The weak and strong laws of large numbers are discussed in Appendix 3.A, along with various notions of convergence for random variables such as convergence in distribution, convergence in probability, almost-sure convergence, and convergence in mean-square. The weak law of large numbers is due to the Swiss mathematician **Jacob Bernoulli (1654–1705)**; it was published posthumously in his book on combinatorics by Bernoulli (1713). The law was known as the "Bernoulli theorem" for many decades and was later referred to as the "law of large numbers" by the French mathematician **Simeon Poisson (1781–1840)** in the work by Poisson (1837). The latter name has since become the common reference to these results. Many other mathematicians followed suit, refining and weakening the conditions required for the conclusions of the law to hold. In particular, the strong version of the law was first proven by the French mathematician **Emile Borel (1871–1956)** in the work by Borel (1909). Some of the weakest conditions for its validity were given later by the Russian mathematician **Andrey Kolmogorov (1903–1987)** in Kolmogorov (1927). A historical account on the laws of large numbers appears in Seneta (2013), in addition to the earlier account given by the Russian mathematician **Andrey Markov (1856–1922)**, which appears in Appendix 1 of Ondar (1981). A useful account on the strong version of the law is given by Prokhorov (2011). For technical details on the laws, the reader may consult Feller (1968, 1971), Billingsley (1986, 1999), Durrett (1996), and Grimmett and Stirzaker (2001).

PROBLEMS[4]

3.1 Refer to the calculations in Example 3.14 on conditional independence. Verify that the marginal pmfs for the variables $\boldsymbol{R}$ and $\boldsymbol{L}$ are given by

$$\mathbb{P}(\boldsymbol{R}=1)=17/36,\quad \mathbb{P}(\boldsymbol{R}=0)=19/36,\quad \mathbb{P}(\boldsymbol{L}=1)=31/60,\quad \mathbb{P}(\boldsymbol{L}=0)=29/60$$

3.2 Refer to the calculations in Example 3.14 on conditional independence. Verify that the conditional pmf of $\boldsymbol{A}$ given $\boldsymbol{R}$ assumes the following values:

$$\begin{aligned}\mathbb{P}(\boldsymbol{A}=1|\boldsymbol{R}=1)=9/17,\quad \mathbb{P}(\boldsymbol{A}=0|\boldsymbol{R}=1)=8/17\\ \mathbb{P}(\boldsymbol{A}=1|\boldsymbol{R}=0)=3/19,\quad \mathbb{P}(\boldsymbol{A}=0|\boldsymbol{R}=0)=16/19\end{aligned}$$

3.3 Conclude from the results of Probs. 3.1 and 3.2 and Example 3.14 that the joint pmf of the variables $\{\boldsymbol{R},\boldsymbol{A},\boldsymbol{L}\}$ factorizes as

$$\mathbb{P}(\boldsymbol{R},\boldsymbol{A},\boldsymbol{L})=\mathbb{P}(\boldsymbol{R})\,\mathbb{P}(\boldsymbol{A}|\boldsymbol{R})\,\mathbb{P}(\boldsymbol{L}|\boldsymbol{A})$$

3.4 Refer to the calculations in Example 3.14 on conditional independence. Verify that the conditional pmf of $\boldsymbol{L}$ given $\boldsymbol{R}$ assumes the following values:

$$\begin{aligned}\mathbb{P}(\boldsymbol{L}=1|\boldsymbol{R}=1)=199/340,\quad \mathbb{P}(\boldsymbol{L}=0|\boldsymbol{R}=1)=141/340\\ \mathbb{P}(\boldsymbol{L}=1|\boldsymbol{R}=0)=173/380,\quad \mathbb{P}(\boldsymbol{L}=0|\boldsymbol{R}=0)=207/380\end{aligned}$$

Use the result of Prob. 3.2 to conclude that the variables $\boldsymbol{L}$ and $\boldsymbol{A}$ are not independent conditioned on $\boldsymbol{R}$ and, hence, the joint pmf of $\{\boldsymbol{R},\boldsymbol{A},\boldsymbol{L}\}$ factors in the form

$$\mathbb{P}(\boldsymbol{R},\boldsymbol{A},\boldsymbol{L})=\mathbb{P}(\boldsymbol{A})\,\mathbb{P}(\boldsymbol{R}|\boldsymbol{A})\,\mathbb{P}(\boldsymbol{L}|\boldsymbol{R},\boldsymbol{A})$$

[4] A couple of problems in this section are adapted from exercises in Sayed (2003, 2008).

where the last factor cannot be replaced by $\mathbb{P}(\boldsymbol{L}|\boldsymbol{R})$. Compare with the factorization in Prob. 3.3.

3.5 Refer to the calculations in Example 3.14 on conditional independence.

Table 3.3 Joint pmf for the variables $\{\boldsymbol{R}, \boldsymbol{L}, \boldsymbol{A}\}$ in Prob. 3.5.

$\boldsymbol{R}$ (rain)	$\boldsymbol{A}$ (accident)	$\boldsymbol{L}$ (late)	$\mathbb{P}(\boldsymbol{R}, \boldsymbol{A}, \boldsymbol{L})$ (joint pmf)
0	0	0	0.40
0	0	1	0.05
0	1	0	0.10
0	1	1	0.10
1	0	0	0.10
1	0	1	0.10
1	1	0	0.05
1	1	1	0.10

Assume instead that the joint pmf of the variables $\{\boldsymbol{R}, \boldsymbol{A}, \boldsymbol{L}\}$ has the values shown in Table 3.3. Repeat the derivation to verify whether the variables $\{\boldsymbol{R}, \boldsymbol{L}\}$ continue to be independent conditioned on knowledge of $\boldsymbol{A}$. Given that the individual arrived late to work, what is the likelihood that there was a traffic accident on the road?

3.6 Consider three continuous random variables $\{\boldsymbol{x}, \boldsymbol{y}, \boldsymbol{z}\}$ and assume their joint pdf factors in the form $f_{\boldsymbol{x},\boldsymbol{y},\boldsymbol{z}}(x, y, z) = f_{\boldsymbol{x}}(x) f_{\boldsymbol{y}|\boldsymbol{x}}(y|x) f_{\boldsymbol{z}|\boldsymbol{y}}(z|y)$. Verify that the variables $\{\boldsymbol{x}, \boldsymbol{z}\}$ are independent of each other conditioned on knowledge of $\boldsymbol{y}$. Verify that the same conclusion applies to discrete random variables.

3.7 Consider three discrete random variables $\{\boldsymbol{x}, \boldsymbol{y}, \boldsymbol{z}\}$ and assume their joint pmf factors in the form $\mathbb{P}(\boldsymbol{x}, \boldsymbol{y}, \boldsymbol{z}) = \mathbb{P}(\boldsymbol{x})\,\mathbb{P}(\boldsymbol{y})\,\mathbb{P}(\boldsymbol{z}|\boldsymbol{x}, \boldsymbol{y})$. Are the variables $\{\boldsymbol{x}, \boldsymbol{y}\}$ independent of each other conditioned on knowledge of $\boldsymbol{z}$? Verify that the same conclusion applies to continuous random variables.

3.8 Let $\boldsymbol{y} = \frac{1}{3}\boldsymbol{x} + \frac{1}{2}\boldsymbol{v}$, where $\boldsymbol{x}$ is uniformly distributed over the interval $[-1, 1]$ and $\boldsymbol{v}$ is a zero-mean Gaussian random variable with variance $1/2$. Both $\boldsymbol{x}$ and $\boldsymbol{v}$ are independent random variables.

(a) Find the mean and variance of $\boldsymbol{y}$.
(b) Find the correlation between $\boldsymbol{y}$ and $\boldsymbol{x}$.
(c) Find the correlation between $\boldsymbol{y}$ and $\boldsymbol{v}$.
(d) How would your answers change if $\boldsymbol{x}$ and $\boldsymbol{v}$ were only uncorrelated rather than independent?

3.9 For what values of the scalar a would the matrix below correspond to the covariance matrix of a 2×1 random vector,

$$R_z = \begin{bmatrix} 1 & a \\ a & 2 \end{bmatrix} \; ?$$

3.10 Consider a column vector $\boldsymbol{y}$ with mean $\bar{y}$ and covariance matrix R_y. What is $\mathbb{E}\,(\boldsymbol{y} \otimes \boldsymbol{y})$? Here, the symbol $\otimes$ refers to the Kronecker product operation.

3.11 If two scalar zero-mean real random variables $\boldsymbol{a}$ and $\boldsymbol{b}$ are uncorrelated, does it follow that $\boldsymbol{a}^2$ and $\boldsymbol{b}^2$ are also uncorrelated?

3.12 Consider the column vector $\boldsymbol{x} = \text{col}\{\boldsymbol{a}, \boldsymbol{b}\}$, where $\boldsymbol{a}$ and $\boldsymbol{b}$ are two scalar random variables with possibly nonzero means. Use the fact that $R_x = \mathbb{E}\,(\boldsymbol{x} - \bar{x})(\boldsymbol{x} - \bar{x})^{\mathsf{T}} \geq 0$ to establish the following Cauchy–Schwarz inequality for random variables:

$$\Big(\mathbb{E}\,(\boldsymbol{a} - \bar{a})(\boldsymbol{b} - \bar{b})\Big)^2 \leq \mathbb{E}\,(\boldsymbol{a} - \bar{a})^2 \times \mathbb{E}\,(\boldsymbol{b} - \bar{b})^2$$

3.13 Problems 3.13–3.18 are adapted from exercises in Sayed (2003, 2008). Consider two scalar random variables $\{\boldsymbol{x}, \boldsymbol{y}\}$ with means $\{\bar{x}, \bar{y}\}$, variances $\{\sigma_x^2, \sigma_y^2\}$, and correlation σ_{xy}. Define the correlation coefficient $\rho_{xy} = \sigma_{xy}/\sigma_x\sigma_y$. Show that it is bounded by 1, i.e., $|\rho_{xy}| \leq 1$.

3.14 A random variable $\boldsymbol{x}_1$ assumes the value $+1$ with probability p and the value -1 with probability $1-p$. A random variable $\boldsymbol{x}_2$ is distributed as follows:

$$\text{if } \boldsymbol{x}_1 = +1 \text{ then } \quad \boldsymbol{x}_2 = \begin{cases} +2 \text{ with probability } q \\ -2 \text{ with probability } 1-q \end{cases}$$

$$\text{if } \boldsymbol{x}_1 = -1 \text{ then } \quad \boldsymbol{x}_2 = \begin{cases} +3 \text{ with probability } r \\ -3 \text{ with probability } 1-r \end{cases}$$

Find the means and variances of $\boldsymbol{x}_1$ and $\boldsymbol{x}_2$.

3.15 Consider a Rayleigh-distributed random variable $\boldsymbol{x}$ with pdf given by (3.26). Show that its mean and variance are given by (3.27).

3.16 Suppose we observe $\boldsymbol{y} = \boldsymbol{x} + \boldsymbol{v}$, where $\boldsymbol{x}$ and $\boldsymbol{v}$ are independent random variables with exponential distributions with parameters λ_1 and λ_2 ($\lambda_1 \neq \lambda_2$). That is, the pdfs of $\boldsymbol{x}$ and $\boldsymbol{v}$ are $f_{\boldsymbol{x}}(x) = \lambda_1 e^{-\lambda_1 x}$ for $x \geq 0$ and $f_{\boldsymbol{v}}(v) = \lambda_2 e^{-\lambda_2 v}$ for $v \geq 0$, respectively.

(a) Using the fact that the pdf of the sum of two independent random variables is the convolution of the individual pdfs, show that

$$f_{\boldsymbol{y}}(y) = \frac{\lambda_1 \lambda_2}{\lambda_2 - \lambda_1} e^{-\lambda_2 y} \left(e^{(\lambda_2 - \lambda_1) y} - 1 \right), \quad y \geq 0$$

(b) Establish that $f_{\boldsymbol{x},\boldsymbol{y}}(x,y) = \lambda_1 \lambda_2 e^{(\lambda_2 - \lambda_1)x - \lambda_2 y}$, for $x \geq 0$ and $y \geq 0$.

3.17 Suppose $\boldsymbol{x}$ is a scalar nonnegative real-valued random variable with pdf $f_{\boldsymbol{x}}(x)$. Show that $\mathbb{P}(\boldsymbol{x} \geq \alpha) \leq \mathbb{E}\,\boldsymbol{x}/\alpha$, for any $\alpha > 0$. This result is known as the *Markov inequality*.

3.18 Consider a scalar real-valued random variable $\boldsymbol{x}$ with mean $\bar{x}$ and variance σ_x^2. Let $\boldsymbol{y} = (\boldsymbol{x} - \bar{x})^2$. Apply Markov inequality to $\boldsymbol{y}$ to establish Chebyshev inequality (3.28).

3.19 Consider a scalar real-valued random variable $\boldsymbol{x}$ with mean $\bar{x}$ and assuming values in the interval $\boldsymbol{x} \in [0,1]$. Apply Markov inequality from Prob. 3.17 to show that, for any real number in the interval $\alpha \in (0,1)$, it holds:

(a) $\mathbb{P}(\boldsymbol{x} > 1-\alpha) \geq (\bar{x} - (1-\alpha))/\alpha$.

(b) $\mathbb{P}(\boldsymbol{x} > \alpha) \geq (\bar{x} - \alpha)/(1-\alpha)$.

3.20 Show that for any positive scalar random variable $\boldsymbol{x}$ with nonzero mean, it holds $1/\mathbb{E}\,\boldsymbol{x} < \mathbb{E}\,(1/\boldsymbol{x})$.

3.21 Suppose $\boldsymbol{x}$ is a scalar real-valued random variable with pdf $f_{\boldsymbol{x}}(x)$ and $\mathbb{E}\,|\boldsymbol{x}|^r < \infty$. Show that, for any $\alpha > 0$ and $r > 2$, $\mathbb{P}(|\boldsymbol{x}| \geq \alpha) \leq \mathbb{E}\,|\boldsymbol{x}|^r/\alpha^r$. This result is a more general version of Markov inequality.

3.22 Consider a real-valued random variable $\boldsymbol{x}$ with mean $\bar{x}$ and variance $\sigma_x^2 < \infty$. Let $a < b$ and $a + b = 2\bar{x}$. Conclude from Chebyshev inequality (3.28) that

$$\mathbb{P}(a < \boldsymbol{x} < b) \geq 1 - \frac{4\sigma_x^2}{(b-a)^2}$$

3.23 Consider a real-valued random variable $\boldsymbol{x}$ with mean $\bar{x}$ and variance $\sigma_x^2 < \infty$. For any real c, conclude from the result of Prob. 3.21 that the following bound also holds:

$$\mathbb{P}(|\boldsymbol{x} - c| \geq \delta) \leq \frac{\sigma_x^2 + (\bar{x} - c)^2}{\delta^2}$$

3.24 Consider a real-valued random variable, $\boldsymbol{x}$, with mean $\bar{x}$ and variance $\sigma_x^2 < \infty$. Apply Markov inequality from Prob. 3.21 to establish the following one-sided versions of Chebyshev inequality, for any $\delta > 0$,

$$\mathbb{P}(\boldsymbol{x} \geq \bar{x} + \delta) \leq \frac{\sigma_x^2}{\sigma_x^2 + \delta^2}, \quad \mathbb{P}(\boldsymbol{x} \leq \bar{x} - \delta) \leq \frac{\sigma_x^2}{\sigma_x^2 + \delta^2}$$

3.25 Consider two real-valued random variables $\boldsymbol{x}$ and $\boldsymbol{y}$. Establish that $\mathbb{E}\left[\mathbb{E}\left(\boldsymbol{x}|\boldsymbol{y}\right)\right] = \mathbb{E}\,\boldsymbol{x}$, where the outermost expectation is over the pdf of $\boldsymbol{y}$ while the innermost expectation is over the conditional pdf $f_{\boldsymbol{x}|\boldsymbol{y}}(x|y)$.

3.26 Consider two random variables $\boldsymbol{x}$ and $\boldsymbol{y}$ and a random vector $\boldsymbol{z} \in \mathbb{R}^M$ that is deterministic conditioned on knowledge of $\boldsymbol{y}$, i.e., $\mathbb{E}\left(\boldsymbol{z}|\boldsymbol{y}\right) = \boldsymbol{z}$. For any deterministic set $\mathcal{S} \subset \mathbb{R}^M$, establish the identity

$$\mathbb{E}\Big\{\mathbb{E}\left(\boldsymbol{x}|\boldsymbol{y}\right) \,|\, \boldsymbol{z} \in \mathcal{S}\Big\} \;=\; \mathbb{E}\left(\boldsymbol{x}|\boldsymbol{z} \in \mathcal{S}\right)$$

where we are further conditioning on the event $\boldsymbol{z} \in \mathcal{S}$. How does this result compare to Prob. 3.25? *Remark*. This result is a special case of the law of total expectations – see, e.g., Billingsley (1986) and Weiss (2005).

3.27 Consider a discrete scalar random variable $\boldsymbol{u} = 0, 1, \ldots, N-1$, and two continuous random vector variables $\boldsymbol{x}$ and $\boldsymbol{y}$. Assume $\boldsymbol{u}$ and $\boldsymbol{y}$ are independent of each other. Verify that

$$\mathbb{E}\left(\boldsymbol{x}|\boldsymbol{y} = y\right) = \sum_{u=0}^{N-1} \mathbb{P}(\boldsymbol{u} = u)\,\mathbb{E}\left(\boldsymbol{x}|\boldsymbol{y} = y, \boldsymbol{u} = u\right)$$

3.28 The following problem is based on an exercise from Sayed (2003, 2008). Consider an $M \times M$ positive-definite symmetric matrix R and introduce its eigen-decomposition, $R = \sum_{m=1}^{M} \lambda_m u_m u_m^{\mathsf{T}}$, where the λ_m are the eigenvalues of R (all positive) and the u_m are the eigenvectors of R. The u_m are orthonormal, i.e., $u_m^{\mathsf{T}} u_k = 0$ for all $m \neq k$ and $u_m^{\mathsf{T}} u_m = 1$. Let $\boldsymbol{h}$ be a random vector with probability distribution $\mathbb{P}(\boldsymbol{h} = u_m) = \lambda_m/\mathrm{Tr}(R)$, where $\mathrm{Tr}(R)$ denotes the trace of R and is equal to the sum of its eigenvalues.

(a) Show that $\mathbb{E}\,\boldsymbol{h}\boldsymbol{h}^{\mathsf{T}} = R/\mathrm{Tr}(R)$ and $\mathbb{E}\,\boldsymbol{h}\boldsymbol{h}^{\mathsf{T}}\boldsymbol{h}\boldsymbol{h}^{\mathsf{T}} = R/\mathrm{Tr}(R)$.

(b) Show that $\mathbb{E}\,\boldsymbol{h}^{\mathsf{T}}R^{-1}\boldsymbol{h} = M/\mathrm{Tr}(R)$ and $\mathbb{E}\,\boldsymbol{h}\boldsymbol{h}^{\mathsf{T}}R^{-1}\boldsymbol{h}\boldsymbol{h}^{\mathsf{T}} = I_M/\mathrm{Tr}(R)$, where I_M is the identity matrix of size $M \times M$.

(c) Show that $\mathbb{E}\,\boldsymbol{h}^{\mathsf{T}}\boldsymbol{h} = 1$ and $\mathbb{E}\,\boldsymbol{h} = \dfrac{1}{\mathrm{Tr}(R)} \sum_{m=1}^{M} \lambda_m u_m$.

3.29 Establish the validity of (3.160).

3.30 Starting from (3.162), verify that:

(a) $\varphi_{\boldsymbol{x}}(0) = 1$.

(b) $|\varphi_{\boldsymbol{x}}(t)| \leq 1$.

(c) $\varphi_{\boldsymbol{x}}(t) = \varphi_{\boldsymbol{x}}^*(-t)$.

(d) Establish the validity of (3.164).

3.31 Assume $\boldsymbol{x}$ is uniformly distributed over the interval $[a, b]$. Show that the characteristic function of $\boldsymbol{x}$ is given by

$$\varphi_{\boldsymbol{x}}(t) = \frac{e^{jtb} - e^{jta}}{jt(b-a)}$$

3.32 Assume $\boldsymbol{x}$ takes the value $x = 1$ with probability p and the value $x = 0$ with probability $1 - p$. What is the characteristic function of $\boldsymbol{x}$? Use the characteristic function to evaluate all moments of $\boldsymbol{x}$.

3.33 What is the mean and variance of a Boltzmann distribution with two states?

3.34 If the probability of a closed ion channel is twice the probability of an open ion channel, what is the relation between the energies of the respective states?

3.35 Let $\Delta E = E_u - E_f$. Verify that the probability of encountering a folded protein can be written in the form

$$\mathbb{P}(\text{protein folded}) \;=\; \frac{1}{1 + e^{-\Delta G/k_B T}}, \quad \text{where } \Delta G \;=\; \Delta E - k_B T \ln L$$

3.36 What is the average energy of N proteins?

3.37 Let $\boldsymbol{x} = \cos\boldsymbol{\theta} + j\sin\boldsymbol{\theta}$, where $\boldsymbol{\theta}$ is uniformly distributed over the interval $[-\pi, \pi]$. Determine the mean and variance of $\boldsymbol{x}$.

3.38 Let

$$\boldsymbol{x} = \begin{bmatrix} 1 + \cos \boldsymbol{\phi} + j \sin \boldsymbol{\phi} \\ \cos \boldsymbol{\phi} + j \sin \boldsymbol{\phi} \end{bmatrix}, \quad \boldsymbol{y} = \begin{bmatrix} 1 + \cos \boldsymbol{\theta} + j \sin \boldsymbol{\theta} \\ \cos \boldsymbol{\theta} + j \sin \boldsymbol{\theta} \end{bmatrix}$$

where $\boldsymbol{\phi}$ and $\boldsymbol{\theta}$ are independent of each other and uniformly distributed over the interval $[-\pi, \pi]$. Determine $\mathbb{E}\,\boldsymbol{x}$, R_x, and R_{xy}.

3.39 Conclude from the axioms of probability (3.209a)–(3.209d) that the probability of any event must be bounded by 1, i.e., $\mathbb{P}(E) \leq 1$.

3.40 Conclude from the axioms of probability (3.209a)–(3.209d) that the probability of the empty event is zero, i.e., $\mathbb{P}(\emptyset) = 0$.

3.41 The pdf of a random variable $\boldsymbol{x} \geq 0$ that is exponentially distributed with parameter $\lambda > 0$ has the form $f_{\boldsymbol{x}}(x) = \lambda e^{-\lambda x}$.

(a) Verify that $\mathbb{E}\,\boldsymbol{x} = 1/\lambda$ and $\sigma_x^2 = 1/\lambda^2$. What is the median of $\boldsymbol{x}$?

(b) Verify that the cumulative distribution function (cdf) of $\boldsymbol{x}$ is given by $F_{\boldsymbol{x}}(x) = 1 - e^{-\lambda x}$. Recall that the cdf at location x is defined as the area under the pdf until that location, i.e., $F_{\boldsymbol{x}}(x) = \int_{-\infty}^{x} f_{\boldsymbol{x}}(x')dx'$.

(c) Consider a sequence of positive-valued random variables $\boldsymbol{x}_n$ with cdf defined as follows:

$$F_{\boldsymbol{x}_n}(x) = 1 - (1 - \lambda/n)^{nx}, \quad x > 0$$

Show that $\boldsymbol{x}_n$ converges to the exponentially distributed random variable $\boldsymbol{x}$ in distribution.

3.42 Consider a sequence of random variables $\boldsymbol{x}_n$ such that $\mathbb{E}\,\boldsymbol{x}_n \to \mu$ and $\sigma_{x_n}^2 \to 0$ as $n \to \infty$. Show that $\boldsymbol{x}_n \xrightarrow{p} \mu$. That is, show that $\boldsymbol{x}_n$ converges to the constant random variable μ in probability. According to definition (3.220a), this is equivalent to showing $\mathbb{P}(|\boldsymbol{x}_n - \mu| \geq \epsilon) \to 0$ for any $\epsilon \geq 0$.

3.43 Consider the random sequence $\boldsymbol{x}_n = \boldsymbol{x} + \boldsymbol{v}_n$, where the perturbation $\boldsymbol{v}_n$ has mean $\mathbb{E}\,\boldsymbol{v}_n = \mu/n^2$ and variance $\sigma_v^2 = \sigma^2/\sqrt{n}$ for some μ and $\sigma^2 > 0$. Show that the sequence $\boldsymbol{x}_n$ converges to $\boldsymbol{x}$ in probability.

3.44 Consider the random sequence $\boldsymbol{x}_n = (1 - \frac{1}{\sqrt{n}})\boldsymbol{x}$, where $\boldsymbol{x}$ is a binary random variable with $\mathbb{P}(\boldsymbol{x} = 0) = p > 0$ and $\mathbb{P}(\boldsymbol{x} = 1) = 1 - p$. Show that the sequence $\boldsymbol{x}_n$ converges to $\boldsymbol{x}$ in probability.

3.45 Consider two random sequences $\{\boldsymbol{x}_n, \boldsymbol{y}_n\}$. Establish the following conclusions, which amount to the statement of Slutsky theorem due to Slutsky (1925) – see also Davidson (1994) and van der Vaart (2000):

(a) $\boldsymbol{x}_n \xrightarrow{d} \boldsymbol{x}$ and $(\boldsymbol{y}_n - \boldsymbol{x}_n) \xrightarrow{p} 0 \;\Rightarrow\; \boldsymbol{y}_n \xrightarrow{d} \boldsymbol{x}$.

(b) $\boldsymbol{x}_n \xrightarrow{p} \boldsymbol{x}$ and $(\boldsymbol{y}_n - \boldsymbol{x}_n) \xrightarrow{p} 0 \;\Rightarrow\; \boldsymbol{y}_n \xrightarrow{p} \boldsymbol{x}$.

(c) $\boldsymbol{x}_n \xrightarrow{a.s.} \boldsymbol{x}$ and $(\boldsymbol{y}_n - \boldsymbol{x}_n) \xrightarrow{a.s.} 0 \Rightarrow \boldsymbol{y}_n \xrightarrow{a.s.} \boldsymbol{x}$.

3.46 Consider a random sequence $\{\boldsymbol{x}_n\}$ that converges in distribution to $\boldsymbol{x}$, and a second random sequence $\{\boldsymbol{y}_n\}$ that converges in probability to a constant c, i.e., $\boldsymbol{x}_n \rightsquigarrow \boldsymbol{x}$ and $\boldsymbol{y}_n \xrightarrow{p} c$. Establish the following consequences of Slutsky theorem:

(a) $\boldsymbol{x}_n + \boldsymbol{y}_n \rightsquigarrow \boldsymbol{x} + c$.

(b) $\boldsymbol{x}_n \boldsymbol{y}_n \rightsquigarrow c\boldsymbol{x}$.

(c) $\boldsymbol{x}_n/\boldsymbol{y}_n \rightsquigarrow \boldsymbol{x}/c, \; c \neq 0$.

3.47 A random variable $\boldsymbol{x}$ is selected uniformly from the interval $[0, 1/2]$. Let $\boldsymbol{x}_n = 1 + 3\boldsymbol{x} + (2\boldsymbol{x})^n$.

(a) Verify that $\boldsymbol{x}_n$ approaches $1 + 3\boldsymbol{x}$ as $n \to \infty$ for any value of $\boldsymbol{x}$ in the semi-open interval $[0, 1/2)$. What happens when $\boldsymbol{x} = 1/2$?

(b) Show that the sequence $\{\boldsymbol{x}_n\}$ converges almost surely. To which random variable?

3.48 Consider a sequence of fair coin tosses with outcome $\boldsymbol{b}_n = 1$ when the coin lands a head at the nth toss or $\boldsymbol{b}_n = 0$ otherwise. We use this sequence to construct $\boldsymbol{x}_n = \prod_{m=1}^{n} \boldsymbol{b}_n$. Show that the sequence $\boldsymbol{x}_n$ converges almost surely to the constant 0.

3.49 Consider a sequence of random variables $\boldsymbol{x}_n$ that are uniformly distributed within the interval $[0 \; \frac{1}{n^2}]$. For what values of $p \geq 1$ does the sequence $\{\boldsymbol{x}_n\}$ converge in the pth mean to $\boldsymbol{x} = 0$?

3.50 Consider a sequence of random variables $\{\boldsymbol{x}_n\}$ that converge in the pth mean to $\boldsymbol{x}$ for some $p \geq 1$. Use the Markov inequality to conclude that the sequence $\{\boldsymbol{x}_n\}$ converges to $\boldsymbol{x}$ in probability, i.e., show that $\boldsymbol{x}_n \xrightarrow{L^p} \boldsymbol{x} \Longrightarrow \boldsymbol{x}_n \xrightarrow{p} \boldsymbol{x}$.

3.51 Show that $\boldsymbol{x}_n \xrightarrow{L^p} \boldsymbol{x} \Longrightarrow \boldsymbol{x}_n \xrightarrow{L^q} \boldsymbol{x}$ for any $p > q \geq 1$. Conclude that convergence in mean-square (for which $p = 2$) implies convergence in mean (for which $p = 1$).

3.52 A biased die is rolled once, resulting in $\mathbb{P}(\text{odd}) = p$ and $\mathbb{P}(\text{even}) = 1 - p$. A sequence of random variables $\{\boldsymbol{x}_n\}$ is constructed as follows:

$$\boldsymbol{x}_n = \begin{cases} \dfrac{2n^2}{n^2 + 1/2}, & \text{when the die roll is even} \\ 2\cos(\pi n), & \text{when the die roll is odd} \end{cases}$$

Verify that $\mathbb{P}(\lim_{n\to\infty} \boldsymbol{x}_n = 2) = 1 - p$. Does the sequence $\{\boldsymbol{x}_n\}$ converge when the result of the die roll is odd?

3.53 Consider the random sequence

$$\boldsymbol{x}_n = \begin{cases} \mu_1, & \text{with probability } 1 - 1/n \\ \mu_2, & \text{with probability } 1/n \end{cases}$$

(a) Show that $\boldsymbol{x}_n$ converges in the mean-square sense to $\boldsymbol{x} = \mu_1$. Does it also converge almost surely to the same limit?

(b) What happens if we change the probabilities to $1 - (1/2)^n$ and $(1/2)^n$?

3.54 Let $\{\boldsymbol{x}_n,\ n = 1, \ldots, N\}$ denote N independent scalar random variables with mean μ, with each variable satisfying $a_n \leq \boldsymbol{x}_n \leq b_n$. Let $\boldsymbol{S}_N = \sum_{n=1}^{N} \boldsymbol{x}_n$ denote the sum of these random variables. Let $\Delta = \sum_{n=1}^{N}(b_n - a_n)^2$ denote the sum of the squared lengths of the respective intervals. A famous inequality known as *Hoeffding inequality* is derived in Appendix 3.B; it asserts that for any $\delta > 0$:

$$\mathbb{P}\Big(|\boldsymbol{S}_N - \mathbb{E}\,\boldsymbol{S}_N| \geq \delta\Big) \leq 2e^{-2\delta^2/\Delta}$$

Introduce the sample average $\widehat{\boldsymbol{\mu}}_N = \frac{1}{N}\sum_{n=1}^{N} \boldsymbol{x}_n$, and assume the bounds $a_n = a$ and $b_n = b$ are uniform over n so that $a \leq \boldsymbol{x}_n \leq b$. Use Hoeffding inequality to justify the following bound, which is independent of the unknown μ:

$$\mathbb{P}\Big(|\widehat{\boldsymbol{\mu}}_N - \mu| \geq \epsilon\Big) \leq 2e^{-2\epsilon^2 N/(b-a)}$$

for any $\epsilon > 0$. Conclude the validity of the weak law of large numbers, namely, the fact that the sample average converges in probability to the actual mean as $N \to \infty$.

3.55 We continue with the setting of Prob. 3.54. Let $\{\boldsymbol{x}_n,\ n = 1, \ldots, N\}$ denote N independent scalar random variables, with each variable lying within the interval $\boldsymbol{x}_n \in [a_n, b_n]$. Introduce the sample mean:

$$\bar{\boldsymbol{x}} \stackrel{\Delta}{=} \frac{1}{N}\sum_{n=1}^{N} \boldsymbol{x}_n$$

(a) Assume $a_n = 0$ and $b_n = 1$, where all random variables lie within the interval $[0, 1]$. Use Hoeffding inequalities (3.232a)–(3.232b) to show that

$$\mathbb{P}\Big(\bar{\boldsymbol{x}} - \mathbb{E}\,\bar{\boldsymbol{x}} \geq \delta\Big) \leq e^{-2N\delta^2}, \quad \mathbb{P}\Big(|\bar{\boldsymbol{x}} - \mathbb{E}\,\bar{\boldsymbol{x}}| \geq \delta\Big) \leq 2e^{-2N\delta^2}$$

(b) Assume we wish to ensure that the likelihood (confidence level) of the sample mean $\bar{\boldsymbol{x}}$ lying within the interval $[\mathbb{E}\,\bar{\boldsymbol{x}} - \delta,\ \mathbb{E}\,\bar{\boldsymbol{x}} + \delta]$ is $1 - \alpha$, for some small significance level α. Show that the number of samples needed to ensure this property is bounded by $N \leq \ln(2/\alpha)/2\delta^2$.

3.56 Consider scalar random variables $\boldsymbol{x}_n \in [0,1]$ and their zero-mean centered versions denoted by $\boldsymbol{x}_{c,n} = \boldsymbol{x}_n - \mathbb{E}\,\boldsymbol{x}_n$. Use the Hoeffding lemma (3.233) to establish the following result, which provides a bound on the expectation of the maximum of a collection of centered random variables:

$$\mathbb{E}\left(\max_{1\leq n\leq N}\left\{\boldsymbol{x}_{c,1},\, \boldsymbol{x}_{c,2}, \ldots,\, \boldsymbol{x}_{c,N}\right\}\right) \leq \sqrt{\frac{1}{2}\ln N}$$

3.57 Consider two scalar random variables $\{\boldsymbol{y}, \boldsymbol{z}\}$ satisfying $\mathbb{E}\,(\boldsymbol{y}|\boldsymbol{z}) = 0$. Assume there exists a function $f(\boldsymbol{z})$ and some constant $c \geq 0$ such that $f(\boldsymbol{z}) \leq \boldsymbol{y} \leq f(\boldsymbol{z}) + c$. Extend the derivation of the Hoeffding lemma (3.233) to verify that the following result also holds for any t:

$$\mathbb{E}\,(e^{t\boldsymbol{y}}|\boldsymbol{z}) \;\leq\; e^{t^2c^2/8}$$

3.58 Consider the problem of multiplying two $N \times N$ matrices A and B to generate $C = AB$. For big data problems, the size of N can be prohibitively largely. Let $\{a_n\}$ denote the $N \times 1$ columns of A, and let $\{b_n^{\mathsf{T}}\}$ denote the $1 \times N$ rows of B. Then, C is a sum of N rank-one products of the form

$$C = \sum_{n=1}^{N} a_n b_n^{\mathsf{T}} \tag{3.214}$$

One simple approximation for computing C employs a *randomized algorithm* and is based on selecting at random R rank-one factors. Indeed, let p_n denote a discrete probability distribution over the indices $1 \leq n \leq N$ with $\sum_{n=1}^{N} p_n = 1$. Select R independent integer indices r from the range $[1, N]$ with $\mathbb{P}(r = n) = p_n$. Denote the set of selected indices by $\mathcal{R}$ and set

$$\widehat{\boldsymbol{C}} = \sum_{r\in\mathcal{R}} \frac{1}{p_r} a_r b_r^{\mathsf{T}}$$

Verify that $\mathbb{E}\,\widehat{\boldsymbol{C}} = C$.

3.59 Consider the same setting of Prob. 3.58. We wish to select the sampling probabilities in order to minimize the mean-square error of the difference between the approximation $\widehat{\boldsymbol{C}}$ and the product AB:

$$\{p_n^o\} \;=\; \underset{\{p_n\}}{\text{argmin}}\; \mathbb{E}\,\|\widehat{\boldsymbol{C}} - AB\|_{\mathrm{F}}^2$$

Let $\|x\|$ denote the Euclidean norm of vector x. Show that the optimal probabilities are given by

$$p_n^o = \frac{\|a_n\|\;\|b_n\|}{\sum_{m=1}^{N} \|a_m\|\;\|b_m\|}$$

Remark. For additional details, the reader may refer to Drineas, Kannan, and Mahoney (2006).

3.60 A drunk person wanders randomly, moving either 10 m to the right or 5 m to the left every minute. Where do you expect to find the person after one hour? Find the expected location and the corresponding standard deviation.

3.61 A particle wanders on average 1 nm every 1 ps with velocity 1000 cm/s. What is the value of the probability p? What is the value of the diffusion coefficient?

3.62 A particle wanders on average 1 nm every 1 ps with velocity -1000 cm/s (the negative sign means that the velocity is in the negative direction of the x-axis). What is the value of the probability p in this case? What is the value of the diffusion coefficient? How does it compare to the case $v = +1000$ cm/s?

3.63 What is the diffusion coefficient of a particle of radius 1 nm diffusing in water?

3.64 If a particle of radius R takes t seconds to wander a distance L, how long does it take a particle of radius $R/2$ to wander the same distance?

3.A CONVERGENCE OF RANDOM VARIABLES

There are several notions of convergence for sequences of random variables, such as convergence in probability, convergence in distribution, convergence in mean, mean-square convergence, and almost-sure convergence (or convergence with probability 1). We review them briefly here for ease of reference. The different notions of convergence vary in how they measure "closeness" between random variables. For additional information and proofs for some of the statements, including illustrative examples and problems, the reader may refer to Feller (1968, 1971), Rohatgi (1976), Billingsley (1986, 1999), Karr (1993), Davidson (1994), Durrett (1996), van der Vaart (2000), Grimmett and Stirzaker (2001), and Dudley (2002).

Convergence in distribution

Consider a sequence of scalar random variables $\{\boldsymbol{x}_n\}$ indexed by the integer n. Each variable is described by a pdf, $f_{\boldsymbol{x}_n}(x)$. By referring to a "sequence $\{\boldsymbol{x}_n\}$" we mean that at each n, the realization for $\boldsymbol{x}_n$ arises from its pdf and the collection of these realizations will constitute the sequence. Consider further a separate random variable $\boldsymbol{x}$ with pdf $f_{\boldsymbol{x}}(x)$. The weakest notion of convergence is *convergence in distribution* (also called weak convergence). Let $F_{\boldsymbol{x}}(x)$ denote the cumulative distribution function (cdf) of the random variable $\boldsymbol{x}$; this is defined as the area under the pdf of $\boldsymbol{x}$ up to location $\boldsymbol{x} = x$:

$$F_{\boldsymbol{x}}(x) \triangleq \int_{-\infty}^{x} f_{\boldsymbol{x}}(x')dx' \;=\; \mathbb{P}(\boldsymbol{x} \leq x) \tag{3.215}$$

Similarly, let $F_{\boldsymbol{x}_n}(x)$ denote the cdf for each $\boldsymbol{x}_n$. The sequence $\{\boldsymbol{x}_n\}$ is said to converge in distribution to the random variable $\boldsymbol{x}$ if the respective cdfs approach each other for large n at all points where $F_{\boldsymbol{x}}(x)$ is continuous, i.e.,

(convergence in distribution I)

$$\boldsymbol{x}_n \xrightarrow{d} \boldsymbol{x} \iff \lim_{n\to\infty} F_{\boldsymbol{x}_n}(x) \;=\; F_{\boldsymbol{x}}(x), \;\; \{\forall\, x \,|\, F_{\boldsymbol{x}}(x) \text{ continuous}\} \tag{3.216}$$

Convergence in distribution is also denoted by the notation $\boldsymbol{x}_n \rightsquigarrow \boldsymbol{x}$. When convergence occurs, then the cdf $F_{\boldsymbol{x}}(x)$ is uniquely defined. This type of convergence depends only on the cdfs (not the actual values of the random variables) and it ensures that, for large n, the likelihoods that $\boldsymbol{x}_n$ and $\boldsymbol{x}$ lie within the same interval are essentially the same:

$$\mathbb{P}(a \leq \boldsymbol{x}_n \leq b) \approx \mathbb{P}(a \leq \boldsymbol{x} \leq b), \quad \text{for large } n \tag{3.217}$$

It also follows from $\boldsymbol{x}_n \rightsquigarrow \boldsymbol{x}$ that, for any continuous function $g(x)$, the sequence $g(\boldsymbol{x}_n)$ converges in distribution to the random variable $g(\boldsymbol{x})$. This result is known as the *continuous mapping theorem.* There are several other equivalent characterizations of convergence in distribution. One useful characterization motivated by the above remark is the following:

(convergence in distribution II)

$$\boldsymbol{x}_n \xrightarrow{d} \boldsymbol{x} \iff \lim_{n\to\infty} \mathbb{P}(\boldsymbol{x}_n \leq x) = \mathbb{P}(\boldsymbol{x} \leq x) \tag{3.218}$$
$$\{\forall\, x \,|\, \mathbb{P}(\boldsymbol{x} \leq x) \text{ continuous}\}$$

A second statement is that

(**convergence in distribution III**)

$$\boldsymbol{x}_n \xrightarrow{d} \boldsymbol{x} \iff \lim_{n\to\infty} \mathbb{E}\, g(\boldsymbol{x}_n) = \mathbb{E}\, g(\boldsymbol{x}) \tag{3.219}$$
$$\forall\, g(x)\text{: bounded and continuous or Lipschitz}$$

where a Lipschitz function is one for which $|g(a) - g(b)| \leq \delta|a - b|$ for all a, b and for some $\delta > 0$. The central limit theorem discussed later in (4.159) is one of the most famous and useful results on convergence in distribution. It is important to note though that convergence in distribution *does not* generally imply convergence of the respective pdfs (i.e., $f_{\boldsymbol{x}_n}(x)$ need not converge to $f_{\boldsymbol{x}}(x)$). Counterexamples can be found to this effect.

Convergence in probability

The second notion we consider is *convergence in probability*, which implies convergence in distribution (the converse is true only when $\boldsymbol{x}$ is the constant random variable). The sequence $\{\boldsymbol{x}_n\}$ is said to converge in probability to the random variable $\boldsymbol{x}$ if there is a high probability that the distance $|\boldsymbol{x}_n - \boldsymbol{x}|$ becomes very small for large n, i.e.,

(**convergence in probability**)

$$\boldsymbol{x}_n \xrightarrow{p} \boldsymbol{x} \iff \lim_{n\to\infty} \mathbb{P}(|\boldsymbol{x}_n - \boldsymbol{x}| \geq \epsilon) = 0, \quad \text{for any } \epsilon > 0 \tag{3.220a}$$
$$\iff \lim_{n\to\infty} \mathbb{P}(|\boldsymbol{x}_n - \boldsymbol{x}| < \epsilon) = 1 \tag{3.220b}$$

This definition is essentially dealing with the convergence of a sequence of probabilities. Note that checking the condition on the right-hand side requires knowledge of the joint distribution of the variables $\{\boldsymbol{x}_n, \boldsymbol{x}\}$. For random vectors $\{\boldsymbol{x}_n\}$, we would replace $|\boldsymbol{x}_n - \boldsymbol{x}|$ in terms of the Euclidean distance $\|\boldsymbol{x}_n - \boldsymbol{x}\|$. Interestingly, although convergence in probability does not imply the stronger notion of almost-sure convergence defined below, it can be shown that convergence in probability of a sequence $\{\boldsymbol{x}_n\}$ to $\boldsymbol{x}$ implies the existence of a subsequence $\{\boldsymbol{x}_{k_n}\}$ that converges almost surely to $\boldsymbol{x}$. The above notion of convergence in probability ensures that, in the limit, $\boldsymbol{x}_n$ will lie with high probability within the disc centered at $\boldsymbol{x}$ and of radius ϵ. The result still does not guarantee "point-wise" convergence of $\boldsymbol{x}_n$ to $\boldsymbol{x}$. That is what almost-sure convergence does.

Almost-sure convergence

The third and strongest notion we consider is *almost-sure convergence*; this implies the other two notions – see Fig. 3.11. The sequence $\{\boldsymbol{x}_n\}$ is said to converge almost surely (or with probability 1) to the random variable $\boldsymbol{x}$ if there is a high probability that $\boldsymbol{x}_n$ approaches $\boldsymbol{x}$ for large n, i.e.,

(**almost-sure convergence**)

$$\boldsymbol{x}_n \xrightarrow{a.s.} \boldsymbol{x} \iff \mathbb{P}\Big(\lim_{n\to\infty} \boldsymbol{x}_n = \boldsymbol{x}\Big) = 1 \tag{3.221}$$

This statement guarantees the convergence of $\boldsymbol{x}_n$ to $\boldsymbol{x}$ except possibly over a set of "measure zero" – see, e.g., Prob. 3.47. In this problem, we construct a particular sequence $\boldsymbol{x}_n$ and define a separate random variable $\boldsymbol{x}$ that is uniformly distributed over the interval $[0, \frac{1}{2}]$. We then verify that $\boldsymbol{x}_n$ converges to $\boldsymbol{x}$ for all points in the semi-open interval $[0, \frac{1}{2})$ but not at the location $x = 1/2$. Since this is a singleton and the probability of $\boldsymbol{x}$ assuming values in the interval $[0, \frac{1}{2})$ is still equal to 1, we are able to conclude that $\boldsymbol{x}_n$ converges almost surely to $\boldsymbol{x}$. This example clarifies the reason for the qualification "almost-surely" since some points in the domain of $\boldsymbol{x}$ may be excluded (i.e., convergence occurs for almost all points). It is not always straightforward to check for almost-sure convergence by applying the definition (3.221). One useful *sufficient* condition is to verify that for any $\epsilon > 0$:

$$\sum_{n=1}^{\infty} \mathbb{P}(|\boldsymbol{x}_n - \boldsymbol{x}| > \epsilon) < \infty \implies \text{almost-sure convergence} \tag{3.222}$$

The *strong law of large numbers* is one of the most famous results illustrating almost-sure convergence. Consider a collection of iid random variables $\{\boldsymbol{x}_n\}$ with mean μ each and bounded absolute first-order moment, $\mathbb{E}\,|\boldsymbol{x}_n| < \infty$. The strong law states that the sample average estimator defined by

$$\widehat{\boldsymbol{\mu}}_N \triangleq \frac{1}{N}\sum_{n=1}^{N} \boldsymbol{x}_n \tag{3.223}$$

converges almost surely to the true mean μ as $N \to \infty$, i.e.,

(**strong law of large numbers**)

$$\widehat{\boldsymbol{\mu}}_N \xrightarrow{a.s.} \mu \iff \mathbb{P}\Big(\lim_{N\to\infty} \widehat{\boldsymbol{\mu}}_N = \mu\Big) = 1 \tag{3.224}$$

In other words, as the number of samples increases, the likelihood that $\widehat{\boldsymbol{\mu}}_N$ will converge to the true value μ tends to 1. In addition, it is possible to specify the rate of convergence of $\widehat{\boldsymbol{\mu}}_N$ toward μ as $N \to \infty$. If the variables $\boldsymbol{x}_n$ have uniform and finite variance, $\mathbb{E}\,(\boldsymbol{x}_n - \mu)^2 = \sigma_x^2 < \infty$, then it is further known that (see, e.g., Durrett (1996, p. 437)):

$$\limsup_{N\to\infty} \left\{ \frac{\widehat{\boldsymbol{\mu}}_N - \mu}{\sigma_x} \times \frac{N^{1/2}}{(2\ln\ln N)^{1/2}} \right\} = 1, \text{ almost surely} \tag{3.225}$$

which indicates that, for large enough N, the difference between $\widehat{\boldsymbol{\mu}}_N$ and μ is on the order of:

$$\widehat{\boldsymbol{\mu}}_N - \mu = O\left(\sqrt{\frac{2\sigma_x^2 \ln\ln N}{N}}\right) \tag{3.226}$$

using the Big-O notation. This notation will be used frequently in our treatment to compare the asymptotic convergence rates of two sequences. Thus, writing $a_n = O(b_n)$ for two sequences $\{a_n, b_n, n \geq 0\}$, with b_n having positive entries, means that there exists some constant $c > 0$ and index n_o such that $|a_n| \leq cb_n$ for all $n > n_o$. This also means that the decay rate of the sequence a_n is at least as fast or faster than b_n. For example, writing $a_n = O(1/n)$ means that the samples of the sequence a_n decay asymptotically at a rate that is comparable to or faster than $1/n$.

The weak version of the law of large numbers is studied in Prob. 3.54; it only ensures convergence in probability, namely, for any $\epsilon > 0$:

(**weak law of large numbers**)

$$\widehat{\boldsymbol{\mu}}_N \xrightarrow{p} \mu \iff \lim_{N\to\infty} \mathbb{P}\Big(|\widehat{\boldsymbol{\mu}}_N - \mu| \geq \epsilon\Big) = 0 \tag{3.227}$$

Remark 3.1. (Inference problems) The weak and strong laws of large numbers provide one useful example of how sequences of random variables arise in inference problems. In future chapters we will describe methods that construct estimators, say, $\widehat{\boldsymbol{\theta}}_n$, for some unknown parameter $\boldsymbol{\theta}$, where n denotes the number of data points used to determine $\widehat{\boldsymbol{\theta}}_n$. We will then be interested in analyzing how well the successive estimators $\widehat{\boldsymbol{\theta}}_n$ approach $\boldsymbol{\theta}$ for increased sample sizes n. In these studies, the notions of convergence in the mean, mean-square, in distribution, and in probability will be very helpful.

■

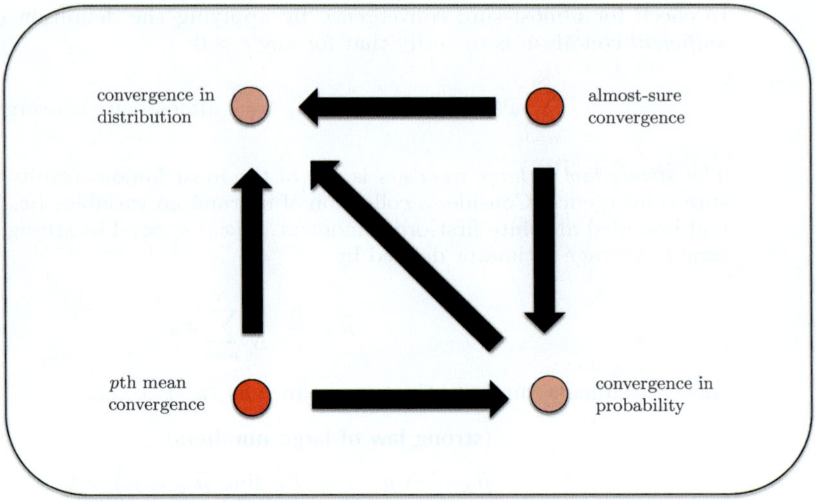

Figure 3.11 Both notions of almost-sure and pth mean convergence are stronger than convergence in probability, which in turn is stronger than convergence in distribution. The direction of an arrow from location A to B means that notion A implies notion B.

Convergence in the pth mean

The last and also strong notion of convergence we consider is convergence in the pth mean; it is stronger than convergence in probability and convergence in distribution – see Fig. 3.11. Consider an exponent $p \geq 1$ and assume the p-th moments of the variables $\boldsymbol{x}_n$ are bounded, i.e., $\mathbb{E}\,|\boldsymbol{x}_n|^p < \infty$. Then, the sequence $\{\boldsymbol{x}_n\}$ is said to converge in the pth mean to the random variable $\boldsymbol{x}$ if

(**convergence in pth mean**)

$$\boldsymbol{x}_n \xrightarrow{L^p} \boldsymbol{x} \iff \lim_{n\to\infty} \mathbb{E}\,|\boldsymbol{x}_n - \boldsymbol{x}|^p = 0 \tag{3.228}$$

The notation L^p above an arrow is used to refer to this notion of convergence. Two special cases are common: $p = 2$ corresponds to mean-square convergence and $p = 1$ corresponds to convergence in the mean. It is easy to verify that convergence in the pth mean implies convergence in probability – see Prob. 3.50.

3.B CONCENTRATION INEQUALITIES

The Markov and Chebyshev bounds are examples of *concentration inequalities*, which help bound the deviation of a random variable (or combinations of random variables) away from certain values (typically their means):

$$\mathbb{P}(\boldsymbol{x} \geq \alpha) \leq \mathbb{E}\,\boldsymbol{x}/\alpha, \quad \boldsymbol{x} \geq 0, \quad \alpha > 0 \qquad (\textbf{Markov inequality}) \tag{3.229a}$$

$$\mathbb{P}(|\boldsymbol{x} - \mathbb{E}\,\boldsymbol{x}| \geq \delta) \;\leq\; \sigma_x^2/\delta^2, \;\; \delta > 0 \qquad (\textbf{Chebyshev inequality}) \tag{3.229b}$$

In this appendix we describe three other famous inequalities, Azuma inequality, Hoeffding inequality, and McDiarmid inequality, which provide bounds on the probability of the sum of a collection of random variables deviating from their mean.

The Hoeffding and McDiarmid inequalities play an important role in the analysis of learning algorithms and will be used in the derivation of generalization bounds in Chapter 64. There are of course other concentration inequalities but we will limit our discussion to those that are most relevant to our treatment in this text.

Hoeffding inequality

We first establish Hoeffding inequality from Prob. 3.54 and the supporting Hoeffding lemma, motivated by arguments from Hoeffding (1963), Serfling (1974), Boucheron, Lugosi, and Bousquet (2004), Massart (2007), Boucheron, Lugosi, and Massart (2013), Mohri, Rostamizadeh, and Talwalkar (2018), Vershynin (2018), and Wainwright (2019).

Hoeffding inequality (Hoeffding (1963)). *Let $\{\boldsymbol{x}_n,\ n=1,\ldots,N\}$ denote N independent scalar random variables, with each variable lying within an interval of the form $a_n \le \boldsymbol{x}_n \le b_n$, with endpoints denoted by $\{a_n, b_n\}$. Let*

$$\boldsymbol{S}_N \triangleq \sum_{n=1}^{N} \boldsymbol{x}_n, \qquad \Delta \triangleq \sum_{n=1}^{N} (b_n - a_n)^2 \tag{3.230}$$

denote the sum of the random variables and the sum of the squared lengths of their respective intervals. The Hoeffding inequality asserts that for any $\delta > 0$:

$$\mathbb{P}\Big(\boldsymbol{S}_N - \mathbb{E}\,\boldsymbol{S}_N \ge \delta \Big) \le e^{-2\delta^2/\Delta} \tag{3.231a}$$

$$\mathbb{P}\Big(\Big|\boldsymbol{S}_N - \mathbb{E}\,\boldsymbol{S}_N\Big| \ge \delta \Big) \le 2e^{-2\delta^2/\Delta} \tag{3.231b}$$

The above inequalities can also be restated in terms of sample means as opposed to sums. It is straightforward to verify that

$$\mathbb{P}\left(\frac{1}{N}\sum_{n=1}^{N} \boldsymbol{x}_n - \frac{1}{N}\sum_{n=1}^{N} \mathbb{E}\,\boldsymbol{x}_n \ge \delta \right) \le e^{-2N^2\delta^2/\Delta} \tag{3.232a}$$

$$\mathbb{P}\left(\left| \frac{1}{N}\sum_{n=1}^{N} \boldsymbol{x}_n - \frac{1}{N}\sum_{n=1}^{N} \mathbb{E}\,\boldsymbol{x}_n \right| \ge \delta \right) \le 2e^{-2N^2\delta^2/\Delta} \tag{3.232b}$$

One of the tools we employ to establish the inequalities is the Hoeffding lemma, which is stated next along with one traditional derivation – see, e.g., Boucheron, Lugosi, and Bousquet (2004), Chung and Lu (2006a, b), Alon and Spencer (2008), Boucheron, Lugosi, and Massart (2013), and Mohri, Rostamizadeh, and Talwalkar (2018).

Hoeffding lemma (Hoeffding (1963)). *Consider a scalar random variable $\boldsymbol{y} \in [a, b]$. Then for any t we have*

$$\mathbb{E}\, e^{t(\boldsymbol{y} - \mathbb{E}\,\boldsymbol{y})} \le e^{t^2(b-a)^2/8} \tag{3.233}$$

Proof: We start by noting that the exponential function $f(x) = e^x$ is convex and, therefore, it holds that

$$e^{t\boldsymbol{y}} \le \left(\frac{b-\boldsymbol{y}}{b-a}\right) e^{ta} + \left(\frac{\boldsymbol{y}-a}{b-a}\right) e^{tb} \tag{3.234}$$

where the nonnegative coefficients $(b-\boldsymbol{y})/(b-a)$ and $(\boldsymbol{y}-a)/(b-a)$ add up to 1. It follows that

$$
\begin{aligned}
e^{-t\mathbb{E}\,\boldsymbol{y}}\,\mathbb{E}\,e^{t\boldsymbol{y}} &\leq e^{-t\mathbb{E}\,\boldsymbol{y}} \times \left\{\left(\frac{b-\mathbb{E}\,\boldsymbol{y}}{b-a}\right)e^{ta} + \left(\frac{\mathbb{E}\,\boldsymbol{y}-a}{b-a}\right)e^{tb}\right\} \\
&= e^{-t\mathbb{E}\,\boldsymbol{y}} \times e^{ta} \times \left\{1-\frac{\mathbb{E}\,\boldsymbol{y}-a}{b-a}+\frac{\mathbb{E}\,\boldsymbol{y}-a}{b-a}e^{t(b-a)}\right\} \\
&= \exp\left\{-t\,(b-a)\,\frac{\mathbb{E}\,\boldsymbol{y}-a}{b-a}\right\} \times \left\{1-\frac{\mathbb{E}\,\boldsymbol{y}-a}{b-a}+\left(\frac{\mathbb{E}\,\boldsymbol{y}-a}{b-a}\right)e^{t(b-a)}\right\} \\
&= \exp\left\{-t\,(b-a)\,\frac{\mathbb{E}\,\boldsymbol{y}-a}{b-a}+\ln\left(1-\frac{\mathbb{E}\,\boldsymbol{y}-a}{b-a}+\frac{\mathbb{E}\,\boldsymbol{y}-a}{b-a}e^{t(b-a)}\right)\right\} \\
&\triangleq \exp\left\{-hp+\ln\left(1-p+pe^{h}\right)\right\} \\
&\triangleq e^{L(h)}
\end{aligned}
\tag{3.235}
$$

where we introduced the quantities:

$$
h \triangleq t(b-a) \geq 0, \qquad p \triangleq \frac{\mathbb{E}\boldsymbol{y}-a}{b-a}, \qquad L(h) \triangleq -hp+\ln(1-p+pe^{h}) \tag{3.236}
$$

We denote the first- and second-order derivatives of $L(h)$ with respect to h by:

$$
L'(h) = -p + pe^{h}\frac{1}{1-p+pe^{h}} \tag{3.237a}
$$

$$
L''(h) = \frac{(1-p)pe^{h}}{(1-p+pe^{h})^2} \tag{3.237b}
$$

and note that $L(0) = L'(0) = 0$ and $L''(h) \leq 1/4$. This last inequality follows from the following equivalent statements:

$$
\begin{aligned}
\frac{(1-p)pe^{h}}{(1-p+pe^{h})^2} \leq \frac{1}{4} &\Longleftrightarrow 4(1-p)pe^{h} \leq (1-p+pe^{h})^2 \\
&\Longleftrightarrow 0 \leq (1-p-pe^{h})^2
\end{aligned}
\tag{3.238}
$$

and the fact that the last statement is obviously true. Now, since $h \geq 0$, we expand $L(h)$ around $h = 0$ and use the mean-value theorem to conclude that there exists a nonnegative value c between 0 and h such that

$$
\begin{aligned}
L(h) &= L(0) + L'(0)h + \frac{L''(c)}{2}h^2 \\
&\leq h^2/8 \\
&= t^2(b-a)^2/8
\end{aligned}
\tag{3.239}
$$

and, consequently,

$$
\mathbb{E}\,e^{t(\boldsymbol{y}-\mathbb{E}\,\boldsymbol{y})} \leq e^{L(h)} \leq e^{t^2(b-a)^2/8} \tag{3.240}
$$

as claimed.

■

We can now return to establish Hoeffding inequality (3.231b).

Proof of Hoeffding inequality (3.231b): To begin with, we note that for any positive scalar $s > 0$, it holds that:

$$\mathbb{P}\Big(\boldsymbol{S}_N - \mathbb{E}\,\boldsymbol{S}_N \geq \delta\Big) = \mathbb{P}\left(e^{s(\boldsymbol{S}_N - \mathbb{E}\,\boldsymbol{S}_N)} \geq e^{s\delta}\right)$$
$$\leq e^{-s\delta}\,\mathbb{E}\left(e^{s(\boldsymbol{S}_N - \mathbb{E}\,\boldsymbol{S}_N)}\right) \tag{3.241}$$

where we used Markov inequality from Prob. 3.17, namely, the fact that for any non-negative real-valued random variable $\boldsymbol{x}$, it holds that $\mathbb{P}(\boldsymbol{x} \geq \alpha) \leq \mathbb{E}\,\boldsymbol{x}/\alpha$. Now, from the definition of $\boldsymbol{S}_N$ in (3.230) and the independence of $\{\boldsymbol{x}(n)\}$ we get:

$$\begin{aligned} e^{-s\delta}\,\mathbb{E}\left(e^{s(\boldsymbol{S}_N - \mathbb{E}\,\boldsymbol{S}_N)}\right) &= e^{-s\delta}\,\mathbb{E}\left(e^{s\left(\sum_{n=1}^{N} \boldsymbol{x}_n - \mathbb{E}\,\boldsymbol{x}_n\right)}\right) \\ &= e^{-s\delta}\,\mathbb{E}\left(\prod_{n=1}^{N} e^{s(\boldsymbol{x}_n - \mathbb{E}\,\boldsymbol{x}_n)}\right) \\ &= e^{-s\delta} \prod_{n=1}^{N} \mathbb{E}\left(e^{s(\boldsymbol{x}_n - \mathbb{E}\,\boldsymbol{x}_n)}\right) \end{aligned} \tag{3.242}$$

so that

$$\mathbb{P}\Big(\boldsymbol{S}_N - \mathbb{E}\,\boldsymbol{S}_N \geq \delta\Big) \leq e^{-s\delta} \prod_{n=1}^{N} \mathbb{E}\left(e^{s(\boldsymbol{x}_n - \mathbb{E}\,\boldsymbol{x}_n)}\right) \tag{3.243}$$

To continue, we call upon result (3.233) from the Hoeffding lemma to conclude that

$$\begin{aligned} \mathbb{P}\Big(\boldsymbol{S}_N - \mathbb{E}\,\boldsymbol{S}_N \geq \delta\Big) &\leq e^{-s\delta} \prod_{n=1}^{N} e^{\frac{s^2}{8}(b_n - a_n)^2} \\ &= e^{-s\delta}\, e^{\frac{s^2}{8}\sum_{n=1}^{N}(b_n - a_n)^2} \\ &= e^{-s\delta}\, e^{\frac{s^2\Delta}{8}} \end{aligned} \tag{3.244}$$

We can tighten the upper bound by selecting the value of s that minimizes the exponent, $-s\delta + s^2\Delta/8$, which is given by $s = 4\delta/\Delta$. Therefore, we get

$$\mathbb{P}\Big(\boldsymbol{S}_N - \mathbb{E}\,\boldsymbol{S}_N \geq \delta\Big) \leq e^{-2\delta^2/\Delta} \tag{3.245}$$

Following similar arguments, we can also get

$$\mathbb{P}\Big(\boldsymbol{S}_N - \mathbb{E}\,\boldsymbol{S}_N \leq -\delta\Big) \;=\; \mathbb{P}\Big(-[\boldsymbol{S}_N - \mathbb{E}\,\boldsymbol{S}_N] \geq \delta\Big) \;\leq\; e^{-2\delta^2/\Delta} \tag{3.246}$$

We then arrive at

$$\begin{aligned} &\mathbb{P}\Big(|\boldsymbol{S}_N - \mathbb{E}\,\boldsymbol{S}_N| \geq \delta\Big) \\ &= \mathbb{P}\Big(\boldsymbol{S}_N - \mathbb{E}\,\boldsymbol{S}_N \leq -\delta\Big) + \mathbb{P}\Big(\boldsymbol{S}_N - \mathbb{E}\,\boldsymbol{S}_N \geq \delta\Big) \\ &\leq 2e^{-2\delta^2/\Delta} \end{aligned} \tag{3.247}$$

■

Azuma and McDiarmid inequalities

The Hoeffding inequalities (3.232a)–(3.232b) provide bounds for the deviation of the sample average function:

$$f(\boldsymbol{x}_1, \boldsymbol{x}_2, \ldots, \boldsymbol{x}_N) \;\stackrel{\Delta}{=}\; \frac{1}{N}\sum_{n=1}^{N} \boldsymbol{x}_n \tag{3.248}$$

away from its mean. The results can be extended to other functions with "*bounded variation.*" These are again functions of the N variables $\{\boldsymbol{x}_1, \boldsymbol{x}_2, \ldots, \boldsymbol{x}_N\}$, except that if any of the variables is changed, say, from $\boldsymbol{x}_m$ to $\boldsymbol{x}'_m$, then the variation in the function remains bounded:

(function with bounded variations)

$$\sup_{x_1,\ldots,x_N,x'_m} \left\{ \left| f(\boldsymbol{x}_{n\neq m}, \boldsymbol{x}_m) - f(\boldsymbol{x}_{n\neq m}, \boldsymbol{x}'_m) \right| \right\} \leq c_m, \;\; \forall\, m = 1, 2, \ldots, N \tag{3.249}$$

For such functions, the Hoeffding inequalities extend to the *McDiarmid inequalities* stated in (3.259a)–(3.259b) further ahead. These results were proven by McDiarmid (1989); see also Ledoux (2001), Chung and Lu (2006a, b), Alon and Spencer (2008), Boucheron, Lugosi, and Massart (2013), Mohri, Rostamizadeh, and Talwalkar (2018), and Wainwright (2019). Motivated by the presentation in these references, we provide one classical derivation that relies on the use of the Azuma inequality, which we motivate first.

Consider two sequences of random variables $\{\boldsymbol{y}_n, \boldsymbol{x}_n\}$ for $n \geq 1$, where each $\boldsymbol{y}_n$ is a function of $\{\boldsymbol{x}_1, \boldsymbol{x}_2, \ldots, \boldsymbol{x}_n\}$. We say that the sequence $\{\boldsymbol{y}_n\}$ is a *martingale difference* relative to the sequence $\{\boldsymbol{x}_n\}$ if the following property holds for every n:

$$\mathbb{E}\left(\boldsymbol{y}_n \,|\, \boldsymbol{x}_1, \boldsymbol{x}_2, \ldots, \boldsymbol{x}_{n-1}\right) = 0 \quad \textbf{(martingale difference)} \tag{3.250}$$

Azuma inequality (Azuma (1967)). *Let $\{\boldsymbol{y}_n,\ n \geq 1\}$ be a martingale difference relative to another sequence $\{\boldsymbol{x}_n,\ n \geq 1\}$, and assume there exist random variables $\{\boldsymbol{z}_n\}$ and constants $\{c_n\}$ such that $\boldsymbol{z}_n \leq \boldsymbol{y}_n \leq \boldsymbol{z}_n + c$ for all n. Let $\Delta' = \sum_{n=1}^N c_n^2$. The Azuma inequality asserts that for any $\delta > 0$:*

$$\mathbb{P}\Big(\sum_{n=1}^N \boldsymbol{y}_n \geq \delta\Big) \leq e^{-2\delta^2/\Delta'} \tag{3.251a}$$

$$\mathbb{P}\Big(\sum_{n=1}^N \boldsymbol{y}_n \leq -\delta\Big) \leq e^{-2\delta^2/\Delta'} \tag{3.251b}$$

Proof: It is sufficient to establish one of the inequalities. We follow an argument similar to Chung and Lu (2006b) and Mohri, Rostamizadeh, and Talwalkar (2018). Introduce the random variable $\boldsymbol{S}_N = \sum_{n=1}^N \boldsymbol{y}_n$, which satisfies $\boldsymbol{S}_N = \boldsymbol{S}_{N-1} + \boldsymbol{y}_N$. For any $s > 0$ we have

$$\begin{aligned}
\mathbb{P}(\boldsymbol{S}_N \geq \delta) &= \mathbb{P}(e^{s\boldsymbol{S}_N} \geq e^{s\delta}) \\
&\leq \frac{\mathbb{E}\, e^{s\boldsymbol{S}_N}}{e^{s\delta}} \qquad \text{(using Markov inequality from Prob. 3.17)} \\
&= e^{-s\delta} \times \mathbb{E}\, e^{s(\boldsymbol{S}_{N-1}+\boldsymbol{y}_N)} \\
&= e^{-s\delta} \times \mathbb{E}\left\{ \mathbb{E}\left(e^{s(\boldsymbol{S}_{N-1}+\boldsymbol{y}_N)} \,|\, \boldsymbol{x}_1, \ldots, \boldsymbol{x}_{N-1} \right) \right\} \\
&\overset{(a)}{=} e^{-s\delta} \times \mathbb{E}\left\{ e^{s\boldsymbol{S}_{N-1}}\, \mathbb{E}\left(e^{s\boldsymbol{y}_N} \,|\, \boldsymbol{x}_1, \ldots, \boldsymbol{x}_{N-1} \right) \right\} \\
&\overset{(b)}{\leq} e^{-s\delta} \times \mathbb{E}\, e^{s\boldsymbol{S}_{N-1}} \times e^{s^2 c_N^2/8}
\end{aligned} \tag{3.252}$$

where step (a) is because $\boldsymbol{S}_{N-1}$ is solely a function of $\{\boldsymbol{x}_1, \boldsymbol{x}_2, \ldots, \boldsymbol{x}_{N-1}\}$, and step (b) uses the result of Prob. 3.57. We therefore arrive at the inequality recursion:

$$\mathbb{E}\, e^{s\boldsymbol{S}_N} \leq \mathbb{E}\, e^{s\boldsymbol{S}_{N-1}} \times e^{s^2 c_N^2/8} \tag{3.253}$$

Iterating starting from $\boldsymbol{S}_0 = 0$ we get

$$\mathbb{E}\, e^{s\boldsymbol{S}_N} \leq e^{\sum_{n=1}^{N} s^2 c_n^2} \tag{3.254}$$

and, hence,

$$\mathbb{P}(\boldsymbol{S}_N \geq \delta) \leq e^{-s\delta} \times e^{s^2 \sum_{n=1}^{N} c_n^2/8} \tag{3.255}$$

We can minimize the upper bound over s and select

$$s = 4\delta \Big/ \sum_{n=1}^{N} c_n^2/8 \tag{3.256}$$

Substituting into the right-hand side of (3.255) gives

$$\mathbb{P}(\boldsymbol{S}_N \geq \delta) \leq e^{-2\delta^2/\sum_{n=1}^{N} c_n^2} \tag{3.257}$$

and the desired result follows. ∎

We are now ready to state the McDiarmid inequality.

McDiarmid inequality (McDiarmid (1989)). *Let $\{\boldsymbol{x}_n,\ n = 1, \ldots, N\}$ denote N independent scalar random variables, and let $f(\boldsymbol{x}_1, \boldsymbol{x}_2, \ldots, \boldsymbol{x}_N)$ be any function with bounded variation as in (3.249). Let*

$$\Delta' \triangleq \sum_{m=1}^{N} c_m^2 \tag{3.258}$$

The McDiarmid inequality asserts that for any $\delta > 0$:

$$\mathbb{P}\Big(f(\boldsymbol{x}_1, \ldots, \boldsymbol{x}_N) - \mathbb{E}\, f(\boldsymbol{x}_1, \ldots, \boldsymbol{x}_N) \geq \delta \Big) \leq e^{-2\delta^2/\Delta'} \tag{3.259a}$$

$$\mathbb{P}\Big(|f(\boldsymbol{x}_1, \ldots, \boldsymbol{x}_N) - \mathbb{E}\, f(\boldsymbol{x}_1, \ldots, \boldsymbol{x}_N)| \geq \delta \Big) \leq 2e^{-2\delta^2/\Delta'} \tag{3.259b}$$

Proof: It is sufficient to establish one of the inequalities. Introduce the following random variables, where the expression for $\boldsymbol{y}_n$ is written in two equivalent forms:

$$\boldsymbol{S} \triangleq f(\boldsymbol{x}_1, \boldsymbol{x}_2, \ldots, \boldsymbol{x}_N) - \mathbb{E}\, f(\boldsymbol{x}_1, \boldsymbol{x}_2, \ldots, \boldsymbol{x}_N) \tag{3.260a}$$

$$\begin{aligned} \boldsymbol{y}_n &\triangleq \mathbb{E}\Big(\boldsymbol{S} \,|\, \boldsymbol{x}_1, \boldsymbol{x}_2, \ldots, \boldsymbol{x}_n\Big) - \mathbb{E}\Big(\boldsymbol{S} \,|\, \boldsymbol{x}_1, \boldsymbol{x}_2, \ldots, \boldsymbol{x}_{n-1}\Big), \quad n \geq 1 \\ &= \mathbb{E}\Big(f(\boldsymbol{x}_1, \ldots, \boldsymbol{x}_N) \,|\, \boldsymbol{x}_1, \boldsymbol{x}_2, \ldots, \boldsymbol{x}_n\Big) - \mathbb{E}\Big(f(\boldsymbol{x}_1, \ldots, \boldsymbol{x}_N) \,|\, \boldsymbol{x}_1, \boldsymbol{x}_2, \ldots, \boldsymbol{x}_{n-1}\Big) \end{aligned} \tag{3.260b}$$

It is clear that $\boldsymbol{S} = \sum_{n=1}^{N} \boldsymbol{y}_n$ and $\mathbb{E}\,(\boldsymbol{y}_n|\boldsymbol{x}_1, \ldots, \boldsymbol{x}_{n-1}) = 0$. The latter result shows that the sequence $\{\boldsymbol{y}_n\}$ defines a martingale difference relative to the sequence $\{\boldsymbol{x}_n\}$. Moreover, the bounded variation property on the function $f(\cdot)$ translates into bounds on each $\boldsymbol{y}_n$ as follows. Let

$$a_n \triangleq \inf_x \left\{ \mathbb{E}\Big(f(\boldsymbol{x}_1, \ldots, \boldsymbol{x}_N) \,|\, \boldsymbol{x}_1, \ldots, \boldsymbol{x}_{n-1}, x\Big) - \mathbb{E}\Big(f(\boldsymbol{x}_1, \ldots, \boldsymbol{x}_N) \,|\, \boldsymbol{x}_1, \ldots, \boldsymbol{x}_{n-1}\Big) \right\} \tag{3.261}$$

Then, each $\boldsymbol{y}_n$ satisfies $a_n \leq \boldsymbol{y}_n \leq a_n + c_n$. We can now apply the Azuma inequality (3.251a) to get (3.259a).

■

REFERENCES

Alon, N. and J. H. Spencer (2008), *The Probabilistic Method*, 3rd ed., Wiley.

Ash, R. B. (2008), *Basic Probability Theory*, Dover Publications.

Azuma, K. (1967), "Weighted sums of certain dependent random variables," *Tôhoku Math. J.*, vol. 19, no. 3, pp. 357–367.

Bass, R. F. (2011), *Stochastic Processes*, Cambridge University Press.

Bayes, T. and R. Price (1763), "An essay towards solving a problem in the doctrine of chances," Bayes's article communicated by R. Price and published posthumously in *Philos. Trans. Roy. Soc. Lond.*, vol. 53, pp. 370–418.

Berg, H. C. (1993), *Random Walks in Biology*, expanded ed., Princeton University Press.

Bernoulli, J. (1713), *Ars Conjectandi*, chapter 4, Thurneysen Brothers. The book was published eight years after the author's death. English translation by E. Sylla available as *The Art of Conjecturing*, Johns Hopkins University Press.

Bernshtein, S. N. (1945), "Chebyshev's work on the theory of probability," in *The Scientific Legacy of P. L. Chebyshev* (in Russian), Akademiya Nauk SSSR, pp. 43–68.

Bienaymé, I. J. (1853), "Considérations al'appui de la découverte de Laplace," *Comptes Rendus de l'Académie des Sciences*, vol. 37, pp. 309–324.

Billingsley, P. (1986), *Probability and Measure*, 2nd ed., Wiley.

Billingsley, P. (1999), *Convergence of Probability Measures*, 2nd ed., Wiley.

Bochner, S. (1955), *Harmonic Analysis and the Theory of Probability*, University of California Press.

Boltzmann, L. (1877), "Uber die beziehung dem zweiten haubtsatze der mechanischen warmetheorie und der wahrscheinlichkeitsrechnung respektive den Satzen uber das warmegleichgewicht," *Wiener Berichte*, vol. 76, pp. 373–435.

Boltzmann, L. (1909), *Wissenschaftliche Abhandlungen*, vols. I, II, and III, F. Hasenöhrl, editor, Barth. Reissued by Chelsea Publishing Company, 1969.

Borel, E. (1909), "Les probabilités dénombrables et leurs applications arithmétique," *Rend. Circ. Mat. Palermo*, vol. 2, no. 27, pp. 247–271.

Boucheron, S., G. Lugosi, and O. Bousquet (2004), "Concentration inequalities," in *Advanced Lectures on Machine Learning*, O. Bousquet, U. von Luxburg, and G. Rätsch, editors, pp. 208–240, Springer.

Boucheron, S., G. Lugosi, and P. Massart (2013), *Concentration Inequalities: A Nonasymptotic Theory of Independence*, Oxford University Press.

Brown, R. (1828), "A brief account of microscopical observations made in the months of June, July, and August, 1827, on the particles contained in the pollen of plants; and on the general existence of active molecules in organic and inorganic bodies," *Edinb. New Philos. J.*, vol. 5, pp. 358–371; reprinted in *Philos. Mag.*, vol. 4, pp. 161–173. Addendum, "Additional remarks on active molecules," appears in *Edinb. J. Sci*, vol. 1, p. 314, 1829.

Brown, R. (1866), *The Miscellaneous Botanical Works of Robert Brown*, vols. 1 and 2, Robert Hardwicke.

Chebyshev, P. (1867), "Des valeurs moyennes," *J. de Mathématiques Pures et Appliquées*, vol. 2, no. 12, pp. 177–184.

Chernoff, H. (1952), "A measure of asymptotic efficiency of tests of a hypothesis based on the sum of observations, " *Ann. of Math. Statist.*, vol. 23, pp. 493–507.

Chung, K. L. (2000), *A Course in Probability Theory*, 2nd ed., Academic Press.

Chung, F. and L. Lu (2006a), *Complex Graphs and Networks*, American Mathematical Society (AMS).

Chung, F. and L. Lu (2006b), "Concentration inequalities and martingale inequalities: A survey," *Internet Math.*, vol. 3, no. 1, pp. 79–127.

Dale, A. I. (2003), *Most Honourable Remembrance: The Life and Work of Thomas Bayes*, Springer.

Davidson, J. (1994), *Stochastic Limit Theory: An Introduction for Econometricians*, Oxford University Press.

Doob, J. L (1996), "The development of rigor in mathematical probability (1900–1950)," *Amer. Math. Monthly*, vol. 103, pp. 586–595.

Drineas, P., R. Kannan, and M. W. Mahoney (2006), "Fast Monte Carlo algorithms for matrices I: Approximating matrix multiplication," *SIAM J. Comput.*, vol. 36, pp. 132–157.

Dubois, J.-M., O. Gilles, and B. Rouzaire-Dubois (2009), "The Boltzmann equation in molecular biology," *Progress Biophys. Molec. Biol.*, vol. 99, pp. 87–93.

Dudley, R. M. (2002), *Real Analysis and Probability*, Cambridge University Press.

Durrett, R. (1996), *Probability: Theory and Examples*, 2nd ed., Duxbury Press. Fifth edition published by Cambridge University Press, 2019.

Edwards, A. W. F. (1986), "Is the reference in Hartley (1749) to Bayesian inference?" *Am. Statistician*, vol. 40, no. 2, pp. 109–110.

Einstein, A. (1905), "Uber die von der molekularkinetischen theorie der warme geforderte bewegung von in ruhenden flussigkeiten suspendierten teilchenm" *Annalen der Physik*, vol. 322, no. 8, pp. 549–560.

Feinberg, S. E. (2003), "When did Bayesian inference become Bayesian?" *Bayesian Analysis*, vol. 1, no. 1, pp. 1–37.

Feller, W. (1968), *An Introduction to Probability Theory and Its Applications*, vol. 1, 3rd ed., Wiley.

Feller, W. (1971), *An Introduction to Probability Theory and Its Applications*, vol. 2, 3rd ed., Wiley.

Fischer, H. (2011), *A History of the Central Limit Theorem*, Springer.

Gallager, R. G. (2014), *Stochastic Processes: Theory for Applications*, Cambridge University Press.

Gibbs, J. W. (1902), *Elementary Principles in Statistical Mechanics*, Charles Scribner's Sons.

Gnedenko, B. V. (1998), *Theory of Probability*, 6th ed., CRC Press.

Grimmett, G. and D. Stirzaker (2001), *Probability and Random Processes*, 3rd ed., Oxford University Press.

Hald, A. (1998), *A History of Mathematical Statistics from 1750 to 1930*, Wiley.

Hardy, G. H., J. E. Littlewood, and G. Pólya (1934), *Inequalities*, Cambridge University Press.

Hill, T. L. (1987), *Statistical Mechanics: Principles and Selected Applications*, Dover Publications.

Hoeffding, W. (1963), "Probability inequalities for sums of bounded random variables," *J. Amer. Stat. Assoc.*, vol. 58, pp. 13–30.

Huang, K. (2005), *Lectures on Statistical Physics and Protein Folding*, World Scientific.

Ingenhousz, J. (1784), "Remarks on the use of the magnifying glass," *Journal de Physique*, vol. II, pp. 122–126. An English translation appears in P. W. van der Pas, "The discovery of the Brownian motion" *Scientiarum Historia*, vol. 13, pp. 27–35.

Karr, A. F. (1993), *Probability*, Springer.

Katznelson, Y. (2004), *An Introduction to Harmonic Analysis*, 3rd ed., Cambridge University Press.

Knuth, D. (1997), *The Art of Computer Programming*, 3rd ed., Addison-Wesley.

Kolmogorov, A. N. (1927) "Sur la loi forte des grands nombres," *C. R. Acad. Sci.*, Ser. I Math, vol. 191, pp. 910–912.

Kolmogorov, A. N. (1931), "Ueber die analytischen methoden der wahrscheinlichkeitsrechnung," *Math. Ann.*, vol. 104, pp. 415–458.

Kolmogorov, A. N. (1933), *Grundbegriffe der Wahrscheinlichkeitsrechnung*, vol. 2, no. 3, Springer. An English translation by N. Morrison exists and is entitled *Foundations of the Theory of Probability*, 2nd ed., Chelsea Publishing Company, 1956.

Kolmogorov, A. N. (1960), *Foundations of the Theory of Probability*, 2nd ed., Chelsea Publishing Company.

Landau, L. D. and E. M. Lifshitz (1980), *Statistical Physics*, 3rd ed., Butterworth-Heinemann.

Laplace, P. S. (1774), "Mémoire sur la probabilité des causes par les événements," *Mém. Acad. R. Sci. de MI (Savants étrangers)*, vol. 4, pp. 621–656. See also *Oeuvres Complètes de Laplace*, vol. 8, pp. 27–65 published by the L'Académie des Sciences, Paris, during the period 1878–1912. Translated by S. M. Sitgler, *Statistical Science*, vol. 1, no. 3, pp. 366–367.

Lawler, G. F. and V. Limic (2010), *Random Walk: A Modern Introduction*, Cambridge University Press.

Ledoux, M. (2001), *The Concentration of Measure Phenomenon*, American Mathematical Society (AMS).

Leon-Garcia, A. (2008), *Probability, Statistics, and Random Processes for Electrical Engineering*, 3rd ed., Prentice Hall.

Lukacs, E. (1970), *Characteristic Functions*, 2nd ed., Charles Griffin & Co.

Markov, A. A. (1884), *On Certain Applications of Algebraic Continued Fractions*, Ph.D. dissertation, St. Petersburg, Russia.

Massart, P. (2007), *Concentration Inequalities and Model Selection*, Springer.

McDiarmid, C. (1989), "On the method of bounded differences," in *Surveys in Combinatorics*, J. Siemons, editor, pp.148–188, Cambridge University Press.

Mohri, M., A. Rostamizadeh, and A. Talwalkar (2018), *Foundations of Machine Learning*, 2nd ed., MIT Press.

Morters, P. and Y. Peres (2010), *Brownian Motion*, Cambridge University Press.

Okamoto, M. (1958), "Some inequalities relating to the partial sum of binomial probabilities," *Ann. Inst. Stat. Math.*, vol. 10, pp. 29–35.

Ondar, K. O. (1981), *The Correspondence Between A. A. Markov and A. A. Chuprov on the Theory of Probability and Mathematical Statistics*, Springer.

Onuchic, J. N., Z. Luthey-Schulten, and P. G. Wolynes (1997), "Theory of protein folding: The energy landscape perspective," *Ann. Rev. Phys. Chem.*, vol. 48, pp. 545–600.

Oppenheim, A. V., R. W. Schafer, and J. R. Buck (2009), *Discrete-Time Signal Processing*, 3rd ed., Prentice Hall.

Papoulis, A. (1991), *Probability, Random Variables, and Stochastic Processes,* 3rd ed., McGraw-Hill.

Pathria, R. K. and P. D. Beale (2011), *Statistical Mechanics*, 3rd ed., Academic Press.

Pearl, J. (1995), "Causal diagrams for empirical research," *Biometrika*, vol. 82, pp. 669–710.

Pearl, J. (2000), *Causality: Models, Reasoning, and Inference*, Cambridge University Press.

Pearle, P., B. Collett, K. Bart, D. Bilderback, D. Newman, and S. Samuels (2010), "What Brown saw and you can too," *Am. J. Phys.*, vol. 78, pp. 1278–1289.

Phillips, R., J. Kondev, J. Theriot, H. Garcia, and J. Kondev (2012), *Physical Biology of the Cell*, 2nd ed., Garland Science.

Picinbono, B. (1993), *Random Signals and Systems*, Prentice Hall.

Poisson, S. D. (1837), *Probabilité des Jugements en Matière Criminelle et en Matière Civile*, Bachelier.

Prokhorov, Y. V. (2011), "Strong law of large numbers," *Encyclopedia of Mathematics*. Available at http://encyclopediaofmath.org.

Rogers, L. C. G. and D. Williams (2000), *Diffusions, Markov Processes, and Martingales*, Cambridge University Press.

Rohatgi, V. K. (1976), *An Introduction to Probability Theory and Mathematical Statistics*, Wiley.

Santana, R., P. Larranaga, and J. A. Lozano (2008), "Protein folding in simplified models with estimation of distribution algorithms," *IEEE Trans. Evolut. Comput.*, vol. 12, no. 4, pp. 418–438.
Sayed, A. H. (2003), *Fundamentals of Adaptive Filtering*, Wiley.
Sayed, A. H. (2008), *Adaptive Filters*, Wiley.
Seneta, E. (2013), "A tricentenary history of the law of large numbers," *Bernoulli*, vol. 19, no. 4, pp. 1088–1121.
Serfling, R. J. (1974), "Probability inequalities for the sum in sampling without replacement," *Ann. Statist.*, vol. 2, no. 1, pp. 39–48.
Shafer, G. and V. Vovk (2006), "The sources of Kolmogorov's Grundbegriffe," *Statist. Sci.*, vol. 21, no. 1, pp. 70–98.
Shiryayev, A. N. (1984), *Probability*, Springer.
Slutsky, E. (1925), "Über stochastische Asymptoten und Grenzwerte," *Metron*, vol. 5, no. 3, pp. 3–89.
Smoluchowski, M. (1906), "Zur kinetischen theorie der Brownschen molekularbewegung und der suspensionen," *Annalen der Physik*, vol. 326, no. 14, pp. 756–780.
Stark, H. and J. W. Woods (1994), *Probability, Random Processes, and Estimation Theory for Engineers*, 2nd ed., Prentice Hall.
Stein, E. M. and R. Shakarchi (2003), *Fourier Analysis: An Introduction*, Princeton University Press.
Stigler, S. M. (1983), "Who discovered Bayes' theorem?" *Am. Statist.*, vol. 37, no. 4, pp. 290–296.
Sutherland, W. (1905), "A dynamical theory of diffusion for non-electrolytes and the molecular mass of albumin," *Phil. Mag.*, vol. 9, pp. 781–785.
Tankard, J. W. (1984), *The Statistical Pioneers*, Schenkman Books.
Tolman, R. C. (2010), *The Principles of Statistical Mechanics*, Dover Publications.
Uffink, J. (2014), "Boltzmann's work in statistical physics," *The Stanford Encyclopedia of Philosophy*, Fall 2014 edition, E. N. Zalta, editor, available at http://plato.stanford.edu/archives/fall2014/entries/statphys-Boltzmann/.
van der Pas, P. W. (1971), "The discovery of the Brownian motion" *Scientiarum Historia*, vol. 13, pp. 27–35.
van der Vaart, A. W. (2000), *Asymptotic Statistics*, Cambridge University Press.
Vershynin, R. (2018), *High-Dimensional Probability*, Cambridge University Press.
Vetterli, M., J. Kovacevic, and V. K. Goyal (2014), *Foundations of Signal Processing*, Cambridge University Press.
Wainwright, M. J. (2019), *High-Dimensional Statistics: A Non-Asymptotic Viewpoint*, Cambridge University Press.
Weiss, N. A. (2005), *A Course in Probability*, Addison-Wesley.

4 Gaussian Distribution

The Gaussian distribution plays a prominent role in inference and learning, especially when we deal with the sum of a large number of samples. In this case, a fundamental result in probability theory, known as the *central limit theorem*, states that under conditions often reasonable in applications, the probability density function (pdf) of the sum of independent random variables approaches that of a Gaussian distribution. It is for this reason that the term "Gaussian noise" generally refers to the combined effect of many independent disturbances. In this chapter we describe the form of the Gaussian distribution for both scalar and vector random variables, and establish several useful properties and integral expressions that will be used throughout our treatment.

4.1 SCALAR GAUSSIAN VARIABLES

We start with the scalar case. Assume $\{\boldsymbol{x}_n, n = 1, 2, \ldots, N\}$ are independent scalar random variables with means $\{\bar{x}_n\}$ and variances $\{\sigma_{x,n}^2\}$ each. Then, as explained in the comments at the end of the chapter, under some technical conditions represented by expressions (4.162)–(4.163), the pdf of the following normalized variable:

$$\boldsymbol{y} \stackrel{\Delta}{=} \frac{\sum_{n=1}^{N}(\boldsymbol{x}_n - \bar{x}_n)}{\left(\sum_{n=1}^{N}\sigma_{x,n}^2\right)^{1/2}} \tag{4.1}$$

can be shown to approach that of a Gaussian distribution with zero mean and unit variance, i.e.,

$$f_{\boldsymbol{y}}(y) = \frac{1}{\sqrt{2\pi}}e^{-y^2/2}, \qquad \text{as } N \to \infty \tag{4.2}$$

or, equivalently,

$$\lim_{N\to\infty} \mathbb{P}(\boldsymbol{y} \leq a) = \frac{1}{\sqrt{2\pi}}\int_{-\infty}^{a} e^{-y^2/2}dy \tag{4.3}$$

More generally, we denote a Gaussian distribution with mean $\bar{y}$ and variance σ_y^2 by the notation $\mathcal{N}_{\boldsymbol{y}}(\bar{y}, \sigma_y^2)$ with the pdf given by

$$\boxed{\boldsymbol{y} \sim \mathcal{N}_{\boldsymbol{y}}(\bar{y}, \sigma_y^2) \iff f_{\boldsymbol{y}}(y) = \frac{1}{\sqrt{2\pi\sigma_y^2}} \exp\left\{ -\frac{1}{2\sigma_y^2}(y-\bar{y})^2 \right\}} \tag{4.4}$$

Figure 4.1 illustrates the form of the Gaussian distribution using $\bar{y} = 2$ and $\sigma_y^2 = 3$. The next example derives three useful integral expressions.

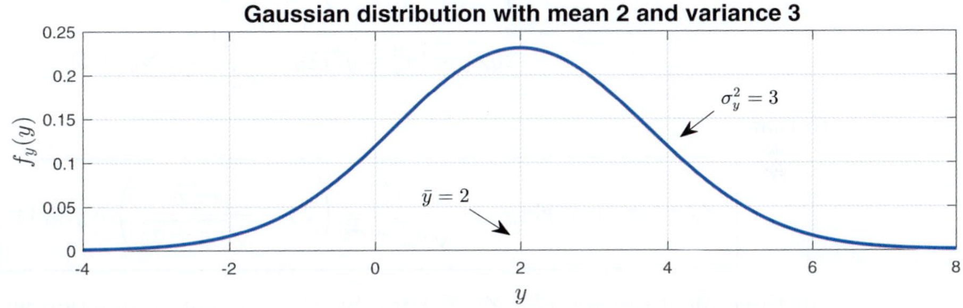

Figure 4.1 Probability density function of a Gaussian distribution with mean $\bar{y} = 2$ and variance $\sigma_y^2 = 3$.

Example 4.1 **(Three useful integral expressions)** There are many results on integrals involving the Gaussian distribution, some of which will appear in our treatment of inference problems. We list some of them here for ease of reference and leave their derivation to the problems.

Let $f_{\boldsymbol{x}}(x)$ denote the standard Gaussian distribution, $\boldsymbol{x} \sim \mathcal{N}_{\boldsymbol{x}}(0, 1)$, i.e.,

$$f_{\boldsymbol{x}}(x) = \frac{1}{\sqrt{2\pi}} e^{-\frac{1}{2}x^2} = \mathcal{N}_{\boldsymbol{x}}(0, 1) \tag{4.5}$$

and introduce its cumulative distribution function (cdf):

$$\Phi(z) \stackrel{\Delta}{=} \frac{1}{\sqrt{2\pi}} \int_{-\infty}^{z} e^{-x^2/2} dx = \int_{-\infty}^{z} \mathcal{N}_{\boldsymbol{x}}(0, 1) dx = \mathbb{P}(\boldsymbol{x} \leq z) \tag{4.6}$$

This function measures the area under $f_{\boldsymbol{x}}(x)$ from $-\infty$ up to location z. Note that $\Phi(z)$ maps real values z to the interval $[0, 1]$. Now, consider a second scalar Gaussian-distributed random variable $\boldsymbol{y}$ with $f_{\boldsymbol{y}}(y) = \mathcal{N}_{\boldsymbol{y}}(\bar{y}, \sigma_y^2)$. One useful integral result is the following identity established in Prob. 4.8 for any a and $\sigma_a > 0$ – a more general result is considered later in Example 4.7:

$$\boxed{Z_0 \stackrel{\Delta}{=} \int_{-\infty}^{\infty} \Phi\left(\frac{y-a}{\sigma_a}\right) \mathcal{N}_{\boldsymbol{y}}(\bar{y}, \sigma_y^2) dy = \Phi(\widehat{y})} \tag{4.7}$$

where

$$\widehat{y} \stackrel{\Delta}{=} \frac{\bar{y}-a}{\sqrt{\sigma_y^2+\sigma_a^2}} \tag{4.8}$$

Differentiating (4.7) once and then twice relative to $\bar{y}$ leads to two other useful results involving multiplication of the integrand by y and y^2 (see Prob. 4.9):

$$\boxed{Z_1 \stackrel{\Delta}{=} \int_{-\infty}^{\infty} y\Phi\Big(\frac{y-a}{\sigma_a}\Big)\,\mathcal{N}_{\boldsymbol{y}}(\bar{y},\sigma_y^2)dy = \bar{y}\,\Phi(\widehat{y}) + \frac{\sigma_y^2}{\sqrt{\sigma_y^2+\sigma_a^2}}\,\mathcal{N}_{\widehat{\boldsymbol{y}}}(0,1)} \tag{4.9}$$

and

$$\begin{aligned} Z_2 &\stackrel{\Delta}{=} \int_{-\infty}^{\infty} y^2\Phi\Big(\frac{y-a}{\sigma_a}\Big)\,\mathcal{N}_{\boldsymbol{y}}(\bar{y},\sigma_y^2)dy \\ &= 2\bar{y}Z_1 + (\sigma_y^2-\bar{y}^2)Z_0 - \frac{\sigma_y^4\,\widehat{y}}{\sigma_y^2+\sigma_a^2}\,\mathcal{N}_{\widehat{\boldsymbol{y}}}(0,1) \end{aligned} \tag{4.10}$$

so that

$$\boxed{Z_2 = (\sigma_y^2+\bar{y}^2)\Phi(\widehat{y}) + \frac{\sigma_y^2}{\sqrt{\sigma_y^2+\sigma_a^2}}\left(2\bar{y} - \frac{\sigma_y^2\,\widehat{y}}{\sqrt{\sigma_y^2+\sigma_a^2}}\right)\mathcal{N}_{\widehat{\boldsymbol{y}}}(0,1)} \tag{4.11}$$

All three identities for $\{Z_0, Z_1, Z_2\}$ involve the cumulative function $\Phi(z)$ in the integrand. We will also encounter integrals involving the sigmoid function

$$\sigma(z) \stackrel{\Delta}{=} \frac{1}{1+e^{-z}} \tag{4.12}$$

This function maps real values z to the same interval $[0,1]$. For these integrals, we will employ the useful approximation

$$\frac{1}{1+e^{-z}} \approx \Phi(bz), \quad \text{where } b^2 = \pi/8 \tag{4.13}$$

4.2 VECTOR GAUSSIAN VARIABLES

Vector Gaussian variables arise frequently in learning and inference problems. We describe next the general form of the pdf for a *vector* Gaussian random variable, and examine several of its properties.

4.2.1 Probability Density Function

We start with a $p \times 1$ random vector $\boldsymbol{x}$ with mean $\bar{x}$ and nonsingular covariance matrix

$$R_x \stackrel{\Delta}{=} \mathbb{E}\,(\boldsymbol{x}-\bar{x})(\boldsymbol{x}-\bar{x})^{\mathsf{T}} > 0 \tag{4.14}$$

We say that $\boldsymbol{x}$ has a Gaussian distribution if its pdf has the form

$$f_{\boldsymbol{x}}(x) = \frac{1}{\sqrt{(2\pi)^p}} \frac{1}{\sqrt{\det R_x}} \exp\left\{-\frac{1}{2}(x-\bar{x})^{\mathsf{T}} R_x^{-1}(x-\bar{x})\right\} \tag{4.15}$$

in terms of the determinant of R_x. Of course, when $p = 1$, the above expression reduces to the pdf considered earlier in (4.4) with R_x replaced by σ_x^2. Figure 4.2 illustrates the form of a two-dimensional Gaussian distribution with

$$\bar{x} = \begin{bmatrix} 1 \\ 2 \end{bmatrix}, \quad R_x = \begin{bmatrix} 1 & 1 \\ 1 & 3 \end{bmatrix} \tag{4.16}$$

The individual entries of $\boldsymbol{x}$ are denoted by $\{\boldsymbol{x}_1, \boldsymbol{x}_2\}$.

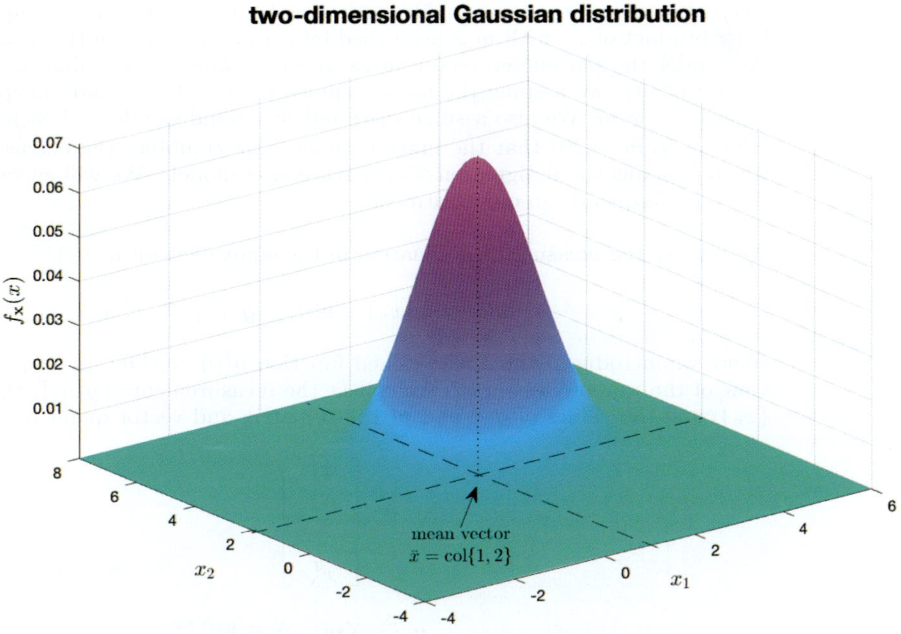

Figure 4.2 Probability density function of a two-dimensional Gaussian distribution with mean vector and covariance matrix given by (4.16).

Example 4.2 **(Noisy measurements of a DC value)** The following example illustrates one possibility by which Gaussian random vectors arise in practice. Consider a collection of N noisy measurements of some unknown constant θ:

$$\boldsymbol{x}(n) = \theta + \boldsymbol{v}(n), \quad n = 1, 2, \ldots, N \tag{4.17}$$

where $\boldsymbol{x}(n)$ is a scalar and $\boldsymbol{v}(n)$ is a zero-mean Gaussian random variable with variance σ_v^2. We assume $\boldsymbol{v}(n)$ and $\boldsymbol{v}(m)$ are independent of each other for all $n \neq m$. Due to the noise, the measurements $\{\boldsymbol{x}(n)\}$ will fluctuate around θ. Each $\boldsymbol{x}(n)$ will be Gaussian distributed with mean θ and variance σ_v^2. We collect the measurements into the vector

$$\boldsymbol{x} \stackrel{\Delta}{=} \text{col}\Big\{\boldsymbol{x}(1), \boldsymbol{x}(2), \ldots, \boldsymbol{x}(N)\Big\} \tag{4.18}$$

Then, the vector $\boldsymbol{x}$ will have a Gaussian distribution with mean $\theta \mathbb{1}_N$ and covariance matrix $R_x = \sigma_v^2 I_N$:

$$\boldsymbol{x} \sim \mathcal{N}_{\boldsymbol{x}}(\theta \mathbb{1}_N, \sigma_v^2 I_N) \tag{4.19}$$

where the notation $\mathbb{1}$ refers to a vector with all its entries equal to 1.

Example 4.3 **(Linear regression model)** Consider next a situation where we observe N noisy scalar measurements $\{\boldsymbol{y}(n)\}$ under Gaussian noise as follows:

$$\boldsymbol{y}(n) = x_n^\mathsf{T} \boldsymbol{w} + \boldsymbol{v}(n), \quad \boldsymbol{v}(n) \sim \mathcal{N}_{\boldsymbol{v}}(0, \sigma_v^2), \quad n = 1, 2, \ldots, N \tag{4.20}$$

where $\{x_n, \boldsymbol{w}\}$ are vectors in $\mathbb{R}^p$ with x_n playing the role of an input vector. The inner product of x_n with $\boldsymbol{w}$ is perturbed by $\boldsymbol{v}(n)$ and results in the measurement $\boldsymbol{y}(n)$. We model the parameter vector $\boldsymbol{w}$ as another Gaussian variable, $\boldsymbol{w} \sim \mathcal{N}_{\boldsymbol{w}}(\bar{w}, R_w)$. For simplicity, we assume the noise samples $\boldsymbol{v}(n)$ and $\boldsymbol{v}(m)$ are independent of each other for $n \neq m$. We also assume $\boldsymbol{v}(n)$ and $\boldsymbol{w}$ are independent of each other for all n. Observe from (4.20) that the inner product $x_n^\mathsf{T} \boldsymbol{w}$ combines the entries of x_n linearly, which explains the designation "linear regression model." We will encounter models of this type frequently in our treatment.

We rewrite the measurement equation in the equivalent form:

$$\boldsymbol{y}(n) = \boldsymbol{g}(x_n) + \boldsymbol{v}(n), \quad \boldsymbol{g}(x_n) \stackrel{\Delta}{=} x_n^\mathsf{T} \boldsymbol{w} \tag{4.21}$$

where we introduced the scalar-valued function $\boldsymbol{g}(x) = x^\mathsf{T} \boldsymbol{w}$; its values are random in view of the randomness in $\boldsymbol{w}$. We collect the measurements $\{\boldsymbol{y}(n)\}$, the input vectors $\{x_n\}$, and the values of $\boldsymbol{g}(\cdot)$ and $\boldsymbol{v}(\cdot)$ into matrix and vector quantities and write:

$$X \stackrel{\Delta}{=} \begin{bmatrix} x_1^\mathsf{T} \\ x_2^\mathsf{T} \\ \vdots \\ x_N^\mathsf{T} \end{bmatrix} \tag{4.22a}$$

$$\boldsymbol{g} \stackrel{\Delta}{=} X\boldsymbol{w}, \quad X \in \mathbb{R}^{N \times p} \tag{4.22b}$$

$$\underbrace{\begin{bmatrix} \boldsymbol{y}(1) \\ \boldsymbol{y}(2) \\ \vdots \\ \boldsymbol{y}(N) \end{bmatrix}}_{\boldsymbol{y}} = \underbrace{\begin{bmatrix} \boldsymbol{g}(x_1) \\ \boldsymbol{g}(x_2) \\ \vdots \\ \boldsymbol{g}(x_N) \end{bmatrix}}_{\boldsymbol{g}} + \underbrace{\begin{bmatrix} \boldsymbol{v}(1) \\ \boldsymbol{v}(2) \\ \vdots \\ \boldsymbol{v}(N) \end{bmatrix}}_{\boldsymbol{v}} \tag{4.22c}$$

$$\boldsymbol{y} = X\boldsymbol{w} + \boldsymbol{v} \tag{4.22d}$$

Note that the input factor $\boldsymbol{g}$ is Gaussian distributed with mean and covariance matrix given by

$$\bar{g} \stackrel{\Delta}{=} \mathbb{E}\boldsymbol{g} = X\bar{w} \tag{4.23a}$$

and

$$\begin{aligned} R_g &= \mathbb{E}\,(\boldsymbol{g}-\bar{g})(\boldsymbol{g}-\bar{g})^{\mathsf{T}} \\ &= \mathbb{E}\,X(\boldsymbol{w}-\bar{w})(\boldsymbol{w}-\bar{w})^{\mathsf{T}}X^{\mathsf{T}} \\ &= XR_wX^{\mathsf{T}} \end{aligned} \tag{4.23b}$$

where each entry of R_g contains the cross-covariance between individual entries of $\boldsymbol{g}$, namely,

$$[R_g]_{m,n} \stackrel{\Delta}{=} \mathbb{E}\Big(\boldsymbol{g}(x_m)-\bar{g}(x_m)\Big)\Big(\boldsymbol{g}(x_n)-\bar{g}(x_n)\Big) \;=\; x_m^{\mathsf{T}}R_wx_n \tag{4.23c}$$

Note further that the mean and covariance matrix of the measurement vector are given by

$$\bar{y} \stackrel{\Delta}{=} \mathbb{E}\,\boldsymbol{y} \;=\; \mathbb{E}\,X\boldsymbol{w}+\mathbb{E}\,\boldsymbol{v} \;=\; X\bar{w}+0 = X\bar{w} \tag{4.24a}$$

and

$$\begin{aligned} R_y &\stackrel{\Delta}{=} \mathbb{E}\,(y-\bar{y})(y-\bar{y})^{\mathsf{T}} \\ &= \mathbb{E}\Big(X(\boldsymbol{w}-\bar{w})+\boldsymbol{v}\Big)\Big(X(\boldsymbol{w}-\bar{w})+\boldsymbol{v}\Big)^{\mathsf{T}} \\ &= \sigma_v^2 I_N + XR_wX^{\mathsf{T}} \end{aligned} \tag{4.24b}$$

Using future result (4.39) that the sum of two independent Gaussian random variables is another Gaussian random variable, we conclude that $\boldsymbol{y}$ is a Gaussian distributed vector with

$$\boldsymbol{y} \sim \mathcal{N}_{\boldsymbol{y}}\Big(X\bar{w},\; \sigma_v^2 I_N + XR_wX^{\mathsf{T}}\Big) \tag{4.25}$$

Example 4.4 **(Fourth-order moment)** We derive a useful result concerning the fourth-order moment of a Gaussian random vector. Thus, let $\boldsymbol{x}$ denote a real-valued Gaussian random column vector with zero mean and a diagonal covariance matrix, say, $\mathbb{E}\,\boldsymbol{x}\boldsymbol{x}^{\mathsf{T}} = \Lambda$. Then, for any symmetric matrix W of compatible dimensions it holds that:

$$\boxed{\mathbb{E}\left\{\boldsymbol{x}\boldsymbol{x}^{\mathsf{T}}W\boldsymbol{x}\boldsymbol{x}^{\mathsf{T}}\right\} = \Lambda\mathrm{Tr}(W\Lambda) + 2\Lambda W\Lambda} \tag{4.26}$$

Proof of (4.26): The argument is based on the fact that uncorrelated Gaussian random variables are also independent (see Prob. 4.4), so that if $\boldsymbol{x}_n$ is the nth element of $\boldsymbol{x}$, then $\boldsymbol{x}_n$ is independent of $\boldsymbol{x}_m$ for $n \neq m$. Now let S denote the desired matrix, i.e., $S = \mathbb{E}\,\boldsymbol{x}\boldsymbol{x}^{\mathsf{T}}W\boldsymbol{x}\boldsymbol{x}^{\mathsf{T}}$, and let S_{nm} denote its (n,m)th element. Assume also that $\boldsymbol{x}$ is p-dimensional. Then

$$S_{nm} = \mathbb{E}\left\{\boldsymbol{x}_n\boldsymbol{x}_m\left(\sum_{i=0}^{p-1}\sum_{j=0}^{p-1}\boldsymbol{x}_iW_{ij}\boldsymbol{x}_j\right)\right\} \tag{4.27}$$

The right-hand side is nonzero only when there are two pairs of equal indices $\{n = m,\, i = j\}$ or $\{n = i,\, m = j\}$ or $\{n = i,\, m = i\}$. Assume first that $n = m$ (which corresponds to the diagonal elements of S). Then, the expectation is nonzero only for $i = j$, i.e.,

$$S_{nn} = \mathbb{E}\left\{\boldsymbol{x}_n^2\sum_{i=0}^{p-1}W_{ii}\boldsymbol{x}_i^2\right\} \;=\; \sum_{i=0}^{p-1}W_{ii}\mathbb{E}\left\{\boldsymbol{x}_n^2\boldsymbol{x}_i^2\right\} \;=\; \lambda_n\mathrm{Tr}(W\Lambda) + 2W_{nn}\lambda_n^2 \tag{4.28}$$

where we used the fact that for a zero-mean *real* scalar-valued Gaussian random variable $\boldsymbol{a}$ we have $\mathbb{E}\,\boldsymbol{a}^4 = 3\big(\mathbb{E}\,\boldsymbol{a}^2\big)^2 = 3\sigma_a^4$, where $\sigma_a^2 = \mathbb{E}\,\boldsymbol{a}^2$ – see Prob. 4.13. We are also denoting the diagonal entries of Λ by $\{\lambda_n\}$.

For the off-diagonal elements of S (i.e., for $n \neq m$), we must have either $n = j$, $m = i$, or $n = i$, $m = j$, so that

$$\begin{aligned} S_{nm} &= \mathbb{E}\left\{\boldsymbol{x}_n \boldsymbol{x}_m \big(\boldsymbol{x}_n W_{nm} \boldsymbol{x}_m\big)\right\} + \mathbb{E}\left\{\boldsymbol{x}_n \boldsymbol{x}_m \big(\boldsymbol{x}_m W_{mn} \boldsymbol{x}_n\big)\right\} \\ &= \Big(W_{nm} + W_{mn}\Big)\mathbb{E}\left\{\boldsymbol{x}_n^2 \boldsymbol{x}_m^2\right\} \\ &= \Big(W_{nm} + W_{mn}\Big)\lambda_n \lambda_m \end{aligned} \tag{4.29}$$

Using the fact that W is symmetric, so that $W_{nm} = W_{mn}$, and collecting the expressions for S_{nm}, in both cases of $n = m$ and $n \neq m$, into matrix form we arrive at the desired result (4.26).

■

We assumed the covariance matrix of $\boldsymbol{x}$ to be diagonal in expression (4.26) to facilitate the derivation. However, it can be verified that the result holds more generally for arbitrary covariance matrices, $R_x = \mathbb{E}\,\boldsymbol{x}\boldsymbol{x}^\mathsf{T}$, and would take the following form with Λ replaced by R_x (see Prob. 4.21):

$$\boxed{\mathbb{E}\left\{\boldsymbol{x}\boldsymbol{x}^\mathsf{T} W \boldsymbol{x}\boldsymbol{x}^\mathsf{T}\right\} = R_x \mathrm{Tr}\big(W R_x\big) + 2 R_x W R_x} \tag{4.30}$$

4.3 USEFUL GAUSSIAN MANIPULATIONS

The fact that the Gaussian pdf is normalized and must integrate to 1 can be exploited to derive a useful multi-dimensional integration result for quadratic forms, as well as useful expressions for integrals involving the product and division of Gaussian distributions.

Multidimensional integral

Consider a $p \times p$ positive-definite matrix A, a $p \times 1$ vector b, a scalar α, and introduce the quadratic form:

$$J(x) \stackrel{\Delta}{=} -\frac{1}{2}x^\mathsf{T} A x + b^\mathsf{T} x + \alpha \tag{4.31}$$

It is straightforward to verify that

$$J(x) = -\frac{1}{2}(x - A^{-1}b)^\mathsf{T} A (x - A^{-1}b) + \alpha + \frac{1}{2}b^\mathsf{T} A^{-1} b \tag{4.32}$$

so that

$$
\begin{aligned}
&\int_{-\infty}^{\infty} e^{J(x)}dx \\
&= \int_{-\infty}^{\infty} \exp\Big\{-\frac{1}{2}(x - A^{-1}b)A(x - A^{-1}b) + \alpha + \frac{1}{2}b^{\mathsf{T}}A^{-1}b\Big\}dx \\
&= \exp\Big\{\alpha + \frac{1}{2}b^{\mathsf{T}}A^{-1}b\Big\} \int_{-\infty}^{\infty} \exp\Big\{-\frac{1}{2}(x - A^{-1}b)A(x - A^{-1}b)\Big\}dx \\
&= \exp\Big\{\alpha + \frac{1}{2}b^{\mathsf{T}}A^{-1}b\Big\} \sqrt{(2\pi)^p}\,\sqrt{\det A^{-1}} \times \\
&\quad \underbrace{\int_{-\infty}^{\infty} \frac{1}{\sqrt{(2\pi)^p}} \frac{1}{\sqrt{\det A^{-1}}} \exp\Big\{-\frac{1}{2}(x - A^{-1}b)A(x - A^{-1}b)\Big\}\,dx}_{=\mathcal{N}_{\boldsymbol{x}}(A^{-1}b,A^{-1})}
\end{aligned}
\tag{4.33}
$$

and, consequently, for $A > 0$,

$$
\boxed{\int_{-\infty}^{\infty} \exp\Big\{-\frac{1}{2}x^{\mathsf{T}}Ax + b^{\mathsf{T}}x + \alpha\Big\}dx \;=\; \sqrt{\frac{(2\pi)^p}{\det A}} \times \exp\Big\{\alpha + \frac{1}{2}b^{\mathsf{T}}A^{-1}b\Big\}}
\tag{4.34}
$$

Sum of Gaussian distributions

The sum of two independent Gaussian distributions is another Gaussian distribution. Specifically, let $\boldsymbol{x} \sim \mathcal{N}_{\boldsymbol{x}}(\bar{x}, R_x)$ and $\boldsymbol{y} \sim \mathcal{N}_{\boldsymbol{y}}(\bar{y}, R_y)$ be two independent Gaussian random variables and introduce their sum $\boldsymbol{z} = \boldsymbol{x} + \boldsymbol{y}$. It is clear that the mean and covariance matrix of $\boldsymbol{z}$ are given by

$$
\bar{z} \overset{\Delta}{=} \mathbb{E}\,\boldsymbol{z} = \bar{x} + \bar{y} \tag{4.35}
$$

$$
R_z \overset{\Delta}{=} \mathbb{E}\,(\boldsymbol{z} - \bar{z})(\boldsymbol{z} - \bar{z})^{\mathsf{T}} \;=\; R_x + R_y \tag{4.36}
$$

The pdf of $\boldsymbol{z}$, on the other hand, is given by the following convolution expression in view of the independence of $\boldsymbol{z}$ and $\boldsymbol{y}$ (recall result (3.160)):

$$
f_{\boldsymbol{z}}(z) = \int_{-\infty}^{\infty} f_{\boldsymbol{x}}(x)\, f_{\boldsymbol{y}}(z - x)dx \tag{4.37}
$$

which involves the integral of the product of two Gaussian distributions. Ignoring the normalization factors we have

$$
\begin{aligned}
&f_{\boldsymbol{z}}(z) \propto \\
&\int_{-\infty}^{\infty} \exp\left\{-\frac{1}{2}(x - \bar{x})^{\mathsf{T}}R_x^{-1}(x - \bar{x})\right\} \exp\left\{-\frac{1}{2}(z - x - \bar{y})^{\mathsf{T}}R_y^{-1}(z - x - \bar{y})\right\} dx
\end{aligned}
\tag{4.38}
$$

The integrand is an exponential function whose exponent is quadratic in x. It can then be verified by using identity (4.34) that the integration leads to a Gaussian pdf with mean $\bar{z} = \bar{x} + \bar{y}$ and covariance matrix $R_z = R_x + R_y$ (see Prob. 4.12):

$$\left.\begin{array}{l} \boldsymbol{x} \sim \mathcal{N}_{\boldsymbol{x}}(\bar{x}, R_x),\ \boldsymbol{y} \sim \mathcal{N}_{\boldsymbol{y}}(\bar{y}, R_y) \\ \boldsymbol{x} \text{ and } \boldsymbol{y} \text{ independent} \end{array}\right\} \Longrightarrow \boldsymbol{z} = \boldsymbol{x} + \boldsymbol{y} \sim \mathcal{N}_{\boldsymbol{z}}(\bar{x} + \bar{y}, R_x + R_y) \tag{4.39}$$

Product of Gaussian distributions

Consider two Gaussian distributions over the *same* random variable $\boldsymbol{x}$, say, $\mathcal{N}_{\boldsymbol{x}}(\bar{x}_a, R_a)$ and $\mathcal{N}_{\boldsymbol{x}}(\bar{x}_b, R_b)$:

$$f_{\boldsymbol{x},a}(x) = \frac{1}{\sqrt{(2\pi)^p}} \frac{1}{\sqrt{\det R_a}} \exp\left\{-\frac{1}{2}(x - \bar{x}_a)^{\mathsf{T}} R_a^{-1} (x - \bar{x}_a)\right\} \tag{4.40a}$$

$$f_{\boldsymbol{x},b}(x) = \frac{1}{\sqrt{(2\pi)^p}} \frac{1}{\sqrt{\det R_b}} \exp\left\{-\frac{1}{2}(x - \bar{x}_b)^{\mathsf{T}} R_b^{-1} (x - \bar{x}_b)\right\} \tag{4.40b}$$

LEMMA 4.1. **(Product of two Gaussians)** *Let $g(x) = f_{\boldsymbol{x},a}(x) f_{\boldsymbol{x},b}(x)$ denote the product of two Gaussian distributions over the same variable $\boldsymbol{x}$, where $f_{\boldsymbol{x},a}(x) \sim \mathcal{N}_{\boldsymbol{x}}(\bar{x}_a, R_a)$ and $f_{\boldsymbol{x},b}(x) \sim \mathcal{N}_{\boldsymbol{x}}(\bar{x}_b, R_b)$. The product is an unnormalized Gaussian distribution given by:*

$$g(x) = Z \times \mathcal{N}_{\boldsymbol{x}}(\bar{x}_c, R_c) \tag{4.41}$$

where

$$R_c^{-1} = R_a^{-1} + R_b^{-1} \tag{4.42a}$$

$$\bar{x}_c = R_c\left(R_a^{-1}\bar{x}_a + R_b^{-1}\bar{x}_b\right) \tag{4.42b}$$

$$Z = \frac{1}{\sqrt{(2\pi)^p}} \frac{1}{\sqrt{\det(R_a + R_b)}} \exp\left\{-\frac{1}{2}(\bar{x}_a - \bar{x}_b)^{\mathsf{T}} (R_a + R_b)^{-1} (\bar{x}_a - \bar{x}_b)\right\} \tag{4.42c}$$

Before proving the result, we recall from Prob. 1.11 that we can rewrite the expression for $\bar{x}_c$ in the equivalent forms:

$$\bar{x}_c = \bar{x}_a + R_c R_b^{-1}(\bar{x}_b - \bar{x}_a) \tag{4.43a}$$

$$= \bar{x}_b + R_c R_a^{-1}(\bar{x}_a - \bar{x}_b) \tag{4.43b}$$

Likewise, we can rewrite the expression for R_c as

$$R_c = R_a - R_a(R_a + R_b)^{-1} R_a \tag{4.44a}$$

$$= R_b - R_b(R_a + R_b)^{-1} R_b \tag{4.44b}$$

Observe further that the expression for Z has the form of a Gaussian distribution, say, over the variable $\bar{\boldsymbol{x}}_a$:

$$Z = \mathcal{N}_{\bar{\boldsymbol{x}}_a}(\bar{x}_b, R_a + R_b) \tag{4.45}$$

Proof of (4.42a)–(4.42c): To begin with, note that

$$g(x) = \frac{1}{\sqrt{(2\pi)^p}} \frac{1}{\sqrt{\det R_a}} \times \frac{1}{\sqrt{(2\pi)^p}} \frac{1}{\sqrt{\det R_b}} \times \\ \exp\left\{-\frac{1}{2}(x-\bar{x}_a)^{\mathsf{T}} R_a^{-1}(x-\bar{x}_a) - \frac{1}{2}(x-\bar{x}_b)^{\mathsf{T}} R_b^{-1}(x-\bar{x}_b)\right\} \tag{4.46}$$

The term in the exponent is quadratic in x. Expanding it gives

$$\begin{aligned}
&(x-\bar{x}_a)^{\mathsf{T}} R_a^{-1}(x-\bar{x}_a) + (x-\bar{x}_b)^{\mathsf{T}} R_b^{-1}(x-\bar{x}_b) \\
&= x^{\mathsf{T}} \underbrace{(R_a^{-1}+R_b^{-1})}_{\triangleq R_c^{-1}} x - 2\underbrace{(R_a^{-1}\bar{x}_a + R_b^{-1}\bar{x}_b)^{\mathsf{T}}}_{\triangleq \bar{x}_c^{\mathsf{T}} R_c^{-1}} x + \underbrace{\bar{x}_a R_a^{-1}\bar{x}_a + \bar{x}_b^{\mathsf{T}} R_b^{-1}\bar{x}_b}_{\triangleq \alpha} \\
&= x^{\mathsf{T}} R_c^{-1} x - 2\bar{x}_c^{\mathsf{T}} R_c^{-1} x + \alpha \\
&= x^{\mathsf{T}} R_c^{-1} x - 2\bar{x}_c^{\mathsf{T}} R_c^{-1} x + \bar{x}_c^{\mathsf{T}} R_c^{-1}\bar{x}_c \underbrace{- \bar{x}_c^{\mathsf{T}} R_c^{-1}\bar{x}_c + \alpha}_{\triangleq \beta} \\
&= (x-\bar{x}_c)^{\mathsf{T}} R_c^{-1}(x-\bar{x}_c) + \beta
\end{aligned} \tag{4.47}$$

where

$$\begin{aligned}
\beta &= \alpha - \bar{x}_c^{\mathsf{T}} R_c^{-1}\bar{x}_c \\
&= \bar{x}_a R_a^{-1}\bar{x}_a + \bar{x}_b^{\mathsf{T}} R_b^{-1}\bar{x}_b - (R_a^{-1}\bar{x}_a + R_b^{-1}\bar{x}_b)^{\mathsf{T}} R_c(R_a^{-1}\bar{x}_a + R_b^{-1}\bar{x}_b) \\
&= \bar{x}_a^{\mathsf{T}}(R_a^{-1} - R_a^{-1}R_cR_a^{-1})\bar{x}_a + \bar{x}_b^{\mathsf{T}}(R_b^{-1} - R_b^{-1}R_cR_b^{-1})\bar{x}_b \\
&\quad -\bar{x}_a^{\mathsf{T}}(R_a+R_b)^{-1}x_b - \bar{x}_b^{\mathsf{T}}(R_a+R_b)^{-1}\bar{x}_a \\
&= \bar{x}_a^{\mathsf{T}}(R_a+R_b)^{-1}\bar{x}_a + \bar{x}_b^{\mathsf{T}}(R_a+R_b)^{-1}\bar{x}_b \\
&\quad -\bar{x}_a^{\mathsf{T}}(R_a+R_b)^{-1}x_b - \bar{x}_b^{\mathsf{T}}(R_a+R_b)^{-1}\bar{x}_a \\
&= (\bar{x}_a - \bar{x}_b)^{\mathsf{T}}(R_a+R_b)^{-1}(\bar{x}_a - \bar{x}_b)
\end{aligned} \tag{4.48}$$

We therefore conclude that

$$g(x) = Z_1 \times \exp\left\{-\frac{1}{2}(x-\bar{x}_c)^{\mathsf{T}} R_c^{-1}(x-\bar{x}_c)\right\} \tag{4.49}$$

where

$$Z_1 = \frac{1}{(2\pi)^p} \frac{1}{\sqrt{\det R_a}} \frac{1}{\sqrt{\det R_b}} \exp\left\{-\frac{1}{2}(\bar{x}_a - \bar{x}_b)^{\mathsf{T}}(R_a+R_b)^{-1}(\bar{x}_a - \bar{x}_b)\right\} \tag{4.50}$$

We can renormalize $g(x)$ to transform its exponential term into a Gaussian distribution as follows. Introduce the block matrix:

$$X \triangleq \begin{bmatrix} R_a + R_b & R_a \\ R_a & R_a \end{bmatrix} \tag{4.51}$$

We can express the determinant of X in two equivalent ways using the Schur complements relative to $R_a + R_b$ (which is equal to R_c) and the Schur complement relative to R_a (which is equal to R_b):

$$\det X = \det(R_a + R_b) \times \det R_c = \det R_a \times \det R_b \tag{4.52}$$

so that

$$\det R_c = \frac{\det R_a \times \det R_b}{\det(R_a + R_b)} \tag{4.53}$$

Using this expression to replace the terms involving $\det R_a$ and $\det R_b$ in (4.50) we arrive at (4.41).

■

We conclude from (4.41) that the product of two Gaussian distributions is an *unnormalized* Gaussian. Equivalently, we obtain a properly normalized Gaussian through scaling by Z as follows:

$$\boxed{\frac{1}{Z} \times \mathcal{N}_{\boldsymbol{x}}(\bar{x}_a, R_a) \times \mathcal{N}_{\boldsymbol{x}}(\bar{x}_b, R_b) \;=\; \mathcal{N}_{\boldsymbol{x}}(\bar{x}_c, R_c)} \tag{4.54}$$

Obviously, the scaling factor Z has the interpretation

$$Z = \int_{-\infty}^{\infty} \mathcal{N}_{\boldsymbol{x}}(\bar{x}_a, R_a) \times \mathcal{N}_{\boldsymbol{x}}(\bar{x}_b, R_b) dx \tag{4.55}$$

Division of Gaussian distributions

Consider the same Gaussian distributions over the random variable $\boldsymbol{x}$. Let now $g(x) = f_{\boldsymbol{x},a}(x)/f_{\boldsymbol{x},b}(x)$ denote their ratio. Repeating the previous arguments we find that

$$\boxed{g(x) = Z_1 \times \exp\left\{-\frac{1}{2}(x - \bar{x}_c)^{\mathsf{T}} R_c^{-1}(x - \bar{x}_c)\right\}} \tag{4.56}$$

where now

$$R_c^{-1} = R_a^{-1} - R_b^{-1} \tag{4.57a}$$

$$\bar{x}_c = R_c(R_a^{-1}\bar{x}_a - R_b^{-1}\bar{x}_b) \tag{4.57b}$$

$$Z_1 = \frac{\sqrt{\det R_b}}{\sqrt{\det R_a}} \exp\left\{-\frac{1}{2}(\bar{x}_a - \bar{x}_b)^{\mathsf{T}}(R_a - R_b)^{-1}(\bar{x}_a - \bar{x}_b)\right\} \tag{4.57c}$$

Observe that in this case, the matrix R_c is not guaranteed to be positive-definite; it can become indefinite. When R_c is positive-definite (which happens when $R_a < R_b$), we observe that $g(x)$ will have the form of an *unnormalized* Gaussian distribution and it can be normalized by noting that

$$\frac{1}{Z_1} \frac{1}{\sqrt{(2\pi)^p}} \frac{1}{\sqrt{\det R_c}} g(x) \;=\; \mathcal{N}_{\boldsymbol{x}}(\bar{x}_c, R_c) \tag{4.58}$$

and, consequently,

$$\boxed{\frac{1}{Z} \frac{\mathcal{N}_{\boldsymbol{x}}(\bar{x}_a, R_a)}{\mathcal{N}_{\boldsymbol{x}}(\bar{x}_b, R_b)} \;=\; \mathcal{N}_{\boldsymbol{x}}(\bar{x}_c, R_c)} \tag{4.59}$$

where

$$Z \;\stackrel{\Delta}{=}\; \sqrt{(2\pi)^p} \sqrt{\det R_c}\; Z_1 \tag{4.60}$$

Let us introduce the block matrix

$$X \;\stackrel{\Delta}{=}\; \begin{bmatrix} R_b - R_a & R_b \\ R_b & R_b \end{bmatrix} \tag{4.61}$$

Its Schur complement relative to $(R_b - R_a)$ is equal to $-R_c$, whereas its Schur complement relative to R_b is $-R_a$. It follows that

$$\det X = \det(R_b - R_a) \times \det(-R_c) = \det R_b \times \det(-R_a) \tag{4.62}$$

Using the fact that $\det(-A) = (-1)^p \det(A)$ for $p \times p$ matrices, we conclude that

$$\det R_c = \frac{\det R_b \det R_a}{\det(R_b - R_a)} \tag{4.63}$$

so that Z admits the following expression as well:

$$\boxed{Z = \sqrt{(2\pi)^p}\,\frac{\det R_b}{\sqrt{\det(R_b - R_a)}}\exp\left\{-\frac{1}{2}(\bar{x}_a - \bar{x}_b)^{\mathsf{T}}(R_a - R_b)^{-1}(\bar{x}_a - \bar{x}_b)\right\}} \tag{4.64}$$

Stein lemma

A useful result pertaining to the evaluation of expectations involving transformations of Gaussian variables is the Stein lemma. We state the result for vector random variables. Let $\boldsymbol{x} \in \mathbb{R}^p$ denote a Gaussian-distributed random variable with mean $\bar{x}$ and covariance matrix R_x:

$$f_{\boldsymbol{x}}(x) = \frac{1}{\sqrt{(2\pi)^p}}\frac{1}{\sqrt{\det R_x}}\exp\left\{-\frac{1}{2}(x - \bar{x})^{\mathsf{T}} R_x^{-1}(x - \bar{x})\right\} \tag{4.65}$$

We will often encounter situations where it is necessary to compute expectations of terms of the form $\boldsymbol{x}g(\boldsymbol{x})$ for some scalar-valued function $g(x)$. This computation is equivalent to evaluating integral expressions of the form

$$\mathbb{E}\,\boldsymbol{x}g(\boldsymbol{x}) = \frac{1}{\sqrt{(2\pi)^p}}\frac{1}{\sqrt{\det R_x}}\int_{-\infty}^{\infty} xg(x)\exp\left\{-\frac{1}{2}(x - \bar{x})^{\mathsf{T}} R_x^{-1}(x - \bar{x})\right\}dx \tag{4.66}$$

Lemma 4.2. (Stein lemma) *Assume the function $g(\boldsymbol{x})$ satisfies the finite expectation conditions $\mathbb{E}\,|\partial g(\boldsymbol{x})/\partial x_m| < \infty$, relative to the individual entries of x. Then, it holds that*

$$\boxed{\mathbb{E}\,(\boldsymbol{x} - \bar{x})g(\boldsymbol{x}) \;=\; R_x\,\mathbb{E}\,\nabla_{\boldsymbol{x}^{\mathsf{T}}}\, g(\boldsymbol{x})} \tag{4.67}$$

For scalar Gaussian random variables $\boldsymbol{x} \sim \mathcal{N}_{\boldsymbol{x}}(\bar{x}, \sigma_x^2)$, the lemma reduces to

$$\mathbb{E}\,(\boldsymbol{x} - \bar{x})g(\boldsymbol{x}) = \sigma_x^2\,\mathbb{E}\,g'(\boldsymbol{x}) \tag{4.68}$$

in terms of the derivative of $g(x)$. Later, in Example 5.2, we extend the Stein lemma to the exponential family of distributions.

Proof: We establish the result for scalar $\boldsymbol{x}$ and defer the vector case to Prob. 4.33 – see also Example 5.2. Thus, note that in the scalar case:

$$\mathbb{E}\,(\boldsymbol{x} - \bar{x})g(\boldsymbol{x}) = \frac{1}{\sqrt{2\pi\sigma_x^2}}\int_{-\infty}^{\infty}(x - \bar{x})g(x)\exp\left\{-\frac{1}{2\sigma_x^2}(x - \bar{x})^2\right\}dx \tag{4.69}$$

We carry out the integration by parts. Let

$$u = g(x) \Longrightarrow du = g'(x)dx \tag{4.70}$$

and

$$dv = (x - \bar{x})\exp\left\{-\frac{1}{2\sigma_x^2}(x-\bar{x})^2\right\}dx \Longrightarrow v = -\sigma_x^2 \exp\left\{-\frac{1}{2\sigma_x^2}(x-\bar{x})^2\right\} \tag{4.71}$$

It follows that

$$\begin{aligned}
\mathbb{E}\,(\boldsymbol{x}-\bar{x})g(\boldsymbol{x}) &= \frac{1}{\sqrt{2\pi\sigma_x^2}}\left\{uv\Big|_{-\infty}^{\infty} - \int_u vdu\right\} \\
&= -\frac{1}{\sqrt{2\pi\sigma_x^2}}g(x)\sigma_x^2\exp\left\{-\frac{1}{2\sigma_x^2}(x-\bar{x})^2\right\}\Bigg|_{-\infty}^{\infty} + \\
&\quad \sigma_x^2\left(\frac{1}{\sqrt{2\pi\sigma_x^2}}\int_{-\infty}^{\infty} g'(x)\exp\left\{-\frac{1}{2\sigma_x^2}(x-\bar{x})^2\right\}dx\right) \\
&= 0 + \sigma_x^2\,\mathbb{E}\,g'(\boldsymbol{x})
\end{aligned} \tag{4.72}$$

as claimed. The mean of $g'(\boldsymbol{x})$ exists in view of the condition $\mathbb{E}\,|g'(\boldsymbol{x})| < \infty$.

■

Example 4.5 **(Fifth-order moment of Gaussian)** Let us apply the Stein lemma to evaluate the fifth-order moment of a Gaussian distribution, $\boldsymbol{x} \sim \mathcal{N}_{\boldsymbol{x}}(\bar{x}, \sigma_x^2)$. Thus, note the following sequence of calculations using (4.68):

$$\begin{aligned}
\mathbb{E}\,\boldsymbol{x}^5 &= \mathbb{E}\,(\boldsymbol{x}-\bar{x}+\bar{x})\boldsymbol{x}^4 \\
&= \mathbb{E}\,(\boldsymbol{x}-\bar{x})\boldsymbol{x}^4 \;+\; \bar{x}\mathbb{E}\,\boldsymbol{x}\boldsymbol{x}^3 \\
&= 4\sigma_x^2\mathbb{E}\,\boldsymbol{x}^3 \;+\; \bar{x}\mathbb{E}\,(\boldsymbol{x}-\bar{x}+\bar{x})\boldsymbol{x}^3 \\
&= 4\sigma_x^2\mathbb{E}\,\boldsymbol{x}^3 \;+\; \bar{x}\mathbb{E}\,(\boldsymbol{x}-\bar{x})\boldsymbol{x}^3 + \bar{x}^2\mathbb{E}\,\boldsymbol{x}^3 \\
&= (4\sigma_x^2+\bar{x}^2)\mathbb{E}\,\boldsymbol{x}^3 \;+\; 3\bar{x}\sigma_x^2\mathbb{E}\,\boldsymbol{x}^2 \\
&= (4\sigma_x^2+\bar{x}^2)\mathbb{E}\,(\boldsymbol{x}-\bar{x}+\bar{x})\boldsymbol{x}^2 \;+\; 3\bar{x}\sigma_x^2\mathbb{E}\,\boldsymbol{x}^2 \\
&= 2(4\sigma_x^2+\bar{x}^2)\sigma_x^2\mathbb{E}\,\boldsymbol{x} + (7\sigma_x^2+\bar{x}^2)\bar{x}\mathbb{E}\,\boldsymbol{x}^2 \\
&= 2(4\sigma_x^2+\bar{x}^2)\sigma_x^2\bar{x} + (7\sigma_x^2+\bar{x}^2)\bar{x}(\sigma_x^2+\bar{x}^2) \\
&= 15\bar{x}\sigma_x^4 + 10\bar{x}^3\sigma_x^2 + \bar{x}^5
\end{aligned} \tag{4.73}$$

4.4 JOINTLY DISTRIBUTED GAUSSIAN VARIABLES

Consider two random vectors $\boldsymbol{x}$ of size $p \times 1$ and $\boldsymbol{y}$ of size $q \times 1$. We denote their respective means by $\{\bar{x}, \bar{y}\}$ and their respective covariance matrices by

$$R_x \overset{\Delta}{=} \mathbb{E}\,(\boldsymbol{x}-\bar{x})(\boldsymbol{x}-\bar{x})^{\mathsf{T}} \tag{4.74a}$$

$$R_y \overset{\Delta}{=} \mathbb{E}\,(\boldsymbol{y}-\bar{y})(\boldsymbol{y}-\bar{y})^{\mathsf{T}} \tag{4.74b}$$

We further let R_{xy} denote the cross-covariance matrix between $\boldsymbol{x}$ and $\boldsymbol{y}$, i.e.,

$$R_{xy} \triangleq \mathbb{E}\,(\boldsymbol{x}-\bar{x})(\boldsymbol{y}-\bar{y})^{\mathsf{T}} = R_{yx}^{\mathsf{T}} \tag{4.75}$$

and introduce the covariance matrix, R, of the aggregate vector $\text{col}\{\boldsymbol{x},\boldsymbol{y}\}$:

$$R \triangleq \mathbb{E}\left(\begin{bmatrix}\boldsymbol{x}\\ \boldsymbol{y}\end{bmatrix}-\begin{bmatrix}\bar{x}\\ \bar{y}\end{bmatrix}\right)\left(\begin{bmatrix}\boldsymbol{x}\\ \boldsymbol{y}\end{bmatrix}-\begin{bmatrix}\bar{x}\\ \bar{y}\end{bmatrix}\right)^{\mathsf{T}} = \begin{bmatrix} R_x & R_{xy}\\ R_{xy}^{\mathsf{T}} & R_y\end{bmatrix} \tag{4.76}$$

We then say that the random variables $\{\boldsymbol{x},\boldsymbol{y}\}$ have a *joint* Gaussian distribution if their joint pdf has the form

$$\begin{aligned} &f_{\boldsymbol{x},\boldsymbol{y}}(x,y) \\ &= \frac{1}{\sqrt{(2\pi)^{p+q}}}\frac{1}{\sqrt{\det R}}\exp\left\{-\frac{1}{2}\begin{bmatrix}(x-\bar{x})^{\mathsf{T}} & (y-\bar{y})^{\mathsf{T}}\end{bmatrix}R^{-1}\begin{bmatrix}x-\bar{x}\\ y-\bar{y}\end{bmatrix}\right\}\end{aligned} \tag{4.77}$$

It can be seen that the joint pdf of $\{\boldsymbol{x},\boldsymbol{y}\}$ is completely determined by the mean, covariances, and cross-covariance of $\{\boldsymbol{x},\boldsymbol{y}\}$, i.e., by the first- and second-order moments $\{\bar{x},\bar{y},R_x,R_y,R_{xy}\}$. It is also straightforward to conclude from (4.77) that uncorrelated Gaussian random vectors are independent – see Prob. 4.4. It takes more effort, though, to show that if $\{\boldsymbol{x},\boldsymbol{y}\}$ are jointly Gaussian-distributed as above, then each of the variables is individually Gaussian-distributed as well, namely, it holds:

$$\boldsymbol{x} \sim \mathbb{N}_{\boldsymbol{x}}(\bar{x},R_x), \quad \boldsymbol{y} \sim \mathbb{N}_{\boldsymbol{y}}(\bar{y},R_y) \tag{4.78}$$

so that

$$f_{\boldsymbol{x}}(x) = \frac{1}{\sqrt{(2\pi)^p}}\frac{1}{\sqrt{\det R_x}}\exp\left\{-\frac{1}{2}(x-\bar{x})^{\mathsf{T}}R_x^{-1}(x-\bar{x})\right\} \tag{4.79a}$$

$$f_{\boldsymbol{y}}(y) = \frac{1}{\sqrt{(2\pi)^q}}\frac{1}{\sqrt{\det R_y}}\exp\left\{-\frac{1}{2}(y-\bar{y})^{\mathsf{T}}R_y^{-1}(y-\bar{y})\right\} \tag{4.79b}$$

LEMMA 4.3. **(Marginal and conditional pdfs)** *Consider two random vectors $\{\boldsymbol{x},\boldsymbol{y}\}$ that are jointly Gaussian distributed as in (4.77), namely,*

$$\begin{bmatrix}\boldsymbol{x}\\ \boldsymbol{y}\end{bmatrix} \sim \mathbb{N}_{\boldsymbol{x},\boldsymbol{y}}\left(\begin{bmatrix}\bar{x}\\ \bar{y}\end{bmatrix},\begin{bmatrix} R_x & R_{xy}\\ R_{xy}^{\mathsf{T}} & R_y\end{bmatrix}\right) \tag{4.80}$$

It follows that the individual marginal distributions are Gaussian, $\boldsymbol{x}\sim\mathbb{N}_{\boldsymbol{x}}(\bar{x},R_x)$ and $\boldsymbol{y}\sim\mathbb{N}_{\boldsymbol{y}}(\bar{y},R_y)$. Moreover, by marginalizing over $\boldsymbol{y}$ and $\boldsymbol{x}$ separately, the resulting conditional pdfs turn out to be Gaussian as well, as listed in Table 4.1.

Table 4.1 Conditional Gaussian distributions.

$f_{\boldsymbol{x}\|\boldsymbol{y}}(x\|y) \sim \mathbb{N}_{\boldsymbol{x}}(\widehat{x},\Sigma_x)$	$f_{\boldsymbol{y}\|\boldsymbol{x}}(y\|x) \sim \mathbb{N}_{\boldsymbol{y}}(\widehat{y},\Sigma_y)$
$\widehat{x} = \bar{x} + R_{xy}R_y^{-1}(y-\bar{y})$	$\widehat{y} = \bar{y} + R_{yx}R_x^{-1}(x-\bar{x})$
$\Sigma_x = R_x - R_{xy}R_y^{-1}R_{yx}$	$\Sigma_y = R_y - R_{yx}R_x^{-1}R_{xy}$

Proof: We start by noting that the block covariance matrix R in (4.76) can be factored into a product of three upper-triangular, diagonal, and lower-triangular matrices, as follows (this can be checked by straightforward algebra or see (1.63b)):

$$R = \begin{bmatrix} I_p & R_{xy}R_y^{-1} \\ 0 & I_q \end{bmatrix} \begin{bmatrix} \Sigma_x & 0 \\ 0 & R_y \end{bmatrix} \begin{bmatrix} I_p & 0 \\ R_y^{-1}R_{yx} & I_q \end{bmatrix} \tag{4.81}$$

where we introduced the Schur complement $\Sigma_x = R_x - R_{xy}R_y^{-1}R_{yx}$. The matrix Σ_x is guaranteed to be positive-definite in view of the assumed positive-definiteness of R itself – recall Example 1.5. It follows that the determinant of R factors into the product

$$\det R = \det \Sigma_x \times \det R_y \tag{4.82}$$

Inverting both sides of (4.81), we find that the inverse of R can be factored as

$$R^{-1} = \begin{bmatrix} I_p & 0 \\ -R_y^{-1}R_{yx} & I_q \end{bmatrix} \begin{bmatrix} \Sigma_x^{-1} & 0 \\ 0 & R_y^{-1} \end{bmatrix} \begin{bmatrix} I_p & -R_{xy}R_y^{-1} \\ 0 & I_q \end{bmatrix} \tag{4.83}$$

where we used the fact that for any matrix A of appropriate dimensions,

$$\begin{bmatrix} I_p & 0 \\ A & I_q \end{bmatrix}^{-1} = \begin{bmatrix} I_p & 0 \\ -A & I_q \end{bmatrix}, \quad \begin{bmatrix} I_p & A \\ 0 & I_q \end{bmatrix}^{-1} = \begin{bmatrix} I_p & -A \\ 0 & I_q \end{bmatrix} \tag{4.84}$$

Then, substituting into the exponent of the joint distribution (4.77) we can rewrite it in the equivalent form:

$$\begin{aligned} &\exp\left\{-\frac{1}{2}\left[\ (x-\bar{x})^{\mathsf{T}} \quad (y-\bar{y})^{\mathsf{T}}\ \right] R^{-1} \begin{bmatrix} x-\bar{x} \\ y-\bar{y} \end{bmatrix}\right\} \\ &= \exp\Big\{-\frac{1}{2}(x-\widehat{x})^{\mathsf{T}}\Sigma_x^{-1}(x-\widehat{x})\Big\}\ \exp\Big\{-\frac{1}{2}(y-\bar{y})^{\mathsf{T}}R_y^{-1}(y-\bar{y})\Big\} \end{aligned} \tag{4.85}$$

where we introduced $\widehat{x} = \bar{x} + R_{xy}R_y^{-1}(y-\bar{y})$. Substituting (4.85) into (4.77) and using (4.82) we find that the joint pdf of $\{\boldsymbol{x}, \boldsymbol{y}\}$ factorizes into the form

$$\begin{aligned} f_{\boldsymbol{x},\boldsymbol{y}}(x,y) = &\frac{1}{\sqrt{(2\pi)^p}}\frac{1}{\sqrt{\det \Sigma_x}} \exp\left\{-\frac{1}{2}(x-\widehat{x})^{\mathsf{T}}\Sigma_x^{-1}(x-\widehat{x})\right\} \times \\ &\frac{1}{\sqrt{(2\pi)^q}}\frac{1}{\sqrt{\det R_y}} \exp\left\{-\frac{1}{2}(y-\bar{y})^{\mathsf{T}}R_y^{-1}(y-\bar{y})\right\} \end{aligned} \tag{4.86}$$

We conclude from the Bayes rule (3.39) that the marginal pdf of $\boldsymbol{y}$ is Gaussian with mean $\bar{y}$ and covariance matrix R_y:

$$\boldsymbol{y} \sim \mathcal{N}_{\boldsymbol{y}}(\bar{y}, R_y) \tag{4.87}$$

and, moreover, the conditional pdf $f_{\boldsymbol{x}|\boldsymbol{y}}(x|y)$ is Gaussian with mean $\widehat{x}$ and covariance matrix Σ_x:

$$f_{\boldsymbol{x}|\boldsymbol{y}}(x|y) \sim \mathcal{N}_{\boldsymbol{x}}(\widehat{x}, \Sigma_x) \tag{4.88a}$$

$$\widehat{x} = \bar{x} + R_{xy}R_y^{-1}(y-\bar{y}) \tag{4.88b}$$

$$\Sigma_x = R_x - R_{xy}R_y^{-1}R_{yx} \tag{4.88c}$$

If we repeat the same argument using instead the following alternative factorization for R (which can again be checked by straightforward algebra or using (1.63a)):

$$R = \begin{bmatrix} I_q & 0 \\ R_{yx}R_x^{-1} & I_p \end{bmatrix} \begin{bmatrix} R_x & 0 \\ 0 & \Sigma_y \end{bmatrix} \begin{bmatrix} I_q & R_x^{-1}R_{xy} \\ 0 & I_p \end{bmatrix} \tag{4.89}$$

where $\Sigma_y = R_y - R_{yx}R_x^{-1}R_{xy}$, then we can similarly conclude that the marginal pdf of $\boldsymbol{x}$ is Gaussian with mean $\bar{x}$ and covariance matrix R_x:

$$\boldsymbol{x} \sim \mathcal{N}_{\boldsymbol{x}}(\bar{x}, R_x) \tag{4.90}$$

Moreover, the reverse conditional pdf $f_{\boldsymbol{y}|\boldsymbol{x}}(y|x)$ is also Gaussian with mean $\widehat{y}$ and covariance matrix Σ_y:

$$f_{\boldsymbol{y}|\boldsymbol{x}}(y|x) \sim \mathcal{N}_{\boldsymbol{y}}(\widehat{y}, \Sigma_y) \tag{4.91a}$$

$$\widehat{y} = \bar{y} + R_{yx}R_x^{-1}(x - \bar{x}) \tag{4.91b}$$

$$\Sigma_y = R_y - R_{yx}R_x^{-1}R_{xy} \tag{4.91c}$$

■

Example 4.6 **(Joint pdf from marginal and conditional pdfs)** Consider two random variables $\boldsymbol{x}$ and $\boldsymbol{y}$ with marginal and conditional pdfs given by

$$f_{\boldsymbol{x}}(x) \sim \mathcal{N}_{\boldsymbol{x}}(\bar{x}, R_x), \quad f_{\boldsymbol{y}|\boldsymbol{x}}(y|x) \sim \mathcal{N}_{\boldsymbol{y}}(Fx, P) \tag{4.92}$$

for some matrices F and $P > 0$. The resulting joint pdf is Gaussian and given by

$$f_{\boldsymbol{x},\boldsymbol{y}}(x, y) \sim \mathcal{N}_{\boldsymbol{x},\boldsymbol{y}}\left(\begin{bmatrix} \bar{x} \\ F\bar{x} \end{bmatrix}, \begin{bmatrix} R_x & R_xF^{\mathsf{T}} \\ FR_x & FR_xF^{\mathsf{T}} + P \end{bmatrix}\right) \tag{4.93}$$

Proof: We rewrite the Gaussian distribution for $\boldsymbol{y}$ conditioned on $\boldsymbol{x}$ in the following form by adding and subtracting $F\bar{x}$ in the second line:

$$\begin{aligned} f_{\boldsymbol{y}|\boldsymbol{x}}(y|x) &\propto \exp\Big\{-\frac{1}{2}(y - Fx)^{\mathsf{T}}P^{-1}(y - Fx)\Big\} \\ &= \exp\Big\{-\frac{1}{2}\Big((y - F\bar{x}) - F(x - \bar{x})\Big)^{\mathsf{T}}P^{-1}\Big((y - F\bar{x}) - F(x - \bar{x})\Big)\Big\} \end{aligned} \tag{4.94}$$

From the Bayes rule (3.39), the joint pdf is given by

$$\begin{aligned} &f_{\boldsymbol{x},\boldsymbol{y}}(x, y) = f_{\boldsymbol{x}}(x)f_{\boldsymbol{y}|\boldsymbol{x}}(y|x) \\ &\propto \exp\Big\{-\frac{1}{2}(x - \bar{x})^{\mathsf{T}}R_x^{-1}(x - \bar{x})\Big\} \times \\ &\quad \exp\Big\{-\frac{1}{2}\Big((y - F\bar{x}) - F(x - \bar{x})\Big)^{\mathsf{T}}P^{-1}\Big((y - F\bar{x}) - F(x - \bar{x})\Big)\Big\} \\ &= \exp\left\{-\frac{1}{2}\begin{bmatrix} (x - \bar{x})^{\mathsf{T}} & (y - F\bar{x})^{\mathsf{T}} \end{bmatrix}\begin{bmatrix} R_x^{-1} + F^{\mathsf{T}}P^{-1}F & -F^{\mathsf{T}}P^{-1} \\ -P^{-1}F & P^{-1} \end{bmatrix}\begin{bmatrix} x - \bar{x} \\ y - F\bar{x} \end{bmatrix}\right\} \\ &\overset{(1.67b)}{=} \exp\left\{-\frac{1}{2}\begin{bmatrix} (x - \bar{x})^{\mathsf{T}} & (y - F\bar{x})^{\mathsf{T}} \end{bmatrix}\begin{bmatrix} R_x & R_xF^{\mathsf{T}} \\ FR_x & FR_xF^{\mathsf{T}} + P \end{bmatrix}^{-1}\begin{bmatrix} x - \bar{x} \\ y - F\bar{x} \end{bmatrix}\right\} \end{aligned} \tag{4.95}$$

from which we conclude that (4.93) holds.

■

Example 4.7 **(Useful integral expressions)** We generalize Example 4.1 to vector Gaussian random variables. Let $f_{\boldsymbol{x}}(x)$ denote the standard N-dimensional Gaussian distribution, $\boldsymbol{x} \sim \mathcal{N}_{\boldsymbol{x}}(0, I_N)$, and introduce its cdf:

$$\Phi(z) \triangleq \int_{-\infty}^{z} \frac{1}{\sqrt{(2\pi)^N}} \exp\Big\{-\frac{1}{2}\|x\|^2\Big\}dx = \int_{-\infty}^{z} \mathcal{N}_{\boldsymbol{x}}(0, I_N)dx \tag{4.96}$$

which is now a multi-dimensional integral since x is a vector. Let $\boldsymbol{y} \sim \mathcal{N}_{\boldsymbol{y}}(\bar{y}, R_y)$ denote an M-dimensional Gaussian distribution. We wish to evaluate an integral expression of the following form involving the product of a Gaussian distribution and the cumulative distribution:

$$\begin{aligned} Z_0 &\stackrel{\Delta}{=} \int_{-\infty}^{\infty} \Phi(Ay+b)\,\mathcal{N}_{\boldsymbol{y}}(\bar{y}, R_y)dy, \quad \text{for some given } A \in \mathbb{R}^{N\times M},\, b \in \mathbb{R}^N \\ &= \frac{1}{\sqrt{(2\pi)^{M+N}}}\frac{1}{\sqrt{\det R_y}} \times \\ &\quad \int_{-\infty}^{\infty}\int_{-\infty}^{Ay+b} \exp\Big\{-\frac{1}{2}\|x\|^2\Big\} \exp\Big\{-\frac{1}{2}(y-\bar{y})R_y^{-1}(y-\bar{y})\Big\}dxdy \end{aligned} \tag{4.97}$$

The evaluation takes some effort. We start with the change of variables

$$w \stackrel{\Delta}{=} y - \bar{y} \in \mathbb{R}^M, \qquad z \stackrel{\Delta}{=} x - Aw \in \mathbb{R}^N \tag{4.98}$$

and replace the integration over x and y by an integration over z and w:

$$\begin{aligned} Z_0 = &\frac{1}{\sqrt{(2\pi)^{M+N}}}\frac{1}{\sqrt{\det R_y}} \times \\ &\int_{-\infty}^{\infty}\int_{-\infty}^{A\bar{y}+b} \exp\Big\{-\frac{1}{2}(z+Aw)^{\mathsf{T}}(z+Aw)\Big\} \exp\Big\{-\frac{1}{2}w^{\mathsf{T}}R_y^{-1}w\Big\}dzdw \end{aligned} \tag{4.99}$$

The advantage of the change of variables is that the limit of the inner integral, $A\bar{y}+b$, is now independent of the variables $\{z, w\}$ over which the integration is performed. The exponent in (4.99) is quadratic in $\{z, w\}$ since

$$(z+Aw)^{\mathsf{T}}(z+Aw) + w^{\mathsf{T}}R_y^{-1}w = \begin{bmatrix} z^{\mathsf{T}} & w^{\mathsf{T}} \end{bmatrix} \begin{bmatrix} I_N & A \\ A^{\mathsf{T}} & R_y^{-1} + A^{\mathsf{T}}A \end{bmatrix} \begin{bmatrix} z \\ w \end{bmatrix} \tag{4.100}$$

This means that the integrand in (4.99) can be written as a joint Gaussian distribution over the extended variable $\text{col}\{z, w\}$ with zero mean and covariance matrix

$$R \stackrel{\Delta}{=} \begin{bmatrix} I_N & A \\ A^{\mathsf{T}} & R_y^{-1} + A^{\mathsf{T}}A \end{bmatrix}^{-1} \stackrel{(1.67b)}{=} \begin{bmatrix} I_N + AR_yA^{\mathsf{T}} & -AR_y \\ -R_yA^{\mathsf{T}} & R_y \end{bmatrix} \tag{4.101}$$

This shows that the covariance matrix of $\boldsymbol{z}$ is

$$R_z \stackrel{\Delta}{=} I_N + AR_yA^{\mathsf{T}} \tag{4.102}$$

Note further that $\det R = \det R_y$ since the Schur complement of R relative to R_y is the identity matrix. It follows that

$$Z_0 = \int_{-\infty}^{A\bar{y}+b} \underbrace{\left(\int_{-\infty}^{\infty} \mathcal{N}_{\boldsymbol{z},\boldsymbol{w}}(0, R)dw\right)}_{\textbf{marginalization}} dz \tag{4.103}$$

The inner integral amounts to marginalizing the joint distribution of $\{\boldsymbol{z}, \boldsymbol{w}\}$ over w so that the result is the marginal distribution for $\boldsymbol{z}$, namely, $f_{\boldsymbol{z}}(z) = \mathcal{N}_{\boldsymbol{z}}(0, R_z)$ and, consequently,

$$\begin{aligned} Z_0 &= \int_{-\infty}^{A\bar{y}+b} \mathcal{N}_{\boldsymbol{z}}(0, R_z)dz \\ &= \frac{1}{\sqrt{(2\pi)^N}}\frac{1}{\sqrt{\det R_z}} \int_{-\infty}^{A\bar{y}+b} \exp\Big\{-\frac{1}{2}z^{\mathsf{T}}R_z^{-1}z\Big\}dz \end{aligned} \tag{4.104}$$

This expression almost has the form of a cumulative distribution calculation except that the Gaussian distribution is not standard (it has covariance matrix R_z rather than the identity matrix). Let X denote a square-root factor for R_z (recall the definition in Section 1.8), namely, X is any invertible square matrix that satisfies $R_z = XX^\mathsf{T}$. We can also use the more explicit notation $R_z^{1/2}$ to refer to X. One choice for X arises from the eigen-decomposition $R_z = U\Lambda U^\mathsf{T}$, where U is orthogonal and Λ is diagonal with positive entries. Using U and Λ, we can select $X = U\Lambda^{1/2}$. Note that

$$R_z = XX^\mathsf{T} \implies \det R_z = (\det X)^2 \tag{4.105}$$

Next, we introduce the change of variables

$$s = X^{-1}z \implies z^\mathsf{T} R_z^{-1} z = s^\mathsf{T} s \tag{4.106}$$

and the $N \times N$ Jacobian matrix J whose entries consist of the partial derivatives:

$$\big[J\big]_{m,n} = \frac{\partial z_m}{\partial s_n} = X \tag{4.107}$$

where z_m is the mth entry of z and s_n is the nth entry of s. We know from the study of multi-dimensional integrals that when a change of variables is used, we need to account for the (absolute value of the) determinant of the Jacobian matrix so that expression (4.104) is replaced by

$$\begin{aligned} Z_0 &= \frac{1}{\sqrt{(2\pi)^N}}\frac{1}{\sqrt{\det R_z}}\int_{-\infty}^{X^{-1}(A\bar{y}+b)} \exp\Big\{-\frac{1}{2}\|s\|^2\Big\}\,|\det X|\,ds \\ &= \underbrace{\frac{|\det X|}{\sqrt{\det R_z}}}_{=1}\int_{-\infty}^{X^{-1}(A\bar{y}+b)} \frac{1}{\sqrt{(2\pi)^N}}\exp\Big\{-\frac{1}{2}\|s\|^2\Big\}ds \\ &\overset{(4.105)}{=} \Phi\Big(X^{-1}(A\bar{y}+b)\Big) \end{aligned} \tag{4.108}$$

In summary, we arrive at the result:

$$\boxed{Z_0 = \int_{-\infty}^{\infty} \Phi(Ay+b)\,\mathcal{N}_{\boldsymbol{y}}(\bar{y}, R_y)dy \;=\; \Phi(\widehat{y})} \tag{4.109}$$

where

$$\widehat{y} \overset{\Delta}{=} R_z^{-1/2}(A\bar{y}+b) \tag{4.110}$$

It is easy to see that the above result reduces to (4.7) with the identifications $A \leftarrow 1/\sigma_a$, $b \leftarrow -a/\sigma_a$, and $R_z \leftarrow 1+\sigma_y^2/\sigma_a^2$. Another useful special case is when A is a row vector and b is a scalar, say, $A = h^\mathsf{T}$ and $b = \alpha$, in which case we get

$$\boxed{Z_0 = \int_{-\infty}^{\infty} \Phi(h^\mathsf{T}y+\alpha)\,\mathcal{N}_{\boldsymbol{y}}(\bar{y}, R_y)dy \;=\; \Phi(\widehat{y})} \tag{4.111}$$

where now

$$\widehat{y} \overset{\Delta}{=} \frac{h^\mathsf{T}\bar{y}+\alpha}{\sqrt{1+h^\mathsf{T}R_y h}} \tag{4.112}$$

If we differentiate (4.111) relative to $\bar{y}$ once and then twice, we arrive at two additional relations where the integrands are further multiplied by y and yy^T (see Prob. 4.34):

$$Z_1 = \int_{-\infty}^{\infty} y\Phi(h^\mathsf{T}y+\alpha)\,\mathcal{N}_{\boldsymbol{y}}(\bar{y}, R_y)dy \;=\; \bar{y}\Phi(\widehat{y}\,) + \frac{R_y h}{\sqrt{1+h^\mathsf{T}R_y h}}\mathcal{N}_{\widehat{\boldsymbol{y}}}(0,1) \tag{4.113}$$

and

$$Z_2 \triangleq \int_{-\infty}^{\infty} yy^{\mathsf{T}}\Phi(h^{\mathsf{T}}y+\alpha)\,\mathcal{N}_y(\bar{y},R_y)dy \tag{4.114}$$

$$= (R_y + \bar{y}\bar{y}^{\mathsf{T}})\Phi(\widehat{y}) + \frac{1}{\sqrt{1+h^{\mathsf{T}}R_yh}}\left(2R_yh\bar{y}^{\mathsf{T}} - \frac{\widehat{y}R_yhh^{\mathsf{T}}R_y}{\sqrt{1+h^{\mathsf{T}}R_yh}}\right)\mathcal{N}_{\widehat{y}}(0,1)$$

4.5 GAUSSIAN PROCESSES

If we consider a p-dimensional Gaussian random vector $\boldsymbol{x} \sim \mathcal{N}_{\boldsymbol{x}}(\bar{x}, R_x)$ with entries $\{\boldsymbol{x}_1, \boldsymbol{x}_2, \ldots, \boldsymbol{x}_p\}$, then we know from Lemma 4.3 that any sub-collection of entries of $\boldsymbol{x}$ will be jointly Gaussian as well. In other words, the Gaussianity property is inherited by any sub-grouping within $\boldsymbol{x}$. For example, $\{\boldsymbol{x}_1, \boldsymbol{x}_2\}$ will be jointly Gaussian with mean vector $\text{col}\{\bar{x}_1, \bar{x}_2\}$ formed from the two top entries of $\bar{x}$ and with covariance matrix formed from the leading 2×2 submatrix of R_x.

The notion of Gaussian processes (GPs) allows us to extend this property to *sequences* of vectors. The concept will be useful when we study learning and inference problems in the kernel domain later in this text. For now, we define a *Gaussian process* as a sequence of random vectors where any finite sub-collection of entries in each vector is jointly Gaussian-distributed. Moreover, entries across vectors can be correlated with each other.

Remark 4.1. (Terminology) We will discuss "random processes" in Chapter 7. We explain there that random processes consist of *sequences* of random variables (or vectors). The sequencing will generally arise from random variables that evolve over time (or space), say, from lower to higher time indices. For the Gaussian processes discussed in this section, the evolution of the random variables is not limited to time or space. The sequential variables can arise from something more abstract, such as repeated experiments involving separate data collections, as the next example and the discussion following it illustrate. In the example, the successive instances of the Gaussian process $\boldsymbol{g}$ arise from repeated data collections.

■

Example 4.8 (Nonlinear transformations) We reconsider the linear regression model from Example 4.3, namely,

$$\boldsymbol{y}(n) = x_n^{\mathsf{T}}\boldsymbol{w} + \boldsymbol{v}(n), \quad \boldsymbol{v}(n) \sim \mathcal{N}_{\boldsymbol{v}}(0, \sigma_v^2), \quad n = 1, 2, \ldots, N \tag{4.115}$$

where $x_n, \boldsymbol{w} \in \mathbb{R}^M$. This expression models the observation $\boldsymbol{y}(n)$ as a noisy measurement of a linear combination of the individual entries $\{x_{n,m}\}$ of x_n, written explicitly as

$$\boldsymbol{y}(n) = \sum_{m=1}^{M} x_{n,m}\boldsymbol{w}_m + \boldsymbol{v}(n) \tag{4.116}$$

In many situations, it will be advantageous to consider more elaborate models for the mapping from the input x_n to the output $\boldsymbol{y}(n)$. One possibility is to replace the

$M \times 1$ vector x_n by a longer $M_\phi \times 1$ vector $\phi(x_n)$, where the notation $\phi(\cdot)$ represents some nonlinear transformation applied to the entries of x_n. For example, if x_n is two-dimensional with individual entries $x_n = [a\ b]$, then one possibility is to use $\phi(x_n) = [a\ b\ a^2\ b^2\ ab]$. For this case, $M = 2$ and $M_\phi = 5$. The M-dimensional weight vector $\boldsymbol{w}$ would also be extended and replaced by some $\boldsymbol{w}^\phi \in \mathbb{R}^{M_\phi}$, in which case the original model would be replaced by

$$\boldsymbol{y}(n) = (\phi(x_n))^{\mathsf{T}} \boldsymbol{w}^\phi + \boldsymbol{v}(n) \tag{4.117}$$

This representation captures more nonlinear dynamics from x_n to $\boldsymbol{y}(n)$. Many other choices for $\phi(\cdot)$ are of course possible. We rewrite the measurement equation in the equivalent form:

$$\boldsymbol{y}(n) = \boldsymbol{g}(x_n) + \boldsymbol{v}(n), \quad \boldsymbol{g}(x_n) \stackrel{\Delta}{=} (\phi(x_n))^{\mathsf{T}} \boldsymbol{w}^\phi \tag{4.118}$$

where the scalar-valued function $\boldsymbol{g}(x)$ now depends on x_n through the transformation $\phi(\cdot)$; it assumes random values because we model $\boldsymbol{w}^\phi$ as Gaussian-distributed:

$$\boldsymbol{w}^\phi \sim \mathcal{N}_{\boldsymbol{w}^\phi}(\bar{w}^\phi, R_w^\phi) \tag{4.119}$$

for some mean vector $\bar{w}^\phi$ and covariance matrix R_w^ϕ. We collect the measurements $\{\boldsymbol{y}(n)\}$, the input vectors $\{x_n\}$, and the values of $\boldsymbol{g}(\cdot)$ and $\boldsymbol{v}(\cdot)$ into matrix and vector quantities and write

$$\Phi \stackrel{\Delta}{=} \begin{bmatrix} (\phi(x_1))^{\mathsf{T}} \\ (\phi(x_2))^{\mathsf{T}} \\ \vdots \\ (\phi(x_N))^{\mathsf{T}} \end{bmatrix} \tag{4.120a}$$

$$\boldsymbol{g} \stackrel{\Delta}{=} \Phi \boldsymbol{w}^\phi, \quad \Phi \in \mathbb{R}^{N \times M_\phi} \tag{4.120b}$$

$$\underbrace{\begin{bmatrix} \boldsymbol{y}(1) \\ \boldsymbol{y}(2) \\ \vdots \\ \boldsymbol{y}(N) \end{bmatrix}}_{\boldsymbol{y}} = \underbrace{\begin{bmatrix} \boldsymbol{g}(x_1) \\ \boldsymbol{g}(x_2) \\ \vdots \\ \boldsymbol{g}(x_N) \end{bmatrix}}_{\boldsymbol{g}} + \underbrace{\begin{bmatrix} \boldsymbol{v}(1) \\ \boldsymbol{v}(2) \\ \vdots \\ \boldsymbol{v}(N) \end{bmatrix}}_{\boldsymbol{v}} \tag{4.120c}$$

$$\boldsymbol{y} = \Phi \boldsymbol{w}^\phi + \boldsymbol{v} \tag{4.120d}$$

where $\boldsymbol{y}$ continues to be Gaussian-distributed with mean and covariance matrix given by

$$\boldsymbol{y} \sim \mathcal{N}_{\boldsymbol{y}}\Big(\Phi \bar{w}^\phi,\ \sigma_v^2 I_N + \Phi R_w^\phi \Phi^{\mathsf{T}}\Big) \tag{4.121}$$

Moreover, the factor $\boldsymbol{g}$ is Gaussian-distributed with mean and covariance matrix given by

$$\bar{g} = \Phi \bar{w}^\phi, \qquad R_g = \Phi R_w^\phi \Phi^{\mathsf{T}} \tag{4.122a}$$

where each entry of R_g corresponds to the cross-covariance:

$$[R_g]_{m,n} = (\phi(x_m))^{\mathsf{T}} R_w^\phi\, \phi(x_n) \tag{4.122b}$$

As explained next, the vector $\boldsymbol{g}$ is an example of a Gaussian process and any sub-collection of entries in $\boldsymbol{g}$ follows a joint Gaussian distribution. The same is true for the vector $\boldsymbol{g}$ defined earlier in (4.22d). However, the addition of the nonlinear mapping $\phi(\cdot)$ to the model enriches the scenario under consideration.

Model (4.120d) involves a finite number of observations in $\boldsymbol{y}$; this observation vector is a perturbed version of the Gaussian process $\boldsymbol{g}$. The mean and covariance matrix of $\boldsymbol{g}$ are described by (4.122a); they both depend on Φ, which in turn is defined in terms of the input vectors $\{x_n\}$. It would appear at first sight that we are dealing with a Gaussian vector with a finite number of elements in it. However, on closer examination, $\boldsymbol{g}$ is a Gaussian process. This is because, in general, we would not know beforehand which input vectors $\{x_n\}$ to expect. For instance, in a second experiment, the observation vector $\boldsymbol{y}$ will be determined by some other collection of input vectors $\{x'_n\}$. In that case, the mean and covariance matrix of this new observation $\boldsymbol{y}'$ would not be given by (4.121) because the matrix Φ will now be different and defined in terms of the $\{x'_n\}$ rather than $\{x_n\}$. Nevertheless, we would still be able to identify the mean and covariance matrix of the new observation vector $\boldsymbol{y}'$ if we define the mean and covariance matrix of the Gaussian process $\boldsymbol{g}$ more broadly, for any possible choice of its arguments $\{x_n\}$. To do so, we proceed as follows.

We let $\boldsymbol{g}(x)$ denote any generic entry of the Gaussian process. Observe that the argument is the M-dimensional vector x; it can assume any value in $\mathbb{R}^M$. We associate with the process $\boldsymbol{g}(\cdot)$ a mean function and a covariance function defined as follows:

$$m(x) \stackrel{\Delta}{=} \mathbb{E}\,\boldsymbol{g}(x) \tag{4.123a}$$

$$K(x,x') \stackrel{\Delta}{=} \mathbb{E}\Big(\boldsymbol{g}(x) - m(x)\Big)\Big(\boldsymbol{g}(x') - m(x')\Big) \tag{4.123b}$$

Using these functions, we can evaluate the mean of $\boldsymbol{g}(x)$ for *any* x, and the cross-covariance between $\boldsymbol{g}(x)$ and $\boldsymbol{g}(x')$ for *any* x, x'. The expectations are over the sources of randomness in $\boldsymbol{g}(x)$. For example, for the case studied above we have

$$m(x) = (\phi(x))^{\mathsf{T}}\bar{w}^{\phi}, \qquad K(x,x') = (\phi(x))^{\mathsf{T}} R_w^{\phi}\, \phi(x') \tag{4.124}$$

so that the mean and covariance matrix that correspond to the particular input vectors $\{x_n\}$ would be constructed as follows (say, for $N = 4$ measurements):

$$\bar{g} = \begin{bmatrix} m(x_1) \\ m(x_2) \\ m(x_3) \\ m(x_4) \end{bmatrix}, \quad R_g = \begin{bmatrix} K(x_1,x_1) & K(x_1,x_2) & K(x_1,x_3) & K(x_1,x_4) \\ K(x_2,x_1) & K(x_2,x_2) & K(x_2,x_3) & K(x_2,x_4) \\ K(x_3,x_1) & K(x_3,x_2) & K(x_3,x_3) & K(x_3,x_4) \\ K(x_4,x_1) & K(x_4,x_2) & K(x_4,x_3) & K(x_4,x_4) \end{bmatrix} \tag{4.125}$$

We will denote a Gaussian process by the notation

$$\boldsymbol{g} \sim \mathcal{GP}_{\boldsymbol{g}}\Big(m(x),\, K(x,x')\Big) \tag{4.126}$$

It is important to note that not any function $K(x,x')$ can be selected as the covariance function for a Gaussian process. This is because when $K(x,x')$ is applied to data to construct matrices like R_g above, these matrices will need to behave like covariance matrices (i.e., they will need to be symmetric and nonnegative-definite). We will examine conditions on $K(x,x')$ in greater detail in Chapter 63 when we study kernel methods. Here, we summarize the main requirement and leave the details to that chapter.

The function $K(x, x')$ will generally be selected to be a kernel, which is a function that maps two vector arguments (x, x') into the *inner-product* of similarly transformed versions of these same vectors, namely, a function that can be written in the form

$$K(x, x') = (\phi(x))^{\mathsf{T}} \phi(x') \tag{4.127}$$

for some mapping $\phi(\cdot)$. Note that the covariance function used in (4.124) is of this form. It is written in terms of a weighted inner product but can be easily recast in the standard form (4.127). For instance, assume for simplicity that R_w^{ϕ} is positive-definite and introduce the eigen-decomposition $R_w^{\phi} = U \Lambda U^{\mathsf{T}}$ (where U is orthogonal and Λ is diagonal with positive entries). Then, we can redefine

$$\phi(x) \leftarrow \Lambda^{1/2} U^{\mathsf{T}} \phi(x) \tag{4.128}$$

and $K(x, x')$ would reduce to the form (4.127). In the above notation, $\Lambda^{1/2}$ is a diagonal matrix with the positive square-roots of the entries of Λ.

Note from definition (4.127) that kernels are symmetric functions since

$$K(x, x') = K(x', x) \tag{4.129}$$

We say that kernel functions induce inner-product operations in the transformed domain. Obviously, not every function, $K(x, x')$, can be expressed in the inner-product form (4.127) and, therefore, not every function is a kernel. A fundamental theorem in functional analysis, known as *Mercer theorem*, clarifies which functions $K(x, x')$ can be expressed in the form (4.127) – see Section 63.2. For any integer N, we introduce the following $N \times N$ Gramian matrix, R_N, which is symmetric:

$$[R_N]_{m,n} \triangleq K(x_m, x_n), \quad m, n = 1, 2, \ldots, N \tag{4.130}$$

(Mercer theorem). *The theorem affirms that a symmetric and square-integrable function $K(x, x')$ is a kernel if, and only if, the Gramian matrix R_N defined by (4.130) is positive semi-definite for any size N and any data $\{x_n\}$.*

There are many popular choices for $K(x, x')$ that satisfy the Mercer condition. One choice is the Gaussian kernel (also called the *radial basis function* or the squared exponential kernel):

$$K(x, x') \triangleq \exp\left\{ -\frac{1}{2\sigma^2} \|x - x'\|^2 \right\} \tag{4.131}$$

for some parameter $\sigma^2 > 0$. This parameter controls the width of the Gaussian pulse. One could also replace the exponent by a weighted squared norm such as

$$K(x, x') \triangleq \exp\left\{ -\frac{1}{2} (x - x')^{\mathsf{T}} W (x - x') \right\} \tag{4.132}$$

with different choices for the positive-definite matrix W, such as

$$W = \frac{1}{\sigma^2} I_N \quad \text{or} \quad W = \text{a diagonal matrix} \tag{4.133}$$

Another kernel choice is the Ornstein–Uhlenbeck kernel:

$$K(x, x') \stackrel{\Delta}{=} \exp\Big\{-\frac{1}{\sigma}\|x - x'\|\Big\}, \quad \sigma > 0 \tag{4.134}$$

defined in terms of the distance between x and x' rather than their squared distance. The parameter σ is referred to as the *length-scale* of the process; it determines how close points x and x' will need to be to each other for a meaningful correlation between them. We can interpret these kernel functions as providing measures of "similarity" between points in space.

We will explain later in Example 63.4 that the Gaussian kernel (4.131) can be written in the inner-product form (4.127) for some function $\phi(\cdot)$; similarly for the Ornstein–Uhlenbeck kernel. Fortunately, explicit knowledge of $\phi(\cdot)$ is unnecessary (and this important fact is what makes kernel methods powerful; as explained later in Chapter 63). Observe that the kernel functions in (4.131)–(4.134) are written directly in terms of the input vectors (x, x') and not their transformed versions $\phi(x)$ or $\phi(x')$. Usually, the mean function of a Gaussian process is taken to be zero. In this way, characterization of the first- and second-order moments of $\boldsymbol{g} \sim \mathcal{GP}_{\boldsymbol{g}}(0, K(x, x'))$ would not require knowledge of the nonlinear mapping $\phi(\cdot)$. Once a kernel function is specified, we are implicitly assuming some nonlinear mapping is applied to the input vectors $\{x_n\}$.

Example 4.9 **(Polynomial kernel)** Let us illustrate the last point by considering a simplified example. Assume $x \in \mathbb{R}^2$ with entries $x = [a \; b]$. We select the polynomial kernel

$$K(x, x') \stackrel{\Delta}{=} (1 + x^{\mathsf{T}} x')^2 \tag{4.135}$$

and verify that it is a kernel function. To do so, and according to definition (4.127), we need to identify a transformation $\phi(x)$ that allows us to express $K(x, x')$ as the inner product $(\phi(x))^{\mathsf{T}} \phi(x')$. Indeed, note that

$$\begin{aligned} K(x, x') &= (1 + aa' + bb')^2 \\ &= (1 + aa')^2 \; + \; b^2 b'^2 \; + \; 2(1 + aa')bb' \\ &= 1 + a^2 a'^2 \; + \; 2aa' \; + \; b^2 b'^2 \; + \; 2bb' \; + \; 2aa'bb' \end{aligned} \tag{4.136}$$

which we can express more compactly as follows. We introduce the transformed vector:

$$\phi(x) = \text{col}\Big\{1, \, \sqrt{2}a, \, \sqrt{2}b, \, \sqrt{2}ab, \, a^2, \, b^2\Big\} \tag{4.137}$$

and note from (4.136) that $K(x, x') = (\phi(x))^{\mathsf{T}} \phi(x')$. In other words, the function (4.135) can be expressed as an inner product between the two transformed vectors $(\phi(x), \phi(x'))$, both of dimension 6×1. Observe further the important fact that the vectors, x and x', have both been transformed in an identical manner.

4.6 CIRCULAR GAUSSIAN DISTRIBUTION[1]

The Gaussian distribution can be extended to complex variables as well. Thus, consider a complex random vector $\boldsymbol{z} = \boldsymbol{x} + j\boldsymbol{y} \in \mathbb{C}^p$. We say $\boldsymbol{z}$ is Gaussian-distributed if its real and imaginary parts are jointly Gaussian (cf. (4.77)), namely, their joint pdf is of the form

$$f_{\boldsymbol{x},\boldsymbol{y}}(x,y) = \frac{1}{(2\pi)^p}\frac{1}{\sqrt{\det R}}\exp\left\{-\frac{1}{2}\left[\begin{array}{cc}(x-\bar{x})^{\mathsf{T}} & (y-\bar{y})^{\mathsf{T}}\end{array}\right]R^{-1}\left[\begin{array}{c}x-\bar{x}\\ y-\bar{y}\end{array}\right]\right\} \tag{4.138}$$

The mean of $\boldsymbol{z}$ is clearly

$$\bar{z} = \mathbb{E}\,\boldsymbol{z} = \bar{x} + j\bar{y} \tag{4.139}$$

while its covariance matrix is

$$R_z \overset{\Delta}{=} \mathbb{E}\,(\boldsymbol{z}-\bar{z})(\boldsymbol{z}-\bar{z})^* = (R_x + R_y) + j(R_{yx} - R_{xy}) \tag{4.140}$$

which is expressed in terms of *both* the covariances and cross-covariance of $\{\boldsymbol{x},\boldsymbol{y}\}$. Note that the variable $\boldsymbol{z}$ can be regarded as a function of the two variables $\{\boldsymbol{x},\boldsymbol{y}\}$ and, therefore, its probabilistic nature is fully characterized by the joint pdf of $\{\boldsymbol{x},\boldsymbol{y}\}$. This joint pdf is defined in terms of the first and second-order moments of $\{\boldsymbol{x},\boldsymbol{y}\}$, i.e., in terms of $\{\bar{x},\bar{y},R_x,R_y,R_{xy}\}$.

It is useful to verify whether it is possible to express the pdf of $\boldsymbol{z}$ *directly* in terms of its own first- and second-order moments, i.e., in terms of $\{\bar{z},R_z\}$. It turns out that this is *not* always possible. This is because knowledge of $\{\bar{z},R_z\}$ alone is not enough to recover the moments $\{\bar{x},\bar{y},R_x,R_y,R_{xy}\}$. More information is needed in the form of a *circularity* condition. To see this, assume we only know $\{\bar{z},R_z\}$. Then, this information is enough to recover $\{\bar{x},\bar{y}\}$ since $\bar{z} = \bar{x} + j\bar{y}$. However, the information is not enough to recover $\{R_x,R_y,R_{xy}\}$. This is because, as we see from (4.140), knowledge of R_z allows us to recover $(R_x + R_y)$ and $(R_{yx} - R_{xy})$ via

$$R_x + R_y = \text{Re}(R_z), \qquad R_{yx} - R_{xy} = \text{Im}(R_z) \tag{4.141}$$

in terms of the real and imaginary parts of R_z. This information is not sufficient to determine the individual covariances $\{R_x,R_y,R_{xy}\}$.

In order to be able to uniquely recover $\{R_x,R_y,R_{xy}\}$ from R_z, it will be further assumed that the random variable $\boldsymbol{z}$ satisfies a *circularity* condition, namely, that

$$\boxed{\mathbb{E}\,(\boldsymbol{z}-\bar{z})(\boldsymbol{z}-\bar{z})^{\mathsf{T}} = 0} \qquad \textbf{(circularity condition)} \tag{4.142}$$

with the transposition symbol T used instead of Hermitian conjugation. Knowledge of R_z, along with circularity, are enough to recover $\{R_x,R_y,R_{xy}\}$ from R_z. Indeed, using the fact that

[1] This section can be skipped on a first reading.

$$\mathbb{E}\,(\boldsymbol{z}-\bar{z})(\boldsymbol{z}-\bar{z})^{\mathsf{T}} \;=\; (R_x - R_y) + j(R_{yx} + R_{xy}) \tag{4.143}$$

we find that, in view of the circularity assumption (4.142), it must now hold that $R_x = R_y$ and $R_{xy} = -R_{yx}$. Consequently, combining with (4.141), we can solve for $\{R_x, R_y, R_{xy}\}$ to get

$$R_x \;=\; R_y \;=\; \frac{1}{2}\,\text{Re}(R_z) \qquad \text{and} \qquad R_{xy} = -R_{yx} = -\frac{1}{2}\,\text{Im}(R_z) \tag{4.144}$$

It follows that the covariance matrix of $\text{col}\{\boldsymbol{x}, \boldsymbol{y}\}$ can be expressed in terms of R_z as

$$R = \frac{1}{2}\begin{bmatrix} \text{Re}(R_z) & -\text{Im}(R_z) \\ \text{Im}(R_z) & \text{Re}(R_z) \end{bmatrix} \tag{4.145}$$

Actually, it also follows that R should have the symmetry structure:

$$R \;=\; \begin{bmatrix} R_x & R_{xy} \\ -R_{xy} & R_x \end{bmatrix} \tag{4.146}$$

with the same matrix R_x appearing on the diagonal, and with R_{xy} and its negative appearing at the off-diagonal locations. Observe further that when $\boldsymbol{z}$ happens to be scalar-valued, then R_{xy} becomes a scalar, say, σ_{xy}, and the condition $R_{xy} = -R_{yx}$ can only hold if $\sigma_{xy} = 0$. That is, the real and imaginary parts of $\boldsymbol{z}$ will need to be independent in the scalar case.

Using result (4.146), we can now verify that the joint pdf of $\{\boldsymbol{x}, \boldsymbol{y}\}$ in (4.138) can be rewritten in terms of $\{\bar{z}, R_z\}$ as shown below – compare with (4.79a) in the real case. Observe in particular that the factors of 2, as well as the square-roots, disappear from the pdf expression in the complex case:

$$\boxed{f_{\boldsymbol{z}}(z) = \frac{1}{\pi^p}\frac{1}{\det R_z}\,\exp\left\{-(z-\bar{z})^* R_z^{-1}(z-\bar{z})\right\}} \tag{4.147}$$

We say that $\boldsymbol{z} \in \mathbb{C}^p$ is a circular or spherically invariant Gaussian random variable. When (4.147) holds, we can check that uncorrelated jointly Gaussian random variables will also be independent; this is one of the main reasons for the assumption of circularity – see Prob. 4.22.

Proof of (4.147): Using (4.146) and the determinantal formula (1.64a), we have

$$\det R = \det(R_x)\,\det(R_x + R_{xy}R_x^{-1}R_{xy}) \tag{4.148}$$

Likewise, using the expression $R_z = 2(R_x - jR_{xy})$, we obtain

$$\begin{aligned}(\det R_z)^2 &= \det(R_z)\,\det(R_z^{\mathsf{T}}) \\ &= 2^{2p}\det(R_x(I - jR_x^{-1}R_{xy}))\,\det(R_x - jR_{xy}^{\mathsf{T}})\end{aligned} \tag{4.149}$$

Noting that

$$R_{xy}^{\mathsf{T}} = R_{yx} \overset{(4.144)}{=} -R_{xy} \tag{4.150}$$

and, for matrices A and B of compatible dimensions, $\det(AB) = \det(BA)$, we get

$$\begin{aligned}(\det R_z)^2 &= 2^{2p} \det R_x \det[(R_x + jR_{xy})(I - jR_x^{-1}R_{xy})] \\ &= 2^{2p} \det(R_x) \det(R_x + R_{xy}R_x^{-1}R_{xy}) \\ &\overset{(4.148)}{=} 2^{2p} \det R \end{aligned} \tag{4.151}$$

so that

$$\frac{1}{(2\pi)^p} \frac{1}{\sqrt{\det R}} = \frac{1}{\pi^p} \frac{1}{\det R_z} \tag{4.152}$$

To conclude the argument, some algebra will show that the exponents in (4.138) and (4.147) are identical – see Prob. 4.23.

■

We can also determine an expression for the fourth-order moment of a circular Gaussian random variable. Following the same argument that led to (4.26), we can similarly verify that if $\boldsymbol{z}$ is a circular Gaussian vector with zero mean and covariance matrix $\mathbb{E}\,\boldsymbol{z}\boldsymbol{z}^* = R_z$ then, for any Hermitian matrix W of compatible dimensions:

$$\boxed{\mathbb{E}\left\{\boldsymbol{z}\boldsymbol{z}^* W \boldsymbol{z}\boldsymbol{z}^*\right\} = R_z \text{Tr}\big(WR_z\big) + R_z W R_z} \tag{4.153}$$

4.7 COMMENTARIES AND DISCUSSION

Gaussian distribution. The origin of the Gaussian distribution is attributed to the German mathematician **Carl Friedrich Gauss (1777–1855)**, who published it in Gauss (1809) while working on two other original ideas, namely, the formulation of the least-squares criterion and the formulation of an early version of the maximum-likelihood criterion. He started from a collection of N independent noisy measurements, $\boldsymbol{y}(n) = \theta + \boldsymbol{v}(n)$, of some unknown parameter θ where the perturbation error was assumed to arise from some unknown pdf, $f_{\boldsymbol{v}}(v)$. Gauss (1809) formulated the problem of estimating the parameter by maximizing the product of the individual likelihoods:

$$\widehat{\theta} = \underset{\theta}{\text{argmax}} \left\{ \prod_{n=1}^{N} f_{\boldsymbol{v}}\Big(y(n) - \theta\Big) \right\} \tag{4.154}$$

He actually worked on a "reverse" problem. He posed the question of determining the form of the noise pdf that would result in an estimate for θ that is equal to the sample mean of the observations, namely, he wanted to arrive at a solution of the form:

$$\widehat{\theta} = \frac{1}{N} \sum_{n=1}^{N} y(n) \tag{4.155}$$

He argued that what we refer to today as the Gaussian distribution is the answer to his inquiry, i.e.,

$$f_{\boldsymbol{v}}(v) = \frac{1}{\sqrt{2\pi\sigma_v^2}} e^{-\frac{v^2}{2\sigma_v^2}} \tag{4.156}$$

For this choice of pdf, the maximum-likelihood formulation (4.154) reduces to the least-squares problem

$$\widehat{\theta} = \underset{\theta}{\text{argmin}} \left\{ \sum_{n=1}^{N} (y(n) - \theta)^2 \right\} \tag{4.157}$$

whose solution is (4.155).

Independently of Gauss, the Irish-American mathematician **Robert Adrain (1775–1843)** also arrived at the Gaussian distribution (as well as the least-squares formulation) in the work by Adrain (1808). He considered a similar estimation problem involving a collection of noisy measurements and postulated that the size of the error in each measurement should be proportional to the size of the measurement itself (namely, larger measurements should contain larger errors). He also postulated that the errors across measurements are independent of each other and moved on to derive the form of the error probability measure that would satisfy these properties, arriving again at the Gaussian distribution – he arrived at the exponential curve e^{-v^2} and referred to it as "*the simplest form of the equation expressing the nature of the curve of probability.*" He used this conclusion to determine the most probable value for the unknown parameter from the noisy observations and arrived again at the sample mean estimate (4.155). In the solution to this problem, he wrote: "*Hence the following rule: Divide the sum of all the observed values by their number, and the quotient will be the most probable value required.*" For further details on the early developments of this branch of statistical analysis, the reader may refer to Stigler (1986).

We will encounter the maximum-likelihood formulation more generally in later chapters. It has become a formidable tool in modern statistical signal analysis, pushed largely by the foundational work of the English statistician **Ronald Fisher (1890–1962)**, who formulated and studied the maximum likelihood approach in its generality in Fisher (1912, 1922, 1925).

Central limit theorem. We explained in the introductory section of this chapter that the Gaussian distribution, also called the *normal distribution*, derives its eminence from the central limit theorem. According to Feller (1945), the name "central limit theorem" is due to Pólya (1920). The earliest formulation of the central limit theorem, and its recognition as a powerful universal approximation law for sums of independent random variables, is due to the French mathematician **Pierre Laplace (1749–1827)** in the treatise by Laplace (1812). He considered a collection of N iid scalar random variables $\{\boldsymbol{x}_n\}$ with mean $\bar{x}$ and finite variance σ_x^2 and showed that the normalized variable

$$\boldsymbol{y} \stackrel{\Delta}{=} \sqrt{N} \left(\frac{1}{N} \sum_{n=1}^{N} (\boldsymbol{x}_n - \bar{x}) \right) \tag{4.158}$$

converges in distribution to $\mathcal{N}_{\boldsymbol{y}}(0, \sigma_x^2)$ as $N \to \infty$, written as

$$\boldsymbol{y} \xrightarrow{d} \mathcal{N}_{\boldsymbol{y}}(0, \sigma_x^2) \tag{4.159}$$

It was not, however, until almost a century later in the works by the Russian and Finnish mathematicians **Aleksandr Lyapunov (1857–1918)** and **Jarl Lindeberg (1876–1932)**, respectively, that the central limit theorem was generalized and placed on a more solid and formal footing – see the treatments by Billingsley (1986) and Fischer (2011). Weaker versions of the theorem were developed where, for example, the requirement of identically distributed random variables was dropped. Both Lyapunov (1901) and Lindeberg (1922) derived sufficient conditions under which the theorem would continue to hold, with the Lindeberg condition being one of the weakest sufficient (and almost necessary) conditions available.

More specifically, let $\{\boldsymbol{x}_n, n = 1, 2, \ldots, N\}$ denote a collection of independent scalar random variables, with possibly different means and variances denoted by $\{\bar{x}_n,\, \sigma^2_{x,n} < \infty\}$. We introduce the sum of variances factor

$$\sigma_N^2 \triangleq \sum_{n=1}^{N} \sigma^2_{x,n} \tag{4.160}$$

and consider the normalized variable

$$\boldsymbol{y} \triangleq \frac{1}{\sigma_N} \sum_{n=1}^{N} (\boldsymbol{x}_n - \bar{x}_n) \tag{4.161}$$

The Lyapunov condition guarantees the convergence of the distribution of $\boldsymbol{y}$ to $\mathcal{N}_{\boldsymbol{y}}(0,1)$ if there exists some $\lambda > 0$ for which

$$\lim_{N\to\infty} \left\{ \frac{1}{\sigma_N^{2+\lambda}} \sum_{n=1}^{N} \mathbb{E}\,(\boldsymbol{x}_n - \bar{x}_n)^{2+\lambda} \right\} = 0 \tag{4.162}$$

A weaker condition is the Lindeberg requirement: It guarantees the convergence of the distribution of $\boldsymbol{y}$ to $\mathcal{N}_{\boldsymbol{y}}(0,1)$ if for every $\epsilon > 0$ it holds that

$$\lim_{N\to\infty} \left\{ \frac{1}{\sigma_N^{2}} \sum_{n=1}^{N} \mathbb{E}\,(\boldsymbol{x}_n - \bar{x}_n)^{2}\, \mathbb{I}\Big[|\boldsymbol{x}_n - \bar{x}_n| > \epsilon\sigma_N \Big] \right\} = 0 \tag{4.163}$$

where the notation $\mathbb{I}[x]$ denotes the indicator function and is defined as follows:

$$\mathbb{I}[x] \triangleq \begin{cases} 1, & \text{when argument } x \text{ is true} \\ 0, & \text{otherwise} \end{cases} \tag{4.164}$$

It can be verified that if condition (4.162) holds then so does (4.163), so that the Lindeberg condition is weaker than the Lyapunov condition. Both conditions essentially amount to requiring the summands that appear in (4.162) and (4.163) to assume small values with high probability.

Stein identity. The Stein lemma (4.68), also known as the Stein identity, is a useful tool in the study of Gaussian-distributed random variables – see Prob. 4.32. The identity is due to Stein (1973, 1981) and was generalized by Hudson (1978) to the family of exponential distributions, as shown in Example 5.2. It was also extended to the vector case by Arnold, Castillo, and Sarabia (2001). The identity is useful in computing moments of transformations of Gaussian random variables. It arises, for example, in the context of the expectation propagation algorithm, which we study in Chapter 34. It has also found applications in many domains, including in finance and asset pricing – see, e.g., Ingersoll (1987) and Cochrane (2001). Several of the other integral identities involving Gaussian distributions derived in Examples 4.1 and 4.7 are motivated by arguments and derivations from Owen (1980), Patel and Read (1996), and Rasmussen and Williams (2006, section 3.9). The proofs of Lemma 4.3 and Example 4.4 are motivated by the derivations from Sayed (2003, 2008).

Gaussian processes. We introduced in Section 4.5 the notion of Gaussian processes and commented on their relation to kernel methods in learning and inference; we will discuss these latter methods in greater detail in Chapter 63. Gaussian processes are a useful modeling tool in the study of learning algorithms, as detailed, for example, in the text by Rasmussen and Williams (2006). We observe from expressions (4.131) and (4.134) for the Gaussian and Ornstein–Uhlenbeck kernels that the "correlation" between two points (x, x') in space decreases as the points move further apart from each other. For this reason, when Gaussian processes are used to model and solve learning and inference problems, it is noted that this property of their kernel translates into the inference decisions being based largely on the closest points in the training data (this behavior is similar to the nearest-neighbor rule discussed in Chapter 52). One early reference on the application of Gaussian processes to statistical inference is the work by O'Hagan (1978). Other notable references on the use of Gaussian processes and kernels for regression and learning applications include Blight and Ott (1975), Poggio and Girosi (1990), Neal (1995, 1996), and Williams and Rasmussen (1995). Similar techniques have also been used in geostatistics under the name of "krigging" – see, e.g., Journel and Huijbregts (1978), Ripley (1981), and Fedorov (1987).

Circular Gaussian distribution. The presentation in Section 4.6 is adapted from Kailath, Sayed, and Hassibi (2000). Expression (4.147) shows the form of a complex Gaussian distribution under the circularity assumption. The original derivation of this form is due to Wooding (1956) – see also Goodman (1963) and the texts by Miller (1974) and Picinbono (1993). This distribution was derived under the circularity condition (4.142), which enables the pdf to be completely characterized by the first- and second-order moments of the complex variable. Under this same condition, uncorrelated Gaussian variables continue to be independent.

PROBLEMS[2]

4.1 Consider two independent and zero-mean real random variables $\{\boldsymbol{u}, \boldsymbol{w}\}$, where $\boldsymbol{u}$ and $\boldsymbol{w}$ are column vectors; both are M-dimensional. The covariance matrices of $\boldsymbol{u}$ and $\boldsymbol{w}$ are defined by $\mathbb{E}\,\boldsymbol{u}\boldsymbol{u}^{\mathsf{T}} = \sigma_u^2 I$ and $\mathbb{E}\,\boldsymbol{w}\boldsymbol{w}^{\mathsf{T}} = C$. Let $\boldsymbol{e}_a = \boldsymbol{u}^{\mathsf{T}}\boldsymbol{w}$.

(a) Show that $\mathbb{E}\,\boldsymbol{e}_a^2 = \sigma_u^2 \text{Tr}(C)$.

(b) Assume $\boldsymbol{u}$ is Gaussian-distributed. Show that $\mathbb{E}\,\boldsymbol{e}_a^2\|\boldsymbol{u}\|^2 = (M+2)\sigma_u^4 \text{Tr}(C)$, where the notation $\|\cdot\|$ denotes the Euclidean norm of its argument.

4.2 Consider K Gaussian distributions with mean μ_k and variance σ_k^2 each. We index these components by $k = 1, 2, \ldots, K$. We select one component k at random with probability π_k. Subsequently, we generate a random variable $\boldsymbol{y}$ according to the selected Gaussian distribution $\mathcal{N}_{\boldsymbol{y}}(\mu_k, \sigma_k^2)$.

(a) What is the pdf of $\boldsymbol{y}$?

(b) What is the mean of $\boldsymbol{y}$?

(c) What is the variance of $\boldsymbol{y}$?

4.3 Let $\boldsymbol{x}$ be a real-valued random variable with pdf $f_{\boldsymbol{x}}(x)$. Define $\boldsymbol{y} = \boldsymbol{x}^2$.

(a) Use the fact that for any nonnegative $\boldsymbol{y}$, the event $\{\boldsymbol{y} \le y\}$ occurs whenever $\{-\sqrt{y} \le \boldsymbol{x} \le \sqrt{y}\}$ to conclude that the pdf of $\boldsymbol{y}$ is given by

$$f_{\boldsymbol{y}}(y) = \frac{1}{2}\frac{f_{\boldsymbol{x}}(\sqrt{y})}{\sqrt{y}} + \frac{1}{2}\frac{f_{\boldsymbol{x}}(-\sqrt{y})}{\sqrt{y}}, \quad y > 0$$

(b) Assume $\boldsymbol{x}$ is Gaussian with zero mean and unit variance. Conclude that $f_{\boldsymbol{y}}(y) = \frac{1}{\sqrt{2\pi y}} e^{-y/2}$ for $y > 0$. *Remark.* The above pdf is known as the chi-square distribution with one degree of freedom. A chi-square distribution with k degrees of freedom is characterized by the pdf:

$$f_{\boldsymbol{y}}(y) = \frac{1}{2^{k/2}\Gamma(k/2)}\, y^{(k-2)/2} e^{-y/2}, \quad y > 0$$

where $\Gamma(\cdot)$ is the so-called gamma function; it is defined by the integral $\Gamma(z) = \int_0^\infty s^{z-1} e^{-s} ds$ for $z > 0$. The function $\Gamma(\cdot)$ has the following useful properties: $\Gamma(1/2) = \sqrt{\pi}$, $\Gamma(z+1) = z\Gamma(z)$ for any $z > 0$, and $\Gamma(n+1) = n!$ for any integer

[2] Some problems in this section are adapted from exercises in Sayed (2003, 2008).

$n \geq 0$. The chi-square distribution with k-degrees of freedom is a special case of the gamma distribution considered in Prob. 5.2 using the parameters $\alpha = k/2$ and $\beta = 1/2$. The mean and variance of $\boldsymbol{y}$ are $\mathbb{E}\,\boldsymbol{y} = k$ and $\sigma_y^2 = 2k$.

(c) Let $\boldsymbol{y} = \sum_{j=1}^{k} \boldsymbol{x}_j^2$ denote the sum of the squares of k independent Gaussian random variables $\boldsymbol{x}_j$, each with zero mean and unit variance. Show that $\boldsymbol{y}$ is chi-square distributed with k degrees of freedom.

4.4 Refer to the pdf expression (4.77) for jointly distributed Gaussian random vectors. Show that if $\boldsymbol{x}$ and $\boldsymbol{y}$ are uncorrelated, then they are also independent.

4.5 Consider the product of three Gaussian distributions over the same random variable $\boldsymbol{x} \in \mathbb{R}^M$:

$$g(x) = \mathcal{N}_{\boldsymbol{x}}(\bar{x}_a, R_a) \times \mathcal{N}_{\boldsymbol{x}}(\bar{x}_b, R_b) \times \mathcal{N}_{\boldsymbol{x}}(\bar{x}_c, R_c)$$

Find an expression for $g(x)$. How should the product be normalized so that $g(x)/Z$ is a Gaussian distribution over x? Find Z.

4.6 Consider a Gaussian distribution over $\boldsymbol{\theta} \sim \mathcal{N}_{\boldsymbol{\theta}}(\bar{\theta}, R_\theta)$, and a second Gaussian distribution over $\boldsymbol{y} \sim \mathcal{N}_{\boldsymbol{y}}(\theta, R_y)$ where the mean is defined in terms of a realization for $\theta \in \mathbb{R}^M$. Find closed-form expressions for the following integrals:

$$I_1 = \int_\theta \theta\, \mathcal{N}_{\boldsymbol{y}}(\theta, R_y)\, \mathcal{N}_\theta(\bar{\theta}, R_\theta) d\theta, \quad I_2 = \int_\theta \theta\theta^{\mathsf{T}}\, \mathcal{N}_{\boldsymbol{y}}(\boldsymbol{\theta}, R_y)\, \mathcal{N}_{\boldsymbol{\theta}}(\bar{\theta}, R_\theta) d\theta$$

4.7 Consider two distributions over the random variables $\boldsymbol{\theta}, \boldsymbol{y} \in \mathbb{R}^M$ of the form:

$$\begin{aligned} f_{\boldsymbol{\theta}}(\theta) &= \mathcal{N}_{\boldsymbol{\theta}}(\bar{\theta}, R_\theta) \\ f_{\boldsymbol{y}|\boldsymbol{\theta}}(y|\theta) &= (1-\alpha)\mathcal{N}_{\boldsymbol{y}}(\theta, R_1) + \alpha\mathcal{N}_{\boldsymbol{y}}(0, R_2), \quad \alpha \in (0,1) \end{aligned}$$

In other words, the conditional pdf of $\boldsymbol{y}$ is parameterized by θ, which appears as the mean of the first Gaussian term. Determine a closed-form expression for the marginal of $\boldsymbol{y}$.

4.8 Let $\boldsymbol{y}$ be a scalar Gaussian-distributed random variable, $\boldsymbol{y} \sim \mathcal{N}_{\boldsymbol{y}}(\bar{y}, \sigma_y^2)$. Establish (4.7). *Remark.* The reader can refer to Owen (1980) and Patel and Read (1996) for a list of similar integral expressions involving Gaussian distributions. Related discussions also appear in Rasmussen and Williams (2006, ch. 3).

4.9 Differentiate both sides of the identity (4.7) once and twice relative to $\bar{y}$ and establish identities (4.9) and (4.11).

4.10 Consider a standard Gaussian distribution with zero mean and unit variance. The error function that is associated with it is denoted by $\text{erf}(z)$ and is defined as the integral

$$\text{erf}(z) \triangleq \frac{2}{\sqrt{\pi}} \int_0^z e^{-x^2} dx$$

The complementary error function is defined by $\text{erfc}(z) = 1 - \text{erf}(z)$.

(a) Verify that $\text{erf}(0) = 0$ and that the function tends to ± 1 as $z \to \pm\infty$.

(b) Comparing with (4.6), verify that $\text{erf}(z) = 2\Phi(\sqrt{2}z) - 1$.

4.11 Consider a Gaussian random vector $\boldsymbol{w} \sim \mathcal{N}_{\boldsymbol{w}}(0, R_w)$ where $\boldsymbol{w} \in \mathbb{R}^M$. Show that

$$\int_{-\infty}^{\infty} \text{erf}\left(h_a^{\mathsf{T}} w\right) \text{erf}\left(h_b^{\mathsf{T}} w\right) \mathcal{N}_{\boldsymbol{w}}(0, R_w) dw = \frac{2}{\pi} \arcsin\left(\frac{2h_a^{\mathsf{T}} R_w h_b}{\sqrt{(1 + 2h_a^{\mathsf{T}} R_w h_a)(1 + 2h_b^{\mathsf{T}} R_w h_b)}}\right)$$

Remark. See Williams (1996) for a related discussion.

4.12 Use identity (4.34) to show that the calculation in (4.38) leads to a Gaussian pdf with mean $\bar{z} = \bar{x} + \bar{y}$ and covariance matrix $R_z = R_x + R_y$.

4.13 Let $\boldsymbol{a}$ denote a real scalar-valued Gaussian random variable with zero mean and variance σ_a^2. Show that $\mathbb{E}\,\boldsymbol{a}^4 = 3\sigma_a^4$.

4.14 Let $\boldsymbol{a}$ denote a complex circular Gaussian random variable with zero mean and variance σ_a^2. Show that $\mathbb{E}\,|\boldsymbol{a}|^4 = 2\sigma_a^4$.

4.15 Assume $\boldsymbol{u}$ is a real Gaussian random column vector with a diagonal covariance matrix Λ. Define $\boldsymbol{z} = \|\boldsymbol{u}\|^2$. What is the variance of $\boldsymbol{z}$?

4.16 Consider two column vectors $\{\boldsymbol{w}, \boldsymbol{z}\}$ that are related via $\boldsymbol{z} = \boldsymbol{w} + \mu \boldsymbol{u}(\boldsymbol{d} - \boldsymbol{u}^{\mathsf{T}} \boldsymbol{w})$, where $\boldsymbol{u}$ is a real Gaussian column vector with a diagonal covariance matrix, $\mathbb{E}\, \boldsymbol{u}\boldsymbol{u}^{\mathsf{T}} = \Lambda$. Moreover, μ is a positive constant and $\boldsymbol{d} = \boldsymbol{u}^{\mathsf{T}} w^o + \boldsymbol{v}$, for some constant vector w^o and random scalar $\boldsymbol{v}$ with variance σ_v^2. The variables $\{\boldsymbol{v}, \boldsymbol{u}, \boldsymbol{w}\}$ are independent of each other. Define $\boldsymbol{e}_a = \boldsymbol{u}^{\mathsf{T}}(w^o - \boldsymbol{w})$, as well as the error vectors $\widetilde{\boldsymbol{z}} = w^o - \boldsymbol{z}$ and $\widetilde{\boldsymbol{w}} = w^o - \boldsymbol{w}$, and denote their covariance matrices by $\{R_{\widetilde{z}}, R_{\widetilde{w}}\}$. Assume $\mathbb{E}\, \boldsymbol{z} = \mathbb{E}\, \boldsymbol{w} = w^o$, while all other random variables are zero-mean.

(a) Verify that $\widetilde{\boldsymbol{z}} = \widetilde{\boldsymbol{w}} - \mu \boldsymbol{u}(\boldsymbol{e}_a + \boldsymbol{v})$.

(b) Show that $R_{\widetilde{z}} = R_{\widetilde{w}} - \mu R_{\widetilde{w}}\Lambda - \mu \Lambda R_{\widetilde{w}} + \mu^2 \left(\Lambda \text{Tr}(R_{\widetilde{w}}\Lambda) + 2\Lambda R_{\widetilde{w}}\Lambda\right) + \mu^2 \sigma_v^2 \Lambda$.

4.17 Consider a collection of N measurements $\{\gamma(n), \boldsymbol{h}_n\}$ where each scalar $\gamma(n)$ is modeled as a noisy perturbation of some Gaussian process $\boldsymbol{g}(h_n)$ defined over the M-dimensional vectors $\{h_n\}$:

$$\gamma(n) = \boldsymbol{g}(h_n) + \boldsymbol{v}(n), \quad \boldsymbol{g} \sim \mathcal{GP}_{\boldsymbol{g}}\Big(0, K(h, h')\Big)$$

Assume the mean function for the Gaussian process $\boldsymbol{g}(\cdot)$ is zero and denote its covariance function by the kernel $K(h, h')$. Assume further that the noise $\boldsymbol{v}(n) \sim \mathcal{N}_{\boldsymbol{v}}(0, \sigma_v^2)$ is white Gaussian with variance σ_v^2 and independent of $\boldsymbol{g}(\cdot)$. Collect the measurements $\{\gamma(n)\}$, the Gaussian process values of $\boldsymbol{g}(\cdot)$, and the perturbations $\{\boldsymbol{v}(n)\}$ into vector quantities:

$$\underbrace{\begin{bmatrix} \gamma(1) \\ \gamma(2) \\ \vdots \\ \gamma(N) \end{bmatrix}}_{\gamma_N} = \underbrace{\begin{bmatrix} \boldsymbol{g}(h_1) \\ \boldsymbol{g}(h_2) \\ \vdots \\ \boldsymbol{g}(h_N) \end{bmatrix}}_{\boldsymbol{g}_N} + \underbrace{\begin{bmatrix} \boldsymbol{v}(1) \\ \boldsymbol{v}(2) \\ \vdots \\ \boldsymbol{v}(N) \end{bmatrix}}_{\boldsymbol{v}_N}$$

so that $\gamma_N = \boldsymbol{g}_N + \boldsymbol{v}_N$. Let R_N denote the covariance matrix of the vector $\boldsymbol{g}_N$ evaluated at the given feature data, $R_N = [K(h_n, h_m)]_{n,m=0}^{N-1}$.

(a) Argue that γ_N has a Gaussian distribution. What is its mean and covariance matrix?

(b) Consider a new vector h and its label γ. What is the conditional pdf of γ given the past data $\{\gamma(n), h_n\}_{n=1}^{N}$?

4.18 Let $\boldsymbol{a}$ and $\boldsymbol{b}$ be scalar real-valued zero-mean jointly Gaussian random variables and denote their correlation by $\rho = \mathbb{E}\, \boldsymbol{a}\boldsymbol{b}$. Price theorem states that for any function $f(\boldsymbol{a}, \boldsymbol{b})$, for which the required derivatives and integrals exist, the following equality due to Price (1958) holds:

$$\frac{\partial^n \mathbb{E}\, f(\boldsymbol{a}, \boldsymbol{b})}{\partial \rho^n} = \mathbb{E}\left(\frac{\partial^{2n} f(\boldsymbol{a}, \boldsymbol{b})}{\partial \boldsymbol{a}^n \partial \boldsymbol{b}^n}\right)$$

in terms of the nth- and $2n$-th order partial derivatives. In simple terms, the Price theorem allows us to move the expectation on the left-hand side outside of the differentiation operation.

(a) Choose $n = 1$ and assume $f(\boldsymbol{a}, \boldsymbol{b})$ has the form $f(\boldsymbol{a}, \boldsymbol{b}) = \boldsymbol{a}g(\boldsymbol{b})$. Verify from the Price theorem that

$$\frac{\partial \mathbb{E}\, \boldsymbol{a}g(\boldsymbol{b})}{\partial \rho} = \mathbb{E}\left(\frac{dg}{dx}\boldsymbol{b}\right)$$

in terms of the derivative of $g(\cdot)$. Integrate both sides over ρ to establish that $\mathbb{E}\, \boldsymbol{a}g(\boldsymbol{b}) = (\mathbb{E}\, \boldsymbol{a}\boldsymbol{b})\, \mathbb{E}\, (dg/dx)\boldsymbol{b}$.

(b) Show further that $\mathbb{E}\,\boldsymbol{b}g(\boldsymbol{b}) = \sigma_b^2\,\mathbb{E}\,(dg/dx)\boldsymbol{b}$ and conclude that the following relation also holds:

$$\mathbb{E}\,\boldsymbol{a}g(\boldsymbol{b}) \;=\; \frac{\mathbb{E}\,\boldsymbol{ab}}{\sigma_b^2}\,\mathbb{E}\,\boldsymbol{b}g(\boldsymbol{b})$$

(c) Assume $g(\boldsymbol{b}) = \text{sign}(\boldsymbol{b})$. Conclude from part (b) that

$$\mathbb{E}\,\boldsymbol{a}\,\text{sign}(\boldsymbol{b}) \;=\; \sqrt{\frac{2}{\pi}}\,\frac{1}{\sigma_b}\,\mathbb{E}\,\boldsymbol{ab}$$

4.19 The Bussgang theorem is a special case of the Price theorem and is due to Bussgang (1952). Let $\{\boldsymbol{a},\boldsymbol{b}\}$ be two real zero-mean Gaussian random variables and define the function

$$g(\boldsymbol{b}) \;\triangleq\; \int_0^{\boldsymbol{b}} e^{-z^2/\sigma^2}\,dz$$

for some $\sigma > 0$. The Bussgang theorem states that

$$\mathbb{E}\,\boldsymbol{a}g(\boldsymbol{b}) \;=\; \frac{1}{\sqrt{\frac{\sigma_b^2}{\sigma^2}+1}}\,\mathbb{E}\,\boldsymbol{ab}$$

The proof of the theorem is as follows. Let $\rho = \mathbb{E}\,\boldsymbol{ab}$. Use the Price general statement from Prob. 4.18 to verify that

$$\frac{\partial \mathbb{E}\,\boldsymbol{a}g(\boldsymbol{b})}{\partial\rho} \;=\; \mathbb{E}\left(\frac{\partial^2 \boldsymbol{a}g(\boldsymbol{b})}{\partial\boldsymbol{a}\partial b}\right) \;=\; \mathbb{E}\left(e^{-\boldsymbol{b}^2/\sigma_z^2}\right)$$

Integrate both sides over ρ to establish the Bussgang theorem.

4.20 Consider two real-valued zero-mean jointly Gaussian random variables $\{\boldsymbol{x},\boldsymbol{y}\}$ with covariance matrix

$$\mathbb{E}\begin{bmatrix}\boldsymbol{x}\\ \boldsymbol{y}\end{bmatrix}\begin{bmatrix}\boldsymbol{x} & \boldsymbol{y}\end{bmatrix} = \begin{bmatrix}1 & \rho\\ \rho & 1\end{bmatrix}$$

That is, $\{\boldsymbol{x},\boldsymbol{y}\}$ have unit variances and correlation ρ. Define the function

$$g(\boldsymbol{x},\boldsymbol{y}) = \frac{2}{\pi\sigma^2}\int_0^{\boldsymbol{x}}\int_0^{\boldsymbol{y}} e^{-\alpha^2/2\sigma^2}e^{-\beta^2/2\sigma^2}\,d\alpha d\beta$$

for some $\sigma > 0$.

(a) Verify that $\partial^2 g(\boldsymbol{x},\boldsymbol{y})/\partial\boldsymbol{x}\partial\boldsymbol{y} = \frac{2}{\pi\sigma^2}e^{-\boldsymbol{x}^2/2\sigma^2}e^{-\boldsymbol{y}^2/2\sigma^2}$, and show that

$$\mathbb{E}\,\frac{\partial^2 g(\boldsymbol{x},\boldsymbol{y})}{\partial\boldsymbol{x}\partial\boldsymbol{y}} \;=\; \frac{2}{\pi}\,\frac{1}{\sqrt{(\sigma^2+1)^2-\rho^2}}$$

(b) Integrate the equality of part (a) over $\rho \in (0,1)$ and conclude that

$$\int_0^1 \mathbb{E}\left(\frac{\partial^2 g(\boldsymbol{x},\boldsymbol{y})}{\partial\boldsymbol{x}\partial\boldsymbol{y}}\right)d\rho \;=\; \frac{2}{\pi}\arcsin\left(\frac{1}{1+\sigma^2}\right)$$

(c) Use the Price identity (cf. Prob. 4.18) to conclude that

$$\mathbb{E}\,g(\boldsymbol{x},\boldsymbol{y}) \;=\; \frac{2}{\pi}\arcsin\left(\frac{1}{1+\sigma^2}\right)$$

4.21 Start from (4.26) and show that result (4.30) holds.

4.22 Refer to the general form (4.147) of a circular Gaussian random vector. Show that uncorrelated Gaussian vectors are also independent.

4.23 Show that the exponents in (4.138) and (4.147) coincide.

4.24 Prove that if condition (4.162) holds then so does condition (4.163).

4.25 Let $\boldsymbol{x}$ denote a Bernoulli random variable, assuming the value 1 with probability p and the value 0 with probability $1-p$. Let $\boldsymbol{S}_N$ denote the sum of N independent Bernoulli experiments. What is the asymptotic distribution of $\boldsymbol{S}_N/N$?

4.26 Let $\boldsymbol{x}_n$ denote a Bernoulli random variable, assuming the value 1 with probability p_n and the value 0 with probability $1-p_n$. Note that we are allowing the probability of success to vary across experiments. Set $\lambda = 1$ and show that the Lyapunov condition (4.162) is satisfied if

$$\lim_{N\to\infty} \sum_{n=1}^{N} p_n(1-p_n) = \infty$$

4.27 Let $\boldsymbol{x}_n$ denote a sequence of iid random variables with mean μ and variance $\sigma_x^2 < \infty$. Show that the Lindeberg condition (4.163) is satisfied.

4.28 Let $\boldsymbol{x}_n$ denote a sequence of iid scalar random variables with mean $\mathbb{E}\,\boldsymbol{x}_n = \mu$ and finite variance, $\sigma_x^2 = \mathbb{E}\,(\boldsymbol{x}(n)-\mu)^2 < \infty$. Introduce the sample average estimator $\widehat{\boldsymbol{\mu}} = \frac{1}{N}\sum_{n=1}^{N} \boldsymbol{x}(n)$ and let $\sigma_{\widehat{\mu}}^2$ denote its variance.

(a) Verify that $\sigma_{\widehat{\mu}}^2 = \sigma_x^2/N$.

(b) Use the Chebyshev inequality (3.28) to conclude the validity of the *weak law of large numbers*, namely, the fact that the sample average converges in probability to the actual mean as $N \to \infty$ – see also future Prob. 3.54:

$$\lim_{N\to\infty} \mathbb{P}\left(|\widehat{\boldsymbol{\mu}} - \mu| \geq \epsilon\right) = 0, \quad \text{for any } \epsilon \geq 0$$

4.29 Let $\{\boldsymbol{x}(n),\ n = 1, \ldots, N\}$ denote N independent realizations with mean μ and and finite variance, $\sigma_x^2 = \mathbb{E}\,(\boldsymbol{x}(n)-\mu)^2 < \infty$. Introduce the *weighted* sample average $\widehat{\boldsymbol{\mu}} = \sum_{n=1}^{N} \alpha(n)\boldsymbol{x}(n)$, where the scalars $\{\alpha(n)\}$ satisfy

$$\alpha(n) \geq 0, \quad \sum_{n=1}^{N} \alpha(n) = 1, \quad \lim_{N\to\infty} \sum_{n=1}^{N} \alpha^2(n) = 0$$

(a) Verify that $\mathbb{E}\,\widehat{\boldsymbol{\mu}} = \mu$ and $\sigma_{\widehat{\mu}}^2 = \sigma_x^2 \left(\sum_{n=1}^{N} \alpha^2(n)\right)$.

(b) Conclude that $\sigma_{\widehat{\mu}}^2 \to 0$ as $N \to \infty$.

4.30 A sequence of $M \times 1$ random vectors $\boldsymbol{x}_n$ converges in distribution to a Gaussian random vector $\boldsymbol{x}$ with zero mean and covariance matrix R_x. A sequence of $M \times M$ random matrices $\boldsymbol{A}_n$ converges in probability to a constant matrix A. What is the asymptotic distribution of the random sequence $\boldsymbol{A}_n\boldsymbol{x}_n$?

4.31 Consider a collection of N iid random variables $\{\boldsymbol{x}_n\}$, each with mean μ and variance σ_x^2. Introduce the sample mean estimator $\widehat{\boldsymbol{\mu}}_N = (1/N)\sum_{n=1}^{N} \boldsymbol{x}_n$, whose mean and variance are given by $\mathbb{E}\,\widehat{\boldsymbol{\mu}} = \mu$ and $\sigma_{\widehat{\mu}}^2 = \sigma_x^2/N$.

(a) Let a and δ be small positive numbers. Use the Chebyshev inequality (3.28) to conclude that at least $N \geq 1/a^2\delta$ samples are needed to ensure that the sample mean lies within the interval $\mu \pm a\sigma_x$ with high likelihood of at least $1-\delta$, namely, $\mathbb{P}(|\widehat{\boldsymbol{\mu}}_N - \mu| < a\sigma_x) \geq 1-\delta$.

(b) Use the central limit theorem (4.158) to find that the conclusion holds by selecting N to satisfy $\delta \leq 2Q(a\sqrt{N})$, where $Q(\cdot)$ denotes the standard Gaussian cdf (i.e., the area under the standard Gaussian distribution $\mathcal{N}(0,1)$):

$$Q(x) \triangleq \frac{1}{\sqrt{2\pi}} \int_{-\infty}^{x} e^{-\frac{1}{2}t^2}\, dt$$

The Q-function is usually tabulated in books on statistics.

(c) Compare the results of parts (a) and (b) to ensure that $\widehat{\boldsymbol{\mu}}$ lies within the interval $\mu \pm 0.05\sigma_x$ with probability larger than or equal to 0.995.

4.32 Consider the same setting of the Stein lemma stated in (4.68). Consider two jointly Gaussian distributed scalar random variables $\boldsymbol{x}$ and $\boldsymbol{y}$. Show that it also holds

$$\mathbb{E}\left(g(\boldsymbol{x}) - \mathbb{E}\,g(\boldsymbol{x})\right)(\boldsymbol{y} - \bar{y}) \;=\; \mathbb{E}\,(\boldsymbol{x} - \bar{x})(\boldsymbol{y} - \bar{y})\mathbb{E}\,g'(\boldsymbol{x})$$

4.33 Repeat the proof of the Stein lemma and establish the validity of (4.67) for vector random variables $\boldsymbol{x} \in \mathbb{R}^M$.

4.34 Refer to expression (4.111) for Z_0. Compute the gradient vector and Hessian matrix of both sides of the equality and establish the validity of results (4.113) and (4.114).

REFERENCES

Adrain, R. (1808), "Research concerning the probabilities of the errors which happen in making observations," *The Analyst*, vol. 1, no. 4, pp. 93–109.

Arnold, B. C., E. Castillo, and J. M. Sarabia (2001), "A multivariate version of Stein's identity with applications to moment calculations and estimation of conditionally specified distributions," *Comm. Statist. Theory Meth.*, vol. 30, pp. 2517–2542.

Billingsley, P. (1986), *Probability and Measure*, 2nd ed., Wiley.

Blight, B. J. N. and L. Ott (1975), "A Bayesian approach to model inadequacy for polynomial regression," *Biometrika*, vol. 62, no. 1, pp. 79–88.

Bussgang, J. J. (1952), *Cross-Correlation Functions of Amplitude Distorted Gaussian Signals*, Technical Report 216, MIT Research Laboratory of Electronics, Cambridge, MA.

Cochrane, J. H. (2001), *Asset Pricing*, Princeton University Press.

Fedorov, V. (1987), "Kriging and other estimators of spatial field characteristics," working paper WP-87-99, International Institute for Applied Systems Analysis (IIASA), Austria.

Feller, W. (1945), "The fundamental limit theorems in probability," *Bull. Amer. Math. Soc.*, vol. 51, pp. 800–832.

Fischer, H. (2011), *A History of the Central Limit Theorem*, Springer.

Fisher, R. A. (1912), "On an absolute criterion for fitting frequency curves," *Messenger Math.*, vol. 41, pp. 155–160.

Fisher, R. A. (1922), "On the mathematical foundations of theoretical statistics," *Philos. Trans. Roy. Soc. London Ser. A.*, vol. 222, pp. 309–368.

Fisher, R. A. (1925), "Theory of statistical estimation," *Proc. Cambridge Philos. Soc.*, vol. 22, pp. 700–725.

Gauss, C. F. (1809), *Theory of the Motion of the Heavenly Bodies Moving about the Sun in Conic Sections*, English translation by C. H. Davis, 1857, Little, Brown, and Company.

Goodman, N. (1963), "Statistical analysis based on a certain multivariate complex Gaussian distribution," *Ann. Math. Stat.*, vol. 34, pp. 152–177.

Hudson, H. M. (1978), "A natural identity for exponential families with application in a multiparameter estimation," *Ann. Statist.*, vol. 6, pp. 473–484.

Ingersoll, J. (1987), *Theory of Financial Decision Making*, Rowman and Littlefield.

Journel, A. G. and C. J. Huijbregts (1978), *Mining Geostatistics*, Academic Press.

Kailath, T., A. H. Sayed, and B. Hassibi (2000), *Linear Estimation*, Prentice Hall.

Laplace, P. S. (1812), *Théorie Analytique des Probabilités*, Paris.

Lindeberg, J. W. (1922), "Eine neue Herleitung des Exponentialgesetzes in der Wahrscheinlichkeitsrechnung," *Mathematische Zeitschrift*, vol. 15, no.1, pp. 211–225.

Lyapunov, A. M. (1901), "Nouvelle forme du théoreme sur la limite de probabilité," *Mémoires de l'Académie de Saint-Petersbourg*, vol. 12, no. 8, pp. 1–24.

Miller, K. (1974), *Complex Stochastic Processes*, Addison-Wesley.

Neal, R. (1995), *Bayesian Learning for Neural Networks*, PhD dissertation, Dept. Computer Science, University of Toronto, Canada.
Neal, R. (1996), *Bayesian Learning for Neural Networks*, Springer.
O'Hagan, A. (1978), "Curve fitting and optimal design for prediction," *J. Roy. Statist. Soc., Ser. B (Methodological)*, vol. 40, no. 1, pp. 1–42.
Owen, D. (1980), "A table of normal integrals" *Commun. Statist. Sim. Comput.*, vol. B9, pp. 389–419.
Patel, J. K. and C. B. Read (1996), *Handbook of the Normal Distribution*, 2nd ed., CRC Press.
Picinbono, B. (1993), *Random Signals and Systems*, Prentice Hall.
Poggio, T. and F. Girosi (1990), "Networks for approximation and learning," *Proc. IEEE*, vol. 78, no. 9, pp. 1481–1497.
Pólya, G. (1920), "Ueber den zentralen Grenzwertsatz der Wahrscheinlichkeitsrechnung und das Momentenproblem," *Math. Zeit.* vol. 8, pp. 171–181.
Price, R. (1958), "A useful theorem for nonlinear devices having Gaussian inputs," *IRE Trans. Inform. Theory*, vol. 4, pp. 69–72.
Rasmussen, C. E. and C. K. I. Williams (2006), *Gaussian Processes for Machine Learning*, MIT Press.
Ripley, B. D. (1981), *Spatial Statistics*, Wiley.
Sayed, A. H. (2003), *Fundamentals of Adaptive Filtering*, Wiley.
Sayed, A. H. (2008), *Adaptive Filters*, Wiley.
Stein, C. M (1973), "Estimation of the mean of a multivariate normal distribution," *Proc. Symp. Asymptotic Statistics*, pp. 345–381, Prague.
Stein, C. M. (1981), "Estimation of the mean of a multivariate normal distribution," *Ann. Statist.*, vol. 9, no. 6, pp. 1135–1151.
Stigler, S. M. (1986), *The History of Statistics: The Measurement of Uncertainty before 1900*, Harvard University Press.
Williams, C. K. I. (1996), "Computing with infinite networks," *Proc. Advances Neural Information Processing Systems* (NIPS), pp. 295–301, Denver, CO.
Williams, C. K. I. and C. E. Rasmussen (1995), "Gaussian processes for regression," *Proc. Advances Neural Information Processing Systems* (NIPS), pp. 514–520, Denver, CO.
Wooding, R. (1956), "The multivariate distribution of complex normal variables," *Biometrika*, vol. 43, pp. 212–215.

5 Exponential Distributions

In this chapter we describe a broad class of probability distributions, known as exponential distributions, which include many important special cases. This family arises frequently in inference and learning problems and deserves close examination.

5.1 DEFINITION

Let $\boldsymbol{y}$ denote a P-dimensional random vector, and consider a vector $\theta \in \mathbb{R}^M$, known as the *natural parameter*, and functions

$$h(y) \geq 0 : \mathbb{R}^P \rightarrow \mathbb{R}, \qquad T(y) : \mathbb{R}^P \rightarrow \mathbb{R}^M, \qquad a(\theta) : \mathbb{R}^M \rightarrow \mathbb{R} \tag{5.1}$$

The variable $\boldsymbol{y}$ is said to follow an exponential distribution (in its *natural* or canonical form) if its probability density function (pdf) has the form:

$$\boxed{f_{\boldsymbol{y}}(y;\theta) \;=\; h(y)e^{\theta^{\mathsf{T}}T(y)-a(\theta)}} \tag{5.2}$$

where, for added emphasis, we are adding the argument θ to the notation $f_{\boldsymbol{y}}(y;\theta)$ to indicate that the pdf depends on θ. The designation "natural" or "canonical" refers to the fact that it is the parameter θ that multiplies $T(y)$ in the exponent. A more general exponential distribution would have the inner product $\theta^{\mathsf{T}}T(y)$ replaced by $(\phi(\theta))^{\mathsf{T}}T(y)$, for some function $\phi(\theta) : \mathbb{R}^M \rightarrow \mathbb{R}^M$. We continue our presentation by focusing on the important case $\phi(\theta) = \theta$, where θ and y appear together *only* in the exponent term $\theta^{\mathsf{T}}T(y)$.

It is clear from (5.2) that the exponential distribution can be rewritten in the equivalent form:

$$f_{\boldsymbol{y}}(y;\theta) \;=\; \frac{1}{Z(\theta)}h(y)e^{\theta^{\mathsf{T}}T(y)} \tag{5.3}$$

where $Z(\theta) = e^{a(\theta)}$ is called the *partition function*; its role is to normalize the area under the curve $h(y)e^{\theta^{\mathsf{T}}T(y)}$ to 1, i.e.,

$$\boxed{Z(\theta) \;=\; \int_{y\in\mathcal{Y}} h(y)e^{\theta^{\mathsf{T}}T(y)}dy} \tag{5.4}$$

where $\mathcal{Y}$ is the domain for the variable y. The function $a(\theta)$ is referred to as the normalization or *log-partition* function (since $a(\theta) = \ln Z(\theta)$). In the factored form (5.3), the function $f_{\boldsymbol{y}}(y)$ is expressed as the product of three terms: one term depends solely on y, a second term depends solely on θ, and a third term involves both θ and y through the inner product $\theta^{\mathsf{T}} T(y)$.

Minimal distribution

An exponential distribution in canonical form is said to be *minimal* if the following two conditions are satisfied:

(a) The entries of $T(y)$ do not satisfy any linear relation among them, i.e., there does not exist any nonzero vector $z \in \mathbb{R}^M$ such that $z^{\mathsf{T}} T(y) = c$ for some constant scalar c. Equivalently, this condition can be stated by saying

$$z^{\mathsf{T}} T(y) = c, \quad \text{for all } y \in \mathcal{Y} \quad \Longrightarrow z = 0 \tag{5.5}$$

(b) The entries of the natural parameter θ do not satisfy a linear relation among them, i.e.,

$$z^{\mathsf{T}} \theta = c, \quad \text{for all } \theta \in \Theta \quad \Longrightarrow z = 0 \tag{5.6}$$

where we are using Θ to denote the domain of θ.

For example, if $T(y) = \text{col}\{y, y\}$ then the inner product $\theta^{\mathsf{T}} T(y)$ can be reduced to a simpler form involving a single parameter:

$$\theta^{\mathsf{T}} T(y) = \theta_1 y + \theta_2 y = (\theta_1 + \theta_2) y \tag{5.7}$$

and we may replace the two-dimensional θ by a new one-dimensional parameter θ. A similar reasoning applies if the entries of θ are related. Minimality avoids these degenerate situations for redundant representations. We will be dealing henceforth with minimal exponential distributions.

Moment parameter

We will encounter a second parameter associated with exponential distributions, known as the *moment parameter* and denoted by η:

$$\boxed{\eta \stackrel{\Delta}{=} \mathbb{E}\, T(\boldsymbol{y})} \qquad \textbf{(moment parameter)} \tag{5.8}$$

The expectation is over the exponential distribution for $\boldsymbol{y}$. We explain later in (5.152) that a one-to-one mapping exists between the two parameters θ and η for minimal distributions so that we can deduce one from the other without loss of information. The mapping from θ to η will be shown to be given by (see expression (5.90)):

$$\boxed{\eta = \nabla_{\theta^{\mathsf{T}}}\, a(\theta)} \tag{5.9}$$

in terms of the gradient of $a(\theta)$. Relation (5.9) maps θ to η. We denote the inverse mapping by the letter ν and write

$$\boxed{\theta = \nu(\eta)} \tag{5.10}$$

We will provide several examples for $\nu(\cdot)$ in the sequel.

5.2 SPECIAL CASES

Many well-known distributions are special cases of the exponential family, as we now verify. The results of this section are summarized in Table 5.1.

Gaussian distribution

Consider a one-dimensional Gaussian distribution of the form

$$f_{\boldsymbol{y}}(y) = \frac{1}{\sqrt{2\pi\sigma^2}} \exp\Big\{-\frac{1}{2\sigma^2}(y-\mu)^2\Big\} \tag{5.11}$$

and note that it can be rewritten as

$$f_{\boldsymbol{y}}(y) = \frac{1}{\sqrt{2\pi}} \exp\left\{-\ln\sigma - \frac{1}{2\sigma^2}y^2 + \frac{\mu}{\sigma^2}y - \frac{\mu^2}{2\sigma^2}\right\} \tag{5.12}$$

Next we introduce the quantities:

$$\theta \stackrel{\Delta}{=} \begin{bmatrix} \mu/\sigma^2 \\ -1/2\sigma^2 \end{bmatrix}, \qquad T(y) \stackrel{\Delta}{=} \begin{bmatrix} y \\ y^2 \end{bmatrix}, \qquad h(y) \stackrel{\Delta}{=} \frac{1}{\sqrt{2\pi}} \tag{5.13}$$

and denote the entries of θ by $\{\theta_1, \theta_2\}$. We further introduce

$$a(\theta) \stackrel{\Delta}{=} \ln\sigma + \frac{\mu^2}{2\sigma^2} = -\frac{1}{2}\ln(-2\theta_2) - \frac{\theta_1^2}{4\theta_2} \tag{5.14}$$

Using these quantities, the Gaussian distribution (5.11) can be written in the form

$$f_{\boldsymbol{y}}(y) = h(y)e^{\theta^{\mathsf{T}}T(y)-a(\theta)} \tag{5.15}$$

which is seen to be of the same form as (5.2). Observe for this example that the moment parameter is given by

$$\eta = \mathbb{E}\,T(\boldsymbol{y}) = \begin{bmatrix} \mathbb{E}\,\boldsymbol{y} \\ \mathbb{E}\,\boldsymbol{y}^2 \end{bmatrix} = \begin{bmatrix} \mu \\ \mu^2+\sigma^2 \end{bmatrix} \tag{5.16}$$

so that the mapping from η to θ is given by

$$\begin{bmatrix} \theta_1 \\ \theta_2 \end{bmatrix} = \begin{bmatrix} \eta_1/(\eta_2-\eta_1^2) \\ -1/2(\eta_2-\eta_1^2) \end{bmatrix} \tag{5.17}$$

Bernoulli distribution

Consider next a random variable $\boldsymbol{y}$ that assumes the value $y = 1$ with probability p and the value $y = 0$ with probability $1 - p$. This situation arises, for example, in an experiment involving the tossing of a coin where the success and failure events correspond to "head" and "tail" outcomes. The pdf of a Bernoulli variable is given by

$$\begin{aligned} f_{\boldsymbol{y}}(y) &= p^y(1-p)^{1-y}, \quad y \in \{0,1\} \\ &= \left(\frac{p}{1-p}\right)^y (1-p) \end{aligned} \tag{5.18}$$

In order to transform a distribution of this form, and of other forms as well, into the exponential form, it is helpful to compute the exponential of the natural logarithm of the given pdf. Applying this step to the above distribution shows that we can rewrite it in the form

$$f_{\boldsymbol{y}}(y) = \exp\left\{ y \ln\left(\frac{p}{1-p}\right) + \ln(1-p)\right\} \tag{5.19}$$

We now introduce the quantities:

$$\theta \stackrel{\Delta}{=} \ln\left(\frac{p}{1-p}\right), \qquad T(y) \stackrel{\Delta}{=} y, \qquad h(y) \stackrel{\Delta}{=} 1 \tag{5.20}$$

as well as

$$a(\theta) \stackrel{\Delta}{=} -\ln(1-p) = \ln(1+e^{\theta}) \tag{5.21}$$

Using these quantities, the Bernoulli distribution (5.18) can be written in the form

$$f_{\boldsymbol{y}}(y) = h(y)e^{\theta T(y) - a(\theta)} \tag{5.22}$$

which is again of the same form as (5.2). Observe further that the moment parameter is given by

$$\eta = \mathbb{E}\, T(\boldsymbol{y}) = p \tag{5.23}$$

and the mapping to θ is

$$\theta = \ln\left(\frac{\eta}{1-\eta}\right) \tag{5.24}$$

Binomial distribution

Consider next a binomial distribution with a known number of experiments, N. Each experiment consists of a Bernoulli trial with success rate p. This situation arises, for example, in an experiment involving the tossing of a coin N times and seeking the probability of observing "heads" or "tails" a certain number of times. Specifically, the probability of y successes occurring in N experiments for a binomial variable is given by

$$
\begin{aligned}
f_{\boldsymbol{y}}(y) &= \binom{N}{y} p^y (1-p)^{N-y}, \quad y = 0, 1, 2, \ldots, N \\
&= \binom{N}{y} \left(\frac{p}{1-p}\right)^y (1-p)^N
\end{aligned} \tag{5.25}
$$

in terms of the notation

$$
\binom{N}{y} \stackrel{\Delta}{=} \frac{N!}{y!(N-y)!} \tag{5.26}
$$

We rewrite (5.25) in the form

$$
f_{\boldsymbol{y}}(y) = \binom{N}{y} \exp\left\{ y \ln\left(\frac{p}{1-p}\right) + N\ln(1-p) \right\} \tag{5.27}
$$

and introduce the quantities:

$$
\theta \stackrel{\Delta}{=} \ln\left(\frac{p}{1-p}\right), \qquad T(y) \stackrel{\Delta}{=} y, \qquad h(y) \stackrel{\Delta}{=} \binom{N}{y} \tag{5.28}
$$

as well as

$$
a(\theta) \stackrel{\Delta}{=} -N\ln(1-p) \;=\; N\ln(1+e^{\theta}) \tag{5.29}
$$

Using these quantities, the binomial distribution (5.25) can be written in the form

$$
f_{\boldsymbol{y}}(y) = h(y) e^{\theta T(y) - a(\theta)} \tag{5.30}
$$

which is of the same form as (5.2). Observe further that the moment parameter is given by

$$
\eta = \mathbb{E}\, T(\boldsymbol{y}) = Np \tag{5.31}
$$

and the mapping to θ is

$$
\theta = \ln\left(\frac{\eta/N}{1-\eta/N}\right) \tag{5.32}
$$

Multinomial distribution

This distribution is an extension of the binomial distribution, where the outcome of each experiment can have more than two possibilities, say, $L \geq 2$ of them. This situation can arise, for instance, if we toss a die with L faces with each face ℓ having a probability p_ℓ of being observed where

$$
\sum_{\ell=1}^{L} p_\ell = 1 \tag{5.33}
$$

The multinomial distribution allows us to assess the likelihood of observing each face a certain number of times in a total of N experiments. Specifically, the probability of observing each face ℓ a number y_ℓ of times is given by

$$f_{\boldsymbol{y}}(y_1, \ldots, y_L) = \frac{N!}{y_1!y_2!\ldots y_L!} p_1^{y_1} p_2^{y_2} \ldots p_L^{y_L} \tag{5.34}$$

where the y_ℓ assume integer values and satisfy

$$y_\ell \in \{0, 1, \ldots, N\}, \qquad \sum_{\ell=1}^{L} y_\ell = N \tag{5.35}$$

We can rewrite the above distribution in the form

$$f_{\boldsymbol{y}}(y) = \frac{N!}{y_1!y_2!\ldots y_L!} \exp\left\{ \sum_{\ell=1}^{L} y_\ell \ln p_\ell \right\} \tag{5.36}$$

This expression suggests that $a(\theta) = 0$ for the multinomial distribution, which is not the case. We need to take into account the fact that the observations $\{y_\ell\}$ satisfy the constraint (5.35); their sum needs to be N. Likewise, the probabilities $\{p_\ell\}$ need to add up to 1 in (5.33). These facts allow us to parameterize one of the observations and one of the probabilities in terms of the remaining observations and probabilities in order to remove redundancy. Thus, note that we can write:

$$\begin{aligned}
\exp\left\{ \sum_{\ell=1}^{L} y_\ell \ln p_\ell \right\} &= \exp\left\{ \sum_{\ell=1}^{L-1} y_\ell \ln p_\ell \; + \; \left(N - \sum_{\ell=1}^{L-1} y_\ell \right) \ln p_L \right\} \\
&= \exp\left\{ \sum_{\ell=1}^{L-1} y_\ell \ln \left(\frac{p_\ell}{p_L} \right) \; + \; N \ln p_L \right\}
\end{aligned} \tag{5.37}$$

We next introduce the quantities:

$$\theta \overset{\Delta}{=} \text{col}\left\{ \ln\left(\frac{p_1}{p_L}\right), \ln\left(\frac{p_2}{p_L}\right), \ldots, \ln\left(\frac{p_{L-1}}{p_L}\right) \right\} \tag{5.38a}$$

$$T(y) \overset{\Delta}{=} \text{col}\{y_1, y_2, \ldots, y_{L-1}\} \tag{5.38b}$$

$$h(y) \overset{\Delta}{=} \frac{N!}{y_1!y_2!\ldots y_L!} \tag{5.38c}$$

and note that we can express the probabilities $\{p_\ell\}$ in terms of the entries $\{\theta_\ell\}$ of θ. Indeed, note first that

$$p_\ell = p_L e^{\theta_\ell}, \quad \ell = 1, 2, \ldots, L-1 \tag{5.39}$$

Adding over ℓ gives

$$\begin{aligned} 1=\sum_{\ell=1}^{L} p_\ell &= p_L+\sum_{\ell=1}^{L-1} p_\ell \\ &= p_L\left(1+\sum_{\ell=1}^{L-1} e^{\theta_\ell}\right) \\ \Longleftrightarrow p_L &= \left(1+\sum_{\ell=1}^{L-1} e^{\theta_\ell}\right)^{-1} \end{aligned} \tag{5.40}$$

and, hence,

$$p_\ell = e^{\theta_\ell}\left(1+\sum_{\ell=1}^{L-1} e^{\theta_\ell}\right)^{-1}, \quad \ell=1,2,\ldots,L-1 \tag{5.41}$$

This leads to the identification

$$a(\theta) \;\stackrel{\Delta}{=}\; -N\ln p_L \;\stackrel{(5.40)}{=}\; N\ln\left(1+\sum_{\ell=1}^{L-1} e^{\theta_\ell}\right) \tag{5.42}$$

Using the quantities $\{\theta, T(y), h(y), a(\theta)\}$, the multinomial distribution (5.34) can then be written in the form

$$f_{\boldsymbol{y}}(y) = h(y)e^{\theta^{\mathsf{T}}T(y)-a(\theta)} \tag{5.43}$$

which is of the same form as (5.2). Observe further that the moment parameter is given by

$$\eta = \mathbb{E}\,T(\boldsymbol{y}) = \operatorname{col}\Big\{Np_1, Np_2, \ldots, Np_{L-1}\Big\} \tag{5.44}$$

and the mapping to θ is

$$\theta = \operatorname{col}\left\{\ln\left(\frac{\eta_\ell/N}{1-\sum_{\ell'=1}^{L-1}\eta_{\ell'}/N}\right)\right\} \tag{5.45}$$

Poisson distribution

Consider next a Poisson distribution with rate $\lambda \geq 0$; this scalar refers to the average number of events expected to occur per unit time (or space). The Poisson distribution is useful to model the likelihood that an integer number of events occurs within a unit time (or unit space). Specifically,

$$f_{\boldsymbol{y}}(y) = \mathbb{P}(y\,\text{events occur}) \;=\; \frac{\lambda^y e^{-\lambda}}{y!}, \quad y=0,1,2,\ldots \tag{5.46}$$

which can be rewritten in the form

$$f_{\boldsymbol{y}}(y) = \frac{1}{y!}\exp\Big\{y\ln\lambda - \lambda\Big\} \tag{5.47}$$

We introduce the quantities:

$$\theta \stackrel{\Delta}{=} \ln\lambda, \qquad T(y) \stackrel{\Delta}{=} y, \qquad h(y) \stackrel{\Delta}{=} 1/y! \tag{5.48}$$

as well as

$$a(\theta) = e^{\theta} \tag{5.49}$$

Using these quantities, the Poisson distribution (5.46) can be written in the form

$$f_{\boldsymbol{y}}(y) = h(y)e^{\theta T(y)-a(\theta)} \tag{5.50}$$

which is again of the same form as (5.2). The moment parameter is given by

$$\eta = \mathbb{E}\,T(\boldsymbol{y}) = \lambda \tag{5.51}$$

and the mapping to θ is

$$\theta = \ln\eta \tag{5.52}$$

Dirichlet distribution

Consider now a Dirichlet distribution of order $L \geq 2$ with positive real parameters $\{s_\ell > 0\}$, namely,

$$\begin{aligned} f_{\boldsymbol{y}}(y) &= K(s)\, y_1^{s_1-1} y_2^{s_2-1} \ldots y_L^{s_L-1} = K(s)\left(\prod_{\ell=1}^{L} y_\ell^{s_\ell-1}\right) \\ &\stackrel{\Delta}{=} \text{Dirichlet}(s) \end{aligned} \tag{5.53}$$

where the scalars $\{y_\ell\}$ are nonnegative and add up to 1:

$$\sum_{\ell=1}^{L} y_\ell = 1, \quad y_\ell \geq 0, \;\; \text{for all } 1 \leq \ell \leq L \tag{5.54}$$

while $K(s)$ is a normalization factor that is a function of $s = \text{col}\{s_1, s_2, \ldots, s_L\}$. This factor is given by the following expression in terms of the gamma function (defined in Prob. 4.3):

$$K(s) = \frac{\Gamma\left(\sum_{\ell=1}^{L} s_\ell\right)}{\prod_{\ell=1}^{L}\Gamma(s_\ell)} \tag{5.55}$$

Each random variable $\boldsymbol{y}_\ell$ lies in the interval $\boldsymbol{y}_\ell \in [0, 1]$. In particular, each $\boldsymbol{y}_\ell$ can be interpreted as the probability of some category ℓ occurring. For this reason, the Dirichlet distribution is useful in generating variables that serve as probability measures (i.e., that are nonnegative and add up to 1). This distribution is commonly used in Bayesian inference to serve as the prior for probability values, such as the ones needed to define a multinomial distribution in (5.34) – see Chapter 37 on topic modeling. The Dirichlet distribution is also a generalization of the beta distribution, which corresponds to the special case $L = 2$, $\boldsymbol{y}_1 = t$, $\boldsymbol{y}_2 = 1 - t$, $s_1 = a$, and $s_2 = b$:

$$f_{\boldsymbol{y}}(y) = \frac{\Gamma(a+b)}{\Gamma(a)\Gamma(b)} t^{a-1}(1-t)^{b-1}, \;\; 0 \leq t \leq 1 \quad \textbf{(beta distribution)} \tag{5.56}$$

When $s_\ell = \alpha$ for all $1 \leq \ell \leq L$ (i.e., when the parameters $\{s_\ell\}$ are uniform and equal to the same value α), the Dirichlet distribution $f_{\boldsymbol{y}}(y)$ is said to be *symmetric*. It reduces to the form

$$f_{\boldsymbol{y}}(y) \;=\; K(\alpha)\left(\prod_{\ell=1}^{L} y_\ell^{\alpha-1}\right), \qquad K(\alpha) = \Gamma(\alpha L)\Big/\Big(\Gamma(\alpha)\Big)^L \tag{5.57}$$

We illustrate the Dirichlet distribution in Fig. 5.1. Each plot in the figure shows $N = 100$ samples from a Dirichlet distribution of order $L = 3$. The three parameters $\{s_1, s_2, s_3\}$ are set equal to 1 in the plot on the left, while the values of s_2 and s_3 are increased to 10 in the middle and rightmost plots. Observe how by controlling the value of the $\{s_\ell\}$ parameters we can concentrate the distribution of the probabilities $\{y_\ell\}$ in different regions within the probability simplex (represented by the triangle connecting the unit probability values on the three axes). We generated the Dirichlet realizations for this figure by exploiting the following useful property for the gamma distribution defined in Prob. 5.2. Assume we generate a collection of L independent gamma-distributed random variables $\boldsymbol{x}_\ell$, where each $\boldsymbol{x}_\ell \sim \text{Gamma}(s_\ell, \beta)$. That is, each gamma distribution has parameters $\alpha = s_\ell$ and arbitrary β (we use $\beta = 1$). We normalize each $\boldsymbol{x}_\ell$ by the sum of the L variables generated in this manner and introduce

$$\boldsymbol{y}_\ell \;\stackrel{\Delta}{=}\; \boldsymbol{x}_\ell\Big/\sum_{\ell'=1}^{L} \boldsymbol{x}'_\ell, \quad \ell = 1, 2, \ldots, L \tag{5.58}$$

Then, it is known that the variables $\boldsymbol{y}_\ell$ will follow a Dirichlet distribution. The proof of this result is left as an exercise – see Prob. 5.4.

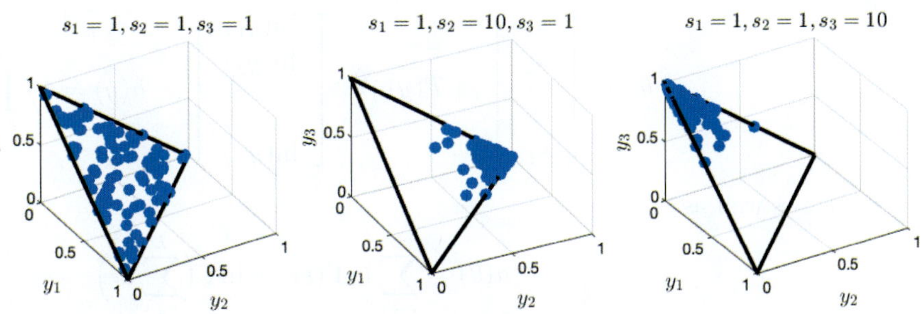

Figure 5.1 Each plot shows $N = 100$ samples from a Dirichlet distribution of order $L = 3$. The parameters $\{s_1, s_2, s_3\}$ are set equal to 1 in the plot on the left, while the values of s_2 and s_3 are increased to 10 in the middle and rightmost plots.

Returning to the general Dirichlet form (5.53), we note that the mean and variance of each $\boldsymbol{y}_\ell$ is given by

$$\bar{y}_\ell \stackrel{\Delta}{=} \mathbb{E}\,\boldsymbol{y}_\ell = s_\ell \Big/ \sum_{\ell'=1}^{L} s_{\ell'}, \qquad \sigma_\ell^2 = \frac{\bar{y}_\ell^2(1-\bar{y}_\ell)}{s_\ell + \bar{y}_\ell} \tag{5.59}$$

Moreover, the mean of $\ln(\boldsymbol{y}_\ell)$ is given by (see Prob. 5.12)

$$\mathbb{E}\ln \boldsymbol{y}_\ell = \psi(s_\ell) - \psi\Big(\sum_{\ell'=1}^{L} s_{\ell'}\Big) \tag{5.60}$$

where $\psi(\cdot)$ is called the *digamma* function and is defined as

$$\psi(x) \stackrel{\Delta}{=} \frac{d}{dx}\ln\Gamma(x) = \frac{\Gamma'(x)}{\Gamma(x)} \tag{5.61}$$

in terms of the ratio of the derivative of the gamma function to the gamma function itself. Based on properties of the gamma function, it is known that

$$\psi(x) \approx -0.577215665 + \sum_{m=0}^{\infty}\left(\frac{1}{1+m} - \frac{1}{x+m}\right) \tag{5.62}$$

This expression can be used to approximate $\psi(x)$ by replacing the infinite series by a finite sum. These results will be useful later when we employ the Dirichlet distribution to define priors for Bayesian inference problems.

To verify that the Dirichlet distribution belongs to the exponential family we first note that

$$\begin{aligned} f_{\boldsymbol{y}}(y) &= \exp\left\{\ln K(s) + \sum_{\ell=1}^{L}(s_\ell - 1)\ln(y_\ell)\right\} \qquad (5.63)\\ &= \exp\left\{\ln\Gamma\Big(\sum_{\ell=1}^{L} s_\ell\Big) - \sum_{\ell=1}^{L}\ln\Gamma(s_\ell) + \sum_{\ell=1}^{L} s_\ell\ln(y_\ell) - \sum_{\ell=1}^{L}\ln(y_\ell)\right\} \end{aligned}$$

Next, we introduce the quantities

$$\theta \stackrel{\Delta}{=} \begin{bmatrix} s_1 \\ s_2 \\ \vdots \\ s_L \end{bmatrix}, \quad T(y) \stackrel{\Delta}{=} \begin{bmatrix} \ln(y_1) \\ \ln(y_2) \\ \vdots \\ \ln(y_L) \end{bmatrix}, \quad h(y) = 1\Big/\prod_{\ell=1}^{L} y_\ell \tag{5.64}$$

as well as

$$a(\theta) = \sum_{\ell=1}^{L}\ln\Gamma(s_\ell) - \ln\Gamma\Big(\sum_{\ell=1}^{L} s_\ell\Big) \tag{5.65}$$

Using these quantities, the Dirichlet distribution (5.53) can be written in the form

$$f_{\boldsymbol{y}}(y) = h(y)e^{\theta^{\mathsf{T}}T(y) - a(\theta)} \tag{5.66}$$

Vector Gaussian distribution

Consider next the vector-valued Gaussian distribution (4.79a), whose form we repeat below for ease of reference:

$$f_{\boldsymbol{y}}(y) = \frac{1}{\sqrt{(2\pi)^P}} \frac{1}{\sqrt{\det R_y}} \exp\left\{-\frac{1}{2}(y-\mu)^\mathsf{T} R_y^{-1}(y-\mu)\right\} \tag{5.67}$$

The parameters for this distribution are the quantities $\{\mu, R_y\}$. Some algebraic manipulations are needed to transform this distribution into a form that fits (5.2). One useful algebraic result we employ to facilitate the derivation is Kronecker property #11 from Table 1.1, namely, for matrices $\{A, C, B\}$ of compatible dimensions it holds that

$$\text{vec}(ACB) = (B^\mathsf{T} \otimes A)\text{vec}(C) \tag{5.68}$$

where the notation vec(A), for a matrix A of size $L \times L$, transforms it into a vector of size $L^2 \times 1$ by stacking the columns of A on top of each other. We denote the reverse operation by unvec: It transforms the $L^2 \times 1$ extended vector back to its $L \times L$ matrix form by unstacking the columns, namely,

$$\text{a} = \text{vec}(A) \iff A = \text{unvec}(a) \tag{5.69}$$

Returning to (5.67), we expand the exponent and write

$$f_{\boldsymbol{y}}(y) = \frac{1}{(2\pi)^{P/2}} \exp\left\{-\frac{1}{2}\ln\det(R_y) - \frac{1}{2}y^\mathsf{T} R_y^{-1} y + \mu^\mathsf{T} R_y^{-1} y - \frac{1}{2}\mu^\mathsf{T} R_y^{-1}\mu\right\} \tag{5.70}$$

Now, using (5.68), we have

$$y^\mathsf{T} R_y^{-1} y = (y^\mathsf{T} \otimes y^\mathsf{T})\text{vec}(R_y^{-1}) \tag{5.71}$$

which motivates us to introduce the quantities:

$$\theta \triangleq \left[\begin{array}{c} R_y^{-1}\mu \\ \hline -\frac{1}{2}\text{vec}(R_y^{-1}) \end{array}\right], \quad T(y) \triangleq \left[\begin{array}{c} y \\ y \otimes y \end{array}\right], \quad h(y) \triangleq \frac{1}{(2\pi)^{P/2}} \tag{5.72}$$

We denote the top and bottom block entries of θ by $\{\theta_1, \theta_2\}$. We further introduce

$$a(\theta) \triangleq \frac{1}{2}\ln\det(R_y) + \frac{1}{2}\mu^\mathsf{T} R_y^{-1}\mu \tag{5.73}$$

and note that

$$R_y^{-1} = \text{unvec}(-2\theta_2) \Rightarrow \frac{1}{2}\ln\det(R_y) = -\frac{1}{2}\ln\det\text{unvec}(-2\theta_2) \tag{5.74}$$

while

$$\begin{aligned} \mu^\mathsf{T} R_y^{-1}\mu &= \mu^\mathsf{T}\theta_1 \\ &= \theta_1^\mathsf{T} R_y \theta_1 \\ &= \theta_1^\mathsf{T}\left(\text{unvec}(-2\theta_2)\right)^{-1}\theta_1 \end{aligned} \tag{5.75}$$

Consequently, we find that

$$a(\theta) \;=\; -\frac{1}{2}\ln\det\mathrm{unvec}(-2\theta_2) - \frac{1}{4}\theta_1^{\mathsf{T}}\left(\mathrm{unvec}(\theta_2)\right)^{-1}\theta_1 \tag{5.76}$$

Using the quantities $\{\theta, T(y), h(y), a(\theta)\}$, the Gaussian distribution (5.67) can then be written as:

$$f_{\boldsymbol{y}}(y) = h(y)e^{\theta^{\mathsf{T}}T(y)-a(\theta)} \tag{5.77}$$

which is of the same form as (5.2). The moment parameter is now given by

$$\eta = \mathbb{E}\,T(\boldsymbol{y}) = \begin{bmatrix} \mu \\ \mathrm{vec}(R_y) + (\mu\otimes\mu) \end{bmatrix} \tag{5.78}$$

and the mapping to θ is

$$\theta = \begin{bmatrix} \Big\{\mathrm{unvec}\Big(\eta_2 - (\eta_1\otimes\eta_1)\Big)\Big\}^{-1}\eta_1 \\ -\frac{1}{2}\Big\{\mathrm{unvec}\Big(\eta_2 - (\eta_1\otimes\eta_1)\Big)\Big\}^{-1} \end{bmatrix} \tag{5.79}$$

In the special case when $R_y = \sigma_y^2 I_P$, it is straightforward to verify that the form (5.77) will hold, albeit with

$$\theta = \begin{bmatrix} \mu/\sigma_y^2 \\ -1/2\sigma_y^2 \end{bmatrix}, \quad T(y) = \begin{bmatrix} y \\ \|y\|^2 \end{bmatrix}, \quad h(y) = \frac{1}{(2\pi)^{P/2}} \tag{5.80}$$

and

$$a(\theta) = \frac{P}{2}\ln(\sigma_y^2) \;+\; \frac{1}{2\sigma_y^2}\|\mu\|^2 \tag{5.81}$$

Moreover, the moment parameter will be

$$\eta = \begin{bmatrix} \mu \\ P\sigma_y^2 + \|\mu\|^2 \end{bmatrix} \tag{5.82}$$

with the mapping to θ given by

$$\theta = \begin{bmatrix} P\eta_1/(\eta_2 - \|\eta_1\|^2) \\ -P/2(\eta_2 - \|\eta_1\|^2) \end{bmatrix} \tag{5.83}$$

Problem 5.10 considers the case when R_y is diagonal.

5.3 USEFUL PROPERTIES

Returning to (5.2), it is useful to note that once the functions $\{h(y), T(y)\}$ are specified, then the function $a(\theta)$ is fully determined by them (i.e., the exponential distribution has fewer degrees of freedom than is apparent). This is because,

Table 5.1 Formulation of popular probability distributions as special cases of the exponential family of distributions (5.2).

Name	pdf	Exponential distribution
Gaussian	$f_{\boldsymbol{y}}(y) = \frac{1}{\sqrt{2\pi\sigma^2}}\, e^{-\frac{1}{2\sigma^2}(y-\mu)^2}$	$\theta = \begin{bmatrix} \mu/\sigma^2 \\ -1/2\sigma^2 \end{bmatrix},\ a(\theta) = -\frac{1}{2}\ln(-2\theta_2) - \frac{\theta_1^2}{4\theta_2}$ $T(y) = \begin{bmatrix} y \\ y^2 \end{bmatrix},\ h(y) = 1/\sqrt{2\pi}$
Bernoulli	$f_{\boldsymbol{y}}(y) = p^y(1-p)^{1-y},\ y \in \{0,1\}$	$\theta = \ln\left(\frac{p}{1-p}\right),\ a(\theta) = \ln(1+e^\theta),\ T(y) = y,\ h(y) = 1$
binomial	$f_{\boldsymbol{y}}(y) = \binom{N}{y} p^y(1-p)^{N-y},\ y = 0,1,2,\ldots,N$	$\theta = \ln\left(\frac{p}{1-p}\right),\ a(\theta) = N\ln(1+e^\theta),\ T(y) = y,\ h(y) = \binom{N}{y}$
multinomial	$f_{\boldsymbol{y}}(y_1,\ldots,y_L) = \frac{N!}{y_1!\ldots y_L!} p_1^{y_1}\ldots p_L^{y_L},\ y_\ell = 0,1,\ldots,N$ $\sum_{\ell=1}^{L} p_\ell = 1,\quad \sum_{\ell=1}^{L} y_\ell = N$	$\theta = \text{col}\left\{\ln\left(\frac{p_1}{p_L}\right),\ldots,\ln\left(\frac{p_{L-1}}{p_L}\right)\right\},\ h(y) = \frac{N!}{y_1!y_2!\ldots y_L!}$ $T(y) = \text{col}\Big\{y_1,y_2,\ldots,y_{L-1}\Big\},\ a(\theta) = N\ln\left(1+\sum_{\ell=1}^{L-1} e^{\theta_\ell}\right)$
Poisson	$f_{\boldsymbol{y}}(y) = \lambda^y e^{-\lambda}/y!,\ y = 0,1,2,\ldots$	$\theta = \ln\lambda,\ a(\theta) = e^\theta,\ T(y) = y,\ h(y) = 1/y!$
Dirichlet	$f_{\boldsymbol{y}}(y) = \frac{\Gamma\left(\sum_{\ell=1}^{L} s_\ell\right)}{\prod_{\ell=1}^{L}\Gamma(s_\ell)} \prod_{\ell=1}^{L} y_\ell^{s_\ell-1},\ \{s_\ell > 0\},\ \sum_{\ell=1}^{L} y_\ell = 1$	$\theta = \text{col}\{s_\ell\},\ a(\theta) = \sum_{\ell=1}^{L}\ln\Gamma(s_\ell) - \ln\Gamma\Big(\sum_{\ell=1}^{L} s_\ell\Big)$ $T(y) = \text{col}\Big\{\ln(y_\ell)\Big\},\ h(y) = 1/\prod_{\ell=1}^{L} y_\ell$
vector Gaussian	$f_{\boldsymbol{y}}(y) = \frac{1}{\sqrt{(2\pi)^P}}\frac{1}{\sqrt{\det R_y}}\exp\left\{-\frac{1}{2}(y-\mu)^{\mathsf{T}} R_y^{-1}(y-\mu)\right\}$	$\theta = \begin{bmatrix} R_y^{-1}\mu \\ \hline -\frac{1}{2}\text{vec}(R_y^{-1}) \end{bmatrix},\ T(y) \overset{\Delta}{=} \begin{bmatrix} y \\ y \otimes y \end{bmatrix},\ h(y) = 1/(2\pi)^{P/2}$ $a(\theta) = -\frac{1}{2}\ln\det\text{unvec}(-2\theta_2) - \frac{1}{4}\theta_1^{\mathsf{T}}\left(\text{unvec}(\theta_2)\right)^{-1}\theta_1$

as befitting of a true pdf, the distribution of $\boldsymbol{y}$ must satisfy the following normalization, where the integration is over the domain of $\boldsymbol{y}$, denoted by $\mathcal{Y}$:

$$\begin{aligned}\int_{y\in\mathcal{Y}} f_{\boldsymbol{y}}(y)dy = 1 \Longleftrightarrow & \int_{y\in\mathcal{Y}} h(y)e^{\theta^{\mathsf{T}}T(y)-a(\theta)}dy = 1 \\ \Longleftrightarrow & \int_{y\in\mathcal{Y}} h(y)e^{\theta^{\mathsf{T}}T(y)}dy \;=\; e^{a(\theta)} = Z(\theta)\end{aligned} \tag{5.84}$$

We let Θ denote the set of all parameter vectors, θ, for which the integral expression on the left-hand side is finite (this is also called the *natural* parameter space):

$$\Theta \;\stackrel{\Delta}{=}\; \left\{\theta\in\mathbb{R}^M \text{ such that } \int_{y\in\mathcal{Y}} h(y)e^{\theta^{\mathsf{T}}T(y)}dy \;<\; \infty\right\} \tag{5.85}$$

When the set Θ is nonempty, we refer to the exponential family of distributions as being *regular*. Then, for such regular distributions and for any $\theta\in\Theta$, it holds that

$$\boxed{a(\theta) \;=\; \ln\left(\int_{y\in\mathcal{Y}} h(y)e^{\theta^{\mathsf{T}}T(y)}dy\right)} \tag{5.86}$$

In other words, the function $a(\theta)$ is the log of a normalization factor determined fully by $h(y)$ and $T(y)$. This relation can be used to establish the following useful properties about Θ and the log-partition function, $a(\theta)$ – the proofs are given in Appendix 5.A:

(a) The set Θ is convex. We will define convex sets in Chapter 8. Here, it suffices to state that convexity means the following: If $\theta_1,\theta_2\in\Theta$, then the combination $\lambda\theta_1+(1-\lambda)\theta_2\in\Theta$ for any scalar $\lambda\in[0,1]$. That is, any point on the segment connecting θ_1 and θ_2 should lie inside Θ as well.

(b) The function $a(\theta)$ is convex. Again, we will define convex functions in Chapter 8. Here, it is sufficient to state that convexity means the following: For any scalar $\lambda\in[0,1]$, it holds that

$$a(\lambda\theta_1+(1-\lambda)\theta_2)\le\lambda a(\theta_1)+(1-\lambda)a(\theta_2) \tag{5.87}$$

The function $a(\theta)$ will be *strictly* convex, with the inequality in (5.87) replaced by strict inequality, if the exponential distribution is *minimal*. We will explain later in (8.30) that convexity and strict convexity of $a(\theta)$ can be characterized in terms of the nonnegative definiteness of its Hessian matrix, namely,

$$\left\{\begin{array}{lll}\text{(a) } \nabla^2_\theta\, a(\theta)\ge 0 \text{ for all } \theta & \Longleftrightarrow & a(\theta) \text{ is convex}\\ \text{(b) } \nabla^2_\theta\, a(\theta)> 0 \text{ for all } \theta & \Longrightarrow & a(\theta) \text{ is strictly convex}\end{array}\right. \tag{5.88}$$

(c) It holds that

$$\boxed{\nabla_{\theta^\mathsf{T}}\, a(\theta) \;=\; \mathbb{E}\, T(\boldsymbol{y})} \tag{5.89}$$

In other words, the expectation of $T(\boldsymbol{y})$ is determined by the gradient of the normalization function, $a(\theta)$. Let us denote the expectation by $\eta = \mathbb{E}\, T(\boldsymbol{y})$, which is the *moment* parameter of the exponential distribution. We therefore get a useful relation between the natural (θ) and moment (η) parameters of the exponential distribution:

$$\boxed{\eta \;=\; \nabla_{\theta^\mathsf{T}}\, a(\theta)} \tag{5.90}$$

This relation maps θ (natural domain) to η (moment domain). When $a(\theta)$ is strictly convex, the above relation is invertible so that we can write

$$\boxed{\theta = \nu(\eta)} \tag{5.91}$$

for some function $\nu(\cdot)$ that now maps η (moment domain) back to θ (natural domain).

(d) It holds that

$$\boxed{\nabla^2_{\theta}\, a(\theta) = R_T \geq 0} \tag{5.92}$$

where R_T refers to the covariance matrix of $T(\boldsymbol{y})$:

$$R_T \;\stackrel{\Delta}{=}\; \mathbb{E}\, T(\boldsymbol{y})\, (T(\boldsymbol{y}))^\mathsf{T} \;-\; (\mathbb{E}\, T(\boldsymbol{y}))\, (\mathbb{E}\, T(\boldsymbol{y}))^\mathsf{T} \tag{5.93}$$

It follows that the Hessian matrix of $a(\theta)$ is nonnegative, which is equivalent to the convexity property (b). Observe that properties (c) and (d) indicate that the first- and second-order moments of $T(\boldsymbol{y})$ can be deduced from $a(\theta)$.

(e) The score function of a pdf, and the associated Fisher information matrix, are denoted by $\mathcal{S}(\theta)$ and $F(\theta)$, respectively, and defined as

$$\mathcal{S}(\theta) \;\stackrel{\Delta}{=}\; \nabla_{\theta^\mathsf{T}}\Big(\ln f_{\boldsymbol{y}}(y;\theta)\Big), \qquad F(\theta) \;=\; \mathbb{E}\, \mathcal{S}(\theta)\mathcal{S}^\mathsf{T}(\theta) \tag{5.94}$$

We will encounter these concepts later in (31.96)–(31.97) and explain their significance in the context of maximum-likelihood estimation at that stage. For now, it is sufficient to define the score matrix and the Fisher matrix as above. Applying these definitions to the exponential distribution (5.2) and using (5.89), we conclude that

$$\mathcal{S}(\theta) = T(\boldsymbol{y}) - \mathbb{E}\, T(\boldsymbol{y}) \tag{5.95}$$

so that the covariance matrix of $T(\boldsymbol{y})$ coincides with the Fisher information matrix for the exponential distribution (5.2), i.e.,

$$\boxed{F(\theta) \;=\; R_T} \tag{5.96}$$

Example 5.1 **(Application to Gaussian and binomial distributions)** We illustrate result (5.89) by considering the Gaussian and binomial distributions from Table 5.1. For the Gaussian case, we have

$$\theta = \begin{bmatrix} \mu/\sigma^2 \\ -1/2\sigma^2 \end{bmatrix}, \quad T(\boldsymbol{y}) = \begin{bmatrix} \boldsymbol{y} \\ \boldsymbol{y}^2 \end{bmatrix}, \quad a(\theta) = -\frac{1}{2}\ln(-2\theta_2) - \frac{\theta_1^2}{4\theta_2} \tag{5.97}$$

Differentiating $a(\theta)$ relative to θ_1 gives

$$\frac{da(\theta)}{d\theta_1} = -\frac{1}{2}\frac{\theta_1}{\theta_2} = \frac{1}{2}\frac{\mu/\sigma^2}{1/2\sigma^2} = \mu \tag{5.98}$$

That is, $\mathbb{E}\,\boldsymbol{y} = \mu$, as expected. For the binomial case we have

$$\theta = \ln\left(\frac{p}{1-p}\right), \quad T(\boldsymbol{y}) = \boldsymbol{y}, \quad a(\theta) = N\ln(1+e^{\theta}) \tag{5.99}$$

Differentiating $a(\theta)$ relative to θ gives

$$\frac{da(\theta)}{d\theta} = \frac{Ne^{\theta}}{1+e^{\theta}} = pN \tag{5.100}$$

since $e^{\theta} = p/(1-p)$. Therefore, $\mathbb{E}\,\boldsymbol{y} = pN$, as expected.

Example 5.2 **(Stein lemma for exponential distributions)** We extend the statement of the Stein lemma (4.67) to the family of exponential distributions. Thus, let the pdf of $\boldsymbol{x} \in \mathbb{R}^M$ take the form

$$f_{\boldsymbol{x}}(x) = h(x)\exp\left\{\theta^{\mathsf{T}}T(x) - a(\theta)\right\}, \quad x \in \mathcal{X} \tag{5.101}$$

with a scalar-valued basis function $h(x)$, a scalar-valued normalization factor $a(\theta)$, and a vector-valued $T(x)$, say,

$$T(x) = \text{col}\left\{T_1(x), T_2(x), \ldots, T_M(x)\right\}, \quad (M \times 1) \tag{5.102}$$

We denote the individual entries of $\boldsymbol{x}$ by $\{\boldsymbol{x}_1, \boldsymbol{x}_2, \ldots, \boldsymbol{x}_M\}$. Consider a function $g(x)$ that satisfies the finite expectation condition $\mathbb{E}\,\|\nabla_x\, g(\boldsymbol{x})\| < \infty$ and, moreover, the product $g(x)f_{\boldsymbol{x}}(x)$ tends to zero as x tends to the boundary of its domain $\mathcal{X}$. Then, it holds that

$$\boxed{\mathbb{E}\left(\frac{\nabla_{x^{\mathsf{T}}}\, h(\boldsymbol{x})}{h(\boldsymbol{x})} + \nabla_{x^{\mathsf{T}}}T(\boldsymbol{x})\,\theta\right)g(\boldsymbol{x}) = -\mathbb{E}\,\nabla_{\boldsymbol{x}^{\mathsf{T}}}\, g(\boldsymbol{x})} \tag{5.103}$$

where, by definition, the gradient of $h(x)$ is a column vector and the Jacobian of $T(x)$ is a square matrix:

$$\nabla_{x^{\mathsf{T}}}\, h(x) = \text{col}\left\{\frac{\partial h(x)}{\partial x_1}, \frac{\partial h(x)}{\partial x_2}, \ldots, \frac{\partial h(x)}{\partial x_M}\right\}, \quad (M \times 1) \tag{5.104a}$$

$$\nabla_{x^{\mathsf{T}}}\, T(x) = \begin{bmatrix} \nabla_{x^{\mathsf{T}}}T_1(x) & \nabla_{x^{\mathsf{T}}}T_2(x) & \ldots & \nabla_{x^{\mathsf{T}}}T_M(x) \end{bmatrix}, \quad (M \times M) \tag{5.104b}$$

Note that the columns of the Jacobian matrix consist of the individual gradient vectors of the functions $T_m(x)$. The proof of (5.103) is as follows. Note first that

$$\mathbb{E}\left(\frac{\nabla_{x^\mathsf{T}} h(\boldsymbol{x})}{h(\boldsymbol{x})} + \nabla_{x^\mathsf{T}} T(\boldsymbol{x})\,\theta\right) g(\boldsymbol{x}) \tag{5.105}$$

$$= \int_{x\in\mathcal{X}} \left(\frac{\nabla_{x^\mathsf{T}} h(x)}{h(x)} + \nabla_{x^\mathsf{T}} T(x)\,\theta\right) g(x)h(x)\exp\left\{\theta^\mathsf{T} T(x) - a(\theta)\right\} dx$$

We can carry out this integration by parts. Let

$$u = g(x) \implies du = \Big(\nabla_{x^\mathsf{T}}\, g(x)\Big) dx \tag{5.106a}$$

$$dv = \left(\frac{\nabla_{x^\mathsf{T}} h(x)}{h(x)} + \nabla_{x^\mathsf{T}} T(x)\,\theta\right) h(x)\exp\left\{\theta^\mathsf{T} T(x) - a(\theta)\right\} dx$$

$$\implies v = h(x)\exp\left\{\theta^\mathsf{T} T(x) - a(\theta)\right\} \;=\; f_{\boldsymbol{x}}(x) \tag{5.106b}$$

It follows that, where we are using b_ℓ and b_u to refer generically to boundaries of the domain of $\mathcal{X}$,

$$\mathbb{E}\left(\frac{\nabla_{x^\mathsf{T}} h(\boldsymbol{x})}{h(\boldsymbol{x})} + \nabla_{x^\mathsf{T}} T(\boldsymbol{x})\,\theta\right) g(\boldsymbol{x}) = uv\Big|_{b_\ell}^{b_u} - \int_{u\in\mathcal{U}} v du \tag{5.107}$$

$$= 0 - \int_{x\in\mathcal{X}} \nabla_{x^\mathsf{T}}\, g(x)\, f_{\boldsymbol{x}}(x) dx \;=\; -\mathbb{E}\,\nabla_{x^\mathsf{T}}\, g(\boldsymbol{x})$$

as claimed. Let us illustrate the result by applying it to a scalar Gaussian distribution with mean μ and variance σ_x^2. We know from (5.14) that this situation corresponds to the case:

$$T(x) = \begin{bmatrix} x \\ x^2 \end{bmatrix}, \quad \theta = \begin{bmatrix} \mu/\sigma_x^2 \\ -1/2\sigma_x^2 \end{bmatrix}, \quad h(x) = 1/\sqrt{2\pi} \tag{5.108}$$

We select $g(x) = x^2$. Then, equality (5.103) leads to

$$\mathbb{E}\left(\frac{\mu}{\sigma_x^2} - \frac{2x}{2\sigma_x^2}\right)\boldsymbol{x}^2 \;=\; -2\mathbb{E}\,\boldsymbol{x} = -2\mu \tag{5.109}$$

Expanding and solving for $\mathbb{E}\,\boldsymbol{x}^3$, we find

$$\mathbb{E}\,\boldsymbol{x}^3 = 3\mu\sigma^2 + \mu^3 \tag{5.110}$$

5.4 CONJUGATE PRIORS

The family of exponential distributions plays a prominent role in statistical Bayesian inference. Often, while seeking optimal inference or classification solutions, simplifications will occur when *conjugate prior* distributions (or conjugate pairs) are used. We will encounter applications of conjugate priors in Chapter 36 in the context of variational inference. Here, we define what the notion means and examine its implications for exponential distributions.

Consider two random variables $\boldsymbol{y}$ and $\boldsymbol{\theta}$. Let $f_{\boldsymbol{\theta}}(\theta)$ denote the pdf of $\boldsymbol{\theta}$, also called the *prior*. Let $f_{\boldsymbol{y}|\boldsymbol{\theta}}(y|\theta)$ denote the conditional pdf of the observation $\boldsymbol{y}$ given $\boldsymbol{\theta}$, which is called the *likelihood* of $\boldsymbol{y}$. Using the Bayes rule (3.39), we can transform the likelihood function into the *posterior* for $\boldsymbol{\theta}$ as follows:

$$f_{\boldsymbol{\theta}|\boldsymbol{y}}(\theta|y) = \frac{f_{\boldsymbol{\theta}}(\theta)\, f_{\boldsymbol{y}|\boldsymbol{\theta}}(y|\theta)}{f_{\boldsymbol{y}}(y)} \tag{5.111}$$

That is,

$$\textbf{posterior} = \frac{\textbf{prior} \times \textbf{likelihood}}{\textbf{evidence}} \tag{5.112}$$

The prior and likelihood distributions that appear in the numerator on the right-hand side, namely, $f_{\boldsymbol{\theta}}(\theta)$ and $f_{\boldsymbol{y}|\boldsymbol{\theta}}(y|\theta)$, are said to be *conjugate pairs* if the posterior, $f_{\boldsymbol{\theta}|\boldsymbol{y}}(\theta|y)$, belongs to the same family of distributions as the prior, $f_{\boldsymbol{\theta}}(\theta)$:

$$\begin{aligned} &f_{\boldsymbol{\theta}}(\theta) \text{ and } f_{\boldsymbol{y}|\boldsymbol{\theta}}(y|\theta) \text{ are conjugate priors} \Longleftrightarrow \\ &f_{\boldsymbol{\theta}}(\theta) \text{ and } f_{\boldsymbol{\theta}|\boldsymbol{y}}(\theta|y) \text{ belong to the same family of distributions} \end{aligned} \tag{5.113a}$$

or, in words,

$$\begin{aligned} &\textbf{prior and likelihood are conjugate priors} \Longleftrightarrow \\ &\textbf{prior and posterior belong to the same family of distributions} \end{aligned} \tag{5.113b}$$

When this happens, we say that the prior distribution is the conjugate prior to the likelihood function.

Example 5.3 **(Illustration of conjugate pairs)**

(a) The Gaussian distribution is its own conjugate prior. Specifically, consider scalar random variables $(\boldsymbol{y}, \boldsymbol{\theta})$ with Gaussian distributions:

$$f_{\boldsymbol{\theta}}(\theta) = \mathcal{N}_{\boldsymbol{\theta}}\left(\bar{\theta}_{\text{prior}}, \sigma^2_{\text{prior}}\right), \quad f_{\boldsymbol{y}|\boldsymbol{\theta}}(y|\theta) = \mathcal{N}_{\boldsymbol{y}|\boldsymbol{\theta}}(\theta, \sigma_y^2) \tag{5.114}$$

for some known variance σ_y^2. In this example, the realization θ defines the mean for the distribution of $\boldsymbol{y}$. Then, starting from the Bayes update (5.111), some straightforward algebra will show that the posterior distribution is also Gaussian (see Prob. 5.13):

$$f_{\boldsymbol{\theta}|\boldsymbol{y}}(\theta|y) = \mathcal{N}_{\boldsymbol{\theta}|\boldsymbol{y}}\left(\bar{\theta}_{\text{post}}, \sigma^2_{\text{post}}\right) \tag{5.115}$$

with its mean and variance satisfying the relations:

$$\frac{1}{\sigma^2_{\text{post}}} = \frac{1}{\sigma^2_{\text{prior}}} + \frac{1}{\sigma_y^2} \tag{5.116a}$$

$$\frac{\bar{\theta}_{\text{post}}}{\sigma^2_{\text{post}}} = \frac{\bar{\theta}_{\text{prior}}}{\sigma^2_{\text{prior}}} + \frac{y}{\sigma_y^2} \tag{5.116b}$$

(b) The beta distribution is the conjugate prior for the binomial distribution. Indeed, assume the likelihood function for a scalar observation $\boldsymbol{y}$ is binomial, i.e.,

$$f_{\boldsymbol{y}|\boldsymbol{\theta}}(y|\theta) = \binom{N}{y}\theta^y(1-\theta)^{N-y}, \quad y = 0, 1, 2, \dots, N \tag{5.117}$$

where N is the number of experiments and θ is the probability of success. Assume also that the prior is the beta distribution:

$$f_{\boldsymbol{\theta}}(\theta; a, b) = \text{Beta}(\theta; a, b) \triangleq \begin{cases} \dfrac{\Gamma(a+b)}{\Gamma(a)\Gamma(b)}\theta^{a-1}(1-\theta)^{b-1}, & 0 \leq \theta \leq 1 \\ 0, & \text{otherwise} \end{cases} \tag{5.118}$$

where $\Gamma(x)$ denotes the gamma function defined earlier in Prob. 4.3. Then, starting from the Bayes update (5.111), some straightforward algebra will show

$$f_{\boldsymbol{\theta}|\boldsymbol{y}}(\theta|y) = \text{Beta}(\theta; a+y, b+N-y) \tag{5.119}$$

That is, the posterior is a beta distribution with parameters $(a+y, b+N-y)$.

(c) The Dirichlet distribution is the conjugate prior for the multinomial distribution. Indeed, let $\boldsymbol{\theta} = \text{col}\{\boldsymbol{\theta}_1, \boldsymbol{\theta}_2, \dots, \boldsymbol{\theta}_L\}$ denote a vector of nonnegative probabilities adding up to 1, $\sum_{\ell=1}^{L} \boldsymbol{\theta}_\ell = 1$. Assume the likelihood function for a collection of observations $\boldsymbol{y} = \text{col}\{\boldsymbol{y}_1, \dots, \boldsymbol{y}_L\}$ over N experiments is the multinomial (recall (5.34)):

$$f_{\boldsymbol{y}|\boldsymbol{\theta}}(y|\theta) = \frac{N!}{y_1!y_2!\dots y_L!}\theta_1^{y_1}\theta_2^{y_2}\dots\theta_L^{y_L} \tag{5.120}$$

where the $\{y_\ell\}$ assume integer values in the interval $0 \leq y_\ell \leq N$ and satisfy:

$$y_\ell \in \{0, 1, \dots, N\}, \qquad \sum_{\ell=1}^{L} y_\ell = N \tag{5.121}$$

Assume further that the prior follows a Dirichlet distribution with parameters $\{s_\ell > 0\}$, namely,

$$f_{\boldsymbol{\theta}}(\theta) = \frac{\Gamma\left(\sum_{\ell=1}^{L} s_\ell\right)}{\prod_{\ell=1}^{L}\Gamma(s_\ell)}\theta_1^{s_1-1}\theta_2^{s_2-1}\dots\theta_L^{s_L-1} = \text{Dirichlet}(s_1, s_2, \dots, s_L) \tag{5.122}$$

Then, starting from the Bayes update (5.111), some straightforward algebra will show:

$$f_{\boldsymbol{\theta}|\boldsymbol{y}}(\theta|y) = \text{Dirichlet}(s_1+y_1, s_2+y_2, \dots, s_L+y_L) \tag{5.123}$$

That is, the posterior is again a Dirichlet distribution with parameters $\{s_\ell + y_\ell\}$.

Let us now examine more generally the form of the conjugate prior for the exponential family of distributions. We return to the case where $\boldsymbol{y}$ is P-dimensional and θ is M-dimensional. Assume the likelihood for an observation $\boldsymbol{y}$ conditioned on $\boldsymbol{\theta}$ has the form of a canonical exponential distribution:

$$\boxed{f_{\boldsymbol{y}|\boldsymbol{\theta}}(y|\theta) = h(y)\,e^{\theta^{\mathsf{T}}T(y)-a(\theta)}} \tag{5.124}$$

Every such exponential form has a conjugate prior. To see this, let us postulate a form for the conjugate prior that is similar to the above likelihood function, say,

$$f_{\boldsymbol{\theta}}(\theta) = k(\theta)\, e^{\lambda_1^\mathsf{T}\theta - \lambda_2 a(\theta) - b(\lambda)} \tag{5.125}$$

for some base function $k(\theta)$, log-normalization factor $b(\lambda)$, and parameters $\lambda_1 \in \mathbb{R}^M$ and $\lambda_2 \in \mathbb{R}$. If we introduce the $(M+1)$-dimensional natural parameter $\lambda = \text{col}\{\lambda_1, \lambda_2\}$ and let $G(\theta) = \text{col}\{\theta, -a(\theta)\}$, then it becomes clear that the above prior has the form of an exponential distribution since

$$f_{\boldsymbol{\theta}}(\theta) = k(\theta)\, e^{\lambda^\mathsf{T} G(\theta) - b(\lambda)} \propto k(\theta) e^{\lambda^\mathsf{T} G(\theta)} \tag{5.126}$$

We now verify that (5.125) is indeed the conjugate prior for (5.124) by showing that the posterior has a similar exponential form.

Proof: From the Bayes rule (5.111) we have, apart from scaling by $f_{\boldsymbol{y}}(y)$,

$$f_{\boldsymbol{\theta}|\boldsymbol{y}}(\theta|y) \propto \left(h(y)\, e^{\theta^\mathsf{T} T(y) - a(\theta)}\right) \left(k(\theta)\, e^{\lambda_1^\mathsf{T}\theta - \lambda_2 a(\theta) - b(\lambda)}\right) \tag{5.127}$$

That is,

$$\begin{aligned} f_{\boldsymbol{\theta}|\boldsymbol{y}}(\theta|y) &\propto k(\theta) \exp\left\{ \Big(\lambda_1 + T(y)\Big)^\mathsf{T} \theta - (\lambda_2 + 1) a(\theta) \right\} \\ &\propto k(\theta) e^{(\lambda')^\mathsf{T} G(\theta)} \end{aligned} \tag{5.128}$$

where we introduced

$$\lambda' \triangleq \begin{bmatrix} \lambda_1 + T(y) \\ \lambda_2 + 1 \end{bmatrix} \tag{5.129}$$

Comparing with (5.126) shows that the posterior for $\boldsymbol{\theta}$ belongs to the same exponential family of distributions as the prior, albeit with parameter λ'.

■

The proportionality constants in (5.126) and (5.128) are independent of the parameter θ. Therefore, conditioned on the observation y, the only major change in moving from the prior distribution to the posterior distribution is replacing the parameter λ by λ':

$$\lambda_1 \text{ is replaced by } \lambda_1 + T(y) \tag{5.130a}$$

$$\lambda_2 \text{ is replaced by } \lambda_2 + 1 \tag{5.130b}$$

$$\lambda = \begin{bmatrix} \lambda_1 \\ \lambda_2 \end{bmatrix} \text{ is replaced by } \lambda' = \begin{bmatrix} \lambda_1 + T(y) \\ \lambda_2 + 1 \end{bmatrix} \tag{5.130c}$$

Example 5.4 **(Collection of independent observations)** Consider a collection of N independent observations $\{\boldsymbol{y}_n\}$, each with an exponential likelihood function:

$$f_{\boldsymbol{y}_n|\boldsymbol{\theta}}(y_n|\theta) = h(y_n)\, e^{\theta^\mathsf{T} T(y_n) - a(\theta)} \tag{5.131}$$

Let $\boldsymbol{y}_{1:N} = \text{col}\{\boldsymbol{y}_1, \ldots, \boldsymbol{y}_N\}$ denote the vector collection of all observations. Due to independence, the aggregate likelihood function for the observations is given by

$$f_{\boldsymbol{y}_{1:N}|\boldsymbol{\theta}}(y_{1:N}|\theta) = \left(\prod_{n=1}^{N} h(y_n)\right) \exp\left\{\theta^{\mathsf{T}}\left(\sum_{n=1}^{N} T(y_n)\right) - Na(\theta)\right\} \tag{5.132}$$

The conjugate prior continues to be given by (5.126) or, equivalently, (5.125). Indeed, repeating the same argument used in the proof leading to the example we can again verify that the posterior in this case is given by

$$\begin{aligned} f_{\boldsymbol{\theta}|\boldsymbol{y}_{1:N}}(\theta|y_{1:N}) &\propto k(\theta)\exp\left\{\left(\lambda_1 + \sum_{n=1}^{N} T(y_n)\right)^{\mathsf{T}} \theta - (\lambda_2 + N)a(\theta)\right\} \\ &= k(\theta)e^{(\lambda')^{\mathsf{T}} G(\theta)} \end{aligned} \tag{5.133}$$

Comparing with (5.125) shows that the posterior for $\boldsymbol{\theta}$ belongs to the same exponential family of distributions as the prior with parameter vector given by

$$\lambda' \triangleq \text{col}\left\{\lambda_1 + \sum_{n=1}^{N} T(y_n),\; N + \lambda_2\right\} \tag{5.134}$$

That is,

$$\lambda_1 \text{ is replaced by } \lambda_1 + \sum_{n=1}^{N} T(y_n) \tag{5.135a}$$

$$\lambda_2 \text{ is replaced by } \lambda_2 + N \tag{5.135b}$$

$$\lambda = \begin{bmatrix} \lambda_1 \\ \lambda_2 \end{bmatrix} \text{ is replaced by } \lambda' = \begin{bmatrix} \lambda_1 + \sum_{n=1}^{N} T(y_n) \\ \lambda_2 + N \end{bmatrix} \tag{5.135c}$$

5.5 COMMENTARIES AND DISCUSSION

Sufficient statistic. The function $T(y)$ that appears in the pdf expression (5.2) for the exponential distribution plays the important role of a *sufficient statistic*. Generally, any function of the observation, say, $S(y)$, is a statistic. However, this notion is most relevant when the statistic happens to be *sufficient*; a concept introduced by **Ronald Fisher (1890–1962)** in the work by Fisher (1922). Recall that $f_{\boldsymbol{y}}(y;\theta)$ is parameterized by θ. Later, starting with Chapter 31 on maximum-likelihood estimation, we will devise methods that allow us to estimate θ from observing a collection of independent realizations $\{y_n\}$ arising from $f_{\boldsymbol{y}}(y;\theta)$. The estimate, denoted by $\widehat{\theta}$, will in general be a function of the observations. For example, assume $f_{\boldsymbol{y}}(y;\theta)$ is a Gaussian distribution with known variance σ_y^2 but unknown mean μ, say, $\boldsymbol{y} \sim \mathcal{N}_{\boldsymbol{y}}(\mu, \sigma_y^2)$. In this case, the mean μ plays the role of the parameter θ. We will find later, when we discuss the maximum-likelihood estimation technique, that one way to estimate μ is to use the sample mean of the observations, i.e.,

$$\widehat{\mu} = \frac{1}{N}\sum_{n=1}^{N} y_n \tag{5.136a}$$

The quantity on the right-hand side (the sample mean) is an example of a statistic: It is a function of the observations:

$$S(y_1, \ldots, y_N) = \frac{1}{N}\sum_{n=1}^{N} y_n \tag{5.136b}$$

The expression for $\widehat{\mu}$ shows that in order to estimate μ, it is not necessary to save the individual observations $\{y_n\}$ but only their sample average. Sufficient statistics allow us to "compress" the observations into a useful summary that can be used to estimate underlying model parameters, θ. While any function of the observations is a statistic, $S(y_{1:N})$ will be *sufficient* for estimating θ if the conditional distribution of $\{\boldsymbol{y}_{1:N}\}$ given $S(\boldsymbol{y}_{1:N})$ does not depend on θ:

$$\begin{aligned} &f_{\boldsymbol{y}_{1:N}|S(\boldsymbol{y}_{1:N})}\Big(y_{1:N}|S(y_{1:N})\Big) \text{ does not depend on } \theta \\ &\Longrightarrow S(y_{1:N}) \text{ is a sufficient statistic for estimating } \theta \end{aligned} \tag{5.137}$$

This definition essentially means that no other statistic computed from the same observations can provide additional information about the parameter of interest, θ. In the next comment, we state the important factorization result (5.138), which allows us to verify whether a statistic $S(\cdot)$ is sufficient. For more details on sufficient statistics, the reader may refer to Zacks (1971), Rao (1973), Cassella and Berger (2002), Cox (2006), and Young and Smith (2010).

Exponential family of distributions. This family of distributions plays a central role in the study of sufficient statistics. According to results from Darmois (1935), Koopman (1936), Pitman (1936), and Barankin and Maitra (1963), and as stated in Andersen (1970, p. 1248), "*under certain regularity conditions on the probability density, a necessary and sufficient condition for the existence of a sufficient statistic of fixed dimension is that the probability density belongs to the exponential family.*" Some useful references on the exponential family of distributions, including coverage of several of their properties as discussed in Section 5.3, are the works by Lehmann (1959), Zacks (1971), Brown (1986), McCullagh and Nelder (1989), Lehmann and Casella (1998), Young and Smith (2010), Nielsen and Garcia (2011), Hogg and McKean (2012), Bickel and Doksum (2015), and Dobson and Barnett (2018). Discussions on conjugate priors for exponential distributions can be found in Diaconis and Ylvisaker (1979) and Bernardo and Smith (2000). A second key result linking the exponential family of distributions to the concept of sufficient statistics is the factorization theorem, which provides a useful mechanism to identify sufficient statistics. The result is known as the Fisher–Neyman factorization theorem – see Prob. 5.11.

Factorization theorem (Fisher (1922), Neyman (1935), Halmos and Savage (1949)): *Consider a pdf, $f_{\boldsymbol{y}}(y;\theta)$, that is parameterized by some vector parameter, θ. A function $S(y)$ is a sufficient statistic for θ if, and only if, the pdf $f_{\boldsymbol{y}}(y;\theta)$ can be factored as the product of two nonnegative functions:*

$$f_{\boldsymbol{y}}(y;\theta) \;=\; h(y)\,g(S(y);\theta) \tag{5.138}$$

where $h(y)$ depends solely on y, while $g(S(y);\theta)$ depends on θ and on the observation through $S(y)$.

The result of the theorem can be applied to a single pdf or to a joint pdf. For example, if we consider a *collection* of N iid observations arising from the exponential distribution $f_{\boldsymbol{y}}(y;\theta)$ defined by (5.2), then the joint pdf is given by

$$f_{\boldsymbol{y}}(y_1, y_2, \ldots, y_N; \theta) = \left(\prod_{n=1}^{N} h(y_n)\right) \exp\left\{\theta^{\mathsf{T}} \left(\sum_{n=1}^{N} T(y_n)\right) - Na(\theta)\right\} \quad (5.139)$$

which is seen to be of the desired factored form (5.138). Therefore, for this case,

$$\text{sufficient statistic} = S(y_1, \ldots, y_N) \triangleq \sum_{n=1}^{N} T(y_n) \quad (5.140)$$

PROBLEMS

5.1 Refer to the expressions in Table 5.1 for $a(\theta)$ for several traditional probability distributions. Use result (5.89) to determine the mean of the variable $T(\boldsymbol{y})$ in each case. Express your results in terms of the original parameters for the various distributions.

5.2 A gamma distribution is defined by the following pdf in terms of two parameters $\alpha > 0$ and $\beta > 0$:

$$f_{\boldsymbol{y}}(y) = \frac{\beta^{\alpha}}{\Gamma(\alpha)} y^{\alpha-1} e^{-\beta y}, \quad y > 0$$

(a) Verify that this distribution belongs to the exponential family with parameter vector $\theta = \text{col}\{-\beta, \alpha - 1\}$, $T(y) = \text{col}\{y, \ln y\}$, and log-partition function $a(\theta) = \ln\Gamma(\alpha) - \alpha \ln \beta$.

(b) Establish that $\mathbb{E}\, \boldsymbol{y} = \alpha/\beta$ and $\mathbb{E} \ln \boldsymbol{y} = \psi(\alpha) - \ln(\beta)$, where $\psi(x)$ refers to the digamma function (5.61).

(c) Show that $\sigma_y^2 = \alpha/\beta^2$ for $\alpha \geq 1$.

5.3 Consider the following generalized logistic distribution:

$$f_{\boldsymbol{y}}(y) = \frac{\theta e^{-y}}{(1 + e^{-y})^{\theta+1}}, \quad \theta > 0$$

(a) Show that this pdf belongs to the exponential family of distributions.

(b) Compute the mean and variance of the random variable $\boldsymbol{x} = \ln(1 + e^{-\boldsymbol{y}})$.

5.4 Show that the variables $\{\boldsymbol{y}_\ell\}$ constructed according to (5.58) follow a Dirichlet distribution with parameters $\{s_\ell\}$.

5.5 One useful property of the exponential family of distributions is that if the pdf of $\boldsymbol{y}$ belongs to this family, then the sufficient statistic $T(\boldsymbol{y})$ will also have a distribution that belongs to the exponential family. Assume y is one-dimensional and refer to the exponential distribution (5.2). Assume $T(y)$ is differentiable relative to y. Show that the distribution of the random variable $T(\boldsymbol{y})$ belongs to the exponential family as well.

5.6 Assume the joint distribution of two variables $\{\boldsymbol{y}, \boldsymbol{z}\}$ belongs to the exponential family of distributions, say, $f_{\boldsymbol{y},\boldsymbol{z}}(y, z) = h(y, z)\exp\left\{\theta^{\mathsf{T}} T(y, z) - a(\theta)\right\}$. Use the Bayes rule to establish that the conditional pdf $f_{\boldsymbol{z}|\boldsymbol{y}}(z|y)$ also belongs to the exponential family of distributions, say, $f_{\boldsymbol{z}|\boldsymbol{y}}(z|y) = h(z)\exp\left\{\eta^{\mathsf{T}} S(z) - r(\eta)\right\}$ for some parameter η, sufficient statistic $S(z)$, and log-normalization factor $r(\eta)$. Repeat when $f_{\boldsymbol{y},\boldsymbol{z}}(y, z) = h(y, z)\exp\left\{(\phi(\theta))^{\mathsf{T}} T(y, z) - a(\theta)\right\}$ for some vector-valued function $\phi(\cdot)$.

5.7 A random variable $\boldsymbol{y}$ is said to follow a categorical distribution if it assumes one of R possible classes with probability p_r each, i.e., $\mathbb{P}(\boldsymbol{y} = r) = p_r$ with $\sum_{r=1}^{R} p_r = 1$.

(a) Argue that the pdf of $\boldsymbol{y}$ can be written in the following form in terms of the indicator function:

$$f_{\boldsymbol{y}}(y) = p_1 \left(\frac{p_2}{p_1}\right)^{\mathbb{I}[y=2]} \left(\frac{p_3}{p_1}\right)^{\mathbb{I}[y=3]} \cdots \left(\frac{p_R}{p_1}\right)^{\mathbb{I}[y=R]}$$

where $\mathbb{I}[x] = 1$ if x is true and is zero otherwise.

(b) Conclude that this distribution belongs to the exponential family of distributions with

$$h(y) = 1, \quad a(\theta) = -\ln p_1, \quad \theta = \begin{bmatrix} \ln(p_2/p_1) \\ \ln(p_3/p_1) \\ \vdots \\ \ln(p_R/p_1) \end{bmatrix}, \quad T(y) = \begin{bmatrix} \mathbb{I}[y=2] \\ \mathbb{I}[y=3] \\ \vdots \\ \mathbb{I}[y=R] \end{bmatrix}$$

(c) Argue that we can also use

$$h(y) = 1, \quad a(\theta) = 0, \quad \theta = \begin{bmatrix} \ln(p_1) \\ \ln(p_2) \\ \vdots \\ \ln(p_R) \end{bmatrix}, \quad T(y) = \begin{bmatrix} \mathbb{I}[y=1] \\ \mathbb{I}[y=2] \\ \vdots \\ \mathbb{I}[y=R] \end{bmatrix}$$

5.8 A random variable $\boldsymbol{y}$ is said to follow a Gibbs (or Boltzmann) distribution if it assumes one of R possible classes with probability p_r each, and where these probabilities are parameterized by a vector $\lambda \in \mathbb{R}^M$ as follows:

$$\mathbb{P}(\boldsymbol{y} = r) = p_r = e^{b_r^\mathsf{T}\lambda} \Big/ \Big(\sum_{r'=1}^{R} e^{b_{r'}^\mathsf{T}\lambda}\Big)^{-1}$$

In this expression, a vector b_r is associated with each class r. Use the result of part (a) from Prob. 5.7 to show that this distribution belongs to the exponential family of distributions with

$$h(y) = 1, \quad a(\theta) = \ln\Big(\sum_{r=1}^{R} e^{(b_r - b_1)^\mathsf{T}\lambda}\Big)$$

$$\theta = \begin{bmatrix} (b_2 - b_1)^\mathsf{T}\lambda \\ (b_3 - b_1)^\mathsf{T}\lambda \\ \vdots \\ (b_R - b_1)^\mathsf{T}\lambda \end{bmatrix}, \quad T(y) = \begin{bmatrix} \mathbb{I}[y=2] \\ \mathbb{I}[y=3] \\ \vdots \\ \mathbb{I}[y=R] \end{bmatrix}$$

5.9 The dispersed exponential family of distributions takes the following form (compare with (5.2)):

$$f_{\boldsymbol{y}}(y;\theta) \;=\; h(y;\sigma^2)\exp\Big\{\frac{1}{\sigma^2}\Big(\theta^\mathsf{T}T(y) - a(\theta)\Big)\Big\}$$

where $\sigma^2 > 0$ is a dispersion factor and $h(\cdot)$ is dependent on σ^2. How does $\nabla_{\theta^\mathsf{T}}\, a(\theta)$ relate to $\mathbb{E}\,T(\boldsymbol{y})$ and to the covariance matrix of $T(\boldsymbol{y})$?

5.10 Refer to the vector Gaussian distribution (5.67) and assume R_y is diagonal, denoted by Σ. Let $\text{diag}(\Sigma^{-1})$ denote the column vector containing the diagonal entries of Σ^{-1}. Show that the exponential form (5.77) will continue to hold albeit with

$$\theta = \begin{bmatrix} \Sigma^{-1}\mu \\ -\frac{1}{2}\text{diag}(\Sigma^{-1}) \end{bmatrix}, \quad T(y) = \begin{bmatrix} y \\ y \odot y \end{bmatrix}, \quad h(y) = \frac{1}{(2\pi)^{P/2}}$$

in terms of the Hadamard (i.e., elementwise) product $y \odot y$ and

$$a(\theta) = \frac{1}{2}\ln\det(\Sigma) \;+\; \frac{1}{2}\mu^\mathsf{T}\Sigma\mu$$

5.11 Assume $\boldsymbol{y}$ is a discrete random variable. Establish the statement of the factorization theorem (5.138). Specifically, establish the following steps:

(a) Assume first that the factorization (5.138) holds. Show that

$$\mathbb{P}(S(\boldsymbol{y}) = t) = g(t;\theta)\left(\sum_{y|S(y)=t} h(y)\right)$$

where the sum is over all realizations y for which $S(y) = t$.

(b) Conclude from part (a) that $\mathbb{P}(\boldsymbol{y} = y|S(\boldsymbol{y}) = t)$ is independent of θ so that $S(y)$ is a sufficient statistic for θ.

(c) Now assume that $S(y)$ is a sufficient statistic. Choose $h(y) = \mathbb{P}(\boldsymbol{y} = y|S(\boldsymbol{y}) = t)$ and $g(S(\boldsymbol{y}) = t;\theta) = \mathbb{P}(S(\boldsymbol{y}) = t;\theta)$. Verify that (5.138) holds for these choices.

5.12 Use property (5.89) to establish expression (5.60) for the mean of the logarithm of a Dirichlet-distributed variable.

5.13 Establish result (5.115) and derive the expressions for $\bar{\theta}_{\text{post}}$ and σ^2_{post}.

5.14 Consider again a Gaussian likelihood in the form $f_{\boldsymbol{y}|\boldsymbol{\theta}}(y|\theta) = \mathcal{N}_{\boldsymbol{y}|\boldsymbol{\theta}}(\theta, \sigma^2)$, where the mean θ is now known whereas the variance σ^2 is a realization from a random distribution with prior $\boldsymbol{\sigma}^2 \sim f_{\boldsymbol{\sigma}^2}(\sigma^2)$. Assume an inverse gamma form for this prior, where an inverse gamma distribution is defined by two parameters (α, β) and takes the form

$$f_{\boldsymbol{\sigma}^2}(\sigma^2;\alpha,\beta) = \frac{\beta^\alpha}{\Gamma(\alpha)}\left(1/\sigma^2\right)^{\alpha+1} e^{-\beta/\sigma^2}, \quad \sigma^2 > 0$$

Show that the posterior, $f_{\boldsymbol{\sigma}^2|\boldsymbol{y}}(\sigma^2|y)$, continues to have an inverse gamma distribution with parameters $\alpha' = \alpha + \frac{1}{2}$ and $\beta' = \beta + \frac{1}{2}(y-\theta)^2$.

5.15 What is the conjugate prior of the Bernoulli distribution?

5.16 What is the conjugate prior of the Poisson distribution?

5.17 Refer to Example 5.4 and consider a more general exponential form for the observations, say,

$$f_{\boldsymbol{y}_n|\boldsymbol{\theta}}(y_n|\theta) = h(y_n)\, e^{\phi(\theta)^\mathsf{T} T(y_n) - a(\theta)}$$

where we are replacing the inner product $\theta^\mathsf{T} T(y_n)$ in the exponent by $\phi(\theta)^\mathsf{T} T(y)$ for some vector function $\phi(\theta)$. Repeat the derivation in the example to show that the posterior function now takes the form

$$f_{\boldsymbol{\theta}|\boldsymbol{y}_{1:N}}(\theta|y_{1:N}) \propto k(\theta) e^{(\lambda')^\mathsf{T} G'(\theta)}$$

where

$$\lambda' \triangleq \text{col}\left\{\lambda_1,\ \sum_{n=1}^{N} T(y_n),\ N + \lambda_2\right\}, \quad G'(\theta) \triangleq \text{col}\left\{\theta,\ \phi(\theta),\ -a(\theta)\right\}$$

5.18 Assume we impose a number of constraints on a pdf, $f_{\boldsymbol{y}}(y)$, such as requiring

$$\mathbb{E}\, T_m(\boldsymbol{y}) = c_m, \quad m = 1, 2, \ldots, M$$

for some functions $T_m(\boldsymbol{y})$ and constants $\{c_m\}$. For example, the $T_m(\cdot)$ could be the functions $\{\boldsymbol{y}, \boldsymbol{y}^2, \boldsymbol{y}^3, \ldots\}$, in which case the constraints would correspond to conditions on the moments of $\boldsymbol{y}$. We wish to determine the pdf $f_{\boldsymbol{y}}(y)$ that solves the following constrained optimization problem:

$$f_{\boldsymbol{y}}(y) \triangleq \underset{f_{\boldsymbol{y}}(\cdot)}{\text{argmin}} \left\{\int_{y\in\mathcal{Y}} f_{\boldsymbol{y}}(y) \ln f_{\boldsymbol{y}}(y) dy\right\}$$

subject to the above conditions and to the normalization

$$\int_{y\in\mathcal{Y}} f_{\boldsymbol{y}}(y) dy = 1, \quad f_{\boldsymbol{y}}(y) \geq 0$$

Based on definition (6.35) for the differential entropy of a probability distribution, the above problem seeks a distribution $f_{\boldsymbol{y}}(y)$ with maximum entropy that satisfies the specified constraints. Note that this is an optimization problem over the space of probability density *functions*. One way to seek the solution is as follows. Introduce the Lagrangian function:

$$\mathcal{L}(f) \triangleq \int_{y\in\mathcal{Y}} f_{\boldsymbol{y}}(y)\ln f_{\boldsymbol{y}}(y)dy \;-\; \alpha\left(\int_{y\in\mathcal{Y}} f_{\boldsymbol{y}}(y)dy - 1\right) - \sum_{m=1}^{M}\theta_m\left(\int_{y\in\mathcal{Y}} T_m(y)f_{\boldsymbol{y}}(y)dy - c_m\right)$$

in terms of real-valued Lagrange multipliers $\{\alpha, \theta_m\}$.

(a) Let $g(y)$ denote any smooth function of y. Write down the expression for the perturbed Lagrangian $\mathcal{L}(f+\epsilon g)$, for small ϵ.

(b) Derive an expression for $\partial\mathcal{L}(f+\epsilon g)/\partial\epsilon$.

(c) Show that the expression derived in part (b) reduces to the following for $\epsilon \to 0$:

$$\lim_{\epsilon\to 0}\left\{\frac{\partial}{\partial\epsilon}\mathcal{L}(f+\epsilon g)\right\} = \int_{y\in\mathcal{Y}}\left(1+\ln(f_{\boldsymbol{y}}(y)) - \alpha - \sum_{m=1}^{M}\theta_m T_m(y)\right)g(y)dy$$

(d) Argue that for the limit in part (c) to be equal to zero for any choice of g, it must hold that

$$1+\ln(f_{\boldsymbol{y}}(y)) - \alpha - \sum_{m=1}^{M}\theta_m T_m(y) = 0$$

Conclude that $f_{\boldsymbol{y}}(y)$ must belong to the exponential family of distributions, namely,

$$f_{\boldsymbol{y}}(y) \propto \exp\left\{\sum_{m=1}^{M}\theta_m T_m(y)\right\}$$

5.A DERIVATION OF PROPERTIES

In this appendix, we establish properties (a)–(e) listed in Section 5.3 by following arguments motivated by the presentations from Zacks (1971), Brown (1986), McCullagh and Nelder (1989), Lehmann and Casella (1998), Young and Smith (2010), Nielsen and Garcia (2011), Bickel and Doksum (2015), and Dobson and Barnett (2018).

To establish property (a) we call upon Hölder inequality, which ensures the following relation for functions $\{r(x), s(x)\}$ and real numbers $p, q \geq 1$ satisfying $1/p + 1/q = 1$:

$$\int_{x\in\mathcal{X}} |r(x)s(x)|dx \;\leq\; \left(\int_{x\in\mathcal{X}} |r(x)|^p dx\right)^{1/p}\left(\int_{x\in\mathcal{X}} |s(x)|^q dx\right)^{1/q} \tag{5.141}$$

where $\mathcal{X}$ denotes the domain of integration. Now let $\theta_1, \theta_2 \in \Theta$ and select $\theta = \lambda\theta_1 + (1-\lambda)\theta_2$. We select $p = 1/\lambda$ and $q = 1/(1-\lambda)$ and note that

$$
\begin{aligned}
\int_{y\in\mathcal{Y}} h(y)e^{\theta^{\mathsf{T}}T(y)}dy &= \int_{y\in\mathcal{Y}} h(y)e^{(\lambda\theta_1+(1-\lambda)\theta_2)^{\mathsf{T}}T(y)}dy \\
&= \int_{y\in\mathcal{Y}} (h(y))^{\lambda}\, e^{\lambda\theta_1^{\mathsf{T}}T(y)}\, (h(y))^{1-\lambda}\, e^{(1-\lambda)\theta_2^{\mathsf{T}}T(y)}dy \\
&\overset{(5.141)}{\leq} \left(\int_{y\in\mathcal{Y}} \left((h(y))^{\lambda}e^{\lambda\theta_1^{\mathsf{T}}T(y)}\right)^{1/\lambda} dy\right)^{\lambda} \times \\
&\qquad \left(\int_{y\in\mathcal{Y}} \left((h(y))^{(1-\lambda)}e^{(1-\lambda)\theta_2^{\mathsf{T}}T(y)}\right)^{1/(1-\lambda)} dy\right)^{1-\lambda} \\
&= \left(\int_{y\in\mathcal{Y}} h(y)e^{\theta_1^{\mathsf{T}}T(y)}dy\right)^{\lambda} \left(\int_{y\in\mathcal{Y}} h(y)e^{\theta_2^{\mathsf{T}}T(y)}dy\right)^{1-\lambda} \\
&< \infty
\end{aligned}
\tag{5.142}
$$

since, by assumption, $\theta_1, \theta_2 \in \Theta$ and, hence,

$$
\int_{y\in\mathcal{Y}} h(y)e^{\theta_1^{\mathsf{T}}T(y)}dy < \infty, \qquad \int_{y\in\mathcal{Y}} h(y)e^{\theta_2^{\mathsf{T}}T(y)}dy < \infty \tag{5.143}
$$

We therefore conclude that $\theta \in \Theta$ so that Θ is a convex set. For properties (c) and (d), let us introduce

$$
Z(\theta) \triangleq \int_{y\in\mathcal{Y}} h(y)e^{\theta^{\mathsf{T}}T(y)}dy \overset{(5.86)}{\Longleftrightarrow} a(\theta) = \ln Z(\theta) \tag{5.144}
$$

Then, it follows from expression (5.86) that

$$
\begin{aligned}
\nabla_{\theta^{\mathsf{T}}}\, a(\theta) &= \frac{1}{Z(\theta)}\nabla_{\theta^{\mathsf{T}}}\, Z(\theta) \\
&= \frac{\int_{y\in\mathcal{Y}} T(y)h(y)e^{\theta^{\mathsf{T}}T(y)}dy}{\int_{y\in\mathcal{Y}} h(y)e^{\theta^{\mathsf{T}}T(y)}dy} \\
&= \frac{\int_{y\in\mathcal{Y}} T(y)h(y)e^{\theta^{\mathsf{T}}T(y)-a(\theta)}dy}{\int_{y\in\mathcal{Y}} h(y)e^{\theta^{\mathsf{T}}T(y)-a(\theta)}dy} \\
&\overset{(a)}{=} \int_{y\in\mathcal{Y}} T(y)h(y)e^{\theta^{\mathsf{T}}T(y)-a(\theta)}dy \\
&= \mathbb{E}\,T(\boldsymbol{y})
\end{aligned}
\tag{5.145}
$$

where step (a) is because the denominator in the previous expression evaluates to 1. In other words, the gradient vector of $a(\theta)$ relative to the parameter θ^{T} is equal to the mean of $T(\boldsymbol{y})$ relative to the distribution of $\boldsymbol{y}$.

We can proceed further and evaluate the Hessian matrix of $a(\theta)$ relative to θ. Using (5.145), this calculation gives:

$$\begin{aligned}
\nabla_\theta^2\, a(\theta) &= \nabla_\theta \left(\nabla_{\theta^\mathsf{T}} a(\theta)\right) \\
&= \nabla_\theta \left(\int_{y \in \mathcal{Y}} h(y) e^{\theta^\mathsf{T} T(y) - a(\theta)} T(y) dy \right) \\
&= \int_{y \in \mathcal{Y}} h(y) e^{\theta^\mathsf{T} T(y) - a(\theta)} T(y) \Big(T^\mathsf{T}(y) - \nabla_\theta\, a(\theta) \Big)\, dy \\
&\overset{(5.145)}{=} \underbrace{\int_{y \in \mathcal{Y}} h(y) e^{\theta^\mathsf{T} T(y) - a(\theta)} T(y) T^\mathsf{T}(y) dy}_{=\mathbb{E}\, T(\boldsymbol{y})(T(\boldsymbol{y}))^\mathsf{T}} \; - \\
&\qquad \underbrace{\left(\int_{y \in \mathcal{Y}} h(y) e^{\theta^\mathsf{T} T(y) - a(\theta)} T(y) dy \right)}_{=\mathbb{E}\, T(\boldsymbol{y})} \mathbb{E}\, T(\boldsymbol{y}) \\
&= \mathbb{E}\, T(\boldsymbol{y})\, (T(\boldsymbol{y}))^\mathsf{T} \; - \; (\mathbb{E}\, T(\boldsymbol{y}))\, (\mathbb{E}\, T(\boldsymbol{y}))^\mathsf{T} \\
&\overset{\Delta}{=} R_T \geq 0
\end{aligned} \tag{5.146}$$

where the notation R_T denotes the covariance matrix of the random variable $T(\boldsymbol{y})$. We therefore find that the Hessian matrix of $a(\theta)$ is equal to the covariance matrix of $T(\boldsymbol{y})$, which in turn implies that this Hessian matrix is necessarily nonnegative definite for any θ. We conclude that property (b) holds, namely, that $a(\theta)$ is a convex function:

$$a(\theta) \text{ is convex over } \theta \tag{5.147}$$

The function $a(\theta)$ will be strictly convex when the exponential distribution is minimal. We verify this fact by contradiction. Assume the exponential distribution is minimal but that $a(\theta)$ is not strictly convex. This means that R_T is singular so that a vector z exists such that $z^\mathsf{T} R_T z = 0$, which implies that

$$z^\mathsf{T} \left\{ \mathbb{E} \Big(T(\boldsymbol{y}) - \mathbb{E}\, T(\boldsymbol{y}) \Big) \Big(T(\boldsymbol{y}) - \mathbb{E}\, T(\boldsymbol{y}) \Big)^\mathsf{T} \right\} z = 0 \tag{5.148}$$

It follows, with probability 1, that

$$z^\mathsf{T} T(\boldsymbol{y}) = z^\mathsf{T} \mathbb{E}\, T(\boldsymbol{y}) \;=\; \text{constant}, \;\; \forall\, \boldsymbol{y} \tag{5.149}$$

so that there exists a linear relation among the entries of $T(\boldsymbol{y})$. This conclusion contradicts the minimality assumption and, hence,

$$a(\theta) \text{ is strictly convex over } \theta \text{ for minimal distributions} \tag{5.150}$$

When $a(\theta)$ is strictly convex, we will establish in relation (8.18) that strict convexity is equivalent to the gradient vector satisfying the following strict monotone property:

$$\Big(\nabla_\theta\, a(\theta_1) - \nabla_\theta\, a(\theta_2) \Big) (\theta_1 - \theta_2) > 0, \;\; \forall\, \theta_1, \theta_2 \in \text{dom}(a) \tag{5.151}$$

Therefore, for two district natural parameters (θ_1, θ_2) with their respective moments (η_1, η_2), it must hold that

$$(\eta_1 - \eta_2)^\mathsf{T} (\theta_1 - \theta_2) > 0 \tag{5.152}$$

We conclude that $\eta_1 \neq \eta_2$ whenever $\theta_1 \neq \theta_2$ so that the mappings from θ to η and from η to θ are both invertible.

For property (e), we use (5.2) to note that

$$\ln f_{\boldsymbol{y}}(y) = \ln h(y) + \theta^\mathsf{T} T(y) - a(\theta) \tag{5.153}$$

so that from definition (5.94), the score function is given by

$$
\begin{aligned}
\mathcal{S}(\theta) &\triangleq \nabla_{\theta^{\mathsf{T}}} \ln f_{\boldsymbol{y}}(y) \\
&\overset{(5.153)}{=} T(\boldsymbol{y}) - \nabla_{\theta^{\mathsf{T}}} a(\theta) \\
&\overset{(5.89)}{=} T(\boldsymbol{y}) - \mathbb{E}\, T(\boldsymbol{y})
\end{aligned} \tag{5.154}
$$

We conclude from (5.94) that the covariance matrix of $T(\boldsymbol{y})$ coincides with the Fisher information matrix of the exponential distribution (5.2).

REFERENCES

Andersen, E. B. (1970), "Sufficiency and exponential families for discrete sample spaces," *J. Amer. Statist. Assoc*, vol. 65, no. 331, pp. 1248–1255.

Barankin, E. W. and A. Maitra (1963), "Generalizations of the Fisher–Darmois–Koopman–Pitman theorem on sufficient statistics," *Sankhya*, A, vol. 25, pp. 217–244.

Bernardo, J. M. and A. F. M. Smith (2000), *Bayesian Theory*, Wiley.

Bickel, P. J. and K. A. Doksum (2015), *Mathematical Statistics: Basic Ideas and Selected Topics*, 2nd ed., Chapman and Hall, CRC Press.

Brown, L. D. (1986), *Fundamentals of Statistical Exponential Families with Applications in Statistical Decision Theory*, Institute of Mathematical Statistics.

Cassella, G. and R. L. Berger (2002), *Statistical Inference*, Duxbury.

Cox, D. R. (2006), *Principles of Statistical Inference*, Cambridge University Press.

Darmois, G. (1935), "Sur les lois de probabilites a estimation exhaustive," *C. R. Acad. Sci. Paris*, vol. 200, pp. 1265–1266.

Diaconis, P. and D. Ylvisaker (1979), "Conjugate priors for exponential families," *Ann. Statist.*, vol. 7, no. 2, pp. 269–281.

Dobson, A. J. and A. G. Barnett (2018), *An Introduction to Generalized Linear Models*, 4th ed., Chapman & Hall.

Fisher, R. A. (1922), "On the mathematical foundations of theoretical statistics," *Philos. Trans. Roy. Soc. London Ser. A.*, vol. 222, pp. 309–368.

Halmos, P. R. and L. J. Savage (1949), "Application of the Radon–Nikodym theorem to the theory of sufficient statistics," *Ann. Math. Statist.*, vol. 20, pp. 225–241.

Hogg, R. V. and J. McKean (2012), *Introduction to Mathematical Statistics*, 7th ed., Pearson.

Koopman, B. (1936), "On distribution admitting a sufficient statistic," *Trans. Amer. Math. Soc.* vol. 39, no. 3, pp. 399–409.

Lehmann, E. L. (1959), *Testing Statistical Hypotheses*, Wiley.

Lehmann, E. L. and G. Casella (1998), *Theory of Point Estimation*, 2nd ed., Springer.

McCullagh, P. and J. A. Nelder (1989), *Generalized Linear Models*, 2nd ed., Chapman & Hall.

Neyman, J. (1935), "Sur un teorema concernente le cosidette statistiche sufficienti," *Giorn. Ist. Ital. Att.*, vol. 6, pp. 320–334.

Nielsen, F. and V. Garcia (2011), "Statistical exponential families: A digest with flash cards," available at arXiv:0911.4863v2.

Pitman, E. (1936), "Sufficient statistics and intrinsic accuracy," *Math. Proc. Cambridge Phil. Soc.*, vol. 32, no. 4, pp. 567–579.

Rao, C. R. (1973), *Linear Statistical Inference and Its Applications*, Wiley.

Young, G. A. and R. L. Smith (2010), *Essentials of Statistical Inference*, Cambridge University Press.

Zacks, S. (1971), *The Theory of Statistical Inference*, Wiley.

6 Entropy and Divergence

Entropy and divergence measures play a prominent role in the solution of inference and learning problems. This chapter provides an overview of several information-theoretic concepts such as the entropy of a random variable, differential entropy, and the Kullback–Leibler (KL) divergence for measuring closeness between probability distributions. The material here will be called upon at different locations in our treatment, for example, when we study variational inference methods and decision trees in future chapters.

6.1 INFORMATION AND ENTROPY

Consider a discrete random variable $\boldsymbol{x}$ and assume we observe an event $\boldsymbol{x} = x$, where x refers to some realization for $\boldsymbol{x}$. We measure the amount of *information* that is conveyed by the event in *bits* and compute it by calculating the quantity:

$$\text{Information}(\boldsymbol{x} = x) \stackrel{\Delta}{=} -\log_2 \mathbb{P}(\boldsymbol{x} = x) \tag{6.1}$$

where the logarithm is relative to base 2. Although the choice of the logarithm base is generally irrelevant, when the base is chosen as 2, the unit of measure for the information is *bits* (where the word "bit" is a shorthand for a binary digit). If $\log_2$ is replaced by the natural logarithm in (6.1), then the unit of measure becomes *nat(s)*. Since probabilities of discrete events are always smaller than or equal to 1, then their information content is always nonnegative.

Consider next a situation in which the random variable $\boldsymbol{x}$ can assume one of three possibilities, say, the colors {green, yellow, red} with probabilities:

$$\mathbb{P}(\boldsymbol{x} = \text{"green"}) = 0.60000 \tag{6.2a}$$

$$\mathbb{P}(\boldsymbol{x} = \text{"yellow"}) = 0.39999 \tag{6.2b}$$

$$\mathbb{P}(\boldsymbol{x} = \text{"red"}) = 0.00001 \tag{6.2c}$$

It is clear that the color "red" is a rare event compared to observing green or yellow since the probability of occurrence of "red" is much smaller. The amount of information contained in each event is:

$$\text{Information}(\boldsymbol{x} = \text{"green"}) = -\log_2(0.60000) = 0.7370\,\text{bits} \tag{6.3a}$$
$$\text{Information}(\boldsymbol{x} = \text{"yellow"}) = -\log_2(0.39999) = 1.3220\,\text{bits} \tag{6.3b}$$
$$\text{Information}(\boldsymbol{x} = \text{"red"}) = -\log_2(0.00001) = 16.6096\,\text{bits} \tag{6.3c}$$

where we see that rare events carry higher levels of information. This is justified for the following reason. If, for example, the color "red" alerts to some catastrophic event in the operation of a manufacturing plant, then observing this rare event would convey a significant amount of information in comparison to the other two colors, which occur more regularly during normal operation.

6.1.1 Entropy

While expression (6.1) conveys the amount of information that is reflected by a single realization for $\boldsymbol{x}$, the *entropy* measures the *average* amount of information that is conveyed by the entire distribution. Specifically, for any discrete random variable, $\boldsymbol{x}$, its entropy is denoted by the symbol $H(\boldsymbol{x})$ and defined as the nonnegative quantity:

$$\boxed{H(\boldsymbol{x}) \stackrel{\Delta}{=} -\mathbb{E}\,\log_2 \mathbb{P}(\boldsymbol{x} = x)} \qquad \textbf{(entropy)} \tag{6.4}$$

where the expectation is over the distribution of $\boldsymbol{x}$. Assuming the domain of $\boldsymbol{x}$ consists of K discrete states, denoted by $\{x_k\}$, we can rewrite $H(\boldsymbol{x})$ more explicitly as

$$H(\boldsymbol{x}) = -\sum_{k=1}^{K} \mathbb{P}(\boldsymbol{x} = x_k) \log_2 \mathbb{P}(\boldsymbol{x} = x_k) \tag{6.5}$$

For simplicity, we also write the above expression in the form:

$$H(\boldsymbol{x}) = -\sum_{x} \mathbb{P}(\boldsymbol{x} = x) \log_2 \mathbb{P}(\boldsymbol{x} = x) \tag{6.6}$$

where the sum over $k = 1, \ldots, K$ is replaced by the simpler notation involving a sum over the discrete values x. For the same $\{\text{green}, \text{yellow}, \text{red}\}$ example mentioned before, we get

$$\begin{aligned} H(\boldsymbol{x}) &= -0.6 \times \log_2(0.6) - 0.39999 \times \log_2(0.39999) - 0.00001 \times \log_2(0.00001) \\ &= 0.9712\,\text{bits} \end{aligned} \tag{6.7}$$

Example 6.1 **(Boolean random variable)** Consider a Boolean (or Bernoulli) random variable $\boldsymbol{x}$ where $\boldsymbol{x}$ is either 0 or 1. Then, expression (6.6) reduces to

$$H(\boldsymbol{x}) = -\mathbb{P}(\boldsymbol{x} = 0) \log_2 \mathbb{P}(\boldsymbol{x} = 0) \;-\; \mathbb{P}(\boldsymbol{x} = 1) \log_2 \mathbb{P}(\boldsymbol{x} = 1) \tag{6.8}$$

or, more compactly, if we let $p = \mathbb{P}(\boldsymbol{x} = 1)$:

$$H(\boldsymbol{x}) = -p \log_2 p - (1-p)\log_2(1-p), \quad p \in [0, 1] \tag{6.9}$$

In this case, the entropy measure is a function of p alone and it is customary to denote it by $H(p)$.

The entropy of a random variable reveals how much uncertainty there is about the variable. For example, if it happens that $p = 1$, then the event $\boldsymbol{x} = 1$ occurs with probability 1 and we would expect to observe it each time a random experiment is performed on $\boldsymbol{x}$. Therefore, this situation corresponds to a case of least uncertainty about $\boldsymbol{x}$ and the corresponding entropy will be $H(\boldsymbol{x}) = 0$, where we use the convention $0 \log_2 0 = 0$ since

$$\lim_{p \to 0+} p \log_2 p = 0 \tag{6.10}$$

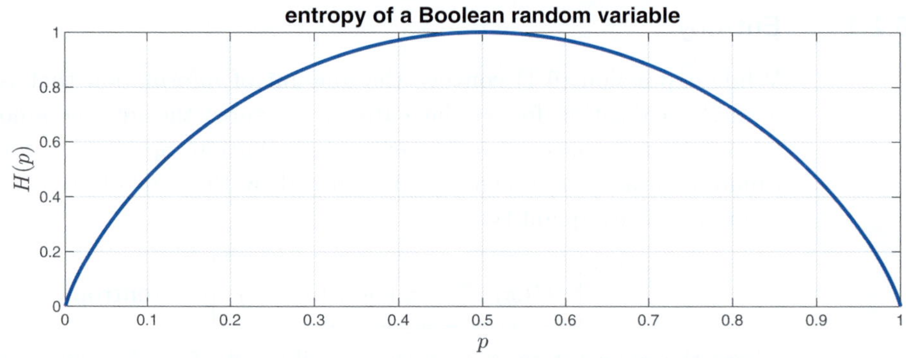

Figure 6.1 Plot of the entropy function (6.9) for a Boolean random variable as a function of $p \in [0, 1]$.

A similar situation occurs when $p = 0$; in this case, we would expect to observe the event $\boldsymbol{x} = 0$ each time a random experiment is performed on $\boldsymbol{x}$. This case again corresponds to a situation of least uncertainty about $\boldsymbol{x}$ and the corresponding entropy is also $H(\boldsymbol{x}) = 0$. The case of most uncertainty about $\boldsymbol{x}$ occurs when $p = 1/2$. In this case, the events $\boldsymbol{x} = 1$ and $\boldsymbol{x} = 0$ are equally likely and the entropy evaluates to $H(\boldsymbol{x}) = 1$. Figure 6.1 plots $H(p)$ versus p; it is seen that the function is concave, attains the value 0 at the locations $p = 0, 1$, and attains the maximum value of 1 at $p = 1/2$.

Example 6.2 **(Source coding theorem)**[1] One fundamental result in information theory is the Shannon source coding theorem. We describe one of its simpler forms. Consider a discrete random variable $\boldsymbol{x} \in \mathcal{X}$ with a probability mass function (pmf) $\mathbb{P}(\boldsymbol{x})$. Assume a code is associated with the variable $\boldsymbol{x}$, and denote the length (in bits) of the representation associated with realization $\boldsymbol{x} = x$ by $C(x)$. Normally, fewer bits will be assigned to more likely realizations. For example, assume $\mathcal{X} = \{x_1, x_2, x_3, x_4, x_5, x_6, x_7, x_8\}$ and all eight realizations are equally probable so that $\mathbb{P}(\boldsymbol{x} = x) = 1/8$. Then, we can associate the bit representations $\{000, 001, 010, 011, 100, 101, 110, 110\}$ with these realizations, in which case $C(x) = 3$ for every x. On the other hand, consider a case where $\mathcal{X} = \{x_1, x_2, x_3\}$ with $\mathbb{P}(\boldsymbol{x} = x_1) = 1/4$, $\mathbb{P}(\boldsymbol{x} = x_2) = 1/4$, and $\mathbb{P}(\boldsymbol{x} = x_3) = 1/2$. In this case, we may represent $\boldsymbol{x} = x_1$ by bits 00, $\boldsymbol{x} = x_2$ by bits 01, and $\boldsymbol{x} = x_3$ by the bit 1. Then, for this code, we have

$$C(x_1) = 2, \quad C(x_2) = 2, \quad C(x_3) = 1 \tag{6.11}$$

[1] This example can be skipped on a first reading.

in which case the average code length is

$$\mathcal{C}_{\rm av} \triangleq \sum_{x\in\mathcal{X}} C(x)\,\mathbb{P}(\boldsymbol{x}=x) = 2\times\frac{1}{4} + 2\times\frac{1}{4} + 1\times\frac{1}{2} = 3/2 \text{ bits} \tag{6.12}$$

The two coding examples considered here amount to *prefix codes*. A prefix code is one where no codeword is a prefix of any other codeword (i.e., no codeword appears as an initial component of other codewords). For such codes, an important result in information theory known as the Kraft inequality states that

$$\boxed{\sum_{x\in\mathcal{X}} 2^{-C(x)} \leq 1} \qquad \textbf{(Kraft inequality)} \tag{6.13}$$

For example, consider the case of a uniform pmf $\mathbb{P}(\boldsymbol{x})$ defined over $\mathcal{X}$ with the number of realizations in $\mathcal{X}$ equal to some power of 2, say, $|\mathcal{X}| = 2^M$ for some integer M. Then, each codeword would consist of $C(x) = M$ bits, and the sum in the above expression would evaluate to:

$$\sum_{m=1}^{2^M} 2^{-M} = 2^{-M} \times 2^M = 1 \tag{6.14}$$

The Kraft inequality is established in Prob. 6.29. The inequality has been extended beyond prefix codes to uniquely decodable codes defined in the comments at the end of the chapter.

Shannon's source coding theorem states that, for any code C defined over $\boldsymbol{x}$, its average code length is lower-bounded by the entropy of $\boldsymbol{x}$, i.e.,

$$\boxed{\mathcal{C}_{\rm av} \geq H(\boldsymbol{x})} \tag{6.15}$$

with equality if, and only if, the code is such that $C(x) = -\log_2 \mathbb{P}(\boldsymbol{x}=x)$. This last conclusion shows that the negative log computation in (6.1) corresponds to the (optimal) bit representation for the event $\boldsymbol{x}=x$.

Proof of (6.15): We use the code length $C(x)$ to define a second pmf over $\boldsymbol{x}$ denoted by $\mathbb{Q}(\boldsymbol{x}=x)$ and whose values are defined as follows:

$$\mathbb{Q}(\boldsymbol{x}=x) = \frac{2^{-C(x)}}{\sum_{x\in\mathcal{X}} 2^{-C(x)}}, \quad \sum_{x\in\mathcal{X}} \mathbb{Q}(\boldsymbol{x}=x) = 1 \tag{6.16}$$

Now, for any two pmfs defined over the same domain, it holds that:

$$\begin{aligned}
\sum_{x\in\mathcal{X}} \mathbb{P}(\boldsymbol{x}=x)\log_2\left\{\frac{\mathbb{Q}(\boldsymbol{x}=x)}{\mathbb{P}(\boldsymbol{x}=x)}\right\} &= \frac{1}{\ln 2}\sum_{x\in\mathcal{X}} \mathbb{P}(\boldsymbol{x}=x)\ln\left\{\frac{\mathbb{Q}(\boldsymbol{x}=x)}{\mathbb{P}(\boldsymbol{x}=x)}\right\} \\
&\overset{(a)}{\leq} \frac{1}{\ln 2}\sum_{x\in\mathcal{X}} \mathbb{P}(\boldsymbol{x}=x)\left\{\frac{\mathbb{Q}(\boldsymbol{x}=x)}{\mathbb{P}(\boldsymbol{x}=x)} - 1\right\} \\
&= \frac{1}{\ln 2}\left\{\underbrace{\sum_{x\in\mathcal{X}} \mathbb{Q}(\boldsymbol{x}=x)}_{=1} - \underbrace{\sum_{x\in\mathcal{X}} \mathbb{P}(\boldsymbol{x}=x)}_{=1}\right\} \\
&= 0
\end{aligned} \tag{6.17}$$

where in step (a) we used the fact that $\ln z \leq z - 1$ for any $z \geq 0$. We thus conclude that

$$\sum_{x \in \mathcal{X}} \mathbb{P}(\boldsymbol{x} = x) \log_2 \left\{ \frac{\mathbb{Q}(\boldsymbol{x} = x)}{\mathbb{P}(\boldsymbol{x} = x)} \right\} \leq 0 \tag{6.18}$$

We replace $\mathbb{Q}(\boldsymbol{x} = x)$ by its definition and find that

$$\begin{aligned}
0 &\geq \sum_{x \in \mathcal{X}} \mathbb{P}(\boldsymbol{x} = x) \log_2 \left\{ \frac{\mathbb{Q}(\boldsymbol{x} = x)}{\mathbb{P}(\boldsymbol{x} = x)} \right\} \\
&= -\sum_{x \in \mathcal{X}} \mathbb{P}(\boldsymbol{x} = x) \log_2 \mathbb{P}(\boldsymbol{x} = x) \; + \; \sum_{x \in \mathcal{X}} \mathbb{P}(\boldsymbol{x} = x) \log_2 \mathbb{Q}(\boldsymbol{x} = x) \\
&= H(\boldsymbol{x}) \; + \; \sum_{x \in \mathcal{X}} \mathbb{P}(\boldsymbol{x} = x) \left\{ -C(x) - \log_2 \sum_{x \in \mathcal{X}} 2^{-C(x)} \right\} \\
&= H(\boldsymbol{x}) \; - \; \sum_{x \in \mathcal{X}} C(x) \mathbb{P}(\boldsymbol{x} = x) \; - \; \left(\log_2 \sum_{x \in \mathcal{X}} 2^{-C(x)} \right) \underbrace{\left(\sum_{x \in \mathcal{X}} \mathbb{P}(\boldsymbol{x} = x) \right)}_{=1} \\
&= H(\boldsymbol{x}) \; - \; \mathcal{C}_{av} \; - \; \log_2 \left(\sum_{x \in \mathcal{X}} 2^{-C(x)} \right)
\end{aligned} \tag{6.19}$$

from which we conclude that $\mathcal{C}_{av} \geq H(\boldsymbol{x})$ in view of Kraft inequality (6.13).

■

6.1.2 Mutual Information

In a similar vein, we define the *conditional* entropy of a random variable, $\boldsymbol{x}$, given an observation of another random variable, $\boldsymbol{y} = y$, by means of the expression:

$$\begin{aligned}
H(\boldsymbol{x}|\boldsymbol{y} = y) &\stackrel{\Delta}{=} -\mathbb{E} \log_2 \mathbb{P}(\boldsymbol{x} = x \,|\, \boldsymbol{y} = y) \\
&= -\sum_{x} \mathbb{P}(\boldsymbol{x} = x \,|\, \boldsymbol{y} = y) \log_2 \mathbb{P}(\boldsymbol{x} = x \,|\, \boldsymbol{y} = y)
\end{aligned} \tag{6.20}$$

This quantity reflects the amount of uncertainty that remains in $\boldsymbol{x}$ after observing $\boldsymbol{y}$. If we average its value over all possible realizations for $\boldsymbol{y}$, then we obtain a measure of the uncertainty that remains in $\boldsymbol{x}$ if $\boldsymbol{y}$ is observable. We denote the result by:

$$\boxed{H(\boldsymbol{x}|\boldsymbol{y}) \stackrel{\Delta}{=} \sum_{y} \mathbb{P}(\boldsymbol{y} = y) H(\boldsymbol{x} \,|\, \boldsymbol{y} = y)} \tag{6.21}$$

where the sum is over the realizations for $\boldsymbol{y}$.

Example 6.3 **(Illustrating conditional entropy)** Consider two random variables $\boldsymbol{x}$ and $\boldsymbol{y}$. The variable $\boldsymbol{y}$ is either "green" or "red" with probabilities:

$$\mathbb{P}(\boldsymbol{y} = \text{"green"}) = 0.9, \quad \mathbb{P}(\boldsymbol{y} = \text{"red"}) = 0.1 \tag{6.22}$$

The state of $\boldsymbol{y}$ indicates whether a machine $\boldsymbol{x}$ is operating normally or not. When $\boldsymbol{y}$ is green, the machine operates efficiently 80% of the time (represented by the state +1) and is inefficient 20% of the time:

$$\mathbb{P}(\boldsymbol{x} = +1|\boldsymbol{y} = \text{“green”}) = 0.8, \quad \mathbb{P}(\boldsymbol{x} = -1|\boldsymbol{y} = \text{“green”}) = 0.2 \tag{6.23}$$

On the other hand, when $\boldsymbol{y}$ is red, the machine operates efficiently only 15% of the time:

$$\mathbb{P}(\boldsymbol{x} = +1|\boldsymbol{y} = \text{“red”}) = 0.15, \quad \mathbb{P}(\boldsymbol{x} = -1|\boldsymbol{y} = \text{“red”}) = 0.85 \tag{6.24}$$

The entropy of $\boldsymbol{y}$ is seen to be

$$\begin{aligned} H(\boldsymbol{y}) &= -\mathbb{P}(\boldsymbol{y} = \text{“green”}) \log_2 \mathbb{P}(\boldsymbol{y} = \text{“green”}) - \mathbb{P}(\boldsymbol{y} = \text{“red”}) \log_2 \mathbb{P}(\boldsymbol{y} = \text{“red”}) \\ &= -0.9 \times \log_2(0.9) - 0.1 \times \log_2(0.1) \\ &= 0.1954 \end{aligned} \tag{6.25}$$

The conditional entropy of $\boldsymbol{x}$ given both states of $\boldsymbol{y}$ is computed as follows:

$$\begin{aligned} H(\boldsymbol{x}|\boldsymbol{y} = \text{“green”}) &= -\mathbb{P}(\boldsymbol{x} = +1|\boldsymbol{y} = \text{“green”}) \log_2 \mathbb{P}(\boldsymbol{x} = +1|\boldsymbol{y} = \text{“green”}) \; - \\ &\qquad \mathbb{P}(\boldsymbol{x} = -1|\boldsymbol{y} = \text{“green”}) \log_2 \mathbb{P}(\boldsymbol{x} = -1|\boldsymbol{y} = \text{“green”}) \\ &= -0.8 \times \log_2(0.8) - 0.2 \times \log_2(0.2) \\ &= 0.7219 \end{aligned} \tag{6.26}$$

and

$$\begin{aligned} H(\boldsymbol{x}|\boldsymbol{y} = \text{“red”}) &= -\mathbb{P}(\boldsymbol{x} = +1|\boldsymbol{y} = \text{“red”}) \log_2 \mathbb{P}(\boldsymbol{x} = +1|\boldsymbol{y} = \text{“red”}) \; - \\ &\qquad \mathbb{P}(\boldsymbol{x} = -1|\boldsymbol{y} = \text{“red”}) \log_2 \mathbb{P}(\boldsymbol{x} = -1|\boldsymbol{y} = \text{“red”}) \\ &= -0.15 \times \log_2(0.15) - 0.85 \times \log_2(0.85) \\ &= 0.6098 \end{aligned} \tag{6.27}$$

Consequently, the entropy of $\boldsymbol{x}$ conditioned on $\boldsymbol{y}$ is

$$\begin{aligned} H(\boldsymbol{x}|\boldsymbol{y}) &= \mathbb{P}(\boldsymbol{y} = \text{“green”}) H(\boldsymbol{x}|\boldsymbol{y} = \text{“green”}) + \mathbb{P}(\boldsymbol{y} = \text{“red”}) H(\boldsymbol{x}|\boldsymbol{y} = \text{“red”}) \\ &= 0.9 \times 0.7219 + 0.1 \times 0.6098 \\ &= 0.7107 \end{aligned} \tag{6.28}$$

If we combine the entropy values $H(\boldsymbol{x})$ and $H(\boldsymbol{x}|\boldsymbol{y})$, we can find by how much the original uncertainty in $\boldsymbol{x}$ is reduced given observations of $\boldsymbol{y}$. This nonnegative quantity is called *mutual information* and is symmetric, meaning it is given by (see Prob. 6.4):

$$I(\boldsymbol{x}; \boldsymbol{y}) \overset{\Delta}{=} H(\boldsymbol{x}) - H(\boldsymbol{x}|\boldsymbol{y}) \tag{6.29a}$$

$$= H(\boldsymbol{y}) - H(\boldsymbol{y}|\boldsymbol{x}) \tag{6.29b}$$

$$= I(\boldsymbol{y}; \boldsymbol{x}) \tag{6.29c}$$

where, by definition,

$$H(\boldsymbol{y}) = -\sum_{y} \mathbb{P}(\boldsymbol{y}=y)\log_2 \mathbb{P}(\boldsymbol{y}=y) \tag{6.30a}$$

$$H(\boldsymbol{y}\,|\,\boldsymbol{x}=x) = -\sum_{y} \mathbb{P}(\boldsymbol{y}=y\,|\,\boldsymbol{x}=x)\log_2 \mathbb{P}(\boldsymbol{y}=y\,|\,\boldsymbol{x}=x) \tag{6.30b}$$

$$H(\boldsymbol{y}|\boldsymbol{x}) = \sum_{x} \mathbb{P}(\boldsymbol{x}=x)H(\boldsymbol{y}|\boldsymbol{x}=x) \tag{6.30c}$$

Equality (6.29b) indicates by how much the entropy of (or uncertainty about) $\boldsymbol{y}$ is reduced if $\boldsymbol{x}$ is observable – see Prob. 6.8 for a numerical example. It is clear from (6.29a) that the mutual information for independent random variables is zero (since, in that case, $H(\boldsymbol{x}|\boldsymbol{y}) = H(\boldsymbol{x})$):

$$I(\boldsymbol{x};\boldsymbol{y}) = 0 \iff \boldsymbol{x} \text{ and } \boldsymbol{y} \text{ are mutually independent} \tag{6.31}$$

Moreover, some straightforward algebra shows that the mutual information of $\boldsymbol{x}$ and $\boldsymbol{y}$ can be written in terms of the joint probability for $\boldsymbol{x}$ and $\boldsymbol{y}$ as well (see Probs. 6.5 and 6.6):

$$I(\boldsymbol{x};\boldsymbol{y}) = \sum_{x}\sum_{y} \mathbb{P}(\boldsymbol{x}=x,\boldsymbol{y}=y)\log_2\left(\frac{\mathbb{P}(\boldsymbol{x}=x,\boldsymbol{y}=y)}{\mathbb{P}(\boldsymbol{x}=x)\mathbb{P}(\boldsymbol{y}=y)}\right) \tag{6.32}$$

and satisfies the relations

$$I(\boldsymbol{x};\boldsymbol{y}) = H(\boldsymbol{x}) + H(\boldsymbol{y}) - H(\boldsymbol{x};\boldsymbol{y}) \tag{6.33a}$$

$$I(\boldsymbol{x};\boldsymbol{y}) = H(\boldsymbol{x};\boldsymbol{y}) - H(\boldsymbol{x}|\boldsymbol{y}) - H(\boldsymbol{y}|\boldsymbol{x}) \tag{6.33b}$$

in terms of the *joint* entropy measure:

$$H(\boldsymbol{x};\boldsymbol{y}) \;\stackrel{\Delta}{=}\; -\sum_{x,y} \mathbb{P}(\boldsymbol{x}=x,\boldsymbol{y}=y)\log_2 \mathbb{P}(\boldsymbol{x}=x,\boldsymbol{y}=y) \tag{6.34}$$

6.1.3 Differential Entropy

When the random variable $\boldsymbol{x}$ is continuous with a probability density function (pdf) $f_{\boldsymbol{x}}(x)$, then its entropy is defined by replacing the sum expression (6.6) with the integral expression:

$$h(\boldsymbol{x}) = -\int_{x\in\mathcal{X}} f_{\boldsymbol{x}}(x)\log_2 f_{\boldsymbol{x}}(x)\,dx \;=\; -\mathbb{E}_{x}\log_2 f_{\boldsymbol{x}}(x) \tag{6.35}$$

where the integration is over the domain of $\boldsymbol{x}$ denoted by $\mathcal{X}$. In this case, entropy is referred to as *differential* or *continuous* entropy and is denoted by $h(\boldsymbol{x})$, using the lower case letter h. It is important to note that, in contrast to the entropy of discrete random variables, differential entropy can be negative. This is because the pdf of $\boldsymbol{x}$ can take values larger than 1. For example, let $\boldsymbol{x}$ be uniformly distributed in the interval $\boldsymbol{x} \in [0, a]$, where $a > 0$. Then, its differential entropy is given by

$$h(\boldsymbol{x}) = -\int_0^a \frac{1}{a}\log_2(1/a)dx \;=\; \log_2(a) \tag{6.36}$$

This result is negative for values of a smaller than 1. It is customary, out of convenience, to employ the natural logarithm in the definition (6.35) instead of the logarithm relative to base 2. We continue with this convention.

Example 6.4 **(Gaussian distribution)** Let $\boldsymbol{x}$ be a Gaussian distributed random variable with mean μ and variance σ^2, i.e.,

$$f_{\boldsymbol{x}}(x) = \frac{1}{\sqrt{2\pi\sigma^2}}\exp\Big\{-\frac{1}{2\sigma^2}(x-\mu)^2\Big\} \tag{6.37}$$

Then, its differential entropy (using the natural logarithm convention) is given by

$$\begin{aligned} h(\boldsymbol{x}) &= -\int_{-\infty}^{\infty} f_{\boldsymbol{x}}(x)\left[-\frac{1}{2}\ln(2\pi\sigma^2) - \frac{1}{2\sigma^2}(x-\mu)^2\right]dx \\ &= \frac{1}{2}\ln(2\pi\sigma^2) + \frac{1}{2\sigma^2}\underbrace{\int_{-\infty}^{\infty}(x-\mu)^2 f_{\boldsymbol{x}}(x)dx}_{=\sigma^2} \\ &= \frac{1}{2}\Big(1+\ln(2\pi\sigma^2)\Big) \\ &= \frac{1}{2}\ln(2\pi e\sigma^2) \end{aligned} \tag{6.38}$$

Note that the entropy expression is only dependent on the variance, σ^2. In other words, all Gaussian distributions with the same variance have the same differential entropy.

We can extend the derivation to M-dimensional vector-valued Gaussian distributions with mean $\bar{x}$ and covariance matrix R_x, namely,

$$f_{\boldsymbol{x}}(x) = \frac{1}{\sqrt{(2\pi)^M}}\,\frac{1}{\sqrt{\det R_x}}\,\exp\left\{-\frac{1}{2}(x-\bar{x})^{\mathsf{T}}R_x^{-1}(x-\bar{x})\right\} \tag{6.39}$$

and find that (see Prob. 6.11):

$$h(x) = \frac{1}{2}\ln\Big((2\pi e)^M\det(R_x)\Big) \tag{6.40}$$

Example 6.5 **(Exponential distribution)** Let $\boldsymbol{x}$ be an exponentially distributed random variable with parameter λ, i.e.,

$$f_{\boldsymbol{x}}(x) = \lambda e^{-\lambda x}, \quad x \in [0,\infty) \tag{6.41}$$

Then, its differential entropy is given by

$$\begin{aligned} h(\boldsymbol{x}) &= -\int_0^{\infty}\lambda e^{-\lambda x}\log_2\Big(\lambda e^{-\lambda x}\Big)dx \\ &= -\log_2(\lambda)\underbrace{\left\{\int_0^{\infty}\lambda e^{-\lambda x}dx\right\}}_{=1} + \lambda\underbrace{\left\{\int_0^{\infty}x\left(\lambda e^{-\lambda x}\right)dx\right\}}_{\mathbb{E}\,\boldsymbol{x}=1/\lambda} \\ &= -\log_2(\lambda) + \lambda\times(1/\lambda) \\ &= 1-\log_2(\lambda) \end{aligned} \tag{6.42}$$

where we used the fact that the mean of the exponential distribution (6.41) is $1/\lambda$.

6.2 KULLBACK–LEIBLER DIVERGENCE

In inference problems, it is often necessary to compare how close two probability distributions are to each other. One useful measure is the Kullback–Leibler (KL) divergence, also called *relative entropy* because it is based on an entropy-like calculation.

6.2.1 Definition

Let $p_{\boldsymbol{x}}(x)$ and $q_{\boldsymbol{x}}(x)$ denote two pdfs defined over the same domain $x \in \mathcal{X}$. The KL divergence relative to $p_{\boldsymbol{x}}(x)$ is denoted by $D_{\text{KL}}(p\|q)$, with p listed first, and is defined as the following mean value (using the natural logarithm for convenience):

$$\begin{aligned} D_{\text{KL}}(p\|q) &\stackrel{\Delta}{=} \mathbb{E}_p\left\{\ln\left(\frac{p_{\boldsymbol{x}}(x)}{q_{\boldsymbol{x}}(x)}\right)\right\} \\ &= \int_{x\in\mathcal{X}} p_{\boldsymbol{x}}(x)\ln\left(\frac{p_{\boldsymbol{x}}(x)}{q_{\boldsymbol{x}}(x)}\right)dx \\ &= \int_{x\in\mathcal{X}} p_{\boldsymbol{x}}(x)\ln p_{\boldsymbol{x}}(x)dx \;-\; \int_{x\in\mathcal{X}} p_{\boldsymbol{x}}(x)\ln q_{\boldsymbol{x}}(x)dx \\ &= \mathbb{E}_p \ln p_{\boldsymbol{x}}(x) \;-\; \mathbb{E}_p \ln q_{\boldsymbol{x}}(x) \end{aligned} \tag{6.43}$$

The expectation is computed relative to the distribution $p_{\boldsymbol{x}}(x)$. For discrete distributions, we use the same notation $p_{\boldsymbol{x}}(x)$ to refer to the probability of event $\boldsymbol{x} = x$, i.e., $p_{\boldsymbol{x}}(x) = \mathbb{P}(\boldsymbol{x} = x)$, and define the KL divergence by replacing integrals by sums:

$$\begin{aligned} D_{\text{KL}}(p\|q) &\stackrel{\Delta}{=} \mathbb{E}_p\left\{\ln\left(\frac{p_{\boldsymbol{x}}(x)}{q_{\boldsymbol{x}}(x)}\right)\right\} \\ &= \sum_x p_{\boldsymbol{x}}(x)\ln\left(\frac{p_{\boldsymbol{x}}(x)}{q_{\boldsymbol{x}}(x)}\right) \\ &= \sum_x p_{\boldsymbol{x}}(x)\ln p_{\boldsymbol{x}}(x) \;-\; \sum_x p_{\boldsymbol{x}}(x)\ln q_{\boldsymbol{x}}(x) \\ &= \mathbb{E}_p \ln p_{\boldsymbol{x}}(x) \;-\; \mathbb{E}_p \ln q_{\boldsymbol{x}}(x) \end{aligned} \tag{6.44}$$

Observe from the third equality in (6.44) that the KL divergence can be interpreted as the difference between the average information contained in $p_{\boldsymbol{x}}(x)$ and $q_{\boldsymbol{x}}(x)$, assuming that the events occur according to the *same* distribution $p_{\boldsymbol{x}}(x)$. We can also interpret the KL divergence as a difference between cross-entropy and entropy terms by rewriting the third equality in the equivalent form:

$$D_{\text{KL}}(p\|q) \stackrel{\Delta}{=} \underbrace{\left(-\sum_x p_{\boldsymbol{x}}(x)\ln q_{\boldsymbol{x}}(x)\right)}_{\textbf{cross-entropy}} - \underbrace{\left(-\sum_x p_{\boldsymbol{x}}(x)\ln p_{\boldsymbol{x}}(x)\right)}_{\textbf{entropy}} \tag{6.45}$$

The previous expressions for the KL divergence assume the following conventions, depending on whether the numerator and/or denominator values are zero:

$$0 \times \ln\left(\frac{0}{0}\right) = 0, \quad 0 \times \ln\left(\frac{0}{q}\right) = 0, \quad p \times \ln\left(\frac{p}{0}\right) = +\infty \tag{6.46}$$

Observe that the KL divergence is not an actual distance function since, for example, it is not symmetric. If we switch the roles of $p_{\boldsymbol{x}}(x)$ and $q_{\boldsymbol{x}}(x)$, then the KL divergence value is changed:

$$D_{\text{KL}}(p\|q) \neq D_{\text{KL}}(q\|p) \tag{6.47}$$

In addition, the KL divergence does not satisfy the usual triangle inequality of norms – see Prob. 6.17. Nevertheless, it still serves as a useful measure of how close two distributions are to each other. An important result in information theory known as the Gibbs inequality ascertains that

$$\boxed{D_{\text{KL}}(p\|q) \geq 0} \tag{6.48a}$$

with

$$D_{\text{KL}}(p\|q) = 0 \text{ if, and only if, } p_{\boldsymbol{x}}(x) = q_{\boldsymbol{x}}(x) \text{ almost everywhere} \tag{6.48b}$$

Proof of the Gibbs inequality (6.48a)–(6.48b): From the definition of the KL divergence:

$$\begin{aligned}
-D_{\text{KL}}(p\|q) &\stackrel{\Delta}{=} -\int_{x\in\mathcal{X}} p_{\boldsymbol{x}}(x) \ln\left(\frac{p_{\boldsymbol{x}}(x)}{q_{\boldsymbol{x}}(x)}\right) dx \\
&= \int_{x\in\mathcal{X}} p_{\boldsymbol{x}}(x) \ln\left(\frac{q_{\boldsymbol{x}}(x)}{p_{\boldsymbol{x}}(x)}\right) dx \\
&\stackrel{(a)}{\leq} \int_{x\in\mathcal{X}} p_{\boldsymbol{x}}(x) \left(\frac{q_{\boldsymbol{x}}(x)}{p_{\boldsymbol{x}}(x)} - 1\right) dx \\
&= \int_{x\in\mathcal{X}} q_{\boldsymbol{x}}(x)dx \; - \int_{x\in\mathcal{X}} p_{\boldsymbol{x}}(x)dx = 0
\end{aligned} \tag{6.49}$$

where in step (a) we used the fact that $\ln(z) \leq z-1$, with equality holding only at $z=1$. It follows that $D_{\text{KL}}(p\|q) \geq 0$ with equality holding if, and only if, $q_{\boldsymbol{x}}(x)/p_{\boldsymbol{x}}(x) = 1$ almost everywhere.

■

Example 6.6 **(KL divergence of two Bernoulli distributions)** Consider two Bernoulli distributions with probabilities of success s_p and s_q. Their respective pdfs are given by

$$p_{\boldsymbol{x}}(x) = s_p^x(1-s_p)^{1-x} \tag{6.50a}$$

$$q_{\boldsymbol{x}}(x) = s_q^x(1-s_q)^{1-x} \tag{6.50b}$$

where $\boldsymbol{x} \in \{0,1\}$ and success refers to the probability of $\boldsymbol{x}$ assuming the value $x=1$ under each distribution. The KL divergence measure of $q_{\boldsymbol{x}}(x)$ relative to $p_{\boldsymbol{x}}(x)$ is given by

$$
\begin{aligned}
D_{\mathrm{KL}}(p\|q) &= \sum_{x=0,1} p_{\boldsymbol{x}}(x)\, \ln\left(\frac{p_{\boldsymbol{x}}(x)}{q_{\boldsymbol{x}}(x)}\right) \\
&= (1-s_p)\ln\left(\frac{1-s_p}{1-s_q}\right) + s_p \ln\left(\frac{s_p}{s_q}\right) \qquad (6.51)
\end{aligned}
$$

Example 6.7 **(KL divergence of two exponential distributions)** Consider two distributions from the exponential family (5.2) with parameters $\{\theta_p, \theta_q\}$, but the same base functions and log-normalization factors, namely,

$$p_{\boldsymbol{x}}(x) = h(x)\exp\left\{\theta_p^{\mathsf{T}} T(x) - a(\theta_p)\right\} \qquad (6.52a)$$

$$q_{\boldsymbol{x}}(x) = h(x)\exp\left\{\theta_q^{\mathsf{T}} T(x) - a(\theta_q)\right\} \qquad (6.52b)$$

The KL divergence measure of $q_{\boldsymbol{x}}(x)$ relative to $p_{\boldsymbol{x}}(x)$ is given by

$$
\begin{aligned}
D_{\mathrm{KL}}(p\|q) &= \int_{x\in\mathcal{X}} p_{\boldsymbol{x}}(x)\, \ln\left(\frac{p_{\boldsymbol{x}}(x)}{q_{\boldsymbol{x}}(x)}\right) dx \\
&= \int_{x\in\mathcal{X}} p_{\boldsymbol{x}}(x)\Big[(\theta_p - \theta_q)^{\mathsf{T}} T(x) + a(\theta_q) - a(\theta_p)\Big] dx \\
&= a(\theta_q) - a(\theta_p) + (\theta_p - \theta_q)^{\mathsf{T}} \mathbb{E}_p\, T(\boldsymbol{x}) \qquad (6.53)
\end{aligned}
$$

where the expectation of the statistic $T(x)$ is relative to the distribution $p_{\boldsymbol{x}}(x)$. Using property (5.89), we know that

$$\nabla_{\theta_p^{\mathsf{T}}}\, a(\theta_p) \;=\; \mathbb{E}_p\, T(\boldsymbol{x}) \qquad (6.54)$$

and, therefore,

$$D_{\mathrm{KL}}(p\|q) = a(\theta_q) - a(\theta_p) + (\theta_p - \theta_q)^{\mathsf{T}} \nabla_{\theta_p^{\mathsf{T}}}\, a(\theta_p) \qquad (6.55)$$

Example 6.8 **(KL divergence of two Gibbs distributions)** Consider two Gibbs (Boltzmann) distributions with parameters $\lambda_p, \lambda_q \in \mathbb{R}^M$ defined along the lines of Prob. 5.8, namely, for one of them:

$$\mathbb{P}(\boldsymbol{x} = r) = e^{b_r^{\mathsf{T}}\lambda_p} \Big/ \Big(\sum_{r'=1}^{R} e^{b_{r'}^{\mathsf{T}}\lambda_p}\Big)^{-1} \qquad (6.56)$$

where a feature vector b_r is associated with each class r. We established in that problem that such distributions belong to the exponential family of distributions with

$$p_{\boldsymbol{x}}(x) = \exp\left\{\theta_p^{\mathsf{T}} T(x) - a(\theta_p)\right\} \qquad (6.57a)$$

$$q_{\boldsymbol{x}}(x) = \exp\left\{\theta_q^{\mathsf{T}} T(x) - a(\theta_q)\right\} \qquad (6.57b)$$

where

$$a(\theta_p) = \ln\Big(\sum_{r=1}^{R} e^{(b_r - b_1)^\mathsf{T}\lambda_p}\Big), \quad a(\theta_q) = \ln\Big(\sum_{r=1}^{R} e^{(b_r - b_1)^\mathsf{T}\lambda_q}\Big) \tag{6.58a}$$

$$\theta_p = \begin{bmatrix} (b_2 - b_1)^\mathsf{T}\lambda_p \\ (b_3 - b_1)^\mathsf{T}\lambda_p \\ \vdots \\ (b_R - b_1)^\mathsf{T}\lambda_p \end{bmatrix}, \quad \theta_q = \begin{bmatrix} (b_2 - b_1)^\mathsf{T}\lambda_q \\ (b_3 - b_1)^\mathsf{T}\lambda_q \\ \vdots \\ (b_R - b_1)^\mathsf{T}\lambda_q \end{bmatrix} \tag{6.58b}$$

$$T(x) = \text{col}\Big\{\mathbb{I}[x=2],\ \mathbb{I}[x=3],\ \ldots,\ \mathbb{I}[x=R]\Big\} \tag{6.58c}$$

Then, using result (6.55) we find that the KL divergence between two Gibbs distributions is given by

$$D_{\text{KL}}(p\|q) = \ln\left(\frac{\sum_{r=1}^{R} e^{(b_r - b_1)^\mathsf{T}\lambda_q}}{\sum_{r=1}^{R} e^{(b_r - b_1)^\mathsf{T}\lambda_p}}\right) + \sum_{r=2}^{R} \frac{e^{(b_r - b_1)^\mathsf{T}\lambda_p}}{\sum_{r'=1}^{R} e^{(b_{r'} - b_1)^\mathsf{T}\lambda_p}}(\lambda_p - \lambda_q)^\mathsf{T}(b_r - b_1) \tag{6.59}$$

Example 6.9 **(KL divergence of two Gaussian distributions)** Consider two Gaussian distributions with mean vectors $\{\bar{x}_p, \bar{x}_q\} \in \mathbb{R}^M$ and positive-definite covariance matrices $\{R_p, R_q\}$. Their respective pdfs are given by

$$p_{\boldsymbol{x}}(x) = \frac{1}{\sqrt{(2\pi)^M}} \frac{1}{\sqrt{\det R_p}} \exp\left\{-\frac{1}{2}(x - \bar{x}_p)^\mathsf{T} R_p^{-1}(x - \bar{x}_p)\right\} \tag{6.60a}$$

$$q_{\boldsymbol{x}}(x) = \frac{1}{\sqrt{(2\pi)^M}} \frac{1}{\sqrt{\det R_q}} \exp\left\{-\frac{1}{2}(x - \bar{x}_q)^\mathsf{T} R_q^{-1}(x - \bar{x}_q)\right\} \tag{6.60b}$$

The KL divergence measure of $q_{\boldsymbol{x}}(x)$ relative to $p_{\boldsymbol{x}}(x)$ is given by

$$\begin{aligned} D_{\text{KL}}(p\|q) &= \int_{-\infty}^{\infty} p_{\boldsymbol{x}}(x)\ \ln\left(\frac{p_{\boldsymbol{x}}(x)}{q_{\boldsymbol{x}}(x)}\right) dx \\ &= \frac{1}{2}\ \ln\left(\frac{\det(R_q)}{\det(R_p)}\right) \\ &\quad - \frac{1}{2}\int_{-\infty}^{\infty} p_{\boldsymbol{x}}(x)\,(x - \bar{x}_p)^\mathsf{T} R_p^{-1}(x - \bar{x}_p)dx \\ &\quad + \frac{1}{2}\int_{-\infty}^{\infty} p_{\boldsymbol{x}}(x)\,(x - \bar{x}_q)^\mathsf{T} R_q^{-1}(x - \bar{x}_q)dx \end{aligned} \tag{6.61}$$

To evaluate the last two terms, we recall that

$$R_p \triangleq \mathbb{E}\,(\boldsymbol{x} - \bar{x}_p)(\boldsymbol{x} - \bar{x}_p)^\mathsf{T} = \int_{-\infty}^{\infty} p_{\boldsymbol{x}}(x)(x - \bar{x}_p)(x - \bar{x}_p)^\mathsf{T} dx \tag{6.62}$$

so that for the second term in (6.61) we have

$$\begin{aligned}
&-\frac{1}{2}\int_{-\infty}^{\infty} p_{\boldsymbol{x}}(x)\,(x-\bar{x}_p)^{\mathsf{T}}R_p^{-1}(x-\bar{x}_p)dx \\
&= -\frac{1}{2}\mathrm{Tr}\left\{\int_{-\infty}^{\infty} p_{\boldsymbol{x}}(x)\,R_p^{-1}(x-\bar{x}_p)(x-\bar{x}_p)^{\mathsf{T}}dx\right\} \\
&= -\frac{1}{2}\mathrm{Tr}\left\{R_p^{-1}\int_{-\infty}^{\infty} p_{\boldsymbol{x}}(x)\,(x-\bar{x}_p)(x-\bar{x}_p)^{\mathsf{T}}dx\right\} \\
&= -\frac{1}{2}\mathrm{Tr}(R_p^{-1}R_p) \\
&= -M/2
\end{aligned} \tag{6.63}$$

In a similar vein, the last term in (6.61) evaluates to

$$\begin{aligned}
&\frac{1}{2}\int_{-\infty}^{\infty} p_{\boldsymbol{x}}(x)\,(x-\bar{x}_q)^{\mathsf{T}}R_q^{-1}(x-\bar{x}_q)dx \\
&= \frac{1}{2}\int_{-\infty}^{\infty} p_{\boldsymbol{x}}(x)\,(x-\bar{x}_p+\bar{x}_p-\bar{x}_q)^{\mathsf{T}}R_q^{-1}(x-\bar{x}_p+\bar{x}_p-\bar{x}_q)dx \\
&= \frac{1}{2}\int_{-\infty}^{\infty} p_{\boldsymbol{x}}(x)\,(x-\bar{x}_p)^{\mathsf{T}}R_q^{-1}(x-\bar{x}_p)dx \\
&\quad+\frac{1}{2}\int_{-\infty}^{\infty} p_{\boldsymbol{x}}(x)\,(\bar{x}_p-\bar{x}_q)^{\mathsf{T}}R_q^{-1}(\bar{x}_p-\bar{x}_q) \\
&\quad+\int_{-\infty}^{\infty} p_{\boldsymbol{x}}(x)\,(x-\bar{x}_p)^{\mathsf{T}}R_q^{-1}(\bar{x}_p-\bar{x}_q)dx \\
&= \frac{1}{2}\mathrm{Tr}(R_q^{-1}R_p)+\frac{1}{2}(\bar{x}_p-\bar{x}_q)^{\mathsf{T}}R_q^{-1}(\bar{x}_p-\bar{x}_q)dx\,+ \\
&\quad\underbrace{\left(\int_{-\infty}^{\infty} p_{\boldsymbol{x}}(x)\,(x-\bar{x}_p)^{\mathsf{T}}dx\right)}_{=0}R_q^{-1}(\bar{x}_p-\bar{x}_q)dx \\
&= \frac{1}{2}\mathrm{Tr}(R_q^{-1}R_p)+\frac{1}{2}(\bar{x}_p-\bar{x}_q)^{\mathsf{T}}R_q^{-1}(\bar{x}_p-\bar{x}_q)
\end{aligned} \tag{6.64}$$

We therefore arrive at the expression

$$D_{\mathrm{KL}}(p\|q) = \frac{1}{2}\left\{\ln\left(\frac{\det(R_q)}{\det(R_p)}\right) - M + \mathrm{Tr}(R_q^{-1}R_p) + (\bar{x}_p-\bar{x}_q)^{\mathsf{T}}R_q^{-1}(\bar{x}_p-\bar{x}_q)\right\} \tag{6.65}$$

For one-dimensional distributions, the above expression simplifies to

$$D_{\mathrm{KL}}(p\|q) = \frac{1}{2}\left\{\ln\left(\sigma_q^2/\sigma_p^2\right) - 1 + (\sigma_p^2/\sigma_q^2) + (\bar{x}_p-\bar{x}_q)^2/\sigma_q^2\right\} \tag{6.66}$$

Example 6.10 **(Relation to cross-entropy)** One popular entropy measure that will be used in later chapters to train learning algorithms, including neural networks, is the cross-entropy between two distributions – see Section 65.7. Consider two pdfs $p_{\boldsymbol{x}}(x)$ and $q_{\boldsymbol{x}}(x)$ over a random variable $\boldsymbol{x}$. Their cross-entropy is denoted by $H(p,q)$ and defined as the value:

$$H(p,q) \stackrel{\Delta}{=} -\mathbb{E}_p \ln q_{\boldsymbol{x}}(x) = -\int_{x\in\mathcal{X}} p_{\boldsymbol{x}}(x)\ln q_{\boldsymbol{x}}(x)dx \tag{6.67}$$

where the expectation is over the distribution $p_{\boldsymbol{x}}(x)$. Comparing with (6.43) we find that

$$\boxed{H(p,q) = D_{\rm KL}(p\|q) + h(\boldsymbol{x})} \tag{6.68}$$

so that the cross-entropy and the KL divergence of $p_{\boldsymbol{x}}(x)$ and $q_{\boldsymbol{x}}(x)$ differ by the (differential) entropy of $\boldsymbol{x}$. In this way, the cross-entropy between two distributions also reflects a measure of closeness between them. Moreover, optimization problems involving the maximization of $D_{\rm KL}(p\|q\|)$ over $q_{\boldsymbol{x}}(x)$ become equivalent to maximizing the cross-entropy over the same distribution, since $h(\boldsymbol{x})$ is independent of $q_{\boldsymbol{x}}(x)$.

6.3 MAXIMUM ENTROPY DISTRIBUTION

The KL divergence can be used to establish important connections to other concepts in inference and learning. In this section, we use it to show that, among all distributions with a fixed variance, the Gaussian distribution attains the largest entropy.

To see this, we let $q_{\boldsymbol{x}}(x)$ denote some Gaussian distribution with mean μ_q and variance σ_q^2:

$$q_{\boldsymbol{x}}(x) = \frac{1}{\sqrt{2\pi\sigma_q^2}} \exp\left\{ -\frac{1}{2\sigma_q^2}(x-\mu_q)^2 \right\} \tag{6.69}$$

where we are using the subscript q in (μ_q, σ_q^2) to emphasize that these are the mean and variance relative to the distribution $q_{\boldsymbol{x}}(x)$, i.e.,

$$\mu_q \stackrel{\Delta}{=} \int_{-\infty}^{\infty} x q_{\boldsymbol{x}}(x)dx, \quad \sigma_q^2 \stackrel{\Delta}{=} \int_{-\infty}^{\infty} (x-\mu_q)^2 q_{\boldsymbol{x}}(x)dx \tag{6.70}$$

We are free to choose any (μ_q, σ_q^2) and will exploit this freedom further ahead.

Next, let $p_x(x)$ denote an arbitrary distribution with differential entropy denoted by $h_p(x)$, i.e.,

$$h_p(x) \stackrel{\Delta}{=} -\int_{-\infty}^{\infty} p_{\boldsymbol{x}}(x) \ln p_{\boldsymbol{x}}(x)dx \tag{6.71}$$

where, without loss of generality, we are using the natural logarithm. Let μ_p and σ_p^2 denote the mean and variance of $\boldsymbol{x}$ relative to this distribution:

$$\mu_p \stackrel{\Delta}{=} \int_{-\infty}^{\infty} x p_{\boldsymbol{x}}(x)dx, \quad \sigma_p^2 \stackrel{\Delta}{=} \int_{-\infty}^{\infty} (x-\mu_p)^2 p_{\boldsymbol{x}}(x)dx \tag{6.72}$$

Now, in view of the Gibbs inequality (6.48a), we have

$$\begin{aligned}
&\int_{-\infty}^{\infty} p_{\boldsymbol{x}}(x)\ln\left(\frac{q_{\boldsymbol{x}}(x)}{p_{\boldsymbol{x}}(x)}\right)dx \leq 0 \\
&\Longleftrightarrow -\int_{-\infty}^{\infty} p_{\boldsymbol{x}}(x)\ln p_{\boldsymbol{x}}(x) \leq -\int_{-\infty}^{\infty} p_{\boldsymbol{x}}(x)\ln q_{\boldsymbol{x}}(x)dx \\
&\Longleftrightarrow \quad h_p(x) \leq -\int_{-\infty}^{\infty} p_{\boldsymbol{x}}(x)\ln q_{\boldsymbol{x}}(x)dx \\
&\overset{(6.69)}{\Longleftrightarrow} \quad h_p(x) \leq -\int_{-\infty}^{\infty} p_{\boldsymbol{x}}(x)\Big(-\frac{1}{2}\ln(2\pi\sigma_q^2) - \frac{1}{2\sigma_q^2}(x-\mu_q)^2\Big)dx \\
&\Longleftrightarrow \quad h_p(x) \leq \frac{1}{2}\ln(2\pi\sigma_q^2) + \frac{1}{2\sigma_q^2}\int_{-\infty}^{\infty}(x-\mu_p+\mu_p-\mu_q)^2 p_{\boldsymbol{x}}(x)dx
\end{aligned} \tag{6.73}$$

Expanding the rightmost term, we get

$$\begin{aligned}
h_p(x) \leq \frac{1}{2}\ln(2\pi\sigma_q^2) + \frac{1}{2\sigma_q^2}\underbrace{\int_{-\infty}^{\infty}(x-\mu_p)^2 p_{\boldsymbol{x}}(x)dx}_{=\sigma_p^2} + \\
\frac{1}{2\sigma_q^2}\underbrace{\int_{-\infty}^{\infty}(\mu_p-\mu_q)^2 p_{\boldsymbol{x}}(x)dx}_{=\,(\mu_p-\mu_q)^2} + \frac{1}{2\sigma_q^2}\underbrace{\int_{-\infty}^{\infty}2(x-\mu_p)(\mu_p-\mu_q)p_{\boldsymbol{x}}(x)dx}_{=\,0}
\end{aligned} \tag{6.74}$$

and we arrive at the general inequality:

$$h_p(x) \leq \frac{1}{2}\ln(2\pi\sigma_q^2) + \frac{\sigma_p^2}{2\sigma_q^2} + \frac{(\mu_p-\mu_q)^2}{2\sigma_q^2} \tag{6.75}$$

This inequality holds for any distribution $p_{\boldsymbol{x}}(x)$ and for any (μ_q, σ_q^2). We can choose these parameters to make the upper bound as tight as possible. For example, if we select $\mu_q = \mu_p$, then the last term in the upper bound disappears and we are left with

$$h_p(x) \leq \frac{1}{2}\ln(2\pi\sigma_q^2) + \frac{\sigma_p^2}{2\sigma_q^2} \tag{6.76}$$

If we minimize the upper bound over σ_q^2 we find that the minimum occurs at $\sigma_q^2 = \sigma_p^2$ (e.g., by setting the derivative relative to σ_q^2 equal to zero and by verifying that the second derivative at the location $\sigma_q^2 = \sigma_p^2$ is positive). Substituting this choice into the upper bound we find that it simplifies to

$$\boxed{h_p(x) \leq \frac{1}{2}\ln(2\pi e\sigma_p^2)} \tag{6.77}$$

This bound is only dependent on the variance of the distribution and is the tightest it can be for any $p_{\boldsymbol{x}}(x)$. Now, we already know from (6.38) that the quantity on the right-hand side is equal to the differential entropy of a Gaussian distribution with variance σ_p^2. Therefore, the bound is achieved for Gaussian distributions. Stated in another way, we conclude that *among all distributions $p_{\boldsymbol{x}}(x)$ with a given variance σ_p^2, the Gaussian distribution has the largest entropy.*

6.4 MOMENT MATCHING

Next, we employ the KL divergence to determine the closest Gaussian approximation to a generic pdf. It turns out that the Gaussian solution needs to match the first- and second-order moments of the given pdf.

Gaussian approximation

Thus, consider a *generic* pdf $p_{\boldsymbol{x}}(x)$ with mean $\bar{x}_p$ and covariance matrix R_p. The parameters $\{\bar{x}_p, R_p\}$ represent the first- and second-order moments of the distribution $p_{\boldsymbol{x}}(x)$:

$$\bar{x} \;=\; \mathbb{E}_p\,\boldsymbol{x}, \qquad R_p = \mathbb{E}_p(\boldsymbol{x}-\bar{x}_p)(\boldsymbol{x}-\bar{x}_p)^\mathsf{T} \tag{6.78}$$

Let $q_{\boldsymbol{x}}(x)$ denote a Gaussian distribution with mean $\bar{x}$ and a positive-definite covariance matrix R_x:

$$\begin{aligned} q_{\boldsymbol{x}}(x) &= \mathcal{N}_{\boldsymbol{x}}(\bar{x}, R_x), \quad x \in \mathbb{R}^M \\ &= \frac{1}{\sqrt{(2\pi)^{M/2}}}\frac{1}{\sqrt{\det R_x}}\exp\left\{-\frac{1}{2}(x-\bar{x})^\mathsf{T}R_x^{-1}(x-\bar{x})\right\} \end{aligned} \tag{6.79}$$

We wish to approximate $p_{\boldsymbol{x}}(x)$ by the closest Gaussian distribution measured in terms of the smallest KL divergence between them, i.e., we wish to solve

$$(\bar{x}_q, R_q) \;\stackrel{\Delta}{=}\; \underset{\bar{x}, R_x}{\text{argmin}}\; D_{\text{KL}}\Big(p_{\boldsymbol{x}}(x)\,\|\,\mathcal{N}_{\boldsymbol{x}}(\bar{x}, R_x)\Big) \tag{6.80}$$

From definition (6.43) we have, apart from some constant terms that do not depend on $(\bar{x}, R_x)$,

$$\begin{aligned} &D_{\text{KL}}(p\|q) \\ &= -h_p(\boldsymbol{x}) + \int_{-\infty}^{\infty}\frac{1}{2}(x-\bar{x})^\mathsf{T}R_x^{-1}(x-\bar{x})p_{\boldsymbol{x}}(x)dx + \frac{1}{2}\ln\det R_x + \text{cte} \end{aligned} \tag{6.81}$$

where $h_p(\boldsymbol{x})$ denotes the differential entropy of $p_{\boldsymbol{x}}(x)$, which is independent of $(\bar{x}, R_x)$. Differentiating the KL divergence relative to $\bar{x}$ gives

$$\begin{aligned} \nabla_{\bar{x}^\mathsf{T}}\Big[D_{\text{KL}}(p\|q)\Big] &= \nabla_{\bar{x}^\mathsf{T}}\left\{\int_{-\infty}^{\infty}\frac{1}{2}(x-\bar{x})^\mathsf{T}R_x^{-1}(x-\bar{x})p_{\boldsymbol{x}}(x)dx\right\} \\ &= \int_{-\infty}^{\infty}\nabla_{\bar{x}^\mathsf{T}}\left\{\frac{1}{2}(x-\bar{x})^\mathsf{T}R_x^{-1}(x-\bar{x})p_{\boldsymbol{x}}(x)\right\}dx \\ &= \int_{-\infty}^{\infty}R_x^{-1}(x-\bar{x})p_{\boldsymbol{x}}(x)dx \end{aligned} \tag{6.82}$$

To arrive at the above expression we had to switch the order of integration and differentiation; this operation is justified under relatively reasonable conditions in most instances in view of the *dominated convergence theorem* discussed later in Appendix 16.A. In essence, the switching is justified here because the function $(x-\bar{x})^{\mathsf{T}}R_x^{-1}(x-\bar{x})$ is continuous and has a continuous gradient over $\bar{x}$. Setting the gradient (6.82) to zero at $\bar{x}=\bar{x}_q$ we conclude that the solution $\bar{x}_q$ satisfies:

$$\bar{x}_q = \int_{-\infty}^{\infty} x p_{\boldsymbol{x}}(x)dx \;=\; \bar{x}_p \tag{6.83}$$

That is, $\bar{x}_q$ should match the mean (i.e., first moment) of $\boldsymbol{x}$ under the distribution $p_{\boldsymbol{x}}(x)$. We next differentiate the KL divergence (6.81) relative to R_x. For this purpose, we appeal to the results from parts (c) and (e) of Probs. 2.11 and 2.12, respectively. Indeed, noting that R is symmetric and invertible, we have

$$\begin{aligned}
&\nabla_{R_x}\Big[D_{\mathrm{KL}}(p\|q)\Big] \\
&= \nabla_{R_x}\left\{\int_{-\infty}^{\infty}\frac{1}{2}\mathrm{Tr}\Big(R_x^{-1}(x-\bar{x})(x-\bar{x})^{\mathsf{T}}\Big)p_{\boldsymbol{x}}(x)dx \;+\; \frac{1}{2}\ln\det R_x\right\} \\
&= -\int_{-\infty}^{\infty}\frac{1}{2}R_x^{-1}(x-\bar{x})(x-\bar{x})^{\mathsf{T}}R_x^{-1}p_{\boldsymbol{x}}(x)dx \;+\; \frac{1}{2}R_x^{-1}
\end{aligned} \tag{6.84}$$

Setting the gradient expression (6.84) to zero at $R_x = R_q$ and solving for it using $\bar{x}=\bar{x}_p$ gives

$$R_q = \int_{-\infty}^{\infty}(x-\bar{x}_p)(x-\bar{x}_p)^{\mathsf{T}}p_x(x)dx \;=\; R_p \tag{6.85}$$

That is, R_q should match the covariance matrix (i.e., second moment) of $\boldsymbol{x}$ under the distribution $p_{\boldsymbol{x}}(x)$. Therefore, we conclude that

$$\boxed{\min_{q_{\boldsymbol{x}}(x)\in\text{ Gaussian}} D_{\mathrm{KL}}\Big(p_{\boldsymbol{x}}(x)\,\|\,q_{\boldsymbol{x}}(x)\Big) \;\Longrightarrow\; q_{\boldsymbol{x}}(x) = \mathcal{N}_{\boldsymbol{x}}(\bar{x}_p, R_p)} \tag{6.86}$$

Approximation using the exponential family

We can generalize this conclusion by allowing $q_{\boldsymbol{x}}(x)$ to belong to the exponential family of distributions, say, to take the form:

$$q_{\boldsymbol{x}}(x) \;=\; h(x)\exp\Big\{\theta^{\mathsf{T}}T(x) - a(\theta)\Big\} \tag{6.87}$$

for some fixed base function $h(x)$, statistic $T(x)$, and log-normalization factor $a(\theta)$. The distribution is parameterized by $\theta\in\mathbb{R}^Q$. We wish to determine θ to minimize the KL divergence to a given distribution $p_{\boldsymbol{x}}(x)$:

$$\widehat{\theta} \;=\; \underset{\theta\in\mathbb{R}^Q}{\mathrm{argmin}}\; D_{\mathrm{KL}}\Big(p_{\boldsymbol{x}}(x)\,\|\,h(x)\exp\Big\{\theta^{\mathsf{T}}T(x)-a(\theta)\Big\}\Big) \tag{6.88}$$

From definition (6.43) we again have (where we are ignoring all terms independent of θ):

$$D_{\text{KL}}(p\|q) = \int_{x\in\mathcal{X}} \Big(\theta^{\mathsf{T}} T(x) - a(\theta)\Big) p_{\boldsymbol{x}}(x)dx + \text{cte} \tag{6.89}$$

Differentiating relative to θ and setting the gradient to zero leads to the result that $\widehat{\theta}$ should satisfy:

$$\nabla_{\theta^{\mathsf{T}}}\, a(\theta)\Big|_{\theta=\widehat{\theta}} = \mathbb{E}_p T(\boldsymbol{x}) \tag{6.90}$$

The term on the right-hand side is the mean of $T(\boldsymbol{x})$ under distribution $p_{\boldsymbol{x}}(x)$. However, we know from property (5.89) for exponential distributions that the gradient on the left-hand side is equal to the mean of $T(\boldsymbol{x})$ under the distribution $q_{\boldsymbol{x}}(x)$. We conclude that the optimal choice for θ should be the parameter that matches the moments under $p_{\boldsymbol{x}}(x)$ and $q_x(x)$:

$$\boxed{\mathbb{E}_p T(\boldsymbol{x}) = \mathbb{E}_q T(\boldsymbol{x})} \tag{6.91}$$

For Gaussian distributions, this equality reduces to matching only the first- and second-order moments.

6.5 FISHER INFORMATION MATRIX

There is another useful connection between the KL divergence and an important concept in statistical signal analysis known as the *Fisher information matrix*. We will encounter this matrix later in Section 31.5.1 when we study the maximum-likelihood estimation problem.

Consider a pdf $p_{\boldsymbol{x}}(x;\theta)$ that is defined by some parameter vector θ (such as a Gaussian distribution and its mean and variance parameters). Let us consider two variants of the pdf, one with parameter θ and another with parameter θ'. Obviously, the two distributions agree when $\theta = \theta'$. Their KL divergence is given by:

$$\begin{aligned} &D_{\text{KL}}\Big(p_{\boldsymbol{x}}(x;\theta)\|p_{\boldsymbol{x}}(x;\theta')\Big) \qquad (6.92)\\ &= \int_{x\in\mathcal{X}} p_{\boldsymbol{x}}(x;\theta)\ln\left(\frac{p_{\boldsymbol{x}}(x;\theta)}{p_{\boldsymbol{x}}(x;\theta')}\right)dx \\ &= \int_{x\in\mathcal{X}} p_{\boldsymbol{x}}(x;\theta)\ln p_{\boldsymbol{x}}(x;\theta)dx - \int_{x\in\mathcal{X}} p_{\boldsymbol{x}}(x;\theta)\ln p_{\boldsymbol{x}}(x;\theta')dx \end{aligned}$$

Differentiating the KL measure with respect to θ' gives

$$\begin{aligned}\nabla_{(\theta')^{\mathsf{T}}} D_{\text{KL}}\Big(p_{\boldsymbol{x}}(x;\theta)\|p_{\boldsymbol{x}}(x;\theta')\Big) &= -\nabla_{(\theta')^{\mathsf{T}}}\left(\int_{x\in\mathcal{X}} p_{\boldsymbol{x}}(x;\theta)\ln p_{\boldsymbol{x}}(x;\theta')dx\right)\\ &= -\int_{x\in\mathcal{X}} p_{\boldsymbol{x}}(x;\theta)\nabla_{(\theta')^{\mathsf{T}}}\Big(\ln p_{\boldsymbol{x}}(x;\theta')\Big)dx\end{aligned} \tag{6.93}$$

Evaluating at $\theta'=\theta$ leads to

$$\begin{aligned}&\nabla_{(\theta')^{\mathsf{T}}} D_{\text{KL}}\Big(p_{\boldsymbol{x}}(x;\theta)\|p_{\boldsymbol{x}}(x;\theta')\Big)\Big|_{\theta'=\theta}\\ &= -\int_{x\in\mathcal{X}} p_{\boldsymbol{x}}(x;\theta)\frac{\nabla_{(\theta)^{\mathsf{T}}}p_{\boldsymbol{x}}(x;\theta)}{p_{\boldsymbol{x}}(x;\theta)}dx\\ &= -\int_{x\in\mathcal{X}} \nabla_{(\theta)^{\mathsf{T}}}p_{\boldsymbol{x}}(x;\theta)dx\\ &= -\nabla_{(\theta)^{\mathsf{T}}}\underbrace{\left\{\int_{x\in\mathcal{X}} p_{\boldsymbol{x}}(x;\theta)dx\right\}}_{=1}\end{aligned} \tag{6.94}$$

and, hence,

$$\boxed{\nabla_{(\theta')^{\mathsf{T}}} D_{\text{KL}}\Big(p_{\boldsymbol{x}}(x;\theta)\|p_{\boldsymbol{x}}(x;\theta')\Big)\Big|_{\theta'=\theta} = 0} \tag{6.95}$$

Differentiating (6.93) again relative to θ' leads to the Hessian matrix of the KL measure and gives:

$$\begin{aligned}\nabla^2_{\theta'} D_{\text{KL}}\Big(p_{\boldsymbol{x}}(x;\theta)\|p_{\boldsymbol{x}}(x;\theta')\Big) &= -\int_{x\in\mathcal{X}} p_{\boldsymbol{x}}(x;\theta)\nabla_{\theta'}\Big(\nabla_{(\theta')^{\mathsf{T}}}\Big[\ln p_{\boldsymbol{x}}(x;\theta')\Big]\Big)dx\\ &= -\int_{x\in\mathcal{X}} p_{\boldsymbol{x}}(x;\theta)\nabla^2_{\theta'}\Big[\ln p_{\boldsymbol{x}}(x;\theta')\Big]dx\end{aligned} \tag{6.96}$$

Evaluating the Hessian matrix at $\theta'=\theta$:

$$\begin{aligned}\nabla^2_{\theta'} D_{\text{KL}}\Big(p_{\boldsymbol{x}}(x;\theta)\|p_{\boldsymbol{x}}(x;\theta')\Big)\Big|_{\theta'=\theta} &= -\int_{x\in\mathcal{X}} p_{\boldsymbol{x}}(x;\theta)\nabla^2_{\theta'}\Big[\ln p_{\boldsymbol{x}}(x;\theta')\Big]\Big|_{\theta'=\theta}dx\\ &= -\int_{x\in\mathcal{X}} p_{\boldsymbol{x}}(x;\theta)\nabla^2_{\theta}\Big[\ln p_{\boldsymbol{x}}(x;\theta)\Big]dx\\ &= -\mathbb{E}_p\Big[\nabla^2_{\theta}\ln p_{\boldsymbol{x}}(x;\theta)\Big]\\ &\stackrel{\Delta}{=} F(\theta)\end{aligned} \tag{6.97}$$

where we introduced the notation $F(\theta)$ to refer to the following matrix:

$$\boxed{F(\theta) \stackrel{\Delta}{=} -\mathbb{E}_p\Big[\nabla^2_{\theta}\ln p_{\boldsymbol{x}}(x;\theta)\Big]} \tag{6.98}$$

This quantity is called the *Fisher information matrix* and will be encountered later in (31.95) in the context of maximum-likelihood estimation. We explain there that this is the same matrix we encountered before in (5.94).

Therefore, result (6.97) is stating that the Hessian matrix of the KL divergence at location θ is equal to the Fisher information matrix for the distribution $p_{\boldsymbol{x}}(x;\theta)$. In summary, we conclude that:

$$\begin{aligned}\nabla^2_{\theta'} D_{\text{KL}}\Big(p_{\boldsymbol{x}}(x;\theta)\|p_{\boldsymbol{x}}(x;\theta')\Big)\Big|_{\theta'=\theta} &= -\mathbb{E}_p\Big[\nabla^2_{\theta}\ln p_{\boldsymbol{x}}(x;\theta)\Big] \\ &\overset{(a)}{=} \mathbb{E}_p\Big[\Big(\nabla_{\theta^{\mathsf{T}}}\ln p_{\boldsymbol{x}}(x;\theta)\Big)\Big(\nabla_{\theta}\ln p_{\boldsymbol{x}}(x;\theta)\Big)\Big] \\ &= F(\theta) \quad \textbf{(Fisher information matrix)}\end{aligned} \tag{6.99}$$

where equality (a) provides an equivalent definition for the Fisher information matrix, which we will establish later in (31.96). Combining results (6.95) and (6.99) we arrive at the following useful second-order Taylor approximation for the KL divergence measure in terms of the Fisher information matrix:

$$D_{\text{KL}}\Big(p_{\boldsymbol{x}}(x;\theta)\|p_{\boldsymbol{x}}(x;\theta+\delta\theta)\Big) = \frac{1}{2}(\delta\theta)^{\mathsf{T}}F(\theta)\,\delta\theta \;+\; o(\|\delta\theta\|^2) \tag{6.100}$$

where the little-o notation $a = o(\|\delta\theta\|^2)$ refers to higher-order terms than $\|\delta\theta\|^2$, namely, $a/\|\delta\theta\|^2 \to 0$ as $\delta\theta \to 0$.

Result (6.100) has one useful interpretation. The distribution $p_{\boldsymbol{x}}(x;\theta)$ is parameterized by $\theta \in \mathbb{R}^M$. If we perturb this parameter, it is generally not clear how the distribution will change. However, since the KL divergence between two distributions serves as a measure of similarity or "closeness" between them, result (6.100) is asserting that the Fisher information matrix defines a weighted Euclidean norm that enables us to assess the effect of the perturbation $\delta\theta$ in the parameter space on the amount of perturbation in the distribution space. For instance, choosing a perturbation $\delta\theta$ along the direction of the eigenvector of $F(\theta)$ corresponding to its largest eigenvalue would cause larger perturbation to the distribution than a similar-size perturbation $\delta\theta$ along the direction of the eigenvector of $F(\theta)$ corresponding to its smallest eigenvalue. The relation between the KL divergence and the Fisher information matrix is helpful in many contexts. For example, we will exploit it later in Chapter 49 to derive policy gradient methods for reinforcement learning purposes.

Example 6.11 **(Gaussian distribution)** Consider a collection of N iid Gaussian random variables with $\boldsymbol{x}_n \sim \mathcal{N}_{\boldsymbol{x}_n}(\theta,\sigma^2)$. We assume σ^2 is fixed and let the Gaussian distribution be parameterized by θ. The joint pdf of the observations $\{\boldsymbol{x}_1,\ldots,\boldsymbol{x}_N\}$, also called their *likelihood*, is given by

$$p_{\boldsymbol{x}_1,\ldots,\boldsymbol{x}_N}(x_1,\ldots,x_N;\theta) = \prod_{n=1}^{N}\frac{1}{\sqrt{2\pi\sigma^2}}e^{-\frac{1}{2\sigma^2}(x_n-\theta)^2} \tag{6.101}$$

so that the log-likelihood function is

$$\ln p_{\boldsymbol{x}_1,\ldots,\boldsymbol{x}_N}(x_1,\ldots,x_N;\theta) = -\frac{N}{2}\ln(2\pi\sigma^2) - \frac{1}{2\sigma^2}\sum_{n=1}^{N}(x_n-\theta)^2 \tag{6.102}$$

It is straightforward to verify by differentiating twice that

$$\frac{\partial^2}{\partial^2\theta}\ln p_{\boldsymbol{x}_1,\ldots,\boldsymbol{x}_N}(x_1,\ldots,x_N;\theta) = -N/\sigma^2 \tag{6.103}$$

so that, in this case, the Fisher information "matrix" is the scalar:

$$F(\theta) = N/\sigma^2 \tag{6.104}$$

In other words, the Hessian of the KL divergence is

$$\frac{\partial^2}{\partial^2\theta}D_{\text{KL}}\Big(p_{\boldsymbol{x}_{1:N}}(x_{1:N};\theta)\,\|\,p_{\boldsymbol{x}_{1:N}}(x_{1:N};\theta')\Big)\bigg|_{\theta=\theta'} = N/\sigma^2 \tag{6.105}$$

where we are writing $x_{1:N}$ to refer to the collection $\{x_1,\ldots,x_N\}$. Alternatively, we can arrive at this same result by working directly with definition (6.43) for the KL divergence. To simplify the notation, we will write $p(\theta)$ to refer to $p_{\boldsymbol{x}_{1:N}}(x_{1:N};\theta)$. Similarly for $p(\theta')$. Then, using (6.43) and (6.102) we get

$$\begin{aligned}
D_{\text{KL}}\Big(p(\theta)\,\|\,p(\theta')\Big) &= \int_{-\infty}^{\infty} p(\theta)\ln p(\theta)dx - \int_{-\infty}^{\infty} p(\theta)\ln p(\theta')dx \\
&= -\frac{N}{2}\ln(2\pi\sigma^2) - \frac{1}{2\sigma^2}\sum_{n=1}^{N}\int_{-\infty}^{\infty}(x_n-\theta)^2 p(\theta)dx + \\
&\quad\ \frac{N}{2}\ln(2\pi\sigma^2) + \frac{1}{2\sigma^2}\sum_{n=1}^{N}\int_{-\infty}^{\infty}(x_n-\theta')^2 p(\theta)dx \\
&= -\frac{1}{2\sigma^2}\sum_{n=1}^{N}\underbrace{\int_{-\infty}^{\infty}(x_n-\theta)^2 p(\theta)dx}_{=\sigma^2} + \frac{1}{2\sigma^2}\sum_{n=1}^{N}\int_{-\infty}^{\infty}(x_n-\theta')^2 p(\theta)dx \\
&= -\frac{N}{2} + \frac{1}{2\sigma^2}\sum_{n=1}^{N}\int_{-\infty}^{\infty}(x_n-\theta+\theta-\theta')^2 p(\theta)dx \\
&= -\frac{N}{2} + \frac{1}{2\sigma^2}\sum_{n=1}^{N}\underbrace{\int_{-\infty}^{\infty}(x_n-\theta)^2 p(\theta)dx}_{=\sigma^2} + \\
&\quad\ \frac{1}{2\sigma^2}\sum_{n=1}^{N}\underbrace{\int_{-\infty}^{\infty}(\theta-\theta')^2 p(\theta)dx}_{(\theta-\theta')^2} + \\
&\quad\ \frac{1}{2\sigma^2}\sum_{n=1}^{N}\underbrace{\int_{-\infty}^{\infty}(x_n-\theta)(\theta-\theta')p(\theta)dx}_{=0} \\
&= \frac{N}{2\sigma^2}(\theta-\theta')^2
\end{aligned} \tag{6.106}$$

Differentiating twice relative to θ' leads back to (6.105), as expected.

6.6 NATURAL GRADIENTS

There is another useful property of the KL divergence that arises in the solution of inference problems, especially in the development of iterative mechanisms.

Newton method

Consider a pdf $p_{\boldsymbol{x}}(x;\theta)$ that is defined by some unknown parameter $\theta \in \mathbb{R}^M$. Let θ_{m-1} denote an estimate for θ at some iteration $m-1$. Assume we desire to update this estimate to $\theta_m = \theta_{m-1} + \delta\theta$ for some correction term $\delta\theta$. We can determine $\delta\theta$ by solving a problem of the form:

$$\delta\theta^o = \underset{\delta\theta\in\mathbb{R}^M}{\operatorname{argmin}}\; J(\theta_{m-1} + \delta\theta) \tag{6.107}$$

where $J(\theta)$ refers to some cost function we wish to minimize. We assume $J(\theta)$ is differentiable. If we perform a second-order Taylor series expansion around $J(\theta_{m-1})$ we get

$$J(\theta_m) \approx J(\theta_{m-1}) + \nabla_\theta J(\theta_{m-1})\delta\theta \;+\; \frac{1}{2}(\delta\theta)^{\mathsf{T}}\left[\nabla_\theta^2 J(\theta_{m-1})\right](\delta\theta) \tag{6.108}$$

To minimize the right-hand side over $\delta\theta$ we set the gradient vector relative to $\delta\theta$ to zero to get

$$\nabla_{\theta^{\mathsf{T}}} J(\theta_{m-1}) + \nabla_\theta^2 J(\theta_{m-1})\delta\theta^o = 0 \tag{6.109}$$

which leads to the gradient descent direction:

$$\boxed{\delta\theta^o = -\left[\nabla_\theta^2 J(\theta_{m-1})\right]^{-1} \nabla_{\theta^{\mathsf{T}}} J(\theta_{m-1})} \qquad \textbf{(Newton correction)} \tag{6.110}$$

Using this correction term we arrive at the standard form of the Newton method, which we will encounter again in (12.197):

$$\theta_m = \theta_{m-1} - \left[\nabla_\theta^2 J(\theta_{m-1})\right]^{-1} \nabla_{\theta^{\mathsf{T}}} J(\theta_{m-1}) \quad \textbf{(Newton method)} \tag{6.111}$$

We refer to (6.110) as the Newton gradient, so that θ_{m-1} is updated to θ_m by moving along this direction. It can be verified that when the cost $J(\theta)$ is quadratic in θ, then the Newton method finds the minimizer in a single step!

Example 6.12 **(Quadratic costs)** Consider a quadratic objective function of the form

$$J(\theta) = \kappa + 2a^{\mathsf{T}}\theta + \theta^{\mathsf{T}}C\theta, \;\; \theta \in \mathbb{R}^M \tag{6.112}$$

with $C > 0$. The gradient vector and Hessian matrix are given by

$$\nabla_{\theta^{\mathsf{T}}} J(\theta) = 2a + 2C\theta, \quad \nabla_\theta^2 J(\theta) = 2C \tag{6.113}$$

Therefore, the Newton recursion (6.111) reduces to

$$\theta_m = \theta_{m-1} - (2C)^{-1}(2a + 2C\theta_{m-1}) \;=\; -C^{-1}a \;=\; \theta^o$$

which shows that the recursion converges to θ^o in a single step.

Example 6.13 **(Line search method)** Consider the same setting as the Newton method. We start from a pdf $p_{\boldsymbol{x}}(x;\theta)$ and let θ_{m-1} denote the estimate for θ at iteration $m-1$. We wish to update θ_{m-1} to θ_m by solving instead the constrained problem:

$$\delta\theta^o = \underset{\delta\theta\in\mathbb{R}^M}{\text{argmin}} \left\{J(\theta_{m-1}+\delta\theta)\right\}, \quad \text{subject to } \frac{1}{2}\|\delta\theta\|^2 \leq \epsilon \tag{6.114}$$

where we are bounding the squared Euclidean norm of the update. We introduce a Lagrange multiplier $\lambda \geq 0$ and consider the unconstrained formulation

$$\delta\theta^o = \underset{\delta\theta\in\mathbb{R}^M}{\text{argmin}} \left\{J(\theta_{m-1}+\delta\theta) \;+\; \lambda\Big(\frac{1}{2}\|\delta\theta\|^2 - \epsilon\Big)\right\} \tag{6.115}$$

To solve the problem, we now introduce a *first-order* (as opposed to second-order) Taylor series expansion around $J(\theta_{m-1})$:

$$J(\theta_m) \approx J(\theta_{m-1}) + \nabla_\theta J(\theta_{m-1})\delta\theta \tag{6.116}$$

so that the cost appearing in (6.115) is approximated by

$$\text{cost} \approx J(\theta_{m-1}) + \nabla_\theta J(\theta_{m-1})\delta\theta + \lambda\Big(\frac{1}{2}\|\delta\theta\|^2 - \epsilon\Big) \tag{6.117}$$

To minimize the right-hand side over $\delta\theta$ we set the gradient vector relative to $\delta\theta$ to zero to get

$$\delta\theta^o = -\frac{1}{\lambda}\nabla_{\theta^{\mathsf{T}}} J(\theta_{m-1}) \tag{6.118}$$

We can determine λ by seeking the "largest" perturbation possible. Specifically, we set

$$\frac{1}{2}\|\delta\theta^o\|^2 = \epsilon \Longrightarrow \lambda = \frac{1}{\sqrt{2\epsilon}}\|\nabla_{\theta^{\mathsf{T}}} J(\theta_{m-1})\| \tag{6.119}$$

and we arrive at the line search update (see Prob. 6.21):

$$\theta_m = \theta_{m-1} - \frac{\sqrt{2\epsilon}}{\|\nabla_{\theta^{\mathsf{T}}} J(\theta_{m-1})\|}\nabla_{\theta^{\mathsf{T}}} J(\theta_{m-1}) \tag{6.120}$$

Change of coordinates

Clearly, for nonquadratic objective functions, convergence of the Newton method will usually take longer than one iteration when (and if) it occurs. It is common practice, especially when dealing with more general objective functions that may not even be convex, to introduce a small step-size parameter $\mu > 0$ into the Newton iteration to control the size of the correction to θ_{m-1}, such as using

$$\boxed{\theta_m = \theta_{m-1} - \mu\left[\nabla_\theta^2 J(\theta_{m-1})\right]^{-1}\nabla_{\theta^{\mathsf{T}}} J(\theta_{m-1})} \tag{6.121}$$

The step size can also change with the iteration index m, as we will discuss at some length in future chapters. Here, it is sufficient to continue with a constant step size.

It is useful to note that sometimes it is possible to introduce a change of coordinate systems and transform an objective function from a less favorable form to a more favorable form. Consider the following example.

Example 6.14 **(Change of coordinates)** If the coordinate system is modified and the parameter θ for the objective function is replaced by a transformed parameter vector θ' in the new space, then it is understandable that the gradient vector of the same objective function $J(\cdot)$ relative to θ' will generally point in a different search direction than the gradient vector relative to θ.

Let $\theta = \text{col}\{x, y\} \in \mathbb{R}^2$ and consider a cost function of the form:

$$J(\theta) = x^2 + (y-1)^2 \implies \nabla_\theta J(\theta) = \begin{bmatrix} \partial J/\partial x & \partial J/\partial y \end{bmatrix} = \begin{bmatrix} 2x & 2(y-1) \end{bmatrix} \tag{6.122}$$

Observe that $J(x, y)$ is quadratic over θ. Applying the Newton recursion (6.111) to $J(x, y)$ would lead to its minimizer $(x^o, y^o) = (0, 1)$ in a single step.

Now assume we perform a change of variables to polar coordinates and let $x = r\cos\phi$ and $y = r\sin\phi$, where $\phi \in [-\pi, \pi]$. Define the new parameter vector $\theta' = \text{col}\{r, \phi\}$. Then, the same cost function can be written as

$$J(\theta') = r^2 + 1 - 2r\sin\phi \tag{6.123}$$

$$\implies \nabla_{\theta'} J(\theta') = \begin{bmatrix} \partial J/\partial r & \partial J/\partial \phi \end{bmatrix} = \begin{bmatrix} 2(r - \sin\phi) & -2r\cos\phi \end{bmatrix}$$

The transformed cost function is not quadratic over θ', and applying Newton recursion (6.111) to $J(\theta')$ will not converge in a single step anymore; it may not even converge.

Figure 6.2 plots the objective function in Cartesian and polar spaces, along with their contour curves. It is evident that a change in the coordinate system leads to a change in the behavior of the objective function. In particular, we can compare the gradient vectors at location $(x, y) = (1, 1)$, which corresponds to $(r, \phi) = (\sqrt{2}, \frac{\pi}{4})$, for both representations to find

$$\nabla_\theta J(\theta)\Big|_{(x,y)=(1,1)} = \begin{bmatrix} 2 & 0 \end{bmatrix} \tag{6.124a}$$

$$\nabla_{\theta'} J(\theta')\Big|_{(r,\phi)=(\sqrt{2},\frac{\pi}{4})} = \begin{bmatrix} \sqrt{2} & -2 \end{bmatrix} \tag{6.124b}$$

Observe how the gradient vectors point in different directions and also have different norms.

The Hessian matrix of $J(\theta')$ is given by

$$\nabla^2_{\theta'} J(\theta') = \begin{bmatrix} 2 & -2\cos\phi \\ -2\cos\phi & 2r\sin\phi \end{bmatrix} \tag{6.125}$$

so that if we were to apply Newton recursion (6.111) to it we get, after simplifications,

$$\begin{bmatrix} r_m \\ \phi_m \end{bmatrix} = \begin{bmatrix} r_{m-1} \\ \phi_{m-1} \end{bmatrix} - \tag{6.126}$$

$$\frac{\mu}{4r_{m-1}\sin\phi_{m-1} - 4\cos^2(\phi_{m-1})} \begin{bmatrix} 4r_{m-1}(r_{m-1}\sin\phi_{m-1} - 1) \\ -2\sin(2\phi_{m-1}) \end{bmatrix}$$

The initial conditions for the parameters (r, ϕ) need to be chosen close enough to the minimizer, which we know occurs at $(r^o, \phi^o) = (1, \pi/2)$. We run the recursion with $\mu = 0.01$, $M = 500$ iterations, and consider two different sets of initial conditions $(r_{-1}, \phi_{-1}) = (0.9, 1.4)$ and $(r_{-1}, \phi_{-1}) = (0.2, 0.5)$. The first choice is close to the minimizer, while the second choice is away from it. We observe from the plots in Fig. 6.3 that convergence occurs in one case but not the other.

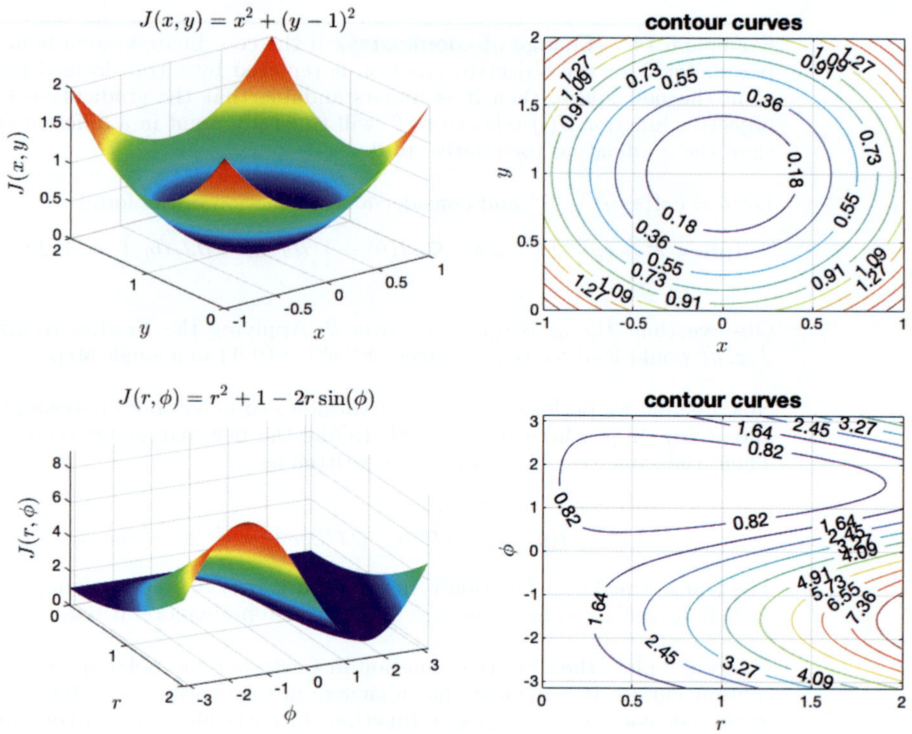

Figure 6.2 The plots in the top row show the objective function $J(\theta) = x^2 + (y-1)^2$ in Cartesian coordinates and its contour curves. The plots in the bottom row show the same objective function in polar coordinates and its contour curves.

There are instances, however, where translation into a new coordinate system can be advantageous. The natural gradient construction takes advantage of this observation.

Definition of natural gradients

The natural gradient recursion can be motivated as follows. Observe that the Newton update (6.111) optimizes $J(\theta)$ over the parameter space $\theta \in \mathbb{R}^M$ without imposing any constraints on the shape of the pdf $p_{\boldsymbol{x}}(x;\theta)$. As a result, two successive iterates θ_{m-1} and θ_m may be close to each other in Euclidean distance while the respective pdfs defined by these parameters may be sufficiently distinct from each other – see top rightmost plot in future Fig. 6.4.

We now describe an alternative approach that leads to a different search direction than Newton's update by requiring the successive pdfs to stay close to each other in the KL divergence sense. Assume that we formulate instead the optimization problem

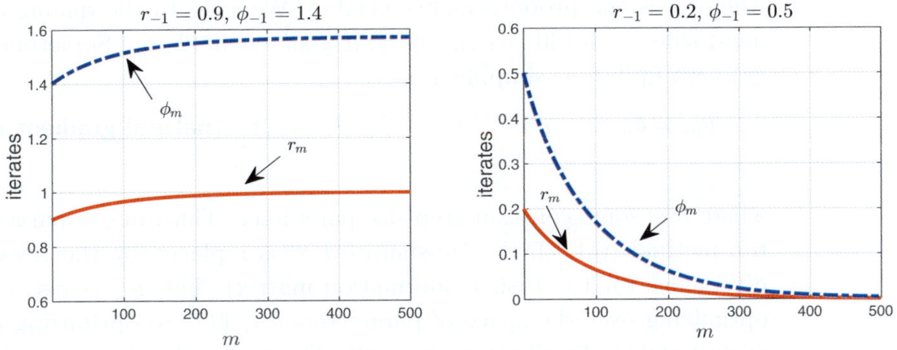

Figure 6.3 Evolution of the iterates (r_m, ϕ_m) for Newton recursion (6.126) for two different choices of the initial conditions. The plots on the left converge to the location of the global minimizer at $(r^o, \phi^o) = (1, \pi/2)$.

$$\begin{aligned} \delta\theta^o = &\underset{\delta\theta \in \mathbb{R}^M}{\text{argmin}}\; J(\theta_{m-1} + \delta\theta) \\ &\text{subject to } D_{\text{KL}}\Big(p_{\boldsymbol{x}}(x;\theta_{m-1}) \| p_{\boldsymbol{x}}(x;\theta_m)\Big) \le \epsilon \end{aligned} \tag{6.127}$$

for some small constant $\epsilon > 0$. One important fact to note is that the KL divergence is an intrinsic measure of similarity between two distributions; it does not matter how these distributions are parameterized. As a result, the solution will be largely invariant to which coordinate system is used to parameterize the distributions.

The natural gradient algorithm is a line search method applied to the solution of an approximate version of (6.127) where, using (6.100), the KL divergence is replaced by a second-order term as follows:

$$\delta\theta^o = \underset{\delta\theta \in \mathbb{R}^M}{\text{argmin}} \Big\{ J(\theta_{m-1} + \delta\theta) \Big\}, \;\; \text{subject to } \frac{1}{2}(\delta\theta)^{\mathsf{T}} F(\theta)\delta\theta \le \epsilon \tag{6.128}$$

Now, repeating the same argument that led to (6.120) gives (see Prob. 6.21):

$$g_{m-1} = \nabla_{\theta^{\mathsf{T}}} J(\theta_{m-1}) \tag{6.129a}$$

$$\theta_m = \theta_{m-1} - \left(\frac{2\epsilon}{\|g_{m-1}\|^2_{F^{-1}(\theta_{m-1})}} \right)^{1/2} F^{-1}(\theta_{m-1}) g_{m-1} \tag{6.129b}$$

where the notation $\|x\|^2_A$ stands for $x^{\mathsf{T}}Ax$. The above expression is in terms of the inverse of the Fisher information matrix, which we already know from (6.99) is related to the inverse of the Hessian matrix of the KL divergence of $p_{\boldsymbol{x}}(x;\theta)$ and $p_{\boldsymbol{x}}(x;\theta')$ as $\theta' \to \theta$. We observe that the search direction is now

$$\boxed{\delta\theta^o \propto F^{-1}(\theta_{m-1}) \nabla_{\theta^{\mathsf{T}}} J(\theta_{m-1})} \quad \textbf{(natural gradient correction)} \tag{6.130}$$

where $\propto$ is the proportionality symbol. We refer to the quantity on the right-hand side of (6.130) as the *natural gradient* at θ_{m-1}. Sometimes, the natural gradient update is simplified to

$$\theta_m = \theta_{m-1} - \mu\, F^{-1}(\theta_{m-1})\, \nabla_{\theta^\mathsf{T}} J(\theta_{m-1}) \quad \textbf{(natural gradient method)} \tag{6.131}$$

where μ is some constant step-size parameter. Thus, in comparison to the Newton method (6.121), the Hessian of $J(\theta)$ is replaced by the Hessian of the KL divergence (or the Fisher information matrix). This is because we moved from optimizing over the space of parameters $\theta \in \mathbb{R}^M$ to optimizing over the space of probability distributions $p_{\boldsymbol{x}}(x;\theta)$. The motivation for using the distribution domain is to ensure that successive iterates of θ maintain the respective distributions close to each other.

Example 6.15 **(Fitting Gaussian distributions)** Consider again a collection of N iid Gaussian random variables with $\boldsymbol{x}_n \sim \mathcal{N}_{\boldsymbol{x}_n}(\mu, \sigma^2)$. We assume in this example that both (μ, σ^2) are unknown and let the Gaussian distribution be parameterized by $\theta = \text{col}\{\mu, \sigma^2\}$. We estimate these parameters by maximizing the log-likelihood function of the observations, which according to (6.102) is defined as the log of the joint pdf of the observations, namely,

$$\theta^o = \underset{\theta \in \mathbb{R}^2}{\text{argmin}}\ \ell(\theta) \tag{6.132}$$

where

$$\begin{aligned} \ell(\theta) = \ell(\mu, \sigma^2) &\overset{\Delta}{=} -\ln\Big\{ p_{\boldsymbol{x}_1, \boldsymbol{x}_2, \ldots, \boldsymbol{x}_N}(x_1, x_2, \ldots, x_N; \theta)\Big\} \\ &= \frac{N}{2}\ln(2\pi\sigma^2) + \frac{1}{2\sigma^2}\sum_{n=1}^{N}(x_n - \mu)^2 \end{aligned} \tag{6.133}$$

We will explain in Chapter 31 that problems of this type arise in the context of maximum-likelihood estimation.

It is straightforward to verify for this example that

$$\nabla_{\theta^\mathsf{T}} \ell(\theta) = \begin{bmatrix} -\dfrac{1}{\sigma^2}\displaystyle\sum_{n=1}^{N}(x_n - \mu) \\ \dfrac{N}{2\sigma^2} - \dfrac{1}{2\sigma^4}\displaystyle\sum_{n=1}^{N}(x_n - \mu)^2 \end{bmatrix} \tag{6.134}$$

$$\nabla_\theta^2 \ell(\theta) = \begin{bmatrix} \dfrac{N}{\sigma^2} & \dfrac{1}{\sigma^4}\displaystyle\sum_{n=1}^{N}(x_n - \mu) \\ \dfrac{1}{\sigma^4}\displaystyle\sum_{n=1}^{N}(x_n - \mu) & -\dfrac{N}{2\sigma^4} + \dfrac{1}{\sigma^6}\displaystyle\sum_{n=1}^{N}(x_n - \mu)^2 \end{bmatrix} \tag{6.135}$$

$$F(\theta) \overset{\Delta}{=} \mathbb{E}_p \nabla_\theta^2\, \ell(\theta) = \begin{bmatrix} \dfrac{N}{\sigma^2} & 0 \\ 0 & \dfrac{N}{2\sigma^4} \end{bmatrix} \tag{6.136}$$

Observe how the Hessian matrix is data-dependent; its entries depend on the samples $\{x_n\}$, while the entries of the Fisher information matrix are data-independent due to the averaging operation. Obviously, by ergodicity, and for N large enough, the Hessian matrix $\nabla_\theta^2 \ell(\theta)$ tends to $F(\theta)$.

We can set the gradient vector of $\ell(\theta)$ to zero at the solution $(\widehat{\mu}, \widehat{\sigma}^2)$ to obtain

$$\widehat{\mu} = \frac{1}{N}\sum_{n=1}^{N} x_n, \quad \widehat{\sigma}^2 = \frac{1}{N}\sum_{n=1}^{N}(x_n - \widehat{\mu})^2 \tag{6.137}$$

However, it can be verified that this estimator for σ^2 is biased, meaning that its mean does not coincide with the true parameter σ^2 (see Prob. 31.1), namely,

$$\mathbb{E}\,\widehat{\boldsymbol{\sigma}}^2 = \frac{N-1}{N}\sigma^2 \tag{6.138}$$

Therefore, it is customary to employ the following alternative expression for estimating the variance of the Gaussian distribution from sample realizations:

$$\widehat{\sigma}^2 = \frac{1}{N-1}\sum_{n=1}^{N}(x_n - \widehat{\mu})^2 \tag{6.139}$$

where we are now dividing by $N-1$. We generate $N = 400$ random realizations $\{x_n\}$ from a Gaussian distribution with its mean chosen randomly at $\mu = 1.3724$ and with variance set to $\sigma^2 = 4$. We used the above expressions to estimate the mean and variance leading to

$$\widehat{\mu} = 1.4406, \quad \widehat{\sigma}^2 = 3.8953 \tag{6.140}$$

Alternatively, in this example, we can minimize $\ell(\theta)$ iteratively by employing the Newton and the natural gradient recursions. For the problem at hand, these recursions take the form

$$\begin{bmatrix} \mu_m \\ \sigma_m^2 \end{bmatrix} = \begin{bmatrix} \mu_{m-1} \\ \sigma_{m-1}^2 \end{bmatrix} - \mu\Big(\nabla_\theta^2\, \ell(\theta_{m-1})\Big)^{-1} \nabla_{\theta^\mathsf{T}}\, \ell(\theta_{m-1}) \tag{6.141}$$

$$\begin{bmatrix} \mu_m \\ \sigma_m^2 \end{bmatrix} = \begin{bmatrix} \mu_{m-1} \\ \sigma_{m-1}^2 \end{bmatrix} - \mu\, F^{-1}(\theta_{m-1}) \nabla_{\theta^\mathsf{T}}\, \ell(\theta_{m-1}) \tag{6.142}$$

The recursions are not always guaranteed to converge toward the true parameter values. We compare their performance in Fig. 6.4 for a case where good convergence is attained. We employed the same $N = 400$ samples along with $\mu = 0.01$, and ran $M = 1000$ iterations. We added ϵI to the Hessian matrix and the Fisher information matrix prior to inversion to avoid singularities, where $\epsilon = 1 \times 10^{-6}$ is a small number. The recursions converged toward $\widehat{\mu} = 1.4405$ (Newton), $\widehat{\mu} = 1.4406$ (natural gradient), $\widehat{\sigma}^2 = 3.8852$ (Newton), and $\widehat{\sigma}^2 = 3.8855$ (natural gradient). Using expression (6.66), with p corresponding to the true pdf with parameters (μ, σ^2) and q corresponding to the estimated pdf with parameters (μ_m, σ_m^2), the figure also plots the evolution of the KL divergence between the true Gaussian pdf and its estimated version using both Newton method and the natural gradient algorithm. It is obvious that the latter leads to smaller KL divergence, as expected by construction.

Remark 6.1. (Unnormalized and normalized histograms) We refer to the two histograms shown in the bottom row of Fig. 6.4. The horizontal axis in both plots is the variable x with values ranging over $[-10, 10]$. This interval is divided into 30 smaller bins in the figure. The vertical axis in the histogram on the left represents the relative frequency of values of x within each bin. For example, if we consider the bin that is centered at $x = 5$, its relative frequency is approximately 0.025, suggesting that 2.5% of

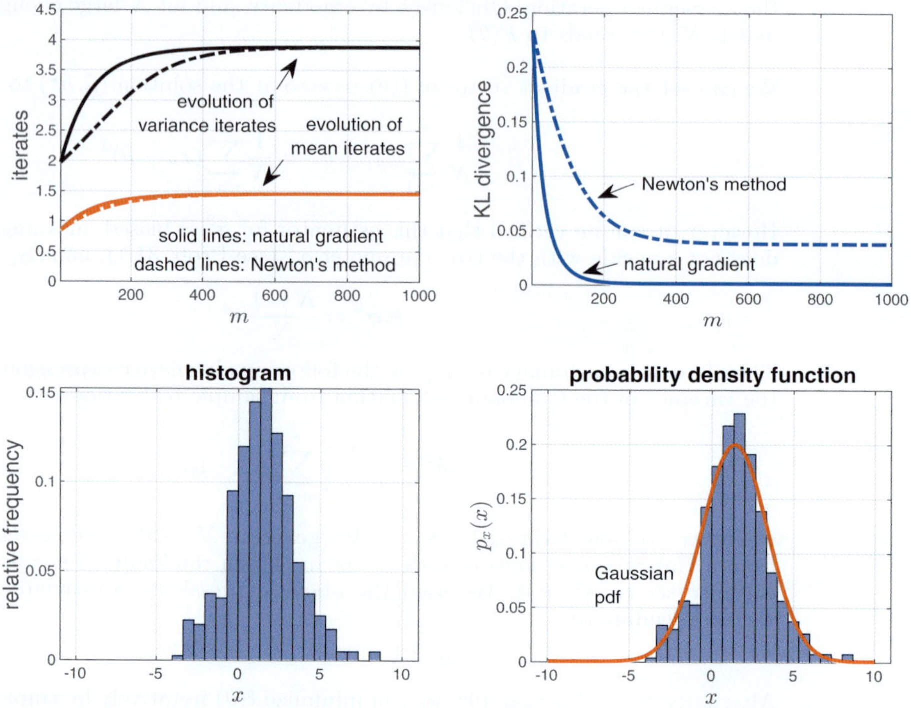

Figure 6.4 Simulation of Newton and natural gradient recursions (6.141)–(6.142) for estimating the mean and variance of a Gaussian distribution from sample measurements. In this simulation, both algorithms converged close to the true values, with the natural gradient algorithm showing faster convergence. The plots in the bottom row show the normalized and unnormalized histograms of the samples that were used.

the values of x fall into this bin. In the histogram on the right, each relative frequency is normalized by the width of the bin. The result is an approximate pdf representation. This normalization is motivated by the following observation. Consider a generic pdf, $p_{\boldsymbol{x}}(x)$. If we integrate it over some interval $[x_0, x_0+\Delta x]$, we obtain a measure for the relative frequency of values of x falling into this range:

$$\int_{x_0}^{x_0+\Delta x} p_{\boldsymbol{x}}(x)dx \;=\; \text{relative frequency within } [x_0, x_0+\Delta x] \tag{6.143}$$

Approximating the area under the integral by $p_{\boldsymbol{x}}(x_0)\Delta x$, we find that the value of the pdf location x_0 is roughly

$$p_{\boldsymbol{x}}(x_0) \approx \frac{\text{relative frequency}}{\Delta x} \tag{6.144}$$

In other words, by dividing relative frequencies by the width of the bins, we obtain an approximation for the pdf of x. We used this method to construct the histogram shown on the right of Fig. 6.4 from the histogram on the left.

■

Example 6.16 (**Natural gradients and Riemannian geometry**) When minimizing over distributions, the natural gradient construction (6.131) ends up employing the inverse of the Fisher information matrix, which we know from (6.99) is related to the curvature (inverse Hessian) of the KL divergence of $p_{\boldsymbol{x}}(x;\theta)$ and $p_{\boldsymbol{x}}(x;\theta')$ as $\theta' \to \theta$.

We can motivate similar natural gradient constructions for more generic optimization problems that need not be limited to the space of probability distributions:

$$\theta^o = \underset{\theta \in \mathbb{R}^M}{\text{argmin}}\ J(\theta) \tag{6.145}$$

In problem (6.127) we sought the optimal solution θ^o by requiring the successive pdfs to lie on a surface of constant KL divergence. We can pursue a similar construction for more general surfaces, other than probability measures. This is because natural gradient implementations are well-suited for optimization over what is known as *Riemannian manifolds*. Manifolds are generalizations of the notion of surfaces to higher dimensions. For example, a sphere or a torus are three-dimensional manifolds. Riemannian manifolds are a particular class of manifolds over which the notions of distances, angles, and curvatures can be defined and measured. We do not need to provide a rigorous treatment of Riemannian manifolds here; it is sufficient to motivate the concept informally.

Thus, recall that in Euclidean space we can measure (and define) the distance between two vectors, say, $\{p, p+\delta p\} \in \mathbb{R}^M$, by computing their Euclidean distance (or the square of it) defined as:

$$d_E^2(p, p+\delta p) \;=\; \sum_{\ell=1}^{M} (\delta p_\ell)^2 \;=\; \|\delta p\|^2 = (\delta p)^{\mathsf{T}} \delta p \tag{6.146}$$

For example, on the plane, this distance metric leads to the convention that the shortest distance between two points is a straight line. However, on curved surfaces, such as on the surface of a sphere, the shortest distance between two points is instead the shortest *arc* connecting them. Riemannian geometry is the branch of geometry that deals with curved spaces of this nature in higher dimensions. The curve with the smallest length that connects two points on a Riemannian manifold is called a *geodesic*.

We associate a metric tensor (also called a *Riemann tensor*) with every point on a Riemannian manifold; loosely, it is a quantity that allows us to measure distances. For Riemannian surfaces in $\mathbb{R}^M$, we denote the metric tensor by $G(q)$ at location q; it is an $M \times M$ positive-definite matrix that is dependent on q. Using the metric tensor, the (squared) distance between two close points q and $q+\delta q$ on the Riemannian manifold is defined by

$$d_R^2(q, q+\delta q) \;=\; \sum_{\ell=1}^{M}\sum_{n=1}^{M} \delta q_\ell G_{\ell n} \delta q_n \;=\; (\delta q)^{\mathsf{T}} G(q) \delta q \tag{6.147}$$

where the $\{G_{\ell n}\}$ denote the entries of the matrix $G(q)$. The natural gradient method for minimizing a cost function $J(\theta)$ over Riemannian geometry replaces the inverse Hessian in the Newton method by the inverse of the curvature matrix, i.e., it uses

$$\boxed{\theta_m = \theta_{m-1} - \mu\Big(G(\theta_{m-1})\Big)^{-1} \nabla J_{\theta^{\mathsf{T}}}(\theta_{m-1}), \quad m \geq 0} \tag{6.148}$$

where $\mu > 0$ is a small step-size parameter. This update ensures that the successive iterates travel over the manifold defined by $G(q)$, just like formulation (6.127) ensures that the successive iterates travel over the manifold of constant KL values. In particular, over the space of probability distributions, we showed in (6.97) that the curvature of the KL measure is given by the Fisher information matrix, which explains its use in (6.131).

Returning to Example 6.14, let us evaluate the curvature matrix of the manifold described by $J(r,\phi)$ and shown in the lower plot of Fig. 6.2. Consider a point $p = \text{col}\{x,y\}$ in Cartesian coordinates and the corresponding point $q = \text{col}\{r,\phi\}$ in polar coordinates, where $r = x\cos\phi$ and $y = r\sin\phi$. Assume q is perturbed slightly to $q+\delta q$ where $\delta q = \text{col}\{\delta r, \delta\phi\}$. The corresponding point p is perturbed to $p+\delta p$ where

$$p = \begin{bmatrix} r\cos(\phi) \\ r\sin(\phi) \end{bmatrix}, \quad p+\delta p \begin{bmatrix} (r+\delta r)\cos(\phi+\delta\phi) \\ (r+\delta r)\sin(\phi+\delta\phi) \end{bmatrix} \tag{6.149}$$

Using the approximations

$$\cos(\delta\phi) \approx 1, \qquad \sin(\delta\phi) \approx \delta\phi \tag{6.150}$$

and ignoring products involving δr and $\delta\phi$, we find

$$\delta p \approx \begin{bmatrix} -r\delta\phi\sin(\phi) + \delta r\cos(\phi) \\ r\delta\phi\cos(\phi) + \delta r\sin(\phi) \end{bmatrix} \tag{6.151}$$

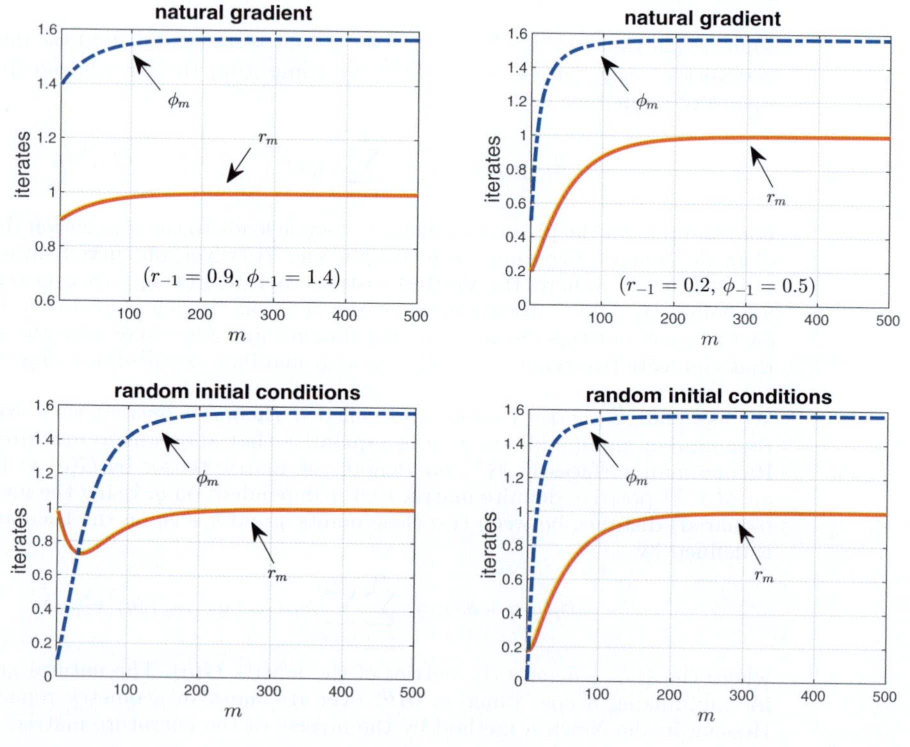

Figure 6.5 Evolution of the iterates (r_m, ϕ_m) for the natural gradient recursion (6.156) for different choices of the initial conditions. The plots converge to the location of the global minimizer at $(r^o, \phi^o) = (1, \pi/2)$.

It follows that the Euclidean distance between p and $p+\delta p$ is given by

$$d_E^2(p, p+\delta p) \approx r^2(\delta\phi)^2 + (\delta r)^2 \tag{6.152}$$

We can determine the curvature tensor at location q by imposing a distance invariance requirement:

$$d_E^2(p, p + \delta p) = d_R^2(q, q + \delta q) \; = \; (\delta q)^{\mathsf{T}} G(q)(\delta q) \tag{6.153}$$

That is, it must hold that

$$r^2(\delta\phi)^2 + (\delta r)^2 \; = \; \begin{bmatrix} \delta r & \delta\phi \end{bmatrix} G(q) \begin{bmatrix} \delta r \\ \delta\phi \end{bmatrix} \tag{6.154}$$

which allows us to identify, for this example,

$$G(q) = \begin{bmatrix} 1 & 0 \\ 0 & r^2 \end{bmatrix} \tag{6.155}$$

In this way, the natural gradient recursion (6.148) reduces to

$$\begin{bmatrix} r_m \\ \phi_m \end{bmatrix} = \begin{bmatrix} r_{m-1} \\ \phi_{m-1} \end{bmatrix} - \mu \begin{bmatrix} 2(r_{m-1} - \sin\phi_{m-1}) \\ -(2/r_{m-1})\cos\phi_{m-1} \end{bmatrix} \tag{6.156}$$

We apply this recursion to the same objective $J(r, \theta)$ from Example 6.14 using again $\mu = 0.01$, $M = 500$ iterations, and the same initial conditions, in addition to other randomly chosen initial conditions. As seen in Fig. 6.5, the iterates now converge to the global minimizer $(r^o, \phi^o) = (1, \pi/2)$ in all simulated cases.

6.7 EVIDENCE LOWER BOUND

We end this chapter by discussing another important use of the KL divergence. This measure is sometimes used to seek approximations for unobservable distributions. We will discuss this application at length in Chapter 36, when we study variational inference methods. Here, we derive a general-purpose bound, known as the evidence lower bound (or ELBO), which lies at the core of these methods.

Consider two random variables $\boldsymbol{z}$ and $\boldsymbol{y}$ where $\boldsymbol{y}$ is observable and $\boldsymbol{z}$ is hidden (i.e., unobservable, also called *latent*). For example, the variable $\boldsymbol{y}$ could correspond to measurements of the heights and weights of individuals and the variable $\boldsymbol{z}$ could correspond to their gender (male or female). By observing only the heights and weights, we would like to infer the gender. In order to infer some latent variable $\boldsymbol{z}$ from observations of another variable $\boldsymbol{y}$, we will learn later in this text that we will need to determine (or approximate) the conditional pdf $p_{\boldsymbol{z}|\boldsymbol{y}}(z|y)$. This distribution is generally unknown or, in many cases, difficult to evaluate in closed form.

Thus, assume we pose the problem of approximating the conditional pdf $p_{\boldsymbol{z}|\boldsymbol{y}}(z|y)$ by some distribution $q_{\boldsymbol{z}}(z)$. To be more precise, we should write $q_{\boldsymbol{z}|\boldsymbol{y}}(z|y)$; but it is sufficient for our purposes to use the lighter notation $q_{\boldsymbol{z}}(z)$ since the approximation is a function of z (though dependent on y as well). One way to determine a good approximation $q_{\boldsymbol{z}}(z)$ is to seek a distribution that is closest to

$p_{\boldsymbol{z}|\boldsymbol{y}}(z|y)$ in terms of the KL divergence:

$$q_{\boldsymbol{z}}^{\star}(z) \stackrel{\Delta}{=} \underset{q_{\boldsymbol{z}}(\cdot)}{\operatorname{argmin}}\; D_{\rm KL}\Big(q_{\boldsymbol{z}}(z) \,\|\, p_{\boldsymbol{z}|\boldsymbol{y}}(z|y)\Big) \tag{6.157}$$

This is a challenging problem to solve because $p_{\boldsymbol{z}|\boldsymbol{y}}(z|y)$ is unknown or hard to compute in closed form even if the joint distribution $p_{\boldsymbol{y},\boldsymbol{z}}(y,z)$ were known. This is because by the Bayes rule:

$$p_{\boldsymbol{z}|\boldsymbol{y}}(z|y) = \frac{p_{\boldsymbol{y},\boldsymbol{z}}(y,z)}{p_{\boldsymbol{y}}(y)} \tag{6.158}$$

and computation of $p_{\boldsymbol{z}|\boldsymbol{y}}(z|y)$ requires the evaluation of the marginal distribution (or evidence) that appears in the denominator. This latter distribution is obtained by marginalizing over z:

$$p_{\boldsymbol{y}}(y) = \int_{z\in\mathcal{Z}} p_{\boldsymbol{y},\boldsymbol{z}}(y,z)dz \tag{6.159}$$

However, the integration is not always straightforward to carry out and may not lead to a closed-form expression; we will provide several examples to this effect in Chapters 33 and 36 when we deal with Bayesian and variational inference problems. It is therefore necessary to pursue an alternative route where the conditional pdf $p_{\boldsymbol{z}|\boldsymbol{y}}(z|y)$ appearing in (6.157) is replaced by the joint distribution $p_{\boldsymbol{y},\boldsymbol{z}}(y,z)$ without the need to evaluate the evidence $p_{\boldsymbol{y}}(y)$.

We now verify that we can manipulate the optimization problem into an equivalent form that is mathematically more tractable. Thus, note the following sequence of equalities (for discrete latent variables $\boldsymbol{z}$, the integral symbols would be replaced by summations over realizations for $\boldsymbol{z}$):

$$\begin{aligned}
& D_{\rm KL}\Big(q_{\boldsymbol{z}}(z)\|p_{\boldsymbol{z}|\boldsymbol{y}}(z|y)\Big) \\
&\stackrel{\Delta}{=} \int_{z\in\mathcal{Z}} q_{\boldsymbol{z}}(z)\ln\left(\frac{q_{\boldsymbol{z}}(z)}{p_{\boldsymbol{z}|\boldsymbol{y}}(z|y)}\right)dz \\
&= -\int_{z\in\mathcal{Z}} q_{\boldsymbol{z}}(z)\ln\left(\frac{p_{\boldsymbol{z}|\boldsymbol{y}}(z|y)}{q_{\boldsymbol{z}}(z)}\right)dz \\
&\stackrel{(a)}{=} -\int_{z\in\mathcal{Z}} q_{\boldsymbol{z}}(z)\ln\left(\frac{p_{\boldsymbol{y},\boldsymbol{z}}(y,z)}{q_{\boldsymbol{z}}(z)p_{\boldsymbol{y}}(y)}\right)dz \\
&= \underbrace{-\int_{z\in\mathcal{Z}} q_{\boldsymbol{z}}(z)\ln\left(\frac{p_{\boldsymbol{y},\boldsymbol{z}}(y,z)}{q_{\boldsymbol{z}}(z)}\right)dz}_{\stackrel{\Delta}{=}\,\mathcal{L}} + \Big(\ln p_{\boldsymbol{y}}(y)\Big)\underbrace{\int_{z\in\mathcal{Z}} q_{\boldsymbol{z}}(z)\,dz}_{=1}
\end{aligned} \tag{6.160}$$

where step (a) uses the Bayes rule. It follows that

$$D_{\rm KL}\Big(q_{\boldsymbol{z}}(z)\|p_{\boldsymbol{z}|\boldsymbol{y}}(z|y)\Big) + \mathcal{L}\Big(q_{\boldsymbol{z}}(z)\|p_{\boldsymbol{y},\boldsymbol{z}}(y,z)\Big) = \ln p_{\boldsymbol{y}}(y) \tag{6.161}$$

where we are writing down explicitly the arguments of the term denoted by $\mathcal{L}$: it depends on the desired approximation $q_{\boldsymbol{z}}(z)$ and on the *joint* distribution

$p_{\boldsymbol{y},\boldsymbol{z}}(y,z)$ (which is assumed known in contrast to the conditional pdf $p_{\boldsymbol{z}|\boldsymbol{y}}(z|y)$ that appears in the expression for the KL divergence). We use a similar notation for $\mathcal{L}$ as the KL divergence due to the similarities in their definitions:

$$\mathcal{L}\Big(q_{\boldsymbol{z}}(z)\|p_{\boldsymbol{y},\boldsymbol{z}}(y,z)\Big) \;\stackrel{\Delta}{=}\; \int_{z\in\mathcal{Z}} q_{\boldsymbol{z}}(z)\,\ln\left(\frac{p_{\boldsymbol{y},\boldsymbol{z}}(y,z)}{q_{\boldsymbol{z}}(z)}\right)dz \tag{6.162}$$

If we drop the arguments, we can rewrite (6.161) more succinctly in the form

$$\boxed{D_{\mathrm{KL}} \;+\; \mathcal{L} \;=\; \ln p_{\boldsymbol{y}}(y)} \tag{6.163}$$

We observe from the expression for $\mathcal{L}$ in (6.162) that it can also be written in the form

$$\boxed{\mathcal{L}(q) \;\stackrel{\Delta}{=}\; \mathbb{E}_q\Big(\ln p_{\boldsymbol{y},\boldsymbol{z}}(y,z)\Big) \;-\; \mathbb{E}_q\Big(\ln q_{\boldsymbol{z}}(z)\Big)} \tag{6.164}$$

where the notation $\mathbb{E}_q$ refers to expectation relative to the distribution $q_{\boldsymbol{z}}(z)$. We are also writing more succinctly $\mathcal{L}(q)$ to highlight that the value of $\mathcal{L}$ changes with the choice of q.

Now recall that the KL divergence is always nonnegative. It follows that (recall that the function $q_{\boldsymbol{z}|\boldsymbol{y}}(z|y)$ on the left-hand side is dependent on y)

$$\mathcal{L}(q) \;\leq\; \ln p_{\boldsymbol{y}}(y) \tag{6.165}$$

which explains why the term $\mathcal{L}$ is referred to as the evidence lower bound, or ELBO(q); it provides a lower bound for the (natural) logarithm of the pdf of the observation $\boldsymbol{y}$; this pdf is called the *evidence*. The general result (6.163) states that the KL divergence and the ELBO should always add up to the (natural) logarithm of the evidence. By changing $q_{\boldsymbol{z}}(z)$, the values of both D_{KL} and $\mathcal{L}$ will change, with one of them increasing and the other decreasing in such a way that their sum remains invariant. We therefore conclude that the minimization problem (6.157) can be replaced by the maximization problem

$$q^{\star}_{\boldsymbol{z}}(z) \;\stackrel{\Delta}{=}\; \underset{q_{\boldsymbol{z}}(\cdot)}{\operatorname{argmax}}\; \mathcal{L}\Big(q_{\boldsymbol{z}}(z)\|p_{\boldsymbol{y},\boldsymbol{z}}(y,z)\Big) \tag{6.166}$$

This is still a challenging problem to solve, as we will explain later in Chapter 36. However, it replaces the *unknown* conditional pdf $p_{\boldsymbol{z}|\boldsymbol{y}}(z|y)$ by the *known* joint pdf $p_{\boldsymbol{y},\boldsymbol{z}}(y,z)$, and this substitution facilitates the search for an optimal approximation $q^{\star}_{\boldsymbol{z}}(z)$. It is worth noting that the ELBO, $\mathcal{L}(q)$, is generally not a convex function of its argument so that multiple local maxima may exist.

Example 6.17 **(Gaussian posterior distribution)** Consider a collection of N iid Gaussian random variables with $\boldsymbol{y}_n \sim \mathcal{N}_{\boldsymbol{y}_n}(\theta, \sigma^2)$, where θ is a realization of another Gaussian distribution, say, $\boldsymbol{\theta} \sim \mathcal{N}_{\boldsymbol{\theta}}(\bar{\theta}, \sigma_\theta^2)$. The variable $\boldsymbol{\theta}$ plays the role of the latent variable $\boldsymbol{z}$ in this problem. We assume the parameters $(\sigma^2, \sigma_\theta^2, \bar{\theta})$ are known.

For this example with Gaussian distributions it is possible to determine a closed-form expression for the conditional pdf $p_{\boldsymbol{\theta}|\boldsymbol{y}_{1:N}}(\theta|y_{1:N})$ and verify that it has a Gaussian form as well – see Prob. 6.19 and also the discussion around future expression (36.10).

Here, we will instead consider the problem of approximating $p_{\boldsymbol{\theta}|\boldsymbol{y}_{1:N}}(\theta|y_{1:N})$ by means of some Gaussian distribution whose mean and variance need to be determined, i.e., assume we postulate

$$q_{\boldsymbol{\theta}}(\theta) = \frac{1}{\sqrt{2\pi\sigma_z^2}} \exp\left\{-\frac{1}{2\sigma_z^2}(\theta-\mu_z)^2\right\} \tag{6.167}$$

for some mean and variance parameters (μ_z, σ_z^2). We will determine these parameters by maximizing the ELBO:

$$\mathcal{L}(q) = \mathbb{E}_q\Big(\ln p_{\boldsymbol{y}_{1:N},\boldsymbol{\theta}}(y_{1:N},\theta)\Big) \;-\; \mathbb{E}_q\Big(\ln q_{\boldsymbol{\theta}}(\theta)\Big) \tag{6.168}$$

We evaluate the two expectations on the right-hand side. Note first that under the assumed model

$$\ln q_{\boldsymbol{\theta}}(\theta) = -\frac{1}{2}\ln(2\pi\sigma_z^2) - \frac{1}{2\sigma_z^2}(\theta-\mu_z)^2 \tag{6.169}$$

so that, computing the expectation over the distribution $q_{\boldsymbol{\theta}}(\theta)$ leads to

$$\mathbb{E}_q\Big(\ln q_{\boldsymbol{\theta}}(\theta)\Big) \;=\; -\frac{1}{2}\ln(2\pi\sigma_z^2) - \frac{1}{2} \tag{6.170}$$

Moreover, from the Bayes rule,

$$\begin{aligned}
&p_{\boldsymbol{y}_{1:N},\boldsymbol{\theta}}(y_{1:N},\theta) \\
&= p_{\boldsymbol{y}_{1:N}|\boldsymbol{\theta}}(y_{1:N}|\theta)\, p_{\boldsymbol{\theta}}(\theta) \\
&= \left(\frac{1}{\sqrt{2\pi\sigma^2}}\right)^N \frac{1}{\sqrt{2\pi\sigma_\theta^2}} \exp\left\{-\frac{1}{2\sigma^2}\sum_{n=1}^{N}(y_n-\theta)^2 - \frac{1}{2\sigma_\theta^2}(\theta-\bar{\theta})^2\right\}
\end{aligned} \tag{6.171}$$

so that

$$\begin{aligned}
&\ln p_{\boldsymbol{y}_{1:N},\boldsymbol{\theta}}(y_{1:N},\theta) \\
&= -\frac{N}{2}\ln(2\pi\sigma^2) - \frac{1}{2}\ln(2\pi\sigma_\theta^2) - \frac{1}{2\sigma^2}\sum_{n=1}^{N}(y_n-\theta)^2 - \frac{1}{2\sigma_\theta^2}(\theta-\bar{\theta})^2
\end{aligned} \tag{6.172}$$

and, hence, computing expectations again over the distribution $q_{\boldsymbol{\theta}}(\theta)$ leads to

$$\begin{aligned}\mathbb{E}_q\Big(\ln p_{\boldsymbol{y}_{1:N},\boldsymbol{\theta}}(y_{1:N},\theta)\Big) &= -\frac{N}{2}\ln(2\pi\sigma^2) - \frac{1}{2}\ln(2\pi\sigma_\theta^2) - \\ &\quad \frac{1}{2\sigma^2}\sum_{n=1}^{N}\Big(y_n^2 + \mathbb{E}_q\boldsymbol{\theta}^2 - 2y_n\,\mathbb{E}_q\boldsymbol{\theta}\Big) - \\ &\quad \frac{1}{2\sigma_\theta^2}\Big(\mathbb{E}_q\boldsymbol{\theta}^2 + \bar{\theta}^2 - 2\bar{\theta}\,\mathbb{E}_q\boldsymbol{\theta}\Big) \\ &= -\frac{N}{2}\ln(2\pi\sigma^2) - \frac{1}{2}\ln(2\pi\sigma_\theta^2) - \\ &\quad \frac{1}{2\sigma^2}\sum_{n=1}^{N}y_n^2 + \frac{\mu_z}{\sigma^2}\sum_{n=1}^{N}y_n - \frac{N(\sigma_z^2+\mu_z^2)}{2\sigma^2} - \\ &\quad \frac{\sigma_z^2 + (\mu_z-\bar{\theta})^2}{2\sigma_\theta^2}\end{aligned} \tag{6.173}$$

We therefore arrive at the following expression for the ELBO in terms of the unknown parameters (μ_z, σ_z^2); we are dropping constant terms that are independent of these parameters:

$$\mathcal{L}(\mu_z,\sigma_z^2) = \frac{\mu_z}{\sigma^2}\sum_{n=1}^{N}y_n - \frac{N(\sigma_z^2+\mu_z^2)}{2\sigma^2} - \frac{\sigma_z^2+(\mu_z-\bar{\theta})^2}{2\sigma_\theta^2} + \frac{1}{2}\ln(2\pi\sigma_z^2) \tag{6.174}$$

Differentiating relative to μ_z and σ_z^2 and setting the derivatives to zero, we get

$$\widehat{\mu}_z = \left(\frac{N}{\sigma^2} + \frac{1}{\sigma_\theta^2}\right)^{-1}\left(\frac{1}{\sigma^2}\sum_{n=1}^{N}y_n + \frac{\bar{\theta}}{\sigma_\theta^2}\right) \tag{6.175a}$$

$$\frac{1}{\widehat{\sigma_z^2}} = \frac{N}{\sigma^2} + \frac{1}{\sigma_\theta^2} \tag{6.175b}$$

6.8 COMMENTARIES AND DISCUSSION

Entropy. The concepts of entropy and mutual information, which are used to quantify the amount of unpredictability (or uncertainty) in random realizations, are due to the American engineer **Claude Shannon (1916–2001)**. He is regarded as the founder of the field of information theory, which was launched with the publication of the seminal papers by Shannon (1948a, b, 1949). According to the account by Tribus and McIrvine (1971), it was the American-Hungarian mathematician **John von Neumann (1903–1957)** who suggested to Shannon the use of the terminology "entropy" for his measure of uncertainty, since the term was already in use in statistical mechanics. Indeed, consider a thermodynamic system that can exist in a multitude of states, with probability p_m for state m. The Gibbs entropy of the system was introduced by the American scientist **J. Willard Gibbs (1839–1903)** and defined as (see Gibbs (1902))

$$G \stackrel{\Delta}{=} k_B\sum_m p_m \ln p_m \tag{6.176}$$

in terms of the Boltzmann constant, $k_B = 1.38 \times 10^{-23}$ J/K (measured in joules/kelvin) – see the discussion by Jaynes (1965). There exist several texts that cover the theoretical foundations of entropy and its application in communications and information theories. The reader may consult, for example, Ash (1990), Cover and Thomas (1991), MacKay (2003), and Gray (2011). The source coding theorem (6.15) and its more common variation with an asymptotic performance guarantee are due to Shannon (1948a, b, 1949). For instance, for N large enough (i.e., as $N \to \infty$), a block of N iid symbols x arising from a random source with entropy $H(\boldsymbol{x})$ can be represented with negligible probability of error by no more than $NH(\boldsymbol{x})$ total bits. An alternative statement is the fact that, given any code with length $C(x)$ bits for realization x, the average code length satisfies

$$H(\boldsymbol{x}) \leq \mathcal{C}_{av} \leq H(\boldsymbol{x}) + \frac{1}{N} \tag{6.177}$$

Proofs for the source coding theorem can be found, for example, in MacKay (2003) and Cover and Thomas (1991). The Kraft inequality (6.13) is due to Kraft (1949); see also the extension in McMillan (1956) for the more general case of *uniquely decodable codes*. A code is said to be *nonsingular* if any two different realizations x and x' are assigned two different codewords. Now consider a collection of N symbols $\{x_1, x_2, \ldots, x_N\}$ and define a block code that concatenates the individual codes of the x_k one after the other. A code is said to be uniquely decodable if this extension is also nonsingular (i.e., two different collections of data will be assigned two different block extensions).

Kullback–Leibler divergence. The KL divergence, or relative entropy, is a measure of closeness between two probability distributions. It first appeared in the article by Kullback and Leibler (1951); see also the book by Kullback (1959). A symmetric form of the KL divergence between two distributions $p_{\boldsymbol{x}}(x)$ and $q_{\boldsymbol{x}}(x)$ is usually defined as:

$$D_{\rm KL}^{\rm sym}(p\|q) \;=\; D_{\rm KL}(p\|q) \;+\; D_{\rm KL}(q\|p) \tag{6.178}$$

The KL divergence is a special case of a broader definition known as the α-divergence, which involves a parameter $\alpha > 0$, $\alpha \neq 1$. The α-divergence, also known as the Renyi divergence, was introduced by Renyi (1961) and defined as

$$\begin{aligned} D_\alpha(p\|q) &\stackrel{\Delta}{=} \frac{1}{\alpha-1}\ln\left(\int_{x\in\mathcal{X}} p_{\boldsymbol{x}}^{\alpha}(x)\, q_{\boldsymbol{x}}^{1-\alpha}(x)dx\right) \\ &= \frac{1}{\alpha-1}\ln\left(\int_{x\in\mathcal{X}} p_{\boldsymbol{x}}(x) p_{\boldsymbol{x}}^{\alpha-1}(x)\, q_{\boldsymbol{x}}^{1-\alpha}(x)dx\right) \\ &= \frac{1}{\alpha-1}\ln\left(\int_{x\in\mathcal{X}} p_{\boldsymbol{x}}(x)\left[\frac{q_{\boldsymbol{x}}(x)}{p_{\boldsymbol{x}}(x)}\right]^{1-\alpha} dx\right) \\ &= \frac{1}{\alpha-1}\ln\left(\mathbb{E}_p\left[\frac{q_{\boldsymbol{x}}(x)}{p_{\boldsymbol{x}}(x)}\right]^{1-\alpha}\right) \end{aligned} \tag{6.179}$$

When $\alpha \to 1$, and using the L'Hospital rule, we recover in the limit the standard KL divergence between $p_{\boldsymbol{x}}(x)$ and $q_{\boldsymbol{x}}(x)$:

$$\lim_{\alpha\to 1} D_\alpha(p\|q) \;=\; D_{\rm KL}(p\|q) \tag{6.180}$$

The works by Minka (2005) and Van Erven and Harremoes (2014) provide useful overviews of the relation between different divergence measures and of the behavior of α-divergence as a function of α. If we use the α-divergence to approximate a distribution $p_{\boldsymbol{x}}(x)$ by another distribution $q_{\boldsymbol{x}}(x)$, then a large negative α value favors distributions $q_{\boldsymbol{x}}(x)$ that approximate well the modes (i.e., peaks) of $p_{\boldsymbol{x}}(x)$. Large positive α values, on the other hand, favor distributions $q_{\boldsymbol{x}}(x)$ that approximate well the entire $p_{\boldsymbol{x}}(x)$.

There is another form for defining α-divergences, which was introduced by Chernoff (1952) and arises frequently in the solution of inference and learning problems. This variation was studied extensively by Amari (1985, 2009), Amari and Nagaoka (2000), and Cichocki and Amari (2010), and takes the form

$$D_\alpha(p\|q) \triangleq \frac{1}{\alpha(1-\alpha)}\left(1 - \int_{x\in\mathcal{X}} p_{\boldsymbol{x}}^{\alpha}(x)\, q_{\boldsymbol{x}}^{1-\alpha}(x)dx\right) \tag{6.181}$$

From this form, we can recover the KL divergence and its reverse form in the limit as $\alpha \to 1$ and $\alpha \to 0$, namely,

$$\lim_{\alpha\to 1} D_\alpha(p\|q) = D_{\text{KL}}(p\|q) \tag{6.182a}$$

$$\lim_{\alpha\to 0} D_\alpha(p\|q) = D_{\text{KL}}(q\|p) \tag{6.182b}$$

Even more generally, the KL and α-divergences are special cases of f-divergence, which is a more general notion introduced independently by Csiszár (1963, 1967), Morimoto (1963), and Ali and Silvey (1966), and defined as

$$D_f(p\|q) = \int_{x\in\mathcal{X}} q_{\boldsymbol{x}}(x) f\left(\frac{p_{\boldsymbol{x}}(x)}{q_{\boldsymbol{x}}(x)}\right) dx \tag{6.183}$$

where $f(\cdot)$ is some continuous convex function satisfying $f(1) = 0$. A useful discussion on properties of the f-divergence appears in Liese and Vajda (2006). The standard KL divergence follows from using $f(x) = x\ln x$, while the α-divergence (6.181) follows from using

$$f(x) = \begin{cases} \dfrac{1}{\alpha(\alpha-1)}(x^\alpha - x), & \text{for } \alpha \neq 1, \alpha \neq 0 \\ x\ln x, & \text{when } \alpha = 1 \\ -\ln x, & \text{when } \alpha = 0 \end{cases} \tag{6.184}$$

Two other useful special cases are the total variation and the Hellinger distances between two distributions obtained by choosing, respectively:

$$f(x) = \frac{1}{2}|x-1| \qquad \text{(used for total variation distance)} \tag{6.185}$$

$$f(x) = \frac{1}{2}\left(\sqrt{x} - 1\right)^2 \quad \text{(used for Hellinger distance)} \tag{6.186}$$

In the first case we get

(total variation distance)

$$D_{\text{TV}} = \frac{1}{2}\int_{\boldsymbol{x}\in\mathcal{X}} |p_{\boldsymbol{x}}(x) - q_{\boldsymbol{x}}(x)|dx \triangleq d_{\text{TV}}(p,q) \tag{6.187}$$

where we are also denoting the total variation distance by $d_{\text{TV}}(p,q)$, while in the second case we get

(square of Hellinger distance)

$$D_H = \frac{1}{2}\int_{\boldsymbol{x}\in\mathcal{X}} \left(\sqrt{p_{\boldsymbol{x}}(x)} - \sqrt{q_{\boldsymbol{x}}(x)}\right)^2 dx \triangleq d_H^2(p,q) \tag{6.188}$$

The above expression provides the *square* of the Hellinger distance between two probability distributions, originally introduced by Hellinger (1909). We are also denoting the distance by $d_H(p,q)$. Later, in Section 8.9, we will introduce another notion of divergence, known as Bregman divergence, which is not limited to probability distributions.

Natural gradients. Some of the first works to recognize the significance of natural gradients for learning purposes are Amari (1985, 1998), followed by contributions by Amari, Cichocki, and Yang (1996) and Amari and Nagaoka (2000). One important feature of the natural gradient is that it scales the gradient of the cost function by the inverse of the Hessian of the KL divergence. The KL measure is an intrinsic measure of similarity between two distributions and it is independent of how the distributions are parameterized – see the overview by Martens (2020). As a result, natural gradient updates lead to algorithms whose performance is largely invariant to which coordinate system is used to parameterize the distributions. Another feature of natural gradient implementations is that they rely on measuring distances between points on a Riemannian manifold. Examples 6.14 and 6.16 illustrate this aspect and are motivated by an example from Amari and Douglas (1998). For more technical discussion on such manifolds and their properties, the reader may refer to Amari (1985), Lee (1997), and Gallot, Hulin, and Lafontaine (2004). We will encounter natural gradient solutions later, mainly in the study of variational inference and reinforcement learning algorithms in Sections 36.6 and 49.7, respectively. One of the main disadvantages of the natural gradient method is that it is computationally expensive (similar to Newton implementation) since it requires inverting the Fisher information matrix and/or the curvature tensor. Moreover, computing these matrices in the first place is challenging. The probability distribution of the underlying signals may not be available to enable computation of the Fisher information matrix. Likewise, determining the curvature tensor is not straightforward in general. One usually needs to resort to approximations to simplify the calculations, which can lead to degradation in performance.

PROBLEMS

6.1 Show that $H(\boldsymbol{x}|\boldsymbol{y}) \leq H(\boldsymbol{x})$ for any two random variables $\boldsymbol{x}$ and $\boldsymbol{y}$.

6.2 Show that $\boldsymbol{x}$ and $\boldsymbol{y}$ are independent if, and only if, $H(\boldsymbol{x}|\boldsymbol{y}) = H(\boldsymbol{y})$.

6.3 Show that $I(\boldsymbol{x},\boldsymbol{y}) \leq \min\{H(\boldsymbol{x}), H(\boldsymbol{y})\}$.

6.4 Show that the mutual information measure (6.29a) is always nonnegative. Establish also the equivalence of expressions (6.29a) and (6.29b).

6.5 Establish the validity of expression (6.32) for the mutual information between two random variables in terms of their joint and marginal pmfs.

6.6 Establish the validity of expressions (6.33a)–(6.33b) for the mutual information between two random variables in terms of their individual and mutual entropy measures.

6.7 Show that two random variables $\boldsymbol{x}$ and $\boldsymbol{y}$ are conditionally independent given $\boldsymbol{z}$ if, and only if, their conditional mutual information is zero where

$$I(\boldsymbol{x};\boldsymbol{y}|\boldsymbol{z}) \stackrel{\Delta}{=} \sum_{x,y,z} \mathbb{P}(\boldsymbol{x}=x,\boldsymbol{y}=y,\boldsymbol{z}=z)\log_2\left(\frac{\mathbb{P}(\boldsymbol{x}=x,\boldsymbol{y}=y|\boldsymbol{z}=z)}{\mathbb{P}(\boldsymbol{x}=x|\boldsymbol{z}=z)\mathbb{P}(\boldsymbol{y}=y|\boldsymbol{z}=z)}\right)$$

6.8 Refer to the setting of Example 6.3 with two random variables $\boldsymbol{x}$ and $\boldsymbol{y}$:

(a) Evaluate $\mathbb{P}(\boldsymbol{x}=+1)$ and $\mathbb{P}(\boldsymbol{x}=-1)$.

(b) Evaluate $\mathbb{P}(\boldsymbol{y}=\text{"green"}|\boldsymbol{x}=+1)$ and $\mathbb{P}(\boldsymbol{y}=\text{"green"}|\boldsymbol{x}=-1)$ using the Bayes rule.

(c) Evaluate $H(\boldsymbol{y})$ and $H(\boldsymbol{y}|\boldsymbol{x})$, and determine the mutual information $I(\boldsymbol{x},\boldsymbol{y})$.

6.9 Refer to definition (6.5) for the Shannon entropy of a discrete-valued random variable, $\boldsymbol{x}$. For any $\alpha \neq 1$ and $\alpha \geq 0$, the Renyi entropy for the same variable is defined as

$$H_\alpha(\boldsymbol{x}) = \frac{1}{1-\alpha}\ln\left(\sum_x \mathbb{P}^\alpha(\boldsymbol{x}=x)\right)$$

Show that $\lim_{\alpha\to 1} H_\alpha(\boldsymbol{x}) = H(\boldsymbol{x})$. Show further that $H_\alpha(\boldsymbol{x})$ is a nonincreasing function of α.

6.10 Consider a continuous random variable $\boldsymbol{x}$ and let $h(\boldsymbol{x})$ denote its differential entropy. Show that:

(a) $h(\alpha\boldsymbol{x}) = h(\boldsymbol{x}) + \log_2|\alpha|$, for any nonzero real constant α.

(b) $h(\boldsymbol{x}+c)=h(\boldsymbol{x})$, for any constant c.
(c) $h(B\boldsymbol{x})=h(\boldsymbol{x})+\log_2|\det(B)|$, for any invertible square matrix B.

6.11 Establish expression (6.40) for the differential entropy of a vector-valued Gaussian distribution.

6.12 Among all distributions $f_{\boldsymbol{x}}(x)$ with a finite support over the interval $x\in[a,b]$, show that the uniform distribution $f_{\boldsymbol{x}}(x)=1/(b-a)$ has the largest entropy. *Remark.* For comparison purposes, recall that we established earlier in Prob. 5.18 that the exponential family of distributions has the largest entropy among all distributions satisfying certain moment-type constraints.

6.13 Among all distributions $f_{\boldsymbol{x}}(x)$ with support over the semi-infinite range $x\in[0,\infty)$ and with finite mean μ, show that the exponential distribution $f_{\boldsymbol{x}}(x)=\frac{1}{\mu}e^{-x/\mu}$ has the largest entropy.

6.14 Consider a random variable $\boldsymbol{x}$ with differential entropy $h(\boldsymbol{x})$. Let $\widehat{\boldsymbol{x}}$ denote any estimator for $\boldsymbol{x}$. Show that

$$\mathbb{E}\,(\boldsymbol{x}-\widehat{\boldsymbol{x}})^2\geq\frac{1}{2\pi e}\,e^{2h(\boldsymbol{x})}$$

6.15 Consider two sequences of positive numbers $\{p_m,q_m\}$ and let $S_p=\sum_{m=1}^{M}p_m$ and $S_q=\sum_{m=1}^{M}q_m$. Show that

$$\sum_{m=1}^{M}p_m\ln(p_m/q_m)\geq S_p\ln(S_p/S_q)$$

6.16 Refer to definition (6.44) for the KL divergence of two discrete distributions. Show that $D_{\rm KL}(p\|q\|)\geq\frac{1}{2}\|p-q\|_1^2$, where the pmfs are collected into the vectors p and q. *Remark.* This result is known as Pinsker inequality and is originally due to Pinsker (1960), and was further refined by Csiszár (1966), Kemperman (1969), and Kullback (1967, 1970).

6.17 Refer to definition (6.43) for the KL divergence. Show by means of an example that it does not satisfy the triangle inequality of norms. In particular, find distributions $p_{\boldsymbol{x}}(x),q_{\boldsymbol{x}}(x),r_{\boldsymbol{x}}(x)$ such that $D_{\rm KL}(p\|r)>D_{\rm KL}(p\|q)+D_{\rm KL}(q\|r)$.

6.18 Let the vectors p and q describe two discrete pmfs of size M each. Let $q=\text{col}\{1/M,1/M,\ldots,1/M\}$ denote the uniform distribution. Show that $D_{\rm KL}(p\|q)\leq\ln(M)$ for any p.

6.19 Refer to the setting of Example 6.17. Determine an expression for the pdf of the observations, i.e., $p_{\boldsymbol{y}_{1:N}}(y_{1:N})$. Use the result and the Bayes rule to determine a closed-form expression for the conditional pdf $p_{\boldsymbol{\theta}|\boldsymbol{y}_{1:N}}(\theta|y_{1:N})$. Compare the resulting mean and variance expressions (6.175a)–(6.175b).

6.20 Refer to Example 6.15 which considers a collection of N iid Gaussian random variables. Assume the Gaussian pdf is parameterized in terms of its mean, μ, and inverse variance, $\tau=1/\sigma^2$. That is, let now $\theta=(\mu,\tau)$.
(a) Express the log-likelihood function in terms of μ and τ.
(b) Determine $\nabla_{\theta^{\sf T}}\,\ell(\theta)$, $\nabla_\theta^2\,\ell(\theta)$, and $F(\theta)$.
(c) Write down Newton recursion and the natural gradient algorithm for estimating μ and τ.

6.21 Refer to the derivation leading to the line search method (6.120) and consider instead the optimization problem

$$\delta\theta^o=\underset{\delta\theta\in{\rm I\!R}^M}{\text{argmin}}\;J(\theta_{m-1}+\delta\theta),\quad\text{subject to}\;\frac{1}{2}(\delta\theta)^{\sf T}A\,\delta\theta\leq\epsilon$$

where $A>0$ is a positive-definite symmetric matrix. Let $g_{m-1}=\nabla_{\theta^{\sf T}}\,J(\theta_{m-1})$. Derive the following line search algorithm for this problem:

$$\theta_m=\theta_{m-1}-\left(\frac{2\epsilon}{g_{m-1}^{\sf T}A^{-1}g_{m-1}}\right)^{1/2}A^{-1}g_{m-1}$$

6.22 Consider a collection of N iid Gaussian random variables with $\boldsymbol{y}_n \sim \mathcal{N}_{\boldsymbol{y}_n}(\theta, \sigma^2)$, where θ is unknown. We seek to estimate the mean parameter θ by maximizing the joint pdf (or the log-likelihood function $\ln p_{\boldsymbol{y}_1,\ldots,\boldsymbol{y}_N}(y_1,\ldots,y_N;\theta)$) over θ.

(a) Show that the optimal choice for θ is given by $\widehat{\theta} = \frac{1}{N}\sum_{n=1}^{N} y_n$.

(b) Show that if we were to employ Newton recursion (with $\mu = 1$) to estimate θ iteratively, the recursion would converge in a single step.

(c) Write down a natural gradient recursion for estimating θ. Does it work?

6.23 Establish property (6.138).

6.24 Refer to definition (6.183) for the f-divergence between two probability distributions. Show that $D_f(p\|q) \geq 0$. Let $\lambda \in [0,1]$ and consider four probability distributions over the random variable $\boldsymbol{x}$, denoted by $\{p_1(x), p_2(x), q_1(x), q_2(x)\}$. Show that the f-divergence satisfies the convexity property:

$$D_f\Big(\lambda p_1(x) + (1-\lambda)p_2(x) \,\|\, \lambda q_1(x) + (1-\lambda)q_2(x)\Big) \leq \lambda D_f(p_1\|q_1) \;+\; (1-\lambda)D_f(p_2\|q_2)$$

6.25 Refer to the definition of the Hellinger distance $d_H(p,q)$ between two probability distributions, whose square value is given by (6.188). Verify that $d_H(p,q)$ is a valid distance metric. In particular, verify that it is symmetric and satisfies a triangle inequality. Verify also that $0 \leq d_H(p,q) \leq 1$.

6.26 Assume we employ the Renyi divergence measure (6.179) in problem (6.157), namely, replace $D_{\rm KL}(q\|p_{\boldsymbol{z}|\boldsymbol{y}})$ by $D_\alpha(q\|p_{\boldsymbol{z}|\boldsymbol{y}})$. Repeat the derivation that led to (6.164) and show that this relation is replaced by $D_\alpha + \mathcal{L}_\alpha = \ln\, p_{\boldsymbol{y}}(y)$, where the ELBO is now given by

$$\mathcal{L}_\alpha \;\stackrel{\Delta}{=}\; \frac{1}{\alpha-1}\ln\left(\mathbb{E}_q\left[\frac{p_{\boldsymbol{z},\boldsymbol{y}}(z,y)}{q_{\boldsymbol{z}}(z)}\right]^{1-\alpha}\right)$$

6.27 Assume data $\boldsymbol{x}$ arises from some unknown distribution $p_{\boldsymbol{x}}(x)$. Let $q_{\boldsymbol{x}}(x;\theta)$ be a parametric approximation that we wish to determine for $p_{\boldsymbol{x}}(x)$; it depends on a parameter θ. Show that $\mathbb{E}_p \ln q_{\boldsymbol{x}}(x;\theta) = -H(\boldsymbol{x}) - D_{\rm KL}(p\|q)$, in terms of the entropy of $\boldsymbol{x}$ and where the first expectation is relative to the distribution $p_{\boldsymbol{x}}(x)$. Conclude that maximizing the average log-likelihood function is equivalent to minimizing the KL divergence between $p_{\boldsymbol{x}}(x)$ and $q_{\boldsymbol{x}}(x)$.

6.28 Assume data $\boldsymbol{x}$ arises from some unknown distribution $p_{\boldsymbol{x}}(x)$. Let $q_{\boldsymbol{x}}(x;\theta)$ be a parametric approximation that we wish to determine for $p_{\boldsymbol{x}}(x)$; it depends on a parameter θ. We may consider maximizing $D_{\rm KL}(p\|q)$ or $D_{\rm KL}(q\|p)$ with the orders of p and q reversed. Argue that the first divergence $D_{\rm KL}(p\|q)$ favors solutions $q_{\boldsymbol{x}}(x;\theta)$ that assign high probability to regions of space where data $\boldsymbol{x}$ are more likely, while the second divergence $D_{\rm KL}(q\|p)$ favors solutions $q_{\boldsymbol{x}}(x;\theta)$ that avoid assigning high probability to regions where data $\boldsymbol{x}$ is almost nonexistent.

6.29 In this problem we motivate one proof for the Kraft inequality (6.13) over prefix codes. These codes can be represented by binary trees, as illustrated in Fig. 6.6. The tree consists of a root node and leaf nodes; the latter are represented by squares. At every node in the tree, branching occurs either to one side or the other (or both). The bit 0 is assigned to branches on the left and the bit 1 is assigned to branches on the right. Codewords are represented by the leaf nodes; each codeword is obtained by reading the sequence of bits that leads to its leaf. Let $C_{\max}$ denote the maximum code length. In the figure, $C_{\max} = 4$.

(a) Consider an arbitrary leaf node X at an arbitrary level $C(x)$, for example, at level 2. Argue that if we were to continue to draw branches from X until we reach the $C_{\max}$ level, then the total number of descendant nodes that X will have at the last level is $2^{C_{\max}-C(x)}$.

(b) Argue that since all descendants at the last level are different from each other, it should hold that $\sum_{x\in\mathcal{X}} 2^{C_{\max}-C(x)} \leq 2^{C_{\max}}$. Conclude the validity of Kraft inequality.

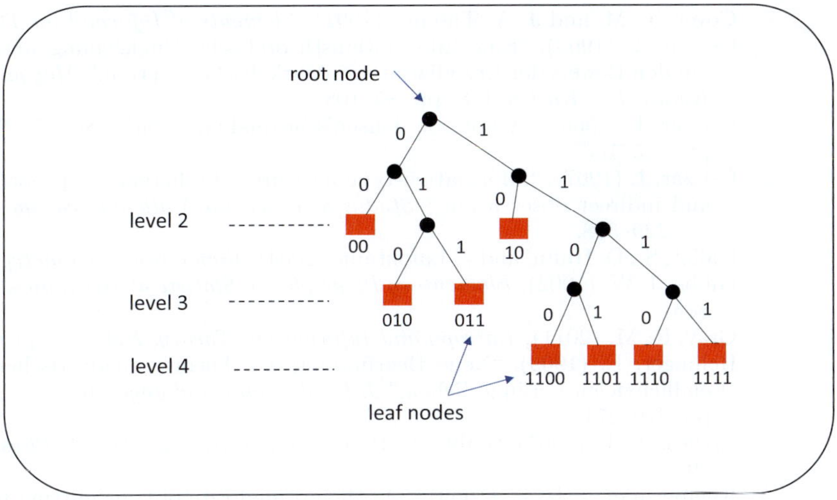

Figure 6.6 Illustration of a binary tree for a prefix code.

6.30 Consider a random variable $\boldsymbol{x} \in \mathcal{X}$ with realizations $\{x_1, x_2, x_3, x_4\}$. Consider the following three code assignments:

$$\begin{array}{lllll} x_1 = 0 & x_2 = 10 & x_3 = 01 & x_4 = 010 & \text{(code A)} \\ x_1 = 00 & x_2 = 10 & x_3 = 110 & x_4 = 11 & \text{(code B)} \\ x_1 = 1 & x_2 = 00 & x_3 = 011 & x_4 = 010 & \text{(code C)} \end{array}$$

Which of the codes is a prefix code? Which one is nonsingular? Which one is uniquely decodable? Verify the Kraft inequality in each case.

REFERENCES

Ali, S. M. and S. D. Silvey (1966), "A general class of coefficients of divergence of one distribution from another," *J. Roy Statist. Soc., Ser. B*, vol. 28, no. 1, pp. 131–142.

Amari, S. (1985), *Differential-Geometrical Methods in Statistics*, Springer.

Amari, S. I. (1998), "Natural gradient works efficiently in learning," *Neur. Comput.*, vol. 10, no. 2, pp. 251–276.

Amari, S. I. (2009), "α-divergence is unique, belonging to both f-divergence and Bregman divergence classes," *IEEE Trans. Inf. Theory*, vol. 55, no. 11, pp. 4925–4931.

Amari, S. I., A. Cichocki, and H. H. Yang (1996), "A new learning algorithm for blind source separation," *Proc. Advances Neural Information Processing Systems* (NIPS), pp. 757–763, Denver, CO.

Amari, S. I. and S. Douglas (1998), "Why natural gradient?" *Proc. IEEE ICASSP*, vol. 2, pp. 1213-1216, Seattle, WA.

Amari, S. and H. Nagaoka (2000), *Methods of Information Geometry*, Oxford University Press.

Ash, R. B. (1990), *Information Theory*, Dover Publications.

Chernoff, H. (1952), "A measure of asymptotic efficiency of tests of a hypothesis based on the sum of observations, " *Ann. Math. Statist.*, vol. 23, pp. 493–507.

Cichocki, A. and S. I. Amari (2010), "Families of alpha- beta- and gamma-divergences: Flexible and robust measures of similarities," *Entropy*, vol. 12, pp. 1532–1568.

Cover, T. M. and J. A. Thomas (1991), *Elements of Information Theory*, Wiley.

Csiszár, I. (1963), "Eine informationstheoretische Ungleichung und ihre Anwendung auf den Beweis der Ergodizitat von Markoffschen Ketten," *Magyar. Tud. Akad. Mat. Kutato Int. Kozl*, vol. 8, pp. 85–108.

Csiszár, I. (1966), "A note on Jensen's inequality," *Studia Sci. Math. Hungar.*, vol. 1. pp. 185–188.

Csiszár, I. (1967), "Information-type measures of difference of probability distributions and indirect observation," *Studia Scientiarum Mathematicarum Hungarica*, vol. 2, pp. 229–318.

Gallot, S., D. Hulin, and J. Lafontaine (2004), *Riemannian Geometry*, 3rd ed., Springer.

Gibbs, J. W. (1902), *Elementary Principles in Statistical Mechanics*, Charles Scribner's Sons.

Gray, R. M. (2011), *Entropy and Information Theory*, 2nd ed., Springer.

Hellinger, E. (1909), "Neue Begründung der Theorie quadratischer Formen von unendlichvielen Veränderlichen," *J. für die reine und angewandte Mathematik*, vol. 136, pp. 210–271.

Jaynes, E. T. (1965), "Gibbs vs Boltzmann entropies," *Am. J. Phys.* vol. 33, pp. 391–398.

Kemperman, J. H. B. (1969), "On the optimal rate of transmitting information," *Ann. Math. Statist.*, vol. 40, pp. 2156–2177.

Kraft, L. G. (1949), *A Device for Quantizing, Grouping, and Coding Amplitude Modulated Pulses*, MS Thesis, Electrical Engineering Department, MIT, MA.

Kullback, S. (1959), *Information Theory and Statistics*, Wiley. Reprinted by Dover Publications, 1978.

Kullback, S. (1967), "A lower bound for discrimination information in terms of variation," *IEEE Trans. Inf. Theory*, vol. 13, pp. 126–127.

Kullback, S. (1970), "Correction to 'A lower bound for discrimination information in terms of variation'," *IEEE Trans. Inf. Theory*, vol. 16, p. 652.

Kullback, S. and R. A. Leibler (1951), "On information and sufficiency," *Ann. Math. Statist.*, vol. 22, no. 1, pp. 79–86, 1951.

Lee, J. M. (1997), *Riemannian Manifolds: An Introduction to Curvature*, Springer.

Liese, F. and I. Vajda (2006), "On divergences and informations in statistics and information theory," *IEEE Trans. Inf. Theory*, vol. 52, no. 10, pp. 4394–4412.

MacKay, D. J. C. (2003), *Information Theory, Inference, and Learning Algorithms*, Cambridge University Press.

Martens, J. (2020), "New insights and perspectives on the natural gradient method," *J. Mach. Learn. Res.*, vol. 21, pp. 1–76.

McMillan, B. (1956), "Two inequalities implied by unique decipherability," *IEEE Trans. Inf. Theory*, vol. 2, no. 4, pp. 115–116.

Minka, T. (2005), "Divergence measures and message passing," *Microsoft Research* Technical Report MSR-TR-2005, Microsoft.

Morimoto, T. (1963), "Markov processes and the H-theorem," *Phys. Soc. Jpn.*, vol. 18, no. 3, pp. 328–331.

Pinsker, M. S. (1960), "Information and information stability of random variables and processes," *Izv. Akad. Nauk*, Moscow, USSR.

Renyi, A. (1961), "On measures of information and entropy," *Proc. 4th Berkeley Symp. Mathematics, Statistics and Probability*, pp. 547–561, Berkeley, CA.

Shannon, C. E. (1948a), "A mathematical theory of communication," *Bell Syst. Tech. J.*, vol. 27, pp. 379–423.

Shannon, C. E. (1948b), "A mathematical theory of communication," *Bell Syst. Tech. J.*, vol. 27, pp. 623–656.

Shannon, C. E. (1949), "Communication in the presence of noise," *Proc. Inst. Radio Eng.*, vol. 37, no. 1, pp. 10–21.

Tribus, M. and E. C. McIrvine (1971), "Energy and information," *Sci. Am.*, vol. 224, pp. 179–188.

Van Erven, T. and P. Harremoes (2014), "Renyi divergence and Kullback–Leibler divergence," *IEEE Trans. Inf. Theory*, vol. 60, no. 7, pp. 3797–3820.

7 Random Processes

Random processes are prevalent in applications. This is because in many instances we will be interested in processing a *sequence* of random variables (also known as a random *process*) rather than a *stand-alone* random variable. In the previous chapters, we explained how to characterize the probability distribution of single random variables and how to define their means and covariance matrices. In this chapter, we elaborate on the extensions that are necessary to study *sequences* of random variables. One key observation is that successive samples in a random sequence will generally be *correlated* with one another and, therefore, individual samples will be able to convey information about their neighboring samples (including past samples or future samples or both). Exploiting the correlation among samples is an important technique to ameliorate the effect of randomness in data and to arrive at more effective inference decisions.

7.1 STATIONARY PROCESSES

The realization of a random process is a sequence of random outcomes. We will be interested in outcomes that are real-valued. We denote the entries of the random process (or sequence) by the notation $\boldsymbol{x}(n)$, which is indexed by the integer n. While this index usually refers to discrete time instants, it may also refer to locations in space or to some other indexing convention. For convenience, we will refer to n as the time index.

For every n, the quantity $\boldsymbol{x}(n)$ corresponds to a standard random variable. As such, when we observe different realizations for the random process, we are in effect observing different realizations for each random variable at the respective point in time n. What makes a random process interesting is that the random variables at different time instants can be correlated with each other.

Covariance function

It is sufficient for our purposes to focus on *scalar-valued* random processes where each $\boldsymbol{x}(n)$ is a real scalar-valued random variable. We associate a mean value and a variance with each variable $\boldsymbol{x}(n)$ and obtain mean and variance sequences over time n, namely,

$$\bar{x}(n) \triangleq \mathbb{E}\,\boldsymbol{x}(n) \tag{7.1a}$$
$$\sigma_x^2(n) \triangleq \mathbb{E}\,(\boldsymbol{x}(n) - \bar{x}(n))^2 \tag{7.1b}$$

The mean of the random process $\{\boldsymbol{x}(n)\}$ is the sequence of means $\{\bar{x}(n)\}$; likewise, the variance of the random process is the sequence of variances $\{\sigma_x^2(n)\}$. In addition, since adjacent random samples can be correlated, we further associate with each random process a *covariance function* that measures the amount of correlation between samples of the random process. Specifically, we let $r_x(n, m)$ denote the covariance between the samples $\boldsymbol{x}(n)$ and $\boldsymbol{x}(m)$ located at instants n and m:

$$\boxed{r_x(n, m) \triangleq \mathbb{E}\,(\boldsymbol{x}(n) - \bar{x}(n))\,(\boldsymbol{x}(m) - \bar{x}(m))} \tag{7.1c}$$

It is generally a two-dimensional sequence, being a function of both n and m. Note in particular that the variance at any time n amounts to the correlation of that sample with itself:

$$\sigma_x^2(n) = r_x(n, n) \tag{7.2}$$

We will generally be dealing with *zero-mean* random processes for which

$$r_x(n, m) = \mathbb{E}\,\boldsymbol{x}(n)\boldsymbol{x}(m) \quad \textbf{(zero-mean case)} \tag{7.3}$$

Wide-sense stationarity

An important type of random processes is the class of wide-sense *stationary* processes, which are characterized by the following properties:

$$\mathbb{E}\,\boldsymbol{x}(n) = \bar{x} \qquad \text{(mean is independent of } n\text{)} \tag{7.4a}$$
$$\mathbb{E}\,(\boldsymbol{x}(n) - \bar{x})^2 = \sigma_x^2 \qquad \text{(variance is independent of } n\text{)} \tag{7.4b}$$
$$r_x(n, m) = r_x(n - m) \qquad \text{(covariance depends on } n - m\text{)} \tag{7.4c}$$

In other words, the mean and variance do not change with time (they assume constant values for all n) and the covariance function is only a function of the time lag between the samples (i.e., it becomes a one-dimensional sequence). In this way, the covariance between samples $\boldsymbol{x}(n)$ and $\boldsymbol{x}(n-2)$ is the same as the covariance between samples $\boldsymbol{x}(n-7)$ and $\boldsymbol{x}(n-9)$, and so forth. It is clear that property (7.4c) implies property (7.4b), so that a wide-sense stationary process is one for which the mean function is constant and the covariance function is only dependent on the time lag. For zero-mean random processes, the defining properties of wide-sense stationarity reduce to the following relations:

$$\mathbb{E}\,\boldsymbol{x}(n) = 0 \tag{7.5a}$$
$$\mathbb{E}\,\boldsymbol{x}^2(n) = \sigma_x^2 \tag{7.5b}$$
$$r_x(n, m) = r_x(n - m) \tag{7.5c}$$

Since $r_x(n,m)$ is a function of the time lag $n-m$, it is customary to denote the covariance function of a stationary random process as follows:

$$\boxed{r_x(k) \stackrel{\Delta}{=} \mathbb{E}\,\boldsymbol{x}(n)\boldsymbol{x}(n-k), \quad k \in \mathbb{Z}} \tag{7.6}$$

and to refer to $r_x(k)$ as the auto-correlation sequence. Clearly, in view of (7.2), we have $\sigma_x^2 = r_x(0)$.

Example 7.1 **(Stationary process)** The auto-correlation sequence of a zero-mean wide-sense stationary random process $\boldsymbol{x}(n)$ is given by $r_x(k) = \rho^{|k|}$, where $0 < \rho < 1$. Let us use this information to determine the covariance matrix of the vector $\boldsymbol{x} = \text{col}\{\boldsymbol{x}(0), \boldsymbol{x}(2), \boldsymbol{x}(5)\}$. Note that since $\boldsymbol{x}$ has zero mean, its covariance matrix is given by

$$R_x = \mathbb{E}\,\boldsymbol{x}\boldsymbol{x}^{\mathsf{T}} = \begin{bmatrix} \mathbb{E}\,\boldsymbol{x}^2(0) & \mathbb{E}\,\boldsymbol{x}(0)\boldsymbol{x}(2) & \mathbb{E}\,\boldsymbol{x}(0)\boldsymbol{x}(5) \\ \mathbb{E}\,\boldsymbol{x}(2)\boldsymbol{x}(0) & \mathbb{E}\,\boldsymbol{x}^2(2) & \mathbb{E}\,\boldsymbol{x}(2)\boldsymbol{x}(5) \\ \mathbb{E}\,\boldsymbol{x}(5)\boldsymbol{x}(0) & \mathbb{E}\,\boldsymbol{x}(5)\boldsymbol{x}(2) & \mathbb{E}\,\boldsymbol{x}^2(5) \end{bmatrix} = \begin{bmatrix} 1 & \rho^2 & \rho^5 \\ \rho^2 & 1 & \rho^3 \\ \rho^5 & \rho^3 & 1 \end{bmatrix} \tag{7.7}$$

We observe from the form of the auto-correlation sequence in this example that the correlation between two samples $\boldsymbol{x}(n)$ and $\boldsymbol{x}(n-k)$ dies out as the lag $|k|$ increases so that closer samples are more strongly correlated with each other than samples that are further apart.

Independent and identically distributed processes

An important sub-class of stationary random processes are processes whose samples are independent of each other but identically distributed (written as iid, for short). What this means is that for every n, the individual random variables $\boldsymbol{x}(n)$ have the same probability density function and, hence, they are identically distributed with the same mean $\bar{x}$ and the same variance σ_x^2. Moreover, any two different samples are statistically independent of each other:

$$\boldsymbol{x}(n) \text{ and } \boldsymbol{x}(m) \text{ are independent for } n \neq m \tag{7.8}$$

It then follows that the covariance sequence in this case is given by

$$r_x(k) = \begin{cases} \sigma_x^2, & k = 0 \\ 0, & k \neq 0 \end{cases} \tag{7.9}$$

That is,

$$r_x(k) = \sigma_x^2\, \delta(k) \tag{7.10}$$

in terms of the Kronecker delta sequence, $\delta(k)$, which assumes the value 1 when $k = 0$ and is 0 otherwise. The scalar σ_x^2 or $r_x(0)$ is referred to as the power of the random process.

White-noise processes

A random process $\boldsymbol{x}(n)$ is said to be a white-noise process if it satisfies the following properties:

$$\mathbb{E}\,\boldsymbol{x}(n) = 0 \qquad \text{(zero mean)} \tag{7.11a}$$

$$\mathbb{E}\,\boldsymbol{x}^2(n) = \sigma_x^2 \qquad \text{(constant variance)} \tag{7.11b}$$

$$r_x(k) = \sigma_x^2\,\delta(k) \qquad \text{(auto-correlation)} \tag{7.11c}$$

so that samples at different time instants are uncorrelated with each other. Thus, note that a zero-mean iid process is a white-noise process since independence among samples implies uncorrelatedness among these same samples, as required by (7.11c).

Example 7.2 **(Sinusoidal stationary process)** Consider the random process

$$\boldsymbol{x}(n) = A\cos(\omega n + \boldsymbol{\theta}) \;+\; \boldsymbol{v}(n) \tag{7.12}$$

where A is a constant, $\boldsymbol{\theta}$ is uniformly distributed in the range $[-\pi, \pi]$, and $\boldsymbol{v}(n)$ is white noise with variance σ_v^2 and independent of $\boldsymbol{\theta}$. The process $\boldsymbol{x}(n)$ can be verified to be wide-sense stationary. First, note that the mean of the term $A\cos(\omega n + \boldsymbol{\theta})$ is zero since

$$\mathbb{E}\,A\cos(\omega n + \boldsymbol{\theta}) = \int_{-\pi}^{\pi} \frac{1}{2\pi} A\cos(\omega n + \theta)d\theta \;=\; \frac{A}{2\pi}\sin(\omega n + \theta)\Big|_{-\pi}^{\pi} \;=\; 0 \tag{7.13}$$

Therefore,

$$\mathbb{E}\,\boldsymbol{x}(n) = \mathbb{E}\,\boldsymbol{v}(n) \;+\; \mathbb{E}\,A\cos(\omega n + \boldsymbol{\theta}) \;=\; 0 \tag{7.14}$$

Moreover,

$$\begin{aligned}
r_x(k) &= \mathbb{E}\,\boldsymbol{x}(n)\boldsymbol{x}(n-k) \\
&= \mathbb{E}\,\big(A\cos(\omega n + \boldsymbol{\theta}) \;+\; \boldsymbol{v}(n)\big)\,\big(A\cos(\omega(n-k) + \boldsymbol{\theta}) \;+\; \boldsymbol{v}(n-k)\big) \\
&= \sigma_v^2\delta(k) \;+\; A^2\mathbb{E}\,\cos(\omega n + \boldsymbol{\theta})\cos(\omega(n-k) + \boldsymbol{\theta}) \\
&= \sigma_v^2\delta(k) \;+\; \frac{A^2}{2}\mathbb{E}\left(\cos(\omega(2n-k) + 2\boldsymbol{\theta}) \;+\; \cos(\omega k)\right) \\
&= \sigma_v^2\delta(k) \;+\; \frac{A^2}{2}\cos(\omega k)
\end{aligned} \tag{7.15}$$

where in the third equality we used the fact that $\boldsymbol{v}(n)$ and $\boldsymbol{v}(n-k)$ are independent of each other (for $k \neq 0$) and of $\boldsymbol{\theta}$, and in the fourth equality we used the trigonometric identity:

$$\cos(a)\cos(b) = \frac{1}{2}\Big(\cos(a+b) + \cos(a-b)\Big) \tag{7.16}$$

Ergodic processes

One useful class of random processes is the family of ergodic processes. An ergodic process is one where the long sample average of *any function* of the random process (also called *time-average*) converges almost everywhere to the actual mean (also called *ensemble average*). In other words, for ergodic processes, time averages tend to the actual signal statistics and, therefore, observations of a sufficiently long realization from a *single* record for the random process help convey important statistical information about the random process itself.

For example, consider a wide-sense stationary random process $\boldsymbol{x}(n)$ with mean $\bar{x}$. Assume we observe a sufficiently large number of samples, say, $\{x(n),\ n = 0, 1, \ldots, N-1\}$ where N is large. If the mean value $\bar{x}$ is unknown, we may consider employing the sample (or time) average of these observations to estimate $\bar{x}$. Thus, let

$$\widehat{\bar{x}}_N = \frac{1}{N}\sum_{n=0}^{N-1} x(n) \qquad \text{(sample-average or time-average)} \tag{7.17}$$

For ergodic processes it will hold that

$$\lim_{N\to\infty} \widehat{\bar{x}}_N = \mathbb{E}\, \boldsymbol{x}(n) \tag{7.18}$$

where it is generally sufficient for the convergence in (7.18) to occur in the mean-square error sense (MSE), i.e., for

$$\lim_{N\to\infty} \mathbb{E}\left(\frac{1}{N}\sum_{n=0}^{N-1} x(n) - \bar{x}\right)^2 = 0 \tag{7.19}$$

or in some other sense including convergence in probability or convergence with probability 1 (see Appendix 3.A for a review of the notion of convergence for random variables). When (7.18) holds, we say that $\boldsymbol{x}(n)$ is first-order ergodic (or ergodic in the mean). For such processes, their mean values can be estimated from averaging a long-enough time sample of the process:

$$\underbrace{\frac{1}{N}\sum_{n=0}^{N-1} x(n)}_{\textbf{sample average}} \quad \overset{N\to\infty}{\longrightarrow} \quad \underbrace{\mathbb{E}\, \boldsymbol{x}(n)}_{\textbf{ensemble average}} \tag{7.20}$$

Consider a second example involving jointly stationary variables $\{\boldsymbol{x}(n), \boldsymbol{y}_n\}$ where the $\boldsymbol{x}(n)$ are scalar random variables and the $\boldsymbol{y}_n \in \mathbb{R}^M$ are random vectors. Assume we observe a large number of realizations $\{x(n), y_n\}$ for these random quantities. Under ergodicity, we will have

$$\underbrace{\frac{1}{N}\sum_{n=0}^{N-1} \left(x(n) - y_n^\mathsf{T} w\right)^2}_{\textbf{sample average}} \quad \overset{N\to\infty}{\longrightarrow} \quad \underbrace{\mathbb{E}\left(\boldsymbol{x}(n) - \boldsymbol{y}_n^\mathsf{T} w\right)^2}_{\textbf{ensemble average}} \tag{7.21}$$

where it is again sufficient for convergence in (7.21) to occur in the MSE sense.

Consider further a third situation involving jointly stationary random variables $\{\boldsymbol{\gamma}(n), \boldsymbol{h}_n\}$ where the $\boldsymbol{\gamma}(n)$ assume the binary values ± 1 randomly and the $\boldsymbol{h}_n \in \mathbb{R}^M$ are random vectors as well. We will encounter in our later studies on logistic regression transformations of the form:

$$\boldsymbol{x}(n) = \ln\left(1 + e^{-\boldsymbol{\gamma}(n)\boldsymbol{h}_n^{\mathsf{T}} w}\right) \tag{7.22}$$

for some deterministic vector w and in terms of the natural logarithm. If we observe a large number of realization pairs $\{\gamma(n), h_n\}$, then under ergodicity of the process $\boldsymbol{x}(n)$ we will have:

$$\underbrace{\frac{1}{N}\sum_{n=0}^{N-1} \ln\left(1 + e^{-\gamma(n)h_n^{\mathsf{T}} w}\right)}_{\textbf{sample average}} \overset{N\to\infty}{\longrightarrow} \underbrace{\mathbb{E}\left\{\ln\left(1 + e^{-\boldsymbol{\gamma}(n)\boldsymbol{h}_n^{\mathsf{T}} w}\right)\right\}}_{\textbf{ensemble average}} \tag{7.23}$$

Example 7.3 (**Nonergodic process**) Consider two boxes A and B. Box A contains balls marked with the values 1 through 5. Box B contains balls marked with the values 6 through 10. One box is chosen at random (with equal probability for A or B). Subsequently, balls are repeatedly selected randomly and with replacement from the chosen box. Let $\boldsymbol{x}(n)$ represent the random value observed at the nth ball selection. The mean of $\boldsymbol{x}(n)$ can be computed as follows. Observe first that if we select balls from box A, then the mean is

$$\bar{x}_A = \frac{1}{5}(1+2+3+4+5) = 3 \tag{7.24}$$

while if we select balls from box B, then the mean is

$$\bar{x}_B = \frac{1}{5}(6+7+8+9+10) = 8 \tag{7.25}$$

Therefore, the mean of $\boldsymbol{x}(n)$ is given by

$$\begin{aligned} \mathbb{E}\,\boldsymbol{x}(n) &= \mathbb{P}(\text{select A}) \times \bar{x}_A \;+\; \mathbb{P}(\text{select B}) \times \bar{x}_B \\ &= \frac{1}{2} \times 3 + \frac{1}{2} \times 8 \;=\; 5.5 \end{aligned} \tag{7.26}$$

On the other hand, the time average of a long sequence of observations $\{x(n)\}$ is expected to converge to $\bar{x}_A = 3$ when box A is selected and to $\bar{x}_B = 8$ when box B is selected. Neither of these values coincides with $\mathbb{E}\,\boldsymbol{x}(n) = 5.5$. Therefore, the process $\boldsymbol{x}(n)$ is not mean ergodic.

7.2 POWER SPECTRAL DENSITY

Consider a zero-mean wide-sense stationary random process $\boldsymbol{x}(n)$ with auto-correlation sequence denoted by $r_x(k)$. This sequence provides one characterization for the random process in the time domain.

Definition

There is an alternative powerful characterization in the transform domain in terms of the so-called *z-spectrum*, which is denoted by $S_x(z)$ and defined as the z-transform of the sequence $\{r_x(k)\}$, i.e.,

$$S_x(z) \triangleq \sum_{k=-\infty}^{\infty} r_x(k) z^{-k} \tag{7.27}$$

where $z \in \mathbb{C}$. This definition replaces the time sequence $\{r_x(n)\}$ by the function $S_x(z)$ whose argument z is complex-valued. Of course, the definition makes sense only for those values of z for which the series (7.27) converges; we refer to this set of values as the *region of convergence* (ROC) of $S_x(z)$. For our purposes, it suffices to assume that $\{r_x(k)\}$ is an exponentially bounded sequence, i.e.,

$$|r_x(k)| \leq \beta\, a^{|k|} \tag{7.28}$$

for some $\beta > 0$ and $0 < a < 1$. In this case, the series (7.27) will be absolutely convergent for all values of z in the annulus:

$$\text{ROC} = \left\{ a < |z| < a^{-1} \right\} \tag{7.29}$$

We thus say that the interval $a < |z| < 1/a$ defines the ROC of $S_x(z)$. Since this ROC includes the unit circle, it follows that the function $S_x(z)$ cannot have poles on the unit circle (poles are locations in the complex plane where $S_x(z)$ becomes unbounded). Evaluating $S_x(z)$ on the unit circle leads to what is called the *power spectrum* (or the power spectral density (PSD) function) of the random process $\boldsymbol{x}(n)$:

$$S_x(e^{j\omega}) \triangleq \sum_{k=-\infty}^{\infty} r_x(k) e^{-j\omega k} \tag{7.30}$$

For readers familiar with the theory of Fourier transforms, the above expression shows that the power spectrum is the discrete-time Fourier transform (DTFT) of the auto-correlation sequence. In particular, it can readily be verified that $r_x(k)$ can be recovered via the following integral expression:

$$r_x(k) = \frac{1}{2\pi} \int_{-\pi}^{\pi} S_x(e^{j\omega}) e^{j\omega k} d\omega \tag{7.31}$$

which amounts to the inverse DTFT formula. We can also recover the signal variance of the random process by evaluating the auto-correlation sequence at lag $k = 0$:

$$\sigma_x^2 \triangleq r_x(0) = \frac{1}{2\pi} \int_{-\pi}^{\pi} S_x(e^{j\omega}) d\omega \tag{7.32}$$

This expression shows that σ_x^2 is a scaled value of the area under the power spectrum over a 2π-wide interval. The variance σ_x^2 or, equivalently, the zeroth-order auto-correlation coefficient, $r_x(0)$, serves as a measure of the *power* of the random process $\boldsymbol{x}(n)$. The assumption of an exponentially bounded auto-correlation sequence ensures $r_x(0) < \infty$, so that the random process $\boldsymbol{x}(n)$ is a finite power process.

Two properties

The power spectrum of a random process satisfies two important properties:

(1) Hermitian symmetry, i.e.,

$$S_x(e^{j\omega}) = \left[S_x(e^{j\omega})\right]^* \tag{7.33a}$$

where the $*$ notation refers to complex conjugation. This property implies that $S_x(e^{j\omega})$ is real-valued for any $0 \leq \omega \leq 2\pi$.

(2) Nonnegativity on the unit circle, i.e.,

$$S_x(e^{j\omega}) \geq 0, \qquad \text{for } 0 \leq \omega \leq 2\pi \tag{7.33b}$$

The first property is easily checked from the definition of $S_x(e^{j\omega})$ since the auto-correlation sequence itself satisfies the symmetry property:

$$r_x(k) = r_x(-k) \tag{7.34}$$

Actually, and more generally, the z-spectrum satisfies the para-Hermitian symmetry property:

$$S_x(z) = \left[S_x\left(1/z^*\right)\right]^* \tag{7.35}$$

That is, if we replace z by $1/z^*$ and conjugate the result, then we recover $S_x(z)$ again. The second claim regarding the nonnegativity of $S_x(e^{j\omega})$ on the unit circle is more demanding to establish; it is proven in Prob. 7.13 under assumption (7.28).

Example 7.4 **(White-noise process)** Consider a white-noise process $\boldsymbol{v}(n)$ with variance σ_v^2. Then, its auto-correlation sequence is given by $r_v(k) = \sigma_v^2\delta(k)$, where $\delta(k)$ is the Kronecker delta: It is equal to 1 at $k = 0$ and is 0 otherwise. The resulting z-spectrum is

$$S_v(z) = \sigma_v^2 \tag{7.36}$$

We say that the power spectrum is flat in frequency since $S_v(e^{j\omega}) = \sigma_v^2$. The same conclusion holds for a process that consists of iid samples.

Example 7.5 **(Exponential correlation)** Consider a stationary process with $r_x(k) = \rho^{|k|}$, where ρ is real and satisfies $|\rho| < 1$. The z-spectrum is given by the z-transform of the sequence $r_x(k)$ and can be verified to evaluate to

$$S_x(z) = \frac{1-\rho^2}{(1-\rho z^{-1})(1-\rho z)}, \qquad |\rho| < |z| < \frac{1}{|\rho|} \tag{7.37a}$$

The power spectrum is obtained by evaluating $S_x(z)$ on the unit circle, i.e., at $z = e^{j\omega}$,

$$S_x(e^{j\omega}) = \frac{1-\rho^2}{(1-\rho e^{j\omega})(1-\rho e^{-j\omega})} = \frac{1-\rho^2}{|1-\rho e^{j\omega}|^2} \tag{7.37b}$$

Filtering of stationary processes

When a stationary random process is filtered by a bounded-input bounded-output (BIBO) (i.e., stable) linear time-invariant (LTI) system, the statistical properties of the output process can be related to the statistical properties of the input process in certain useful ways. LTI systems are described by transfer functions $H(z)$, which map the z-transform of input sequences $\{x(n)\}$ to the z-transform of the corresponding output sequences $\{y(n)\}$:

$$Y(z) = H(z)X(z) \tag{7.38}$$

The transfer function $H(z)$ is the z-transform of the impulse response sequence $h(n)$ of the LTI system; this is the sequence that relates $x(n)$ and $y(n)$ through the convolution expression:

$$y(n) = \sum_{k=-\infty}^{\infty} x(k)h(n-k) = \sum_{k=-\infty}^{\infty} h(k)x(n-k) \tag{7.39}$$

BIBO stability means that bounded input sequences are mapped to bounded output sequences. This can be shown to be equivalent to requiring $h(n)$ to be absolutely summable:

$$\sum_{n=-\infty}^{\infty} |h(n)| < \infty \tag{7.40}$$

which in turn is equivalent to requiring the ROC of $H(z)$ to include the unit circle:

$$\Big\{|z| = 1\Big\} \subset \mathrm{ROC}\Big(H(z)\Big) \tag{7.41}$$

Now, assume we feed a zero-mean wide-sense stationary random process $\boldsymbol{x}(n)$ with z-spectrum $S_x(z)$ into the filter $H(z)$. Let $S_y(z)$ denote the z-spectrum of the random process $\boldsymbol{y}(n)$ at the output of the filter – see Fig. 7.1.

Relating z-spectra

The output process $\boldsymbol{y}(n)$ will also be zero-mean and wide-sense stationary. Moreover, its z-spectrum will be given by

$$S_y(z) \;=\; H(z)\,S_x(z)\,\left[H\left(\frac{1}{z^*}\right)\right]^* \tag{7.42a}$$

and the power spectra of the input and output processes will be related as

$$S_y(e^{j\omega}) \;=\; \left|H(e^{j\omega})\right|^2\,S_x(e^{j\omega}) \tag{7.42b}$$

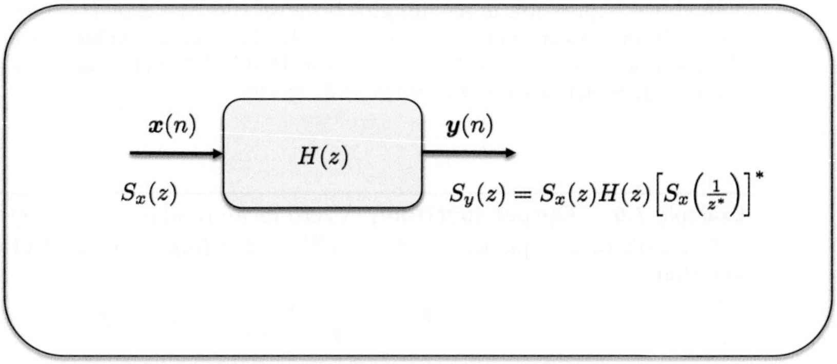

Figure 7.1 Filtering of a wide-sense stationary random process $\{\boldsymbol{x}(n)\}$ by a stable LTI system $H(z)$.

Proof: The BIBO stability of $H(z)$ ensures that the impulse response sequence is absolutely summable. Now, the input–output relation is given by the convolution sum:

$$\boldsymbol{y}(n) = \sum_{j=-\infty}^{\infty} \boldsymbol{x}(j)h(n-j) = \sum_{j=-\infty}^{\infty} h(j)\boldsymbol{x}(n-j) \tag{7.43}$$

from which we conclude that $\mathbb{E}\,\boldsymbol{y}(n) = 0$ since $\mathbb{E}\,\boldsymbol{x}(j) = 0$ for all j. Hence, the random process $\boldsymbol{y}(n)$ has zero mean. Moreover, for any integer k,

$$\begin{aligned}\mathbb{E}\,\boldsymbol{y}(n)\boldsymbol{y}(n-k) &= \mathbb{E}\left(\sum_{j=-\infty}^{\infty} h(j)\boldsymbol{x}(n-j)\right) \times \left(\sum_{m=-\infty}^{\infty} h(m)\boldsymbol{x}(n-k-m)\right)\\ &= \sum_{j=-\infty}^{\infty}\sum_{m=-\infty}^{\infty} h(j)h(m)r_x(k+m-j) \triangleq r_y(k)\end{aligned} \tag{7.44}$$

The expression on the right-hand side in the second equality depends only on k (and not on both k and n). For this reason, we denote it by $r_y(k)$ to highlight that the auto-correlation sequence of $\boldsymbol{y}(n)$ is only a function of the time lag, k. Therefore, the random process $\boldsymbol{y}(n)$ is wide-sense stationary.

Finally, the z-spectrum of $\{\boldsymbol{y}(n)\}$ is given by

$$\begin{aligned}S_y(z) &= \sum_{k=-\infty}^{\infty} r_y(k)z^{-k}\\ &= \sum_{k=-\infty}^{\infty}\sum_{j=-\infty}^{\infty}\sum_{m=-\infty}^{\infty} h(j)h(m)r_x(k+m-j)z^{-k}\\ &= \sum_{p=-\infty}^{\infty}\sum_{j=-\infty}^{\infty}\sum_{m=-\infty}^{\infty} h(j)h(m)r_x(p)z^{-p+m-j}\\ &= \left(\sum_{j=-\infty}^{\infty} h(j)z^{-j}\right) \times \left(\sum_{p=-\infty}^{\infty} r_x(p)z^{-p}\right) \times \left(\sum_{m=-\infty}^{\infty} h(m)z^{+m}\right)\\ &= H(z)\,S_x(z)\left[H\left(\frac{1}{z^*}\right)\right]^*\end{aligned} \tag{7.45}$$

The above expression is meaningful because the ROCs of $H(z)$ and $S_x(z)$ have a nontrivial intersection. This is because the ROC of $H(z)$ includes the unit circle (due to the assumed stability of $H(z)$), and the ROC of $S_x(z)$ is of the form (7.29) for some $0 < a < 1$, which also includes the unit circle.

■

Example 7.6 **(Output spectrum)** A zero-mean random process $\boldsymbol{x}(n)$ with exponential auto-correlation sequence, $r_x(k) = \rho^{|k|}$, is fed into a stable LTI filter with transfer function

$$H(z) = \frac{1}{1 - \frac{1}{2}z^{-1}}, \quad |z| > 1/2 \tag{7.46a}$$

It is assumed that $0 < \rho < 1$. Let us determine the z-spectrum of the output process, $\boldsymbol{y}(n)$. We already know from Example 7.5 that the z-spectrum of the process $\boldsymbol{x}(n)$ is given by

$$S_x(z) = \frac{1 - \rho^2}{(1 - \rho z^{-1})(1 - \rho z)}, \quad |\rho| < |z| < \frac{1}{|\rho|} \tag{7.46b}$$

Using (7.42a) we get

$$S_y(z) = \frac{1}{1 - \frac{1}{2}z^{-1}} \frac{1 - \rho^2}{(1 - \rho z^{-1})(1 - \rho z)} \frac{1}{1 - \frac{1}{2}z} \tag{7.46c}$$

with

$$\text{ROC of } S_y(z) = \max\{|\rho|, 1/2\} < |z| < \frac{1}{\max\{|\rho|, 1/2\}} \tag{7.46d}$$

Cross–spectrum

Let $r_{xy}(n, m)$ denote the cross-correlation sequence between any two arbitrary zero-mean wide-sense stationary processes $\{\boldsymbol{x}(n), \boldsymbol{y}(n)\}$, and which is defined as

$$r_{xy}(n, m) \stackrel{\Delta}{=} \mathbb{E}\,\boldsymbol{x}(n)\boldsymbol{y}(m) \tag{7.47}$$

When the random processes $\{\boldsymbol{x}(n), \boldsymbol{y}(n)\}$ are related, as in Fig. 7.1, it turns out that $r_{xy}(n, m)$ becomes a function of the difference $n - m$, i.e.,

$$\boxed{r_{xy}(k) \stackrel{\Delta}{=} \mathbb{E}\,\boldsymbol{x}(n)\boldsymbol{y}(n-k)} \tag{7.48}$$

Indeed, in that case, it holds that:

$$\begin{aligned} r_{xy}(n, m) &= \mathbb{E}\,\boldsymbol{x}(n)\boldsymbol{y}(m) \\ &= \mathbb{E}\,\boldsymbol{x}(n)\left(\sum_{k=-\infty}^{\infty} h(m)\boldsymbol{x}(m-k)\right) \\ &= \sum_{k=-\infty}^{\infty} h(m)r_x(n-m+k) \\ &\stackrel{\Delta}{=} r_{xy}(n-m) \end{aligned} \tag{7.49}$$

since the quantity on the right-hand side of the third equality is a function of $n - m$. When this happens, we say that the processes $\{\boldsymbol{x}(n), \boldsymbol{y}(n)\}$ are jointly wide-sense stationary. In this case, we can associate a z-cross spectrum with their cross-correlation sequence; it is denoted by $S_{xy}(z)$ and is defined as the z-transform of $r_{xy}(k)$:

$$S_{xy}(z) \triangleq \sum_{k=-\infty}^{\infty} r_{xy}(k) z^{-k} \tag{7.50}$$

We can again conclude from the setting of Fig. 7.1 that the above z-cross-spectrum is given by

$$S_{xy}(z) = S_x(z) \left[H\left(\frac{1}{z^*}\right)\right]^* \tag{7.51}$$

Proof: First, note that

$$\begin{aligned} r_{xy}(k) &= \mathbb{E}\, \boldsymbol{x}(n)\boldsymbol{y}(n-k) \\ &= \mathbb{E}\, \boldsymbol{x}(n) \left(\sum_{j=-\infty}^{\infty} h(j)\boldsymbol{x}(n-k-j) \right) \\ &= \sum_{j=-\infty}^{\infty} h(j) r_x(k+j) \end{aligned} \tag{7.52}$$

Accordingly,

$$\begin{aligned} S_{xy}(z) &= \sum_{k=-\infty}^{\infty} r_{xy}(k) z^{-k} \\ &= \sum_{k=-\infty}^{\infty} \sum_{j=-\infty}^{\infty} h(k) r_x(k+j) z^{-k} \\ &= \sum_{p=-\infty}^{\infty} \sum_{j=-\infty}^{\infty} h(j) r_x(p) z^{-p+j} \\ &= \left(\sum_{p=-\infty}^{\infty} r_x(p) z^{-p} \right) \left(\sum_{j=-\infty}^{\infty} h(j) z^{j} \right) \\ &= S_x(z) \left[H\left(\frac{1}{z^*}\right)\right]^* \end{aligned} \tag{7.53}$$

■

Example 7.7 (Cross-spectrum and correlations) A white-noise process $\boldsymbol{v}(n)$ with power σ_v^2 is fed into a stable LTI system with transfer function

$$H(z) = \frac{1}{1 - \frac{1}{2}z^{-1}}, \quad |z| > 1/2 \tag{7.54}$$

The z-spectrum of the input sequence is $S_v(z) = \sigma_v^2$. Using (7.42a), we find that the z-spectrum of the output sequence $\boldsymbol{y}(n)$ is

$$S_y(z) = \frac{\sigma_v^2}{\left(1 - \frac{1}{2}z^{-1}\right)\left(1 - \frac{1}{2}z\right)}, \quad \frac{1}{2} < |z| < 2$$

Observe that the output process is not white even though the input process is white. We thus say that passing a white-noise process through a stable LTI system colors the noise. In particular, using (7.37a), observe by inverse-transforming $S_y(z)$ that the auto-correlation sequence of the process $\boldsymbol{y}(n)$ is given by

$$r_y(k) = \frac{4\sigma_v^2}{3}\left(\frac{1}{2}\right)^{|k|} \tag{7.55}$$

We can further determine the z-cross-spectrum between the input and output processes $\boldsymbol{v}(n)$ and $\boldsymbol{y}(n)$ by using (7.51) so that

$$S_{vy}(z) = \frac{\sigma_v^2}{1 - \frac{1}{2}z}, \quad |z| < 2 \tag{7.56}$$

The region $|z| < 2$ ensures that the unit circle is included in the ROC of $S_{vy}(z)$. By inverse-transforming $S_{vy}(z)$ we find the cross-correlation sequence between the processes $\boldsymbol{v}(n)$ and $\boldsymbol{y}(n)$:

$$\begin{aligned} r_{vy}(k) &\triangleq \mathbb{E}\,\boldsymbol{v}(n)\boldsymbol{y}(n-k) \\ &= \begin{cases} 0, & k > 0 \\ \sigma_v^2, & k = 0 \\ \sigma_v^2 2^k, & k < 0 \end{cases} \end{aligned} \tag{7.57}$$

7.3 SPECTRAL FACTORIZATION

Spectral factorization is one important tool in the design of optimal filters to ameliorate the effect of distortions and randomness in data. It can be motivated as follows. Consider a zero-mean wide-sense stationary process $\boldsymbol{x}(n)$ with z-spectrum $S_x(z)$. We assume from now on that $S_x(z)$ is a proper *rational* function in z and z^{-1} and that it does not have zeros on the unit-circle so that

$$S_x(e^{j\omega}) > 0, \quad \text{for all} \quad -\pi \leq \omega \leq \pi \tag{7.58}$$

Then, using the para-Hermitian symmetry property (7.35), it is easy to verify that for every pole (or zero) at a point ξ, there must exist a pole (respectively a zero) at the point $1/\xi^*$. In addition, it follows from the fact that $S_x(z)$ does not have poles and zeros on the unit circle that any such rational function $S_x(z)$ can be factored as

$$S_x(z) = \gamma \times \frac{\prod_{k=1}^{m}(z - z_k)(z^{-1} - z_k^*)}{\prod_{k=1}^{n}(z - p_k)(z^{-1} - p_k^*)} \tag{7.59}$$

for some positive scalar γ and for poles and zeros satisfying $|z_k| < 1$ and $|p_k| < 1$. We will establish later in (7.72) that the scalar γ has the interpretation of a variance measure. The spectral factorization of $S_x(z)$ is defined as a factorization of the special form:

$$\boxed{S_x(z) = \gamma \times L(z) \times \left[L\left(\frac{1}{z^*}\right)\right]^*} \tag{7.60}$$

where $\{\gamma, L(z)\}$ satisfy the following conditions:

(1) γ is a positive scalar.

(2) $L(z)$ is normalized to unity at infinity, i.e., $L(\infty) = 1$.

(3) $L(z)$ and $L^{-1}(z)$ are stable rational minimum-phase functions (i.e., their poles and zeros lie inside the unit circle) and their ROCs include the unit circle.

(4) $[L(1/z^*)]^*$ and its inverse are stable rational maximum-phase functions (i.e., their poles and zeros lie outside the unit circle) and their ROCs include the unit circle.

The normalization $L(\infty) = 1$ makes the choice of $L(z)$ unique since otherwise infinitely many choices for $\{\gamma, L(z)\}$ would exist. In order to determine $L(z)$ from (7.59) we just have to extract the poles and zeros that are inside the unit circle. However, in order to meet the normalization condition $L(\infty) = 1$, we take

$$\boxed{L(z) = z^{n-m} \times \frac{\prod_{k=1}^{m}(z - z_k)}{\prod_{k=1}^{n}(z - p_k)} = \frac{\prod_{k=1}^{m}(1 - z_k z^{-1})}{\prod_{k=1}^{n}(1 - p_k z^{-1})}} \tag{7.61}$$

Therefore, to obtain the spectral factor $L(z)$ we keep the stable poles and zeros of $S_x(z)$ and then multiply by z^{n-m} in order to meet the normalization condition $L(\infty) = 1$. The factor $L(z)$ so defined is called the *canonical spectral factor* of $S_x(z)$.

Since the poles of $L(z)$ lie inside the unit circle, and since $L(z)$ is stable, it follows that the ROC of $L(z)$ is the outside of a circular region and includes the unit circle. It also follows that $L(z)$ is the transfer function of a causal system (one with a causal impulse response sequence $\{\ell(n)\}$ with zero samples for all $n < 0$). Likewise, the ROC of $L^{-1}(z)$ is the outside of a circular region that includes the unit circle. Therefore, $L^{-1}(z)$ is also the transfer function of a causal system. We sometimes refer to the fact that both $L(z)$ and $L^{-1}(z)$ are stable causal systems by saying that

$$L(z) \textbf{ is causal and causally invertible} \tag{7.62}$$

or that $L(z)$ and $L^{-1}(z)$ have causal inverse transforms (meaning that their time-domain sequences are causal sequences and therefore have zero samples for all $n < 0$). In a similar vein, we say that

$$\left[L\left(\frac{1}{z^*}\right)\right]^* \quad \textbf{and its inverse have anti-causal inverse transforms} \tag{7.63}$$

which means that the inverse transform of $[L(1/z^*)]^*$ or the inverse transform of its inverse are anti-causal sequences (their samples are zero over $n > 0$). We illustrate the above construction by means of examples.

Example 7.8 **(Spectral factorization)** Consider a zero-mean wide-sense stationary process with z-spectrum $S_x(z) = (2z+1)(2z^{-1}+1)$. To determine its spectral factorization we first express $S_x(z)$ in the form (7.59):

$$S_y(z) = 4\left(z + \frac{1}{2}\right)\left(z^{-1} + \frac{1}{2}\right) \tag{7.64}$$

It then follows that $\gamma = 4$. Moreover, since $m = 1$ and $n = 0$, expression (7.61) gives

$$L(z) = z^{-1}\left(z + \frac{1}{2}\right) = 1 + \frac{1}{2}z^{-1} \tag{7.65}$$

Example 7.9 **(Spectral factor)** Consider now a stationary process with auto-correlation sequence $r_x(k) = \rho^{|k|}$, where ρ is real and satisfies $|\rho| < 1$. We already know from Example 7.5 that its z-spectrum is given by

$$S_x(z) = \frac{1-\rho^2}{(1-\rho z^{-1})(1-\rho z)}, \qquad |\rho| < |z| < \frac{1}{|\rho|} \tag{7.66}$$

It follows that the canonical spectral factor is $L(z) = 1/(1-\rho z^{-1})$ with $\gamma = 1 - \rho^2$.

Example 7.10 **(ARMA process)** Consider a stable causal LTI system that is described by an auto-regressive moving-average (ARMA) difference equation:

$$\boldsymbol{y}(n+1) = a_o\boldsymbol{y}(n) + a_1\boldsymbol{y}(n-1) + \boldsymbol{x}(n) + b\boldsymbol{x}(n-1) \tag{7.67}$$

with $|b| < 1$. Assume the excitation started in the remote past so that the output process is stationary. Assume further that $\boldsymbol{x}(n)$ is a white-noise process with variance σ_x^2. The transfer function of the filter is

$$H(z) = \frac{z+b}{z^2 - a_0 z - a_1} \tag{7.68}$$

The poles of $H(z)$ lie inside the unit circle since the system is assumed to be stable and causal. Now, since $S_x(z) = \sigma_x^2$, we find that the z-spectrum of the output process is

$$S_y(z) = \sigma_x^2 \times \left(\frac{z+b}{z^2 - a_0 z - a_1}\right) \times \left(\frac{z^{-1}+b}{z^{-2} - a_0 z^{-1} - a_1}\right) \tag{7.69}$$

But since $|b| < 1$, by assumption, we conclude that the spectral factor of $S_y(z)$ is given by

$$L(z) = \frac{z(z+b)}{z^2 - a_0 z - a_1} \tag{7.70}$$

Modeling and whitening filters

The spectral factor $1/L(z)$ has a useful interpretation in view of result (7.42a) and Fig. 7.1. Assume $\boldsymbol{x}(n)$ is fed through the stable LTI filter $1/L(z)$ and let $\boldsymbol{w}(n)$ denote the resulting output process. Then, the z-spectrum of $\boldsymbol{w}(n)$ is

$$S_w(z) = \frac{1}{L(z)} \times S_x(z) \times \frac{1}{\left[L\left(\frac{1}{z^*}\right)\right]^*} = \gamma \tag{7.71}$$

Consequently, the auto-correlation sequence of $\boldsymbol{w}(n)$ is

$$r_w(k) = \gamma\, \delta(k) \tag{7.72}$$

In other words, the process $\{\boldsymbol{w}(\cdot)\}$ becomes a white-noise process with variance equal to γ. For this reason, the filter $1/L(z)$ is called the *whitening filter*.

On the other hand, if a white-noise process with variance γ is fed into the filter $L(z)$, then the z-spectrum of the output sequence will be $S_x(z)$. For this reason, the filter $L(z)$ is called the *modeling filter*. The modeling and whitening filters can therefore be used as follows:

(a) The modeling filter $L(z)$ can be used to construct a wide-sense stationary process with a desired z-spectrum, $S_x(z)$. This is achieved by feeding a white-noise process with variance γ through $L(z)$.

(b) The whitening filter $1/L(z)$ can be used to whiten a process with z-spectrum $S_x(z)$ and obtain a white-noise process with variance γ.

7.4 COMMENTARIES AND DISCUSSION

Stationarity, ergodicity, and spectral factorization. The exposition in Sections 7.2 and 7.3 is motivated by the presentation in Kailath, Sayed, and Hassibi (2000). Section 7.2 assumes some basic knowledge of random processes, stationarity, and ergodicity. If additional help is needed, some useful references are Halmos (1956), Feller (1968, 1971), Petersen (1983), Papoulis (1991), Durrett (1996), and Leon-Garcia (2008). We explained in Section 7.3 that the spectral factorization result (7.60) is an important tool in the design of optimal filters, smoothers, and predictors. We will encounter it later in Section 30.5 when we examine the steady-state behavior of the Kalman filter. It is clear from the representation (7.59) that determination of the spectral factor, $L(z)$, requires computing the poles and zeros of the numerator and denominator polynomials for rational z-spectra, $S_x(z)$. Such factorizations are straightforward to perform for low-order polynomials, but require more systematic procedures for higher-order polynomials. It is no wonder then that a variety of techniques have been developed to determine $L(z)$, and the normalization constant, γ, in an iterative manner. In Section 30.5 we explain how the Kalman filter performs this calculation by means of a Riccati recursion. But there are other methods that can be used for this purpose as well, such as the famed Levinson and Schur algorithms. The article by Sayed and Kailath (2001) provides a survey of these iterative techniques, including other solution methods – see also Goodman *et al.* (1997).

Paley–Wiener condition. In our treatment, we made the stronger than needed assumption in (7.28) that the auto-correlation sequence, $\{r_x(k)\}$, is exponentially bounded and that the z-spectrum, $S_x(z)$, is rational. However, as already explained in Sayed and Kailath (2001) and repeated here, some of the iterative techniques for spectral factorization, such as the Levinson and Schur algorithms, can still be applied to non-rational spectra provided we weaken the definition of the canonical spectral factor and provided $S_x(z)$ has finite power and satisfies a certain Paley–Wiener condition. Following Doob (1953) and Grenander and Szegö (1958), for general (not necessarily rational) z-spectra, $S_x(z)$, a canonical spectral factorization with $\gamma > 0$ and a unique function $L(z)$ with the following two properties exist:

(a) $L(z)$ and $1/L(z)$ are analytic in the region $|z| > 1$ (i.e., they are differentiable at every point in $|z| > 1$, which also means that they admit a convergent power series representation in a neighborhood around any point in $|z| > 1$), and
(b) the impulse response sequence of $L(z)$ is square-summable, i.e., $\sum_{n=1}^{\infty} |\ell(n)|^2 < \infty$

if, and only if, $S_x(z)$ has finite power, namely,

$$r_x(0) \triangleq \frac{1}{2\pi}\int_{-\pi}^{\pi} S_x(e^{j\omega})d\omega < \infty \tag{7.73}$$

and satisfies the Paley–Wiener condition:

$$\frac{1}{2\pi}\int_{-\pi}^{\pi} \ln S_x(e^{j\omega})d\omega > -\infty \tag{7.74}$$

When these conditions are met, the normalization factor γ is given by the expression

$$\gamma = \exp\left[\frac{1}{2\pi}\int_{-\pi}^{\pi} \ln S_x(e^{j\omega})d\omega\right] \tag{7.75}$$

In order for $S_x(z)$ to satisfy the Paley–Wiener condition (7.74), it cannot have zeros over a set of nonzero measure on the unit circle. Thus, for example, random processes with band-limited power spectra will not qualify. For our discussions we have made the stronger assumption that $S_x(z)$ is never zero on the unit circle. We remark that because of the finite-power condition (7.73), the Paley–Wiener condition (7.74) is equivalent to requiring the absolute integrability of $\ln S_x(e^{j\omega})$.

Condition (7.74) is due to Paley and Wiener (1934). The result is usually stated under a more general setting that helps answer the following important design question (see, e.g., Papoulis (1962, 1977) and Paarmann (2001)). Given a desired magnitude response, $|H(e^{j\omega})|$ for a filter, can a causal LTI system be determined with that magnitude response? By a causal system we mean one where the output signal at any time n can only depend on the input signal up to that time instant. The answer is given by the following statement for discrete-time systems; a proof of the result for continuous-time systems can be found in Papoulis (1962).

Paley–Wiener condition (Paley and Wiener (1934)) *Consider a square-integrable frequency response, $H(e^{j\omega})$, i.e.,*

$$\int_{-\pi}^{\pi} |H(e^{j\omega})|^2 d\omega < \infty \tag{7.76}$$

Then, $H(e^{j\omega})$ is the frequency response of a causal LTI system if, and only if,

$$\int_{-\pi}^{\pi} \Big| \ln |H(e^{j\omega})| \Big| d\omega < \infty \tag{7.77}$$

Therefore, given a square-integrable *magnitude* response, $|H(e^{j\omega})|$, that satisfies condition (7.77), the theorem guarantees that a phase response, $\phi(w)$, can be determined such that the resulting frequency response

$$H(e^{j\omega}) = |H(e^{j\omega})| \times \exp\Big\{j\phi(w)\Big\} \tag{7.78}$$

corresponds to a causal LTI system.

Ergodicity. We encountered the law of large numbers earlier in Appendix 3.A. It establishes the almost-sure convergence of the sample mean of a collection of *independent and identically distributed* (iid) random variables $\{\boldsymbol{x}(n)\}$ to their actual mean μ:

(strong law of large numbers)

$$\lim_{N\to\infty} \left\{ \frac{1}{N} \sum_{n=1}^{N} \boldsymbol{x}(n) \right\} \overset{a.s.}{\longrightarrow} \mu \tag{7.79}$$

Ergodic processes exhibit similar behavior where convergence of the sample mean toward the actual mean is guaranteed even for samples $\{\boldsymbol{x}(n)\}$ that are not necessarily iid, as well as for transformations of these samples. These conclusions follow from different versions of the so-called *ergodic theorem*, starting with Birkhoff (1931) and von Neumann (1932b). One version is the following statement.

Mean ergodic theorem (von Neumann (1932b)) *Let $g(x)$ denote a square-integrable function, evaluated at realizations $x(n)$ of a stationary process whose samples arise from the probability distribution $\boldsymbol{x} \sim f_{\boldsymbol{x}}(x)$. Then, under mild technical conditions, it holds that*

$$\lim_{N\to\infty} \frac{1}{N} \sum_{n=1}^{N} g(\boldsymbol{x}(n)) \overset{L^2}{\to} \int_{x\in\mathcal{X}} g(x) f_{\boldsymbol{x}}(x) d(x) = \mathbb{E}_f\, g(\boldsymbol{x}) \tag{7.80}$$

where convergence is in the mean-square sense (cf. definition (3.228)).

Result (7.80) ensures that every infinitely long trajectory $\{g(\boldsymbol{x}(n))\}$ admits the same time average, which is the quantity given on the right-hand side. We can interpret the sample mean on the left-hand side of the equation as an *average over time*, and the integral expression on the right-hand side of the same equation as an *average over space* (or over the domain $x \in \mathcal{X}$). The notation $\mathbb{E}_f\, g(\boldsymbol{x})$ represents the mean value of $g(\boldsymbol{x})$ relative to the probability distribution $f_{\boldsymbol{x}}(x)$. The Birkhoff version of the theorem guarantees that if $g(x)$ is absolutely integrable, then the time-average exists almost always and converges to the constant value $\mathbb{E}_f\, g(\boldsymbol{x})$. For more rigorous statements of the theorems, along with their proofs, the reader may refer to Billingsley (1965), Cornfeld, Fomin, and Sinai (1982), Walters (1982), and Petersen (1983).

PROBLEMS

7.1 Assume $\boldsymbol{A}$ is Gaussian with zero mean and variance σ_a^2. Assume $\boldsymbol{\theta}$ is uniformly distributed over $[-\pi, \pi]$ and is independent of $\boldsymbol{A}$. Show that $\boldsymbol{x}(n) = \boldsymbol{A}\cos(\omega n + \boldsymbol{\theta})$ is a wide-sense stationary process.

7.2 Consider a random process $\boldsymbol{x}(n)$ that consists of two tones embedded in noise, say, of the form

$$\boldsymbol{x}(n) = A_1\cos(\omega_1 n + \boldsymbol{\theta}_1) \;+\; A_2\cos(\omega_2 n + \boldsymbol{\theta}_2) \;+\; \boldsymbol{v}(n)$$

where A_1 and A_2 are constants, $\boldsymbol{\theta}_1$ and $\boldsymbol{\theta}_2$ are independent random variables that are uniformly distributed in the range $[-\pi, \pi]$, and $\boldsymbol{v}(n)$ is additive white noise with power

σ_v^2 and independent of both $\boldsymbol{\theta}_1$ and $\boldsymbol{\theta}_2$. Verify that $\boldsymbol{x}(n)$ is a wide-sense stationary process with zero mean and auto-correlation sequence given by

$$r_x(\ell) = \frac{A_1^2}{2}\cos(\omega_1 \ell) \; + \; \frac{A_2^2}{2}\cos(\omega_2 \ell) \; + \; \sigma_v^2 \delta(\ell)$$

7.3 Let $\boldsymbol{a}$ denote a random variable with probability distribution $f_{\boldsymbol{a}}(a)$. One realization a is picked for $\boldsymbol{a}$. Let $\boldsymbol{z}(n)$ denote a discrete random process whose samples are defined as $\boldsymbol{z}(n) = a$ for all n. Is this a stationary process? Is it ergodic?
7.4 True or false? An ergodic random process must be stationary.
7.5 Consider a collection of N_T samples $\{x(n)\}$ for a random process $\boldsymbol{x}(n)$ over the interval $0 \leq n \leq N_T - 1$. Using these samples, we estimate the auto-correlation sequence of $\boldsymbol{x}(n)$ by using the following expression:

$$\widehat{r}_x(\ell) = \frac{1}{N_T}\sum_{n=\ell}^{N_T-1} x(n)x(n-\ell), \quad \ell = 0, 1, \ldots, N_T - 1$$

where the lower limit of the summation ensures that only samples of $x(n)$ within the available interval of time are included in the computation of $\widehat{r}_x(\ell)$. The values of $\widehat{r}_x(\ell)$ for $\ell < 0$ are obtained by imposing the symmetry relation $\widehat{r}_x(-\ell) = \widehat{r}_x(\ell)$. Verify that the random variable $\widehat{\boldsymbol{r}}_x(\ell)$ is a biased estimator for $r_x(\ell)$ and that its mean is given by

$$\mathbb{E}\,\widehat{\boldsymbol{r}}_x(\ell) \; = \; \frac{N_T - \ell}{N_T} r_x(\ell)$$

7.6 Let $S_{xy}(z)$ denote the z-cross spectrum between two zero-mean wide-sense stationary processes $\{\boldsymbol{x}(n), \boldsymbol{y}(n)\}$. What is the z-cross spectrum between the random processes $\boldsymbol{x}(n-d)$ and $\boldsymbol{y}(n)$, for some integer delay $d > 0$?
7.7 Consider $\boldsymbol{y}(n) = a\boldsymbol{x}(n-1) + \boldsymbol{v}(n)$ where all processes are zero-mean and wide-sense stationary. Moreover, $|a| < 1$ and $\boldsymbol{v}(n)$ is a white-noise process with power σ_v^2 and its samples are independent of all samples of $\boldsymbol{x}(n)$. The z-spectrum of $\boldsymbol{x}(n)$ is

$$S_x(z) = \frac{3/4}{\left(1 - \frac{1}{2}z^{-1}\right)\left(1 - \frac{1}{2}z\right)}, \quad 1/2 < |z| < 2$$

(a) What is the power of $\boldsymbol{x}(n)$?
(b) What is the z-spectrum of $\boldsymbol{y}(n)$?
(c) Find the power of $\boldsymbol{y}(n)$.
(d) Find the auto-correlation sequence of $\boldsymbol{y}(n)$.
7.8 Consider the stable linear time-invariant (LTI) system $\boldsymbol{y}(n) = \frac{1}{2}\boldsymbol{y}(n-1) + \frac{1}{2}\boldsymbol{x}(n)$, where $\boldsymbol{x}(n)$ is a zero-mean random process with auto-correlation sequence $r_x(\ell) = \left(\frac{1}{2}\right)^{|2\ell|}$. Find $r_y(\ell)$, $S_y(z)$, $r_{xy}(\ell)$, and $S_{xy}(z)$.
7.9 A white-noise process $\boldsymbol{w}(n)$ with variance σ_w^2 is fed into a bounded-input bounded-output (BIBO) stable filter with transfer function

$$H(z) = \frac{1 + bz^{-1}}{1 + az^{-1}}$$

with real coefficients (a, b). Determine the z-spectrum of the output process, $\boldsymbol{y}(n)$. Determine also the spectral factorization of $S_y(z)$ for the following scenarios:
(a) $|a| < 1$ and $|b| < 1$.
(b) $|a| < 1$ and $|b| > 1$.
(c) $|a| > 1$ and $|b| < 1$.
(d) $|a| > 1$ and $|b| > 1$.
7.10 A unit-variance white-noise random process $\boldsymbol{s}(n)$ is fed into a first-order auto-regressive model with transfer function $\sqrt{1 - a^2}/(1 - az^{-1})$, where a is real. The output process is denoted by $\boldsymbol{u}(n)$. Assume $|a| < 1$. Show that the auto-correlation sequence of $\boldsymbol{u}(n)$ is given by $r_u(\ell) = a^{|\ell|}$ for all integer values ℓ.

7.11 Let

$$S_y(z) = \frac{1}{3} \times \frac{1+\frac{1}{2}z^{-1}}{1+\frac{1}{4}z^{-1}} \times \frac{1+\frac{1}{2}z}{1+\frac{1}{4}z}$$

Determine the corresponding modeling and whitening filters.

7.12 Consider a zero-mean random process with auto-correlation sequence $r_x(\ell) = \left(\frac{1}{4}\right)^{|2\ell|}$. Determine the following quantities:

(a) The signal power.

(b) $\frac{1}{2\pi}\int_{-\pi}^{\pi} S_x(e^{j\omega})d\omega$.

(c) $S_x(z)$.

7.13 Let $r_x(\ell) = \mathbb{E}\,\boldsymbol{x}(n)\boldsymbol{x}(n-\ell)$ and assume the auto-correlation sequence is exponentially bounded as in (7.28). Introduce the corresponding power spectrum (7.30). Let R_M be the $M \times M$ symmetric Toeplitz covariance matrix whose first column is $\text{col}\{r_x(0), \ldots, r_x(M-1)\}$; a Toeplitz matrix has equal entries along its diagonals. Pick any finite scalar α and define

$$b = \text{col}\left\{\alpha, \alpha e^{-j\omega}, \ldots, \alpha e^{-j(M-1)\omega}\right\}$$

where $j = \sqrt{-1}$. Show that

$$0 \leq \frac{1}{M} b^* R_M b = |\alpha|^2 \left[\sum_{k=-(M-1)}^{M-1} r_x(k)e^{-j\omega k}\right] - \frac{|\alpha|^2}{M} \sum_{k=-(M-1)}^{M-1} |k|\, r_x(k)e^{-j\omega k}$$

Now take the limit as $M \to \infty$ to conclude that $S_x(e^{j\omega}) \geq 0$. *Remark.* This problem is adapted from Kailath, Sayed, and Hassibi (2000). The reader may consult this reference for additional discussion on power spectra and spectral factorization.

REFERENCES

Billingsley, P. (1965), *Ergodic Theory and Information*, Wiley.

Birkhoff, G. D. (1931), "Proof of the ergodic theorem," *Proc. Nat. Acad. Sci.*, vol. 17, no. 12, pp. 656–660.

Cornfeld, I., S. Fomin, and Y. G. Sinai (1982), *Ergodic Theory*, Springer.

Doob, J. L. (1953), *Stochastic Processes*, Wiley.

Durrett, R. (1996), *Probability: Theory and Examples*, 2nd ed., Duxbury Press. Fifth edition published by Cambridge University Press, 2019.

Feller, W. (1968), *An Introduction to Probability Theory and Its Applications*, vol. 1, 3rd ed., Wiley.

Feller, W. (1971), *An Introduction to Probability Theory and Its Applications*, vol. 2, 3rd ed., Wiley.

Goodman, T. N. T., C. A. Micchelli, G. Rodriguez, and S. Seatzu (1997), "Spectral factorization of Laurent polynomials," *Adv. Comput. Math.*, vol. 7, pp. 429–445.

Grenander, U. and G. Szegö (1958), *Toeplitz Forms and their Applications*, University of California Press.

Halmos, P. R. (1956), *Lectures on Ergodic Theory*, Chelsea Publishing Company. Republished by Cambridge University Press, 1990.

Kailath, T., A. H. Sayed, and B. Hassibi (2000), *Linear Estimation*, Prentice Hall.

Leon-Garcia, A. (2008), *Probability, Statistics, and Random Processes for Electrical Engineering,* 3rd ed., Prentice Hall.

Paarmann, L. D. (2001), *Design and Analysis of Analog Filters: A Signal Processing Perspective*, Kluwer Academic Publishers.

Paley, R. and N. Wiener (1934), *Fourier Transforms in the Complex Domain*, vol. 19, American Mathematical Society Colloquium Publishing.
Papoulis, A. (1962), *The Fourier Integral and Its Applications*, McGraw-Hill.
Papoulis, A. (1977), *Signal Analysis*, McGraw-Hill.
Papoulis, A. (1991), *Probability, Random Variables, and Stochastic Processes,* 3rd ed., McGraw-Hill.
Petersen, K. E. (1983), *Ergodic Theory*, Cambridge University Press.
Sayed, A. H. and T. Kailath (2001), "A survey of spectral factorization methods," *Numer. Linear Algebra Appl.*, vol. 8, pp. 467–496.
von Neumann, J. (1932b), "Proof of the quasi-ergodic hypothesis," *Proc. Nat. Acad. Sci.*, vol. 18, no. 1, pp. 70–82.
Walters, P. (1982), *An Introduction to Ergodic Theory*, Springer.

8 Convex Functions

Convex functions are prevalent in inference and learning, where optimization problems involving convex risks are commonplace. In this chapter, we review basic properties of smooth and nonsmooth convex functions and introduce the concept of subgradient vectors. In the next chapter, we discuss projections onto convex sets and the solution of convex optimization problems by duality arguments.

8.1 CONVEX SETS

Let $g(z) : \mathbb{R}^M \to \mathbb{R}$ denote a *real-valued* function of a possibly vector argument, $z \in \mathbb{R}^M$. We consider initially the case in which $g(z)$ is at least first- and second-order differentiable, meaning that the gradient vector, $\nabla_z\, g(z)$, and the Hessian matrix, $\nabla_z^2\, g(z)$, exist and are well-defined at all points in the domain of the function. Later we comment on the case of nonsmooth convex functions, which are not differentiable at some locations.

To begin with, a set $\mathcal{S} \subset \mathbb{R}^M$ is said to be convex if for any pair of points $z_1, z_2 \in \mathcal{S}$, all points that lie on the line segment connecting z_1 and z_2 also belong to $\mathcal{S}$. Specifically,

$$\forall\, z_1, z_2 \in \mathcal{S} \text{ and } 0 \le \alpha \le 1 \implies \alpha z_1 + (1-\alpha) z_2 \in \mathcal{S} \tag{8.1}$$

Figure 8.1 illustrates this definition by showing two convex sets and one nonconvex set in its first row. In the rightmost set, a segment is drawn between two points inside the set and it is seen that some of the points on the segment lie outside the set.

We will regularly deal with *closed* sets, including closed convex sets. A set $\mathcal{S}$ in any metric space (i.e., in a space with a distance measure between its elements) is said to be *closed* if any converging sequence of points in $\mathcal{S}$ converges to a point in $\mathcal{S}$. This characterization is equivalent to stating that the complement of $\mathcal{S}$ is an open set or that any point outside $\mathcal{S}$ has a neighborhood around it that is disjoint from $\mathcal{S}$. For example, the segment $[-1, 1]$ on the real line is a closed set, while $[-1, 1)$ is an open set.

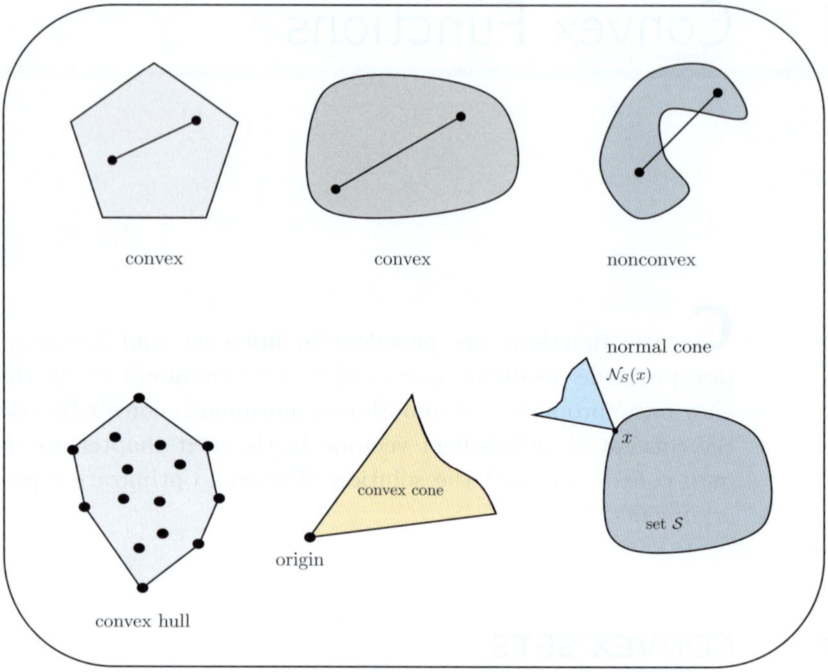

Figure 8.1 The two sets on the left in the first row are examples of convex sets, while the set on the right is nonconvex. The bottom row shows examples of a convex hull on the left, a convex cone in the middle, and a normal cone on the right. The curved line in the figure for the cones is meant to indicate that the cone extends indefinitely.

Example 8.1 (**Convex hull, convex cone, and conic hull**) Consider an arbitrary set $\mathcal{S} \subset \mathbb{R}^M$ that is not necessarily convex. The *convex hull* of $\mathcal{S}$, denoted by $\text{conv}(\mathcal{S})$, is the set of all convex combinations of elements in $\mathcal{S}$. Intuitively, the convex hull of $\mathcal{S}$ is the smallest convex set that contains $\mathcal{S}$. This situation is illustrated in the leftmost plot in the bottom row of Fig. 8.1. The dark circles represent the elements of $\mathcal{S}$, and the connected lines define the contour of the smallest convex set that contains $\mathcal{S}$.

A set $\mathcal{C} \subset \mathbb{R}^M$ is said to be a *cone* if for any element $z \in \mathcal{C}$ it holds that $tz \in \mathcal{C}$ for any $t \geq 0$. The cone is convex if $\alpha z_1 + (1-\alpha)z_2 \in \mathcal{C}$ for any $\alpha \in [0,1]$ and $z_1, z_2 \in \mathcal{C}$. One useful example of a convex cone is the *normal cone*. Consider a closed convex set $\mathcal{S} \subset \mathbb{R}^M$ and pick an arbitrary point $x \in \mathcal{S}$. We define the normal cone at point x, denoted by $\mathcal{N}_S(x)$, by considering all vectors y that satisfy

$$\mathcal{N}_S(x) = \left\{ y \,|\, \text{such that } y^{\mathsf{T}}(s-x) \leq 0, \text{ for all } s \in \mathcal{S} \right\} \quad \textbf{(normal cone)} \tag{8.2}$$

If x happens to lie in the interior of $\mathcal{S}$, then $\mathcal{N}_S(x) = \{0\}$.

The conic hull of a set $\mathcal{S} \subset \mathbb{R}^M$ is the set of all combinations of elements of $\mathcal{S}$ of the form $\alpha_1 z_1 + \alpha_2 z_2 + \ldots + \alpha_m z_m$ for any finite m and any $\alpha_m \geq 0$. These are called conic combinations because the result is a cone. The conic hull is a convex set – see Prob. 8.4.

8.2 CONVEXITY

Let dom(g) denote the domain of $g(z)$, namely, the set of values z where $g(z)$ is well-defined (i.e., finite). The function $g(z)$ is said to be convex if its domain is a convex set and, for any points $z_1, z_2 \in \text{dom}(g)$ and for any scalar $0 \leq \alpha \leq 1$, it holds that

$$g\Big(\alpha z_1 + (1-\alpha)z_2\Big) \leq \alpha g(z_1) + (1-\alpha)g(z_2) \tag{8.3}$$

In other words, all points belonging to the line segment connecting $g(z_1)$ to $g(z_2)$ lie on or above the graph of $g(z)$ – see the plot on the left side in Fig. 8.2. We will be dealing primarily with *proper* convex functions, meaning that the function has a finite value for at least one location z in its domain and, moreover, it is bounded from below, i.e., $g(z) > -\infty$ for all $z \in \text{dom}(g)$. We will also be dealing with *closed* functions. A general function $g(z) : \mathbb{R}^M \to \mathbb{R}$ is said to be closed if for every scalar $c \in \mathbb{R}$, the sublevel set defined by the points $\{z \in \text{dom}(g) \,|\, g(z) \leq c\}$ is a closed set. It is easy to verify that if $g(z)$ is continuous in z and dom(g) is a closed set, then $g(z)$ is a closed function.

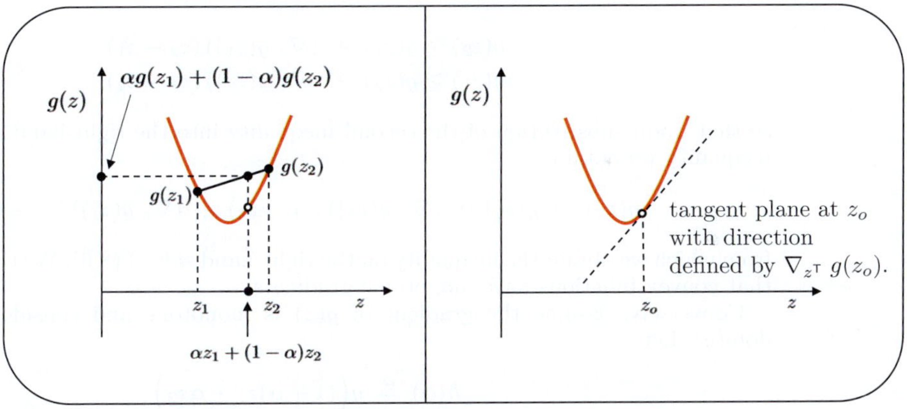

Figure 8.2 Two equivalent characterizations of convexity for *differentiable* functions $g(z)$ as defined by (8.3) and (8.4).

An equivalent characterization of the convexity definition (8.3) under first-order differentiability is that for any $z_o, z \in \text{dom}(g)$:

$$\boxed{g(z) \geq g(z_o) + (\nabla_z\, g(z_o))\,(z - z_o)} \tag{8.4}$$

in terms of the inner product between the gradient vector at z_o and the difference $(z - z_o)$; recall from (2.2) that, by definition, the gradient vector $\nabla_z\, g(z_o)$ is a *row vector*. Condition (8.4) means that the tangent plane at z_o lies beneath the graph of the function – see the plot on the right side of Fig. 8.2. For later reference, we rewrite (8.4) in the alternative form

$$g(z) \geq g(z_o) + \left(\nabla_{z^\mathsf{T}} g(z_o)\right)^\mathsf{T} (z - z_o) \tag{8.5}$$

in terms of the gradient of $g(z)$ relative to z^T at location z_o.

A useful property of every convex function is that, when a minimum exists, it can only be a global minimum; there can be multiple global minima but no local minima. That is, any stationary point at which the gradient vector of $g(z)$ is annihilated will correspond to a global minimum of the function; the function cannot have local maxima, local minima, or saddle points. A second useful property of convex functions, and which follows from characterization (8.4), is that for any $z_1, z_2 \in \text{dom}(g)$:

$$\boxed{g(z) \text{ convex} \iff \left(\nabla_z g(z_2) - \nabla_z g(z_1)\right)(z_2 - z_1) \geq 0} \tag{8.6}$$

in terms of the inner product between two differences: the difference in the gradient vectors and the difference in the vectors themselves. The inequality on the right-hand side in (8.6) is equivalent to saying that the gradient function is *monotone*.

Proof of (8.6): One direction is straightforward. Assume $g(z)$ is convex. Using (8.4) we have

$$g(z_2) \geq g(z_1) + \left(\nabla_z g(z_1)\right)(z_2 - z_1) \tag{8.7}$$

$$g(z_1) \geq g(z_2) + \left(\nabla_z g(z_2)\right)(z_1 - z_2) \tag{8.8}$$

so that upon substitution of the second inequality into the right-hand side of the first inequality we obtain

$$g(z_2) \geq g(z_2) + \left(\nabla_z g(z_2)\right)(z_1 - z_2) + \left(\nabla_z g(z_1)\right)(z_2 - z_1) \tag{8.9}$$

from which we obtain the inequality on the right-hand side of (8.6). We therefore showed that convex functions have monotonic gradients.

Conversely, assume the gradient of $g(z)$ is monotone and consider any $z_1, z_2 \in \text{dom}(g)$. Let

$$h(\alpha) \stackrel{\Delta}{=} g\Big((1-\alpha)z_1 + \alpha z_2\Big) \tag{8.10}$$

for any $0 \leq \alpha \leq 1$. Differentiating $h(\alpha)$ with respect to α gives

$$h'(\alpha) = \Big(\nabla_z g((1-\alpha)z_1 + \alpha z_2)\Big)(z_2 - z_1) \tag{8.11}$$

In particular, it holds that

$$h'(0) = \nabla_z g(z_1)(z_2 - z_1) \tag{8.12}$$

From the assumed monotonicity for the gradient vector we have, for $\alpha \neq 0$,

$$\Big\{\nabla_z g\Big((1-\alpha)z_1 + \alpha z_2\Big) - \nabla_z g(z_1)\Big\}(z_2 - z_1) \geq 0 \tag{8.13}$$

which implies that

$$h'(\alpha) \geq h'(0), \quad \forall\, 0 \leq \alpha \leq 1 \tag{8.14}$$

We know from the fundamental theorem of calculus that

$$h(1) - h(0) = \int_0^1 h'(\alpha)d\alpha \tag{8.15}$$

and, hence,

$$\begin{aligned} g(z_2) &= h(1) \\ &= h(0) + \int_0^1 h'(\alpha)d\alpha \\ &\geq h(0) + h'(0), \quad \text{(in view of (8.14))} \\ &= g(z_1) + \nabla_z\, g(z_1)(z_2 - z_1) \end{aligned} \tag{8.16}$$

so that $g(z)$ is convex from (8.4).

■

Example 8.2 **(Some useful operations that preserve convexity)** It is straightforward to verify from definition (8.3) that the following operations preserve convexity:

(1) If $g(z)$ is convex then $h(z) = g(Az + b)$ is also convex for any constant matrix A and vector b. That is, affine transformations of z do not destroy convexity.

(2) If $g_1(z)$ and $g_2(z)$ are convex functions, then $h(z) = \max\{g_1(z), g_2(z)\}$ is convex. That is, pointwise maximization does not destroy convexity.

(3) If $g_1(z)$ and $g_2(z)$ are convex functions, then $h(z) = a_1 g_1(z) + a_2 g_2(z)$ is also convex for any nonnegative coefficients a_1 and a_2.

(4) If $h(z)$ is convex and nondecreasing, and $g(z)$ is convex, then the composite function $f(z) = h(g(z))$ is convex.

(5) If $h(z)$ is convex and nonincreasing, and $g(z)$ is concave (i.e., $-g(z)$ is convex), then the composite function $f(z) = h(g(z))$ is convex.

8.3 STRICT CONVEXITY

The function $g(z)$ is said to be *strictly* convex if the inequalities in (8.3) or (8.4) are replaced by *strict* inequalities. More specifically, for any $z_1 \neq z_2 \in \text{dom}(g)$ and $0 < \alpha < 1$, a strictly convex function should satisfy:

$$g\Big(\alpha z_1 + (1-\alpha)z_2\Big) \;<\; \alpha g(z_1) \;+\; (1-\alpha)g(z_2) \tag{8.17}$$

A useful property of every strictly convex function is that, when a minimum exists, then it is both *unique* and also the global minimum for the function. A second useful property replaces (8.6) by the following statement with a strict inequality for any $z_1 \neq z_2 \in \text{dom}(g)$:

$$\boxed{g(z) \text{ strictly convex} \iff \left(\nabla_z\, g(z_2) - \nabla_z\, g(z_1)\right)(z_2 - z_1) > 0} \tag{8.18}$$

The inequality on the right-hand side in (8.18) is equivalent to saying that the gradient function is now *strictly monotone*.

8.4 STRONG CONVEXITY

The function $g(z)$ is said to be *strongly* convex (or, more specifically, ν-strongly convex) if it satisfies the following stronger condition for any $0 \leq \alpha \leq 1$:

$$g\Big(\alpha z_1 + (1-\alpha)z_2\Big) \;\leq\; \alpha g(z_1) \;+\; (1-\alpha)g(z_2) \;-\; \frac{\nu}{2}\alpha(1-\alpha)\|z_1 - z_2\|^2 \tag{8.19}$$

for some scalar $\nu > 0$, and where the notation $\|\cdot\|$ denotes the Euclidean norm of its vector argument; other norms can be used – see Prob. 8.60. Comparing (8.19) with (8.17) we conclude that strong convexity implies strict convexity. Therefore, every strongly convex function has a unique global minimum as well. Nevertheless, strong convexity is a stronger requirement than strict convexity so that functions exist that are strictly convex but not necessarily strongly convex. For example, for scalar arguments z, the function $g(z) = z^4$ is strictly convex but not strongly convex. On the other hand, the function $g(z) = z^2$ is strongly convex – see Fig. 8.3. In summary, it holds that:

$$\boxed{\textsf{\textbf{strong convexity}} \implies \textsf{\textbf{strict convexity}} \implies \textsf{\textbf{convexity}}} \tag{8.20}$$

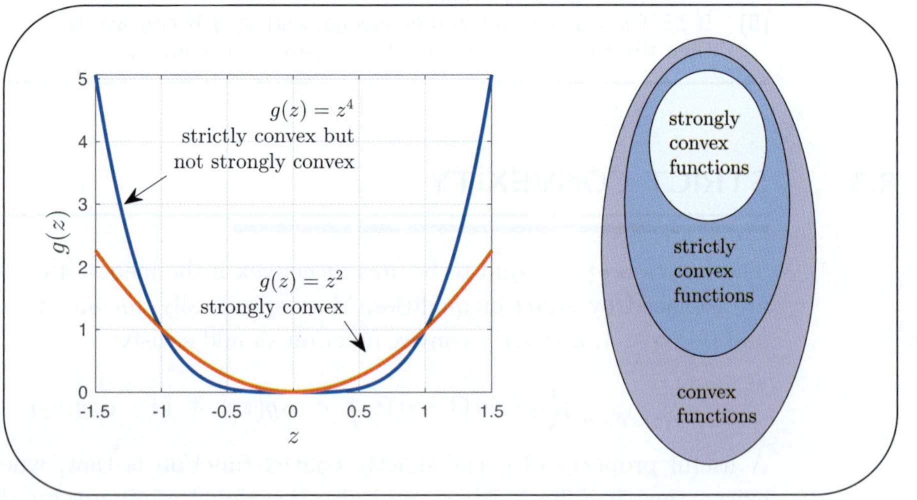

Figure 8.3 The function $g(z) = z^4$ is strictly convex but not strongly convex, while the function $g(z) = z^2$ is strongly convex. Observe how $g(z) = z^4$ is more flat around its global minimizer and moves away from it more slowly than in the quadratic case.

A useful property of strong convexity is that there exists a quadratic lower bound on the function since an equivalent characterization of strong convexity is that for any $z_o, z \in \text{dom}(g)$:

$$\boxed{g(z) \geq g(z_o) + (\nabla_z\, g(z_o))\,(z - z_o) + \frac{\nu}{2}\|z - z_o\|^2} \tag{8.21}$$

This means that the graph of $g(z)$ is strictly above the tangent plane at location z_o and moreover, for any z, the distance between the graph and the corresponding point on the tangent plane is at least as large as the quadratic term $\frac{\nu}{2}\|z-z_o\|^2$. In particular, if we specialize (8.21) to the case in which z_o is selected to correspond to the global minimizer of $g(z)$, i.e.,

$$z_o = z^o, \quad \text{where} \quad \nabla_z\, g(z^o) = 0 \tag{8.22}$$

then we conclude that every strongly convex function satisfies the following useful property for every z:

$$g(z) - g(z^o) \geq \frac{\nu}{2}\|z - z^o\|^2 \qquad (z^o \text{ is global minimizer}) \tag{8.23}$$

This property is illustrated in Fig. 8.4. Another useful property that follows from (8.21) is that for any z_1, z_2:

$$\boxed{g(z) \text{ strongly convex} \Longleftrightarrow (\nabla_z\, g(z_2) - \nabla_z\, g(z_1))\,(z_2 - z_1) \geq \nu\|z_2 - z_1\|^2} \tag{8.24}$$

The inequality on the right-hand side in (8.24) is equivalent to saying that the gradient function is *strongly monotone*. Strong monotonicity is also called *coercivity*. This monotonicity property, along with the earlier conclusions (8.6) and (8.18), are important properties of convex functions. We summarize them in Table 8.1 for ease of reference.

Table 8.1 Useful monotonicity properties implied by the convexity, strict convexity, or strong convexity of a real-valued function $g(z) \in \mathbb{R}$ of a *real* argument $z \in \mathbb{R}^M$.

$g(z)$ **convex**	$\Longleftrightarrow$	$\Big(\nabla_z\, g(z_2) - \nabla_z\, g(z_1)\Big)(z_2 - z_1) \geq 0$
$g(z)$ **strictly convex**	$\Longleftrightarrow$	$\Big(\nabla_z\, g(z_2) - \nabla_z\, g(z_1)\Big)(z_2 - z_1) > 0$
$g(z)$ ν**-strongly convex**	$\Longleftrightarrow$	$\Big(\nabla_z\, g(z_2) - \nabla_z\, g(z_1)\Big)(z_2 - z_1) \geq \nu\|z_2 - z_1\|^2$

Inequality (8.23) provides a bound from below for the difference $g(z) - g(z^o)$, where z^o is the global minimizer. We can establish a second bound from above, which will be useful in the analysis of learning algorithms later in our treatment. Referring to the general property (8.21) for ν-strongly convex functions, we can write for any $z_2, z_1 \in \text{dom}(g)$:

$$g(z_2) \geq g(z_1) + (\nabla_z\, g(z_1))\,(z_2 - z_1) + \frac{\nu}{2}\|z_2 - z_1\|^2 \tag{8.25}$$

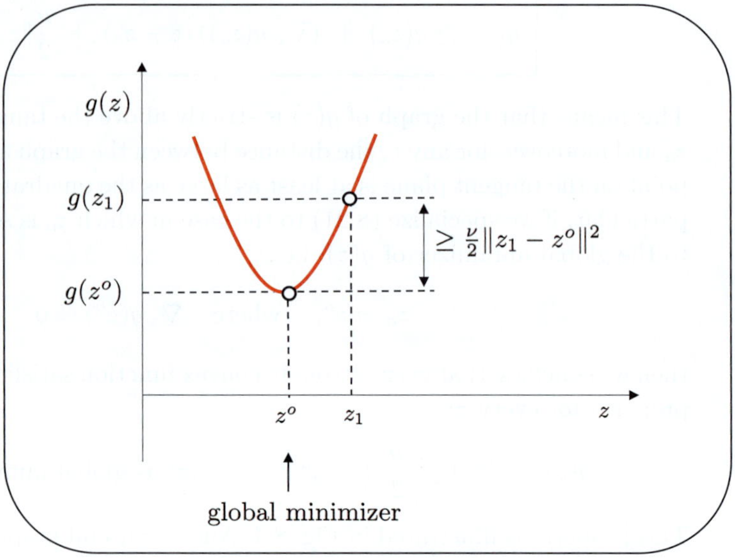

Figure 8.4 For ν-strongly convex functions, the increment $g(z_1) - g(z^o)$ grows at least as fast as the quadratic term $\frac{\nu}{2}\|z_1 - z^o\|^2$, as indicated by (8.23) and where z^o is the global minimizer of $g(z)$.

The right-hand side is quadratic in z_2; its minimum value occurs at

$$\nabla_{z^\mathsf{T}}\, g(z_1) + \nu(z_2 - z_1) = 0 \implies (z_2 - z_1) = -\frac{1}{\nu}\nabla_{z^\mathsf{T}}\, g(z_1) \tag{8.26}$$

Substituting into the right-hand side of (8.25) gives

$$g(z_2) \geq g(z_1) - \frac{1}{2\nu}\|\nabla_z\, g(z_1)\|^2 \tag{8.27}$$

Selecting $z_1 = z$ and $z_2 = z^o$ (the global minimizer) leads to

$$g(z) - g(z^o) \leq \frac{1}{2\nu}\|\nabla_z\, g(z)\|^2 \tag{8.28}$$

Combining with (8.23) we arrive at the following useful lower and upper bounds for ν-strongly convex functions:

$$\boxed{\frac{\nu}{2}\|z - z^o\|^2 \;\leq\; g(z) - g(z^o) \;\leq\; \frac{1}{2\nu}\|\nabla_z\, g(z)\|^2} \tag{8.29}$$

8.5 HESSIAN MATRIX CONDITIONS

When $g(z)$ is twice differentiable, the properties of convexity, strict convexity, and strong convexity can be inferred directly from the inertia of the Hessian matrix of $g(z)$ as follows:

$$\begin{cases} \text{(a)} \;\; \nabla_z^2\, g(z) \geq 0 \text{ for all } z & \Longleftrightarrow g(z) \text{ is convex.} \\ \text{(b)} \;\; \nabla_z^2\, g(z) > 0 \text{ for all } z & \Longrightarrow g(z) \text{ is strictly convex.} \\ \text{(c)} \;\; \nabla_z^2\, g(z) \geq \nu I_M > 0 \text{ for all } z & \Longleftrightarrow g(z) \text{ is } \nu\text{-strongly convex.} \end{cases} \tag{8.30}$$

where, by definition,

$$\nabla_z^2\, g(z) \;\stackrel{\Delta}{=}\; \nabla_{z^\mathsf{T}}\Big(\nabla_z\, g(z)\Big) \tag{8.31}$$

Observe from (8.30) that the positive-definiteness of the Hessian matrix is only a sufficient condition for strict convexity (for example, the function $g(z) = z^4$ is strictly convex even though its second-order derivative is not strictly positive for all z). One of the main advantages of working with strongly convex functions is that their Hessian matrices are sufficiently bounded away from zero.

Example 8.3 **(Strongly convex functions)** The following is a list of useful strongly convex functions that appear in applications involving inference and learning:

(1) Consider the quadratic function

$$g(z) \;=\; \kappa + 2a^\mathsf{T} z + z^\mathsf{T} C z, \quad a, z \in \mathbb{R}^M, \; \kappa \in \mathbb{R} \tag{8.32}$$

with a symmetric positive-definite matrix C. The Hessian matrix is $\nabla_z^2\, g(z) \;=\; 2C$, which is sufficiently bounded away from zero for all z since

$$\nabla_z^2\, g(z) \;\geq\; 2\lambda_{\min}(C)\, I_M > 0 \tag{8.33}$$

in terms of the smallest eigenvalue of C. Therefore, such quadratic functions are strongly convex.

The top row in Fig. 8.5 shows a surface plot for the quadratic function (8.32) for $z \in \mathbb{R}^2$ along with its contour lines for the following (randomly generated) parameter values:

$$C = \begin{bmatrix} 3.3784 & 0 \\ 0 & 3.4963 \end{bmatrix}, \quad a = \begin{bmatrix} 0.4505 \\ 0.0838 \end{bmatrix}, \quad \kappa = 0.5 \tag{8.34}$$

The minimum of the corresponding $g(z)$ occurs at location:

$$z^o \approx \begin{bmatrix} 0.1334 \\ 0.0240 \end{bmatrix}, \quad \text{with} \;\; g(z^o) \approx 0.4379 \tag{8.35}$$

The individual entries of z are denoted by $z = \text{col}\{z_1, z_2\}$. Recall that a contour line of a function $g(z)$ is a curve along which the value of the function remains invariant. In this quadratic case, the location of the minimizer z^o can be determined in closed form and is given by $z^o = C^{-1}a$. In the plot, the surface curve is determined by evaluating $g(z)$ on a dense grid with values of (z_1, z_2) varying in the range $[-2, 2]$ in small steps of size 0.01. The location of z^o is approximated by determining the grid location where the surface attains its smallest value. This approximate numerical evaluation is applied to the other two examples below involving logistic and hinge functions where closed-form expressions for z^o are not readily available. In later chapters, we are going to introduce recursive algorithms, of the gradient-descent type, and also of the subgradient and proximal gradient type, which will allow us to seek the minimizers of strongly convex functions in a more systematic manner.

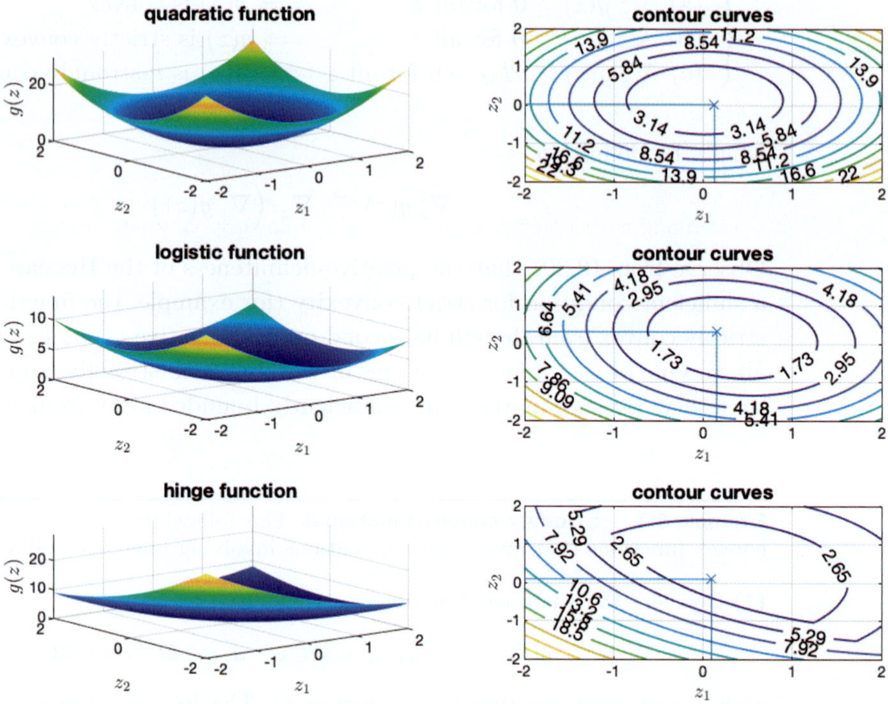

Figure 8.5 Examples of three strongly convex functions $g(z) : \mathbb{R}^2 \to \mathbb{R}$ with their contour lines. (*Top*) Quadratic function, (*Middle*) regularized logistic function; (*Bottom*) regularized hinge function. The locations of the minimizers are indicated by the $\times$ notation with horizontal and vertical lines emanating from them in the plots on the right.

(2) Consider next the regularized logistic (or log-)loss function:

$$g(z) = \ln\left(1 + e^{-\gamma h^{\mathsf{T}} z}\right) + \frac{\rho}{2}\|z\|^2, \quad z \in \mathbb{R}^M \tag{8.36}$$

with a scalar γ, column vector h, and $\rho > 0$. This function is also strongly convex, as can be seen from examining its Hessian matrix:

$$\nabla_z^2\, g(z) = \rho I_M + \left(\frac{e^{-\gamma h^{\mathsf{T}} z}}{(1 + e^{-\gamma h^{\mathsf{T}} z})^2}\right) hh^{\mathsf{T}} \geq \rho I_M > 0 \tag{8.37}$$

The middle row in Fig. 8.5 shows a surface plot for the logistic function (8.36) for $z \in \mathbb{R}^2$ along with its contour lines for the following parameter values:

$$\gamma = 1, \quad h = \begin{bmatrix} 1 \\ 2 \end{bmatrix}, \quad \rho = 2 \tag{8.38}$$

The minimum of the corresponding $g(z)$ occurs roughly at location:

$$z^o \approx \begin{bmatrix} 0.1568 \\ 0.3135 \end{bmatrix}, \quad \text{with } g(z^o) \approx 0.4990 \tag{8.39}$$

(3) Now consider the regularized hinge loss function:

$$g(z) = \max\left\{0, 1 - \gamma h^{\mathsf{T}} z\right\} + \frac{\rho}{2}\|z\|^2 \tag{8.40}$$

with a scalar γ, column vector h, and $\rho > 0$ is also strongly convex, although nondifferentiable. This result can be verified by noting that the function $\max\left\{0, 1 - \gamma h^{\mathsf{T}} z\right\}$ is convex in z while the regularization term $\frac{\rho}{2}\|z\|^2$ is ρ-strongly convex in z – see Prob. 8.23. The bottom row in Fig. 8.5 shows a surface plot for the hinge function (8.40) for $z \in \mathbb{R}^2$ along with its contour lines for the following parameter values:

$$\gamma = 1, \quad h = \begin{bmatrix} 5 \\ 5 \end{bmatrix}, \quad \rho = 2 \tag{8.41}$$

The minimum of the corresponding $g(z)$ occurs roughly at location:

$$z^o \approx \begin{bmatrix} 0.1000 \\ 0.1000 \end{bmatrix}, \quad \text{with} \quad g(z^o) \approx 0.0204 \tag{8.42}$$

Figure 8.6 shows an enlarged surface plot for the same regularized hinge function from a different view angle, where it is possible to visualize the locations of nondifferentiability in $g(z)$; these consist of all points z where $1 = \gamma h^{\mathsf{T}} z$ or, more explicitly, $z_1 + z_2 = 1/5$ by using the assumed numerical values for γ and h.

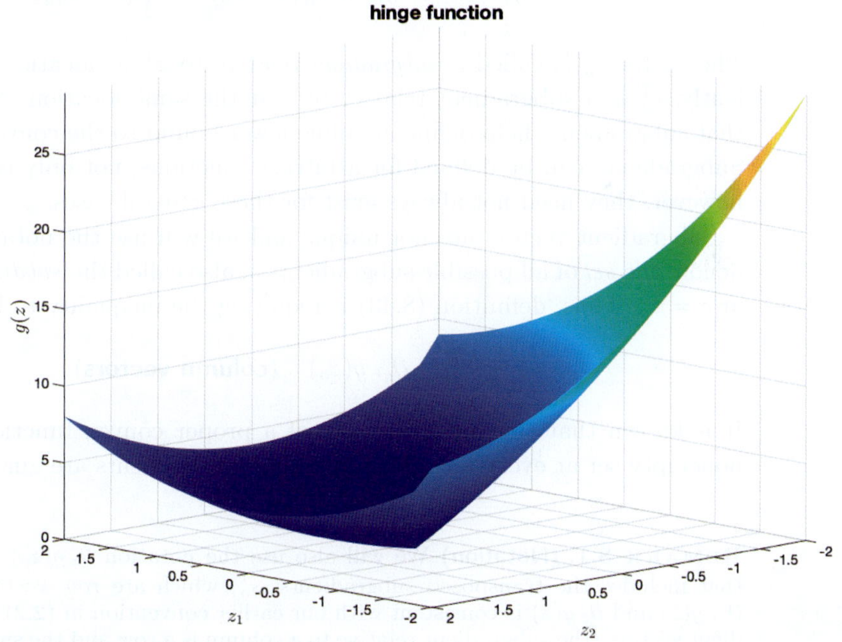

Figure 8.6 Surface plot for the same regularized hinge function from Fig. 8.5, albeit from a different viewpoint. The points of nondifferentiability occur at the locations satisfying $z_1 + z_2 = 1/5$.

8.6 SUBGRADIENT VECTORS

The characterization of convexity in (8.4) is stated in terms of the gradient vector for $g(z)$. This gradient exists because we have assumed so far that the function $g(z)$ is differentiable. There are, however, many situations of interest where the function $g(z)$ need not be differentiable at all points. For example, for scalar arguments z, the function $g(z) = |z|$ is convex but is not differentiable at $z = 0$. For such nondifferentiable convex functions, the characterizations (8.4) or (8.5) will need to be adjusted and replaced by the statement that the function $g(z)$ is convex if, and only if, for every z_o, a *column* vector s_o (dependent on z_o) exists such that

$$g(z) \geq g(z_o) + s_o^{\mathsf{T}}(z - z_o), \quad \text{for all } z_o, z \in \text{dom}(g) \tag{8.43}$$

Expression (8.43) is in terms of the inner product between s_o and the difference $(z - z_o)$. Similarly, the characterization of strong convexity in (8.21) is replaced by

$$g(z) \geq g(z_o) + s_o^{\mathsf{T}}(z - z_o) + \frac{\nu}{2}\|z - z_o\|^2 \tag{8.44}$$

The vector s_o is called a *subgradient* relative to z^{T} at location $z = z_o$; equivalently, s_o^{T} is a subgradient relative to z at the same location. Note from (8.43) that subgradients help define an affine lower bound to the convex function $g(z)$. Subgradients can be defined for arbitrary functions, not only convex functions; however, they need not always exist for these general cases.

Subgradient vectors are not unique and we will use the notation $\partial_{z^{\mathsf{T}}}\, g(z_o)$ to denote the *set* of all possible subgradients s, also called the *subdifferential* of $g(z)$, at $z = z_o$. Thus, definition (8.43) is requiring the inequality to hold for some

$$s_o \in \partial_{z^{\mathsf{T}}}\, g(z_o) \quad \textbf{(column vectors)} \tag{8.45}$$

It is known that the subdifferential of a proper convex function is a bounded nonempty set at every location z_o, so that subgradients are guaranteed to exist.

Remark 8.1. (Notation) We will also use the notation $\partial_z\, g(z_o)$ to denote the set that includes the transposed subgradients, s_o^{T}, which are *row* vectors. The notation $\partial_{z^{\mathsf{T}}}\, g(z)$ and $\partial_z\, g(z)$ is consistent with our earlier convention in (2.2) and (2.5) for gradient vectors (the subgradient relative to a column is a row and the subgradient relative to a row is a column). Sometimes, for compactness, we may simply write $\partial\, g(z)$ to refer to the subdifferential $\partial_{z^{\mathsf{T}}}\, g(z)$, where every element in $s \in \partial\, g(z)$ is a column vector.

■

The concept of subgradients is illustrated in Fig. 8.7. When $g(z)$ is differentiable at z_o, then there exists a unique subgradient at that location and it coincides with $\nabla_{z^{\mathsf{T}}}\, g(z_o)$. In this case, statement (8.43) reduces to (8.4) or (8.5).

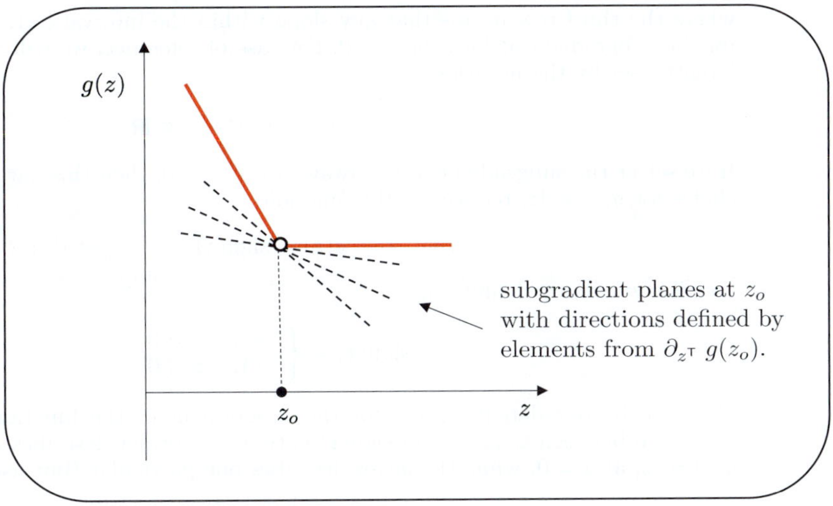

Figure 8.7 A nondifferentiable convex function admits a multitude of subgradient directions at every point of nondifferentiability.

One useful property that follows from (8.43) is that for any $z_1, z_2 \in \text{dom}(g)$:

$$\boxed{g(z) \text{ convex} \implies (s_2 - s_1)^\mathsf{T}(z_2 - z_1) \;\geq\; 0} \tag{8.46}$$

where $\{s_1, s_2\}$ correspond to subgradient vectors relative to z^T at locations $\{z_1, z_2\}$, respectively. A second useful property of subgradient vectors is the following condition for the global minimum of a convex function (see Prob. 8.41):

$$\begin{cases} g(z) \text{ differentiable at } z^o\text{:} & z^o \text{ is a minimum} \iff 0 = \nabla_z\, g(z^o) \\ g(z) \text{ nondifferentiable at } z^o\text{:} & z^o \text{ is a minimum} \iff 0 \in \partial_z\, g(z^o) \end{cases} \tag{8.47}$$

The second condition states that the set of subgradients at z^o must include the zero vector. This condition reduces to the first statement when $g(z)$ is differentiable at z^o.

Example 8.4 **(Absolute value function)** Let $z \in \mathbb{R}$ and consider the function

$$\boxed{g(z) = |z|} \tag{8.48}$$

This function is differentiable everywhere except at $z = 0$ – see Fig. 8.8 (*left*). The slope of the function is $+1$ over $z > 0$ and -1 over $z < 0$. At $z = 0$, any line passing through the origin with slope in the range $[-1, 1]$ can serve as a valid subgradient direction. Therefore, we find that

$$\partial_z\, g(z) = \begin{cases} +1, & z > 0 \\ -1, & z < 0 \\ [-1, +1], & z = 0 \end{cases} \tag{8.49}$$

where the third row means that any slope within the interval $[-1, 1]$ is a valid choice for the subgradient at location $z = 0$. For ease of reference, we will denote this subdifferential set by the notation:

$$\mathbb{G}_{\text{abs}}(z) \stackrel{\Delta}{=} \partial_z |z|, \quad z \in \mathbb{R} \tag{8.50a}$$

If we select the subgradient to be always $+1$ at $z = 0$, then this particular subgradient choice for $g(z) = |z|$ reduces to the function:

$$s(z) = \text{sign}(z) \tag{8.50b}$$

for all z where, by definition,

$$\text{sign}(z) = \begin{cases} +1, & z \geq 0 \\ -1, & z < 0 \end{cases} \tag{8.50c}$$

This will be our default choice for the subgradient of the function $g(z) = |z|$. The difference between $\mathbb{G}_{\text{abs}}(z)$ and $\text{sign}(z)$ is that the former describes *all* subgradients of $g(z) = |z|$ at $z = 0$, while the latter describes one particular (but useful) choice.

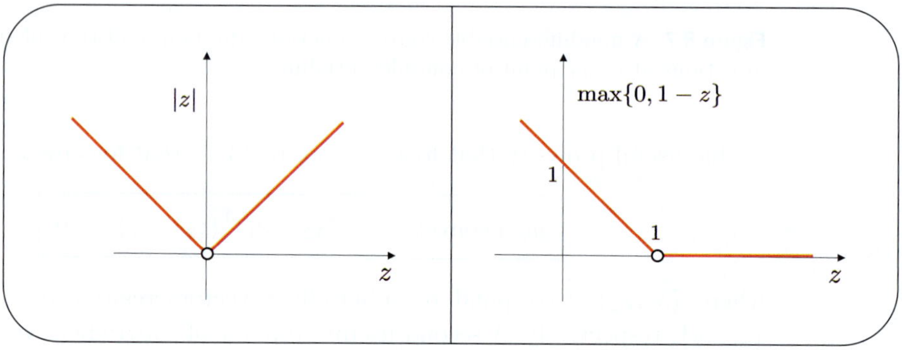

Figure 8.8 (*Left*) Absolute value function, $g(z) = |z|$. (*Right*) Hinge function, $g(z) = \max\{0, 1 - z\}$.

Consider next the case in which z is an M-dimensional vector and

$$\boxed{g(z) = \|z\|_1 = \sum_{m=1}^{M} |z_m|} \tag{8.51}$$

where the $\{z_m\}$ denote the individual entries of z. The function is not differentiable at all locations $z \in \mathbb{R}^M$ with at least one zero entry z_m. It follows that the subdifferential of $g(z)$, which consists of vectors of size $M \times 1$, can be constructed as follows:

$$\mathbb{G}(z) \stackrel{\Delta}{=} \begin{bmatrix} \mathbb{G}_{\text{abs}}(z_1) \\ \mathbb{G}_{\text{abs}}(z_2) \\ \vdots \\ \mathbb{G}_{\text{abs}}(z_M) \end{bmatrix}, \quad (M \times 1) \tag{8.52a}$$

where each $\mathbb{G}_{\text{abs}}(z_m)$ is given by (8.49). One particular subgradient for $g(z)$ relative to z^{T} is then

$$s(z) = \text{sign}(z) \tag{8.52b}$$

where the sign function now returns an $M \times 1$ vector consisting of ± 1 entries corresponding to the signs of the individual entries of z.

Example 8.5 **(Hinge function)** Consider the hinge function

$$\boxed{g(z) = \max\{0, 1 - z\}} \tag{8.53}$$

shown in Fig. 8.8 (*right*). This function is differentiable everywhere except at $z = 1$. The slope of the function is -1 over $z < 1$ and 0 over $z > 1$. At $z = 1$, any line passing through this point with slope in the range $[-1, 0]$ can serve as a valid subgradient direction. Therefore, we find that

$$\partial_z\, g(z) \;=\; \left\{ \begin{array}{cl} 0, & z > 1 \\ -1, & z < 1 \\ [-1, 0], & z = 1 \end{array} \right. \tag{8.54}$$

For ease of reference, we denote this subdifferential set by the notation:

$$\mathbb{G}_1(z) \;\stackrel{\Delta}{=}\; \partial_z \max\{0, 1 - z\}, \quad z \in \mathrm{I\!R} \tag{8.55}$$

If we select the subgradient to be always -1 at $z = 1$, then this particular subgradient choice reduces to the (negative of the) indicator function:

$$s(z) = -\mathbb{I}[z \leq 1] \tag{8.56}$$

for all z where, by definition,

$$\mathbb{I}[a] = \left\{ \begin{array}{ll} 1, & \text{if statement } a \text{ is true} \\ 0, & \text{otherwise} \end{array} \right. \tag{8.57}$$

This will be our default choice for the subgradient of the function $g(z) = \max\{0, 1-z\}$. The difference between $\mathbb{G}_1(z)$ and $-\mathbb{I}[z \leq 1]$ is that the former describes *all* subgradients of $g(z) = \max\{0, 1 - z\}$ at location $z = 1$, while the latter describes one particular (but useful) choice.

Consider next a slight adjustment where the argument z is scaled by a nonzero constant β, say,

$$\boxed{g(z) = \max\{0, 1 - \beta z\}, \quad z \in \mathrm{I\!R}, \quad \beta \neq 0} \tag{8.58}$$

Using similar arguments, it can be verified that the subdifferential for $g(z)$ relative to z is now given by

$$\mathbb{G}_\beta(z) \;=\; \left\{ \begin{array}{cl} 0, & \beta z > 1 \\ -\beta, & \beta z < 1 \\ [-\beta, 0], & \beta z = 1, \;\; \beta > 0 \\ [0, -\beta], & \beta z = 1, \;\; \beta < 0 \end{array} \right. \tag{8.59a}$$

where we added β as a subscript in $\mathbb{G}_\beta(z)$. Moreover, one particular choice for the subgradient is

$$s(z) = -\beta\, \mathbb{I}[\beta z \leq 1] \tag{8.59b}$$

In the degenerate case when $\beta = 0$ in (8.58), we get $g(z) = 1$. Its derivative is zero everywhere so that the above expression for $s(z)$ continues to hold in this situation.

Consider a third example involving the hinge function where z is now an M-dimensional vector:

$$\boxed{g(z) = \max\Big\{0, 1 - h^{\mathsf{T}} z\Big\}} \tag{8.60}$$

for some given $h \in \mathbb{R}^M$. Let $\{h_m, z_m\}$ denote the individual entries of $\{h, z\}$. The function is not differentiable at all locations $z \in \mathbb{R}^M$ where $h^\mathsf{T} z = 1$. It follows that the subdifferential of $g(z)$ consists of vectors of size $M \times 1$ with entries constructed as follows:

$$\mathbb{G}(z) \triangleq \begin{bmatrix} \mathbb{A}_{h_1}(z_1) \\ \mathbb{A}_{h_2}(z_2) \\ \vdots \\ \mathbb{A}_{h_M}(z_M) \end{bmatrix}, \quad \mathbb{A}_{h_m}(z_m) \triangleq \begin{cases} 0, & h^\mathsf{T} z > 1 \\ -h_m, & h^\mathsf{T} z < 1 \\ [-h_m, 0], & h^\mathsf{T} z = 1, \;\; h_m \geq 0 \\ [0, -h_m], & h^\mathsf{T} z = 1, \;\; h_m < 0 \end{cases} \tag{8.61a}$$

where each $\mathbb{A}_{h_m}(z_m)$ is defined as above using h_m and $h^\mathsf{T} z$. One particular subgradient for $g(z)$ relative to z^T is

$$s(z) = -h\,\mathbb{I}[h^\mathsf{T} z \leq 1] \tag{8.61b}$$

Subgradients and subdifferentials will arise frequently in our study of inference methods and optimization problems. They possess several useful properties, some of which are collected in Table 8.2 for ease of reference. These properties are established in the problems at the end of the chapter; the last column in the table provides the relevant reference.

Table 8.2 Some useful properties of subgradients and subdifferentials for convex functions $g(z)$.

	Property	Prob.
1.	$\partial_z\, \alpha g(z) = \alpha\, \partial_z\, g(z), \;\; \alpha \geq 0$	8.29
2.	$\partial_{z^\mathsf{T}}\, g(Az+b) = A^\mathsf{T}\, \partial_{z^\mathsf{T}}\, g(z)\Big\vert_{z \leftarrow Az+b}$	8.30
3.	$\partial_{z^\mathsf{T}}\, g_1(z) + \partial_{z^\mathsf{T}}\, g_2(z) \subset \partial_{z^\mathsf{T}}\Big(g_1(z) + g_2(z)\Big)$	8.31
4.	$\partial_{z^\mathsf{T}}\, \Vert z\Vert_2 = \begin{cases} z/\Vert z\Vert_2, & z \neq 0 \\ \{a \mid \Vert a\Vert_2 \leq 1\}, & z = 0 \end{cases}$	8.32
5.	$\partial_{z^\mathsf{T}}\, \Vert z\Vert_q = \underset{\Vert y\Vert_p \leq 1}{\text{argmax}} \Big\{z^\mathsf{T} y\Big\}, \;\; p, q \geq 1, \; 1/p + 1/q = 1$	8.33
6.	$\partial_{z^\mathsf{T}}\, \mathbb{I}_{C,\infty}[z] = \mathcal{N}_C(z)$ (normal cone to convex set $\mathcal{C}$ at z)	8.34

For example, the first row in the table states that the subdifferential of the scaled function $\alpha g(z)$ consists of all elements in the subdifferential of $g(z)$ scaled by α. The second row in the table shows what happens to the subdifferential set when the argument of $g(z)$ is replaced by the affine transformation $Az + b$. The result shows that the subdifferential of $g(z)$ should be evaluated at the transformations $Az + b$ and subsequently scaled by A^T. The third row shows that the subdifferential of the sum of two functions is *not* equal to the sum of the individual subdifferentials; it is a larger set. We will use this result in the following manner. Assume we wish to seek a minimizer z^o for the sum of two convex functions, $g_1(z) + g_2(z)$. We know that the zero vector must satisfy

$$0 \in \partial_{z^\mathsf{T}} \Big(g_1(z) + g_2(z)\Big)\Bigg\vert_{z=z^o} \tag{8.62}$$

That is, the zero vector must belong to the subdifferential of the sum evaluated at $z = z^o$. We will seek instead a vector z^o that ensures

$$0 \in \left\{ \partial_{z^\mathsf{T}}\, g_1(z)\Big|_{z=z^o} + \partial_{z^\mathsf{T}}\, g_2(z)\Big|_{z=z^o} \right\} \tag{8.63}$$

If this step is successful then the zero vector will satisfy (8.62) by virtue of the property in the third row of the table. The next example provides more details on the subdifferential of sums of convex functions.

Example 8.6 **(Subdifferential of sums of convex functions)** We will encounter in later chapters functions that are expressed in the form of empirical averages of convex components such as

$$g(z) \triangleq \frac{1}{L}\sum_{\ell=1}^{L} g_\ell(z), \quad z \in \mathbb{R}^M \tag{8.64}$$

where each $g_\ell(z)$ is convex. The subdifferential set for $g(z)$ will be characterized *fully* by the relation:

$$\partial_{z^\mathsf{T}}\, g(z) = \left\{ \frac{1}{L}\sum_{\ell=1}^{L} \partial_{z^\mathsf{T}}\, g_\ell(z) \right\} \tag{8.65}$$

under some conditions:

(a) First, from the third row in the table we know that whenever we combine subdifferentials of individual convex functions we obtain a subgradient for $g(z)$ so that

$$\left\{ \frac{1}{L}\sum_{\ell=1}^{L} \partial_{z^\mathsf{T}}\, g_\ell(z) \right\} \subset \partial_{z^\mathsf{T}}\, g(z) \tag{8.66}$$

(b) The converse statement is more subtle, meaning that we should be able to express every subgradient for $g(z)$ in the same sample average form and ensure

$$\partial_{z^\mathsf{T}}\, g(z) \subset \left\{ \frac{1}{L}\sum_{\ell=1}^{L} \partial_{z^\mathsf{T}}\, g_\ell(z) \right\} \tag{8.67}$$

We provide a counterexample in Prob. 8.31 to show that this direction is not always true, as already anticipated by the third row of Table 8.2. However, we explain in the comments at the end of the chapter that equality of both sets is possible under condition (8.113). The condition requires the domains of the individual functions $\{g_\ell(z)\}$ to have a nonempty intersection. This situation will be satisfied in most cases of interest since the individual functions will have the same form over z. For this reason, we will regularly assume that expression (8.65) describes all subgradients for $g(z)$. At the same time we remark that in most applications, we will not need to characterize the full subdifferential for $g(z)$; it will be sufficient to find one particular subgradient for it and this subgradient can be obtained by adding individual subgradients for $\{g_\ell(z)\}$.

Example 8.7 **(Sum of hinge functions)** Consider the convex function

$$\boxed{g(z) = \frac{1}{L}\sum_{\ell=1}^{L} \max\{0, 1 - \beta_\ell z\}, \quad z \in \mathbb{R}, \quad \beta_\ell \neq 0} \tag{8.68}$$

which involves a sum of individual hinge functions, $g_\ell(z) = \max\{0, 1 - \beta_\ell z\}$. We know from (8.59a) how to characterize the subdifferential of each of these terms:

$$\mathbb{G}_{\beta_\ell}(z) \;=\; \begin{cases} 0, & \beta_\ell z > 1 \\ -\beta_\ell, & \beta_\ell z < 1 \\ [-\beta_\ell, 0], & \beta_\ell z = 1, \;\; \beta_\ell > 0 \\ [0, -\beta_\ell], & \beta_\ell z = 1, \;\; \beta_\ell < 0 \end{cases} \tag{8.69}$$

Moreover, one subgradient for each term can be chosen as $s_\ell(z) = -\beta_\ell\, \mathbb{I}[\beta_\ell z \leq 1]$. Using the conclusions from parts (a) and (b) in Example 8.6, we find that the subdifferential for $g(z)$ is given by

$$\partial_z g(z) \;=\; \frac{1}{L}\sum_{\ell=1}^{L} \mathbb{G}_{\beta_\ell}(z) \tag{8.70a}$$

while a subgradient for it can be chosen as

$$s(z) \;=\; -\frac{1}{L}\sum_{\ell=1}^{L} \beta_\ell\, \mathbb{I}_\ell[\beta_\ell z \leq 1] \tag{8.70b}$$

Consider next a situation in which z is M-dimensional:

$$\boxed{g(z) = \frac{1}{L}\sum_{\ell=1}^{L} \max\left\{0, 1 - h_\ell^{\mathsf{T}} z\right\}, \quad z, h_\ell \in \mathbb{R}^M} \tag{8.71}$$

which again involves a sum of individual hinge functions, $g_\ell(z) = \max\{0, 1 - h_\ell^{\mathsf{T}} z\}$. We know from (8.61a) how to characterize the subdifferential for each of these terms:

$$\mathbb{G}_\ell(z) \;=\; \begin{bmatrix} \mathbb{A}_{h_{\ell,1}}(z_1) \\ \mathbb{A}_{h_{\ell,2}}(z_2) \\ \vdots \\ \mathbb{A}_{h_{\ell,M}}(z_M) \end{bmatrix} \tag{8.72a}$$

in terms of the individual entries $\{h_{\ell,m}\}$ of h_ℓ, and where each $\mathbb{A}_{h_{\ell,m}}(z_m)$ is defined according to (8.61a). The subdifferential for $g(z)$ is then given by

$$\partial_{z^{\mathsf{T}}}\, g(z) = \frac{1}{L}\sum_{\ell=1}^{L} \mathbb{G}_\ell(z) \tag{8.72b}$$

One particular subgradient for $g(z)$ relative to z^{T} is then

$$s(z) = -\frac{1}{L}\sum_{\ell=1}^{L} h_\ell\, \mathbb{I}[h_\ell^{\mathsf{T}} z \leq 1] \tag{8.72c}$$

The last three rows in Table 8.2 provide some useful subdifferential expressions for the ℓ_2-norm, ℓ_q-norm, and for the indicator function of a convex set. In particular, recall from the discussion on dual norms in Section 1.10 that the maximum of $z^{\mathsf{T}} y$ over the ball $\|y\|_p \leq 1$ is equal to the ℓ_q-norm, $\|z\|_q$. The result in the table is therefore stating that the subdifferential of $\|z\|_q$ consists of the vectors y within the ball $\|y\|_p \leq 1$ that attain this maximum value (i.e., that attain the dual norm). The last row in the table deals with the subdifferential of the indicator function of a convex set, denoted by $\mathbb{I}_{C,\infty}[z]$. Given a set $\mathcal{C}$, this function indicates whether a point z lies in $\mathcal{C}$ or not as follows:

$$\mathbb{I}_{C,\infty}[z] \triangleq \begin{cases} 0, & \text{if } z \in \mathcal{C} \\ \infty, & \text{otherwise} \end{cases} \tag{8.73}$$

The result in the table describes the subdifferential of the indicator function in terms of the normal cone at location z; this conclusion is illustrated geometrically in Fig. 8.9, where the normal cone is shown at one of the corner points.

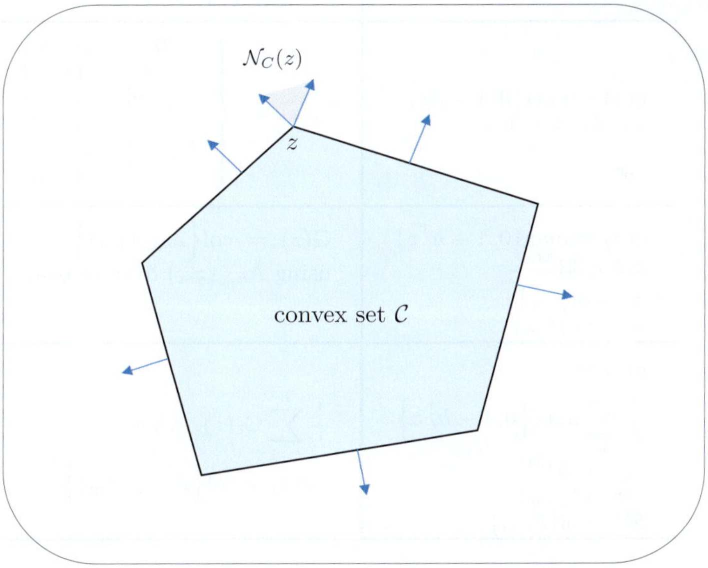

Figure 8.9 Geometric illustration of the subdifferential for the indicator function of a convex set at location z.

We collect, for ease of reference, in Table 8.3 some useful subdifferential and subgradient expressions derived in the earlier examples for a couple of convex functions that will arise in our study of learning problems.

8.7 JENSEN INEQUALITY

There are several variations and generalizations of the Jensen inequality, which is a useful result associated with convex functions. One form is the following. Let $\{z_k \in \mathbb{R}^M, k = 1, 2, \ldots, N\}$ denote a collection of N column vectors that lie in the domain of a real-valued convex function $g(z)$. Let $\{\alpha_k\}$ denote a collection

Table 8.3 Some useful subdifferentials and subgradients for convex functions $g(z)$.

Function, $g(z)$	**Subdifferential**, $\partial_{z^\mathsf{T}} g(z)$	**Subgradient**, $s(z)$
$g(z) = \|z\|,\ z \in \mathbb{R}$	$\mathbb{G}_{\text{abs}}(z) = \begin{cases} +1, & z > 0 \\ -1, & z < 0 \\ [-1, +1], & z = 0 \end{cases}$	$\text{sign}(z)$
$g(z) = \\|z\\|_1,\ z \in \mathbb{R}^M$ $z = \text{col}\{z_m\}$	$\mathbb{G}(z) = \text{col}\Big\{\mathbb{G}_{\text{abs}}(z_m)\Big\}$	$\text{sign}(z)$
$g(z) = \max\{0, 1 - z\}$ $z \in \mathbb{R}$	$\mathbb{G}_1(z) = \begin{cases} 0, & z > 1 \\ -1, & z < 1 \\ [-1, 0], & z = 1 \end{cases}$	$-\mathbb{I}[z \leq 1]$
$g(z) = \max\{0, 1 - \beta z\}$ $z \in \mathbb{R}, \beta \neq 0$	$\mathbb{G}_\beta(z) = \begin{cases} 0, & \beta z > 1 \\ -\beta, & \beta z < 1 \\ [-\beta, 0], & \beta z = 1 \\ & \beta > 0 \\ [0, -\beta], & \beta z = 1 \\ & \beta < 0 \end{cases}$	$-\beta\, \mathbb{I}[\beta z \leq 1]$
$g(z) = \max\{0, 1 - h^\mathsf{T} z\}$ $z, h \in \mathbb{R}^M$ $z = \text{col}\{z_m\}$ $h = \text{col}\{h_m\}$	$\mathbb{G}(z) = \text{col}\Big\{\mathbb{A}_{h_m}(z_m)\Big\}$ using $\mathbb{A}_{h_m}(z_m)$ from (8.61a)	$-h\, \mathbb{I}[h^\mathsf{T} z \leq 1]$
$g(z) =$ $\frac{1}{L}\sum_{\ell=1}^{L} \max\Big\{0, 1 - h_\ell^\mathsf{T} z\Big\}$ $z, h_\ell \in \mathbb{R}^M$ $z = \text{col}\{z_m\}$ $h_\ell = \text{col}\{h_{\ell,m}\}$	$\frac{1}{L}\sum_{\ell=1}^{L} \mathbb{G}_\ell(z)$, where $\mathbb{G}_\ell(z) = \text{col}\Big\{\mathbb{A}_{h_{\ell,m}}(z_m)\Big\}$	$-\frac{1}{L}\sum_{\ell=1}^{L} h_\ell\, \mathbb{I}[h_\ell^\mathsf{T} z \leq 1]$

of nonnegative real coefficients that add up to 1:

$$\sum_{k=1}^{N} \alpha_k = 1, \qquad 0 \leq \alpha_k \leq 1 \tag{8.74}$$

The Jensen inequality states that

$$\boxed{g\left(\sum_{k=1}^{N} \alpha_k z_k\right) \leq \sum_{k=1}^{N} \alpha_k g(z_k)} \tag{8.75}$$

and equality holds if, and only if, $z_1 = z_2 = \ldots = z_N$. For example, if we select $g(z) = \|z\|^2$ in terms of the squared Euclidean norm of z, then it follows from (8.75) that

$$\left\|\sum_{k=1}^{N} \alpha_k z_k\right\|^2 \leq \sum_{k=1}^{N} \alpha_k \|z_k\|^2 \tag{8.76}$$

There is also a stochastic version of Jensen inequality. If $\boldsymbol{a} \in \mathbb{R}^M$ is a real-valued random variable, then it holds that

$$g(\mathbb{E}\,\boldsymbol{a}) \leq \mathbb{E}\,(g(\boldsymbol{a})) \quad \text{(when } g(z) \in \mathbb{R} \text{ is convex)} \tag{8.77}$$

$$g(\mathbb{E}\,\boldsymbol{a}) \geq \mathbb{E}\,(g(\boldsymbol{a})) \quad \text{(when } g(z) \in \mathbb{R} \text{ is concave)} \tag{8.78}$$

where it is assumed that $\boldsymbol{a}$ and $g(\boldsymbol{a})$ have bounded expectations. We remark that a function $g(z)$ is said to be concave if, and only if, $-g(z)$ is convex.

Example 8.8 **(Vector norm)** For any vectors $a, b, c \in \mathbb{R}^M$, we know from the triangle inequality of norms that

$$\|a+b+c\| \;\leq\; \|a\| + \|b\| + \|c\| \tag{8.79}$$

Using the Jensen inequality (8.75), we can determine an upper bound for the quantity $\|a+b+c\|^4$. For this purpose, we consider the convex function $g(z) = \|z\|^4$ and note that

$$\begin{aligned}
\|a+b+c\|^4 &= \left\| 3\left(\frac{1}{3}a + \frac{1}{3}b + \frac{1}{3}c\right)\right\|^4 \\
&= 81\left\|\frac{1}{3}a + \frac{1}{3}b + \frac{1}{3}c\right\|^4 \\
&\overset{(8.75)}{\leq} 81\left(\frac{1}{3}\|a\|^4 + \frac{1}{3}\|b\|^4 + \frac{1}{3}\|c\|^4\right) \\
&= 27\left(\|a\|^4 + \|b\|^4 + \|c\|^4\right)
\end{aligned} \tag{8.80}$$

Example 8.9 **(Value at averaged arguments)** Consider a convex function $g(z)$ with vector argument $z \in \mathbb{R}^M$, and assume we are able to establish that its average value at a collection of points $\{z_n\}$ is upper-bounded by some value β:

$$\frac{1}{N}\sum_{n=1}^{N} g(z_n) \;\leq \beta \tag{8.81}$$

From Jensen inequality (8.75), it follows that

$$g\Big(\frac{1}{N}\sum_{n=1}^{N} z_n\Big) \;\leq\; \frac{1}{N}\sum_{n=1}^{N} g(z_n) \;\leq\; \beta \tag{8.82}$$

so that the value of the function at the averaged arguments is also bounded by β.

8.8 CONJUGATE FUNCTIONS

Conjugate functions play an important role in the solution of optimization problems. In this section, we define them, list several of their properties, and provide some intuition for their role in convex analysis.

Consider a convex function $h(w)$ defined over M-dimensional vectors w. We denote its *conjugate function* (also called the Fenchel conjugate) by the notation $h^\star(x)$ and define it as follows:

$$h^\star(x) \triangleq \sup_{w} \left\{ x^\mathsf{T} w - h(w) \right\}, \quad x \in \mathcal{X} \tag{8.83}$$

where $\mathcal{X}$ denotes the set of all x where the supremum operation is finite. It can be verified that $h^\star(x)$ is always a closed convex function regardless of whether $h(w)$ itself is convex or not. This is because, for every fixed w, the function $x^\mathsf{T} w - h(w)$ is linear in x (and, hence, convex) and the supremum of a set of convex functions is convex. Likewise, the set $\mathcal{X}$ is a convex set – see Prob. 8.47. The transformation from $h(w)$ to $h^\star(x)$ is useful in many domains and appears frequently in optimization problems. We provide some intuition next.

Interpretation

Assume w and x are *scalar* variables. The situation is illustrated in Fig. 8.10 for some arbitrary function $h(w)$. In the figure, the term $x^\mathsf{T} w$ corresponds to a line passing through the origin with slope x. For the situation illustrated in the figure, the difference $x^\mathsf{T} w - h(w)$ is negative for all w, and the supremum will occur at the location of minimal distance between the line $x^\mathsf{T} w$ and the function. That distance is the value $-h^\star(x)$. If we move the line $x^\mathsf{T} w$ up by that amount it will become tangent to the function $h(w)$. The tangent is the dotted line in the figure; it is characterized by the pair $(x, h^\star(x))$: the value of x determines its slope and the value $-h^\star(x)$ determines its offset (i.e., the point where it crosses the vertical axis). We can repeat this construction for many other values of x. We find that the conjugate function provides an alternative characterization for $h(w)$: it identifies all lines $(x, h^\star(x))$ that serve as tangents to $h(w)$.

More generally, when x and w are vector-valued, we can interpret $x^\mathsf{T} w$ as representing a hyperplane passing through the origin. The normal direction of the plane is the vector x. The term $x^\mathsf{T} w - h(w)$ measures the difference between the convex function $h(w)$ and the hyperplane. For each x, the conjugate function is finding the largest possible difference between the hyperplane and the function. And the value $-h^\star(x)$ will correspond to the amount of offset that needs to be added to the hyperplane $x^\mathsf{T} w$ to make it tangent to $h(w)$. For this reason, we can interpret $h^\star(x)$ as a mapping from normal directions x to offset values $h^\star(x)$ so that the pairs $(x, h^\star(x))$ define tangent hyperplanes to $h(w)$.

Conjugate functions also arise in finance and economics in the form of conjugate utility or profit functions. In this context, $h(w)$ measures the cost of producing an amount w of some product. The variable x represents the market price per unit so that $x^\mathsf{T} w$ is the total expected market price. The difference $x^\mathsf{T} w - h(w)$ measures the profit that is expected if w items are produced. For a fixed market price x, the conjugate value $h^\star(x)$ then indicates the maximal profit at this price level.

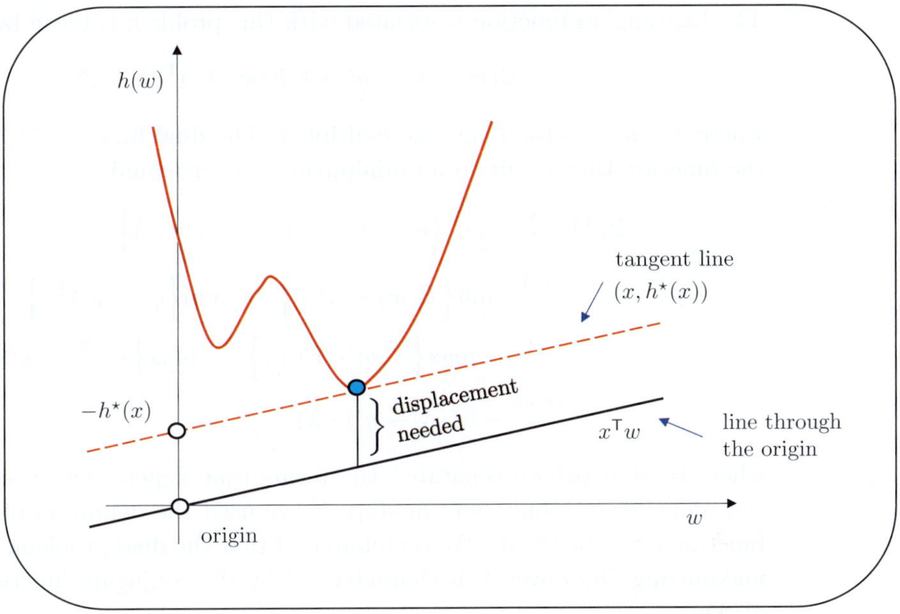

Figure 8.10 Illustration of the concept of a conjugate function for the case in which x and w are scalars. In this case, x represents the slope of the line $x^\mathsf{T} w$ passing through the origin. The conjugate value $-h^\star(x)$ is the amount of displacement needed for this line to become tangent to the function $h(w)$. The tangent line is characterized by the pair $(x, h^\star(x))$: its slope is x and its offset is $h^\star(x)$.

Relation to optimization problems

Conjugate functions are useful for the solution of optimization problems, as will be illustrated in greater detail in Example 51.6. Here we motivate the procedure and provide a couple of motivating examples.

Consider first a problem involving the unconstrained optimization of the sum of two convex functions, say,

$$\boxed{\min_{w \in \mathbb{R}^M} \Big\{ q(w) + E(w) \Big\} \qquad \textbf{(primal problem)}} \tag{8.84}$$

Problems of this type are commonplace when solving inference and learning problems with regularization, as will be discussed in later chapters, where the term $q(w)$ will play the role of the regularizer. We can replace problems of the above form by an equivalent formulation that involves working instead with conjugate functions as follows. First, we transform the problem into a constrained formulation by introducing a dummy variable $z \in \mathbb{R}^M$ to write:

$$\min_{w, z \in \mathbb{R}^M} \Big\{ q(z) + E(w) \Big\}, \quad \text{subject to } z = w \tag{8.85}$$

The Lagrangian function associated with this problem is given by

$$\mathcal{L}(w, z, \lambda) = q(z) + E(w) + \lambda^{\mathsf{T}}(z - w) \tag{8.86}$$

where $\lambda \in \mathbb{R}^M$ is the Lagrange multiplier. The dual function $\mathcal{D}(\lambda)$ is defined as the function that results from minimizing $\mathcal{L}$ over w and z:

$$\begin{aligned}
\mathcal{D}(\lambda) &\stackrel{\Delta}{=} \min_{w,z} \left\{q(z) + E(w) + \lambda^{\mathsf{T}}(z - w)\right\} \\
&\stackrel{(a)}{=} \min_{w}\left\{E(w) - \lambda^{\mathsf{T}} w\right\} + \min_{z}\left\{q(z) + \lambda^{\mathsf{T}} z\right\} \\
&= -\max_{w}\left\{\lambda^{\mathsf{T}} w - E(w)\right\} - \max_{z}\left\{-\lambda^{\mathsf{T}} z - q(z)\right\} \\
&\stackrel{(8.83)}{=} -E^{\star}(\lambda) - q^{\star}(-\lambda)
\end{aligned} \tag{8.87}$$

where in step (a) we separated the terms that depend on w only from those that depend on z only, and in step (b) we used the definition of the conjugate function given by (8.83). We therefore find that the dual problem, which involves maximizing $\mathcal{D}(\lambda)$ over λ, is characterized by the conjugate functions $E^{\star}(\lambda)$ and $q^{\star}(\lambda)$:

$$\boxed{\max_{\lambda \in \mathbb{R}^M} \left\{-q^{\star}(-\lambda) - E^{\star}(\lambda)\right\} \quad \textbf{(dual problem)}} \tag{8.88}$$

We will exploit this duality result later in Section 51.4.2, when we study sparsity-inducing regularization problems.

A second application in the context of optimization problems is the following. Consider a closed convex function $h(w)$ and its conjugate $h^{\star}(x)$. Assume we are interested in solving the optimization problem:

$$w^{\star} = \underset{w \in \mathbb{R}^M}{\operatorname{argmin}}\ h(w) \tag{8.89}$$

Then, we know that the solution $w^{\star}$ must satisfy

$$0 \in \partial_{w^{\mathsf{T}}}\, h(w^{\star}) \tag{8.90}$$

One challenge is that it is not always possible to solve this equation directly to determine $w^{\star}$. Nevertheless, in Prob. 8.46 we establish one useful property that explains how subgradients of $h(w)$ are related to subgradients of its conjugate function $h^{\star}(x)$, namely,

$$v \in \partial_{w^{\mathsf{T}}} h(w) \iff w \in \partial_{x^{\mathsf{T}}} h^{\star}(v) \tag{8.91}$$

Applying this property to (8.90) we conclude that $w^{\star}$ should satisfy

$$\boxed{w^{\star} \in \partial_{v^{\mathsf{T}}} h^{\star}(0)} \tag{8.92}$$

In other words, $w^{\star}$ should belong to the subdifferential of $h^{\star}(x)$ at the origin.

Relation to subdifferentials

Another useful application of conjugate functions arises in the characterization of the subdifferential of convex functions. Thus, consider a convex function $h(w)$: $\mathbb{R}^M \rightarrow \mathbb{R}$. Its subdifferential at any point z is the set of all vectors $s \in \mathbb{R}^M$ such that

$$\begin{aligned}
\partial_{z^\mathsf{T}} h(z) &= \Big\{ s \,|\, h(w) \geq h(z) + s^\mathsf{T}(w - z), \;\; \forall\, w \in \text{dom}(h) \Big\} \\
&\Longleftrightarrow \Big\{ s \,|\, s^\mathsf{T} w - h(w) \leq s^\mathsf{T} z - h(z), \;\; \forall\, w \in \text{dom}(h) \Big\} \\
&\Longleftrightarrow \Big\{ s \,|\, \sup_{w \in \text{dom}(h)} \Big(s^\mathsf{T} w - h(w) \Big) = s^\mathsf{T} z - h(z) \Big\}
\end{aligned} \tag{8.93}$$

The upper bound is attained by selecting $w = z$ in the sup operation. It follows that the subdifferential of $h(z)$ at a point z consists of the set of all vectors s where the conjugate function evaluates to the following:

$$\partial_{z^\mathsf{T}} h(z) = \Big\{ s \,|\, h^\star(s) = s^\mathsf{T} z - h(z) \Big\} \tag{8.94}$$

or, stated equivalently,

$$\boxed{s \in \partial_{z^\mathsf{T}} h(z) \Longleftrightarrow h^\star(s) = s^\mathsf{T} z - h(z)} \tag{8.95}$$

Properties

Conjugate functions have several useful properties. We list them in Table 8.4 for ease of reference and leave the proofs to the problems. The last column in the table provides the relevant references.

8.9 BREGMAN DIVERGENCE

The Kullback–Leibler (KL) divergence studied earlier in Section 6.2 is a special case of what is known as Bregman divergence, which serves as a measure of "distance" or "similarity" and is not limited to probability density functions (pdfs). Its definition and properties rely on the notions of convexity and conjugate functions, which explains our treatment of Bregman divergence at this location in the text.

Definition

Consider a closed convex set Γ and let $\phi(w) : \Gamma \rightarrow \mathbb{R}$ be a differentiable and *strictly convex* function. Let p and q be two points in Γ. The Bregman divergence between p and q is defined as the difference:

$$D_\phi(p, q) \stackrel{\Delta}{=} \phi(p) - \Big(\phi(q) + \nabla_w\, \phi(q)\, (p - q) \Big) \tag{8.96}$$

where $\nabla_w\, \phi(q)$ refers to the gradient of $\phi(w)$ relative to w and evaluated at $w = q$. Note that the Bregman divergence measures the difference between the

Table 8.4 Some useful properties of conjugate functions.

	Given conditions or name	Property	Prob.
1.	closed convex function, $h(w)$	$v \in \partial h(w) \Longleftrightarrow w \in \partial h^\star(v)$	8.46
2.	closed ν-strongly convex, $h(w)$	$h^\star(x)$ is differentiable everywhere with $1/\nu$-Lipschitz gradients and $\nabla_{x^\mathsf{T}} h^\star(x) = \underset{w\in\mathbb{R}^M}{\text{argmax}}\left\{x^\mathsf{T} w - h(w)\right\}$	8.47
3.	$h(w)+c$ $\alpha h(w),\ \alpha>0$ $h(\alpha w),\ \alpha\neq 0$ $h(w-w_o)$ $h(Aw)$, A invertible $h(w)+z^\mathsf{T} w$	$h^\star(x)-c$ $\alpha h^\star(x/\alpha)$ $h^\star(x/\alpha)$ $h^\star(x)+x^\mathsf{T} w_o$ $h^\star(A^{-\mathsf{T}}x)$ $h^\star(x-z)$	8.49
4.	Fenchel–Young inequality	$h(w)+h^\star(x) \geq w^\mathsf{T} x$ with equality when $x\in\partial_{w^\mathsf{T}} h(w)$ or $w \in \partial_{x^\mathsf{T}} h^\star(x)$	8.48
5.	$g(w_1,w_2) = h(w_1)+h(w_2)$ (separable function)	$g^\star(x_1,x_2) = h^\star(x_1)+h^\star(x_2)$	8.50
6.	$h(w)=\frac{1}{2}\\|w\\|_A^2,\ A>0$	$h^\star(x)=\frac{1}{2}\\|x\\|^2_{A^{-1}}$	8.51
7.	$h(w)=\frac{1}{2}w^\mathsf{T} Aw + b^\mathsf{T} w + c$ $A>0$	$h^\star(x)=\frac{1}{2}(x-b)^\mathsf{T} A^{-1}(x-b)-c$	8.52
8.	$h(w)=\\|w\\|_1$	$h^\star(x)=\mathbb{I}_{C,\infty}[x],\ \mathcal{C}=\{x \mid \\|x\\|_\infty \leq 1\}$	8.55
9.	$h(w)=\frac{\nu}{2}\\|w\\|_1^2$	$h^\star(x)=\frac{1}{2\nu}\\|x\\|^2_\infty$	8.53
10.	$h(w)=\frac{1}{2}\\|w\\|_p^2,\ p\geq 1$	$h^\star(x)=\frac{1}{2}\\|x\\|_q^2,\ \frac{1}{p}+\frac{1}{q}=1$	8.54
11.	$h(w)=\\|w\\|$	$h^\star(x)=\mathbb{I}_{C,\infty}[x],\ \mathcal{C}=\{x \mid \\|x\\|_\star \leq 1\}$	8.55
12.	$h(w)=\mathbb{I}_{C,\infty}[w]$	$h^\star(x)=\sup_{w\in\mathcal{C}}\left\{x^\mathsf{T} w\right\}$	8.56
13.	$\begin{cases} h(w)=\sum_{m=1}^{M} w_m \ln w_m \\ w_m \geq 0\end{cases}$	$h^\star(x)=\sum_{m=1}^{M} e^{x_m-1}$	8.61
14.	$h(w)=-\sum_{m=1}^{M}\ln w_m,\ w_m>0$	$h^\star(x)=-\sum_{m=1}^{M}\ln(-x_m)-M$	8.61
15.	$\begin{cases} h(w)=\sum_{m=1}^{M} w_m(\ln w_m -1) \\ w_m \geq 0\end{cases}$	$h^\star(x)=\sum_{m=1}^{M} e^{x_m}$	8.61
16.	$\begin{cases} h(w)=\sum_{m=1}^{M} w_m \ln w_m \\ w_m \geq 0,\ \sum_{m=1}^{M} w_m = 1\end{cases}$	$h^\star(x)=\ln\left(\sum_{m=1}^{M} e^{x_m}\right)$	8.61
17.	$h(W)=-\ln\det(W)$ $W\in\mathbb{R}^{M\times M},\ W>0$	$h^\star(X)=-\ln\det(-X)-M$	8.59

value of the function $\phi(\cdot)$ at $w = p$ and a first-order Taylor expansion around point $w = q$. In this way, the divergence reflects the gap between the convex function $\phi(p)$ and the tangent plane at $w = q$ – see Fig. 8.11. Since $\phi(w)$ is strictly convex, it will lie above the tangent plane and the difference will always be nonnegative:

$$D_\phi(p, q) \geq 0, \quad \forall\, p, q \in \text{dom}(\phi) \tag{8.97}$$

Equality to zero will hold if, and only if, $p = q$. However, the Bregman divergence is not symmetric in general, meaning that $D_\phi(p, q)$ and $D_\phi(q, p)$ need not agree with each other.

It is clear from the definition that $D_\phi(p, q)$ is strictly convex over p since $\phi(p)$ is strictly convex by choice and $\nabla_w \phi(q)\,(p - q)$ is linear in p. Note that if we were to approximate $\phi(p)$ by a second-order Taylor expansion around the same point $w = q$, we would get

$$\phi(p) \approx \phi(q) + \nabla_w \phi(q)(p - q) + \frac{1}{2}(p - q)^{\mathsf{T}} \nabla_w^2\, \phi(q)(p - q) \tag{8.98}$$

so that by substituting into (8.96) we will find the Bregman divergence can be interpreted as a locally weighted squared-Euclidean distance between p and q:

$$D_\phi(p, q) \approx \|p - q\|^2_{\frac{1}{2}\nabla_w^2\, \phi(q)} \tag{8.99}$$

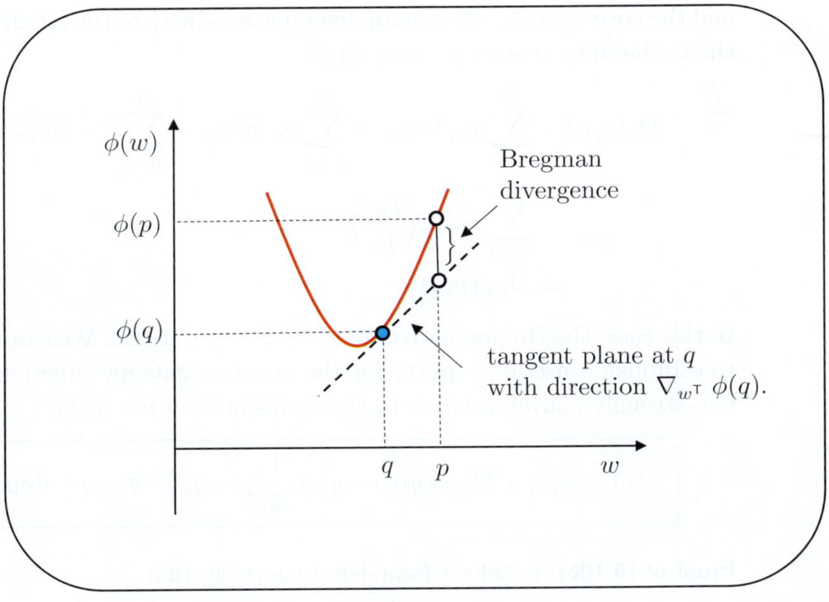

Figure 8.11 The Bregman divergence measures the gap between the function $\phi(p)$ at $w = p$ and its tangent plane at $w = q$.

Two examples

Consider the space of M-dimensional vectors and select

$$\phi(w) = \frac{1}{2}\|w\|^2 \tag{8.100}$$

Then, for any two vectors $p, q \in \mathbb{R}^M$:

$$D_\phi(p,q) = \frac{1}{2}\|p\|^2 - \frac{1}{2}\|q\|^2 - q^\mathsf{T}(p-q) \; = \; \frac{1}{2}\|p-q\|^2 \tag{8.101}$$

which shows that the squared Euclidean distance between two vectors is a Bregman distance. In this case, the Bregman divergence is symmetric. Consider next two probability mass functions (pmfs), with probability values $\{p_m, q_m\}$, defined over the simplex:

$$\Gamma = \left\{ w \in \mathbb{R}^M \;\middle|\; w_m \geq 0 \text{ and } \sum_{m=1}^{M} w_m = 1 \right\} \tag{8.102}$$

Choose $\phi(w)$ as the (negative) entropy of $\{w_m\}$, which is the convex function:

$$\phi(w) = \sum_{m=1}^{M} w_m \ln(w_m) \tag{8.103}$$

Then, the gradient vector is given by

$$\nabla_{w^\mathsf{T}}\, \phi(w) = \text{col}\Big\{1 + \ln w_1,\; 1 + \ln w_2, \ldots, 1 + \ln w_M\Big\} \tag{8.104}$$

and the corresponding Bregman divergence reduces to the KL divergence between the probability vectors p and q since

$$\begin{aligned} D_\phi(p,q) &= \sum_{m=1}^{M} p_m \ln p_m - \sum_{m=1}^{M} q_m \ln q_m - \sum_{m=1}^{M} (1 + \ln q_m)(p_m - q_m) \\ &= \sum_{m=1}^{M} p_m \ln\left(\frac{p_m}{q_m}\right) \\ &= D_{\text{KL}}(p,q) \end{aligned} \tag{8.105}$$

In this case, the Bregman divergence is not symmetric. We can use result (8.105) to establish a useful property for the negative entropy function, namely, that it is ν-strongly convex relative to the ℓ_1-norm with $\nu = 1$, i.e.,

$$\boxed{\phi(p) \geq \phi(q) + \nabla_w\, \phi(q)(p-q) + \frac{1}{2}\|p-q\|_1^2, \;\; \forall\, p,q \in \text{dom}(\phi)} \tag{8.106}$$

Proof of (8.106): It follows from definition (8.96) that

$$\begin{aligned} \phi(p) \;&=\; \phi(q) + \nabla_w\, \phi(q)\,(p-q) + D_\phi(p,q) \\ &\overset{(8.105)}{=} \phi(q) + \nabla_w\, \phi(q)\,(p-q) + D_{\text{KL}}(p\|q) \\ &\overset{(a)}{\geq} \phi(q) + \nabla_w\, \phi(q)\,(p-q) + \frac{1}{2}\|p-q\|_1^2 \end{aligned} \tag{8.107}$$

where in step (a) we used the result of Prob. 6.16, which showed that the KL-divergence of two distributions is lower-bounded by $\frac{1}{2}\|p-q\|_1^2$.

■

Some properties

The Bregman divergence has several useful properties, which facilitate the development of inference methods. We list some of them in this section and leave the arguments to the problems. One first property is the following interesting interpretation.

Theorem 8.1. (Average Bregman divergence) *Let $\boldsymbol{u} \sim p_{\boldsymbol{u}}(u)$ be a random variable defined over a domain $u \in \mathcal{U}$ with pdf $p_{\boldsymbol{u}}(u)$. Let $D_\phi(u,x)$ denote the Bregman divergence between any points $u, x \in \mathcal{U}$. Then, the solution to the following optimization problem:*

$$\bar{u} \stackrel{\Delta}{=} \underset{x\in\mathcal{U}}{\text{argmin}}\; \mathbb{E}_{\boldsymbol{u}}\, D_\phi(\boldsymbol{u}, x) \tag{8.108}$$

is the mean value:

$$\bar{u} = \mathbb{E}\boldsymbol{u} = \int_{u\in\mathcal{U}} u p_{\boldsymbol{u}}(u) du \tag{8.109}$$

In other words, the mean of the distribution $p_{\boldsymbol{u}}(u)$ is the point that minimizes the average Bregman divergence to all points $u \in \mathcal{U}$.

Proof: Denote the cost function by $P(x) = \mathbb{E}_{\boldsymbol{u}}\, D_\phi(\boldsymbol{u}, x)$. Then,

$$\begin{aligned}
&P(x) - P(\bar{u}) \\
&= \int_{u\in\mathcal{U}} p_{\boldsymbol{u}}(u) D_\phi(u,x)du \;-\; \int_{u\in\mathcal{U}} p_{\boldsymbol{u}}(u) D_\phi(u,\bar{u})du \\
&= \int_{u\in\mathcal{U}} p_{\boldsymbol{u}}(u)\Big[D_\phi(u,x) \;-\; D_\phi(u,\bar{u})\Big]du \\
&= \int_{u\in\mathcal{U}} p_{\boldsymbol{u}}(u)\Big[\cancel{\phi(u)} - \phi(x) - \nabla_x\,\phi(x)(u-x) - \cancel{\phi(u)} + \phi(\bar{u}) + \nabla_x\,\phi(\bar{u})(u-\bar{u})\Big]du \\
&= \int_{u\in\mathcal{U}} p_{\boldsymbol{u}}(u)\Big[\phi(\bar{u}) - \phi(x) - \nabla_x\,\phi(x)(u-x) + \nabla_x\,\phi(\bar{u})(u-\bar{u})\Big]du \\
&= \phi(\bar{u}) - \phi(x) - \nabla_x\,\phi(x)\Big(\int_{u\in\mathcal{U}} u p_{\boldsymbol{u}}(u)du - x\Big) + \nabla_x\,\phi(\bar{u})\Big(\int_{u\in\mathcal{U}} u p_{\boldsymbol{u}}(u)du - \bar{u}\Big) \\
&= \phi(\bar{u}) - \phi(x) - \nabla_x\,\phi(x)(\bar{u}-x) + \nabla_x\,\phi(\bar{u})(\bar{u}-\bar{u}) \\
&= \phi(\bar{u}) - \phi(x) - \nabla_x\,\phi(x)(\bar{u}-x) \\
&= D_\phi(\bar{u}, x) \\
&\geq 0
\end{aligned} \tag{8.110}$$

It follows that $P(\bar{u}) \leq P(x)$ for all $x \in \mathcal{U}$ with equality only when $x = \bar{u}$.

■

We collect in Table 8.5 other useful properties, which are established in the problems. The last column in the table provides the relevant reference. Observe for the result in the first row of the table that the Bregman divergences are computed relative to ϕ and its conjugate $\phi^\star$, and that the order of the arguments are reversed. The last two rows of the table extend the Bregman divergence to matrix arguments. In Section 9.4 we describe the use of Bregman divergences in the context of projections onto convex sets.

Table 8.5 Some useful properties of the Bregman divergence where $\phi(x)$ is a differentiable and strictly convex function and $\phi^\star(x)$ is its conjugate function.

	Property	Reference
1.	$D_\phi(p,q) = D_{\phi^\star}\Big(\nabla_w\phi(q),\, \nabla_w\phi(p)\Big)$ (duality)	Prob. 8.64
2.	$D_\phi(r,p) + D_\phi(p,q) = D_\phi(r,q) + \Big(\nabla\phi_w(q) - \nabla_w\phi(p)\Big)(r-p)$ (generalized triangle inequality)	Prob. 8.65
3.	$D_\phi(p,q) = \frac{1}{2}\Vert p-q\Vert_Q^2,\ \phi(w) = \frac{1}{2}\Vert w\Vert_Q^2,\ Q>0$ (Mahalanobis distance)	Prob. 8.66
4.	$D_\phi(p,q) = D_{\rm KL}(p,q) = \sum_{m=1}^{M} p_m \ln\left(\frac{p_m}{q_m}\right)$ $\phi(w) = \sum_{m=1}^{M} w_m \ln(w_m),\ w_m>0,\ \sum_{m=1}^{M} w_m = 1$ (negative entropy)	Eq. (8.105)
5.	$D_\phi(P,Q) = \text{Tr}(P\ln P - P\ln Q - P + Q)$ $\phi(W) = \text{Tr}(W\ln W), W>0$ (von Neumann divergence)	Prob. 8.67
6.	$D_\phi(P,Q) = \text{Tr}(PQ^{-1} - I_M) - \ln\det(PQ^{-1})$ $\phi(W) = -\ln\det(W),\ W>0,\ W\in\mathbb{R}^{M\times M}$	Prob. 8.68

8.10 COMMENTARIES AND DISCUSSION

Convex functions. Excellent references on convex analysis are the texts by Rockafellar (1970), Polyak (1987), Hiriart-Urruty and Lemaréchal (2001), Bertsekas (2003), Boyd and Vandenberghe (2004), and Nesterov (2004). Useful accounts on the history of convexity, dating back to the development of Greek geometry, appear in Fenchel (1983) and Dwilewicz (2009). According to the latter reference and also Heath (1912, p. 8), the first definition of convexity was apparently given by the ancient Greek mathematician **Archimedes of Syracuse (ca 287 BC–212 BC)** in the work by Archimedes (225 BC) – see the exposition by Dunham (1990). Result (8.28) for ν-strongly convex functions is often referred to as the *Polyak–Lojasiewicz bound* due to Polyak (1963) and Lojasiewicz (1963); it is useful in the study of the convergence behavior of gradient descent algorithms – see, e.g., Example 12.10 and the proof of Theorem 12.3.

Subgradients. In future chapters we will encounter optimization problems that involve nonsmooth functions with nondifferentiable terms. In these cases, iterative algorithms for minimizing these functions will be constructed by replacing traditional gradient vectors by subgradients whenever necessary. The idea of employing subgradient vectors was proposed by Shor (1962) in his work on maximizing piecewise linear concave functions. The method was well-received at the time and generated tremendous interest due to its simplicity and effectiveness. Some of the earliest works that helped solidify the theoretical foundations of the method were published by Ermoliev (1966, 1969, 1983a, b), Ermoliev and Shor (1967), and Polyak (1967, 1969), culminating with the manuscripts by Ermoliev (1976) and Shor (1979). Useful surveys on the history and development of subgradient methods are given by Shor (1991) and Goffin (2012). Some additional references include Rockafellar (1970), Bertsekas (1973), Held, Wolfe, and Crowder (1974), Clarke (1983), Nemirovsky and Yudin (1983), Kiwiel (1985), Polyak (1987), Shor (1998, 2012), Bertsekas, Nedic, and Ozdaglar (2003), Nesterov (2004), Shalev-Shwartz *et al.* (2011), Duchi, Hazan, and Singer (2011), Duchi, Bartlett, and Wainwright (2012), Shamir and Zhang (2013), and Ying and Sayed (2018).

Subgradients of sums of convex functions. It will be common in our treatment of inference and learning methods in this text to encounter objective functions that are expressed as the sum of a finite number of convex functions such as

$$g(z) = \ell_1(z) + \ldots + \ell_N(z) \tag{8.111}$$

in terms of individual convex terms $\ell_n(z)$. These individual terms need not be differentiable. Let $\partial_z \ell_n(z)$ denote the subdifferential set for $\ell_n(z)$ at location z. Then, the result of Prob. 8.31 indicates that we can construct a subgradient for the sum $g(z)$ by adding individual subgradients for the $\{\ell_n(z)\}$. This is because

$$\Big\{\partial_z \ell_1(z) + \partial_z \ell_2(z) + \ldots + \partial_z \ell_N(z)\Big\} \subset \partial_z g(z) \tag{8.112}$$

A useful question is whether *all* elements of the subdifferential set of $g(z)$ can be constructed from the sum of individual subgradient vectors for the $\{\ell_n(z)\}$, i.e., whether the two sets in (8.112) are actually *equal* to each other. We provide a counterexample in Prob. 8.31 to show that these two sets are not generally equal. While establishing property (8.112) is relatively straightforward, and is left as an exercise in Prob. 8.31, the study of conditions under which both sets coincide is more challenging and can be found, for example, in Rockafellar (1963). In particular, it is shown there that both sets will coincide when the domains of the individual functions satisfy the following condition:

$$\bigcap_{n=1}^{N} \text{ri}\Big(\text{dom}(\ell_n(z))\Big) \neq \emptyset \tag{8.113}$$

in terms of the *relative interior* (ri) of the domains of the individual functions. Condition (8.113) requires the domains of all individual functions to have a nonempty intersection. This situation will be satisfied in most cases of interest to us since the individual functions will have the same form over z, or their domains will generally be $\mathbb{R}^M$. In these situations, the following two directions will hold:

$$\Big\{\partial_z \ell_1(z) + \partial_z \ell_2(z) + \ldots + \partial_z \ell_N(z)\Big\} \subset \partial_z g(z) \tag{8.114a}$$

$$\partial_z g(z) \subset \Big\{\partial_z \ell_1(z) + \partial_z \ell_2(z) + \ldots + \partial_z \ell_N(z)\Big\} \tag{8.114b}$$

To explain the notion of the relative interior, consider the segment $\{-1 \leq x \leq 1\}$ on the real axis. The interior of this set consists of all points $\{-1 < x < 1\}$. Recall that a point is in the interior of a set $\mathcal{S}$ if a small ϵ-size open interval around the point continues to be in $\mathcal{S}$. Now, let us take the same interval $\{-1 \leq x \leq 1\}$ and view it as a set in the higher-dimensional space $\mathbb{R}^2$. In this space, this interval does *not* have an interior anymore. This is because, for any point in the interval, if we draw a small circle of radius ϵ around it, the circle will contain points outside the interval no matter how small ϵ is. Therefore, the interval $\{-1 \leq x \leq 1\}$ does not have an interior in $\mathbb{R}^2$. However, one can extend the notion of interiors to allow for such intervals to have interiors in higher-dimensional spaces. This is what the notion of *relative interior* does. Loosely, under this concept, to check whether a set $\mathcal{S}$ has an interior, we limit our examination to the subspace where the set lies. For any set $\mathcal{S}$, we define its *affine hull* as the collection of all points resulting from any *affine* combination of elements of $\mathcal{S}$:

$$\text{affine}(\mathcal{S}) \triangleq \left\{ \sum_{p=1}^{P} a_p s_p \,\middle|\, \text{for any integer } P > 0, s_p \in \mathcal{S} \right\} \tag{8.115a}$$

$$a_p \in \mathbb{R}, \; \sum_{p=1}^{P} a_p = 1 \tag{8.115b}$$

where the combination coefficients $\{a_p\}$ are real numbers and required to add up to 1; if these combination coefficients were further restricted to being nonnegative, then the affine hull would become the convex hull of the set $\mathcal{S}$. For example, for the interval $\mathcal{S} = \{-1 \leq x \leq 1\}$, its affine hull will be a line along the x-axis containing the interval. Once affine($\mathcal{S}$) is identified, we then determine whether the interval $\mathcal{S}$ has an interior *within* this affine hull, which we already know is true and given by $\{-1 < x < 1\}$. This interior is referred to as the *relative interior* of the set in $\mathbb{R}^2$; the qualification "relative" is referring to the fact that the interior is defined *relative* to the affine hull space and not the entire $\mathbb{R}^2$ space where the interval lies.

Jensen inequality. We described deterministic and stochastic forms of Jensen inequality in Section 8.7. They are useful in various contexts in probability theory, information theory, and statistics. Inequality (8.75) is generally attributed to Jensen (1906), although in an addendum on page 192 of his article, Jensen acknowledges that he discovered an earlier instance of his inequality in the work by Hölder (1889). In this latter reference, the inequality appears in the following form:

$$g\left(\frac{\sum_{k=1}^{N} \beta_k z_k}{\sum_{\ell=1}^{N} \beta_\ell} \right) \leq \frac{\sum_{k=1}^{N} \beta_k g(z_k)}{\sum_{\ell=1}^{N} \beta_\ell} \tag{8.116}$$

where $g(z)$ is a convex function and the $\{\beta_k\}$ are positive scalars. If we redefine

$$\alpha_k \triangleq \frac{\beta_k}{\sum_{\ell=1}^{N} \beta_\ell} \tag{8.117}$$

then the $\{\alpha_k\}$ become convex combination coefficients and the result reduces to (8.75). More information on Jensen and Hölder inequalities can be found in Hardy, Littlewood, and Pólya (1934) and Abramowitz and Stegun (1965).

Conjugate functions. These functions are also known as *Fenchel conjugates* after Fenchel (1949); they play an important supporting role in the solution of optimization problems through duality. For more details on their mathematical properties, the reader may refer to Rockafellar (1970, 1974), Boyd and Vandenberghe (2004), and Bertsekas (2009). We explained in Section 8.8 that the transformation from $h(w)$ to $h^\star(x)$ defined by (8.83) is useful in many scenarios, including in the solution of optimization problems. We also indicated that conjugate functions arise in finance and economics in the form of conjugate utility or profit functions – see, e.g., Eatwell, Newman, and Milgate (1987). We will explain later in the commentaries to Chapter 11 the close relationship that exists between conjugate functions and proximal operators in the form of the Moreau decomposition established by Moreau (1965).

Bregman divergence. The KL divergence is a special case of the Bregman divergence defined in (8.96) and introduced by Bregman (1967). As explained in the body of the chapter, this divergence serves as a measure of "distance" or "similarity" and is not limited to pdfs. However, when the arguments p and q correspond to pmfs and $\phi(\cdot)$ is selected as the negative entropy function (8.103), we get

$$D_\phi(p,q) = D_{\text{KL}}(p,q) \tag{8.118}$$

The important result in Theorem 8.1 is due to Banerjee, Gou, and Wang (2005). It states that for a collection of points randomly distributed within some space $\mathcal{U}$, their mean is the point that minimizes the average Bregman divergence to all of them. For further details on Bregman divergences, the reader may refer to Censor and Zenios (1998), Azoury and Warmuth (2001), Banerjee *et al.* (2005), Chen, Chen, and Rao (2008), Adamcík (2014), Harremoës (2017), and Siahkamari *et al.* (2020). In Chapter 15 we will exploit properties of the Bregman divergence in the derivation of mirror-descent learning algorithms.

PROBLEMS

8.1 Is the column span of any matrix $A \in \mathbb{R}^{N\times M}$ a convex set? What about its row span? What about its nullspace?

8.2 Consider a convex function $g(z) : \mathbb{R}^M \to \mathbb{R}$. Denote the individual entries of z by $\{z_m\}$ for $m = 1, 2 \ldots, M$. Select an arbitrary z_m and fix all other entries. Is $g(z)$ convex over z_m?

8.3 Show that the intersection of convex sets is a convex set.

8.4 Show that the zero vector belongs to the conic hull of a set $\mathcal{S} \subset \mathbb{R}^M$. Show also that the conic hull is a convex set.

8.5 Show that the normal cone defined by (8.2) is always a convex cone regardless of the nature of the set $\mathcal{S}$.

8.6 Verify that each of the following sets is convex:

(a) The nonnegative orthant denoted by $\mathbb{R}^M_+$, which consists of all M-dimensional vectors with nonnegative entries.

(b) Any affine subspace consisting of all vectors $x \in \mathbb{R}^M$ satisfying $Ax = b$, where $A \in \mathbb{R}^{N\times M}$ and $b \in \mathbb{R}^N$.

(c) The halfspace consisting of all vectors $x \in \mathbb{R}^M$ satisfying $a^\mathsf{T} x \le \alpha$, where a is a vector and α is a scalar.

(d) Any polyhedron consisting of all vectors $x \in \mathbb{R}^M$ satisfying $Ax \preceq b$, where $A \in \mathbb{R}^{N\times M}$, $b \in \mathbb{R}^N$, and the notation $x \preceq y$ refers to component-wise inequalities for the individual elements of the vectors $\{x, y\}$.

(e) The set of symmetric and nonnegative-definite matrices, $A \in \mathbb{R}^{N\times N}$.

8.7 Consider two convex functions $h(z)$ and $g(z)$. Is their composition $f(z) = g(h(z))$ convex?

8.8 Given any x_o and a square matrix $A \ge 0$, show that the ellipsoid consisting of all vectors $x \in \mathbb{R}^M$ such that $(x - x_o)^\mathsf{T} A(x - x_o) \le 1$ is a convex set.

8.9 Show that the probability simplex defined by all vectors $p \in \mathbb{R}^M$ with entries p_m satisfying $p_m \ge 0$ and $\sum_{m=1}^M p_m = 1$ is a convex set.

8.10 Consider a convex function $g(z)$ with $z \in \mathbb{R}^M$. The α-sublevel set of $g(z)$ is defined as the set of all vectors $z \in \text{dom}(g)$ that satisfy $g(z) \le \alpha$. Show that α-sublevel sets are convex.

8.11 Consider a collection of convex functions, $\{g_\ell(z),\ \ell = 1, \ldots, L\}$, and introduce the weighted combination (also called conic combination) $g(z) = \sum_{\ell=1}^L a_\ell g_\ell(z)$, where $a_\ell \ge 0$. Show that $g(z)$ is convex.

8.12 Show that definitions (8.3) and (8.4) are equivalent characterizations of convexity when $g(z)$ is differentiable.

8.13 A continuous function $g(z) : \mathbb{R}^M \to \mathbb{R}$ is said to be midpoint convex if for any $z_1, z_2 \in \text{dom}(g)$, it holds that $g\left(\frac{1}{2}(z_1 + z_2)\right) \le \frac{1}{2}(g(z_1)+g(z_2))$. Show that a real-valued continuous function $g(z)$ is convex if, and only if, it is midpoint convex.

8.14 Establish the three properties listed in Example 8.2.
8.15 Consider the indicator function (8.73) for some set $\mathcal{C}$. Show that $\mathbb{I}_{C,\infty}[z]$ is a convex function if, and only if, the set $\mathcal{C}$ is convex.
8.16 Consider a function $g(z;a) : \mathbb{R}^M \to \mathbb{R}$ that is parameterized by a vector a in some set $\mathcal{A}$. Show that if $g(z;a)$ is convex in z for every $a \in \mathcal{A}$, then the following function is also convex in z:

$$g(z) \triangleq \max_{a\in\mathcal{A}} g(z;a)$$

8.17 Let $z \in \mathbb{R}^M$ be a vector in the probability simplex and denote its entries by $\{z_m \geq 0\}$. Assume the convention $0 \times \ln 0 = 0$. Consider the negative entropy function $g(z) = \sum_{m=1}^{M} z_m \ln z_m$. Verify that $g(z)$ is convex. Show further that $g(z)$ is ν-strongly convex relative to the ℓ_1-norm, i.e., it satisfies the following relation with $\nu = 1$ for any (z, z_0):

$$g(z) \geq g(z_0) + \nabla_z g(z_0)\,(z - z_0) + \frac{\nu}{2}\|z - z_0\|_1^2$$

8.18 Consider a function $g(z) : \mathbb{R}^M \to \mathbb{R}$. Pick any $z \in \text{dom}(g)$ and any scalar t and vector w such that $z + tw \in \text{dom}(g)$. Show that $g(z)$ is convex in z if, and only if, the function $h(t) = g(z + tw)$ is convex in t. In other words, a function $g(z)$ is convex if, and only if, its restriction to *any* line in its domain, namely, $g(z + tw)$, is also convex.
8.19 Let z^o denote the global minimizer of a ν-strongly convex function $g(z)$. Use (8.24) to show that $\nu\|z - z^o\| \leq \|\nabla_z\, g(z)\|$.
8.20 Consider a ν-strongly convex function $g(z) : \mathbb{R}^M \to \mathbb{R}$ satisfying (8.19) . Denote the individual entries of z by $\{z_m\}$ for $m = 1, 2, \ldots, M$. Select an arbitrary z_m and fix all other entries. Is $g(z)$ ν-strongly convex over z_m?
8.21 True or false. Refer to definition (8.19). A function $g(z)$ is ν-strongly convex if, and only if, the function $g(z) - \frac{\nu}{2}\|z\|^2$ is convex.
8.22 Let $z \in \mathbb{R}^M$. Show that $g(z) = \|z\|^4$ is strictly convex.
8.23 Show that the regularized hinge loss function (8.40) is strongly convex.
8.24 Establish (8.21) as an equivalent characterization for ν-strong convexity for differentiable functions $g(z) : \mathbb{R}^M \to \mathbb{R}$.
8.25 Establish property (8.24) for ν-strongly convex functions $g(z) : \mathbb{R}^M \to \mathbb{R}$.
8.26 Let $z \in \mathbb{R}^M$ and consider a full-rank matrix $A \in \mathbb{R}^{N\times M}$ with $N \geq M$. Examine the convexity, strict convexity, and strong convexity of the function $g(z) = \|Az\|^\alpha$ for all values of α in the range $\alpha \in [1, \infty)$. How would your answers change if A were nonzero but rank-deficient?
8.27 Consider a ν-strongly convex function $g(z) : \mathbb{R}^M \to \mathbb{R}$, as defined by (8.19). Relation (8.21) provides a useful property for such functions when they are differentiable. Assume now that the function is not necessarily differentiable. For any arbitrary points z and z_o, let s and s_o denote subgradients for $g(z)$ at locations z and z_o, respectively, i.e., $s \in \partial_{z^\mathsf{T}}\, g(z)$ and $s_o \in \partial_{z^\mathsf{T}}\, g(z_o)$. Establish the validity of the following properties:

$$g(z) \geq g(z_o) \;+\; s_o^\mathsf{T}(z - z_o) \;+\; \frac{\nu}{2}\|z - z_o\|^2$$

$$g(z) \leq g(z_o) \;+\; s_o^\mathsf{T}(z - z_o) \;+\; \frac{1}{2\nu}\|s - s_o\|^2$$

$$\nu\|z - z_o\|^2 \leq (s - s_o)^\mathsf{T}(z - z_o) \leq \frac{1}{\nu}\|s - s_o\|^2$$

8.28 Let $g(z)$ be a strictly convex function and consider two distinct points z_1 and z_2 in its domain. Show that $\partial_z g(z_1) \cap \partial_z g(z_2) = \emptyset$.
8.29 For any $\alpha \geq 0$, show that $\partial_z\, \alpha g(z) = \alpha\, \partial_z\, g(z)$.
8.30 Let $g(z)$ be a convex function and introduce the transformation $h(z) = g(Az+b)$ where $z \in \mathbb{R}^M$, $A \in \mathbb{R}^{N\times M}$, and $b \in \mathbb{R}^N$. Show that $\partial_{z^\mathsf{T}}\, h(z) = A^\mathsf{T}\, \partial_{z^\mathsf{T}}\, g(z)|_{z\leftarrow Az+b}$.

8.31 Let $g_1(z)$ and $g_2(z)$ be two convex functions with $z \in \mathbb{R}^M$. Show that

$$\partial_{z^\mathsf{T}}\, g_1(z) + \partial_{z^\mathsf{T}}\, g_2(z) \subset \partial_{z^\mathsf{T}}\Big(g_1(z) + g_2(z)\Big)$$

To verify that these two sets are not identical in general, consider the following two functions with $z \in \mathbb{R}$:

$$g_1(z) = \begin{cases} z, & z \geq 0 \\ +\infty, & z < 0 \end{cases} \qquad\qquad g_2(z) = \begin{cases} +\infty, & z > 0 \\ -z, & z \leq 0 \end{cases}$$

(a) Let $g(z) = g_1(z) + g_2(z)$. What is $g(z)$?
(b) Determine $\partial_{z^\mathsf{T}}\, g_1(0), \partial_{z^\mathsf{T}}\, g_2(0)$, and $\partial_{z^\mathsf{T}}\, g(0)$.

8.32 Let $g(z) = \|z\|$, in terms of the Euclidean norm of $z \in \mathbb{R}^M$. Show that the subdifferential of $g(z)$ is given by the following expression where $a \in \mathbb{R}^M$:

$$\partial_{z^\mathsf{T}}\, g(z) = \begin{cases} z/\|z\|, & z \neq 0 \\ \{a \mid \|a\| \leq 1\}, & z = 0 \end{cases}$$

8.33 Refer to the definition of the dual norm in (1.157). Let $g(z) = \|z\|_q$ denote the q-norm of vector $z \in \mathbb{R}^M$ and define p through $1/p + 1/q = 1$ for $p, q \geq 1$. Show that

$$\partial_{z^\mathsf{T}}\, \|z\|_q = \underset{\|y\|_p \leq 1}{\text{argmax}}\; z^\mathsf{T} y$$

Explain how this characterization leads to the same conclusion in Prob. 8.32.

8.34 Consider a convex set $\mathcal{C}$ and its indicator function $\mathbb{I}_{C,\infty}[z]$: it is equal to zero when $z \in$ and $+\infty$ otherwise. Show that $\partial_z\, \mathbb{I}_{C,\infty}[z] = \mathcal{N}_C(z)$ in terms of the normal cone at location z. The result is illustrated geometrically in Fig. 8.9, where the normal cone is shown at one of the corner points.

8.35 Let $g(z) = \|z\|_\infty$ in terms of the ∞-norm of the vector $z \in \mathbb{R}^M$. What is $\partial_{z^\mathsf{T}}\, g(0)$?

8.36 Let

$$g(z) \;=\; \max_{1 \leq n \leq N} \left\{a_n^\mathsf{T} z + \alpha(n)\right\}, \;\; a_n, z \in \mathbb{R}^M, \;\; \alpha(n) \in \mathbb{R}$$

and assume the maximum is attained at some index n_o. Show that $a_{n_o} \in \partial_{z^\mathsf{T}}\, g(z)$.

8.37 For any convex function $g(z)$ that is nondifferentiable at some location z_o, show that its subdifferential at this location is a convex set.

8.38 Consider L convex functions $g_\ell(z)$ for $\ell = 1, 2, \ldots, L$ and define the pointwise maximum function

$$g(z) = \max_{\ell=1,\ldots,L} \left\{g_\ell(z)\right\}$$

At any point z_1, let $g_{\ell^o}(z_1)$ be one of the functions for which $g_{\ell^o}(z_1) = g(z_1)$. There may exist more than one function attaining the value $g(z_1)$. It is sufficient to consider one of them. Show that if $s \in \partial_{z^\mathsf{T}}\, g_{\ell^o}(z_1)$, then $s \in \partial_{z^\mathsf{T}}\, g(z_1)$. That is, show that a subgradient for $g_{\ell^o}(z)$ at z_1 can serve as a subgradient for $g(z)$ at the same location. More generally, show that the subdifferential of $g(z)$ is given by the following convex hull:

$$\partial_{z^\mathsf{T}}\, g(z) \;=\; \text{conv}\left\{ \bigcup_{g_\ell(z) = g(z)} \partial_{z^\mathsf{T}}\, g_\ell(z) \right\}$$

8.39 Consider two differentiable convex functions $\{g_1(z), g_2(z)\}$ and define $g(z) = \max\{g_1(z), g_2(z)\}$. Show that

$$\partial_z g(z) \;=\; \begin{cases} \nabla_z\, g_1(z), & \text{if } g_1(z) > g_2(z) \\ \nabla_z\, g_2(z), & \text{if } g_2(z) > g_1(z) \\ \alpha \nabla_z\, g_1(z) + (1-\alpha)\nabla_z\, g_2(z), & \text{if } g_1(z) = g_2(z) \end{cases}$$

where $\alpha \in [0,1]$. The last condition amounts to selecting any point on the segment linking the gradient vectors of $g_1(z)$ and $g_2(z)$.

8.40 Let $g(z) : \mathbb{R}^M \rightarrow \mathbb{R}$ denote a convex function and consider subgradient vectors $s_1 \in \partial_{z^\mathsf{T}}\, g(z_1)$ and $s_2 \in \partial_{z^\mathsf{T}}\, g(z_2)$ at locations z_1 and z_2. Establish the following inner product inequality:

$$(s_2 - s_1)^\mathsf{T}(z_2 - z_1) \geq 0$$

Let z^o denote the global minimizer of $g(z)$. Conclude that $(\partial_{z^\mathsf{T}}\, g(z))^\mathsf{T}\,(z - z^o) \geq 0$ for any subgradient vector at location z.

8.41 For a convex function $g(z)$, show that z^o is a minimum if, and only if, $0 \in \partial_{z^\mathsf{T}}\, g(z^o)$.

8.42 Let $g(z) = \sum_{n=1}^{N} |\gamma(n) - h_n^\mathsf{T} z|$, where $z, h_n \in \mathbb{R}^M$ and $\gamma(n) \in \mathbb{R}$. Show that a subgradient for $g(z)$ is given by

$$\sum_{n=1}^{N} -h_n \text{sign}\left(\gamma(n) - h_n^\mathsf{T} z\right) \in \partial_{z^\mathsf{T}}\, g(z)$$

where $\text{sign}(x) = +1$ if $x \geq 0$ and $\text{sign}(x) = -1$ if $x < 0$.

8.43 Let $g(z) = \max_{1 \leq n \leq N} \left(\gamma(n) - h_n^\mathsf{T} z\right)$, where $z, h_n \in \mathbb{R}^M$ and $\gamma(n) \in \mathbb{R}$. Show that the subdifferential of $g(z)$ is given by

$$\partial_{z^\mathsf{T}}\, g(z) = \sum_{n=1}^{N} -\alpha(n) h_n$$

where the scalars $\{\alpha(n)\}$ satisfy the conditions

$$\alpha(n) \geq 0, \quad \sum_{n=1}^{N} \alpha(n) = 1, \quad \alpha(m) = 0 \text{ if } (\gamma(m) - h_m^\mathsf{T} z) < g(z)$$

8.44 Consider the set of points $(x, y, 0) \in \mathbb{R}^3$ satisfying $2x^2 + y^2 \leq 1$. Does this set of points have an interior? Does it have a relative interior? If so, identify it.

8.45 What is the affine hull of two points in $\mathbb{R}^3$?

8.46 Consider a closed convex function $h(w)$ and its Fenchel conjugate $h^\star(x)$ as defined by (8.83). Show that the subgradients of $h(w)$ and $h^\star(x)$ are related as follows:

$$v \in \partial_{w^\mathsf{T}} h(w) \iff w \in \partial h^\star_{x^\mathsf{T}}(v)$$

8.47 Refer to definition (8.83) for the conjugate function of $h(w)$. Show that the set $\mathcal{X}$ is a convex set. Furthermore, assume $h(w)$ is ν-strongly convex and closed. Show that in this case $\mathcal{X} = \mathbb{R}^M$ so that $\text{dom}(h^\star) = \mathbb{R}^M$ and, moreover, $h^\star(x)$ is differentiable everywhere with the gradient vector given by

$$\nabla_{x^\mathsf{T}}\, h^\star(x) \;=\; \underset{w \in \mathbb{R}^M}{\text{argmax}} \left\{x^\mathsf{T} w - h(w)\right\}$$

and satisfies the $1/\nu$-Lipschitz condition

$$\|\nabla_{x^\mathsf{T}}\, h^\star(x_1) - \nabla_{x^\mathsf{T}}\, h^\star(x_2)\| \leq \frac{1}{\nu}\|x_1 - x_2\|$$

8.48 Refer to definition (8.83) for the conjugate function of $h(w)$. Show that for any function $h(w)$ and its conjugate, the so-called Fenchel–Young inequality holds:

$$h(w) + h^\star(x) \geq \; w^\mathsf{T} x, \quad \text{for any } w, x$$

Show that the inequality becomes an equality when $x \in \partial_{w^\mathsf{T}}\, h(w)$, i.e., when x belongs to the subdifferential set of $h(\cdot)$ at location w (or, alternatively, when $w \in \partial_{x^\mathsf{T}} h^\star(x)$). Conclude that if $h^\star(x)$ is differentiable, then equality holds when $w = \nabla_{x^\mathsf{T}} h^\star(x)$.

8.49 Refer to definition (8.83) for the conjugate function of $h(w)$. Establish the properties listed in Table 8.6.

Table 8.6 List of properties for conjugate pairs $(h(w), h^\star(x))$.

Function transformation	Conjugate function
$h(w) = g(w) + c$	$h^\star(x) = g^\star(x) - c$ (c is a constant)
$h(w) = \alpha g(w),\ \alpha > 0$	$h^\star(x) = \alpha g^\star(x/\alpha)$
$h(w) = g(\alpha w),\ \alpha \neq 0$	$h^\star(x) = g^\star(x/\alpha)$
$h(w) = g(w - w_o)$	$h^\star(x) = g^\star(x) + x^\mathsf{T} w_o$
$h(w) = g(Aw)$, A invertible	$h^\star(x) = g^\star(A^{-\mathsf{T}} x)$
$h(w) = g(w) + z^\mathsf{T} w$	$h^\star(x) = g^\star(x - z)$

8.50 Refer to definition (8.83) for the conjugate function of $h(w)$ and consider the separable sum $g(w_1, w_2) = h(w_1) + h(w_2)$. Show that $g^\star(x_1, x_2) = h^\star(x_1) + h^\star(x_2)$.
8.51 Let $h(w) = \frac{1}{2}\|w\|_A^2$, where $w \in \mathbb{R}^M$ and $A > 0$. Show that $h^\star(x) = \frac{1}{2}\|x\|_{A^{-1}}^2$.
8.52 Let $h(w) = \frac{1}{2}w^\mathsf{T} A w + b^\mathsf{T} w + c$, where $w \in \mathbb{R}^M$ and $A > 0$. Show that $h^\star(x) = \frac{1}{2}(x - b)^\mathsf{T} A^{-1}(x - b) - c$.
8.53 Let $h(w) = \frac{\nu}{2}\|w\|_1^2$. Show that $h^\star(x) = \frac{1}{2\nu}\|x\|_\infty^2$.
8.54 Let $h(w) = \frac{1}{2}\|w\|_p^2$. Show that $h^\star(x) = \frac{1}{2}\|x\|_q^2$, where $1/p + 1/q = 1$.
8.55 Let $h(w) = \|w\|_1$. Show that $h^\star(x) = 0$ if $\|x\|_\infty \leq 1$ and ∞ otherwise. That is, the conjugate function is the indicator function that verifies whether x belongs to the convex set $\|x\|_\infty \leq 1$. More generally, let $h(w) = \|w\|$ denote an arbitrary norm over $w \in \mathbb{R}^M$ and let $\|\cdot\|_\star$ denote the dual norm defined as $\|x\|_\star = \sup_w\{x^\mathsf{T} w \mid \|w\| \leq 1\}$ – recall (1.157). Show that $h^\star(x) = 0$ if $\|x\|_\star \leq 1$ and ∞ otherwise.
8.56 Consider a convex set $\mathcal{C}$ and the indicator function $h(w) = \mathbb{I}_{\mathcal{C},\infty}[w]$ defined as follows: its value is zero if $w \in \mathcal{C}$ and $+\infty$ otherwise. Show that the conjugate function $h^\star(x)$ is given by $h^\star(x) = \sup_{w\in\mathcal{C}} x^\mathsf{T} w$. The function $h^\star(x)$ is called the support function of the set $\mathcal{C}$. Show that the support function is convex over x.
8.57 Let $h(w) = \alpha w + \beta$ where $\{\alpha, \beta, w\}$ are all scalars. Show that its conjugate function is given by

$$h^\star(x) = \begin{cases} -\beta, & x = \alpha \\ \infty, & \text{otherwise} \end{cases}$$

8.58 Let $h(w) = \max\{0, 1 - w\}$, where w is scalar. Show that its conjugate function is given by

$$h^\star(x) = \begin{cases} x, & x \in [-1, 0] \\ +\infty, & \text{otherwise} \end{cases}$$

8.59 For matrix arguments W, we define the conjugate function using

$$h^\star(X) = \sup_W \Big\{ \mathrm{Tr}(X^\mathsf{T} W) - h(W) \Big\}$$

Consider the matrix function $h(W) = -\ln\det(W)$ for positive-definite $W \in \mathbb{R}^{M\times M}$. Show that $h^\star(X) = -\ln\det(-X) - M$.
8.60 The characterization (8.21) for a ν-strongly convex function relied on the use of the squared-Euclidean norm term, $\|z - z_o\|^2$. We indicated then that other vector norms can be used as well. For instance, the same function will also be ν_1-strongly convex relative to the squared ℓ_1-norm, namely, it will satisfy for any (z_2, z):

$$g(z_2) \geq g(z) + (\nabla_z\, g(z))\,(z_2 - z) + \frac{\nu_1}{2}\|z_2 - z\|_1^2$$

for some parameter $\nu_1 > 0$.

(a) Maximize the right-hand side over z_2 and use the result of Prob. 8.53 to conclude that at the minimizer z^o (compare with the upper bound in (8.29)):

$$g(z^o) \geq g(z) - \frac{1}{2\nu_1}\|\nabla_z\, g(z)\|_\infty^2$$

(b) For vectors $x \in \mathbb{R}^M$, use the known norm bounds $\frac{1}{\sqrt{M}}\|x\|_1 \leq \|x\|_2 \leq \|x\|_1$ to conclude that the strong convexity constants (ν, ν_1) can be selected to satisfy $\frac{\nu}{M} \leq \nu_1 \leq \nu$. *Remark.* See Nutini *et al.* (2015) for a related discussion.

8.61 Let $w \in \mathbb{R}^M$ with entries $\{w_m\}$. Establish the conjugate pairs (with the convention $0 \times \ln 0 = 0$) listed in Table 8.7.

Table 8.7 List of conjugate pairs $(h(w), h^\star(x))$.

Original function, $h(w)$	**Conjugate function,** $h^\star(x)$
$\sum_{m=1}^{M} w_m \ln w_m,\quad w_m \geq 0$	$\sum_{m=1}^{M} e^{x_m - 1}$
$\sum_{m=1}^{M} w_m\Big(\ln w_m - 1\Big),\quad w_m \geq 0$	$\sum_{m=1}^{M} e^{x_m}$
$-\sum_{m=1}^{M} \ln w_m,\quad w_m > 0$	$-\sum_{m=1}^{M} \ln(-x_m) - M$
$\sum_{m=1}^{M} w_m \ln w_m,\quad w_m \geq 0,\ \sum_{m=1}^{M} w_m = 1$	$\ln\Big(\sum_{m=1}^{M} e^{x_m}\Big)$

8.62 Refer to definition (8.96) for the Bregman divergence. Show that $\phi(w)$ is ν-strongly convex with respect to some norm $\|\cdot\|$ if, and only if, $D_\phi(p,q) \geq \frac{\nu}{2}\|p-q\|^2$ for any $p, q \in \text{dom}(\phi)$.

8.63 Refer to definition (8.96) for the Bregman divergence. The function $\phi(w)$ is said to be α-strongly smooth relative to some norm $\|\cdot\|$ if it is differentiable and $D_\phi(p,q) \leq \frac{\alpha}{2}\|p-q\|^2$ for all $p, q \in \text{dom}(\phi)$, where $\alpha \geq 0$. Let $\phi^\star(x)$ denote the conjugate function of some closed convex $\phi(w)$. Show that

$$\begin{aligned} &\phi(w) \text{ is } \nu\text{-strongly convex relative to some norm } \|\cdot\| \iff \\ &\phi^\star(x) \text{ is } \tfrac{1}{\nu}\text{-strongly smooth relative to the dual norm } \|\cdot\|_\star \end{aligned}$$

Argue from the differentiability of $\phi^\star(x)$ and the result of Prob. 8.48 that the equality $\phi^\star(x) = x^\mathsf{T} w - \phi(w)$ holds when $w = \nabla_x\, \phi^\star(x)$. Conclude that

$$\nabla_{x^\mathsf{T}}\, \phi^\star(x) \;=\; \underset{w}{\text{argmax}}\; \Big\{x^\mathsf{T} w - \phi(w)\Big\}$$

Remark. The reader may refer to Zalinescu (2002) and Shalev-Shwartz (2011) for a related discussion.

8.64 We continue with definition (8.96) for the Bregman divergence. Let $\phi(w)$ be a differentiable and strictly convex closed function and consider its conjugate function, $\phi^\star(x)$. Show that

$$D_\phi(p,q) = D_{\phi^\star}\Big(\nabla_{w^\mathsf{T}}\phi(q),\ \nabla_{w^\mathsf{T}}\phi(p)\Big)$$

where the Bregman divergences are computed relative to ϕ and $\phi^\star$ and the arguments are swapped.

8.65 Refer to definition (8.96) for the Bregman divergence. Show that it satisfies

$$D_\phi(p,q) + D_\phi(r,p) - D_\phi(r,q) \;=\; \Big(\nabla_w\phi(p) - \nabla_w\phi(q)\Big)(p-r)$$

8.66 Refer to definition (8.96) for the Bregman divergence. Let $\phi(w) = \frac{1}{2}w^\mathsf{T} Qw$ where $Q > 0$ is symmetric and $w \in \mathbb{R}^M$. Verify that the Bregman divergence in this case reduces to the weighted Euclidean distance shown below, also called the squared Mahalanobis distance:

$$D_\phi(p, q) = \frac{1}{2}(p - q)^\mathsf{T} Q(p - q), \quad p, q \in \mathbb{R}^M$$

8.67 We can extend definition (8.96) for the Bregman divergence to matrix arguments P and Q as follows:

$$D_\phi(P, Q) \triangleq \phi(P) \; - \; \phi(Q) - \mathrm{Tr}\Big(\nabla_{W^\mathsf{T}} \phi(Q)\,(P - Q)\Big)$$

Let $\phi(W) = \mathrm{Tr}(W \ln W - W)$, where W is symmetric positive-definite. If $W = U \Lambda W^\mathsf{T}$ is the eigen-decomposition for W, then $\ln(W)$ is defined as $\ln(W) = U \ln(\Lambda) V^\mathsf{T}$. Show that the resulting Bregman divergence (also called the von Neumann divergence in this case) is given by

$$D_\phi(P, Q) = \mathrm{Tr}\Big(P \ln P - P \ln Q - P + Q\Big)$$

8.68 Continuing with Prob. 8.67, choose now $\phi(W) = -\ln\det(W)$ where $W > 0$ is $M \times M$ symmetric. Show that

$$D_\phi(P, Q) = \mathrm{Tr}(PQ^{-1} - I_M) - \ln\det(PQ^{-1})$$

8.69 Consider a proper convex function $f(w) : \mathbb{R}^M \to \mathbb{R}$ and a closed convex set $\mathcal{C}$ such that $\mathcal{C} \subset \mathrm{dom}(f)$. Consider the optimization problem for a given w_{n-1}:

$$w_n \triangleq \operatorname*{argmin}_{w \in \mathcal{C}} \Big\{ f(w) + D_\phi(w, w_{n-1}) \Big\}$$

Show that

$$f(c) + D_\phi(c, w_{n-1}) \; \geq \; f(w_n) + D_\phi(w_n, w_{n-1}) + D_\phi(c, w_n), \;\; \forall c \in \mathcal{C}$$

8.70 Determine the Bregman divergences corresponding to the choices $\phi(w) = 1/w$ and $\phi(w) = e^w$.

REFERENCES

Abramowitz, M. and I. Stegun (1965), *Handbook of Mathematical Functions*, Dover Publications.

Adamcík, M. (2014), "The information geometry of Bregman divergences and some applications in multi-expert reasoning," *Entropy*, vol. 16, no. 12, pp. 6338–6381.

Archimedes (225 BC), *On the Sphere and Cylinder*, 2 volumes, Greece. See also Heath (1912).

Azoury, K. S. and M. K. Warmuth (2001), "Relative loss bounds for on-line density estimation with the exponential family of distributions," *Mach. Learn.*, vol. 43, pp. 211–246.

Banerjee, A., X. Gou, and H. Wang (2005), "On the optimality of conditional expectation as a Bregman predictor," *IEEE Trans. Inf. Theory*, vol. 51, no. 7, pp. 2664–2669.

Banerjee, A., S. Merugu, I. S. Dhillon, and J. Ghosh (2005), "Clustering with Bregman divergences," *J. Mach. Learn. Res.*, vol. 6, pp. 1705–1749.

Bertsekas, D. P. (1973), "Stochastic optimization problems with nondifferentiable cost functionals," *J. Optim. Theory Appl.*, vol. 12, no. 2, pp. 218–231.

Bertsekas, D. P. (2003), *Convex Analysis and Optimization*, Athena Scientific.

Bertsekas, D. P. (2009), *Convex Optimization Theory*, Athena Scientific.

Bertsekas, D. P., A. Nedic, and A. Ozdaglar (2003), *Convex Analysis and Optimization*, 2nd ed., Athena Scientific.

Boyd, S. and L. Vandenberghe (2004), *Convex Optimization*, Cambridge University Press.

Bregman, L. M. (1967), "The relaxation method of finding the common points of convex sets and its application to the solution of problems in convex programming," *USSR Comput. Math. Math. Phys.*, vol. 7, no. 3, pp. 200–217.

Censor, Y. and S. Zenios (1998), *Parallel Optimization: Theory, Algorithms, and Applications*, Oxford University Press.

Chen, P., Y. Chen, and M. Rao (2008), "Metrics defined by Bregman divergences," *Comm. Math. Sci.*, vol. 6, no. 4, pp. 915–926.

Clarke, F. H. (1983), *Optimization and Nonsmooth Analysis*, Wiley.

Duchi, J. C., P. L. Bartlett, and M. J. Wainwright (2012), "Randomized smoothing for stochastic optimization," *SIAM J. Optim.*, vol. 22, no. 2, pp. 674–701.

Duchi, J., E. Hazan, and Y. Singer (2011), "Adaptive subgradient methods for online learning and stochastic optimization," *J. Mach. Learn. Res.*, vol. 12, pp. 2121–2159.

Dunham, W. (1990), *Journey Through Genius*, Wiley.

Dwilewicz, R. J. (2009), "A history of convexity," *Differ. Geom. Dyn. Syst.*, vol. 11, pp. 112–129.

Eatwell, J., P. Newman, and M. Milgate (1987), *The New Palgrave: A Dictionary of Economics*, Groves Dictionaries.

Ermoliev, Y. M. (1966), "Methods of solutions of nonlinear extremal problems," *Cybernetics*, vol. 2, no. 4, pp. 1–16.

Ermoliev, Y. M. (1969), "On the stochastic quasi-gradient method and stochastic quasi-Feyer sequences," *Kibernetika*, vol. 2, pp. 72–83.

Ermoliev, Y. M. (1976), *Stochastic Programming Methods*, Nauka.

Ermoliev, Y. M. (1983a), "Stochastic quasigradient methods and their application to system optimization," *Stochastic*, vol. 9, pp. 1–36.

Ermoliev, Y. M. (1983b), "Stochastic quasigradient methods," in *Numerical Techniques for Stochastic Optimization*, Y. M. Ermoliev and R.J-.B. Wets, editors, pp. 141–185, Springer.

Ermoliev, Y. M. and N. Z. Shor (1967), "On the minimization of nondifferentable functions," *Cybernetics*, vol. 3, no. 1, pp. 101–102.

Fenchel, W. (1949), "On conjugate convex functions," *Canad. J. Math.*, vol. 1, pp. 73–77.

Fenchel, W. (1983), "Convexity through the ages," in *Convexity and Its Applications*, P. M. Gruber *et al.*, editors, pp. 120–130, Springer.

Goffin, J.-L. (2012), "Subgradient optimization in nonsmooth optimization," *Documenta Mathematica*, Extra Volume ISMP, pp. 277–290.

Hardy, G. H., J. E. Littlewood, and G. Pólya (1934), *Inequalities*, Cambridge University Press.

Harremoës, P. (2017), "Divergence and sufficiency for convex optimization," *Entropy*, vol. 19, no. 5, pp. 1–27.

Heath, J. L. (1912), *The Works of Archimedes*, Dover Publications.

Held, M., P. Wolfe, and H. P. Crowder (1974), "Validation of subgradient optimization," *Math. Program.*, vol. 6, pp. 62–88.

Hiriart-Urruty, J.-B. and C. Lemaréchal (2001), *Fundamentals of Convex Analysis*, Springer.

Hölder, O. L. (1889), "Ueber einen Mittelwertsatz" *Nachrichten von der Königl. Gesellschaft der Wissenschaften und der Georg-Augusts-Universität zu Göttingen* (in German), vol. 1889 no. 2, pp. 38–47.

Jensen, J. (1906), "Sur les fonctions convexes et les inégalités entre les valeurs moyennes," *Acta Mathematica*, vol. 30, no. 1, pp. 175–193.

Kiwiel, K. (1985), *Methods of Descent for Non-differentiable Optimization*, Springer.

Lojasiewicz, S. (1963), "A topological property of real analytic subsets," *Coll. du CNRS,* Les équations aux dérivés partielles, pp. 87–89.

Moreau, J. J. (1965), "Proximité et dualité dans un espace hilbertien," *Bull. Soc. Math. de France*, vol. 93, pp. 273–299.

Nemirovsky, A. S. and D. B. Yudin (1983), *Problem Complexity and Method Efficiency in Optimization*, Wiley.

Nesterov, Y. (2004), *Introductory Lectures on Convex Optimization*, Springer.

Nutini, J., M. Schmidt, I. Laradji, M. Friedlander, and H. Koepke (2015), "Coordinate descent converges faster with the Gauss–Southwell rule than random selection," *Proc. Int. Conf. Machine Learning* (ICML), pp. 1632–1641, Lille.

Polyak, B. T. (1963), "Gradient methods for minimizing functionals," *Zh. Vychisl. Mat. Mat. Fiz.*, vol. 3, no. 4 pp. 643–653.

Polyak, B. T. (1967), "A general method of solving extremal problems," *Soviet Math. Doklady*, vol. 8, pp. 593–597.

Polyak, B. T. (1969), "Minimization of nonsmooth functionals," *Zhurn. Vychisl. Matem. i Matem. Fiz.*, vol. 9, no. 3, pp. 509–521.

Polyak, B. T. (1987), *Introduction to Optimization*, Optimization Software.

Rockafellar, R. T. (1963), *Convex Functions and Dual Extremum Problems*, Ph.D. dissertation, Harvard University, Cambridge, MA.

Rockafellar, R. T. (1970), *Convex Analysis*, Princeton University Press.

Rockafellar, R. T. (1974), *Conjugate Duality and Optimization*, SIAM.

Shalev-Shwartz, S. (2011), "Online learning and online convex optimization," *Found. Trends Mach. Learn.*, vol. 4, no. 2, pp. 107–194.

Shalev-Shwartz, S., Y. Singer, N. Srebro, and A. Cotter (2011b), "Pegasos: Primal estimated sub-gradient solver for SVM," *Math. Program.*, Ser. B, vol. 127, no. 1, pp. 3–30.

Shamir O. and T. Zhang (2013), "Stochastic gradient descent for non-smooth optimization: Convergence results and optimal averaging schemes," *Proc. Int. Conf. Machine Learning* (PMLR), vol. 28, no. 1, pp. 71–79, Atlanta, GA.

Shor, N. Z. (1962), "Application of the method of gradient descent to the solution of the network transportation problem," in *Materialy Naucnovo Seminara po Teoret i Priklad. Voprosam Kibernet. i Issted. Operacii, Nucnyi Sov. po Kibernet*, Akad. Nauk Ukrain. SSSR, pp. 9–17 (in Russian).

Shor, N. Z. (1979), *Minimization Methods for Non-differentiable Functions and Their Applications*, Naukova Dumka.

Shor, N. Z. (1991), "The development of numerical methods for nonsmooth optimization in the USSR," in *History of Mathematical Programming*, J. K. Lenstra, A. H. G. Rinnoy Kan, and A. Shrijver, editors, pp. 135–139, North-Holland.

Shor, N. Z. (1998), *Nondifferentiable Optimization and Polynomial Problems*, Kluwer.

Shor, N. Z. (2012), *Minimization Methods for Non-differentiable Functions*, Springer.

Siahkamari, A., X. Xia, V. Saligrama, D. Castanon, and B. Kulis (2020), "Learning to approximate a Bregman divergence," *Proc. Advances. Neural Information Processing Systems* (NIPS), pp. 1–10, Vancouver.

Ying, B. and A. H. Sayed (2018), "Performance limits of stochastic sub-gradient learning, part I: Single-agent case," *Signal Process.*, vol. 144, pp. 271–282.

Zalinescu, C. (2002), *Convex Analysis in General Vector Spaces*, World Scientific.

9 Convex Optimization

We build on the theory of convex functions from the last chapter and introduce two useful tools: (a) the use of duality arguments to solve convex optimization problems with inequality and equality constraints, and (b) the computation of projections onto convex sets.

9.1 CONVEX OPTIMIZATION PROBLEMS

We will encounter in our treatment several optimization problems involving convex cost functions subject to linear or convex constraints. In this section, we explain how these problems can be solved by duality arguments. Although the methodology holds more generally and can be applied to optimization problems involving cost functions and/or constraints that are not necessarily convex, it is sufficient for our purposes here to focus on convex optimization problems.

9.1.1 Constrained Optimization

Consider a *convex* scalar-valued and differentiable cost function $P(w) \in \mathbb{R}^M \rightarrow \mathbb{R}$, and a general constrained optimization problem involving N inequality constraints and L equality constraints:

$$\left\{\begin{array}{lll} w^{\star} = \underset{w\in\mathbb{R}^M}{\text{argmin}} & P(w) & \\ \text{subject to} & g_n(w) \leq 0, & n = 1, 2, \ldots, N \\ & h_\ell(w) = 0, & \ell = 1, 2, \ldots, L \end{array}\right. \qquad \left(\begin{array}{c}\textbf{primal}\\ \textbf{problem}\end{array}\right) \tag{9.1}$$

The scalar-valued functions $\{g_n(w)\}$ define the inequality constraints and they are assumed to be differentiable and *convex* in w; later we comment on the case of nonsmooth $g_n(w)$. Likewise, the scalar-valued functions $\{h_\ell(w)\}$ define the equality constraints and are assumed to be *affine* in w. Recall that a generic *affine function* in w takes the form

$$h_\ell(w) = a_\ell^{\mathsf{T}} w - \kappa_\ell \tag{9.2}$$

for some column vector a_ℓ and scalar κ_ℓ. In this case, the equality constraints can be grouped together into the linear equation:

$$Aw = b, \quad \text{where} \quad A = \begin{bmatrix} a_1^\mathsf{T} \\ a_2^\mathsf{T} \\ \vdots \\ a_L^\mathsf{T} \end{bmatrix}, \quad b = \begin{bmatrix} \kappa_1 \\ \kappa_2 \\ \vdots \\ \kappa_L \end{bmatrix} \tag{9.3}$$

The variable w in (9.1) is called the *primal variable*. The domain of the optimization problem is the set of all points in the domain of $P(w)$ and the domains of $\{g_n(w)\}$, written as:

$$\text{dom(opt)} = \text{dom}(P) \cap \left\{ \bigcap_{n=1}^{N} \text{dom}(g_n) \right\} \tag{9.4}$$

A point $w \in \text{dom(opt)}$ is said to be primal *feasible* if it satisfies the inequality and equality constraints in (9.1).

Problems of the form (9.1) are called *convex optimization problems*: They are constrained problems that involve minimizing a convex function $P(w)$ over a convex set. To see that problem (9.1) is indeed of this form observe first that for each n, the set of vectors w that satisfy $g_n(w) \leq 0$ is a convex set (for this to hold, it is critical that the direction of the inequality be $g_n(w) \leq 0$ and not $g_n(w) \geq 0$). Then, the set of vectors w that satisfy all N conditions $g_n(w) \leq 0$, for $n = 1, \ldots, N$, is convex as well since the intersection of convex sets is itself convex. Likewise, the set of vectors w that satisfy the affine conditions, $h_\ell(w) = 0$ for all ℓ, is convex. Therefore, problem (9.1) is of the form:

$$\begin{cases} w^\star = \underset{w \in \mathbb{R}^M}{\text{argmin}} \ P(w) \\ \text{subject to } w \in \text{some convex set} \end{cases} \tag{9.5}$$

It is straightforward to verify that convex optimization problems of this form can only have *global* minima, denoted by $w^\star$.

Proof: Assume w_1 is a local minimum while $w^\star \neq w_1$ is a global minimum. Then, $P(w^\star) < P(w_1)$. The local minimum condition means that there exists a small neighborhood around w_1, say, $\|w - w_1\| \leq \delta$ for some $\delta > 0$, such that $P(w_1) < P(w)$ within this region. Clearly, point $w^\star$ is outside this region so that $\|w_1 - w^\star\| > \delta$. Introduce the positive scalar:

$$\theta \triangleq \frac{\delta}{2\|w_1 - w^\star\|} \tag{9.6}$$

which is smaller than 1, and select the convex combination vector:

$$w_2 = \theta w^\star + (1 - \theta) w_1 \tag{9.7}$$

Then, note that

$$\begin{aligned} \|w_1 - w_2\| &= \|w_1 - \theta w^\star - (1-\theta) w_1\| \\ &= \|\theta(w_1 - w^\star)\| \\ &= \delta/2 < \delta \end{aligned} \tag{9.8}$$

Moreover, it follows from the convexity of $P(w)$ and the assumed conditions that

$$\begin{aligned} P(w_2) &\leq \theta P(w^\star) + (1-\theta)P(w_1) \\ &= \theta(P(w^\star) - P(w_1)) + P(w_1) \\ &< P(w_1) \end{aligned} \tag{9.9}$$

since $\theta(P(w^\star) - P(w_1)) < 0$. Therefore, we found a point w_2 inside the neighborhood around w_1, i.e., $\|w_1 - w_2\| < \delta$, and satisfying $P(w_2) < P(w_1)$. This conclusion contradicts the fact that w_1 is a local minimum within this neighborhood.

■

Moreover, when $P(w)$ happens to be strictly convex, then the optimization problem (9.1) will have a *unique* global solution – see Prob. 9.22. One special case of (9.1) is problems with only equality constraints, such as

$$\min_{w\in\mathbb{R}^M} P(w), \quad \text{subject to } h_\ell(w) = 0, \;\; \ell = 1, 2, \ldots, L \tag{9.10}$$

The construction in the next section is applicable to these cases as well (it reduces to the traditional Lagrange multiplier approach).

Example 9.1 **(Linear and quadratic programs)** A special case of (9.1) arises when the cost function and the inequality constraints are *affine*, resulting in what is known as a *linear program*. More specifically, a linear program is a convex optimization problem of the following form:

$$\left\{ \begin{array}{lll} w^\star = \underset{w\in\mathbb{R}^M}{\text{argmin}} \; c^\mathsf{T} w & & \\ \text{subject to} & g_n^\mathsf{T} w - \alpha_n \leq 0, & n = 1, 2, \ldots, N \\ & a_\ell^\mathsf{T} w - \kappa_\ell = 0, & \ell = 1, 2, \ldots, L \end{array} \right. \tag{9.11}$$

where the $\{c, g_n, a_\ell\}$ are column vectors and the $\{\alpha_n, \kappa_\ell\}$ are scalars. If we introduce the matrix and vector quantities:

$$A = \begin{bmatrix} a_1^\mathsf{T} \\ a_2^\mathsf{T} \\ \vdots \\ a_L^\mathsf{T} \end{bmatrix}, \quad b = \begin{bmatrix} \kappa_1 \\ \kappa_2 \\ \vdots \\ \kappa_L \end{bmatrix}, \quad G = \begin{bmatrix} g_1^\mathsf{T} \\ g_2^\mathsf{T} \\ \vdots \\ g_N^\mathsf{T} \end{bmatrix}, \quad d = \begin{bmatrix} \alpha_1 \\ \alpha_2 \\ \vdots \\ \alpha_N \end{bmatrix} \tag{9.12}$$

then we can rewrite (9.11) more compactly as

$$\left\{ \begin{array}{ll} w^\star = \underset{w\in\mathbb{R}^M}{\text{argmin}} & c^\mathsf{T} w \\ \text{subject to} & Gw \preceq d \\ & Aw = b \end{array} \right. \qquad \left(\textbf{linear program} \right) \tag{9.13}$$

where the notation $x \preceq y$ for two vectors amounts to elementwise inequalities. In the same vein, a *quadratic program* is one of the form:

$$\left\{ \begin{array}{ll} w^\star = \underset{w\in\mathbb{R}^M}{\text{argmin}} & \left\{ w^\mathsf{T} Q w + c^\mathsf{T} w \right\}, \;\; Q \geq 0 \\ \text{subject to} & Gw \preceq d \\ & Aw = b \end{array} \right. \qquad \left(\textbf{quadratic program} \right) \tag{9.14}$$

Example 9.2 **(Other problems)** Many formulations can be recast as linear programs. We provide one useful example and show that the problem listed on the left (also known as *basis pursuit* and studied later in Section 51.4.2 and Prob. 51.7) is equivalent to the linear program listed on the right:

$$\left\{\begin{array}{l} w^\star = \underset{w \in \mathbb{R}^M}{\text{argmin}}\ \|w\|_1 \\ \text{subject to } Aw = b \end{array}\right\} \iff \left\{\begin{array}{ll} (w^o, z^o) = & \underset{z, w \in \mathbb{R}^M}{\text{argmin}}\ \mathbb{1}^\mathsf{T} z \\ & e_m^\mathsf{T} w \leq z_m \\ \text{subject to} & -e_m^\mathsf{T} w \leq z_m \\ & Aw = b \end{array}\right\} \tag{9.15}$$

where the $\{e_m\}$ denote the basis vectors in $\mathbb{R}^M$ and the $\{z_m\}$ are the individual entries of z. The first two conditions on w on the right-hand side require the entries of w to lie within the interval $w_m \in [-z_m, z_m]$; this requirement is stated in the form of two inequalities as required by the standard notation in (9.13). We prove that both problems are equivalent, so that $w^\star = w^o$, as follows.

Proof of (9.15): Let $w^\star$ denote a solution to the problem on the left (the one involving minimizing the ℓ_1-norm of w), and introduce $z^\star = |w^\star|$. This notation is used to mean that the vector $z^\star$ consists of the absolute values of the individual entries of $w^\star$. Then, the pair $(w^\star, z^\star)$ is feasible for the linear program on the right-hand side (meaning that it satisfies the constraints). Moreover, note that

$$\|w^\star\|_1 \overset{(a)}{=} \mathbb{1}^\mathsf{T} z^\star \overset{(b)}{\geq} \mathbb{1}^\mathsf{T} z^o \overset{(c)}{\geq} \|w^o\|_1 \tag{9.16}$$

where step (a) is by construction of $z^\star$, step (b) is because z^o is an optimal solution for the linear program, and step (c) is because $-z^o_m \leq w^o_m \leq z^o_m$ from the constraints in the linear program. The above inequality means that w^o must also solve the ℓ_1-optimization problem on the left. In other words, any solution to the linear program on the right must also be a solution to the ℓ_1-optimization problem on the left. Conversely, note that

$$\mathbb{1}^\mathsf{T} z^\star \overset{(a)}{=} \|w^\star\|_1 \overset{(b)}{\leq} \|w^o\|_1 \overset{(c)}{\leq} \mathbb{1}^\mathsf{T} z^o \tag{9.17}$$

where step (a) is again by construction of $z^\star$, step (b) is because $w^\star$ is an optimal solution for the ℓ_1-optimization problem, and step (c) is because $-z^o_m \leq w^o_m \leq z^o_m$ from the constraints in the linear program. The above inequality means that $(w^\star, z^\star)$ must solve the linear program on the right.

■

In a similar vein, we can show that the following ℓ_1-regularized least-squares problem with $\alpha > 0$ can be reformulated as a quadratic program (see Prob. 9.20):

$$w^\star = \underset{w \in \mathbb{R}^M}{\text{argmin}} \left\{\alpha\|w\|_1 + \|d - Hw\|^2\right\} \iff$$

$$\left\{\begin{array}{ll} (w^o, z^o) = & \underset{z, w \in \mathbb{R}^M}{\text{argmin}} \left\{w^\mathsf{T} H^\mathsf{T} H w - 2d^\mathsf{T} H w + \alpha \mathbb{1}^\mathsf{T} z\right\} \\ \text{subject to} & \begin{array}{l} e_m^\mathsf{T} w \leq z_m \\ -e_m^\mathsf{T} w \leq z_m \end{array} \end{array}\right. \tag{9.18}$$

9.1.2 Solution by Duality Arguments

Returning to (9.1), one useful solution technique is by means of Lagrange duality arguments. We motivate the method here and delay the justifications to the next section.

We start by introducing the Lagrangian function, which is a function of three parameters as follows:

$$\boxed{\mathcal{L}(w, \lambda_n, \beta_\ell) \stackrel{\Delta}{=} P(w) + \sum_{n=1}^{N} \lambda_n g_n(w) + \sum_{\ell=1}^{L} \beta_\ell h_\ell(w)} \tag{9.19}$$

where the $\{\lambda_n \geq 0\}$ are *nonnegative* scalar coefficients and the $\{\beta_\ell \in \mathbb{R}\}$ are *real* scalar coefficients; the former variables are associated with the inequality constraints while the latter variables are associated with the equality constraints. The scalars $\{\lambda_n, \beta_\ell\}$ are called by various names, including Lagrange multipliers, Karush–Kuhn–Tucker (or KKT) multipliers, or more broadly *dual variables.* Note that we are requiring the $\{\lambda_n\}$ to be nonnegative scalars while the $\{\beta_\ell\}$ are not restricted. It is clear that

$$\mathcal{L}(w, \lambda_n, \beta_\ell) \leq P(w), \quad \text{at any feasible point } w \tag{9.20}$$

One of the advantages of working with the Lagrangian function (9.19) is that it allows the conversion of the constrained optimization problem into an unconstrained form. Thus, consider now the problem of minimizing the Lagrangian function over w:

$$w^o = \underset{w \in \mathbb{R}^M}{\text{argmin}}\; \mathcal{L}(w, \lambda_n, \beta_\ell) \tag{9.21}$$

We denote the potential solution by w^o, which should satisfy

$$\nabla_w \mathcal{L}(w, \lambda_n, \beta_\ell)\Big|_{w=w^o} = 0 \tag{9.22}$$

This condition is equivalent to

$$\nabla_w P(w^o) + \sum_{n=1}^{N} \lambda_n \nabla_w g_n(w^o) + \sum_{\ell=1}^{L} \beta_\ell \nabla_w h_\ell(w^o) = 0 \tag{9.23}$$

Obviously, the solution w^o will depend on the choices for the scalars $\{\lambda_n, \beta_\ell\}$. We can denote this dependency explicitly by writing $w^o(\lambda_n, \beta_\ell)$. Once we determine w^o and substitute back into expression (9.19) for $\mathcal{L}(w, \lambda_n, \beta_\ell)$, we arrive at the *dual function* denoted by:

$$\boxed{\mathcal{D}(\lambda_n, \beta_\ell) \stackrel{\Delta}{=} \mathcal{L}(w^o, \lambda_n, \beta_\ell)} \quad (\textbf{dual function}) \tag{9.24}$$

This function is only dependent on the *dual* variables $\{\lambda_n, \beta_\ell\}$; the dependency on w is eliminated. One useful property of the dual function is that it is *always* concave in its arguments $\{\lambda_n, \beta_\ell\}$ even if the original cost $P(w)$ is not convex.

This is because the dual function is constructed as the pointwise minimum over w of the Lagrangian $\mathcal{L}(w, \lambda_n, \beta_\ell)$:

$$\mathcal{D}(\lambda_n, \beta_\ell) \;=\; \min_{w\in \mathbb{R}^M} \; \mathcal{L}(w, \lambda_n, \beta_\ell) \tag{9.25}$$

Since $\mathcal{L}(w, \lambda_n, \beta_\ell)$ is affine and, hence, concave over each of the variables $\{\lambda_n, \beta_\ell\}$, it follows that $\mathcal{D}(\lambda_n, \beta_\ell)$ will be concave over these same parameters. This is because the minimum of a collection of concave functions is concave. Now, let $\{\lambda_n^o, \beta_\ell^o\}$ denote solutions to the *dual* problem defined as:

$$\boxed{\{\lambda_n^o, \beta_\ell^o\} = \underset{\lambda_n \geq 0, \beta_\ell}{\text{argmax}} \; \mathcal{D}(\lambda_n, \beta_\ell)} \qquad (\textbf{dual problem}) \tag{9.26}$$

Since the β_ℓ variables are unconstrained, the solutions $\{\beta_\ell^o\}$ should satisfy:

$$\left.\frac{\partial \mathcal{D}(\lambda_n, \beta_\ell)}{\partial \beta_\ell}\right|_{\beta=\beta_\ell^o} = 0, \;\; \ell = 1, 2\ldots, L \tag{9.27}$$

Once the optimal dual variables $\{\lambda_n^o, \beta_n^o\}$ are determined, they define the optimal vector $w^o(\lambda_n^o, \beta_\ell^o)$. The question of interest is whether this vector relates to the desired optimizer $w^\star$ for the original problem (9.1).

9.1.3 Karush–Kuhn–Tucker Conditions

Necessary conditions for the vector $w^o(\lambda_n^o, \beta_\ell^o)$ found in this manner to coincide with $w^\star$ are given by the Karush–Kuhn–Tucker (KKT) conditions, namely, it should hold for every $n = 1, 2, \ldots, N$ and $\ell = 1, 2, \ldots, L$ that (see the derivation further ahead):

$$g_n(w^o) \leq 0 \qquad \text{(feasibility of primal problem)} \tag{9.28a}$$

$$h_\ell(w^o) = 0 \qquad \text{(feasibility of primal problem)} \tag{9.28b}$$

$$\lambda_n^o \geq 0 \qquad \text{(feasibility of dual problem)} \tag{9.28c}$$

$$\lambda_n^o \, g_n(w^o) = 0 \qquad \text{(complementary condition)} \tag{9.28d}$$

$$\left.\left(\nabla_w \, P(w) \; + \; \sum_{n=1}^{N} \lambda_n^o \, \nabla_w \, g_n(w) \; + \; \sum_{\ell=1}^{L} \beta_\ell^o \, \nabla_w \, h_\ell(w)\right)\right|_{w=w^o} = 0 \tag{9.28e}$$

where the last condition follows from (9.22) and amounts to requiring w^o to be a stationary point for the Lagrangian function. The first two conditions (9.28a)–(9.28b) are required by the statement of the original optimization problem; they amount to feasibility conditions for the primal problem. The third condition (9.28c) is due to the inequality constraints. The last condition (9.28d) is known as the KKT *complementary condition*; it shows that zero values for λ_n^o will occur for constraints $g_n(w^o)$ that are satisfied strictly, i.e.,

$$\begin{cases} g_n(w^o) < 0 \;\; \Longrightarrow \;\; \lambda_n^o = 0 \\ \quad \lambda_n^o > 0 \;\;\; \Longrightarrow \;\; g_n(w^o) = 0 \end{cases} \tag{9.29}$$

The KKT conditions are in general only necessary but they are *both* necessary and sufficient when strong duality holds (e.g., under the Slater condition stated in (9.58a)).

Example 9.3 **(Water-filling strategy)** Let us employ the KKT conditions to arrive at the water-filling strategy for the allocation of power among several communication channels. Let $h_m \in \mathbb{R}$ denote the gain of a channel over which iid data with power p_m is transmitted – see Fig. 9.1. The power of the received signal over each channel is $|h_m|^2 p_m$.

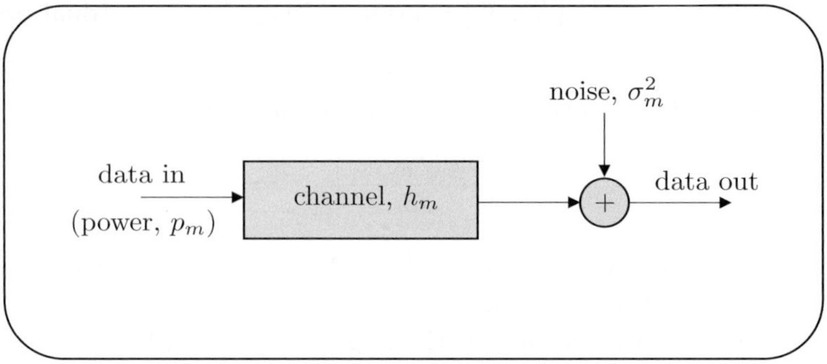

Figure 9.1 Independent and identically distributed data with power p_m is scaled by h_m when transmitted through a channel under additive noise with power σ_m^2. The power of the signal at the output of the channel is $|h_m|^2 p_m$ prior to corruption by the noise component.

The channel capacity is a measure of how much information can be transmitted reliably in the presence of additive noise; it is measured in bits/second/Hz and its value is proportional to

$$C_m \propto \ln\left(1 + \frac{|h_m|^2 p_m}{\sigma_m^2}\right) \tag{9.30}$$

where σ_m^2 is the noise power at the receiver. The ratio $|h_m|^2 p_m/\sigma_m^2$ is the resulting signal-to-noise ratio at the receiver. Now assume we have M channels, $\{h_m\}$ for $m = 1, 2, \ldots, M$, and a total power budget of $p > 0$. We would like to allocate the power distribution among the channels by solving the following constrained optimization problem:

$$\begin{aligned} &\max_{\{p_m\}} \sum_{m=1}^{M} \ln\left(1 + \frac{|h_m|^2 p_m}{\sigma_m^2}\right) \\ &\text{subject to } p_m \geq 0, \; m = 1, 2, \ldots, M \\ &\qquad\qquad \sum_{m=1}^{M} p_m = p \end{aligned} \tag{9.31}$$

We can find closed-form expressions for the optimal power allocations, $\{p_m^o\}$, across all M channels by appealing to the KKT conditions (9.28a)–(9.28d). Observe first that the above optimization problem involves maximization rather than minimization. This can be easily handled by minimizing the negative of the cost function. Observe further that the inequality constraints $p_m \geq 0$ can be replaced by $-p_m \leq 0$. In this way, the above problem now fits into the description (9.1) with

$$\begin{aligned}
w &\triangleq \text{col}\{p_1, p_2, \ldots, p_M\} && (9.32a)\\
P(w) &\triangleq -\sum_{m=1}^{M} \ln\left(1 + \frac{|h_m|^2 e_m^{\mathsf T} w}{\sigma_m^2}\right) && (9.32b)\\
h(w) &\triangleq p - \mathbb{1}^{\mathsf T} w && (9.32c)\\
g_m(w) &\triangleq -e_m^{\mathsf T} w \le 0,\ m = 1, 2 \ldots, M && (9.32d)
\end{aligned}$$

where the $\{e_m\}$ denote basis vectors in $\mathbb{R}^M$. We are therefore dealing with a convex optimization problem involving affine inequality and equality constraints; strong duality (defined further ahead) holds for this problem and the KKT conditions become both necessary and sufficient for optimality. For simplicity of presentation, we continue to work with the individual entries $\{p_m\}$ of w rather than with w itself. Let $c_m = \sigma_m^2/|h_m|^2$. We start by introducing the Lagrangian function:

$$\mathcal{L}(p_m, \lambda_n, \beta_\ell) = -\sum_{m=1}^{M} \ln\left(1 + \frac{p_m}{c_m}\right) - \sum_{m=1}^{M} \lambda_m p_m + \beta\Big(p - \sum_{m=1}^{M} p_m\Big) \tag{9.33}$$

Differentiating relative to each p_m gives

$$\left.\frac{\partial \mathcal{L}}{\partial p_m}\right|_{p_m = p_m^o} = 0 \implies p_m^o + c_m = -1/(\beta + \lambda_m) \tag{9.34}$$

Since $c_m > 0$, this implies that $\beta < -\lambda_m$, which also implies that $\beta < 0$. Next, we could in principle substitute the expression for p_m^o into the Lagrangian function $\mathcal{L}(p_m, \lambda_m, \beta)$ to determine the dual function, $\mathcal{D}(\lambda_m, \beta)$. By differentiating $\mathcal{D}$ relative to the dual variables $\{\lambda_m, \beta\}$ we would then determine their optimal values $\{\lambda_m^o, \beta^o\}$. We follow an alternative simpler route by appealing to the KKT conditions.

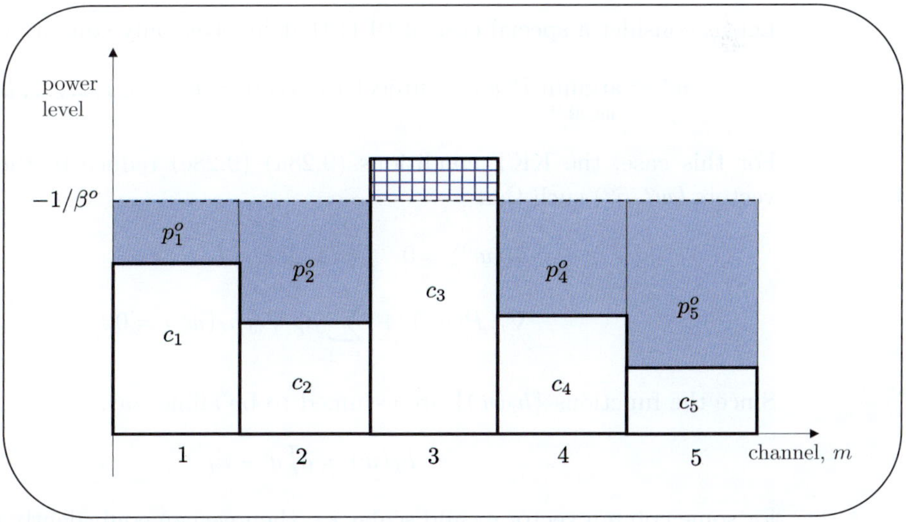

Figure 9.2 Illustration of the water-filling procedure (9.37). Each channel has a starting level dictated by c_m. This value is complemented until it reaches the level $-1/\beta^o$. In this illustration, the power assigned to channel $m = 3$ is $p_3^o = 0$.

First, we know from the complementary condition (9.28d) that the following property should hold:

$$\lambda_m^o p_m^o = 0 \Longrightarrow \lambda_m^o = 0 \text{ when } p_m^o > 0 \tag{9.35}$$

and, therefore, we conclude from (9.34) that

$$p_m^o = \begin{cases} -\frac{1}{\beta^o} - c_m, & \text{when } -1/\beta^o > c_m \\ 0, & \text{otherwise} \end{cases} \tag{9.36}$$

In other words,

$$p_m^o = \max\left\{0, -\frac{1}{\beta^o} - c_m\right\} \tag{9.37}$$

Next, using the equality constraint we must have

$$p = \sum_{m=1}^{M} \max\left\{0, -\frac{1}{\beta^o} - c_m\right\} \tag{9.38}$$

This relation can be solved for β^o, from which p_m^o can be determined. Figure 9.2 illustrates the result and explains why this construction is referred to as a *water-filling* solution. Each channel has a starting power level dictated by c_m. This value is then complemented until it reaches the level $-1/\beta^o$. Channels that happen to be above $-1/\beta^o$ do not receive any power assignment.

9.2 EQUALITY CONSTRAINTS

Let us consider a special case of (9.1) that involves only equality constraints:

$$w^\star = \underset{w\in\mathbb{R}^M}{\text{argmin}}\ P(w), \quad \text{subject to } h_\ell(w) = 0, \ \ \ell = 1, 2, \ldots, L \tag{9.39}$$

For this case, the KKT conditions (9.28a)–(9.28e) reduce to the existence of vectors (w^o, β^o) such that

$$h_\ell(w^o) = 0, \ \ \ell = 1, 2, \ldots, L \tag{9.40a}$$

$$\nabla_w P(w^o) + \sum_{\ell=1}^{L} \beta_\ell^o \nabla_w h_\ell(w^o) = 0 \tag{9.40b}$$

Since the functions $\{h_\ell(w)\}$ are assumed to be affine, say,

$$h_\ell(w) = a_\ell^\mathsf{T} w - \kappa_\ell \tag{9.41}$$

for some column vector a_ℓ and scalar κ_ℓ, then we can equivalently replace (9.39) by

$$\boxed{w^\star = \underset{w\in\mathbb{R}^M}{\text{argmin}}\ P(w), \quad \text{subject to } Aw = b} \tag{9.42}$$

where the rows of A are the vectors $\{a_\ell^\mathsf{T}\}$ and the entries of b are the $\{\kappa_\ell\}$ with $A \in \mathbb{R}^{L\times M}$ and $b \in \mathbb{R}^L$. The KKT conditions (9.40a)–(9.40b) translate into determining vectors (w^o, β^o) such that

$$Aw^o - b = 0 \tag{9.43a}$$

$$\nabla_{w^\mathsf{T}} P(w^o) + A^\mathsf{T}\beta^o = 0 \tag{9.43b}$$

where $\beta^o \overset{\Delta}{=} \text{col}\{\beta_\ell^o\} \in \mathbb{R}^L$. It turns out that these KKT conditions are both necessary and sufficient for the solution of (9.42). We also know from the second row in Table 9.1 that these are the same conditions for solving the *saddle point problem*:

$$\boxed{\min_{w\in\mathbb{R}^M}\ \max_{\beta\in\mathbb{R}^L}\ \left\{P(w) + \beta^\mathsf{T}(Aw - b)\right\}} \tag{9.44}$$

Lemma 9.1. (Existence of optimal solutions) *Assume the convex optimization problem (9.42) is feasible. A vector $w^\star$ solves (9.42) if, and only if, a pair $(w^\star, \beta^\star)$ exists that satisfies*

$$Aw^\star - b = 0 \tag{9.45a}$$

$$\nabla_{w^\mathsf{T}} P(w^\star) + A^\mathsf{T}\beta^\star = 0 \tag{9.45b}$$

Proof: We refer to the result in Prob. 9.30, which establishes that $w^\star$ is a solution to (9.42) if, and only if,

$$\nabla_w P(w^\star)\, v = 0, \quad \text{for any } v \in \mathcal{N}(A) \tag{9.46}$$

Let w be any feasible solution satisfying $Aw = b$. Using the fact that $Aw^\star = b$, we conclude that the difference $(w - w^\star)$ must lie in the nullspace of A, written as

$$w - w^\star \in \mathcal{N}(A) \tag{9.47}$$

It follows from (9.46) that $\nabla_{w^\mathsf{T}} P(w^\star)$ is orthogonal to $w - w^\star$. But since, for any matrix A, it holds that vectors in the range space of A^T, written as $\mathcal{R}(A^\mathsf{T})$, are orthogonal to $\mathcal{N}(A)$, we conclude that $\nabla_{w^\mathsf{T}} P(w^\star)$ must belong to the range space of A^T, which establishes condition (9.45b) for some vector $\beta^\star$.

■

Consider next situations in which the solution $w^\star$ of (9.42) is *unique*. In that case, even under this uniqueness condition, there can exist multiple $\beta^\star$ that satisfy (9.45b) since A need not have full row rank. Nevertheless, among these possibilities there exists a unique dual solution $\beta^\star$ that happens to belong to the range space of A (i.e., $\beta^\star \perp \mathcal{N}(A^\mathsf{T})$). This solution will be useful later in Section 26.1 when we study primal–dual iterative algorithms for the solution of saddle point problems.

Lemma 9.2. (One particular dual variable) *Consider the same setting of Lemma 9.1 and assume $w^\star$ is unique. There exists a unique dual variable, denoted by $\beta_d^\star \in \mathcal{R}(A)$, that satisfies (9.45b).*

Proof: Let $\beta^\star$ denote any dual variable that satisfies (9.45b). We split it into $\beta^\star = \beta_d^\star + \beta_n^\star$ where $\beta_d^\star \in \mathcal{R}(A)$ and $\beta_n^\star \in \mathcal{N}(A^\mathsf{T})$. It follows that if the pair $(w^\star, \beta^\star)$ satisfies (9.45a)–(9.45b), then so does the pair $(w^\star, \beta_d^\star)$. We next establish that $\beta_d^\star$ is unique by contradiction. Assume, otherwise, that there exist two distinct dual solutions in the range space of A, say, $\beta_{d1}^\star = Ax_1$ and $\beta_{d2}^\star = Ax_2$ for some distinct vectors (x_1, x_2). Then, substituting each of these values into the second relation (9.45b) and subtracting both relations, we find that $A^\mathsf{T} A(x_1 - x_2) = 0$. Multiplying from the left by $(x_1 - x_2)^\mathsf{T}$ we get $\|A(x_1 - x_2)\|^2 = 0$, which in turn implies that $Ax_1 = Ax_2$ or, equivalently, $\beta_{d1}^\star = \beta_{d2}^\star$.

■

The proof of the lemma shows that all possible dual solutions $\beta^\star$ share the *same* component $\beta_d^\star$ in the range space of A and vary only in the component $\beta_n^\star$ that lies in the nullspace of A^T. Since $\beta_d^\star$ lies in the range space of A, we can appeal to the result of Prob. 1.48 to conclude that it satisfies

$$\|A^\mathsf{T} \beta_d^\star\|^2 \;\geq\; \sigma_{\min}^2(A)\, \|\beta_d^\star\|^2 \tag{9.48}$$

where $\sigma_{\min}^2(A)$ denotes the smallest nonzero singular value of A. We will use this result in a future section.

9.3 MOTIVATING THE KKT CONDITIONS

We now return to motivate the solution method by duality and justify the KKT conditions. We will examine two problems in this section, namely, a primal problem and its dual version stated as follows:

$$\min_{w \in \mathbb{R}^M} \left\{ \max_{\lambda_n \geq 0, \beta_\ell} \; \mathcal{L}(w, \lambda_n, \beta_\ell) \right\} \qquad \textbf{(primal problem)} \tag{9.49a}$$

$$\max_{\lambda_n \geq 0, \beta_\ell} \left\{ \min_{w \in \mathbb{R}^M} \; \mathcal{L}(w, \lambda_n, \beta_\ell) \right\} \qquad \textbf{(dual problem)} \tag{9.49b}$$

where the order of the minimization and maximization steps is swapped. Two observations are key to the duality argument. The first observation is the fact that the first problem (9.49a) is related to the original constrained optimization problem (9.1), and the second observation is that, under some technical conditions, one can swap the order of the minimization and maximization steps and solve instead the dual problem (9.49b) to arrive at the solution for the original problem. The arguments are as follows.

To begin with, observe that the inner function in the dual problem (9.49b) is the dual function $\mathcal{D}(\lambda_n, \beta_\ell)$ defined earlier in (9.25). We already know that this dual function is *concave* in the arguments $\{\lambda_n, \beta_\ell\}$. We let d^o denote the optimal value of the dual problem, i.e.,

$$d^o \stackrel{\Delta}{=} \mathcal{D}(\lambda_n^o, \beta_\ell^o) \tag{9.50a}$$

where $\{\lambda_n^o, \beta_\ell^o\}$ denote the optimal dual variables:

$$\{\lambda_n^o, \beta_\ell^o\} = \underset{\lambda_n \geq 0, \beta_\ell}{\text{argmax}}\; \mathcal{D}(\lambda_n, \beta_\ell) \tag{9.50b}$$

Next, consider the primal problem (9.49a). We denote its inner function by

$$\mathcal{K}(w) \stackrel{\Delta}{=} \max_{\lambda_n \geq 0, \beta_\ell} \mathcal{L}\,(w,\, \lambda_n,\, \beta_\ell) \tag{9.51}$$

and refer to it as the *primal* function. It is straightforward to verify that $\mathcal{K}(w)$ is *convex* in w. This is because, by definition,

$$\mathcal{K}(w) = \max_{\lambda_n \geq 0, \beta_\ell} \left\{ P(w) + \sum_{n=1}^{N} \lambda_n g_n(w) + \sum_{\ell=1}^{L} \beta_\ell h_\ell(w) \right\} \tag{9.52}$$

where each $g_n(w), h_\ell(w)$, and $P(w)$ are convex in w. Since $\lambda_n \geq 0$, we have that $\lambda_n g_n(w)$ is also convex. Likewise, since $h_\ell(w)$ is affine we have that $\beta_\ell h_\ell(w)$ is convex as well. Therefore, the sum between brackets in the expression for $\mathcal{K}(w)$ is convex in w. And we know that the maximum of a collection of convex functions is again convex. Therefore, $\mathcal{K}(w)$ is convex.

It is important to note that minimizing $\mathcal{K}(w)$ over w excludes infeasible points w (i.e., it excludes points that violate the constraints $g_n(w) \leq 0$ and $h_\ell(w) = 0$). This is because if some inequality constraint is violated, say, $g_{n_o}(w) > 0$ for some index n_o, then $\mathcal{K}(w)$ can be made unbounded by setting the corresponding λ_{n_o} in the maximization problem (9.52) to $\lambda_{n_o} \to \infty$. A similar observation holds if some equality constraint is violated. On the other hand, when w is feasible, optimal values for λ_n are $\lambda_n = 0$; this is because when $g_n(w) < 0$, the value of λ_n cannot be positive since otherwise it decreases the value of the term between brackets in the maximization (9.52). We conclude that

$$\mathcal{K}(w) = \begin{cases} P(w), & \text{for all feasible } w \\ \infty, & \text{otherwise} \end{cases} \tag{9.53}$$

and, therefore, $\mathcal{K}(w)$ is minimized at the global solution $w^\star$ for the original constrained problem (9.1). Let p^o denote the optimal value for this primal problem, i.e.,

$$p^o \stackrel{\Delta}{=} \mathcal{K}(w^\star) \tag{9.54a}$$

where

$$w^\star = \underset{w \in \mathbb{R}^M}{\text{argmin}}\; \mathcal{K}(w) \tag{9.54b}$$

It is now straightforward to verify that

$$\boxed{d^o \leq p^o} \qquad \textbf{(weak duality)} \tag{9.55}$$

This is because

$$\begin{aligned}
d^o &\stackrel{\Delta}{=} \mathcal{D}(\lambda_n^o, \beta_\ell^o) \\
&\stackrel{\Delta}{=} \min_{w \in \mathbb{R}^M} \mathcal{L}(w, \lambda_n^o, \beta_\ell^o) \\
&\leq \mathcal{L}(w^\star, \lambda_n^o, \beta_\ell^o) \\
&\leq \max_{\lambda_n \geq 0, \beta_\ell} \mathcal{L}(w^\star, \lambda_n, \beta_\ell) \\
&\leq \max_{\lambda_n \geq 0, \beta_\ell} \left\{ P(w^\star) + \sum_{n=1}^{N} \lambda_n g_n(w^\star) + \sum_{\ell=1}^{L} \beta_\ell h_\ell(w^\star) \right\} \\
&= \mathcal{K}(w^\star) \\
&= p^o
\end{aligned} \tag{9.56}$$

We conclude that, for any pair of primal–dual problems of the form (9.49a)–(9.49b), *weak duality* holds. We also say that a duality gap exists between both problems.

Assume next that *strong duality* holds, that is,

$$\boxed{d^o = p^o} \qquad \textbf{(strong duality)} \tag{9.57}$$

There are known conditions on the convex functions $\{P(w), g_n(w), h_\ell(w)\}$ to ensure strong duality. These conditions are not of interest here, except perhaps to mention that when $g_n(w)$ and $h_\ell(w)$ are both affine functions of w and $P(w)$ is convex and differentiable, then strong duality will hold. We will encounter these situations in future studies in this text. Strong duality also holds under the following Slater condition.

(Slater condition) *If there exists at least one primal strictly feasible vector $\bar{w}$ satisfying the constraints in (9.1) with*

$$g_n(\bar{w}) < 0, \;\; n = 1, 2, \ldots, N \tag{9.58a}$$

$$h_\ell(\bar{w}) = 0, \;\; \ell = 1, 2, \ldots, L \tag{9.58b}$$

then strong duality holds.

When strong duality is in effect, then the complementary condition (9.28d) must hold:

$$\begin{array}{c}\textbf{(complementary condition)} \\ \text{strong duality} \Longrightarrow \left\{ \lambda_n^o \, g_n(w^\star) = 0, \; \forall \, n \right\}\end{array} \tag{9.59}$$

Proof of (9.59): To see this, note that

$$\begin{aligned} p^o = d^o &\stackrel{\Delta}{=} D(\lambda_n^o, \beta_\ell^o) \\ &= \min_{w \in \mathbb{R}^M} \mathcal{L}(w, \lambda_n^o, \beta_\ell^o) \\ &\leq \mathcal{L}(w^\star, \lambda_n^o, \beta_\ell^o) \\ &= P(w^\star) + \sum_{n=1}^N \lambda_n^o g_n(w^\star) + \sum_{\ell=1}^L \beta_\ell^o h_\ell(w^\star) \\ &= P(w^\star) + \sum_{n=1}^N \lambda_n^o g_n(w^\star), \quad \text{since } h_\ell^o(w^\star) = 0 \\ &\leq P(w^\star), \quad \text{since } \lambda_n^o \geq 0,\ g_n(w^\star) \leq 0 \\ &\stackrel{(9.53)}{=} p^o \end{aligned} \tag{9.60}$$

It follows that all inequalities in this sequence of arguments must be equalities and, hence, it must hold that

$$\sum_{n=1}^N \lambda_n^o\, g_n(w^\star) = 0 \tag{9.61}$$

This is a sum of nonpositive terms. Therefore, each one of the terms must be zero.

■

We summarize the main relations between the primal and dual problems in Table 9.1 for ease of reference.

Nondifferentiable functions

The KKT conditions can be extended to functions $\{P(w), g_n(w), h_\ell(w)\}$ that are not necessarily differentiable. In this case, conditions (9.28a)–(9.28d) will continue to hold. However, expression (9.28e) for determining w^o will employ subgradient vectors instead. Using characterization (8.47), we replace condition (9.28e) by

$$0 \in \partial_w \left(P(w^o) + \sum_{n=1}^N \lambda_n^o\, g_n(w^o) + \sum_{\ell=1}^L \beta_\ell^o\, h_\ell(w^o) \right) \tag{9.62}$$

9.4 PROJECTION ONTO CONVEX SETS

One important concept that arises frequently in the study of inference methods is that of projection onto convex sets. Let $\mathcal{C} \subset \mathbb{R}^M$ denote a closed convex set and let x be any point outside $\mathcal{C}$. We define the distance from x to $\mathcal{C}$ as the quantity:

$$\text{dist}(x, \mathcal{C}) \stackrel{\Delta}{=} \min_{c \in \mathcal{C}} \|x - c\|_2, \qquad \textbf{(distance to } \mathcal{C}\textbf{)} \tag{9.63}$$

which measures the smallest Euclidean distance from x to any point in $\mathcal{C}$. We can of course employ other norms in place of the Euclidean norm. The distance

Table 9.1 Primal and dual optimization problems where the functions $\{P(w), g_n(w), h_\ell(w)\}$ are convex.

<table>
<tr><th>Primal domain</th><th>Dual domain</th></tr>
<tr><td colspan="2">(Lagrangian function): $\mathcal{L} \triangleq P(w) + \sum_{n=1}^{N} \lambda_n g_n(w) + \sum_{\ell=1}^{L} \beta_\ell h_\ell(w)$</td></tr>
<tr><td>(primal problem)
$\begin{cases} w^\star = \underset{w \in \mathbb{R}^M}{\text{argmin}}\ P(w) \\ \text{s.t. } g_n(w) \leq 0,\ n = 1, \ldots, N \\ \qquad h_\ell(w) = 0,\ \ell = 1, \ldots, L \end{cases}$</td><td>(dual problem)
$\begin{cases} \text{(dual function)} \\ \mathcal{D}(\lambda_n, \beta_\ell) \triangleq \min_{w \in \mathbb{R}^M} \mathcal{L}(w, \lambda_n, \beta_\ell) \\ \{\lambda_n^o, \beta_\ell^o\} = \underset{\lambda_n \geq 0, \beta_\ell}{\text{argmax}}\ \mathcal{D}(\lambda_n, \beta_\ell) \end{cases}$</td></tr>
<tr><td>(equivalent primal problem)
$w^\star =$
$\underset{w \in \mathbb{R}^M}{\text{argmin}} \left\{ \max_{\lambda_n \geq 0, \beta_\ell} \mathcal{L}\left(w, \lambda_n, \beta_\ell\right) \right\}$</td><td>(equivalent dual problem)
$\{\lambda_n^o, \beta_\ell^o\} =$
$\underset{\lambda_n \geq 0, \beta_\ell}{\text{argmax}} \left\{ \min_{w \in \mathbb{R}^M} \mathcal{L}\left(w, \lambda_n, \beta_\ell\right) \right\}$</td></tr>
<tr><td>primal variable, w</td><td>dual variables, $\{\lambda_n, \beta_\ell\}$</td></tr>
<tr><td>optimal solution: $w^\star$</td><td>optimal solution: $\{w^o, \lambda_n^o, \beta_\ell^o\}$</td></tr>
<tr><td>optimal cost: $p^o \triangleq P(w^\star)$</td><td>optimal cost: $d^o \triangleq \mathcal{D}(\lambda_n^o, \beta_\ell^o)$</td></tr>
<tr><td colspan="2">weak duality always holds: $d^o \leq p^o$.</td></tr>
<tr><td colspan="2">$w^\star = w^o$ under the Slater condition (9.58a)–(9.58b),
in which case strong duality will hold: $d^o = p^o$</td></tr>
<tr><td colspan="2">necessary conditions for $w^\star = w^o$ (KKT conditions):
$\begin{cases} g_n(w^o) \leq 0 \\ h_\ell(w^o) = 0 \\ \lambda_n^o \geq 0 \\ \lambda_n^o\, g_n(w^o) = 0 \\ \nabla_w\, \mathcal{L}(w^o, \lambda_n^o, \beta_\ell^o) = 0 \end{cases}$</td></tr>
<tr><td colspan="2">KKT conditions are both necessary and sufficient
for $w^\star = w^o$ when strong duality holds</td></tr>
<tr><td colspan="2">strong duality holds when only
equality constraints are involved</td></tr>
</table>

will be zero for points x that lie inside or on the boundary of $\mathcal{C}$. When x is outside $\mathcal{C}$, we define its projection onto $\mathcal{C}$ as the point in $\mathcal{C}$ that is closest to x, i.e., as the point that attains the minimal distance defined by (9.63). We denote the projection by the following notation:

$$\widehat{x} = \mathcal{P}_C(x) \triangleq \underset{c \in \mathcal{C}}{\text{argmin}}\ \|x - c\|_2 \tag{9.64}$$

This is illustrated in Fig. 9.3. The projection is guaranteed to exist and is unique.

Proof: If $x \in \mathcal{C}$, then clearly $\widehat{x} = x$. Assume $x \notin \mathcal{C}$. The squared distance function $f(c) = \|x - c\|^2$ is strongly convex over $c \in \mathcal{C}$ (its Hessian matrix is equal to $2I_M$

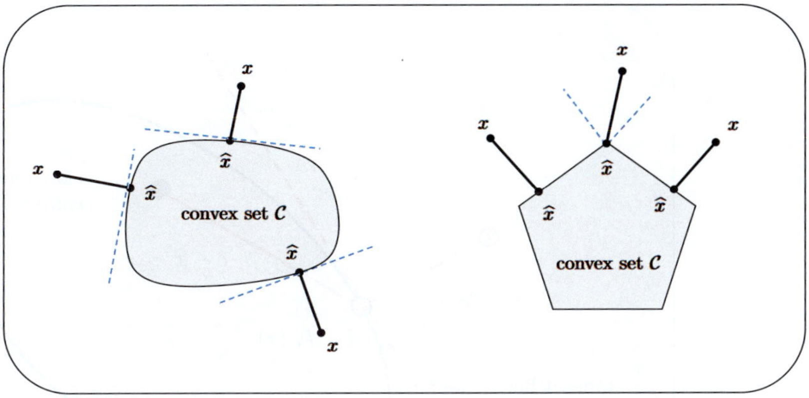

Figure 9.3 Projections of points $x \notin \mathcal{C}$ for two types of convex sets; the one on the right includes sharp edges.

for all c). Therefore, $f(c)$ has a unique global minimum at some point $c^o \in \mathcal{C}$, which would correspond to the projection of x. Another way to establish uniqueness is to assume that there are two distinct projections, $\widehat{x}_1$ and $\widehat{x}_2$, both of them in $\mathcal{C}$ and both of them attaining $\text{dist}(x, \mathcal{C})$. Then, the convex combination $\frac{1}{2}(\widehat{x}_1 + \widehat{x}_2)$ is also in $\mathcal{C}$ and, by invoking the strict convexity of $f(c)$:

$$f\left(\frac{1}{2}(\widehat{x}_1 + \widehat{x}_2)\right) < \frac{1}{2}f(\widehat{x}_1) + \frac{1}{2}f(\widehat{x}_2) = \text{dist}(x, \mathcal{C}) \tag{9.65}$$

This result would imply that point $\frac{1}{2}(\widehat{x}_1 + \widehat{x}_2)$ attains a smaller distance than $\widehat{x}_1$ and $\widehat{x}_2$, which contradicts the fact that $\widehat{x}_1$ and $\widehat{x}_2$ are projections. Therefore, the projection needs to be unique.

■

The projection onto a convex set is characterized by an important inequality:

$$\boxed{\widehat{x} = \mathcal{P}_C(x) \iff (x - \widehat{x})^\mathsf{T}(c - \widehat{x}) \leq 0, \ \ \forall c \in \mathcal{C}} \tag{9.66}$$

In other words, the inner product between the residual, $x - \widehat{x}$, and the vector linking $\widehat{x}$ to any point in $\mathcal{C}$ is necessarily nonpositive; we provide a graphical interpretation for this result in Fig. 9.4.

Proof of (9.66): We need to establish both directions. Assume first that $\widehat{x}$ is the projection of x onto $\mathcal{C}$. Consider any point $c \in \mathcal{C}$ that is different from $\widehat{x}$. For any scalar $\alpha \in (0, 1)$, the convex combination $\alpha c + (1 - \alpha)\widehat{x} \in \mathcal{C}$ by definition. Then, since $\widehat{x}$ is the closest point in $\mathcal{C}$ to x, we have

$$\begin{aligned}
\|x - \widehat{x}\|^2 &\leq \|x - \alpha c - (1 - \alpha)\widehat{x}\|^2 \\
&= \|x - \widehat{x} - \alpha(c - \widehat{x})\|^2 \\
&= \|x - \widehat{x}\|^2 + \alpha^2\|c - \widehat{x}\|^2 - 2\alpha(x - \widehat{x})^\mathsf{T}(c - \widehat{x})
\end{aligned} \tag{9.67}$$

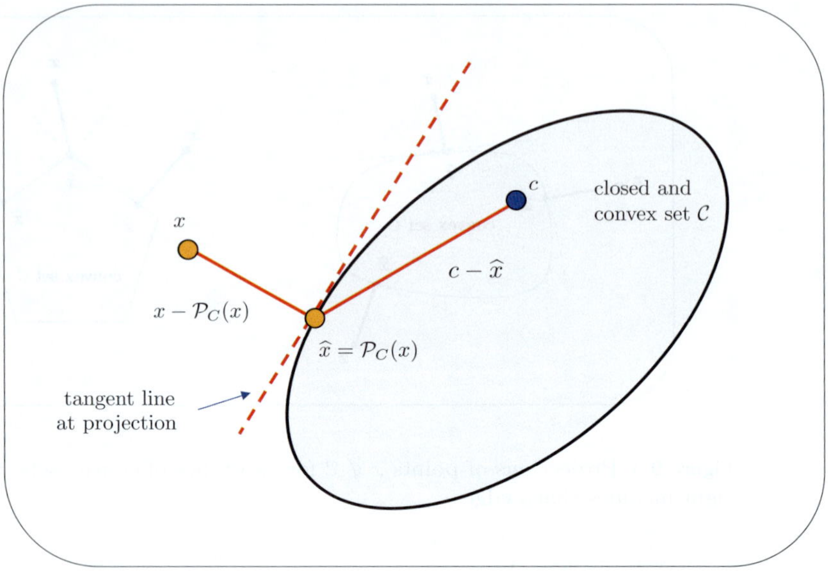

Figure 9.4 The segment connecting x to its projection $\widehat{x}$ is orthogonal to the tangent to $\mathcal{C}$ at $\widehat{x}$. The segment linking this projection to any $c \in \mathcal{C}$ has a negative inner product with the orthogonal segment arriving from x.

from which we conclude that

$$(x-\widehat{x})^{\mathsf{T}}(c-\widehat{x}) \leq \frac{\alpha}{2}\|c-\widehat{x}\|^2 \tag{9.68}$$

The right-hand side can be made arbitrarily small as α approaches zero. It follows that $(x-\widehat{x})^{\mathsf{T}}(c-\widehat{x}) \leq 0$. Conversely, assume a point $\widehat{x} \in \mathcal{C}$ satisfies $(x-\widehat{x})^{\mathsf{T}}(c-\widehat{x}) \leq 0$. Let us verify that it has to coincide with the projection of x onto $\mathcal{C}$. Indeed, note that, for any $c \in \mathcal{C}$ that is different from $\widehat{x}$, we have:

$$\begin{aligned}
\|x-c\|^2 - \|x-\widehat{x}\|^2 &= \|x-\widehat{x}+\widehat{x}-c\|^2 - \|x-\widehat{x}\|^2 \\
&= \underbrace{\|\widehat{x}-c\|^2}_{>0} - 2\underbrace{(x-\widehat{x})^{\mathsf{T}}(c-\widehat{x})}_{\leq 0} \\
&> 0
\end{aligned} \tag{9.69}$$

so that $\|x-\widehat{x}\|^2 < \|x-c\|^2$ for any $c \neq \widehat{x}$ in $\mathcal{C}$.

■

The projection operator satisfies several useful properties. For example, it is *nonexpansive*, meaning that:

$$\boxed{\|\mathcal{P}_C(x) - \mathcal{P}_C(y)\| \leq \|x-y\|, \quad \forall\, x, y \in \mathcal{C}} \tag{9.70}$$

Proof: Let $\widehat{x} = \mathcal{P}_C(x)$ and $\widehat{y} = \mathcal{P}_C(y)$. Let also $\widetilde{x} = x - \widehat{x}$ and $\widetilde{y} = y - \widehat{y}$ denote the projection errors. Then, note that

$$\begin{aligned}\|x-y\|^2 &= \|x+\widehat{x}-\widehat{x}-y+\widehat{y}-\widehat{y}\|^2\\ &= \|(\widehat{x}-\widehat{y})+(\widetilde{x}-\widetilde{y})\|^2\\ &= \|\widehat{x}-\widehat{y}\|^2+\underbrace{\|\widetilde{x}-\widetilde{y}\|^2}_{\geq 0}-2\underbrace{\widetilde{x}^{\mathsf{T}}(\widehat{y}-\widehat{x})}_{\leq 0}-2\underbrace{\widetilde{y}^{\mathsf{T}}(\widehat{x}-\widehat{y})}_{\leq 0}\end{aligned} \tag{9.71}$$

where we applied property (9.66) for the two rightmost terms, from which we conclude that

$$\|\widehat{x}-\widehat{y}\|^2 \leq \|x-y\|^2 \tag{9.72}$$

as claimed.

■

Moreover, we establish in Prob. 9.1 a second useful property for the projection operator, namely, it helps characterize the gradient vector of the distance function to $\mathcal{C}$:

$$\boxed{\nabla_{x^{\mathsf{T}}}\,\text{dist}(x,\mathcal{C}) \;=\; \frac{x-\widehat{x}}{\|x-\widehat{x}\|_2},\quad \widehat{x}=\mathcal{P}_C(x),\quad x\neq\widehat{x}} \tag{9.73}$$

If $x\in\mathcal{C}$, then $\widehat{x}=x$ and the distance of x to $\mathcal{C}$ is zero.

Now, given a convex set $\mathcal{C}$, it is not always possible to compute the projection of some point x onto $\mathcal{C}$ in closed form. There are some important exceptions, however. We collect in Table 9.2 expressions for several useful projection operations and leave the verification of their validity to the problems. The last column in the table gives the relevant reference. Row 9 in the table uses the soft-thresholding function $\mathbb{T}_\lambda(x)$, which is defined as follows:

$$\mathbb{T}_\lambda(x) = \begin{cases} x-\lambda, & \text{if } x\geq\lambda\\ 0, & \text{if } -\lambda<x<\lambda\\ x+\lambda, & \text{if } x\leq-\lambda\end{cases} \tag{9.74}$$

where $\lambda>0$. In other words, values of x within the interval $(-\lambda,\lambda)$ are annihilated, while values outside this interval have their magnitudes decreased by an amount equal to λ.

Bregman projections

We can extend the notion of projections onto convex sets by relying instead on the use of the Bregman divergence (8.96) in definition (9.64) as follows. Let $\mathcal{C}\subset\mathbb{R}^M$ denote a closed convex set. The projection of a point $z\in\mathbb{R}^M$ onto $\mathcal{C}$ is now defined by (observe that z is the second argument in $D_\phi(c,z)$):

$$\boxed{\widehat{z}=\mathcal{P}_{C,\phi}(z) \;\stackrel{\Delta}{=}\; \underset{c\in\mathcal{C}}{\text{argmin}}\; D_\phi(c,z)} \tag{9.75}$$

where we are using the notation $\mathcal{P}_{C,\phi}(z)$, with an additional subscript ϕ to refer to the role played by $\phi(w)$ in determining the Bregman divergence. The minimizer $\widehat{z}$

Table 9.2 List of some useful convex sets along with expressions for computing projections onto them.

	Convex set, $\mathcal{C}$	Projection, $\mathcal{P}_{C}(x)$	Prob.
1.	$\mathcal{C} = \mathcal{R}(A)$, $A \in \mathbb{R}^{N\times M}$ orthonormal columns $\{a_m\}$	$\widehat{x} = \sum_{m=1}^{M} (a_m^{\mathsf{T}} x) a_m$	9.2
2.	$\mathcal{C} = \mathcal{R}(A)$, $A \in \mathbb{R}^{N\times M}$ $M \leq N$, full rank	$\widehat{x} = A(A^{\mathsf{T}}A)^{-1}A^{\mathsf{T}}x$	9.2
3.	$\mathcal{C} = \{h \mid h^{\mathsf{T}}w - \theta = 0\}$ (hyperplane)	$\widehat{x} = x - \frac{(x^{\mathsf{T}}w - \theta)}{\\|w\\|^2} w$	9.3
4.	$\mathcal{C} = \{h \mid Wh - b = 0\}$ $W = \text{col}\{w_1^{\mathsf{T}}, \ldots, w_L^{\mathsf{T}}\}$ $b = \text{col}\{\theta_1, \ldots, \theta_L\}$ (intersection of hyperplanes)	$\widehat{x} = x - W^{\dagger}(Wx - b)$	9.4
5.	$\mathcal{C} = \{h \mid h^{\mathsf{T}}w - \theta \leq 0\}$ (halfspace)	$\widehat{x} = x - \frac{w}{\\|w\\|^2} \max\left\{0, x^{\mathsf{T}}w - \theta\right\}$	9.5
6.	$\mathcal{C} = \mathcal{R}(y)$, $y \in \mathbb{R}^M$	$\widehat{x} = \frac{y^{\mathsf{T}}x}{\\|y\\|^2} y$	9.6
7.	$\mathcal{C} = \{h \mid \\|h - a\\|_2 \leq r\}$ (Euclidean ball of radius r)	$\widehat{x} = \begin{cases} x, \text{ if } \\|x - a\\|_2 \leq r \\ a + r\frac{x - a}{\\|x - a\\|}, \text{ otherwise} \end{cases}$	9.7
8.	$\mathcal{B} = \{h \mid \\|h - a\\|_\infty \leq r\}$ (∞-ball of radius r)	$[\widehat{x}]_m = \begin{cases} x_m, \text{ if } \lvert x_m - a_m\rvert \leq r \\ a_m + r\,\text{sign}(x_m - a_m), \\ \text{otherwise} \end{cases}$	9.8
9.	$\mathcal{B} = \{h \mid \\|h\\|_1 \leq 1\}$ (ℓ_1-ball of unit radius)	• $\lambda = 0$ if $\\|x\\|_1 \leq 1$, otherwise • solve $\sum_{m=1}^{M}(\lvert x_m\rvert - \lambda)_+ = 1$. • $\widehat{x} = \mathbb{T}_\lambda(x)$	9.10
10.	$\mathcal{R} = \{h \mid \ell_m \leq h_m \leq u_m\}$ $h = \text{col}\{h_m\}$ (rectangular slab)	$[\widehat{x}]_m = \begin{cases} \ell_m, \text{ if } x_m \leq \ell_m \\ x_m, \text{ if } \ell_m \leq x_m \leq u_m \\ u_m, \text{ if } x_m \geq u_m \end{cases}$	9.9
11.	$\mathcal{S} = \{s \mid s_m \geq 0,\ \mathbb{1}^{\mathsf{T}}s = 1\}$ $s = \text{col}\{s_m\}$ (simplex)	• order $x_1 \geq x_2 \geq \ldots \geq x_M$. • $b_m = x_m + \frac{1}{m}(1 - \sum_{j=1}^{m} x_j)$. • largest m^o for which $b_{m^o} > 0$. • $\lambda^o = \frac{1}{m^o}(1 - \sum_{j=1}^{m_o} x_j)$. • $[\widehat{x}]_m = \max\{0, x_m + \lambda^o\}$	9.11

exists and is unique since, as we already know, the divergence $D_\phi(c, z)$ is strictly convex over c. Moreover, in a manner similar to the arguments used to establish inequality (9.66), we can similarly verify that (see Prob. 9.14):

$$\boxed{\widehat{z} = \mathcal{P}_{C,\phi}(z) \iff \Big(\nabla_x \phi(z) - \nabla_x \phi(\widehat{z})\Big)(c - \widehat{z}) \leq 0, \;\; \forall c \in \mathcal{C}} \tag{9.76}$$

and that this characterization is further equivalent to the following:

$$\widehat{z} = \mathcal{P}_{C,\phi}(z) \iff D_\phi(c, z) \geq D_\phi(c, \widehat{z}) + D_\phi(\widehat{z}, z), \;\; \forall c \in \mathcal{C} \tag{9.77}$$

This last inequality amounts to a generalization of the classical Pythagoras relation in Euclidean space. It also holds that

$$D_\phi(c, z) \geq D_\phi(c, \widehat{z}), \; \forall \, c \in \mathcal{C} \tag{9.78}$$

Example 9.4 **(Projection under negative entropy function)** We illustrate Bregman projections by considering the case of the negative entropy function (8.103). Consider the simplex set:

$$\mathcal{C} \;=\; \left\{ w \in \mathbb{R}^M \;\middle|\; w_m \geq 0, \;\; \sum_{m=1}^{M} w_m = 1 \right\} \tag{9.79}$$

which is a closed convex set, and the positive orthant:

$$\mathcal{C}_\phi \;=\; \left\{ w \in \mathbb{R}^M \;\middle|\; w_m > 0 \right\} \tag{9.80}$$

which is an open set. The function $\phi(w)$ is well-defined over $\mathcal{C}_\phi$. Pick a point $z \in \mathcal{C}_\phi \backslash \mathcal{C}$, i.e., a vector z with positive entries that do not add up to 1. We wish to compute the Bregman projection of z onto the intersection $\mathcal{C} \cap \mathcal{C}_\phi$ using the definition:

$$\widehat{z} = \underset{c \in \mathcal{C} \cap \mathcal{C}_\phi}{\text{argmin}} \; D_\phi(c, z) \tag{9.81}$$

It is easy to see in this case that the Bregman divergence is given by

$$\begin{aligned} D_\phi(c, z) &= \sum_{m=1}^{M} c_m \ln c_m - \sum_{m=1}^{M} z_m \ln z_m - \sum_{m=1}^{M} (1 + \ln z_m)(c_m - z_m) \\ &= \sum_{m=1}^{M} \Big\{ c_m \ln(c_m/z_m) - c_m + z_m \Big\} \\ &= \sum_{m=1}^{M} \Big\{ c_m \ln(c_m/z_m) + z_m \Big\} - 1 \end{aligned} \tag{9.82}$$

since $\sum_{m=1}^{M} c_m = 1$ for all $c \in \mathcal{C} \cap \mathcal{C}_\phi$. It follows that

$$\underset{c \in \mathcal{C} \cap \mathcal{C}_\phi}{\text{argmin}} \Big\{ D_\phi(c, z) \Big\} = \underset{c \in \mathcal{C} \cap \mathcal{C}_\phi}{\text{argmin}} \left\{ \sum_{m=1}^{M} c_m \ln(c_m/z_m) \right\} \tag{9.83}$$

One can solve this constrained optimization problem using a Lagrangian argument, where the constraint arises from the positive entries of c_m adding up to 1. Let us ignore the fact that the entries need to be positive and focus on their sum evaluating to 1. We introduce the Lagrangian function

$$\mathcal{L}(c, \lambda) = \sum_{m=1}^{M} c_m \ln(c_m/z_m) + \lambda\Big(\sum_{m=1}^{M} c_m - 1 \Big) \tag{9.84}$$

Differentiating relative to c_m and setting the derivative to zero gives

$$\ln(c_m/z_m) + 1 + \lambda = 0 \Longrightarrow c_m = z_m e^{-(\lambda+1)} \tag{9.85}$$

Using $\sum_{m=1}^{M} c_m = 1$ we find

$$e^{-(\lambda+1)} = \left(\sum_{m=1}^{M} z_m\right)^{-1} = 1/\|z\|_1 \tag{9.86}$$

Consequently, $c_m = z_m/\|z\|_1$ (which are positive) and we conclude that the Bregman projection is given by

$$\boxed{\widehat{z} = z/\|z\|_1} \tag{9.87}$$

9.5 COMMENTARIES AND DISCUSSION

Convex optimization. The intent of the chapter is to provide a concise introduction to the topic of convex optimization without being exhaustive, with particular attention to concepts that are sufficient for our treatment in this text. There is already an extensive literature on convex optimization problems, along with expansive treatments on various effective techniques that are available for the solution of such problems. Excellent references on convex analysis and projections onto convex sets are the texts by Rockafellar (1970), Polyak (1987), Hiriart-Urruty and Lemaréchal (2001), Bertsekas (2003, 2009), Boyd and Vandenberghe (2004), Nesterov (2004), and Dattoro (2016). The material in the body of the chapter focused on the interplay between primal and dual problems, the corresponding KKT conditions, and the duality gap. This interplay is often called upon to transform an optimization problem to a "simpler" form in the dual domain and vice-versa. One prominent example of the use of the duality theory in the context of learning methods will arise in the design of support vector machines later in Chapter 61.

Duality and KKT conditions. We introduced the KKT conditions in Section 9.1 and illustrated their application to a water-filling problem in Example 9.3 by adapting a similar discussion from Boyd and Vandenberghe (2004). We limited our discussion to convex optimization problems with a convex cost subject to convex inequality constraints and affine equality constraints. These are the types of problems we will encounter most often in our treatment. According to the historical accounts by Kuhn (1976), Kjeldsen (2000), and Cottle (2012), the KKT conditions appeared originally in the 1939 unpublished master's thesis of William Karush at the University of Chicago, and were rediscovered independently almost a decade later in the conference article by Kuhn and Tucker (1950). The article by Kuhn (1976) includes an appendix that presents the unpublished results of Karush (1939). Another similar contribution noted in these accounts is the work by John (1948). The account by Kuhn (2002) indicates that Kuhn and Tucker were made aware of John's work during their preparation of the galley proofs for their article and included a reference to his work in their article. According to the same account, they became aware of Karush's work only later in 1974 through a reference in the book by Takayama (1974). The Slater condition appeared originally in Slater (1950). For further results on the KKT conditions, and on the general field of nonlinear programming and the concept of duality in optimization, readers may consult the texts by Dantzig (1963), Rockafellar (1970), Fletcher (1987), Bertsekas (1995),

Avriel (2003), Boyd and Vandenberghe (2004), Ruszczynski (2006), and Luenberger and Ye (2008).

PROBLEMS

9.1 Let $\mathcal{C} \subset \mathbb{R}^M$ denote a closed convex set and let x be any point outside $\mathcal{C}$. The distance from x to $\mathcal{C}$ is defined as $\text{dist}(x, \mathcal{C}) = \min_{c \in \mathcal{C}} \|x - c\|_2$, with the projection of x onto $\mathcal{C}$ given by $\widehat{x} = \mathcal{P}_{\mathcal{C}}(x) = \arg\min_{c \in \mathcal{C}} \|x - c\|_2$. Verify that $\text{dist}(x, \mathcal{C})$ is differentiable with gradient vector given by (recall Prob. 8.32):

$$\nabla_{x^\mathsf{T}} \text{dist}(x, \mathcal{C}) = \frac{x - \widehat{x}}{\|x - \widehat{x}\|_2}, \quad \widehat{x} = \mathcal{P}_{\mathcal{C}}(x)$$

9.2 Consider an $N \times M$ matrix A with orthonormal columns $\{a_m\}$ and $M \leq N$. Is $\mathcal{R}(A)$ convex? Pick an arbitrary point $x \in \mathbb{R}^M$. We wish to project it onto the range space of A. Show that

$$\widehat{x} = \mathcal{P}_{\mathcal{R}(A)}(x) = \sum_{m=1}^{M} (a_m^\mathsf{T} x) a_m$$

Next, assume the columns of A are not necessarily orthonormal but that A has full column rank. Show that

$$\widehat{x} = \mathcal{P}_{\mathcal{R}(A)}(x) = A(A^\mathsf{T} A)^{-1} A^\mathsf{T} x$$

9.3 Consider a hyperplane defined by $\mathcal{H} = \{h \,|\, h^\mathsf{T} w - \theta = 0\}$ for some normal direction vector $w \in \mathbb{R}^M$ and scalar offset parameter θ. Is $\mathcal{H}$ convex? Pick an arbitrary point $x \in \mathbb{R}^M$. Show that the projection of x onto $\mathcal{H}$ is given by

$$\widehat{x} = \mathcal{P}_{\mathcal{H}}(x) = (I_M - ww^\dagger)x + ww^\dagger z = x - \frac{(x^\mathsf{T} w - \theta)}{\|w\|^2} w$$

where $w^\dagger = w^\mathsf{T}/\|w\|^2$ and z is any vector satisfying $z^\mathsf{T} w = \theta$.

9.4 Consider a collection of $L \leq M$ hyperplanes defined by $\mathcal{H}_\ell = \{h \,|\, h^\mathsf{T} w_\ell - \theta_\ell = 0\}$ for some normal direction vectors $w_\ell \in \mathbb{R}^M$ and scalar offset parameters θ_ℓ. We collect the $\{w_\ell, \theta_\ell\}$ into the matrix and vector quantities

$$W \triangleq \begin{bmatrix} w_1^\mathsf{T} \\ w_2^\mathsf{T} \\ \vdots \\ w_L^\mathsf{T} \end{bmatrix}, \quad b \triangleq \begin{bmatrix} \theta_1 \\ \theta_2 \\ \vdots \\ \theta_L \end{bmatrix}$$

The intersection of all hyperplanes is then described by the set $\mathcal{H} = \{h \,|\, Wh - b = 0\}$. Is $\mathcal{H}$ a convex set? Pick an arbitrary point $x \in \mathbb{R}^M$. Show that the projection of x onto $\mathcal{H}$ is given by

$$\widehat{x} = \mathcal{P}_{\mathcal{H}}(x) = (I_M - W^\dagger W)x + W^\dagger W z = x - W^\dagger(Wx - b)$$

where $W^\dagger$ denotes the pseudo-inverse of W and z is any vector satisfying $Wz = b$.

9.5 Consider the halfspace defined as the set of all points $h \in \mathbb{R}^M$ satisfying $\mathcal{H} = \{h \,|\, h^\mathsf{T} w - \theta \leq 0\}$ for some normal direction vector $w \in \mathbb{R}^M$ and scalar θ. Pick an arbitrary point $x \in \mathbb{R}^M$. Is $\mathcal{H}$ a convex set? Show that the projection of x onto $\mathcal{H}$ is given by

$$\widehat{x} = \mathcal{P}_{\mathcal{H}}(x) \;=\; x - \frac{w}{\|w\|^2}\max\Big\{0, x^{\mathsf{T}}w - \theta\Big\}$$

9.6 Consider two vectors $x, y \in \mathbb{R}^M$. We wish to project x onto the direction of y. Show that

$$\widehat{x} = \mathcal{P}_y(x) = \frac{y^{\mathsf{T}}x}{\|y\|^2}y$$

9.7 Consider the closed Euclidean ball $\mathcal{B} = \{h \,|\, \|h - a\| \leq r\}$ for some center location $a \in \mathbb{R}^M$ and radius r. Pick an arbitrary point $x \in \mathbb{R}^M$. Show that the projection of x onto $\mathcal{B}$ is given by

$$\widehat{x} = \mathcal{P}_{\mathcal{B}}(x) \;=\; \begin{cases} x, & \text{if } \|x - a\| \leq r \\ a + r\dfrac{x-a}{\|x-a\|}, & \text{if } \|x - a\| > r \end{cases}$$

Verify that the projection can also be written in the form

$$\widehat{x} = \mathcal{P}_{\mathcal{B}}(x) = a + \frac{r}{\max\{\|x-a\|, r\}}(x - a)$$

9.8 Consider the closed ∞-norm ball $\mathcal{B} = \{h \,|\, \|h - a\|_\infty \leq r\}$ for some center location $a \in \mathbb{R}^M$ and radius r. Pick an arbitrary point $x \in \mathbb{R}^M$. Show that the projection of x onto $\mathcal{B}$ is a vector with individual entries constructed as follows:

$$[\,\widehat{x}\,]_m = \Big[\mathcal{P}_{\mathcal{B}}(x)\Big]_m \;=\; \begin{cases} x_m, & \text{if } |x_m - a_m| \leq r \\ a_m + r\,\text{sign}(x_m - a_m), & \text{if } |x_m - a_m| > r \end{cases}$$

9.9 Consider the closed rectangular region defined by the box constraints $\mathcal{R} = \{h |\, \ell_m \leq h_m \leq u_m\}$, where the $\{h_m\}$ refer to the individual entries of $h \in \mathbb{R}^M$ and $\{\ell_m, u_m\}$ are lower and upper bounds. Show that the projection of $x \in \mathbb{R}^M$ onto $\mathcal{R}$ is given by

$$[\,\widehat{x}\,]_m = \begin{cases} \ell_m, & \text{if } x_m \leq \ell_m \\ x_m, & \text{if } \ell_m \leq x_m \leq u_m \\ u_m, & \text{if } x_m \geq u_m \end{cases}$$

9.10 Consider the closed ℓ_1-norm ball $\mathcal{B} = \{h \,|\, \|h\|_1 \leq 1\}$. Pick an arbitrary point $x \in \mathbb{R}^M$. Show that the projection of x onto $\mathcal{B}$ is given by the soft-thresholding operator $\widehat{x} = \mathbb{T}_\lambda(x)$, where $\lambda = 0$ if $\|x\|_1 \leq 1$; otherwise, λ is chosen as a solution to the equation $\sum_{m=1}^{M}(|x_m| - \lambda)_+ = 1$. *Remark.* For a related discussion and implementation of the solution, the reader may refer to Duchi *et al.* (2008).

9.11 Consider the simplex $\mathcal{S} = \{s \,|\, s_m \geq 0,\; \mathbb{1}^{\mathsf{T}}s = 1\}$, where the $\{s_m\}$ denote the entries of vector $s \in \mathbb{R}^M$. We wish to project an arbitrary vector $x \in \mathbb{R}^M$ onto $\mathcal{S}$, i.e.,

$$\widehat{x} = \mathcal{P}_{\mathcal{S}}(x) = \underset{s \in \mathbb{R}^M}{\text{argmin}}\; \|s - x\|^2, \;\; \text{subject to } s_m \geq 0,\;\; \mathbb{1}^{\mathsf{T}}s = 1$$

(a) Use the KKT conditions to show that the individual entries of $\widehat{x}$ are given by

$$\widehat{x}_m = \max\{0, x_m + \lambda^o\}$$

where λ^o is related to the maximizer of the dual function that is associated with the above constrained problem.

(b) Explain that λ^o can be determined as follows:

(b.1) Order the entries of x in decreasing order, say, $x_1 \geq x_2 \geq \ldots \geq x_M$.

(b.2) Compute the scalars $b_m = x_m + \frac{1}{m}(1 - \sum_{j=1}^{m} x_j)$, for $m = 1, 2, \ldots, M$.

(b.3) Determine the largest m^o for which $b_{m^o} > 0$.
(b.4) Set $\lambda^o = \frac{1}{m^o}(1 - \sum_{j=1}^{m_o} x_j)$.

Remark. For a related discussion and implementation of the solution, the reader may refer to Pardalos and Kovoor (1990), Shalev-Shwartz and Singer (2006), and Duchi *et al.* (2008).

9.12 Consider a convex set $\mathcal{S}$ and translate each vector $s \in \mathcal{S}$ to $s+a$ for some constant vector a. Show that $\mathcal{P}_{\mathcal{S}+a}(x+a) = a + \mathcal{P}_{\mathcal{S}}(x)$.

9.13 Consider a finite collection of closed convex sets $\{\mathcal{C}_\ell,\ \ell = 1, 2, \ldots, L\}$ and define the following optimization problem, whose minimum value we denote by $J^\star$ and whose minimizer we denote by $x^\star$:

$$J^\star = \min_{x \in \mathbb{R}^M} \left\{ \max_{1 \leq \ell \leq L} \text{dist}(x, \mathcal{C}_\ell) \right\}$$

Show that provided the intersection is nonempty, $J^\star = 0$ if, and only if, the minimizer $x^\star$ belongs to the intersection of the convex sets:

$$J^\star = 0 \iff x^\star \in \left\{ \bigcap_{\ell=1}^{L} \mathcal{C}_\ell \right\}$$

9.14 Let $\mathcal{C} \subset \mathbb{R}^M$ denote a closed convex set. The projection of a point $z \in \mathbb{R}^M$ onto $\mathcal{C}$ is defined by (9.75). Show the following:

(a) $\widehat{z} = \mathcal{P}_{\mathcal{C}}(z) \iff \Big(\nabla_x \phi(z) - \nabla_x \phi(\widehat{z})\Big)(c - \widehat{z}) \leq 0, \ \forall c \in \mathcal{C}$.
(b) $D_\phi(c, z) \geq D_\phi(c, \widehat{z}), \ \forall c \in \mathcal{C}$.
(c) $D_\phi(c, z) \geq D_\phi(c, \widehat{z}) + D_\phi(\widehat{z}, z)$.
(d) Verify that conditions (a) and (c) are equivalent to each other.

9.15 It turns out that partial optimization need not destroy convexity. Assume $g(x, y)$ is convex over both x and y. Let $\mathcal{Y}$ be a convex set and define $f(x) = \min_{y \in \mathcal{Y}} g(x, y)$. Show that $f(x)$ is convex over x.

9.16 Use the result of Prob. 9.15 to justify the following procedure. Consider a constrained convex optimization problem of the form:

$$\min_{x, y \in \mathbb{R}^M} \ g(x, y), \quad \text{subject to } a(x) \leq 0 \text{ and } b(y) \leq 0$$

under convex inequality constraints. Argue that the problem can be solved in two steps:

$$\text{(Step I): solve } f(x) = \min_y \ g(x, y), \quad \text{subject to } b(y) \leq 0$$

$$\text{(Step II): solve } \min_x \ f(x), \quad \text{subject to } a(x) \leq 0$$

9.17 Consider vectors $c, z \in \mathbb{R}^M$ with positive entries denoted by $\{c_m, z_m\}$. Moreover, the entries of c add up to 1 so that $\sum_{m=1}^{M} c_m = 1$. Introduce the optimization problem

$$\widehat{z} = \underset{c \in \mathbb{R}^M}{\text{argmin}} \left\{ \sum_{m=1}^{M} c_m \ln(c_m / z_m) \right\}, \quad \text{subject to } c_m > 0, \ \sum_{m=1}^{M} c_m = 1$$

Use the Jensen inequality to argue that the solution is $\widehat{z} = z / \|z\|_1$.

9.18 Consider the optimization problem

$$\min_{w \in \mathbb{R}^M} \left\{ \alpha \|w\|_1 + \frac{1}{N} \|d - Hw\|^2 \right\}, \quad \alpha > 0$$

where $d \in \mathbb{R}^N$ and $H \in \mathbb{R}^{N \times M}$. Show that $w^\star$ is a minimizer if, and only if, there exists a vector $a \in \partial_{w^\mathsf{T}} \|w\|_1$ such that $H^\mathsf{T}(d - Hw) = \frac{\alpha N}{2} a$. Conclude that in the special case $H = I$, the solution is given by $w^\star = \mathbb{T}_{\alpha \frac{N}{2}}(d)$, where the soft-thresholding function $\mathbb{T}_\lambda(x)$ is defined by (9.74).

9.19 Consider the problem of minimizing a differentiable convex function $g(z)$ over a convex set, $z \in \mathcal{C}$. Let $\mathcal{N}_C(z)$ denote the normal cone for set $\mathcal{C}$ at location z. Show that

$$z^o \in \left\{\nabla_{z^\mathsf{T}}\, g(z^o) + \mathcal{N}_C(z^o)\right\} \implies z^o \text{ is a minimizer for } g(z) \text{ over } \mathcal{C}$$

9.20 Extend the argument in Example 9.2 to establish the equivalence shown in (9.18) between an ℓ_1-regularized least-squares problem and a quadratic program.

9.21 Let Δ denote the probability simplex in $\mathbb{R}^M$ consisting of the set of all probability vectors $\pi \in \mathbb{R}^M$ with nonnegative entries $\{\pi_m\}$ that add up to 1, namely, $\Delta = \{\pi \in \mathbb{R}^M \,|\, \mathbb{1}^\mathsf{T}\pi = 1,\ \pi_m \geq 0\}$. Each vector $\pi \in \Delta$ defines a discrete probability distribution. We denote its entropy by $\mathcal{H}_\pi = -\sum_{m=1}^{M} \pi_m \ln \pi_m$. Consider an arbitrary vector $q \in \mathbb{R}^M$ with entries $\{q_m\}$ and a scalar $\lambda > 0$. Solve the following constrained optimization problem:

$$F(q,\lambda) \stackrel{\Delta}{=} \max_{\pi \in \Delta} \left\{\pi^\mathsf{T} q + \lambda\, \mathcal{H}_\pi\right\}$$

(a) Show that the solution is given by the log-sum exponential function

$$F(q,\lambda) = \lambda \ln\left[\sum_{m=1}^{M} \exp\Big(q_m/\lambda\Big)\right]$$

(b) Show further that the minimizing vector π is unique and its entries are given by

$$\pi_m = e^{q_m/\lambda} \Big/ \sum_{m'=1}^{M} e^{q_{m'}/\lambda}$$

Verify that the optimal π satisfies $\pi = \nabla_{q^\mathsf{T}} F(q,\lambda)$.

(c) Verify that the optimal value $F(q,\lambda)$ is also equal to $F(q,\lambda) = q_m - \lambda \ln \pi_m$, for all m.

Remark. See the appendix in Nachum *et al.* (2017) for a useful discussion and application in the context of reinforcement learning.

9.22 Assume $P(w)$ in (9.1) is strictly convex. Show that the constrained optimization problem has a unique solution.

9.23 Let $A \in \mathbb{R}^{N,M}$. Show that the second problem below is dual to the first problem:

$$\text{(P1)}: \quad \min_{w \in \mathbb{R}^M} \|w\|_1 \quad \text{subject to } Aw = b$$

$$\text{(P2)}: \quad \max_{z \in \mathbb{R}^N} b^\mathsf{T} z \quad \text{subject to } \|A^\mathsf{T} z\|_\infty \leq 1$$

9.24 Consider the quadratic optimization problem:

$$\min_{w} \left\{\alpha + a^\mathsf{T} w + \frac{1}{2} w^\mathsf{T} A w\right\}, \quad \text{subject to } Bw \succeq b, \;\; w \in \mathbb{R}^M, \;\; B \in \mathbb{R}^{N \times M}$$

where $A \geq 0$ is symmetric, $\alpha \in \mathbb{R}$, $a \in \mathbb{R}^M$, and $b \in \mathbb{R}^N$. Moreover, the notation $\succeq$ denotes elementwise comparison. Apply the KKT conditions to conclude that w^o is a solution if, and only if, there exist scalars $\{\lambda_n^o\}$, collected into a column vector λ^o, such that (i) $Aw^o + a = B^\mathsf{T}\lambda^o$, (ii) $Bw^o \succeq b$, (iii) $\lambda^o \succeq 0$, and (iv) $(\lambda^o)^\mathsf{T}(Bw^o - b) = 0$.

9.25 Consider the quadratic optimization problem

$$\min_{w \in \mathbb{R}^M} \left\{a^\mathsf{T} w + \frac{1}{2} w^\mathsf{T} A w\right\}, \quad \text{subject to } Bw = b, \;\; w \succeq 0$$

where A is symmetric and a is a column vector.

(a) Assume first that $A > 0$. Show that the dual function is given by

$$D(\lambda, \beta) = -b^{\mathsf{T}}\lambda - \frac{1}{2}\|a - \lambda + B^{\mathsf{T}}\beta\|^2_{A^{-1}}$$

where the notation $\|x\|^2_C$ stands for $x^{\mathsf{T}}Cx$.

(b) Assume $A \geq 0$. Show that the dual function is given by

$$D(\lambda, \beta) = \begin{cases} -b^{\mathsf{T}}\lambda - \frac{1}{2}\|a - \lambda + B^{\mathsf{T}}\beta\|^2_{A^{\dagger}}, & \text{if } a - \lambda + B^{\mathsf{T}}\beta \in \mathcal{R}(A) \\ -\infty, & \text{otherwise} \end{cases}$$

9.26 Consider a constrained optimization problem of the form (9.1) where the functions $\{P(w), g_n(w), h_\ell(w)\}$ are not necessarily convex. Let w^o denote a feasible solution. Introduce the Lagrangian function (9.19) where the λ_n are nonnegative. Establish the weak duality result

$$\min_{w} \mathcal{L}(w, \lambda_n, \beta_\ell) \leq P(w^o)$$

9.27 Consider problem (9.1) where the functions $\{P(w), g_n(w), h_\ell(w)\}$ are not necessarily convex. Assume that there exist $\{w^o, \lambda_n^o, \beta_\ell^o\}$ that satisfy the following five conditions for all n and ℓ: (i) $\mathcal{L}(w^o, \lambda_n^o, \beta_\ell^o) = \min_w \mathcal{L}(w, \lambda_n^o, \beta_\ell^o)$, (ii) $g_n(w^o) \leq 0$, (iii) $h_\ell(w^o) = 0$, (iv) $\lambda_n^o \geq 0$, and (v) $\lambda_n^o g_n(w^o) = 0$. Show that w^o is a feasible solution to (9.1). In other words, establish that conditions (i)–(v) are necessary for optimality.

9.28 Consider the setting of Prob. 9.27 and assume further that the functions $\{g_n(w)\}$ are convex and continuously differentiable, while the $\{h_\ell(w)\}$ are affine. Assume also that there exists a strictly feasible point $\bar{w}$ where $g_n(\bar{w}) < 0$ for all n. That is, the inequality conditions are satisfied with strict inequality at $w = \bar{w}$. Show that the KKT conditions stated in Prob. 9.27 become both sufficient *and* necessary for optimality. Argue further that condition (i) reduces to requiring

$$\nabla_w P(w^o) + \sum_{n=0}^{N-1} \lambda_n^o \nabla_w g_n(w^o) + \sum_{\ell=0}^{N-1} \beta_\ell^o \nabla_w h_\ell(w^o) = 0$$

9.29 Show that the following two convex optimization problems are equivalent, where $\{P(w), g_n(w)\}$ are all convex functions, A_ℓ are matrices, and b_ℓ are vectors for $n = 1, 2, \ldots, N$ and $\ell = 1, 2, \ldots, L$. The second formulation on the right is sometimes referred to as the epigraph form of the standard formulation on the left:

$$\left\{ \begin{array}{ll} \min\limits_{w \in \mathbb{R}^M} & P(w) \\ \text{subject to} & g_n(w) \leq 0, \\ & A_\ell w = b_\ell \end{array} \right\} \iff \left\{ \begin{array}{ll} \min\limits_{t, w \in \mathbb{R}^M} & t \\ \text{subject to} & P(w) - t \leq 0 \\ & g_n(w) \leq 0 \\ & A_\ell w = b_\ell \end{array} \right\}$$

9.30 Refer to the constrained optimization problem (9.1) and let $\mathcal{W}$ denote the feasible set consisting of all vectors w that satisfy the constraints:

$$\mathcal{W} = \left\{ w \,\middle|\, g_n(w) \leq 0, h_\ell(w) = 0, \begin{array}{l} n = 1, 2, \ldots, N \\ \ell = 1, 2, \ldots, L \end{array} \right\}$$

Show that $w^\star$ is an optimal solution if, and only if, $w^\star \in \mathcal{W}$ and

$$\nabla_w P(w^\star)(w - w^\star) \geq 0, \quad \text{for any } w \in \mathcal{W}$$

Explain that for an optimization problem involving *only* equality constraints, the above condition reduces to

$$\nabla_w P(w^\star)v = 0, \quad \text{for any } v \in \mathcal{N}(A)$$

Remark. For more details on such convex optimization problems, see the discussion in Boyd and Vandenberghe (2004, section 4.2.3). We use the above result in the proof of Lemma 9.1 following the arguments from this same reference.

9.31 Refer to the primal problem (8.84) and its dual (8.88). Assume strong duality holds and $E(w)$ is differentiable. Show that (w^o, λ^o) are optimal solutions to these problems if, and only if,

$$\lambda^o = \nabla_{w^\mathsf{T}} E(w^o), \quad -\lambda^o = \partial_{w^\mathsf{T}} q(w^o)$$

Remark. The result explains why the gradient vector of a function is interpreted as belonging to a dual space; it is seen here that it is equal to the dual variable at the optimal solution(s).

REFERENCES

Avriel, M. (2003), *Nonlinear Programming: Analysis and Methods*, Dover Publications.

Bertsekas, D. P. (1995), *Nonlinear Programming*, Athena Scientific.

Bertsekas, D. P. (2003), *Convex Analysis and Optimization*, Athena Scientific.

Bertsekas, D. P. (2009), *Convex Optimization Theory*, Athena Scientific.

Boyd, S. and L. Vandenberghe (2004), *Convex Optimization*, Cambridge University Press.

Cottle, R. W. (2012), "William Karush and the KKT theorem," *Documenta Mathematica*, extra vol. ISMP, pp. 255–269. Available at www.math.uni-bielefeld.de/documenta/vol-ismp/41_cottle-richard.pdf

Dantzig, G. B. (1963), *Linear Programming and Extensions*, Princeton University Press.

Dattoro, J. (2016), *Convex Optimization and Euclidean Distance Geometry*, Meboo Publishing.

Duchi, J., S. Shalev-Shwartz, Y. Singer, and T. Chandra (2008), "Efficient projections onto the ℓ_1-ball for learning in high dimensions," *Proc. Intern. Conf. Machine Learning* (ICML), pp. 272–279, Helsinki, Finland.

Fletcher, R. (1987), *Practical Methods of Optimization*, 2nd ed., Wiley.

Hiriart-Urruty, J.-B. and C. Lemaréchal (2001), *Fundamentals of Convex Analysis*, Springer.

John, F. (1948), "Extremum problems with inequalities as subsidiary conditions," in *Studies and Essays*, presented to R. Courant on his 60th Birthday, pp. 187–204, Interscience.

Karush, W. (1939), *Minima of Functions of Several Variables with Inequalities as Side Conditions*, Master's thesis, Department of Mathematics, University of Chicago, IL.

Kjeldsen, T. H. (2000), "A contextualized historical analysis of the Kuhn–Tucker Theorem in nonlinear programming: The impact of World War II," *Historica Mathematica*, vol. 27, pp. 331–361.

Kuhn, H. W. (1976), "Nonlinear programming: A historical view," *SIAM-AMS Proc.*, vol. 9, pp. 1–26.

Kuhn, H. W. (2002), "Being in the right place at the right time," *Oper. Res.*, vol. 50, no. 1, pp. 132–134.

Kuhn, H. W. and A. W. Tucker (1950), "Nonlinear programming," in *Proc. Berkeley Symp. Mathematical Statistics and Probability*, J. Neyman, editor, pp. 481–492, Berkeley, CA.

Luenberger, D. G. and Y. Ye (2008), *Linear and Nonlinear Programming*, Springer.

Nachum, O., M. Norouzi, K. Xu, and D. Schuurmans (2017), "Bridging the gap between value and policy based reinforcement learning," *Proc. Advances Neural Information Processing Systems* (NIPS), pp. 2772–2782, Long Beach, CA.

Nesterov, Y. (2004), *Introductory Lectures on Convex Optimization*, Springer.

Pardalos, P. M. and N. Kovoor (1990), "An algorithm for a singly constrained class of quadratic programs subject to upper and lower bounds," *Math. Prog.*, vol. 46, no. 1–3, pp. pp. 321–328.
Polyak, B. T. (1987), *Introduction to Optimization*, Optimization Software.
Rockafellar, R. T. (1970), *Convex Analysis*, Princeton University Press.
Ruszczynski, A. (2006), *Nonlinear Optimization*, Princeton University Press.
Shalev-Shwartz, S. and Y. Singer (2006), "Efficient learning of label ranking by soft projections onto polyhedra," *J. Mach. Learn. Res.*, vol. 7, pp. 1567–1599.
Slater, M. (1950), "Lagrange multipliers revisited," Cowles Commission Discussion Paper No. 403. Article reprinted in G. Giorgi and T. Hoff, editors, *Traces and Emergence of Nonlinear Programming*, pp. 293–306, Birkhauser, 2014.
Takayama, A. (1974), *Mathematical Economics*, Drydale Press.

10 Lipschitz Conditions

In this chapter, we review some useful integral equalities that involve increments of a function $g(z)$ and its gradient vector. The equalities will involve extensions of the classical mean-value theorem from single-variable calculus to functions of several variables. We also establish useful lower and upper bounds on functions that satisfy certain Lipschitz conditions. Lipschitz continuity will arise frequently in the study of inference and learning problems, and the bounds derived in this chapter will prove useful in several instances.

10.1 MEAN-VALUE THEOREM

Let $g(z) : \mathbb{R}^M \rightarrow \mathbb{R}$ denote a real-valued function of a possibly vector argument z. We assume that $g(z)$ is differentiable whenever necessary. Let z_o and Δz denote arbitrary M-dimensional vectors, and introduce the following real-valued and differentiable function of the scalar variable $t \in [0, 1]$:

$$f(t) \stackrel{\Delta}{=} g(z_o + t\,\Delta z) \tag{10.1}$$

Then, it holds that

$$f(0) = g(z_o), \qquad f(1) = g(z_o + \Delta z) \tag{10.2}$$

Using the fundamental theorem of calculus we have:

$$f(1) - f(0) = \int_0^1 \frac{df(t)}{dt} dt \tag{10.3}$$

Moreover, from definition (10.1) we have

$$\frac{df(t)}{dt} = \frac{d}{dt} g(z_o + t\,\Delta z) = \nabla_z\, g(z_o + t\,\Delta z)\,\Delta z \tag{10.4}$$

in terms of the inner product computation on the far right, where $\nabla_z\, g(z)$ denotes the (row) gradient vector of $g(z)$ with respect to z. The notation $\nabla_z\, g(z_o + t\,\Delta z)$ refers to the gradient vector $\nabla_z\, g(z)$ evaluated at $z = z_o + t\Delta z$. Substituting (10.4) into (10.3) we arrive at the first mean-value theorem result:

$$g(z_o+\Delta z) - g(z_o) = \left(\int_0^1 \nabla_z\, g(z_o+t\Delta z)dt\right)\Delta z \tag{10.5}$$

This result holds for any differentiable (*not necessarily convex*) real-valued function $g(z)$. The expression on the right-hand side is an inner product between the column vector Δz and the result of the integration, which is a row vector. Expression (10.5) tells us how the increment of the function $g(z)$ in moving from $z = z_o$ to $z = z_o + \Delta z$ is related to the integral of the gradient vector of $g(z)$ over the segment $z_o + t\,\Delta z$ as t varies within the interval $t \in [0, 1]$.

We can derive a similar relation for increments of the gradient vector itself. To do so, we introduce the column vector function

$$h(z) \stackrel{\Delta}{=} \nabla_{z^{\mathsf{T}}}\, g(z) \tag{10.6}$$

and apply (10.5) to its individual entries to conclude that

$$h(z_o+\Delta z) - h(z_o) = \left(\int_0^1 \nabla_z\, h(z_o+r\,\Delta z)dr\right)\Delta z \tag{10.7}$$

where $\nabla_z\, h(z)$ denotes the Jacobian of $h(z)$ relative to z. Replacing $h(z)$ by its definition, and transposing both sides of the equality, we arrive at a second mean-value theorem result:

$$\nabla_z\, g(z_o+\Delta z) - \nabla_z\, g(z_o) = \Delta z^{\mathsf{T}}\left(\int_0^1 \nabla_z^2\, g(z_o+r\,\Delta z)dr\right) \tag{10.8}$$

This expression tells us how increments in the gradient vector in moving from $z = z_o$ to $z = z_o + \Delta z$ are related to the integral of the Hessian matrix of $g(z)$ over the segment $z_o + r\,\Delta z$ as r varies within the interval $r \in [0, 1]$. In summary, we arrive at the following statement.

Lemma 10.1. (Mean-value theorem) *Consider a real-valued and twice-differentiable function $g(z) : \mathbb{R}^M \to \mathbb{R}$. Then, for any M-dimensional vectors z_o and Δz, the following increment equalities hold:*

$$g(z_o+\Delta z) - g(z_o) = \left(\int_0^1 \nabla_z\, g(z_o+t\,\Delta z)dt\right)\Delta z \tag{10.9}$$

$$\nabla_z\, g(z_o+\Delta z) - \nabla_z\, g(z_o) = (\Delta z)^{\mathsf{T}}\left(\int_0^1 \nabla_z^2\, g(z_o+r\,\Delta z)dr\right) \tag{10.10}$$

10.2 δ-SMOOTH FUNCTIONS

Let $g(z) : \mathbb{R}^M \rightarrow \mathbb{R}$ be a δ-smooth function, also called a function with a δ-Lipschitz gradient or a function with a Lipschitz continuous gradient, namely, there exists a constant $\delta \geq 0$ (referred to as the Lipschitz constant) such that:

$$\boxed{\|\nabla_z g(z_2) - \nabla_z g(z_1)\| \leq \delta \|z_2 - z_1\|} \quad (\delta\text{-Lipschitz gradients}) \tag{10.11}$$

for any $z_1, z_2 \in \text{dom}(g)$, and where $\|\cdot\|$ denotes the Euclidean norm of its vector argument. It readily follows from the Cauchy–Schwarz inequality for the inner product of two vectors that

$$\begin{aligned}(\nabla_z g(z_2) - \nabla_z g(z_1))(z_2 - z_1) &\leq \|\nabla_z g(z_2) - \nabla_z g(z_1)\| \, \|z_2 - z_1\| \\ &\leq \delta \|z_2 - z_1\|^2\end{aligned} \tag{10.12}$$

Quadratic upper bound

We will use this observation to show that the Lipschitz condition (10.11) implies a quadratic upper bound on any function $g(z)$. Namely, for any $z, z_1 \in \text{dom}(g)$:

$$\boxed{g(z) \leq g(z_1) + (\nabla_z g(z_1))(z - z_1) + \frac{\delta}{2}\|z - z_1\|^2} \tag{10.13}$$

This result is sometimes referred to as the *descent lemma*. Observe that the upper bound on $g(z)$ is quadratic in z and it holds irrespective of whether $g(z)$ is convex or not. When $g(z)$ happens to be convex, then the lower bound in (8.4) will hold as well, namely,

$$g(z) \geq g(z_1) + (\nabla_z g(z_1))(z - z_1) \tag{10.14}$$

The difference between both bounds is the quadratic term $\frac{\delta}{2}\|z - z_1\|^2$ – see Prob. 10.12.

Proof of (10.13): Introduce the function

$$h(\alpha) = g((1-\alpha)z_1 + \alpha z) \tag{10.15}$$

for any $0 \leq \alpha \leq 1$. Differentiating $h(\alpha)$ with respect to α gives

$$h'(\alpha) = \Big(\nabla_z g((1-\alpha)z_1 + \alpha z)\Big)(z - z_1) \tag{10.16}$$

and, hence,

$$\begin{aligned}h'(\alpha) - h'(0) &= \Big(\nabla_z g((1-\alpha)z_1 + \alpha z) - \nabla_z g(z_1)\Big)(z - z_1) \\ &\leq \|\nabla_z g((1-\alpha)z_1 + \alpha z) - \nabla_z g(z_1)\| \, \|z - z_1\| \\ &= \alpha\delta \|z - z_1\|^2\end{aligned} \tag{10.17}$$

It follows from the fundamental theorem of calculus that

$$\begin{aligned} h(1) - h(0) &= \int_0^1 h'(\alpha)d\alpha \\ &\leq \int_0^1 \left(h'(0) + \alpha\delta\|z - z_1\|^2\right)d\alpha \\ &= h'(0) + \frac{\delta}{2}\|z - z_1\|^2 \end{aligned} \tag{10.18}$$

Using

$$h(1) = g(z), \quad h(0) = g(z_1), \quad h'(0) = (\nabla_z\, g(z_1))(z - z_1) \tag{10.19}$$

we obtain (10.13). Actually, statements (10.12) and (10.13) are equivalent. If we write (10.13) with the roles of z and z_1 switched, and combine the resulting relation with the original (10.13), we deduce that (10.12) holds.

■

Upper and lower bounds

We continue with the same δ-smooth function $g(z)$. Let z^o be a minimizer for $g(z)$. We now establish useful lower and upper bounds for the increment $g(z) - g(z^o)$. Specifically, for any function $g(z)$ (that is *not* necessarily convex) it holds that:

(**functions with δ-Lipschitz gradients**)

$$\frac{1}{2\delta}\|\nabla_z\, g(z)\|^2 \;\leq\; g(z) - g(z^o) \;\leq\; \frac{\delta}{2}\|z - z^o\|^2 \tag{10.20}$$

These bounds are important in the analysis of learning algorithms.

Proof of (10.20): The upper bound follows directly from (10.13) by selecting $z_1 \leftarrow z^o$ since $\nabla_z\, g(z^o) = 0$. The lower bound follows by minimizing the quadratic term on the right-hand side of (10.13) over z. The minimum occurs at

$$\nabla_{z^{\mathsf{T}}}\, g(z_1) + \delta(z - z_1) = 0 \implies (z - z_1) = -\frac{1}{\delta}\nabla_{z^{\mathsf{T}}}\, g(z_1) \tag{10.21}$$

Substituting into the right-hand side of (10.13) gives

$$g(z) \leq g(z_1) - \frac{1}{2\delta}\|\nabla_z\, g(z_1)\|^2 \tag{10.22}$$

Selecting $z \leftarrow z^o$ and $z_1 \leftarrow z$ gives

$$g(z) - g(z^o) \geq \frac{1}{2\delta}\|\nabla_z\, g(z)\|^2 \tag{10.23}$$

which is the desired lower bound.

■

The bounds in (10.20) do not require convexity. Earlier, in (8.29), we established alternative bounds when $g(z)$ happens to be ν-strongly convex, namely,

(**ν-strongly convex functions**)

$$\frac{\nu}{2}\|z - z^o\|^2 \;\leq\; g(z) - g(z^o) \;\leq\; \frac{1}{2\nu}\|\nabla_z\, g(z)\|^2 \tag{10.24}$$

Combining these bounds we obtain the following statement.

LEMMA 10.2. (Perturbation bound) *Consider a ν-strongly convex function $g(z) \in \mathbb{R}$ with δ-Lipschitz gradients. Let $z^o \in \mathbb{R}^M$ denote its global minimizer. Combining (10.20) and (10.24) we conclude that, for any Δz, the function increments are bounded by the squared Euclidean norm of Δz as follows:*

$$\frac{\nu}{2}\,\|\Delta z\|^2 \;\leq g(z^o+\Delta z) \;-\; g(z^o) \;\leq\; \frac{\delta}{2}\,\|\Delta z\|^2 \tag{10.25}$$

One useful conclusion from (10.25) is that $g(z)$ can be sandwiched between two quadratic functions, namely,

$$g(z^o)+\frac{\nu}{2}\,\|z-z^o\|^2 \;\leq\; g(z) \;\leq\; g(z^o)+\frac{\delta}{2}\,\|z-z^o\|^2 \tag{10.26}$$

A second useful conclusion follows if we further assume that $g(z)$ is second-order differentiable with a Hessian matrix that is locally Lipschitz continuous in a small neighborhood around $z = z^o$, namely,

$$\left\|\nabla_z^2\, g(z^o+\Delta z) - \nabla_z^2\, g(z^o)\right\| \;\leq\; \kappa\,\|\Delta z\| \tag{10.27}$$

for sufficiently small values $\|\Delta z\| \leq \epsilon$ and for some constant $\kappa > 0$. This condition implies that we can write

$$\nabla_z^2\, g(z^o+\Delta z) \;=\; \nabla_z^2\, g(z^o) \;+\; O(\|\Delta z\|) \tag{10.28}$$

where the notation $O(\|\Delta z\|)$ means that this term decays to zero at the same or comparable rate as $\|\Delta z\|$. Combining the mean-value theorem results (10.9) and (10.10) we know that

$$g(z^o+\Delta z) - g(z^o) = (\nabla z)^{\mathsf{T}}\left(\int_0^1\int_0^1 t\nabla_z^2\, g(z^o+tr\Delta z)drdt\right)\Delta z \tag{10.29}$$

It then follows that, for sufficiently small Δz:

$$\begin{aligned} g(z^o+\Delta z) \;-\; g(z^o) &= (\Delta z)^{\mathsf{T}}\left(\frac{1}{2}\nabla_z^2\, g(z^o)\right)\Delta z \;+\; O(\|\Delta z\|^3) \\ &\approx (\Delta z)^{\mathsf{T}}\left(\frac{1}{2}\nabla_z^2\, g(z^o)\right)\Delta z \\ &= \|\Delta z\|^2_{\frac{1}{2}\nabla_z^2\, g(z^o)} \end{aligned} \tag{10.30}$$

where the symbol $\approx$ in the second line is used to indicate that higher-order powers in $\|\Delta z\|$ are being ignored. Moreover, for any symmetric positive-definite weighting matrix $W > 0$, the notation $\|x\|^2_W$ refers to the weighted square Euclidean norm $x^{\mathsf{T}}Wx$.

We conclude from (10.30) that the increment in the value of the function in a small neighborhood around $z = z^o$ can be well approximated by means of a weighted square Euclidean norm; the weighting matrix in this case is equal to the Hessian matrix of $g(z)$ evaluated at $z = z^o$ and scaled by $1/2$. The error in this approximate evaluation is on the order of $\|\Delta z\|^3$.

Lemma 10.3. (Perturbation approximation) *Consider the same setting in the statement leading to (10.25) and assume additionally that the Hessian matrix is locally Lipschitz continuous in a small neighborhood around $z = z^o$ as defined by (10.27). It then follows that the increment in the value of the function $g(z)$ for sufficiently small variations around $z = z^o$ can be well approximated by*

$$g(z^o + \Delta z) \; - \; g(z^o) \; \approx \; \Delta z^{\mathsf{T}} \left(\frac{1}{2} \nabla_z^2 \, g(z^o) \right) \Delta z \tag{10.31}$$

where the approximation error is on the order of $O(\|\Delta z\|^3)$.

Bounded Hessian

For twice-differentiable convex functions, the condition of δ-Lipschitz gradients is equivalent to bounding the Hessian matrix by $\nabla_z^2 \, g(z) \leq \delta I_M$ for all z. Although both conditions are equivalent, one advantage of using the δ-Lipschitz condition is that the function $g(z)$ would not need to be twice-differentiable.

Lemma 10.4. (Lipschitz and bounded Hessian matrix) *Consider a real-valued and twice-differentiable convex function $g(z) : \mathbb{R}^M \to \mathbb{R}$. Then, the following two conditions are equivalent:*

$$\begin{aligned} &\nabla_z^2 g(z) \leq \delta \, I_M, \text{ for all } z \Longleftrightarrow \\ &\|\nabla_z \, g(z_2) - \nabla_z \, g(z_1)\| \leq \delta \, \|z_2 - z_1\|, \text{ for all } z_1, z_2 \end{aligned} \tag{10.32}$$

Proof: Assume first that the Hessian matrix, $\nabla_z^2 \, g(z)$, is uniformly upper-bounded by $\delta \, I_M$ for all z; we know that it is nonnegative-definite since $g(z)$ is convex and, therefore, $\nabla_z^2 \, g(z)$ is lower-bounded by zero. We pick any z_1 and z_2 and introduce the column vector function $h(z) = \nabla_{z^{\mathsf{T}}} \, g(z)$. Applying (10.9) to $h(z)$ gives

$$h(z_2) \; - \; h(z_1) \; = \; \left(\int_0^1 \nabla_z \, h(z_1 + t(z_2 - z_1)) dt \right) (z_2 - z_1) \tag{10.33}$$

so that using $0 \leq \nabla_z^2 \, g(z) \leq \delta \, I_M$, we get

$$\|\nabla_{z^{\mathsf{T}}} \, g(z_2) \; - \; \nabla_{z^{\mathsf{T}}} \, g(z_1)\| \; \leq \; \left(\int_0^1 \delta dt \right) \|z_2 - z_1\| \tag{10.34}$$

and we arrive at the Lipschitz condition shown in (10.32).

Conversely, assume the δ-Lipschitz gradient condition in (10.32) holds, and introduce the column vector function $f(t) = \nabla_{z^{\mathsf{T}}} \, g(z + t\Delta z)$ defined in terms of a scalar real parameter t. Then,

$$\frac{df(t)}{dt} \; = \; \left(\nabla_z^2 \, g(z + t\Delta z) \right) \Delta z \tag{10.35}$$

Now, for any Δt and in view of the Lipschitz condition, it holds that

$$\begin{aligned} \|f(t + \Delta t) \; - \; f(t)\| &= \|\nabla_{z^{\mathsf{T}}} \, g(z + (t + \Delta t)\Delta z) \; - \; \nabla_{z^{\mathsf{T}}} \, g(z + t\Delta z)\| \\ &\leq \delta \, |\Delta t| \, \|\Delta z\| \end{aligned} \tag{10.36}$$

so that

$$\underbrace{\lim_{\Delta t \to 0} \left\{ \frac{\|f(t + \Delta t) \; - \; f(t)\|}{|\Delta t|} \right\}}_{= \|df(t)/dt\|} \; \leq \; \delta \, \|\Delta z\| \tag{10.37}$$

Using (10.35) we conclude that

$$\|(\nabla_z^2\, g(z+t\Delta z))\;\Delta z\| \;\leq\; \delta\,\|\Delta z\|, \quad \text{for any } t, z, \text{ and } \Delta z \tag{10.38}$$

Setting $t=0$, squaring both sides, and recalling that the Hessian matrix is symmetric, we obtain

$$(\Delta z)^{\mathsf T}\left(\nabla_z^2\, g(z)\right)^2 \Delta z \;\leq\; \delta^2\|\Delta z\|^2, \quad \text{for any } z, \Delta z \tag{10.39}$$

from which we conclude that $\nabla_z^2\, g(z) \leq \delta\, I_M$ for all z, as desired.

■

Lipschitz-continuous functions

We provide one final useful result relating the Lipschitz condition on a convex function (smooth or not) to bounds on its gradient (or subgradient) vectors. Thus, assume now that $g(z)$ is a convex function that is Lipschitz continuous (note that the Lipschitz condition is imposed on the function itself):

$$\boxed{|g(z_2)-g(z_1)| \;\leq\; \delta\,\|z_2-z_1\|} \qquad \textbf{(Lipschitz continuous)} \tag{10.40}$$

for some $\delta \geq 0$ and all $z_1, z_2 \in \text{dom}(g)$. The function $g(z)$ is not required to be differentiable at all points in its domain; it can have points of nondifferentiability. The following result applies to gradient and subgradient vectors.

Lemma 10.5. **(Lipschitz and bounded subgradients)** *Consider a real-valued convex and Lipschitz-continuous function $g(z) : \mathbb{R}^M \to \mathbb{R}$ satisfying (10.40). Let $\|\cdot\|_\star$ denote the dual norm that corresponds to the norm $\|\cdot\|$ used in (10.40); recall the definition of the dual norm from (1.157). In particular, if the 2-norm is used, then the dual norm is the 2-norm itself. The following two conditions are equivalent:*

$$|g(z_2)-g(z_1)| \;\leq\; \delta\,\|z_2-z_1\|,\; \forall z_1, z_2 \iff \|s\|_\star \leq \delta, \;\; \forall\, s \in \partial_{z^{\mathsf T}}\, g(z) \tag{10.41}$$

Proof: Assume first that $g(z)$ is Lipschitz continuous. Pick any subgradient vector $s \in \partial_{z^{\mathsf T}}\, g(z)$ and let

$$v^o = \underset{\|v\|=1}{\text{argmax}}\; s^{\mathsf T} v \tag{10.42}$$

We know from the definition of the dual norm in (1.157) that $\|s\|_\star = w^{\mathsf T} v^o$. Moreover, from the definition of subgradients and in view of the convexity of $g(z)$ we have

$$g(z+v^o) - g(z) \geq s^{\mathsf T} v^o \;=\; \|s\|_\star \tag{10.43}$$

On the other hand, from the Lipschitz condition:

$$g(z+v^o)-g(z) \leq |g(z+v^o)-g(z)| \leq \delta\|v^o\| = \delta, \quad \text{since } \|v^o\|=1 \tag{10.44}$$

Combining the last two inequalities we conclude that $\|s\|_\star \leq \delta$.

Conversely, assume $\|s\|_\star \leq \delta$ for any $s \in \partial_{z^{\mathsf T}}\, g(z)$ and let us verify that $g(z)$ needs to be Lipschitz continuous. For any $z_1, z_2 \in \text{dom}(g)$, we know from the definition of subgradients that

$$g(z_2)-g(z_1) \geq s^{\mathsf T}(z_2-z_1), \quad \text{for some subgradient } s \text{ at } z_1 \tag{10.45}$$

Reversing the inequality and using the property $x^{\mathsf{T}}y \leq \|y\|\,\|x\|_\star$ for dual norms from (1.159) we get

$$g(z_1) - g(z_2) \leq s^{\mathsf{T}}(z_1 - z_2) \;\leq\; \|z_1 - z_2\|\,\|s\|_\star \leq \delta\|z_1 - z_2\| \tag{10.46}$$

Similarly, starting from

$$g(z_1) - g(z_2) \geq (s')^{\mathsf{T}}(z_1 - z_2), \quad \text{for some subgradient } s' \text{ at } z_2 \tag{10.47}$$

we get

$$g(z_2) - g(z_1) \leq \delta\|z_1 - z_2\| \tag{10.48}$$

Combining (10.46) and (10.48) we conclude that (10.40) holds.

■

10.3 COMMENTARIES AND DISCUSSION

Mean-value theorem. The increment formula (10.5) holds for any differentiable (*not necessarily convex*) real-valued functions, $g(z)$; it is also valid for scalar or vector arguments z. The equality can be viewed as a useful alternative to the Taylor series expansion of $g(z)$ around the point $z = z_o$, where the series expansion is replaced by an integral representation. For more discussion on mean-value theorems in differential calculus, the reader may refer to Binmore and Davies (2007) and Larson and Edwards (2009). For discussion in the context of optimization, adaptation, and learning theories, the reader may consult Polyak (1987) and Sayed (2014a); the latter reference contains extensions to the case of complex arguments z.

For scalar-valued functions $g(z) : \mathbb{R} \to \mathbb{R}$, the traditional version of the mean-value theorem is due to mathematicians **Bernard Bolzano (1781–1848)** and **Augustin Cauchy (1789–1857)**, who established versions of the result in Bolzano (1817) and Cauchy (1821). The result states that there exists a point $c \in (z_o, z_o + \Delta z)$ such that

$$g'(c) \;=\; \frac{g(z_o + \Delta z) - g(z_o)}{\Delta z} \tag{10.49}$$

where we are writing $g'(c)$ to refer to the derivative of $g(z)$ at $z = c$. This relation can be rewritten equivalently as

$$g(z_o + \Delta z) \;=\; g(z_o) \;+\; g'(c)\Delta z \tag{10.50}$$

The analogue of this relation for vector arguments z is the statement that there exists a scalar $t_o \in (0, 1)$ such that

$$g(z_o + \Delta z) \;=\; g(z_o) \;+\; \Big(\nabla_{z^{\mathsf{T}}}\, g(z_o + t_o\Delta z)\Big)^{\mathsf{T}} \Delta z \tag{10.51}$$

Comparing this statement with (10.9) we find that the gradient vector $\nabla_z\, g(z_o + t\Delta z)$ is replaced by the average value over the interval $(z_o, z_o + \Delta z)$. These are two useful forms of the mean-value theorem.

Lipschitz conditions. The property of Lipschitz continuity is prevalent in the study of differentiable and/or continuous functions; it provides a midterm characterization that is weaker than continuous differentiability and stronger than uniform continuity. We say that a real-valued function $h(z) : \mathbb{R}^M \to \mathbb{R}$, defined over some domain $\mathcal{Z} \subset \mathbb{R}^M$, is δ-Lipschitz over $\mathcal{Z}$ if

$$|h(z_2) - h(z_1)| \;\leq\; \delta\,\|z_2 - z_1\| \tag{10.52}$$

for all $z_1, z_2 \in \mathcal{Z}$. The scalar δ is referred to as the Lipschitz constant. The function $h(z)$ is said to be *continuously differentiable* if its derivative exists at every point in $\mathcal{Z}$ (and, therefore, $h(z)$ is continuous over $\mathcal{Z}$) and, in addition, the derivative function itself is also continuous over the same domain. Likewise, the function $h(z)$ is said to be *uniformly continuous* if for any $\epsilon > 0$, there exists a $\delta > 0$ such that for any $z_1, z_2 \in \mathcal{Z}$:

$$|z_1 - z_2| \leq \delta \implies |h(z_1) - h(z_2)| \leq \epsilon \tag{10.53}$$

This definition is stronger than regular continuity in that the value of δ is independent of the values of z_1 and z_2. Accordingly, every uniformly continuous function is also continuous. It can be verified that

$$h(z) \text{ continuously differentiable} \implies \text{Lipschitz continuous} \implies \text{uniformly continuous} \tag{10.54}$$

According to the account by Yushkevich (1981), the concept of Lipschitz continuity is named after the German mathematician **Rudolf Lipschitz (1832–1903)**, who introduced it in Lipschitz (1854) as a condition for the uniqueness of solutions to first-order ordinary differential equations with initial conditions of the form:

$$\frac{dz}{dt} = h(t, z(t)), \quad z(t_o) = z_o \tag{10.55}$$

for differentiable functions $h(t, \cdot)$. In particular, if $h(t, z)$ is δ-Lipschitz, i.e.,

$$|h(t, z_2) - h(t, z_1)| \;\leq\; \delta|z_2 - z_1| \tag{10.56}$$

for any t and z_1, z_2 in a domain $\mathcal{Z}$, then the above differential equation has a unique solution over $\mathcal{Z}$. For further discussion on Lipschitz conditions and ordinary differential equations, the reader may consult, for example, Coddington and Levinson (1955), Ince (1956), Arnold (1978), Tenenbaum and Pollard (1985), Coddington (1989), Farlow (1993), Zill (2000), and Boyce and DiPrima (2012). For discussion of Lipschitz conditions in the context of convexity, the reader may refer to Polyak (1987), Bertsekas (2003, 2009), Nesterov (2004), Boyd and Vandenberghe (2004), and Sayed (2014a). The argument used in the proof of Lemma 10.5 is from Shalev-Shwartz (2011).

PROBLEMS

10.1 Let $z = \text{col}\{z_m\} \in \mathbb{R}^M$. Assume a convex function $g(z)$ is δ-Lipschitz relative to each z_m, namely, for any $z \in \mathbb{R}^M$ and $\beta \in \mathbb{R}$ (compare with (10.11)):

$$\left|\frac{\partial}{\partial z_m} g(z + \beta e_m) - \frac{\partial}{\partial z_m} g(z)\right| \leq \delta|\beta|$$

where e_m denotes the mth basis vector in $\mathbb{R}^M$. Repeat the argument that led to (10.13) and (10.20) to conclude that for any (z, z_1) with entries $\{z_m, z_{1,m}\}$:

$$g(z) \leq g(z_1) + \frac{\partial g(z_1)}{\partial z_m}(z_m - z_{1,m}) + \frac{\delta}{2}(z_m - z_{1,m})^2$$

$$\frac{1}{2\delta}\left|\partial g(z)/\partial z_m\right|^2 \leq g(z) - g(z^o) \;\leq \frac{\delta}{2}(z_m - z_m^o)^2$$

where z^o is a global minimizer.

10.2 Consider any real-valued function $g(z) : \mathbb{R}^M \to \mathbb{R}$ (which is not necessarily convex). Assume its gradient is δ-Lipschitz. Show that its Hessian matrix satisfies $-\delta I \leq \nabla_z^2\, g(z) \leq \delta I$. *Hint.* Introduce the function $f(t) = \nabla_{z^\mathsf{T}}\, g(z + t\Delta z)$ and establish $\lim_{\Delta \to 0} \|f(t + \Delta t) - f(t)\|/|\Delta t| \leq \delta\|\Delta z\|$. Use this result to conclude the argument.

10.3 Consider a convex function $g(z) : \mathbb{R}^M \to \mathbb{R}$ with $\text{dom}(g) = \mathbb{R}^M$. Assume $g(z)$ has δ-Lipschitz gradients. Show that the function $f(z) = \frac{\delta}{2}\|z\|^2 - g(z)$ is convex.

10.4 Consider a convex function $g(z) : \mathbb{R}^M \to \mathbb{R}$ with $\text{dom}(g) = \mathbb{R}^M$. Assume $g(z)$ has δ-Lipschitz gradients. Establish the co-coercivity property:

$$\Big(\nabla_{z^\mathsf{T}}\, g(z_2) - \nabla_{z^\mathsf{T}}\, g(z_1)\Big)^\mathsf{T}(z_2 - z_1) \geq \frac{1}{\delta}\,\|\nabla_z\, g(z_2) - \nabla_z\, g(z_1)\|^2$$

Remark. This result is a special case of the bound in the next problem for $\nu = 0$. See also the text by Nesterov (2004).

10.5 Consider a ν-strongly convex function $g(z) : \mathbb{R}^M \to \mathbb{R}$ with $\text{dom}(g) = \mathbb{R}^M$. Assume $g(z)$ has δ-Lipschitz gradients. Establish the co-coercivity property:

$$\Big(\nabla_{z^\mathsf{T}}\, g(z_2) - \nabla_{z^\mathsf{T}}\, g(z_1)\Big)^\mathsf{T}(z_2 - z_1) \geq \frac{\nu\delta}{\nu+\delta}\|z_2 - z_1\|^2 + \frac{1}{\nu+\delta}\|\nabla_z\, g(z_2) - \nabla_z\, g(z_1)\|^2$$

10.6 Let $s(z)$ denote a subgradient of the convex function $g(z)$ relative to z^T. Assume $s(z)$ is bounded so that $\|s(z)\| \leq d$ for all z and for some $d \geq 0$. Show that this boundedness condition is equivalent to $g(z)$ satisfying the following Lipschitz condition: $|g(z_1) - g(z_2)| \leq d\,\|z_1 - z_2\|$ for any $z_1, z_2 \in \text{dom}(g)$.

10.7 Consider $g(z) = \|z\|^4 + \|z\|^2$, where $z \in \mathbb{R}^M$. Let z^o denote a stationary point of $g(z)$ and introduce $\widetilde{z} = z^o - z$. Use the mean-value theorem (10.9) to express the difference $g(z) - g(z^o)$ in terms of $\widetilde{z}$. Evaluate the integral expressions whenever possible.

10.8 Use the mean-value theorem (10.50) to establish (10.51) when $z \in \mathbb{R}^M$.

10.9 Problems 10.9 and 10.11 are motivated by useful properties derived from Polyak (1987, ch. 1). Consider a convex function $g(z) : \mathbb{R}^M \to \mathbb{R}$ with a δ-Lipschitz gradient. Let z^o denote the location of a global minimum for $g(z)$ and define $\widetilde{z} = z^o - z$. Show that $(\nabla_{z^\mathsf{T}}\, g(z))^\mathsf{T}\,\widetilde{z} \leq -(1/\delta)\,\|\nabla_z\, g(z)\|^2$. Provide one interpretation for this result.

10.10 Let $g(z)$ be a real-valued twice-differentiable function with $z \in \mathbb{R}^M$. Use the mean-value theorem statement (10.9)–(10.10) to conclude that

$$g(z_o + \Delta z) = g(z_o) + (\nabla_{z^\mathsf{T}}\, g(z_o))^\mathsf{T}\,\Delta z \;+\; (\Delta z)^\mathsf{T}\left(\int_0^1\!\!\int_0^1 t\nabla_z^2\, g(z_o + tr\Delta z)drdt\right)\Delta z$$

10.11 Let $g(z)$ be a real-valued twice-differentiable function with $z \in \mathbb{R}^M$. Refer to the mean-value theorem (10.9).

(a) Assume the gradient of $g(z)$ is δ-Lipschitz continuous for $z \in [z_o, z_o + \Delta z]$ so that $\|\nabla_z\, g(z_o + t\Delta z) - \nabla_z\, g(z_o)\| \leq \delta\, t\,\|\Delta z\|$ for any $t \in [0,1]$. Show that

$$\Big\|g(z_o + \Delta z) - g(z_o) - (\nabla_{z^\mathsf{T}}\, g(z_o))^\mathsf{T}\,\Delta z\Big\| \;\leq\; \frac{\delta}{2}\|\Delta z\|^2$$

(b) Assume instead that the Hessian matrix of $g(z)$ is δ-Lipschitz continuous in the interval $z \in [z_o, z_o + \Delta z]$, i.e., $\|\nabla_z^2\, g(z_o + t\Delta z) - \nabla_z^2\, g(z_o)\| \leq \delta\, t\,\|\Delta z\|$ for any $t \in [0,1]$. Show that

$$\Big\|g(z_o + \Delta z) - g(z_o) - (\nabla_{z^\mathsf{T}}\, g(z_o))^\mathsf{T}\,\Delta z - \frac{1}{2}\,(\Delta z)^\mathsf{T}\,\big(\nabla_z^2\, g(z_o)\big)\,\Delta z\Big\| \;\leq\; \frac{\delta}{6}\|\Delta z\|^3$$

10.12 We established in (10.13) that any function $g(z)$ with δ-Lipschitz gradients satisfies the quadratic upper bound (10.13). Assume instead that the Hessian matrix of $g(z)$ is ρ-Lipschitz, meaning that $\|\nabla_z^2\, g(z_1) - \nabla_z^2\, g(z_2)\| \leq \rho\|z_1 - z_2\|$, for any (z_1, z_2). Establish the tighter cubic bound:

$$g(z) \leq g(z_1) + (\nabla_z\, g(z_1))^\mathsf{T}(z - z_1) + \frac{1}{2}(z - z_1)^\mathsf{T}\nabla_z^2\, g(z_1)\,(z - z_1) + \frac{\rho}{6}\|z - z_1\|^3$$

Remark. See Nesterov and Polyak (2006) for more discussion on this result.

10.13 Consider column vectors $h, z \in \mathbb{R}^M$ and $\rho > 0$. Introduce the logistic function:

$$g(z) = \frac{\rho}{2}\|z\|^2 + \ln\left(1 + e^{-h^\mathsf{T} z}\right)$$

(a) Show that $g(z)$ is strongly convex. Let z^o denote its global minimizer.
(b) Show that, for any z, the Hessian matrix function of $g(z)$ satisfies a Lipschitz condition of the form $\left\|\nabla_z^2\, g(z) - \nabla_z^2\, g(z^o)\right\| \leq \kappa\, \|z^o - z\|$, for some $\kappa \geq 0$. Determine an expression for κ in terms of h.

REFERENCES

Arnold, V. I. (1978), *Ordinary Differential Equations*, MIT Press.

Bertsekas, D. P. (2003), *Convex Analysis and Optimization*, Athena Scientific.

Bertsekas, D. P. (2009), *Convex Optimization Theory*, Athena Scientific.

Binmore, K. and J. Davies (2007), *Calculus Concepts and Methods*, Cambridge University Press.

Bolzano, B. (1817), "Rein analytischer Beweis des Lehrsatzes dass zwischen je zwey Werthen, die ein entgegengesetztes Resultat gewaehren, wenigstens eine reele Wurzel der Gleichung liege," Prague. English translation by S. B. Russ, "A translation of Bolzano's paper on the intermediate value theorem," *Hist. Math.*, vol. 7, pp. 156–185.

Boyce, W. E. and R. C. DiPrima (2012), *Elementary Differential Equations and Boundary Value Problems*, 10th ed., Wiley.

Boyd, S. and L. Vandenberghe (2004), *Convex Optimization*, Cambridge University Press.

Cauchy, A.-L. (1821), *Cours d'Analyse de l'École Royale Polytechnique, Ire Partie: Analyse Algébrique*, Chez Debure frères, Libraires du Roi et de la Bibliothèque du Roi. Edition republished by Cambridge University Press, 2009.

Coddington, E. A. (1989), *An Introduction to Ordinary Differential Equations*, Dover Publications.

Coddington, E. A. and N. Levinson (1955), *Theory of Ordinary Differential Equations*, McGraw-Hill.

Farlow, S. J. (1993), *Partial Differential Equations for Scientists and Engineers*, Dover Publications.

Ince, E. L. (1956), *Ordinary Differential Equations*, reprint edition, Dover.

Larson, R. and B. H. Edwards (2009), *Calculus*, Thomson Brooks Cole.

Lipschitz, R. (1854), "De explicatione per series trigonometricas insttuenda functionum unius variablis arbitrariarum, et praecipue earum, quae per variablis spatium finitum valorum maximorum et minimorum numerum habent infintum disquisitio," *J. Reine Angew. Math.*, vol. 63, pp. 296–308.

Nesterov, Y. (2004), *Introductory Lectures on Convex Optimization*, Springer.

Nesterov, Y. and B. T. Polyak (2006), "Cubic regularization of Newton method and its global performance," *Math. Program.*, vol. 108, no. 1, pp. 177–205.

Polyak, B. T. (1987), *Introduction to Optimization*, Optimization Software.

Sayed, A. H. (2014a), *Adaptation, Learning, and Optimization over Networks*, Foundations and Trends in Machine Learning, NOW Publishers, vol. 7, no. 4–5, pp. 311–801.

Shalev-Shwartz, S. (2011), "Online learning and online convex optimization," *Found. Trends Mach. Learn.*, vol. 4, no. 2, pp. 107–194.

Tenenbaum, M. and H. Pollard (1985), *Ordinary Differential Equations*, Dover.

Yushkevich, A. P. (1981), "Sur les origines de la méthode Cauchy-Lipschitz dans la théorie des equations différentielles," *Revue d'Histoire des Sciences*, vol. 34, pp. 209–215.

Zill, D. G. (2000), *A First Course in Differential Equations*, 5th ed., Thomson Brooks Cole.

11 Proximal Operator

Proximal projection is a useful procedure for the minimization of nonsmooth convex functions. Its main power lies in transforming the minimization of convex functions into the equivalent problem of determining fixed points for contractive operators. The purpose of this chapter is to introduce proximal operators, highlight some of their properties, and explain the role that soft-thresholding or shrinkage plays in this context.

11.1 DEFINITION AND PROPERTIES

Let $h(w) : \mathbb{R}^M \rightarrow \mathbb{R}$ denote a convex function of real arguments, $w \in \mathbb{R}^M$. Since $h(w)$ is convex, it can only have minimizers and all of them will be global minimizers. We are interested in locating a minimizer for $h(w)$. The function $h(w)$ may be nondifferentiable at some locations.

11.1.1 Definition

Let $z \in \mathbb{R}^M$ denote some given vector. We add a quadratic term to $h(w)$ that measures the squared distance from w to z and define:

$$h_p(w) \stackrel{\Delta}{=} h(w) + \frac{1}{2\mu}\|w - z\|^2, \quad \forall\, w \in \text{dom}(h) \tag{11.1}$$

for some positive scalar $\mu > 0$ chosen by the designer. The proximal operator of $h(w)$, also called the *proximity operator*, is denoted by the notation $\text{prox}_{\mu h}(z)$ and defined as the mapping that transforms z into the vector $\widehat{w}$ computed as follows:

$$\widehat{w} \stackrel{\Delta}{=} \underset{w \in \mathbb{R}^M}{\text{argmin}}\ h_p(w) \tag{11.2}$$

We will refer to $\widehat{w}$ as the proximal projection of z relative to h. The function $h_p(w)$ is strongly convex in w since $h(w)$ is convex and $\|w - z\|^2$ is strongly convex. It follows that $h_p(w)$ has a unique global minimizer and, therefore, the proximal projection, $\widehat{w}$, exists and is unique. Obviously, the vector $\widehat{w}$ is a function of z and we express the transformation from z to $\widehat{w}$ by writing

$$\widehat{w} = \text{prox}_{\mu h}(z) \tag{11.3}$$

where the role of $h(w)$ is highlighted in the subscript notation, and where the proximal operator is defined by

$$\boxed{\operatorname{prox}_{\mu h}(z) \;\stackrel{\Delta}{=}\; \underset{w\in\mathbb{R}^M}{\operatorname{argmin}} \left\{ h(w) + \frac{1}{2\mu}\|w-z\|^2 \right\}} \tag{11.4}$$

In the trivial case when $h(w) = 0$, we get $\widehat{w} = z$ so that

$$\operatorname{prox}_0(z) = z \tag{11.5}$$

Using (11.3), the minimum value of $h_p(w)$, when w is replaced by $\widehat{w}$, is called the *Moreau envelope*, written as

$$\mathcal{M}_{\mu h}(z) \;\stackrel{\Delta}{=}\; h(\widehat{w}) + \frac{1}{2\mu}\|\widehat{w}-z\|^2 \tag{11.6}$$

Intuitively, the proximal construction (11.3) approximates z by the vector $\widehat{w}$ that is "close" to it under the squared Euclidean norm and subject to the "penalty" $h(w)$ on the "size" of w. The Moreau value at $\widehat{w}$ serves as a measure of a "generalized" distance between z and its proximal projection, $\widehat{w}$. The reason for the qualification "projection" is because the proximal operation (11.3) can be interpreted as a generalization of the notion of projection, as the following example illustrates for a particular choice of $h(w)$.

Example 11.1 **(Projection onto a convex set)** Let $\mathcal{C}$ denote some convex set and introduce the indicator function:

$$\mathbb{I}_{C,\infty}[w] \;\stackrel{\Delta}{=}\; \begin{cases} 0, & w \in \mathcal{C} \\ \infty, & \text{otherwise} \end{cases} \tag{11.7}$$

That is, the function assumes the value zero whenever w belongs to the set $\mathcal{C}$ and is infinite otherwise. It is straightforward to verify that when $h(w) = \mathbb{I}_{C,\infty}[w]$, the definition

$$\operatorname{prox}_{\mu\,\mathbb{I}_{C,\infty}}(z) = \underset{w\in\mathbb{R}^M}{\operatorname{argmin}} \left\{ \mathbb{I}_{C,\infty}(w) + \frac{1}{2\mu}\|w-z\|^2 \right\} \tag{11.8}$$

reduces to

$$\operatorname{prox}_{\mu\,\mathbb{I}_{C,\infty}}(z) = \underset{w\in\mathcal{C}}{\operatorname{argmin}}\, \|w-z\|^2 \tag{11.9}$$

In other words, the proximal operator (11.8) corresponds to projecting onto the set $\mathcal{C}$ and determining the closest element in $\mathcal{C}$ to the vector z. We express the result in the form:

$$\operatorname{prox}_{\mu\,\mathbb{I}_{C,\infty}}(z) = \mathcal{P}_C(z) \tag{11.10}$$

If we compare (11.8) with the general definition (11.4), it becomes clear why the proximal operation is viewed as a generalization of the concept of projection onto convex sets. The generalization results from replacing the indicator function, $\mathbb{I}_{C,\infty}[w]$, by an arbitrary convex function, $h(w)$.

Optimality condition

When $h(w)$ is differentiable, the minimizer $\widehat{w}$ of (11.4) should satisfy

$$\nabla_{w^\mathsf{T}} h(\widehat{w}) + \frac{1}{\mu}(\widehat{w} - z) = 0 \iff \widehat{w} = z - \mu \nabla_{w^\mathsf{T}} h(\widehat{w}) \tag{11.11}$$

On the other hand, when $h(w)$ is not differentiable, the minimizer $\widehat{w}$ should satisfy

$$0 \in \partial_{w^\mathsf{T}} h(\widehat{w}) \; + \; \frac{1}{\mu}(\widehat{w} - z) \iff (z - \widehat{w}) \in \mu \, \partial_{w^\mathsf{T}} h(\widehat{w}) \tag{11.12}$$

This is a critical property and we rewrite it more generically as follows for ease of reference in terms of two vectors a and b:

$$\boxed{a = \text{prox}_{\mu h}(b) \iff (b - a) \in \mu \, \partial_{w^\mathsf{T}} h(a)} \tag{11.13}$$

11.1.2 Soft Thresholding

One useful choice for the proximal function is

$$h(w) = \alpha \|w\|_1 \tag{11.14}$$

in terms of the ℓ_1-norm of the vector w and where $\alpha > 0$. In this case, the function $h(w)$ is nondifferentiable at $w = 0$ and the function $h_p(w)$ becomes

$$h_p(w) = \alpha \|w\|_1 \; + \; \frac{1}{2\mu} \|w - z\|^2 \tag{11.15}$$

It turns out that a closed-form expression exists for the corresponding proximal operator:

$$\widehat{w} \; \stackrel{\Delta}{=} \; \text{prox}_{\mu\alpha\|w\|_1}(z) \tag{11.16}$$

We show below that $\widehat{w}$ is obtained by applying a soft-thresholding (or shrinkage) operation to z as follows. Let z_m denote the m-th entry of $z \in \mathbb{R}^M$. Then, the corresponding entry of $\widehat{w}$, denoted by $\widehat{w}_m$, is found by shrinking z_m in the following manner:

$$\widehat{w}_m \;=\; \mathbb{T}_{\mu\alpha}(z_m), \quad m = 1, 2, \ldots, M \tag{11.17}$$

where the soft-thresholding function, denoted by $\mathbb{T}_\beta(x) : \mathbb{R} \to \mathbb{R}$, with threshold $\beta \geq 0$, is defined as

$$\mathbb{T}_\beta(x) \; \stackrel{\Delta}{=} \; \begin{cases} x - \beta, & \text{if } x \geq \beta \\ 0, & \text{if } -\beta < x < \beta \\ x + \beta, & \text{if } x \leq -\beta \end{cases} \tag{11.18}$$

This can also be written in the alternative form:

$$\begin{aligned} \mathbb{T}_\beta(x) &= \text{sign}(x) \times \max\Big\{0, |x| - \beta\Big\} \\ &= \text{sign}(x) \Big(|x| - \beta\Big)_+ \end{aligned} \tag{11.19}$$

where $(a)_+ \triangleq \max\{0, a\}$. Figure 11.1 plots the function $\mathbb{T}_\beta(x)$ defined by (11.18); observe how values of x outside the interval $(-\beta, \beta)$ have their magnitudes reduced by the amount β, while values of x within this interval are set to zero.

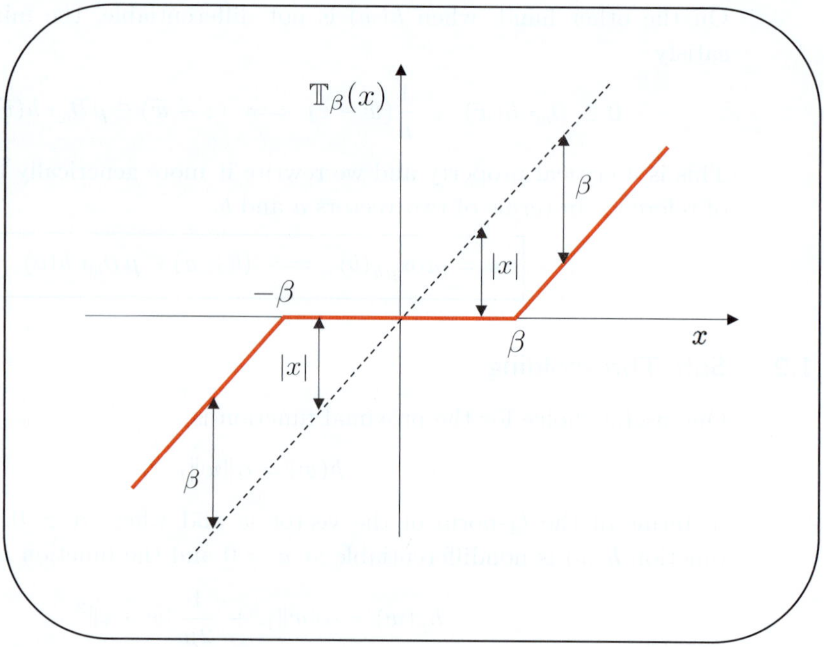

Figure 11.1 The soft-thresholding function, $\mathbb{T}_\beta(x)$, reduces the value of x gradually. Small values of x within the interval $(-\beta, \beta)$ are set to zero, while values of x outside this interval have their size reduced by an amount equal to β. The dotted curve corresponds to the line $y = x$, where y is the vertical coordinate.

We will replace notation (11.17) by the more compact representation:

$$\widehat{w} = \mathbb{T}_{\mu\alpha}(z) \tag{11.20}$$

in terms of the vector arguments $\{\widehat{w}, z\}$, with the understanding that $\mathbb{T}_{\mu\alpha}(\cdot)$ is applied to the individual entries of z to construct the corresponding entries of $\widehat{w}$ according to (11.18).

Proof of (11.17): To determine $\widehat{w}$ we need to minimize the function $h_p(w)$ defined by (11.15). We rewrite (11.15) in terms of the individual entries $\{w_m, z_m\}$ as follows:

$$h_p(w) = \sum_{m=1}^{M} \alpha|w_m| + \sum_{m=1}^{M} \frac{1}{2\mu}(w_m - z_m)^2 \tag{11.21}$$

It follows from this expression that the minimization of $h_p(w)$ over w decouples into M separate minimization problems:

$$\widehat{w}_m \triangleq \underset{w_m \in \mathbb{R}}{\text{argmin}} \left\{\alpha|w_m| + \frac{1}{2\mu}(w_m - z_m)^2\right\} \tag{11.22}$$

For ease of reference, we denote the function that appears on the right-hand side by

$$h_m(w_m) \;\overset{\Delta}{=}\; \alpha|w_m| + \frac{1}{2\mu}(w_m - z_m)^2 \tag{11.23}$$

This cost is convex in w_m but is not differentiable at $w_m = 0$. We can arrive at a closed-form expression for the minimizer by examining the behavior of the function separately over the ranges $w_m \geq 0$ and $w_m \leq 0$:

(1) $\underline{w_m \geq 0}$: In this case, we can use the expression for $h_m(w_m)$ to write

$$\begin{aligned} 2\mu h_m(w_m) &= 2\mu\alpha w_m + (w_m - z_m)^2 \\ &= \big(w_m - (z_m - \mu\alpha)\big)^2 + z_m^2 \; - \; (z_m - \mu\alpha)^2 \end{aligned} \tag{11.24}$$

It follows that the minimizer of $h_m(w_m)$ over the range $w_m \geq 0$ is given by

$$w_m^+ \;\overset{\Delta}{=}\; \underset{w_m \geq 0}{\text{argmin}}\; h_m(w_m) = \begin{cases} 0, & \text{if } \; z_m < \mu\alpha \\ z_m - \mu\alpha, & \text{if } \; z_m \geq \mu\alpha \end{cases} \tag{11.25}$$

(2) $\underline{w_m \leq 0}$: In this case, we get

$$\begin{aligned} 2\mu h_m(w_m) &= -2\mu\alpha w_m + (w_m - z_m)^2 \\ &= \big(w_m - (z_m + \mu\alpha)\big)^2 + z_m^2 \; - \; (z_m + \mu\alpha)^2 \end{aligned} \tag{11.26}$$

It follows that the minimizer of $h_m(w_m)$ over the range $w_m \leq 0$ is given by

$$w_m^- \;\overset{\Delta}{=}\; \underset{w_m \leq 0}{\text{argmin}}\; h_m(w_m) = \begin{cases} 0, & \text{if } \; z_m > -\mu\alpha \\ z_m + \mu\alpha, & \text{if } \; z_m \leq -\mu\alpha \end{cases} \tag{11.27}$$

We conclude that the optimal value for each w_m is given by

$$\widehat{w}_m = \begin{cases} z_m - \mu\alpha, & \text{if } \; z_m \geq \mu\alpha \\ 0, & \text{if } \; -\mu\alpha < z_m < \mu\alpha \\ z_m + \mu\alpha, & \text{if } \; z_m \leq -\mu\alpha \end{cases} \tag{11.28}$$

which agrees with (11.17).

■

Example 11.2 **(A second useful choice)** Let

$$h(w) = \alpha\|w\|_1 + \frac{\rho}{2}\|w\|^2, \quad \alpha > 0, \quad \rho > 0 \tag{11.29}$$

which involves a combination of ℓ_1- and ℓ_2-norms. For this choice of $h(w)$, a similar derivation leads to the expression (see Prob. 11.4):

$$\widehat{w} \;=\; \mathbb{T}_{\frac{\mu\alpha}{1+\mu\rho}}\left(\frac{z}{1+\mu\rho}\right) \tag{11.30}$$

Compared with (11.20), we see that the value of z is scaled by $(1+\mu\rho)$ and the threshold value is also scaled by the same amount.

Example 11.3 **(General statement I)** It is useful to state the main conclusion that follows from the derivation leading to (11.17). This conclusion is of general interest and will be called upon later in different contexts, especially when we study regularization problems. Thus, consider a generic optimization problem of the following form (compare with (11.15)):

$$\widehat{w} \triangleq \underset{w\in\mathbb{R}^M}{\text{argmin}} \left\{\alpha\|w\|_1 + \frac{1}{2\mu}\|w-z\|^2 + \phi\right\} \tag{11.31}$$

for some constants $\alpha \geq 0$, $\mu > 0$, ϕ, and vector $z \in \mathbb{R}^M$. Then, the solution is unique and given by the following expression:

$$\widehat{w} = \mathbb{T}_{\mu\alpha}(z) = \text{sign}(z) \odot \Big(|z| - \mu\alpha\,\mathbb{1}\Big)_+ \tag{11.32}$$

in terms of the soft-thresholding function whose entry-wise operation is defined by (11.18), and where $\odot$ denotes the Hadamard elementwise product.

Example 11.4 **(General statement II)** Consider a diagonal scaling matrix

$$D = \text{diag}\Big\{\sigma_1^2, \sigma_2^2, \ldots, \sigma_M^2\Big\} \tag{11.33}$$

with $\sigma_m^2 > 0$ and replace (11.31) by

$$\widehat{w} \triangleq \underset{w\in\mathbb{R}^M}{\text{argmin}} \left\{\alpha\|Dw\|_1 + \frac{1}{2\mu}\|w-z\|^2 + \phi\right\} \tag{11.34}$$

Repeating the arguments that led to (11.32), it can be verified that (see Prob. 11.11):

$$\widehat{w} = \text{sign}(z) \odot \Big(|z| - \mu\alpha D\,\mathbb{1}\Big)_+ \tag{11.35}$$

where the operations $\text{sign}(x)$, $|x|$, and $(a)_+$ are applied elementwise.

We collect in Table 11.1 several proximal operators established before and in the problems at the end of the chapter. In the table, the notation $(a)_+ = \max\{0, a\}$ and $\mathbb{I}[x] = 1$ when statement x is true; otherwise, it is equal to zero.

11.1.3 Fixed Points

One main motivation for introducing proximal operators is the fact that fixed points for the proximal mapping coincide with global minimizers for the convex function $h(w)$. Specifically, if we let w^o denote any global minimum for $h(w)$, i.e., a point where

$$0 \in \partial_w h(w^o) \tag{11.36}$$

then w^o will be a fixed point for $\text{prox}_{\mu h}(z)$, namely, it holds that:

$$\underbrace{w^o = \text{prox}_{\mu h}(w^o)}_{\textbf{fixed point}} \iff \underbrace{0 \in \partial_w h(w^o)}_{\textbf{global minimum}} \tag{11.37}$$

Table 11.1 Some useful proximal operators along with some properties.

	Convex function, $h(w)$	Proximal operator, $\widehat{w} = \text{prox}_{\mu h}(z)$	Reference
1.	$\mathbb{I}_{C,\infty}[w]$ (C convex set)	$\widehat{w} = \mathcal{P}_C(z)$ (projection operator)	Eq. (11.10)
2.	$\alpha\|w\|_1$, $\alpha > 0$	$\widehat{w} = \mathbb{T}_{\mu\alpha}(z)$ (soft thresholding)	Eq. (11.20)
3.	$\alpha\|w\|_1 + \frac{\rho}{2}\|w\|^2$	$\widehat{w} = \mathbb{T}_{\frac{\mu\alpha}{1+\mu\rho}}\left(\frac{z}{1+\mu\rho}\right)$	Eq. (11.30)
4.	$\alpha\|w\|$	$\widehat{w} = \left(1 - \frac{\mu\alpha}{\|z\|}\right)_+ z$	Prob. 11.7
5.	$\alpha\|w\|_0$	$\widehat{w} = z\,\mathbb{I}\left[\,\|z\| > \sqrt{2\mu\alpha\,\mathbb{1}}\,\right]$	Prob. 11.8
6.	$g(w) = h(w) + c$	$\text{prox}_{\mu g}(z) = \text{prox}_{\mu h}(z)$	Prob. 11.1
7.	$g(w) = h(\alpha w + b)$	$\text{prox}_g(z) = \frac{1}{\alpha}\left(\text{prox}_{\alpha^2 h}(\alpha z + b) - b\right)$	Prob. 11.2
8.	$g(w) = h(w) + \frac{\rho}{2}\|w\|^2$	$\text{prox}_{\mu g}(z) = \text{prox}_{\frac{\mu h}{1+\mu\rho}}\left(\frac{z}{1+\mu\rho}\right)$	Prob. 11.6

This fact follows immediately from property (11.13). Consequently, iterative procedures for finding fixed points of $\text{prox}_{\mu h}(z)$ can be used to find minimizers for $h(w)$. Although unnecessary, we will generally be dealing with the case when there is a unique minimizer, w^o, for $h(w)$ and, correspondingly, a unique fixed point for $\text{prox}_{\mu h}(z)$.

11.2 PROXIMAL POINT ALGORITHM

We now exploit property (11.37) to motivate iterative procedures that converge to global minima w^o for $h(w)$. We know from (11.37) that these minima are fixed points for the proximal operator, i.e., they satisfy

$$w^o = \text{prox}_{\mu h}(w^o) \tag{11.38}$$

There are many iterative constructions that can be used to seek fixed points for operators of this type. We examine in this section the *proximal point algorithm*.

Let $w = f(z) : \mathbb{R}^M \rightarrow \mathbb{R}^M$ denote some generic strictly contractive operator, i.e., a mapping from z to w that satisfies

$$\|f(z_1) - f(z_2)\| < \lambda\,\|z_1 - z_2\|, \text{ for some } 0 \leq \lambda < 1 \tag{11.39}$$

with strict inequality for any vectors $z_1, z_2 \in \text{dom}(f)$. Then, a result known as the *Banach fixed-point theorem* ensures that such operators have unique fixed points, i.e., a unique w^o satisfying $w^o = f(w^o)$ and, moreover, this point can be determined by iterating:

$$w_n = f(w_{n-1}), \quad n \geq 0 \tag{11.40}$$

starting from any initial condition w_{-1}. Then, $w_n \to w^o$ as $n \to \infty$ – see Prob. 11.18.

Remark 11.1. (Convention for initial conditions) Throughout our treatment, most recursive implementations will run over $n \geq 0$, as in (11.40), which means that the initial condition for the recursion will be specified at $n = -1$. This is simply a matter of convention; one can of course run recursions for $n > 0$ and specify the initial condition at $n = 0$. We adopt $n = -1$ as the time instant for the initial condition, w_{-1}.

■

The main challenge in using (11.40) for proximal operators is that the function $\text{prox}_{\mu h}(z)$ is *not* strictly contractive. It is shown in Prob. 11.19 that $\text{prox}_{\mu h}(z)$ satisfies

$$\|\text{prox}_{\mu h}(z_1) - \text{prox}_{\mu h}(z_2)\| \leq \|z_1 - z_2\| \tag{11.41}$$

with *inequality* rather than strict inequality as required by (11.39). In this case, we say that the operator $\text{prox}_{\mu h}(z)$ is *nonexpansive*. Nevertheless, an iterative procedure can still be developed for determining fixed points for $\text{prox}_{\mu h}(z)$ by exploiting the fact that the operator is, in addition, *firmly* nonexpansive. This means that $\text{prox}_{\mu h}(z)$ satisfies the stronger property (see again Prob. 11.19):

$$\|\text{prox}_{\mu h}(z_1) - \text{prox}_{\mu h}(z_2)\|^2 \leq (z_1 - z_2)^{\mathsf{T}}\Big(\text{prox}_{\mu h}(z_1) - \text{prox}_{\mu h}(z_2)\Big) \tag{11.42}$$

This relation implies (11.41); it also implies the following conclusion, in view of the Cauchy–Schwarz relation for inner products:

$$\begin{aligned} \|z_1 - z_2\| &= \|\text{prox}_{\mu h}(z_1) - \text{prox}_{\mu h}(z_2)\| \\ \overset{(11.42)}{\Longleftrightarrow} \; z_1 - z_2 &= \text{prox}_{\mu h}(z_1) - \text{prox}_{\mu h}(z_2) \end{aligned} \tag{11.43}$$

By using this fact, along with (11.41), it is shown in Prob. 11.20 that for such (firmly nonexpansive) operators, an iteration similar to (11.40) will converge to a fixed point w^o of $\text{prox}_{\mu h}(z)$, namely,

$$\boxed{w_n = \text{prox}_{\mu h}(w_{n-1}), \quad n > 0} \qquad \textbf{(proximal iteration)} \tag{11.44}$$

Recursion (11.44) is known as the *proximal point algorithm*, which is summarized in (11.45).

Proximal point algorithm for minimizing $h(w)$.

function $h(w)$ is convex;
given the proximal operator for $h(w)$;
start from an arbitrary initial condition w_{-1}.
repeat over $n \geq 0$ until convergence:
$\quad w_n = \text{prox}_{\mu h}(w_{n-1})$
end
return minimizer $w^o \leftarrow w_n$. (11.45)

By appealing to cases where the proximal projection has a closed-form expression in terms of the entries of w_{n-1}, we arrive at a *realizable* implementation for (11.45). For example, when $h(w) = \alpha\|w\|_1$, we already know from (11.20) that the proximal step in (11.45) can be replaced by:

$$w_n = \mathbb{T}_{\mu\alpha}(w_{n-1}), \quad n > 0 \tag{11.46}$$

On the other hand, when $h(w) = \alpha\|w\|_1 + \frac{\rho}{2}\|w\|^2$, we would use instead:

$$w_n = \mathbb{T}_{\frac{\mu\alpha}{1+\mu\rho}}\left(\frac{w_{n-1}}{1+\mu\rho}\right), \quad n > 0 \tag{11.47}$$

based on the result of Prob. 11.6. Procedures of the form (11.46) and (11.47) are called *iterative soft-thresholding algorithms* for obvious reasons.

11.3 PROXIMAL GRADIENT ALGORITHM

The proximal iteration (11.45) is useful for seeking global minimizers of stand-alone convex functions $h(w)$. However, the algorithm requires the proximal operator for $h(w)$, which is not always available in closed form. We now consider an extension of the method for situations where the objective function $h(w)$ can be expressed as the sum of *two* components, and where the proximal operator for one of the components is available in closed form. The extension leads to the *proximal gradient algorithm*. In preparation for the notation used in subsequent chapters where the optimization objective is denoted by $P(w)$, we will henceforth replace the notation $h(w)$ by $P(w)$ and seek to minimize $P(w)$.

Thus, assume that we are faced with an optimization problem of the form:

$$\min_{w\in\mathbb{R}^M} \left\{P(w) \stackrel{\Delta}{=} q(w) + E(w)\right\} \tag{11.48}$$

involving the sum of two convex components, $E(w)$ and $q(w)$. Usually, $E(w)$ is differentiable and $q(w)$ is nonsmooth. For example, we will encounter in later chapters, while studying the LASSO and basis pursuit problems, optimization problems of the following form:

$$\min_{w\in \mathbb{R}^M} \left\{\alpha\|w\|_1 \;+\; \|d - Hw\|^2\right\} \tag{11.49}$$

where $d \in \mathbb{R}^{N\times 1}$ and $H \in \mathbb{R}^{N\times M}$. For this case, we have $q(w) = \alpha\|w\|_1$ and $E(w) = \|d - Hw\|^2$. At first sight, we could consider applying the proximal iteration (11.44) directly to the aggregate function $P(w)$ and write

$$w_n \;=\; \text{prox}_{\mu P}(w_{n-1}), \quad n \geq 0 \tag{11.50}$$

The difficulty with this approach is that, in general, the proximal operator for $P(w)$ may not be available in closed form. We now derive an alternative procedure for situations when the form of $\text{prox}_{\mu q}(z)$ is known.

Thus, note first that a minimizer for $P(w)$ is a fixed point for the operator $\text{prox}_{\mu q}(z - \mu\nabla_{w^{\mathsf{T}}} E(z))$. Indeed, let w^o denote a fixed point for this operator so that

$$w^o \;=\; \text{prox}_{\mu q}\Big(w^o - \mu\nabla_{w^{\mathsf{T}}} E(w^o)\Big) \tag{11.51a}$$

Then, we know from property (11.13) that

$$\Big(w^o \;-\; \mu\nabla_{w^{\mathsf{T}}} E(w^o) - w^o\Big) \;\in\; \mu\,\partial_{w^{\mathsf{T}}} q(w^o) \tag{11.51b}$$

from which we conclude that

$$0 \;\in\; \nabla_{w^{\mathsf{T}}} E(w^o) + \partial_{w^{\mathsf{T}}} q(w^o) \tag{11.51c}$$

or, equivalently,

$$0 \;\in\; \partial_{w^{\mathsf{T}}} P(w^o) \tag{11.51d}$$

We therefore find, as claimed, that the fixed point w^o is a minimizer for $P(w)$. We can then focus on finding fixed points for $\text{prox}_{\mu q}(z - \mu\nabla_{w^{\mathsf{T}}} E(z))$ by using the recursion (cf. (11.44)):

$$\boxed{w_n \;=\; \text{prox}_{\mu q}\Big(w_{n-1} - \mu\nabla_{w^{\mathsf{T}}} E(w_{n-1})\Big), \quad n \geq 0} \tag{11.52}$$

We refer to this implementation as the *proximal gradient algorithm*, which we rewrite in expanded form in (11.53) by introducing an intermediate variable z_n. The algorithm involves two steps: a gradient-descent step on the differentiable component $E(w)$ to obtain z_n, followed by a proximal projection step relative to the nonsmooth component $q(w)$.

Proximal gradient algorithm for minimizing $P(w) = q(w) + E(w)$.

$q(w)$ and $E(w)$ are convex functions;
given the proximal operator for the nonsmooth component, $q(w)$;
given the gradient operator for the smooth component, $E(w)$;
start from an arbitrary initial condition w_{-1}. (11.53)
repeat over $n \geq 0$ until convergence:
$z_n = w_{n-1} - \mu \nabla_{w^\mathsf{T}} E(w_{n-1})$
$w_n = \text{prox}_{\mu q}(z_n)$
end
return minimizer $w^o \leftarrow w_n$.

Remark 11.2. (Forward–backward splitting) In view of property (11.13), the proximal step in (11.53) implies the following relation:

$$w_n = \text{prox}_{\mu q}(z_n) \iff (z_n - w_n) \in \mu\, \partial_{w^\mathsf{T}} q(w_n) \tag{11.54}$$

which in turn implies that we can rewrite the algorithm in the form:

$$\text{(forward–backward splitting)} \quad \begin{cases} z_n &= w_{n-1} - \mu \nabla_{w^\mathsf{T}} E(w_{n-1}) \\ w_n &= z_n - \mu\, \partial_{w^\mathsf{T}} q(w_n) \end{cases} \tag{11.55}$$

where $\partial_{w^\mathsf{T}} q(w_n)$ denotes *some* subgradient for $q(w)$ at location w_n. In this form, the first step corresponds to a forward update step moving from w_{n-1} to z_n, while the second step corresponds to a backward (or implicit) update since it involves w_n on both sides. For these reasons, recursion (11.53) is sometimes referred to as a *forward–backward splitting* implementation.

■

Example 11.5 (Two useful cases) Let us consider two special instances of formulation (11.48). When $q(w) = \alpha \|w\|_1$, the proximal gradient algorithm reduces to

$$\begin{cases} z_n &= w_{n-1} - \mu \nabla_{w^\mathsf{T}} E(w_{n-1}) \\ w_n &= \mathbb{T}_{\mu\alpha}(z_n) \end{cases} \tag{11.56}$$

and when $q(w) = \alpha \|w\|_1 + \frac{\rho}{2}\|w\|^2$, we get

$$\begin{cases} z_n &= w_{n-1} - \mu \nabla_{w^\mathsf{T}} E(w_{n-1}) \\ w_n &= \mathbb{T}_{\frac{\mu\alpha}{1+\mu\rho}}\left(\dfrac{z_n}{1+\mu\rho}\right) \end{cases} \tag{11.57}$$

Example 11.6 (Logistic cost function) Consider a situation where $E(w)$ is chosen as the logistic loss function, i.e.,

$$E(w) = \ln\left(1 + e^{-\gamma h^\mathsf{T} w}\right) \tag{11.58}$$

where $\gamma \in \mathbb{R}$ and $h \in \mathbb{R}^M$. Note that we are using here the notation h to refer to the column vectors that appear in the exponent expression; this is in line with our future notation for feature vectors in subsequent chapters. Let

$$q(w) = \alpha \|w\|_1 + \frac{\rho}{2}\|w\|^2 \tag{11.59}$$

so that the function that we wish to minimize is

$$P(w) = \alpha\|w\|_1 + \frac{\rho}{2}\|w\|^2 + \ln\left(1 + e^{-\gamma h^\mathsf{T} w}\right) \tag{11.60}$$

The function $E(w)$ is differentiable everywhere with

$$\nabla_{w^\mathsf{T}} E(w) \;=\; -\gamma h \times \frac{1}{1 + e^{\gamma h^\mathsf{T} w}} \tag{11.61}$$

and we find that the proximal gradient recursion (11.53) reduces to

$$\left\{\begin{array}{l} z_n = w_{n-1} \;+\; \mu\gamma h \times \frac{1}{1+e^{\gamma h^\mathsf{T} w_{n-1}}} \\ w_n = \mathbb{T}_{\frac{\mu\alpha}{1+\mu\rho}}\left(\dfrac{z_n}{1+\mu\rho}\right) \end{array}\right. \tag{11.62}$$

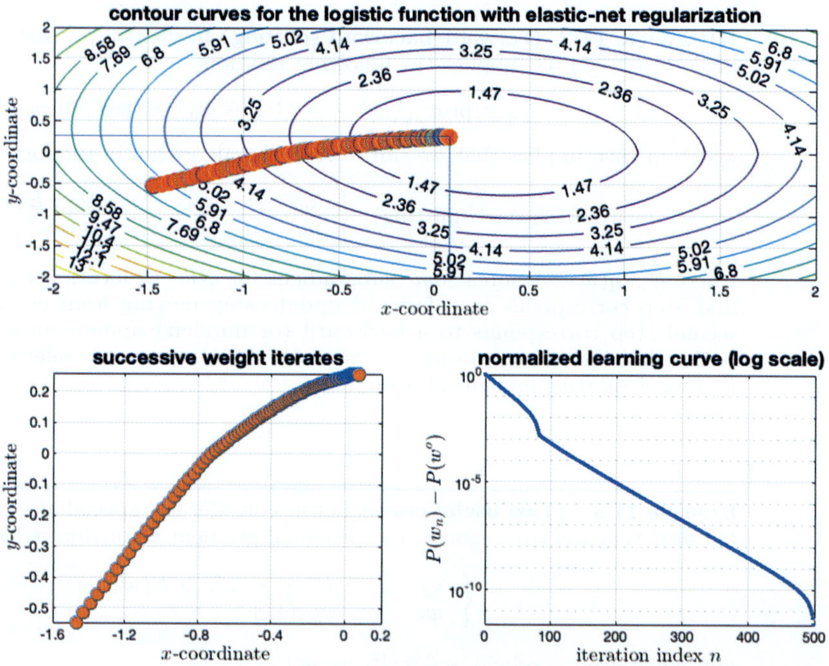

Figure 11.2 (*Top*) Contour curves of a regularized logistic function $P(w)$ of the form (11.60) with $\alpha = 0.2$, $\rho = 2$, $\gamma = 1$, and $h = \text{col}\{1, 2\}$. The successive locations of the weight iterates generated by the proximal gradient recursion (11.62) are shown in circles moving from left to right toward the minimizer location of $P(w)$. (*Bottom left*) Trajectory of the successive weight iterates in $\mathbb{R}^2$, moving from left to right, as they approach the location of the minimizer w^o of $P(w)$. (*Bottom right*) Normalized learning curve in logarithmic scale. The curve illustrates the expected "linear convergence" rate for the successive iterate values toward $P(w^o)$.

Figure 11.2 plots the contour curves for the regularized logistic function $P(w) : \mathbb{R}^2 \to \mathbb{R}$ with parameters

$$\gamma = 1, \;\; h = \begin{bmatrix} 1 \\ 2 \end{bmatrix}, \;\; \alpha = 0.2, \;\;\; \rho = 2 \tag{11.63}$$

The location of the minimizer w^o for $P(w)$ and the corresponding minimum value are determined to be approximately:

$$w^o \approx \begin{bmatrix} 0.0782 \\ 0.2564 \end{bmatrix}, \quad P(w^o) \approx 0.5795 \tag{11.64}$$

These values are obtained by running the proximal gradient recursion (11.62) for 500 iterations starting from a random initial condition w_{-1} and using $\mu = 0.01$. The top plot in the figure illustrates the trajectory of the successive weight iterates, moving from left to right, in relation to the contour curves of $P(w)$; this same trajectory is shown in the lower left plot of the same figure. The lower right plot shows the evolution of the learning curve $P(w_n) - P(w^o)$ over the iteration index n in logarithmic scale for the vertical axis. Specifically, the figure shows the evolution of the *normalized* quantity

$$\ln\left(\frac{P(w_n) - P(w^o)}{\max_n\{P(w_n) - P(w^o)\}}\right) \tag{11.65}$$

where the curve is normalized by its maximum value so that the peak value in the logarithmic scale is zero. This will be our standing assumption in plots for learning curves in the logarithmic scale; the peak values will be normalized to zero. It is clear from the figure, due to its decaying "linear form" that the convergence of $P(w_n)$ toward $P(w^o)$ occurs at an exponential rate, as predicted further ahead by result (11.74).

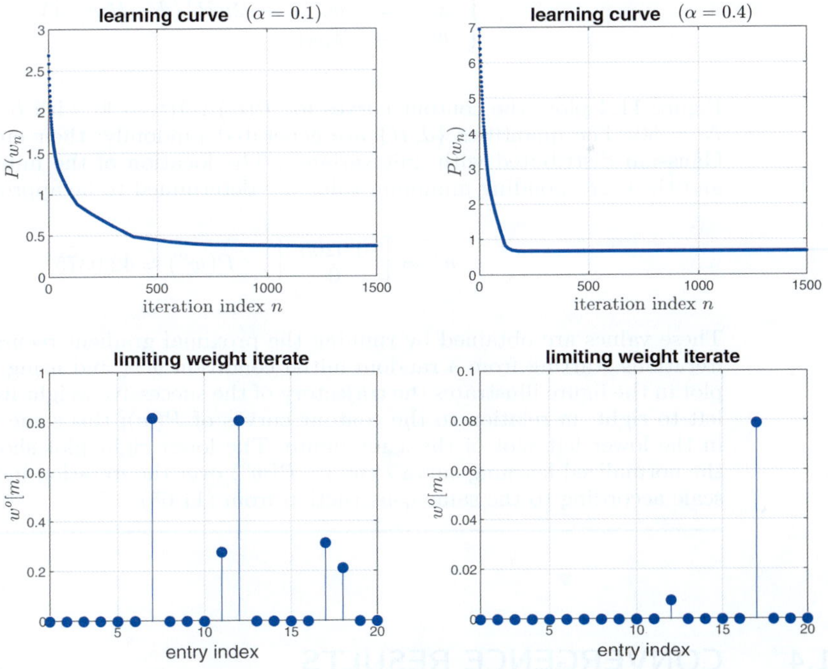

Figure 11.3 (*Top*) Learning curves for the proximal gradient recursion (11.62) using $\mu = 0.03$ when applied to $P(w) = \alpha\|w\|_1 + \ln(1 + e^{-\gamma h^\mathsf{T} w})$ using $\alpha = 0.1$ and $\alpha = 0.4$. We use $\gamma = 1$ and generate a random h with entries selected uniformly from within the interval $[-1, 1]$. (*Bottom*) Form of resulting minimizers, w^o, for both choices of α. Observe how the minimizer w^o is more sparse for $\alpha = 0.4$ than for $\alpha = 0.1$.

Figure 11.3 illustrates the influence of the parameter α on the sparsity of the minimizer w^o. We set $\rho = 0$ and plot the learning curves and the resulting minimizers for $P(w)$ with $w \in \mathbb{R}^{20}$, i.e., $M = 20$. We continue to use $\gamma = 1$ but generate a random h with entries selected uniformly from within the interval $[-1, 1]$. The curves in the figure are the result of running the proximal gradient recursion (11.62) for 1500 iterations using $\mu = 0.03$. The learning curves tend toward the minimum values $P(w^o) \approx 0.3747$ for $\alpha = 0.1$ and $P(w^o) \approx 0.6924$ for $\alpha = 0.4$. Observe how the minimizer w^o is more sparse for $\alpha = 0.4$ than for $\alpha = 0.1$.

Example 11.7 **(LASSO or basis pursuit)** Consider the optimization problem

$$w^o = \underset{w \in \mathbb{R}^M}{\operatorname{argmin}} \left\{ P(w) \triangleq \alpha \|w\|_1 + \|d - Hw\|^2 \right\} \tag{11.66}$$

where $d \in \mathbb{R}^{N \times 1}$ and $H \in \mathbb{R}^{N \times M}$. We will encounter this problem later when we study sparsity-inducing solutions and, in particular, the LASSO and basis pursuit algorithms. The function $E(w) = \|d - Hw\|^2$ is differentiable everywhere with

$$\nabla_{w^{\mathsf{T}}} E(w) = -2H^{\mathsf{T}}(d - Hw) \tag{11.67}$$

so that the proximal gradient recursion (11.53) reduces to

$$\begin{cases} z_n = w_{n-1} + 2\mu H^{\mathsf{T}}(d - Hw_{n-1}) \\ w_n = \mathbb{T}_{\mu\alpha}(z_n) \end{cases} \tag{11.68}$$

Figure 11.4 plots the contour curves for $P(w) : \mathbb{R}^2 \to \mathbb{R}$ with $\alpha = 0.5$, $M = 2$, and $N = 50$. The quantities $\{d, H\}$ are generated randomly; their entries are zero-mean Gaussian distributed with unit variance. The location of the minimizer w^o for $P(w)$ and the corresponding minimum value are determined to be approximately

$$w^o \approx \begin{bmatrix} 0.0205 \\ 0 \end{bmatrix}, \quad P(w^o) \approx 42.0375 \tag{11.69}$$

These values are obtained by running the proximal gradient recursion (11.68) for 400 iterations starting from a random initial condition w_{-1} and using $\mu = 0.001$. The top plot in the figure illustrates the trajectory of the successive weight iterates, moving from left to right, in relation to the contour curves of $P(w)$; this same trajectory is shown in the lower left plot of the same figure. The lower right plot shows the evolution of the normalized learning curve $P(w_n) - P(w^o)$ over the iteration index n in logarithmic scale according to the same construction from (11.65).

11.4 CONVERGENCE RESULTS

We list two convergence results for the proximal gradient algorithm (11.53) for small enough μ. We examine two cases: **(a)** $E(w)$ is convex and **(b)** $E(w)$ is strongly convex. The latter case has faster convergence rate. The proof of the following first result appears in Appendix 11.A.

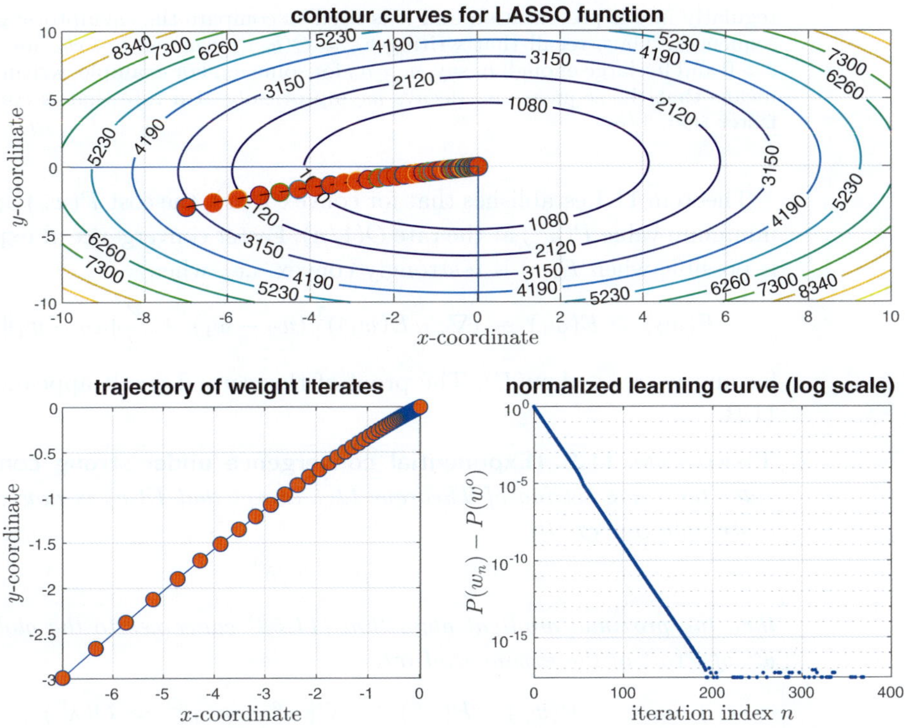

Figure 11.4 (*Top*) Contour curves of a regularized LASSO function $P(w)$ of the form (11.66) with $\alpha = 0.5$, $M = 2$, and $N = 50$. The quantities (d, H) are generated randomly; their entries are zero-mean Gaussian distributed with unit variance. The successive locations of the weight iterates generated by the proximal gradient recursion (11.68) are shown in circles moving from left to right toward the minimizer location of $P(w)$. (*Bottom left*) Trajectory of the successive weight iterates in $\mathbb{R}^2$, moving from left to right, as they approach the location of the minimizer w^o of $P(w)$. (*Bottom right*) Normalized learning curve in logarithmic scale.

Theorem 11.1. (Convergence of proximal gradient algorithm) *Consider the problem of minimizing $P(w) = q(w) + E(w)$ where $q(w)$ and $E(w)$ are both convex functions, with $E(w)$ differentiable and having δ-Lipschitz gradients. If*

$$\mu < 1/\delta \tag{11.70}$$

then the proximal gradient algorithm (11.53) converges to a global minimizer w^o of $P(w)$ at the following rate

$$P(w_n) - P(w^o) \leq \frac{1}{2(n+1)\mu}\|w^o - w_{-1}\|^2 \;=\; O(1/n) \tag{11.71}$$

where w_{-1} is an arbitrary initial condition.

Remark 11.3. (Big-O notation) Statement (11.71) uses the big-O notation, which we already encountered in the earlier expression (3.226). This notation will appear regularly in our presentation and it is used to compare the asymptotic growth rates of sequences. Thus, recall that writing $a_n = O(b_n)$ means $|a_n| \leq cb_n$ for some constant $c > 0$ and for large enough n, say, $n > n_o$ for some n_o. For example, writing $a_n = O(1/n)$ means that the sequence a_n decays asymptotically at a rate that is comparable to or faster than $1/n$.

■

Theorem 11.1 establishes that for convex $E(w)$, the cost $P(w_n)$ approaches the minimum value $P(w^o)$ at the rate $O(1/n)$. Faster convergence at exponential rate is possible when $E(w)$ is ν-strongly convex, i.e., when

$$E(w_2) \geq E(w_1) + \left(\nabla_{w^{\mathsf{T}}} E(w_1)\right)^{\mathsf{T}} (w_2 - w_1) + \frac{\nu}{2}\|w_2 - w_1\|^2 \tag{11.72}$$

for any $w_1, w_2 \in \text{dom}(E)$. The proof of this second result appears in Appendix 11.B.

THEOREM 11.2. (Exponential convergence under strong convexity) *Consider the same setting of Theorem 11.1 except that $E(w)$ is now assumed to be ν-strongly convex. If*

$$\mu < 2\nu/\delta^2 \tag{11.73}$$

then the proximal gradient algorithm (11.53) converges to the global minimizer w^o of $P(w)$ at the exponential rate

$$P(w_n) - P(w^o) \leq \beta\lambda^n \|w^o - w_{-1}\|^2 = O(\lambda^n) \tag{11.74}$$

for some constant β and where

$$\lambda \stackrel{\Delta}{=} 1 - 2\mu\nu + \mu^2\delta^2 \in [0, 1) \tag{11.75}$$

REMARK 11.4. (A more relaxed bound on μ) The result of Theorem 11.2 establishes the exponential convergence of the excess risk to zero for sufficiently small step sizes, μ. In most instances, this result is sufficient for our purposes since our objective will generally be to verify whether the iterative algorithms approach their desired limit. This conclusion is established in Theorem 11.2 under the bound $\mu < 2\nu/\delta^2$. We can relax the bound and show that convergence will continue to occur for the more relaxed bound $\mu < 2/\delta$ at the rate $O((\lambda')^n)$, where $\lambda' = 1 - 2\mu\nu + \mu^2\nu\delta$. We explain in the same Appendix 11.B that this can be achieved by exploiting a certain *co-coercivity* property that is satisfied by convex functions with δ-Lipschitz gradients – see Example 11.8.

■

11.5 DOUGLAS–RACHFORD ALGORITHM

We continue with the optimization problem (11.48) except that we now allow for both functions $q(w)$ and $E(w)$ to be nonsmooth and present a splitting algorithm for minimizing their aggregate sum. This second algorithm will involve the proximal operators for *both* functions and is therefore suitable when these proximal operators can be determined beforehand. There are several variations of the Douglas–Rachford algorithm – see Probs. 11.22 and 11.23. We list one form in (11.76).

Douglas–Rachford algorithm for minimizing $P(w) = q(w) + E(w)$.

$q(w)$ and $E(w)$ are (possibly nonsmooth) convex functions;
given the proximal operator for $q(w)$;
given the proximal operator for $E(w)$;
start from an arbitrary initial condition z_{-1};
repeat over $n \geq 0$ until convergence:

$$\begin{aligned} w_n &= \text{prox}_{\mu q}(z_{n-1}) \\ t_n &= \text{prox}_{\mu E}(2w_n - z_{n-1}) \\ z_n &= t_n - w_n + z_{n-1} \end{aligned} \tag{11.76}$$

end
return minimizer $w^o \leftarrow w_n$.

The main motivation for the algorithm lies in the fact that the mapping from z_{n-1} to z_n in (11.76) can be shown to be firmly nonexpansive and that fixed points for this mapping determine the desired minimizer(s) for (11.48). Let us comment on the second property first. For this purpose, we write down the fixed-point relations:

$$\left\{ \begin{array}{rcl} w^o &=& \text{prox}_{\mu q}(z^o) \\ t^o &=& \text{prox}_{\mu E}(2w^o - z^o) \\ z^o &=& t^o - w^o + z^o \end{array} \right. \tag{11.77}$$

where we replaced the variables $\{w_n, w_{n-1}, z_n, z_{n-1}, t_n\}$ by fixed-point values $\{w^o, z^o, t^o\}$. Using property (11.13) for proximal projections, these relations translate into:

$$\left\{ \begin{array}{rcl} (z^o - w^o) &\in& \mu\, \partial_{w^\mathsf{T}}\, q(w^o) \\ (2w^o - z^o - t^o) &\in& \mu\, \partial_{w^\mathsf{T}}\, E(t^o) \\ w^o &=& t^o \end{array} \right. \tag{11.78}$$

which imply that

$$(w^o - z^o) \in \mu\, \partial_{w^\mathsf{T}} E(w^o) \;\; \text{and} \;\; (z^o - w^o) \in \mu\, \partial_{w^\mathsf{T}} q(w^o) \tag{11.79}$$

Using the result of Prob. 8.31 we conclude that

$$0 \in \partial_{w^\mathsf{T}} \Big(E(w^o) + q(w^o) \Big) \tag{11.80}$$

which confirms that w^o is a minimizer for the sum $E(w)+q(w)$, as claimed. This argument establishes that if z^o is a fixed point for the mapping from z_{n-1} to z_n in (11.76), then $w^o = \text{prox}_{\mu q}(z^o)$ is a minimizer for the $P(w)$. We still need to establish that the mapping from z_{n-1} to z_n is firmly nonexpansive, in which case algorithm (11.76) would correspond to a fixed-point iteration applied to this mapping. If we group the three relations appearing in (11.76) we find that the mapping from z_{n-1} to z_n is given by

$$z_n = z_{n-1} + \text{prox}_{\mu E}\Big(2\text{prox}_{\mu q}(z_{n-1}) - z_{n-1}\Big) \; - \; \text{prox}_{\mu q}(z_{n-1}) \qquad (11.81)$$

which we denote more compactly by writing $z_n = R(z_{n-1})$ with the mapping $R(z)$ defined by

$$R(z) \;\stackrel{\Delta}{=}\; z + \text{prox}_{\mu E}\Big(2\text{prox}_{\mu q}(z) - z\Big) \; - \; \text{prox}_{\mu q}(z) \qquad (11.82)$$

By exploiting the fact that proximal operators are themselves firmly nonexpansive, it is shown in Prob. 11.21 that $R(z)$ is also firmly nonexpansive, meaning that it satisfies:

$$\|R(z_1) - R(z_2)\|^2 \le \; (z_1 - z_2)^{\mathsf{T}} \left(R(z_1) - R(z_2)\right) \qquad (11.83)$$

for any z_1, z_2. In this case, recursion (11.81) can be viewed as a fixed-point iteration for this mapping and implementation (11.76) amounts to unfolding this iteration into three successive steps.

11.6 COMMENTARIES AND DISCUSSION

Proximal operators. The proximal projection of a convex function $h(w)$ was defined in (11.4) as the mapping that transforms a vector $z \in \mathbb{R}^M$ into the vector:

$$\widehat{w} \;=\; \text{prox}_{\mu h}(z) \;\stackrel{\Delta}{=}\; \underset{w\in\mathbb{R}^M}{\text{argmin}} \; \left\{ h(w) \; + \; \frac{1}{2\mu}\|w - z\|^2 \right\} \qquad (11.84)$$

with the resulting minimum value given by its Moreau envelope from (11.6):

$$\mathcal{M}_{\mu h}(z) \;\stackrel{\Delta}{=}\; \min_{w\in\mathbb{R}^M} \; \left\{ h(w) \; + \; \frac{1}{2\mu}\|w - z\|^2 \right\} \qquad (11.85)$$

This envelope is also called the Moreau–Yosida envelope in recognition of the contributions by Moreau (1965) and Yosida (1968). Intuitively, the proximal construction approximates z by some vector, $\widehat{w}$, that is close to it under the squared Euclidean norm but subject to the penalty $h(w)$. The Moreau value at $\widehat{w}$ serves as a measure of a (generalized) distance between z and its proximal projection.

We discussed several properties of proximal projections in Section 11.1. The form of the soft-thresholding operator (11.18) appeared in the works by Donoho and Johnstone (1994, 1995) on the recovery of signals embedded in additive Laplace-distributed noise. Other useful interpretations and properties of proximal operators can be found in the treatments by Lemaire (1989a, b) Rockafellar and Wets (1998), Combettes and Pesquet (2011), and Parikh and Boyd (2013). In Prob. 11.7 we highlight one useful connection of proximal operators to the Huber function, which is a popular tool in robust statistics used to reduce the effect of data outliers – see Huber (1981). We explain in that problem that the Moreau envelope that corresponds to the choice $h(w) = \|w\|$ is the Huber loss function, denoted by

$$H_\mu(z) = \begin{cases} \frac{1}{2\mu}\|z\|^2, & \|z\| \le \mu \\ \|z\| - \frac{\mu}{2}, & \|z\| > \mu \end{cases} \qquad (11.86)$$

where $z, w \in \mathbb{R}^M$. The Huber function is linear in $\|z\|$ over the range $\|z\| > \mu$, for some parameter $\mu > 0$ and, therefore, it penalizes less drastically large values for $\|z\|$ in comparison to the quadratic loss, $\|z\|^2$.

Moreau decomposition. The concept of the proximal operator (11.84) and its envelope were introduced and studied in a series of works by Moreau (1962, 1963a, b, 1965). One of the main driving themes in the work by Moreau (1965) was to establish the following interesting decomposition. Consider a convex function $h(w)$ defined over $w \in \mathbb{R}^M$, and let $h^\star(x)$ denote its conjugate function defined earlier by (8.83), i.e.,

$$h^\star(x) \triangleq \sup_w \left(x^\mathsf{T} w - h(w)\right), \quad x \in \mathcal{X} \tag{11.87}$$

where $\mathcal{X}$ denotes the set of all x where the supremum operation is finite. It is shown in Prob. 8.47 that $\mathcal{X}$ is a convex set. Then, any vector $z \in \mathbb{R}^M$ can be decomposed as (see Prob. 11.24):

$$z = \text{prox}_h(z) \; + \; \text{prox}_{h^\star}(z) \tag{11.88a}$$

$$\frac{1}{2}\|z\|^2 = \mathcal{M}_h(z) + \mathcal{M}_{h^\star}(z) \tag{11.88b}$$

These expressions provide an interesting generalization of the orthogonal projection decomposition that is familiar from Euclidean geometry, namely, for any closed vector space $\mathcal{C} \subset \mathbb{R}^M$, every vector z can be decomposed as

$$z = \mathcal{P}_C(z) \; + \; \mathcal{P}_{C^\perp}(z) \tag{11.89a}$$

$$\|z\|^2 = \|\mathcal{P}_C(z)\|^2 \; + \; \|\mathcal{P}_{C^\perp}(z)\|^2 \tag{11.89b}$$

where $\mathcal{P}_C(z)$ denotes the orthogonal projection of z onto $\mathcal{C}$, and similarly $\mathcal{P}_{C^\perp}(z)$ denotes the orthogonal projection of z onto the orthogonal complement space, $\mathcal{C}^\perp$, namely,

$$\mathcal{P}_C(z) \triangleq \min_{w \in \mathcal{C}} \|w - z\|^2 \tag{11.90}$$

where for any $x \in \mathcal{C}$ and $y \in \mathcal{C}^\perp$, it holds that $x^\mathsf{T} y = 0$.

Two other critical properties established by Moreau (1965) are that **(a)** the proximal operator is a firmly nonexpansive mapping (cf. (11.42)) and **(b)** the Moreau envelope is *differentiable* over z, regardless of whether the original function $h(w)$ is differentiable or not over w. Actually, the gradient vector of the Moreau envelope is given by (see Prob. 11.25):

$$\nabla_{z^\mathsf{T}} \mathcal{M}_{\mu h}(z) \; = \; \frac{1}{\mu}\left(z - \text{prox}_{\mu h}(z)\right) \tag{11.91}$$

It is then clear from property (11.37) relating minimizers of $h(w)$ to fixed points of $\text{prox}_{\mu h}(z)$ that it also holds

$$\underbrace{w^o = \text{prox}_{\mu h}(w^o)}_{\textbf{fixed point}} \iff \underbrace{0 \in \partial_w h(w^o)}_{\textbf{global minimum}} \iff \nabla_z \mathcal{M}_{\mu h}(w^o) = 0 \tag{11.92}$$

In this way, the problem of determining the minimizer(s) of a possibly nonsmooth (i.e., nondifferentiable) convex function $h(w)$ can be reduced to the equivalent problems of determining the fixed point(s) of a firmly nonexpansive proximal operator or the stationary point(s) of a smooth (i.e., differentiable) Moreau envelope. This observation is further reinforced by noting that we can rewrite (11.91) as

$$\text{prox}_{\mu h}(z) \; = \; z - \mu \nabla_{z^\mathsf{T}} \mathcal{M}_{\mu h}(z) \tag{11.93}$$

which shows that the proximal operation amounts to performing a gradient-descent step over the Moreau envelope.

Fixed-point iterations. In future chapters we will derive several algorithms for online learning by relying on the use of proximal projections to deal with optimization problems that involve nonsmooth components. There are two key properties that make proximal operators particularly suitable for solving nonsmooth optimization problems. The first property is the fact, established in (11.37), that their fixed points coincide with the minimizers of the functions defining them, namely,

$$\underbrace{w^o = \text{prox}_{\mu h}(w^o)}_{\textbf{fixed point}} \iff \underbrace{0 \in \partial_w h(w^o)}_{\textbf{global minimum}} \tag{11.94}$$

The second property is the fact that proximal operators are firmly nonexpansive, meaning that they satisfy property (11.42). Consequently, as shown by (11.44), a convergent iteration can be used to determine their fixed points, which leads to the proximal point algorithm:

$$w_n = \text{prox}_{\mu h}(w_{n-1}), \quad n \geq 0 \tag{11.95}$$

These observations have motivated broad research efforts into constructing new families of algorithms for the solution of nonsmooth convex optimization problems by exploiting the theory of fixed-point iterations for firmly nonexpansive operators – see, e.g., the works by Minty (1962), Browder (1965, 1967), Bruck and Reich (1977), and Combettes (2004), as well as the texts by Brezis (1973), Granas and Dugundji (2003), and Bauschke and Combettes (2011). One of the earliest contributions in this regard is the proximal point algorithm (11.95). It was proposed by Martinet (1970, 1972) as a way to construct iterates for minimizing a convex function $h(w)$ by solving successive problems of the type (using our notation):

$$w_n = \underset{w\in\mathbb{R}^M}{\text{argmin}} \left\{ h(w) + \frac{1}{2\mu}\|w - w_{n-1}\|^2 \right\} \tag{11.96}$$

where the variable z is replaced by the prior iterate w_{n-1}. According to (11.84), this leads to $w_n = \text{prox}_{\mu h}(w_{n-1})$, which is the proximal iteration (11.95). Two other early influential works in the area of proximal operators for nonsmooth optimization, with stronger convergence results, are the articles by Rockafellar (1976a, b) on the proximal point algorithm and generalizations. Two early works on the proximal gradient method (11.52) are Sibony (1970) and Mercier (1979). Since then, several important advances have occurred in the development of techniques for the optimization of nonsmooth problems. The presentation and convergence analysis in the chapter and appendices are motivated by the useful overviews given by Polyak (1987), Combettes and Pesquet (2011), Parikh and Boyd (2013), Polson, Scott, and Willard (2015), and Beck (2017), in addition to the contributions by Luo and Tseng (1992b, 1993), Nesterov (2004, 2005), Combettes and Wajs (2005), Figueiredo, Bioucas-Dias, and Nowak (2007), and Beck and Teboulle (2009a, 2012).

Resolvents and splitting techniques. Motivated by expression (11.12) for $\widehat{w}$, it is customary in the literature to express the proximal operator of a convex function $h(w)$ by writing

$$\text{prox}_{\mu h}(z) = (I + \mu\,\partial_{w^\mathsf{T}} h)^{-1}(z) \tag{11.97}$$

The notation on the right-hand side means that the point z is mapped to its proximal projection $\widehat{w}$. The operation that performs this mapping is denoted either by $\text{prox}_{\mu h}(z)$, which is our standard notation, or more broadly by the operator notation $(I+\mu\partial_{w^\mathsf{T}} h)^{-1}$. This latter notation is called the *resolvent* of the operator $\mu\partial_{w^\mathsf{T}} h$. Observe that although the subdifferential of a function at any particular location is not uniquely defined (i.e., it generally maps one point in space to multiple points since there can be many choices for the subgradient vector), the result of the resolvent operation is always unique since we already know that proximal projections are unique. In this way, notation (11.97) maps z to a unique point $\widehat{w}$.

We can employ the resolvent notation to motivate splitting algorithms, along the lines of Lions and Mercier (1979) and Eckstein and Bertsekas (1992). For instance, one other way to motivate the Douglas–Rachford splitting procedure (11.76) is to consider initially the simpler but related problem of determining vectors $w \in \mathbb{R}^M$ that lie in the nullspace of the sum of two nonnegative-definite matrices, i.e., vectors w that satisfy

$$(A+B)w = 0 \iff Aw + Bw = 0 \tag{11.98}$$

for some matrices $A \geq 0, B \geq 0$. If we consider the equivalent form

$$w = \mu Aw + \mu Bw + w \tag{11.99}$$

for some $\mu > 0$, or

$$w = \big(I + \mu(A+B)\big)^{-1} w \tag{11.100}$$

then the solution can be pursued by considering a fixed-point iteration of the form:

$$w_n = \big(I + \mu(A+B)\big)^{-1} w_{n-1} \tag{11.101}$$

This recursion requires inverting a matrix consisting of the sum $A + B$. We would like to replace it by an alternative procedure that requires inverting the terms $(I + \mu A)$ and $(I + \mu B)$ separately. This can be achieved by introducing auxiliary variables as follows. From (11.99), the vector w is a fixed point for the equation:

$$2w = (I + \mu A)w + (I + \mu B)w \tag{11.102}$$

We next introduce the variable:

$$z \stackrel{\Delta}{=} (I + \mu B)w \iff w = (I + \mu B)^{-1} z \tag{11.103}$$

and note from (11.102) that

$$2w - z = (I + \mu A)w \iff w = (I + \mu A)^{-1}(2w - z) \tag{11.104}$$

We therefore have two separate expressions in (11.103)–(11.104) involving the inverses $(I + \mu A)^{-1}$ and $(I + \mu B)^{-1}$. Now observe from the trivial equality:

$$z = z + w - w \tag{11.105}$$

that we can write using (11.103)–(11.104):

$$\begin{aligned} z &= z \;+\; (I + \mu A)^{-1}(2w - z) \;-\; (I + \mu B)^{-1} z \\ &= z \;+\; (I + \mu A)^{-1}\big(2(I + \mu B)^{-1} z - z\big) \;-\; (I + \mu B)^{-1} z \end{aligned} \tag{11.106}$$

The mapping on the right-hand side has a form similar to (11.82) if we make the identifications:

$$(I + \mu A)^{-1} \leftarrow (I + \mu \partial_{w^{\mathsf{T}}} E)^{-1} \;=\; \text{prox}_{\mu E}(\cdot) \tag{11.107}$$

$$(I + \mu B)^{-1} \leftarrow (I + \mu \partial_{w^{\mathsf{T}}} q)^{-1} \;=\; \text{prox}_{\mu q}(\cdot) \tag{11.108}$$

Indeed, the same argument can be repeated to arrive at (11.82) by noting that minimizers of (11.48) should satisfy, for any nonzero μ,

$$0 \in \mu \partial_{w^{\mathsf{T}}} E(w) + \mu \partial_{w^{\mathsf{T}}} q(w) \tag{11.109}$$

or, equivalently,

$$2w \in (I + \mu \partial_{w^{\mathsf{T}}} E)(w) + (I + \mu \partial_{w^{\mathsf{T}}} q)(w) \tag{11.110}$$

If we now introduce the variable:

$$z \stackrel{\Delta}{=} (I + \mu \partial_{w^{\mathsf{T}}} q)(w) \iff w = (I + \mu \partial_{w^{\mathsf{T}}} q)^{-1}(z) \;=\; \text{prox}_{\mu q}(z) \tag{11.111}$$

and note from (11.110) that

$$2w - z = (I + \mu\partial_{w^\mathsf{T}} E)(w) \iff w = (I + \mu\partial_{w^\mathsf{T}} E)^{-1}(2w - z) = \text{prox}_{\mu E}(2w - z) \tag{11.112}$$

then the same argument ends up leading to the mapping (11.82).

Discussions on, and variations of, the forward–backward proximal splitting technique (11.55) can be found in Combettes and Wajs (2005), Bach *et al.* (2012), Combettes and Pesquet (2011), and the many references therein, as well as in the articles by Lions and Mercier (1979), Passty (1979), Fukushima and Mine (1981), Guler (1991), Tseng (1991), Chen and Rockafellar (1997), Daubechies, Defrise, and De Mol (2004), Hale, Yin, and Zhang (2008), Bredies (2009), Beck and Teboulle (2009a, b), and Duchi and Singer (2009). The proximal Douglas–Rachford splitting algorithm (11.76) was introduced by Lions and Mercier (1979), who were motivated by the original work of Douglas and Rachford (1956) on a numerical discretized solution for the heat conduction problem. More discussion on this algorithm can be found in Eckstein and Bertsekas (1992), Combettes (2004), Combettes and Pesquet (2011), and O'Connor and Vandenberghe (2014), as well as in the lecture notes by Vandenberghe (2010).

PROBLEMS

11.1 Consider two convex functions $h(w)$ and $f(w)$ related by $h(w) = f(w) + c$, for some constant c. Show that $\text{prox}_{\mu h}(z) = \text{prox}_{\mu f}(z)$ for any $z \in \mathbb{R}^M$ and $\mu > 0$.

11.2 Let $h(w) : \mathbb{R}^M \to \mathbb{R}$ denote a convex function and introduce the transformation $g(w) = h(\alpha w + b)$, where $\alpha \neq 0$ and $b \in \mathbb{R}^M$. Show that

$$\text{prox}_g(z) = \frac{1}{\alpha}\Big(\text{prox}_{\alpha^2 h}(\alpha z + b) - b\Big)$$

11.3 Let $h(w) = \alpha\|w\|$. Show that

$$\text{prox}_{\mu h}(z) = \begin{cases} \left(1 - \frac{\mu\alpha}{\|z\|}\right) z, & \text{if } \|z\| \geq \mu\alpha \\ 0, & \text{otherwise} \end{cases}$$

11.4 Establish the validity of expression (11.30).

11.5 Show that the soft-thresholding function (11.18) satisfies the property $\mathbb{T}_{\rho\beta}(\rho x) = \rho\mathbb{T}_\beta(x)$, for any scalars $\rho > 0, \beta > 0$.

11.6 Let $h(w) : \mathbb{R}^M \to \mathbb{R}$ denote a convex function and introduce the transformation $g(w) = h(w) + \frac{\rho}{2}\|w\|^2$, where $\rho \neq 0$. Show that

$$\text{prox}_{\mu g}(z) = \text{prox}_{\frac{\mu h}{1+\mu\rho}}\left(\frac{z}{1+\mu\rho}\right)$$

Conclude that the proximal operator of $f(w) = \alpha\|w\|_1 + \frac{\rho}{2}\|w\|^2$ is given by

$$\text{prox}_{\mu f}(z) = \mathbb{T}_{\frac{\mu\alpha}{1+\mu\rho}}\left(\frac{z}{1+\mu\rho}\right)$$

11.7 Assume we select $h(w) = \|w\|$, where $w \in \mathbb{R}^M$. Show that

$$\text{prox}_{\mu h}(z) = \left(1 - \frac{\mu}{\|z\|}\right)_+ z, \qquad \mathcal{M}_{\mu h}(z) = \begin{cases} \frac{1}{2\mu}\|z\|^2, & \|z\| \leq \mu \\ \|z\| - \frac{\mu}{2}, & \|z\| > \mu \end{cases}$$

where $(x)_+ = \max\{x, 0\}$. Verify that when $M = 1$, the above expression for the proximal projection reduces to the soft-thresholding operation (11.18); the corresponding Moreau envelope will be the Huber function.

11.8 Let $h(w) = \alpha\|w\|_0$, where the notation $\|x\|_0$ counts the number of nonzero elements in vector x. Show that $\text{prox}_h(z) = z\,\mathbb{I}[\,|z| > \sqrt{2\alpha}\,]$. That is, all values of z larger in magnitude than $\sqrt{2\alpha}$ are retained otherwise they are set to zero. This function is sometimes referred to as the *hard-thresholding* mapping as opposed to soft-thresholding.

11.9 Let $w \in \mathbb{R}^M$ with entries $\{w(m)\}$ and M even. Consider the function

$$h(w) = |w(1) - w(2)| + |w(3) - w(4)| + \ldots + |w(M-1) - w(M)|$$

Introduce the $\frac{M}{2} \times M$ matrix (e.g., for $M = 10$)

$$D = \begin{bmatrix} 1 & -1 & 0 & 0 & 0 & 0 & 0 & 0 & 0 & 0 \\ 0 & 0 & 1 & -1 & 0 & 0 & 0 & 0 & 0 & 0 \\ 0 & 0 & 0 & 0 & 1 & -1 & 0 & 0 & 0 & 0 \\ 0 & 0 & 0 & 0 & 0 & 0 & 1 & -1 & 0 & 0 \\ 0 & 0 & 0 & 0 & 0 & 0 & 0 & 0 & 1 & -1 \end{bmatrix}$$

Verify that $DD^{\mathsf{T}} = 2I_{M/2}$ and $h(w) = \|Dw\|_1$. Show that the proximal operator can be expressed in terms of the soft-thresholding operator as follows:

$$\text{prox}_{\mu h}(z) = z + \frac{1}{2\mu} D^{\mathsf{T}}\Big(\text{prox}_{2\mu^2\|w\|_1}(\mu D z) - \mu D z\Big)$$

Remark. For more information on this problem and the next, the reader may refer to Beck (2017, ch. 6).

11.10 Let $w \in \mathbb{R}^M$ with entries $\{w(m)\}$ and M even. Consider the function

$$h(w) = |\sqrt{2}w(1) - 1| + |w(2) - w(3)| + |w(4) - w(5)| + \ldots + |w(M-2) - w(M-1)|$$

Introduce the $\frac{M}{2} \times M$ matrix D and basis vector $e_1 \in \mathbb{R}^{M/2}$ (e.g., for $M = 10$)

$$D = \begin{bmatrix} \sqrt{2} & 0 & 0 & 0 & 0 & 0 & 0 & 0 & 0 & 0 \\ 0 & 1 & -1 & 0 & 0 & 0 & 0 & 0 & 0 & 0 \\ 0 & 0 & 0 & 1 & -1 & 0 & 0 & 0 & 0 & 0 \\ 0 & 0 & 0 & 0 & 0 & 1 & -1 & 0 & 0 & 0 \\ 0 & 0 & 0 & 0 & 0 & 0 & 0 & 1 & -1 & 0 \end{bmatrix}, \quad e_1 = \begin{bmatrix} 1 \\ 0 \\ 0 \\ 0 \\ 0 \end{bmatrix}$$

Verify that $DD^{\mathsf{T}} = 2I_{M/2}$ and $h(w) = \|Dw - e_1\|_1$. Show that the proximal operator can be expressed in terms of the soft-thresholding operator as follows:

$$\text{prox}_{\mu h}(z) = z + \frac{1}{2\mu} D^{\mathsf{T}}\Big(\text{prox}_{2\mu^2\|w\|_1}(\mu D z - \mu e_1) - \mu(Dz - e_1)\Big)$$

11.11 Establish the validity of expression (11.35).

11.12 Consider a matrix $W \in \mathbb{R}^{N \times M}$ and introduce the following matrix-based proximal function definition:

$$\text{prox}_{\mu}(Z) \stackrel{\Delta}{=} \underset{W \in \mathbb{R}^{N\times M}}{\text{argmin}} \left\{ \alpha\,\|W\|_\star + \frac{1}{2}\|W - Z\|_{\mathrm{F}}^2 \right\}, \quad \alpha, \mu > 0$$

where $\|W\|_\star$ denotes the nuclear norm of W (sum of its nonzero singular values). Introduce the singular value decomposition $Z = U\Sigma V^{\mathsf{T}}$, and replace Σ by a new matrix Σ_α whose diagonal entries are computed from the nonzero singular values in Σ as follows:

$$[\Sigma_\alpha]_{kk} = \max\Big\{0, \Sigma_{kk} - \alpha\Big\}$$

That is, nonzero singular values in Σ larger than α are reduced by α, while nonzero singular values smaller than α are set to zero. Show that the proximal solution is given by

$$\widehat{W} = \text{prox}(Z) \; = \; U\Sigma_\alpha V^\mathsf{T} \;\stackrel{\Delta}{=}\; \mathbb{S}_\alpha(Z)$$

where we are also using the notation $\mathbb{S}_\alpha(Z)$ to refer to the singular value soft-thresholding operation defined above. *Remark.* This result appears in Cai, Candes, and Shen (2010), Mazumder, Hastie, and Tibshirani (2010), and Ma, Goldfarb, and Chen (2011).

11.13 Consider a full rank matrix $H \in \mathbb{R}^{N\times M}$ with $N \geq M$ and a vector $d \in \mathbb{R}^M$. We introduce the least-squares problem

$$\widehat{w} \; = \; \underset{w\in\mathbb{R}^M}{\text{argmin}} \; \left\{ \|w - \bar{w}\|^2 \; + \; \|d - Hw\|^2 \right\}$$

where $\bar{w} \in \mathbb{R}^M$ is some given vector.

(a) Determine the solution $\widehat{w}$.

(b) Determine the proximal projection of $\bar{w}$ using $h(w) = \frac{1}{2}\|d - Hw\|^2$ and $\mu = 1$. Show that the result agrees with the solution to part (a).

11.14 Consider the proximal projection problem

$$\widehat{w} = \underset{w\in\mathbb{R}^M}{\text{argmin}} \; \left\{ \frac{1}{2} w^\mathsf{T} A w \; + \; \frac{1}{2}\|w - z\|^2 \right\}$$

where $A \geq 0$. Show that the solution is given by $\widehat{w} = (I_M + A)^{-1} z$.

11.15 Consider the proximal projection problem

$$\underset{w^e\in\mathbb{R}^{M+1}}{\text{argmin}} \; \left\{ \frac{1}{2} (w^e)^\mathsf{T} A w^e \; + \; \frac{1}{2}\|w^e - z^e\|^2 \right\}$$

where $A = \text{diag}\{0, \rho I_M\}$ with $\rho > 0$, $w^e = \text{col}\{-\theta, w\}$, and $z^e = \text{col}\{-\phi, z\}$. Both w^e and z^e are extended vectors of size $M+1$ each, and $\{\theta, \phi\}$ are scalars. Show that the solution is given by

$$\widehat{\theta} = \phi, \quad \widehat{w} = \frac{z}{1+\rho}$$

11.16 Consider the proximal projection problem

$$\underset{w^e\in\mathbb{R}^{M+1}}{\text{argmin}} \; \left\{ \alpha\|Aw^e\|_1 \; + \; \frac{\rho}{2} (w^e)^\mathsf{T} A w^e \; + \; \frac{1}{2}\|w^e - z^e\|^2 \right\}$$

where $A = \text{diag}\{0, \rho I_M\}$ with $\rho > 0$, $w^e = \text{col}\{-\theta, w\}$, and $z^e = \text{col}\{-\phi, z\}$. Both w^e and z^e are extended vectors of size $M+1$ each, and $\{\theta, \phi\}$ are scalars. Show that the solution is given by

$$\widehat{\theta} = \phi, \quad \widehat{w} = \mathbb{T}_{\frac{\alpha}{1+\rho}}\left(\frac{z}{1+\rho}\right)$$

11.17 True or false. A firmly nonexpansive operator is also nonexpansive.

11.18 Refer to the fixed-point iteration (11.40) for a strictly contractive operator, $f(z)$. Show that $f(z)$ has a unique fixed point, z^o, and that $z_n \rightarrow z^o$ as $n \rightarrow \infty$.

11.19 Let $h(w) : \mathbb{R}^M \rightarrow \mathbb{R}$ denote a convex function. Establish the following properties for the proximal operator of $h(w)$, for any vectors $a, b \in \mathbb{R}^M$:

(a) $\|\text{prox}_{\mu h}(a) - \text{prox}_{\mu h}(b)\| \leq \|a - b\|$.

(b) $\|\text{prox}_{\mu h}(a) - \text{prox}_{\mu h}(b)\|^2 \leq (a-b)^\mathsf{T}\left(\text{prox}_{\mu h}(a) - \text{prox}_{\mu h}(b)\right)$.

(c) $\|\text{prox}_{\mu h}(a) - \text{prox}_{\mu h}(b)\|^2 + \|(a - \text{prox}_{\mu h}(a)) - (b - \text{prox}_{\mu h}(b))\|^2 \leq \|a - b\|^2$.

(d) $\|a - b\| = \|\text{prox}_{\mu h}(a) - \text{prox}_{\mu h}(b)\| \iff a - b = \text{prox}_{\mu h}(a) - \text{prox}_{\mu h}(b)$.

Property (a) means that the proximal operator is nonexpansive. Property (b) means that the operator is firmly nonexpansive. Property (c) is equivalent to (b). Property (d) follows from (c).

11.20 Refer to the proximal iteration (11.44).

(a) Let w^o denote a fixed point for the proximal operator, i.e., $w^o = \text{prox}_{\mu h}(w^o)$. Show that $\|w^o - w_n\| \leq \|w^o - w_{n-1}\|$.

(b) Let $a(n) = \|w^o - w_n\|$. Since the sequence $\{a(n)\}$ is bounded from below, conclude from the monotone convergence theorem that $a(n)$ converges to some limit value $\bar{a}$ as $n \to \infty$. Conclude further that $\|\text{prox}_{\mu h}(w^o) - \text{prox}_{\mu h}(w_{n-1})\|$ converges to the same value $\bar{a}$.

(c) Use the analogue of property (11.43) for $\mu q(w)$ and the result of part (b) to conclude that w_n converges to a fixed point of $\text{prox}_{\mu h}(w)$.

11.21 Refer to the Douglas–Rachford algorithm (11.76) and the mapping $R(z)$ defined by (11.82). The purpose of this problem is to establish that $R(z)$ is firmly nonexpansive. For any z_1, z_2, introduce the variables:

$$w_1 = \text{prox}_{\mu q}(z_1), \qquad t_1 = \text{prox}_{\mu E}(2w_1 - z_1)$$
$$w_2 = \text{prox}_{\mu q}(z_2), \qquad t_2 = \text{prox}_{\mu E}(2w_2 - z_2)$$

(a) Use the fact that proximal operators are firmly nonexpansive to conclude that

$$\|w_1 - w_2\|^2 \leq (z_1 - z_2)^\mathsf{T}(w_1 - w_2)$$
$$\|t_1 - t_2\|^2 \leq (2w_1 - z_1 - 2w_2 + z_2)^\mathsf{T}(t_1 - t_2)$$

(b) Verify that

$$\begin{aligned}(z_1 - z_2)^\mathsf{T}\left(R(z_1) - R(z_2)\right) \geq (z_1 - z_2)^\mathsf{T}(t_1 - w_1 + z_1 - t_2 + w_2 - z_2) +\\ \|w_1 - w_2\|^2 - (z_1 - z_2)^\mathsf{T}(w_1 - w_2)\end{aligned}$$

(c) Simplify the expression in part (b) to verify that

$$\begin{aligned}(z_1 - z_2)^\mathsf{T}\left(R(z_1) - R(z_2)\right) \geq \|R(z_1) - R(z_2)\|^2 +\\ (2w_1 - z_1 - 2w_2 + z_2)^\mathsf{T}(t_1 - t_2) - \|t_1 - t_2\|^2\end{aligned}$$

Conclude that $R(z)$ is firmly nonexpansive.

11.22 Consider the following variation of the Douglas–Rachford algorithm, starting from any w_{-1} and z_{-1}:

$$\begin{cases} t_n &= \text{prox}_{\mu E}(2w_{n-1} - z_{n-1}) \\ z_n &= t_n + z_{n-1} - w_{n-1} \\ w_n &= \text{prox}_{\mu q}(z_n) \end{cases}$$

Show that fixed points of the mapping from w_{n-1} to w_n are minimizers of the aggregate cost in (11.48). Show further that the mapping from w_{n-1} to w_n is firmly nonexpansive.

11.23 Consider the following variation of the Douglas–Rachford algorithm:

$$\begin{cases} w_n &= \text{prox}_{\mu q}(z_{n-1}) \\ t_n &= \text{prox}_{\mu E}(2w_n - z_{n-1}) \\ z_n &= (1-\rho)z_{n-1} + \rho(t_n - w_n + z_{n-1}) \end{cases}$$

where $0 < \rho < 2$ is called a relaxation parameter. Comparing with (11.76), we see that the last step is now a linear combination of z_{n-1} with the original quantity $t_n - w_n + z_{n-1}$ with the combination coefficients adding up to 1. Show that w_n converges to a minimizer of (11.48).

11.24 Establish the validity of decomposition (11.88a)–(11.88b). More generally, show that

$$z = \text{prox}_{\mu h}(z) + \mu \, \text{prox}_{\frac{1}{\mu} h^\star}(z/\mu)$$

11.25 Refer to the definition of the Moreau envelope in (11.85). Is the Moreau envelope a convex function over z? Show that $\mathcal{M}_{\mu h}(z)$ is differentiable with respect to z and that its gradient vector is given by expression (11.91).

11.26 Consider a convex function $h(w)$ and its Fenchel conjugate $h^\star(x)$ as defined by (8.83). Show that $\text{prox}_h(z) = \nabla_{z^\mathsf{T}} \mathcal{M}_{h^\star}(z)$.

11.27 Consider a convex function $h(w)$ and its Fenchel conjugate $h^\star(x)$ as defined by (8.83). Show that the Moreau envelope satisfies:

$$\mathcal{M}_{\mu h}(z) = \left(h^\star(w) + \frac{\mu}{2}\|w\|^2\right)^\star$$

That is, the Moreau envelope is obtained by perturbing the Fenchel conjugate of $h(w)$ by a quadratic term and then computing the Fenchel conjugate of the result.

11.28 Using the Bregman divergence, assume we extend the definition of the proximal operator by replacing the quadratic measure in the original definition (11.4) by

$$\text{prox}_{\mu h}(z) \;\stackrel{\Delta}{=}\; \underset{w \in \mathbb{R}^M}{\text{argmin}} \left\{ h(w) \;+\; \frac{1}{\mu} D_\phi(w, z) \right\}$$

(a) Verify that the optimality condition (11.13) is replaced by

$$a = \text{prox}_{\mu h}(b) \iff \Big(\nabla_{w^\mathsf{T}} \phi(a) - \nabla_{w^\mathsf{T}} \phi(b)\Big) \in \partial_{w^\mathsf{T}} h(a)$$

(b) Let $\mathcal{C}$ denote some closed convex set and introduce its indicator function $\mathbb{I}_{C,\infty}[w]$, which is equal to zero when $w \in \mathcal{C}$ and $+\infty$ otherwise. Verify that

$$\text{prox}_{\mathbb{I}_C}(z) = \underset{w \in \mathcal{C}}{\text{argmin}} \; D_\phi(w, z)$$

which amounts to finding the projection of z onto $\mathcal{C}$ using the Bregman measure.

11.A CONVERGENCE UNDER CONVEXITY

The convergence analyses in the two appendices to this chapter benefit from the presentations in Polyak (1987), Combettes and Wajs (2005), Combettes and Pesquet (2011), Polson, Scott, and Willard (2015), and Beck (2017). In this first appendix, we establish the statement of Theorem 11.1, which relates to the convergence of the proximal gradient algorithm under convexity of the smooth component, $E(w)$. In preparation for the argument we establish three useful facts. First, we rewrite algorithm (11.53) in the form:

$$w_n = w_{n-1} - \mu g_\mu(w_{n-1}) \tag{11.113}$$

where

$$g_\mu(w) \;\stackrel{\Delta}{=}\; \frac{1}{\mu}\Big(w - \text{prox}_{\mu q}(w - \mu \nabla_{w^\mathsf{T}} E(w))\Big) \tag{11.114}$$

Form (11.113) shows that the proximal gradient algorithm adjusts w_{n-1} by adding a correction along the direction of $-g_\mu(w_{n-1})$; the size of the correction is modulated by μ. Observe in particular that evaluating $g_\mu(w)$ at a minimizer w^o for $P(w)$ we get

$$g_\mu(w^o) = 0 \tag{11.115}$$

This is because w^o is a fixed point for the proximal operator by (11.51a).

Second, we have from (11.114) that

$$w - \mu\, g_\mu(w) \;=\; \text{prox}_{\mu q}\Big(w - \mu\, \nabla_{w^\mathsf{T}} E(w)\Big) \tag{11.116}$$

Therefore, using (11.13) with the identifications

$$a \leftarrow w - \mu\, g_\mu(w), \quad b \leftarrow w - \mu\, \nabla_{w^\mathsf{T}} E(w) \tag{11.117}$$

we find that

$$g_\mu(w) - \nabla_{w^\mathsf{T}} E(w) \;\in \partial_{w^\mathsf{T}}\, q(w - \mu g_\mu(w)) \tag{11.118}$$

Third, the smooth component $E(w) : \mathbb{R}^M \rightarrow \mathbb{R}$ is assumed to be convex differentiable with δ-Lipschitz gradients, i.e.,

$$\|\nabla_w\, E(a) - \nabla_w\, E(b)\| \;\leq\; \delta\, \|a - b\| \tag{11.119}$$

for any $a, b \in \text{dom}(E)$ and some $\delta \geq 0$. It then follows from property (10.13) that

$$E(a) \leq E(b) + \nabla_w\, E(b)\,(a - b) + \frac{\delta}{2}\|a - b\|^2 \tag{11.120}$$

We are now ready to establish Theorem 11.1.

Proof of Theorem 11.1: The argument involves several steps:

(**Step 1**) We use property (11.120) and select $a \leftarrow w_n$ and $b \leftarrow w_{n-1}$ to write

$$\begin{aligned} E(w_n) \;\; &\leq \;\; E(w_{n-1}) + \nabla_w\, E(w_{n-1})(w_n - w_{n-1}) + \frac{\delta}{2}\|w_n - w_{n-1}\|^2 \\ &\overset{(11.113)}{=} E(w_{n-1}) - \mu \nabla_w\, E(w_{n-1})\, g_\mu(w_{n-1}) + \frac{\mu^2\delta}{2}\|g_\mu(w_{n-1})\|^2 \\ &\overset{(11.70)}{\leq} E(w_{n-1}) - \mu \nabla_w\, E(w_{n-1})\, g_\mu(w_{n-1}) + \frac{\mu}{2}\|g_\mu(w_{n-1})\|^2 \end{aligned} \tag{11.121}$$

(**Step 2**) We use the inequality from Step 1 to establish that, for any $z \in \text{dom}(h)$,

$$P(w_n) \leq P(z) + (g_\mu(w_{n-1}))^\mathsf{T}\,(w_{n-1} - z) - \frac{\mu}{2}\|g_\mu(w_{n-1})\|^2 \tag{11.122}$$

Indeed, note that

$$\begin{aligned} P(w_n) \;\; &\overset{\Delta}{=} \;\; q(w_n) + E(w_n) \\ &\overset{(11.121)}{\leq} q(w_n) + E(w_{n-1}) - \mu \nabla_w\, E(w_{n-1})\, g_\mu(w_{n-1}) + \frac{\mu}{2}\|g_\mu(w_{n-1})\|^2 \\ &\overset{(a)}{\leq} q(z) + (s(w_n))^\mathsf{T}(w_n - z) + E(z) + \nabla_w\, E(w_{n-1})(w_{n-1} - z) - \\ &\qquad \mu \nabla_w E(w_{n-1}) g_\mu(w_{n-1}) + \frac{\mu}{2}\|g_\mu(w_{n-1})\|^2 \end{aligned} \tag{11.123}$$

where in step (a) we used the convexity of $q(w)$ and $E(w)$ and properties (8.4) and (8.43). Moreover, the notation $s(w_n)$ refers to a subgradient of $q(w)$ relative to w^T:

$$s(w_n) \;\in \partial_{w^\mathsf{T}}\, q(w_n) \tag{11.124}$$

Now, appealing to (11.118) and letting $w = w_{n-1}$, we find that

$$g_\mu(w_{n-1}) - \nabla_{w^\mathsf{T}} E(w_{n-1}) \;\in \partial_{w^\mathsf{T}} q(w_n) \tag{11.125}$$

where the argument of $q(\cdot)$ becomes w_n under (11.113). This relation implies that we can select the subgradient $s(w_n)$ as the difference on the left so that

$$\nabla_{w^{\mathsf{T}}} E(w_{n-1}) = g_\mu(w_{n-1}) - s(w_n) \tag{11.126}$$

We can now evaluate three terms appearing in (11.123) as follows:

$$\begin{aligned}(s(w_n))^{\mathsf{T}}(w_n - z) &= (s(w_n))^{\mathsf{T}}(w_{n-1} - \mu g_\mu(w_{n-1}) - z) \\ &= (s(w_n))^{\mathsf{T}}(w_{n-1} - z) \;-\; \mu(s(w_n))^{\mathsf{T}} g_\mu(w_{n-1})\end{aligned} \tag{11.127a}$$

and

$$\begin{aligned}\nabla_w E(w_{n-1})(w_{n-1} - z) &= \Big(g_\mu(w_{n-1}) - s(w_n)\Big)^{\mathsf{T}}(w_{n-1} - z) \\ &= \Big(g_\mu(w_{n-1})\Big)^{\mathsf{T}}(w_{n-1} - z) \;-\; (s(w_n))^{\mathsf{T}}(w_{n-1} - z)\end{aligned} \tag{11.127b}$$

and

$$\begin{aligned}-\mu\nabla_w E(w_{n-1}) g_\mu(w_{n-1}) &= -\mu\Big(\big(g_\mu(w_{n-1}) - s(w_n)\big)^{\mathsf{T}} g_\mu(w_{n-1})\Big) \\ &= -\mu\|g_\mu(w_{n-1})\|^2 \;+\; \mu(s(w_n))^{\mathsf{T}} g_\mu(w_{n-1})\end{aligned} \tag{11.127c}$$

Substituting into (11.123) and simplifying gives

$$\begin{aligned}P(w_n) &\le q(z) + E(z) + (g_\mu(w_{n-1}))^{\mathsf{T}}(w_{n-1} - z) - \frac{\mu}{2}\|g_\mu(w_{n-1})\|^2 \\ &= P(z) + (g_\mu(w_{n-1}))^{\mathsf{T}}(w_{n-1} - z) - \frac{\mu}{2}\|g_\mu(w_{n-1})\|^2\end{aligned} \tag{11.128}$$

(**Step 3**) Using $z = w_{n-1}$ in the last inequality we get

$$P(w_n) \le P(w_{n-1}) - \frac{\mu}{2}\|g_\mu(w_{n-1})\|^2 \tag{11.129}$$

which shows that $P(w_n)$ is a nonincreasing sequence. If we use instead $z = w^o$ in (11.128) we get

$$\begin{aligned}0 \le P(w_n) - P(w^o) \;&\le\; (g_\mu(w_{n-1}))^{\mathsf{T}}(w_{n-1} - w^o) - \frac{\mu}{2}\|g_\mu(w_{n-1})\|^2 \\ &\overset{(a)}{=} \frac{1}{2\mu}\Big(\|w_{n-1} - w^o\|^2 - \|w_{n-1} - w^o - \mu g_\mu(w_{n-1})\|^2\Big) \\ &\overset{(11.113)}{=} \frac{1}{2\mu}(\|w_{n-1} - w^o\|^2 - \|w_n - w^o\|^2)\end{aligned} \tag{11.130}$$

where step (a) follows by expanding the terms in the second line and noting that they coincide with those in the first line. It follows that

$$\|w^o - w_n\| \le \|w^o - w_{n-1}\| \tag{11.131}$$

so that $\|w^o - w_n\|$ is a nonincreasing sequence.

(**Step 4**) Adding from $n = 0$ up to some $N-1$ gives

$$\begin{aligned}\sum_{n=0}^{N-1}(P(w_n) - P(w^o)) &\le \frac{1}{2\mu}\sum_{n=0}^{N-1}(\|w^o - w_{n-1}\|^2 - \|w^o - w_n\|^2) \\ &= \frac{1}{2\mu}(\|w^o - w_{-1}\|^2 - \|w^o - w_{N-1}\|^2) \\ &\le \frac{1}{2\mu}\|w^o - w_{-1}\|^2\end{aligned} \tag{11.132}$$

Now since $P(w_n)$ is nonincreasing we get

$$P(w_{N-1}) - P(w^o) \leq \frac{1}{N}\sum_{n=0}^{N-1}(P(w_n) - P(w^o)) \leq \frac{1}{2\mu N}\|w^o - w_{-1}\|^2 \tag{11.133}$$

11.B CONVERGENCE UNDER STRONG CONVEXITY

In this appendix we establish the statement of Theorem 11.2, which relates to the convergence of the proximal gradient algorithm under strong convexity of the smooth component, $E(w)$. Theorem 11.1 shows that under a *convexity* condition on $E(w)$, the cost value $P(w_n)$ approaches the minimum $P(w^o)$ at the rate $O(1/n)$. Faster convergence at an exponential rate is possible when $E(w)$ is ν-strongly convex.

Proof of Theorem 11.2: From the proximal gradient recursion (11.53) and the fixed-point relation (11.51a) we have

$$w_n = \text{prox}_{\mu q}\Big(w_{n-1} - \mu\nabla_{w^\mathsf{T}} E(w_{n-1})\Big) \tag{11.134a}$$

$$w^o = \text{prox}_{\mu q}\Big(w^o - \mu\nabla_{w^\mathsf{T}} E(w^o)\Big) \tag{11.134b}$$

Applying the nonexpansive property (11.41) of the proximal operator we get

$$\begin{aligned}
&\|w^o - w_n\|^2 \\
&= \left\|\text{prox}_{\mu q}(w^o - \mu\nabla_{w^\mathsf{T}} E(w^o)) - \text{prox}_{\mu q}(w_{n-1} - \mu\nabla_{w^\mathsf{T}} E(w_{n-1}))\right\|^2 \\
&\overset{(11.41)}{\leq} \|w^o - \mu\nabla_{w^\mathsf{T}} E(w^o) - w_{n-1} + \mu\nabla_{w^\mathsf{T}} E(w_{n-1})\|^2 \\
&= \|w^o - w_{n-1}\|^2 - \\
&\quad 2\mu\left(\nabla_{w^\mathsf{T}} E(w^o) - \nabla_{w^\mathsf{T}} E(w_{n-1})\right)^\mathsf{T}(w^o - w_{n-1}) + \\
&\quad \mu^2\|\nabla_{w^\mathsf{T}} E(w^o) - \nabla_{w^\mathsf{T}} E(w_{n-1})\|^2 \\
&\overset{(a)}{\leq} \|w^o - w_{n-1}\|^2 - 2\mu\nu\|w^o - w_{n-1}\|^2 + \mu^2\delta^2\|w^o - w_{n-1}\|^2 \\
&= (1 - 2\mu\nu + \mu^2\delta^2)\|w^o - w_{n-1}\|^2 \\
&\overset{\Delta}{=} \lambda^{n+1}\|w^o - w_{-1}\|^2
\end{aligned} \tag{11.135}$$

where λ is defined by (11.75) and step (a) is because of the Lipschitz condition (11.119) and the assumed strong convexity of $E(w)$, which implies, in view of property (8.24), the relation:

$$\Big(\nabla_{w^\mathsf{T}} E(w^o) - \nabla_{w^\mathsf{T}} E(w_{n-1})\Big)^\mathsf{T}(w^o - w_{n-1}) \geq \nu\|w^o - w_{n-1}\|^2 \tag{11.136}$$

We conclude from (11.135) that the squared error vector converges exponentially fast to zero at a rate dictated by λ. To verify that condition (11.73) ensures $0 \leq \lambda < 1$, we refer to Fig. 11.5 where we plot the coefficient $\lambda(\mu)$ as a function of μ. The minimum value of $\lambda(\mu)$, which occurs at the location $\mu = \nu/\delta^2$ and is equal to $1 - \nu^2/\delta^2$, is nonnegative since $0 < \nu \leq \delta$. It is clear from the figure that $0 \leq \lambda < 1$ for $\mu \in (0, \frac{2\nu}{\delta^2})$.

We next establish the exponential convergence of the excess cost value as shown by (11.74). For this purpose, we first note from the convexity of $P(w)$ and property (8.43) for nonsmooth convex functions that

$$P(w_n) - P(w^o) \leq (s(w_n))^\mathsf{T}(w_n - w^o) \tag{11.137}$$

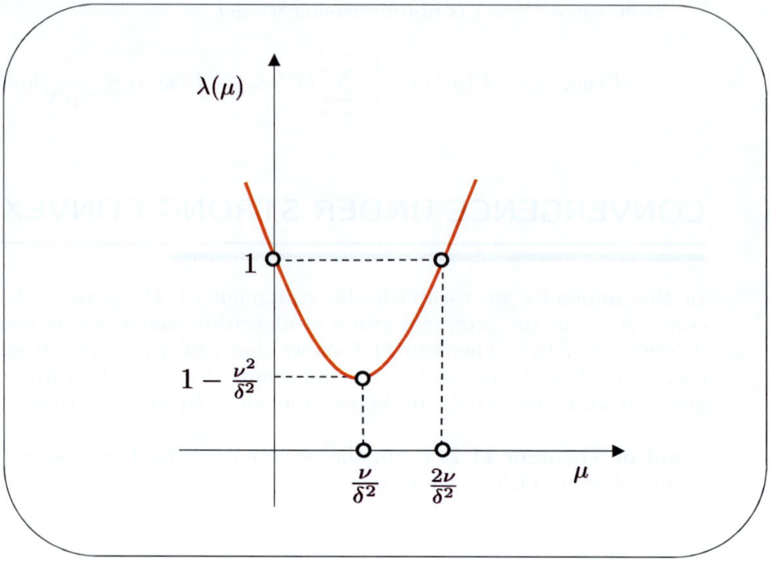

Figure 11.5 Plot of the function $\lambda(\mu) = 1 - 2\nu\mu + \mu^2\delta^2$ given by (11.75). It shows that the function $\lambda(\mu)$ assumes values below 1 in the range $0 < \mu < 2\nu/\delta^2$.

where $s(w_n)$ denotes a subgradient vector for $P(w)$ relative to w^T at location w_n. Using (11.54) and the fact that $P(w) = q(w) + E(w)$, we know that this subgradient vector can be chosen as:

$$s(w_n) = \frac{1}{\mu}(z_n - w_n) \; + \; \nabla_{w^\mathsf{T}} E(w_n) \tag{11.138}$$

Substituting into (11.137) gives

$$\begin{aligned} &P(w_n) - P(w^o) \\ &\leq \left(\frac{1}{\mu}(z_n - w_n) \; + \; \nabla_{w^\mathsf{T}} E(w_n)\right)^\mathsf{T} (w_n - w^o) \\ &= \left(\frac{1}{\mu}(z_n - \text{prox}_{\mu q}(z_n)) \; + \; \nabla_{w^\mathsf{T}} E(w_n)\right)^\mathsf{T} (w_n - w^o) \\ &= \left(\frac{1}{\mu}(z_n - \text{prox}_{\mu q}(z_n)) + \nabla_{w^\mathsf{T}} E(w^o) + \nabla_{w^\mathsf{T}} E(w_n) - \nabla_{w^\mathsf{T}} E(w^o)\right)^\mathsf{T} (w_n - w^o) \end{aligned} \tag{11.139}$$

so that using (11.119):

$$\begin{aligned} &P(w_n) - P(w^o) \\ &\leq \left(\frac{1}{\mu}(z_n - \text{prox}_{\mu q}(z_n)) + \nabla_{w^\mathsf{T}} E(w^o)\right)^\mathsf{T} (w_n - w^o) + \delta\|w_n - w^o\|^2 \\ &\leq \left\|\frac{1}{\mu}(z_n - \text{prox}_{\mu q}(z_n)) + \nabla_{w^\mathsf{T}} E(w^o)\right\| \|w_n - w^o\| + \delta\|w_n - w^o\|^2 \end{aligned} \tag{11.140}$$

To continue, we substitute $\nabla_{w^\mathsf{T}} E(w^o)$ by an equivalent expression as follows. We know from the first part of the proof that w_n converges to w^o, which satisfies the fixed-point relation (11.134b). Let

$$z^o \triangleq w^o - \mu\nabla_{w^\mathsf{T}} E(w^o) \tag{11.141a}$$
$$w^o = \text{prox}_{\mu q}(z^o) \tag{11.141b}$$

Combining these two relations we get

$$\nabla_{w^\mathsf{T}} E(w^o) = -\frac{1}{\mu}(z^o - w^o) = -\frac{1}{\mu}\left(z^o - \text{prox}_{\mu q}(z^o)\right) \tag{11.142}$$

Substituting into (11.140) we obtain

$$\begin{aligned}
&P(w_n) - P(w^o)\\
&\leq \frac{1}{\mu}\left\| z_n - \text{prox}_{\mu q}(z_n) - z^o + \text{prox}_{\mu q}(z^o)\right\| \|w_n - w^o\| + \delta\|w_n - w^o\|^2\\
&\overset{(11.41)}{\leq} \frac{2}{\mu}\|z_n - z^o\| \, \|w_n - w^o\| + \delta\|w_n - w^o\|^2\\
&\overset{(a)}{\leq} \frac{2}{\mu}\|w_{n-1} - w^o\| \, \|w_n - w^o\| + 2\|\nabla_{w^\mathsf{T}} E(w_{n-1}) - \nabla_{w^\mathsf{T}} E(w^o)\| \, \|w_n - w^o\| +\\
&\qquad \delta\|w_n - w^o\|^2\\
&\overset{(11.119)}{\leq} \frac{2}{\mu}\|w_{n-1} - w^o\| \, \|w_n - w^o\| + 2\delta\|w_{n-1} - w^o\| \, \|w_n - w^o\| + \delta\|w_n - w^o\|^2\\
&\overset{(11.135)}{\leq} \frac{2\sqrt{\lambda}}{\mu}\|w_{n-1} - w^o\|^2 + 2\sqrt{\lambda}\delta\|w_{n-1} - w^o\|^2 + \lambda\delta\|w_{n-1} - w^o\|^2\\
&= \underbrace{\left(\frac{2(1+\mu\delta)}{\mu\sqrt{\lambda}} + \delta\right)\lambda}_{\triangleq\, \beta} \times \|w_{n-1} - w^o\|^2\\
&\overset{(b)}{=} \beta\|w_{n-1} - w^o\|^2\\
&\overset{(11.135)}{\leq} \beta\lambda^n\|w^o - w_{-1}\|^2
\end{aligned} \tag{11.143}$$

where in step (a) we used the relation

$$z_n - z^o = (w_{n-1} - \mu\nabla_{w^\mathsf{T}} E(w_{n-1})) - (w^o - \mu\nabla_{w^\mathsf{T}} E(w^o)) \tag{11.144}$$

and in step (b) we introduced the scalar β.

■

Example 11.8 **(A more relaxed bound on μ)** We revisit Remark 11.4 and explain that the bound on μ for convergence can be relaxed to $\mu < 2/\delta$. For this purpose, we exploit the *co-coercivity* property of convex functions with δ-Lipschitz gradients. We know from Prob. 10.4 that:

$$\Big(\nabla_{w^\mathsf{T}} E(w_2) - \nabla_{w^\mathsf{T}} E(w_1)\Big)^\mathsf{T}(w_2 - w_1) \geq \frac{1}{\delta}\|\nabla_w E(w_2) - \nabla_w E(w_1)\|^2 \tag{11.145}$$

We use this inequality in the third line of (11.135) as follows:

$$\begin{aligned}
\|\widetilde{w}_n\|^2 &\overset{(11.145)}{\leq} \|\widetilde{w}_{n-1}\|^2 - 2\mu\Big(\nabla_{w^{\mathsf{T}}}E(w^o) - \nabla_{w^{\mathsf{T}}}E(w_{n-1})\Big)^{\mathsf{T}}\widetilde{w}_{n-1} + \\
&\qquad + \mu^2\delta\Big(\nabla_{w^{\mathsf{T}}}E(w^o) - \nabla_{w^{\mathsf{T}}}E(w_{n-1})\Big)^{\mathsf{T}}\widetilde{w}_{n-1} \\
&= \|\widetilde{w}_{n-1}\|^2 - (2\mu - \mu^2\delta)\Big(\nabla_{w^{\mathsf{T}}}E(w^o) - \nabla_{w^{\mathsf{T}}}E(w_{n-1})\Big)^{\mathsf{T}}\widetilde{w}_{n-1} \\
&\overset{(11.136)}{\leq} \|\widetilde{w}_{n-1}\|^2 - (2\mu - \mu^2\delta)\nu\|\widetilde{w}_{n-1}\|^2 \\
&= \underbrace{(1 - 2\mu\nu + \mu^2\nu\delta)}_{\triangleq\,\lambda'}\|\widetilde{w}_{n-1}\|^2 \\
&= (\lambda')^{n+1}\|w^o - w_{-1}\|^2
\end{aligned} \tag{11.146}$$

This result is consistent with (11.135) since $\lambda' \leq \lambda$ in view of $\nu \leq \delta$. Working with λ' instead of λ and repeating the argument leading to (11.143), we will arrive at the bound $0 < \mu < 2/\delta$ for stability with convergence occurring at $O((\lambda')^n)$.

REFERENCES

Bach, F., R. Jenatton, J. Mairal, and G. Obozinski (2012), "Optimization with sparsity-inducing penalties," *Found. Trends Mach. Learn.*, vol. 4, no. 1, pp. 1–106.

Bauschke, H. H. and P. L. Combettes (2011), *Convex Analysis and Monotone Operator Theory in Hilbert Spaces*, Springer.

Beck, A. (2017), *First-Order Methods in Optimization*, SIAM.

Beck, A. and M. Teboulle (2009a), "A fast iterative shrinkage-thresholding algorithm for linear inverse problems," *SIAM J. Imaging Sci.*, vol. 2, no. 1, pp. 183–202.

Beck, A. and M. Teboulle (2009b), "Fast gradient-based algorithms for constrained total variation image denoising and deblurring problems," *IEEE Trans. Image Process.*, vol. 18, no. 11, pp. 2419–2434.

Beck, A. and M. Teboulle (2012), "Smoothing and first order methods: A unified framework," *SIAM J. Optim.*, vol. 22, no. 2, pp. 557–580.

Bredies, K. (2009), "A forward–backward splitting algorithm for the minimization of non-smooth convex functionals in Banach space," *Inverse Problems*, vol. 25, Art. 015005.

Brezis, H. (1973), *Opérateurs Maximaux Monotones et Semi-Groupes de Contractions dans les Espaces de Hilbert*, North-Holland.

Browder, F. (1965), "Nonlinear monotone operators and convex sets in Banach spaces," *Bull. Amer. Math. Soc.*, vol. 71, no. 5, pp. 780–785.

Browder, F. (1967), "Convergence theorems for sequences of nonlinear operators in Banach spaces," *Mathematische Zeitschrift*, vol. 100, no. 3, pp. 201–225.

Bruck, R. E. and S. Reich (1977), "Nonexpansive projections and resolvents of accretive operators in Banach spaces," *Houston J. Math.*, vol. 3, pp. 459–470.

Cai, J., E. J. Candes, and Z. Shen (2010), "A singular value thresholding algorithm for matrix completion," *SIAM J. Optim.*, vol. 20, no. 4, pp. 1956–1982.

Chen, G. H. G. and R. T. Rockafellar (1997), "Convergence rates in forward–backward splitting," *SIAM J. Optim.* vol. 7, pp. 421–444.

Combettes, P. L. (2004), "Solving monotone inclusions via compositions of nonexpansive averaged operators," *Optimization*, vol. 53, no. 5–6, pp. 475–504.

Combettes, P. L. and J.-C. Pesquet (2011), "Proximal splitting methods in signal processing," in *Fixed-Point Algorithms for Inverse Problems in Science and Engineering*, H. H. Bauschke *et al.*, editors, pp. 185–212, Springer.

Combettes, P. L. and V. R. Wajs (2005), "Signal recovery by proximal forward–backward splitting," *Multiscale Model. Simul.*, vol. 4, no. 4, pp. 1168–1200.

Daubechies, I., M. Defrise, and C. De Mol (2004), "An iterative thresholding algorithm for linear inverse problems with a sparsity constraint," *Commun. Pure App. Math.*, vol. LVII, pp. 1413–1457.

Donoho, D. L. and I. M. Johnstone (1994), "Ideal spatial adaptation via wavelet shrinkage," *Biomefrika*, vol. 81, pp. 425–455.

Donoho, D. L. and I. M. Johnstone (1995), "Adapting to unknown smoothness via wavelet shrinkage," *J. Amer. Stat. Assoc.*, vol. 90, pp. 1200–1224.

Douglas, J. and H. H. Rachford (1956), "On the numerical solution of heat conduction problems in two and three space variables," *Trans. Amer. Math. Soc.*, vol. 82, pp. 421–439.

Duchi, J. and Y. Singer (2009), "Efficient online and batch learning using forward backward splitting,' *J. Mach. Learn. Res.*, vol. 10, pp. 2873–2908.

Eckstein, J. and D. P. Bertsekas (1992), "On the Douglas–Rachford splitting method and the proximal point algorithm for maximal monotone operators," *Math. Program.*, vol. 55, pp. 293–318.

Figueiredo, M., J. Bioucas-Dias, and R. Nowak (2007), "Majorization–minimization algorithms for wavelet-based image restoration," *IEEE Trans. Image Process.*, vol. 16, no. 12, pp. 2980–2991.

Fukushima, M. and H. Mine (1981), "A generalized proximal point algorithm for certain non-convex minimization problems," *Inter. J. Syst. Sci.*, vol. 12, no. 8, pp. 989–1000.

Granas, A. and J. Dugundji (2003), *Fixed Point Theory*, Springer.

Guler, O. (1991), "On the convergence of the proximal point algorithm for convex minimization," *SIAM J. Control Optim.*, vol. 20, pp. 403–419.

Hale, E. T., M. Yin, and Y. Zhang (2008), "Fixed-point continuation for ℓ_1-minimization: Methodology and convergence," *SIAM J. Optim.*, vol. 19, pp. 1107–1130.

Huber, P. J. (1981), *Robust Statistics*, Wiley.

Lemaire, B. (1989a), "The proximal algorithm," *Intern. Series Numer. Math.*, vol. 87, pp. 73–87.

Lemaire, B. (1989b), "New methods in optimization and their industrial uses," *Inter. Ser. Numer. Math.*, vol. 87, pp. 73–87.

Lions, P. and B. Mercier (1979), "Splitting algorithms for the sum of two nonlinear operators," *SIAM J. Numer. Anal.*, vol. 16, pp. 964–979.

Luo, Z. Q. and P. Tseng (1992b), "On the linear convergence of descent methods for convex essentially smooth minimization," *SIAM J. Control Optim.*, vol. 30, pp. 408–425.

Luo, Z. Q. and P. Tseng (1993), "Error bounds and convergence analysis of feasible descent methods: A general approach," *Ann. Oper. Res.*, vol. 46, pp. 157–178.

Ma, S., D. Goldfarb, and L. Chen (2011), "Fixed point and Bregman iterative methods for matrix rank minimization," *Math. Program.*, vol. 128, pp. 321–353.

Martinet, B. (1970), "Régularisation d'inéquations variationnelles par approximations successives," *Rev. Francaise Informat. Recherche Opérationnelle*, vol. 4, no. 3, pp. 154–158.

Martinet, B. (1972), "Determination approchtfe d'un point fixe d'une application pseudo-contractante," *Comptes Rendus de l'Académie des Sciences de Paris*, vol. 274, pp. 163–165.

Mazumder, R., T. Hastie, and R. Tibshirani (2010), "Spectral regularization algorithms for learning large incomplete matrices," *J. Mach. Learn. Res.*, vol. 11, pp. 2287–2322.

Mercier, B. (1979), *Topics in Finite Element Solution of Elliptic Problems*, Tata Institute of Fundamental Research.

Minty, G. J. (1962), "Monotone (nonlinear) operators in Hilbert space," *Duke Math. J.*, vol. 29, pp. 341–346.
Moreau, J. J. (1962), "Fonctions convexes duales et points proximaux dans un espace hilbertien," *Comptes Rendus de l'Académie des Sciences de Paris*, vol. A255, pp. 2897–2899.
Moreau, J. J. (1963a), "Propriétés des applications prox," *Comptes Rendus de l'Académie des Sciences de Paris*, vol. A256, pp. 1069–1071.
Moreau, J. J. (1963b), "Fonctionnelles sous-différentiables," *Comptes Rendus de l'Académie des Sciences de Paris*, vol. A257, pp. 4117–4119.
Moreau, J. J. (1965), "Proximité et dualité dans un espace hilbertien," *Bull. Soc. Math. de France*, vol. 93, pp. 273–299.
Nesterov, Y. (2004), *Introductory Lectures on Convex Optimization*, Springer.
Nesterov, Y. (2005), "Smooth minimization of non-smooth functions," *Math. Program.*, vol. 103, no. 1, pp. 127–152.
O'Connor, D. and L. Vandenberghe (2014), "Primal–dual decomposition by operator splitting and applications to image deblurring," *SIAM J. Imag. Sci.*, vol. 7, no. 3, pp. 1724–1754.
Parikh, N. and S. Boyd (2013), "Proximal algorithms," *Found. Trends Optim.*, vol. 1, no. 3, pp. 127–239.
Passty, G. (1979), "Ergodic convergence to a zero of the sum of monotone operators in Hilbert space," *J. Math. Anal. Appl.*, vol. 72, no. 2, pp. 383–390.
Polson, N. G., J. G. Scott, and B. T. Willard (2015), "Proximal algorithms in statistics and machine learning," *Statist. Sci.*, vol. 30, no. 4, pp. 559–581.
Polyak, B. T. (1987), *Introduction to Optimization*, Optimization Software.
Rockafellar, R. T. (1976a), "Monotone operators and the proximal point algorithm," *SIAM J. Control Optim.*, vol. 14, no. 5, pp. 877–898.
Rockafellar, R. T. (1976b), "Augmented Lagrangians and applications of the proximal point algorithm in convex programming," *Math. Oper. Res.*, vol. 1, no. 2, pp. 97–116.
Rockafellar, R. T. and R. Wets (1998), *Variational Analysis*, Springer.
Sibony, M. (1970), "Méthodes itératives pour les équations et inéquations aux dérivées partielles non linéaires de type monotone," *Calcolo*, vol. 7, pp. 65–183.
Tseng, P. (1991), "Applications of a splitting algorithm to decomposition in convex programming and variational inequalities," *SIAM J. Control Optim.*, vol. 29, pp. 119–138.
Vandenberghe, L. (2010), "Optimization methods for large-scale systems," unpublished lecture notes for the graduate course EE236C, University of California, Los Angeles (UCLA).
Yosida, K. (1968), *Functional Analysis*, Springer.

12 Gradient-Descent Method

The gradient-descent method is the backbone of learning algorithms. It is a powerful iterative procedure that allows us to approach minimizers of objective functions when closed-form expressions for these minimizers are not possible. Several variations will be described in this and the following chapters. We focus initially on objective functions that are *first-order differentiable*. In subsequent chapters we consider nonsmooth functions that may have points of nondifferentiability and introduce subgradient and proximal algorithms for their minimization. Although gradient-descent algorithms can be applied to both convex and nonconvex functions, we will focus largely on convex objectives and examine their convergence properties. Later, in Chapter 24, we consider nonconvex optimization problems.

12.1 EMPIRICAL AND STOCHASTIC RISKS

We consider an optimization problem of the following generic form:

$$w^\star \stackrel{\Delta}{=} \underset{w \in \mathbb{R}^M}{\operatorname{argmin}} \; P(w) \tag{12.1}$$

where $P(w)$ refers to the objective function that we wish to minimize, $w \in \mathbb{R}^M$ is the independent variable, and $w^\star$ denotes a minimizing argument. In the context of learning algorithms, objective functions are called *risks* because they provide a measure of how much error or risk is incurred in using a solution w to make inference decisions.

12.1.1 Empirical Risks

The results in this chapter are applicable to convex risk functions, $P(w)$. In learning problems, $P(w)$ will generally be some function of N data points denoted by the notation $\{\gamma(m), h_m,\ m = 0, 1, \ldots, N-1\}$, where $\gamma \in \mathbb{R}$ is a scalar referred to as the *target* or *label* variable and $h \in \mathbb{R}^M$ is a vector referred to as the *feature* vector. In particular, $P(w)$ will often take the form of a *sample average* over this data, written as

$$P(w) = \frac{1}{N}\sum_{m=0}^{N-1} Q\Big(w;\gamma(m),h_m\Big), \quad \gamma(m)\in \mathbb{R},\quad h_m \in \mathbb{R}^M \tag{12.2}$$

for some convex function $Q(w;\cdot,\cdot)$, referred to as the *loss*. The value $Q(w;\gamma(m), h_m)$ represents the loss at the mth data pair $(\gamma(m),h_m)$. When $P(w)$ has the sample average form (12.2), we refer to it as an *empirical* risk; one that is defined directly from data measurements. In this way, problem (12.1) becomes an empirical risk minimization (ERM) problem of the form:

$$\boxed{w^\star \triangleq \underset{w\in\mathbb{R}^M}{\text{argmin}}\ \frac{1}{N}\sum_{m=0}^{N-1} Q\Big(w;\gamma(m),h_m\Big)} \quad \textbf{(empirical risk minimization)} \tag{12.3}$$

We will encounter several choices for the loss function in future chapters, such as:

$$Q\Big(w;\gamma(m),h_m\Big) = \begin{cases} q(w) \ + \ (\gamma(m)-h_m^{\mathsf T}w)^2 & \text{(quadratic)}\\ q(w) \ + \ \ln\Big(1+e^{-\gamma(m)h_m^{\mathsf T}w}\Big) & \text{(logistic)}\\ q(w) \ + \ \max\Big\{0,-\gamma(m)h_m^{\mathsf T}w\Big\} & \text{(perceptron)}\\ q(w) \ + \ \max\Big\{0,1-\gamma(m)h_m^{\mathsf T}w\Big\} & \text{(hinge)}\end{cases} \tag{12.4}$$

The (also convex) function $q(w)$ is called the *regularization* factor and it usually takes one of several forms, such as:

$$q(w) \ = \begin{cases} 0 & \text{(no regularization)}\\ \rho\|w\|^2 & (\ell_2\text{-regularization})\\ \alpha\|w\|_1 & (\ell_1\text{-regularization})\\ \alpha\|w\|_1+\rho\|w\|^2 & \text{(elastic-net regularization)}\end{cases} \tag{12.5}$$

where $\alpha > 0$ and $\rho > 0$. Other choices for $q(w)$ are possible. We explain in Chapter 51 that the choice of $q(w)$ plays an important role in determining the form of the minimizer $w^\star$, such as forcing it to have a small norm or forcing it to be sparse and have many zero entries. Table 12.1 lists the empirical risk functions described so far.

Note that all loss functions in (12.4) depend on $\{h_m, w\}$ through the inner product $h_m^{\mathsf T}w$. Although unnecessary, this property will hold for most loss functions of interest in our treatment. It is customary to interpret this inner product as an estimate or prediction for $\gamma(m)$, written as

$$\widehat{\gamma}(m) = h_m^{\mathsf T}w \qquad \textbf{(prediction)} \tag{12.6}$$

In this way, the loss functions in (12.4) can be interpreted as measuring the discrepancy between the labels $\{\gamma(m)\}$ and their predictions $\{\widehat{\gamma}(m)\}$. By seeking a minimizer $w^\star$ in (12.3), we are in effect seeking a model that "best" matches the $\{\widehat{\gamma}(m)\}$ to the $\{\gamma(m)\}$.

Table 12.1 Examples of empirical risks based on N data pairs $\{\gamma(m), h_m\}$, and where $q(w)$ denotes a convex regularization factor.

Name	Empirical risk, $P(w)$
least-squares	$q(w) + \frac{1}{N}\sum_{m=0}^{N-1}\left(\gamma(m) - h_m^{\mathsf{T}} w\right)^2$
logistic	$q(w) + \frac{1}{N}\sum_{m=0}^{N-1}\ln\left(1 + e^{-\gamma(m)h_m^{\mathsf{T}} w}\right)$
perceptron	$q(w) + \frac{1}{N}\sum_{m=0}^{N-1}\max\left\{0,\ -\gamma(m)h_m^{\mathsf{T}} w\right\}$
hinge	$q(w) + \frac{1}{N}\sum_{m=0}^{N-1}\max\left\{0,\ 1 - \gamma(m)h_m^{\mathsf{T}} w\right\}$

12.1.2 Stochastic Risks

In many instances, the objective function $P(w)$ will not have an empirical form but will instead be *stochastic* in nature. In these cases, $P(w)$ will be defined as the expectation of the loss function:

$$P(w) = \mathbb{E}\, Q(w; \boldsymbol{\gamma}, \boldsymbol{h}) \tag{12.7}$$

Here, the expectation operator $\mathbb{E}$ is relative to the distribution of the data $\{\boldsymbol{\gamma}, \boldsymbol{h}\}$, now assumed to be randomly distributed according to some joint probability density function (pdf), $f_{\boldsymbol{\gamma},\boldsymbol{h}}(\gamma, h)$. In this way, the optimization problem (12.1) becomes one of the form:

$$\boxed{w^o \stackrel{\Delta}{=} \underset{w\in\mathbb{R}^M}{\text{argmin}}\ \mathbb{E}\, Q(w; \boldsymbol{\gamma}, \boldsymbol{h})} \qquad \textbf{(stochastic risk minimization)} \tag{12.8}$$

where we are denoting the minimizing argument by w^o. In analogy with Table 12.1, we list examples of stochastic risks in Table 12.2. Note that all loss functions in the table depend again on $\{\boldsymbol{h}, w\}$ through the inner product $\boldsymbol{h}^{\mathsf{T}} w$, which we also interpret as a prediction for $\boldsymbol{\gamma}$, written as $\widehat{\boldsymbol{\gamma}} = \boldsymbol{h}^{\mathsf{T}} w$. In this way, the loss functions in Table 12.2 measure the average discrepancy between the label $\boldsymbol{\gamma}$ and its prediction $\widehat{\boldsymbol{\gamma}}$ over the distribution of the data. By seeking the minimizer w^o in (12.8), we are in effect seeking a model w^o that "best" matches $\widehat{\boldsymbol{\gamma}}$ to $\boldsymbol{\gamma}$ in some average loss sense.

Remark 12.1. (Notation for minimizers) We will employ the following convention throughout our treatment to distinguish between the empirical and stochastic scenarios. We will denote the minimizer of an empirical risk by $w^\star$ and the minimizer of a stochastic risk by w^o:

$$w^\star : \text{ minimizers for empirical risks} \tag{12.9a}$$
$$w^o : \text{ minimizers for stochastic risks} \tag{12.9b}$$

In general, though, when we are dealing with a generic optimization problem where $P(w)$ can refer to either an empirical or stochastic risk, we will denote the minimizing argument generically by $w^\star$, as was already done in (12.1).

■

Table 12.2 Examples of stochastic risks defined over the joint distribution of the data $\{\boldsymbol{\gamma}, \boldsymbol{h}\}$, and where $q(w)$ denotes the regularization factor.

Name	**Stochastic risk**, $P(w)$
mean-square error(MSE)	$q(w) + \mathbb{E}\left(\boldsymbol{\gamma} - \boldsymbol{h}^\mathsf{T} w\right)^2$
logistic	$q(w) + \mathbb{E} \ln\left(1 + e^{-\boldsymbol{\gamma}\boldsymbol{h}^\mathsf{T} w}\right)$
perceptron	$q(w) + \mathbb{E}\max\left\{0, -\boldsymbol{\gamma}\boldsymbol{h}^\mathsf{T} w\right\}$
hinge	$q(w) + \mathbb{E}\max\left\{0, 1 - \boldsymbol{\gamma}\boldsymbol{h}^\mathsf{T} w\right\}$

One difficulty that arises in the minimization of stochastic risks of the form (12.8) is that the joint distribution of the data $\{\boldsymbol{\gamma}, \boldsymbol{h}\}$ is rarely known beforehand. This means that the expectation in (12.7) cannot be computed, which in turn means that the risk function $P(w)$ itself is not known! This situation is different from the empirical risk case (12.2) where $P(w)$ is defined in terms of N data pairs $\{\gamma(m), h_m\}$ and is therefore known. However, motivated by the ergodicity property (7.18), we can approximate the expectation in (12.7) and replace it by a sample average computed over a good number of data samples $\{\gamma(m), h_m\}$ arising from the unknown distribution. Using these samples, we can approximate the *stochastic* risk (12.7) by the *empirical* risk (12.2). For this reason, gradient-descent methods for minimizing empirical risks are equally applicable to the minimization of stochastic risks, as our presentation will reveal.

12.1.3 Generalization

The models $\{w^\star, w^o\}$ that result from minimizing empirical or stochastic risks will be used to perform inference on new feature vectors. If we denote a generic feature vector by h, then $w^\star$ can be used to predict its target or label by using $\widehat{\gamma} = h^\mathsf{T} w^\star$; likewise, for w^o. One important distinction arises in the performance of the two models $\{w^\star, w^o\}$:

(a) An empirical risk formulation of the form (12.3) determines the optimizer $w^\star$ that is implied by the *given* collection of N data points, $\{\gamma(m), h_m\}$. As such, the performance of the empirical model $w^\star$ in predicting future labels will be strongly dependent on how representative the original dataset $\{\gamma(m), h_m\}$

is of the space from which features and labels arise. This issue relates to the important question of "*generalization*," and will be discussed in greater detail in Chapter 64. Intuitively, one model w_a is said to generalize better than another model w_b if w_a is able to perform more accurate predictions than w_b for new feature data. The concept of "generalization" is also referred to as *inductive inference* or *inductive reasoning* because it endows models with the ability to reason about new feature data based on experience learned from training data.

(b) In contrast, the stochastic risk formulation (12.8) seeks the optimizer w^o that is defined by the joint probability distribution of the data $\{\boldsymbol{\gamma}, \boldsymbol{h}\}$, and not by any finite collection of data points arising from this distribution. This is because the optimization criterion seeks to minimize the *average loss* over the joint pdf. The resulting model w^o is expected to perform better on average in predicting new labels. The challenge, however, as we are going to see, is that it is not possible to minimize stochastic risks directly because they require knowledge of the joint pdf of the data, and this information is rarely available. For this reason, solutions for the stochastic risk problem will often involve a step that reduces it to an empirical risk problem through an ergodic approximation, which is then minimized from a collection of data points. The bottom line is that, either way, whether we are dealing with empirical or stochastic risks, it is important to examine how well inference models generalize. We defer the technical details to Chapter 64.

12.2 CONDITIONS ON RISK FUNCTION

Three observations are warranted at this stage:

(a) First, in many cases of interest in this and subsequent chapters, the risk $P(w)$ will have a *unique* global minimizer $w^\star$ since $P(w)$ will generally be strongly convex. This is because the addition of regularization factors will often ensure strong convexity. We will examine this case in some detail. We will also comment on the case when $P(w)$ is only convex, as well as study nonconvex risks in Chapter 24.

(b) Second, the development in this chapter is not limited to the risks and losses shown in the previous tables.

(c) Third, the risk function $P(w)$ need not be smooth (i.e., it need not be differentiable everywhere). For example, for the logistic risk in Table 12.1 we have

$$P(w) = q(w) + \frac{1}{N}\sum_{m=0}^{N-1} \ln\left(1 + e^{-\gamma(m)h_m^\mathsf{T} w}\right) \tag{12.10}$$

This function is differentiable for all w when $q(w) = \rho\|w\|^2$ but is not differentiable at $w = 0$ when $q(w) = \alpha\|w\|_1$ or $q(w) = \alpha\|w\|_1 + \rho\|w\|^2$. Likewise, for the hinge risk in Table 12.1 we have

$$P(w) = q(w) + \frac{1}{N}\sum_{m=0}^{N-1} \max\left\{0,\, 1 - \gamma(m)h_m^{\mathsf{T}} w\right\} \tag{12.11}$$

This function is not differentiable at all points w satisfying $1 = \gamma(m)h_m^{\mathsf{T}} w$. The function is also not differentiable at $w = 0$ when $q(w) = \alpha\|w\|_1$ or $q(w) = \alpha\|w\|_1 + \rho\|w\|^2$. Observe that nondifferentiability can arise either from the regularization term or from the unregularized component of the risk. The recursive techniques for determining $w^\star$ will need to account for the possibility of points of nondifferentiability. We focus in this chapter on the case in which $P(w)$ is first-order differentiable, and defer the case of nonsmooth risks to future chapters.

Motivated by these considerations, we will consider in this chapter optimization problems of the form (12.1) where the risk function $P(w)$ satisfies two conditions:

(A1) (**Strong convexity**). $P(w)$ is ν-strongly convex and first-order differentiable at all w so that, from definition (8.21),

$$P(w_2) \geq P(w_1) + \nabla_w P(w_1)(w_2 - w_1) + \frac{\nu}{2}\|w_2 - w_1\|^2 \tag{12.12a}$$

for every $w_1, w_2 \in \text{dom}(P)$ and some $\nu > 0$.

(A2) (**δ-Lipschitz gradients**). The gradient vectors of $P(w)$ are δ-Lipschitz:

$$\|\nabla_w P(w_2) - \nabla_w P(w_1)\| \leq \delta\,\|w_2 - w_1\| \tag{12.12b}$$

for any $w_1, w_2 \in \text{dom}(P)$, and where $\|\cdot\|$ denotes the Euclidean norm of its vector argument.

For reference, we know from the earlier results (8.29) and (10.20) derived for strongly convex and δ-Lipschitz functions that conditions **A1** and **A2** imply, respectively:

$$(\textbf{A1}) \implies \frac{\nu}{2}\|\widetilde{w}\|^2 \leq P(w) - P(w^\star) \leq \frac{1}{2\nu}\|\widetilde{w}\|^2 \tag{12.13a}$$

$$(\textbf{A2}) \implies \frac{1}{2\delta}\|\widetilde{w}\|^2 \leq P(w) - P(w^\star) \leq \frac{\delta}{2}\|\widetilde{w}\|^2 \tag{12.13b}$$

where $\widetilde{w} = w^\star - w$. The upper bounds in both expressions indicate that whenever we bound $\|\widetilde{w}\|^2$ we will also be automatically bounding the excess risk, $P(w) - P(w^\star)$.

Example 12.1 (**Second-order differentiability**) Conditions (12.12a)–(12.12b) only require $P(w)$ to be first-order differentiable since the conditions are stated in terms of the gradient of the risk function. However, if $P(w)$ happens to be *second-order* differentiable over w, then we can combine both conditions into a single statement involving the Hessian matrix of $P(w)$. Recall from property (8.30) that strong convexity is equivalent to $P(w)$ having a Hessian matrix that is uniformly bounded from *below* by ν, i.e.,

$$0 < \nu I_M \leq \nabla_w^2 P(w), \quad \forall\, w \in \mathrm{dom}(P) \tag{12.14a}$$

We also know from (10.32) that the δ-Lipschitz condition (12.12b) is equivalent to the Hessian matrix being uniformly bounded from *above* by δ, i.e.,

$$\nabla_w^2 P(w) \;\leq\; \delta I_M, \quad \forall\, w \in \mathrm{dom}(P) \tag{12.14b}$$

Therefore, combining (12.14a) and (12.14b) we find that under second-order differentiability of $P(w)$, the two conditions (12.12a)–(12.12b) are equivalent to requiring the Hessian matrix of $P(w)$ to be uniformly bounded from below *and* from above as follows:

$$\boxed{0 < \nu I_M \leq \nabla_w^2 P(w) \;\leq\; \delta I_M} \tag{12.15}$$

Clearly, condition (12.15) requires $\nu \leq \delta$. Several risks from Table 12.1 satisfy property (12.15). Here is one example (see also Prob. 12.2).

Example 12.2 (Logistic empirical risk) Consider the ℓ_2-regularized logistic risk from Table 12.1, namely,

$$P(w) = \rho\|w\|^2 \;+\; \frac{1}{N}\sum_{m=0}^{N-1} \ln\left(1 + e^{-\gamma(m)h_m^{\mathsf{T}} w}\right) \tag{12.16}$$

It can be verified that

$$\nabla_w^2\, P(w) = 2\rho I_M \;+\; \frac{1}{N}\sum_{m=0}^{N-1} h_m h_m^{\mathsf{T}} \underbrace{\frac{e^{-\gamma(m)h_m^{\mathsf{T}} w}}{\left(1 + e^{-\gamma(m)h_m^{\mathsf{T}} w}\right)^2}}_{\leq 1} \tag{12.17}$$

from which we conclude that

$$0 \;<\; \underbrace{2\rho}_{\triangleq\, \nu}\, I_M \;\leq\; \nabla_w^2\, P(w) \;\leq\; \underbrace{2\rho I_M + \lambda_{\max}\left(\frac{1}{N}\sum_{m=0}^{N-1} h_m h_m^{\mathsf{T}}\right) I_M}_{\triangleq\, \delta} \tag{12.18}$$

where the notation $\lambda_{\max}(\cdot)$ denotes the maximum eigenvalue of its symmetric matrix argument.

12.3 CONSTANT STEP SIZES

We are now ready to motivate the gradient-descent method. We consider first the case in which a constant step size is employed in the implementation of the algorithm. In a future section, we examine the case of iteration-dependent step sizes.

12.3.1 Derivation of Algorithm

When $P(w)$ is first-order differentiable and strongly convex, its unique global minimizer $w^\star$ satisfies:

$$\nabla_{w^\mathsf{T}} P(w)\Big|_{w=w^\star} = 0 \tag{12.19}$$

We are differentiating relative to w^T and not w in order to be consistent with our earlier convention from Chapter 2 that differentiation relative to a row vector results in a column vector. Equality (12.19) does not change if we scale the gradient vector by any positive scalar $\mu > 0$ and add and subtract $w^\star$, so that it also holds

$$w^\star = w^\star - \mu \nabla_{w^\mathsf{T}} P(w)\Big|_{w=w^\star} \tag{12.20}$$

This relation indicates that we can view the solution $w^\star$ as a *fixed point* for the mapping $f(w) : \mathbb{R}^M \rightarrow \mathbb{R}^M$ defined by

$$f(w) \stackrel{\Delta}{=} w - \mu \nabla_{w^\mathsf{T}} P(w) \tag{12.21}$$

The idea of the gradient-descent method is based on transforming the fixed-point equality (12.20) into a recursion, written as:

$$\boxed{w_n = w_{n-1} - \mu \nabla_{w^\mathsf{T}} P(w_{n-1}), \quad n \geq 0} \tag{12.22}$$

where the $w^\star$ on the left-hand side of (12.20) is replaced by w_n, while the $w^\star$ on the right-hand side of the same expression is replaced by w_{n-1}. The vectors $\{w_{n-1}, w_n\}$ represent two successive iterates that serve as estimates for $w^\star$. The scalar $\mu > 0$ is known as the *step-size* parameter and it is usually a small number. The gradient-descent algorithm is listed in (12.24). It is iterated over n until a maximum number of iterations is reached, or until the change in the weight iterate is small, or until the norm of the gradient vector is small:

$$n \leq n_{\max} \tag{12.23a}$$

$$\|w_n - w_{n-1}\|^2 \leq \epsilon, \quad \text{for some small } \epsilon \tag{12.23b}$$

$$\|\nabla_{w^\mathsf{T}} P(w_n)\| \leq \epsilon', \quad \text{for some small } \epsilon' \tag{12.23c}$$

Gradient-descent method for minimizing $P(w)$.

given gradient operator, $\nabla_{w^\mathsf{T}} P(w)$;
given a small step-size parameter $\mu > 0$;
start from an arbitrary initial condition, w_{-1}; (12.24)
repeat until convergence over $n \geq 0$**:**
$\quad w_n = w_{n-1} - \mu \nabla_{w^\mathsf{T}} P(w_{n-1})$
end
return $w^\star \leftarrow w_n$.

Recursion (12.24) starts from some initial condition, denoted by w_{-1} (usually the zero vector), and updates the iterate w_{n-1} along the negative direction of the gradient vector of $P(w)$ at w_{n-1}. The reason for the negative sign in front of μ in (12.22) is to ensure that the update to w_{n-1} is in the direction of the minimizer $w^\star$. This is because, by definition, the gradient vector of a function points in the direction toward which the function is increasing and, hence, the negative gradient points in the opposite direction. This is illustrated in Fig. 12.1. The panel on the left shows the mechanics of one update step, while the panel on the right shows the result of several successive steps.

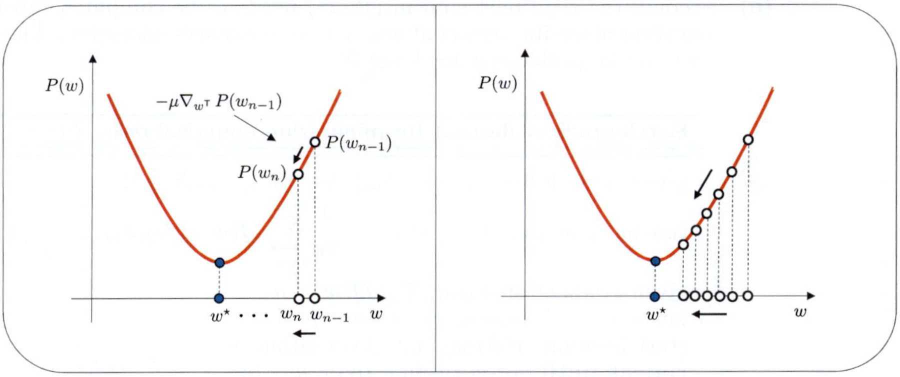

Figure 12.1 The panel on the left shows the mechanics of one update step where w_{n-1} is updated in the direction of the minimizer $w^\star$. The panel on the right shows the result of several successive steps with the iterates approaching $w^\star$.

Example 12.3 (Quadratic approximation) There are many ways by which the gradient-descent method can be motivated. For example, we can motivate the same gradient-descent recursion (12.22) by minimizing a quadratic approximation to $P(w)$. Let w_{n-1} denote an estimate for $w^\star$ that is available at iteration $n-1$ and approximate the Hessian matrix by $\nabla_w^2 P(w_{n-1}) \approx \frac{1}{\mu} I_M$, for some $\mu > 0$. We consider a second-order expansion for $P(w)$ around w_{n-1} and pose the problem of updating w_{n-1} to w_n by solving:

$$w_n = \underset{w\in\mathbb{R}^M}{\text{argmin}} \left\{ P(w_{n-1}) + \nabla_w P(w_{n-1})(w - w_{n-1}) + \frac{1}{2\mu}\|w - w_{n-1}\|^2 \right\} \tag{12.25}$$

Differentiating the right-hand side relative to w we find that the solution w_n is given by the relation:

$$w_n = w_{n-1} - \mu \nabla_{w^\mathsf{T}} P(w_{n-1}) \tag{12.26}$$

Example 12.4 (Batch gradient-descent) If we apply the gradient-descent algorithm (12.22) to *empirical risk* functions of the form (12.2), then the gradient vector will have the form of a sample average expression:

$$\nabla_{w^\mathsf{T}} P(w) = \frac{1}{N}\sum_{m=0}^{N-1} \nabla_{w^\mathsf{T}} Q(w;\gamma(m),h_m) \tag{12.27}$$

In this case, we can be more explicit about the description of the gradient-descent method and write it in the form shown in (12.28). The reason for the designation "batch algorithm" is because each iteration of (12.28) employs the *entire* set of data, i.e., all N data pairs $\{\gamma(m),h_m\}$. Moreover, this dataset is used repeatedly until the algorithm approaches its limiting behavior. There are at least two disadvantages for these types of batch implementations:

(a) First, the entire set of N data pairs $\{\gamma(m),h_m\}$ needs to be available beforehand to be used at every iteration. For this reason, batch implementations cannot respond to streaming data, i.e., to data that arrive sequentially at every time instant.

(b) Second, the rightmost sum in (12.28) needs to be computed repeatedly at every iteration since its argument w_{n-1} is continuously changing. The computational cost can be prohibitive for large N.

Batch gradient-descent for minimizing empirical risks, $P(w)$.

given N data pairs $\{\gamma(m),h_m\}, m=0,1,\ldots,N-1$;

risk has empirical form $P(w) = \frac{1}{N}\sum_{m=0}^{N-1} Q(w;\gamma(m),h_m)$;

given gradient operator, $\nabla_{w^\mathsf{T}}Q(w;\gamma,h)$;

given a small step-size parameter $\mu>0$; (12.28)

start from an arbitrary initial condition, w_{-1};

repeat until convergence over $n\geq 0$**:**

$$w_n = w_{n-1} - \mu\left(\frac{1}{N}\sum_{m=0}^{N-1}\nabla_{w^\mathsf{T}} Q(w_{n-1};\gamma(m),h_m)\right)$$

end

return $w^\star \leftarrow w_n$.

We will explain in Chapter 16 how difficulties **(a)** and **(b)** can be addressed by resorting to *stochastic* gradient algorithms. In one implementation, the gradient sum in (12.27) is approximated by a *single* term, $\nabla_{w^\mathsf{T}} Q(w_{n-1},\boldsymbol{\gamma}(n),\boldsymbol{h}_n)$, where the pair $(\boldsymbol{\gamma}(n),\boldsymbol{h}_n)$ is selected at random from the dataset. In a second implementation, the gradient sum is approximated by a *mini-batch* where a small *subset* of the N-long data $\{\gamma(m),h_m\}$ is selected at random at every iteration and used in the update from w_{n-1} to w_n.

Example 12.5 **(Batch logistic regression)** Consider the ℓ_2-regularized logistic empirical risk from Table 12.1 along with its gradient vector:

$$P(w) = \rho\|w\|^2 + \frac{1}{N}\sum_{m=0}^{N-1}\ln\left(1+e^{-\gamma(m)h_m^\mathsf{T} w}\right) \tag{12.29a}$$

$$\nabla_{w^\mathsf{T}} P(w) = 2\rho w - \frac{1}{N}\sum_{m=0}^{N-1}\frac{\gamma(m)h_m}{1+e^{\gamma(m)h_m^\mathsf{T} w}} \tag{12.29b}$$

The corresponding gradient-descent recursion (12.22) is given by

$$w_n = (1-2\mu\rho)\,w_{n-1} + \mu\left(\frac{1}{N}\sum_{m=0}^{N-1}\frac{\gamma(m)h_m}{1+e^{\gamma(m)h_m^\mathsf{T} w_{n-1}}}\right),\quad n\geq 0 \tag{12.30}$$

The reason for the designation "logistic" is because the logistic loss in (12.29a) will arise when we study logistic regression problems in Chapters 28 and 59. When ρ is zero, the above recursion simplifies to

$$w_n = w_{n-1} + \mu\left(\frac{1}{N}\sum_{m=0}^{N-1}\frac{\gamma(m)h_m}{1+e^{\gamma(m)h_m^{\mathsf{T}}w_{n-1}}}\right), \quad n \geq 0 \tag{12.31}$$

Example 12.6 (**MSE stochastic risk**) Consider next an example involving a stochastic risk, say, one of the form:

$$P(w) = \rho\|w\|^2 + \mathbb{E}\,(\boldsymbol{\gamma} - \boldsymbol{h}^{\mathsf{T}}w)^2 \tag{12.32}$$

where $\boldsymbol{\gamma} \in \mathbb{R}$ and $\boldsymbol{h} \in \mathbb{R}^M$ are assumed to have zero means with second-order moments denoted by $\sigma_\gamma^2 = \mathbb{E}\,\boldsymbol{\gamma}^2$, $R_h = \mathbb{E}\,\boldsymbol{h}\boldsymbol{h}^{\mathsf{T}}$, and $r_{h\gamma} = \mathbb{E}\,\boldsymbol{h}\boldsymbol{\gamma}$. Then, it holds that

$$P(w) = \rho\|w\|^2 + \sigma_\gamma^2 - 2r_{h\gamma}^{\mathsf{T}}w + w^{\mathsf{T}}R_h w \tag{12.33a}$$

$$\nabla_{w^{\mathsf{T}}} P(w) = 2\rho w - 2r_{h\gamma} + 2R_h w \tag{12.33b}$$

and the gradient-descent recursion (12.22) leads to

$$w_n = (1 - 2\mu\rho)\,w_{n-1} + 2\mu\,(r_{h\gamma} - R_h w_{n-1}), \quad n \geq 0 \tag{12.34}$$

Example 12.7 (**Batch least-squares**) Consider the ℓ_2-regularized least-squares empirical risk from Table 12.1, where

$$P(w) = \rho\|w\|^2 + \frac{1}{N}\sum_{m=0}^{N-1}(\gamma(m) - h_m^{\mathsf{T}}w)^2 \tag{12.35a}$$

$$\nabla_{w^{\mathsf{T}}} P(w) = 2\rho w - \frac{2}{N}\sum_{m=0}^{N-1}h_m(\gamma(m) - h_m^{\mathsf{T}}w) \tag{12.35b}$$

and the gradient-descent method reduces to

$$w_n = (1 - 2\mu\rho)\,w_{n-1} + 2\mu\left(\frac{1}{N}\sum_{m=0}^{N-1}h_m(\gamma(m) - h_m^{\mathsf{T}}w_{n-1})\right), \quad n \geq 0 \tag{12.36}$$

where $\mu > 0$ is a small step size.

Example 12.8 (**Batch least-squares with offset**) In many inference problems, there will be a need to incorporate an offset parameter θ into the problem formulation. We illustrate this fact by considering a variation of the ℓ_2-regularized least-squares risk from the previous example:

$$(w^\star, \theta^\star) \triangleq \underset{w\in\mathbb{R}^M,\theta\in\mathbb{R}}{\text{argmin}} \left\{\rho\|w\|^2 + \frac{1}{N}\sum_{m=0}^{N-1}\left(\gamma(m) - h_m^{\mathsf{T}}w + \theta\right)^2\right\} \tag{12.37}$$

where θ represents the scalar offset parameter. In this case, the prediction for the target γ corresponding to a feature h is computed by means of the affine relation

$\widehat{\gamma} = h^{\mathsf{T}} w - \theta$. Observe that regularization is applied to w only and not to θ. We now have two parameters $\{w, \theta\}$ and, therefore,

$$P(w,\theta) = \rho\|w\|^2 + \frac{1}{N}\sum_{m=0}^{N-1}(\gamma(m) - h_m^{\mathsf{T}} w)^2 \tag{12.38a}$$

$$\nabla_{w^{\mathsf{T}}} P(w) = 2\rho w \;-\; \frac{2}{N}\sum_{m=0}^{N-1} h_m(\gamma(m) - h_m^{\mathsf{T}} w + \theta) \tag{12.38b}$$

$$\partial P(w,\theta)/\partial\theta = \frac{2}{N}\sum_{m=0}^{N-1}(\gamma(m) - h_m^{\mathsf{T}} w + \theta) \tag{12.38c}$$

The batch iteration (12.28) then becomes

$$\theta(n) = \theta(n-1) - 2\mu\left(\frac{1}{N}\sum_{m=0}^{N-1}\Big(\gamma(m) - h_m^{\mathsf{T}} w_{n-1} + \theta(n-1)\Big)\right) \tag{12.39a}$$

$$w_n = (1 - 2\mu\rho)\, w_{n-1} + 2\mu\left(\frac{1}{N}\sum_{m=0}^{N-1} h_m\Big(\gamma(m) - h_m^{\mathsf{T}} w_{n-1} + \theta(n-1)\Big)\right) \tag{12.39b}$$

We can combine the two recursions into a single relation by introducing the augmented variables of size $M+1$ each:

$$w' \triangleq \begin{bmatrix} -\theta \\ w \end{bmatrix}, \quad h' \triangleq \begin{bmatrix} 1 \\ h \end{bmatrix} \tag{12.40}$$

and writing

$$w'_n = \begin{bmatrix} 1 & \\ & (1-2\mu\rho)\, I_M \end{bmatrix} w'_{n-1} + 2\mu\left(\frac{1}{N}\sum_{m=0}^{N-1} h'_m\Big(\gamma(m) - (h'_m)^{\mathsf{T}} w'_{n-1}\Big)\right) \tag{12.41}$$

12.3.2 Convergence Analysis

The size of the step taken in (12.22) along the (negative) gradient direction is determined by μ. A small μ helps the iterates $\{w_n\}$ approach $w^\star$ in small steps, while a large μ can result in unstable behavior with the iterates bouncing back and forth around $w^\star$. Most convergence analyses specify bounds on how large μ can be to ensure the convergence of w_n to $w^\star$ as $n \to \infty$.

Theorem 12.1. (Convergence under constant step sizes) *Consider the gradient-descent recursion (12.22) for minimizing a first-order differentiable risk function $P(w)$, where $P(w)$ is ν-strongly convex with δ-Lipschitz gradients according to (12.12a)–(12.12b). Introduce the error vector $\widetilde{w}_n = w^\star - w_n$, which measures the difference between the nth iterate and the global minimizer of $P(w)$. If the step size μ satisfies (i.e., is small enough)*

$$0 < \mu < 2\nu/\delta^2 \tag{12.42}$$

then w_n and the excess risk converge exponentially fast in the following sense:

$$\|\widetilde{w}_n\|^2 \leq \lambda\,\|\widetilde{w}_{n-1}\|^2, \quad n \geq 0 \tag{12.43a}$$

$$P(w_n) - P(w^\star) \leq \frac{\delta}{2}\lambda^{n+1}\|\widetilde{w}_{-1}\|^2 \;=\; O(\lambda^n), \quad n \geq 0 \tag{12.43b}$$

where

$$\lambda \;\stackrel{\Delta}{=}\; 1 - 2\mu\nu + \mu^2\delta^2 \;\in [0,1) \tag{12.44}$$

Proof: We subtract $w^\star$ from both sides of (12.22) to get

$$\widetilde{w}_n \;=\; \widetilde{w}_{n-1} \;+\; \mu\,\nabla_{w^\mathsf{T}} P(w_{n-1}) \tag{12.45}$$

We compute the squared Euclidean norms (or energies) of both sides of the above equality and use the fact that $\nabla_{w^\mathsf{T}} P(w^\star) = 0$ to write

$$\begin{aligned}
&\|\widetilde{w}_n\|^2 \\
&= \|\widetilde{w}_{n-1}\|^2 + 2\mu\left(\nabla_{w^\mathsf{T}} P(w_{n-1})\right)^\mathsf{T}\widetilde{w}_{n-1} + \mu^2\,\|\nabla_{w^\mathsf{T}} P(w_{n-1})\|^2 \\
&= \|\widetilde{w}_{n-1}\|^2 + 2\mu\left(\nabla_{w^\mathsf{T}} P(w_{n-1})\right)^\mathsf{T}\widetilde{w}_{n-1} + \mu^2\,\|\nabla_{w^\mathsf{T}} P(w^\star) - \nabla_{w^\mathsf{T}} P(w_{n-1})\|^2 \\
&\stackrel{(12.12b)}{\leq} \|\widetilde{w}_{n-1}\|^2 + 2\mu\left(\nabla_{w^\mathsf{T}} P(w_{n-1})\right)^\mathsf{T}\widetilde{w}_{n-1} + \mu^2\delta^2\|\widetilde{w}_{n-1}\|^2
\end{aligned} \tag{12.46}$$

We appeal to the strong-convexity property (12.12a) and use $w_2 = w^\star$, $w_1 = w_{n-1}$ in step (a) below and $w_2 = w_{n-1}$, $w_1 = w^\star$ in step (b) to find that

$$\begin{aligned}
\left(\nabla_{w^\mathsf{T}} P(w_{n-1})\right)^\mathsf{T}\widetilde{w}_{n-1} &\stackrel{(a)}{\leq} P(w^\star) - P(w_{n-1}) - \frac{\nu}{2}\|\widetilde{w}_{n-1}\|^2 \\
&\stackrel{(b)}{\leq} -\frac{\nu}{2}\|\widetilde{w}_{n-1}\|^2 \;-\; \frac{\nu}{2}\|\widetilde{w}_{n-1}\|^2 \\
&= -\nu\|\widetilde{w}_{n-1}\|^2
\end{aligned} \tag{12.47}$$

Substituting into (12.46) gives

$$\|\widetilde{w}_n\|^2 \;\leq\; (1 - 2\mu\nu + \mu^2\delta^2)\,\|\widetilde{w}_{n-1}\|^2 \tag{12.48}$$

which coincides with (12.43a)–(12.44). Iterating, we find that

$$\|\widetilde{w}_n\|^2 \;\leq\; \lambda^{n+1}\,\|\widetilde{w}_{-1}\|^2 \tag{12.49}$$

which highlights the exponential convergence of $\|\widetilde{w}_n\|^2$ to zero. We next verify that condition (12.42) ensures $0 \leq \lambda < 1$ using the same argument from Fig. 11.5. We plot the coefficient $\lambda(\mu)$ as a function of μ in Fig. 12.2. The minimum value of $\lambda(\mu)$ occurs at location $\mu = \nu/\delta^2$ and is equal to $1 - \nu^2/\delta^2$. This value is nonnegative since $0 < \nu \leq \delta$. It is clear from the figure that $0 \leq \lambda < 1$ for $\mu \in (0, \frac{2\nu}{\delta^2})$.

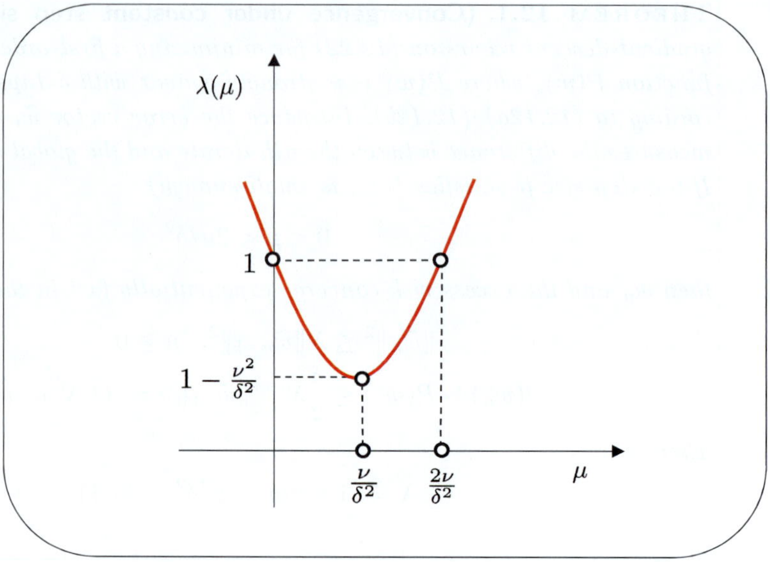

Figure 12.2 Plot of the function $\lambda(\mu) = 1 - 2\nu\mu + \mu^2\delta^2$ given by (12.44). It shows that the function $\lambda(\mu)$ assumes values below 1 in the range $0 < \mu < 2\nu/\delta^2$.

To establish (12.43b), we first note that $P(w_n) \geq P(w^\star)$ since $w^\star$ is the minimizer of $P(w)$. Using the upper bound (12.13b) we have

$$0 \leq P(w_n) - P(w^\star) \leq \frac{\delta}{2}\|\widetilde{w}_n\|^2 \overset{(12.48)}{\leq} \frac{\delta}{2}\lambda^{n+1}\|\widetilde{w}_{-1}\|^2 \tag{12.50}$$

■

Remark 12.2. (Exponential or linear convergence) Recursions evolving according to a dynamics of the form (12.43a), such as $a(n) \leq \lambda\, a(n-1)$ for some $\lambda \in [0, 1)$, are said to converge *exponentially fast* since, by iterating, we get $a(n) \leq \lambda^{n+1}a(-1)$. This expression shows that $a(n)$ decays to zero exponentially at the rate λ^n. This mode of convergence is also referred to as *linear* convergence because, when plotted on a semi-log scale, the curve $\ln a(n) \times n$ will be linear in n with slope $\ln\lambda$, namely,

$$\ln a(n) \leq (n+1)\ln\lambda + \text{cte} \tag{12.51}$$

■

Remark 12.3. (Big-O, little-o, and big-Θ notation) The statement (12.43b) uses the big-O notation. In other locations, we will employ the little-o notation. We therefore compare their meanings in this remark. We already explained in the earlier Remark 11.3 that the big-O notation is used to compare the asymptotic growth rate of two sequences. Thus, writing $a_n = O(b_n)$, with a big O for a sequence b_n with positive entries, means that there exists some constant $c > 0$ and index n_o such that $|a_n| \leq cb_n$ for all $n > n_o$. This also means that the decay rate of a_n is at least as fast or faster than b_n. For example, writing $a_n = O(1/n)$ means that the samples of the sequence a_n decay asymptotically at a rate that is comparable to or faster than $1/n$. Sometimes, one may use the big-Θ notation, which is more specific than the big-O notation in that it bounds the sequence $|a(n)|$ both from above and from below. Thus, writing $a_n = \Theta(b_n)$ now means that there exist two constants $c_1 > 0$ and $c_2 > 0$, and an index n_o, such that $c_1 b_n \leq |a_n| \leq c_2 b_n$ for all $n > n_o$. This means that the decay rate of the sequence a_n is comparable to the decay rate of b_n. For instance, writing $a_n = \Theta(1/n)$ would now mean that the samples of the sequence a_n decay asymptotically at the rate $1/n$.

On the other hand, the little-o notation, $a_n = o(b_n)$, means that, asymptotically, the sequence a_n decays faster than the sequence b_n so that it should hold $|a_n|/b_n \to 0$ as $n \to \infty$. In this case, the notation $a_n = o(1/n)$ implies that the samples of a_n decay at a faster rate than $1/n$. Table 12.3 summarizes these definitions.

Table 12.3 Interpretation of the big-O, little-o, and big-Θ notation.

Notation	Interpretation
$a_n = O(b_n)$	$\|a_n\| \leq cb_n,\ n > n_o$
$a_n = \Theta(b_n)$	$c_1 b_n \leq \|a_n\| \leq c_2 b_n,\ n > n_o$
$a_n = o(b_n)$	$\|a_n\|/b_n \to 0$ as $n \to \infty$

■

Example 12.9 **(A more relaxed bound on μ)** The result of Theorem 12.1 establishes the exponential convergence of the squared weight error, $\|\widetilde{w}_n\|^2$, and the excess risk, $P(w_n) - P(w^\star)$, toward zero for sufficiently small step sizes, μ. In most instances, these results are sufficient since our objective is often to verify whether the iterative algorithms approach their desired limits. This conclusion is established in Theorem 12.1 under the bound $\mu < 2\nu/\delta^2$. We can relax the result and show that convergence will continue to occur for $\mu < 2/\delta$. We do this by exploiting a certain *co-coercivity* property that is satisfied by convex functions with δ-Lipschitz gradients. Specifically, we know from the result of Prob. 10.4 that:

$$\Big(\nabla_{w^\mathsf{T}} P(w_2) - \nabla_{w^\mathsf{T}} P(w_1)\Big)^\mathsf{T}(w_2 - w_1) \geq \frac{1}{\delta}\,\|\nabla_w P(w_2) - \nabla_w P(w_1)\|^2 \qquad (12.52)$$

We use this inequality in (12.46) as follows:

$$\begin{aligned}
&\|\widetilde{w}_n\|^2 \\
&= \|\widetilde{w}_{n-1}\|^2 - 2\mu\Big(\nabla_{w^\mathsf{T}} P(w^\star) - \nabla_{w^\mathsf{T}} P(w_{n-1})\Big)^\mathsf{T}\widetilde{w}_{n-1} + \mu^2\,\|\nabla_{w^\mathsf{T}} P(w_{n-1})\|^2 \\
&\overset{(12.52)}{\leq} \|\widetilde{w}_{n-1}\|^2 - 2\mu\Big(\nabla_{w^\mathsf{T}} P(w^\star) - \nabla_{w^\mathsf{T}} P(w_{n-1})\Big)^\mathsf{T}\widetilde{w}_{n-1} \\
&\qquad\quad + \mu^2\delta\Big(\nabla_{w^\mathsf{T}} P(w^\star) - \nabla_{w^\mathsf{T}} P(w_{n-1})\Big)^\mathsf{T}\widetilde{w}_{n-1} \\
&= \|\widetilde{w}_{n-1}\|^2 - (2\mu - \mu^2\delta)\Big(\nabla_{w^\mathsf{T}} P(w^\star) - \nabla_{w^\mathsf{T}} P(w_{n-1})\Big)^\mathsf{T}\widetilde{w}_{n-1} \\
&= \|\widetilde{w}_{n-1}\|^2 + (2\mu - \mu^2\delta)(\nabla_{w^\mathsf{T}} P(w_{n-1}))^\mathsf{T}\widetilde{w}_{n-1} \\
&\overset{(12.47)}{\leq} \|\widetilde{w}_{n-1}\|^2 - (2\mu - \mu^2\delta)\nu\|\widetilde{w}_{n-1}\|^2 \\
&= \underbrace{(1 - 2\mu\nu + \mu^2\nu\delta)}_{\triangleq\,\lambda'}\,\|\widetilde{w}_{n-1}\|^2
\end{aligned} \qquad (12.53)$$

This result is consistent with (12.48) since $\lambda' \leq \lambda$ in view of $\nu \leq \delta$. Working with λ', we obtain the bound $0 < \mu < 2/\delta$ for stability with convergence occurring at $O((\lambda')^n)$.

Example 12.10 **(Convergence analysis based on excess risk)** The convergence analysis used to establish Theorem 12.1 was based on examining the evolution of the squared error, $\|\widetilde{w}_n\|^2$, and from there we were able to conclude how the excess risk term evolves with time. We will adopt this approach uniformly throughout our presentation. However, we remark here that we can arrive at similar conclusions by working directly with the risk function. To do so, we exploit two properties of the risk function: its strong convexity and the fact that it has Lipschitz gradients. These properties were shown before to induce certain bounds on the risk.

For instance, using the ν-strong convexity of $P(w)$, we use property (8.29) to deduce that

$$P(w^\star) \geq P(w_{n-1}) - \frac{1}{2\nu}\|\nabla_w P(w_{n-1})\|^2 \tag{12.54}$$

On the other hand, from the δ-Lipschitz property on the gradients of $P(w)$, we use result (10.13) to write

$$\begin{aligned} P(w_n) &\leq P(w_{n-1}) + (\nabla_w P(w_{n-1}))(w_n - w_{n-1}) + \frac{\delta}{2}\|w_n - w_{n-1}\|^2 \\ &\overset{(12.22)}{=} P(w_{n-1}) - \mu\Big(1 - \frac{\mu\delta}{2}\Big)\|\nabla_w P(w_{n-1})\|^2 \end{aligned} \tag{12.55}$$

Subtracting $P(w^\star)$ from both sides of this inequality and using (12.54) we obtain

$$P(w_n) - P(w^\star) \leq \underbrace{(1 - 2\mu\nu + \mu^2\nu\delta)}_{\lambda'}\Big(P(w_{n-1}) - P(w^\star)\Big) \tag{12.56}$$

This result is consistent with (12.53) and convergence again occurs for $0 < \mu < 2/\delta$ at the rate $O((\lambda')^n)$.

Regret analysis

The statement of Theorem 12.1 examines the convergence behavior of the squared weight error, $\|\widetilde{w}_n\|^2$, and the risk value $P(w_n)$. Another common performance measure for learning algorithms is the *average regret*. It is defined over a window of N iterations and computes the deviation of the accumulated risk relative to the minimal risk:

$$\boxed{\mathcal{R}(N) \triangleq \frac{1}{N}\sum_{n=0}^{N-1} P(w_{n-1}) - P(w^\star)} \qquad \textbf{(average regret)} \tag{12.57}$$

The sum involves all risk values over the first N iterations. Using (12.43b) we find that the regret decays at the rate of $1/N$ since

$$\begin{aligned} \mathcal{R}(N) &\leq \frac{1}{N}\frac{\delta\|\widetilde{w}_{-1}\|^2}{2}\sum_{n=0}^{N-1}\lambda^{n+1} \\ &= \frac{1}{N}\frac{\delta\|\widetilde{w}_{-1}\|^2}{2}\frac{(1-\lambda^N)\lambda}{1-\lambda} \\ &= O(1/N) \end{aligned} \tag{12.58}$$

This calculation shows that we can transform bounds on the excess risk $P(w_n) - P(w^\star)$ into bounds on the average regret. For this reason, we will continue to derive excess risk bounds throughout our analysis of learning algorithms, with the understanding that they can be easily transformed into regret bounds.

Remark 12.4. (Regret analysis and convexity) There is another useful bound for the average regret for convex risk functions. Using property (8.4) we have

$$P(w_{n-1}) - P(w^\star) \leq -(\nabla_{w^\mathsf{T}} P(w_{n-1}))^\mathsf{T} \widetilde{w}_{n-1} \tag{12.59}$$

so that we can bound (12.57) by

$$\mathcal{R}(N) \leq -\frac{1}{N}\sum_{n=0}^{N-1} (\nabla_{w^\mathsf{T}} P(w_{n-1}))^\mathsf{T} \widetilde{w}_{n-1} \quad \textbf{(for convex risks)} \tag{12.60}$$

For this reason, it is also customary to study $\mathcal{R}(N)$ by bounding the inner product $(\nabla_{w^\mathsf{T}} P(w_{n-1}))^\mathsf{T} \widetilde{w}_{n-1}$ and its cumulative sum. Examples to this effect will be encountered later in Section 16.5 and Appendix 17.A in the context of stochastic optimization algorithms.

■

Example 12.11 (Dependence of convergence on problem dimension) Result (12.58) may suggest at first sight that the regret bound is not dependent on the parameter dimension M. However, the bound is scaled by the Lipschitz constant δ and this constant is implicitly dependent on M. This is because the value of δ depends on the norm used in (12.12b). For most of our treatment, we will be working with the Euclidean norm, but there are important cases where the gradient Lipschitz property will hold for other norms. For example, assume for the sake of argument that the gradients of the risk function $P(w)$ happen to be δ-Lipschitz relative to some other norm, such as the ℓ_∞-norm. In this case, expression (12.12b) will be replaced by

$$\|\nabla_w P(w_2) - \nabla_w P(w_1)\|_\infty \;\leq\; \delta\, \|w_2 - w_1\|_\infty \tag{12.61}$$

Using the norm inequalities:

$$\|x\|_2 \leq \sqrt{M}\, \|x\|_\infty, \quad \|x\|_\infty \leq \|x\|_2 \tag{12.62}$$

for any $x \in \mathbb{R}^M$, relation (12.61) can be transformed into an inequality involving the ℓ_2-norm, as in (12.12b):

$$\|\nabla_w P(w_2) - \nabla_w P(w_1)\| \;\leq\; \sqrt{M}\delta\, \|w_2 - w_1\| \tag{12.63}$$

with a new δ value that is scaled by $\sqrt{M}$. If we were to write the regret and performance bounds derived so far in the chapter using this new δ, then the results will be scaled by $\sqrt{M}$ and become dependent on the problem dimension. This fact is problematic for large dimensional inference problems. Later, in Section 15.3, we will motivate the *mirror-descent* algorithm, which addresses this problem for a class of constrained optimization problems and leads to performance bounds that are independent of M.

Convexity versus strong convexity

The statement of Theorem 12.1 assumes *strongly convex* risk functions $P(w)$ with δ-Lipschitz gradients satisfying (12.12a)–(12.12b). The theorem establishes in (12.43a) the exponential convergence of $\|\widetilde{w}_n\|^2$ to zero at the rate λ^n. It also establishes in (12.43b) that $P(w_n)$ converges exponentially at the same rate to $P(w^\star)$. We will express these conclusions by adopting the following notation:

$$\begin{cases} \|\widetilde{w}_n\|^2 \;\leq\; O(\lambda^n) \\ P(w_n) - P(w^\star) \;\leq\; O(\lambda^n) \end{cases} \quad \textbf{(for strongly convex } P(w)\textbf{)} \qquad (12.64a)$$

This means that $O(\ln(1/\epsilon))$ iterations are needed for the risk value $P(w_n)$ to get ϵ-close to $P(w^\star)$. We will be dealing largely with strongly convex risks $P(w)$, especially since regularization will ensure strong convexity in many cases of interest. Nevertheless, when $P(w)$ happens to be only convex (but not necessarily strongly convex) then, following an argument similar to the derivation of (11.71), we can establish that convergence in this case will be *sublinear* (rather than linear). Specifically, it will hold for $\mu < 1/\delta$ that the successive risk values approach the minimum value at the slower rate of $1/n$:

$$P(w_n) - P(w^\star) \;\leq\; O(1/n) \quad \textbf{(for convex } P(w)\textbf{)} \qquad (12.64b)$$

This result is established in Prob. 12.13. In this case, $O(1/\epsilon)$ iterations will be needed for the risk value $P(w_n)$ to get ϵ-close to $P(w^\star)$.

12.4 ITERATION-DEPENDENT STEP SIZES

Although recursion (12.22) employs a *constant* step size μ, one can also consider iteration-dependent step sizes, denoted by $\mu(n)$, and write:

$$\boxed{w_n = w_{n-1} - \mu(n)\nabla_{w^\mathsf{T}} P(w_{n-1}), \quad n \geq 0} \qquad (12.65)$$

The ability to vary the step size with n provides an opportunity to control the size of the gradient step, for example, by using larger steps during the initial stages of learning and smaller steps later. There are several ways by which the step-size sequence $\mu(n)$ can be selected.

12.4.1 Vanishing Step Sizes

The convergence analysis in Theorem 12.2 assumes step sizes that satisfy either one of the following two conditions:

$$\textbf{(Condition I)} \qquad \sum_{n=0}^{\infty} \mu^2(n) < \infty \quad \text{and} \quad \sum_{n=0}^{\infty} \mu(n) = \infty \qquad (12.66a)$$

$$\textbf{(Condition II)} \qquad \lim_{n\to\infty} \mu(n) = 0 \quad \text{and} \quad \sum_{n=0}^{\infty} \mu(n) = \infty \qquad (12.66b)$$

Clearly, any sequence that satisfies (12.66a) also satisfies (12.66b). In either case, the step-size sequence vanishes asymptotically but the rate of decay of $\mu(n)$ to zero should not be too fast (so that the sequence is not absolutely summable). For example, step-size sequences of the form:

$$\mu(n) = \frac{\tau}{(n+1)^c}, \quad \text{for any } \tau > 0 \text{ and } \tfrac{1}{2} < c \leq 1 \tag{12.67}$$

satisfy (12.66b). The choice $c = 1$ is common. There are other choices for $\mu(n)$, besides sequences that satisfy (12.66a) or (12.66b), that can ensure convergence of the gradient-descent method. We will illustrate this fact further ahead when we examine *backtracking* in Section 12.4.2. There, we will introduce another sufficient requirement on $\mu(n)$ to guarantee convergence known as the *Armijo condition.* Other examples of convergent gradient-descent methods with iteration-dependent step sizes include the alternating projection algorithm from Section 12.6 and the Kaczmarz method from Prob. 12.34. For now, we continue with the popular conditions (12.66a)–(12.66b). The following result shows that the convergence rate is not exponential any longer and is slower than under constant step sizes.

THEOREM 12.2. (Convergence under vanishing step sizes) *Consider the gradient-descent recursion (12.65) for minimizing a first-order differentiable risk function $P(w)$, where $P(w)$ is ν-strongly convex with δ-Lipschitz gradients according to (12.12a)–(12.12b). If the step-size sequence $\mu(n)$ satisfies either (12.66a) or (12.66b), then w_n converges to the global minimizer, $w^\star$. In particular, when the step-size sequence is chosen as $\mu(n) = \tau/(n+1)$, the convergence rate is on the order of*

$$\|\widetilde{w}_n\|^2 \leq O(1/n^{2\nu\tau}) \tag{12.68a}$$

$$P(w_n) - P(w^\star) \leq O(1/n^{2\nu\tau}) \tag{12.68b}$$

for large enough n.

Proof: The argument that led to (12.48) will similarly lead to

$$\|\widetilde{w}_n\|^2 \leq \lambda(n)\, \|\widetilde{w}_{n-1}\|^2 \tag{12.69}$$

where now $\lambda(n) = 1 - 2\nu\mu(n) + \delta^2\mu^2(n)$. We split $2\nu\mu(n)$ into the sum of two factors and write

$$\lambda(n) = 1 - \nu\mu(n) - \nu\mu(n) + \delta^2\mu^2(n) \tag{12.70}$$

Now, since $\mu(n) \to 0$ under (12.66a) or (12.66b), we conclude that for large enough $n > n_o$, the value of $\mu^2(n)$ will be smaller than $\mu(n)$. Therefore, a large enough time index, n_o, exists such that the following two conditions are satisfied:

$$\nu\mu(n) \geq \delta^2\mu^2(n), \qquad 0 < 1 - \nu\mu(n) \leq 1, \qquad n > n_o \tag{12.71}$$

It follows that

$$\lambda(n) \;\leq\; 1 - \nu\mu(n), \qquad n > n_o \tag{12.72}$$

and, hence,

$$\|\widetilde{w}_n\|^2 \;\leq\; (1 - \nu\mu(n))\, \|\widetilde{w}_{n-1}\|^2, \qquad n > n_o \tag{12.73}$$

Iterating over n we can write (assuming a finite n_o exists for which $\|\widetilde{w}_{n_o}\| \neq 0$, otherwise the algorithm would have converged):

$$\lim_{n\to\infty} \left(\frac{\|\widetilde{w}_n\|^2}{\|\widetilde{w}_{n_o}\|^2}\right) \leq \prod_{n=n_o+1}^{\infty} (1-\nu\mu(n)) \tag{12.74}$$

or, equivalently,

$$\lim_{n\to\infty} \ln\left(\frac{\|\widetilde{w}_n\|^2}{\|\widetilde{w}_{n_o}\|^2}\right) \leq \sum_{n=n_o+1}^{\infty} \ln(1-\nu\mu(n)) \tag{12.75}$$

Now, using the following property for the natural logarithm function:

$$\ln(1-y) \;\leq\; -y, \qquad \text{for all } 0 \leq y < 1 \tag{12.76}$$

and letting $y = \nu\mu(n)$, we have that

$$\ln(1-\nu\mu(n)) \;\leq\; -\nu\mu(n), \qquad n > n_o \tag{12.77}$$

so that

$$\sum_{n=n_o+1}^{\infty} \ln(1-\nu\mu(n)) \;\leq\; -\sum_{n=n_o+1}^{\infty} \nu\mu(n) \;=\; -\nu\left(\sum_{n=n_o+1}^{\infty} \mu(n)\right) \;=\; -\infty \tag{12.78}$$

since the step-size series is assumed to be divergent under (12.66a) or (12.66b). We conclude that

$$\lim_{n\to\infty} \ln\left(\frac{\|\widetilde{w}_n\|^2}{\|\widetilde{w}_{n_o}\|^2}\right) = -\infty \tag{12.79}$$

so that $\widetilde{w}_n \to 0$ as $n \to \infty$.

We next examine the rate at which this convergence occurs for step-size sequences of the form $\mu(n) = \tau/(n+1)$. Note first that these sequences satisfy the following two conditions

$$\sum_{n=0}^{\infty} \mu(n) = \infty, \qquad \sum_{n=0}^{\infty} \mu^2(n) \;=\; \tau^2\left(\sum_{n=1}^{\infty} \frac{1}{n^2}\right) \;=\; \beta\tau^2 < \infty \tag{12.80}$$

for $\beta = \pi^2/6$. Again, since $\mu(n) \to 0$ and $\mu^2(n)$ decays faster than $\mu(n)$, we know that for some large enough $n > n_1$, it will hold that

$$2\nu\mu(n) \geq \delta^2\mu^2(n) \tag{12.81}$$

and, hence,

$$0 < \lambda(n) \leq 1, \qquad n > n_1 \tag{12.82}$$

We can now repeat the same steps up to (12.79) using $y = 2\nu\mu(n) - \delta^2\mu^2(n)$ to conclude that

$$
\begin{aligned}
\ln\left(\frac{\|\widetilde{w}_n\|^2}{\|\widetilde{w}_{n_1}\|^2}\right) &\le \sum_{m=n_1+1}^{n} \ln\left(1-2\nu\mu(m)+\delta^2\mu^2(m)\right) \\
&\le -\sum_{m=n_1+1}^{n} \left(2\nu\mu(m)-\delta^2\mu^2(m)\right) \\
&= -2\nu\left(\sum_{m=n_1+1}^{n}\mu(m)\right)+\delta^2\left(\sum_{m=n_1+1}^{n}\mu^2(m)\right) \\
&\le -2\nu\left(\sum_{m=n_1+1}^{n}\mu(m)\right)+\beta\tau^2\delta^2 \\
&= -2\nu\tau\left(\sum_{m=n_1+2}^{n+1}\frac{1}{m}\right)+\beta\tau^2\delta^2 \\
&\overset{(a)}{\le} -2\nu\tau\left(\int_{n_1+2}^{n+2}\frac{1}{x}dx\right)+\beta\tau^2\delta^2 \\
&= 2\nu\tau\ln\left(\frac{n_1+2}{n+2}\right)+\beta\tau^2\delta^2 \\
&= \ln\left(\frac{n_1+2}{n+2}\right)^{2\nu\tau}+\beta\tau^2\delta^2
\end{aligned}
\tag{12.83}
$$

where in step (a) we used the following integral bound, which reflects the fact that the area under the curve $f(x)=1/x$ over the interval $x\in[n_1+2,n+2]$ is upper-bounded by the sum of the areas of the rectangles shown in Fig. 12.3:

$$
\int_{n_1+2}^{n+2}\frac{1}{x}dx \le \sum_{m=n_1+2}^{n+1}\frac{1}{m}
\tag{12.84}
$$

We conclude from (12.83) that

$$
\begin{aligned}
\|\widetilde{w}_n\|^2 &\le \left\{e^{\left(\ln\left(\frac{n_1+2}{n+2}\right)^{2\nu\tau}+\beta\tau^2\delta^2\right)}\right\}\|\widetilde{w}_{n_1}\|^2, \quad i>i_1 \\
&= e^{\beta\tau^2\delta^2}\|\widetilde{w}_{n_1}\|^2\left(\frac{n_1+2}{n+2}\right)^{2\nu\tau} \\
&= O(1/n^{2\nu\tau})
\end{aligned}
\tag{12.85}
$$

as claimed. Result (12.68b) follows by noting from (12.50) that

$$
0\le P(w_n)-P(w^\star)\le\frac{\delta}{2}\|\widetilde{w}_n\|^2
\tag{12.86}
$$

■

Convexity versus strong convexity

The statement of Theorem 12.2 assumes *strongly convex* risks $P(w)$ with δ-Lipschitz gradients. In Prob. 12.15 we relax these conditions and limit $P(w)$ to being convex (as opposed to strongly convex) and Lipschitz as opposed to gradient-Lipschitz, i.e.,

$$
\|P(w_1)-P(w_2)\|\le\delta\|w_1-w_2\|, \quad \forall\, w_1,w_2\in\text{dom}(P)
\tag{12.87}
$$

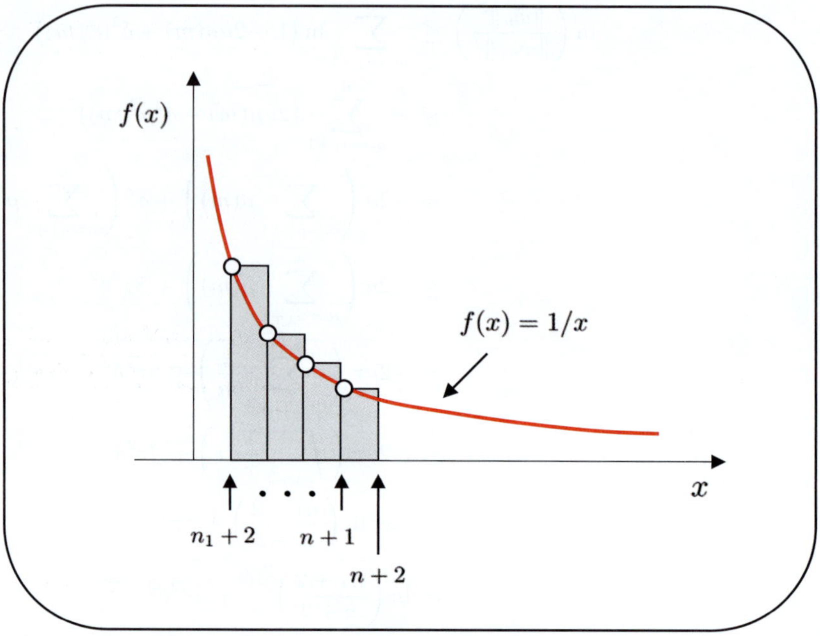

Figure 12.3 The area under the curve $f(x) = 1/x$ over the interval $x \in [n_1 + 2, n + 2]$ is upper-bounded by the sum of the areas of the rectangles shown in the figure.

We know from property (10.41) that the condition of a Lipschitz function translates into bounded gradient vectors, so that we are in effect requiring $\|\nabla_w P(w)\| \leq \delta$. Assume we run the gradient-descent recursion (12.65) for N iterations using a decaying step-size sequence of the form $\mu(n) = c/\sqrt{n+1}$ for some positive constant c, and let w^{best} denote the iterate that results in the smallest risk value, namely,

$$w^{\text{best}} \stackrel{\Delta}{=} \underset{0 \leq n \leq N-1}{\text{argmin}} \; P(w_n) \tag{12.88}$$

Then, we show in Prob. 12.15 that

$$P(w^{\text{best}}) - P(w^\star) = O\Big(\ln(N)/N\Big) \tag{12.89}$$

Example 12.12 **(Steepest-descent algorithm)** We motivate another choice for the step-size sequence $\mu(n)$ by seeking the steepest-descent direction along which the update of w_{n-1} should be performed. We use the same reasoning from Example 6.13, which dealt with the line search method.

Starting from an iterate w_{n-1}, our objective is to determine a small adjustment to it, say, $w_n = w_{n-1} + \delta w$, by solving

$$\delta w^o = \underset{\delta w \in \mathbb{R}^M}{\text{argmin}} \Big\{ P(w_{n-1} + \delta w) \Big\}, \quad \text{subject to } \frac{1}{2}\|\delta w\|^2 \leq \epsilon \tag{12.90}$$

We introduce a Lagrange multiplier $\lambda \geq 0$ and consider the unconstrained formulation

$$\delta w^o = \underset{\delta w \in \mathbb{R}^M}{\text{argmin}} \left\{ P(w_{n-1} + \delta w) \;+\; \lambda\Big(\frac{1}{2}\|\delta w\|^2 - \epsilon\Big) \right\} \tag{12.91}$$

To solve the problem, we introduce the *first-order* Taylor series expansion:

$$P(w_n) \approx P(w_{n-1}) + \nabla_w P(w_{n-1})\delta w \tag{12.92}$$

so that the cost appearing in (12.91) is approximated by

$$\text{cost} \approx P(w_{n-1}) + \nabla_w P(w_{n-1})\delta w + \lambda\Big(\frac{1}{2}\|\delta w\|^2 - \epsilon\Big) \tag{12.93}$$

To minimize the right-hand side over δw, and to find λ, we repeat the argument from Example 6.13 to arrive at the same conclusion:

$$w_n = w_{n-1} - \underbrace{\frac{\sqrt{2\epsilon}}{\|\nabla_{w^\mathsf{T}} P(w_{n-1})\|}}_{\triangleq\, \mu(n)} \nabla_{w^\mathsf{T}} P(w_{n-1}) \tag{12.94}$$

The term multiplying $\nabla_{w^\mathsf{T}} P(w_{n-1})$ plays the role of an iteration-dependent step size. In this case, the step size is chosen to result in the "largest" descent possible per iteration.

Example 12.13 **(Comparing constant and vanishing step sizes)** We return to the logistic algorithm (12.31) and simulate its performance under both constant and vanishing step sizes. Figure 12.4 plots a learning curve for the algorithm using parameters

$$\rho = 2, M = 10, N = 200, \quad \mu = 0.001 \tag{12.95}$$

For this simulation, the data $\{\gamma(m), h_m\}$ are generated randomly as follows. First, a random parameter model $w^a \in \mathbb{R}^{10}$ is selected, and a random collection of feature vectors $\{h_m\}$ are generated, say, with zero-mean and unit-variance Gaussian entries. Then, for each h_m, the label $\gamma(m)$ is set to either $+1$ or -1 according to the following construction:

$$\gamma(m) = +1 \text{ if } \left(\frac{1}{1 + e^{-h_m^\mathsf{T} w^a}}\right) \geq 0.5; \text{ otherwise } \gamma(m) = -1 \tag{12.96}$$

We will explain in expression (59.5a) that construction (12.96) amounts to generating data $\{\gamma(m), h_m\}$ that satisfy a logistic probability model. The gradient-descent recursion (12.30) is run for 2000 iterations on the data $\{\gamma(m), h_m\}$. The resulting weight iterate, denoted by $w^\star$, is shown in the bottom plot of the figure and the value of the risk function at this weight iterate is found to be

$$P(w^\star) \approx 0.6732 \tag{12.97}$$

The two plots in the top row display the learning curve $P(w_n)$ relative to the minimum value $P(w^\star)$, both in linear scale (on the left) and in normalized logarithmic scale on the right (according to construction (11.65)). The plot on the right in the top row reveals the linear convergence of $P(w_n)$ toward $P(w^\star)$ under constant step sizes, as anticipated by result (12.43b).

Figure 12.5 repeats the simulation using the same logistic data $\{\gamma(m), h_m\}$, albeit with a decaying step-size sequence of the form:

$$\mu(n) = \tau/(n+1), \quad \tau = 0.1 \tag{12.98}$$

The gradient-descent recursion (12.65) is now repeated for 4000 iterations with μ replaced by $\mu(n)$, and the resulting learning curve is compared against the curve generated under the constant step-size regime from the previous simulation. The plot on the left

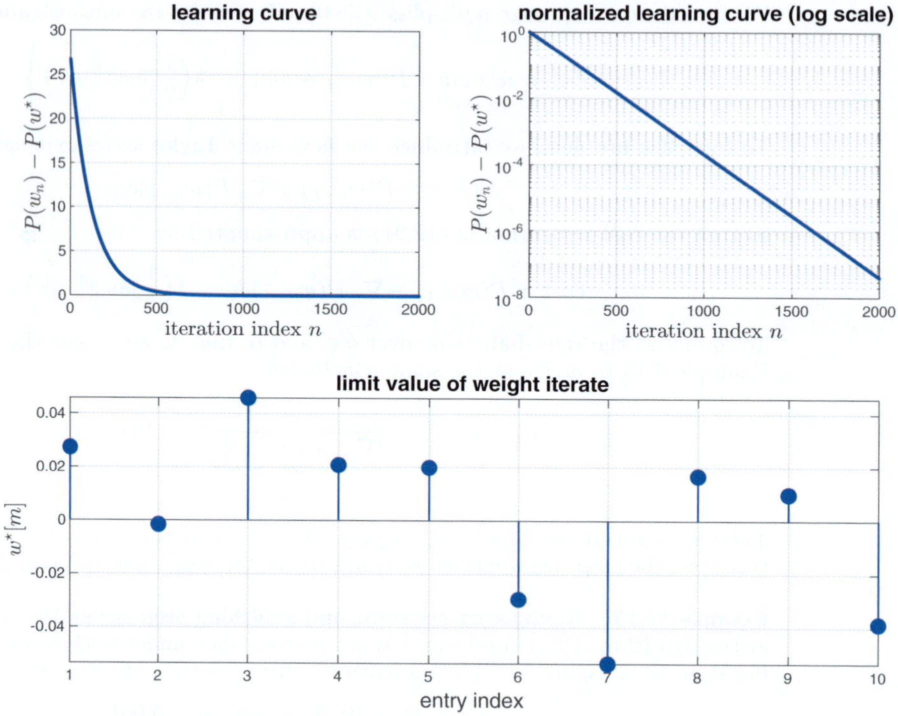

Figure 12.4 (*Top*) Learning curves $P(w_n)$ relative to the minimum risk value $P(w^\star)$ in linear scale (on the left) and in normalized logarithmic scale (on the right). This latter plot confirms the linear convergence of the risk value toward $P(w^\star)$. (*Bottom*) Limiting value of the weight iterate w_n, which tends to the minimizer $w^\star$ according to result (12.43a).

shows the learning curves in normalized logarithmic scale; it is clear that the convergence rate under decaying step sizes is much slower (it starts converging faster but ultimately becomes slower). The plot on the right illustrates this effect; it shows the limiting value $w^\star$ that was determined under constant step-size learning in Fig. 12.4 after 2000 iterations along with the weight iterate that is obtained under the decaying step size after 4000 iterations. It is clear that convergence has not been attained yet in the latter case, and many more iterations would be needed; this is because $\mu(n)$ becomes vanishingly small as n increases.

12.4.2 Backtracking Line Search

There are other methods to select the step-size sequence $\mu(n)$, besides (12.66a) or (12.66b). One method is the *backtracking line search* technique. Recall that the intent is to move from w_{n-1} to w_n in a manner that reduces the risk function at $w_{n-1} + \delta w$; it is also desirable to take "larger" steps when possible. Motivated by these considerations, at every iteration n, the backtracking method runs a

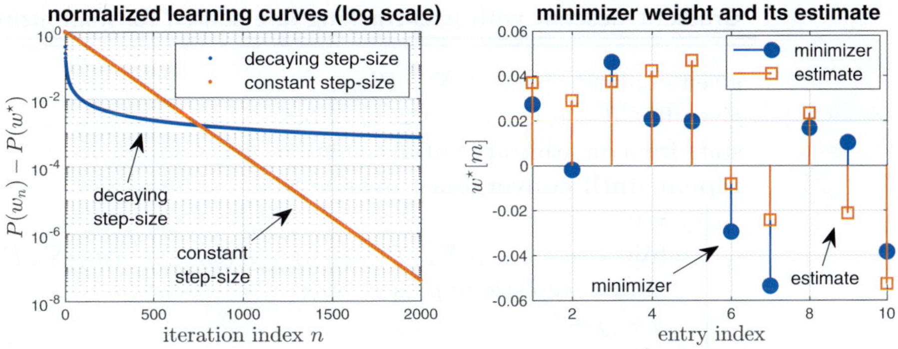

Figure 12.5 (*Left*) Learning curves $P(w_n)$ relative to the minimum risk value $P(w^\star)$ in normalized logarithmic scale for both cases of constant and decaying step sizes. (*Right*) After 4000 iterations, the weight iterate w_n in the decaying step-size implementation has not converged yet.

separate search to select $\mu(n)$ for that iteration. It starts from some large initial value and repeatedly shrinks it until a convenient value is found. The procedure is motivated as follows.

Starting from the iterate w_{n-1}, we introduce a first-order Taylor series approximation for the risk function around w_{n-1}:

$$P(w) \approx P(w_{n-1}) + \nabla_w P(w_{n-1})(w - w_{n-1}) \tag{12.99}$$

If we take a gradient-descent step from w_{n-1} to w_n with some generic step-size value μ, i.e.,

$$w_n = w_{n-1} - \mu \nabla_{w^\mathsf{T}} P(w_{n-1}) \tag{12.100}$$

then, substituting into (12.99), the new risk value is approximately

$$P(w_n) \approx P(w_{n-1}) - \mu \left\| \nabla_w P(w_{n-1}) \right\|^2 \tag{12.101}$$

The backtracking line search method selects μ to ensure that the decrease in the risk value is at least a fraction of the amount suggested above, say,

$$P\big(w_{n-1} - \mu \nabla_{w^\mathsf{T}} P(w_{n-1})\big) \;\leq\; P(w_{n-1}) - \alpha\mu \|\nabla_w P(w_{n-1})\|^2 \tag{12.102}$$

for some $0 < \alpha \leq \frac{1}{2}$. This is achieved as shown in listing (12.103). Typical values for the parameters are $\beta = 0.2$, $\alpha = 0.01$, and $\mu_0 = 1$.

Gradient-descent with backtracking line search for minimizing $P(w)$.

given gradient operator, $\nabla_{w^{\mathsf{T}}} P(w)$;
given $0 < \beta < 1$, $0 < \alpha < 1/2$, $\mu_0 > 0$;
start from an arbitrary initial condition, w_{-1};
repeat until convergence over $n \geq 0$:
  $j = 0$;
  while $P(w_{n-1} - \mu_j \nabla_{w^{\mathsf{T}}} P(w_{n-1})) > P(w_{n-1}) - \alpha\mu_j \|\nabla_w P(w_{n-1})\|^2$:
    shrink step size to $\mu_{j+1} = \beta\mu_j$;
    $j \leftarrow j + 1$;
  end
  set $\mu(n) = \mu_j$;
  $w_n = w_{n-1} - \mu(n)\nabla_{w^{\mathsf{T}}} P(w_{n-1})$;
end
return $w^\star \leftarrow w_n$.

(12.103)

Once $\mu(n)$ is selected at step n (say, $\mu(n) = \mu$ for some value μ), the risk function will satisfy

$$P(w_n) \leq P(w_{n-1}) - \alpha\mu \|\nabla_w P(w_{n-1})\|^2 \tag{12.104}$$

This result is known as the *Armijo condition*, which is usually stated in the following more abstract form. Consider an update step δw and introduce the function:

$$\phi(\mu) \stackrel{\Delta}{=} P(w + \mu\, \delta w) \tag{12.105}$$

where μ is some step-size parameter that we wish to determine in order to update w to $w + \mu\, \delta w$. The Armijo condition chooses μ to satisfy:

(**Armijo condition**)
$$\phi(\mu) \leq \phi(0) + \alpha\mu\, \phi'(0), \quad \text{for some } 0 < \alpha < 1/2 \tag{12.106}$$

where $\phi'(\mu)$ denotes the derivative of $\phi(\mu)$ relative to μ. It is easy to verify that this condition reduces to (12.104) for the case of the gradient-descent algorithm where $\delta w = -\nabla_{w^{\mathsf{T}}} P(w)$. The step-sizes $\mu(n)$ that result from the backtracking procedure (12.103) satisfy the above Armijo condition at every step, n. We show next that the Armijo condition is sufficient to ensure convergence of the gradient-descent algorithm.

THEOREM 12.3. (Convergence under Armijo condition) *Consider the gradient-descent recursion (12.65) for minimizing a first-order differentiable risk function $P(w)$, where $P(w)$ is ν-strongly convex with δ-Lipschitz gradients according to (12.12a)–(12.12b). If the step-size sequence $\mu(n)$ is chosen to satisfy the Armijo condition (12.104) at every step, then the excess risk converges exponentially fast, namely,*

$$P(w_n) - P(w^\star) \leq O(\lambda^n) \quad \textit{(for strongly convex } P(w)) \tag{12.107}$$

for some $\lambda \in [0, 1)$. If $P(w)$ is only convex, then

$$P(w_n) - P(w^\star) \leq O(1/n) \quad \textit{(for convex } P(w)) \tag{12.108}$$

Proof: First, we call upon property (10.13) for the δ-Lipschitz gradient of $P(w)$, which allows us to write (using $z \leftarrow w_n$, $z_1 \leftarrow w_{n-1}$, and $z - z_1 \leftarrow -\mu_j \nabla_{w^\mathsf{T}} P(w_{n-1})$):

$$P(w_{n-1} - \mu_j \nabla_{w^\mathsf{T}} P(w_{n-1})) \leq P(w_{n-1}) - \mu_j \left(1 - \frac{\delta \mu_j}{2}\right) \|\nabla_w P(w_{n-1}\|^2 \tag{12.109}$$

According to the backtracking construction (12.103), the search for the step-size parameter will stop when

$$P(w_{n-1} - \mu_j \nabla_{w^\mathsf{T}} P(w_{n-1})) \leq P(w_{n-1}) - \alpha \mu_j \|\nabla_w P(w_{n-1})\|^2 \tag{12.110}$$

Combining with (12.109), we find that for the search to stop it is sufficient to require

$$\mu_j \left(1 - \frac{\delta \mu_j}{2}\right) > \alpha \mu_j \tag{12.111}$$

Since, by choice, $\alpha < 1/2$, the search is guaranteed to stop when

$$\mu_j \left(1 - \frac{\delta \mu_j}{2}\right) > \frac{1}{2} \mu_j \iff \mu_j < 1/\delta \tag{12.112}$$

This argument shows that the exit condition for the backtracking construction will be satisfied whenever $\mu_j < 1/\delta$. Using this condition in (12.109), and noting that the argument of $P(\cdot)$ on the left-hand side becomes w_n at the exit point, we find that at that point:

$$P(w_n) \leq P(w_{n-1}) - \frac{\mu_j}{2} \|\nabla_w P(w_{n-1}\|^2 \tag{12.113}$$

On the other hand, using the ν-strong convexity of $P(w)$, we apply the upper bound from property (8.29) to deduce that:

$$P(w^\star) \geq P(w_{n-1}) - \frac{1}{2\nu} \|\nabla_w P(w_{n-1})\|^2 \tag{12.114}$$

Subtracting $P(w^\star)$ from both sides of inequality (12.113) and using (12.114), we obtain

$$P(w_n) - P(w^\star) \leq (1 - \mu_j \nu)\Big(P(w_{n-1}) - P(w^\star)\Big) \tag{12.115}$$

Now recall that we launch the backtracking search from the initial condition $\mu_0 = 1$. Two scenarios are possible: either $1/\delta > 1$ or $1/\delta \leq 1$. In the first case, the backtracking search will stop right away at $\mu_j = \mu_0 = 1$ since the condition $\mu_j < 1/\delta$ will be met. In the second case, the step size will be scaled down repeatedly by β until the first time it goes below $1/\delta$, at which point the search stops. In this case, the final μ_j will satisfy $\mu_j \geq \beta/\delta$. Therefore, it holds that

$$\mu_j \geq \min\{1, \beta/\delta\} \tag{12.116}$$

Substituting into (12.115) we obtain

$$P(w_n) - P(w^\star) \leq \underbrace{\Big(1 - \min\{\nu, \nu\beta/\delta\}\Big)}_{\triangleq\, \lambda} \Big(P(w_{n-1}) - P(w^\star)\Big) \tag{12.117}$$

from which we deduce exponential convergence of $P(w_n)$ to $P(w^\star)$. For convex risk functions $P(w)$, we can establish a conclusion similar to (12.64b) by following an argument similar to the derivation of (11.71). This result is established in Prob. 12.16.

■

12.5 COORDINATE-DESCENT METHOD

The gradient-descent algorithm described in the earlier sections minimizes the risk function $P(w)$ over the *entire* vector $w \in \mathbb{R}^M$. One alternative technique is the *coordinate-descent* approach, which optimizes $P(w)$ over a *single* entry of w at a time while keeping all other entries fixed. The individual entries of w are called *coordinates* and, hence, the designation "coordinate-descent."

12.5.1 Derivation of Algorithm

In its traditional form, the coordinate-descent technique writes $P(w)$ as an explicit function of the individual entries of $w = \text{col}\{w_m\}$ for $m = 1, 2, \ldots, M$:

$$P(w) = P(w_1, \ldots, w_m, \ldots, w_M) \tag{12.118}$$

and minimizes it over each argument separately. Listing (12.119) describes the algorithm when the minimization can be carried out in closed form. The weight iterate at iteration $n-1$ is denoted by w_{n-1} and its coordinates by $\{w_{n-1,m}\}$. At every iteration n, the algorithm cycles through the coordinates and updates each $w_{n-1,m}$ to $w_{n,m}$ by minimizing $P(w)$ over w_m while keeping all other coordinates of indices $m' \neq m$ fixed at their most *recent* values. Observe in particular that once the first coordinate w_1 is updated to $w_{n,1}$ in the first step, this new value is used as an argument in the second step that updates the second coordinate to $w_{n,2}$. The process continues in this manner by using the updated coordinates from the previous steps as arguments in subsequent steps.

Traditional coordinate-descent for minimizing $P(w)$.

let $w = \text{col}\{w_m\}$, $m = 1, 2, \ldots, M$;
start from an arbitrary initial condition $w_{-1} = \text{col}\{w_{m,-1}\}$.
repeat until convergence over $n \geq 0$:
$w_{n-1} = \text{col}\{w_{n-1,m}\}$ is available at start of iteration;
for each coordinate $m = 1, 2, \ldots, M$ compute:

$$\begin{aligned}
w_{n,1} &= \underset{w_1 \in \mathbb{R}}{\text{argmin}}\; P(\underline{w_1}, w_{n-1,2}, w_{n-1,3}, \ldots, w_{n-1,M}) \\
w_{n,2} &= \underset{w_2 \in \mathbb{R}}{\text{argmin}}\; P(w_{n,1}, \underline{w_2}, w_{n-1,3}, w_{n-1,4}, \ldots, w_{n-1,M}) \\
w_{n,3} &= \underset{w_3 \in \mathbb{R}}{\text{argmin}}\; P(w_{n,1}, w_{n,2}, \underline{w_3}, w_{n-1,4}, \ldots, w_{n-1,M}) \\
&\bullet \\
&\bullet \\
w_{n,M} &= \underset{w_M \in \mathbb{R}}{\text{argmin}}\; P(w_{n,1}, w_{n,2}, \ldots, w_{n,M-1}, \underline{w_M})
\end{aligned} \tag{12.119}$$

end
end
$w_n = \text{col}\{w_{n,m}\}_{m=1}^{M}$
return $w^\star \leftarrow w_n$.

The coordinate-descent procedure can be motivated as follows. Consider a convex risk $P(w)$ and let $w^\star$ denote a global minimizer so that $\nabla_w P(w^\star) = 0$. This also means that

$$\left.\frac{\partial P(w)}{\partial w_m}\right|_{w_m = w_m^\star} = 0 \tag{12.120}$$

so that $P(w)$ is minimized over each coordinate. Specifically, for any step λ and basis vector $e_m \in \mathbb{R}^M$, it will hold that

$$P(w^\star + \lambda e_m) \geq P(w^\star), \quad m = 1, 2, \ldots, M \tag{12.121}$$

which justifies searching for $w^\star$ by optimizing separately over the coordinates of w. Unfortunately, this property is lost when $P(w)$ is not differentiable – see Prob. 12.27. This fact highlights one of the weaknesses of the coordinate-descent method. We will revisit this issue in the next chapter and explain for what type of nondifferentiable risks the coordinate-descent construction will continue to work.

Example 12.14 **(Coordinate-descent for ℓ_2-regularized least-squares)** Consider the regularized least-squares problem:

$$w^\star \stackrel{\Delta}{=} \underset{w\in\mathbb{R}^M}{\text{argmin}} \left\{\rho\|w\|^2 + \frac{1}{N}\sum_{\ell=0}^{N-1}\Big(\gamma(\ell) - h_\ell^\mathsf{T} w\Big)^2\right\} \tag{12.122}$$

We are using the subscript ℓ to index the data points $\{\gamma(\ell), h_\ell\}$ to avoid confusion with the subscript m used to index the individual coordinates of w. We denote the individual entries of h_ℓ by $\text{col}\{h_{\ell,m}\}$ for $m = 1, 2, \ldots, M$. We also use the notation w_{-m} and $h_{\ell,-m}$ to refer to the vectors w and h_ℓ with their mth entries excluded. Then, as a function of w_m, the risk can be written in the form:

$$\begin{aligned}
P(w) &= \rho\|w\|^2 + \frac{1}{N}\sum_{\ell=0}^{N-1}\Big(\gamma(\ell) - h_{\ell,-m}^\mathsf{T} w_{-m} - h_{\ell,m}w_m\Big)^2 \\
&\stackrel{(a)}{=} \rho\, w_m^2 + \underbrace{\left(\frac{1}{N}\sum_{\ell=0}^{N-1} h_{\ell,m}^2\right)}_{\stackrel{\Delta}{=}\, a_m} w_m^2 - \\
&\qquad \underbrace{\frac{2}{N}\left(\sum_{\ell=0}^{N-1} h_{\ell,m}\Big(\gamma(\ell) - h_{\ell,-m}^\mathsf{T} w_{-m}\Big)\right)}_{\stackrel{\Delta}{=}\, 2c_m} w_m + \text{cte} \\
&\stackrel{(b)}{=} (\rho + a_m)w_m^2 - 2c_m w_m + \text{cte}
\end{aligned} \tag{12.123}$$

where terms independent of w_m are collected into the constant factor in step (a). In step (b), we introduced the scalars $a_m \geq 0$ and c_m for compactness of notation. Minimizing $P(w)$ over w_m we get

$$\widehat{w}_m = c_m/(\rho + a_m) \tag{12.124}$$

and arrive at listing (12.125).

Traditional coordinate-descent algorithm for solving (12.122).

given N data points $\{\gamma(\ell), h_\ell\}$, $\ell = 0, 1, \ldots, N-1$;
start from an arbitrary initial condition $w_{-1} = 0$.
repeat until convergence over $n \geq 0$:
  iterate is $w_{n-1} = \text{col}\{w_{n-1,m}\}_{m=1}^M$
  repeat for each coordinate $m = 1, 2, \ldots, M$:

$$a_m = \frac{1}{N}\sum_{\ell=0}^{N-1} h_{\ell,m}^2$$

$$c_m = \frac{1}{N}\sum_{\ell=0}^{N-1} h_{\ell,m}\Big(\gamma(\ell) - h_{\ell,-m}^\mathsf{T} w_{n-1,-m}\Big) \tag{12.125}$$

    $w_{n,m} = c_m/(\rho + a_m)$
    $w_{n-1,m} \leftarrow w_{n,m}$ (use updated coordinate in next step)
  end
end
$w_n = \text{col}\{w_{n,m}\}_{m=1}^M$
return $w^\star \leftarrow w_n$.

Observe that the expressions for a_m and c_m depend on all data points. Moreover, at each iteration n, all coordinates of w_n are updated. We discuss next some simplifications.

12.5.2 Randomized Implementation

In practice, the minimization of $P(w)$ over the coordinates $\{w_m\}$ is often difficult to solve in closed form. In these cases, it is customary to replace the minimization in (12.119) by a gradient-descent step of the form:

$$w_{n,m} = w_{n-1,m} - \mu \left.\frac{\partial P(w)}{\partial w_m}\right|_{w=w_{n-1}}, \quad m = 1, 2, \ldots, M \tag{12.126}$$

While implementation (12.119) cycles through *all* coordinates of w at each iteration n, there are popular variants that limit the update to a single coordinate per iteration. The coordinate may be selected in different ways, for example, uniformly at random or as the coordinate corresponding to the maximal absolute gradient value. Description (12.127) is the randomized version of coordinate-descent.

Randomized coordinate-descent for minimizing $P(w)$.

let $w = \text{col}\{w_m\}$, $m = 1, 2, \ldots, M$;
start from an arbitrary initial condition w_{-1}.
repeat until convergence over $n \geq 0$:
  $w_{n-1} = \text{col}\{w_{n-1,m}\}$ is available at start of iteration
  select an index m^o at random within $1 \leq m \leq M$ (12.127)
  update $w_{n,m^o} = w_{n-1,m^o} - \mu \left.\frac{\partial P(w)}{\partial w_{m^o}}\right|_{w=w_{n-1}}$
  keep $w_{n,m} = w_{n-1,m}$, for all $m \neq m^o$
end
$w_n = \text{col}\{w_{n,m}\}_{m=1}^{M}$
return $w^\star \leftarrow w_n$.

This implementation can be viewed as a variation of gradient descent. At every iteration n, we select some basis vector e_{m^o} at random with probability $1/M$ and use it to construct the scaling diagonal matrix $D_n = \text{diag}\{e_{m^o}\}$, with a single unit entry on its diagonal at the m^oth location. Note that D_n is a random matrix, and therefore we will write $\boldsymbol{D}_n$ using the boldface notation to highlight this fact. The selection of $\boldsymbol{D}_n$ at iteration n is performed independently of any other variables in the optimization problem. Then, the update generated by the algorithm can be written in vector form as follows:

$$\boldsymbol{w}_n = \boldsymbol{w}_{n-1} - \mu \boldsymbol{D}_n \nabla_{w^\mathsf{T}} P(\boldsymbol{w}_{n-1}) \tag{12.128}$$

where the variables $\{\boldsymbol{w}_n, \boldsymbol{w}_{n-1}\}$ are also *random* in view of the randomness in $\boldsymbol{D}_n$. In particular, observe that on average:

$$\mathbb{E}\,\boldsymbol{D}_n = \frac{1}{M}\sum_{m=1}^{M} \text{diag}\{e_m\} \;=\; \frac{1}{M} I_M \;=\; \mathbb{E}\,\boldsymbol{D}_n^2 \tag{12.129}$$

The following statement establishes the convergence of the randomized algorithm. In contrast to the earlier arguments in this chapter on the convergence of the gradient-descent implementation, we now need to take the randomness of the weight iterates into account.

THEOREM 12.4. (Convergence of randomized coordinate-descent) *Consider the randomized coordinate-descent algorithm (12.127) for minimizing a first-order differentiable risk function $P(w)$, where $P(w)$ is ν-strongly convex with δ-Lipschitz gradients according to (12.12a)–(12.12b). Introduce the error vector $\widetilde{\boldsymbol{w}}_n = w^\star - \boldsymbol{w}_n$, which measures the difference between the nth iterate and the global minimizer of $P(w)$. If the step size μ satisfies (i.e., is small enough)*

$$0 < \mu < 2\nu/\delta^2 \tag{12.130}$$

then the MSE, $\mathbb{E}\,\|\widetilde{\boldsymbol{w}}_n\|^2$, and the average excess risk converge exponentially fast in the sense that

$$\mathbb{E}\,\|\widetilde{\boldsymbol{w}}_n\|^2 \;\le\; \lambda\,\mathbb{E}\,\|\widetilde{\boldsymbol{w}}_{n-1}\|^2, \;\; n \ge 0 \tag{12.131a}$$

$$\mathbb{E}\,P(\boldsymbol{w}_n) - P(w^\star) \;\le\; \frac{\delta}{2}\lambda^{n+1}\|\widetilde{w}_{-1}\|^2 \;=\; O(\lambda^n) \tag{12.131b}$$

where

$$\lambda \;=\; 1 - \frac{2\mu\nu}{M} + \frac{\mu^2\delta^2}{M} \;\in [0,1) \tag{12.132}$$

Proof: We subtract $w^\star$ from both sides of (12.128) to get

$$\widetilde{\boldsymbol{w}}_n \;=\; \widetilde{\boldsymbol{w}}_{n-1} \;+\; \mu\,\boldsymbol{D}_n\,\nabla_{w^\mathsf{T}}\,P(\boldsymbol{w}_{n-1}) \tag{12.133}$$

We compute the squared Euclidean norms (or energies) of both sides and use the fact that $\nabla_{w^\mathsf{T}}\,P(w^\star) = 0$ to write

$$\|\widetilde{\boldsymbol{w}}_n\|^2 = \|\widetilde{\boldsymbol{w}}_{n-1}\|^2 + 2\mu\left(\nabla_{w^\mathsf{T}}\,P(\boldsymbol{w}_{n-1})\right)^\mathsf{T} \boldsymbol{D}_n \widetilde{\boldsymbol{w}}_{n-1} + \mu^2\,\|\nabla_{w^\mathsf{T}}\,P(\boldsymbol{w}_{n-1})\|^2_{\boldsymbol{D}_n^2} \tag{12.134}$$

where the notation $\|x\|_A^2$ stands for $x^\mathsf{T} A x$. Conditioning on $\widetilde{\boldsymbol{w}}_{n-1}$ and taking expectations of both sides gives

$$
\begin{aligned}
&\mathbb{E}\left(\|\widetilde{\boldsymbol{w}}_n\|^2 \,|\, \widetilde{\boldsymbol{w}}_{n-1}\right) \\
&= \|\widetilde{w}_{n-1}\|^2 + 2\mu\left(\nabla_{w^\mathsf{T}} P(w_{n-1})\right)^\mathsf{T} (\mathbb{E}\,\boldsymbol{D}_n)\widetilde{w}_{n-1} + \mu^2 \left\|\nabla_{w^\mathsf{T}} P(w_{n-1})\right\|^2_{\mathbb{E}\,\boldsymbol{D}_n^2} \\
&\overset{(12.129)}{=} \|\widetilde{w}_{n-1}\|^2 + \frac{2\mu}{M}\left(\nabla_{w^\mathsf{T}} P(w_{n-1})\right)^\mathsf{T} \widetilde{w}_{n-1} + \frac{\mu^2}{M}\left\|\nabla_{w^\mathsf{T}} P(w_{n-1})\right\|^2 \\
&\overset{(12.47)}{\leq} \|\widetilde{w}_{n-1}\|^2 - \frac{2\mu\nu}{M}\|\widetilde{w}_{n-1}\|^2 + \frac{\mu^2\delta^2}{M}\|\widetilde{w}_{n-1}\|^2
\end{aligned}
\tag{12.135}
$$

Taking expectations again to eliminate the conditioning over $\widetilde{\boldsymbol{w}}_{n-1}$ we arrive at

$$
\mathbb{E}\,\|\widetilde{\boldsymbol{w}}_n\|^2 \;\leq\; \left(1 - \frac{2\mu\nu}{M} + \frac{\mu^2\delta^2}{M}\right)\mathbb{E}\,\|\widetilde{\boldsymbol{w}}_{n-1}\|^2 \tag{12.136}
$$

Comparing with (12.48) we find that the recursion is in terms of the MSE. The structure of the above recursion is similar to (12.43a) and we arrive at the conclusions stated in the theorem.

■

We conclude that the weight iterate $\boldsymbol{w}_n$ converges in the MSE sense to $w^\star$. Referring to the earlier diagram from Fig. 3.11 on the convergence of random sequences, we conclude that this fact implies that $\boldsymbol{w}_n$ converges to $w^\star$ in probability. Moreover, comparing expression (12.132) for λ with (12.44) in the gradient-descent case, we find that

$$
\lambda \approx 1 - 2\mu\nu \qquad \text{(for gradient-descent)} \tag{12.137a}
$$

$$
\lambda \approx 1 - \frac{2\mu\nu}{M} \qquad \text{(for randomized coordinate-descent)} \tag{12.137b}
$$

which suggests that randomized coordinate descent converges at a slower pace. This is understandable because only one entry of the weight iterate is updated at every iteration. However, if we compare the performance of both algorithms by considering one iteration for gradient descent against M iterations for randomized coordinate-descent, then the decay of the squared weight error vectors will occur at comparable rates.

Remark 12.5. (Bound on step-size) If we follow the argument from Example 12.10 based on the risk function, we can relax the upper bound on μ in (12.130) to $\mu < 2/\delta$ and replace λ by

$$
\lambda \;=\; 1 - \frac{2\mu\nu}{M} + \frac{\mu^2\nu\delta}{M} \;\in [0,1) \tag{12.138}
$$

This fact is exploited in the proof of convergence for the Gauss–Southwell variant in the next section.

■

Example 12.15 **(Randomized coordinate-descent for regularized least-squares)** Consider the ℓ_2-regularized least-squares problem (12.122). In this case, we can determine the optimal coordinate $w^o_{n,m}$ in closed form at every iteration n. Using

$$
\partial P(w)/\partial w_m = 2(\rho + a_m)w_m - 2c_m, \quad m = 1,2,\ldots,M \tag{12.139}
$$

we find that the corresponding randomized coordinate-descent implementation is given by listing (12.140).

Randomized coordinate-descent for solving (12.122).

given N data points $\{\gamma(\ell), h_\ell\}$, $\ell = 0, 1, \ldots, N-1$;
start from an arbitrary initial condition $w_{-1} = 0$.
repeat until convergence over $n \geq 0$:
  iterate is $w_{n-1} = \text{col}\{w_{n-1,m}\}_{m=1}^{M}$
  select an index m^o at random within $1 \leq m \leq M$

$$a_{m^o} = \frac{1}{N}\sum_{\ell=0}^{N-1} h_{\ell,m^o}^2 \tag{12.140}$$

$$c_{m^o} = \frac{1}{N}\sum_{\ell=0}^{N-1} h_{\ell,m^o}\Big(\gamma(\ell) - h_{\ell,-m^o}^{\mathsf{T}} w_{n-1,-m^o}\Big)$$

  update $w_{n,m^o} = c_{m^o}/(\rho + a_{m^o})$
  keep $w_{n,m} = w_{n-1,m}$, for all $m \neq m^o$
end
$w_n = \text{col}\{w_{n,m}\}_{m=1}^{M}$
return $w^\star \leftarrow w_n$.

We can describe the algorithm in vector form by introducing the vector and matrix quantities:

$$\gamma_N \triangleq \begin{bmatrix} \gamma(0) \\ \gamma(1) \\ \vdots \\ \gamma(N-1) \end{bmatrix}, \quad H_N = \begin{bmatrix} h_0^{\mathsf{T}} \\ h_1^{\mathsf{T}} \\ \vdots \\ h_{N-1}^{\mathsf{T}} \end{bmatrix} \tag{12.141}$$

where γ_N is $N \times 1$ and H_N is $N \times M$. We let x_{m^o} denote the column of index m^o in H_N and write $H_{N,-m^o}$ to refer to the data matrix H_N with its m^oth column excluded. That is, $H_{N,-m^o}$ has dimensions $N \times (M-1)$. Then, it can be verified that (see Prob. 12.30):

$$w_{n,m^o} = \frac{1}{\rho + \frac{1}{N}\|x_{m^o}\|^2} \times \frac{1}{N} x_{m^o}^{\mathsf{T}}\Big(\gamma_N - H_{N,-m^o}\, w_{n-1,-m^o}\Big) \tag{12.142}$$

We illustrate the operation of the algorithm by generating a random model $w^o \in \mathbb{R}^{10}$ with $M = 10$, and a collection of $N = 200$ random feature vectors $\{h_n\}$. The entries of w^o are selected randomly from a Gaussian distribution with mean zero and unit variance; likewise for the entries of the feature vectors. We also generate noisy target signals:

$$\gamma(n) = h_n^{\mathsf{T}} w^o + v(n) \tag{12.143}$$

where $v(n)$ are realizations of zero-mean Gaussian noise with variance $\sigma_v^2 = 0.01$. We set the step-size parameter to $\mu = 0.01$ and the regularization parameter to $\rho = 2/N$. If we differentiate the risk function (12.122) relative to w, it is straightforward to determine that the minimizer is given by:

$$w^\star = (\rho N I_M + H_N^{\mathsf{T}} H_N)^{-1} H_N^{\mathsf{T}} \gamma_N \tag{12.144}$$

Substituting $w^\star$ into the risk function we find the minimal risk value, $P(w^\star) = 0.0638$. The learning curves in Fig. 12.6 are plotted relative to this value; the curves in the right plot in the first row are normalized by the maximum value of $P(w_n) - P(w^\star)$ so that they start from the value 1. The learning curves for the coordinate-descent implementation are downsampled by a factor $M = 10$ since, on average, it takes 10 iterations for all entries of the weight vector to be updated (whereas, under the gradient-descent implementation, all entries are updated at every single iteration). The downsampling allows for a fair comparison of the convergence rates of the two methods. It is observed from the results in the figure that both methods are able to estimate $w^\star$ and that their learning curves practically coincide with each other.

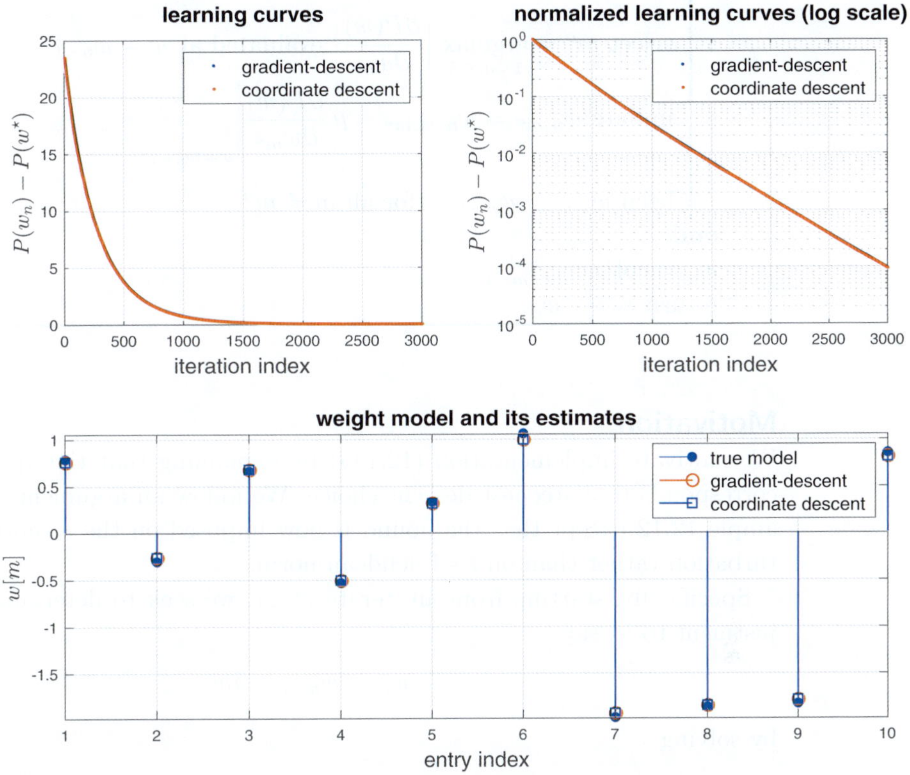

Figure 12.6 (*Top*) Learning curves $P(w_n)$ relative to the minimum risk value $P(w^\star)$ in regular and normalized logarithmic scales for gradient-descent and randomized coordinate-descent; the learning curve for the latter is downsampled and plotted every $M = 10$ iterations. (*Bottom*) The original and estimated parameter models.

12.5.3 Gauss–Southwell Implementation

We consider next a coordinate-descent implementation where at every iteration n, the coordinate corresponding to the maximal absolute gradient is updated. This variant is known as the Gauss–Southwell (GS) rule.

Gauss–Southwell coordinate-descent for minimizing $P(w)$.

let $w = \text{col}\{w_m\}$, $m = 1, 2, \ldots, M$;
start from an arbitrary initial condition w_{-1}.
repeat until convergence over $n \geq 0$**:**

$w_{n-1} = \text{col}\{w_{n-1,m}\}$ is available at start of iteration

$$\text{select } m^o = \underset{1\leq m\leq M}{\text{argmax}} \left|\frac{\partial P(w)}{\partial w_m}\right| \text{ evaluated at } w = w_{n-1} \tag{12.145}$$

$$\text{update } w_{n,m^o} = w_{n-1,m^o} - \mu \left.\frac{\partial P(w)}{\partial w_{m^o}}\right|_{w=w_{n-1}}$$

keep $w_{n,m} = w_{n-1,m}$, for all $m \neq m^o$

end

$w_n = \text{col}\{w_{n,m}\}_{m=1}^{M}$
return $w^\star \leftarrow w_n$.

Motivation

We motivate implementation (12.145) by explaining that the update direction corresponds to a steepest-descent choice. We follow an argument similar to Example 12.12 except that the bound is now imposed on the ℓ_1-norm of the perturbation rather than on its Euclidean norm.

Specifically, starting from an iterate w_{n-1}, we seek to determine a small adjustment to it, say,

$$w_n = w_{n-1} + \delta w \tag{12.146}$$

by solving

$$\delta w^o = \underset{\delta w\in \mathbb{R}^M}{\text{argmin}} \Big\{P(w_{n-1} + \delta w)\Big\}, \quad \text{subject to } \|\delta w\|_1 \leq \mu' \tag{12.147}$$

for some $\mu' > 0$. We introduce the *first-order* Taylor series expansion:

$$P(w_n) \approx P(w_{n-1}) + \nabla_w P(w_{n-1})\delta w \tag{12.148}$$

and approximate the problem by solving

$$y^o = \underset{y\in \mathbb{R}^M}{\text{argmin}} \Big\{\mu' \,\nabla_w P(w_{n-1}) y\Big\}, \quad \text{subject to } \|y\|_1 \leq 1 \tag{12.149}$$

where we introduced the change of variables $y = \delta w/\mu'$. We recall from the result of part (c) in Prob. 1.26 that for any vectors $\{x, y\}$ of matching dimensions, the ℓ_1- and ℓ_∞-norms satisfy:

$$\|x\|_\infty = \sup_{\|y\|_1\leq 1} \Big\{x^\mathsf{T} y\Big\} \tag{12.150}$$

Problem (12.149) has the same form (if we negate the argument and replace max by min):

$$y^o = \underset{\|y\|_1 \leq 1}{\text{argmax}} \left\{ -\mu' \nabla_w P(w_{n-1}) y \right\} \tag{12.151}$$

It follows that the optimal value of (12.151) is equal to $\mu' \|\nabla_w P(w_{n-1})\|_\infty$, which is the maximum absolute entry of the gradient vector scaled by μ'. Let m^o denote the index of this maximum absolute entry. Then, the maximal value of (12.151) is attained if we select

$$y^o = -\, e_{m^o} \,\text{sign}\left(\frac{\partial P(w_{n-1})}{\partial w_{m^o}} \right) \tag{12.152}$$

Taking a step along this direction leads to the update

$$w_{n,m^o} = w_{n-1,m^o} - \mu' \,\text{sign}\left(\frac{\partial P(w_{n-1})}{\partial w_{m^o}} \right) \tag{12.153}$$

which updates w_{n-1} along the descent direction determined by the maximal absolute entry of the gradient vector, in a manner "similar" to (12.145).

Convergence

The GS implementation (12.145) can again be viewed as a variation of gradient-descent. At every iteration n, we construct the diagonal matrix $D_n = \text{diag}\{e_{m^o}\}$, where m^o is the index of the entry in the gradient vector at w_{n-1} with the largest absolute value. The matrix D_n is not random anymore, as was the case in the randomized coordinate-descent implementation. Instead, its value depends on w_{n-1} and the update can be written in vector form as follows:

$$w_n = w_{n-1} + \mu D_n \nabla_{w^\mathsf{T}} P(w_{n-1}) \tag{12.154}$$

In the analysis for the randomized algorithm we were able to remove the effect of the matrix $\boldsymbol{D}_n$ through expectation. This is not possible here because D_n is now deterministic and dependent on w_{n-1}. Nevertheless, a similar convergence analysis is applicable with one minor adjustment. We continue to assume that the risk function $P(w)$ is ν-strongly convex as in (12.12a), but require $P(w)$ to have δ-Lipschitz gradients relative to each coordinate, namely,

(1) (**Strong convexity**). $P(w)$ is ν-strongly convex and first-order differentiable:

$$P(w_2) \geq P(w_1) + \nabla_{w^\mathsf{T}} P(w_1)(w_2 - w_1) + \frac{\nu}{2}\|w_2 - w_1\|^2 \tag{12.155a}$$

for every $w_1, w_2 \in \text{dom}(P)$ and some $\nu > 0$.

(2) (**δ-Lipschitz gradients relative to each coordinate**). The gradient vectors of $P(w)$ are δ-Lipschitz relative to each coordinate, meaning that:

$$\left| \frac{\partial}{\partial w_m} P(w + \alpha e_m) - \frac{\partial}{\partial w_m} P(w) \right| \leq \delta |\alpha| \tag{12.155b}$$

for any $w \in \text{dom}(P)$, $\alpha \in \mathbb{R}$, and where e_m denotes the mth basis vector in $\mathbb{R}^M$.

Theorem 12.5. (Convergence of Gauss–Southwell coordinate-descent) *Consider the GS coordinate-descent algorithm (12.145) for minimizing a first-order differentiable risk function $P(w)$, where $P(w)$ is ν-strongly convex with δ-Lipschitz gradients relative to each coordinate according to (12.155a)–(12.155b). Introduce the error vector $\widetilde{w}_n = w^\star - w_n$, which measures the difference between the nth iterate and the global minimizer of $P(w)$. If the step size μ satisfies (i.e., is small enough)*

$$0 < \mu < 2/\delta \tag{12.156}$$

then the risk value converges exponentially fast as follows:

$$P(w_n) - P(w^\star) \;\leq\; \lambda\Big(P(w_{n-1}) - P(w^\star)\Big) \tag{12.157}$$

where

$$\lambda = 1 - \frac{2\mu\nu}{M} + \frac{\mu^2\nu\delta}{M} \in [0,1) \tag{12.158}$$

Proof: We follow an argument similar to Example 12.10 based on the risk function. In view of the ν-strong convexity of $P(w)$, we first use property (8.29) to deduce that

$$P(w^\star) \geq P(w_{n-1}) - \frac{1}{2\nu}\|\nabla_w P(w_{n-1})\|^2 \tag{12.159}$$

Next, using the coordinate-wide δ-Lipschitz property (12.155b) and the result of Prob. 10.1, we write

$$\begin{aligned}
P(w_n) \;&\leq\; P(w_{n-1}) + \frac{\partial P(w_{n-1})}{\partial w_{m^o}}(w_{n,m^o} - w_{n-1,m^o}) + \frac{\delta}{2}(w_{n,m^o} - w_{n-1,m^o})^2 \\
&\overset{(12.127)}{=} P(w_{n-1}) - \mu\left(\frac{\partial P(w_{n-1})}{\partial w_{m^o}}\right)^2 + \frac{\mu^2\delta}{2}\left(\frac{\partial P(w_{n-1})}{\partial w_{m^o}}\right)^2 \\
&= P(w_{n-1}) - \mu\Big(1 - \frac{\mu\delta}{2}\Big)\left(\frac{\partial P(w_{n-1})}{\partial w_{m^o}}\right)^2
\end{aligned} \tag{12.160}$$

Now note the bound

$$\left(\frac{\partial P(w_{n-1})}{\partial w_{m^o}}\right)^2 \overset{(a)}{=} \|\nabla_w P(w_{n-1})\|_\infty^2 \overset{(b)}{\geq} \frac{1}{M}\|\nabla_w P(w_{n-1})\|_2^2 \tag{12.161}$$

where step (a) is by construction and step (b) is the property of norms $\|x\|_2 \leq \sqrt{M}\|x\|_\infty$ for any M-dimensional vector x. Subtracting $P(w^\star)$ from both sides of (12.160) and using (12.159) we obtain, after grouping terms:

$$P(w_n) - P(w^\star) \leq \Big(1 - \frac{2\mu\nu}{M} + \frac{\mu^2\nu\delta}{M}\Big)\Big(P(w_{n-1}) - P(w^\star)\Big) \tag{12.162}$$

■

12.6 ALTERNATING PROJECTION ALGORITHM[1]

We end this chapter with one application of the gradient-descent methodology to the derivation of a popular *alternating projection* algorithm. This method can be used to check whether two convex sets have a nontrivial intersection and to retrieve points from that intersection. It can also be used to verify whether a convex optimization problem with constraints has feasible solutions.

Consider two closed convex sets $\mathcal{C}_1$ and $\mathcal{C}_2$ in $\mathbb{R}^M$ and assume we are interested in determining a point $w^\star$ in their intersection, $\mathcal{C}_1 \cap \mathcal{C}_2$. Let w denote some arbitrary point. The distance from w to any of the sets is denoted by $\text{dist}(w, \mathcal{C})$ and defined as the smallest Euclidean distance to the elements in $\mathcal{C}$:

$$\text{dist}(w, \mathcal{C}) \stackrel{\Delta}{=} \min_{c \in \mathcal{C}} \|c - w\|_2 \tag{12.163}$$

To determine $w^\star$, we formulate the optimization problem (see Prob. 12.22):

$$P(w) \stackrel{\Delta}{=} \max\left\{\text{dist}(w, \mathcal{C}_1),\ \text{dist}(w, \mathcal{C}_2)\right\} \tag{12.164a}$$

$$w^\star = \underset{w \in \mathbb{R}^M}{\text{argmin}}\ P(w) \tag{12.164b}$$

For every w, the cost function $P(w)$ measures its distance to the set that is furthest away from it. The minimization seeks a point w with the smallest maximal distance. This formulation is motivated by the result of Prob. 9.13, which showed that, provided the sets intersect:

$$P(w^\star) = 0 \iff w^\star \in \mathcal{C}_1 \cap \mathcal{C}_2 \tag{12.165}$$

That is, we will succeed in finding a point in the intersection if, and only if, the minimal cost value turns out to be zero.

Now, for any w outside the intersection set, it will generally be further away from one of the sets. Let ℓ be the index of this set so that $P(w) = \text{dist}(w, \mathcal{C}_\ell)$. Let $\mathcal{P}_{\mathcal{C}_\ell}(w)$ denote the projection of w onto $\mathcal{C}_\ell$. We can now evaluate the gradient of $P(w)$ by using the result of Prob. 8.32, which shows that the gradient of $\text{dist}(w, \mathcal{C}_\ell)$ relative to w is given by

$$\nabla_{w^\mathsf{T}} \text{dist}(w, \mathcal{C}_\ell) = \frac{w - \mathcal{P}_{C_\ell}(w)}{\|w - \mathcal{P}_{C_\ell}(w)\|_2} \tag{12.166}$$

[1] This section can be skipped on a first reading.

We can therefore use this expression to write the following gradient-descent recursion to minimize $P(w)$ with an iteration-dependent step size:

$$
\begin{array}{l}
\textbf{for each iteration } n \textbf{ do:}\\
\quad w_{n-1} \text{ is the iterate at step } n-1;\\
\quad \text{let } \ell \text{ denote the index of the convex set furthest from it;}\\
\quad \text{let } \mathcal{P}_{C_\ell}(w_{n-1}) \text{ denote the projection of } w_{n-1} \text{ onto this set;}\\
\quad \text{set the step size to } \mu(n) = \|w_{n-1} - \mathcal{P}_{C_\ell}(w_{n-1})\| \; (=\text{distance to } \mathcal{C}_\ell);\\
\quad \text{update } w_n = w_{n-1} - \mu(n)\, \nabla_{w^\mathsf{T}} \text{dist}(w_{n-1}, \mathcal{C}_\ell);\\
\textbf{end}
\end{array}
\tag{12.167}
$$

We can simplify the last step as follows:

$$
\begin{aligned}
w_n &= w_{n-1} - \mu(n)\, \nabla_{w^\mathsf{T}} \text{dist}(w_{n-1}, \mathcal{C}_\ell)\\
&= w_{n-1} \; - \; \|w_{n-1} - \mathcal{P}_{C_\ell}(w_{n-1})\| \times \frac{w_{n-1} - \mathcal{P}_{C_\ell}(w_{n-1})}{\|w_{n-1} - \mathcal{P}_{C_\ell}(w_{n-1})\|}\\
&= w_{n-1} - \Big(w_{n-1} - \mathcal{P}_{C_\ell}(w_{n-1})\Big)
\end{aligned}
\tag{12.168}
$$

That is, the gradient-descent step reduces to the following projection step:

$$
\boxed{w_n = \mathcal{P}_{C_\ell}(w_{n-1})}
\tag{12.169}
$$

Thus, for example, if $\mathcal{C}_1$ happens to be the convex set that is furthest from w_{n-1}, then w_n will be its projection onto $\mathcal{C}_1$. For the next iteration, $\mathcal{C}_2$ will be the set that is furthest away from w_n and we will project onto $\mathcal{C}_2$. In this way, we arrive at a procedure that involves projecting onto the two convex sets alternately until $w^\star$ is attained. This construction is illustrated in Fig. 12.7.

We can describe the alternating projection procedure as generating two sequences of vectors, $\{a_n, b_n\}$, one in $\mathcal{C}_1$ and the other in $\mathcal{C}_2$. Assume we start from an initial condition $a_{-1} \in \mathcal{C}_1$; if the initial condition is outside $\mathcal{C}_1$, we can always project it onto $\mathcal{C}_1$ first and take that projection as the initial condition. Then, we can alternate as shown in listing (12.171) for $n \geq 0$ or, equivalently,

$$
a_n = \mathcal{P}_{\mathcal{C}_1}\Big(\mathcal{P}_{C_2}(a_{n-1})\Big), \quad n \geq 0
\tag{12.170}
$$

In this way, the sequence of vectors $\{a_n,\; n \geq -1\}$ will belong to $\mathcal{C}_1$ while the sequence of vectors $\{b_n,\; n \geq 0\}$ will belong to $\mathcal{C}_2$.

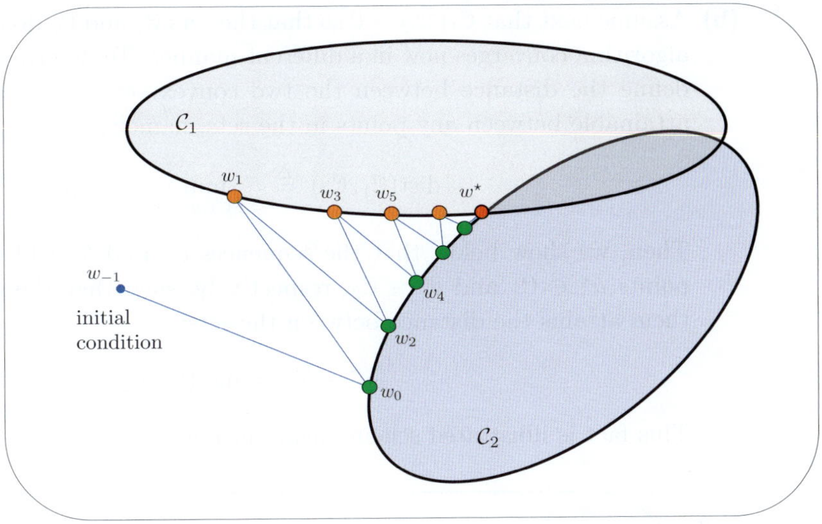

Figure 12.7 Illustration of the alternating projection procedure over two convex sets. Starting from an initial condition, the algorithm successively alternates the projections on the sets.

Alternating projection algorithm.

---

given two closed convex sets $\mathcal{C}_1$ and $\mathcal{C}_2$;
given projection operators onto $\mathcal{C}_1$ and $\mathcal{C}_2$;
start from arbitrary $a_{-1} \in \mathcal{C}_1$;
objective:
  if $\mathcal{C}_1 \cap \mathcal{C}_2 \neq \emptyset$: find a point $w^\star$ in the intersection;
  else find points $\{a^\star \in \mathcal{C}_1, b^\star \in \mathcal{C}_2\}$ closest to each other; (12.171)
repeat until convergence over $n \geq 0$**:**
  $b_n = \mathcal{P}_{C_2}(a_{n-1})$
  $a_n = \mathcal{P}_{C_1}(b_n)$
end
if $\|a_n - b_n\|$ small, return $w^\star \leftarrow a_n$;
  else return $a^\star \leftarrow a_n$, $b^\star \leftarrow b_n$.
end

We examine next the convergence of the algorithm; its behavior will depend on whether the sets $\mathcal{C}_1$ and $\mathcal{C}_2$ have a nontrivial intersection:

(a) Assume first that $\mathcal{C}_1 \cap \mathcal{C}_2 \neq \emptyset$. Then, we will verify that the sequences a_n and b_n will converge to the *same* limit point $w^\star \in \mathcal{C}_1 \cap \mathcal{C}_2$. That is, both sequences will converge to a point in the intersection set.

(b) Assume next that $\mathcal{C}_1 \cap \mathcal{C}_2 = \emptyset$ so that the sets $\mathcal{C}_1$ and $\mathcal{C}_2$ do not intersect. The algorithm converges now in a different manner. To describe the behavior, we define the distance between the two convex sets as the smallest distance attainable between any points in the sets, namely,

$$\text{dist}(\mathcal{C}_1, \mathcal{C}_2) \stackrel{\Delta}{=} \min_{x \in \mathcal{C}_1, y \in \mathcal{C}_2} \|x - y\| \tag{12.172}$$

Then, we show below that the sequences a_n and b_n will converge to limit points $a^\star \in \mathcal{C}_1$ and $b^\star \in \mathcal{C}_2$, respectively, such that the distance between them attains the distance between the sets:

$$\|a^\star - b^\star\| = \text{dist}(\mathcal{C}_1, \mathcal{C}_2) \tag{12.173}$$

This fact is illustrated schematically in Fig. 12.8.

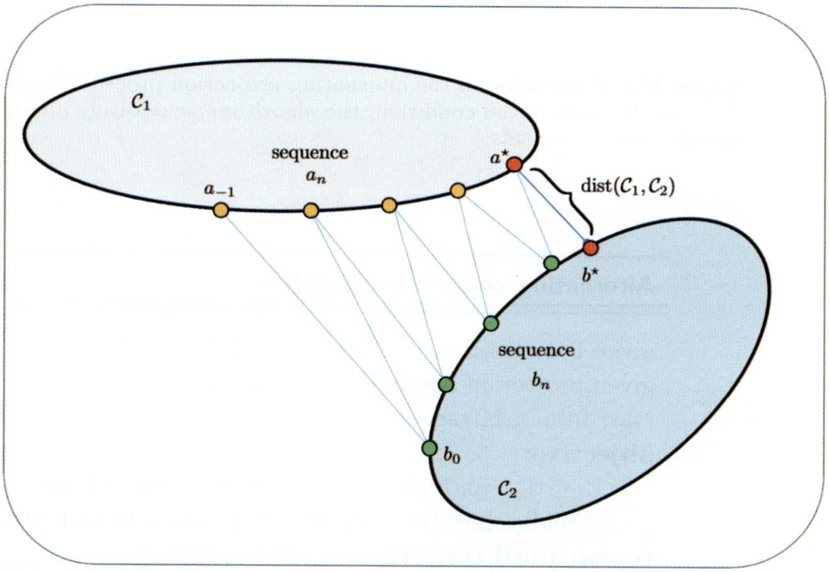

Figure 12.8 The sequences $\{a_n, b_n\}$ generated by the alternating projection algorithm converge to limit points $\{a^\star, b^\star\}$ that are closest to each other from both convex sets.

THEOREM 12.6. (Convergence of alternating projection) *Consider the alternating projection algorithm (12.171) and two closed convex sets $\mathcal{C}_1$ and $\mathcal{C}_2$:*

(a) When $\mathcal{C}_1 \cap \mathcal{C}_2 \neq \emptyset$, the sequences $\{a_n, b_n\}$ converge to the same limit point $w^\star \in \mathcal{C}_1 \cap \mathcal{C}_2$.

(b) When $\mathcal{C}_1 \cap \mathcal{C}_2 = \emptyset$, the sequences $\{a_n, b_n\}$ converge to limit points $\{a^\star \in \mathcal{C}_1, b^\star \in \mathcal{C}_2\}$ that are closest to each other.

Proof: Assume initially that the sets intersect and let $w^\star$ denote some arbitrary point in their intersection. Then, obviously, projecting $w^\star$ onto either set results in $w^\star$ again:

$$w^\star = \mathcal{P}_{\mathcal{C}_1}(w^\star), \quad w^\star = \mathcal{P}_{\mathcal{C}_2}(w^\star) \tag{12.174}$$

We now call upon the nonexpansive property (9.70) of projection operators, namely, the fact that, for any convex set $\mathcal{C}$:

$$\|\mathcal{P}_{\mathcal{C}}(x) - \mathcal{P}_{\mathcal{C}}(y)\| \leq \|x - y\|, \quad \forall\, x, y \in \mathcal{C} \tag{12.175}$$

Applying this property to the sequences $\{a_n, b_n\}$ generated by (12.171) we get (using $x = a_{n-1}$ in the first line and $x = b_n$ in the second line while $y = w^\star$ in both lines)

$$\|\mathcal{P}_{\mathcal{C}_2}(a_{n-1}) - w^\star\| = \|b_n - w^\star\| \leq \|a_{n-1} - w^\star\| \tag{12.176a}$$

$$\|\mathcal{P}_{\mathcal{C}_1}(b_n) - w^\star\| = \|a_n - w^\star\| \leq \|b_n - w^\star\| \tag{12.176b}$$

It follows that a_n is closer to $w^\star$ than b_n, and b_n is closer to $w^\star$ than a_{n-1}. More importantly, by combining both inequalities we observe that the sequence of squared distances $\|a_n - w^\star\|^2$ is decreasing and bounded from below by zero since it satisfies

$$0 \leq \|a_n - w^\star\|^2 \leq \|a_{n-1} - w^\star\|^2 \leq \ldots \leq \|a_{-1} - w^\star\|^2 \tag{12.177}$$

We conclude that the sequence of projections $a_n \in \mathcal{C}_1$ converges to some limit point denoted by $a^\star$. Since $\mathcal{C}_1$ is closed by assumption, this point belongs to $\mathcal{C}_1$, i.e., $a^\star \in \mathcal{C}_1$. A similar argument shows that

$$0 \leq \|b_n - w^\star\|^2 < \|b_{n-1} - w^\star\|^2 < \ldots < \|b_0 - w^\star\| < \|a_{-1} - w^\star\|^2 \tag{12.178}$$

so that the sequence of projections $b_n \in \mathcal{C}_2$ converges to some limit point denoted by $b^\star \in \mathcal{C}_2$. The limit points satisfy

$$a^\star = \mathcal{P}_{\mathcal{C}_1}(b^\star), \quad b^\star = \mathcal{P}_{\mathcal{C}_2}(a^\star) \tag{12.179}$$

We next apply the inner-product property (9.66) for projections, namely, the fact that

$$(x - \mathcal{P}_{\mathcal{C}}(x))^\mathsf{T}(c - \mathcal{P}_{\mathcal{C}}(x)) \leq 0, \quad \forall\, c \in \mathcal{C} \tag{12.180}$$

Therefore, the limit points $\{a^\star, b^\star\}$ satisfy

$$(b^\star - a^\star)^\mathsf{T}(c_1 - a^\star) \leq 0, \quad \forall\, c_1 \in \mathcal{C}_1 \tag{12.181a}$$

$$(a^\star - b^\star)^\mathsf{T}(c_2 - b^\star) \leq 0, \quad \forall\, c_2 \in \mathcal{C}_2 \tag{12.181b}$$

Adding gives

$$\|b^\star - a^\star\|^2 \leq (b^\star - a^\star)^\mathsf{T}(c_2 - c_1) \overset{(a)}{\leq} \|b^\star - a^\star\|\, \|c_2 - c_1\| \tag{12.182}$$

where step (a) is by Cauchy–Schwarz. It follows that

$$\|b^\star - a^\star\| \leq \|c_2 - c_1\|, \quad \forall\, c_1 \in \mathcal{C}_1,\ c_2 \in \mathcal{C}_2 \tag{12.183}$$

When the sets $\mathcal{C}_1$ and $\mathcal{C}_2$ have a nontrivial intersection, the right-hand side can be made equal to zero by selecting $c_1 = c_2 = w^\star$, from which we conclude that $b^\star = a^\star$. But since $b^\star \in \mathcal{C}_2$ and $a^\star \in \mathcal{C}_1$ by construction, the limit point satisfying $b^\star = a^\star$ must belong to the intersection of both sets. On the other hand, when the intersection is an empty set, we conclude that $\|b^\star - a^\star\|$ attains the smallest distance between any two points in $\mathcal{C}_1$ and $\mathcal{C}_2$.

■

The alternating projection method suffers from one inconvenience when the intersection set $\mathcal{C}_1 \cap \mathcal{C}_2$ has more than one point. Starting from an initial vector

a_{-1}, the method generates two sequences $\{a_n \in \mathcal{C}_1\}$ and $\{b_n \in \mathcal{C}_2\}$ that are only guaranteed to converge to some *arbitrary* point in the intersection. We describe a modification in the comments at the end of the chapter, known as the *Dykstra method*, which allows the algorithm to converge to the point $w^\star$ in the intersection that is closest to the initial condition a_{-1} – see listing (12.208).

12.7 COMMENTARIES AND DISCUSSION

Method of gradient descent. In Section 12.3 we motivated the gradient-descent recursion (12.22) for minimizing differentiable convex functions. The method is credited to the French mathematician **Augustine Cauchy (1789–1857)**, who proposed it as an iterative procedure for locating the roots of a function. Consider a function $P(x, y, z)$ of three scalar parameters (x, y, z), and assume that $P(x, y, z)$ has a unique root at some location $(x^\star, y^\star, z^\star)$ where $P(x^\star, y^\star, z^\star) = 0$. The objective is to identify this location. Cauchy (1847) worked with nonnegative and continuous functions $P(x, y, z)$ so that finding their roots corresponds to finding minimizers for the function. He argued that starting from some initial guess for $(x^\star, y^\star, z^\star)$, one can repeatedly move to new locations (x, y, z) where the values of the function continue to decrease – a description of Cauchy's argument is given by Lemaréchal (2012). Cauchy identified the direction of the update in terms of the negative gradient of the function. If we let (P_x, P_y, P_z) denote the partial derivatives of $P(x, y, z)$ relative to its individual arguments, then Cauchy proposed the following recursive form:

$$x(n) = x(n-1) - \mu\, P_x\Big(x(n-1), y(n-1), z(n-1)\Big) \tag{12.184a}$$

$$y(n) = y(n-1) - \mu\, P_y\Big(x(n-1), y(n-1), z(n-1)\Big) \tag{12.184b}$$

$$z(n) = z(n-1) - \mu\, P_z\Big(x(n-1), y(n-1), z(n-1)\Big) \tag{12.184c}$$

where μ is a small positive parameter. If we introduce the vector notation $w = \text{col}\{x, y, z\}$, this construction can be rewritten as

$$w_n = w_{n-1} - \mu\, \nabla_{w^{\mathsf{T}}} P(w_{n-1}) \tag{12.185}$$

which is the gradient-descent step we considered in (12.22). Cauchy did not analyze the convergence of his procedure. Convergence studies appeared later, e.g., in works by Curry (1944) and Goldstein (1962). For further discussion on gradient and *steepest-descent* methods, the reader may refer to Polyak (1987), Fletcher (1987), Nash and Sofer (1996), Luenberger and Ye (2008), Bertsekas (1995), Bertsekas and Tsitsiklis (1997), and Sayed (2014a). The convergence analysis given in the text for gradient-descent algorithms for both constant and decaying step sizes follows the presentations by Polyak (1987) and Sayed (2014a). The argument in Section 12.4.2 for the convergence of the backtracking method is based on the analysis in Boyd and Vandenberghe (2004); see also Curry (1944), Wolfe (1969, 1971), Goldstein (1966), and Kelley (1996). The Armijo condition (12.106) is from Armijo (1966). Some further applications of gradient-descent to batch algorithms appear in Bottou (1998), Bottou and LeCun (2004), Le Roux, Schmidt, and Bach (2012), and Cevher, Becker, and Schmidt (2014).

Momentum acceleration methods. In Probs. 12.9 and 12.11 we describe two popular methods to accelerate the convergence of gradient-descent methods by incorporating *momentum* terms into the update recursion. One method is known as the *heavy-ball* implementation or *Polyak momentum acceleration* and is due to Polyak (1964, 1987) and Polyak and Juditsky (1992). This method modifies the gradient-descent recursion

(12.185) by adding a driving term that is proportional to the difference of the last two iterates, namely,

$$w_n = w_{n-1} - \mu \nabla_{w^\mathsf{T}} P(w_{n-1}) + \beta(w_{n-1} - w_{n-2}), \quad n \geq 0 \tag{12.186}$$

The scalar $0 \leq \beta < 1$ is called the *momentum parameter.* It is shown in Prob. 12.10 that recursion (12.186) can be described in the equivalent form:

(**Polyak momentum acceleration**)

$$\begin{cases} b_n = \nabla_{w^\mathsf{T}} P(w_{n-1}) \\ \bar{b}_n = \beta \bar{b}_{n-1} + b_n, \quad \bar{b}_{-1} = 0 \\ w_n = w_{n-1} - \mu \bar{b}_n \end{cases} \tag{12.187}$$

which helps clarify the role of the momentum term. In this description, the gradient vector is denoted by b_n. It is seen that b_n is smoothed over time into $\bar{b}_n$, and the smoothed direction $\bar{b}_n$ is used to update the weight iterate from w_{n-1} to w_n. By doing so, momentum helps reinforce search directions with more pronounced progress toward the location of the sought-after minimizer.

A second momentum method is known as *Nesterov momentum acceleration* and is due to Nesterov (1983, 2004, 2005). It modifies the gradient-descent recursion in the following manner:

$$w_n = w_{n-1} - \mu \nabla_{w^\mathsf{T}} P\Big(w_{n-1} + \beta(w_{n-1} - w_{n-2})\Big) + \beta(w_{n-1} - w_{n-2}), \quad n \geq 0 \tag{12.188}$$

Compared with Polyak momentum (12.186), we find that the main difference is that the gradient vector is evaluated at the intermediate iterate $w_{n-1} + \beta(w_{n-1} - w_{n-2})$. It is shown in Prob. 12.12 that recursion (12.188) can be described in the equivalent form:

(**Nesterov momentum acceleration**)

$$\begin{cases} w'_{n-1} = w_{n-1} - \mu\beta \bar{b}_{n-1} \\ b'_n = \nabla_{w^\mathsf{T}} P(w'_{n-1}) \\ \bar{b}_n = \beta \bar{b}_{n-1} + b'_n, \quad \bar{b}_{-1} = 0 \\ w_n = w_{n-1} - \mu \bar{b}_n \end{cases} \tag{12.189}$$

That is, we first adjust w_{n-1} to the intermediate value w'_{n-1} and denote the gradient at this location by b'_n. We smooth this gradient over time and use the smoothed direction $\bar{b}_n$ to update the weight iterate.

When the risk function $P(w)$ is ν-strongly convex and has δ-Lipschitz gradients, both momentum methods succeed in accelerating the gradient-descent method to attain a faster exponential convergence rate, and this rate has been proven to be optimal for problems with smooth $P(w)$ and cannot be attained by the standard gradient-descent method – see Polyak (1987) and Nesterov (2004). Specifically, it is shown in these references that the convergence of the squared error $\|\widetilde{w}_n\|^2$ to zero occurs for these acceleration methods at the rate (see Prob. 12.9):

$$\|\widetilde{w}_n\|^2 \leq \left(\frac{\sqrt{\delta} - \sqrt{\nu}}{\sqrt{\delta} + \sqrt{\nu}}\right)^2 \|\widetilde{w}_{n-1}\|^2 \tag{12.190}$$

In contrast, in theorem 2.1.15 of Nesterov (2005) and theorem 4 in section 1.4 of Polyak (1987), the fastest rate for the gradient-descent method is shown to be (see Prob. 12.5):

$$\|\widetilde{w}_n\|^2 \leq \left(\frac{\delta - \nu}{\delta + \nu}\right)^2 \|\widetilde{w}_{n-1}\|^2 \tag{12.191}$$

It can be verified that

$$\frac{\sqrt{\delta}-\sqrt{\nu}}{\sqrt{\delta}+\sqrt{\nu}} < \frac{\delta-\nu}{\delta+\nu} \tag{12.192}$$

for $\nu < \delta$. This inequality confirms that the momentum algorithms can achieve faster rates in minimizing strongly convex risks $P(w)$ with Lipschitz gradients and that this faster rate cannot be attained by standard gradient descent.

Newton and quasi-Newton methods. In Section 12.3 we motivated the gradient-descent recursion for minimizing a first-order differentiable convex risk function. This technique is a *first-order method* since it relies solely on gradient calculations. There are other iterative techniques that can be used for the same purpose. If we examine the derivation that led to the gradient-descent recursion (12.22) starting from (12.20), we observe that the search direction may be scaled by any positive-definite matrix, say, by A^{-1}, so that the corresponding iteration (12.22) would become

$$w_n = w_{n-1} - \mu\, A^{-1}\, \nabla_{w^{\mathsf{T}}}\, P(w_{n-1}), \quad n \geq 0 \tag{12.193}$$

The step-size parameter μ can be incorporated into A if desired. This construction is referred to as the *quasi-Newton method.* If A is diagonal with individual positive entries, $A = \text{diag}\{a(1), a(2), \ldots, a(M)\}$, one for each entry of w, then we end up with a gradient-descent implementation where a separate step size is used for each individual entry of the weight iterate. The value of A can also vary with n. A second popular procedure is the Newton method (also called Newton–Raphson) described below, which is a *second-order method* since it uses the Hessian matrix of the risk function. A third procedure is the *natural gradient method* encountered in (6.131), which replaces the Hessian matrix by the Fisher information matrix. We elaborate here on the Newton method.

The Newton–Raphson method is named after the English mathematician and physicist **Isaac Newton (1643–1727)**, whose contribution appeared in Wallis (1685), and also after Newton's contemporary Raphson (1697). Both Newton and Raphson were interested in finding roots of polynomials. According to the account by Kollerstrom (1992), Newton's original method was not iterative, while Raphson's method was not expressed in differential form. It was the British mathematician **Thomas Simpson (1710–1761)** who introduced the current form of the method in Simpson (1740) – see the account by Christensen (1996). Thus, consider a ν-strongly convex second-order differentiable risk function, $P(w)$. Let δw denote a small perturbation vector. We approximate the value $P(w + \delta w)$ in terms of a second-order Taylor series expansion of $P(w)$ around w as follows:

$$P(w+\delta w) \approx P(w) + \Big(\nabla_{w^{\mathsf{T}}} P(w)\Big)^{\mathsf{T}} \delta w + \frac{1}{2}\,(\delta w)^{\mathsf{T}}\, \nabla_w^2 P(w) \delta w \tag{12.194}$$

We select δw to minimize the difference $P(w + \delta w) - P(w)$. We differentiate the expression on the right-hand side with respect to δw and set the result to zero, leading to:

$$\nabla_{w^{\mathsf{T}}} P(w) + \nabla_w^2 P(w) \delta w = 0 \tag{12.195}$$

When $P(w)$ is ν-strongly convex, its Hessian matrix satisfies $\nabla_w^2 P(w) \geq \nu I$ and is therefore positive-definite and invertible. It follows that the optimal perturbation is given by

$$\delta w^o = -\left(\nabla_w^2 P(w)\right)^{-1} \nabla_{w^{\mathsf{T}}}\, P(w) \tag{12.196}$$

Starting from the iterate w_{n-1}, the Newton method updates it to $w_n = w_{n-1} + \delta w^o$, leading to listing (12.197). Often, a small correction ϵI_M is added to the Hessian matrix to avoid degenerate situations and to ensure the inverse operation is valid.

Newton method for minimizing a risk function $P(w)$.

given gradient operator, $\nabla_{w^{\mathsf{T}}} P(w)$;
given Hessian operator, $\nabla_w^2 P(w)$;
given small $\epsilon > 0$;
start from arbitrary w_{-1}.
repeat until convergence over $n \geq 0$**:**
  $A_{n-1} = \epsilon I_M + \nabla_w^2 P(w_{n-1})$
  $w_n = w_{n-1} - A_{n-1}^{-1} \nabla_{w^{\mathsf{T}}} P(w_{n-1})$
end
return $w^\star \leftarrow w_n$.

(12.197)

We can verify that δw^o plays the role of a "descent" direction by noting that (see also Prob. 12.7):

$$\begin{aligned}
&P(w + \delta w^o) - P(w) \\
&= \left(\nabla_{w^{\mathsf{T}}} P(w)\right)^{\mathsf{T}} \delta w^o + \frac{1}{2} (\delta w)^{\mathsf{T}} \nabla_w^2 P(w) \delta w^o \\
&= -\left(\nabla_{w^{\mathsf{T}}} P(w)\right)^{\mathsf{T}} \left(\nabla_w^2 P(w)\right)^{-1} \nabla_w P(w) + \\
&\quad \frac{1}{2} \left(\nabla_{w^{\mathsf{T}}} P(w)\right)^{\mathsf{T}} \left(\nabla_w^2 P(w)\right)^{-1} \nabla_w^2 P(w) \left(\nabla_w^2 P(w)\right)^{-1} \nabla_{w^{\mathsf{T}}} P(w) \\
&= -\left(\nabla_{w^{\mathsf{T}}} P(w)\right)^{\mathsf{T}} \left(\nabla_w^2 P(w)\right)^{-1} \nabla_w P(w) + \\
&\quad \frac{1}{2} \left(\nabla_{w^{\mathsf{T}}} P(w)\right)^{\mathsf{T}} \left(\nabla_w^2 P(w)\right)^{-1} \nabla_{w^{\mathsf{T}}} P(w) \\
&= -\frac{1}{2} \left(\nabla_{w^{\mathsf{T}}} P(w)\right)^{\mathsf{T}} \left(\nabla_w^2 P(w)\right)^{-1} \nabla_w P(w) \\
&< 0
\end{aligned} \tag{12.198}$$

BFGS method. The Newton method (12.197) selects the update term δw^o in terms of the inverse of the Hessian matrix of the risk function, namely,

$$\delta w^o = -\left(\nabla_w^2 P(w_{n-1})\right)^{-1} \nabla_{w^{\mathsf{T}}} P(w_{n-1}) \tag{12.199}$$

The BFGS method, named after the initials of its independent developers, Broyden (1970), Fletcher (1970), Goldfarb (1970), and Shanno (1970), is a quasi-Newton method that approximates the Hessian matrix. It employs instead

$$\delta w^o = -B_{n-1}^{-1} \nabla_{w^{\mathsf{T}}} P(w_{n-1}) \tag{12.200}$$

for some positive-definite matrix B_{n-1} updated recursively:

$$B_n = B_{n-1} + \alpha a_n a_n^{\mathsf{T}} + \beta b_n b_n^{\mathsf{T}} \tag{12.201}$$

through the addition of two rank-one matrices to B_{n-1}. The scalars $\{\alpha, \beta\}$ are chosen to enforce a certain constraint on the successive matrices $\{B_n\}$ as follows. Introduce first the vectors:

$$a_n \triangleq \nabla_{w^{\mathsf{T}}} P(w_n) - \nabla_{w^{\mathsf{T}}} P(w_{n-1}) \tag{12.202a}$$

$$z_n \triangleq w_n - w_{n-1} \tag{12.202b}$$

$$b_n \triangleq B_{n-1} z_n \tag{12.202c}$$

Note that a_n is the difference between two successive gradient vectors, while z_n is the difference between two successive iterates. The vector b_n is the result of transforming

z_n by B_{n-1}. The scalars $\{\alpha, \beta\}$ are selected to ensure the following constraint (also called the *secant* equation):

$$B_n(w_n - w_{n-1}) = \nabla_{w^\mathsf{T}} P(w_n) - \nabla_{w^\mathsf{T}} P(w_{n-1}) \tag{12.203}$$

That is, $B_n z_n = a_n$. This condition is reminiscent of the classical *secant method* for finding the roots of a function $f(x)$ over $x \in \mathbb{R}$, which is described by the recursion

$$x_{n+1} = x_n - f(x_n) \times \frac{x_n - x_{n-1}}{f(x_n) - f(x_{n-1})} \tag{12.204}$$

The BFGS method seeks a root for the equation $\nabla_{w^\mathsf{T}} P(w) = 0$. It does so by extending construction (12.204) to the vector case through the imposition of (12.203). It is easily verified by multiplying the update relation for B_n from the left by z_n, that (12.203) is satisfied for the choices:

$$\alpha = \frac{1}{z_n^\mathsf{T} a_n}, \quad \beta = -\frac{1}{z_n^\mathsf{T} B_{n-1} z_n} \tag{12.205}$$

We arrive at listing (12.206) where the initial matrix can be selected as $B_{-1} = I_M$. Problem 12.35 derives an expression that updates B_{n-1}^{-1} to B_n^{-1} directly and shows that the successive matrices B_n are positive-definite. For more details, see Fletcher (1987), Kelley (1996), and Nocedal and Wright (2006).

BFGS method for minimizing a risk function $P(w)$.

given gradient operator, $\nabla_{w^\mathsf{T}} P(w)$;
given small step size, $\mu > 0$;
start from arbitrary w_{-1} and $B_{-1} > 0$.
repeat until convergence over $n \geq 0$**:**

$$\begin{aligned}
w_n &= w_{n-1} - \mu B_{n-1}^{-1} \nabla_{w^\mathsf{T}} P(w_{n-1}) \\
z_n &= w_n - w_{n-1} \\
a_n &= \nabla_{w^\mathsf{T}} P(w_n) - \nabla_{w^\mathsf{T}} P(w_{n-1}) \\
b_n &= B_{n-1} z_n \\
B_n &= B_{n-1} + \frac{1}{z_n^\mathsf{T} a_n} a_n a_n^\mathsf{T} - \frac{1}{z_n^\mathsf{T} B_{n-1} z_n} b_n b_n^\mathsf{T}
\end{aligned} \tag{12.206}$$

end
return $w^\star \leftarrow w_n$.

Method of coordinate-descent. The coordinate-descent method is a simple and effective technique for the solution of optimization problems; it relies on a sequence of coordinate-wise steps to reduce an M-dimensional problem to M one-dimensional problems. In each step, the optimization is carried out over a single entry (or coordinate) of the parameter vector w, while holding all other entries fixed at their current estimated values. Ideally, when the optimization can be carried out in closed form, the coordinate-descent construction minimizes $P(w)$ over its individual coordinates in sequence before repeating the iteration, as was shown in listing (12.119). In practice, though, the optimization step is generally difficult to compute analytically and it is replaced by a gradient-descent step. The classical implementation of coordinate-descent cycles through all coordinates, while more popular variants update one coordinate per iteration. This coordinate is selected either uniformly at random, which is one of the most popular variants, or as the coordinate that corresponds to the maximal absolute gradient value. For more details, the reader may refer to Nesterov (2012), which studies the case of smooth functions under both convexity and strong convexity conditions. The work by Richtárik and Takác (2011) simplifies the results and considers the case of smooth plus separable risks. There are many variants of coordinate-descent implementations, including the important case where blocks of coordinates (rather than a single coordinate) are updated at the same time – see, e.g., the convergence analysis in Tseng (2001), where the future separable form (14.123) for nonsmooth risks is studied in some detail. In the block case, the parameter w is divided into sub-blocks, say, $w = \text{blkcol}\{w_1, w_2, \ldots, w_B\}$ where each w_b is now a sub-vector with multiple entries in it. Then, the same constructions we described in the body of the chapter will apply working with block sub-vectors rather than single coordinates as shown, for example, in listing (12.207).

Block coordinate-descent for minimizing a risk function $P(w)$.

let $w = \text{blkcol}\{w_1, w_2, \ldots, w_B\}$;
start from an arbitrary initial condition w_{-1}.
repeat until convergence over $n \geq 0$:
  w_{n-1} with B blocks $\text{blkcol}\{w_{n-1,b}\}$ is available at start of iteration;
  for each block $b = 1, 2, \ldots, B$ compute:

$$w_{n,b} = \underset{w_b}{\text{argmin}}\ P\Big(w_{n,1}, \ldots, \boxed{w_b}, \ldots, w_{n-1,B}\Big) \tag{12.207}$$

  end
end
return $w^\star \leftarrow w_n$.

The idea of estimating one component at a time while fixing all other components at their current values appears already in the classical Gauss–Seidel approach to solving linear systems of equations – see, e.g., Golub and Van Loan (1996). The Gauss–Seidel approach is accredited to the German mathematicians **Carl Friedrich Gauss (1777–1855)** and Seidel (1874). Gauss described the method 50 years prior to Seidel in a correspondence from 1823 – see the collection of works by Gauss (1903). In more recent times, some of the earliest references on the use of coordinate-descent in optimization include Hildreth (1957), where block coordinate descent was first introduced, in addition to Warga (1963) and Ortega and Rheinboldt (1970). The Gauss–Southwell (GS) rule that amounts to selecting the gradient component with the largest absolute value in (12.145) is also due to Gauss (1903) and Southwell (1940). Some analysis on comparing the convergence rates of the GS and randomized rules appear in Nutini *et al.* (2015), where it is argued that the GS rule leads to improved convergence – see Prob. 12.19; this argument is consistent with the steepest-descent derivation of the GS rule in Section 12.5.3 and is related to the derivation used in the proof of Theorem 12.5. More recent applications of coordinate-descent in computer tomography, machine learning, statistics, and multi-agent optimization appear, for example, in Luo and Tseng (1992a), Sauer and Bouman (1993), Fu (1998), Daubechies, Defrise, and De Mol (2004), Friedman *et al.* (2007), Wu and Lange (2008), Chang, Hsieh, and Lin (2008), Tseng and Yun (2009, 2010), Beck and Tetruashvili (2013), Lange, Chi, and Zhou (2014), Wright (2015), Shi *et al.* (2017), Wang *et al.* (2018), Fercoq and Bianchi (2019), and the many references therein.

Alternating projection method. We described in Section 12.6 the *alternating projection algorithm* (12.171) for finding points in the intersection of two closed convex sets, $\mathcal{C}_1$ and $\mathcal{C}_2$; the technique is also known as the *successive projection method.* It has found applications in a wide range of fields, including statistics, optimization, medical imaging, machine learning, and finance. The algorithm was originally developed by the Hungarian-American mathematician **John von Neumann (1903–1957)** in 1933 in unpublished lecture notes, which appeared later in press in the works by von Neumann (1949, 1950) on operator theory. von Neumann's work focused on the intersection of two affine subspaces in Hilbert space (such as hyperplanes), and it was subsequently extended by Halperin (1962) to the intersection of multiple affine subspaces and by Bregman (1965) to the intersection of multiple closed convex sets. For the benefit of the reader, a set $\mathcal{S} \subset \mathbb{R}^M$ is an affine subspace if every element $s \in \mathcal{S}$ can be written as $s = p + v$, for some fixed p and where $v \in \mathcal{V}$ denotes an arbitrary element from a vector space. For example, the set $\mathcal{S} = \{x \,|\, a^\mathsf{T} x = b\}$ is an affine subspace. If we let $\widehat{x}$ denote any solution to $a^\mathsf{T} x = b$, then any element $s \in \mathcal{S}$ can be written as $s = \widehat{x} + v$ for any $v \in \mathcal{N}(a)$.

The proof of Theorem 12.6 assumes initially that the convex sets have a nontrivial intersection. When the sets do not intersect, the argument reveals through expression (12.183), as was discovered by Cheney and Goldstein (1959), that the alternating projection algorithm converges to points that are closest to each other from both sets. This fact was illustrated schematically in Fig. 12.8 and it has many useful applications. In particular, the result shows that the alternating projection method can be used to check whether a collection of convex sets intersect or not (such as checking whether the constraints in a convex optimization problem of the form (9.1) admit feasible solutions). It can also be used to determine the minimum distance between two convex sets. Moreover, we can use any hyperplane that is orthogonal to the segment connecting $a^\star$ and $b^\star$ to separate the two convex sets from each other: one set will be on one side of the hyperplane while the other set will be on the other side. This observation is useful for classification problems, as we will explain in later chapters when we discuss linearly separable datasets. The presentation in the chapter benefited from useful overviews on the alternating projection method, including proofs for its convergence properties, given in the works by Cheney and Goldstein (1959), Bregman (1967), Gubin, Polyak, and Raik (1967), Combettes (1993), Bauschke and Borwein (1996), Escalante and Raydan (2011), and Dattoro (2016). Although we have focused on finding points in the intersection of *two* convex sets, the algorithm can be applied to a larger number of sets by projecting sequentially onto the sets, one at a time, with minimal adjustments to the arguments, leading to what is known as the *cyclic projection algorithm* – see Prob. 12.34.

The alternating projection method suffers from one inconvenience when the intersection set $\mathcal{C}_1 \cap \mathcal{C}_2$ has more than one point. Starting from an initial vector a_{-1}, the method generates two sequences $\{a_n \in \mathcal{C}_1\}$ and $\{b_n \in \mathcal{C}_2\}$ that are only guaranteed to converge to some *arbitrary* point in the intersection. An elegant variation is the Dykstra algorithm listed in (12.208), which was developed by Dykstra (1983); see also Boyle and Dykstra (1986). The same method was rediscovered by Han (1988).

Dykstra alternating projection algorithm.

given two closed convex sets $\mathcal{C}_1$ and $\mathcal{C}_2$;
given projection operators onto $\mathcal{C}_1$ and $\mathcal{C}_2$;
start from $a_{-1} \in \mathcal{C}_1$, and set $c_{-1} = d_0 = 0$;
objective:
  if $\mathcal{C}_1 \cap \mathcal{C}_2 \neq \emptyset$: find projection $w^\star$ of a_{-1} onto the intersection;
  else find points $\{a^\star \in \mathcal{C}_1, b^\star \in \mathcal{C}_2\}$ closest to each other;
repeat until convergence over $n \geq 0$: (12.208)
  $b_n = \mathcal{P}_{C_2}(a_{n-1} + c_{n-1})$
  $c_n = a_{n-1} + c_{n-1} - b_n$ (residual)
  $a_{n+1} = \mathcal{P}_{C_1}(b_n + d_n)$
  $d_{n+1} = b_n + d_n - a_{n+1}$ (residual)
end
if $\|a_n - b_n\|$ small, return $w^\star \leftarrow a_n$;
  else return $a^\star \leftarrow a_n$, $b^\star \leftarrow b_n$.
end

The Dykstra method ensures convergence to the point $w^\star$ in the intersection that is closest to a_{-1}, namely,

$$\|a_{-1} - w^\star\| \leq \|a_{-1} - w\|, \quad \text{for any } w \in \mathcal{C}_1 \cap \mathcal{C}_2 \tag{12.209}$$

That is, the method ends up determining the projection of a_{-1} onto $\mathcal{C}_1 \cap \mathcal{C}_2$. Compared with the classical formulation (12.171), the Dykstra method (12.208) introduces two auxiliary vectors $\{c_n, d_n\}$ and performs the calculations shown in (12.208) repeatedly, starting from the same initial vector $a_{-1} \in \mathcal{C}_1$ and using $c_{-1} = d_0 = 0$.

Using the traditional alternating projection recursions, and the Dykstra variation, we are able to solve the following two types of problems:

$$\textbf{(feasibility problem)} \text{ finding } w^\star \in \mathcal{C}_1 \cap \mathcal{C}_2 \tag{12.210}$$

$$\textbf{(projection problem)} \text{ finding } \mathcal{P}_{C_1 \cap C_2}(x), \text{ for a given } x \tag{12.211}$$

Both problems are solved by working independently with the projection operators $\mathcal{P}_{\mathcal{C}_1}(x)$ and $\mathcal{P}_{\mathcal{C}_2}(x)$, and by applying them alternately starting from x.

Zeroth-order optimization. The gradient-descent algorithm described in the body of the chapter is an example of a first-order optimization method, which requires the availability of the gradient information, $\nabla_w P(w)$, in order to perform the update (12.22). In some situations of interest, it may not be possible to evaluate the gradient function either because the risk $P(w)$ may not have a closed analytical form or is unknown to the designer altogether – see, e.g., Brent (2002) and Conn, Scheinberg, and Vicente (2009) for examples. The latter situation arises, for example, in adversarial learning scenarios, studied in Chapter 71, where the designer wishes to misguide the operation of a learning algorithm and is only able to perform function evaluations of the risk function, $P(w)$, at different data samples. We provide a brief review of zeroth-order optimization, also known as *derivative-free optimization* in Appendix 12.A. This technique enables the designer to approximate the gradient vector by relying solely on function evaluations. One of the earliest references on the use of finite difference approximations for gradient evaluations is Kiefer and Wolfowitz (1952). We explain in the appendix how gradient vectors can be approximated by means of two function evaluations in a manner that satisfies the useful unbiasedness property (12.218). The proof given for this latter property follows Nesterov and Spokoiny (2017) and Flaxman, Kalai, and McMahan (2005). There have been several works in more recent years on the performance of optimization algorithms based on such constructions. They are slower to converge than traditional gradient descent and have been shown to require at least M times more iterations to converge. The slowdown in performance is due to the error variance in estimating the true gradient vector. Results along these lines can be found in Wibisono *et al.* (2012), Nesterov and Spokoiny (2017), and Liu *et al.* (2018). Overviews of gradient-free optimization appear in Rios and Sahinidis (2013), Duchi *et al.* (2015), Larson, Menickelly, and Wild (2019), and Liu *et al.* (2020).

PROBLEMS

12.1 Establish that for ν-strongly convex risk functions $P(w)$ with δ-Lipschitz gradients it holds that

$$\frac{2}{\delta}\left(P(w) - P(w^\star)\right) \le \|\widetilde{w}\|^2 \le \frac{2}{\nu}\left(P(w) - P(w^\star)\right)$$

12.2 Show that the ℓ_2-regularized least-squares risk listed in Table 12.1 satisfies condition (12.15). Determine the values for $\{\nu, \delta\}$.

12.3 Assume all we know is that $P(w)$ is twice-differentiable over w and satisfies condition (12.15). Establish the validity of Theorem 12.1.

12.4 Assume that $P(w)$ is twice-differentiable over w and satisfies (12.15).

(a) Use the mean-value relation (10.10) to show that the error vector satisfies a recursion of the form $\widetilde{w}_n = (I_M - \mu H_{n-1})\widetilde{w}_{n-1}$ where

$$H_{n-1} \triangleq \int_0^1 \nabla_w^2\, P(w^\star - t\widetilde{w}_{n-1})dt$$

(b) Show that the conclusions of Theorem 12.1 continue to hold over the wider step-size interval $\mu < 2/\delta$, which is independent of ν.

12.5 Problems 12.5–12.9 are motivated by results from Polyak (1987, chs. 1, 3). Consider the same setting as Prob. 12.4. Show that convergence of $\|\widetilde{w}_n\|^2$ to zero also occurs at an exponential rate that is given by $\lambda_2 = \max\{(1-\mu\delta)^2, (1-\mu\nu)^2\}$. Conclude that the convergence rate is fastest when the step size is chosen as $\mu^o = 2/(\nu+\delta)$ for which $\lambda_2^o = (\delta-\nu)^2/(\delta+\nu)^2$. Roughly how many iterations are needed for the squared error, $\|\widetilde{w}_n\|^2$, to fall below a small threshold value, ϵ?

12.6 Let $P(w)$ be a real-valued first-order differentiable risk function whose gradient vector satisfies the δ-Lipschitz condition (12.12b). The risk $P(w)$ is *not* assumed convex. Instead, we assume that it is lower-bounded, namely, $P(w) \geq L$ for all w and for some finite value L. Consider the gradient-descent algorithm (12.22). Show that if the step size μ satisfies $\mu < 2/\delta$, then the sequence of iterates $\{w_n\}$ satisfies the following two properties:

(a) $P(w_n) \leq P(w_{n-1})$.

(b) $\lim_{n\to\infty} \nabla_w P(w_n) = 0$.

12.7 Let $P(w)$ denote a real-valued ν-strongly convex and twice-differentiable cost function with $w \in \mathbb{R}^M$. Assume the Hessian matrix of $P(w)$ is δ-Lipschitz continuous, i.e., $\left\|\nabla_w^2 P(w_2) - \nabla_w^2 P(w_1)\right\| \leq \delta\|w_2 - w_1\|$. The global minimizer of $P(w)$ is sought by means of the Newton method:

$$w_n = w_{n-1} - \left(\nabla_w^2\, P(w_{n-1})\right)^{-1} \nabla_{w^\mathsf{T}}\, P(w_{n-1}), \quad n \geq 0$$

which employs the inverse of the Hessian matrix. The initial condition is denoted by w_{-1}. Let $\lambda \triangleq \left(\delta/2\nu^2\right)^2 \|\nabla_w P(w_{-1})\|^2$, and assume $\lambda < 1$. Show that $\|\widetilde{w}_n\|^2$ converges to zero at the rate

$$\|\widetilde{w}_n\|^2 \leq \left(\frac{2\nu^2}{\delta}\right)\lambda^{2^n}$$

Conclude that the convergence rate is now dependent on the quality of the initial condition.

12.8 Consider the gradient-descent recursion (12.65) where the step-size sequence is selected as

$$\mu(n) = \frac{\tau}{(n+1)^q}, \qquad 1/2 < q \leq 1, \quad \tau > 0$$

(a) Verify that the step-size sequence satisfies conditions (12.66b).

(b) Follow the proof of Theorem 12.2 to determine the rate of convergence of $\|\widetilde{w}_n\|^2$ to zero.

(c) For a fixed τ, which value of q in the range $1/2 < q \leq 1$ results in the fastest convergence rate?

12.9 Let $P(w)$ be a real-valued risk function, assumed ν-strongly convex, first-order differentiable at all $w \in \text{dom}(P)$, and with δ-Lipschitz gradients as in (12.12a)–(12.12b). Consider a *heavy-ball* implementation of the gradient-descent algorithm, which is a form of *momentum acceleration*, also known as the Polyak momentum method, and given by

$$w_n = w_{n-1} - \mu\nabla_{w^\mathsf{T}}\, P(w_{n-1}) + \beta(w_{n-1} - w_{n-2}), \quad n \geq 0$$

where the past iterate w_{n-2} is also used in the update equation, and $0 \leq \beta < 1$ is called the momentum parameter. Assume the initial conditions w_{-1} and w_{-2} lie sufficiently close to $w^\star$, i.e., $\|\widetilde{w}_{-1}\|^2 < \epsilon$ and $\|\widetilde{w}_{-2}\|^2 < \epsilon$ for some small enough ϵ.

(a) Show that if $0 < \mu < 2(1+\beta)/\delta$, then $\|\widetilde{w}_n\|^2$ converges to zero at the exponential rate $O(\lambda_3^n)$. Identify λ_3 and show that optimal values for $\{\mu, \beta, \lambda_3\}$ are

$$\mu^o = \frac{4}{\left(\sqrt{\delta}+\sqrt{\nu}\right)^2}, \qquad \beta^o = \left(\frac{\sqrt{\delta}-\sqrt{\nu}}{\sqrt{\delta}+\sqrt{\nu}}\right)^2, \qquad \lambda_3^o = \beta^o$$

(b) Let $\kappa = \delta/\nu$. Large values for κ indicate ill-conditioned Hessian matrices, $\nabla_w^2 P(w)$, since their spectra will lie over wider intervals. Let λ_2^o denote the optimal rate of convergence when $\beta = 0$. We already know from Prob. 12.5 that $\lambda_2^o = (\delta - \nu)^2/(\delta+\nu)^2$. Argue that for large κ:

$$\lambda_2^o \approx 1 - 2/\kappa, \qquad \lambda_3^o \approx 1 - 2/\sqrt{\kappa}$$

Compare the number of iterations that are needed for $\|\widetilde{w}_n\|^2$ to fall below a threshold ϵ for both cases of $\beta = 0$ and $\beta = \beta^o$.

12.10 Show that the momentum method that updates w_{n-1} to w_n in Prob. 12.9 can be described in the equivalent form:

$$\begin{aligned} b_n &\stackrel{\Delta}{=} \nabla_{w^\mathsf{T}} P(w_{n-1}) \\ \bar{b}_n &= \beta \bar{b}_{n-1} + b_n, \quad \bar{b}_{-1} = 0 \\ w_n &= w_{n-1} - \mu \bar{b}_n \end{aligned}$$

Remark. We will encounter this construction later in Section 17.5 when we study adaptive gradient methods with momentum acceleration.

12.11 Consider the same setting of Prob. 12.9. A second momentum implementation is the Nesterov momentum method, which is given by the following recursion:

$$w_n = w_{n-1} - \mu \nabla_{w^\mathsf{T}} P\Big(w_{n-1} + \beta(w_{n-1} - w_{n-2})\Big) + \beta(w_{n-1} - w_{n-2}), \quad n \geq 0$$

Compared with Polyak momentum from Prob. 12.10, we find that the main difference is that the gradient vector is evaluated at the intermediate iterate $w_{n-1} + \beta(w_{n-1} - w_{n-2})$. Study the convergence properties of the Nesterov method in a manner similar to Prob. 12.9. *Remark.* For more details on this implementation and its convergence properties, see Nesterov (1983, 2004) and Yu, Jin, and Yang (2019).

12.12 Show that the Nesterov momentum method that updates w_{n-1} to w_n in Prob. 12.11 can be described in the equivalent form:

$$\begin{aligned} w'_{n-1} &\stackrel{\Delta}{=} w_{n-1} - \mu\beta \bar{b}_{n-1} \\ b'_n &\stackrel{\Delta}{=} \nabla_{w^\mathsf{T}} P(w'_{n-1}) \\ \bar{b}_n &= \beta \bar{b}_{n-1} + b'_n, \quad \bar{b}_{-1} = 0 \\ w_n &= w_{n-1} - \mu \bar{b}_n \end{aligned}$$

Remark. We will encounter this construction later in Section 17.5 when we study adaptive gradient methods with momentum acceleration.

12.13 Refer to the gradient-descent recursion (12.22) and assume that $P(w)$ is only convex (but not necessarily strongly convex) with a δ-Lipschitz gradient satisfying (12.12b). Let $\mu < 1/\delta$.

(a) Use property (11.120) for convex functions with δ-Lipschitz gradients to argue that

$$P(w_n) \leq P(w_{n-1}) - \frac{\mu}{2} \|\nabla_w P(w_{n-1})\|^2$$

Conclude that $P(w_n)$ is nonincreasing.

(b) Use part (a) to show that, for any $z \in \mathbb{R}^M$, it holds

$$P(w_n) \leq P(z) + \Big(\nabla_{w^\mathsf{T}} P(w_{n-1})\Big)(w_{n-1} - z) - \frac{\mu}{2}\|\nabla_w P(w_{n-1})\|^2$$

(c) Show that $P(w_n) - P(w^\star) \leq \frac{1}{2\mu}(\|\widetilde{w}_{n-1}\|^2 - \|\widetilde{w}_n\|^2)$.

(d) Conclude that $P(w_n) - P(w^\star) \leq \frac{1}{2\mu n}\|\widetilde{w}_0\|^2$ so that (12.64b) holds.

12.14 Refer to the gradient-descent recursion (12.22) and assume that $P(w)$ is only convex and δ-Lipschitz, namely,

$$\|P(w_1) - P(w_2)\| \leq \delta\|w_1 - w_2\|, \quad \forall w_1, w_2 \in \text{dom}(P)$$

Observe that we are now assuming that $P(w)$ itself is Lipschitz rather than its gradient. We know from property (10.41) that the condition of a Lipschitz function translates into bounded gradient vectors, so that $\|\nabla_w P(w)\| \leq \delta$. Let $w^\star$ be a minimizer for $P(w)$ and assume $\|\widetilde{w}_{-1}\| \leq W$, where w_{-1} is the initial condition for the gradient-descent recursion.

(a) Use the convexity of $P(w)$, the gradient-descent recursion, and the bounded gradients to verify that

$$P(w_{n-1}) - P(w^\star) \leq \frac{\mu\delta^2}{2} + \frac{1}{2\mu}\|\widetilde{w}_{n-1}\|^2 - \frac{1}{2\mu}\|\widetilde{w}_n\|^2$$

(b) Sum over the first N iterations to verify that

$$\frac{1}{N}\sum_{n=0}^{N-1} P(w_n) - P(w^\star) \leq \frac{\mu\delta^2}{2} + \frac{W^2}{2N\mu}$$

(c) Show that the upper bound is minimized for $\mu^o = \frac{W}{\delta\sqrt{N}}$ and apply the Jensen inequality to $P(w)$ to conclude that

$$P\Big(\frac{1}{N}\sum_{n=0}^{N-1} w_n\Big) - P(w^\star) \leq W\delta/\sqrt{N}$$

In other words, the risk value evaluated at the average iterate approaches the minimal value $P(w^\star)$ at the rate $O(1/\sqrt{N})$.

(d) Let w^{best} denote the iterate value that results in the smallest risk from among all iterates, namely, $w^{\text{best}} = \underset{0\leq n\leq N-1}{\text{argmin}}\ P(w_n)$. Conclude further that

$$P(w^{\text{best}}) - P(w^\star) \leq W\delta/\sqrt{N}$$

12.15 Refer to the gradient-descent algorithm (12.65) with decaying step sizes. Repeat the argument from parts (a) and (b) of Prob. 12.14 to establish that

$$\sum_{n=0}^{N-1} \mu(n)\Big(P(w_n) - P(w^\star)\Big) \leq \frac{\delta}{2}\sum_{n=0}^{N-1}\mu^2(n) + \frac{W^2}{2}$$

Select $\mu(n) = \frac{c}{\sqrt{n+1}}$ for some constant c and show that

$$P(w^{\text{best}}) - P(w^\star) = O\Big(\ln(N)/\sqrt{N}\Big)$$

12.16 Refer to the backtracking line search method (12.103) and assume now that $P(w)$ is only convex. Repeat the argument of Prob. 12.13 to establish that

$$P(w_n) - P(w^\star) \leq \frac{1}{2\mu_{\text{bt}}\, n}\|\widetilde{w}_0\|^2$$

where $\mu_{\text{bt}} = \min\{1, \beta/\delta\}$.

12.17 Show that step-size sequences of the form (12.67) satisfy condition (12.66b).

12.18 Consider the gradient-descent recursion (12.65) with the step-size sequence selected as $\mu(n) = \tau/(n+1)^q$ where $\frac{1}{2} < q \leq 1$ and $\tau > 0$.

(a) Verify that the step-size sequence satisfies condition (12.66b).
(b) Follow the proof of Theorem 12.2 to determine the rate of convergence of $\|\widetilde{w}_n\|^2$ to zero.
(c) For a fixed τ, which value of q results in the fastest rate of convergence?

12.19 Refer to the statement of Theorem 12.5 for the Gauss–Southwell coordinate-descent algorithm. The value of $\lambda \in [0,1)$ determines the convergence rate of the algorithm. We can establish a tighter result with a smaller $\lambda' \in [0,1)$, which would imply a faster rate than the one suggested by the theorem as follows. We know from the result of Prob. 8.60 that $P(w)$ is also ν_1-strongly convex relative to the infinity norm, i.e., expression (12.155a) implies

$$P(w_2) \geq P(w_1) + \Big(\nabla_{w^\mathsf{T}} P(w_1)\Big)^\mathsf{T}(w_2 - w_1) + \frac{\nu_1}{2}\|w_2 - w_1\|_\infty^2$$

where ν_1 satisfies $\frac{\nu}{M} \leq \nu_1 \leq \nu$. Repeat the argument in the proof of Theorem 12.5 to show that the algorithm continues to be stable for $\mu < 2/\delta$ with the excess risk now evolving according to

$$P(w_n) - P(w^\star) \;\leq\; \lambda'\Big(P(w_n) - P(w^\star)\Big)$$

where $\lambda' \stackrel{\Delta}{=} 1 - 2\mu\nu_1 + \mu^2\nu_1\delta$. Verify that $\lambda' < \lambda$ when $\nu_1 > \nu/M$, and conclude that this result suggests a faster rate of convergence for the Gauss–Southwell coordinate descent recursion than the randomized coordinate-descent recursion. *Remark.* The reader may see Nutini *et al.* (2015) for a related discussion.

12.20 Consider a risk function $P(w)$ and let w_{n-1} denote an estimate for the minimizer of $P(w)$ at iteration $n-1$. Assume we are able to construct a function $G(w, w_{n-1})$ that satisfies the two conditions:

$$G(w, w_{n-1}) = P(w), \quad G(w, w_{n-1}) \geq P(w_{n-1}), \; \forall w$$

We say that $G(w, w_{n-1})$ "majorizes" $P(w)$ by bounding it from above. Note that the definition of $G(w)$ depends on w_{n-1}. Now set w_n to the minimizer of $G(w, w_{n-1})$, i.e.,

$$w_n = \underset{w\in\mathbb{R}^M}{\text{argmin}}\; G(w, w_{n-1})$$

Show that $P(w_n) \leq P(w_{n-1})$. That is, the updates constructed in this manner lead to nonincreasing risks. This method of design is referred to as *majorization–minimization*; one example is encountered in Example 58.2 in the context of the multiplicative update algorithm for nonnegative matrix factorization.

12.21 Consider the ℓ_2-regularized MSE risk:

$$w^o = \underset{w\in\mathbb{R}^M}{\text{argmin}} \left(\rho\|w\|^2 + \mathbb{E}\,(\boldsymbol{\gamma} - \boldsymbol{h}^\mathsf{T} w)^2\right)$$

(a) Denote the risk in the above optimization problem by $P(w)$. Verify that $P(w)$ is quadratic in w and given by $P(w) = \sigma_\gamma^2 - 2r_{\gamma h}^\mathsf{T} w + w^\mathsf{T}(\rho I + R_h)w$.
(b) Show that $P(w)$ is ν-strongly convex and find a value for ν.
(c) Show that $P(w)$ has δ-Lipschitz gradients and find a value for δ.
(d) Show that $w^o = (\rho I + R_h)^{-1} r_{h\gamma}$.
(e) In this case we know a closed-form expression for the optimal solution w^o. Still, let us determine its value iteratively. Show that the gradient-descent recursion (12.22) applied to the above $P(w)$ leads to

$$w_n = (1 - 2\rho\mu)w_{n-1} + 2\mu(r_{h\gamma} - R_h w_{n-1}), \;\; n \geq 0$$

(f) Let $\widetilde{w}_n = w^o - w_n$. Verify that $\widetilde{w}_n = ((1 - 2\rho\mu)I_M - 2\mu R_h)\widetilde{w}_{n-1}$.
(g) Show that $\widetilde{w}_n$ converges to zero for step sizes μ satisfying $\mu < 2/(\rho + \lambda_{\max})$, where $\lambda_{\max}$ denotes the maximum eigenvalue of R_h (which is also equal to its spectral radius).

12.22 Is the objective function $P(w)$ defined by (12.164a) convex in w?

12.23 Refer to Newton recursion (12.197) for the minimization of $P(w)$ with $\epsilon = 0$. Introduce the change of variables $w = Az$ and write down the Newton method for minimizing the function $Q(z) = P(Az)$ over z, namely,

$$z_n = z_{n-1} - \left(\nabla_w^2 Q(z_{n-1})\right)^{-1} \nabla_{z^\mathsf{T}} Q(z_{n-1})$$

Multiply by A from the left and show that the result reduces to Newton's original recursion over w_n for minimizing $P(w)$. Conclude that the Newton method is invariant to affine scaling.

12.24 Consider the optimization problem (9.42) with linear equality constraints and the corresponding saddle point formulation (9.44). Apply gradient-descent on w and gradient-ascent on the dual variable β by using

$$\begin{aligned} w_n &= w_{n-1} - \mu_w \nabla_{w^\mathsf{T}} P(w_{n-1}) - \mu_w B^\mathsf{T} \beta_{n-1}, \; n \geq 0 \\ \beta_n &= \beta_{n-1} + \mu_\beta (B w_n - c) \end{aligned}$$

where $\mu_w > 0$ and $\mu_\beta > 0$ are step-size parameters and w_{-1} and β_{-1} are arbitrary initial conditions satisfying $\beta_{-1} = Bw_{-1}$ or $\beta_{-1} = 0$. Note that the updated iterate w_n is used in the recursion for β_n instead of w_{n-1}; we say that we are implementing the recursions in an *incremental* form. Assume $P(w)$ is ν-strongly convex and has δ-Lipschitz gradients, i.e., for any w_1, w_2:

$$\|\nabla_{w^\mathsf{T}} P(w_1) - \nabla_{w^\mathsf{T}} P(w_2)\| \leq \delta \|w_1 - w_2\|$$

Consider the unique saddle point $(w^\star, \beta_d^\star)$ defined by Lemma 9.2 and introduce the error quantities $\widetilde{w}_n = w^\star - w_n$ and $\widetilde{\beta}_n = \beta_d^\star - \beta_n$. Show that, for $\mu_w < 1/\delta$ and $\mu_\beta < \nu/\sigma_{\max}^2(B)$, the above recursions converge linearly to $(w^\star, \beta_d^\star)$, namely,

$$\|\widetilde{w}_n\|_{c_w}^2 + \|\widetilde{\beta}_n\|_{c_\beta}^2 \;\leq\; \rho\left(\|\widetilde{w}_{n-1}\|_{c_w}^2 + \|\widetilde{\beta}_{n-1}\|_{c_\beta}^2\right)$$

where $c_\beta = \mu_w/\mu_\beta > 0$, $c_w = 1 - \mu_w \mu_\beta \sigma_{\max}^2(B) > 0$, and

$$\rho \overset{\Delta}{=} \max\left\{1 - \mu_w \nu (1 - \mu_w \delta),\; 1 - \mu_w \mu_\beta \sigma_{\min}^2(B)\right\} < 1$$

where $\sigma_{\min}(B)$ is the smallest nonzero singular value of B. *Remark.* For more details, the reader may refer to Alghunaim and Sayed (2020) and the discussion therein.

12.25 Consider the same setting of Prob. 12.24 except that the saddle point $(w^\star, \beta_d^\star)$ is now sought by means of the following recursions

$$\begin{aligned} w_n &= w_{n-1} - \mu_w \nabla_{w^\mathsf{T}} P(w_{n-1}) - \mu_w B^\mathsf{T} \beta_{n-1}, \; n \geq 0 \\ \beta_n &= \beta_{n-1} + \mu_\beta (B w_{n-1} - c) \end{aligned}$$

The main difference is that w_{n-1} is used in the recursion for β_n instead of w_n. The above recursions, with w_{n-1} instead of w_n, are due to Arrow and Hurwicz (1956) and are known as the *Arrow–Hurwicz algorithm*. We will encounter an instance of it in Example 44.10 when studying Markov decision processes. We will also comment on the history of the algorithm in the concluding remarks of that chapter.

(a) Introduce the modified cost function $P_a(w) = P(w) - \frac{\mu_\beta}{2}\|Bw - c\|^2$. Write down the incremental recursions of Prob. 12.24 that would correspond to the problem of minimizing $P_a(w)$ subject to $Bw = c$. Verify that these recursions can be transformed into the Arrow–Hurwicz algorithm for $P(w)$ given above.

(b) Use the result of Prob. 12.24 to conclude that the Arrow–Hurwicz recursions also converge linearly to the unique saddle point $(w^\star, \beta_d^\star)$ for $\mu_w < 1/(\delta + \mu_\beta \sigma_{\max}^2(B))$ and $\mu_\beta < \nu/2\sigma_{\max}^2(B)$.

12.26 Consider a ν-strongly convex risk function $P(w)$ and apply the following gradient-descent algorithm for its minimization:

$$\boldsymbol{w}_n = \boldsymbol{w}_{n-1} - \mu \boldsymbol{D}_n \nabla_{w^{\mathsf{T}}} P(\boldsymbol{w}_{n-1})$$

where $\boldsymbol{D}_n$ is a diagonal matrix; each of its diagonal entries is either 1 with probability p or 0 with probability $1-p$. Repeat an analysis similar to the proof of Theorem 12.4 to determine conditions for the convergence of this scheme.

12.27 Consider a nonsmooth convex risk function $P(w) : \mathbb{R}^M \to \mathbb{R}$ for which a point $w^\star$ is found that satisfies (12.121). Does it follow that $w^\star$ is a global minimizer for $P(w)$? Consider the risk $P(w) = \|w\|^2 + |w_1 - w_2|$ and the point $(w_1, w_2) = (0.5, 0.5)$.

12.28 The following example is from Powell (1973). Consider a risk function over $\mathbb{R}^3$ of the form

$$P(w) = -w_1 w_2 - w_2 w_3 - w_1 w_3 + \sum_{k=1}^{3} \Big(|w_k| - 1\Big)_+^2$$

where the notation $(x)_+ = x$ if $x \geq 0$ and zero otherwise.

(a) Is $P(w)$ convex over w?

(b) Verify that $P(w)$ has two minimizers at locations $\text{col}\{1,1,1\}$ and $\text{col}\{-1,-1,-1\}$ within the cube defined by $\mathbb{C} = \{w \in \mathbb{R}^3 \mid -1 \leq w_k \leq 1\}$, where the $\{w_k\}$ denote the individual entries of w for $k = 1, 2, 3$.

(c) Choose an initial condition w_{-1} close to the vertices of $\mathbb{C}$, but distinct from the minimizers. Write down the corresponding coordinate-descent algorithm (12.119). Does it converge?

12.29 Consider the optimization problem

$$W^\star = \underset{W}{\text{argmin}}\ \|X - WZ\|_{\text{F}}^2$$

where X is $M \times N$, W is $M \times K$, and Z is $K \times N$.

(a) Verify that the cost function can be rewritten as

$$P(W) = \text{Tr}\Big\{(X - WZ)^{\mathsf{T}}(X - WZ)\Big\}$$

(b) Let $B = XZ^{\mathsf{T}}$ and $A = ZZ^{\mathsf{T}}$. Let also W_m denote the estimate for W at iteration m. Follow a coordinate-descent argument similar to the one leading to (12.125) to estimate the individual columns of W and show that the resulting algorithm involves an update of the form:

$$W_m = W_{m-1} + (B - W_{m-1}A)\text{diag}(A^{-1})$$

12.30 Refer to expression (12.142) and assume $\rho = 0$. Show that it can be rewritten in the equivalent form:

$$w_{n,m^o} \;=\; w_{n-1,m^o} + \frac{1}{\|x_{m^o}\|^2}\, x_{m^o}^{\mathsf{T}} \Big(\gamma_N - H_N w_{n-1}\Big)$$

12.31 The alternating projection algorithm (12.171) can be written in the equivalent form $a_n = \mathcal{P}_{\mathcal{C}_1}(\mathcal{P}_{\mathcal{C}_2}(a_{n-1}))$. Show that the cascade projection operator $\mathcal{P}(x) = \mathcal{P}_{\mathcal{C}_1}(\mathcal{P}_{\mathcal{C}_2}(x))$ is nonexpansive, i.e.,

$$\|\mathcal{P}(x) - \mathcal{P}(y)\| \;\leq\; \|x - y\|, \quad \forall x, y \in \mathcal{C}_1$$

12.32 The method of *averaged projections* is an alternative to the alternating projection algorithm (12.171). It starts from an initial vector w_{-1} and updates it recursively as follows by averaging its projections on the two convex sets:

$$w_n = \frac{1}{2}\Big(\mathcal{P}_{\mathcal{C}_1}(w_{n-1}) \;+\; \mathcal{P}_{\mathcal{C}_2}(w_{n-1})\Big), \quad n \geq 0$$

Assume the intersection $\mathcal{C}_1 \cap \mathcal{C}_2$ is nonempty. Establish convergence of the sequence w_n to some point $w^\star$ in this intersection.

12.33 The Dykstra method (12.208) is a variation of the alternating projection algorithm. Show that, starting from an initial vector a_{-1}, it converges to the unique point $w^\star \in \mathcal{C}_1 \cap \mathcal{C}_2$ that is closest to a_{-1} from within the intersection set. *Remark.* The reader may refer to Dykstra (1983), Boyle and Dykstra (1986), Combettes and Pesquet (2011), and Dattoro (2016) for a related discussion.

12.34 We wish to determine a solution to the linear system of equations $Hw = d$, where $H \in \mathbb{R}^{N\times M}$. This objective can be recast as the problem of determining the intersection of a collection of affine subspaces. We denote the rows of H by $\{h_k^\mathsf{T}\}$ and the entries of d by $\{\theta_k\}$ for $k = 1, 2, \ldots, N$, and consider the hyperplanes $\mathcal{H}_k = \{z \,|\, h_k^\mathsf{T} z - \theta_k = 0\}$. Starting from an initial condition w_{-1}, assume we apply a cyclic projection algorithm and project successively onto the $\mathcal{H}_k$ and keep repeating the procedure. Use the result of Prob. 9.3 for projections onto hyperplanes to verify that this construction leads to the following so-called Kaczmarz algorithm:

$$\left\{\begin{array}{l} \text{current projection is } w_{n-1} \\ \text{current selected row of } H \text{ is } h_k \text{ and selected entry of } d \text{ is } \theta_k \\ \text{update } w_n = w_{n-1} - \dfrac{(h_k^\mathsf{T} w_{n-1} - \theta_k)}{\|w_{n-1}\|^2} w_{n-1} \end{array}\right.$$

Remark. These recursions correspond to the classical method of Kaczmarz (1937) for the solution of linear systems of equations. We will encounter a randomized version later in Prob. 16.7 and apply it to the solution of least-squares problems.

12.35 Refer to the BFGS algorithm (12.206). Verify that $z_n^\mathsf{T} a_n > 0$ for strictly convex risks $P(w)$. Use the matrix inversion lemma to show that

$$B_n^{-1} = \left(I_M - \frac{1}{z_n^\mathsf{T} a_n} z_n a_n^\mathsf{T}\right) B_{n-1}^{-1} \left(I_M - \frac{1}{z_n^\mathsf{T} a_n} z_n a_n^\mathsf{T}\right)^\mathsf{T} + \frac{1}{z_n^\mathsf{T} a_n} z_n z_n^\mathsf{T}$$

Conclude that $B_n > 0$ when $P(w)$ is strictly convex and $z_n \neq 0$.

12.36 Continuing with the BFGS algorithm (12.206), introduce the second-order approximation for the risk function around w_n:

$$\widehat{P}(w) = P(w_n) + \nabla_w P(w_n)(w - w_n) + \frac{1}{2}(w - w_n)^\mathsf{T} B_n (w - w_n)$$

Use the secant condition to show that the gradient vectors of $P(w)$ and its approximation coincide at the locations w_n and w_{n-1}, i.e.,

$$\nabla_w \widehat{P}(w_n) = \nabla_w P(w_n), \quad \nabla_w \widehat{P}(w_{n-1}) = \nabla_w P(w_{n-1})$$

12.37 The material in Probs. 12.37–12.39 is motivated by results from Nesterov and Spokoiny (2017). Refer to the smoothed function (12.216) relative to the Gaussian distribution $f_{\boldsymbol{x}}(x) = \mathcal{N}_{\boldsymbol{x}}(0, I_M)$. Verify that:
(a) $P_\alpha(w) \geq P(w)$ for any $\alpha > 0$.
(b) $P_\alpha(w)$ is convex when $P(w)$ is convex.
(c) If $P(w)$ is δ-Lipschitz then $P_\alpha(w)$ is δ_α-Lipschitz with $\delta_\alpha \leq \delta$.
(d) If $P(w)$ has δ-Lipschitz gradients then $P_\alpha(w)$ has δ_α-Lipschitz gradients with $\delta_\alpha \leq \delta$.

12.38 Refer to the smoothed function (12.216) relative to the Gaussian distribution $f_{\boldsymbol{x}}(x) = \mathcal{N}_{\boldsymbol{x}}(0, I_M)$. Assume $P(w)$ is δ-Lipschitz, i.e., $|P(w_1) - P(w_2)| \leq \delta\|w_1 - w_2\|$ for all $w_1, w_2 \in \text{dom}(P)$. Show that $P_\alpha(w)$ is first-order differentiable and its gradient vector is δ_α-Lipschitz, i.e.,

$$\|\nabla_w P_\alpha(w_1) - \nabla_w P_\alpha(w_2)\| \leq \delta_\alpha \|w_1 - w_2\|$$

with $\delta_\alpha = \delta\sqrt{M}/\alpha$.

12.39 Refer to the smoothed function (12.216) relative to the Gaussian distribution $f_{\boldsymbol{x}}(x) = \mathbb{N}_{\boldsymbol{x}}(0, I_M)$. Assume $P(w)$ has δ-Lipschitz gradients, i.e.,

$$\|\nabla_w P(w_1) - \nabla_w P(w_2)\| \leq \delta\|w_1 - w_2\|$$

for all $w_1, w_2 \in \text{dom}(P)$. Establish the following inequalities:
(a) $|P_\alpha(w) - P(w)| \leq \alpha^2 M\delta/2$.
(b) $\|\nabla_w P_\alpha(w) - \nabla_w P(w)\| \leq \alpha\delta(M+3)^{3/2}/2$.
(c) $\|\nabla_w P(w)\|^2 \leq 2\|\nabla_w P_\alpha(w)\|^2 + \alpha^2\delta^2(M+6)^3/2$.
Show further that the second-order moment of the gradient approximation (12.214) satisfies

$$\mathbb{E}_{\boldsymbol{u}}\|\widehat{\nabla_w P}(w)\|^2 \leq 2(M+4)\|\nabla_w P(w)\|^2 + \alpha^2\delta^2(M+6)^3/2$$

Conclude that the error variance in estimating the gradient vector is bounded by the sum of two components: one varies with α^2 and decays with α, while the other is independent of α but depends on the problem dimension M.
12.40 Verify equality (12.226), which relates the integral over a ball to the integral over its spherical surface.
12.41 Consider the following one-point estimate for the gradient vector in place of (12.214):

$$\widehat{\nabla_{w^\mathsf{T}} P}(w) = \frac{\beta}{\alpha} P(w + \alpha u)u$$

Verify that the unbiasedness property (12.218) continues to hold.

12.A ZEROTH-ORDER OPTIMIZATION

The gradient-descent algorithm described in the body of the chapter is an example of a first-order optimization method, which requires the availability of the gradient information, $\nabla_w P(w)$, in order to perform the update:

$$w_n = w_{n-1} - \mu\nabla_{w^\mathsf{T}} P(w_{n-1}), \quad n \geq 0 \tag{12.212}$$

where we are assuming a constant step size for illustration purposes. In some situations of interest it may not be possible to evaluate the gradient function either because the risk $P(w)$ may not have a closed analytical form or is unknown to the designer altogether. This latter situation will arise, for example, in adversarial learning scenarios, studied in Chapter 71, where the designer wishes to misguide the operation of a learning algorithm but is only able to perform function evaluations, $P(w)$. Zeroth-order optimization is a technique that enables the designer to approximate the gradient vector by relying on function evaluations. We provide a brief overview of the methodology in this appendix.

Two-point gradient estimate

One way to approximate $\nabla_{w^\mathsf{T}} P(w)$ is to construct a *two-point estimate* as follows. Let $u \in \mathbb{R}^M$ denote a realization for a random vector that is selected according to some predefined distribution. Two choices are common:

$$\boldsymbol{u} \sim \mathbb{N}_{\boldsymbol{u}}(0, I_M) \quad \textbf{(Gaussian distribution)} \tag{12.213a}$$

$$\boldsymbol{u} \sim \mathcal{U}(\mathbb{S}) \quad \textbf{(uniform distribution on the unit sphere, } \mathbb{S}\textbf{)} \tag{12.213b}$$

where $\mathbb{S}$ denotes the unit sphere in $\mathbb{R}^M$ of radius 1 and centered at the origin. In both cases, the variable $\boldsymbol{u}$ has zero mean. Let $\alpha > 0$ denote a small parameter known as the *smoothing factor*. Then, we construct

$$\boxed{\widehat{\nabla_{w^{\mathsf{T}}} P(w)} = \frac{\beta}{\alpha}\Big(P(w+\alpha\, u) - P(w)\Big)u} \tag{12.214}$$

where the value of the scalar β depends on which mechanism is used to generate the directional vector u:

$$\beta = \begin{cases} 1, & \text{if } \boldsymbol{u} \sim \mathcal{N}_{\boldsymbol{u}}(0, I_M) \\ M, & \text{if } \boldsymbol{u} \sim \mathcal{U}(\mathbb{S}) \end{cases} \tag{12.215}$$

Note that expression (12.214) requires two function evaluations to approximate the gradient vector.

Smoothed risk function

Construction (12.214) has one useful property. We introduce the following smoothed version of $P(w)$, which is dependent on α and where the integration is over the domain of the variable x:

$$P_\alpha(w) \triangleq \int_{x\in\mathcal{X}} P(w+\alpha x) f_{\boldsymbol{x}}(x)dx \;=\; \mathbb{E}_{\boldsymbol{x}}\Big\{P(w+\alpha \boldsymbol{x})\Big\} \tag{12.216}$$

The distribution of $\boldsymbol{x} \in \mathbb{R}^M$ depends on the mechanism used to generate $\boldsymbol{u}$, namely,

$$f_{\boldsymbol{x}}(x) \triangleq \begin{cases} \mathcal{N}_{\boldsymbol{x}}(0, I_M) & \textbf{(Gaussian distribution)} \\ \mathcal{U}(\mathbb{B}) & \textbf{(unit ball)} \end{cases} \tag{12.217}$$

where the symbol $\mathbb{B}$ denotes the unit ball in $\mathbb{R}^M$ of radius 1 and centered at the origin (all vectors $x \in \mathbb{B}$ will have $\|x\| \leq 1$). The sphere $\mathbb{S}$ is the surface of this ball, and the ball is the interior of the sphere. The smoothed function $P_\alpha(w)$ is differentiable (even when $P(w)$ is not) with its gradient vector satisfying:

$$\boxed{\nabla_{w^{\mathsf{T}}} P_\alpha(w) \;=\; \mathbb{E}_{\boldsymbol{u}}\Big\{\widehat{\nabla_{w^{\mathsf{T}}} P(w)}\Big\}} \tag{12.218}$$

where the expectation is over the distribution used to generate $\boldsymbol{u}$. This result shows that construction (12.214) provides an unbiased estimate for the gradient of $P_\alpha(w)$. Additional properties for the smoothed function are listed in Prob. 12.37.

Proof of (12.218): Consider first the Gaussian case. Then, it holds that

$$\begin{aligned} P_\alpha(w) &\overset{(12.216)}{=} \int_{x\in\mathcal{X}} P(w+\alpha x)\frac{1}{\sqrt{(2\pi)^M}}\exp\Big\{-\frac{1}{2}\|x\|^2\Big\}dx \\ &= \frac{1}{\alpha}\frac{1}{\sqrt{(2\pi)^M}}\int_{y\in\mathcal{Y}} P(y)\exp\Big\{-\frac{1}{2\alpha^2}\|y-w\|^2\Big\}dy, \quad \text{using } y = w+\alpha x \end{aligned} \tag{12.219}$$

Differentiating relative to w gives

$$\begin{aligned}
\nabla_{w^\mathsf{T}} P_\alpha(w) &= \frac{1}{\alpha}\frac{1}{\sqrt{(2\pi)^M}}\int_{y\in\mathcal{Y}} P(y)\exp\Big\{-\frac{1}{2\alpha^2}\|y-w\|^2\Big\}\times\frac{1}{\alpha^2}(y-w)dy \\
&= \frac{1}{\sqrt{(2\pi)^M}}\int_{x\in\mathcal{X}} P(w+\alpha x)\exp\Big\{-\frac{1}{2}\|x\|^2\Big\}\times\frac{1}{\alpha^2}(\alpha x)dx \\
&= \frac{1}{\alpha}\int_{x\in\mathcal{X}} P(w+\alpha x)x\,\mathcal{N}_{\boldsymbol{x}}(0,I_M)dx \\
&= \frac{1}{\alpha}\mathbb{E}_{\boldsymbol{x}}\Big\{P(w+\alpha\boldsymbol{x})\boldsymbol{x}\Big\} \\
&= \frac{1}{\alpha}\mathbb{E}_{\boldsymbol{x}}\Big\{P(w+\alpha\boldsymbol{x})\boldsymbol{x}\Big\} - \underbrace{\frac{1}{\alpha}\mathbb{E}_{\boldsymbol{x}}\Big\{P(w)\boldsymbol{x}\Big\}}_{=0} \\
&= \mathbb{E}_{\boldsymbol{x}}\left\{\frac{1}{\alpha}\Big(P(w+\alpha\boldsymbol{x})-P(w)\Big)\boldsymbol{x}\right\} \\
&\overset{(12.214)}{=} \mathbb{E}_{\boldsymbol{u}}\Big\{\widehat{\nabla_{w^\mathsf{T}}P(w)}\Big\}
\end{aligned} \tag{12.220}$$

as claimed, where the last equality is because $\boldsymbol{u}$ and $\boldsymbol{x}$ have the same Gaussian distribution.

Consider next the case in which $\boldsymbol{u}$ is selected uniformly from the unit sphere, $\mathbb{S}$. Then,

$$\begin{aligned}
P_\alpha(w) &\overset{(12.216)}{=} \mathbb{E}_{\boldsymbol{x}\in\mathbb{B}}\Big\{P(w+\alpha\boldsymbol{x})\Big\} \\
&= \mathbb{E}_{\boldsymbol{y}\in\alpha\mathbb{B}}\Big\{P(w+\boldsymbol{y})\Big\}, \text{ using change of variables } \boldsymbol{y}=\alpha\boldsymbol{x} \\
&= \frac{1}{\text{vol}(\alpha\mathbb{B})}\int_{y\in\alpha\mathbb{B}} P(w+y)dy
\end{aligned} \tag{12.221}$$

where $\text{vol}(\alpha\mathbb{B})$ denotes the volume of the ball of radius α; we recall that the volume of a ball in $\mathbb{R}^M$ centered at the origin with radius α is given by

$$\text{vol}(\alpha\mathbb{B}) = \frac{\pi^{M/2}\,\alpha^M}{\Gamma(\frac{M}{2}+1)} \tag{12.222}$$

in terms of the gamma function. Moreover, it also holds that

$$\begin{aligned}
\mathbb{E}_{\boldsymbol{u}\in\mathbb{S}}\Big\{P(w+\alpha\boldsymbol{u})\boldsymbol{u}\Big\} &= \mathbb{E}_{\boldsymbol{z}\in\alpha\mathbb{S}}\Big\{P(w+\boldsymbol{z})\frac{\boldsymbol{z}}{\alpha}\Big\}, \text{ using change of variables } \boldsymbol{z}=\alpha\boldsymbol{u} \\
&= \mathbb{E}_{\boldsymbol{z}\in\alpha\mathbb{S}}\Big\{P(w+\boldsymbol{z})\frac{\boldsymbol{z}}{\|\boldsymbol{z}\|}\Big\}, \text{ since } \|\boldsymbol{z}\|=\alpha \\
&= \frac{1}{\text{surf}(\alpha\mathbb{S})}\int_{z\in\alpha\mathbb{S}} P(w+z)\frac{z}{\|z\|}dz
\end{aligned} \tag{12.223}$$

where $\text{surf}(\alpha\mathbb{S})$ denotes the surface area of the ball of radius α; we recall that the surface area of a ball in $\mathbb{R}^M$ centered at the origin with radius α is given by

$$\text{surf}(\alpha\mathbb{S}) = \frac{2\pi^{M/2}\,\alpha^{M-1}}{\Gamma(\frac{M}{2})} \tag{12.224}$$

Using the property $\Gamma(z+1)=z\Gamma(z)$ for gamma functions, we conclude that

$$\frac{\text{surf}(\alpha\mathbb{S})}{\text{vol}(\alpha\mathbb{B})} = \frac{M}{\alpha} \tag{12.225}$$

From the divergence theorem in calculus, which allows us to relate integration over a volume to the integral over its surface, it can be verified that (see Prob. 12.40):

$$\nabla_{w^\mathsf{T}}\left\{\int_{y\in\alpha\mathbb{B}} P(w+y)dy\right\} = \int_{z\in\alpha\mathbb{S}} P(w+z)\frac{z}{\|z\|}dz \tag{12.226}$$

Collecting terms we conclude that (12.218) holds.

■

An alternative two-point estimate for the gradient in place of (12.214) is the symmetric version

$$\widehat{\nabla_{w^\mathsf{T}} P}(w) = \frac{\beta}{2\alpha}\Big(P(w+\alpha u) - P(w-\alpha u)\Big)u \tag{12.227}$$

where the arguments of the risk function are $w \pm \alpha u$ and the factor in the denominator is 2α. In Prob. 12.41 we consider another example.

Zeroth-order algorithm

We can now list a zeroth-order algorithm for minimizing a risk function $P(w) : \mathbb{R}^M \to \mathbb{R}$. The original gradient-descent recursion (12.24) is replaced by (12.228). In the listing, we denote the distribution from which the directional vectors $\boldsymbol{u}$ are sampled by $f_{\boldsymbol{u}}(u)$; it can refer either to the Gaussian distribution $\mathcal{N}_{\boldsymbol{u}}(0, I_M)$ or the uniform distribution $\mathcal{U}(\mathbb{S})$, as described by (12.213a)–(12.213b).

Zeroth-order gradient-based method for minimizing $P(w)$.

given a small step-size parameter $\mu > 0$;
given a small smoothing factor $\alpha > 0$;
select the sampling distribution $f_{\boldsymbol{u}}(u)$ and set $\beta \in \{1, M\}$;
start from an arbitrary initial condition, w_{-1}.
repeat until sufficient convergence over $n \geq 0$:
  sample $u_n \sim f_{\boldsymbol{u}}(u)$
  $\widehat{\nabla_{w^\mathsf{T}} P}(w_{n-1}) = \frac{\beta}{\alpha}\Big(P(w_{n-1}+\alpha u_n) - P(w_{n-1})\Big)u_n$
  $w_n = w_{n-1} - \mu\,\widehat{\nabla_{w^\mathsf{T}} P}(w_{n-1})$
end
return $w^\star \leftarrow w_n$.
(12.228)

REFERENCES

Alghunaim, S. A. and A. H. Sayed (2020), "Linear convergence of primal–dual gradient methods and their performance in distributed optimization," *Automatica*, vol. 117, article 109003.

Armijo, L. (1966), "Minimization of functions having Lipschitz continuous first partial derivatives," *Pacific J. Math.*, vol. 16, no. 1, pp. 1–3.

Arrow, K. J. and L. Hurwicz (1956), "Reduction of constrained maxima to saddle-point problems," *Proc. 3rd Berkeley Symp. Mathematical Statistics and Probability*, pp. 1–20.

Bauschke, H. H. and Borwein, J. M. (1996), "On projection algorithms for solving convex feasibility problems," *SIAM Rev.*, vol. 38, no. 3, pp. 367–426.

Beck, A. and L. Tetruashvili (2013), "On the convergence of block coordinate descent type methods," *SIAM J. Optim.*, vol. 23, no. 4, pp. 2037–2060.

Bertsekas, D. P. (1995), *Nonlinear Programming*, Athena Scientific.

Bertsekas, D. P. and J. N. Tsitsiklis (1997), *Parallel and Distributed Computation: Numerical Methods*, Athena Scientific.

Bottou, L. (1998), "Online algorithms and stochastic approximations," in *Online Learning and Neural Networks*, D. Saad, editor, Cambridge University Press.

Bottou, L. and Y. LeCun (2004), "Large scale online learning," *Proc. Advances Neural Information Processing Systems* (NIPS), vol. 16, pp. 217–224, Cambridge, MA.

Boyd, S. and L. Vandenberghe (2004), *Convex Optimization*, Cambridge University Press.

Boyle, J. P. and R. L. Dykstra (1986), "A method for finding projections onto the intersection of convex sets in Hilbert spaces," *Lect. Notes Statist.*, vol. 37, pp. 28–47.

Bregman, L. M. (1965), "The method of successive projection for finding a common point of convex sets," *Soviet Math.*, vol. 6, pp. 688–692.

Bregman, L. M. (1967), "The relaxation method of finding the common points of convex sets and its application to the solution of problems in convex programming," *USSR Comput. Math. Math. Phys.*, vol. 7, no. 3, pp. 200–217.

Brent, R. (2002), *Algorithms for Minimization without Derivatives*, Prentice Hall.

Broyden, C. G. (1970), "The convergence of a class of double-rank minimization algorithms," *J. Institu. Math. Appl.*, vol. 6, pp. 76–90.

Cauchy, A.-L. (1847), "Methode générale pour la résolution des systems déquations simultanes," *Comptes Rendus Hebd. Séances Acad. Sci.*, vol. 25, pp. 536–538.

Cevher, V., S. Becker, and M. Schmidt (2014), "Convex optimization for big data: Scalable, randomized, and parallel algorithms for big data analytics," *IEEE Signal Process. Mag.*, vol. 31, no. 5, pp. 32–43.

Chang, K. W., C. J. Hsieh, and C. J. Lin (2008), "Coordinate descent method for large-scale L2-loss linear SVM," *J. Mach. Learn. Res.*, vol. 9, pp. 1369–1398.

Cheney, W. and A. Goldstein (1959), "Proximity maps for convex sets," *Proc. AMS*, vol. 10, pp. 448–450.

Christensen, C. (1996), "Newton's method for resolving affected equations," *College Math. J.* vol. 27, no. 5, pp. 330–340.

Combettes, P. L. (1993), "The foundations of set theoretic estimation," *Proc. IEEE*, vol. 81, no. 2, pp. 182–208.

Combettes, P. L. and J.-C. Pesquet (2011), "Proximal splitting methods in signal processing," in *Fixed-Point Algorithms for Inverse Problems in Science and Engineering*, H. H. Bauschke *et al.*, editors, pp. 185–212, Springer.

Conn, A. R., K. Scheinberg, and L. N. Vicente (2009), *Introduction to Derivative-Free Optimization*, SIAM.

Curry, H. B. (1944), "The method of steepest descent for nonlinear minimization problems," *Q. J. Mech. App. Math.*, vol. 2, pp. 258–261.

Dattoro, J. (2016), *Convex Optimization and Euclidean Distance Geometry*, Meboo Publishing.

Daubechies, I., M. Defrise, and C. De Mol (2004), "An iterative thresholding algorithm for linear inverse problems with a sparsity constraint," *Commun. Pure App. Math.*, vol. LVII, pp. 1413–1457.

Duchi, J. C., M. I. Jordan, M. J. Wainwright, and A. Wibisono (2015), "Optimal rates for zero-order convex optimization: The power of two function evaluations," *IEEE Trans. Inf. Theory*, vol. 61, no. 5, pp. 2788–2806.

Dykstra, R. L. (1983), "An algorithm for restricted least squares regression," *J. Amer. Statist. Assoc.*, vol. 78, no. 384, pp. 837–842.

Escalante, R. and M. Raydan (2011), *Alternating Projection Methods*, SIAM.

Fercoq O. and P. Bianchi (2019), "A coordinate-descent primal–dual algorithm with large step size and possibly nonseparable functions," *SIAM J. Optim.*, vol. 29, no. 1, pp. 100–134.

Flaxman, A. D., A. T. Kalai, and H. B. McMahan (2005), "Online convex optimization in the bandit setting: Gradient descent without a gradient," *Proc. Ann. ACM-SIAM Symp. Discrete Algorithms*, pp. 385–394, Vancouver, BC.

Fletcher, R. (1970), "A new approach to variable metric algorithms," *Comput. J.*, vol. 13, no. 3, pp. 317–322.

Fletcher, R. (1987), *Practical Methods of Optimization*, 2nd ed., Wiley.

Friedman, J. H., T. Hastie, H. Höfling, and R. Tibshirani (2007), "Pathwise coordinate optimization," *Ann. App. Statist.*, vol. 1, no. 2, pp. 302–332.

Fu, W. J. (1998), "Penalized regressions: The bridge versus the Lasso," *J. Comput. Graph. Statist.*, vol. 7, no. 3, pp. 397–416.

Gauss, C. F. (1903), *Carl Friedrich Gauss Werke,* Akademie der Wissenschaften.

Goldfarb, D. (1970), "A family of variable metric updates derived by variational means," *Math. Comput.*, vol. 24, no. 109, pp. 23–26.

Goldstein, A. A. (1962), "Cauchy's method of minimization," *Numer. Math.*, vol. 4, no. 2, pp. 146–150.

Goldstein, A. A. (1966), "Minimizing functionals on normed-linear spaces," *SIAM J. Control*, vol. 4, pp. 91–89.

Golub, G. H. and C. F. Van Loan (1996), *Matrix Computations*, 3rd ed., John Hopkins University Press.

Gubin, L. G., B. T. Polyak, and E. V. Raik (1967), "The method of projections for finding the common point of convex sets," *USSR Comput. Math. Math. Phys.*, vol. 7, no. 6, pp. 1–24.

Halperin, I. (1962), "The product of projection operators," *Acta. Sci. Math.*, vol. 23, pp. 96–99.

Han, S.-P. (1988), "A successive projection method," *Math. Program.*, vol. 40, pp. 1–14.

Hildreth, C. (1957), "A quadratic programming procedure," *Naval Res. Logist. Q.*, vol. 4, pp. 79–85. See also erratum in same volume on page 361.

Kaczmarz, S. (1937), "Angenäherte Auflösung von Systemen linearer Gleichungen," *Bull. Int. Acad. Polon. Sci. Lett.* A, vol. 35, pp. 335–357.

Kelley, C. T. (1996), *Iterative Methods for Optimization*, SIAM.

Kiefer, J. and J. Wolfowitz (1952), "Stochastic estimation of the maximum of a regression function," *Ann. Math. Statist.*, vol. 23, no. 3, pp. 462–466.

Kollerstrom, N. (1992), "Thomas Simpson and Newton's method of approximation: An enduring myth," *Br. J. Hist. Sci.*, vol. 25, no. 3, pp. 347–354.

Lange, K., E. C. Chi, and H. Zhou (2014), "A brief survey of modern optimization for statisticians," *Int. Statist. Rev.*, vol. 82, no. 1, pp. 46–70.

Larson, J., M. Menickelly, and S. M. Wild (2019), "Derivative-free optimization methods," *Acta Numer.*, vol. 28, pp. 287–404.

Lemaréchal, C. (2012), "Cauchy and the gradient method," *Documenta Mathematica*, Extra Volume ISMP, pp. 251–254.

Le Roux, N., M. Schmidt, and F. Bach (2012), "A stochastic gradient method with an exponential convergence rate for finite training sets," *Proc. Advances Neural Information Processing Systems* (NIPS), pp. 2672–2680, Lake Tahoe.

Liu, S., P.-Y. Chen, B. Kailkhura, G. Zhang, A. Hero, and P. Varshney (2020), "A primer on zeroth-order optimization in signal processing and machine learning," *IEEE Signal Process. Mag.*, vol 37, no. 5, pp. 43–54.

Liu, S., B. Kailkhura, P.-Y. Chen, P. Ting, S. Chang, and L. Amini (2018), "Zeroth-order stochastic variance reduction for nonconvex optimization,"*Proc. Advances Neural Information Processing Systems* (NIPS), pp. 3727–3737.

Luenberger, D. G. and Y. Ye (2008), *Linear and Nonlinear Programming*, Springer.

Luo, Z. Q. and P. Tseng (1992a), "On the convergence of the coordinate descent method for convex differentiable minimization," *J. Optim. Theory Appl.*, vol. 72, pp. 7–35.

Nash, S. G. and A. Sofer (1996), *Linear and Nonlinear Programming*, McGraw-Hill.

Nesterov, Y. (1983), "A method for unconstrained convex minimization problem with the rate of convergence $O(1/k^2)$," *Doklady AN USSR*, vol. 269, pp. 543–547.

Nesterov, Y. (2004), *Introductory Lectures on Convex Optimization*, Springer.

Nesterov, Y. (2005), "Smooth minimization of non-smooth functions," *Math. Program.*, vol. 103, no. 1, pp. 127–152.

Nesterov, Y. (2012), "Efficiency of coordinate descent methods on huge-scale optimization problems," *SIAM J. Optim.*, vol. 22, no. 2, pp. 341–362.

Nesterov, Y. and V. Spokoiny (2017), "Random gradient-free minimization of convex functions," *Found. Comput. Math.*, vol. 17, no. 2, pp. 527–566.

Nocedal, J. and S. J. Wright (2006), *Numerical Optimization*, Springer, NY.

Nutini, J., M. Schmidt, I. Laradji, M. Friedlander, and H. Koepke (2015), "Coordinate descent converges faster with the Gauss–Southwell rule than random selection," *Proc. Int. Conf. Machine Learning* (ICML), pp. 1632–1641, Lille, France.

Ortega, J. M. and W. Rheinboldt (1970), *Iterative Solution of Nonlinear Equations in Several Variables*, Academic Press.

Polyak, B. T. (1964), "Some methods of speeding up the convergence of iteration methods," *USSR Comput. Math. Math. Phys.*, vol. 4, no. 5, pp. 1–17.

Polyak, B. T. (1987), *Introduction to Optimization*, Optimization Software.

Polyak, B. T. and A. Juditsky (1992), "Acceleration of stochastic approximation by averaging," *SIAM J. Control Optim.*, vol. 30, no. 4, pp. 838–855.

Powell, M. J. D. (1973), "On search directions for minimization algorithms," *Math. Program.*, vol. 4, pp. 193–201.

Raphson, J. (1697), *Analysis aequationum universalis seu ad aequationes algebraicas resolvendas methodus generalis, and expedita, ex nova infinitarum serierum methodo, deducta ac demonstrata,* publisher Th. Braddyll.

Richtárik, P. and M. Takác (2011), "Iteration complexity of randomized block-coordinate descent methods for minimizing a composite function," *Math. Program.*, Series A, 144, no. 1–2, pp. 1–38.

Rios, L. M. and N. V. Sahinidis (2013), "Derivative-free optimization: A review of algorithms and comparison of software implementations," *J. Global Optim.*, vol. 56, no. 3, pp. 1247–1293.

Sauer, K. and C. Bouman (1993), "A local update strategy for iterative reconstruction from projections," *IEEE Trans. Signal Process.*, vol. 41, no. 2, pp. 534–548.

Sayed, A. H. (2014a), *Adaptation, Learning, and Optimization over Networks,* Foundations and Trends in Machine Learning, NOW Publishers, vol. 7, no. 4–5, pp. 311–801.

Seidel, L. (1874), "Über ein Verfahren, die Gleichungen, auf welche die Methode der kleinsten Quadrate führt, sowie lineare Gleichungen überhaupt, durch sukzessive Anŋnäherung au fzulösen," *Abh. Bayer. Akad. Wiss.*, vol. 11, no. 3, pp. 81–108.

Shanno, D. F. (1970), "Conditioning of quasi-Newton methods for function minimization," *Math. Comput.*, vol. 24, no. 111, pp. 647–656.

Shi, H.-J. M., S. Tu, Y. Xu, and W. Yin (2017), "A primer on coordinate descent algorithms," available at arXiv:1610.00040.

Simpson, T. (1740), *Essays on Several Curious and Useful Subjects in Speculative and Mix'd Mathematicks*, London.

Southwell, R. V. (1940), *Relaxation Methods in Engineering Science: A Treatise on Approximate Computation*, Oxford University Press.

Tseng, P. (2001), "Convergence of a block coordinate descent method for nondifferentiable minimization," *J. Optim. Theory Appl.*, vol. 109, pp. 475–494.

Tseng, P. and S. Yun (2009), "A coordinate gradient descent method for nonsmooth separable minimization," *Math. Program.*, vol. 117, pp. 387–423.

Tseng, P. and S. Yun (2010), "A coordinate gradient descent method for linearly constrained smooth optimization and support vector machines training," *Comput. Optim. Appl.*, vol. 47, pp. 179–206.

von Neumann, J. (1949), "On rings of operators: Reduction theory," *Ann. Math.*, vol. 50, no. 2, pp. 401–485.

von Neumann, J. (1950), *Functional Operators II: The Geometry of Orthogonal Spaces*, vol. 22, *Annal. Math. Studies.* Reprinted from lecture notes first distributed in 1933.

Wallis, J. (1685), *A Treatise of Algebra, both Historical and Practical. Shewing the Original, Progress, and Advancement thereof, from time to time, and by what Steps it hath attained to the Heighth at which it now is*, printed by John Playford, London.

Wang, C., Y. Zhang, B. Ying, and A. H. Sayed (2018), "Coordinate-descent diffusion learning by networked agents," *IEEE Trans. Signal Process.*, vol. 66, no. 2, pp. 352–367.

Warga, J. (1963), "Minimizing certain convex functions," *SIAM J. App. Math.,* vol. 11, pp. 588–593.

Wibisono, A., M. J Wainwright, M. I. Jordan, and J. C. Duchi (2012), "Finite sample convergence rates of zero-order stochastic optimization methods," *Proc. Advances Neural Information Processing Systems* (NIPS), pp. 1439–1447.

Wolfe, P. (1969), "Convergence conditions for ascent methods," *SIAM Rev.*, vol. 11, no. 2, pp. 226–235.

Wolfe, P. (1971), "Convergence conditions for ascent methods II: Some corrections," *SIAM Rev.,* vol. 13, pp. 185–188.

Wright, S. J. (2015), "Coordinates descent algorithms," *Math. Program.*, vol. 151, pp. 3–34.

Wu, T. T. and K. Lange (2008), "Coordinate descent algorithms for LASSO penalized regression," *Ann. App. Statist.*, vol. 2, no. 1, pp. 224–244.

Yu, H., R. Jin, and S. Yang (2019), "On the linear speedup analysis of communication efficient momentum SGD for distributed non-convex optimization," *Proc. Mach. Learn. Res.* (PMLR), vol. 97, pp. 7184–7193.

13 Conjugate Gradient Method

Before discussing extensions of the gradient-descent method to deal with non-smooth risk functions, such as the subgradient and proximal gradient methods, we pause to introduce another class of iterative algorithms for smooth functions, known as the *conjugate gradient method*. This technique was originally developed for the solution of linear systems of equations with symmetric positive-definite coefficient matrices, but was subsequently extended to nonlinear optimization problems as well. The method is based on the idea of updating the successive weight iterates along directions that are "orthogonal" to each other. As such, the method involves steps to determine these update directions in comparison to gradient descent implementations. We first explain how the conjugate gradient solves linear systems of equations, and later extend it to nonlinear optimization problems.

13.1 LINEAR SYSTEMS OF EQUATIONS

Consider linear systems of equations with symmetric and positive-definite coefficient matrices, such as

$$Aw = b, \quad \text{where } A = A^{\mathsf{T}} \in \mathbb{R}^{M\times M},\ A > 0 \tag{13.1}$$

Since A is invertible, the solution is unique and given by $w^\star = A^{-1}b$. The conjugate gradient method will determine $w^\star$ in an alternative manner, without the need to invert A. The method is motivated as follows.

13.1.1 *A*-Orthogonality

First, we observe that $w^\star$ is the unique minimizer to the following quadratic risk:

$$w^\star = \underset{w\in\mathbb{R}^M}{\operatorname{argmin}} \left\{ P(w) \overset{\Delta}{=} \frac{1}{2}w^{\mathsf{T}}Aw - b^{\mathsf{T}}w \right\} \tag{13.2}$$

Next, we assume that we have available a collection of M nonzero vectors $q_m \in \mathbb{R}^M$, for $m = 0, 1, \ldots, M-1$, satisfying the property:

$$q_k^{\mathsf{T}} A q_\ell = 0, \quad \forall k \neq \ell \tag{13.3}$$

We will explain further ahead how these vectors can be determined. For now, we assume they are available and will explain how the solution $w^\star$ can be determined from knowledge of the $\{q_m\}$. In view of (13.3), we say that the vectors $\{q_m\}$ are A-orthogonal or A-conjugate (or simply conjugate) to each other relative to the inner product induced by the matrix A, namely,

$$\langle x, y\rangle_A \stackrel{\Delta}{=} x^\mathsf{T} A y \qquad \textbf{(weighted inner product)} \tag{13.4}$$

for any two vectors $\{x, y\}$ of compatible dimensions. It is straightforward to verify that this definition satisfies the properties of inner products and, moreover (see Prob. 13.1):

$$\langle x, x\rangle_A = 0 \iff x = 0 \tag{13.5}$$

Using property (13.3) we can verify that the vectors $\{q_m\}$ are *linearly independent.* To see this, we collect them into the $M \times M$ matrix:

$$Q = \left[\begin{array}{cccc} q_0 & q_1 & \ldots & q_{M-1} \end{array}\right] \tag{13.6}$$

and introduce the diagonal matrix:

$$D \stackrel{\Delta}{=} \text{diag}\Big\{\|q_0\|_A^2,\ \|q_1\|_A^2,\ \ldots,\ \|q_{M-1}\|_A^2\Big\} \tag{13.7}$$

where the notation $\|x\|_A^2$ stands for the weighted squared Euclidean norm, $x^\mathsf{T} A x$. The matrix D has positive entries since the $\{q_m\}$ are nonzero vectors and $A > 0$. We conclude from the orthogonality conditions (13.3) that

$$Q^\mathsf{T} A Q = D \tag{13.8}$$

Now assume, to the contrary, that the vectors $\{q_m\}$ are linearly dependent. This means that there exists a nonzero vector a such that $Qa = 0$. Multiplying (13.8) by a^T and a from left and right gives

$$a^\mathsf{T} D a = 0 \tag{13.9}$$

which is not possible since $D > 0$. It follows that the A-orthogonal vectors $\{q_m\}$ must be linearly independent, which also means that they span the range space of A:

$$\text{span}\Big\{q_0, q_1, \ldots, q_{M-1}\Big\} = \mathcal{R}(A) = \mathbb{R}^M \tag{13.10}$$

13.1.2 Line Search

Let w_{-1} denote an arbitrary initial guess for the solution $w^\star$. Our first step is to update this iterate to the next estimate w_0. We set $w_0 = w_{-1} + \delta w$ and search for δw by solving:

$$\delta w^o = \underset{\alpha \in \mathbb{R}}{\text{argmin}} \Big\{P(w_{-1} + \delta w)\Big\}, \quad \text{subject to } \delta w = \alpha q_0 \tag{13.11}$$

for some $\alpha \in \mathbb{R}$. This means that we are limiting the update δw to be along the direction of the first vector q_0. This formulation amounts to performing a *line search* along q_0 to determine the "size" of the update to w_{-1}. Due to the quadratic form of $P(w)$, the problem can be solved in closed form:

$$\begin{aligned}
&P(w_{-1} + \delta w) \\
&= P(w_{-1} + \alpha q_0) \\
&= \frac{1}{2}(w_{-1} + \alpha q_0)^{\mathsf{T}} A (w_{-1} + \alpha q_0) - b^{\mathsf{T}}(w_{-1} + \alpha q_0) \\
&= \frac{1}{2}\alpha^2 q_0^{\mathsf{T}} A q_0 + \alpha(Aw_{-1} - b)^{\mathsf{T}} q_0 + \text{terms indep. of } \alpha
\end{aligned} \tag{13.12}$$

Differentiating relative to α and setting the derivative to zero at $\alpha = \alpha_0$ leads to

$$r_{-1} \stackrel{\Delta}{=} b - Aw_{-1} \quad \textbf{(residual vector)} \tag{13.13}$$

$$\alpha_0 = \frac{r_{-1}^{\mathsf{T}} q_0}{q_0^{\mathsf{T}} A q_0} \tag{13.14}$$

and, consequently, the first update takes the form:

$$\boxed{w_0 = w_{-1} + \frac{r_{-1}^{\mathsf{T}} q_0}{q_0^{\mathsf{T}} A q_0} q_0} \tag{13.15}$$

Note that the residual vector r_{-1} is the negative of the gradient of $P(w)$ evaluated at $w = w_{-1}$ since

$$\nabla_{w^{\mathsf{T}}} P(w) = Aw - b \tag{13.16}$$

so that the update (13.15) can also be written in the form

$$w_0 = w_{-1} - \frac{\nabla_w P(w_{-1}) q_0}{q_0^{\mathsf{T}} A q_0} q_0 \tag{13.17}$$

Relation (13.15) reveals three useful geometric properties. First, the updated residual $r_0 = b - Aw_0$ will be orthogonal to the update direction q_0 (relative to the standard inner product without weighting by A). Indeed, using

$$r_0 \stackrel{\Delta}{=} b - Aw_0 \quad \textbf{(new residual vector)} \tag{13.18}$$

and taking inner products we get

$$\begin{aligned}
q_0^{\mathsf{T}} r_0 &= q_0^{\mathsf{T}}(b - Aw_0) \\
&= q_0^{\mathsf{T}} b - q_0^{\mathsf{T}} A w_0 \\
&\stackrel{(13.15)}{=} q_0^{\mathsf{T}} b - q_0^{\mathsf{T}} A w_{-1} - \frac{r_{-1}^{\mathsf{T}} q_0}{\cancel{q_0^{\mathsf{T}} A q_0}} \cancel{q_0^{\mathsf{T}} A q_0} \\
&= q_0^{\mathsf{T}} b - q_0^{\mathsf{T}} A w_{-1} - (b - Aw_{-1})^{\mathsf{T}} q_0 \\
&= 0
\end{aligned} \tag{13.19}$$

Second, the error vector $\widetilde{w}_0 = w^\star - w_0$ will be A-orthogonal to the update direction q_0. To see this, we use the fact that $Aw^\star = b$ and note that

$$
\begin{aligned}
q_0^{\mathsf{T}} A \widetilde{w}_0 &= q_0^{\mathsf{T}} A(w^\star - w_0) \\
&= q_0^{\mathsf{T}} A w^\star - q_0^{\mathsf{T}} A\Big(w_{-1} + \frac{r_{-1}^{\mathsf{T}} q_0}{q_0^{\mathsf{T}} A q_0} q_0\Big) \\
&= q_0^{\mathsf{T}} b - q_0^{\mathsf{T}} A w_{-1} - r_{-1}^{\mathsf{T}} q_0 \\
&= q_0^{\mathsf{T}} b - q_0^{\mathsf{T}} A w_{-1} - (b - A w_{-1})^{\mathsf{T}} q_0 \\
&= 0
\end{aligned} \tag{13.20}
$$

Third, if we subtract $w^\star$ from both sides of (13.15) and use $Aw^\star = b$ we obtain the weight-error update

$$
\widetilde{w}_0 = \widetilde{w}_{-1} - \frac{\widetilde{w}_{-1}^{\mathsf{T}} A q_0}{q_0^{\mathsf{T}} A q_0} q_0 \tag{13.21}
$$

The rightmost term has the interpretation of projecting $\widetilde{w}_{-1}$ onto the direction q_0 in the space equipped with the weighted inner product $\langle \cdot, \cdot \rangle_A$. More specifically, consider two generic vectors $q, z \in \mathbb{R}^M$. These vectors need not be aligned. The projection of z onto the direction of q is determined by solving

$$
\widehat{z} = \underset{\lambda \in \mathbb{R}}{\operatorname{argmin}} \left\{ \|z - \lambda q\|_A^2 \; = \; \lambda^2 \|q\|_A^2 - 2\lambda z^{\mathsf{T}} A q + \|z\|_A^2 \right\} \tag{13.22}
$$

Differentiating relative to λ gives the optimal value $\lambda^o = z^{\mathsf{T}} Aq / q^{\mathsf{T}} Aq$ so that the projection of z onto q is given by $\widehat{z} = \lambda^o q$, i.e.,

$$
\widehat{z} = \left(\frac{z^{\mathsf{T}} A q}{q^{\mathsf{T}} A q} \right) q \tag{13.23}
$$

One useful property of projections of the form (13.23) is that their residual, $\widetilde{z} = z - \widehat{z}$, is A-orthogonal to the vector over which the projection was performed. This is because

$$
q^{\mathsf{T}} A \widetilde{z} = q^{\mathsf{T}} A z - \frac{z^{\mathsf{T}} A q}{\cancel{q^{\mathsf{T}} A q}} \cancel{q^{\mathsf{T}} A q} \; = \; 0 \tag{13.24}
$$

Comparing expression (13.23) with the last term in (13.21) we observe that this last term corresponds to the projection of $\widetilde{w}_{-1}$ onto q_0, i.e.,

$$
\widehat{\widetilde{w}_{-1}} = \frac{\widetilde{w}_{-1}^{\mathsf{T}} A q_0}{q_0^{\mathsf{T}} A q_0} q_0 \tag{13.25}
$$

In this way, the result $\widetilde{w}_0$ is the residual that remains from this projection and we already know from (13.20) that it is A-orthogonal to q_0. In summary, we arrive at the following three geometric conclusions for the first step of updating w_{-1} to w_0:

$$
\boxed{\left\{ \begin{array}{l} q_0^{\mathsf{T}} r_0 = 0 \\ q_0^{\mathsf{T}} A \widetilde{w}_0 = 0 \\ \widetilde{w}_0 = \text{error from projecting } \widetilde{w}_{-1} \text{ onto } q_0 \end{array} \right.} \tag{13.26}
$$

13.1.3 Geometric Properties

We can repeat the argument for a generic step and examine its geometric properties as well. Let w_{m-1} denote an estimate for the solution $w^\star$ at a generic iteration $m-1$. We wish to update it to $w_m = w_{m-1} + \delta w$ by solving the problem:

$$\delta w^o = \underset{\alpha \in \mathbb{R}}{\text{argmin}} \left\{ P(w_{m-1} + \delta w) \right\}, \quad \text{subject to } \delta w = \alpha q_m \tag{13.27}$$

for some $\alpha \in \mathbb{R}$. This again involves a line search along q_m. Repeating the same derivation we find that

$$\boxed{\begin{cases} r_{m-1} = b - Aw_{m-1} \\ \alpha_m = \dfrac{r_{m-1}^\mathsf{T} q_m}{q_m^\mathsf{T} A q_m} \\ w_m = w_{m-1} + \alpha_m q_m \end{cases}} \tag{13.28}$$

where, again,

$$r_{m-1} = -\nabla_{w^\mathsf{T}} P(w)\Big|_{w=w_{m-1}} \tag{13.29}$$

If desired, the residual vector can be updated recursively as follows:

$$r_m \triangleq b - Aw_m = b - A(w_{m-1} + \alpha_m q_m) \tag{13.30}$$

so that

$$\boxed{r_m = r_{m-1} - \alpha_m A q_m} \tag{13.31}$$

Some stronger geometric properties will hold for $m > 0$. To begin with, the same argument applied to $m = 0$ would lead to:

$$\boxed{\begin{cases} q_m^\mathsf{T} r_m = 0 \\ q_m^\mathsf{T} A \widetilde{w}_m = 0 \\ \widetilde{w}_m = \text{error from projecting } \widetilde{w}_{m-1} \text{ onto } q_m \end{cases}} \tag{13.32}$$

However, we can strengthen these conclusions for $m > 0$. In particular, it will hold that the residual r_m is orthogonal to *all* prior directions and not just q_m, i.e., $q_j^\mathsf{T} r_m = 0$ for all $j \le m$. We already know that the result holds for $j = m$ from (13.32). Consider now $j = m - 1$ and note that

$$\begin{aligned}
q_{m-1}^{\mathsf{T}} r_m &= q_{m-1}^{\mathsf{T}}(b - Aw_m) \\
&= q_{m-1}^{\mathsf{T}}(b - Aw_{m-1} - \alpha_m A q_m) \\
&= q_{m-1}^{\mathsf{T}} b - q_{m-1}^{\mathsf{T}} A w_{m-1} - \frac{r_{m-1}^{\mathsf{T}} q_m}{q_m^{\mathsf{T}} A q_m} \underbrace{q_{m-1}^{\mathsf{T}} A q_m}_{=0} \\
&= q_{m-1}^{\mathsf{T}}(b - Aw_{m-1}) \\
&= q_{m-1}^{\mathsf{T}} r_{m-1} \\
&\overset{(13.32)}{=} 0
\end{aligned} \tag{13.33}$$

and the argument can be repeated for other values $j < m-1$. Therefore, we find that

$$\boxed{q_j^{\mathsf{T}} r_m = 0, \quad j = 0, 1, \ldots, m} \tag{13.34}$$

The same is true for $\widetilde{w}_m$ being A-orthogonal to all prior directions and not just q_m. To see this, note first that the weight error vector satisfies

$$\widetilde{w}_m = \widetilde{w}_{m-1} - \left(\frac{\widetilde{w}_{m-1}^{\mathsf{T}} A q_m}{q_m^{\mathsf{T}} A q_m} \right) q_m \tag{13.35}$$

We know from (13.28) that at any generic iteration m, the algorithm projects the error $\widetilde{w}_{m-1}$ onto q_m and that the updated $\widetilde{w}_m$ is A-orthogonal to q_m. Now consider $j = m-1$. Then,

$$\begin{aligned}
q_{m-1}^{\mathsf{T}} A \widetilde{w}_m &= q_{m-1}^{\mathsf{T}} A \left\{ \widetilde{w}_{m-1} - \left(\frac{\widetilde{w}_{m-1}^{\mathsf{T}} A q_m}{q_m^{\mathsf{T}} A q_m} \right) q_m \right\} \\
&= \underbrace{q_{m-1}^{\mathsf{T}} A \widetilde{w}_{m-1}}_{=0} - \frac{\widetilde{w}_{m-1}^{\mathsf{T}} A q_m}{q_m^{\mathsf{T}} A q_m} \underbrace{q_{m-1}^{\mathsf{T}} A q_m}_{=0} \\
&= 0
\end{aligned} \tag{13.36}$$

We can continue recursively. For example,

$$\begin{aligned}
q_{m-2}^{\mathsf{T}} A \widetilde{w}_m &= q_{m-2}^{\mathsf{T}} A \left\{ \widetilde{w}_{m-1} - \left(\frac{\widetilde{w}_{m-1}^{\mathsf{T}} A q_m}{q_m^{\mathsf{T}} A q_m} \right) q_m \right\} \\
&= \underbrace{q_{m-2}^{\mathsf{T}} A \widetilde{w}_{m-1}}_{=0} - \frac{\widetilde{w}_{m-1}^{\mathsf{T}} A q_m}{q_m^{\mathsf{T}} A q_m} \underbrace{q_{m-2}^{\mathsf{T}} A q_m}_{=0} \\
&= 0
\end{aligned} \tag{13.37}$$

where the first zero in the second line is because of result (13.36) applied at the previous time instant. Repeating the argument for other values of j, we conclude that

$$\boxed{q_j^{\mathsf{T}} A \widetilde{w}_m = 0, \quad j = 0, 1, \ldots, m} \tag{13.38}$$

as claimed. We therefore find that $\widetilde{w}_m$ is A-orthogonal to q_m and to all previous projection directions, $\{q_j, j < m\}$. This means that $\widetilde{w}_m$ belongs to the linear span of the remaining directions (see Prob. 13.4):

$$\boxed{\widetilde{w}_m \in \text{span}\Big\{q_{m+1}, q_{m+2}, \ldots, q_{M-1}\Big\}} \tag{13.39}$$

This is consistent with the fact that the next step of the algorithm now projects the error $\widetilde{w}_m$ onto q_{m+1} to give:

$$\widetilde{w}_{m+1} = \widetilde{w}_m - \left(\frac{\widetilde{w}_m^{\mathsf{T}} A q_{m+1}}{q_{m+1}^{\mathsf{T}} A q_{m+1}}\right) q_{m+1} \tag{13.40}$$

where now

$$\widetilde{w}_{m+1} \in \text{span}\Big\{q_{m+2}, \ldots, q_{M-1}\Big\} \tag{13.41}$$

and the algorithm continues in this manner until the error $\widetilde{w}$ is reduced to zero at iteration $M-1$, i.e.,

$$\boxed{\widetilde{w}_{M-1} = 0} \tag{13.42}$$

The reason this happens is the following. The initial weight error $\widetilde{w}_{-1}$ lies in $\mathbb{R}^M$. This space is spanned by the directions $\{q_m\}$, which are independent of each other. Recursion (13.35) starts by projecting $\widetilde{w}_{-1}$ onto q_0. The error that remains will not have any component left along q_0 and, therefore, $\widetilde{w}_0$ will be projected onto q_1 in the next step. The error that remains from this projection will not have any component in the span of $\{q_1, q_2\}$ and $\widetilde{w}_1$ will be projected onto q_2, and so forth until all components of the original error vector have been determined along the M independent directions specified by the $\{q_m\}$. Since the $\{q_m\}$ span the space $\mathbb{R}^M$, there will be no error left after M projections and (13.42) will hold. We can establish the same conclusion algebraically as follows. We express the arbitrary initial error in the form:

$$\widetilde{w}_{-1} = \sum_{j'=0}^{M-1} \beta_{j'} q_{j'} \tag{13.43}$$

for some coefficients $\beta_{j'}$. This is always possible since the $\{q_j\}$ span $\mathbb{R}^M$. The error recursion (13.35) can be written in the form

$$\widetilde{w}_m = \left(I_M - \frac{q_m q_m^{\mathsf{T}} A}{q_m^{\mathsf{T}} A q_m}\right) \widetilde{w}_{m-1} \tag{13.44}$$

so that iterating we get

$$\begin{aligned} \widetilde{w}_{M-1} &= \left\{ \prod_{j=0}^{M-1} \left(I_M - \frac{q_j q_j^{\mathsf{T}} A}{q_j^{\mathsf{T}} A q_j} \right) \right\} \widetilde{w}_{-1} \\ &= \left\{ \prod_{j=0}^{M-1} \left(I_M - \frac{q_j q_j^{\mathsf{T}} A}{q_j^{\mathsf{T}} A q_j} \right) \right\} \times \left(\sum_{j'=0}^{M-1} \beta_{j'} q_{j'} \right) \\ &= 0 \end{aligned} \tag{13.45}$$

where the last equality follows from the fact that $q_j^{\mathsf{T}} A q_{j'} = 0$ for any $j \neq j'$. We therefore conclude that recursion (13.35) is able to find $w^\star$ in M iterations so that

$$w_{M-1} = w^\star \tag{13.46}$$

It also follows by iterating (13.28) that

$$\boxed{w^\star = w_{-1} + \sum_{m=0}^{M-1} \alpha_m q_m} \tag{13.47}$$

For ease of reference, we collect in (13.48) some of the key relations derived so far.

Relations for conjugate gradient for solving $Aw = b$, $A = A^{\mathsf{T}} > 0$.

given an arbitrary initial condition, $w_{-1} \in \mathbb{R}^M$;
given M conjugate vectors $\{q_m\}$ satisfying $q_m^{\mathsf{T}} A q_\ell = 0$ for $k \neq \ell$;

relations

$$\begin{aligned} & r_{m-1} = b - A w_{m-1} \\ & \alpha_m = r_{m-1}^{\mathsf{T}} q_m / q_m^{\mathsf{T}} A q_m \\ & w_m = w_{m-1} + \alpha_m q_m \\ & r_m = r_{m-1} - \alpha_m A q_m \end{aligned} \tag{13.48}$$

properties

$$\begin{aligned} & q_j^{\mathsf{T}} A \widetilde{w}_m = 0, \quad j \leq m \\ & q_j^{\mathsf{T}} r_m = 0, \quad j \leq m \\ & \widetilde{w}_m = \text{error from } A\text{-projecting } \widetilde{w}_{m-1} \text{ onto } q_m \\ & \widetilde{w}_m \in \text{span}\{q_{m+1}, q_{m+2}, \ldots, q_{M-1}\} \\ & \widetilde{w}_{M-1} = 0 \end{aligned}$$

13.1.4 Conjugate Directions

To complete the argument we still need to explain how to select the conjugate directions $\{q_m\}$ that are used by the algorithm. These can be obtained from a Gram–Schmidt orthogonalization procedure, which we already encountered in Section 1.6, except that **(a)** inner products will now need to be weighted by the matrix A and **(b)** the vectors will only need to be orthogonal but not necessarily orthonormal. The Gram–Schmidt procedure starts from an arbitrary collection of M independent vectors, denoted by $\{h_0, h_1, \ldots, h_{M-1}\}$, and transforms them into A-orthogonal vectors $\{q_0, q_1, \ldots, q_{M-1}\}$ as follows. It sets

$$q_0 = h_0 \tag{13.49}$$

and repeats for $m = 1, \ldots, M-1$:

$$q_m = h_m - \sum_{j=0}^{m-1} \left(\frac{h_m^{\mathsf{T}} A q_j}{q_j^{\mathsf{T}} A q_j} \right) q_j \tag{13.50}$$

Each term inside the sum on the right-hand side corresponds to the projection of h_m onto q_j using result (13.23). In this way, the vector q_m is generated by projecting h_m onto each of the directions $\{q_j, j < m\}$ and keeping the projection error. It follows that q_m is A-orthogonal to all prior directions $\{q_\ell,\ \ell < m\}$ since

$$\begin{aligned}
q_\ell^{\mathsf{T}} A q_m &= q_\ell^{\mathsf{T}} A \left\{ h_m - \sum_{j=0}^{m-1} \left(\frac{h_m^{\mathsf{T}} A q_j}{q_j^{\mathsf{T}} A q_j} \right) q_j \right\} \\
&= q_\ell^{\mathsf{T}} A h_m - \sum_{j=0}^{m-1} \left(\frac{h_m^{\mathsf{T}} A q_j}{q_j^{\mathsf{T}} A q_j} \right) q_\ell^{\mathsf{T}} A q_j \\
&= q_\ell^{\mathsf{T}} A h_m - \left(\frac{h_m^{\mathsf{T}} A q_\ell}{q_\ell^{\mathsf{T}} A q_\ell} \right) q_\ell^{\mathsf{T}} A q_\ell, \quad \text{setting } j = \ell \\
&= 0
\end{aligned} \tag{13.51}$$

We conclude that the vectors $\{q_m,\ m = 0, 1, \ldots, M-1\}$ generated by means of (13.50) are A-orthogonal to each other, as desired. This construction therefore shows how to transform an initial collection of M arbitrary vectors $\{h_m\}$ into a collection of A-orthogonal vectors $\{q_m\}$ – see Prob. 13.7.

The conjugate gradient method selects the $\{h_m\}$ in a particular manner to simplify the computations involved in (13.50) and reduce the sum of m terms to computing a single term. It sets them to the residual vectors:

$$h_m \stackrel{\Delta}{=} r_{m-1} = b - A w_{m-1} \quad \textbf{(residual vector)} \tag{13.52}$$

This choice is motivated by property (13.34), namely, the fact that $q_j^{\mathsf{T}} r_m = 0$ for $j \le m$. Using this property we can verify that only one term in the sum appearing in (13.50) will be nonzero, while all other terms will be zero. First, using (13.52), the Gram–Schmidt construction becomes

$$q_m = r_{m-1} - \sum_{j=0}^{m-1} \left(\frac{r_{m-1}^{\mathsf{T}} A q_j}{q_j^{\mathsf{T}} A q_j} \right) q_j \tag{13.53}$$

Computing the inner product of q_m with r_{m-1} and using the orthogonality condition (13.34) gives

$$r_{m-1}^{\mathsf{T}} q_m = \|r_{m-1}\|^2 \tag{13.54}$$

so that the expression for α_m in (13.28) can be written as

$$\alpha_m = \frac{\|r_{m-1}\|^2}{q_m^{\mathsf{T}} A q_m} \tag{13.55}$$

In a similar vein, if we compute the inner product of q_m in (13.53) with r_ℓ for any $\ell \geq m-1$, we conclude using the orthogonality condition (13.34) that

$$r_\ell^{\mathsf{T}} q_m = r_\ell^{\mathsf{T}} r_{m-1} \;=\; \begin{cases} \|r_{m-1}\|^2, & \ell = m-1 \\ 0, & \ell > m-1 \text{ by (13.34)} \end{cases} \tag{13.56}$$

This result can now be used to show that only one term inside the summation (13.53) will prevail while all other terms will be zero. Computing the inner product of (13.31) written for $m = j$ with r_{m-1} we get

$$r_{m-1}^{\mathsf{T}} r_j = r_{m-1}^{\mathsf{T}} r_{j-1} - \alpha_j r_{m-1}^{\mathsf{T}} A q_j \tag{13.57}$$

so that

$$\begin{aligned} r_{m-1}^{\mathsf{T}} A q_j &= \frac{1}{\alpha_j}\left(r_{m-1}^{\mathsf{T}} r_{j-1} - r_{m-1}^{\mathsf{T}} r_j \right) \\ &\overset{(13.56)}{=} \begin{cases} \|r_{m-1}\|^2/\alpha_m, & j = m \\ -\|r_{m-1}\|^2/\alpha_{m-1}, & j = m-1 \\ 0, & j < m-1 \end{cases} \end{aligned} \tag{13.58}$$

Substituting this conclusion into (13.53) and using (13.55) leads to the update relation:

$$\boxed{\begin{cases} \beta_{m-1} = \dfrac{\|r_{m-1}\|^2}{\|r_{m-2}\|^2} \\ q_m = r_{m-1} + \beta_{m-1}\, q_{m-1} \end{cases}} \tag{13.59}$$

where the scalar β_{m-1} arises from the calculation

$$\beta_{m-1} = -\frac{r_{m-1}^{\mathsf{T}} A q_{m-1}}{q_{m-1} A q_{m-1}} \overset{(13.58)}{=} \frac{1}{\alpha_{m-1}} \frac{\|r_{m-1}\|^2}{q_{m-1} A q_{m-1}} \overset{(13.55)}{=} \frac{\|r_{m-1}\|^2}{\|r_{m-2}\|^2} \tag{13.60}$$

For later use, we note that β_{m-1} also admits the representation:

$$\beta_{m-1} = -\frac{r_{m-1}^{\mathsf{T}} A q_{m-1}}{q_{m-1} A q_{m-1}} \overset{(13.58)}{=} -\frac{1}{\alpha_{m-1}} \frac{r_{m-1}^{\mathsf{T}}(r_{m-2} - r_{m-1})}{q_{m-1} A q_{m-1}} \tag{13.61}$$

which upon using (13.55) leads to

$$\boxed{\beta_{m-1} = \frac{r_{m-1}^{\mathsf{T}}(r_{m-1} - r_{m-2})}{\|r_{m-2}\|^2}} \tag{13.62}$$

This expression reduces to (13.60) since $r_{m-1}^{\mathsf{T}} r_{m-2} = 0$ by (13.56); however, the two expressions will not be equivalent when the conjugate gradient method is applied to the minimization of nonquadratic risk functions, as we will discuss in the next section. A third expression for β_{m-1} follows by using (13.54) in (13.60) to get

$$\boxed{\beta_{m-1} = \frac{r_{m-1}^{\mathsf{T}} q_m}{r_{m-2}^{\mathsf{T}} q_{m-1}}} \tag{13.63}$$

Yet a fourth expression for β_{m-1} follows from replacing the denominator in the middle expression of (13.60) by

$$\alpha_{m-1} q_{m-1} A q_{m-1} = q_{m-1}(r_{m-2} - r_{m-1}) \tag{13.64}$$

so that

$$\boxed{\beta_{m-1} = \frac{\|r_{m-1}\|^2}{q_{m-1}(r_{m-2} - r_{m-1})}} \tag{13.65}$$

A fifth expression for β_{m-1} follows from

$$\beta_{m-1} = -\frac{r_{m-1}^{\mathsf{T}} A q_{m-1}}{q_{m-1} A q_{m-1}} \overset{(13.58)}{=} -\frac{1}{\alpha_{m-1}} \frac{r_{m-1}^{\mathsf{T}}(r_{m-2} - r_{m-1})}{q_{m-1} A q_{m-1}} \tag{13.66}$$

which upon using (13.64) in the denominator leads to

$$\boxed{\beta_{m-1} = -\frac{r_{m-1}^{\mathsf{T}}(r_{m-1} - r_{m-2})}{q_{m-1}^{\mathsf{T}}(r_{m-1} - r_{m-2})}} \tag{13.67}$$

In summary, we arrive at listing (13.68) for the conjugate gradient algorithm. Note that the algorithm involves only computations of inner products and one matrix-vector multiplication; it does not require any matrix inversion. Recall that at every iteration the algorithm orthogonalizes the update direction relative to all prior directions before carrying out the update. In this way, the method ensures that the current update is done along a direction over which none of the prior updates have progressed.

Conjugate gradient algorithm for solving $Aw = b, A = A^{\mathsf{T}} > 0$.

start from an arbitrary initial condition, $w_{-1} \in \mathbb{R}^M$;
set $r_{-1} = b - Aw_{-1}$, $\beta_{-1} = 0$, $q_{-1} = 0$;
repeat for $m = 0, 1, 2, \ldots, M-1$:

$$\begin{aligned}
q_m &= r_{m-1} + \beta_{m-1} q_{m-1} \\
\alpha_m &= \|r_{m-1}\|^2 / q_m^{\mathsf{T}} A q_m \\
w_m &= w_{m-1} + \alpha_m q_m \\
r_m &= r_{m-1} - \alpha_m A q_m \\
\beta_m &= \|r_m\|^2 / \|r_{m-1}\|^2
\end{aligned} \tag{13.68}$$

end
return $w^\star \leftarrow w_{M-1}$.

Example 13.1 **(Preconditioned conjugate gradient)** Preconditioning is a useful tool to speed up the convergence of solvers for linear systems of equations. The conjugate gradient method converges in M steps to the exact solution $w^\star$. It is useful to examine the rate at which $\widetilde{w}_m$ approaches zero, especially for large-scale linear systems of equations. It is mentioned in (13.134b) in the comments at the end of the chapter that the convergence rate is dependent on the condition number of A (the ratio of its largest to smallest eigenvalues). The result suggests that matrices A with clustered eigenvalues (small condition numbers) lead to faster rates of decay. We exploit this property to motivate *preconditioning*.

Assume we are interested in solving $Aw = b$ where we continue to assume that $A = A^{\mathsf{T}} > 0$ and $w \in \mathbb{R}^M$. We introduce an invertible transformation

$$w = Tx \tag{13.69}$$

for some $M \times M$ matrix T chosen by the designer. Then, solving $Aw = b$ is equivalent to solving

$$\underbrace{T^{\mathsf{T}} A T}_{\triangleq A'} x = \underbrace{T^{\mathsf{T}} b}_{\triangleq b'} \tag{13.70}$$

where the new matrix A' is congruent (but not similar) to the original matrix A. Therefore, A and A' have different eigenvalue distributions. The intention is to choose T to result in matrices A' with clustered eigenvalues. Once T is chosen, as illustrated further ahead, we can apply conjugate gradient to the pair (A', b') and write (where we use the prime notation to refer to the corresponding variables):

$$\begin{cases} r'_{-1} = b' - A'x_{-1}, \;\; \beta'_{-1} = 0, \;\; q'_{-1} = 0 \\ q'_m = r'_{m-1} + \beta'_{m-1}\, q'_{m-1} \\ \alpha'_m = \dfrac{\|r'_{m-1}\|^2}{(q'_m)^{\mathsf{T}} A' q'_m} \\ x_m = x_{m-1} + \alpha'_m q'_m \\ r'_m = r'_{m-1} - \alpha'_m A' q'_m \\ \beta'_m = \dfrac{\|r'_m\|^2}{\|r'_{m-1}\|^2} \end{cases} \tag{13.71}$$

The calculations can be simplified if we redefine

$$q_m \triangleq Tq'_m \tag{13.72a}$$

$$w_m \triangleq Tx_m \tag{13.72b}$$

$$r_m \triangleq (T^{-1})^{\mathsf{T}} r'_m \tag{13.72c}$$

$$z_m \triangleq Tr'_m \tag{13.72d}$$

in which case the above equations lead to the preconditioned conjugate gradient algorithm shown in listing (13.73). This implementation does not require knowledge of T but rather of $P = TT^{\mathsf{T}}$; we refer to P as the *preconditioning matrix*.

Preconditioned conjugate gradient for solving $Aw = b, A = A^{\mathsf{T}} > 0.$

given an $M \times M$ preconditioning matrix P;
start from an arbitrary initial condition, $w_{-1} \in \mathbb{R}^M$;
set $r_{-1} = b - Aw_{-1}, \;\; z_{-1} = Pr_{-1}, \;\; \beta_{-1} = 0, \;\; q_{-1} = 0.$
repeat for $m = 0, 1, 2, \ldots, M-1$**:**

$$\begin{aligned} q_m &= z_{m-1} + \beta_{m-1}\, q_{m-1} \\ \alpha_m &= r_{m-1}^{\mathsf{T}} z_{m-1} / q_m^{\mathsf{T}} A q_m \\ w_m &= w_{m-1} + \alpha_m q_m \\ r_m &= r_{m-1} - \alpha_m A q_m \\ z_m &= P r_m \\ \beta_m &= r_m^{\mathsf{T}} z_m / r_{m-1}^{\mathsf{T}} z_{m-1} \end{aligned} \tag{13.73}$$

end
return $w^\star \leftarrow w_{M-1}$.

There are several options for choosing P in practice. One approximation is to set it to the diagonal matrix:

$$P = \text{diag}\Big\{1/a_{mm}\Big\}_{m=1}^{M} \tag{13.74}$$

where the $\{a_{mm}\}$ refer to the diagonal entries of A. Observe from the listing of the algorithm that P appears only in the expression that computes z_m, as well as in the initial condition z_{-1}.

13.2 NONLINEAR OPTIMIZATION

The discussion in the previous section focused on the solution of linear systems of equations, $Ax = b$ for $A = A^{\mathsf{T}} > 0$. We can apply similar ideas and develop conjugate gradient algorithms for the solution of general optimization problems of the form:

$$w^\star = \underset{w\in\mathbb{R}^M}{\text{argmin}}\ P(w) \tag{13.75}$$

where $P(w)$ is not necessarily quadratic anymore. The main difference in relation to the quadratic case is that the residual vector $r_m = b - Aw_m$ will be replaced by the (negative of the) gradient vector, $r_m = -\nabla_{w^{\mathsf{T}}} P(w_m)$. We motivate the algorithm as follows.

Let $\{q_m\}$ denote a collection of M-dimensional vectors for $m = 0, 1, \ldots, M-1$. These vectors will serve as update directions and will be constructed in the sequel. For now, let w_{m-1} denote the estimate that is available for $w^\star$ at iteration $m-1$. We wish to update it to $w_m = w_{m-1} + \delta w^o$ by solving:

$$\delta w^o = \underset{\alpha\in\mathbb{R}}{\text{argmin}} \Big\{P(w_{m-1} + \delta w)\Big\}, \quad \text{subject to } \delta w = \alpha q_m \tag{13.76}$$

for some $\alpha \in \mathbb{R}$ and where the update direction is along q_m. Differentiating the objective function $P(w + \alpha q_m)$ relative to α we observe that the optimal α, denoted by α_m, should guarantee:

$$\boxed{\nabla_w P(w_m) q_m = 0} \qquad \textbf{(orthogonality condition)} \tag{13.77}$$

However, solving for α_m is now challenging. In the case studied in the previous section for quadratic $P(w)$, we were able to determine a closed-form expression for α_m. Here we need to resort to some approximations.

13.2.1 Line Search Methods

One first method introduces a second-order Taylor series expansion for $P(w+\delta w)$ around $w = w_{m-1}$:

$$P(w + \delta w) \approx P(w_{m-1}) + \alpha\, \nabla_w P(w_{m-1}) q_m + \frac{1}{2}\alpha^2\, q_m^{\mathsf{T}} \nabla_w^2 P(w_{m-1}) q_m \tag{13.78}$$

and minimizes over α to get

$$r_{m-1} \triangleq -\nabla_{w^{\mathsf{T}}} P(w_{m-1}) \quad \text{(negative gradient vector)} \tag{13.79a}$$

$$A_{m-1} \triangleq \nabla_w^2 P(w_{m-1}) \quad \text{(Hessian matrix)} \tag{13.79b}$$

$$\alpha_m = \frac{r_{m-1}^{\mathsf{T}} q_m}{q_m^{\mathsf{T}} A_{m-1} q_m} \tag{13.79c}$$

This solution requires that we evaluate the Hessian matrix of $P(w)$ at the successive iterates.

Alternatively, we can estimate α_m by resorting to the *backtracking* search method described earlier in Section 12.4.2. We select two parameters $0 < \beta < 1$ and $0 < \lambda < 1/2$. Typical values are $\beta = 0.2$ and $\lambda = 0.01$. We also select a large initial value $\alpha_0 > 0$ (e.g., $\alpha_0 = 1$). At every step $j \geq 0$, we perform the following:

$$\left\{\begin{array}{l} \textbf{while } P(w_{m-1} + \alpha_j q_m) \; > P(w_{m-1}) + \lambda \alpha_j \nabla_w P(w_{m-1}) q_m \\ \quad \text{shrink } \alpha \text{ to } \alpha_{j+1} = \beta \alpha_j; \\ \quad j \leftarrow j+1; \\ \textbf{end} \\ \text{set } \alpha_m = \alpha_j; \\ \text{update } w_m = w_{m-1} + \alpha_m q_m \end{array}\right. \tag{13.80}$$

This method finds α_m in an iterative manner and it does not require calculation of the Hessian of $P(w)$. Once α_m is determined (we denote its value by α), this construction will guarantee the Armijo condition (12.104) to hold, namely,

$$P(w_m) \leq P(w_{m-1}) + \lambda \alpha \, \nabla_w P(w_{m-1}) q_m \tag{13.81}$$

We can restate this condition in terms of the α-function:

$$\phi(\alpha) \;\stackrel{\Delta}{=}\; P(w + \alpha \, \delta w) \tag{13.82}$$

by writing

$$\phi(\alpha) \leq \phi(0) + \lambda \alpha \, \phi'(0), \quad \text{for some } 0 < \lambda < 1/2 \tag{13.83}$$

It was shown in Section 12.4.2 that the Armijo condition is sufficient to ensure the convergence of traditional gradient-descent algorithms, where the $\{q_m\}$ were chosen as $q_m = -\nabla_{w^\mathsf{T}} P(w_{m-1})$. The same conclusion will hold here if the $\{q_m\}$ are constructed as descent directions for all m, i.e., if they satisfy $\nabla_w P(w_{m-1}) q_m < 0$ – see Prob. 13.12. It turns out that this property will hold if we are able to select the $\{\alpha_m\}$ to satisfy the following two conditions combined (and not just (13.83)), either in their strong form:

(**strong Wolfe conditions**)

$$\phi(\alpha) \leq \phi(0) + \lambda \alpha \, \phi'(0), \quad \text{for some } 0 < \lambda < 1 \tag{13.84a}$$

$$|\phi'(\alpha)| \leq \eta |\phi'(0)|, \quad \text{for some } 0 < \lambda < \eta < 1 \tag{13.84b}$$

or weaker form:

(**weak Wolfe conditions**)

$$\phi(\alpha) \leq \phi(0) + \lambda \alpha \; \phi'(0), \quad \text{for some } 0 < \lambda < 1 \tag{13.85a}$$

$$\phi'(\alpha) \geq \eta \phi'(0), \quad \text{for some } 0 < \lambda < \eta < 1 \tag{13.85b}$$

Condition (13.84a) guarantees a sufficient decrease in the risk value, while condition (13.84b) ensures that the $\{q_m\}$ constructed below remain descent directions (as is explained in Lemma 13.1). Typical values are $\lambda = 10^{-4}$ and $\eta = 0.1$. For well-behaved differentiable risk functions $P(w)$, it is always guaranteed that an α can be found that satisfies the weak and strong Wolfe conditions – see Prob. 13.15.

One approximate method to determine an α that "meets" the weak Wolfe conditions, is to mimic the backtracking line search method (13.80). We select a parameter $0 < \kappa < 1$ (say, $\kappa = 0.2$) and a large initial value $\alpha_0 > 0$ (e.g., $\alpha_0 = 1$). At every step $j \geq 0$, we perform the following steps until a suitable α is found or a maximum number of iterations is reached:

$$\left\{\begin{array}{l} \textbf{while } \Big(\phi(\alpha_j) > \phi(0) + \lambda\alpha_j\phi'(0)\Big) \text{ and } \Big(\phi'(\alpha_j) < \eta\phi'(0)\Big) \\ \quad \text{shrink } \alpha \text{ to } \alpha_{j+1} = \kappa\alpha_j; \\ \quad j \leftarrow j+1; \\ \textbf{end} \\ \text{set } \alpha_m = \alpha_j; \\ \text{update } w_m = w_{m-1} + \alpha_m q_m \end{array}\right. \tag{13.86}$$

A more accurate approach for selecting an α that satisfies the Wolfe conditions is the Moré–Thuente line search method; it is considerably more complex than (13.86) and is mentioned in the references listed at the end of the chapter.

13.2.2 Fletcher–Reeves Algorithm

Once α_m is chosen, the remaining relations in the conjugate gradient method remain similar and the resulting algorithm, known as the Fletcher–Reeves method, is listed in (13.87). The algorithm does not converge anymore in M steps to the exact minimizer $w^\star$. Often, it is run continually or is restarted by using the iterate w_{M-1} after M iterations as the initial condition for the next run and resetting β_m to zero at the beginning of each run.

Fletcher–Reeves algorithm for minimizing a risk function $P(w)$.

start from an arbitrary initial condition, w_{-1};
set $r_{-1} = -\nabla_{w^\mathsf{T}} P(w_{-1})$, $\beta_{-1} = 0$, $q_{-1} = 0$.
repeat until convergence for $m \geq 0$:

$$\begin{array}{l} q_m = r_{m-1} + \beta_{m-1}\, q_{m-1} \\ \text{find } \alpha_m \text{ to satisfy the weak or strong Wolfe conditions;} \\ w_m = w_{m-1} + \alpha_m q_m \\ r_m = -\nabla_{w^\mathsf{T}} P(w_m) \\ \beta_m = \|r_m\|^2/\|r_{m-1}\|^2 \end{array} \tag{13.87}$$

end
return $w^\star \leftarrow w_m$.

Polak–Ribière algorithm

An alternative implementation known as the Polak–Ribière method uses the second expression (13.62) for β_m, namely,

$$\beta_m = \frac{r_m^\mathsf{T}(r_m - r_{m-1})}{\|r_{m-1}\|^2} \triangleq \beta_m^{\text{PR}} \tag{13.88}$$

For general optimization problems, it does not hold anymore that $r_m^\mathsf{T} r_{m-1} = 0$ as was the case for quadratic risks and, therefore, the two expressions for β_m used by Fletcher–Reeves and Polak–Ribière are not equivalent any longer. In particular, observe that while β^{FR} is always nonnegative, the above expression for β_m^{PR} can become negative. As a result, and as indicated in the references at the end of the chapter, the Polak–Ribière method can get stuck in loops indefinitely and need not converge.

Polak–Ribière–Powell algorithm

A variation is to use the Polak–Ribière–Powell method for which convergence is guaranteed under exact and inexact line searches by limiting β_m to being positive:

$$\beta_m = \max\left\{ \frac{r_m^\mathsf{T}(r_m - r_{m-1})}{\|r_{m-1}\|^2}, 0 \right\} \triangleq \beta_m^{\text{PRP}} \tag{13.89}$$

This construction is one of the preferred forms in practice. Yet another variation is to select β_m as follows:

$$\beta_m = \begin{cases} -\beta_m^{\text{FR}}, & \text{when } \beta_m^{\text{PR}} < \beta^{\text{FR}} \\ -\beta_m^{\text{PR}}, & \text{when } -\beta^{\text{FR}} \leq \beta_m^{\text{PR}} \leq \beta^{\text{FR}} \\ \beta_m^{\text{FR}}, & \text{when } \beta_m^{\text{PR}} > \beta^{\text{FR}} \end{cases} \tag{13.90}$$

where β_m^{FR} is the parameter used by Fletcher–Reeves. The above construction ensures that the resulting β_m satisfies $|\beta_m| \leq \beta_m^{\text{FR}}$. The reason for doing so is because convergence is guaranteed for this choice of β_m – see Prob. 13.17.

Example 13.2 **(Application to logistic regression)** We illustrate the operation of the Fletcher–Reeves algorithm by considering the ℓ_2-regularized logistic regression risk function:

$$P(w) = \rho\|w\|^2 + \frac{1}{N}\sum_{m=0}^{N-1} \ln\left(1 + e^{-\gamma(m)h_m^\mathsf{T} w}\right) \tag{13.91}$$

For this simulation, the data $\{\gamma(m), h_m\}$ are generated *randomly* as follows. First, a 10th-order random parameter model $w^a \in \mathbb{R}^{10}$ is selected, and $N = 200$ random feature vectors $\{h_m\}$ are generated, say, with zero-mean unit-variance Gaussian entries. Then, for each h_m, the label $\gamma(m)$ is set to either $+1$ or -1 according to the following construction:

$$\gamma(m) = +1 \text{ if } \left(\frac{1}{1 + e^{-h_m^\mathsf{T} w^a}}\right) \geq 0.5; \text{ otherwise } \gamma(m) = -1 \tag{13.92}$$

We will explain in future expression (59.5a) that construction (13.92) amounts to generating data $\{\gamma(m), h_m\}$ that satisfy a logistic probability model. The Fletcher–Reeves algorithm (13.87) is run for 100 iterations on the data $\{\gamma(m), h_m\}$ using parameters

$$\rho = 2, \quad \mu = 0.001, \quad \lambda = 1 \times 10^{-4}, \quad \eta = 0.1, \quad \kappa = 0.2 \tag{13.93}$$

The maximum number of iterations for the line search procedure (13.86) is set to 20. Every $M = 10$ iterations, we reset the q-variable to zero. The minimizer $w^\star$ and the resulting weight iterate after $L = 100$ iterations are shown in the bottom plot of Fig. 13.1; the minimal risk value is found to be

$$P(w^\star) \approx 0.6818 \tag{13.94}$$

The two plots in the top row display the learning curve $P(w_m)$ relative to the minimum risk value $P(w^\star)$, both in linear scale on the left and in normalized logarithmic scale on the right (according to construction (11.65)).

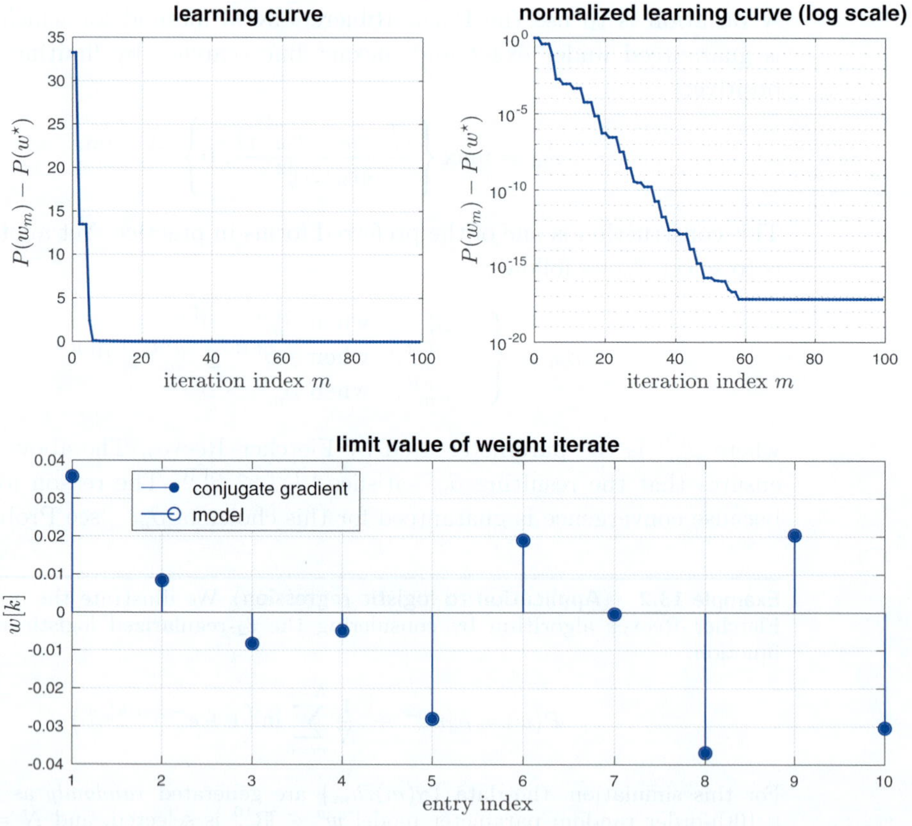

Figure 13.1 (*Top*) Learning curves $P(w_m)$ relative to the minimum risk value $P(w^\star)$ in linear scale (on the left) and in normalized logarithmic scale (on the right) generated by the Fletcher–Reeves algorithm. (*Bottom*) Limiting value of the weight iterate w_m, which tends to the minimizer $w^\star$.

Example 13.3 **(Relation to momentum methods)** The gradient-descent method for minimizing a convex function $P(w)$ updates w_{m-1} to w_m along the negative direction of the gradient vector:

$$w_m = w_{m-1} - \mu \nabla_{w^{\mathsf{T}}} P(w_{m-1}) \tag{13.95}$$

We introduced a variation of this approach, known as the *momentum method*, in Prob. 12.9, where the update includes an additional term proportional to the difference of the last two iterates:

$$w_m = w_{m-1} - \mu \nabla_{w^{\mathsf{T}}} P(w_{m-1}) + \beta(w_{m-1} - w_{m-2}) \tag{13.96}$$

for some nonnegative scalar β. We will discuss momentum methods in greater detail in Section 17.5, where we will explain that they help enhance the convergence behavior of the algorithms by adjusting the search direction. We verify here that the conjugate-gradient method (13.87) is one example of a momentum method. To see this, note first that

$$w_m = w_{m-1} + \alpha_m q_m \Longrightarrow q_m = \frac{1}{\alpha_m}(w_m - w_{m-1}) \tag{13.97}$$

Consequently,

$$\begin{aligned} w_m &= w_{m-1} + \alpha q_m \\ &= w_{m-1} + \alpha_m (r_{m-1} + \beta_{m-1} q_{m-1}) \\ &= w_{m-1} + \alpha_m r_{m-1} + \frac{\alpha_m \beta_{m-1}}{\alpha_{m-1}}(w_{m-1} - w_{m-2}) \\ &= w_{m-1} - \alpha_m \nabla_{w^{\mathsf{T}}} P(w_{m-1}) + \frac{\alpha_m \beta_{m-1}}{\alpha_{m-1}}(w_{m-1} - w_{m-2}) \end{aligned} \tag{13.98}$$

This recursion is of the same general form as (13.96); the main difference is that the parameters $\{\alpha_m, \beta_m\}$ are updated over m.

13.3 CONVERGENCE ANALYSIS

The convergence analysis of conjugate gradient methods for nonquadratic risks is influenced by the *inexact* line search method used to compute the $\{\alpha_m\}$. We establish the convergence of the Fletcher–Reeves method in the sequel and defer extensions to the comments at the end of the chapter and to Probs. 13.14–13.17.

To begin with, the Fletcher–Reeves and Polak–Ribière implementations require the computation of α_m to be sufficiently accurate. To see this, starting from the recursion

$$q_m = -\nabla_{w^{\mathsf{T}}} P(w_{m-1}) + \beta_{m-1} q_{m-1} \tag{13.99}$$

and computing the inner product with the gradient vector, we get:

$$\nabla_w P(w_{m-1}) q_m = -\|\nabla_{w^{\mathsf{T}}} P(w_{m-1})\|^2 + \beta_{m-1} \underbrace{\nabla_{w^{\mathsf{T}}} P(w_{m-1}) q_{m-1}}_{\overset{(13.77)}{=} 0} \tag{13.100}$$

The last term is zero if α_{m-1} is exact, in which case the inner product between q_m and $\nabla_w P(w_{m-1})$ on the left will be negative. This in turn means that q_m

will correspond to a descent direction for $P(w)$ at $w = w_{m-1}$, which is a desirable property. However, when the successive $\{\alpha_m\}$ are approximated by means of a line search method, the last term in (13.100) will be nonzero and can become dominant due to errors. As a result, the inner product $\nabla_w P(w_{m-1})q_m$ can become positive and q_m will not be a descent direction anymore. For this reason, convergence analyses for conjugate gradient methods for nonlinear optimization problems focus on verifying whether choosing the $\{\alpha_m\}$ to satisfy the weak or strong Wolfe conditions would ensure vectors $\{q_m\}$ that remain descent directions.

Convergence guarantees are available for the Fletcher–Reeves method under exact and inexact line searches. The same is not true for Polak–Ribière, which need not converge even under exact line searches since the corresponding vectors $\{q_m\}$ are not guaranteed to be descent directions. The Polak–Ribière–Powell method, on the other hand, has guaranteed convergence under exact and inexact line searches; it also has superior performance to Fletcher–Reeves.

To establish the convergence of the Fletcher–Reeves method we proceed as follows. Suppose the $\{\alpha_m\}$ are selected to satisfy the strong Wolfe conditions (13.84a)–(13.84b) with parameters $0 < \lambda < \eta < 1/2$. This means that when we exit the line search procedure, the value of α_m satisfies:

$$P(w_m) \leq P(w_{m-1}) + \lambda\alpha_m \nabla_w P(w_{m-1})q_m \tag{13.101a}$$

$$|\nabla_w P(w_m)q_m| \leq \eta|\nabla_w P(w_{m-1})q_m| \tag{13.101b}$$

We first establish two auxiliary results. The first lemma only uses the fact that α_m satisfies the second Wolfe condition (13.101b) to show that the successive vectors $\{q_m\}$ will correspond to descent directions, meaning that the inner product of each q_m with the corresponding gradient vector is negative, $\nabla_w P(w_{m-1})q_m < 0$. The first Wolfe condition (13.101a) will be used in the subsequent convergence proof. The statement below assumes $\nabla_w P(w_{m-1}) \neq 0$; otherwise, the algorithm would have converged.

LEMMA 13.1. (Risk descent directions) *Assume the $\{\alpha_m\}$ are selected to satisfy (13.101a)–(13.101b) for $0 < \lambda < \eta < 1/2$, then it holds for any $m \geq 0$ that*

$$-\frac{1}{1-\eta} \leq \frac{\nabla_w P(w_{m-1})q_m}{\|\nabla_w P(w_{m-1})\|^2} \leq \frac{2\eta - 1}{1-\eta} < 0 \tag{13.102}$$

and, hence, the successive $\{q_m\}$ generated by Fletcher–Reeves are descent directions.

Proof: First note that the function $f(\eta) = (2\eta-1)/(1-\eta)$ is increasing over the interval $\eta \in (0, 1/2)$ with $f(0) = -1$ and $f(1/2) = 0$. Therefore, the upper bound satisfies

$$-1 < \frac{2\eta - 1}{1-\eta} < 0 \tag{13.103}$$

On the other hand, the function $g(\eta) = -1/(1-\eta)$ is decreasing over $\eta \in (0, 1/2)$ with $g(0) = -1$ and $g(1/2) = -2$. Next, we establish (13.102) by induction. For $m = 0$, we have $q_0 = -\nabla_w P(w_{-1})$ so that the ratio in the middle is equal to -1 and both sides of the inequality are satisfied. Now suppose the inequality holds for iteration m and let us verify that it holds for iteration $m+1$. It follows that q_m is a descent direction so that condition (13.101b) becomes

$$|\nabla_w P(w_m)q_m| \leq -\eta\nabla_w P(w_{m-1})q_m \tag{13.104}$$

which is equivalent to

$$\eta\nabla_w P(w_{m-1})q_m \leq \nabla_w P(w_m)q_m \leq -\eta\nabla_w P(w_{m-1})q_m \tag{13.105}$$

Now note from recursions (13.87) that

$$\begin{aligned}
\frac{\nabla_w P(w_m)q_{m+1}}{\|\nabla_w P(w_m)\|^2} &= \frac{\nabla_w P(w_m)(r_m + \beta_m q_m)}{\|\nabla_w P(w_m)\|^2} \\
&= -1 + \beta_m \frac{\nabla_w P(w_m)q_m}{\|\nabla_w P(w_m)\|^2}, \quad \text{since } r_m = -\nabla_{w^\mathsf{T}} P(w_m) \\
&= -1 + \frac{\|r_m\|^2}{\|r_{m-1}\|^2}\frac{\nabla_w P(w_m)q_m}{\|\nabla_w P(w_m)\|^2} \\
&= -1 + \frac{\nabla_w P(w_m)q_m}{\|\nabla_w P(w_{m-1})\|^2}
\end{aligned} \tag{13.106}$$

Using (13.105) we obtain

$$-1 + \eta\frac{\nabla_w P(w_{m-1})q_m}{\|\nabla_w P(w_{m-1})\|^2} \leq -1 + \frac{\nabla_w P(w_m)q_m}{\|\nabla_w P(w_{m-1})\|^2} \leq -1 - \eta\frac{\nabla_w P(w_{m-1})q_m}{\|\nabla_w P(w_{m-1})\|^2} \tag{13.107}$$

Applying the lower bound from (13.102) to the leftmost and rightmost terms and using (13.106) we get

$$-1 - \frac{\eta}{1-\eta} \leq \frac{\nabla_w P(w_m)q_{m+1}}{\|\nabla_w P(w_m)\|^2} \leq -1 + \frac{\eta}{1-\eta} \tag{13.108}$$

which establishes the validity of (13.102) for $m+1$.

■

The next result is a powerful property for updates of the general form $w_m = w_{m-1} + \alpha_m q_m$, of which gradient descent algorithms are a special case. The result is known as the *Zoutendijk condition* and it holds irrespective of whether the risk function $P(w)$ is convex or not. Although the statement below assumes the $\{q_m\}$ are constructed according to the Fletcher–Reeves method (13.87), so that the $\{q_m\}$ are descent directions by Lemma 13.1, the result is actually more general and holds for any algorithm where the q_m are guaranteed to be descent directions; it also holds for $\{\alpha_m\}$ that satisfy the weaker Wolfe conditions; this extension is studied in Prob. 13.14.

Lemma 13.2. (Zoutendijk condition) *Assume the risk function $P(w) : \mathbb{R}^M \rightarrow \mathbb{R}$ is first-order differentiable with δ-Lipschitz gradients and is bounded from below; the risk function is not required to be convex. Assume the $\{\alpha_m\}$ are selected to satisfy (13.101a)–(13.101b) for $0 < \lambda < \eta < 1/2$ so that the $\{q_m\}$ generated by Fletcher–Reeves are descent directions by Lemma 13.1. The iterate w_{m-1} is updated to $w_m = w_{m-1} + \alpha_m q_m$. Let θ_m denote the angle between q_m and the negative of the gradient vector at w_{m-1}:*

$$\cos(\theta_m) \triangleq \frac{-\nabla_w P(w_{m-1})q_m}{\|\nabla_w P(w_{m-1})\|\, \|q_m\|} \tag{13.109}$$

It then holds that

$$\sum_{m=0}^{\infty} \cos^2(\theta_m)\|\nabla_w P(w_{m-1})\|^2 < +\infty \tag{13.110a}$$

which is equivalent to

$$\sum_{m=0}^{\infty} \frac{(\nabla_w P(w_{m-1})q_m)^2}{\|q_m\|^2} < +\infty \tag{13.110b}$$

Moreover, $P(w_m)$ is nonincreasing, meaning that $P(w_m) \le P(w_{m-1})$.

Proof: Since q_m is a descent direction, we conclude from the second Wolfe condition (13.101b) that

$$\nabla_w P(w_m)q_m \;\ge\; \eta \nabla_w P(w_{m-1})q_m \tag{13.111}$$

Subtracting $\nabla_w P(w_{m-1})q_m$ from both sides gives

$$\Big(\nabla_w P(w_m) - \nabla_w P(w_{m-1})\Big)q_m \;\ge\; (\eta - 1)\nabla_w P(w_{m-1})q_m \tag{13.112}$$

From the δ-Lipschitz condition on the gradient of $P(w)$ we have

$$\|\nabla_w P(w_m) - \nabla_w P(w_{m-1})\| \le \delta \|w_m - w_{m-1}\| \tag{13.113}$$

and, hence, by Cauchy–Schwarz:

$$\begin{aligned}\Big(\nabla_w P(w_m) - \nabla_w P(w_{m-1})\Big)q_m &\le \|\nabla_w P(w_m) - \nabla_w P(w_{m-1})\|\, \|q_m\| \\ &\le \delta \|w_m - w_{m-1}\|\, \|q_m\| \\ &= \delta \alpha_m \|q_m\|^2, \quad \text{using } w_m = w_{m-1} + \alpha q_m\end{aligned} \tag{13.114}$$

Combining (13.112) and (13.114) shows that α_m is lower-bounded by

$$\alpha_m \ge \frac{(\eta - 1)}{\delta}\, \frac{\nabla_w P(w_{m-1})q_m}{\|q_m\|^2} \tag{13.115}$$

where the term on the right-hand side is positive since q_m is a descent direction and $\eta < 1$. Substituting this conclusion into the first Wolfe condition (13.101a), and recalling that $\nabla_w P(w_{m-1})q_m < 0$, we find that

$$P(w_m) \le P(w_{m-1}) + \underbrace{\lambda \frac{(\eta - 1)}{\delta}}_{\triangleq\, -c} \frac{(\nabla_w P(w_{m-1})q_m)^2}{\|q_m\|^2} \tag{13.116}$$

where we introduced the positive constant $c = \lambda(1-\eta)/\delta$. In other words, in terms of the angles $\{\theta_m\}$ we have that

$$P(w_m) \leq P(w_{m-1}) - c\,\cos^2(\theta_m)\,\|\nabla_w P(w_{m-1})\|^2 \tag{13.117}$$

This relation shows that $P(w_m)$ is nonincreasing. Summing over m gives

$$\sum_{m=0}^{\infty} \cos^2(\theta_m)\,\|\nabla_w P(w_{m-1})\|^2 \;\leq\; \frac{1}{c}\Big(P(w_{-1}) - \lim_{m\to\infty} P(w_m)\Big) \tag{13.118}$$

Since, by assumption, the risk function is bounded from below, the term on the right-hand side is bounded by some positive constant and, therefore, conclusion (13.110a) holds.

■

We conclude from the previous lemma that

$$\lim_{m\to\infty} \Big\{\cos^2(\theta_m)\|\nabla_w P(w_{m-1})\|^2\Big\} = 0 \tag{13.119}$$

This shows that if the update directions $\{q_m\}$ are selected to ensure that $\cos(\theta_m)$ is uniformly bounded away from zero, say, $\cos(\theta_m) \geq \sigma > 0$ (i.e., if the search directions are not too close to being orthogonal to the gradient directions), then $\|\nabla_w P(w_{m-1})\|^2 \to 0$ and convergence to a stationary point is attained. We use this observation to establish convergence of the Fletcher–Reeves method (13.87) to a stationary point $w^\star$ of (13.75); this point will be a minimizer when $P(w)$ is convex. One useful corollary of the previous two lemmas follows if we multiply (13.102) by $\|\nabla_w P(w_{m-1})\|/\|q_m\|$ and use (13.110a) to conclude that the following condition must also hold:

$$\sum_{m=0}^{\infty} \frac{\|\nabla_w P(w_{m-1})\|^4}{\|q_m\|^2} < +\infty \tag{13.120}$$

THEOREM 13.1. (Convergence of Fletcher–Reeves) *Assume the risk function $P(w) : \mathbb{R}^M \to \mathbb{R}$ is first-order differentiable, has $\delta-$Lipschitz gradients, and is bounded from below; the risk function is not required to be convex. The $\{\alpha_m\}$ are selected to satisfy the Wolfe conditions (13.101a)–(13.101b) for $0 < \lambda < \eta < 1/2$ so that the $\{q_m\}$ generated by Fletcher–Reeves are descent directions by Lemma 13.1. Then, it holds that*

$$\liminf_{m\to\infty} \|\nabla_w P(w_m)\| = 0 \tag{13.121}$$

This implies that the sequence of gradient vectors $\{\nabla_w P(w_m)\}$ contains a subsequence that converges to zero.

Proof: Since q_m is a descent direction, we conclude from the second Wolfe condition (13.101b) that

$$|\nabla_w P(w_m) q_m| \;\leq\; -\eta \nabla_w P(w_{m-1}) q_m \;\overset{(13.102)}{\leq}\; \frac{\eta}{1-\eta}\|\nabla_w P(w_{m-1})\|^2 \tag{13.122}$$

It follows that

$$\begin{aligned}
\|q_{m+1}\|^2 &= \|r_m + \beta_m q_m\|^2 \\
&\leq \|\nabla_w P(w_m)\|^2 + 2\beta_m |\nabla_w P(w_m) q_m| + \beta_m^2 \|q_m\|^2 \\
&= \|\nabla_w P(w_m)\|^2 + 2\frac{\|\nabla_w P(w_m)\|^2}{\|\nabla_w P(w_{m-1})\|^2} |\nabla_w P(w_m) q_m| + \beta_m^2 \|q_m\|^2 \\
&\overset{(13.122)}{\leq} \|\nabla_w P(w_m)\|^2 + \frac{2\eta}{1-\eta} \|\nabla_w P(w_m)\|^2 + \beta_m^2 \|q_m\|^2 \\
&= \underbrace{\frac{1+\eta}{1-\eta}}_{\triangleq\, c_2 > 1} \|\nabla_w P(w_m)\|^2 \; + \; \beta_m^2 \|q_m\|^2
\end{aligned} \tag{13.123}$$

Iterating we get

$$\begin{aligned}
\|q_{m+1}\|^2 &\leq c_2 \|\nabla_w P(w_m)\|^2 + c_2 \frac{\|\nabla_w P(w_m)\|^4}{\|\nabla_w P(w_{m-1})\|^2} + c_2 \frac{\|\nabla_w P(w_m)\|^4}{\|\nabla_w P(w_{m-2})\|^2} + \ldots \\
&= c_2 \|\nabla_w P(w_m)\|^4 \left\{ \frac{1}{\|\nabla_w P(w_m)\|^2} + \frac{1}{\|\nabla_w P(w_{m-1})\|^2} + \ldots \right\} \\
&= c_2 \|\nabla_w P(w_m)\|^4 \sum_{j=0}^{m+1} \frac{1}{\|\nabla_w P(w_{j-1})\|^2}
\end{aligned} \tag{13.124}$$

We can establish the validity of (13.121) by contradiction. Assume it does not hold. This means that the norm of $\nabla_w P(w)$ is bounded from below for all m, say,

$$\|\nabla_w P(w_m)\| \geq c_3, \quad \text{for some } c_3 > 0 \text{ and for all } m > 0 \tag{13.125}$$

Using this bound we have

$$\sum_{j=0}^{m+1} \frac{1}{\|\nabla_w P(w_{j-1})\|^2} \leq \frac{1}{c_3^2} \times (m+2) \tag{13.126}$$

so that (13.124) gives

$$\frac{\|\nabla_w P(w_m)\|^4}{\|q_{m+1}\|^2} \geq \frac{c_3^2}{c_2} \times \frac{1}{m+2} \tag{13.127}$$

Summing over m leads to

$$\sum_{m=0}^{\infty} \frac{\|\nabla_w P(w_{m-1})\|^4}{\|q_m\|^2} \geq \frac{c_3^2}{c_2} \sum_{m=0}^{\infty} \frac{1}{m+1} \tag{13.128}$$

which is not bounded. This conclusion contradicts the Zoutendijk condition (13.110a) or its corollary (13.120). We conclude that condition (13.121) is valid. It follows that a sequence of indices $\{n_1, n_2, n_3, \ldots\}$ exists such that $\nabla_w P(w_{n_j}) \to 0$ as $j \to \infty$. One selection for the indices $\{n_j\}$ is as follows. Let $\epsilon_1 > \epsilon_2 > \epsilon_3 > \ldots$ be a sequence of strictly positive numbers with $\epsilon_j \to 0$ as $j \to \infty$, and let

$$n_1 : \underset{n \geq 0}{\text{argmin}} \left\{ \|\nabla_w P(w_n)\| \leq \epsilon_1 \right\} \tag{13.129a}$$

$$n_j : \underset{n > n_{j-1}}{\text{argmin}} \left\{ \|\nabla_w P(w_n)\| \leq \epsilon_j \right\}, \; j = 2, 3, \ldots \tag{13.129b}$$

■

Remark 13.1. (Strongly convex risks) Assume $P(w)$ is ν-strongly convex with δ-Lipschitz gradients. Using the δ-Lipschitz property, we established in (13.117) that

$$P(w_n) \le P(w_{n-1}) - \cos^2(\theta_m)\, \|\nabla_w P(w_{n-1})\|^2 \tag{13.130}$$

On the other hand, using the ν-strong convexity of $P(w)$, we apply property (8.29) to deduce that:

$$P(w^\star) \ge P(w_{n-1}) - \frac{1}{2\nu}\|\nabla_w P(w_{n-1})\|^2 \tag{13.131}$$

Subtracting $P(w^\star)$ from both sides of (13.130) and using (13.131) we obtain

$$P(w_n) - P(w^\star) \le (1 - 2\nu\cos^2(\theta_m))(P(w_{n-1}) - P(w^\star)) \tag{13.132}$$

If the sequence $\{\cos^2(\theta_m)\}$ is uniformly bounded from below, say, $0 < \sigma^2 \le \cos^2(\theta_m) \le 1$, then

$$P(w_n) - P(w^\star) \le (1 - 2\nu\sigma^2)(P(w_{n-1}) - P(w^\star)) \tag{13.133}$$

from which we deduce exponential convergence of $P(w_n)$ to $P(w^\star)$.

■

13.4 COMMENTARIES AND DISCUSSION

Conjugate gradient method. The conjugate gradient algorithm was developed independently by Hestenes (1951) and Stiefel (1952); they collaborated on a joint publication by Hestenes and Stiefel (1952). The method helps improve the convergence rate of iterative implementations by moving along A-orthogonal (or conjugate) directions derived from the residual vectors. Its complexity is higher than gradient descent but more affordable than the Newton method since the latter requires inverting a Hessian matrix. The conjugate gradient method was originally developed to solve $M \times M$ linear systems of equations with symmetric positive-definite coefficient matrices; it can solve such systems exactly in M iterations even though fewer iterations tend to be sufficient when M is large. However, the performance can degrade in finite arithmetic due to rounding errors that can cause the conjugate vectors to lose the required A-orthogonality property. It is worth noting that the use of conjugate (A-orthogonal) directions to represent the solution of equations $Aw = b$ in the form shown in Prob. 13.2 precedes the conjugate gradient algorithm and was already used by Fox, Huskey, and Wilkinson (1948) in their study of linear systems of equations.

The convergence rate of the conjugate gradient method is sensitive to the condition number of A. Assume we order the eigenvalues of A as $\lambda_1 \le \lambda_2 \le \ldots \le \lambda_M$. It was shown by Kaniel (1966), Luenberger (1973), and van der Sluis and van der Vorst (1986) that the weighted Euclidean norm of the error vector evolves according to the relations:

$$\|\widetilde{w}_m\|_A^2 \le \left(\frac{\lambda_{M-m} - \lambda_1}{\lambda_{M-m} + \lambda_1}\right)^2 \|\widetilde{w}_{-1}\|_A^2 \tag{13.134a}$$

$$\|\widetilde{w}_m\|_A \le 2\left(\frac{\sqrt{\kappa} - 1}{\sqrt{\kappa} + 1}\right)^{m+1} \|\widetilde{w}_{-1}\|_A \tag{13.134b}$$

where

$$\kappa \triangleq \lambda_{\max}(A)/\lambda_{\min}(A) \tag{13.135}$$

These relations reveal that matrices A with clustered eigenvalues lead to faster convergence. The use of preconditioning, as explained in Example 13.1, is meant to replace

A by a congruent matrix with better eigenvalue distribution to help improve the convergence rate.

Nonlinear optimization. The nonlinear versions of the conjugate gradient algorithm are from Fletcher and Reeves (1964) and Polak and Ribière (1969). The Moré–Thuente line search method is due to Moré and Thuente (1994). The strong and weak Wolfe conditions (13.84a)–(13.84b) and (13.85a)–(13.85b) are from Wolfe (1969, 1971). Convergence analyses for these nonlinear variants appear, for example, in Daniel (1967, 1970, 1971), Cohen (1972), Zoutendijk (1970), Al-Baali (1985), Gilbert and Nocedal (1992), Sun and Zhang (2001), Dai (2002), and Nocedal and Wright (2006). The derivations and proofs given in Section 13.3 for the Fletcher–Reeves algorithm are adapted from the arguments in Powell (1984) and Al-Baali (1985), and more closely from the presentation in Nocedal and Wright (2006); this last reference provides an excellent treatment of conjugate gradient techniques for optimization problems. Theorem 13.1 on the convergence of Fletcher–Reeves under inexact linear searches is due to Al-Baali (1985). The work by Zoutendijk (1970) established that the Fletcher–Reeves method converges to a stationary point under the assumption that the optimal parameters $\{\alpha_m\}$ are found exactly by the line search method. Al-Baali (1985) extended the convergence guarantee to inexact line searches. The Polak–Ribière method does not have the same convergence guarantees as Fletcher–Reeves; it was shown by Powell (1984) that Polak–Ribière can get stuck in loops indefinitely without converging to a stationary point even under exact line searches. The Polak–Ribière–Powell method, on the other hand, is from Powell (1986) and it was shown by Gilbert and Nocedal (1992) to converge even under inexact line searches. Two other nonlinear versions are due to Dai and Yuan (1999) and Hestenes and Stiefel (1952) where the β-parameter is computed based on expressions (13.65) and (13.67), respectively. The former reference provides convergence guarantees under its β_m while the Hestenes–Stiefel construction behaves similarly to Polak–Ribière. For additional discussion on the conjugate gradient method, the reader may refer to Fletcher (1987), Shewchuk (1994), Kelley (1995), Golub and O'Leary (1989), Golub and Van Loan (1996), van der Vorst (2003), and Nocedal and Wright (2006).

PROBLEMS

13.1 Let A be a symmetric positive-definite matrix. Establish that $\langle x, y\rangle_A = x^{\mathsf T} A y$, which maps vectors from $\mathbb{R}^M$ to scalars in $\mathbb{R}$, satisfies the standard inner product properties over the field of real numbers, namely,
(a) $\langle x, y\rangle_A = \langle y, x\rangle_A$.
(b) $\langle \alpha x, y\rangle_A = \alpha\langle x, y\rangle_A$.
(c) $\langle x+z, y\rangle_A = \langle x, y\rangle_A + \langle z, y\rangle_A$.
(d) $\langle x, x\rangle_A > 0, \forall x \neq 0$.
(e) $\langle x, x\rangle_A = 0 \iff x = 0$.

13.2 Consider any collection of M-dimensional A-orthogonal vectors $\{q_m\}$ where A is symmetric positive-definite. Let $w^\star$ denote the unique solution to the linear system of equations $Aw = b$. Show that

$$w^\star = \sum_{m=0}^{M-1} \left(\frac{q_m^{\mathsf T} b}{q_m^{\mathsf T} A q_m}\right) q_m$$

13.3 Refer to expression (13.59) for β_{m-1}. Show that $\beta_{m-1} = \|r_{m-1}\|^2 / q_{m-1}^{\mathsf T} r_{m-2}$.

13.4 Refer to the conjugate gradient relations (13.28). Let $\mathcal{R}_{m-1}$ denote the linear span of the conjugate vectors $\{q_0, q_1, \ldots, q_{m-1}\}$. Show that w_m can be interpreted as the solution to the following problem:

$$w_m = \underset{w \in \mathbb{R}^M}{\text{argmin}} \left\{ P(w) = \frac{1}{2} w^\mathsf{T} A w - b^\mathsf{T} w \right\}$$
$$\text{subject to } (w - w_{-1}) \in \mathcal{R}_{m-1}$$

13.5 Continuing with Prob. 13.4, argue that w_m can be interpreted as the solution to the following problem:

$$w_m = \underset{w \in \mathbb{R}^M}{\text{argmin}} \; \|\widetilde{w}\|_A^2, \quad \text{subject to } (w - w_{-1}) \in \mathcal{R}_{m-1}$$

13.6 The Krylov subspace of order $m-1$ that is associated with a vector b and a matrix A is denoted by $\mathcal{K}_{m-1}$ and defined as the following linear span:

$$\mathcal{K}_{m-1} \triangleq \text{span}\{b,\, Ab,\, A^2 b,\, \ldots,\, A^{m-1} b\}$$

Consider the linear system of equations $Aw = b$ where $A > 0$ and $w \in \mathbb{R}^M$. Refer to the conjugate gradient relations (13.28).

(a) Show that w_m can be interpreted as the solution to the following problem (we say that the $\{w_m\}$ forms a Krylov sequence):

$$w_m = \underset{w \in \mathbb{R}^M}{\text{argmin}} \left\{ P(w) = \frac{1}{2} w^\mathsf{T} A w - b^\mathsf{T} w \right\}$$
$$\text{subject to } (w - w_{-1}) \in \mathcal{K}_{m-1}$$

(b) Show further that

$$\begin{aligned} \text{span}\{r_{-1}, r_0, r_1, \ldots, r_{m-2}\} &= \mathcal{K}_{m-1} \oplus Aw_{-1} \\ \text{span}\{q_0, q_1, q_2, \ldots, q_{m-1}\} &= \mathcal{K}_{m-1} \oplus Aw_{-1} \end{aligned}$$

where the $\oplus$ notation means that for each m, it holds $(r_{m-2} - Aw_{-1}) \in \mathcal{K}_{m-1}$ and $(q_{m-1} - Aw_{-1}) \in \mathcal{K}_{m-1}$.

13.7 Construction (13.50) transforms a collection of M vectors $\{h_m\}$ into A-orthogonal vectors $\{q_m\}$. Show that these collections have the same linear span (i.e., they span the same space): $\text{span}\{h_0, h_1, \ldots, h_{M-1}\} = \text{span}\{q_0, q_1, \ldots, q_{M-1}\}$.

13.8 How would you use the conjugate gradient method (13.68) to solve the least-squares problem:

$$w^\star = \underset{w \in \mathbb{R}^M}{\text{argmin}} \; \|d - Hw\|^2$$

where $H \in \mathbb{R}^{N \times M}$ and $d \in \mathbb{R}^N$ with $N \geq M$? Moreover, H has full rank.

13.9 Refer to the relations and properties associated with the conjugate gradient method listed in (13.48). Which of these properties would break down, if any, when A is only invertible but not necessarily positive-definite?

13.10 Refer to the Fletcher–Reeves algorithm (13.87). Write down its equations for the case of the logistic empirical risk (13.91), namely,

$$P(w) = \rho\|w\|^2 + \frac{1}{N}\sum_{m=0}^{N-1} \ln\left(1 + e^{-\gamma(m) h_m^\mathsf{T} w}\right)$$

13.11 Refer to the Fletcher–Reeves recursions (13.87). Verify that

$$\|q_m\|^2 = 2r_{m-1}^\mathsf{T} q_m - \|r_{m-1}\|^2 + \beta_{m-1}^2 \|q_{m-1}\|^2$$

13.12 The proof of Theorem 12.3 for the backtracking line search method is based on selecting the update direction as $-\nabla_{w^\mathsf{T}} P(w_{n-1})$. More generally, assume w_{n-1} is updated to $w_n = w_{n-1} + \mu q_n$, where q_n denotes any descent direction, i.e., any vector that satisfies $\nabla_{w^\mathsf{T}} P(w_{n-1}) q_n < 0$. Can you extend the proof to conclude that the excess risk $P(w_n) - P(w^\star)$ converges to zero under the same conditions imposed on the risk function?

13.13 Assume the risk function $P(w) : \mathbb{R}^M \rightarrow \mathbb{R}$ is differentiable, has δ-Lipschitz gradients, and is bounded from below; the risk function is not required to be convex. Assume further that the level set $\mathcal{L} = \{w \,|\, P(w) \leq P(w_{-1})\}$ is bounded. The $\{\alpha_m\}$ are selected to satisfy the strong Wolfe conditions (13.101a)–(13.101b) for $0 < \lambda < \eta < 1$. Assume the update directions $\{q_m\}$ are selected as descent directions, namely, $\nabla_w P(w_{m-1})q_m < 0$. Show that the convergence result (13.121) holds if $\sum_{m=0}^{\infty} 1/\|q_m\|^2 = +\infty$. *Remark.* The reader may consult Zoutendijk (1970) and Nocedal and Wright (2006, ch. 3) for a related discussion.

13.14 This problem relaxes some of the assumptions in Lemma 13.2 leading to a Zoutendijk condition for general updates of the form $w_n = w_{n-1} + \mu q_n$ of which gradient descent is a special case (i.e., we are not limiting the algorithms to Fletcher–Reeves anymore). Thus, consider a differentiable risk function $P(w) : \mathbb{R}^M \rightarrow \mathbb{R}$ with δ-Lipschitz gradients and assume it is bounded from below. The risk function is not required to be convex. Let q_n be any descent direction at location w_{m-1} and assume the step size μ is chosen to satisfy the weak (rather than strong) Wolfe conditions:

$$
\begin{aligned}
&\phi(\mu) \overset{\Delta}{=} P(w + \mu \delta w) \\
&\phi(\mu) \leq \phi(0) + \alpha\mu\, \phi'(0), \quad \text{for some } 0 < \alpha < 1 \\
&\phi'(\mu) \geq \eta\phi'(0), \quad \text{for some } 0 < \alpha < \eta < 1
\end{aligned}
$$

with the bounds on α and η extended to 1 from 1/2. Show that the conclusion of Lemma 13.2 continues to hold. *Remark.* The reader may consult Zoutendijk (1970) and Nocedal and Wright (2006, ch. 3) for a related discussion.

13.15 Consider a differentiable risk function $P(w) : \mathbb{R}^M \rightarrow \mathbb{R}$ and an iterate $w_{m-1} \in \mathbb{R}^M$. Let q_m be a descent direction at w_{m-1}, i.e., $\nabla_w P(w_{m-1})q_m < 0$. Assume $P(w)$ is bounded from below on the set $\{w_{m-1} + \alpha q_m\}$ for any $\alpha > 0$. Assume $0 < \lambda < \eta < 1$. Show that there exist nonempty intervals of steps α that satisfy both the strong and weak Wolfe conditions (13.84a)–(13.84b) and (13.85a)–(13.85b). *Remark.* The reader may consult Nocedal and Wright (2006, ch. 3) for a related discussion.

13.16 Refer to the Fletcher–Reeves algorithm (13.87) for nonlinear optimization and assume the $\{\alpha_m\}$ are selected to satisfy the strong Wolfe conditions (13.84a)–(13.84b) with $\eta < 1/2$.

(a) Let θ_m denote the angle between q_m and the negative of the gradient at w_{n-1}:

$$
\cos(\theta_m) = \frac{-\nabla_w P(w_{m-1})q_m}{\|\nabla_w P(w_{m-1})\|\, \|q_m\|}
$$

Assume q_m is a poor direction for which $\cos(\theta_m) \approx 0$. Argue that $\beta_m^{\text{FR}} \approx 1$ and $q_{m+1} \approx q_m$ so that there will be little improvement in the search direction. Conclude that once the Fletcher–Reeves method produces a "bad" direction q_m, it is likely to produce a succession of bad directions.

(b) Under the same conditions of item (a), show for the Polak–Ribière method that $\beta_m^{\text{PR}} \approx 0$ and q_{m+1} will be close to $-\nabla_{w^\mathsf{T}} P(w_m)$. Conclude that the Polak–Ribière method is able to recover from bad directions.

Remark. The reader may consult Nocedal and Wright (2006, ch. 5) for a related discussion.

13.17 Show that the conclusion of Theorem 13.1 holds for any sequence β_m satisfying $|\beta_m| \leq \beta_m^{\text{FR}}$.

REFERENCES

Al-Baali, M. (1985), "Descent property and global convergence of the Fletcher–Reeves method with inexact line search," *IMA J. Numer. Anal.*, vol. 5, pp. 121–124.

Cohen, A. (1972), "Rate of convergence of several conjugate gradient algorithms," *SIAM J. Numer. Anal.*, vol. 9, pp. 248–259.

Dai, Y. H. (2002), "Conjugate gradient methods with Armijo-type line searches," *Acta Mathematicae Applicatae Sinica*, English Series, vol. 18, no. 1, pp. 123–130.

Dai, Y. H. and Y. Yuan (1999), "A nonlinear conjugate gradient method with a strong global convergence property," *SIAM J. Optim.*, vol. 10, no. 1, pp. 177–182.

Daniel, J. W. (1967), "Convergence of the conjugate gradient method with computationally convenient modifications,"*Numerische Mathematik*, vol. 10, pp. 125–131.

Daniel, J. W. (1970), "Correction concerning the convergence rate for the conjugate gradient method," *Numerische Mathematik*, vol. 7, pp. 277–280.

Daniel, J. W. (1971), *The Approximate Minimization of Functionals*, Prentice-Hall.

Fletcher, R. (1987), *Practical Methods of Optimization*, 2nd ed., Wiley.

Fletcher, R. and C. M. Reeves (1964), "Function minimization by conjugate gradients," *Comput. J.*, vol. 7, pp. 149–154.

Fox, L., H. D. Huskey, and J. H. Wilkinson (1948), "Notes on the solution of algebraic linear simultaneous equations," *Q. J. Mech. App. Math.*, vol. 1, pp. 149–173.

Gilbert, J. C. and J. Nocedal (1992), "Global convergence properties of conjugate gradient methods for optimization," *SIAM J. Optim.*, vol. 2, no. 1, pp. 21–42.

Golub, G. H. and D. P. O'Leary (1989), "Some history of the conjugate gradient and Lanczos algorithms: 1948–1976," *SIAM Rev.*, vol. 31, no. 1, pp. 50–102.

Golub, G. H. and C. F. Van Loan (1996), *Matrix Computations*, 3rd ed., John Hopkins University Press.

Hestenes, M. R. (1951), "Iterative methods for solving linear equations," *NAML Report 52-9*, National Bureau of Standards, USA. Published subsequently in *J. Optim. Theory Appl.*, vol. 11, no. 4, pp. 323–334, 1973.

Hestenes, M. R. and E. Stiefel (1952), "Methods of conjugate gradients for solving linear systems," *J. Res. Nat. Bur. Standards*, vol. 49, no. 6, pp. 409–436.

Kaniel, S. (1966), "Estimates for some computational techniques in linear algebra," *Math. Comput.*, vol. 20, pp. 369–378.

Kelley, C. T. (1995), *Iterative Methods for Linear and Nonlinear Equations*, SIAM.

Luenberger, D. G. (1973), *Introduction to Linear and Nonlinear Programming*, Addison-Wesley.

Moré, J. J. and D. J. Thuente (1994), "Line search algorithms with guaranteed sufficient decrease," *ACM Trans. Mathe. Softw.*, vol. 20, pp. 286–307.

Nocedal, J. and S. J. Wright (2006), *Numerical Optimization*, Springer.

Polak, E. and G. Ribière (1969), "Note sur la convergence de méthodes de directions conjuguées," *Rev. Francaise Informat Recherche Operationelle*, vol. 3, no. 1, pp. 35–43.

Powell, M. J. D. (1984), "Nonconvex minimization calculations and the conjugate gradient method," in *Lecture Notes in Mathematics*, vol. 1066, pp. 122–141, Springer.

Powell, M. J. D. (1986), "Convergence properties of algorithms for nonlinear optimization," *SIAM Rev.*, vol. 28, no. 4, pp. 487–500.

Shewchuk, J. R. (1994), "An introduction to the conjugate gradient method without the agonizing pain," Technical Report 10.5555/865018, Carnegie Mellon University, PA, USA.

Stiefel, E. (1952), "Ueber einige methoden der relaxationsrechnung," *ZAMP Zeitschrift für angewandte Mathematik und Physik*, vol. 3, no. 1, pp. 1–33.

Sun, J. and J. Zhang (2001), "Global convergence of conjugate gradient methods," *Ann. Oper. Res.*, vol. 103, pp. 161–173.

van der Sluis, A. and H. A. van der Vorst (1986), "The rate of convergence of conjugate gradients," *Numerische Mathematik*, vol. 48, no. 5, pp. 543–560.

van der Vorst, H. A. (2003), *Iterative Krylov Methods for Large Linear Systems*, Cambridge University Press.

Wolfe, P. (1969), "Convergence conditions for ascent methods," *SIAM Rev.*, vol. 11, no. 2, pp. 226–235.

Wolfe, P. (1971), "Convergence conditions for ascent methods II: Some corrections," *SIAM Rev.* vol. 13, pp. 185–188.

Zoutendijk, G. (1970), "Nonlinear programming, computational methods," in *Integer and Nonlinear Programming*, J. Abadie, editor, pp. 37–86, North-Holland.

14 Subgradient Method

The gradient-descent method from Chapter 12 assumes the risk functions are first-order differentiable and uses the gradient information to carry out the successive updates. However, in many situations of interest, the risk functions are nonsmooth and include points of nondifferentiability. In this chapter, we explain how to replace gradient calculations by subgradients in order to minimize nonsmooth convex risks. Some loss in performance will ensue, with the successive iterations of the subgradient method not converging to the exact minimizer anymore, as was the case with gradient descent, but rather to a small neighborhood around it.

14.1 SUBGRADIENT ALGORITHM

Consider again an optimization problem of the generic form:

$$w^\star \stackrel{\Delta}{=} \underset{w\in\mathbb{R}^M}{\text{argmin}}\; P(w) \tag{14.1}$$

where $P(w)$ is the risk function to be minimized, which could be represented either in empirical or stochastic form as was already explained in Section 12.1. We showed earlier that when $P(w)$ is first-order differentiable, the minimizer $w^\star$ can be found by means of the gradient-descent recursion (recall (12.22)):

$$w_n = w_{n-1} - \mu \nabla_{w^\mathsf{T}} P(w_{n-1}), \quad n \geq 0 \tag{14.2}$$

in terms of the gradient vector of $P(w)$ evaluated at $w = w_{n-1}$. We are now interested in situations when $P(w)$ need not be differentiable at $w = w_{n-1}$. In this case, one option is to replace the gradient vector appearing in (14.2) by a *subgradient*, thus leading to what we will refer to as the *subgradient algorithm* with an update of the form:

$$\boxed{w_n = w_{n-1} - \mu\, s(w_{n-1}), \quad n \geq 0} \tag{14.3}$$

In this description, the notation $s(w_{n-1})$ refers either to the gradient of $P(w)$ at $w = w_{n-1}$ or to a subgradient at the same location depending on whether $P(w)$ is differentiable or not:

$$s(w_{n-1}) = \begin{cases} \nabla_{w^\mathsf{T}} P(w_{n-1}), & \text{if } P(w) \text{ is differentiable at } w_{n-1} \\ \in \partial_{w^\mathsf{T}} P(w_{n-1}), & \text{otherwise} \end{cases} \tag{14.4}$$

The second line states that $s(w_{n-1})$ belongs to the subdifferential of $P(w)$ at w_{n-1} (i.e., it can be selected as any of the elements within this set). We already know from the discussion in Section 8.6 that when $P(w)$ is differentiable at w_{n-1}, then its subdifferential at that location will consist of a single element that coincides with the gradient at w_{n-1}. Therefore, we can combine both choices in (14.4) into a single statement by writing

$$\boxed{s(w_{n-1}) \in \partial_{w^\mathsf{T}} P(w_{n-1})} \tag{14.5}$$

so that recursion (14.3) requires that we select at each iteration a subgradient vector from the subdifferential set of $P(w)$ at $w = w_{n-1}$. In summary, we arrive at listing (14.6) for the subgradient method.

Subgradient method for minimizing $P(w)$.

given a construction for subgradient vectors, $s(w)$;
given a small step-size parameter $\mu > 0$;
start from an arbitrary initial condition, w_{-1}. (14.6)
repeat over $n \geq 0$:
  $w_n = w_{n-1} - \mu\, s(w_{n-1})$
end
return $w^\star \leftarrow w_n$.

The next examples illustrate the subgradient construction – other examples are considered in Prob. 14.1.

Example 14.1 (LASSO or ℓ_1-regularized least-squares) Consider an ℓ_1-regularized least-squares risk, which corresponds to the risk function for LASSO or basis pursuit problems (which we will encounter in Chapter 51):

$$P(w) = \alpha \|w\|_1 + \frac{1}{N}\sum_{m=0}^{N-1}(\gamma(m) - h_m^\mathsf{T} w)^2, \quad \alpha > 0 \tag{14.7}$$

Using the result from the second row in Table 8.3, we know that one subgradient choice for $P(w)$ at an arbitrary location w is

$$s(w) = \alpha\, \text{sign}(w) - \frac{2}{N}\sum_{m=0}^{N-1} h_m(\gamma(m) - h_m^\mathsf{T} w) \tag{14.8}$$

in which case the subgradient recursion (14.3) becomes

$$w_n = w_{n-1} - \mu\alpha\, \text{sign}(w_{n-1}) + \mu\left(\frac{2}{N}\sum_{m=0}^{N-1} h_m(\gamma(m) - h_m^\mathsf{T} w_{n-1})\right), \quad n \geq 0 \tag{14.9}$$

This implementation is in batch form since it employs all data pairs $\{\gamma(m), h_m\}$ at every iteration.

Example 14.2 (**ℓ_2-regularized hinge risk**) Consider next the ℓ_2-regularized hinge risk from Table 12.1, which will appear in our study of support vector machines in Chapter 61:

$$P(w) = \rho\|w\|^2 + \frac{1}{N}\sum_{m=0}^{N-1} \max\left\{0, 1-\gamma(m)h_m^{\mathsf{T}}w\right\}, \quad w \in \mathbb{R}^M \tag{14.10}$$

This risk is strongly convex but not differentiable at any w for which $\gamma(m)h_m^{\mathsf{T}}w = 1$. Using the result from the last row in Table 8.3, we find that one subgradient for $P(w)$ at an arbitrary location w is given by

$$s(w) = 2\rho w - \frac{1}{N}\sum_{m=0}^{N-1}\gamma(m)h_m\,\mathbb{I}\left[\gamma(m)h_m^{\mathsf{T}}w \le 1\right] \tag{14.11}$$

in which case the subgradient recursion (14.3) becomes

$$w_n = (1-2\mu\rho)w_{n-1} + \mu\left(\frac{1}{N}\sum_{m=0}^{N-1}\gamma(m)h_m\,\mathbb{I}\left[\gamma(m)h_m^{\mathsf{T}}w_{n-1} \le 1\right]\right), \quad n \ge 0 \tag{14.12}$$

Given that $P(w)$ in (14.10) has the form of an empirical risk, the above recursion is in batch form since it employs all data pairs $\{\gamma(m), h_m\}$ at every iteration.

Example 14.3 (**Batch perceptron**) Consider similarly the ℓ_2-regularized empirical perceptron risk from Table 12.1, namely,

$$P(w) = \rho\|w\|^2 + \frac{1}{N}\sum_{m=0}^{N-1}\max\left\{0, -\gamma(m)h_m^{\mathsf{T}}w\right\} \tag{14.13}$$

This risk is strongly convex but not differentiable at any w for which $\gamma(m)h_m^{\mathsf{T}}w = 0$. Similar to (14.11), we can construct a subgradient for $P(w)$ as follows:

$$s(w) = 2\rho w - \frac{1}{N}\sum_{m=0}^{N-1}\gamma(m)h_m\,\mathbb{I}\left[\gamma(m)h_m^{\mathsf{T}}w \le 0\right] \tag{14.14}$$

in which case the subgradient recursion (14.3) becomes

$$w_n = (1-2\mu\rho)\,w_{n-1} + \mu\left(\frac{1}{N}\sum_{m=0}^{N-1}\gamma(m)h_m\,\mathbb{I}\left[\gamma(m)h_m^{\mathsf{T}}w_{n-1} \le 0\right]\right) \tag{14.15}$$

The reason for the designation "perceptron" is because this type of construction will appear later when we study the perceptron algorithm for classification purposes in Chapter 60.

Example 14.4 (**Batch subgradient method**) More generally, consider *empirical risk* functions of the form

$$P(w) = \frac{1}{N}\sum_{m=0}^{N-1} Q\Big(w;\gamma(m), h_m\Big), \quad \gamma(m) \in \mathbb{R}, \quad h_m \in \mathbb{R}^M \tag{14.16}$$

for some convex loss function $Q(w;\cdot,\cdot)$. Let $s_Q(w;\gamma,h)$ denote a subgradient construction for the loss at location w. Then, one choice for a subgradient of $P(w)$ is the sample average:

$$s(w) = \frac{1}{N}\sum_{m=0}^{N-1} s_Q(w;\gamma(m),h_m) \tag{14.17}$$

which leads to listing (14.18). The reason for the designation "batch algorithm" is because each iteration of (14.18) employs the *entire* set of data, i.e., all N data pairs $\{\gamma(m),h_m\}$. Moreover, this set is used repeatedly until the algorithm approaches its limiting behavior.

Batch subgradient method for minimizing empirical risks, $P(w)$.

given N data pairs $\{\gamma(m),h_m\}, m=0,1,\ldots,N-1$;
risk has empirical form $P(w)=\frac{1}{N}\sum_{m=0}^{N-1}Q(w;\gamma(m),h_m)$;
given subgradient construction for loss, $s_Q(w;\gamma,h)$;
given a small step-size parameter $\mu>0$;
start from an arbitrary initial condition, w_{-1}.
repeat until convergence over $n\geq 0$**:**

$$w_n = w_{n-1} - \mu\left(\frac{1}{N}\sum_{m=0}^{N-1}s_Q(w_{n-1};\gamma(m),h_m)\right)$$

end
return $w^\star \leftarrow w_n$.

(14.18)

Example 14.5 **(Empirical risks with offset parameters)** We continue with the empirical hinge risk (14.10) and include a scalar offset parameter $\theta\in\mathbb{R}$ in the following manner:

$$P(w,\theta) = \rho\|w\|^2 + \frac{1}{N}\sum_{m=0}^{N-1}\max\left\{0,\, 1-\gamma(m)\left(h_m^{\mathsf{T}}w-\theta\right)\right\} \tag{14.19}$$

This function is now dependent on two parameters, $\{w,\theta\}$; it continues to be strongly convex relative to these parameters but is nondifferentiable at all locations w satisfying $\gamma(m)(h_m^{\mathsf{T}}w-\theta)=1$. If we now compute subgradients relative to θ and w and write down the corresponding subgradient algorithm we get:

$$\theta(n) = \theta(n-1) - \mu\left(\frac{1}{N}\sum_{m=0}^{N-1}\gamma(m)\,\mathbb{I}\left[\gamma(m)(h_m^{\mathsf{T}}w_{n-1}-\theta(n-1))\leq 1\right]\right) \tag{14.20a}$$

$$w_n = (1-2\mu\rho)\,w_{n-1} + \mu\left(\frac{1}{N}\sum_{m=0}^{N-1}\gamma(m)h_m\,\mathbb{I}\left[\gamma(m)(h_m^{\mathsf{T}}w_{n-1}-\theta(n-1))\leq 1\right]\right) \tag{14.20b}$$

We can combine these iterations by introducing the augmented $(M+1)$-dimensional quantities:

$$w' \triangleq \begin{bmatrix}-\theta\\ w\end{bmatrix},\quad h' \triangleq \begin{bmatrix}1\\ h\end{bmatrix} \tag{14.21}$$

to arrive at the following batch implementation:

$$w'_n = \begin{bmatrix} 1 & 0 \\ 0 & (1-2\mu\rho)I_M \end{bmatrix} w'_{n-1} + \mu\left(\frac{1}{N}\sum_{m=0}^{N-1}\gamma(m)h'_m\,\mathbb{I}\left[\gamma(m)(h'_m)^\mathsf{T}w'_{n-1}\le 1\right]\right) \tag{14.22}$$

We illustrate the behavior of the algorithm by means of a numerical simulation. The top row of Fig. 14.1 plots the learning curve for the subgradient recursion (14.22) in linear and normalized logarithmic scales using the parameters:

$$\mu = 0.002,\quad \rho = 1,\quad M = 10,\quad N = 200. \tag{14.23}$$

The logarithmic curve plots

$$\ln\left(\frac{P(w_n,\theta(n)) - P(w^\star,\theta^\star)}{\max_n\{P(w_n,\theta(n)) - P(w^\star,\theta^\star)\}}\right) \tag{14.24}$$

where $P(w_n,\theta(n))$ denotes the risk values at the successive iterates $(w_n,\theta(n))$ and $P(w^\star,\theta^\star)$ is the minimum risk. The subgradient recursion is run for 12,000 iterations at the end of which the limit quantities $(w_n,\theta(n))$ are used as approximations for $(w^\star,\theta^\star)$ with the corresponding risk value taken as the approximate minimum risk. In order to simplify the plot of the logarithmic curve, we sample it down by a factor of 100 and place markers every 100 samples.

The $N = 200$ data points $\{\gamma(m), h_m\}$ are generated according to a logistic model. Specifically, a random parameter model $w^a \in \mathbb{R}^{10}$ is selected, and a random collection of feature vectors $\{h_m\}$ are generated with zero-mean and unit-variance Gaussian entries. Then, for each h_m, the label $\gamma(m)$ is set to either $+1$ or -1 according to the following construction:

$$\gamma(m) = +1 \text{ if } \left(\frac{1}{1+e^{-h_m^\mathsf{T}w^a}}\right) \ge 0.5, \text{ otherwise } \gamma(m) = -1 \tag{14.25}$$

The resulting weight iterate for recursion (14.22), denoted by $w^\star$, at the end of the 12,000 iterations is shown in the bottom left plot of the figure. The evolution of the θ-iterate is shown in the bottom right plot for the first 5000 iterations; it approaches the value

$$\theta^\star \approx 0.8214 \tag{14.26}$$

The minimum value of the risk function at the limiting points of these iterates is found to be

$$P(w^\star,\theta^\star) \approx 0.6721 \tag{14.27}$$

The plot in the log scale for the learning curve helps illustrate the linear convergence of $P(w_n,\theta(n))$ toward its state value, as anticipated by result (14.50).

14.2 CONDITIONS ON RISK FUNCTION

Before we examine the convergence behavior of the subgradient method, we need to introduce conditions on the risk function $P(w)$ in a manner similar to what we did for the gradient-descent method in Section 12.2. Since $P(w)$ is now possibly nonsmooth (i.e., nondifferentiable at some locations), the conditions of strong convexity and δ-Lipschitz gradients will need to be restated in terms of

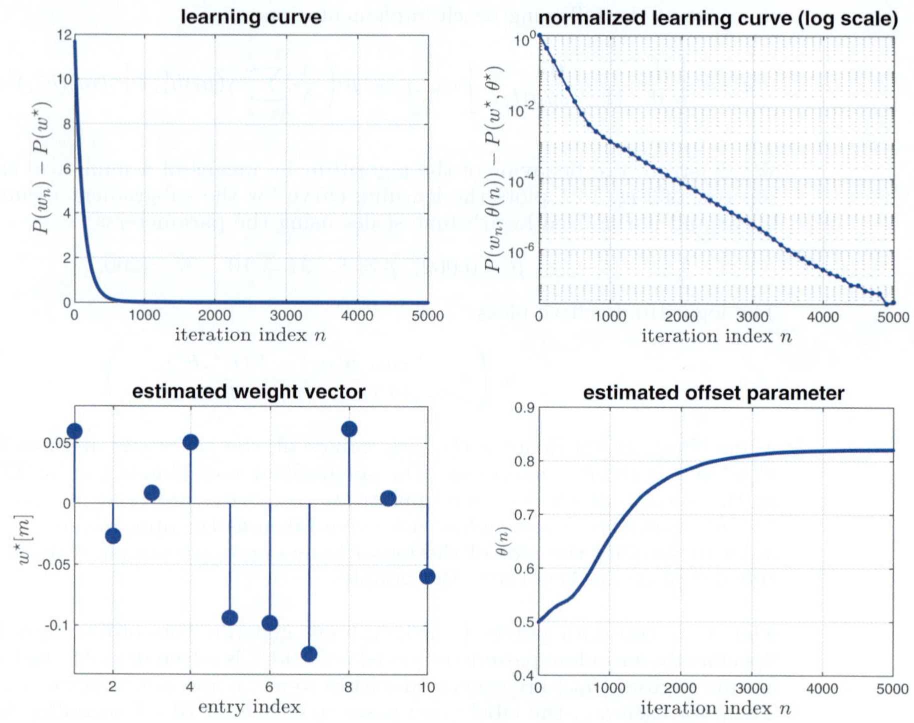

Figure 14.1 (*Top*) Learning curves $P(w_n, \theta(n))$ relative to the minimum risk value $P(w^\star, \theta^\star)$ over the first 5000 iterations in linear scale (on the left) and in normalized logarithmic scale (on the right) for the ℓ_2-regularized hinge risk (14.19). The rightmost plot on top illustrates the linear convergence of the risk value toward its limiting behavior. (*Bottom*) The limiting value of the weight iterate w_n, is shown on the left and the evolution of the offset iterate $\theta(n)$ is shown on the right.

subgradients. The following two conditions replace (12.12a)–(12.12b) used earlier in the smooth case:

(A1) (**Strong convexity**). $P(w)$ is ν-strongly convex so that, from definition (8.44) and for every $w_1, w_2 \in \text{dom}(P)$, there should exist subgradient vectors at w_1 such that

$$P(w_2) \geq P(w_1) + (s(w_1))^\mathsf{T}(w_2 - w_1) + \frac{\nu}{2}\|w_2 - w_1\|^2 \qquad (14.28a)$$

for some $\nu > 0$.

(A2) (**Affine-Lipschitz subgradients**). The subgradient vectors of $P(w)$ satisfy an *affine*-Lipschitz condition. Specifically, there should exist nonnegative constants $\{\delta, \delta_2\}$ such that the subgradient construction $s(w)$ used in recursion (14.3) satisfies

$$\|s(w_2) - s'(w_1)\| \leq \delta\|w_2 - w_1\| + \delta_2 \qquad (14.28b)$$

for any $w_1, w_2 \in \text{dom}(P)$ and *any* $s'(w) \in \partial_{w^\mathsf{T}} P(w)$. Condition (14.28b) is stated in terms of the *particular* subgradient construction $s(w)$ used in recursion (14.3) and in terms of *any* of the possible subgradients $s'(w)$ from the subdifferential set of $P(w)$ (these other subgradients could have the same construction as $s(w)$ or other possible constructions). Clearly, when $P(w)$ is differentiable everywhere, then condition (12.12b) implies (14.28b) by setting $\delta_2 = 0$. For later use, we note that condition (14.28b) implies the following relation, which involves the squared norms as opposed to the actual norms:

$$\|s(w_2) - s'(w_1)\|^2 \; \leq \; 2\delta^2 \|w_2 - w_1\|^2 + 2\delta_2^2 \tag{14.29}$$

This result follows from the inequality $(a+b)^2 \leq 2a^2 + 2b^2$ for any scalars $\{a, b\}$.

One useful conclusion that follows from conditions (14.28a)–(14.28b) is that

$$\boxed{\nu \leq \delta} \tag{14.30}$$

Proof of (14.30): We show in part (b) of Prob. 14.3 that subgradients for ν-strongly convex risks satisfy:

$$\|s(w)\| \; \geq \; \nu \|w - w^\star\|, \quad \forall\, w \in \text{dom}(P) \tag{14.31}$$

Moreover, since $w^\star$ is the minimizer for $P(w)$, we know that $0 \in \partial_{w^\mathsf{T}} P(w^\star)$. In other words, there exists a subgradient construction such that $s'(w^\star) = 0$. Using (14.31) along with (14.28b) we find that subgradients for $P(w)$ should satisfy:

$$\nu \|w - w^\star\| \; \leq \; \|s(w)\| \; \leq \; \delta \|w - w^\star\| + \delta_2 \tag{14.32}$$

Since these inequalities must hold for all w, it will follow that inequality (14.30) is valid. We can verify this statement by contradiction. Assume, to the contrary, that $\nu > \delta$. Pick a vector w_1 that satisfies

$$(\nu - \delta)\|w_1 - w^\star\| > \delta_2 \tag{14.33}$$

which is possible since the quantity $(\nu - \delta)$ is positive by assumption. For instance, let $w_1 = w^\star + \alpha e_m$, where e_m is some arbitrary basis vector in $\mathbb{R}^M$ and α is a positive scalar to be selected. Then, $\|w_1 - w^\star\| = \alpha$ and we simply need to select

$$\alpha > \frac{\delta_2}{\nu - \delta} \tag{14.34}$$

in order to satisfy (14.33). However, condition (14.33) leads to a contradiction since it implies that

$$\nu \|w_1 - w^\star\| \; > \; \delta \|w_1 - w^\star\| + \delta_2 \tag{14.35}$$

which violates (14.32).

■

Several of the risks listed in Table 12.1 satisfy condition (14.28b). Here are two examples; others are treated in Prob. 14.2.

Example 14.6 (LASSO or ℓ_1-regularized least-squares) Consider the ℓ_1-regularized least-squares risk (14.7), written using a subscript ℓ to index the data $\{\gamma(\ell), h_\ell\}$ instead of m:

$$P(w) = \alpha\|w\|_1 + \frac{1}{N}\sum_{\ell=0}^{N-1}(\gamma(\ell) - h_\ell^\mathsf{T} w)^2, \quad w = \text{col}\{w_m\} \in \mathbb{R}^M \tag{14.36}$$

We determined one particular subgradient construction for $P(w)$ in (14.8), namely,

$$s(w) = \alpha\,\text{sign}(w) - \frac{2}{N}\sum_{\ell=0}^{N-1} h_\ell\left(\gamma(\ell) - h_\ell^\mathsf{T} w\right) \tag{14.37}$$

From the second row in Table 8.3 we know that all subgradients for $P(w)$ are described by the following expression:

$$s'(w) = \alpha\,\mathbb{G}(w) - \frac{2}{N}\sum_{\ell=0}^{N-1} h_\ell\left(\gamma(\ell) - h_\ell^\mathsf{T} w\right) \tag{14.38a}$$

$$\mathbb{G}(w) = \text{col}\left\{\mathbb{G}_{\text{abs}}(w_m)\right\} \tag{14.38b}$$

$$\mathbb{G}_{\text{abs}}(x) \triangleq \begin{cases} +1, & x > 0 \\ -1, & x < 0 \\ [-1, +1], & x = 0 \end{cases} \tag{14.38c}$$

Then, for any w_1, w_2 we have:

$$\begin{aligned} &\|s(w_2) - s'(w_1)\| \\ &\leq \alpha\,\|\text{sign}(w_2) - \mathbb{G}(w_1)\| + \frac{2}{N}\sum_{\ell=0}^{N-1}\|h_\ell h_\ell^\mathsf{T}\|\,\|w_2 - w_1\| \\ &\overset{(a)}{\leq} \underbrace{\left(\frac{2}{N}\sum_{\ell=0}^{N-1}\|h_\ell\|^2\right)}_{\triangleq\,\delta}\|w_2 - w_1\| + \underbrace{2\alpha\sqrt{M}}_{\triangleq\,\delta_2} \end{aligned} \tag{14.39}$$

where in step (a) we used the fact that the difference between any of the entries of $\text{sign}(w_2)$ and $\mathbb{G}(w_1)$ cannot be larger than 2 in magnitude so that

$$\|\text{sign}(w_2) - \mathbb{G}(w_1)\|^2 \leq \|\text{col}\{2, 2, \ldots, 2\}\|^2 = 4M \tag{14.40}$$

Observe how the factor δ_2 arises from the *nonsmooth* component in the risk function, $P(w)$.

Example 14.7 **(ℓ_2-regularized hinge risk)** Consider next the ℓ_2-regularized hinge risk (14.10), written using a subscript ℓ to index the data $\{\gamma(\ell), h_\ell\}$ instead of m:

$$P(w) = \rho\|w\|^2 + \frac{1}{N}\sum_{\ell=0}^{N-1}\max\left\{0, 1 - \gamma(\ell)h_\ell^\mathsf{T} w\right\}, \quad w = \text{col}\{w_m\} \in \mathbb{R}^M \tag{14.41}$$

We already determined one particular subgradient construction for this risk function in (14.11), namely,

$$s(w) = 2\rho w - \frac{1}{N}\sum_{\ell=0}^{N-1}\gamma(\ell)h_\ell\,\mathbb{I}\left[\gamma(\ell)h_\ell^\mathsf{T} w \leq 1\right] \tag{14.42}$$

From the last row in Table 8.3 we know that all subgradients for $P(w)$ are described by the expression:

$$
\begin{aligned}
s'(w) &= 2\rho w \;-\; \frac{1}{N}\sum_{\ell=0}^{N-1} \mathbb{G}_\ell(w) && \text{(14.43a)}\\
\mathbb{G}_\ell(w) &= \text{col}\Big\{\mathbb{A}_{\gamma(\ell)h_{\ell,m}}(w_m)\Big\} && \text{(14.43b)}\\
\mathbb{A}_{\gamma h_m}(w_m) &\triangleq \begin{cases} 0, & \gamma h^{\mathsf{T}} w > 1\\ -\gamma h_m, & \gamma h^{\mathsf{T}} w < 1\\ [-\gamma h_m, 0], & \gamma h^{\mathsf{T}} w = 1,\;\; \gamma h_m \geq 0\\ [0, -\gamma h_m], & \gamma h^{\mathsf{T}} w = 1,\;\; \gamma h_m < 0 \end{cases} && \text{(14.43c)}
\end{aligned}
$$

where the $\{h_{\ell,m}\}$ denote the individual entries of h_ℓ, and the last expression for $\mathbb{A}_{\gamma h_m}(w_m)$ is defined for a generic $\gamma \in \mathbb{R}$ and vector $h \in \mathbb{R}^M$ with individual entries $\{h_m\}$. Using the triangle inequality of norms we then get:

$$
\begin{aligned}
&\|s(w_2) - s'(w_1)\|\\
&\leq 2\rho\|w_2 - w_1\| + \left\|\frac{1}{N}\sum_{\ell=0}^{N-1}\gamma(\ell)h_\ell\,\mathbb{I}\left[\gamma(\ell)h_\ell^{\mathsf{T}} w_2 \leq 1\right]\right\| + \left\|\frac{1}{N}\sum_{\ell=0}^{N-1}\mathbb{G}_\ell(w_1)\right\|\\
&\leq \underbrace{2\rho}_{\triangleq\,\delta}\,\|w_2 - w_1\| + \underbrace{\frac{2}{N}\sum_{\ell=0}^{N-1}\|\gamma(\ell)h_\ell\|}_{\triangleq\,\delta_2} && \text{(14.44)}
\end{aligned}
$$

since $\mathbb{I}[a]$ is bounded by 1 and $\|\mathbb{G}_\ell(w_1)\|$ is bounded by $\|\gamma(\ell)h_\ell\|$. Observe again how the factor δ_2 arises from the *nonsmooth* component in the risk function, $P(w)$.

14.3 CONVERGENCE BEHAVIOR

While the gradient-descent algorithm (14.2) was shown to generate iterates w_n that converge exactly to the minimizer $w^\star$ of $P(w)$ for small enough μ, the same is *not* generally true for the subgradient algorithm (14.3). This is because, unlike gradient vectors, negative subgradient vectors do *not* necessarily correspond to descent directions. We illustrate this situation in Prob. 14.7 for a particular example. This means that it can now occur that $P(w_n) > P(w_{n-1})$ no matter how small μ is. For this reason, it will not be possible any longer to guarantee the exact convergence of w_n to $w^\star$. It will instead hold that w_n will approach a small neighborhood around $w^\star$ whose size depends on μ. And, moreover, in the limit, the iterate values w_n may end up oscillating around $w^\star$ within this neighborhood as the following example shows.

Example 14.8 **(Oscillations can occur)** Assume w is one-dimensional and let us apply the subgradient algorithm (14.3) to the minimization of the strongly convex risk function $P(w) = |w| + \frac{1}{2}w^2$. This function is not differentiable at $w = 0$. In this case, the subgradient algorithm (14.3) takes the following form with a small step size $\mu > 0$:

$$
w_n = w_{n-1} - \mu\,\text{sign}(w_{n-1}) - \mu w_{n-1} \qquad \text{(14.45)}
$$

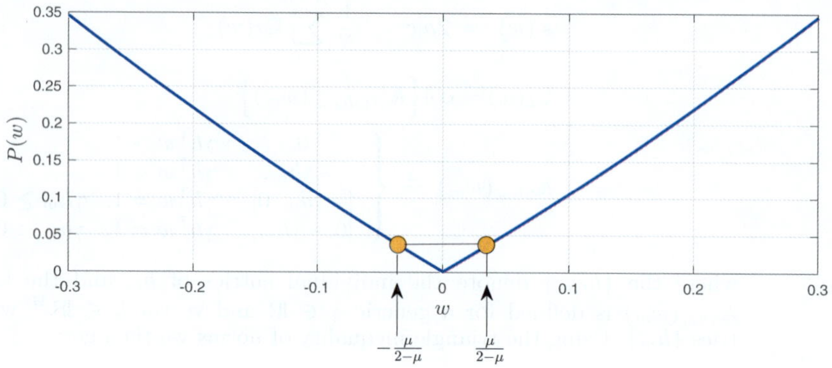

Figure 14.2 Starting from $w_{n-1} = -\mu/(2-\mu)$, the subgradient iteration (14.3) applied to $P(w) = |w| + \frac{1}{2}w^2$ oscillates between $-\mu/(2-\mu)$ and $\mu/(2-\mu)$.

Assume that at some instant in time we arrive at the value $w_{n-1} = -\mu/(2-\mu)$ (or we could consider starting from the initial condition $w_{-1} = -\mu/(2-\mu)$). Then, iteration (14.45) will give $w_n = \mu/(2-\mu)$. At this new location, iteration (14.45) will move to $w_{n+1} = -\mu/(2-\mu)$ and the process will now repeat. In this way, the iterates will end up oscillating around the optimal solution $w^\star = 0$ and the size of the error (distance from the iterate to $w^\star$) remains at $\mu/(2-\mu) = O(\mu)$ for small μ, and it does not get smaller. The situation is illustrated in Fig. 14.2.

Theorem 14.1. (Convergence with constant step size) *Consider the subgradient recursion (14.3) for minimizing a risk function $P(w)$, where $P(w)$ is ν-strongly convex with affine-Lipschitz subgradients satisfying (14.28a)–(14.28b). If the step size μ satisfies*

$$0 < \mu < \nu/\delta^2 \tag{14.46}$$

then $\|\widetilde{w}_n\|^2$ converges exponentially fast according to the dynamics:

$$\|\widetilde{w}_n\|^2 \leq \lambda\, \|\widetilde{w}_{n-1}\|^2 + 2\mu^2\delta_2^2, \quad n \geq 0 \tag{14.47}$$

where

$$\lambda = 1 - 2\mu\nu + 2\mu^2\delta^2 \, \in [0, 1) \tag{14.48}$$

By iterating, it follows for n large enough that $\|\widetilde{w}_n\|^2$ approaches an $O(\mu)$-neighborhood at the rate λ^n. We represent this behavior by writing:

$$\|\widetilde{w}_n\|^2 \leq O(\lambda^n) \, + \, O(\mu) \tag{14.49}$$

Moreover, the risk value $P(w_n)$ converges exponentially at the rate $O(\lambda^{n/2})$ to an $O(\sqrt{\mu})$-neighborhood of $P(w^\star)$ for $0 < \mu < \min\{\nu/\delta^2, 1\}$:

$$P(w_n) - P(w^\star) \, \leq \, O(\lambda^{n/2}) \, + \, O(\sqrt{\mu}) \tag{14.50}$$

Proof: We subtract $w^\star$ from both sides of (14.3) to get

$$\widetilde{w}_n \;=\; \widetilde{w}_{n-1} \;+\; \mu\, s(w_{n-1}) \tag{14.51}$$

which gives upon squaring

$$\begin{aligned}
\|\widetilde{w}_n\|^2 \;&=\; \|\widetilde{w}_{n-1}\|^2 \;+\; 2\mu(s(w_{n-1}))^\mathsf{T}\widetilde{w}_{n-1} \;+\; \mu^2\,\|s(w_{n-1})\|^2 \\
&\overset{(a)}{=}\; \|\widetilde{w}_{n-1}\|^2 \;+\; 2\mu(s(w_{n-1}))^\mathsf{T}\widetilde{w}_{n-1} \;+\; \mu^2\,\|s'(w^\star) - s(w_{n-1})\|^2 \\
&\overset{(14.29)}{\le}\; \|\widetilde{w}_{n-1}\|^2 + 2\mu(s(w_{n-1}))^\mathsf{T}\widetilde{w}_{n-1} + \mu^2\,(2\delta^2\|\widetilde{w}_{n-1}\|^2 + 2\delta_2^2)
\end{aligned} \tag{14.52}$$

where in step (a) we used the fact that, since $w^\star$ is the minimizer for $P(w)$, there must exist some subgradient construction such that $s'(w^\star) = 0$ – recall (8.47). Now, we appeal to the strong convexity condition (14.28a) and use $w_2 \leftarrow w^\star$, $w_1 \leftarrow w_{n-1}$ in step (b) below and $w_2 \leftarrow w_{n-1}$, $w_1 \leftarrow w^\star$ in step (c) to find that

$$\begin{aligned}
(s(w_{n-1}))^\mathsf{T}\widetilde{w}_{n-1} \;&\overset{(b)}{\le}\; P(w^\star) - P(w_{n-1}) - \frac{\nu}{2}\|\widetilde{w}_{n-1}\|^2 \\
&\overset{(c)}{\le}\; -\frac{\nu}{2}\|\widetilde{w}_{n-1}\|^2 \;-\; \frac{\nu}{2}\|\widetilde{w}_{n-1}\|^2 \\
&=\; -\nu\|\widetilde{w}_{n-1}\|^2
\end{aligned} \tag{14.53}$$

Substituting into (14.52) gives

$$\|\widetilde{w}_n\|^2 \le \underbrace{(1 - 2\nu\mu + 2\mu^2\delta^2)}_{\triangleq\,\lambda}\,\|\widetilde{w}_{n-1}\|^2 \;+\; 2\mu^2\delta_2^2 \tag{14.54}$$

which agrees with (14.47). The same argument that was used earlier in Fig. 12.2 can be repeated here to verify that condition (14.46) ensures $0 \le \lambda < 1$. Iterating (14.54) gives

$$\begin{aligned}
\|\widetilde{w}_n\|^2 \;&\le\; \lambda^{n+1}\|\widetilde{w}_{-1}\|^2 \;+\; 2\mu^2\delta_2^2\left(\frac{1-\lambda^{n+1}}{1-\lambda}\right) \\
&\le\; \lambda^{n+1}\|\widetilde{w}_{-1}\|^2 \;+\; 2\mu^2\delta_2^2\left(\frac{1}{1-\lambda}\right) \\
&\overset{(14.48)}{=}\; \lambda^{n+1}\|\widetilde{w}_{-1}\|^2 \;+\; \underbrace{\frac{\mu\delta_2^2}{\nu - \mu\delta^2}}_{\triangleq\,\phi^2}
\end{aligned} \tag{14.55}$$

where we introduced the positive constant ϕ^2, which is dependent on μ. We readily conclude that

$$\lim_{n\to\infty}\|\widetilde{w}_n\|^2 \;\le\; \phi^2 \tag{14.56}$$

It is easy to check that ϕ^2 does not exceed $a\mu\delta_2^2/\nu$, for any positive constant $a > 1$, for values of the step size satisfying

$$\mu < \frac{a-1}{a}\frac{\nu}{\delta^2} \tag{14.57}$$

This condition is satisfied by step sizes that are strictly smaller than the upper bound in (14.46). It follows that $\phi^2 = O(\mu)$ so that $\|\widetilde{w}_n\|^2 \rightarrow O(\mu)$. Note further, for use below, that

$$\begin{aligned}
\|\widetilde{w}_n\|^2 &\leq \lambda^{n+1}\|\widetilde{w}_{-1}\|^2 + \phi^2 \\
&\leq \lambda^{n+1}\|\widetilde{w}_{-1}\|^2 + \phi^2 + 2\phi\lambda^{(n+1)/2}\|\widetilde{w}_{-1}\| \\
&\leq \left(\lambda^{(n+1)/2}\|\widetilde{w}_{-1}\| + \phi\right)^2
\end{aligned} \tag{14.58}$$

and, therefore,

$$\|\widetilde{w}_n\| \leq \lambda^{(n+1)/2}\|\widetilde{w}_{-1}\| + \phi \tag{14.59}$$

Next, to establish (14.50), we first note that $P(w_n) \geq P(w^\star)$ since $w^\star$ is the minimizer of $P(w)$. Using (14.28a) again with $w_2 \leftarrow w^\star$ and $w_1 \leftarrow w_n$ we get

$$\begin{aligned}
0 &\leq P(w_n) - P(w^\star) \\
&\leq -\left(s(w_n)\right)^{\mathsf{T}}\widetilde{w}_n \\
&\leq \|s(w_n)\|\,\|\widetilde{w}_n\| \quad \text{(by Cauchy–Schwarz)} \\
&\overset{(a)}{=} \|s(w_n) - s'(w^\star)\|\,\|\widetilde{w}_n\| \\
&\overset{(14.28b)}{\leq} \left(\delta\|\widetilde{w}_n\| + \delta_2\right)\|\widetilde{w}_n\| \\
&= \delta\|\widetilde{w}_n\|^2 + \delta_2\|\widetilde{w}_n\| \\
&\overset{(14.58)}{=} \delta\lambda^{n+1}\|\widetilde{w}_{-1}\|^2 + \delta\phi^2 + \delta_2\lambda^{(n+1)/2}\|\widetilde{w}_{-1}\| + \delta_2\phi \\
&\leq \left(\delta\|\widetilde{w}_{-1}\|^2\lambda^{(n+1)/2} + \delta_2\|\widetilde{w}_{-1}\|\right)\lambda^{(n+1)/2} + O(\delta\mu) + O(\delta_2\sqrt{\mu}) \\
&\overset{(b)}{\leq} \left(\delta\|\widetilde{w}_{-1}\|^2 + \delta_2\|\widetilde{w}_{-1}\|\right)\lambda^{(n+1)/2} + O(\delta\mu) + O(\delta_2\sqrt{\mu}) \\
&\overset{(c)}{\leq} O(\lambda^{n/2}) + O(\sqrt{\mu})
\end{aligned} \tag{14.60}$$

where in step (a) we used again the fact that, since $w^\star$ is the minimizer of $P(w)$, there exists some subgradient vector such that $s'(w^\star) = 0$. In step (b) we used the fact that $\lambda^{n/2} < 1$ and in step (c) we assumed $\mu < 1$, which is generally the case, in which case the term $O(\sqrt{\mu})$ dominates $O(\mu)$ for small μ. It follows that, in the limit, as $n \rightarrow \infty$, the risk value $P(w_n)$ approaches an $O(\sqrt{\mu})$-neighborhood around the optimal value $P(w^\star)$ at the rate $O(\lambda^{n/2})$.

■

Regret analysis

The statement of Theorem 14.1 examines the convergence behavior of the squared weight error, $\|\widetilde{w}_n\|^2$, and of the risk value $P(w_n)$. We can transform the bound on the excess risk into a bound on the average regret defined by

$$\mathcal{R}(N) \overset{\Delta}{=} \frac{1}{N}\sum_{n=0}^{N-1} P(w_{n-1}) - P(w^\star) \quad \textbf{(average regret)} \tag{14.61}$$

Indeed, using (14.50) we find that the regret decays at the rate of $1/N$ toward $O(\sqrt{\mu})$ since

$$\begin{aligned}
\mathcal{R}(N) &\triangleq \frac{1}{N}\sum_{n=0}^{N-1}\Big(P(w_{n-1}) - P(w^\star)\Big) \\
&\leq \frac{1}{N}O\left(\sum_{n=0}^{N-1}\lambda^{n/2}\right) + O(\sqrt{\mu}) \\
&= \frac{1}{N}O\left(\frac{1-\lambda^{N/2}}{1-\lambda^{1/2}}\right) + O(\sqrt{\mu}) \\
&= O(1/N) + O(\sqrt{\mu})
\end{aligned} \tag{14.62}$$

14.4 POCKET VARIABLE

Theorem 14.1 establishes the exponential convergence of $\|\widetilde{w}_n\|^2$ to a region of size $O(\mu)$ at the rate of λ^n, as well as the exponential convergence of $P(w_n)$ at the slower rate $\lambda^{n/2}$ to an $O(\sqrt{\mu})$-ball around $P(w^\star)$. We express these conclusions in the convenient form:

(**for strongly convex** $P(w)$)

$$\left\{\begin{array}{rcl}
\|\widetilde{w}_n\|^2 & \leq & O(\lambda^n) + O(\mu) \\
P(w_n) - P(w^\star) & \leq & O(\lambda^{n/2}) + O(\sqrt{\mu})
\end{array}\right. \tag{14.63}$$

We can improve on the second result and obtain faster convergence for the risk at the rate of λ^n toward a smaller $O(\mu)$-neighborhood around $P(w^\star)$ by introducing the concept of a *pocket variable*.

Let w_n^{best} denote the iterate value that results in the smallest risk from among all iterates up to iteration n, namely,

$$w_n^{\text{best}} \triangleq \underset{0\leq m\leq n}{\text{argmin}}\; P(w_m) \tag{14.64}$$

This best iterate is called the *pocket* variable because its value is saved into a "pocket" denoted by w_p. At every iteration n, the subgradient algorithm compares w_n against the stored pocket variable and updates the latter as follows:

$$w_p \leftarrow \left\{\begin{array}{ll}
w_n, & \text{if } P(w_n) < P(w_p) \\
w_p, & \text{otherwise}
\end{array}\right. \tag{14.65}$$

The modified subgradient algorithm with the inclusion of a pocket variable is listed in (14.66).

Subgradient method with pocket variable for minimizing $P(w)$.

given a construction for subgradient vectors, $s(w)$;
given a small step-size parameter $\mu > 0$;
start from an arbitrary initial condition, w_{-1};
given a pocket variable $w_p = w_{-1}$.
repeat over $n \geq 0$:
  $w_n = w_{n-1} - \mu\, s(w_{n-1})$
  if $P(w_n) < P(w_p)$
    set $w_p \leftarrow w_n$
    set $P(w_p) \leftarrow P(w_n)$
  end
end
return $w^\star \leftarrow w_p$.

(14.66)

THEOREM 14.2. (Convergence with pocket variable) *Consider the subgradient algorithm (14.66) for minimizing a risk function $P(w)$, where $P(w)$ is ν-strongly convex with affine-Lipschitz subgradients satisfying (14.28a)–(14.28b). If the step size μ satisfies the tighter bound:*

$$0 < \mu < \nu/2\delta^2 \tag{14.67}$$

then it holds that for n large enough:

$$\|w^\star - w_n^{\text{best}}\|^2 \leq O(\lambda^n) + O(\mu) \tag{14.68a}$$

$$P(w_n^{\text{best}}) - P(w^\star) \;\leq\; O(\lambda^n) + O(\mu) \tag{14.68b}$$

where now

$$\lambda \;=\; 1 - \nu\mu + 2\mu^2\delta^2 \;\in\; [0,1) \tag{14.69}$$

That is, $P(w_n^{\text{best}})$ converges toward an $O(\mu)$-neighborhood around $P(w^\star)$ at the exponential rate λ^n.

Proof: We subtract $w^\star$ from both sides of (14.3) to get

$$\widetilde{w}_n \;=\; \widetilde{w}_{n-1} \;+\; \mu\, s(w_{n-1}) \tag{14.70}$$

which gives upon squaring

$$\|\widetilde{w}_n\|^2 \;=\; \|\widetilde{w}_{n-1}\|^2 \;+\; 2\mu (s(w_{n-1}))^{\mathsf{T}} \widetilde{w}_{n-1} \;+\; \mu^2\, \|s(w_{n-1})\|^2 \tag{14.71}$$

Now, from the strong convexity condition (14.28a), it holds that

$$(s(w_{n-1}))^{\mathsf{T}} \widetilde{w}_{n-1} \;\leq\; P(w^\star) - P(w_{n-1}) - \frac{\nu}{2}\|\widetilde{w}_{n-1}\|^2 \tag{14.72}$$

so that substituting into (14.71) gives

$$\|\widetilde{w}_n\|^2 \leq \|\widetilde{w}_{n-1}\|^2 \;+\; 2\mu\,(P(w^\star) - P(w_{n-1})) - \nu\mu\|\widetilde{w}_{n-1}\|^2 + \mu^2\,\|s(w_{n-1})\|^2 \tag{14.73}$$

Referring to (14.29), if we set $w_1 \leftarrow w_{n-1}$ and $w_2 \leftarrow w^\star$, and use the fact that there exists one particular subgradient satisfying $s'(w^\star) = 0$, we obtain

$$\|s(w_{n-1})\|^2 \;\leq\; 2\delta^2 \,\|\widetilde{w}_{n-1}\|^2 \;+\; 2\delta_2^2 \tag{14.74}$$

Substituting into (14.73), we get

$$\|\widetilde{w}_n\|^2 \leq (1 - \nu\mu + 2\delta^2\mu^2)\|\widetilde{w}_{n-1}\|^2 \;+\; 2\mu\,(P(w^\star) - P(w_{n-1})) \;+\; 2\delta_2^2\mu^2 \tag{14.75}$$

which we rewrite as

$$2\mu\,(P(w_{n-1}) - P(w^\star)) \leq (1 - \nu\mu + 2\delta^2\mu^2)\|\widetilde{w}_{n-1}\|^2 - \|\widetilde{w}_n\|^2 + 2\delta_2^2\mu^2 \tag{14.76}$$

To proceed, we simplify the notation and introduce the scalars

$$a(n) = P(w_{n-1}) - P(w^\star) \tag{14.77a}$$
$$b(n) = \|\widetilde{w}_n\|^2 \tag{14.77b}$$
$$\lambda = 1 - \nu\mu + 2\delta^2\mu^2 \tag{14.77c}$$
$$\tau^2 = 2\delta_2^2 \tag{14.77d}$$

Since $w^\star$ is the global minimizer for $P(w)$, it holds that $a(n) \geq 0$ for all n. Note that $a(n)$ represents the excess risk at w_{n-1}. Furthermore, it holds that $\lambda \in (0,1)$. This can be seen as follows. First, observe from expression (14.77c) that λ is a function of μ and that $\lambda(\mu)$ is quadratic in μ. This function attains its minimum at the location $\mu_o = \nu/4\delta^2$. For any μ, the value of $\lambda(\mu)$ is larger than the minimum value of the function at μ_o, i.e., it holds that

$$\lambda \;\geq\; 1 - \frac{\nu^2}{8\delta^2} \tag{14.78}$$

We know from (14.30) that $\nu \leq \delta$ and, hence, $\lambda > 0$. Moreover, it holds that $\lambda < 1$ under (14.46). We conclude that $\lambda \in (0,1)$, as claimed.

Using the quantities $\{a(n), b(n), \lambda, \tau^2\}$, we rewrite (14.76) more compactly as

$$2\mu a(n) \leq \lambda b(n-1) - b(n) + \mu^2\tau^2 \tag{14.79}$$

Since subgradients are not necessarily descent directions, we cannot guarantee $a(n) < a(n-1)$. However, we can still arrive at a useful conclusion by employing the pocket variable, $w_n^{\rm best}$, defined by (14.64). Iterating (14.79) over $0 \leq n \leq L$, for some duration L, gives

$$\sum_{n=0}^{L} \lambda^{L-n}\left(2\mu a(n) - \mu^2\tau^2\right) \;\leq\; \lambda^{L+1}b(-1) - b(L) \;\leq\; \lambda^{L+1}b(-1) \tag{14.80}$$

Let

$$a^{\rm best}(L) \;\stackrel{\Delta}{=}\; \min_{0\leq n\leq L}\; a(n) \tag{14.81}$$

The vector $w_{L-1}^{\rm best}$ will be the corresponding iterate where $a^{\rm best}(L)$ is attained. Replacing $a(n)$ by $a^{\rm best}(L)$ in (14.80) gives

$$\left(2\mu a^{\rm best}(L) - \mu^2\tau^2\right) \leq \lambda^{L+1}b(-1)\left(\sum_{n=0}^{L}\lambda^{L-n}\right)^{-1} \;=\; b(-1)\,\frac{\lambda^{L+1}(1-\lambda)}{1-\lambda^{L+1}} \tag{14.82}$$

or, equivalently,

$$2\mu a^{\rm best}(L) \;\leq\; \mu^2\tau^2 \;+\; b(-1)\,\frac{\lambda^{L+1}(1-\lambda)}{1-\lambda^{L+1}} \tag{14.83}$$

Using the definitions of $a^{\text{best}}(L)$ and $b(-1)$ we can rewrite the above in the form:

$$P(w_{L-1}^{\text{best}}) - P(w^\star) \leq \frac{\lambda^{L+1}(1-\lambda)}{2\mu(1-\lambda^{L+1})}\|\widetilde{w}_{-1}\|^2 + \mu\delta_2^2 \tag{14.84}$$

This result leads to (14.68b) by noting that $1-\lambda^{L+1} \geq 1/2$ for L large enough. We can also characterize how far w_L^{best} is from $w^\star$ by recalling that the strong convexity property (14.28a) implies:

$$\|w_L^{\text{best}} - w^\star\|^2 \leq \frac{2}{\nu}\left(P(w_L^{\text{best}}) - P(w^\star)\right) \tag{14.85}$$

■

Convexity versus strong convexity

Theorems 14.1 and 14.2 consider *strongly convex* risks $P(w)$ with affine-Lipschitz subgradients satisfying (14.28a)–(14.28b) and establish that

(**for strongly convex** $P(w)$)

$$\left\{\begin{array}{rcl} \|\widetilde{w}_n\|^2 & \leq & O(\lambda^n) + O(\mu) \\ P(w_n^{\text{best}}) - P(w^\star) & \leq & O(\lambda^n) + O(\mu) \end{array}\right. \tag{14.86a}$$

where, for small μ, the parameter $\lambda = 1 - O(\mu)$. If $P(w)$ happens to be only convex (but not necessarily strongly convex), then the argument in part (b) of Prob. 14.4 shows that for risks $P(w)$ with *bounded* subgradients, the convergence rate will become sublinear:

(**for convex** $P(w)$**; bounded subgradients**)

$$P(w_n^{\text{best}}) - P(w^\star) \leq O(1/n) + O(\mu) \tag{14.86b}$$

14.5 EXPONENTIAL SMOOTHING

The convergence result (14.68b) shows that the risk value at the pocket variable, w_n^{best}, is able to approach the global minimum within $O(\mu)$. However, keeping track of the pocket variable is not computationally feasible. This is because we need to evaluate the risk value $P(w_n)$ at every iteration and compare it against $P(w_p)$. For empirical risks, this calculation will require processing the entire dataset at every iteration. A more practical construction is possible and is already suggested by the arguments used in the last convergence proof.

Motivation

Let us introduce the geometric sum, for some integer L:

$$S_L \triangleq \sum_{n=0}^{L} \lambda^{L-n} = \frac{1-\lambda^{L+1}}{1-\lambda} \tag{14.87}$$

and note that

$$S_L = \lambda S_{L-1} + 1, \;\; S_0 = 1 \tag{14.88}$$

We also introduce the convex combination coefficients

$$r_L(n) \stackrel{\Delta}{=} \frac{\lambda^{L-n}}{S_L}, \quad n = 0, 1, 2, \ldots, L \tag{14.89}$$

which depend on both L and n. Using these coefficients, we define a smoothed iterate as follows:

$$\begin{aligned} \bar{w}_L &\stackrel{\Delta}{=} \sum_{n=0}^{L} r_L(n) w_{n-1} \\ &= \frac{1}{S_L}\left(\lambda^L w_{-1} + \lambda^{L-1} w_0 + \ldots + \lambda w_{L-2} + w_{L-1}\right) \end{aligned} \tag{14.90}$$

where the $\{w_n\}$ are the iterates generated by the subgradient algorithm (14.3). In this calculation, recent iterates are given more weight that iterates from the remote past. Observe that, in contrast to the pocket variable w_L^{best}, the smoothed variable $\bar{w}_L$ is computed directly from the $\{w_m\}$ and does not require evaluation of the risk values $\{P(w_m)\}$. Observe further that $\bar{w}_L$ can be updated recursively as follows:

$$\bar{w}_L = \left(1 - \frac{1}{S_L}\right)\bar{w}_{L-1} + \frac{1}{S_L} w_{L-1}, \quad \bar{w}_0 = w_{-1}, \; L \geq 0 \tag{14.91}$$

Moreover, since $P(w)$ is a convex function, it holds that

$$P(\bar{w}_L) = P\left(\sum_{n=0}^{L} r_L(n) w_{n-1}\right) \leq \sum_{n=0}^{L} r_L(n) P(w_{n-1}) \tag{14.92}$$

Using this fact, we will verify that the smoothed iterate $\bar{w}_L$ satisfies

$$\lim_{L\to\infty}\Big(P(\bar{w}_L) - P(w^\star)\Big) \leq O(\lambda^L) + O(\mu) \tag{14.93}$$

which is similar to (14.68b). In other words, the smoothed iterate is able to deliver the same performance guarantee, but without the need for a pocket variable and for the evaluation of the risk function at every iteration.

Proof of (14.93): We start from (14.80) and divide both sides by S_L to get

$$\sum_{n=0}^{L} \frac{\lambda^{L-n}}{S_L}\left(2\mu a(n) - \mu^2\tau^2\right) \leq \frac{\lambda^{L+1}}{S_L} b(-1) \tag{14.94}$$

or, equivalently,

$$\sum_{n=0}^{L} 2\mu\, r_L(n)\Big(P(w_{n-1}) - P(w^\star)\Big) \leq \frac{\lambda^{L+1}(1-\lambda)}{1-\lambda^{L+1}} b(-1) + \mu^2\tau^2 \tag{14.95}$$

Appealing to the convexity property (14.92), we conclude that

$$2\mu\Big(P(\bar{w}_L) - P(w^\star)\Big) \leq \frac{\lambda^{L+1}(1-\lambda)}{1-\lambda^{L+1}} b(-1) + \mu^2\tau^2 \tag{14.96}$$

Taking the limit as $L \to \infty$ leads to the desired conclusion (14.93).

■

Incorporating smoothing

We still have one challenge. Using λ as a scaling weight in (14.90) is problematic because its value depends on (μ, ν, δ) and the parameters (ν, δ) may not be known beforehand. The analysis, however, suggests that we may replace λ by a design parameter, denoted by κ and satisfying $\lambda \le \kappa < 1$. We therefore introduce a new smoothed variable (we continue to denote it by $\bar{w}_L$ to avoid a proliferation of symbols; we also continue to denote the scaling coefficients by $r_L(n)$):

$$\bar{w}_L \stackrel{\Delta}{=} \sum_{n=0}^{L} r_L(n) w_{n-1} \tag{14.97}$$

where now

$$r_L(n) \stackrel{\Delta}{=} \frac{\kappa^{L-n}}{S_L} \tag{14.98a}$$

$$S_L \stackrel{\Delta}{=} \sum_{n=0}^{L} \kappa^{L-n} = \frac{1-\kappa^{L+1}}{1-\kappa} \tag{14.98b}$$

The resulting algorithm is listed in (14.99), where κ is chosen close to but smaller than 1 (since the value of λ is usually unknown beforehand). For this procedure, the convergence property (14.93) will continue to hold with λ replaced by κ, as we proceed to verify.

Subgradient method with smoothing for minimizing $P(w)$.

given a construction for subgradient vectors, $s(w)$;
given a small step-size parameter $\mu > 0$;
start from an arbitrary initial condition, w_{-1};
select a positive scalar $\kappa \in [\lambda, 1)$;
set $S_0 = 0$, $\bar{w}_0 = w_{-1}$.
repeat until convergence over $n \ge 0$: (14.99)

$$w_n = w_{n-1} - \mu\, s(w_{n-1})$$
$$S_{n+1} = \kappa S_n + 1$$
$$\bar{w}_{n+1} = \left(1 - \frac{1}{S_{n+1}}\right)\bar{w}_n + \frac{1}{S_{n+1}} w_n$$

end
return $w^\star \leftarrow \bar{w}_{n+1}$.

Theorem 14.3. (Convergence with exponential smoothing) *Consider the subgradient algorithm (14.99) for minimizing a risk function $P(w)$, where $P(w)$ is ν-strongly convex with affine-Lipschitz subgradients satisfying (14.28a)–(14.28b). Assume the step size μ satisfies the same bound (14.67) so that the parameter λ defined by (14.69) lies in the interval $\lambda \in [0,1)$. Select a scalar κ satisfying $\lambda \leq \kappa < 1$. Then, it holds that the smoothed iterates $\bar{w}_n$ generated by algorithm (14.99) satisfy for large enough n:*

$$\|w^\star - \bar{w}_n\|^2 \leq O(\kappa^n) + O(\mu) \tag{14.100a}$$

$$P(\bar{w}_n) - P(w^\star) \leq O(\kappa^n) + O(\mu) \tag{14.100b}$$

Proof: We start from (14.79) but unlike the previous derivation in (14.80), we now use κ to get

$$\sum_{n=0}^{L} \kappa^{L-n}\left(2\mu a(n) - \mu^2\tau^2\right) \leq \sum_{n=0}^{L} \kappa^{L-n}\left(\lambda b(n-1) - b(n)\right) \overset{(a)}{\leq} \kappa^L b(-1) \tag{14.101}$$

where in step (a) we used the fact that $\lambda \leq \kappa < 1$. We can now proceed from here and complete the argument as before to arrive at (14.100b). We can also characterize how far $\bar{w}_L$ is from $w^\star$ by recalling that the strong convexity property (14.28a) implies

$$\|\bar{w}_L - w^\star\|^2 \leq \frac{2}{\nu}\left(P(\bar{w}_L) - P(w^\star)\right) \tag{14.102}$$

■

14.6 ITERATION-DEPENDENT STEP SIZES

We can also consider iteration-dependent step sizes and use the subgradient recursion:

$$\boxed{w_n = w_{n-1} - \mu(n)\, s(w_{n-1}), \quad n \geq 0} \tag{14.103}$$

There are several ways to select the step-size sequence $\mu(n)$. The convergence analysis in the sequel assumes step sizes that satisfy either one of the following two conditions:

$$\textbf{(Condition I)} \qquad \sum_{n=0}^{\infty} \mu^2(n) < \infty \quad \text{and} \quad \sum_{n=0}^{\infty} \mu(n) = \infty \tag{14.104a}$$

$$\textbf{(Condition II)} \qquad \lim_{n\to\infty} \mu(n) = 0 \quad \text{and} \quad \sum_{n=0}^{\infty} \mu(n) = \infty \tag{14.104b}$$

The next result establishes that the iterates w_n are now able to converge to the *exact* minimizer $w^\star$ albeit at the slower *sublinear* rate of $O(1/n)$.

THEOREM 14.4. (Convergence under vanishing step sizes) *Consider the subgradient recursion (14.103) for minimizing a risk function $P(w)$, where $P(w)$ is ν-strongly convex with affine-Lipschitz gradients satisfying (14.28a)–(14.28b). If the step-size sequence $\mu(n)$ satisfies either (14.104a) or (14.104b), then w_n converges to the global minimizer, $w^\star$. In particular, when the step-size sequence is chosen as*

$$\mu(n) = \frac{\tau}{n+1} \tag{14.105}$$

then three convergence rates are possible depending on how the factor $\nu\tau$ compares to the value 1. Specifically, for large enough n, it holds that:

$$\begin{cases} \|\widetilde{w}_n\|^2 \leq \left(\frac{2\tau^2\delta_2^2}{\nu\tau-1}\right)\frac{1}{n} + o\left(\frac{1}{n}\right), & \nu\tau > 1 \\ \|\widetilde{w}_n\|^2 = O\left(\frac{\log n}{n}\right), & \nu\tau = 1 \\ \|\widetilde{w}_n\|^2 = O\left(\frac{1}{n^{\nu\tau}}\right), & \nu\tau < 1 \end{cases} \tag{14.106}$$

The fastest convergence rate occurs when $\nu\tau > 1$ (i.e., for large enough τ) and is on the order of $O(1/n)$. Likewise, $P(w_n)$ converges to $P(w^\star)$ at the same rate as $\|\widetilde{w}_n\|$ so that the fastest rate of convergence is

$$P(w_n) - P(w^\star) \leq O(1/\sqrt{n}) \tag{14.107}$$

Proof: The argument that led to (14.54) will similarly lead to

$$\|\widetilde{w}_n\|^2 \leq \lambda(n)\|\widetilde{w}_{n-1}\|^2 + 2\mu^2(n)\delta_2^2 \tag{14.108}$$

with $\lambda(n) = 1-2\nu\mu(n)+2\delta^2\mu^2(n)$. We split the term $2\nu\mu(n)$ into the sum of two terms and write

$$\lambda(n) = 1 - \nu\mu(n) - \nu\mu(n) + 2\delta^2\mu^2(n) \tag{14.109}$$

And since $\mu(n) \to 0$, we conclude that for large enough $n > n_o$, the value of $\mu^2(n)$ is smaller than $\mu(n)$ so that

$$\nu\mu(n) \geq 2\delta^2\mu^2(n), \quad 0 \leq \nu\mu(n) < 1, \quad n > n_o \tag{14.110}$$

Consequently,

$$\lambda(n) \leq 1 - \nu\mu(n), \quad n > n_o \tag{14.111}$$

Then, inequalities (14.108) and (14.111) imply that

$$\|\widetilde{w}_n\|^2 \leq (1-\nu\mu(n))\,\|\widetilde{w}_{n-1}\|^2 + 2\mu^2(n)\delta_2^2, \quad n > n_o \tag{14.112}$$

We can now compare this result with recursion (14.136) in the appendix and make the identifications:

$$u(n+1) \leftarrow \|\widetilde{w}_n\|^2 \tag{14.113a}$$

$$a(n) \leftarrow \nu\mu(n) \tag{14.113b}$$

$$b(n) \leftarrow 2\delta_2^2\mu^2(n) \tag{14.113c}$$

We know from conditions (14.104a) or (14.104b) on $\mu(n)$ that for $n > n_o$:

$$0 \leq a(n) < 1, \quad b(n) \geq 0, \quad \sum_{n=0}^{\infty} a(n) = \infty, \quad \lim_{n\to\infty}\frac{b(n)}{a(n)} = 0 \tag{14.114}$$

Therefore, conditions (14.137) in the appendix are met and we conclude that $\|\widetilde{w}_n\|^2 \to 0$. We can be more specific and quantify the rate at which the squared error $\|\widetilde{w}_n\|^2$ converges for step-size sequences of the form $\mu(n) = \tau/(n+1)$. To establish (14.106), we use (14.112) and the assumed form for $\mu(n)$ to write

$$\|\widetilde{w}_n\|^2 \leq \left(1 - \frac{\nu\tau}{n+1}\right) \|\widetilde{w}_{n-1}\|^2 + \frac{2\tau^2\delta_2^2}{(n+1)^2}, \quad n > n_o \tag{14.115}$$

This recursion has the same form as recursion (14.138) in the appendix, with the identifications:

$$a(n) \leftarrow \frac{\nu\tau}{n+1}, \quad b(n) \leftarrow \frac{2\tau^2\delta_2^2}{(n+1)^2}, \quad p = 1 \tag{14.116}$$

The rates of convergence (14.106) then follow from part (b) of Lemma 14.1 in the appendix. To establish (14.107) we recall (14.60), which shows that

$$0 \leq P(w_n) - P(w^\star) \leq \delta\|\widetilde{w}_n\|^2 + \delta_2\|\widetilde{w}_n\| \tag{14.117}$$

so that the convergence of $P(w_n)$ to $P(w^\star)$ follows the same rate as $\|\widetilde{w}_n\|$.

■

Example 14.9 (LASSO or ℓ_1-regularized least-squares) We revisit the LASSO formulation from Example 14.1, where we considered the ℓ_1-regularized least-squares risk:

$$P(w) = \alpha\|w\|_1 + \frac{1}{N}\sum_{m=0}^{N-1}(\gamma(m) - h_m^{\mathsf{T}} w)^2 \tag{14.118}$$

From expression (14.9) we know that a subgradient implementation for the LASSO risk is given by

$$w_n = w_{n-1} - \mu\alpha\,\text{sign}(w_{n-1}) + \mu\left(\frac{2}{N}\sum_{m=0}^{N-1} h_m\left(\gamma(m) - h_m^{\mathsf{T}} w_{n-1}\right)\right), \quad n \geq 0 \tag{14.119}$$

We illustrate the behavior of the algorithm by means of a numerical simulation. The top row of Fig. 14.3 plots the learning curves with and without exponential smoothing using the parameters:

$$\mu = 0.002, \quad \alpha = 0.7, \quad \kappa = 0.95, \quad M = 10, \quad N = 200 \tag{14.120}$$

The smoothed implementation follows construction (14.99) where the weight iterates are smoothed to $\bar{w}_n$. The regular subgradient recursion (14.119) is run for 5000 iterations at the end of which the limit quantity w_n is used as an approximation for $w^\star$ with the corresponding risk value taken as $P(w^\star)$. The smoothed version of the recursion is also run for the same number of iterations, and we know from the convergence analysis that $\bar{w}_n$ will approach $w^\star$, albeit with a smaller risk value.

The $N = 200$ data points $\{\gamma(m), h_m\}$ are generated according to a linear model of the form

$$\gamma(m) = h_m^{\mathsf{T}} w^a + v(m) \tag{14.121}$$

where the $\{v(m)\}$ are realizations of a zero-mean Gaussian noise with variance 10^{-4}, while the entries of each h_m are generated from a Gaussian distribution with zero mean and unit variance. The entries of the model $w^a \in \mathbb{R}^{10}$ are also generated from a Gaussian distribution with zero mean and unit variance.

The resulting weight iterates, obtained under the regular subgradient algorithm (without smoothing) and under the smoothed version, are shown on the left in the bottom

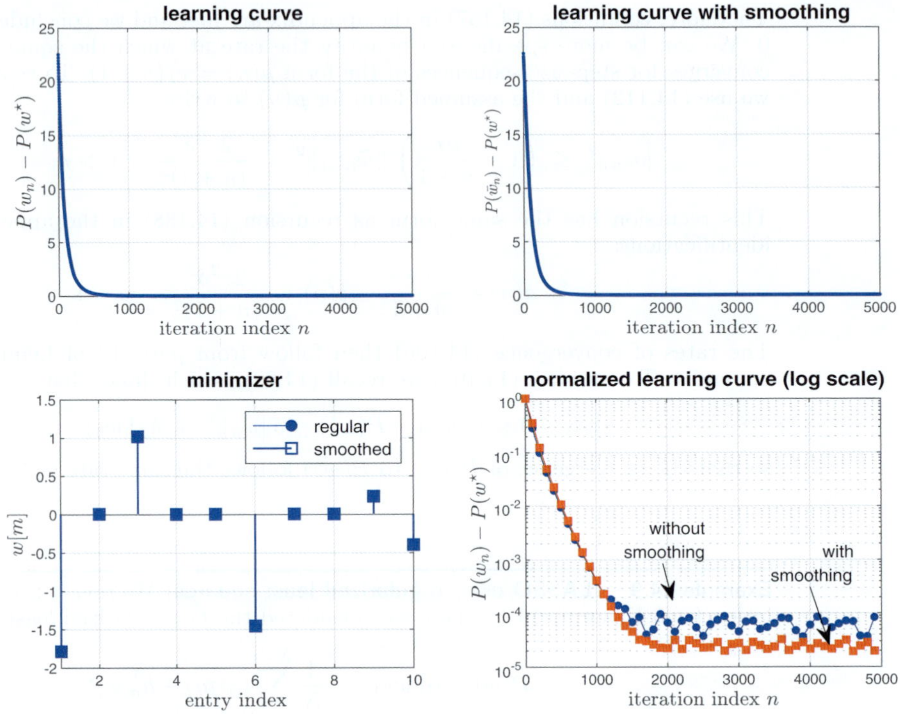

Figure 14.3 (*Top*) Learning curves $P(w_n)$ relative to the minimum risk value $P(w^\star)$ for the regular LASSO subgradient recursion (14.119) (*left*) and for the smoothed version (*right*) over the first 5000 iterations. The lower plot on the right compares these curves in logarithmic scale and illustrates how the smoothed version has superior "steady-state" behavior. The lower plot on the left shows the limiting minimizers obtained by applying the regular subgradient recursion (14.119) or its smoothed version according to construction (14.99). The limit values of these minimizers approach each other.

plot of the figure. It is seen that they approach the same value. The bottom-right plot compares the learning curves in normalized logarithmic scale. In order to simplify the plot, the learning curves are downsampled by a factor of 100, meaning that one marker is shown for every 100 values over a run of 5000 iterations. It is seen that the smoothed version of the algorithm tends to a lower "steady-state" level, as anticipated by result (14.93).

Recall that result (14.93) ensures that the smoothed iterates $\bar{w}_n$ of the subgradient recursion generate risk values $P(\bar{w}_n)$ that converge toward an $O(\mu)$-neighborhood of $P(w^\star)$, while the regular subgradient recursion (14.119) generates risk values $P(w_n)$ that approach a larger $O(\sqrt{\mu})$-neighborhood of $w^\star$. This latter effect is illustrated in Fig. 14.4, where we repeat the simulation for different values of the step-size parameter, namely,

$$\mu \in \{0.0001, 0.0002, 0.0005, 0.001, 0.002, 0.005, 0.007\} \tag{14.122}$$

For each step size, the regular and smoothed recursions are run for 50,000 iterations. The average of the last 50 risk values, $P(w_n)$ or $P(\bar{w}_n)$, are used as approximations for the limiting risk values, namely, $P(w_n)$ and $P(\bar{w}_n)$ as $n \to \infty$. The figure plots the values of $P(w_\infty) - P(w^\star)$ and $P(\bar{w}_\infty) - P(w^\star)$ as functions of the step-size parameter. The plot on the left uses a linear scale while the plot on the right uses a normalized logarithmic scale for both axes. It is evident that smoothing improves the steady-state performance level.

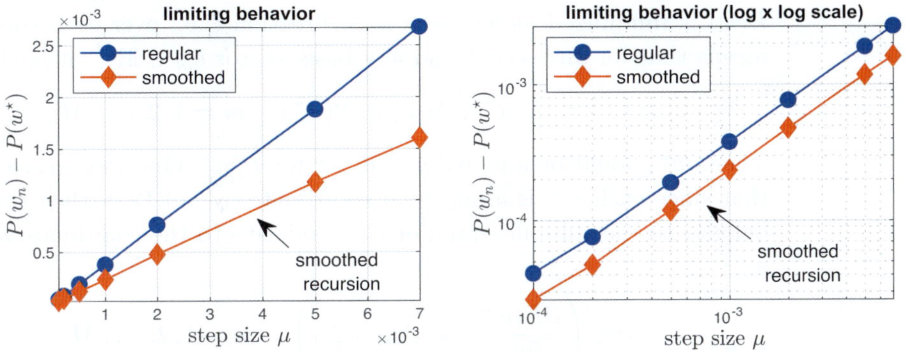

Figure 14.4 (*Left*) Values of $P(w_\infty) - P(w^\star)$ and $P(\bar{w}_\infty) - P(w^\star)$ as functions of the step-size parameter using a linear scale. (*Right*) Values of $P(w_\infty) - P(w^\star)$ and $P(\bar{w}_\infty) - P(w^\star)$ as functions of the step-size parameter using a normalized logarithmic scale for both axes. The plots illustrate the expected $O(\mu)$ versus $O(\sqrt{\mu})$ behavior for the smoothed and nonsmoothed subgradient LASSO recursion.

14.7 COORDINATE-DESCENT ALGORITHMS

We end the chapter by revisiting the coordinate-descent method from Section 12.5 and explain how it can be used to minimize nonsmooth risks as well. Recall that we motivated the method by showing in (12.121) that coordinate-wise minimization leads to global minimization for convex smooth risks $P(w)$. We remarked though that this conclusion does not hold for nonsmooth functions and illustrated this fact by means of an example in Prob. 12.27. Nevertheless, coordinate-wise minimization will continue to work for a sub-class of nonsmooth functions, namely, those that can be expressed as the sum of two components: one is convex and differentiable while the other is separable over the coordinates. Specifically, let us assume $P(w)$ can be expressed in the form:

$$P(w) \;=\; E(w) + \sum_{m=1}^{M} q_m(w_m) \tag{14.123}$$

where the convex function $E(w)$ is at least first-order differentiable while the rightmost convex component is possibly nonsmooth but separates over the individual entries of w denoted by $\{w_m\}$. For example, consider the ℓ_1-regularized least-squares risk (14.118). It has the above form since we can write

$$P(w) = \underbrace{\frac{1}{N}\sum_{n=0}^{N-1}(\gamma(n) - h_n^{\mathsf{T}}w)^2}_{=E(w)} + \underbrace{\alpha\sum_{m=1}^{M}|w_m|}_{=q(w)} \tag{14.124}$$

Returning to (14.123), assume that we succeed in finding a point $w^\star$ such that $P(w)$ is minimized along every axis direction (i.e., over each coordinate). This means that for any step λ and any basis vector $e_m \in \mathbb{R}^M$, it holds that

$$P(w^\star + \lambda e_m) \geq P(w^\star), \quad m = 1, 2, \ldots, M \tag{14.125}$$

Then, $w^\star$ should be a global minimizer for $P(w)$. One way to see this is to note that since each $w_m^\star$ is an optimal coordinate by (14.125), then the number zero belongs to the subdifferential of $P(w)$ relative to the coordinate w_m at location $w_m^\star$:

$$0 \in \left(\frac{\partial E(w^\star)}{\partial w_m} + \partial q_m(w_m^\star)\right), \quad m = 1, 2, \ldots, M \tag{14.126}$$

It follows that $0_M \in \partial P(w^\star)$. This is because a subgradient for $P(w)$ at location $w^\star$ can be constructed by concatenating individual subgradients of $E(w) + q_m(w_m)$, at the separate coordinates, on top of each other. The resulting algorithm is listed in (14.127) assuming a randomized selection of the coordinates. In this listing, the notation $s_m(a)$ refers to a subgradient construction for $q_m(a)$ at location a.

Randomized coordinate-descent for minimizing (14.123).

let $w = \text{col}\{w_m\}$, $m = 1, 2, \ldots, M$;
given first-order differentiable convex $E(w)$;
given nonsmooth separable components $\{q_m(w_m)\}$;
given a subgradient construction, $s_m(\cdot)$ for each $q_m(\cdot)$;
start from an arbitrary initial condition w_{-1};
repeat until convergence over $n \geq 0$:
  $w_{n-1} = \text{col}\{w_{n-1,m}\}$ is available at start of iteration (14.127)
  select index m^o uniformly at random within $1 \leq m \leq M$
  update $w_{n,m^o} = w_{n-1,m^o} - \mu\left\{\dfrac{\partial E(w_{n-1})}{\partial w_{m^o}} + s_{m^o}(w_{n-1,m^o})\right\}$
  keep $w_{n,m} = w_{n-1,m}$ for all $m \neq m^o$
  end
end
return $w^\star \leftarrow w_n$.

We showed earlier in the proof of Theorem 12.4 how the convergence argument for the gradient-descent algorithm can be extended to randomized coordinate descent by working with the mean-square error (MSE) $\mathbb{E}\,\|\widetilde{\boldsymbol{w}}_n\|^2$. A similar analysis can be repeated to extend the derivation to the nonsmooth case under the separable condition (14.123). We leave this extension to Probs. 14.15 and 14.16.

Example 14.10 (**Coordinate-descent LASSO**) Consider the regularized problem:

$$w^\star = \underset{w\in\mathbb{R}^M}{\text{argmin}} \left\{ \alpha\|w\|_1 + \rho\|w\|^2 \ + \ \frac{1}{N}\sum_{\ell=0}^{N-1}(\gamma(\ell) - h_\ell^\mathsf{T} w)^2 \right\} \tag{14.128}$$

We denote the individual entries of w and h_ℓ by $w = \text{col}\{w_m\}$ and $h_\ell = \text{col}\{h_{\ell,m}\}$, respectively, with $m = 1, 2, \ldots, M$. We also use the notation w_{-m} and $h_{\ell,-m}$ to refer to the vectors w and h_ℓ with their mth entries excluded. We are using the subscript ℓ to index the data $\{\gamma(\ell), h_\ell\}$ in order to avoid confusion with the index m used for the coordinates of w. Then, as a function of w_m, the risk can be written in the form:

$$\begin{aligned} P(w) &= \alpha\|w\|_1 + \rho\|w\|^2 \ + \ \frac{1}{N}\sum_{\ell=0}^{N-1}\Big(\gamma(\ell) - h_{\ell,-m}^\mathsf{T} w_{-m} - h_{\ell,m}w_m\Big)^2 \\ &\overset{(a)}{=} \alpha\,|w_m| + \rho\, w_m^2 \ + \underbrace{\left(\frac{1}{N}\sum_{\ell=0}^{N-1} h_{\ell,m}^2\right)}_{\triangleq\, a_m} w_m^2 \ - \\ &\qquad \underbrace{\frac{2}{N}\left(\sum_{\ell=0}^{N-1} h_{\ell,m}\Big(\gamma(\ell) - h_{\ell,-m}^\mathsf{T} w_{-m}\Big)\right)}_{\triangleq\, 2c_m} w_m + \text{cte} \end{aligned} \tag{14.129}$$

where terms independent of w_m are collected into the constant factor in step (a). Thus, we have

$$\begin{aligned} P(w) &\overset{(b)}{=} \alpha\,|w_m| + (\rho + a_m)w_m^2 - 2c_m w_m + \text{cte} \\ &= (\rho + a_m)\left\{\frac{\alpha}{\rho + a_m}|w_m| + \left(w_m - \frac{c_m}{\rho + a_m}\right)^2\right\} + \text{cte} \end{aligned} \tag{14.130}$$

In step (b) we introduced the scalars $a_m \geq 0$ and c_m for compactness of notation. We conclude from the earlier result (11.31)–(11.32) that the minimizer over w_m is given by

$$\widehat{w}_m = \mathbb{T}_{\alpha/(\rho+a_m)}\left(\frac{c_m}{\rho + a_m}\right) \ = \ \frac{1}{\rho + a_m}\mathbb{T}_\alpha(c_m) \tag{14.131}$$

in terms of the soft-thresholding operator. We then arrive at listing (14.132) for a coordinate-descent algorithm (without randomization) for solving the LASSO problem (14.128).

Coordinate-descent for solving the LASSO problem (14.128).

given N data pairs $\{\gamma(\ell), h_\ell\}$;
start from an arbitrary initial condition w_{-1}.
repeat until convergence over $n \geq 0$**:**
  iterate at start of iteration is $w_{n-1} = \text{col}\{w_{n-1,m}\}_{m=1}^{M}$
  repeat for each entry $m = 1, 2, \ldots, M$**:**

$$
\begin{aligned}
a_m &= \frac{1}{N}\sum_{\ell=0}^{N-1} h_{\ell,m}^2 \\
c_m &= \frac{1}{N}\sum_{\ell=0}^{N-1} h_{\ell,m}\Big(\gamma(\ell) - h_{\ell,-m}^{\mathsf{T}} w_{n-1,-m}\Big) \\
w_{n,m} &= \frac{1}{\rho + a_m}\mathbb{T}(c_m)
\end{aligned}
\tag{14.132}
$$

  end
end
return $w^\star \leftarrow w_n$.

14.8 COMMENTARIES AND DISCUSSION

Subgradient method. The idea of replacing gradients by subgradients was proposed by Shor (1962) in his work on maximizing piecewise linear concave functions. The method was well-received at the time and generated tremendous interest due to its simplicity and effectiveness. Subsequent analyses and contributions were published by Ermoliev (1966, 1969, 1983a, b), Ermoliev and Shor (1967), and Polyak (1967, 1987), culminating with the manuscripts by Ermoliev (1976) and Shor (1979). Useful surveys on the history and development of subgradient algorithms are given by Shor (1991) and Goffin (2012). Some earlier works that have elements of subgradient optimization in them, although not mentioned explicitly, are the back-to-back publications by Agmon (1954) and Motzkin and Schoenberg (1954) on relaxation methods for finding feasible solutions to systems of linear inequalities; these 1954 articles proposed iterative solutions that rely on selecting update directions that can now be interpreted as corresponding to subgradient vectors. The works dealt with a specific application and were not interested in developing a framework for general nonsmooth optimization problems. That framework had to wait until Shor (1962) and the subsequent developments. Among other early proponents of the concept of subgradients and subdifferentials for nonsmooth functions are the studies by Moreau (1963a) and Rockafellar (1963). Further discussions on the use of subgradient methods for nonsmooth optimization problems can be found in the works by Rockafellar (1970), Bertsekas (1973), Held, Wolfe, and Crowder (1974), Clarke (1983), Nemirovski and Yudin (1983), Kiwiel (1985), Polyak (1987), Shor (1998, 2012), Bertsekas, Nedic, and Ozdaglar (2003), Nesterov (2004), Shalev-Shwartz *et al.* (2011), Duchi, Hazan, and Singer (2011), Duchi, Bartlett, and Wainwright (2012), Shamir and Zhang (2013), and Ying and Sayed (2018).

Affine-Lipschitz subgradient condition. The convergence analysis in the chapter follows the approach by Ying and Sayed (2018), which relies on the affine-Lipschitz condition (14.28b). This condition is weaker than the assumption of bounded subgradient vectors commonly employed in the literature, e.g., in Bertsekas (1999), Nemirovski *et al.* (2009), Nedic and Ozdaglar (2009), Ram, Nedic, and Veeravalli (2010), Srivastava and Nedic (2011), and Agarwal *et al.* (2012), namely, the assumption that

$$\|s_Q(w;\gamma,h)\| \leq G \tag{14.133}$$

for some constant $G \geq 0$ and for all subgradients in the subdifferential set of $Q(w,\cdot)$. We explain in Prob. 14.3 that (14.133) is in direct conflict with the ν-strong convexity assumption on $P(w)$. The affine-Lipschitz condition (14.28b) is automatically satisfied by several cases of interest including LASSO (ℓ_1-regularized quadratic losses), support vector machines (ℓ_2-regularized hinge losses), and total-variation denoising problems – see, for instance, Examples 14.6 and 14.7 as well as Prob. 14.10. In comparison, these applications do not satisfy the bounded subgradient assumption (14.133) and, therefore, conclusions derived under this assumption are not directly applicable to these problems.

Convergence analysis. The convergence analysis in the body of the chapter for subgradient algorithms with pocket and exponentially smoothed variables follows the presentation by Ying and Sayed (2018). The performance result (14.68b) for the subgradient implementation with a pocket variable, namely,

$$\lim_{n\to\infty} \Big(P(w_n^{\text{best}}) - P(w^\star)\Big) \;\leq\; O(\lambda^n) \;+\; O(\mu) \tag{14.134a}$$

complements other results by Nedic (2002) and Agarwal *et al.* (2012), which established the following upper and lower bounds, respectively:

$$\lim_{n\to\infty} \Big(P(w_n^{\text{best}}) - P(w^\star)\Big) \;\leq\; O(1/n) \;+\; O(\mu) \tag{14.134b}$$

$$\lim_{n\to\infty} \Big(P(w_n^{\text{best}}) - P(w^\star)\Big) \;\geq O(1/n) \tag{14.134c}$$

Expression (14.134c) shows that the convergence of $P(w_n^{\text{best}})$ toward $P(w^\star)$ cannot be better than a sublinear rate. In comparison, result (14.134a) shows that $P(w_n^{\text{best}})$ approaches a small $O(\mu)$-neighborhood around $P(w^\star)$ exponentially fast at the rate λ^n. The results (14.134a) and (14.134c) do not contradict each other. On the contrary, they provide complementary views on the convergence behavior (from below and from above). However, the analysis used by Agarwal *et al.* (2012) imposes a stronger condition on the risk function than used in the chapter; they require $P(w)$ to be Lipschitz continuous. In comparison, we require the subgradient (and not the risk function) to be affine-Lipschitz, which makes the results applicable to a broader class of problems. Likewise, result (14.134b) only guarantees a sublinear rate of convergence toward an $O(\mu)$-neighborhood around $P(w^\star)$, while (14.134a) shows that convergence actually occurs at a linear rate. Result (14.134b) was derived by assuming the subgradient vectors are bounded, which is a stronger condition than the affine-Lipschitz condition used in the text – see Prob. 14.3. It is worth noting that, as shown by Nesterov (2004), the optimal convergence rate for subgradient methods cannot be faster than $O(1/\sqrt{n})$ for convex risks – see Prob. 14.4.

Pocket variable and smoothing. Discussions on the role of pocket variables in the study of subgradient learning appear in the texts by Shor (1979, 2012), Kiwiel (1985), Bertsekas (1995), and Nesterov (2004). However, as the analysis in the chapter revealed, the pocket method is not practical because its implementation and performance guarantees rely on knowledge of the convergence rate λ, which is not known beforehand. We followed Ying and Sayed (2018) and employed an exponentially weighted scheme. The analysis showed that exponential smoothing does not degrade convergence but retains the desired smoothing effect. There is also an alternative smoothing construction based on the use of *averaged iterates*, which was proposed independently by Ruppert (1988) and Polyak and Juditsky (1992). If we refer to the subgradient algorithm (14.3), the

Polyak–Ruppert averaging (smoothing) procedure is a post-processing calculation that relies on the use of an averaged iterate, denoted by $\bar{w}_n$, as follows:

$$w_n = w_{n-1} - \mu\, s(w_{n-1}), \quad n \geq 0 \tag{14.135a}$$

$$\bar{w}_n = \frac{1}{n - n_o} \sum_{m=n_o}^{n+1} w_{m-1} \tag{14.135b}$$

where $n_o \geq 0$ denotes the index at which averaging starts. Clearly, such averaged constructions can be applied as well to gradient-descent implementations with differentiable risks.

PROBLEMS

14.1 Consider the regularized risks listed in Table 12.1. Determine subgradient expressions for the ℓ_1-regularized logistic risk, ℓ_2-regularized Perceptron risk, and ℓ_2-regularized hinge risk.

14.2 Show that the perceptron and hinge empirical risks listed in Table 12.1, under ℓ_2 or elastic-net regularization, satisfy the affine-Lipschitz condition (14.28b). Determine for each case the respective values for $\{\delta, \delta_2\}$. In which of these cases is the regularized risk strongly convex? Determine the corresponding ν factor.

14.3 In lieu of the affine-Lipschitz condition (14.28b), some works in the literature on subgradient algorithms assume instead that the subgradient vector is bounded, namely, $\|s(w)\| \leq G$ for all w and some constant G. This assumption is in conflict with strong convexity. To see this, assume $P(w)$ is ν-strongly convex with global minimizer $w^\star$.

(a) Deduce from (14.28a) that

$$P(w) \geq P(w^\star) + \frac{\nu}{2}\|w - w^\star\|^2$$
$$P(w^\star) \geq P(w) + (s(w))^{\mathsf{T}}(w^\star - w) + \frac{\nu}{2}\|w - w^\star\|^2$$

(b) Conclude that $\|s(w)\| \geq \nu\|w - w^\star\|$, which contradicts the boundedness assumption on $s(w)$.

14.4 Consider the subgradient algorithm (14.3) for minimizing a possibly nonsmooth regularized convex risk $P(w)$ over $w \in \mathbb{R}^M$. Let $w^\star$ denote the minimizer. Introduce the quantities

$$\widetilde{w}_n \;\stackrel{\Delta}{=}\; w^\star - w_n, \quad P^{\text{best}}(n) \;\stackrel{\Delta}{=}\; \min_{0 \leq k \leq n} P(w_k)$$

We assume in this problem that $P(w)$ has bounded subgradient vectors, i.e., $\|s(w)\| \leq G$ for all $s(w) \in \partial_{w^{\mathsf{T}}} P(w)$ and for some finite constant G. We know from (10.41) that this condition is equivalent to assuming that $P(w)$ is G-Lipschitz continuous. We also assume that $\|\widetilde{w}_{-1}\| \leq W$, for some constant W (that is, the initial condition is within a bounded range from the optimal solution).

(a) Verify that $\|\widetilde{w}_n\|^2 \leq \|\widetilde{w}_{n-1}\|^2 + \mu^2\|s(w_{n-1})\|^2 - 2\mu\,(P(w_{n-1}) - P(w^\star))$. Conclude by iterating that

$$0 \leq \|\widetilde{w}_n\|^2 \leq W^2 + n\mu^2 G^2 - 2\mu n\left(P(w_{n-1}^{\text{best}}) - P(w^\star)\right)$$

(b) Show that $P(w_{n-1}^{\text{best}}) - P(w^\star) \leq (W^2 + nG^2\mu^2)/2n\mu$. Conclude that under constant step sizes it holds $P(w_n^{\text{best}}) - P(w^\star) \leq O(1/n) + O(\mu)$, for large n.

(c) Show that $\mu^o(n) = W/G\sqrt{n}$ minimizes the upper bound in part (b); note that this step-size value is iteration-dependent. Conclude that $P(w_{n-1}^{\text{best}}) - P(w^\star) \leq WG/\sqrt{n}$ and that the subgradient algorithm converges at the rate of $O(1/\sqrt{n})$.

14.5 Consider the same setting of Prob. 14.4 except that the constant step size μ is replaced by an iteration-dependent step size $\mu(n)$ that satisfies (14.104a).

(a) Show that

$$P^{\text{best}}(n-1) - P(w^\star) \leq \frac{W^2 + G^2\left(\sum_{m=1}^{n} \mu^2(m)\right)}{2\left(\sum_{m=1}^{n} \mu(m)\right)}$$

(b) Conclude that $P^{\text{best}}(n) \to P(w^\star)$ as $n \to \infty$.

14.6 We consider the same setting as Prob. 14.4 except that we now constrain the solution w to lie in a closed convex set, $\mathcal{C}$:

$$\min_{w} P(w), \quad \text{subject to } w \in \mathcal{C}$$

For this purpose, we rely on the projection subgradient method:

$$\begin{cases} z_n &= w_{n-1} - \mu s(w_{n-1}) \\ w_n &= \mathcal{P}_C[z_n] \end{cases}$$

where $\mathcal{P}_C[z]$ denotes the unique projection of z onto $\mathcal{C}$. We continue to denote the minimizer of $P(w)$ by $w^\star \in C$.

(a) Verify that the projection operator satisfies for any $z_1, z_2 \in \mathbb{R}^M$:

$$\|\mathcal{P}_C[z_1] - \mathcal{P}_C[z_2]\|^2 \leq (z_1 - z_2)^\mathsf{T}\left(\mathcal{P}_C[z_1] - \mathcal{P}_C[z_2]\right)$$

(b) Use part (a) to show that

$$\|\widetilde{w}_n\|^2 \leq \|\widetilde{w}_{n-1}\|^2 + \mu^2\|s(w_{n-1})\|^2 - 2\mu\left(P(w_{n-1}) - P(w^\star)\right)$$

(c) Conclude that for large enough n it holds that $P^{\text{best}}(n) - P(w^\star) \leq O(\mu)$.

14.7 Consider the two-dimensional function $g(z) = |a| + 2|b|$, where $z \in \mathbb{R}^2$ and has entries denoted by $\{a, b\}$.

(a) Verify that $g(z)$ is a convex function.

(b) Show that the row vector $[1\ 1]$ is a valid subgradient at location $z_o = \text{col}\{1, 0\}$.

(c) Assume we start from location $z_o = \text{col}\{1, 0\}$ and perform a subgradient update of the form $z_1 = z_o - \mu\partial_{z^\mathsf{T}} g(z_o)$, using the subgradient vector found in part (b) and a small $\mu > 0$. Compare the values of $g(z_o)$ and $g(z_1)$. What do you observe?

14.8 Use the projection subgradient method of Prob. 14.6 to solve the ℓ_1-regularized least-squares (or LASSO) problem:

$$\min_{w} \left\{\alpha\|w\|_1 + \frac{1}{N}\sum_{m=0}^{N-1}\left(\gamma(m) - h_m^\mathsf{T} w\right)^2\right\}, \quad \text{subject to } \|w\|_\infty \leq 1$$

14.9 Consider the ℓ_1-regularized MSE risk:

$$w^o = \underset{w \in \mathbb{R}^M}{\text{argmin}} \left\{\alpha\|w\|_1 + \mathbb{E}\,(\gamma - \boldsymbol{h}^\mathsf{T} w)^2\right\}$$

(a) Denote the risk by $P(w)$. Verify that $P(w) = \alpha\|w\|_1 + \sigma_\gamma^2 - 2r_{\gamma h}^\mathsf{T} w + w^\mathsf{T} R_h w$.

(b) Assume $R_h > 0$. Show that $P(w)$ is strongly convex.

(c) Show that $P(w)$ has affine-Lipschitz subgradients.

(d) Show that the subgradient recursion (14.3) reduces to

$$w_n = w_{n-1} - \alpha\, \text{sign}(w_{n-1}) + 2\mu(r_{h\gamma} - R_h w_{n-1}), \quad n \geq 0$$

14.10 The total-variation denoising problem introduced by Rudin, Osher, and Fatemi (19
involves solving a regularized least-squares problem of the following form:

$$\min_{w} \left\{ \alpha \sum_{m'=1}^{M-1} \Big| w[m'+1] - w[m'] \Big| + \frac{1}{N}\sum_{m=0}^{N-1} \left(\gamma(m) - h_m^{\mathsf{T}} w\right)^2 \right\}$$

where the $\{w[m']\}$, for $m' = 1, 2, \ldots, M$, denote the individual entries of $w \in \mathbb{R}^M$. Derive a subgradient batch solution for this problem.

14.11 The fused LASSO problem introduced by Tibshirani *et al.* (2005) adds ℓ_1-regularization to the total variation formulation in Prob. 14.10 and considers instead

$$\min_{w} \left\{ \alpha_1 \|w\|_1 + \alpha_2 \sum_{m'=1}^{M-1} \Big| w[m'+1] - w[m'] \Big| + \frac{1}{N}\sum_{m=0}^{N-1} \left(\gamma(m) - h_m^{\mathsf{T}} w\right)^2 \right\}$$

Derive a subgradient batch solution for this problem.

14.12 The group LASSO problem introduced by Yuan and Lin (2006) involves solving a regularized least-squares problem of the following form. We partition each observation vector into K sub-vectors, say, $h_m = \text{col}\{h_{mk}\}$ for $k = 1, 2, \ldots, K$. We similarly partition the weight vector into K sub-vectors, $w = \text{col}\{w_k\}$, of similar dimensions to h_{mk}. Now consider the problem:

$$\min_{w} \left\{ \alpha \sum_{k=1}^{K} \|w_k\| + \frac{1}{N}\sum_{m=0}^{N-1} \left(\gamma(m) - \sum_{k=1}^{K} h_{mk}^{\mathsf{T}} w_k\right)^2 \right\}$$

Derive a subgradient batch solution for this problem.

14.13 Consider a regularized empirical risk optimization problem of the form:

$$w^\star = \arg\min_{w \in \mathbb{R}^M} \left\{ P(w) = \alpha \|w\|_1 + \rho \|w\|^2 + \frac{1}{N}\sum_{n=1}^{N} |\gamma(n) - h_n^{\mathsf{T}} w| \right\}$$

where $\alpha \geq 0, \rho > 0$.

(a) Is the objective function strongly convex? Can you find a value for ν?

(b) Write down a stochastic subgradient algorithm with constant step size μ.

(c) What is the convergence rate of $\mathbb{E}\|\widetilde{\boldsymbol{w}}_n\|$ when $\alpha > 0$?

(d) What is the convergence rate when $\alpha = 0$?

(e) What is the convergence rate when μ is replaced by $\mu(n) = \nu^{-1}/(n+1)$?

(f) What is the convergence rate of $\mathbb{E}P(\boldsymbol{w}_n)$ toward $P(w^\star)$?

14.14 Derive a randomized coordinate-descent algorithm for the regularized logistic regression problem with risk function:

$$P(w) = \alpha \|w\|_1 + \rho \|w\|^2 + \frac{1}{N}\sum_{\ell=0}^{N-1} \ln\left(1 + e^{-\gamma(\ell) h_\ell^{\mathsf{T}} w}\right)$$

14.15 Extend the proof of Theorem 14.1 to the randomized coordinate-descent algorithm (14.127) under the separable condition (14.123).

14.16 Extend the argument used in the proof of Theorem 12.5 to the Gauss–Southwell coordinate-descent implementation where m^o in (14.127) is selected as the index corresponding to the entry in the subgradient vector with the largest absolute value. Assume the separable condition (14.123).

14.A DETERMINISTIC INEQUALITY RECURSION

The following useful statement is from Polyak (1987, pp. 45–46). The result is helpful in studying the convergence of recursive learning algorithms, as already illustrated in the body of this chapter.

LEMMA 14.1. (Deterministic recursion) *Let $u(n) \geq 0$ denote a scalar deterministic (i.e., nonrandom) sequence that satisfies:*

$$u(n+1) \;\leq\; (1-a(n))u(n) \;+\; b(n), \quad n \geq 0 \tag{14.136}$$

(a) *When the scalar sequences $\{a(n), b(n)\}$ satisfy the four conditions:*

$$0 \leq a(n) < 1, \qquad b(n) \geq 0, \qquad \sum_{n=0}^{\infty} a(n) = \infty, \qquad \lim_{n\to\infty} \frac{b(n)}{a(n)} = 0 \tag{14.137}$$

it holds that $\lim_{n\to\infty} u(n) = 0$.

(b) *When the scalar sequences $\{a(n), b(n)\}$ are of the form*

$$a(n) = \frac{c}{n+1}, \qquad b(n) = \frac{d}{(n+1)^{p+1}}, \quad c > 0, \;\; d > 0, \;\; p > 0 \tag{14.138}$$

it holds that, for large enough n, the sequence $u(n)$ converges to zero at one of the following rates depending on the value of c:

$$\begin{cases} u(n) \;\leq\; \left(\dfrac{d}{c-p}\right)\dfrac{1}{n^p} \;+\; o\left(1/n^p\right), & c > p \\ u(n) \;=\; O\left(\log n/n^p\right), & c = p \\ u(n) \;=\; O\left(1/n^c\right), & c < p \end{cases} \tag{14.139}$$

The fastest convergence rate occurs when $c > p$ and is on the order of $1/n^p$.

Proof: Given that the lemma is used frequently in the convergence arguments, we include its proof for completeness by following the elegant derivation from Polyak (1987, pp. 45–46) adjusted to our notation and conventions.

(a) By iterating we get

$$u(n+1) \leq \left(\prod_{j=0}^{n}(1-a(j))\right) u(0) \;+\; \sum_{j=0}^{n}\left\{\left(\prod_{k=j+1}^{n}(1-a(k))\right) b(j)\right\} \tag{14.140}$$

Introduce the auxiliary variable:

$$s(n+1) \;\stackrel{\Delta}{=}\; u(n+1) - \sum_{j=0}^{n}\left\{\left(\prod_{k=j+1}^{n}(1-a(k))\right) b(j)\right\} \tag{14.141}$$

Then, it holds that

$$s(n+1) \leq \left(\prod_{j=0}^{n}(1-a(j))\right) u(0) \tag{14.142}$$

If $u(0) = 0$, then $s(n+1) = 0$. When $u(0) \neq 0$, we can conclude that $s(n+1) \to 0$ as $n \to \infty$ based on the following argument. We have

$$\frac{s(n+1)}{u(0)} \leq \prod_{j=0}^{n}(1-a(j)) \tag{14.143}$$

so that

$$\lim_{n\to\infty}\left(\frac{s(n+1)}{u(0)}\right)\leq\prod_{j=0}^{\infty}(1-a(j)) \tag{14.144}$$

where the $\{a(j)\}$ satisfy $0<1-a(j)\leq 1$. Repeating the argument following (12.74) we conclude that $s(n+1)\to 0$ as $n\to\infty$. Consequently,

$$\lim_{n\to\infty} u(n+1) \;=\; \lim_{n\to\infty}\sum_{j=0}^{n}\left\{\left(\prod_{k=j+1}^{n}(1-a(k))\right)b(j)\right\} \tag{14.145}$$

Let us now examine the limit of the quantity on the right-hand side and show that it also goes to zero. For that purpose, we introduce the auxiliary nonnegative variable:

$$r(n)\;\triangleq\;\left(\prod_{k=j+1}^{n}(1-a(k))\right)b(j) \tag{14.146}$$

This variable can be zero if the $b(j)$ appearing in it is zero. Otherwise, we can show that the limit of $r(n)$ tends to zero as $n\to\infty$. Indeed, assume $b(j)>0$. Then,

$$\ln r(n)\;\triangleq\;\ln b(j)\;+\;\sum_{k=j+1}^{n}\ln(1-a(k)) \tag{14.147}$$

Using property (12.76), it holds that

$$\ln r(n)\leq\ln b(j)\;-\;\sum_{k=j+1}^{n}a(k) \tag{14.148}$$

which implies that

$$\lim_{n\to\infty}\ln r(n)\leq\ln b(j)\;-\;\sum_{k=j}^{\infty}a(k)\;=\;-\infty \tag{14.149}$$

since $b(j)/a(j)\to 0$ as $j\to\infty$ implies that for $j\geq j_o$ large enough, $b(j)\leq a(j)<1$ (so that $\ln b(j)$ is bounded). Hence, $r(n)\to 0$ as $n\to\infty$ and we conclude that $u(n+1)\to 0$ as $n\to\infty$.

(b) Using the assumed forms for $a(n)$ and $b(n)$ we have

$$u(n+1)\leq\left(1-\frac{c}{n+1}\right)u(n)\;+\;\frac{d}{(n+1)^{p+1}} \tag{14.150}$$

Consider first the case $c>p$ and introduce the variable

$$\nu(n+1)\;\triangleq\;(n+2)^{p}u(n+1)-\frac{d}{c-p} \tag{14.151}$$

Then, from the first equation:

$$\begin{aligned}\nu(n+1) &\le (n+2)^p\left\{\left(1-\frac{c}{n+1}\right)u(n) + \frac{d}{(n+1)^{p+1}}\right\} - \frac{d}{c-p}\\ &= \frac{(n+1)^p}{(n+1)^p}(n+2)^p\left\{\left(1-\frac{c}{n+1}\right)u(n) + \frac{d}{(n+1)^{p+1}}\right\} - \frac{d}{c-p}\\ &= (n+1)^p u(n)\cdot\left(1+\frac{1}{n+1}\right)^p\left(1-\frac{c}{n+1}\right) + \\ &\quad \left(1+\frac{1}{n+1}\right)^p\frac{d}{n+1} - \frac{d}{c-p}\end{aligned} \tag{14.152}$$

Using the expansion

$$\left(1+\frac{1}{n+1}\right)^p = 1+\frac{p}{n+1}+o\left(\frac{1}{n+1}\right) \tag{14.153}$$

we get

$$\begin{aligned}\nu(n+1) &\le \left(\nu(n)+\frac{d}{c-p}\right)\left(1-\frac{c}{n+1}+\frac{p}{n+1}-\frac{pc}{(n+1)^2}+o\left(\frac{1}{n+1}\right)\right) + \\ &\quad \frac{d}{n+1}\left(1+\frac{p}{n+1}+o\left(\frac{1}{n+1}\right)\right) - \frac{d}{c-p}\\ &= \left(\nu(n)+\frac{d}{c-p}\right)\left(1-\frac{c-p}{n+1}+o\left(\frac{1}{n+1}\right)\right) + \\ &\quad \frac{d}{n+1}\left(1+\frac{p}{n+1}+o\left(\frac{1}{n+1}\right)\right) - \frac{d}{c-p}\\ &= \nu(n)\left(1-\frac{c-p}{n+1}+o\left(\frac{1}{n+1}\right)\right) + \frac{dp}{(n+1)^2}+o\left(\frac{1}{(n+1)^2}\right)\end{aligned} \tag{14.154}$$

From an argument similar to part (a) we conclude that (note that now $\nu(n)$ is not necessarily nonnegative)

$$\limsup_{n\to\infty}\ \nu(n) \le 0 \tag{14.155}$$

and consequently,

$$\limsup_{n\to\infty}\ (n+1)^p u(n) - \frac{d}{c-p} \le 0 \tag{14.156}$$

which leads to

$$u(n) \le \left(\frac{d}{c-p}\right)\left(\frac{1}{n+1}\right)^p + o\left(\frac{1}{(n+1)^p}\right), \quad \text{as } n\to\infty \tag{14.157}$$

which also implies the following for large enough n:

$$u(n) \le \left(\frac{d}{c-p}\right)\left(\frac{1}{n^p}\right) + o\left(\frac{1}{n^p}\right) \tag{14.158}$$

as claimed.

Next, let us consider the case $\underline{p \ge c}$. In this case, we introduce instead

$$\nu(n+1) \triangleq (n+2)^c u(n+1) \tag{14.159}$$

Then, it holds that

$$\nu(n+1) \leq \left\{\left(1 - \frac{c}{n+1}\right) u(n) + \frac{d}{(n+1)^{p+1}}\right\} (n+2)^c \tag{14.160}$$

Using the expansion

$$\begin{aligned}(n+2)^c &= (n+1)^c \left(1 + \frac{1}{n+1}\right)^c \\ &= (n+1)^c \left\{1 + \frac{c}{n+1} + \frac{c^2}{2(n+1)^2} + o\left(\frac{1}{(n+1)^2}\right)\right\}\end{aligned} \tag{14.161}$$

we obtain

$$\begin{aligned}&\nu(n+1) \\ &\leq \left(1 - \frac{c}{n+1}\right) u(n)(n+1)^c \left\{1 + \frac{c}{n+1} + \frac{c^2}{2(n+1)^2} + o\left(\frac{1}{(n+1)^2}\right)\right\} + \\ &\frac{d}{(n+1)^{p-c+1}} \left\{1 + \frac{c}{n+1} + \frac{c^2}{2(n+1)^2} + o\left(\frac{1}{(n+1)^2}\right)\right\} \\ &= \nu(n) \left\{1 - \frac{c^2}{(n+1)^2} + o\left(\frac{1}{(n+1)^2}\right)\right\} + \frac{d}{(n+1)^{p-c+1}} \left\{1 + o\left(\frac{1}{n+1}\right)\right\} \\ &\leq \nu(n) + \frac{d'}{(n+1)^{p-c+1}}, \quad \text{for } n \text{ large enough and for some } d'\end{aligned} \tag{14.162}$$

By iterating we obtain

$$\nu(n+1) \leq \nu(0) + \sum_{j=1}^{n+1} \frac{d'}{j^{p-c+1}} \tag{14.163}$$

When $p > c$ we have that the series on the right-hand side converges to a bounded value since

$$\sum_{j=1}^{\infty} \frac{1}{j^{\alpha}} < \infty \quad \text{for any } \alpha > 1 \tag{14.164}$$

It follows that, in this case,

$$\limsup_{n\to\infty} \nu(n+1) \leq d'' \tag{14.165}$$

for some finite constant d'', from which we conclude that

$$\limsup_{n\to\infty} u(n+1) \leq d''/(n+2)^c \tag{14.166}$$

For large enough n, this result implies that

$$u(n) = O\left(1/n^c\right), \quad \text{when } p > c \text{ and for large enough } n \tag{14.167}$$

Finally, for the case $p = c$, we have that

$$\sum_{j=1}^{n+1} \frac{1}{j} = O\left(\ln(n+1)\right) \tag{14.168}$$

Consequently,

$$\limsup_{n\to\infty} \nu(n+1) \leq \nu(0) + O\left(\log(n+1)\right) \tag{14.169}$$

so that

$$\limsup_{n\to\infty} u(n+1) \le \frac{\nu(0)}{(n+2)^p} + O\left(\frac{\log(n+1)}{(n+2)^p}\right) \tag{14.170}$$

which for large enough n implies that

$$\limsup_{n\to\infty} u(n) = O\left(\frac{\log n}{n^p}\right) \tag{14.171}$$

■

REFERENCES

Agarwal, A., P. L. Bartlett, P. Ravikumar, and M. J. Wainwright (2012), "Information-theoretic lower bounds on the oracle complexity of convex optimization," *IEEE Trans. Inf. Theory*, vol. 58, no. 5, pp. 3235–3249.

Agmon, S. (1954), "The relaxation method for linear inequalities," *Canad. J. Math.*, vol. 6, no. 3, pp. 382–392.

Bertsekas, D. P. (1973), "Stochastic optimization problems with nondifferentiable cost functionals," *J. Optim. Theory Appl.*, vol. 12, no. 2, pp. 218–231.

Bertsekas, D. P. (1995), *Nonlinear Programming*, Athena Scientific.

Bertsekas, D. P. (1999), *Nonlinear Programming*, 2nd ed., Athena Scientific.

Bertsekas, D. P., A. Nedic, and A. Ozdaglar (2003), *Convex Analysis and Optimization*, 2nd ed., Athena Scientific.

Clarke, F. H. (1983), *Optimization and Nonsmooth Analysis*, Wiley.

Duchi, J. C., P. L. Bartlett, and M. J. Wainwright (2012), "Randomized smoothing for stochastic optimization," *SIAM J. Optim.*, vol. 22, no. 2, pp. 674–701.

Duchi, J., E. Hazan, and Y. Singer (2011), "Adaptive subgradient methods for online learning and stochastic optimization," *J. Mach. Learn. Res.*, vol. 12, pp. 2121–2159.

Ermoliev, Y. M. (1966), "Methods of solutions of nonlinear extremal problems," *Cybernetics*, vol. 2, no. 4, pp. 1–16.

Ermoliev, Y. M. (1969), "On the stochastic quasi-gradient method and stochastic quasi-Feyer sequences," *Kibernetika*, no. 2, pp. 72–83.

Ermoliev, Y. M. (1976), *Stochastic Programming Methods*, Nauka.

Ermoliev, Y. M. (1983a), "Stochastic quasigradient methods and their application to system optimization," *Stochastic*, vol. 9, pp. 1–36.

Ermoliev, Y. M. (1983b), "Stochastic quasigradient methods," in *Numerical Techniques for Stochastic Optimization*, Y. M. Ermoliev and R.J-.B. Wets, editors, pp. 141–185, Springer.

Ermoliev, Y. M. and N. Z. Shor (1967), "On the minimization of nondifferentiable functions," *Cybernetics*, vol. 3, no. 1, pp. 101–102.

Goffin, J.-L. (2012), "Subgradient optimization in nonsmooth optimization," *Documenta Mathematica*, Extra Volume ISMP, pp. 277–290.

Held, M., P. Wolfe, and H. P. Crowder (1974), "Validation of subgradient optimization," *Math. Program.*, vol. 6, pp. 62–88.

Kiwiel, K. (1985), *Methods of Descent for Non-differentiable Optimization*, Springer.

Moreau, J. J. (1963a), "Propriétés des applications prox," *Comptes Rendus de l'Académie des Sciences de Paris*, vol. A256, pp. 1069–1071.

Motzkin, T. and I. J. Schoenberg (1954), "The relaxation method for linear inequalities," *Canad. J. Math.*, vol. 6, no. 3, pp. 393–404.

Nedic, A. (2002), *Subgradient Methods for Convex Minimization*, Ph.D. dissertation, MIT.

Nedic, A. and A. Ozdaglar (2009), "Distributed subgradient methods for multiagent optimization," *IEEE Trans. Aut. Control*, vol. 54, no. 1, pp. 48–61.

Nemirovski, A. S., A. Juditsky, G. Lan, and A. Shapiro (2009), "Robust stochastic approximation approach to stochastic programming," *SIAM J. Optim.*, vol. 19, no. 4, pp. 1574–1609.

Nemirovski, A. S. and D. B. Yudin (1983), *Problem Complexity and Method Efficiency in Optimization*, Wiley.

Nesterov, Y. (2004), *Introductory Lectures on Convex Optimization*, Springer.

Polyak, B. T. (1967), "A general method of solving extremal problems," *Soviet Math. Doklady*, vol. 8, pp. 593–597.

Polyak, B. T. (1987), *Introduction to Optimization*, Optimization Software.

Polyak, B. T. and A. Juditsky (1992), "Acceleration of stochastic approximation by averaging," *SIAM J. Control Optim.*, vol. 30, no. 4, pp. 838–855.

Ram, S. S., A. Nedic, and V. V. Veeravalli (2010), "Distributed stochastic subgradient projection algorithms for convex optimization," *J. Optim. Theory Appl.*, vol. 147, no. 3, pp. 516–545.

Rockafellar, R. T. (1963), *Convex Functions and Dual Extremum Problems*, Ph.D. dissertation, Harvard University.

Rockafellar, R. T. (1970), *Convex Analysis*, Princeton University Press.

Rudin, L. I., S. Osher, and E. Fatemi (1992), "Nonlinear total variation based noise removal algorithms," *Physica D.*, vol. 60, no. 1–4, pp. 259–268.

Ruppert, D. (1988), "Efficient Estimation from a Slowly Convergent Robbins-Monro Process," Technical Report 781, Cornell University, School of Operations Research and Industrial Engineering.

Shalev-Shwartz, S., Y. Singer, N. Srebro, and A. Cotter (2011), "Pegasos: Primal estimated sub-gradient solver for SVM," *Math. Program.*, Ser. B, vol. 127, no. 1, pp. 3–30.

Shamir O. and T. Zhang (2013), "Stochastic gradient descent for nonsmooth optimization: Convergence results and optimal averaging schemes," *Proc. Int. Conf. Machine Learning* (PMLR) vol. 28, no.1, pp. 71–79, Atlanta, GA.

Shor, N. Z. (1962), "Application of the method of gradient descent to the solution of the network transportation problem," in *Materialy Naucnovo Seminara po Teoret i Priklad. Voprosam Kibernet. i Issted. Operacii, Nucnyi Sov. po Kibernet*, pp. 9–17, Akad. Nauk Ukrain. SSSR (in Russian).

Shor, N. Z. (1979), *Minimization Methods for Non-differentiable Functions and Their Applications*, Naukova Dumka.

Shor, N. Z. (1991), "The development of numerical methods for nonsmooth optimization in the USSR," in *History of Mathematical Programming*, J. K. Lenstra, A. H. G. Rinnoy Kan, and A. Shrijver, editors, pp. 135–139, CWI, North-Holland.

Shor, N. Z. (1998), *Nondifferentiable Optimization and Polynomial Problems*, Kluwer.

Shor, N. Z. (2012), *Minimization Methods for Non-differentiable Functions*, Springer.

Srivastava, K. and A. Nedic (2011), "Distributed asynchronous constrained stochastic optimization," *IEEE J. Sel. Topics. Signal Process.*, vol. 5, no. 4, pp. 772–790.

Tibshirani, R., M. Saunders, S. Rosset, J. Zhu, and K. Knight (2005), "Sparsity and smoothness via the fused LASSO," *J. Roy. Statist. Soc.*, Series B (Statistical Methodology), vol. 67, no. 1, pp. 91–108.

Ying, B. and A. H. Sayed (2018), "Performance limits of stochastic sub-gradient learning, part I: Single-agent case," *Signal Process.*, vol. 144, pp. 271–282.

Yuan, M. and Y. Lin (2006), "Model selection and estimation in regression with grouped variables," *J. Roy. Statist. Soc.*, Series B (Statistical Methodology), vol. 68, no. 1, pp. 49–67.

15 Proximal and Mirror-Descent Methods

The subgradient method is the natural extension of the gradient-descent method to nonsmooth risks. It nevertheless suffers from the fact that subgradient vectors are not descent directions in general and, therefore, exact convergence to the minimizer of the risk function is not guaranteed. Instead, convergence occurs toward a small neighborhood around the minimizer. In this chapter, we describe three other methods for the minimization of nonsmooth convex risks: proximal gradient, projection gradient, and mirror descent. These methods are less general than the subgradient method in that they impose certain requirements on the structure of the risk function. They do, however, guarantee exact convergence.

15.1 PROXIMAL GRADIENT METHOD

We consider first the proximal gradient method, which we already encountered in Section 11.3. Here we provide an alternative motivation and discuss both cases of constant and vanishing step-size implementations.

15.1.1 Motivation

Consider the optimization problem:

$$w^{\star} \triangleq \underset{w\in \mathbb{R}^M}{\text{argmin}}\; P(w) \tag{15.1a}$$

where it is now assumed that the risk function $P(w)$ can be split into the sum of two convex components:

$$P(w) = q(w) + E(w) \tag{15.1b}$$

where $E(w)$ is at least first-order differentiable anywhere while $q(w)$ may have points of nondifferentiability. For convenience, we will say that $E(w)$ is smooth while $q(w)$ is nonsmooth. The splitting is often suggested naturally, as is evident from the empirical risks listed in Table 12.1. For example, for the hinge risk with elastic-net regularization we have

$$P(w) = \alpha\|w\|_1 + \rho\|w\|^2 + \frac{1}{N}\sum_{m=0}^{N-1}\max\Big\{0, 1-\gamma(m)h_m^{\mathsf T}w\Big\} \tag{15.2}$$

for which

$$q(w) \triangleq \alpha\|w\|_1 \tag{15.3a}$$

$$E(w) \triangleq \rho\|w\|^2 + \frac{1}{N}\sum_{m=0}^{N-1}\max\Big\{0, 1-\gamma(m)h_m^{\mathsf T}w\Big\} \tag{15.3b}$$

or, if it is desired to keep the regularization terms together,

$$q(w) \triangleq \alpha\|w\|_1 + \rho\|w\|^2 \tag{15.4a}$$

$$E(w) \triangleq \frac{1}{N}\sum_{m=0}^{N-1}\max\Big\{0, 1-\gamma(m)h_m^{\mathsf T}w\Big\} \tag{15.4b}$$

Likewise, for the ℓ_1-regularized least-squares risk we have

$$P(w) = \alpha\|w\|_1 + \frac{1}{N}\sum_{m=0}^{N-1}(\gamma(m) - h_m^{\mathsf T}w)^2 \tag{15.5a}$$

$$q(w) \triangleq \alpha\|w\|_1 \tag{15.5b}$$

$$E(w) \triangleq \frac{1}{N}\sum_{m=0}^{N-1}(\gamma(m) - h_m^{\mathsf T}w)^2 \tag{15.5c}$$

Now, if we were to apply the subgradient method to seek the minimizer of (15.1a)–(15.1b) directly, the iteration would take the form:

$$w_n = w_{n-1} - \mu\nabla_{w^{\mathsf T}}E(w_{n-1}) - \mu s_q(w_{n-1}) \tag{15.6}$$

in terms of a subgradient vector for $q(w)$ at location w_{n-1}, which we are denoting by $s_q(w_{n-1})$. We already know from our study of subgradient algorithms in the previous chapter that the iterates w_n generated by (15.6) will not converge to $w^\star$, but will only approach a small neighborhood around it. We will explain in this chapter how to exploit proximal operators to improve performance and guarantee exact convergence. We derived the proximal gradient algorithm earlier in (11.53) by exploiting connections to fixed-point theory. We will provide here a second motivation for the same method from first principles by building it up from the subgradient implementation (15.6).

Observe first that we can rewrite recursion (15.6) in the equivalent form:

$$\textbf{(subgradient method)}\quad \left\{\begin{array}{lcl} z_n & \triangleq & w_{n-1} - \mu\nabla_{w^{\mathsf T}}E(w_{n-1}) \\ w_n & = & z_n - \mu\, s_q(w_{n-1}) \end{array}\right. \tag{15.7}$$

where we carry out the computation in two steps and introduce the intermediate variable z_n. The first step uses the gradient of $E(w)$ to update w_{n-1} to z_n, while the second step uses a subgradient for $q(w)$ to update z_n to w_n. This type of splitting is reminiscent of *incremental* techniques where calculations are performed in an incremental manner. Actually, the incremental technique goes a step further and replaces w_{n-1} in the second equation by z_n. This is because z_n is an updated version of w_{n-1} and is expected to be a "better" estimate for $w^\star$ than w_{n-1}:

$$\text{(incremental method)} \quad \left\{ \begin{array}{rcl} z_n & = & w_{n-1} \; - \; \mu \nabla_{w^\mathsf{T}} E(w_{n-1}) \\ w_n & = & z_n \; - \; \mu \, s_q(z_n) \end{array} \right. \tag{15.8}$$

If we move even further and evaluate the subgradient at the "better" estimate w_n rather than z_n, we would arrive at the *same* proximal gradient algorithm derived earlier in (11.53):

$$\text{(proximal gradient)} \quad \left\{ \begin{array}{rcl} z_n & = & w_{n-1} \; - \; \mu \nabla_{w^\mathsf{T}} E(w_{n-1}) \\ w_n & = & z_n \; - \; \mu \, s_q(w_n) \end{array} \right. \tag{15.9}$$

The second line in (15.9) is perplexing at first sight since w_n appears on both sides of the equation. However, we know from the earlier result (11.55) that the second step in (15.9) corresponds to the proximal projection of z_n relative to $q(\cdot)$. As such, we can rewrite (15.9) in the form of (15.10), which agrees with the proximal gradient recursion (11.53). Observe how the nonsmooth component $q(w)$ enters into the algorithm through the proximal step only. Observe also that the algorithm reduces to the standard gradient-descent iteration when $q(w) = 0$, and to the proximal point algorithm (11.44) when $E(w) = 0$.

Proximal gradient algorithm for minimizing $q(w) + E(w)$.

---

given smooth convex function $E(w)$;
given nonsmooth convex function $q(w)$;
given gradient operator $\nabla_{w^\mathsf{T}} E(w)$;
start from any initial condition, w_{-1}. (15.10)
repeat until convergence over $n \geq 0$:
  $z_n = w_{n-1} \; - \; \mu \nabla_{w^\mathsf{T}} E(w_{n-1})$
  $w_n = \text{prox}_{\mu q}(z_n)$
end
return $w^\star \leftarrow w_n$.

Example 15.1 **(Oscillations removed)** Let us reconsider the strongly convex risk function from Example 14.8:

$$P(w) = |w| + \frac{1}{2}w^2, \quad w \in \mathbb{R} \tag{15.11}$$

where w is a scalar. This function has a global minimum at $w = 0$. We split $P(w)$ into $q(w) + E(w)$ where $q(w) = |w|$ and $E(w) = \frac{1}{2}w^2$. Writing down the proximal gradient algorithm (15.10) for this case we get

$$\begin{cases} z_n &= (1-\mu)w_{n-1} \\ w_n &= \text{prox}_{\mu|w|}(z_n) \end{cases} \tag{15.12}$$

We know from the second row in Table 11.1 that $\text{prox}_{\mu|w|}(z) = \mathbb{T}_\mu(z)$ in terms of the soft-thresholding operator applied to z. Therefore, the proximal gradient recursion simplifies to

$$w_n = \mathbb{T}_\mu\Big((1-\mu)w_{n-1}\Big) = (1-\mu)\mathbb{T}_{\frac{\mu}{1-\mu}}(w_{n-1}) \tag{15.13}$$

Introduce the interval $(-a, a)$ of width $2\mu/(1-2\mu)$ where $a = \mu/(1-\mu)$. Then, expression (15.13) shows that

$$w_n = \begin{cases} (1-\mu)(w_{n-1} - a), & \text{if } w_{n-1} \geq a \\ 0, & \text{if } w_{n-1} \in (-a, a) \\ (1-\mu)(w_{n-1} + a), & \text{if } w_{n-1} \leq -a \end{cases} \tag{15.14}$$

Assume we start from some initial condition to the right of a. Then, the value of the successive w_n are decreased by a and scaled by $(1-\mu)$ until they enter the interval $(-a, a)$ where w_n stays at zero. Similarly, if we start from some initial condition to the left of a, then the successive w_n are increased by a and scaled by $(1-\mu)$ until they enter $(-a, a)$ where w_n stays at zero.

Example 15.2 **(Quadratic approximation)** We explained earlier in Example 12.3 that the gradient-descent method can be motivated by minimizing a quadratic approximation for $P(w)$. We can apply a similar argument for the case in which $P(w) = q(w) + E(w)$, where only $E(w)$ is smooth, and recover the proximal gradient algorithm. Thus, assume we approximate $\nabla^2_w E(w_{n-1}) \approx \frac{1}{\mu} I_M$ and introduce the second-order expansion of $E(w)$ around w_{n-1}:

$$E(w) \approx E(w_{n-1}) + \nabla_w E(w_{n-1})(w - w_{n-1}) + \frac{1}{2\mu}\|w - w_{n-1}\|^2 \tag{15.15}$$

We consider the problem of updating w_{n-1} to w_n by solving

$$w_n = \underset{w \in \mathbb{R}^M}{\text{argmin}} \left\{ q(w) + E(w_{n-1}) + \nabla_w E(w_{n-1})(w - w_{n-1}) + \frac{1}{2\mu}\|w - w_{n-1}\|^2 \right\} \tag{15.16}$$

Completing the squares on the right-hand side allows us to rewrite the problem in the equivalent form:

$$w_n = \underset{w \in \mathbb{R}^M}{\text{argmin}} \left\{ q(w) + \frac{1}{2\mu}\left\| w - \Big(w_{n-1} - \mu \nabla_{w^\mathsf{T}} E(w_{n-1})\Big) \right\|^2 \right\} \tag{15.17}$$

Using the definition of the proximal operator from (11.1) we conclude that

$$w_n = \text{prox}_{\mu q}\Big(w_{n-1} - \mu \nabla_{w^\mathsf{T}} E(w_{n-1})\Big) \tag{15.18}$$

which is the same recursion in (15.10).

Example 15.3 **(Iterated soft-thresholding algorithm (ISTA))** Consider the ℓ_1-regularized least-squares risk corresponding to the LASSO problem (which we will encounter again in Chapter 51):

$$P(w) = \alpha\|w\|_1 + \frac{1}{N}\sum_{m=0}^{N-1}(\gamma(m) - h_m^\mathsf{T} w)^2 \tag{15.19}$$

for which

$$q(w) = \alpha\|w\|_1, \quad E(w) = \frac{1}{N}\sum_{m=0}^{N-1}(\gamma(m) - h_m^\mathsf{T} w)^2 \tag{15.20}$$

The function $E(w)$ is differentiable and its gradient vector is given by

$$\nabla_{w^\mathsf{T}} E(w) = -\frac{2}{N}\sum_{m=0}^{N-1} h_m(\gamma(m) - h_m^\mathsf{T} w) \tag{15.21}$$

so that the proximal iteration (15.10) becomes

$$\begin{cases} z_n = w_{n-1} + \dfrac{2\mu}{N}\displaystyle\sum_{m=0}^{N-1} h_m(\gamma(m) - h_m^\mathsf{T} w_{n-1}) \\ w_n = \mathbb{T}_{\mu\alpha}(z_n) \end{cases}, \quad n \geq 0 \tag{15.22}$$

where we appealed to result (11.20) to replace $\text{prox}_{\mu q}(z_n)$ by the soft-thresholding operation $\mathbb{T}_{\mu\alpha}(z_n)$. We can rewrite the algorithm more compactly as

$$w_n = \mathbb{T}_{\mu\alpha}\left(w_{n-1} + \frac{2\mu}{N}\sum_{m=0}^{N-1}\gamma(m)h_m(\gamma(m) - h_m^\mathsf{T} w_{n-1})\right) \tag{15.23}$$

This recursion is known as the iterated soft-thresholding algorithm (ISTA).

Example 15.4 **(Proximal logistic regression)** Consider the ℓ_1-regularized logistic risk:

$$P(w) = \alpha\|w\|_1 + \frac{1}{N}\sum_{m=0}^{N-1}\ln\left(1 + e^{-\gamma(m)h_m^\mathsf{T} w}\right) \tag{15.24}$$

for which

$$q(w) = \alpha\|w\|_1, \quad E(w) = \frac{1}{N}\sum_{m=0}^{N-1}\ln\left(1 + e^{-\gamma(m)h_m^\mathsf{T} w}\right) \tag{15.25}$$

The function $E(w)$ is differentiable with

$$\nabla_{w^\mathsf{T}} E(w) = -\frac{1}{N}\sum_{m=0}^{N-1}\gamma(m)h_m\left(\frac{1}{1 + e^{\gamma(m)h_m^\mathsf{T} w}}\right) \tag{15.26}$$

so that the proximal iteration (15.10) becomes

$$\begin{cases} z_n = w_{n-1} + \dfrac{\mu}{N}\displaystyle\sum_{m=0}^{N-1}\frac{\gamma(m)h_m}{1 + e^{\gamma(m)h_m^\mathsf{T} w_{n-1}}} \\ w_n = \mathbb{T}_{\mu\alpha}(z_n) \end{cases}, \quad n \geq 0 \tag{15.27}$$

where we appealed to result (11.20) to replace $\text{prox}_{\mu q}(z_n)$ by the soft-thresholding operation $\mathbb{T}_{\mu\alpha}(z_n)$.

Example 15.5 **(Batch proximal gradient algorithm)** Consider the case in which the smooth component $E(w)$ is expressed in the form of an empirical average, say, as

$$E(w) = \frac{1}{N}\sum_{m=0}^{N-1} Q_u(w; \gamma(m), h_m) \tag{15.28}$$

for some loss function $Q_u(w; \cdot)$. We encountered this situation in the two previous examples involving least-squares and logistic regression smooth components, $E(w)$.

The proximal gradient iteration (15.10) leads in this case to the batch algorithm listed in (15.29).

Batch proximal gradient algorithm for minimizing $q(w) + E(w)$.

N data pairs $\{\gamma(m), h_m\}$;

given smooth convex function $E(w) = \frac{1}{N}\sum_{m=0}^{N-1} Q_u(w; \gamma(m), h_m)$;

given nonsmooth convex function $q(w)$;
given gradient operator $\nabla_{w^\mathsf{T}} Q_u(w, \cdot)$;
start from any initial condition, w_{-1}. (15.29)
repeat until convergence over $n \geq 0$**:**

$$z_n = w_{n-1} - \mu\left(\frac{1}{N}\sum_{m=0}^{N-1} \nabla_{w^\mathsf{T}} Q_u(w_{n-1}; \gamma(m), h_m)\right)$$

$$w_n = \text{prox}_{\mu q}(z_n)$$

end
return $w^\star \leftarrow w_n$.

The batch solution (15.29) continues to suffer from the same disadvantages mentioned before, namely, the entire set of N data pairs $\{\gamma(m), h_m\}$ needs to be available beforehand, and the sum of gradients (15.29) needs to be computed repeatedly at every iteration. We will explain in Chapter 16 how to resolve these difficulties by resorting to *stochastic* proximal implementations.

Example 15.6 (Proximal randomized coordinate descent) Let us denote the individual entries of w by $\{w_m\}$ and assume that the nonsmooth component $q(w)$ is separable over these coordinates, say,

$$q(w) = \sum_{m=1}^{M} q_m(w_m) \tag{15.30}$$

Then, we can readily derive the following proximal randomized coordinate-descent algorithm for minimizing $P(w) = q(w) + E(w)$ (see Prob. 15.8).

Proximal randomized coordinate descent for minimizing $q(w) + E(w)$.

---

given smooth convex function $E(w)$;
let $w = \text{col}\{w_m\}_{m=1}^M$ denote the individual entries of w;
given nonsmooth convex function $q(w) = \sum_{m=1}^M q_m(w_m)$;
given partial derivatives $\partial E(w)/\partial w_m$;
start from any initial condition, w_{-1}.
repeat until convergence for $n = 0, 1, 2, \ldots$**:** (15.31)
  iterate at start of iteration is $w_{n-1} = \text{col}\{w_{n-1,m}\}_{m=1}^M$
  select index m^o uniformly at random within $1 \leq m \leq M$
  $z_{n,m^o} = w_{n-1,m^o} - \mu\, \partial E(w_{n-1})/\partial w_{m^o}$
  $w_{n,m^o} = \text{prox}_{\mu q_{m^o}}(z_{n,m^o})$
  keep $w_{n,m} = w_{n-1,m}$ for all $m \neq m^o$
end
return $w^\star \leftarrow w_n$.

15.1.2 Convergence Analysis

We already established the convergence of the proximal gradient algorithm (15.10) earlier in Theorem 11.2. We repeat the statement here for ease of reference. The result assumes the following conditions on the risk function:

(A1) (**Splitting**). The risk function is assumed to split into the sum of two convex components,

$$P(w) = q(w) + E(w) \tag{15.32a}$$

where $q(w)$ is possibly nonsmooth while $E(w)$ is smooth.

(A2) (**Strong convexity**). The smooth component $E(w)$ is ν-strongly convex and first-order differentiable, namely,

$$E(w_2) \geq E(w_1) + \Big(\nabla_{w^\mathsf{T}} E(w_1)\Big)^\mathsf{T}(w_2 - w_1) + \frac{\nu}{2}\|w_2 - w_1\|^2 \tag{15.32b}$$

for any $w_1, w_1 \in \text{dom}(E)$ and some $\nu > 0$.

(A3) (**δ-Lipschitz gradients**). The gradients of $E(w)$ are δ-Lipschitz:

$$\|\nabla_w E(w_2) - \nabla_w E(w_1)\| \leq \delta\,\|w_2 - w_1\| \tag{15.32c}$$

for any $w_1, w_1 \in \text{dom}(E)$ and some $\delta \geq 0$.

Theorem 15.1. (Convergence of proximal gradient) *Consider convex risks $P(w)$ satisfying conditions (15.32a)–(15.32c). If the step size μ satisfies*

$$0 < \mu < 2\nu/\delta^2 \tag{15.33}$$

then the iterates w_n generated by the proximal gradient algorithm (15.10), as well as the risk values $P(w_n)$, converge exponentially at the rate λ^n to their respective optimal values, i.e.,

$$\|\widetilde{w}_n\|^2 \leq \lambda\,\|\widetilde{w}_{n-1}\|^2, \quad n \geq 0 \tag{15.34a}$$

$$0 \leq P(w_n) - P(w^\star) \leq O(\lambda^n), \quad n \geq 0 \tag{15.34b}$$

where

$$\lambda = 1 - 2\mu\nu + \mu^2\delta^2 \in [0,1) \tag{15.35}$$

Regret analysis

The statement of Theorem 15.1 examines the convergence behavior of the squared weight error, $\|\widetilde{w}_n\|^2$, and the risk value $P(w_n)$. Another useful performance measure is the *average regret*, defined by

$$\mathcal{R}(N) \triangleq \frac{1}{N}\sum_{n=0}^{N-1} P(w_{n-1}) - P(w^\star) \qquad (\textbf{average regret}) \tag{15.36}$$

Using (15.34b) we find that the regret decays at the rate of $1/N$ since

$$\begin{aligned}
\mathcal{R}(N) &\triangleq \frac{1}{N}\sum_{n=0}^{N-1}\Big(P(w_{n-1}) - P(w^\star)\Big) \\
&\leq O\Big(\frac{1}{N}\sum_{n=0}^{N-1}\lambda^n\Big) \\
&= O\Big(\frac{1}{N}\,\frac{1-\lambda^N}{1-\lambda}\Big) \\
&= O(1/N)
\end{aligned} \tag{15.37}$$

Convexity versus strong convexity

Theorem 15.1 considers risk functions $P(w) = q(w) + E(w)$ where $E(w)$ is *strongly convex* with δ-Lipschitz gradients. The theorem establishes the exponential convergence of $\|\widetilde{w}_n\|^2$ to zero at the rate λ^n. It also establishes that $P(w_n)$ converges exponentially to $P(w^\star)$ at the same rate λ^n. We thus write:

$$\left\{\begin{array}{rcl} \|\widetilde{w}_n\|^2 &\leq& O(\lambda^n) \\ P(w_n) - P(w^\star) &\leq& O(\lambda^n) \end{array}\right. \quad (\textbf{for strongly convex } E(w)) \tag{15.38a}$$

We will be dealing mainly with strongly convex components $E(w)$. Nevertheless, it is useful to note that if $E(w)$ happens to be only convex (but not necessarily strongly convex) then we showed earlier in Theorem 11.1 that $P(w_n)$ approaches $P(w^\star)$ at a slower *sublinear* rate:

$$P(w_n) - P(w^\star) \leq O(1/n) \quad (\textbf{for convex } E(w)) \tag{15.38b}$$

Iteration-dependent step sizes

We can also employ iteration-dependent step sizes in the proximal gradient iteration (15.10) and write

$$(\textbf{proximal gradient}) \quad \left\{\begin{array}{rcl} z_n &=& w_{n-1} - \mu(n)\nabla_{w^\mathsf{T}} E(w_{n-1}) \\ w_n &=& \text{prox}_{\mu(n)q}(z_n) \end{array}\right. \tag{15.39}$$

where the sequence $\mu(n)$ satisfies either one of the following two conditions:

$$(\textbf{Condition I}) \quad \sum_{n=0}^{\infty}\mu^2(n) < \infty \quad \text{and} \quad \sum_{n=0}^{\infty}\mu(n) = \infty \tag{15.40a}$$

$$(\textbf{Condition II}) \quad \lim_{n\to\infty}\mu(n) = 0 \quad \text{and} \quad \sum_{n=0}^{\infty}\mu(n) = \infty \tag{15.40b}$$

Theorem 15.2. (Convergence under vanishing step sizes) *Consider convex risks $P(w)$ satisfying conditions (15.32a)–(15.32c). If the step-size sequence $\mu(n)$ satisfies either (15.40a) or (15.40b), then the iterate w_n generated by the proximal gradient recursion (15.39) converges to the global minimizer, $w^\star$. In particular, when the step-size sequence is chosen as $\mu(n) = \tau/(n+1)$, the convergence rates are on the order of:*

$$\|\widetilde{w}_n\|^2 \leq O(1/n^{2\nu\tau}) \tag{15.41a}$$

$$P(w_n) - P(w^\star) \leq O(1/n^{2\nu\tau-1}), \quad \text{when } 2\nu\tau > 1 \tag{15.41b}$$

for large enough n.

Proof: The same argument used in Appendix 11.B to arrive at (11.135) can be repeated here to find that

$$\|\widetilde{w}_n\|^2 \leq \lambda(n)\, \|\widetilde{w}_{n-1}\|^2 \tag{15.42}$$

where $\lambda(n) = 1 - 2\nu\mu(n) + \delta^2\mu^2(n)$. The derivation can now continue as in the proof of Theorem 12.2 in the gradient-descent case to arrive at (15.41a). In a similar vein, repeating the argument from Appendix 11.B that led to (11.143) we get

$$0 \leq P(w_n) - P(w^\star) \leq \left(\frac{2\,(1+\mu(n)\delta)}{\mu(n)\sqrt{\lambda(n)}} + \delta\right) \lambda(n) \times \|\widetilde{w}_{n-1}\|^2 \tag{15.43}$$

Taking the limit as $n \to \infty$ and using $\mu(n) = \tau/(n+1)$, we conclude that (15.41b) holds.

■

15.2 PROJECTION GRADIENT METHOD

We move on to study constrained optimization problems where the parameter w is required to lie within some given convex set, $\mathcal{C}$, say,

$$w^\star = \underset{w \in \mathbb{R}^M}{\text{argmin}}\ E(w), \quad \text{subject to } w \in \mathcal{C} \tag{15.44}$$

where $E(w)$ is a smooth convex function; usually, ν-strongly convex. One popular method for the solution of such problems is the *projection gradient algorithm* (also referred to as the *projected* gradient algorithm), which is motivated as follows.

15.2.1 Motivation

We can embed the constraint into the problem formulation by adding the nonsmooth term $q(w) = \mathbb{I}_{C,\infty}[w]$ to $E(w)$:

$$w^\star = \underset{w \in \mathbb{R}^M}{\text{argmin}} \left\{E(w) + \mathbb{I}_{C,\infty}[w]\right\} \tag{15.45}$$

Recall that $\mathbb{I}_{C,\infty}[w]$ is the indicator function for the set $\mathcal{C}$: Its value is equal to zero if $w \in \mathcal{C}$ and $+\infty$ otherwise. Statement (15.45) fits into the formulation we used in the previous section to derive the proximal gradient algorithm with

$$P(w) = E(w) + \mathbb{I}_{C,\infty}[w] \tag{15.46}$$

Recursion (15.10) then leads to

$$w_n = \text{prox}_{\mu \mathbb{I}_{C,\infty}}\Big(w_{n-1} - \mu \nabla_{w^\mathsf{T}} E(w_{n-1})\Big) \tag{15.47}$$

We know from the earlier result (11.9) that the proximal operator of the indicator function is nothing but the projection operator onto $\mathcal{C}$, i.e.,

$$\text{prox}_{\mu \mathbb{I}_{C,\infty}}(x) = \mathcal{P}_C(x) \tag{15.48}$$

where the projection is defined as follows using Euclidean norms:

$$\widehat{x} = \mathcal{P}_C(x) \iff \widehat{x} = \underset{z \in \mathcal{C}}{\text{argmin}} \; \|x - z\| \tag{15.49}$$

In this way, recursion (15.47) reduces to what is known as the *projection gradient algorithm* listed in (15.51). Clearly, for algorithm (15.51) to be useful, we should be able to evaluate the projection in a tractable manner. This is possible in several instances, as shown before in Table 9.2. For example, assume $\mathcal{C}$ is defined as the intersection of hyperplanes, $\mathcal{C} = \{w \,|\, Aw - b = 0\}$ where $A \in \mathbb{R}^{N \times M}$. Using the result from the fourth row in that table, we know that

$$\mathcal{P}_\mathcal{C}(x) = x - A^\dagger(Ax - b) \tag{15.50}$$

in terms of the pseudo-inverse of A.

Projection gradient algorithm for minimizing $E(w)$ over $w \in \mathcal{C}$.

given convex function $E(w)$;
given convex set $\mathcal{C}$ given gradient operator $\nabla_{w^\mathsf{T}} E(w)$;
start from any initial condition, w_{-1}. (15.51)
repeat until convergence over $n \geq 0$:
$\quad w_n = \mathcal{P}_C\Big(w_{n-1} - \mu \nabla_{w^\mathsf{T}} E(w_{n-1})\Big)$
end
return $w^\star \leftarrow w_n$.

15.2.2 Convergence Analysis

We motivated the projection gradient algorithm as a special instance of the proximal gradient method (15.47) and, hence, the convergence result of Theorem 15.1 applies, namely, under

$$0 < \mu < 2\nu/\delta^2 \tag{15.52}$$

and for large enough n, it holds that:

$$\|\widetilde{w}_n\|^2 \le \lambda \|\widetilde{w}_{n-1}\|^2, \quad n \ge 0 \tag{15.53a}$$

$$0 \le E(w_n) - E(w^\star) \le O(\lambda^n), \quad n \ge 0 \tag{15.53b}$$

where

$$\lambda = 1 - 2\mu\nu + \mu^2\delta^2 \in [0,1) \tag{15.54}$$

The theorem assumes a ν-strongly convex function $E(w)$ with δ-Lipschitz gradients as in (15.32b)–(15.32c).

Convexity versus strong convexity

We relax the strong-convexity requirement and assume $E(w)$ is only convex. We will also allow it to be *nonsmooth*, in which case the gradient of $E(w)$ in the update relation (15.51) will be replaced by a subgradient vector denoted by $s_E(w_{n-1}) \in \partial_{w^\mathsf{T}} E(w_{n-1})$. In this case, we rewrite algorithm (15.51) using an auxiliary variable z_n:

$$z_n = w_{n-1} - \mu\, s_E(w_{n-1}) \tag{15.55a}$$

$$w_n = \mathcal{P}_{\mathcal{C}}(z_n) \tag{15.55b}$$

The convergence analysis in the sequel assumes the following conditions:

(A1) (**Convex and nonsmooth**). The risk function $E(w)$ is convex but can be nonsmooth.

(A2) (**δ-Lipschitz function**). The risk $E(w)$ is δ-Lipschitz itself (rather than its gradients), namely, there exists some $\delta \ge 0$ such that

$$\|E(w_2) - E(w_1)\| \le \delta\|w_2 - w_1\|, \quad \forall\, w_1, w_2 \in \text{dom}(E) \tag{15.56}$$

relative to the Euclidean norm. We know from property (10.41) that this is equivalent to stating that the subgradient vectors of $E(w)$ are bounded by δ relative to the same norm, namely,

$$\|s_E(w)\| \le \delta \tag{15.57}$$

The convergence analysis further assumes that the projected iterates w_n resulting from (15.55b) are smoothed in a manner similar to the Polyak–Ruppert averaging procedure (14.135b) by defining the variable:

$$\bar{w}_{N-2} \stackrel{\Delta}{=} \frac{1}{N}\sum_{n=0}^{N-1} w_{n-1} \tag{15.58}$$

It is straightforward to verify that this quantity can be computed recursively by means of the construction:

$$\bar{w}_N = \left(1 - \frac{1}{N+2}\right)\bar{w}_{N-1} + \frac{1}{N+2} w_N, \quad \bar{w}_{-1} = w_{-1}, \quad N \ge 0 \tag{15.59}$$

Theorem 15.3. (Convergence under convexity) *Consider a convex and possibly nonsmooth risk $E(w)$ satisfying the δ-Lipschitz condition (15.56). Assume w_{-1} is selected from within the region $\|w^\star - w_{-1}\| \leq W$ for some constant $W \geq 0$. Then, if we run the projected gradient algorithm (15.55a)–(15.55b) over N iterations using $\mu = W/\delta\sqrt{N}$ and smooth the iterates using (15.59), it will hold that*

$$E(\bar{w}_{N-2}) - E(w^\star) = O\big(1/\sqrt{N}\big) \tag{15.60}$$

That is, the risk function evaluated at the averaged iterate approaches the minimal value at the rate of $O(1/\sqrt{N})$.

Proof: The argument extends the steps from Prob. 14.4 to the projection gradient method. Using the convexity of $E(w)$ we have

$$\begin{aligned} E(w_{n-1}) - E(w^\star) &\leq (s_E(w_{n-1}))^\mathsf{T}(w_{n-1} - w^\star) \\ &\overset{(15.55a)}{=} \frac{1}{\mu}(w_{n-1} - z_n)^\mathsf{T}(w_{n-1} - w^\star) \end{aligned} \tag{15.61}$$

Introduce the error vectors

$$\widetilde{w}_{n-1} = w^\star - w_{n-1}, \qquad \widetilde{z}_n = w^\star - z_{n-1} \tag{15.62}$$

and recall the useful algebraic result for any two vectors $x, y \in \mathbb{R}^M$

$$2x^\mathsf{T}y = \|x\|^2 + \|y\|^2 - \|x - y\|^2 \tag{15.63}$$

Applying this result to the right-hand side of (15.61) we get

$$\begin{aligned} E(w_{n-1}) - E(w^\star) &\leq \frac{1}{2\mu}\Big(\|w_{n-1} - z_n\|^2 + \|\widetilde{w}_{n-1}\|^2 - \|\widetilde{z}_n\|^2\Big) \\ &\overset{(15.55a)}{\leq} \frac{1}{2\mu}\Big(\mu^2\|s_E(w_{n-1})\|^2 + \|\widetilde{w}_{n-1}\|^2 - \|\widetilde{z}_n\|^2\Big) \\ &\overset{(a)}{\leq} \frac{\mu\delta^2}{2} + \frac{1}{2\mu}\Big(\|\widetilde{w}_{n-1}\|^2 - \|\widetilde{z}_n\|^2\Big) \end{aligned} \tag{15.64}$$

where step (a) uses (15.57). Now using properties of the projection operator we have

$$\begin{aligned} \|\widetilde{z}_n\|^2 &= \|(w^\star - w_n) - (z_n - w_n)\|^2 \\ &= \|\widetilde{w}_n\|^2 + \|w_n - z_n\|^2 \underbrace{-2(z_n - w_n)^\mathsf{T}\widetilde{w}_n}_{\leq 0} \\ &\overset{(a)}{\geq} \|\widetilde{w}_n\|^2 \end{aligned} \tag{15.65}$$

where step (a) uses the fact that $w_n = \mathcal{P}(z_n)$ along with property (9.66) for projections onto convex sets. Substituting (15.65) into (15.64) gives

$$E(w_{n-1}) - E(w^\star) \leq \frac{\mu\delta^2}{2} + \frac{1}{2\mu}\Big(\|\widetilde{w}_{n-1}\|^2 - \|\widetilde{w}_n\|^2\Big) \tag{15.66}$$

The last difference contributes a telescoping sum. Summing over n leads to

$$\frac{1}{N}\sum_{n=0}^{N-1} E(w_{n-1}) - E(w^\star) \leq \frac{\mu\delta^2}{2} + \frac{W^2}{2N\mu} \tag{15.67}$$

The right-hand side is minimized at $\mu = W/\delta\sqrt{N}$ so that

$$\frac{1}{N}\sum_{n=0}^{N-1} E(w_{n-1}) - E(w^\star) \leq \frac{W\delta}{\sqrt{N}} \tag{15.68}$$

Recall that $E(w)$ is convex. Applying property (8.82) we conclude

$$E\Big(\frac{1}{N}\sum_{n=0}^{N-1} w_{n-1}\Big) - E(w^\star) \leq \frac{W\delta}{\sqrt{N}} \tag{15.69}$$

as claimed.

■

15.3 MIRROR-DESCENT METHOD

We discuss next a second useful method for the solution of the constrained optimization problem (15.44). One main motivation for the method is that its performance guarantees will be insensitive to the dimension M of the parameter space.

15.3.1 Motivation

In most of our discussions so far we have assumed that the risk function is ν-strongly convex and its gradients are δ-Lipschitz. The value of δ affects the range of step sizes for which the pertinent algorithms converge. For instance, for the projection gradient method, we listed the following condition on μ in (15.52):

$$0 < \mu < 2\nu/\delta^2 \tag{15.70}$$

In this case, the weight-error vector $\widetilde{w}_n$ was shown to converge to zero at an exponential rate λ^n with the value of λ given by (15.54), which is seen to be dependent on δ:

$$\lambda \;=\; 1 - 2\mu\nu + \mu^2\delta^2 \;\in\; [0,1) \tag{15.71}$$

In some situations, however, the risk function $E(w)$ may have δ-Lipschitz gradients relative to some *different norm*, other than the Euclidean norm, say, the ℓ_∞-norm. For example, assume we encounter a scenario where the following Lipschitz gradient condition holds:

$$\|\nabla_w E(w_2) - \nabla_w E(w_1)\|_\infty \;\leq\; \delta\, \|w_2 - w_1\|_\infty \tag{15.72}$$

If we were to use this condition to deduce a range for μ for convergence based on our previous results, then we would first need to transform the bound in (15.72) into one that involves the Euclidean norm and not the ℓ_∞-norm. This can be done by using the following properties of norms for any vector $x \in \mathbb{R}^M$:

$$\|x\|_2 \leq \sqrt{M}\,\|x\|_\infty, \quad \|x\|_\infty \leq \|x\|_2 \tag{15.73}$$

which allow us to deduce from (15.72) that:

$$\|\nabla_w E(w_2) - \nabla_w E(w_1)\| \leq \sqrt{M}\delta \|w_2 - w_1\| \tag{15.74}$$

Observe, however, that the δ factor appears multiplied by $\sqrt{M}$, where M is the problem dimension. If we use this new scaled value of δ in expressions (15.70) and (15.71), we will find that the range of stability and the convergence rate will both be affected by the problem dimension. In particular, the regret bound (15.37) will be replaced by $O(\sqrt{M}/N)$, leading to performance deterioration for large-size problems – recall the earlier discussion in Example 12.11. The mirror descent technique helps ameliorate this difficulty, as we proceed to explain.

We refer again to the constrained optimization problem (15.44) involving a convex and smooth function $E(w)$:

$$w^\star = \underset{w\in\mathrm{IR}^M}{\operatorname{argmin}}\ E(w), \quad \text{subject to } w \in \mathcal{C} \tag{15.75}$$

To motivate the mirror-descent algorithm we start from the projection gradient update (15.51) and rewrite it in the equivalent form:

$$w_n = \underset{w\in\mathcal{C}}{\operatorname{argmin}} \left\| w - \Big(w_{n-1} - \mu\nabla_{w^\mathsf{T}} E(w_{n-1})\Big)\right\|^2 \tag{15.76}$$

Expanding the right-hand side and ignoring terms that are independent of w, the above minimization is equivalent to:

$$w_n = \underset{w\in\mathcal{C}}{\operatorname{argmin}} \left\{ \Big(\nabla_w E(w_{n-1})\Big) w + \frac{1}{2\mu}\|w - w_{n-1}\|^2 \right\} \tag{15.77}$$

The first term inside the parentheses involves the inner product between w and the gradient vector of the risk at w_{n-1}; this term embeds information about the descent direction. The second term involves a squared Euclidean distance between w and w_{n-1}; its purpose is to ensure that the update from w_{n-1} to w_n takes "small" steps and keeps w_n close to w_{n-1}. The mirror-descent method replaces this second term by a more general "measure of similarity" between points in $\mathcal{C}$. But first we need to introduce the concept of *mirror functions.*

15.3.2 Mirror Function

Consider a differentiable, closed, and ν_ϕ-*strongly convex* function $\phi(x) : \mathcal{C}_\phi \to \mathrm{IR}$. Its domain $\mathcal{C}_\phi$ is an open set whose closure subsumes the convex set $\mathcal{C}$ over which the optimization is performed, i.e.,

$$\mathcal{C} \subseteq \bar{\mathcal{C}}_\phi \tag{15.78}$$

By definition, the closure of the open set $\mathcal{C}_\phi$, denoted by $\bar{\mathcal{C}}_\phi$, consists of all points $\mathcal{C}_\phi$ in addition to all limit points of converging sequences in $\mathcal{C}_\phi$. Note that we are requiring $\phi(x)$ to be defined over the same convex domain $\mathcal{C}$ over which $E(w)$ is being minimized. For example, assume $\mathcal{C}$ is chosen as the simplex:

$$\mathcal{C} = \left\{ x \in \mathbb{R}^M \,\middle|\, x_m \geq 0, \; \sum_{m=1}^{M} x_m = 1 \right\} \quad \textbf{(simplex)} \tag{15.79}$$

where the $\{x_m\}$ denote the individual entries of x. This set arises when dealing with probability distributions where the vector x plays the role of a probability mass function (pmf). One choice for $\mathcal{C}_\phi$ would be the positive orthant, defined by

$$\mathcal{C}_\phi = \left\{ x \in \mathbb{R}^M \,|\, x_m > 0 \right\} \quad \textbf{(positive orthant)} \tag{15.80}$$

This set is open and its closure includes points with $x_m = 0$. Therefore, the sets $\{\mathcal{C}, \mathcal{C}_\phi\}$ chosen in this manner satisfy requirement (15.78).

We refer to $\phi(x)$ as the *mirror function* or mirror map. We associate a *dual* space with $\phi(x)$, which is defined as the space spanned by its gradient operator. It consists of all vectors y that can be generated from the gradient transformation:

$$\text{dual space} \stackrel{\Delta}{=} \left\{ y \in \mathbb{R}^M \,|\, y = \nabla_{x^\mathsf{T}} \phi(x), \; \forall\, x \in \mathcal{C}_\phi \right\} \stackrel{\Delta}{=} \text{dual}(\phi) \tag{15.81}$$

We use the notation dual(ϕ) to refer to the dual space, and assume that

$$\text{dual}(\phi) = \mathbb{R}^M \tag{15.82}$$

Moreover, since, by choice, the function $\phi(x)$ is closed and strongly convex, we know from the second row of Table 8.4 that its conjugate function $\phi^\star(y)$ will be differentiable everywhere. We can then appeal to Prob. 8.46, which relates the gradient of a function to the gradient of its conjugate function, to conclude that:

$$\boxed{y = \nabla_{x^\mathsf{T}} \phi(x) \iff x = \nabla_{y^\mathsf{T}} \phi^\star(y), \;\; y \in \text{dual}(\phi), \;\; x \in \text{dom}(\phi)} \tag{15.83}$$

These important relations imply that the mapping $y = \nabla_{x^\mathsf{T}} \phi(x)$ is a bijection so that x and y determine each other uniquely. The bijective mapping between the primal and dual spaces is illustrated in Fig. 15.1. The primal space consists of all vectors $x \in \text{dom}(\phi)$, while the dual space consists of all vectors generated by the gradient operator, $y = \nabla_{x^\mathsf{T}} \phi(x)$. The inverse mapping is given by the gradient of the conjugate function, $x = \nabla_{y^\mathsf{T}} \phi^\star(y)$.

Example 15.7 **(Negative entropy function)** Consider the negative entropy function:

$$\phi(x) \stackrel{\Delta}{=} \sum_{m=1}^{M} x_m \ln x_m, \quad x_m > 0 \tag{15.84}$$

where the $\{x_m\}$ denote the individual entries of x. The domain of this function is the positive orthant:

$$\text{dom}(\phi) = \mathcal{C}_\phi = \left\{ x \in \mathbb{R}^M \,|\, x_m > 0 \right\} \tag{15.85}$$

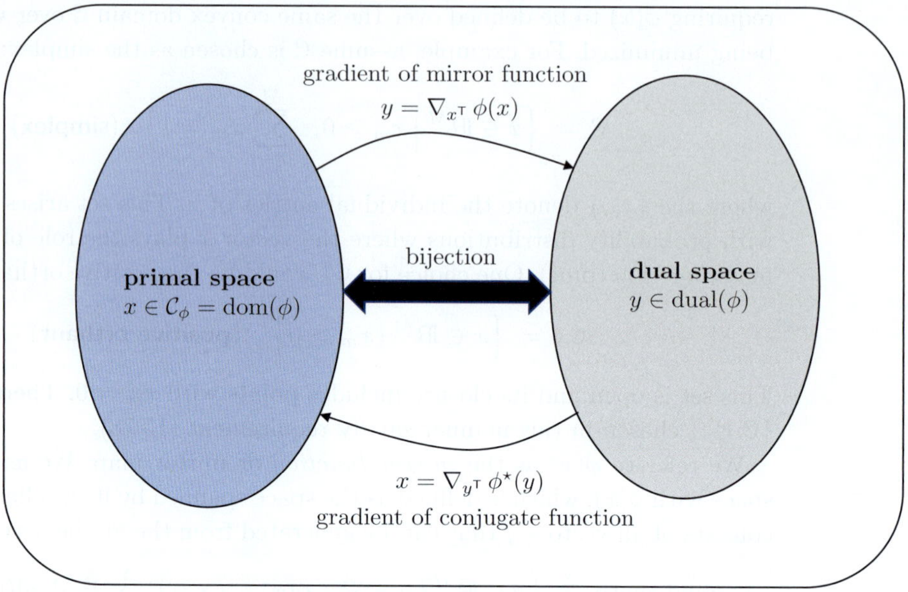

Figure 15.1 The primal space consists of all vectors $x \in \text{dom}(\phi)$, while the dual space consists of all vectors generated by the gradient operator, $y = \nabla_{x^\mathsf{T}}\, \phi(x)$. The inverse mapping is given by the gradient of the conjugate function, $x = \nabla_{y^\mathsf{T}} \phi^\star(y)$.

The function $\phi(x)$ is ν_ϕ-strongly convex relative to the ℓ_1-norm, as already established in (8.106) with $\nu_\phi = 1$:

$$\phi(x) \geq \phi(y) + \nabla_x\, \phi(y)\,(x-y) + \frac{1}{2}\|x-y\|_1^2, \quad x, y \in \mathbb{R}^M \tag{15.86}$$

The entries of the gradient vector of $\phi(x)$ are given by

$$\partial\phi(x)/\partial x_m \;=\; 1 + \ln x_m \tag{15.87}$$

Observe that these entries can generate any value in $\mathbb{R}$. Indeed, if we set $y_m = 1 + \ln x_m$, for an arbitrary y_m, then $x_m = e^{y_m - 1} > 0$. It follows that

$$\text{dual}(\phi) = \mathbb{R}^M \tag{15.88}$$

and, moreover, the mapping between the domain of $\phi(x)$ and its dual space is given by

$$y = \mathbb{1}_M + \ln(x) \iff x = e^{y-\mathbb{1}} \tag{15.89}$$

where x and y are M-dimensional vectors, and the operations $\ln(x)$ and $e^{y-\mathbb{1}}$ are defined elementwise. Expression (15.89) is consistent with the form of the conjugate function for $\phi(x)$, which we know from Table 8.4 to be

$$\phi^\star(y) = \sum_{m=1}^{M} e^{y_m - 1} \tag{15.90}$$

15.3.3 Bregman Divergence

Now, given a mirror function $\phi(x)$, the Bregman distance between any vectors $w, w_{n-1} \in \mathbb{R}^M$ is defined as the difference (*cf.* Section 8.9):

$$D_\phi(w, w_{n-1}) \stackrel{\Delta}{=} \phi(w) - \Big(\phi(w_{n-1}) + \nabla_w\, \phi(w_{n-1})\, (w - w_{n-1})\Big) \tag{15.91}$$

Since $\phi(\cdot)$ is convex, we already know from (8.4) that $\phi(w)$ will lie above the tangent hyperplane at w_{n-1} so that the difference (15.91) is nonnegative for any w. In other words, $D_\phi(w, w_{n-1}) \geq 0$ for all $\{w, w_{n-1}\}$ and equality holds if, and only if, $w = w_{n-1}$. Moreover, $D_\phi(w, w_{n-1})$ is strongly convex over w since $\phi(w)$ is strongly convex and $\nabla\phi(w_{n-1})\,(w - w_{n-1})$ is affine in w. Note further that the gradient of $D_\phi(\cdot)$ relative to w is the following difference of gradients:

$$\nabla_{w^\mathsf{T}} D_\phi(w, w_{n-1}) \;=\; \nabla_{w^\mathsf{T}}\, \phi(w) - \nabla_{w^\mathsf{T}}\, \phi(w_{n-1}) \tag{15.92}$$

15.3.4 Derivation of Algorithm

The mirror-descent method uses the Bregman divergence between w and w_{n-1} to replace the term $\frac{1}{2}\|w - w_{n-1}\|^2$ in (15.77) leading to (see also Prob. 15.9):

$$\boxed{w_n = \underset{w \in \mathcal{C}\cap\mathcal{C}_\phi}{\text{argmin}} \left\{ \Big(\nabla_w E(w_{n-1})\Big) w \;+\; \frac{1}{\mu}\, D_\phi(w, w_{n-1}) \right\}} \tag{15.93}$$

where we also replaced the constraint set $w \in \mathcal{C}$ by $w \in \mathcal{C} \cap \mathcal{C}_\phi$ to make sure all terms in the risk function are well defined. Recall that since $\mathcal{C} \subseteq \bar{\mathcal{C}}_\phi$, this restriction may only exclude points from the boundary of $\mathcal{C}$. For simplicity of notation, we let

$$\mathcal{C}' \stackrel{\Delta}{=} \mathcal{C} \cap \mathcal{C}_\phi \tag{15.94}$$

In expression (15.93), the Bregman distance plays the role of a regularization term. We verify next that we can solve this optimization in two steps.

THEOREM 15.4. (Two-step solution of (15.93)) *Consider a convex and first-order differentiable risk $E(w)$, and a ν_ϕ-strongly convex and first-order differentiable mirror function $\phi(x)$. The solution of (15.93) can be attained in two steps involving an unconstrained optimization problem followed by a Bregman projection, namely, it holds that*

$$z_n = \underset{w \in \mathbb{R}^M}{\text{argmin}} \left\{ \Big(\nabla_w E(w_{n-1})\Big) w \;+\; \frac{1}{\mu}\, D_\phi(w, w_{n-1}) \right\} \tag{15.95a}$$

$$w_n = \mathcal{P}_{\mathcal{C}',\phi}(z_n) \stackrel{\Delta}{=} \underset{w \in \mathcal{C}'}{\text{argmin}}\; D_\phi(w, z_n) \tag{15.95b}$$

where the projection is carried out using the Bregman distance measure.

Proof: A solution z_n for the first step (15.95a) can be obtained by setting the gradient vector to zero at z_n, i.e.,

$$\mu \nabla_{w^\mathsf{T}} E(w_{n-1}) + \nabla_{w^\mathsf{T}} \phi(z_n) - \nabla_{w^\mathsf{T}} \phi(w_{n-1}) = 0 \tag{15.96}$$

Let

$$b_n \stackrel{\Delta}{=} \nabla_{w^\mathsf{T}} \phi(z_n) \tag{15.97}$$

Then, we deduce from (15.96) that

$$b_n = \nabla_{w^\mathsf{T}} \phi(w_{n-1}) - \mu \nabla_{w^\mathsf{T}} E(w_{n-1}) \tag{15.98}$$

We can solve for z_n by appealing to the result of Prob. 8.46, which shows how the gradient of a function relates to the gradient of its conjugate, namely,

$$z_n = \nabla_{x^\mathsf{T}} \phi^\star(b_n) \tag{15.99}$$

where, by definition,

$$\phi^\star(x) = \sup_{w \in \mathbb{R}^M} \left\{ w^\mathsf{T} x - \phi(w) \right\} \tag{15.100}$$

Since the function $\phi(w)$ is closed and ν_ϕ-strongly convex over $w \in \mathcal{C}_\phi$, we know from the second row of Table 8.4 dealing with properties of conjugate functions that $\phi^\star(x)$ will be differentiable everywhere. For this reason, we are using the gradient notation in expression (15.99) for z_n. We therefore find that the variables $\{z_n, b_n\}$ define each other via

$$z_n = \nabla_{x^\mathsf{T}} \phi^\star(b_n), \quad b_n = \nabla_{w^\mathsf{T}} \phi(z_n) \tag{15.101}$$

Using these expressions, we can now verify that problems (15.93) and (15.95a)–(15.95b) are equivalent. Starting from the cost in (15.93), we can write

$$\begin{aligned}
w_n &= \underset{w \in \mathcal{C}'}{\text{argmin}} \left\{ \Big(\mu \nabla_w E(w_{n-1})\Big) w + D_\phi(w, w_{n-1}) \right\} \\
&= \underset{w \in \mathcal{C}'}{\text{argmin}} \left\{ \Big(\mu \nabla_w E(w_{n-1})\Big) w + \phi(w) - \phi(w_{n-1}) - \nabla_w \phi(w_{n-1})(w - w_{n-1}) \right\} \\
&= \underset{w \in \mathcal{C}'}{\text{argmin}} \left\{ \Big(\mu \nabla_w E(w_{n-1})\Big) w + \phi(w) - \nabla_w \phi(w_{n-1}) w \right\} \\
&= \underset{w \in \mathcal{C}'}{\text{argmin}} \left\{ \phi(w) - \Big[\nabla_w \phi(w_{n-1}) - \mu \nabla_w E(w_{n-1})\Big] w \right\} \\
&\stackrel{(15.98)}{=} \underset{w \in \mathcal{C}'}{\text{argmin}} \left\{ \phi(w) - b_n^\mathsf{T} w \right\} \\
&\stackrel{(15.101)}{=} \underset{w \in \mathcal{C}'}{\text{argmin}} \left\{ \phi(w) - \nabla_w \phi(z_n) w \right\} \\
&= \underset{w \in \mathcal{C}'}{\text{argmin}} \left\{ \phi(w) - \phi(z_n) - \nabla_w \phi(z_n)(w - z_n) \right\} \\
&= \underset{w \in \mathcal{C}'}{\text{argmin}} \; D_\phi(w, z_n)
\end{aligned} \tag{15.102}$$

which agrees with the cost in the second step (15.95b).

■

In this way, we arrive at the listing of the online mirror-descent (OMD) algorithm shown in (15.103). For cases when $E(w)$ happens to be nonsmooth, it is possible to replace $\nabla_w E(w_{n-1})$ by a subgradient vector $s_E \in \partial_w E(w_{n-1})$.

Online mirror-descent for minimizing $E(w)$ over $w \in \mathcal{C}$

given convex and smooth function $E(w)$.
choose ν_ϕ-strongly convex mirror function $\phi(w) : \mathcal{C}_\phi \rightarrow \mathrm{IR}$;
let $\phi^\star(x) = \sup_w \{w^\mathsf{T} x - \phi(w)\}$ denote its conjugate function;
given convex set $\mathcal{C} \subset \bar{\mathcal{C}}_\phi$;
given gradient operator $\nabla_{w^\mathsf{T}} E(w)$;
given gradient operator $\nabla_{w^\mathsf{T}} \phi(w)$;
let $\mathcal{C}' = \mathcal{C} \cap \mathcal{C}_\phi$; (15.103)
start from an initial condition, $w_{-1} \in \mathcal{C}'$.
repeat until convergence over $n \geq 0$:
$b_n = \nabla_{w^\mathsf{T}} \phi(w_{n-1}) - \mu \nabla_{w^\mathsf{T}} E(w_{n-1})$
$z_n = \nabla_{x^\mathsf{T}} \phi^\star(b_n)$
$w_n = \mathcal{P}_{C',\phi}(z_n)$ (Bregman projection)
end
return $w^\star \leftarrow w_n$.

Examining the recursions in (15.103) we observe that they admit the following interpretation. Starting from an iterate $w_{n-1} \in \mathcal{C}'$, we perform the following steps in succession (see Fig. 15.2):

(1) Map w_{n-1} to the dual space by computing first $y_{n-1} = \nabla_{w^\mathsf{T}} \phi(w_{n-1})$.

(2) Perform a gradient update in the dual domain, $b_n = y_{n-1} - \mu \nabla_{w^\mathsf{T}} E(w_{n-1})$.

(3) Map the result back to the primal domain, $z_n = \nabla_{x^\mathsf{T}} \phi^\star(b_n)$.

(4) Project z_n onto $\mathcal{C}'$ to get w_n and ensure the constraint is satisfied.

(5) Continue the process in this manner.

The reason for the name "mirror-descent" is because the algorithm maps w_{n-1} into the dual space in step (1) and then performs the gradient update in step (2). By doing so, the method allows one to move away from Euclidean geometry to other types of geometries, defined by the choice of the mirror function $\phi(x)$, and to use the Bregman divergence to measure "distances." The gradient update is performed in the dual space and, as the convergence analysis in the next section shows, this mapping helps avoid the deterioration in convergence rate that would result from the modified Lipschitz constant in (15.74) if we were to insist on working in the original domain – see the explanation in the paragraph following expression (15.150).

Example 15.8 (Bregman projection using the negative entropy function) The last step in algorithm (15.103) involves a Bregman projection. We illustrate how this step

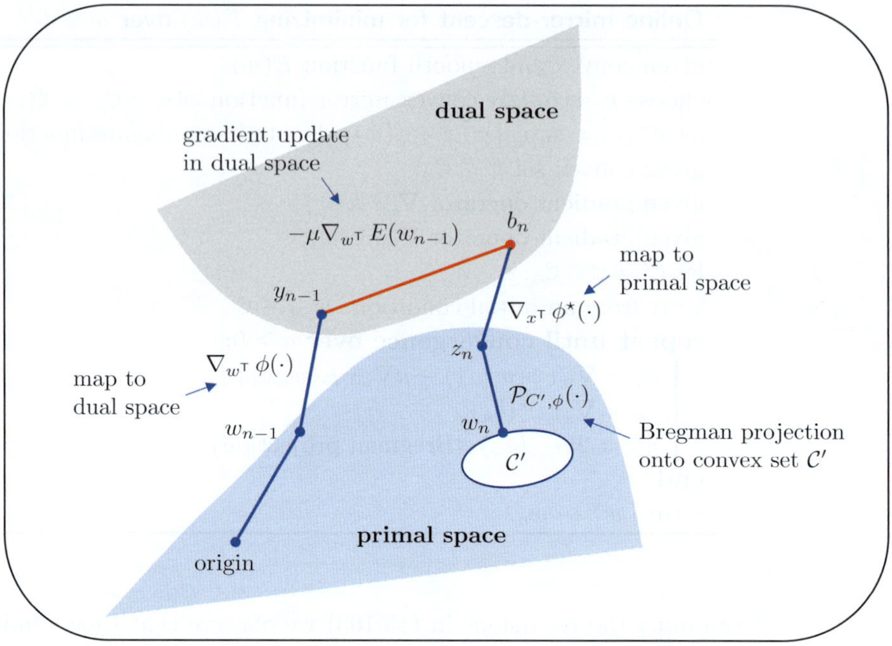

Figure 15.2 Graphical interpretation of the steps carried out by the mirror-descent algorithm (15.103). Starting at w_{n-1} on the left, the iterate is mapped by $\nabla_{w^\mathsf{T}} \phi(\cdot)$ into the dual variable y_{n-1}, where a gradient update is performed to generate b_n. The result is mapped back by $\nabla_{x^\mathsf{T}} \phi^\star(\cdot)$ to z_n in the primal space and projected onto the convex set $\mathcal{C}'$ to get w_n.

can be computed for the negative entropy mirror function (15.84). Let $x \in \mathbb{R}^M$ with individual entries $\{x_m\}$, and consider the simplex:

$$\mathcal{C} = \left\{ x \in \mathbb{R}^M \;\middle|\; x_m \geq 0, \; \sum_{m=1}^{M} x_m = 1 \right\} \tag{15.104}$$

which is a closed convex set, as well as the positive orthant:

$$\mathcal{C}_\phi = \left\{ x \in \mathbb{R}^M \;\middle|\; x_m > 0 \right\} \tag{15.105}$$

which is an open convex set. The negative entropy mirror function defined by (15.84) is well-defined over $\mathcal{C}_\phi$. Pick any point $z \in \mathcal{C}_\phi \backslash \mathcal{C}$, i.e., a vector z with positive entries that do not add up to 1. We wish to compute the Bregman projection of z onto the intersection $\mathcal{C} \cap \mathcal{C}_\phi$ using the definition:

$$\widehat{z} = \underset{c \in \mathcal{C} \cap \mathcal{C}_\phi}{\text{argmin}} \; D_\phi(c, z) \tag{15.106}$$

In this case, the Bregman divergence is given by

$$\begin{aligned} D_\phi(c,z) &= \sum_{m=1}^{M} c_m \ln c_m - \sum_{m=1}^{M} z_m \ln z_m - \sum_{m=1}^{M}(1+\ln z_m)(c_m - z_m) \\ &= \sum_{m=1}^{M}\Big\{c_m \ln(c_m/z_m) - c_m + z_m\Big\} \\ &= \sum_{m=1}^{M}\Big\{c_m \ln(c_m/z_m) + z_m\Big\} - 1 \end{aligned} \tag{15.107}$$

since $\sum_{m=1}^{M} c_m = 1$ for all $c \in \mathcal{C} \cap \mathcal{C}_\phi$. It follows that

$$\underset{c\in\mathcal{C}\cap\mathcal{C}_\phi}{\text{argmin}}\Big\{D_\phi(c,z)\Big\} = \underset{c\in\mathcal{C}\cap\mathcal{C}_\phi}{\text{argmin}}\left\{\sum_{m=1}^{M} c_m \ln(c_m/z_m)\right\} \tag{15.108}$$

One can solve this constrained optimization problem using a Lagrangian argument, as was done in Example 9.4, to find that

$$\boxed{\widehat{z} = z/\|z\|_1} \tag{15.109}$$

Example 15.9 **(Mirror descent under negative entropy)** Combining the results of Examples 15.7 and 15.8, we can simplify the mirror-descent recursion (15.103) for the case when the mirror function is the negative entropy function (15.84) and when the sets $\mathcal{C}$ and $\mathcal{C}_\phi$ correspond to the simplex and the positive orthant, respectively. The result is shown in listing (15.110), where the notation $\ln(x)$ and $\exp(x)$ for a vector x refers to vectors where the ln and exp functions are applied elementwise.

Online mirror-descent for minimizing $E(w)$ subject to $w \in \mathcal{C}$ using a negative entropy mirror function.

given convex and smooth function $E(w)$;
let $\phi(w) : \mathcal{C}_\phi \to \mathbb{R}$ be the negative entropy function (15.84);
let $\mathcal{C}$ be the simplex (15.104) and $\mathcal{C}_\phi$ the positive orthant (15.105);
given gradient operator $\nabla_{w^\mathsf{T}} E(w)$;
start from an initial condition, $w_{-1} \in \mathcal{C}' = \mathcal{C} \cap \mathcal{C}_\phi$. (15.110)
repeat until convergence over $n \geq 0$:

$b_n = \mathbb{1}_M + \ln(w_{n-1}) - \mu\nabla_{w^\mathsf{T}} E(w_{n-1})$
$z_n = \exp\{b_n - \mathbb{1}_M\}$
$w_n = z_n/\|z_n\|_1$

end
return $w^\star \leftarrow w_n$.

We can simplify the expression for z_n in (15.110) by eliminating b_n and noting that

$$\begin{aligned} z_n &= \exp\Big\{\ln(w_{n-1}) - \mu\nabla_{w^\mathsf{T}} E(w_{n-1})\Big\} \\ &= w_{n-1} \odot \exp\Big\{-\mu\nabla_{w^\mathsf{T}} E(w_{n-1})\Big\} \end{aligned} \tag{15.111}$$

in terms of the Hadamard (elementwise) product. If we denote the individual entries of w_n by $w_{n,m}$, we find that the mapping from the entries $\{w_{n-1,m}\}$ to $\{w_{n,m}\}$ in the mirror-descent construction (15.110) is given by

$$w_{n,m} = \frac{w_{n-1,m}\exp\Big\{-\mu\,\partial E(w_{n-1})/\partial w_m\Big\}}{\sum_{m'=1}^{M} w_{n-1,m'}\exp\Big\{-\mu\,\partial E(w_{n-1})/\partial w_{m'}\Big\}} \tag{15.112}$$

for $m = 1, 2, \ldots, M$.

We illustrate the operation of the online mirror-descent algorithm (15.110) and the projected gradient algorithm (15.51) by considering the ℓ_2-regularized logistic regression risk function:

$$E(w) = \rho\|w\|^2 + \frac{1}{N}\sum_{m=0}^{N-1} \ln\left(1 + e^{-\gamma(m)h_m^{\mathsf{T}} w}\right) \tag{15.113}$$

For this simulation, the data $\{\gamma(m), h_m\}$ are generated randomly as follows. First, a 10th-order random parameter model $w^a \in \mathbb{R}^{10}$ is selected, and $N = 200$ random feature vectors $\{h_m\}$ are generated, say, with zero-mean unit-variance Gaussian entries. Then, for each h_m, the label $\gamma(m)$ is set to either $+1$ or -1 according to the following construction:

$$\gamma(m) = +1 \;\text{ if }\; \left(\frac{1}{1 + e^{-h_m^{\mathsf{T}} w^a}}\right) \geq 0.5; \;\text{ otherwise } \gamma(m) = -1 \tag{15.114}$$

The algorithms are run for 10,000 iterations on the data $\{\gamma(m), h_m\}$ using parameters

$$\rho = 2, \quad \mu = 0.001 \tag{15.115}$$

The projection onto the simplex, which is necessary for running the projected gradient algorithm, is implemented according to the description in Prob. 9.11. The resulting weight iterates are shown in the bottom plot of Fig. 15.3; the minimal risk is found to be

$$E(w^\star) \approx 0.6903 \tag{15.116}$$

The two plots in the top row display the learning curves $E(w_n)$ relative to the minimum value $E(w^\star)$, both in linear scale on the left and in normalized logarithmic scale on the right (according to construction (11.65)). The results indicate faster convergence of the projected gradient method for this simulation; nevertheless, each step of the algorithm requires the evaluation of a projection onto the simplex whereas the computations involved in the mirror-descent implementation (15.110) are more straightforward.

Example 15.10 **(Mirror-descent under quadratic mirror function)** Let A be a symmetric positive-definite matrix of size $M \times M$. Consider the mirror function $\phi(w) = \frac{1}{2}w^{\mathsf{T}}Aw$, whose domain is $\mathcal{C}_\phi = \mathbb{R}^M$. Then, for any two vectors (p, q):

$$\begin{aligned} D_\phi(p,q) &= \frac{1}{2}p^{\mathsf{T}}Ap - \frac{1}{2}q^{\mathsf{T}}Aq - q^{\mathsf{T}}A^{\mathsf{T}}(p-q) \\ &= \frac{1}{2}(p-q)^{\mathsf{T}}A(p-q) \\ &= \frac{1}{2}\|p-q\|_A^2 \end{aligned} \tag{15.117}$$

That is, the weighted Euclidean distance between two vectors is a Bregman distance. The mirror function $\phi(w)$ is ν-strongly convex relative to the Euclidean norm with $\nu = \lambda_{\min}(A)$, and

$$\nabla_{w^{\mathsf{T}}}\phi(w) = Aw \tag{15.118}$$

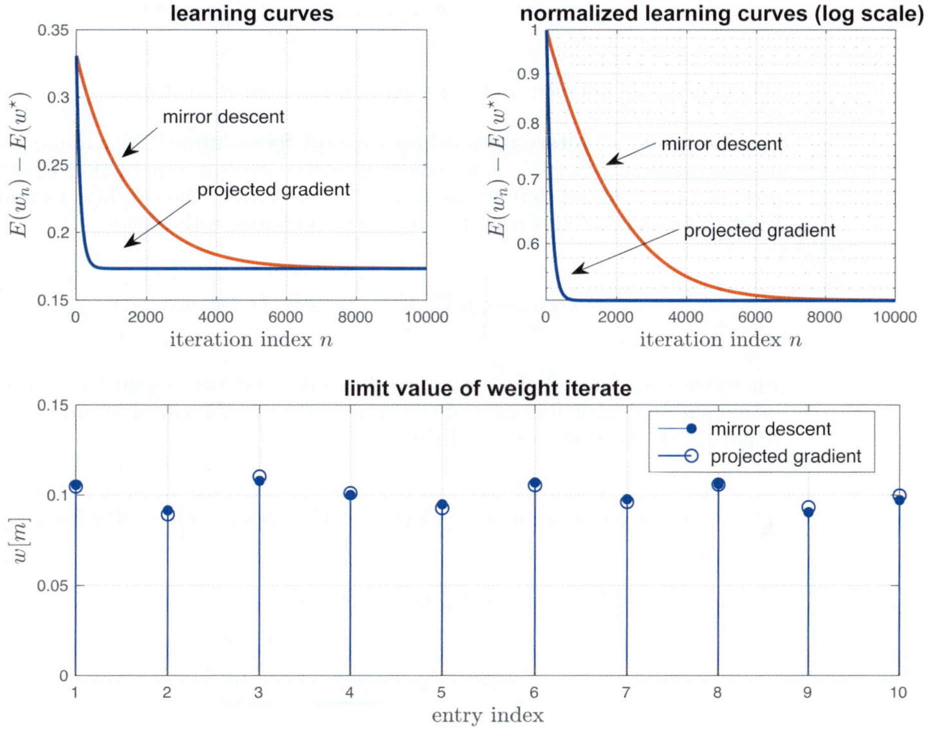

Figure 15.3 (*Top*) Learning curves $E(w_n)$ relative to the minimum risk value $E(w^\star)$ in linear scale (on the left) and in normalized logarithmic scale (on the right) generated by the projected gradient and mirror-descent algorithms (15.51) and (15.110), respectively. (*Bottom*) Limiting value of the weight iterate w_n, which tends to the minimizer $w^\star$.

We know from Prob. 8.51 that $\phi^\star(x) = \frac{1}{2}x^\mathsf{T} A^{-1}x$ so that $\nabla_{x^\mathsf{T}}\phi^\star(x) = A^{-1}x$. In other words, the mapping between the primal and dual spaces is determined by the matrix A and its inverse. In this way, the mirror-descent recursion (15.103) reduces to (15.119).

Online mirror-descent for minimizing $E(w)$ subject to $w \in \mathcal{C}$ using a quadratic mirror function.

given convex and smooth function $E(w)$;
let $\phi(w) = \frac{1}{2}w^\mathsf{T} Aw$, $A > 0$, and $\mathcal{C}_\phi = \mathrm{I\!R}^M$;
let $\mathcal{C} \subset \mathrm{I\!R}^M$ be a closed convex set;
given gradient operator $\nabla_{w^\mathsf{T}} E(w)$;
start from an initial condition, $w_{-1} \in \mathcal{C}$.
repeat until convergence over $n \geq 0$**:**

$$b_n = Aw_{n-1} - \mu\nabla_{w^\mathsf{T}} E(w_{n-1})$$
$$z_n = A^{-1}b_n$$
$$w_n = \mathcal{P}_C(z_n) \text{ using (15.120)}$$

end
return $w^\star \leftarrow w_n$.

(15.119)

In this listing, the notation $\mathcal{P}_C(z)$ refers to the weighted projection onto $\mathcal{C}$:

$$\widehat{z} = \mathcal{P}_C(z) \Longleftrightarrow \underset{c \in \mathcal{C}}{\text{argmin}} \; \|c - z\|_A^2 \tag{15.120}$$

When $A = I_M$, we recover the projection gradient algorithm.

Example 15.11 **(Alternative mirror-descent formulation)** We can motivate the mirror-descent method (15.103) by following an alternative argument that avoids the two-step procedure (15.95a)–(15.95b) and works directly with the original formulation (15.93). First, note that we can rewrite this latter problem in the form:

$$w_n = \underset{w}{\text{argmin}} \left\{ \mu \nabla_w E(w_{n-1}) w + D_\phi(w, w_{n-1}) + \mathbb{I}_{C',\infty}[w] \right\}$$

where we transformed the constrained problem into an unconstrained one by incorporating the indicator function of the convex set, $\mathcal{C}'$. We can determine a solution w_n by requiring the stationarity condition:

$$0 \in \left\{ \mu \nabla_{w^\mathsf{T}} E(w_{n-1}) + \Big(\nabla_{w^\mathsf{T}} \phi(w_n) - \nabla_{w^\mathsf{T}} \phi(w_{n-1}) \Big) \; + \; \partial_{w^\mathsf{T}} \mathbb{I}_{C',\infty}[w_n] \right\} \tag{15.121}$$

This is satisfied by selecting w_n to ensure that

$$b_n \in \partial_{w^\mathsf{T}} \Big(\underbrace{\phi(w) + \mathbb{I}_{C',\infty}[w]}_{\triangleq\, g(w)} \Big) \bigg|_{w = w_n} \tag{15.122}$$

for the same vector b_n defined in (15.98). To continue, we need to evaluate the subdifferential of the function $g(w) = \phi(w) + \mathbb{I}_{C',\infty}[w]$ that appears in (15.122) in order to arrive at an expression for w_n. For this purpose, we appeal to the result of Prob. 8.46, which shows that w_n should satisfy

$$w_n \in \partial_{x^\mathsf{T}} \, g^\star(x) \Big|_{x = b_n} \tag{15.123}$$

in terms of the conjugate function:

$$\begin{aligned} g^\star(x) &= \sup_{w \in \mathbb{R}^M} \left\{ w^\mathsf{T} x - g(w) \right\} \\ &= \sup_{w \in \mathbb{R}^M} \left\{ w^\mathsf{T} x - \phi(w) \; - \; \mathbb{I}_{C',\infty}[w] \right\} \\ &= \sup_{w \in \mathcal{C}'} \left\{ w^\mathsf{T} x - \phi(w) \right\} \end{aligned} \tag{15.124}$$

Note that we are limiting the maximization over $w \in \mathcal{C}'$ in the last line. Since the function $g(w)$ is closed and ν-strongly convex over $w \in \mathcal{C}'$, we know from Prob. 8.47 that its conjugate function $g^\star(x)$ is differentiable everywhere. For this reason, we can replace the subdifferential in (15.123) by an actual gradient vector and write

$$w_n \;=\; \nabla_{x^\mathsf{T}} \, g^\star(x) \Big|_{x \,=\, b_n} \tag{15.125}$$

In this way, we arrive at the listing shown in (15.103).

Alternative mirror-descent for minimizing $E(w)$ over $w \in \mathcal{C}$.

given convex and smooth function $E(w)$;
choose ν_ϕ-strongly convex mirror function $\phi(w) : \mathcal{C}_\phi \rightarrow \mathbb{R}$;
let $g^\star(x) = \sup_{w\in\mathcal{C}'}\{w^\mathsf{T} x - \phi(w)\}$;
given convex set $\mathcal{C} \subset \bar{\mathcal{C}}_\phi$;
given gradient operator $\nabla_{w^\mathsf{T}} E(w)$;
given gradient operator $\nabla_{w^\mathsf{T}} \phi(w)$; (15.126)
let $\mathcal{C}' = \mathcal{C} \cap \mathcal{C}_\phi$;
start from an initial condition, $w_{-1} \in \mathcal{C}'$.
repeat until convergence over $n \geq 0$:
$\quad b_n = \nabla_{w^\mathsf{T}} \phi(w_{n-1}) - \mu \nabla_{w^\mathsf{T}} E(w_{n-1})$
$\quad w_n = \nabla_{x^\mathsf{T}} g^\star(b_n)$
end
return $w^\star \leftarrow w_n$.

Example 15.12 (**Quadratic approximation**) We can also motivate the mirror-descent algorithm (15.103) by following the same heuristic quadratic approximation argument from Example 12.3 for the classical gradient-descent method. There, we approximated $\nabla_w^2 E(w_{n-1}) \approx \frac{1}{\mu} I_M$ and introduced the following second-order expansion for $E(w)$ around w_{n-1}:

$$E(w) \approx E(w_{n-1}) + \nabla_w E(w_{n-1})(w - w_{n-1}) + \frac{1}{2\mu}\|w - w_{n-1}\|^2 \tag{15.127}$$

Then, we minimized the right-hand side over w and arrived at

$$w_n = w_{n-1} - \mu \nabla_{w^\mathsf{T}} E(w_{n-1}) \tag{15.128}$$

Here, we consider instead the approximation:

$$E(w) \approx E(w_{n-1}) + \nabla_w E(w_{n-1})(w - w_{n-1}) + \frac{1}{\mu} D_\phi(w, w_{n-1}) \tag{15.129}$$

where the squared distance between w and w_{n-1} is replaced by the Bregman distance. We then minimize the right-hand side of (15.129) over $w \in \mathcal{C}'$; this objective can be accomplished in two steps for the same reason explained before to justify (15.95a)–(15.95b): First we ignore the constraint and determine the vector z_n that minimizes the right-hand, and subsequently project z_n onto $\mathcal{C}'$. The first step leads to

$$\begin{aligned} 0 &= \nabla_{w^\mathsf{T}} E(w_{n-1}) + \frac{1}{\mu}\nabla_{w^\mathsf{T}} D_\phi(w, w_{n-1}) \\ &= \nabla_{w^\mathsf{T}} E(w_{n-1}) + \frac{1}{\mu}\Big(\nabla_{w^\mathsf{T}} \phi(w) - \nabla_{w^\mathsf{T}} \phi(w_{n-1})\Big) \end{aligned} \tag{15.130}$$

Setting the gradient to zero at $w = z_n$ gives

$$\nabla_{w^\mathsf{T}} \phi(z_n) = \nabla_{w^\mathsf{T}} \phi(w_{n-1}) - \mu \nabla_{w^\mathsf{T}} E(w_{n-1}) \stackrel{\Delta}{=} b_n \tag{15.131}$$

Using the conjugate mapping that takes b_n back to z_n we obtain the mirror-descent recursions:

$$b_n = \nabla_{w^\mathsf{T}} \phi(w_{n-1}) - \mu \nabla_{w^\mathsf{T}} E(w_{n-1}) \tag{15.132a}$$
$$z_n = \nabla_{x^\mathsf{T}} \phi^\star(b_n) \tag{15.132b}$$
$$w_n = \mathcal{P}_{C',\phi}(z_n) \tag{15.132c}$$

where the last projection is added to ensure $w_n \in \mathcal{C}'$.

15.3.5 Convergence Analysis

We are ready to examine the convergence behavior of the mirror-descent algorithm (15.103). We will consider the possibility of nonsmooth risks $E(w)$ as well, in which case the gradient vector $\nabla_w E(w_{n-1})$ in the statement of the algorithm is replaced by a subgradient vector, $s_E \in \partial_{w^\mathsf{T}} E(w_{n-1})$. We consider the optimization problem

$$w^\star \overset{\Delta}{=} \underset{w\in\mathcal{C}\cap\mathcal{C}_\phi}{\text{argmin}}\ E(w) \tag{15.133}$$

where $\mathcal{C}$ is a closed convex set included in the closure of the open convex set $\mathcal{C}_\phi$. The convergence analysis assumes the following conditions:

(A1) (**Convex and nonsmooth**). The risk function $E(w)$ is convex but can be nonsmooth.

(A2) (**δ-Lipschitz function**). The risk $E(w)$ is δ-Lipschitz itself (rather than its gradient), namely, there exists some $\delta \geq 0$ such that

$$\|E(w_2) - E(w_1)\| \leq \delta\|w_2 - w_1\|, \quad \forall\, w_1, w_2 \in \text{dom}(E) \tag{15.134}$$

relative to some arbitrary norm. Let $\|\cdot\|_\star$ denote the dual of the norm $\|\cdot\|$ used in (15.134). We know from property (10.41) that the Lipschitz condition (15.134) implies bounded subgradients:

$$\|s_E\|_\star \leq \delta, \quad \forall\, s_E \in \partial_{w^\mathsf{T}} E(w) \tag{15.135}$$

(A3) (**Strong convexity of mirror function**). The mirror function $\phi(w) : \mathcal{C}_\phi \to \mathbb{R}$ is first-order differentiable and ν_ϕ-strongly convex relative to the same norm used for the Lipschitz bound in $E(w)$, i.e.,

$$\phi(w) \geq \phi(z) + \nabla_w\, \phi(z)(w - z) + \frac{\nu}{2}\|w - z\|^2 \tag{15.136}$$

(A4) (**Domain of optimization**). The closed convex set $\mathcal{C}$ over which optimization is carried out is included in the closure of $\mathcal{C}_\phi$.

The analysis below further assumes that the projected iterates w_n resulting from the mirror-descent algorithm (15.103) are smoothed in a manner similar to the Polyak–Ruppert averaging procedure (14.135b) by defining

$$\bar{w}_{N-2} \overset{\Delta}{=} \frac{1}{N}\sum_{n=0}^{N-1} w_{n-1} \tag{15.137}$$

This iterate can be computed recursively as follows:

$$\bar{w}_N = \left(1 - \frac{1}{N+2}\right)\bar{w}_{N-1} + \frac{1}{N+2}w_N, \quad \bar{w}_{-1} = w_{-1}, \quad N \geq 0 \tag{15.138}$$

THEOREM 15.5. (Convergence of mirror descent) *Refer to the mirror-descent algorithm (15.103) and assume the four conditions listed prior to the statement of the theorem hold. Let $w_{-1} \in \mathcal{C} \cap \mathcal{C}_\phi$ denote the initial condition for the algorithm and define*

$$W^2 \triangleq \sup_{w \in \mathcal{C} \cap \mathcal{C}_\phi} \Big\{ D_\phi(w, w_{-1}) \Big\} \tag{15.139}$$

Set the step-size parameter to

$$\mu = \frac{W}{\delta} \sqrt{\frac{2\nu}{N}} \tag{15.140}$$

Then, the averaged iterate generated by (15.138) over a window of N iterations satisfies

$$0 \leq E(\bar{w}_{N-2}) - E(w^\star) \leq O(1/\sqrt{N}) \tag{15.141}$$

Proof: The minimal risk value is given by $E(w^\star)$. For any iterate w_{n-1} we have

$$\begin{aligned}
0 \leq E(w_{n-1}) - E(w^\star) &\overset{(a)}{\leq} \big(s_E(w_{n-1})\big)^\mathsf{T}(w_{n-1} - w^\star) \\
&\overset{(b)}{=} \frac{1}{\mu}\Big(\nabla_{w^\mathsf{T}}\, \phi(w_{n-1} - b_n)\Big)^\mathsf{T}(w_{n-1} - w^\star) \\
&\overset{(c)}{=} \frac{1}{\mu}\Big(\nabla_{w}\, \phi(w_{n-1}) - \nabla_w\, \phi(z_n)\Big)(w_{n-1} - w^\star) \\
&\overset{(d)}{=} \frac{1}{\mu}\Big(D_\phi(w^\star, w_{n-1}) - D_\phi(w^\star, z_n) + D_\phi(w_{n-1}, z_n)\Big)
\end{aligned} \tag{15.142}$$

where step (a) is by convexity of $E(w)$, step (b) follows from the first step of the mirror-descent algorithm (15.103), step (c) uses $b_n = \nabla_{w^\mathsf{T}}\phi(z_n)$, and step (d) can be verified by substituting the Bregman divergences by their definitions. Next, we know that w_n results from the Bregman projection of z_n onto $\mathcal{C} \cap \mathcal{C}_\phi$. Using the result of part (c) of Prob. 9.14 that such projections satisfy the Pythagorean inequality:

$$D_\phi(w^\star, z_n) \geq D_\phi(w^\star, w_n) + D_\phi(w_n, z_n) \tag{15.143}$$

we substitute into (15.142) to get

$$\begin{aligned}
0 &\leq E(w_{n-1}) - E(w^\star) \\
&\leq \frac{1}{\mu}\Big(\underbrace{D_\phi(w^\star, w_{n-1}) - D_\phi(w^\star, w_n)}_{=\text{ telescoping sequence}} + \underbrace{D_\phi(w_{n-1}, z_n) - D_\phi(w_n, z_n)}_{\triangleq\, B}\Big)
\end{aligned} \tag{15.144}$$

where the second term evaluates to

$$
\begin{aligned}
B &\overset{\Delta}{=} D_\phi(w_{n-1}, z_n) - D_\phi(w_n, z_n) \\
&\overset{(a)}{=} \phi(w_{n-1}) - \phi(w_n) - \nabla_w \phi(z_n)(w_{n-1} - w_n) \\
&\overset{(b)}{\leq} \nabla_w \phi(w_{n-1})(w_{n-1} - w_n) - \frac{\nu}{2}\|w_n - w_{n-1}\|^2 - \nabla_w \phi(z_n)(w_{n-1} - w_n) \\
&= \Big(\nabla_w \phi(w_{n-1}) - \nabla_w \phi(z_n)\Big)(w_{n-1} - w_n) - \frac{\nu}{2}\|w_n - w_{n-1}\|^2 \\
&\overset{(c)}{=} \Big(\nabla_{w^\mathsf{T}} \phi(w_{n-1}) - b_n\Big)^\mathsf{T}(w_{n-1} - w_n) - \frac{\nu}{2}\|w_n - w_{n-1}\|^2 \\
&\overset{(d)}{=} \mu(s_E(w_{n-1}))^\mathsf{T}(w_{n-1} - w_n) - \frac{\nu}{2}\|w_n - w_{n-1}\|^2
\end{aligned} \tag{15.145}
$$

where step (a) is obtained by replacing the Bregman divergences by their definitions, step (b) uses the strong-convexity condition (15.136), step (c) uses $b_n = \nabla_{w^\mathsf{T}} \phi(z_n)$, and step (d) uses the first step of the mirror-descent algorithm (15.103).

Next we call upon the Cauchy–Schwarz property $|x^\mathsf{T} y| \leq \|x\|_\star \|y\|$ in terms of the original and dual norms. Using this property we have

$$
|(s_E(w_{n-1}))^\mathsf{T}(w_{n-1} - w_n)| \leq \|s_E(w_{n-1})\|_\star \|w_{n-1} - w_n\| \overset{(15.135)}{\leq} \delta\|w_{n-1} - w_n\| \tag{15.146}
$$

Substituting into (15.145), we get

$$
B \leq \mu\delta \|w_{n-1} - w_n\| - \frac{\nu}{2}\|w_n - w_{n-1}\|^2 \tag{15.147}
$$

Consider the quadratic function $f(x) = \mu\delta x - \frac{\nu}{2}x^2$. Its maximum occurs at $x = \mu\delta/\nu$ and the maximum value is $(\mu\delta)^2/2\nu$. Therefore, we can bound B by

$$
B \leq (\mu\delta)^2/2\nu \tag{15.148}
$$

Returning to (15.144), summing from $n = 0$ to $n = N - 1$, and dividing by N, we get

$$
\begin{aligned}
\frac{1}{N}\sum_{n=0}^{N-1} E(w_{n-1}) \; - \; E(w^\star) &\leq \frac{1}{N\mu} D_\phi(w^\star, w_{-1}) + \frac{\mu\delta^2}{2\nu} \\
&\overset{(a)}{\leq} \frac{W^2}{N\mu} + \frac{\mu\delta^2}{2\nu} \\
&\overset{(b)}{\leq} W\delta\sqrt{\frac{2}{N\nu}}
\end{aligned} \tag{15.149}
$$

where step (a) uses (15.139) and step (b) uses (15.140). Applying property (8.82) to the convex function $E(w)$ we conclude that (15.141) holds. An alternative convergence argument is suggested in Prob. 15.10.

■

Comparison with projection gradient method

We established the following convergence rate for the projection gradient algorithm in (15.69):

$$
E(\bar{w}_{N-2}) - E(w^\star) \leq \frac{W\delta}{\sqrt{N}} \tag{15.150}
$$

where the value of δ is the Lipschitz constant relative to the *Euclidean norm.* If $E(w)$ happens to be Lipschitz relative to some other norm, say, the ℓ_∞-norm, then, as explained earlier in (15.74), the upper bound in the above expression will become $O(\sqrt{M/N})$. In contrast, the δ-parameter that appears in result (15.149) for the mirror-descent algorithm is relative to whichever norm was used to establish the Lipschitz property of $E(w)$, and the upper bound is independent of the problem size, M:

$$\textit{under (15.72), excess risk improves from } O(\sqrt{M/N}) \textit{ to } O(\,1/\sqrt{N}) \tag{15.151}$$

Convexity versus strong convexity

The convergence rate of the mirror-descent method improves when either smoothness or strong convexity of the risk function holds.

(1) (**Smooth vs. nonsmooth risks**). The $O(1/\sqrt{N})$ convergence rate in (15.141) for mirror descent can be improved to $O(1/N)$ when $E(w)$ is first-order differentiable with δ-Lipschitz gradients. It is shown in Prob. 15.13 that for this case:

$$0 \le E(\bar{w}_{N-2}) - E(w^\star) \le \frac{W^2\delta}{N\nu} = O(1/N) \tag{15.152}$$

(2) (**Convex vs. strongly convex risks**). If, on the other hand, $E(w)$ is ρ-strongly convex relative to the Bregman measure, say, if it satisfies:

$$E(w) \ge E(z) + (s_E(z))^\mathsf{T}(w-z) + \rho\, D_\phi(w,z), \quad \forall\, s \in \partial_{z^\mathsf{T}} E(z), \;\; \rho > 0 \tag{15.153}$$

and has δ-Lipschitz gradients, then it is verified in Prob. 15.14 that the $O(1/\sqrt{N})$ convergence rate in (15.141) improves to $O(\ln(N)/N)$:

$$0 \le E(\bar{w}_{N-2}) - E(w^\star) \le \frac{\delta^2}{2\nu\rho}\frac{O(\ln N)}{N} = O(\ln(N)/N) \tag{15.154}$$

We summarize the main convergence properties for the mirror-descent method in Table 15.1 under varied conditions on the risk function $E(w)$.

Example 15.13 (**Lazy mirror descent**) The mirror-descent algorithm (15.103) maps w_{n-1} to the dual space using $\nabla_{w^\mathsf{T}}\phi(\cdot)$ and maps b_n back to the primal space using $\nabla_{x^\mathsf{T}}\phi^\star(\cdot)$. The lazy variant of mirror descent does away with the first mapping and implements instead the recursions shown in (15.156). The first recursion over b_n essentially averages the gradient vectors over time, which is why the algorithm is also sometimes referred to as a *dual averaging* method. By repeating an argument similar to the proof of Theorem 15.5 under the same conditions stated in the theorem, it can

Table 15.1 Convergence properties of the mirror-descent algorithm (15.103).

Conditions on risk $E(w)$	Asymptotic convergence property
ν-strongly convex as in (15.153) δ-Lipschitz gradients	$E(\bar{w}_{n-2}) - E(w^\star) \leq O(\ln(n)/n)$ Eq. (15.154)
smooth convex function δ-Lipschitz gradients	$E(\bar{w}_{n-2}) - E(w^\star) \leq O(1/n)$ Eq. (15.152)
nonsmooth, convex, and δ-Lipschitz function	$E(\bar{w}_{n-2}) - E(w^\star) \leq O(1/\sqrt{n})$ Thm. 15.5

be verified that for $\mu = (W/\delta)\sqrt{\nu/(2N)}$, the averaged iterate over a window of N iterations will satisfy (see Prob. 15.16):

$$0 \leq E(\bar{w}_{N-2}) - E(w^\star) \leq 2W\delta\sqrt{2/N\nu} \tag{15.155}$$

Lazy mirror descent for minimizing $E(w)$ subject to $w \in \mathcal{C}$ (15.44).

given convex and smooth function $E(w)$;
choose ν_ϕ-strongly convex mirror function $\phi(w) : \mathcal{C}_\phi \rightarrow \mathrm{IR}$;
let $\phi^\star(x) = \sup_w \{w^\mathsf{T} x - \phi(w)\}$ denote its conjugate function;
given convex set $\mathcal{C} \subset \bar{\mathcal{C}}_\phi$;
given gradient operator $\nabla_{w^\mathsf{T}} E(w)$;
given gradient operator $\nabla_{w^\mathsf{T}} \phi(w)$;
let $\mathcal{C}' = \mathcal{C} \cap \mathcal{C}_\phi$; (15.156)
start from an initial condition, $b_{-1} = w_{-1} \in \mathcal{C}'$.
repeat until convergence over $n \geq 0$:
  $b_n = b_{n-1} - \mu\nabla_{w^\mathsf{T}} E(w_{n-1})$
  $z_n = \nabla_{x^\mathsf{T}} \phi^\star(b_n)$
  $w_n = \mathcal{P}_{C',\phi}(z_n)$
end
return $w^\star \leftarrow w_n$.

Example 15.14 (**Mirror prox algorithm**) Another variant of mirror descent applies two "mirror steps" in the manner described in listing (15.158). The first mirror step results in the intermediate variable w_n', which is used to assess the gradient of the risk $E(w)$ in the second mirror step. This second step is the original mirror-descent recursion *except* that it evaluates the gradient at w_n' rather than w_{n-1}. Assume $E(w)$ has δ-Lipschitz gradients and set $\mu = \nu/\delta$. By repeating an argument similar to the proof of Theorem 15.5, it can be verified that (see Prob. 15.17):

$$0 \leq E\left(\frac{1}{N}\sum_{n=0}^{N-1} w_{n-1}'\right) - E(w^\star) \leq \frac{W^2\delta}{N\nu} = O(1/N) \tag{15.157}$$

Observe that the rate of convergence is now improved to $O(1/N)$ from $O(1/\sqrt{N})$ for convex functions with δ-Lipschitz gradients.

Mirror prox method for minimizing $E(w)$ subject to $w \in \mathcal{C}$.

given convex and smooth function $E(w)$;
choose ν_ϕ-strongly convex mirror function $\phi(w) : \mathcal{C}_\phi \rightarrow \mathbb{R}$;
let $\phi^\star(x) = \sup_w \{w^\mathsf{T} x - \phi(w)\}$ denote its conjugate function;
given convex set $\mathcal{C} \subset \bar{\mathcal{C}}_\phi$;
given gradient operator $\nabla_{w^\mathsf{T}} E(w)$;
given gradient operator $\nabla_{w^\mathsf{T}} \phi(w)$;
let $\mathcal{C}' = \mathcal{C} \cap \mathcal{C}_\phi$;
start from an initial condition, $w_{-1} \in \mathcal{C}'$.
repeat until convergence over $n \geq 0$: (15.158)

$$
\begin{aligned}
b_n &= \nabla_{w^\mathsf{T}} \phi(w_{n-1}) - \mu \nabla_{w^\mathsf{T}} E(w_{n-1}) \\
z_n &= \nabla_{x^\mathsf{T}} \phi^\star(b_n) \\
w_n' &= \mathcal{P}_{C',\phi}(z_n) \\
b_n' &= \nabla_{w^\mathsf{T}} \phi(w_{n-1}) - \mu \nabla_{w^\mathsf{T}} E(w_n') \\
z_n' &= \nabla_{x^\mathsf{T}} \phi^\star(b_n') \\
w_n &= \mathcal{P}_{C',\phi}(z_n')
\end{aligned}
$$

end
return $w^\star \leftarrow w_n'$.

In the mirror-prox implementation, the algorithm looks "forward" and anticipates the iterate w_n' that would result from the mirror-descent step; this information is then used to generate w_n. It can be shown that this construction amounts to transforming the constrained convex optimization problem (of minimizing $E(w)$ over $w \in \mathcal{C}$) into the solution of a related convex–concave saddle point problem of the general form (see Probs. 15.18 and 15.19):

$$
w^\star = \underset{w \in \mathcal{C}}{\operatorname{argmin}} \left\{ E(w) \right\} \iff w^\star = \underset{w \in \mathcal{C}}{\operatorname{argmin}} \max_{w' \in \mathcal{W}} \left\{ S(w, w') \right\} \tag{15.159}
$$

over variables $\{w, w'\}$ constrained to two closed convex sets $\{\mathcal{C}, \mathcal{W}\}$ and some smooth function $S(w, w')$ with δ-Lipschitz gradients such that:

$$
\text{for each } w \in \mathcal{C}, \text{ the function } S(w, w') \text{ is concave over } w'; \tag{15.160a}
$$

$$
\text{for each } w' \in \mathcal{W}, \text{ the function } S(w, w') \text{ is convex over } w; \tag{15.160b}
$$

This suggests that we can seek $w^\star$ by performing a gradient-ascent step on $S(w, w')$ over w' followed by a gradient-descent step over w, or vice-versa. The gradient update steps can be replaced by "mirror descent" steps, leading to the mirror-prox algorithm.

15.4 COMPARISON OF CONVERGENCE RATES

For ease of reference, Tables 15.2 and 15.3 summarize the main results derived so far in the text for the asymptotic behavior of various algorithms under constant and vanishing step sizes, along with the assumed conditions on the risk functions. The summary for the mirror-descent algorithm is already given in Table 15.1.

Table 15.2 Convergence properties of gradient-descent, subgradient, and proximal algorithms with *constant* step-sizes.

Algorithm	Conditions on risk $P(w)$	Asymptotic convergence property	Reference
gradient descent algorithm (12.22)	ν-strongly convex δ-Lipschitz gradients (12.12a)–(12.12b)	$P(w_n) - P(w^\star) \leq O(\lambda^n)$	Thm. 12.1
subgradient algorithm (14.3)	ν-strongly convex affine-Lipschitz subgradient (14.28a)–(14.28b)	$P(w_n) - P(w^\star) \leq O(\lambda^{n/2}) + O(\sqrt{\mu})$	Thm. 14.1
subgradient algorithm with smoothed iterate (14.99)	ν-strongly convex affine-Lipschitz subgradients (14.28a)–(14.28b)	$P(\bar{w}_n) - P(w^\star) \leq O(\kappa^n) + O(\mu)$	Eq. (14.96)
proximal gradient algorithm (15.10)	$P(w) = q(w) + E(w)$ $E(w)$: ν-strongly convex $E(w)$: δ-Lipschitz gradients (15.32b)–(15.32c)	$P(w_n) - P(w^\star) \leq O(\lambda^n)$	Thm. 15.1
projected gradient algorithm (15.51)	$P(w) = \mathbb{I}_{C,\infty}[w] + E(w)$ $E(w)$: ν-strongly convex $E(w)$: δ-Lipschitz gradients (15.32b)–(15.32c)	$E(w_n) - E(w^\star) \leq O(\lambda^n)$	Thm. 15.1
gradient descent algorithm (12.22)	convex, δ-Lipschitz gradients	$P(w_n) - P(w^\star) \leq O(1/n)$	Prob. 12.13
subgradient algorithm (14.3)	convex, bounded subgradients	$P(w_n^{\text{best}}) - P(w^\star) \leq O(1/n) + O(\mu)$	Prob. 14.4(b)
proximal gradient algorithm (15.10)	$P(w) = q(w) + E(w)$ $E(w)$: convex, δ-Lipschitz gradients	$P(w_n) - P(w^\star) \leq O(1/n)$	Thm. 11.1
projected gradient algorithm (15.10)	$P(w) = \mathbb{I}_{C,\infty}[w] + E(w)$ $E(w)$: convex nonsmooth $E(w)$: δ-Lipschitz	$E(\bar{w}_n) - E(w^\star) \leq O(1/\sqrt{n})$	Thm. 15.3

Table 15.3 Convergence properties of gradient-descent, subgradient, and proximal algorithms with decaying step sizes, $\mu(n) = \tau/(n+1)$, $\tau > 0$.

Algorithm	Conditions on risk $P(w)$	Asymptotic convergence property
gradient descent (12.22)	ν-strongly convex δ-Lipschitz gradients (12.12a)–(12.12b)	$P(w_n) - P(w^\star) \leq O(1/n^{2\nu\tau})$ Thm. 12.2
subgradient method (14.3)	ν-strongly convex affine-Lipschitz subgradients (14.28a)–(14.28b)	$P(w_n) - P(w^\star) \leq O(1/\sqrt{n})$ Thm. 14.4
proximal gradient (15.10)	$P(w) = q(w) + E(w)$ $E(w)$: ν-strongly convex $E(w)$: δ-Lipschitz gradients (15.32b)–(15.32c)	$P(w_n) - P(w^\star) \leq O(1/n^{2\nu\tau-1})$ Thm. 15.2
projected gradient (15.10)	$P(w) = \mathbb{I}_{C,\infty}[w] + E(w)$ $E(w)$: ν-strongly convex $E(w)$: δ-Lipschitz gradients (15.32b)–(15.32c)	$E(w_n) - E(w^\star) \leq O(1/n^{2\nu\tau-1})$ Thm. 15.2

15.5 COMMENTARIES AND DISCUSSION

Proximal gradient algorithm. We commented extensively on the proximal operator, its properties, and the construction of proximal gradient algorithms in the comments at the end of Chapter 10. Briefly, the proximal operator provides a powerful tool for solving optimization problems with nonsmooth components. We explained in that chapter that there are two key properties that make proximal operators particularly suitable:

(a) According to (11.37), the fixed points of proximal operators coincide with the minimizers of the functions defining them, namely,

$$\underbrace{w^o = \text{prox}_{\mu h}(w^o)}_{\textbf{fixed point}} \iff \underbrace{0 \in \partial_w h(w^o)}_{\textbf{global minimum}} \tag{15.161}$$

(b) Proximal operators are firmly nonexpansive, meaning that they satisfy property (11.42). Consequently, as shown by (11.44), a convergent iteration can be used to determine their fixed points, which led to the proximal point algorithm:

$$w_n = \text{prox}_{\mu h}(w_{n-1}), \quad n \geq 0 \tag{15.162}$$

These properties motivated expansive research efforts on new families of algorithms for the solution of nonsmooth convex optimization problems, including the proximal gradient algorithm (15.10) – see, e.g., the works by Minty (1962), Browder (1965, 1967), Bruck and Reich (1977), and Combettes, (2004), as well as the texts by Brezis (1973), Granas and Dugundji (2003), and Bauschke and Combettes (2011). One of the earliest contributions in this regard is the proximal point algorithm (15.162). It was proposed by Martinet (1970, 1972) as a way to construct iterates for minimizing a convex function $h(w)$ by solving successive problems of the type (using our notation):

$$w_n = \underset{w \in \mathbb{R}^M}{\text{argmin}} \left\{ h(w) + \frac{1}{2\mu}\|w - w_{n-1}\|^2 \right\} \tag{15.163}$$

The solution is given by $w_n = \text{prox}_{\mu h}(w_{n-1})$. Two other early influential works in the area of proximal operators for nonsmooth optimization, with stronger convergence results, are the articles by Rockafellar (1976a, b). Since then, several important advances have occurred in the development of smooth techniques for the optimization of nonsmooth problems – see, e.g., the works by Polyak (1987), Combettes and Pesquet (2011), Parikh and Boyd (2013), and Cevher, Becker, and Schmidt (2014), in addition to the contributions by Luo and Tseng (1992b, 1993), Nesterov (2004, 2005), Combettes and Wajs (2005), Figueiredo Bioucas-Dias, and Nowak (2007), and Beck and Teboulle (2009a, 2012).

Projection gradient algorithm. The projection gradient method (15.51) was proposed by Goldstein (1964) and Levitin and Polyak (1966). Goldstein (1964) provided a concise convergence proof for twice-differentiable functions $E(w)$ in a short two-page article; the same argument can be applied to the proximal gradient algorithm. Further discussions on the projected gradient technique, including convergence properties, can be found in the works by McCormick and Tapia (1972), Polak (1973), Bruck (1975), Bertsekas (1976), and Dunn (1981), as well as in the texts by Goldstein (1967), Demyanov and Rubinov (1971), Daniel (1971), Polyak (1987), Bertsekas and Tsitsiklis (1997, 2000), Bauschke and Combettes (2011), and Bach *et al.* (2012). There are, of course, other methods, besides the projection gradient algorithm, for the solution of constrained convex optimization problems of the form (15.44), namely,

$$w^\star = \operatorname*{argmin}_{w\in \mathbb{R}^M} E(w), \quad \text{subject to } w \in \mathcal{C} \tag{15.164}$$

One such method is the *Frank–Wolfe algorithm* devised by Frank and Wolfe (1956) and generalized by Demyanov and Rubinov (1967, 1971). It is an iterative procedure that relies on minimizing successive first-order Taylor approximations for the objective function $E(w)$. Assume we start from an initial condition $w_{-1} \in \mathcal{C}$. Let w_{n-1} denote the iterate at instant $n-1$ and introduce the first-order approximation around $w = w_{n-1}$:

$$E(w) \approx E(w_{n-1}) + \Big(\nabla_{w^\mathsf{T}} E(w_{n-1})\Big)^\mathsf{T}(w - w_{n-1}) \tag{15.165}$$

The Frank–Wolfe method constructs an update direction by minimizing the expression on the right-hand side over $w \in \mathcal{C}$. We denote the resulting minimizer by s_{n-1}, i.e., we set

$$s_{n-1} \stackrel{\Delta}{=} \operatorname*{argmin}_{w\in\mathcal{C}} \Big(\nabla_{w^\mathsf{T}} E(w_{n-1})\Big)^\mathsf{T} w \tag{15.166}$$

and use it to update w_{n-1} to w_n by using the convex combination (see Prob. 15.22):

$$\begin{aligned} w_n &= (1-\mu_n)w_{n-1} + \mu_n s_{n-1} \\ &= w_{n-1} + \mu_n(s_{n-1} - w_{n-1}) \end{aligned} \tag{15.167}$$

where the update direction is seen to be $p_{n-1} = s_{n-1} - w_{n-1}$. The step-size parameter μ_n is determined from the line search procedure:

$$\mu_n \stackrel{\Delta}{=} \operatorname*{argmin}_{0\le\mu\le 1} E(w_{n-1} + \mu p_{n-1}) \tag{15.168}$$

The method is also known as the *conditional gradient algorithm*, and it was shown by Frank and Wolfe (1956) and Dunn and Harshbarger (1978) that its excess risk satisfies $E(w_n) - E(w^\star) = O(1/n)$ for convex $E(w)$. Observe that the constrained optimization step (15.166) involves a linear objective function, as happens in linear programming problems. Good overviews are given by Polak (1973) and Jaggi (2013).

Mirror-descent algorithm. The mirror-descent method was introduced in the book by Nemirovski and Yudin (1983) for the solution of large-scale optimization problems. The method was shown to be effective and was applied to the solution of a convex problem in tomography involving millions of variables by Ben-Tal, Margalit, and Nemirovski (2001). The derivation used in Example 15.11 to arrive at listing (15.126) is motivated by the approach from Beck and Teboulle (2003), based on using Bregman divergence to replace the squared-Euclidean distance. The mirror-prox algorithm is from

Nemirovski (2004); the algorithm is motivated by transforming a constrained convex optimization problem into the solution of a convex–concave saddle point problem. The argument in the proof of Theorem 15.5 is based on the presentation in Bubeck (2015). This reference provides an excellent overview of mirror-descent methods and variations thereof. Additional discussion and overviews can be found in the works by Juditsky and Nemirovski (2011a, b), Beck (2017), Hazan (2019), and Lei and Zhou (2020).

PROBLEMS

15.1 Under the same conditions of Theorem 15.1 for the proximal gradient algorithm, show that $\|w_n - w_{n-1}\|^2 \leq \lambda \|w_{n-1} - w_{n-2}\|^2$.

15.2 Consider the same setting as Prob. 14.9. Show that, for this case, the proximal gradient recursion (15.10) reduces to

$$\begin{cases} z_n &= w_{n-1} + 2\mu(r_{h\gamma} - R_h w_{n-1}) \\ w_n &= \mathbb{T}_{\mu\alpha}(z_n) \end{cases}$$

15.3 Consider the ℓ_1-regularized least-squares problem:

$$\min_w \left\{ \alpha\|w\|_1 + \frac{1}{N}\sum_{m=0}^{N-1}\left(\gamma(m) - h_m^\mathsf{T} w\right)^2 \right\}$$

Verify that algorithm (15.10) reduces the following proximal batch LASSO recursion:

$$\begin{cases} z_n &= w_{n-1} + 2\mu\left(\dfrac{1}{N}\displaystyle\sum_{m=0}^{N-1} h_m(\gamma(m) - h_m^\mathsf{T} w_{n-1})\right) \\ w_n &= \mathbb{T}_{\mu\alpha}(z_n) \end{cases}$$

How would the algorithm change if elastic-net regularization were used?

15.4 The total-variation denoising problem introduced by Rudin, Osher, and Fatemi (1992) involves solving a regularized least-squares problem of the following form:

$$\min_w \left\{ \alpha\sum_{m'=1}^{M-1}\Big|w[m'+1] - w[m']\Big| + \frac{1}{N}\sum_{m=0}^{N-1}\left(\gamma(m) - h_m^\mathsf{T} w\right)^2 \right\}$$

where the $\{w[m']\}$, for $m' = 1, 2, \ldots, M$, denote the individual entries of $w \in \mathbb{R}^M$. Derive a proximal solution for this problem.

15.5 The fused LASSO problem introduced by Tibshirani *et al.* (2005) adds ℓ_1-regularization to the total variation formulation in Prob. 15.4 and considers instead

$$\min_w \left\{ \alpha_1\|w\|_1 + \alpha_2\sum_{m'=1}^{M-1}\Big|w[m'+1] - w[m']\Big| + \frac{1}{N}\sum_{m=0}^{N-1}\left(\gamma(m) - h_m^\mathsf{T} w\right)^2 \right\}$$

Derive a proximal solution for this problem.

15.6 The group LASSO problem introduced by Yuan and Lin (2006) involves solving a regularized least-squares problem of the following form. We partition each observation vector into K sub-vectors, say, $h_m = \text{col}\{h_{mk}\}$ for $k = 1, 2, \ldots, K$. We similarly partition the weight vector into K sub-vectors, $w = \text{col}\{w_k\}$, of similar dimensions to h_{mk}. Now consider the problem:

$$\min_w \left\{ \alpha\sum_{k=1}^{K}\|w_k\| + \frac{1}{N}\sum_{m=0}^{N-1}\left(\gamma(m) - \sum_{k=1}^{K} h_{mk}^\mathsf{T} w_k\right)^2 \right\}$$

Derive a proximal solution for this problem.

15.7 Consider a partially observed matrix $R \in \mathbb{R}^{N\times M}$ with missing entries. We denote the set of all indices (i,j) for which R_{ij} are known by $\mathcal{R}$. Matrices like this appear, for example, in the study of recommender systems and matrix completion problems – see future Example 16.7. We wish to approximate R by a completed matrix $W \in \mathbb{R}^{N\times M}$ of rank r by solving the optimization problem:

$$\widehat{W} \triangleq \underset{W\in\mathbb{R}^{N\times M}}{\text{argmin}} \left\{ \alpha\|W\|_\star + \sum_{(i,j)\in\mathcal{R}} (W_{ij} - R_{ij})^2 \right\}, \quad \text{rank}(W) = r$$

where $\|\cdot\|_\star$ denotes the nuclear norm of its matrix argument. Introduce the projection operation onto the set of indices $\mathcal{R}$:

$$\mathcal{P}_{\mathcal{R}}(W) = \begin{cases} W_{ij}, & \text{if } (i,j)\in\mathcal{R} \\ 0, & \text{otherwise} \end{cases}$$

Justify the following proximal gradient algorithm for seeking $\widehat{W}$ in terms of the matrix soft-thresholding operation $\mathbb{T}_\alpha(Z)$ from Prob. 11.12:

$$W_n = \mathbb{T}_\alpha\Big(W_{n-1} + \mu\big[\mathcal{P}_{\mathcal{R}}(R) - \mathcal{P}_{\mathcal{R}}(W_{n-1})\big]\Big)$$

Remark. This result is the *soft-impute* algorithm from Mazumder, Hastie, and Tibshirani (2010).

15.8 In a manner similar to the proof of Theorem 15.1, establish the convergence of the proximal randomized coordinate descent algorithm (15.31).

15.9 Verify that an equivalent restatement of problem (15.93) is

$$w_n = \underset{w\in\mathcal{C}\cap\mathcal{C}_\phi}{\text{argmin}} \left\{ E(w_{n-1}) + \nabla_w E(w_{n-1})(w - w_{n-1}) + \frac{1}{\mu} D_\phi(w, w_{n-1}) \right\}$$

where the first two terms on the right-hand side amount to a first-order approximation for $E(w)$ around w_{n-1}.

15.10 We can examine the convergence of mirror-descent constructions in a manner that extends the approach of Prob. 14.4 where squared Euclidean distances are replaced by Bregman divergences. To see this, we refer to the formulation (15.93), whose solution is w_n. Use the result of Prob. 8.69 to show that

$$D_\phi(w^\star, w_n) - D_\phi(w^\star, w_{n-1}) + D_\phi(w_n, w_{n-1}) \le \mu\Big(E(w^\star) - E(w_{n-1})\Big)$$

15.11 Continuing with Prob. 15.10, use the δ-Lipschitz property of $E(w)$ to show that

$$0 \le \mu\Big(E(w_{n-1}) - E(w^\star)\Big) \le D_\phi(w^\star, w_{n-1}) - D_\phi(w^\star, w_n) + \frac{\mu^2\delta^2}{2\nu}$$

15.12 Continuing with the mirror-descent construction, assume $E(w)$ is not necessarily smooth but continues to be convex over $\mathcal{C}$ and δ-Lipschitz relative to some norm $\|\cdot\|$. Consider a mirror function $\phi(w) : \mathcal{C}_\phi \to \mathbb{R}$ that is ν-strongly convex relative to the same norm, and where $\mathcal{C} \subset \bar{\mathcal{C}}_\phi$. Repeat the argument used in the proof of Theorem 15.5 to show that

$$0 \le \frac{1}{N}\sum_{n=0}^{N-1} E(w_{n-1}) - E(w^\star) \le \frac{1}{N\mu} D_\phi(w^\star, w_{-1}) + \frac{\mu}{2N\nu}\sum_{n=0}^{N-1}\|s_n\|_\star^2$$

where s_n is a subgradient for $E(w)$ at location w_{n-1} and $\|\cdot\|_\star$ is the dual norm. Conclude that

$$0 \le \frac{1}{N}\sum_{n=0}^{N-1} E(w_{n-1}) - E(w^\star) \le \frac{1}{N\mu} D_\phi(w^\star, w_{-1}) + \frac{\mu}{2\nu}\max_n \|s_n\|_\star^2$$

15.13 Assume $\phi(w)$ is differentiable and strictly convex over $\mathcal{C}_\phi$, while the risk function $E(w)$ is convex over $\mathcal{C} \subset \bar{\mathcal{C}}_\phi$ with δ-Lipschitz gradients. Establish (15.152).

15.14 Assume $\phi(w)$ is differentiable and strictly convex over $\mathcal{C}_\phi$, while the risk function $E(w)$ is ρ-strongly convex over $\mathcal{C} \subset \bar{\mathcal{C}}_\phi$ relative to the Bregman measure as defined by (15.153). Establish (15.154).

15.15 Consider the set of nonnegative-definite symmetric matrices whose trace is equal to 1, $\mathcal{C} = \{X \in \mathbb{R}^{M\times M} \mid X \geq 0,\ \text{Tr}(X) = 1\}$. This set is referred to as the *spectrahedron*. Consider also the set of positive-definite matrices $\mathcal{C}_\phi = \{X \in \mathbb{R}^{M\times M} \mid X > 0\}$. Introduce the von Neumann mirror map $\phi(X) : \sum_{m=1}^{M} \lambda_m(X) \ln \lambda_m(X)$, in terms of the eigenvalues of X.

(a) Are the sets $\{\mathcal{C}, \mathcal{C}_\phi\}$ convex?

(b) Show that $\phi(X)$ is $\frac{1}{2}$-strongly convex relative to the nuclear norm, $\|X\|_\star = \sum_{m=1}^{M} \lambda_m(X)$.

(c) Verify that $d_{\max} = \ln M$, where $d_{\max} = \sup_{X\in\mathcal{C}\cap\mathcal{C}_\phi} D_\phi(X, X^\star)$ and $X^\star$ is a solution to the optimization problem

$$X^\star = \underset{X\in\mathcal{C}\cap\mathcal{C}_\phi}{\text{argmin}}\ E(X)$$

where $E(W)$ is convex over $\mathcal{C}$ and δ-Lipschitz relative to the same nuclear norm.

(d) Show that the mirror descent algorithm in this case reduces to

$$Z_n = \exp\Big\{\ln X_{n-1} - \mu \nabla_X E(X_{n-1})\Big\}$$
$$X_n = Z_n/\text{Tr}(Z_n)$$

Remark. The reader may refer to Bubeck (2015) for a related discussion.

15.16 Establish the convergence result (15.155) for the lazy mirror-descent algorithm under the same conditions stated in Theorem 15.5.

15.17 Establish the convergence result (15.157) for the mirror-prox algorithm using an argument similar to Theorem 15.5. *Remark.* The reader may refer to Bubeck (2015) for a related discussion.

15.18 Consider the risk function $E(w) = \frac{1}{N}\sum_{m=0}^{N-1} \max\{0, 1-\gamma(m)h_m^\mathsf{T} w\}$. Show that the minimizer of $E(w)$ over some closed convex set $\mathcal{C}$ can be found by solving the convex–concave saddle point problem:

$$w^\star = \underset{w\in\mathcal{C}}{\text{argmin}}\ \max_{0\leq w'_m\leq 1} \left\{\frac{1}{N}\sum_{m=0}^{N-1} w'_m \max\{0, 1-\gamma(m)h_m^\mathsf{T} w\}\right\}$$

15.19 Consider the risk function $E(w) = \|d - Hw\|_p$, where $d \in \mathbb{R}^N$, $H \in \mathbb{R}^{N\times M}$, and $p \geq 1$. Let $1/p + 1/q = 1$. Show that the minimizer of $E(w)$ over some closed convex set $\mathcal{C}$ can be found by solving the convex–concave saddle point problem:

$$w^\star = \underset{w\in\mathcal{C}}{\text{argmin}}\ \max_{\|w'\|_q\leq 1} \Big\{(d - Hw)^\mathsf{T} w'\Big\}$$

15.20 Refer to the constrained optimization problem (15.164) and assume $E(w)$ is convex and first-order differentiable. Show that a necessary condition for a point $w^\star$ to be the global minimizer is

$$\Big(\nabla_{w^\mathsf{T}} E(w^\star)\Big)^\mathsf{T}(w - w^\star) \geq 0,\ \ \forall\, w \in \mathcal{C}$$

When $E(w)$ is not necessarily convex, argue that this same condition is necessary for $w^\star$ to be a local minimizer.

15.21 Refer to the same constrained optimization problem (15.164) and assume $E(w)$ is convex and first-order differentiable. Show that a necessary and sufficient condition for a point $w^\star$ to be the global minimizer is

$$\left(\nabla_{w^\mathsf{T}} E(w^\star)\right)^\mathsf{T}(w - w^\star) = 0, \quad \forall\, w \in \mathcal{C}$$

When $E(w)$ is not necessarily convex, argue that this same condition is only necessary but not sufficient.

15.22 Refer to the Frank–Wolfe algorithm (15.166)–(15.168). Argue that $E(w_n) \le E(w_{n-1})$. Let $L(w)$ denote the first-order approximation that appears on the right-hand side of (15.165), i.e.,

$$L(w) = E(w_{n-1}) + \left(\nabla_{w^\mathsf{T}} E(w_{n-1})\right)^\mathsf{T}(w - w_{n-1})$$

Argue also that, at every iteration n, it holds: $L(s_{n-1}) \le E(w^\star) \le E(w_{n-1})$. Conclude that one criterion to stop the algorithm is when the width of the interval defined by $[L(s_{n-1}), E(w_{n-1})]$ becomes sufficiently small.

REFERENCES

Bach, F., R. Jenatton, J. Mairal, and G. Obozinski (2012), "Optimization with sparsity-inducing penalties," *Found. Trends Mach. Learn.*, vol. 4, no. 1, pp. 1–106.

Bauschke, H. H. and P. L. Combettes (2011), *Convex Analysis and Monotone Operator Theory in Hilbert Spaces*, Springer.

Beck, A. (2017), *First-Order Methods in Optimization*, SIAM.

Beck, A. and M. Teboulle (2003), "Mirror descent and nonlinear projected subgradient methods for convex optimization," *Oper. Res. Lett.*, vol. 31, pp. 167–175.

Beck, A. and M. Teboulle (2009a), "A fast iterative shrinkage-thresholding algorithm for linear inverse problems," *SIAM J. Imaging Sci.*, vol. 2, no. 1, pp. 183–202.

Beck, A. and M. Teboulle (2012), "Smoothing and first order methods: A unified framework," *SIAM J. Optim.*, vol. 22, no. 2, pp. 557–580.

Ben-Tal, A., T. Margalit, and A. Nemirovski (2001), "The ordered subsets mirror descent optimization method with applications to tomography," *SIAM J. Optim.*, vol. 12, pp. 79–108.

Bertsekas, D. P. (1976), "On the Goldstein–Levitin–Polyak gradient projection method," *IEEE Trans. Aut. Control*, vol. 21, no. 2, pp. 174–184.

Bertsekas, D. P. and J. N. Tsitsiklis (1997), *Parallel and Distributed Computation: Numerical Methods*, Athena Scientific.

Bertsekas, D. P. and J. N. Tsitsiklis (2000), "Gradient convergence in gradient methods with errors," *SIAM J. Optim.*, vol. 10, no. 3, pp. 627–642.

Brezis, H. (1973), *Opérateurs Maximaux Monotones et Semi-Groupes de Contractions dans les Espaces de Hilbert*, North-Holland.

Browder, F. (1965), "Nonlinear monotone operators and convex sets in Banach spaces," *Bull. Amer. Math. Soc.*, vol. 71, no. 5, pp. 780–785.

Browder, F. (1967), "Convergence theorems for sequences of nonlinear operators in Banach spaces," *Mathematische Zeitschrift*, vol. 100, no. 3, pp. 201–225.

Bruck, R. E. (1975), "An iterative solution of a variational inequality for certain monotone operators in a Hilbert space," *Bull. Amer. Math. Soc.*, vol. 81, no. 5, pp. 890–892.

Bruck, R. E. and S. Reich (1977), "Nonexpansive projections and resolvents of accretive operators in Banach spaces," *Houston J. Math.*, vol. 3, pp. 459–470.

Bubeck, S. (2015), *Convex Optimization: Algorithms and Complexity*, Foundations and Trends in Machine Learning, NOW Publishers, vol. 8, no. 3–4, pp. 231–357.

Cevher, V., S. Becker, and M. Schmidt (2014), "Convex optimization for big data: Scalable, randomized, and parallel algorithms for big data analytics," *IEEE Signal Process. Mag.*, vol. 31, no. 5, pp. 32–43.

Combettes, P. L. (2004), "Solving monotone inclusions via compositions of nonexpansive averaged operators," *Optimization*, vol. 53, no. 5–6, 2004.

Combettes, P. L. and J.-C. Pesquet (2011), "Proximal splitting methods in signal processing," in *Fixed-Point Algorithms for Inverse Problems in Science and Engineering*, H. H. Bauschke *et al.*, editors, pp. 185–212, Springer.

Combettes, P. L. and V. R. Wajs (2005), "Signal recovery by proximal forward–backward splitting," *Multiscale Model. Simul.*, vol. 4, no. 4., pp. 1168–1200.

Daniel, J. W. (1971), *The Approximate Minimization of Functionals*, Prentice-Hall.

Demyanov, V. F. and A. M. Rubinov (1967), "The minimization of a smooth convex functional on a convex set," *SIAM J. Control*, vol. 5, no. 2, pp. 280–294.

Demyanov, V. F. and A. M. Rubinov (1971), *Approximate Methods in Optimization Problems*, Elsevier.

Dunn, J. C. (1981), "Global and asymptotic convergence rate estimates for a class of projected gradient processes," *SIAM J. Control Optim.*, vol. 19, pp. 368–400.

Dunn, J. C. and S. Harshbarger (1978), "Conditional gradient algorithms with open loop step size rules," *J. Math. Anal. Appl.*, vol. 62, no. 2, pp. 432–444.

Figueiredo, M., J. Bioucas-Dias, and R. Nowak (2007), "Majorization–minimization algorithms for wavelet-based image restoration," *IEEE Trans. Image Process.*, vol. 16, no. 12, pp. 2980–2991.

Frank, M. and P. Wolfe (1956), "An algorithm for quadratic programming," *Naval Res. Logist. Q.*, vol. 3, pp. 95–110.

Goldstein, A. A. (1964), "Convex programming in Hilbert space," *Bull. Amer. Math. Soc.*, vol. 70, pp. 709–710.

Goldstein, A. A. (1967), *Constructive Real Analysis*, Harper & Row.

Granas, A. and J. Dugundji (2003), *Fixed Point Theory*, Springer.

Hazan, E. (2019), *Introduction to Online Convex Optimization*, available at arXiv:1909.05207.

Jaggi, M. (2013), "Revisiting Frank–Wolfe: Projection-free sparse convex optimization," *J. Mach. Learn. Res.*, vol. 28, no. 1, pp. 427–435.

Juditsky, A. and A. Nemirovski (2011a), "First order methods for nonsmooth convex large-scale optimization, Part I: General purpose methods," in *Optimization for Machine Learning*, S. Sra, S. Nowozin, and S. J. Wright, editors, pp. 121–146, MIT Press.

Juditsky, A. and A. Nemirovski (2011b), "First order methods for nonsmooth convex large-scale optimization, Part II: Utilizing problem's structure," in *Optimization for Machine Learning*, S. Sra, S. Nowozin, and S. J. Wright, editors, pp. 149–183, MIT Press.

Lei, Y. and D.-X. Zhou (2020), "Convergence of online mirror descent," *Appl. Comput. Harmonic Anal.*, vol. 48, no. 1, pp. 343–373.

Levitin, E. S. and B. T. Polyak (1966), "Constrained minimization methods," *Zh. Vychisl. Mat. Mat. Fiz.*, vol. 6, no. 5, pp. 787–823 (in Russian).

Luo, Z. Q. and P. Tseng (1992b), "On the linear convergence of descent methods for convex essentially smooth minimization," *SIAM J. Control Optim.*, vol. 30, pp. 408–425.

Luo, Z. Q. and P. Tseng (1993), "Error bounds and convergence analysis of feasible descent methods: A general approach," *Ann. Oper. Res.*, vol. 46, pp. 157–178.

Martinet, B. (1970), "Régularisation d'inéquations variationnelles par approximations successives," *Rev. Francaise Informat. Recherche Opérationnelle*, vol. 4, no. 3, pp. 154–158.

Martinet, B. (1972), "Determination approchtfe d'un point fixe d'une application pseudo-contractante," *Comptes Rendus de l'Académie des Sciences de Paris*, vol. 274, pp. 163–165.

Mazumder, R., T. Hastie, and R. Tibshirani (2010), "Spectral regularization algorithms for learning large incomplete matrices," *J. Mach. Learn. Res.*, vol. 11, pp. 2287–2322.

McCormick, G. P. and R. A. Tapia (1972), "The gradient projection method under mild differentiability conditions," *SIAM J. Control Optim.*, vol. 10, pp. 93–98.

Minty, G. J. (1962), "Monotone (nonlinear) operators in Hilbert space," *Duke Math. J.*, vol. 29, pp. 341–346.

Nemirovski, A. S. (2004), "Prox-method with rate of convergence o(1/t) for variational inequalities with Lipschitz continuous monotone operators and smooth convex–concave saddle point problems," *SIAM J. Optim.*, vol. 15, no. 1, pp. 229–251.

Nemirovski, A. S. and D. B. Yudin (1983), *Problem Complexity and Method Efficiency in Optimization*, Wiley.

Nesterov, Y. (2004), *Introductory Lectures on Convex Optimization*, Springer.

Nesterov, Y. (2005), "Smooth minimization of non-smooth functions," *Math. Program.*, vol. 103, no. 1, pp. 127–152.

Parikh, N. and S. Boyd (2013), "Proximal algorithms," *Found. Trends Optim.*, vol. 1, no. 3, pp. 127–239.

Polak, E. (1973), "An historical survey of computational methods in optimal control," *SIAM Rev.*, vol. 15, no. 2, pp. 553–584.

Polyak, B. T. (1987), *Introduction to Optimization*, Optimization Software.

Rockafellar, R. T. (1976a), "Monotone operators and the proximal point algorithm," *SIAM J. Control Optim.*, vol. 14, no. 5, pp. 877–898.

Rockafellar, R. T. (1976b), "Augmented Lagrangians and applications of the proximal point algorithm in convex programming," *Math. Oper. Res.*, vol. 1, no. 2, pp. 97–116.

Rudin, L. I., S. Osher, and E. Fatemi (1992), "Nonlinear total variation based noise removal algorithms," *Physica D.*, vol. 60, no. 1–4, pp. 259–268.

Tibshirani, R., M. Saunders, S. Rosset, J. Zhu, and K. Knight (2005), "Sparsity and smoothness via the fused LASSO," *J. Roy. Statist. Soc.*, Series B (Statistical Methodology), vol. 67, no. 1, pp. 91–108.

Yuan, M. and Y. Lin (2006), "Model selection and estimation in regression with grouped variables," *J. Roy. Statist. Soc.*, Series B (Statistical Methodology), vol. 68, no. 1, pp. 49–67.

16 Stochastic Optimization

We examined several types of algorithms in the last chapters, including gradient-descent, coordinate-descent, subgradient, proximal-gradient, projection-gradient, and mirror-descent algorithms. We applied the methods to the solution of general convex optimization problems of the form:

$$w^\star = \underset{w\in\mathbb{R}^M}{\text{argmin}}\ P(w) \tag{16.1}$$

with and without constraints on w, and for both smooth and nonsmooth risks. In this chapter, we are going to exploit the structure of $P(w)$ and the fact that it may correspond to an empirical or stochastic risk, in which case (16.1) takes either form:

$$w^\star \overset{\Delta}{=} \underset{w\in\mathbb{R}^M}{\text{argmin}} \left\{ P(w) \overset{\Delta}{=} \frac{1}{N}\sum_{m=0}^{N-1} Q(w;\gamma(m),h_m)\right\} \tag{16.2a}$$

$$w^o \overset{\Delta}{=} \underset{w\in\mathbb{R}^M}{\text{argmin}} \left\{ P(w) \overset{\Delta}{=} \mathbb{E}\,Q(w;\boldsymbol{\gamma},\boldsymbol{h})\right\} \tag{16.2b}$$

where $Q(w;\cdot,\cdot)$ is some convex loss function, $\gamma(m)\in\mathbb{R}$ are target signals, and $h_m\in\mathbb{R}^M$ are observed vectors (also called feature vectors). The expectation in the second line is over the joint distribution of the data $\{\boldsymbol{\gamma},\boldsymbol{h}\}$. The exposition will address two main challenges:

(a) (**Amount of data**). The size N of the dataset in empirical risk minimization problems can be large, which leads to a serious computational burden on the optimization methods devised so far in our treatment.

(b) (**Unknown statistics**). The joint probability density function (pdf) of $\{\boldsymbol{\gamma},\boldsymbol{h}\}$ for stochastic risks is generally unknown and, therefore, the risk $P(w)$ is unavailable. As a result, it is not possible to evaluate gradients or subgradients of $P(w)$ and many of the methods devised in the previous chapters will not be applicable.

These two issues are recurrent in the context of inference and learning problems. We will provide ways to alleviate their effect by appealing to *stochastic approximation*. Under this approach, and independent of whether we are dealing with empirical or stochastic risks, the gradient or subgradient vectors of $P(w)$ will be approximated from a small amount of data samples, using instantaneous or mini-batch calculations. The (sub)gradient estimates will then be used to drive the updates of the weight iterates. Some degradation in performance will ensue from the gradient approximations, but it is generally tolerable. The resulting *stochastic approximation algorithms* form the backbone of most data-driven inference and learning methods: They do not require knowledge of the underlying signal statistics, and do not need to process the entire dataset repeatedly at every iteration.

16.1 STOCHASTIC GRADIENT ALGORITHM

We start our treatment by considering the gradient-descent method where the risk function is smooth in order to illustrate the main steps of the stochastic approximation method. Recall that we use the qualification "smooth" to refer to risks that are at least first-order differentiable everywhere.

Applying the gradient-descent method to the solution of either (16.2a) or (16.2b) leads to update relations of the form:

$$\textbf{(empirical risk)} \quad w_n = w_{n-1} - \mu \times \left(\frac{1}{N}\sum_{m=0}^{N-1} \nabla_{w^\mathsf{T}} Q(w_{n-1}; \gamma(m), h_m)\right) \tag{16.3a}$$

$$\textbf{(stochastic risk)} \quad w_n = w_{n-1} - \mu \times \Big(\mathbb{E}\, \nabla_{w^\mathsf{T}} Q(w_{n-1}, \boldsymbol{\gamma}, \boldsymbol{h})\Big) \tag{16.3b}$$

where in the second line we exchanged the order of the differentiation and expectation operators, i.e., we used

$$\nabla_{w^\mathsf{T}} P(w) \stackrel{\Delta}{=} \nabla_{w^\mathsf{T}} \mathbb{E}\, Q(w; \boldsymbol{\gamma}, \boldsymbol{h}) = \mathbb{E}\, \nabla_{w^\mathsf{T}} Q(w; \boldsymbol{\gamma}, \boldsymbol{h}) \tag{16.4}$$

Example 16.1 (Switching gradient and expectation operators) We will often encounter situations that require switching gradient and expectation operators, as already seen in (16.4). We explain in Appendix 16.A on the *dominated convergence theorem* that the switching is possible under some conditions that are generally valid for our cases of interest. For instance, the switching will be possible when the loss function $Q(w;\cdot)$ and its gradient are continuous functions of w. The following example considers a situation where we can verify the validity of the switching operation directly from first principles. Consider the quadratic loss:

$$Q(w; \boldsymbol{\gamma}, \boldsymbol{h}) = (\boldsymbol{\gamma} - \boldsymbol{h}^\mathsf{T} w)^2, \ \boldsymbol{\gamma} \in \mathbb{R}, \ \ \boldsymbol{h} \in \mathbb{R}^M \tag{16.5a}$$

$$\nabla_{w^\mathsf{T}} Q(w; \boldsymbol{\gamma}, \boldsymbol{h}) = -2\boldsymbol{h}(\boldsymbol{\gamma} - \boldsymbol{h}^\mathsf{T} w) \tag{16.5b}$$

and assume $(\boldsymbol{\gamma}, \boldsymbol{h})$ have zero means with second-order moments $\mathbb{E}\,\boldsymbol{\gamma}^2 = \sigma_\gamma^2$, $\mathbb{E}\,\boldsymbol{h}\boldsymbol{\gamma} = r_{h\gamma}$, and $\mathbb{E}\,\boldsymbol{h}\boldsymbol{h}^\mathsf{T} = R_h$. In this case, direct calculations show that

$$\mathbb{E}\, Q(w; \boldsymbol{\gamma}, \boldsymbol{h}) \;=\; \sigma_\gamma^2 - 2(r_{h\gamma})^\mathsf{T} w + 2w^\mathsf{T} R_h w \tag{16.6a}$$

$$\nabla_{w^\mathsf{T}} \mathbb{E}\, Q(w; \boldsymbol{\gamma}, \boldsymbol{h}) \;=\; -2r_{h\gamma} + 2R_h w \tag{16.6b}$$

$$\mathbb{E}\, \nabla_{w^\mathsf{T}} Q(w; \boldsymbol{\gamma}, \boldsymbol{h}) \stackrel{(16.5b)}{=} -2r_{h\gamma} + 2R_h w \tag{16.6c}$$

from which we conclude, by comparing the last two lines, that the switching operation is justified.

Returning to the gradient-descent recursions (16.3a)–(16.3b), we observe first that implementation (16.3a) in the empirical case employs the *entire* dataset $\{\gamma(m), h_m\}$ at every iteration n. This means that the data needs to be available beforehand, prior to running the algorithm. This also means that this particular implementation cannot be used in streaming scenarios where data pairs $\{\gamma(n), h_n\}$ arrive sequentially, one pair at every iteration n. Moreover, the same dataset is used repeatedly by the algorithm, from one iteration to the other, until sufficient convergence is attained. The data is needed to compute the gradient of $P(w)$ at w_{n-1}, and this calculation can be costly for large N.

For implementation (16.3b) in the stochastic case, the situation is different but related. The main problem here is that the gradient of $P(w)$ is not known because the statistical distribution of the data $\{\boldsymbol{\gamma}, \boldsymbol{h}\}$ is unavailable and, hence, the risk $P(w)$ itself is not known. Even if the joint distribution of the data were known, evaluation of $P(w)$ and its gradient in closed form can still be difficult. Therefore, this second implementation suffers from the unavailability of statistical information. And even if this information is available, calculations can still be analytically intractable.

16.1.1 Stochastic Approximation

Stochastic approximation helps address these challenges for both cases of empirical and stochastic risks. Interestingly, the method will lead to the same implementation for both types of risks (empirical or stochastic). We motivate the approach by treating both risks separately initially for the benefit of the reader, until it becomes clear that one can proceed thereafter by using a unified presentation.

Empirical risks

First, in the empirical case (16.2a), the true gradient vector is given by

$$\nabla_{w^\mathsf{T}} P(w) = \frac{1}{N} \sum_{m=0}^{N-1} \nabla_{w^\mathsf{T}} Q(w; \gamma(m), h_m) \tag{16.7}$$

The gradient in this case is *known* but costly to compute and involves the entire dataset. Two popular ways to approximate it are as follows:

(a) We can select one data pair $(\boldsymbol{\gamma}(n), \boldsymbol{h}_n)$ *at random* from the N-size dataset and use it to approximate the true gradient by the expression:

$$\text{(\textbf{instantaneous approximation})}$$
$$\widehat{\nabla_{w^\mathsf{T}} P(w)} = \nabla_{w^\mathsf{T}} Q(w; \boldsymbol{\gamma}(n), \boldsymbol{h}_n) \tag{16.8}$$

In other words, we select a single term from the right-hand side of (16.7) to approximate $\nabla_{w^\mathsf{T}} P(w)$. There are many ways by which the data sample can be selected from the dataset (e.g., by sampling with or without replacement). We will discuss sampling strategies in the sequel.

(b) More generally, we can select more than one data pair at random, say, a *mini-batch* of size B, and use them to approximate the true gradient by means of a sample average:

(**mini-batch approximation**)

$$\widehat{\nabla_{w^\mathsf{T}} P}(w) = \frac{1}{B}\sum_{b=0}^{B-1} \nabla_{w^\mathsf{T}} Q(w;\gamma(b),\boldsymbol{h}_b) \tag{16.9}$$

Again, there are different ways by which the samples in the mini-batch can be selected from the given dataset. The value of B is usually a power of 2.

Stochastic risks

For the case of stochastic risks in (16.2b), the situation is different but related. The true gradient vector of the risk function now has the form:

$$\nabla_{w^\mathsf{T}} P(w) = \mathbb{E}\, \nabla_{w^\mathsf{T}} Q(w;\boldsymbol{\gamma},\boldsymbol{h}) \tag{16.10}$$

The main difficulty is that this gradient is generally unknown because the expectation on the right-hand side cannot be computed either because the statistical distribution of the data is unavailable or because the computation is analytically intractable. Stochastic approximation provides one useful way out:

(a') We assume we can sample (or observe) one data pair $(\gamma(n),\boldsymbol{h}_n)$ from the underlying joint distribution for $\{\boldsymbol{\gamma},\boldsymbol{h}\}$, and use this data sample to approximate the true gradient by the expression:

(**instantaneous approximation**)

$$\widehat{\nabla_{w^\mathsf{T}} P}(w) = \nabla_{w^\mathsf{T}} Q(w;\gamma(n),\boldsymbol{h}_n) \tag{16.11}$$

Comparing with (16.10), we observe that we are effectively dropping the expectation operator $\mathbb{E}$ altogether and evaluating the gradient of the loss at one realization for the data. We say that we are replacing the expectation $\mathbb{E}\, \nabla_{w^\mathsf{T}} Q(w;\boldsymbol{\gamma},\boldsymbol{h})$ by the *instantaneous approximation* $\nabla_{w^\mathsf{T}} Q(w;\gamma(n),\boldsymbol{h}_n)$. The resulting approximation (16.11) has the same form as in the empirical case, shown in (16.8).

(b') We can alternatively sample (or observe) more than one data pair from the underlying joint distribution for $\{\boldsymbol{\gamma},\boldsymbol{h}\}$, say, a *mini-batch* of size B, and use the samples to compute:

(**mini-batch approximation**)

$$\widehat{\nabla_{w^\mathsf{T}} P}(w) = \frac{1}{B}\sum_{b=0}^{B-1} \nabla_{w^\mathsf{T}} Q(w;\gamma(b),\boldsymbol{h}_b) \tag{16.12}$$

Again, this expression has the same form as the approximation (16.9) used in the empirical case – compare with (16.9).

Comparison

The gradient approximations in the empirical and stochastic risk cases have the same form, whether a single data point is used or a mini-batch of data samples. The difference between the constructions lies in their interpretation. In the empirical case, the data samples are extracted from the already given dataset, whereas in the stochastic case the data samples stream in and correspond to realizations from the underlying distribution for $\{\boldsymbol{\gamma}, \boldsymbol{h}\}$. Either way, for empirical or stochastic risks, the resulting stochastic gradient algorithm (often denoted by the letters SGA) takes the following form for instantaneous approximations:

(**stochastic gradient algorithm**)
for every iteration $n \geq 0$**:**
  select or receive a random data pair $(\boldsymbol{\gamma}(n), \boldsymbol{h}_n)$ (16.13a)
  update $\boldsymbol{w}_n = \boldsymbol{w}_{n-1} - \mu \nabla_{w^{\mathsf{T}}} Q(\boldsymbol{w}_{n-1}; \boldsymbol{\gamma}(n), \boldsymbol{h}_n)$
end

or, in the mini-batch case,

(**mini-batch stochastic gradient algorithm**)
for every iteration $n \geq 0$**:**
  select or receive B random data pairs $(\boldsymbol{\gamma}(b), \boldsymbol{h}_b)$ (16.13b)
  update $\boldsymbol{w}_n = \boldsymbol{w}_{n-1} - \mu \times \left(\dfrac{1}{B} \displaystyle\sum_{b=0}^{B-1} \nabla_{w^{\mathsf{T}}} Q(\boldsymbol{w}_{n-1}; \boldsymbol{\gamma}(b), \boldsymbol{h}_b) \right)$
end

Obviously, the first implementation (16.13a) is a special case of the mini-batch algorithm when $B = 1$. Note that in these listings, we are denoting the samples $\{\boldsymbol{\gamma}(b), \boldsymbol{h}_b\}$ as well as the iterates $\{\boldsymbol{w}_{n-1}, \boldsymbol{w}_n\}$ in boldface. This is because they are now random variables. Indeed, assume we run either algorithm for N iterations and obtain the sequence of realizations $\{w_0, w_1, w_2, \ldots, w_{N-1}\}$. If we re-run the same algorithm again, and even if we start from the same initial condition w_{-1}, the sequence of realizations that will result for the iterates $\{\boldsymbol{w}_n\}$ will generally be different from the previous run. This is because the samples $\{\boldsymbol{\gamma}(b), \boldsymbol{h}_b\}$ are selected at random in both runs. However, as the analysis will reveal, all the random trajectories for $\{\boldsymbol{w}_n\}$ will continue to converge close enough to $w^\star$ in some meaningful sense.

We will refer to algorithms that employ sample-based approximations for the true gradients (or subgradients) as *stochastic optimization* methods. We will also refer to these methods as online learning algorithms or simply *online algorithms* because they respond to streaming data.

16.1.2 Convergence Questions

The randomness in (16.13a) or (16.13b) raises several interesting questions. For example, we established earlier in Theorem 12.1 that the gradient-descent implementation (16.3a) in the empirical case generates iterates w_n that converge at some exponential rate λ^n to the true minimizer $w^\star$ of $P(w)$. Now, however, the true gradient vector of $P(w)$ is replaced by an *approximation* based on a randomly selected data point (or on a mini-batch of randomly selected data points). Some loss in performance is expected to occur due to the gradient approximation. Specifically, the sequence of iterates $\boldsymbol{w}_n$ (and the corresponding risk values $P(\boldsymbol{w}_n)$) need not converge any longer to the exact minimizer $w^\star$ (and the corresponding minimum value $P(w^\star)$). Even the convergence question becomes more subtle because we are not dealing anymore with a *deterministic* sequence of iterates $\{w_n\}$ converging towards a limit point $w^\star$. Instead, we are now dealing with a *random* sequence $\{\boldsymbol{w}_n\}$. Each run of the stochastic algorithm generates a trajectory for this random process, and these trajectories will generally be different over different runs. A well-designed stochastic algorithm should be able to ensure that these trajectories will approach $w^\star$ in some useful probabilistic sense. Using the *random* error vector:

$$\widetilde{\boldsymbol{w}}_n \stackrel{\Delta}{=} w^\star - \boldsymbol{w}_n \tag{16.14}$$

we will examine in future chapters convergence questions such as the following:

$$\begin{cases} \text{does } \mathbb{E}\,\|\widetilde{\boldsymbol{w}}_n\|^2 \text{ approach zero?} & (\textbf{mean-square-error convergence}) \\ \text{does } \lim\limits_{n\to\infty} \mathbb{P}(\|\widetilde{\boldsymbol{w}}_n\|^2 > \epsilon) = 0? & (\textbf{convergence in probability}) \\ \text{does } \mathbb{P}\left(\lim\limits_{n\to\infty} \|\widetilde{\boldsymbol{w}}_n\|^2 = 0\right) = 1? & (\textbf{almost-sure convergence}) \end{cases} \tag{16.15}$$

We will often examine the limiting value of the mean-square error (MSE; also called mean-square deviation or MSD):

$$\limsup_{n\to\infty} \mathbb{E}\,\|\widetilde{\boldsymbol{w}}_n\|^2 \tag{16.16}$$

from which we will be able to comment on the behavior of the algorithms in probability. This is because MSE convergence implies convergence in probability in view of Markov inequality (recall the discussion in Appendix 3.A):

$$\mathbb{P}\Big(\|\widetilde{\boldsymbol{w}}_n\|^2 \geq \epsilon\Big) \;\leq\; \mathbb{E}\,\|\widetilde{\boldsymbol{w}}_n\|^2/\epsilon, \quad \text{for any } \epsilon > 0 \tag{16.17}$$

The convergence analysis of stochastic optimization methods is demanding due to the degradation that is introduced by the stochastic gradient (or subgradient) approximations. Nevertheless, the conclusions will be reassuring in that the methods will be shown to perform well for small-enough step sizes.

16.1.3 Sample Selection

In stochastic risk minimization, the samples $(\gamma(n), h_n)$ stream in and the stochastic gradient algorithm or its mini-batch version respond to them accordingly. However, in empirical risk minimization, when a collection of N data samples $\{\gamma(m), h_m\}$ is already available, it is necessary to devise strategies to sample from this dataset. There are several ways by which the random samples can be selected.

For generality, we let $\boldsymbol{\sigma}$ denote a random integer index from within the range $0 \leq \boldsymbol{\sigma} \leq N-1$. The value of $\boldsymbol{\sigma}$ determines the data pair that is used by the algorithm at iteration n, namely, $(\boldsymbol{\gamma}(\boldsymbol{\sigma}), \boldsymbol{h}_{\boldsymbol{\sigma}})$. Clearly, the value of $\boldsymbol{\sigma}$ varies with the iteration index n, which means that, in principle, we should be writing $\boldsymbol{\sigma}(n)$. We lighten the notation and write $\boldsymbol{\sigma}$. Using this variable, we rewrite (16.13a) more explicitly in the equivalent form:

$$
\begin{array}{l}
(\textbf{stochastic gradient algorithm})\\
\textbf{for every iteration } n \geq 0\textbf{:}\\
\quad \left| \begin{array}{l}
\text{select a random index } \boldsymbol{\sigma} \text{ from within } 0 \leq \boldsymbol{\sigma} \leq N-1\\
\text{consider the random data pair } (\boldsymbol{\gamma}(\boldsymbol{\sigma}), \boldsymbol{h}_{\boldsymbol{\sigma}})\\
\text{update } \boldsymbol{w}_n = \boldsymbol{w}_{n-1} - \mu \nabla_{w^\mathsf{T}} Q(\boldsymbol{w}_{n-1}; \boldsymbol{\gamma}(\boldsymbol{\sigma}), \boldsymbol{h}_{\boldsymbol{\sigma}})
\end{array} \right.\\
\textbf{end}
\end{array}
\tag{16.18}
$$

The random index $\boldsymbol{\sigma}$ can be selected in a number of ways:

(a) (**Uniform sampling**). In this case, $\boldsymbol{\sigma}$ is selected uniformly from the discrete set of indices $\{0, 1, \ldots, N-1\}$ so that, for each integer m in this set:

$$
\mathbb{P}(\boldsymbol{\sigma} = m) = \frac{1}{N}, \quad m \in \{0, 1, 2, \ldots, N-1\} \tag{16.19}
$$

This mode of operation amounts to sampling from the N data pairs $\{\gamma(m), h_m\}$ *with replacement*, which means that some sample points may be selected multiple times during the operation of the algorithm.

(b) (**Random reshuffling**). In this case, we sample from the N data pairs *without replacement*. Another way to describe the sampling process is to randomly reshuffle the N data points first, and then process the samples *sequentially*, one after the other, from the reshuffled data.

(c) (**Importance sampling**). In this case, a probability value p_m is assigned to every index $m \in \{0, 1, \ldots, N-1\}$ with their sum adding up to 1:

$$
\sum_m p_m = 1 \tag{16.20}
$$

The probabilities $\{p_m\}$ need not be uniform. At every iteration n, the random index $\boldsymbol{\sigma}$ is selected according to this distribution, i.e.,

$$
\mathbb{P}(\boldsymbol{\sigma} = m) = p_m, \quad m \in \{0, 1, 2, \ldots, N-1\} \tag{16.21}
$$

This mode of operation amounts to sampling from the N-data pairs *with* replacement, albeit one where some data points are more or less likely to be selected according to the probabilities $\{p_m\}$. This is in contrast to (16.19), where all data points are equally likely to be selected. While this description assumes the $\{p_m\}$ are known, there are ways for stochastic algorithms to learn what values to use for the $\{p_m\}$ – see Section 19.6.

Likewise, for the mini-batch implementation (16.13b), the B samples can be chosen with or without replacement:

(i) We can sample *with replacement* one data point $(\boldsymbol{\gamma}(b), \boldsymbol{h}_b)$ at a time until all B samples have been selected. In this mode of operation, the samples within each mini-batch are selected independently of each other, although some samples may appear repeated.

(ii) We can sample *without replacement* one point $(\boldsymbol{\gamma}(b), \boldsymbol{h}_b)$ at a time until all B samples have been selected. In this case, the samples within each mini-batch will be different. However, the samples are not independent anymore because the selection of one sample is dependent on the previously selected samples.

Example 16.2 **(Bias under importance sampling)** It is useful to remark that we can always transform a stochastic implementation that relies on importance sampling into an equivalent implementation that applies *uniform* sampling to a larger dataset. The original dataset has N samples $\{\gamma(m), h_m\}$. We extend it to size $N' > N$ as follows. We repeat each sample $(\gamma(m), h_m)$ a number N_m of times such that its proportion, measured by N_m/N', in the new dataset becomes equal to p_m. In this way, selecting *uniformly* from the N'-long dataset will correspond to selecting with probability p_m from the original N-long dataset. Clearly, when this is done, the empirical risk that we will be minimizing will not be (16.3a) any longer, which is defined over the original N data samples, but rather the modified risk:

$$P'(w) \triangleq \frac{1}{N'} \sum_{m'=0}^{N'-1} Q\Big(w; \gamma(m'), h_{m'}\Big) \tag{16.22}$$

where the sum is over the enlarged N' data samples (which include sample repetitions). This modified risk function can be rewritten in a weighted form in terms of the original data samples:

$$P'(w) \triangleq \sum_{m=0}^{N-1} p_m \, Q(w; \gamma(m), h_m) \tag{16.23}$$

where the loss values are scaled by the respective probabilities, $\{p_m\}$. This observation means that, under importance sampling, the stochastic gradient recursion, using either instantaneous or mini-batch gradient approximations, will be converging toward the minimizer of the weighted risk $P'(w)$ and not toward the desired minimizer $w^\star$ of the original risk $P(w)$. We say that importance sampling biases the solution. For this reason, in implementations that employ importance sampling, the instantaneous and mini-batch approximations for the gradient vector are usually redefined by incorporating an additional scaling by $1/Np_\sigma$:

$$\textbf{(instantaneous)}\ \widehat{\nabla_{w^\mathsf{T}} P}(w) = \frac{1}{Np_\sigma}\nabla_{w^\mathsf{T}} Q(w;\gamma(\boldsymbol{\sigma}),\boldsymbol{h_\sigma}) \tag{16.24a}$$

$$\textbf{(mini-batch)}\ \widehat{\nabla_{w^\mathsf{T}} P}(w) = \frac{1}{B}\sum_{b=0}^{B-1}\frac{1}{Np_b}\nabla_{w^\mathsf{T}} Q(w;\boldsymbol{\gamma}(b),\boldsymbol{h}_b) \tag{16.24b}$$

The role of the scaling in removing the bias is explained in the next chapter.

16.1.4 Data Runs

Besides random sample selection, there is another element that arises in the implementation of stochastic approximation algorithms for *empirical risk minimization* when a collection of N data points $\{\gamma(m), h_m\}$ is available. It is customary to perform repeated *runs* (also called *epochs*) over the N data points to enhance performance and smooth out the effect of errors due to the gradient approximations. During each run k, the algorithm starts from some initial condition, denoted by w_{-1}^k, and carries out N iterations by sampling from the data and performing stochastic gradient updates. At the end of the run, the final iterate is w_{N-1}^k. This iterate is chosen as the initial condition for the next run:

$$\underbrace{\boldsymbol{w}_{-1}^{k+1}}_{\substack{\text{start of}\\ \text{run } k+1}} = \underbrace{\boldsymbol{w}_{N-1}^{k}}_{\substack{\text{end of}\\ \text{run } k}} \tag{16.25}$$

and the process repeats. The algorithm starts from $\boldsymbol{w}_{-1}^{k+1}$, and carries out N iterations by sampling from the same data again and performing stochastic gradient updates. At the end of the run, the final iterate is $\boldsymbol{w}_{N-1}^{k+1}$, which is set to $\boldsymbol{w}_{-1}^{k+2}$ and so forth. The sampling of the data is random during each run so that the data is generally covered in different orders between runs. Within each run, the random samples are selected either with or without replacement.

We can describe the multiple runs explicitly by using the superscript $k \geq 1$ to index the iterates within the kth run, such as writing $\boldsymbol{w}_n^k$. Using this notation, and incorporating multiple runs, the stochastic gradient algorithm (16.18) applied to the minimization of an empirical risk of the form (16.2a) can be as shown in (16.26); we can similarly write a mini-batch version:

(stochastic gradient algorithm with multiple runs)
given N data pairs $\{\gamma(m), h_m\}, m = 0, 1, \ldots, N-1$;
given a desired number of epochs, K;
start with an arbitrary initial condition $\boldsymbol{w}_{N-1}^0$.
for each epoch $k = 1, 2, \ldots, K$**:**
  set initial condition to $\boldsymbol{w}_{-1}^k = \boldsymbol{w}_{N-1}^{k-1}$;
  repeat for $n = 0, 1, 2, \ldots, N-1$**:**
    select a random index $0 \leq \boldsymbol{\sigma} \leq N-1$;
    $\boldsymbol{w}_n^k = \boldsymbol{w}_{n-1}^k - \mu \nabla_{w^\mathsf{T}} Q(\boldsymbol{w}_{n-1}^k; \gamma(\boldsymbol{\sigma}), \boldsymbol{h}_{\boldsymbol{\sigma}})$
  end
end
return $w^\star \leftarrow \boldsymbol{w}_{N-1}^K$.   (16.26)

Since each run starts from the iterate obtained at the end of the previous run then, for all practical purposes, the above implementation can be described more succinctly as one standard iteration running continuously over the data as listed in (16.27) and (16.28) for instantaneous and mini-batch gradient approximations. The notation $(\boldsymbol{\gamma}(n), \boldsymbol{h}_n)$ in listing (16.27) refers to the data sample that is selected at iteration n. For example, in empirical risk minimization this sample is selected at random from the N-size dataset $\{\gamma(m), h_m\}$ according to some selection policy (e.g., uniform sampling, random reshuffling) and then used to update $\boldsymbol{w}_{n-1}$ to $\boldsymbol{w}_n$. Similar remarks apply to the mini-batch version (16.28). From the discussion in the previous section, we already know that *the same algorithm is applicable to the minimization of stochastic risks* of the form (16.2b). The main difference is that the dataset $\{\gamma(m), h_m\}$ will not be available beforehand. Instead, the samples $(\boldsymbol{\gamma}(n), \boldsymbol{h}_n)$ will stream in successively over time. The description (16.27) accounts for this possibility as well, and is applicable to the minimization of both empirical and stochastic risks.

Stochastic gradient algorithm for minimizing (16.2a) or (16.2b).

given dataset $\{\gamma(m), h_m\}_{m=0}^{N-1}$ or streaming data $(\gamma(n), h_n)$;
start from an arbitrary initial condition, $\boldsymbol{w}_{-1}$.
repeat until convergence over $n \geq 0$**:**
  select at random or receive a sample $(\boldsymbol{\gamma}(n), \boldsymbol{h}_n)$ at iteration n;
  $\boldsymbol{w}_n = \boldsymbol{w}_{n-1} - \mu \nabla_{w^\mathsf{T}} Q(\boldsymbol{w}_{n-1}; \boldsymbol{\gamma}(n), \boldsymbol{h}_n)$
end
return $w^\star \leftarrow \boldsymbol{w}_n$.   (16.27)

Mini-batch stochastic gradient algorithm for minimizing (16.2a) or (16.2b).

given dataset $\{\gamma(m), h_m\}_{m=0}^{N-1}$ or streaming data $(\gamma(n), h_n)$;
given a mini-batch size, B;
start from an arbitrary initial condition, $\boldsymbol{w}_{-1}$.
repeat until convergence over $n \geq 0$**:**
select at random or receive B samples $\{\boldsymbol{\gamma}(b), \boldsymbol{h}_b\}_{b=0}^{B-1}$ at iteration n;

$$\boldsymbol{w}_n = \boldsymbol{w}_{n-1} - \mu \left(\frac{1}{B} \sum_{b=0}^{B-1} \nabla_{w^\mathsf{T}} Q(\boldsymbol{w}_{n-1}; \boldsymbol{\gamma}(b), \boldsymbol{h}_b) \right)$$

end
return $w^\star \leftarrow \boldsymbol{w}_n$.

(16.28)

Example 16.3 **(Delta rule and adaline)** Consider the ℓ_2-regularized least-squares risk:

$$P(w) = \rho \|w\|^2 + \frac{1}{N} \sum_{m=0}^{N-1} (\gamma(m) - h_m^\mathsf{T} w)^2 \tag{16.29}$$

For an arbitrary data point (γ, h) we have:

$$Q(w; \gamma, h) = \rho \|w\|^2 + (\gamma - h^\mathsf{T} w)^2 \tag{16.30a}$$

$$\nabla_{w^\mathsf{T}} Q(w; \gamma, h) = 2\rho w - 2h(\gamma - h^\mathsf{T} w) \tag{16.30b}$$

so that the stochastic gradient iteration (16.27) reduces to:

$$\boldsymbol{w}_n = (1 - 2\mu\rho)\, \boldsymbol{w}_{n-1} + 2\mu \boldsymbol{h}_n(\boldsymbol{\gamma}(n) - \boldsymbol{h}_n^\mathsf{T} \boldsymbol{w}_{n-1}), \quad n \geq 0 \tag{16.31}$$

where $\mu > 0$ is a small step size. This recursion is known as the leaky LMS (least-mean-squares) algorithm in the adaptive filtering literature – the reason for the designation "least-mean-squares" is explained in the next example, where we re-derive the same algorithm as the solution to a stochastic risk-minimizing problem involving the MSE criterion. When $\rho = 0$, the recursion reduces to the plain LMS algorithm:

$$\boldsymbol{w}_n = \boldsymbol{w}_{n-1} + 2\mu \boldsymbol{h}_n(\boldsymbol{\gamma}(n) - \boldsymbol{h}_n^\mathsf{T} \boldsymbol{w}_{n-1}), \quad n \geq 0 \tag{16.32}$$

If we focus on a single entry of the weight vector, say, $\boldsymbol{w}_n[m']$ for the m'th entry, and if we introduce the quantities

$$\delta \boldsymbol{w}_n[m'] \stackrel{\Delta}{=} \boldsymbol{w}_n[m'] - \boldsymbol{w}_{n-1}[m'] \quad \text{(change due to update)} \tag{16.33}$$

$$\boldsymbol{e}(n) \stackrel{\Delta}{=} \boldsymbol{\gamma}(n) - \boldsymbol{h}_n^\mathsf{T} \boldsymbol{w}_{n-1} \tag{16.34}$$

then recursion (16.32) gives, for the individual entries of the weight vector:

$$\delta \boldsymbol{w}_n[m'] = 2\mu \boldsymbol{e}(n) \boldsymbol{h}_n[m'] \tag{16.35}$$

In other words, the change in the individual entries of the weight vector is proportional to the observation entry, $\boldsymbol{h}_n[m']$, scaled by the error signal and step size. This form of the LMS recursion is known as the *delta rule* in the machine learning and neural network literature. In the particular case when the $\boldsymbol{\gamma}(n)$ are binary variables assuming the values ± 1, recursion (16.32) is also referred to as the *adaline* algorithm, where "adaline" stands for "adaptive linear" solution.

Example 16.4 **(Delta rule from stochastic risk minimization)** We re-derive the same delta rule by considering instead the stochastic risk minimization problem:

$$P(w) = \rho\|w\|^2 + \mathbb{E}\,(\boldsymbol{\gamma} - \boldsymbol{h}^\mathsf{T} w)^2 \tag{16.36}$$

where the loss function is now given by

$$Q(w;\gamma,\boldsymbol{h}) = \rho\|w\|^2 + (\gamma - \boldsymbol{h}^\mathsf{T} w)^2 \tag{16.37a}$$

$$\nabla_{w^\mathsf{T}} Q(w;\gamma,\boldsymbol{h}) = 2\rho w - 2\boldsymbol{h}(\gamma - \boldsymbol{h}^\mathsf{T} w) \tag{16.37b}$$

In this case, the stochastic gradient iteration (16.27) leads to the same delta rule:

$$\boldsymbol{w}_n = (1 - 2\mu\rho)\,\boldsymbol{w}_{n-1} + 2\mu\boldsymbol{h}_n(\boldsymbol{\gamma}(n) - \boldsymbol{h}_n^\mathsf{T}\boldsymbol{w}_{n-1}) \tag{16.38}$$

We illustrate the performance of the algorithm in Fig. 16.1, which shows the learning curve in linear scale using $\rho = 0.5$, $\mu = 0.001$, and $M = 10$. The simulation generates random pairs of data $\{\gamma(m), h_m\}$ according to a linear model. First, a random parameter model $w^a \in \mathbb{R}^{10}$ is selected, and a random collection of feature vectors $\{h_m\}$ are generated with zero-mean unit-variance Gaussian entries. Likewise, a collection of N independent noise Gaussian entries $\{v(m)\}$ with zero mean and variance $\sigma_v^2 = 0.0001$ is generated. Then, each $\gamma(m)$ is set to

$$\gamma(m) = h_m^\mathsf{T} w^a + v(m) \tag{16.39}$$

The minimizer w^o for the risk (16.36) can be determined in closed form and is given by

$$w^o = (\rho I_M + R_h)^{-1} r_{h\gamma} \tag{16.40}$$

where, for the simulated data,

$$R_h = \mathbb{E}\,\boldsymbol{h}\boldsymbol{h}^\mathsf{T} = I_M \quad \text{(by construction)} \tag{16.41a}$$

$$r_{h\gamma} \stackrel{\Delta}{=} \mathbb{E}\,\boldsymbol{\gamma}\boldsymbol{h} = \mathbb{E}\,(\boldsymbol{h}^\mathsf{T} w^a + \boldsymbol{v})\boldsymbol{h} = R_h w^a = w^a \tag{16.41b}$$

In the stochastic gradient implementation (16.38), data $(\boldsymbol{\gamma}(n), \boldsymbol{h}_n)$ stream in, one pair at a time. The *learning curve* of the stochastic algorithm is denoted by $P(n)$ and defined as

$$P(n) \stackrel{\Delta}{=} \mathbb{E}\,P(\boldsymbol{w}_{n-1}) \qquad \textbf{(learning curve)} \tag{16.42}$$

where the expectation is over the randomness in the weight iterates. The learning curve shows how the risk value evolves on average over time. We can simulate the learning curve by using repeated experiments, with each experiment having its own data and starting from the same initial condition. We construct the learning curve in this example as follows. We generate L replicas of $N = 3000$-long data sequences $\{\gamma(m), h_m\}$ arising from the same linear model to ensure they have the *same* statistical distribution. We subsequently run the stochastic gradient algorithm on each of these datasets, always starting from the *same* initial condition w_{-1}. In the simulations we perform $L = 500$ experiments. Each experiment ℓ results in a realization for a risk curve of the form:

$$\widehat{P}(w_{n-1}^\ell) = \rho\|w_{n-1}^\ell\|^2 + \Big(\gamma(n) - h_n^\mathsf{T} w_{n-1}^\ell\Big)^2, \quad 0 \le n \le N-1 \tag{16.43}$$

This curve is evaluated at the successive weight iterates during the ℓth experiment. By averaging the curves over all L experiments we obtain an ensemble average approximation for the true risk value as follows:

$$P(n) \approx \frac{1}{L}\sum_{\ell=1}^{L} \widehat{P}(w_{n-1}^\ell), \quad 0 \le n \le N-1 \tag{16.44}$$

where $P(n)$ denotes the estimate for the risk value $P(w)$ at the nth iteration of the algorithm. This is the curve that is shown in the left plot of Fig. 16.1.

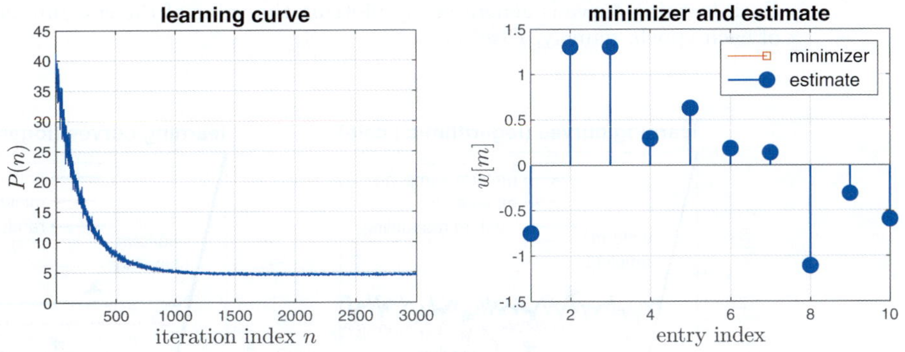

Figure 16.1 (*Left*) Ensemble-average learning curve $P(n)$ for the stochastic gradient implementation (16.27) in linear scale obtained by averaging over repeated experiments. (*Right*) Comparison of the minimizer w^o and the limit iterate w_n obtained at the end of one experiment.

Example 16.5 **(Logistic regression)** Consider the ℓ_2-regularized logistic regression empirical risk:

$$P(w) = \rho\|w\|^2 + \frac{1}{N}\sum_{m=0}^{N-1} \ln\left(1 + e^{-\gamma(m)h_m^{\mathsf{T}} w}\right) \tag{16.45}$$

for which

$$Q(w;\gamma(n),h_n) = \rho\|w\|^2 \;+\; \ln\left(1 + e^{-\gamma(n)h_n^{\mathsf{T}} w}\right) \tag{16.46a}$$

$$\nabla_{w^{\mathsf{T}}} Q(w;\gamma(n),h_n) = 2\rho w \;-\; \frac{\gamma(n)h_n}{1 + e^{\gamma(n)h_n^{\mathsf{T}} w}} \tag{16.46b}$$

The stochastic gradient iteration (16.27) becomes

$$\boldsymbol{w}_n \;=\; (1 - 2\mu\rho)\,\boldsymbol{w}_{n-1} + \mu \frac{\gamma(n)\boldsymbol{h}_n}{1 + e^{\gamma(n)\boldsymbol{h}_n^{\mathsf{T}} \boldsymbol{w}_{n-1}}}, \quad n \geq 0 \tag{16.47}$$

We illustrate the performance of this algorithm in Fig. 16.2, which shows the normalized learning curves in logarithmic scale under uniform sampling and random reshuffling, in addition to the learning curve for a mini-batch implementation.

The simulation uses $\rho = 1$, $\mu = 0.0001$, and $M = 10$. It generates $N = 500$ random pairs of data $\{\gamma(m), h_m\}$ according to a logistic model. First, a random parameter model $w^a \in \mathbb{R}^{10}$ is selected, and a random collection of feature vectors $\{h_m\}$ are generated, say, with zero-mean unit-variance Gaussian entries. Then, for each h_m, the label $\gamma(m)$ is set to either $+1$ or -1 according to the following construction:

$$\gamma(m) = +1 \;\text{ if }\; \left(\frac{1}{1 + e^{-h_m^{\mathsf{T}} w^a}}\right) \geq 0.5;\; \text{otherwise } \gamma(m) = -1 \tag{16.48}$$

A total of $K = 500$ epochs are run over the data. The learning curves are plotted in normalized logarithmic scale in line with construction (11.65), namely,

$$\ln\left(\frac{P(w_n) - P(w^\star)}{\max_n\{P(w_n) - P(w^\star)\}}\right) \tag{16.49}$$

where $w^\star$ is approximated by the limit value of the weight iterate after sufficient convergence. The mini-batch implementation employs mini-batches of size $B = 5$ samples.

Each learning curve is generated by plotting the value of the risk function at the start of each epoch, namely, $P(w^k_{-1})$.

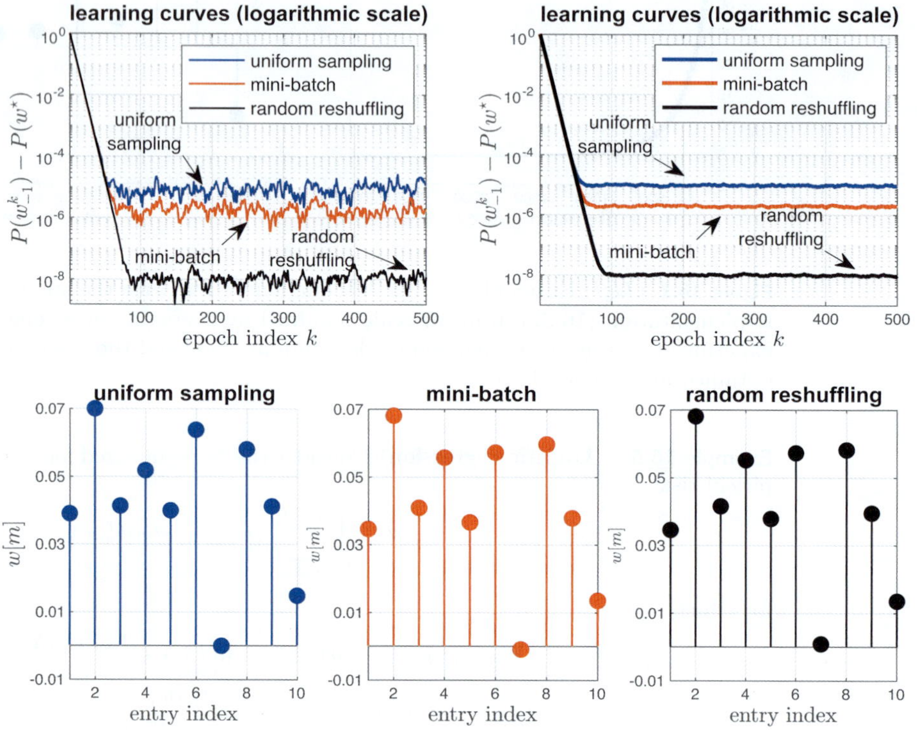

Figure 16.2 (*Top left*) Learning curves $P(w^k_{-1})$ relative to the minimum risk value $P(w^\star)$ in normalized logarithmic scale for the stochastic gradient implementation (16.47) under uniform sampling and random reshuffling, in addition to the learning curve for the mini-batch implementation. (*Top right*) The learning curves are smoothed over $L = 100$ experiments. (*Bottom*) Limiting values for the weight iterates under three data sampling policies.

The plot on the left in the top row of the figure shows the evolution of these values relative to the minimum risk value $P(w^\star)$ for one experiment using $K = 500$ epochs. The noisy variations in these learning curves are a reflection of the stochastic nature of the updates. We repeat this experiment for a total of $L = 100$ times and average the learning curves, with each experiment starting from the same initial condition $\boldsymbol{w}_{-1}$. The result leads to the smoother curves shown in the plot on the right in the top row of the figure. The curves illustrate the improved performance that is delivered by the mini-batch and random reshuffling versions of the stochastic gradient algorithm; these observations will be established analytically in our future derivations – see Table 16.2. The plots in the bottom row show the limiting value of the weight iterates under the three data sampling policies at the end of one experiment involving $K = 500$ runs.

Example 16.6 (**Polyak–Ruppert averaging**) Sometimes a running average step is coupled with the stochastic gradient implementation (16.27) in order to smooth the iterates. Let

$$\bar{w}_{n-2} \triangleq \frac{1}{n}\sum_{m=0}^{n-1} w_{m-1}, \quad n \geq 0 \tag{16.50}$$

which averages all iterates up to w_{n-1}. The smoothed variable $\bar{w}_n$ can be computed recursively as follows:

$$\bar{w}_n = \bar{w}_{n-1} + \frac{1}{n+2}(w_n - \bar{w}_{n-1}), \quad \bar{w}_{-1} = w_{-1} \tag{16.51}$$

and the stochastic gradient recursion (16.27) is adjusted to

(stochastic gradient algorithm with Polyak–Ruppert averaging)

$$\begin{cases} w_n = w_{n-1} - \mu\nabla_{w^\mathsf{T}} Q(w_{n-1}; \gamma(n), h_n), \quad n \geq 0 \\ \bar{w}_n = \bar{w}_{n-1} + \dfrac{1}{n+2}(w_n - \bar{w}_{n-1}) \end{cases} \tag{16.52}$$

where $\bar{w}_n$ is taken as the output variable.

Example 16.7 **(Recommender systems and matrix factorization)** We provide an example with a nonconvex risk function, which is widely used in the design of *recommender systems*. These are automated systems used to suggest recommendations to users for products based on their past preferences and the preferences of other similar users. Such systems are widely used by online business sites. We motivate the approach by considering the example of a streaming movie service.

Assume there are U users, labeled $u = 1, 2, \ldots, U$, and I items (i.e., movies), labeled $i = 1, 2, \ldots, I$. Based on past interactions between the users and the service provider, the users have provided ratings for different movies (say, on a numerical scale from 1= poor to 5 = excellent). Table 16.1 shows one example of the type of information that is available. Each row corresponds to a user u, and each column corresponds to a movie item, i. The table shows data for nine movies and seven users. Usually, users rate only some of the movies; they may not have watched all movies and they may not provide feedback on all movies they watch. For this reason, some entries in the table are marked with question marks to indicate that these ratings are missing.

Table 16.1 Ratings provided by users for some movie items from the service provider.

user	M1	M2	M3	M4	M5	M6	M7	M8	M9
U1	3	?	4	?	1	5	?	?	1
U2	3	4	?	?	2	?	4	4	?
U3	?	5	2	1	1	2	4	5	3
U4	?	?	?	4	?	?	?	2	5
U5	3	2	4	?	3	3	3	3	?
U6	?	3	3	2	2	?	1	?	1
U7	?	1	?	?	?	3	?	2	4

We collect all ratings into a $U \times I$ user–item (or ratings) matrix $R = [r_{ui}]$; it contains all the scores from the table, with r_{ui} representing the score given by user u for item i. Some entries in the matrix R will be missing and we mark them by a question mark. We denote the set of available entries in R by $\mathcal{R}$. Thus, when we write $(u, i) \in \mathcal{R}$ we mean that entry r_{ui} has a valid ratings score. The objective of a recommender system is to predict what ratings users are likely to provide in place of the question marks. For example, referring to the table, we would like to know what ratings user U1 is likely to provide to movies M2, M4, M7, and M8. Based on these predictions, the service provider will then recommend some movies to the user. There are several methods that can be used to predict the numerical values for the missing entries (i.e., to perform what is known as *imputation* or matrix completion). Here, we follow one approach known as *collaborative filtering*. It is based on exploiting relationships between users and using a convenient matrix factorization.

We assume that each item i can be represented by a feature vector $h_i \in \mathbb{R}^M$; the entries of this vector are called *latent* variables because they are hidden and will need to be discovered or learned. For example, for the case of movies, the entries of h_i could be some explicit attributes that relate to the type of movie (comedy, action, thriller, etc.), the duration of the movie, if the movie has won any awards, or other more implicit attributes. Since the latent variables will be discovered by the matrix factorization approach, they may not relate directly to explicit attributes.

We further assume that each user u employs a weight vector $w_u \in \mathbb{R}^M$ to arrive at its ratings. The entries of this vector scale different attributes for the item (or movie) differently. Some users prefer comedy movies over suspense movies, or shorter movies over longer movies. If we happen to know the feature representation h_i for some item i, then we model the rating process used by user u as computing the inner product:

$$r_{ui} \approx h_i^{\mathsf{T}} w_u - \theta \tag{16.53}$$

where θ models some bias term. For example, some users may provide consistently higher-than-average ratings, or some items (movies) may be perceived consistently as being superior to other movies. These perceptions can bias the rating process. To be more specific, we should split the bias into two sources: one arises from user-related biases (users behave differently) and the second arises from item-related biases (some items elicit different types of reactions from users, perhaps because they are promoted more strongly than other movies). For this reason, it is common to replace the above ratings generation model by one of the form:

$$r_{ui} \approx h_i^{\mathsf{T}} w_u - \theta_u - \alpha_i, \quad u = 1, 2, \ldots, U, \quad i = 1, 2, \ldots I \tag{16.54}$$

with two scalar bias terms $\{\theta_u, \alpha_i\}$: one by the user and the other by the item. If we collect the feature vectors into an $M \times I$ matrix H and all user models into a $U \times M$ matrix W, a $U \times 1$ vector θ_U, and an $I \times 1$ vector α:

$$H = \begin{bmatrix} h_1 & h_2 & \ldots & h_I \end{bmatrix}, \; W = \begin{bmatrix} w_1^{\mathsf{T}} \\ w_2^{\mathsf{T}} \\ \vdots \\ w_U^{\mathsf{T}} \end{bmatrix}, \; \theta = \begin{bmatrix} \theta_1 \\ \theta_2 \\ \vdots \\ \theta_U \end{bmatrix}, \; \alpha = \begin{bmatrix} \alpha_1 \\ \alpha_2 \\ \vdots \\ \alpha_I \end{bmatrix} \tag{16.55}$$

then expression (16.54) amounts to assuming that the ratings matrix R is generated according to the model:

$$R \approx WH - \theta \mathbb{1}_I^{\mathsf{T}} - \mathbb{1}_U \alpha^{\mathsf{T}} = \begin{bmatrix} W & -\theta & \mathbb{1}_U \end{bmatrix} \begin{bmatrix} H \\ \mathbb{1}_I^{\mathsf{T}} \\ -\alpha^{\mathsf{T}} \end{bmatrix} \tag{16.56}$$

This expression factors R into the product of two matrices; the first has $M+2$ columns and the second has $M+2$ rows. If we succeed in determining the quantities $\{W, H, \theta, \alpha\}$, then we can use relation (16.56) to predict the ratings at all locations in R. For this purpose, we will minimize the following regularized least-squares risk function:

$$\left\{\widehat{w}_u, \widehat{h}_i, \widehat{\theta}_u, \widehat{\alpha}_i\right\} = \underset{\{w_u, h_i, \theta_u, \alpha_i\}}{\text{argmin}} \left\{ \sum_{u=1}^{U} \rho \|w_u\|^2 + \sum_{i=1}^{I} \rho \|h_i\|^2 + \sum_{(u,i)\in\mathcal{R}} \left(r_{ui} - h_i^{\mathsf{T}} w_u + \theta_u + \alpha_i\right)^2 \right\} \tag{16.57}$$

where the last sum is over the valid indices $(u, i) \in \mathcal{R}$. The above risk function is nonconvex because of the products $h_i^{\mathsf{T}} w_u$. We can approximate the solution by means of a stochastic gradient implementation, which takes the form shown in listing (16.58). The entries of the initial iterates $\boldsymbol{w}_{u,-1}$ and $\boldsymbol{h}_{i,-1}$ are selected at random from a uniform distribution in the range $[0, 1/\sqrt{M}]$.

Stochastic gradient algorithm applied to recommender problem (16.57).

given valid ratings in locations $(u, i) \in \mathcal{R}$;
start from arbitrary $\{\boldsymbol{w}_{u,-1}, \boldsymbol{h}_{i,-1}, \boldsymbol{\theta}_u(-1), \boldsymbol{\alpha}_i(-1)\}$.

repeat until convergence over $m \geq 0$**:**

select a random entry $(\boldsymbol{u}, \boldsymbol{i}) \in \mathcal{R}$

$$\begin{aligned}
\boldsymbol{e}(m) &= \boldsymbol{r_{ui}} - \boldsymbol{h}_{\boldsymbol{i},m-1}^{\mathsf{T}} \boldsymbol{w}_{\boldsymbol{u},m-1} + \boldsymbol{\theta}_{\boldsymbol{u}}(m-1) + \boldsymbol{\alpha}_{\boldsymbol{i}}(m-1) \\
\boldsymbol{w}_{\boldsymbol{u},m} &= (1 - 2\mu\rho)\boldsymbol{w}_{\boldsymbol{u},m-1} + 2\mu \boldsymbol{h}_{\boldsymbol{i},m-1}\boldsymbol{e}(m) \\
\boldsymbol{h}_{\boldsymbol{i},m} &= (1 - 2\mu\rho)\boldsymbol{h}_{\boldsymbol{i},m-1} + 2\mu \boldsymbol{w}_{\boldsymbol{u},m-1}\boldsymbol{e}(m) \\
\boldsymbol{\theta}_{\boldsymbol{u}}(m) &= \boldsymbol{\theta}_{\boldsymbol{u}}(m-1) - 2\mu \boldsymbol{e}(m) \\
\boldsymbol{\alpha}_{\boldsymbol{i}}(m) &= \boldsymbol{\alpha}_{\boldsymbol{i}}(m-1) - 2\mu \boldsymbol{e}(m)
\end{aligned} \tag{16.58}$$

end

return $\{w_u^\star, h_i^\star, \theta_u^\star, \alpha_i^\star\}$.

In the above listing, the term $w_{u,m}$ represents the estimate for w_u at iteration m; likewise for $h_{i,m}$, $\theta_u(m)$, and $\alpha_i(m)$. It is useful to note that although the recursions for updating θ_u and α_i look similar, these variables will generally be updated at different instants. This is because the same i will appear under different u values, and the same u will appear with different i values. Later, in Example 50.6, we revisit this problem and solve it by applying an alternating least-squares solution. We also revisit the same problem in Example 68.2 and solve it by employing variational autoencoders.

We simulate recursions (16.58) by generating a random ratings matrix R with $U = 10$ users and $I = 10$ items. The scores are integer numbers in the range $1 \leq r \leq 5$, and unavailable scores are indicated by the symbol ?:

$$R = \begin{bmatrix}
5 & 3 & 2 & 2 & ? & 3 & 4 & ? & 3 & 3 \\
5 & 4 & 1 & 3 & 1 & 4 & 4 & ? & 3 & ? \\
3 & 5 & ? & 2 & 1 & 5 & 4 & 1 & 4 & 1 \\
? & 2 & 3 & 4 & 4 & 5 & 2 & 5 & 1 & 1 \\
2 & 1 & 2 & 2 & 1 & 5 & 1 & 4 & 1 & ? \\
? & 2 & 1 & 3 & ? & ? & 5 & 3 & 3 & 5 \\
3 & 4 & ? & 2 & 5 & 5 & 3 & 2 & ? & 4 \\
4 & 5 & 3 & 4 & 2 & 2 & 1 & ? & 5 & 5 \\
2 & 4 & 2 & 5 & ? & 1 & 1 & 3 & 1 & 4 \\
? & 1 & 4 & 4 & 3 & ? & 5 & 2 & 4 & 3
\end{bmatrix} \tag{16.59}$$

We set $M = 5$ (feature vectors of size 5) and generate uniform random initial conditions for the variables $\{\boldsymbol{w}_{u,-1}, \boldsymbol{h}_{i,-1}, \boldsymbol{\theta}_u(-1), \boldsymbol{\alpha}_i(-1)\}$ in the open interval $(0, 1)$. We set $\mu = 0.0001$ and $\rho = 0$ (no regularization). We normalize the entries of R to lie in the range $[0,1]$ by replacing each numerical entry r by the value $r \leftarrow (r-1)/4$ where the denominator is the score range (highest value minus lowest value) and the lowest score is subtracted from the numerator. We run a large number of iterations until sufficient convergence is attained. Specifically, we run $K = 50,000$ epochs with the data randomly reshuffled at the beginning of each run. At the end of the simulation, we use the parameters $\{w_u^\star, h_i^\star, \theta_u^\star, \alpha_i^\star\}$ to estimate each entry of R using

$$\widehat{r}_{ui} = (h_i^\star)^\mathsf{T} w_u^\star - \theta_u^\star - \alpha_i^\star \tag{16.60}$$

We undo the normalization by replacing each of these predicted values by $\widehat{r}_{ui} \leftarrow 4\,\widehat{r}_{ui}+1$ and rounding each value to the closest integer; scores above 5 are saturated at 5 and scores below 1 are fixed at 1. The result is the matrix $\widehat{R}$ shown below, where we indicate the scores predicted for the unknown entries in red; we also indicate in blue those locations where the estimated scores differ by one level from the original scores:

$$\widehat{R} = \begin{bmatrix} 5 & 3 & 2 & 2 & \mathbf{2} & 3 & 4 & \mathbf{2} & 3 & 3 \\ 5 & 4 & 1 & 3 & 1 & 4 & 4 & \mathbf{3} & 3 & \mathbf{3} \\ 3 & \mathbf{4} & \mathbf{4} & 2 & 1 & 5 & \mathbf{3} & 1 & 4 & 1 \\ \mathbf{2} & 2 & 3 & 4 & 4 & 5 & 2 & 5 & 1 & 1 \\ 2 & \mathbf{2} & 2 & 2 & 1 & 5 & \mathbf{2} & 4 & 1 & \mathbf{1} \\ \mathbf{4} & 2 & 1 & 3 & \mathbf{5} & \mathbf{3} & 5 & 3 & 3 & 5 \\ 3 & 4 & \mathbf{2} & 2 & 5 & 5 & 3 & 2 & \mathbf{4} & 4 \\ \mathbf{3} & 5 & 3 & 4 & 2 & 2 & \mathbf{2} & \mathbf{1} & 5 & 5 \\ 2 & 4 & 2 & 5 & \mathbf{2} & 1 & 1 & 3 & 1 & 4 \\ \mathbf{4} & 1 & 4 & 4 & 3 & \mathbf{1} & 5 & 2 & 4 & 3 \end{bmatrix} \tag{16.61}$$

Figure 16.3 provides a color-coded representation of the entries of the original matrix R on the left, with the locations of the missing entries highlighted by red squares, and the recovered matrix $\widehat{R}$ on the right.

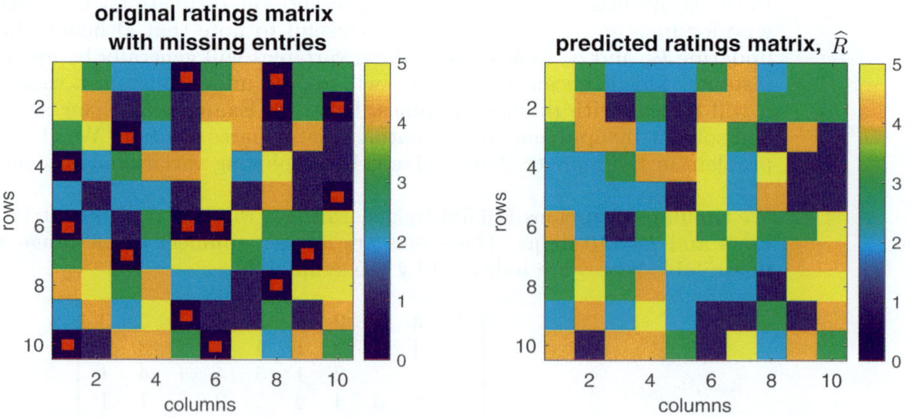

Figure 16.3 Color-coded representation of the entries of the original matrix R with missing entries (*left*) and the recovered matrix $\widehat{R}$ (*right*).

We further denote the risk value at the start of each epoch of index k by

$$P(k) \triangleq \sum_{u=1}^{U} \rho\|w_u\|^2 + \sum_{i=1}^{I} \rho\|h_i\|^2 + \sum_{(u,i)\in\mathcal{R}} \Big(r_{ui} - h_i^\mathsf{T} w_u + \theta_u + \alpha_i\Big)^2 \tag{16.62}$$

where the parameters on the right-hand side are set to the values at the start of epoch k. Figure 16.4 plots the evolution of the risk curve (normalized by its maximum value so that its peak value is set to 1.

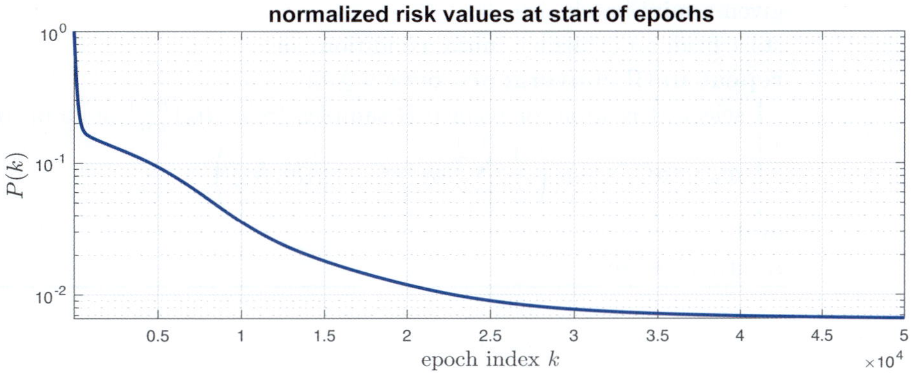

Figure 16.4 Evolution of the risk curve (16.62) with its peak value normalized to 1.

16.2 STOCHASTIC SUBGRADIENT ALGORITHM

The same arguments used for the derivation of the stochastic gradient algorithm and its mini-batch version can be applied to nonsmooth risks (i.e., risks with points of nondifferentiability) to arrive at the *stochastic subgradient algorithm*. The main difference is that gradients will now be replaced by subgradients:

$$\nabla_{w^{\mathsf{T}}} Q(w;\gamma,h) \;\; \text{replaced by} \;\; s_Q(w;\gamma,h) \tag{16.63}$$

where $s_Q(w;\gamma,h)$ refers to a subgradient construction for the loss function $Q(w,\cdot)$ evaluated at the data point (γ,h). The substitution leads to listings (16.64) and (16.65) for the minimization of empirical or stochastic risks using instantaneous or mini-batch subgradient approximations.

Stochastic subgradient algorithm for minimizing (16.2a) or (16.2b).

given dataset $\{\gamma(m),h_m\}_{m=0}^{N-1}$ or streaming data $(\gamma(n),h_n)$;
start from an arbitrary initial condition, $\boldsymbol{w}_{-1}$.
repeat until convergence over $n \geq 0$**:**
  select at random or receive a sample $(\boldsymbol{\gamma}(n),\boldsymbol{h}_n)$ at iteration n;
  $\boldsymbol{w}_n = \boldsymbol{w}_{n-1} - \mu\, s_Q(\boldsymbol{w}_{n-1};\boldsymbol{\gamma}(n),\boldsymbol{h}_n)$
end
return $w^\star \leftarrow \boldsymbol{w}_n$.

(16.64)

Mini-batch stochastic subgradient algorithm for minimizing (16.2a) or (16.2b).

given dataset $\{\gamma(m), h_m\}_{m=0}^{N-1}$ or streaming data $(\gamma(n), h_n)$;
given a mini-batch size, B;
start from an arbitrary initial condition, $\boldsymbol{w}_{-1}$.
repeat until convergence over $n \geq 0$**:**
 select at random or receive B samples $\{\boldsymbol{\gamma}(b), \boldsymbol{h}_b\}_{b=0}^{B-1}$ at iteration n;

$$\boldsymbol{w}_n = \boldsymbol{w}_{n-1} - \mu \left(\frac{1}{B} \sum_{b=0}^{B-1} s_Q(\boldsymbol{w}_{n-1}; \boldsymbol{\gamma}(b), \boldsymbol{h}_b) \right)$$

end
return $w^\star \leftarrow \boldsymbol{w}_n$.

(16.65)

If desired, and as was explained earlier in (14.99), we can incorporate exponential smoothing into the operation of the stochastic subgradient algorithm. The result is shown in listing (16.66), where the value of the positive scalar κ is smaller than but close to 1.

Mini-batch stochastic subgradient algorithm with exponential smoothing for minimizing (16.2a) or (16.2b).

given dataset $\{\gamma(m), h_m\}_{m=0}^{N-1}$ or streaming data $(\gamma(n), h_n)$;
given a mini-batch size, B;
select a positive scalar κ close to but smaller than 1;
start from an arbitrary initial condition, $\boldsymbol{w}_{-1}$;
start from $\bar{\boldsymbol{w}}_0 = \boldsymbol{w}_{-1}$;
start from $S_0 = 0$;
repeat until convergence over $n \geq 0$**:**
 select at random or receive B samples $\{\boldsymbol{\gamma}(b), \boldsymbol{h}_b\}_{b=0}^{B-1}$ at iteration n;

$$\boldsymbol{w}_n = \boldsymbol{w}_{n-1} - \mu \left(\frac{1}{B} \sum_{b=0}^{B-1} s_Q(\boldsymbol{w}_{n-1}; \boldsymbol{\gamma}(b), \boldsymbol{h}_b) \right)$$

$$S_{n+1} = \kappa S_n + 1;$$

$$\bar{\boldsymbol{w}}_{n+1} = \left(1 - \frac{1}{S_{n+1}}\right) \bar{\boldsymbol{w}}_n + \frac{1}{S_{n+1}} \boldsymbol{w}_n$$

end
return $w^\star \leftarrow \bar{\boldsymbol{w}}_{n+1}$.

(16.66)

Example 16.8 **(ℓ_2-regularized hinge loss)** Consider the ℓ_2-regularized hinge loss function:

$$Q(w; \gamma, h) = \rho\|w\|^2 \; + \; \max\left\{0, 1 - \gamma h^\mathsf{T} w\right\} \tag{16.67}$$

We already know that one subgradient construction for it is

$$s_Q(w; \gamma, h) \;=\; 2\rho w - \gamma h \, \mathbb{I}\left[\gamma h^\mathsf{T} w \leq 1\right] \tag{16.68}$$

Substituting into (16.64), we arrive at the stochastic subgradient implementation:

$$\boldsymbol{w}_n = (1 - 2\mu\rho)\boldsymbol{w}_{n-1} + \mu\,\boldsymbol{\gamma}(n)\boldsymbol{h}_n \, \mathbb{I}\left[\boldsymbol{\gamma}(n)\boldsymbol{h}_n^\mathsf{T}\boldsymbol{w}_{n-1} \leq 1\right] \tag{16.69}$$

We illustrate the performance of algorithm (16.69) in Fig. 16.5, which shows the normalized learning curves in logarithmic scale under both random reshuffling and uniform sampling with and without smoothing, in addition to the mini-batch implementation.

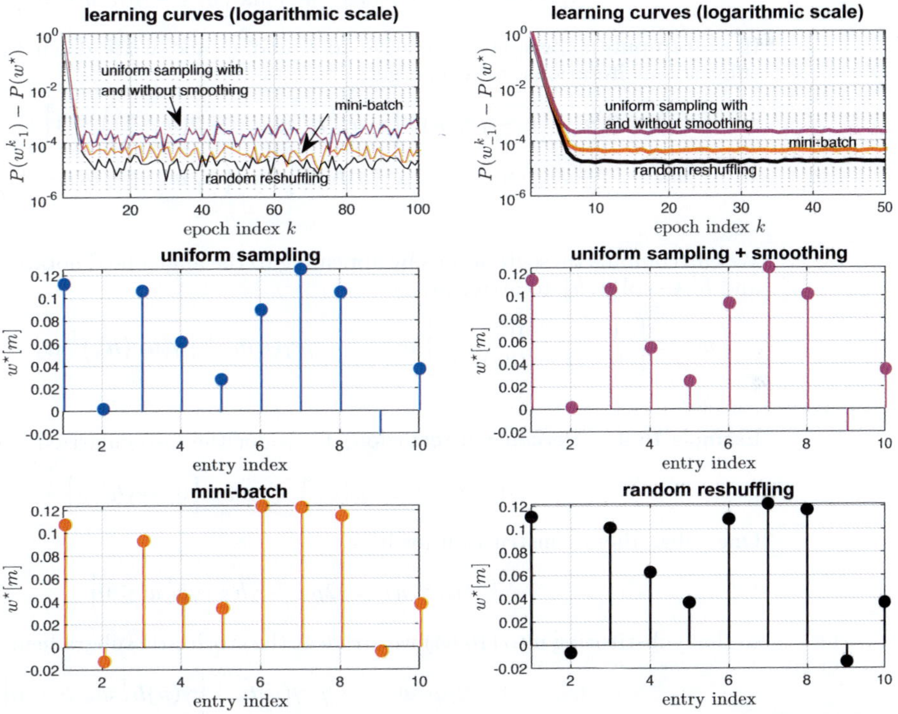

Figure 16.5 (*Top row*) Learning curves $P(w_{-1}^k)$ relative to the minimum risk value $P(w^\star)$ in normalized logarithmic scale for the stochastic subgradient implementation (16.64) under random reshuffling and uniform sampling with and without smoothing, in addition to a mini-batch implementation. (*Lower rows*) Limit values for the weight iterates obtained under different data sampling policies.

The simulation uses $\rho = 1$, $\mu = 0.001$, $\kappa = 0.95$, and $M = 10$. It generates $N = 500$ random pairs of data $\{\gamma(m), h_m\}$ according to the logistic model described earlier in Example 16.5. A total of $K = 500$ epochs are run over the data. The learning curves are plotted in normalized logarithmic scale in line with construction (16.49), where $w^\star$ is approximated by the limit value of the weight iterate after sufficient convergence. The mini-batch implementation employs mini-batches of size $B = 5$ samples. Each learning curve is generated by plotting the values of the risk function at the start of each epoch, namely, $P(w_{-1}^k)$. The plot on the left in the top row of the figure shows the evolution of these values relative to the minimum risk value $P(w^\star)$ for one run of the algorithm over the first 100 epochs. The noisy variations in these learning curves are a reflection of the stochastic nature of the updates. We repeat this experiment for a total of $L = 100$ times and average the learning curves over these experiments. Each experiment starts from the same initial condition w_{-1}. The result leads to the smoother curves shown in the plot on the right in the top row of the figure (for the first 50 epochs). The curves illustrate that the mini-batch and random reshuffling versions of the stochastic subgradient algorithm lead to improved steady-state performance. The plots in the bottom two rows show the limiting value of the weight iterates under four data sampling policies.

Consider next a variation of the hinge loss that incorporates an offset parameter θ:

$$Q(w, \theta; \gamma, h) = \rho\|w\|^2 + \max\left\{0,\, 1 - \gamma(h^{\mathsf T} w - \theta)\right\} \tag{16.70}$$

If we compute subgradients relative to θ and w and write down the corresponding subgradient iterations, we would get:

$$\boldsymbol{\theta}(n) = \boldsymbol{\theta}(n-1) - \mu\boldsymbol{\gamma}(n)\,\mathbb{I}\left[\boldsymbol{\gamma}(n)\left(\boldsymbol{h}_n^{\mathsf T}\boldsymbol{w}_{n-1} - \boldsymbol{\theta}(n-1)\right) \le 1\right] \tag{16.71a}$$

$$\boldsymbol{w}_n = (1 - 2\mu\rho)\,\boldsymbol{w}_{n-1} + \mu\boldsymbol{\gamma}(n)\boldsymbol{h}_n\,\mathbb{I}\left[\boldsymbol{\gamma}(n)\left(\boldsymbol{h}_n^{\mathsf T}\boldsymbol{w}_{n-1} - \boldsymbol{\theta}(n-1)\right) \le 1\right] \tag{16.71b}$$

We can combine these iterations by appealing to the augmented notation $w' \leftarrow \text{col}\{-\theta, w\}$ and $h' \leftarrow \text{col}\{1, h\}$ to arrive at:

$$\boldsymbol{w}_n' = \begin{bmatrix} 1 & 0 \\ 0 & (1-2\mu\rho)I_M \end{bmatrix} \boldsymbol{w}_{n-1}' + \mu\boldsymbol{\gamma}(n)\boldsymbol{h}_n'\,\mathbb{I}\left[\boldsymbol{\gamma}(n)\left(\boldsymbol{h}_n'\right)^{\mathsf T}\boldsymbol{w}_{n-1}' \le 1\right] \tag{16.72}$$

Example 16.9 **(Perceptron recursion)** Consider the ℓ_2-regularized perceptron loss:

$$Q(w; \boldsymbol{\gamma}, \boldsymbol{h}) = \rho\|w\|^2 + \max\left\{0,\, -\boldsymbol{\gamma}\boldsymbol{h}^{\mathsf T} w\right\} \tag{16.73}$$

One subgradient construction for it is

$$s_Q(w; \gamma, h) = 2\rho w - \gamma h\,\mathbb{I}\left[\gamma h^{\mathsf T} w \le 0\right] \tag{16.74}$$

so that substituting into (16.64) we arrive at the stochastic subgradient implementation:

$$\boldsymbol{w}_n = (1-2\mu\rho)\boldsymbol{w}_{n-1} + \mu\,\boldsymbol{\gamma}(n)\boldsymbol{h}_n\,\mathbb{I}\left[\boldsymbol{\gamma}(n)\boldsymbol{h}_n^{\mathsf T}\boldsymbol{w}_{n-1} \le 0\right] \tag{16.75}$$

We can also consider a variation with an offset parameter:

$$Q(w, \theta; \boldsymbol{\gamma}, \boldsymbol{h}) = \rho\|w\|^2 + \max\left\{0,\, -\boldsymbol{\gamma}(\boldsymbol{h}^{\mathsf T} w - \theta)\right\} \tag{16.76}$$

If we compute subgradients relative to θ and w and write down the corresponding subgradient iterations, we get:

$$\boldsymbol{\theta}(n) = \boldsymbol{\theta}(n-1) - \mu\boldsymbol{\gamma}(n)\,\mathbb{I}\left[\boldsymbol{\gamma}(n)\left(\boldsymbol{h}_n^{\mathsf T}\boldsymbol{w}_{n-1} - \boldsymbol{\theta}(n-1)\right) \le 0\right] \tag{16.77a}$$

$$\boldsymbol{w}_n = (1 - 2\mu\rho)\,\boldsymbol{w}_{n-1} + \mu\boldsymbol{\gamma}(n)\boldsymbol{h}_n\,\mathbb{I}\left[\boldsymbol{\gamma}(n)\left(\boldsymbol{h}_n^{\mathsf T}\boldsymbol{w}_{n-1} - \boldsymbol{\theta}(n-1)\right) \le 0\right] \tag{16.77b}$$

We can combine these iterations by appealing to the augmented notation $w' \leftarrow \text{col}\{-\theta, w\}$ and $h' \leftarrow \text{col}\{1, h\}$ to arrive at:

$$\boldsymbol{w}'_n = \begin{bmatrix} 1 & 0 \\ 0 & (1-2\mu\rho)I_M \end{bmatrix} \boldsymbol{w}'_{n-1} + \mu\boldsymbol{\gamma}(n)\boldsymbol{h}'_n \, \mathbb{I}\left[\boldsymbol{\gamma}(n)\left(\boldsymbol{h}'_n\right)^\mathsf{T}\boldsymbol{w}'_{n-1} \leq 0\right] \tag{16.78}$$

Example 16.10 **(LASSO or basis pursuit)** Consider the ℓ_1-regularized quadratic loss:

$$Q(w;\gamma,h) = \alpha\|w\|_1 + (\gamma - h^\mathsf{T} w)^2 \tag{16.79}$$

One subgradient construction for it is

$$s_Q(w;\gamma,h) = \alpha\,\text{sign}(w) - 2h(\gamma - h^\mathsf{T} w) \tag{16.80}$$

so that substituting into (16.64), we arrive at:

$$\boldsymbol{w}_n = \boldsymbol{w}_{n-1} - \mu\,\text{sign}(\boldsymbol{w}_{n-1}) + 2\mu\boldsymbol{h}_n(\boldsymbol{\gamma}(n) - \boldsymbol{h}_n^\mathsf{T}\boldsymbol{w}_{n-1}) \tag{16.81}$$

Example 16.11 **(Switching expectation and sub-differentiation)** The stochastic algorithms of this section are applicable to empirical and stochastic risks, as in (16.3a) and (16.3b). In the latter case, we need to justify switching the order of the expectation and sub-differentiation operators in order to write (in a manner similar to (16.4)):

$$\partial_w P(w) = \partial_w\left(\mathbb{E}\,Q(w;\boldsymbol{\gamma},\boldsymbol{h})\right) \overset{(a)}{=} \mathbb{E}\left(\partial_w Q(w;\boldsymbol{\gamma},\boldsymbol{h})\right) \tag{16.82}$$

Step (a) is possible under conditions that are generally valid for our cases of interest – see the explanation in Lemma 16.1 in the appendix. In essence, the switching is possible when the loss function $Q(w;\cdot)$ is convex and bounded in neighborhoods where the subgradients are evaluated.

16.3 STOCHASTIC PROXIMAL GRADIENT ALGORITHM

We motivate next stochastic approximations for the proximal gradient method by following similar arguments to those used for gradient descent and the subgradient method in the last two sections. First, recall from the discussion in Section 15.1 that the proximal gradient algorithm is suitable for minimizing risk functions $P(w)$ that can be split into the sum of two convex components:

$$P(w) = q(w) + E(w) \tag{16.83}$$

where $E(w)$ is first-order differentiable and $q(w)$ is nonsmooth. In the empirical case, the component $E(w)$ is expressed as the sample average

$$E(w) = \frac{1}{N}\sum_{m=0}^{N-1} Q_u\Big(w;\gamma(m),h_m\Big) \tag{16.84}$$

in terms of some convex loss function $Q_u(w,\cdot)$. The proximal gradient algorithm was listed in (15.10) and had the following form:

$$\left\{ \begin{array}{lcl} z_n & = & w_{n-1} - \mu \nabla_{w^\mathsf{T}} E(w_{n-1}) \\ w_n & = & \text{prox}_{\mu q}(z_n) \end{array} \right. \tag{16.85a}$$

where

$$\nabla_{w^\mathsf{T}} E(w) \stackrel{\Delta}{=} \frac{1}{N} \sum_{m=0}^{N-1} \nabla_{w^\mathsf{T}} Q_u\Big(w; \gamma(m), h_m\Big) \tag{16.85b}$$

When the size N of the dataset is large, we observe again that it becomes impractical to evaluate the gradient of $E(w)$. We therefore resort to *stochastic* approximations, where this gradient is approximated either by an instantaneous value or by a mini-batch calculation. In the first case, we arrive at the stochastic proximal gradient algorithm listed in (16.87), and in the second case we arrive at (16.88). The same listings apply to the minimization of stochastic risks when $E(w)$ is defined instead as

$$E(w) = \mathbb{E}\, Q_u(w; \boldsymbol{\gamma}, \boldsymbol{h}) \tag{16.86}$$

Stochastic proximal gradient for minimizing $P(w) = q(w) + E(w)$.

given dataset $\{\gamma(m), h_m\}_{m=0}^{N-1}$ or streaming data $(\gamma(n), h_n)$;
start from an arbitrary initial condition, $\boldsymbol{w}_{-1}$.
repeat until convergence over $n \geq 0$**:** (16.87)
  select at random or receive a sample $(\boldsymbol{\gamma}(n), \boldsymbol{h}_n)$ at iteration n;
  $\boldsymbol{z}_n = \boldsymbol{w}_{n-1} - \mu \nabla_{w^\mathsf{T}} Q_u(\boldsymbol{w}_{n-1}; \boldsymbol{\gamma}(n), \boldsymbol{h}_n)$
  $\boldsymbol{w}_n = \text{prox}_{\mu q}(\boldsymbol{z}_n)$
end
return $w^\star \leftarrow \boldsymbol{w}_n$.

Mini-batch stochastic proximal gradient for minimizing $P(w) = q(w) + E(w)$.

given dataset $\{\gamma(m), h_m\}_{m=0}^{N-1}$ or streaming data $(\gamma(n), h_n)$;
given a mini-batch size, B;
start from an arbitrary initial condition, $\boldsymbol{w}_{-1}$.
repeat until convergence over $n \geq 0$**:**
  select at random or receive B samples $\{\boldsymbol{\gamma}(b), \boldsymbol{h}_b\}_{b=0}^{B-1}$ at iteration n;
  $\boldsymbol{z}_n = \boldsymbol{w}_{n-1} - \mu \left(\frac{1}{B} \sum_{b=0}^{B-1} \nabla_{w^\mathsf{T}} Q_u(\boldsymbol{w}_{n-1}; \boldsymbol{\gamma}(b), \boldsymbol{h}_b) \right)$
  $\boldsymbol{w}_n = \text{prox}_{\mu q}(\boldsymbol{z}_n)$
end
return $w^\star \leftarrow \boldsymbol{w}_n$.

(16.88)

Example 16.12 **(LASSO or basis pursuit)** Consider the quadratic risk with elastic-net regularization:

$$P(w) = \alpha\|w\|_1 + \rho\|w\|^2 + \frac{1}{N}\sum_{m=0}^{N-1}\left(\gamma(m) - h_m^{\mathsf{T}} w\right)^2, \quad w \in \mathbb{R}^M \tag{16.89}$$

so that

$$q(w) = \alpha\|w\|_1, \quad Q_u(w;\gamma(n),h_n) = \rho\|w\|^2 + (\gamma(n) - h_n^{\mathsf{T}} w)^2 \tag{16.90}$$

In this case, the stochastic proximal recursion (16.87) becomes

$$\boldsymbol{z}_n = (1-2\mu\rho)\boldsymbol{w}_{n-1} + 2\mu \boldsymbol{h}_n(\boldsymbol{\gamma}(n) - \boldsymbol{h}_n^{\mathsf{T}}\boldsymbol{w}_{n-1}) \tag{16.91a}$$

$$\boldsymbol{w}_n = \mathbb{T}_{\mu\alpha}(\boldsymbol{z}_n) \tag{16.91b}$$

in terms of the soft-thresholding operator applied to $\boldsymbol{z}_n$.

We illustrate the performance of the algorithm in Fig. 16.6, which shows the normalized learning curve in logarithmic scale under uniform sampling using $\rho = 0$, $\alpha = 1$, $\mu = 0.001$, and $M = 10$. The simulation generates $N = 500$ random pairs of data $\{\gamma(m), h_m\}$ according to the same linear model (16.39). The minimizer $w^\star$ for the risk (16.89) is estimated by running the batch proximal algorithm (16.85a), which employs the full gradient vector of $P(w)$, for a sufficient number of iterations. The plot on the left in the top row of the figure shows the normalized learning curve in logarithmic scale, where $P(w)$ is evaluated at the start of each epoch. The plot on the right in the first row averages these learning curves over $L = 100$ experiments to generate a smoother curve. The lower plot in the figure shows the limit value of w_n resulting from (16.91a)–(16.91b) and obtained after running $K = 300$ epochs over the data. It is seen that w_n approaches $w^\star$ and that it also exhibits sparsity.

Example 16.13 **(Logistic regression)** Consider an empirical logistic regression risk with elastic-net regularization:

$$P(w) = \alpha\|w\|_1 + \rho\|w\|^2 + \frac{1}{N}\sum_{m=0}^{N-1}\ln\left(1 + e^{-\gamma(m)h_m^{\mathsf{T}} w}\right), \quad w \in \mathbb{R}^M \tag{16.92}$$

so that

$$q(w) = \alpha\|w\|_1, \quad Q_u(w;\gamma(n),h_n) = \rho\|w\|^2 + \ln\left(1 + e^{-\gamma(n)h_n^{\mathsf{T}} w}\right) \tag{16.93}$$

In this case, the stochastic proximal recursion (16.87) becomes

$$\boldsymbol{z}_n = (1-2\mu\rho)\boldsymbol{w}_{n-1} + \mu\frac{\boldsymbol{\gamma}(n)\boldsymbol{h}_n}{1 + e^{\boldsymbol{\gamma}(n)\boldsymbol{h}_n^{\mathsf{T}}\boldsymbol{w}_{n-1}}} \tag{16.94a}$$

$$\boldsymbol{w}_n = \mathbb{T}_{\mu\alpha}(\boldsymbol{z}_n) \tag{16.94b}$$

in terms of the soft-thresholding operator applied to $\boldsymbol{z}_n$.

Example 16.14 **(Stochastic projection gradient)** Consider the constrained optimization problem:

$$w^\star = \underset{w\in\mathbb{R}^M}{\text{argmin}}\left\{E(w) \triangleq \frac{1}{N}\sum_{m=0}^{N-1} Q_u(w;\gamma(m),h_m)\right\}, \quad \text{subject to } w \in \mathcal{C} \tag{16.95}$$

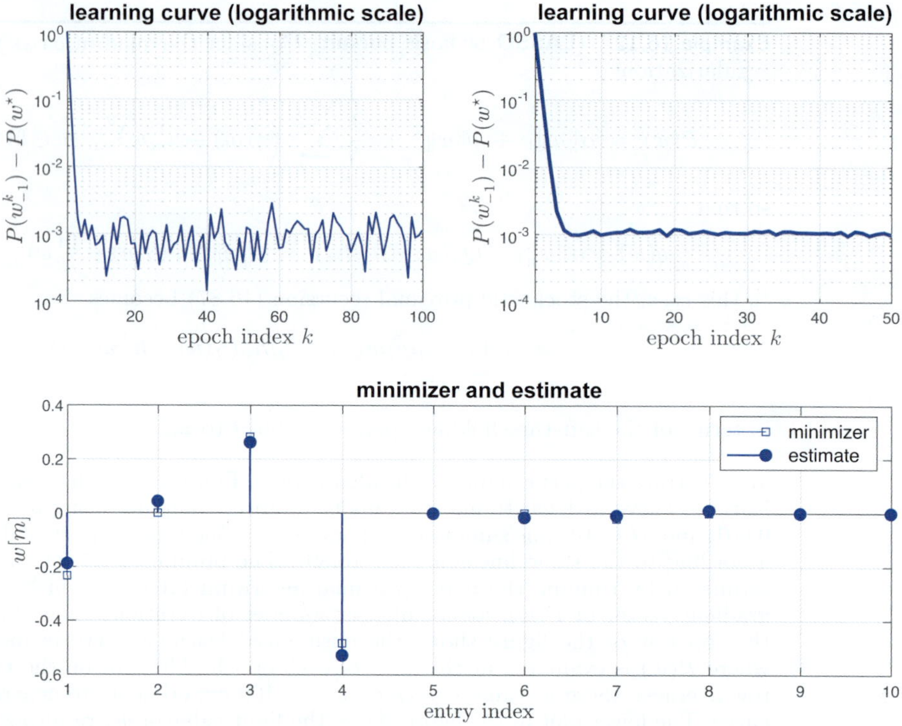

Figure 16.6 (*Top left*) Learning curve $P(w^k_{-1})$ relative to the minimum risk value $P(w^\star)$ in normalized logarithmic scale for the stochastic proximal gradient implementation (16.91a)–(16.91b) under uniform sampling. (*Top right*) Learning curve obtained by averaging over 100 experiments. (*Bottom*) The limiting value of the weight iterate in comparison to the minimizer $w^\star$.

where $\mathcal{C}$ is a closed convex set. The projection gradient algorithm (15.51) was shown to be a special case of the proximal gradient method and it uses the gradient of $E(w)$. We can approximate this gradient by using either an instantaneous sample or a mini-batch calculation. The former case leads to listing (16.96), where $\mathcal{P}_C(x)$ denotes the projection of $x \in \mathbb{R}^M$ onto $\mathcal{C}$.

Stochastic projection gradient algorithm for solving (16.95) and (16.97).

given dataset $\{\gamma(m), h_m\}_{m=0}^{N-1}$ or streaming data $(\gamma(n), h_n)$;
start from any initial condition, w_{-1}.
repeat until convergence over $n \geq 0$:
  select at random or receive $(\boldsymbol{\gamma}(n), \boldsymbol{h}_n)$
  $\boldsymbol{w}_n = \mathcal{P}_C\Big(\boldsymbol{w}_{n-1} - \mu \nabla_{w^\mathsf{T}} Q_u(\boldsymbol{w}_{n-1}; \boldsymbol{\gamma}(n), \boldsymbol{h}_n)\Big)$
end
return $w^\star \leftarrow \boldsymbol{w}_n$.

(16.96)

This algorithm can also handle stochastic risks of the form

$$w^o = \underset{w\in\mathbb{R}^M}{\text{argmin}} \left\{ E(w) \stackrel{\Delta}{=} \mathbb{E}\, Q_u(w;\boldsymbol{\gamma},\boldsymbol{h}) \right\}, \quad \text{subject to } w \in \mathcal{C} \tag{16.97}$$

In this case, the data samples $(\gamma(n), h_n)$ would stream in successively over time.

Example 16.15 (**Stochastic mirror descent**) Consider the same constrained optimization problems (16.95) or (16.97). The mirror-descent algorithm (15.103) also relies on the gradient of $E(w)$. We can again approximate this gradient by using either an instantaneous sample or a mini-batch calculation. The former case leads to listing (16.98). The listing can be specialized for particular choices of the mirror function $\phi(x)$, such as choosing it as the negative entropy function or as a quadratic function. In a similar manner, we can write down stochastic versions for the lazy mirror-descent and mirror-prox algorithms.

Stochastic mirror-descent for solving (16.95) or (16.97).

given dataset $\{\gamma(m), h_m\}_{m=0}^{N-1}$ or streaming data $(\gamma(n), h_n)$;
choose ν_ϕ-strongly convex mirror function $\phi(w) : \mathcal{C}_\phi \rightarrow \mathbb{R}$;
let $\phi^\star(x) = \sup_w\{w^\mathsf{T} x - \phi(w)\}$ denote its conjugate function;
let $\mathcal{C}' = \mathcal{C} \cap \mathcal{C}_\phi$;
start from an initial condition, $\boldsymbol{w}_{-1} \in \mathcal{C}'$.
repeat until convergence over $n \geq 0$**:** (16.98)
  select at random or receive $(\boldsymbol{\gamma}(n), \boldsymbol{h}_n)$
  $\boldsymbol{b}_n = \nabla_{w^\mathsf{T}}\, \phi(\boldsymbol{w}_{n-1}) - \mu\nabla_{w^\mathsf{T}}\, Q_u(\boldsymbol{w}_{n-1}; \boldsymbol{\gamma}(n), \boldsymbol{h}_n)$
  $\boldsymbol{z}_n = \nabla_{x^\mathsf{T}}\, \phi^\star(\boldsymbol{b}_n)$
  $\boldsymbol{w}_n = \mathcal{P}_{C',\phi}(\boldsymbol{z}_n)$ (Bregman projection)
end
return $w^\star \leftarrow \boldsymbol{w}_n$.

Example 16.16 (**Stochastic coordinate descent**) The same stochastic approximation approach can be applied to coordinate-descent algorithms. It is sufficient to illustrate the construction by considering the randomized proximal version listed earlier in (15.31); similar constructions apply to other variants of coordinate descent. Thus, consider an empirical risk $P(w)$ that separates into the form:

$$P(w) \;=\; E(w) + \sum_{m=1}^{M} q_m(w_m) \tag{16.99a}$$

where the second component is separable over the individual coordinates of w denoted by $\{w_m\}$. Moreover, the smooth component $E(w)$ is expressed as the sample average of loss values:

$$E(w) = \frac{1}{N}\sum_{m=0}^{N-1} Q_u(w;\gamma_m, h_m) \tag{16.99b}$$

We approximate the gradient of $E(w)$ by using either an instantaneous approximation or a mini-batch calculation. The former case leads to (16.101), which is applicable to both empirical risks as in (16.99a)–(16.99b), or to stochastic risks where

$$E(w) = \mathbb{E}\, Q_u(w;\boldsymbol{\gamma},\boldsymbol{h}) \tag{16.100}$$

The main difference is that $(\boldsymbol{\gamma}(n), \boldsymbol{h}_n)$ will now stream in successively over time.

Stochastic randomized proximal coordinate descent for minimizing (16.99a).

given dataset $\{\gamma(m), h_m\}_{m=0}^{N-1}$ or streaming data $(\gamma(n), h_n)$;
start with an arbitrary initial condition $\boldsymbol{w}_{-1}$.
repeat until convergence over $n \geq 0$**:**
  iterate is $\boldsymbol{w}_{n-1} = \text{col}\{\boldsymbol{w}_{n-1,m}\}_{m=1}^{M}$
  select at random or receive $(\boldsymbol{\gamma}(n), \boldsymbol{h}_n)$;
  select a random index $1 \leq m^o \leq M$;
  $\boldsymbol{z}_{n,m^o} = \boldsymbol{w}_{n-1,m^o} - \mu\, \partial Q_u(\boldsymbol{w}_{n-1}; \boldsymbol{\gamma}(n), \boldsymbol{h}_n)/\partial w_{m^o}$
  $\boldsymbol{w}_{n,m^o} = \text{prox}_{\mu q_{m^o}}(\boldsymbol{z}_{n,m^o})$
  keep $\boldsymbol{w}_{n,m} = \boldsymbol{w}_{n-1,m}$ for all $m \neq m^o$
end
return $w^\star \leftarrow \boldsymbol{w}_n$. (16.101)

Example 16.17 **(Stochastic conjugate gradient)** We can similarly devise a stochastic implementation for the Fletcher–Reeves conjugate gradient algorithm (13.87) by approximating the gradient of $P(w)$ using an instantaneous or mini-batch calculation. The former case is listed in (16.102), which can handle both empirical and stochastic risk minimization. In the latter case, the data samples $(\gamma(n), h_n)$ stream in successively over time.

Stochastic Fletcher–Reeves algorithm for minimizing (16.2a) or (16.2b).

given dataset $\{\gamma(m), h_m\}_{m=0}^{N-1}$ or streaming data $(\gamma(n), h_n)$;
start with an arbitrary initial condition $\boldsymbol{w}_{-1}$;
set $\boldsymbol{q}_{-1} = 0$;
repeat until convergence over $n \geq 0$**:**
  select at random or receive $(\boldsymbol{\gamma}(n), \boldsymbol{h}_n)$;
  $\boldsymbol{r}_{n-1} = -\nabla_{w^{\mathsf{T}}} Q(\boldsymbol{w}_{n-1}; \boldsymbol{\gamma}(n), \boldsymbol{h}_n)$
  if $n = 0$ then $\boldsymbol{\beta}_{-1} = 0$
    else $\boldsymbol{\beta}_{n-1} = \|\boldsymbol{r}_{n-1}\|^2/\|\boldsymbol{r}_{n-2}\|^2$
  end
  $\boldsymbol{q}_n = \boldsymbol{r}_{n-1} + \boldsymbol{\beta}_{n-1}\, \boldsymbol{q}_{n-1}$
  find $\boldsymbol{\alpha}_n$ by solving $\min_{\alpha \in \mathbb{R}} Q(\boldsymbol{w}_{n-1} + \alpha \boldsymbol{q}_n)$ using line search.
  $\boldsymbol{w}_n = \boldsymbol{w}_{n-1} + \boldsymbol{\alpha}_n \boldsymbol{q}_n$
end
return $w^\star \leftarrow \boldsymbol{w}_n$. (16.102)

16.4 GRADIENT NOISE

In all stochastic algorithms studied in this chapter, the desired gradient or subgradient search direction is approximated by using either instantaneous or mini-batch calculations. For example, when $P(w)$ is smooth, we used approximations of the form:

$$\text{(instantaneous)} \quad \widehat{\nabla_{w^\mathsf{T}} P}(w) = \nabla_{w^\mathsf{T}} Q(w; \boldsymbol{\gamma}, \boldsymbol{h}) \tag{16.103a}$$

$$\text{(mini-batch)} \quad \widehat{\nabla_{w^\mathsf{T}} P}(w) = \frac{1}{B} \sum_{b=0}^{B-1} \nabla_{w^\mathsf{T}} Q\Big(w; \boldsymbol{\gamma}(b), \boldsymbol{h}_b\Big) \tag{16.103b}$$

and when $P(w)$ is nonsmooth we replaced the gradients of $Q(w; \cdot)$ by subgradients, $s_Q(w; \boldsymbol{\gamma}, \boldsymbol{h})$. We continue with the smooth case for illustration purposes. The difference between the true gradient and its approximation is called *gradient noise* and denoted by

$$\boldsymbol{g}(w) \stackrel{\Delta}{=} \widehat{\nabla_{w^\mathsf{T}} P}(w) - \nabla_{w^\mathsf{T}} P(w) \tag{16.104}$$

The presence of this noise alters the dynamics of optimization algorithms. To see this, the following two relations highlight the difference between the original gradient-descent method and its stochastic version:

$$\text{(gradient descent)} \quad w_n = w_{n-1} - \mu \nabla_{w^\mathsf{T}} P(w_{n-1}) \tag{16.105a}$$

$$\text{(stochastic version)} \quad \boldsymbol{w}_n = \boldsymbol{w}_{n-1} - \mu \nabla_{w^\mathsf{T}} P(\boldsymbol{w}_{n-1}) - \mu \boldsymbol{g}(\boldsymbol{w}_{n-1}) \tag{16.105b}$$

where the gradient noise process appears as a driving term in the second recursion. The noise seeps into the operation of the algorithm and some degradation in performance is expected. For instance, while we were able to show in a previous chapter that the gradient-descent implementation (16.105a) converges to the exact minimizer $w^\star$ for sufficiently small step sizes, we will discover that the stochastic version (16.105b) can only approach a small neighborhood around $w^\star$. Specifically, we will prove in later chapters that, for this case, the MSD $\mathbb{E}\,\|\widetilde{\boldsymbol{w}}_n\|^2$ approaches $O(\mu)$:

$$\limsup_{n\to\infty} \mathbb{E}\,\|\widetilde{\boldsymbol{w}}_n\|^2 = O(\mu) \tag{16.106}$$

Obviously, the smaller μ is, the smaller the size of the limiting error will be. However, small step sizes slow down convergence and there is a need to strike a balance between convergence rate and error size. One way to reduce the size of the error is to employ decaying step sizes; in this case, future results will show that it is possible for stochastic algorithms to converge to the exact minimizer even in the presence of gradient noise. Nevertheless, vanishing step sizes reduce the ability of an optimization algorithm to continue to update and learn, which is problematic for scenarios involving drifting minimizers. Another way to reduce the limiting error, even with constant step sizes, is to resort to variance-reduction techniques. We will discuss these various elements in future chapters and establish performance limits. The derivations are demanding and are not necessary at this stage.

For illustration and comparison purposes, we collect in Tables 16.2 and 16.3 some of the results that will be derived in future analyses for strongly convex risks. The first table assumes constant step sizes, while the second table assumes a vanishing step size of the form $\mu(k) = \tau/k$, with its value decaying with the epoch index $k \geq 1$. The tables also list the conditions under which the results will be derived; these conditions are generally satisfied for our problems of interest. The tables further indicate the locations in the text where the results can be found. Observe that all algorithms deliver exponential (i.e., linear) convergence rates under strong-convexity conditions, and that the size of the limiting neighborhood depends on several factors, including whether uniform sampling or random reshuffling is used, whether the risk function is smooth or not, and on whether mini-batch processing is employed. More results will be discussed in Chapters 19–22.

Remark 16.1. (Big-O and little-o notation) The results in the tables employ the Big-O notation, with its argument being either a sequence, such as λ^k, or a function of the step-size parameter μ. We already explained in Remark 12.3 what the notation means for sequences, namely, it compares the asymptotic growth rate of two sequences. Thus, writing $a_k = O(b_k)$, with a big O for a sequence b_k with positive entries, means that there exists some constant $c > 0$ and index k_o such that $|a_k| \leq cb_k$ for $k > k_o$. This also means that the decay rate of a_k is at least as fast or faster than b_k. On the other hand, the notation $f(\mu) \in O(g(\mu))$ for some positive function $g(\mu)$ means that there exists a constant c independent of μ such that $\lim_{\mu\to 0} |f(\mu)|/g(\mu) \leq c$. In a similar vein, using the little-o notation and writing $f(\mu) \in o(g(\mu))$ means that $\lim_{\mu\to 0} |f(\mu)|/g(\mu) = 0$. Using these definitions, we note for example that

$$\mu \in O(\sqrt{\mu}), \quad 10\mu \in O(\mu), \quad \mu \in o(\sqrt{\mu}), \quad \mu^2 \in o(\mu), \quad \mu^{-1/2} \in o(\mu^{-1}) \tag{16.107}$$

We will use the compact symbols $O(g(\mu))$ and $o(g(\mu))$ as placeholders for the more explicit notation $f(\mu) \in O(g(\mu))$ and $f(\mu) \in o(g(\mu))$, respectively. Also, writing $a(\mu) \leq b(\mu) + o(g(\mu))$ means that there exists $f(\mu) \in o(g(\mu))$ such that $a(\mu) \leq b(\mu) + f(\mu)$. ■

16.5 REGRET ANALYSIS

We will examine the convergence performance of stochastic approximation algorithms in some detail in future chapters by studying the behavior of the mean-square-error, $\mathbb{E}\,\|\widetilde{\boldsymbol{w}}_n\|^2$, and the MSE, $\mathbb{E}\,P(\boldsymbol{w}_n) - P(w^\star)$. In this section, we comment on another approach for the study of these algorithms, which is based on regret analysis. We explain the value of this type of analysis and also some of its limitations.

We have already encountered the regret measure in earlier chapters – see, e.g., expression (12.58) for the gradient-descent algorithm. The definition of the regret needs to be adjusted for *stochastic approximation methods*. It is sufficient to illustrate the construction for the case of smooth loss functions and for the stochastic gradient algorithm.

Table 16.2 Convergence properties of stochastic optimization algorithms with *constant* step sizes. The risk $P(w)$ is ν-strongly convex and the loss is $Q(w;\cdot,\cdot)$. For proximal implementations, we assume $P(w) = q(w) + E(w)$ and denote the loss associated with $E(w)$ by $Q_u(w;\cdot,\cdot)$.

Algorithm	Conditions on risk and loss functions	Asymptotic convergence property	Reference
stochastic gradient algorithm (16.27) (uniform sampling)	$P(w)$: ν-strongly convex $Q(w)$: δ-Lipschitz gradients (18.10a)–(18.10b)	$\mathbb{E}\,P(\boldsymbol{w}_n^k) - P(w^\star) \leq O(\lambda^k) + O(\mu)$ $k \geq 1$: epoch index $n \geq 0$: iteration index	Thm. 19.1
stochastic gradient algorithm (16.27); (random reshuffling)	same as above	$\mathbb{E}\,P(\boldsymbol{w}_n^k) - P(w^\star) \leq O(\lambda^k) + O(\mu^2)$	Thm. 19.5
stochastic gradient algorithm (16.27); (importance sampling)	same as above	$\mathbb{E}\,P(\boldsymbol{w}_n^k) - P(w^\star) \leq O(\lambda^k) + O(\mu)$	Thm. 19.1 Sec. 19.6
stochastic gradient algorithm (16.27) (data streaming)	$P(w)$: ν-strongly convex $Q(w)$: δ-Lipschitz gradients in mean-square sense (18.13a)–(18.13b)	$\mathbb{E}\,P(\boldsymbol{w}_n^k) - P(w^o) \leq O(\lambda^k) + O(\mu)$	Thm. 19.1
stochastic subgradient algorithm with smoothing (16.66), $B = 1$ (uniform sampling)	$P(w)$: ν-strongly convex $Q(w)$: affine-Lipschitz subgradients (18.66a)–(18.66b)	$\mathbb{E}\,P(\boldsymbol{w}_n^k) - P(w^\star) \leq O(\kappa^k) + O(\mu)$ (using κ close to 1)	Eq. (20.68)
stochastic proximal gradient (16.87) (uniform sampling)	$E(w)$: ν-strongly convex $Q_u(w)$: δ-Lipschitz gradients (21.11a)–(21.11b) and (21.13) $P(w) = q(w) + E(w)$	$\mathbb{E}\,P(\boldsymbol{w}_n^k) - P(w^\star) \leq O(\lambda^{k/2}) + O(\sqrt{\mu})$	Thm. 21.1
mini-batch stochastic gradient (16.28)	$P(w)$: ν-strongly convex $Q(w)$: δ-Lipschitz gradients (18.10a)–(18.10b)	$\mathbb{E}\,P(\boldsymbol{w}_n) - P(w^\star) \leq O(\lambda^n) + O(\mu/B)$	Thm. 19.2
mini-batch stochastic subgradient algorithm with smoothing (16.66)	$P(w)$: ν-strongly convex $Q(w)$: affine-Lipschitz subgradients (18.66a)–(18.66b)	$\mathbb{E}\,P(\boldsymbol{w}_n) - P(w^\star) \leq O(\kappa^n) + O(\mu/B)$ (using κ close to 1)	Prob. 20.2
mini-batch proximal gradient (16.88)	$E(w)$: ν-strongly convex $Q_u(w)$: δ-Lipschitz gradients (21.11a)–(21.11b) and (21.13) $P(w) = q(w) + E(w)$	$\mathbb{E}\,P(\boldsymbol{w}_n) - P(w^\star) \leq O(\lambda^{n/2}) + O(\sqrt{\mu/B})$	Thm. 21.2

Table 16.3 Convergence properties of various stochastic optimization algorithms with step size $\mu(k) = \tau/k$, which decays with the epoch index, k. In the listing, n refers to the iteration index and $k \geq 1$ refers to the epoch index. In this table, the risk $P(w)$ is assumed to be ν-strongly convex and the loss function is denoted by $Q(w;\cdot,\cdot)$. For proximal implementations, we assume the risk is split as $q(w) + E(w)$ and denote the loss associated with $E(w)$ by $Q_u(w;\cdot,\cdot)$.

Algorithm	Conditions on risk and loss functions	Asymptotic convergence property	Reference
stochastic gradient algorithm (16.27) (uniform sampling) (decaying step size)	$P(w)$: ν-strongly convex $Q(w)$: δ-Lipschitz gradients (18.10a)–(18.10b) $\mu(k) = \tau/k$	$\mathbb{E}\, P(\boldsymbol{w}_n^k) - P(w^\star) \leq O(1/k)$ $k \geq 1$: epoch index $n \geq 0$: iteration index	Thm. 19.4
stochastic subgradient algorithm (16.64) (uniform sampling) (decaying step size)	$P(w)$: ν-strongly convex $Q(w)$: affine-Lipschitz subgradients (18.66a)–(18.66b) $\mu(k) = \tau/k$	$\mathbb{E}\, P(\boldsymbol{w}_n^k) - P(w^\star) \leq O(1/\sqrt{k}\,)$	Thm. 20.4
stochastic proximal gradient (16.87) (uniform sampling) (decaying step size)	$E(w)$: ν-strongly convex $Q_u(w)$: δ-Lipschitz gradients (21.11a)–(21.11b) and (21.13) $\mu(k) = \tau/k$	$\mathbb{E}\, P(\boldsymbol{w}_n^k) - P(w^\star) \leq O(1/\sqrt{k}\,)$	Thm. 21.4

Consider the empirical risk minimization problem:

$$w^\star \triangleq \underset{w\in\mathcal{C}}{\text{argmin}} \left\{ P(w) \triangleq \frac{1}{N}\sum_{m=0}^{N-1} Q(w;\gamma(m),h_m) \right\} \tag{16.108}$$

where $Q(w,\cdot)$ is a smooth convex function over w, and where we are now adding a constraint by requiring w to belong to some convex set $\mathcal{C}$. The reason for the addition of this constraint will become evident soon. One way to solve constrained problems of this type is to resort to the stochastic projection method (16.96):

$$\boldsymbol{w}_n = \mathcal{P}_C\Big\{\boldsymbol{w}_{n-1} \;-\; \mu(n)\,\nabla_{w^\mathsf{T}}\, Q(\boldsymbol{w}_{n-1};\boldsymbol{\gamma}(n),\boldsymbol{h}_n)\Big\} \tag{16.109}$$

where, for generality, we are allowing for an iteration-dependent step size, $\mu(n)$. In this description, the operator $\mathcal{P}_C(x)$ projects the vector $x \in \mathbb{R}^M$ onto the convex set $\mathcal{C}$. We already know from the nonexpansive property (9.70) of projection operators that

$$\|w - \mathcal{P}_C(x)\| \;\le\; \|w - x\|, \quad \text{for any } x \in \mathbb{R}^M, w \in \mathcal{C} \tag{16.110}$$

in terms of Euclidean distances.

Observe next that the gradient vector in (16.109) changes with n. We simplify the notation and introduce

$$Q_n(w) \;\triangleq\; Q(w;\gamma(n),h_n) \tag{16.111}$$

so that the stochastic algorithm can be rewritten more compactly as

$$\boxed{\boldsymbol{w}_n = \mathcal{P}_C\Big\{\boldsymbol{w}_{n-1} \;-\; \mu(n)\,\nabla_{w^\mathsf{T}}\, Q_n(\boldsymbol{w}_{n-1})\Big\}, \quad n \ge 0} \tag{16.112}$$

For such *stochastic* optimization procedures, the average *regret* is defined directly in terms of the loss function $Q_n(w;\cdot)$; this is in contrast to the earlier definition in the gradient-descent case (12.57), where the regret was defined in terms of the risk function itself. This is because the updates there involved the gradient of the risk function, whereas the updates here involve the gradient of the loss function and *sampled* data points $(\gamma(n), h_n)$. Thus, for stochastic optimization algorithms we define the average regret that is based on a dataset of size N as the difference between the accumulated *loss* and the smallest possible value for it:

$$\boxed{\mathcal{R}(N) \;\triangleq\; \frac{1}{N}\sum_{n=0}^{N-1} Q_n(\boldsymbol{w}_{n-1}) - \min_{w\in\mathcal{C}}\left\{\frac{1}{N}\sum_{n=0}^{N-1} Q_n(w)\right\}} \tag{16.113}$$

where the iterate $\boldsymbol{w}_{n-1}$ changes with the iteration index within the first sum. In a manner similar to (16.108), we denote the minimizer for the rightmost term in the above expression by $w^\star$. Observe that this term involves the samples

$\{\gamma(n), h_n\}$ that were used during the N iterations of the stochastic algorithm. We then have

$$\mathfrak{R}(N) \stackrel{\Delta}{=} \frac{1}{N}\sum_{n=0}^{N-1}\Big\{Q_n(\boldsymbol{w}_{n-1}) - Q_n(w^\star)\Big\} \tag{16.114}$$

The boldface notation for $\mathfrak{R}$ is meant to reflect the random nature of the regret due to its dependence on the random iterates, $\{\boldsymbol{w}_{n-1}\}$. We will show below that, in the process of bounding the regret, the weight iterates will disappear and we will be able to bound the regret by some deterministic value.

The purpose of regret analysis is to examine how the regret evolves with N, for any sequence of iterates $\{\boldsymbol{w}_{n-1}\}$. For example, if it can be shown that the average regret decays to zero at the rate $\mathfrak{R}(N) = O(1/\sqrt{N})$, then this would imply the desirable conclusion that, asymptotically, the average accumulated loss by the algorithm is able to approach the smallest possible risk value. We can bound the regret as follows.

Regret bound

First, we rewrite the stochastic gradient algorithm (16.112) in the following equivalent form involving an intermediate variable $\boldsymbol{z}_n$:

$$\left\{\begin{array}{l} \boldsymbol{z}_n = \boldsymbol{w}_{n-1} \; - \; \mu(n)\nabla_{w^\mathsf{T}} Q_n(\boldsymbol{w}_{n-1}) \\ \boldsymbol{w}_n = \mathcal{P}_C(\boldsymbol{z}_n) \end{array}\right. \tag{16.115}$$

where, using property (16.110), it holds that

$$\|w - \boldsymbol{w}_n\| \leq \|w - \boldsymbol{z}_n\|, \quad \text{for any } w \in \mathcal{C} \tag{16.116}$$

Subtracting $w^\star$ from both sides of the first relation in (16.115) gives

$$(w^\star - \boldsymbol{z}_n) = (w^\star - \boldsymbol{w}_{n-1}) + \mu(n)\nabla_{w^\mathsf{T}} Q_n(\boldsymbol{w}_{n-1}) \tag{16.117}$$

Let $\widetilde{\boldsymbol{w}}_n = w^\star - \boldsymbol{w}_n$. Squaring both sides of the above equation and using (16.116) leads to

$$\begin{aligned} \|\widetilde{\boldsymbol{w}}_n\|^2 \leq \|\widetilde{\boldsymbol{w}}_{n-1}\|^2 + 2\mu(n)\Big(\nabla_{w^\mathsf{T}}\, Q(\boldsymbol{w}_{n-1})\Big)^\mathsf{T} \widetilde{\boldsymbol{w}}_{n-1} + \\ \mu^2(n)\|\nabla_{w^\mathsf{T}} Q_n(\boldsymbol{w}_{n-1})\|^2 \end{aligned} \tag{16.118}$$

Invoking the convexity of $Q(w, \cdot)$ over w and using property (8.5) for convex functions we have

$$Q_n(\boldsymbol{w}_{n-1}) - Q_n(w^\star) \leq -\Big(\nabla_{w^\mathsf{T}}\, Q_n(\boldsymbol{w}_{n-1})\Big)^\mathsf{T} \widetilde{\boldsymbol{w}}_{n-1} \tag{16.119}$$

Substituting into the regret expression (16.114), we obtain

$$\begin{aligned}
N\mathfrak{R}(N) &\leq -\sum_{n=0}^{N-1}\Big(\nabla_{w^{\mathsf{T}}} Q_n(\boldsymbol{w}_{n-1})\Big)^{\mathsf{T}} \widetilde{\boldsymbol{w}}_{n-1} \\
&\overset{(16.118)}{\leq} \sum_{n=0}^{N-1}\left\{\frac{1}{2\mu(n)}\Big(\|\widetilde{\boldsymbol{w}}_{n-1}\|^2 - \|\widetilde{\boldsymbol{w}}_n\|^2\Big) + \frac{\mu(n)}{2}\|\nabla_{w^{\mathsf{T}}} Q_n(\boldsymbol{w}_{n-1})\|^2\right\} \\
&= \frac{1}{2\mu(0)}\|\widetilde{\boldsymbol{w}}_{-1}\|^2 - \frac{1}{2\mu(N-1)}\|\widetilde{\boldsymbol{w}}_{N-1}\|^2 + \\
&\quad \frac{1}{2}\sum_{n=1}^{N-1}\Big(\frac{1}{\mu(n)} - \frac{1}{\mu(n-1)}\Big)\|\widetilde{\boldsymbol{w}}_{n-1}\|^2 + \\
&\quad \frac{1}{2}\sum_{n=0}^{N-1}\mu(n)\|\nabla_{w^{\mathsf{T}}} Q_n(\boldsymbol{w}_{n-1})\|^2
\end{aligned} \tag{16.120}$$

Introduce the two constants:

$$d \overset{\Delta}{=} \max_{x,y\in\mathcal{C}} \|x-y\| \tag{16.121a}$$

$$c \overset{\Delta}{=} \max_{w\in\mathcal{C},0\leq n\leq N-1} \|\nabla_{w^{\mathsf{T}}} Q_n(w)\| \tag{16.121b}$$

where d is the largest distance between any two points in the convex set $\mathcal{C}$, and c is the largest norm of the gradient of the loss function over both $\mathcal{C}$ and the data. Then, we can simplify the bound (16.120) on the regret function by noting that

$$N\mathfrak{R}(N) \leq \frac{d^2}{2}\left\{\frac{1}{\mu(0)} + \sum_{n=1}^{N-1}\left(\frac{1}{\mu(n)} - \frac{1}{\mu(n-1)}\right)\right\} + \frac{c^2}{2}\sum_{n=0}^{N-1}\mu(n) \tag{16.122}$$

and, consequently,

$$\boxed{\mathfrak{R}(N) \leq \frac{1}{N}\left\{\frac{d^2}{2\mu(N-1)} + \frac{c^2}{2}\sum_{n=0}^{N-1}\mu(n)\right\}} \tag{16.123}$$

Vanishing step size

For illustration purposes, assume the step-size sequence decays as $\mu(n) = 1/\sqrt{n+1}$. Then,

$$\sum_{n=0}^{N-1}\mu(n) = \sum_{n=1}^{N}\frac{1}{\sqrt{n}} \leq 1 + \int_1^N \frac{1}{\sqrt{x}}dx = 2\sqrt{N} - 1 \tag{16.124}$$

and we arrive at

$$\mathfrak{R}(N) \leq \frac{1}{N}\left\{\frac{d^2\sqrt{N}}{2} + \frac{c^2}{2}\Big(2\sqrt{N}-1\Big)\right\} \tag{16.125}$$

It follows that the average regret converges to zero as $N \to \infty$ at the rate

$$\boxed{\mathcal{R}(N) \leq O(1/\sqrt{N}\,)} \tag{16.126}$$

Constant step size

On the other hand, when the step size is constant, say, $\mu(n) = \mu$, then the bound (16.123) leads to

$$\boxed{\mathcal{R}(N) \leq O(1/N\,) \;+\; O(\mu)} \tag{16.127}$$

which shows that the average regret approaches a small value on the order of μ and the convergence rate toward this region is $O(1/N)$. We will encounter a more detailed example of regret analysis in Appendix 17.A when we apply it to examine the performance of adaptive gradient algorithms.

There are at least two differences in the regret analysis approach in comparison to the MSE analysis performed in future chapters; see also Prob. 16.2. First, the regret argument relies on a worst-case scenario in the sense that the effect of the random trajectory $\{\widetilde{\boldsymbol{w}}_{-1}, \widetilde{\boldsymbol{w}}_0, \ldots, \widetilde{\boldsymbol{w}}_{N-1}\}$ is removed completely by replacing their norms by d. Second, the size of the constants d and c can be large. For example, if we were to remove the constraint $w \in \mathcal{C}$ and replace $\mathcal{C}$ by $\mathbb{R}^M$, then the above argument would not carry through since d or c will become unbounded.

16.6 COMMENTARIES AND DISCUSSION

Stochastic approximation theory. The idea of using data realizations to approximate actual gradient or subgradient vectors is at the core of *stochastic approximation theory*. According to Tsypkin (1971, p. 70) and Lai (2003), the pioneering work in the field is the landmark paper by Robbins and Monro (1951), which developed a recursive method for finding roots of functions, i.e., points $w^\star$ where $P(w^\star) = 0$. Their procedure was a variation of a scheme developed two decades earlier by von Mises and Pollaczek-Geiringer (1929), and it can be succinctly described as follows. Consider a risk function $P(w)$, of a *scalar* parameter w, and assume $P(w)$ is represented as the mean of some loss function, say, in the form:

$$P(w) \stackrel{\Delta}{=} \mathbb{E}\, Q(w; \boldsymbol{x}) \tag{16.128}$$

Robbins and Monro (1951) argued that the root $w^\star$ can be approximated by evaluating the loss function at successive realizations x_n and employing the update relation:

$$\boldsymbol{w}_n = \boldsymbol{w}_{n-1} - \mu(n) Q(\boldsymbol{w}_{n-1}; \boldsymbol{x}_n), \quad n \geq 0 \tag{16.129}$$

where $\mu(n)$ is a step-size sequence that satisfies:

$$\sum_{n=0}^{\infty} \mu(n) = \infty, \quad \sum_{n=0}^{\infty} \mu^2(n) < \infty \tag{16.130}$$

They showed that the algorithm converges in the MSE sense, and also in probability, namely, $\mathbb{E}\,\widetilde{\boldsymbol{w}}_n^2 \to 0$ and, for any $\epsilon > 0$:

$$\lim_{n\to\infty} \mathbb{P}\Big(|\widetilde{\boldsymbol{w}}_n| \geq \epsilon\Big) = 0 \tag{16.131}$$

Stronger almost-sure convergence results were later given by Blum (1954), Dvoretzky (1956), and Gladyshev (1965), among others, showing that under certain technical conditions:

$$\mathbb{P}\left(\lim_{n\to\infty} \widetilde{\boldsymbol{w}}_n = 0\right) = 1 \tag{16.132}$$

The same construction can be extended from root-finding to the solution of minimization problems. Assume $P(w)$ is convex and has a unique minimum at some location $w^\star$. Then, finding $w^\star$ is equivalent to finding the root of $dP(w)/dw = 0$, which suggests using the stochastic gradient recursion:

$$\boldsymbol{w}_n = \boldsymbol{w}_{n-1} - \mu(n)\left.\frac{dQ(w;\boldsymbol{x}_n)}{dw}\right|_{w=\boldsymbol{w}_{n-1}} \tag{16.133}$$

Motivated by the work of Robbins and Monro (1951), an alternative stochastic construction was proposed by Kiefer and Wolfowitz (1952) to solve minimization problems from noisy measurements. Their procedure relies on the same general concept of using data to approximate unknown quantities, but took a different form, namely, they proposed using the recursion:

$$\boldsymbol{w}_{n-1}^{+} = \boldsymbol{w}_{n-1} + \tau(n) \tag{16.134a}$$

$$\boldsymbol{w}_{n-1}^{-} = \boldsymbol{w}_{n-1} - \tau(n) \tag{16.134b}$$

$$\Delta Q(n) = Q(\boldsymbol{w}_{n-1}^{+};\boldsymbol{x}_n) - Q(\boldsymbol{w}_{n-1}^{-};\boldsymbol{x}_n) \tag{16.134c}$$

$$\boldsymbol{w}_n = \boldsymbol{w}_{n-1} - \mu(n)\,\frac{\Delta Q(n)}{\tau(n)} \tag{16.134d}$$

In this recursion, a first-order finite difference calculation is used to approximate the derivative of $Q(w)$ with $\tau(n)$ denoting the interval width. The nonnegative sequences $\{\mu(n), \tau(n)\}$ are chosen to tend asymptotically to zero and to satisfy the conditions:

$$\sum_{n=0}^{\infty} \mu(n) = \infty, \qquad \sum_{n=0}^{\infty}\left(\frac{\mu(n)}{\tau(n)}\right)^2 < \infty \tag{16.135}$$

The work by Robbins and Monro (1951) generated tremendous interest in the statistical, optimization, and engineering literature and led to many subsequent studies and extensions. While their work dealt primarily with a *scalar* weight w, Blum (1954) and Schmetterer (1961) extended the procedure to weight *vectors*. A description of these developments can be found in the book by Wetherhill (1966). Further discussions on stochastic approximation methods, including some detailed treatments of their convergence properties, can be found in the works by Albert and Gardner (1967), Wasan (1969), Mendel and Fu (1970), Tsypkin (1971), Ljung (1977), Kushner and Clark (1978), Kushner (1984), Polyak (1987), Benveniste, Métivier, and Priouret (1987), Bertsekas and Tsitsiklis (1997, 2000), Kushner and Yin (2003), Spall (2003), Marti (2005), and Sayed (2003, 2008, 2014a). The use of averaged iterates as in (16.52) was proposed independently by Ruppert (1988) and Polyak and Juditsky (1992).

Stochastic gradient and subgradient algorithms have become commonplace in online inference and learning solutions for at least three reasons. First, the explosive interest in large-scale and big data scenarios favors the use of simple algorithmic structures, of which these methods are prime examples. Second, as shown in future chapters, these algorithms are able to deliver solid performance guarantees, with the MSE $\mathbb{E}\,\|\widetilde{\boldsymbol{w}}_n\|^2$ and the excess risk $\mathbb{E}\,P(\boldsymbol{w}_n) - P(w^\star)$ approaching small neighborhoods on the order of

$O(\mu)$. Third, and importantly, it is increasingly evident that employing more sophisticated optimization techniques does not necessarily ensure improved performance – see, e.g., Bottou and Bousquet (2008) and Bottou (2012). This is because the assumed data models or risk functions do not always capture faithfully the underlying data structure anyway. In addition, the presence of noise in the data generally implies that a solution that may be perceived to be optimal is actually suboptimal due to the perturbations in the data and models.

Adaline and perceptron. During the 1950s, stochastic approximation theory did not receive much attention in the engineering community until the landmark work by Widrow and Hoff (1960) in which they developed the delta or adaline recursion (16.32). Using a target sequence $\gamma(n)$, the algorithm enabled adjusting the weight parameter $\boldsymbol{w}_n$ in order to close the gap between the target signal and its prediction given by $\widehat{\gamma}(n) = \boldsymbol{h}_n^\mathsf{T} \boldsymbol{w}_{n-1}$. Their filter launched the design of adaptive systems with adjustable structures and has found applications in a remarkable range of areas – see, e.g., the treatments by Widrow and Stearns (1985), Haykin (2001), and Sayed (2003, 2008). A useful interpretation for the LMS algorithm (or delta rule) as the solution to a min–max optimization problem was given by Hassibi, Sayed, and Kailath (1994a, 1996). The algorithm was further studied by Sayed and Rupp (1996) and Sayed (2003, 2008) using energy arguments to establish several robustness properties.

Besides adaline and LMS, there have been other notable works on stochastic gradient algorithms in the early 1960s. One example is the perceptron algorithm (16.75), which was developed by Rosenblatt (1957, 1958, 1962) for pattern classification problems and which we will study in greater detail in Chapter 60.

Early examples of stochastic approximation structures. There are other examples of adjustable designs from the 1950s that bear resemblance to stochastic approximation constructions. One such contribution is the works by Mattson (1959a, b) in the context of pattern classifiers. According to Widrow and Hoff (1960, p. 97), these works were among the first to apply adjustable structures to classification problems. However, unlike Rosenblatt (1957), the construction proposed in Mattson (1959b) was a trial-and-error procedure based on varying the weight entries and a threshold value until satisfactory performance is attained. There was no explicit optimization problem guiding the design procedure. This is reflected in the description by Mattson (1959b), where it is stated that *"it is the purpose of this paper to define a model for a self-organizing logical system which determines, by an iterative trial-and-error procedure, the proper Boolean function for a process."*

According to the presentation in Sayed (2003, 2008), another example of early work on stochastic algorithms is a procedure that minimizes the MSE between an input signal and a reference signal, developed by Gabor, Wilby, and Woodcock (1961); their filter is described in Tsypkin (1971, p. 156). This latter reference also contains on pages 172–173 commentaries on works on adaptation and learning during the early 1960s, including a description of a stochastic gradient algorithm by Sefl (1960) that is the continuous-time counterpart of the delta or LMS rule; it employs a differential update equation of the form

$$\frac{d\boldsymbol{w}(t)}{dt} = 2\mu(t)\boldsymbol{h}(t)\Big(\gamma(t) - (\boldsymbol{h}(t))^\mathsf{T}\,\boldsymbol{w}(t)\Big) \tag{16.136}$$

with continuous-time vector variables $\{\boldsymbol{w}(t), \boldsymbol{h}(t)\}$ and scalar variables $\{\gamma(t), \mu(t)\}$. Other noteworthy works on stochastic gradient algorithms in the 1960s are those by Applebaum (1966) and Widrow *et al.* (1967), on adaptive antenna arrays, and Amari (1967), on pattern classification. In Applebaum (1966), a stochastic gradient algorithm is derived that is based on maximizing a signal-to-noise ratio measure, while Widrow *et al.* (1967) focus on MSE performance and use the LMS algorithm. The work by Amari (1967) uses a stochastic gradient recursion to learn the weight vector in a pattern classification problem.

For further readings and discussion on online learning techniques, the reader may consult the books by Sayed (2003, 2008, 2014a), Cesa-Bianchi and Lugosi (2006), Shalev-Shwartz (2011), and Theodoridis (2015), and the articles by Bottou (1998, 2012), Cesa-Bianchi, Conconi, and Gentile (2004), Bottou and Bousquet (2008), Bach and Moulines (2011), and Agarwal *et al.* (2012).

Zeroth-order learning algorithms. We can also develop stochastic versions for the zeroth-order optimization algorithms described earlier in Appendix 12.A. In this case, we would sample directional vectors u in order to approximate the gradient vector of the loss (rather than the risk) function by using either instantaneous or mini-batch calculations. For instance, the stochastic gradient algorithm (16.27) would be replaced by listing (16.137) for instantaneous gradient approximations or by listing (16.138) for mini-batch approximations. In either listing, we denote the distribution from which the directional vectors $\boldsymbol{u}$ are sampled by $f_{\boldsymbol{u}}(u)$; it can refer to either the Gaussian distribution $\mathcal{N}_{\boldsymbol{u}}(0, I_M)$ or the uniform distribution $\mathcal{U}(\mathbb{S})$, as described by (12.213a)–(12.213b).

Zeroth-order stochastic gradient algorithm for minimizing (16.2a) or (16.2b).

given dataset $\{\gamma(m), h_m\}_{m=0}^{N-1}$ or streaming data $(\gamma(n), h_n)$;
given a small smoothing factor $\alpha > 0$;
select the sampling distribution $f_{\boldsymbol{u}}(u)$ and set $\beta \in \{1, M\}$;
start from an arbitrary initial condition, $\boldsymbol{w}_{-1}$.
repeat until sufficient convergence over $n \geq 0$**:**
  select at random or receive a sample $(\boldsymbol{\gamma}(n), \boldsymbol{h}_n)$ at iteration n;
  sample $\boldsymbol{u} \sim f_{\boldsymbol{u}}(u)$;

$$\widehat{\nabla_{w^{\mathsf{T}}} Q}(\boldsymbol{w}_{n-1}; \boldsymbol{u}) = \frac{\beta}{\alpha}\left\{Q\Big(\boldsymbol{w}_{n-1} + \alpha \boldsymbol{u}; \boldsymbol{\gamma}(n), \boldsymbol{h}_n\Big) - Q\Big(\boldsymbol{w}_{n-1}; \boldsymbol{\gamma}(n), \boldsymbol{h}_n\Big)\right\}\boldsymbol{u}$$

$$\boldsymbol{w}_n = \boldsymbol{w}_{n-1} - \mu\, \widehat{\nabla_{w^{\mathsf{T}}} Q}(\boldsymbol{w}_{n-1}; \boldsymbol{u})$$

end
return $w^\star \leftarrow \boldsymbol{w}_n$.

(16.137)

Zeroth-order mini-batch stochastic algorithm for minimizing (16.2a) or (16.2b).

given dataset $\{\gamma(m), h_m\}_{m=0}^{N-1}$ or streaming data $(\gamma(n), h_n)$;
given a mini-batch size, B;
given a small smoothing factor $\alpha > 0$;
select the sampling distribution $f_{\boldsymbol{u}}(u)$ and set $\beta \in \{1, M\}$;
start from an arbitrary initial condition, $\boldsymbol{w}_{-1}$.
repeat until sufficient convergence over $n \geq 0$**:**
  select at random or receive B samples $\{\boldsymbol{\gamma}(b), \boldsymbol{h}_b\}_{b=0}^{B-1}$ at iteration n;
  sample direction $\boldsymbol{u} \sim f_{\boldsymbol{u}}(u)$

$$\widehat{\nabla_{w^{\mathsf{T}}} Q}(\boldsymbol{w}_{n-1}; \boldsymbol{u}, b) \stackrel{\Delta}{=} \frac{\beta}{\alpha}\Big(Q(\boldsymbol{w}_{n-1} + \alpha \boldsymbol{u}; \boldsymbol{\gamma}(b), \boldsymbol{h}_b) - Q(\boldsymbol{w}_{n-1}; \boldsymbol{\gamma}(b), \boldsymbol{h}_b)\Big)\boldsymbol{u}$$

$$\boldsymbol{w}_n = \boldsymbol{w}_{n-1} - \mu\left(\frac{1}{B}\sum_{b=0}^{B-1} \widehat{\nabla_{w^{\mathsf{T}}} Q}(\boldsymbol{w}_{n-1}; \boldsymbol{u}, b)\right)$$

end
return $w^\star \leftarrow \boldsymbol{w}_n$.

(16.138)

There are many variations by which the gradient vector of the loss function can be approximated. The mini-batch listing assumes that, for each iteration n, a single direction u is sampled and used for all entries in the mini-batch. Alternatively, we could consider sampling J directions u_j, computing the mini-batch gradient approximation for each one of them, and then averaging over the J directions, namely,

$$\left\{\begin{array}{l}
\text{sample } J \text{ directions } \boldsymbol{u}_j \sim f_{\boldsymbol{u}}(u),\ j=0,1,\ldots,J-1;\\
\textbf{for each } j\textbf{, compute } B \textbf{ gradients:}\\
\quad\left|\begin{array}{l}\widehat{\nabla_{w^\mathsf{T}} Q}(\boldsymbol{w}_{n-1};\boldsymbol{u}_j,b) = \dfrac{\beta}{\alpha}\Big\{Q\Big(\boldsymbol{w}_{n-1}+\alpha\boldsymbol{u}_j;\boldsymbol{\gamma}(b),\boldsymbol{h}_b\Big) - Q\Big(\boldsymbol{w}_{n-1};\boldsymbol{\gamma}(b),\boldsymbol{h}_b\Big)\Big\}\boldsymbol{u}_j\\
b=0,1,\ldots,B-1\end{array}\right.\\
\textbf{end}\\
\text{set } \widehat{\nabla_{w^\mathsf{T}} Q}(\boldsymbol{w}_{n-1}) = \dfrac{1}{J}\displaystyle\sum_{j=0}^{J-1}\left\{\dfrac{1}{B}\sum_{b=0}^{B-1}\widehat{\nabla_{w} Q}(\boldsymbol{w}_{n-1};\boldsymbol{u}_j,b)\right\};\\
\text{update } \boldsymbol{w}_n = \boldsymbol{w}_{n-1} - \mu\widehat{\nabla_{w^\mathsf{T}} Q}(\boldsymbol{w}_{n-1}).
\end{array}\right. \tag{16.139}$$

Other constructions are possible.

Collaborative filtering. The stochastic gradient implementation (16.58) for solving the matrix factorization problem (16.57) is motivated by the works of Funk (2006), Paterek (2007), Takacs *et al.* (2007), and Koren (2008). The alternating least-squares version described later in Example 50.6 is motivated by Bell and Koren (2007a), Hu, Koren, and Volinsky (2008), Zhou *et al.* (2008), and Pilaszy, Zibriczky, and Tikk (2010). Most of these works were driven by the Netflix prize challenge, which was an open competition during 2006–2009 offering a prize of US$1 million for the best collaborative filtering solution to predict user ratings of movies – overviews appear in Bell, Koren, and Volinsky (2007) and Bell and Koren (2007b). Netflix provided training data consisting of over 100 million ratings from close to 500,000 users for about 18,000 movies. For more details on the Netflix challenge and on matrix factorization methods, readers may consult the tutorial by Koren, Bell, and Volinsky (2009) and the text by Symeonidis and Zioupos (2017). The stochastic gradient algorithm (16.58) was also among the top solutions in the KDDCup 2011 challenge, which dealt with the problem of recommending music items to users from the Yahoo music dataset. This challenge released about 250 million ratings from over 1 million anonymized users – see the account by Dror *et al.* (2012).

PROBLEMS

16.1 Consider an empirical risk minimization problem with uniform data sampling. Given a finite number of data samples $\{(\gamma(0),h_0),(\gamma(1),h_1),\ldots,(\gamma(N-1),h_{N-1})\}$, we define discrete random variables $\{\boldsymbol{\gamma},\boldsymbol{h}\}$ that are generated according to the following probability distribution:

$$\mathbb{P}(\boldsymbol{\gamma}=\gamma,\boldsymbol{h}=h) = \begin{cases} 1/N, & \text{if } \gamma=\gamma(0), h=h_0\\ 1/N, & \text{if } \gamma=\gamma(1), h=h_1\\ \vdots & \vdots\\ 1/N, & \text{if } \gamma=\gamma(N-1), h=h_{N-1}\end{cases}$$

Verify that, under this construction, the following holds:

$$\mathbb{E}\,Q(w;\boldsymbol{\gamma},\boldsymbol{h}) \;=\; \frac{1}{N}\sum_{m=0}^{N-1} Q(w;\gamma(m),h_m)$$

Conclude that the solutions to the following stochastic and empirical risk problems coincide:

$$w^o \;\stackrel{\Delta}{=}\; \underset{w\in\mathbb{R}^M}{\operatorname{argmin}}\;\mathbb{E}\,Q(w;\boldsymbol{\gamma},\boldsymbol{h}) \;=\; \underset{w\in\mathbb{R}^M}{\operatorname{argmin}}\;\frac{1}{N}\sum_{m=0}^{N-1} Q(w;\gamma(m),h_m) \;\stackrel{\Delta}{=}\; w^\star$$

16.2 Refer to the empirical risk minimization problem (16.2a) with minimizer denoted by $w^\star$. Assume the minimization is restricted over a bounded convex set, $w \in \mathcal{C}$. Refer further to the rightmost term in (16.113) in the definition of the average regret. Assuming a stochastic gradient implementation, how does the minimizer of this rightmost term compare to $w^\star$?

16.3 Consider the ℓ_1-regularized MSE risk:

$$w^o = \underset{w\in\mathbb{R}^M}{\operatorname{argmin}}\;\left\{\alpha\|w\|_1 \;+\; \mathbb{E}\,(\boldsymbol{\gamma}-\boldsymbol{h}^\mathsf{T} w)^2\right\}$$

(a) Write down a stochastic subgradient algorithm for its solution.
(b) Write down a stochastic proximal algorithm for the same problem.
(c) Write down a stochastic coordinate descent solution.

16.4 The total-variation denoising problem involves solving a regularized least-squares problem of the following form:

$$\min_{w}\left\{\alpha\sum_{m'=1}^{M-1}\Big|w[m'+1]-w[m']\Big| \;+\; \frac{1}{N}\sum_{m=0}^{N-1}\Big(\gamma(m)-h_m^\mathsf{T} w\Big)^2\right\}$$

where the $\{w[m']\}$, for $m' = 1,2,\ldots,M$, denote the individual entries of $w \in \mathbb{R}^M$.
(a) Derive a stochastic subgradient solution for this problem.
(b) Derive a stochastic proximal solution for the same problem.

16.5 The fused LASSO problem adds ℓ_1-regularization to the total variation formulation in Prob. 16.4 and considers instead

$$\min_{w}\left\{\alpha_1\|w\|_1+\alpha_2\sum_{m'=1}^{M-1}\Big|w[m'+1]-w[m']\Big| \;+\; \frac{1}{N}\sum_{m=0}^{N-1}\Big(\gamma(m)-h_m^\mathsf{T} w\Big)^2\right\}$$

(a) Derive a stochastic subgradient solution for this problem.
(b) Derive a stochastic proximal solution for the same problem.

16.6 The group LASSO problem involves solving a regularized least-squares problem of the following form. We partition each observation vector into K sub-vectors, say, $h_m = \operatorname{col}\{h_{mk}\}$ for $k = 1,2,\ldots,K$. We similarly partition the weight vector into K sub-vectors, $w = \operatorname{col}\{w_k\}$, of similar dimensions to h_{mk}. Now consider the problem:

$$\min_{w}\left\{\alpha\sum_{k=1}^{K}\|w_k\| \;+\; \frac{1}{N}\sum_{m=0}^{N-1}\left(\gamma(m)-\sum_{k=1}^{K}h_{mk}^\mathsf{T} w_k\right)^2\right\}$$

(a) Derive a stochastic subgradient solution for this problem.
(b) Derive a stochastic proximal solution for the same problem.

16.7 Consider a collection of N-data pairs $\{\gamma(m),h_m\}$ where $\gamma(m)\in\mathbb{R}$ and $h_m \in \mathbb{R}^M$, and formulate the least-squares problem:

$$w^\star = \underset{w\in\mathbb{R}^M}{\operatorname{argmax}}\left\{\frac{1}{N}\sum_{m=0}^{N-1}\Big(\gamma(m)-h_m^\mathsf{T} w\Big)^2\right\}$$

One stochastic gradient method for minimizing this risk function can be devised as follows (this method is known as a *randomized* Kaczmarz method):

$$\begin{cases} \text{select an index } 0 \le n \le N-1 \text{ at random} \\ \text{consider the corresponding data pair } \{\gamma(n), \boldsymbol{h}_n\} \\ \text{update } \boldsymbol{w}_n = \boldsymbol{w}_{n-1} + \dfrac{\boldsymbol{h}_n}{\|\boldsymbol{h}_n\|^2}(\gamma(n) - \boldsymbol{h}_n^\mathsf{T}\boldsymbol{w}_{n-1}) \end{cases}$$

Collect all vectors $\{h_m\}$ into the $N \times M$ matrix $H = \text{row}\{h_m^\mathsf{T}\}$ and all target values into the $N \times 1$ vector $d = \text{col}\{\gamma(m)\}$. Assume H has full rank. Assume also the random index $\boldsymbol{n}$ is selected according to the following importance sampling procedure $\mathbb{P}(\boldsymbol{n} = m) = \|h_m\|^2/\|H\|_\text{F}^2$. Note that the vectors $\boldsymbol{h}_n$ are selected independently of each other and of any other random variable in the problem. Let $w^\star$ denote the solution to the least-squares problem, i.e., it satisfies $H^\mathsf{T}Hw^\star = H^\mathsf{T}d$, and introduce the weight-error vector $\widetilde{\boldsymbol{w}}_n = w^\star - \boldsymbol{w}_n$. Show that

$$\mathbb{E}\,\|\widetilde{\boldsymbol{w}}_n\|^2 \le \left(1 - \frac{\sigma_{\min}^2(H)}{\|H\|_\text{F}^2}\right)\mathbb{E}\,\|\widetilde{\boldsymbol{w}}_{n-1}\|^2, \quad n \ge 0$$

in terms of the smallest singular value of H. *Remark.* See the work by Strohmer and Vershynin (2009), who studied this randomized version of a popular method by Kaczmarz (1937) for the solution of linear systems of equations. The traditional Kaczmarz method studied earlier in Prob. 12.34 cycles through the rows of H, whereas the randomized version described here samples the rows of H at random, as described above.

16.8 Continuing with Prob. 16.7, let $w^\star$ denote the minimum-norm solution in the over-parameterized case when $N < M$. Show that $\mathbb{E}\,\|\widetilde{\boldsymbol{w}}_n\|^2$ converges to zero.

16.9 Let $\{\boldsymbol{x}(n),\ n = 1, \ldots, N\}$ denote N independent realizations with mean μ and and finite variance, $\sigma_x^2 = \mathbb{E}\,(\boldsymbol{x}(n) - \mu)^2 < \infty$. Introduce the sample average $\widehat{\boldsymbol{\mu}}_N = \frac{1}{N}\sum_{n=1}^N \boldsymbol{x}(n)$.

(a) Verify that $\widehat{\boldsymbol{\mu}}_N = \widehat{\boldsymbol{\mu}}_{N-1} + \frac{1}{N}\left(\boldsymbol{x}(N) - \widehat{\boldsymbol{\mu}}_{N-1}\right)$ with $\widehat{\boldsymbol{\mu}}_0 = 0$ and $N \ge 1$.

(b) Assume that the sample average is approximated recursively as follows (written now as $\boldsymbol{\mu}_N$ to distinguish it from $\widehat{\boldsymbol{\mu}}_N$):

$$\boldsymbol{\mu}_N = \boldsymbol{\mu}_{N-1} + \alpha(N)\left(\boldsymbol{x}(N) - \boldsymbol{\mu}_{N-1}\right), \quad \boldsymbol{\mu}_0 = 0,\ N \ge 1$$

where the scalar sequence $\{\alpha(n)\}$ satisfies

$$0 \le \alpha(N) < 1, \quad \lim_{N\to\infty} \alpha(N) = 0, \quad \lim_{N\to\infty} \sum_{n=1}^N \alpha(n) = \infty$$

We want to verify that $\boldsymbol{\mu}_N$ tends to μ in probability. Let σ_N^2 denote the variance of $\boldsymbol{\mu}_N$, i.e., $\sigma_N^2 = \mathbb{E}\,(\boldsymbol{\mu}_N - \mathbb{E}\,\boldsymbol{\mu}_N)^2$.

(b.1) Verify that $(\mathbb{E}\,\boldsymbol{\mu}_N - \mu) = (1 - \alpha(N))\,(\mathbb{E}\,\boldsymbol{\mu}_{N-1} - \mu)$, $N \ge 1$.

(b.2) Show that σ_N^2 satisfies $\sigma_N^2 = (1 - \alpha(N))^2\sigma_{N-1}^2 + \alpha^2(N)\sigma_x^2$.

(b.3) Compare the recursion in (b.2) with (14.136) and conclude that $\sigma_N^2 \to 0$ as $N \to \infty$. Conclude also that $\mathbb{E}\,\boldsymbol{\mu}_N \to \mu$ as $N \to \infty$.

16.10 Consider the regularized logistic risk

$$P(w) = \rho\|w\|^2 + \mathbb{E}\left\{\ln\left(1 + e^{-\boldsymbol{\gamma}\boldsymbol{h}^\mathsf{T}w}\right)\right\}$$

where $\boldsymbol{\gamma}$ is a binary random variable assuming the values ± 1 and $R_h = \mathbb{E}\,\boldsymbol{h}\boldsymbol{h}^\mathsf{T}$. Let w^o denote the minimizer of $P(w)$. Show that

(a) $\|w^o\| \le \mathbb{E}\,\|\boldsymbol{h}\|/2\rho$.

(b) $\|w^o\|^2 \le \text{Tr}(R_h)/4\rho^2$.

16.11 Consider the MSE cost $P(w) = \mathbb{E}\,(\boldsymbol{\gamma}(n) - \boldsymbol{h}_n^\mathsf{T} w)^2$, where $\boldsymbol{\gamma}(n)$ denotes a streaming sequence of zero-mean random variables with variance $\sigma_\gamma^2 = \mathbb{E}\,\boldsymbol{\gamma}^2(n)$ and $\boldsymbol{h}_n \in \mathbb{R}^M$ is a streaming sequence of independent zero-mean Gaussian random vectors with covariance matrix $R_h = \mathbb{E}\,\boldsymbol{h}_n \boldsymbol{h}_n^\mathsf{T} > 0$. Both processes $\{\boldsymbol{\gamma}(n), \boldsymbol{h}_n\}$ are assumed to be jointly wide-sense stationary. The cross-covariance vector between $\boldsymbol{\gamma}(n)$ and $\boldsymbol{h}_n$ is denoted by $r_{h\gamma} = \mathbb{E}\,\boldsymbol{\gamma}(n)\boldsymbol{h}_n$. The data $\{\boldsymbol{\gamma}(n), \boldsymbol{h}_n\}$ are assumed to be related via a linear regression model of the form $\boldsymbol{\gamma}(n) = \boldsymbol{h}_n^\mathsf{T} w^\bullet + \boldsymbol{v}(n)$, for some unknown parameter vector $w^\bullet$, and where $\boldsymbol{v}(n)$ is a zero-mean white-noise process with power $\sigma_v^2 = \mathbb{E}\,\boldsymbol{v}^2(n)$ and assumed independent of $\boldsymbol{h}_m$ for all n, m.

(a) Let w^o denote the minimizer for $P(w)$. Show that $w^o = w^\bullet$.

(b) Consider the stochastic gradient algorithm for estimating w^o from streaming data, $\boldsymbol{w}_n = \boldsymbol{w}_{n-1} + 2\mu \boldsymbol{h}_n(\boldsymbol{\gamma}(n) - \boldsymbol{h}_n^\mathsf{T} \boldsymbol{w}_{n-1})$. Let $\widetilde{\boldsymbol{w}}_n = w^o - \boldsymbol{w}_n$. Verify that

$$\widetilde{\boldsymbol{w}}_n = (1 - 2\mu \boldsymbol{h}_n \boldsymbol{h}_n^\mathsf{T})\widetilde{\boldsymbol{w}}_{n-1} + 2\mu \boldsymbol{h}_n \boldsymbol{v}(n)$$

(c) Determine a necessary and sufficient condition on μ to ensure convergence in the mean, i.e., for $\mathbb{E}\,\widetilde{\boldsymbol{w}}_n \to 0$ as $n \to \infty$.

(d) Determine a recursion for $\mathbb{E}\,\|\widetilde{\boldsymbol{w}}_n\|^2$.

(e) Find a necessary and sufficient condition on μ to ensure that $\mathbb{E}\,\|\widetilde{\boldsymbol{w}}_n\|^2$ converges. How does this condition compare to the one in part (c)?

(f) Find an expression for the limiting value of $\mathbb{E}\,\|\widetilde{\boldsymbol{w}}_n\|^2$ (also referred to as the mean-square-deviation (MSD) of the algorithm).

16.12 Consider the same setting of Prob. 16.11, albeit with $R_h = \sigma_h^2 I_M$. Assume the limits exist and define the MSD and excess mean-square error (EMSE) figures of merit:

$$\text{MSD} = \lim_{n\to\infty} \mathbb{E}\,\|\widetilde{\boldsymbol{w}}_n\|^2, \qquad \text{EMSE} = \lim_{n\to\infty} \mathbb{E}\,|\boldsymbol{h}_n^\mathsf{T} \widetilde{\boldsymbol{w}}_{n-1}|^2$$

(a) Verify that $P(w^o) = \sigma_v^2$ and $R(w_{n-1}) = R(w^o) + \text{EMSE}$. Justify the name "excess mean-square error."

(b) Determine expressions for the MSD and EMSE.

(c) Define the convergence time, $\mathcal{K}$, as the number of iterations it takes for the MSE, $P(w_{n-1})$, to be within $\epsilon\%$ of its steady-state value. Find a closed-form expression for $\mathcal{K}$.

16.13 Consider the regret bound (16.123) and assume $\mu(n) = 1/(n+1)$. At what rate does the regret approach its limiting behavior?

16.14 Consider a stochastic gradient recursion of the form:

$$\boldsymbol{w}_n = \boldsymbol{w}_{n-1} + \boldsymbol{\mu}(n)\boldsymbol{h}_n\left(\boldsymbol{\gamma}(n) - \boldsymbol{h}_n^\mathsf{T} \boldsymbol{w}_{n-1}\right)$$

where $\boldsymbol{\gamma}(n) \in \mathbb{R}$ and $\boldsymbol{h}_n \in \mathbb{R}^M$. The step size $\boldsymbol{\mu}(n)$ is an iid random process with mean $\bar{\mu}$ and variance σ_μ^2. The feature vectors $\{\boldsymbol{h}_n\}$ are iid Gaussian with zero mean and covariance matrix $R_h = \sigma_h^2 I_M > 0$. Moreover, the data $\{\boldsymbol{\gamma}(n), \boldsymbol{h}_n\}$ is assumed to arise from the stationary data model $\boldsymbol{\gamma}(n) = \boldsymbol{h}_n^\mathsf{T} w^o + \boldsymbol{v}(n)$, where $\boldsymbol{h}_n$ and $\boldsymbol{v}(m)$ are independent of each other for all n and m. The variance of the zero-mean process $\boldsymbol{v}(n)$ is denoted by σ_v^2. In addition, the step-size variable $\boldsymbol{\mu}(n)$ is assumed to be independent of all random variables in the learning algorithm for any time instant.

(a) Determine conditions to ensure mean convergence of $\boldsymbol{w}_n$ toward w^o.

(b) Determine a recursion for $\mathbb{E}\,\|\widetilde{\boldsymbol{w}}_n\|^2$, where $\widetilde{\boldsymbol{w}}_n = w^o - \boldsymbol{w}_n$.

(c) Determine conditions to ensure the convergence of $\mathbb{E}\,\|\widetilde{\boldsymbol{w}}_n\|^2$ to a steady-state value.

(d) Use the recursion of part (b) to determine an exact closed-form expression for the limiting MSD of the algorithm, which is defined as the limiting value of $\mathbb{E}\,\|\widetilde{\boldsymbol{w}}_n\|^2$ as $n \to \infty$.

(e) Determine an approximation for the MSD metric to first order in $\bar{\mu}$.

(f) Determine an approximation for the convergence rate to first order in $\bar{\mu}$.

(g) Assume $\boldsymbol{\mu}(n)$ is Bernoulli and assumes the values μ and 0 with probabilities p and $1-p$, respectively. What are the values of $\bar{\mu}$ and σ_μ^2 in this case? Consider the alternative stochastic gradient implementation with $\boldsymbol{\mu}(n)$ replaced by a constant value μ. How do the MSD values for these two implementations, with $\boldsymbol{\mu}(n)$ and μ, compare to each other?

16.15 Consider the stochastic mirror-descent algorithm and apply it to the solution of the following optimization problem:

$$w^\star = \underset{w \in \mathcal{C} \cap \mathcal{C}_\phi}{\text{argmin}} \frac{1}{N} \sum_{n=0}^{N-1} Q_n(w)$$

Here, each loss term $Q_n(w) : \mathbb{R}^M \to \mathbb{R}$ is convex over $\mathcal{C}$ and δ-Lipschitz relative to some norm $\|\cdot\|$. Observe that in this case the objective function is the average of several loss values changing with n. Consider a mirror function $\phi(w) : \mathcal{C}_\phi \to \mathbb{R}$ that is ν-strongly convex relative to the same norm, and where $\mathcal{C} \subset \bar{\mathcal{C}}_\phi$. Repeat the argument used in the proof of Theorem 15.5 to show that

$$0 \leq \frac{1}{N} \sum_{n=0}^{N-1} \Big(Q_n(w_{n-1}) - Q_n(w^\star) \Big) \leq \frac{1}{N\mu} D_\phi(w^\star, w_{-1}) + \frac{\mu}{2N\nu} \sum_{n=0}^{N-1} \|g_n\|_\star^2$$

where g_n is a subgradient for $Q_n(w)$ at w_{n-1} and $\|\cdot\|_\star$ is the dual norm.

16.16 Continuing with the setting of Prob. 16.15, assume $\|\cdot\|$ is the ℓ_1-norm and $\delta = 1$. Choose $\phi(x)$ as the negative entropy function for which we already know from the discussion in the body of the chapter that $\nu = 1$. Choose $w_{-1} = \frac{1}{M}\mathbb{1}$ and $\mu = \sqrt{(2/N) \ln M}$. Conclude that

$$0 \leq \frac{1}{N} \sum_{n=0}^{N-1} \Big(Q_n(w_{n-1}) - Q_n(w^\star) \Big) \leq \sqrt{\frac{2 \ln M}{N}}$$

16.A SWITCHING EXPECTATION AND DIFFERENTIATION

We encountered in the body of the chapter instances where it is necessary to switch the order of the expectation and differentiation operators. The switching can be justified by appealing to the *dominated convergence theorem* from measure theory, which we state under conditions that are sufficient for our purposes.

Dominated convergence theorem (e.g., Rudin (1976), Royden (1988)). *Assume $R_n(x)$ is a sequence of real-valued functions parameterized by n and converging pointwise to a limit as $n \to \infty$. Assume that $R_n(x)$ is dominated by another function independent of n, i.e., $|R_n(x)| \leq a(x)$ for all $x \in \mathcal{D}$ in the domain of $R_n(x)$ and where $a(x)$ is integrable, i.e., $\int_\mathcal{D} a(x)dx < \infty$. Then, it holds that*

$$\lim_{n\to\infty} \left(\int_\mathcal{D} R_n(x)dx \right) = \int_\mathcal{D} \left(\lim_{n\to\infty} R_n(x) \right) dx \tag{16.140}$$

That is, we can switch the limit and integral signs.

We can use the above result to justify exchanging derivatives (which are limit operations) with expectations (which are integral operations). We provide three statement variations that lead to similar conclusions under related but different conditions. In the proofs, we follow the same line of reasoning that is used to establish the classical Leibniz integral rule from calculus for the derivative of an integral expression by means of the dominated convergence theorem – see, e.g., Natanson (1961), Buck (1965), Hewitt and Stromberg (1969), Dieudonné (1969), Apostol (1974), Lewin (1987), Bartle (1995), Troutman (1996), or Norris (2013).

Theorem 16.1. (Switching expectation and gradient operations I) *Consider a function $Q(w;\boldsymbol{x}) : \mathbb{R}^M \times \mathbb{R}^P \to \mathbb{R}$, where $\boldsymbol{x}$ is a random vector. Assume Q is first-order differentiable with respect to w, and that for any $w \in \text{dom}\, Q(w;x)$ there exists a function $b(w;x) : \mathbb{R}^M \times \mathbb{R}^P \to [0,\infty)$ satisfying $\mathbb{E}\, b(w;\boldsymbol{x}) < \infty$ and*

$$\|\nabla_w Q(w+\delta w;x)\| \;\le\; b(w;x), \quad \textit{for any } \|\delta w\| \le \epsilon \tag{16.141}$$

Then, assuming the expectations over the distribution of $\boldsymbol{x}$ exist, it holds that

$$\nabla_w\Big(\mathbb{E}\, Q(w;\boldsymbol{x})\Big) \;=\; \mathbb{E}\Big(\nabla_w Q(w;\boldsymbol{x})\Big) \tag{16.142}$$

Proof: The argument is motivated by the discussion in Hewitt and Stromberg (1969, pp. 172–173), Bartle (1995, corollary 5.7), and also by the derivation used in the proof of Theorem 3.5.1 from Norris (2013) extended to vector variables. Let $w[m]$ denote the mth entry of $w \in \mathbb{R}^M$. Then, from the definition of the gradient vector of a multivariable function we know that:

$$\nabla_{w^\mathsf{T}} \mathbb{E}\, Q(w;\boldsymbol{x}) \;\stackrel{\Delta}{=}\; \text{col}\left\{ \frac{\partial \mathbb{E}\, Q(w;\boldsymbol{x})}{\partial w[1]}, \frac{\partial \mathbb{E}\, Q(w;\boldsymbol{x})}{\partial w[2]}, \ldots, \frac{\partial \mathbb{E}\, Q(w;\boldsymbol{x})}{\partial w[M]} \right\} \tag{16.143}$$

We establish result (16.142) by considering the individual entries of the gradient vector. Let $\alpha > 0$ denote a small positive scalar and let e_m denote the mth basis vector in $\mathbb{R}^M$ with all its entries equal to zero except for the mth entry, which is equal to 1. According to the definition of the differentiation operation, we have

$$\begin{aligned}
\frac{\partial \mathbb{E}\, Q(w;\boldsymbol{x})}{\partial w[m]} &= \lim_{\alpha\to 0} \frac{1}{\alpha}\Big(\mathbb{E}\, Q(w+\alpha e_m;\boldsymbol{x}) \;-\; \mathbb{E}\, Q(w;\boldsymbol{x})\Big) \\
&= \lim_{\alpha\to 0} \frac{1}{\alpha}\mathbb{E}\Big(Q(w+\alpha e_m;\boldsymbol{x}) \;-\; Q(w;\boldsymbol{x})\Big) \\
&\stackrel{(a)}{=} \lim_{\alpha\to 0} \frac{1}{\alpha}\,\mathbb{E}\Big(\nabla_w Q(w+t_m\alpha e_m;\boldsymbol{x})\Big)^\mathsf{T} \alpha e_m \\
&\stackrel{(b)}{=} \lim_{\alpha\to 0}\, \mathbb{E}\left(\frac{\partial Q(w+t_m\alpha e_m;\boldsymbol{x})}{\partial w[m]}\right)
\end{aligned} \tag{16.144}$$

for some constant $t_m \in (0,1)$. Equality (a) holds because of the mean-value theorem, while equality (b) holds because the elements in e_m are all zero except for the mth entry. We next introduce the function

$$g_m(w;x) \;\stackrel{\Delta}{=}\; \frac{\partial Q(w;x)}{\partial w[m]} \tag{16.145}$$

so that (16.144) becomes

$$\frac{\partial \mathbb{E}\, Q(w;\boldsymbol{x})}{\partial w[m]} \;=\; \lim_{\alpha\to 0}\, \mathbb{E}\, g_m(w+t_m\alpha e_m;\boldsymbol{x}) \;=\; \lim_{\alpha\to 0}\left(\int_{x\in\mathcal{D}} g_m(w+t_m\alpha e_m;\boldsymbol{x}) f_{\boldsymbol{x}}(x)dx\right) \tag{16.146}$$

where $\mathcal{D}$ is the domain of $\boldsymbol{x}$. Now consider any scalar sequence $\alpha^{(n)} \to 0$. We define $\alpha^{(n)}e_m \in \mathbb{R}^M$ as the vector in which all the elements are zero except for the mth entry set to $\alpha^{(n)}$. With this notation, expression (16.144) becomes

$$\frac{\partial \mathbb{E}\, Q(w;\boldsymbol{x})}{\partial w[m]} \;=\; \lim_{n\to+\infty}\left(\int_{x\in\mathcal{D}} g_m\Big(w+t_m\alpha^{(n)}e_m;\boldsymbol{x}\Big)f_{\boldsymbol{x}}(x)dx\right) \tag{16.147}$$

Next, we define

$$R_n(x;w) \overset{\Delta}{=} g_m\Big(w+t_m\alpha^{(n)}e_m;x\Big)f_{\boldsymbol{x}}(x), \quad w\in\text{dom}(Q) \tag{16.148}$$

In $R_n(x;w)$, the symbol x is the variable and w is a parameter. The function $R_n(x;w)$ is dominated. Indeed, using condition (16.141) we have

$$\begin{aligned}\left|g_m\Big(w+t_m\alpha^{(n)}e_m;x\Big)\right| &= \left|\frac{\partial Q\Big(w+t_m\alpha^{(n)}e_m;x\Big)}{\partial w[m]}\right| \\ &\le \left\|\nabla_w\, Q\Big(w+t_m\alpha^{(n)}e_m;x\Big)\right\| \\ &\le b(w;x)\end{aligned} \tag{16.149}$$

for any $n\ge N_o$ where N_o is sufficiently large. Therefore, for any $x\in\mathcal{D}$ and $w\in\text{dom}(Q)$, it holds that

$$|R_n(x;w)| \;\le\; b(w;x)f_{\boldsymbol{x}}(x) \tag{16.150}$$

Since, by assumption, $\mathbb{E}\, b(w;\boldsymbol{x}) < +\infty$, we know that $b(w;x)f_{\boldsymbol{x}}(x)$ is integrable. Finally, applying the dominated convergence theorem, we know that for any $w\in\text{dom}\,(Q)$, it holds that

$$\begin{aligned}\lim_{n\to+\infty}\int_{x\in\mathcal{D}} R_n(x;w)dx &= \int_{x\in\mathcal{D}}\lim_{n\to+\infty} R_n(x;w)dx \\ &= \int_{x\in\mathcal{D}}\frac{\partial Q(w;x)}{\partial w[m]}f_{\boldsymbol{x}}(x)dx \\ &= \mathbb{E}\left(\frac{\partial Q(w;\boldsymbol{x})}{\partial w[m]}\right)\end{aligned} \tag{16.151}$$

which also implies from (16.147) that

$$\frac{\partial \mathbb{E}\, Q(w;\boldsymbol{x})}{\partial w[m]} \;=\; \mathbb{E}\left(\frac{\partial Q(w;\boldsymbol{x})}{\partial w[m]}\right) \tag{16.152}$$

From (16.143) and (16.152) we arrive at (16.142).

■

We state a second related variation of the theorem, which holds when $\nabla_w Q(w;\boldsymbol{x})$ is continuous in w and the distribution of $\boldsymbol{x}$ is such that the means of the absolute entries of $\nabla_w Q(w;\boldsymbol{x})$ are bounded – see, e.g., the statement for the dominated convergence theorem for bounded functions in Natanson (1961), Luxemburg (1971), Lewin (1987), and the brief note by Ene (1999). We continue with the same notation used in the proof of the previous version.

Theorem 16.2. (Switching expectation and gradient operations II) *Consider a function $Q(w;\boldsymbol{x}) : \mathbb{R}^M \times \mathbb{R}^P \rightarrow \mathbb{R}$, where $\boldsymbol{x}$ is a random vector. Assume Q is first-order differentiable with respect to w, $\nabla_w Q(w;x)$ is continuous over w, and*

$$\mathbb{E}\left|\frac{\partial Q(w;\boldsymbol{x})}{\partial w[m]}\right| < +\infty, \quad m = 1,2,\ldots,M \tag{16.153}$$

Then, it holds that

$$\nabla_w\Big(\mathbb{E}\, Q(w;\boldsymbol{x})\Big) = \mathbb{E}\Big(\nabla_w\, Q(w;\boldsymbol{x})\Big) \tag{16.154}$$

where the expectations are over the distribution of $\boldsymbol{x}$.

Proof: We repeat similar arguments. We start from (16.146):

$$\frac{\partial \mathbb{E}\, Q(w;\boldsymbol{x})}{\partial w[m]} = \lim_{\alpha\rightarrow 0} \mathbb{E}\, g_m(w + t_m\alpha e_m;\boldsymbol{x}) = \lim_{\alpha\rightarrow 0}\left(\int_{x\in\mathcal{D}} g_m(w + t_m\alpha e_m;\boldsymbol{x}) f_{\boldsymbol{x}}(x)dx\right) \tag{16.155}$$

Since $\nabla_w Q(w;x)$ is continuous with respect to w, we know that $g_m(w;x)$ is continuous with respect to w. Using the same sequence $\alpha^{(n)}$ defined in the previous proof, this property implies that

$$\lim_{\alpha\rightarrow 0} g_m\Big(w + t_m\alpha^{(n)}e_m;x\Big) = g_m(w;x) \tag{16.156}$$

which also means that for any $\epsilon > 0$, there exists a positive integer N_o such that for all $n > N_o$, it holds that

$$\Big|g_m\Big(w + t_m\alpha^{(n)}e_m;x\Big) - g_m(w;x)\Big| < \epsilon \tag{16.157}$$

As a result, for any $n > N_o$, we have

$$\begin{aligned}
\Big|g_m\Big(w + t_m\alpha^{(n)}e_m;x\Big)\Big| &= \Big|g_m\Big(w + t_m\alpha^{(n)}e_m;x\Big) - g_m(w;x) + g_m(w;x)\Big| \\
&= \Big|g_m\Big(w + t_m\alpha^{(n)}e_m;x\Big) - g_m(w;x)\Big| + |g_m(w;x)| \\
&\overset{(16.157)}{\leq} |g_m(w;x)| + \epsilon
\end{aligned} \tag{16.158}$$

Next, we define, for any $w \in \text{dom}(Q)$,

$$R_n(x;w) \triangleq g_m\Big(w + t_m\alpha^{(n+N)}e_m;x\Big) f_{\boldsymbol{x}}(x) \tag{16.159a}$$

$$b(w;x) \triangleq (|g_m(w;x)| + \epsilon)\, f_{\boldsymbol{x}}(x) \tag{16.159b}$$

It follows from (16.158) and the fact that $f_{\boldsymbol{x}}(x) \geq 0$ that $|R_n(x)| \leq b(w;x)$ for all n. Moreover, under assumption (16.153):

$$\begin{aligned}
\int_{x\in\mathcal{D}} b(w;x)dx &= \int_{x\in\mathcal{D}} (|g_m(w;x)| + \epsilon)\, f_{\boldsymbol{x}}(x)dx \\
&= \epsilon + \int_{x\in\mathcal{D}} |g_m(w;x)|\, f_{\boldsymbol{x}}(x)dx \\
&= \epsilon + \mathbb{E}\left|\frac{\partial Q(w;\boldsymbol{x})}{\partial w[m]}\right| < +\infty
\end{aligned} \tag{16.160}$$

which implies that $b(w;x)$ is integrable. We conclude that $R_n(x)$ is dominated by an integrable function $b(w;x)$. Using (16.155) and the dominated convergence theorem, we know that for any $w \in \text{dom}\, Q$, it holds that

$$\begin{aligned}
\frac{\partial \mathbb{E}\, Q(w;\boldsymbol{x})}{\partial w[m]} &= \lim_{n\to+\infty} \int_{x\in\mathcal{D}} R_n(x;w)dx \\
&= \int_{x\in\mathcal{D}} \lim_{n\to+\infty} R_n(x;w)dx \\
&= \int_{x\in\mathcal{D}} \lim_{n\to+\infty} g_m\left(w + t_m\alpha^{(n+N)}e_m; x\right) f_{\boldsymbol{x}}(x)dx \\
&= \int_{x\in\mathcal{D}} g_m(w;x) f_{\boldsymbol{x}}(x)dx \\
&= \int_{x\in\mathcal{D}} \frac{\partial Q(w;x)}{\partial w[m]} f_{\boldsymbol{x}}(x)dx \\
&= \mathbb{E}\left(\frac{\partial Q(w;\boldsymbol{x})}{\partial w[m]}\right)
\end{aligned} \tag{16.161}$$

From this result and (16.143) we arrive at (16.154).

■

A straightforward corollary follows if the random variable $\boldsymbol{x}$ takes values in a *compact* (i.e., closed and bounded) set.

Corollary 16.1. (Switching expectation and gradient operations III) *Consider a function $Q(w;\boldsymbol{x}) : \mathbb{R}^M \times \mathbb{R}^P \to \mathbb{R}$, where $\boldsymbol{x}$ is a random vector variable taking on values in a compact set, $\mathcal{D}$. Assume Q is first-order differentiable with respect to w and $\nabla_w Q(w;\boldsymbol{x})$ is continuous with respect to w and $\boldsymbol{x}$, respectively. Then, it holds that*

$$\nabla_w \Big(\mathbb{E}\, Q(w;\boldsymbol{x})\Big) = \mathbb{E}\Big(\nabla_w Q(w;\boldsymbol{x})\Big) \tag{16.162}$$

where the expectations are over the distribution of $\boldsymbol{x}$.

Proof: Recall that $g_m(w;x) = \partial Q(w;x)/\partial w[m]$, which is continuous with respect to x since we are assuming that $\nabla_w Q(w;x)$ is continuous with respect to x. Now, given that x is defined over a compact set, we know that

$$|g_m(w;x)| < C, \quad \forall\, x \in \mathcal{D} \tag{16.163}$$

Therefore, it holds that

$$\int_{x\in\mathcal{D}} |g_m(w;x)| f_{\boldsymbol{x}}(x)dx \le C\left(\int_{x\in\mathcal{D}} f_{\boldsymbol{x}}(x)dx\right) = C < +\infty. \tag{16.164}$$

In other words, the expectation $\mathbb{E}\,|g_m(w;\boldsymbol{x})| = \mathbb{E}\,|\partial Q(w;x)/\partial w[m]|$ exists for any $m \in \{1,\ldots,M\}$. Since all conditions of Theorem 16.2 are satisfied, we conclude that (16.162) holds.

■

There are similar results allowing the exchange of expectation and *subgradient* operations when the function $Q(w;\boldsymbol{x})$ is nondifferentiable at some locations. The proof of the following statement is given in Rockafellar and Wets (1981, eq. 20) and also Wets (1989, prop. 2.10).

LEMMA 16.1. (Switching expectation and subgradient operations) *Consider a convex function* $Q(w;\boldsymbol{x}) : \mathbb{R}^M \times \mathbb{R}^P \rightarrow \mathbb{R}$*, where* $\boldsymbol{x}$ *is a random vector variable, and assume* $\mathbb{E}\,Q(w;\boldsymbol{x})$ *is finite in a neighborhood of* w *where the subgradient is computed. Then, it holds that*

$$\partial_w \Big(\mathbb{E}\,Q(w;\boldsymbol{x})\Big) = \mathbb{E}\,\Big(\partial_w\, Q(w;\boldsymbol{x})\Big) \tag{16.165}$$

where ∂_w *refers to the subdifferential operator.*

REFERENCES

Agarwal, A., P. L. Bartlett, P. Ravikumar, and M. J. Wainwright (2012), "Information-theoretic lower bounds on the oracle complexity of convex optimization," *IEEE Trans. Inf. Theory*, vol. 58, no. 5, pp. 3235–3249.

Albert, A. E. and L. A. Gardner (1967), *Stochastic Approximation and Nonlinear Regression*, MIT Press.

Amari, S. I. (1967), "A theory of adaptive pattern classifiers," *IEEE Trans. Elec. Comput.*, vol. 16, pp. 299–307.

Apostol, T. (1974), *Mathematical Analysis*, 2nd ed., Addison-Wesley.

Applebaum, S. P. (1966), *Adaptive Arrays*, Rep. SPLTR 66-1, Syracuse University Research Corporation.

Bach, F. and E. Moulines (2011), "Non-asymptotic analysis of stochastic approximation algorithms for machine learning," *Proc. Advances Neural Information Processing Systems* (NIPS), pp. 451–459, Granada.

Bartle, R. G. (1995), *The Elements of Integration and Lebesgue Measure*, Wiley.

Bell, R. and Y. Koren (2007a), "Scalable collaborative filtering with jointly derived neighborhood interpolation weights," *Proc. IEEE Int. Conf. Data Mining* (ICDM), pp. 43–52, Omaha, NE.

Bell, R. and Y. Koren (2007b), "Lessons from the Netflix prize challenge," *ACM SIGKDD Explorations Newsletter*, vol. 9, no. 2, pp. 75–79.

Bell, R., Y. Koren, and C. Volinsky (2007), "The BellKor solution to the Netflix Prize," available at www.netflixprize.com/assets/ProgressPrize2007_KorBell.pdf

Benveniste, A., M. Métivier, and P. Priouret (1987), *Adaptive Algorithms and Stochastic Approximations*, Springer.

Bertsekas, D. P. and J. N. Tsitsiklis (1997), *Parallel and Distributed Computation: Numerical Methods*, Athena Scientific.

Bertsekas, D. P. and J. N. Tsitsiklis (2000), "Gradient convergence in gradient methods with errors," *SIAM J. Optim.*, vol. 10, no. 3, pp. 627–642.

Blum, J. R. (1954), "Multidimensional stochastic approximation methods," *Ann. Math. Stat.*, vol. 25, pp. 737–744.

Bottou, L. (1998), "Online algorithms and stochastic approximations," in *Online Learning and Neural Networks*, D. Saad, editor, Cambridge University Press.

Bottou, L. (2012), "Stochastic gradient descent tricks," in *Neural Networks: Tricks of the Trade*, 2nd ed., G. Montavon, G. B. Orr, and K.-R. Muller, editors, pp. 421–436, Springer.

Bottou, L. and O. Bousquet (2008), "The tradeoffs of large scale learning," *Proc. Advances Neural Information Processing Systems* (NIPS), vol. 20, pp. 161–168, Vancouver.

Buck, R. C. (1965), *Advanced Calculus*, McGraw-Hill.

Cesa-Bianchi, N., A. Conconi, and C. Gentile (2004), "On the generalization ability of on-line learning algorithms," *IEEE Trans. Inf. Theory*, vol. 50, no. 9, pp. 2050–2057.

Cesa-Bianchi, N. and G. Lugosi (2006), *Prediction, Learning, and Games*, Cambridge University Press.

Dieudonné, J. (1969), *Foundations of Modern Analysis*, vol. 1, Academic Press.

Dror, G., N. Koenigstein, Y. Koren, and M. Weimer (2012), "The Yahoo! music dataset and KDDCup 11," *J. Mach. Learn. Res.*, vol. 18, pp. 8–18.

Dvoretzky, A. (1956), "On stochastic approximation," *Proc. 3rd Berkeley Symp. Mathematical Statistics and Probability*, vol. 1, pp. 39–56, Berkeley, CA.

Ene, V. (1999), "Some queries concerning convergence theorems," *Real Anal. Exchange*, vol. 25, no. 2, pp. 955–958.

Funk, S. (2006), "Netflix update: Try this at home," available at https://sifter.org/ simon/journal/20061211.html

Gabor, D., W. P. Z. Wilby, and R. Woodcock (1961), "An universal nonlinear filter, predictor, and simulator which optimizes itself by a learning process," *Proc. IEE*, vol. 108, no. 40, pp. 422–436.

Gladyshev, E. G. (1965), "On stochastic approximations," *Theory Prob. Appl.*, vol. 10, pp. 275–278.

Hassibi, B., A. H. Sayed, and T. Kailath (1994a), "$\mathcal{H}^\infty$-optimality criteria for LMS and backpropagation," *Proc. Advances Neural Information Processing Systems* (NIPS), vol. 6, pp. 351–358, Denver, CO.

Hassibi, B., A. H. Sayed, and T. Kailath (1996), "$\mathcal{H}^\infty$-optimality of the LMS algorithm," *IEEE Trans. Signal Process.*, vol. 44, no. 2, pp. 267–280.

Haykin, S. (2001), *Adaptive Filter Theory*, 4th ed., Prentice Hall.

Hewitt, E. and K. Stromberg (1969), *Real and Abstract Analysis*, Springer.

Hu, Y. F., Y. Koren, and C. Volinsky (2008), "Collaborative filtering for implicit feedback datasets," *Proc. IEEE Int. Conf. Data Mining* (ICDM), pp. 263–272, Pisa, Italy.

Kaczmarz, S. (1937), "Angenäherte Auflösung von Systemen linearer Gleichungen," *Bull. Int. Acad. Polon. Sci. Lett.* A, pp. 335–357.

Kiefer, J. and J. Wolfowitz (1952), "Stochastic estimation of the maximum of a regression function," *Ann. Math. Statist.*, vol. 23, no. 3, pp. 462–466.

Koren, Y. (2008), "Factorization meets the neighborhood: A multifaceted collaborative filtering model," *Proc. ACM SIGKDD Int. Conf. Knowledge Discovery and Data Mining*, pp. 426–434, Las Vegas, NV.

Koren, Y., R. Bell, and C. Volinsky (2009), "Matrix factorization techniques for recommender systems," *Computer*, vol. 42, no. 8, pp. 30–37.

Kushner, H. J. (1984), *Approximation and Weak Convergence Methods for Random Processes, with Applications to Stochastic System Theory*, MIT Press.

Kushner, H. J. and D. S. Clark (1978), *Stochastic Approximation for Constrained and Unconstrained Systems*, Springer.

Kushner, H. J. and G. G. Yin (2003), *Stochastic Approximation and Recursive Algorithms and Applications*, Springer.

Lai, T. L. (2003), "Stochastic approximation," *Ann. Statist.*, vol. 31, no. 2, pp. 391–406.

Lewin, J. W. (1987), "Some applications of the bounded convergence theorem for an introductory course in analysis," *Amer. Math. Monthly*, vol. 94, no. 10, pp. 988–993.

Ljung, L. (1977), "Analysis of recursive stochastic algorithms," *IEEE Trans. Aut. Control*, vol. 22, pp. 551–575.

Luxemburg, W. A. J. (1971), "Arzelà's dominated convergence theorem for the Riemann integral," *Amer. Math. Monthly*, vol. 78, pp. 970–979.

Marti, K. (2005), *Stochastic Optimization Methods*, Springer.

Mattson, R. L. (1959a), *The Design and Analysis of an Adaptive System for Statistical Classification*, S. M. Thesis, MIT.

Mattson, R. L. (1959b), "A self-organizing binary system," *Proc. Eastern Joint IRE-AIEE-ACM Computer Conf.*, pp. 212–217, Boston, MA.

Mendel, J. M. and K. S. Fu (1970), *Adaptive, Learning, and Pattern Recognition Systems: Theory and Applications*, Academic Press.

Natanson, I. P. (1961), *Theory of Functions of a Real Variable*, 2nd ed., Frederick Ungar Publishing Co.

Norris, J. R. (2013), "Probability and measure," unpublished notes, available at www.statslab.cam.ac.uk/~james/Lectures/pm.pdf

Paterek, A. (2007), "Improving regularized singular value decomposition for collaborative filtering," *Proc. KDD Cup and Workshop*, pp. 39–42, ACM Press.

Pilaszy, I., D. Zibriczky, and D. Tikk (2010), "Fast ALS-based matrix factorization for explicit and implicit feedback datasets," *Proc. ACM Conf. Recommender Systems*, pp. 71–78, Barcelona.

Polyak, B. T. (1987), *Introduction to Optimization*, Optimization Software.

Polyak, B. T. and A. Juditsky (1992), "Acceleration of stochastic approximation by averaging," *SIAM J. Control Optim.*, vol. 30, no. 4, pp. 838–855.

Robbins, H. and S. Monro (1951), "A stochastic approximation method," *Ann. Math. Stat.*, vol. 22, pp. 400–407.

Rockafellar, R. T. and R. Wets (1981), "On the interchange of subdifferentiation and conditional expectation for convex functionals," International Institute for Applied Systems Analysis, IIASA working paper WP-81-089, Laxenburg, Austria.

Rosenblatt, F. (1957), *The Perceptron: A Perceiving and Recognizing Automaton*, Technical Report 85-460-1, Project PARA, Cornell Aeronautical Lab.

Rosenblatt, F. (1958), "The Perceptron: A probabilistic model for information storage and organization in the brain," *Psychol. Rev.*, vol. 65, no. 6, pp. 386–408.

Rosenblatt, F. (1962), *Principles of Neurodynamics: Perceptrons and the Theory of Brain Mechanisms*, Spartan Press.

Royden, H. L. (1988), *Real Analysis*, Prentice Hall.

Rudin, W. (1976), *Principles of Mathematical Analysis*, 3rd ed., McGraw-Hill.

Ruppert, D. (1988), *Efficient Estimation from a Slowly Convergent Robbins–Monro Process*, Technical Report 781, Cornell University, School of Operations Research and Industrial Engineering.

Sayed, A. H. (2003), *Fundamentals of Adaptive Filtering*, Wiley.

Sayed, A. H. (2008), *Adaptive Filters*, Wiley.

Sayed, A. H. (2014a), *Adaptation, Learning, and Optimization over Networks,* Foundations and Trends in Machine Learning, NOW Publishers, vol. 7, no. 4–5, pp. 311–801.

Sayed, A. H. and M. Rupp (1996), "Error energy bounds for adaptive gradient algorithms," *IEEE Trans. Signal Process.*, vol. 44, no. 8, pp. 1982–1989.

Schmetterer, L. (1961), "Stochastic approximation," *Proc. Berkeley Symp. Math. Statist. Probab.*, pp. 587–609, Berkeley, CA.

Sefl, O. (1960), "Filters and predictors which adapt their values to unknown parameters of the input process," *Trans. 2nd Conf. Information Theory,* Czechoslovak Academy of Sciences, Prague.

Shalev-Shwartz, S. (2011), "Online learning and online convex optimization," *Found. Trends Mach. Learn.*, vol. 4, no. 2, pp. 107–194.

Spall, J. C. (2003), *Introduction to Stochastic Search and Optimization: Estimation, Simulation and Control*, Wiley.

Strohmer, T. and R. Vershynin (2009), "A randomized Kaczmarz algorithm with exponential convergence," *J. Fourier Anal. Appl.*, vol. 15, no. 2, pp. 262–278.

Symeonidis, P. and A. Zioupos (2017), *Matrix and Tensor Factorization Techniques for Recommender Systems*, Springer.

Takacs, G., I. Pilaszy, B. Nemeh, and D. Tikk (2007), "Major components of the gravity recommendation system," *SIGKDD Explorations*, vol. 9, pp. 80–84.

Theodoridis, S. (2015), *Machine Learning: A Bayesian and Optimization Perspective*, Academic Press.

Troutman, J. L. (1996), *Variational Calculus and Optimal Control*, Springer.

Tsypkin, Y. Z. (1971), *Adaptation and Learning in Automatic Systems*, Academic Press.

von Mises, R. and H. Pollaczek-Geiringer (1929), "Praktische verfahren der gleichungs-auflösung," *Z. Agnew. Math. Mech.*, vol. 9.

Wasan, M. T. (1969), *Stochastic Approximation*, Cambridge University Press.

Wetherhill, G. B. (1966), *Sequential Methods in Statistics,* Methuen.

Wets, R. (1989), "Stochastic programming," in *Handbook for Operations Research and Management Sciences*, G. Nemhauser and A. Rinnnooy Kan, editors, vol. 1, pp. 573–629.

Widrow, B. and M. E. Hoff (1960), "Adaptive switching circuits," *IRE WESCON Conv. Rec.*, pt. 4, pp. 96–104.

Widrow, B., P. Mantey, L. J. Griffiths, and B. Goode (1967), "Adaptive antenna systems," *Proc. IEEE*, vol. 55, no. 12, pp. 2143–2159.

Widrow, B. and S. D. Stearns (1985), *Adaptive Signal Processing*, Prentice Hall.

Zhou, Y., D. Wilkinson, R. Schreiber, and R. Pan (2008), "Large-scale parallel collaborative filtering for the Netflix prize," in *Algorithmic Aspects in Information and Management*, Rudolf Fleischer and Jinhui Xu, editors, pp. 337–348, Springer.

17 Adaptive Gradient Methods

In this chapter we examine variations of stochastic approximation methods where the learning rates (or step sizes) and even the update directions are continually adjusted in a manner that is dependent on the data. The intention is to enhance convergence performance and move away from unfavorable limit points, such as saddle points in nonconvex optimization scenarios. We describe several *adaptive gradient methods*, also known as methods with *adaptive learning rates*. They can be broadly categorized into three classes: One group adjusts the learning rate, a second group adjusts the update direction, and a third group adjusts both the learning rate and the update direction. It is sufficient for our purposes to illustrate the algorithms for smooth loss functions and for empirical risk minimization problems. We already know from the discussion in the previous chapter that we would simply replace gradients by subgradients for nonsmooth losses. We can also incorporate proximal steps and apply the same algorithms to both empirical and stochastic risk optimization.

17.1 MOTIVATION

Consider the empirical risk minimization problem:

$$w^\star = \underset{w\in\mathbb{R}^M}{\text{argmin}} \left\{ P(w) \triangleq \frac{1}{N}\sum_{m=0}^{N-1} Q(w;\gamma(m),h_m) \right\} \tag{17.1}$$

with a first-order differentiable loss function, $Q(w;\cdot)$. The stochastic gradient algorithm for seeking its minimizer is given by

$$\left\{ \begin{array}{l} \text{select a random data point } (\boldsymbol{\gamma}(n),\boldsymbol{h}_n) \\ \boldsymbol{w}_n = \boldsymbol{w}_{n-1} - \mu\,\nabla_{w^\mathsf{T}} Q(\boldsymbol{w}_{n-1};\boldsymbol{\gamma}(n),\boldsymbol{h}_n), \quad n\geq 0 \end{array} \right. \tag{17.2}$$

where we can also consider mini-batch versions of the recursion, if desired. We will establish in a future chapter that the mean-square error (MSE) performance of this algorithm tends asymptotically to a region of size $O(\mu)$ around the minimizer, namely,

$$\limsup_{n\to\infty} \mathbb{E}\,\|\widetilde{\boldsymbol{w}}_n\|^2 = O(\mu) \tag{17.3}$$

where $\widetilde{\boldsymbol{w}}_n = w^\star - \boldsymbol{w}_n$. This means that the smaller μ is, the closer $\boldsymbol{w}_n$ will be to $w^\star$ in the MSE sense. We will also establish that the rate at which $\boldsymbol{w}_n$ approaches this limiting region depends on $1 - O(\mu)$. In this case, smaller values for μ slow down convergence; they also slow down the response of the algorithm to drifts in $w^\star$ that may occur due to changes in the statistical properties of the data. Therefore, a balance needs to be attained between performance and convergence guarantees through the proper selection of the learning rate μ.

One common approach in the literature is to replace μ by a time-dependent sequence of the form:

$$\mu(n) = \begin{cases} \left(1 - \dfrac{n}{N_b}\right)\mu_a + \dfrac{n}{N_b}\mu_b, & n \le N_b \\ \mu_b, & n > N_b \end{cases} \tag{17.4}$$

with positive terms μ_a, μ_b and where μ_b is much smaller than μ_a, say, $\mu_b = 0.01\mu_a$. The above expression is plotted in Fig. 17.1. The step-size sequence starts at value μ_a at $n = 0$, decays to value μ_b at $n = N_b$, and stays at this smaller value afterwards. This construction combines larger and smaller step sizes; it enables faster convergence during the initial stages of the algorithm and smaller MSE in later stages.

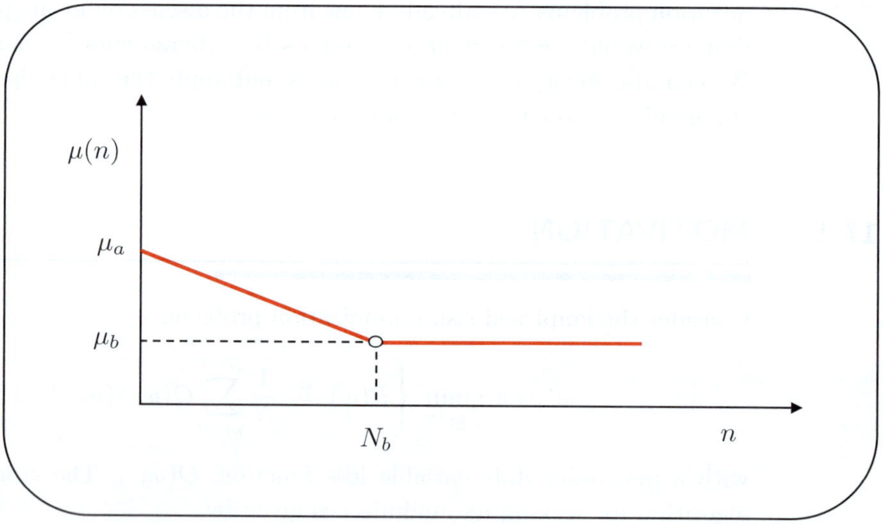

Figure 17.1 Linear decay of the step-size sequence from $\mu(n) = \mu_a$ at $n = 0$ to $\mu(n) = \mu_b$ at $n = N_b$. The step size stays at the value μ_b for $n \ge N_b$.

Adaptive learning rates

Observe from (17.2) that all entries of the gradient vector, $\nabla_{w^\mathsf{T}} Q(\boldsymbol{w}_{n-1}, \cdot)$, are scaled by the *same* step size μ. In many situations, it is desirable to scale different entries of the gradient vector by different amounts because some of these entries may be more or less relevant to the learning task than other entries. For example, some entries in the gradient vector may vary faster from one iteration to another, while other entries may vary slower. Some entries may also be more expressive than other entries (e.g., in classification problems, some attributes within the feature vectors $\{h_m\}$ may be more revealing than other attributes). Furthermore, the gradient vector may be sparse, with some entries assuming negligible values compared to the remaining entries in the vector. It is therefore useful to devise algorithms that are able to exploit on a finer scale the contributions of the individual entries of the gradient vector.

One way to develop such algorithms is to scale the gradient direction by some $M \times M$ positive-definite matrix, A^{-1}, and replace the update (17.2) by

$$\boldsymbol{w}_n = \boldsymbol{w}_{n-1} - \mu\, A^{-1}\, \nabla_{w^{\mathsf{T}}}\, Q(\boldsymbol{w}_{n-1}; \boldsymbol{\gamma}(n), \boldsymbol{h}_n), \quad n \geq 0 \tag{17.5}$$

This recursion corresponds to a stochastic implementation of the *quasi-Newton* procedure encountered earlier in (12.193). For simplicity, we will limit the choice of A to diagonal matrices, say, with entries denoted by

$$A = \text{diag}\Big\{a_1, a_2, \ldots, a_M\Big\} \tag{17.6}$$

In this way, the mth entry in the gradient vector is scaled by μ/a_m. In the commentaries at the end of Chapter 12 we listed other popular choices for A, which are not necessarily diagonal, such as the Newton method (12.197), where A is chosen as the Hessian matrix of $P(w)$ evaluated at $w = \boldsymbol{w}_{n-1}$, and the natural gradient method (6.131), where A is chosen as the Fisher information matrix.

In this chapter, we will employ scaling diagonal matrices A. We will describe three choices for A, known as AdaGrad, ADAM, and RMSprop. In all three cases, the matrix A will be both iteration- and data-dependent. For this reason, we will denote it more explicitly by the notation $\boldsymbol{A}_n$ at iteration n; the boldface letter is because the entries of $\boldsymbol{A}_n$ will be constructed directly from the data and will therefore be random. The methods described in the following will differ in their performance, with some being more successful than others. It is difficult to provide a general statement about which method is preferred, even in comparison to the vanilla stochastic gradient method (17.2) because it has been observed in practice that this method performs reasonably well in many situations of interest.

Saddle points

There is a second motivation for adaptive gradient methods, which is helpful for nonconvex optimization problems. Although we have focused mainly on *convex* risks in our treatment so far in the text, the various stochastic algorithms from the previous chapter and this chapter can also be applied to the "minimization" of nonconvex risks (which we will study in greater detail in Chapter 24). Some challenges arise in this case, such as the presence of local minima, local maxima, and saddle points. These situations are illustrated in Figs. 17.2 and 17.3 for two-dimensional weight vectors, $w \in \mathbb{R}^2$, with entries $\{w_1, w_2\}$.

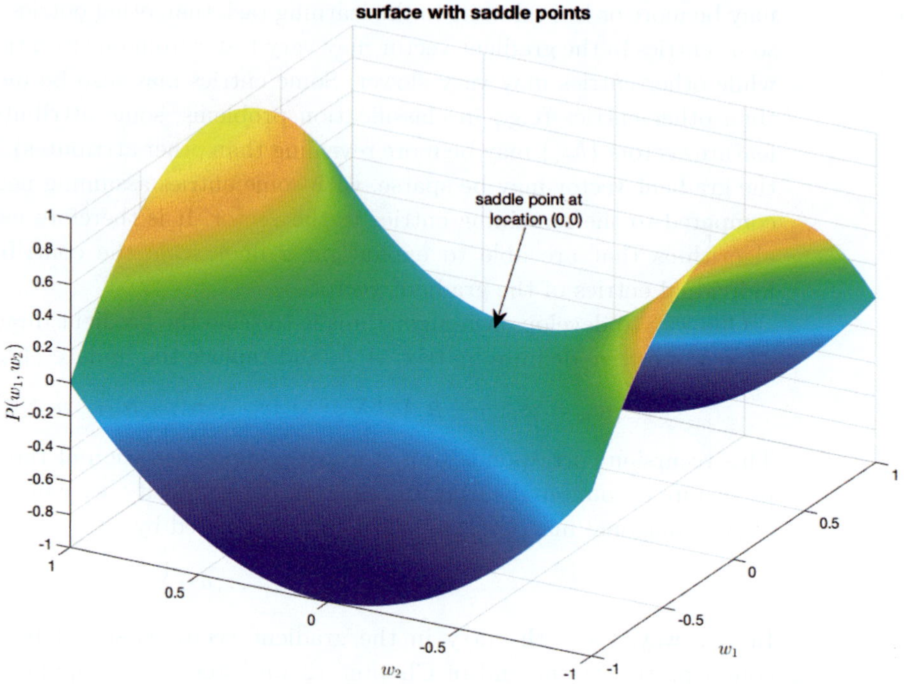

Figure 17.2 Plot of the function $P(w) = w_1^2 - w_2^2$, which has a saddle point at location $w = (0, 0)$.

Figure 17.2 plots the risk function $P(w) = w_1^2 - w_2^2$, which has a saddle point at location $w = (0, 0)$; a saddle point is neither a local maximum nor a local minimum for the function. For instance, it is seen in Fig. 17.2 that the location $w = (0, 0)$ is a maximum along the w_1-direction and a minimum along the w_2-direction. The gradient vector of the function will be zero at a saddle point, but the Hessian matrix will be indefinite. Figure 17.3 shows a risk function with global and local maxima, global and local minima, as well as saddle points (such as the saddle point appearing between two "hills" on the surface of the plot). Since the algorithms described in this chapter will differ in the search directions they employ, some will be more successful than others in moving away from saddle points. For example, referring to the plot in Fig. 17.2, if a gradient-descent implementation approaches the saddle point by taking steps along the w_2-direction, then it will end up assuming that it has reached a minimizer. The fact that the search directions are noisy will help inject some element of randomness into the update direction and consequently reduce the possibility of the algorithm continually updating along the w_2-direction toward the saddle point.

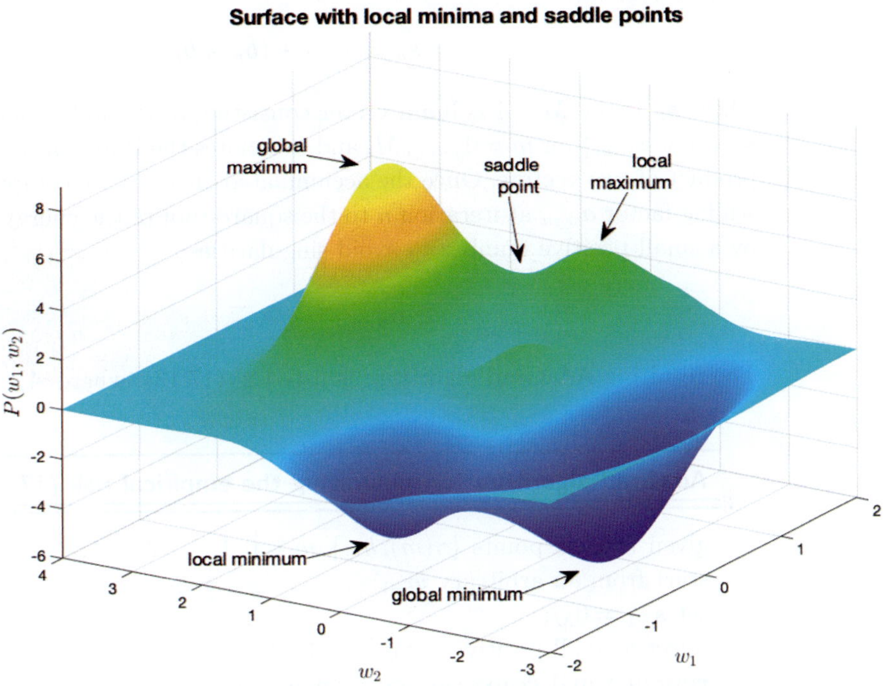

Figure 17.3 Plot of a risk function with global and local maxima, global and local minima, as well as saddle points.

17.2 ADAGRAD ALGORITHM

We start our exposition of adaptive gradient methods by considering the AdaGrad (adaptive gradient) algorithm. Its update relation takes the form:

$$\boldsymbol{w}_n = \boldsymbol{w}_{n-1} - \mu\, \boldsymbol{A}_n^{-1} \nabla_{w^\mathsf{T}} Q(\boldsymbol{w}_{n-1}; \boldsymbol{\gamma}(n), \boldsymbol{h}_n), \quad n \geq 0 \tag{17.7}$$

for a particular choice of $\boldsymbol{A}_n$ constructed as follows. To simplify the notation, we denote the $M \times 1$ gradient vector of the loss function at iteration n by:

$$\boldsymbol{b}_n \stackrel{\Delta}{=} \nabla_{w^\mathsf{T}} Q(\boldsymbol{w}_{n-1}; \boldsymbol{\gamma}(n), \boldsymbol{h}_n) = \text{col}\Big\{\boldsymbol{b}_{n,1}, \boldsymbol{b}_{n,2}, \ldots, \boldsymbol{b}_{n,M}\Big\} \tag{17.8}$$

with individual entries $\{\boldsymbol{b}_{n,m}\}$. For each entry of index m, we compute its *accumulated* energy over time:

$$\boldsymbol{s}_{n,m} \stackrel{\Delta}{=} \sum_{n'=0}^{n} \boldsymbol{b}_{n',m}^2 = \boldsymbol{b}_{0,m}^2 + \boldsymbol{b}_{1,m}^2 + \boldsymbol{b}_{2,m}^2 + \boldsymbol{b}_{3,m}^2 + \ldots + \boldsymbol{b}_{n,m}^2 \tag{17.9}$$

which can also be evaluated recursively by using

$$\boldsymbol{s}_{n,m} = \boldsymbol{s}_{n-1,m} + \boldsymbol{b}_{n,m}^2, \quad \boldsymbol{s}_{-1,m} = 0, \;\; m = 1, \ldots, M \tag{17.10}$$

or, in vector form,

$$\boldsymbol{s}_n = \boldsymbol{s}_{n-1} + (\boldsymbol{b}_n \odot \boldsymbol{b}_n) \tag{17.11}$$

where $\boldsymbol{s}_n$ is the $M \times 1$ column vector consisting of the individual energy terms, $\boldsymbol{s}_n = \text{col}\{\boldsymbol{s}_{n,m}\}$ for $m = 1, \ldots, M$, and $\odot$ denotes the Hadamard (or elementwise) product of two vectors. Once the accumulated energies are computed, we set the scaling factor $\boldsymbol{a}_{n,m}$ at iteration n to the square-root of the energy term, adjusted by a small positive number to avoid singularities:

$$\boldsymbol{a}_{n,m} \stackrel{\Delta}{=} \sqrt{\boldsymbol{s}_{n,m}} + \epsilon \tag{17.12}$$

The resulting AdaGrad algorithm is listed in (17.13), where $\boldsymbol{A}_n = \text{diag}\{\boldsymbol{a}_{n,m}\}$.

AdaGrad algorithm for minimizing the empirical risk (17.1).

given N data points $\{\gamma(m), h_m\}, m = 0, 1, \ldots, N-1$;
start from an arbitrary $\boldsymbol{w}_{-1}$;
set $\boldsymbol{s}_{-1} = 0_M$;
select a small positive ϵ, say, $\epsilon = 10^{-6}$.
repeat until convergence over $n \geq 0$:
select at random a data sample $(\boldsymbol{\gamma}(n), \boldsymbol{h}_n)$
$\boldsymbol{b}_n = \nabla_{w^\mathsf{T}} Q(\boldsymbol{w}_{n-1}; \boldsymbol{\gamma}(n), \boldsymbol{h}_n)$
$\boldsymbol{s}_n = \boldsymbol{s}_{n-1} + (\boldsymbol{b}_n \odot \boldsymbol{b}_n)$
denote the entries of $\boldsymbol{s}_n$ by $\{\boldsymbol{s}_{n,m}\}$
$\boldsymbol{A}_n = \text{diag}\left\{\sqrt{\boldsymbol{s}_{n,m}}\right\}_{m=1}^{M} + \epsilon I_M$
$\boldsymbol{w}_n = \boldsymbol{w}_{n-1} - \mu\, \boldsymbol{A}_n^{-1}\, \boldsymbol{b}_n$
end
return $w^\star \leftarrow \boldsymbol{w}_n$.

(17.13)

Observe that each entry of $\boldsymbol{w}_n$ is updated using an individual step size, which is equal to $\mu/\boldsymbol{a}_{n,m}$. In this way, AdaGrad is an adaptive learning rate method, one that adjusts the learning rate (or step size) in accordance with the energy of the gradient vector. Note in particular that since the accumulated energy increases for frequently occurring gradient entries, the size of the update step will be smaller for these entries. It has been observed in practice that AdaGrad works best in situations involving sparse gradient vectors and convex risk functions.

Derivation

We justify the form of the scaling matrix $\boldsymbol{A}_n$ used by AdaGrad by appealing to the regret analysis arguments from Section 16.5; the same choice can be motivated using MSE arguments instead – see future Prob. 19.24.

Consider the empirical minimization problem:

$$w^\star \triangleq \underset{w\in\mathbb{R}^M}{\text{argmin}} \left\{ P(w) \triangleq \frac{1}{N}\sum_{m=0}^{N-1} Q(w;\gamma(m),h_m) \right\} \tag{17.14}$$

where $Q(w,\cdot)$ is a first-order differentiable convex function over w. We wish to seek $w^\star$ by employing a stochastic gradient construction of the following form, modified to include a scaling positive-definite diagonal matrix A^{-1}:

$$w_n = w_{n-1} \; - \; \mu\, A^{-1}\, \nabla_{w^\mathsf{T}}\, Q(w_{n-1};\gamma(n),h_n) \tag{17.15}$$

We are dropping the boldface notation for convenience to indicate that we are working with specific realizations for the data and iterates. Observe that the gradient vector changes with n. We simplify the notation and introduce:

$$Q_n(w) \triangleq Q(w;\gamma(n),h_n) \tag{17.16}$$

so that the stochastic algorithm can be rewritten as

$$w_n = w_{n-1} \; - \; \mu\, A^{-1}\nabla_{w^\mathsf{T}}\, Q_n(w_{n-1}), \quad n \geq 0 \tag{17.17}$$

Subtracting $w^\star$ from both sides gives

$$\widetilde{w}_n = \widetilde{w}_{n-1} + \mu\, A^{-1}\, \nabla_{w^\mathsf{T}} Q_n(w_{n-1}) \tag{17.18}$$

where $\widetilde{w}_n = w^\star - w_n$. Equating the weighted squared Euclidean norms leads to

$$\|\widetilde{w}_n\|^2_A = \|\widetilde{w}_{n-1}\|^2_A + 2\mu(\nabla_{w^\mathsf{T}}\, Q(w_{n-1}))^\mathsf{T}\widetilde{w}_{n-1} + \mu^2\|\nabla_{w^\mathsf{T}} Q_n(w_{n-1})\|^2_{A^{-1}} \tag{17.19}$$

where the notation $\|x\|^2_A$ stands for $x^\mathsf{T} Ax$. Invoking the convexity of $Q(w,\cdot)$ over w and using property (8.5) for convex functions, we have

$$Q_n(w_{n-1}) - Q_n(w^\star) \leq -(\nabla_{w^\mathsf{T}}\, Q_n(w_{n-1}))^\mathsf{T}\widetilde{w}_{n-1} \tag{17.20}$$

We use the last two relations to derive a bound on the average regret. This regret, which is based on the dataset of size N, was defined by (16.114):

$$\mathcal{R}(N) \triangleq \frac{1}{N}\sum_{n=0}^{N-1}\Big\{Q_n(w_{n-1}) - Q_n(w^\star)\Big\} \tag{17.21}$$

Note that the losses involved in the definition of $\mathcal{R}(N)$ are those used in attaining the final iterate w_{N-1} since the sum goes up to $N-1$ and the last loss value included in the definition of $\mathcal{R}(N)$ is $Q_{N-1}(w_{N-2})$, which is used in the update:

$$w_{N-1} = w_{N-2} - \mu A^{-1}\nabla_{w^\mathsf{T}}\, Q\Big(w_{N-2};\gamma(N-1),h_{N-1}\Big) \tag{17.22}$$

Now, the same argument that led to (16.120) would establish the upper bound (see Prob. 17.3):

$$\boxed{N\mathfrak{R}(N) \leq \frac{1}{2\mu}\|\widetilde{w}_{-1}\|_A^2 + \frac{\mu}{2}\sum_{n=0}^{N-1}\|\nabla_{w^\mathsf{T}} Q_n(w_{n-1})\|_{A^{-1}}^2} \tag{17.23}$$

We assume the initial condition w_{-1} is chosen to ensure $\|\widetilde{w}_{-1}\|_A^2$ is bounded. Ignoring this bounded term, the above expression suggests one criterion for selecting A by reducing the size of the rightmost term to its smallest value. The objective then is to minimize over the diagonal entries $\{a_m\}$ the following expression:

$$J(a_1, \ldots, a_M) \stackrel{\Delta}{=} \sum_{n=0}^{N-1} b_n^\mathsf{T} A^{-1} b_n \; = \; \sum_{m=1}^{M} \frac{s_{N-1,m}}{a_m} \tag{17.24}$$

where $s_{N-1,m}$ denotes the cumulative energy of each entry $b_{n,m}$ until time $N-1$, i.e.,

$$s_{N-1,m} \stackrel{\Delta}{=} \sum_{n'=0}^{N-1} b_{n',m}^2 \tag{17.25}$$

The subscript $N-1$ is added to the solution $\{a_{N-1,m}\}$ because these entries will correspond to the diagonal of A_{N-1}. For well-posedness, we impose the constraint $\text{Tr}(A) \leq \alpha$, for some positive constant α, so that the entries of A do not grow unbounded. We then formulate the problem of optimally selecting each entry a_m by solving:

$$\{a_{N-1,m}\} \stackrel{\Delta}{=} \underset{\{a_m>0\}}{\text{argmin}} \left\{\sum_{m=1}^{M} \frac{s_{N-1,m}}{a_m}\right\}, \;\text{subject}\; \sum_{m=1}^{M} a_m \leq \alpha \tag{17.26}$$

To solve this problem, we appeal to the techniques used to solve constrained optimization problems of the general form (9.1). We introduce the Lagrangian function:

$$\mathcal{L}(A, \beta, \lambda_m) \stackrel{\Delta}{=} \sum_{m=1}^{M} \frac{s_{N-1,m}}{a_m} + \beta\Big(\sum_{m=1}^{M} a_m - \alpha\Big) - \sum_{m=1}^{M} \lambda_m a_m \tag{17.27}$$

where $\beta \geq 0$ and $\lambda_m \geq 0$ in view of the inequality constraints $\text{Tr}(A) - \alpha \leq 0$ and $-a_m < 0$. Differentiating relative to a_m and setting the gradient to zero at $a_{N-1,m}$ gives

$$-\frac{s_{N-1,m}}{(a_{N-1,m})^2} + \beta^o - \lambda_m^o = 0 \tag{17.28}$$

We know from the complementary requirement (9.28d) under the Karush–Kuhn–Tucker (KKT) conditions that the $\{\lambda_m^o\}$ must be zero since the $\{a_{N-1,m}\}$ are required to be positive. It follows that

$$a_{N-1,m} = (s_{N-1,m}/\beta^o)^{1/2} \tag{17.29}$$

Substituting this value into the Lagrangian function leads to the dual function

$$\mathcal{D}(\beta^o) = \left(2\sum_{m=1}^{M} s_{N-1,m}\right)\sqrt{\beta^o} - \beta^o\alpha \tag{17.30}$$

Differentiating relative to β^o and setting the gradient to zero leads to

$$(\beta^o)^{1/2} = \sum_{m=1}^{M} s_{N-1,m}/\alpha \tag{17.31}$$

Consequently, substituting into (17.29) we find that the optimal coefficients satisfy

$$a_{N-1,m} \propto \sqrt{s_{N-1,m}} \tag{17.32}$$

The important conclusion is that $a_{N-1,m}$ is proportional to $\sqrt{s_{N-1,m}}$. AdaGrad uses this construction for every iteration index n. An equivalent way to write the above result is to consider the second-order "moment" matrix:

$$S_{N-1} \triangleq \sum_{n=0}^{N-1} b_n b_n^{\mathsf{T}} = \sum_{n=0}^{N-1} \nabla_{w^{\mathsf{T}}} Q_n(w_{n-1}) \nabla_w Q_n(w_{n-1}) \tag{17.33}$$

and then select

$$A_{N-1} \propto \Big(\text{diag}(S_{N-1})\Big)^{1/2} \tag{17.34}$$

Remark 17.1. (Nondiagonal scaling) If we remove the restriction of a diagonal A and allow it to be any positive-definite matrix, then we can replace (17.26) by

$$A_{N-1} \triangleq \underset{A>0}{\text{argmin}} \left\{\sum_{n=0}^{N-1} b_n^{\mathsf{T}} A^{-1} b_n\right\}, \;\; \text{subject to } A > 0 \text{ and } \text{Tr}(A) \leq \alpha \tag{17.35}$$

This is a more challenging problem to solve. It can be shown that the optimal A_{N-1} is given by the following expression (see Prob. 17.4). Let $S_{N-1}^{1/2}$ denote a symmetric square-root for S_{N-1}, i.e., a square matrix such that $S_{N-1} = (S_{N-1}^{1/2})^2$. Then, it holds that

$$A_{N-1} = \alpha\, S_{N-1}^{1/2}/\text{Tr}(S_{N-1}^{1/2}) \tag{17.36}$$

■

Remark 17.2. (Minimization over bounded convex sets) It is customary to seek the minimizer of the empirical risk function (17.1) over bounded closed convex sets, denoted by $w \in \mathcal{C}$. In this case, the last step in (17.13) would be modified to include a projection step onto $\mathcal{C}$, such as writing

$$\boldsymbol{w}_n = \mathcal{P}_c(\boldsymbol{w}_{n-1} - \mu\, \boldsymbol{A}_n^{-1}\, \boldsymbol{b}_n) \tag{17.37}$$

This adjustment can be applied to all subsequent algorithms in this chapter.

■

17.3 RMSPROP ALGORITHM

Unfortunately, AdaGrad suffers from one serious limitation: The energy factors $\{s_{n,m}\}$ grow with n. The growth causes the scaling factors $1/a_{n,m}$ to decay close to zero until the training eventually stops. As a result, the update from w_{n-1} to w_n becomes less effective over time. One way to alleviate this difficulty is to introduce a forgetting factor into the evaluation of the energy terms, $\{s_{n,m}\}$. This is the approach followed by the RMSprop (root-mean-square propagation) method, where the energy values are smoothed through a moving average filter as follows:

$$s_{n,m} = \beta s_{n-1,m} + (1-\beta) b_{n,m}^2, \quad m = 1, 2, \ldots, M \tag{17.38}$$

with boundary condition $s_{-1,m} = 0$. Here, the scalar $\beta \in (0, 1)$ is a forgetting factor close to 1, say, $\beta = 0.95$. Observe by iterating (17.38) that now

$$s_{n,m} = (1-\beta)\Big\{ b_{n,m}^2 + \beta b_{n-1,m}^2 + \ldots + \beta^n b_{0,m}^2 \Big\} \tag{17.39}$$

which shows that energies from gradient vectors in the remote past are weighted down by larger powers of β. The resulting algorithm is listed in (17.40).

RMSprop algorithm for minimizing the empirical risk (17.1).

given N data points $\{\gamma(m), h_m\}, m = 0, 1, \ldots, N-1$;
start from an arbitrary w_{-1};
set $s_{-1} = 0_M$;
set $\beta = 0.95$ $(0 \ll \beta < 1)$;
select a small positive ϵ, say, $\epsilon = 10^{-6}$.
repeat until convergence over $n \geq 0$:
  select at random a data sample $(\gamma(n), h_n)$
  $b_n = \nabla_{w^\mathsf{T}} Q(w_{n-1}; \gamma(n), h_n)$
  $s_n = \beta s_{n-1} + (1-\beta)(b_n \odot b_n)$
  denote the entries of s_n by $\{s_{n,m}\}$
  $A_n = \text{diag}\Big\{ \sqrt{s_{n,m}} \Big\}_{m=1}^{M} + \epsilon I_M$
  $w_n = w_{n-1} - \mu A_n^{-1} b_n$
end
return $w^\star \leftarrow w_n$. (17.40)

Example 17.1 (**AdaDelta algorithm**) We motivate another adaptive gradient algorithm by building upon the RMSprop formulation; one notable feature of this new variant is that it does not involve a step-size parameter μ.

Let c_n denote the $M \times 1$ vector difference between two successive weight iterates:

$$c_n \stackrel{\Delta}{=} w_n - w_{n-1} \stackrel{\Delta}{=} \text{col}\{c_{n,m}\} \tag{17.41}$$

with individual entries $\{\boldsymbol{c}_{n,m}\}$, and let us track the energy of its entries as well, say,

$$\boldsymbol{\delta}_{n,m} = \beta\boldsymbol{\delta}_{n-1,m} + (1-\beta)\boldsymbol{c}^2_{n,m}, \quad m = 1, 2, \ldots, M \tag{17.42}$$

We further introduce the $M \times M$ diagonal matrix:

$$\boldsymbol{D}_n = \text{diag}\Big\{ \sqrt{\boldsymbol{\delta}_{n,1}}, \sqrt{\boldsymbol{\delta}_{n,2}}, \ldots, \sqrt{\boldsymbol{\delta}_{n,M}} \Big\} + \epsilon I_M \tag{17.43}$$

where a small ϵ is added to assist with the starting phase of the algorithm when the $\boldsymbol{\delta}_{n,m}$ are zero. The so-called AdaDelta algorithm modifies RMSprop by replacing the step-size parameter μI_M by the matrix $\boldsymbol{D}_{n-1}$, as shown in listing (17.44).

AdaDelta algorithm for minimizing the empirical risk (17.1).

given N data points $\{\gamma(m), h_m\}, m = 0, 1, \ldots, N-1$;
start from an arbitrary $\boldsymbol{w}_{-1}$;
set $\boldsymbol{s}_{-1} = \boldsymbol{\delta}_{-1} = 0_M$;
set $\beta = 0.95$ $(0 \ll \beta < 1)$;
select a small positive ϵ, say, $\epsilon = 10^{-6}$.
repeat until convergence over $n \geq 0$**:**
  select at random a data sample $(\boldsymbol{\gamma}(n), \boldsymbol{h}_n)$
  $\boldsymbol{b}_n = \nabla_{w^{\mathsf{T}}} Q(\boldsymbol{w}_{n-1}; \boldsymbol{\gamma}(n), \boldsymbol{h}_n)$
  $\boldsymbol{s}_n = \beta\boldsymbol{s}_{n-1} + (1-\beta)(\boldsymbol{b}_n \odot \boldsymbol{b}_n)$
  denote the entries of $\boldsymbol{s}_n$ by $\{\boldsymbol{s}_{n,m}\}$
  $\boldsymbol{A}_n = \text{diag}\Big\{ \sqrt{\boldsymbol{s}_{n,m}} \Big\}_{m=1}^{M} + \epsilon I_M$
  denote the entries of $\boldsymbol{\delta}_{n-1}$ by $\{\boldsymbol{\delta}_{n-1,m}\}$
  $\boldsymbol{D}_{n-1} = \text{diag}\Big\{ \sqrt{\boldsymbol{\delta}_{n-1,m}} \Big\}_{m=1}^{M} + \epsilon I_M$
  $\boldsymbol{w}_n = \boldsymbol{w}_{n-1} - \boldsymbol{D}_{n-1}\boldsymbol{A}_n^{-1}\boldsymbol{b}_n$
  $\boldsymbol{c}_n = \boldsymbol{w}_n - \boldsymbol{w}_{n-1}$
  $\boldsymbol{\delta}_n = \beta\boldsymbol{\delta}_{n-1} + (1-\beta)(\boldsymbol{c}_n \odot \boldsymbol{c}_n)$
end
return $w^\star \leftarrow \boldsymbol{w}_n$.

(17.44)

We motivate the substitution of μI_M by $\boldsymbol{D}_{n-1}$ as follows. Consider the stochastic gradient iteration with μ:

$$\boldsymbol{w}_n = \boldsymbol{w}_{n-1} - \mu\boldsymbol{A}_n^{-1}\boldsymbol{b}_n \iff \boldsymbol{c}_{n,m} = -\mu\,\frac{\boldsymbol{b}_{n,m}}{\boldsymbol{a}_{n,m}} \tag{17.45}$$

We know that $\boldsymbol{a}^2_{n,m}$ approximates the energy in $\boldsymbol{b}_{n,m}$ and write

$$\mathbb{E}\,\boldsymbol{a}^2_{n,m} \approx \mathbb{E}\,\boldsymbol{b}^2_{n,m} \tag{17.46}$$

This approximation can be deduced by applying an argument similar to future expression (17.50). Using the expression for $\boldsymbol{c}_{n,m}$ in (17.45), we set

$$\mu^2 \approx \mathbb{E}\,\boldsymbol{c}^2_{n,m} \tag{17.47}$$

This means that if we are able to estimate $\sqrt{\mathbb{E}\,\boldsymbol{c}^2_{n,m}}$, then we can use it in place of μ. The AdaDelta algorithm (17.44) uses a smoothing filter to estimate $\mathbb{E}\,\boldsymbol{c}^2_{n,m}$ by means of the $\boldsymbol{\delta}_{n,m}$ parameters and replaces μI_M by $\boldsymbol{D}_{n-1}$.

17.4 ADAM ALGORITHM

Another method to alleviate the decay in the learning rate of AdaGrad is the ADAM (adaptive moment) algorithm. In contrast to the algorithms derived so far, ADAM adjusts *both* the learning rate *and* the update direction. It does so by estimating the first- and second-order moments of the entries of the gradient vector.

Again, we let $\boldsymbol{b}_n$ denote the gradient vector at iteration n as defined by (17.8), with individual entries $\{\boldsymbol{b}_{n,m}\}$ for $m = 1, 2, \ldots, M$. For each entry $\boldsymbol{b}_{n,m}$, we evaluate its first-order moment, $\mathbb{E}\,\boldsymbol{b}_{n,m}$, and second-order moment, $\mathbb{E}\,\boldsymbol{b}^2_{n,m}$, by means of two separate first-order smoothing (or moving average) filters of the following form:

$$\bar{\boldsymbol{b}}_{n,m} = \beta_1 \bar{\boldsymbol{b}}_{n-1,m} + (1-\beta_1)\boldsymbol{b}_{n,m}, \quad \bar{\boldsymbol{b}}_{-1,m} = 0 \tag{17.48a}$$

$$\boldsymbol{s}_{n,m} = \beta_2 \boldsymbol{s}_{n-1,m} + (1-\beta_2)\boldsymbol{b}^2_{n,m}, \quad \boldsymbol{s}_{-1,m} = 0 \tag{17.48b}$$

Typical values for the $\beta \in (0,1)$ are $\beta_1 = 0.9$ and $\beta_2 = 0.999$. The filter for $\boldsymbol{s}_{n,m}$ is the same one used for RMSprop in (17.38). Unfortunately, the estimates that result from the above two recursions are *biased* and ADAM corrects for the bias. The presence of the bias can be seen as follows. Note by iterating the first recursion that:

$$\bar{\boldsymbol{b}}_{n,m} = (1-\beta_1)\sum_{n'=0}^{n} \beta_1^{n-n'} \boldsymbol{b}_{n',m} \tag{17.49}$$

so that

$$\begin{aligned}
\mathbb{E}\,\bar{\boldsymbol{b}}_{n,m} &= (1-\beta_1)\sum_{n'=0}^{n} \beta_1^{n-n'} \Big(\mathbb{E}\,\boldsymbol{b}_{n',m}\Big) \\
&\overset{(a)}{=} (1-\beta_1)\sum_{n'=0}^{n} \beta_1^{n-n'} \Big(\mathbb{E}\,\boldsymbol{b}_{n,m} + x_{n',m}\Big) \\
&\overset{(b)}{=} \mathbb{E}\,\boldsymbol{b}_{n,m}\left\{(1-\beta_1)\sum_{n'=0}^{n} \beta_1^{n-n'}\right\} + x \\
&= (1-\beta_1^{n+1})\mathbb{E}\,\boldsymbol{b}_{n,m} + x
\end{aligned} \tag{17.50}$$

where in step (a) we expressed $\mathbb{E}\,\boldsymbol{b}_{n',m}$ in terms of the mean $\mathbb{E}\,\boldsymbol{b}_{n,m}$ at the boundary $n' = n$ plus a correction term, denoted by $x_{n',m}$. In step (b) we collected all extra terms into x. Result (17.50) highlights two points. First, the scalar β_1 acts as a forgetting factor: It scales data from the past with increasing powers of β_1. Second, the estimator $\bar{\boldsymbol{b}}_{n,m}$ is biased due to the scaling by $(1-\beta_1^{n+1})$; this coefficient disappears and approaches 1 as n increases. We can assume the x term is small for large n; this is because we expect the mean of $\boldsymbol{b}_{n',m}$ to get closer to the mean of $\boldsymbol{b}_{n,m}$ at later iterations as the algorithm approaches convergence. This discussion suggests that we should scale $\bar{\boldsymbol{b}}_{n,m}$ by $(1-\beta_1^{n+1})$ to reduce the bias. A similar analysis applies to the second-moment $\boldsymbol{s}_{n,m}$, which would need to be scaled by $(1-\beta_2^{n+1})$:

$$\bar{\boldsymbol{b}}_{n,m} \leftarrow \bar{\boldsymbol{b}}_{n,m}/(1-\beta_1^{n+1}) \tag{17.51a}$$

$$\boldsymbol{s}_{n,m} \leftarrow \boldsymbol{s}_{n,m}/(1-\beta_2^{n+1}) \tag{17.51b}$$

Using these values, ADAM introduces two modifications into the AdaGrad algorithm; the gradient vector $\boldsymbol{b}_n$ is replaced by the bias-corrected vector $\bar{\boldsymbol{b}}_n$ and the entries of $\boldsymbol{A}_n$ are defined in terms of the bias-corrected moments $\boldsymbol{s}_{n,m}$. We write the resulting algorithm in the form shown in (17.52), where we have blended the bias correction into a single scalar multiplying μ.

ADAM algorithm for minimizing the empirical risk (17.1).

given N data points $\{\gamma(m), h_m\}, m = 0, 1, \ldots, N-1$;
start from an arbitrary $\boldsymbol{w}_{-1}$;
set $\boldsymbol{s}_{-1} = \bar{\boldsymbol{b}}_{-1} = 0_M$;
set $\beta_1 = 0.9$, $\beta_2 = 0.999$ $(0 \ll \beta_1, \beta_2 < 1)$;
select a small positive ϵ, say, $\epsilon = 10^{-6}$.
repeat until convergence over $n \geq 0$**:**
  select at random a data sample $(\boldsymbol{\gamma}(n), \boldsymbol{h}_n)$
  $\boldsymbol{b}_n = \nabla_{w^\mathsf{T}} Q(\boldsymbol{w}_{n-1}; \boldsymbol{\gamma}(n), \boldsymbol{h}_n)$
  $\bar{\boldsymbol{b}}_n = \beta_1 \bar{\boldsymbol{b}}_{n-1} + (1-\beta_1)\boldsymbol{b}_n$
  $\boldsymbol{s}_n = \beta_2 \boldsymbol{s}_{n-1} + (1-\beta_2)(\boldsymbol{b}_n \odot \boldsymbol{b}_n)$
  denote the entries of $\boldsymbol{s}_n$ by $\{\boldsymbol{s}_{n,m}\}$
  $\boldsymbol{A}_n = \text{diag}\left\{\sqrt{\boldsymbol{s}_{n,m}}\right\}_{m=1}^{M} + \epsilon I_M$
  $\boldsymbol{w}_n = \boldsymbol{w}_{n-1} - \mu \frac{\sqrt{1-\beta_2^{n+1}}}{1-\beta_1^{n+1}} \boldsymbol{A}_n^{-1} \bar{\boldsymbol{b}}_n$
end
return $w^\star \leftarrow \boldsymbol{w}_n$.

(17.52)

For later reference, especially when we discuss the Nadam algorithm in Example 17.4, we rewrite the ADAM recursions in a form that scales $\bar{\boldsymbol{b}}_n(m)$ and $\boldsymbol{s}_n(m)$ separately, namely, as:

$$\left\{\begin{array}{l}
\boldsymbol{b}_n = \nabla_{w^\mathsf{T}} Q(\boldsymbol{w}_{n-1}; \boldsymbol{\gamma}(n), \boldsymbol{h}_n) \\
\bar{\boldsymbol{b}}_n = \beta_1 \bar{\boldsymbol{b}}_{n-1} + (1-\beta_1)\boldsymbol{b}_n \\
\bar{\boldsymbol{b}}_n \leftarrow \bar{\boldsymbol{b}}_n/(1-\beta_1^{n+1}) \\
\boldsymbol{s}_n = \beta_2 \boldsymbol{s}_{n-1} + (1-\beta_2)(\boldsymbol{b}_n \odot \boldsymbol{b}_n) \\
\boldsymbol{s}_n \leftarrow \boldsymbol{s}_n/(1-\beta_2^{n+1}) \\
\boldsymbol{A}_n = \text{diag}\left\{\sqrt{\boldsymbol{s}_{n,m}}\right\}_{m=1}^{M} + \epsilon I_M \\
\boldsymbol{w}_n = \boldsymbol{w}_{n-1} - \mu \boldsymbol{A}_n^{-1} \bar{\boldsymbol{b}}_n
\end{array}\right. \tag{17.53}$$

The replacement of the gradient vector $\boldsymbol{b}_n$ by the smoothed vector $\bar{\boldsymbol{b}}_n$ is a form of *momentum acceleration*, which we describe in the next section.

Example 17.2 **(AMSGrad algorithm)** The ADAM implementation (17.52) uses moving averages of the entries of the gradient vector and their squared values. It has been observed in practice that its convergence performance can be improved by endowing the algorithm with "longer-term memory" of past gradients. One way to achieve this objective is to define $\boldsymbol{A}_n$ in terms of the *maximal* squared entries until time n (denoted by $\{\widehat{\boldsymbol{s}}_{n,m}\}$ in the listing below), thus leading to what is known as the AMSGrad algorithm.

AMSGrad algorithm for minimizing the empirical risk (17.1).

given N data points $\{\gamma(m), h_m\}, m = 0, 1, \ldots, N-1$;
start from an arbitrary $\boldsymbol{w}_{-1}$;
set $\widehat{\boldsymbol{s}}_{-1} = \boldsymbol{s}_{-1} = \bar{\boldsymbol{b}}_{-1} = 0_M$;
set $\beta_1 = 0.9$, $\beta_2 = 0.999$ $(0 \ll \beta_1, \beta_2 < 1)$;
select a small positive ϵ, say, $\epsilon = 10^{-6}$.
repeat until convergence over $n \geq 0$:

select at random a data sample $(\boldsymbol{\gamma}(n), \boldsymbol{h}_n)$

$$\begin{aligned}
\boldsymbol{b}_n &= \nabla_{w^{\mathsf{T}}} Q(\boldsymbol{w}_{n-1}; \boldsymbol{\gamma}(n), \boldsymbol{h}_n)\\
\bar{\boldsymbol{b}}_n &= \beta_1 \bar{\boldsymbol{b}}_{n-1} + (1-\beta_1)\boldsymbol{b}_n\\
\boldsymbol{s}_n &= \beta_2 \boldsymbol{s}_{n-1} + (1-\beta_2)(\boldsymbol{b}_n \odot \boldsymbol{b}_n)\\
\widehat{\boldsymbol{s}}_n &= \max\{\widehat{\boldsymbol{s}}_{n-1}, \boldsymbol{s}_n\} \quad \text{(elementwise operation)}
\end{aligned} \tag{17.54}$$

denote the entries of $\widehat{\boldsymbol{s}}_n$ by $\{\widehat{\boldsymbol{s}}_{n,m}\}$

$$\boldsymbol{A}_n = \text{diag}\left\{\sqrt{\widehat{\boldsymbol{s}}_{n,m}}\right\}_{m=1}^{M} + \epsilon I_M$$

$$\boldsymbol{w}_n = \boldsymbol{w}_{n-1} - \mu\, \frac{\sqrt{1-\beta_2^{n+1}}}{1-\beta_1^{n+1}}\, \boldsymbol{A}_n^{-1}\, \bar{\boldsymbol{b}}_n$$

end
return $w^\star \leftarrow \boldsymbol{w}_n$.

We carry out a detailed regret analysis for the AMSGrad algorithm in Appendix 17.A and show that, under constant step-size learning, it holds

$$\mathcal{R}(N) = O(1/N) + O(\mu) \tag{17.55}$$

which shows that the regret will approach a small value on the order of μ and the convergence rate toward this region is $O(1/N)$.

Example 17.3 **(AdaMax algorithm)** There are other variations of ADAM that use different norms in evaluating the energy terms $\boldsymbol{s}_{n,m}$, such as using pth power:

$$\boldsymbol{s}_{n,m} = \beta_2^p\, \boldsymbol{s}_{n-1,m} + (1-\beta_2^p)\, |\boldsymbol{b}_{n,m}|^p \tag{17.56}$$

Here, p can be any positive real number. In this computation, we are also raising the forgetting factor β_2 to the same power p. The construction for the scalars $\boldsymbol{a}_{n,m}$ becomes

$$\boldsymbol{a}_{n,m} = (\boldsymbol{s}_{n,m})^{1/p} + \epsilon, \quad m = 1, 2, \ldots, M \tag{17.57}$$

It has been observed in practice that the choice $p \to \infty$ leads to more stable behavior for the algorithm. We can evaluate the value $(\boldsymbol{s}_{n,m})^{1/p}$ in the limit over p as follows:

$$\begin{aligned} \boldsymbol{u}_{n,m} &\stackrel{\Delta}{=} \lim_{p\to\infty} (\boldsymbol{s}_{n,m})^{1/p} \\ &\stackrel{(17.56)}{=} \lim_{p\to\infty} (1-\beta_2^p)^{1/p} \left\{ \sum_{n'=0}^{n} \beta_2^{p(n-n')} |\boldsymbol{b}_{n',m}|^p \right\}^{1/p} \\ &\stackrel{(a)}{=} \lim_{p\to\infty} \left\{ \sum_{n'=0}^{n} \left(\beta_2^{(n-n')} |\boldsymbol{b}_{n',m}| \right)^p \right\}^{1/p} \\ &\stackrel{(b)}{=} \max\Big\{ |\boldsymbol{b}_{n,m}|,\ \beta_2 |\boldsymbol{b}_{n-1,m}|,\ \ldots,\ \beta_2^n |\boldsymbol{b}_{0,m}| \Big\} \end{aligned} \tag{17.58}$$

where in step (a) we used the result

$$\lim_{p\to\infty} (1-x^p)^{1/p} = 1, \quad x \in [0,1) \tag{17.59}$$

and in step (b) we used the property of p-norms for vectors $x \in \mathbb{R}^L$ with entries $\{x_\ell\}$:

$$\max_{1\leq \ell \leq L} |x_\ell| \stackrel{\Delta}{=} \|x\|_\infty = \lim_{p\to\infty} \|x\|_p = \lim_{p\to\infty} \left(\sum_{\ell=1}^{L} |x_\ell|^p \right)^{1/p} \tag{17.60}$$

Result (17.58) allows us to write the following recursive formula for updating $\boldsymbol{u}_n(m)$ as the maximum between two values:

$$\boldsymbol{u}_{n,m} = \max\Big\{ \beta_2\, \boldsymbol{u}_{n-1,m},\ |\boldsymbol{b}_{n,m}| \Big\} \tag{17.61}$$

We therefore arrive at the AdaMax listing shown in (17.62).

AdaMax algorithm for minimizing the empirical risk (17.1).

given N data points $\{\gamma(m), h_m\}, m = 0, 1, \ldots, N-1$;
start from an arbitrary $\boldsymbol{w}_{-1}$;
set $\boldsymbol{u}_{-1} = \bar{\boldsymbol{b}}_{-1} = 0_M$;
set $\beta_1 = 0.9$, $\beta_2 = 0.999$ $(0 \ll \beta_1, \beta_2 < 1)$;
select a small positive ϵ, say, $\epsilon = 10^{-6}$.
repeat until convergence over $n \geq 0$**:**
  select at random a data sample $(\boldsymbol{\gamma}(n), \boldsymbol{h}_n)$
  $\boldsymbol{b}_n = \nabla_{w^{\mathsf{T}}} Q(\boldsymbol{w}_{n-1}; \boldsymbol{\gamma}(n), \boldsymbol{h}_n)$ (17.62)
  $\bar{\boldsymbol{b}}_n = \beta_1 \bar{\boldsymbol{b}}_{n-1} + (1-\beta_1)\boldsymbol{b}_n$
  denote the entries of $\boldsymbol{b}_n$ by $\{\boldsymbol{b}_{n,m}\}$
  $\boldsymbol{u}_{n,m} = \max\Big\{ \beta_2\, \boldsymbol{u}_{n-1,m},\ |\boldsymbol{b}_{n,m}| \Big\},\ m = 1, \ldots, M$
  $\boldsymbol{A}_n = \text{diag}\Big\{ \boldsymbol{u}_{n,m} \Big\}_{m=1}^{M} + \epsilon I_M$
  $\boldsymbol{w}_n = \boldsymbol{w}_{n-1} - \mu \frac{1}{1-\beta_1^{n+1}} \boldsymbol{A}_n^{-1} \bar{\boldsymbol{b}}_n$
end
return $w^\star \leftarrow \boldsymbol{w}_n$.

Note that the update for $\boldsymbol{u}_{n,m}$ is keeping track of the "largest" absolute value for the mth gradient entry over time; this tracking mechanism involves a forgetting factor β_2 so that the influence of entries from the remote past is limited. Since the $\boldsymbol{u}_{n,m}$ track the largest absolute entry value, there is no need for scaling it by $(1-\beta_2^{n+1})^{1/2}$.

17.5 MOMENTUM ACCELERATION METHODS

We describe next two useful techniques for enhancing the convergence of stochastic gradient methods by altering the update direction, while keeping the step size (or learning rate) fixed at μ. The techniques are based on using *momentum acceleration* to improve convergence, reduce oscillations close to local minima (which is a problem for nonconvex risks), and keep the weight iterates moving along preferred directions. We encountered these techniques earlier in the concluding remarks of Chapter 12 and in Probs. 12.9 and 12.11 while studying gradient-descent methods.

17.5.1 Polyak Momentum Acceleration

In the momentum method, the gradient vector $\boldsymbol{b}_n$ is smoothed to a vector $\bar{\boldsymbol{b}}_n$ (also called a *velocity vector*) through a moving average filter with some positive coefficient β smaller than but close to 1 (known as the momentum parameter) as follows:

$$\bar{\boldsymbol{b}}_n = \beta \bar{\boldsymbol{b}}_{n-1} + \boldsymbol{b}_n \tag{17.63}$$

This update is similar to the smoothing filter (17.48a) used by ADAM; the main difference is the removal of the scaling of $\boldsymbol{b}_n$ by $(1-\beta)$. Iterating (17.63) we get

$$\bar{\boldsymbol{b}}_n = \boldsymbol{b}_n + \beta \boldsymbol{b}_{n-1} + \beta^2 \boldsymbol{b}_{n-2} + \beta^3 \boldsymbol{b}_{n-3} + \ldots \tag{17.64}$$

which shows that $\bar{\boldsymbol{b}}_n$ is an exponentially weighted combination of present and past gradient vectors; the effect of past directions dies out with increasing powers of β. The momentum method helps reinforce directions along which more progress has been made in the search space, de-emphasizes slow directions or undesirable regions in the same space, and helps smooth out the effect of gradient noise through the filtering operation. The resulting algorithm is listed in (17.65). Setting $\beta = 0$ recovers the stochastic gradient iteration (17.2).

On the other hand, when $\beta > 0$, we can expand the last relation in (17.65) to observe that $\boldsymbol{w}_n = \boldsymbol{w}_{n-1} - \mu \boldsymbol{b}_n - \mu\beta \bar{\boldsymbol{b}}_{n-1}$. This expression shows that the update to $\boldsymbol{w}_{n-1}$ happens along a direction that combines that gradient vector $\boldsymbol{b}_n$, as is common in traditional stochastic gradient implementations, and the previous smoothed gradient value $\bar{\boldsymbol{b}}_{n-1}$.

Stochastic gradient algorithm with Polyak momentum acceleration for minimizing the empirical risk (17.1).

given N data points $\{\gamma(m), h_m\}, m = 0, 1, \ldots, N-1$;
start from an arbitrary $\boldsymbol{w}_{-1}$;
set $\bar{\boldsymbol{b}}_{-1} = 0$;
set $\beta = 0.95$ $(0 \ll \beta < 1)$. (17.65)
repeat until convergence over $n \geq 0$**:**

  select at random a data sample $(\boldsymbol{\gamma}(n), \boldsymbol{h}_n)$
  $\boldsymbol{b}_n = \nabla_{w^{\mathsf{T}}} Q(\boldsymbol{w}_{n-1}; \boldsymbol{\gamma}(n), \boldsymbol{h}_n)$
  $\bar{\boldsymbol{b}}_n = \beta \bar{\boldsymbol{b}}_{n-1} + \boldsymbol{b}_n$
  $\boldsymbol{w}_n = \boldsymbol{w}_{n-1} - \mu \, \bar{\boldsymbol{b}}_n$

end
return $w^\star \leftarrow \boldsymbol{w}_n$.

17.5.2 Nesterov Momentum Acceleration

Expanding the weight update from (17.65) we find that the net effect is:

$$\begin{aligned} \boldsymbol{w}_n &= \boldsymbol{w}_{n-1} - \mu \, \bar{\boldsymbol{b}}_n \\ &= \boldsymbol{w}_{n-1} - \mu(\beta \bar{\boldsymbol{b}}_{n-1} + \boldsymbol{b}_n) \\ &= \underbrace{\boldsymbol{w}_{n-1} - \mu \beta \bar{\boldsymbol{b}}_{n-1}}_{\triangleq \, \boldsymbol{w}'_{n-1}} - \mu \nabla_{w^{\mathsf{T}}} Q(\boldsymbol{w}_{n-1}; \boldsymbol{\gamma}(n), \boldsymbol{h}_n) \end{aligned} \tag{17.66}$$

We group the first two terms into the variable $\boldsymbol{w}'_{n-1}$ so that

$$\boldsymbol{w}_n = \boldsymbol{w}'_{n-1} - \mu \nabla_{w^{\mathsf{T}}} Q(\boldsymbol{w}_{n-1}; \boldsymbol{\gamma}(n), \boldsymbol{h}_n) \tag{17.67}$$

Observe that this relation is a "distorted" stochastic gradient recursion. This is because the starting point is $\boldsymbol{w}'_{n-1}$, while the gradient of the loss is evaluated at $\boldsymbol{w}_{n-1}$. This suggests that we should evaluate the gradient of the loss at $\boldsymbol{w}'_{n-1}$ rather than $\boldsymbol{w}_{n-1}$, which in turn suggests that we should smooth this adjusted gradient over time. Applying the same construction (17.65), and taking this adjustment into consideration, leads to the Nesterov accelerated gradient (NAG) method listed in (17.68). In this method, the algorithm first takes a *look-ahead* step and updates $\boldsymbol{w}_{n-1}$ to $\boldsymbol{w}'_{n-1}$. We denote the gradient vector at this location by $\boldsymbol{b}'_n$ and smooth it over time. The smoothed direction is then used to update the weight iterate.

In this way, the NAG implementation is expected to improve the responsiveness of the stochastic gradient algorithm by allowing it to anticipate the behavior of the loss surface around the current iterate location. This is because NAG first computes a rough prediction for the location of the next iterate, and then uses the smoothed gradient of the loss at this location to drive the update rule.

Stochastic gradient algorithm with Nesterov momentum acceleration for minimizing the empirical (17.1) – VERSION I.

$$\begin{array}{l}
\text{given } N \text{ data points } \{\gamma(m), h_m\}, m = 0, 1, \ldots, N-1; \\
\text{start from an arbitrary } \boldsymbol{w}_{-1}; \\
\text{set } \bar{\boldsymbol{b}}_{-1} = 0; \\
\text{set } \beta = 0.95 \ (0 \ll \beta < 1). \\
\textbf{repeat until convergence over } n \geq 0\textbf{:} \\
\quad \text{select at random a data sample } (\boldsymbol{\gamma}(n), \boldsymbol{h}_n) \\
\quad \boldsymbol{w}'_{n-1} = \boldsymbol{w}_{n-1} - \mu\beta\bar{\boldsymbol{b}}_{n-1} \qquad \text{(look-ahead step)} \\
\quad \boldsymbol{b}'_n = \nabla_{w^\mathsf{T}} Q(\boldsymbol{w}'_{n-1}; \boldsymbol{\gamma}(n), \boldsymbol{h}_n) \\
\quad \bar{\boldsymbol{b}}_n = \beta\bar{\boldsymbol{b}}_{n-1} + \boldsymbol{b}'_n \\
\quad \boldsymbol{w}_n = \boldsymbol{w}_{n-1} - \mu\,\bar{\boldsymbol{b}}_n \\
\textbf{end} \\
\text{return } w^\star \leftarrow \boldsymbol{w}_n.
\end{array} \tag{17.68}$$

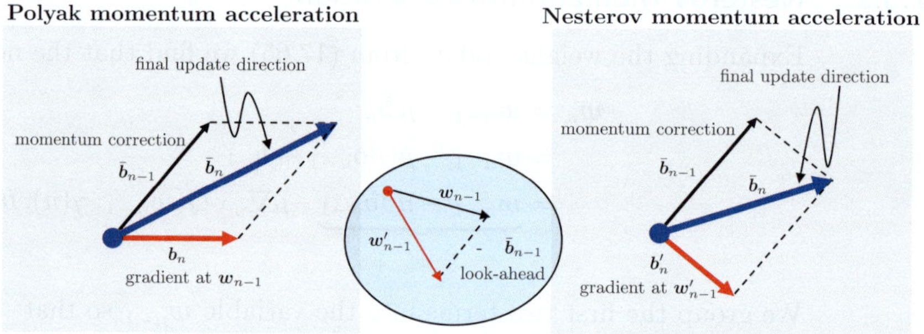

Figure 17.4 Comparison of the update directions $\bar{\boldsymbol{b}}_n$ for the momentum and accelerated momentum algorithms.

Figure 17.4 compares the update directions $\bar{\boldsymbol{b}}_n$ for Polyak and Nesterov acceleration methods. In the panel on the left, we show the gradient vector $\boldsymbol{b}_n$ at $\boldsymbol{w}_{n-1}$ in red. It is added to the momentum correction $\beta\bar{\boldsymbol{b}}_{n-1}$ to generate the final update direction $\bar{\boldsymbol{b}}_n$ for the Polyak implementation. In the center panel, the current iterate $\boldsymbol{w}_{n-1}$ is updated to $\boldsymbol{w}'_{n-1}$ by employing the look-ahead term $-\mu\beta\bar{\boldsymbol{b}}_{n-1}$. In the panel on the right, we show the gradient vector $\boldsymbol{b}'_n$ at this updated location $\boldsymbol{w}'_{n-1}$ in red. It is added to the momentum correction $\beta\bar{\boldsymbol{b}}_{n-1}$ to generate the final update direction $\bar{\boldsymbol{b}}_n$ for the Nesterov momentum implementation.

A variation of the algorithm follows by replacing the argument $\boldsymbol{w}_{n-1}$ in the gradient of the loss on the right-hand side of (17.67) by $\boldsymbol{w}'_{n-1}$ to get:

$$\boldsymbol{w}_n = \boldsymbol{w}'_{n-1} - \mu\nabla_{w^\mathsf{T}} Q(\boldsymbol{w}'_{n-1}; \boldsymbol{\gamma}(n), \boldsymbol{h}_n) \tag{17.69}$$

This leads to listing (17.70).

Stochastic gradient algorithm with Nesterov momentum acceleration for minimizing the empirical (17.1) – VERSION II.

$$
\begin{array}{l}
\text{given } N \text{ data points } \{\gamma(m), h_m\}, m = 0, 1, \ldots, N-1; \\
\text{start from an arbitrary } \boldsymbol{w}_{-1}; \\
\text{set } \bar{\boldsymbol{b}}_{-1} = 0; \\
\text{set } \beta = 0.95 \ (0 \ll \beta < 1). \\
\textbf{repeat until convergence over } n \geq 0\textbf{:} \\
\quad \left| \begin{array}{l}
\text{select at random a data sample } (\boldsymbol{\gamma}(n), \boldsymbol{h}_n) \\
\boldsymbol{w}'_{n-1} = \boldsymbol{w}_{n-1} - \mu\beta\bar{\boldsymbol{b}}_{n-1}, \qquad \text{(look-ahead step)} \\
\boldsymbol{b}'_n = \nabla_{w^\mathsf{T}} Q(\boldsymbol{w}'_{n-1}; \boldsymbol{\gamma}(n), \boldsymbol{h}_n) \\
\bar{\boldsymbol{b}}_n = \beta\bar{\boldsymbol{b}}_{n-1} + \boldsymbol{b}'_n \\
\boldsymbol{w}_n = \boldsymbol{w}'_{n-1} - \mu\, \boldsymbol{b}'_n
\end{array} \right. \\
\textbf{end} \\
\text{return } w^\star \leftarrow \boldsymbol{w}_n.
\end{array}
\tag{17.70}
$$

There is a third way to adjust the update recursion (17.66). The middle equation is repeated here for ease of reference:

$$
\boldsymbol{w}_n = \boldsymbol{w}_{n-1} - \mu\beta\bar{\boldsymbol{b}}_{n-1} - \mu\boldsymbol{b}_n \tag{17.71}
$$

Note that $\boldsymbol{b}_n$ depends on $\boldsymbol{w}_{n-1}$ since $\boldsymbol{b}_n = \nabla_{w^\mathsf{T}} Q(\boldsymbol{w}_{n-1}; \boldsymbol{\gamma}(n), \boldsymbol{h}_n)$, while $\bar{\boldsymbol{b}}_{n-1}$ does not depend on $\boldsymbol{w}_{n-1}$ but only on previous iterates occurring before $\boldsymbol{w}_{n-1}$. We may consider replacing $\bar{\boldsymbol{b}}_{n-1}$ by the look-ahead variable $\bar{\boldsymbol{b}}_n$, which is now dependent on $\boldsymbol{w}_{n-1}$, and write:

$$
\begin{aligned}
\boldsymbol{w}_n &\approx \boldsymbol{w}_{n-1} - \mu\beta\bar{\boldsymbol{b}}_n - \mu\boldsymbol{b}_n \\
&= \boldsymbol{w}_{n-1} - \mu\underbrace{(\beta\bar{\boldsymbol{b}}_n + \boldsymbol{b}_n)}_{\triangleq\, \widehat{\boldsymbol{b}}_n}
\end{aligned}
\tag{17.72}
$$

In this case, the last two terms on the right-hand side become dependent on the same starting point, $\boldsymbol{w}_{n-1}$, for the iteration. This adjustment leads to listing (17.73). Observe that this implementation does not require $\boldsymbol{w}'_{n-1}$ any longer.

Stochastic gradient algorithm with Nesterov momentum acceleration for minimizing the empirical (17.1) – VERSION III.

given N data points $\{\gamma(m), h_m\}, m = 0, 1, \ldots, N-1$;
start from an arbitrary $\boldsymbol{w}_{-1}$;
set $\bar{\boldsymbol{b}}_{-1} = 0$;
set $\beta = 0.95$ $(0 \ll \beta < 1)$.
repeat until convergence over $n \geq 0$: (17.73)

$$
\begin{aligned}
&\text{select at random a data sample } (\boldsymbol{\gamma}(n), \boldsymbol{h}_n)\\
&\boldsymbol{b}_n = \nabla_{w^\mathsf{T}} Q(\boldsymbol{w}_{n-1}; \boldsymbol{\gamma}(n), \boldsymbol{h}_n)\\
&\bar{\boldsymbol{b}}_n = \beta \bar{\boldsymbol{b}}_{n-1} + \boldsymbol{b}_n\\
&\widehat{\boldsymbol{b}}_n = \beta \bar{\boldsymbol{b}}_n + \boldsymbol{b}_n \quad \text{(look-ahead step)}\\
&\boldsymbol{w}_n = \boldsymbol{w}_{n-1} - \mu\, \widehat{\boldsymbol{b}}_n
\end{aligned}
$$

end
return $w^\star \leftarrow \boldsymbol{w}_n$.

Example 17.4 (**Nadam algorithm**) We can incorporate momentum acceleration into the operation of the ADAM algorithm (17.53) by employing a similar look-ahead idea to implementation (17.73). The resulting construction leads to the Nadam algorithm (ADAM with Nesterov acceleration), which is listed in (17.74).

Nadam algorithm for minimizing the empirical risk (17.1).

given N data points $\{\gamma(m), h_m\}, m = 0, 1, \ldots, N-1$;
start from an arbitrary $\boldsymbol{w}_{-1}$;
set $\boldsymbol{s}_{-1} = \bar{\boldsymbol{b}}_{-1} = 0_M$;
set $\beta_1 = 0.9$, $\beta_2 = 0.999$ $(0 \ll \beta_1, \beta_2 < 1)$;
select a small positive ϵ, say, $\epsilon = 10^{-6}$.
repeat until convergence over $n \geq 0$:

$$
\begin{aligned}
&\text{select at random a data sample } (\boldsymbol{\gamma}(n), \boldsymbol{h}_n)\\
&\boldsymbol{b}_n = \nabla_{w^\mathsf{T}} Q(\boldsymbol{w}_{n-1}; \boldsymbol{\gamma}(n), \boldsymbol{h}_n)\\
&\bar{\boldsymbol{b}}_n = \beta_1 \bar{\boldsymbol{b}}_{n-1} + (1-\beta_1)\boldsymbol{b}_n\\
&\widehat{\boldsymbol{b}}_n = \beta_1 \bar{\boldsymbol{b}}_n + (1-\beta_1)\boldsymbol{b}_n\\
&\widehat{\boldsymbol{b}}_n \leftarrow \widehat{\boldsymbol{b}}_n/(1-\beta_1^{n+1})\\
&\boldsymbol{s}_n = \beta_2 \boldsymbol{s}_{n-1} + (1-\beta_2)(\boldsymbol{b}_n \odot \boldsymbol{b}_n)\\
&\boldsymbol{s}_n \leftarrow \boldsymbol{s}_n/(1-\beta_2^{n+1})\\
&\text{denote the entries of } \boldsymbol{s}_n \text{ by } \{\boldsymbol{s}_{n,m}\}\\
&\boldsymbol{A}_n = \text{diag}\left\{\sqrt{\boldsymbol{s}_{n,m}}\right\} + \epsilon I_M\\
&\boldsymbol{w}_n = \boldsymbol{w}_{n-1} - \mu\, \boldsymbol{A}_n^{-1} \widehat{\boldsymbol{b}}_n
\end{aligned}
\tag{17.74}
$$

end
return $w^\star \leftarrow \boldsymbol{w}_n$.

In the algorithm we apply the look-ahead step to $\bar{\boldsymbol{b}}_n$, correct for the bias, and use the adjusted gradient vector $\widehat{\boldsymbol{b}}_n$ to update $\boldsymbol{w}_{n-1}$ to $\boldsymbol{w}_n$.

Example 17.5 **(Logistic regression)** We repeat the simulations from Example 16.5 and illustrate the operation of various adaptive gradient algorithms. Thus, consider again the ℓ_2-regularized logistic regression empirical risk for which the gradient of the loss function is given by

$$\nabla_{w^{\mathsf{T}}} Q(w;\gamma(n),h_n) = 2\rho w \; - \; \frac{\gamma(n)h_n}{1+e^{\gamma(n)h_n^{\mathsf{T}}w}} \;\stackrel{\Delta}{=}\; b_n \tag{17.75}$$

We generate normalized learning curves in logarithmic scale for a variety of algorithms, including the vanilla stochastic gradient implementation, its momentum accelerated versions, and the adaptive gradient variations AdaGrad, RMSprop, ADAM, AdaMax, AdaDelta, and Nadam. We employ uniform sampling and set the parameters to (it is recommended that the ϵ-variable for AdaDelta be chosen between 1×10^{-2} and 1×10^{-5}):

$$\rho = 1, \beta = 0.95, \beta_1 = 0.9, \beta_2 = 0.999, M = 10 \tag{17.76a}$$

$$\mu = 0.005, \;\; \mu_{\text{AdaGrad}} = 0.1, \;\; \epsilon = 10^{-8} \tag{17.76b}$$

$$\epsilon = 1\times 10^{-8}, \;\; \epsilon_{\text{AdaDelta}} = 1\times 10^{-4} \tag{17.76c}$$

with a larger step size for AdaGrad (otherwise, its convergence will be slow). The simulation generates $N = 100$ random pairs of data $\{\gamma(m), h_m\}$ according to the same logistic model explained earlier in Example 16.5. A total of $K = 50$ epochs are run over the data. We repeat this experiment $L = 100$ times and average the corresponding learning curves, with each experiment starting from the same initial condition $\boldsymbol{w}_{-1}$. The resulting learning curves are plotted in normalized logarithmic scale in line with construction (11.65), namely,

$$\ln\left(\frac{P(w_n) - P(w^\star)}{\max_n\{P(w_n) - P(w^\star)\}}\right) \tag{17.77}$$

where $w^\star$ is approximated by the limit value of the weight iterate after sufficient convergence. Each learning curve is generated by plotting the values of the risk function at the start of each epoch, namely, $P(w_{-1}^k)$. The plot in the top row of Fig. 17.5 shows the evolution of these values relative to the minimum risk $P(w^\star)$ over the first 50 epochs. It is seen from the plot in the top-left corner how the incorporation of momentum accelerates the convergence of the stochastic gradient iteration; the steady-state risk error stays largely unaltered. The plot in the top-right corner shows how AdaGrad, RMSprop, and ADAM reduce the steady-state error in comparison with the stochastic gradient implementation; the convergence rate is, however, affected. The plot in the bottom-left corner compares different variations of ADAM, Nadam, AdaMax, and AdaDelta. The first two exhibit similar performance in this example, while AdaMax converges faster but tends to a higher error level. The plot in the bottom-right corner compares the performance of Nadam to accelerated versions of the stochastic gradient algorithm.

17.6 FEDERATED LEARNING

We end this chapter by describing an application of stochastic gradient algorithms, where any of the methods described in this chapter or the previous one can be used to drive the learning process. Specifically, we discuss the *federated*

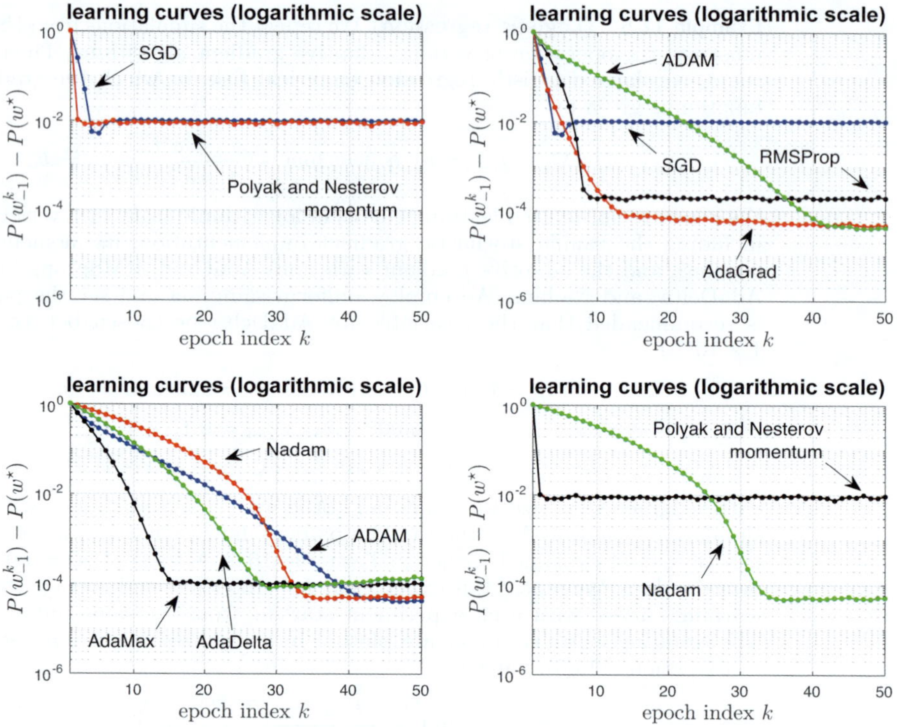

Figure 17.5 (*Top-left*) Learning curves $P(w_{-1}^k)$ relative to the minimum risk value $P(w^\star)$ in normalized logarithmic scale for the stochastic-gradient implementation and its momentum accelerated versions. (*Top-right*) Learning curves for the stochastic gradient implementation (SGD), AdaGrad, RMSprop, and ADAM. (*Bottom-left*) Learning curves for ADAM, AdaMax, AdaDelta, and Nadam. (*Bottom-right*) Learning curves for the stochastic gradient with momentum accelerated versions as well.

learning approach, which is an architecture that takes advantage of local processing at distributed agents to reduce communication cost and enhance privacy protection at the agents.

It is sufficient to illustrate the structure of federated learning for smooth loss functions and empirical risk minimization problems of the form:

$$w^\star = \underset{w \in \mathbb{R}^M}{\text{argmin}} \left\{ P(w) \triangleq \frac{1}{N} \sum_{m=0}^{N-1} Q(w; \gamma(m), h_m) \right\} \tag{17.78}$$

with a total of N data points $\{\gamma(m), h_m\}$. Assume there are L computing devices or agents (also called *clients*) spread over some geographical area (such as smart phones or autonomous vehicles). These devices collect data continually and the totality of the N data points $\{\gamma(m), h_m\}$ correspond to the collection of data available from across all agents. We index the agents by $\ell = 1, 2, \ldots, L$, and assume that agent ℓ contributes N_ℓ data samples so that

$$N = \sum_{\ell=1}^{L} N_\ell \tag{17.79}$$

If the data points from the agents could be transferred from their locations to some powerful central processor (or *server*), then the server would have access to the entire dataset and could run its preferred stochastic gradient algorithm to estimate the solution $w^\star$. However, in situations when L is large, as well as when the amount of data N_ℓ at the agents is large, transferring all data to the central processor poses at least two challenges. First, the transfer would entail significant amounts of communication and energy consumption. Second, the individual agents may be reluctant to share their raw data directly with a remote server due to privacy or security concerns. Federated learning is one approach to alleviate these problems. In this approach, the agents perform local processing on their data and share only the result of this operation with the central server. The construction for federated learning can be motivated as follows.

We associate with each agent a local risk function based on its data and denoted by:

$$P_\ell(w) \stackrel{\Delta}{=} \frac{1}{N_\ell} \sum_{m=0}^{N_\ell-1} Q(w;\gamma(m),h_m) \tag{17.80}$$

where the $\{\gamma(m), h_m\}$ here refer to data points at location ℓ. We could have used a superscript to indicate their location explicitly, such as writing $\{\gamma^{(\ell)}(m), h_m^{(\ell)}\}$. We prefer to avoid cluttering the notation at this stage; it is understood from the context that the data associated with $P_\ell(w)$ relate to agent ℓ. We next express the global risk $P(w)$ in terms of these local risks by noting that

$$P(w) = \frac{1}{N} \sum_{\ell=1}^{L} N_\ell P_\ell(w) \tag{17.81}$$

where the contribution from agent ℓ appears scaled by the amount of data available at that location. Now, let w_{n-1} denote the global estimate for $w^\star$ that is available at the central server at time $n-1$. If the server were to employ a gradient-descent algorithm to update it to w_n, the recursion would take the form

$$\begin{aligned} w_n &= w_{n-1} - \mu \nabla_{w^\mathsf{T}} P(w_{n-1})) \\ &= w_{n-1} - \mu\left(\frac{1}{N}\sum_{\ell=1}^{L} N_\ell \nabla_{w^\mathsf{T}} P_\ell(w_{n-1})\right) \end{aligned} \tag{17.82}$$

in terms of the gradients of the local risks. The dispersed agents could assist with this calculation by performing local computations that update w_{n-1} to some local iterates denoted by $w_{\ell,n}$ at agent ℓ, namely,

$$\begin{aligned} w_{1,n} &= w_{n-1} - \mu \nabla_{w^{\mathsf{T}}} P_1(w_{n-1}) && \text{(17.83a)} \\ w_{2,n} &= w_{n-1} - \mu \nabla_{w^{\mathsf{T}}} P_2(w_{n-1}) && \text{(17.83b)} \\ \vdots &= \vdots \\ w_{L,n} &= w_{n-1} - \mu \nabla_{w^{\mathsf{T}}} P_L(w_{n-1}) && \text{(17.83c)} \end{aligned}$$

If the agents share their iterates with the central server, then it is easy to see from the above updates and (17.82) that the $\{w_{\ell,n}\}$ should be combined in the following manner to recover w_n:

$$w_n = \sum_{\ell=1}^{L} \frac{N_\ell}{N}\, w_{\ell,n} \tag{17.84}$$

Note that this construction follows because the global update (17.82) employs the full gradient vector, while the local updates (17.83a)–(17.83c) employ the gradient vectors of their local risks.

Now, in a *stochastic* gradient implementation the local agents replace their actual gradients by approximate gradients. For example, agent ℓ may employ a mini-batch approximation of the form:

$$\nabla_{w^{\mathsf{T}}} P_\ell(w_{n-1}) \approx \frac{1}{B}\sum_{b=0}^{B-1} \nabla_{w^{\mathsf{T}}} Q(w_{n-1}; \gamma(b), h_b) \tag{17.85}$$

where the batch samples are selected at random from the N_ℓ samples that are available at location ℓ. It is common for the local agents to run *multiple* mini-batch iterations on $\boldsymbol{w}_{n-1}$ (say, R_ℓ of them) before generating the iterates $\{\boldsymbol{w}_{\ell,n}\}$ that they would share with the server. For example, agent ℓ would apply the following construction repeatedly over its data:

$$\left\{\begin{array}{l} \text{start from condition } \boldsymbol{w}_{\ell,-1} = \boldsymbol{w}_{n-1} \text{ received from server} \\ \textbf{repeat for } n' = 0, 1, \ldots, R_\ell - 1\textbf{:} \\ \quad \left| \begin{array}{l} \text{select } B \text{ random samples } \{\boldsymbol{\gamma}(b), \boldsymbol{h}_b\} \text{ from the data at agent } \ell; \\ \text{set } \boldsymbol{w}_{\ell,n'} = \boldsymbol{w}_{\ell,n'-1} - \dfrac{\mu}{R_\ell}\left(\dfrac{1}{B}\displaystyle\sum_{b=0}^{B-1} \nabla_{w^{\mathsf{T}}} Q(\boldsymbol{w}_{\ell,n'-1}; \boldsymbol{\gamma}(b), \boldsymbol{h}_b)\right) \end{array}\right. \\ \textbf{end} \\ \text{return } \boldsymbol{w}_{\ell,n} = \boldsymbol{w}_{\ell,R_\ell-1}. \end{array}\right. \tag{17.86}$$

Observe that we are scaling the step size by R_ℓ. The reason is as follows. At the end of the recursions, the final iterate $\boldsymbol{w}_{\ell,n}$ will be given by:

$$\boldsymbol{w}_{\ell,n} = \boldsymbol{w}_{n-1} - \mu \underbrace{\left\{\frac{1}{R_\ell}\sum_{n'=0}^{R_\ell-1}\left(\frac{1}{B}\sum_{b=0}^{B-1} \nabla_{w^{\mathsf{T}}} Q(\boldsymbol{w}_{\ell,n'-1}; \boldsymbol{\gamma}(b), \boldsymbol{h}_b)\right)\right\}}_{\triangleq\, \boldsymbol{g}_\ell} \tag{17.87}$$

Observe that, due to the appearance of the sum over n', the scaling by R_ℓ helps ensure that the vector $\boldsymbol{g}_\ell$ serves as a sample approximation for the true gradient

$\nabla_{w^\mathsf{T}} P_\ell(\cdot)$ at agent ℓ. Even more, agents may run repeated epochs over the data, say, K_ℓ epochs at agent ℓ. For this reason, in the listing below, the step size is further scaled by K_ℓ for the same reason.

We are now in a position to describe the federated learning procedure in some detail. Thus, let $\boldsymbol{w}_{n-1}$ denote the global estimate for $w^\star$ that is available at the central server at the end of cycle $n-1$. At cycle n, the server selects a random subset $\mathcal{L}_n$ from the set of agents $\mathcal{L} = \{1, 2, \ldots, L\}$ to perform data processing (note that not all agents participate in the learning process at every cycle):

$$\mathcal{L}_n \;=\; \text{random subset of } \{1, 2, \ldots, L\} \text{ at cycle } n \tag{17.88a}$$

The server transmits $\boldsymbol{w}_{n-1}$ to the agents $\ell \in \mathcal{L}_n$ in the selected subset, which will use $\boldsymbol{w}_{n-1}$ as the initial condition for their local iterative algorithms. Each selected agent ℓ runs multiple epochs, say K_ℓ of them, of a mini-batch stochastic gradient algorithm as shown in (17.88b).

Mini-batch processing at each agent $\ell \in \mathcal{L}_n$ during the nth cycle.

set initial condition for epoch $k = 0$ to $\boldsymbol{w}^0_{\ell, R_\ell - 1} = \boldsymbol{w}_{n-1}$.
repeat for each epoch $k = 1, 2, \ldots, K_\ell$**:**
  set $\boldsymbol{w}^k_{\ell,-1} = \boldsymbol{w}^{k-1}_{\ell, R_\ell - 1}$
  repeat for $n' = 0, 1, 2, \ldots, R_\ell - 1$**:**
    select B random samples $\{\boldsymbol{\gamma}(b), \boldsymbol{h}_b\}$ from the data at agent ℓ

$$\boldsymbol{w}^k_{\ell,n'} = \boldsymbol{w}^k_{\ell,n'-1} - \frac{\mu}{K_\ell R_\ell}\left(\frac{1}{B}\sum_{b=0}^{B-1} \nabla_{w^\mathsf{T}} Q\Big(\boldsymbol{w}^k_{\ell,n'-1}; \boldsymbol{\gamma}(b), \boldsymbol{h}_b\Big)\right)$$

  end
end
$\boldsymbol{w}_{\ell,n} \leftarrow \boldsymbol{w}^K_{\ell, R_\ell - 1}$
send $\boldsymbol{w}_{\ell,n}$ to server.

(17.88b)

In this way, each agent $\ell \in \mathcal{L}_n$ receives $\boldsymbol{w}_{n-1}$ from the server, runs a mini-batch procedure for K_ℓ epochs over its local data starting from $\boldsymbol{w}_{n-1}$, and returns $\boldsymbol{w}_{\ell,n}$ to the server. The server in turn fuses the iterates it receives from the agents in $\mathcal{L}_n$ in order to update $\boldsymbol{w}_{n-1}$ to $\boldsymbol{w}_n$ by means of the convex combination:

$$\boldsymbol{w}_n = \sum_{\ell \in \mathcal{L}_n} \frac{R_\ell}{R_\mathcal{L}} \boldsymbol{w}_{\ell,n}, \quad \text{where } R_\mathcal{L} \triangleq \sum_{\ell' \in \mathcal{L}_n} R_{\ell'} \tag{17.88c}$$

In this expression, each local iterate is weighted by the relative number of mini-batches used to generate it. Expressions (17.88a)–(17.88c) describe the federated learning algorithm, also known as federated averaging. We collect these expressions into listing (17.89). The construction is illustrated in Fig. 17.6. In the next example, we motivate the fusion formula (17.88c).

Federated averaging algorithm for minimizing the empirical risk (17.78).

given N data points $\{\gamma(m), h_m\}$, $m = 0, 1, \ldots, N-1$;
select random initial vector $\boldsymbol{w}_{-1}$;
repeat for each cycle $n \geq 0$**:**
  select a random subset of agents $\mathcal{L}_n \subset \{1, 2, \ldots, L\}$;
  for each agent $\ell \in \mathcal{L}_n$**:**
    set initial condition to $\boldsymbol{w}^0_{\ell, R_\ell - 1} = \boldsymbol{w}_{n-1}$
    repeat for each epoch $k = 1, 2, \ldots, K_\ell$**:**
      set $\boldsymbol{w}^k_{\ell,-1} = \boldsymbol{w}^{k-1}_{\ell, R_\ell - 1}$
      repeat for $n' = 0, 1, 2, \ldots, R_\ell - 1$**:**
        select B random samples $\{\gamma(b), \boldsymbol{h}_b\}$ from the data at agent ℓ

$$\boldsymbol{w}^k_{\ell,n'} = \boldsymbol{w}^k_{\ell,n'-1} - \frac{\mu}{K_\ell R_\ell}\left(\frac{1}{B}\sum_{b=0}^{B-1} \nabla_{w^\mathsf{T}} Q(\boldsymbol{w}^k_{\ell,n'-1}; \gamma(b), \boldsymbol{h}_b)\right)$$

      end
    end
    $\boldsymbol{w}_{\ell,n} \leftarrow \boldsymbol{w}^K_{\ell, R_\ell - 1}$
    send $\boldsymbol{w}_{\ell,n}$ to server
  end

$$R_{\mathcal{L}} = \sum_{\ell' \in \mathcal{L}_n} R_{\ell'}$$

$$\boldsymbol{w}_n = \sum_{\ell \in \mathcal{L}_n} \frac{R_\ell}{R_{\mathcal{L}}} \boldsymbol{w}_{\ell,n}$$

end
return $w^\star \leftarrow \boldsymbol{w}_n$.

(17.89)

Example 17.6 **(Data fusion simplified)** Let $w^\star \in \mathbb{R}^M$ denote some unknown parameter vector that we wish to estimate. Assume we have L noisy estimates for it, denoted by $\{w_1, w_2, \ldots, w_L\}$. We express each one of these estimates as some perturbed version of $w^\star$, say, as:

$$w_\ell = w^\star + v_\ell \tag{17.90}$$

where v_ℓ denotes the perturbation. For simplicity, we model v_ℓ as a realization of a noise process with some diagonal covariance matrix, $\sigma_\ell^2 I_M$. Clearly, the smaller the value of σ_ℓ^2 is, the "closer" we expect the estimate w_ℓ to be to $w^\star$. That is, smaller noise variances imply "cleaner" estimates.

One way to fuse the L estimates $\{w_\ell\}$ to obtain a "better" estimate for $w^\star$ is to smooth the available estimates by solving:

$$\widehat{w} = \underset{w \in \mathbb{R}^M}{\text{argmin}} \left\{ \sum_{\ell=1}^{L} \frac{1}{\sigma_\ell^2} \|w - w_\ell\|^2 \right\} \tag{17.91}$$

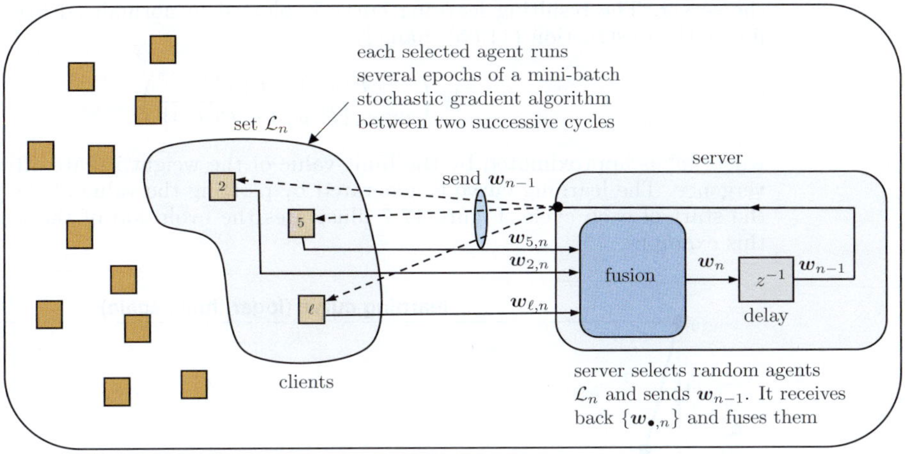

Figure 17.6 At every cycle, the server selects a random subset of agents $\mathcal{L}_n$, and shares $\boldsymbol{w}_{n-1}$ with them. Each of the agents runs multiple epochs of a mini-batch stochastic gradient algorithm and returns its iterate $\boldsymbol{w}_{\ell,n}$. These iterates are fused at the server to generate $\boldsymbol{w}_n$ and the process continues in this manner.

Differentiating relative to w and setting the gradient to zero leads to

$$\widehat{w} = \sum_{\ell=1}^{L} \left(\frac{1/\sigma_\ell^2}{\sum_{\ell'=1}^{L} 1/\sigma_{\ell'}^2} \right) w_\ell \tag{17.92}$$

Later, in Chapter 19, when we study the performance of stochastic gradient algorithms for generating estimates w_ℓ, we will discover that their noise variance σ_ℓ^2 is proportional to a quantity denoted by σ_g^2 (see expression (19.26)). The value of σ_g^2 will be inversely proportional to the amount of data N_ℓ used to run the stochastic gradient algorithm (see expression (18.35b)). In other words, when the estimates $\{w_\ell\}$ are generated by means of stochastic gradient algorithms, it is reasonable to expect that $\sigma_\ell^2 \propto 1/N_\ell$. Substituting into (17.92) leads to

$$\widehat{w} = \sum_{\ell=1}^{L} \left(\frac{N_\ell}{\sum_{\ell'=1}^{L} N_{\ell'}} \right) w_\ell \tag{17.93}$$

which agrees with the fusion expression (17.88c). For more details on data fusion in the context of linear estimation models, the reader may refer to the discussion in Appendix 29.4, and in particular to Lemma 29.2.

Example 17.7 **(Logistic regression)** We illustrate the operation of the federated averaging algorithm by repeating the simulations of Example 17.5 involving an ℓ_2-regularized logistic regression empirical risk problem. We employ uniform sampling and set the parameters to:

$$\rho = 1, \mu = 1, M = 10, L = 20, N_\ell = 200 = R_\ell, N = 4000, B = 5 \tag{17.94}$$

We also assume that at each cycle n the server selects four random clients (i.e., the cardinality of the subset $\mathcal{L}_n$ is $|\mathcal{L}_n| = 4$). We assume each agent runs $K_\ell = 50$ epochs, and we repeat the simulation over 50 cycles (i.e., for $n = 1, 2, \ldots, 50$). In order to generate a smooth learning curve, we average the risk error function over five experiments, with each experiment starting from the same initial condition $\boldsymbol{w}_{-1}$ for the weight at the server. The resulting learning curve is plotted in normalized logarithmic scale in line with construction (11.65), namely,

$$\ln\left(\frac{P(w_n) - P(w^\star)}{\max_n\{P(w_n) - P(w^\star)\}}\right) \tag{17.95}$$

where $w^\star$ is approximated by the limit value of the weight iterate after sufficient convergence. The learning curve is generated by plotting the value of the risk function at the start of each cycle. Figure 17.7 illustrates the evolution of the learning curve for this example.

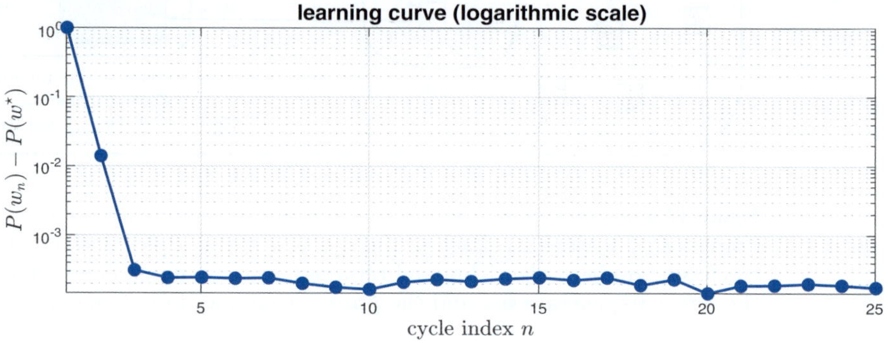

Figure 17.7 Learning curve $P(w_n)$ relative to the minimum risk value $P(w^\star)$ in normalized logarithmic scale over the first 25 cycles for the federated averaging algorithm.

17.7 COMMENTARIES AND DISCUSSION

Adaptive gradient algorithms. These methods adjust the step-size parameters continually and scale the entries of the gradient vectors individually. The first method we presented was the AdaGrad (adaptive gradient) scheme (17.13) introduced by Duchi, Hazan, and Singer (2011). The method scales the search direction by the inverse of a diagonal matrix, A_n^{-1}, and computes the entries of this matrix from the cumulative energies of the entries of the gradient of the loss function. Overall, the method performs reasonably well for problems involving sparse gradient vectors and convex risk function. However, one major difficulty with the method is the fact that the entries of A_n increase continually, which eventually leads to loss of adaptation as A_n^{-1} approaches zero. The RMSprop (root-mean-square propagation) method (17.40) proposes a simple fix by smoothing the computation of the energy terms by means of a first-order moving average filter. This algorithm is an unpublished result proposed by Tieleman and Hinton (2012) in one of their online video lectures. The AdaDelta method (17.44) was introduced by Zeiler (2012) to alleviate the problem with the decreasing learning rate in AdaGrad. It can be viewed as a modification to RMSprop, where it additionally tracks the energies in the weight differences and uses $D_{n-1}A_n^{-1}$ as the scaling matrix, instead of only A_n^{-1}. By doing so, AdaDelta forgoes the need to select a step-size parameter μ since it is eliminated from the final recursions. The original derivation of AdaDelta justifies the incorporation of the matrix D_{n-1} by arguing that units need to be matched in the stochastic gradient recursion. We followed an alternative argument in Example 17.1.

The ADAM (adaptive moment) method is among the most popular stochastic gradient algorithms with adaptive learning rates. It resolves the difficulty with AdaGrad by smoothing the computation of the energy terms through a moving average filter, as was the case with RMSprop and AdaDelta. It also performs bias correction. Additionally, it employs a momentum acceleration step by smoothing the gradient vector of the loss and using this smoothed value, after bias correction, in the update of the weight vectors. The ADAM method was introduced by Kingma and Ba (2015); the authors present in section 7 of their paper the AdaMax version (17.62) – we followed the same arguments in motivating the algorithm. Subsequent work by Dozat (2016) incorporated Nesterov momentum acceleration into ADAM, resulting in the Nadam algorithm. The original convergence analysis in Kingma and Ba (2015) assumes a decaying factor β_1 but suffers from a technical error identified by Reddi, Kale, and Kumar (2018). This latter reference showed that the ADAM recursion can diverge and proposed the modification known as AMSGrad listed in (17.54). Convergence results for these algorithms based on regret analysis appear in Reddi, Kale, and Kumar (2018) and Alacaoglu *et al.* (2020). In particular, motivated by arguments from these works, we show in Appendix 17.A that the average regret for the AMSGrad algorithm under constant step-size learning satisfies

$$\mathcal{R}(N) = O(1/N) + O(\mu) \tag{17.96}$$

That is, the regret approaches a small value on the order of μ and the convergence rate toward this region is $O(1/N)$.

It is worth noting that there exists a rich history in the adaptive filtering literature on the design of "optimal" step-size sequences and also on the scaling of individual entries of the gradient vector for enhanced performance – see, e.g., the summary in Sayed (2003, p. 255) and Sayed (2008, p. 207) for examples of variable step-size LMS (least mean squares) adaptive algorithms, and the overview by Paleologu, Benesty, and Ciochina (2011, ch. 5) on proportionate LMS algorithms in the context of echo cancellation, and the many other references therein. In these problems, the risk function is generally quadratic in nature.

Momentum acceleration. For additional discussion on Polyak and Nesterov acceleration methods, readers may refer to Polyak (1987), Polyak and Juditsky (1992), and Nesterov (1983, 2004) – see also the comments at the end of Chapter 12 and Probs. 19.21 and 19.22. Motivated by the useful acceleration properties of momentum implementations when applied to gradient descent methods, many subsequent works were motivated to introduce momentum terms into stochastic optimization algorithms as well, as was described in the body of the chapter – see, e.g., Tugay and Tanik (1989), Roy and Shynk (1990), Sharma, Sethares, and Bucklew (1998), Xiao (2010), Ghadimi and Lan (2012), Sutskever *et al.* (2013), and Zareba *et al.* (2015). The analysis by Yuan, Ying, and Sayed (2016) shows that the advantages of the momentum technique for gradient-descent problems do not always carry over to stochastic implementations due to the presence of gradient noise. It is shown in this last work that for sufficiently small step sizes, and for a momentum parameter not too close to 1, the advantages brought forth by momentum terms can be achieved by staying with the original stochastic gradient algorithm and adjusting its step size to a larger value.

Federated learning. In Section 17.6 we described the federated learning formulation, which distributes data processing over a federation of dispersed agents and fuses their iterates regularly at a central server. This architecture is motivated by the explosion in wireless smart devices, the growth in autonomous systems, and the proliferation of Internet-of-Things (IoT) solutions, all of which will be able to collect large and continuous amounts of raw data. Rather than aggregate the data at a central processor, which would require significant amounts of communication, federated learning allows the dispersed agents to perform local processing and to share their local iterates with the server. The framework of federated learning was introduced in Konecny *et al.* (2016) and McMahan *et al.* (2017). A useful performance evaluation appears in Nilsson *et al.* (2018). Analysis of federated learning in dynamic environments under model drifts is given by Rizk, Vlaski, and Sayed (2020).

Observe that in the federated learning architecture, clients share their local information $\boldsymbol{w}_{\ell,n}$ regularly with the server. In applications such as this one, involving the sharing of information, it is often desirable to ensure some level of privacy for the agents. For example, the sharing of weight vector $\boldsymbol{w}_{\ell,n}$ with the central server should not reveal any private information about user ℓ or other users to the server. One useful notion of privacy is *ϵ-differential privacy*, which we describe briefly for the benefit of the reader.

Differential privacy. The concept of differential privacy, introduced by Dwork (2006) and Dwork *et al.* (2006), is a mechanism that allows sharing aggregate statistics about a dataset without revealing information about individual elements in the dataset. It is a method to limit the disclosure of private information.

Let $\mathcal{A}$ denote some generic *randomized* algorithm that operates on a dataset $\mathcal{D}$ and generates some output $\mathcal{S}$. By "randomized" we mean that the way the algorithm processes the data involves some element of randomness, as happens for example with stochastic gradient algorithms where samples or mini-batches are chosen at random or are randomly reshuffled. The dataset usually consists of feature vectors $\{h_n\}$ that contain, for example, personal information about each patient n such as their age, sex, height, weight, habits, and medical conditions. This information may have been supplied by individuals participating in a study, and algorithm $\mathcal{A}$ could have been designed to infer whether certain habits are more likely to lead to a particular medical condition. In situations like these, the individuals participating in the study would prefer to keep their personal information private. For instance, if the algorithm were to release information about which habits are more likely to cause the medical condition, then this information should not enable others to identify specific individuals from the dataset that are likely to develop the medical condition. One way to embed a layer of privacy onto the operation of the algorithm is by adding differential privacy, which is defined as follows.

Algorithm $\mathcal{A}$ can generate multiple pieces of information about the dataset, such as the percentage of individuals having a particular habit, the average weight and height of individuals, the percentage of male and female participants, which habits are more likely to cause the medical condition under study, and so forth. We refer to the output generated by algorithm $\mathcal{A}$ as its image and denote it by the letter $\mathcal{S}$. Now consider any two subsets of $\mathcal{D}$, denoted by $D_1 \subset \mathcal{D}$ and $D_2 \subset \mathcal{D}$, which are assumed to differ by a single entry, i.e., the individuals contributing to both sets are the same except for one individual in each. We refer to these sets as *neighbors*. Algorithm $\mathcal{A}$ will be said to be ϵ-differentially private if it satisfies the condition:

$$\frac{\mathbb{P}\Big(\mathcal{A}(D_1) \in \mathcal{S}\Big)}{\mathbb{P}\Big(\mathcal{A}(D_2) \in \mathcal{S}\Big)} \leq e^{\epsilon}, \quad \text{for some small } \epsilon > 0 \tag{17.97}$$

for *any* two neighboring subsets D_1, D_2, and where the probability is computed over the randomness by the algorithm. The above definition assumes that the output of the algorithm is discrete for simplicity. Since the definition should hold for any D_1 and D_2, we can reverse their order in the numerator and denominator to conclude that it also implies

$$e^{-\epsilon} \leq \frac{\mathbb{P}\Big(\mathcal{A}(D_1) \in \mathcal{S}\Big)}{\mathbb{P}\Big(\mathcal{A}(D_2) \in \mathcal{S}\Big)} \leq e^{\epsilon} \tag{17.98}$$

Using $e^\epsilon \approx 1 + \epsilon$, we can write

$$1 - \epsilon \leq \frac{\mathbb{P}\Big(\mathcal{A}(D_1) \in \mathcal{S}\Big)}{\mathbb{P}\Big(\mathcal{A}(D_2) \in \mathcal{S}\Big)} \leq 1 + \epsilon \tag{17.99}$$

This requirement essentially means that the probability of observing the same output for the algorithm, whether acting on one set or the other, is largely invariant. In other words, if we apply the algorithm to two neighboring sets (D_1, D_2) that differ by a single entry, it is not possible to observe any noticeable difference in the output (behavior) of the algorithm. In this way, every individual in the dataset $\mathcal{D}$ is afforded the same privacy whether they are included or removed from the dataset. The parameter ϵ is called the *privacy loss* parameter since it controls the level of privacy loss in response to differential changes in the dataset: The smaller ϵ is, the higher the level of privacy protection will be. However, as we are going to see, decreasing ϵ leads to a loss in accuracy (such as classification accuracy by algorithm $\mathcal{A}$); this is because decreasing ϵ would entail adding more noise to the computations.

One way to achieve differential privacy is by means of a Laplace mechanism, which corresponds to adding Laplacian noise (i.e., noise arising from a Laplace distribution) to the output of the algorithm – see Dwork *et al.* (2006). Recall that if $\boldsymbol{v}$ is Laplace-distributed with zero mean and variance σ_v^2, then its pdf is given by:

$$f_{\boldsymbol{v}}(v) = \frac{1}{\sqrt{2\sigma_v^2}} \exp\left\{-\frac{\sqrt{2}|v|}{\sqrt{\sigma_v^2}}\right\} \triangleq \mathcal{L}\text{ap}(0, \sigma_v^2) \tag{17.100}$$

To see how adding Laplacian noise assists with privacy, we first introduce the notion of the *sensitivity* of a function $g(D) : \mathcal{D} \rightarrow \mathbb{R}^S$. This function acts on subsets of the dataset $\mathcal{D}$ and generates some output vector. Algorithm $\mathcal{A}$ is one example of such a function. Let D_1 and D_2 be any neighboring subsets in $\mathcal{D}$ that differ by only one element. The sensitivity of algorithm $\mathcal{A}$ is denoted by Δ and is defined as

$$\Delta \triangleq \max_{D_1, D_2 \subset \mathcal{D}} \Big\|\mathcal{A}(D_1) - \mathcal{A}(D_2)\Big\|_1 \tag{17.101}$$

in terms of the ℓ_1-norm. This expression is measuring the largest possible change that can occur to $\mathcal{A}$ over all possible choices for $\{D_1, D_2\}$. The more sensitive an algorithm is, the larger the change in its output for small changes in its input. One method to embed privacy into an algorithm is to add Laplacian noise whose variance is dependent on the level of sensitivity of the algorithm.

Laplace mechanism (Dwork *et al.* (2006)) *Let $\mathcal{A}(D)$ denote the unperturbed output of an algorithm with sensitivity Δ. Assume the output of the algorithm is perturbed to $\mathcal{A}'(D)$ by adding Laplacian noise as follows:*

$$\boxed{\mathcal{A}'(D) = \mathcal{A}(D) + v, \quad v \sim \mathcal{L}\text{ap}(0, \sigma_v^2 I_S), \quad \sigma_v^2 = 2\Delta^2/\epsilon^2} \tag{17.102}$$

It follows that $\mathcal{A}'(D)$ is ϵ-differentially private.

Proof: Let D_1 and D_2 be two neighboring sets in $\mathcal{D}$ differing by only one element. Algorithm $\mathcal{A}'$ results in continuous real-valued outputs due to the addition of v. We therefore need to assess the likelihood that $\mathcal{A}'(D_1)$ assumes values in a small infinitesimal interval, say, of the form $[s, s + ds]$:

$$\mathbb{P}\Big(\mathcal{A}'(D_1) \in [s, s + ds]\Big) = \int_s^{s+ds} f_{\boldsymbol{v}}(x - \mathcal{A}(D_1))dx \approx f_{\boldsymbol{v}}(s - \mathcal{A}(D_1))ds \tag{17.103}$$

and compare it to the probability that $\mathcal{A}'(D_2)$ assumes values in the same small infinitesimal interval:

$$\mathbb{P}\Big(\mathcal{A}'(D_2) \in [s, s+ds]\Big) = \int_s^{s+ds} f_{\boldsymbol{v}}(x - \mathcal{A}(D_2))dx \approx f_{\boldsymbol{v}}(s - \mathcal{A}(D_2))ds \qquad (17.104)$$

The above two expressions are written in terms of the pdf of the vector $\boldsymbol{v}$, which we are denoting by $f_{\boldsymbol{v}}(v)$. Let s_j and $\mathcal{A}_j(\cdot)$ denote the jth entries of the S-dimensional vectors s and $\mathcal{A}(\cdot)$, respectively. It follows that

$$\begin{aligned}
\frac{\mathbb{P}(\mathcal{A}'(D_1) \in [s, s+ds])}{\mathbb{P}(\mathcal{A}'(D_2) \in [s, s+ds])} &\approx \frac{\prod_{j=1}^S \exp\Big\{-\epsilon|s_j - \mathcal{A}_j(D_1)|/\Delta\Big\}}{\prod_{j=1}^S \exp\Big\{-\epsilon|s_j - \mathcal{A}_j(D_2)|/\Delta\Big\}} \\
&= \exp\left\{\frac{\epsilon}{\Delta}\sum_{j=1}^S \Big(|s_j - \mathcal{A}_j(D_2)| - |s_j - \mathcal{A}_j(D_1)|\Big)\right\} \\
&\overset{(a)}{\leq} \exp\left\{\frac{\epsilon}{\Delta}\sum_{j=1}^S \Big|\mathcal{A}_j(D_2) - \mathcal{A}_j(D_1)\Big|\right\} \\
&= \exp\left\{\frac{\epsilon}{\Delta}\Big\|\mathcal{A}(D_2) - \mathcal{A}(D_1)\Big\|_1\right\} \\
&\overset{(b)}{\leq} \exp\left\{\frac{\epsilon}{\Delta}\Delta\right\} \\
&= e^{\epsilon}
\end{aligned} \qquad (17.105)$$

where step (a) is from the triangle inequality and step (b) is from definition (17.101). We conclude that the Laplace mechanism (17.102) ensures ϵ-differential privacy.

■

Observe from (17.102) that the more privacy is desired (i.e., the smaller ϵ is), the larger the noise variance will need to be. Also, for a fixed privacy level ϵ, algorithms with higher sensitivity will require higher levels of noise to deliver ϵ-differential privacy. A useful overview of differential privacy appears in Dwork and Roth (2014). Applications to empirical risk optimization and convex optimization problems appear in Chaudhuri, Monteleoni, and Sarwate (2011) and Kifer, Smith, and Thakurta (2012), respectively.

PROBLEMS

17.1 Consider the LASSO formulation with elastic-net regularization:

$$\min_{w\in\mathbb{R}^M} \left\{P(w) \triangleq \alpha\|w\|_1 + \rho\|w\|^2 + \frac{1}{N}\sum_{m=0}^{N-1}(\gamma(m) - h_m^{\mathsf{T}}w)^2\right\}$$

where the $\{h_m \in \mathbb{R}^M\}$ are feature vectors and the $\{\gamma(m)\}$ are scalar target signals.

(a) Derive a subgradient stochastic gradient algorithm for the minimization of the empirical risk, $P(w)$.

(b) Derive a stochastic proximal gradient algorithm for the same purpose.

(c) Write down ADAM recursion for solving the same problem.

17.2 Introduce the projection operator $\mathcal{P}_{\mathcal{C},A}(x)$, which projects a vector $x \in \mathbb{R}^M$ onto the convex set $\mathcal{C}$ by using a weighted Euclidean norm (also known as the Mahalanobis norm):

$$\widehat{x} = \mathcal{P}_{\mathcal{C},A}(x) \iff \widehat{x} = \underset{x'\in\mathcal{C}}{\operatorname{argmin}}\ \|x - x'\|_A^2$$

where $\|y\|_A^2 = y^\mathsf{T} A y$ denotes the weighted squared Euclidean norm of vector y, and $A > 0$. Show that such projection operations satisfy the nonexpansive property:

$$\|w - \mathcal{P}_{\mathcal{C},A}(x)\|_A^2 \leq \|w - x\|_A^2, \quad \text{for any } x \in \mathbb{R}^M \text{ and } w \in \mathcal{C}$$

17.3 Derive the regret bound (17.23).

17.4 Solve the constrained optimization problem (17.36) over symmetric positive-definite matrices A with bounded trace. *Remark.* See lemma 15 in the work by Duchi, Hazan, and Singer (2011).

17.5 Refer to the AMSGrad algorithm (17.54) and replace μ by a decaying step-size sequence, $\mu(n) = O(1/\sqrt{n})$. Repeat the derivation leading to (17.151) in the appendix to conclude that

$$\sum_{n=0}^{N-1} \mu(n)\|\bar{b}_n\|_{B_n}^2 = O(\sqrt{N}\,)$$

17.6 For the same setting of Prob. 17.5, verify that it also holds

$$\sum_{n=0}^{N-1} \mu(n)\|\bar{b}_n\|_{B_n}^2 = O\left(\sqrt{1 + \ln N} \times \sum_{m=1}^{M}\left\{\sum_{n=0}^{N-1} b_{n,m}^2\right\}^{1/2}\right)$$

Remark. See lemma 4 in Alacaoglu *et al.* (2020) for a related discussion.

17.7 Refer to the AMSGrad algorithm (17.54) and replace μ by a decaying step-size sequence, $\mu(n) = O(1/\sqrt{n})$. Repeat the derivation in Appendix 17.A to conclude that the average regret satisfies $\mathcal{R}(N) = O(1/\sqrt{N})$. *Remark.* See the works by Kingma and Ba (2015), Reddi, Kale, and Kumar (2018), and Alacaoglu *et al.* (2020) for a related discussion.

17.8 Conclude from algorithm (17.70) that $\boldsymbol{w}_n = \boldsymbol{w}_{n-1} - \mu\bar{\boldsymbol{b}}_n$.

17.9 Verify that algorithm (17.65) with Polyak acceleration can be written as:

$$\boldsymbol{w}_n = \boldsymbol{w}_{n-1} - \mu\nabla_{w^\mathsf{T}} Q(\boldsymbol{w}_{n-1}; \boldsymbol{\gamma}(n), \boldsymbol{h}_n) + \beta(\boldsymbol{w}_{n-1} - \boldsymbol{w}_{n-2}), \quad n \geq 0$$

17.10 Verify that algorithm (17.68) with Nesterov acceleration can be written as

$$\begin{aligned}\boldsymbol{w}_n = \boldsymbol{w}_{n-1} &+ \beta(\boldsymbol{w}_{n-1} - \boldsymbol{w}_{n-2}) - \\ &\mu\nabla_{w^\mathsf{T}} Q\Big(\boldsymbol{w}_{n-1} + \beta(\boldsymbol{w}_{n-1} - \boldsymbol{w}_{n-2}); \boldsymbol{\gamma}(n), \boldsymbol{h}_n\Big)\end{aligned}$$

17.11 Verify that algorithm (17.73) with Nesterov acceleration can be written as

$$\begin{aligned}\boldsymbol{w}_n = \boldsymbol{w}_{n-1} &+ \beta(\boldsymbol{w}_{n-1} - \boldsymbol{w}_{n-2}) - \mu\nabla_{w^\mathsf{T}} Q(\boldsymbol{w}_{n-1}; \boldsymbol{\gamma}(n), \boldsymbol{h}_n) + \\ &\mu\beta\Big(\nabla_{w^\mathsf{T}} Q(\boldsymbol{w}_{n-2}; \boldsymbol{\gamma}(n), \boldsymbol{h}_n) - \nabla_{w^\mathsf{T}} Q(\boldsymbol{w}_{n-1}; \boldsymbol{\gamma}(n), \boldsymbol{h}_n)\Big)\end{aligned}$$

17.12 Consider the following stochastic gradient formulation with two momentum parameters $\beta_1, \beta_2 \in [0, 1)$:

$$\begin{cases} \boldsymbol{z}_n = \boldsymbol{w}_{n-1} + \beta_1(\boldsymbol{w}_{n-1} - \boldsymbol{w}_{n-2}) \\ \boldsymbol{w}_n = \boldsymbol{z}_n - \mu\nabla_{w^\mathsf{T}} Q(\boldsymbol{z}_n, \boldsymbol{\gamma}(n), \boldsymbol{h}_n) + \beta_2(\boldsymbol{z}_n - \boldsymbol{z}_{n-1}) \end{cases}$$

Verify that the Polyak and Nesterov momentum acceleration methods from Probs. 17.9 and 17.10 are special cases by selecting $(\beta_1, \beta_2) = (0, \beta)$ in one case and $(\beta_1, \beta_2) = (\beta, 0)$ in the other case. *Remark.* See the work by Yuan, Ying, and Sayed (2016) for analysis of the behavior of this stochastic construction.

17.13 Refer to the discussion on federated learning in Section 17.6. Associate the same local risk function $P_\ell(w)$ with each agent ℓ and introduce the global risk function $P(w) = \sum_{\ell=1}^{L} p_\ell P_\ell(w)$, where the $\{p_\ell\}$ are nonnegative weighting scalars that add up to 1. Repeat the arguments in that section to derive the corresponding federated learning algorithm and explain that the local iterates $\boldsymbol{w}_{\ell,n}$ can now be fused as follows:

$$p_{\mathcal{L}} \stackrel{\Delta}{=} \sum_{\ell' \in \mathcal{L}_n} p_{\ell'}, \quad \boldsymbol{w}_n = \sum_{\ell \in \mathcal{L}_n} \frac{p_\ell}{p_{\mathcal{L}}} \boldsymbol{w}_{\ell,n}$$

Compare the situations corresponding to the choices $p_\ell = 1/L$ and $p_\ell = N_\ell/N$.

17.14 Consider a dataset of N scalar binary points $D = \{a_n\}$, where each $a_n \in \{0, 1\}$. Let $g(D)$ denote their sum, $g(D) = \sum_{n=1}^{N} a_n$. Let $\mathcal{A}(D)$ refer to an algorithm whose output is a Laplace-perturbed version of $g(D)$, namely, $\mathcal{A}(D) = g(D) + \boldsymbol{v}$, where $\boldsymbol{v}$ is Laplace-distributed with mean zero and variance $2/\epsilon^2$.

(a) Show that the sensitivity of the function g is $\Delta = 1$.

(b) Show that $\mathcal{A}(D)$ is ϵ-differentially private.

(c) Show that $\mathbb{E}\,\boldsymbol{\mathcal{A}}(D) = g(D)$.

17.15 Consider again a dataset of N scalar data points $D = \{a_n\}$, where each $a_n \in [0, \alpha]$. Let $g(D)$ denote their average, $g(D) = (1/N)\sum_{n=1}^{N} a_n$. Let $\mathcal{A}(D)$ refer to an algorithm whose output is a Laplace-perturbed version of $g(D)$, namely, $\mathcal{A}(D) = g(D) + \boldsymbol{v}$, where $\boldsymbol{v}$ is Laplace-distributed with zero mean and variance $2\alpha^2/N^2\epsilon^2$.

(a) Show that the sensitivity of the function g is $\Delta = \alpha/N$.

(b) Show that $\mathcal{A}(D)$ is ϵ-differentially private.

17.A REGRET ANALYSIS FOR ADAM

In this appendix we examine the regret performance of the AMSGrad algorithm (17.54), which is an improved variant of ADAM; similar derivations can be applied to other forms of adaptive gradient methods discussed in the body of the chapter. The presentation given here combines and adjusts arguments from Kingma and Ba (2015), Reddi, Kale, and Kumar (2018), and Alacaoglu *et al.* (2020).

As was explained in Remark 17.2, it is customary to seek the minimizer of (17.1) over bounded closed convex sets, denoted by $w \in \mathcal{C}$. In this case, the last step in (17.54) is modified to include a projection step onto $\mathcal{C}$. For simplicity, we list the main equations of the algorithm using slightly redefined symbols as follows:

$$\left\{\begin{aligned} \bar{b}_n &= \beta_1 \bar{b}_{n-1} + (1-\beta_1) b_n \\ s_n &= \beta_2 s_{n-1} + (1-\beta_2)(b_n \odot b_n) \\ \widehat{s}_n &= \max\{\widehat{s}_{n-1}, s_n\} \\ b'_n &\stackrel{\Delta}{=} (1-\beta_2^{n+1})^{1/2} \times \bar{b}_n \\ B_n &\stackrel{\Delta}{=} \left\{(1-\beta_1^{n+1}) \times \left(\epsilon I_M + \text{diag}(\widehat{s}_n^{1/2})\right)\right\}^{-1} \\ w_n &= \mathcal{P}_C\Big(w_{n-1} - \mu B_n b'_n\Big) \end{aligned}\right. \tag{17.106}$$

where we introduced the scaled quantities $\{b'_n, B_n\}$. The notation $s^{1/2}$, when applied to a vector s with nonnegative entries, refers to another vector consisting of the square-roots of the individual entries of s. Moreover, by definition,

$$b_n \stackrel{\Delta}{=} \nabla_{w^{\mathsf{T}}} Q(w_{n-1}; \gamma(n), h_n), \quad \text{(gradient of loss function)} \tag{17.107}$$

$$\beta_1 = 0.9, \quad \beta_2 = 0.999, \quad \epsilon: \text{ a small positive number} \tag{17.108}$$

$$\bar{b}_{-1} = s_{-1} = \widehat{s}_{-1} = 0, \quad \text{(zero initial conditions)} \tag{17.109}$$

Observe from the definition of $\widehat{s}_n$ and the expression for the *diagonal* matrix B_n that $(B_{-1})^{-1} = 0$ and

$$B_n \leq B_{n-1}, \quad \forall\, n \geq 0 \tag{17.110}$$

Average regret

We know from (16.113) that the regret measure for stochastic optimization algorithms is defined by

$$\begin{aligned}
\mathcal{R}(N) &\triangleq \frac{1}{N}\sum_{n=0}^{N-1} Q(w_{n-1};\gamma(n),h_n) - \min_{w\in\mathcal{C}}\left\{\frac{1}{N}\sum_{n=0}^{N-1} Q(w;\gamma(n),h_n)\right\} \\
&\overset{(a)}{=} \frac{1}{N}\sum_{n=0}^{N-1} Q(w_{n-1};\gamma(n),h_n) - \frac{1}{N}\sum_{n=0}^{N-1} Q(w^\star;\gamma(n),h_n) \\
&\overset{(b)}{\leq} -\frac{1}{N}\sum_{n=0}^{N-1}\Big(\nabla_{w^\mathsf{T}} Q(w_{n-1};\gamma(n),h_n)\Big)^\mathsf{T}\widetilde{w}_{n-1} \\
&\overset{(c)}{=} -\frac{1}{N}\sum_{n=0}^{N-1} b_n^\mathsf{T}\widetilde{w}_{n-1}
\end{aligned} \tag{17.111}$$

where $\widetilde{w}_{n-1} = w^\star - w_{n-1}$. In step (a) we are denoting the minimizer from the first line by $w^\star$, in step (b) we are using the convexity of the loss function and Remark 12.4, and in step (c) we are replacing the gradient vector by the simpler notation b_n from (17.106). Expression (17.111) shows that one way to bound the regret is by bounding the inner product terms $b_n^\mathsf{T}\widetilde{w}_{n-1}$. However, the recursion for w_n involves the filtered and scaled vectors b_n' instead of b_n. Therefore, some analysis is necessary to arrive at the desired bounds.

Inner product terms

We deduce from the first relation in (17.106) that

$$b_n = \frac{1}{1-\beta_1}\bar{b}_n - \frac{\beta_1}{1-\beta_1}\bar{b}_{n-1} \tag{17.112}$$

so that

$$\begin{aligned}
&b_n^\mathsf{T}\widetilde{w}_{n-1} \\
&= \frac{1}{1-\beta_1}\bar{b}_n^\mathsf{T}\widetilde{w}_{n-1} - \frac{\beta_1}{1-\beta_1}\bar{b}_{n-1}^\mathsf{T}\widetilde{w}_{n-1} \\
&= \frac{1}{1-\beta_1}\bar{b}_n^\mathsf{T}\widetilde{w}_{n-1} - \frac{\beta_1}{1-\beta_1}\bar{b}_{n-1}^\mathsf{T}(\underbrace{w^\star - w_{n-1}}_{=\widetilde{w}_{n-1}} + \underbrace{w_{n-2} - w_{n-2}}_{=0}) \\
&= \frac{1}{1-\beta_1}\bar{b}_n^\mathsf{T}\widetilde{w}_{n-1} - \frac{\beta_1}{1-\beta_1}\bar{b}_{n-1}^\mathsf{T}\widetilde{w}_{n-2} - \frac{\beta_1}{1-\beta_1}\bar{b}_{n-1}^\mathsf{T}(w_{n-2} - w_{n-1}) \\
&= \bar{b}_{n-1}^\mathsf{T}\widetilde{w}_{n-2} - \frac{\beta_1}{1-\beta_1}\bar{b}_{n-1}^\mathsf{T}(w_{n-2} - w_{n-1}) + \frac{1}{1-\beta_1}\left\{\bar{b}_n^\mathsf{T}\widetilde{w}_{n-1} - \bar{b}_{n-1}^\mathsf{T}\widetilde{w}_{n-2}\right\}
\end{aligned} \tag{17.113}$$

This result expresses each term $b_n^\mathsf{T}\widetilde{w}_{n-1}$ in terms of inner products involving the filtered gradient vectors $\{\bar{b}_n, \bar{b}_{n-1}\}$. Substituting into (17.111) gives an expression with three terms marked by the circled numbers:

$$
\begin{aligned}
\mathcal{R}(N) = & -\frac{1}{N}\sum_{n=0}^{N-1} \bar{b}_{n-1}^{\mathsf{T}}\widetilde{w}_{n-2} & \text{①} \\
& +\frac{\beta_1}{1-\beta_1}\frac{1}{N}\sum_{n=0}^{N-1} \bar{b}_{n-1}^{\mathsf{T}}(w_{n-2}-w_{n-1}) & \text{②} \\
& -\frac{1}{1-\beta_1}\frac{1}{N}\sum_{n=0}^{N-1}\left\{\bar{b}_n^{\mathsf{T}}\widetilde{w}_{n-1} - \bar{b}_{n-1}^{\mathsf{T}}\widetilde{w}_{n-2}\right\} & \text{③}
\end{aligned}
\tag{17.114}
$$

We proceed to bound each of the three terms appearing in the above inequality.

Useful bounds

First, note that

$$
\begin{aligned}
\text{②} & \triangleq \bar{b}_{n-1}^{\mathsf{T}}(w_{n-2}-w_{n-1}) \\
& \overset{(a)}{=} \bar{b}_{n-1}^{\mathsf{T}} B_{n-1}^{1/2} B_{n-1}^{-1/2}(w_{n-2}-w_{n-1}) \\
& = \left(B_{n-1}^{1/2}\bar{b}_{n-1}\right)^{\mathsf{T}} B_{n-1}^{-1/2}(w_{n-2}-w_{n-1}) \\
& \leq \|B_{n-1}^{1/2}\bar{b}_{n-1}\| \times \|B_{n-1}^{-1/2}(w_{n-2}-w_{n-1})\| \qquad \text{(by Cauchy–Schwarz)} \\
& = \sqrt{\|\bar{b}_{n-1}\|^2_{B_{n-1}}} \times \sqrt{\|w_{n-2}-w_{n-1}\|^2_{B_{n-1}^{-1}}}
\end{aligned}
\tag{17.115}
$$

where in step (a) we inserted the identity matrix $B_{n-1}^{1/2}B_{n-1}^{-1/2}$, and in the last line we are using the notation $\|x\|_B^2 = x^{\mathsf{T}}Bx$. Next, since $w_{n-2} \in \mathcal{C}$, we can use the update iteration from (17.106) to write

$$
\|w_{n-2}-w_{n-1}\|^2_{B_{n-1}^{-1}} = \|\mathcal{P}_C(w_{n-2}) - \mathcal{P}_C(w_{n-2}-\mu B_{n-1}b'_{n-1})\|^2_{B_{n-1}^{-1}} \tag{17.116}
$$

Using the nonexpansive property of the projection operator, namely, for any positive-definite weighting matrix B:

$$
\|\mathcal{P}_C(x) - \mathcal{P}_C(y)\|_B^2 \;\leq\; \|x-y\|_B^2, \;\; \forall\, x, y \in \mathbb{R}^M \tag{17.117}
$$

we readily conclude that

$$
\begin{aligned}
\|w_{n-2}-w_{n-1}\|^2_{B_{n-1}^{-1}} & \leq \|\mu B_{n-1}b'_{n-1}\|^2_{B_{n-1}^{-1}} \\
& = \|\mu b'_{n-1}\|^2_{B_{n-1}} \\
& = \mu^2(1-\beta_2^{n+1})\|\bar{b}_{n-1}\|^2_{B_{n-1}} \\
& \leq \mu^2\|\bar{b}_{n-1}\|^2_{B_{n-1}}
\end{aligned}
\tag{17.118}
$$

since $1-\beta_2^{n+1} \leq 1$. Substituting into (17.115) we get

$$
\boxed{\frac{\beta_1}{1-\beta_1}\frac{1}{N}\sum_{n=0}^{N-1}\bar{b}_{n-1}^{\mathsf{T}}(w_{n-2}-w_{n-1}) \;\leq\; \frac{\mu\beta_1}{1-\beta_1}\frac{1}{N}\sum_{n=0}^{N-1}\|\bar{b}_{n-1}\|^2_{B_{n-1}}} \tag{17.119}
$$

We consider next another term from (17.114). Subtracting $w^\star$ from both sides of the update iteration in (17.106) and using $w^\star = \mathcal{P}_C(w^\star)$ we have

$$
\widetilde{w}_n = \mathcal{P}_C(w^\star) - \mathcal{P}_C(w_{n-1}-\mu B_n b'_n) \tag{17.120}
$$

so that from the same nonexpansive property (17.117):

$$\begin{aligned}\|\widetilde{w}_n\|^2_{B_n^{-1}} &\leq \|\widetilde{w}_{n-1} + \mu B_n b'_n\|^2_{B_n^{-1}} \\ &\leq \|\widetilde{w}_{n-1}\|^2_{B_n^{-1}} + \mu^2\|b'_n\|^2_{B_n} + 2\mu(b'_n)^{\mathsf{T}}\widetilde{w}_{n-1} \\ &\leq \|\widetilde{w}_{n-1}\|^2_{B_n^{-1}} + \mu^2(1-\beta_2^{n+1})\|\bar{b}_n\|^2_{B_n} + 2\mu(1-\beta_2^{n+1})^{1/2}\bar{b}_n^{\mathsf{T}}\widetilde{w}_{n-1}\end{aligned} \tag{17.121}$$

and, hence, by rearranging terms:

$$\begin{aligned}\textcircled{1} &\stackrel{\Delta}{=} -\bar{b}_n^{\mathsf{T}}\widetilde{w}_{n-1} \\ &\leq \frac{1}{2\mu(1-\beta_2^{n+1})^{1/2}}\left\{\|\widetilde{w}_{n-1}\|^2_{B_n^{-1}} - \|\widetilde{w}_n\|^2_{B_n^{-1}}\right\} + \frac{\mu(1-\beta_2^{n+1})^{1/2}}{2}\|\bar{b}_n\|^2_{B_n} \\ &\leq \frac{1}{2\mu(1-\beta_2)^{1/2}}\left\{\|\widetilde{w}_{n-1}\|^2_{B_n^{-1}} - \|\widetilde{w}_n\|^2_{B_n^{-1}}\right\} + \frac{\mu}{2}\|\bar{b}_n\|^2_{B_n}\end{aligned} \tag{17.122}$$

where we used the fact that $1-\beta_2^{n+1} < 1$ and $1-\beta_2^{n+1} \geq 1-\beta_2$ for $n \geq 0$. Let $\mu' = \mu(1-\beta_2)^{1/2}$. Let also d denote the largest distance between any two points in the bounded convex set $\mathcal{C}$:

$$d \stackrel{\Delta}{=} \max_{x,y\in\mathcal{C}} \|x-y\| \tag{17.123}$$

Using μ' and d, the above inequality can be manipulated as follows:

$$\begin{aligned}\textcircled{1} &\stackrel{\Delta}{=} -\bar{b}_n^{\mathsf{T}}\widetilde{w}_{n-1} \\ &\leq \frac{1}{2\mu'}\left\{\|\widetilde{w}_{n-1}\|^2_{B_n^{-1}} - \|\widetilde{w}_n\|^2_{B_n^{-1}}\right\} + \frac{\mu}{2}\|\bar{b}_n\|^2_{B_n} \\ &= \frac{1}{2\mu'}\left\{\|\widetilde{w}_{n-1}\|^2_{B_n^{-1}} + \underbrace{\|\widetilde{w}_{n-1}\|^2_{B_{n-1}^{-1}} - \|\widetilde{w}_{n-1}\|^2_{B_{n-1}^{-1}}}_{=0} - \|\widetilde{w}_n\|^2_{B_n^{-1}}\right\} + \frac{\mu}{2}\|\bar{b}_n\|^2_{B_n} \\ &= \underbrace{\frac{1}{2\mu'}\left\{\|\widetilde{w}_{n-1}\|^2_{B_{n-1}^{-1}} - \|\widetilde{w}_n\|^2_{B_n^{-1}}\right\}}_{\text{telescoping sequence}} + \frac{1}{2\mu'}\left\{\|\widetilde{w}_{n-1}\|^2_{(B_n^{-1}-B_{n-1}^{-1})}\right\} + \frac{\mu}{2}\|\bar{b}_n\|^2_{B_n} \\ &\stackrel{(17.123)}{\leq} \underbrace{\frac{1}{2\mu'}\left\{\|\widetilde{w}_{n-1}\|^2_{B_{n-1}^{-1}} - \|\widetilde{w}_n\|^2_{B_n^{-1}}\right\}}_{\text{telescoping sequence}} + \frac{d^2}{2\mu'}\underbrace{\mathrm{Tr}(B_n^{-1} - B_{n-1}^{-1})}_{\text{telescoping sequence}} + \frac{\mu}{2}\|\bar{b}_n\|^2_{B_n}\end{aligned} \tag{17.124}$$

It follows that

$$\begin{aligned}&-\frac{1}{N}\sum_{n=0}^{N-1}\bar{b}_n^{\mathsf{T}}\widetilde{w}_{n-1} \leq \\ &\frac{1}{2\mu' N}\left\{\cancelto{0}{\|\widetilde{w}_{-1}\|^2_{B_{-1}^{-1}}} - \|\widetilde{w}_{N-1}\|^2_{B_{N-1}^{-1}}\right\} + \frac{d^2}{2\mu' N}\mathrm{Tr}\left(B_{N-1}^{-1} - \cancelto{0}{B_{-1}^{-1}}\right) + \frac{\mu}{2N}\sum_{n=0}^{N-1}\|\bar{b}_n\|^2_{B_n}\end{aligned} \tag{17.125}$$

and, therefore,

$$-\frac{1}{N}\sum_{n=0}^{N-1}\bar{b}_n^{\mathsf{T}}\widetilde{w}_{n-1} \le \frac{d^2}{2\mu' N}\mathrm{Tr}(B_{N-1}^{-1}) + \frac{\mu}{2N}\sum_{n=0}^{N-1}\|\bar{b}_n\|_{B_n}^2 \tag{17.126}$$

Finally, the last term in (17.114) leads again to a telescoping sequence so that

$$\begin{aligned}\sum_{n=0}^{N-1}\left\{\bar{b}_n^{\mathsf{T}}\widetilde{w}_{n-1} - \bar{b}_{n-1}^{\mathsf{T}}\widetilde{w}_{n-2}\right\} &= \bar{b}_{N-1}^{\mathsf{T}}\widetilde{w}_{N-2}, \;\text{ since } \bar{b}_{-1}=0\\ &= \bar{b}_{N-1}^{\mathsf{T}}B_{N-1}^{1/2}B_{N-1}^{-1/2}\widetilde{w}_{N-2}\\ &= \left(B_{N-1}^{1/2}\bar{b}_{N-1}\right)^{\mathsf{T}}B_{N-1}^{-1/2}\widetilde{w}_{N-2}\end{aligned} \tag{17.127}$$

where we inserted the identity matrix $B_{N-1}^{1/2}B_{N-1}^{-1/2}$. We now appeal to a famous inequality known as the *Young inequality*, which states that for any two vectors $x, y \in \mathbb{R}^M$ and any scalar λ, it holds that (see Hardy, Littlewood, and Pólya (1934)):

$$|x^{\mathsf{T}}y| \le \frac{\lambda^2}{2}\|x\|^2 + \frac{1}{2\lambda^2}\|y\|^2 \qquad (\textbf{Young inequality}) \tag{17.128}$$

Applying this inequality to (17.127) and using $\lambda^2/2 = \mu$ gives

$$\begin{aligned}\left|\left(B_{N-1}^{1/2}\bar{b}_{N-1}\right)^{\mathsf{T}}B_{N-1}^{-1/2}\widetilde{w}_{N-2}\right| &\le \mu\|\bar{b}_{N-1}\|_{B_{N-1}}^2 + \frac{1}{4\mu}\|\widetilde{w}_{N-2}\|_{B_{N-1}^{-1}}^2\\ &\overset{(17.123)}{\le} \mu\|\bar{b}_{N-1}\|_{B_{N-1}}^2 + \frac{d^2}{4\mu}\mathrm{Tr}(B_{N-1}^{-1})\end{aligned} \tag{17.129}$$

and we conclude that

$$\begin{aligned}&\left|\frac{1}{1-\beta_1}\frac{1}{N}\sum_{n=0}^{N-1}\left\{\bar{b}_n^{\mathsf{T}}\widetilde{w}_{n-1} - \bar{b}_{n-1}^{\mathsf{T}}\widetilde{w}_{n-2}\right\}\right|\\ &\le \frac{1}{N(1-\beta_1)}\left\{\mu\|\bar{b}_{N-1}\|_{B_{N-1}}^2 + \frac{d^2}{4\mu}\mathrm{Tr}(B_{N-1}^{-1})\right\}\end{aligned} \tag{17.130}$$

Grouping the bounds obtained this far for the three terms from (17.114) we find that

$$\begin{aligned}\mathcal{R}(N) \le \;&\frac{d^2}{2\mu' N}\mathrm{Tr}(B_{N-1}^{-1}) + \frac{\mu}{2N}\sum_{n=0}^{N-1}\|\bar{b}_n\|_{B_n}^2 +\\ &\frac{\mu\beta_1}{1-\beta_1}\frac{1}{N}\sum_{n=0}^{N-1}\|\bar{b}_{n-1}\|_{B_{n-1}}^2 +\\ &\frac{1}{N(1-\beta_1)}\left\{\mu\|\bar{b}_{N-1}\|_{B_{N-1}}^2 + \frac{d^2}{4\mu}\mathrm{Tr}(B_{N-1}^{-1})\right\}\end{aligned} \tag{17.131}$$

Using the fact that $\mu' < \mu$ and $\beta_1 < 1$, we group the two terms involving the trace of B_{N-1}^{-1} and bound them by

$$\frac{d^2}{2\mu' N}\mathrm{Tr}(B_{N-1}^{-1}) + \frac{1}{N(1-\beta_1)}\frac{d^2}{4\mu}\mathrm{Tr}(B_{N-1}^{-1}) \le \frac{1}{N}\times\frac{3d^2}{4\mu'(1-\beta_1)}\times\mathrm{Tr}(B_{N-1}^{-1}) \tag{17.132}$$

Likewise, we group the three terms involving $\|\bar{b}_n\|^2_{B_n}$ in (17.131) and bound them as follows:

$$
\begin{aligned}
&\frac{\mu}{2N}\sum_{n=0}^{N-1}\|\bar{b}_n\|^2_{B_n} + \frac{\mu\beta_1}{N(1-\beta_1)}\sum_{n=0}^{N-1}\|\bar{b}_{n-1}\|^2_{B_{n-1}} + \frac{\mu}{N(1-\beta_1)}\|\bar{b}_{N-1}\|^2_{B_{N-1}} \\
&\leq \frac{\mu}{2N}\sum_{n=0}^{N-1}\|\bar{b}_n\|^2_{B_n} + \frac{\mu}{N(1-\beta_1)}\sum_{n=0}^{N-1}\|\bar{b}_{n-1}\|^2_{B_{n-1}} + \frac{\mu}{N(1-\beta_1)}\|\bar{b}_{N-1}\|^2_{B_{N-1}} \\
&= \frac{\mu}{2N}\sum_{n=0}^{N-1}\|\bar{b}_n\|^2_{B_n} + \frac{\mu}{N(1-\beta_1)}\sum_{n=0}^{N-1}\|\bar{b}_n\|^2_{B_n} \\
&\leq \frac{\mu}{2N} \times \frac{3}{1-\beta_1} \times \sum_{n=0}^{N-1}\|\bar{b}_n\|^2_{B_n}
\end{aligned}
\tag{17.133}
$$

Substituting back into (17.131) we find that

$$
\mathcal{R}(N) \leq \frac{1}{N}\frac{3}{(1-\beta_1)}\left\{\frac{d^2}{4\mu(1-\beta_2)}\text{Tr}(B_{N-1}^{-1}) + \frac{\mu}{2}\sum_{n=0}^{N-1}\|\bar{b}_n\|^2_{B_n}\right\}
\tag{17.134}
$$

Bounding the filtered gradient vector

To conclude the argument we still need to bound the term $\|\bar{b}_n\|^2_{B_n}$, which appears on the right-hand side of the last expression. The following derivation bounds this term by the entries of the original gradient vectors, b_n, as shown further ahead in expression (17.148). Assuming the gradient vectors are bounded, as is often customary in regret analysis, we will then end up concluding from (17.134) that the average regret for the AMSgrad algorithm under constant step sizes is

$$
\mathcal{R}(N) = O(1/N) + O(\mu) \qquad \textbf{(for constant step sizes)}
\tag{17.135}
$$

which shows that the regret approaches a small value on the order of μ and the convergence rate toward this region is $O(1/N)$.

To arrive at (17.148) we first note from the definitions of $\widehat{s}_n$ and B_n in (17.106) that $\epsilon > 0$ and $1-\beta_2^{n+1}$ is lower-bounded by $1-\beta_2$ for all $n \geq 0$ so that we can write

$$
B_n \leq \frac{1}{1-\beta_2} \times \left(\text{diag}(\,s_n^{1/2}\,)\right)^{-1}
\tag{17.136}
$$

We denote the individual entries of the vectors $\{\bar{b}_n, b_n, s_n\}$ by $\{\bar{b}_{n,m}, b_{n.m}, s_{n,m}\}$ for $m = 1, 2, \ldots, M$, and the diagonal entries of B_n by $[B_n]_{m,m}$. From the definitions (17.48a)–(17.48b) we have

$$
\bar{b}_{n,m} = (1-\beta_1)\sum_{n'=0}^{n}\beta_1^{n-n'}b_{n',m}
\tag{17.137a}
$$

$$
s_{n,m} = (1-\beta_2)\sum_{n'=0}^{n}\beta_2^{n-n'}b^2_{n',m}
\tag{17.137b}
$$

and, consequently,

$$
[B_n]_{m.m} \leq \frac{1}{1-\beta_2} \times \frac{1}{(1-\beta_2)^{1/2}(\sum_{n'=0}^{n}\beta_2^{n-n'}b^2_{n',m})^{1/2}}
\tag{17.138}
$$

It follows that

$$\|\bar{b}_n\|^2_{B_n} \triangleq \sum_{m=1}^{M} [B_n]_{m,m}\, \bar{b}^2_{n,m}$$
$$= \frac{(1-\beta_1)^2}{(1-\beta_2)^{3/2}} \times \sum_{m=1}^{M} \frac{\left(\sum_{n'=0}^{n} \beta_1^{n-n'} b_{n',m}\right)^2}{\left(\sum_{n'=0}^{n} \beta_2^{n-n'} b^2_{n',m}\right)^{1/2}} \tag{17.139}$$

At this point, we follow Alacaoglu *et al.* (2020) and appeal to the generalized Hölder inequality (see Hardy, Littlewood, and Pólya (1934) and Beckenbach and Bellman (1961)), which states that for any positive sequences a_ℓ, b_ℓ, c_ℓ and positive numbers p, q, and r satisfying $1/p + 1/q + 1/r = 1$, it holds that

$$\sum_{\ell=1}^{L} a_\ell b_\ell c_\ell \leq \|a\|_p\, \|b\|_q\, \|c\| \quad \textbf{(generalized Holder inequality)} \tag{17.140}$$

where $\{a, b, c\}$ are the L-dimensional vectors with entries $\{a_\ell, b_\ell, c_\ell\}$ for $\ell = 1, 2, \ldots, L$, and $\|\cdot\|_p$ denotes the pth vector norm. Using this inequality, we select:

$$p = q = 4, \quad r = 2 \tag{17.141}$$

and define the sequences:

$$a_{n'} = \beta_2^{\frac{n-n'}{4}} \times |b_{n',m}|^{1/2} \tag{17.142}$$
$$b_{n'} = \left(\beta_1 \beta_2^{-1/2}\right)^{\frac{n-n'}{2}} \tag{17.143}$$
$$c_{n'} = \left(\, \beta_1^{n-n'} |b_{n',m}|\, \right)^{1/2} \tag{17.144}$$

Then, referring to the term that appears in the numerator of (17.139), we can bound it as

$$\begin{aligned}
&\left\{ \sum_{n'=0}^{n} \beta_1^{n-n'} b_{n',m} \right\}^2 \\
&\leq \left\{ \sum_{n'=0}^{n} \beta_2^{\frac{n-n'}{4}} |b_{n',m}|^{1/2} \times \left(\beta_1\beta_2^{-1/2}\right)^{\frac{n-n'}{2}} \times \left(\, \beta_1^{n-n'} |b_{n',m}|\, \right)^{1/2} \right\}^2 \\
&\leq \left\{ \left(\sum_{n'=0}^{n} \left[\beta_2^{\frac{n-n'}{4}} |b_{n',m}|^{1/2} \right]^4 \right)^{1/4} \times \left(\sum_{n'=0}^{n} \left[\left(\beta_1\beta_2^{-1/2}\right)^{\frac{n-n'}{2}} \right]^4 \right)^{1/4} \times \right. \\
&\qquad \left. \left(\sum_{n'=0}^{n} \left[\left(\beta_1^{n-n'} |b_{n',m}|\right)^{1/2} \right]^2 \right)^{1/2} \right\}^2
\end{aligned} \tag{17.145}$$

Let

$$\kappa \triangleq \beta_1^2/\beta_2 < 1 \tag{17.146}$$

Then, the second term in the last expression is a convergent geometric series:

$$\left(\sum_{n'=0}^{n} \left[\left(\beta_1\beta_2^{-1/2}\right)^{\frac{n-n'}{2}} \right]^4 \right)^{1/4} = \left(\sum_{n'=0}^{n} \kappa^{n-n'} \right)^{1/2} \leq \left(\frac{1}{1-\kappa} \right)^{1/2} \tag{17.147}$$

Substituting into (17.145) we get

$$\left\{\sum_{n'=0}^{n}\beta_1^{n-n'}b_{n',m}\right\}^2 \leq \left(\frac{1}{1-\kappa}\right)^{1/2}\times\left(\sum_{n'=0}^{n}\left(\beta_2^{n-n'}b_{n',m}^2\right)\right)^{1/2}\times\sum_{n'=0}^{n}\beta_1^{n-n'}|b_{n',m}|$$

Substituting into (17.139) we arrive at

$$\|\bar{b}_n\|_{B_n}^2 \leq \frac{(1-\beta_1)^2}{(1-\beta_2)^{3/2}}\times\left(\frac{1}{1-\kappa}\right)^{1/2}\times\sum_{m=1}^{M}\sum_{n'=0}^{n}\beta_1^{n-n'}|b_{n',m}| \tag{17.148}$$

Assume the gradient vectors are bounded and introduce

$$c \triangleq \max_{w\in\mathcal{C},0\leq n\leq N-1}\left\{\|\nabla_{w^\mathsf{T}}Q(w;\gamma(n),h_n)\|\right\} \tag{17.149}$$

so that each entry $|b_{n',m}|$ is bounded by c. Then,

$$\begin{aligned}\|\bar{b}_n\|_{B_n}^2 &\leq \frac{(1-\beta_1)^2}{(1-\beta_2)^{3/2}}\times\left(\frac{1}{1-\kappa}\right)^{1/2}\times\sum_{m=1}^{M}\sum_{n'=0}^{n}c\,\beta_1^{n-n'}\\ &\leq \frac{(1-\beta_1)^2}{(1-\beta_2)^{3/2}}\times\left(\frac{1}{1-\kappa}\right)^{1/2}\times\frac{Mc}{1-\beta_1}\\ &\leq \frac{(1-\beta_1)}{(1-\beta_2)^{3/2}}\times\left(\frac{1}{1-\kappa}\right)^{1/2}\times Mc\end{aligned} \tag{17.150}$$

Note further that

$$\begin{aligned}\frac{1}{1-\kappa} &\overset{(17.146)}{=} \frac{\beta_2}{\beta_2-\beta_1^2}\\ &\leq \frac{\beta_2}{\beta_2-\beta_1\beta_2},\quad \text{since } \beta_1^2<\beta_1\beta_2\\ &= \frac{1}{1-\beta_1}\end{aligned}$$

and, hence,

$$\|\bar{b}_n\|_{B_n}^2 \leq \frac{cM(1-\beta_1)^{1/2}}{(1-\beta_2)^{3/2}} \tag{17.151}$$

Substituting into (17.134) leads to

$$\mathcal{R}(N) \leq \frac{1}{N}\frac{3}{(1-\beta_1)}\left\{\frac{d^2}{4\mu(1-\beta_2)}\text{Tr}(B_{N-1}^{-1}) + \frac{\mu}{2}\frac{cMN(1-\beta_1)^{1/2}}{(1-\beta_2)^{3/2}}\right\} \tag{17.152}$$

and, therefore, $\mathcal{R}(N) = O(1/N) + O(\mu)$, as claimed before.

REFERENCES

Alacaoglu, A., Y. Malitsky, P. Mertikopoulos, and V. Cevher (2020), "A new regret analysis for Adam-type algorithms," *Proc. Int. Conf. Machine Learning*, PMLR, vol. 119, pp. 202–210.

Beckenbach, E. F. and R. E. Bellman (1961), *Inequalities*, vol. 30, Springer.

Chaudhuri, K., C. Monteleoni, and A. D. Sarwate (2011), "Differentially private empirical risk minimization," *J. Mach. Learn. Res.*, vol. 12, pp. 1069–1109.

Dozat, T. (2016), "Incorporating Nesterov momentum into Adam," *Proc. Int. Conf. Learning Representations* (ICLR), pp. 2013–2016, San Jose, Puerto Rico.

Duchi, J., E. Hazan, and Y. Singer (2011), "Adaptive subgradient methods for online learning and stochastic optimization," *J. Mach. Learn. Res.*, vol. 12, pp. 2121–2159.

Dwork, C. (2006), "Differential privacy," in *Proc. Int. Colloquium Automata, Languages, and Programming* (ICALP), pp. 1–12, Venice, Italy.

Dwork, C., F. McSherry, K. Nissim, and A. Smith (2006), "Calibrating noise to sensitivity in private data analysis," *Proc. Conf. Theory of Cryptography*, pp. 265–284, New York.

Dwork, C. and A. Roth (2014), *The Algorithmic Foundations of Differential Privacy*, Foundations and Trends in Theoretical Computer Science, NOW Publishers, vol. 9, no. 3–4, pp. 211–407.

Ghadimi, S. and G. Lan (2012), "Optimal stochastic approximation algorithms for strongly convex stochastic composite optimization I: A generic algorithmic framework," *SIAM J. Optim.*, vol. 22, no. 4, pp. 1469–1492.

Hardy, G. H., J. E. Littlewood, and G. Pólya (1934), *Inequalities*, Cambridge University Press.

Kifer, D., A. Smith, and A. Thakurta (2012), "Private convex optimization for empirical risk minimization with applications to high-dimensional regression," *J. Mach. Learn. Res.*, vol. 23, pp. 25.1–25.40.

Kingma, D. P. and J. L. Ba (2015), "Adam: A method for stochastic optimization," *Proc. Int. Conf. Learning Representations* (ICLR), pp. 1–13, San Diego, CA.

Konecny, J., H. B. McMahan, D. Ramage, and P. Richtarik (2016), "Federated optimization: Distributed machine learning for on-device intelligence," available at arXiv:1610.02527.

McMahan, H. B., E. Moore, D. Ramage, S.Hampson, B. Aguera y Arcas (2017), "Communication-efficient learning of deep networks from decentralized data," *Artif. Intelli. Statist.*, vol. 54, pp. 1273–1282.

Nesterov, Y. (1983), "A method for unconstrained convex minimization problem with the rate of convergence $O(1/k^2)$," *Doklady AN USSR*, vol. 269, pp. 543–547.

Nesterov, Y. (2004), *Introductory Lectures on Convex Optimization*, Springer.

Nilsson, A., S. Smith, G. Ulm, E. Gustavsson, and M. Jirstrand (2018), "A performance evaluation of federated learning algorithms," *Proc. Workshop Distributed Infrastructures for Deep Learning* (DIDL), pp. 1–8, Rennes, France.

Paleologu, C., J. Benesty, and S. Ciochina (2011), *Sparse Adaptive Filters for Echo Cancellation*, Morgan and Claypool.

Polyak, B. T. (1987), *Introduction to Optimization*, Optimization Software.

Polyak, B. T. and A. Juditsky (1992), "Acceleration of stochastic approximation by averaging," *SIAM J. Control Optim.*, vol. 30, no. 4, pp. 838–855.

Reddi, S. J., S. Kale, and S. Kumar (2018), "On the convergence of ADAM and beyond," *Proc. Int. Conf. Learning Representations* (ICLR), pp. 1–23, Vancouver, BC. Also available at arXiv:1904.09237.

Rizk, E., S. Vlaski, and A. H. Sayed (2020), "Dynamic federated learning," *Proc. IEEE Workshop Signal Process. Advances in Wireless Comm.* (SPAWC), pp. 1–5.

Roy, S. and J. J. Shynk (1990), "Analysis of the momentum LMS algorithm," *IEEE Trans. Acoustics, Speech Signal Process.*, vol. 38, no. 12, pp. 2088–2098.

Sayed, A. H. (2003), *Fundamentals of Adaptive Filtering*, Wiley.

Sayed, A. H. (2008), *Adaptive Filters*, Wiley.

Sharma, R., W. A. Sethares, and J. A. Bucklew (1998), "Analysis of momentum adaptive filtering algorithms," *IEEE Trans. Signal Process.*, vol. 46, no. 5, pp. 1430–1434.

Sutskever, I., J. Martens, G. Dahl, and G. Hinton (2013), "On the importance of initialization and momentum in deep learning," in *Proc. Int. Conf. Machine Learning* (ICML), pp. 1139–1147, Atlanta, GA.

Tieleman, T. and G. Hinton. (2012), *Neural Networks for Machine Learning*, online course available at www.cs.toronto.edu and www.youtube.com, lecture 6.5.

Tugay, M. A. and Y. Tanik (1989), "Properties of the momentum LMS algorithm," *Signal Process.*, vol. 18, no. 2, pp. 117–127.

Xiao, L. (2010), "Dual averaging methods for regularized stochastic learning and online optimization," *J. Mach. Learn. Res.*, vol. 11, pp. 2543–2596.

Yuan, K., B. Ying, and A. H. Sayed (2016), "On the influence of momentum acceleration on online learning," *J. Mach. Learn. Res.*, vol. 17, no. 192, pp. 1–66.

Zareba, S., A. Gonczarek, J. M. Tomczak, and J. Swiatek (2015), "Accelerated learning for restricted Boltzmann machine with momentum term," in *Proc. Int. Conf. Systems Engineering*, pp. 187–192, Coventry, UK.

Zeiler, M. D. (2012), "AdaDelta: An adaptive learning rate method," available at arXiv:1212.5701.

18 Gradient Noise

The purpose of this chapter is to study the gradient noise process more closely, for cases of both smooth and nonsmooth risk functions, and to derive expressions for its first- and second-order moments (i.e., mean and variance). The results will then be exploited in the subsequent chapters to assess how gradient noise affects the convergence behavior of various stochastic approximation algorithms. The presentation in the chapter prepares the ground for the detailed convergence analyses given in the next chapters. Throughout this chapter, we will use the terminology "*smooth*" functions to refer to risks that are at least first-order differentiable everywhere in their domain, and apply the qualification "*nonsmooth*" functions to risks that are not differentiable at some points in their domains.

18.1 MOTIVATION

We examined several stochastic optimization algorithms in the previous chapters for the solution of convex optimization problems of the form:

$$w^\star = \underset{w \in \mathbb{R}^M}{\text{argmin}} \; P(w) \tag{18.1}$$

with and without constraints on w, for both smooth and nonsmooth risks, as well as for empirical and stochastic risks, namely,

$$w^\star \triangleq \underset{w \in \mathbb{R}^M}{\text{argmin}} \left\{ P(w) \triangleq \frac{1}{N} \sum_{m=0}^{N-1} Q(w; \gamma(m), h_m) \right\} \tag{18.2a}$$

$$w^o \triangleq \underset{w \in \mathbb{R}^M}{\text{argmin}} \left\{ P(w) \triangleq \mathbb{E}\, Q(w; \boldsymbol{\gamma}, \boldsymbol{h}) \right\} \tag{18.2b}$$

In these expressions, $Q(w, \cdot)$ denotes some convex loss function, $\{\gamma(m), h_m\}$ refer to a collection of N data points with $\gamma(m) \in \mathbb{R}$ and $h_m \in \mathbb{R}^M$, and the expectation in the second line is over the joint distribution of $\{\boldsymbol{\gamma}, \boldsymbol{h}\}$. In most algorithms, the desired gradient or subgradient search direction was approximated by using either instantaneous or mini-batch calculations. For example, for smooth risk functions $P(w)$ we used approximations of the form:

$$\text{(instantaneous)} \quad \widehat{\nabla_{w^\mathsf{T}} P}(w) = \nabla_{w^\mathsf{T}} Q(w; \boldsymbol{\gamma}, \boldsymbol{h}) \tag{18.3a}$$

$$\text{(mini-batch)} \quad \widehat{\nabla_{w^\mathsf{T}} P}(w) = \frac{1}{B}\sum_{b=0}^{B-1} \nabla_{w^\mathsf{T}} Q\Big(w; \boldsymbol{\gamma}(b), \boldsymbol{h}_b\Big) \tag{18.3b}$$

where the boldface notation $(\boldsymbol{\gamma}, \boldsymbol{h})$ or $(\boldsymbol{\gamma}(b), \boldsymbol{h}_b)$ refers to data samples selected at random from the dataset $\{\gamma(m), h_m\}$ in empirical risk minimization, or assumed to stream in independently over time in stochastic risk minimization. When $P(w)$ happens to be nonsmooth, the gradient vectors of $Q(w; \cdot)$ are replaced by subgradients, denoted by $s_Q(w; \boldsymbol{\gamma}, \boldsymbol{h})$. The difference between the true gradient and its approximation is *gradient noise* and is denoted by

$$\boldsymbol{g}(w) \overset{\Delta}{=} \widehat{\nabla_{w^\mathsf{T}} P}(w) - \nabla_{w^\mathsf{T}} P(w) \tag{18.4}$$

We explained in Section 16.4 that the presence of this noise source alters the dynamics of the optimization algorithms. For example, the following two relations highlight the difference between the original gradient-descent method and its stochastic version for smooth risks:

$$\text{(gradient descent)} \quad w_n = w_{n-1} - \mu \nabla_{w^\mathsf{T}} P(w_{n-1}) \tag{18.5a}$$

$$\begin{aligned} \text{(stochastic version)} \quad \boldsymbol{w}_n &= \boldsymbol{w}_{n-1} - \mu \widehat{\nabla_{w^\mathsf{T}} P}(\boldsymbol{w}_{n-1}) \\ &= \boldsymbol{w}_{n-1} - \mu \nabla_{w^\mathsf{T}} P(\boldsymbol{w}_{n-1}) - \mu\, \boldsymbol{g}(\boldsymbol{w}_{n-1}) \end{aligned} \tag{18.5b}$$

The gradient noise appears as a driving perturbation in the second recursion. This is illustrated in Fig. 18.1, where the block with z^{-1} represents a unit delay element. The panel on top shows the dynamics of (18.5a), while the panel in the bottom shows the dynamics of the perturbed update (18.5b). The gradient noise seeps into the operation of the algorithm and some degradation in performance is expected. While we were able to show in a previous chapter that the gradient-descent implementation (18.5a) converges to the exact minimizer $w^\star$ of $P(w)$ for sufficiently small step sizes, we will discover in future chapters that the stochastic version (18.5b) can only approach a small neighborhood around $w^\star$ of size $\mathbb{E}\,\|\widetilde{\boldsymbol{w}}_n\|^2 = O(\mu)$ as $n \to \infty$.

Example 18.1 (**Gradient noise for quadratic risks**) We illustrate the concept of gradient noise by considering two quadratic risks: one empirical and the other stochastic. Consider first the empirical risk:

$$P(w) \;=\; \rho\|w\|^2 \;+\; \frac{1}{N}\sum_{m=0}^{N-1}(\gamma(m) - h_m^\mathsf{T} w)^2, \quad \rho > 0 \tag{18.6}$$

In this case, the gradient vector and its instantaneous approximation are given by

$$\nabla_w P(w) = 2\rho w - \frac{2}{N}\sum_{m=0}^{N-1} h_m(\gamma(m) - h_m^\mathsf{T} w) \tag{18.7a}$$

$$\widehat{\nabla_w P}(w) = 2\rho w - 2\boldsymbol{h}_n(\boldsymbol{\gamma}(n) - \boldsymbol{h}_n^\mathsf{T} w) \tag{18.7b}$$

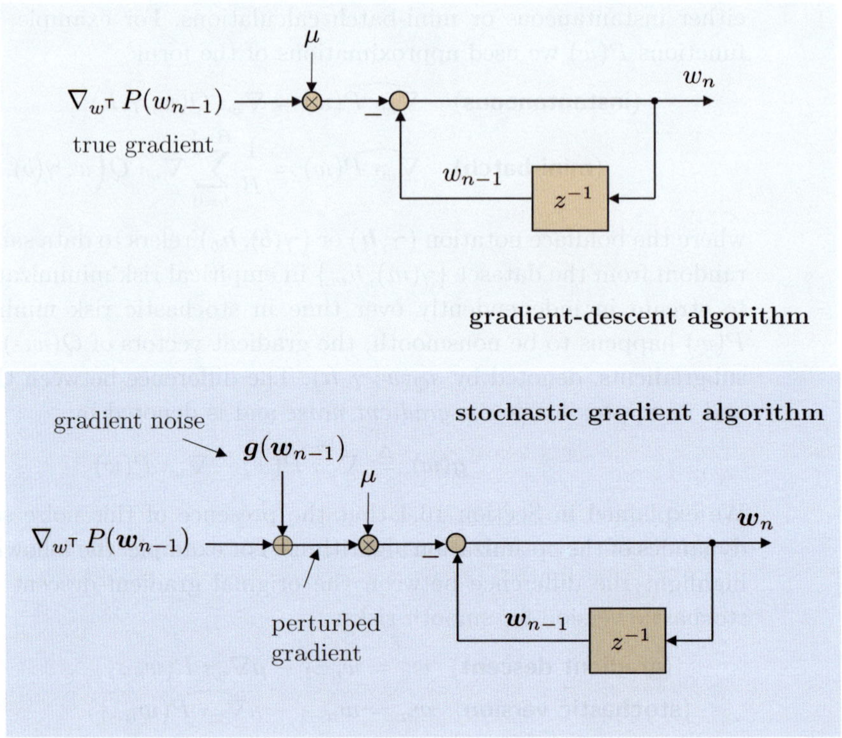

Figure 18.1 The panel on top shows the dynamics of the original gradient-descent recursion (18.5a), while the panel at the bottom shows the dynamics of the stochastic version (18.5b). The true gradient vector is perturbed by gradient noise, which seeps into the operation of the algorithm. The block with z^{-1} represents a unit delay element.

where $\{\gamma(n), \boldsymbol{h}_n\}$ refer to the random sample selected at iteration n by the stochastic gradient implementation. The resulting gradient noise process is then given by

$$\boldsymbol{g}(w) = \frac{2}{N}\sum_{m=0}^{N-1} h_m(\gamma(m) - h_m^\mathsf{T} w) \; - \; 2\boldsymbol{h}_n(\boldsymbol{\gamma}(n) - \boldsymbol{h}_n^\mathsf{T} w) \tag{18.7c}$$

Observe that $\boldsymbol{g}(w)$ depends on the sample $\{\boldsymbol{\gamma}(n), \boldsymbol{h}_n\}$ and, therefore, in principle, we should be writing $\boldsymbol{g}_n(w)$ with a subscript n to highlight its dependency on n.

Consider next the stochastic risk:

$$\begin{aligned} P(w) &= \rho\|w\|^2 \; + \; \mathbb{E}\,(\boldsymbol{\gamma} - \boldsymbol{h}^\mathsf{T} w)^2 \\ &= \sigma_\gamma^2 - 2r_{h\gamma}^\mathsf{T} w + w^\mathsf{T}(\rho I_M + R_h)w \end{aligned} \tag{18.8}$$

which we expanded in terms of the second-order moments $\sigma_\gamma^2 = \mathbb{E}\,\boldsymbol{\gamma}^2$, $r_{h\gamma} = \mathbb{E}\,\boldsymbol{h}\boldsymbol{\gamma}$, and $R_h = \mathbb{E}\,\boldsymbol{h}\boldsymbol{h}^\mathsf{T}$. The random variables $\{\boldsymbol{\gamma}, \boldsymbol{h}\}$ are assumed to have zero means. In this case, the gradient of $P(w)$ and its instantaneous approximation are given by

$$\nabla_w P(w) = 2\rho w - 2(r_{h\gamma} - R_h w) \tag{18.9a}$$

$$\widehat{\nabla_w P}(w) = 2\rho w - 2\boldsymbol{h}_n(\boldsymbol{\gamma}(n) - \boldsymbol{h}_n^\mathsf{T} w) \tag{18.9b}$$

so that the corresponding gradient noise process is now

$$\boldsymbol{g}(w) = 2(r_{h\gamma} - R_h w) - 2\boldsymbol{h}_n(\boldsymbol{\gamma}(n) - \boldsymbol{h}_n^{\mathsf{T}} w) \tag{18.9c}$$

Observe again that $\boldsymbol{g}(w)$ depends on the streaming sample $\{\boldsymbol{\gamma}(n), \boldsymbol{h}_n\}$.

18.2 SMOOTH RISK FUNCTIONS

To facilitate the analysis and presentation, we will treat smooth and nonsmooth risks separately, although we will end up with the same ultimate conclusion about the gradient noise for both cases. We start with smooth risks and describe the conditions that are normally imposed on the risk and loss functions, $P(w)$ and $Q(w,\cdot)$. The conditions listed here are satisfied by several risk and loss functions of interest, as illustrated in the problems at the end of the chapter.

Empirical risks

Consider smooth empirical risks of the form (18.2a). We will assume that the risk and loss functions satisfy the two conditions listed below. Compared with the earlier conditions (12.12a)–(12.12b) in the gradient-descent case, we see that we now need to take the loss function into consideration since its gradients are the ones used in the stochastic implementation:

(A1) (**Strongly convex risk**). $P(w)$ is ν-strongly convex and first-order differentiable, namely, for every $w_1, w_2 \in \text{dom}(P)$:

$$P(w_2) \geq P(w_1) + (\nabla_{w^{\mathsf{T}}} P(w_1))^{\mathsf{T}}(w_2 - w_1) + \frac{\nu}{2}\|w_2 - w_1\|^2 \tag{18.10a}$$

for some $\nu > 0$.

(A2) (**δ-Lipschitz loss gradients**). The gradient vectors of $Q(w,\cdot)$ are δ-Lipschitz regardless of the data argument, i.e.,

$$\|\nabla_w Q(w_2; \gamma(k), h_k) - \nabla_w Q(w_1; \gamma(\ell), h(\ell))\| \leq \delta\,\|w_2 - w_1\| \tag{18.10b}$$

for any $w_1, w_2 \in \text{dom}(Q)$, any $0 \leq k, \ell \leq N-1$, and with $\delta \geq \nu$ (this latter requirement can always be met by enlarging δ). Condition (18.10b) is equivalent to saying the loss function is δ-smooth. It is easy to verify from the triangle inequality of norms that (18.10b) implies that the gradient of $P(w)$ is itself δ-Lipschitz:

$$\|\nabla_w P(w_2) - \nabla_w P(w_1)\| \leq \delta\|w_2 - w_1\| \tag{18.11}$$

Moreover, if it happens that $P(w)$ is twice-differentiable, then we already know from (12.15) that conditions (18.10a) and (18.11) combined are equivalent to:

$$0 < \nu I_M \leq \nabla_w^2 P(w) \leq \delta I_M \tag{18.12}$$

in terms of the Hessian matrix of $P(w)$.

Stochastic risks

For stochastic risks of the form (18.2b), we continue to assume that $P(w)$ is ν-strongly convex but that the loss function has gradients that are δ-Lipschitz in the *mean-square sense*:

(A1) (**Strongly convex risk**). $P(w)$ is ν-strongly convex and first-order differentiable, namely, for every $w_1, w_2 \in \text{dom}(P)$:

$$P(w_2) \geq P(w_1) + (\nabla_{w^\mathsf{T}} P(w_1))^\mathsf{T}(w_2 - w_1) + \frac{\nu}{2}\|w_2 - w_1\|^2 \quad (18.13a)$$

for some $\nu > 0$.

(A2') (**Mean-square δ-Lipschitz loss gradients**). The gradient vectors of $Q(w, \cdot)$ satisfy the mean-square bound:

$$\mathbb{E}\,\|\nabla_w Q(w_2; \boldsymbol{\gamma}, \boldsymbol{h}) - \nabla_w Q(w_1; \boldsymbol{\gamma}, \boldsymbol{h})\|^2 \leq \delta^2\|w_2 - w_1\|^2 \quad (18.13b)$$

for any $w_1, w_2 \in \text{dom}(Q)$ and with $\delta \geq \nu$. The expectation is over the joint distribution of the random data $\{\boldsymbol{\gamma}, \boldsymbol{h}\}$. Using the fact that for any scalar random variable $\boldsymbol{x}$ it holds that $(\mathbb{E}\,\boldsymbol{x})^2 \leq \mathbb{E}\,\boldsymbol{x}^2$, we conclude from condition (18.13b) that the gradient vectors of the loss function are also δ-Lipschitz on *average*, namely,

$$\mathbb{E}\,\|\nabla_w Q(w_2; \boldsymbol{\gamma}, \boldsymbol{h}) - \nabla_w Q(w_1; \boldsymbol{\gamma}, \boldsymbol{h})\| \leq \delta\|w_2 - w_1\| \quad (18.14)$$

By further applying Jensen inequality (8.77) that $f(\mathbb{E}\,\boldsymbol{x}) \leq \mathbb{E}\,f(\boldsymbol{x})$ for the convex function $f(x) = \|x\|$, we can conclude from (18.14) that the gradients of $P(w)$ are themselves δ-Lipschitz as well:

$$\|\nabla_w P(w_2) - \nabla_w P(w_1)\| \leq \delta\|w_2 - w_1\| \quad (18.15)$$

Proof of (18.15): Note that

$$\begin{aligned}
\|\nabla_w P(w_2) - \nabla_w P(w_1)\| &\overset{\Delta}{=} \|\nabla_w \mathbb{E}\,Q(w_2; \boldsymbol{\gamma}, \boldsymbol{h}) - \nabla_w \mathbb{E}\,Q(w_1; \boldsymbol{\gamma}, \boldsymbol{h})\| \\
&\overset{(a)}{=} \left\|\mathbb{E}\left(\nabla_w Q(w_2; \boldsymbol{\gamma}, \boldsymbol{h}) - \nabla_w Q(w_1; \boldsymbol{\gamma}, \boldsymbol{h})\right)\right\| \\
&\leq \mathbb{E}\,\|\nabla_w Q(w_2; \boldsymbol{\gamma}, \boldsymbol{h}) - \nabla_w Q(w_1; \boldsymbol{\gamma}, \boldsymbol{h})\| \\
&\overset{(18.14)}{\leq} \delta\|w_2 - w_1\| \quad (18.16)
\end{aligned}$$

Step (a) switches the order of the expectation and differentiation operators, which is possible under certain conditions that are generally valid for our cases of interest – recall the explanation in Appendix 16.A on the *dominated convergence theorem*. In particular, the switching is possible when the loss function $Q(w; \cdot, \cdot)$ and its gradient are continuous functions of w.

■

For ease of reference, we collect in Table 18.1 the main relations and conditions described so far for smooth empirical and stochastic risk minimization.

Table 18.1 The main relations and conditions used for *smooth* empirical and stochastic risk minimization problems.

Quantity	Empirical risk minimization	Stochastic risk minimization
optimization problem	$w^\star = \underset{w\in\mathbb{R}^M}{\text{argmin}} \left\{ P(w) \stackrel{\Delta}{=} \frac{1}{N}\sum_{m=0}^{N-1} Q(w;\gamma(m),h_m) \right\}$	$w^o = \underset{w\in\mathbb{R}^M}{\text{argmin}} \left\{ P(w) \stackrel{\Delta}{=} \mathbb{E}\, Q(w;\boldsymbol{\gamma},\boldsymbol{h}) \right\}$
true gradient vector	$\nabla_{w^\mathsf{T}} P(w) = \frac{1}{N}\sum_{m=0}^{N-1} \nabla_{w^\mathsf{T}} Q(w;\gamma(m),h_m)$	$\nabla_{w^\mathsf{T}} P(w) = \nabla_{w^\mathsf{T}} \mathbb{E}\, Q(w;\boldsymbol{\gamma},\boldsymbol{h})$
instantaneous approximation	$\widehat{\nabla_{w^\mathsf{T}} P}(w) = \nabla_{w^\mathsf{T}} Q(w;\boldsymbol{\gamma}(n),\boldsymbol{h}_n)$ $(\boldsymbol{\gamma}(n),\boldsymbol{h}_n)$ selected at random	$\widehat{\nabla_{w^\mathsf{T}} P}(w) = \nabla_{w^\mathsf{T}} Q(w;\boldsymbol{\gamma}(n),\boldsymbol{h}_n)$ $(\boldsymbol{\gamma}(n),\boldsymbol{h}_n)$ streaming in
mini-batch approximation	$\widehat{\nabla_{w^\mathsf{T}} P}(w) = \frac{1}{B}\sum_{b=0}^{B-1} \nabla_{w^\mathsf{T}} Q(w;\boldsymbol{\gamma}(b),\boldsymbol{h}_b)$ $\{\boldsymbol{\gamma}(b),\boldsymbol{h}_b\}$ selected at random	$\widehat{\nabla_{w^\mathsf{T}} P}(w) = \frac{1}{B}\sum_{b=0}^{B-1} \nabla_{w^\mathsf{T}} Q(w;\boldsymbol{\gamma}(b),\boldsymbol{h}_b)$ $\{\boldsymbol{\gamma}(b),\boldsymbol{h}_b\}$ streaming in
conditions on risk and loss functions	(18.10a)–(18.10b) $P(w)$ ν-strongly convex $\nabla_{w^\mathsf{T}} Q(w;\gamma,h)$ δ-Lipschitz	(18.13a)–(18.13b) $P(w)$ ν-strongly convex $\nabla_{w^\mathsf{T}} Q(w;\boldsymbol{\gamma},\boldsymbol{h})$ δ-Lipschitz in mean-square sense

18.3 GRADIENT NOISE FOR SMOOTH RISKS

Using the δ-Lipschitz conditions on the gradient of the loss function alone, we will now derive expressions for the first- and second-order moments of the gradient noise. For the instantaneous and mini-batch constructions (18.3a)–(18.3b), the gradient noise at iteration n is given by

(**instantaneous approximation**)

$$\boldsymbol{g}(\boldsymbol{w}_{n-1}) = \nabla_{w^{\mathsf{T}}} Q(\boldsymbol{w}_{n-1}; \boldsymbol{\gamma}(n), \boldsymbol{h}_n) - \nabla_{w^{\mathsf{T}}} P(\boldsymbol{w}_{n-1}) \tag{18.17a}$$

for instantaneous gradient approximations, and by

(**mini-batch approximation**)

$$\boldsymbol{g}(\boldsymbol{w}_{n-1}) = \frac{1}{B}\sum_{b=0}^{B-1} \nabla_{w^{\mathsf{T}}} Q(\boldsymbol{w}_{n-1}; \boldsymbol{\gamma}(b), \boldsymbol{h}_b) - \nabla_{w^{\mathsf{T}}} P(\boldsymbol{w}_{n-1}) \tag{18.17b}$$

for mini-batch approximations, where $(\boldsymbol{\gamma}(n), \boldsymbol{h}_n)$ and $\{\boldsymbol{\gamma}(b), \boldsymbol{h}_b\}$ denote the random data samples used at the nth iteration while updating $\boldsymbol{w}_{n-1}$ to $\boldsymbol{w}_n$. It is important to recognize that the gradient noise is *random* in nature because its calculation depends on the random data samples. For this reason, we are denoting it in boldface. Moreover, the gradient noise is dependent on the iteration index n because its calculation depends on $\boldsymbol{w}_{n-1}$ and on the data samples used at that iteration. For added clarity, we will often write $\boldsymbol{g}_n(\boldsymbol{w}_{n-1})$ instead of just $\boldsymbol{g}(\boldsymbol{w}_{n-1})$, with an added subscript n, in order to emphasize that we are referring to the gradient noise computed at iteration n.

The main conclusion of this section (and actually of this chapter) will be to show that the conditional second-order moment of the gradient noise is bounded as follows:

$$\boxed{\mathbb{E}\left(\|\boldsymbol{g}_n(\boldsymbol{w}_{n-1})\|^2 \,|\, \boldsymbol{w}_{n-1}\right) \leq \beta_g^2 \|\widetilde{w}_{n-1}\|^2 + \sigma_g^2} \tag{18.18}$$

for some nonnegative constants (β_g^2, σ_g^2) that will be independent of the error $\widetilde{w}_{n-1} = w^\star - w_{n-1}$ (here, we are using $w^\star$ to refer generically to the minimizer of the risk function $P(w)$, whether empirical or stochastic in nature). The conditioning on $\boldsymbol{w}_{n-1}$ in (18.18) could have been written more explicitly as

$$\mathbb{E}\left(\|\boldsymbol{g}_n(\boldsymbol{w}_{n-1})\|^2 \,|\, \boldsymbol{w}_{n-1} = w_{n-1}\right) \tag{18.19}$$

to indicate that the conditioning is based on an actual realization for $\boldsymbol{w}_{n-1}$. For convenience, we will be using the simpler notation shown in (18.18) throughout our presentation, where the conditioning is written relative to the random variable. Result (18.18) shows that the second-order moment of the gradient noise is upper-bounded by a quadratic term that involves *two* factors: one factor is dependent on $\|\widetilde{w}_{n-1}\|^2$ and, therefore, gets smaller as the quality of the iterate

w_{n-1} improves, while the second factor is a *constant* term σ_g^2. This latter term is persistent and continues to exist even if $\|\widetilde{w}_{n-1}\|^2$ approaches zero.

It is important to remark that result (18.18) is only dependent on how the approximate gradient vector is constructed; the result does not depend on the particular stochastic approximation algorithm used to update the successive iterates from $\boldsymbol{w}_{n-1}$ to $\boldsymbol{w}_n$. By conditioning on $\boldsymbol{w}_{n-1}$, we are in effect stating that the bound holds regardless of how this iterate is generated. Once its value is given and used to compute the gradient approximation, then the resulting gradient noise will satisfy (18.18).

Before establishing (18.18), it is worth recalling the types of sampling strategies that can be employed by a stochastic approximation algorithm to select its random samples.

18.3.1 Sampling Strategies

For *mini-batch* implementations, the B samples can be chosen with or without replacement or they can be streaming in:

(a) (**Sampling with replacement**). In this case, we sample *with replacement* one data point $(\gamma(b), \boldsymbol{h}_b)$ at a time from the N dataset $\{\gamma(m), h_m\}$ until B samples have been selected. All samples within the mini-batch are selected independently of each other, but some samples may appear repeated.

(b) (**Sampling without replacement**). We can also sample *without replacement*, one data point $(\gamma(b), \boldsymbol{h}_b)$ at a time from the dataset $\{\gamma(m), h_m\}$ until B samples have been selected. The samples within the mini-batch will be different but not independent any more. The process is repeated to construct the successive mini-batches by sampling from the original dataset.

(c) (**Streaming data**). For stochastic risk minimization, the samples $\{\boldsymbol{\gamma}(b), \boldsymbol{h}_b\}$ used in the mini-batch will be streaming in independently of each other.

(d) (**Importance sampling**). In this case, a probability value p_m is assigned to each sample $(\gamma(m), h_m)$ in the dataset, and the mini-batch samples are selected at random (with replacement) from the dataset according to this distribution. We explained in Example 16.2 that the approximation for the gradient vector will need to be adjusted to include an additional scaling by $1/Np_b$ – compare with (18.3b):

$$\widehat{\nabla_{w^\mathsf{T}} P}(w) = \frac{1}{B} \sum_{b=0}^{B-1} \frac{1}{Np_b} \nabla_{w^\mathsf{T}} Q(w; \boldsymbol{\gamma}(b), \boldsymbol{h}_b) \tag{18.20}$$

We clarify in the following how this scaling corrects an inherent bias that is present under importance sampling – see argument (18.30).

For implementations with *instantaneous gradient approximations*, the random sample can also be selected with or without replacement or it can stream in:

(a') (**Sampling with replacement**). In this case, the sample $(\boldsymbol{\gamma}(n), \boldsymbol{h}_n)$ at iteration n is selected uniformly at random from the dataset $\{\gamma(m), h_m\}$ with replacement. Some sample points may be selected multiple times.

(b') (**Sampling without replacement**). In this case, the sample $(\boldsymbol{\gamma}(n), \boldsymbol{h}_n)$ at iteration n is selected at random from the same dataset but without replacement.

(c') (**Streaming data**). For stochastic risk minimization, the samples $(\boldsymbol{\gamma}(n), \boldsymbol{h}_n)$ stream in independently of each other.

(d') (**Importance sampling**). In this case, a probability value p_m is assigned to each sample $(\gamma(m), h_m)$ in the dataset, and the sample $(\boldsymbol{\gamma}(n), \boldsymbol{h}_n)$ is selected at random according to this distribution. We also explained in Example 16.2 that the approximation for the gradient vector will need to be adjusted to include an additional scaling by $1/Np_n$ – compare with (18.3a):

$$\widehat{\nabla_{w^{\mathsf{T}}} P}(w) = \frac{1}{Np_n} \nabla_{w^{\mathsf{T}}} Q(w; \boldsymbol{\gamma}(n), \boldsymbol{h}_n) \tag{18.21}$$

where p_n is the probability with which sample $(\boldsymbol{\gamma}(n), \boldsymbol{h}_n)$ is selected. We clarify in the sequel how the scaling corrects the bias that arises under importance sampling – see argument (18.30).

The derivations in the remainder of this section are meant to establish the following main conclusion.

LEMMA 18.1. **(Gradient noise under smooth risks)** *Consider the empirical or stochastic risk optimization problems (18.2a)–(18.2b) and assume the risk and loss functions are first-order differentiable with the gradients of the loss function satisfying the δ-Lipschitz conditions (18.10b) or (18.13b). The first- and second-order moments of the gradient noise process will satisfy:*

$$\mathbb{E}\left(\boldsymbol{g}_n(\boldsymbol{w}_{n-1}) \,|\, \boldsymbol{w}_{n-1}\right) = 0 \tag{18.22a}$$

$$\mathbb{E}\left(\|\boldsymbol{g}_n(\boldsymbol{w}_{n-1})\|^2 \,|\, \boldsymbol{w}_{n-1}\right) \leq \beta_g^2 \|\widetilde{w}_{n-1}\|^2 + \sigma_g^2 \tag{18.22b}$$

for some nonnegative constants $\{\beta_g^2, \sigma_g^2\}$ that are independent of $\widetilde{w}_{n-1}$.

Results (18.22a)–(18.22b) hold for instantaneous and mini-batch gradient approximations, regardless of whether the samples are streaming in independently of each other, sampled uniformly with replacement, sampled without replacement, or selected under importance sampling. *The only exception is that the zero-mean property (18.22a) will not hold for the instantaneous gradient implementation when the samples are selected without replacement.* This exception is not of major consequence for the convergence analyses in the next chapters. When property (18.22a) does not hold, the convergence argument will need to be adjusted (and becomes more demanding) but will continue to lead to the same conclusion.

To establish properties (18.22a)–(18.22b), we proceed by examining each sampling procedure separately and then show that they all lead to the same result. We consider the zero-mean property (18.22a) first.

18.3.2 First-Order Moment

We verify in this section that for almost all cases of interest, the gradient noise process has zero mean conditioned on the previous iterate, i.e.,

$$\mathbb{E}\left(\boldsymbol{g}_n(\boldsymbol{w}_{n-1})\,|\,\boldsymbol{w}_{n-1}\right) \;=\; 0 \tag{18.23}$$

Sampling with replacement

Consider first the case of an instantaneous gradient approximation where a single sample is chosen at each iteration n. Let $\boldsymbol{\sigma}$ denote the index of the data sample selected at that iteration so that

$$\mathbb{P}(\boldsymbol{\sigma}=m) = 1/N, \quad m \in \{0,1,2,\ldots,N-1\} \tag{18.24}$$

In this case, the approximate search direction is unbiased since, by conditioning on $\boldsymbol{w}_{n-1}$, we get

$$\begin{aligned}
\mathbb{E}\left(\widehat{\nabla_{w^\mathsf{T}} P}(\boldsymbol{w}_{n-1})\,|\,\boldsymbol{w}_{n-1}\right) &= \mathbb{E}\left(\nabla_{w^\mathsf{T}} Q(\boldsymbol{w}_{n-1};\boldsymbol{\gamma}(\boldsymbol{\sigma}),\boldsymbol{h}_{\boldsymbol{\sigma}})\,|\,\boldsymbol{w}_{n-1}\right)\\
&= \frac{1}{N}\sum_{\sigma=0}^{N-1} \nabla_{w^\mathsf{T}} Q(w_{n-1};\gamma(\sigma),h_\sigma)\\
&\overset{(18.2a)}{=} \nabla_{w^\mathsf{T}} P(w_{n-1})
\end{aligned} \tag{18.25}$$

where in the second equality we used the fact that the loss function assumes each of the values $Q(w_{n-1};\gamma(\sigma),h_\sigma)$ with probability $1/N$. We conclude that (18.23) holds. This is a reassuring conclusion because it means that, on average, the approximation we are using for the gradient vector agrees with the actual gradient.

The gradient noise process continues to have zero conditional mean in the mini-batch implementation. This is because the approximate search direction is again unbiased:

$$\begin{aligned}
\mathbb{E}\left(\widehat{\nabla_{w^\mathsf{T}} P}(\boldsymbol{w}_{n-1})\,|\,\boldsymbol{w}_{n-1}\right) &= \mathbb{E}\left(\frac{1}{B}\sum_{b=0}^{B-1} \nabla_{w^\mathsf{T}} Q(\boldsymbol{w}_{n-1};\boldsymbol{\gamma}(b),\boldsymbol{h}_b)\,|\,\boldsymbol{w}_{n-1}\right)\\
&= \frac{1}{B}\sum_{b=0}^{B-1} \mathbb{E}\left(\nabla_{w^\mathsf{T}} Q(\boldsymbol{w}_{n-1};\boldsymbol{\gamma}(b),\boldsymbol{h}_b)\,|\,\boldsymbol{w}_{n-1}\right)\\
&\overset{(a)}{=} \mathbb{E}\left(\nabla_{w^\mathsf{T}} Q(\boldsymbol{w}_{n-1};\boldsymbol{\gamma}(\boldsymbol{\sigma}),\boldsymbol{h}_{\boldsymbol{\sigma}})\,|\,\boldsymbol{w}_{n-1}\right)\\
&= \frac{1}{N}\sum_{\sigma=0}^{N-1} \nabla_{w^\mathsf{T}} Q(w_{n-1};\gamma(\sigma),h_\sigma)\\
&= \nabla_{w^\mathsf{T}} P(w_{n-1})
\end{aligned} \tag{18.26}$$

where in step (a) we used the fact that the data samples $(\gamma(b), \boldsymbol{h}_b)$ are selected independently of each other.

Sampling without replacement

When the data point $(\gamma(n), \boldsymbol{h}_n)$ is sampled *without* replacement from the dataset $\{\gamma(m), h_m\}$ and used to compute an instantaneous gradient approximation, we find that

$$\begin{aligned}\mathbb{E}\left(\widehat{\nabla_{w^\mathsf{T}} P}(\boldsymbol{w}_{n-1})\,|\,\boldsymbol{w}_{n-1}\right) &= \mathbb{E}\left(\nabla_{w^\mathsf{T}} Q(\boldsymbol{w}_{n-1}; \gamma(\boldsymbol{\sigma}), h_{\boldsymbol{\sigma}})\,|\,\boldsymbol{w}_{n-1}\right)\\ &\neq \frac{1}{N}\sum_{m=0}^{N-1} \nabla_{w^\mathsf{T}} Q(w_{n-1}; \gamma(m), h_m)\\ &= \nabla_{w^\mathsf{T}} P(w_{n-1})\end{aligned} \tag{18.27}$$

where the first line is not equal to the second line because $\boldsymbol{\sigma}$ cannot be selected uniformly with probability $1/N$ when conditioned on $\boldsymbol{w}_{n-1}$. This is due to the fact that knowledge of $\boldsymbol{w}_{n-1}$ carries with it information about the samples that were selected in the previous iterations leading to $\boldsymbol{w}_{n-1}$. As a result, the gradient noise process under random reshuffling is *biased*. For this reason, we will need to adjust the convergence arguments for algorithms employing random reshuffling in comparison to uniform sampling.

A different conclusion holds for mini-batch implementations where the $B > 1$ samples within each batch are selected randomly *without* replacement, with the successive mini-batches constructed by sampling from the original dataset. In this case, the zero-mean property for the gradient noise will continue to hold. To see this, observe first that collecting B samples sequentially, one at a time without replacement, is equivalent to choosing B data points at once from the original N-long dataset. The number of possible choices for this mini-batch of data is given by the combinatorial expression:

$$C_N^B \triangleq \binom{N}{B} = \frac{N!}{B!(N-B)!} \triangleq L \tag{18.28}$$

which we are denoting by L. We number the L possible choices for the mini-batch by $\mathcal{B}_1, \mathcal{B}_2, \ldots, \mathcal{B}_L$, and each one of them can be selected with equal probability $1/L$. Assuming that some random mini-batch ℓ is selected at iteration n, we can write

$$
\begin{aligned}
\mathbb{E}\left(\widehat{\nabla_{w^{\mathsf{T}}} P}(\boldsymbol{w}_{n-1})\,|\,\boldsymbol{w}_{n-1}\right) &= \mathbb{E}\left(\frac{1}{B}\sum_{b\in\mathcal{B}_\ell}\nabla_{w^{\mathsf{T}}} Q(\boldsymbol{w}_{n-1};\boldsymbol{\gamma}(b),\boldsymbol{h}_b)\,|\,\boldsymbol{w}_{n-1}\right) \\
&= \frac{1}{L}\sum_{\ell=1}^{L}\left(\frac{1}{B}\sum_{b\in\mathcal{B}_\ell}\nabla_{w^{\mathsf{T}}} Q(\boldsymbol{w}_{n-1};\boldsymbol{\gamma}(b),\boldsymbol{h}_b)\,|\,\boldsymbol{w}_{n-1}\right) \\
&\overset{(a)}{=} \frac{C_{N-1}^{B-1}}{LB}\sum_{m=0}^{N-1}\nabla_{w^{\mathsf{T}}} Q(w_{n-1};\gamma(m),h_m) \\
&= \frac{1}{N}\sum_{m=0}^{N-1}\nabla_{w^{\mathsf{T}}} Q(w_{n-1};\gamma(m),h_m) \\
&= \nabla_{w^{\mathsf{T}}} P(w_{n-1})
\end{aligned}
\tag{18.29}
$$

where the expectation in the first line is relative to the randomness in the mini-batch selections, and step (a) uses result (18.117) from the appendix. Observe that the mini-batches $\{\mathcal{B}_\ell\}$ in the second line will generally contain some common samples. Equality (a) accounts for these repetitions and rewrites the equality only in terms of the original samples within $0 \leq m \leq N-1$ without any repetitions. We therefore conclude that the gradient noise process continues to have zero mean in this case.

Importance sampling

Under importance sampling, the scaling by $1/Np_n$ of the gradient approximation renders the search directions unbiased since

$$
\begin{aligned}
\mathbb{E}\left(\widehat{\nabla_{w^{\mathsf{T}}} P}(\boldsymbol{w}_{n-1})\,|\,\boldsymbol{w}_{n-1}\right) &= \mathbb{E}\left(\frac{1}{Np_{\boldsymbol{\sigma}}}\nabla_{w^{\mathsf{T}}} Q(\boldsymbol{w}_{n-1};\gamma(\boldsymbol{\sigma}),h_{\boldsymbol{\sigma}})\,|\,\boldsymbol{w}_{n-1}\right) \\
&\overset{(a)}{=} \sum_{\sigma=0}^{N-1} p_\sigma\left(\frac{1}{Np_\sigma}\nabla_{w^{\mathsf{T}}} Q(w_{n-1};\gamma(\sigma),h_\sigma)\right) \\
&= \frac{1}{N}\sum_{\sigma=0}^{N-1}\nabla_{w^{\mathsf{T}}} Q(w_{n-1};\gamma(\sigma),h_\sigma) \\
&= \nabla_{w^{\mathsf{T}}} P(w_{n-1})
\end{aligned}
\tag{18.30}
$$

where in step (a) we used the fact that each $(\gamma(\boldsymbol{\sigma}), h_{\boldsymbol{\sigma}})$ is selected with probability $p_{\boldsymbol{\sigma}}$. The same unbiasedness result holds for the mini-batch version.

Streaming data

Under stochastic risk minimization, the data samples stream in *independently* of each other. As a result, the approximate search direction continues to be unbiased conditioned on the prior weight iterate since, now,

$$\begin{aligned}
\mathbb{E}\left(\widehat{\nabla_{w^{\mathsf{T}}} P}(\boldsymbol{w}_{n-1})\,|\,\boldsymbol{w}_{n-1}\right) &= \mathbb{E}\left(\nabla_{w^{\mathsf{T}}} Q(\boldsymbol{w}_{n-1};\boldsymbol{\gamma}(n),\boldsymbol{h}_n)\,|\,\boldsymbol{w}_{n-1}\right)\\
&\overset{(a)}{=} \nabla_{w^{\mathsf{T}}}\left(\mathbb{E}\, Q(\boldsymbol{w}_{n-1};\boldsymbol{\gamma}(n),\boldsymbol{h}_n)\,|\,\boldsymbol{w}_{n-1}\right)\\
&\overset{(b)}{=} \nabla_{w^{\mathsf{T}}}\,\mathbb{E}\, Q(w_{n-1};\boldsymbol{\gamma}(n),\boldsymbol{h}_n)\\
&\overset{(18.2b)}{=} \nabla_{w^{\mathsf{T}}} P(w_{n-1}) \qquad (18.31)
\end{aligned}$$

Step (a) switches the order of the expectation and differentiation operators which, as explained earlier, is possible in most cases of interest since the loss $Q(w;\cdot,\cdot)$ and its gradient will generally be continuous functions of w. Step (b) is because the samples $(\boldsymbol{\gamma}(n),\boldsymbol{h}_n)$ are independent over time and therefore independent of $\boldsymbol{w}_{n-1}$ (which is a function of previous data samples). The conditioning on $\boldsymbol{w}_{n-1}$ that appears in step (a) can therefore be removed in step (b). It follows that the gradient noise has zero mean conditioned on $\boldsymbol{w}_{n-1}$, and result (18.23) continues to hold.

18.3.3 Second-Order Moment

We examine next the second-order moment of the gradient noise process under different sampling procedures and verify that it satisfies

$$\mathbb{E}\left(\|\boldsymbol{g}_n(\boldsymbol{w}_{n-1})\|^2\,|\,\boldsymbol{w}_{n-1}\right) \;\le\; \beta_g^2\|\widetilde{w}_{n-1}\|^2+\sigma_g^2 \qquad (18.32)$$

for some constants (β_g^2,σ_g^2) independent of $\widetilde{w}_{n-1}$.

Sampling with replacement

Consider first the case of an instantaneous gradient approximation where a single sample is chosen at each iteration n. Let $\boldsymbol{\sigma}$ denote the index of the random data sample selected at that iteration. The squared Euclidean norm of the gradient noise is given by

$$\begin{aligned}
\|\boldsymbol{g}_n(\boldsymbol{w}_{n-1})\|^2 &\overset{\Delta}{=} \left\|\widehat{\nabla_{w^{\mathsf{T}}} P}(\boldsymbol{w}_{n-1}) - \nabla_{w^{\mathsf{T}}} P(\boldsymbol{w}_{n-1})\right\|^2\\
&= \left\|\nabla_{w^{\mathsf{T}}} Q(\boldsymbol{w}_{n-1};\gamma(\boldsymbol{\sigma}),h_{\boldsymbol{\sigma}}) - \nabla_{w^{\mathsf{T}}} P(\boldsymbol{w}_{n-1})\right\|^2\\
&\overset{(a)}{=} \|\nabla_{w^{\mathsf{T}}} Q(\boldsymbol{w}_{n-1}) - \nabla_{w^{\mathsf{T}}} P(\boldsymbol{w}_{n-1})\|^2 \qquad (18.33)
\end{aligned}$$

where we are removing the data argument $(\gamma(\boldsymbol{\sigma}),h_{\boldsymbol{\sigma}})$ from $Q(w;\cdot,\cdot)$ in step (a) to simplify the notation. Adding and subtracting the same term $\nabla_{w^{\mathsf{T}}} Q(w^{\star})$ gives

$$\begin{aligned}
&\|\boldsymbol{g}_n(\boldsymbol{w}_{n-1})\|^2 \\
&= \|\nabla_{w^\mathsf{T}} Q(\boldsymbol{w}_{n-1}) - \nabla_{w^\mathsf{T}} Q(w^\star) + \nabla_{w^\mathsf{T}} Q(w^\star) - \nabla_{w^\mathsf{T}} P(\boldsymbol{w}_{n-1})\|^2 \\
&\overset{(b)}{\leq} 2\,\|\nabla_{w^\mathsf{T}} Q(\boldsymbol{w}_{n-1}) - \nabla_{w^\mathsf{T}} Q(w^\star) - \nabla_{w^\mathsf{T}} P(\boldsymbol{w}_{n-1})\|^2 + 2\,\|\nabla_{w^\mathsf{T}} Q(w^\star)\|^2 \\
&\overset{(c)}{\leq} 4\,\|\nabla_{w^\mathsf{T}} Q(\boldsymbol{w}_{n-1}) - \nabla_{w^\mathsf{T}} Q(w^\star)\|^2 + 2\,\|\nabla_{w^\mathsf{T}} Q(w^\star)\|^2 + \\
&\qquad 4\,\|\nabla_{w^\mathsf{T}} P(w^\star) - \nabla_{w^\mathsf{T}} P(\boldsymbol{w}_{n-1})\|^2 \\
&\overset{(18.10b)}{\leq} 4\delta^2\|\widetilde{\boldsymbol{w}}_{n-1}\|^2 + 4\delta^2\|\widetilde{\boldsymbol{w}}_{n-1}\|^2 + 2\,\|\nabla_{w^\mathsf{T}} Q(w^\star)\|^2
\end{aligned} \tag{18.34}$$

In step (b) we applied the Jensen inequality $\|a+b\|^2 \leq 2\|a\|^2 + 2\|b\|^2$ for any vectors (a,b), and in step (c) we added $\nabla_{w^\mathsf{T}} P(w^\star) = 0$ and applied the Jensen inequality again. Conditioning on $\boldsymbol{w}_{n-1}$ and taking expectations over the randomness in data selection, we conclude that (18.32) holds for the following parameters:

$$\beta_g^2 = 8\delta^2 \tag{18.35a}$$

$$\begin{aligned}
\sigma_g^2 &= 2\,\mathbb{E}\left(\|\nabla_{w^\mathsf{T}} Q(w^\star; \boldsymbol{\gamma}(\boldsymbol{\sigma}), \boldsymbol{h}_{\boldsymbol{\sigma}})\|^2\right) \\
&\overset{(18.24)}{=} \frac{2}{N}\sum_{\sigma=0}^{N-1}\left\|\nabla_{w^\mathsf{T}} Q(w^\star; \gamma(\sigma), h_\sigma)\right\|^2
\end{aligned} \tag{18.35b}$$

If desired, the value for β_g^2 can be tightened to $\beta_g^2 = 2\delta^2$ – see Prob. 18.3. It is sufficient for our purposes to know that a bound of the form (18.32) exists; the specific values for the parameters $\{\beta_g^2, \sigma_g^2\}$ are not relevant at this stage.

For mini-batch implementations, the gradient noise process continues to satisfy relation (18.32) albeit with the parameters (β_g^2, σ_g^2) scaled by B. Indeed, since the data points $\{\boldsymbol{\gamma}(b), \boldsymbol{h}_b\}$ are now sampled with replacement and are independent of each other, we have

$$\begin{aligned}
&\mathbb{E}\left(\|\boldsymbol{g}_n(\boldsymbol{w}_{n-1})\|^2 \,|\, \boldsymbol{w}_{n-1}\right) \\
&\overset{\Delta}{=} \mathbb{E}\left(\left\|\widehat{\nabla_{w^\mathsf{T}} P}(\boldsymbol{w}_{n-1}) - \nabla_{w^\mathsf{T}} P(\boldsymbol{w}_{n-1})\right\|^2 \,\Big|\, \boldsymbol{w}_{n-1}\right) \\
&= \mathbb{E}\left(\left\|\frac{1}{B}\sum_{b=0}^{B-1}\nabla_{w^\mathsf{T}} Q(\boldsymbol{w}_{n-1}; \boldsymbol{\gamma}(b), \boldsymbol{h}_b) - \nabla_{w^\mathsf{T}} P(\boldsymbol{w}_{n-1})\right\|^2 \,\Big|\, \boldsymbol{w}_{n-1}\right) \\
&= \mathbb{E}\left\|\frac{1}{B}\sum_{b=0}^{B-1}\Big(\nabla_{w^\mathsf{T}} Q(w_{n-1}; \boldsymbol{\gamma}(b), \boldsymbol{h}_b) - \nabla_{w^\mathsf{T}} P(w_{n-1})\Big)\right\|^2 \\
&\overset{(a)}{=} \frac{1}{B^2}\sum_{b=0}^{B-1}\mathbb{E}\,\|\nabla_{w^\mathsf{T}} Q(w_{n-1}; \boldsymbol{\gamma}(b), \boldsymbol{h}_b) - \nabla_{w^\mathsf{T}} P(w_{n-1})\|^2
\end{aligned} \tag{18.36}$$

where step (a) is because the samples within the mini-batch are independent of each other. Using the same bound from argument (18.34) we get

$$\mathbb{E}\left(\|\boldsymbol{g}_n(\boldsymbol{w}_{n-1})\|^2 \,|\, \boldsymbol{w}_{n-1}\right) \leq \frac{1}{B^2}\sum_{b=0}^{B-1}\left(\beta_g^2\|\widetilde{w}_{n-1}\|^2 + \sigma_g^2\right)$$
$$= \frac{1}{B}\left(\beta_g^2\|\widetilde{w}_{n-1}\|^2 + \sigma_g^2\right) \tag{18.37}$$

and step (a) uses the same bound that would result from argument (18.34).

Sampling without replacement

If we repeat the same argument for the implementation with instantaneous gradient approximation, we will similarly find that the same relation (18.32) continues to hold, albeit with parameters

$$\beta_g^2 = 8\delta^2 \tag{18.38a}$$
$$\sigma_g^2 = 2\,\mathbb{E}\left(\|\nabla_{w^\mathsf{T}} Q(w^\star;\gamma(\boldsymbol{\sigma}), h_{\boldsymbol{\sigma}}\|^2 \,|\, \boldsymbol{w}_{n-1}\right) \tag{18.38b}$$

where the expression for σ_g^2 involves an inconvenient conditioning on $\boldsymbol{w}_{n-1}$. We can remove the conditioning as follows. Let $\boldsymbol{\sigma}$ denote the data index selected at iteration n. We know that $\boldsymbol{\sigma}$ is not necessarily chosen uniformly when we condition on $\boldsymbol{w}_{n-1}$ due to data sampling with replacement. Let us introduce, for the sake of argument, the conditional probabilities:

$$\kappa_m \;\stackrel{\Delta}{=}\; \mathbb{P}\Big(\boldsymbol{\sigma} = m \,|\, \boldsymbol{w}_{n-1}\Big), \quad \kappa_m \geq 0, \quad \sum_{m=0}^{N-1}\kappa_m = 1 \tag{18.39}$$

That is, κ_m is the likelihood of selecting index $\boldsymbol{\sigma} = m$ at iteration n conditioned on knowledge of $\boldsymbol{w}_{n-1}$. Then, substituting into (18.38b), we get

$$\sigma_g^2 = 2\sum_{m=0}^{N-1}\kappa_m\,\|\nabla_{w^\mathsf{T}} Q(w^\star;\gamma(m), h_m)\|^2 \;\leq\; 2\sum_{m=0}^{N-1}\|\nabla_{w^\mathsf{T}} Q(w^\star;\gamma(m), h_m)\|^2 \tag{18.40}$$

which is independent of $\boldsymbol{w}_{n-1}$. This result can be used as the expression for σ_g^2 in (18.32).

A similar conclusion holds for the mini-batch gradient approximation where the B samples are randomly selected *without* replacement. To establish this result, we need to appeal to the auxiliary Lemma 18.3 from the appendix. First, for any iterate value w_{n-1}, we introduce the auxiliary vectors:

$$x_m \;\stackrel{\Delta}{=}\; \nabla_{w^\mathsf{T}} Q\Big(w_{n-1};\gamma(m), h_m\Big) - \nabla_{w^\mathsf{T}} P(w_{n-1}), \quad m = 0, 1, \ldots, N-1 \tag{18.41}$$

It is clear from the definition of the empirical risk function (18.2a) that

$$\frac{1}{N}\sum_{m=0}^{N-1} x_m = 0 \tag{18.42}$$

which means that the vectors $\{x_m\}$ satisfy condition (18.129) required by the lemma. At iteration n, the mini-batch implementation selects B vectors $\{\boldsymbol{x}_b\}$ at random without replacement. Let

$$\boldsymbol{g}_n(\boldsymbol{w}_{n-1}) = \frac{1}{B}\sum_{b=0}^{B-1} \boldsymbol{x}_b \tag{18.43}$$

Then, result (18.130) in the appendix implies that

$$\begin{aligned}
&\mathbb{E}\left(\|\boldsymbol{g}_n(\boldsymbol{w}_{n-1})\|^2 \,|\, \boldsymbol{w}_{n-1}\right) \\
&= \frac{1}{B^2}\frac{B(N-B)}{N(N-1)}\sum_{m=0}^{N-1} \|\nabla_w^{\mathsf{T}} Q(w_{n-1};\gamma(m),h_m) - \nabla_{w^{\mathsf{T}}} P(w_{n-1})\|^2 \\
&\overset{(a)}{\leq} \frac{1}{B}\frac{(N-B)}{N(N-1)}\sum_{m=0}^{N-1}\left(8\delta^2\|\widetilde{w}_{n-1}\|^2 + 2\,\|\nabla_{w^{\mathsf{T}}}\, Q(w^\star;\gamma(m),h_m)\|^2\right) \\
&\overset{(b)}{=} \frac{1}{B}\frac{(N-B)}{(N-1)}\,(\beta_g^2\|\widetilde{w}_{n-1}\|^2 + \sigma_g^2)
\end{aligned} \tag{18.44}$$

where step (a) uses the same bound that would result from argument (18.34), and step (b) uses (18.35a)–(18.35b). We can group the results for the mini-batch implementations under sampling with and without replacement into a single statement as follows:

$$\mathbb{E}\left(\|\boldsymbol{g}_n(\boldsymbol{w}_{n-1})\|^2 \,|\, \boldsymbol{w}_{n-1}\right) \;\leq\; \frac{1}{\tau_B}\,(\beta_g^2\|\widetilde{w}_{n-1}\|^2 + \sigma_g^2) \tag{18.45}$$

where the factor τ_B is chosen as:

$$\tau_B \;\overset{\Delta}{=}\; \begin{cases} B, & \text{when samples are selected } \textit{with} \text{ replacement} \\ B\dfrac{N-1}{N-B}, & \text{when samples are selected } \textit{without} \text{ replacement} \end{cases} \tag{18.46}$$

Observe from the second line that $\tau_B \approx B$ for N large enough.

Importance sampling

Under importance sampling, the same bound (18.32) holds for both cases of instantaneous and mini-batch gradient approximations, as can be seen from the following argument.

For the instantaneous gradient implementations we have:

$$
\begin{aligned}
&\|\boldsymbol{g}_n(\boldsymbol{w}_{n-1})\|^2 \\
&\stackrel{\Delta}{=} \left\|\widehat{\nabla_{w^\mathsf{T}} P}(\boldsymbol{w}_{n-1}) - \nabla_{w^\mathsf{T}} P(\boldsymbol{w}_{n-1})\right\|^2 \\
&= \left\|\frac{1}{Np_n}\nabla_{w^\mathsf{T}} Q(\boldsymbol{w}_{n-1};\boldsymbol{\gamma}(n),\boldsymbol{h}_n) - \nabla_{w^\mathsf{T}} P(\boldsymbol{w}_{n-1})\right\|^2 \\
&\stackrel{(a)}{=} \left\|\frac{1}{Np_n}\nabla_{w^\mathsf{T}} Q(\boldsymbol{w}_{n-1}) - \nabla_{w^\mathsf{T}} P(\boldsymbol{w}_{n-1})\right\|^2 \\
&\stackrel{(b)}{=} \left\|\frac{1}{Np_n}\nabla_{w^\mathsf{T}} Q(\boldsymbol{w}_{n-1}) - \frac{1}{Np_n}\nabla_{w^\mathsf{T}} Q(w^\star) + \frac{1}{Np_n}\nabla_{w^\mathsf{T}} Q(w^\star) - \nabla_{w^\mathsf{T}} P(\boldsymbol{w}_{n-1})\right\|^2
\end{aligned} \tag{18.47}
$$

where step (a) removes the data arguments $(\boldsymbol{\gamma}(n),\boldsymbol{h}_n)$ from $\nabla_{w^\mathsf{T}} Q(w;\cdot,\cdot)$ for convenience, and step (b) adds and subtracts the same quantity $\nabla_{w^\mathsf{T}} Q(w^\star)$. We therefore get

$$
\begin{aligned}
&\|\boldsymbol{g}_n(\boldsymbol{w}_{n-1})\|^2 \\
&\leq 2\left\|\frac{1}{Np_n}\nabla_{w^\mathsf{T}} Q(\boldsymbol{w}_{n-1}) - \frac{1}{Np_n}\nabla_{w^\mathsf{T}} Q(w^\star) - \nabla_{w^\mathsf{T}} P(\boldsymbol{w}_{n-1})\right\|^2 + \\
&\quad\; 2\left\|\frac{1}{Np_n}\nabla_{w^\mathsf{T}} Q(w^\star)\right\|^2 \\
&\stackrel{(c)}{\leq} 4\left\|\frac{1}{Np_n}\nabla_{w^\mathsf{T}} Q(\boldsymbol{w}_{n-1}) - \frac{1}{Np_n}\nabla_{w^\mathsf{T}} Q(w^\star)\right\|^2 + \\
&\quad\; 4\left\|\nabla_{w^\mathsf{T}} P(w^\star) - \nabla_{w^\mathsf{T}} P(\boldsymbol{w}_{n-1})\right\|^2 + 2\left\|\frac{1}{Np_n}\nabla_{w^\mathsf{T}} Q(w^\star)\right\|^2 \\
&\stackrel{(18.10b)}{\leq} 4\delta^2\|\widetilde{\boldsymbol{w}}_{n-1}\|^2 + 4\left\|\frac{1}{Np_n}\nabla_{w^\mathsf{T}} Q(\boldsymbol{w}_{n-1}) - \frac{1}{Np_n}\nabla_{w^\mathsf{T}} Q(w^\star)\right\|^2 + \\
&\quad\; 2\left\|\frac{1}{Np_n}\nabla_{w^\mathsf{T}} Q(w^\star)\right\|^2
\end{aligned} \tag{18.48}
$$

where in step (c) we applied the Jensen inequality $\|a+b\|^2 \leq 2\|a\|^2 + 2\|b\|^2$ for any two vectors (a,b) and added $\nabla_{w^\mathsf{T}} P(w^\star) = 0$. Next, we need to condition on $\boldsymbol{w}_{n-1}$ and take expectations. For that purpose, we note that

$$
\begin{aligned}
\mathbb{E}\left\{2\left\|\frac{1}{Np_n}\nabla_{w^\mathsf{T}} Q(w^\star)\right\|^2 \,\middle|\, \boldsymbol{w}_{n-1}\right\} &= \sum_{m=0}^{N-1} p_m \frac{2}{N^2 p_m^2}\|\nabla_{w^\mathsf{T}} Q(w^\star;\gamma(m),h_m)\|^2 \\
&= \frac{2}{N^2}\sum_{m=0}^{N-1}\frac{1}{p_m}\|\nabla_{w^\mathsf{T}} Q(w^\star;\gamma(m),h_m)\|^2
\end{aligned} \tag{18.49}
$$

while

$$\mathbb{E}\left\{4\left\|\frac{1}{Np_n}\nabla_{w^\mathsf{T}}Q(\boldsymbol{w}_{n-1})-\frac{1}{Np_n}\nabla_{w^\mathsf{T}}Q(w^\star)\right\|^2\,\Big|\,\boldsymbol{w}_{n-1}\right\}$$
$$=4\sum_{m=0}^{N-1}p_m\frac{1}{N^2p_m^2}\left\|\nabla_{w^\mathsf{T}}Q(w_{n-1};\gamma(m),h_m)-\nabla_{w^\mathsf{T}}Q(w^\star)\right\|^2$$
$$\overset{(18.10b)}{\leq}\frac{4\delta^2}{N^2}\sum_{m=0}^{N-1}\frac{1}{p_m}\|\widetilde{w}_{n-1}\|^2 \tag{18.50}$$

Substituting into (18.48) we conclude that

$$\mathbb{E}\left(\|\boldsymbol{g}_n(\boldsymbol{w}_{n-1})\|^2\,|\,\boldsymbol{w}_{n-1}\right)\;\leq\;\beta_g^2\|\widetilde{w}_{n-1}\|^2+\sigma_g^2 \tag{18.51}$$

where the parameters $\{\beta_g^2,\sigma_g^2\}$ are given by

$$\beta_g^2=4\delta^2\left(1+\frac{1}{N^2}\sum_{m=0}^{N-1}\frac{1}{p_m}\right) \tag{18.52a}$$
$$\sigma_g^2=\frac{2}{N^2}\sum_{m=0}^{N-1}\frac{1}{p_m}\left\|\nabla_{w^\mathsf{T}}Q(w^\star;\gamma(m),h_m)\right\|^2 \tag{18.52b}$$

A similar bound holds with the above parameters $\{\beta_g^2,\sigma_g^2\}$ divided by B for the mini-batch version – see Prob. 18.13.

Streaming data

Under stochastic risk minimization, the gradient noise process continues to satisfy relation (18.32), as can be seen from the following sequence of inequalities:

$$\mathbb{E}\left(\|\boldsymbol{g}_n(\boldsymbol{w}_{n-1})\|^2\,|\,\boldsymbol{w}_{n-1}\right)$$
$$\triangleq\mathbb{E}\left(\left\|\widehat{\nabla_{w^\mathsf{T}}P}(\boldsymbol{w}_{n-1})-\nabla_{w^\mathsf{T}}P(\boldsymbol{w}_{n-1})\right\|^2\,\Big|\,\boldsymbol{w}_{n-1}\right)$$
$$=\mathbb{E}\left(\|\nabla_{w^\mathsf{T}}Q(\boldsymbol{w}_{n-1};\boldsymbol{\gamma}(n),\boldsymbol{h}_n)-\nabla_{w^\mathsf{T}}P(\boldsymbol{w}_{n-1})\|^2\,|\,\boldsymbol{w}_{n-1}\right)$$
$$\overset{(a)}{=}\mathbb{E}\left(\|\nabla_{w^\mathsf{T}}Q(\boldsymbol{w}_{n-1})-\nabla_{w^\mathsf{T}}P(\boldsymbol{w}_{n-1})\|^2\,|\,\boldsymbol{w}_{n-1}\right)$$
$$\overset{(b)}{=}\mathbb{E}\left(\left\|\nabla_{w^\mathsf{T}}Q(\boldsymbol{w}_{n-1})-\nabla_{w^\mathsf{T}}Q(w^o)+\nabla_{w^\mathsf{T}}Q(w^o)-\nabla_{w^\mathsf{T}}P(\boldsymbol{w}_{n-1})\right\|^2\,|\,\boldsymbol{w}_{n-1}\right) \tag{18.53}$$

In step (a) we removed the data argument $(\boldsymbol{\gamma}(n),\boldsymbol{h}_n)$ from $Q(w;\cdot,\cdot)$ to simplify the notation, and in step (b) we added and subtracted the same quantity

$\nabla_{w^\mathsf{T}} Q(w^o)$. It follows that

$$\begin{aligned}
&\mathbb{E}\left(\|\boldsymbol{g}_n(\boldsymbol{w}_{n-1})\|^2 \,|\, \boldsymbol{w}_{n-1}\right) \\
&\overset{(c)}{\leq} 2\,\mathbb{E}\left(\|\nabla_{w^\mathsf{T}} Q(\boldsymbol{w}_{n-1}) - \nabla_{w^\mathsf{T}} Q(w^o) - \nabla_{w^\mathsf{T}} P(\boldsymbol{w}_{n-1})\|^2 \,|\, \boldsymbol{w}_{n-1}\right) + \\
&\quad 2\,\mathbb{E}\left(\|\nabla_{w^\mathsf{T}} Q(w^o)\|^2\right) \\
&\overset{(d)}{\leq} 4\,\mathbb{E}\left(\|\nabla_{w^\mathsf{T}} Q(\boldsymbol{w}_{n-1}) - \nabla_{w^\mathsf{T}} Q(w^o)\|^2 \,|\, \boldsymbol{w}_{n-1}\right) + 2\,\mathbb{E}\left(\|\nabla_{w^\mathsf{T}} Q(w^o)\|^2\right) + \\
&\quad 4\,\|\nabla_{w^\mathsf{T}} P(w^o) - \nabla_{w^\mathsf{T}} P(w_{n-1})\|^2 \\
&\overset{(18.14)}{\leq} 4\delta^2\|\widetilde{w}_{n-1}\|^2 + 4\delta^2\|\widetilde{w}_{n-1}\|^2 + 2\,\mathbb{E}\,\|\nabla_{w^\mathsf{T}} Q(w^o)\|^2 \qquad (18.54)
\end{aligned}$$

In step (c) we applied the Jensen inequality $\|a+b\|^2 \leq 2\|a\|^2 + 2\|b\|^2$ for any vectors (a,b), and in step (d) we added $\nabla_{w^\mathsf{T}} P(w^o) = 0$ and applied the Jensen inequality again. We conclude that relation (18.32) holds with parameters:

$$\beta_g^2 = 8\delta^2 \qquad (18.55a)$$

$$\sigma_g^2 = 2\,\mathbb{E}\left(\|\nabla_{w^\mathsf{T}} Q(w^o;\boldsymbol{\gamma},\boldsymbol{h})\|^2\right) \qquad (18.55b)$$

where the expectation in σ_g^2 is over the joint distribution of the data $\{\boldsymbol{\gamma},\boldsymbol{h}\}$.

REMARK 18.1. **(Variance-reduced techniques)** We will discover through expressions (19.18a) and (19.26) that the constant factor σ_g^2 in (18.32) is a source of performance degradation. It will prevent the iterates $\boldsymbol{w}_n$ from converging exactly to $w^\star$. In Chapter 22 we will introduce the class of *variance-reduced* algorithms for empirical risk minimization. These algorithms adjust the gradient approximation in such a way that the constant driving term, σ_g^2, will end up disappearing from the variance expression (18.32). By doing so, we will be able to recover the exact convergence of $\boldsymbol{w}_n$ to $w^\star$.

■

REMARK 18.2. **(Bound on second-order moment for gradient noise)** The derivation of the bound (18.32) relied almost exclusively on the assumption that the loss function has δ-Lipschitz gradients either in the deterministic sense (18.10b) or in the mean-square sense (18.13b). The convergence analyses in future chapters will continue to hold if one assumes (or imposes) from the start that the gradient noise satisfies (18.32) for some nonnegative constants (β_g^2, σ_g^2).

■

18.4 NONSMOOTH RISK FUNCTIONS

We consider next the case of nonsmooth risk functions where gradient vectors are replaced by subgradients and these are in turn approximated by using either instantaneous or mini-batch versions, say as,

$$\textbf{(instantaneous)} \quad \widehat{s}(w) = s_Q(w;\boldsymbol{\gamma},\boldsymbol{h}) \qquad (18.56a)$$

$$\textbf{(mini-batch)} \quad \widehat{s}(w) = \frac{1}{B}\sum_{b=0}^{B-1} s_Q(w;\boldsymbol{\gamma}(b),\boldsymbol{h}_b) \qquad (18.56b)$$

In this notation, $s(w)$ denotes a subgradient construction for $P(w)$, and $s_Q(w;\boldsymbol{\gamma},\boldsymbol{h})$ refers to a subgradient of the loss function at the same location. For example, for the empirical risk minimization case (18.2a), a subgradient construction for $P(w)$ can be chosen as

$$s(w)=\frac{1}{N}\sum_{m=0}^{N-1}s_Q(w;\gamma(m),h_m) \tag{18.57}$$

in terms of individual subgradients of the loss function parameterized by the data points $(\gamma(m),h_m)$. The instantaneous approximation (18.56a) selects one subgradient vector at random, while the mini-batch approximation (18.56b) selects B subgradient vectors at random. The difference between the original subgradient construction and its approximation is again called the *gradient noise*:

$$\boldsymbol{g}(w)\overset{\Delta}{=}\widehat{s}(w)-s(w) \tag{18.58}$$

The following two relations highlight the difference between the original subgradient method and its stochastic version:

$$\textbf{(subgradient method)}\quad w_n=w_{n-1}-\mu s(w_{n-1}) \tag{18.59a}$$

$$\begin{aligned}\textbf{(stochastic version)}\quad \boldsymbol{w}_n&=\boldsymbol{w}_{n-1}-\mu\widehat{s}(\boldsymbol{w}_{n-1})\\&=\boldsymbol{w}_{n-1}-\mu s(\boldsymbol{w}_{n-1})-\mu\,\boldsymbol{g}(\boldsymbol{w}_{n-1})\end{aligned} \tag{18.59b}$$

Example 18.2 **(Gradient noise for a nonsmooth quadratic risk)** We illustrate the form of the gradient noise process for two ℓ_1-regularized quadratic risks: an empirical risk and a stochastic risk. Consider first the empirical case:

$$P(w)\;=\;\alpha\|w\|_1\;+\;\frac{1}{N}\sum_{m=0}^{N-1}(\gamma(m)-h_m^{\mathsf T}w)^2 \tag{18.60}$$

Subgradient constructions for $P(w)$ and its loss function can be chosen as

$$s_Q(w;\boldsymbol{\gamma}(n),\boldsymbol{h}_n)=\alpha\,\text{sign}(w)-2\boldsymbol{h}_n(\boldsymbol{\gamma}(n)-\boldsymbol{h}_n^{\mathsf T}w) \tag{18.61a}$$

$$s(w)=\alpha\,\text{sign}(w)-\frac{2}{N}\sum_{m=0}^{N-1}h_m(\gamma(m)-h_m^{\mathsf T}w) \tag{18.61b}$$

with the resulting gradient noise vector given by

$$\boldsymbol{g}(w)=\frac{2}{N}\sum_{m=0}^{N-1}h_m(\gamma(m)-h_m^{\mathsf T}w)\;-\;2\boldsymbol{h}_n(\boldsymbol{\gamma}(n)-\boldsymbol{h}_n^{\mathsf T}w) \tag{18.62}$$

Observe that $\boldsymbol{g}(w)$ depends on the data $(\boldsymbol{\gamma}(n),\boldsymbol{h}_n)$ and, hence, as explained before, we could have written instead $\boldsymbol{g}_n(w)$ to highlight this dependency. Consider next the stochastic risk

$$\begin{aligned}P(w)&=\alpha\|w\|_1\;+\;\mathbb{E}\,(\boldsymbol{\gamma}-\boldsymbol{h}^{\mathsf T}w)^2\\&=\alpha\|w\|_1+\sigma_\gamma^2-2r_{h\gamma}^{\mathsf T}w+w^{\mathsf T}R_hw\end{aligned} \tag{18.63}$$

Subgradient constructions for $P(w)$ and its loss function can be chosen as

$$s(w) = \alpha\, \text{sign}(w) - 2(r_{h\gamma} - R_h w) \tag{18.64a}$$

$$s_Q(w; \boldsymbol{\gamma}(n), \boldsymbol{h}_n) = \alpha\, \text{sign}(w) - 2\boldsymbol{h}_n(\boldsymbol{\gamma}(n) - \boldsymbol{h}_n^\mathsf{T} w) \tag{18.64b}$$

with the resulting gradient noise vector given by

$$\boldsymbol{g}(w) = 2(r_{h\gamma} - R_h w) - 2\boldsymbol{h}_n(\boldsymbol{\gamma}(n) - \boldsymbol{h}_n^\mathsf{T} w) \tag{18.65}$$

Observe again that $\boldsymbol{g}(w)$ depends on n.

We are again interested in characterizing the first- and second-order moments of the gradient noise process. For this purpose, we describe below the conditions that are normally imposed on the risk and loss functions, $P(w)$ and $Q(w, \cdot)$. The conditions listed here are satisfied by several risk and loss functions of interest, as illustrated in the next example and in the problems at the end of the chapter.

Empirical risks

Consider initially the case of nonsmooth empirical risks of the form (18.2a). We will assume that the risk and loss functions satisfy the two conditions listed below. Compared with the earlier conditions (14.28a)–(14.28b), we see that we are now taking the loss function into consideration since its subgradients are the ones used in the stochastic approximation implementation:

(1) (**Strongly convex risk**). $P(w)$ is ν-strongly convex, namely, for every $w_1, w_2 \in \text{dom}(P)$, there exists a subgradient $s(w_1)$ relative to w^T such that

$$P(w_2) \geq P(w_1) + (s(w_1))^\mathsf{T}(w_2 - w_1) + \frac{\nu}{2}\|w_2 - w_1\|^2 \tag{18.66a}$$

for some $\nu > 0$.

(2) (**Affine-Lipschitz loss subgradients**). The loss function $Q(w, \cdot)$ is convex over w and its subgradients are affine-Lipschitz, i.e., there exist nonnegative constants $\{\delta, \delta_2\}$ such that, independently of the data samples,

$$\|s_Q(w_2; \gamma(\ell), h_\ell) - s'_Q(w_1; \gamma(k), h_k)\| \leq \delta\|w_2 - w_1\| + \delta_2 \tag{18.66b}$$

for any $w_1, w_2 \in \text{dom}(Q)$, any indices $0 \leq \ell, k \leq N-1$, and for *any* subgradients:

$$s'_Q(w; \gamma, h) \in \partial_{w^\mathsf{T}} Q(w; \gamma, h) \tag{18.67}$$

Observe that condition (18.66b) is stated in terms of the *particular* subgradient construction $s_Q(w; \gamma, w)$ used by the stochastic implementation and *any* of the subgradients $s'_Q(w; \gamma, h)$ from the subdifferential set of $Q(w; \gamma, h)$. For later use, it is useful to note that condition (18.66b) implies the following relation, which involves the squared norms as opposed to the actual norms:

$$\|s_Q(w_2; \gamma(\ell), h_\ell) - s'_Q(w_1; \gamma(k), h_k)\|^2 \leq 2\delta^2\|w_2 - w_1\| + 2\delta_2^2 \tag{18.68}$$

This result follows from the inequality $(a+b)^2 \leq 2a^2 + 2b^2$ for any (a, b).

Based on the explanation given in Example 8.6 on the subdifferential for sums of convex functions, we characterize all subgradients for $P(w)$ by writing

$$s'(w) = \frac{1}{N}\sum_{m=0}^{N-1} s'_Q(w;\gamma(m),h_m) \tag{18.69}$$

in terms of any subdifferential $s'_Q(w,\cdot)$ for $Q(w,\cdot)$. It readily follows from the triangle inequality of norms that subgradient vectors for the risk function $P(w)$ also satisfy affine-Lipschitz conditions, namely, for all $w_1, w_2 \in \text{dom}(P)$:

$$\|s(w_2) - s'(w_1)\| \;\leq\; \delta\|w_2 - w_1\| + \delta_2 \tag{18.70a}$$

$$\|s(w_2) - s'(w_1)\|^2 \;\leq\; 2\delta^2\|w_2 - w_1\|^2 + 2\delta_2^2 \tag{18.70b}$$

Example 18.3 (ℓ_2-**regularized hinge risk**) We illustrate condition (18.66b) by considering the following ℓ_2-regularized hinge risk, written using a subscript ℓ to index the data $\{\gamma(\ell), h_\ell\}$ instead of m:

$$P(w) = \rho\|w\|^2 + \frac{1}{N}\sum_{\ell=0}^{N-1} \max\left\{0, 1-\gamma(\ell)h_\ell^{\mathsf{T}}w\right\}, \quad w = \text{col}\{w_m\} \in \mathbb{R}^M \tag{18.71}$$

The corresponding loss function is given by

$$Q(w;\gamma(\ell),h_\ell) \;=\; \rho\|w\|^2 + \max\left\{0, 1-\gamma(\ell)h_\ell^{\mathsf{T}}w\right\} \tag{18.72}$$

We know from the earlier results (8.59a) and (8.72a) how to characterize the subdifferential set of $Q(w,\gamma(\ell),h_\ell)$. Let $h_\ell = \text{col}\{h_{\ell,m}\}$ denote the individual entries of h_ℓ. Then, it holds that

$$s'_Q(w;\gamma(\ell),h_\ell) = 2\rho w + \text{col}\left\{\mathbb{A}_{\gamma(\ell)h_{\ell,m}}(w_m)\right\} \tag{18.73a}$$

where each $\mathbb{A}_{\gamma h_m}(w_m)$ is defined for a generic $\gamma \in \mathbb{R}$ and vector $h \in \mathbb{R}^M$ with individual entries $\{h_m\}$ as follows:

$$\mathbb{A}_{\gamma h_m}(w_m) \triangleq \begin{cases} 0, & \gamma h^{\mathsf{T}}w > 1 \\ -\gamma h_m, & \gamma h^{\mathsf{T}}w < 1 \\ [-\gamma h_m, 0], & \gamma h^{\mathsf{T}}w = 1,\;\; \gamma h_m \geq 0 \\ [0, -\gamma h_m], & \gamma h^{\mathsf{T}}w = 1,\;\; \gamma h_m < 0 \end{cases} \tag{18.73b}$$

Moreover, one particular subgradient construction for $Q(w;\cdot)$ is given by

$$s_Q(w;\gamma(\ell),h_\ell) \;=\; 2\rho w - \gamma(\ell)h_\ell\,\mathbb{I}[\gamma(\ell)h_\ell w \leq 1] \tag{18.73c}$$

Using the triangle inequality of norms we get:

$$\begin{aligned}
&\|s_Q(w_2;\gamma(\ell),h_\ell) - s'_Q(w_1;\gamma(k),h_k)\| \\
&\le 2\rho\|w_2-w_1\| + \left\|\gamma(\ell)h_\ell\,\mathbb{I}[\gamma(\ell)h_\ell^\mathsf{T} w_2 \le 1]\right\| + \left\|\text{col}\left\{\mathbb{A}_{\gamma(k)h_{k,m}}(w_m)\right\}\right\| \\
&\le \underbrace{2\rho}_{\triangleq\,\delta}\|w_2-w_1\| + \|\gamma(\ell)h_\ell\| + \|\gamma(k)h_k\| \\
&\le \underbrace{2\rho}_{\triangleq\,\delta}\|w_2-w_1\| + \underbrace{\max_{0\le \ell\le N-1} 2\|\gamma(\ell)h_\ell\|}_{\triangleq\,\delta_2} \\
&= \delta\|w_2-w_1\| + \delta_2
\end{aligned} \tag{18.74}$$

since $\mathbb{I}[a]$ is bounded by 1 and $\|\text{col}\{\mathbb{A}_{\gamma(k)h_{k,m}}(w_m)\}\|$ is bounded by $\|\gamma(k)h_k\|$. Observe how the factor δ_2 arises from the *nonsmooth* component in $P(w)$.

Stochastic risks

For stochastic risks of the form (18.2b), we continue to assume that $P(w)$ is ν-strongly convex but that the loss function has subgradients that are affine-Lipschitz in the *mean-square sense*:

(1') (**Strongly convex risk**). $P(w)$ is ν-strongly convex, namely, for every $w_1, w_2 \in$ dom(P) there exists a subgradient vector $s(w_1)$ relative to w^T such that

$$P(w_2) \ge P(w_1) + (s(w_1))^\mathsf{T}(w_2-w_1) + \frac{\nu}{2}\|w_2-w_1\|^2 \tag{18.75a}$$

for some $\nu > 0$.

(2') (**Mean-square affine-Lipschitz loss gradients**). The loss function $Q(w,\cdot)$ is convex over w and its subgradient vectors satisfy

$$\mathbb{E}\,\|s_Q(w_2;\boldsymbol{\gamma},\boldsymbol{h}) - s'_Q(w_1;\boldsymbol{\gamma},\boldsymbol{h})\|^2 \le \delta^2\|w_2-w_1\|^2 + \delta_2^2 \tag{18.75b}$$

for any $w_1, w_2 \in$ dom(Q) and for *any*

$$s'_Q(w;\gamma,h) \in \partial_{w^\mathsf{T}} Q(w;\gamma,h) \tag{18.76}$$

Again, condition (18.75b) is stated in terms of the *particular* subgradient construction $s_Q(w;\cdot,\cdot)$ used by the stochastic optimization algorithm and *any* of the possible subgradients $s'_Q(w;\cdot,\cdot)$ from the subdifferential set of $Q(w;\cdot,\cdot)$. Note from (18.75b) that

$$\begin{aligned}
&\mathbb{E}\,\|s_Q(w_2;\boldsymbol{\gamma},\boldsymbol{h}) - s'_Q(w_1;\boldsymbol{\gamma},\boldsymbol{h})\|^2 \\
&\le \delta^2\|w_2-w_1\|^2 + \delta_2^2 + 2\delta\delta_2\|w_2-w_1\| \\
&= (\delta\|w_2-w_1\| + \delta_2)^2
\end{aligned} \tag{18.77}$$

Now using the fact that for any scalar random variable $\boldsymbol{x}$ it holds that $(\mathbb{E}\,\boldsymbol{x})^2 \le \mathbb{E}\,\boldsymbol{x}^2$, we conclude that the subgradient vectors are also affine-Lipschitz on average, namely,

$$\mathbb{E}\,\|s_Q(w_2;\boldsymbol{\gamma},\boldsymbol{h}) - s'_Q(w_1;\boldsymbol{\gamma},\boldsymbol{h})\| \le \delta\|w_2-w_1\| + \delta_2 \tag{18.78}$$

Moreover, the constructions

$$s(w) \stackrel{\Delta}{=} \mathbb{E}\, s_Q(w;\boldsymbol{\gamma},\boldsymbol{h}) \tag{18.79a}$$

$$s'(w) \stackrel{\Delta}{=} \mathbb{E}\, s'_Q(w;\boldsymbol{\gamma},\boldsymbol{h}) \tag{18.79b}$$

correspond to subgradient vectors for the risk function $P(w)$ and we can also conclude that they satisfy similar affine-Lipschitz conditions:

$$\|s(w_2) - s'(w_1)\| \;\leq\; \delta\|w_2 - w_1\| + \delta_2 \tag{18.80a}$$

$$\|s(w_2) - s'(w_1)\|^2 \;\leq\; \delta^2\|w_2 - w_1\|^2 + \delta_2^2 \tag{18.80b}$$

for any $w_1, w_2 \in \text{dom}(P)$. Expressions (18.79a)–(18.79b) are justified by switching the order of the expectation and subdifferentiation operators to write:

$$\partial_w\, P(w) = \partial_w \Big(\mathbb{E}\, Q(w;\boldsymbol{\gamma},\boldsymbol{h})\Big) \stackrel{(a)}{=} \mathbb{E}\Big(\partial_w\, Q(w;\boldsymbol{\gamma},\boldsymbol{h})\Big) \tag{18.81}$$

Step (a) is possible under conditions that are generally valid for our cases of interest – as was already explained in Lemma 16.1. In particular, the switching is possible whenever the loss function $Q(w;\cdot)$ is convex and bounded in neighborhoods where the subgradients are evaluated.

Proof of (18.80a)–(18.80b): Note that

$$\begin{aligned}
\|s(w_2) - s'(w_1)\|^2 &= \|\mathbb{E}\, s_Q(w_2;\boldsymbol{\gamma},\boldsymbol{h}) \;-\; \mathbb{E}\, s'_Q(w_1;\boldsymbol{\gamma},\boldsymbol{h})\|^2 \\
&\leq \mathbb{E}\,\|s_Q(w_2;\boldsymbol{\gamma},\boldsymbol{h}) - s'_Q(w_1;\boldsymbol{\gamma},\boldsymbol{h})\|^2 \\
&\stackrel{(18.75b)}{\leq} \delta^2\|w_2 - w_1\|^2 + \delta_2^2 \\
&\leq \delta^2\|w_2 - w_1\|^2 + \delta_2^2 + 2\delta\delta_2\|w_2 - w_1\| \\
&\leq (\delta\|w_2 - w_1\| + \delta_2)^2
\end{aligned} \tag{18.82}$$

■

For ease of reference, we collect in Table 18.2 the main relations and conditions described so far for nonsmooth empirical and stochastic risk minimization.

18.5 GRADIENT NOISE FOR NONSMOOTH RISKS

Using the affine-Lipschitz conditions on the subgradients of the convex loss function alone, we will now derive expressions for the first- and second-order moments of the gradient noise. For the instantaneous and mini-batch constructions (18.56a)–(18.56b), the gradient noise at iteration n is given by

(**instantaneous approximation**)

$$\boldsymbol{g}(\boldsymbol{w}_{n-1}) = s_Q(\boldsymbol{w}_{n-1};\boldsymbol{\gamma}(n),\boldsymbol{h}_n) - s(\boldsymbol{w}_{n-1}) \tag{18.83a}$$

Table 18.2 Main relations and conditions used for *nonsmooth* empirical and stochastic risk minimization problems.

Quantity	Empirical risk minimization	Stochastic risk minimization
optimization problem	$w^\star = \underset{w\in\mathbb{R}^M}{\text{argmin}} \left\{ P(w) \stackrel{\Delta}{=} \frac{1}{N}\sum_{m=0}^{N-1} Q(w;\gamma(m),h_m) \right\}$	$w^o = \underset{w\in\mathbb{R}^M}{\text{argmin}} \left\{ P(w) \stackrel{\Delta}{=} \mathbb{E}\, Q(w;\boldsymbol{\gamma},\boldsymbol{h}) \right\}$
subgradient vector	$s(w) = \frac{1}{N}\sum_{m=0}^{N-1} s_Q(w;\gamma(m),h_m)$	$s(w) = \mathbb{E}\, s_Q(w;\boldsymbol{\gamma},\boldsymbol{h})$
instantaneous approximation	$\widehat{s}(w) = s_Q(w;\boldsymbol{\gamma}(n),\boldsymbol{h}_n)$ $(\boldsymbol{\gamma}(n),\boldsymbol{h}_n)$ selected at random	$\widehat{s}(w) = s_Q(w;\boldsymbol{\gamma}(n),\boldsymbol{h}_n)$ $(\boldsymbol{\gamma}(n),\boldsymbol{h}_n)$ streaming in
mini-batch approximation	$\widehat{s}(w) = \frac{1}{B}\sum_{b=0}^{B-1} s_Q(w;\boldsymbol{\gamma}(b),\boldsymbol{h}_b)$ $\{\boldsymbol{\gamma}(b),\boldsymbol{h}_b\}$ selected at random	$\widehat{s}(w) = \frac{1}{B}\sum_{b=0}^{B-1} s_Q(w;\boldsymbol{\gamma}(b),\boldsymbol{h}_b)$ $\{\boldsymbol{\gamma}(b),\boldsymbol{h}_b\}$ streaming in
conditions on risk and loss functions	(18.66a)–(18.66b) $P(w)$ ν-strongly convex, $Q(w,\cdot)$ convex $s_Q(w;\gamma,h)$ affine-Lipschitz	(18.75a)–(18.75b) $P(w)$ ν-strongly convex, $Q(w,\cdot)$ convex $s_Q(w;\boldsymbol{\gamma},\boldsymbol{h})$ affine-Lipschitz in mean-square sense

for instantaneous subgradient approximations or by

$$\textbf{(mini-batch approximation)}$$
$$\boldsymbol{g}(\boldsymbol{w}_{n-1}) = \frac{1}{B}\sum_{b=0}^{B-1} s_Q(\boldsymbol{w}_{n-1};\boldsymbol{\gamma}(b),\boldsymbol{h}_b) - s(\boldsymbol{w}_{n-1}) \tag{18.83b}$$

for mini-batch approximations, where $(\boldsymbol{\gamma}(n),\boldsymbol{h}_n)$ and $\{\boldsymbol{\gamma}(b),\boldsymbol{h}_b\}$ denote the random data samples used at the nth iteration while updating $\boldsymbol{w}_{n-1}$ to $\boldsymbol{w}_n$. The main conclusion of this section is again to show that the second-order moment of the gradient noise is bounded in the same manner as in (18.18), i.e., as follows:

$$\boxed{\mathbb{E}\left(\|\boldsymbol{g}_n(\boldsymbol{w}_{n-1})\|^2\,|\,\boldsymbol{w}_{n-1}\right) \leq \beta_g^2\|\widetilde{w}_{n-1}\|^2 + \sigma_g^2} \tag{18.84}$$

for some nonnegative constants $\{\beta_g^2,\sigma_g^2\}$ that will be independent of the error $\widetilde{w}_{n-1} = w^\star - w_{n-1}$. More specifically, the derivations in the remainder of this section are meant to establish the following conclusion, which is similar to the statement of Lemma 18.1 for smooth risks except for the condition on the subgradients of the loss function.

Lemma 18.2. (Gradient noise under nonsmooth risks) *Consider the empirical or stochastic risk optimization problems (18.2a)–(18.2b) and assume the subgradients of the convex loss function satisfy the affine-Lipschitz conditions (18.66b) or (18.75b). The first- and second-order moments of the gradient process will satisfy:*

$$\mathbb{E}\left(\boldsymbol{g}_n(\boldsymbol{w}_{n-1})\,|\,\boldsymbol{w}_{n-1}\right) = 0 \tag{18.85a}$$
$$\mathbb{E}\left(\|\boldsymbol{g}_n(\boldsymbol{w}_{n-1})\|^2\,|\,\boldsymbol{w}_{n-1}\right) \leq \beta_g^2\|\widetilde{w}_{n-1}\|^2 + \sigma_g^2 \tag{18.85b}$$

for some nonnegative constants $\{\beta_g^2,\sigma_g^2\}$ that are independent of $\widetilde{w}_{n-1}$.

Results (18.85a)–(18.85b) hold for instantaneous or mini-batch gradient approximations, regardless of whether the samples are streaming in independently of each other, sampled uniformly with replacement, sampled without replacement, or selected under importance sampling. Again, the *only exception* is that the zero-mean property (18.85a) will not hold for the instantaneous gradient implementation when the samples are selected without replacement.

To establish properties (18.85a)–(18.85b), we proceed by examining each sampling procedure separately and then show that they all lead to the same result. We consider the zero-mean property (18.85a) first. Since the arguments are similar to what we have done in the smooth case, we will be brief.

18.5.1 First-Order Moment

We verify again, as was the case with smooth risks, that the gradient noise process has zero mean conditioned on the previous iterate, i.e.,

$$\mathbb{E}\left(\boldsymbol{g}_n(\boldsymbol{w}_{n-1})\,|\,\boldsymbol{w}_{n-1}\right) = 0 \tag{18.86}$$

where the expectation is over the randomness in sample selection.

Sampling with replacement

Consider first the case of empirical risk minimization where samples are selected uniformly from the given dataset. Let $\boldsymbol{\sigma}$ denote the sample index that is selected at iteration n with

$$\mathbb{P}(\boldsymbol{\sigma} = m) = \frac{1}{N}, \quad m \in \{0, 1, 2, \ldots, N-1\} \tag{18.87}$$

It follows that

$$\begin{aligned} \mathbb{E}\left(\widehat{s}(\boldsymbol{w}_{n-1})\,|\,\boldsymbol{w}_{n-1}\right) &= \mathbb{E}\left(s_Q(\boldsymbol{w}_{n-1}; \boldsymbol{\gamma}(\boldsymbol{\sigma}), \boldsymbol{h}_{\boldsymbol{\sigma}})\,|\,\boldsymbol{w}_{n-1}\right) \\ &= \frac{1}{N}\sum_{m=0}^{N-1} s_Q(w_{n-1}; \gamma(m), h_m) \\ &\overset{(18.57)}{=} s(w_{n-1}) \end{aligned} \tag{18.88}$$

where in the first equality we used the fact that the loss function assumes each of the values $s_Q(w_{n-1}; \gamma(m), h_m)$ with probability $1/N$. It follows that (18.86) holds for instantaneous subgradient approximations.

The gradient noise process continues to have zero conditional mean in the mini-batch implementation. This is because the approximate search direction is again unbiased:

$$\begin{aligned} \mathbb{E}\left(\widehat{s}(\boldsymbol{w}_{n-1})\,|\,\boldsymbol{w}_{n-1}\right) &= \mathbb{E}\left(\frac{1}{B}\sum_{b=0}^{B-1} s_Q(\boldsymbol{w}_{n-1}; \boldsymbol{\gamma}(b), \boldsymbol{h}_b)\,|\,\boldsymbol{w}_{n-1}\right) \\ &= \frac{1}{B}\sum_{b=0}^{B-1} \mathbb{E}\left(s_Q(\boldsymbol{w}_{n-1}; \boldsymbol{\gamma}(b), \boldsymbol{h}_b)\,|\,\boldsymbol{w}_{n-1}\right) \\ &\overset{(a)}{=} \mathbb{E}\left(s_Q(\boldsymbol{w}_{n-1}; \boldsymbol{\gamma}(\boldsymbol{\sigma}), \boldsymbol{h}_{\boldsymbol{\sigma}})\,|\,\boldsymbol{w}_{n-1}\right) \\ &= \frac{1}{N}\sum_{\sigma=0}^{N-1} s_Q(w_{n-1}; \gamma(\sigma), h_\sigma) \\ &\overset{(18.57)}{=} s(w_{n-1}) \end{aligned} \tag{18.89}$$

where in step (a) we used the fact that the data samples $(\boldsymbol{\gamma}(b), \boldsymbol{h}_b)$ are selected independently of each other.

Sampling without replacement

When samples are selected at random without replacement, we obtain for the instantaneous subgradient approximation:

$$\begin{aligned}\mathbb{E}\left(\widehat{s}(\boldsymbol{w}_{n-1})\,|\,\boldsymbol{w}_{n-1}\right) &= \mathbb{E}\left(s_Q(\boldsymbol{w}_{n-1};\boldsymbol{\gamma}(\boldsymbol{\sigma}),\boldsymbol{h}_{\boldsymbol{\sigma}})\,|\,\boldsymbol{w}_{n-1}\right)\\ &\neq \frac{1}{N}\sum_{m=0}^{N-1} s_Q(w_{n-1};\gamma(m),h_m)\\ &= s(w_{n-1})\end{aligned} \tag{18.90}$$

where the first line is not equal to the second line because, conditioned on $\boldsymbol{w}_{n-1}$, the sample index $\boldsymbol{\sigma}$ cannot be selected uniformly. As a result, the gradient noise process under random reshuffling is *biased* and does *not* have zero mean anymore.

A different conclusion holds for mini-batch implementations where the B samples in the batch are selected randomly *without* replacement. In this case, the zero-mean property for the gradient noise continues to hold, as can be verified by repeating the argument that led to (18.29) in the smooth case.

Importance sampling

Under importance sampling, the instantaneous and mini-batch subgradient approximations would be scaled as follows:

$$\textbf{(instantaneous)}\quad \widehat{s}(w) = \frac{1}{Np_{\boldsymbol{\sigma}}}s_Q(w;\boldsymbol{\gamma}(\boldsymbol{\sigma}),\boldsymbol{h}_{\boldsymbol{\sigma}}) \tag{18.91a}$$

$$\textbf{(mini-batch)}\quad \widehat{s}(w) = \frac{1}{B}\sum_{b=0}^{B-1}\frac{1}{Np_b}s_Q(w;\boldsymbol{\gamma}(b),\boldsymbol{h}_b) \tag{18.91b}$$

The scalings render these search directions unbiased so that (18.23) continues to hold. The argument is similar to the one used to establish (18.30) in the smooth case.

Streaming data

Under stochastic risk minimization, the data samples stream in *independently* of each other. As a result, the gradient noise process continues to satisfy relation (18.86) since

$$\mathbb{E}\left(s_Q(\boldsymbol{w}_{n-1};\boldsymbol{\gamma}(n),\boldsymbol{h}_n)\,|\,\boldsymbol{w}_{n-1}\right) \overset{(a)}{=} \mathbb{E}\,s_Q(w_{n-1};\boldsymbol{\gamma}(n),\boldsymbol{h}_n) \overset{(18.79a)}{=} s(w_{n-1}) \tag{18.92}$$

Step (a) is because the samples $(\boldsymbol{\gamma}(n),\boldsymbol{h}_n)$ are independent over time and, hence, they are also independent of $\boldsymbol{w}_{n-1}$ (which is a function of all previous data samples).

18.5.2 Second-Order Moment

We examine next the second-order moment of the gradient noise process under different sampling procedures and verify that it satisfies:

$$\mathbb{E}\left(\|\boldsymbol{g}_n(\boldsymbol{w}_{n-1})\|^2\,|\,\boldsymbol{w}_{n-1}\right) \;\leq\; \beta_g^2\|\widetilde{w}_{n-1}\|^2 + \sigma_g^2 \tag{18.93}$$

for some constants (β_g^2, σ_g^2) independent of $\widetilde{w}_{n-1}$. The arguments are similar to the ones used in the smooth case and, therefore, we shall be brief again.

Sampling with replacement

Consider first the case of empirical risk minimization where samples are selected uniformly from the given dataset $\{\gamma(m), h_m\}$. It follows for instantaneous subgradient approximations that:

$$\begin{aligned}
&\|\boldsymbol{g}(\boldsymbol{w}_{n-1})\|^2\\
&\leq 2\left\|\frac{1}{Np_n}s_Q(\boldsymbol{w}_{n-1}) - \frac{1}{Np_n}s_Q(w^\star) - s(\boldsymbol{w}_{n-1})\right\|^2 + 2\left\|\frac{1}{Np_n}s_Q(w^\star)\right\|^2\\
&\overset{(c)}{\leq} 4\left\|\frac{1}{Np_n}s_Q(\boldsymbol{w}_{n-1}) - \frac{1}{Np_n}s_Q(w^\star)\right\|^2 + 2\left\|\frac{1}{Np_n}s_Q(w^\star)\right\|^2\\
&\quad 4\,\|s'(w^\star) - s(\boldsymbol{w}_{n-1})\|^2\\
&\overset{(18.70b)}{\leq} 8\delta^2\|\widetilde{\boldsymbol{w}}_{n-1}\|^2 + 8\delta_2^2 + 4\left\|\frac{1}{Np_n}\Big(s_Q(\boldsymbol{w}_{n-1}) - s_Q(w^\star)\Big)\right\|^2 + 2\left\|\frac{1}{Np_n}s_Q(w^\star)\right\|^2
\end{aligned} \tag{18.94}$$

In step (a) we removed the data argument $(\gamma(\boldsymbol{\sigma}), h_{\boldsymbol{\sigma}})$ from $s_Q(w;\cdot)$ to simplify the notation and added and subtracted $s_Q(w^\star)$. In step (b) we applied Jensen inequality $\|a+b\|^2 \leq 2\|a\|^2 + 2\|b\|^2$ for any vectors (a, b), and in step (c) we added and subtracted $s'(w^\star) = 0$. We know from property (8.47) that a subgradient vector $s'(w^\star)$ for $P(w)$ exists that evaluates to zero at the minimizer. We conclude that (18.93) holds with

$$\beta_g^2 = 16\delta^2 \tag{18.95a}$$

$$\begin{aligned}
\sigma_g^2 &= 16\delta_2^2 + 2\,\mathbb{E}\left(\|s_Q(w^\star; \gamma(\boldsymbol{\sigma}), h_{\boldsymbol{\sigma}})\|^2\,|\,\boldsymbol{w}_{n-1}\right)\\
&= 16\delta_2^2 + 2\,\mathbb{E}\,\|s_Q(w^\star; \gamma(\boldsymbol{\sigma}), h_{\boldsymbol{\sigma}})\|^2\\
&\overset{(18.87)}{=} 16\delta_2^2 + \frac{2}{N}\sum_{m=0}^{N-1}\|s_Q(w^\star; \gamma(m), h_m)\|^2
\end{aligned} \tag{18.95b}$$

A similar conclusion holds for the mini-batch version with the parameters (β_g^2, σ_g^2) divided by B.

Sampling without replacement

If we repeat the same argument leading to (18.94), we will conclude that the same relation (18.93) holds, albeit with parameters:

$$\beta_g^2 = 16\delta^2 \tag{18.96a}$$

$$\sigma_g^2 = 16\delta_2^2 + 2\,\mathbb{E}\left(\|\nabla_{w^\mathsf{T}} Q(w^\star;\gamma(\boldsymbol{\sigma}),\boldsymbol{h}_{\boldsymbol{\sigma}}\|^2\,|\,\boldsymbol{w}_{n-1}\right) \tag{18.96b}$$

where the expression for σ_g^2 still involves an inconvenient conditioning on $\boldsymbol{w}_{n-1}$. We can remove the conditioning as explained earlier in (18.40) to arrive at

$$\sigma_g^2 \leq 16\delta_2^2 + 2\sum_{m=0}^{N-1} \|\nabla_{w^\mathsf{T}} Q(w^\star;\gamma(m),h_m)\|^2 \tag{18.97}$$

which is independent of $\boldsymbol{w}_{n-1}$. This result can be used as the expression for σ_g^2 in (18.93).

A similar conclusion holds for the mini-batch subgradient approximation where the B samples are randomly selected *without* replacement. The same argument leading to (18.45) will continue to hold and lead to

$$\mathbb{E}\left(\|\boldsymbol{g}_n(\boldsymbol{w}_{n-1})\|^2\,|\,\boldsymbol{w}_{n-1}\right) \;\leq\; \frac{1}{\tau_B}\left(\beta_g^2\|\widetilde{w}_{n-1}\|^2 + \sigma_g^2\right) \tag{18.98}$$

where

$$\tau_B \;\stackrel{\Delta}{=}\; \begin{cases} B, & \text{when samples are selected } \textit{with} \text{ replacement} \\ B\dfrac{N-1}{N-B}, & \text{when samples are selected } \textit{without} \text{ replacement} \end{cases} \tag{18.99}$$

We see from the second line that $\tau_B \approx B$ for N large enough.

Importance sampling

Under importance sampling, we have for implementations with instantaneous subgradient approximations:

$$\begin{aligned} &\|\boldsymbol{g}_n(\boldsymbol{w}_{n-1})\|^2 \\ &\stackrel{\Delta}{=} \|\widehat{s}(\boldsymbol{w}_{n-1}) - s(\boldsymbol{w}_{n-1})\|^2 \\ &= \left\|\frac{1}{Np_n}s_Q(\boldsymbol{w}_{n-1};\boldsymbol{\gamma}(n),\boldsymbol{h}_n) - s(\boldsymbol{w}_{n-1})\right\|^2 \\ &\stackrel{(a)}{=} \left\|\frac{1}{Np_n}s_Q(\boldsymbol{w}_{n-1}) - s(\boldsymbol{w}_{n-1})\right\|^2 \\ &\stackrel{(b)}{=} \left\|\frac{1}{Np_n}s_Q(\boldsymbol{w}_{n-1}) - \frac{1}{Np_n}s_Q(w^\star) + \frac{1}{Np_n}s_Q(w^\star) - s(\boldsymbol{w}_{n-1})\right\|^2 \end{aligned} \tag{18.100}$$

where step (a) removes the data arguments $(\boldsymbol{\gamma}(n),\boldsymbol{h}_n)$ from $s_Q(w;\cdot,\cdot)$ for convenience, and step (b) adds and subtracts the same quantity $s_Q(w^\star)$. We therefore have

$$
\begin{aligned}
&\|\boldsymbol{g}(\boldsymbol{w}_{n-1})\|^2 \\
&\leq 2\left\|\frac{1}{Np_n}s_Q(\boldsymbol{w}_{n-1}) - \frac{1}{Np_n}s_Q(w^\star) - s(\boldsymbol{w}_{n-1})\right\|^2 + 2\left\|\frac{1}{Np_n}s_Q(w^\star)\right\|^2 \\
&\overset{(c)}{\leq} 4\left\|\frac{1}{Np_n}s_Q(\boldsymbol{w}_{n-1}) - \frac{1}{Np_n}s_Q(w^\star)\right\|^2 + 2\left\|\frac{1}{Np_n}s_Q(w^\star)\right\|^2 \\
&\quad 4\left\|s'(w^\star) - s(\boldsymbol{w}_{n-1})\right\|^2 \\
&\overset{(18.70b)}{\leq} 8\delta^2\|\widetilde{\boldsymbol{w}}_{n-1}\|^2 + 8\delta_2^2 + 4\left\|\frac{1}{Np_n}\left(s_Q(\boldsymbol{w}_{n-1}) - s_Q(w^\star)\right)\right\|^2 \\
&\qquad +2\left\|\frac{1}{Np_n}s_Q(w^\star)\right\|^2
\end{aligned} \tag{18.101}
$$

where in step (c) we applied Jensen inequality $\|a+b\|^2 \leq 2\|a\|^2 + 2\|b\|^2$ for any two vectors (a, b), and used the fact that there exists a subgradient for $P(w)$ such that $s'(w^\star) = 0$. Next, we need to condition on $\boldsymbol{w}_{n-1}$ and take expectations. For that purpose, we note that

$$
\mathbb{E}\left\{2\left\|\frac{1}{Np_n}s_Q(w^\star)\right\|^2 \,\Big|\, \boldsymbol{w}_{n-1}\right\} = \frac{2}{N^2}\sum_{m=0}^{N-1}\frac{1}{p_m}\|s_Q(w^\star;\gamma(m),h_m)\|^2 \tag{18.102}
$$

and

$$
\mathbb{E}\left\{\frac{4}{N^2p_n^2}\,\|s_Q(\boldsymbol{w}_{n-1}) - s_Q(w^\star)\|^2 \,\Big|\, \boldsymbol{w}_{n-1}\right\} \leq \frac{8}{N^2}\sum_{m=0}^{N-1}\frac{1}{p_m}(\delta^2\|\widetilde{w}_{n-1}\|^2 + \delta_2^2) \tag{18.103}
$$

Substituting into (18.101) we conclude that

$$
\mathbb{E}\left(\|\boldsymbol{g}_n(\boldsymbol{w}_{n-1})\|^2 \,|\, \boldsymbol{w}_{n-1}\right) \;\leq\; \beta_g^2\|\widetilde{w}_{n-1}\|^2 + \sigma_g^2 \tag{18.104}
$$

where the parameters $\{\beta_g^2, \sigma_g^2\}$ are given by

$$
\beta_g^2 = 8\delta^2\left(1 + \frac{1}{N^2}\sum_{m=0}^{N-1}\frac{1}{p_m}\right) \tag{18.105a}
$$

$$
\sigma_g^2 = 8\delta_2^2 + \frac{2}{N^2}\sum_{m=0}^{N-1}\frac{1}{p_m}\left(\|s_Q(w^\star;\gamma(m),h_m)\|^2 + 4\delta_2^2\right) \tag{18.105b}
$$

A similar bound holds with the above parameters $\{\beta_g^2, \sigma_g^2\}$ divided by B for the mini-batch version – see Prob. 18.13.

Streaming data

Under stochastic risk minimization, the data samples stream in *independently* of each other. As a result, the gradient noise process continues to satisfy relation (18.93), as can be seen from the following sequence of inequalities:

$$
\begin{aligned}
&\mathbb{E}\left(\|\boldsymbol{g}(\boldsymbol{w}_{n-1})\|^2 \,|\, \boldsymbol{w}_{n-1}\right) \\
&\stackrel{\Delta}{=} \mathbb{E}\left(\|s_Q(\boldsymbol{w}_{n-1};\boldsymbol{\gamma}(n),\boldsymbol{h}_n) - s(\boldsymbol{w}_{n-1})\|^2 \,|\, \boldsymbol{w}_{n-1}\right) \\
&\stackrel{(a)}{=} \mathbb{E}\left(\|s_Q(\boldsymbol{w}_{n-1}) - s_Q(w^o) + s_Q(w^o) - s(\boldsymbol{w}_{n-1})\|^2 \,|\, \boldsymbol{w}_{n-1}\right) \\
&\stackrel{(b)}{\leq} 2\,\mathbb{E}\left(\|s_Q(\boldsymbol{w}_{n-1}) - s_Q(w^o) - s(w_{n-1})\|^2 \,|\, \boldsymbol{w}_{n-1}\right) + 2\,\mathbb{E}\,\|s_Q(w^o)\|^2 \\
&\stackrel{(c)}{\leq} 4\,\mathbb{E}\left(\|s_Q(\boldsymbol{w}_{n-1}) - s_Q(w^o)\|^2 \,|\, \boldsymbol{w}_{n-1}\right) + 2\,\mathbb{E}\,\|s_Q(w^o)\|^2 \\
&\qquad 4\,\|s(w_{n-1}) - s'(w^o)\|^2 \\
&\stackrel{(18.75b)}{\leq} 4\delta^2\|\widetilde{w}_{n-1}\|^2 + 4\delta_2^2 + 4\delta^2\|\widetilde{w}_{n-1}\|^2 + 4\delta_2^2 + 2\,\mathbb{E}\,\|s_Q(w^o;\gamma,h)\|^2
\end{aligned}
\tag{18.106}
$$

In step (a) we removed the argument $(\boldsymbol{\gamma}(n), \boldsymbol{h}_n)$ from $s_Q(w;\cdot,\cdot)$ to simplify the notation and added and subtracted $s_Q(w^o)$. In step (b) we applied Jensen inequality $\|a+b\|^2 \leq 2\|a\|^2 + 2\|b\|^2$ for any vectors (a,b), and in step (c) we added and subtracted $s'(w^o) = 0$. We know from property (8.47) that a subgradient vector $s'(w^o)$ exists for $P(w)$ that evaluates to zero at the minimizer. We conclude that a relation of the form (18.93) continues to hold with

$$\beta_g^2 = 8\delta^2 \tag{18.107a}$$

$$\sigma_g^2 = 8\delta_2^2 + 2\,\mathbb{E}\left(\|s_Q(w^o;\boldsymbol{\gamma},\boldsymbol{h})\|^2\right) \tag{18.107b}$$

where the expectation in σ_g^2 is over the joint distribution of the data $(\boldsymbol{\gamma}, \boldsymbol{h})$.

18.6 COMMENTARIES AND DISCUSSION

Moments of gradient noise. We established in the body of the chapter that under certain δ-Lipschitz or affine-Lipschitz conditions on the gradients or subgradients of the loss function, the second-order moment of the gradient noise process in stochastic implementations satisfies the bound

$$\mathbb{E}\left(\|\boldsymbol{g}_n(\boldsymbol{w}_{n-1})\|^2 \,|\, \boldsymbol{w}_{n-1}\right) \leq \beta_g^2\|\widetilde{w}_{n-1}\|^2 + \sigma_g^2 \tag{18.108}$$

for some parameters (β_g^2, σ_g^2) that are independent of the error vector, $\widetilde{w}_{n-1}$. It is verified in the problems that this bound holds for many cases of interest involving risks that arise frequently in inference and learning problems with and without regularization. The bound (18.108) is similar to conditions used earlier in the optimization literature. In Polyak (1987), the term involving $\|\widetilde{w}_{n-1}\|^2$ is termed the "relative noise component," while the term involving σ_g^2 is termed the "absolute noise component." In Polyak and Tsypkin (1973) and Bertsekas and Tsitsiklis (2000) it is assumed instead that the gradient noise satisfies a condition of the type:

$$\mathbb{E}\left(\|\boldsymbol{g}_n(\boldsymbol{w}_{n-1})\|^2 \,|\, \boldsymbol{w}_{n-1}\right) \leq \alpha\left(1 + \|\nabla_{w^{\mathsf{T}}} P(w_{n-1})\|^2\right) \tag{18.109}$$

for some positive constant α. One main difference is that (18.109) is introduced as an assumption in these works, whereas we established the validity of (18.108) in the chapter. We can verify that, for strongly convex risks, conditions (18.108) and (18.109) are equivalent. One direction follows from the earlier result (10.20) for δ-smooth risks that

$$\frac{1}{2\delta}\|\nabla_w P(w)\|^2 \leq P(w) - P(w^\star) \leq \frac{\delta}{2}\|\widetilde{w}\|^2 \tag{18.110}$$

where $\widetilde{w} = w^\star - w$. Substituting into the right-hand side of (18.109) we get

$$\mathbb{E}\left(\|\boldsymbol{g}_n(\boldsymbol{w}_{n-1})\|^2 \,|\, \boldsymbol{w}_{n-1}\right) \leq \alpha + \alpha\delta^2\|\widetilde{\boldsymbol{w}}_{n-1}\|^2 \tag{18.111}$$

The other direction follows similarly from property (8.29) for ν-strongly convex functions, namely,

$$\frac{\nu}{2}\|\widetilde{w}\|^2 \leq P(w) - P(w^\star) \leq \frac{1}{2\nu}\|\nabla_w P(w)\|^2 \tag{18.112}$$

Absolute and relative noise terms. The presence of *both* relative and absolute terms in the bound (18.108) is necessary in most cases of interest – see, e.g., Chen and Sayed (2012a) and Sayed (2014a). An example to this effect is treated in Prob. 18.9. Consider the quadratic stochastic risk optimization problem:

$$w^o = \underset{w\in\mathbb{R}^M}{\text{argmin}} \left\{\mathbb{E}\,(\boldsymbol{\gamma} - \boldsymbol{h}^\mathsf{T} w)^2\right\} \tag{18.113}$$

Assume the streaming data $\{\boldsymbol{\gamma}(n), \boldsymbol{h}_n\}$ arises from a linear regression model of the form $\boldsymbol{\gamma}(n) = \boldsymbol{h}_n^\mathsf{T} w^\bullet + \boldsymbol{v}(n)$, for some model $w^\bullet \in \mathbb{R}^M$, and where $\boldsymbol{h}_n$ and $\boldsymbol{v}(n)$ are zero-mean uncorrelated processes. Let $R_h = \mathbb{E}\,\boldsymbol{h}_n\boldsymbol{h}_n^\mathsf{T} > 0$, $r_{h\gamma} = \mathbb{E}\,\boldsymbol{h}_n\boldsymbol{\gamma}(n)$, and $\sigma_v^2 = \mathbb{E}\,\boldsymbol{v}^2(n)$. Moreover, $\boldsymbol{v}(n)$ is a white-noise process that is independent of all other random variables. It is shown in the problem that $w^o = w^\bullet$, which means that the solution to the optimization problem is able to recover the underlying model $w^\bullet$. A stochastic gradient algorithm with instantaneous gradient approximation can then be used to estimate $w^\bullet$ and it is verified in the same problem that the gradient noise process in this case will satisfy

$$\mathbb{E}\left(\|\boldsymbol{g}(\boldsymbol{w}_{n-1})\|^2 | \boldsymbol{w}_{n-1}\right) \leq \beta_g^2\|\widetilde{\boldsymbol{w}}_{n-1}\|^2 + \sigma_g^2 \tag{18.114}$$

where

$$\sigma_g^2 = 4\sigma_v^2\,\text{Tr}(R_h), \quad \beta_g^2 = 4\mathbb{E}\,\|R_h - \boldsymbol{h}_n\boldsymbol{h}_n^\mathsf{T}\|^2 \tag{18.115}$$

We observe that even in this case, dealing with a quadratic risk function, the upper bound includes both relative and absolute noise terms.

Affine-Lipschitz conditions. For nonsmooth risks, the affine-Lipschitz conditions (18.66b)–(18.75b) are from the work by Ying and Sayed (2018). It is customary in the literature to use a more restrictive condition that assumes the subgradient vectors $s_Q(w,\cdot)$ are uniformly bounded either in the absolute sense or in the mean-square sense depending on whether one is dealing with empirical or stochastic minimization problems – see, e.g., Bertsekas (1999), Nemirovski *et al.* (2009), Nedic and Ozdaglar (2009), Ram, Nedic, and Veeravalli (2010), Srivastava and Nedic (2011), and Agarwal *et al.* (2010). That is, in these works, it is generally imposed that

$$\|s_Q(w;\gamma,h)\| \leq G \quad \text{or} \quad \mathbb{E}\,\|s_Q(w;\boldsymbol{\gamma},\boldsymbol{h})\|^2 \leq G \tag{18.116}$$

for some constant $G \geq 0$ and for all subgradients in the subdifferential set of $Q(w,\cdot)$. We know from the result of Prob. 14.3 that the bounded subgradient assumption, $\|s_Q(w;\gamma,h\cdot)\| \leq G$, is in conflict with the ν-strong convexity assumption on $P(w)$; the latter condition implies that the subgradient norm cannot be bounded. One common way to circumvent the difficulty with the bounded requirement on the subgradients is to restrict the domain of $P(w)$ to some bounded convex set, say, $w \in \mathcal{W}$, in order to bound its subgradient vectors, and then employ a projection-based subgradient implementation. This approach can still face challenges. First, projections onto $\mathcal{W}$ may not be straightforward to perform unless the set $\mathcal{W}$ is simple enough and, second, the bound G that results on the subgradient vectors by limiting w to $\mathcal{W}$ can be loose. In our presentation, we established and adopted the more relaxed affine-Lipschitz conditions (18.66b) or (18.75b).

PROBLEMS

18.1 Consider the ℓ_2-regularized quadratic and logistic losses defined by

$$Q(w;\boldsymbol{\gamma},\boldsymbol{h}) = \begin{cases} \rho\|w\|^2 \;+\; (\boldsymbol{\gamma}-\boldsymbol{h}^{\mathsf{T}}w)^2 & \text{(quadratic)} \\ \rho\|w\|^2 \;+\; \ln\left(1+e^{-\boldsymbol{\gamma}\boldsymbol{h}^{\mathsf{T}}w}\right) & \text{(logistic)} \end{cases}$$

Verify that these losses satisfy the mean-square δ-Lipschitz condition (18.13b) for zero-mean random variables $\{\boldsymbol{\gamma},\boldsymbol{h}\}$.

18.2 Verify that the mini-batch gradient approximation (18.20) is unbiased under importance sampling conditioned on $\boldsymbol{w}_{n-1}$.

18.3 If desired, we can tighten the bound in (18.32) to $\beta_g^2 = 2\delta^2$ as follows. Use the fact that, for any scalar random variable $\boldsymbol{x}$, we have $\mathbb{E}\,(\boldsymbol{x}-\mathbb{E}\,\boldsymbol{x})^2 \le \mathbb{E}\,\boldsymbol{x}^2$ to show that

$$\mathbb{E}\left(\|\boldsymbol{g}_n(\boldsymbol{w}_{n-1})\|^2 \,|\, \boldsymbol{w}_{n-1}\right) \le \mathbb{E}\left(\|\nabla_{w^{\mathsf{T}}} Q(\boldsymbol{w}_{n-1})\|^2 \,|\, \boldsymbol{w}_{n-1}\right)$$

where we are not showing the data arguments of $Q(w,\cdot)$ for convenience. Conclude that $\mathbb{E}\left(\|\boldsymbol{g}_n(\boldsymbol{w}_{n-1})\|^2 \,|\, \boldsymbol{w}_{n-1}\right) \le 2\delta^2\|\widetilde{\boldsymbol{w}}_{n-1}\|^2 + \sigma_g^2$.

18.4 Repeat the argument that led to the second-order moment bound (18.32) for both cases of empirical and stochastic risks and establish that the fourth-order moment of the gradient noise process satisfies a similar relation, namely,

$$\mathbb{E}\left(\|\boldsymbol{g}_n(\boldsymbol{w}_{n-1})\|^4 \,|\, \boldsymbol{w}_{n-1}\right) \;\le\; \beta_{g4}^4\|\widetilde{\boldsymbol{w}}_{n-1}\|^4 + \sigma_{g4}^4$$

for some nonnegative constants $(\beta_{g4}^4, \sigma_{g4}^4)$. Show further that if the above bound on the fourth-order moment of the gradient noise process holds, then it automatically implies that the following bound on the second-order moment also holds:

$$\mathbb{E}\left(\|\boldsymbol{g}_n(\boldsymbol{w}_{n-1})\|^2 \,|\, \boldsymbol{w}_{n-1}\right) \;\le\; \beta_g^2\|\widetilde{\boldsymbol{w}}_{n-1}\|^2 + \sigma_g^2$$

where $\beta_g^2 = (\beta_{g4}^4)^{1/2}$ and $\sigma_g^2 = (\sigma_{g4}^4)^{1/2}$.

18.5 Assume the bound given in Prob. 18.4 holds for the fourth-order moment of the gradient noise process generated by a stochastic gradient algorithm with instantaneous gradient approximation. Consider instead a mini-batch implementation for smooth risks. Show that the gradient noise satisfies

$$\mathbb{E}\left(\|\boldsymbol{g}_n(\boldsymbol{w}_{n-1})\|^4 \,|\, \boldsymbol{w}_{n-1}\right) \;\le\; \frac{C_B}{B^2}\left(\beta_{g4}^4\|\widetilde{\boldsymbol{w}}_{n-1}\|^4 + \sigma_{g4}^4\right)$$

where $C_B = 3 - \frac{2}{B} \le 3$. Conclude that a B^2-fold decrease occurs in the mean-fourth moment of the gradient noise.

18.6 Consider a stochastic gradient implementation with instantaneous gradient approximation using data sampling without replacement. Show that $\frac{1}{N}\sum_{n=0}^{N-1}\boldsymbol{g}_n(w) = 0$.

18.7 Consider a stochastic gradient implementation with instantaneous gradient approximation. Assume multiple epochs are run using random reshuffling at the start of each epoch. Show that the conditional mean of the gradient noise at the *beginning* of every kth epoch satisfies $\mathbb{E}\left(\boldsymbol{g}_0(\boldsymbol{w}_{-1}^k)\,|\,\boldsymbol{w}_{-1}^k\right) = 0$.

18.8 Let $\boldsymbol{\gamma}(n)$ be a streaming sequence of binary random variables assuming the values ± 1, and let $\boldsymbol{h}_n \in \mathbb{R}^M$ be a streaming sequence of real random vectors with $R_h = \mathbb{E}\,\boldsymbol{h}_n\boldsymbol{h}_n^\mathsf{T} > 0$. Assume the random processes $\{\boldsymbol{\gamma}(n), \boldsymbol{h}_n\}$ are jointly wide-sense stationary and zero-mean. Consider the regularized logistic risk function:

$$P(w) = \frac{\rho}{2}\|w\|^2 + \mathbb{E}\ln\left(1+e^{-\boldsymbol{\gamma}\boldsymbol{h}^\mathsf{T}w}\right)$$

(a) Write down the expression for the gradient noise process, $\boldsymbol{g}_n(\boldsymbol{w}_{n-1})$, that would result from using a constant step-size stochastic gradient algorithm with instantaneous gradient approximation.

(b) Verify from first principles that this noise process satisfies

$$\mathbb{E}\left(\boldsymbol{g}_n(\boldsymbol{w}_{n-1})\,|\,\boldsymbol{w}_{n-1}\right) = 0$$
$$\mathbb{E}\left(\|\boldsymbol{g}_n(\boldsymbol{w}_{n-1})\|^2\,|\,\boldsymbol{w}_{n-1}\right) \le \beta_g^2\,\|\widetilde{\boldsymbol{w}}_{n-1}\|^2 + \sigma_g^2$$

for some nonnegative constants β_g^2 and σ_g^2.

(c) Verify also that the fourth-order moment of the gradient noise process satisfies

$$\mathbb{E}\left(\|\boldsymbol{g}_n(\boldsymbol{w}_{n-1})\|^4\,|\,\boldsymbol{w}_{n-1}\right) \le \beta_{g4}^4\,\|\widetilde{\boldsymbol{w}}_{n-1}\|^4 + \sigma_{g4}^4$$

for some nonnegative constants β_{g4}^4 and σ_{g4}^4. What conditions on the moments of the data are needed to ensure this result?

(d) Define $R_{g,n}(\boldsymbol{w}) = \mathbb{E}\left(\boldsymbol{g}_n(\boldsymbol{w})\boldsymbol{g}_n^\mathsf{T}(\boldsymbol{w})\,|\,\boldsymbol{w}_{n-1}\right)$, which denotes the conditional second-order moment of the gradient noise process. Show that

$$\left\|\nabla_w^2\, P(w^o+\Delta w) - \nabla_w^2\, P(w^o)\right\| \le \kappa_1\,\|\Delta w\|$$
$$\|R_{g,n}(w^o+\Delta w) - R_{g,n}(w^o)\| \le \kappa_2\,\|\Delta w\|^\alpha$$

for small perturbations $\|\Delta w\| \le \epsilon$ and for some constants $\kappa_1 \ge 0$, $\kappa_2 \ge 0$, and positive exponent α. What conditions on the moments of the data are needed to ensure these results?

18.9 Consider the quadratic stochastic risk optimization problem:

$$w^o = \underset{w\in\mathbb{R}^M}{\text{argmin}}\ \mathbb{E}\,(\boldsymbol{\gamma} - \boldsymbol{h}^\mathsf{T}w)^2$$

Assume the streaming data $\{\boldsymbol{\gamma}(n), \boldsymbol{h}_n\}$ arise from a linear regression model of the form $\boldsymbol{\gamma}(n) = \boldsymbol{h}_n^\mathsf{T}w^\bullet + \boldsymbol{v}(n)$, for some model parameter $w^\bullet \in \mathbb{R}^M$ and where $\boldsymbol{h}_n$ and $\boldsymbol{v}_n$ are zero-mean uncorrelated processes. Let $R_h = \mathbb{E}\,\boldsymbol{h}_n\boldsymbol{h}_n^\mathsf{T} > 0$, $r_{h\gamma} = \mathbb{E}\,\boldsymbol{h}_n\boldsymbol{\gamma}(n)$, and $\sigma_v^2 = \mathbb{E}\,\boldsymbol{v}^2(n)$. Moreover, $\boldsymbol{v}(n)$ is a white-noise process that is independent of all other random variables.

(a) Show that $w^o = w^\bullet$. That is, show that the optimal solution w^o is able to recover the underlying model $w^\bullet$.

(b) Verify that the gradient noise is $\boldsymbol{g}_n(\boldsymbol{w}_{n-1}) = 2(R_h - \boldsymbol{h}_n\boldsymbol{h}_n^\mathsf{T})\widetilde{\boldsymbol{w}}_{n-1} - 2\boldsymbol{h}_n\boldsymbol{v}(n)$.

(c) Show that $\mathbb{E}\,(\boldsymbol{g}_n(\boldsymbol{w}_{n-1})\,|\,\boldsymbol{w}_{n-1}) = 0$.

(d) Show that $\mathbb{E}\,(\|\boldsymbol{g}(\boldsymbol{w}_{n-1})\|^2|\boldsymbol{w}_{n-1}) \le \beta_g^2\|\widetilde{\boldsymbol{w}}_{n-1}\|^2 + \sigma_g^2$ where $\sigma_g^2 = 4\sigma_v^2\,\text{Tr}(R_h)$ and $\beta_g^2 = 4\mathbb{E}\,\|R_h - \boldsymbol{h}_n\boldsymbol{h}_n^\mathsf{T}\|^2$.

18.10 Consider the ℓ_1-regularized quadratic and logistic losses defined by

$$Q(w;\gamma,h) = \begin{cases} \alpha\|w\|_1 + (\gamma - \boldsymbol{h}^\mathsf{T}w)^2 & \text{(quadratic)} \\ \alpha\|w\|_1 + \ln\left(1+e^{-\gamma\boldsymbol{h}^\mathsf{T}w}\right) & \text{(logistic)} \end{cases}$$

Verify that these losses satisfy the mean-square affine-Lipschitz condition (18.75b) for some (δ, δ_2). *Remark.* For a related discussion, see Ying and Sayed (2018).

18.11 Consider the ℓ_2-regularized hinge loss

$$Q(w;\boldsymbol{\gamma},\boldsymbol{h}) = \rho\|w\|^2 + \max\{0, 1-\boldsymbol{\gamma}\boldsymbol{h}^{\mathsf{T}}w\}$$

Verify that this loss satisfies the mean-square affine-Lipschitz condition (18.75b) for some (δ,δ_2). *Remark.* For a related discussion, see Ying and Sayed (2018).

18.12 Consider the quadratic, perceptron, and hinge losses defined by

$$Q(w;\gamma(m),h_m) = \begin{cases} q(w) + (\gamma(n) - h_m^{\mathsf{T}}w)^2 & \text{(quadratic)} \\ q(w) + \max\{0, -\gamma(m)h_m^{\mathsf{T}}w\} & \text{(perceptron)} \end{cases}$$

Show that these losses satisfy the affine-Lipschitz condition (18.66b) under ℓ_1, ℓ_2, or elastic-net regularization. Determine for each case the respective values for $\{\delta,\delta_2\}$.

18.13 Refer to the bound (18.104) derived for a stochastic gradient implementation under importance sampling. Repeat the derivation assuming instead a mini-batch implementation where the B samples are selected with replacement. Show that the same bound holds with $\{\beta_g^2,\sigma_g^2\}$ divided by B.

18.14 Refer to the statement of Lemma 18.3 and let β be any nonnegative constant. Verify that

$$\mathbb{E}\left\|\sum_{j=1}^{B}\beta^{B-j}\boldsymbol{x}_{\boldsymbol{\sigma}(j)}\right\|^2 = \frac{1}{N(N-1)}\times\left(N\sum_{j=0}^{B-1}\beta^{2j} - \left(\sum_{j=0}^{B-1}\beta^j\right)^2\right)\times\sum_{n=0}^{N-1}\|x_n\|^2$$

18.A AVERAGING OVER MINI-BATCHES

In this appendix we establish the validity of the third step in the argument leading to conclusion (18.29). To do so, we need to validate the equality:

$$\sum_{\ell=1}^{L}\left(\frac{1}{B}\sum_{b\in\mathcal{B}_\ell}\nabla_{w^{\mathsf{T}}}Q(w;\gamma(b),h_b)\right) = \frac{C_{N-1}^{B-1}}{B}\sum_{m=0}^{N-1}\nabla_{w^{\mathsf{T}}}Q(w;\gamma(m),h_m) \tag{18.117}$$

where C_{N-1}^{B-1} is the combinatorial coefficient for choosing $B-1$ data points out of $N-1$ total samples. We simplify the notation and denote the data point $(\gamma(m),h_m)$ by the letter x_m. We also introduce the symbols

$$q(w;x_m) \triangleq \nabla_{w^{\mathsf{T}}}Q(w;x_m) \tag{18.118a}$$

$$q^{\mathcal{B}_\ell}(w) \triangleq \frac{1}{B}\sum_{b\in B_\ell}\nabla_{w^{\mathsf{T}}}Q(w;x_b) = \frac{1}{B}\sum_{b\in B_\ell}q(w;x_b) \tag{18.118b}$$

so that we are interested in establishing the identity:

$$\sum_{\ell=1}^{L}q^{\mathcal{B}_\ell}(w) = \frac{C_{N-1}^{B-1}}{B}\sum_{m=0}^{N-1}q(w;x_m) \tag{18.119}$$

Consider first a few illustrative examples. Assume there are $N=3$ data samples x_0, x_1, and x_2 and that the mini-batch size is $B=2$. Then, there are $L=3$ candidate mini-batches $\{\mathcal{B}_1,\mathcal{B}_2,\mathcal{B}_3\}$ and, for this case,

$$q^{\mathcal{B}_1}(w) = \frac{1}{2}\Big(q(w;x_0)+q(w;x_1)\Big) \tag{18.120a}$$

$$q^{\mathcal{B}_2}(w) = \frac{1}{2}\Big(q(w;x_0)+q(w;x_2)\Big) \tag{18.120b}$$

$$q^{\mathcal{B}_3}(w) = \frac{1}{2}\Big(q(w;x_1)+q(w;x_2)\Big) \tag{18.120c}$$

As a result, it holds that

$$\sum_{\ell=1}^{L} q^{\mathcal{B}_\ell}(w) = q(w;x_0) + q(w;x_1) + q(w;x_2) = \sum_{m=0}^{N-1} q(w;x_m) \tag{18.121}$$

which satisfies (18.119). Assume next that there are $N = 4$ data samples x_0, x_1, x_2 and x_3 with the size of the mini-batch still at $B = 2$. Then, there are $L = C_4^2 = 6$ candidate mini-batches with:

$$q^{\mathcal{B}_1}(w) = \frac{1}{2}\Big(q(w;x_0) + q(w;x_1)\Big) \tag{18.122a}$$

$$q^{\mathcal{B}_2}(w) = \frac{1}{2}\Big(q(w;x_0) + q(w;x_2)\Big) \tag{18.122b}$$

$$q^{\mathcal{B}_3}(w) = \frac{1}{2}\Big(q(w;x_0) + q(w;x_3)\Big) \tag{18.122c}$$

$$q^{\mathcal{B}_4}(w) = \frac{1}{2}\Big(q(w;x_1) + q(w;x_2)\Big) \tag{18.122d}$$

$$q^{\mathcal{B}_5}(w) = \frac{1}{2}\Big(q(w;x_1) + q(w;x_3)\Big) \tag{18.122e}$$

$$q^{\mathcal{B}_6}(w) = \frac{1}{2}\Big(q(w;x_2) + q(w;x_3)\Big) \tag{18.122f}$$

As a result, it holds that

$$\sum_{\ell=1}^{L} q^{\mathcal{B}_\ell}(w) = \frac{1}{2}\Big(3q(w;x_0) + 3q(w;x_1) + 3q(w;x_2) + 3q(w;x_3)\Big) = \frac{3}{2}\sum_{m=0}^{N-1} q(w;x_m) \tag{18.123}$$

which again satisfies (18.119). In the third example, we assume there are $N = 4$ data samples x_0, x_1, x_2, and x_3 and increase the mini-batch size to $B = 3$. Then, there are $L = C_4^3 = 4$ candidate mini-batches with:

$$q^{\mathcal{B}_1}(w) = \frac{1}{3}\Big(q(w;x_0) + q(w;x_1) + q(w;x_2)\Big) \tag{18.124a}$$

$$q^{\mathcal{B}_2}(w) = \frac{1}{3}\Big(q(w;x_0) + q(w;x_1) + q(w;x_3)\Big) \tag{18.124b}$$

$$q^{\mathcal{B}_3}(w) = \frac{1}{3}\Big(q(w;x_0) + q(w;x_2) + q(w;x_3)\Big) \tag{18.124c}$$

$$q^{\mathcal{B}_4}(w) = \frac{1}{3}\Big(q(w;x_1) + q(w;x_2) + q(w;x_3)\Big) \tag{18.124d}$$

It follows that the following result holds, which satisfies (18.119):

$$\sum_{\ell=1}^{L} q^{\mathcal{B}_\ell}(w) = \frac{1}{3}\Big(3q(w;x_0) + 3q(w;x_1) + 3q(w;x_2) + 3q(w;x_3)\Big) = \sum_{m=0}^{N-1} q(w;x_m) \tag{18.125}$$

Let us consider next the general scenario with N data samples and mini-batches of size B. Then, there are

$$L = C_N^B = \frac{N!}{B!(N-B)!} \tag{18.126}$$

candidate batches. We denote these batches by $\mathcal{B}_1, \mathcal{B}_2, \ldots, \mathcal{B}_L$. It then holds that

$$\sum_{\ell=1}^{L} q^{\mathcal{B}_\ell}(w) = \sum_{\ell=1}^{L}\left(\frac{1}{B}\sum_{b\in B_\ell} q(w;x_b)\right) \tag{18.127a}$$

$$= \frac{1}{B}\sum_{\ell=1}^{L}\sum_{\ell=1}^{B} q(w;x_b)$$

$$= \frac{1}{B}\Big(\alpha_0 q(w;x_0) + \alpha_1 q(w;x_1) + \cdots + \alpha_{N-1} q(w;x_{N-1})\Big)$$

where the $\{\alpha_m\}$ are integers; each α_m counts how many times the term $q(w;x_m)$ appears in (18.127a). A critical observation here is that α_m is equal to the number of mini-batches that involve the data sample x_m, as is evident from the previous examples. Thus, suppose x_m is already selected. Then, the number of mini-batches that will contain x_m can be determined by counting in how many ways $B-1$ data samples (that exclude x_m) can be selected from the remaining $N-1$ data samples. This number is given by C_{N-1}^{B-1}. That is,

$$\alpha_0 = \cdots = \alpha_{N-1} = C_{N-1}^{B-1} = \frac{(N-1)!}{(B-1)!(N-B)!} \tag{18.128}$$

from which we conclude that (18.117) holds.

18.B AUXILIARY VARIANCE RESULT

In this appendix we establish the following result.

Lemma 18.3. (Variance expression) *Consider N vectors $\{x_0, x_1, \ldots, x_{N-1}\}$ satisfying*

$$\frac{1}{N}\sum_{n=0}^{N-1} x_n = 0 \tag{18.129}$$

Assume we sample B of the vectors without replacement and obtain the random sequence $\{\boldsymbol{x}_{\boldsymbol{\sigma}(1)}, \boldsymbol{x}_{\boldsymbol{\sigma}(2)}, \ldots, \boldsymbol{x}_{\boldsymbol{\sigma}(B)}\}$. Then, it holds that:

$$\mathbb{E}\left\|\sum_{j=1}^{B} \boldsymbol{x}_{\boldsymbol{\sigma}(j)}\right\|^2 = \frac{B(N-B)}{N(N-1)}\sum_{n=0}^{N-1}\|x_n\|^2 \tag{18.130}$$

Proof: The proof employs mathematical induction and follows the derivation from Ying *et al.* (2018). We introduce the notation:

$$f(B) \triangleq \mathbb{E}\left\|\sum_{j=1}^{B} \boldsymbol{x}_{\boldsymbol{\sigma}(j)}\right\|^2 \tag{18.131}$$

and note that for any single sample selected at random from the collection of N samples:

$$f(1) = \mathbb{E}\,\|\boldsymbol{x}_{\boldsymbol{\sigma}(j)}\|^2 = \frac{1}{N}\sum_{n=0}^{N-1}\|x_n\|^2 \triangleq \text{var}(x) \tag{18.132}$$

where we are using the notation var(x) to refer to the average squared value of the samples. It follows that (18.130) holds for $B = 1$. Next we assume result (18.130) holds up to B and establish that it also holds at $B + 1$. Indeed, note that, by definition,

$$\begin{aligned}
f(B+1) &= \mathbb{E}\left\|\sum_{j=1}^{B+1} \boldsymbol{x}_{\boldsymbol{\sigma}(j)}\right\|^2 \\
&= \mathbb{E}\left\|\sum_{j=1}^{B} x_{\boldsymbol{\sigma}(j)} + x_{\boldsymbol{\sigma}(B+1)}\right\|^2 \\
&= \mathbb{E}\left\|\sum_{j=1}^{B} x_{\boldsymbol{\sigma}(j)}\right\|^2 + \mathbb{E}\,\|x_{\boldsymbol{\sigma}(B+1)}\|^2 + 2\,\mathbb{E}\left(\sum_{j=1}^{B} x_{\boldsymbol{\sigma}(j)}\right)^{\mathsf{T}} x_{\boldsymbol{\sigma}(B+1)} \\
&= f(B) + \text{var}(x) + 2\,\mathbb{E}\left(\sum_{j=1}^{B} x_{\boldsymbol{\sigma}(j)}\right)^{\mathsf{T}} x_{\boldsymbol{\sigma}(B+1)}
\end{aligned} \tag{18.133}$$

where we used

$$\mathbb{E}\,\|x_{\boldsymbol{\sigma}(B+1)}\|^2 \overset{(18.132)}{=} \text{var}(x) \tag{18.134}$$

We introduce the notation $\boldsymbol{\sigma}(1\colon B)$ to denote the collection of sample indices selected during steps 1 through B. To evaluate the last cross term in (18.133), we exploit the conditional mean property $\mathbb{E}\,\boldsymbol{a} = \mathbb{E}\,(\mathbb{E}\,(\boldsymbol{a}|\boldsymbol{b}))$ for any two random variables $\boldsymbol{a}$ and $\boldsymbol{b}$ to write:

$$\begin{aligned}
&\mathbb{E}\left(\sum_{j=1}^{B} x_{\boldsymbol{\sigma}(j)}\right)^{\mathsf{T}} x_{\boldsymbol{\sigma}(B+1)} \\
&= \mathbb{E}_{\boldsymbol{\sigma}(1:B)}\left[\mathbb{E}_{\boldsymbol{\sigma}(B+1)}\left\{\left(\sum_{j=1}^{B} x_{\boldsymbol{\sigma}(j)}\right)^{\mathsf{T}} x_{\boldsymbol{\sigma}(B+1)} \,\middle|\, \boldsymbol{\sigma}(1:B)\right\}\right] \\
&\overset{(a)}{=} \mathbb{E}_{\boldsymbol{\sigma}(1:B)}\left[\left(\sum_{j=1}^{B} x_{\boldsymbol{\sigma}(j)}\right)^{\mathsf{T}}\left(\frac{1}{N-B}\sum_{j'\notin\boldsymbol{\sigma}(1:B)} x_{j'}\right)\right] \\
&= \frac{1}{N-B}\mathbb{E}_{\boldsymbol{\sigma}(1:B)}\left[\left(\sum_{j=1}^{B} x_{\boldsymbol{\sigma}(j)}\right)^{\mathsf{T}}\left(\sum_{j'=0}^{N-1} x_{j'} - \sum_{j'=1}^{B} x_{\boldsymbol{\sigma}(j')}\right)\right] \\
&\overset{(b)}{=} -\frac{1}{N-B}\mathbb{E}_{\boldsymbol{\sigma}(1:B)}\left[\left(\sum_{j=1}^{B} x_{\boldsymbol{\sigma}(j)}\right)^{\mathsf{T}}\sum_{j'=1}^{B} x_{\boldsymbol{\sigma}(j')}\right] \\
&= -\frac{1}{N-B}\mathbb{E}_{\boldsymbol{\sigma}(1:B)}\left(\sum_{j=1}^{B}\|x_{\boldsymbol{\sigma}(j)}\|^2\right) - \\
&\qquad \frac{1}{N-B}\mathbb{E}_{\boldsymbol{\sigma}(1:B)}\left[\sum_{j=1}^{B}\left(\sum_{j'=1,j'\neq j}^{B} x_{\boldsymbol{\sigma}(j)}^{\mathsf{T}} x_{\boldsymbol{\sigma}(j')}\right)\right] \\
&\overset{(18.132)}{=} -\frac{B}{N-B}\text{var}(x) - \frac{1}{N-B}\mathbb{E}_{\boldsymbol{\sigma}(1:B)}\left[\sum_{j=1}^{B}\left(\sum_{j'=1,j'\neq j}^{B} x_{\boldsymbol{\sigma}(j)}^{\mathsf{T}} x_{\boldsymbol{\sigma}(j')}\right)\right]
\end{aligned} \tag{18.135}$$

where in step (a) we used the fact that

$$\mathbb{E}\left(x_{\boldsymbol{\sigma}(B+1)}\,|\,\boldsymbol{\sigma}(1:B)\right) = \frac{1}{N-B}\sum_{j'\notin\boldsymbol{\sigma}(1:B)} x_{j'} \tag{18.136}$$

since the expectation is over the distribution of $x_{\boldsymbol{\sigma}(B+1)}$, and in step (b) we used the condition

$$\sum_{j'=0}^{N-1} x_{j'} = 0 \tag{18.137}$$

We continue with (18.135). Without loss of generality, we assume $j < j'$ in the following argument. If $j > j'$, exchanging the places of $x_{\boldsymbol{\sigma}(j)}$ and $x_{\boldsymbol{\sigma}(j')}$ leads to the same conclusion:

$$\begin{aligned}
\mathbb{E}_{\boldsymbol{\sigma}(1:B)}\left(x^{\mathsf{T}}_{\boldsymbol{\sigma}(j)}x_{\boldsymbol{\sigma}(j')}\right) &= \mathbb{E}_{\boldsymbol{\sigma}(j),\boldsymbol{\sigma}(j')}\left(x^{\mathsf{T}}_{\boldsymbol{\sigma}(j)}x_{\boldsymbol{\sigma}(j')}\right)\\
&= \mathbb{E}_{\boldsymbol{\sigma}(j)}\left\{x^{\mathsf{T}}_{\boldsymbol{\sigma}(j)}\left(\mathbb{E}_{\boldsymbol{\sigma}(j')}[x_{\boldsymbol{\sigma}(j')}\,|\,\boldsymbol{\sigma}(j)]\right)\right\}\\
&= \mathbb{E}_{\boldsymbol{\sigma}(j)}\left\{x^{\mathsf{T}}_{\boldsymbol{\sigma}(j)}\left(\frac{1}{N-1}\sum_{j'\neq j}^{N-1} x_n\right)\right\}\\
&= \mathbb{E}_{\boldsymbol{\sigma}(j)}\left\{x^{\mathsf{T}}_{\boldsymbol{\sigma}(j)}\left(\frac{1}{N-1}\left[\sum_{j'=1}^{N-1} x_{j'} - x_{\boldsymbol{\sigma}(j)}\right]\right)\right\}\\
&\overset{(18.137)}{=} -\frac{1}{N-1}\mathbb{E}_{\boldsymbol{\sigma}(j)}\|x_{\boldsymbol{\sigma}(j)}\|^2\\
&= -\frac{1}{N-1}\mathrm{var}(x)
\end{aligned} \tag{18.138}$$

Substituting (18.138) into (18.135), we obtain:

$$\mathbb{E}\left(\sum_{j=1}^{B} x_{\boldsymbol{\sigma}(j)}\right)^{\mathsf{T}} x_{\boldsymbol{\sigma}(B+1)} = -\frac{B}{N-1}\mathrm{var}(x) \tag{18.139}$$

Combining (18.133), (18.134), and (18.139), we get:

$$\begin{aligned}
f(B+1) &= f(B) + \mathrm{var}(x) - \frac{2B}{N-1}\mathrm{var}(x)\\
&\overset{(a)}{=} \frac{B(N-B)}{N-1}\mathrm{var}(x) + \mathrm{var}(x) - \frac{2B}{N-1}\mathrm{var}(x)\\
&= \left(\frac{(B+1)(N-B-1)}{N-1}\right)\mathrm{var}(x)
\end{aligned} \tag{18.140}$$

where in step (a) we used the induction assumption on $f(B)$ and form (18.130). The same form turns out to be valid for $f(B+1)$ and we conclude that (18.130) is valid. ∎

REFERENCES

Agarwal, A., P. L. Bartlett, P. Ravikumar, and M. J. Wainwright (2012), "Information-theoretic lower bounds on the oracle complexity of convex optimization," *IEEE Trans. Inf. Theory*, vol. 58, no. 5, pp. 3235–3249.

Bertsekas, D. P. (1999), *Nonlinear Programming*, 2nd ed., Athena Scientific.
Bertsekas, D. P. and J. N. Tsitsiklis (2000), "Gradient convergence in gradient methods with errors," *SIAM J. Optim.*, vol. 10, no. 3, pp. 627–642.
Chen, J. and A. H. Sayed (2012a), "Diffusion adaptation strategies for distributed optimization and learning over networks," *IEEE Trans. Signal Process.*, vol. 60, no. 8, pp. 4289–4305.
Nedic, A. and A. Ozdaglar (2009), "Distributed subgradient methods for multiagent optimization," *IEEE Trans. Aut. Control*, vol. 54, no. 1, pp. 48–61.
Nemirovski, A. S., A. Juditsky, G. Lan, and A. Shapiro (2009), "Robust stochastic approximation approach to stochastic programming," *SIAM J. Optim.*, vol. 19, no. 4, pp. 1574–1609.
Polyak, B. T. (1987), *Introduction to Optimization*, Optimization Software.
Polyak, B. T. and Y. Z. Tsypkin (1973), "Pseudogradient adaptation and training algorithms," *Aut. Remote Control*, vol. 12, pp. 83–94.
Ram, S. S., A. Nedic, and V. V. Veeravalli (2010), "Distributed stochastic subgradient projection algorithms for convex optimization," *J. Optim. Theory Appl.*, vol. 147, no. 3, pp. 516–545.
Sayed, A. H. (2014a), *Adaptation, Learning, and Optimization over Networks,* Foundations and Trends in Machine Learning, NOW Publishers, vol. 7, no. 4–5, pp. 311–801.
Srivastava, K. and A. Nedic (2011), "Distributed asynchronous constrained stochastic optimization," *IEEE J. Sel. Topics. Signal Process.*, vol. 5, no. 4, pp. 772–790.
Ying, B. and A. H. Sayed (2018), "Performance limits of stochastic sub-gradient learning, part I: Single-agent case," *Signal Process.*, vol. 144, pp. 271–282.
Ying, B., K. Yuan, S. Vlaski, and A. H. Sayed (2018), "Stochastic learning under random reshuffling with constant step-sizes," *IEEE Trans. Signal Process.*, vol. 67, no. 2, pp. 474–489.

19 Convergence Analysis I: Stochastic Gradient Algorithms

We are ready to examine the convergence behavior of the stochastic gradient algorithm for smooth risks under various operation modes. We will consider updates with constant and vanishing step sizes. We will also consider data sampling with and without replacement, as well as under importance sampling. We will further consider instantaneous and mini-batch gradient approximations. In all cases, the main conclusion will be that the mean-square error (MSE), $\mathbb{E}\,\|\widetilde{\boldsymbol{w}}_n\|^2$, is guaranteed to approach a small $O(\mu)$-neighborhood, while exact convergence of $\boldsymbol{w}_n$ to $w^\star$ can be guaranteed for some vanishing step-size sequences. These are reassuring results in that the deterioration due to the stochastic gradient approximations remains small, which explains in large part the explosive success of stochastic approximation methods in inference and learning.

19.1 PROBLEM SETTING

We start our exposition by recalling the problem formulation, and the conditions imposed on the risk and loss functions. We also recall the constructions for the gradient approximations, and the first- and second-order moment results derived in the last chapter for the gradient noise process.

19.1.1 Risk Minimization Problems

To begin with, in this chapter, we are interested in examining the convergence behavior of the stochastic gradient implementation:

$$\boxed{\boldsymbol{w}_n = \boldsymbol{w}_{n-1} - \mu\,\widehat{\nabla_{w^\mathsf{T}} P}(\boldsymbol{w}_{n-1}),\; n \geq 0} \tag{19.1}$$

with constant μ, or even decaying step sizes $\mu(n)$, for the solution of convex optimization problems of the form:

$$w^\star = \underset{w \in \mathbb{R}^M}{\operatorname{argmin}}\; P(w) \tag{19.2}$$

where $P(w)$ is a first-order differentiable empirical or stochastic risk, i.e., for the solution of:

$$w^\star \triangleq \underset{w\in\mathbb{R}^M}{\text{argmin}} \left\{ P(w) \triangleq \frac{1}{N}\sum_{m=0}^{N-1} Q(w;\gamma(m),h_m) \right\} \tag{19.3a}$$

$$w^o \triangleq \underset{w\in\mathbb{R}^M}{\text{argmin}} \left\{ P(w) \triangleq \mathbb{E}\, Q(w;\boldsymbol{\gamma},\boldsymbol{h}) \right\} \tag{19.3b}$$

Observe that we use $w^\star$ to refer to the minimizer in the empirical case, and w^o for the minimizer in the stochastic case. Often, when there is no room for confusion, we will use $w^\star$ to refer generically to the minimizer of $P(w)$ independent of whether it represents an empirical or stochastic risk. In the above expressions, $Q(w,\cdot)$ denotes the loss function, $\{\gamma(m),h_m\}$ refers to a collection of N data points with $\gamma(m)\in\mathbb{R}$ and $h_m\in\mathbb{R}^M$, and the expectation in the second line is over the joint distribution of $\{\boldsymbol{\gamma},\boldsymbol{h}\}$.

19.1.2 Gradient Vector Approximations

The gradient search direction will be approximated by using either instantaneous or mini-batch calculations, namely,

(*approximations under sampling with and without replacement*)

$$\textbf{(instantaneous)}\quad \widehat{\nabla_{w^\mathsf{T}} P}(w) = \nabla_{w^\mathsf{T}} Q(w;\boldsymbol{\gamma},\boldsymbol{h}) \tag{19.4a}$$

$$\textbf{(mini-batch)}\quad \widehat{\nabla_{w^\mathsf{T}} P}(w) = \frac{1}{B}\sum_{b=0}^{B-1} \nabla_{w^\mathsf{T}} Q(w;\boldsymbol{\gamma}(b),\boldsymbol{h}_b) \tag{19.4b}$$

where the boldface notation $(\boldsymbol{\gamma},\boldsymbol{h})$ or $(\boldsymbol{\gamma}(b),\boldsymbol{h}_b)$ refers to data samples selected at random (with or without replacement) from the given dataset $\{\gamma(m),h_m\}$ in empirical risk minimization, or assumed to stream in independently over time in stochastic risk minimization. The difference between the true gradient and its approximation is *gradient noise*, denoted by

$$\boldsymbol{g}(w) \triangleq \widehat{\nabla_{w^\mathsf{T}} P}(w) - \nabla_{w^\mathsf{T}} P(w) \tag{19.5}$$

When the stochastic gradient algorithm operates under importance sampling, the gradient approximations are further scaled by $1/Np_b$, where p_b is the probability of selecting sample $(\boldsymbol{\gamma}(b),\boldsymbol{h}_b)$:

(*approximations under importance sampling*)

$$\textbf{(instantaneous)}\quad \widehat{\nabla_{w^\mathsf{T}} P}(w) = \frac{1}{Np}\nabla_{w^\mathsf{T}} Q(w;\boldsymbol{\gamma},\boldsymbol{h}) \tag{19.6a}$$

$$\textbf{(mini-batch)}\quad \widehat{\nabla_{w^\mathsf{T}} P}(w) = \frac{1}{B}\sum_{b=0}^{B-1} \frac{1}{Np_b}\nabla_{w^\mathsf{T}} Q(w;\boldsymbol{\gamma}(b),\boldsymbol{h}_b) \tag{19.6b}$$

19.1.3 Conditions on Risk and Loss Functions

In Section 18.2 we introduced the following conditions for empirical risk minimization problems of the form (19.3a):

(A1) (**Strongly convex risk**). $P(w)$ is ν-strongly convex and first-order differentiable, namely, for every $w_1, w_2 \in \text{dom}(P)$:

$$P(w_2) \geq P(w_1) + (\nabla_{w^\mathsf{T}} P(w_1))^\mathsf{T}(w_2 - w_1) + \frac{\nu}{2}\|w_2 - w_1\|^2 \quad (19.7\text{a})$$

(A2) (**δ-Lipschitz loss gradients**). The gradient vectors of $Q(w, \cdot)$ are δ-Lipschitz regardless of the data argument, i.e.,

$$\|\nabla_w Q(w_2; \gamma(k), h_k) - \nabla_w Q(w_1; \gamma(\ell), h_\ell)\| \leq \delta\,\|w_2 - w_1\| \quad (19.7\text{b})$$

for any $w_1, w_2 \in \text{dom}(Q)$, any $0 \leq k, \ell \leq N-1$, and with $\delta \geq \nu$. We explained that condition (19.7b) implies that the gradient of $P(w)$ is itself δ-Lipschitz:

$$\|\nabla_w P(w_2) - \nabla_w P(w_1)\| \leq \delta\|w_2 - w_1\| \quad (19.8)$$

On the other hand, for stochastic risk minimization problems of the form (19.3b), we continue to assume the strong convexity of $P(w)$ but replace (**A2**) by the requirement that the gradients of the loss are now δ-Lipschitz in the mean-square sense:

(A2') (**Mean-square δ-Lipschitz loss gradients**). The gradient vectors of $Q(w, \cdot)$ satisfy the mean-square bound:

$$\mathbb{E}\,\|\nabla_w Q(w_2; \boldsymbol{\gamma}, \boldsymbol{h}) - \nabla_w Q(w_1; \boldsymbol{\gamma}, \boldsymbol{h})\|^2 \leq \delta^2\|w_2 - w_1\|^2 \quad (19.9)$$

for any $w_1, w_2 \in \text{dom}(Q)$ and with $\delta \geq \nu$. We further showed that condition (19.9) implies that the gradients of $P(w)$ are δ-Lipschitz as well:

$$\|\nabla_w P(w_2) - \nabla_w P(w_1)\| \leq \delta\|w_2 - w_1\| \quad (19.10)$$

which is the same condition (19.8) under empirical risk minimization.

Note that under conditions (**A1**,**A2**) for empirical risk minimization or (**A1**,**A2'**) for stochastic risk minimization, the following two conditions hold:

(**P1**) ν-strong convexity of $P(w)$ as in (19.7a) (19.11a)

(**P2**) δ-Lipschitz gradients for $P(w)$ as in (19.8) and (19.10) (19.11b)

Moreover, we know from the earlier results (8.29) and (10.20) derived for strongly convex and δ-smooth functions that conditions (**P1**) and (**P2**) imply, respectively:

$$(\textbf{P1}) \quad \Longrightarrow \quad \frac{\nu}{2}\|\widetilde{w}\|^2 \leq P(w) - P(w^\star) \leq \frac{1}{2\nu}\|\nabla_w P(w)\|^2 \quad (19.12\text{a})$$

$$(\textbf{P2}) \quad \Longrightarrow \quad \frac{1}{2\delta}\|\nabla_w P(w)\|^2 \leq P(w) - P(w^\star) \leq \frac{\delta}{2}\|\widetilde{w}\|^2 \quad (19.12\text{b})$$

where $\widetilde{w} = w^\star - w$. The bounds in both expressions affirm that whenever we bound $\|\widetilde{w}\|^2$ we will also be automatically bounding the excess risk, $P(w) - P(w^\star)$.

19.1.4 Gradient Noise

We further showed in Section 18.3 that, as a result of the Lipschitz conditions (**A2**) or (**A2'**), the first- and second-order moments of the gradient noise process satisfy the following two properties denoted by **G1** and **G2** for ease of reference:

$$\textbf{(G1)}\quad \mathbb{E}\left(\boldsymbol{g}_n(\boldsymbol{w}_{n-1})\,|\,\boldsymbol{w}_{n-1}\right) = 0 \tag{19.13a}$$

$$\textbf{(G2)}\quad \mathbb{E}\left(\|\boldsymbol{g}_n(\boldsymbol{w}_{n-1})\|^2\,|\,\boldsymbol{w}_{n-1}\right) \leq \beta_g^2\|\widetilde{\boldsymbol{w}}_{n-1}\|^2 + \sigma_g^2 \tag{19.13b}$$

for some nonnegative constants $\{\beta_g^2, \sigma_g^2\}$ that are independent of $\widetilde{\boldsymbol{w}}_{n-1}$. Condition (**G1**) refers to the zero-mean property of the gradient noise while (**G2**) refers to its bounded "variance." We explained in the previous chapter that results (19.13a)–(19.13b) hold for instantaneous and mini-batch gradient approximations, regardless of whether the samples are streaming in independently of each other, sampled uniformly with replacement, sampled without replacement, or selected under importance sampling. The *only exception* is that the zero-mean property (19.13a) will *not* hold for the instantaneous gradient implementation when the samples are selected without replacement. This exception is not of major consequence for the convergence results in this chapter. When property (19.13a) does not hold, the convergence argument will need to be adjusted (and becomes more demanding) but will continue to lead to the same conclusion.

In summary, we find that conditions (**A1**,**A2**) for empirical risk minimization or (**A1**,**A2'**) for stochastic risk minimization imply the validity of conditions (**P1**,**P2**) on the risk function and conditions (**G1**,**G2**) on the gradient noise:

$$(\textbf{A1},\textbf{A2}) \text{ or } (\textbf{A1},\textbf{A2'}) \implies (\textbf{P1},\textbf{P2},\textbf{G1},\textbf{G2}) \tag{19.14}$$

Remark 19.1. (Conditions on risk and loss functions) The δ-Lipschitz conditions **A2** and **A2'** on the loss function were shown in the previous chapter to lead to the gradient noise properties **G1**,**G2**. They also imply the δ-Lipschitz property **P2**. The convergence analyses in the following will rely largely on **G2** and **P2**, which relate to the bound on the second-order moment of the gradient noise and to the δ-Lipschitz condition on the gradients of $P(w)$. While the two properties (**G2**,**P2**) follow from (**A2**,**A2'**), they are sometimes introduced on their own as starting assumptions for the convergence analysis.

■

19.2 CONVERGENCE UNDER UNIFORM SAMPLING

We are ready to examine the convergence behavior of the stochastic gradient recursion (19.1) in the MSE sense under conditions (**A1**,**A2**) or (**A1**,**A2'**).

19.2.1 Mean-Square-Error Convergence

The first result shows that the MSE, denoted by $\mathbb{E}\,\|\widetilde{\boldsymbol{w}}_n\|^2$, does not converge to zero but rather to a small neighborhood of size $O(\mu)$. Specifically, results (19.18b) and (19.18c) mean that the behavior of $\mathbb{E}\,\|\widetilde{\boldsymbol{w}}_n\|^2$ and the excess-risk $\mathbb{E}\,P(\boldsymbol{w}_n) - P(w^\star)$ can be described by the combined effect of two terms: One term $O(\lambda^n)$ decays exponentially at the rate λ^n and a second term $O(\mu)$ describes the size of the steady-state value that is left after sufficient iterations so that

$$\limsup_{n\to\infty}\ \mathbb{E}\,\|\widetilde{\boldsymbol{w}}_n\|^2 = O(\mu) \tag{19.15a}$$

$$\limsup_{n\to\infty}\ \Big(\mathbb{E}\,P(\boldsymbol{w}_n) - P(w^\star)\Big) = O(\mu) \tag{19.15b}$$

in terms of the *limit superior* of the variables involved. The limit superior of a sequence corresponds to the smallest upper bound for the limiting behavior of that sequence; this concept is useful when a sequence is not necessarily convergent but tends toward a small bounded region. This situation is illustrated schematically in Fig. 19.1 for $\mathbb{E}\,\|\widetilde{\boldsymbol{w}}_n\|^2$. If a sequence happens to be convergent, then the limit superior will coincide with its normal limiting value.

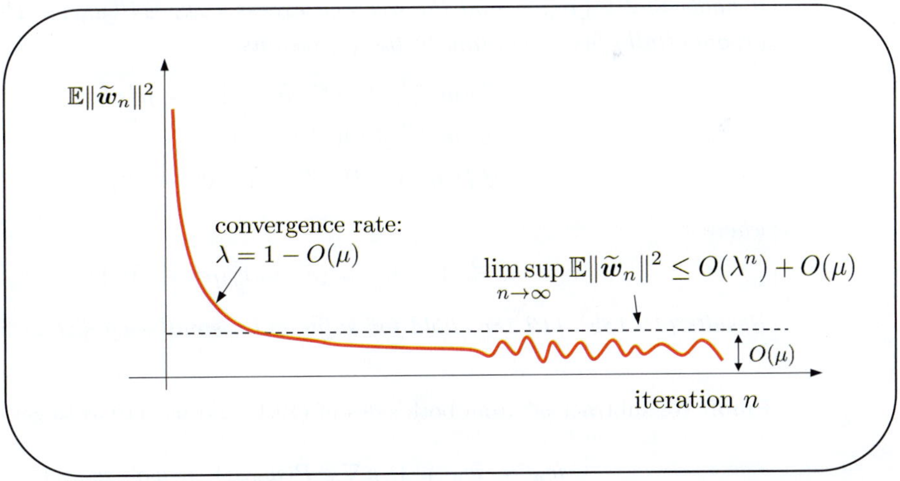

Figure 19.1 Exponential decay of the MSE described by expression (19.18a) to a level that is bounded by $O(\mu)$ and at a rate that is on the order of λ^n where $\lambda = 1 - O(\mu)$.

It further follows from the proof of Theorem 19.1 that, for sufficiently small step sizes, the size of the $O(\mu)$ limiting region in the above expressions is actually dependent on σ_g^2 (the absolute noise term) since it will hold that (see the arguments after (19.26) and (19.28)):

$$\limsup_{n\to\infty} \mathbb{E}\,\|\widetilde{\boldsymbol{w}}_n\|^2 = O(\mu\sigma_g^2/2\nu) \tag{19.16a}$$

$$\limsup_{n\to\infty} \Big(\mathbb{E}\,P(\boldsymbol{w}_n) - P(w^\star)\Big) = O(\mu\sigma_g^2/4) \tag{19.16b}$$

where ν is the strong-convexity factor. Thus, observe from the statement of the theorem that the parameters (β_g^2, σ_g^2), which define the bound on the second-order moment of the gradient noise, affect performance in different ways. The value of β_g^2 affects both stability (by defining the bound on μ for convergence) and the rate of convergence λ, whereas σ_g^2 affects the size of the limiting region (i.e., the size of the steady-state error). This observation holds for all convergence results in this chapter.

THEOREM 19.1. (MSE convergence under constant step sizes) *Consider the stochastic gradient recursion (19.1) with the instantaneous gradient approximation (19.4a) under uniform data sampling or streaming data, used to seek the minimizers of empirical or stochastic risks. The risk and loss functions are assumed to satisfy conditions* (**A1,A2**) *or* (**A1,A2'**). *For step-size values satisfying (i.e., for μ small enough):*

$$\mu < \frac{2\nu}{\delta^2 + \beta_g^2} \stackrel{\Delta}{=} \mu_o \tag{19.17}$$

it holds that $\mathbb{E}\,\|\widetilde{\boldsymbol{w}}_n\|^2$ and the average excess risk, $\mathbb{E}\,P(\boldsymbol{w}_n) - P(w^\star)$, converge exponentially fast according to the recursions:

$$\mathbb{E}\,\|\widetilde{\boldsymbol{w}}_n\|^2 \le \lambda\,\mathbb{E}\,\|\widetilde{\boldsymbol{w}}_{n-1}\|^2 + \mu^2\sigma_g^2 \tag{19.18a}$$

$$\mathbb{E}\,\|\widetilde{\boldsymbol{w}}_n\|^2 \le O(\lambda^n) + O(\mu) \tag{19.18b}$$

$$\mathbb{E}\,P(\boldsymbol{w}_n) - P(w^\star) \le O(\lambda^n) + O(\mu) \tag{19.18c}$$

where

$$\lambda \stackrel{\Delta}{=} 1 - 2\nu\mu + (\delta^2 + \beta_g^2)\mu^2 \in [0,1) \tag{19.19}$$

Results (19.18b)–(19.18c) hold for sufficiently small step sizes.

Proof: We subtract $w^\star$ from both sides of (19.1) and use (19.5) to get

$$\widetilde{\boldsymbol{w}}_n = \widetilde{\boldsymbol{w}}_{n-1} + \mu\nabla_{w^\mathsf{T}}P(\boldsymbol{w}_{n-1}) + \mu\,\boldsymbol{g}_n(\boldsymbol{w}_{n-1}) \tag{19.20}$$

We will be squaring this expression. In preparation for that step, we first use the fact that $\nabla_w P(w^\star) = 0$ to note that for the first two terms on the right-hand side:

$$\begin{aligned}
&\|\widetilde{\boldsymbol{w}}_{n-1} + \mu\nabla_{w^\mathsf{T}}P(\boldsymbol{w}_{n-1})\|^2\\
&= \|\widetilde{\boldsymbol{w}}_{n-1}\|^2 + 2\mu\left(\nabla_{w^\mathsf{T}}P(\boldsymbol{w}_{n-1})\right)^\mathsf{T}\widetilde{\boldsymbol{w}}_{n-1} + \mu^2\|\nabla_{w^\mathsf{T}}P(\boldsymbol{w}_{n-1})\|^2\\
&= \|\widetilde{\boldsymbol{w}}_{n-1}\|^2 + 2\mu\left(\nabla_{w^\mathsf{T}}P(\boldsymbol{w}_{n-1})\right)^\mathsf{T}\widetilde{\boldsymbol{w}}_{n-1} + \mu^2\|\nabla_{w^\mathsf{T}}P(w^\star) - \nabla_{w^\mathsf{T}}P(\boldsymbol{w}_{n-1})\|^2\\
&\stackrel{(\mathbf{P2})}{\le} \|\widetilde{\boldsymbol{w}}_{n-1}\|^2 + 2\mu\left(\nabla_{w^\mathsf{T}}P(\boldsymbol{w}_{n-1})\right)^\mathsf{T}\widetilde{\boldsymbol{w}}_{n-1} + \mu^2\delta^2\|\widetilde{\boldsymbol{w}}_{n-1}\|^2
\end{aligned} \tag{19.21}$$

Next, we appeal to the strong convexity property (19.7a) to get

$$\begin{aligned}(\nabla_{w^{\mathsf T}} P(\boldsymbol{w}_{n-1}))^{\mathsf T}\, \widetilde{\boldsymbol{w}}_{n-1} \;&\leq\; P(w^\star) - P(\boldsymbol{w}_{n-1}) - \frac{\nu}{2}\|\widetilde{\boldsymbol{w}}_{n-1}\|^2 \\ &\overset{(8.23)}{\leq}\; -\frac{\nu}{2}\|\widetilde{\boldsymbol{w}}_{n-1}\|^2 \;-\; \frac{\nu}{2}\|\widetilde{\boldsymbol{w}}_{n-1}\|^2 \\ &=\; -\nu\|\widetilde{\boldsymbol{w}}_{n-1}\|^2 \end{aligned} \tag{19.22}$$

Substituting into (19.21) gives

$$\boxed{\|\widetilde{\boldsymbol{w}}_{n-1} + \mu\,\nabla_{w^{\mathsf T}} P(\boldsymbol{w}_{n-1})\|^2 \leq (1 - 2\mu\nu + \mu^2\delta^2)\|\widetilde{\boldsymbol{w}}_{n-1}\|^2} \tag{19.23}$$

which is a useful intermediate result. Returning to (19.20), squaring both sides, conditioning on $\boldsymbol{w}_{n-1}$, and taking expectations we obtain

$$\begin{aligned}&\mathbb{E}\left(\|\widetilde{\boldsymbol{w}}_n\|^2 \,|\, \boldsymbol{w}_{n-1}\right) \\ &= \mathbb{E}\Big(\|\widetilde{\boldsymbol{w}}_{n-1} \;+\; \mu\,\nabla_{w^{\mathsf T}} P(\boldsymbol{w}_{n-1}) \;+\; \mu\, \boldsymbol{g}_n(\boldsymbol{w}_{n-1})\|^2 \,|\, \boldsymbol{w}_{n-1}\Big) \\ &\overset{(a)}{=} \mathbb{E}\left(\|\widetilde{\boldsymbol{w}}_{n-1} \;+\; \mu\,\nabla_{w^{\mathsf T}} P(\boldsymbol{w}_{n-1})\|^2 \,|\, \boldsymbol{w}_{n-1}\right) \;+ \\ &\qquad \mu^2\, \mathbb{E}\left(\|\boldsymbol{g}_n(\boldsymbol{w}_{n-1})\|^2 \,|\, \boldsymbol{w}_{n-1}\right) \\ &\overset{(19.23)}{\leq} (1 - 2\mu\nu + \mu^2\delta^2)\|\widetilde{w}_{n-1}\|^2 \;+\; \mu^2\,\mathbb{E}\left(\|\boldsymbol{g}_n(\boldsymbol{w}_{n-1})\|^2 \,|\, \boldsymbol{w}_{n-1}\right) \\ &\overset{(\mathbf{G2})}{\leq} (1 - 2\mu\nu + \mu^2\delta^2)\|\widetilde{w}_{n-1}\|^2 + \mu^2\left(\beta_g^2\|\widetilde{w}_{n-1}\|^2 + \sigma_g^2\right) \\ &= (1 - 2\mu\nu + \mu^2(\delta^2 + \beta_g^2))\|\widetilde{w}_{n-1}\|^2 + \mu^2\sigma_g^2 \end{aligned} \tag{19.24}$$

where the cross term in (a) is zero because of the zero-mean property **G1**; it is at this step that the zero-mean property of the gradient noise process is used. Taking expectations of both sides again removes the conditioning on $\boldsymbol{w}_{n-1}$ and leads to (19.18a), where λ is defined by (19.19). The same argument used in Fig. 12.2 can be repeated here to show that condition (19.17) ensures $0 \leq \lambda < 1$. By further iterating recursion (19.18a) we obtain

$$\mathbb{E}\,\|\widetilde{\boldsymbol{w}}_n\|^2 \leq \lambda^{n+1}\,\mathbb{E}\,\|\widetilde{\boldsymbol{w}}_{-1}\|^2 \;+\; \frac{\mu^2\sigma_g^2}{1-\lambda} \tag{19.25}$$

which proves that $\mathbb{E}\,\|\widetilde{\boldsymbol{w}}_n\|^2$ converges exponentially toward a region that is upper-bounded by:

$$\limsup_{n\to\infty}\, \mathbb{E}\,\|\widetilde{\boldsymbol{w}}_n\|^2 \leq \frac{\mu^2\sigma_g^2}{1-\lambda} \;=\; \frac{\mu\sigma_g^2}{2\nu - \mu(\delta^2 + \beta_g^2)} \tag{19.26}$$

It is easy to check that the upper bound does not exceed $\mu\sigma_g^2/\nu$ for any step size $\mu < \mu_o/2$. If, on the other hand, $\mu \ll \mu_o$ so that the denominator is approximately 2ν, then the upper bound is on the order of $\mu\sigma_g^2/2\nu$. We conclude that (19.18b) holds for sufficiently small step sizes.

To establish (19.18c), we use (19.12b) to get

$$0 \leq \mathbb{E}\,P(\boldsymbol{w}_n) - P(w^\star) \leq \frac{\delta}{2}\,\mathbb{E}\,\|\widetilde{\boldsymbol{w}}_n\|^2 \tag{19.27}$$

so that from (19.26)

$$\limsup_{n\to\infty}\, \Big(\mathbb{E}\,P(\boldsymbol{w}_n) - P(w^\star)\Big) \leq \frac{\delta\mu\sigma_g^2}{2(2\nu - \mu(\delta^2 + \beta_g^2))} \tag{19.28}$$

where the upper bound does not exceed $\mu\delta\sigma_g^2/2\nu$ for any $\mu < \mu_o/2$. If, on the other hand, $\mu \ll \mu_o$ so that the denominator is approximately 4ν, and since ν and δ are of the same order, then the upper bound is on the order of $\mu\sigma_g^2/4$. Either way, we conclude from (19.27) that $\mathbb{E}\, P(\boldsymbol{w}_n)$ converges toward an $O(\mu)$-neighborhood around $P(w^\star)$ at the same exponential rate as $\mathbb{E}\,\|\widetilde{\boldsymbol{w}}_n\|^2$, which is λ^n.

■

Observe that we can rewrite (19.18a) in the equivalent form

$$\left(\mathbb{E}\,\|\widetilde{\boldsymbol{w}}_n\|^2 - \frac{\mu^2\sigma_g^2}{1-\lambda}\right) \leq \lambda\left(\mathbb{E}\,\|\widetilde{\boldsymbol{w}}_{n-1}\|^2 - \frac{\mu^2\sigma_g^2}{1-\lambda}\right) \tag{19.29}$$

where the steady-state bound is subtracted from both sides. It is clear from this representation that λ determines the rate of decay of the MSE toward its steady-state bound – refer again to Fig. 19.1.

Example 19.1 **(Randomized coordinate descent)** A similar convergence analysis can be applied to a randomized version of coordinate descent using stochastic gradient approximations – see listing (19.30).

Stochastic randomized coordinate descent for solving (19.3a) or (19.3b).

given dataset $\{\gamma(m), h_m\}_{m=0}^{N-1}$ or streaming data $(\gamma(n), h_n)$;
start with an arbitrary initial condition $\boldsymbol{w}_{-1}$.
repeat until convergence over $n \geq 0$**:**
  iterate is $\boldsymbol{w}_{n-1} = \text{col}\{\boldsymbol{w}_{n-1,m}\}_{m=1}^{M}$
  select at random or receive $(\boldsymbol{\gamma}(n), \boldsymbol{h}_n)$; (19.30)
  select a random index $1 \leq m^o \leq M$;
  $\boldsymbol{w}_{n,m^o} = \boldsymbol{w}_{n-1,m^o} - \mu \dfrac{\partial Q_u(\boldsymbol{w}_{n-1};\boldsymbol{\gamma}(n),\boldsymbol{h}_n)}{\partial w_{m^o}}$
  keep $\boldsymbol{w}_{n,m} = \boldsymbol{w}_{n-1,m}$ for all $m \neq m^o$
end
return $w^\star \leftarrow \boldsymbol{w}_n$.

The same argument used in the proof of Theorem 19.1 can be repeated to establish convergence conditions for (19.30). We leave the analysis to Prob. 19.7.

19.2.2 Regret Analysis

For empirical risks minimized by stochastic gradient algorithms, the regret value over a window of N iterations is defined as the deviation of the accumulated *loss* from the minimal risk value:

$$\begin{aligned}\mathcal{R}(N) &\triangleq \frac{1}{N}\sum_{n=0}^{N-1} Q(\boldsymbol{w}_{n-1};\boldsymbol{\gamma}(n),\boldsymbol{h}_n) - \min_{w\in\mathbb{R}^M}\left(\frac{1}{N}\sum_{n=0}^{N-1} Q(w;\gamma(n),h_n)\right)\\ &= \frac{1}{N}\sum_{n=0}^{N-1} Q(\boldsymbol{w}_{n-1};\boldsymbol{\gamma}(n),\boldsymbol{h}_n) - P(w^\star)\end{aligned} \tag{19.31}$$

where the arguments $(\boldsymbol{w}_{n-1}, \boldsymbol{\gamma}(n), \boldsymbol{h}_n)$ are random due to uniform sampling and gradient noise. For this reason, we are denoting the regret variable in boldface to highlight its random nature as well. We already encountered this definition earlier in Section 17.2 while discussing the AdaGrad algorithm.

We may compare the above expression with the earlier definition (12.57) used for the gradient-descent case when the actual gradient of $P(w)$ is employed in the update recursion. We observe that in the above expression for the regret, the risk function $P(w)$ from (12.57) is replaced by the loss function $Q(w;\cdot)$. If we evaluate the conditional expectation of $\boldsymbol{\mathcal{R}}(N)$ over the trajectory of weight iterates, and use the unbiasedness property $\mathbb{E}\, Q(w, \boldsymbol{\gamma}, \boldsymbol{h}) = P(w)$, we arrive at the following expression for the conditional regret:

$$\begin{aligned} \mathcal{R}(N) &\stackrel{\Delta}{=} \mathbb{E}\left(\boldsymbol{\mathcal{R}}(N)\,|\,\boldsymbol{w}_{-1}, \boldsymbol{w}_0, \ldots, \boldsymbol{w}_{N-1}\right) \\ &= \frac{1}{N}\sum_{n=0}^{N-1} P(w_{n-1}) \;-\; P(w^\star) \end{aligned} \tag{19.32}$$

which agrees with the earlier definition (12.57). In stochastic optimization implementations, it is common to employ the regret (19.32) as a performance measure as well. Using (19.18c) and the same argument that led to (12.58), we can readily find that the regret for the stochastic gradient algorithm (19.1) under uniform sampling satisfies (see Prob. 19.5):

$$\mathbb{E}\,\boldsymbol{\mathcal{R}}(N) \;\leq\; O(1/N) \;+\; O(\mu) \tag{19.33}$$

19.3 CONVERGENCE OF MINI-BATCH IMPLEMENTATION

Let (β_g^2, σ_g^2) denote the parameters that characterize the bound in (19.13b) for the stochastic gradient algorithm based on *instantaneous* gradient approximations. We showed in (18.45) that these parameters get scaled down by a factor τ_B when a mini-batch implementation is used since the second-order moment of the gradient noise will then satisfy

$$\mathbb{E}\left(\|\boldsymbol{g}_n(\boldsymbol{w}_{n-1})\|^2\,|\,\boldsymbol{w}_{n-1}\right) \;\leq\; \frac{1}{\tau_B}\left(\beta_g^2\|\widetilde{w}_{n-1}\|^2 + \sigma_g^2\right) \tag{19.34}$$

The value of τ_B depends on whether the mini-batch samples are selected with or without replacement:

$$\tau_B \;\stackrel{\Delta}{=}\; \begin{cases} B, & \text{sampling } \textit{with} \text{ replacement} \\ B\dfrac{N-1}{N-B}, & \text{sampling } \textit{without} \text{ replacement} \end{cases} \tag{19.35}$$

Observe from the second line that $\tau_B \approx B$ for N large enough. The same analysis used to establish Theorem 19.1 will lead to a similar conclusion apart from the scaling of β_g^2 by τ_B – see Prob. 19.1. Observe Theorem 19.2 how the mini-batch size B influences the performance expressions. In particular, the size of the steady-state neighborhood is reduced from $O(\mu)$ to $O(\mu/\tau_B)$.

Theorem 19.2. (MSE convergence of mini-batch implementation) *Consider the stochastic gradient recursion (19.1) with the mini-batch gradient approximation (19.4b) under random sampling with or without replacement or streaming data, used to seek the minimizers of empirical or stochastic risks. The risk and loss functions are assumed to satisfy conditions* (**A1,A2**) *or* (**A1,A2'**). *For step-size values satisfying (i.e., for μ small enough):*

$$\mu < \frac{2\nu}{\delta^2 + \frac{\beta_g^2}{\tau_B}} \stackrel{\Delta}{=} \mu_o \tag{19.36}$$

it holds that $\mathbb{E}\,\|\widetilde{\boldsymbol{w}}_n\|^2$ and the average excess risk, $\mathbb{E}\,P(\boldsymbol{w}_n) - P(w^\star)$, converge exponentially fast according to the recursions:

$$\mathbb{E}\,\|\widetilde{\boldsymbol{w}}_n\|^2 \;\leq\; \lambda\,\mathbb{E}\,\|\widetilde{\boldsymbol{w}}_{n-1}\|^2 \;+\; \frac{\mu^2\sigma_g^2}{\tau_B} \tag{19.37a}$$

$$\mathbb{E}\,\|\widetilde{\boldsymbol{w}}_n\|^2 \leq O(\lambda^n) + O(\mu/\tau_B) \tag{19.37b}$$

$$\mathbb{E}\,P(\boldsymbol{w}_n) - P(w^\star) \leq O(\lambda^n) + O(\mu/\tau_B) \tag{19.37c}$$

where

$$\lambda \stackrel{\Delta}{=} 1 - 2\nu\mu + \left(\delta^2 + \frac{\beta_g^2}{\tau_B}\right)\mu^2 \;\in\; [0,1) \tag{19.38}$$

Results (19.37b) and (19.37c) hold for sufficiently small step sizes.

19.4 CONVERGENCE UNDER VANISHING STEP SIZES

We observe from (19.18b) that under constant step-size learning, the MSE value $\mathbb{E}\,\|\widetilde{\boldsymbol{w}}_n\|^2$ converges to a small neighborhood of size $O(\mu)$; the smaller the value of μ is, the smaller the size of this neighborhood will be. However, small step sizes affect the convergence rate of the algorithm because they cause the value of λ to approach 1. One way to reduce the size of the limiting region to zero is to employ a decaying step size $\mu(n)$ in place of the constant μ. Doing so allows us to employ larger step-size values during the initial stages of the algorithm to speed up convergence and smaller step-size values during the latter stages to improve steady-state performance. It is common to choose the sequence $\mu(n) > 0$ to satisfy either of the following two conditions:

$$(\textbf{Condition I}) \qquad \sum_{n=0}^{\infty} \mu^2(n) < \infty \quad \text{and} \quad \sum_{n=0}^{\infty} \mu(n) = \infty \tag{19.39a}$$

$$(\textbf{Condition II}) \qquad \lim_{n\to\infty} \mu(n) = 0 \quad \text{and} \quad \sum_{n=0}^{\infty} \mu(n) = \infty \tag{19.39b}$$

Clearly, any sequence that satisfies the stronger condition (19.39a) also satisfies (19.39b). Recursion (19.1) would then be replaced by

$$\boxed{\boldsymbol{w}_n = \boldsymbol{w}_{n-1} - \mu(n)\,\widehat{\nabla_{w^\mathsf{T}} P}(\boldsymbol{w}_{n-1}), \quad n \geq 0} \tag{19.40}$$

The decaying step size helps annihilate the effect of gradient noise and ensures convergence of $\boldsymbol{w}_n$ to $w^\star$. Specifically, we will show below that $\boldsymbol{w}_n$ will now converge to $w^\star$ in the MSE sense under both choices (19.39a) or (19.39b), i.e., $\mathbb{E}\,\|\widetilde{\boldsymbol{w}}_n\|^2 \to 0$ as $n \to \infty$. Based on the discussion from Appendix 3.A on the convergence of random variables, this conclusion implies convergence in probability so that

$$\lim_{n\to\infty} \mathbb{P}\left(\|\widetilde{\boldsymbol{w}}_n\|^2 > \epsilon\right) = 0, \quad \text{for any small } \epsilon > 0 \tag{19.41}$$

We will actually establish below the stronger result that under (19.39a), $\boldsymbol{w}_n$ converges to $w^\star$ almost surely, i.e., with probability 1:

$$\mathbb{P}\left(\lim_{n\to\infty} \boldsymbol{w}_n = w^\star\right) = 1 \tag{19.42}$$

While decaying step-size sequences of the form (19.39a)–(19.39b) provide favorable convergence properties toward $w^\star$, and assist in countering the effect of gradient noise, they nevertheless force the step size to approach zero. This is problematic for applications requiring continuous learning from streaming data because the algorithm will update more slowly and become less effective at tracking drifts in the location of $w^\star$ due to changes in the statistical properties of the data.

Theorem 19.3. (Convergence under vanishing step sizes) *Consider the stochastic gradient recursion (19.1) with the instantaneous gradient approximation (19.4a) under uniform data sampling or streaming data, used to seek the minimizers of empirical or stochastic risks. The risk and loss functions are assumed to satisfy conditions (***A1***,***A2***) or (***A1***,***A2'***). Then, the following convergence properties hold:*

(a) *If the step-size sequence $\mu(n)$ satisfies (19.39a), then $\boldsymbol{w}_n$ converges almost surely to $w^\star$, written as $\boldsymbol{w}_n \to w^\star$ a.s.*

(b) *If the step-size sequence $\mu(n)$ satisfies (19.39b), then $\boldsymbol{w}_n$ converges in the MSE sense to $w^\star$, i.e., $\mathbb{E}\,\|\widetilde{\boldsymbol{w}}_n\|^2 \to 0$, which in turn implies convergence in probability according to (19.41).*

Proof: The same argument leading to (19.24) for constant step sizes continues to hold, giving the inequality:

$$\mathbb{E}\left(\|\widetilde{\boldsymbol{w}}_n\|^2 \,|\, \boldsymbol{w}_{n-1}\right) \leq \lambda(n)\,\|\widetilde{w}_{n-1}\|^2 + \mu^2(n)\sigma_g^2 \tag{19.43}$$

with μ replaced by $\mu(n)$ and where now

$$\lambda(n) \triangleq 1 - 2\nu\mu(n) + (\delta^2 + \beta_g^2)\mu^2(n) \tag{19.44}$$

We split the term $2\nu\mu(n)$ into the sum of two terms and write

$$\lambda(n) \;=\; 1-\nu\mu(n)-\nu\mu(n)+(\delta^2+\beta_g^2)\mu^2(n) \tag{19.45}$$

Now, since $\mu(n)\rightarrow 0$, we conclude that for large enough $n>n_o$, the value of $\mu^2(n)$ is smaller than $\mu(n)$. Therefore, a large enough time index, n_o, exists such that the following two conditions are satisfied:

$$\nu\mu(n)\geq(\delta^2+\beta_g^2)\mu^2(n), \qquad 0\leq\nu\mu(n)<1, \qquad n>n_o \tag{19.46}$$

Consequently,

$$\lambda(n)\;\leq\;1-\nu\mu(n), \qquad n>n_o \tag{19.47}$$

Then, inequalities (19.43) and (19.47) imply that

$$\mathbb{E}\left(\|\widetilde{\boldsymbol{w}}_n\|^2\,|\,\boldsymbol{w}_{n-1}\right)\;\leq\;(1-\nu\mu(n))\,\|\widetilde{w}_{n-1}\|^2\;+\;\mu^2(n)\sigma_g^2, \qquad n>n_o \tag{19.48}$$

Due to the Markovian property of recursion (19.40), where $\boldsymbol{w}_n$ is solely dependent on the most recent iterate $\boldsymbol{w}_{n-1}$, we can also write that

$$\mathbb{E}\left(\|\widetilde{\boldsymbol{w}}_n\|^2\,|\,\boldsymbol{w}_{n-1},\ldots,\boldsymbol{w}_0,\boldsymbol{w}_{-1}\right)\;\leq\;(1-\nu\mu(n))\,\|\widetilde{w}_{n-1}\|^2\;+\;\mu^2(n)\sigma_g^2, \qquad n>n_o \tag{19.49}$$

where the conditioning on the left-hand side is now relative to the entire trajectory. For compactness of notation, let

$$\boldsymbol{u}(n+1)\;\stackrel{\Delta}{=}\;\|\widetilde{\boldsymbol{w}}_n\|^2 \tag{19.50}$$

Then, inequality (19.49) implies

$$\mathbb{E}\Big(\boldsymbol{u}(n+1)|\,\boldsymbol{u}(0),\boldsymbol{u}(1),\ldots,\boldsymbol{u}(n)\Big)\;\leq\;(1-\nu\mu(n))\,u(n)\;+\;\mu^2(n)\sigma_g^2, \quad n>n_o \tag{19.51}$$

We now call upon the useful result (19.158) from Appendix 19.A and make the identifications:

$$a(n)\leftarrow\nu\mu(n), \quad b(n)\leftarrow\mu^2(n)\sigma_g^2 \tag{19.52}$$

These sequences satisfy conditions (19.159) in the appendix in view of assumption (19.39a) on the step-size sequence and the second condition in (19.46). We then conclude that $\boldsymbol{u}(n)\rightarrow 0$ almost surely and, hence, $\boldsymbol{w}_n\rightarrow w^\star$ almost surely.

Finally, taking expectations of both sides of (19.51) leads to

$$\mathbb{E}\,\boldsymbol{u}(n+1)\;\leq\;(1-\nu\mu(n))\,\mathbb{E}\,\boldsymbol{u}(n)\;+\;\mu^2(n)\sigma_g^2, \qquad n>n_o \tag{19.53}$$

with the expectation operator appearing on both sides of the inequality. Then, we conclude from the earlier result (14.136), under conditions (19.39b), that $\mathbb{E}\,\|\widetilde{\boldsymbol{w}}_n\|^2\rightarrow 0$ so that $\boldsymbol{w}_n$ converges to $w^\star$ in the MSE sense.

■

We can be more specific and quantify the rate at which the error variance $\mathbb{E}\,\|\widetilde{\boldsymbol{w}}_n\|^2$ converges to zero for step-size sequences of the form:

$$\mu(n)=\frac{\tau}{n+1}, \qquad \tau>0 \tag{19.54}$$

which satisfy both conditions (19.39a) and (19.39b). This particular form for $\mu(n)$ is motivated in the next example. In contrast to the previous result (12.68a) on the convergence rate of gradient-descent algorithms, which was seen to be on the order of $O(1/n^{2\nu\tau})$, the next statement indicates that three rates of convergence are now possible, depending on how $\nu\tau$ compares to the value 1.

THEOREM 19.4. (Rates of convergence under (19.54)) *Consider the stochastic gradient recursion (19.1) with the instantaneous gradient approximation (19.4a) under uniform data sampling or streaming data, used to seek the minimizers of empirical or stochastic risks. The risk and loss functions are assumed to satisfy conditions (**A1,A2**) or (**A1,A2'**). Assume the step-size sequence is selected according to (19.54). Then, three convergence rates are possible depending on how the factor $\nu\tau$ compares to the value 1. Specifically, for large enough n, it holds that:*

$$\begin{cases} \mathbb{E}\,\|\widetilde{\boldsymbol{w}}_n\|^2 \leq O\left(\frac{1}{n}\right), & \nu\tau > 1 \\ \mathbb{E}\,\|\widetilde{\boldsymbol{w}}_n\|^2 = O\left(\frac{\log n}{n}\right), & \nu\tau = 1 \\ \mathbb{E}\,\|\widetilde{\boldsymbol{w}}_n\|^2 = O\left(\frac{1}{n^{\nu\tau}}\right), & \nu\tau < 1 \end{cases} \tag{19.55}$$

The fastest convergence rate occurs when $\nu\tau > 1$ (i.e., for large enough τ) and is on the order of $O(1/n)$. The risk values follow a similar convergence behavior as $\mathbb{E}\,\|\widetilde{\boldsymbol{w}}_n\|^2$, namely,

$$\begin{cases} \mathbb{E}\,P(\boldsymbol{w}_n) - P(w^\star) \leq O\left(\frac{1}{n}\right), & \nu\tau > 1 \\ \mathbb{E}\,P(\boldsymbol{w}_n) - P(w^\star) = O\left(\frac{\log n}{n}\right), & \nu\tau = 1 \\ \mathbb{E}\,P(\boldsymbol{w}_n) - P(w^\star) = O\left(\frac{1}{n^{\nu\tau}}\right), & \nu\tau < 1 \end{cases} \tag{19.56}$$

The fastest convergence rate again occurs when $\nu\tau > 1$ and is on the order of $O(1/n)$.

Proof: We use (19.53) and the assumed form for $\mu(n)$ in (19.54) to write

$$\mathbb{E}\,\boldsymbol{u}(n+1) \leq \left(1 - \frac{\nu\tau}{n+1}\right)\mathbb{E}\,\boldsymbol{u}(n) + \frac{\tau^2\sigma_g^2}{(n+1)^2}, \quad n > n_o \tag{19.57}$$

This recursion has the same form as (14.136), with the identifications:

$$a(n) \leftarrow \frac{\nu\tau}{n+1}, \quad b(n) \leftarrow \frac{\tau^2\sigma_g^2}{(n+1)^2}, \quad p \leftarrow 1 \tag{19.58}$$

The above rates of convergence then follow from the statement in part (b) of Lemma 14.1 from Appendix 14.A. Result (19.56) follows from (19.27).

■

Example 19.2 (Motivating step-size sequences of the form (19.54)) We refer to expression (19.44) for $\lambda(n)$ and notice that the following relation holds whenever $\mu(n) < \nu/(\delta^2 + \beta_g^2)$ (this condition is possible for decaying step sizes and large enough n):

$$1 - 2\nu\mu(n) + (\delta^2 + \beta_g^2)\mu^2(n) < 1 - \mu(n)\nu \tag{19.59}$$

Then, for large n and sufficiently small $\mu(n)$,

$$
\begin{aligned}
1 - 2\nu\mu(n)\nu + (\delta^2 + \beta_g^2)\mu^2(n) &< 1 - \mu(n)\nu \\
&\leq 1 - \mu(n)\nu + \frac{\mu^2(n)\nu^2}{4} \\
&= \left(1 - \frac{\mu(n)\nu}{2}\right)^2 \\
&\leq 1 - \frac{\mu(n)\nu}{2}
\end{aligned}
\tag{19.60}
$$

so that taking expectations of both sides of (19.43):

$$
\mathbb{E}\,\|\widetilde{\boldsymbol{w}}_n\|^2 \leq \left(1 - \frac{\mu(n)\nu}{2}\right)\mathbb{E}\,\|\widetilde{\boldsymbol{w}}_{n-1}\|^2 + \mu^2(n)\sigma_g^2 \tag{19.61}
$$

We can select $\mu(n)$ to tighten the upper bound. By minimizing over $\mu(n)$ we arrive at the choice:

$$
\mu^o(n) = \frac{\nu}{4\sigma_g^2}\,\mathbb{E}\,\|\widetilde{\boldsymbol{w}}_{n-1}\|^2 \tag{19.62}
$$

We now verify that this choice leads to $\mathbb{E}\,\|\widetilde{\boldsymbol{w}}_{n-1}\|^2 = O(1/n)$ so that the step-size sequence itself satisfies $\mu^o(n) = O(1/n)$. Indeed, substituting into (19.61) gives

$$
\mathbb{E}\,\|\widetilde{\boldsymbol{w}}_n\|^2 \leq \mathbb{E}\,\|\widetilde{\boldsymbol{w}}_{n-1}\|^2\left(1 - \frac{\nu^2}{16\sigma_g^2}\mathbb{E}\,\|\widetilde{\boldsymbol{w}}_{n-1}\|^2\right) \tag{19.63}
$$

Inverting both sides we obtain a linear recursion for the inverse quantity $1/\mathbb{E}\,\|\widetilde{\boldsymbol{w}}_n\|^2$:

$$
\begin{aligned}
\frac{1}{\mathbb{E}\,\|\widetilde{\boldsymbol{w}}_n\|^2} &\geq \frac{1}{\mathbb{E}\,\|\widetilde{\boldsymbol{w}}_{n-1}\|^2}\left(1 - \frac{\nu^2}{16\sigma_g^2}\mathbb{E}\,\|\widetilde{\boldsymbol{w}}_{n-1}\|^2\right)^{-1} \\
&\overset{(a)}{\geq} \frac{1}{\mathbb{E}\,\|\widetilde{\boldsymbol{w}}_{n-1}\|^2}\left(1 + \frac{\nu^2}{16\sigma_g^2}\mathbb{E}\,\|\widetilde{\boldsymbol{w}}_{n-1}\|^2\right) \\
&= \frac{1}{\mathbb{E}\,\|\widetilde{\boldsymbol{w}}_{n-1}\|^2} + \frac{\nu^2}{16\sigma_g^2}
\end{aligned}
\tag{19.64}
$$

where in step (a) we used the fact that for any small enough scalar $x^2 < 1$, it holds that $1 - x^2 \leq 1$ and

$$
(1-x)(1+x) \leq 1 \implies (1-x)^{-1} \geq (1+x) \tag{19.65}
$$

Iterating (19.64) gives a bound on the value of the MSE:

$$
\frac{1}{\mathbb{E}\,\|\widetilde{\boldsymbol{w}}_n\|^2} \geq \frac{1}{\mathbb{E}\,\|\widetilde{\boldsymbol{w}}_{-1}\|^2} + \frac{(n+1)\nu^2}{16\sigma_g^2} \tag{19.66}
$$

Substituting into (19.62) we find that

$$
\mu^o(n) \leq \frac{\nu}{4\sigma_g^2}\left(\frac{1}{\mathbb{E}\,\|\widetilde{\boldsymbol{w}}_{-1}\|^2} + \frac{n\nu^2}{16\sigma_g^2}\right)^{-1} = O(1/n) \tag{19.67}
$$

Example 19.3 **(Comparing workloads)** Theorem 19.4 reveals that the stochastic gradient algorithm (19.1) is able to converge to the exact minimizer at the rate of $O(1/n)$. This means that the algorithm will need on the order of $1/\epsilon$ iterations for the average risk value $\mathbb{E}\,P(\boldsymbol{w}_n)$ to get ϵ-close to the optimal value $P(w^\star)$. Since each iteration requires the computation of a single gradient vector, we say that the workload (or computing time) that is needed is proportional to $1/\epsilon$:

$$\textbf{workload or computing time} = \text{number of iterations} \times \text{gradient computations per iteration} \tag{19.68}$$

The result of the theorem is equally applicable to *mini-batch* stochastic implementations – see Prob. 19.2. Therefore, a total of $1/\epsilon$ iterations will again be necessary for $\mathbb{E}\,P(\boldsymbol{w}_n)$ to get ϵ-close to $P(w^\star)$. Now, however, for mini-batches of size B, it is necessary to evaluate B gradient vectors per iteration so that the workload is increased to B/ϵ.

If we were to rely instead on the *full-batch* gradient-descent implementation (12.28) for the minimization of the same empirical risk, then we know from the statement after (12.64a) that a smaller number of $\ln(1/\epsilon)$ iterations will be needed for the risk value $P(w_n)$ to get ϵ-close to $P(w^\star)$. The workload in this case will become $N\ln(1/\epsilon)$. Table 19.1 summarizes the conclusions.

Table 19.1 Computation time or workload needed for the risk value of each algorithm to get ϵ-close to the optimal value.

	Algorithm for empirical risk minimization	Workload or computing time
1.	stochastic gradient with decaying step size	$1/\epsilon$
2.	mini-batch stochastic gradient with decaying step size and mini-batch size B	B/ϵ
3.	full-batch gradient-descent algorithm (12.28)	$N\ln(1/\epsilon)$

Example 19.4 **(Comparing generalization abilities)** Each of the algorithms considered in the previous example is concerned with the solution of the empirical risk minimization problem (19.3a). For added clarity, in this example alone, we will refer to the risk function by writing $P_{\text{emp}}(w)$, where the subscript is meant to emphasize its empirical nature. Thus, these three algorithms are solving:

$$w^\star \stackrel{\Delta}{=} \underset{w\in\mathbb{R}^M}{\text{argmin}} \left\{ P_{\text{emp}}(w) \stackrel{\Delta}{=} \frac{1}{N}\sum_{m=0}^{N-1} Q(w;\gamma(m),h_m) \right\} \tag{19.69}$$

We explained earlier in Section 12.1.3 when discussing the concept of "generalization" that, ideally, we would like the solutions by these algorithms to serve as good approximations for the minimizer of the following stochastic optimization problem:

$$w^o \stackrel{\Delta}{=} \underset{w\in\mathbb{R}^M}{\text{argmin}} \left\{ P(w) \stackrel{\Delta}{=} \mathbb{E}\,Q(w;\boldsymbol{\gamma},\boldsymbol{h}) \right\} \tag{19.70}$$

Assume we run each algorithm for a sufficient number of L iterations and let ϵ denote the resulting gap between the empirical risk at $\boldsymbol{w}_L$ and its optimal value, i.e.,

$$\epsilon \stackrel{\Delta}{=} \mathbb{E}\Big(P_{\text{emp}}(\boldsymbol{w}_L) - P_{\text{emp}}(w^\star)\Big) \qquad \textbf{(empirical excess risk)} \tag{19.71}$$

Clearly, the value of ϵ is dependent on the algorithm. We can derive an expression that reveals how close $P(\boldsymbol{w}_L)$ gets to $P(w^o)$, namely, how close the *stochastic* risk at $\boldsymbol{w}_L$ gets to the optimal value. For this purpose, we first note that

$$\begin{aligned}
&\mathbb{E}\left(P(\boldsymbol{w}_L) - P(w^o)\right) \\
&= \mathbb{E}\left(P(\boldsymbol{w}_L) - P_{\rm emp}(\boldsymbol{w}_L)\right) + \underbrace{\mathbb{E}\left(P_{\rm emp}(\boldsymbol{w}_L) - P_{\rm emp}(w^\star)\right)}_{=\epsilon} + \\
&\quad \underbrace{\mathbb{E}\left(P_{\rm emp}(w^\star) - P_{\rm emp}(w^o)\right)}_{\leq 0} + \mathbb{E}\left(P_{\rm emp}(w^o) - P(w^o)\right) \\
&\leq \epsilon + \mathbb{E}\left(P(\boldsymbol{w}_L) - P_{\rm emp}(\boldsymbol{w}_L)\right) + \mathbb{E}\left(P_{\rm emp}(w^o) - P(w^o)\right)
\end{aligned} \tag{19.72}$$

so that

(stochastic excess risk)

$$\mathbb{E}\left(P(\boldsymbol{w}_L) - P(w^o)\right) = \epsilon + O\left(\sqrt{\frac{2\ln\ln N}{N}}\right) \tag{19.73}$$

where we used result (3.226) to approximate the differences between the empirical and true risk values in the last two terms appearing in (19.72). Result (19.73) shows that, for N large enough,

$$\left(\begin{array}{c}\textbf{stochastic}\\ \textbf{excess risk}\end{array}\right) = \left(\begin{array}{c}\textbf{empirical}\\ \textbf{excess risk}\end{array}\right) + O\left(\sqrt{\frac{2\ln\ln N}{N}}\right) \tag{19.74}$$

This expression reveals how well algorithms generalize. It shows that the stochastic excess risk depends on two factors: the sample size N *and* the empirical excess risk (19.71).

Assume we fix the computational (or workload) budget at a maximum value $\mathcal{C}_{\max}$ for each of the algorithms. Consider first the stochastic gradient implementation (19.1). From the first row in Table 19.1 we know that this algorithm will lead to an empirical excess risk on the order of $\epsilon = 1/\mathcal{C}_{\max}$ and, moreover, this excess is independent of N. Therefore, we conclude from (19.73) that increasing the sample size N will help reduce the stochastic excess risk and improve generalization.

In contrast, consider next the full-batch gradient-descent algorithm (12.28). From the third row in Table 19.1 we know that this algorithm will lead to an empirical excess risk on the order of $\epsilon = e^{-\mathcal{C}_{\max}/N}$, which *depends* on N. In other words, both factors on the right-hand side of (19.73) will now be dependent on N and an optimal choice for N can be selected.

19.5 CONVERGENCE UNDER RANDOM RESHUFFLING

We return to the stochastic gradient algorithm (19.1) with the instantaneous gradient approximation (19.4a) and constant step size μ. We now examine its convergence behavior under random reshuffling (i.e., sampling without replacement) for *empirical risk minimization*. In this case, the gradient noise process does not have zero mean, and we need to adjust the convergence argument. The analysis will require that we make explicit the multiple epochs (or runs) over the data. Thus, let k denote the epoch index. Before the start of an epoch, the N-size data $\{\gamma(m), h_m\}$ is reshuffled at random. During the run, we select samples sequentially from the reshuffled dataset. We describe the algorithm in listing (19.75).

Stochastic gradient algorithm with random reshuffling for solving the empirical risk minimization problem (19.3a).

given dataset $\{\gamma(m), h_m\}_{m=0}^{N-1}$;
start from an arbitrary initial condition $\boldsymbol{w}_{N-1}^{0}$.
for each run $k = 1, 2, \ldots, K$**:**
  set $\boldsymbol{w}_{-1}^{k} = \boldsymbol{w}_{N-1}^{k-1}$;
  reshuffle the dataset;
  repeat for $n = 0, 1, 2, \ldots, N-1$**:**
    $\boldsymbol{w}_n^k = \boldsymbol{w}_{n-1}^k - \mu\, \nabla_{w^{\mathsf{T}}} Q(\boldsymbol{w}_{n-1}^k; \boldsymbol{\gamma}(n), \boldsymbol{h}_n)$
  end
end
return $w^\star \leftarrow \boldsymbol{w}_{N-1}^{K}$. (19.75)

In this description, the notation $(\boldsymbol{\gamma}(n), \boldsymbol{h}_n)$ denotes the random sample that is selected at iteration n of the kth run. The initial iterate for each run is the value that is attained at the end of the previous run:

$$\boldsymbol{w}_{-1}^{k} = \boldsymbol{w}_{N-1}^{k-1} \tag{19.76}$$

Since operation under random reshuffling corresponds to sampling *without* replacement, it is clear that no sample points are repeated during each run of the algorithm. This is in contrast to uniform sampling *with replacement*, where some data points may be repeated during the same run of the algorithm. The argument will show that this simple adjustment to the operation of the algorithm, using data sampling *without* as opposed to *with* replacement, results in performance *improvement*. The MSE value $\mathbb{E}\,\|\widetilde{\boldsymbol{w}}_n\|^2$ will now be reduced to $O(\mu^2)$ in comparison to the earlier $O(\mu)$ value shown in (19.18b). The proof of the following theorem appears in Appendix 19.B. The statement assumes a fixed number N of data samples and repeated epochs, indexed by k.

THEOREM 19.5. (Convergence under random reshuffling) *Consider the stochastic gradient recursion (19.1) with the instantaneous gradient approximation (19.4a) under data sampling without replacement, used to seek the minimizers of empirical risks. The risk and loss functions are assumed to satisfy conditions (**A1**,**A2**) or (**A1**,**A2'**). For step-size values satisfying:*

$$\mu < \frac{\nu}{\sqrt{24N\delta^2}} \tag{19.77}$$

it holds that $\mathbb{E}\,\|\widetilde{\boldsymbol{w}}_n^k\|^2$ converges exponentially over the epoch index k at the rate λ^k where

$$\lambda = 1 - \frac{\mu}{2}\nu N \tag{19.78}$$

and, for any $0 \le n \le N-1$:

$$\mathbb{E}\,\|\widetilde{\boldsymbol{w}}_n^k\|^2 \le O(\lambda^k) + O(\mu^2) \tag{19.79a}$$

$$\mathbb{E}\,P(\boldsymbol{w}_n^k) - P(w^\star) \le O(\lambda^k) + O(\mu^2) \tag{19.79b}$$

Observe that the results in the theorem are expressed in terms of the epoch index k tending to $+\infty$. We can also examine the behavior of random reshuffling when an *epoch-dependent* step size is used, say, of the form

$$\mu(k) = \tau/k, \quad k \ge 1, \quad \tau > 0 \tag{19.80}$$

By repeating the arguments from Appendix 19.B and the technique used to establish Theorem 19.4, we can similarly verify that three convergence rates are possible depending on how the factor $\nu\tau N/2$ compares to the value 1. Specifically, for any $0 \le n \le N-1$ and large enough k, it holds that (see Prob. 19.9):

$$\begin{cases} \mathbb{E}\,\|\widetilde{\boldsymbol{w}}_n^k\|^2 \le O\left(\frac{1}{k}\right), & \nu\tau N > 2 \\ \mathbb{E}\,\|\widetilde{\boldsymbol{w}}_n^k\|^2 = O\left(\frac{\log k}{k}\right), & \nu\tau N = 2 \\ \mathbb{E}\,\|\widetilde{\boldsymbol{w}}_n^k\|^2 = O\left(\frac{1}{k^{\nu\tau N/2}}\right), & \nu\tau N < 2 \end{cases} \tag{19.81}$$

The fastest convergence rate occurs when $\nu\tau N > 2$ (i.e., for large enough τ) and is on the order of $O(1/k)$. The risk values follow a similar convergence behavior as $\mathbb{E}\,\|\widetilde{\boldsymbol{w}}_n^k\|^2$, namely,

$$\begin{cases} \mathbb{E}\,P(\boldsymbol{w}_n^k) - P(w^\star) \le O\left(\frac{1}{k}\right), & \nu\tau N > 2 \\ \mathbb{E}\,P(\boldsymbol{w}_n^k) - P(w^\star) = O\left(\frac{\log k}{k}\right), & \nu\tau N = 2 \\ \mathbb{E}\,P(\boldsymbol{w}_n^k) - P(w^\star) = O\left(\frac{1}{k^{\nu\tau N/2}}\right), & \nu\tau N < 2 \end{cases} \tag{19.82}$$

The fastest convergence rate again occurs when $\nu\tau N > 2$ and is on the order of $O(1/k)$.

Example 19.5 (Simulating random reshuffling) We compare the performance of the stochastic gradient algorithm (19.1) with constant step size under both uniform sampling and random reshuffling for the instantaneous gradient approximation (19.4a).

The objective is to illustrate the superior steady-state performance under random reshuffling. According to result (19.79b), if we plot the steady-state deviation value $P(\boldsymbol{w}_n^k) - P(w^\star)$, as $k \to \infty$, versus the step-size parameter μ in a log–log scale, the slope of the resulting line should be at least 2 since

$$\log_{10}\Big(\mathbb{E}\,P(\boldsymbol{w}_n^k) - P(w^\star)\Big) \;\leq\; 2\log_{10}(\mu), \quad k \to \infty \tag{19.83}$$

In other words, if the step size is reduced by a factor of 10 from μ to $\mu/10$, then the risk deviation should be reduced by at least a factor of 100. In comparison, under uniform sampling, we know from (19.18c) that the risk deviation will be reduced by at least the same factor 10. We illustrate this behavior by means of a simulation. Consider the ℓ_2-regularized logistic empirical risk:

$$P(w) = \rho\|w\|^2 + \frac{1}{N}\sum_{m=0}^{N-1} \ln\left(1 + e^{-\gamma(m)h_m^{\mathsf{T}} w}\right), \quad w \in \mathbb{R}^M \tag{19.84}$$

with $\rho = 0.1$ and $M = 10$. The step-size parameter is varied between 10^{-4} and 10^{-3}. The simulation generates $N = 1000$ random pairs of data $\{\gamma(m), h_m\}$ according to a logistic model. First, a random parameter model $w^a \in \mathbb{R}^{10}$ is selected, and a random collection of feature vectors $\{h_m\}$ are generated, say, with zero-mean unit-variance Gaussian entries. Then, for each h_m, the label $\gamma(m)$ is set to either $+1$ or -1 according to the following construction:

$$\gamma(m) = +1 \;\text{ if }\; \left(\frac{1}{1 + e^{-h_m^{\mathsf{T}} w^a}}\right) \geq 0.5; \;\text{ otherwise } \gamma(m) = -1 \tag{19.85}$$

A total of $K = 2000$ epochs are run over the data. In one simulation, we evaluate the risk value $P(w_{-1}^k)$ at the beginning of each epoch and subsequently average these values over all epochs to approximate the average deviation

$$\mathbb{E}\,P(\boldsymbol{w}_n^k) - P(w^\star) \;\approx\; \frac{1}{K}\sum_{k'=1}^{K} P(w_{-1}^{k'}) - P(w^\star) \tag{19.86}$$

In a second simulation, we use the risk value $P(w_{-1}^K)$ at the last epoch as the approximation for $\mathbb{E}\,P(\boldsymbol{w}_{-1}^k)$, i.e.,

$$\mathbb{E}\,P(\boldsymbol{w}_{-1}^k) - P(w^\star) \;\approx\; P(w_{-1}^K) - P(w^\star) \tag{19.87}$$

Both approximations lead to similar results. We plot the variation of the risk deviations in the logarithmic scale against $\log_{10}(\mu)$ in Fig. 19.2. The plot shows the simulated values for these risk deviations against the step-size parameter. The vertical and horizontal scales are logarithmic. The dotted lines are the fitted regression lines, which provide an estimate of the slope variations for the measurements. The slopes of the lines for uniform sampling and random reshuffling are found in this simulation to be 1.3268 and 2.8512, respectively.

19.6 CONVERGENCE UNDER IMPORTANCE SAMPLING

We now examine the convergence behavior of the stochastic gradient algorithm (19.1) under *importance sampling* for empirical risk minimization. In this implementation, a probability value p_m is assigned to each sample $(\gamma(m), h_m)$ in the dataset, and the samples are selected at random according to this distribution.

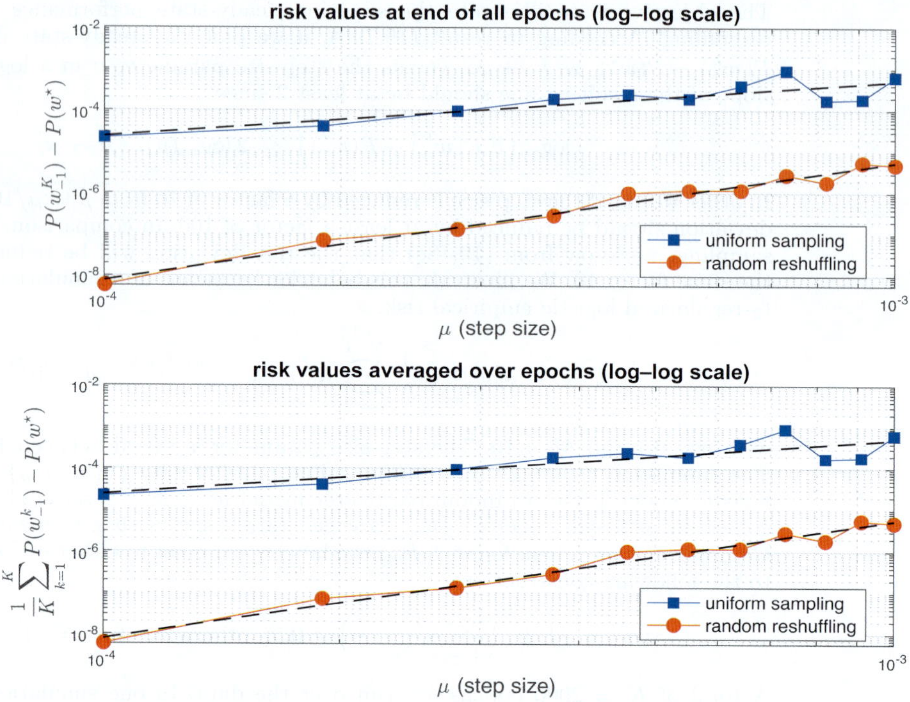

Figure 19.2 Random reshuffling has better risk deviation performance than uniform sampling. The plot shows the simulated values for these risk deviations against the step-size parameter. The vertical and horizontal scales are logarithmic. The dotted lines are the fitted regression lines, which provide estimates for the slopes.

We explained earlier that the approximations for the gradient vector will need to be adjusted and scaled as shown in (19.6a)–(19.6b).

The result of Theorem 19.1 can be extended to the scaled gradient approximations (19.6a)–(19.6b). We will therefore leave the analysis to the problems – see Prob. 19.17, where it is shown that the limiting MSE will continue to be $O(\mu\sigma_g^2)$, where the expression for σ_g^2 was derived earlier in (18.52b):

$$\sigma_g^2 = \frac{2}{N^2}\sum_{m=0}^{N-1}\frac{1}{p_m}\left\|\nabla_{w^\mathsf{T}} Q(w^\star;\gamma(m),h_m)\right\|^2 \tag{19.88}$$

for instantaneous gradient approximations. A similar expression holds with σ_g^2 divided by B for the mini-batch version. Observe that the expression for σ_g^2 involves the selection probabilities $\{p_m\}$.

19.6.1 Optimal Importance Sampling

One important question that is relevant for importance sampling implementations is the choice of the selection probabilities $\{p_m\}$. One possibility is to minimize σ_g^2 over the $\{p_m\}$ in order to reduce the size of the $O(\mu\sigma_g^2)$-limiting neighborhood. Thus, we consider the following constrained optimization problem:

$$\{p_m^o\} \stackrel{\Delta}{=} \underset{\{p_m\}}{\text{argmin}} \left\{ \sum_{m=0}^{N-1} \frac{1}{p_m} \left\| \nabla_{w^\mathsf{T}} Q(w^\star; \gamma(m), h_m) \right\|^2 \right\} \tag{19.89a}$$

subject to

$$0 \le p_m \le 1, \;\; \sum_{m=0}^{N-1} p_m = 1 \tag{19.89b}$$

This problem has a closed-form solution. Let us ignore for the moment the constraint $0 \le p_m \le 1$ and solve the remaining constrained problem by introducing the Lagrangian function:

$$\mathcal{L}(p_m, \alpha) \stackrel{\Delta}{=} \sum_{m=0}^{N-1} \frac{1}{p_m} \left\| \nabla_{w^\mathsf{T}} Q(w^\star; \gamma(m), h_m) \right\|^2 + \alpha \left(\sum_{m=0}^{N-1} p_m - 1 \right) \tag{19.90}$$

where α is a Lagrange multiplier. Differentiating relative to p_m and setting the derivative to zero gives an expression for p_m^o in terms of α:

$$p_m^o = \frac{1}{\sqrt{\alpha}} \left\| \nabla_{w^\mathsf{T}} Q(w^\star; \gamma(m), h_m) \right\| \tag{19.91}$$

Since the sum of the $\{p_m^o\}$ must be 1, we find that

$$\sqrt{\alpha} = \sum_{m=0}^{N-1} \left\| \nabla_{w^\mathsf{T}} Q(w^\star; \gamma(m), h_m) \right\| \tag{19.92}$$

and, hence,

$$p_m^o = \frac{\left\| \nabla_{w^\mathsf{T}} Q(w^\star; \gamma(m), h_m) \right\|}{\sum_{m=0}^{N-1} \left\| \nabla_{w^\mathsf{T}} Q(w^\star; \gamma(m), h_m) \right\|} \tag{19.93}$$

This solution satisfies the constraint $0 \le p_m \le 1$ and leads to an optimal sampling strategy. However, this particular strategy is not practical for two reasons: the values of p_m^o depend on the unknown $w^\star$, and the denominator involves a sum over all N data samples.

19.6.2 Adaptive Importance Sampling

We can address the first difficulty, at every iteration n, by replacing $w^\star$ by the estimate that is available at the start of that iteration, namely, w_{n-1}. This leads to an *adaptive* importance sampling procedure with:

$$p_{m,n}^o = \frac{\left\| \nabla_{w^\mathsf{T}} Q(w_{n-1}; \gamma(m), h_m) \right\|}{\sum_{m=0}^{N-1} \left\| \nabla_{w^\mathsf{T}} Q(w_{n-1}; \gamma(m), h_m) \right\|}, \quad m = 0, 1, \ldots, N-1 \tag{19.94}$$

where we are adding a subscript n to indicate that the probabilities $\{p_{m,n}^o\}$ are the ones used at iteration n while updating $\boldsymbol{w}_{n-1}$ to $\boldsymbol{w}_n$. Expression (19.94) is still inefficient because of the sum in the denominator, which involves all data samples and N can be large. We address this second difficulty by devising a recursive scheme to update the denominator.

We introduce an auxiliary vector variable $\psi_n \in \mathbb{R}^N$, whose size is equal to the number of data samples. One entry of $\boldsymbol{\psi}_n$ is updated at each iteration n. Let σ denote the index of the sample $(\boldsymbol{\gamma}, \boldsymbol{h})$ that is selected for use at iteration n. Then, only the σ-th entry of ψ_n is updated at that iteration:

$$\psi_{n,\sigma} = \beta\psi_{n-1,\sigma} + (1-\beta)\|\nabla_{w^\mathsf{T}} Q(w_{n-1}; \gamma(\sigma), h_\sigma)\| \tag{19.95}$$

where $\beta \in (0,1)$ is a design parameter. All other entries of ψ_n stay identical to the entries from the previous instant ψ_{n-1}. We can express this update in vector form as follows. Let D_σ denote the $N \times N$ diagonal matrix with β at location (σ, σ) and 1s at all other diagonal entries:

$$D_\sigma = \text{diag}\Big\{1, \ldots, 1, \beta, 1, \ldots, 1\Big\}, \quad (N \times N) \tag{19.96}$$

Let also e_σ denote the basis vector in $\mathbb{R}^N$ with a unit entry at location σ. Then,

$$\boxed{\psi_n = D_\sigma \psi_{n-1} + (1-\beta)\|\nabla_{w^\mathsf{T}} Q(w_{n-1}; \gamma(\sigma), h_\sigma)\|\, e_\sigma, \quad n \geq 0} \tag{19.97}$$

We initialize ψ_{-1} to large positive values. Note that at iteration n, only one entry of ψ_n is updated, and hence this update is computationally inexpensive. Moreover, each entry of index σ in ψ_n corresponds to a smooth running estimate of the norm $\|\nabla_{w^\mathsf{T}} Q(w_{n-1}; \gamma(\sigma), h_\sigma)\|$ (which is the quantity that appears in the numerator of (19.94)).

We introduce a second auxiliary scalar quantity, denoted by τ_n for iteration n, in order to keep track of the sum of the entries of ψ; this sum (and, hence, τ) will serve as the approximation for the quantity appearing in the denominator of $p^o_{m,n}$:

$$\tau_n \stackrel{\Delta}{=} \|\psi_n\|_1 = \sum_{m=0}^{N-1} \psi_{n,m} = \sum_{m=0}^{N-1} \psi_{n-1,m} + (\psi_{n,\sigma} - \psi_{n-1,\sigma}) \tag{19.98}$$

and, hence, using (19.97):

$$\boxed{\tau_n = \tau_{n-1} + (1-\beta)\Big(\|\nabla_{w^\mathsf{T}} Q(w_{n-1}; \gamma(\sigma), h_\sigma)\| - \psi_{n-1,\sigma}\Big), \quad n \geq 0} \tag{19.99}$$

with $\tau_{-1} = \|\psi_{-1}\|_1$. Note that each update of τ only requires $O(1)$ operations, which is also inexpensive. This construction leads to a procedure that automatically learns an "optimal" sampling strategy. The algorithm is listed below using instantaneous gradient approximations. In the listing, the vector $r_n \in \mathbb{R}^N$ contains the values of the probabilities $\{p_{m,n}\}$ used at iteration n:

$$r_n \stackrel{\Delta}{=} \{p_{m,n}\}, \quad m = 1, 2, \ldots, M \tag{19.100}$$

Stochastic gradient algorithm with adaptive importance sampling for minimizing the empirical risk (19.3a).

given dataset $\{\gamma(m), h_m\}_{m=0}^{N-1}$;
given a scalar $\beta \in (0,1)$;
start from an arbitrary initial condition $\boldsymbol{w}_{-1} \in \mathbb{R}^M$;
initialize $\boldsymbol{\psi}_{-1} \in \mathbb{R}^N$ to large positive entries;
set $\boldsymbol{\tau}_{-1} = \|\boldsymbol{\psi}_{-1}\|_1$;
set $\boldsymbol{r}_0 = \frac{1}{N}\mathbb{1}_N$; (uniform distribution).
repeat until convergence over $n \geq 0$:
  entries of $\boldsymbol{r}_n$ are the probabilities $\{p_{m,n}\}_{m=0}^{N-1}$ at iteration n; (19.101)
  select an index $0 \leq \sigma \leq N-1$ according to probabilities $\{p_{m,n}\}$;
  let $\boldsymbol{x}_n = \nabla_{w^\mathsf{T}} Q(\boldsymbol{w}_{n-1}; \boldsymbol{\gamma}(\sigma), \boldsymbol{h}_\sigma)$; (approximate gradient)
  $\boldsymbol{w}_n = \boldsymbol{w}_{n-1} - \dfrac{\mu}{N p_{\sigma,n}} \boldsymbol{x}_n$
  $D_\sigma = \text{diag}\{1, \ldots, 1, \beta, 1, \ldots, 1\}$; ($\beta$ at σ-th location)
  $\boldsymbol{\psi}_n = D_\sigma \boldsymbol{\psi}_{n-1} + (1-\beta)\|\boldsymbol{x}_n\|\, e_\sigma$
  $\boldsymbol{\tau}_n = \boldsymbol{\tau}_{n-1} + (1-\beta)(\|\boldsymbol{x}_n\| - \boldsymbol{\psi}_{n-1,\sigma})$
  $\boldsymbol{r}_{n+1} = \boldsymbol{\psi}_n / \boldsymbol{\tau}_n$
end
return $w^\star \leftarrow \boldsymbol{w}_n$.

Example 19.6 (**Simulating importance sampling**) We illustrate the results by considering the regularized logistic regression problem:

$$P(w) = \rho\|w\|^2 + \frac{1}{N}\sum_{m=0}^{N-1} \ln\left(1 + e^{-\gamma(m)h_m^\mathsf{T} w}\right) \tag{19.102}$$

where $h_m \in \mathbb{R}^{10}$ and $\gamma(m) \in \{\pm 1\}$. In the simulation, we generate a random dataset $\{h_m, \gamma(m)\}$ with $N = 500$ using the same logistic model from Example 19.5. We set $\rho = 0.01$, $\mu = 0.001$, and $\beta = 0.25$. We also set the initial condition $\psi_{-1} = 1000\,\mathbb{1}_N$. We run algorithm (19.101) over $K = 200$ epochs and compute the risk values $P(w_{-1}^k)$ at the start of each epoch. This leads to a risk deviation curve $P(w_{-1}^k) - P(w^\star)$ over the epoch index k. We repeat this simulation over $L = 100$ trials and average the deviation curves. The results for uniform sampling and importance sampling are shown in Fig. 19.3.

Table 19.2 summarizes the performance results obtained in this chapter for various stochastic gradient algorithms under uniform sampling, importance sampling, random reshuffling, data streaming, and also for mini-batch implementations. Results are shown for strongly convex risks under both constant and vanishing step sizes.

Table 19.2 Convergence properties of the stochastic gradient algorithm (19.1) with *constant* and *vanishing* step sizes, where $P(w)$ denotes either an empirical risk with minimizer $w^\star$ or a stochastic risk with minimizer w^o.

Mode of operation	Conditions on risk and loss functions	Asymptotic convergence property	Reference
instantaneous gradient approximation (uniform sampling)	$P(w)$: ν-strongly convex $Q(w)$: δ-Lipschitz gradients conditions (**A1**,**A2**)	$\mathbb{E}\,P(\boldsymbol{w}_n) - P(w^\star) \le O(\lambda^n) + O(\mu)$	Thm. 19.1
instantaneous gradient approximation (data streaming)	$P(w)$: ν-strongly convex $Q(w)$: δ-Lipschitz gradients in mean-square sense conditions (**A1**,**A2'**)	$\mathbb{E}\,P(\boldsymbol{w}_n) - P(w^o) \le O(\lambda^n) + O(\mu)$	Thm. 19.1
instantaneous gradient approximation (random reshuffling)	$P(w)$: ν-strongly convex $Q(w)$: δ-Lipschitz gradients conditions (**A1**,**A2**)	$\mathbb{E}\,P(\boldsymbol{w}_n^k) - P(w^\star) \le O(\lambda^k) + O(\mu^2)$ $k \ge 1$: epoch index $n \ge 0$: iteration index	Thm. 19.5
instantaneous gradient approximation (importance sampling)	$P(w)$: ν-strongly convex $Q(w)$: δ-Lipschitz gradients conditions (**A1**,**A2**)	$\mathbb{E}\,P(\boldsymbol{w}_n) - P(w^\star) \le O(\lambda^n) + O(\mu)$	Prob. 19.17 Sec. 19.6
mini-batch gradient approximation (sampling with or without) (replacement or data streaming)	$P(w)$: ν-strongly convex $Q(w)$: δ-Lipschitz gradients in deterministic or mean-square sense conditions (**A1**,**A2**) or (**A1**,**A2'**)	$\mathbb{E}\,P(\boldsymbol{w}_n) - P(w^\star) \le O(\lambda^n) + O(\mu/B)$	Thm. 19.2
instantaneous gradient approximation (uniform sampling or data streaming) (decaying step size)	$P(w)$: ν-strongly convex $Q(w)$: δ-Lipschitz gradients in deterministic or mean-square sense conditions (**A1**,**A2**) or (**A1**,**A2'**) $\mu(n) = \tau/(n+1)$	$\mathbb{E}\,P(\boldsymbol{w}_n) - P(w^\star) \le O(1/n)$	Thm. 19.4
instantaneous gradient approximation (random reshuffling) (decaying step size)	$P(w)$: ν-strongly convex $Q(w)$: δ-Lipschitz gradients conditions (**A1**,**A2**) $\mu(k) = \tau/k$	$\mathbb{E}\,P(\boldsymbol{w}_n^k) - P(w^\star) \le O(1/k)$	Eq. (19.82)

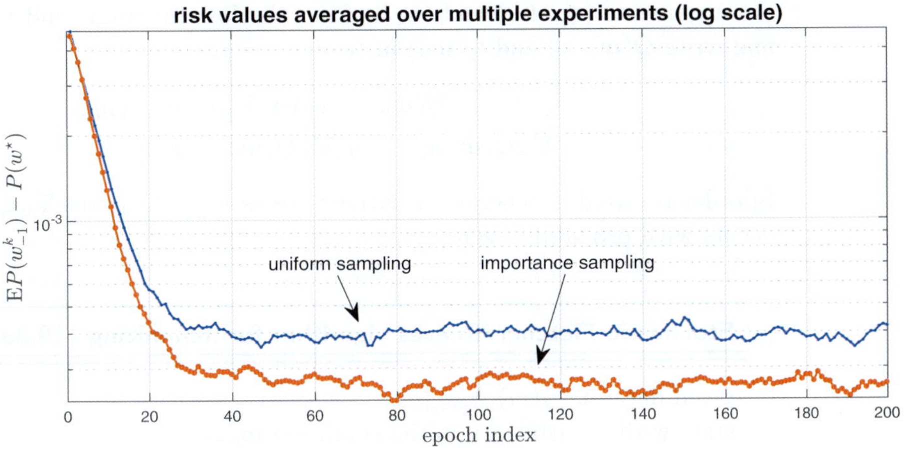

Figure 19.3 Adaptive importance sampling for a regularized logistic regression problem.

19.7 CONVERGENCE OF STOCHASTIC CONJUGATE GRADIENT

We examine in this last section the convergence of the *stochastic* version of the Fletcher–Reeves algorithm listed earlier in (16.102) and repeated in (19.105) for ease of reference. The data is assumed to be sampled uniformly with replacement from a dataset $\{\gamma(m), h_m\}$ for $m = 0, 1, \ldots, N-1$, and the algorithm is used to solve an empirical minimization problem of the form (19.3a).

The convergence analysis given here extends the arguments from Section 13.2 to the stochastic case where the gradient of the risk function is replaced by the gradient of the loss function evaluated at a random data point $(\boldsymbol{\gamma}(n), \boldsymbol{h}_n)$. Several of the steps in the argument are similar and we will therefore be brief. Recall that the arguments in Section 13.2 dealt with general nonlinear optimization problems *without restricting* $P(w)$ to being convex; they established convergence toward a stationary point. We consider the same scenario.

We assume that a line search procedure is used at each iteration n to select parameters $\{\boldsymbol{\alpha}_n\}$ that satisfy the following variation of the Wolfe conditions for some $0 < \lambda < \eta < 1/2$ (compare with (13.101a)–(13.101b)):

$$Q(\boldsymbol{w}_n; \boldsymbol{\gamma}, \boldsymbol{h}) \leq Q(\boldsymbol{w}_{n-1}; \boldsymbol{\gamma}_n, \boldsymbol{h}_n) + \lambda \boldsymbol{\alpha}_n \nabla_w Q(\boldsymbol{w}_{n-1}; \boldsymbol{\gamma}_n, \boldsymbol{h}_n) \boldsymbol{q}_n \tag{19.103a}$$

$$|\nabla_w Q(\boldsymbol{w}_n; \boldsymbol{\gamma}, \boldsymbol{h}) \boldsymbol{q}_n| \leq \eta \, |\nabla_w Q(\boldsymbol{w}_{n-1}; \boldsymbol{\gamma}_n, \boldsymbol{h}_n) \boldsymbol{q}_n| \tag{19.103b}$$

Here, the notation $(\boldsymbol{\gamma}(n), \boldsymbol{h}_n)$ refers to the data sample selected at iteration n, and $(\boldsymbol{\gamma}, \boldsymbol{h})$ denotes the random sample for iteration $n+1$. For simplicity, we drop the arguments $(\gamma_n, \boldsymbol{h}_n)$ and $(\gamma, \boldsymbol{h})$ from the loss functions and their gradients and write $Q(\boldsymbol{w}_{n-1})$ and $Q(\boldsymbol{w}_n)$ instead:

$$Q(\boldsymbol{w}_n) \leq Q(\boldsymbol{w}_{n-1}) + \lambda \boldsymbol{\alpha}_n \nabla_w Q(\boldsymbol{w}_{n-1})\boldsymbol{q}_n \tag{19.104a}$$

$$|\nabla_w Q(\boldsymbol{w}_n)\boldsymbol{q}_n| \leq \eta\, |\nabla_w Q(\boldsymbol{w}_{n-1})\boldsymbol{q}_n| \tag{19.104b}$$

In order to avoid degenerate situations, we assume the event $\nabla_w Q(\boldsymbol{w}; \boldsymbol{\gamma}, \boldsymbol{h}) = 0$ occurs with probability zero.

Stochastic Fletcher–Reeves algorithm for minimizing (19.3a).

given dataset $\{\gamma(m), h_m\}_{m=0}^{N-1}$;
start with an arbitrary initial condition $\boldsymbol{w}_{-1}$;
set $\boldsymbol{q}_{-1} = 0$;
repeat until convergence over $n \geq 0$:
  select at random $(\boldsymbol{\gamma}(n), \boldsymbol{h}_n)$;
  $\boldsymbol{r}_{n-1} = -\nabla_{w^\mathsf{T}} Q(\boldsymbol{w}_{n-1}; \boldsymbol{\gamma}(n), \boldsymbol{h}_n)$
  if $n = 0$ then $\boldsymbol{\beta}_{-1} = 0$
    else $\boldsymbol{\beta}_{n-1} = \|\boldsymbol{r}_{n-1}\|^2 / \|\boldsymbol{r}_{n-2}\|^2$
  end
  $\boldsymbol{q}_n = \boldsymbol{r}_{n-1} + \boldsymbol{\beta}_{n-1}\boldsymbol{q}_{n-1}$
  find $\boldsymbol{\alpha}_n$ using line search: $\min\limits_{\alpha \in \mathbb{R}} Q(\boldsymbol{w}_{n-1} + \alpha \boldsymbol{q}_n)$
  $\boldsymbol{w}_n = \boldsymbol{w}_{n-1} + \boldsymbol{\alpha}_n \boldsymbol{q}_n$
end
return $w^\star \leftarrow \boldsymbol{w}_n$.

(19.105)

Lemma 19.1. (Loss descent directions) *Assume the* $\{\boldsymbol{\alpha}_n\}$ *are selected to satisfy (19.104a)–(19.104b) for* $0 < \lambda < \eta < 1/2$, *then it holds for any* $n \geq 0$ *that*

$$-\frac{1}{1-\eta} \leq \frac{\nabla_w Q(\boldsymbol{w}_{n-1})\boldsymbol{q}_n}{\|\nabla_w Q(\boldsymbol{w}_{n-1})\|^2} \leq \frac{2\eta - 1}{1-\eta} < 0 \tag{19.106}$$

and, hence, the successive $\{\boldsymbol{q}_n\}$ *generated by Fletcher–Reeves are descent directions relative to the loss function* $Q(w; \cdot)$.

Proof: The argument is similar to the one used to establish Lemma 13.1. For $n = 0$, we have $\boldsymbol{q}_0 = -\nabla_w Q(\boldsymbol{w}_{-1})$ so that the ratio in the middle is equal to -1 and both sides of the inequality are satisfied. Now suppose the inequality holds for iteration n and let us verify that it holds for iteration $n+1$. It follows that $\boldsymbol{q}_n$ is a descent direction so that condition (19.104b) becomes

$$|\nabla_w Q(\boldsymbol{w}_n)\boldsymbol{q}_n| \leq -\eta \nabla_w Q(\boldsymbol{w}_{n-1})\boldsymbol{q}_n \tag{19.107}$$

which is equivalent to

$$\eta \nabla_w Q(\boldsymbol{w}_{n-1})\boldsymbol{q}_n \leq \nabla_w Q(\boldsymbol{w}_n)\boldsymbol{q}_n \leq -\eta \nabla_w Q(\boldsymbol{w}_{n-1})\boldsymbol{q}_n \tag{19.108}$$

Now note from recursions (19.105) that

$$\begin{aligned}
\frac{\nabla_w Q(\boldsymbol{w}_n)\boldsymbol{q}_{n+1}}{\|\nabla_w Q(\boldsymbol{w}_n)\|^2} &= \frac{\nabla_w Q(\boldsymbol{w}_n)(\boldsymbol{r}_n + \boldsymbol{\beta}_n \boldsymbol{q}_n)}{\|\nabla_w Q(\boldsymbol{w}_n)\|^2} \\
&= -1 + \boldsymbol{\beta}_n \frac{\nabla_w Q(\boldsymbol{w}_n)\boldsymbol{q}_n}{\|\nabla_w Q(\boldsymbol{w}_n)\|^2}, \quad \text{since } \boldsymbol{r}_n = -\nabla_{w^\mathsf{T}} Q(\boldsymbol{w}_n) \\
&= -1 + \frac{\|\boldsymbol{r}_n\|^2}{\|\boldsymbol{r}_{n-1}\|^2} \frac{\nabla_w Q(\boldsymbol{w}_n)\boldsymbol{q}_n}{\|\nabla_w Q(\boldsymbol{w}_n)\|^2} \\
&= -1 + \frac{\nabla_w Q(\boldsymbol{w}_n)\boldsymbol{q}_n}{\|\nabla_w Q(\boldsymbol{w}_{n-1})\|^2}
\end{aligned} \tag{19.109}$$

Using (19.108) we obtain

$$-1 + \eta \frac{\nabla_w Q(\boldsymbol{w}_{n-1})\boldsymbol{q}_n}{\|\nabla_w Q(\boldsymbol{w}_{n-1})\|^2} \le -1 + \frac{\nabla_w Q(\boldsymbol{w}_n)\boldsymbol{q}_n}{\|\nabla_w Q(\boldsymbol{w}_{n-1})\|^2} \le -1 - \eta \frac{\nabla_w Q(\boldsymbol{w}_{n-1})\boldsymbol{q}_n}{\|\nabla_w Q(\boldsymbol{w}_{n-1})\|^2} \tag{19.110}$$

and applying the lower bound from (19.106) to the leftmost and rightmost terms we get

$$-1 - \frac{\eta}{1-\eta} \le \frac{\nabla_w Q(\boldsymbol{w}_n)\boldsymbol{q}_{n+1}}{\|\nabla_w Q(\boldsymbol{w}_n)\|^2} \le -1 + \frac{\eta}{1-\eta} \tag{19.111}$$

which establishes the validity of (19.106) for $n+1$.

■

The next result extends the Zoutendijk condition to the stochastic case. Again, neither the loss nor the risk function are required to be convex in this statement.

LEMMA 19.2. (Stochastic Zoutendijk condition) *Consider an empirical risk minimization problem of the form (19.3a), where $P(w)$ is bounded from below (but not necessarily convex) and the loss function $Q(w, \cdot)$ is first-order differentiable with δ-Lipschitz gradients as in (19.7b). The data is sampled uniformly with replacement from $\{\gamma(m), h_m\}$. Assume the $\{\boldsymbol{\alpha}_n\}$ are selected to satisfy (19.104a)–(19.104b) for $0 < \lambda < \eta < 1/2$ so that the $\{\boldsymbol{q}_n\}$ generated by the stochastic Fletcher–Reeves procedure (19.105) are descent directions relative to $Q(w; \cdot)$ by Lemma 19.1. The iterate $\boldsymbol{w}_{n-1}$ is updated to $\boldsymbol{w}_n = \boldsymbol{w}_{n-1} + \boldsymbol{\alpha}_n \boldsymbol{q}_n$. Let $\boldsymbol{\theta}_n$ denote the angle defined by:*

$$\cos(\boldsymbol{\theta}_n) \overset{\Delta}{=} \frac{-\nabla_w Q(\boldsymbol{w}_{n-1})\boldsymbol{q}_n}{\|\nabla_w Q(\boldsymbol{w}_{n-1})\| \, \|\boldsymbol{q}_n\|} \tag{19.112}$$

It then holds that

$$\sum_{n=0}^{\infty} \mathbb{E}\left(\cos^2(\boldsymbol{\theta}_n)\|\nabla_w Q(\boldsymbol{w}_{n-1})\|^2\right) < +\infty \tag{19.113}$$

where the expectation is over the randomness of the data. Moreover, $\mathbb{E}\,P(\boldsymbol{w}_n)$ is nonincreasing, meaning that $\mathbb{E}\,P(\boldsymbol{w}_n) \le \mathbb{E}\,P(\boldsymbol{w}_{n-1})$.

Proof: The argument is similar to the one used to establish Lemma 13.2. Since $\boldsymbol{q}_n$ is a descent direction relative to the loss function, we conclude from the second Wolfe condition (19.104b) or from (19.108) that

$$\nabla_w Q(\boldsymbol{w}_n)\boldsymbol{q}_n \;\geq\; \eta\,\nabla_w Q(\boldsymbol{w}_{n-1})\boldsymbol{q}_n \tag{19.114}$$

Subtracting $\nabla_w Q(\boldsymbol{w}_{n-1})\boldsymbol{q}_n$ from both sides gives

$$\Big(\nabla_w Q(\boldsymbol{w}_n) - \nabla_w Q(\boldsymbol{w}_{n-1})\Big)\boldsymbol{q}_n \;\geq\; (\eta-1)\nabla_w Q(\boldsymbol{w}_{n-1})\boldsymbol{q}_n \tag{19.115}$$

From the δ-Lipschitz condition (19.7b) on the gradient of $Q(w;\cdot)$ we have by Cauchy–Schwarz:

$$\begin{aligned}\Big(\nabla_w Q(\boldsymbol{w}_n) - \nabla_w Q(\boldsymbol{w}_{n-1})\Big)\boldsymbol{q}_n &\leq \|\nabla_w Q(\boldsymbol{w}_n) - \nabla_w Q(\boldsymbol{w}_{n-1})\|\,\|\boldsymbol{q}_n\| \\ &\leq \delta\|\boldsymbol{w}_n - \boldsymbol{w}_{n-1}\|\,\|\boldsymbol{q}_n\| \\ &= \delta\boldsymbol{\alpha}_n\|\boldsymbol{q}_n\|^2, \quad \text{since } \boldsymbol{w}_n = \boldsymbol{w}_{n-1} + \boldsymbol{\alpha}_n\boldsymbol{q}_n\end{aligned} \tag{19.116}$$

Combining (19.115) and (19.116) shows that $\boldsymbol{\alpha}_n$ is lower-bounded by

$$\boldsymbol{\alpha}_n \geq \frac{(\eta-1)}{\delta}\,\frac{\nabla_w Q(\boldsymbol{w}_{n-1})\boldsymbol{q}_n}{\|\boldsymbol{q}_n\|^2} \tag{19.117}$$

where the term on the right-hand side is positive since $\boldsymbol{q}_n$ is a descent direction and $\eta < 1$. Substituting this conclusion into the first Wolfe condition (19.104a) we find that

$$Q(\boldsymbol{w}_n) \leq Q(\boldsymbol{w}_{n-1}) + \underbrace{\lambda\frac{(\eta-1)}{\delta}}_{\triangleq\, -c}\,\frac{\left(\nabla_w Q(\boldsymbol{w}_{n-1})\boldsymbol{q}_n\right)^2}{\|\boldsymbol{q}_n\|^2} \tag{19.118}$$

where we introduced the positive constant $c = \lambda(1-\eta)/\delta$. In other words, in terms of the angles $\{\boldsymbol{\theta}_n\}$ we have that

$$Q(\boldsymbol{w}_n) \leq Q(\boldsymbol{w}_{n-1}) - c\,\cos^2(\boldsymbol{\theta}_n)\,\|\nabla_w Q(\boldsymbol{w}_{n-1})\|^2 \tag{19.119}$$

Taking expectations over the randomness in the data we get

$$\mathbb{E}\,P(\boldsymbol{w}_n) \leq \mathbb{E}\,P(\boldsymbol{w}_{n-1}) - c\,\mathbb{E}\left(\cos^2(\boldsymbol{\theta}_n)\,\|\nabla_w Q(\boldsymbol{w}_{n-1})\|^2\right) \tag{19.120}$$

which shows that $\mathbb{E}\,P(\boldsymbol{w}_m)$ is nonincreasing. Summing over n gives

$$\sum_{n=0}^{\infty}\mathbb{E}\left(\cos^2(\boldsymbol{\theta}_n)\,\|\nabla_w Q(\boldsymbol{w}_{n-1})\|^2\right) \;\leq\; \frac{1}{c}\Big(\mathbb{E}\,P(\boldsymbol{w}_{-1}) - \lim_{n\to\infty}\mathbb{E}\,P(\boldsymbol{w}_n)\Big) \tag{19.121}$$

Since, by assumption, the risk function is bounded from below, the term on the right-hand side is bounded by some positive constant, and conclusion (19.113) holds.

■

One useful corollary of the previous two lemmas follows if we multiply (19.106) by $\|\nabla_w Q(\boldsymbol{w}_{n-1})\|/\|\boldsymbol{q}_n\|$ and use (19.113) to conclude that the following condition must also hold:

$$\sum_{n=0}^{\infty}\mathbb{E}\left(\frac{\|\nabla_w Q(\boldsymbol{w}_{n-1})\|^4}{\|\boldsymbol{q}_n\|^2}\right) < +\infty \tag{19.122}$$

Theorem 19.6. (Convergence of stochastic Fletcher-Reeves) *Consider the same setting of Lemma 19.2 and assume further that the* $\{\boldsymbol{\beta}_n\}$ *are uniformly bounded almost surely, say,* $\boldsymbol{\beta}_n \leq B$ *a.s., for some finite* B. *Then, a sequence of random indices* $\{\boldsymbol{m}_1, \boldsymbol{m}_2, \boldsymbol{m}_3, \ldots\}$ *exists such that*

$$\lim_{k\to\infty} \nabla_w P(\boldsymbol{w}_{\boldsymbol{m}_k}) \overset{\text{a.s.}}{=} 0 \tag{19.123}$$

Proof: Since $\boldsymbol{q}_m$ is a descent direction relative to the loss, we conclude from the second Wolfe condition (19.104b) that

$$|\nabla_w Q(\boldsymbol{w}_n)\boldsymbol{q}_n| \leq -\eta \nabla_w Q(\boldsymbol{w}_{n-1})\boldsymbol{q}_n \overset{(19.106)}{\leq} \frac{\eta}{1-\eta}\|\nabla_w Q(\boldsymbol{w}_{n-1})\|^2 \tag{19.124}$$

It follows that

$$\begin{aligned}
\|\boldsymbol{q}_{n+1}\|^2 &= \|\boldsymbol{r}_n + \boldsymbol{\beta}_n \boldsymbol{q}_n\|^2 \\
&\leq \|\nabla_w Q(\boldsymbol{w}_n)\|^2 + 2\boldsymbol{\beta}_n |\nabla_w Q(\boldsymbol{w}_n)\boldsymbol{q}_n| + \boldsymbol{\beta}_n^2 \|\boldsymbol{q}_n\|^2 \\
&= \underbrace{\frac{1+\eta}{1-\eta}}_{\triangleq\, c_2 > 1} \|\nabla_w Q(\boldsymbol{w}_n)\|^2 \;+\; \boldsymbol{\beta}_n^2 \|\boldsymbol{q}_n\|^2
\end{aligned} \tag{19.125}$$

Iterating we get

$$\|\boldsymbol{q}_{n+1}\|^2 \leq c_2 \|\nabla_w Q(\boldsymbol{w}_n)\|^4 \sum_{j=0}^{n+1} \frac{1}{\|\nabla_w Q(\boldsymbol{w}_{j-1})\|^2} \tag{19.126}$$

In a manner similar to the proof of Theorem 13.1 we can establish that the limit inferior of $\|\nabla_w Q(\boldsymbol{w}_n)\|$ is zero for almost every trajectory $\{\boldsymbol{w}_{-1}, \boldsymbol{w}_0, \boldsymbol{w}_1, \boldsymbol{w}_2, \ldots\}$:[1]

$$\liminf_{n\to\infty} \|\nabla_w Q(\boldsymbol{w}_n)\| \overset{\text{a.s.}}{=} 0 \tag{19.127}$$

Indeed, assume, by contradiction, that this is not the case. Then, since by assumption the event $\|\nabla_w Q(\boldsymbol{w}_n)\| = 0$ occurs with probability 0, the norm $\|\nabla_w Q(\boldsymbol{w}_n)\|$ should be bounded from below for all n, say,

$$\|\nabla_w Q(\boldsymbol{w}_n)\| \geq \boldsymbol{c}_3 > 0, \quad \text{for all } n > 0 \tag{19.128}$$

where $\boldsymbol{c}_3$ is some random variable with $\mathbb{E}\,\boldsymbol{c}_3^2 > 0$. Applying this bound to the rightmost sum term in (19.126), taking expectations, and summing over n leads to

$$\sum_{n=0}^{\infty} \mathbb{E}\left(\frac{\|\nabla_w Q(\boldsymbol{w}_n)\|^4}{\|\boldsymbol{q}_{n+1}\|^2}\right) \geq \frac{\mathbb{E}\,\boldsymbol{c}_3^2}{c_2} \sum_{n=0}^{\infty} \frac{1}{n+2} \tag{19.129}$$

which is not bounded. This conclusion contradicts (19.122). We conclude that (19.127) should hold almost surely. This in turn implies that a subsequence of nonnegative random indices $\{\boldsymbol{n}_1, \boldsymbol{n}_2, \boldsymbol{n}_3, \ldots\}$ exists such that

$$\lim_{j\to\infty} \|\nabla_w Q(\boldsymbol{w}_{\boldsymbol{n}_j})\| \overset{\text{a.s.}}{=} 0 \tag{19.130}$$

[1] The author gratefully acknowledges the contribution of Dr. Vincenzo Matta from the University of Salerno, Italy, to the remainder of the argument in this proof.

One selection for the indices $\{\boldsymbol{n}_j\}$ is as follows. Let $\epsilon_1 > \epsilon_2 > \epsilon_3 > \ldots$ be a sequence of strictly positive numbers with $\epsilon_j \rightarrow 0$ as $j \rightarrow \infty$, and let

$$\boldsymbol{n}_1\text{: } \underset{n \geq 0}{\text{argmin}} \left\{ \|\nabla_w Q(\boldsymbol{w}_n; \boldsymbol{\gamma}(n+1), \boldsymbol{h}_{n+1})\| \leq \epsilon_1 \right\} \tag{19.131}$$

$$\boldsymbol{n}_j\text{: } \underset{n > \boldsymbol{n}_{j-1}}{\text{argmin}} \left\{ \|\nabla_w Q(\boldsymbol{w}_n; \boldsymbol{\gamma}(n+1), \boldsymbol{h}_{n+1})\| \leq \epsilon_j \right\}, \; j = 2, 3, \ldots \tag{19.132}$$

Since the indices $\{\boldsymbol{n}_j\}$ are data-dependent random quantities, the distribution of the data $\{\boldsymbol{\gamma}(n+1), \boldsymbol{h}_{n+1}\}$ in the above expressions does not coincide with the original distribution for $\{\boldsymbol{\gamma}, \boldsymbol{h}\}$. Let us, however, shift $\boldsymbol{n}_j$ by one time lag and consider $\nabla_w Q(\boldsymbol{w}_{\boldsymbol{n}_j+1}; \boldsymbol{\gamma}(\boldsymbol{n}_j+2), \boldsymbol{h}_{\boldsymbol{n}_j+2})$. Using $\boldsymbol{\beta}_{\boldsymbol{n}_j+1} \leq B$ a.s., we find from (19.130) that

$$\lim_{j \rightarrow \infty} \|\nabla_w Q(\boldsymbol{w}_{\boldsymbol{n}_j+1}; \boldsymbol{\gamma}(\boldsymbol{n}_j+2), \boldsymbol{h}_{\boldsymbol{n}_j+2})\| \overset{\text{a.s.}}{=} 0 \tag{19.133}$$

It follows from the uniform boundedness of $\|\nabla_w Q(\boldsymbol{w}_{\boldsymbol{n}_j})\|$ constructed via (19.131)–(19.132), and from the dominated convergence theorem (*cf.* Appendix 16.A), that

$$\lim_{j \rightarrow \infty} \mathbb{E}\, \|\nabla_w Q(\boldsymbol{w}_{\boldsymbol{n}_j+1}; \boldsymbol{\gamma}(\boldsymbol{n}_j+2), \boldsymbol{h}_{\boldsymbol{n}_j+2})\| = 0 \tag{19.134}$$

where the expectation is relative to the randomness in the data and weight vectors. But since $\boldsymbol{n}_j$ is now independent of $\{\boldsymbol{\gamma}(\boldsymbol{n}_j+2), \boldsymbol{h}_{\boldsymbol{n}_j+2}\}$ we have that the distribution of $\{\boldsymbol{\gamma}(\boldsymbol{n}_j+2), \boldsymbol{h}_{\boldsymbol{n}_j+2}\}$ coincides with the original distribution for $\{\boldsymbol{\gamma}, \boldsymbol{h}\}$ so that

$$\begin{aligned}
\mathbb{E}_{\,w,\gamma,h} \|\nabla_w Q(\boldsymbol{w}_{\boldsymbol{n}_j+1})\| &= \mathbb{E}_{\,w} \mathbb{E}_{\,\gamma,h|w} \|\nabla_w Q(\boldsymbol{w}_{\boldsymbol{n}_j+1}; \boldsymbol{\gamma}(\boldsymbol{n}_j+2), \boldsymbol{h}_{\boldsymbol{n}_j+2})\| \\
&\geq \mathbb{E}_{\,w} \|\nabla_w \mathbb{E}_{\,\gamma,h|w} Q(\boldsymbol{w}_{\boldsymbol{n}_j+1}; \boldsymbol{\gamma}(\boldsymbol{n}_j+2), \boldsymbol{h}_{\boldsymbol{n}_j+2})\| \\
&= \mathbb{E}_{\,w} \|\nabla_w P(\boldsymbol{w}_{\boldsymbol{n}_j+1})\|
\end{aligned} \tag{19.135}$$

Combining with (19.134) we get $\lim_{j \rightarrow \infty} \mathbb{E}\, \|\nabla_w P(\boldsymbol{w}_{\boldsymbol{n}_j+1})\| = 0$, so that $\nabla_w P(\boldsymbol{w}_{\boldsymbol{n}_j+1})$ converges to 0 in probability. This in turm implies (*cf.* Appendix 3.A) that a subsequence $\{\boldsymbol{m}_k \overset{\Delta}{=} \boldsymbol{n}_{j_k} + 1\}$ exists such that $\nabla_w P(\boldsymbol{w}_{\boldsymbol{m}_k})$ converges to zero almost surely.

■

19.8 COMMENTARIES AND DISCUSSION

Stochastic gradient algorithms. There are extensive works in the literature on stochastic gradient algorithms and their convergence behavior, including by Albert and Gardner (1967), Wasan (1969), Mendel and Fu (1970), Tsypkin (1971), Ljung (1977), Kushner and Clark (1978), Kushner (1984), Polyak (1987), Benveniste, Métivier, and Priouret (1987), Bertsekas and Tsitsiklis (1997, 2000), Bottou (1998, 2010, 2012), Kushner and Yin (2003), Spall (2003), Marti (2005), Sayed (2003, 2008, 2014a), Shalev-Shwartz and Ben-David (2014), and Bottou, Curtis, and Nocedal (2018). The proof of Theorem 19.1 for operation under constant step sizes follows the argument from Sayed (2014a). The convergence result (19.158) in the appendix for a stochastic inequality recursion is from Polyak (1987, pp. 49–50). The result is useful in characterizing the convergence rates of stochastic approximation algorithms with diminishing step sizes, as was shown in the proof of Theorem 19.3 following Polyak (1987). Some of the earlier studies on regret analysis for stochastic optimization algorithms are the works by Gordon (1999) and Zinkevich (2003) – see also the treatment by Shalev-Shwartz (2011) and the references therein. Analysis of the convergence behavior of the stochastic gradient algorithm under Polyak and Nesterov momentum acceleration schemes appear in Yuan, Ying, and Sayed (2016) using the general model described earlier in Prob. 17.12 and arguments similar to those employed in this chapter. The discussion in Example 19.2 motivating the choice $\mu(n) = O(1/n)$ for the decaying step-size sequence is in line with the analysis and conclusions from Robbins and Monro (1951) and Nemirovski *et al.* (2009). The presentation in Examples 19.3–19.4 is motivated by arguments from

Bottou, Curtis, and Nocedal (2018). In Section 19.7 we examined the convergence of the *stochastic*version of the Fletcher–Reeves algorithm by extending the derivation from Section 13.2 to the stochastic case (19.105). The analysis and proofs follow arguments similar to Zoutendijk (1970), Powell (1984, 1986), Al-Baali (1985), and more closely the presentation from Nocedal and Wright (2006).

Mean-square deviation and excess risk measures. Theorem 19.1 characterizes the size of the limiting region for the MSE, namely, $\mathbb{E}\,\|\widetilde{\boldsymbol{w}}_n\|^2 \to O(\mu)$ for sufficiently small μ. If desired, one can pursue a more detailed MSE analysis and quantify *more accurately* the size of the constant multiplying μ in the $O(\mu)$-result. Consider a stochastic gradient implementation that is based on instantaneous gradient approximations and uniform sampling of the data. Let $w^\star$ denote the minimizer of the risk function $P(w)$, which can be an empirical or stochastic risk. Let $\boldsymbol{g}_n(w^\star)$ denote the gradient noise at location $w = w^\star$, i.e.,

$$\boldsymbol{g}_n(w^\star) \;=\; \widehat{\nabla_{w^\mathsf{T}} P}(w^\star) - \nabla_{w^\mathsf{T}} P(w^\star) \;=\; \nabla_{w^\mathsf{T}} Q(w^\star; \boldsymbol{\gamma}(n), \boldsymbol{h}_n) \tag{19.136}$$

and denote its steady-state covariance matrix, assumed to exist, by

$$R_g \;\overset{\Delta}{=}\; \lim_{n\to\infty} \mathbb{E}\,\boldsymbol{g}_n(w^\star)\boldsymbol{g}_n^\mathsf{T}(w^\star) \tag{19.137}$$

where the expectation is over the randomness in the data. The above expression assumes that the covariance matrix approaches a stationary value R_g. We know from (19.13b) that $\mathrm{Tr}(R_g) \leq \sigma_g^2$. Assume further that $P(w)$ is twice-differentiable and denote its Hessian matrix at $w = w^\star$ by

$$H \;\overset{\Delta}{=}\; \nabla_w^2\, P(w^\star) \tag{19.138}$$

Since $P(w)$ is ν-strongly convex, we know that $H \geq \nu I_M$. Then, it can be verified under (**A1**,**A2**) or (**A1**,**A2'**) and by exploiting the bound on the fourth-order moment of the gradient noise process established in Prob. 18.4, that (see the derivation in Sayed (2014a, ch.4)):

$$(\textbf{MSD})\quad \limsup_{n\to\infty} \mathbb{E}\,\|\widetilde{\boldsymbol{w}}_n\|^2 \;=\; \frac{\mu}{2}\mathrm{Tr}(H^{-1}R_g) + O(\mu^{3/2}) \tag{19.139a}$$

$$(\textbf{ER})\quad \limsup_{n\to\infty} \Big(\mathbb{E}\,P(\boldsymbol{w}_n) - P(w^\star)\Big) \;=\; \frac{\mu}{4}\mathrm{Tr}(R_g) + O(\mu^{3/2}) \tag{19.139b}$$

where the notation MSD and ER stands for "mean-square deviation" and "excess-risk," respectively. Extensions of these results to sampling *without* replacement appear in table I of Ying *et al.* (2019). The above expressions are consistent with the bounds derived in the body of the chapter. For example, if we replace the upper bound $H^{-1} \leq \nu I_M$ into (19.139a) we find that $\mathrm{MSD} = O(\mu\sigma_g^2/2\nu)$, which is consistent with (19.16a). Likewise, using $\mathrm{Tr}(R_g) \leq \sigma_g^2$ in (19.139b) we find $\mathrm{ER} = O(\mu\sigma_g^2/4)$, which is consistent with (19.16b).

A simplified justification for (19.139a) is given further ahead under the remarks on the "long-term model." Consider, for illustration purposes, the following special case involving a quadratic stochastic risk:

$$P(w) = \mathbb{E}\,(\boldsymbol{\gamma} - \boldsymbol{h}^\mathsf{T} w)^2 \;=\; \sigma_\gamma^2 - 2r_{h\gamma}^\mathsf{T} w + w^\mathsf{T} R_h w \tag{19.140}$$

which we expanded in terms of the second-order moments $\sigma_\gamma^2 = \mathbb{E}\,\boldsymbol{\gamma}^2$, $r_{h\gamma} = \mathbb{E}\,\boldsymbol{h}\boldsymbol{\gamma}$, and $R_h = \mathbb{E}\,\boldsymbol{h}\boldsymbol{h}^\mathsf{T} > 0$. The random variables $\{\boldsymbol{\gamma}, \boldsymbol{h}\}$ are assumed to have zero means. The Hessian matrix of $P(w)$ is $H = 2R_h$ for all w. If we differentiate $P(w)$ relative to w and set the gradient vector to zero, we find that the minimizer occurs at location

$$R_h w^\star = r_{h\gamma} \quad\Longleftrightarrow\quad w^\star = R_h^{-1} r_{h\gamma} \tag{19.141}$$

Assume the streaming data $\{\boldsymbol{\gamma}(n), \boldsymbol{h}_n\}$ arise from a linear regression model of the form $\boldsymbol{\gamma}(n) = \boldsymbol{h}_n^\mathsf{T} w^\bullet + \boldsymbol{v}(n)$, for some model $w^\bullet \in \mathbb{R}^M$, and where $\boldsymbol{h}_n$ and $\boldsymbol{v}(n)$ are zero-mean uncorrelated processes. Moreover, $\boldsymbol{v}(n)$ is a white-noise process that is independent of all other random variables and has variance denoted by $\sigma_v^2 = \mathbb{E}\,\boldsymbol{v}^2(n)$. We showed in Prob. 18.9 that $w^\star = w^\bullet$, which means that the minimizer $w^\star$ is able to recover the underlying model $w^\bullet$ and, hence, it also holds that

$$\boldsymbol{v}(n) = \boldsymbol{\gamma}(n) - \boldsymbol{h}_n^\mathsf{T} w^\star \tag{19.142}$$

The gradient noise for instantaneous gradient approximations is given by

$$\boldsymbol{g}_n(w) = 2(r_{h\gamma} - R_h \boldsymbol{w}_{n-1}) - 2\boldsymbol{h}_n(\boldsymbol{\gamma}(n) - \boldsymbol{h}_n^\mathsf{T} w) \tag{19.143}$$

Evaluating at $w = w^\star$ gives

$$\boldsymbol{g}_n(w^\star) = -2\boldsymbol{h}_n(\boldsymbol{\gamma}(n) - \boldsymbol{h}_n^\mathsf{T} w^\star) = -2\boldsymbol{h}_n \boldsymbol{v}(n) \tag{19.144}$$

whose covariance matrix is

$$R_g \stackrel{\Delta}{=} \mathbb{E}\,\boldsymbol{g}_n(w^\star)\boldsymbol{g}_n^\mathsf{T}(w^\star) \;=\; 4\sigma_v^2 R_h \tag{19.145}$$

Substituting into (19.139a)–(19.139b) we arrive at the famous expressions for the performance of the least-mean-squares (LMS) algorithm (see, e.g., Widrow and Stearns (1985), Haykin (2001), Sayed (2003, 2008)):

$$\text{MSD}^{\text{LMS}} \approx \mu M \sigma_v^2 \tag{19.146a}$$

$$\text{ER}^{\text{LMS}} \approx \mu \sigma_v^2 \text{Tr}(R_h) \tag{19.146b}$$

Long-term model. Consider the stochastic gradient algorithm (19.1) and assume an implementation with an instantaneous gradient approximation (similar remarks will hold for the mini-batch version):

$$\begin{aligned} \boldsymbol{w}_n &= \boldsymbol{w}_{n-1} - \mu \nabla_{w^\mathsf{T}} Q(\boldsymbol{w}_{n-1}; \boldsymbol{\gamma}(n), \boldsymbol{h}_n)) \\ &= \boldsymbol{w}_{n-1} - \mu \nabla_{w^\mathsf{T}} P(\boldsymbol{w}_{n-1}) - \mu \boldsymbol{g}_n(\boldsymbol{w}_{n-1}) \end{aligned} \tag{19.147}$$

where the second equality is in terms of the true gradient vector and the gradient noise. We analyzed the convergence behavior of this recursion in the body of the chapter and discovered that $\mathbb{E}\,\|\widetilde{\boldsymbol{w}}_n\|^2 \to O(\mu)$ for sufficiently small step sizes. Recursion (19.147) describes a stochastic system, with its state vector $\boldsymbol{w}_n$ evolving randomly over time due to the randomness in the data samples and the resulting gradient noise. In many instances, it is useful to introduce an approximate model, with constant dynamics, that could serve as a good approximation for the evolution of the state vector for large time instants. We motivate this *long-term model* as follows – see Sayed (2014a, b).

Assume $P(w)$ is twice-differentiable and that its Hessian matrix is τ-Lipschitz relative to the minimizer $w^\star$, meaning that

$$\|\nabla_w^2 P(w) - \nabla_w^2 P(w^\star)\| \;\leq\; \tau\,\|\widetilde{w}\|, \quad \widetilde{w} = w^\star - w \tag{19.148}$$

Since $P(w)$ is ν-strongly convex and its gradient vectors are δ-Lipschitz, we also know that $\nu I_M \leq \nabla_w^2\, P(w) \leq \delta I_M$. Using the mean-value theorem (10.8) we can write

$$\nabla_{w^\mathsf{T}} P(\boldsymbol{w}_{n-1}) = -\underbrace{\left(\int_0^1 \nabla_w^2 P(w^\star - t\widetilde{\boldsymbol{w}}_{n-1})dt\right)}_{\stackrel{\Delta}{=}\, \boldsymbol{H}_{n-1}} \widetilde{\boldsymbol{w}}_{n-1} \tag{19.149}$$

where $\boldsymbol{H}_{n-1} \leq \delta I_M$ is a symmetric and random matrix changing with the time index n. Subtracting $w^\star$ from both sides of (19.147) and using (19.149) gives

$$\widetilde{\boldsymbol{w}}_n = (I_M - \mu \boldsymbol{H}_{n-1})\widetilde{\boldsymbol{w}}_{n-1} + \mu \boldsymbol{g}_n(\boldsymbol{w}_{n-1}) \tag{19.150}$$

This is a nonlinear stochastic recursion in the error vector. Let $H = \nabla_w^2 P(w^\star)$ denote the Hessian matrix at the minimizer and introduce the deviation relative to it:

$$\widetilde{\boldsymbol{H}}_{n-1} \stackrel{\Delta}{=} H - \boldsymbol{H}_{n-1} \tag{19.151}$$

Then, recursion (19.150) can be rewritten as

$$\begin{aligned} \boldsymbol{c}_{n-1} &\stackrel{\Delta}{=} \widetilde{\boldsymbol{H}}_{n-1}\widetilde{\boldsymbol{w}}_{n-1} && (19.152a)\\ \widetilde{\boldsymbol{w}}_n &= (I_M - \mu H)\widetilde{\boldsymbol{w}}_{n-1} + \mu \boldsymbol{g}_n(\boldsymbol{w}_{n-1}) + \mu \boldsymbol{c}_{n-1} && (19.152b) \end{aligned}$$

Using (19.148), we have that $\|\boldsymbol{c}_{n-1}\| \leq \tau \|\widetilde{\boldsymbol{w}}_{n-1}\|^2$. Now since $\mathbb{E}\,\|\widetilde{\boldsymbol{w}}_n\|^2 \to O(\mu)$, we conclude that, for large n, the weight-error vector evolves according to the dynamics:

$$\boxed{\widetilde{\boldsymbol{w}}_n = (I_M - \mu H)\widetilde{\boldsymbol{w}}_{n-1} + \mu \boldsymbol{g}_n(\boldsymbol{w}_{n-1}) + O(\mu^2)} \tag{19.153}$$

Working with this long-term model is helpful because its dynamics are driven by the constant matrix H, as opposed to the random matrix $\boldsymbol{H}_{n-1}$. Also, the driving $O(\mu^2)$ term can be ignored for small enough μ. Using this model, we can justify the first term in the MSD expression (19.139a). Indeed, computing the weighted Euclidean norm of both sides of (19.153) using H^{-1} as the weighting matrix we get

$$\begin{aligned} &\widetilde{\boldsymbol{w}}_n^\mathsf{T} H^{-1} \widetilde{\boldsymbol{w}}_n && (19.154)\\ &= \widetilde{\boldsymbol{w}}_{n-1}^\mathsf{T}(I - \mu H)H^{-1}(I - \mu H)\widetilde{\boldsymbol{w}}_{n-1} + \mu^2 \boldsymbol{g}_n^\mathsf{T}(\boldsymbol{w}_{n-1})H^{-1}\boldsymbol{g}_n(\boldsymbol{w}_{n-1}) + \text{cross term}\\ &\approx \widetilde{\boldsymbol{w}}_{n-1}^\mathsf{T}(H^{-1} - 2\mu I_M)\widetilde{\boldsymbol{w}}_{n-1} + \mu^2 \boldsymbol{g}_n^\mathsf{T}(\boldsymbol{w}_{n-1})H^{-1}\boldsymbol{g}_n(\boldsymbol{w}_{n-1}) + \text{cross term} \end{aligned}$$

where we are ignoring the term $\mu^2 \widetilde{\boldsymbol{w}}_{n-1}^\mathsf{T} H \widetilde{\boldsymbol{w}}_{n-1}$, which is on the order of μ^3 as $n \to \infty$. Under expectation, the cross term is zero since $\mathbb{E}\left(\boldsymbol{g}_n(\boldsymbol{w}_{n-1})|\boldsymbol{w}_{n-1}\right) = 0$. Taking expectations and letting $n \to \infty$ we get

$$\mathbb{E}\,\|\widetilde{\boldsymbol{w}}_n\|^2 \to \frac{\mu}{2}\mathrm{Tr}(H^{-1}R_g) \tag{19.155}$$

as claimed.

Random reshuffling. We established in the body of the chapter (see, e.g., the summary in Table 19.2) that the performance of stochastic gradient algorithms differs under sampling *with* and *without* replacement. In the first case, the steady-state MSE, $\mathbb{E}\,\|\widetilde{\boldsymbol{w}}_n\|^2$, approaches a neighborhood of size $O(\mu)$, while in the second case under random reshuffling the neighborhood size is reduced to $O(\mu^2)$, where μ is the small step-size parameter. This is a remarkable conclusion showing that the manner by which the same data points are processed by the algorithm can have a nontrivial effect on performance. It has been noted in several studies, e.g., by Bottou (2009), Recht and Re (2012), Gürbüzbalaban, Ozdaglar, and Parrilo (2015b), and Shamir (2016) that incorporating random reshuffling into the operation of a stochastic gradient algorithm helps improve performance. The last three works pursued justifications for the enhanced behavior of the algorithm by examining the convergence rate of the learning process under vanishing step sizes. Some of the justifications rely on loose bounds or their conclusions are dependent on the sample size. Also, some of the results only establish that random reshuffling will not degrade performance relative to uniform sampling. In the body of the chapter, and specifically the arguments used in Appendix 19.B, we follow the approach by Ying *et al.* (2019). The contribution in this work provided a detailed analysis justifying analytically the improved performance from $O(\mu)$ to $O(\mu^2)$ under *constant* step-size operation.

Importance sampling. The derivation of the optimal and adaptive sampling strategies in Section 19.6 follows the approach proposed by Yuan *et al.* (2016). There are of course other sampling strategies in the literature. For example, in some works, condition (19.7b) on the loss function is stated instead in the form:

$$\|\nabla_w Q(w_2;\gamma(m),h_m) \;-\; \nabla_w Q(w_1;\gamma(m),h_m)\| \;\leq\; \delta_m\,\|w_2 - w_1\| \tag{19.156}$$

with a separate Lipschitz constant δ_m for each sample $m = 0, 1, \ldots, N-1$. One sampling strategy proposed by Needell, Ward, and Srebro (2014) and Zhao and Zhang (2015) measures the importance of each sample according to its Lipschitz constant and selects the assignment probabilities according to

$$p_m = \frac{\delta_m}{\sum_{m=0}^{N-1}\delta_m} \tag{19.157}$$

This construction is not the result of an optimized design and it requires knowledge of the Lipschitz constants, which are generally not available in advance. One feature of the adaptive sampling strategy described in (19.101) is that it relies solely on the available data.

PROBLEMS

19.1 Repeat the steps in the proof of Theorem 19.1 to establish Theorem 19.2 for the mini-batch stochastic gradient implementation.

19.2 Extend Theorem 19.4 for decaying step sizes to the mini-batch stochastic gradient implementation.

19.3 Consider a stochastic gradient implementation with instantaneous gradient approximations. Assume an empirical risk minimization problem where the N data points $\{\gamma(m), h_m\}$ are randomly reshuffled at the start of each run. Let $(\gamma(n), \boldsymbol{h}_n)$ denote generically the sample that is selected at iteration n in the kth run. Let $\boldsymbol{\sigma}(0:n-1)$ denote the history of all sample selections before the nth iteration.

(a) Show that, conditioned on $\boldsymbol{w}_{n-1}$ and $\boldsymbol{\sigma}(0:n-1)$, it holds that

$$\mathbb{E}\left(\|\boldsymbol{g}(\boldsymbol{w}_{n-1})\|^2\,|\,\boldsymbol{w}_{n-1},\boldsymbol{\sigma}(0:n-1)\right) \leq$$
$$8\delta^2\|\widetilde{w}_{n-1}\|^2 + 2\,\mathbb{E}\left(\|\nabla_{w^{\mathsf{T}}} Q(w^\star;\boldsymbol{\gamma}(n),\boldsymbol{h}_n\|^2\,|\,\boldsymbol{w}_{n-1},\boldsymbol{\sigma}(0:n-1)\right)$$

(b) Conclude that $\mathbb{E}\left(\|\boldsymbol{g}_n(\boldsymbol{w}_{n-1})\|^2\,|\,\boldsymbol{w}_{n-1}\right) \leq \beta_g^2\|\widetilde{w}_{n-1}\|^2 + \sigma_g^2$, where $\beta_g^2 = 8\delta^2$ and $\sigma_g^2 = \max_{0\leq m\leq N-1}\|\nabla_{w^{\mathsf{T}}} Q(w^\star;\gamma(m),h_m)\|^2$.

19.4 Refer to the intermediate result (19.23). Show that it also holds

$$\|\widetilde{\boldsymbol{w}}_{n-1} + \mu\,\nabla_{w^{\mathsf{T}}} P(\boldsymbol{w}_{n-1})\|^2 \leq \left(1 - \frac{\mu\nu}{2}\right)^2\|\widetilde{\boldsymbol{w}}_{n-1}\|^2$$

19.5 Establish bound (19.33) on the average regret for the stochastic gradient algorithm with constant step size.

19.6 How would the convergence rates shown in (19.55) change for step-size sequences of the form $\mu(n) = \tau/(n+1)^q$ for $\frac{1}{2} < q \leq 1$ and $\tau > 0$?

19.7 Extend the proof of Theorem 19.1 to the stochastic coordinate descent recursion (19.30) to derive conditions on the step size for convergence. Assess the limiting behavior of the algorithm: its convergence rate and limiting MSE performance.

19.8 Assume the search direction in a stochastic gradient implementation is scaled by a diagonal positive-definite matrix A as follows:

$$\boldsymbol{w}_n = \boldsymbol{w}_{n-1} - \mu\,A^{-1}\,\nabla_{w^{\mathsf{T}}} Q(\boldsymbol{w}_{n-1};\boldsymbol{\gamma}(n),\boldsymbol{h}_n),\quad n\geq 0$$

where $A \triangleq \text{diag}\{a(1), a(2), \ldots, a(M)\}$, $0 < a(m) \leq 1$, and $\mu > 0$.

(a) Extend the result of Theorem 19.1 to this case.

(b) Extend the result of Theorem 19.3 to this case when μ is replaced by $\mu(n)$.

19.9 Establish result (19.81).

19.10 Refer to the stochastic gradient recursion (19.1) and assume that the step size is a random parameter with mean $\mathbb{E}\,\boldsymbol{\mu} = \bar{\mu}$ and variance σ_μ^2. Assume $\boldsymbol{\mu}$ is independent of all other random variables. Follow the arguments used in the proof of Theorem 19.1 and show how the results of the theorem would need to be adjusted. *Remark.* For a related discussion, the reader may refer to Zhao and Sayed (2015a, b) and Sayed and Zhao (2018).

19.11 Refer to the stochastic gradient recursion (19.1) and assume that the step size $\boldsymbol{\mu}$ is a Bernoulli random variable that is equal to μ with probability p and zero with probability $1-p$. That is, the recursion is active p fraction of the times. Assume $\boldsymbol{\mu}$ is independent of all other random variables. Follow the arguments used in the proof of Theorem 19.1 and show how the results of the theorem would need to be adjusted. *Remark.* For a related discussion, the reader may refer to Zhao and Sayed (2015a, b) and Sayed and Zhao (2018).

19.12 Assume $P(w)$ is only convex (but not necessarily strongly convex) with a loss function whose gradients are δ-Lipschitz satisfying (18.10b). Consider the stochastic gradient recursion (19.1). Show that

$$\frac{1}{N}\sum_{n=0}^{N-1}\|\widetilde{\boldsymbol{w}}_{n-1}\|^2 \;\geq\; \frac{1}{N\mu^2\delta^2}\left(\|\widetilde{\boldsymbol{w}}_{N-1}\|^2 - \|\widetilde{\boldsymbol{w}}_{-1}\|^2\right)$$

How would the result change if $P(w)$ is ν-strongly-convex?

19.13 This problem extends the result of Prob. 12.13 to the stochastic gradient scenario. Thus, refer to the stochastic gradient recursion (19.1) and assume $P(w)$ is only convex (but not necessarily strongly convex) with a loss function whose gradients are δ-Lipschitz satisfying (18.10b). Let $\mu < 1/\delta$.

(a) Use property (11.120) for convex functions with δ-Lipschitz gradients to argue that the average risk value, $\mathbb{E}\,P(\boldsymbol{w}_n)$, increases by at most $O(\mu^2)$ per iteration. Specifically, verify that $\mathbb{E}\,P(\boldsymbol{w}_n) \leq \mathbb{E}\,P(\boldsymbol{w}_{n-1}) - \frac{\mu}{2}\mathbb{E}\,\|\nabla_w P(\boldsymbol{w}_{n-1})\|^2 + \frac{1}{2}\mu^2\delta\sigma_g^2$.

(b) Show that

$$\mathbb{E}\,P(\boldsymbol{w}_n) - P(w^\star) \leq \frac{1}{2\mu}\left(\mathbb{E}\,\|\widetilde{\boldsymbol{w}}_{n-1}\|^2 - \mathbb{E}\,\|\widetilde{\boldsymbol{w}}_n\|^2\right) + \mu\left(\beta_g^2\mathbb{E}\,\|\nabla_w P(\boldsymbol{w}_{n-1})\|^2 + \sigma_g^2\right)$$

(c) Conclude that $\frac{1}{n}\sum_{k=1}^{n}\mathbb{E}\,P(\boldsymbol{w}_k) - P(w^\star) \leq O(1/n) + O(\mu)$.

19.14 Refer to the stochastic gradient algorithm (19.75) under random reshuffling and assume an epoch-dependent step size $\mu(k) = \tau/k$, for $k \geq 1$, is used. Repeat the arguments in Appendix 19.B and the technique used to derive Lemma 19.4 to establish the convergence rates (19.81)–(19.82).

19.15 The proof technique used to establish the convergence properties in Theorem 19.1 exploits the fact that the gradient noise process has zero mean conditioned on the past iterate $\boldsymbol{w}_{n-1}$. Motivated by the arguments used in Appendix 19.B, assume we follow now a similar proof technique to avoid the reliance on the zero-mean property for the gradient noise.

(a) Let $0 < t < 1$ be any scalar that we are free to choose. Subtract $w^\star$ from both sides of (19.1) and establish the result

$$\|\widetilde{\boldsymbol{w}}_n\|^2 \leq \frac{1}{t}\left(1 - 2\mu\nu + \mu^2\delta^2\right)\|\widetilde{\boldsymbol{w}}_{n-1}\|^2 \;+\; \frac{\mu^2}{1-t}\,\|\boldsymbol{g}_n(\boldsymbol{w}_{n-1})\|^2$$

(b) Verify that $1 - 2\mu\nu + \mu^2\delta^2 \leq \left(1 - \frac{\mu\nu}{2}\right)^2$ for $\mu < \nu/\delta^2$. Select $t = 1 - \frac{\mu\nu}{2}$ and show that $\mathbb{E}\,\|\widetilde{\boldsymbol{w}}_n\|^2 \leq \lambda\,\mathbb{E}\,\|\widetilde{\boldsymbol{w}}_{n-1})\|^2 + 2\mu\sigma_g^2/\nu$, where $\lambda = 1 - \mu(\frac{\nu}{2} - \frac{2\beta_g^2}{\nu})$. Does $\lambda \in (0,1)$?

(c) Are you able to conclude from the recursion in part (b) that the mean-square deviation $\mathbb{E}\,\|\widetilde{\boldsymbol{w}}_n\|^2$ approaches a neighborhood of size $O(\mu)$?

19.16 Show that the convergence rates in Lemma 19.4 continue to hold for the mini-batch stochastic gradient implementation.

19.17 Consider an empirical risk minimization problem and apply a stochastic gradient algorithm with importance sampling using either (19.6a) or (19.6b) to approximate the gradient direction. Extend the proof of Theorem 19.1 to show that the limiting MSE region will continue to be $O(\mu\sigma_g^2)$.

19.18 Probs. 19.18–19.20 are motivated by the discussion in Bottou, Curtis, and Nocedal (2018). Refer to the stochastic gradient recursion (19.1) and assume the risk function $P(w)$ has δ-Lipschitz gradients as in (19.8) or (19.10). Use property (10.13) for δ-smooth functions to establish the following inequality regardless of how the stochastic gradient is constructed:

$$\begin{aligned}&\mathbb{E}\Big(P(\boldsymbol{w}_n)|\boldsymbol{w}_{n-1}\Big) - P(w_{n-1})\\&\leq -\mu\Big(\nabla_{w^\mathsf{T}} P(w_{n-1})\Big)^\mathsf{T}\,\mathbb{E}\,\widehat{\nabla_{w^\mathsf{T}} P}(w_{n-1}) + \frac{\mu^2\delta}{2}\mathbb{E}\,\|\widehat{\nabla_{w^\mathsf{T}} P}(w_{n-1})\|^2\end{aligned}$$

where the expectation operator $\mathbb{E}$ is over the statistical distribution of the data $\{\boldsymbol{\gamma}, \boldsymbol{h}\}$ conditioned on the past iterate $\boldsymbol{w}_{n-1}$. Conclude that if the gradient approximation is unbiased then

$$\mathbb{E}\Big(P(\boldsymbol{w}_n)|\boldsymbol{w}_{n-1}\Big) - P(w_{n-1}) \leq -\mu\|\nabla_{w^\mathsf{T}} P(w_{n-1})\|^2 + \frac{\mu^2\delta}{2}\mathbb{E}\,\|\widehat{\nabla_{w^\mathsf{T}} P}(w_{n-1})\|^2$$

19.19 Continuing with Prob. 19.18, assume the stochastic gradient approximation satisfies the following three conditions in terms of the squared Euclidean norm:

$$\begin{aligned}&\text{(i) } \Big(\nabla_{w^\mathsf{T}} P(w_{n-1})\Big)^\mathsf{T}\,\mathbb{E}\,\widehat{\nabla_{w^\mathsf{T}} P}(w_{n-1}) \geq a\,\|\nabla_{w^\mathsf{T}} P(w_{n-1})\|^2\\&\text{(ii) } \|\mathbb{E}\,\widehat{\nabla_{w^\mathsf{T}} P}(w_{n-1})\| \leq b\,\|\nabla_{w^\mathsf{T}} P(w_{n-1})\|\\&\text{(iii) } \text{var}\Big(\widehat{\nabla_{w^\mathsf{T}} P}(w_{n-1})\Big) \leq \alpha + \beta\|\nabla_{w^\mathsf{T}} P(w_{n-1})\|^2\end{aligned}$$

for some constants $b \geq a > 0$ and $\alpha, \beta \geq 0$ and where, by definition,

$$\text{var}\Big(\widehat{\nabla_{w^\mathsf{T}} P}(w_{n-1})\Big) \stackrel{\Delta}{=} \mathbb{E}\,\|\widehat{\nabla_{w^\mathsf{T}} P}(w_{n-1})\|^2 - \|\mathbb{E}\,\widehat{\nabla_{w^\mathsf{T}} P}(w_{n-1})\|^2$$

(a) Let $\beta_1 = \beta + b^2$. Verify that

$$\mathbb{E}\,\|\widehat{\nabla_{w^\mathsf{T}} P}(w_{n-1})\|^2 \leq \alpha + \beta_1\|\nabla_{w^\mathsf{T}} P(w_{n-1})\|^2$$

(b) Conclude that

$$\mathbb{E}\Big(P(\boldsymbol{w}_n)|\boldsymbol{w}_{n-1}\Big) - P(w_{n-1}) \leq -\Big(a - \frac{1}{2}\delta\mu\beta_1\Big)\mu\|\nabla_{w^\mathsf{T}} P(w_{n-1})\|^2 + \frac{\mu^2}{2}\alpha\delta$$

19.20 Continuing with Prob. 19.19, assume now that $P(w)$ is a ν-strongly convex risk that is bounded from below. Verify that for $\mu \leq a/\delta\beta_1$ we have

$$\mathbb{E}\,P(\boldsymbol{w}_n) - P(w^\star) \leq \frac{\mu\delta\alpha}{2\nu a} + (1-\mu\nu a)^{n+1} \times \Big(P(w_{-1}) - P(w^\star) - \frac{\mu\delta\alpha}{2\nu a}\Big)$$

and conclude that

$$\limsup_{n\to\infty}\Big(\mathbb{E}\,P(\boldsymbol{w}_n) - P(w^\star)\Big) \leq \frac{\mu\delta\alpha}{2\nu a} = O(\mu)$$

19.21 The stochastic gradient algorithm can be implemented with Polyak–Ruppert averaging, as shown earlier in (16.52), i.e.,

$$\begin{cases} \boldsymbol{w}_n = \boldsymbol{w}_{n-1} - \mu \nabla_{w^\mathsf{T}} Q(\boldsymbol{w}_{n-1}; \boldsymbol{\gamma}(n), \boldsymbol{h}_n), \quad n \geq 0 \\ \bar{\boldsymbol{w}}_n = \bar{\boldsymbol{w}}_{n-1} + \dfrac{1}{n+2}(\boldsymbol{w}_n - \bar{\boldsymbol{w}}_{n-1}) \end{cases}$$

Extend the result of Theorem 19.1 to this case. *Remark.* For more discussion on this technique, the reader may refer to Ruppert (1988) and Polyak and Juditsky (1992).

19.22 A variation of the Polyak–Ruppert averaging algorithm of Prob. 19.21 is to generate $\bar{\boldsymbol{w}}_n$ by means of a convex combination, say

$$\begin{aligned} \boldsymbol{w}_n &= \boldsymbol{w}_{n-1} - \mu \nabla_{\boldsymbol{w}^\mathsf{T}} Q(\boldsymbol{w}_{n-1}; \boldsymbol{\gamma}(n), \boldsymbol{h}_n) \\ \bar{\boldsymbol{w}}_n &= \beta \bar{\boldsymbol{w}}_{n-1} + (1-\beta)\boldsymbol{w}_n, \quad \bar{\boldsymbol{w}}_{-1} = \boldsymbol{w}_{-1} = 0 \end{aligned}$$

where $\beta \in [0,1]$. Extend the result of Theorem 19.1 to this case.

19.23 Refer to the stochastic Nesterov momentum method (17.73) and examine its convergence properties. *Remark.* For a related discussion, refer to Yu, Jin, and Yang (2019).

19.24 In this problem we seek to re-derive the AdaGrad algorithm (17.13) by relying on the same MSE analysis used in the proof of Theorem 19.1. Thus, consider a stochastic gradient recursion of the form

$$\boldsymbol{w}_n = \boldsymbol{w}_{n-1} - \mu A^{-1} \nabla_{w^\mathsf{T}} Q(\boldsymbol{w}_{n-1}; \boldsymbol{\gamma}(n), \boldsymbol{h}_n)$$

with a constant step size μ and a scaling symmetric and positive-definite matrix A^{-1}. Let $\sigma_{\max}(A^{-1})$ denote the maximum singular value of A. If A is restricted to being diagonal with positive entries, then $\sigma_{\max}(A^{-1}) = 1/a_{\min}$, where $a_{\min}$ is the smallest entry in A. Introduce the gradient noise vector

$$\boldsymbol{g}(w) \triangleq A^{-1}\widehat{\nabla_{w^\mathsf{T}} P}(w) - A^{-1}\nabla_{w^\mathsf{T}} P(w)$$

(a) Verify that under uniform sampling:

$$\mathbb{E}\Big(\boldsymbol{g}_n(\boldsymbol{w}_{n-1}) \,|\, \boldsymbol{w}_{n-1}\Big) = 0, \quad \mathbb{E}\Big(\|\boldsymbol{g}_n(\boldsymbol{w}_{n-1})\|_A^2 \,|\, \boldsymbol{w}_{n-1}\Big) \leq \beta_g^2 \|\widetilde{w}_{n-1}\|_A^2 + \sigma_g^2$$

$$\beta_g^2 = 8\delta^2/\sigma_{\min}^2(A)$$

$$\sigma_g^2 = \frac{2}{N}\sum_{m=0}^{N-1} \|\nabla_{w^\mathsf{T}} Q(w^\star; \gamma(m), h_m)\|_{A^{-1}}^2$$

(b) Repeat the argument leading to (19.24) and verify that the relation now becomes

$$\mathbb{E}\,\|\widetilde{\boldsymbol{w}}_n\|_A^2 \leq \left(1 - \frac{2\mu\nu}{\sigma_{\max}(A)} + \frac{9\mu^2\delta^2}{\sigma_{\min}^2(A)}\right)\mathbb{E}\,\|\widetilde{\boldsymbol{w}}_{n-1}\|_A^2 + \mu^2\sigma_g^2$$

(c) Argue that selecting A to minimize σ_g^2 under the condition $\text{Tr}(A) \leq c$, leads to the same optimization problem (17.35) obtained from the regret analysis.

19.25 Refer to the stochastic Fletcher–Reeves algorithm (19.105). Assume the parameters $\boldsymbol{\beta}_n$ are bounded for all n, say, $\boldsymbol{\beta}_n \leq \beta$ for some $\beta > 0$.

(a) Use an argument similar to (13.123) to show that

$$\mathbb{E}\,\|\boldsymbol{q}_{n+1}\|^2 \leq \beta^2 \mathbb{E}\,\|\boldsymbol{q}_n\|^2 + \frac{(1+\eta)}{1-\eta}\mathbb{E}\,\|\nabla_w P(\boldsymbol{w}_{n-1})\|^2$$

(b) Let $c = (1+\eta)/(1-\eta)$. Iterate part (a) to conclude that

$$\mathbb{E}\,\|\boldsymbol{q}_{n+1}\|^2 \leq c\beta^{n+1}\left(\frac{1-\beta^{n+3}}{1-\beta}\right)\mathbb{E}\,\|\nabla_w P(\boldsymbol{w}_{-1})\|^2$$

(c) Assume the $\{\boldsymbol{\alpha}_m\}$ are limited to the bounded interval $\boldsymbol{\alpha}_m = (\alpha_\ell, \alpha_u)$ where $0 < \alpha_\ell < \alpha_u$. Assume each loss term $Q(w;\cdot)$ is ν-strongly convex and has δ-Lipschitz gradients. Show that the average excess risk evolves according to

$$\mathbb{E}\, P(\boldsymbol{w}_n) - P(w^\star) \;\leq\; \rho^n \Big(\mathbb{E}\, P(\boldsymbol{w}_{n-1}) - P(w^\star)\Big)$$

for some positive factor $\rho < 1$.

Remark. The reader may refer to Jin *et al.* (2019) for a related discussion.

19.26 Refer to the stochastic Fletcher–Reeves algorithm (19.105). Assume the parameters $\{\boldsymbol{\alpha}_n\}$ are generated as follows:

$$\boldsymbol{\alpha}_n = -\rho \times \frac{\nabla_w Q(\boldsymbol{w}_{n-1})\boldsymbol{q}_n}{\boldsymbol{q}_n^\mathsf{T}\Sigma_n \boldsymbol{q}_n}$$

where $\rho \in (0, \nu_{\min}/\delta)$ and Σ_n is a given deterministic sequence of matrices satisfying $\nu_{\min}\|x\|^2 \leq x^\mathsf{T}\Sigma_n x \leq \nu_{\max}\|x\|^2$ for any x and where $\nu_{\min}$ and $\nu_{\max}$ are positive. Assume $P(w)$ is ν-strongly convex with δ-Lipschitz gradients. Establish that $\liminf_{n\to\infty} \|\mathbb{E}\,\nabla_w P(\boldsymbol{w}_n)\| = 0$. *Remark.* See the work by Sun and Zhang (2001) for a related discussion in the nonstochastic case.

19.A STOCHASTIC INEQUALITY RECURSION

The following useful result from Polyak (1987, p. 49) is originally from Gladyshev (1965) and deals with the convergence of stochastic inequality recursions; it is the stochastic analogue of the earlier deterministic recursion (14.136).

Lemma 19.3. (Stochastic recursion) *Let $\boldsymbol{u}(n) \geq 0$ denote a scalar sequence of nonnegative random variables satisfying $\mathbb{E}\,\boldsymbol{u}(0) < \infty$ and consider the stochastic recursion:*

$$\mathbb{E}\Big(\boldsymbol{u}(n+1)|\, \boldsymbol{u}(0), \boldsymbol{u}(1), \ldots, \boldsymbol{u}(n)\Big) \;\leq\; (1-a(n))\boldsymbol{u}(n) \,+\, b(n), \quad n \geq 0 \tag{19.158}$$

where the scalar deterministic sequences $\{a(n), b(n)\}$ satisfy the five conditions:

$$0 \leq a(n) < 1, \quad b(n) \geq 0, \quad \sum_{n=0}^{\infty} a(n) = \infty, \quad \sum_{n=0}^{\infty} b(n) < \infty, \quad \lim_{n\to\infty} \frac{b(n)}{a(n)} = 0 \tag{19.159}$$

Then, it holds that

$$\lim_{n\to\infty} \boldsymbol{u}(n) = 0, \quad \textit{almost surely} \tag{19.160a}$$

$$\lim_{n\to\infty} \mathbb{E}\,\boldsymbol{u}(n) = 0 \tag{19.160b}$$

Proof: For completeness, we establish the lemma by following the same argument from Polyak (1987, pp. 49–50). First, observe by taking expectations of both sides of (19.158) that the recursion reduces to the same form covered by Lemma 14.1, namely,

$$\mathbb{E}\,\boldsymbol{u}(n+1) \leq (1-a(n))\mathbb{E}\,\boldsymbol{u}(n) \,+\, b(n) \tag{19.161}$$

and, therefore, $\mathbb{E}\,\boldsymbol{u}(n) \to 0$ as $n \to \infty$. Next, introduce the auxiliary variable:

$$\boldsymbol{s}(n) \triangleq \boldsymbol{u}(n) + \sum_{j=n}^{\infty} b(j) \tag{19.162}$$

We know from the conditions $0 \leq a(n) < 1$ and $b(n) \geq 0$ that $\boldsymbol{s}(n) \geq 0$. Moreover, we also get $\mathbb{E}\,\boldsymbol{s}(0) < \infty$ since

$$\mathbb{E}\,\boldsymbol{s}(0) = \mathbb{E}\,\boldsymbol{u}(0) + \sum_{j=0}^{\infty} b(j) < \infty \tag{19.163}$$

Computing the conditional expectation of $\boldsymbol{s}(n+1)$ relative to $\{\boldsymbol{s}(0), \boldsymbol{s}(1), \ldots, \boldsymbol{s}(n)\}$ we get

$$\begin{aligned}
&\mathbb{E}\left(\boldsymbol{s}(n+1) |\, \boldsymbol{s}(0), \boldsymbol{s}(1), \ldots, \boldsymbol{s}(n)\right) \\
&= \mathbb{E}\left(\boldsymbol{u}(n+1)|\boldsymbol{u}(0), \boldsymbol{u}(1), \ldots, \boldsymbol{u}(n)\right) + \sum_{j=n+1}^{\infty} b(j) \\
&\leq (1 - a(n))\boldsymbol{u}(n) + b(n) + \sum_{j=n+1}^{\infty} b(j) \\
&= (1 - a(n))\boldsymbol{u}(n) + \sum_{j=n}^{\infty} b(j) \\
&\leq \boldsymbol{u}(n) + \sum_{j=n}^{\infty} b(j) \\
&= \boldsymbol{s}(n)
\end{aligned} \tag{19.164}$$

In other words, we established that

$$\mathbb{E}\left(\boldsymbol{s}(n+1) |\, \boldsymbol{s}(0), \boldsymbol{s}(1), \ldots, \boldsymbol{s}(n)\right) \leq \boldsymbol{s}(n) \tag{19.165}$$

This property means that $\boldsymbol{s}(n) \geq 0$ is a semi-martingale process, which also satisfies $\mathbb{E}\,\boldsymbol{s}(0) < \infty$. For such processes, it is known that there exists a random variable $\boldsymbol{s} \geq 0$ such that $\boldsymbol{s}(n) \to \boldsymbol{s}$ almost surely (see, e.g., Lipster and Shiryayev (1989), Williams (1991), and He, Wang, and Yan (1992)). Now note that, by construction,

$$\boldsymbol{u}(n) = \boldsymbol{s}(n) - \sum_{j=n}^{\infty} b(j) \tag{19.166}$$

so that, as $n \to \infty$,

$$\begin{aligned}
\mathbb{P}\left(\lim_{n\to\infty} \boldsymbol{u}(n) = \boldsymbol{s}\right) &= \mathbb{P}\left(\lim_{n\to\infty} \boldsymbol{s}(n) - \sum_{j=n}^{\infty} b(j) = \boldsymbol{s}\right) \\
&= \mathbb{P}\left(\lim_{n\to\infty} \boldsymbol{s}(n) = \boldsymbol{s}\right) \\
&= 1
\end{aligned} \tag{19.167}$$

and we conclude that $\boldsymbol{u}(n)$ also tends almost surely to $\boldsymbol{s} \geq 0$. We showed earlier that $\mathbb{E}\,\boldsymbol{u}(n) \to 0$ as $n \to \infty$. It follows that $\boldsymbol{s} = 0$ so that $\boldsymbol{u}(n)$ converges in probability to zero.

■

19.B PROOF OF THEOREM 19.5

In this appendix we follow the derivation from Ying *et al.* (2019) to establish the performance results (19.79a)–(19.79b) for operation under random reshuffling.

To begin with, note that recursion (19.75) shows how to move from one iterate to another within the same run k. The argument below will deduce from this recursion a similar relation that shows how to move from the initial iterate $\boldsymbol{w}^{k-1}_{-1}$ for run $k-1$ to the initial iterate $\boldsymbol{w}^{k}_{-1}$ for run k. That is, we first transform the description of the algorithm from iterations within the same run to iterations across epochs. Doing so will enable us to exploit a useful property of the random reshuffling mechanism, as explained in (19.170). Once this new recursion across epochs is derived, we will then use it to establish (19.79a)–(19.79b).

Proof: Subtracting $w^\star$ from both sides of (19.75) gives

$$\widetilde{\boldsymbol{w}}^k_n = \widetilde{\boldsymbol{w}}^k_{n-1} + \mu\, \nabla_{w^{\mathsf T}} Q(\boldsymbol{w}^k_{n-1}; \boldsymbol{\gamma}(n), \boldsymbol{h}_n) \tag{19.168}$$

where the notation $(\boldsymbol{\gamma}(n), \boldsymbol{h}_n)$ denotes the random sample that is selected at iteration n of the kth epoch. Iterating gives, where we are now dropping the data samples as arguments for $Q(w;\cdot,\cdot)$ for simplicity (we will restore them when necessary),

$$\begin{aligned}
\widetilde{\boldsymbol{w}}^{k+1}_{-1} &\overset{\Delta}{=} \widetilde{\boldsymbol{w}}^{k}_{N-1}\\
&= \widetilde{\boldsymbol{w}}^{k}_{-1} + \mu \sum_{n=0}^{N-1} \nabla_{w^{\mathsf T}} Q(\boldsymbol{w}^k_{n-1})\\
&\overset{(a)}{=} \widetilde{\boldsymbol{w}}^{k}_{-1} + \mu \sum_{n=0}^{N-1} \nabla_{w^{\mathsf T}} Q(\boldsymbol{w}^k_{n-1}) + \mu \sum_{n=0}^{N-1} \nabla_{w^{\mathsf T}} Q(\boldsymbol{w}^k_{-1}) - \mu \sum_{n=0}^{N-1} \nabla_{w^{\mathsf T}} Q(\boldsymbol{w}^k_{-1})\\
&\overset{(b)}{=} \widetilde{\boldsymbol{w}}^{k}_{-1} + \mu N \nabla_{w^{\mathsf T}} P(\boldsymbol{w}^k_{-1}) + \mu \sum_{n=0}^{N-1} \Big(\nabla_{w^{\mathsf T}} Q(\boldsymbol{w}^k_{n-1}) - \nabla_{w^{\mathsf T}} Q(\boldsymbol{w}^k_{-1}) \Big)
\end{aligned} \tag{19.169}$$

where in step (a) we added and subtracted the same quantity, and in step (b) we used the fact that under random reshuffling:

$$\frac{1}{N}\sum_{n=0}^{N-1} Q(\boldsymbol{w}^k_{-1}; \boldsymbol{\gamma}(n), \boldsymbol{h}_n) = \frac{1}{N}\sum_{m=0}^{N-1} Q(\boldsymbol{w}^k_{-1}; \gamma(m), h_m) \;=\; P(\boldsymbol{w}^k_{-1}) \tag{19.170}$$

The first equality in (19.170) is because each data pair is sampled once under random reshuffling. Observe that this property would not hold under uniform sampling *with* replacement.

Now, let $0 < t < 1$ be any scalar that we are free to choose. Continuing with (19.169), we square both sides and note that

$$\begin{aligned}
\|\widetilde{\boldsymbol{w}}_{-1}^{k+1}\|^2 &= \left\| \frac{t}{t}\left(\widetilde{\boldsymbol{w}}_{-1}^k + \mu N \nabla_{w^\mathsf{T}} P(\boldsymbol{w}_{-1}^k)\right) + \right. \\
&\qquad \left. \frac{1-t}{1-t}\mu \sum_{n=0}^{N-1}\left(\nabla_{w^\mathsf{T}} Q(\boldsymbol{w}_{n-1}^k) - \nabla_{w^\mathsf{T}} Q(\boldsymbol{w}_{-1}^k)\right)\right\|^2 \\
&\overset{(a)}{\leq} t\left\| \frac{1}{t}\left(\widetilde{\boldsymbol{w}}_{-1}^k + \mu N \nabla_{w^\mathsf{T}} P(\boldsymbol{w}_{-1}^k)\right)\right\|^2 + \\
&\qquad (1-t)\left\| \frac{\mu}{1-t} \sum_{n=0}^{N-1}\left(\nabla_{w^\mathsf{T}} Q(\boldsymbol{w}_{n-1}^k) - \nabla_{w^\mathsf{T}} Q(\boldsymbol{w}_{-1}^k)\right)\right\|^2 \\
&= \frac{1}{t}\left\|\widetilde{\boldsymbol{w}}_{-1}^k + \mu N \nabla_{w^\mathsf{T}} P(\boldsymbol{w}_{-1}^k)\right\|^2 + \\
&\qquad \frac{\mu^2}{1-t}\left\| \sum_{n=0}^{N-1}\left(\nabla_{w^\mathsf{T}} Q(\boldsymbol{w}_{n-1}^k) - \nabla_{w^\mathsf{T}} Q(\boldsymbol{w}_{-1}^k)\right)\right\|^2 \\
&\overset{(b)}{\leq} \frac{1}{t}\left\|\widetilde{\boldsymbol{w}}_{-1}^k + \mu N \nabla_{w^\mathsf{T}} P(\boldsymbol{w}_{-1}^k)\right\|^2 + \\
&\qquad \frac{\mu^2 N}{1-t} \sum_{n=0}^{N-1}\left\|\nabla_{w^\mathsf{T}} Q(\boldsymbol{w}_{n-1}^k) - \nabla_{w^\mathsf{T}} Q(\boldsymbol{w}_{-1}^k)\right\|^2
\end{aligned} \tag{19.171}$$

where step (a) uses the Jensen inequality (8.76) and step (b) uses the same inequality again to justify the following property for any vectors $\{x_n\}$:

$$\left\|\sum_{n=1}^{N-1} x_n\right\|^2 = N^2\left\|\sum_{n=1}^{N-1}\frac{1}{N}x_n\right\|^2 \overset{(8.76)}{\leq} N\sum_{n=1}^{N-1}\|x_n\|^2 \tag{19.172}$$

Let us now examine the two terms on the right-hand side of (19.171). First note that

$$\begin{aligned}
&\left\|\widetilde{\boldsymbol{w}}_{-1}^k + \mu N \nabla_{w^\mathsf{T}} P(\boldsymbol{w}_{-1}^k)\right\|^2 \\
&= \|\widetilde{\boldsymbol{w}}_{-1}^k\|^2 + 2\mu N\left(\nabla_{w^\mathsf{T}} P(\boldsymbol{w}_{-1}^k)\right)^\mathsf{T}\widetilde{\boldsymbol{w}}_{-1}^k + \mu^2N^2\|\nabla_{w^\mathsf{T}} P(\boldsymbol{w}_{-1}^k)\|^2 \\
&= \|\widetilde{\boldsymbol{w}}_{-1}^k\|^2 + 2\mu N\left(\nabla_{w^\mathsf{T}} P(\boldsymbol{w}_{-1}^k)\right)^\mathsf{T}\widetilde{\boldsymbol{w}}_{-1}^k + \mu^2N^2\|\underbrace{\nabla_{w^\mathsf{T}} P(w^\star)}_{=0} - \nabla_{w^\mathsf{T}} P(\boldsymbol{w}_{-1}^k)\|^2 \\
&\overset{(\textbf{P2})}{\leq} \|\widetilde{\boldsymbol{w}}_{-1}^k\|^2 + 2\mu N\left(\nabla_{w^\mathsf{T}} P(\boldsymbol{w}_{-1}^k)\right)^\mathsf{T}\widetilde{\boldsymbol{w}}_{-1}^k + \mu^2N^2\delta^2\|\widetilde{\boldsymbol{w}}_{-1}^k\|^2
\end{aligned} \tag{19.173}$$

Next, we appeal to the strong convexity property (18.10a) to find that

$$\begin{aligned}
\left(\nabla_{w^\mathsf{T}} P(\boldsymbol{w}_{-1}^k)\right)^\mathsf{T}\widetilde{\boldsymbol{w}}_{-1}^k &\leq P(w^\star) - P(\boldsymbol{w}_{-1}^k) - \frac{\nu}{2}\|\widetilde{\boldsymbol{w}}_{-1}^k\|^2 \\
&\overset{(8.23)}{\leq} -\frac{\nu}{2}\|\widetilde{\boldsymbol{w}}_{-1}^k\|^2 - \frac{\nu}{2}\|\widetilde{\boldsymbol{w}}_{-1}^k\|^2 \\
&= -\nu\|\widetilde{\boldsymbol{w}}_{-1}^k\|^2
\end{aligned} \tag{19.174}$$

Substituting into (19.173) gives

$$\|\widetilde{\boldsymbol{w}}_{-1}^k + \mu N\,\nabla_{w^\mathsf{T}} P(\boldsymbol{w}_{-1}^k)\|^2 \leq (1 - 2\mu\nu N + \mu^2\delta^2N^2)\|\widetilde{\boldsymbol{w}}_{-1}^k\|^2 \tag{19.175}$$

Note that for

$$\mu < \frac{2\nu}{3N\delta^2} \tag{19.176}$$

we have

$$
\begin{aligned}
1 - 2\mu\nu N + \mu^2\delta^2 N^2 &\le 1 - \frac{4\mu\nu N}{3} \\
&\le 1 - \frac{4\mu\nu N}{3} + \frac{4\mu^2\nu^2 N^2}{9} \\
&\le \left(1 - \frac{2\mu\nu N}{3}\right)^2
\end{aligned}
\tag{19.177}
$$

which has the form of a perfect square. It follows from (19.175) that

$$
\|\widetilde{\boldsymbol{w}}_{-1}^k + \mu N\, \nabla_{w^\mathsf{T}} P(\boldsymbol{w}_{-1}^k)\|^2 \le \left(1 - \frac{2\mu\nu N}{3}\right)^2 \|\widetilde{\boldsymbol{w}}_{-1}^k\|^2 \tag{19.178}
$$

Consider now the second term on the right-hand side of (19.171) and note that

$$
\begin{aligned}
\sum_{n=0}^{N-1} \left\| \nabla_{w^\mathsf{T}} Q(\boldsymbol{w}_{n-1}^k) - \nabla_{w^\mathsf{T}} Q(\boldsymbol{w}_{-1}^k) \right\|^2 &\overset{(18.10b)}{\le} \delta^2 \sum_{n=0}^{N-1} \left\| \boldsymbol{w}_{n-1}^k - \boldsymbol{w}_{-1}^k \right\|^2 \\
&\overset{(a)}{=} \delta^2 \sum_{n=0}^{N-1} \left\| \sum_{m=0}^{n-1} \boldsymbol{w}_m^k - \boldsymbol{w}_{m-1}^k \right\|^2 \\
&\overset{(19.172)}{\le} \delta^2 \sum_{n=0}^{N-1} n \sum_{m=0}^{n-1} \left\| \boldsymbol{w}_m^k - \boldsymbol{w}_{m-1}^k \right\|^2 \\
&\overset{(b)}{=} \delta^2 \sum_{m=0}^{N-2} \left\| \boldsymbol{w}_m^k - \boldsymbol{w}_{m-1}^k \right\|^2 \left(\sum_{n=m+1}^{N-1} n \right) \\
&\overset{(c)}{\le} \frac{\delta^2 N^2}{2} \sum_{m=0}^{N-2} \left\| \boldsymbol{w}_m^k - \boldsymbol{w}_{m-1}^k \right\|^2
\end{aligned}
\tag{19.179}
$$

where step (a) uses the telescoping sum

$$
\boldsymbol{w}_{n-1}^k - \boldsymbol{w}_{-1}^k = \sum_{m=0}^{n-1} \boldsymbol{w}_m^k - \boldsymbol{w}_{m-1}^k \tag{19.180}
$$

Step (b) uses the easily verified property

$$
\sum_{n=0}^{N-1} \sum_{m=0}^{n-1} a_{nm} = \sum_{m=0}^{N-2} \sum_{n=m+1}^{N-1} a_{nm} \tag{19.181}
$$

and step (c) uses

$$
\sum_{n=m+1}^{N-1} n \le \sum_{n=0}^{N-1} n = \frac{N(N-1)}{2} \le \frac{N^2}{2} \tag{19.182}
$$

Continuing with (19.179), we appeal to the stochastic gradient recursion to observe that for each term in the sum:

$$
\begin{aligned}
\left\|\boldsymbol{w}_m^k - \boldsymbol{w}_{m-1}^k\right\|^2 &\overset{(19.75)}{=} \mu^2 \left\|\nabla_{w^\mathsf{T}} Q(\boldsymbol{w}_{m-1}^k)\right\|^2 \\
&= \mu^2 \left\|\nabla_{w^\mathsf{T}} Q(\boldsymbol{w}_{m-1}^k) + \nabla_{w^\mathsf{T}} Q(w^\star) - \nabla_{w^\mathsf{T}} Q(w^\star)\right\|^2 \\
&\le 2\mu^2 \left\|\nabla_{w^\mathsf{T}} Q(\boldsymbol{w}_{m-1}^k) + \nabla_{w^\mathsf{T}} Q(w^\star)\right\|^2 + 2\mu^2 \left\|\nabla_{w^\mathsf{T}} Q(w^\star)\right\|^2 \\
&\overset{(18.10b)}{\le} 2\mu^2\delta^2 \|\widetilde{\boldsymbol{w}}_{m-1}^k\|^2 + 2\mu^2 \left\|\nabla_{w^\mathsf{T}} Q(w^\star)\right\|^2
\end{aligned}
\tag{19.183}
$$

Therefore, we have

$$
\begin{aligned}
\left\|\boldsymbol{w}_m^k - \boldsymbol{w}_{m-1}^k\right\|^2 &\le 2\mu^2\delta^2 \|w^\star - \boldsymbol{w}_{-1}^k + \boldsymbol{w}_{-1}^k - \boldsymbol{w}_{m-1}^k\|^2 + 2\mu^2 \left\|\nabla_{w^\mathsf{T}} Q(w^\star)\right\|^2 \\
&\le 4\mu^2\delta^2 \|\widetilde{\boldsymbol{w}}_{-1}^k\|^2 + 4\mu^2\delta^2 \|\boldsymbol{w}_{-1}^k - \boldsymbol{w}_{m-1}^k\|^2 + 2\mu^2 \left\|\nabla_{w^\mathsf{T}} Q(w^\star)\right\|^2
\end{aligned}
\tag{19.184}
$$

Introduce the average loss value

$$
Q_{\text{av}} \overset{\Delta}{=} \frac{1}{N} \sum_{m=0}^{N-1} \left\|\nabla_{w^\mathsf{T}} Q(w^\star; \gamma(m), h_m)\right\|^2 \tag{19.185}
$$

We know from (18.40) that

$$
Q_{\text{av}} = O(\sigma_g^2) \tag{19.186}
$$

i.e., it is on the order of the factor σ_g^2 that bounds the second-order moment of the gradient noise process. Adding (19.184) over m gives

$$
\begin{aligned}
&\sum_{m=0}^{N-1} \left\|\boldsymbol{w}_m^k - \boldsymbol{w}_{m-1}^k\right\|^2 \\
&\le 4\mu^2\delta^2 N \|\widetilde{\boldsymbol{w}}_{-1}^k\|^2 + 2\mu^2 N Q_{\text{av}} + 4\mu^2\delta^2 \sum_{m=0}^{N-1} \|\boldsymbol{w}_{m-1}^k - \boldsymbol{w}_{-1}^k\|^2 \\
&\overset{(a)}{=} 4\mu^2\delta^2 N \|\widetilde{\boldsymbol{w}}_{-1}^k\|^2 + 2\mu^2 N Q_{\text{av}} + 4\mu^2\delta^2 \sum_{m=0}^{N-1} \left\|\sum_{n=0}^{m-1} \boldsymbol{w}_n^k - \boldsymbol{w}_{n-1}^k\right\|^2 \\
&\overset{(b)}{\le} 4\mu^2\delta^2 N \|\widetilde{\boldsymbol{w}}_{-1}^k\|^2 + 2\mu^2 N Q_{\text{av}} + 4\mu^2\delta^2 \sum_{m=0}^{N-1} \sum_{n=0}^{m-1} m \left\|\boldsymbol{w}_n^k - \boldsymbol{w}_{n-1}^k\right\|^2 \\
&\overset{(c)}{=} 4\mu^2\delta^2 N \|\widetilde{\boldsymbol{w}}_{-1}^k\|^2 + 2\mu^2 N Q_{\text{av}} + 4\mu^2\delta^2 \sum_{n=0}^{N-2} \left\|\boldsymbol{w}_n^k - \boldsymbol{w}_{n-1}^k\right\|^2 \left(\sum_{m=n+1}^{N-1} m\right) \\
&\overset{(d)}{\le} 4\mu^2\delta^2 N \|\widetilde{\boldsymbol{w}}_{-1}^k\|^2 + 2\mu^2 N Q_{\text{av}} + 2\mu^2\delta^2 N^2 \sum_{n=0}^{N-2} \left\|\boldsymbol{w}_n^k - \boldsymbol{w}_{n-1}^k\right\|^2 \\
&\le 4\mu^2\delta^2 N \|\widetilde{\boldsymbol{w}}_{-1}^k\|^2 + 2\mu^2 N Q_{\text{av}} + 2\mu^2\delta^2 N^2 \sum_{n=0}^{N-1} \left\|\boldsymbol{w}_n^k - \boldsymbol{w}_{n-1}^k\right\|^2
\end{aligned}
\tag{19.187}
$$

where in step (a) we used again a telescoping sum representation, in step (b) we used property (19.172), in step (c) we appealed again to (19.181), and in step (d) we used

(19.182). In the last step, we increased the upper limit on the summation on the right-hand side to $N-1$. It follows that

$$\sum_{m=0}^{N-1}\left\|\boldsymbol{w}_m^k - \boldsymbol{w}_{m-1}^k\right\|^2 \leq \frac{1}{1-2\mu^2\delta^2N^2}\left(4\mu^2\delta^2 N\,\|\widetilde{\boldsymbol{w}}_{-1}^k\|^2 \;+\; 2\mu^2 N Q_{\rm av}\right) \tag{19.188}$$

Combining (19.178), (19.179), and (19.188) into (19.171), we arrive at

$$\begin{aligned}\|\widetilde{\boldsymbol{w}}_{-1}^{k+1}\|^2 &\leq \frac{1}{t}\left(1-\frac{2\mu\nu N}{3}\right)^2\|\widetilde{\boldsymbol{w}}_{-1}^k\|^2 + \\ &\qquad \frac{\mu^2\delta^2N^3}{2(1-t)(1-2\mu^2\delta^2N^2)}\left(4\mu^2\delta^2N\,\|\widetilde{\boldsymbol{w}}_{-1}^k\|^2 \;+\; 2\mu^2NQ_{\rm av}\right)\end{aligned} \tag{19.189}$$

We select $t = 1 - \frac{2\mu\nu N}{3}$ so that

$$\begin{aligned}\|\widetilde{\boldsymbol{w}}_{-1}^{k+1}\|^2 &\leq \left(1-\frac{2\mu\nu N}{3}\right)\|\widetilde{\boldsymbol{w}}_{-1}^k\|^2 + \\ &\qquad \frac{3\mu\delta^2N^2}{4\nu(1-2\mu^2\delta^2N^2)}\left(4\mu^2\delta^2N\,\|\widetilde{\boldsymbol{w}}_{-1}^k\|^2 \;+\; 2\mu^2NQ_{\rm av}\right) \\ &\leq \left(1-\frac{2\mu\nu N}{3} + \frac{3\mu^3\delta^4N^3}{\nu(1-2\mu^2\delta^2N^2)}\right)\|\widetilde{\boldsymbol{w}}_{-1}^k\|^2 + \\ &\qquad \frac{3\mu^3\delta^2N^3}{2\nu(1-2\mu^2\delta^2N^2)}Q_{\rm av}\end{aligned} \tag{19.190}$$

Assume again that μ is small enough such that

$$1-2\mu^2\delta^2N^2 > \frac{3}{4} \iff \mu < \frac{1}{\sqrt{8}N\delta} \tag{19.191}$$

Since $\nu \leq \delta$, this condition is met by any

$$\mu < \frac{\nu}{\sqrt{8}N\delta^2} \tag{19.192}$$

Then, we have

$$\|\widetilde{\boldsymbol{w}}_{-1}^{k+1}\|^2 \leq \left(1-\frac{2\mu\nu N}{3}+\frac{4\mu^3\delta^4N^3}{\nu}\right)\|\widetilde{\boldsymbol{w}}_{-1}^k\|^2 + \frac{2\mu^3\delta^2N^3}{\nu}Q_{\rm av} \tag{19.193}$$

Assume further that μ is small enough such that

$$1-\frac{2\mu\nu N}{3}+\frac{4\mu^3\delta^4N^3}{\nu} < 1-\frac{\mu}{2}\nu N \tag{19.194}$$

which is equivalent to

$$\mu < \frac{\nu}{\sqrt{24}N\delta^2} \tag{19.195}$$

Conditions (19.176), (19.192), and (19.195) are met by (19.77). Then, it follows that

$$\|\widetilde{\boldsymbol{w}}_{-1}^{k+1}\|^2 \leq \left(1-\frac{\mu}{2}\nu N\right)\|\widetilde{\boldsymbol{w}}_{-1}^k\|^2 + \frac{2\mu^3\delta^2N^3}{\nu}Q_{\rm av} \tag{19.196}$$

or, by taking expectations of both sides,

$$\mathbb{E}\,\|\widetilde{\boldsymbol{w}}_{-1}^{k+1}\|^2 \leq \left(1-\frac{\mu}{2}\nu N\right)\mathbb{E}\,\|\widetilde{\boldsymbol{w}}_{-1}^{k}\|^2 + \frac{2\mu^3\delta^2N^3}{\nu}Q_{\rm av} \tag{19.197}$$

and, hence,

$$\mathbb{E}\,\|\widetilde{\boldsymbol{w}}_{-1}^{k}\|^2 \;\leq\; O(\lambda^k)+O(\mu^2) \tag{19.198}$$

with $\lambda = 1-\frac{\mu\nu N}{2}$. Finally, note that for any n we have

$$\begin{aligned}
\|\widetilde{\boldsymbol{w}}_n^k\|^2 &= \|\widetilde{\boldsymbol{w}}_n^k - \widetilde{\boldsymbol{w}}_{-1}^k + \widetilde{\boldsymbol{w}}_{-1}^k\|^2 \\
&\leq 2\|\widetilde{\boldsymbol{w}}_{-1}^k\|^2 + 2\|\widetilde{\boldsymbol{w}}_n^k - \widetilde{\boldsymbol{w}}_{-1}^k\|^2 \\
&= 2\|\widetilde{\boldsymbol{w}}_{-1}^k\|^2 + 2\|\boldsymbol{w}_n^k - \boldsymbol{w}_{-1}^k\|^2 \\
&\overset{(a)}{=} 2\|\widetilde{\boldsymbol{w}}_{-1}^k\|^2 + 2\left\|\sum_{m=0}^{n}\boldsymbol{w}_m^k - \boldsymbol{w}_{m-1}^k\right\|^2 \\
&\overset{(19.172)}{\leq} 2\|\widetilde{\boldsymbol{w}}_{-1}^k\|^2 + 2(n+1)\sum_{m=0}^{n}\left\|\boldsymbol{w}_m^k - \boldsymbol{w}_{m-1}^k\right\|^2 \\
&\overset{(b)}{\leq} O(\lambda^k)+O(\mu^2)+O(\mu^2), \quad \text{large } k \\
&= O(\lambda^k)+O(\mu^2)
\end{aligned} \tag{19.199}$$

where in step (a) we used a telescoping series representation and in step (b) we used (19.188) and (19.198). We therefore arrive at (19.79a). To establish (19.79b), we use (19.12b) to note that

$$0 \leq \mathbb{E}\,P(\boldsymbol{w}_n^k) - P(w^\star) \leq \frac{\delta}{2}\,\mathbb{E}\,\|\widetilde{\boldsymbol{w}}_n^k\|^2 \tag{19.200}$$

■

REFERENCES

Al-Baali, M. (1985), "Descent property and global convergence of the Fletcher–Reeves method with inexact line search," *IMA J. Numer. Anal.*, vol. 5, pp. 121–124.

Albert, A. E. and L. A. Gardner (1967), *Stochastic Approximation and Nonlinear Regression*, MIT Press.

Benveniste, A., M. Métivier, and P. Priouret (1987), *Adaptive Algorithms and Stochastic Approximations*, Springer.

Bertsekas, D. P. and J. N. Tsitsiklis (1997), *Parallel and Distributed Computation: Numerical Methods*, Athena Scientific.

Bertsekas, D. P. and J. N. Tsitsiklis (2000), "Gradient convergence in gradient methods with errors," *SIAM J. Optim.*, vol. 10, no. 3, pp. 627–642.

Bottou, L. (1998), "Online algorithms and stochastic approximations," in *Online Learning and Neural Networks*, D. Saad, editor, pp. 9–42, Cambridge University Press.

Bottou, L. (2009), "Curiously fast convergence of some stochastic gradient descent algorithms," in *Proc. Symp. Learning and Data Science*, pp. 1–4, Paris.

Bottou, L. (2010), "Large-scale machine learning with stochastic gradient descent," *Proc. Int. COnf. Computational Statistics*, pp. 177–186, Paris.

Bottou, L. (2012), "Stochastic gradient descent tricks," in *Neural Networks: Tricks of the Trade*, G. Montavon, G. B. Orr, and K.-R. Muller, editors, pp. 421–436, Springer.

Bottou, L., F. E. Curtis, and J. Nocedal (2018), "Optimization methods for large-scale machine learning," *SIAM Rev.*, vol. 60, no. 2, pp. 223–311.

Gladyshev, E. G. (1965), "On stochastic approximations," *Theory Probab. Appl.*, vol. 10, pp. 275–278.

Gordon, G. J. (1999), "Regret bounds for prediction problems," *Proc. Ann. Conf. Computational Learning Theory* (COLT), pp. 29–40, Santa Cruz, CA.

Gürbüzbalaban, M., A. Ozdaglar, and P. Parrilo (2015b), "Why random reshuffling beats stochastic gradient descent," available at arXiv:1510.08560.

Haykin, S. (2001), *Adaptive Filter Theory*, 4th ed., Prentice Hall.

He, S., J. Wang, and J. Yan (1992), *Semimartingale Theory and Stochastic Calculus*, CRC Press.

Jin, X.-B., X.-Y. Zhang, K. Huang, and G.-G. Geng (2019), "Stochastic conjugate gradient algorithm with variance reduction," *IEEE Trans. Neural Netw. Learn. Syst.*, vol. 30, no. 5, pp. 1360–1369.

Kushner, H. J. (1984), *Approximation and Weak Convergence Methods for Random Processes, with Applications to Stochastic System Theory*, MIT Press.

Kushner, H. J. and D. S. Clark (1978), *Stochastic Approximation for Constrained and Unconstrained Systems*, Springer.

Kushner, H. J. and G. G. Yin (2003), *Stochastic Approximation and Recursive Algorithms and Applications*, Springer.

Lipster, R. and A. N. Shiryayev (1989), *Theory of Martingales*, Springer.

Ljung, L. (1977), "Analysis of recursive stochastic algorithms," *IEEE Trans. Aut. Control*, vol. 22, pp. 551–575.

Marti, K. (2005), *Stochastic Optimization Methods*, Springer.

Mendel, J. M. and K. S. Fu (1970), *Adaptive, Learning, and Pattern Recognition Systems: Theory and Applications*, Academic Press.

Needell, D., R. Ward, and N. Srebro (2014), "Stochastic gradient descent, weighted sampling, and the randomized Kaczmarz algorithm," *Proc. Advances Neural Information Processing Systems* (NIPS), pp. 1017–1025, Montreal.

Nemirovski, A. S., A. Juditsky, G. Lan, and A. Shapiro (2009), "Robust stochastic approximation approach to stochastic programming," *SIAM J. Optim.*, vol. 19, no. 4, pp. 1574–1609.

Nocedal, J. and S. J. Wright (2006), *Numerical Optimization*, Springer.

Polyak, B. T. (1987), *Introduction to Optimization*, Optimization Software.

Polyak, B. T. and A. Juditsky (1992), "Acceleration of stochastic approximation by averaging," *SIAM J. Control Optim.*, vol. 30, no. 4, pp. 838–855.

Powell, M. J. D. (1984), "Nonconvex minimization calculations and the conjugate gradient method," *Lecture Notes Math.*, vol. 1066, pp. 122–141, Springer.

Powell, M. J. D. (1986), "Convergence properties of algorithms for nonlinear optimization," *SIAM Rev.*, vol. 28, no. 4, pp. 487–500.

Recht, B. and C. Re (2012), "Toward a noncommutative arithmetic-geometric mean inequality: Conjectures, case-studies, and consequences," in *Proc. Conf. on Learning Theory* (COLT), pp. 1–11, Edinburgh.

Robbins, H. and S. Monro (1951), "A stochastic approximation method," *Ann. Math. Statist.*, vol. 22, pp. 400–407.

Ruppert, D. (1988), *Efficient Estimation from a Slowly Convergent Robbins–Monro Process*, Technical Report 781, Cornell University, School of Operations Research and Industrial Engineering.

Sayed, A. H. (2003), *Fundamentals of Adaptive Filtering*, Wiley.

Sayed, A. H. (2008), *Adaptive Filters*, Wiley.

Sayed, A. H. (2014a), *Adaptation, Learning, and Optimization over Networks*, Foundations and Trends in Machine Learning, NOW Publishers, vol. 7, no. 4–5, pp. 311–801.

Sayed, A. H. (2014b), "Adaptive networks," *Proc. IEEE*, vol. 102, no. 4, pp. 460–497.

Sayed, A. H. and X. Zhao (2018), "Asynchronous adaptive networks," in *Cooperative and Graph Signal Processing*, P. Djuric and C. Richard, editors, pp. 3–68, Elsevier. Also available at https://arxiv.org/abs/1511.09180.

Shalev-Shwartz, S. (2011), "Online learning and online convex optimization," *Foundations and Trends in Machine Learning*, vol. 4, no. 2, pp. 107–194.

Shalev-Shwartz, S. and S. Ben-David (2014), *Understanding Machine Learning: From Theory to Algorithms*, Cambridge University Press.

Shamir, O. (2016), "Without-replacement sampling for stochastic gradient methods: Convergence results and application to distributed optimization," *Proc. Advances Neural Information Processing Systems* (NIPS), pp. 46–54, Barcelona.

Spall, J. C. (2003), *Introduction to Stochastic Search and Optimization: Estimation, Simulation and Control*, Wiley.

Sun, J. and J. Zhang (2001), "Global convergence of conjugate gradient methods," *Ann. Oper. Res.*, vol. 103, pp. 161–173.

Tsypkin, Y. Z. (1971), *Adaptation and Learning in Automatic Systems*, Academic Press.

Wasan, M. T. (1969), *Stochastic Approximation*, Cambridge University Press.

Widrow, B. and S. D. Stearns (1985), *Adaptive Signal Processing*, Prentice Hall.

Williams, D. (1991), *Probability with Martingales*, Cambridge University Press.

Ying, B., K. Yuan, S. Vlaski, and A. H. Sayed (2019), "Stochastic learning under random reshuffling with constant step-sizes," *IEEE Trans. Signal Process.*, vol. 67, no. 2, pp. 474–489.

Yu, H., R. Jin, and S. Yang (2019), "On the linear speedup analysis of communication efficient momentum SGD for distributed non-convex optimization," *Proc. Mach. Learn. Res.* (PMLR), vol. 97, pp. 7184–7193.

Yuan, K., B. Ying, S. Vlaski, and A. H. Sayed (2016), "Stochastic gradient descent with finite sample sizes," *Proc. IEEE Int. Workshop on Machine Learning for Signal Processing* (MLSP), pp. 1–6, Salerno.

Zhao, X. and A. H. Sayed (2015a), "Asynchronous adaptation and learning over networks – part I: Modeling and stability analysis," *IEEE Trans. Signal Process.*, vol. 63, no. 4, pp. 811–826.

Zhao, X. and A. H. Sayed (2015b), "Asynchronous adaptation and learning over networks – part II: Performance analysis," *IEEE Trans. Signal Process.*, vol. 63, no. 4, pp. 827–842.

Zhao, P. and T. Zhang (2015), "Stochastic optimization with importance sampling for regularized loss minimization," in *Proc. Int. Conf. Machine Learning* (ICML), pp. 1355–1363, Lille.

Zinkevich, M. (2003), "Online convex programming and generalized infinitesimal gradient ascent," *Proc. Int. Conf. Machine Learning* (ICML), pp. 928–936, Washington, DC.

Zoutendijk, G. (1970), "Nonlinear programming, computational methods," in *Integer and Nonlinear Programming*, J. Abadie, editor, pp. 37–86, North-Holland.

20 Convergence Analysis II: Stochastic Subgradient Algorithms

We extend the convergence analysis from the previous chapter to stochastic subgradient algorithms for nonsmooth risks. We consider updates with instantaneous and mini-batch gradient approximations, for both cases of constant and vanishing step sizes. We also examine the incorporation of a pocket variable and exponential smoothing into the operation of the algorithms to improve convergence and steady-state performance. The main conclusion will be that the mean-square error (MSE), $\mathbb{E}\,\|\widetilde{\boldsymbol{w}}_n\|^2$, is guaranteed to approach a small $O(\mu)$-neighborhood at an exponential rate λ^n, while exact convergence of $\boldsymbol{w}_n$ to $w^\star$ can be guaranteed for some vanishing step-size sequences. On the other hand, the excess risk $\mathbb{E}\,P(\boldsymbol{w}_n) - P(w^\star)$ is guaranteed to approach a larger $O(\sqrt{\mu})$-neighborhood at the slower rate $\lambda^{n/2}$. The use of a pocket variable and exponential smoothing ensures convergence of the excess risk to the smaller $O(\mu)$-neighborhood and recovers the λ^n rate.

20.1 PROBLEM SETTING

We recall the problem formulation, and the conditions imposed on the risk and loss functions. We also recall the constructions for the subgradient approximations, and the first- and second-order moment results derived for the gradient noise process from Chapter 18.

20.1.1 Risk Minimization Problems

Let $s(w)$ denote a subgradient construction for the risk function $P(w)$ at location w. In this chapter, we are interested in examining the convergence behavior of the stochastic subgradient implementation:

$$\boxed{\boldsymbol{w}_n = \boldsymbol{w}_{n-1} - \mu\,\widehat{s}(\boldsymbol{w}_{n-1}),\ n \geq 0} \tag{20.1}$$

where $\widehat{s}(w)$ denotes an approximation for $s(w)$. We will be using either a constant or decaying step size for the solution of convex optimization problems of the form:

$$w^\star = \underset{w\in\mathbb{R}^M}{\operatorname{argmin}}\ P(w) \tag{20.2}$$

where $P(w)$ is a nonsmooth empirical or stochastic risk, i.e., for the solution of:

$$w^\star \stackrel{\Delta}{=} \underset{w\in\mathbb{R}^M}{\text{argmin}} \left\{ P(w) \stackrel{\Delta}{=} \frac{1}{N}\sum_{m=0}^{N-1} Q(w;\gamma(m),h_m) \right\} \tag{20.3a}$$

$$w^o \stackrel{\Delta}{=} \underset{w\in\mathbb{R}^M}{\text{argmin}} \left\{ P(w) \stackrel{\Delta}{=} \mathbb{E}\, Q(w;\boldsymbol{\gamma},\boldsymbol{h}) \right\} \tag{20.3b}$$

In these expressions, $Q(w;\cdot)$ denotes a convex loss function, $\{\gamma(m),h_m\}$ refer to a collection of N data points with $\gamma(m)\in\mathbb{R}$ and $h_m\in\mathbb{R}^M$, and the expectation in the second line is over the joint distribution of $\{\boldsymbol{\gamma},\boldsymbol{h}\}$.

20.1.2 Subgradient Vector Approximations

The search direction will be approximated by using either instantaneous or mini-batch calculations, namely,

$$\textbf{(instantaneous)}\quad \widehat{s}(w) = s_Q(w;\boldsymbol{\gamma},\boldsymbol{h}) \tag{20.4a}$$

$$\textbf{(mini-batch)}\quad \widehat{s}(w) = \frac{1}{B}\sum_{b=0}^{B-1} s_Q(w;\boldsymbol{\gamma}(b),\boldsymbol{h}_b) \tag{20.4b}$$

where the notation $s_Q(w;\boldsymbol{\gamma},\boldsymbol{h})$ denotes a subgradient construction for the loss function $Q(w;\cdot)$ evaluated at the data sample $(\boldsymbol{\gamma},\boldsymbol{h})$. For example, for the empirical risk minimization case (20.3a), a subgradient construction for $P(w)$ can be chosen as

$$s(w) = \frac{1}{N}\sum_{m=0}^{N-1} s_Q(w;\gamma(m),h_m) \tag{20.5}$$

in terms of individual subgradients for the loss function parameterized by the data points $\{\gamma(m),h_m\}$. The instantaneous approximation (20.4a) selects one subgradient vector at random, while the mini-batch approximation (20.4b) selects B subgradient vectors at random. Likewise, for the stochastic risk minimization case (20.3b), a subgradient construction for $P(w)$ can be chosen as

$$s(w) = \mathbb{E}\, s_Q(w;\boldsymbol{\gamma},\boldsymbol{h}) \tag{20.6}$$

in terms of the expectation of the subgradient for the loss function. The instantaneous approximation (20.4a) selects one realization for the subgradient, while the mini-batch approximation (20.4b) averages B realizations.

The difference between the original subgradient construction and its approximation is the *gradient noise*:

$$\boldsymbol{g}(w) \stackrel{\Delta}{=} \widehat{s}(w) - s(w) \tag{20.7}$$

The boldface notation $(\boldsymbol{\gamma}, \boldsymbol{h})$ or $(\boldsymbol{\gamma}(b), \boldsymbol{h}_b)$ refers to data samples selected at random from the given dataset $\{\gamma(m), h_m\}$ in empirical risk minimization, or assumed to stream in independently over time in stochastic risk minimization.

20.1.3 Conditions on Risk and Loss Functions

In Section 18.4 we introduced the following conditions for empirical risk minimization problems of the form (20.3a):

(A1) (**Strongly convex risk**). $P(w)$ is ν-strongly convex and first-order differentiable, namely, for every $w_1, w_2 \in \text{dom}(P)$, there exists a subgradient $s(w_1)$ relative to w^{T} such that

$$P(w_2) \geq P(w_1) + (s(w_1))^{\mathsf{T}}(w_2 - w_1) + \frac{\nu}{2}\|w_2 - w_1\|^2 \tag{20.8a}$$

for all $w_1, w_2 \in \text{dom}(P)$ and some $\nu > 0$.

(A2) (**Affine-Lipschitz loss gradients**). The loss function $Q(w, \cdot)$ is convex over w and its subgradients are affine-Lipschitz, i.e., there exist nonnegative constants $\{\delta, \delta_2\}$ such that, independently of the data samples,

$$\|s_Q(w_2; \gamma(\ell), h_\ell) - s'_Q(w_1; \gamma(k), h_k)\| \leq \delta\|w_2 - w_1\| + \delta_2 \tag{20.8b}$$

for any $w_1, w_2 \in \text{dom}(Q)$, any indices $0 \leq \ell, k \leq N-1$, and for *any* subgradients $s'_Q(w; \gamma, h) \in \partial_{w^{\mathsf{T}}} Q(w; \gamma, h)$. We explained that condition (20.8b) implies that the subgradient vectors for the risk function $P(w)$ satisfy the same affine-Lipschitz conditions, namely, for all $w_1, w_2 \in \text{dom}(P)$:

$$\|s(w_2) - s'(w_1)\| \leq \delta\|w_2 - w_1\| + \delta_2 \tag{20.9a}$$

$$\|s(w_2) - s'(w_1)\|^2 \leq 2\delta^2\|w_2 - w_1\| + 2\delta_2^2 \tag{20.9b}$$

On the other hand, for stochastic risk minimization problems of the form (20.3b), we continue to assume the strong convexity of $P(w)$ but replace (**A2**) by the requirement that the subgradients of the loss are now affine-Lipschitz in the mean-square sense:

(A2') (**Mean-square δ-Lipschitz loss gradients**). The gradient vectors of $Q(w, \cdot)$ satisfy the mean-square bound:

$$\mathbb{E}\,\|s_Q(w_2; \boldsymbol{\gamma}, \boldsymbol{h}) - s'_Q(w_1; \boldsymbol{\gamma}, \boldsymbol{h})\|^2 \leq \delta^2\|w_2 - w_1\|^2 + \delta_2^2 \tag{20.10}$$

for any $w_1, w_2 \in \text{dom}(Q)$ and for *any* $s'_Q(w; \gamma, h) \in \partial_{w^{\mathsf{T}}} Q(w; \gamma, h)$. We explained that condition (20.10) implies that the subgradients of $P(w)$ satisfy similar affine-Lipschitz conditions:

$$\|s(w_2) - s'(w_1)\| \leq \delta\|w_2 - w_1\| + \delta_2 \tag{20.11a}$$

$$\|s(w_2) - s'(w_1)\|^2 \leq \delta^2\|w_2 - w_1\|^2 + \delta_2^2 \tag{20.11b}$$

for any $w_1, w_2 \in \text{dom}(P)$. In this case, the constructions $s(w) = \mathbb{E}\, s_Q(w;\gamma,\boldsymbol{h})$ and $s'(w) = \mathbb{E}\, s'_Q(w;\gamma,\boldsymbol{h})$ correspond to subgradients for $P(w)$.

Remark 20.1. (Subgradients of risk function) Note that for stochastic risks, the Lipschitz property (20.11b) has factors (δ^2, δ_2^2) rather than $(2\delta^2, 2\delta 2^2)$, which appear in the empirical case (20.9b). The additional scaling by 2 is not relevant because we can replace δ^2 by $2\delta^2$ and δ_2^2 by $2\delta_2^2$ in (20.11b) and the bound will continue to hold. Therefore, we will pursue our arguments by treating both the empirical and stochastic cases similarly and assuming a bound of the form (20.9b) with factors $(2\delta^2, 2\delta_2^2)$ for both cases:

$$\|s(w_2) - s'(w_1)\|^2 \;\le\; 2\delta^2\|w_2 - w_1\|^2 + 2\delta_2^2 \tag{20.12}$$

This condition will be used in the convergence proofs in the remainder of this chapter. Thus, note that under conditions (**A1**,**A2**) for empirical risk minimization or (**A1**,**A2'**) for stochastic risk minimization, the following two conditions hold:

$$\begin{aligned}
&(\textbf{P1})\ \nu\text{-strong convexity of } P(w) \text{ as in (20.8a)} && \text{(20.13a)}\\
&(\textbf{P2})\ \text{affine-Lipschitz gradients for } P(w) \text{ as in (20.12)} && \text{(20.13b)}
\end{aligned}$$

■

Example 20.1 **(ℓ_1-regularized MSE risk)** We illustrate condition (20.12) by considering the following ℓ_1-regularized quadratic risk:

$$P(w) = \alpha\|w\|_1 \;+\; \mathbb{E}\,(\gamma - \boldsymbol{h}^\mathsf{T} w)^2, \quad w \in \mathbb{R}^M \tag{20.14}$$

We denote the covariance matrix of $\boldsymbol{h}$ by $R_h = \mathbb{E}\,\boldsymbol{h}\boldsymbol{h}^\mathsf{T}$. The corresponding loss function is given by

$$Q(w;\gamma,\boldsymbol{h}) \;=\; \alpha\|w\|_1 \;+\; (\gamma - \boldsymbol{h}^\mathsf{T} w)^2 \tag{20.15}$$

We know from the earlier results (8.49) and (8.52a) how to characterize the subdifferential set of $Q(w,\gamma,\boldsymbol{h})$. Let $w = \text{col}\{w_m\}$ denote the individual entries of the parameter vector. Then, it holds that

$$s'_Q(w_1;\gamma,\boldsymbol{h}) = \alpha\,\mathbb{G}(w_1) \;-\; 2\boldsymbol{h}^\mathsf{T}(\gamma - \boldsymbol{h}^\mathsf{T} w_1) \tag{20.16a}$$

where $\mathbb{G}(w_1) = \text{col}\{\mathbb{G}_m(w_{1,m})\}$ and each $\mathbb{G}_m(w_{1,m})$ is defined by

$$\mathbb{G}_m(w_{1,m}) \;=\; \begin{cases} +1, & w_{1,m} > 0\\ -1, & w_{1,m} < 0\\ [-1,+1], & w_{1,m} = 0 \end{cases} \tag{20.16b}$$

Moreover, one particular subgradient construction for $Q(w;\gamma,\boldsymbol{h})$ is given by

$$s_Q(w_2;\gamma,\boldsymbol{h}) \;=\; \alpha\,\text{sign}(w_2) \;-\; 2\boldsymbol{h}(\gamma - \boldsymbol{h}^\mathsf{T} w_2) \tag{20.16c}$$

Using the these results, we can write down a particular subgradient for $P(w)$, as well as describe the elements from the subdifferential set of $P(w)$ as follows:

$$\begin{aligned}
s(w_2) &= \alpha\,\text{sign}(w_2) \;-\; 2\mathbb{E}\,\boldsymbol{h}(\gamma - \boldsymbol{h}^\mathsf{T} w_2) && \text{(20.17a)}\\
s'(w_1) &= \alpha\,\mathbb{G}(w_1) - 2\mathbb{E}\,\boldsymbol{h}(\gamma - \boldsymbol{h}^\mathsf{T} w_1) && \text{(20.17b)}
\end{aligned}$$

Using the Jensen inequality that $\|a+b\|^2 \leq 2\|a\|^2 + 2\|b\|^2$ for any vectors a, b, we get

$$\begin{aligned}\|s(w_2) - s'(w_1)\|^2 &= \left\|\alpha\left(\text{sign}(w_2) - \mathbb{G}(w_1)\right) + 2\mathbb{E}\,\boldsymbol{h}\boldsymbol{h}^\mathsf{T}(w_2 - w_1)\right\|^2 \\ &\leq 2\alpha^2\,\|\text{sign}(w_2) - \mathbb{G}(w_1)\|^2 + 8\|\mathbb{E}\,\boldsymbol{h}\boldsymbol{h}^\mathsf{T}(w_2 - w_1)\|^2 \\ &\leq 8\alpha^2 M + 8\|\mathbb{E}\,\boldsymbol{h}\boldsymbol{h}^\mathsf{T}\|^2\,\|w_2 - w_1\|^2 \\ &= 8\alpha^2 M + 8\|R_h\|^2\,\|w_2 - w_1\|^2 \end{aligned} \tag{20.18}$$

which has the same form as (20.12) with

$$\delta = 2\|R_h\|, \quad \delta_2 = 2\alpha\sqrt{M} \tag{20.19}$$

Observe that, for this example, the value of δ_2 is proportional to the regularization factor α.

20.1.4 Gradient Noise

We further showed in Section 18.5 that, as a result of the Lipschitz conditions (**A2**) or (**A2'**), the first- and second-order moments of the gradient noise process satisfy the following two properties denoted by **G1** and **G2** for ease of reference:

$$(\textbf{G1})\ \ \mathbb{E}\left(\boldsymbol{g}_n(\boldsymbol{w}_{n-1})\,|\,\boldsymbol{w}_{n-1}\right) = 0 \tag{20.20a}$$

$$(\textbf{G2})\ \ \mathbb{E}\left(\|\boldsymbol{g}_n(\boldsymbol{w}_{n-1})\|^2\,|\,\boldsymbol{w}_{n-1}\right) \leq \beta_g^2\|\widetilde{w}_{n-1}\|^2 + \sigma_g^2 \tag{20.20b}$$

for some nonnegative constants $\{\beta_g^2, \sigma_g^2\}$ that are independent of $\widetilde{w}_{n-1}$. Condition (**G1**) refers to the zero-mean property of the gradient noise while (**G2**) refers to its bounded "variance." Results (19.13a)–(19.13b) hold for instantaneous and mini-batch gradient approximations, regardless of whether the samples are streaming in independently of each other, sampled uniformly with replacement, sampled without replacement, or selected under importance sampling. The *only exception* is that the zero-mean property (19.13a) will *not* hold for the instantaneous gradient implementation when the samples are selected without replacement.

In summary, we find that conditions (**A1**,**A2**) for empirical risk minimization or (**A1**,**A2'**) for stochastic risk minimization imply the validity of the following conditions on the risk function and gradient noise denoted by:

$$(\textbf{A1},\textbf{A2}) \text{ or } (\textbf{A1},\textbf{A2'}) \Longrightarrow (\textbf{P1},\textbf{P2},\textbf{G1},\textbf{G2}) \tag{20.21}$$

REMARK 20.2. **(Conditions on risk and loss functions)** The affine-Lipschitz conditions **A2** and **A2'** on the loss function were shown in Chapter 18 to lead to the gradient noise properties **G1**,**G2**. They also imply the δ-Lipschitz property **P2**. The convergence analyses in the following will rely largely on **G2** and **P2**, which relate to the bound on the second-order moment of the gradient noise and the δ-Lipschitz condition on the subgradients of $P(w)$. While the properties (**G2**,**P2**) follow from (**A2**,**A2'**), they can be introduced on their own as starting assumptions for the convergence analysis.

■

20.2 CONVERGENCE UNDER UNIFORM SAMPLING

We examine in this section the convergence behavior of the stochastic subgradient recursion (20.1) in the MSE sense under conditions (**A1**,**A2**) or (**A1**,**A2'**). Throughout this chapter, we will focus on uniform sampling (i.e., sampling with replacement) with and without pocket variables.

20.2.1 Mean-Square-Error Convergence

The next result shows that the MSE, denoted by $\mathbb{E}\,\|\widetilde{\boldsymbol{w}}_n\|^2$, does not converge to zero but to a small neighborhood of size $O(\mu)$, namely,

$$\limsup_{n\to\infty} \mathbb{E}\,\|\widetilde{\boldsymbol{w}}_n\|^2 \;=\; O(\mu) \tag{20.22}$$

The convergence occurs at an exponential rate λ^n. The same result shows that the excess risk approaches a larger neighborhood of size:

$$\limsup_{n\to\infty} \Big(\mathbb{E}\,P(\boldsymbol{w}_n) - P(w^\star)\Big) \;=\; O(\sqrt{\mu}\,) \tag{20.23}$$

where convergence in this case occurs at the slower rate $\lambda^{n/2}$. In the next section, we will explain how the convergence limit (20.23) for the excess risk can be improved to $O(\mu)$ at the faster rate λ^n by means of a pocket variable or exponential smoothing. The proof of the theorem will further reveal that the size of the limiting region for the MSE is given by (see the argument after (20.35))

$$\limsup_{n\to\infty} \mathbb{E}\,\|\widetilde{\boldsymbol{w}}_n\|^2 \;=\; O\Big(\mu(\sigma_g^2 + 2\delta_2^2)/2\nu\Big) \tag{20.24}$$

Observe in particular that the parameters (β_g^2, σ_g^2), which define the bound on the second-order moment of the gradient noise, affect performance in different ways. The value of β_g^2 affects stability (i.e., it defines a bound on μ for convergence) and the rate of convergence λ, while σ_g^2 affects the size of the limiting neighborhood (i.e., the size of the steady-state error).

Theorem 20.1. (MSE convergence under constant step sizes) *Consider the stochastic subgradient recursion (20.1) with the instantaneous subgradient approximation (20.4a) under uniform sampling with replacement or streaming data, used to seek minimizers of empirical or stochastic risks. The risk and loss functions are assumed to satisfy conditions (**A1,A2**) or (**A1,A2'**). For step-size values satisfying (i.e., for μ small enough):*

$$\mu \;<\; \frac{2\nu}{2\delta^2+\beta_g^2} \;\stackrel{\Delta}{=}\; \mu_o \tag{20.25}$$

it holds that $\mathbb{E}\,\|\widetilde{\boldsymbol{w}}_n\|^2$ and the average excess risk, $\mathbb{E}\,P(\boldsymbol{w}_n)-P(w^\star)$, converge exponentially fast according to the recursions:

$$\mathbb{E}\,\|\widetilde{\boldsymbol{w}}_n\|^2 \;\leq\; \lambda\,\mathbb{E}\,\|\widetilde{\boldsymbol{w}}_{n-1}\|^2 \;+\; \mu^2(\sigma_g^2+2\delta_2^2) \tag{20.26a}$$

$$\mathbb{E}\,\|\widetilde{\boldsymbol{w}}_n\|^2 \leq O(\lambda^n)+O(\mu) \tag{20.26b}$$

$$\mathbb{E}\,P(\boldsymbol{w}_n)-P(w^\star) \leq O(\lambda^{n/2})+O(\sqrt{\mu}) \tag{20.26c}$$

where

$$\lambda \;=\; 1-2\nu\mu+(2\delta^2+\beta_g^2)\mu^2 \;\in\; [0,1) \tag{20.27}$$

Results (20.26b)–(20.26c) hold for sufficiently small step sizes. Result (20.26c) holds whenever $\delta_2\neq 0$. Otherwise, the convergence of the excess risk occurs at the rate λ^n toward $O(\mu)$.

Proof: We subtract $w^\star$ from both sides of (20.1) and use the definition of the gradient noise to get

$$\widetilde{\boldsymbol{w}}_n \;=\; \widetilde{\boldsymbol{w}}_{n-1} \;+\; \mu\, s(\boldsymbol{w}_{n-1}) \;+\; \mu\, \boldsymbol{g}_n(\boldsymbol{w}_{n-1}) \tag{20.28}$$

We will be squaring this expression. In preparation for that step, we first recall that since $w^\star$ is the minimizer of $P(w)$, there should exist a subgradient vector $s'(w)$ at $w=w^\star$ such that

$$s'(w^\star)=0 \tag{20.29}$$

It follows that

$$\begin{aligned}
&\|\widetilde{\boldsymbol{w}}_{n-1}+\mu\, s(\boldsymbol{w}_{n-1})\|^2 \\
&= \|\widetilde{\boldsymbol{w}}_{n-1}\|^2+2\mu\,\left(s(\boldsymbol{w}_{n-1})\right)^{\mathsf T}\widetilde{\boldsymbol{w}}_{n-1}+\mu^2\|s(\boldsymbol{w}_{n-1})\|^2 \\
&= \|\widetilde{\boldsymbol{w}}_{n-1}\|^2+2\mu\,\left(s(\boldsymbol{w}_{n-1})\right)^{\mathsf T}\widetilde{\boldsymbol{w}}_{n-1}+\mu^2\|s'(w^\star)-s(\boldsymbol{w}_{n-1})\|^2 \\
&\stackrel{(20.12)}{\leq} \|\widetilde{\boldsymbol{w}}_{n-1}\|^2+2\mu\,\left(s(\boldsymbol{w}_{n-1})\right)^{\mathsf T}\widetilde{\boldsymbol{w}}_{n-1}+2\mu^2\delta^2\|\widetilde{\boldsymbol{w}}_{n-1}\|^2+2\mu^2\delta_2^2
\end{aligned} \tag{20.30}$$

Next, we appeal to the strong convexity property (20.8a) to find that

$$\begin{aligned}
\left(s(\boldsymbol{w}_{n-1})\right)^{\mathsf T}\widetilde{\boldsymbol{w}}_{n-1} \;&\leq\; P(w^\star)-P(\boldsymbol{w}_{n-1})-\frac{\nu}{2}\|\widetilde{\boldsymbol{w}}_{n-1}\|^2 \\
&\stackrel{(8.23)}{\leq}\; -\frac{\nu}{2}\|\widetilde{\boldsymbol{w}}_{n-1}\|^2 \;-\; \frac{\nu}{2}\|\widetilde{\boldsymbol{w}}_{n-1}\|^2 \\
&=\; -\nu\|\widetilde{\boldsymbol{w}}_{n-1}\|^2
\end{aligned} \tag{20.31}$$

Substituting into (20.30) gives

$$\boxed{\|\widetilde{\boldsymbol{w}}_{n-1}+\mu\, s(\boldsymbol{w}_{n-1})\|^2 \leq (1-2\mu\nu+2\mu^2\delta^2)\|\widetilde{\boldsymbol{w}}_{n-1}\|^2+2\mu^2\delta_2^2} \tag{20.32}$$

which is a useful intermediate result. Now returning to (20.28), squaring both sides, conditioning on $\boldsymbol{w}_{n-1}$, and taking expectations we obtain

$$
\begin{aligned}
&\mathbb{E}\left(\|\widetilde{\boldsymbol{w}}_n\|^2 \mid \boldsymbol{w}_{n-1}\right) \\
&= \mathbb{E}\left(\|\widetilde{\boldsymbol{w}}_{n-1} + \mu\, s(\boldsymbol{w}_{n-1}) + \mu\, \boldsymbol{g}_n(\boldsymbol{w}_{n-1})\|^2 \mid \boldsymbol{w}_{n-1}\right) \\
&\overset{(a)}{=} \mathbb{E}\left(\|\widetilde{\boldsymbol{w}}_{n-1} + \mu\, s(\boldsymbol{w}_{n-1})\|^2 \mid \boldsymbol{w}_{n-1}\right) + \mu^2\, \mathbb{E}\left(\|\boldsymbol{g}_n(\boldsymbol{w}_{n-1})\|^2 \mid \boldsymbol{w}_{n-1}\right) \\
&\overset{(20.32)}{\leq} (1 - 2\mu\nu + 2\mu^2\delta^2)\|\widetilde{w}_{n-1}\|^2 + 2\mu^2\delta_2^2 + \mu^2\, \mathbb{E}\left(\|\boldsymbol{g}_n(\boldsymbol{w}_{n-1})\|^2 \mid \boldsymbol{w}_{n-1}\right) \\
&\overset{(20.20b)}{\leq} (1 - 2\mu\nu + 2\mu^2\delta^2)\|\widetilde{w}_{n-1}\|^2 + 2\mu^2\delta_2^2 + \mu^2\left(\beta_g^2\|\widetilde{w}_{n-1}\|^2 + \sigma_g^2\right) \\
&= (1 - 2\mu\nu + \mu^2(2\delta^2 + \beta_g^2))\|\widetilde{w}_{n-1}\|^2 + \mu^2\sigma_g^2 + 2\mu^2\delta_2^2 \\
&= \lambda\|\widetilde{w}_{n-1}\|^2 + \mu^2(\sigma_g^2 + 2\delta_2^2)
\end{aligned}
\tag{20.33}
$$

where the cross term in step (a) is zero because of the zero-mean property (20.20a). Taking expectations of both sides again removes the conditioning on $\boldsymbol{w}_{n-1}$ and leads to (20.26a) where λ is defined by (20.27). The same argument used in Fig. 12.2 can be repeated here to show that condition (20.25) ensures $0 \leq \lambda < 1$. By further iterating recursion (20.26a) we obtain

$$
\mathbb{E}\,\|\widetilde{\boldsymbol{w}}_n\|^2 \leq \lambda^{n+1}\,\mathbb{E}\,\|\widetilde{\boldsymbol{w}}_{-1}\|^2 + \frac{\mu^2(\sigma_g^2 + 2\delta_2^2)}{1 - \lambda}
\tag{20.34}
$$

which proves that $\mathbb{E}\,\|\widetilde{\boldsymbol{w}}_n\|^2$ converges exponentially fast to a region that is upper-bounded by

$$
\limsup_{n\to\infty}\ \mathbb{E}\,\|\widetilde{\boldsymbol{w}}_n\|^2 \leq \frac{\mu^2(\sigma_g^2 + 2\delta_2^2)}{1 - \lambda} = \frac{\mu(\sigma_g^2 + 2\delta_2^2)}{2\nu - \mu(2\delta^2 + \beta_g^2)}
\tag{20.35}
$$

It is easy to check that the upper bound does not exceed $\mu(\sigma_g^2 + 2\delta_2^2)/\nu$ for any step size $\mu < \mu_o$. If, on the other hand, the step size is sufficiently small so that the denominator is approximately 2ν, then the size of the limiting region is $O(\mu(\sigma_g^2 + 2\delta_2^2)/2\nu)$. Either way, we conclude that

$$
\limsup_{n\to\infty}\ \mathbb{E}\,\|\widetilde{\boldsymbol{w}}_n\|^2 \leq O(\mu)
\tag{20.36}
$$

We show in Prob. 20.1 that the norm of the weight-error vector evolves on average according to the relation:

$$
\mathbb{E}\,\|\widetilde{\boldsymbol{w}}_n\| \leq \sqrt{\lambda}\,\mathbb{E}\,\|\widetilde{\boldsymbol{w}}_{n-1}\| + O(\mu)
\tag{20.37}
$$

To establish (20.26c), we use (14.60) to note that

$$
0 \leq \mathbb{E}\,P(\boldsymbol{w}_n) - P(w^\star) \leq 2\delta\,\mathbb{E}\,\|\widetilde{\boldsymbol{w}}_n\|^2 + 2\delta_2\,\mathbb{E}\,\|\widetilde{\boldsymbol{w}}_n\|
\tag{20.38}
$$

so that using (20.36) and (20.37):

$$
\limsup_{n\to\infty}\ \left(\mathbb{E}\,P(\boldsymbol{w}_n) - P(w^\star)\right) \leq O(\sqrt{\mu})
\tag{20.39}
$$

with the decay of $\mathbb{E}\,P(\boldsymbol{w}_n)$ toward $P(w^\star)$ occurring at the same rate as $\mathbb{E}\,\|\widetilde{\boldsymbol{w}}_n\|$, which is $\lambda^{n/2}$.

■

20.2.2 Regret Analysis

For empirical risks minimized by stochastic algorithms, the regret value over a window of N iterations is defined as the deviation of the accumulated *loss* relative from the minimal risk value:

$$\boldsymbol{\mathcal{R}}(N) \stackrel{\Delta}{=} \frac{1}{N}\sum_{n=0}^{N-1} Q(\boldsymbol{w}_{n-1};\boldsymbol{\gamma}(n),\boldsymbol{h}_n) \; - \; P(w^\star) \tag{20.40}$$

where the arguments $(\boldsymbol{w}_{n-1},\boldsymbol{\gamma}(n),\boldsymbol{h}_n)$ of the loss function are random due to sampling and gradient noise. As was explained earlier following (19.31), if we evaluate the conditional expectation of $\boldsymbol{\mathcal{R}}(N)$ over a trajectory of the weight iterates, and use the unbiasedness property $\mathbb{E}\,Q(w,\boldsymbol{\gamma},\boldsymbol{h}) = P(w)$, we arrive at the following expression for the average regret:

$$\mathcal{R}(N) \stackrel{\Delta}{=} \frac{1}{N}\sum_{n=0}^{N-1} P(w_{n-1}) \; - \; P(w^\star) \tag{20.41}$$

Using (20.26c) and the same argument that led to (12.58), we will be able to conclude that the regret for the stochastic subgradient algorithm (20.1) under uniform sampling satisfies:

$$\mathcal{R}(N) \leq O(1/N) + O(\sqrt{\mu}) \tag{20.42}$$

20.3 CONVERGENCE WITH POCKET VARIABLES

The statement of Theorem 20.1 establishes the exponential convergence of $\mathbb{E}\,\|\widetilde{\boldsymbol{w}}_n\|^2$ to a region of size $O(\mu)$ at the rate of λ^n. It also establishes that $\mathbb{E}\,P(\boldsymbol{w}_n)$ converges exponentially at the slower rate $\lambda^{n/2}$ to an $O(\sqrt{\mu})$-neighborhood around $P(w^\star)$. We express both conclusions in the following convenient form:

$$\begin{cases} \mathbb{E}\,\|\widetilde{\boldsymbol{w}}_n\|^2 \leq O(\lambda^n) + O(\mu) \\ \mathbb{E}\,P(\boldsymbol{w}_n) - P(w^\star) \leq O(\lambda^{n/2}) + O(\sqrt{\mu}) \end{cases} \tag{20.43}$$

We can improve these results and obtain faster convergence for the risk value at the rate λ^n toward a smaller $O(\mu)$-neighborhood around $P(w^\star)$ by relying on the use of a pocket variable defined as:

$$w_n^{\text{best}} \stackrel{\Delta}{=} \underset{0\leq m\leq n}{\operatorname{argmin}} \left\{P(w_m)\right\} \tag{20.44}$$

Determination of w_n^{best} requires the ability to compute $P(w_m)$, which can be costly for large datasets in empirical minimization or due to the unknown data distribution in stochastic minimization. We ignore this complexity for now and establish an initial result that shows the improvement in performance that is attained by the use of pocket variables. In the next section we address the computational challenge and resort to exponential smoothing to achieve the desired improvement in performance.

LEMMA 20.1. (Convergence with a pocket variable) *Consider the same setting of Theorem 20.1. If the step size μ satisfies*

$$0 < \mu < \frac{\nu}{2\delta^2 + \beta_g^2} = \mu_o/2 \tag{20.45}$$

then it holds that

$$\mathbb{E}\, P(\boldsymbol{w}_n^{\text{best}}) - P(w^\star) \leq O(\lambda^n) \;+\; O(\mu) \tag{20.46}$$

where now

$$\lambda \;=\; 1 - \nu\mu + \mu^2(2\delta^2 + \beta_g^2) \;\in\; [0, 1) \tag{20.47}$$

Proof: Squaring the weight-error recursion (20.28), conditioning on $\boldsymbol{w}_{n-1}$, and taking expectations:

$$\mathbb{E}\,\left(\|\widetilde{\boldsymbol{w}}_n\|^2\,|\,\boldsymbol{w}_{n-1}\right) = \|\widetilde{\boldsymbol{w}}_{n-1}\|^2 \;+\; 2\mu(s(\boldsymbol{w}_{n-1}))^{\mathsf{T}}\widetilde{\boldsymbol{w}}_{n-1} \;+\; \mu^2\,\|s(\boldsymbol{w}_{n-1})\|^2 + \mu^2\mathbb{E}\,\left(\boldsymbol{g}_n(\boldsymbol{w}_{n-1})\,|\,\boldsymbol{w}_{n-1}\right) \tag{20.48}$$

Now, from the strong-convexity condition (20.8a), it holds that

$$(s(\boldsymbol{w}_{n-1}))^{\mathsf{T}}\widetilde{\boldsymbol{w}}_{n-1} \;\leq\; P(w^\star) - P(\boldsymbol{w}_{n-1}) - \frac{\nu}{2}\|\widetilde{\boldsymbol{w}}_{n-1}\|^2 \tag{20.49}$$

so that substituting into (20.48) gives

$$\mathbb{E}\,\left(\|\widetilde{\boldsymbol{w}}_n\|^2\,|\,\boldsymbol{w}_{n-1}\right) \leq \|\widetilde{\boldsymbol{w}}_{n-1}\|^2 \;+\; 2\mu\,(P(w^\star) - P(\boldsymbol{w}_{n-1})) - \nu\mu\|\widetilde{\boldsymbol{w}}_{n-1}\|^2 \;+\; \mu^2\,\|s(\boldsymbol{w}_{n-1})\|^2 + \mu^2\mathbb{E}\,\left(\boldsymbol{g}_n(\boldsymbol{w}_{n-1}\,|\,\boldsymbol{w}_{n-1})\right) \tag{20.50}$$

Referring to (18.68), if we set $w_1 \leftarrow \boldsymbol{w}_{n-1}$ and $w_2 \leftarrow w^\star$, and use the fact that there exists one particular subgradient for $P(w)$ satisfying $s'(w^\star) = 0$, we obtain

$$\|s(\boldsymbol{w}_{n-1})\|^2 \;\leq\; 2\delta^2\,\|\widetilde{\boldsymbol{w}}_{n-1}\|^2 \;+\; 2\delta_2^2 \tag{20.51}$$

Substituting into (20.50) and using (18.85b), we get

$$\mathbb{E}\,\left(\|\widetilde{\boldsymbol{w}}_n\|^2\,|\,\boldsymbol{w}_{n-1}\right) \leq (1 - \nu\mu + 2\delta^2\mu^2)\|\widetilde{\boldsymbol{w}}_{n-1}\|^2 \;+\; 2\mu\,(P(w^\star) - P(\boldsymbol{w}_{n-1})) \;+\; 2\delta_2^2\mu^2 + \mu^2(\beta_g^2\|\widetilde{\boldsymbol{w}}_{n-1}\|^2 + \sigma_g^2) \tag{20.52}$$

Taking expectations again we remove the conditioning on $\boldsymbol{w}_{n-1}$ and obtain

$$\mathbb{E}\,\|\widetilde{\boldsymbol{w}}_n\|^2 \leq (1 - \nu\mu + 2\delta^2\mu^2)\,\mathbb{E}\,\|\widetilde{\boldsymbol{w}}_{n-1}\|^2 \;+\; 2\mu\,(P(w^\star) - \mathbb{E}\,P(\boldsymbol{w}_{n-1})) \;+\; 2\delta_2^2\mu^2 + \mu^2(\beta_g^2\mathbb{E}\,\|\widetilde{\boldsymbol{w}}_{n-1}\|^2 + \sigma_g^2) \tag{20.53}$$

which we rewrite as

$$2\mu\,(\mathbb{E}\,P(\boldsymbol{w}_{n-1}) - P(w^\star)) \leq (1 - \nu\mu + \mu^2(2\delta^2 + \beta_g^2))\,\mathbb{E}\,\|\widetilde{\boldsymbol{w}}_{n-1}\|^2 \;-\; \mathbb{E}\,\|\widetilde{\boldsymbol{w}}_n\|^2 + \mu^2(2\delta_2^2 + \sigma_g^2) \tag{20.54}$$

To proceed, we simplify the notation and introduce the scalars

$$a(n) = \mathbb{E}\,P(\boldsymbol{w}_{n-1}) - P(w^\star) \tag{20.55a}$$

$$b(n) = \mathbb{E}\,\|\widetilde{\boldsymbol{w}}_n\|^2 \tag{20.55b}$$

$$\lambda = 1 - \nu\mu + \mu^2(2\delta^2 + \beta_g^2) \tag{20.55c}$$

$$\tau^2 = 2\delta_2^2 + \sigma_g^2 \tag{20.55d}$$

Since $w^\star$ is the global minimizer for $P(w)$, it holds that $a(n) \geq 0$ for all n. Note that $a(n)$ represents the excess risk at iteration $n-1$. Using the above scalars we rewrite (20.54) more compactly as

$$2\mu a(n) \leq \lambda b(n-1) - b(n) + \mu^2\tau^2 \tag{20.56}$$

This recursion has the same form as the earlier recursion (14.79) derived while studying the convergence of subgradient algorithms with pocket variables. Therefore, from this point onwards, the same arguments apply and we would similarly conclude that, over any interval $L \geq 1$,

$$\mathbb{E}\, P(\boldsymbol{w}_{L-1}^{\rm best}) - P(w^\star) \;\leq\; \frac{\lambda^{L+1}(1-\lambda)}{2\mu(1-\lambda^{L+1})}\,\mathbb{E}\,\|\widetilde{\boldsymbol{w}}_{-1}\|^2 + \frac{1}{2}\mu(2\delta_2^2+\sigma_g^2) \tag{20.57}$$

This result leads to (20.46) by noting that $1-\lambda^L \geq 1/2$ for L large enough. We conclude that

$$\mathbb{E}\, P(\boldsymbol{w}_L^{\rm best}) - P(w^\star) \;\leq\; O(\lambda^L) \;+\; O(\mu) \tag{20.58}$$

■

20.4 CONVERGENCE WITH EXPONENTIAL SMOOTHING

The convergence result (20.46) shows that the average risk value at the pocket variable, $\boldsymbol{w}_n^{\rm best}$, is able to approach the global minimum within $O(\mu)$. This is a reassuring conclusion. However, determination of the pocket variable is a challenging task. A more practical construction is possible motivated by the arguments in the convergence proof.

Let us introduce the geometric sum, for some integer L:

$$S_L \;\stackrel{\Delta}{=}\; \sum_{n=0}^{L} \lambda^{L-n} \;=\; \frac{1-\lambda^{L+1}}{1-\lambda} \tag{20.59}$$

and note that

$$S_L = \lambda S_{L-1} + 1 \tag{20.60}$$

We also introduce the convex combination coefficients:

$$r_L(n) \;\stackrel{\Delta}{=}\; \frac{\lambda^{L-n}}{S_L}, \quad n = 0, 1, 2, \ldots, L \tag{20.61}$$

Using these coefficients, we define a smoothed iterate as follows:

$$\begin{aligned}\bar{\boldsymbol{w}}_L \;&\stackrel{\Delta}{=}\; \sum_{n=0}^{L} r_L(n)\boldsymbol{w}_{n-1} \\ &= \frac{1}{S_L}\left(\lambda^L \boldsymbol{w}_{-1} + \lambda^{L-1}\boldsymbol{w}_0 + \ldots + \lambda \boldsymbol{w}_{L-2} + \boldsymbol{w}_{L-1}\right)\end{aligned} \tag{20.62}$$

where the $\{\boldsymbol{w}_n\}$ are the iterates generated by the stochastic subgradient algorithm (20.1). Observe that, in contrast to the pocket variable $\boldsymbol{w}_L^{\rm best}$, the smoothed variable $\bar{w}_L$ is computed directly from the iterates $\{w_m\}$ and not from the risk values $\{P(w_m)\}$. Observe further that $\bar{w}_L$ satisfies the recursive construction:

$$\bar{w}_L = \left(1 - \frac{1}{S_L}\right)\bar{w}_{L-1} + \frac{1}{S_L}w_{L-1}, \quad \bar{w}_0 = w_{-1}, \quad L \geq 0 \tag{20.63}$$

Moreover, since $P(w)$ is a convex function, it holds that

$$P(\bar{w}_L) = P\left(\sum_{n=0}^{L} r_L(n) w_{n-1}\right) \leq \sum_{n=0}^{L} r_L(n) P(w_{n-1}) \tag{20.64}$$

Using this fact, and following an argument similar to the one that led to (20.58), we can establish for the smoothed iterate that

$$\mathbb{E}\, P(\bar{w}_L) - P(w^\star) \leq O(\lambda^L) + O(\mu) \tag{20.65}$$

so that convergence continues to occur at the rate λ^n.

Proof of (20.65): Iterating (20.56) and dividing both sides by S_L we get

$$\sum_{n=0}^{L} \frac{\lambda^{L-n}}{S_L}\left(2\mu a(n) - \mu^2\tau^2\right) \leq \frac{\lambda^{L+1}}{S_L} b(-1) \tag{20.66}$$

or, equivalently,

$$\sum_{n=0}^{L} 2\mu r_L(n)\Big(\mathbb{E}\, P(w_n) - P(w^\star)\Big) \leq \frac{\lambda^{L+1}(1-\lambda)}{1-\lambda^{L+1}} b(-1) + \mu^2\tau^2 \tag{20.67}$$

Appealing to the convexity property (20.64), we conclude that

$$2\mu\Big(\mathbb{E}\, P(\bar{w}_L) - P(w^\star)\Big) \leq \frac{\lambda^{L+1}(1-\lambda)}{1-\lambda^{L+1}} b(-1) + \mu^2\tau^2 \tag{20.68}$$

from which we arrive at (20.65).

■

Incorporating smoothing

Using λ as a scaling weight in (20.62) is problematic because its value depends on (μ, ν, δ) and the parameters (ν, δ) may not be known beforehand. The analysis, however, suggests that we may replace λ by a design parameter, denoted by κ and satisfying $\lambda \leq \kappa < 1$. We therefore introduce a new smoothed variable (we continue to denote it by $\bar{w}_L$ to avoid a proliferation of symbols; we also continue to denote the scaling coefficients by $r_L(n)$):

$$\bar{w}_L \triangleq \sum_{n=0}^{L} r_L(n) w_{n-1} \tag{20.69}$$

where now

$$r_L(n) \triangleq \frac{\kappa^{L-n}}{S_L} \tag{20.70a}$$

$$S_L \triangleq \sum_{n=0}^{L} \kappa^{L-n} = \frac{1-\kappa^{L+1}}{1-\kappa} \tag{20.70b}$$

The resulting algorithm is listed in (20.71), where κ is chosen close to but smaller than 1.

Stochastic subgradient algorithm with exponential smoothing for minimizing (20.3a)–(20.3b).

---

given dataset $\{\gamma(m), h_m\}_{m=0}^{N-1}$ or streaming data $(\gamma(n), h_n)$;
select a positive scalar $\kappa = 1 - O(\mu)$ close to but smaller than 1;
start from an arbitrary initial condition, $\boldsymbol{w}_{-1}$.
start from $\bar{\boldsymbol{w}}_0 = \boldsymbol{w}_{-1}$;
start from $S_0 = 0$;
repeat until convergence over $n \geq 0$**:** (20.71)
  select at random or receive $(\boldsymbol{\gamma}(n), \boldsymbol{h}_n)$;
  $\boldsymbol{w}_n = \boldsymbol{w}_{n-1} - \mu\, s_Q(\boldsymbol{w}_{n-1}; \boldsymbol{\gamma}(n), \boldsymbol{h}_n)$
  $S_{n+1} = \kappa S_n + 1$;
  $\bar{\boldsymbol{w}}_{n+1} = \left(1 - \frac{1}{S_{n+1}}\right)\bar{\boldsymbol{w}}_n + \frac{1}{S_{n+1}}\boldsymbol{w}_n$
end
return $w^\star \leftarrow \bar{\boldsymbol{w}}_{n+1}$.

For this procedure, the convergence property (14.93) will continue to hold with λ replaced by κ.

Lemma 20.2. **(Convergence with smoothing)** *Consider the same setting of Theorem 20.1, and choose a smoothing parameter κ satisfying $\lambda \leq \kappa < 1$. For the same bound (20.45) on the step size μ, it holds that*

$$\mathbb{E}\,\|w^\star - \bar{\boldsymbol{w}}_n\|^2 \leq O(\kappa^n) + O(\mu) \tag{20.72a}$$

$$\mathbb{E}\,P(\bar{\boldsymbol{w}}_n) - P(w^\star) \leq O(\kappa^n) + O(\mu) \tag{20.72b}$$

and the steady-state regions are given by

$$\limsup_{n\to\infty} \mathbb{E}\,\|w^\star - \bar{\boldsymbol{w}}_n\|^2 \leq O(\mu\tau^2/\nu) \tag{20.73a}$$

$$\limsup_{n\to\infty} \Big(\mathbb{E}\,P(\bar{\boldsymbol{w}}_n) - P(w^\star)\Big) \leq O(\mu\tau^2/2) \tag{20.73b}$$

where $\tau^2 = 2\delta_2^2 + \sigma_g^2$.

Proof: The argument requires some adjustment relative to what we did before. We scale each term in (20.56) by κ^{L-n} and iterate from $n = 0$ to $n = L$ to get

$$\begin{aligned}\sum_{n=0}^{L}\kappa^{L-n}\left(2\mu a(n) - \mu^2\tau^2\right) &\leq \sum_{n=0}^{L}\kappa^{L-n}\left(\lambda b(n-1) - b(n)\right)\\ &= \kappa^L\lambda b(-1) - b(L) + (\lambda-\kappa)\sum_{n=0}^{L-1}\kappa^{L-n-1}b(n)\\ &\leq \kappa^L\lambda b(-1)\end{aligned} \tag{20.74}$$

where in the last inequality we used the fact that $\kappa \geq \lambda$. We can now proceed from here and complete the argument as before to arrive at

$$2\mu\left(\mathbb{E}\,P(\bar{\boldsymbol{w}}_{L-1}) - P(w^\star)\right) \leq \frac{\lambda\kappa^L(1-\kappa)}{1-\kappa^{L+1}}b(-1) + \mu^2\tau^2 \tag{20.75}$$

We can also characterize how far $\bar{\boldsymbol{w}}_L$ is from $w^\star$ by recalling the strong-convexity definition (8.44), which implies that

$$\mathbb{E}\,\|\bar{\boldsymbol{w}}_L - w^\star\|^2 \leq \frac{2}{\nu}\Big(\mathbb{E}\,P(\bar{\boldsymbol{w}}_L) - P(w^\star)\Big) \tag{20.76}$$

■

Example 20.2 (**ℓ_1-regularized MSE risk**) We illustrate the behavior of the stochastic subgradient algorithm, with and without exponential smoothing, by applying it to the minimization of an ℓ_1-regularized MSE risk of the form:

$$w^o \stackrel{\Delta}{=} \underset{w\in\mathbb{R}^M}{\text{argmin}} \left\{P(w) \stackrel{\Delta}{=} \alpha\|w\|_1 + \mathbb{E}\,(\boldsymbol{\gamma} - \boldsymbol{h}^\mathsf{T} w)^2\right\} \tag{20.77}$$

The streaming data samples $(\boldsymbol{\gamma}(n), \boldsymbol{h}_n)$ are assumed to arise from the linear model:

$$\boldsymbol{\gamma}(n) = \boldsymbol{h}_n^\mathsf{T} w^\bullet + \boldsymbol{v}(n) \tag{20.78}$$

for some parameter $w^\bullet \in \mathbb{R}^M$ and where $\boldsymbol{v}(n)$ is white noise with zero mean and variance σ_v^2; the noise is independent of all other random variables. We assume that the data $\{\boldsymbol{\gamma}, \boldsymbol{h}\}$ have zero means with second-order moments given by

$$\sigma_\gamma^2 \stackrel{\Delta}{=} \mathbb{E}\,\boldsymbol{\gamma}^2, \quad r_{h\gamma} \stackrel{\Delta}{=} \mathbb{E}\,\boldsymbol{h}\boldsymbol{\gamma}, \quad R_h \stackrel{\Delta}{=} \mathbb{E}\,\boldsymbol{h}\boldsymbol{h}^\mathsf{T} = \sigma_h^2 I_M \tag{20.79}$$

If we multiply both sides of (20.78) by $\boldsymbol{h}_n$ and take expectations, we find that $w^\bullet$ satisfies

$$r_{h\gamma} = \sigma_h^2\, w^\bullet \tag{20.80}$$

Expanding (20.77) we get

$$\begin{aligned} P(w) &= \alpha\|w\|_1 + \mathbb{E}\,(\boldsymbol{\gamma} - \boldsymbol{h}^\mathsf{T} w)^2 \\ &= \alpha\|w\|_1 + \mathbb{E}\,(\boldsymbol{h}^\mathsf{T} w^\bullet + \boldsymbol{v} - \boldsymbol{h}^\mathsf{T} w)^2 \\ &= \alpha\|w\|_1 + \sigma_h^2\|w - w^\bullet\|^2 + \sigma_v^2 \end{aligned} \tag{20.81}$$

Comparing with (11.31)–(11.32) we can write down a closed-form expression for the minimizer w^o of $P(w)$ in terms of the soft-thresholding operator defined by (11.18):

$$w^o = \mathbb{T}_{\alpha/2\sigma_h^2}(w^\bullet) \tag{20.82}$$

We know from result (20.19) that the affine factor δ_2 for the subgradients of $P(w)$ is given by $\delta_2 = 2\alpha\sqrt{M}$. Moreover, subtracting (20.16c) and (20.17a), with both expressions evaluated at $w_2 \leftarrow \boldsymbol{w}_{n-1}$ and at the data sample $(\boldsymbol{\gamma}(n), \boldsymbol{h}_n)$, we find that the gradient noise is given by

$$\begin{aligned} &\boldsymbol{g}_n(\boldsymbol{w}_{n-1}) \\ &= 2(r_{h\gamma} - \sigma_h^2\boldsymbol{w}_{n-1}) - 2\boldsymbol{h}_n(\boldsymbol{\gamma}(n) - \boldsymbol{h}_n^\mathsf{T}\boldsymbol{w}_{n-1}) \\ &\overset{(20.80)}{=} 2\sigma_h^2(w^\bullet - \boldsymbol{w}_{n-1}) - 2\boldsymbol{h}_n\boldsymbol{h}_n^\mathsf{T}(w^\bullet + \boldsymbol{v}(n) - \boldsymbol{w}_{n-1}) \\ &\overset{(a)}{=} 2\sigma_h^2(w^\bullet - w^o + w^o - \boldsymbol{w}_{n-1}) - 2\boldsymbol{h}_n\boldsymbol{h}_n^\mathsf{T}(w^\bullet + \boldsymbol{v}(n) - w^o + w^o - \boldsymbol{w}_{n-1}) \\ &\overset{(b)}{=} 2(\sigma_h^2 I_M - \boldsymbol{h}_n\boldsymbol{h}_n^\mathsf{T})\widetilde{\boldsymbol{w}}_{n-1} - 2\boldsymbol{h}_n\boldsymbol{h}_n^\mathsf{T}\boldsymbol{v}(n) + 2(\boldsymbol{h}_n\boldsymbol{h}_n^\mathsf{T} - \sigma_h^2 I_M)\widetilde{w}^\bullet \end{aligned} \tag{20.83}$$

where in step (a) we added and subtracted the minimizer w^o, and in step (b) we used $\widetilde{\boldsymbol{w}}_{n-1} = w^o - \boldsymbol{w}_{n-1}$ and $\widetilde{w}^\bullet = w^o - w^\bullet$. Conditioning on $\boldsymbol{w}_{n-1}$, squaring both sides, and taking expectations we arrive at the following bound for the second-order moment of the gradient noise:

$$\mathbb{E}\left(\|\boldsymbol{g}_n(\boldsymbol{w}_{n-1})\|^2 \,|\, \boldsymbol{w}_{n-1}\right) \leq \beta_g^2\|\widetilde{w}_{n-1}\|^2 + \delta_g^2 \tag{20.84a}$$

$$\beta_g^2 \overset{\Delta}{=} 4\,\mathbb{E}\,\|\sigma_h^2 I_M - \boldsymbol{h}\boldsymbol{h}^\mathsf{T}\|^2 \tag{20.84b}$$

$$\sigma_g^2 = 4\sigma_v^2\sigma_h^2 M + \beta_g^2\|\widetilde{w}^\bullet\|^2 \tag{20.84c}$$

We deduce from (20.73b) that the excess risk for the stochastic subgradient algorithm with exponential smoothing will be within the following region:

$$\limsup_{n\to\infty}\left(\mathbb{E}\,P(\bar{\boldsymbol{w}}_n) - P(w^\star)\right) = O\left(\mu(2\delta_2^2 + \sigma_g^2)/2\right) \tag{20.85}$$

in terms of the parameters (δ_2^2, σ_g^2) that we have identified for the current problem.

We simulate the following scenario. We set $M = 50$, $\alpha = 0.002$, $\mu = 0.001$, $\sigma_v^2 = 0.01$, $\sigma_h^2 = 1$, and $\kappa = 0.9980$. The entries of $\boldsymbol{h}$ and $\boldsymbol{v}$ are Gaussian distributed with zero mean. We generate a random *sparse* model $w^\bullet$ with only two nonzero entries set equal to 1, while all other entries are set to 0. For each run of the algorithms listed below, we generate $N = 10{,}000$ data realizations $\{\gamma(n), h_n\}$. We run the stochastic subgradient algorithm with and without exponential smoothing, namely,

$$w_n = w_{n-1} + 2\mu h_n\left(\gamma(n) - h_n^\mathsf{T} w_{n-1}\right) - \mu\,\alpha\,\text{sign}(w_{n-1}) \tag{20.86}$$

and

$$\begin{cases} w_n = w_{n-1} + 2\mu h_n\left(\gamma(n) - h_n^\mathsf{T} w_{n-1}\right) - \mu\,\alpha\,\text{sign}(w_{n-1}) \\ S_{n+1} = \kappa S_n + 1 \\ \bar{w}_{n+1} = \left(1 - \frac{1}{S_{n+1}}\right)\bar{w}_n + \frac{1}{S_{n+1}} w_n \end{cases} \tag{20.87}$$

For comparison purposes, we also apply the traditional stochastic gradient algorithm without any regularization to estimate $w^\bullet$ by using

$$w_n = w_{n-1} + 2\mu h_n\left(\gamma(n) - h_n^\mathsf{T} w_{n-1}\right) \tag{20.88}$$

For each run, we use expression (20.81) to evaluate the excess-risk curve:

$$P(w) - P(w^o) = \alpha\left(\|w\|_1 - \|w^o\|_1\right) + \sigma_h^2\|w - w^\bullet\|^2 - \sigma_h^2\|\widetilde{w}^\bullet\|^2 \tag{20.89}$$

and average the curves over 50 experiments. The results are shown in Fig. 20.1 using the logarithmic scale; each curve shows the evolution of $10\log_{10}(P(w_n) - P(w^o))$. We will learn in a future chapter when discussing the addition of regularization terms such as $\alpha\|w\|_1$ to the risk function, that the resulting minimizer w^o will tend to be sparse. For this reason, stochastic subgradient implementations that exploit this information in their update relations deliver superior performance in comparison to the plain stochastic gradient algorithm (20.88), which does not use the regularization information. Observe further that the use of exponential smoothing reduces the convergence rate of the algorithm but leads to improved steady-state performance.

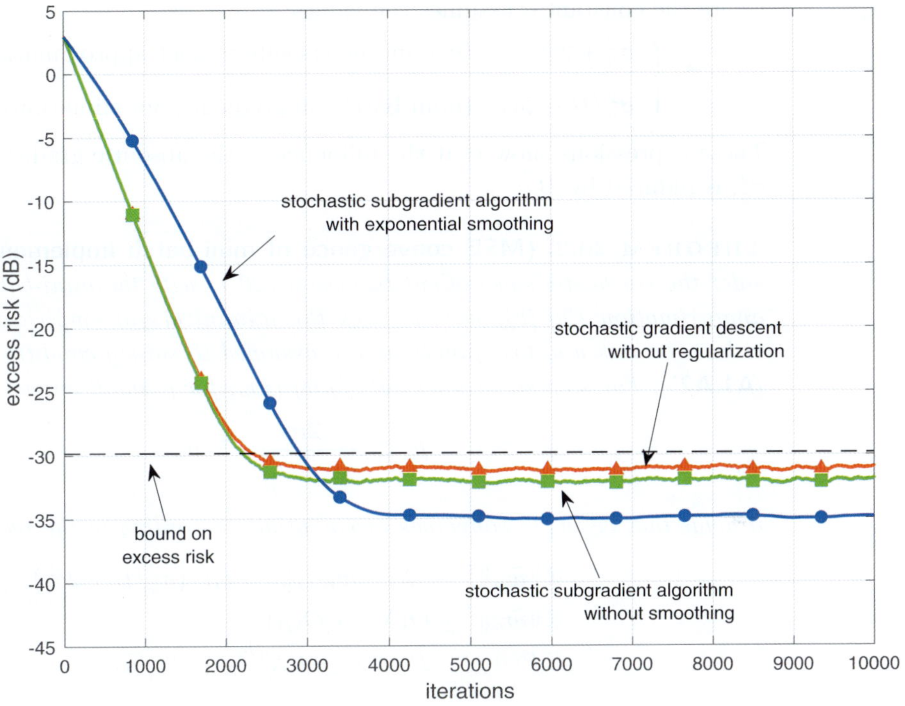

Figure 20.1 Learning curves in logarithmic scale for the stochastic gradient algorithm (20.88), and for the stochastic subgradient algorithms (20.86)–(20.87) with and without smoothing. The horizontal line shows the value of the bound on the right-hand side of (20.85).

20.5 CONVERGENCE OF MINI-BATCH IMPLEMENTATION

Let (β_g^2, σ_g^2) denote the parameters that characterize the bound in (20.20b) for the stochastic subgradient algorithm based on *instantaneous* subgradient approximations. We explained in (18.98) that these parameters get scaled down by a factor B when a mini-batch implementation is used with uniform sampling. Now recall that we showed in Theorem 19.2 that this scaling reduces the size of the steady-state MSE from $O(\mu)$ to $O(\mu/B)$ for *smooth* risks. This conclusion will not hold any longer for nonsmooth risks where the limiting neighborhood will continue to be $O(\mu)$. This is due to the Lipschitz factor δ_2, which arises from the *nonsmoothness* of $P(w)$. Still, if we compare the mean-square error performance expressions under instantaneous and mini-batch subgradient implementations we find from (20.34) and (20.98) that

$$\mathbb{E}\,\|\widetilde{\boldsymbol{w}}_n\|^2 \leq O(\lambda^n) + O(\mu a) \tag{20.90}$$

where the constant a assumes the values:

$$a \triangleq \begin{cases} \sigma_g^2 + 2\delta_2^2 & \text{(instantaneous subgradient approximation)} \\ \sigma_g^2/B + 2\delta_2^2 & \text{(mini-batch subgradient approximation)} \end{cases} \tag{20.91}$$

These expressions show that the influence of the absolute gradient noise factor, σ_g^2, is reduced by B.

Theorem 20.2. (MSE convergence of mini-batch implementation) *Consider the stochastic subgradient recursion (20.1) with the mini-batch subgradient approximation (20.4b), used to seek the minimizers of empirical or stochastic risks. The risk and loss functions are assumed to satisfy conditions* (**A1**,**A2**) *or* (**A1**,**A2'**)*. For step-size values satisfying (i.e., for μ small enough):*

$$\mu < \frac{2\nu}{2\delta^2 + \frac{\beta_g^2}{B}} \triangleq \mu_o \tag{20.92}$$

it holds that $\mathbb{E}\,\|\widetilde{\boldsymbol{w}}_n\|^2$ converges exponentially according to the recursion

$$\mathbb{E}\,\|\widetilde{\boldsymbol{w}}_n\|^2 \leq \lambda\,\mathbb{E}\,\|\widetilde{\boldsymbol{w}}_{n-1}\|^2 + \mu^2\left(\sigma_g^2/B + 2\delta_2^2\right) \tag{20.93a}$$

$$\mathbb{E}\,\|\widetilde{\boldsymbol{w}}_n\|^2 \leq O(\lambda^n) + O(\mu) \tag{20.93b}$$

$$\mathbb{E}\,P(\boldsymbol{w}_n) - P(w^\star) \leq O(\lambda^{n/2}) + O(\sqrt{\mu}) \tag{20.93c}$$

where

$$\lambda = 1 - 2\nu\mu + \left(2\delta^2 + \beta_g^2/B\right)\mu^2 \in [0, 1) \tag{20.94}$$

Results (20.93b)–(20.93c) hold for sufficiently small step sizes.

Proof: We subtract $w^\star$ from both sides of (20.1) and use the definition of the gradient noise to get

$$\widetilde{\boldsymbol{w}}_n = \widetilde{\boldsymbol{w}}_{n-1} + \mu\, s(\boldsymbol{w}_{n-1}) + \mu\, \boldsymbol{g}_n(\boldsymbol{w}_{n-1}) \tag{20.95}$$

We already know from (20.32) that

$$\|\widetilde{\boldsymbol{w}}_{n-1} + \mu\, s(\boldsymbol{w}_{n-1})\|^2 \leq (1 - 2\mu\nu + 2\mu^2\delta^2)\|\widetilde{\boldsymbol{w}}_{n-1}\|^2 + 2\mu^2\delta_2^2 \tag{20.96}$$

Squaring both sides of (20.95), conditioning on $\boldsymbol{w}_{n-1}$, and taking expectations we obtain

$$\begin{aligned} &\mathbb{E}\left(\|\widetilde{\boldsymbol{w}}_n\|^2 \,|\, \boldsymbol{w}_{n-1}\right) \\ &= \mathbb{E}\left(\|\widetilde{\boldsymbol{w}}_{n-1} + \mu\, s(\boldsymbol{w}_{n-1}) + \mu\, \boldsymbol{g}_n(\boldsymbol{w}_{n-1})\|^2 \,|\, \boldsymbol{w}_{n-1}\right) \\ &\overset{(a)}{=} \mathbb{E}\left(\|\widetilde{\boldsymbol{w}}_{n-1} + \mu\, s(\boldsymbol{w}_{n-1})\|^2 \,|\, \boldsymbol{w}_{n-1}\right) + \mu^2\,\mathbb{E}\left(\|\boldsymbol{g}_n(\boldsymbol{w}_{n-1})\|^2 \,|\, \boldsymbol{w}_{n-1}\right) \\ &= \Big(1 - 2\mu\nu + \mu^2\left(2\delta^2 + \beta_g^2/B\right)\Big)\|\widetilde{w}_{n-1}\|^2 + \mu^2\sigma_g^2/B + 2\mu^2\delta_2^2 \end{aligned} \tag{20.97}$$

where the cross term in step (a) is zero because of the zero-mean property (18.85a). Taking expectations of both sides removes the conditioning on $\boldsymbol{w}_{n-1}$ and leads to (20.93a), where λ is defined by (20.94). The same argument used in Fig. 12.2 can be repeated here to show that condition (20.92) ensures $0 \leq \lambda < 1$. By further iterating recursion (20.93a) we obtain

$$\mathbb{E}\,\|\widetilde{\boldsymbol{w}}_n\|^2 \leq \lambda^{n+1}\,\mathbb{E}\,\|\widetilde{\boldsymbol{w}}_{-1}\|^2 + \frac{\mu^2\left(\sigma_g^2/B + 2\delta_2^2\right)}{1-\lambda} \tag{20.98}$$

which, in a manner similar to the earlier arguments, can again be used to show that $\mathbb{E}\,\|\widetilde{\boldsymbol{w}}_n\|^2$ converges exponentially to a region that is upper-bounded by

$$\mathbb{E}\,\|\widetilde{\boldsymbol{w}}_n\|^2 \leq O(\lambda^n) \;+\; O(\mu) \tag{20.99}$$

Observe that due to the presence of a nonzero δ_2 factor, which arises from the non-smoothness in the risk function $P(w)$, the size of the limiting neighborhood is now $O(\mu)$ instead of $O(\mu/B)$ as was the case for smooth risks. To establish (20.93c), we use (20.38) to note that

$$0 \leq \mathbb{E}\,P(\boldsymbol{w}_n) - P(w^\star) \leq 2\delta\,\mathbb{E}\,\|\widetilde{\boldsymbol{w}}_n\|^2 \;+\; 2\delta_2\,\mathbb{E}\,\|\widetilde{\boldsymbol{w}}_n\| \tag{20.100}$$

and, hence,

$$\mathbb{E}\,P(\boldsymbol{w}_n) - P(w^\star) \leq O(\lambda^{n/2}) + O(\sqrt{\mu}) \tag{20.101}$$

■

Observe from (20.101) that $\mathbb{E}\,P(\boldsymbol{w}_n)$ converges exponentially at the rate $\lambda^{n/2}$ to an $O(\sqrt{\mu})$-neighborhood around $P(w^\star)$. In a manner similar to what was done in the previous two sections, we can improve on this result and obtain faster convergence at the rate λ^n to a smaller $O(\mu)$-neighborhood around $P(w^\star)$ by smoothing the iterates $\boldsymbol{w}_n$ as follows:

$$\left\{\begin{array}{l} \text{select } B \text{ samples } \{\boldsymbol{\gamma}(b), \boldsymbol{h}_b\} \text{ at random} \\ \boldsymbol{w}_n = \boldsymbol{w}_{n-1} - \mu\left(\dfrac{1}{B}\displaystyle\sum_{b=0}^{B-1} s_Q\Big(\boldsymbol{w}_{n-1}; \boldsymbol{\gamma}(b), \boldsymbol{h}_b\Big)\right) \\ S_{n+1} = \kappa S_n + 1, \;\; S_{-1} = 0 \\ \bar{\boldsymbol{w}}_{n+1} = \left(1 - \dfrac{1}{S_{n+1}}\right)\bar{\boldsymbol{w}}_n + \dfrac{1}{S_{n+1}}\boldsymbol{w}_n \end{array}\right. \tag{20.102}$$

where $\lambda \leq \kappa < 1$ is a number close to but smaller than 1. Doing so leads to the performance result (see Prob. 20.2):

$$\mathbb{E}\,P(\bar{\boldsymbol{w}}_n) - P(w^\star) \leq O(\kappa^n) \;+\; O(\mu) \tag{20.103}$$

20.6 CONVERGENCE UNDER VANISHING STEP SIZES

We observe from (20.26b) that under constant step-size learning, the MSE $\mathbb{E}\,\|\widetilde{\boldsymbol{w}}_n\|^2$ converges to a small neighborhood of size $O(\mu)$; the smaller the value of μ is, the smaller the size of this neighborhood will be. However, small step sizes affect the convergence rate of the algorithm because they cause the value of λ to approach 1. One way to reduce the size of the limiting region to zero is to employ a decaying step size $\mu(n)$ in place of the constant μ. Doing so allows us to employ larger step-size values during the initial stages of the algorithm to speed up convergence and smaller step-size values during the latter stages to improve steady-state performance. It is common to choose the sequence $\mu(n) > 0$ to satisfy either of the following two conditions:

$$\text{(Condition I)} \qquad \sum_{n=0}^{\infty} \mu^2(n) < \infty \quad \text{and} \quad \sum_{n=0}^{\infty} \mu(n) = \infty \tag{20.104a}$$

$$\text{(Condition II)} \qquad \lim_{n\to\infty} \mu(n) = 0 \quad \text{and} \quad \sum_{n=0}^{\infty} \mu(n) = \infty \tag{20.104b}$$

Clearly, any sequence that satisfies the stronger condition (20.104a) also satisfies (20.104b). In either case, recursion (20.1) would be replaced by:

$$\boxed{\boldsymbol{w}_n \;=\; \boldsymbol{w}_{n-1} \;-\; \mu(n)\, s_Q(\boldsymbol{w}_{n-1}; \boldsymbol{\gamma}(n), \boldsymbol{h}_n), \quad n \geq 0} \tag{20.105}$$

The decaying step size helps annihilate the effect of gradient noise and ensures convergence of $\boldsymbol{w}_n$ to $w^\star$. Specifically, we will show below that $\boldsymbol{w}_n$ will now converge to $w^\star$ in the mean-square sense under both choices (20.104a) or (20.104b), i.e., $\mathbb{E}\,\|\widetilde{\boldsymbol{w}}_n\|^2 \to 0$ as $n \to \infty$. From the discussion in Appendix 3.A on the convergence of random variables we know that this conclusion implies convergence in probability as well, so that

$$\lim_{n\to\infty} \mathbb{P}\left(\|\widetilde{\boldsymbol{w}}_n\|^2 > \epsilon\right) = 0, \quad \text{for any small } \epsilon > 0 \tag{20.106}$$

We will actually establish below the stronger conclusion that, under (20.104a), $\boldsymbol{w}_n$ converges to $w^\star$ almost surely (i.e., with probability 1):

$$\mathbb{P}\left(\lim_{n\to\infty} \boldsymbol{w}_n = w^\star\right) = 1 \tag{20.107}$$

THEOREM 20.3. (Convergence under vanishing step sizes) *Consider the stochastic subgradient recursion (20.1) with the instantaneous subgradient approximation (20.4a), used to seek minimizers of empirical or stochastic risks. The risk and loss functions are assumed to satisfy conditions* (**A1,A2**) *or* (**A1,A2'**). *Then, the following convergence properties hold:*

(a) *If the step-size sequence $\mu(n)$ satisfies (20.104a), then $\boldsymbol{w}_n$ converges almost surely to $w^\star$, written as $\boldsymbol{w}_n \to w^\star$ a.s.*

(b) *If the step-size sequence $\mu(n)$ satisfies (20.104b), then $\boldsymbol{w}_n$ converges in the MSE sense to $w^\star$, i.e., $\mathbb{E}\,\|\widetilde{\boldsymbol{w}}_n\|^2 \to 0$, which in turn implies convergence in probability according to (20.106).*

Proof: The same argument leading to (20.33) from the proof of Theorem 20.1 in the constant step-size case continues to hold, leading to the inequality:

$$\mathbb{E}\left(\|\widetilde{\boldsymbol{w}}_n\|^2 \,|\, \boldsymbol{w}_{n-1}\right) \leq \lambda(n)\, \|\widetilde{w}_{n-1}\|^2 + \mu^2(n)(\sigma_g^2 + 2\delta_2^2) \tag{20.108}$$

with μ replaced by $\mu(n)$ and where

$$\lambda(n) \stackrel{\Delta}{=} 1 - 2\nu\mu(n) + (2\delta^2 + \beta_g^2)\mu^2(n) \tag{20.109}$$

We split the term $2\nu\mu(n)$ into the sum of two terms and write

$$\lambda(n) \;=\; 1 - \nu\mu(n) - \nu\mu(n) + (2\delta^2 + \beta_g^2)\mu^2(n) \tag{20.110}$$

And since $\mu(n) \to 0$, we conclude that for large enough $n > n_o$, the value $\mu^2(n)$ is smaller than $\mu(n)$. Therefore, a large enough time index, n_o, exists such that the following two conditions are satisfied:

$$\nu\mu(n) \geq (2\delta^2 + \beta_g^2)\mu^2(n), \quad 0 \leq \nu\mu(n) < 1, \quad n > n_o \tag{20.111}$$

Consequently,

$$\lambda(n) \;\leq\; 1 - \nu\mu(n), \quad n > n_o \tag{20.112}$$

Then, inequalities (20.108) and (20.112) imply that

$$\mathbb{E}\left(\|\widetilde{\boldsymbol{w}}_n\|^2 \,|\, \boldsymbol{w}_{n-1}\right) \;\leq\; (1 - \nu\mu(n))\, \|\widetilde{w}_{n-1}\|^2 \;+\; \mu^2(n)(\sigma_g^2 + 2\delta_2^2), \quad n > n_o \tag{20.113}$$

Due to the Markovian property of recursion (20.105), where $\boldsymbol{w}_n$ is solely dependent on the most recent iterate $\boldsymbol{w}_{n-1}$, we conclude that for $n > n_o$:

$$\mathbb{E}\left(\|\widetilde{\boldsymbol{w}}_n\|^2 \,|\, \boldsymbol{w}_{n-1}, \ldots, \boldsymbol{w}_0, \boldsymbol{w}_{-1}\right) \;\leq\; (1 - \nu\mu(n))\, \|\widetilde{w}_{n-1}\|^2 \;+\; \mu^2(n)(\sigma_g^2 + 2\delta_2^2) \tag{20.114}$$

For convenience of notation, let

$$\boldsymbol{u}(n+1) \;\stackrel{\Delta}{=}\; \|\widetilde{\boldsymbol{w}}_n\|^2 \tag{20.115}$$

Then, inequality (20.114) implies for $n > n_o$:

$$\mathbb{E}\left(\boldsymbol{u}(n+1) |\, \boldsymbol{u}(0), \boldsymbol{u}(1), \ldots, \boldsymbol{u}(n)\right) \;\leq\; (1 - \nu\mu(n))\; u(n) \;+\; \mu^2(n)(\sigma_g^2 + 2\delta_2^2) \tag{20.116}$$

We now call upon the useful result (19.158) from Appendix 19.A and make the identifications:

$$a(n) \leftarrow \nu\mu(n), \quad b(n) \leftarrow \mu^2(n)(\sigma_g^2 + 2\delta_2^2) \tag{20.117}$$

These sequences satisfy conditions (19.159) in the appendix in view of assumption (20.104a) on the step-size sequence and the second condition in (20.111). We then conclude that $\boldsymbol{u}(n) \to 0$ almost surely and, hence, $\boldsymbol{w}_n \to w^\star$ almost surely.

Finally, taking expectations of both sides of (20.116) leads to

$$\mathbb{E}\,\boldsymbol{u}(n+1) \;\leq\; (1 - \nu\mu(n))\; \mathbb{E}\,\boldsymbol{u}(n) \;+\; \mu^2(n)(\sigma_g^2 + 2\delta_2^2), \quad n > n_o \tag{20.118}$$

with the expectation operator appearing on both sides of the inequality. We conclude from result (14.136) in Appendix 19.A, under conditions (20.104b), that $\mathbb{E}\,\|\widetilde{\boldsymbol{w}}_n\|^2 \to 0$ so that $\boldsymbol{w}_n$ converges to $w^\star$ in the MSE sense.

■

We can be more specific and quantify the rate at which the error variance $\mathbb{E}\,\|\widetilde{\boldsymbol{w}}_n\|^2$ converges toward zero for step-size sequences of the form:

$$\mu(n) = \frac{\tau}{n+1}, \quad \tau > 0 \tag{20.119}$$

which satisfy both conditions (20.104a) and (20.104b).

Theorem 20.4. (Rates of convergence under (20.119)) *Consider the stochastic subgradient recursion (20.1) with the instantaneous subgradient approximation (20.4a), used to seek minimizers of empirical or stochastic risks. The risk and loss functions are assumed to satisfy conditions* (**A1,A2**) *or* (**A1,A2'**)*. Then, three convergence rates are possible, depending on how the factor $\nu\tau$ compares to the value 1. Specifically, for large enough n, it holds that:*

$$\begin{cases} \mathbb{E}\,\|\widetilde{\boldsymbol{w}}_n\|^2 \leq O\left(\frac{1}{n}\right), & \nu\tau > 1 \\ \mathbb{E}\,\|\widetilde{\boldsymbol{w}}_n\|^2 = O\left(\frac{\log n}{n}\right), & \nu\tau = 1 \\ \mathbb{E}\,\|\widetilde{\boldsymbol{w}}_n\|^2 = O\left(\frac{1}{n^{\nu\tau}}\right), & \nu\tau < 1 \end{cases} \tag{20.120}$$

The fastest convergence rate occurs when $\nu\tau > 1$ (i.e., for large enough τ) and is on the order of $O(1/n)$. The risk values follow a similar convergence behavior as $\mathbb{E}\,\|\widetilde{\boldsymbol{w}}_n\|$, namely,

$$\begin{cases} \mathbb{E}\,P(\boldsymbol{w}_n) - P(w^\star) \leq O\left(\frac{1}{\sqrt{n}}\right), & \nu\tau > 1 \\ \mathbb{E}\,P(\boldsymbol{w}_n) - P(w^\star) = O\left(\frac{(\log n)^{1/2}}{\sqrt{n}}\right), & \nu\tau = 1 \\ \mathbb{E}\,P(\boldsymbol{w}_n) - P(w^\star) = O\left(\frac{1}{n^{\nu\tau/2}}\right), & \nu\tau < 1 \end{cases} \tag{20.121}$$

The fastest convergence rate again occurs when $\nu\tau > 1$ and is on the order of $O(1/\sqrt{n})$.

Proof: We use (20.118) and the assumed form for $\mu(n)$ in (20.119) to write

$$\mathbb{E}\,\boldsymbol{u}(n+1) \leq \left(1 - \frac{\nu\tau}{n+1}\right)\mathbb{E}\,\boldsymbol{u}(n) + \frac{\tau^2(\sigma_g^2 + 2\delta_2^2)}{(n+1)^2}, \quad n > n_o \tag{20.122}$$

This recursion has the same form as recursion (14.136) with the identifications:

$$a(n) \leftarrow \frac{\nu\tau}{n+1}, \quad b(n) \leftarrow \frac{\tau^2(\sigma_g^2 + 2\delta_2^2)}{(n+1)^2}, \quad p \leftarrow 1 \tag{20.123}$$

The above rates of convergence then follow from Lemma 14.1. Result (20.121) follows from (20.38).

■

Table 20.1 summarizes the performance results obtained in this chapter for the stochastic subgradient algorithm under uniform sampling (i.e., sampling with replacement) and data streaming, and also for mini-batch implementations. Results are shown for strongly convex risks under both constant and decaying step sizes.

20.7 COMMENTARIES AND DISCUSSION

Stochastic subgradient algorithms. There is extensive work on subgradient algorithms for both deterministic and stochastic optimization problems in the literature, including in the contributions by Rockafellar (1970), Bertsekas (1973, 1999), Clarke (1983), Nemirovski and Yudin (1983), Kiwiel (1985), Polyak (1987), Shor (2012), Bertsekas, Nedic, and Ozdaglar (2003), Nesterov (2004), Duchi, Hazan, and Singer (2011), Duchi,

Table 20.1 Convergence properties of stochastic subgradient algorithms with *constant* and *vanishing* step sizes, where $P(w)$ denotes either an empirical risk (with minimizer $w^\star$) or a stochastic risk (with minimizer w^o).

Mode of operation	Conditions on risk and loss functions	Asymptotic convergence property	Reference
instantaneous subgradient approximation (uniform sampling)	$P(w)$: ν-strongly convex $Q(w)$: affine-Lipschitz subgradients conditions (**A1**,**A2**)	$\mathbb{E}\, P(\boldsymbol{w}_n) - P(w^\star) \leq O(\lambda^{n/2}) + O(\sqrt{\mu})$	Thm. 20.1
instantaneous subgradient approximation (data streaming)	$P(w)$: ν-strongly convex $Q(w)$: affine-Lipschitz subgradients in mean-square sense conditions (**A1**,**A2'**)	$\mathbb{E}\, P(\boldsymbol{w}_n) - P(w^o) \leq O(\lambda^{n/2}) + O(\sqrt{\mu})$	Thm. 20.1
instantaneous subgradient approximation with smoothing (uniform sampling)	$P(w)$: ν-strongly convex $Q(w)$: affine-Lipschitz subgradients conditions (**A1**,**A2**)	$\mathbb{E}\, P(\boldsymbol{w}_n) - P(w^\star) \leq O(\kappa^n) + O(\mu)$ using $\lambda \leq \kappa < 1$	Lemma 20.2
instantaneous subgradient approximation with smoothing (data streaming)	$P(w)$: ν-strongly convex $Q(w)$: affine-Lipschitz subgradients in mean-square sense conditions (**A1**,**A2'**)	$\mathbb{E}\, P(\boldsymbol{w}_n) - P(w^o) \leq O(\kappa^n) + O(\mu)$ using $\lambda \leq \kappa < 1$	Lemma 20.2
mini-batch subgradient approximation with smoothing (uniform sampling or data streaming)	$P(w)$: ν-strongly convex $Q(w)$: affine-Lipschitz subgradients in deterministic or mean-square sense conditions (**A1,A2**) or (**A1**,**A2'**)	$\mathbb{E}\, P(\boldsymbol{w}_n) - P(w^\star) \leq O(\kappa^n) + O(\mu)$ using $\lambda \leq \kappa < 1$	Prob. 20.2
instantaneous subgradient approximation (uniform sampling) (decaying step size)	$P(w)$: ν-strongly convex $Q(w)$: affine-Lipschitz subgradients conditions (**A1**,**A2**) $\mu(n) = \tau/(n+1)$	$\mathbb{E}\, P(\boldsymbol{w}_n) - P(w^\star) \leq O(1/\sqrt{n})$	Thm. 20.4
instantaneous subgradient approximation (data streaming) (decaying step size)	$R(w)$: ν-strongly convex $Q(w)$: affine-Lipschitz subgradients in mean-square sense conditions (**A1**,**A2'**) $\mu(n) = \tau/(n+1)$	$\mathbb{E}\, P(\boldsymbol{w}_n) - P(w^o) \leq O(1/\sqrt{n})$	Thm. 20.4

Bartlett, and Wainwright (2012), and Shamir and Zhang (2013). The convergence analyses in the chapter follow closely the approach by Ying and Sayed (2018), which relies on the affine-Lipschitz conditions **A2** and **A2'** on the loss function. As was explained in the concluding remarks of Chapter 18, these affine-Lipschitz conditions are weaker than earlier conditions employed in the literature, e.g., in Bertsekas (1999), Nemirovski *et al.* (2009), Nedic and Ozdaglar (2009), Ram, Nedic, and Veeravalli (2010), Srivastava and Nedic (2011), and Agarwal *et al.* (2012), such as bounding the subgradient vectors in the absolute or MSE sense:

$$\|s_Q(w;\gamma,h\cdot)\| \leq G \quad \text{or} \quad \mathbb{E}\,\|s_Q(w;\boldsymbol{\gamma},\boldsymbol{h})\|^2 \leq G \tag{20.124}$$

for some constant $G \geq 0$ and for all subgradients in the subdifferential set of $Q(w,\cdot)$. We explained in Prob. 14.3 that the bounded subgradient condition is in conflict with the ν-strong-convexity assumption on $P(w)$.

Effect of nonsmoothness on performance. We examined two performance measures for stochastic subgradient algorithms, involving the mean-square error and the average excess risk. We showed in Theorem 20.1 that for the traditional implementation, the steady-state performance metrics satisfy:

$$\limsup_{n\to\infty} \mathbb{E}\,\|\widetilde{\boldsymbol{w}}_n\|^2 = O(\mu) \tag{20.125a}$$

$$\limsup_{n\to\infty} \Big(\mathbb{E}\,P(\boldsymbol{w}_n) - P(w^\star)\Big) = O(\sqrt{\mu}\,) \tag{20.125b}$$

where the neighborhood for the excess risk is larger and on the order of $O(\sqrt{\mu})$. Motivated by the concept of pocket variables, we subsequently incorporated exponential smoothing and showed in Lemma 20.2 that this step helps recover the $O(\mu)$ performance:

$$\limsup_{n\to\infty} \Big(\mathbb{E}\,P(\boldsymbol{w}_n) - P(w^\star)\Big) = O(\mu\,) \tag{20.126}$$

If we examine the derivation of these results in the body of the chapter, we will find that the $O(\mu)$ term could be written more explicitly as

$$\limsup_{n\to\infty} \mathbb{E}\,\|\widetilde{\boldsymbol{w}}_n\|^2 = O\Big(\mu(\sigma_g^2 + 2\delta_2^2)/\nu\Big) \tag{20.127}$$

where μ appears scaled by the sum $\sigma_g^2 + 2\delta_2^2$, and ν is the strong-convexity factor. The factor σ_g^2 refers to the absolute noise term from the gradient noise bound (20.20b), whereas the factor δ_2^2 arises from the nonsmoothness of the loss and risk functions, as is evident from conditions **A2** and **A2'**. In comparison, for smooth risks, we established in the previous chapter that the MSE reaches a neighborhood of size

$$\limsup_{n\to\infty} \mathbb{E}\,\|\widetilde{\boldsymbol{w}}_n\|^2 = O(\mu\sigma_g^2/2\nu) \tag{20.128}$$

Observe from (20.127) how the nonsmoothness factor δ_2 in the loss and risk functions degrades performance in comparison to (20.128) by enlarging the size of the steady-state region.

On the other hand, when a mini-batch implementation of size B is used, the performance in the smooth case is scaled down by B, whereas the performance in the nonsmooth case was shown in Theorem 20.2 to scale in the following manner:

$$\limsup_{n\to\infty} \mathbb{E}\,\|\widetilde{\boldsymbol{w}}_n\|^2 = O\Big(\mu(\sigma_g^2/B + 2\delta_2^2)/\nu\Big) \tag{20.129}$$

In other words, only the effect of the absolute gradient noise component is scaled down by B, while the effect of the nonsmoothness in the risk function (captured by δ_2^2) remains unchanged.

PROBLEMS

20.1 Use the fact that $(\mathbb{E}\,\boldsymbol{x})^2 \leq \mathbb{E}\,\boldsymbol{x}^2$ for any random variable $\boldsymbol{x}$ to conclude from (20.33) that $\mathbb{E}\,(\|\widetilde{\boldsymbol{w}}_n\| \,|\, \boldsymbol{w}_{n-1}) \leq \sqrt{\lambda}\,\|\widetilde{w}_{n-1}\| + O(\mu)$.

20.2 Consider the mini-batch subgradient implementation (20.102). Extend the argument of Sections 20.3 and 20.4 to show that $\mathbb{E}\,P(\bar{\boldsymbol{w}}_n) - P(w^\star) \leq O(\kappa^n) + O(\mu)$.

20.3 This problem extends the result of Prob. 14.4 to stochastic subgradient algorithms. Consider the stochastic subgradient recursion (20.1) for minimizing a possibly nonsmooth regularized convex risk $P(w)$ over $w \in \mathbb{R}^M$. Let $w^\star$ denote the minimizer. Introduce the quantities:

$$\widetilde{\boldsymbol{w}}_n \stackrel{\Delta}{=} w^\star - \boldsymbol{w}_n, \quad \boldsymbol{P}^{\text{best}}(n) \stackrel{\Delta}{=} \min_{k\leq n} P(\boldsymbol{w}_k), \quad P_1^{\text{best}}(n) = \min_{k\leq n} \mathbb{E}\,P(\boldsymbol{w}_k)$$

where $\boldsymbol{P}^{\text{best}}(n)$ is now a random variable. We assume in this problem that $Q(w;\cdot)$ has bounded subgradient vectors, i.e., $\|s_Q(w;\cdot)\| \leq G$ for all $s_Q(w;\cdot) \in \partial_{w^\mathsf{T}} Q(w;\cdot)$ and for some finite constant G. We also assume that $\mathbb{E}\,\|\widetilde{\boldsymbol{w}}_{-1}\| \leq W$, for some constant W (that is, the initial condition is within a bounded range from the optimal solution). We know from the zero-mean property of the gradient noise that $\mathbb{E}\,s_Q(\boldsymbol{w}_{n-1};\boldsymbol{\gamma},\boldsymbol{h}\,|\,\boldsymbol{w}_{n-1}) = s(w_{n-1})$, where $s(w_{n-1})$ denotes a subgradient construction for $P(w)$ at location w_{n-1}.

(a) Verify that the subgradients for $P(w)$, denoted by $s(w)$, are also bounded by G.

(b) Conclude that the gradient noise process satisfies $\mathbb{E}\,\left(\|\boldsymbol{g}_n(\boldsymbol{w}_{n-1})\|^2 | \boldsymbol{w}_{n-1}\right) \leq 4G^2$.

(c) Verify that

$$\mathbb{E}\,\|\widetilde{\boldsymbol{w}}_n\|^2 \leq \mathbb{E}\,\|\widetilde{\boldsymbol{w}}_{n-1}\|^2 + \mu^2 \mathbb{E}\,\|s(\boldsymbol{w}_{n-1};\boldsymbol{\gamma}(n),\boldsymbol{h}_n)\|^2 - 2\mu\Big(\mathbb{E}\,P(\boldsymbol{w}_{n-1}) - P(w^\star)\Big)$$

Conclude by iterating that

$$0 \leq \mathbb{E}\,\|\widetilde{\boldsymbol{w}}_n\|^2 \leq W^2 \;+\; n\mu^2 G^2 \;-\; 2n\mu\left(P_1^{\text{best}}(n-1) - P(w^\star)\right)$$

(d) Show that $P_1^{\text{best}}(n-1) - P(w^\star) \leq (W^2 + nG^2\mu^2)/2n\mu$.

(e) Verify that $\mathbb{E}\,\boldsymbol{P}^{\text{best}}(n) \leq P_1^{\text{best}}(n)$. Conclude that under constant step sizes:

$$\mathbb{E}\,\boldsymbol{P}^{\text{best}}(n) - P(w^\star) \leq O(1/n) + O(\mu), \quad \text{for large } n$$

(f) Show that $\mu^o(n) = W/G\sqrt{n}$ minimizes the upper bound in part (d); note that this step size is iteration-dependent. Conclude that $\mathbb{E}\,\boldsymbol{P}^{\text{best}}(n-1) - P(w^\star) \leq O(1/\sqrt{n})$.

20.4 Consider a regularized stochastic risk optimization problem of the form

$$w^o = \underset{w\in\mathbb{R}^M}{\text{argmin}} \left\{ P(w) \stackrel{\Delta}{=} \alpha\|w\|_1 + \rho\|w\|^2 \;+\; \mathbb{E}\,Q(w;\boldsymbol{\gamma},\boldsymbol{h}) \right\}$$

where $\alpha \geq 0$, $\rho > 0$, and $Q(w;\boldsymbol{\gamma},\boldsymbol{h}) = |\boldsymbol{\gamma} - \boldsymbol{h}^\mathsf{T} w|$.

(a) Is the risk function $P(w)$ strongly convex? Can you find a value for ν?

(b) Write down a stochastic subgradient algorithm with constant step size μ.

(c) What is the convergence rate of $\mathbb{E}\,\|\widetilde{\boldsymbol{w}}_n\|^2$ when $\alpha > 0$?

(d) What is the convergence rate when $\alpha = 0$?

(e) What is the convergence rate when μ is replaced by $\mu(n) = \nu^{-1}/(n+1)$?

(f) What is the convergence rate of $\mathbb{E}\,P(\boldsymbol{w}_n)$ toward $P(w^o)$?

20.5 Consider a regularized empirical risk optimization problem of the form

$$w^\star = \underset{w\in\mathbb{R}^M}{\text{argmin}} \left\{ P(w) \stackrel{\Delta}{=} \alpha\|w\|_1 + \rho\|w\|^2 \;+\; \frac{1}{N}\sum_{n=1}^{N} |\gamma(n) - h_n^\mathsf{T} w| \right\}$$

where $\alpha \geq 0$, $\rho > 0$.

(a) Is the risk function $P(w)$ strongly convex? Can you find a value for ν?
(b) Write down a stochastic subgradient algorithm with constant step size μ.
(c) What is the convergence rate of $\mathbb{E}\,\|\widetilde{\boldsymbol{w}}_n\|^2$ when $\alpha > 0$?
(d) What is the convergence rate when $\alpha = 0$?
(e) What is the convergence rate when μ is replaced by $\mu(n) = \nu^{-1}/(n+1)$?
(f) What is the convergence rate of $\mathbb{E}\,P(\boldsymbol{w}_n)$ toward $P(w^\star)$?

20.6 Refer to the stochastic subgradient recursion (20.1) and assume that the step size is a random parameter with mean $\mathbb{E}\,\boldsymbol{\mu} = \bar{\mu}$ and variance σ_μ^2. Assume $\boldsymbol{\mu}$ is independent of all other random variables. Follow the arguments used in the proof of Theorem 20.1 and show how the results of the theorem would need to be adjusted. For a related discussion, the reader may refer to Zhao and Sayed (2015a, b) and Sayed and Zhao (2018).

20.7 Refer to the stochastic subgradient recursion (20.1) and assume that the step size $\boldsymbol{\mu}$ is a Bernoulli random variable that is equal to μ with probability p and is zero with probability $1-p$. That is, the recursion is active p fraction of the times. Assume $\boldsymbol{\mu}$ is independent of all other random variables. Follow the arguments used in the proof of Theorem 20.1 and show how the results of the theorem would need to be adjusted. For a related discussion, the reader may refer to Zhao and Sayed (2015a, b) and Sayed and Zhao (2018).

20.8 Consider the empirical risk minimization problem (20.3a) and associate a probability p_m with each sample $(\gamma(m), h_m)$. Consider further the stochastic subgradient algorithm (20.1) with instantaneous gradient approximations and adjusted as

$$\boldsymbol{w}_n = \boldsymbol{w}_{n-1} - \frac{\mu}{Np_n}\,\widehat{s}(\boldsymbol{w}_{n-1}),\ n \geq 0$$

where p_n denotes the probability with which sample $(\boldsymbol{\gamma}(n), \boldsymbol{h}_n)$ is selected from the given dataset $\{\gamma(m), h_m\}$. The data points are selected at random according to the distribution defined by $\{p_m\}$. Study the convergence behavior of the algorithm under this importance sampling scheme. Derive a condition on μ for mean-square convergence and characterize the size of the limiting MSE region.

20.9 Let $\mathcal{C}$ denote a convex set and $\mathcal{P}_{\mathcal{C}}(z)$ the projection of $z \in \mathbb{R}^M$ onto $\mathcal{C}$. The projection subgradient algorithm replaces (20.1) by

$$\boldsymbol{w}_n = \mathcal{P}_{\mathcal{C}}\Big(\boldsymbol{w}_{n-1} - \mu\,\widehat{s}(\boldsymbol{w}_{n-1})\Big)$$

Extend the results of Theorem 20.1 to this case.

REFERENCES

Agarwal, A., P. L. Bartlett, P. Ravikumar, and M. J. Wainwright (2012), "Information-theoretic lower bounds on the oracle complexity of convex optimization," *IEEE Trans. Inf. Theory*, vol. 58, no. 5, pp. 3235–3249.

Bertsekas, D. P. (1973), "Stochastic optimization problems with nondifferentiable cost functionals," *J. Optim. Theory Appl.* (JOTA), vol. 12, no. 2, pp. 218–231.

Bertsekas, D. P. (1999), *Nonlinear Programming*, 2nd ed., Athena Scientific.

Bertsekas, D. P., A. Nedic, and A. Ozdaglar (2003), *Convex Analysis and Optimization*, 2nd ed., Athena Scientific.

Clarke, F. H. (1983), *Optimization and Nonsmooth Analysis*, Wiley.

Duchi, J. C., P. L. Bartlett, and M. J. Wainwright (2012), "Randomized smoothing for stochastic optimization," *SIAM J. Optim.*, vol. 22, no. 2, pp. 674–701.

Duchi, J., E. Hazan, and Y. Singer (2011), "Adaptive subgradient methods for online learning and stochastic optimization," *J. Mach. Learn. Res.*, vol. 12, pp. 2121–2159.

Kiwiel, K. (1985), *Methods of Descent for Non-differentiable Optimization*, Springer.

Nedic, A. and A. Ozdaglar (2009), "Distributed subgradient methods for multiagent optimization,"*IEEE Trans. Aut. Control*, vol. 54, no. 1, pp. 48–61.

Nemirovski, A. S., A. Juditsky, G. Lan, and A. Shapiro (2009), "Robust stochastic approximation approach to stochastic programming," *SIAM J. Optim.*, vol. 19, no. 4, pp. 1574–1609.

Nemirovski, A. S. and D. B. Yudin (1983), *Problem Complexity and Method Efficiency in Optimization*, Wiley.

Nesterov, Y. (2004), *Introductory Lectures on Convex Optimization*, Springer.

Polyak, B. T. (1987), *Introduction to Optimization*, Optimization Software.

Ram, S. S., A. Nedic, and V. V. Veeravalli (2010), "Distributed stochastic subgradient projection algorithms for convex optimization," *J. Optim. Theory Appl.*, vol. 147, no. 3, pp. 516–545.

Rockafellar, R. T. (1970), *Convex Analysis*, Princeton University Press.

Sayed, A. H. and X. Zhao (2018), "Asynchronous adaptive networks," in *Cooperative and Graph Signal Processing*, P. Djuric and C. Richard, editors, pp. 3–68, Elsevier. Also available at https://arxiv.org/abs/1511.09180.

Shamir, O. and T. Zhang (2013), "Stochastic gradient descent for non-smooth optimization: Convergence results and optimal averaging schemes," *Proc. Int. Conf. Machine Learning* (PMLR) vol. 28, no.1, pp. 71–79, Atlanta, GA.

Shor, N. Z. (2012), *Minimization Methods for Non-differentiable Functions*, Springer.

Srivastava, K. and A. Nedic (2011), "Distributed asynchronous constrained stochastic optimization," *IEEE J. Sel. Topics. Signal Process.*, vol. 5, no. 4, pp. 772–790.

Ying, B. and A. H. Sayed (2018), "Performance limits of stochastic sub-gradient learning, part I: Single-agent case," *Signal Process.*, vol. 144, pp. 271–282.

Zhao, X. and A. H. Sayed (2015a), "Asynchronous adaptation and learning over networks – part I: Modeling and stability analysis," *IEEE Trans. Signal Process.*, vol. 63, no. 4, pp. 811–826.

Zhao, X. and A. H. Sayed (2015b), "Asynchronous adaptation and learning over networks – part II: Performance analysis," *IEEE Trans. Signal Process.*, vol. 63, no. 4, pp. 827–842.

21 Convergence Analysis III: Stochastic Proximal Algorithms

We examine in this chapter the convergence behavior of stochastic proximal algorithms, which are particularly well-suited for risk functions that can be expressed as the sum of two convex components: usually, one component is a nonsmooth regularization term while the second component is a smooth function expressed either in empirical or stochastic form. We will consider updates with constant and vanishing step sizes. The main conclusion will be that the mean-square error (MSE), $\mathbb{E}\,\|\widetilde{\boldsymbol{w}}_n\|^2$, is guaranteed to approach a small $O(\mu)$-neighborhood at an exponential rate λ^n, while exact convergence of $\boldsymbol{w}_n$ to $w^\star$ can be guaranteed for some vanishing step-size sequences. On the other hand, the excess risk $\mathbb{E}\,P(\boldsymbol{w}_n) - P(w^\star)$ is guaranteed to approach a larger $O(\sqrt{\mu})$-neighborhood at the slower rate $\lambda^{n/2}$.

21.1 PROBLEM SETTING

We recall the problem formulation, and the conditions imposed on the risk and loss functions. We also recall the constructions for the gradient approximations, and the first- and second-order moment results derived for the gradient noise process from an earlier chapter.

21.1.1 Risk Minimization Problems

In this chapter we are interested in examining the convergence behavior of the stochastic proximal implementation:

$$\begin{cases} \boldsymbol{z}_n = \boldsymbol{w}_{n-1} - \mu\, \widehat{\nabla_{w^\mathsf{T}} E}(\boldsymbol{w}_{n-1}),\ n \geq 0 \\ \boldsymbol{w}_n = \text{prox}_{\mu q}(\boldsymbol{z}_n) \end{cases} \tag{21.1}$$

with constant μ or decaying step sizes, $\mu(n)$, for the solution of convex optimization problems of the form:

$$w^\star = \underset{w \in \mathbb{R}^M}{\text{argmin}} \left\{ q(w) + E(w) \right\} \tag{21.2}$$

where $q(w)$ is possibly nonsmooth, usually a regularization term such as

$$q(w) \;=\; \alpha\|w\|_1 + \rho\|w\|^2, \;\; \alpha > 0, \;\; \rho \geq 0 \tag{21.3}$$

and $E(w)$ is a smooth empirical or stochastic risk. In other words, we are interested in solving either of the following two problems:

$$w^\star \;\stackrel{\Delta}{=}\; \underset{w\in\mathbb{R}^M}{\operatorname{argmin}} \left\{ P(w) \;\stackrel{\Delta}{=}\; q(w) + \frac{1}{N}\sum_{m=0}^{N-1} Q_u(w;\gamma(m),h_m) \right\} \tag{21.4a}$$

$$w^o \;\stackrel{\Delta}{=}\; \underset{w\in\mathbb{R}^M}{\operatorname{argmin}} \left\{ P(w) \;\stackrel{\Delta}{=}\; q(w) + \mathbb{E}\,Q_u(w;\boldsymbol{\gamma},\boldsymbol{h}) \right\} \tag{21.4b}$$

We continue to use $w^\star$ to refer to minimizers of empirical risks, and w^o for the minimizers of stochastic risks. Often, when there is no room for confusion, we will use $w^\star$ to refer generically to the minimizer of $P(w)$ independent of whether it represents an empirical or stochastic risk. In the above expressions, $Q_u(w;\cdot)$ refers to the loss function that is associated with the smooth component $E(w)$, $\{\gamma(m), h_m\}$ refer to a collection of N data points with $\gamma(m) \in \mathbb{R}$ and $h_m \in \mathbb{R}^M$, and the expectation in the second line is over the joint distribution of $\{\boldsymbol{\gamma}, \boldsymbol{h}\}$.

21.1.2 Gradient Vector Approximations

The gradient search direction will be approximated by using either instantaneous or mini-batch calculations, namely,

$$(\textbf{instantaneous}) \;\; \widehat{\nabla_{w^\mathsf{T}} E}(w) = \nabla_{w^\mathsf{T}} Q_u(w;\boldsymbol{\gamma},\boldsymbol{h}) \tag{21.5a}$$

$$(\textbf{mini-batch}) \;\; \widehat{\nabla_{w^\mathsf{T}} E}(w) = \frac{1}{B}\sum_{b=0}^{B-1} \nabla_{w^\mathsf{T}} Q_u\Big(w;\boldsymbol{\gamma}(b),\boldsymbol{h}_b\Big) \tag{21.5b}$$

where the boldface notation $(\boldsymbol{\gamma},\boldsymbol{h})$ or $(\boldsymbol{\gamma}(b),\boldsymbol{h}_b)$ refers to data samples selected uniformly at random with replacement from the given dataset $\{\gamma(m), h_m\}$ in empirical risk minimization, or assumed to stream in independently over time in stochastic risk minimization. The difference between the true gradient and its approximation is *gradient noise* and is denoted by

$$\boldsymbol{g}(w) \;\stackrel{\Delta}{=}\; \widehat{\nabla_{w^\mathsf{T}} E}(w) - \nabla_{w^\mathsf{T}} E(w) \tag{21.6}$$

Example 21.1 **(Gradient noise for quadratic risks)** Consider the empirical risk

$$P(w) \;=\; \alpha\|w\|_1 \;+\; \frac{1}{N}\sum_{m=0}^{N-1}(\gamma(m) - h_m^\mathsf{T} w)^2 \tag{21.7}$$

In this case we have

$$E(w) = \frac{1}{N}\sum_{m=0}^{N-1}(\gamma(m) - h_m^{\mathsf{T}} w)^2 \tag{21.8a}$$

$$Q_u(w;\gamma(m),h_m) = (\gamma(m) - h_m^{\mathsf{T}} w)^2 \tag{21.8b}$$

so that

$$\nabla_{w^{\mathsf{T}}} E(w) = -\frac{2}{N}\sum_{m=0}^{N-1} h_m(\gamma(m) - h_m^{\mathsf{T}} w) \tag{21.8c}$$

$$\nabla_{w^{\mathsf{T}}} Q_u(w;\boldsymbol{\gamma}(n),\boldsymbol{h}_n) = -2\boldsymbol{h}_n(\boldsymbol{\gamma}(n) - \boldsymbol{h}_n^{\mathsf{T}} w) \tag{21.8d}$$

and, therefore,

$$\boldsymbol{g}(w) = \frac{1}{N}\sum_{m=0}^{N-1} 2h_m(\gamma(m) - h_m^{\mathsf{T}} w) - 2\boldsymbol{h}_n(\boldsymbol{\gamma}(n) - \boldsymbol{h}_n^{\mathsf{T}} w) \tag{21.8e}$$

Consider next the stochastic risk

$$\begin{aligned} P(w) &= \alpha\|w\|_1 \; + \; \mathbb{E}\,(\boldsymbol{\gamma} - \boldsymbol{h}^{\mathsf{T}} w)^2 \\ &= \alpha\|w\|_1 + \sigma_\gamma^2 - 2r_{h\gamma}^{\mathsf{T}} w + w^{\mathsf{T}} R_h w \end{aligned} \tag{21.9}$$

where $\sigma_\gamma^2 = \mathbb{E}\,\boldsymbol{\gamma}^2$, $r_{h\gamma} = \mathbb{E}\,\boldsymbol{h\gamma}$, and $R_h = \mathbb{E}\,\boldsymbol{hh}^{\mathsf{T}}$. In this case we have

$$\begin{aligned} E(w) &= \mathbb{E}\,(\boldsymbol{\gamma} - \boldsymbol{h}^{\mathsf{T}} w)^2 \\ Q_u(w;\boldsymbol{\gamma},\boldsymbol{h}) &= (\boldsymbol{\gamma} - \boldsymbol{h}^{\mathsf{T}} w)^2 \end{aligned} \tag{21.10a}$$

so that

$$\nabla_{w^{\mathsf{T}}} E(w) = -2(r_{h\gamma} - R_h w) \tag{21.10b}$$

$$\nabla_{w^{\mathsf{T}}} Q_u(w;\boldsymbol{\gamma}(n),\boldsymbol{h}_n) = -2\boldsymbol{h}_n(\boldsymbol{\gamma}(n) - \boldsymbol{h}_n^{\mathsf{T}} w) \tag{21.10c}$$

and, therefore,

$$\boldsymbol{g}(w) = 2(r_{h\gamma} - R_h w) - 2\boldsymbol{h}_n(\boldsymbol{\gamma}(n) - \boldsymbol{h}_n^{\mathsf{T}} w) \tag{21.10d}$$

21.1.3 Conditions on Risk and Loss Functions

In a manner similar to what was done in Section 18.2, we introduce the following conditions for empirical risk minimization problems of the form (21.4a):

(A1) (**Strongly convex risk**). $E(w)$ is ν-strongly convex and first-order differentiable, namely, for every $w_1, w_2 \in \text{dom}(E)$ and some $\nu > 0$:

$$E(w_2) \;\geq E(w_1) + (\nabla_{w^{\mathsf{T}}} E(w_1))^{\mathsf{T}}(w_2 - w_1) \; + \; \frac{\nu}{2}\|w_2 - w_1\|^2 \tag{21.11a}$$

(A2) (**δ-Lipschitz loss gradients**). The gradient vectors of $Q_u(w,\cdot)$ are δ-Lipschitz regardless of the data argument, i.e.,

$$\|\nabla_w Q_u(w_2;\gamma(k),h_k) - \nabla_w Q_u(w_1;\gamma(\ell),h_\ell)\| \;\leq\; \delta\,\|w_2 - w_1\| \tag{21.11b}$$

for any $w_1, w_2 \in \text{dom}(Q_u)$, any $0 \leq k, \ell \leq N-1$, and with $\delta \geq \nu$. We already know that condition (21.11b) implies that the gradient of $E(w)$ is itself δ-Lipschitz:

$$\|\nabla_w E(w_2) - \nabla_w E(w_1)\| \leq \delta \|w_2 - w_1\| \tag{21.12}$$

(A3) (**Affine-Lipschitz subgradients for regularizer**). The regularization term $q(w)$ has affine-Lipschitz subgradients, namely, there exist constants (δ_1, δ_2) such that for any $w_1, w_2 \in \text{dom}(q)$:

$$\|s_q(w_2) - s_q'(w_1)\| \;\leq\; \delta_1 \, \|w_2 - w_1\| + \delta_2 \tag{21.13}$$

where $s_q'(w)$ denotes any subgradient from the subdifferential set $\partial_{w^{\mathsf{T}}}\, q(w)$, while $s_q(w)$ denotes the particular subgradient construction used by the algorithm.

On the other hand, for stochastic risk minimization problems of the form (21.4b), we continue to assume the strong convexity of $E(w)$ but replace (**A2**) by the requirement that the gradients of the loss are now δ-Lipschitz in the mean-square sense:

(A2') (**Mean-square δ-Lipschitz loss gradients**). The gradient vectors of $Q_u(w, \cdot)$ satisfy the mean-square bound:

$$\mathbb{E}\, \|\nabla_w Q_u(w_2; \boldsymbol{\gamma}, \boldsymbol{h}) - \nabla_w Q_u(w_1; \boldsymbol{\gamma}, \boldsymbol{h})\|^2 \leq \delta^2 \|w_2 - w_1\|^2 \tag{21.14}$$

for any $w_1, w_2 \in \text{dom}(Q_u)$ and with $\delta \geq \nu$. We also know that condition (21.14) implies that the gradients of $E(w)$ are δ-Lipschitz as well:

$$\|\nabla_w E(w_2) - \nabla_w E(w_1)\| \leq \delta \|w_2 - w_1\| \tag{21.15}$$

which is the same condition (21.12) under empirical risk minimization.

Note that under conditions (**A1**,**A2**) for empirical risk minimization or (**A1**,**A2'**) for stochastic risk minimization, the following two conditions hold:

(**P1**) ν-strong convexity of $E(w)$ as in (21.11a) (21.16a)

(**P2**) δ-Lipschitz gradients for $E(w)$ as in (21.12) and (21.15) (21.16b)

Moreover, we know from the earlier results (8.29) and (10.20) derived for strongly convex and δ-smooth functions that conditions **P1** and **P2** imply, respectively:

$$\textbf{(P1)} \implies \frac{\nu}{2}\|\widetilde{w}\|^2 \leq E(w) - E(w^\star) \leq \frac{1}{2\nu}\|\nabla w\, E(w)\|^2 \tag{21.17a}$$

$$\textbf{(P2)} \implies \frac{1}{2\delta}\|\nabla_w E(w)\|^2 \leq E(w) - E(w^\star) \leq \frac{\delta}{2}\|\widetilde{w}\|^2 \tag{21.17b}$$

where $\widetilde{w} = w^\star - w$. The upper bounds in both expressions indicate that whenever we bound $\|\widetilde{w}\|^2$ we will also be automatically bounding the excess risk, $E(w) - E(w^\star)$.

Example 21.2 **(Affine-Lipschitz regularizer)** Condition (**A3**) is introduced in view of the two separate steps in the stochastic proximal implementation: the first step depends on the gradient of $Q_u(w,\cdot)$, while the second step depends on the proximal operator relative to $q(w)$.

The affine-Lipschitz condition on $q(w)$ is satisfied by most cases of interest. Consider, for example, the elastic-net regularizer (21.3). One particular subgradient construction for $q(w)$ is given by

$$s_q(w) = \alpha\,\text{sign}(w) + 2\rho w \tag{21.18}$$

while the elements from the subdifferential set $\partial_{w^\mathsf{T}} q(w)$ are described by (recall (8.52a)):

$$s_q'(w) = 2\rho w + \alpha\,\mathbb{G}(w) \;=\; 2\rho w + \alpha\,\text{col}\Big\{\mathbb{G}_m(w_m)\Big\} \tag{21.19}$$

where the $\{w_m\}$ denote the individual entries of w and each $\mathbb{G}_m(w_m)$ is defined by

$$\mathbb{G}_m(w_m) = \begin{cases} +1, & w_m > 0 \\ -1, & w_m < 0 \\ [-1,+1], & w_m = 0 \end{cases} \tag{21.20}$$

It follows that

$$\begin{aligned} \|s_q(w_2) - s_q'(w_1)\| &= \|\alpha\,(\text{sign}(w_2) - \mathbb{G}(w_1)) + 2\rho(w_2 - w_1)\| \\ &\le \alpha\,\|\text{sign}(w_2) - \mathbb{G}(w_1)\| \;+ 2\rho\,\|w_2 - w_1\| \\ &\le 2\alpha\sqrt{M} + 2\rho\,\|w_2 - w_1\| \end{aligned} \tag{21.21}$$

which satisfies (21.13) with $\delta_1 = 2\rho$ and $\delta_2 = 2\alpha\sqrt{M}$.

21.1.4 Gradient Noise

In a manner similar to Section 18.3 we can again verify that, as a result of the Lipschitz conditions (**A2**) or (**A2'**), the first- and second-order moments of the gradient noise process satisfy the following two properties denoted by **(G1)** and **(G2)** for ease of reference (see Probs. 21.1 and 21.2):

$$\textbf{(G1)}\;\; \mathbb{E}\left(\boldsymbol{g}_n(\boldsymbol{w}_{n-1}) \,|\, \boldsymbol{w}_{n-1}\right) = 0 \tag{21.22a}$$

$$\textbf{(G2)}\;\; \mathbb{E}\left(\|\boldsymbol{g}_n(\boldsymbol{w}_{n-1})\|^2 \,|\, \boldsymbol{w}_{n-1}\right) \le \beta_g^2\|\widetilde{\boldsymbol{w}}_{n-1}\|^2 + \sigma_g^2 \tag{21.22b}$$

for some nonnegative constants $\{\beta_g^2, \sigma_g^2\}$ that are independent of $\widetilde{\boldsymbol{w}}_{n-1}$. Condition (**G1**) refers to the zero-mean property of the gradient noise while (**G2**) refers to its bounded "variance." Results (21.22a)–(21.22b) hold for instantaneous and mini-batch gradient approximations, regardless of whether the samples are streaming in independently of each other, sampled uniformly with replacement, sampled without replacement, or selected under importance sampling. The *only exception* is that the zero-mean property (21.22a) will *not* hold for the instantaneous gradient implementation when the samples are selected without replacement.

In summary, we find that conditions (**A1**,**A2**) for empirical risk minimization or (**A1**,**A2'**) for stochastic risk minimization imply the validity of the following conditions on the risk function and gradient noise denoted by

$$(\textbf{A1},\textbf{A2}) \text{ or } (\textbf{A1},\textbf{A2'}) \implies (\textbf{P1},\textbf{P2}, \textbf{G1},\textbf{G2}) \tag{21.23}$$

21.2 CONVERGENCE UNDER UNIFORM SAMPLING

We are ready to examine the convergence behavior of the stochastic proximal gradient algorithm for both cases of empirical and stochastic risk optimization. We will focus on operation under *uniform sampling* (i.e., sampling with replacement) for empirical risks and data streaming for stochastic risks. The dynamics of the algorithms in both scenarios are identical and the same convergence properties will apply.

THEOREM 21.1. **(MSE convergence under constant step sizes)** *Consider the stochastic proximal recursion (21.1) with the instantaneous gradient approximation (21.5a) under sampling with replacement or streaming data, used to seek minimizers of empirical or stochastic risks. The risk and loss functions are assumed to satisfy conditions (***A1**,**A2**,**A3***) or (***A1**,**A2'**,**A3***). For step-size values satisfying (i.e., for small enough μ):*

$$\mu < \frac{2\nu}{\delta^2+\beta_g^2} \overset{\Delta}{=} \mu_o \tag{21.24}$$

it holds that $\mathbb{E}\,\|\widetilde{\boldsymbol{w}}_n\|^2$ and the average excess risk, $\mathbb{E}\,P(\boldsymbol{w}_n)-P(w^\star)$, converge exponentially fast according to the recursions:

$$\mathbb{E}\,\|\widetilde{\boldsymbol{w}}_n\|^2 \;\leq\; \lambda\,\mathbb{E}\,\|\widetilde{\boldsymbol{w}}_{n-1}\|^2 \;+\; \mu^2\sigma_g^2 \tag{21.25a}$$

$$\mathbb{E}\,\|\widetilde{\boldsymbol{w}}_n\|^2 \leq O(\lambda^n) + O(\mu) \tag{21.25b}$$

$$\mathbb{E}\,P(\boldsymbol{w}_n) - P(w^\star) \leq O(\lambda^{n/2}) + O(\sqrt{\mu}) \tag{21.25c}$$

where

$$\lambda \;=\; 1-2\nu\mu+(\delta^2+\beta_g^2)\mu^2 \;\in\; [0,1) \tag{21.26}$$

Results (21.25b)–(21.25c) hold for sufficiently small step sizes. Result (21.25c) holds whenever $\delta_2 \neq 0$. Otherwise, the convergence of the excess risk occurs at the rate λ^n toward $O(\mu)$.

Proof: We know from the results that led to (11.51a) that the minimizer $w^\star$ is a fixed point for the operator $\text{prox}_{\mu q}(z - \mu\nabla_{w^{\mathsf{T}}} E(z))$, i.e., it satisfies

$$w^\star \;=\; \text{prox}_{\mu q}\Big(w^\star - \mu\nabla_{w^{\mathsf{T}}} E(w^\star)\Big) \tag{21.27}$$

We exploit next the firmly nonexpansive property of proximal operators to arrive at the desired convergence result. Thus, note that

$$\begin{aligned}
\|\widetilde{\boldsymbol{w}}_n\|^2 &\stackrel{\Delta}{=} \|w^\star - \boldsymbol{w}_n\|^2 \\
&\stackrel{(21.27)}{=} \left\|\text{prox}_{\mu q}(w^\star - \mu\nabla_{w^\mathsf{T}} E(w^\star)) - \text{prox}_{\mu q}(\boldsymbol{z}_n)\right\|^2 \\
&\stackrel{(11.41)}{\leq} \|w^\star - \mu\nabla_{w^\mathsf{T}} E(w^\star) - \boldsymbol{z}_n\|^2 \\
&\stackrel{(21.1)}{=} \|w^\star - \mu\nabla_{w^\mathsf{T}} E(w^\star) - \boldsymbol{w}_{n-1} + \mu\nabla_{w^\mathsf{T}} Q_u(\boldsymbol{w}_{n-1})\|^2 \\
&= \|\widetilde{\boldsymbol{w}}_{n-1} - \mu\nabla_{w^\mathsf{T}} E(w^\star) + \mu\nabla_{w^\mathsf{T}} Q_u(\boldsymbol{w}_{n-1})\|^2 \\
&= \|\widetilde{\boldsymbol{w}}_{n-1} - \mu\nabla_{w^\mathsf{T}} E(w^\star) + \mu\nabla_{w^\mathsf{T}} E(\boldsymbol{w}_{n-1}) + \mu\boldsymbol{g}_n(\boldsymbol{w}_{n-1})\|^2
\end{aligned} \tag{21.28}$$

Conditioning on $\boldsymbol{w}_{n-1}$, expanding and taking expectations while exploiting the fact that the conditional mean of the gradient noise process is zero, we get

$$\begin{aligned}
\mathbb{E}\left(\|\widetilde{\boldsymbol{w}}_n\|^2 \,|\, \boldsymbol{w}_{n-1}\right) &= \|\widetilde{w}_{n-1}\|^2 + \mu^2\|\nabla_{w^\mathsf{T}} E(w^\star) - \nabla_{w^\mathsf{T}} E(w_{n-1})\|^2 + \\
&\qquad \mu^2\mathbb{E}\left(\|\boldsymbol{g}_n(\boldsymbol{w}_{n-1})\|^2 \,|\, \boldsymbol{w}_{n-1}\right) - \\
&\qquad 2\mu\left(\nabla_{w^\mathsf{T}} E(w^\star) - \nabla_{w^\mathsf{T}} E(w_{n-1})\right)^\mathsf{T} \widetilde{w}_{n-1} \\
&\leq \|\widetilde{w}_{n-1}\|^2 + \mu^2\delta^2\|\widetilde{w}_{n-1}\|^2 + \mu^2(\beta_g^2\|\widetilde{w}_{n-1}\|^2 + \sigma_g^2) - \\
&\qquad 2\mu\left(\nabla_{w^\mathsf{T}} E(w^\star) - \nabla_{w^\mathsf{T}} E(w_{n-1})\right)^\mathsf{T} \widetilde{w}_{n-1} \\
&\stackrel{(8.24)}{\leq} \|\widetilde{w}_{n-1}\|^2 + \mu^2\delta^2\|\widetilde{w}_{n-1}\|^2 + \mu^2(\beta_g^2\|\widetilde{w}_{n-1}\|^2 + \sigma_g^2) - \\
&\qquad 2\mu\nu\|\widetilde{w}_{n-1}\|^2 \\
&= \lambda\|\widetilde{w}_{n-1}\|^2 + \mu^2\sigma_g^2
\end{aligned} \tag{21.29}$$

where λ is defined by (21.26). Taking expectation again we remove the conditioning on $\widetilde{\boldsymbol{w}}_{n-1}$ and arrive at (21.25a). We know from the argument used in Fig. 12.2 that condition (21.24) ensures $0 \leq \lambda < 1$.

By iterating recursion (21.29) we obtain

$$\limsup_{n\to\infty} \mathbb{E}\,\|\widetilde{\boldsymbol{w}}_n\|^2 \leq \frac{\mu^2\sigma_g^2}{1-\lambda} = \frac{\mu\sigma_g^2}{2\nu - \mu(\delta^2 + \beta_g^2)} \tag{21.30}$$

It is easy to check that the upper bound does not exceed $\mu\sigma_g^2/\nu$ for any step size $\mu < \mu_o/2$. If, on the other hand, $\mu \ll \mu_o$ so that the denominator is approximately 2ν, then the upper bound is on the order of $\mu\sigma_g^2/2\nu$. We conclude that (21.25b) holds for sufficiently small step sizes.

The same argument from Prob. 20.1 can be repeated to show that the norm of the weight-error vector evolves on average according to the relation

$$\mathbb{E}\,\|\widetilde{\boldsymbol{w}}_n\| \;\leq\; \sqrt{\lambda}\,\mathbb{E}\,\|\widetilde{\boldsymbol{w}}_{n-1}\| \;+\; O(\mu) \tag{21.31}$$

To establish (21.25c), we proceed as follows. First, the risk $P(w) = q(w) + E(w)$ is a convex function but not necessarily differentiable. According to definition (8.43), this means that there exists a subgradient of $P(w)$ at location $\boldsymbol{w}_n$, denoted by $s_P(\boldsymbol{w}_n)$, such that

$$P(\boldsymbol{w}_n) - P(w^\star) \;\leq -\Big(s_P(\boldsymbol{w}_n)\Big)^\mathsf{T} \widetilde{\boldsymbol{w}}_n \tag{21.32}$$

where $s_P(w) = s_q(w) + \nabla_{w^\mathsf{T}} E(w)$, for some subgradient of $q(w)$ relative to w^T. Consequently,

$$\begin{aligned}
0 &\le P(\boldsymbol{w}_n) - P(w^\star) \\
&\le -\Big(s_q(\boldsymbol{w}_n) + \nabla_{w^\mathsf{T}} E(\boldsymbol{w}_n)\Big)^\mathsf{T} \widetilde{\boldsymbol{w}}_n \\
&\overset{(a)}{=} -\Big(s_q(\boldsymbol{w}_n) + \nabla_{w^\mathsf{T}} E(\boldsymbol{w}_n) - s_q'(w^\star) - \nabla_{w^\mathsf{T}} E(w^\star)\Big)^\mathsf{T} \widetilde{\boldsymbol{w}}_n \\
&\le \|s_q(\boldsymbol{w}_n) - s_q'(w^\star)\| \, \|\widetilde{\boldsymbol{w}}_n\| + \|\nabla_{w^\mathsf{T}} E(\boldsymbol{w}_n) - \nabla_{w^\mathsf{T}} E(w^\star)\| \, \|\widetilde{\boldsymbol{w}}_n\| \\
&\overset{(b)}{=} (\delta_1 + \delta)\, \|\widetilde{\boldsymbol{w}}_n\|^2 + \delta_2 \, \|\widetilde{\boldsymbol{w}}_n\|
\end{aligned} \tag{21.33}$$

where in step (a) we used the fact that there should exist a subgradient $s_P'(w)$ that evaluates to zero at the minimizer $w = w^\star$. This subgradient is written as $s_P'(w) = s_q'(w) + \nabla_{w^\mathsf{T}} E(w)$ for some subgradient $s_q'(w)$ of $q(w)$. In step (b) we used the affine-Lipschitz condition on the subgradients of $q(w)$ and the δ-Lipschitz condition on the gradients of $E(w)$. Under expectation we get

$$\mathbb{E}\, P(\boldsymbol{w}_n) - P(w^\star) \le (\delta_1 + \delta)\, \mathbb{E}\, \|\widetilde{\boldsymbol{w}}_n\|^2 + \delta_2 \, \mathbb{E}\, \|\widetilde{\boldsymbol{w}}_n\| \tag{21.34}$$

so that using (21.31) we get (21.25c) whenever $\delta_2 \neq 0$.

■

Example 21.3 **(ℓ_1-regularized MSE risk)** We illustrate the behavior of the stochastic proximal algorithm by reconsidering the same setting from Example 20.2 where we illustrated the behavior of the stochastic subgradient algorithm with and without smoothing.

We consider the same ℓ_1-regularized MSE risk minimization problem:

$$w^o \triangleq \underset{w \in \mathbb{R}^M}{\text{argmin}} \left\{ P(w) \triangleq \alpha \|w\|_1 + \mathbb{E}\, (\boldsymbol{\gamma} - \boldsymbol{h}^\mathsf{T} w)^2 \right\} \tag{21.35}$$

where the streaming data samples $(\boldsymbol{\gamma}(n), \boldsymbol{h}_n)$ are assumed to arise from the linear model:

$$\boldsymbol{\gamma}(n) = \boldsymbol{h}_n^\mathsf{T} w^\bullet + \boldsymbol{v}(n) \tag{21.36}$$

for some parameter $w^\bullet \in \mathbb{R}^M$ and where $\boldsymbol{v}(n)$ is white noise with zero mean and variance σ_v^2; the noise is independent of all other random variables. We assume that the data $(\boldsymbol{\gamma}, \boldsymbol{h})$ have zero means with second-order moments given by

$$\sigma_\gamma^2 \triangleq \mathbb{E}\, \boldsymbol{\gamma}^2, \quad r_{h\gamma} \triangleq \mathbb{E}\, \boldsymbol{h}\boldsymbol{\gamma}, \quad R_h = \mathbb{E}\, \boldsymbol{h}\boldsymbol{h}^\mathsf{T} \triangleq \sigma_h^2 I_M \tag{21.37}$$

We already know from (20.81) and (20.82) that

$$P(w) = \alpha \|w\|_1 + \sigma_h^2 \|w - w^\bullet\|^2 + \sigma_v^2 \tag{21.38a}$$

$$w^o = \mathbb{T}_{\alpha/2\sigma_h^2}(w^\bullet) \tag{21.38b}$$

We simulate the following scenario. We set $M = 25$, $\alpha = 0.002$, $\mu = 0.001$, $\sigma_v^2 = 0.01$, $\sigma_h^2 = 1$, and $\kappa = 0.9980$ for subgradient smoothing. The entries of $\boldsymbol{h}$ and $\boldsymbol{v}$ are Gaussian distributed with zero mean. We generate a random sparse model $w^\bullet$ with only two nonzero entries set equal to 1, while all other entries are set to 0. For each run of the algorithms listed below, we generate $N = 10{,}000$ data realizations $\{\gamma(m), h_m\}$. We run the stochastic subgradient algorithm with and without exponential smoothing, namely,

$$w_n = w_{n-1} + 2\mu h_n(\gamma(n) - h_n^\mathsf{T} w_{n-1}) - \mu\, \alpha\, \text{sign}(w_{n-1}) \tag{21.39}$$

and

$$\begin{cases} w_n = w_{n-1} + 2\mu h_n(\gamma(n) - h_n^\mathsf{T} w_{n-1}) - \mu\,\alpha\,\text{sign}(w_{n-1}) \\ S_{n+1} = \kappa S_n + 1 \\ \bar{w}_{n+1} = \left(1 - \frac{1}{S_{n+1}}\right)\bar{w}_n + \frac{1}{S_{n+1}} w_n \end{cases} \tag{21.40}$$

We also run the stochastic proximal gradient algorithm:

$$\begin{cases} z_n = w_{n-1} + 2\mu h_n(\gamma(n) - h_n^\mathsf{T} w_{n-1}) \\ w_n = \text{prox}_{\mu\alpha\|w\|_1}(z_n) \;=\; \mathbb{T}_{\mu\alpha}(z_n) \end{cases} \tag{21.41}$$

For each run, we use expression (21.38a) to evaluate the excess-risk curve:

$$P(w) - P(w^o) = \alpha\Big(\|w\|_1 - \|w^o\|_1\Big) + \sigma_h^2\|w - w^\bullet\|^2 - \sigma_h^2\|\widetilde{w}^\bullet\|^2 \tag{21.42}$$

and average the curves over 50 experiments. The results are shown in Fig. 21.1 using the logarithmic scale; each curve is showing the evolution of $10\log_{10}(P(w_n) - P(w^o))$ (normalized to start from 0 dB). It is observed that the stochastic subgradient and proximal algorithms have similar performance, while smoothing enhances steady-state performance.

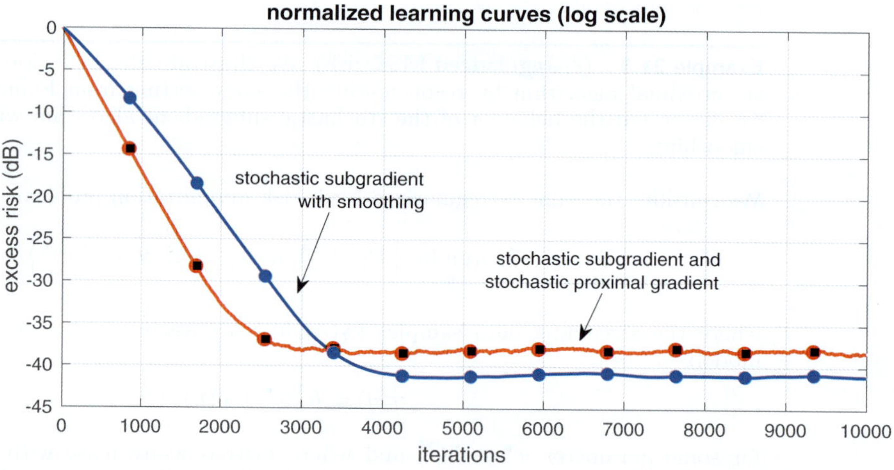

Figure 21.1 Learning curves in logarithmic scale for the stochastic subgradient algorithms (20.86)–(20.87) with and without smoothing and for the stochastic proximal gradient algorithm (21.41).

In a second simulation, we vary the step size in the range $10^{-4} \leq \mu \leq 10^{-3}$ and assess the steady-state excess risk for the three algorithms; the steady-state values are obtained by averaging the last 50 samples of each excess-risk curve. The results are plotted in Fig. 21.2. It is again observed that the stochastic subgradient and proximal algorithms have similar performance, while smoothing enhances performance.

Example 21.4 (**Stochastic coordinate descent**) A similar convergence analysis can be applied to the randomized version of coordinate descent listed in (16.101), namely,

$$\begin{cases} \text{select a random index } 1 \leq m^o \leq M; \\ \boldsymbol{z}_{n,m^o} = \boldsymbol{w}_{n-1,m^o} \;-\; \mu\,\partial Q_u(\boldsymbol{w}_{n-1};\boldsymbol{\gamma}(n),\boldsymbol{h}_n)/\partial w_{m^o} \\ \boldsymbol{w}_{n,m^o} = \text{prox}_{\mu q_{m^o}}(\boldsymbol{z}_{n,m^o}) \\ \boldsymbol{w}_{n,m} = \boldsymbol{w}_{n-1,m},\; m \neq m^o \end{cases} \tag{21.43}$$

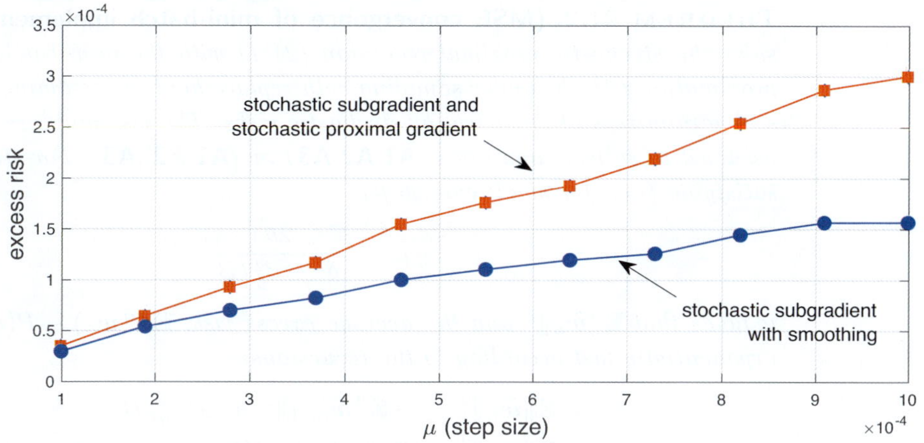

Figure 21.2 Excess risk values as function of the step size μ for the stochastic subgradient algorithms (20.86)–(20.87) with and without smoothing and for the stochastic proximal gradient algorithm (21.41).

The regularization term in this case is assumed to separate over the $\{w_m\}$, as shown earlier in (14.123), i.e.,

$$P(w) \;=\; E(w) + \sum_{m=1}^{M} q_m(w_m) \tag{21.44}$$

We leave the convergence analysis to Prob. 21.8.

21.3 CONVERGENCE OF MINI-BATCH IMPLEMENTATION

Let (β_g^2, σ_g^2) denote the parameters that characterize the bound in (21.22b) for the stochastic gradient algorithm based on *instantaneous* gradient approximations. We showed in (18.37) that these parameters get scaled down by a factor B when a mini-batch implementation is used with uniform sampling. The second-order moment of the gradient noise will then satisfy

$$\mathbb{E}\left(\|\boldsymbol{g}_n(\boldsymbol{w}_{n-1})\|^2 \,|\, \boldsymbol{w}_{n-1}\right) \;\le\; \frac{1}{B}\left(\beta_g^2\|\widetilde{w}_{n-1}\|^2 + \sigma_g^2\right) \tag{21.45}$$

The same analysis used to establish Theorem 21.1 will lead to a similar conclusion apart from the scaling of β_g^2 and σ_g^2 by B – see Prob. 21.3. As a result, the size of the steady-state neighborhood will be reduced from $O(\mu)$ to $O(\mu/B)$.

Theorem 21.2. (MSE convergence of mini-batch implementation) *Consider the stochastic proximal recursion (21.1) with the mini-batch gradient approximation (21.5b) under sampling with replacement or streaming data, used to seek minimizers of empirical or stochastic risks. The risk and loss functions are assumed to satisfy conditions (**A1,A2,A3**) or (**A1,A2',A3**). For step-size values satisfying (i.e., for small enough μ):*

$$\mu < \frac{2\nu}{\delta^2 + \beta_g^2/B} \tag{21.46}$$

it holds that $\mathbb{E}\,\|\widetilde{\boldsymbol{w}}_n\|^2$ and the average excess risk, $\mathbb{E}\,P(\boldsymbol{w}_n) - P(w^\star)$, converge exponentially fast according to the recursions:

$$\mathbb{E}\,\|\widetilde{\boldsymbol{w}}_n\|^2 \;\leq\; \lambda\,\mathbb{E}\,\|\widetilde{\boldsymbol{w}}_{n-1}\|^2 \;+\; \mu^2\sigma_g^2/B \tag{21.47a}$$

$$\mathbb{E}\,\|\widetilde{\boldsymbol{w}}_n\|^2 \leq O(\lambda^n) + O(\mu/B) \tag{21.47b}$$

$$\mathbb{E}\,P(\boldsymbol{w}_n) - P(w^\star) \leq O(\lambda^{n/2}) + O\Big(\sqrt{\mu/B}\Big) \tag{21.47c}$$

where

$$\lambda \;=\; 1 - 2\nu\mu + \left(\delta^2 + \beta_g^2/B\right)\mu^2 \;\in\; [0,1) \tag{21.48}$$

Results (21.47b)–(21.47c) hold for sufficiently small step sizes. Result (21.47c) holds whenever $\delta_2 \neq 0$. Otherwise, the convergence of the excess risk occurs at the rate λ^n toward $O(\mu/B)$.

21.4 CONVERGENCE UNDER VANISHING STEP SIZES

We observe from (21.25b) that under constant step-size learning, the MSE $\mathbb{E}\,\|\widetilde{\boldsymbol{w}}_n\|^2$ converges to a small neighborhood of size $O(\mu)$; the smaller the value of μ is, the smaller the size of this neighborhood will be. However, small step sizes affect the convergence rate of the algorithm because they cause the value of λ to approach 1. One way to reduce the size of the limiting region to zero is to employ a decaying step size $\mu(n)$ instead of the constant μ. It is common to choose the sequence $\mu(n) > 0$ to satisfy either of the following two conditions:

$$\textbf{(Condition I)} \qquad \sum_{n=0}^{\infty} \mu^2(n) < \infty \quad \text{and} \quad \sum_{n=0}^{\infty} \mu(n) = \infty \tag{21.49a}$$

$$\textbf{(Condition II)} \qquad \lim_{n\to\infty} \mu(n) = 0 \quad \text{and} \quad \sum_{n=0}^{\infty} \mu(n) = \infty \tag{21.49b}$$

Clearly, any sequence that satisfies the stronger condition (21.49a) also satisfies (21.49b). In either case, recursion (21.1) would be replaced by

$$\begin{cases} z_n = w_{n-1} - \mu(n)\,\widehat{\nabla_{w^\mathsf{T}} E}(w_{n-1}),\ n \geq 0 \\ w_n = \text{prox}_{\mu q}(z_n) \end{cases} \tag{21.50}$$

Theorem 21.3. (MSE convergence under vanishing step sizes) *Consider the stochastic proximal recursion (21.1) with the instantaneous gradient approximation (21.5a) under sampling with replacement or streaming data, used to seek minimizers of empirical or stochastic risks. The risk and loss functions are assumed to satisfy conditions* **(A1,A2,A3)** *or* **(A1,A2',A3)**. *Then, the following convergence properties hold:*

(a) If the step-size sequence $\mu(n)$ satisfies (21.49a), then w_n converges almost surely to $w^\star$, written as $w_n \to w^\star$ a.s.

(b) If the step-size sequence $\mu(n)$ satisfies (21.49b), then w_n converges in the MSE sense to $w^\star$, i.e., $\mathbb{E}\,\|\widetilde{w}_n\|^2 \to 0$.

Proof: The same argument leading to (21.29) in the constant step-size case continues to hold, leading to the inequality:

$$\mathbb{E}\left(\|\widetilde{w}_n\|^2 \,|\, w_{n-1}\right) \leq \lambda(n)\,\|\widetilde{w}_{n-1}\|^2 + \mu^2(n)\delta_g^2 \tag{21.51}$$

with μ replaced by $\mu(n)$ and where

$$\lambda(n) \overset{\Delta}{=} 1 - 2\nu\mu(n) + (\delta^2 + \beta_g^2)\mu^2(n) \tag{21.52}$$

We split the term $2\nu\mu(n)$ into the sum of two terms and write

$$\lambda(n) \;=\; 1 - \nu\mu(n) - \nu\mu(n) + (\delta^2 + \beta_g^2)\mu^2(n) \tag{21.53}$$

And since $\mu(n) \to 0$, we conclude that for large enough $n > n_o$, the value $\mu^2(n)$ is smaller than $\mu(n)$. Therefore, a large enough time index, n_o, exists such that the following two conditions are satisfied:

$$\nu\mu(n) \geq (\delta^2 + \beta_g^2)\mu^2(n), \quad 0 \leq \nu\mu(n) < 1, \quad n > n_o \tag{21.54}$$

Consequently,

$$\alpha(n) \;\leq\; 1 - \nu\mu(n), \quad n > n_o \tag{21.55}$$

Then, inequalities (21.51) and (21.55) imply that

$$\mathbb{E}\left(\|\widetilde{w}_n\|^2 \,|\, w_{n-1}\right) \;\leq\; (1 - \nu\mu(n))\,\|\widetilde{w}_{n-1}\|^2 \;+\; \mu^2(n)\sigma_g^2, \quad n > n_o \tag{21.56}$$

Due to the Markovian property of recursion (21.1), where w_n is solely dependent on the most recent iterate w_{n-1}, we also conclude that for $n > n_o$:

$$\mathbb{E}\left(\|\widetilde{w}_n\|^2, w_{n-1}, \ldots, w_0, w_{-1}\right) \;\leq\; (1 - \nu\mu(n))\,\|\widetilde{w}_{n-1}\|^2 \;+\; \mu^2(n)\sigma_g^2 \tag{21.57}$$

For convenience of notation, let

$$u(n+1) \overset{\Delta}{=} \|\widetilde{w}_n\|^2 \tag{21.58}$$

Then, inequality (21.56) gives

$$\mathbb{E}\left(\boldsymbol{u}(n+1)|\,\boldsymbol{u}(0),\boldsymbol{u}(1),\ldots,\boldsymbol{u}(n)\right) \;\leq\; (1-\nu\mu(n))\,u(n) \;+\; \mu^2(n)\sigma_g^2, \quad n>n_o \tag{21.59}$$

We now call upon the useful result (19.158) and make the identifications:

$$a(n) \leftarrow \nu\mu(n), \quad b(n) \leftarrow \mu^2(n)\sigma_g^2 \tag{21.60}$$

These sequences satisfy conditions (19.159) in view of assumption (21.49a) on the step-size sequence and the second condition in (21.54). We conclude that $\boldsymbol{u}(n)\to 0$ almost surely and, hence, $\boldsymbol{w}_n \to w^\star$ almost surely.

Finally, taking expectations of both sides of (21.59) leads to

$$\mathbb{E}\,\boldsymbol{u}(n+1) \;\leq\; (1-\nu\mu(n))\,\mathbb{E}\,\boldsymbol{u}(n) \;+\; \mu^2(n)\sigma_g^2, \qquad n>n_o \tag{21.61}$$

with the expectation operator appearing on both sides of the inequality. Then, we conclude from result (14.136), under conditions (21.49b), that $\mathbb{E}\,\|\widetilde{\boldsymbol{w}}_n\|^2 \to 0$ so that $\boldsymbol{w}_n$ converges to $w^\star$ in the MSE sense.

■

We can be more specific and quantify the rate at which the error variance $\mathbb{E}\,\|\widetilde{\boldsymbol{w}}_n\|^2$ converges toward zero for step-size sequences of the form:

$$\mu(n) = \frac{\tau}{n+1}, \qquad \tau>0 \tag{21.62}$$

which satisfy both conditions (21.49a) and (21.49b).

THEOREM 21.4. (Rates of convergence under (21.62)) *Consider the stochastic proximal recursion (21.1) with the instantaneous gradient approximation (21.5a) under sampling with replacement or streaming data, used to seek minimizers of empirical or stochastic risks. The risk and loss functions are assumed to satisfy conditions (**A1,A2,A3**) or (**A1,A2',A3**). Then, three convergence rates are possible depending on how the factor $\nu\tau$ compares to the value 1. Specifically, for large enough n, it holds that:*

$$\begin{cases} \mathbb{E}\,\|\widetilde{\boldsymbol{w}}_n\|^2 \leq O\left(\frac{1}{n}\right), & \nu\tau>1 \\ \mathbb{E}\,\|\widetilde{\boldsymbol{w}}_n\|^2 = O\left(\frac{\log n}{n}\right), & \nu\tau=1 \\ \mathbb{E}\,\|\widetilde{\boldsymbol{w}}_n\|^2 = O\left(\frac{1}{n^{\nu\tau}}\right), & \nu\tau<1 \end{cases} \tag{21.63}$$

The fastest convergence rate occurs when $\nu\tau>1$ (i.e., for large enough τ) and is on the order of $O(1/n)$. When the regularization term is of the form $q(w)=\alpha\|w\|_1+\rho\|w\|^2$, with $\alpha>0$, the risk values exhibit similar convergence behavior as $\mathbb{E}\,\|\widetilde{\boldsymbol{w}}_n\|$, namely,

$$\begin{cases} \mathbb{E}\,P(\boldsymbol{w}_n)-P(w^\star) \leq O\left(\frac{1}{\sqrt{n}}\right), & \nu\tau>1 \\ \mathbb{E}\,P(\boldsymbol{w}_n)-P(w^\star) = O\left(\frac{(\log n)^{1/2}}{\sqrt{n}}\right), & \nu\tau=1 \\ \mathbb{E}\,P(\boldsymbol{w}_n)-P(w^\star) = O\left(\frac{1}{n^{\nu\tau/2}}\right), & \nu\tau<1 \end{cases} \tag{21.64}$$

The fastest convergence rate again occurs when $\nu\tau>1$ and is on the order of $O(1/\sqrt{n})$.

Proof: We use (21.61) and the assumed form for $\mu(n)$ in (21.62) to write

$$\mathbb{E}\,\boldsymbol{u}(n+1) \;\leq\; \left(1 - \frac{\nu\tau}{n+1}\right)\mathbb{E}\,\boldsymbol{u}(n) \;+\; \frac{\tau^2\sigma_g^2}{(n+1)^2}, \quad n > n_o \tag{21.65}$$

This recursion has the same form as recursion (14.136) with the identifications:

$$a(n) \leftarrow \frac{\nu\tau}{n+1}, \quad b(n) \leftarrow \frac{\tau^2\sigma_g^2}{(n+1)^2}, \quad p \leftarrow 1 \tag{21.66}$$

The above rates of convergence then follow from Lemma 14.1. Result (21.64) follows from (21.34).

■

Table 21.1 summarizes the performance results obtained thus far in the chapter for the stochastic proximal algorithm. Results are shown for strongly convex risks under both constant and vanishing step sizes.

21.5 STOCHASTIC PROJECTION GRADIENT

The stochastic projection gradient algorithm was listed earlier in (15.51) and was shown to be a special case of the stochastic proximal algorithm. It corresponds to choosing the regularizer term as the indicator function for a convex set $\mathcal{C}$, $q(w) = \mathbb{I}_{C,\infty}[w]$. This indicator function assumes the value zero when $w \in \mathcal{C}$ and the value $+\infty$ otherwise. The algorithm is meant to solve problems of the form:

$$w^\star = \underset{w\in\mathbb{R}^M}{\text{argmin}}\; E(w), \quad \text{subject to } w \in \mathcal{C} \tag{21.67}$$

where $E(w)$ can take the form of an empirical or stochastic risk, namely,

$$E(w) \;\stackrel{\Delta}{=}\; \frac{1}{N}\sum_{m=0}^{N-1} Q_u(w;\gamma(m),h_m) \quad \text{or} \quad E(w) \;\stackrel{\Delta}{=}\; \mathbb{E}\,Q_u(w;\boldsymbol{\gamma},\boldsymbol{h}) \tag{21.68}$$

We showed in (15.51) that the stochastic version of the projection gradient construction takes the following form (compare with (21.1)):

$$\boxed{\left\{\begin{array}{l} \boldsymbol{z}_n = \boldsymbol{w}_{n-1} - \mu\,\widehat{\nabla_{w^\mathsf{T}} E}(\boldsymbol{w}_{n-1}),\; n \geq 0 \\ \boldsymbol{w}_n = \mathcal{P}_C(\boldsymbol{z}_n) \end{array}\right.} \tag{21.69}$$

where $\mathcal{P}_\mathcal{C}(x)$ is the projection operator onto $\mathcal{C}$. If we examine the convergence proof of Theorem 21.1 we will find that most of the arguments go through. The only modification is in relation to the argument leading to the bound on the excess risk in (21.34). Here, since $E(w)$ is strongly convex, we can use property (21.17b) to write

$$0 \leq \mathbb{E}\,E(\boldsymbol{w}_n) - E(w^\star) \leq \frac{\delta}{2}\,\mathbb{E}\,\|\widetilde{\boldsymbol{w}}_n\|^2 \tag{21.70}$$

and arrive at the following conclusion for instantaneous gradient approximations.

Table 21.1 Convergence properties of the stochastic proximal gradient algorithm with *constant* and *vanishing* step sizes. The risk function is split as $P(w) = q(w) + E(w)$ and the smooth component $E(w)$ can either be an empirical or stochastic risk. In both cases, the loss function associated with $E(w)$ is denoted by $Q_u(w; \cdot)$.

Mode of operation	Conditions on risk and loss functions	Asymptotic convergence property	Reference
instantaneous gradient approximation (uniform sampling)	$E(w)$: ν-strongly convex $Q_u(w)$: δ-Lipschitz gradients (**A1**,**A2**,**A3**)	$\mathbb{E}\,P(\boldsymbol{w}_n) - P(w^\star) \leq O(\lambda^{n/2}) + O(\sqrt{\mu})$	Thm. 21.1
instantaneous gradient approximation (data streaming)	$E(w)$: ν-strongly convex $Q_u(w)$: δ-Lipschitz gradients in mean-square sense (**A1**,**A2'**,**A3**)	$\mathbb{E}\,P(\boldsymbol{w}_n) - P(w^o) \leq O(\lambda^{n/2}) + O(\sqrt{\mu})$	Thm. 21.1
instantaneous gradient approximation (uniform sampling) (vanishing step size)	$E(w)$: ν-strongly convex $Q_u(w)$: δ-Lipschitz gradients (**A1**,**A2**,**A3**) $\mu(n) = \tau/(n+1)$	$\mathbb{E}\,P(\boldsymbol{w}_n) - P(w^\star) \leq O(1/\sqrt{n})$	Thm. 21.4
instantaneous gradient approximation (data streaming) (vanishing step size)	$E(w)$: ν-strongly convex $Q_u(w)$: δ-Lipschitz gradients in mean-square sense (**A1**,**A2'**,**A3**) $\mu(n) = \tau/(n+1)$	$\mathbb{E}\,P(\boldsymbol{w}_n) - P(w^o) \leq O(1/\sqrt{n})$	Thm. 21.4
mini-batch gradient approximation (uniform sampling or data streaming)	$E(w)$: ν-strongly convex $Q_u(w)$: δ-Lipschitz gradients in deterministic or mean-square sense (**A1**,**A2**,**A3**) or (**A1**,**A2'**,**A3**)	$\mathbb{E}\,P(\boldsymbol{w}_n) - P(w^\star) \leq O(\lambda^{n/2}) + O(\sqrt{\mu/B})$	Thm. 21.2

Theorem 21.5. (MSE convergence under constant step sizes) *Consider the stochastic projection recursion (21.69) with the instantaneous gradient approximation (21.5a) under sampling with replacement or streaming data, used to seek minimizers of empirical or stochastic risks over a closed convex set $w \in \mathcal{C}$. The risk and loss functions are assumed to satisfy conditions (**A1**,**A2**) or (**A1**,**A2'**). For step-size values satisfying (i.e., for small enough μ):*

$$\mu < \frac{2\nu}{\delta^2 + \beta_g^2} \stackrel{\Delta}{=} \mu_o \tag{21.71}$$

it holds that $\mathbb{E}\,\|\widetilde{\boldsymbol{w}}_n\|^2$ and the average excess risk, $\mathbb{E}\,E(\boldsymbol{w}_n) - E(w^\star)$, converge exponentially fast according to the recursions:

$$\mathbb{E}\,\|\widetilde{\boldsymbol{w}}_n\|^2 \;\leq\; \lambda\,\mathbb{E}\,\|\widetilde{\boldsymbol{w}}_{n-1}\|^2 + \;\mu^2\sigma_g^2 \tag{21.72a}$$

$$\mathbb{E}\,\|\widetilde{\boldsymbol{w}}_n\|^2 \leq O(\lambda^n) + O(\mu) \tag{21.72b}$$

$$\mathbb{E}\,E(\boldsymbol{w}_n) - E(w^\star) \leq O(\lambda^n) + O(\mu) \tag{21.72c}$$

where

$$\lambda \;=\; 1 - 2\nu\mu + (\delta^2 + \beta_g^2)\mu^2 \;\in\; [0, 1) \tag{21.73}$$

Results (21.72b)–(21.72c) hold for sufficiently small step sizes.

21.6 MIRROR-DESCENT ALGORITHM

We described the stochastic version of the mirror-descent algorithm in listing (16.98), which is repeated here for ease of reference.

Stochastic mirror descent for solving the constrained problem (21.67).

given dataset $\{\gamma(m), h_m\}_{m=0}^{N-1}$ or streaming data $(\gamma(n), h_n)$;
choose ν_ϕ-strongly convex mirror function $\phi(w) : \mathcal{C}_\phi \to \mathbb{R}$;
let $\phi^\star(x) = \sup_w\{w^\mathsf{T}x - \phi(w)\}$ denote its conjugate function;
let $\mathcal{C}' = \mathcal{C} \cap \mathcal{C}_\phi$;
start from an initial condition, $\boldsymbol{w}_{-1} \in \mathcal{C}'$.
repeat until convergence over $n \geq 0$**:**
  select at random or receive $(\boldsymbol{\gamma}(n), \boldsymbol{h}_n)$
  $\boldsymbol{b}_n = \nabla_{w^\mathsf{T}}\,\phi(\boldsymbol{w}_{n-1}) - \mu\nabla_{w^\mathsf{T}}\,Q_u(\boldsymbol{w}_{n-1}; \boldsymbol{\gamma}(n), \boldsymbol{h}_n)$
  $\boldsymbol{z}_n = \nabla_{x^\mathsf{T}}\,\phi^\star(\boldsymbol{b}_n)$
  $\boldsymbol{w}_n = \mathcal{P}_{C',\phi}(\boldsymbol{z}_n)$ (Bregman projection)
end
return $w^\star \leftarrow \boldsymbol{w}_n$.

(21.74)

We examine the convergence behavior of the algorithm under the following relaxed conditions on $E(w)$:

(M1) (**Convex but possibly nonsmooth risk**). We assume $E(w)$ is only convex and allow it to be nonsmooth, in which case the gradient vector $\nabla_{w^\mathsf{T}} Q_u(w;\gamma,h)$ is replaced by a subgradient, denoted by $s_u(w;\gamma,h) \in \partial_{w^\mathsf{T}} Q_u(w;\gamma,h)$.

(M2) (**δ-Lipschitz loss function**). We assume $Q_u(w;\gamma,h)$ is itself δ-Lipschitz relative to some *arbitrary* norm, i.e., it satisfies for any (γ,γ',h,h'):

$$\|Q_u(w_2;\gamma,h) - Q_u(w_1;\gamma',h')\| \leq \delta\,\|w_2 - w_1\| \tag{21.75}$$

for any $w_1, w_2 \in \text{dom}(Q_u)$. Observe that we are requiring $Q_u(w;\cdot)$ to be Lipschitz rather than its subgradient. Let $\|\cdot\|_\star$ denote the dual norm that corresponds to the norm $\|\cdot\|$ used in (21.75). For example, if the 2-norm is used, then the dual norm is the 2-norm itself. Then, we know from the result of Lemma 10.5 that the Lipschitz condition (21.75) implies bounded subgradients:

$$\|s_u(w;\gamma,h)\|_\star \leq \delta \tag{21.76}$$

(M3) (**Strongly convex mirror function**). We assume the mirror function $\phi(w)$ is ν-strongly convex relative to the *same* norm used in (21.75) for the Lipschitz bound on $Q_u(w;\cdot)$, i.e.,

$$\phi(w) \geq \phi(z) + \nabla_w\,\phi(z)(w-z) + \frac{\nu}{2}\|w-z\|^2 \tag{21.77}$$

We denote the domain of $\phi(w)$ by $\mathcal{C}_\phi$.

THEOREM 21.6. (**Convergence of stochastic mirror descent**) *Consider the stochastic mirror-descent algorithm (21.74) used to seek the minimizer of the constrained problem (21.67) for an empirical or stochastic risk over the closed convex set $w \in \mathcal{C}$. The risk, loss, and mirror functions are assumed to satisfy conditions (**M1**,**M2**,**M3**). Set the step-size parameter to*

$$\mu = \frac{W}{\delta}\sqrt{\frac{2\nu}{L}} \tag{21.78}$$

in terms of the maximal average Bregman distance from the initial condition to the convex set, namely,

$$W^2 \triangleq \sup_{w\in\mathcal{C}\cap\mathcal{C}_\phi} \Big\{\mathbb{E}\,D_\phi(w,\boldsymbol{w}_{-1})\Big\} \tag{21.79}$$

Then, the averaged iterate over L iterations satisfies

$$0 \leq \mathbb{E}\left\{E\left(\frac{1}{L}\sum_{n=0}^{L-1}\boldsymbol{w}_{n-1}\right) - E(w^\star)\right\} \leq W\delta\sqrt{\frac{2}{L\nu}} \tag{21.80}$$

which shows that the risk value at the averaged iterate converges to $E(w^\star)$ at the rate of $O(1/\sqrt{L})$.

Proof: The minimal risk value is $E(w^\star)$. Let $s_E(\boldsymbol{w}_{n-1})$ denote a subgradient for $E(w)$ at location $\boldsymbol{w}_{n-1}$ and introduce the gradient noise process:

$$\boldsymbol{g}_n(\boldsymbol{w}_{n-1}) \triangleq s_u(\boldsymbol{w}_{n-1};\boldsymbol{\gamma}(n),\boldsymbol{h}_n) - s_E(\boldsymbol{w}_{n-1}) \tag{21.81}$$

Here, the data point $(\boldsymbol{\gamma}(n),\boldsymbol{h}_n)$ is either selected uniformly at random from the dataset $\{\gamma(m),h_m\}$ in empirical risk minimization, or streams in over time independently of other samples. Either way, it is easy to verify that

$$\mathbb{E}\left(\boldsymbol{g}_n(\boldsymbol{w}_{n-1})\,|\,\boldsymbol{w}_{n-1}\right) = 0 \tag{21.82}$$

Now, since $E(w)$ is convex, there exists a subgradient $s_E(\boldsymbol{w}_{n-1})$ such that

$$\begin{aligned} 0 \le E(\boldsymbol{w}_{n-1}) - E(w^\star) \;&\le\; \left(s_E(\boldsymbol{w}_{n-1})\right)^{\mathsf{T}}(\boldsymbol{w}_{n-1}-w^\star) \\ &\overset{(21.81)}{=} \left(s_u(\boldsymbol{w}_{n-1};\boldsymbol{\gamma}(n),\boldsymbol{h}_n) - \boldsymbol{g}_n(\boldsymbol{w}_{n-1})\right)^{\mathsf{T}}(\boldsymbol{w}_{n-1}-w^\star) \end{aligned} \tag{21.83}$$

It then follows from (21.74) that

$$\begin{aligned} &0 \le E(\boldsymbol{w}_{n-1}) - E(w^\star) \\ &\overset{(a)}{=} \frac{1}{\mu}\left(\nabla_{w^{\mathsf{T}}}\phi(\boldsymbol{w}_{n-1}) - \boldsymbol{b}_n\right)^{\mathsf{T}}(\boldsymbol{w}_{n-1}-w^\star) + \boldsymbol{g}_n(\boldsymbol{w}_{n-1})\widetilde{\boldsymbol{w}}_{n-1} \\ &\overset{(b)}{=} \frac{1}{\mu}\left(\nabla_{w}\phi(\boldsymbol{w}_{n-1}) - \nabla_w\phi(\boldsymbol{z}_n)\right)(\boldsymbol{w}_{n-1}-w^\star) + \boldsymbol{g}_n(\boldsymbol{w}_{n-1})\widetilde{\boldsymbol{w}}_{n-1} \\ &\overset{(c)}{=} \frac{1}{\mu}\left(D_\phi(w^\star,\boldsymbol{w}_{n-1}) - D_\phi(w^\star,\boldsymbol{z}_n) + D_\phi(\boldsymbol{w}_{n-1},\boldsymbol{z}_n)\right) + \boldsymbol{g}_n(\boldsymbol{w}_{n-1})\widetilde{\boldsymbol{w}}_{n-1} \end{aligned} \tag{21.84}$$

where step (a) follows from the first step of the algorithm, step (b) uses $\boldsymbol{b}_n = \nabla_{w^{\mathsf{T}}}\phi(\boldsymbol{z}_n)$, and step (c) can be verified by substituting the Bregman divergences by their definitions.

Next, we know that $\boldsymbol{w}_n$ results from the Bregman projection of $\boldsymbol{z}_n$ onto $\mathcal{C}\cap\mathcal{C}_\phi$. We know from the result of part (c) of Prob. 9.14 that such projections satisfy the Pythagorean inequality:

$$D_\phi(w^\star,\boldsymbol{z}_n) \;\ge\; D_\phi(w^\star,\boldsymbol{w}_n) + D_\phi(\boldsymbol{w}_n,\boldsymbol{z}_n) \tag{21.85}$$

Substituting into (21.84) gives

$$\begin{aligned} &E(\boldsymbol{w}_{n-1}) - E(w^\star) \;\le\; \boldsymbol{g}_n(\boldsymbol{w}_{n-1})\widetilde{\boldsymbol{w}}_{n-1} + \\ &\frac{1}{\mu}\Big(\underbrace{D_\phi(w^\star,\boldsymbol{w}_{n-1}) - D_\phi(w^\star,\boldsymbol{w}_n)}_{=\text{ telescoping sequence}} + \underbrace{D_\phi(\boldsymbol{w}_{n-1},\boldsymbol{z}_n) - D_\phi(\boldsymbol{w}_n,\boldsymbol{z}_n)}_{\triangleq\, B}\Big) \end{aligned} \tag{21.86}$$

where, similar to (15.145), the second term evaluates to

$$B \triangleq \mu\left(s_u(\boldsymbol{w}_{n-1};\boldsymbol{\gamma}(n),\boldsymbol{h}_n)\right)^{\mathsf{T}}(\boldsymbol{w}_{n-1}-\boldsymbol{w}_n) - \frac{\nu}{2}\|\boldsymbol{w}_n-\boldsymbol{w}_{n-1}\|^2 \tag{21.87}$$

Next we call upon the Cauchy–Schwarz property $|x^{\mathsf{T}}y| \le \|x\|_\star\,\|y\|$ from Prob. 1.24 in terms of the original and dual norms. Using this property we have

$$\begin{aligned} &\left|\left(s_u(\boldsymbol{w}_{n-1};\boldsymbol{\gamma}(n),\boldsymbol{h}_n)\right)^{\mathsf{T}}(\boldsymbol{w}_{n-1}-\boldsymbol{w}_n)\right| \\ &\le \|s_u(\boldsymbol{w}_{n-1};\boldsymbol{\gamma},\boldsymbol{h}_n)\|_\star\,\|\boldsymbol{w}_{n-1}-\boldsymbol{w}_n\| \overset{(21.76)}{\le} \delta\|\boldsymbol{w}_{n-1}-\boldsymbol{w}_n\| \end{aligned} \tag{21.88}$$

Substituting into (21.87) we get

$$B \le \mu\delta \, \|\boldsymbol{w}_{n-1} - \boldsymbol{w}_n\| - \frac{\nu}{2}\|\boldsymbol{w}_n - \boldsymbol{w}_{n-1}\|^2 \tag{21.89}$$

Consider the quadratic function $f(x) = \mu\delta x - \frac{\nu}{2}x^2$. Its maximum occurs at $x = \mu\delta/\nu$ and the maximum value is $(\mu\delta)^2/2\nu$. Therefore, we can bound B by

$$B \le (\mu\delta)^2/2\nu \tag{21.90}$$

and (21.86) becomes

$$\begin{aligned} E(\boldsymbol{w}_{n-1}) - E(w^\star) \;\le\; & \boldsymbol{g}_n(\boldsymbol{w}_{n-1})\widetilde{\boldsymbol{w}}_{n-1} + (\mu\delta)^2/2\nu \; + \\ & \frac{1}{\mu}\underbrace{\Big(D_\phi(w^\star, \boldsymbol{w}_{n-1}) - D_\phi(w^\star, \boldsymbol{w}_n)\Big)}_{=\text{ telescoping sequence}} \end{aligned} \tag{21.91}$$

Conditioning on $\boldsymbol{w}_{n-1}$, computing expectations, and using (21.82) we get

$$\begin{aligned} & \mathbb{E}\left(E(\boldsymbol{w}_{n-1}) \,|\, \boldsymbol{w}_{n-1}\right) - E(w^\star) \\ & \le\; (\mu\delta)^2/2\nu \;+\; \frac{1}{\mu}\mathbb{E}\left(D_\phi(w^\star, \boldsymbol{w}_{n-1}) - D_\phi(w^\star, \boldsymbol{w}_n) \,|\, \boldsymbol{w}_{n-1}\right) \end{aligned} \tag{21.92}$$

Taking expectations again we remove the conditioning on $\boldsymbol{w}_{n-1}$ to obtain

$$\mathbb{E}\left(E(\boldsymbol{w}_{n-1}) - E(w^\star)\right) \;\le\; (\mu\delta)^2/2\nu \;+ \frac{1}{\mu}\mathbb{E}\left(D_\phi(w^\star, \boldsymbol{w}_{n-1}) - D_\phi(w^\star, \boldsymbol{w}_n)\right) \tag{21.93}$$

Summing from $n = 0$ to $n = L-1$, and dividing by L we find

$$\begin{aligned} \mathbb{E}\left(\frac{1}{L}\sum_{n=0}^{L-1} E(\boldsymbol{w}_{n-1})\right) \;-\; E(w^\star) \;&\le\; \frac{1}{L\mu}\mathbb{E}\, D_\phi(w^\star, \boldsymbol{w}_{-1}) + \frac{\mu\delta^2}{2\nu} \\ &\overset{(a)}{\le} \frac{W^2}{L\mu} + \frac{\mu\delta^2}{2\nu} \\ &\overset{(b)}{\le} W\delta\sqrt{\frac{2}{L\nu}} \end{aligned} \tag{21.94}$$

where step (a) uses (21.79) and step (b) uses (21.78). Applying property (8.82) to the convex function $E(w)$ we conclude that (21.80) holds.

■

21.7 COMMENTARIES AND DISCUSSION

Stochastic proximal and mirror-descent algorithms. There is an extensive literature on proximal gradient and mirror-descent methods, including on their stochastic versions along with convergence analyses and variations to enhance their performance. Some works that deal with the stochastic proximal gradient algorithm and accelerated versions include Bottou and LeCun (2005), Kwok, Hu, and Pan (2009), Juditsky and Nesterov (2010), Shalev-Shwartz *et al.* (2011), Juditsky and Nemirovski (2011a, b), Ghadimi and Lan (2012), Shalev-Shwartz and Zhang (2013), Xiao and Zhang (2014), Nassif *et al.* (2016), Schmidt, Le Roux, and Bach (2017), Atchadé, Fort, and Moulines (2017), Rosasco, Villa, and Vu (2020), and Alghunaim *et al.* (2021). Other works that deal with the stochastic mirror-descent algorithm and variations thereof include Nemirovski and Yudin (1983), Nemirovski *et al.* (2009), Xiao (2010), Juditsky, Nemirovski, and Tauvel (2011), Juditsky and Nemirovski (2011a, b), Lan (2012), Duchi *et al.* (2012), Cesa-Bianchi *et al.* (2012), Shamir and Zhang (2013), Zhou *et al.* (2017), Mertikopoulos and Staudigl (2018), and Azizan and Hassibi (2019). The proof of Theorem 21.6 is an extension to the stochastic case of the argument from Bubeck (2015).

Comparison of performance. The results from this chapter and the previous one show that under instantaneous gradient approximations, the *stochastic* subgradient and proximal algorithms deliver similar performance under strong-convexity conditions, namely, the average excess risk evolves according to the dynamics (see Theorems 20.1 and 21.1):

(*using instantaneous gradient approximations and constant step size*)

$$\text{(\textbf{stochastic subgradient})}\quad \mathbb{E}\,P(\boldsymbol{w}_n) - P(w^\star) \;\le\; O(\lambda^{n/2}) + O(\sqrt{\mu}) \tag{21.95a}$$

$$\text{(\textbf{stochastic proximal})}\quad \mathbb{E}\,P(\boldsymbol{w}_n) - P(w^\star) \;\le\; O(\lambda^{n/2}) + O(\sqrt{\mu}) \tag{21.95b}$$

with the same convergence rate and the size of the limiting region being $O(\sqrt{\mu})$ in both cases. When a vanishing step size of the form $\mu(n) = \tau/(n+1)$ is used, both algorithms converge at the rate of $1/\sqrt{n}$, as was shown in Theorems 20.4 and 21.4:

(*using instantaneous gradient approximations and vanishing step size*)

$$\text{(\textbf{stochastic subgradient})}\quad \mathbb{E}\,P(\boldsymbol{w}_n) - P(w^\star) \;\le\; O(1/\sqrt{n}) \tag{21.96a}$$

$$\text{(\textbf{stochastic proximal})}\quad \mathbb{E}\,P(\boldsymbol{w}_n) - P(w^\star) \;\le\; O(1/\sqrt{n}) \tag{21.96b}$$

The results (21.95a)–(21.95b) under constant step sizes are different from the behavior derived for both algorithms when the actual gradients are used instead of their approximations. In this latter case, the proximal gradient algorithm was shown to exhibit superior performance. Again under strong-convexity conditions, we determined in Theorems 14.1 and 15.1 that

(*using true gradients and constant step size*)

$$\text{(\textbf{subgradient algorithm})}\quad P(w_n) - P(w^\star) \;\le\; O(\lambda^{n/2}) \;+\; O(\sqrt{\mu}) \tag{21.97a}$$

$$\text{(\textbf{proximal gradient})}\quad P(w_n) - P(w^\star) \;\le\; O(\lambda^{n}) \tag{21.97b}$$

PROBLEMS

21.1 Consider the empirical risk minimization problem (21.4a) and apply the stochastic proximal algorithm (21.1) with instantaneous gradient approximations. The samples $(\boldsymbol{\gamma}(n), \boldsymbol{h}_n)$ are selected uniformly at random from the dataset $\{\gamma(m), h_m\}$. Show that the second-order moment of the gradient noise satisfies (21.22b) with

$$\beta_g^2 = 12\delta^2, \qquad \sigma_g^2 \le \frac{10}{N}\sum_{m=0}^{N-1} \|\nabla_{w^{\mathsf{T}}} Q_u(w^\star; \gamma(m), h_m)\|^2$$

21.2 Consider the stochastic risk minimization problem (21.4b) and apply the stochastic proximal algorithm (21.1) with instantaneous gradient approximations. The samples $(\boldsymbol{\gamma}(n), \boldsymbol{h}_n)$ stream in independently of each other. Show that the second-order moment of the gradient noise satisfies (21.22b) with

$$\beta_g^2 = 12\delta^2, \qquad \sigma_g^2 \le 8\|\nabla_{w^{\mathsf{T}}} E(w^o)\|^2 + 2\,\mathbb{E}\,\|\nabla_{w^{\mathsf{T}}} Q_u(w^o; \boldsymbol{\gamma}, \boldsymbol{h})\|^2$$

where the expectation is over the distribution of $\{\boldsymbol{\gamma}, \boldsymbol{h}\}$.

21.3 Establish the convergence results stated in Theorem 21.2 for the mini-batch implementation of the stochastic proximal algorithm.

21.4 Consider the stochastic projection algorithm (21.69). We relax the conditions on $E(w)$ and assume only that it is convex. We also allow it to be nonsmooth, in which case the gradient of $Q_u(w;\cdot)$ is replaced by a subgradient vector denoted by $s_u(w;\gamma,h) \in \partial_{w^\mathsf{T}} Q_u(w;\gamma,h)$. The algorithm becomes

$$\begin{cases} \boldsymbol{z}_n = \boldsymbol{w}_{n-1} - \mu\, s_u(\boldsymbol{w}_{n-1};\boldsymbol{\gamma}(n),\boldsymbol{h}_n),\ n \geq 0 \\ \boldsymbol{w}_n = \mathcal{P}_C(\boldsymbol{z}_n) \end{cases}$$

Assume further that $Q_u(w;\cdot)$ is Lipschitz, i.e., for any (γ,γ',h,h'):

$$\|Q_u(w_2;\gamma,h) - Q_u(w_1,;\gamma',h')\| \leq \delta\|w_2 - w_1\|, \quad w_1, w_2 \in \text{dom}(E)$$

relative to the Euclidean norm. We know from property (10.41) that this is equivalent to $\|s_u(w;\cdot)\| \leq \delta$. Assume the initial condition satisfies $\mathbb{E}\,\|\widetilde{\boldsymbol{w}}_{-1}\| \leq W^2$. Adjust the argument that led to (15.69) to show that if we run the algorithm for L iterations using $\mu = W/\delta\sqrt{L}$, then

$$\mathbb{E}\left\{E\left(\frac{1}{L}\sum_{n=0}^{L-1}\boldsymbol{w}_n\right)\right\} - E(w^\star) = O(1/\sqrt{L})$$

21.5 Consider the following modification of the stochastic proximal algorithm (21.1) where a smoothing step is added after the proximal step:

$$\begin{cases} \boldsymbol{z}_n = \boldsymbol{w}_{n-1} - \mu\,\widehat{\nabla_{w^\mathsf{T}} E}(\boldsymbol{w}_{n-1}),\ n \geq 0 \\ \boldsymbol{w}_n = \text{prox}_{\mu q}(\boldsymbol{z}_n) \\ \bar{\boldsymbol{w}}_n = \eta\bar{\boldsymbol{w}}_{n-1} + (1-\eta)\boldsymbol{w}_n \end{cases}$$

where $\eta \in (0,1)$ and $\bar{\boldsymbol{w}}_n$ is the output. Derive expressions for the convergence rate of the algorithm and the limiting steady-state MSE and average excess risk.

21.6 Refer to the stochastic proximal recursion (21.1) and assume that the step size is a random parameter with mean $\mathbb{E}\,\boldsymbol{\mu} = \bar{\mu}$ and variance σ_μ^2. Assume $\boldsymbol{\mu}$ is independent of all other random variables. Follow the arguments used in the proof of Theorem 21.1 and show how the results of the theorem would need to be adjusted.

21.7 Refer to the stochastic proximal recursion (21.1) and assume that the step size $\boldsymbol{\mu}$ is a Bernoulli random variable that is equal to μ with probability p and is zero with probability $1-p$. That is, the recursion is active p fraction of the times. Assume $\boldsymbol{\mu}$ is independent of all other random variables. Follow the arguments used in the proof of Theorem 21.1 and show how the results of the theorem would need to be adjusted.

21.8 Extend the arguments used in the proof of Theorem 21.1 to establish the convergence of the randomized coordinate-descent recursion (21.43) under the separability condition (21.44).

21.9 Write down a stochastic version for the lazy mirror-descent algorithm based on instantaneous subgradient approximations. Repeat the arguments used in the proof of Theorem 21.6 to examine its convergence behavior.

21.10 Write down a stochastic version for the mirror prox algorithm based on instantaneous subgradient approximations. Repeat the arguments used in the proof of Theorem 21.6 to examine its convergence behavior.

REFERENCES

Alghunaim, S. A., E. K. Ryu, K. Yuan, and A. H. Sayed (2021), "Decentralized proximal gradient algorithms with linear convergence rates," *IEEE Trans. Aut. Control*, vol. 66, no. 6, pp. 2787–2794.

Atchadé, Y. F., G. Fort, and E. Moulines (2017), "On perturbed proximal gradient algorithms," *J. Mach. Learn. Res.*, vol. 18, pp. 1–33.

Azizan, N. and B. Hassibi (2019), "A characterization of stochastic mirror descent algorithms and their convergence properties," *Proc. IEEE ICASSP*, pp. 5167–5171, Brighton.

Bottou, L. and Y. LeCun (2005), "On-line learning for very large data sets," *App. Stoch. Model. Bus. Ind.*, vol. 21, no. 2, pp. 137–151.

Bubeck, S. (2015), *Convex Optimization: Algorithms and Complexity*, Foundations and Trends in Machine Learning, NOW Publishers, vol. 8, no. 3–4, pp. 231–357.

Cesa-Bianchi, N., P. Gaillard, G. Lugosi, and G. Stoltz (2012), "Mirror descent meets fixed share (and feels no regret)," *Proc. Advances Neural Information Processing Systems* (NIPS), pp. 980–988, Lake Tahoe, NV.

Duchi, J., A. Agarwal, M. Johansson, and M. I. Jordan (2012), "Ergodic mirror descent," *SIAM J. Optim.*, vol. 22, pp. 1549–1578.

Ghadimi, S. and G. Lan (2012), "Optimal stochastic approximation algorithms for strongly convex stochastic composite optimization I: A generic algorithmic framework," *SIAM J. Optim.*, vol. 22, no. 4, pp. 1469–1492.

Juditsky, A. and A. S. Nemirovski (2011a), "First order methods for nonsmooth convex large-scale optimization, part I: General purpose methods," in *Optimization for Machine Learning*, S. Sra, S. Nowozin, and S. J. Wright, editors, pp. 121–146, MIT Press.

Juditsky, A. and A. S. Nemirovski (2011b), "First order methods for nonsmooth convex large-scale optimization, part II: Utilizing problem's structure," in *Optimization for Machine Learning*, S. Sra, S. Nowozin, and S. J. Wright, editors, pp. 149–183, MIT Press.

Juditsky, A., A. S. Nemirovski, and C. Tauvel (2011), "Solving variational inequalities with stochastic mirror-prox algorithm," *Stoch. Syst.*, vol. 1, no. 1, pp. 17–58.

Juditsky, A. and Y. Nesterov (2010), "Primal–dual subgradient methods for minimizing uniformly convex functions," available at https://hal.archives-ouvertes.fr/hal-00508933. Also available at https://arxiv.org/abs/1401.1792.

Kwok, J. T., C. Hu, and W. Pan (2009), "Accelerated gradient methods for stochastic optimization and online learning," *Proc. Advances Neural Information Processing Systems*, pp. 781–789, Vancouver, BC.

Lan, G. (2012), "An optimal method for stochastic composite optimization," *Math. Program.*, Ser. A, vol. 133, nos. 1–2, pp. 365–397.

Mertikopoulos, P. and M. Staudigl (2018), "On the convergence of gradient-like flows with noisy gradient input," *SIAM J. Optim.*, vol. 28, pp. 163–197.

Nassif, R., A. Ferrari, C. Richard, and A. H. Sayed (2016), "Proximal multitask learning over networks with sparsity-inducing coregularization," *IEEE Trans. Signal Process.*, vol. 64, no. 23, pp. 6329–6344.

Nemirovski, A. S., A. Juditsky, G. Lan, and A. Shapiro (2009), "Robust stochastic approximation approach to stochastic programming," *SIAM J. Optim.*, vol. 19, no. 4, pp. 1574–1609.

Nemirovski, A. S. and D. B. Yudin (1983), *Problem Complexity and Method Efficiency in Optimization*, Wiley.

Rosasco, L., S. Villa, and B. C. Vu (2020), "Convergence of stochastic proximal gradient algorithm," *Appl. Math. Optim.*, vol. 82, pp. 891–917.

Schmidt, M., N. Le Roux, and F. Bach (2017), "Convergence rates of inexact proximal gradient methods for convex optimization." *Proc. Advances Neural Information Processing Systems* (NIPS), pp. 1458–1466, Long Beach, CA.

Shalev-Shwartz, S., Y. Singer, N. Srebro, and A. Cotter (2011), "Pegasos: Primal estimated sub-gradient solver for SVM," *Math. Program.*, Ser. B, vol. 127, no. 1, pp. 3–30.

Shalev-Shwartz, S. and T. Zhang (2013), "Stochastic dual coordinate ascent methods for regularized loss minimization," *J. Mach. Learn. Res.*, vol. 14, pp. 567–599.

Shamir, O. and T. Zhang (2013), "Stochastic gradient descent for non-smooth optimization: Convergence results and optimal averaging schemes," *Proc. Int. Conf. Machine Learning* (PMLR) vol. 28, no.1, pp. 71–79, Atlanta, GA.

Xiao, L. (2010), "Dual averaging methods for regularized stochastic learning and online optimization," *J. Mach. Learn. Res.*, vol. 11, pp. 2543–2596.

Xiao, L. and T. Zhang (2014), "A proximal stochastic gradient method with progressive variance reduction," *SIAM J. Optim.*, vol. 24, no. 4, pp. 2057–2075.

Zhou, Z., P. Mertikopoulos, N. Bambos, S. Boyd, and P. W. Glynn (2017), "Stochastic mirror descent in variationally coherent optimization problems," *Proc. Advances Neural Information Processing Systems* (NIPS), pp. 7043–7052, Long Beach, CA.

22 Variance-Reduced Methods I: Uniform Sampling

Stochastic optimization methods replace true gradients by instantaneous or mini-batch approximations, which introduce gradient noise into the operation of the algorithms. We explained in the previous chapters that the variance of the noise process, denoted by $\mathbb{E}\,(\|\boldsymbol{g}_n(\boldsymbol{w}_{n-1}\|^2 \,|\, \boldsymbol{w}_{n-1})$, is bounded by two components: a relative noise component $\beta_g^2\|\widetilde{w}_{n-1}\|^2$ and an absolute noise component σ_g^2. The presence of this last factor causes the mean-square error (MSE), $\mathbb{E}\,\|\widetilde{\boldsymbol{w}}_n\|^2$, and the mean excess risk, $\mathbb{E}\,P(\boldsymbol{w}_n) - P(w^\star)$, to converge to small neighborhoods, usually of size $O(\mu\sigma_g^2)$, rather than to zero under constant step-size learning. In this and the next chapter, we introduce the class of variance-reduced methods. These algorithms use different approximations for the gradient (or subgradient) vectors and ensure convergence of the iterates $\boldsymbol{w}_n$ to the *exact* minimizer, $w^\star$. The techniques are suitable for *empirical* risk optimization problems under both uniform sampling (i.e., sampling with replacement) and random reshuffling (i.e., sampling without replacement). We focus on uniform sampling in this chapter and study random reshuffling in the next chapter. In all, we will study three types of variance-reduction algorithms known as SAGA (stochastic average-gradient algorithm), SVRG (stochastic variance-reduced gradient algorithm), and AVRG (amortized variance-reduced gradient algorithm). We also consider both cases of smooth and nonsmooth risk functions.

22.1 PROBLEM SETTING

We motivate variance-reduced methods by focusing first, at some length, on smooth (first-order differentiable) risk functions. Later, in a future section, we explain how the constructions and derivations extend to nonsmooth risks.

Consider the traditional stochastic gradient implementation:

$$\boldsymbol{w}_n = \boldsymbol{w}_{n-1} - \mu\,\widehat{\nabla_{w^\mathsf{T}} P}(\boldsymbol{w}_{n-1}),\ n \geq 0 \tag{22.1}$$

with constant μ for the solution of convex optimization problems in empirical risk form:

$$w^\star \stackrel{\Delta}{=} \underset{w\in\mathbb{R}^M}{\text{argmin}} \left\{ P(w) \stackrel{\Delta}{=} \frac{1}{N}\sum_{m=0}^{N-1} Q(w;\gamma(m),h_m) \right\} \tag{22.2}$$

where $P(w)$ is a smooth risk, $Q(w,\cdot)$ is a convex loss, and $\{\gamma(m), h_m\}$ refer to the collection of N data points with $\gamma(m) \in \mathbb{R}$ and $h_m \in \mathbb{R}^M$.

22.1.1 Conditions on Risk and Loss Functions

In Section 18.2 we introduced the following conditions for empirical risk minimization problems of the form (22.2):

(A1) (**Strongly convex risk**). $P(w)$ is ν-strongly convex and first-order differentiable, namely, for every $w_1, w_2 \in \text{dom}(P)$ and some $\nu > 0$:

$$P(w_2) \geq P(w_1) + (\nabla_{w^\mathsf{T}} P(w_1))^\mathsf{T}(w_2 - w_1) + \frac{\nu}{2}\|w_2 - w_1\|^2 \quad (22.3a)$$

(A2) (**δ-Lipschitz loss gradients**). The gradient vectors of $Q(w,\cdot)$ are δ-Lipschitz regardless of the data argument, i.e.,

$$\|\nabla_w Q(w_2; \gamma(k), h_k) - \nabla_w Q(w_1; \gamma(\ell), h_\ell)\| \leq \delta\,\|w_2 - w_1\| \quad (22.3b)$$

for any $w_1, w_2 \in \text{dom}(Q)$, any $0 \leq k, \ell \leq N-1$, and with $\delta \geq \nu$. We explained that condition (22.3b) implies that the gradient of $P(w)$ is itself δ-Lipschitz:

$$\|\nabla_w P(w_2) - \nabla_w P(w_1)\| \leq \delta\|w_2 - w_1\| \quad (22.4)$$

Note that under conditions (**A1**,**A2**), the following two conditions hold:

$$\textbf{(P1)}\ \nu\text{-strong convexity of } P(w) \text{ as in (22.3a)} \quad (22.5a)$$

$$\textbf{(P2)}\ \delta\text{-Lipschitz gradients for } P(w) \text{ as in (22.4)} \quad (22.5b)$$

Moreover, we know from the earlier results (8.29) and (10.20) derived for strongly convex and δ-smooth functions that conditions **P1** and **P2** imply respectively:

$$\textbf{(P1)} \implies \frac{\nu}{2}\|\widetilde{w}\|^2 \leq P(w) - P(w^\star) \leq \frac{1}{2\nu}\|\nabla_w P(w)\|^2 \quad (22.6a)$$

$$\textbf{(P2)} \implies \frac{1}{2\delta}\|\nabla_w P(w)\|^2 \leq P(w) - P(w^\star) \leq \frac{\delta}{2}\|\widetilde{w}\|^2 \quad (22.6b)$$

where $\widetilde{w} = w^\star - w$. The upper bounds in both expressions indicate that whenever we bound $\|\widetilde{w}\|^2$ we will also be automatically bounding the excess risk, $P(w) - P(w^\star)$.

22.1.2 Traditional Gradient Noise

In previous chapters we considered constructions where the gradient search direction was approximated by using either instantaneous or mini-batch calculations, namely,

$$\textbf{(instantaneous)}\ \widehat{\nabla_{w^{\mathsf{T}}} P}(w) = \nabla_{w^{\mathsf{T}}} Q(w;\boldsymbol{\gamma},\boldsymbol{h}) \tag{22.7a}$$

$$\textbf{(mini-batch)}\ \widehat{\nabla_{w^{\mathsf{T}}} P}(w) = \frac{1}{B}\sum_{b=0}^{B-1} \nabla_{w^{\mathsf{T}}} Q(w;\boldsymbol{\gamma}(b),\boldsymbol{h}_b) \tag{22.7b}$$

where the boldface notation $(\boldsymbol{\gamma},\boldsymbol{h})$ or $(\boldsymbol{\gamma}(b),\boldsymbol{h}_b)$ refers to data samples selected at random (with or without replacement) from the given dataset $\{\gamma(m),h_m\}$. The difference between the true gradient and its approximation was referred to as *gradient noise* and denoted by

$$\boldsymbol{g}(w) \triangleq \widehat{\nabla_{w^{\mathsf{T}}} P}(w) - \nabla_{w^{\mathsf{T}}} P(w) \tag{22.8}$$

We will focus almost exclusively on instantaneous gradient approximations in this chapter, although the discussion can be extended to mini-batch solutions in a straightforward manner, as already shown on multiple occasions in the previous chapters.

We showed in Section 18.3 that, as a result of the Lipschitz condition **A2**, the first- and second-order moments of the gradient noise process (22.8) satisfy the following two properties denoted by **G1** and **G2** for ease of reference:

$$\textbf{(G1)}\ \mathbb{E}\left(\boldsymbol{g}_n(\boldsymbol{w}_{n-1})\,|\,\boldsymbol{w}_{n-1}\right) = 0 \tag{22.9a}$$

$$\textbf{(G2)}\ \mathbb{E}\left(\|\boldsymbol{g}_n(\boldsymbol{w}_{n-1})\|^2\,|\,\boldsymbol{w}_{n-1}\right) \leq \beta_g^2\|\widetilde{w}_{n-1}\|^2 + \sigma_g^2 \tag{22.9b}$$

for some nonnegative constants $\{\beta_g^2,\sigma_g^2\}$ that are independent of $\widetilde{w}_{n-1}$. Condition (**G1**) refers to the zero-mean property of the gradient noise while (**G2**) refers to its bounded "variance." We explained earlier that results (22.9a)–(22.9b) hold for instantaneous and mini-batch gradient approximations, regardless of whether the samples are streaming in independently of each other, sampled uniformly with replacement, sampled without replacement, or selected under importance sampling. The *only exception* is that the zero-mean property (22.9a) will *not* hold for the instantaneous gradient implementation when the samples are selected without replacement.

22.1.3 Size of Error Variance

For operation under uniform sampling (i.e., sampling with replacement), we established in (19.16a)–(19.16b) that

$$\limsup_{n\to\infty}\ \mathbb{E}\,\|\widetilde{\boldsymbol{w}}_n\|^2 = O(\mu\sigma_g^2/2\nu) \tag{22.10a}$$

$$\limsup_{\mathbb{E}\,n\to\infty}\ \Big(\mathbb{E}\,P(\boldsymbol{w}_n) - P(w^\star)\Big) = O(\mu\sigma_g^2/4) \tag{22.10b}$$

where ν is the strong-convexity factor. Observe that $\mathbb{E}\,\|\widetilde{\boldsymbol{w}}_n\|^2$ does not converge to zero but rather approaches a small neighborhood of size $O(\mu)$ at an exponential rate λ^n. Similarly, the average risk value $\mathbb{E}\,P(\boldsymbol{w}_n)$ converges at the same exponential rate to a small $O(\mu)$-neighborhood around $P(w^\star)$. In both cases, the size of the neighborhood is dependent on the absolute gradient noise component, σ_g^2. The smaller the value of σ_g^2 is, the closer $\mathbb{E}\,\|\widetilde{\boldsymbol{w}}_n\|^2$ and $\mathbb{E}\,P(\boldsymbol{w}_n)$ will be to their optimal values. A similar conclusion holds for operation under random reshuffling (i.e., for sampling without replacement) – recall (19.186) and (19.197).

22.2 NAÏVE STOCHASTIC GRADIENT ALGORITHM

Let us examine the origin of the $O(\mu)$ steady-state bias and the absolute gradient noise term σ_g^2 more closely under *uniform sampling*. In the next chapter we consider operation under random reshuffling. Assuming instantaneous gradient approximations, we rewrite (22.1) in the following form, where we make explicit the random nature of the sample index chosen at iteration n:

$$\begin{array}{l}
\textbf{repeat until convergence over } n \geq 0\textbf{:} \\
\quad \text{select a random index } \boldsymbol{u} \text{ uniformly from } 0 \leq u \leq N-1 \\
\quad \text{consider the data sample } (\gamma(\boldsymbol{u}), h_{\boldsymbol{u}}) \\
\quad \text{update } \boldsymbol{w}_n = \boldsymbol{w}_{n-1} - \mu\, \nabla_{w^{\mathsf{T}}} Q(\boldsymbol{w}_{n-1}; \gamma(\boldsymbol{u}), h_{\boldsymbol{u}}) \\
\textbf{end}
\end{array} \tag{22.11}$$

For convenience, we will be using the letter $\boldsymbol{u}$ in this chapter to refer to the random index selected at iteration n (we used $\boldsymbol{\sigma}$ in previous chapters). Thus, the notation $(\gamma(\boldsymbol{u}), h_{\boldsymbol{u}})$ denotes the data point sampled uniformly at random (with replacement) from the given dataset $\{\gamma(m), h_m\}$ of size N. To examine the origin of the factor σ_g^2, we refer back to the derivation (18.34) that led to expression (18.35b) for σ_g^2. Note from the argument leading to (18.34) that in the first step we added and subtracted $\nabla_{w^{\mathsf{T}}} Q(w^\star)$. This step led to the appearance of the term $2\|\nabla_{w^{\mathsf{T}}} Q(w^\star)\|^2$ in (18.34) and ultimately to the expression for σ_g^2 in (18.35b). One main feature of *variance-reduced techniques* is that they will employ other gradient approximations in (22.11) in such a way that the driving term σ_g^2 will end up disappearing from the second-order moment of the gradient noise process.

Naïve gradient approximation

We illustrate one possibility in this section, which we refer to as a *naïve solution* because it will only serve as motivation for more sophisticated constructions. This naïve solution is not realizable since it requires knowledge of the unknown minimizer $w^\star$, but is helpful to drive the discussion. Thus, assume we modify (22.11) in the following manner:

(Naïve stochastic gradient algorithm; unrealizable)

repeat until convergence over $n \geq 0$**:**

$$\left|\begin{array}{l}\text{select a random index } \boldsymbol{u} \text{ uniformly from } 0 \leq u \leq N-1 \\ \text{consider the data sample } (\gamma(\boldsymbol{u}), h_{\boldsymbol{u}}) \\ \text{update } \boldsymbol{w}_n = \boldsymbol{w}_{n-1} - \mu\Big(\nabla_{w^\mathsf{T}} Q(\boldsymbol{w}_{n-1}) - \nabla_{w^\mathsf{T}} Q(w^\star)\Big)\end{array}\right. \tag{22.12}$$

end

In this description, we are not showing all the arguments for the loss function to keep the notation compact. Specifically, we should have written

$$Q(\boldsymbol{w}_{n-1}) \stackrel{\Delta}{=} Q(\boldsymbol{w}_{n-1}; \gamma(\boldsymbol{u}), h_{\boldsymbol{u}}), \qquad Q(w^\star) \stackrel{\Delta}{=} Q(w^\star; \gamma(\boldsymbol{u}), h_{\boldsymbol{u}}) \tag{22.13}$$

Construction (22.12) corresponds to centering the original gradient approximation around the loss value at $w^\star$ and using the following instantaneous gradient approximation in (22.1):

$$\widehat{\nabla_{w^\mathsf{T}} P}(\boldsymbol{w}_{n-1}) = \nabla_{w^\mathsf{T}} Q(\boldsymbol{w}_{n-1}; \gamma(\boldsymbol{u}), h_{\boldsymbol{u}}) - \nabla_{w^\mathsf{T}} Q(w^\star; \gamma(\boldsymbol{u}), h_{\boldsymbol{u}}) \tag{22.14}$$

Removal of absolute noise term

By doing so, the resulting gradient noise process continues to have zero mean, but its variance expression will change in a favorable way. Note first that since $\boldsymbol{u}$ takes integer values in the range $0 \leq u \leq N-1$ with equal probability, then

$$\begin{aligned}
&\mathbb{E}\left(\widehat{\nabla_{w^\mathsf{T}} P}(\boldsymbol{w}_{n-1}) \,|\, \boldsymbol{w}_{n-1}\right) \\
&= \mathbb{E}\left(\nabla_{w^\mathsf{T}} Q(\boldsymbol{w}_{n-1}; \gamma(\boldsymbol{u}), h_{\boldsymbol{u}}) \,|\, \boldsymbol{w}_{n-1}\right) - \mathbb{E}\left(\nabla_{w^\mathsf{T}} Q(w^\star; \gamma(\boldsymbol{u}), h_{\boldsymbol{u}}) \,|\, \boldsymbol{w}_{n-1}\right) \\
&= \frac{1}{N}\sum_{u=0}^{N-1} \nabla_{w^\mathsf{T}} Q(w_{n-1}; \gamma(u), h_u) - \frac{1}{N}\sum_{u=0}^{N-1} \nabla_{w^\mathsf{T}} Q(w^\star; \gamma(u), h_u) \\
&= \nabla_{w^\mathsf{T}} P(w_{n-1}) - \underbrace{\nabla_{w^\mathsf{T}} P(w^\star)}_{=0} \\
&= \nabla_{w^\mathsf{T}} P(w_{n-1})
\end{aligned} \tag{22.15}$$

where in the second equality we used the fact that the loss function assumes each of the values $Q(w_{n-1}; \gamma(u), h_u)$ with probability $1/N$ under uniform sampling. It follows that the zero-mean property (**G1**) continues to hold. On the other hand, if we repeat the derivation leading to (18.34), we find that we do not need to add and subtract $\nabla_{w^\mathsf{T}} Q(w^\star)$ anymore in the first step of that argument since we now have:

$$
\begin{aligned}
\|\boldsymbol{g}_n(\boldsymbol{w}_{n-1})\|^2 &\stackrel{\Delta}{=} \left\|\widehat{\nabla_{w^\mathsf{T}} P}(\boldsymbol{w}_{n-1}) - \nabla_{w^\mathsf{T}} P(\boldsymbol{w}_{n-1})\right\|^2 \\
&\stackrel{(22.14)}{=} \|\nabla_{w^\mathsf{T}} Q(\boldsymbol{w}_{n-1}) - \nabla_{w^\mathsf{T}} Q(w^\star) - \nabla_{w^\mathsf{T}} P(\boldsymbol{w}_{n-1})\|^2 \\
&\stackrel{(a)}{\leq} 2\,\|\nabla_{w^\mathsf{T}} Q(\boldsymbol{w}_{n-1}) - \nabla_{w^\mathsf{T}} Q(w^\star)\|^2 + 2\,\|\nabla_{w^\mathsf{T}} P(\boldsymbol{w}_{n-1})\|^2 \\
&\stackrel{(22.3b)}{\leq} 2\delta^2\|\widetilde{\boldsymbol{w}}_{n-1}\|^2 + 2\,\|\nabla_{w^\mathsf{T}} P(w^\star) - \nabla_{w^\mathsf{T}} P(\boldsymbol{w}_{n-1})\|^2 \\
&\stackrel{(22.3b)}{\leq} 2\delta^2\|\widetilde{\boldsymbol{w}}_{n-1}\|^2 + 2\delta^2\|\widetilde{\boldsymbol{w}}_{n-1}\|^2 \\
&= 4\delta^2\|\widetilde{\boldsymbol{w}}_{n-1}\|^2
\end{aligned}
\tag{22.16}
$$

where $\widetilde{\boldsymbol{w}}_n = w^\star - \boldsymbol{w}_n$. In the second equality we removed the arguments $(\boldsymbol{\gamma}(n), \boldsymbol{h}_n)$ to simplify the notation, and in step (a) we applied the Jensen inequality that $\|a+b\|^2 \leq 2\|a\|^2 + 2\|b\|^2$ for any vectors (a, b). We conclude therefore that the variance property (**G2**) is modified to the following (see Prob. 22.1):

$$
\boxed{\textbf{(G2')} \quad \mathbb{E}\left(\|\boldsymbol{g}_n(\boldsymbol{w}_{n-1})\|^2 \,|\, \boldsymbol{w}_{n-1}\right) \;\leq\; \beta_g^2\|\widetilde{w}_{n-1}\|^2}
\tag{22.17}
$$

where $\beta_g^2 = 4\delta^2$. Observe that in comparison to (**G2**), the driving factor σ_g^2 is now zero! We can then repeat the proof of Theorem 19.1 to establish the following convergence result.

THEOREM 22.1. (Convergence of naïve algorithm under uniform sampling) *Consider the naïve stochastic gradient recursion (22.12) used to seek the minimizer for the empirical risk minimization problem (22.2). The risk and loss functions are assumed to satisfy conditions (**A1**,**A2**). For step-size values satisfying (i.e., for μ small enough):*

$$
\mu < \frac{2\nu}{\delta^2 + \beta_g^2}
\tag{22.18}
$$

it holds that $\mathbb{E}\,\|\widetilde{\boldsymbol{w}}_n\|^2$ and $\mathbb{E}\,P(\boldsymbol{w}_n)$ converge to zero and $P(w^\star)$, respectively, at an exponential rate, according to the recursions

$$
\mathbb{E}\,\|\widetilde{\boldsymbol{w}}_n\|^2 \leq O(\lambda^n)
\tag{22.19a}
$$

$$
\mathbb{E}\,P(\boldsymbol{w}_n) - P(w^\star) \leq O(\lambda^n)
\tag{22.19b}
$$

where

$$
\lambda = 1 - 2\nu\mu + (\delta^2 + \beta_g^2)\mu^2 \in [0, 1)
\tag{22.20}
$$

Proof: See Prob. 22.2. ∎

22.3 STOCHASTIC AVERAGE-GRADIENT ALGORITHM (SAGA)

The main difficulty with the naïve solution (22.12) is that the correction term $\nabla_{w^\mathsf{T}} Q(w^\star)$ requires knowledge of the minimizer $w^\star$, which is of course unavailable. One path forward is to replace $w^\star$ by an approximation for it. We will denote the approximation that is used at iteration n by $\boldsymbol{\phi}_{n-1} \in \mathbb{R}^M$, i.e.,

$$\boldsymbol{\phi}_{n-1} \stackrel{\Delta}{=} \widehat{w^\star} \quad \text{(approximation used at } n\text{th iteration)} \tag{22.21}$$

We explain in the following that the value used for $\boldsymbol{\phi}_{n-1}$ will depend on which data sample $\boldsymbol{u}$ is selected at the nth iteration. For this reason, we will be writing $\boldsymbol{\phi}_{n-1,\boldsymbol{u}}$, with an additional index $\boldsymbol{u}$, to highlight the dependency on $\boldsymbol{u}$. In addition, moving forward, we will incorporate multiple runs (epochs) into the description of the variance-reduced algorithms. We will use the superscript k to refer to the kth run (or epoch). In this way, the notation $\boldsymbol{\phi}^k_{n-1,\boldsymbol{u}}$ will be referring to the estimate used at iteration n of the kth run and its value will depend on which data index $\boldsymbol{u}$ was selected at that iteration. For reasons explained in the following, we will refer to the ϕ-vectors as *memory* variables.

22.3.1 SAGA Gradient Approximation

Using this notation, we replace (22.12) by the following form where the starting condition for one epoch is the iterate obtained at the end of the previous epoch:

$$\begin{array}{l}
\textbf{(preliminary or incomplete SAGA algorithm – version I)}\\
\textbf{repeat over epochs } k \geq 1\textbf{:}\\
\quad \text{set } \boldsymbol{w}^k_{-1} = \boldsymbol{w}^{k-1}_{N-1}\\
\quad \textbf{repeat for } n = 0, 1, \ldots, N-1\textbf{:}\\
\quad\quad \text{select a random index } \boldsymbol{u} \text{ uniformly from } 0 \leq u \leq N-1\\
\quad\quad \text{consider the data sample } (\gamma(\boldsymbol{u}), h_{\boldsymbol{u}})\\
\quad\quad \text{pick the memory variable } \boldsymbol{\phi}^k_{n-1,\boldsymbol{u}} \text{ (as explained ahead)}\\
\quad\quad \text{update } \boldsymbol{w}^k_n = \boldsymbol{w}^k_{n-1} - \mu\Big(\nabla_{w^\mathsf{T}} Q(\boldsymbol{w}^k_{n-1}) - \nabla_{w^\mathsf{T}} Q(\boldsymbol{\phi}^k_{n-1,\boldsymbol{u}})\Big)\\
\quad \textbf{end}\\
\textbf{end}
\end{array} \tag{22.22}$$

Again, we are not showing all the arguments for the loss function to keep the notation compact. More explicitly, if desired, the loss values in the above expression are given by

$$Q(\boldsymbol{w}^k_{n-1}) \stackrel{\Delta}{=} Q\Big(\boldsymbol{w}^k_{n-1}; \gamma(\boldsymbol{u}), h_{\boldsymbol{u}}\Big) \tag{22.23a}$$

$$Q(\boldsymbol{\phi}^k_{n-1}) \stackrel{\Delta}{=} Q\Big(\boldsymbol{\phi}^k_{n-1,\boldsymbol{u}}; \gamma(\boldsymbol{u}), h_{\boldsymbol{u}}\Big) \tag{22.23b}$$

The above implementation amounts to using the following gradient approximation in (22.1):

$$\widehat{\nabla_{w^\mathsf{T}} P}(\boldsymbol{w}_{n-1}^k) = \nabla_{w^\mathsf{T}} Q(\boldsymbol{w}_{n-1}^k) - \nabla_{w^\mathsf{T}} Q(\boldsymbol{\phi}_{n-1,\boldsymbol{u}}^k) \tag{22.24}$$

We still need to explain how to select $\boldsymbol{\phi}_{n-1,\boldsymbol{u}}^k$; in the process of doing so, we will discover that we will need to modify (22.22) by adding yet another term to the gradient approximation. That is the reason why we are referring to the above listing as a "preliminary" (or incomplete) version of the SAGA algorithm.

22.3.2 Memory Variables

Observe from (22.22) that each time an index value $\boldsymbol{u}$ is selected, during any iteration in any run of the algorithm, the same data pair $(\gamma(\boldsymbol{u}), h_{\boldsymbol{u}})$ will be used during that step. For example, if an index value u_1 is selected at iteration n during run k, then that index would result in updating $\boldsymbol{w}_{n-1}^k$ to $\boldsymbol{w}_n^k$ based on the data pair $(\gamma(u_1), h_{u_1})$. If the *same* index u_1 is selected at some other iteration n' during some other run k', then that same index would result in updating $\boldsymbol{w}_{n'-1}^{k'}$ to $\boldsymbol{w}_{n'}^{k'}$ based on the *same* data $(\gamma(u_1), h_{u_1})$.

Therefore, we can view $\boldsymbol{\phi}_{n-1,\boldsymbol{u}}^k$ as a *memory* variable: At each iteration n, there will be a total of N such vector variables; one for each possible data index, u. The variable $\boldsymbol{\phi}_{n-1,\boldsymbol{u}}^k$ stores the last w-iterate that was computed using the data $(\gamma(u), h_u)$; this value changes with time as the algorithm moves from one iteration to another. Specifically, the memory variable will be updated as follows:

$$\boldsymbol{\phi}_{n,u}^k = \begin{cases} \boldsymbol{w}_n^k, & \text{if } \boldsymbol{u} = u \\ \boldsymbol{\phi}_{n-1,u}^k, & \text{otherwise, i.e., no change} \end{cases} \tag{22.25}$$

In other words, when an update is performed and $\boldsymbol{w}_n^k$ is generated, this w-iterate will be stored in the memory location corresponding to index $\boldsymbol{u}$. In this way, the memory variable $\boldsymbol{\phi}_{n,u}^k$ either stays invariant when u is not selected or is updated. Note that since the data is sampled with replacement, then during any particular run, the memory variable for a particular u-location may be updated more than once during that run. It may also not be updated at all. With this in mind, we rewrite (22.22) in the more detailed form (22.27) where we update the memory variables; this version is still incomplete and will be adjusted in the following. We can describe this update more succinctly by writing:

$$\begin{cases} \boldsymbol{w}_n^k = \boldsymbol{w}_{n-1}^k - \mu\Big(\nabla_{w^\mathsf{T}} Q(\boldsymbol{w}_{n-1}^k) - \nabla_{w^\mathsf{T}} Q(\boldsymbol{\phi}_{n-1,\boldsymbol{u}}^k)\Big) \\ \boldsymbol{\phi}_{n,\boldsymbol{u}}^k = \boldsymbol{w}_n^k \\ \boldsymbol{\phi}_{n,u'}^k = \boldsymbol{\phi}_{n-1,u'}^k, \quad u' \neq \boldsymbol{u} \end{cases} \tag{22.26}$$

where the last expression is simply stating that all other memory variables, associated with the other data points, remain unchanged.

$$
\begin{array}{l}
\textbf{(preliminary or incomplete SAGA algorithm – version II)}\\
\textbf{repeat over epochs } k \geq 1\textbf{:}\\
\quad \text{set } \boldsymbol{w}_{-1}^{k} = \boldsymbol{w}_{N-1}^{k-1}\\
\quad \textbf{repeat for } n = 0, 1, \ldots, N-1\textbf{:}\\
\quad\quad \text{select a random index } \boldsymbol{u} \text{ uniformly from } 0 \leq u \leq N-1\\
\quad\quad \text{consider the data sample } (\gamma(\boldsymbol{u}), h_{\boldsymbol{u}})\\
\quad\quad \text{read the } \boldsymbol{u}\text{th memory location, } \boldsymbol{\phi}_{n-1,\boldsymbol{u}}^{k}\\
\quad\quad \text{update } \boldsymbol{w}_{n}^{k} = \boldsymbol{w}_{n-1}^{k} - \mu\Big(\nabla_{w^{\mathsf{T}}} Q(\boldsymbol{w}_{n-1}^{k}) - \nabla_{w^{\mathsf{T}}} Q(\boldsymbol{\phi}_{n-1,\boldsymbol{u}}^{k})\Big)\\
\quad\quad \text{update the } \boldsymbol{u}\text{th memory location: } \boldsymbol{\phi}_{n,\boldsymbol{u}}^{k} \leftarrow \boldsymbol{w}_{n}^{k}\\
\quad \textbf{end}\\
\textbf{end}
\end{array}
\tag{22.27}
$$

Observe that this implementation needs to store and continually update information related to N memory vectors $\{\boldsymbol{\phi}_{n,u}^{k}\}$ of size $M \times 1$ each. The indices n and k in this notation refer to the fact that the values stored in these memory variables are continually updated and change with the epoch index k and the iteration index n. For simplicity, let us denote the memory vector simply by ϕ_u with $0 \leq u \leq N-1$. These vectors are indexed by u and there is one vector location for each data index u:

$$
\Big\{\phi_u \in \mathbb{R}^M,\ u = 0, 1, \ldots, N-1\Big\} \tag{22.28}
$$

We collect the memory vectors $\{\phi_u\}$ into an $M \times N$ matrix Φ consisting of N columns, one for each index $0 \leq u \leq N-1$; a snapshot of the memory matrix is shown in Fig. 22.1. The uth column in this matrix saves the value of the last w-iterate that was updated using the uth data sample. We will introduce a more efficient algorithm in a later section with a smaller storage requirement. Note from (22.26) that it is sufficient to store the values of the gradients of the loss function at these memory vectors, rather than the memory vectors themselves, i.e.,

$$
\Big\{\nabla_{w^{\mathsf{T}}} Q(\phi_{n,u}^{k}; \gamma(u), h_u),\ u = 0, 1, \ldots, N-1\Big\} \tag{22.29}
$$

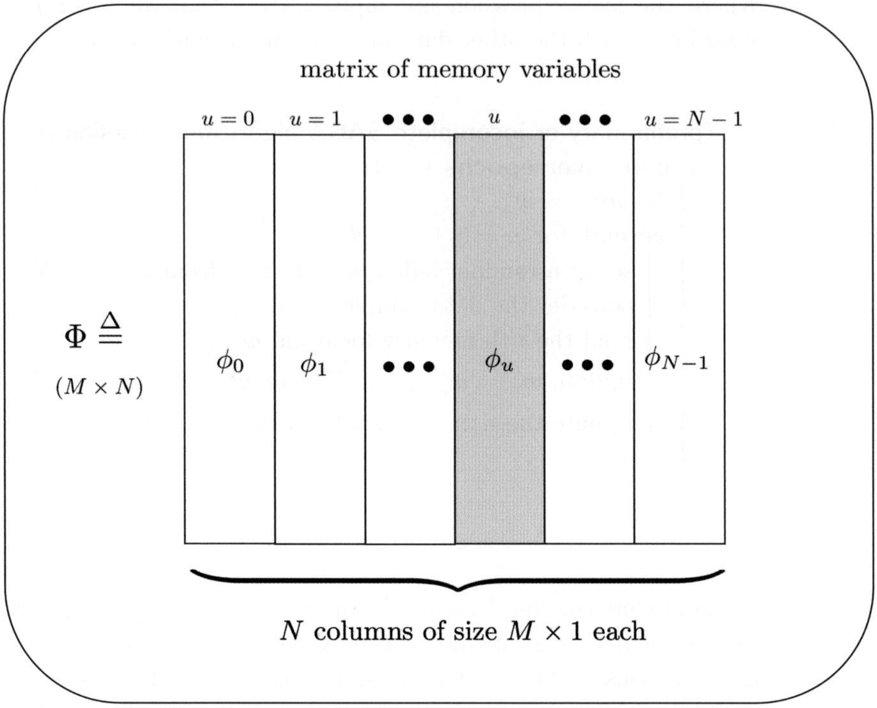

Figure 22.1 A snapshot of the memory matrix Φ, which consists of N columns $\{\phi_u\}$ of size $M \times 1$ each. The uth column saves the value of the last w-iterate that was updated using the uth data pair $(\gamma(u), h_u)$. The values stored in this matrix are continually updated over epoch k and iteration n.

Unfortunately, the gradient noise process that would result from the preliminary construction (22.27) will not have zero mean. To see this, note first that the gradient approximation that corresponds to this implementation is given by

$$\widehat{\nabla_{w^{\mathsf{T}}} P}(\boldsymbol{w}_{n-1}^k) = \nabla_{w^{\mathsf{T}}} Q(\boldsymbol{w}_{n-1}^k) - \nabla_{w^{\mathsf{T}}} Q(\boldsymbol{\phi}_{n-1,\boldsymbol{u}}^k) \tag{22.30}$$

so that for the kth run:

$$\begin{aligned}
&\mathbb{E}\left(\widehat{\nabla_{w^{\mathsf{T}}} P}(\boldsymbol{w}_{n-1}^k) \,|\, \boldsymbol{w}_{n-1}^k\right) \\
&= \mathbb{E}\left(\nabla_{w^{\mathsf{T}}} Q(\boldsymbol{w}_{n-1}^k; \gamma(\boldsymbol{u}), h_{\boldsymbol{u}}) | \boldsymbol{w}_{n-1}^k\right) - \mathbb{E}\left(\nabla_{w^{\mathsf{T}}} Q(\boldsymbol{\phi}_{n-1,\boldsymbol{u}}^k; \gamma(\boldsymbol{u}), h_{\boldsymbol{u}}) \,|\, \boldsymbol{w}_{n-1}^k\right) \\
&\overset{(a)}{=} \frac{1}{N} \sum_{u=0}^{N-1} \nabla_{w^{\mathsf{T}}} Q(w_{n-1}^k; \gamma(u), h_u) - \frac{1}{N} \sum_{u=0}^{N-1} \nabla_{w^{\mathsf{T}}} Q(\phi_{n-1,u}^k; \gamma(u), h_u) \\
&= \nabla_{w^{\mathsf{T}}} P(w_{n-1}^k) - \frac{1}{N} \sum_{u=0}^{N-1} \nabla_{w^{\mathsf{T}}} Q(\phi_{n-1,u}^k; \gamma(u), h_u)
\end{aligned} \tag{22.31}$$

where the rightmost expression in step (a) is because $\boldsymbol{u}$ is selected uniformly within the interval $0 \le u \le N-1$ and also independently of $\boldsymbol{w}_{n-1}^k$. Expression

(22.31) is not zero generally and, therefore, the gradient noise will not have zero mean under the construction proposed so far. For this reason, we modify the SAGA description as follows to arrive at the final listing for the algorithm.

22.3.3 SAGA Algorithm

The above argument suggests a modification to (22.26) that would lead to a zero-mean gradient noise process. Specifically, we replace the weight recursion in (22.26) by the following update, which is known as the stochastic *average-gradient* algorithm (SAGA):

$$\begin{aligned} \boldsymbol{w}_n^k = \boldsymbol{w}_{n-1}^k - \mu \Big\{ & \nabla_{w^\mathsf{T}} Q(\boldsymbol{w}_{n-1}^k; \gamma(\boldsymbol{u}), h_{\boldsymbol{u}}) - \nabla_{w^\mathsf{T}} Q(\boldsymbol{\phi}_{n-1,\boldsymbol{u}}^k; \gamma(\boldsymbol{u}), h_{\boldsymbol{u}}) \\ & + \frac{1}{N}\sum_{u=0}^{N-1} \nabla_{w^\mathsf{T}} Q(\phi_{n-1,u}^k; \gamma(u), h_u) \Big\} \end{aligned} \tag{22.32}$$

The algorithm is shown in listing (22.33).

SAGA algorithm for solving the empirical risk problem (22.2).

given N data points $\{\gamma(m), h_m\}$, $m = 0, 1, \ldots, N-1$;
start with arbitrary $\boldsymbol{w}_{N-1}^0$;
start with $\boldsymbol{\phi}_{N-1,u}^0 = 0$, $u = 0, 1, \ldots, N-1$;
start with $\boldsymbol{b}_{N-1}^0 = \frac{1}{N}\sum_{u=0}^{N-1} \nabla_{w^\mathsf{T}} Q(\boldsymbol{\phi}_{N-1,u}^0; \gamma(u), h_u)$.
for each run $k = 1, 2, \ldots, K$:
  set $\boldsymbol{w}_{-1}^k = \boldsymbol{w}_{N-1}^{k-1}$;
  set $\boldsymbol{\phi}_{-1,u}^k = \boldsymbol{\phi}_{N-1,u}^{k-1}$, $u = 0, 1, \ldots, N-1$;
  set $\boldsymbol{b}_{-1}^k = \boldsymbol{b}_{N-1}^{k-1}$
  repeat for $n = 0, 1, \ldots, N-1$:
    select $\boldsymbol{u}$ uniformly from the set $\{0, 1, \ldots, N-1\}$;
    select the data $(\gamma(\boldsymbol{u}), h_{\boldsymbol{u}})$;
    read the $\boldsymbol{u}$th memory variable, $\boldsymbol{\phi}_{n-1,\boldsymbol{u}}^k$;
    $\boldsymbol{w}_n^k = \boldsymbol{w}_{n-1}^k - \mu\Big(\nabla_{w^\mathsf{T}} Q(\boldsymbol{w}_{n-1}^k) - \nabla_{w^\mathsf{T}} Q(\boldsymbol{\phi}_{n-1,\boldsymbol{u}}^k) + \boldsymbol{b}_{n-1}^k\Big)$;
    $\boldsymbol{\phi}_{n,\boldsymbol{u}}^k \leftarrow \boldsymbol{w}_n^k$;
    $\boldsymbol{\phi}_{n,u'}^k = \boldsymbol{\phi}_{n-1,u'}^k$, $u' \neq \boldsymbol{u}$;
    $\boldsymbol{b}_n^k = \boldsymbol{b}_{n-1}^k + \frac{1}{N}\Big(\nabla_{w^\mathsf{T}} Q(\boldsymbol{\phi}_{n,\boldsymbol{u}}^k) - \nabla_{w^\mathsf{T}} Q(\boldsymbol{\phi}_{n-1,\boldsymbol{u}}^k)\Big)$;
  end
end
return $w^\star \leftarrow \boldsymbol{w}_{N-1}^K$.

(22.33)

Observe the following from examining the update relation (22.32):

(a) Each step involves retrieving the memory vector $\boldsymbol{\phi}^k_{n-1,\boldsymbol{u}}$ (or the gradient loss value at this location) for the $\boldsymbol{u}$th data pair.

(b) Each step also involves computing the last term, which corresponds to a sample average of the gradient of the loss function over *all* memory variables. We denote this term by

$$b^k_{n-1} \stackrel{\Delta}{=} \frac{1}{N}\sum_{u=0}^{N-1} \nabla_{w^{\mathsf{T}}} Q(\phi^k_{n-1,m};\gamma(u),h_u) \tag{22.34}$$

The variable b^k_n can be updated in an efficient manner by noting that, at every iteration n, *only one* of the memory locations is updated from $\boldsymbol{\phi}^k_{n-1,\boldsymbol{u}}$ to $\boldsymbol{\phi}^k_{n,\boldsymbol{u}} = \boldsymbol{w}^k_n$. This is the location that corresponds to the selected sample index $\boldsymbol{u}$. All other values $\boldsymbol{\phi}^k_{n,u'}$ stay equal to their previous values $\boldsymbol{\phi}^k_{n-1,u'}$ for all $u' \neq \boldsymbol{u}$. Therefore, the expression for b^k_n can be updated recursively as shown in the listing of the algorithm.

We remarked earlier in (22.29) that rather than store the memory variables $\{\phi^k_{n,u}\}$ directly, it is more efficient to store the gradient values at these locations, namely, $\nabla_{w^{\mathsf{T}}} Q(\phi^k_{n,u};\gamma(u),h_u)$. This is because evaluation of the sample average b^k_n requires these gradient values. In this way, the SAGA recursion requires $O(N)$ storage locations, one for each index u. For each epoch, the total number of gradient calculations that are needed are on the order of $O(N)$ as well. This is because, for each iteration n, there is one gradient calculation needed for $\nabla_{w^{\mathsf{T}}} Q(\boldsymbol{w}^k_{n-1};\gamma(\boldsymbol{u}),h_{\boldsymbol{u}})$.

22.3.4 SAGA Gradient Noise

The gradient noise process for the SAGA implementation (22.32) is given by

$$\begin{aligned} \boldsymbol{g}_n(\boldsymbol{w}^k_{n-1}) &\stackrel{\Delta}{=} \widehat{\nabla_{w^{\mathsf{T}}} P}(\boldsymbol{w}^k_{n-1}) - \nabla_{w^{\mathsf{T}}} P(\boldsymbol{w}^k_{n-1}) \\ &\stackrel{(22.32)}{=} \nabla_{w^{\mathsf{T}}} Q(\boldsymbol{w}^k_{n-1};\gamma(\boldsymbol{u}),h_{\boldsymbol{u}}) - \nabla_{w^{\mathsf{T}}} Q(\boldsymbol{\phi}^k_{n-1,\boldsymbol{u}};\gamma(\boldsymbol{u}),h_{\boldsymbol{u}}) \\ &\quad + \frac{1}{N}\sum_{u=0}^{N-1} \nabla_{w^{\mathsf{T}}} Q(\phi^k_{n-1,u};\gamma(u),h_u) - \nabla_{w^{\mathsf{T}}} P(\boldsymbol{w}^k_{n-1}) \end{aligned} \tag{22.35}$$

We will now examine its first- and second-order moment properties. For this purpose, we consider the following collection of random variables at iteration $n-1$ and epoch k:

$$\boldsymbol{\mathcal{F}}^k_{n-1} = \left\{\boldsymbol{w}^k_{n-1},\ \boldsymbol{\phi}^k_{n-1,u}\right\},\ u = 0,1,\ldots,N-1 \tag{22.36}$$

Conditioning on $\mathfrak{F}_{n-1}^k$ and computing expectations, we will be able to verify that the first-order moment continues to be zero since

$$\begin{aligned}\mathbb{E}\left(\boldsymbol{g}_n(\boldsymbol{w}_{n-1}^k)|\mathfrak{F}_{n-1}^k\right) &\overset{(22.31)}{=} \nabla_{w^\mathsf{T}} P(w_{n-1}^k) - \frac{1}{N}\sum_{u=0}^{N-1}\nabla_{w^\mathsf{T}} Q(\phi_{n-1,u}^k;\gamma(u),h_u) \\ &\quad +\frac{1}{N}\sum_{u=0}^{N-1}\nabla_{w^\mathsf{T}} Q(\phi_{n-1,u}^k;\gamma(u),h_u) \\ &\quad -\nabla_{w^\mathsf{T}} P(w_{n-1}^k) \;=\; 0 \end{aligned} \tag{22.37}$$

and, hence, under uniform sampling *with* replacement:

$$\boxed{\mathbb{E}\left(\boldsymbol{g}_n(\boldsymbol{w}_{n-1}^k)\,|\,\mathfrak{F}_{n-1}^k\right) \;=\; 0} \tag{22.38}$$

Let us examine next the second-order moment of the gradient noise process (22.35). Thus, note that

$$\mathbb{E}\left(\|\boldsymbol{g}(\boldsymbol{w}_{n-1}^k)\|^2\,|\,\mathfrak{F}_{n-1}^k\right) \overset{\Delta}{=} \mathbb{E}\left(\|\widehat{\nabla_{w^\mathsf{T}} P}(\boldsymbol{w}_{n-1}^k) - \nabla_{w^\mathsf{T}} P(\boldsymbol{w}_{n-1}^k)\|^2\,|\,\mathfrak{F}_{n-1}^k\right) \tag{22.39}$$

from which we get

$$\begin{aligned}
&\mathbb{E}\left(\|\boldsymbol{g}(\boldsymbol{w}_{n-1}^k)\|^2\,|\,\mathfrak{F}_{n-1}^k\right) \\
&= \mathbb{E}\left(\left\|\nabla_{w^\mathsf{T}} Q(\boldsymbol{w}_{n-1}^k;\gamma(\boldsymbol{u}),h_{\boldsymbol{u}}) - \nabla_{w^\mathsf{T}} Q(\boldsymbol{\phi}_{n-1,\boldsymbol{u}}^k;\gamma(\boldsymbol{u}),h_{\boldsymbol{u}}) \;+\right.\right. \\
&\qquad \left.\left.\frac{1}{N}\sum_{u=0}^{N-1}\nabla_{w^\mathsf{T}} Q(\phi_{n-1,u}^k;\gamma(u),h_u) - \nabla_{w^\mathsf{T}} P(\boldsymbol{w}_{n-1}^k)\right\|^2\,|\,\mathfrak{F}_{n-1}^k\right) \\
&\overset{(a)}{=} \mathbb{E}\left(\left\|\nabla_{w^\mathsf{T}} Q(\boldsymbol{w}_{n-1}^k) - \nabla_{w^\mathsf{T}} Q(w^\star) \;+\; \nabla_{w^\mathsf{T}} Q(w^\star) - \nabla_{w^\mathsf{T}} Q(\boldsymbol{\phi}_{n-1,\boldsymbol{u}}^k) \;+\right.\right. \\
&\qquad \frac{1}{N}\sum_{u=0}^{N-1}\nabla_{w^\mathsf{T}} Q(\phi_{n-1,\boldsymbol{u}}^k) - \frac{1}{N}\sum_{m=0}^{N-1}\nabla_{w^\mathsf{T}} Q(w^\star) + \\
&\qquad \left.\left.\nabla_{w^\mathsf{T}} P(w^\star) - \nabla_{w^\mathsf{T}} P(\boldsymbol{w}_{n-1}^k)\right\|^2\,|\,\mathfrak{F}_{n-1}^k\right) \\
&\overset{(b)}{\leq} 8\delta^2\|\widetilde{w}_{n-1}^k\|^2 + 4\delta^2\mathbb{E}\left(\|\widetilde{\boldsymbol{\phi}}_{n-1,\boldsymbol{u}}^k\|^2\,|\,\mathfrak{F}_{n-1}^k\right) + \frac{4\delta^2}{N}\sum_{u=0}^{N-1}\|\widetilde{\phi}_{n-1,u}^k\|^2 \\
&\leq 8\delta^2\|\widetilde{w}_{n-1}^k\|^2 + \frac{4\delta^2}{N}\sum_{u=0}^{N-1}\|\widetilde{\phi}_{n-1,u}^k\|^2 + \frac{4\delta^2}{N}\sum_{u=0}^{N-1}\|\widetilde{\phi}_{n-1,u}^k\|^2 \\
&= 8\delta^2\|\widetilde{w}_{n-1}^k\|^2 + \frac{8\delta^2}{N}\sum_{u=0}^{N-1}\|\widetilde{\phi}_{n-1,u}^k\|^2
\end{aligned} \tag{22.40}$$

where in step (a) we removed the data arguments from the loss functions to simplify the notation, and added and subtracted the same terms $\nabla_{w^\mathsf{T}} Q(w^\star)$ and $\nabla_{w^\mathsf{T}} P(w^\star)$. Step (b) uses the Lipschitz conditions (22.3b) and (22.4), the Jensen inequality that $\|a+b+c+d\|^2 \leq 4\|a\|^2 + 4\|b\|^2 + 4\|c\|^2 + 4\|d\|^2$ for any vectors $\{a, b, c, d\}$ of compatible dimensions, and property (19.172). We also introduce the error quantity

$$\widetilde{\boldsymbol{\phi}}^k_{n-1,\boldsymbol{u}} \stackrel{\Delta}{=} w^\star - \boldsymbol{\phi}^k_{n-1,\boldsymbol{u}} \tag{22.41}$$

Comparing the bound in (22.40) with (22.16) where the variance-reduced algorithm employed $w^\star$ rather than its approximation $\boldsymbol{\phi}^k_{n-1,\boldsymbol{u}}$, we observe two facts:

(a) As before, the bound (22.40) does not contain the constant driving term σ_g^2 and is only dependent on weight-error quantities.

(b) However, and unlike (22.16), there is one deterioration in the bound in that it also depends on the sum of the squared-error memory terms from the previous run. This dependency appears in the last term of (22.42).

We therefore conclude that for the SAGA algorithm (22.32), the second-order moment of the gradient noise satisfies a bound of the form:

$$\boxed{\mathbb{E}\left(\|\boldsymbol{g}_n(\boldsymbol{w}^k_{n-1})\|^2 \,|\, \boldsymbol{\mathcal{F}}^k_{n-1}\right) \leq \beta_g^2 \left(\|\widetilde{w}^k_{n-1}\|^2 + \frac{1}{N}\sum_{u=0}^{N-1} \|\widetilde{\phi}^k_{n-1,u}\|^2\right)} \tag{22.42}$$

for some finite nonnegative constant $\beta_g^2 = 8\delta^2$; in the next proof we will use the fact that $\nu^2 \leq \delta^2 \leq \beta_g^2$.

22.3.5 Convergence under Uniform Sampling

We are now ready to establish the *exact* convergence of the SAGA algorithm.

Theorem 22.2. (Convergence of SAGA under uniform sampling) *Consider the SAGA algorithm (22.33) used to seek the minimizer for the empirical risk minimization problem (22.2). The risk and loss functions are assumed to satisfy conditions* (**A1**,**A2**). *For step-size values satisfying (i.e., for μ small enough):*

$$\mu < \frac{\nu}{\delta^2 + 2\beta_g^2 + 2N\nu^2} \tag{22.43}$$

it holds that $\mathbb{E}\,\|\widetilde{\boldsymbol{w}}^k_{-1}\|^2$ and $\mathbb{E}\,P(\boldsymbol{w}^k_{-1})$ converge to zero and $P(w^\star)$, respectively, at an exponential rate, namely,

$$\mathbb{E}\,\|\widetilde{\boldsymbol{w}}^k_{-1}\|^2 \leq O(\lambda^k) \tag{22.44a}$$

$$\mathbb{E}\,P(\boldsymbol{w}^k_{-1}) - P(w^\star) \leq O(\lambda^k) \tag{22.44b}$$

where

$$\lambda = \frac{1-\mu\nu}{1-\frac{\mu^2\beta_g^2}{1-\mu N\nu}} = 1 - O(\mu) \in [0,1) \tag{22.45}$$

The notation $\boldsymbol{w}^k_{-1}$ refers to the weight iterate at the start of the kth epoch (which is identical to the iterate at the end of the previous epoch).

Proof: See Appendix 22.A. ■

22.4 STOCHASTIC VARIANCE-REDUCED GRADIENT ALGORITHM (SVRG)

One inconvenience of the SAGA implementation (22.33) is its high storage requirement since it needs to save the history variables $\{\boldsymbol{\phi}^k_{n,u}\}$ or their gradients $\{\nabla_{w^\mathsf{T}} Q(\boldsymbol{\phi}^k_{n-1,u}; \gamma(u), h_u),\ u = 0, \ldots, N-1\}$ for use in the algorithm. This amounts to storing $O(N)$ variables, one for each training data pair $(\gamma(u), h_u)$. Sometimes, the size of N is prohibitive. An alternative method is the stochastic variance-reduced gradient (SVRG) algorithm. This method replaces the history variables $\{\boldsymbol{\phi}^k_{n,u}\}$ for SAGA by a *fixed* value for each epoch, denoted by $\boldsymbol{\phi}^k_{-1}$. This vector stays invariant during the *entire* epoch. At the end of the kth run, we update its value to

$$\boldsymbol{\phi}^{k+1}_{-1} = \boldsymbol{w}^k_{N-1} \tag{22.46}$$

That is, we fix $\boldsymbol{\phi}^{k+1}_{-1}$ to the iterate value at the end of the previous run. This simplification greatly reduces the storage requirement. However, each epoch in SVRG will now need to be preceded by an aggregation step to compute the variable

$$b^k = \frac{1}{N}\sum_{u=0}^{N-1} \nabla_{w^\mathsf{T}} Q(\phi^k_{-1}, \gamma(u), h_u) \tag{22.47}$$

This computation can be time-consuming for large datasets. It also causes the operation of SVRG to become *unbalanced*, with a larger amount of time possibly needed before each epoch starts and a shorter time needed within the epoch itself. Observe further that for each epoch, the total number of gradient calculations that are needed are on the order of $O(2N)$. This is because, at the start of each epoch, there are N gradient calculations needed for b^k and, moreover, for each iteration n, there is one gradient calculation needed for $\nabla_{w^\mathsf{T}} Q(\boldsymbol{w}^k_{n-1}; \boldsymbol{\gamma}(\boldsymbol{u}), \boldsymbol{h}_{\boldsymbol{u}})$. Since n runs over $n = 0, \ldots, N-1$, we arrive at the estimated figure of $O(2N)$ gradient calculations per epoch.

SVRG algorithm for solving the empirical risk problem (22.2).

given N data points $\{\gamma(m), h_m\}$, $m = 0, 1, \ldots, N-1$;
start with arbitrary $\boldsymbol{w}^0_{N-1}$;
for each run $k = 1, 2, \ldots, K$**:**
set $\boldsymbol{w}^k_{-1} = \boldsymbol{w}^{k-1}_{N-1}$;
set $\boldsymbol{\phi}^k_{-1} = \boldsymbol{w}^{k-1}_{N-1}$;

$$\boldsymbol{b}^k = \frac{1}{N}\sum_{u=0}^{N-1} \nabla_{w^\mathsf{T}} Q(\boldsymbol{\phi}^k_{-1}, \gamma(u), h_u); \tag{22.48}$$

repeat for $n = 0, 1, \ldots, N-1$**:**
select $\boldsymbol{u}$ uniformly from the set $\{0, 1, \ldots, N-1\}$;
select the data $(\boldsymbol{\gamma}(\boldsymbol{u}), \boldsymbol{h}_{\boldsymbol{u}})$;

$$\boldsymbol{w}^k_n = \boldsymbol{w}^k_{n-1} - \mu\Big(\nabla_{w^\mathsf{T}} Q(\boldsymbol{w}^k_{n-1}) - \nabla_{w^\mathsf{T}} Q(\boldsymbol{\phi}^k_{-1}) + \boldsymbol{b}^k\Big);$$

end
end
return $w^\star \leftarrow \boldsymbol{w}^K_{N-1}$.

22.4.1 SVRG Gradient Noise

The gradient noise process for the SVRG implementation (22.48) is given by

$$\begin{aligned}
\boldsymbol{g}_n(\boldsymbol{w}^k_{n-1}) &\overset{\Delta}{=} \widehat{\nabla_{w^\mathsf{T}} P}(\boldsymbol{w}^k_{n-1}) - \nabla_{w^\mathsf{T}} P(\boldsymbol{w}^k_{n-1}) \qquad (22.49)\\
&\overset{(22.48)}{=} \nabla_{w^\mathsf{T}} Q(\boldsymbol{w}^k_{n-1}; \boldsymbol{\gamma}(\boldsymbol{u}), \boldsymbol{h}_{\boldsymbol{u}}) - \nabla_{w^\mathsf{T}} Q(\boldsymbol{\phi}^k_{-1}; \boldsymbol{\gamma}(\boldsymbol{u}), \boldsymbol{h}_{\boldsymbol{u}})\\
&\quad + \frac{1}{N}\sum_{u=0}^{N-1} \nabla_{w^\mathsf{T}} Q(\phi^k_{-1}; \gamma(u), h_u) - \nabla_{w^\mathsf{T}} P(\boldsymbol{w}^k_{n-1})
\end{aligned}$$

We will again examine its first- and second-order moments. We consider the following collection of random variables at iteration $n-1$ and epoch k:

$$\boldsymbol{\mathfrak{F}}^k_{n-1} = \Big\{\boldsymbol{w}^k_{n-1}, \boldsymbol{\phi}^k_{-1}\Big\} \tag{22.50}$$

where, in comparison with (22.36), the filtration $\boldsymbol{\mathfrak{F}}_{n-1}^k$ now contains only $\boldsymbol{\phi}_{-1}^k$ instead of the entire memory variables $\{\boldsymbol{\phi}_{n-1,\boldsymbol{u}}^k\}$. Conditioning on $\boldsymbol{\mathfrak{F}}_{n-1}^k$ and computing expectations we will find that the gradient noise continues to have zero mean since

$$\begin{aligned}\mathbb{E}\left(\boldsymbol{g}_n(\boldsymbol{w}_{n-1}^k)\,|\,\boldsymbol{\mathfrak{F}}_{n-1}^k\right) \overset{(22.31)}{=} & \nabla_{w^\mathsf{T}} P(w_{n-1}^k) - \frac{1}{N}\sum_{u=0}^{N-1} \nabla_{w^\mathsf{T}} Q(\phi_{-1}^k;\gamma(u),h_u) \\ & +\frac{1}{N}\sum_{u=0}^{N-1} \nabla_{w^\mathsf{T}} Q(\phi_{-1}^k;\gamma(u),h_u) \\ & -\nabla_{w^\mathsf{T}} P(w_{n-1}^k) \;=\; 0 \end{aligned} \tag{22.51}$$

and, hence, under uniform sampling *with* replacement:

$$\boxed{\mathbb{E}\left(\boldsymbol{g}_n(\boldsymbol{w}_{n-1}^k)\,|\,\boldsymbol{\mathfrak{F}}_{n-1}^k\right) \;=\; 0} \tag{22.52}$$

Let us examine next the second-order moment of the modified gradient noise process (22.49). Thus, note that

$$\mathbb{E}\left(\|\boldsymbol{g}(\boldsymbol{w}_{n-1}^k)\|^2\,|\,\boldsymbol{\mathfrak{F}}_{n-1}^k\right) \overset{\Delta}{=} \mathbb{E}\left(\|\widehat{\nabla_{w^\mathsf{T}} P}(\boldsymbol{w}_{n-1}^k) - \nabla_{w^\mathsf{T}} P(\boldsymbol{w}_{n-1}^k\|^2\,|\,\boldsymbol{\mathfrak{F}}_{n-1}^k\right) \tag{22.53}$$

so that

$$\begin{aligned}&\mathbb{E}\left(\|\boldsymbol{g}(\boldsymbol{w}_{n-1}^k)\|^2|\boldsymbol{\mathfrak{F}}_{n-1}^k\right)\\ &\overset{(22.49)}{=} \mathbb{E}\Bigg(\Big\|\nabla_{w^\mathsf{T}} Q(\boldsymbol{w}_{n-1}^k;\gamma(\boldsymbol{u}),h_{\boldsymbol{u}}) - \nabla_{w^\mathsf{T}} Q(\boldsymbol{\phi}_{-1}^k;\gamma(\boldsymbol{u}),h_{\boldsymbol{u}}) \;+\\ &\qquad \frac{1}{N}\sum_{u=0}^{N-1} \nabla_{w^\mathsf{T}} Q(\phi_{-1}^k;\gamma(u),h_u) - \nabla_{w^\mathsf{T}} P(\boldsymbol{w}_{n-1}^k)\Big\|^2\,|\,\boldsymbol{\mathfrak{F}}_{n-1}^k\Bigg)\\ &\overset{(a)}{=} \mathbb{E}\Bigg(\Big\|\nabla_{w^\mathsf{T}} Q(\boldsymbol{w}_{n-1}^k) - \nabla_{w^\mathsf{T}} Q(w^\star) \;+\\ &\qquad \nabla_{w^\mathsf{T}} Q(w^\star) - \nabla_{w^\mathsf{T}} Q(\boldsymbol{\phi}_{-1}^k;\gamma(\boldsymbol{u}),h_{\boldsymbol{u}}) \;+\\ &\qquad \frac{1}{N}\sum_{u=0}^{N-1} \nabla_{w^\mathsf{T}} Q(\boldsymbol{\phi}_{-1}^k;\gamma(u),h_u) - \frac{1}{N}\sum_{m=0}^{N-1} \nabla_{w^\mathsf{T}} Q(w^\star) \;+\\ &\qquad \nabla_{w^\mathsf{T}} P(w^\star) - \nabla_{w^\mathsf{T}} P(\boldsymbol{w}_{n-1}^k)\Big\|^2\,|\,\boldsymbol{\mathfrak{F}}_{n-1}^k\Bigg)\\ &\overset{(b)}{\le} 8\delta^2\|\widetilde{w}_{n-1}^k\|^2 + 4\delta^2\|\widetilde{\phi}_{-1}^k\|^2 + \frac{4\delta^2}{N}\sum_{u=0}^{N-1}\|\widetilde{\phi}_{-1}^k\|^2\\ &= 8\delta^2\|\widetilde{w}_{n-1}^k\|^2 + 4\delta^2\|\widetilde{\phi}_{-1}^k\|^2 + 4\delta^2\|\widetilde{\phi}_{-1}^k\|^2\\ &= 8\delta^2\|\widetilde{w}_{n-1}^k\|^2 + 8\delta^2\|\widetilde{\phi}_{-1}^k\|^2 \end{aligned} \tag{22.54}$$

where in step (a) we removed the data arguments from the loss functions to simplify the notation and added and subtracted the same terms $\nabla_{w^\mathsf{T}} Q(w^\star)$ and $\nabla_{w^\mathsf{T}} P(w^\star)$. Step ($b$) uses the Lipschitz conditions (22.3b) and (22.4), the Jensen inequality that $\|a+b+c+d\|^2 \leq 4\|a\|^2+4\|b\|^2+4\|c\|^2+4\|d\|^2$ for any vectors $\{a,b,c,d\}$ of compatible dimensions, and property (19.172). We also introduced the error quantity

$$\widetilde{\phi}_{-1}^k \stackrel{\Delta}{=} w^\star - \phi_{-1}^k \tag{22.55}$$

Comparing the bound in (22.54) with (22.16), where the variance-reduced algorithm employed $w^\star$ rather than its approximation ϕ_{-1}^k, we observe two facts:

(a) As before, the bound (22.54) does not contain a constant driving term σ_g^2 and is only dependent on weight-error quantities.

(b) However, and unlike (22.16), there is some deterioration in the bound in that it is also dependent on the squared-error term $\|\widetilde{\phi}_{-1}^k\|^2$.

We therefore conclude that for the SVRG algorithm, the second-order moment of the gradient noise satisfies:

$$\boxed{\mathbb{E}\left(\|\boldsymbol{g}_n(\boldsymbol{w}_{n-1}^k)\|^2|\boldsymbol{\mathcal{F}}_{n-1}^k\right) \leq \beta_g^2\left(\|\widetilde{\boldsymbol{w}}_{n-1}^k\|^2 + \|\widetilde{\phi}_{-1}^k\|^2\right)} \tag{22.56}$$

for some finite nonnegative constant $\beta_g^2 = 8\delta^2$; in the next proof we will use the fact that $\nu^2 \leq \delta^2 \leq \beta_g^2$.

22.4.2 Convergence under Uniform Sampling

We are now ready to establish the *exact* convergence of the SVRG algorithm. The argument is simpler than in the case of the SAGA implementation.

THEOREM 22.3. **(Convergence of SVRG under uniform sampling)** *Consider the SVRG algorithm (22.48) used to seek the minimizer for the empirical risk minimization problem (22.2). The risk and loss functions are assumed to satisfy conditions (***A1**,**A2***). For step-size values satisfying (i.e., for μ small enough):*

$$\mu < \frac{1}{N\nu} \tag{22.57}$$

and for N large enough satisfying $N > 2\beta_g^2/\nu^2$, it holds that $\mathbb{E}\,\|\widetilde{\boldsymbol{w}}_{-1}^k\|^2$ and $\mathbb{E}\,P(\boldsymbol{w}_{-1}^k)$ converge to zero and $P(w^\star)$, respectively, at an exponential rate, according to

$$\mathbb{E}\,\|\widetilde{\boldsymbol{w}}_{-1}^k\|^2 \leq O(\lambda^k) \tag{22.58a}$$

$$\mathbb{E}\,P(\boldsymbol{w}_{-1}^k) - P(w^\star) \leq O(\lambda^k) \tag{22.58b}$$

where

$$\lambda = 1 - \frac{\mu}{2\nu}\left(N - 2\beta_g^2/\nu\right) \in [0,1) \tag{22.59}$$

The notation $\boldsymbol{w}_{-1}^k$ refers to the weight iterate at the start of the kth epoch (which is identical to the iterate at the end of the previous epoch).

Proof: See Appendix 22.B. ■

Example 22.1 **(Illustrating performance of SAGA and SVRG)** We illustrate the convergence performance of the SAGA and SVRG variance-reduction techniques by numerical simulations. We consider the following regularized logistic empirical risk problem:

$$\min_{w\in\mathbb{R}^M}\left\{P(w)\ \stackrel{\Delta}{=}\ \rho\|w\|^2+\frac{1}{N}\sum_{m=0}^{N-1}\ln\Big(1+e^{-\gamma(m)h_m^{\mathsf{T}}w}\Big)\right\} \tag{22.60}$$

where $h_m\in\mathbb{R}^M$ and $\gamma(m)\in\{\pm1\}$. In all experiments, we set $\rho=0.5$ and $M=25$. We generate $N=400$ random data pairs $\{\gamma(m),h_m\}$ according to a logistic model. First, a random parameter model $w^a\in\mathbb{R}^M$ is selected, and a random collection of vectors $\{h_m\}$ are generated with zero-mean unit-variance Gaussian entries. Then, for each h_m, the label $\gamma(m)$ is set to either $+1$ or -1 according to the following construction:

$$\gamma(m)=+1\ \text{ if }\ \left(\frac{1}{1+e^{-h_m^{\mathsf{T}}w^a}}\right)\geq 0.5;\ \text{ otherwise }\gamma(m)=-1 \tag{22.61}$$

A total of $K=40$ epochs are run over the data using $\mu=0.001$.

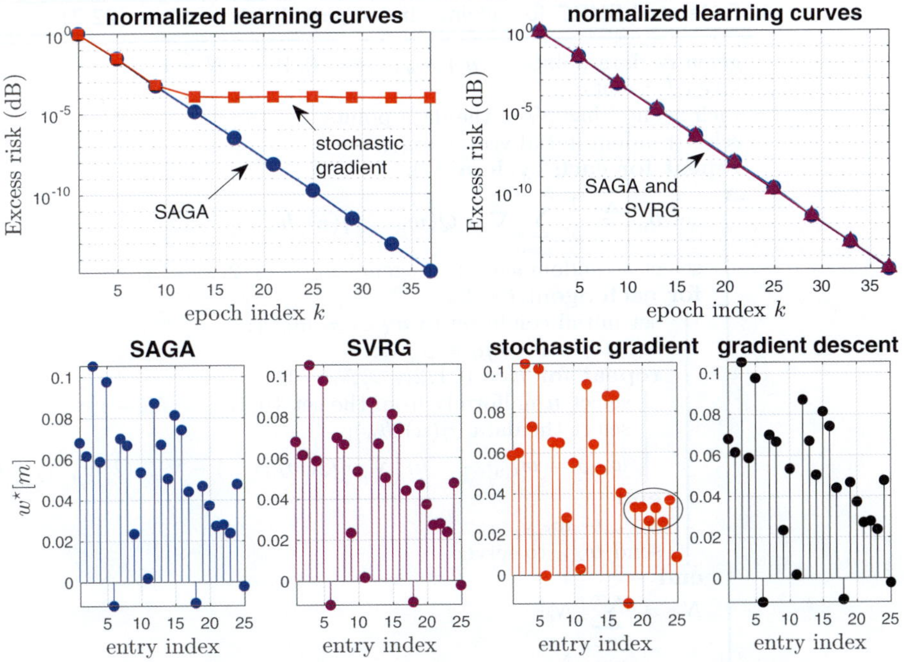

Figure 22.2 The top plot shows the learning curves for the SAGA recursion (22.33) and the stochastic gradient recursion (16.47) in normalized log scale. The bottom row plots the limiting weight iterates that are obtained by the SAGA recursion (22.33), the stochastic gradient recursion (16.47), and also the logistic gradient-descent implementation (12.30).

Figure 22.2 compares the performance of the SAGA and SVRG recursions with the traditional stochastic gradient algorithm (16.47). The plot in the top row compares the learning curves in normalized log scale. Observe how the learning curves for SAGA and SVRG continue to decrease while the learning curve for stochastic gradient tends toward a steady-state level. The learning curves shown in the figure reflect the values of the risk function at the start of each epoch, namely, the successive values $P(w_{-1}^k)$. The plots in the lower row of the figure show the limiting weight iterates that are obtained by SAGA, SVRG, the stochastic gradient recursion (16.47), and also the logistic gradient-descent implementation (12.30); the latter employs the full gradient vector of the risk function. Observe that, for the number of epochs used, the weight iterate by the SAGA and SVRG recursions are already close to the weight iterate attained by the gradient-descent implementation. In contrast, the weight iterate by the stochastic gradient implementation (the third plot) still needs additional iterations to converge, as is evident from the encircled weight values.

Example 22.2 **(Federated learning using SVRG)** We described the federated learning architecture earlier in Section 17.6, along with the corresponding federated averaging algorithm (17.89). The algorithm relied on the use of a mini-batch stochastic gradient procedure. In this example, we describe a second variation for federated learning that employs the SVRG variance-reduction technique. In this version, the algorithm requires at the beginning of each cycle that the global gradient vector be computed centrally using the entire set of data; this is the vector denoted by b_n in listing (22.62). To determine b_n, all dispersed agents (or clients) will receive $\boldsymbol{w}_{n-1}$, evaluate the gradients of the loss function at their data, and share these gradients back with the server.

Federated SVRG for solving the empirical risk problem (22.2).

given N data points $\{\gamma(m), h_m\}$, $m = 0, 1, \ldots, N-1$;
given L agents;
each agent ℓ has N_ℓ of the data points;
select random initial vector $\boldsymbol{w}_{-1}$;
repeat for each cycle $n \geq 0$**:**

set $\boldsymbol{b}_n \triangleq \frac{1}{N}\sum_{u=0}^{N-1} \nabla_{w^{\mathsf{T}}} Q(\boldsymbol{w}_{n-1}; \gamma(u), \boldsymbol{h}_u)$;
select a random subset of agents $\mathcal{L}_n \subset \{1, 2, \ldots, L\}$;
for each agent $\ell \in \mathcal{L}_n$
  set initial condition to $\boldsymbol{w}_{\ell,-1} = \boldsymbol{w}_{n-1}$;
  set step size to $\mu_\ell = \mu/N_\ell$;
  repeat for $n' = 0, 1, 2, \ldots, N_\ell - 1$**:**
    select $\boldsymbol{u}$ uniformly from the set $\{0, 1, , \ldots, N_\ell - 1\}$;
    select the data $(\gamma(\boldsymbol{u}), h_{\boldsymbol{u}})$;
    $\boldsymbol{w}_{\ell,n'} = \boldsymbol{w}_{\ell,n'-1} - \mu_\ell\Big(\nabla_{w^{\mathsf{T}}} Q(\boldsymbol{w}_{\ell,n'-1}) - \nabla_{w^{\mathsf{T}}} Q(\boldsymbol{w}_{n-1}) + \boldsymbol{b}_n\Big)$;
  end
  $\boldsymbol{w}_{\ell,n} \leftarrow \boldsymbol{w}_{\ell,N_\ell-1}$;
  send $\boldsymbol{w}_{\ell,n}$ to server
end
$N_{\mathcal{L}} = \sum_{\ell' \in \mathcal{L}_n} N_{\ell'}$
$\boldsymbol{w}_n = \sum_{\ell \in \mathcal{L}_n} \frac{N_\ell}{N_{\mathcal{L}}} \boldsymbol{w}_{\ell,n}$

end
return $w^\star \leftarrow \boldsymbol{w}_n$. (22.62)

Observe in the listing that we are also scaling the step size μ by the amount of data at each agent, namely, $\mu_\ell = \mu/N_\ell$, so that agents with different amounts of data progress similarly in their updates.

22.5 NONSMOOTH RISK FUNCTIONS

We have focused so far on the case in which the empirical risk function $P(w)$ in (22.2) and the corresponding loss function $Q(w;\cdot)$ are first-order differentiable. In this section we consider nonsmooth risks. It is sufficient to focus on the SAGA implementation since similar arguments can be used to extend the results to SVRG. In particular, we will illustrate how to incorporate a *proximal projection* step into SAGA in order to deal with the minimization of the nonsmooth risk.

We consider a scenario similar to what we have studied in the previous chapter for stochastic proximal algorithms. We replace (22.2) by the optimization problem

$$w^\star = \underset{w\in\mathbb{R}^M}{\text{argmin}} \left\{q(w) + E(w)\right\} \tag{22.63}$$

where $q(w)$ is possibly nonsmooth, usually a regularization term such as

$$q(w) \;=\; \alpha\|w\|_1 + \rho\|w\|^2, \;\; \alpha > 0, \;\; \rho \geq 0 \tag{22.64}$$

and $E(w)$ is a smooth empirical risk so that (22.63) is effectively a problem of the form

$$w^\star \;\stackrel{\Delta}{=}\; \underset{w\in\mathbb{R}^M}{\text{argmin}} \left\{P(w) \;\stackrel{\Delta}{=}\; q(w) + \frac{1}{N}\sum_{m=0}^{N-1} Q_u(w;\gamma(m),h_m)\right\} \tag{22.65}$$

In this expression, $Q_u(w;\cdot)$ is the loss function that is associated with the smooth component $E(w)$.

22.5.1 Conditions on Risk and Loss Functions

In a manner similar to what was done in Section 21.1.3, we introduce the following conditions for empirical risk minimization problems of the form (22.65):

(A1) (**Strongly convex risk**). $E(w)$ is ν-strongly convex and first-order differentiable, namely, for every $w_1, w_2 \in \text{dom}(E)$ and some $\nu > 0$:

$$E(w_2) \geq E(w_1) + (\nabla_{w^{\mathsf{T}}} E(w_1))^{\mathsf{T}}(w_2 - w_1) \;+\; \frac{\nu}{2}\|w_2 - w_1\|^2 \tag{22.66a}$$

(A2) (**δ-Lipschitz loss gradients**). The gradient vectors of $Q_u(w,\cdot)$ are δ-Lipschitz regardless of the data argument, i.e.,

$$\|\nabla_w Q_u(w_2;\gamma(k),h_k) - \nabla_w Q_u(w_1;\gamma(\ell),h_\ell)\| \;\leq\; \delta\,\|w_2 - w_1\| \tag{22.66b}$$

for any $w_1, w_2 \in \text{dom}(Q_u)$, any $0 \leq k, \ell \leq N-1$, and with $\delta \geq \nu$. We already know that condition (22.66b) implies that the gradient of $E(w)$ is itself δ-Lipschitz:

$$\|\nabla_w E(w_2) - \nabla_w E(w_1)\| \leq \delta\|w_2 - w_1\| \tag{22.67}$$

(A3) (**Affine-Lipschitz subgradients for regularizer**). The regularization term $q(w)$ has affine-Lipschitz subgradients, namely, there exist constants (δ_1, δ_2) such that for any $w_1, w_2 \in \text{dom}(q)$:

$$\|s_q(w_2) - s'_q(w_1)\| \;\leq\; \delta_1 \|w_2 - w_1\| + \delta_2 \tag{22.68}$$

where $s'_q(w)$ denotes any subgradient from the subdifferential set $\partial_{w^\mathsf{T}} q(w)$, while $s_q(w)$ denotes the particular subgradient construction used by the algorithm.

Note that under conditions (**A1**,**A2**), the following two conditions hold:

$$(\textbf{P1})\ \nu\text{-strong convexity of } E(w) \text{ as in (22.66a)} \tag{22.69a}$$

$$(\textbf{P2})\ \delta\text{-Lipschitz gradients for } E(w) \text{ as in (22.67)} \tag{22.69b}$$

Moreover, we know from the earlier results (8.29) and (10.20) derived for strongly convex and δ-smooth functions that conditions **P1** and **P2** imply respectively:

$$(\textbf{P1}) \quad \Longrightarrow \quad \frac{\nu}{2}\|\widetilde{w}\|^2 \leq E(w) - E(w^\star) \leq \frac{1}{2\nu}\|\widetilde{w}\|^2 \tag{22.70a}$$

$$(\textbf{P2}) \quad \Longrightarrow \quad \frac{1}{2\delta}\|\widetilde{w}\|^2 \leq E(w) - E(w^\star) \leq \frac{\delta}{2}\|\widetilde{w}\|^2 \tag{22.70b}$$

where $\widetilde{w} = w^\star - w$. The upper bounds in both expressions indicate that whenever we bound $\|\widetilde{w}\|^2$ we will also be automatically bounding the excess risk, $E(w) - E(w^\star)$.

22.5.2 Traditional Gradient Noise

In the previous chapter we considered constructions where the gradient search direction was approximated by using either instantaneous or mini-batch calculations, namely,

$$(\textbf{instantaneous})\ \widehat{\nabla_{w^\mathsf{T}} E}(w) = \nabla_{w^\mathsf{T}} Q_u(w; \boldsymbol{\gamma}, \boldsymbol{h}) \tag{22.71a}$$

$$(\textbf{mini-batch})\ \widehat{\nabla_{w^\mathsf{T}} E}(w) = \frac{1}{B}\sum_{b=0}^{B-1} \nabla_{w^\mathsf{T}} Q_u(w; \boldsymbol{\gamma}(b), \boldsymbol{h}_b) \tag{22.71b}$$

where the boldface notation $(\boldsymbol{\gamma}, \boldsymbol{h})$ or $(\boldsymbol{\gamma}(b), \boldsymbol{h}_b)$ refers to data samples selected uniformly at random with replacement from the given dataset $\{\gamma(m), h_m\}$. The difference between the true gradient and its approximation was referred to as *gradient noise* and denoted by

$$\boldsymbol{g}(w) \;\stackrel{\Delta}{=}\; \widehat{\nabla_{w^\mathsf{T}} E}(w) - \nabla_{w^\mathsf{T}} E(w) \tag{22.72}$$

We indicated in the previous chapter that, as a result of the Lipschitz condition (**A2**), the first- and second-order moments of the gradient noise process satisfy the following two properties:

$$\textbf{(G1)}\ \mathbb{E}\left(\boldsymbol{g}_n(\boldsymbol{w}_{n-1})\,|\,\boldsymbol{w}_{n-1}\right)=0 \qquad (22.73a)$$

$$\textbf{(G2)}\ \mathbb{E}\left(\|\boldsymbol{g}_n(\boldsymbol{w}_{n-1})\|^2\,|\,\boldsymbol{w}_{n-1}\right)\leq\beta_g^2\|\widetilde{w}_{n-1}\|^2+\sigma_g^2 \qquad (22.73b)$$

for some nonnegative constants $\{\beta_g^2,\sigma_g^2\}$ that are independent of $\widetilde{w}_{n-1}$.

For operation under uniform sampling (i.e., sampling with replacement), we established in (21.25b), in view of these properties on the gradient noise process, that

$$\limsup_{n\to\infty}\ \mathbb{E}\,\|\widetilde{\boldsymbol{w}}_n\|^2=O(\mu\sigma_g^2/2\nu) \qquad (22.74)$$

where ν is the strong-convexity factor. Observe that $\mathbb{E}\,\|\widetilde{\boldsymbol{w}}_n\|^2$ does not converge to zero but rather approaches a small neighborhood of size $O(\mu)$ at an exponential rate λ^n. Moreover, the size of the neighborhood is dependent on σ_g^2. The variance-reduction techniques will employ different approximations for the gradient of $E(w)$ and reduce the value of σ_g^2 to zero.

22.5.3 Proximal SAGA Algorithm

Using the proximal operator, we show in (22.75) how the earlier SAGA implementation (22.33) can be modified to account for a nondifferentiable $q(w)$ through the incorporation of a proximal projection step. The main difference is the replacement of the earlier loss function $Q(w;\cdot,\cdot)$ for $P(w)$ by the loss function $Q_u(w;\cdot,\cdot)$ for $E(w)$, and the introduction of the intermediate variable $\boldsymbol{z}_n^k$, which is passed through a proximal operator to generate $\boldsymbol{w}_n^k$. The subscript u in $Q_u(w;\cdot)$ should not be confused with the index u used to index the random sample by the algorithm. The notation $Q_u(w;\cdot)$ is borrowed from the previous chapters, where it was used to denote the loss function associated with the "unregularized" component $E(w)$ in the composite $P(w)=q(w)+E(w)$.

Proximal SAGA for solving the empirical risk problem (22.65).

given N data points $\{\gamma(m), h_m\}$, $m = 0, 1, \ldots, N-1$;
start with arbitrary $\boldsymbol{w}_{N-1}^0$;
start with $\boldsymbol{\phi}_{N-1,u}^0 = 0$, $u = 0, 1, \ldots, N-1$;
start with $b_{N-1}^0 = \frac{1}{N}\sum_{m=0}^{N-1} \nabla_{w^\mathsf{T}} Q_u(\phi_{N-1,m}^0; \gamma(m), h_m)$.
for each run $k = 1, 2, \ldots, K$:
  set $\boldsymbol{w}_{-1}^k = \boldsymbol{w}_{N-1}^{k-1}$;
  set $\boldsymbol{\phi}_{-1,m}^k = \boldsymbol{\phi}_{N-1,u}^{k-1}$, $u = 0, 1, \ldots, N-1$
  set $\boldsymbol{b}_{-1}^k = \boldsymbol{b}_{N-1}^{k-1}$;
  repeat for $n = 0, 1, \ldots, N-1$:
    select $\boldsymbol{u}$ uniformly from the set $\{0, 1, \ldots, N-1\}$;
    select the data $(\gamma(\boldsymbol{u}), h_{\boldsymbol{u}})$;
    read the $\boldsymbol{u}$th memory variable, $\boldsymbol{\phi}_{n-1,\boldsymbol{u}}^k$;
    $\boldsymbol{z}_n^k = \boldsymbol{w}_{n-1}^k - \mu\Big(\nabla_{w^\mathsf{T}} Q_u(\boldsymbol{w}_{n-1}^k) - \nabla_{w^\mathsf{T}} Q_u(\boldsymbol{\phi}_{n-1,\boldsymbol{u}}^k) + \boldsymbol{b}_{n-1}^k\Big)$;
    $\boldsymbol{w}_n^k = \text{prox}_{\mu q}(\boldsymbol{z}_n^k)$;
    $\boldsymbol{\phi}_{n,\boldsymbol{u}}^k \leftarrow \boldsymbol{w}_n^k$;
    $\boldsymbol{\phi}_{n,u'}^k = \boldsymbol{\phi}_{n-1,u'}^k$, $u' \neq \boldsymbol{u}$;
    $\boldsymbol{b}_n^k = \boldsymbol{b}_{n-1}^k + \frac{1}{N}\Big(\nabla_{w^\mathsf{T}} Q_u(\boldsymbol{\phi}_{n,\boldsymbol{u}}^k) - \nabla_{w^\mathsf{T}} Q_u(\boldsymbol{\phi}_{n-1,\boldsymbol{u}}^k)\Big)$;
  end
end
return $w^\star \leftarrow \boldsymbol{w}_{N-1}^K$.

(22.75)

22.5.4 Proximal SAGA Gradient Noise

The gradient noise process for the proximal SAGA implementation (22.75) is given by

$$\begin{aligned} \boldsymbol{g}_n(\boldsymbol{w}_{n-1}^k) &\stackrel{\Delta}{=} \widehat{\nabla_{w^\mathsf{T}} E}(\boldsymbol{w}_{n-1}^k) - \nabla_{w^\mathsf{T}} E(\boldsymbol{w}_{n-1}^k) \qquad (22.76)\\ &= \nabla_{w^\mathsf{T}} Q_u(\boldsymbol{w}_{n-1}^k; \gamma(\boldsymbol{u}), h_{\boldsymbol{u}}) - \nabla_{w^\mathsf{T}} Q_u(\boldsymbol{\phi}_{n-1,\boldsymbol{u}}^k; \gamma(\boldsymbol{u}), h_{\boldsymbol{u}})\\ &\quad + \frac{1}{N}\sum_{u=0}^{N-1} \nabla_{w^\mathsf{T}} Q_u(\phi_{n-1,u}^k; \gamma(u), h_u\Big) - \nabla_{w^\mathsf{T}} E(\boldsymbol{w}_{n-1}^k) \end{aligned}$$

We will examine its first- and second-order moment properties. For this purpose, we consider the following collection of random variables at iteration $n-1$ and epoch k:

$$\boldsymbol{\mathcal{F}}_{n-1}^k = \Big\{\boldsymbol{w}_{n-1}^k, \boldsymbol{\phi}_{n-1,u}^k\Big\}, \; u = 0, 1, \ldots, N-1 \qquad (22.77)$$

Conditioning on $\boldsymbol{\mathcal{F}}_{n-1}^k$ and computing expectations we get

$$\begin{aligned}\mathbb{E}\left(\boldsymbol{g}_n(\boldsymbol{w}_{n-1}^k)\,|\,\boldsymbol{\mathcal{F}}_{n-1}^k\right) &= \nabla_{w^\mathsf{T}} E(w_{n-1}^k) - \frac{1}{N}\sum_{u=0}^{N-1}\nabla_{w^\mathsf{T}} Q_u(\phi_{n-1,u}^k;\gamma(u),h_u)\\ &\quad +\frac{1}{N}\sum_{u=0}^{N-1}\nabla_{w^\mathsf{T}} Q_u(\phi_{n-1,u}^k;\gamma(u),h_u)\\ &\quad -\nabla_{w^\mathsf{T}}E(w_{n-1}^k) \;=\; 0\end{aligned} \tag{22.78}$$

so that the gradient noise process has zero mean under uniform sampling *with* replacement:

$$\boxed{\mathbb{E}\left(\boldsymbol{g}_n(\boldsymbol{w}_{n-1}^k)\,|\,\boldsymbol{\mathcal{F}}_{n-1}^k\right) \;=\; 0} \tag{22.79}$$

Let us examine next the second-order moment of the modified gradient noise process. Thus, note that

$$\begin{aligned}\|\boldsymbol{g}_n(\boldsymbol{w}_{n-1}^k)\|^2 &\overset{\Delta}{=} \|\widehat{\nabla_{w^\mathsf{T}} E}(\boldsymbol{w}_{n-1}^k) - \nabla_{w^\mathsf{T}} E(\boldsymbol{w}_{n-1}^k)\|^2\\ &\overset{(22.76)}{=} \Big\|\nabla_{w^\mathsf{T}} Q_u(\boldsymbol{w}_{n-1}^k;\gamma(\boldsymbol{u}),h_{\boldsymbol{u}}) - \nabla_{w^\mathsf{T}} Q_u(\boldsymbol{\phi}_{n-1,\boldsymbol{u}}^k;\gamma(\boldsymbol{u}),h_{\boldsymbol{u}}) \;+\\ &\qquad \frac{1}{N}\sum_{u=0}^{N-1}\nabla_{w^\mathsf{T}} Q_u(\phi_{n-1,u}^k;\gamma(u),h_u) - \nabla_{w^\mathsf{T}} E(\boldsymbol{w}_{n-1}^k)\Big\|^2\end{aligned} \tag{22.80}$$

and, hence,

$$\begin{aligned}\|\boldsymbol{g}_n(\boldsymbol{w}_{n-1}^k)\|^2 &\overset{(a)}{=} \Big\|\nabla_{w^\mathsf{T}} Q_u(\boldsymbol{w}_{n-1}^k) - \nabla_{w^\mathsf{T}} Q_u(w^\star) \;+\\ &\qquad \nabla_{w^\mathsf{T}} Q_u(w^\star) - \nabla_{w^\mathsf{T}} Q_u(\boldsymbol{\phi}_{n-1,\boldsymbol{u}}^k) \;+\\ &\qquad \frac{1}{N}\sum_{u=0}^{N-1}\nabla_{w^\mathsf{T}} Q_u(\boldsymbol{\phi}_{n-1,\boldsymbol{u}}^k) - \frac{1}{N}\sum_{u=0}^{N-1}\nabla_{w^\mathsf{T}} Q_u(w^\star) +\\ &\qquad \nabla_{w^\mathsf{T}} E(w^\star) - \nabla_{w^\mathsf{T}} E(\boldsymbol{w}_{n-1}^k)\Big\|^2\\ &\overset{(b)}{\leq} 8\delta^2\|\widetilde{\boldsymbol{w}}_{n-1}^k\|^2 + 4\delta^2\|\widetilde{\boldsymbol{\phi}}_{n-1,\boldsymbol{u}}^k\|^2| + \frac{4\delta^2}{N}\sum_{u=0}^{N-1}\|\widetilde{\boldsymbol{\phi}}_{n-1,u}^k\|^2\\ &\leq 8\delta^2\|\widetilde{\boldsymbol{w}}_{n-1}^k\|^2 + \frac{4\delta^2}{N}\sum_{u=0}^{N-1}\|\widetilde{\boldsymbol{\phi}}_{n-1,u}^k\|^2 + \frac{4\delta^2}{N}\sum_{u=0}^{N-1}\|\widetilde{\boldsymbol{\phi}}_{n-1,u}^k\|^2\\ &= 8\delta^2\|\widetilde{\boldsymbol{w}}_{n-1}^k\|^2 + \frac{8\delta^2}{N}\sum_{u=0}^{N-1}\|\widetilde{\boldsymbol{\phi}}_{n-1,u}^k\|^2\end{aligned} \tag{22.81}$$

where in step (a) we removed the data arguments from the loss functions to simplify the notation, and added and subtracted the same terms $\nabla_{w^\mathsf{T}} Q_u(w^\star)$ and $\nabla_{w^\mathsf{T}} E(w^\star)$. Step (b) uses the Lipschitz conditions (22.66b) and (22.67), the Jensen inequality that $\|a+b+c+d\|^2 \leq 4\|a\|^2 + 4\|b\|^2 + 4\|c\|^2 + 4\|d\|^2$ for any vectors $\{a, b, c, d\}$ of compatible dimensions, and property (19.172). We therefore conclude that for the SAGA algorithm operating under nonsmooth risks, the second-order moment of the gradient noise continues to satisfy:

$$\boxed{\mathbb{E}\left(\|\boldsymbol{g}_n(\boldsymbol{w}_{n-1}^k)\|^2 \,|\, \boldsymbol{\mathcal{F}}_{n-1}^k\right) \;\leq\; \beta_g^2 \left(\|\widetilde{\boldsymbol{w}}_{n-1}^k\|^2 \;+\; \frac{1}{N}\sum_{u=0}^{N-1} \|\widetilde{\boldsymbol{\phi}}_{n-1,u}^k\|^2\right)} \tag{22.82}$$

for some finite nonnegative constant $\beta_g^2 = 8\delta^2$; in the next proof we will use the fact that $\nu^2 \leq \delta^2 \leq \beta_g^2$.

22.5.5 Convergence under Uniform Sampling

We are now ready to establish the *exact* convergence of the SAGA algorithm for nonsmooth risks.

THEOREM 22.4. **(Convergence of proximal SAGA under uniform sampling)** *Consider the proximal SAGA algorithm (22.75) used to seek the minimizer for the empirical risk minimization problem (22.65). The risk and loss functions are assumed to satisfy conditions* (**A1**,**A2**,**A3**)*. For step-size values satisfying (i.e., for μ small enough):*

$$\mu < \frac{\nu}{\delta^2 + 2\beta_g^2 + 2N\nu^2} \tag{22.83}$$

it holds that $\mathbb{E}\,\|\widetilde{\boldsymbol{w}}_{-1}^k\|^2$ and $\mathbb{E}\,P(\boldsymbol{w}_{-1}^k)$ converge to zero and $P(w^\star)$, respectively, at an exponential rate, namely,

$$\mathbb{E}\,\|\widetilde{\boldsymbol{w}}_{-1}^k\|^2 \leq O(\lambda^k) \tag{22.84a}$$

$$\mathbb{E}\,P(\boldsymbol{w}_{-1}^k) - P(w^\star) \leq O(\lambda^{k/2}) \tag{22.84b}$$

where

$$\lambda = \frac{1-\mu\nu}{1 - \frac{\mu^2\beta_g^2}{1-\mu N\nu}} = O(1-\mu\nu) \in [0,1) \tag{22.85}$$

Condition (22.84b) holds when $\delta_2 \neq 0$; otherwise, the excess risk converges to zero at the exponential rate λ^k. The notation $\boldsymbol{w}_{-1}^k$ refers to the weight iterate at the start of the kth epoch (which is identical to the iterate at the end of the previous epoch).

Proof: From the proximal SAGA recursion (22.75) we first note that

$$\begin{aligned}
\boldsymbol{w}_n^k &= \text{prox}_{\mu q}\left(\boldsymbol{w}_{n-1}^k - \mu\left(\nabla_{w^\mathsf{T}} Q_u(\boldsymbol{w}_{n-1}^k) - \nabla_{w^\mathsf{T}} Q_u(\boldsymbol{\phi}_{n-1,\boldsymbol{u}}^k) + \boldsymbol{b}_n^k\right)\right) \\
&= \text{prox}_{\mu q}\left(\boldsymbol{w}_{n-1}^k - \mu\widehat{\nabla_{w^\mathsf{T}} E}(\boldsymbol{w}_{n-1}^k)\right) \\
&= \text{prox}_{\mu q}\left(\boldsymbol{w}_{n-1}^k - \mu\nabla_{w^\mathsf{T}} E(\boldsymbol{w}_{n-1}^k) - \mu\boldsymbol{g}_n(\boldsymbol{w}_{n-1}^k)\right)
\end{aligned} \tag{22.86}$$

Next we know from result (11.51a) that if $w^\star$ minimizes $P(w) = q(w) + E(w)$, then it satisfies

$$w^\star = \text{prox}_{\mu q}\left(w^\star - \mu\nabla_{w^\mathsf{T}} E(w^\star)\right) \tag{22.87}$$

Using this relation and subtracting $w^\star$ from both sides of (22.86) we get

$$\begin{aligned}
\widetilde{\boldsymbol{w}}_n^k = \;& \text{prox}_{\mu q}\left(w^\star - \mu\nabla_{w^\mathsf{T}} E(w^\star)\right) \\
& -\text{prox}_{\mu q}\left(\boldsymbol{w}_{n-1}^k - \mu\nabla_{w^\mathsf{T}} E(\boldsymbol{w}_{n-1}^k) - \mu\boldsymbol{g}_n(\boldsymbol{w}_{n-1}^k)\right)
\end{aligned} \tag{22.88}$$

Squaring both sides, conditioning on $\boldsymbol{\mathcal{F}}_{n-1}^k$, and taking expectations we get

$$\begin{aligned}
&\mathbb{E}\left(\|\widetilde{\boldsymbol{w}}_n^k\|^2 | \boldsymbol{\mathcal{F}}_{n-1}^k\right) \\
&\overset{(a)}{\leq} \mathbb{E}\left(\left\|\widetilde{\boldsymbol{w}}_{n-1}^k - \mu\left(\nabla_{w^\mathsf{T}} E(w^\star) - \nabla_{w^\mathsf{T}} E(\boldsymbol{w}_{n-1}^k)\right) + \mu\boldsymbol{g}_n(\boldsymbol{w}_{n-1}^k)\right\|^2 | \boldsymbol{\mathcal{F}}_{n-1}^k\right) \\
&= \left\|\widetilde{w}_{n-1}^k - \mu\left(\nabla_{w^\mathsf{T}} E(w^\star) - \nabla_{w^\mathsf{T}} E(w_{n-1}^k)\right)\right\|^2 + \mu^2\mathbb{E}\left(\boldsymbol{g}_n(\boldsymbol{w}_{n-1}^k)\|^2 | \boldsymbol{\mathcal{F}}_{n-1}^k\right) \\
&\overset{(b)}{\leq} (1 - 2\mu\nu + \mu^2\delta^2)\|\widetilde{w}_{n-1}^k\|^2 + \mu^2\mathbb{E}\left(\|\boldsymbol{g}_n(\boldsymbol{w}_{n-1}^k)\|^2 | \boldsymbol{\mathcal{F}}_{n-1}^k\right) \\
&\overset{(22.82)}{\leq} (1 - 2\mu\nu + \mu^2(\delta^2 + \beta_g^2))\|\widetilde{w}_{n-1}^k\|^2 + \frac{\mu^2\beta_g^2}{N}\sum_{u=0}^{N-1}\|\widetilde{\phi}_{n-1,u}^k\|^2
\end{aligned} \tag{22.89}$$

where step (a) is because the proximal operator is nonexpansive – recall property (11.41), and step (b) is similar to (19.22) under the ν-strong convexity of $E(w)$ and is established in Prob. 22.8. From this point onward the proof continues in a manner similar to (22.99) in the appendix until the end of the proof of Theorem 22.2, except that to establish (22.84b) we refer to property (21.34).

■

22.5.6 Proximal SVRG Algorithm

We can pursue a similar analysis for proximal SVRG, which is shown in listing (22.90). We leave the details for the problems.

SVRG algorithm for solving the empirical risk problem (22.65).

given N data samples $\{\gamma(m), h_m\}$, $m = 0, 1, \ldots, N-1$;
start with arbitrary $\boldsymbol{w}_{N-1}^0$;
for each run $k = 1, 2, \ldots, K$:
  set $\boldsymbol{w}_{-1}^k = \boldsymbol{w}_{N-1}^{k-1}$;
  set $\boldsymbol{\phi}_{-1}^k = \boldsymbol{w}_{N-1}^{k-1}$;
  set $\boldsymbol{b}^k = \frac{1}{N}\sum_{u=0}^{N-1} \nabla_{w^\mathsf{T}} Q_u(\boldsymbol{\phi}_{-1}^k, \gamma(u), h_u)$;
  repeat for $n = 0, 1, \ldots, N-1$:
    select $\boldsymbol{u}$ uniformly from the set $\{0, 1, \ldots, N-1\}$;
    select the data $(\gamma(\boldsymbol{u}), h_{\boldsymbol{u}})$;
    $\boldsymbol{z}_n^k = \boldsymbol{w}_{n-1}^k - \mu\Big(\nabla_{w^\mathsf{T}} Q_u(\boldsymbol{w}_{n-1}^k) - \nabla_{w^\mathsf{T}} Q_u(\boldsymbol{\phi}_{-1}^k) + \boldsymbol{b}^k\Big)$;
    $\boldsymbol{w}_n^k = \text{prox}_{\mu q}(\boldsymbol{z}_n^k)$;
  end
end
return $w^\star \leftarrow w_{N-1}^K$. (22.90)

22.6 COMMENTARIES AND DISCUSSION

Variance-reduced techniques. There are several variations of variance-reduced techniques. In this chapter we discussed SAGA (stochastic average-gradient algorithm) and SVRG (stochastic variance-reduced gradient algorithm). These variants are sufficient to convey the main properties of this class of algorithms. The SVRG algorithm was proposed by Johnson and Zhang (2013), while SAGA was proposed by Defazio, Bach, and Lacoste-Julien (2014). There are other variance-reduction techniques such as Finito by Defazio, Domke, and Caetano (2014), and SAG (stochastic average gradient) by Le Roux, Schmidt, and Bach (2012). Some other useful references on variance-reduced methods include the works by Shalev-Shwartz and Zhang (2013), Xiao and Zhang (2014), Gürbüzbalaban, Ozdaglar, and Parrilo (2015a), Reddi *et al.* (2016a,b), Schmidt, Le Roux, and Bach (2017), Shang *et al.* (2018), Poon, Liang, and Schoenlieb (2018), and Dubois-Taine *et al.* (2021). A discussion on federated learning and SVRG appears in Konecny *et al.* (2016). The derivations in the body of the chapter and the proofs in Appendices 22.A and 22.B follow the work by Ying, Yuan, and Sayed (2018, 2020).

The SAGA and SVRG implementations are able to guarantee *exact* convergence of the weight iterates $\boldsymbol{w}_n$ to the minimizer $w^\star$. They differ nevertheless in one important feature. We remarked in the body of the chapter that rather than store the memory variables $\{\phi_{n,u}^k\}$ directly, it is more efficient for SAGA to store the gradient values at these locations, namely, the quantities $\nabla_{w^\mathsf{T}} Q(\phi_{n,u}^k; \gamma(u), h_u)$. This is because evaluation of the sample average b_n^k requires these gradient values. In this way, the SAGA recursion (22.33) requires $O(N)$ storage locations, one for each index $0 \leq u \leq N-1$. For each epoch, the total number of gradient calculations that are needed is on the order of $O(N)$. This is because, for each iteration n, there is one gradient calculation for $\nabla_{w^{\mathsf{T}}} Q(\boldsymbol{w}_{n-1}^k; \gamma(\boldsymbol{u}), \boldsymbol{h}_{\boldsymbol{u}})$.

Since the size of N can be prohibitive for large datasets, the SVRG implementation (22.48) seeks to reduce the storage requirement. It replaces the history variables $\{\boldsymbol{\phi}_{n,u}^k\}$ for SAGA by a *fixed* value computed at the start of each epoch, and denoted by $\boldsymbol{\phi}_{-1}^k$. This vector stays invariant during the *entire* epoch. Although this simplification greatly reduces the storage needs, it however requires a demanding aggregation step prior to the start of each epoch to compute the variable:

$$\boldsymbol{b}^k = \frac{1}{N}\sum_{u=0}^{N-1} \nabla_{w^{\mathsf{T}}} Q(\boldsymbol{\phi}_{-1}^k, \gamma(u), h_u) \tag{22.91}$$

This computation can be time-consuming for large datasets. As a result, the operation of SVRG becomes *unbalanced*, with a larger amount of time needed before each epoch and a shorter time needed within the epoch itself. Observe further that for each epoch, the total number of gradient calculations that are needed are on the order of $O(2N)$. This is because, at the start of each epoch, there are N gradient calculations needed for $\boldsymbol{b}^k$ and, moreover, for each iteration n, there is one gradient calculation needed for $\nabla_{w^{\mathsf{T}}} Q(\boldsymbol{w}_{n-1}^k; \gamma(\boldsymbol{u}), \boldsymbol{h}_{\boldsymbol{u}})$. Since n runs over $n = 0, \ldots, N-1$, we arrive at the estimated figure of $O(2N)$ gradient calculations per epoch:

$$\textsf{(SAGA)}\;\; O(N) \text{ gradient calculations per epoch} \tag{22.92a}$$

$$\textsf{(SVRG)}\;\; O(2N) \text{ gradient calculations per epoch} \tag{22.92b}$$

Taylor series. In the proof of convergence for the SVRG algorithm we employed the following form of the Taylor series expansion. Consider a differentiable function $f(x)$: $\mathbb{R} \rightarrow \mathbb{R}$ and let us expand it around some point $a \leq x$. Then,

$$f(x) = f(a) + (x-a)f'(a) + \frac{1}{2!}(x-a)^2 f''(a) + \ldots + \frac{1}{M!}(x-a)^M f^{(M)}(a) + R_M(a) \tag{22.93}$$

where the notation $f'(a), f''(a)$, and $f^{(M)}(a)$ denotes the first-, second-, and Mth-order derivatives of $f(x)$ at location $x = a$. Moreover, M is an integer and the term $R_M(a)$ denotes the residual that is left if we employ M terms in the Taylor series expansion. It is known that the residual term can be expressed in the following form – see, e.g., Abramowitz and Stegun (1965) and Whittaker and Watson (1927):

$$R_M(a) = \frac{1}{(M+1)!}(x-a)^{M+1} f^{(M+1)}(x_a) \tag{22.94}$$

for some point $x_a \in (a, x)$. For example, when $M = 1$, we would obtain

$$f(x) = f(a) + (x-a)f'(a) + \frac{1}{2}(x-a)^2 f''(x_a) \tag{22.95}$$

Applying this form to the function $f(x) = (1-x)^N$ leads to (22.141).

PROBLEMS

22.1 If desired, we can tighten the bound in (22.17) by showing that $\beta_g^2 = 4\delta^2$ can be replaced by $\beta_g^2 = \delta^2$ as follows. Use the fact that, for any scalar random variable $\boldsymbol{x}$, we have $\mathbb{E}\,(\boldsymbol{x} - \mathbb{E}\,\boldsymbol{x})^2 \leq \mathbb{E}\,\boldsymbol{x}^2$ to show that

$$\mathbb{E}\left(\|\boldsymbol{g}_n(\boldsymbol{w}_{n-1})\|^2 \,|\, \boldsymbol{w}_{n-1}\right) \leq \mathbb{E}\left(\|\widehat{\nabla_{w^\mathsf{T}} P}(\boldsymbol{w}_{n-1})\|^2 \,|\, \boldsymbol{w}_{n-1}\right)$$

Conclude that $\mathbb{E}\,(\|\boldsymbol{g}_n(\boldsymbol{w}_{n-1})\|^2 \,|\, \boldsymbol{w}_{n-1}) \leq \delta^2 \|\widetilde{w}_{n-1}\|^2$.

22.2 Extend the arguments from Theorem 19.1 to establish the validity of Theorem 22.1 for the convergence of the naïve stochastic gradient algorithm with instantaneous gradient approximation.

22.3 Consider a mini-batch version of the naïve stochastic gradient algorithm (22.12) where the B samples in the mini-batch can be chosen either uniformly (i.e., with replacement) or without replacement. Extend the analysis from Theorem 19.2 to this case to establish the convergence of the naïve implementation to the exact minimizer.

22.4 Refer to the SAGA algorithm (22.33) and consider a mini-batch version in the manner listed below (with the updates written for the variables within the kth epoch, and with the letter p used to index the mini-batch samples instead of b to avoid confusion with the variable $\boldsymbol{b}_{n-1}^k$). Examine the convergence behavior of the algorithm.

repeat for $n = 0, 1, \ldots, N-1$**:**
  select B samples uniformly with replacement from the set $\{0, 1, \ldots, N-1\}$;
  select the mini-batch samples $\{\boldsymbol{\gamma}(p), \boldsymbol{h}_p\}$, $p = 0, 1, \ldots, B-1$;
  read the B memory variables, $\{\boldsymbol{\phi}_{n-1,p}^k\}$, $p = 0, 1, \ldots, B-1$;
  compute $\boldsymbol{a}_{n,1} = \dfrac{1}{B}\displaystyle\sum_{p=0}^{B-1} \nabla_{w^\mathsf{T}} Q(\boldsymbol{w}_{n-1}^k; \boldsymbol{\gamma}(p), \boldsymbol{h}_p)$;
  compute $\boldsymbol{a}_{n,2} = \dfrac{1}{B}\displaystyle\sum_{p=0}^{B-1} \nabla_{w^\mathsf{T}} Q(\boldsymbol{\phi}_{n-1,p}^k; \boldsymbol{\gamma}(p), \boldsymbol{h}_p)$;
  $\boldsymbol{w}_n^k = \boldsymbol{w}_{n-1}^k - \mu(\boldsymbol{a}_{n,1} - \boldsymbol{a}_{n,2} + \boldsymbol{b}_{n-1}^k)$;
  $\boldsymbol{\phi}_{n,p}^k \leftarrow \boldsymbol{w}_n^k$, $p = 0, 1, \ldots, B-1$;
  $\boldsymbol{\phi}_{n,u'}^k = \boldsymbol{\phi}_{n-1,u'}^k$, $u' \notin B$ selected samples;
  $\boldsymbol{b}_n^k = \boldsymbol{b}_{n-1}^k + \dfrac{B}{N}(\boldsymbol{a}_{2,n} - \boldsymbol{a}_{2,n-1})$;
end

22.5 Continuing with Prob. 22.4, assume we adjust the algorithm by applying the mini-batch construction to $\boldsymbol{a}_{n,1}$ only so that the update for the weight vector becomes

$$\boldsymbol{w}_n^k = \boldsymbol{w}_{n-1}^k - \mu(\boldsymbol{a}_{n,1} - \nabla_{w^\mathsf{T}} Q(\boldsymbol{\phi}_{n-1,\boldsymbol{u}}^k) + \boldsymbol{b}_{n-1}^k)$$

where $\boldsymbol{u}$ is selected at random from the set $\{0, 1, \ldots, N-1\}$, and only $\boldsymbol{\phi}_{n,\boldsymbol{u}}^k$ is updated with $\boldsymbol{w}_n^k$. Moreover, the variable $\boldsymbol{b}_n^k$ continues to be updated as

$$\boldsymbol{b}_n^k = \boldsymbol{b}_{n-1}^k + \frac{1}{N}\Big(\nabla_{w^\mathsf{T}} Q(\boldsymbol{\phi}_{n,\boldsymbol{u}}^k) - \nabla_{w^\mathsf{T}} Q(\boldsymbol{\phi}_{n-1,\boldsymbol{u}}^k)\Big)$$

Examine the convergence behavior of this implementation.

22.6 Refer to the SVRG algorithm (22.48) and consider a mini-batch version in the manner listed below (with the updates written for the variables within the kth epoch, and with the letter p used to index the mini-batch samples instead of b to avoid confusion with the variable $\boldsymbol{b}^k$):

repeat for $n = 0, 1, \ldots, N-1$:
select B samples uniformly with replacement from the set $\{0, 1, \ldots, N-1\}$;
select the mini-batch samples $\{\boldsymbol{\gamma}(p), \boldsymbol{h}_p\}$, $p = 0, 1, \ldots, B-1$;
read the B memory variables, $\{\boldsymbol{\phi}^k_{n-1,p}\}$, $p = 0, 1, \ldots, B-1$;
compute $\boldsymbol{a}_{n,1} = \dfrac{1}{B}\sum_{p=0}^{B-1} \nabla_{w^\mathsf{T}} Q(\boldsymbol{w}^k_{n-1}; \boldsymbol{\gamma}(b), \boldsymbol{h}_b)$;
compute $\boldsymbol{a}_{n,2} = \dfrac{1}{B}\sum_{p=0}^{B-1} \nabla_{w^\mathsf{T}} Q(\boldsymbol{\phi}^k_{-1}; \boldsymbol{\gamma}(p), \boldsymbol{h}_p)$;
$\boldsymbol{w}^k_n = \boldsymbol{w}^k_{n-1} - \mu(\boldsymbol{a}_{n,1} - \boldsymbol{a}_{n,2} + \boldsymbol{b}^k)$;
$\boldsymbol{\phi}^k_{n,p} \leftarrow \boldsymbol{w}^k_n$, $p = 0, 1, \ldots, B-1$;
$\boldsymbol{\phi}^k_{n,u'} = \boldsymbol{\phi}^k_{n-1,u'}$, $u' \notin B$ selected samples;
end

22.7 Continuing with Prob. 22.6, assume again that we adjust the algorithm by applying the mini-batch construction to $\boldsymbol{a}_{n,1}$ only so that the update for the weight vector becomes

$$\boldsymbol{w}^k_n = \boldsymbol{w}^k_{n-1} - \mu(\boldsymbol{a}_{n,1} - \nabla_{w^\mathsf{T}} Q(\boldsymbol{\phi}^k_{n-1,\boldsymbol{u}}) + \boldsymbol{b}^k_{n-1})$$

where $\boldsymbol{u}$ is selected at random from the set $\{0, 1, \ldots, N-1\}$, and only $\boldsymbol{\phi}^k_{n,\boldsymbol{u}}$ is updated with $\boldsymbol{w}^k_n$. Examine the convergence behavior of this implementation.

22.8 Refer to step (b) in (22.89). Show that

$$(\widetilde{\boldsymbol{w}}^k_{n-1})^\mathsf{T}\Big(\nabla_{w^\mathsf{T}} E(\boldsymbol{w}^k_{n-1}) - \nabla_{w^\mathsf{T}} E(w^\star)\Big) \leq -\nu\|\widetilde{\boldsymbol{w}}^k_{n-1}\|^2$$

22.9 Refer to the discussion in Section 22.5 on nonsmooth risk functions. Extend the analysis to examine the convergence behavior of the proximal SVRG algorithm (22.90) under uniform sampling and nonsmooth risks satisfying conditions (**A1**,**A2**,**A3**).

22.10 Consider the empirical risk minimization problem (22.2). The SAG (stochastic average gradient) algorithm by Le Roux, Schmidt, and Bach (2012) can be described as follows. A memory matrix Φ of size $N \times M$ is maintained with columns denoted by $\{\phi_u\}$. Each column ϕ_u saves the last gradient calculation that used the data sample $(\gamma(u), h_u)$. The notation $\phi_{n,u}$ refers to the value of the uth column saved at iteration n. The algorithm is described as follows:

repeat until convergence over $n \geq 0$:
select an index $\boldsymbol{u}$ uniformly at random from the set $\{0, 1, \ldots, N-1\}$;
select the sample $(\gamma(\boldsymbol{u}), h_{\boldsymbol{u}}\}$;
update the $\boldsymbol{u}$th column, $\boldsymbol{\phi}_{n,\boldsymbol{u}} = \nabla_{w^\mathsf{T}} Q(\boldsymbol{w}_{n-1}; \gamma(\boldsymbol{u}), h_{\boldsymbol{u}})$;
retain the other columns: $\boldsymbol{\phi}_{n,u'} = \boldsymbol{\phi}_{n-1,u'}$, $u' \neq \boldsymbol{u}$;
compute $\boldsymbol{a}_n = \dfrac{1}{N}\sum_{u=0}^{N-1} \boldsymbol{\phi}_{n,u}$;
update $\boldsymbol{w}_n = \boldsymbol{w}_{n-1} - \mu\, \boldsymbol{a}_n$;
end

(a) Verify that $\boldsymbol{a}_n$ can be updated recursively using $\boldsymbol{a}_n = \boldsymbol{a}_{n-1} + \frac{1}{N}(\boldsymbol{\phi}_{n,\boldsymbol{u}} - \boldsymbol{\phi}_{n-1,\boldsymbol{u}})$.

(b) Assume the risk and loss functions satisfy conditions (**A1**,**A2**) in the smooth case. Examine the first- and second-order moments of the resulting gradient noise process.

(c) Study the convergence behavior of the algorithm.

22.A PROOF OF THEOREM 22.2

The argument in this appendix is based on the derivations from Ying, Yuan, and Sayed (2018, 2020) and Yuan *et al.* (2019). From the SAGA recursion (22.33) and expression (22.35) for the gradient noise process, we have

$$\boldsymbol{w}_n^k = \boldsymbol{w}_{n-1}^k - \mu\left(\nabla_{w^\mathsf{T}} P(\boldsymbol{w}_{n-1}^k) + \boldsymbol{g}_n(\boldsymbol{w}_{n-1}^k)\right) \tag{22.96}$$

Subtracting $w^\star$ from both sides gives the error recursion:

$$\widetilde{\boldsymbol{w}}_n^k = \widetilde{\boldsymbol{w}}_{n-1}^k + \mu\left(\nabla_{w^\mathsf{T}} P(\boldsymbol{w}_{n-1}^k) + \boldsymbol{g}_n(\boldsymbol{w}_{n-1}^k)\right) \tag{22.97}$$

Squaring both sides, conditioning on $\boldsymbol{\mathcal{F}}_{n-1}^k$, and taking expectations we get

$$\begin{aligned}
&\mathbb{E}\left(\|\widetilde{\boldsymbol{w}}_n^k\|^2 \,|\, \boldsymbol{\mathcal{F}}_{n-1}^k\right) \qquad\qquad (22.98)\\
&= \|\widetilde{w}_{n-1}^k + \mu\nabla_{w^\mathsf{T}} P(w_{n-1}^k)\|^2 + \mu^2\mathbb{E}\left(\|\boldsymbol{g}_n(\boldsymbol{w}_{n-1}^k)\|^2 \,|\, \boldsymbol{\mathcal{F}}_{n-1}^k\right)\\
&\overset{(19.23)}{\leq} (1-2\mu\nu+\mu^2\delta^2)\|\widetilde{w}_{n-1}^k\|^2 + \mu^2\mathbb{E}\left(\|\boldsymbol{g}_n(\boldsymbol{w}_{n-1}^k)\|^2 \,|\, \boldsymbol{\mathcal{F}}_{n-1}^k\right)\\
&\overset{(22.42)}{\leq} (1-2\mu\nu+\mu^2(\delta^2+\beta_g^2))\|\widetilde{w}_{n-1}^k\|^2 + \frac{\mu^2\beta_g^2}{N}\sum_{u=0}^{N-1}\|\widetilde{\phi}_{n-1,u}^k\|^2
\end{aligned}$$

By taking expectation again, we remove the conditional calculation and obtain

$$\begin{aligned}
\mathbb{E}\,\|\widetilde{\boldsymbol{w}}_n^k\|^2 &\leq (1-2\mu\nu+\mu^2(\delta^2+\beta_g^2))\,\mathbb{E}\,\|\widetilde{\boldsymbol{w}}_{n-1}^k\|^2 + \frac{\mu^2\beta_g^2}{N}\sum_{u=0}^{N-1}\mathbb{E}\,\|\widetilde{\boldsymbol{\phi}}_{n-1,u}^k\|^2\\
&\leq (1-\mu\nu)\mathbb{E}\,\|\widetilde{\boldsymbol{w}}_{n-1}^k\|^2 + \frac{\mu^2\beta_g^2}{N}\sum_{u=0}^{N-1}\mathbb{E}\,\|\widetilde{\boldsymbol{\phi}}_{n-1,u}^k\|^2
\end{aligned} \tag{22.99}$$

where the last inequality holds whenever

$$\mu \leq \frac{\nu}{\delta^2+\beta_g^2} \tag{22.100}$$

Relation (22.99) is the first inequality we need in our convergence analysis. We establish a second inequality.

Note that at iteration n at epoch k, the update for $\boldsymbol{\phi}_{n,\boldsymbol{u}}^k$ is

$$\begin{cases} \boldsymbol{\phi}_{n,u}^k = \boldsymbol{w}_n^k, & \text{if } \boldsymbol{u} = u\\ \boldsymbol{\phi}_{n,u'}^k = \boldsymbol{\phi}_{n-1,u'}^k, & \text{for } \boldsymbol{u} \neq u' \end{cases} \tag{22.101}$$

The first event occurs with probability $1/N$ while the second event occurs with probability $(N-1)/N$. It follows that

$$\begin{aligned}
&\mathbb{E}\left(\|\widetilde{\boldsymbol{\phi}}^k_{n,\boldsymbol{u}}\|^2 \,|\, \boldsymbol{\mathfrak{F}}^k_{n-1}\right)\\
&\overset{(a)}{=} \sum_{u=0}^{N-1} \mathbb{P}(\boldsymbol{u}=u|\boldsymbol{\mathfrak{F}}^k_{n-1})\,\mathbb{E}\left(\|\widetilde{\boldsymbol{\phi}}^k_{n,\boldsymbol{u}}\|^2 \,|\, \boldsymbol{\mathfrak{F}}^k_{n-1}, \boldsymbol{u}=u\right)\\
&\overset{(b)}{=} \sum_{u=0}^{N-1} \mathbb{P}(\boldsymbol{u}=u)\,\mathbb{E}\left(\|\widetilde{\boldsymbol{\phi}}^k_{n,\boldsymbol{u}}\|^2 \,|\, \boldsymbol{\mathfrak{F}}^k_{n-1}, \boldsymbol{u}=u\right)\\
&= \frac{1}{N}\mathbb{E}\left(\|\widetilde{\boldsymbol{w}}^k_n\|^2 \,|\, \boldsymbol{\mathfrak{F}}^k_{n-1}, \boldsymbol{u}=u\right) + \frac{N-1}{N}\mathbb{E}\left(\|\widetilde{\boldsymbol{\phi}}^k_{n-1,\boldsymbol{u}}\|^2 \,|\, \boldsymbol{\mathfrak{F}}^k_{n-1}, \boldsymbol{u}\neq u\right)\\
&= \frac{1}{N}\mathbb{E}\left(\|\widetilde{\boldsymbol{w}}^k_n\|^2 \,|\, \boldsymbol{\mathfrak{F}}^k_{n-1}\right) + \frac{N-1}{N}\mathbb{E}\,\|\widetilde{\boldsymbol{\phi}}^k_{n-1,\boldsymbol{u}}\|^2
\end{aligned} \tag{22.102}$$

where step (a) uses the result of Prob. 3.27 and step (b) is because $\boldsymbol{u}$ is independent of $\boldsymbol{\mathfrak{F}}^k_{n-1}$. By taking expectations again and summing over the index u, we get

$$\frac{1}{N}\sum_{u=0}^{N-1}\mathbb{E}\,\|\widetilde{\boldsymbol{\phi}}^k_{n,u}\|^2 = \frac{1}{N}\mathbb{E}\,\|\widetilde{\boldsymbol{w}}^k_n\|^2 + \frac{N-1}{N}\left(\frac{1}{N}\sum_{u=0}^{N-1}\mathbb{E}\,\|\widetilde{\boldsymbol{\phi}}^k_{n-1,u}\|^2\right) \tag{22.103}$$

In summary, under (22.100), we arrive at the following two inequalities:

$$\mathbb{E}\,\|\widetilde{\boldsymbol{w}}^k_n\|^2 \overset{(22.99)}{\leq} (1-\mu\nu)\mathbb{E}\,\|\widetilde{\boldsymbol{w}}^k_{n-1}\|^2 + \mu^2\beta_g^2 G^k_{n-1} \tag{22.104a}$$

$$G^k_n \overset{(22.103)}{=} (1-1/N)G^k_{n-1} + \frac{1}{N}\mathbb{E}\,\|\widetilde{\boldsymbol{w}}^k_n\|^2 \tag{22.104b}$$

where we introduced the variable

$$G^k_n \triangleq \frac{1}{N}\sum_{u=0}^{N-1}\mathbb{E}\,\|\widetilde{\boldsymbol{\phi}}^k_{n,u}\|^2 \tag{22.105}$$

We now use these inequalities to establish the convergence of an energy function of the form:

$$V^k_n \triangleq \mathbb{E}\,\|\widetilde{\boldsymbol{w}}^k_n\|^2 + \gamma G^k_n \tag{22.106}$$

where γ is some positive constant whose value will be decided later (see (22.112)). Using (22.104a) and (22.104b), we have

$$\mathbb{E}\,\|\widetilde{\boldsymbol{w}}^k_n\|^2 + \gamma G^k_n \leq (1-\mu\nu)\mathbb{E}\,\|\widetilde{\boldsymbol{w}}^k_{n-1}\|^2 + \frac{\gamma}{N}\mathbb{E}\,\|\widetilde{\boldsymbol{w}}^k_n\|^2 + \left(\mu^2\beta_g^2 + \gamma(1-1/N)\right)G^k_{n-1} \tag{22.107}$$

which can be rearranged into

$$(1-\gamma/N)\mathbb{E}\,\|\widetilde{\boldsymbol{w}}^k_n\|^2 + \gamma G^k_n \leq (1-\mu\nu)\mathbb{E}\,\|\widetilde{\boldsymbol{w}}^k_{n-1}\|^2 + \left(\mu^2\beta_g^2 + \gamma(1-1/N)\right)G^k_{n-1} \tag{22.108}$$

If γ satisfies

$$1-\gamma/N > 0 \tag{22.109}$$

then relation (22.108) is equivalent to

$$\begin{aligned}
\mathbb{E}\,\|\widetilde{\boldsymbol{w}}^k_n\|^2 + \frac{\gamma}{1-\gamma/N}G^k_n &\leq \frac{1-\mu\nu}{1-\gamma/N}\mathbb{E}\,\|\widetilde{\boldsymbol{w}}^k_{n-1}\|^2 + \frac{\mu^2\beta_g^2+\gamma(1-1/N)}{1-\gamma/N}G^k_{n-1}\\
&= \left(\frac{1-\mu\nu}{1-\gamma/N}\right)\left(\mathbb{E}\,\|\widetilde{\boldsymbol{w}}^k_{n-1}\|^2 + \frac{\mu^2\beta_g^2+\gamma(1-1/N)}{1-\mu\nu}G^k_{n-1}\right)
\end{aligned} \tag{22.110}$$

Note that $0 < 1 - \gamma/N < 1$ and, hence, the above inequality also implies that

$$\mathbb{E}\,\|\widetilde{\boldsymbol{w}}_n^k\|^2 + \gamma G_n^k \leq \left(\frac{1-\mu\nu}{1-\gamma/N}\right)\left(\mathbb{E}\,\|\widetilde{\boldsymbol{w}}_{n-1}^k\|^2 + \frac{\mu^2\beta_g^2 + \gamma(1-1/N)}{1-\mu\nu} G_{n-1}^k\right) \tag{22.111}$$

If we select

$$\mu < \frac{1}{\nu N} \quad \text{and} \quad \gamma = \frac{\mu^2\beta_g^2}{1/N - \mu\nu} \tag{22.112}$$

then γ satisfies

$$\gamma = \frac{\mu^2\beta_g^2 + \gamma(1-1/N)}{1-\mu\nu} \tag{22.113}$$

and relation (22.111) becomes

$$\mathbb{E}\,\|\widetilde{\boldsymbol{w}}_n^k\|^2 + \gamma G_n^k \leq \lambda\left(\mathbb{E}\,\|\widetilde{\boldsymbol{w}}_{n-1}^k\|^2 + \gamma G_{n-1}^k\right), \quad \text{where} \quad \lambda = \frac{1-\mu\nu}{1-\gamma/N} \tag{22.114}$$

Using the expression for γ from (22.112), we have

$$\lambda = \frac{1-\mu\nu}{1 - \frac{\mu^2\beta_g^2}{1-\mu N\nu}} \tag{22.115}$$

and it holds that

$$\lambda < 1 \iff \mu < \frac{\nu}{\beta_g^2 + N\nu^2} \tag{22.116}$$

It is also easy to see that requirement (22.109) is satisfied by selecting

$$\mu < \min\left\{\frac{1}{2N\nu}, \frac{1}{\sqrt{2}\beta_g}\right\} \tag{22.117}$$

This is because

$$\gamma = \frac{\mu^2\beta_g^2}{1/N - \mu\nu} \overset{(22.117)}{<} \frac{\mu^2\beta_g^2}{1/2N} = 2N\mu^2\beta_g^2 \overset{(22.117)}{<} N \tag{22.118}$$

Note further from $\nu \leq \beta_g$ that the condition $\mu < 1/\sqrt{2}\beta_g$ is satisfied since

$$\mu \overset{(22.43)}{<} \frac{\nu}{2\beta_g^2} < \frac{\nu}{2\beta_g\nu} < \frac{\nu}{\sqrt{2}\beta_g\nu} < \frac{1}{\sqrt{2}\beta_g} \tag{22.119}$$

Therefore, the four conditions on μ given by (22.100), (22.112), (22.116), and (22.117) are satisfied by a small enough μ such that

$$\mu < \frac{\nu}{\delta^2 + 2\beta_g^2 + 2N\nu^2} \tag{22.120}$$

Under this condition we get

$$\mathbb{E}\,\|\widetilde{\boldsymbol{w}}_n^k\|^2 + \gamma G_n^k \leq \lambda\left(\mathbb{E}\,\|\widetilde{\boldsymbol{w}}_{n-1}^k\|^2 + \gamma G_{n-1}^k\right), \quad \text{where} \quad \lambda < 1 \tag{22.121}$$

Iterating over $n = 0, 1, \ldots, N-1$ and noting that $\widetilde{\boldsymbol{w}}_{-1}^{k+1} = \widetilde{\boldsymbol{w}}_{N-1}^k$ and $\boldsymbol{\phi}_{-1}^{k+1} = \boldsymbol{\phi}_{N-1}^k$ (so that $G_{-1}^{k+1} = G_{N-1}^k$), we find that

$$\mathbb{E}\,\|\widetilde{\boldsymbol{w}}_{-1}^{k+1}\|^2 + \gamma G_{-1}^{k+1} \leq \lambda^N\left(\mathbb{E}\,\|\widetilde{\boldsymbol{w}}_{-1}^k\|^2 + \gamma G_{-1}^k\right) \tag{22.122}$$

As iterating gives

$$\mathbb{E}\,\|\widetilde{\boldsymbol{w}}_{-1}^{k+1}\|^2 + \gamma G_{-1}^{k+1} \le \lambda^{kN}\left(\mathbb{E}\,\|\widetilde{\boldsymbol{w}}_{-1}^{1}\|^2 + \gamma G_{-1}^{1}\right) \tag{22.123}$$

from which we conclude that

$$\mathbb{E}\,\|\widetilde{\boldsymbol{w}}_{-1}^{k+1}\|^2 \le C\lambda^{kN} \tag{22.124}$$

where $C = \mathbb{E}\,\|\widetilde{\boldsymbol{w}}_{-1}^{1}\|^2 + \gamma G_{-1}^{1}$. In other words, SAGA converges to the exact solution exponentially fast. To establish (22.44b), we use (22.6b) to note that

$$0 \le \mathbb{E}\,P(\boldsymbol{w}_n^k) - P(w^\star) \le \frac{\delta}{2}\,\mathbb{E}\,\|\widetilde{\boldsymbol{w}}_n^k\|^2 \;\le\; O(\lambda^{(k-1)N}) \tag{22.125}$$

22.B PROOF OF THEOREM 22.3

The argument in this appendix is based on the derivations from Ying, Yuan, and Sayed (2018, 2020) and Yuan *et al.* (2019). From the SVRG recursion (22.48) and expression (22.49) for the gradient noise process, we have

$$\boldsymbol{w}_n^k = \boldsymbol{w}_{n-1}^k - \mu\left(\nabla_{w^\mathsf{T}} P(\boldsymbol{w}_{n-1}^k) + \boldsymbol{g}_n(\boldsymbol{w}_{n-1}^k)\right) \tag{22.126}$$

Subtracting $w^\star$ from both sides gives the error recursion:

$$\widetilde{\boldsymbol{w}}_n^k = \widetilde{\boldsymbol{w}}_{n-1}^k + \mu\left(\nabla_{w^\mathsf{T}} P(\boldsymbol{w}_{n-1}^k) + \boldsymbol{g}_n(\boldsymbol{w}_{n-1}^k)\right) \tag{22.127}$$

Squaring both sides, conditioning on $\boldsymbol{\mathcal{F}}_{n-1}^k$, and taking expectations we get

$$\begin{aligned}
\mathbb{E}\left(\|\widetilde{\boldsymbol{w}}_n^k\|^2 \,|\, \boldsymbol{\mathcal{F}}_{n-1}^k\right) &= \left\|\widetilde{w}_{n-1}^k + \mu\nabla_{w^\mathsf{T}} P(w_{n-1}^k)\right\|^2 + \mu^2\mathbb{E}\left(\|\boldsymbol{g}_n(\boldsymbol{w}_{n-1}^k)\|^2 \,|\, \boldsymbol{\mathcal{F}}_{n-1}^k\right)\\
&\overset{(19.22)}{\le} (1 - 2\mu\nu + \mu^2\delta^2)\|\widetilde{w}_{n-1}^k\|^2 + \mu^2\mathbb{E}\left(\|\boldsymbol{g}_n(\boldsymbol{w}_{n-1}^k)\|^2 \,|\, \boldsymbol{\mathcal{F}}_{n-1}^k\right)\\
&\overset{(22.42)}{\le} (1 - 2\mu\nu + \mu^2(\delta^2 + \beta_g^2))\|\widetilde{w}_{n-1}^k\|^2 + \mu^2\beta_g^2\|\widetilde{\phi}_{-1}^k\|^2
\end{aligned} \tag{22.128}$$

By taking expectations again, we remove the conditional calculation and obtain

$$\begin{aligned}
\mathbb{E}\,\|\widetilde{\boldsymbol{w}}_n^k\|^2 &\le (1 - 2\mu\nu + \mu^2(\delta^2 + \beta_g^2))\,\mathbb{E}\,\|\widetilde{\boldsymbol{w}}_{n-1}^k\|^2 + \mu^2\beta_g^2\mathbb{E}\,\|\widetilde{\boldsymbol{\phi}}_{-1}^k\|^2\\
&\le (1 - \mu\nu)\,\mathbb{E}\,\|\widetilde{\boldsymbol{w}}_{n-1}^k\|^2 + \mu^2\beta_g^2\,\mathbb{E}\,\|\widetilde{\boldsymbol{\phi}}_{-1}^k\|^2
\end{aligned} \tag{22.129}$$

where the second inequality holds when

$$\mu \le \frac{\nu}{\delta^2 + \beta_g^2} \tag{22.130}$$

Noting that, by construction, $\widetilde{\boldsymbol{\phi}}_{-1}^k = \widetilde{\boldsymbol{w}}_{N-1}^{k-1} = \widetilde{\boldsymbol{w}}_{-1}^k$, we rewrite the above inequality as

$$\mathbb{E}\,\|\widetilde{\boldsymbol{w}}_n^k\|^2 \le (1 - \mu\nu)\,\mathbb{E}\,\|\widetilde{\boldsymbol{w}}_{n-1}^k\|^2 + \mu^2\beta_g^2\,\mathbb{E}\,\|\widetilde{\boldsymbol{w}}_{-1}^k\|^2 \tag{22.131}$$

Iterating (22.131) over $n = 0, 1, \ldots, N-1$, we get

$$\begin{aligned}
\mathbb{E}\,\|\widetilde{\boldsymbol{w}}_{N-1}^k\|^2 &\le (1 - \mu\nu)^N\mathbb{E}\,\|\widetilde{\boldsymbol{w}}_{-1}^k\|^2 + \mu^2\beta_g^2\mathbb{E}\,\|\widetilde{\boldsymbol{w}}_{-1}^k\|^2\sum_{n=0}^{N-1}(1 - \mu\nu)^n\\
&\le (1 - \mu\nu)^N\mathbb{E}\,\|\widetilde{\boldsymbol{w}}_{-1}^k\|^2 + \frac{\mu\beta_g^2}{\nu}\mathbb{E}\,\|\widetilde{\boldsymbol{w}}_{-1}^k\|^2
\end{aligned} \tag{22.132}$$

under the condition $\mu < 1/\nu$ so that $0 < 1 - \mu\nu < 1$. We verify at the end of this proof that when the step size satisfies

$$\mu < 1/N\nu \tag{22.133}$$

then it holds that

$$(1-\mu\nu)^N \;\leq 1 - \frac{\mu\nu N}{2} \tag{22.134}$$

Substituting into (22.132) gives

$$\mathbb{E}\,\|\widetilde{\boldsymbol{w}}^k_{N-1}\|^2 \leq \left(1 - \frac{\mu N\nu}{2} + \frac{\mu\beta_g^2}{\nu}\right)\mathbb{E}\,\|\widetilde{\boldsymbol{w}}^k_{-1}\|^2 \tag{22.135}$$

It is obvious that when N is large enough such that

$$\frac{N\nu}{2} > \frac{\beta_g^2}{\nu} \iff N > \frac{2\beta_g^2}{\nu^2} \tag{22.136}$$

it holds that

$$\lambda \triangleq 1 - \frac{\mu N\nu}{2} + \frac{\mu\beta_g^2}{\nu} < 1 \tag{22.137}$$

Also, since μ satisfies (22.133), we have

$$\mu < \frac{2}{N\nu} < \frac{1}{N\nu/2} < \frac{1}{N\nu/2 - \beta_g^2/\nu} \iff \lambda > 0 \tag{22.138}$$

In summary, when $\mu \leq 1/N\nu$ and N is large enough such that $N > 2\beta_g^2/\nu^2$, it holds that

$$\mathbb{E}\,\|\widetilde{\boldsymbol{w}}^{k+1}_{-1}\|^2 \;=\; \mathbb{E}\,\|\widetilde{\boldsymbol{w}}^k_{N-1}\|^2 \;\leq\; \lambda\,\mathbb{E}\,\|\widetilde{\boldsymbol{w}}^k_{-1}\|^2 \;\leq\; C\lambda^k \tag{22.139}$$

where $C = \mathbb{E}\,\|\widetilde{\boldsymbol{w}}^1_{-1}\|^2$. To establish (22.58b), we use (22.6b) to note that

$$0 \leq \mathbb{E}\,P(\boldsymbol{w}^k_{-1}) - P(w^\star) \leq \frac{\delta}{2}\,\mathbb{E}\,\|\widetilde{\boldsymbol{w}}^k_{-1}\|^2 \;\leq\; \delta C\lambda^k \tag{22.140}$$

We still need to establish (22.134) under (22.133). To do so, we first examine the form $(1-x)^N$ for any $x \in (0,1)$. Using the Taylor theorem, we can write the expansion (see relation (22.95) in the comments at the end of the chapter):

$$(1-x)^N = 1 - Nx + \frac{1}{2}N(N-1)(1-\tau)^{N-2}x^2 \tag{22.141}$$

for some constant $\tau \in (0, x)$ and, hence, $\tau < 1$. To ensure $(1-x)^N \leq 1 - \frac{1}{2}Nx$ we require

$$1 - Nx + \frac{1}{2}N(N-1)(1-\tau)^{N-2}x^2 \leq 1 - \frac{Nx}{2} \tag{22.142}$$

which is equivalent to

$$x \leq \frac{1}{(N-1)(1-\tau)^{N-2}} \tag{22.143}$$

Note that

$$\frac{1}{N} < \frac{1}{N-1} < \frac{1}{(N-1)(1-\tau)^{N-2}} \tag{22.144}$$

If we choose $x \leq 1/N$, then it will also satisfy (22.143). If we now let $x = \mu\nu$, we conclude that (22.134) holds under (22.133).

REFERENCES

Abramowitz, M. and I. Stegun (1965), *Handbook of Mathematical Functions*, Dover Publications.

Defazio, A., F. Bach, and S. Lacoste-Julien (2014), "SAGA: A fast incremental gradient method with support for non-strongly convex composite objectives," *Proc. Advances Neural Information Processing Systems* (NIPS), pp. 1646–1654, Montreal.

Defazio, A., J. Domke, and T. S. Caetano (2014), "Finito: A faster, permutable incremental gradient method for big data problems," in *Proc. Int. Conf. Machine Learning* (ICML), pp. 1125–1133, Beijing.

Dubois-Taine, B., S. Vaswani, R. Babanezhad, M. Schmidt, and S. Lacoste-Julien (2021), "SVRG meets AdaGrad: Painless variance reduction," available at arXiv:2102.09645.

Gürbüzbalaban, M., A. Ozdaglar, and P. Parrilo (2015a), "A globally convergent incremental Newton method," *Math. Program.*, vol. 151, no. 1, pp. 283–313.

Johnson, R. and T. Zhang (2013), "Accelerating stochastic gradient descent using predictive variance reduction," *Proc. Advances Neural Information Processing Systems* (NIPS), pp. 315–323, Lake Tahoe, NV.

Konecny, J., H. B. McMahan, D. Ramage, and P. Richtarik (2016), "Federated optimization: Distributed machine learning for on-device intelligence," available at arXiv:1610.02527.

Le Roux, N., M. Schmidt, and F. Bach (2012), "A stochastic gradient method with an exponential convergence rate for finite training sets," *Proc. Advances Neural Information Processing Systems* (NIPS), pp. 2672–2680, Lake Tahoe, NV.

Poon, C., J. Liang, and C. Schoenlieb (2018), "Local convergence properties of SAGA/Prox-SVRG and acceleration," *Proc. Int. Conf. Machine Learning* (ICML), vol. 80, pp. 4124–4132.

Reddi, S. J., A. Hefny, S. Sra, B. Póczos, and A. Smola (2016a), "On variance reduction in stochastic gradient descent and its asynchronous variants," *Proc. Advances Neural Information Processing Systems* (NIPS), pp. 2647–2655, Barcelona.

Reddi, S. J., A. Hefny, S. Sra, B. Póczos, and A. Smola (2016b), "Stochastic variance reduction for nonconvex optimization" *Proc. Int. Conf. Machine Learning* (ICML), pp. 324–323, New York, NY.

Schmidt, M., N. Le Roux, and F. Bach (2017), "Convergence rates of inexact proximal gradient methods for convex optimization." *Proc. Advances Neural Information Processing Systems* (NIPS), pp. 1458–1466, Long Beach, CA.

Shalev-Shwartz, S. and T. Zhang (2013), "Stochastic dual coordinate ascent methods for regularized loss minimization," *J. Mach. Learn. Res.*, vol. 14, pp. 567–599.

Shang, F., L. Jiao, K. Zhou, J. Cheng, Y. Ren, and Y. Jin (2018), "ASVRG: Accelerated proximal SVRG," *Proc. Machine Learning Research* (PMLR), vol. 95, pp. 815–830, Beijing.

Whittaker, E. T. and G. N. Watson (1927), *A Course of Modern Analysis*, 4th ed., Cambridge University Press.

Xiao, L. and T. Zhang (2014), "A proximal stochastic gradient method with progressive variance reduction," *SIAM J. Optim.*, vol. 24, no. 4, pp. 2057–2075.

Ying, B., K. Yuan, and A. H. Sayed (2018), "Convergence of variance-reduced learning under random reshuffling," *Proc. IEEE ICASSP*, pp. 2286–2290, Calgary.

Ying, B., K. Yuan, and A. H. Sayed (2020), "Variance-reduced stochastic learning under random reshuffling," *IEEE Trans. Signal Process.*, vol. 68, pp. 1390–1408.

Yuan, K., B. Ying, J. Liu, and A. H. Sayed (2019), "Variance-reduced stochastic learning by networked agents under random reshuffling," *IEEE Trans. Signal Process.*, vol. 67, no. 2, pp. 351–366.

23 Variance-Reduced Methods II: Random Reshuffling

We described two variance-reduced techniques in the previous chapter, namely, SAGA and SVRG, and established their convergence to the exact minimizer of the empirical risk under uniform sampling. These two procedures can also operate under random reshuffling, where the dataset is reshuffled at the start of each epoch. One main difference in relation to uniform sampling is that during each run over the data, all data points will now be used in updating the weight iterates. This is because sampling will be performed without replacement. In this chapter, we examine the convergence behavior of variance-reduced methods under random reshuffling and establish that exact convergence continues to occur. However, the convergence analysis is more demanding. For this reason, we place most of the derivations and proofs in the appendices and list only the main conclusions in the body of the chapter. We continue with the same assumptions and notation from the previous chapter for both smooth and nonsmooth risks. For this reason, we shall be brief.

23.1 AMORTIZED VARIANCE-REDUCED GRADIENT ALGORITHM (AVRG)

One inconvenience of the SVRG algorithm (22.48) is that it leads to an *unbalanced* implementation: each epoch is preceded by an aggregation step to compute the variable $\boldsymbol{b}^k$. This step requires the algorithm to spend a larger amount of processing time before each epoch starts than is usually needed within the epoch itself, especially for large datasets. We consider now a third variation that relies on an *amortized* computation of the variable $\boldsymbol{b}^k$. The resulting algorithm is referred to as the *amortized variance-reduced gradient* (AVRG) method. It continues to use a *fixed* variable $\boldsymbol{\phi}_{-1}^k$ for each epoch. However, it removes the initial aggregation step involving the computation of $\boldsymbol{b}^k$ from SVRG, and replaces it by a running update *within* the epoch loop over the data, namely,

$$\boldsymbol{b}^{k+1} \leftarrow \boldsymbol{b}^k + \frac{1}{N}\nabla_{w^\mathsf{T}} Q(\boldsymbol{w}_{n-1}^k; \gamma(\boldsymbol{u}), h_{\boldsymbol{u}}) \tag{23.1}$$

One useful feature of this calculation is that it *reuses* the gradient vector, $\nabla_{w^\mathsf{T}} Q(\boldsymbol{w}_{n-1}^k; \gamma(\boldsymbol{u}), h_{\boldsymbol{u}})$, which is already used in the update from $\boldsymbol{w}_{n-1}^k$ to $\boldsymbol{w}_n^k$.

The variable $\boldsymbol{u}$ so far has been used to refer to the sample index that is selected uniformly from within the range $\{0, 1, \ldots, N-1\}$. However, for AVRG, *random reshuffling* is used so that $\boldsymbol{u}$ will now be selected randomly *without* replacement from the same range. In this way, during each run all data points $\{\gamma(u), h_u\}$ for $u = 0, 1, 2, \ldots, N-1$ will contribute to the computation of $\boldsymbol{b}^k$. In order to incorporate random reshuffling into the description of the algorithm, we will let $\boldsymbol{\sigma}^k(\cdot)$ denote the reshuffling permutation function for the kth epoch. For each epoch k, this function reshuffles the integers $0 \leq n \leq N-1$ at random and, for each n, provides a random mapping $\boldsymbol{u} = \boldsymbol{\sigma}^k(n)$ so that the sample of index n in the original dataset becomes the sample of index $\boldsymbol{u}$ in the reshuffled dataset.

AVRG algorithm for solving the empirical risk problem (22.2).

given N data points $\{\gamma(m), h_m\}$, $m = 0, 1, \ldots, N-1$.
start with arbitrary $\boldsymbol{w}_{N-1}^0$.
for each run $k = 1, 2, \ldots, K$**:**
  set $\boldsymbol{w}_{-1}^k = \boldsymbol{w}_{N-1}^{k-1}$;
  set $\boldsymbol{\phi}_{-1}^k = \boldsymbol{w}_{N-1}^{k-1}$;
  $\boldsymbol{b}^k = 0$;
  randomly reshuffle the dataset $\{\gamma(m), h_m\}$;
  let $\boldsymbol{\sigma}^k(\cdot)$ denote the random permutation function; (23.2)
  repeat for $n = 0, 1, \ldots, N-1$**:**
    select $\boldsymbol{u} = \boldsymbol{\sigma}^k(n)$;
    select the data $(\gamma(\boldsymbol{u}), h_{\boldsymbol{u}})$;
    $\boldsymbol{w}_n^k = \boldsymbol{w}_{n-1}^k - \mu\Big(\nabla_{w^\mathsf{T}} Q(\boldsymbol{w}_{n-1}^k) - \nabla_{w^\mathsf{T}} Q(\boldsymbol{\phi}_{-1}^k) + \boldsymbol{b}^k\Big)$
    $\boldsymbol{b}^k \leftarrow \boldsymbol{b}^k + \dfrac{1}{N}\nabla_{w^\mathsf{T}} Q(\boldsymbol{w}_{n-1}^k)$
  end
end
return $w^\star \leftarrow \boldsymbol{w}_{N-1}^K$.

We are removing the arguments from the loss function for compactness of notation. If needed, the gradient vectors can be written more explicitly as follows:

$$\nabla_{w^\mathsf{T}} Q(\boldsymbol{w}_{n-1}^k) = \nabla_{w^\mathsf{T}} Q\Big(\boldsymbol{w}_{n-1}^k; \gamma(\boldsymbol{u}), h_{\boldsymbol{u}}\Big) \tag{23.3a}$$

$$\nabla_{w^\mathsf{T}} Q(\boldsymbol{\phi}_{-1}^k) = \nabla_{w^\mathsf{T}} Q\Big(\boldsymbol{\phi}_{-1}^k; \gamma(\boldsymbol{u}), h_{\boldsymbol{u}}\Big) \tag{23.3b}$$

Observe from listing (23.2) that the storage requirement for AVRG for each epoch is just the variables $\{\boldsymbol{b}^k, \boldsymbol{\phi}_{-1}^k\}$; this requirement is similar to SVRG and considerably less than SAGA. Observe also that every iteration n of (23.2) only requires two gradients to be evaluated, namely, $\nabla_{w^\mathsf{T}} Q(\boldsymbol{w}_{n-1}^k)$ and $\nabla_{w^\mathsf{T}} Q(\boldsymbol{\phi}_{-1}^k)$.

Thus, the effective computation of gradients per epoch is $O(2N)$.

REMARK 23.1. **(SAGA and SVRG with random reshuffling)** We can similarly list the SAGA and SVRG algorithms using random reshuffling. Their earlier listings in (22.33) and (22.48) assumed that $\boldsymbol{u}$ is selected uniformly (i.e., *with* replacement) from the set $\{0, 1, \ldots, N-1\}$. We can modify the listings to incorporate random reshuffling by writing $\boldsymbol{u} = \boldsymbol{\sigma}^k(n)$ in the same manner that we did for the AVRG algorithm, where $\boldsymbol{\sigma}^k(\cdot)$ denotes the random perturbation function for the kth epoch.

■

It turns out that all three algorithms, SAGA, SVRG, and AVRG, continue to converge to the exact minimizer $w^\star$ under random reshuffling. The arguments, however, are more challenging than in the uniform sampling case. This is because the gradient noise process will not have zero mean anymore. We examine the convergence question in the next section.

23.2 EVOLUTION OF MEMORY VARIABLES

We will carry out the convergence analysis in steps. First, we examine the evolution of the memory variables $\{\boldsymbol{\phi}_{n,u}^k\}$ under SAGA. Then, we establish some useful properties for these variables and, finally, we examine convergence of SAGA under random reshuffling before extending the analysis to SVRG and AVRG.

23.2.1 Updates under Random Reshuffling

Refer to the SAGA implementation (22.33), albeit one where the samples $\gamma(\boldsymbol{u})$ and $h_{\boldsymbol{u}}$ are selected under *random reshuffling*. That is, for each run k, the original data $\{\gamma(m), h_m\}_{m=0}^{N-1}$ is first randomly reshuffled so that the sample of index n becomes the sample of index $\boldsymbol{u} = \boldsymbol{\sigma}^k(n)$. To facilitate the understanding of the algorithms under random reshuffling, we recall the memory matrix Φ from Fig. 22.1; its columns consist of the memory vectors $\{\phi_u \in \mathbb{R}^M\}$. The memory matrix evolves with time and its entries vary with k and n. We can make the dependency on k and n explicit and write Φ_n^k instead of Φ, with the notation Φ_n^k referring to the state of the memory matrix at the nth iteration of the kth epoch. This matrix is only introduced for visualization purposes.

Now, consider a particular epoch of index k and let us examine the evolution of the entries of Φ_n^k as the iteration index runs from $n = 0$ to $N-1$ within that epoch. This situation is illustrated in Fig. 23.1.

The cells in the figure play the role of memory variables. Their values are updated as follows during epoch k. At iteration $n = 0$, the cells in the first block row Φ_{-1}^k will contain a randomly reshuffled version of all iterates generated during the previous epoch, namely, $\{\boldsymbol{w}_0^{k-1}, \boldsymbol{w}_1^{k-1}, \ldots, \boldsymbol{w}_{N-1}^{k-1}\}$. During the first iteration $n = 0$, a random index $\boldsymbol{u} = \boldsymbol{\sigma}^k(0)$ is selected. Assume this value turns out to be $\boldsymbol{u} = 1$. Then, as indicated in the highlighted cell in the second block row in the

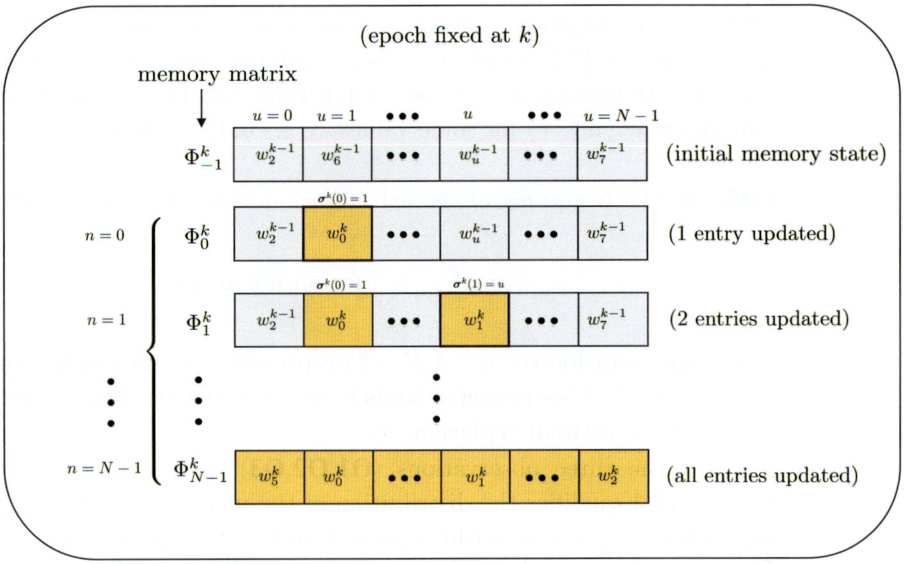

Figure 23.1 Illustration of the evolution of the memory matrix Φ under random reshuffling for an epoch k. During each iteration n, a random cell of index $\boldsymbol{u}$ is selected and its value updated to $\boldsymbol{w}_n^k$. This is indicated by the highlighted rectangles in the figure. Since sampling is performed *without* replacement, a different cell is selected at each iteration n. By the time all N iterations have been completed, all cells in the last row would have been updated.

figure, the second cell of Φ_0^k is updated to $\boldsymbol{w}_0^k$ while all other cells in this row remain invariant. Moving to iteration $n = 1$, a new random index $\boldsymbol{u} = \boldsymbol{\sigma}^k(1)$ is selected. Assume this value turns out to be $\boldsymbol{u} = u$. Then, as indicated again in the third block row in the figure, the uth cell of Φ_1^k is updated to $\boldsymbol{w}_1^k$ while all other cells in this row remain invariant. The process continues in this manner, by populating the cell corresponding to location $\boldsymbol{u}$ in the nth block row. By the end of iteration $N-1$, all cells of Φ_{N-1}^k would have been populated by the iterates $\{\boldsymbol{w}_n^k\}$ generated during the kth run. Observe that, since sampling *without* replacement is used, then *all* weight iterates $\{\boldsymbol{w}_n^k\}$ from $n = 0$ to $n = N-1$ will appear in Φ_{N-1}^k. These iterates appear randomly shuffled in the last row in the figure and they constitute the initial state for Φ_{-1}^{k+1} for the next run.

23.2.2 Properties of Memory Variables

Three useful observations can be drawn from Fig. 23.1. These properties are useful for the convergence proof in subsequent sections:

(**O1**): At the start of each epoch k, the components $\{\phi_{-1,u}^k\}_{u=0}^{N-1}$ in the first block row correspond to a random permutation of the weight iterates from the previous run, $\{\boldsymbol{w}_n^{k-1}\}_{n=0}^{N-1}$.

(**O2**): At the beginning of the nth iteration of an epoch k, all components of indices $\{\boldsymbol{\sigma}^k(m)\}_{m=0}^{n-1}$ would have been updated during the kth run to $\{\boldsymbol{w}_m^k\}_{m=0}^{n-1}$, while the remaining $N-n$ history positions will have values from the previous run, namely, $\{\boldsymbol{w}_{k_n}^{k-1}\}$ for some values $k_n \in \{0, 1, \ldots, N-1\}$.

(**O3**): At the beginning of the nth iteration of an epoch k, it holds that

$$\boldsymbol{\phi}_{n,\boldsymbol{u}}^k = \boldsymbol{\phi}_{-1,\boldsymbol{u}}^k, \qquad \text{for all } \boldsymbol{u} \in \boldsymbol{\sigma}^k(n+1{:}N-1) \tag{23.4}$$

where the notation $\boldsymbol{\sigma}^k(n+1{:}N-1)$ represents the indices for future iterations $n+1$ to $N-1$. This property holds because, under random reshuffling, sampling is performed without replacement.

Using the three observations (**O1**,**O2**,**O3**), two results can be established. Lemma 23.1 qualifies the distribution of the memory variable $\boldsymbol{\phi}_{n,\boldsymbol{u}}^k$. Assume the algorithm is operating within epoch k and that n updates have already passed. The remaining iterations within that epoch will select the random index $\boldsymbol{u}$ from within the set $\{\boldsymbol{\sigma}^k(n+1),\ \boldsymbol{\sigma}^k(n+2), \ldots, \boldsymbol{\sigma}^k(N-1)\}$, which we write more compactly as

$$\boldsymbol{u} \in \boldsymbol{\sigma}^k(n+1 : N-1) \tag{23.5}$$

Each choice for $\boldsymbol{u}$ will define a memory variable $\boldsymbol{\phi}_{n,\boldsymbol{u}}^k$. Lemma 23.2 shows that $\boldsymbol{\phi}_{n,\boldsymbol{u}}^k$ is equally likely to be any of the weight iterates from the previous epoch.

LEMMA 23.1. **(Distribution of memory variables)** *Conditioned on the previous $k-1$ epochs, each memory variable $\boldsymbol{\phi}_{n,\boldsymbol{u}}^k$ in the SAGA implementation (22.33) under random reshuffling has the following uniform probability distribution at the beginning of the nth iteration for epoch k:*

$$\mathbb{P}\left(\boldsymbol{\phi}_{n,\boldsymbol{u}}^k = x \,|\, \boldsymbol{\mathcal{F}}_{-1}^k\right) = \begin{cases} 1/N, & \text{if } x = w_0^{k-1} \\ 1/N, & \text{if } x = w_1^{k-1} \\ \quad\vdots & \\ 1/N, & \text{if } x = w_{N-1}^{k-1} \end{cases}, \quad \textit{for } \boldsymbol{u} \in \boldsymbol{\sigma}^k(n+1{:}N-1) \tag{23.6}$$

The notation $\boldsymbol{\mathcal{F}}_{-1}^k$ denotes all information that is available at the start of epoch k, namely all information from the previous $k-1$ epochs (their weight iterates and memory variables):

$$\boldsymbol{\mathcal{F}}_{-1}^k = \left\{\boldsymbol{w}_n^\ell, \boldsymbol{\phi}_{n,u}^\ell\right\}, \quad 0 \le u, n \le N-1, \;\; \ell < k \tag{23.7}$$

Proof: For any $\boldsymbol{u} = \boldsymbol{\sigma}^k(n+1{:}N-1)$ and any $\boldsymbol{w}_n^{k-1}$, $n = 0, 1, \ldots, N-1$, it holds that

$$\begin{aligned}
\mathbb{P}\Big(\boldsymbol{\phi}_{n,\boldsymbol{u}}^k = w_n^{k-1} \,|\, \boldsymbol{\mathcal{F}}_{-1}^k\Big) &= \sum_{\sigma^k} \mathbb{P}(\boldsymbol{\sigma}^k = \sigma^k)\mathbb{P}\Big(\boldsymbol{\phi}_{n,\boldsymbol{u}}^k = w_n^{k-1} \,|\, \boldsymbol{\mathcal{F}}_{-1}^k, \boldsymbol{\sigma}^k\Big) \\
&= \sum_{\sigma^k} \frac{1}{N!}\mathbb{P}\Big(\boldsymbol{\phi}_{n,\boldsymbol{u}}^k = w_n^{k-1} \,|\, \boldsymbol{\mathcal{F}}_{-1}^k, \boldsymbol{\sigma}^k\Big) \\
&= \sum_{\sigma^k} \frac{1}{N!}\mathbb{P}\Big(\boldsymbol{\phi}_{-1,\boldsymbol{u}}^k = w_n^{k-1} \,|\, \boldsymbol{\mathcal{F}}_{-1}^k, \boldsymbol{\sigma}^k\Big) \\
&= \frac{1}{N!}\sum_{\sigma^k} \mathbb{I}\left[\boldsymbol{\phi}_{-1,\boldsymbol{u}}^k = w_n^{k-1} \,|\, \boldsymbol{\mathcal{F}}_{-1}^k, \boldsymbol{\sigma}^k\right]
\end{aligned} \tag{23.8}$$

The first equality is a sum over all possible permutation functions σ^k in the kth epoch. The second equality is because there are $N!$ permutations and they are equally probable. The third equality applies observation (**O3**). The last equality follows from noting that, given $\boldsymbol{\mathcal{F}}_{-1}^k$ and $\boldsymbol{\sigma}^k$, the quantity $\boldsymbol{\phi}_{-1,\boldsymbol{u}}^k$ becomes a deterministic variable. In this case, the probability $\mathbb{P}(\boldsymbol{\phi}_{-1,\boldsymbol{u}}^k = w_n^{k-1}|\boldsymbol{\mathcal{F}}_{-1}^k, \boldsymbol{\sigma}^k)$ is either 1 or 0. We therefore express it in terms of the indicator function, where the notation $\mathbb{I}[a] = 1$ when the statement a is true and is zero otherwise. Next, note that there are $(N-1)!$ permutations $\boldsymbol{\sigma}^k$ with the $\boldsymbol{u}$th position storing $\boldsymbol{w}_n^{k-1}$ so that

$$\sum_{\sigma^k} \mathbb{I}\left[\boldsymbol{\phi}_{-1,\boldsymbol{u}}^k = w_n^{k-1} \,|\, \boldsymbol{\mathcal{F}}_{-1}^k, \boldsymbol{\sigma}^k\right] \;=\; (N-1)! \tag{23.9}$$

Substituting into (23.8), we get

$$\mathbb{P}\Big(\boldsymbol{\phi}_{n,\boldsymbol{u}}^k = w_n^{k-1} \,|\, \boldsymbol{\mathcal{F}}_{-1}^k\Big) = \frac{(N-1)!}{N!} \;=\; \frac{1}{N} \tag{23.10}$$

■

The next result uses the uniform distribution from Lemma 23.1 to determine an expression for the second-order moment of the aggregate memory variables.

LEMMA 23.2. (Second-order moment of memory variables) *The aggregate second-order moment of the memory variables from the nth iteration within epoch k in the SAGA implementation (22.33) under random reshuffling is given by*

$$\mathbb{E}\left(\sum_{u=0}^{N-1} \|\boldsymbol{\phi}_{n,u}^k\|^2\right) = \sum_{n'=0}^{n} \mathbb{E}\,\|\boldsymbol{w}_{n'}^k\|^2 + \frac{N-n-1}{N}\sum_{n=0}^{N-1} \mathbb{E}\,\|\boldsymbol{w}_n^{k-1}\|^2 \tag{23.11}$$

Proof: We split the aggregate sum into two terms: one involving updates up to iteration n and the other involving the remaining updates. Specifically, conditioned on the information from the past epochs:

$$
\begin{aligned}
&\mathbb{E}\left[\left(\sum_{u=0}^{N-1}\|\boldsymbol{\phi}_{n,u}^k\|^2\right)\Big|\,\boldsymbol{\mathcal{F}}_{-1}^k\right]\\
&=\mathbb{E}\left[\left(\sum_{u\in\boldsymbol{\sigma}^k(0:n)}\|\boldsymbol{\phi}_{n,u}^k\|^2\right)\Big|\,\boldsymbol{\mathcal{F}}_{-1}^k\right]+\mathbb{E}\left[\left(\sum_{u\notin\boldsymbol{\sigma}^k(0:n)}\|\boldsymbol{\phi}_{n,u}^k\|^2\right)\Big|\,\boldsymbol{\mathcal{F}}_{-1}^k\right]\\
&=\mathbb{E}\left[\left(\sum_{n'=0}^{n}\|\boldsymbol{w}_{n'}^k\|^2\right)\Big|\,\boldsymbol{\mathcal{F}}_{-1}^k\right]+\mathbb{E}\left[\left(\sum_{n'=n+1}^{N-1}\|\boldsymbol{\phi}_{n,\boldsymbol{\sigma}^k(n')}^k\|^2\right)\Big|\,\boldsymbol{\mathcal{F}}_{-1}^k\right]\\
&=\sum_{n'=0}^{n}\mathbb{E}\left(\|\boldsymbol{w}_{n'}^k\|^2\,|\,\boldsymbol{\mathcal{F}}_{-1}^k\right)+\sum_{n'=n+1}^{N-1}\mathbb{E}\left(\|\boldsymbol{\phi}_{n,\boldsymbol{\sigma}^k(n')}^k\|^2\,|\,\boldsymbol{\mathcal{F}}_{-1}^k\right)\\
&\overset{(23.6)}{=}\sum_{n'=0}^{n}\mathbb{E}\left(\|\boldsymbol{w}_{n'}^k\|^2\,|\,\boldsymbol{\mathcal{F}}_{-1}^k\right)+\sum_{n'=n+1}^{N-1}\frac{1}{N}\sum_{n=0}^{N-1}\|\boldsymbol{w}_n^{k-1}\|^2\\
&=\sum_{n'=0}^{n}\mathbb{E}\left(\|\boldsymbol{w}_{n'}^k\|^2\,|\,\boldsymbol{\mathcal{F}}_{-1}^k\right)+\frac{N-n-1}{N}\sum_{n=0}^{N-1}\|\boldsymbol{w}_n^{k-1}\|^2
\end{aligned}
\tag{23.12}
$$

Taking expectation over $\boldsymbol{\mathcal{F}}_{-1}^k$, we arrive at (23.11).

■

For comparison purposes, the properties and results discussed so far under random reshuffling for SAGA do not hold under uniform sampling where the data is sampled *with* replacement. For example, from the earlier result (22.102) under uniform sampling we would get instead:

$$
\mathbb{E}\left(\sum_{u=0}^{N-1}\|\boldsymbol{\phi}_{n,u}^k\|^2\right)=\mathbb{E}\,\|\widetilde{\boldsymbol{w}}_n^k\|^2+\frac{N-1}{N}\sum_{u=0}^{N-1}\mathbb{E}\,\|\widetilde{\boldsymbol{\phi}}_{n-1,u}^k\|^2 \tag{23.13}
$$

This result is similar to (23.11) only for $n=0$. However, observe that (23.13) involves memory variables $\{\widetilde{\boldsymbol{\phi}}_{n-1,u}^k\}$ on the right-hand side, instead of the iterates $\{\boldsymbol{w}_n^{k-1}\}$ that appear in (23.11). This is because random reshuffling updates every history variable during each run, while uniform sampling may leave some variables $\boldsymbol{\phi}_{n-1,u}^k$ untouched. This fact also helps explain why SAGA under random reshuffling tends to have faster convergence rate.

23.3 CONVERGENCE OF SAGA

It turns out that, under random reshuffling, the gradient noise process in SAGA does not have zero mean any longer. This fact will require that we adjust the convergence proof for SAGA relative to what we did under uniform sampling in the previous chapter.

23.3.1 Biased Gradient Noise

To verify the nonzero-mean property, recall that the gradient noise is given by (22.35):

$$\begin{aligned} \boldsymbol{g}_n(\boldsymbol{w}_{n-1}^k) &\stackrel{\Delta}{=} \widehat{\nabla_{w^\mathsf{T}} P}(\boldsymbol{w}_{n-1}^k) - \nabla_{w^\mathsf{T}} P(\boldsymbol{w}_{n-1}^k) \\ &\stackrel{(22.32)}{=} \nabla_{w^\mathsf{T}} Q(\boldsymbol{w}_{n-1}^k; \gamma(\boldsymbol{u}), h_{\boldsymbol{u}}) - \nabla_{w^\mathsf{T}} Q(\boldsymbol{\phi}_{n-1,\boldsymbol{u}}^k; \gamma(\boldsymbol{u}), h_{\boldsymbol{u}}) \\ &\quad + \frac{1}{N}\sum_{u=0}^{N-1} \nabla_{w^\mathsf{T}} Q(\phi_{n-1,u}^k; \gamma(u), h_u) - \nabla_{w^\mathsf{T}} P(\boldsymbol{w}_{n-1}^k) \end{aligned} \tag{23.14}$$

Consider again the collection $\boldsymbol{\mathcal{F}}_{n-1}^k = \{\boldsymbol{w}_{n-1}^k, \boldsymbol{\phi}_{n-1,u}^k\}$, $u = 0, 1, \ldots, N-1$, which was defined earlier in (22.36). Conditioning on $\boldsymbol{\mathcal{F}}_{n-1}^k$ and computing expectations we now get

$$\begin{aligned} &\mathbb{E}\left(\boldsymbol{g}_n(\boldsymbol{w}_{n-1}^k) \,|\, \boldsymbol{\mathcal{F}}_{n-1}^k\right) \\ &= \frac{1}{N-n} \sum_{u \notin \boldsymbol{\sigma}^k(0:n)} \Big(\nabla_{w^\mathsf{T}} Q(\boldsymbol{w}_{n-1}^k; \gamma(u), h_u) - \nabla_{w^\mathsf{T}} Q(\boldsymbol{\phi}_{n-1,u}^k; \gamma(u), h_u)\Big) \\ &\quad + \frac{1}{N}\sum_{u=0}^{N-1} \nabla_{w^\mathsf{T}} Q(\phi_{n-1,u}^k; \gamma(u), h_u) - \nabla_{w^\mathsf{T}} P(w_{n-1}^k) \end{aligned} \tag{23.15}$$

which is generally not equal to zero. The first equality is because, at iteration n and under sampling *without* replacement, only the data $\{\gamma(\boldsymbol{u}), h_{\boldsymbol{u}}\}$ for $\boldsymbol{u} \notin \boldsymbol{\sigma}^k(0{:}n)$ are available for sampling. Consequently, the gradient estimate that is employed by SAGA under random reshuffling is not an unbiased estimator for the true gradient vector. Nevertheless, we will be able to establish two useful facts in the following sections. First, the gradient estimate becomes *asymptotically unbiased* as the algorithm converges. Second, the biased gradient estimation does not harm the convergence rate because, as we are going to observe later, SAGA under random reshuffling actually converges faster than SAGA under uniform sampling. Similar remarks are applicable to SVRG and AVRG.

23.3.2 Convergence to the Exact Minimizer

The convergence analysis for SAGA employs two supporting lemmas. To begin with, note that recursion (22.33) shows how to move from one iterate to another within the *same* epoch k. The argument below will derive from this recursion a similar relation that shows how to move from the initial iterate $\boldsymbol{w}_{-1}^{k-1}$ for epoch $k-1$ to the initial iterate $\boldsymbol{w}_{-1}^k$ for epoch k. That is, we will first transform the update from an iteration *within* runs to an iteration *across* runs, in a manner similar to what we did earlier for stochastic gradient algorithms in Appendix 19.B. Doing so will enable us to exploit a useful property of the random reshuffling mechanism. Once this new recursion across epochs is derived, we will then use it to establish convergence.

We start by replacing the sample index $\boldsymbol{u}$ by the more explicit notation $\boldsymbol{u}_n^k$ to indicate that it is the random sample that is selected at iteration n of epoch k, namely,

$$\boldsymbol{u}_n^k = \boldsymbol{\sigma}^k(n) \tag{23.16}$$

Next, we note by iterating the SAGA recursion (22.33) that

$$\begin{aligned} \boldsymbol{w}_{-1}^{k+1} &= \boldsymbol{w}_{N-1}^k \\ &= \boldsymbol{w}_{-1}^k - \mu \sum_{n=0}^{N-1} \nabla_{w^\mathsf{T}} Q(\boldsymbol{w}_{n-1}^k; \gamma(\boldsymbol{u}_n^k), h_{\boldsymbol{u}_n^k}) \\ &\quad + \mu \sum_{n=0}^{N-1} \nabla_{w^\mathsf{T}} Q(\boldsymbol{\phi}_{n-1,\boldsymbol{u}_n^k}^k; \gamma(\boldsymbol{u}_n^k), h_{\boldsymbol{u}_n^k}) \\ &\quad - \frac{\mu}{N} \sum_{n=0}^{N-1} \sum_{u=0}^{N-1} \nabla_{w^\mathsf{T}} Q(\boldsymbol{\phi}_{n-1,u}^k; \gamma(u), h_u) \end{aligned} \tag{23.17}$$

As already alluded to, one main difficulty in the analysis is the fact that the gradient estimate is biased. For this reason, we will compare against the gradient at the start of the epoch. Since under random reshuffling, each index is selected only once during each epoch, it holds that

$$P(\boldsymbol{w}_{-1}^k) = \frac{1}{N} \sum_{n=0}^{N-1} Q\Big(\boldsymbol{w}_{-1}^k; \gamma(\boldsymbol{u}_n^k), h_{\boldsymbol{u}_n^k}\Big) \tag{23.18}$$

Moreover, for the same reason that each index value is selected only once during each epoch, it holds that

$$\boldsymbol{\phi}_{n-1,\boldsymbol{u}_n^k}^k = \boldsymbol{\phi}_{-1,\boldsymbol{u}_n^k}^k \tag{23.19}$$

That is, the value of the memory variable $\boldsymbol{\phi}$ corresponding to index $\boldsymbol{u}_n^k$ coincides with its initial value. This is because when $\boldsymbol{u}_n^k$ is selected, it will be the first time that this index is chosen and the value of $\boldsymbol{\phi}_{n-1,\boldsymbol{u}_n^k}^k$ would not have been overwritten yet during the operation of algorithm (22.33). For the same reason, it is also easy to see that

$$\sum_{n=0}^{N-1} \nabla_{w^\mathsf{T}} Q\Big(\boldsymbol{\phi}_{-1,\boldsymbol{u}_n^k}^k; \gamma(\boldsymbol{u}_n^k), h_{\boldsymbol{u}_n^k}\Big) = \sum_{u=0}^{N-1} \nabla_{w^\mathsf{T}} Q(\boldsymbol{\phi}_{-1,u}^k; \gamma(u), h_u) \tag{23.20}$$

Using these two remarks, adding and subtracting the similar terms that are underlined in the equation below, we obtain an initial recursion over epochs:

$$
\begin{aligned}
&\boldsymbol{w}_{-1}^{k+1} \\
&\overset{(a)}{=} \boldsymbol{w}_{-1}^{k} - \underline{\mu N \nabla_{w^{\mathsf{T}}} P(\boldsymbol{w}_{-1}^{k})} \\
&\quad -\mu \sum_{n=0}^{N-1} \left[\nabla_{w^{\mathsf{T}}} Q(\boldsymbol{w}_{n-1}^{k}; \gamma(\boldsymbol{u}_n^k), h_{\boldsymbol{u}_n^k}) - \underline{\nabla_{w^{\mathsf{T}}} Q(\boldsymbol{w}_{-1}^{k}; \gamma(\boldsymbol{u}_n^k), h_{\boldsymbol{u}_n^k})} \right] \\
&\quad +\mu \sum_{n=0}^{N-1} \left[\nabla_{w^{\mathsf{T}}} Q(\boldsymbol{\phi}_{-1,\boldsymbol{u}_n^k}^{k}; \gamma(\boldsymbol{u}_n^k), h_{\boldsymbol{u}_n^k}) - \underline{\underline{\frac{1}{N}\sum_{u=0}^{N-1} \nabla_{w^{\mathsf{T}}} Q(\boldsymbol{\phi}_{-1,u}^{k}; \gamma(u), h_u)}} \right] \\
&\quad -\frac{\mu}{N} \sum_{n=0}^{N-1}\sum_{u=0}^{N-1} \left[\nabla_{w^{\mathsf{T}}} Q(\boldsymbol{\phi}_{n-1,u}^{k}; \gamma(u), h_u) - \underline{\underline{\nabla_{w^{\mathsf{T}}} Q(\boldsymbol{\phi}_{-1,u}^{k}; \gamma(u), h_u)}} \right] \\
&\overset{(b)}{=} \boldsymbol{w}_{-1}^{k} - \mu N \nabla_{w^{\mathsf{T}}} P(\boldsymbol{w}_{-1}^{k}) \\
&\quad -\mu \sum_{n=0}^{N-1} \left[\nabla_{w^{\mathsf{T}}} Q(\boldsymbol{w}_{n-1}^{k}; \gamma(\boldsymbol{u}_n^k), h_{\boldsymbol{u}_n^k}) - \nabla_{w^{\mathsf{T}}} Q(\boldsymbol{w}_{-1}^{k}; \gamma(\boldsymbol{u}_n^k), h_{\boldsymbol{u}_n^k}) \right] \\
&\quad -\frac{\mu}{N} \sum_{n=0}^{N-1}\sum_{u=0}^{N-1} \left[\nabla_{w^{\mathsf{T}}} Q(\boldsymbol{\phi}_{n-1,u}^{k}; \gamma(u), h_u) - \nabla_{w^{\mathsf{T}}} Q(\boldsymbol{\phi}_{-1,u}^{k}; \gamma(u), h_u) \right]
\end{aligned} \tag{23.21}
$$

where in step (b) we removed the third term from step (a) because of (23.20).

We will also need to appeal to a second recursion *within* the same epoch k. By moving $\boldsymbol{w}_{n-1}^k$ in (22.33) to the left-hand side and computing the squared norm, we obtain (where we are removing the data arguments to compress the notation):

$$
\begin{aligned}
&\|\boldsymbol{w}_n^k - \boldsymbol{w}_{n-1}^k\|^2 \\
&= \mu^2 \left\| \nabla_{w^{\mathsf{T}}} Q(\boldsymbol{w}_{n-1}^k) - \nabla_{w^{\mathsf{T}}} Q(\boldsymbol{\phi}_{n-1,\boldsymbol{u}_n^k}^k) + \frac{1}{N}\sum_{u=0}^{N-1} \nabla_{w^{\mathsf{T}}} Q(\boldsymbol{\phi}_{n-1,u}^k) \right\|^2 \\
&\overset{(a)}{\leq} 3\mu^2 \left\| \nabla_{w^{\mathsf{T}}} Q(\boldsymbol{w}_{n-1}^k) - \nabla_{w^{\mathsf{T}}} Q(\boldsymbol{w}_{-1}^k) \right\|^2 \\
&\quad + 3\mu^2 \left\| \nabla_{w^{\mathsf{T}}} Q(\boldsymbol{\phi}_{n-1,\boldsymbol{u}_n^k}^k) - \nabla_{w^{\mathsf{T}}} Q(\boldsymbol{w}_{N-1}^{k-1}) \right\|^2 \\
&\quad + 3\mu^2 \left\| \frac{1}{N}\sum_{u=0}^{N-1} \Big(\nabla_{w^{\mathsf{T}}} Q(\boldsymbol{\phi}_{n-1,u}^k) - \nabla_{w^{\mathsf{T}}} Q(w^\star) \Big) \right\|^2
\end{aligned} \tag{23.22}
$$

which implies that

$$
\begin{aligned}
\|\boldsymbol{w}_n^k - \boldsymbol{w}_{n-1}^k\|^2 &\overset{(b)}{\leq} 3\delta^2\mu^2 \|\boldsymbol{w}_{n-1}^k - \boldsymbol{w}_{-1}^k\|^2 + 3\delta^2\mu^2 \left\| \boldsymbol{w}_{N-1}^{k-1} - \boldsymbol{\phi}_{n-1,\boldsymbol{u}_n^k}^k \right\|^2 \\
&\quad + \frac{3\delta^2\mu^2}{N} \sum_{u=0}^{N-1} \|\boldsymbol{\phi}_{n-1,u}^k - w^\star\|^2
\end{aligned} \tag{23.23}
$$

where in step (a) we added and subtracted $\nabla_{w^{\mathsf{T}}} Q(\boldsymbol{w}_{-1}^{k})$ and used the fact that $\boldsymbol{w}_{-1}^{k} = \boldsymbol{w}_{N-1}^{k-1}$. We also used the equality

$$\frac{1}{N}\sum_{u=0}^{N-1} \nabla_{w^{\mathsf{T}}} Q(w^{\star}; \gamma(u), h_u) = 0 \tag{23.24}$$

at the optimal solution $w^{\star}$, and applied the Jensen inequality that

$$\|a+b+c\|^2 \leq 3\|a\|^2 + 3\|b\|^2 + 3\|c\|^2 \tag{23.25}$$

for any vectors a, b, c of compatible dimensions. In step (b) we used the Lipschitz conditions (22.3a)–(22.3b) and property (19.172). Introducing the error quantity $\widetilde{\boldsymbol{w}}_n^k = w^{\star} - \boldsymbol{w}_n^k$, we can now establish two auxiliary lemmas. Lemma 23.3 shows how the second-order moment of the error vector at the start of each epoch evolves over time.

LEMMA 23.3. (Mean-square-error recursion) *The mean-square error at the start of each epoch satisfies the following inequality recursion for step sizes $\mu \leq 1/N\nu$:*

$$\begin{aligned}\mathbb{E}\,\|\widetilde{\boldsymbol{w}}_{-1}^{k+1}\|^2 \leq{}& \left(1 - \frac{\mu\nu N - \mu^2 N^2 \delta^2}{1-\mu N \nu}\right)\mathbb{E}\,\|\widetilde{\boldsymbol{w}}_{-1}^{k}\|^2 \\ &+ 4\mu\frac{\delta^2}{\nu}\left(\sum_{n=0}^{N-1}\mathbb{E}\,\|\boldsymbol{w}_n^k - \boldsymbol{w}_{-1}^k\|^2 + \sum_{n'=0}^{N-1}\mathbb{E}\,\|\boldsymbol{w}_{N-1}^{k-1} - \boldsymbol{w}_{n'}^{k-1}\|^2\right)\end{aligned} \tag{23.26}$$

Proof: See Appendix 23.A.

■

Roughly, the above result shows that the mean-square error across epochs evolves according to a dynamics that is determined by a scaling factor (smaller than 1 for small μ) in addition to two driving terms:

$$\sum_{n=0}^{N-1}\mathbb{E}\,\|\boldsymbol{w}_n^k - \boldsymbol{w}_{-1}^k\|^2 \qquad \text{(forward inner difference term)} \tag{23.27a}$$

$$\sum_{n'=0}^{N-1}\mathbb{E}\,\|\boldsymbol{w}_{N-1}^{k-1} - \boldsymbol{w}_{n'}^{k-1}\|^2 \qquad \text{(backward inner difference term)} \tag{23.27b}$$

Lemma 23.4 derives bounds for these inner difference terms.

LEMMA 23.4. **(Inner differences)** *The forward inner difference satisfies:*

$$\sum_{n=0}^{N-1} \mathbb{E}\,\|\boldsymbol{w}_n^k - \boldsymbol{w}_{-1}^k\|^2 \le 5\delta^2\mu^2N^3\mathbb{E}\,\|\widetilde{\boldsymbol{w}}_{-1}^k\|^2 + \tag{23.28}$$
$$7\delta^2\mu^2N^2\left(\sum_{n=0}^{N-1}\mathbb{E}\,\|\boldsymbol{w}_n^k - \boldsymbol{w}_{-1}^k\|^2 + \sum_{n=0}^{N-1}\mathbb{E}\,\|\boldsymbol{w}_{N-1}^{k-1} - \boldsymbol{w}_n^{k-1}\|^2\right)$$

while the backward inner difference satisfies:

$$\sum_{n=0}^{N-1} \mathbb{E}\,\|\boldsymbol{w}_{N-1}^{k-1} - \boldsymbol{w}_n^{k-1}\|^2 \le 5\delta^2\mu^2N^3\mathbb{E}\,\|\widetilde{\boldsymbol{w}}_{-1}^{k-1}\|^2 + \tag{23.29}$$
$$7\delta^2\mu^2N^2\left(\sum_{n=0}^{N-1}\mathbb{E}\|\boldsymbol{w}_n^{k-1} - \boldsymbol{w}_{-1}^{k-1}\|^2 + \sum_{n=0}^{N-1}\mathbb{E}\|\boldsymbol{w}_{N-1}^{k-2} - \boldsymbol{w}_n^{k-2}\|^2\right)$$

Proof: See Appendix 23.B. ■

Combining the above lemmas, we arrive at the following theorem.

THEOREM 23.1. **(Convergence of SAGA under random reshuffling)** *Consider the SAGA algorithm (22.33) operating under random reshuffling and used to seek the minimizer for the empirical risk minimization problem (22.2). The risk and loss functions are assumed to satisfy conditions (***A1***,***A2***) in the smooth case from the previous chapter, namely, conditions (22.3a)–(22.3b). For* $\mu < \nu/11\delta^2N$, *it holds that*

$$\limsup_{k\to\infty} \mathbb{E}\,\|\widetilde{\boldsymbol{w}}_{-1}^k\|^2 \le O(\lambda^k) \tag{23.30}$$

where

$$\lambda = \frac{1-\mu\nu N/4}{1-33\delta^4\mu^3N^3/\nu} = 1 - O(\mu) < 1 \tag{23.31}$$

Proof: See Appendix 23.C. ■

23.4 CONVERGENCE OF AVRG

The same approach used to establish the convergence of SAGA under random reshuffling is also suitable for AVRG. For this reason, we can be brief. First, similar to (23.17) and using the relations:

$$\boldsymbol{\phi}_{-1}^k = \boldsymbol{w}_{-1}^k, \quad \boldsymbol{b}^k = \frac{1}{N}\sum_{n=0}^{N-1}\nabla_{w^\mathsf{T}}Q\Big(\boldsymbol{w}_{n-1}^{k-1};\gamma(\boldsymbol{u}_n^{k-1}), h_{\boldsymbol{u}_n^{k-1}}\Big) \tag{23.32}$$

we can establish the following recursion for one epoch (see Prob. 23.2):

$$
\begin{aligned}
\boldsymbol{w}_{-1}^{k+1} &= \boldsymbol{w}_{-1}^{k} - \mu N \nabla_{w^\mathsf{T}} P(\boldsymbol{w}_{-1}^{k}) \\
&\quad - \mu \sum_{n=0}^{N-1} \Big[\nabla_{w^\mathsf{T}} Q(\boldsymbol{w}_{n-1}^{k}; \gamma(\boldsymbol{u}_n^k), h_{\boldsymbol{u}_n^k}) - \nabla_{w^\mathsf{T}} Q(\boldsymbol{w}_{-1}^{k}; \gamma(\boldsymbol{u}_n^k), h_{\boldsymbol{u}_n^k}) \Big] \\
&\quad + \mu \sum_{n=0}^{N-1} \Big[\nabla_{w^\mathsf{T}} Q(\boldsymbol{w}_{-1}^{k}; \gamma(\boldsymbol{u}_n^k), h_{\boldsymbol{u}_n^k}) - \nabla_{w^\mathsf{T}} Q(\boldsymbol{w}_{n-1}^{k-1}; \gamma(\boldsymbol{u}_n^{k-1}), h_{\boldsymbol{u}_n^{k-1}}) \Big]
\end{aligned}
\tag{23.33}
$$

Second, similar to (23.23), and using $\boldsymbol{w}_{-1}^{k} = \boldsymbol{w}_{N-1}^{k-1}$, we derive the inner difference recursion (where we are removing the data arguments from the loss functions for simplicity):

$$
\begin{aligned}
&\|\boldsymbol{w}_n^k - \boldsymbol{w}_{n-1}^k\|^2 \\
&= \mu^2 \left\| \nabla_{w^\mathsf{T}} Q(\boldsymbol{w}_{n-1}^k) - \nabla_{w^\mathsf{T}} Q(\boldsymbol{w}_{-1}^k) + \boldsymbol{b}^k \right\|^2 \\
&= \mu^2 \left\| \nabla_{w^\mathsf{T}} Q(\boldsymbol{w}_{n-1}^k) - \nabla_{w^\mathsf{T}} Q(\boldsymbol{w}_{-1}^k) \; + \frac{1}{N} \sum_{n=0}^{N-1} \Big(\nabla_{w^\mathsf{T}} \underline{Q(\boldsymbol{w}_{-1}^k)} - Q(w^\star) \Big) \right. \\
&\qquad \left. + \frac{1}{N} \sum_{n=0}^{N-1} \Big(\nabla_{w^\mathsf{T}} Q(\boldsymbol{w}_{n-1}^{k-1}) - \nabla_{w^\mathsf{T}} \underline{Q(\boldsymbol{w}_{N-1}^{k-1})} \Big) \right\|^2 \\
&\leq 3\mu^2\delta^2 \left(\|\boldsymbol{w}_{n-1}^k - \boldsymbol{w}_{-1}^k\|^2 + \frac{1}{N} \sum_{n=0}^{N-1} \|\boldsymbol{w}_{n-1}^{k-1} - \boldsymbol{w}_{N-1}^{k-1}\|^2 + \|\widetilde{\boldsymbol{w}}_{-1}^k\|^2 \right)
\end{aligned}
\tag{23.34}
$$

Next, we establish recursions related to $\widetilde{\boldsymbol{w}}_{-1}^k$, and the forward and backward difference terms.

LEMMA 23.5. (Recursions for AVRG analysis) *The mean-square error at the start of each epoch satisfies the following inequality recursion for $\mu \le 1/N\nu$:*

$$\mathbb{E}\,\|\widetilde{\boldsymbol{w}}_{-1}^{k+1}\|^2 \le \left(1 - \frac{\mu\nu N - \mu^2 N^2 \delta^2}{1-\mu N \nu}\right)\mathbb{E}\,\|\widetilde{\boldsymbol{w}}_{-1}^{k}\|^2 + \frac{2\mu\delta^2}{\nu}\left(\sum_{n=0}^{N-1}\mathbb{E}\,\|\boldsymbol{w}_n^k - \boldsymbol{w}_{-1}^k\|^2 + \sum_{n=0}^{N-1}\mathbb{E}\,\|\boldsymbol{w}_{N-1}^{k-1} - \boldsymbol{w}_{n-1}^{k-1}\|^2\right) \tag{23.35}$$

Moreover, the forward inner difference satisfies:

$$\sum_{n=0}^{N-1}\mathbb{E}\,\|\boldsymbol{w}_n^k - \boldsymbol{w}_{-1}^k\|^2 \le 3\mu^2\delta^2 N^2 \sum_{n=0}^{N-1}\mathbb{E}\,\|\boldsymbol{w}_n^k - \boldsymbol{w}_{-1}^k\|^2 + \mu^2\delta^2 N^2\left(\sum_{n=0}^{N-1}\mathbb{E}\,\|\boldsymbol{w}_{N-1}^{k-1} - \boldsymbol{w}_{n-1}^{k-1}\|^2 + N\mathbb{E}\,\|\widetilde{\boldsymbol{w}}_{-1}^{k}\|^2\right) \tag{23.36}$$

while the backward inner difference satisfies:

$$\sum_{n=0}^{N-1}\mathbb{E}\,\|\boldsymbol{w}_{N-1}^k - \boldsymbol{w}_{n}^k\|^2 \le 3\mu^2\delta^2 N^2 \sum_{n=0}^{N-1}\mathbb{E}\,\|\boldsymbol{w}_n^k - \boldsymbol{w}_{-1}^k\|^2 + 3\mu^2\delta^2 N^2\left(\sum_{n=0}^{N-1}\mathbb{E}\,\|\boldsymbol{w}_{n-1}^{k-1} - \boldsymbol{w}_{N-1}^{k-1}\|^2 + N\mathbb{E}\,\|\widetilde{\boldsymbol{w}}_{-1}^{k}\|^2\right) \tag{23.37}$$

Proof: See Appendix 23.D.

■

Using these relations we can establish the convergence of AVRG.

THEOREM 23.2. (Convergence of AVRG under random reshuffling) *Consider the AVRG algorithm (23.2) used to seek the minimizer for the empirical risk minimization problem (22.2). The risk and loss functions are assumed to satisfy conditions (**A1**,**A2**) in the smooth case from the previous chapter, namely, conditions (22.3a)–(22.3b). For $\mu \le \nu/9\delta^2 N$, it holds that*

$$\limsup_{k\to\infty}\mathbb{E}\,\|\widetilde{\boldsymbol{w}}_{-1}^{k}\|^2 \le O(\lambda^k) \tag{23.38}$$

where

$$\lambda = \frac{1-\mu\nu N/4}{1-18\delta^4\mu^3N^3/\nu} = 1 - O(\mu) < 1 \tag{23.39}$$

Proof: See Appendix 23.E.

■

23.5 CONVERGENCE OF SVRG

The same approach used to establish the convergence of AVRG in the previous section can be simplified to establish the convergence of SVRG. For instance, similar to (23.17) and using the relations:

$$\phi_{-1}^{k} = w_{-1}^{k}, \quad b^{k} = \frac{1}{N}\sum_{m=0}^{N-1} \nabla_{w^{\mathsf{T}}} Q(w_{-1}^{k}; \gamma(m), h_m) \tag{23.40}$$

we can establish the following recursion for one epoch (see Prob. 23.3):

$$\begin{aligned} w_{-1}^{k+1} &= w_{-1}^{k} - \mu N \nabla_{w^{\mathsf{T}}} P(w_{-1}^{k}) \\ &\quad - \mu \sum_{n=0}^{N-1} \Big[\nabla_{w^{\mathsf{T}}} Q\Big(w_{n-1}^{k}; \gamma(u_n^k), h_{u_n^k}\Big) - \nabla_{w^{\mathsf{T}}} Q\Big(w_{-1}^{k}; \gamma(u_n^k), h_{u_n^k}\Big)\Big] \end{aligned} \tag{23.41}$$

Likewise, similar to (23.23) and using $w_{-1}^{k} = w_{N-1}^{k-1}$, we can derive the inner difference recursion (where we are removing the data arguments from the loss functions for simplicity):

$$\begin{aligned} &\|w_n^k - w_{n-1}^k\|^2 \\ &= \mu^2 \left\| \nabla_{w^{\mathsf{T}}} Q(w_{n-1}^{k}) - \nabla_{w^{\mathsf{T}}} Q(w_{-1}^{k}) + b^k \right\|^2 \\ &= \mu^2 \left\| \nabla_{w^{\mathsf{T}}} Q(w_{n-1}^{k}) - \nabla_{w^{\mathsf{T}}} Q(w_{-1}^{k}) + \frac{1}{N}\sum_{n=0}^{N-1} \left(\nabla_{w^{\mathsf{T}}} Q(w_{-1}^{k}) - Q(w^{\star})\right) \right\|^2 \\ &\le 2\mu^2\delta^2 \Big(\|w_{n-1}^{k} - w_{-1}^{k}\|^2 + \|\widetilde{w}_{-1}^{k}\|^2 \Big) \end{aligned} \tag{23.42}$$

The argument can now be continued in a manner similar to AVRG and is left as an exercise – see Prob. 23.1.

Example 23.1 **(Comparing performance of algorithms)** We illustrate the convergence performance of SAGA, SVRG, and AVRG on simulated data. We consider the same regularized logistic empirical risk problem from (22.60), namely,

$$\min_{w \in \mathbb{R}^M} \left\{ P(w) \triangleq \rho\|w\|^2 + \frac{1}{N}\sum_{m=0}^{N-1} \ln\Big(1 + e^{-\gamma(m) h_m^{\mathsf{T}} w}\Big) \right\} \tag{23.43}$$

where $h_m \in \mathbb{R}^M$ and $\gamma(m) \in \{\pm 1\}$. In all experiments, we set $\rho = 0.5$ and $M = 25$. We run two sets of simulations: one involving $N = 500$ samples and another involving $N = 2000$ samples. Since the convergence rates under random reshuffling are on the order of $\lambda = 1 - \mu\nu N/4$, we keep the value of the product μN constant at $\mu = 0.5$. This corresponds to simulating with the step sizes $\mu = 0.001$ in one case and $\mu = 0.00025$ in the other case. We generate N random data pairs $\{\gamma(m), h_m\}$ according to a logistic model. First, a random parameter model $w^a \in \mathbb{R}^M$ is selected, and a random collection of vectors $\{h_m\}$ are generated with zero-mean unit-variance Gaussian entries. Then, for each h_m, the label $\gamma(m)$ is set to either $+1$ or -1 according to the following construction:

$$\gamma(m) = +1 \text{ if } \left(\frac{1}{1+e^{-h_m^\mathsf{T} w^a}}\right) \geq 0.5; \text{ otherwise } \gamma(m) = -1 \tag{23.44}$$

A total of $K = 30$ epochs are run over the data. Figure 23.2 compares the learning curves for SAGA, SVRG, and AVRG under uniform sampling and random reshuffling in normalized logarithmic scale; all normalized curves start at the value 1 and they are obtained by averaging over the $K = 30$ epochs.

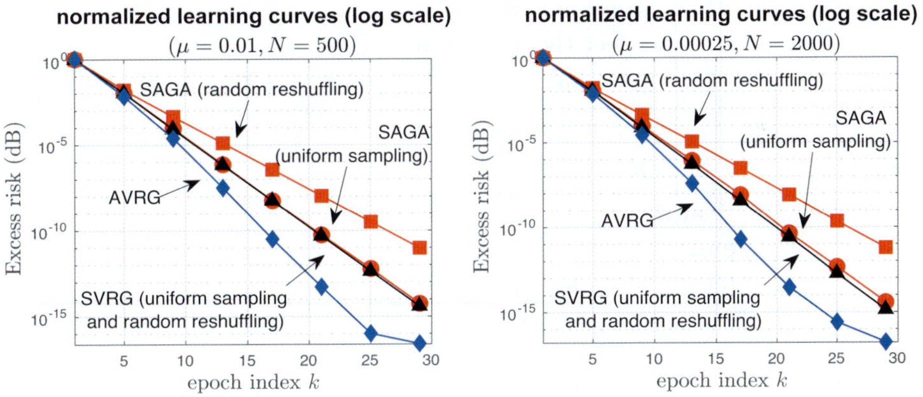

Figure 23.2 (*Left*) Learning curves for $\mu = 0.001$ and $N = 500$ data points. (*Right*) Learning curves for $\mu = 0.00025$ and $N = 2000$ data points.

23.6 NONSMOOTH RISK FUNCTIONS

We have focused so far on the case in which the empirical risk function $P(w)$ in (22.2) and the corresponding loss function $Q(w; \cdot)$ are first-order differentiable. As was done in Section 22.5, we can consider extensions to nonsmooth risks by incorporating *proximal projection* steps. For example, we list in (23.45) the proximal AVRG algorithm.

Proximal AVRG for solving the empirical risk problem (22.65).

$$
\begin{array}{l}
\text{given } N \text{ data points } \{\gamma(m), h_m\},\ m = 0, 1, \ldots, N-1. \\
\text{start with arbitrary } \boldsymbol{w}^0_{N-1}. \\
\textbf{for each run } k = 1, 2, \ldots, K\textbf{:} \\
\quad \text{set } \boldsymbol{w}^k_{-1} = \boldsymbol{w}^{k-1}_{N-1}; \\
\quad \text{set } \boldsymbol{\phi}^k_{-1} = \boldsymbol{w}^{k-1}_{N-1}; \\
\quad \boldsymbol{b}^k = 0; \\
\quad \text{randomly reshuffle the dataset } \{\gamma(m), h_m\}; \\
\quad \text{let } \boldsymbol{\sigma}^k(\cdot) \text{ denote the random permutation function;} \\
\quad \textbf{repeat for } n = 0, 1, \ldots, N-1\textbf{:} \\
\qquad \text{select } \boldsymbol{u} = \boldsymbol{\sigma}^k(n); \\
\qquad \text{select the data } (\gamma(\boldsymbol{u}), h_{\boldsymbol{u}}); \\
\qquad \boldsymbol{z}^k_n = \boldsymbol{w}^k_{n-1} - \mu\left(\nabla_{w^{\mathsf{T}}} Q(\boldsymbol{w}^k_{n-1}) - \nabla_{w^{\mathsf{T}}} Q(\boldsymbol{\phi}^k_{-1}) + \boldsymbol{b}^k\right) \\
\qquad \boldsymbol{w}^k_n = \text{prox}_{\mu q}(\boldsymbol{z}^k_n); \\
\qquad \boldsymbol{b}^k \leftarrow \boldsymbol{b}^k + \dfrac{1}{N}\nabla_{w^{\mathsf{T}}} Q(\boldsymbol{w}^k_{n-1}) \\
\quad \textbf{end} \\
\textbf{end} \\
\text{return } w^\star \leftarrow \boldsymbol{w}^K_{N-1}.
\end{array}
\tag{23.45}
$$

23.7 COMMENTARIES AND DISCUSSION

Variance-reduced techniques under random reshuffling. In the current and previous chapter we described some of the main algorithms for variance reduction in stochastic optimization solutions. The SVRG algorithm was proposed by Johnson and Zhang (2013), while SAGA was proposed by Defazio, Bach, and Lacoste-Julien (2014), and AVRG was proposed by Ying, Yuan, and Sayed (2018, 2020). There are of course other variations as explained in the comments at the end of the previous chapter. That chapter focused on analyzing the performance of variance-reduced algorithms under uniform sampling, whereas the current chapter analyzes their performance under random reshuffling.

It was remarked earlier in the comments of Chapter 19 that several works in the literature have observed that stochastic gradient algorithms tend to deliver improved performance under random reshuffling, such as the works by Bottou (2009), Recht and Re (2012), Gürbüzbalaban, Ozdaglar, and Parrilo (2015b), and Shamir (2016). A detailed analytical justification for this improved performance was provided by Ying *et al.* (2019), where it was shown that random reshuffling (i.e., sampling *without* replacement) enables convergence of stochastic gradient implementations toward the smaller $O(\mu^2)$-size neighborhood around the optimal solution. This is in contrast to the traditional $O(\mu)$-performance delivered by uniform sampling (i.e., sampling *with* replacement). We included this justification and discussion in Section 19.5. In this chapter, we followed the approach from Ying, Yuan, and Sayed (2018, 2020) and Yuan *et al.* (2019) for variance-reduced methods where convergence analyses under random reshuffling was provided for SAGA and AVRG. Apparently, prior to these latter works, no formal guarantees of exact convergence existed for variance-reduced algorithms under random reshuffling. In De and Goldstein (2016), another variance-reduction algorithm is proposed under reshuffling; however, no proof of convergence is provided either. The closest attempts at proofs are the arguments given in Gürbüzbalaban, Ozdaglar, and Parrilo (2015c) and Shamir (2016); the first work deals with the case of incremental aggregated gradients, which corresponds to a deterministic version of random reshuffling for SAGA, while the second work deals with SVRG in the context of ridge regression problems using regret analysis.

Observe from listing (23.2) that the storage requirement for AVRG for each epoch is just the variables $\{\boldsymbol{b}^k, \boldsymbol{\phi}^k_{-1}\}$; this requirement is similar to SVRG and considerably less than SAGA. Observe also that every iteration n of (23.2) only requires two gradients to be evaluated, namely, $\nabla_{w^{\mathsf{T}}} Q(\boldsymbol{w}^k_{n-1})$ and $\nabla_{w^{\mathsf{T}}} Q(\boldsymbol{\phi}^k_{-1})$. Thus, the effective computation of gradients per epoch is $O(2N)$. However, these computations are distributed in a more balanced manner over the epoch in contrast to SVRG, where a costly aggregation step is necessary at the start of the epoch. Combining with results (22.92a)–(22.92b) we find that

$$\text{(\textbf{SAGA})}\;\; O(N) \text{ gradient calculations per epoch} \tag{23.46a}$$
$$\text{(\textbf{SVRG})}\;\; O(2N) \text{ gradient calculations per epoch (unbalanced)} \tag{23.46b}$$
$$\text{(\textbf{AVRG})}\;\; O(2N) \text{ gradient calculations per epoch (balanced)} \tag{23.46c}$$

If we further compare the expressions for the convergence rates of SAGA, SVRG, and AVRG from the previous and current chapter we arrive at the summary shown in Table 23.1. In this summary, we ignore terms involving higher-order powers in μ and assume μ is sufficiently small. Observe that the convergence rates under random reshuffling are sensitive to N; the expressions suggest that μ should be decreased for larger values of N.

Table 23.1 Approximate convergence rates for SAGA, SVRG, and AVRG under uniform sampling and random reshuffling assuming small step sizes.

Algorithm	Uniform sampling	Random reshuffling
SAGA	$\lambda = 1 - \mu\nu$	$\lambda = 1 - \mu\nu N/4$
SVRG	$\lambda = 1 - \mu N/2\nu$	$\lambda = 1 - \mu\nu N/4$
AVRG	not applicable	$\lambda = 1 - \mu\nu N/4$

PROBLEMS

23.1 Follow arguments similar to those used in Section 23.4 to establish the convergence of the SVRG algorithm under random reshuffling for smooth risks that satisfy conditions (**A1**,**A2**) from the previous chapter.

23.2 Establish the weight recursion (23.33) over epochs for the AVRG algorithm.

23.3 Establish the weight recursion (23.41) over epochs for the SVRG algorithm.

23.4 Establish the validity of inequality (23.56).

23.5 Establish the second-order moment relation (23.65) for SAGA operating under random reshuffling.

23.6 Establish inequality (23.106) for the AVRG algorithm.

23.7 Examine the convergence behavior of the proximal AVRG algorithm under nonsmooth risks satisfying conditions (**A1**,**A2**,**A3**) from the previous chapter.

23.8 Refer to the discussion in Sections 22.5 and 23.6 on nonsmooth risk functions. Examine the convergence behavior of the proximal SAGA algorithm under uniform sampling and nonsmooth risks satisfying conditions (**A1**,**A2**,**A3**) from the previous chapter.

23.9 Refer to the description of the SAG algorithm in Prob. 22.10. We rewrite it by incorporating random reshuffling into its operation using k as an index for the epoch:

randomly reshuffle the data $\{\gamma(m), h_m\}$;
let $\boldsymbol{\sigma}^k(\cdot)$ denote the random permutation function for epoch k;
repeat for $n = 0, 1, \ldots, N-1$:
  select $\boldsymbol{u} = \boldsymbol{\sigma}^k(n)$;
  select the sample $(\gamma(\boldsymbol{u}), h_{\boldsymbol{u}}\}$;
  update the $\boldsymbol{u}$th column, $\boldsymbol{\phi}_{n,\boldsymbol{u}} = \nabla_{w^\mathsf{T}} Q(\boldsymbol{w}_{n-1}; \gamma(\boldsymbol{u}), h_{\boldsymbol{u}})$;
  retain the other columns: $\boldsymbol{\phi}_{n,u'} = \boldsymbol{\phi}_{n-1,u'},\ u' \neq \boldsymbol{u}$;
  compute $\boldsymbol{a}_n = \dfrac{1}{N}\displaystyle\sum_{u=0}^{N-1} \boldsymbol{\phi}_{n,u}$;
  update $\boldsymbol{w}_n = \boldsymbol{w}_{n-1} - \mu\, \boldsymbol{a}_n$;
end

Examine the convergence behavior of the algorithm.

23.A PROOF OF LEMMA 23.3

The proofs in this and the following appendices are based on the arguments from Ying, Yuan, and Sayed (2018, 2020). Subtracting $w^\star$ from both sides of (23.21) we get the error recursion:

$$\begin{aligned}
\widetilde{\boldsymbol{w}}_{-1}^{k+1} = \widetilde{\boldsymbol{w}}_{-1}^{k} &+ \mu N \nabla_{w^\mathsf{T}} P(\boldsymbol{w}_{-1}^k) \\
&+ \mu \left[\sum_{n=0}^{N-1} \Big(\nabla_{w^\mathsf{T}} Q(\boldsymbol{w}_{n-1}^k; \gamma(\boldsymbol{u}_n^k), h_{\boldsymbol{u}_n^k}) - \nabla_{w^\mathsf{T}} Q(\boldsymbol{w}_{-1}^k; \gamma(\boldsymbol{u}_n^k), h_{\boldsymbol{u}_n^k}) \Big) \right. \\
&\left. + \frac{1}{N} \sum_{u=0}^{N-1} \Big(\nabla_{w^\mathsf{T}} Q(\boldsymbol{\phi}_{n-1,u}^k; \gamma(u), h_u) - \nabla_{w^\mathsf{T}} Q(\boldsymbol{\phi}_{-1,u}^k; \gamma(u), h_u) \Big) \right]
\end{aligned} \tag{23.47}$$

Recall that the notation $\boldsymbol{\mathcal{F}}_{-1}^k$ denotes the collection of all information before iteration $n = 0$ at epoch k, i.e., all information from the previous $k-1$ epochs. Computing the conditional mean-square error of both sides of (23.47) gives:

$$\begin{aligned}
&\mathbb{E}\left(\left\| \widetilde{\boldsymbol{w}}_{-1}^{k+1} \right\|^2 \mid \boldsymbol{\mathcal{F}}_{-1}^k \right) \\
&\overset{(a)}{\leq} \frac{1}{1-t} \left\| \widetilde{w}_{-1}^k + \mu N \nabla_{w^\mathsf{T}} P(w_{-1}^k) \right\|^2 \\
&\quad + \frac{\mu^2}{t} \mathbb{E} \left\{ \left\| \sum_{n=0}^{N-1} \Big(\nabla_{w^\mathsf{T}} Q(\boldsymbol{w}_{n-1}^k; \gamma(\boldsymbol{u}_n^k), h_{\boldsymbol{u}_n^k}) - \nabla_{w^\mathsf{T}} Q(\boldsymbol{w}_{-1}^k; \gamma(\boldsymbol{u}_n^k), h_{\boldsymbol{u}_n^k}) \Big) \right. \right. \\
&\quad \left. \left. + \frac{1}{N} \sum_{u=0}^{N-1} \Big(\nabla_{w^\mathsf{T}} Q(\boldsymbol{\phi}_{n-1,u}^k; \gamma(u), h_u) - \nabla_{w^\mathsf{T}} Q(\boldsymbol{\phi}_{-1,u}^k; \gamma(u), h_u) \Big) \right\|^2 \mid \boldsymbol{\mathcal{F}}_{-1}^k \right\}
\end{aligned} \tag{23.48}$$

where step (a) follows from the Jensen inequality and t can be chosen arbitrarily in the open interval $t \in (0,1)$. Continuing, we have

$$
\begin{aligned}
&\mathbb{E}\left(\left\|\widetilde{\boldsymbol{w}}_{-1}^{k+1}\right\|^2 \mid \boldsymbol{\mathcal{F}}_{-1}^k\right) \\
&\overset{(b)}{\leq} \frac{1}{1-t}\left\|\widetilde{w}_{-1}^k + \mu N \nabla_{w^\mathsf{T}} P(w_{-1}^k)\right\|^2 \\
&+\frac{\mu^2 N}{t}\left\{\sum_{n=0}^{N-1}\mathbb{E}\left(\left\|\left(\nabla_{w^\mathsf{T}} Q(\boldsymbol{w}_{n-1}^k;\gamma(\boldsymbol{u}_n^k),h_{\boldsymbol{u}_n^k}) - \nabla_{w^\mathsf{T}} Q(\boldsymbol{w}_{-1}^k;\gamma(\boldsymbol{u}_n^k),h_{\boldsymbol{u}_n^k})\right)\right.\right.\right. \\
&\left.\left.\left.+\frac{1}{N}\sum_{u=0}^{N-1}\left(\nabla_{w^\mathsf{T}} Q(\boldsymbol{\phi}_{n-1,u}^k;\gamma(u),h_u) - \nabla_{w^\mathsf{T}} Q(\boldsymbol{\phi}_{-1,u}^k;\gamma(u),h_u)\right)\right\|^2 \mid \boldsymbol{\mathcal{F}}_{-1}^k\right)\right\} \\
&\overset{(c)}{\leq} \frac{1}{1-t}\left\|\widetilde{w}_{-1}^k + \mu N \nabla_{w^\mathsf{T}} P(w_{-1}^k)\right\|^2 \\
&+\frac{2\mu^2 N}{t}\sum_{n=0}^{N-1}\mathbb{E}\left(\left\|\nabla_{w^\mathsf{T}} Q(\boldsymbol{w}_{n-1}^k;\gamma(\boldsymbol{u}_n^k),h_{\boldsymbol{u}_n^k}) - \nabla_{w^\mathsf{T}} Q(\boldsymbol{w}_{-1}^k;\gamma(\boldsymbol{u}_n^k),h_{\boldsymbol{u}_n^k})\right\|^2 \mid \boldsymbol{\mathcal{F}}_{-1}^k\right) \\
&+\frac{2\mu^2 N}{t}\sum_{n=0}^{N-1}\mathbb{E}\left(\left\|\frac{1}{N}\sum_{u=0}^{N-1}\nabla_{w^\mathsf{T}} Q(\boldsymbol{\phi}_{n-1,u}^k;\gamma(u),h_u) - \nabla_{w^\mathsf{T}} Q(\boldsymbol{\phi}_{-1,u}^k;\gamma(u),h_u)\right\|^2 \mid \boldsymbol{\mathcal{F}}_{-1}^k\right) \\
&\overset{(d)}{=} \frac{1}{1-t}\left\|\widetilde{w}_{-1}^k + \mu N \nabla_{w^\mathsf{T}} P(w_{-1}^k)\right\|^2 \\
&+\frac{2\mu^2 N}{t}\sum_{n=1}^{N-1}\mathbb{E}\left(\left\|\nabla_{w^\mathsf{T}} Q(\boldsymbol{w}_{n-1}^k;\gamma(\boldsymbol{u}_n^k),h_{\boldsymbol{u}_n^k}) - \nabla_{w^\mathsf{T}} Q(\boldsymbol{w}_{-1}^k;\gamma(\boldsymbol{u}_n^k),h_{\boldsymbol{u}_n^k})\right\|^2 \mid \boldsymbol{\mathcal{F}}_{-1}^k\right) \\
&+\frac{2\mu^2 N}{t}\sum_{n=1}^{N-1}\mathbb{E}\left(\left\|\frac{1}{N}\sum_{u=0}^{N-1}\left(\nabla_{w^\mathsf{T}} Q(\boldsymbol{\phi}_{n-1,u}^k;\gamma(u),h_u) - \nabla_{w^\mathsf{T}} Q(\boldsymbol{\phi}_{-1,u}^k;\gamma(u),h_u)\right)\right\|^2 \mid \boldsymbol{\mathcal{F}}_{-1}^k\right)
\end{aligned}
\tag{23.49}
$$

where steps (b) and (c) follow from the following corollary to the Jensen inequality:

$$
\left\|\sum_{n=0}^{N-1} y_n\right\|^2 = N^2\left\|\sum_{n=0}^{N-1}\frac{1}{N}y_n\right\|^2 \leq N\sum_{n=0}^{N-1}\|y_n\|^2 \tag{23.50}
$$

In step (d), the summations start from $n=1$ since the terms corresponding to $n=0$ are equal to zero.

We further know from the Lipschitz condition (22.3b) that:

$$
\mathbb{E}\left(\left\|\nabla_{w^\mathsf{T}} Q(\boldsymbol{w}_{n-1}^k) - \nabla_{w^\mathsf{T}} Q(\boldsymbol{w}_{-1}^k)\right\|^2 \mid \boldsymbol{\mathcal{F}}_{-1}^k\right) \leq \delta^2 \mathbb{E}\left(\|\boldsymbol{w}_{n-1}^k - \boldsymbol{w}_{-1}^k\|^2 \mid \boldsymbol{\mathcal{F}}_{-1}^k\right) \tag{23.51}
$$

and, likewise,

$$\begin{aligned}
&\mathbb{E}\left(\left\|\frac{1}{N}\sum_{u=0}^{N-1}\left(\nabla_{w^\mathsf{T}} Q(\phi^k_{n-1,u};\gamma(u),h_u) - \nabla_{w^\mathsf{T}} Q(\phi^k_{-1,u};\gamma(u),h_u)\right)\right\|^2 \,|\, \boldsymbol{\mathcal{F}}^k_{-1}\right)\\
&\overset{(a)}{=} \mathbb{E}\left(\left\|\frac{1}{N}\sum_{m=0}^{n-1}\nabla_{w^\mathsf{T}} Q(\boldsymbol{w}^k_m;\gamma(\boldsymbol{u}^k_m),h_{\boldsymbol{u}^k_m}) - \nabla_{w^\mathsf{T}} Q(\boldsymbol{\phi}^k_{-1,\boldsymbol{u}^k_m};\gamma(\boldsymbol{u}^k_m),h_{\boldsymbol{u}^k_m})\right\|^2 \,|\, \boldsymbol{\mathcal{F}}^k_{-1}\right)\\
&\overset{(b)}{\le} \frac{n\delta^2}{N^2}\sum_{m=0}^{n-1}\mathbb{E}\left(\|\boldsymbol{w}^k_m - \boldsymbol{\phi}^k_{-1,\boldsymbol{u}^k_m}\|^2 \,|\, \boldsymbol{\mathcal{F}}^k_{-1}\right)\\
&= \frac{n\delta^2}{N^2}\sum_{m=0}^{n-1}\mathbb{E}\left(\|\boldsymbol{w}^k_m - \boldsymbol{w}^k_{-1} + \boldsymbol{w}^{k-1}_{N-1} - \boldsymbol{\phi}^k_{-1,\boldsymbol{u}^k_m}\|^2 \,|\, \boldsymbol{\mathcal{F}}^k_{-1}\right)\\
&\overset{(c)}{\le} \frac{n\delta^2}{N^2}\sum_{m=0}^{n-1}\left(2\mathbb{E}\left(\|\boldsymbol{w}^k_m - \boldsymbol{w}^k_{-1}\|^2 \,|\, \boldsymbol{\mathcal{F}}^k_{-1}\right) + 2\mathbb{E}\left(\|\boldsymbol{w}^{k-1}_{N-1} - \boldsymbol{\phi}^k_{-1,\boldsymbol{u}^k_m}\|^2 \,|\, \boldsymbol{\mathcal{F}}^k_{-1}\right)\right)\\
&\overset{(d)}{=} \frac{n\delta^2}{N^2}\sum_{m=0}^{n-1}\left(2\mathbb{E}\left(\|\boldsymbol{w}^k_m - \boldsymbol{w}^k_{-1}\|^2 \,|\, \boldsymbol{\mathcal{F}}^k_{-1}\right) + \frac{2}{N}\sum_{n'=0}^{N-1}\|\boldsymbol{w}^{k-1}_{N-1} - \boldsymbol{w}^{k-1}_{n'}\|^2\right)
\end{aligned} \tag{23.52}$$

where step (a) holds because of observation (**O2**) (namely, up to iteration n, only the memory variables $\boldsymbol{\phi}^k_{m,\cdot}$ for $m = 0, \ldots, n-1$ would have been updated to $\boldsymbol{w}^k_m$ while the remaining memory variables continue with values $\boldsymbol{\phi}^k_{-1,\cdot}$). Steps (b) and (c) apply the Jensen inequality, and step (d) is because of (23.6).

Next, in a manner similar to (19.22), we use the strong-convexity property of the empirical risk $P(w)$ to note that

$$\begin{aligned}
&\left\|\widetilde{w}^k_{-1} + \mu N\nabla_{w^\mathsf{T}} P(w^k_{-1})\right\|^2\\
&= \|\widetilde{w}^k_{-1}\| + \mu^2N^2\|\nabla_{w^\mathsf{T}} P(w^k_{-1})\|^2 + 2\mu N(\widetilde{w}^k_{-1})^\mathsf{T}\nabla_{w^\mathsf{T}} P(w^k_{-1})\\
&\overset{(19.22)}{\le} \|\widetilde{w}^k_{-1}\| + \mu^2N^2\|\nabla_{w^\mathsf{T}} P(w^k_{-1})\|^2 - 2\mu\nu N\|\widetilde{w}^k_{-1})\|^2\\
&\overset{(19.22)}{=} \|\widetilde{w}^k_{-1}\| + \mu^2N^2\|\nabla_{w^\mathsf{T}} P(w^k_{-1}) - \nabla_{w^\mathsf{T}} P(w^\star)\|^2 - 2\mu\nu N\|\widetilde{w}^k_{-1})\|^2\\
&\overset{(22.4)}{\le} \|\widetilde{w}^k_{-1}\| + \mu^2\delta^2N^2\|\widetilde{w}^k_{-1})\|^2 - 2\mu\nu N\|\widetilde{w}^k_{-1})\|^2\\
&\le (1 - 2\mu\nu N + \mu^2N^2\delta^2)\|\widetilde{w}^k_{-1}\|^2
\end{aligned} \tag{23.53}$$

Substituting (23.51), (23.52), and (23.53) into (23.49), choosing $t = \mu N\nu$, and assuming $\mu \le 1/(N\nu)$ so that $t < 1$, we get

$$
\begin{aligned}
&\mathbb{E}\left(\|\widetilde{w}_{-1}^{k+1}\|^2 \,|\, \boldsymbol{\mathcal{F}}_{-1}^k\right) \\
&\le \left(\frac{1-2\mu\nu N+\mu^2N^2\delta^2}{1-\mu N\nu}\right)\|\widetilde{w}_{-1}^k\|^2 + 2\mu\frac{\delta^2}{\nu}\sum_{n=0}^{N-1}\mathbb{E}\left(\|\boldsymbol{w}_{n-1}^k - \boldsymbol{w}_{-1}^k\|^2 \,|\, \boldsymbol{\mathcal{F}}_{-1}^k\right) \\
&\quad + 2\mu\frac{\delta^2}{\nu}\sum_{n=0}^{N-1}\frac{n}{N^2}\sum_{m=0}^{n-1}\left(2\mathbb{E}\left(\|\boldsymbol{w}_m^k - \boldsymbol{w}_{-1}^k\|^2 \,|\, \boldsymbol{\mathcal{F}}_{-1}^k\right) + \frac{2}{N}\sum_{n'=0}^{N-1}\|\boldsymbol{w}_{N-1}^{k-1} - \boldsymbol{w}_{n'}^{k-1}\|^2\right) \\
&\overset{(a)}{=} \left(1-\frac{\mu\nu N-\mu^2N^2\delta^2}{1-\mu N\nu}\right)\|\widetilde{w}_{-1}^k\|^2 + 2\mu\frac{\delta^2}{\nu}\sum_{n=0}^{N-1}\mathbb{E}\left(\|\boldsymbol{w}_{n-1}^k - \boldsymbol{w}_{-1}^k\|^2 \,|\, \boldsymbol{\mathcal{F}}_{-1}^k\right) \\
&\quad + 2\mu\frac{\delta^2}{\nu}\sum_{m=0}^{N-1}\sum_{n=m+1}^{N-1}\frac{n}{N^2}\left(2\mathbb{E}\left(\|\boldsymbol{w}_m^k - \boldsymbol{w}_{-1}^k\|^2 \,|\, \boldsymbol{\mathcal{F}}_{-1}^k\right) + \frac{2}{N}\sum_{n'=0}^{N-1}\|\boldsymbol{w}_{N-1}^{k-1} - \boldsymbol{w}_{n'}^{k-1}\|^2\right) \\
&\overset{(b)}{\le} \left(1-\frac{\mu\nu N-\mu^2N^2\delta^2}{1-\mu N\nu}\right)\|\widetilde{w}_{-1}^k\|^2 + 2\mu\frac{\delta^2}{\nu}\sum_{n=0}^{N-1}\mathbb{E}\left(\|\boldsymbol{w}_n^k - \boldsymbol{w}_{-1}^k\|^2 \,|\, \boldsymbol{\mathcal{F}}_{-1}^k\right) \\
&\quad + 2\mu\frac{\delta^2}{\nu}\sum_{m=0}^{N-1}\frac{1}{2}\left(2\mathbb{E}\left(\|\boldsymbol{w}_m^k - \boldsymbol{w}_{-1}^k\|^2 \,|\, \boldsymbol{\mathcal{F}}_{-1}^k\right) + \frac{2}{N}\sum_{n'=0}^{N-1}\|\boldsymbol{w}_{N-1}^{k-1} - \boldsymbol{w}_{n'}^{k-1}\|^2\right) \\
&\le \left(1-\frac{\mu\nu N-\mu^2N^2\delta^2}{1-\mu N\nu}\right)\|\widetilde{w}_{-1}^k\|^2 \\
&\quad + 4\mu\frac{\delta^2}{\nu}\left(\sum_{n=0}^{N-1}\mathbb{E}\left(\|\boldsymbol{w}_n^k - \boldsymbol{w}_{-1}^k\|^2 \,|\, \boldsymbol{\mathcal{F}}_{-1}^k\right) + \sum_{n'=0}^{N-1}\|\boldsymbol{w}_{N-1}^{k-1} - \boldsymbol{w}_{n'}^{k-1}\|^2\right)
\end{aligned}
\tag{23.54}
$$

where in step (a) and in several similar steps later, we use the equality:

$$
\sum_{n=0}^{N-1}\sum_{m=0}^{n-1} f(n,m) \equiv \sum_{m=0}^{N-1}\sum_{n=m+1}^{N-1} f(n,m)
\tag{23.55}
$$

and in step (b), the factor $1/2$ appears because, for any $0 \le m \le N-1$ (see Prob. 23.4):

$$
\sum_{n=m+1}^{N-1} n \;=\; \frac{1}{2}(N-m-1)(N+m) < \frac{N^2}{2}
\tag{23.56}
$$

and, consequently,

$$
\sum_{n=m+1}^{N-1}\frac{n}{N^2} \;<\; \frac{N^2}{2N^2} = 1/2
\tag{23.57}
$$

Taking the expectation of (23.54) over the past history $\boldsymbol{\mathcal{F}}_{-1}^k$ leads to (23.26).

23.B PROOF OF LEMMA 23.4

Using (23.23), we can establish an upper bound for any inner difference based on $\boldsymbol{w}_{-1}^k$ as follows:

$$\begin{aligned}
\|\boldsymbol{w}_n^k - \boldsymbol{w}_{-1}^k\|^2 &= \|\boldsymbol{w}_n^k - \boldsymbol{w}_{n-1}^k + \boldsymbol{w}_{n-1}^k - \cdots + \cdots - \boldsymbol{w}_{-1}^k\|^2 \\
&= (n+1)^2 \left\| \frac{1}{n+1}\left(\boldsymbol{w}_n^k - \boldsymbol{w}_{n-1}^k + \boldsymbol{w}_{n-1}^k - \cdots - \boldsymbol{w}_{-1}^k\right)\right\|^2 \\
&\leq (n+1)\sum_{m=0}^{n} \|\boldsymbol{w}_m^k - \boldsymbol{w}_{m-1}^k\|^2 \\
&\overset{(23.23)}{\leq} 3\delta^2\mu^2(n+1)\sum_{m=0}^{n}\Big(\|\boldsymbol{w}_{m-1}^k - \boldsymbol{w}_{-1}^k\|^2 + \|\boldsymbol{w}_{N-1}^{k-1} - \boldsymbol{\phi}_{m-1,\boldsymbol{u}_m^k}^k\|^2 \\
&\qquad + \frac{1}{N}\sum_{u=0}^{N-1}\|\widetilde{\boldsymbol{\phi}}_{m-1,u}^k\|^2\Big)
\end{aligned} \tag{23.58}$$

where $\widetilde{\boldsymbol{\phi}}_{m,u}^k \triangleq w^\star - \boldsymbol{\phi}_{m,u}^k$. It is important to recall that $\boldsymbol{u}_n^k = \boldsymbol{\sigma}^k(n)$ so that the index $\boldsymbol{u}_n^k$ is selected at iteration n of epoch k. Summing over n, we have:

$$\begin{aligned}
&\sum_{n=0}^{N-1}\|\boldsymbol{w}_n^k - \boldsymbol{w}_{-1}^k\|^2 \\
&\leq 3\delta^2\mu^2\sum_{n=0}^{N-1}(n+1)\sum_{m=0}^{n-1}\left(\|\boldsymbol{w}_{m-1}^k - \boldsymbol{w}_{-1}^k\|^2 + \left\|\boldsymbol{w}_{N-1}^{k-1} - \boldsymbol{\phi}_{m-1,\boldsymbol{u}_m^k}^k\right\|^2 + \right. \\
&\qquad\qquad \left. \frac{1}{N}\sum_{u=0}^{N-1}\|\widetilde{\boldsymbol{\phi}}_{m-1,u}^k\|^2\right) \\
&\overset{(23.55)}{=} 3\delta^2\mu^2\sum_{m=0}^{N-1}\sum_{n=m+1}^{N-1}(n+1)\left(\|\boldsymbol{w}_{m-1}^k - \boldsymbol{w}_{-1}^k\|^2 + \left\|\boldsymbol{w}_{N-1}^{k-1} - \boldsymbol{\phi}_{m-1,\boldsymbol{u}_m^k}^k\right\|^2 + \right. \\
&\qquad\qquad \left. \frac{1}{N}\sum_{u=0}^{N-1}\|\widetilde{\boldsymbol{\phi}}_{m-1,u}^k\|^2\right) \\
&\overset{(a)}{\leq} \frac{9}{4}\delta^2\mu^2N^2\sum_{m=0}^{N-1}\left(\|\boldsymbol{w}_{m-1}^k - \boldsymbol{w}_{-1}^k\|^2 + \left\|\boldsymbol{w}_{N-1}^{k-1} - \boldsymbol{\phi}_{m-1,\boldsymbol{u}_m^k}^k\right\|^2 + \frac{1}{N}\sum_{u=0}^{N-1}\|\widetilde{\boldsymbol{\phi}}_{m-1,u}^k\|^2\right) \\
&= \frac{9}{4}\delta^2\mu^2N^2\left\{\sum_{m=0}^{N-1}\|\boldsymbol{w}_{m-1}^k - \boldsymbol{w}_{-1}^k\|^2 + \sum_{m=0}^{N-1}\left\|\boldsymbol{w}_{N-1}^{k-1} - \boldsymbol{\phi}_{m-1,\boldsymbol{u}_m^k}^k\right\|^2 \right. \\
&\qquad\qquad \left. + \sum_{m=0}^{N-1}\frac{1}{N}\sum_{u=0}^{N-1}\|\widetilde{\boldsymbol{\phi}}_{m-1,u}^k\|^2\right\} \\
&\triangleq X
\end{aligned} \tag{23.59}$$

where step (a) is because

$$\sum_{n=m+1}^{N-1}(n+1) \leq N + \sum_{n=m+1}^{N-1} n \leq N^2 + \frac{N^2}{2} \leq \frac{3N^2}{2} \tag{23.60}$$

The quantity X is further equal to

$$X \overset{(b)}{=} \frac{9}{4}\delta^2\mu^2 N^2 \left\{ \sum_{m=0}^{N-1} \|\boldsymbol{w}_{m-1}^k - \boldsymbol{w}_{-1}^k\|^2 + \sum_{m'=0}^{N-1} \|\boldsymbol{w}_{N-1}^{k-1} - \boldsymbol{w}_{m'}^{k-1}\|^2 + \sum_{m=0}^{N-1} \frac{1}{N} \sum_{u=0}^{N-1} \|\widetilde{\boldsymbol{\phi}}_{m-1,u}^k\|^2 \right\}$$

$$= \frac{9}{4}\delta^2\mu^2 N^2 \left\{ \sum_{n=0}^{N-1} \|\boldsymbol{w}_{n-1}^k - \boldsymbol{w}_{-1}^k\|^2 + \sum_{n=0}^{N-1} \|\boldsymbol{w}_{N-1}^{k-1} - \boldsymbol{w}_{n}^{k-1}\|^2 + \sum_{n=0}^{N-1} \frac{1}{N} \sum_{u=0}^{N-1} \|\widetilde{\boldsymbol{\phi}}_{n-1,u}^k\|^2 \right\} \tag{23.61}$$

where step (b) is because the index $\boldsymbol{u}_m^k$ is selected at iteration m, which is posterior to $m-1$, and, hence,

$$\sum_{m=0}^{N-1} \left\| \boldsymbol{w}_{N-1}^{k-1} - \boldsymbol{\phi}_{m-1,\boldsymbol{u}_m^k}^k \right\|^2 = \sum_{m'=0}^{N-1} \|\boldsymbol{w}_{N-1}^{k-1} - \boldsymbol{w}_{m'}^{k-1}\|^2 \tag{23.62}$$

since the memory variables will be filled with past weight iterates from epoch $k-1$. Then, computing the conditional expectation, we get:

$$\sum_{n=0}^{N-1} \mathbb{E}\left(\|\boldsymbol{w}_n^k - \boldsymbol{w}_{-1}^k\|^2 \,|\, \boldsymbol{\mathcal{F}}_{-1}^k \right)$$
$$\leq \frac{9}{4}\delta^2\mu^2 N^2 \left\{ \sum_{n=0}^{N-1} \mathbb{E}\left(\|\boldsymbol{w}_{n-1}^k - \boldsymbol{w}_{-1}^k\|^2 \,|\, \boldsymbol{\mathcal{F}}_{-1}^k \right) + \sum_{n=0}^{N-1} \|\boldsymbol{w}_{N-1}^{k-1} - \boldsymbol{w}_n^{k-1}\|^2 + \sum_{n=0}^{N-1} \frac{1}{N} \sum_{u=0}^{N-1} \mathbb{E}\left(\|\widetilde{\boldsymbol{\phi}}_{n-1,u}^k\|^2 \,|\, \boldsymbol{\mathcal{F}}_{-1}^k \right) \right\} \tag{23.63}$$

To bound the last term, we first separate it into two quantities:

$$\mathbb{E}\left(\|\widetilde{\boldsymbol{\phi}}_{n-1,u}^k\|^2 \,|\, \boldsymbol{\mathcal{F}}_{-1}^k \right) = \mathbb{E}\left(\|\widetilde{\boldsymbol{\phi}}_{n-1,u}^k - \widetilde{\boldsymbol{w}}_{-1}^k + \widetilde{\boldsymbol{w}}_{-1}^k\|^2 \,|\, \boldsymbol{\mathcal{F}}_{-1}^k \right)$$
$$\leq 2\mathbb{E}\left(\|\boldsymbol{\phi}_{n-1,u}^k - \boldsymbol{w}_{-1}^k\|^2 \,|\, \boldsymbol{\mathcal{F}}_{-1}^k \right) + 2\|\widetilde{\boldsymbol{w}}_{-1}^k\|^2 \tag{23.64}$$

Using an argument similar to the proof of Lemma 23.2, and the fact that $\boldsymbol{w}_{-1}^k = \boldsymbol{w}_{N-1}^{k-1}$, we can verify that (see Prob. 23.5):

$$\mathbb{E}\left(\sum_{u=0}^{N-1} \|\boldsymbol{\phi}_{n-1,u}^k - \boldsymbol{w}_{-1}^k\|^2 \,|\, \boldsymbol{\mathcal{F}}_{-1}^k \right) \tag{23.65}$$
$$= \sum_{n'=0}^{n-1} \mathbb{E}\left(\|\boldsymbol{w}_{n'}^k - \boldsymbol{w}_{-1}^k\|^2 \,|\, \boldsymbol{\mathcal{F}}_{-1}^k \right) + \frac{N-n}{N} \sum_{n=0}^{N-1} \|\boldsymbol{w}_{N-1}^{k-1} - \boldsymbol{w}_n^{k-1}\|^2$$

Combining results (23.64) and (23.65), we can bound the last term of (23.63):

$$\begin{aligned}
&\sum_{n=0}^{N-1}\frac{1}{N}\sum_{u=0}^{N-1}\mathbb{E}\left(\|\widetilde{\boldsymbol{\phi}}_{n-1,u}^{k}\|^2\,|\,\boldsymbol{\mathcal{F}}_{-1}^{k}\right)\\
&\leq \sum_{n=0}^{N-1}\frac{2}{N}\left(\sum_{n'=0}^{n-1}\mathbb{E}\left(\|\boldsymbol{w}_{n'}^{k}-\boldsymbol{w}_{-1}^{k}\|^2\,|\,\boldsymbol{\mathcal{F}}_{-1}^{k}\right)+\frac{N-n}{N}\sum_{n=0}^{N-1}\|\boldsymbol{w}_{N-1}^{k-1}-\boldsymbol{w}_{n}^{k-1}\|^2\right)\\
&\quad +2N\|\widetilde{\boldsymbol{w}}_{-1}^{k}\|^2\\
&\overset{(a)}{=}\frac{2}{N}\sum_{n=0}^{N-1}\sum_{n'=0}^{n}\mathbb{E}\left(\|\boldsymbol{w}_{n'}^{k}-\boldsymbol{w}_{-1}^{k}\|^2\,|\,\boldsymbol{\mathcal{F}}_{-1}^{k}\right)+\frac{N+1}{N}\sum_{n=0}^{N-1}\|\boldsymbol{w}_{N-1}^{k-1}-\boldsymbol{w}_{n}^{k-1}\|^2\\
&\quad +2N\|\widetilde{\boldsymbol{w}}_{-1}^{k}\|^2\\
&\overset{(b)}{\leq} 2\sum_{n=0}^{N-1}\mathbb{E}\left(\|\boldsymbol{w}_{n}^{k}-\boldsymbol{w}_{-1}^{k}\|^2\,|\,\boldsymbol{\mathcal{F}}_{-1}^{k}\right)+2\sum_{n=0}^{N-1}\|\boldsymbol{w}_{N-1}^{k-1}-\boldsymbol{w}_{n}^{k-1}\|^2+2N\|\widetilde{\boldsymbol{w}}_{-1}^{k}\|^2
\end{aligned}\tag{23.66}$$

where step (a) is because

$$\begin{aligned}
\sum_{n=0}^{N-1}\frac{2}{N}\frac{N-n}{N}&=\frac{2}{N^2}\left(\sum_{n=0}^{N-1}N-\sum_{n=0}^{N-1}n\right)\\
&=\frac{2}{N^2}\left(N^2-\frac{N(N-1)}{2}\right)=\frac{N+1}{N}
\end{aligned}\tag{23.67}$$

and step (b) is because

$$\sum_{n=0}^{N-1}\sum_{n'=0}^{n}\mathbb{E}\left(\|\boldsymbol{w}_{n'}^{k}-\boldsymbol{w}_{-1}^{k}\|^2\right)\leq N\sum_{n=0}^{N-1}\mathbb{E}\left(\|\boldsymbol{w}_{n}^{k}-\boldsymbol{w}_{-1}^{k}\|^2\right)\tag{23.68}$$

and

$$\frac{N+1}{N}\leq 2,\quad \text{for } N\geq 1\tag{23.69}$$

Substituting back into (23.63), we get

$$\begin{aligned}
&\sum_{n=0}^{N-1}\mathbb{E}\left(\|\boldsymbol{w}_{n}^{k}-\boldsymbol{w}_{-1}^{k}\|^2\,|\,\boldsymbol{\mathcal{F}}_{-1}^{k}\right)\\
&\leq\frac{9}{4}\delta^2\mu^2N^2\left(\sum_{n=0}^{N-1}\mathbb{E}\left(\|\boldsymbol{w}_{n-1}^{k}-\boldsymbol{w}_{-1}^{k}\|^2\,|\,\boldsymbol{\mathcal{F}}_{-1}^{k}\right)+\sum_{n=0}^{N-1}\|\boldsymbol{w}_{N-1}^{k-1}-\boldsymbol{w}_{n}^{k-1}\|^2\right.\\
&\qquad\left.+2\sum_{n=0}^{N-1}\mathbb{E}\left(\|\boldsymbol{w}_{n}^{k}-\boldsymbol{w}_{-1}^{k}\|^2\,|\,\boldsymbol{\mathcal{F}}_{-1}^{k}\right)+2\sum_{n=0}^{N-1}\|\boldsymbol{w}_{N-1}^{k-1}-\boldsymbol{w}_{n}^{k-1}\|^2+2N\|\widetilde{\boldsymbol{w}}_{-1}^{k}\|^2\right)\\
&\leq\frac{9}{4}\delta^2\mu^2N^2\left\{\sum_{n=0}^{N-1}\mathbb{E}\left(\|\boldsymbol{w}_{n}^{k}-\boldsymbol{w}_{-1}^{k}\|^2\,|\,\boldsymbol{\mathcal{F}}_{-1}^{k}\right)+\sum_{n=0}^{N-1}\|\boldsymbol{w}_{N-1}^{k-1}-\boldsymbol{w}_{n}^{k-1}\|^2\right.\\
&\qquad\left.+2\sum_{n=0}^{N-1}\mathbb{E}\left(\|\boldsymbol{w}_{n}^{k}-\boldsymbol{w}_{-1}^{k}\|^2\,|\,\boldsymbol{\mathcal{F}}_{-1}^{k}\right)+2\sum_{n=0}^{N-1}\|\boldsymbol{w}_{N-1}^{k-1}-\boldsymbol{w}_{n}^{k-1}\|^2+2N\|\widetilde{\boldsymbol{w}}_{-1}^{k}\|^2\right\}\\
&\overset{\Delta}{=} Y
\end{aligned}\tag{23.70}$$

where Y is further bounded by

$$
\begin{aligned}
Y \leq\ & \frac{27}{4}\delta^2\mu^2N^2\left\{\sum_{n=0}^{N-1}\mathbb{E}\left(\|\boldsymbol{w}_n^k-\boldsymbol{w}_{-1}^k\|^2\,|\,\boldsymbol{\mathcal{F}}_{-1}^k\right)+\sum_{n=0}^{N-1}\|\boldsymbol{w}_{N-1}^{k-1}-\boldsymbol{w}_n^{k-1}\|^2\right\}\\
&+\frac{9}{2}\delta^2\mu^2N^3\|\widetilde{\boldsymbol{w}}_{-1}^k\|^2\\
\leq\ & 7\delta^2\mu^2N^2\left\{\sum_{n=0}^{N-1}\mathbb{E}\left(\|\boldsymbol{w}_n^k-\boldsymbol{w}_{-1}^k\|^2\,|\,\boldsymbol{\mathcal{F}}_{-1}^k\right)+\sum_{n=0}^{N-1}\|\boldsymbol{w}_{N-1}^{k-1}-\boldsymbol{w}_n^{k-1}\|^2\right\}\\
&+5\delta^2\mu^2N^3\|\widetilde{\boldsymbol{w}}_{-1}^k\|^2 \qquad (23.71)
\end{aligned}
$$

Taking expectation over the filtration leads to (23.28).

Next, following similar arguments, we have the following for the backward inner difference term:

$$
\begin{aligned}
&\|\boldsymbol{w}_{N-1}^{k-1}-\boldsymbol{w}_n^{k-1}\|^2\\
&=\|\boldsymbol{w}_{N-1}^{k-1}-\boldsymbol{w}_{N-2}^{k-1}+\boldsymbol{w}_{N-2}^{k-1}-\cdots+\cdots-\boldsymbol{w}_n^{k-1}\|^2\\
&\leq(N-n-1)\sum_{m=n+1}^{N-1}\|\boldsymbol{w}_m^{k-1}-\boldsymbol{w}_{m-1}^{k-1}\|^2\\
&\overset{(23.23)}{\leq}3\delta^2\mu^2(N-n-1)\sum_{m=n+1}^{N-1}\left(\|\boldsymbol{w}_{m-1}^{k-1}-\boldsymbol{w}_{-1}^{k-1}\|^2+\left\|\boldsymbol{w}_{N-1}^{k-2}-\boldsymbol{\phi}_{m-1,\boldsymbol{u}_m^{k-1}}^{k-1}\right\|^2\right.\\
&\qquad\qquad\left.+\frac{1}{N}\sum_{u=0}^{N-1}\|\widetilde{\boldsymbol{\phi}}_{m-1,u}^{k-1}\|^2\right) \qquad (23.72)
\end{aligned}
$$

Summing over n, we have

$$
\begin{aligned}
&\sum_{n=0}^{N-1}\|\boldsymbol{w}_{N-1}^{k-1}-\boldsymbol{w}_n^{k-1}\|^2\\
&\leq 3\delta^2\mu^2\sum_{n=0}^{N-1}(N-n-1)\sum_{m=n+1}^{N-1}\left(\|\boldsymbol{w}_{m-1}^{k-1}-\boldsymbol{w}_{-1}^{k-1}\|^2+\|\boldsymbol{w}_{N-1}^{k-2}-\boldsymbol{\phi}_{m-1,\boldsymbol{u}_m^{k-1}}^{k-1}\|^2\right.\\
&\qquad\qquad\left.+\frac{1}{N}\sum_{u=0}^{N-1}\|\widetilde{\boldsymbol{\phi}}_{m-1,u}^{k-1}\|^2\right)\\
&=3\delta^2\mu^2\sum_{m=0}^{N-1}\sum_{n=0}^{m-1}(N-n-1)\left(\|\boldsymbol{w}_{m-1}^{k-1}-\boldsymbol{w}_{-1}^{k-1}\|^2+\|\boldsymbol{w}_{N-1}^{k-2}-\boldsymbol{\phi}_{m-1,\boldsymbol{u}_m^{k-1}}^{k-1}\|^2\right.\\
&\qquad\qquad\left.+\frac{1}{N}\sum_{u=0}^{N-1}\|\widetilde{\boldsymbol{\phi}}_{m-1,u}^{k-1}\|^2\right)\\
&\overset{(a)}{\leq}\frac{3}{2}\delta^2\mu^2N^2\sum_{m=0}^{N-1}\left(\|\boldsymbol{w}_{m-1}^{k-1}-\boldsymbol{w}_{-1}^{k-1}\|^2+\|\boldsymbol{w}_{N-1}^{k-2}-\boldsymbol{\phi}_{m-1,\boldsymbol{u}_m^{k-1}}^{k-1}\|^2+\frac{1}{N}\sum_{u=0}^{N-1}\|\widetilde{\boldsymbol{\phi}}_{m-1,u}^{k-1}\|^2\right)\\
&\overset{(b)}{=}\frac{3}{2}\delta^2\mu^2N^2\left\{\sum_{m=0}^{N-1}\|\boldsymbol{w}_{m-1}^{k-1}-\boldsymbol{w}_{-1}^{k-1}\|^2+\sum_{m'=0}^{N-1}\|\boldsymbol{w}_{N-1}^{k-2}-\boldsymbol{w}_{m'}^{k-2}\|^2+\sum_{m=0}^{N-1}\frac{1}{N}\sum_{u=0}^{N-1}\|\widetilde{\boldsymbol{\phi}}_{n,u}^{t-1}\|^2\right\}
\end{aligned}
$$

(23.73)

where step (a) is because

$$\sum_{n=0}^{m-1}(N-n-1) \leq \sum_{n=0}^{N-1}(N-n-1) = \sum_{n=1}^{N-1} n = \frac{N(N-1)}{2} \leq \frac{N^2}{2} \tag{23.74}$$

and step (b) is because

$$\sum_{m=0}^{N-1}\left\|\boldsymbol{w}_{N-1}^{k-2} - \boldsymbol{\phi}_{m-1,\boldsymbol{u}_m^{k-1}}^{k-1}\right\|^2 = \sum_{m'=0}^{N-1}\|\boldsymbol{w}_{N-1}^{k-2} - \boldsymbol{w}_{m'}^{k-2}\|^2 \tag{23.75}$$

since the index $\boldsymbol{u}_m^{k-1}$ depends on m, which is posterior to $m-1$; in this case, the memory variables will be filled with past weight iterates from epoch $k-2$. The above result is similar to (23.61) with k replaced by $k-1$. Therefore, the same procedure can now be followed to arrive at the upper bound in (23.29) where we are using the same scaling coefficients (5 and 7) as in (23.28) for convenience.

23.C PROOF OF THEOREM 23.1

To simplify the notation, we introduce the symbols:

$$a_k^2 \triangleq \frac{1}{N}\sum_{n=0}^{N-1}\mathbb{E}\,\|\boldsymbol{w}_n^k - \boldsymbol{w}_{-1}^k\|^2, \quad b_{k-1}^2 \triangleq \frac{1}{N}\sum_{n=0}^{N-1}\mathbb{E}\,\|\boldsymbol{w}_{N-1}^{k-1} - \boldsymbol{w}_n^{k-1}\|^2 \tag{23.76}$$

Then, the results of Lemmas 23.3 and 23.4 can be combined into the form:

$$\mathbb{E}\,\|\widetilde{\boldsymbol{w}}_{-1}^{k+1}\|^2 \leq \left(1 - \frac{\mu\nu N - \mu^2 N^2\delta^2}{1-\mu N\nu}\right)\mathbb{E}\|\widetilde{\boldsymbol{w}}_{-1}^k\|^2 + 4\mu N\frac{\delta^2}{\nu}(a_k^2 + b_{k-1}^2) \tag{23.77a}$$

$$a_{k+1}^2 \leq 7\delta^2\mu^2N^2(a_{k+1}^2 + b_k^2) + 5\delta^2\mu^2N^2\mathbb{E}\,\|\widetilde{\boldsymbol{w}}_{-1}^{k+1}\|^2 \tag{23.77b}$$

$$b_k^2 \leq 7\delta^2\mu^2N^2(a_k^2 + b_{k-1}^2) + 5\delta^2\mu^2N^2\mathbb{E}\,\|\widetilde{\boldsymbol{w}}_{-1}^k\|^2 \tag{23.77c}$$

We can simplify these relations by recognizing certain bounds. To begin with, note that

$$\begin{aligned}
1 - \frac{\mu\nu N - \mu^2N^2\delta^2}{1-\mu N\nu} &= 1 - \frac{\mu\nu N - \mu^2N^2\nu^2 + \mu^2N^2\nu^2 - \mu^2N^2\delta^2}{1-\mu N\nu} \\
&= 1 - \mu\nu N + \frac{\mu^2N^2\delta^2 - \mu^2N^2\nu^2}{1-\mu N\nu} \\
&= 1 - \frac{3\mu\nu N}{4} - \left(\frac{\mu\nu N}{4} - \frac{\mu^2N^2\delta^2 - \mu^2N^2\nu^2}{1-\mu N\nu}\right) \\
&\leq 1 - \frac{3\mu\nu N}{4}
\end{aligned} \tag{23.78}$$

where the last inequality holds when

$$\begin{aligned}
&1-\mu N\nu > 0, \; \frac{\mu\nu N}{4} - \frac{\mu^2N^2\delta^2 - \mu^2N^2\nu^2}{1-\mu N\nu} \geq 0 \\
\Longleftrightarrow \;& \mu \leq \min\left\{\frac{1}{N\nu}, \frac{\nu}{N(4\delta^2 - 3\nu^2)}\right\}
\end{aligned} \tag{23.79}$$

Since $\nu \leq \delta$, we can replace (23.79) by the sufficient condition

$$\boxed{\text{condition \#1}: \; \mu < \frac{\nu}{4\delta^2 N}} \tag{23.80}$$

Under this condition, and substituting (23.78) into (23.77a), we get

$$\mathbb{E}\,\|\widetilde{\boldsymbol{w}}_{-1}^{k+1}\|^2 \leq \left(1 - \frac{3\mu\nu N}{4}\right)\mathbb{E}\,\|\widetilde{\boldsymbol{w}}_{-1}^{k}\|^2 + 4\mu N\frac{\delta^2}{\nu}(a_k^2 + b_{k-1}^2) \tag{23.81}$$

Let γ denote an arbitrary positive scalar that we are free to choose. Multiplying relations (23.77b) and (23.77c) by γ and adding to (23.81) we obtain:

$$\begin{aligned}
&\mathbb{E}\,\|\widetilde{\boldsymbol{w}}_{-1}^{k+1}\|^2 + \gamma(a_{k+1}^2 + b_k^2)\\
&\leq \left(1 - \frac{3}{4}\mu\nu N\right)\mathbb{E}\,\|\widetilde{\boldsymbol{w}}_{-1}^{k}\|^2 + 4\mu N\frac{\delta^2}{\nu}(a_k^2 + b_{k-1}^2)\\
&\quad + 7\gamma\delta^2\mu^2N^2(a_{k+1}^2 + b_k^2) + 5\gamma\delta^2\mu^2N^2\mathbb{E}\,\|\widetilde{\boldsymbol{w}}_{-1}^{k+1}\|^2\\
&\quad + 7\gamma\delta^2\mu^2N^2(a_k^2 + b_{k-1}^2) + 5\gamma\delta^2\mu^2N^2\mathbb{E}\,\|\widetilde{\boldsymbol{w}}_{-1}^{k}\|^2
\end{aligned} \tag{23.82}$$

which simplifies to

$$\begin{aligned}
&(1 - 5\gamma\delta^2\mu^2N^2)\mathbb{E}\,\|\widetilde{\boldsymbol{w}}_{-1}^{k+1}\|^2 + \gamma(1 - 7\delta^2\mu^2N^2)(a_{k+1}^2 + b_k^2)\\
&\leq \left(1 - \frac{3}{4}\mu\nu N + 5\gamma\delta^2\mu^2N^2\right)\mathbb{E}\,\|\widetilde{\boldsymbol{w}}_{-1}^{k}\|^2 + \left(4\mu N\frac{\delta^2}{\nu} + 7\gamma\delta^2\mu^2N^2\right)(a_k^2 + b_{k-1}^2)
\end{aligned} \tag{23.83}$$

Under the condition $1 - 5\gamma\delta^2\mu^2N^2 > 0$, which is equivalent to

$$\boxed{\text{condition \#2}:\ \mu^2\gamma < \frac{1}{5\delta^2N^2}} \tag{23.84}$$

it holds that

$$\begin{aligned}
&\mathbb{E}\,\|\widetilde{\boldsymbol{w}}_{-1}^{k+1}\|^2 + \gamma\frac{1 - 7\delta^2\mu^2N^2}{1 - 5\gamma\delta^2\mu^2N^2}(a_{k+1}^2 + b_k^2) \qquad (23.85)\\
&\leq \frac{1 - 3\mu\nu N/4 + 5\gamma\delta^2\mu^2N^2}{1 - 5\gamma\delta^2\mu^2N^2}\mathbb{E}\,\|\widetilde{\boldsymbol{w}}_{-1}^{k}\|^2 + \frac{4\mu N\delta^2/\nu + 7\gamma\delta^2\mu^2N^2}{1 - 5\gamma\delta^2\mu^2N^2}(a_k^2 + b_{k-1}^2)\\
&= \frac{1 - 3\mu\nu N/4 + 5\gamma\delta^2\mu^2N^2}{1 - 5\gamma\delta^2\mu^2N^2}\left(\mathbb{E}\,\|\widetilde{\boldsymbol{w}}_{-1}^{k}\|^2 + \frac{4\mu N\delta^2/\nu + 7\gamma\delta^2\mu^2N^2}{1 - 3\mu\nu N/4 + 5\gamma\delta^2\mu^2N^2}(a_k^2 + b_{k-1}^2)\right)
\end{aligned}$$

Since $0 < 1 - 5\gamma\delta^2\mu^2N^2 < 1$, this relation in turn implies that

$$\begin{aligned}
&\mathbb{E}\,\|\widetilde{\boldsymbol{w}}_{-1}^{k+1}\|^2 + \gamma(1 - 7\delta^2\mu^2N^2)(a_{k+1}^2 + b_k^2) \qquad (23.86)\\
&\leq \frac{1 - 3\mu\nu N/4 + 5\gamma\delta^2\mu^2N^2}{1 - 5\gamma\delta^2\mu^2N^2}\left(\mathbb{E}\,\|\widetilde{\boldsymbol{w}}_{-1}^{k}\|^2 + \frac{4\mu N\delta^2/\nu + 7\gamma\delta^2\mu^2N^2}{1 - 3\mu\nu N/4 + 5\gamma\delta^2\mu^2N^2}(a_k^2 + b_{k-1}^2)\right)
\end{aligned}$$

We can again simplify the result by noting that

$$\begin{aligned}
1 - 3\mu\nu N/4 + 5\gamma\delta^2\mu^2N^2 &= 1 - \mu\nu N/4 - (\mu\nu N/2 - 5\gamma\delta^2\mu^2N^2)\\
&\leq 1 - \mu\nu N/4
\end{aligned} \tag{23.87}$$

where the inequality holds when $\mu\nu N/2 - 5\gamma\delta^2\mu^2N^2 \geq 0$, i.e., under the sufficient condition

$$\boxed{\text{condition \#3}:\ \mu\gamma < \frac{\nu}{10\delta^2N}} \tag{23.88}$$

In addition, we have the lower bound

$$1-\frac{3}{4}\mu\nu N+5\gamma\delta^2\mu^2N^2 \geq 1-\frac{3}{4}\mu\nu N \tag{23.89}$$

Using condition #1 from (23.80) and the fact that $\nu \leq \delta$, we have

$$1-\frac{3}{4}\mu\nu N \geq 1-\frac{3\nu^2}{16\delta^2} \geq 1-\frac{3}{16} \geq \frac{13}{16} \tag{23.90}$$

In a similar manner, using condition #3 from (23.88) and $\nu \leq \delta$, we have

$$\begin{aligned}4\mu N\frac{\delta^2}{\nu}+7\gamma\delta^2\mu^2N^2 &\leq 4\mu N\frac{\delta^2}{\nu}+\frac{7}{10}\mu N\nu\\ &\leq 4\mu N\frac{\delta^2}{\nu}+\mu N\nu\\ &\leq 4\mu N\frac{\delta^2}{\nu}+\mu N\delta\\ &\leq 4\mu N\frac{\delta^2}{\nu}+\mu N\frac{\delta^2}{\nu}\\ &= 5\mu N\frac{\delta^2}{\nu}\end{aligned} \tag{23.91}$$

Substituting (23.87), (23.90), and (23.91) into (23.86), we find that

$$\begin{aligned}&\mathbb{E}\,\|\widetilde{\boldsymbol{w}}_{-1}^{k+1}\|^2+\gamma(1-7\delta^2\mu^2N^2)(a_{k+1}^2+b_k^2)\\ &\leq \frac{1-\mu\nu N/4}{1-5\gamma\delta^2\mu^2N^2}\left(\mathbb{E}\,\|\widetilde{\boldsymbol{w}}_{-1}^{k}\|^2+\frac{16}{13}5\mu N\frac{\delta^2}{\nu}(a_k^2+b_{k-1}^2)\right)\end{aligned} \tag{23.92}$$

Under condition #1 in (23.80) and using $\nu \leq \delta$, we have

$$1-7\delta^2\mu^2N^2 \geq 1-7\delta^2N^2\frac{\nu^2}{16\delta^4N^2} \geq 1-\frac{7}{16} \geq \frac{9}{16} \tag{23.93}$$

and, hence,

$$\begin{aligned}&\mathbb{E}\,\|\widetilde{\boldsymbol{w}}_{-1}^{k+1}\|^2+\frac{9}{16}\gamma(a_{k+1}^2+b_k^2)\\ &\leq \frac{1-\mu\nu N/4}{1-53\gamma\delta^2\mu^2N^2}\left(\mathbb{E}\,\|\widetilde{\boldsymbol{w}}_{-1}^{k}\|^2+\frac{80}{13}\mu N\frac{\delta^2}{\nu}(a_k^2+b_{k-1}^2)\right)\\ &\leq \frac{1-\mu\nu N/4}{1-53\gamma\delta^2\mu^2N^2}\left(\mathbb{E}\,\|\widetilde{\boldsymbol{w}}_{-1}^{k}\|^2+\frac{99}{16}\mu N\frac{\delta^2}{\nu}(a_k^2+b_{k-1}^2)\right)\end{aligned} \tag{23.94}$$

where we replaced 80/13 by the larger number 99/16 for convenience. Recall that we are free to choose γ, so assume we choose it to satisfy

$$\frac{9}{16}\gamma=\frac{99}{16}\mu N\frac{\delta^2}{\nu} \implies \gamma=11\mu N\frac{\delta^2}{\nu} \tag{23.95}$$

It then follows that

$$\begin{aligned}\mathbb{E}\,\|\widetilde{\boldsymbol{w}}_{-1}^{k+1}\|^2+\frac{9}{16}\gamma(a_{k+1}^2+b_k^2) &\leq \frac{1-\mu\nu N/4}{1-5\gamma\delta^2\mu^2N^2}\left(\mathbb{E}\,\|\widetilde{\boldsymbol{w}}_{-1}^{k}\|^2+\frac{9}{16}\gamma(a_k^2+b_{k-1}^2)\right)\\ &\overset{\Delta}{=} \lambda\left(\mathbb{E}\,\|\widetilde{\boldsymbol{w}}_{-1}^{k}\|^2+\frac{9}{16}\gamma(a_t^2+b_{t-1}^2)\right)\end{aligned} \tag{23.96}$$

where we introduced the positive parameter

$$\lambda \overset{\Delta}{=} \frac{1-\mu\nu N/4}{1-5\gamma\delta^2\mu^2N^2} \tag{23.97}$$

We can rewrite (23.96) in terms of the energy function V_k defined by:

$$\begin{aligned} &V_{k+1} \\ &\triangleq \mathbb{E}\,\|\widetilde{\boldsymbol{w}}_{-1}^{k+1}\|^2 + \frac{9}{16}\gamma\left(\frac{1}{N}\sum_{n=0}^{N-1}\mathbb{E}\,\|\boldsymbol{w}_n^{k+1} - \boldsymbol{w}_{-1}^{k+1}\|^2 + \frac{1}{N}\sum_{n=0}^{N-1}\mathbb{E}\,\|\boldsymbol{w}_{N-1}^{k} - \boldsymbol{w}_n^{k}\|^2\right) \end{aligned} \tag{23.98}$$

as follows:

$$V_{k+1} \;\leq\; \lambda V_k \tag{23.99}$$

The parameter λ controls the speed of convergence. It will hold that $\lambda < 1$ when

$$\frac{1-\mu\nu N/4}{1-5\gamma\delta^2\mu^2N^2} \overset{(23.95)}{=} \frac{1-\mu\nu N/4}{1-33\delta^4\mu^3N^3/\nu} < 1 \iff \mu < \sqrt{\frac{1}{132}\frac{\nu}{\delta^2 N}} \tag{23.100}$$

Let us now re-examine conditions #1 through #3, along with (23.100) when γ is chosen according to (23.95). In this case, conditions #1 through #3 become

$$\text{conditions \#1 to \#3}: \mu < \frac{\nu}{4\delta^2 N},\ \mu^3 < \frac{\nu}{55\delta^4 N^3},\ \mu^2 < \frac{\nu^2}{110\delta^4 N^2} \tag{23.101}$$

which can be met by

$$\mu \leq \frac{\nu}{4\delta^2 N},\quad \mu < \frac{1}{\delta N}\left(\frac{\nu}{55\delta}\right)^{1/3},\quad \mu \leq \sqrt{\frac{1}{110}\frac{\nu}{\delta^2 N}} \tag{23.102}$$

All three conditions and condition (23.100) can be satisfied by the following single sufficient bound on the step-size parameter (since $11^2 > 108$):

$$\mu < \frac{\nu}{11\delta^2 N} \tag{23.103}$$

23.D PROOF OF LEMMA 23.5

Subtracting $w^\star$ from both sides of (23.33), we obtain:

$$\begin{aligned} \widetilde{\boldsymbol{w}}_{-1}^{k+1} &= \widetilde{\boldsymbol{w}}_{-1}^{k} + \mu N \nabla_{w^\mathsf{T}} P(\boldsymbol{w}_{-1}^{k}) \\ &\quad + \mu\sum_{n=0}^{N-1}\left[\nabla_{w^\mathsf{T}} Q\Big(\boldsymbol{w}_{n-1}^{k};\gamma(\boldsymbol{u}_n^{k}),h_{\boldsymbol{u}_n^{k}}\Big) - \nabla_{w^\mathsf{T}} Q\Big(\boldsymbol{w}_{-1}^{k};\gamma(\boldsymbol{u}_n^{k}),h_{\boldsymbol{u}_n^{k}}\Big)\right] \\ &\quad - \mu\sum_{n=0}^{N-1}\left[\nabla_{w^\mathsf{T}} Q\Big(\boldsymbol{w}_{-1}^{k};\gamma(\boldsymbol{u}_n^{k}),h_{\boldsymbol{u}_n^{k}}\Big) - \nabla_{w^\mathsf{T}} Q\Big(\boldsymbol{w}_{n-1}^{k-1};\gamma(\boldsymbol{u}_n^{k-1}),h_{\boldsymbol{u}_n^{k-1}}\Big)\right] \end{aligned} \tag{23.104}$$

Then, taking the squared norm, applying the Jensen inequality, and using $\boldsymbol{w}_{-1}^{k} = \boldsymbol{w}_{N-1}^{k-1}$, we establish the first recursion for any $t \in (0,1)$:

$$
\begin{aligned}
&\|\widetilde{\boldsymbol{w}}_{-1}^{k+1}\|^2 \\
&\leq \frac{1}{t}\left\|\widetilde{\boldsymbol{w}}_{-1}^{k} + \mu N \nabla_{w^{\mathsf{T}}} P(\boldsymbol{w}_{-1}^{k})\right\|^2 \\
&\quad + \frac{2\mu^2}{1-t}\left\| \sum_{n=0}^{N-1} \left[\nabla_{w^{\mathsf{T}}} Q(\boldsymbol{w}_{n-1}^{k}; \gamma(\boldsymbol{u}_n^k), h_{\boldsymbol{u}_n^k}) - \nabla_{w^{\mathsf{T}}} Q(\boldsymbol{w}_{-1}^{k}; \gamma(\boldsymbol{u}_n^k), h_{\boldsymbol{u}_n^k})\right] \right\|^2 \\
&\quad + \frac{2\mu^2}{1-t}\left\| \sum_{n=0}^{N-1} \left[\nabla_{w^{\mathsf{T}}} Q(\boldsymbol{w}_{-1}^{k}; \gamma(\boldsymbol{u}_n^k), h_{\boldsymbol{u}_n^k}) - \nabla_{w^{\mathsf{T}}} Q(\boldsymbol{w}_{n-1}^{k-1}; \gamma(\boldsymbol{u}_n^{k-1}), h_{\boldsymbol{u}_n^{k-1}})\right] \right\|^2 \\
&\leq \frac{1}{t}\|\widetilde{\boldsymbol{w}}_{-1}^{k} + \mu N \nabla_{w^{\mathsf{T}}} P(\boldsymbol{w}_{-1}^{k})\|^2 + \frac{2\mu^2\delta^2 N}{1-t}\sum_{n=0}^{N-1}\|\boldsymbol{w}_{n-1}^{k} - \boldsymbol{w}_{-1}^{k}\|^2 \\
&\qquad + \frac{2\mu^2\delta^2 N}{1-t}\sum_{n=0}^{N-1}\|\boldsymbol{w}_{N-1}^{k-1} - \boldsymbol{w}_{n-1}^{k-1}\|^2
\end{aligned} \tag{23.105}
$$

Using an argument similar to (23.53), letting $t = 1 - \mu N \nu$, and assuming $\mu \leq 1/(N\nu)$, we obtain (see Prob. 23.6):

$$
\begin{aligned}
\|\widetilde{\boldsymbol{w}}_{-1}^{k+1}\|^2 \leq{}& \left(1 - \frac{\mu\nu N - \mu^2 N^2 \delta^2}{1 - \mu N \nu}\right)\|\widetilde{\boldsymbol{w}}_{-1}^{k}\|^2 \\
&+ \frac{2\mu\delta^2}{\nu}\left(\sum_{n=0}^{N-1}\|\boldsymbol{w}_{n}^{k} - \boldsymbol{w}_{-1}^{k}\|^2 + \sum_{n=0}^{N-1}\|\boldsymbol{w}_{N-1}^{k-1} - \boldsymbol{w}_{n-1}^{k-1}\|^2\right)
\end{aligned} \tag{23.106}
$$

Taking the expectation of both sides, we establish (23.35). The forward inner difference recursion can be obtained by following the same procedure as in (23.58):

$$
\begin{aligned}
&\|\boldsymbol{w}_{n}^{k} - \boldsymbol{w}_{-1}^{k}\|^2 \\
&\leq (n+1)\sum_{m=0}^{n}\|\boldsymbol{w}_{m}^{k} - \boldsymbol{w}_{m-1}^{k}\|^2 \\
&\overset{(23.34)}{\leq} 3\mu^2\delta^2(n+1)\sum_{m=0}^{n}\left(\|\boldsymbol{w}_{m-1}^{k} - \boldsymbol{w}_{-1}^{k}\|^2 + \frac{1}{N}\sum_{n'=0}^{N-1}\|\boldsymbol{w}_{n'-1}^{k-1} - \boldsymbol{w}_{N-1}^{k-1}\|^2 + \|\widetilde{\boldsymbol{w}}_{-1}^{k}\|^2\right) \\
&= 3\mu^2\delta^2(n+1)\sum_{m=0}^{n}\|\boldsymbol{w}_{m-1}^{k} - \boldsymbol{w}_{-1}^{k}\|^2 \\
&\quad + 3\mu^2\delta^2(n+1)^2\left(\frac{1}{N}\sum_{n'=0}^{N-1}\|\boldsymbol{w}_{n'-1}^{k-1} - \boldsymbol{w}_{N-1}^{k-1}\|^2 + \|\widetilde{\boldsymbol{w}}_{-1}^{k}\|^2\right)
\end{aligned} \tag{23.107}
$$

Summing over n, we have

$$\begin{aligned}
\sum_{n=0}^{N-1} \|\boldsymbol{w}_n^k - \boldsymbol{w}_{-1}^k\|^2 \leq\; & 3\mu^2\delta^2 \left\{ \sum_{n=0}^{N-1} (n+1) \sum_{m=0}^{n} \|\boldsymbol{w}_{m-1}^k - \boldsymbol{w}_{-1}^k\|^2 \right. \\
& \left. + \sum_{n=0}^{N-1} (n+1)^2 \left(\frac{1}{N} \sum_{n'=0}^{N-1} \|\boldsymbol{w}_{n'-1}^{k-1} - \boldsymbol{w}_{N-1}^{k-1}\|^2 + \|\widetilde{\boldsymbol{w}}_{-1}^k\|^2 \right) \right\} \\
\overset{(a)}{=}\; & 3\mu^2\delta^2 \left\{ \sum_{m=0}^{N-1} \sum_{n=m}^{N-1} (n+1) \|\boldsymbol{w}_{m-1}^k - \boldsymbol{w}_{-1}^k\|^2 \right. \\
& \left. + \sum_{n=0}^{N-1} (n+1)^2 \left(\frac{1}{N} \sum_{n'=0}^{N-1} \|\boldsymbol{w}_{n'-1}^{k-1} - \boldsymbol{w}_{N-1}^{k-1}\|^2 + \|\widetilde{\boldsymbol{w}}_{-1}^k\|^2 \right) \right\} \\
\overset{(b)}{\leq}\; & 3\mu^2\delta^2 N^2 \sum_{m=0}^{N-1} \|\boldsymbol{w}_{m-1}^k - \boldsymbol{w}_{-1}^k\|^2 \\
& + \mu^2\delta^2 N^2 \left(\sum_{n'=0}^{N-1} \|\boldsymbol{w}_{N-1}^{k-1} - \boldsymbol{w}_{n'-1}^{k-1}\|^2 + N \|\widetilde{\boldsymbol{w}}_{-1}^k\|^2 \right) \\
=\; & 3\mu^2\delta^2 N^2 \sum_{n=0}^{N-1} \|\boldsymbol{w}_{n-1}^k - \boldsymbol{w}_{-1}^k\|^2 \\
& + \mu^2\delta^2 N^2 \left(\sum_{n'=0}^{N-1} \|\boldsymbol{w}_{N-1}^{k-1} - \boldsymbol{w}_{n'}^{k-1}\|^2 + N \|\widetilde{\boldsymbol{w}}_{-1}^k\|^2 \right)
\end{aligned} \tag{23.108}$$

where step (a) is because:

$$\sum_{n=0}^{N-1} \sum_{m=0}^{n} f(n,m) = \sum_{m=0}^{N-1} \sum_{n=m}^{N-1} f(n,m) \tag{23.109}$$

and step (b) is because

$$\sum_{n=m}^{N-1} (n+1) \leq \sum_{n=0}^{N-1} (n+1) \leq \sum_{n=1}^{N} n = \frac{N(N+1)}{2} \leq N^2 \tag{23.110}$$

and

$$\begin{aligned}
\sum_{n=0}^{N-1} (n+1)^2 = \sum_{n=1}^{N} n^2 &= \frac{1}{6} N(N+1)(2N+1) \\
&= \frac{1}{6}(2N^3 + 3N^2 + N) \\
&\leq \frac{1}{6}(2N^3 + 3N^3 + N^3) \\
&= N^3
\end{aligned} \tag{23.111}$$

Lastly, we establish the backward inner difference term using the same argument as in (23.72):

$$\begin{aligned}
\|\boldsymbol{w}_{N-1}^k - \boldsymbol{w}_n^k\|^2 &= \|\boldsymbol{w}_{N-1}^k - \boldsymbol{w}_{N-2}^k + \boldsymbol{w}_{N-2}^k - \cdots + \boldsymbol{w}_{n+1}^k - \boldsymbol{w}_n^k\|^2 \\
&\leq (N-n-1) \sum_{m=n+1}^{N-1} \|\boldsymbol{w}_m^k - \boldsymbol{w}_{m-1}^k\|^2 \\
&\overset{(23.34)}{\leq} 3\mu^2\delta^2(N-n-1) \sum_{m=n+1}^{N-1} \Bigg(\|\boldsymbol{w}_{m-1}^k - \boldsymbol{w}_{-1}^k\|^2 \\
&\qquad + \frac{1}{N} \sum_{n'=0}^{N-1} \|\boldsymbol{w}_{n'-1}^{k-1} - \boldsymbol{w}_{N-1}^{k-1}\|^2 + \|\widetilde{\boldsymbol{w}}_{-1}^k\|^2 \Bigg) \\
&\leq 3\mu^2\delta^2(N-n-1) \sum_{m=n+1}^{N-1} \|\boldsymbol{w}_{m-1}^k - \boldsymbol{w}_{-1}^k\|^2 \\
&\qquad + 3\mu^2\delta^2 \frac{(N-n-1)^2}{N} \sum_{n'=0}^{N-1} \|\boldsymbol{w}_{n'-1}^{k-1} - \boldsymbol{w}_{N-1}^{k-1}\|^2 \\
&\qquad + 3\mu^2\delta^2(N-n-1)^2 \|\widetilde{\boldsymbol{w}}_{-1}^k\|^2
\end{aligned} \tag{23.112}$$

Summing over n we get:

$$\begin{aligned}
&\sum_{n=0}^{N-1} \|\boldsymbol{w}_{N-1}^k - \boldsymbol{w}_n^k\|^2 \\
&\leq 3\mu^2\delta^2 \sum_{n=0}^{N-1} (N-n-1) \sum_{m=n+1}^{N-1} \|\boldsymbol{w}_{m-1}^k - \boldsymbol{w}_{-1}^k\|^2 \\
&\quad + 3\mu^2\delta^2 \sum_{n=0}^{N-1} (N-n-1)^2 \left(\frac{1}{N} \sum_{n'=0}^{N-1} \|\boldsymbol{w}_{n'-1}^{k-1} - \boldsymbol{w}_{N-1}^{k-1}\|^2 + \|\widetilde{\boldsymbol{w}}_{-1}^k\|^2 \right) \\
&\overset{\Delta}{=} Z
\end{aligned} \tag{23.113}$$

where Z is further bounded by

$$\begin{aligned}
Z = {} & 3\mu^2\delta^2 \sum_{m=0}^{N-1} \sum_{n=0}^{m-1} (N-n-1) \|\boldsymbol{w}_{m-1}^k - \boldsymbol{w}_{-1}^k\|^2 \\
& + 3\mu^2\delta^2 \frac{(N-1)N(2N-1)}{6} \left(\frac{1}{N} \sum_{n'=0}^{N-1} \|\boldsymbol{w}_{n'-1}^{k-1} - \boldsymbol{w}_{N-1}^{k-1}\|^2 + \|\widetilde{\boldsymbol{w}}_{-1}^k\|^2 \right) \\
\leq {} & 3\mu^2\delta^2 N^2 \sum_{m=0}^{N-1} \|\boldsymbol{w}_{m-1}^k - \boldsymbol{w}_{-1}^k\|^2 \\
& + 3\mu^2\delta^2 N^2 \left(\sum_{n'=0}^{N-1} \|\boldsymbol{w}_{n'-1}^{k-1} - \boldsymbol{w}_{N-1}^{k-1}\|^2 + N\|\widetilde{\boldsymbol{w}}_{-1}^k\|^2 \right) \\
= {} & 3\mu^2\delta^2 N^2 \sum_{n=0}^{N-1} \|\boldsymbol{w}_{n-1}^k - \boldsymbol{w}_{-1}^k\|^2 \\
& + 3\mu^2\delta^2 N^2 \left(\sum_{n=0}^{N-1} \|\boldsymbol{w}_{n-1}^{k-1} - \boldsymbol{w}_{N-1}^{k-1}\|^2 + N\|\widetilde{\boldsymbol{w}}_{-1}^k\|^2 \right)
\end{aligned} \tag{23.114}$$

23.E PROOF OF THEOREM 23.2

We let

$$a_k^2 \triangleq \frac{1}{N}\sum_{n=0}^{N-1}\mathbb{E}\,\|\boldsymbol{w}_n^k - \boldsymbol{w}_{-1}^k\|^2, \quad b_k^2 \triangleq \frac{1}{N}\sum_{n=0}^{N-1}\mathbb{E}\,\|\boldsymbol{w}_{N-1}^k - \boldsymbol{w}_{n-1}^k\|^2 \tag{23.115}$$

The recursions available for AVRG are then given by:

$$\mathbb{E}\,\|\widetilde{\boldsymbol{w}}_{-1}^{k+1}\|^2 \leq \left(1 - \frac{\mu\nu N - \mu^2N^2\delta^2}{1-\mu N\nu}\right)\mathbb{E}\,\|\widetilde{\boldsymbol{w}}_{-1}^k\|^2 + \frac{2\mu\delta^2 N}{\nu}\,(a_k^2 + b_{k-1}^2) \tag{23.116}$$

$$a_{k+1}^2 \leq 3\mu^2\delta^2N^2a_{k+1}^2 + \mu^2\delta^2N^2b_k^2 + \mu^2\delta^2N^2\mathbb{E}\,\|\widetilde{\boldsymbol{w}}_{-1}^{k+1}\|^2 \tag{23.117}$$

$$b_k^2 \leq 3\mu^2\delta^2N^2(a_k^2 + b_{k-1}^2) + 3\mu^2\delta^2N^2\mathbb{E}\,\|\widetilde{\boldsymbol{w}}_{-1}^k\|^2 \tag{23.118}$$

which have the same form as recursions (23.77a)–(23.77c) except for the coefficients. To simplify the argument, we replace (23.117) by:

$$a_{k+1}^2 \leq 3\mu^2\delta^2N^2(a_{k+1}^2 + b_k^2) + \mu^2\delta^2N^2\mathbb{E}\,\|\widetilde{\boldsymbol{w}}_{-1}^{k+1}\|^2 \tag{23.119}$$

Similar to the derivation of (23.82), we have:

$$\begin{aligned}
&(1-\gamma\mu^2\delta^2N^2)\mathbb{E}\,\|\widetilde{\boldsymbol{w}}_{-1}^{k+1}\|^2 + \gamma(1-3\mu^3\delta^2N^2)(a_{k+1}^2 + b_k^2)\\
&\leq \left(1-\frac{3\mu\nu N}{4}\right)\mathbb{E}\,\|\widetilde{\boldsymbol{w}}_{-1}^k\|^2 + \frac{2\mu\delta^2N}{\nu}\,(a_k^2 + b_{k-1}^2)\\
&\quad+\gamma\Big(3\mu^2\delta^2N^2(a_k^2+b_{k-1}^2) + 3\mu^2\delta^2N^2\mathbb{E}\,\|\widetilde{\boldsymbol{w}}_{-1}^k\|^2\Big)\\
&= \left(1-\frac{3\mu\nu N}{4} + 3\gamma\mu^2\delta^2N^2\right)\mathbb{E}\,\|\widetilde{\boldsymbol{w}}_{-1}^k\|^2 + \left(\frac{2\mu\delta^2N}{\nu} + 3\gamma\mu^2\delta^2N^2\right)(a_k^2 + b_{k-1}^2)
\end{aligned} \tag{23.120}$$

under

$$\boxed{\text{condition \#1}: \ \mu < \frac{\nu}{4\delta^2 N}} \tag{23.121}$$

Under the condition $1-\gamma\delta^2\mu^2N^2 > 0$, which is equivalent to

$$\boxed{\text{condition \#2}: \ \mu^2\gamma < \frac{1}{\delta^2 N^2}} \tag{23.122}$$

it holds that

$$\begin{aligned}
&\mathbb{E}\,\|\widetilde{\boldsymbol{w}}_{-1}^{t+1}\|^2 + \gamma(1-3\mu^2\delta^2N^2)(a_{k+1}^2 + b_k^2)\\
&\leq \frac{1-3\mu\nu N/4 + 3\gamma\mu^2\delta^2N^2}{1-\gamma\delta^2\mu^2N^2}\left(\mathbb{E}\,\|\widetilde{\boldsymbol{w}}_{-1}^k\|^2 + \frac{2\mu\delta^2N/\nu + 3\gamma\mu^2\delta^2N^2}{1-3\mu\nu N/4 + 3\gamma\mu^2\delta^2N^2}\,(a_k^2 + b_{k-1}^2)\right)
\end{aligned} \tag{23.123}$$

Note that the numerator $1-3\mu\nu N/4+3\gamma\mu^2\delta^2N^2$ is similar to SAGA in (23.87). Thus, under condition:

$$\boxed{\text{condition \#3}: \ \mu\gamma < \frac{\nu}{6\delta^2 N}} \tag{23.124}$$

we have

$$\frac{11}{16} \leq 1 - 3\mu\nu N/4 + 3\gamma\delta^2\mu^2 N^2 \leq 1 - \mu\nu N/4 \tag{23.125}$$

Lastly, we can verify that

$$\frac{2\mu\delta^2 N}{\nu} + 3\gamma\mu^2\delta^2 N^2 \leq \frac{2\mu\delta^2 N}{\nu} + \frac{\mu\delta^2 N}{\nu} \leq \frac{3\mu\delta^2 N}{\nu} \tag{23.126}$$

where the last inequality holds when $\mu\gamma < 1/3\nu N$, which is always valid under condition #3. Now, collecting the results, we have

$$\begin{aligned}\mathbb{E}\,\|\widetilde{\boldsymbol{w}}_{-1}^{k+1}\|^2 + \gamma\frac{13}{16}(a_{k+1}^2 + b_k^2) &\leq \frac{1-\mu\nu N/4}{1-\gamma\mu^2\delta^2N^2}\left(\mathbb{E}\,\|\widetilde{\boldsymbol{w}}_{-1}^{k}\|^2 + 3\frac{16}{11}\mu N\frac{\delta^2}{\nu}\left(a_k^2 + b_{k-1}^2\right)\right)\\ &\leq \frac{1-\mu\nu N/4}{1-\gamma\mu^2\delta^2N^2}\left(\mathbb{E}\,\|\widetilde{\boldsymbol{w}}_{-1}^{k}\|^2 + 3\frac{26}{16}\mu N\frac{\delta^2}{\nu}\left(a_k^2 + b_{k-1}^2\right)\right)\end{aligned} \tag{23.127}$$

Assume we choose γ such that

$$\gamma\frac{13}{16} = 3\frac{26}{16}\mu N\frac{\delta^2}{\nu} \quad\Longrightarrow\quad \gamma = 6\mu N\frac{\delta^2}{\nu} \tag{23.128}$$

It then follows that

$$\begin{aligned}\mathbb{E}\,\|\widetilde{\boldsymbol{w}}_{-1}^{k+1}\|^2 + \frac{13}{16}\gamma(a_{k+1}^2 + b_k^2) \;&\leq\; \frac{1-\mu\nu N/4}{1-3\gamma\delta^2\mu^2N^2}\left(\mathbb{E}\,\|\widetilde{\boldsymbol{w}}_{-1}^{k}\|^2 + \frac{13}{16}\gamma(a_k^2 + b_{k-1}^2)\right)\\ &\stackrel{\Delta}{=}\; \lambda\left(\mathbb{E}\,\|\widetilde{\boldsymbol{w}}_{-1}^{k}\|^2 + \frac{13}{16}\gamma(a_k^2 + b_{k-1}^2)\right)\end{aligned} \tag{23.129}$$

where we introduced the positive parameter:

$$\lambda \stackrel{\Delta}{=} \frac{1-\mu\nu N/4}{1-3\gamma\delta^2\mu^2N^2} \tag{23.130}$$

We can rewrite (23.129) in terms of the energy function defined by

$$\begin{aligned}V_{k+1} \stackrel{\Delta}{=}\; &\mathbb{E}\,\|\widetilde{\boldsymbol{w}}_{-1}^{k+1}\|^2 + \\ &\frac{13}{16}\gamma\left(\frac{1}{N}\sum_{n=0}^{N-1}\mathbb{E}\,\|\boldsymbol{w}_n^{k+1} - \boldsymbol{w}_{-1}^{k+1}\|^2 + \frac{1}{N}\sum_{n=0}^{N-1}\mathbb{E}\,\|\boldsymbol{w}_{N-1}^{k} - \boldsymbol{w}_n^{k}\|^2\right)\end{aligned} \tag{23.131}$$

as follows:

$$V_{k+1} \leq \lambda V_k \tag{23.132}$$

The parameter λ satisfies $\lambda < 1$ for step sizes such that

$$\frac{1-\mu\nu N/4}{1-3\gamma\delta^2\mu^2N^2} = \frac{1-\mu\nu N/4}{1-18\delta^4\mu^3N^3/\nu} < 1 \iff \mu < \sqrt{\frac{1}{72}\frac{\nu}{\delta^2 N}} \tag{23.133}$$

We re-examine conditions #1 through #3 when γ is chosen according to (23.128). In this case, these conditions become

$$\text{conditions \#1 to \#3} : \mu < \frac{\nu}{4\delta^2 N},\; \mu^3 < \frac{\nu}{6\delta^4 N^3},\; \mu^2 < \frac{\nu^2}{36\delta^4 N^2} \tag{23.134}$$

which can be met by

$$\mu < \frac{\nu}{4\delta^2 N},\; \mu < \frac{1}{2\delta N}\left(\frac{\nu}{\delta}\right)^{1/3},\; \mu < \frac{\nu}{6\delta^2 N} \tag{23.135}$$

All three conditions and the condition for $\lambda < 1$ can be satisfied by the following single sufficient bound on the step-size parameter:

$$\mu < \frac{\nu}{9\delta^2 N} \tag{23.136}$$

REFERENCES

Bottou, L. (2009), "Curiously fast convergence of some stochastic gradient descent algorithms," in *Proc. Symp. Learning and Data Science*, pp. 1–4, Paris.

De, S. and T. Goldstein (2016), "Efficient distributed SGD with variance reduction," in *Proc. IEEE Int. Conf. Data Mining* (ICDM), pp. 111–120, Barcelona.

Defazio, A., F. Bach, and S. Lacoste-Julien (2014), "SAGA: A fast incremental gradient method with support for non-strongly convex composite objectives," *Proc. Advances Neural Information Processing Systems* (NIPS), pp. 1646–1654, Montreal.

Gürbüzbalaban, M., A. Ozdaglar, and P. Parrilo (2015b), "Why random reshuffling beats stochastic gradient descent," available at arXiv:1510.08560.

Gürbüzbalaban, M., A. Ozdaglar, and P. Parrilo (2015c), "Convergence rate of incremental gradient and Newton methods," available at arXiv:1510.08562.

Johnson, R. and T. Zhang (2013), "Accelerating stochastic gradient descent using predictive variance reduction," *Proc. Advances Neural Information Processing Systems* (NIPS), pp. 315–323, Lake Tahoe, NV.

Recht, B. and C. Re (2012), "Toward a noncommutative arithmetic-geometric mean inequality: Conjectures, case-studies, and consequences," *Proc. Conf. Learning Theory* (COLT), pp. 1–11, Edinburgh.

Shamir, O. (2016), "Without-replacement sampling for stochastic gradient methods: Convergence results and application to distributed optimization," *Proc. Advances Neural Information Processing Systems* (NIPS), pp. 46–54, Long Beach, CA.

Ying, B., K. Yuan, and A. H. Sayed (2018), "Convergence of variance-reduced learning under random reshuffling," *Proc. IEEE ICASSP*, pp. 2286–2290, Calgary.

Ying, B., K. Yuan, and A. H. Sayed (2020), "Variance-reduced stochastic learning under random reshuffling," *IEEE Trans. Signal Process.*, vol. 68, pp. 1390–1408.

Ying, B., K. Yuan, S. Vlaski, and A. H. Sayed (2019), "Stochastic learning under random reshuffling with constant step-sizes," *IEEE Trans. Signal Process.*, vol. 67, no. 2, pp. 474–489.

Yuan, K., B. Ying, J. Liu, and A. H. Sayed (2019), "Variance-reduced stochastic learning by networked agents under random reshuffling," *IEEE Trans. Signal Process.*, vol. 67, no. 2, pp. 351–366.

24 Nonconvex Optimization

The analyses in the previous chapters focused on the convergence behavior of stochastic optimization algorithms for convex risks, $P(w)$. There are, however, many instances in practice where nonconvex risks are commonplace, as happens for example in applications involving the training of neural networks. This is because advances in data collection and processing capabilities have allowed for the use of increasingly complex models, and these naturally give rise to nonconvex formulations. Surprisingly, some relatively simple algorithms can still lead to good performance results even in the nonconvex setting. One prominent example is the success of the backpropagation algorithm for the training of neural networks, which we will study in future chapters. Several works in the literature have pursued rigorous analytical justification for this phenomenon by studying the structure of nonconvex optimization problems and establishing that simple algorithms, such as stochastic gradient and its variants, are able to converge towards local minima and avoid saddle points. One important insight from these studies is that *gradient perturbations* play a critical role in allowing the algorithms to efficiently distinguish desirable from undesirable stationary points and to escape from the latter. In this chapter, we examine the convergence behavior of stochastic gradient algorithms in nonconvex environments, and reveal first- and second-order guarantees for their convergence behavior toward desirable stationary points.

24.1 FIRST- AND SECOND-ORDER STATIONARITY

We initiate our presentation by explaining the nature of the stationary points that arise in nonconvex optimization problems.

24.1.1 Gradient Norm and Excess Risk

A desirable feature for automated optimization (or learning) algorithms is the ability to learn models directly from data with minimal intervention by the designer. This objective is usually achieved by parameterizing a family of models of sufficient explanatory power through a set of parameters $w \in \mathbb{R}^M$ and subsequently searching for the "optimal" choice that fits the data "well" by minimizing either an empirical or stochastic risk function such as

$$w^\star \stackrel{\Delta}{=} \underset{w\in\mathbb{R}^M}{\text{argmin}} \left\{ P(w) = \frac{1}{N}\sum_{m=0}^{N-1} Q(w;\gamma(m),h_m) \right\} \tag{24.1a}$$

$$w^o \stackrel{\Delta}{=} \underset{w\in\mathbb{R}^M}{\text{argmin}} \left\{ P(w) = \mathbb{E}\,Q(w;\boldsymbol{\gamma},\boldsymbol{h}) \right\} \tag{24.1b}$$

for some loss function $Q(w;\cdot)$. In the above expressions, $\boldsymbol{\gamma} \in \mathbb{R}$ refers to the target variable and $\boldsymbol{h} \in \mathbb{R}^M$ refers to the observation vector. In the first formulation, the pairs $(\gamma(m), h_m)$ are data realizations arising from a joint distribution for the random variables $\{\boldsymbol{\gamma}, \boldsymbol{h}\}$. In the second formulation, the expectation is over this distribution. We remain consistent with the notation from the earlier chapters and use $w^\star$ to denote minimizers for *empirical* risks and w^o to denote minimizers for *stochastic* risks. We hasten to add, though, that it will be sufficient for our exposition in this chapter to refer to a generic risk $P(w)$ and its minimizer $w^\star$ regardless of whether we are dealing with an empirical or stochastic minimization problem. Thus, we will consider a generic optimization problem of the form:

$$w^\star \stackrel{\Delta}{=} \underset{w\in\mathbb{R}^M}{\text{argmin}}\; P(w) \tag{24.2}$$

For simplicity of presentation, we will assume that the risk and loss functions are *first-* and *second*-order differentiable. We have shown in previous chapters how problems involving nonsmooth functions can be handled by appealing to subgradient calculations or proximal steps. Similar arguments can be applied in the nonconvex context with proper adjustments. We focus on conveying the main ideas by considering the differentiable case.

The objective of any (stochastic) optimization algorithm is to produce "high-quality" estimates for the minimizer $w^\star$ in (24.2). When the risk $P(w)$ is strongly convex, there is little ambiguity in assessing the quality of the estimate. This is because for ν-strongly convex risks we know from property (8.29) that

$$\frac{\nu}{2}\|w-w^\star\|^2 \leq P(w) - P(w^\star) \leq \frac{1}{2\nu}\|\nabla_w\, P(w)\|^2 \tag{24.3}$$

If the risk function additionally has δ-Lipschitz gradients, we further know from property (10.20) that:

$$\frac{1}{2\delta}\|\nabla_w\, P(w)\|^2 \leq P(w) - P(w^\star) \leq \frac{\delta}{2}\|w-w^\star\|^2 \tag{24.4}$$

By inspecting these two inequalities we find that all three measures of optimality, namely, the squared deviation from the minimizer $\|w - w^\star\|^2$, the excess risk $P(w) - P(w^\star)$, and the squared gradient norm $\|\nabla_w\, P(w^\star)\|^2$, are equivalent up to constants that depend on the strong-convexity and Lipschitz parameters ν and δ, respectively. This means that, as long as the convex optimization problem is reasonably well-conditioned, meaning that the fraction δ/ν does not grow too large, the choice of the performance measure is not particularly relevant, since accurate performance in one measure necessarily implies accurate performance in both other measures. In other words, any point $w \in \mathbb{R}^M$ with a small gradient norm $\|\nabla_w P(w)\|^2$, for strongly convex problems, will be globally optimal in the sense that both the excess risk $P(w) - P(w^\star)$ and distance to the minimizer $\|w - w^\star\|^2$ will be small as well. Unfortunately, in the nonconvex setting and, hence, in the absence of property (24.3), this is no longer the case as we explain in the following.

24.1.2 First-Order Stationarity

We start with the following definition. We say that a point $w \in \mathbb{R}^M$ is $O(\mu)$-*first-order stationarity* if:

$$\boxed{\|\nabla_w P(w)\|^2 \leq O(\mu)} \qquad (\boldsymbol{O(\mu)}\textbf{-first-order stationarity}) \tag{24.5}$$

where μ is a small parameter (it will correspond to the step-size parameter in the stochastic gradient algorithm). The above expression requires the squared norm of the gradient vector at w to be small and on the order of the step size μ. Points w satisfying (24.5) are only *approximately* first-order stationary because *exact* first-order stationarity requires $\nabla_w P(w) = 0$. Moreover, points w satisfying (24.5) can be "close" to any type of stationary points such as local or global minima, local or global maxima, or saddle points where the gradient vector is zero. These possibilities are illustrated in Fig. 24.1, which depicts a risk function in two-dimensional space with global and local maxima, global and local minima, as well as saddle points (such as the saddle point indicated in the figure and appearing between two "hills" on the surface of the plot). A saddle point is neither a local maximum nor a local minimum for the function. While the gradient vector of the risk is zero at a saddle location, its Hessian matrix will be indefinite.

In light of relation (24.4), for risks with δ-Lipschitz gradients, $O(\mu)$-first-order stationarity is a *necessary condition* to have both $P(w) - P(w^\star) \leq O(\mu)$ and $\|w - w^\star\|^2 \leq O(\mu)$ since

$$\|w - w^\star\|^2 \leq O(\mu) \Longrightarrow P(w) - P(w^\star) \leq O(\mu) \Longrightarrow \|\nabla_w P(w)\|^2 \leq O(\mu) \tag{24.6}$$

That is, if the excess risk $P(w) - P(w^\star)$ is small, then w is $O(\mu)$-first-order stationary. However, the converse is not true in general. Unless the risk $P(w)$ is assumed to be additionally strongly convex, $O(\mu)$-first-order stationarity *is not sufficient* to guarantee that the point w has small excess risk $P(w) - P(w^\star)$ or small distance to the minimizer $\|w - w^\star\|^2$. This is because establishing sufficiency

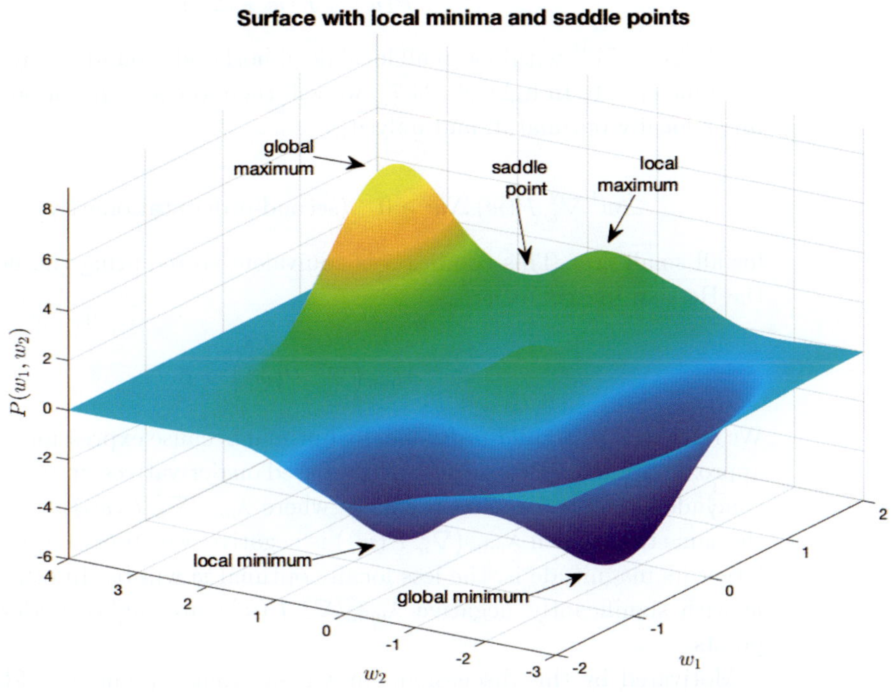

Figure 24.1 Plot of a risk function with global and local maxima, global and local minima, and saddle points.

requires (24.3), which only holds for *strongly convex* costs. This conclusion is consistent with the fact that the set of $O(\mu)$-first-order stationary points for nonconvex risk functions includes the set of local minima, maxima, and saddle points.

24.1.3 Second-Order Stationarity

We can formulate a stronger notion of stationarity, referred to as *second-order stationarity*. It involves a condition on the Hessian matrix of the risk function – see (24.10).

Let w denote some first-order stationary point according to (24.5), where we already know that $\nabla_{w^\mathsf{T}} P(w) \approx 0$. We introduce the second-order Taylor series expansion around w:

$$\begin{aligned} P(w+\Delta w) - P(w) &\approx (\nabla_{w^\mathsf{T}} P(w))^\mathsf{T} \Delta w + \frac{1}{2}\Delta w^\mathsf{T} \nabla_w^2 P(w)\, \Delta w \\ &\approx \frac{1}{2}\Delta w^\mathsf{T}\, \nabla_w^2 P(w)\, \Delta w \end{aligned} \tag{24.7}$$

and drop the first (linear) term since $\nabla w^\mathsf{T} P(w) \approx 0$. Now, recall that the objective is to converge toward first-order stationary points $w \in \mathbb{R}^M$ that are local minima, i.e., that satisfy

$$P(w) \leq P(w + \Delta w) \tag{24.8}$$

for all $\Delta w \in \mathbb{R}^M$ within a small local neighborhood around w, say, for $\|\Delta w\| \leq \epsilon_1$ for some $\epsilon_1 > 0$. In light of (24.7), we will therefore say that a point w is *second-order* locally optimal if, and only if,

$$\Delta w^{\mathsf{T}} \nabla_w^2 P(w) \Delta w \geq 0 \quad \text{(second-order stationarity)} \tag{24.9a}$$

for all small Δw. This condition is equivalent to requiring the nonnegativity of the Hessian matrix at w:

$$\lambda_{\min}\Big(\nabla_w^2 P(w)\Big) \geq 0 \tag{24.9b}$$

We refer to w as *second-order* locally optimal because expression (24.7) is only an approximation for $P(w+\Delta w) - P(w)$ based on derivatives up to second order. We conclude that approaching points w where $\lambda_{\min}(\nabla_w^2 P(w)) \geq 0$ is desirable. At the same time, when $\lambda_{\min}\left(\nabla_w^2 P(w)\right)$ is negative, we observe from (24.7) that the larger its magnitude is, the less locally optimal w will be. In other words, points w with significantly negative $\lambda_{\min}\left(\nabla_w^2 P(w)\right)$ are highly undesirable limiting points.

Motivated by this discussion, we will say that a point $w \in \mathbb{R}^M$ is *τ-second-order stationary* if it is $O(\mu)$-first-order stationary *and* additionally, for some small $\tau > 0$,

$$\boxed{\lambda_{\min}\Big(\nabla_w^2 P(w)\Big) > -\tau} \qquad (\boldsymbol{\tau}\textbf{-second-order stationarity}) \tag{24.10}$$

We will be focusing on the case when τ is *small*. Intuitively, points w that satisfy condition (24.10) are either local minima (e.g., when all eigenvalues of the Hessian matrix are positive) or they are "weak" saddle points that are close to local minima (when the smallest eigenvalue is negative but only by a small amount). Returning to (24.7), we find that every τ-second-order stationary point w satisfies the approximate inequality:

$$P(w) \lessapprox P(w + \Delta w) + \tau \|\Delta w\|^2 \tag{24.11}$$

Note that, as $\tau \to 0$, the definition of τ-second-order stationarity corresponds to the definition of local optimality (24.8). The freedom to set any $\tau > 0$, rather than requiring $\tau \to 0$, allows us to set an expectation of local optimality in the sense of (24.11). The quantity τ will not appear as a parameter in the optimization algorithm, but will appear in the expressions on the convergence time as $O(1/\tau)$, meaning that a higher expectation of local optimality (i.e., smaller τ) will require longer running time for the algorithm, which conforms with intuition – see Theorems 24.2 and 24.3. We conclude that any τ-second-order stationary point is *almost* locally optimal for small τ in the sense of (24.11).

24.1.4 Strict Saddle Points

By choosing τ small enough in (24.10) we are able to exclude first-order stationary points where the smallest eigenvalue of the Hessian matrix is negative and bounded away from zero. We will refer to these undesirable points as strict saddle points (which include both local maxima and saddle points). Specifically, we say that a point $w \in \mathbb{R}^M$ is a *τ-strict saddle point* if it is $O(\mu)$-first-order stationary but satisfies

$$\boxed{\lambda_{\min}\Big(\nabla_w^2 P(w)\Big) \leq -\tau} \qquad (\boldsymbol{\tau}\text{-}\textbf{strict saddle point}) \qquad (24.12)$$

Note that the only difference in relation to definition (24.10) is the reversal of the inequality. In other words, the set of τ-strict saddle points is the complement of the set of τ-second-order stationary points.

Note that, depending on the choice of the parameter τ, not all saddle points of the risk $P(w)$ need to be τ-strict saddle points. If $P(w)$ happens to have a saddle point where $-\tau \leq \lambda_{\min}\left(\nabla_w^2 P(w)\right) < 0$, then this particular saddle point would not be τ-strict and would instead fall under definition (24.10) of a τ-second-order stationary point. Nevertheless, so long as τ is small, such saddle points can intuitively be viewed as "weak" saddle points in the sense that they are *almost* locally optimal according to (24.11).

Moreover, under definition (24.12), the set of strict saddle points includes local maxima. In fact, if *all* eigenvalues of $\nabla_w^2 P(w)$ are bounded from above by $-\tau$, then w would be a *local maximum*. The set of strict saddle points, however, is larger than the set of local maxima, since *only one* eigenvalue of the Hessian is required to be bounded from above by $-\tau$, while other eigenvalues are unrestricted. Hence, the incorporation of second-order information into the definition of stationarity allows us to distinguish between τ-second-order stationary points and τ-strict saddle points.

Figure 24.2 provides a graphical representation of the sets of first-order stationary, second-order stationary, and strict saddle points for a specific numerical example involving a nonconvex risk with two local minima and one saddle point. The plots in the lower row in the figure illustrate in color the regions of second-order stationary points (yellow) and strict saddle points (green) that result from applying a stochastic gradient algorithm (described in the next section) to seek the local minima of the risk function shown in the top row. Two values for τ are used: $\tau = 0.1$ and $\tau = 0.01$. Observe how the smaller value for τ allows for a finer resolution of the sets $\mathcal{D}$ and $\mathcal{U}$, defined in the next diagram.

Figure 24.3 illustrates the notions of first-order stationary, second-order stationary, and strict saddle points. The set of first-order stationary points is denoted by $\mathcal{S}$ and consists of all points w with *small gradient* sizes according to (24.5). This set is split into the union of two *disjoint sets* denoted by $\mathcal{D}$ and $\mathcal{U}$: the former consists of all second-order stationary points (with smallest

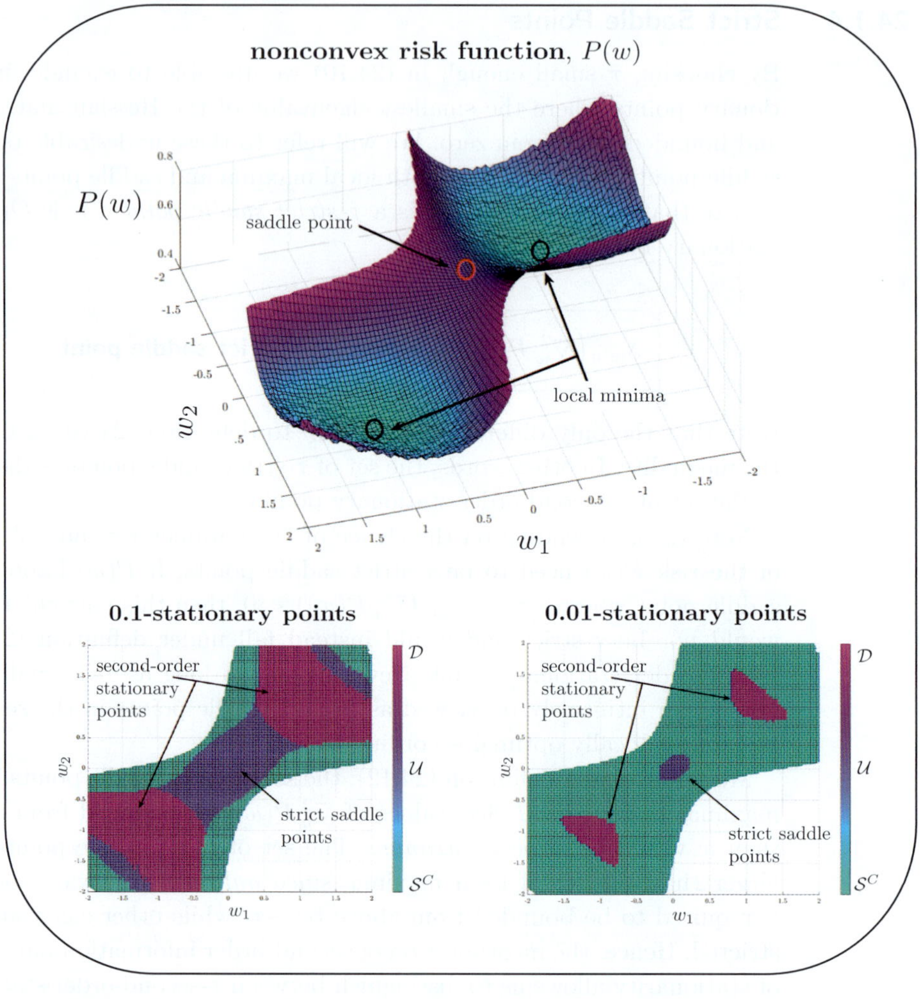

Figure 24.2 A visual representation of first-order stationary, second-order stationary, and strict saddle points for a representative nonconvex risk surface with two local minimizers and one saddle point. The risk surface is shown in the top row for a two-dimensional vector w with entries $\{w_1, w_2\}$; this surface originates from the data in Example 24.2. The plots in the bottom row correspond to the two values $\tau = 0.1$ (*left*) and $\tau = 0.01$ (*right*). The color bar identifies the regions in $\mathbb{R}^2$ corresponding to the sets $\mathcal{D}$ and $\mathcal{U}$ defined in the next diagram. The parts in yellow correspond to desirable second-order stationary points in $\mathcal{D}$ while the parts in green correspond to undesirable strict saddle points in $\mathcal{U}$.

eigenvalue of Hessian beyond $-\tau$) and the latter consists of the remaining points (with smallest eigenvalue of Hessian matrix below $-\tau$):

$$\mathcal{S} = \mathcal{D} \cup \mathcal{U} \tag{24.13}$$

Note that $\mathcal{D}$ consists of all *desirable* limit points (hence the designation $\mathcal{D}$), while $\mathcal{U}$ consists of all *undesirable* limit points (hence the designation $\mathcal{U}$). The complement of $\mathcal{S}$, denoted by $\mathcal{S}^C$, consists of all points w with "large" gradient sizes (which, therefore, cannot correspond to stationary points):

$$\mathrm{IR}^M = \mathcal{S} \cup \mathcal{S}^C \tag{24.14}$$

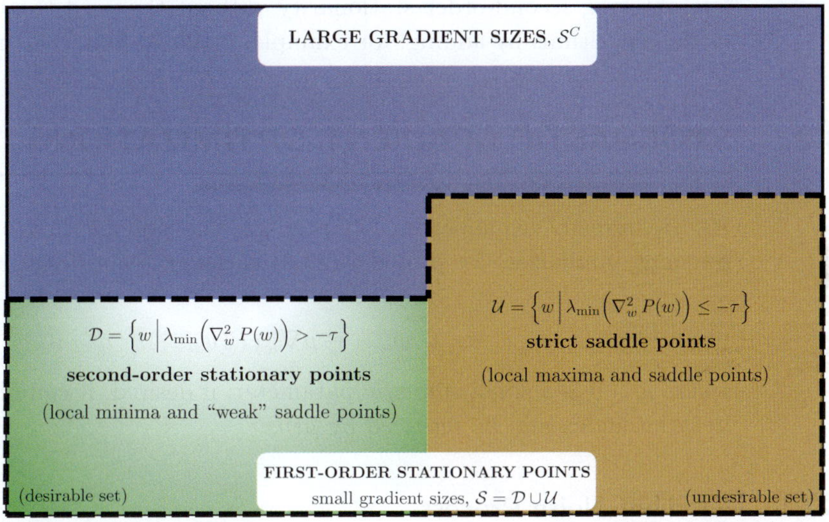

Figure 24.3 The set of first-order stationary points is denoted by $\mathcal{S}$ and consists of all points w with small gradient sizes according to (24.5). This set is split into the union of two disjoint sets denoted by $\mathcal{D}$ and $\mathcal{U}$: The former consists of all (desirable) second-order stationary points and the latter consists of the remaining (undesirable) points. The complement of $\mathcal{S}$, denoted by $\mathcal{S}^C$, consists of all points w with "large" gradient sizes.

Now, the types of convergence analyses for optimization algorithms in nonconvex settings can be broadly categorized into two groups:

(a) *First-order stationarity analyses*: These studies only establish convergence toward first-order stationary points, i.e., toward points within the set $\mathcal{S}$ where the norm of the gradient of the risk function is small. First-order convergence guarantees can only ensure that algorithms do not return points in the complement of $\mathcal{S}$ where the norm of the gradient is large.

(b) *Second-order stationarity analyses*: These studies provide finer results and establish convergence toward *desirable* second-order stationary points within the set $\mathcal{D}$. That is, these studies are able to discriminate between the two disjoint sets $\mathcal{D}$ and $\mathcal{U}$ within $\mathcal{S}$, and establish that the optimization algorithm is able to escape from saddle points within $\mathcal{U}$ and approach desirable limiting points within $\mathcal{D}$.

The analysis in the following will focus on stochastic gradient strategies; it will show that these techniques benefit from the presence of perturbations in the update direction (due to *gradient noise*) to escape from saddle points. The perturbations in the gradient direction will be shown to be sufficient to guarantee *efficient* escape from saddle points, meaning that the escape-time will be bounded by quantities that scale favorably with problem dimension and parameters, resulting in simple yet effective solutions for escaping saddle points and guaranteeing second-order stationarity without the need to alter the operation of the algorithms by adding, for example, artificial (external) noise.

24.2 STOCHASTIC GRADIENT OPTIMIZATION

As was already explained in Chapter 12, one popular first-order approach to pursuing minimizers for problem (24.2) is the gradient-descent recursion:

$$w_n = w_{n-1} - \mu \nabla_{w^\mathsf{T}} P(w_{n-1}), \quad n \geq 0 \tag{24.15}$$

where $\mu > 0$ is a small step-size parameter; if desired, it can also depend on the iteration index and be replaced by $\mu(n)$.

24.2.1 Gradient Noise Process

The main limitation of this recursion, however, is the need to compute the exact gradient of $P(w)$, either due to the lack of statistical information about the data in the stochastic formulation (24.1b) or due to the complexity in evaluating an aggregate gradient over all data points in the empirical formulation (24.1a). As was explained in Chapter 16 on stochastic gradient methods, the most common remedy for this challenge is to employ a stochastic approximation for the true gradient. Independent of the nature of $P(w)$ (whether empirical or stochastic), the stochastic gradient algorithm takes the following form:

$$\boldsymbol{w}_n = \boldsymbol{w}_{n-1} - \mu \, \widehat{\nabla_{w^\mathsf{T}} P}(\boldsymbol{w}_{n-1}) \tag{24.16}$$

which employs an approximation for the true gradient of $P(w)$ evaluated at $w = \boldsymbol{w}_{n-1}$. For both cases of empirical and stochastic risks, we will construct the gradient approximation by relying on the gradient of the loss function evaluated at some random data pair $(\boldsymbol{\gamma}(n), \boldsymbol{h}_n)$, i.e.,

$$\widehat{\nabla_{w^\mathsf{T}} P}(\boldsymbol{w}_{n-1}) \;=\; \nabla_{w^\mathsf{T}} Q(\boldsymbol{w}_{n-1}; \boldsymbol{\gamma}(n), \boldsymbol{h}_n) \tag{24.17}$$

The sample pair $(\boldsymbol{\gamma}(n), \boldsymbol{h}_n)$ is indicated in boldface to highlight its random nature, with the index set to n. This notation means that, at each iteration n, a data pair $(\boldsymbol{\gamma}(n), \boldsymbol{h}_n)$ is selected (or streams in) at random. The selection depends on the mode of operation. For example, in empirical risk minimization where we are already given a collection of N data pairs $\{\gamma(m), h_m\}$ for $m = 0, 1, 2, \ldots, N-1$, the random sample $(\boldsymbol{\gamma}(n), \boldsymbol{h}_n)$ is chosen according to some selection policy such as uniform sampling (i.e., sampling with replacement). In stochastic risk minimization, on the other hand, the random sample $(\boldsymbol{\gamma}(n), \boldsymbol{h}_n)$ is assumed to arise from data that stream in independently of each other, with one data pair arriving at each iteration. Either way, the stochastic gradient recursion (24.16) takes the form:

$$\boxed{\boldsymbol{w}_n = \boldsymbol{w}_{n-1} - \mu\, \nabla_{w^{\mathsf{T}}} Q(\boldsymbol{w}_{n-1}; \boldsymbol{\gamma}(n), \boldsymbol{h}_n), \quad n \geq 0} \tag{24.18}$$

which runs continually over the data. At each iteration n, some random data pair $(\boldsymbol{\gamma}(n), \boldsymbol{h}_n)$ is made available and used to update $\boldsymbol{w}_{n-1}$ to $\boldsymbol{w}_n$. Observe that we are denoting $\boldsymbol{w}_n$ in bold font to emphasize the fact that, by utilizing a stochastic approximation $\widehat{\nabla_w P}(\cdot)$ based on *realizations* of the data in place of the true gradient $\nabla_w P(\cdot)$, the resulting iterates $\boldsymbol{w}_n$ will become stochastic themselves.

We refer to the difference between the true and approximate gradient as *gradient noise* and denote it by the notation:

$$\boldsymbol{g}(w) \stackrel{\Delta}{=} \widehat{\nabla_{w^{\mathsf{T}}} P}(w) \;-\; \nabla_{w^{\mathsf{T}}} P(w) \tag{24.19}$$

It is useful to emphasize that the gradient noise is a function of the iteration index n because its calculation depends on the data at that iteration. Therefore, for added clarity, we should write $\boldsymbol{g}_n(w)$ instead of $\boldsymbol{g}(w)$. The gradient noise $\boldsymbol{g}_n(w)$ is also a *random* process, which is why we are denoting it in boldface; its value varies randomly with the data $(\boldsymbol{\gamma}(n), \boldsymbol{h}_n)$ and with the weight iterate $\boldsymbol{w}_{n-1}$ used to evaluate $\boldsymbol{g}_n(\boldsymbol{w}_{n-1})$. The statistical properties of the gradient noise process (i.e., how good the gradient approximation is) influence the performance of the stochastic gradient algorithm in important ways.

24.2.2 Gradient Noise Conditions

We will establish convergence guarantees for the stochastic gradient recursion (24.18) under some conditions on the risk function and the gradient noise process.

First, we assume that the risk function $P(w)$ is first- and second-order differentiable with both Lipschitz gradient and Hessian. Specifically, we assume there exist constants $\delta \geq 0$ and $\delta_2 \geq 0$ such that, for any w_2, w_1:

$$\|\nabla_w P(w_2) - \nabla_w P(w_1)\| \leq \; \delta \|w_2 - w_1\| \tag{24.20a}$$

$$\|\nabla_w^2 P(w_2) - \nabla_w^2 P(w_1)\| \leq \delta_2 \|w_2 - w_1\| \tag{24.20b}$$

We encountered condition (24.20a) in the study of convex optimization problems (e.g., in Section 18.2); this condition will be used to establish *first-order* optimality guarantees. The Lipschitz condition (24.20b) on the Hessian matrix will be used to establish *second-order* guarantees and escape from saddle points; this Hessian condition was not needed in the case of (strongly) convex risks.

Second, we assume bounds on the first- and fourth-order moments of the gradient noise process (compare with Prob. 18.4):

$$\mathbb{E}\left(\boldsymbol{g}_n(\boldsymbol{w}_{n-1})\,|\,\boldsymbol{w}_{n-1}\right) = 0 \tag{24.21a}$$

$$\mathbb{E}\left(\|\boldsymbol{g}_n(\boldsymbol{w}_{n-1})\|^4\,|\,\boldsymbol{w}_{n-1}\right) \le \beta^4\|\nabla_w P(w_{n-1})\|^4 + \sigma^4 \tag{24.21b}$$

for some nonnegative constants β^4, σ^4. Relation (24.21a) requires the gradient approximation $\widehat{\nabla_w P}(\cdot)$ to be unbiased. We already know from the analyses in the previous chapters that this condition is satisfied for instantaneous gradient approximations when the data samples $(\boldsymbol{\gamma}(n), \boldsymbol{h}_n)$ are selected uniformly at random or stream in independently of each other. Condition (24.21b), on the other hand, imposes a bound on the fourth-order moment of the gradient noise, but allows the bound to grow with the fourth norm of the gradient $\|\nabla_w P(w_{n-1})\|^4$. Note that, in light of Jensen inequality and the sub-additivity of the square root operation, condition (24.21b) implies and is slightly stronger than:

$$\mathbb{E}\left(\|\boldsymbol{g}_n(\boldsymbol{w}_{n-1})\|^2\,|\,\boldsymbol{w}_{n-1}\right) \le \beta^2\|\nabla_w P(w_{n-1})\|^2 + \sigma^2 \tag{24.22}$$

REMARK 24.1. (Bound in terms of gradient of risk function) In previous chapters, when dealing with convex risks, we established that the second-order moment of the gradient noise process satisfies a bound of the form:

$$\mathbb{E}\left(\|\boldsymbol{g}_n(\boldsymbol{w}_{n-1})\|^2\,|\,\boldsymbol{w}_{n-1}\right) \le \beta_g^2\|\widetilde{w}_{n-1}\|^2 + \sigma_g^2 \tag{24.23}$$

for some parameters (β_g^2, σ_g^2). This bound is in terms of the squared weight error, $\|\widetilde{w}_{n-1}\|^2$, whereas the assumed bound in (24.22) for nonconvex settings is expressed in terms of the squared norm of the gradient of the risk, $\|\nabla_w P(w_{n-1})\|^2$. Obviously, both conditions are equivalent for strongly convex and δ-smooth risks in view of properties (24.3) and (24.4). This is not true anymore for nonconvex risks. In that case, condition (24.22) is more strict since it implies (24.23), but not the other way around.

■

We also impose a condition on the covariance matrix of the gradient noise process, denoted by

$$R_g(\boldsymbol{w}_{n-1}) \triangleq \mathbb{E}\left(\boldsymbol{g}_n(\boldsymbol{w}_{n-1})\boldsymbol{g}_n^{\mathsf{T}}(\boldsymbol{w}_{n-1})\,|\,\boldsymbol{w}_{n-1}\right) \tag{24.24a}$$

Specifically, we assume that the gradient noise process has a Lipschitz covariance matrix in the following sense:

$$\|R_g(w_2) - R_g(w_1)\| \le \beta_R\|w_2 - w_1\|^{\eta} \tag{24.24b}$$

for any w_2, w_1, some $\beta_R \ge 0$ and $0 < \eta \le 4$. Note from the definition of the gradient noise covariance (24.24a) that the distribution of the gradient noise process is a function of the iterate $\boldsymbol{w}_{n-1}$. The perturbations introduced into the stochastic recursion (24.18) are not necessarily identically distributed over time; this fact introduces challenges into the study of the cumulative effect of these perturbations on the performance of the optimization algorithm. Condition

(24.24b) ensures that the covariance $R_g(\boldsymbol{w}_{n-1})$ is sufficiently smooth in space, leading to a more tractable analysis.

We need to introduce one final condition on the gradient noise process, which is motivated as follows. Assume $w \in \mathcal{U}$ is some τ-strict saddle point according to definition (24.12); that is, it is a point with a small gradient norm but is undesirable because it may correspond to a local maximum or a saddle point location. Let $R_g(w)$ denote the covariance matrix of the gradient noise at this particular location. We introduce the eigen-decomposition of the Hessian matrix of $P(w)$ at the same location, namely,

$$\nabla_w^2\, P(w) \;\stackrel{\Delta}{=}\; V \Lambda V^{\mathsf{T}} \tag{24.25}$$

where V is an orthogonal matrix and Λ is a diagonal matrix; their values depend on w. We order the entries of Λ with its nonnegative entries coming first followed by its negative entries:

$$\Lambda \;\stackrel{\Delta}{=}\; \begin{bmatrix} \Lambda^{\geq 0} & 0 \\ 0 & \Lambda^{<0} \end{bmatrix} \tag{24.26}$$

where $\Lambda^{\geq 0} \geq 0$ and $\Lambda^{<0} < 0$, i.e., these diagonal submatrices contain the nonnegative and negative eigenvalues of Λ, respectively. We also partition the columns of V accordingly as

$$V \;\stackrel{\Delta}{=}\; \begin{bmatrix} V^{\geq 0} & V^{<0} \end{bmatrix} \tag{24.27}$$

In this partitioning, the columns of $V^{<0}$ correspond to eigenvectors of $\nabla_w^2\, P(w)$ with negative eigenvalues. If we again consider a Taylor series expansion of the form (24.7) around the strict saddle location $w \in \mathcal{U}$, we get

$$\begin{aligned} P(w + \Delta w) &\approx P(w) + (\nabla_{w^{\mathsf{T}}} P(w))^{\mathsf{T}} \Delta w + \frac{1}{2} \Delta w^{\mathsf{T}}\, \nabla_w^2 P(w)\, \Delta w \\ &\approx P(w) + \frac{1}{2} \Delta w^{\mathsf{T}}\, \nabla_w^2 P(w)\, \Delta w \end{aligned} \tag{24.28}$$

since at strict saddle points $\nabla_w\, P(w) \approx 0$. For every small perturbation Δw in the range of $V^{<0}$, i.e., $\Delta w = V^{<0} x$ for some vector x, we have

$$\Delta w^{\mathsf{T}}\, \nabla_w^2 P(w)\, \Delta w = x^{\mathsf{T}} \Big((V^{<0})^{\mathsf{T}}\, \nabla_w^2 P(w) V^{<0} \Big) x < 0 \tag{24.29}$$

and, hence, we end up with the approximate inequality:

$$P(w + \Delta w) \lessapprox P(w), \quad \text{for any small } \Delta w \in \mathcal{R}(V^{<0}) \tag{24.30}$$

where $\mathcal{R}(V^{<0})$ refers to the range space of V^o. We conclude that the space spanned by $V^{<0}$ around the strict saddle point w corresponds to local descent directions. For this reason, we will assume that the covariance matrix of the

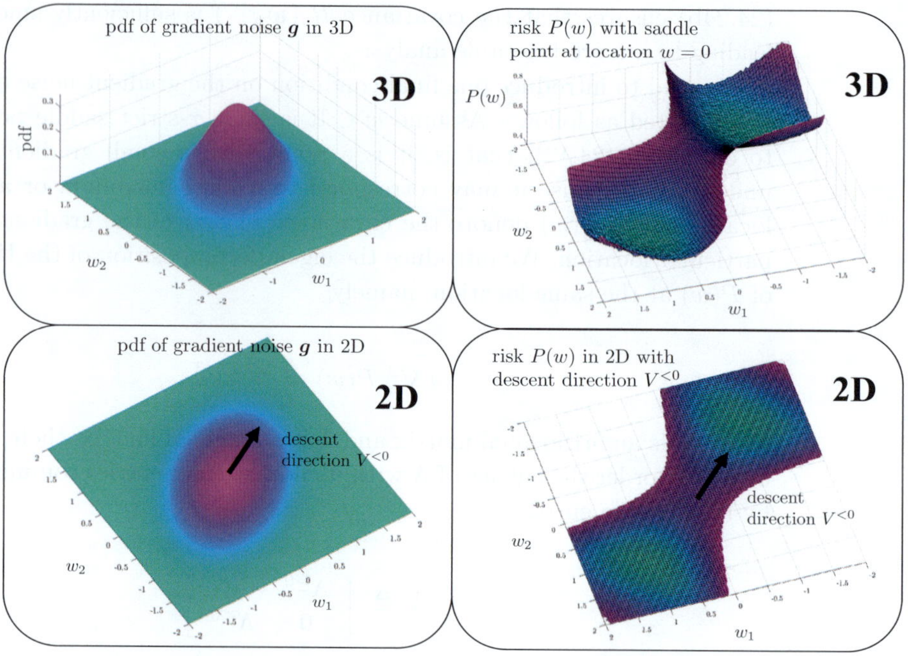

Figure 24.4 Visual illustration of condition (24.31). Examples in 3D of a probability density function (pdf) for the gradient noise process $\boldsymbol{g}_n(\boldsymbol{w}_{n-1})$ (*top-left*) and risk function $P(w)$ (*bottom-left*). The risk $P(w)$ exhibits a strict saddle point at $w = 0$. A local descent direction is shown by the red arrows in the right column. Condition (24.31) requires the presence of noise components along this direction.

gradient noise process satisfies the following condition at all strict saddle points:

$$\lambda_{\min}\left\{\left(V^{<0}\right)^{\mathsf{T}} R_g(w)\, V^{<0}\right\} \geq \sigma_\ell^2 > 0, \quad \forall\, w \in \mathcal{U} \ \text{(strict saddle points)}$$

(**persistent gradient noise components along directions in** $\mathcal{R}(V^o)$) (24.31)

for some uniform σ_ℓ^2. This condition requires the presence of *persistent* gradient noise components along local descent directions spanned by $V^{<0}$ in the vicinity of strict saddle points – see Fig. 24.4. Condition (24.31) is not merely a technical requirement but is necessary for the algorithm to be able to escape saddle points.

Example 24.1 **(Perturbing the stochastic update)** In the absence of prior knowledge that there are gradient noise components along a descent direction for every strict saddle point, as required by condition (24.31), we can guarantee (24.31) by adding a small perturbation term $\boldsymbol{v}_n$ with positive-definite covariance matrix $R_v \triangleq \mathbb{E}\boldsymbol{v}\boldsymbol{v}^{\mathsf{T}} > 0$ to construct:

$$\widehat{\nabla_w P}(\boldsymbol{w}_{n-1}) = \nabla_w Q(\boldsymbol{w}_{n-1}; \boldsymbol{\gamma}(n), \boldsymbol{h}_n) + \boldsymbol{v}_n \qquad (24.32)$$

The gradient noise covariance will then be replaced by $R_g(\boldsymbol{w}_{n-1}) + R_v > 0$. More elaborate constructions, such as adding an additional perturbation only when the iterate $\boldsymbol{w}_{n-1}$ is suspected to be near a first-order stationary point, is also possible.

For ease of reference, we collect the modeling assumptions listed in this section into the following statement.

Assumption 24.1. (Modeling assumptions) *The convergence analysis of the stochastic gradient recursion (24.18) for nonconvex risks $P(w)$i assumes the following conditions:*

$$(\mathbb{A}1)\quad P(w) \text{ is first- and second-order differentiable} \tag{24.33a}$$

$$(\mathbb{A}2)\quad \|\nabla_w P(w_2) - \nabla_w P(w_1)\| \leq \delta\|w_2 - w_1\| \tag{24.33b}$$

$$(\mathbb{A}3)\quad \|\nabla_w^2 P(w_2) - \nabla_w^2 P(w_1)\| \leq \delta_2\|w_2 - w_1\| \tag{24.33c}$$

and

$$(\mathbb{A}4)\quad \mathbb{E}\left(\boldsymbol{g}_n(\boldsymbol{w}_{n-1})\,|\,\boldsymbol{w}_{n-1}\right) = 0 \tag{24.34a}$$

$$(\mathbb{A}5)\quad \mathbb{E}\left(\|\boldsymbol{g}_n(\boldsymbol{w}_{n-1})\|^4\,|\,\boldsymbol{w}_{n-1}\right) \leq \beta^4\|\nabla_w P(w_{n-1})\|^4 + \sigma^4 \tag{24.34b}$$

$$(\mathbb{A}6)\quad \|R_g(w_2) - R_g(w_1)\| \leq \beta_R\|w_2 - w_1\|^\eta \tag{24.34c}$$

and

$$(\mathbb{A}7)\quad \lambda_{\min}\left\{\left(V^{<0}\right)^\mathsf{T} R_g(w)\, V^{<0}\right\} \geq \sigma_\ell^2 > 0,\ \ \forall w \in \mathcal{U} \tag{24.35}$$

The first set of conditions ($\mathbb{A}1$–$\mathbb{A}3$) deals with the risk function $P(w)$, while the second set of conditions ($\mathbb{A}4$–$\mathbb{A}6$) deals with the quality of the gradient approximation and the gradient noise process. The last condition ($\mathbb{A}7$) links the behavior of the risk function and the gradient noise process (the latter needs to have persistent presence along the descent directions in $V^{<0}$, which are determined by the Hessian of $P(w)$). We noted in (24.22) that condition ($\mathbb{A}5$) implies

$$(\mathbb{A}5')\quad \mathbb{E}\left(\|\boldsymbol{g}_n(\boldsymbol{w}_{n-1})\|^2|\boldsymbol{w}_{n-1}\right) \leq \beta^2\|\nabla_w P(w_{n-1})\|^2 + \sigma^2 \tag{24.36}$$

24.3 CONVERGENCE BEHAVIOR

We now list the main results concerning the convergence behavior of the stochastic gradient recursion (24.18) for nonconvex risk functions. We defer the proofs and derivations to the appendices. We begin by formalizing the space decomposition $\{\mathcal{S}, \mathcal{S}^C, \mathcal{D}, \mathcal{U}\}$ introduced earlier.

24.3.1 Space Decomposition

The set $\mathcal{S}$ consists of all locations $w \in \mathbb{R}^M$ with "small" gradient sizes. More formally, we define:

$$\mathcal{S} \stackrel{\Delta}{=} \left\{ w \in \mathbb{R}^M : \|\nabla_w P(w)\|^2 < \mu \frac{c_2}{c_1}\left(1 + \frac{1}{\epsilon}\right) = O(\mu) \right\} \tag{24.37}$$

"locations w with 'small' gradient sizes bounded by $O(\mu)$"

where $0 < \epsilon < 1$ is a small parameter to be chosen by the designer (specified further ahead in Theorem 24.3), and for some constants $\{c_1, c_2\}$ defined as follows (these constants will be used in the derivations in the appendices to establish the conclusions listed in this section):

$$c_1 \stackrel{\Delta}{=} 1 - \frac{\mu\delta}{2}(1 + \beta^2) = O(1) \tag{24.38}$$

$$c_2 \stackrel{\Delta}{=} \frac{\delta\sigma^2}{2} = O(1) \tag{24.39}$$

The constant c_1 is guaranteed to remain positive when the step size μ is small enough, satisfying:

$$\mu < \frac{2}{\delta(1 + \beta^2)} \tag{24.40}$$

We denote the complement space to $\mathcal{S}$ by $\mathcal{S}^C$: It consists of all locations w with "large" gradient sizes:

$$\mathcal{S}^C \stackrel{\Delta}{=} \left\{ w \in \mathbb{R}^M : \|\nabla_w P(w)\|^2 \geq \mu \frac{c_2}{c_1}\left(1 + \frac{1}{\epsilon}\right) = O(\mu) \right\} \tag{24.41}$$

"locations w with 'large' gradient sizes"

We further decompose the set $\mathcal{S}$ of first-order stationary points into the union of two disjoint sets, $\mathcal{S} = \mathcal{D} \cup \mathcal{U}$. The symbol $\mathcal{D}$ refers to the *desirable* space of second-order stationary points consisting of all local minima and "weak" saddle locations:

$$\mathcal{D} \stackrel{\Delta}{=} \left\{ w \in \mathbb{R}^M : w \in \mathcal{S}, \lambda_{\min}\left(\nabla_w^2 P(w)\right) > -\tau \right\} \tag{24.42}$$

"desirable locations: local minima, weak saddle points"

for some small $\tau > 0$, while the symbol $\mathcal{U}$ refers to the *undesirable* locations of local maxima and strict saddle points:

$$\mathcal{U} \stackrel{\Delta}{=} \left\{ w \in \mathbb{R}^M : w \in \mathcal{S}, \lambda_{\min}\left(\nabla_w^2 P(w)\right) \leq -\tau \right\} \tag{24.43}$$

"undesirable locations: local maxima, strict saddle points"

For visualization of the spaces $\{\mathcal{S}, \mathcal{S}^C, \mathcal{D}, \mathcal{U}\}$, we refer back to Fig. 24.3. Points in both $\mathcal{S}^C$ and $\mathcal{U}$ are "undesirable" limiting points in the sense that they will still have local directions of descent. Our objective is to show that for iterates within both sets, the stochastic gradient algorithm (24.18) will continue to descend toward locations in $\mathcal{D}$.

24.3.2 First-Order Convergence

The first result below guarantees convergence of the stochastic gradient algorithm (24.18) to first-order stationary points, i.e., to points $w \in \mathcal{S}$ with small gradient norms. The statement does not discriminate between the subsets $\mathcal{D}$ and $\mathcal{U}$ within $\mathcal{S}$. To be able to ascertain whether the limiting point is ultimately within the desirable subset $\mathcal{D}$, we will need to resort to a second-order analysis, which is performed in the next section. In this first statement, we can only guarantee convergence toward the set $\mathcal{S}$ of first-order stationary points.

THEOREM 24.1. (Descent in the large-gradient regime) *For sufficiently small step sizes satisfying*

$$\mu \leq \frac{2}{\delta\,(1+\beta^2)} \tag{24.44}$$

and when the gradient of the risk $P(w)$ at the iterate $w = \boldsymbol{w}_{n-1}$ happens to be sufficiently large, i.e., $\boldsymbol{w}_{n-1} \in \mathcal{S}^C$, the stochastic gradient recursion (24.18) will yield descent of the risk value under expectation in a single iteration, namely, it will hold that:

$$\mathbb{E}\left(P(\boldsymbol{w}_n)\,|\,\boldsymbol{w}_{n-1} \in \mathcal{S}^C\right) \leq \mathbb{E}\left(P(\boldsymbol{w}_{n-1})\,|\,\boldsymbol{w}_{n-1} \in \mathcal{S}^C\right) - \mu^2\frac{c_2}{\epsilon} \tag{24.45}$$

This conclusion holds under conditions ($\mathbb{A}1$), ($\mathbb{A}2$), ($\mathbb{A}4$), *and* ($\mathbb{A}5'$).

Proof: See Appendix 24.A.

∎

Starting from an iterate $\boldsymbol{w}_{n-1}$ with a "large" risk gradient value, i.e., $\boldsymbol{w}_{n-1} \in \mathcal{S}^C$, result (24.45) shows that the stochastic gradient algorithm (24.18) behaves in a desirable manner and updates $\boldsymbol{w}_{n-1}$ to a new location $\boldsymbol{w}_n$ where the risk value is smaller on average, i.e., $\mathbb{E}\,P(\boldsymbol{w}_n) < \mathbb{E}\,P(\boldsymbol{w}_{n-1})$ since the positive term $\mu^2 c_2/\epsilon$ is subtracted from $\mathbb{E}\,P(\boldsymbol{w}_{n-1})$ on the right-hand side. Moreover, this decrease in the average risk value occurs over a *single* iteration. This is a reassuring result because it means that, starting from iterates within the large gradient space $\mathcal{S}^C$, the stochastic gradient algorithm will move these iterates closer to locations with smaller risk values. That is, the iterates will move away from $\mathcal{S}^C$ and toward $\mathcal{S}$.

The proof of Theorem 24.1 in the appendix relies solely on conditions ($\mathbb{A}1$), ($\mathbb{A}2$), ($\mathbb{A}4$), and ($\mathbb{A}5'$) listed under the modeling Assumption 24.1. Specifically, the following conditions are sufficient to ensure the first-order convergence result:

$$\begin{aligned}
&(\mathbb{A}1)\;\; P(w) \text{ is first-order differentiable} && (24.46a)\\
&(\mathbb{A}2)\;\; \|\nabla_w P(w_2) - \nabla_w P(w_1)\| \leq \delta\|w_2 - w_1\| && (24.46b)\\
&(\mathbb{A}4)\;\; \mathbb{E}\left(\boldsymbol{g}_n(\boldsymbol{w}_{n-1}) \,|\, \boldsymbol{w}_{n-1}\right) = 0 && (24.46c)\\
&(\mathbb{A}5')\;\; \mathbb{E}\left(\|\boldsymbol{g}_n(\boldsymbol{w}_{n-1})\|^2 \,|\, \boldsymbol{w}_{n-1}\right) \leq \beta^2\|\nabla_w P(w_{n-1})\|^2 + \sigma^2 && (24.46d)
\end{aligned}$$

24.3.3 Second-Order Convergence

Now that we are assured of convergence toward the set $\mathcal{S}$ of first-order stationary points (i.e., small gradient norms), let us examine more closely how the algorithm evolves within this set and, in particular, in the vicinity of strict saddle points in the set $\mathcal{U}$. As the analysis will reveal, the two sets $\mathcal{S}^C$ and $\mathcal{U}$ are distinguished by the fact that for points in $\mathcal{S}^{\mathbb{C}}$, the gradient norm $\|\nabla_w P(w)\|^2$ is large enough for a single (stochastic) gradient step to be sufficient to guarantee descent in expectation (as was shown by Theorem 24.1), while points in $\mathcal{U}$ (i.e., strict saddle points) have small gradient norms and a single gradient step will not be sufficient anymore to guarantee descent. The arguments in the appendices show that it is the cumulative effect of the gradient noise perturbations that ends up pushing the iterates from $\mathcal{U}$ toward $\mathcal{D}$.

Theorem 24.2. (Descent away from strict saddle points) *Starting from any strict saddle point $\boldsymbol{w}_{n-1} \in \mathcal{U}$ and iterating for $n^s = O(1/\mu\tau)$ additional iterations after $n-1$ with*

$$n^s = \frac{\log\left(2M\dfrac{\sigma^2}{\sigma_\ell^2} + 1 + O(\mu)\right)}{\log(1+2\mu\tau)} = O\left(\frac{1}{\mu\tau}\right) \tag{24.47}$$

guarantees

$$\mathbb{E}\Big(P(\boldsymbol{w}_{n+n^s}) \,|\, \boldsymbol{w}_{n-1} \in \mathcal{U}\Big) \leq \mathbb{E}\Big(P(\boldsymbol{w}_{n-1}) \,|\, \boldsymbol{w}_{n-1} \in \mathcal{U}\Big) - \frac{\mu}{2}M\sigma^2 + o(\mu) \tag{24.48}$$

This conclusion holds under conditions ($\mathbb{A}1$–$\mathbb{A}7$) listed under the modeling Assumption 24.1.

Proof: The argument is demanding and involves several steps. It appears in Appendix 24.C.

■

Theorem 24.2 ensures that, even when the norm of the gradient at $\boldsymbol{w}_{n-1} \in \mathcal{U}$ is too small to carry sufficient information about descent directions, the persistent gradient noise at $\boldsymbol{w}_{n-1}$ guaranteed by condition (24.31) and the negative local curvature of the risk surface guaranteed by (24.12) are both sufficient to ensure descent in n^s iterations, where the escape-time n^s scales favorably with problem parameters. For example, the escape-time scales logarithmically with the dimension M, implying that we can expect fast evasion of saddle points even in high dimensions. Moreover, the escape-time scales inversely with $1/\tau$, meaning that the larger the value of τ is, the faster the escape. The proof of Theorem 24.2 in the appendix takes advantage of all conditions ($\mathbb{A}1$–$\mathbb{A}7$) listed under the modeling Assumption 24.1.

Having established descent from both the large-gradient regime ($\boldsymbol{w}_{n-1} \in \mathcal{S}^C$) and strict saddle point regime ($\boldsymbol{w}_{n-1} \in \mathcal{U}$), we can combine the results to conclude eventual second-order stationarity. Theorem 24.3 states that, starting from an initial condition w_{-1} for the stochastic gradient algorithm, it will reach the desirable set $\mathcal{D}$ in at most $O(1/\mu^3\tau)$ iterations.

THEOREM 24.3. (Second-order convergence guarantee) *Assume the risk function $P(w)$ is bounded from below, say, $P(w) \geq P^o$ for all w. Let $\epsilon \in (0,1)$ denote a small positive number selected by the designer. Then, there exist sufficiently small step sizes μ such that, starting from an arbitrary initial condition w_{-1} and with high probability $1-2\epsilon$, the iterate $\boldsymbol{w}_{n^o}$ will belong to the desirable set $\mathcal{D}$, i.e., $\|\nabla_w P(\boldsymbol{w}_{n^o})\|^2 \leq O(\mu)$ and $\lambda_{\min}\left(\nabla_w^2 P(\boldsymbol{w}_{n^o})\right) > -\tau$ in at most n^o iterations, where*

$$n^o \leq \frac{(P(w_{-1}) - P^o)}{\mu^2 c_2} n^s = O(1/\mu^3\tau) \tag{24.49}$$

This conclusion holds under conditions ($\mathbb{A}1$–$\mathbb{A}7$).

Proof: See Appendix 24.D.

■

The statement of the theorem guarantees convergence to a point $\boldsymbol{w}_{n^o} \in \mathcal{D}$ within a bounded number of iterations n^o.

Example 24.2 (Training a classifier with a sigmoidal output) We illustrate the operation of the stochastic gradient recursion (24.18) by applying it to the structure shown in Fig. 24.5. The input vector $h \in \mathbb{R}^M$ is multiplied by a weighting matrix W to generate $y = Wh \in \mathbb{R}^J$. The inner product of y and w generates the scalar z, which is fed into a sigmoidal function to produce the output signal $\widehat{\gamma} = f(z)$. The input–output mapping shown in the figure can be described by the relations:

$$\begin{cases} y = Wh \\ z = w^\mathsf{T} y = w^\mathsf{T} Wh \\ \widehat{\gamma} = f(z) \end{cases} \tag{24.50}$$

where $f(z)$ is the sigmoid activation function:

$$\widehat{\gamma} = \frac{1}{1+e^{-z}} \tag{24.51}$$

In order to learn the weight parameters $\{w, W\}$, we will employ the following cross-entropy empirical risk:

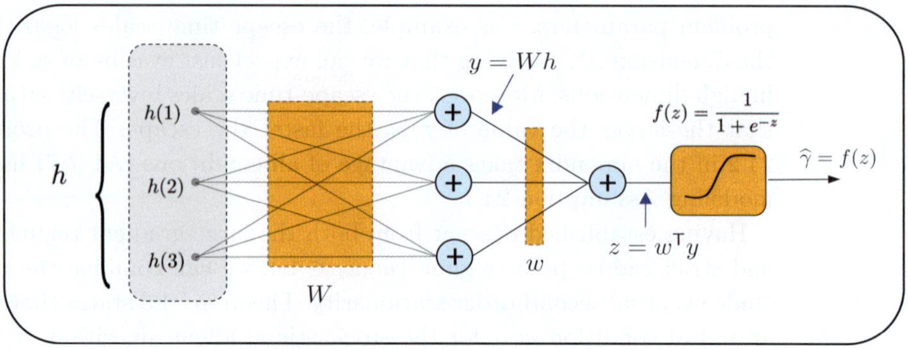

Figure 24.5 A classifier structure with a sigmoidal output node.

$$\mathcal{P}(w,W) \;\stackrel{\Delta}{=}\; \rho\|w\|_2^2 + \rho\|W_1\|_{\rm F}^2 - \frac{1}{N}\sum_{m=0}^{N}\Big\{\gamma(m)\ln(\widehat{\gamma}(m)) + (1-\gamma(m))\ln(1-\widehat{\gamma}(m))\Big\} \tag{24.52}$$

where we assume a collection of training points $\{\gamma(m), h_m\}$ with $h_m \in \mathbb{R}^M$ representing feature vectors and $\gamma(m) \in \{0,1\}$ representing the corresponding binary labels. We rewrite the rightmost term in the risk function in an equivalent form using the sigmoidal function as follows. We first note that

$$\gamma\ln(\widehat{\gamma}) + (1-\gamma)\ln(1-\widehat{\gamma}) = \begin{cases} -\ln(1+e^{-w^{\mathsf T}Wh}), & \gamma = 1 \\ -\ln(1+e^{w^{\mathsf T}Wh}), & \gamma = 0 \end{cases} \tag{24.53}$$

so that we can rewrite the empirical risk (24.52) in the alternative form:

$$\boxed{\mathcal{P}(w,W) = \rho\|w\|_2^2 + \rho\|W_1\|_{\rm F}^2 + \frac{1}{N}\sum_{m=0}^{N}\ln\Big(1+e^{-(2\gamma(m)-1)w^{\mathsf T}Wh_m}\Big)} \tag{24.54}$$

From relations (24.50) we see that the mapping from h to z is linear since $z = a^{\mathsf T}h$, for some vector a. However, the vector a is being represented in factored form, as the product $a = w^{\mathsf T}W$. This representation causes the empirical function $P(w,W)$ to become nonconvex over the parameters $\{w,W\}$ since they appear multiplied together.

We evaluate the gradients of $\mathcal{P}(w,W)$ with respect to w and W. For the matrix W, we recall that the entries in the notation $\partial\mathcal{P}(w,W)/\partial W$ consist of the partial derivatives of $\mathcal{P}(\cdot)$ relative to the individual entries of W. In this way, we have

$$\begin{bmatrix} \partial\mathcal{P}(w,W)/\partial W \\ \partial\mathcal{P}(w,W)/\partial w \end{bmatrix} = \begin{bmatrix} 2\rho W - \dfrac{1}{N}\displaystyle\sum_{m=0}^{N-1}\frac{(2\gamma(m)-1)wh_m^{\mathsf T}}{1+e^{(2\gamma(m)-1)w^{\mathsf T}Wh_m}} \\ 2\rho w^{\mathsf T} - \dfrac{1}{N}\displaystyle\sum_{m=0}^{N-1}\frac{(2\gamma(m)-1)(Wh_m)^{\mathsf T}}{1+e^{(2\gamma(m)-1)w^{\mathsf T}Wh_m}} \end{bmatrix} \tag{24.55}$$

These expressions provide the *true* gradients relative to the weight parameters. We can employ stochastic approximations for both gradients and update the parameter estimates by using:

$$
\begin{aligned}
&\textbf{select a pair } (\gamma(n), \boldsymbol{h}_n) \textbf{ at random}\\
&\boldsymbol{W}_n = (1-2\mu\rho)\boldsymbol{W}_{n-1} + \mu\frac{(2\gamma(n)-1)\boldsymbol{w}_{n-1}\boldsymbol{h}_n^\mathsf{T}}{1+e^{(2\gamma(n)-1)\boldsymbol{w}_{n-1}^\mathsf{T}\boldsymbol{W}_{n-1}\boldsymbol{h}_n}} \quad (24.56a)\\
&\boldsymbol{w}_n = (1-2\mu\rho)\boldsymbol{w}_{n-1} + \mu\frac{(2\gamma(n)-1)\boldsymbol{W}_{n-1}\boldsymbol{h}_n}{1+e^{(2\gamma(n)-1)\boldsymbol{w}_{n-1}^\mathsf{T}\boldsymbol{W}_{n-1}\boldsymbol{h}_n}} \quad (24.56b)
\end{aligned}
$$

Next, the Hessian matrix of $\mathcal{P}(w, W)$ at the origin ($w = 0, W = 0$) is denoted by $\mathcal{H}$ and is given by

$$
\mathcal{H} = \left[\begin{array}{c|c} 2\rho I_M & -\dfrac{1}{2N}\displaystyle\sum_{m=0}^{N-1}(2\gamma(m)-1)h_m \\ \hline -\dfrac{1}{2N}\displaystyle\sum_{m=0}^{N-1}(2\gamma(m)-1)h_m^\mathsf{T} & 2\rho \end{array}\right] \tag{24.57}
$$

The $(1, 1)$ block is positive-definite and equal to $2\rho I_M$. The Schur complement relative to this block is given by

$$
\Delta = 2\rho - \frac{1}{2\rho}\left(\frac{1}{4N^2}\sum_{m'=0}^{N-1}\sum_{m=0}^{N-1}(2\gamma(m')-1)(2\gamma(m)-1)h_{m'}^\mathsf{T}h_m\right) \tag{24.58}
$$

This Schur complement will be negative-definite for small enough regularization parameter ρ chosen to satisfy

$$
4\rho^2 \leq \frac{1}{4N^2}\sum_{m'=0}^{N-1}\sum_{m=0}^{N-1}(2\gamma(m')-1)(2\gamma(m)-1)h_{m'}^\mathsf{T}h_m \tag{24.59}
$$

Assume that ρ satisfies this condition. Then, the Hessian matrix of $\mathcal{P}(w, W)$ will be indefinite at the location ($w = 0, W = 0$), having both negative and positive eigenvalues. It follows that this location will be a strict saddle point with some descent directions associated with it. However, we observe from (24.55) and (24.56a)–(24.56b) that both the true and approximate gradients are zero at ($w = 0, W = 0$). This implies that the gradient noise is also zero at this location, which means that the gradient noise induced by the stochastic gradient construction will not have a component in any of the descent directions at the origin. Hence, initializing the algorithm at ($w = 0, W = 0$) would cause it to remain there with probability 1. This suggests that condition (24.31) is not merely a technical requirement but indeed necessary. To satisfy the assumption, we will instead run the stochastic gradient recursions (24.56a)–(24.56b) by adding a small noise perturbation to each gradient approximation.

We illustrate these conclusions by running a contrived example in order to enable visualization of the trajectories. We let $M = 1$ and $J = 1$ so that the input feature h_n is a scalar and both w and W are also scalars. The stochastic gradient recursions (24.56a)–(24.56b) can be written as:

$$
\begin{bmatrix} \boldsymbol{W}_n \\ \boldsymbol{w}_n \end{bmatrix} = (1-2\mu\rho)\begin{bmatrix} \boldsymbol{W}_{n-1} \\ \boldsymbol{w}_{n-1} \end{bmatrix} + \frac{\mu(2\gamma(n)-1)\boldsymbol{h}_n}{1+e^{(2\gamma(n)-1)\boldsymbol{w}_{n-1}\boldsymbol{W}_{n-1}\boldsymbol{h}_n}}\begin{bmatrix} \boldsymbol{w}_{n-1} \\ \boldsymbol{W}_{n-1} \end{bmatrix} \tag{24.60}
$$

We set $\rho = 0.05$ and generate random labels $\gamma(n) \in \{0, 1\}$ with equal probability. We also generate the corresponding feature vectors from a Gaussian distribution as follows:

$$
\begin{cases} \boldsymbol{h}_n \sim \mathcal{N}_{\boldsymbol{h}_n}(+1, 1), & \text{if } \gamma(n) = 1 \\ \boldsymbol{h}_n \sim \mathcal{N}_{\boldsymbol{h}_n}(-1, 1), & \text{if } \gamma(n) = 0 \end{cases} \tag{24.61}
$$

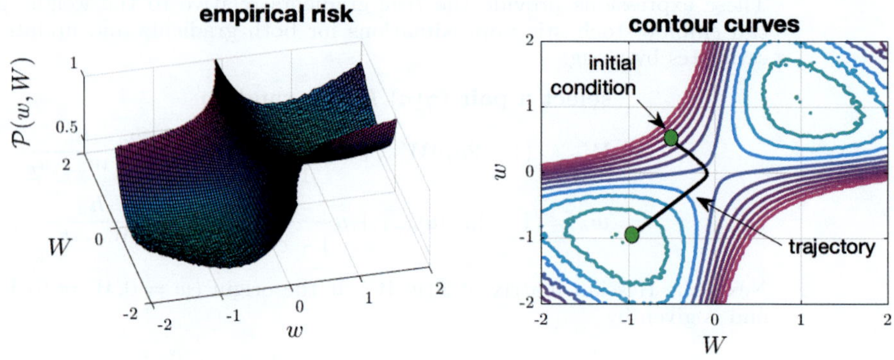

Figure 24.6 The risk surface on the left has one saddle point at the origin and two local minima in the positive and negative quadrants. The plot on the right shows the contour curves and a trajectory converging toward a local minimum.

Under these conditions, it is clear that $\mathbb{E}\,(2\boldsymbol{\gamma}(n)-1)\boldsymbol{h}_n = 1$ so that by ergodicity and for large sample size N:

$$\frac{1}{2N}\sum_{m=0}^{N-1}(2\gamma(m)-1)h_m \rightarrow \frac{1}{2}, \quad N\rightarrow\infty \tag{24.62}$$

and, hence, the Hessian matrix at the origin becomes

$$\mathcal{H} \approx \begin{bmatrix} 0.1 & -0.5 \\ -0.5 & 0.1 \end{bmatrix} \tag{24.63}$$

which has an eigenvalue at -0.4 with corresponding eigenvector col$\{1,1\}$. This implies that $w = W = 0$ is a strict saddle point with local descent direction col$\{1,1\}$. To satisfy condition (24.31), we perturb the update direction in (24.60) by adding $\mu\boldsymbol{v}\,\text{col}\{1,1\}$ where $\boldsymbol{v}\sim\mathbb{N}_{\boldsymbol{v}}(0,1)$. This step ensures that gradient noise is present in the descent direction around the strict saddle point at $w = W = 0$.

The risk surface is depicted on the left in Fig. 24.6. It can be observed from the figure, and also analytically verified, that $\mathcal{P}(w,W)$ has two local minima in the positive and negative quadrants, respectively, and a single saddle point at $w = W = 0$. One realization of the trajectory (w_n, W_n) is shown in the same figure on the right converging toward a local minimum. The plots are generated using $\mu = 0.0001$ and $N = 10{,}000$ data points. The stochastic gradient recursions are run for 100000 iterations.

24.4 COMMENTARIES AND DISCUSSION

Nonconvex optimization. The presentation in the body of the chapter follows closely the overview by Vlaski and Sayed (2020), while the derivations and proofs in the appendices follow Vlaski and Sayed (2019, 2020, 2021a, b). There are many important scenarios in learning and inference that give rise to nonconvex optimization problems, as happens for example in the training of neural network architectures. These problems can be arbitrarily difficult to solve in general in the sense that even verifying that a given point in space is a local minimum can be NP-hard according to Murty and Kabadi (1987). Still, as the analysis in the chapter has shown, simple algorithms such as stochastic gradient learning can perform well in converging toward local minima and avoiding saddle points. One important insight from the analysis is that *gradient perturbations* play a critical role in allowing local descent algorithms to efficiently distinguish desirable from undesirable stationary points and escape from the latter.

First-order stationarity. Many works in the literature have been successful in establishing performance guarantees for stochastic gradient algorithms and showing that their limiting points are approximately first-order stationary using variations of definition (24.5), such as the works by Bertsekas and Tsitsiklis (2000), Nesterov (2004), and Reddi *et al.* (2016a, b). These results are reassuring since first-order stationarity is a necessary condition for local optimality, and hence any algorithm that does not produce a first-order stationary point will necessarily not produce a point with small excess risk, or small distance to the minimizer. However, these results cannot ensure that the limiting first-order stationary point will not correspond to a saddle point, which has been identified by Choromanska *et al.* (2015) as a bottleneck in many nonconvex problems of interest. This observation, following the works by Nesterov and Polyak (2006), Ge *et al.* (2015), and Lee *et al.* (2016), has motivated us to consider the stronger notion of second-order optimality in the body of the chapter.

Second-order stationarity. For many risk functions commonly found in inference and learning problems, such as tensor decomposition in Ge *et al.* (2015), matrix completion in Ge, Lee, and Ma (2016), low-rank recovery in Ge, Jin, and Zheng (2017), and some deep learning formulations in Kawaguchi (2016), *all* saddle points and local maxima have been shown to have a significant negative eigenvalue in the Hessian, and can hence be excluded from the set of second-order stationary points (24.10) for sufficiently small but finite τ. For such risk functions, *all* τ-second-order stationary points for some small but finite τ correspond to local, or even global, minima. This observation has motivated a number of works to pursue higher-order stationarity guarantees of local descent algorithms by means of second-order information in Nesterov and Polyak (2006), Curtis, Robinson, and Samadi (2017), and Tripuraneni *et al.* (2018), intermediate searches for the negative curvature direction in Fang *et al.* (2018), Allen-Zhu (2018), and Allen-Zhu and Li (2018), perturbations in the initialization in Lee *et al.* (2016) and Du *et al.* (2017), or perturbations to the update direction in Gelfand and Mitter (1991), Ge *et al.* (2015), Jin *et al.* (2017), Daneshmand *et al.* (2018), Fang, Lin, and Zhang (2019), and Vlaski and Sayed (2019, 2020, 2021a, b).

Modeling conditions. We explained in Section 18.2 that the Lipschitz condition (24.20a) on the gradient of the risk function appears frequently in the study of first-order optimality guarantees for (stochastic) gradient algorithms in the context of convex optimization problems. For these problems, the Lipschitz condition (24.20b) on the Hessian matrix is not necessary to establish convergence, but it can be used to more accurately quantify the size of the deviation from the minimizer in the steady state, as was done in Sayed (2014a), or to establish the escape from saddle points, as was done in the body of the chapter and in Ge *et al.* (2015), Ge, Jin, and Zheng (2017), Daneshmand *et al.* (2018), and Vlaski and Sayed (2019, 2020, 2021a, b). The second-order moment condition (24.22) is sufficient to establish limiting first-order stationarity, as was shown in Bertsekas and Tsitsiklis (2000), while the fourth-order moment condition (24.21b) allows us to more carefully analyze the dynamics of (24.18) around first-order stationary points and establish escape from saddle points, resulting in second-order guarantees, as is done in the body of the chapter and in Vlaski and Sayed (2019, 2020, 2021a, b). Note from the definition of the gradient noise covariance (24.24a) that the distribution of the gradient noise process is a function of the iterate $\boldsymbol{w}_{n-1}$. The fact that the perturbations introduced into the stochastic recursion (24.18) are not necessarily identically distributed over time introduces challenges into the study of their cumulative effect on the performance of the algorithm. Condition (24.24b) ensures that the covariance $R_g(\boldsymbol{w}_{n-1})$ is sufficiently smooth in localized regions in space around strict saddle points, thus resulting in essentially identically distributed gradient noise perturbations and leading to a more tractable analysis. The condition has been exploited to derive accurate steady-state performance expressions in the strongly convex setting in Sayed (2014a).

Persistent gradient noise components. Condition (24.31) is a notable deviation from the assumptions typically imposed in the *convex* setting. The condition (24.31) is relevant to the study of stochastic gradient algorithms in the vicinity of saddle points in nonconvex optimization, as shown in Daneshmand *et al.* (2018) and Vlaski and Sayed (2019, 2021a, b). Condition (24.31) indicates that the presence of gradient noise, especially along the directions of local descent in the vicinity of saddle points, is beneficial for the ultimate escape from saddle locations. This conclusion may appear puzzling at first sight since the presence of gradient perturbations is generally perceived as detrimental to the performance of stochastic gradient algorithms in convex settings. Nevertheless, when generalizing to nonconvex risks, as the analyses in Ge *et al.* (2015), Jin *et al.* (2017), Daneshmand *et al.* (2018), and Vlaski and Sayed (2019, 2021a, b) have shown, the persistent presence of gradient perturbations allows the algorithm to efficiently escape from saddle points and arrive at local minima. In this sense, condition (24.31) allows the algorithm to distinguish stable local minima from unstable saddle points, both of which are first-order stationary points. In the absence of prior knowledge that there are gradient noise components in the descent direction for every strict saddle point, as required by condition (24.31), one can guarantee (24.31) by adding a small perturbation term $\boldsymbol{v}_n$ with positive-definite covariance matrix as explained in Example 24.1 – see, e.g., Ge *et al.* (2015) and Jin *et al.* (2019).

PROBLEMS

24.1 Refer to the definition of the gradient noise covariance matrix in (24.24a). Show that $\|R_g(\boldsymbol{w}_{n-1})\| \leq \beta\|\nabla_w P(\boldsymbol{w}_{n-1}\|^2 + \sigma^2$ in terms of the same (β^2, σ^2) parameters appearing in (24.22).

24.2 For positive T, μ, δ satisfying $\mu < 1/\delta$ and integers $k \in \mathbb{Z}_+$, show that

$$\lim_{\mu\to 0}\left(\frac{(1+\mu\delta)^k + O(\mu^2)}{(1-\mu\delta)^{k-1}}\right)^{\frac{T}{\mu}} = e^{T\delta(2k-1)} = O(1)$$

Remark. The reader may refer to Vlaski and Sayed (2019, 2021a).

24.3 Assume $\mu \leq 2/(\delta\left(1+\beta^2\right))$. Show that the following result holds, which bounds the negative effect of the gradient noise in the vicinity of local minima in $\mathcal{D}$:

$$\mathbb{E}\left(P(\boldsymbol{w}_n) \,|\, \boldsymbol{w}_{n-1} \in \mathcal{D}\right) \leq \mathbb{E}\left(P(\boldsymbol{w}_{n-1}) \,|\, \boldsymbol{w}_{n-1} \in \mathcal{D}\right) + \mu^2 c_2$$

24.4 Establish bound (24.81) on the second-order moment of the state variable for the short-term model.

24.5 Establish bound (24.82) on the third-order moment of the state variable for the short-term model, i.e., $\mathbb{E}\left(\|\boldsymbol{z}_m^{n-1}\|^3 \,|\, \boldsymbol{w}_{n-1} \in \mathcal{U}\right) \leq O(\mu^{3/2})$.

24.6 Iterate recursion (24.108) and establish bound (24.109) for the difference between the true iterate and the one approximated by the short-term model.

24.7 Consider the argument leading to (24.158). Verify that

$$\lambda_{\max}\left\{\left(\boldsymbol{V}_{n-1}^{>0}\right)^{\mathsf{T}} R_g\left(\boldsymbol{w}_{n-1}\right) \boldsymbol{V}_{n-1}^{>0}\right\} \leq \beta^2\|\nabla_{w^{\mathsf{T}}} P(\boldsymbol{w}_{n-1})\|^2 + \sigma^2$$

24.8 We consider a simplified version of the federated learning algorithm from Section 17.6 used to minimize the aggregate risk function (17.81), namely, $P(w) = \sum_{\ell=1}^{L} p_\ell P_\ell(w)$, where the $\{p_\ell\}$ are positive convex weights that add up to 1 and satisfy $p_\ell = N_\ell/N$. Moreover, the individual risks are given by $P_\ell(w) = \frac{1}{N_\ell}\sum_{m=0}^{N_\ell-1} Q(w;\gamma(m),h_m)$, where the $\{\gamma(m),h_m\}$ refer to data points at location ℓ. At every cycle n we select a subset $\bar{\mathcal{L}}$ from the set of agents $\mathcal{L}=\{1,2,\dots,L\}$. Agent $\ell\in\bar{\mathcal{L}}$ then performs a single mini-batch update as follows (we are setting $R_\ell = 1$ in (17.86)):

$$\boldsymbol{w}_{\ell,n} = \boldsymbol{w}_{n-1} - \mu\left\{\frac{1}{B}\sum_{b=0}^{B-1}\nabla_{w^\mathsf{T}} Q\Big(\boldsymbol{w}_{n-1};\boldsymbol{\gamma}^{(\ell)}(b),\boldsymbol{h}_b^{(\ell)}\Big)\right\}$$

where the $\{\boldsymbol{\gamma}^{(\ell)}(b),\boldsymbol{h}_b^{(\ell)}\}$, with superscript (ℓ), refer to the mini-batch data within agent ℓ. The server aggregates the iterates $\{\boldsymbol{w}_{\ell,n}\}$ by using $\boldsymbol{w}_n = \frac{1}{|\bar{\mathcal{L}}|}\sum_{\ell\in\bar{\mathcal{L}}}\boldsymbol{w}_{\ell,n}$.

(a) Introduce the indicator function $\mathbb{I}_{\ell,n} = 1$ if agent ℓ is selected at cycle n and zero otherwise. Verify that the update from $\boldsymbol{w}_{n-1}$ to $\boldsymbol{w}_n$ at the server can be written as

$$\boldsymbol{w}_n = \boldsymbol{w}_{n-1} - \mu\left\{\frac{1}{|\bar{\mathcal{L}}|}\sum_{\ell=1}^{L}\mathbb{I}_{\ell,n}\left(\frac{1}{B}\sum_{b=0}^{B-1}\nabla_{w^\mathsf{T}} Q\Big(\boldsymbol{w}_{n-1};\boldsymbol{\gamma}^{(\ell)}(b),\boldsymbol{h}_b^{(\ell)}\Big)\right)\right\}$$

(b) Assume the gradients of the local risks satisfy the bounded disagreement condition $\|\nabla_w P_\ell(w) - \nabla_w P_k(w)\| \le G$, for some constant G and any (ℓ,k). Conclude that the deviation from the aggregate gradient of the risk is also bounded, namely, it holds that $\|\nabla_w P_\ell(w) - \nabla_w P(w)\| \le G$ for all ℓ.

24.9 Continuing with Prob. 24.8, introduce the approximate gradient vector used in part (a) of that problem:

$$\widehat{\nabla_{w^\mathsf{T}} P}(\boldsymbol{w}_{n-1}) \triangleq \frac{1}{|\bar{\mathcal{L}}|}\sum_{\ell=1}^{L}\mathbb{I}_{\ell,n}\left(\frac{1}{B}\sum_{b=1}^{B}\nabla_{w^\mathsf{T}} Q\Big(\boldsymbol{w}_{n-1};\boldsymbol{\gamma}^{(\ell)}(b),\boldsymbol{h}_b^{(\ell)}\Big)\right)$$

as well as the gradient noise process $\boldsymbol{g}_n(\boldsymbol{w}_{n-1}) = \widehat{\nabla_{w^\mathsf{T}} P}(\boldsymbol{w}_{n-1}) - \nabla_{w^\mathsf{T}} P(\boldsymbol{w}_{n-1})$.

(a) Verify that $\mathbb{E}\left(\boldsymbol{g}_n(\boldsymbol{w}_{n-1})\,|\,\boldsymbol{w}_{n-1}\right) = 0$.

(b) Verify that $\mathbb{E}\left(\|\boldsymbol{g}_n(\boldsymbol{w}_{n-1})\|^4\,|\,\boldsymbol{w}_{n-1}\right) \le \beta_{\text{Fed}}^4\,\|\nabla_{w^\mathsf{T}} P(w_{n-1})\|^4 + \sigma_{\text{Fed}}^4$, for some constants that depend on the participation rate $|\bar{\mathcal{L}}|/L$ and the mini-batch size B.

(c) Conclude that, over nonconvex environments and for sufficiently small step sizes, the federated learning algorithm reaches a point w^a within the desirable set $\mathcal{D}$ of local minima satisfying $\|\nabla_w P(w^a)\| \le O(\mu)$ and $\lambda_{\min}(\nabla_w^2 P(w^x)) > -\tau$ in at most $O(1/\mu^3\tau)$ iterations.

Remark. The reader may refer to Vlaski and Sayed (2020) and Vlaski, Rizk, and Sayed (2020).

24.10 Refer to the stochastic gradient recursion (24.16) with an iteration-dependent step-size sequence $\mu(n)$ that satisfies:

$$\sum_{n=0}^{\infty}\mu(n) = \infty, \qquad \sum_{n=0}^{\infty}\mu^2(n) < \infty$$

Assume the risk function $P(w)$ is bounded from below, say, $P(w) \ge P^o$, and has δ-Lipschitz gradients as in (24.20a). Assume further the stochastic gradient approximation satisfies the following three conditions in terms of the squared Euclidean norm:

$$\text{(i) } \Big(\nabla_{w^\mathsf{T}} P(w_{n-1})\Big)^\mathsf{T}\,\mathbb{E}\,\widehat{\nabla_{w^\mathsf{T}} P}(w_{n-1}) \ge a\,\|\nabla_{w^\mathsf{T}} P(w_{n-1})\|^2$$

$$\text{(ii) } \|\mathbb{E}\,\widehat{\nabla_{w^\mathsf{T}} P}(w_{n-1})\| \le b\,\|\nabla_{w^\mathsf{T}} P(w_{n-1})\|$$

$$\text{(iii) } \text{var}\Big(\widehat{\nabla_{w^\mathsf{T}} P}(w_{n-1})\Big) \le \alpha + \beta\|\nabla_{w^\mathsf{T}} P(w_{n-1})\|^2$$

for some constants $b \geq a > 0$ and $\alpha, \beta \geq 0$ and where, by definition,

$$\text{var}\left(\widehat{\nabla_{w^\mathsf{T}} P}(w_{n-1})\right) \triangleq \mathbb{E}\,\|\widehat{\nabla_{w^\mathsf{T}} P}(w_{n-1})\|^2 - \|\mathbb{E}\,\widehat{\nabla_{w^\mathsf{T}} P}(w_{n-1})\|^2$$

In the above relations, the expectation operator $\mathbb{E}$ is over the statistical distribution of the data $\{\gamma, \boldsymbol{h}\}$ conditioned on the past iterate $\boldsymbol{w}_{n-1}$. Let $S_N = \sum_{n=0}^{N-1} \mu(n)$ and $\beta_1 = \beta + b^2$. Repeat arguments similar to Probs. 19.18–19.20 to show that for $\mu(n) \leq a/\delta\beta_1$:

$$\lim_{N\to\infty} \mathbb{E}\left\{\sum_{n=0}^{N-1} \mu(n)\|\nabla_{w^\mathsf{T}} P(\boldsymbol{w}_n)\|^2\right\} < \infty$$

$$\lim_{N\to\infty} \mathbb{E}\left\{\frac{1}{S_N}\sum_{n=0}^{N-1} \mu(n)\|\nabla_{w^\mathsf{T}} P(\boldsymbol{w}_n)\|^2\right\} = 0$$

Conclude that

$$\liminf_{n\to\infty} \mathbb{E}\,\|\nabla_{w^\mathsf{T}} P(\boldsymbol{w}_n)\|^2 = 0$$

Remark. See Bottou, Curtis, and Nocedal (2018) for a related discussion.

24.11 Consider a nonconvex but δ-smooth risk function $P(w)$ (i.e., its gradient is δ-Lipschitz). Apply the gradient-descent recursion for N iterations:

$$w_n = w_{n-1} - \mu \nabla_{w^\mathsf{T}} P(w_{n-1}), \quad n \geq 0$$

Show that there must exist at least one iterate w_{n^o} with small gradient satisfying $\|\nabla_w P(w_{n^o})\|^2 \leq O(1/N)$, for some $n^o \leq N$.

24.A DESCENT IN THE LARGE GRADIENT REGIME

The derivation in these appendices follow closely the arguments from Vlaski and Sayed (2019, 2020, 2021a, b). In this first appendix we establish Theorem 24.1, which ensures convergence toward first-order stationary locations.

To begin with, since the risk function $P(w)$ has δ-Lipschitz gradients, we know from property (10.13) that

$$P(\boldsymbol{w}_n) \leq P(\boldsymbol{w}_{n-1}) + (\nabla_{w^\mathsf{T}} P(\boldsymbol{w}_{n-1}))^\mathsf{T}(\boldsymbol{w}_n - \boldsymbol{w}_{n-1}) + \frac{\delta}{2}\|\boldsymbol{w}_n - \boldsymbol{w}_{n-1}\|^2 \tag{24.64}$$

Using the stochastic gradient recursion (24.16) and the definition of the gradient noise from (24.19) we have:

$$\begin{aligned} P(\boldsymbol{w}_n) &\leq P(\boldsymbol{w}_{n-1}) + (\nabla_{w^\mathsf{T}} P(\boldsymbol{w}_{n-1}))^\mathsf{T}\left(-\mu\widehat{\nabla_{w^\mathsf{T}} P}(\boldsymbol{w}_{n-1})\right) + \frac{\delta}{2}\left\|-\mu\widehat{\nabla_{w^\mathsf{T}} P}(\boldsymbol{w}_{n-1})\right\|^2 \\ &\leq P(\boldsymbol{w}_{n-1}) - \mu(\nabla_{w^\mathsf{T}} P(\boldsymbol{w}_{n-1}))^\mathsf{T}\nabla_{w^\mathsf{T}} P(\boldsymbol{w}_{n-1}) - \mu(\nabla_{w^\mathsf{T}} P(\boldsymbol{w}_{n-1}))^\mathsf{T}\boldsymbol{g}_n(\boldsymbol{w}_{n-1}) \\ &\quad + \mu^2\frac{\delta}{2}\|\nabla_{w^\mathsf{T}} P(\boldsymbol{w}_{n-1}) + \boldsymbol{g}_n(\boldsymbol{w}_{n-1})\|^2 \end{aligned} \tag{24.65}$$

Conditioning on $\boldsymbol{w}_{n-1}$, taking expectations, and using the conditions on the mean and second-order moment of the gradient noise process from (24.46c)–(24.46d) gives:

$$
\begin{aligned}
&\mathbb{E}\Big(P(\boldsymbol{w}_n)|\boldsymbol{w}_{n-1}\Big)\\
&\leq P(\boldsymbol{w}_{n-1}) - \mu\left(1-\frac{\mu\delta}{2}\right)\|\nabla_w P(\boldsymbol{w}_{n-1})\|^2 + \mu^2\frac{\delta}{2}\mathbb{E}\left(\|\boldsymbol{g}_n(\boldsymbol{w}_{n-1})\|^2|\boldsymbol{w}_{n-1}\right)\\
&\leq P(\boldsymbol{w}_{n-1}) - \mu\left(1-\frac{\mu\delta}{2}(1+\beta^2)\right)\|\nabla_w P(\boldsymbol{w}_{n-1})\|^2 + \mu^2\frac{\delta}{2}\sigma^2\\
&\stackrel{(a)}{=} P(\boldsymbol{w}_{n-1}) - \mu c_1\|\nabla_w P(\boldsymbol{w}_{n-1})\|^2 + \mu^2 c_2
\end{aligned}
\tag{24.66}
$$

where step (a) follows from (24.38)–(24.39). Taking expectations when $\boldsymbol{w}_{n-1} \in \mathcal{S}^C$, and using the result of the earlier Prob. 3.26 for conditional expectations, we find:

$$
\begin{aligned}
&\mathbb{E}\Big(P(\boldsymbol{w}_n)\,|\,\boldsymbol{w}_{n-1}\in\mathcal{S}^C\Big)\\
&\leq \mathbb{E}\Big(P(\boldsymbol{w}_{n-1})\,|\,\boldsymbol{w}_{n-1}\in\mathcal{S}^C\Big) - \mu c_1\mathbb{E}\Big(\|\nabla_w P(\boldsymbol{w}_{n-1})\|^2\,|\,\boldsymbol{w}_{n-1}\in\mathcal{S}^C\Big) + \mu^2 c_2\\
&\leq \mathbb{E}\Big(P(\boldsymbol{w}_{n-1})\,|\,\boldsymbol{w}_{n-1}\in\mathcal{S}^C\Big) - \mu c_1\cdot\mu\frac{c_2}{c_1}\left(1+\frac{1}{\epsilon}\right) + \mu^2 c_2\\
&= \mathbb{E}\Big(P(\boldsymbol{w}_{n-1})\,|\,\boldsymbol{w}_{n-1}\in\mathcal{S}^C\Big) - \mu^2\frac{c_2}{\epsilon}
\end{aligned}
\tag{24.67}
$$

24.B INTRODUCING A SHORT-TERM MODEL

This appendix introduces a short-term approximation model and establishes two preliminary results on the accuracy of the model. These results are then used in Appendix 24.C to establish Theorem 24.2.

Motivating the short-term model

Starting from some strict saddle point $\boldsymbol{w}_{n-1} \in \mathcal{U}$ and writing the stochastic gradient recursion (24.16) for step $n+m$ where $m \geq 0$:

$$
\boldsymbol{w}_{n+m} = \boldsymbol{w}_{n+m-1} - \mu\nabla_{w^\mathsf{T}}P(\boldsymbol{w}_{n+m-1}) - \mu\boldsymbol{g}_{n+m}(\boldsymbol{w}_{n+m-1}) \tag{24.68}
$$

Subtracting this relation from $\boldsymbol{w}_{n-1}$, we find:

$$
\boldsymbol{w}_{n-1} - \boldsymbol{w}_{n+m} = \boldsymbol{w}_{n-1} - \boldsymbol{w}_{n+m-1} + \mu\nabla_{w^\mathsf{T}}P(\boldsymbol{w}_{n+m-1}) + \mu\boldsymbol{g}_{n+m}(\boldsymbol{w}_{n+m-1}) \tag{24.69}
$$

We will study the evolution of the deviation $\boldsymbol{w}_{n-1} - \boldsymbol{w}_{n+m}$ over several iterations m. For brevity, we introduce the deviation vector starting from $\boldsymbol{w}_{n-1}$:

$$
\boxed{\boldsymbol{d}_m^{n-1} \triangleq \boldsymbol{w}_{n-1} - \boldsymbol{w}_{n+m}} \quad \text{(deviation vector after } m \text{ steps)} \tag{24.70}
$$

so that (24.69) becomes:

$$
\boldsymbol{d}_m^{n-1} = \boldsymbol{d}_{m-1}^{n-1} + \mu\nabla_{w^\mathsf{T}}P(\boldsymbol{w}_{n+m-1}) + \mu\boldsymbol{g}_{n+m}(\boldsymbol{w}_{n+m-1}) \tag{24.71}
$$

From the mean-value theorem (10.10) we have

$$
\nabla_{w^\mathsf{T}}P(\boldsymbol{w}_{n+m-1}) - \nabla_{w^\mathsf{T}}P(\boldsymbol{w}_{n-1}) \stackrel{(24.70)}{=} -\boldsymbol{H}_{n+m-1}\,\boldsymbol{d}_{m-1}^{n-1} \tag{24.72}
$$

where we are introducing the random matrix:

$$
\boldsymbol{H}_{n+m-1} \triangleq \int_0^1 \nabla_w^2 P\Big((1-t)\boldsymbol{w}_{n+m-1} + t\boldsymbol{w}_{n-1}\Big)dt \tag{24.73}
$$

so that (24.71) can be reformulated to:

$$\boxed{\boldsymbol{d}_m^{n-1} = (I - \mu \boldsymbol{H}_{n+m-1})\, \boldsymbol{d}_{m-1}^{n-1} + \mu \nabla_{w^{\mathsf{T}}} P(\boldsymbol{w}_{n-1}) + \mu \boldsymbol{g}_{n+m}(\boldsymbol{w}_{n+m-1})} \tag{24.74}$$

This recursion is over m; it involves a random and time-variant coefficient matrix multiplying the deviation vector $\boldsymbol{d}_{m-1}^{n-1}$ on the right-hand side. In a manner similar to the analysis performed in Chen and Sayed (2013) and Sayed (2014a), we replace $\boldsymbol{H}_{n+m-1}$ by the Hessian matrix $\nabla_w^2 P(\boldsymbol{w}_{n-1})$ evaluated at the starting point $n-1$. This substitution obviously leads to an approximate recursion in place of (24.74); we will denote its state vector by $\boldsymbol{z}_m^{n-1}$ instead of $\boldsymbol{d}_m^{n-1}$, as seen below in (24.75). The point is that while the Hessian matrix $\nabla_w^2 P(\boldsymbol{w}_{n-1})$ is random and depends on the time instant $n-1$, it is nevertheless constant over m and, moreover, becomes deterministic and constant when conditioned on $\boldsymbol{w}_{n-1}$. We thus arrive at the following recursion, which we shall refer to as the *short-term* model in lieu of (24.74):

$$\boxed{\boldsymbol{z}_m^{n-1} = \big(I - \mu \nabla_w^2 P(\boldsymbol{w}_{n-1})\big)\, \boldsymbol{z}_{n-1}^{m-1} + \mu \nabla_{w^{\mathsf{T}}} P(\boldsymbol{w}_{n-1}) + \mu \boldsymbol{g}_{n+m}(\boldsymbol{w}_{n+m-1})} \tag{24.75}$$

This is again a recursion over m. Motivated by definition (24.70), we denote the state iterates that result from using the approximate deviation $\boldsymbol{z}_m^{n-1}$ by

$$\boldsymbol{w}'_{n+m} \stackrel{\Delta}{=} \boldsymbol{w}_{n-1} - \boldsymbol{z}_m^{n-1} \tag{24.76}$$

This iterate approximates $\boldsymbol{w}_{n+m}$. The fact that the driving matrix $I - \mu \nabla_w^2 P(\boldsymbol{w}_{n-1})$ is constant for all m ensures that (24.75) is a more tractable recursion to study than (24.74). In order for this model to be useful, however, we need to ensure that the risk $P(\boldsymbol{w}'_{n+m})$ is close to the actual value $P(\boldsymbol{w}_{n+m})$. We begin by establishing a set of deviation bounds over a finite time horizon. These ensure that the iterates $\boldsymbol{w}'_{n+m}$ and $\boldsymbol{w}_{n+m}$ remain close to each other for a bounded number of iterations, which will allow us to relate $P(\boldsymbol{w}'_{n+m})$ and $P(\boldsymbol{w}_{n+m})$ further below in (24.83).

LEMMA 24.1. (Useful deviation bounds) *Starting from an iterate $\boldsymbol{w}_{n-1} \in \mathcal{U}$ in the undesirable set of strict saddle points and iterating the stochastic gradient recursion (24.16) for m steps leads to deviation vectors $\{\boldsymbol{d}_m^{n-1}\}$ whose moments satisfy the following bounds for sufficiently small step size μ:*

$$\mathbb{E}\left(\|\boldsymbol{d}_m^{n-1}\|^2 \,|\, \boldsymbol{w}_{n-1} \in \mathcal{U}\right) \leq O(\mu) \tag{24.77}$$

$$\mathbb{E}\left(\|\boldsymbol{d}_m^{n-1}\|^3 \,|\, \boldsymbol{w}_{n-1} \in \mathcal{U}\right) \leq O(\mu^{3/2}) \tag{24.78}$$

$$\mathbb{E}\left(\|\boldsymbol{d}_m^{n-1}\|^4 \,|\, \boldsymbol{w}_{n-1} \in \mathcal{U}\right) \leq O(\mu^2) \tag{24.79}$$

$$\mathbb{E}\left(\|\boldsymbol{d}_m^{n-1} - \boldsymbol{z}_m^{n-1}\|^2 \,|\, \boldsymbol{w}_{n-1} \in \mathcal{U}\right) \leq O(\mu^2) \tag{24.80}$$

$$\mathbb{E}\left(\|\boldsymbol{z}_m^{n-1}\|^2 \,|\, \boldsymbol{w}_{n-1} \in \mathcal{U}\right) \leq O(\mu) \tag{24.81}$$

$$\mathbb{E}\left(\|\boldsymbol{z}_m^{n-1}\|^3 \,|\, \boldsymbol{w}_{n-1} \in \mathcal{U}\right) \leq O(\mu^{3/2}) \tag{24.82}$$

for any $m + 1 \leq T/\mu$, where T denotes some arbitrary constant that is independent of the step size μ. The result of this lemma holds under conditions ($\mathbb{A}$1–$\mathbb{A}$5).

We will establish the lemma shortly. But, first, we comment that these deviation bounds will allow us to establish later the following corollary.

COROLLARY 24.1. (Short-term model accuracy) *Starting from an iterate $\boldsymbol{w}_{n-1} \in \mathcal{U}$ in the undesirable set of strict saddle points and iterating the stochastic gradient recursion (24.16) for m steps, the short-term model (24.75) is accurate over every finite horizon $m+1 \leq T/\mu$, i.e.,*

$$\mathbb{E}\left(P(\boldsymbol{w}_{n+m}) \,|\, \boldsymbol{w}_{n-1} \in \mathcal{U}\right) \leq \mathbb{E}\left(P(\boldsymbol{w}'_{n+m}) \,|\, \boldsymbol{w}_{n-1} \in \mathcal{U}\right) + O(\mu^{3/2}) \tag{24.83}$$

The above result shows that $P(w)$ evaluated at the true iterate $\boldsymbol{w}_{n+m}$ is upper-bounded by $P(w)$ evaluated at the approximate short-term iterate $\boldsymbol{w}'_{n+m}$ up to an approximation error on the order of $O(\mu^{3/2})$; this error is negligible for small step-sizes. This conclusion holds as long as both recursions (the actual recursion and the short-term approximation) are initialized at strict saddle points $\boldsymbol{w}_{n-1} \in \mathcal{U}$.

Bounds on deviation vectors

We now establish the bounds listed in Lemma 24.1. Let $\boldsymbol{\mathcal{F}}_k$ denote the filtration generated by the random processes $\{\boldsymbol{w}_j\}$ for all $j \leq k$. Informally, $\boldsymbol{\mathcal{F}}_k$ captures all information that is available about the stochastic processes $\{\boldsymbol{w}_j\}$ up to iteration k.

We refer to (24.74), condition on $\boldsymbol{\mathcal{F}}_{n+m-1}$, and note that:

$$\begin{aligned}
&\mathbb{E}\left(\|\boldsymbol{d}_m^{n-1}\|^2 \,|\, \boldsymbol{\mathcal{F}}_{n+m-1}\right) &(24.84)\\
&\overset{(24.74)}{=} \mathbb{E}\left\{\left\|(I-\mu\boldsymbol{H}_{n+m-1})\,\boldsymbol{d}_{m-1}^{n-1} + \mu\nabla_{w^\mathsf{T}} P(\boldsymbol{w}_{n-1}) + \mu\boldsymbol{g}_{n+m}\right\|^2 \,|\, \boldsymbol{\mathcal{F}}_{n+m-1}\right\}\\
&\overset{(a)}{=} \left\|(I-\mu\boldsymbol{H}_{n+m-1})\,\boldsymbol{d}_{m-1}^{n-1} + \mu\nabla_{w^\mathsf{T}} P(\boldsymbol{w}_{n-1})\right\|^2 + \mu^2\mathbb{E}\left(\|\boldsymbol{g}_{n+m}\|^2 \,|\, \boldsymbol{\mathcal{F}}_{n+m-1}\right)\\
&\overset{(b)}{\leq} \frac{1}{1-\mu\delta}\left\|(I-\mu\boldsymbol{H}_{n+m-1})\,\boldsymbol{d}_{m-1}^{n-1}\right\|^2 + \frac{\mu}{\delta}\|\nabla_{w^\mathsf{T}} P(\boldsymbol{w}_{n-1})\|^2\\
&\qquad + \mu^2\mathbb{E}\left(\|\boldsymbol{g}_{n+m}\|^2 \,|\, \boldsymbol{\mathcal{F}}_{n+m-1}\right)\\
&\overset{(c)}{\leq} \frac{(1+\mu\delta)^2}{1-\mu\delta}\|\boldsymbol{d}_{m-1}^{n-1}\|^2 + \frac{\mu}{\delta}\|\nabla_{w^\mathsf{T}} P(\boldsymbol{w}_{n-1})\|^2 + \mu^2\mathbb{E}\left(\|\boldsymbol{g}_{n+m}\|^2 \,|\, \boldsymbol{\mathcal{F}}_{n+m-1}\right)
\end{aligned}$$

where step (a) follows from the conditional zero-mean property (24.21a) for the gradient noise process, step (b) follows from the Jensen inequality:

$$\|a+b\|^2 \leq \frac{1}{\alpha}\|a\|^2 + \frac{1}{1-\alpha}\|b\|^2 \tag{24.85}$$

with α chosen as $\alpha = \mu\delta < 1$, and step (c) follows from the sub-multiplicative property of norms along with $-\delta I \leq \nabla_w^2 P(\boldsymbol{w}_{n-1}) \leq \delta I$, which follows from the Lipschitz gradient condition and the result of Prob. 10.2. It follows that

$$\begin{aligned}
&\mathbb{E}\left(\|\boldsymbol{d}_m^{n-1}\|^2 \,|\, \boldsymbol{\mathcal{F}}_{n+m-1}\right)\\
&\overset{(d)}{\leq} \frac{(1+\mu\delta)^2}{1-\mu\delta}\|\boldsymbol{d}_{m-1}^{n-1}\|^2 + \frac{\mu}{\delta}\|\nabla_{w^\mathsf{T}} P(\boldsymbol{w}_{n-1})\|^2 + \mu^2\beta^2\|\nabla_{w^\mathsf{T}} P(\boldsymbol{w}_{n+m-1})\|^2 + \mu^2\sigma^2\\
&= \frac{(1+\mu\delta)^2}{1-\mu\delta}\|\boldsymbol{d}_{m-1}^{n-1}\|^2 + \frac{\mu}{\delta}\|\nabla_{w^\mathsf{T}} P(\boldsymbol{w}_{n-1})\|^2\\
&\quad + \mu^2\beta^2\|\nabla_{w^\mathsf{T}} P(\boldsymbol{w}_{n-1}) + \nabla_{w^\mathsf{T}} P(\boldsymbol{w}_{n+m-1}) - \nabla_{w^\mathsf{T}} P(\boldsymbol{w}_{n-1})\|^2 + \mu^2\sigma^2\\
&\overset{(e)}{\leq} \frac{(1+\mu\delta)^2}{1-\mu\delta}\|\boldsymbol{d}_{m-1}^{n-1}\|^2 + \frac{\mu}{\delta}\|\nabla_{w^\mathsf{T}} P(\boldsymbol{w}_{n-1})\|^2 + 2\mu^2\beta^2\|\nabla_{w^\mathsf{T}} P(\boldsymbol{w}_{n-1})\|^2\\
&\quad + 2\mu^2\beta^2\|\nabla_{w^\mathsf{T}} P(\boldsymbol{w}_{n+m-1}) - \nabla_{w^\mathsf{T}} P(\boldsymbol{w}_{n-1})\|^2 + \mu^2\sigma^2\\
&\overset{(f)}{\leq} \frac{(1+\mu\delta)^2 + (1-\mu\delta)2\mu^2\beta^2\delta^2}{1-\mu\delta}\|\boldsymbol{d}_{m-1}^{n-1}\|^2 + \mu\left(\frac{1}{\delta} + 2\mu\beta^2\right)\|\nabla_{w^\mathsf{T}} P(\boldsymbol{w}_{n-1})\|^2 + \mu^2\sigma^2\\
&\leq \frac{(1+\mu\delta)^2 + O(\mu^2)}{1-\mu\delta}\|\boldsymbol{d}_{m-1}^{n-1}\|^2 + O(\mu)\|\nabla_{w^\mathsf{T}} P(\boldsymbol{w}_{n-1})\|^2 + \mu^2\sigma^2
\end{aligned} \tag{24.86}$$

where step (d) uses the bound (24.22) on the second-order moment of the gradient noise. Step (e) follows from $\|a+b\|^2 \leq 2\|a\|^2 + 2\|b\|^2$, and step (f) applies the Lipschitz bound on the gradient of the risk function.

We can now take expectations over $\boldsymbol{w}_{n-1} \in \mathcal{U}$ to obtain:

$$\begin{aligned}
\mathbb{E}\left(\|\boldsymbol{d}_m^{n-1}\|^2 \,|\, \boldsymbol{w}_{n-1} \in \mathcal{U}\right) &\leq \frac{(1+\mu\delta)^2 + O(\mu^2)}{1-\mu\delta}\mathbb{E}\left\{\|\boldsymbol{d}_{m-1}^{n-1}\|^2 \,|\, \boldsymbol{w}_{n-1} \in \mathcal{U}\right\}\\
&\quad + O(\mu)\mathbb{E}\left(\|\nabla_{w^\mathsf{T}} P(\boldsymbol{w}_{n-1})\|^2 \,|\, \boldsymbol{w}_{n-1} \in \mathcal{U}\right) + O(\mu^2)\\
&\overset{(a)}{\leq} \frac{(1+\mu\delta)^2 + O(\mu^2)}{1-\mu\delta}\mathbb{E}\left(\|\boldsymbol{d}_{m-1}^{n-1}\|^2 \,|\, \boldsymbol{w}_{n-1} \in \mathcal{U}\right) + O(\mu^2)
\end{aligned} \tag{24.87}$$

where (a) follows from the definition of the set $\mathcal{U}$ in (24.43) where the squared norm of the gradient of $P(w)$ is $O(\mu)$. Note from definition (24.70) that $\boldsymbol{d}_{-1}^{n-1} = \boldsymbol{w}_{n-1} - \boldsymbol{w}_{n-1} = 0$ and, hence, the initial deviation is zero, by definition. Iterating (24.87) starting from $m = 0$ yields:

$$
\begin{aligned}
&\mathbb{E}\left(\left\|\boldsymbol{d}_m^{n-1}\right\|^2 \mid \boldsymbol{w}_{n-1} \in \mathcal{H}\right) \\
&\leq O(\mu^2) \times \left(\sum_{j=0}^{m} \left(\frac{(1+\mu\delta)^2 + O(\mu^2)}{1-\mu\delta}\right)^j\right) \\
&= O(\mu^2) \times \left(\frac{1 - \left(\frac{(1+\mu\delta)^2 + O(\mu^2)}{1-\mu\delta}\right)^{m+1}}{1 - \frac{(1+\mu\delta)^2 + O(\mu^2)}{1-\mu\delta}}\right) \\
&= O(\mu^2) \times \frac{\left(\left(\frac{(1+\mu\delta)^2 + O(\mu^2)}{1-\mu\delta}\right)^{m+1} - 1\right)(1-\mu\delta)}{1 + 2\mu\delta + \mu^2\delta^2 - 1 + \mu\delta} \\
&\leq O(\mu) \times \frac{\left(\left(\frac{(1+\mu\delta)^2 + O(\mu^2)}{1-\mu\delta}\right)^{\frac{T}{\mu}} - 1\right)(1-\mu\delta)}{3\delta + \mu\delta^2} \\
&\leq O(\mu) \times \left(\left(\frac{(1+\mu\delta)^2 + O(\mu^2)}{1-\mu\delta}\right)^{\frac{T}{\mu}} - 1\right)\frac{(1-\mu\delta)}{3\delta} \\
&\leq O(\mu) \times \left(\frac{(1+\mu\delta)^2 + O(\mu^2)}{1-\mu\delta}\right)^{\frac{T}{\mu}} \frac{1}{3\delta} \\
&= O(\mu)
\end{aligned}
\tag{24.88}
$$

where the last line follows from Prob. 24.2 for small step sizes. We therefore established (24.77). We proceed to establish a bound on the fourth-order moment. Using the following easily verifiable inequality for any two vectors a and b:

$$
\|a+b\|^4 \leq \|a\|^4 + 3\|b\|^4 + 8\|a\|^2\|b\|^2 + 4\|a\|^2(a^{\mathsf{T}}b) \tag{24.89}
$$

we have

$$
\begin{aligned}
\mathbb{E}\left(\left\|\boldsymbol{d}_m^{n-1}\right\|^4 \mid \boldsymbol{\mathcal{F}}_{n+m-1}\right) \leq\ & \left\|(I - \mu\boldsymbol{H}_{n+m-1})\,\boldsymbol{d}_{m-1}^{n-1} + \mu\nabla_{w^{\mathsf{T}}} P(\boldsymbol{w}_{n-1})\right\|^4 \\
& + 3\mu^4\mathbb{E}\left(\left\|\boldsymbol{g}_{n+m}\right\|^4 \mid \boldsymbol{\mathcal{F}}_{n+m-1}\right) \\
& + 8\mu^2\left\|(I - \mu\boldsymbol{H}_{n+m-1})\,\boldsymbol{d}_{m-1}^{n-1} + \mu\nabla_{w^{\mathsf{T}}} P(\boldsymbol{w}_{n-1})\right\|^2 \\
& \quad \times \mathbb{E}\left(\left\|\boldsymbol{g}_{n+m}\right\|^2 \mid \boldsymbol{\mathcal{F}}_{n+m-1}\right) \\
& + 4\mu\left\|(I - \mu\boldsymbol{H}_{n+m-1})\,\boldsymbol{d}_{m-1}^{n-1} + \mu\nabla_{w^{\mathsf{T}}} P(\boldsymbol{w}_{n-1})\right\|^2 \\
& \quad \times \left((I - \mu\boldsymbol{H}_{n+m-1})\,\boldsymbol{d}_{m-1}^{n-1} + \mu\nabla_{w^{\mathsf{T}}} P(\boldsymbol{w}_{n-1})\right)^{\mathsf{T}} \\
& \quad \times \left(\mathbb{E}\left(\boldsymbol{g}_{n+m} \middle| \boldsymbol{\mathcal{F}}_{n+m-1}\right)\right)
\end{aligned}
\tag{24.90}
$$

That is,

$$
\begin{aligned}
\mathbb{E}\left(\|\boldsymbol{d}_m^{n-1}\|^4 \,|\, \boldsymbol{\mathcal{F}}_{n+m-1}\right) &\overset{(a)}{=} \left\|(I-\mu\boldsymbol{H}_{n+m-1})\,\boldsymbol{d}_{m-1}^{n-1}+\mu\nabla_{w^\mathsf{T}}P(\boldsymbol{w}_{n-1})\right\|^4 \\
&\quad +3\mu^4\mathbb{E}\left(\|\boldsymbol{g}_{n+m}\|^4 \,|\, \boldsymbol{\mathcal{F}}_{i+j}\right) \\
&\quad +8\mu^2\left\|(I-\mu\boldsymbol{H}_{n+m-1})\,\boldsymbol{d}_{m-1}^{n-1}+\mu\nabla_{w^\mathsf{T}}P(\boldsymbol{w}_{n-1})\right\|^2 \\
&\qquad \times\mathbb{E}\left(\|\boldsymbol{g}_{n+m}\|^2 \,|\, \boldsymbol{\mathcal{F}}_{n+m-1}\right) \\
&\overset{(b)}{\leq} \left\|(I-\mu\boldsymbol{H}_{n+m-1})\,\boldsymbol{d}_{m-1}^{n-1}+\mu\nabla_{w^\mathsf{T}}P(\boldsymbol{w}_{n-1})\right\|^4 \\
&\quad +3\mu^4\left(\beta^4\|\nabla_{w^\mathsf{T}}P(\boldsymbol{w}_{n+m-1})\|^4+\sigma^4\right) \\
&\quad +8\mu^2\left\|(I-\mu\boldsymbol{H}_{n+m-1})\,\boldsymbol{d}_{m-1}^{n-1}+\mu\nabla_{w^\mathsf{T}}P(\boldsymbol{w}_{n-1})\right\|^2 \\
&\qquad \times\left(\beta\|\nabla_{w^\mathsf{T}}P(\boldsymbol{w}_{n+m-1})\|^2+\sigma^2\right)
\end{aligned}
\tag{24.91}
$$

where in step (a) we dropped cross terms due to the conditional zero-mean property (24.21a) of the gradient noise, and step (b) follows from the fourth- and second-order conditions (24.21b) and (24.22) on the gradient noise. We next bound each of the terms appearing in (24.91). From the Jensen inequality, we have for any $0 < \alpha < 1$ and vectors (a, b):

$$
\|a+b\|^4 = \frac{1}{\alpha^3}\,\|a\|^4 + \frac{1}{(1-\alpha)^3}\,\|b\|^4 \tag{24.92}
$$

and hence, using $\alpha = 1-\mu\delta$ and $0 < \mu < 1/\delta$, we obtain for the first term on the right-hand side of (24.91):

$$
\begin{aligned}
&\left\|(I-\mu\boldsymbol{H}_{n+m-1})\,\boldsymbol{d}_{m-1}^{n-1}+\mu\nabla_{w^\mathsf{T}}P(\boldsymbol{w}_{n-1})\right\|^4 \\
&\leq \frac{(1+\mu\delta)^4}{(1-\mu\delta)^3}\left\|\boldsymbol{d}_{m-1}^{n-1}\right\|^4 + O(\mu)\|\nabla_{w^\mathsf{T}}P(\boldsymbol{w}_{n-1})\|^4
\end{aligned}
\tag{24.93}
$$

After taking expectations conditioned on $\boldsymbol{w}_{n-1}\in\mathcal{U}$, we find:

$$
\begin{aligned}
&\mathbb{E}\left(\left\|(I-\mu\boldsymbol{H}_{n+m-1})\,\boldsymbol{d}_{m-1}^{n-1}+\mu\nabla_{w^\mathsf{T}}P(\boldsymbol{w}_{n-1})\right\|^4 \,|\, \boldsymbol{w}_{n-1}\in\mathcal{U}\right) \\
&\leq \frac{(1+\mu\delta)^4}{(1-\mu\delta)^3}\mathbb{E}\left(\left\|\boldsymbol{d}_{m-1}^{n-1}\right\|^4 \,|\, \boldsymbol{w}_{n-1}\in\mathcal{U}\right) + O(\mu)\mathbb{E}\left(\|\nabla_{w^\mathsf{T}}P(\boldsymbol{w}_{n-1})\|^4 \,|\, \boldsymbol{w}_{n-1}\in\mathcal{U}\right) \\
&\leq \frac{(1+\mu\delta)^4}{(1-\mu\delta)^3}\mathbb{E}\left\{\left\|\boldsymbol{d}_{m-1}^{n-1}\right\|^4 \,|\, \boldsymbol{w}_{n-1}\in\mathcal{H}\right\} + O(\mu^3)
\end{aligned}
\tag{24.94}
$$

where in the last line we used the fact that the square norm of the gradient of the risk function is bounded by $O(\mu)$ in the set $\mathcal{U}$.

For the second term in (24.91) we have, again from (24.92) with $\alpha = 1/2$:

$$
\begin{aligned}
&3\mu^4\left(\beta^4\|\nabla_{w^\mathsf{T}}P(\boldsymbol{w}_{n+m-1})\|^4+\sigma^4\right) \\
&= 3\mu^4\left(\beta^4\|\nabla_{w^\mathsf{T}}P(\boldsymbol{w}_{n-1})+\nabla_{w^\mathsf{T}}P(\boldsymbol{w}_{n+m-1})-\nabla_{w^\mathsf{T}}P(\boldsymbol{w}_{n-1})\|^4+\sigma^4\right) \\
&\overset{(24.92)}{\leq} 3\mu^4\left(8\beta^4\|\nabla_{w^\mathsf{T}}P(\boldsymbol{w}_{n-1})\|^4+8\beta^4\|\nabla_{w^\mathsf{T}}P(\boldsymbol{w}_{n+m-1})-\nabla_{w^\mathsf{T}}P(\boldsymbol{w}_{n-1})\|^4+\sigma^4\right) \\
&\overset{(a)}{\leq} \mu^4\left(8\beta^4\|\nabla_{w^\mathsf{T}}P(\boldsymbol{w}_{n-1})\|^4+8\beta^4\delta^4\left\|\boldsymbol{d}_{m-1}^{n-1}\right\|^4+\sigma^4\right) \\
&= O(\mu^4)\|\nabla_{w^\mathsf{T}}P(\boldsymbol{w}_{n-1})\|^4+O(\mu^4)\left\|\boldsymbol{d}_{m-1}^{n-1}\right\|^4+O(\mu^4)
\end{aligned}
\tag{24.95}
$$

where step (a) uses the δ-Lipschitz condition on the gradient of $P(w)$. After taking expectations over $\boldsymbol{w}_{n-1} \in \mathcal{H}$ we have

$$\begin{aligned}
&\mathbb{E}\left(3\mu^4\left(\beta^4\|\nabla_{w^\mathsf{T}} P(\boldsymbol{w}_{i+j})\|^4+\sigma^4\right) \,\middle|\, \boldsymbol{w}_{n-1} \in \mathcal{U}\right) \\
&\leq O(\mu^4)\mathbb{E}\left(\|\nabla_{w^\mathsf{T}} P(\boldsymbol{w}_{n-1})\|^4 \,\middle|\, \boldsymbol{w}_{n-1} \in \mathcal{U}\right)+O(\mu^4)\mathbb{E}\left(\|\boldsymbol{d}_{m-1}^{n-1}\|^4 \,\middle|\, \boldsymbol{w}_{n-1} \in \mathcal{U}\right)+O(\mu^4) \\
&\leq O(\mu^4)\mathbb{E}\left(\|\boldsymbol{d}_{m-1}^{n-1}\|^4 \,\middle|\, \boldsymbol{w}_{n-1} \in \mathcal{H}\right)+O(\mu^4)
\end{aligned} \tag{24.96}$$

since the squared norm of the gradient of the risk function is bounded by $O(\mu)$ in the set $\mathcal{U}$. For the last term in (24.91), we have

$$\begin{aligned}
&8\mu^2\left\|(I-\mu\boldsymbol{H}_{n+m-1})\,\boldsymbol{d}_{m-1}^{n-1}+\mu\nabla_{w^\mathsf{T}} P(\boldsymbol{w}_{n-1})\right\|^2\left(\beta\|\nabla_{w^\mathsf{T}} P(\boldsymbol{w}_{n+m-1})\|^2+\sigma^2\right) \\
&= 8\beta\mu^2\left\|(I-\mu\boldsymbol{H}_{n+m-1})\,\boldsymbol{d}_{m-1}^{n-1}+\mu\nabla_{w^\mathsf{T}} P(\boldsymbol{w}_{n-1})\right\|^2\|\nabla_{w^\mathsf{T}} P(\boldsymbol{w}_{n+m-1})\|^2 \\
&\quad +8\sigma^2\mu^2\left\|(I-\mu\boldsymbol{H}_{n+m-1})\,\boldsymbol{d}_{m-1}^{n-1}+\mu\nabla_{w^\mathsf{T}} P(\boldsymbol{w}_{n-1})\right\|^2 \\
&\overset{(24.85)}{\leq} 8\beta^2\mu^2\left(\frac{(1+\mu\delta)^2}{1-\mu\delta}\|\boldsymbol{d}_{m-1}^{n-1}\|^2+\frac{\mu}{\delta}\|\nabla_{w^\mathsf{T}} P(\boldsymbol{w}_{n-1})\|^2\right)\|\nabla_{w^\mathsf{T}} P(\boldsymbol{w}_{n+m-1})\|^2 \\
&\quad +8\sigma^2\mu^2\left(\frac{(1+\mu\delta)^2}{1-\mu\delta}\|\boldsymbol{d}_{m-1}^{n-1}\|^2+\frac{\mu}{\delta}\|\nabla_{w^\mathsf{T}} P(\boldsymbol{w}_{n-1})\|^2\right) \\
&= 8\beta^2\mu^2\left(\frac{(1+\mu\delta)^2}{1-\mu\delta}\|\boldsymbol{d}_{m-1}^{n-1}\|^2+\frac{\mu}{\delta}\|\nabla_{w^\mathsf{T}} P(\boldsymbol{w}_{n-1})\|^2\right) \\
&\quad \times\|\nabla_{w^\mathsf{T}} P(\boldsymbol{w}_{n-1})+\nabla_{w^\mathsf{T}} P(\boldsymbol{w}_{n+m-1})-\nabla_{w^\mathsf{T}} P(\boldsymbol{w}_{n-1})\|^2 \\
&\quad +8\sigma^2\mu^2\left(\frac{(1+\mu\delta)^2}{1-\mu\delta}\|\boldsymbol{d}_{m-1}^{n-1}\|^2+\frac{\mu}{\delta}\|\nabla_{w^\mathsf{T}} P(\boldsymbol{w}_{n-1})\|^2\right)
\end{aligned} \tag{24.97}$$

so that applying the Jensen inequality to the second term we get

$$\begin{aligned}
&8\mu^2\left\|(I-\mu\boldsymbol{H}_{n+m-1})\,\boldsymbol{d}_{m-1}^{n-1}+\mu\nabla_{w^\mathsf{T}} P(\boldsymbol{w}_{n-1})\right\|^2\left(\beta\|\nabla_{w^\mathsf{T}} P(\boldsymbol{w}_{n+m-1})\|^2+\sigma^2\right) \\
&\leq 8\beta^2\mu^2\left(\frac{(1+\mu\delta)^2}{1-\mu\delta}\|\boldsymbol{d}_{m-1}^{n-1}\|^2+\frac{\mu}{\delta}\|\nabla_{w^\mathsf{T}} P(\boldsymbol{w}_{n-1})\|^2\right) \\
&\quad \times\left(2\|\nabla_{w^\mathsf{T}} P(\boldsymbol{w}_{n-1})\|^2+2\|\nabla_{w^\mathsf{T}} P(\boldsymbol{w}_{n+m-1})-\nabla_{w^\mathsf{T}} P(\boldsymbol{w}_{n-1})\|^2\right) \\
&\quad +8\sigma^2\mu^2\left(\frac{(1+\mu\delta)^2}{1-\mu\delta}\|\boldsymbol{d}_{m-1}^{n-1}\|^2+\frac{\mu}{\delta}\|\nabla_{w^\mathsf{T}} P(\boldsymbol{w}_{n-1})\|^2\right) \\
&\leq 8\beta^2\mu^2\left(\frac{(1+\mu\delta)^2}{1-\mu\delta}\|\boldsymbol{d}_{m-1}^{n-1}\|^2+\frac{\mu}{\delta}\|\nabla_{w^\mathsf{T}} P(\boldsymbol{w}_{n-1})\|^2\right) \\
&\quad \times\left(2\|\nabla_{w^\mathsf{T}} P(\boldsymbol{w}_{n-1})\|^2+2\delta^2\|\boldsymbol{d}_{m-1}^{n-1}\|^2\right) \\
&\quad +8\sigma^2\mu^2\left(\frac{(1+\mu\delta)^2}{1-\mu\delta}\|\boldsymbol{d}_{m-1}^{n-1}\|^2+\frac{\mu}{\delta}\|\nabla_{w^\mathsf{T}} P(\boldsymbol{w}_{n-1})\|^2\right) \\
&= O(\mu^2)\|\boldsymbol{d}_{m-1}^{n-1}\|^4+O(\mu^3)\|\nabla_{w^\mathsf{T}} P(\boldsymbol{w}_{n-1})\|^4+O(\mu^3)\|\nabla_{w^\mathsf{T}} P(\boldsymbol{w}_{n-1})\|^2 \\
&\quad +O(\mu^2)\|\nabla_{w^\mathsf{T}} P(\boldsymbol{w}_{n-1})\|^2\|\boldsymbol{d}_{m-1}^{n-1}\|^2 O(\mu^2)+O(\mu^2)\|\boldsymbol{d}_{m-1}^{n-1}\|^2
\end{aligned} \tag{24.98}$$

After taking conditional expectations:

$$\begin{aligned}
&\mathbb{E}\Big\{(8\mu^2\big\|(I-\mu \boldsymbol{H}_{n+m-1})\,\boldsymbol{d}_{m-1}^{n-1}+\mu\nabla_{w^\mathsf{T}}P(\boldsymbol{w}_{n-1})\big\|^2 \\
&\qquad\times\left(8\|\nabla_{w^\mathsf{T}}P(\boldsymbol{w}_{n+m-1})\|^2+\sigma^2\right)\mid \boldsymbol{w}_{n-1}\in\mathcal{U}\Big\} \\
&\le O(\mu^2)\mathbb{E}\left(\|\boldsymbol{d}_{m-1}^{n-1}\|^4\mid \boldsymbol{w}_{n-1}\in\mathcal{U}\right)+O(\mu^3)\mathbb{E}\left(\|\nabla_{w^\mathsf{T}}P(\boldsymbol{w}_{n-1})\|^4\mid \boldsymbol{w}_{n-1}\in\mathcal{U}\right) \\
&\quad+O(\mu^2)\mathbb{E}\left(\|\nabla_{w^\mathsf{T}}P(\boldsymbol{w}_{n-1})\|^2\|\boldsymbol{d}_{m-1}^{n-1}\|^2\mid \boldsymbol{w}_{n-1}\in\mathcal{U}\right) \\
&\quad+O(\mu^2)\mathbb{E}\left(\|\boldsymbol{d}_{m-1}^{n-1}\|^2\mid \boldsymbol{w}_{n-1}\in\mathcal{U}\right)+O(\mu^3)\mathbb{E}\left(\|\nabla_{w^\mathsf{T}}P(\boldsymbol{w}_{n-1})\|^2\mid \boldsymbol{w}_{n-1}\in\mathcal{U}\right) \\
&\le O(\mu^2)\mathbb{E}\left(\|\boldsymbol{d}_{m-1}^{n-1}\|^4\mid \boldsymbol{w}_{n-1}\in\mathcal{U}\right)+O(\mu^3)\cdot O(\mu^2) \\
&\quad+O(\mu^3)\mathbb{E}\left(\|\boldsymbol{d}_{m-1}^{n-1}\|^2\mid \boldsymbol{w}_{n-1}\in\mathcal{U}\right)+O(\mu^2)\cdot O(\mu)+O(\mu^3)\cdot O(\mu) \\
&\le O(\mu^2)\mathbb{E}\left(\|\boldsymbol{d}_{m-1}^{n-1}\|^4\mid \boldsymbol{w}_{n-1}\in\mathcal{U}\right)+O(\mu^3)
\end{aligned} \tag{24.99}$$

Returning to (24.91), after taking expectations over $\boldsymbol{w}_{n-1}\in\mathcal{U}$ on both sides and grouping terms, we find:

$$\mathbb{E}\left(\|\boldsymbol{d}_{m}^{n-1}\|^4\mid \boldsymbol{w}_{n-1}\in\mathcal{U}\right)\le\frac{(1+\mu\delta)^4+O(\mu^2)}{(1-\mu\delta)^3}\mathbb{E}\left(\|\boldsymbol{d}_{m-1}^{n-1}\|^4\mid \boldsymbol{w}_{n-1}\in\mathcal{U}\right)+O(\mu^3) \tag{24.100}$$

Recall again that $d_{-1}^{n-1}=0$ and therefore iterating (24.100) yields:

$$\begin{aligned}
\mathbb{E}\left(\|\boldsymbol{d}_{m}^{n-1}\|^4\mid \boldsymbol{w}_{n-1}\in\mathcal{U}\right)&\le\left(\sum_{j=0}^{m}\left(\frac{(1+\mu\delta)^4+O(\mu^2)}{(1-\mu\delta)^3}\right)^j\right)O(\mu^3) \\
&=\frac{1-\left(\frac{(1+\mu\delta)^4+O(\mu^2)}{(1-\mu\delta)^3}\right)^{m+1}}{1-\frac{(1+\mu\delta)^4+O(\mu^2)}{(1-\mu\delta)^3}}O(\mu^3) \\
&=\frac{\left(\left(\frac{(1+\mu\delta)^4+O(\mu^2)}{(1-\mu\delta)^3}\right)^{m+1}-1\right)(1-\mu\delta)^3}{(1+\mu\delta)^4+O(\mu^2)-(1-\mu\delta)^3}O(\mu^3) \\
&\le\frac{\left(\frac{(1+\mu\delta)^4+O(\mu^2)}{(1-\mu\delta)^3}\right)^{m+1}-1}{(1+\mu\delta)^4+O(\mu^2)-(1-\mu\delta)^3}O(\mu^3) \\
&\le\frac{\left(\frac{(1+\mu\delta)^4+O(\mu^2)}{(1-\mu\delta)^3}\right)^{j}}{(1+\mu\delta)^4+O(\mu^2)-(1-\mu\delta)^3}O(\mu^3)
\end{aligned} \tag{24.101}$$

Using the expansion

$$\begin{aligned}
&(1+\mu\delta)^4+O(\mu^2)-(1-\mu\delta)^3 \\
&=1+4\mu\delta+O(\mu^2)-1+3\mu\delta-O(\mu^2)=O(\mu)
\end{aligned} \tag{24.102}$$

we get

$$\begin{aligned}
\mathbb{E}\left(\left\|\boldsymbol{d}_m^{n-1}\right\|^4 \mid \boldsymbol{w}_{n-1} \in \mathcal{U}\right) &\leq \frac{\left(\frac{(1+\mu\delta)^4+O(\mu^2)}{(1-\mu\delta)^3}\right)^{m+1}}{O(\mu)} O(\mu^3) \\
&= \left(\frac{(1+\mu\delta)^4+O(\mu^2)}{(1-\mu\delta)^3}\right)^{m+1]} O(\mu^2) \\
&\leq \left(\frac{(1+\mu\delta)^4+O(\mu^2)}{(1-\mu\delta)^3}\right)^{\frac{T}{\mu}} O(\mu^2) \\
&\leq O(\mu^2)
\end{aligned} \tag{24.103}$$

where the last step follows from Prob. 24.2 for small step sizes. We therefore established bound (24.79). Equation (24.78) then follows from the Jensen inequality by noting that

$$\mathbb{E}\left(\left\|\boldsymbol{d}_{m-1}^{n-1}\right\|^3 \,\middle|\, \boldsymbol{w}_{n-1} \in \mathcal{U}\right) \leq \left(\mathbb{E}\left(\left\|\boldsymbol{d}_{m-1}^{n-1}\right\|^4 \,\middle|\, \boldsymbol{w}_{n-1} \in \mathcal{U}\right)\right)^{3/4} = \left(O(\mu^2)\right)^{3/4} = O(\mu^{3/2}) \tag{24.104}$$

We next establish bound (24.80). We start by examining the difference between the short-term model (24.75) and the true recursion (24.74). Subtracting both recursions gives:

$$\begin{aligned}
\boldsymbol{w}_{n+m} - \boldsymbol{w}'_{n+m} &= \boldsymbol{z}_m^{n-1} - \boldsymbol{d}_m^{n-1} \\
&= \left(I - \mu\nabla_w^2 P(\boldsymbol{w}_{n-1})\right)\boldsymbol{z}_{m-1}^{n-1} + \mu\nabla_{w^{\mathsf{T}}} P(\boldsymbol{w}_{n-1}) + \mu\boldsymbol{g}_{n+m} \\
&\quad - \left(I - \mu\boldsymbol{H}_{n+m-1}\right)\boldsymbol{d}_{m-1}^{n-1} - \mu\nabla_{w^{\mathsf{T}}} P(\boldsymbol{w}_{n-1}) - \mu\boldsymbol{g}_{n+m} \\
&= \left(I - \mu\nabla_w^2 P(\boldsymbol{w}_{n-1})\right)\boldsymbol{z}_{m-1}^{n-1} - \left(I - \mu\boldsymbol{H}_{n+m-1}\right)\boldsymbol{d}_{m-1}^{n-1} \\
&= \left(I - \mu\nabla_w^2 P(\boldsymbol{w}_{n-1})\right)\left(\boldsymbol{w}_{n+m-1} - \boldsymbol{w}'_{n+m-1}\right) \\
&\quad + \mu\left(\boldsymbol{H}_{n+m-1} - \nabla_w^2 P(\boldsymbol{w}_{n-1})\right)\boldsymbol{d}_{m-1}^{n-1}
\end{aligned} \tag{24.105}$$

Before proceeding, note that the difference between the "Hessian matrices" in the last term can be bounded as:

$$\begin{aligned}
&\left\|\nabla_w^2 P(\boldsymbol{w}_{n-1}) - \boldsymbol{H}_{n+m-1}\right\| \\
&= \left\|\nabla_w^2 P(\boldsymbol{w}_{n-1}) - \int_0^1 \nabla_w^2 P\left((1-t)\boldsymbol{w}_{n+m-1} + t\boldsymbol{w}_{n-1}\right) dt\right\| \\
&= \left\|\int_0^1 \left(\nabla_w^2 P(\boldsymbol{w}_{n-1}) - \nabla_w^2 P\left((1-t)\boldsymbol{w}_{n+m-1} + t\boldsymbol{w}_{n-1}\right)\right) dt\right\| \\
&\leq \int_0^1 \left\|\nabla_w^2 P(\boldsymbol{w}_{n-1}) - \nabla_w^2 P\left((1-t)\boldsymbol{w}_{n+m-1} + t\boldsymbol{w}_{n-1}\right)\right\| dt \\
&\overset{(a)}{\leq} \delta_2 \int_0^1 \left\|(1-t)\boldsymbol{w}_{n-1} - (1-t)\boldsymbol{w}_{n+m-1}\right\| dt \\
&= \delta_2 \left\|\boldsymbol{d}_{m-1}^{n-1}\right\| \int_0^1 (1-t) dt = \frac{\delta_2}{2}\left\|\boldsymbol{d}_{m-1}^{n-1}\right\|
\end{aligned} \tag{24.106}$$

where in step (a) we used the δ_2-Lipschitz condition (24.20b) on the Hessian of the risk function. Returning to (24.105) and taking norms yields:

$$\begin{aligned}
&\|\boldsymbol{w}_{n+m} - \boldsymbol{w}'_{n+m}\|^2 \\
&= \Big\| \left(I - \mu \nabla_w^2 P(\boldsymbol{w}_{n-1})\right) \left(\boldsymbol{w}_{n+m-1} - \boldsymbol{w}'_{n+m-1}\right) + \mu \left(\boldsymbol{H}_{n+m-1} - \nabla_w^2 P(\boldsymbol{w}_{n-1})\right) \boldsymbol{d}_{m-1}^{n-1} \Big\|^2 \\
&\overset{(a)}{\leq} \frac{1}{1-\mu\delta} \left\| \left(I - \mu \nabla_w^2 P(\boldsymbol{w}_{n-1})\right) \left(\boldsymbol{w}_{n+m-1} - \boldsymbol{w}'_{n+m-1}\right) \right\|^2 \\
&\quad + \frac{\mu^2}{\mu\delta} \left\| \left(\boldsymbol{H}_{n+m-1} - \nabla_w^2 P(\boldsymbol{w}_{n-1})\right) \boldsymbol{d}_{m-1}^{n-1} \right\|^2 \\
&\overset{(24.106)}{\leq} \frac{(1+\mu\delta)^2}{1-\mu\delta} \|\boldsymbol{w}_{n+m-1} - \boldsymbol{w}'_{n+m-1}\|^2 + \frac{\mu}{\delta}\frac{\delta_2}{2} \|\boldsymbol{d}_{m-1}^{n-1}\|^4
\end{aligned} \tag{24.107}$$

where (a) again follows from the Jensen inequality (24.85) with $\alpha = 1 - \mu\delta$. Taking expectations over $\boldsymbol{w}_{n-1} \in \mathcal{U}$ yields:

$$\begin{aligned}
&\mathbb{E}\left(\|\boldsymbol{w}_{n+m} - \boldsymbol{w}'_{n+m}\|^2 \,\Big|\, \boldsymbol{w}_{n-1} \in \mathcal{U} \right) \\
&\leq \frac{(1+\mu\delta)^2}{1-\mu\delta} \mathbb{E}\left(\|\boldsymbol{w}_{n+m-1} - \boldsymbol{w}'_{n+m-1}\|^2 \,\Big|\, \boldsymbol{w}_{n-1} \in \mathcal{U} \right) \\
&\quad + \frac{\mu}{\delta}\frac{\delta_2}{2} \mathbb{E}\left(\|\boldsymbol{d}_{m-1}^{n-1}\|^4 \,\Big|\, \boldsymbol{w}_{n-1} \in \mathcal{U} \right) \\
&\overset{(24.103)}{\leq} \frac{(1+\mu\delta)^2}{1-\mu\delta} \mathbb{E}\left(\|\boldsymbol{w}_{n+m-1} - \boldsymbol{w}'_{n+m-1}\| \,\Big|\, \boldsymbol{w}_{n-1} \in \mathcal{U} \right)^2 + O(\mu^3)
\end{aligned} \tag{24.108}$$

Since both the true and the short-term models are initialized at the same location $\boldsymbol{w}_{n-1} \in \mathcal{U}$, iterating the above recursion and using the same argument as before leads to (see Prob. 24.6):

$$\mathbb{E}\,\|\boldsymbol{w}_{n+m} - \boldsymbol{w}'_{n+m}\|^2 \leq O(\mu^2) \tag{24.109}$$

which is (24.80). The proof of (24.81) and (24.82) is left as an exercise.

Accuracy of the short-term model

We now establish Corollary 24.1. Since the risk function $P(w)$ has δ-Lipschitz gradients, we know from property (10.13) that

$$\begin{aligned}
P(\boldsymbol{w}_{n+m}) \leq\; & P(\boldsymbol{w}'_{n+m}) + \left(\nabla_{w^{\mathsf{T}}} P\left(\boldsymbol{w}'_{n+m}\right)\right)^{\mathsf{T}} \left(\boldsymbol{w}_{n+m} - \boldsymbol{w}'_{n+m}\right) \\
& + \frac{\delta}{2} \|\boldsymbol{w}_{n+m} - \boldsymbol{w}'_{n+m}\|^2
\end{aligned} \tag{24.110}$$

In the vicinity of saddle points, we can refine the upper bound (24.110) by taking expectations conditioned on $\boldsymbol{w}_{n-1} \in \mathcal{U}$:

$$\begin{aligned}
&\mathbb{E}\left(P(\boldsymbol{w}_{n+m}) \,|\, \boldsymbol{w}_{n-1} \in \mathcal{U}\right) \\
&\leq \mathbb{E}\left(P(\boldsymbol{w}'_{n+m}) \,|\, \boldsymbol{w}_{n-1} \in \mathcal{U}\right) \\
&\quad + \mathbb{E}\left(\left(\nabla_{w^\mathsf{T}} P\left(\boldsymbol{w}'_{n+m}\right)\right)^\mathsf{T} \left(\boldsymbol{w}_{n+m} - \boldsymbol{w}'_{n+m}\right) \,|\, \boldsymbol{w}_{n-1} \in \mathcal{U}\right) \\
&\quad + \frac{\delta}{2}\mathbb{E}\left(\left\|\boldsymbol{w}_{n+m} - \boldsymbol{w}'_{n+m}\right\|^2 \,|\, \boldsymbol{w}_{n-1} \in \mathcal{U}\right) \\
&\overset{(a)}{\leq} \mathbb{E}\left(P(\boldsymbol{w}'_{n+m}) \,|\, \boldsymbol{w}_{n-1} \in \mathcal{U}\right) \\
&\quad + \sqrt{\mathbb{E}\left(\left\|\nabla_w P\left(\boldsymbol{w}'_{n+m}\right)\right\|^2 \,|\, \boldsymbol{w}_{n-1} \in \mathcal{U}\right)} \\
&\quad \times \sqrt{\mathbb{E}\left(\left\|\boldsymbol{w}_{n+m} - \boldsymbol{w}'_{n+m}\right\|^2 \,|\, \boldsymbol{w}_{n-1} \in \mathcal{U}\right)} \\
&\quad + \frac{\delta}{2}\mathbb{E}\left(\left\|\boldsymbol{w}_{n+m} - \boldsymbol{w}'_{n+m}\right\|^2 \,|\, \boldsymbol{w}_{n-1} \in \mathcal{U}\right)
\end{aligned} \tag{24.111}$$

That is,

$$\begin{aligned}
&\mathbb{E}\left(P(\boldsymbol{w}_{n+m}) \,|\, \boldsymbol{w}_{n-1} \in \mathcal{U}\right) \\
&\overset{(b)}{\leq} \mathbb{E}\left(P(\boldsymbol{w}'_{n+m}) \,|\, \boldsymbol{w}_{n-1} \in \mathcal{U}\right) \\
&\quad + \sqrt{\mathbb{E}\left(2\|\nabla_w P\left(\boldsymbol{w}_{n-1}\right)\|^2 + 2\delta^2\left\|\boldsymbol{z}_m^{n-1}\right\|^2 \,|\, \boldsymbol{w}_{n-1} \in \mathcal{U}\right)} \\
&\quad \times \sqrt{\mathbb{E}\left(\left\|\boldsymbol{w}_{n+m} - \boldsymbol{w}'_{n+m}\right\|^2 \,|\, \boldsymbol{w}_{n-1} \in \mathcal{U}\right)} \\
&\quad + \frac{\delta}{2}\mathbb{E}\left(\left\|\boldsymbol{w}_{n+m} - \boldsymbol{w}'_{n+m}\right\|^2 \,|\, \boldsymbol{w}_{n-1} \in \mathcal{U}\right) \\
&\overset{(c)}{\leq} \mathbb{E}\left(P(\boldsymbol{w}'_{n+m}) \,|\, \boldsymbol{w}_{n-1} \in \mathcal{U}\right) \\
&\quad + O\left(\mu^{1/2}\right)\sqrt{\mathbb{E}\left(\left\|\boldsymbol{w}_{n+m} - \boldsymbol{w}'_{n+m}\right\|^2 \,|\, \boldsymbol{w}_{n-1} \in \mathcal{U}\right)} \\
&\quad + \frac{\delta}{2}\mathbb{E}\left(\left\|\boldsymbol{w}_{n+m} - \boldsymbol{w}'_{n+m}\right\|^2 \,|\, \boldsymbol{w}_{n-1} \in \mathcal{U}\right) \\
&\overset{(d)}{\leq} \mathbb{E}\left(P(\boldsymbol{w}'_{n+m}) \,|\, \boldsymbol{w}_{n-1} \in \mathcal{U}\right) + O(\mu^{3/2})
\end{aligned} \tag{24.112}$$

where step (a) follows from Cauchy–Schwarz inequality and step (b) from:

$$\begin{aligned}
\left\|\nabla_w P\left(\boldsymbol{w}'_{n+m}\right)\right\|^2 &= \left\|\nabla_w P\left(\boldsymbol{w}_{n-1}\right) + \nabla_w P\left(\boldsymbol{w}'_{n+m}\right) - \nabla_w P\left(\boldsymbol{w}_{n-1}\right)\right\|^2 \\
&\leq 2\|\nabla_w P\left(\boldsymbol{w}_{n-1}\right)\|^2 + 2\left\|\nabla_w P\left(\boldsymbol{w}'_{n+m}\right) - \nabla_w P\left(\boldsymbol{w}_{n-1}\right)\right\|^2 \\
&\leq 2\|\nabla_w P\left(\boldsymbol{w}_{n-1}\right)\|^2 + 2\delta^2\left\|\boldsymbol{w}'_{n+m} - \boldsymbol{w}_{n-1}\right\|^2 \\
&\leq 2\|\nabla_w P\left(\boldsymbol{w}_{n-1}\right)\|^2 + 2\delta^2\|\boldsymbol{z}_m^{n-1}\|^2
\end{aligned} \tag{24.113}$$

Moreover, step (c) uses (24.81) and the fact that the squared norm of the gradient of the risk function within $\mathcal{U}$ is bounded by $O(\mu)$. Step (d) follows from (24.80).

24.C DESCENT AWAY FROM STRICT SADDLE POINTS

We are now ready to establish Theorem 24.2. The argument relies on the results from Appendix 24.B on the short-term model and its accuracy. The reader is advised to study that appendix first before continuing.

From Corollary 24.1, we have:

$$\mathbb{E}\left(P(\boldsymbol{w}_{n+m})\,|\,\boldsymbol{w}_{n-1}\in\mathcal{U}\right)\leq\mathbb{E}\left(P(\boldsymbol{w}'_{n+m})\,|\,\boldsymbol{w}_{n-1}\in\mathcal{U}\right)+O(\mu^{3/2}) \tag{24.114}$$

for any $m+1\leq T/\mu$. We will therefore proceed by studying the behavior of the approximate risk $\mathbb{E}\,(P(\boldsymbol{w}'_{n+m-1})\,|\,\boldsymbol{w}_{n-1}\in\mathcal{U})$ and then add the approximation error $O(\mu^{3/2})$ to the end result.

From the result of the earlier Prob. 10.12 for functions with δ_2-Lipschitz Hessian matrices we can write:

$$P(\boldsymbol{w}'_{n+m})\leq P(\boldsymbol{w}_{n-1})-(\nabla_{w^{\mathsf{T}}}P(\boldsymbol{w}_{n-1}))^{\mathsf{T}}\boldsymbol{z}_m^{n-1}+\frac{1}{2}\|\boldsymbol{z}_m^{n-1}\|^2_{\nabla_w^2 P(\boldsymbol{w}_{n-1})}+\frac{\delta_2}{6}\|\boldsymbol{z}_m^{n-1}\|^3 \tag{24.115}$$

where the notation $\|a\|_A^2$ stands for $a^{\mathsf{T}}Aa$. We will bound each term appearing on the right-hand side of (24.115). From (24.75) we find after conditioning on $\boldsymbol{\mathcal{F}}_{m+n-1}$:

$$\begin{aligned}
&\mathbb{E}\left(\boldsymbol{z}_m^{n-1}\,|\,\boldsymbol{\mathcal{F}}_{m+n-1}\right)\\
&=\left(I-\mu\nabla_w^2P(\boldsymbol{w}_{n-1})\right)\boldsymbol{z}_{m-1}^{n-1}+\mu\nabla_{w^{\mathsf{T}}}P(\boldsymbol{w}_{n-1})+\mu\mathbb{E}\left(\boldsymbol{g}_{n+m}\,|\,\boldsymbol{\mathcal{F}}_{m+n-1}\right)\\
&\overset{(24.21a)}{=}\left(I-\mu\nabla_w^2P(\boldsymbol{w}_{n-1})\right)\boldsymbol{z}_{m-1}^{n-1}+\mu\nabla_{w^{\mathsf{T}}}P(\boldsymbol{w}_{n-1})
\end{aligned} \tag{24.116}$$

Note that $\boldsymbol{\mathcal{F}}_{m+n-1}$ captures the information in all random variables $\{\boldsymbol{w}_j\}$ up to time $j\leq m+n-1$. In a similar manner, the filtration $\boldsymbol{\mathcal{F}}_{n-1}$ captures the information available up to time $n-1$. Hence, we can write:

$$\boldsymbol{\mathcal{F}}_{m+n-1}=\boldsymbol{\mathcal{F}}_{n-1}\cup\text{filtration}\Big(\boldsymbol{w}_n,\ldots,\boldsymbol{w}_{n+m-1}\Big) \tag{24.117}$$

Taking expectation again of (24.116) conditioned on $\boldsymbol{\mathcal{F}}_{n-1}$ removes the elements that appear within filtration$(\boldsymbol{w}_n,\ldots,\boldsymbol{w}_{n+m-1})$ and yields:

$$\mathbb{E}\left(\boldsymbol{z}_m^{n-1}\,|\,\boldsymbol{\mathcal{F}}_{n-1}\right)=\left(I-\mu\nabla_w^2\,P(\boldsymbol{w}_{n-1})\right)\mathbb{E}\left(\boldsymbol{z}_{m-1}^{n-1}\,|\,\boldsymbol{\mathcal{F}}_{n-1}\right)+\mu\nabla_{w^{\mathsf{T}}}P(\boldsymbol{w}_{n-1}) \tag{24.118}$$

Since $\boldsymbol{z}_{-1}^{n-1}=0$, iterating starting at $m=0$ yields:

$$\mathbb{E}\left(\boldsymbol{z}_m^{n-1}\,|\,\boldsymbol{\mathcal{F}}_{n-1}\right)=\mu\left(\sum_{j=0}^{m}\left(I-\mu\nabla_w^2\,P(\boldsymbol{w}_{n-1})\right)^j\right)\nabla_{w^{\mathsf{T}}}P(\boldsymbol{w}_{n-1}) \tag{24.119}$$

This result allows us to evaluate the linear term appearing in (24.115) as:

$$\begin{aligned}
&-\mathbb{E}\left((\nabla_{w^{\mathsf{T}}}P(\boldsymbol{w}_{n-1}))^{\mathsf{T}}\boldsymbol{z}_m^{n-1}\,|\,\boldsymbol{\mathcal{F}}_{n-1}\right)\\
&=-(\nabla_{w^{\mathsf{T}}}P(\boldsymbol{w}_{n-1}))^{\mathsf{T}}\mathbb{E}\left(\boldsymbol{z}_m^{n-1}\,|\,\boldsymbol{\mathcal{F}}_{n-1}\right)\\
&\overset{(24.119)}{=}-\mu(\nabla_{w^{\mathsf{T}}}P(\boldsymbol{w}_{n-1}))^{\mathsf{T}}\left(\sum_{j=0}^{m}\left(I-\mu\nabla_w^2P(\boldsymbol{w}_{n-1})\right)^j\right)\nabla_{w^{\mathsf{T}}}P(\boldsymbol{w}_{n-1})\\
&=-\mu\|\nabla_{w^{\mathsf{T}}}P(\boldsymbol{w}_{n-1})\|^2_{\boldsymbol{Z}_m^{n-1}}
\end{aligned} \tag{24.120}$$

with the weighting matrix denoted by

$$\boldsymbol{Z}_m^{n-1} \triangleq \sum_{j=0}^{m} \left(I - \mu\nabla_w^2 P(\boldsymbol{w}_{n-1})\right)^j \tag{24.121}$$

To study the quadratic term in (24.115), we introduce the eigenvalue decomposition of the Hessian matrix around the iterate at time $n-1$:

$$\nabla_w^2 P(\boldsymbol{w}_{n-1}) \triangleq \boldsymbol{V}_{n-1}\boldsymbol{\Lambda}_{n-1}\boldsymbol{V}_{n-1}^{\mathsf{T}} \tag{24.122}$$

where $\boldsymbol{V}_{n-1}$ is orthogonal and $\boldsymbol{\Lambda}_{n-1}$ is diagonal. The decomposition motivates the transformations:

$$\overline{\boldsymbol{z}}_m^{n-1} \triangleq \boldsymbol{V}_{n-1}^{\mathsf{T}}\boldsymbol{z}_m^{n-1} \tag{24.123a}$$

$$\overline{\boldsymbol{g}}_{n+m} \triangleq \boldsymbol{V}_{n-1}^{\mathsf{T}}\boldsymbol{g}_{n+m} \tag{24.123b}$$

$$\overline{\nabla_{w^{\mathsf{T}}} P}(\boldsymbol{w}_{n-1}) \triangleq \boldsymbol{V}_{n-1}^{\mathsf{T}}\Big(\nabla_{w^{\mathsf{T}}} P(\boldsymbol{w}_{n-1})\Big) \tag{24.123c}$$

$$\overline{\boldsymbol{Z}}_m^{n-1} \triangleq \boldsymbol{V}_{n-1}^{\mathsf{T}}\boldsymbol{Z}_m^{n-1}\boldsymbol{V}_{n-1} = \sum_{j=0}^{m} (I - \mu\boldsymbol{\Lambda}_{n-1})^j \tag{24.123d}$$

so that

$$\left\|\boldsymbol{z}_m^{n-1}\right\|^2_{\nabla_w^2 P(\boldsymbol{w}_{n-1})} = \left\|\overline{\boldsymbol{z}}_m^{n-1}\right\|^2_{\boldsymbol{\Lambda}_{n-1}} \tag{24.124}$$

and recursion (24.75) is diagonalized into

$$\overline{\boldsymbol{z}}_m^{n-1} = (I - \mu\boldsymbol{\Lambda}_{n-1})\,\overline{\boldsymbol{z}}_{m-1}^{n-1} + \mu\overline{\nabla_{w^{\mathsf{T}}} P}(\boldsymbol{w}_{n-1}) + \mu\overline{\boldsymbol{g}}_{n+m} \tag{24.125}$$

Applying the same transformation to the conditional mean recursion (24.118) gives

$$\mathbb{E}\left(\overline{\boldsymbol{z}}_m^{n-1} \,|\, \boldsymbol{\mathcal{F}}_{n-1}\right) = (I - \mu\boldsymbol{\Lambda}_{n-1})\,\mathbb{E}\left(\overline{\boldsymbol{z}}_{m-1}^{n-1} \,|\, \boldsymbol{\mathcal{F}}_{n-1}\right) + \mu\overline{\nabla_{w^{\mathsf{T}}} P}(\boldsymbol{w}_{n-1}) \tag{24.126}$$

Subtracting from (24.125) leads to the centered recursion

$$\begin{aligned} &\overline{\boldsymbol{z}}_m^{n-1} - \mathbb{E}\left(\overline{\boldsymbol{z}}_m^{n-1} \,|\, \boldsymbol{\mathcal{F}}_{n-1}\right) \\ &= (I - \mu\boldsymbol{\Lambda}_{n-1})\left(\overline{\boldsymbol{z}}_{m-1}^{n-1} - \mathbb{E}\left(\overline{\boldsymbol{z}}_{m-1}^{n-1} \,|\, \boldsymbol{\mathcal{F}}_{n-1}\right)\right) + \mu\overline{\boldsymbol{g}}_{n+m} \end{aligned} \tag{24.127}$$

where the driving term involving the gradient of the risk function has been canceled. For compactness of notation, we introduce the (conditionally) zero-mean centered random variable:

$$\check{\boldsymbol{z}}_m^{n-1} \triangleq \overline{\boldsymbol{z}}_m^{n-1} - \mathbb{E}\left(\overline{\boldsymbol{z}}_m^{n-1} \,|\, \boldsymbol{\mathcal{F}}_{n-1}\right) \tag{24.128}$$

so that

$$\boxed{\check{\boldsymbol{z}}_m^{n-1} = (I - \mu\boldsymbol{\Lambda}_{n-1})\,\check{\boldsymbol{z}}_{m-1}^{n-1} + \mu\overline{\boldsymbol{g}}_{n+m}} \tag{24.129}$$

Note further that the weighted second-order moment of the centered variable satisfies:

$$\mathbb{E}\left(\left\|\check{\boldsymbol{z}}_m^{n-1}\right\|^2_{\boldsymbol{\Lambda}_{n-1}} \,|\, \boldsymbol{\mathcal{F}}_{n-1}\right) = \mathbb{E}\left(\left\|\overline{\boldsymbol{z}}_m^{n-1}\right\|^2_{\boldsymbol{\Lambda}_{n-1}} \,|\, \boldsymbol{\mathcal{F}}_{n-1}\right) - \left\|\mathbb{E}\left(\overline{\boldsymbol{z}}_m^{n-1} \,|\, \boldsymbol{\mathcal{F}}_{n-1}\right)\right\|^2_{\boldsymbol{\Lambda}_{n-1}} \tag{24.130}$$

This result is simply a manifestation of the general property that $\sigma_x^2 = \mathbb{E}\,\boldsymbol{x}^2 - (\mathbb{E}\,\boldsymbol{x})^2$ for any random variable $\boldsymbol{x}$. Hence, we have

$$\begin{aligned}\mathbb{E}\left(\left\|\boldsymbol{z}_m^{n-1}\right\|^2_{\nabla_w^2 P(\boldsymbol{w}_{n-1})} \,|\, \boldsymbol{\mathcal{F}}_{n-1}\right) &= \mathbb{E}\left(\left\|\overline{\boldsymbol{z}}_m^{n-1}\right\|^2_{\boldsymbol{\Lambda}_{n-1}} \,|\, \boldsymbol{\mathcal{F}}_{n-1}\right) \\ &= \mathbb{E}\left(\left\|\overline{\boldsymbol{z}}_m^{n-1}\right\|^2_{\boldsymbol{\Lambda}_{n-1}} \,|\, \boldsymbol{\mathcal{F}}_{n-1}\right) + \left\|\mathbb{E}\left(\overline{\boldsymbol{z}}_m^{n-1} \,|\, \boldsymbol{\mathcal{F}}_{n-1}\right)\right\|^2_{\boldsymbol{\Lambda}_{n-1}}\end{aligned} \tag{24.131}$$

We examine the two terms on the right-hand side. Note first that

$$\begin{aligned}&\left\|\mathbb{E}\left(\overline{\boldsymbol{z}}_m^{n-1} \,|\, \boldsymbol{\mathcal{F}}_{n-1}\right)\right\|^2_{\boldsymbol{\Lambda}_{n-1}} \\ &= \left\|\mathbb{E}\left(\boldsymbol{V}_{n-1}^{\mathsf{T}}\boldsymbol{z}_m^{n-1} \,|\, \boldsymbol{\mathcal{F}}_{n-1}\right)\right\|^2_{\boldsymbol{\Lambda}_{n-1}} \\ &\overset{(24.119)}{=} \mu^2\left\|\boldsymbol{V}_{n-1}^{\mathsf{T}}\boldsymbol{Z}_m^{n-1}\nabla_{w^{\mathsf{T}}} P(\boldsymbol{w}_{n-1})\right\|^2_{\boldsymbol{\Lambda}_{n-1}} \\ &= \mu^2\left\|\overline{\boldsymbol{Z}}_m^{n-1}\overline{\nabla_{w^{\mathsf{T}}} P(\boldsymbol{w}_{n-1})}\right\|^2_{\boldsymbol{\Lambda}_{n-1}} \\ &= \mu^2\left(\overline{\nabla_{w^{\mathsf{T}}} P(\boldsymbol{w}_{n-1})}\right)^{\mathsf{T}}\, \overline{\boldsymbol{Z}}_m^{n-1}\boldsymbol{\Lambda}_{n-1}\overline{\boldsymbol{Z}}_m^{n-1}\, \overline{\nabla_{w^{\mathsf{T}}} P(\boldsymbol{w}_{n-1})}\end{aligned} \tag{24.132}$$

We order the eigenvalues of $\nabla_w^2 P(\boldsymbol{w}_{n-1})$ such that its eigen-decomposition has a block structure of the form:

$$\boldsymbol{V}_{n-1} = \begin{bmatrix} \boldsymbol{V}_{n-1}^{>0} & \boldsymbol{V}_{n-1}^{=0} & \boldsymbol{V}_{n-1}^{<0} \end{bmatrix} \tag{24.133a}$$

$$\boldsymbol{\Lambda}_{n-1} = \begin{bmatrix} \boldsymbol{\Lambda}_{n-1}^{>0} & 0 & 0 \\ 0 & \boldsymbol{\Lambda}_{n-1}^{=0} & 0 \\ 0 & 0 & \boldsymbol{\Lambda}_{n-1}^{<0} \end{bmatrix} \tag{24.133b}$$

with $\delta I \geq \boldsymbol{\Lambda}_{n-1}^{>0} \geq 0$, $\boldsymbol{\Lambda}_{n-1}^{<0} < 0$, and $\boldsymbol{\Lambda}_{n-1}^{=0} = 0$. We are collecting the zero eigenvalues into $\boldsymbol{\Lambda}_{n-1}^{=0}$. Note that since $\nabla_w^2 P(\boldsymbol{w}_{n-1})$ is random, the decomposition itself is random as well. Nevertheless, it exists with probability 1. We also decompose the transformed gradient vector with appropriate dimensions:

$$\overline{\nabla_{w^{\mathsf{T}}} P(\boldsymbol{w}_{n-1})} = \begin{bmatrix} \overline{\nabla_{w^{\mathsf{T}}} P(\boldsymbol{w}_{n-1})}^{>0} \\ \overline{\nabla_{w^{\mathsf{T}}} P(\boldsymbol{w}_{n-1})}^{=0} \\ \overline{\nabla_{w^{\mathsf{T}}} P(\boldsymbol{w}_{n-1})}^{<0} \end{bmatrix} \tag{24.134}$$

and the transformed matrix

$$\begin{aligned}\overline{\boldsymbol{Z}}_m^{n-1} &\triangleq \begin{bmatrix} (\overline{\boldsymbol{Z}}_m^{n-1})^{>0} & 0 & 0 \\ 0 & (\overline{\boldsymbol{Z}}_m^{n-1})^{=0} & 0 \\ 0 & 0 & (\overline{\boldsymbol{Z}}_m^{n-1})^{<0} \end{bmatrix} \\ &= \begin{bmatrix} \sum_{j=0}^{m}(I - \mu\boldsymbol{\Lambda}_{n-1}^{>0})^j & 0 & 0 \\ 0 & I & 0 \\ 0 & 0 & \sum_{j=0}^{m}(I - \mu\boldsymbol{\Lambda}_{n-1}^{<0})^j \end{bmatrix}\end{aligned} \tag{24.135}$$

We can then decompose (24.132):

$$
\begin{aligned}
&\left\|\mathbb{E}\left(\overline{\boldsymbol{z}}_m^{n-1}\,|\,\boldsymbol{\mathfrak{F}}_{n-1}\right)\right\|^2_{\boldsymbol{\Lambda}_{n-1}}\\
&= \mu^2\left(\overline{\nabla_{w^\mathsf{T}} P(\boldsymbol{w}_{n-1})}^{>0}\right)^\mathsf{T}\left(\overline{\boldsymbol{Z}}_m^{n-1}\right)^{>0}\boldsymbol{\Lambda}_{n-1}^{>0}\left(\overline{\boldsymbol{Z}}_m^{n-1}\right)^{>0}\overline{\nabla_{w^\mathsf{T}} P(\boldsymbol{w}_{n-1})}^{>0}\\
&\quad+\mu^2\left(\overline{\nabla_{w^\mathsf{T}} P(\boldsymbol{w}_{n-1})}^{<0}\right)^\mathsf{T}\left(\overline{\boldsymbol{Z}}_m^{n-1}\right)^{<0}\boldsymbol{\Lambda}_{n-1}^{<0}\left(\overline{\boldsymbol{Z}}_m^{n-1}\right)^{<0}\overline{\nabla_{w^\mathsf{T}} P(\boldsymbol{w}_{n-1})}^{<0}\\
&\overset{(a)}{\leq} \mu^2\left(\overline{\nabla_{w^\mathsf{T}} P(\boldsymbol{w}_{n-1})}^{>0}\right)^\mathsf{T}\left(\overline{\boldsymbol{Z}}_m^{n-1}\right)^{>0}\boldsymbol{\Lambda}_{n-1}^{>0}\left(\overline{\boldsymbol{Z}}_m^{n-1}\right)^{>0}\overline{\nabla_{w^\mathsf{T}} P(\boldsymbol{w}_{n-1})}^{>0}\\
&\overset{(b)}{\leq} \mu^2\left(\overline{\nabla_{w^\mathsf{T}} P(\boldsymbol{w}_{n-1})}^{>0}\right)^\mathsf{T}\left(\overline{\boldsymbol{Z}}_\infty^{n-1}\right)^{>0}\boldsymbol{\Lambda}_{n-1}^{>0}\left(\overline{\boldsymbol{Z}}_m^{n-1}\right)^{>0}\overline{\nabla_{w^\mathsf{T}} P(\boldsymbol{w}_{n-1})}^{>0}
\end{aligned}
\tag{24.136}
$$

That is,

$$
\begin{aligned}
&\left\|\mathbb{E}\left(\overline{\boldsymbol{z}}_m^{n-1}\,|\,\boldsymbol{\mathfrak{F}}_{n-1}\right)\right\|^2_{\boldsymbol{\Lambda}_{n-1}}\\
&\overset{(c)}{=} \mu^2\left(\overline{\nabla_{w^\mathsf{T}} P(\boldsymbol{w}_{n-1})}^{>0}\right)^\mathsf{T}\left(\mu\boldsymbol{\Lambda}_{n-1}^{>0}\right)^{-1}\boldsymbol{\Lambda}_{n-1}^{>0}\left(\overline{\boldsymbol{Z}}_m^{n-1}\right)^{>0}\overline{\nabla_{w^\mathsf{T}} P(\boldsymbol{w}_{n-1})}^{>0}\\
&= \mu\left(\overline{\nabla_{w^\mathsf{T}} P(\boldsymbol{w}_{n-1})}^{>0}\right)^\mathsf{T}\left(\overline{\boldsymbol{Z}}_m^{n-1}\right)^{>0}\overline{\nabla_{w^\mathsf{T}} P(\boldsymbol{w}_{n-1})}^{>0}\\
&\overset{(d)}{\leq} \mu\left(\overline{\nabla_{w^\mathsf{T}} P(\boldsymbol{w}_{n-1})}^{>0}\right)^\mathsf{T}\left(\overline{\boldsymbol{Z}}_m^{n-1}\right)^{>0}\overline{\nabla_{w^\mathsf{T}} P(\boldsymbol{w}_{n-1})}^{>0}\\
&\quad+\mu\left(\overline{\nabla_{w^\mathsf{T}} P(\boldsymbol{w}_{n-1})}^{<0}\right)^\mathsf{T}\left(\overline{\boldsymbol{Z}}_m^{n-1}\right)^{<0}\overline{\nabla_{w^\mathsf{T}} P(\boldsymbol{w}_{n-1})}^{<0}\\
&\leq \mu\overline{\nabla_{w^\mathsf{T}} P(\boldsymbol{w}_{n-1})}^\mathsf{T}\,\overline{\boldsymbol{Z}}_m^{n-1}\,\overline{\nabla_{w^\mathsf{T}} P(\boldsymbol{w}_{n-1})}\\
&= \mu\left\|\overline{\nabla_{w^\mathsf{T}} P(\boldsymbol{w}_{n-1})}\right\|^2_{\overline{\boldsymbol{Z}}_m^{n-1}}
\end{aligned}
\tag{24.137}
$$

where step (a) follows from $\boldsymbol{\Lambda}_{n-1}^{<0} < 0$, and step (b) follows from:

$$
\sum_{j=0}^{m}\left(I-\mu\boldsymbol{\Lambda}_{n-1}^{>0}\right)^j \leq \sum_{j=0}^{\infty}\left(I-\mu\boldsymbol{\Lambda}_{n-1}^{>0}\right)^j \tag{24.138}
$$

for $\mu < 1/\delta$. Step (c) follows from the formula for the geometric matrix series, and step (d) follows from:

$$
\left(\overline{\nabla_{w^\mathsf{T}} P(\boldsymbol{w}_{n-1})}^{<0}\right)^\mathsf{T}\left(\overline{\boldsymbol{Z}}_m^{n-1}\right)^{<0}\overline{\nabla_{w^\mathsf{T}} P(\boldsymbol{w}_{n-1})}^{<0} \geq 0 \tag{24.139}
$$

Comparing (24.137) to (24.120), we find that we can bound:

$$
-\mathbb{E}\left(\left(\nabla_{w^\mathsf{T}} P(\boldsymbol{w}_{n-1})\right)^\mathsf{T}\boldsymbol{z}_m^{n-1}\,|\,\boldsymbol{\mathfrak{F}}_{n-1}\right)+\left\|\mathbb{E}\left(\overline{\boldsymbol{z}}_m^{n-1}\,|\,\boldsymbol{\mathfrak{F}}_{n-1}\right)\right\|^2_{\boldsymbol{\Lambda}_{n-1}} \leq 0 \tag{24.140}
$$

so that from (24.115) and (24.130) we obtain:

$$
\begin{aligned}
&\mathbb{E}\left(P(\boldsymbol{w}'_{n+m})\,|\,\boldsymbol{\mathfrak{F}}_{n-1}\right)\\
&\leq P(\boldsymbol{w}_{n-1})+\frac{1}{2}\mathbb{E}\left(\left\|\overline{\boldsymbol{z}}_m^{n-1}\right\|^2_{\boldsymbol{\Lambda}_{n-1}}\,|\,\boldsymbol{\mathfrak{F}}_{n-1}\right)+\frac{\delta_2}{6}\mathbb{E}\left(\left\|\boldsymbol{z}_m^{n-1}\right\|^3\,|\,\boldsymbol{\mathfrak{F}}_{n-1}\right)
\end{aligned}
\tag{24.141}
$$

We proceed with the now simplified quadratic term. We square both sides of (24.129) under an arbitrary diagonal weighting matrix $\boldsymbol{\Sigma}_m$ (chosen to be deterministic when conditioned on $\boldsymbol{w}_{n-1}$ and $\boldsymbol{w}_{n+m-1}$):

$$\begin{aligned}\left\|\breve{\boldsymbol{z}}_m^{n-1}\right\|_{\boldsymbol{\Sigma}_m}^2 &= \left\|(I-\mu\boldsymbol{\Lambda}_{n-1})\,\breve{\boldsymbol{z}}_{m-1}^{n-1}+\mu\overline{\boldsymbol{g}}_{n+m}\right\|_{\boldsymbol{\Sigma}_m}^2\\ &= \left\|(I-\mu\boldsymbol{\Lambda}_{n-1})\,\breve{\boldsymbol{z}}_{m-1}^{n-1}\right\|_{\boldsymbol{\Sigma}_m}^2+\mu^2\left\|\overline{\boldsymbol{g}}_{n+m}\right\|_{\boldsymbol{\Sigma}_m}^2\\ &\quad+2\mu(\breve{\boldsymbol{z}}_{m-1}^{n-1})^{\mathsf{T}}\,(I-\mu\boldsymbol{\Lambda}_{n-1})\,\boldsymbol{\Sigma}_m\overline{\boldsymbol{g}}_{n+m}\end{aligned} \tag{24.142}$$

Note that upon conditioning on $\boldsymbol{\mathcal{F}}_{m+n-1}$, all elements of the cross term, aside from $\overline{\boldsymbol{g}}_{n+m}$, become deterministic, and as such the term disappears when taking expectations. We obtain using the covariance matrix of the gradient noise process:

$$\begin{aligned}&\mathbb{E}\left(\left\|\breve{\boldsymbol{z}}_m^{n-1}\right\|_{\boldsymbol{\Sigma}_m}^2\,|\,\boldsymbol{\mathcal{F}}_{m+n-1}\right)\\ &=\left\|(I-\mu\boldsymbol{\Lambda}_{n-1})\,\breve{\boldsymbol{z}}_{m-1}^{n-1}\right\|_{\boldsymbol{\Sigma}_m}^2+\mu^2\mathbb{E}\left(\left\|\overline{\boldsymbol{g}}_{n+m}\right\|_{\boldsymbol{\Sigma}_m}^2\,|\,\boldsymbol{\mathcal{F}}_{m+n-1}\right)\\ &=\left\|\breve{\boldsymbol{z}}_{m-1}^{n-1}\right\|_{\boldsymbol{\Sigma}_m-2\mu\boldsymbol{\Lambda}_{n-1}\boldsymbol{\Sigma}_m+\mu^2\boldsymbol{\Lambda}_{n-1}\boldsymbol{\Sigma}_m\boldsymbol{\Lambda}_{n-1}}^2\\ &\quad+\mu^2\mathrm{Tr}\left(\boldsymbol{V}_{n-1}\boldsymbol{\Sigma}_m\boldsymbol{V}_{n-1}^{\mathsf{T}}R_g\left(\boldsymbol{w}_{n+m-1}\right)\right)\\ &=\left\|\breve{\boldsymbol{z}}_{m-1}^{n-1}\right\|_{\boldsymbol{\Sigma}_m-2\mu\boldsymbol{\Lambda}_{n-1}\boldsymbol{\Sigma}_m}^2+\mu^2\mathrm{Tr}\left(\boldsymbol{V}_{n-1}\boldsymbol{\Sigma}_m\boldsymbol{V}_{n-1}^{\mathsf{T}}R_g\left(\boldsymbol{w}_{n-1}\right)\right)\\ &\quad+\mu^2\mathrm{Tr}\left(\boldsymbol{V}_{n-1}\boldsymbol{\Sigma}_m\boldsymbol{V}_{n-1}^{\mathsf{T}}\left(R_g\left(\boldsymbol{w}_{n+m-1}\right)-R_g\left(\boldsymbol{w}_{n-1}\right)\right)\right)\\ &\quad+\mu^2\left\|\breve{\boldsymbol{z}}_{m-1}^{n-1}\right\|_{\boldsymbol{\Lambda}_{n-1}\boldsymbol{\Sigma}_m\boldsymbol{\Lambda}_{n-1}}^2\end{aligned} \tag{24.143}$$

We proceed to bound the terms in (24.143). First, we have:

$$\begin{aligned}&\mathrm{Tr}\left(\boldsymbol{V}_{n-1}\boldsymbol{\Sigma}_m\boldsymbol{V}_{n-1}^{\mathsf{T}}\left(R_g\left(\boldsymbol{w}_{n+m-1}\right)-R_g\left(\boldsymbol{w}_{n-1}\right)\right)\right)\\ &\overset{(a)}{\leq}\left\|\boldsymbol{V}_{n-1}\boldsymbol{\Sigma}_m\boldsymbol{V}_{n-1}^{\mathsf{T}}\right\|\left\|R_g\left(\boldsymbol{w}_{n+m-1}\right)-R_g\left(\boldsymbol{w}_{n-1}\right)\right\|\\ &\overset{(b)}{\leq}\rho\left(\boldsymbol{\Sigma}_m\right)\beta_R\left\|\boldsymbol{d}_{m-1}^{n-1}\right\|^{\eta}\end{aligned} \tag{24.144}$$

where step (a) follows from the Cauchy–Schwarz inequality since $\mathrm{Tr}(A^{\mathsf{T}}B)$ is an inner product over the space of symmetric matrices and, hence, $|\mathrm{Tr}(A^{\mathsf{T}}B)|\leq\|A\|\|B\|$, and step (b) follows from the Lipschitz condition (24.24b). Moreover, the notation $\rho(A)$ denotes the spectral radius of matrix A. For the last term in (24.143) we have:

$$\begin{aligned}\left\|\breve{\boldsymbol{z}}_{m-1}^{n-1}\right\|_{\boldsymbol{\Lambda}_{n-1}\boldsymbol{\Sigma}_m\boldsymbol{\Lambda}_{n-1}}^2&\leq\rho\left(\boldsymbol{\Lambda}_{n-1}\boldsymbol{\Sigma}_m\boldsymbol{\Lambda}_{n-1}\right)\left\|\breve{\boldsymbol{z}}_{m-1}^{n-1}\right\|^2\\ &\leq\delta^2\rho\left(\boldsymbol{\Sigma}_m\right)\left\|\breve{\boldsymbol{z}}_{m-1}^{n-1}\right\|^2\end{aligned} \tag{24.145}$$

since $-\delta I\leq\boldsymbol{\Lambda}_{n-1}\leq\delta I$. We conclude that

$$\begin{aligned}&\mathbb{E}\left(\left\|\breve{\boldsymbol{z}}_m^{n-1}\right\|_{\boldsymbol{\Sigma}_m}^2\,|\,\boldsymbol{\mathcal{F}}_{n-1}\right)\\ &\leq\mathbb{E}\left(\left\|\breve{\boldsymbol{z}}_{m-1}^{n-1}\right\|_{\boldsymbol{\Sigma}_m-2\mu\boldsymbol{\Lambda}_{n-1}\boldsymbol{\Sigma}_m}^2\,|\,\boldsymbol{\mathcal{F}}_{n-1}\right)+\mu^2\mathrm{Tr}\left(\boldsymbol{V}_{n-1}\boldsymbol{\Sigma}_m\boldsymbol{V}_{n-1}^{\mathsf{T}}R_g\left(\boldsymbol{w}_{n-1}\right)\right)\\ &\quad+\mu^2\rho\left(\boldsymbol{\Sigma}_m\right)\mathbb{E}\left(\boldsymbol{q}_{m-1}^{n-1}\,|\,\boldsymbol{\mathcal{F}}_{n-1}\right)\end{aligned} \tag{24.146}$$

where we introduced

$$q_{m-1}^{n-1} \triangleq \beta_R \left\| d_{m-1}^{n-1} \right\|^{\eta} + \delta^2 \left\| \check{z}_{m-1}^{n-1} \right\|^2 \tag{24.147}$$

For brevity, we introduce the quantities:

$$D \triangleq I - 2\mu \Lambda_{n-1} \tag{24.148}$$

$$Y \triangleq V_{n-1}^{\mathsf{T}} R_g\left(w_{n-1}\right) V_{n-1} \tag{24.149}$$

and write:

$$\begin{aligned} &\mathbb{E}\left(\left\|\check{z}_m^{n-1}\right\|_{\Sigma_m}^2 \mid \mathcal{F}_{n-1}\right) \\ &= \mathbb{E}\left(\left\|\check{z}_{m-1}^{n-1}\right\|_{D\Sigma_m}^2 \mid \mathcal{F}_{n-1}\right) + \mu^2 \mathrm{Tr}\left(\Sigma_m Y\right) + \mu^2 \rho\left(\Sigma_m\right) \mathbb{E}\left(q_{m-1}^{n-1} \mid \mathcal{F}_{n-1}\right) \end{aligned} \tag{24.150}$$

where $\check{z}_{-1}^{n-1} = 0$. For convenience, we rewrite this relation by replacing the running index m by j:

$$\begin{aligned} &\mathbb{E}\left(\left\|\check{z}_j^{n-1}\right\|_{\Sigma_j}^2 \mid \mathcal{F}_{n-1}\right) \\ &= \mathbb{E}\left(\left\|\check{z}_{j-1}^{n-1}\right\|_{D\Sigma_j}^2 \mid \mathcal{F}_{n-1}\right) + \mu^2 \mathrm{Tr}\left(\Sigma_j Y\right) + \mu^2 \rho\left(\Sigma_j\right) \mathbb{E}\left(q_{j-1}^{n-1} \mid \mathcal{F}_{n-1}\right) \end{aligned} \tag{24.151}$$

We are free to select the diagonal weighting matrices Σ_j at every iteration j. We wish to run the recursion up to $j = m$. We select the weighting matrices as follows:

$$\Sigma_0 = D^m \Lambda_{n-1} \tag{24.152a}$$

$$\Sigma_1 = D^{m-1} \Lambda_{n-1} \tag{24.152b}$$

$$\Sigma_2 = D^{m-2} \Lambda_{n-1} \tag{24.152c}$$

$$\vdots = \vdots$$

$$\Sigma_m = \Lambda_{n-1} \tag{24.152d}$$

That is, we are setting $\Sigma_j = D^{m-j} \Lambda_{n-1}$ for any $j \le m$. Observe that the matrices D and Λ commute so that it also holds $\Sigma_j = \Lambda_{n-1} D^{m-j}$. Note further that, under this construction, it holds that

$$D\Sigma_j = \Sigma_{j-1} \tag{24.153}$$

Iterating (24.151) over $j \ge 0$ and until $j = m$ gives (we are removing the conditioning on $\mathcal{F}_{n-1}$ to lighten the notation)

$$\begin{aligned} \mathbb{E}\left(\|\check{z}_0^{n-1}\|_{\Sigma_0}^2\right) &= \mu^2 \mathrm{Tr}\left(\Sigma_0 Y\right) + \mu^2 \rho\left(\Sigma_0\right) \mathbb{E}\, q_{-1}^{n-1} \\ &= \mu^2 \mathrm{Tr}\left(D^m \Lambda_{n-1} Y\right) + \mu^2 \rho\left(D^m \Lambda_{n-1}\right) \mathbb{E}\, q_{-1}^{n-1} \end{aligned} \tag{24.154}$$

and

$$\begin{aligned} \mathbb{E}\left(\|\check{z}_1^{n-1}\|_{\Sigma_1}^2\right) &= \mathbb{E}\left(\|\check{z}_0^{n-1}\|_{D\Sigma_1}^2\right) + \mu^2 \mathrm{Tr}\left(\Sigma_1 Y\right) + \mu^2 \rho\left(\Sigma_1\right) \mathbb{E}\, q_0^{n-1} \\ &\overset{(24.153)}{=} \mathbb{E}\left(\|\check{z}_0^{n-1}\|_{\Sigma_0}^2\right) + \mu^2 \mathrm{Tr}\left(\Sigma_1 Y\right) + \mu^2 \rho\left(\Sigma_1\right) \mathbb{E}\, q_0^{n-1} \\ &\overset{(24.154)}{=} \mu^2 \mathrm{Tr}\left(D^m \Lambda_{n-1} Y\right) + \mu^2 \rho\left(D^m \Lambda_{n-1}\right) \mathbb{E}\, q_{-1}^{n-1} \\ &\quad + \mu^2 \mathrm{Tr}\left(D^{m-1} \Lambda_{n-1} Y\right) + \mu^2 \rho\left(D^{m-1} \Lambda_{n-1}\right) \mathbb{E}\, q_0^{n-1} \end{aligned} \tag{24.155}$$

and so on. Continuing in this manner until $j = m$ gives the expression:

$$\begin{aligned}
&\mathbb{E}\left(\|\check{z}_m^{n-1}\|^2_{\Lambda_{n-1}} \,|\, \mathcal{F}_{n-1}\right) \\
&= \mu^2 \sum_{j=0}^{m} \text{Tr}\left(\Lambda_{n-1} D^{m-j} Y\right) + \mu^2 \sum_{j=0}^{m} \rho\left(\Lambda_{n-1} D^{m-j}\right) \cdot \mathbb{E}\left(q_{j-1}^{n-1} \,|\, \mathcal{F}_{n-1}\right) \\
&= \mu^2 \text{Tr}\left\{\Lambda_{n-1}\left(\sum_{j=0}^{m} D^{m-j}\right) Y\right\} + \mu^2 \sum_{j=0}^{m} \rho\left(\Lambda_{n-1} D^{m-j}\right) \cdot \mathbb{E}\left(q_{j-1}^{n-1} \,|\, \mathcal{F}_{n-1}\right) \\
&= \mu^2 \text{Tr}\left\{\Lambda_{n-1}\left(\sum_{j=0}^{m} D^{j}\right) Y\right\} + \mu^2 \sum_{j=0}^{m} \rho\left(\Lambda_{n-1} D^{j}\right) \cdot \mathbb{E}\left(q_{m-j-1}^{n-1} \,|\, \mathcal{F}_{n-1}\right)
\end{aligned} \tag{24.156}$$

where we employed a change of variables from $m - j$ to j in the last step. Our objective is to show that the first term on the right-hand side yields sufficient descent (i.e., will be sufficiently negative), while the second term is small enough to be negligible. To this end, we again make use of the eigen-decomposition (24.122). We have:

$$\begin{aligned}
&\mu^2 \text{Tr}\left\{\Lambda_{n-1}\left(\sum_{j=0}^{m} D^{j}\right) V_{n-1}^{\mathsf{T}} R_g\left(w_{n-1}\right) V_{n-1}\right\} \\
&= \mu^2 \text{Tr}\left\{\Lambda_{n-1}^{>0}\left(\sum_{j=0}^{m}\left(I - 2\mu\Lambda_{n-1}^{>0}\right)^j\right)\left(V_{n-1}^{>0}\right)^{\mathsf{T}} R_g\left(w_{n-1}\right) V_{n-1}^{>0}\right\} \\
&\quad + \mu^2 \text{Tr}\left\{\Lambda_{n-1}^{<0}\left(\sum_{j=0}^{m}\left(I - 2\mu\Lambda_{n-1}^{<0}\right)^j\right)\left(V_{n-1}^{<0}\right)^{\mathsf{T}} R_g\left(w_{n-1}\right) V_{n-1}^{<0}\right\} \\
&= \mu^2 \text{Tr}\left\{\Lambda_{n-1}^{>0}\left(\sum_{j=0}^{m}\left(I - 2\mu\Lambda_{n-1}^{>0}\right)^j\right)\left(V_{n-1}^{>0}\right)^{\mathsf{T}} R_g\left(w_{n-1}\right) V_{n-1}^{>0}\right\} \\
&\quad - \mu^2 \text{Tr}\left\{\left(-\Lambda_{n-1}^{<0}\right)\left(\sum_{j=0}^{m}\left(I - 2\mu\Lambda_{n-1}^{<0}\right)^j\right) \times \left(V_{n-1}^{<0}\right)^{\mathsf{T}} R_g\left(w_{n-1}\right) V_{n-1}^{<0}\right\}
\end{aligned} \tag{24.157}$$

Using the property $\text{Tr}(A)\lambda_{\min}(B) \leq \text{Tr}(AB) \leq \text{Tr}(A)\lambda_{\max}(B)$, which holds for $A = A^{\mathsf{T}}, B = B^{\mathsf{T}} \geq 0$ as shown earlier in Prob. 1.8, we get

$$\begin{aligned}
&\mu^2 \text{Tr}\left\{\Lambda_{n-1}\left(\sum_{j=0}^{m} D^{j}\right) V_{n-1}^{\mathsf{T}} R_g\left(w_{n-1}\right) V_{n-1}\right\} \\
&\leq \mu^2 \text{Tr}\left\{\Lambda_{n-1}^{>0}\left(\sum_{j=0}^{m}\left(I - 2\mu\Lambda_{n-1}^{>0}\right)^j\right)\right\} \times \lambda_{\max}\left(\left(V_{n-1}^{>0}\right)^{\mathsf{T}} R_g\left(w_{n-1}\right) V_{n-1}^{>0}\right) \\
&\quad - \mu^2 \text{Tr}\left\{\left(-\Lambda_{n-1}^{<0}\right)\left(\sum_{j=0}^{m}\left(I - 2\mu\Lambda_{n-1}^{<0}\right)^j\right)\right\} \times \lambda_{\min}\left(\left(V_{n-1}^{<0}\right)^{\mathsf{T}} R_g\left(w_{n-1}\right) V_{n-1}^{<0}\right) \\
&\overset{(a)}{\leq} \mu^2 \text{Tr}\left(\Lambda_{n-1}^{>0}\left(\sum_{j=0}^{m}\left(I - 2\mu\Lambda_{n-1}^{>0}\right)^j\right)\right)\left(\beta^2\|\nabla_{w^{\mathsf{T}}} P(w_{n-1})\|^2 + \sigma^2\right) \\
&\quad - \mu^2 \text{Tr}\left\{\left(-\Lambda_{n-1}^{<0}\right)\left(\sum_{j=0}^{m}\left(I - 2\mu\Lambda_{n-1}^{<0}\right)^j\right)\right\} \sigma_\ell^2
\end{aligned} \tag{24.158}$$

where step (a) uses the persistent noise condition (24.31) and the result of Probs. 24.1 and 24.7.

For the first term on the right-hand side of (24.158), we have

$$\begin{aligned}
&\mu^2 \mathrm{Tr}\left\{\boldsymbol{\Lambda}_{n-1}^{>0}\left(\sum_{j=0}^{m}(I-2\mu\boldsymbol{\Lambda}_{n-1}^{>0})^j\right)\right\}(\beta^2\|\nabla_{w^\mathsf{T}}P(\boldsymbol{w}_{n-1})\|^2+\sigma^2)\\
&\overset{(a)}{\le}\mu^2 \mathrm{Tr}\left\{\boldsymbol{\Lambda}_{n-1}^{>0}\left(\sum_{j=0}^{\infty}(I-2\mu\boldsymbol{\Lambda}_{n-1}^{>0})^n\right)\right\}(\beta^2\|\nabla_{w^\mathsf{T}}P(\boldsymbol{w}_{n-1})\|^2+\sigma^2)\\
&\overset{(b)}{\le}\mu^2 \mathrm{Tr}\left\{\boldsymbol{\Lambda}_{n-1}^{>0}(2\mu\boldsymbol{\Lambda}_{n-1}^{>0})^{-1}\right\}(\beta^2\|\nabla_{w^\mathsf{T}}P(\boldsymbol{w}_{n-1})\|^2+\sigma^2)\\
&\le\frac{\mu}{2}M\left(\beta^2\|\nabla_{w^\mathsf{T}}P(\boldsymbol{w}_{n-1})\|^2+\sigma^2\right)
\end{aligned}\tag{24.159}$$

where step (a) exploits the fact that the matrix $I-2\mu\boldsymbol{\Lambda}_{n-1}^{>0}$ is elementwise nonnegative for $\mu \le 1/2\delta$, and step (b) follows from the geometric series formula $\sum_{n=0}^{\infty}A^n = (I-A)^{-1}$ for stable A. Hence, under expectation:

$$\begin{aligned}
&\mu^2\mathbb{E}\left[\mathrm{Tr}\left\{\boldsymbol{\Lambda}_{n-1}^{>0}\left(\sum_{j=0}^{m}(I-2\mu\boldsymbol{\Lambda}_{n-1}^{>0})^j\right)\right\}(\beta^2\|\nabla_{w^\mathsf{T}}P(\boldsymbol{w}_{n-1})\|^2+\sigma^2)\mid \boldsymbol{w}_{n-1}\in\mathcal{U}\right]\\
&\le\frac{\mu}{2}M\left(\beta^2\mathbb{E}\left(\|\nabla_{w^\mathsf{T}}P(\boldsymbol{w}_{n-1})\|^2\mid \boldsymbol{w}_{n-1}\in\mathcal{U}\right)+\sigma^2\right)\\
&\le\frac{\mu}{2}M\left(\beta^2\cdot O(\mu)+\sigma^2\right)\\
&\le\frac{\mu}{2}M\sigma^2+O(\mu^2)
\end{aligned}\tag{24.160}$$

where the last line uses the fact that the squared norm of the gradient of the risk function is $O(\mu)$ within the set $\mathcal{U}$.

For the second term on the right-hand side of (24.158), we have under expectation conditioned on $\boldsymbol{w}_{n-1}\in\mathcal{U}$:

$$\begin{aligned}
&\mathbb{E}\left[\sigma_\ell^2\,\mathrm{Tr}\left\{(-\boldsymbol{\Lambda}_{n-1}^{<0})\left(\sum_{j=0}^{m}(I-2\mu\boldsymbol{\Lambda}_{n-1}^{<0})^j\right)\right\}\mid \boldsymbol{w}_{n-1}\in\mathcal{U}\right]\\
&\overset{(a)}{\ge}\mathbb{E}\left(\sigma_\ell^2\,\tau\left(\sum_{j=0}^{m}(1+2\mu\tau)^j\right)\mid \boldsymbol{w}_{n-1}\in\mathcal{U}\right)\\
&\overset{(b)}{=}\sigma_\ell^2\,\tau\left(\sum_{j=0}^{m}(1+2\mu\tau)^j\right)\\
&=\sigma_\ell^2\,\tau\frac{1-(1+2\mu\tau)^{m+1}}{1-(1+2\mu\tau)}\\
&=\frac{\sigma_\ell^2}{2\mu}\left((1+2\mu\tau)^{m+1}-1\right)
\end{aligned}\tag{24.161}$$

Step (a) exploits the fact that the trace of a diagonal matrix with nonnegative entries can be lower-bounded by any of its diagonal elements:

$$\mathrm{Tr}\left((-\boldsymbol{\Lambda}_{n-1}^{<0})\left(\sum_{j=0}^{m}(I-2\mu\boldsymbol{\Lambda}_{n-1}^{<0})^j\right)\right)\ge\tau\left(\sum_{j=0}^{m}(1+2\mu\tau)^j\right)\tag{24.162}$$

The lower bound is because at least one entry of $\Lambda_{n-1}^{<0}$ is less than $-\tau$ over the set $\mathcal{U}$. In step (b) we dropped the expectation since the expression is no longer random. We return to the full expression (24.158) and find:

$$\begin{aligned}
&\mu^2 \mathbb{E}\left\{ \mathrm{Tr}\left(\boldsymbol{\Lambda}_{n-1} \left(\sum_{j=0}^{m} \boldsymbol{D}^j \right) \boldsymbol{V}_{n-1}^{\mathsf{T}} R_g\left(\boldsymbol{w}_{n-1}\right) \boldsymbol{V}_{n-1} \right) \mid \boldsymbol{w}_{n-1} \in \mathcal{U} \right\} \\
&\leq \frac{\mu}{2} M\sigma^2 + O(\mu^2) - \frac{\sigma_\ell^2 \mu}{2}\left((1+2\mu\tau)^{m+1} - 1 \right) \\
&\overset{(a)}{\leq} -\frac{\mu}{2} M\sigma^2
\end{aligned} \tag{24.163}$$

where (a) holds if, and only if,

$$\begin{aligned}
&\frac{\mu}{2} M\sigma^2 + O(\mu^2) - \frac{\sigma_\ell^2 \mu}{2}\left((1+2\mu\tau)^{m+1} - 1 \right) \leq -\frac{\mu}{2} M\sigma^2 \\
&\Longleftrightarrow\ 2M\frac{\sigma^2}{\sigma_\ell^2} + O(\mu) + 1 \leq (1+2\mu\tau)^{m+1} \\
&\Longleftrightarrow\ \log\left(2M\frac{\sigma^2}{\sigma_\ell^2} + 1 + O(\mu) \right) \leq (m+1)\log\left(1+2\mu\tau\right) \\
&\Longleftrightarrow\ \frac{\log\left(2M\frac{\sigma^2}{\sigma_\ell^2} + 1 + O(\mu) \right)}{\log\left(1+2\mu\tau\right)} \leq m+1 \\
&\Longleftrightarrow\ \log\left(2M\frac{\sigma^2}{\sigma_\ell^2} + 1 + O(\mu) \right) O\left(\frac{1}{\mu\tau}\right) \leq m+1
\end{aligned} \tag{24.164}$$

where the last line follows from $\lim_{x\to 0} \frac{1}{x}\log(1+x) = 1$. Choosing n^s to be the smallest integer m satisfying (24.164) we get:

$$\mu^2 \mathbb{E}\left\{ \mathrm{Tr}\left(\boldsymbol{\Lambda}_{n-1} \left(\sum_{j=0}^{n^s} \boldsymbol{D}^n \right) \boldsymbol{V}_{n-1}^{\mathsf{T}} R_g\left(\boldsymbol{w}_{n-1}\right) \boldsymbol{V}_{n-1} \right) \right\} \leq -\frac{\mu}{2} M\sigma^2 \tag{24.165}$$

Applying this relation to (24.156) and taking expectations over $\boldsymbol{w}_{n-1} \in \mathcal{U}$, we obtain:

$$\begin{aligned}
&\mathbb{E}\left(\left\| \check{\boldsymbol{z}}_{n^s}^{n-1} \right\|^2_{\boldsymbol{\Lambda}_{n-1}} \mid \boldsymbol{w}_{n-1} \in \mathcal{U} \right) \\
&\leq \mu^2 \sum_{j=0}^{n^s} \mathbb{E}\left\{ \left(\rho\left(\boldsymbol{\Lambda}_{n-1} \boldsymbol{D}^j \right) \cdot \mathbb{E}\left(\boldsymbol{q}_{n^s-j-1}^{n-1} \mid \boldsymbol{\mathcal{F}}_{n-1} \right) \right) \mid \boldsymbol{w}_{n-1} \in \mathcal{U} \right\} - \frac{\mu}{2} M\sigma^2
\end{aligned} \tag{24.166}$$

We now bound the first term on the right-hand side:

$$\begin{aligned}
&\mu^2 \sum_{j=0}^{n^s} \mathbb{E}\left\{\left(\rho\left(\boldsymbol{\Lambda}_{n-1}\boldsymbol{D}^j\right)\cdot \mathbb{E}\left(\boldsymbol{q}_{n^s-j-1}^{n-1} \,|\, \boldsymbol{\mathcal{F}}_{n-1}\right)\right) \,|\, \boldsymbol{w}_{n-1} \in \mathcal{U}\right\} \\
&\le \mu^2 \sum_{j=0}^{n^s} \left(\delta(1+2\mu\delta)^j \cdot \mathbb{E}\left\{\boldsymbol{q}_{n^s-j-1}^{n-1} \,|\, \boldsymbol{w}_{n-1} \in \mathcal{U}\right\}\right) \\
&\overset{(24.147)}{=} \mu^2 \sum_{j=0}^{n^s} \delta(1+2\mu\delta)^j \cdot \mathbb{E}\left(\beta_R \left\|\boldsymbol{d}_{n^s-j-1}^{n-1}\right\|^{\eta} + \delta^2 \left\|\widetilde{\boldsymbol{z}}_{n^s-j-1}^{n-1}\right\|^2 \,|\, \boldsymbol{w}_{n-1} \in \mathcal{U}\right)\Big) \\
&\overset{(a)}{\le} \mu^2 \sum_{j=0}^{n^s} \delta(1+2\mu\delta)^j \cdot \left(O(\mu^{\eta/2}) + O(\mu)\right) \\
&\le \delta \left(\sum_{j=0}^{n^s} (1+2\mu\delta)^j\right) O(\mu^{2+\gamma/2}) \\
&\overset{(b)}{\le} O(\mu^{1+\gamma/2}) = o(\mu)
\end{aligned} \tag{24.167}$$

where step (a) follows from the bounds in Lemma 24.1 and step (b) follows from the result of Prob. 24.2. We conclude that

$$\mathbb{E}\left(\|\widetilde{\boldsymbol{z}}_{n^s}^{n-1}\|^2_{\boldsymbol{\Lambda}_{n-1}} \,|\, \boldsymbol{w}_{n-1} \in \mathcal{U}\right) \le -\frac{\mu}{2} M\sigma^2 + o(\mu) \tag{24.168}$$

Returning to (24.141), we find:

$$\begin{aligned}
&\mathbb{E}\left(P(\boldsymbol{w}'_{n+n^s}) \,|\, \boldsymbol{w}_{n-1} \in \mathcal{U}\right) \\
&\le \mathbb{E}\left(P(\boldsymbol{w}_{n-1}) \,|\, \boldsymbol{w}_{n-1} \in \mathcal{U}\right) + \frac{1}{2}\mathbb{E}\left(\left\|\widetilde{\boldsymbol{z}}_{n^s}^{n-1}\right\|^2_{\boldsymbol{\Lambda}_{n-1}} \,|\, \boldsymbol{w}_{n-1} \in \mathcal{U}\right) \\
&\quad + \frac{\delta_2}{6}\mathbb{E}\left(\left\|\boldsymbol{z}_{n^s}^{n-1}\right\|^3 \,|\, \boldsymbol{w}_{n-1} \in \mathcal{U}\right) \\
&\le \mathbb{E}\left(P(\boldsymbol{w}_{n-1}) \,|\, \boldsymbol{w}_{n-1} \in \mathcal{U}\right) - \frac{\mu}{2} M\sigma^2 + o(\mu)
\end{aligned} \tag{24.169}$$

where we further used the result of Prob. 24.5. If we now recall relation (24.114), we arrive at the desired result (24.48).

24.D SECOND-ORDER CONVERGENCE GUARANTEE

We finally establish Theorem 24.3. We introduce the probabilities of an iterate $\boldsymbol{w}_n$ belonging to one of the sets $\{\mathcal{S}, \mathcal{S}^C, \mathcal{D}, \mathcal{U}\}$:

$$\pi_n^{\mathcal{S}^C} \triangleq \mathbb{P}\left(\boldsymbol{w}_n \in \mathcal{S}^C\right) \tag{24.170a}$$

$$\pi_n^{\mathcal{S}} \triangleq \mathbb{P}(\boldsymbol{w}_n \in \mathcal{S}) \tag{24.170b}$$

$$\pi_n^{\mathcal{D}} \triangleq \mathbb{P}(\boldsymbol{w}_n \in \mathcal{D}) \tag{24.170c}$$

$$\pi_n^{\mathcal{U}} \triangleq \mathbb{P}(\boldsymbol{w}_n \in \mathcal{U}) \tag{24.170d}$$

Obviously, for all n, we have

$$\pi_n^{\mathcal{S}} = \pi_n^{\mathcal{D}} + \pi_n^{\mathcal{U}} \tag{24.171a}$$

$$\pi_n^{\mathcal{S}^C} + \pi_n^{\mathcal{D}} + \pi_n^{\mathcal{U}} = 1 \tag{24.171b}$$

Next, we introduce the scalar stochastic process:

$$\mathbf{t}(k+1) = \begin{cases} \mathbf{t}(k)+1, & \text{if } \boldsymbol{w}_{\mathbf{t}(k)} \in \mathcal{S}^C, \\ \mathbf{t}(k)+1 & \text{if } \boldsymbol{w}_{\mathbf{t}(k)} \in \mathcal{D}, \\ \mathbf{t}(k)+n^s & \text{if } \boldsymbol{w}_{\mathbf{t}(k)} \in \mathcal{U} \end{cases} \tag{24.172}$$

where $\mathbf{t}(0) = -1$. Here, $\boldsymbol{t}(k)$ is a random index; it is updated by one step if we start from an iterate within the sets $\mathcal{S}^C$ or $\mathcal{D}$, whereas it is updated by n^s steps if we start from an iterate within $\mathcal{U}$. We will be examining the weight iterate $\boldsymbol{w}_n$ at two successive values of $n = \boldsymbol{t}(k)$ and $n = \boldsymbol{t}(k+1)$.

Assume first that at the iteration of index $\boldsymbol{t}(k)$, the iterate $\boldsymbol{w}_{\boldsymbol{t}(k)}$ happens to be within $\mathcal{S}^C$. In this case, the updated iterate of index $\boldsymbol{t}(k+1) = \boldsymbol{t}(k)+1$ will satisfy, in view of (24.45), the following relation:

$$\mathbb{E}\left(P(\boldsymbol{w}_{\mathbf{t}(k)}) - P(\boldsymbol{w}_{\mathbf{t}(k)+1}) \,|\, \boldsymbol{w}_{\mathbf{t}(k)} \in \mathcal{S}^C\right) \geq \mu^2 \frac{c_2}{\epsilon} \tag{24.173}$$

Since result (24.45) holds for every n, it follows that it holds for every realization of $\boldsymbol{t}(k)$ as long as $\boldsymbol{w}_{\mathbf{t}(k)} \in \mathcal{S}^C$. We then have with probability 1:

$$\mathbb{E}\left(P(\boldsymbol{w}_{\mathbf{t}(k)}) - P(\boldsymbol{w}_{\mathbf{t}(k)+1}) \,|\, \boldsymbol{t}(k), \boldsymbol{w}_{\mathbf{t}(k)} \in \mathcal{S}^C\right) \geq \mu^2 \frac{c_2}{\epsilon} \tag{24.174}$$

Similarly, if the iterate $\boldsymbol{w}_{\boldsymbol{t}(k)}$ at index $\boldsymbol{t}(k)$ happens to be within the set $\mathcal{D}$, then the updated iterate of index $\boldsymbol{t}(k+1) = \boldsymbol{t}(k)+1$ will satisfy, in view of Prob. 24.3:

$$\mathbb{E}\left(P(\boldsymbol{w}_{\mathbf{t}(k)}) - P(\boldsymbol{w}_{\mathbf{t}(k)+1}) \,|\, \boldsymbol{t}(k), \boldsymbol{w}_{\mathbf{t}(k)} \in \mathcal{D}\right) \geq -\mu^2 c_2 \tag{24.175}$$

Finally, if the iterate $\boldsymbol{w}_{\boldsymbol{t}(k)}$ at index $\boldsymbol{t}(k)$ happens to be within the set $\mathcal{U}$, then the updated iterate of index $\boldsymbol{t}(k+1) = \boldsymbol{t}(k)+n^s$ will satisfy, in view of Theorem 24.2:

$$\mathbb{E}\left(P(\boldsymbol{w}_{\mathbf{t}(k)}) - P(\boldsymbol{w}_{\mathbf{t}(k)+n^s}) \,|\, \boldsymbol{t}(k), \boldsymbol{w}_{\mathbf{t}(k)} \in \mathcal{U}\right) \geq \frac{\mu}{2} M\sigma^2 - o(\mu) \tag{24.176}$$

Combining these three conclusions, we find that the risk values at the iterates of indices $\boldsymbol{t}(k)$ and $\boldsymbol{t}(k+1)$ satisfy

$$
\begin{aligned}
\mathbb{E}\Big(P(\boldsymbol{w}_{\mathbf{t}(k)}) - P(\boldsymbol{w}_{\mathbf{t}(k+1)})\Big)[][] &= \mathbb{E}\left\{\mathbb{E}\Big(P(\boldsymbol{w}_{\mathbf{t}(k)}) - P(\boldsymbol{w}_{\mathbf{t}(k+1)})\,|\,\boldsymbol{t}(k)\Big)\right\} \\
&= \mathbb{E}\left\{\mathbb{E}\Big(P(\boldsymbol{w}_{\mathbf{t}(k)}) - P(\boldsymbol{w}_{\mathbf{t}(k+1)})\,|\,\boldsymbol{t}(k), \boldsymbol{w}_{\mathbf{t}(k)} \in \mathcal{S}^C\Big)\cdot \pi^{\mathcal{S}^C}_{\mathbf{t}(k)}\right. \\
&\quad + \mathbb{E}\Big(P(\boldsymbol{w}_{\mathbf{t}(k)}) - P(\boldsymbol{w}_{\mathbf{t}(k+1)})\,|\,\boldsymbol{t}(k), \boldsymbol{w}_{\mathbf{t}(k)} \in \mathcal{U}\Big)\cdot \pi^{\mathcal{U}}_{\mathbf{t}(k)} \\
&\quad \left. + \mathbb{E}\Big(P(\boldsymbol{w}_{\mathbf{t}(k)}) - P(\boldsymbol{w}_{\mathbf{t}(k+1)})\,|\,\boldsymbol{t}(k), \boldsymbol{w}_{\mathbf{t}(k)} \in \mathcal{D}\Big)\cdot \pi^{\mathcal{D}}_{\mathbf{t}(k)}\right\} \\
&\geq \mu^2\frac{c_2}{\epsilon}\cdot \mathbb{E}\,\pi^{\mathcal{S}^C}_{\mathbf{t}(k)} + \Big(\frac{\mu}{2}M\sigma^2 - o(\mu)\Big)\cdot \mathbb{E}\,\pi^{\mathcal{U}}_{\mathbf{t}(k)} - \mu^2 c_2\cdot \mathbb{E}\,\pi^{\mathcal{D}}_{\mathbf{t}(k)} \\
&\geq \min\Big\{\mu^2\frac{c_2}{\epsilon},\, \frac{\mu}{2}M\sigma^2 - o(\mu)\Big\}\Big(\mathbb{E}\,\pi^{\mathcal{S}^C}_{\mathbf{t}(k)} + \mathbb{E}\,\pi^{\mathcal{U}}_{\mathbf{t}(k)}\Big) - \mu^2 c_2\,\mathbb{E}\,\pi^{\mathcal{D}}_{\mathbf{t}(k)} \\
&\overset{(a)}{\geq} \mu^2\frac{c_2}{\epsilon}\Big(\mathbb{E}\,\pi^{\mathcal{S}^C}_{\mathbf{t}(k)} + \mathbb{E}\,\pi^{\mathcal{U}}_{\mathbf{t}(k)}\Big) - \mu^2 c_2\,\mathbb{E}\,\pi^{\mathcal{D}}_{\mathbf{t}(k)} \\
&\geq \mu^2\frac{c_2}{\epsilon}\Big(\mathbb{E}\,\pi^{\mathcal{S}^C}_{\mathbf{t}(k)} + \mathbb{E}\,\pi^{\mathcal{U}}_{\mathbf{t}(k)}\Big) - \mu^2 c_2
\end{aligned}
\tag{24.177}
$$

where step (a) holds for sufficiently small μ. We then have by telescoping (recall that, by construction $\boldsymbol{t}(0) = -1$ so that $P(w_{\mathbf{t}(0)}) = P(w_{-1})$ and, moreover, $P^o \leq P(w)$ for any w):

$$
\begin{aligned}
P(w_{-1}) - P^o \geq\; & \mathbb{E}\,P(w_{\mathbf{t}(0)}) - \mathbb{E}\,P(\boldsymbol{w}_{\mathbf{t}(k)}) \\
=\; & \mathbb{E}\,P(w_{\mathbf{t}(0)}) - \mathbb{E}\,P(\boldsymbol{w}_{\mathbf{t}(1)}) \\
& + \mathbb{E}\,P(\boldsymbol{w}_{\mathbf{t}(1)}) - \mathbb{E}\,P(\boldsymbol{w}_{\mathbf{t}(2)}) \\
& \bullet \\
& \bullet \\
& + \mathbb{E}\,P(\boldsymbol{w}_{\mathbf{t}(k-1)}) - \mathbb{E}\,P(\boldsymbol{w}_{\mathbf{t}(k)}) \\
\geq\; & \mu^2\frac{c_2}{\epsilon}\sum_{j=0}^{k}\Big(\mathbb{E}\,\pi^{\mathcal{S}^C}_{\mathbf{t}(j)} + \mathbb{E}\,\pi^{\mathcal{U}}_{\mathbf{t}(j)}\Big) - \mu^2 c_2 k
\end{aligned}
\tag{24.178}
$$

Rearranging terms we get

$$
\frac{1}{k}\sum_{j=0}^{k}\Big(\mathbb{E}\,\pi^{\mathcal{S}^C}_{\mathbf{t}(j)} + \mathbb{E}\,\pi^{\mathcal{U}}_{\mathbf{t}(j)}\Big) \leq \frac{(P(w_{-1}) - P^o)\epsilon}{\mu^2 c_2 k} + \epsilon \overset{(b)}{\leq} 2\epsilon \tag{24.179}
$$

where step (b) holds whenever

$$
k \geq k^o \;\overset{\Delta}{=}\; \frac{P(w_{-1}) - P^o}{\mu^2 c_2} \tag{24.180}
$$

Let us select $k = k^o$. Since the average on the left-hand side of (24.179) is bounded by 2ϵ and each element is nonnegative, it follows that at least one element in the average is bounded by 2ϵ. We conclude that there exists $j \leq k^o$ such that

$$
\mathbb{E}\,\pi^{\mathcal{S}^C}_{\mathbf{t}(j)} + \mathbb{E}\,\pi^{\mathcal{U}}_{\mathbf{t}(j)} \leq 2\epsilon \iff \mathbb{E}\,\pi^{\mathcal{D}}_{\mathbf{t}(j)} \geq 1 - 2\epsilon \tag{24.181}
$$

Now, by definition of the stochastic process $\boldsymbol{t}(k)$:

$$
n^o = \boldsymbol{t}(k^o) \;\leq\; k^o n^s \;=\; \frac{P(w_{-1}) - P^o}{\mu^2 c_2} n^s \tag{24.182}
$$

From (24.181) we conclude that $\pi^{\mathcal{D}}_{\mathbf{t}(j)} \geq 1 - 2\epsilon$ with a nonzero probability and, hence, there must exist an $n \leq n^o$ such that $\pi^{\mathcal{D}}_{n} \geq 1 - 2\epsilon$. This concludes the proof.

REFERENCES

Allen-Zhu, Z. (2018), "Natasha 2: Faster non-convex optimization than SGD," *Proc. Advances Neural Information Processing Systems* (NIPS), pp. 2675–2686, Montreal.

Allen-Zhu, Z. and Y. Li (2018), "NEON2: Finding local minima via first-order oracles," *Proc. Advances Neural Information Processing Systems* (NIPS), pp. 3716–3726, Montreal.

Bertsekas, D. P. and J. N. Tsitsiklis (2000), "Gradient convergence in gradient methods with errors," *SIAM J. Optim.*, vol. 10, no. 3, pp. 627–642.

Bottou, L., F. E. Curtis, and J. Nocedal (2018), "Optimization methods for large-scale machine learning," *SIAM Rev.*, vol. 60, no. 2, pp. 223–311.

Chen, J. and A. H. Sayed (2013), "Distributed Pareto optimization via diffusion strategies," *IEEE J. Sel. Topics Signal Process.*, vol. 7, no. 2, pp. 205–220.

Choromanska, A., M. Hena, M. Mathieu, G. B. Arous, and Y. LeCun (2015), "The loss surfaces of multilayer networks," *Proc. Int. Conf. Artificial Intelligence and Statistics*, pp. 192–204, San Diego, CA.

Curtis, F. E., D. P. Robinson, and M. Samadi (2017), "A trust region algorithm with a worst-case iteration complexity of $o(\epsilon^{-3/2})$ for nonconvex optimization," *Math. Program.*, vol. 162, pp. 1–32.

Daneshmand, H., J. Kohler, A. Lucchi, and T. Hofmann (2018), "Escaping saddles with stochastic gradients," available at arXiv:1803.05999.

Du S. S., C. Jin, J. D. Lee, M. I. Jordan, B. Póczos, and A. Singh (2017), "Gradient descent can take exponential time to escape saddle points," *Proc. Advances Neural Information Processing Systems* (NIPS), pp. 1067–1077, Long Beach, CA.

Fang, C., C. J. Li, Z. Lin, and T. Zhang (2018), "SPIDER: Near-optimal non-convex optimization via stochastic path-integrated differential estimator," *Proc. Neural Information Processing Systems* (NIPS), pp. 689–699, Montreal.

Fang, C., Z. Lin and T. Zhang (2019), "Sharp analysis for nonconvex SGD escaping from saddle points," available at arXiv:1902.00247.

Ge, R., F. Huang, C. Jin, and Y. Yuan (2015), "Escaping from saddle points: Online stochastic gradient for tensor decomposition," *Proc. Conf. Learning Theory* (COLT), pp. 797–842, Paris.

Ge, R., C. Jin, and Y. Zheng (2017), "No spurious local minima in nonconvex low rank problems: A unified geometric analysis," *Proc. Int. Conf. Machine Learning* (ICML), pp. 1233–1242, Sydney.

Ge, R., J. D. Lee, and T. Ma (2016), "Matrix completion has no spurious local minimum," *Proc. Advances Neural Information Processing Systems* (NIPS), pp. 2973–2981, Barcelona.

Gelfand, S. and S. Mitter (1991), "Recursive stochastic algorithms for global optimization in $\mathbb{R}^d$," *SIAM J. Control Optim.*, vol. 29, no. 5, pp. 999–1018.

Jin, C., R. Ge, P. Netrapalli, S. M. Kakade, and M. I. Jordan (2017), "How to escape saddle points efficiently," *Proc. Int. Conf. Machine Learning* (ICML), pp. 1724–1732, Sydney.

Jin, X.-B., X.-Y. Zhang, K. Huang, and G.-G. Geng (2019), "Stochastic conjugate gradient algorithm with variance reduction," *IEEE Trans. Neural Netw. Learn. Syst.*, vol. 30, no. 5, pp. 1360–1369.

Kawaguchi, K. (2016), "Deep learning without poor local minima," *Proc. Advances Neural Information Processing Systems* (NIPS), pp. 586–594, Barcelona.

Lee, J. D., M. Simchowitz, M. I. Jordan, and B. Recht (2016), "Gradient descent only converges to minimizers," *Proc. Conf. Learning Theory* (COLT), pp. 1246–1257, New York, NY.

Murty, K. G. and S. N. Kabadi (1987), "Some NP-complete problems in quadratic and nonlinear programming," *Math. Program.*, vol. 39, no. 2, pp. 117–129.

Nesterov, Y. (2004), *Introductory Lectures on Convex Optimization*, Springer.

Nesterov, Y. and B. T. Polyak (2006), "Cubic regularization of Newton method and its global performance," *Math. Program.*, vol. 108, no. 1, pp. 177–205.

Reddi, S. J., A. Hefny, S. Sra, B. Póczos, and A. Smola (2016a), "On variance reduction in stochastic gradient descent and its asynchronous variants," *Proc. Advances Neural Information Processing Systems* (NIPS), pp. 2647–2655, Barcelona.

Reddi, S. J., A. Hefny, S. Sra, B. Póczos, and A. Smola (2016b), "Stochastic variance reduction for nonconvex optimization" *Proc. Int. Conf. Machine Learning* (ICML), pp. 324–323, New York, NY.

Sayed, A. H. (2014a), *Adaptation, Learning, and Optimization over Networks,* Foundations and Trends in Machine Learning, NOW Publishers, vol. 7, no. 4–5, pp. 311–801.

Tripuraneni, N., M. Stern, C. Jin, J. Regier, and M. I. Jordan (2018), "Stochastic cubic regularization for fast nonconvex optimization," *Proc. Advances Neural Information Processing Systems* (NIPS), pp. 2904–2913, Montreal.

Vlaski, S., E. Rizk, and A. H. Sayed (2020), "Second-order guarantees in federated learning," *Proc. Asilomar Conf. Signals, Systems and Computers*, pp. 1–8, Pacific Grove, CA.

Vlaski, S. and A. H. Sayed (2019), "Second-order guarantees of stochastic gradient descent in nonconvex optimization," available at arXiv:1908.07023. Accepted for publication in *IEEE Trans. Aut. Control*, doi: 10.1109/TAC.2021.3131963.

Vlaski, S. and A. H. Sayed (2020), "Second-order guarantees in centralized, federated and decentralized nonconvex optimization," *Commun. Inf. Syst.*, vol. 20, no. 3, pp. 353–388.

Vlaski, S. and A. H. Sayed (2021a), "Distributed learning in non-convex environments, part I: Agreement at a linear rate," *IEEE Trans. Signal Process.*, vol. 69, pp. 1242–1256.

Vlaski, S. and A. H. Sayed (2021b) "Distributed learning in non-convex environments, part II: Polynomial escape from saddle-points," *IEEE Trans. Signal Process.*, vol. 69, pp. 1257–1270.

25 Decentralized Optimization I: Primal Methods

Most of the treatment in the earlier chapters focused on *single-agent* optimization and online learning methods, where a single agent (or learner) responds to training data $\{\gamma(m), h_m\}$ by using gradient, subgradient, or proximal methods and variations thereof to approach the minimizers of certain empirical or stochastic risks. For various reasons explained below, it is useful in many applications to consider scenarios where the optimization and learning efforts are split among *multiple* dispersed agents or learners. In these cases, each individual agent or learner will have access to *part* of the data, and agents will be linked by some graph topology that allows neighboring agents to share information with each other. We will derive in this chapter and the next various algorithms that allow the connected agents, through continuous interactions with their neighbors, to attain the same level of performance that would be possible if there were a single powerful centralized processor with access to all the data from across all agents in the graph.

We will refer to these methods as *decentralized* algorithms since all processing is performed locally and there is no central processor involved in fusing data from the agents. Thus, in the context of decentralized algorithms, *both* the data *and* the processing are localized, with the data being collected locally by individual agents and the processing being performed locally across neighborhoods. We will make a distinction between *decentralized* and *distributed* algorithms. In the latter case, the data is collected locally at the agents, and some local processing may also be performed at the agents, but there is a *fusion center* that receives processed data from the agents, fuses the data, and shares the result back with the dispersed agents. The architecture of *federated learning* studied earlier in Section 17.6 is one example of a distributed algorithm. Sometimes the terms *decentralized* and *distributed* are used interchangeably in the literature, with some authors using the qualification "distributed" when they mean "decentralized" or vice-versa. We will maintain a distinction between the terms in the manner described above for clarity.

There are many reasons for the interest in decentralized implementations:

(1) A centralized solution with a fusion center is costly to maintain and has a serious point of failure. If the central processor fails, the entire solution method becomes moot.

(2) Decentralized solutions are robust to link and agent failures. If one agent fails, or if a link in the graph topology is dropped, the remaining agents remain connected and the learning process can continue.

(3) Decentralized solutions provide the individual agents with more privacy and secrecy; agents keep their raw data and do not share it globally. In many situations involving sensitive data, agents may not be comfortable sharing their raw personal data with remote fusion centers in centralized implementations or even with their neighboring agents. Decentralized solutions provide one way out where agents only share some processed statistic with their neighbors.

(4) The data may already be available in dispersed locations, as happens with cloud computing, and it can be costly to transfer the data to a central location for processing. Decentralized learning procedures offer an attractive approach to dealing with such large datasets.

In summary, decentralized solutions are more robust than centralized solutions, more resilient to failure, offer more privacy, and avoid the need to coalesce large datasets in a single location. At the same time, decentralized mechanisms can serve as important enablers for the design of cooperative multi-agent systems, such as robotic swarms, consisting of multiple agents moving in coordination, which can assist, for example, in the exploration of disaster areas.

This chapter focuses on the class of *primal methods* for decentralized optimization, both in the deterministic and stochastic settings. These methods result from minimizing a regularized risk function where the regularization term is constructed by penalizing deviations away from agreement among the agents – see expression (25.113). The penalty term introduces some tolerable bias into the operation of the algorithms on the order of $O(\mu^2)$ for the squared-weight error in deterministic optimization, and on the order of $O(\mu)$ for the mean-square error (MSE) in stochastic optimization. In the next chapter, we describe the class of *primal–dual methods*, which result from solving *constrained* optimization problems and lead to the exact solution without bias for deterministic optimization. These latter methods can also be applied to stochastic optimization; however, their MSE performance will continue to be similar to primal methods and on the order of $O(\mu)$. We will also describe dual methods in the next chapter, and discuss performance for nonconvex optimization problems as well.

25.1 GRAPH TOPOLOGY

We will be using the terms "agent" and "learner" interchangeably in our presentation, and will derive in the sequel several decentralized strategies that can be used by a collection of interacting learners. These learners will be assumed to be interconnected by some graph topology. We describe the graph model in this section in preparation for future discussions.

25.1.1 Neighborhoods

Figure 25.1 shows a graph consisting of K agents, labeled $k = 1, 2, \ldots, K$. The agents are represented by vertices in the graph with edges linking them. An edge that connects an agent to itself is called a *self-loop*. The neighborhood of agent k is denoted by $\mathcal{N}_k$ and consists of all agents that are connected to k by an edge, in addition to agent k itself. Any two neighboring agents k and ℓ have the ability to share information over the edge connecting them. Each agent k will be assigned a risk function, denoted by $P_k(w)$.

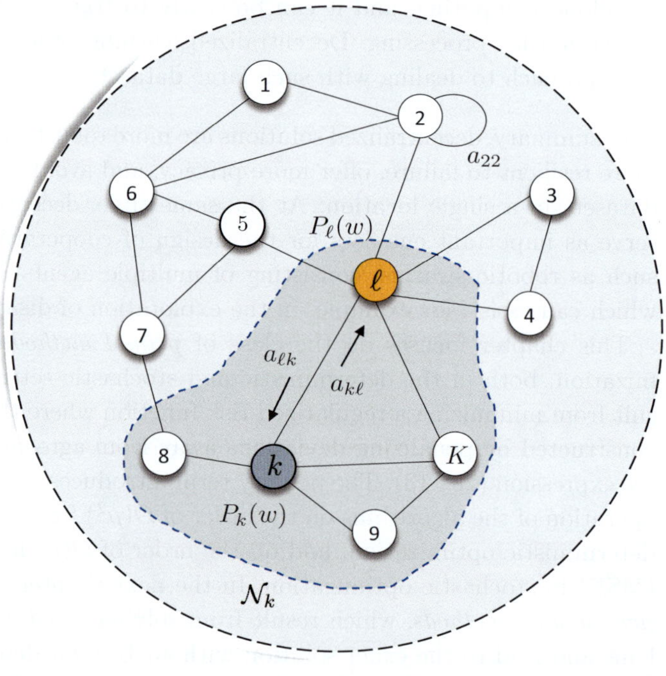

Figure 25.1 Agents are linked by a graph topology and can share information over their shared edges. The neighborhood of an agent is the collection of all agents linked to it. The neighborhood of agent k is marked by the highlighted area and denoted by $\mathcal{N}_k$. Each agent k is assigned a local risk function, denoted by $P_k(w)$.

We will assume an *undirected* graph, meaning that if agent k is a neighbor of agent ℓ, then agent ℓ is also a neighbor of agent k so that information is able to flow both ways between the agents. We assign a pair of nonnegative scaling weights, $\{a_{k\ell}, a_{\ell k}\}$, to the edge connecting k and ℓ. The scalar $a_{\ell k}$ is used by agent k to scale data it receives from agent ℓ; this scaling can be interpreted as a measure of the confidence level that agent k assigns to its interaction with agent ℓ. Likewise, $a_{k\ell}$ is used by agent ℓ to scale the data it receives from agent k. Whether agents linked by edges will exchange information, and whether the exchange will be unidirectional, bidirectional, or nonexistent, will depend on the values of the weights $\{a_{k\ell}, a_{\ell k}\}$. In principle, the weights $\{a_{k\ell}, a_{\ell k}\}$ can be different so that the exchange of information between the neighboring agents $\{k, \ell\}$ need not be symmetrical. One or both weights can also be zero. Although many of the decentralized algorithms described in this chapter can operate under such asymmetric conditions, it is generally assumed that $a_{k\ell} = a_{\ell k}$.

25.1.2 Strong Connectedness

The graph is said to be *connected* if paths with *nonzero* scaling weights can be found linking any two distinct agents in *both* directions, either directly when they are neighbors or by passing through intermediate agents when they are not neighbors. In this way, information can flow in both directions between any two agents in the graph, although the forward path from an agent k to some other agent ℓ need not be the same as the backward path from ℓ to k. A *strongly connected* network is a connected network with at least one nontrivial *self-loop*, meaning that $a_{kk} > 0$ for some agent k. Intuitively, this means that there exists at least one agent in the network that trusts its own information and will assign some positive weight to it. If $a_{kk} = 0$ for all k, then this means that all agents will be ignoring their individual information and will be relying instead on information received from other agents. The terminology of "strongly connected networks" is a bit excessive because it may convey the wrong impression that the graph has more connectivity than is actually necessary. In fact, a strongly connected graph can be sparsely connected.

Observe that we are defining connectivity over graphs not in terms of whether paths can be found connecting their vertices but in terms of whether these paths allow for the *meaningful* exchange of information between the vertices. This is because of the requirement that all *scaling weights* over the edges must be *positive* over at least one of the paths connecting any two distinct vertices. The assumption of a connected graph ensures that information will be flowing between any two arbitrary agents in the network and that this flow of information is bidirectional: Information flows from k to ℓ and from ℓ to k, although the paths over which the flows occur need not be the same and the manner by which information is scaled over these paths can also be different.

The strong connectivity of a graph translates into a useful property on the combination weights. Assume we collect the coefficients $\{a_{\ell k}\}$ into a $K \times K$ matrix $A = [a_{\ell k}]$, such that the entries on the kth column of A contain the coefficients used by agent k to scale data arriving from its neighbors $\ell \in \mathcal{N}_k$; we set $a_{\ell k} = 0$ if $\ell \notin \mathcal{N}_k$. In this way, the row index in (ℓ, k) designates the *source* agent and the column index designates the *sink* agent (or destination). We refer to A as the *combination* matrix or policy. It turns out that combination matrices that correspond to strongly connected networks are *primitive* – recall from definition (1.196) that a $K \times K$ matrix A with nonnegative entries is said to be primitive if there exists some finite integer $n_o > 0$ such that all entries of A^{n_o} are strictly positive:

$$[A^{n_o}]_{\ell,k} > 0, \quad \text{uniformly for all } (\ell, k) \tag{25.1}$$

Lemma 25.1. (Primitive combination matrix) *The $K \times K$ combination matrix of a strongly connected graph with K vertices is a primitive matrix.*

Proof: See Appendix. 25.A.

■

25.1.3 Graph Laplacian

We associate a Laplacian matrix with every graph; it is a symmetric matrix with useful properties that reveal the connectedness of the graph.

Consider again a graph consisting of K vertices and L edges connecting these vertices to each other. We only need to focus on the edges that connect distinct nodes to each other; we can ignore any self-loops that may exist in the graph. In other words, when we refer to the L edges of the graph, we are excluding self-loops from this set; but we are still allowing loops of at least length 2 (i.e., loops generated by paths covering at least two edges). Figure 25.2 shows a graph with $K = 7$ nodes and $L = 9$ edges.

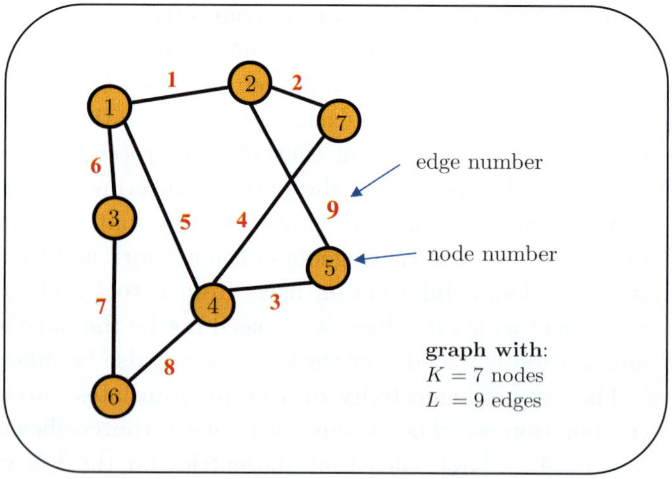

Figure 25.2 A graph with $K = 7$ vertices and $L = 9$ edges. The vertices are numbered 1 through 7 and the edges are numbered 1 through 9.

The degree of a node k, which we denote by n_k, is defined as the size of its neighborhood, $\mathcal{N}_k$:

$$n_k \triangleq |\mathcal{N}_k| \tag{25.2}$$

Since by convention $k \in \mathcal{N}_k$, it holds that $n_k \geq 1$. The entries of the $K \times K$ Laplacian matrix, denoted by $\mathcal{L}$, are defined by:

$$[\mathcal{L}]_{k\ell} = \begin{cases} n_k - 1, & \text{if } k = \ell \\ -1, & \text{if } k \neq \ell \text{ and nodes } k \text{ and } \ell \text{ are neighbors} \\ 0, & \text{otherwise} \end{cases} \tag{25.3}$$

In this expression, the value $n_k - 1$ corresponds to the number of edges that are incident on node k, and the locations of the -1s on the kth row indicate the nodes that are connected to node k. For the graph shown in Fig. 25.2, the Laplacian matrix has dimensions 7×7 and is given by

$$\mathcal{L} = \begin{bmatrix} 3 & -1 & -1 & -1 & 0 & 0 & 0 \\ -1 & 3 & 0 & 0 & -1 & 0 & -1 \\ -1 & 0 & 2 & 0 & 0 & -1 & 0 \\ -1 & 0 & 0 & 4 & -1 & -1 & -1 \\ 0 & -1 & 0 & -1 & 2 & 0 & 0 \\ 0 & 0 & -1 & -1 & 0 & 2 & 0 \\ 0 & -1 & 0 & -1 & 0 & 0 & 2 \end{bmatrix} \quad (7 \times 7) \tag{25.4}$$

We also associate with the graph a $K \times L$ *incidence matrix*, denoted by $\mathcal{I}$. The entries of $\mathcal{I}$ are defined as follows. Every column of $\mathcal{I}$ represents one edge in the graph. Each edge connects two nodes and its column will display two nonzero entries at the rows corresponding to these nodes: one entry will be $+1$ and the other entry will be -1. For directed graphs, the choice of which entry is positive or negative can be used to identify the nodes from which edges emanate (source nodes) and the nodes to which edges arrive (sink nodes). Since we are dealing with undirected graphs, we will simply assign positive values to lower indexed nodes and negative values to higher indexed nodes:

$$[\mathcal{I}]_{ke} = \begin{cases} +1, & \text{if node } k \text{ is the lower-indexed node connected to edge } e \\ -1, & \text{if node } k \text{ is the higher-indexed node connected to edge } e \\ 0, & \text{otherwise} \end{cases} \tag{25.5}$$

For the graph shown in Fig. 25.2, the incidence matrix has dimensions 7×9 and is given by

$$\mathcal{I} = \begin{bmatrix} +1 & 0 & 0 & 0 & +1 & +1 & 0 & 0 & 0 \\ -1 & +1 & 0 & 0 & 0 & 0 & 0 & 0 & +1 \\ 0 & 0 & 0 & 0 & 0 & -1 & +1 & 0 & 0 \\ 0 & 0 & +1 & +1 & -1 & 0 & 0 & +1 & 0 \\ 0 & 0 & -1 & 0 & 0 & 0 & 0 & 0 & -1 \\ 0 & 0 & 0 & 0 & 0 & 0 & -1 & -1 & 0 \\ 0 & -1 & 0 & -1 & 0 & 0 & 0 & 0 & 0 \end{bmatrix} \quad (7 \times 9) \tag{25.6}$$

Consider, for example, column 6 in the matrix. This column corresponds to edge 6, which links nodes 1 and 3. Therefore, at location $\mathcal{I}_{1,6}$ we have $+1$ and at location $\mathcal{I}_{3,6}$ we have -1.

It can be easily verified that the Laplacian and incidence matrices of a graph are related via

$$\mathcal{L} = \mathcal{I}\,\mathcal{I}^{\mathsf{T}} \tag{25.7}$$

which means that $\mathcal{L}$ is always symmetric and nonnegative-definite. If we let $\theta_1 \geq \theta_2 \geq \ldots \geq \theta_K$ denote the ordered eigenvalues of $\mathcal{L}$, then it holds that

$$\theta_k \geq 0, \quad k = 1, 2, \ldots, K \tag{25.8}$$

Moreover, from the definition of $\mathcal{L}$, the entries on each of its rows add up to zero so that

$$\mathcal{L}\mathbb{1} = 0 \tag{25.9}$$

This means that $\mathbb{1}$ is a right eigenvector for $\mathcal{L}$ corresponding to the eigenvalue at zero. This also means that the smallest eigenvalue of $\mathcal{L}$ is always zero, i.e.,

$$\theta_K = 0 \tag{25.10}$$

The second smallest eigenvalue, θ_{K-1}, is called the *algebraic connectivity* of the graph and it conveys important information about the topology.

LEMMA 25.2. (Laplacian and graph connectivity) *Let $\theta_1 \geq \theta_2 \geq \ldots \geq \theta_K$ denote the ordered eigenvalues of $\mathcal{L}$. Then, the graph is connected if, and only if, its algebraic connectivity is nonzero (i.e., $\theta_{K-1} \neq 0$). Moreover, the number of times that zero appears as an eigenvalue of $\mathcal{L}$ (i.e., its multiplicity) is equal to the number of connected subgraphs.*

Proof: If the algebraic connectivity is nonzero then the graph must be connected. This statement can be verified by contradiction. Assume, to the contrary, that $\theta_{K-1} \neq 0$ and the graph is disconnected. Then, the Laplacian matrix of the graph will have a block diagonal structure, with one block for each subgraph. As a result, the algebraic multiplicity of zero as an eigenvalue for $\mathcal{L}$ will be larger than 1, which implies that θ_{K-1} will need to be zero. This conclusion contradicts the fact that $\theta_{K-1} \neq 0$.

Conversely, assume the graph is connected and let us verify that we must have $\theta_{K-1} \neq 0$. Let x denote an arbitrary eigenvector for the Laplacian matrix $\mathcal{L}$ corresponding to the eigenvalue at zero, i.e., $\mathcal{L}x = 0$. We already know from (25.9) that $\mathcal{L}\mathbb{1} = 0$. This means that in order to establish that the algebraic multiplicity of the zero eigenvalue is 1, we need to verify that x is proportional to the vector $\mathbb{1}$. Let us denote the individual entries of x by x_k. Multiplying $\mathcal{L}x = 0$ by x^{T} from the left we get $x^{\mathsf{T}}\mathcal{L}x = 0$. This equality in turn implies that over the neighborhood of each node k:

$$\sum_{\ell \in \mathcal{N}_k} (x_k - x_\ell)^2 = 0 \tag{25.11}$$

from which we conclude that

$$x_\ell = x_k, \quad \forall\, \ell \in \mathcal{N}_k \tag{25.12}$$

That is, the entries of x within each neighborhood have uniform values. But since the graph is connected, we conclude that all entries of x must be equal to each other. It follows that the eigenvector x is proportional to the vector $\mathbb{1}$, as desired.

For the second part of the lemma, assume the graph consists of S separate connected subgraphs. Then, the Laplacian matrix will have a block diagonal structure, say, of the form $\mathcal{L} = \text{diag}\{\mathcal{L}_1, \mathcal{L}_2, \ldots, \mathcal{L}_S\}$, where the individual $\mathcal{L}_s$ are the Laplacian matrices for the smaller subgraphs. The smallest eigenvalue for each of these Laplacian matrices will be zero and unique, so that the multiplicity of zero as an eigenvalue of $\mathcal{L}$ is equal to S.

■

25.2 WEIGHT MATRICES

Most of the time, the weights $\{a_{\ell k}\}$ will be nonnegative scalars that satisfy the following conditions for each agent $k = 1, 2, \ldots, K$:

$$a_{\ell k} \geq 0, \quad a_{\ell k} = a_{k\ell}, \quad \sum_{\ell=1}^{K} a_{\ell k} = 1, \quad \text{and} \quad a_{\ell k} = 0 \text{ if } \ell \notin \mathcal{N}_k \tag{25.13}$$

Condition (25.13) implies that the combination matrix $A = [a_{\ell k}]$ is symmetric and satisfies $A^{\mathsf{T}}\mathbb{1} = \mathbb{1}$, where $\mathbb{1}$ denotes the vector with all entries equal to 1. It then follows that A is *doubly stochastic* since the entries on each of its rows and on each of its columns add up to 1:

$$A^{\mathsf{T}}\mathbb{1} = \mathbb{1}, \quad A = A^{\mathsf{T}} \qquad \textbf{(doubly stochastic)} \tag{25.14}$$

A doubly stochastic matrix is not required to be symmetric, although this is the case here under the symmetry condition in (25.13). If, on the other hand, A is not symmetric (i.e., if condition $a_{\ell k} = a_{k\ell}$ does not hold), then the sum in (25.13) would imply that A is a *left-stochastic* matrix with the entries on each of its columns adding up to 1 – recall Section 1.12:

$$A^{\mathsf{T}}\mathbb{1} = \mathbb{1} \qquad \textbf{(left-stochastic)} \tag{25.15}$$

We will be dealing mainly with doubly stochastic matrices in this chapter.

25.2.1 Stochastic Matrices

There are many choices for the matrix A. The following are some examples for graphs with K nodes:

(1) (**Averaging rule**). It has a left-stochastic combination matrix with entries

$$a_{\ell k} = \begin{cases} 1/n_k, & \text{if } k \neq \ell \text{ are neighbors or } k = \ell \\ 0, & \text{otherwise} \end{cases} \tag{25.16}$$

where n_k denotes the degree of node k. Under this rule, each node averages the data from their neighbors.

(2) (**Laplacian rule**). It has a doubly stochastic combination matrix, defined generally as follows:

$$A = I_K - \beta\mathcal{L} \tag{25.17}$$

where $\beta > 0$ is some positive constant and $\mathcal{L}$ is the Laplacian matrix defined earlier in (25.3). One important special case corresponds to the choice $\beta = 1/n_{\max}$, where $n_{\max}$ refers to the maximum degree across the graph:

$$n_{\max} \stackrel{\Delta}{=} \max_{1 \leq k \leq K} n_k \tag{25.18}$$

In this case, the expression for A reduces to

$$a_{\ell k} = \begin{cases} 1/n_{\max}, & \text{if } k \neq \ell \text{ are neighbors} \\ 1 - (n_k - 1)/n_{\max}, & k = \ell \\ 0, & \text{otherwise} \end{cases} \tag{25.19}$$

A second special case occurs when $\beta = 1/K$ and leads to the maximum-degree rule:

$$a_{\ell k} = \begin{cases} 1/K, & \text{if } k \neq \ell \text{ are neighbors} \\ 1 - (n_k - 1)/K, & k = \ell \\ 0, & \text{otherwise} \end{cases} \tag{25.20}$$

Obviously, since $n_{\max} \leq K$, the choice $\beta = 1/n_{\max}$ assigns larger weights to neighbors than the choice $\beta = 1/K$.

(3) (**Metropolis rule**). It has a doubly stochastic combination matrix with entries

$$a_{\ell k} = \begin{cases} 1/\max\{n_k, n_\ell\}, & \text{if } k \neq \ell \text{ are neighbors} \\ 1 - \displaystyle\sum_{\ell \in \mathcal{N}_k \setminus \{k\}} a_{\ell k}, & \text{when } k = \ell \\ 0, & \text{otherwise} \end{cases} \tag{25.21}$$

Stochastic weight matrices A have some useful properties, which we will exploit in later sections. We establish them here for ease of reference.

LEMMA 25.3. (Spectral norm of stochastic matrices) *Let A be a $K \times K$ left- (i.e., $A^{\mathsf{T}}\mathbb{1} = \mathbb{1}$), right- (i.e., $A\mathbb{1} = \mathbb{1}$), or doubly stochastic (i.e., $A^{\mathsf{T}}\mathbb{1} = A\mathbb{1} = \mathbb{1}$) matrix. Let $\rho(A)$ denote the spectral radius of A. Then, $\rho(A) = 1$ and, therefore, all eigenvalues of A lie inside the unit disc, i.e., $|\lambda(A)| \leq 1$.*

Proof: We prove the result for right-stochastic matrices; a similar argument applies to left- or doubly stochastic matrices. Let A be right-stochastic. Then, $A\mathbb{1} = \mathbb{1}$ so that $\lambda = 1$ is one of the eigenvalues of A. Moreover, for any matrix A, it holds that $\rho(A) \leq \|A\|_\infty$, where $\|\cdot\|_\infty$ denotes the maximum absolute row sum of its matrix argument. But since all entries on each row of A are nonnegative and add up to 1, we have $\|A\|_\infty = 1$. Therefore, $\rho(A) \leq 1$. And since we already know that A has an eigenvalue at $\lambda = 1$, we conclude that $\rho(A) = 1$.

■

The above result asserts that the spectral radius of a stochastic matrix is equal to 1 and that A has an eigenvalue at $\lambda = 1$. The result, however, does not rule out the possibility of multiple eigenvalues at $\lambda = 1$, or even other (complex) eigenvalues with magnitude equal to 1. If the stochastic matrix A happens to be *primitive* (which happens when the underlying graph is strongly connected by

Lemma 25.1), then the *Perron–Frobenius theorem* leads to a stronger characterization of the eigen-structure of A. We encountered this theorem earlier in the commentary at the end of Chapter 1. In particular, it asserts that A has a single eigenvalue at 1 with multiplicity 1, while all other eigenvalues have magnitude strictly less than 1. We reproduce some elements of the theorem here. When A is stochastic (left, right, or doubly) and *primitive*, then:

(a) A will have a *single* eigenvalue at 1.
(b) All other eigenvalues of A are *strictly inside* the unit circle so that $\rho(A) = 1$.
(c) With proper sign scaling, all entries of the right-eigenvector of A corresponding to the single eigenvalue at 1 are *positive*. Let p denote this right-eigenvector, with its entries $\{p_k\}$ normalized to add up to 1, i.e.,

$$\boxed{Ap = p, \quad \mathbb{1}^{\mathsf{T}} p = 1, \quad p_k > 0, \; k = 1, 2, \ldots, K} \tag{25.22}$$

We refer to p as the *Perron vector* or eigenvector of A; its kth entry represents the *centrality* of agent k. All other eigenvectors of A associated with the other eigenvalues will have at least one negative or complex entry.

(d) It holds that (see Prob. 25.6):

$$\boxed{\lim_{n\to\infty} A^n \; = \; p\mathbb{1}^{\mathsf{T}}} \tag{25.23}$$

25.2.2 Doubly Stochastic Matrices

In particular, when A is doubly stochastic, then its Perron vector p will have identical entries and will be given by

$$p = \frac{1}{K}\mathbb{1} = \text{col}\Big\{1/K,\, 1/K, \ldots,\, 1/K\Big\} \tag{25.24}$$

We list additional useful results for doubly stochastic matrices, which will be called upon in later parts of our development.

Lemma 25.4. (Some properties of doubly stochastic matrices) *Let A be a $K \times K$ doubly stochastic matrix. Then, since A is also symmetric, the following properties hold:*

(a) $\rho(A) = 1$.
(b) $I_K - A$ *is symmetric and nonnegative-definite.*
(c) A^2 *is doubly stochastic.*
(d) $\rho(A^2) = 1$.
(e) The eigenvalues of A^2 *are real and lie inside the interval* $[0, 1]$.
(f) $I_K - A^2 \geq 0$.

Proof: Part (a) follows from Lemma 25.3. For part (b), since A is symmetric and $\rho(A) = 1$, we have that $\lambda(A) \in [-1, 1]$. That is, the eigenvalues of A are real and in the range $[-1, 1]$. It follows that $I - A$ is symmetric and also nonnegative-definite since its eigenvalues are in the range $[0, 2]$. For part (c), note that A^2 is symmetric and $A^2\mathbb{1} = A\mathbb{1} = \mathbb{1}$. Therefore, A^2 is doubly stochastic. Part (d) follows from part (a) since A^2 is doubly stochastic. For part (e), note that A^2 is symmetric and nonnegative-definite. Therefore, its eigenvalues are real and nonnegative. But since $\rho(A^2) = 1$, we must have $\lambda(A^2) \in [0, 1]$. Part (f) follows from part (d).

■

When A happens to be additionally primitive, we get the following useful characterization of the nullspace of $I - A$ (in addition to (25.30) further ahead).

LEMMA 25.5. (Nullspace of $I - A$) *Let A be a $K \times K$ doubly stochastic and primitive matrix. Then, the nullspace of the matrix $I_K - A$ is spanned by $\mathbb{1}_K$:*

$$\text{nullspace}(I_K - A) = \text{span}(\mathbb{1}_K) \tag{25.25}$$

This means that every nonzero vector x such that $(I_K - A)x = 0$ is of the form $x = \alpha\mathbb{1}_K$, for some nonzero scalar α. Moreover, it also holds that

$$\text{nullspace}(I_K - A) \;=\; \text{nullspace}(I_K - A^2) \tag{25.26}$$

Proof: Let x be a vector in the nullspace of $I_K - A$ so that $(I_K - A)x = 0$, which is equivalent to $Ax = x$. Since A is primitive and doubly stochastic, it has a single eigenvalue at 1 and $A\mathbb{1} = \mathbb{1}$. It follows that $x = \alpha\mathbb{1}$, for some α so that (25.25) is justified.

To establish (25.26), consider again any vector x satisfying $(I_K - A)x = 0$. Then,

$$Ax = x \Longrightarrow A^2x = Ax = x \Longrightarrow (I_K - A^2)x = 0 \Longrightarrow x \in \mathcal{N}(I_K - A^2) \tag{25.27}$$

Conversely, the matrix A^2 is also doubly stochastic and primitive. It therefore has a single eigenvalue at 1 and $A^2\mathbb{1} = \mathbb{1}$. Let x be any vector in the nullspace of $(I_K - A^2)$. Then, the equality $A^2x = x$ implies that $x = \alpha\mathbb{1}$ for some scalar α. This implies that x is also an eigenvector for A with eigenvalue 1, which means $Ax = x$ so that $(I_K - A)x = 0$ and x lies in the nullspace of $(I_K - A)$.

■

Since the matrix $(I_K - A)$ is symmetric and nonnegative definite, we introduce the eigen-decomposition:

$$V \;\stackrel{\Delta}{=}\; c(I_K - A) = U\Lambda U^\mathsf{T}, \qquad V^{1/2} \;\stackrel{\Delta}{=}\; U\Lambda^{1/2}U^\mathsf{T} \tag{25.28}$$

for any scalar $c > 0$, and where U is a $K \times K$ orthogonal matrix ($UU^\mathsf{T} = U^\mathsf{T}U = I_K$) and Λ is a $K \times K$ diagonal matrix with nonnegative entries. Moreover, $\Lambda^{1/2}$ is a diagonal matrix with the nonnegative square-roots of the entries of Λ. We use the notation $V^{1/2}$ to refer to the "square-root" of the matrix V since $V^{1/2}(V^{1/2})^\mathsf{T} = V$. Observe that $V^{1/2}$ is a symmetric square-root since $V^{1/2} = (V^{1/2})^\mathsf{T}$ and, hence,

$$V = (V^{1/2})^2 \tag{25.29}$$

Since V has the same nullspace as $I_K - A$, it follows that $V^{1/2}$ and $(I_K - A)$ also have the same nullspace, i.e.,

$$\boxed{(I_K - A)x = 0 \iff Vx = 0 \iff V^{1/2}x = 0} \tag{25.30}$$

Proof of (25.30): Consider first any $x \in \mathcal{N}(V^{1/2})$. Then,

$$V^{1/2}x = 0 \Longrightarrow V^{1/2}\left(V^{1/2}x\right) = 0 \Longrightarrow Vx = 0 \Longrightarrow x \in \mathcal{N}(V) \tag{25.31}$$

Conversely, consider any $x \in \mathcal{N}(V)$. Then,

$$\begin{aligned} Vx = 0 &\Longrightarrow x^{\mathsf{T}}Vx = 0 \\ &\Longrightarrow x^{\mathsf{T}}V^{1/2}(V^{1/2}x) = 0 \\ &\Longrightarrow \|V^{1/2}x\|^2 = 0 \\ &\Longrightarrow x \in \mathcal{N}(V^{1/2}) \end{aligned} \tag{25.32}$$

■

25.3 AGGREGATE AND LOCAL RISKS

We motivate decentralized optimization solutions by considering first an empirical risk minimization problem of the form:

$$w^\star = \underset{w \in \mathbb{R}^M}{\operatorname{argmin}} \left\{ P(w) \triangleq \frac{1}{N}\sum_{m=0}^{N-1} Q(w;\gamma(m), h_m) \right\}, \quad \gamma(m) \in \mathbb{R}, \quad h_m \in \mathbb{R}^M \tag{25.33}$$

with a total of N data pairs $\{(\gamma(m), h_m)\}$ and where $Q(w;\cdot)$ is some loss function. For example, some useful choices for the loss function include:

$$Q(w;\gamma,h) = \begin{cases} q(w) \;+\; (\gamma - h^{\mathsf{T}}w)^2 & \text{(quadratic)} \\ q(w) \;+\; \ln(1+e^{-\gamma h^{\mathsf{T}}w}) & \text{(logistic)} \\ q(w) \;+\; \max\{0, -\gamma h^{\mathsf{T}}w\} & \text{(perceptron)} \\ q(w) \;+\; \max\{0, 1-\gamma h^{\mathsf{T}}w\} & \text{(hinge)} \end{cases} \tag{25.34}$$

where $q(w)$ is a regularization term such as

$$q(w) = \begin{cases} \kappa_2\|w\|^2 & (\ell_2\text{-regularization}) \\ \kappa_1\|w\|_1 & (\ell_1\text{-regularization}) \\ \kappa_1\|w\|_1 + \kappa_2\|w\|^2 & \text{(elastic-net regularization)} \end{cases} \tag{25.35}$$

with $\kappa_1 \geq 0$ and $\kappa_2 \geq 0$.

We continue our discussion by assuming *differentiable* loss and risk functions for convenience of presentation. We already know from the extensive discussion in prior chapters that we can easily extend the resulting algorithms by replacing gradients by subgradients or by employing proximal projections, to deal with nondifferentiable losses or risk functions.

25.3.1 Aggregate Risk Function

Assume the amount of data N is large and let us split it into K *disjoint* sets, of size N_k each:

$$N = N_1 + N_2 + \ldots + N_K \tag{25.36}$$

We denote the kth set of samples by $\mathcal{D}_k$ and place it at vertex k of a graph. We associate with this vertex the *local* risk function:

$$P_k(w) \stackrel{\Delta}{=} \frac{1}{N_k} \sum_{m \in \mathcal{D}_k} Q(w; \gamma_k(m), h_{k,m}) \tag{25.37}$$

where we are adding the subscript k to the data $(\gamma_k(m), h_{k,m})$ to indicate that they are the data samples placed at the kth agent. The risk $P_k(w)$ averages the losses over the data available at agent k only. The global risk $P(w)$ from (25.33) can then be rewritten in the equivalent form:

$$P(w) = \frac{1}{N} \sum_{k=1}^{K} N_k P_k(w) \tag{25.38}$$

We are therefore motivated to consider optimization problems of the following aggregate form:

$$\boxed{w^\star = \underset{w \in \mathbb{R}^M}{\text{argmin}} \left\{ P(w) \stackrel{\Delta}{=} \sum_{k=1}^{K} \alpha_k P_k(w) \right\}} \tag{25.39}$$

where $P(w)$ is expressed as a combination of local risks and the $\{\alpha_k\}$ are positive convex combination coefficients that add up to 1:

$$\sum_{k=1}^{K} \alpha_k = 1, \quad \alpha_k > 0, \quad k = 1, 2, \ldots, K \tag{25.40}$$

If the data size N happens to be split equally among the K agents so that $N_k = N/K$, then $\alpha_k = 1/K$ and the aggregate risk would take the form

$$P(w) = \frac{1}{K} \sum_{k=1}^{K} P_k(w) \tag{25.41}$$

In general, the $\{\alpha_k\}$ need not be uniform across the agents because some agents may have access to more data than other agents or may be more relevant to the solution of the optimization task. For this reason, we will continue to assume possibly nonuniform values for the $\{\alpha_k\}$ for generality.

The derivation leading to (25.39) assumed that the local risks $P_k(w)$ are *empirical* in nature and defined by sample-average expressions of the form (25.37). We can also consider the same optimization problem (25.39) where the local risks are instead *stochastic* in nature and defined as:

$$P_k(w) \stackrel{\Delta}{=} \mathbb{E}_k\, Q(w; \boldsymbol{\gamma}, \boldsymbol{h}) \tag{25.42}$$

where the expectation is over the distribution of the data at the kth agent. This distribution may vary across the agents, which is why we added a subscript k to the expectation symbol. We will continue our derivation by assuming empirical risks of the form (25.37) and will comment later (see Section 25.4.4) on how the resulting algorithms are *equally* applicable to stochastic risks of the form (25.42). Thus, the problem we wish to address is the following.

(Decentralized optimization problem) *Consider a risk function $P(w)$ that is expressed as the (convex) weighted combination of individual risk functions $\{P_k(w)\}$, as in (25.39). The individual risks and their respective data are available at K agents linked by some graph topology. We wish to develop decentralized algorithms that enable all agents in the graph to approach $w^\star$ by relying solely on their local datasets $\{\mathcal{D}_k\}$ and on information exchanges with their immediate neighbors. At any point in time, no single agent will have access to all the data that is available across the graph or to other agents beyond those that are directly linked to them.*

Example 25.1 **(Interpretation as Pareto solution)** The unique vector $w^\star$ that solves (25.39) can be interpreted as corresponding to a Pareto optimal solution for the collection of convex functions $\{\alpha_k P_k(w)\}$. To explain why this is the case, let us review first the concept of *Pareto optimality.*

Let $w_k^\star$ denote a minimizer for $\alpha_k P_k(w)$. In general, the minimizers $\{w_k^\star,\ k = 1, 2, \ldots, K\}$ are distinct from each other. The agents would like to cooperate to seek some common vector, denoted by $w^\star$, that is "optimal" in some sense for the entire network. One useful notion of optimality is *Pareto optimality.* A solution $w^\star$ will be Pareto optimal for all K agents if there does not exist any other vector, $w^\bullet$, that dominates $w^\star$, i.e., that satisfies the following two conditions:

$$P_k(w^\bullet) \leq P_k(w^\star), \quad \text{for all } k \in \{1, 2, \ldots, K\} \tag{25.43}$$

$$P_{k^o}(w^\bullet) < P_{k^o}(w^\star), \quad \text{for at least one } k^o \in \{1, 2, \ldots, K\} \tag{25.44}$$

In other words, there does not exist any other vector $w^\bullet$ that improves one of the risks, say, $P_{k^o}(w^\bullet) < P_{k^o}(w^\star)$, without degrading the performance of the other agents. In this way, solutions $w^\star$ that are Pareto optimal are such that no agent in a cooperative network can have its performance improved by moving away from $w^\star$ without degrading the performance of some other agent.

To illustrate this concept, let us consider an example with $K = 2$ agents with the argument $w \in \mathbb{R}$ being real-valued and scalar. Let the set

$$\mathcal{S} \triangleq \Big\{\alpha_1 P_1(w),\ \alpha_2 P_2(w)\Big\} \subset \mathbb{R}^2 \tag{25.45}$$

denote the achievable risk values over all feasible choices of $w \in \mathbb{R}$; each point $S \in \mathcal{S}$ belongs to the two-dimensional space $\mathbb{R}^2$ and represents values attained by the risk functions $\{\alpha_1 P_1(w), \alpha_2 P_2(w)\}$ for a particular w. The shaded areas in Fig. 25.3 represent the set $\mathcal{S}$ for two situations of interest. The plot on the left represents the situation in which the two risk functions $\alpha_1 P_1(w)$ and $\alpha_2 P_2(w)$ achieve their minima at the *same* location,

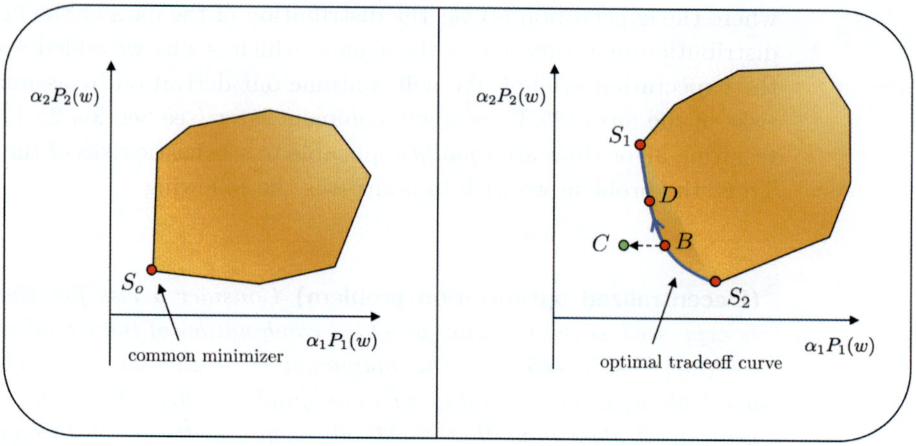

Figure 25.3 Pareto optimal points for the case $K = 2$. In the figure on the left, point S_o denotes the optimal point where both risk functions are minimized simultaneously. In the figure on the right, all points that lie on the heavy boundary curve are Pareto optimal solutions.

namely, $w_1^\star = w_2^\star$. This location is indicated by the point $S_o = \{\alpha_1 P_1(w_1^\star); \alpha_2 P_2(w_2^\star)\}$ where $w_1^\star = w_2^\star$ denotes the common minimizer. In comparison, the plot on the right represents the situation in which the two risk functions $\alpha_1 P_1(w)$ and $\alpha_2 P_2(w)$ achieve their minima at two distinct locations, $w_1^\star$ and $w_2^\star$. Point S_1 in the figure indicates the location where $\alpha_1 P_1(w)$ attains its minimum value, while point S_2 indicates the location where $\alpha_2 P_2(w)$ attains its minimum value. In this case, the two risk functions do not have a common minimizer. It is easy to verify that all points that lie on the heavy curve between points S_1 and S_2 are Pareto optimal solutions for $\{\alpha_1 P_1(w), \alpha_2 P_2(w)\}$. For example, starting at some arbitrary point B on the curve, if we want to reduce the value of $\alpha_1 P_1(w)$ without increasing the value of $\alpha_2 P_2(w)$, then we will need to move out of the achievable set $\mathcal{S}$ toward point C, which is not feasible. The alternative choice to reducing the value of $\alpha_1 P_1(w)$ is to move from B on the curve to another Pareto optimal point, such as point D. This move, while feasible, would increase the value of $\alpha_2 P_2(w)$. In this way, we would need to trade the value of $\alpha_2 P_2(w)$ for $\alpha_1 P_1(w)$. For this reason, the curve from S_1 to S_2 is called the optimal trade-off curve (or *optimal trade-off surface* when $K > 2$).

Pareto optimal solutions are generally nonunique. One useful method to determine a Pareto solution is a *scalarization* technique, whereby an *aggregate* risk function is first formed as the weighted sum of the component convex risk functions as follows:

$$P^{\text{glob},\pi}(w) \triangleq \sum_{k=1}^{N} \pi_k \Big(\alpha_k P_k(w)\Big) \tag{25.46}$$

where the $\{\pi_k\}$ are positive scalars. It is known that the unique minimizer for the above aggregate risk, which we denote by w^π, corresponds to a Pareto optimal solution for the collection $\{\alpha_k P_k(w),\ k = 1, 2, \ldots, K\}$. Moreover, by varying the values of $\{\pi_k\}$, we determine different Pareto optimal solutions. If we compare expression (25.46) with the earlier aggregate risk in (25.39), we conclude that the solution $w^\star$ can be interpreted as the Pareto optimal solution corresponding to $\pi_k = 1$.

25.3.2 Conditions on Risk and Loss Functions

In the following, and for convenience of presentation, we will assume that each individual risk function $P_k(w)$ is *convex* with δ-Lipschitz gradients, while the aggregate risk $P(w)$ is *ν-strongly convex*. This latter condition implies that the minimizer $w^\star$ is unique. We hasten to add that these requirements can be relaxed; for example, it is not necessary for all local risks to be convex. Nevertheless, we will examine the convergence behavior of the various distributed algorithms derived in this chapter under the stated convexity assumptions to keep the presentation more tractable.

More specifically, and in a manner similar to what we did in the earlier chapters, we will assume the following conditions on the risk and loss functions for both cases of empirical and stochastic risk minimization problems. For empirical risk minimization problems of the form (25.33), we assume:

(A1) (**Strongly convex risk**). $P(w)$ is ν-strongly convex and first-order differentiable, namely, for every $w_1, w_2 \in \text{dom}(P)$ and some $\nu > 0$:

$$P(w_2) \geq P(w_1) \;+\; (\nabla_{w^\mathsf{T}} P(w_1))^\mathsf{T}(w_2 - w_1) \;+\; \frac{\nu}{2}\|w_2 - w_1\|^2 \qquad (25.47)$$

(A2) (**Convex individual risks**). We also assume that the individual risks $P_k(w)$ are convex and first-order differentiable.

(A3) (**δ-Lipschitz loss gradients**). The gradient vectors of $Q(w;\cdot)$ are δ-Lipschitz regardless of the data argument, i.e.,

$$\|\nabla_w Q(w_2; \gamma(k), h_k) - \nabla_w Q(w_1; \gamma(\ell), h_\ell)\| \;\leq\; \delta\, \|w_2 - w_1\| \qquad (25.48)$$

for any $w_1, w_2 \in \text{dom}(Q)$, any $0 \leq k, \ell \leq N-1$, and with $\delta \geq \nu$. It can be verified that condition (25.48) implies that the gradients of $P(w)$ and of each $P_k(w)$ are δ-Lipschitz as well, namely,

$$\|\nabla_w P(w_2) - \nabla_w P(w_1)\| \;\; \leq \delta\|w_2 - w_1\| \qquad (25.49a)$$

$$\|\nabla_w P_k(w_2) - \nabla_w P_k(w_1)\| \leq \delta\|w_2 - w_1\| \qquad (25.49b)$$

If $P(w)$ happens to be twice-differentiable, then we also know from (12.15) that conditions (25.47) and (25.49a) are equivalent to

$$0 < \nu I_M \leq \nabla_w^2 P(w) \;\leq\; \delta I_M \qquad (25.50)$$

in terms of the Hessian matrix of $P(w)$.

On the other hand, for stochastic risk minimization problems of the form (25.39) and (25.42), we continue to assume the strong convexity of $P(w)$ and the convexity of the individual $P_k(w)$ but replace (**A3**) by the requirement that the gradients of the loss are now δ-Lipschitz in the mean-square sense:

(A3') (**Mean-square δ-Lipschitz loss gradients**). The gradient vectors of $Q(w;\cdot)$ satisfy the mean-square bound:

$$\mathbb{E}\,\|\nabla_w\, Q(w_2;\boldsymbol{\gamma},\boldsymbol{h}) - \nabla_w\, Q(w_1;\boldsymbol{\gamma},\boldsymbol{h})\|^2 \leq \delta^2\|w_2 - w_1\|^2 \qquad (25.51)$$

for any $w_1, w_2 \in \text{dom}(Q)$. Using the fact that for any scalar random variable $\boldsymbol{x}$ it holds that $(\mathbb{E}\,\boldsymbol{x})^2 \leq \mathbb{E}\,\boldsymbol{x}^2$, we conclude from (25.51) that the gradient vectors of the loss function are also δ-Lipschitz on average, namely,

$$\mathbb{E}\,\|\nabla_w\, Q(w_2;\boldsymbol{\gamma},\boldsymbol{h}) - \nabla_w\, Q(w_1;\boldsymbol{\gamma},\boldsymbol{h})\| \leq \delta\|w_2 - w_1\| \qquad (25.51)$$

By further applying Jensen inequality (8.77) that $f(\mathbb{E}\,\boldsymbol{x}) \leq \mathbb{E}\,f(\boldsymbol{x})$ for the convex function $f(x) = \|x\|$, we conclude from (25.3.2) that the gradient vectors of the risk functions are also δ-Lipschitz (recall the argument used to establish (18.15)):

$$\|\nabla_w\, P(w_2) - \nabla_w\, P(w_1)\| \leq \delta\|w_2 - w_1\| \qquad (25.52a)$$

$$\|\nabla_w\, P_k(w_2) - \nabla_w\, P_k(w_1)\| \leq \delta\|w_2 - w_1\| \qquad (25.52b)$$

Note that under conditions (**A1**,**A2**,**A3**) for empirical risk minimization or conditions (**A1**,**A2**,**A3'**) for stochastic risk minimization, the following conditions hold:

(**P1**) ν-strong convexity of $P(w)$ as in (25.47) (25.53a)

(**P2**) convexity of all individual risks, $\{P_k(w)\}$ (25.53b)

(**P3**) δ-Lipschitz gradients for $P(w)$ as in (25.49a) and (25.52a) (25.53c)

(**P4**) δ-Lipschitz gradients for $P_k(w)$ as in (25.49b) and (25.52b) (25.53d)

25.4 INCREMENTAL, CONSENSUS, AND DIFFUSION

There are several decentralized strategies that can be used to seek the minimizer of (25.39). We motivate the incremental, consensus, and diffusion strategies in this initial section before pursuing a more formal derivation by solving a primal optimization problem. In the next chapter, we derive other methods of the primal–dual type.

25.4.1 Incremental Strategy

To begin with, if all data was available at a centralized location (say, if all agents transmit their datasets $\{\mathcal{D}_k\}$ to a central processor), then a centralized gradient-descent algorithm could be employed at this central location by using:

$$w_n = w_{n-1} - \mu\sum_{k=1}^{K}\alpha_k\,\nabla_{w^{\mathsf{T}}}\,P_k(w_{n-1}), \quad n \geq 0 \qquad (25.54)$$

in terms of the gradients of the local risk functions evaluated at w_{n-1}, and where $\mu > 0$ is a small step-size parameter. We can use this algorithm to motivate the incremental strategy as follows (refer to Fig. 25.4).

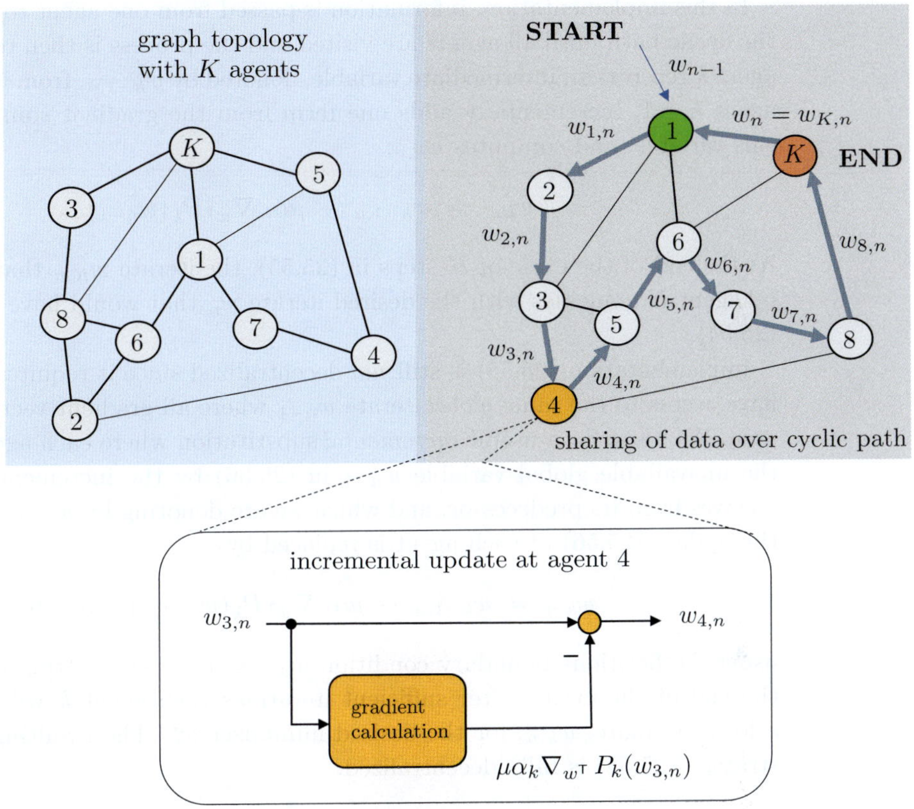

Figure 25.4 The plot on the left shows the original numbering in the given graph topology involving K agents. A cyclic path visiting all agents is found in the plot on the right and the agents are renumbered 1 through K from start to finish. The diagram at the bottom shows the incremental update that is performed at agent 4.

We are given K agents that are linked by some *connected* graph. Starting from this graph, we first determine a *cyclic* trajectory that visits all agents in the graph in succession. Once a cycle is identified, we renumber the agents along the trajectory from 1 to K, with 1 corresponding to the starting location for the cyclic path and K corresponding to the last location. Then, at every iteration n, the centralized update (25.54) can be split into K consecutive *incremental* steps, with each step performed locally at one of the agents:

$$\begin{cases} w_{1,n} &= \boxed{w_{n-1}} - \mu\alpha_1 \nabla_{w^\mathsf{T}} P_1(w_{n-1}) \\ w_{2,n} &= w_{1,n} - \mu\alpha_2 \nabla_{w^\mathsf{T}} P_2(w_{n-1}) \\ \vdots &= \vdots \\ \boxed{w_n} &= w_{K-1,n} - \mu\alpha_K \nabla_{w^\mathsf{T}} P_K(w_{n-1}) \end{cases} \tag{25.55}$$

In this implementation, information is passed from one agent to the next over the cyclic path until all agents are visited and the process is then repeated. Each agent k receives an intermediate variable, denoted by $w_{k-1,n}$, from its predecessor agent $k-1$, incrementally adds one term from the gradient sum in (25.54) to this variable, and computes $w_{k,n}$:

$$w_{k,n} = w_{k-1,n} - \mu\alpha_k \nabla_{w^\mathsf{T}} P_k(w_{n-1}) \tag{25.56}$$

At the end of the cycle of K steps in (25.55), the iterate $w_{K,n}$ that is computed by agent K coincides with the desired iterate w_n that would have resulted from (25.54).

Implementation (25.55) is still *not* decentralized since it requires all agents to have access to the same *global* iterate w_{n-1} where all gradient vectors are evaluated. We resort to a useful *incremental* substitution where each agent k replaces the unavailable global variable w_{n-1} in (25.56) by the incremental variable it receives from its predecessor, and which we are denoting by $w_{k-1,n}$. In this way, the update (25.56) at each agent is replaced by

$$w_{k,n} = w_{k-1,n} - \mu\alpha_k \nabla_{w^\mathsf{T}} P_k(w_{k-1,n}), \quad n \geq 0 \tag{25.57}$$

using the fictitious boundary condition $w_{0,n} = w_{n-1}$ and setting $w_n = w_{K,n}$ at the end of the cycle. After sufficient iterations each agent k will end up with a local estimate, $w_{k,n}$, for the desired minimizer $w^\star$. The resulting incremental strategy (25.58) is fully decentralized.

Incremental strategy for solving problem (25.39).

given a connected graph linking K agents;
determine a cyclic path over the agents;
number the agents from 1 to K from start to finish on the path;
start from any initial condition, $w_{K,-1} = w_{-1}$.
repeat until convergence over $n \geq 0$**:** (25.58)
  $w_{0,n} \leftarrow w_{K,n-1}$
  repeat over agents $k = 1, 2, \ldots, K$**:**
    $w_{k,n} = w_{k-1,n} - \mu\alpha_k \nabla_{w^\mathsf{T}} P_k(w_{k-1,n})$
  end
end
return $w^\star \leftarrow w_{K,n}$.

The incremental strategy (25.58) suffers from a number of drawbacks for online learning over networks. First, the strategy is sensitive to agent or link failures. Second, determining a cyclic path that visits all agents is generally an NP-hard problem. Third, cooperation between agents is limited since each agent receives data from only its preceding agent and shares data with only its successor agent. Fourth, for every iteration n it is necessary to perform K incremental steps and to visit all agents in order to update w_{n-1} to w_n before the next cycle begins. As a result, the processing at the agents will need to be performed at a faster pace than the rate of arrival of the data.

We next motivate two other decentralized strategies based on consensus and diffusion techniques that do not suffer from these limitations. These techniques take advantage of the following flexibility:

(a) First, there is no reason why agents should only receive information from one neighbor at a time and pass information to only one other neighbor.

(b) Second, there is also no reason why the global variable w_{n-1} in (25.54) cannot be replaced by some other choice, other than $w_{k-1,n}$, to attain decentralization.

(c) Third, there is no reason why agents cannot learn simultaneously with other agents rather than wait for each cycle to complete.

25.4.2 Consensus Strategy

The iterate $w_{k-1,n}$ appears twice on the right-hand side of the incremental update (25.57). The first $w_{k-1,n}$ represents the information that agent k receives from its preceding agent. In the consensus strategy, this term is replaced by a convex combination of the iterates that are available at the neighbors of agent k – see the first term on the right-hand side of (25.59). With regards to the second $w_{k-1,n}$ on the right-hand side of (25.57), it is replaced by $w_{k,n-1}$; this quantity is the iterate that is already available at agent k from a previous time instant. In this manner, the resulting consensus iteration at each agent k becomes:

$$w_{k,n} = \sum_{\ell \in \mathcal{N}_k} a_{\ell k}\, w_{\ell,n-1} \; - \; \mu\alpha_k \,\nabla_{w^\mathsf{T}} P_k(w_{k,n-1}) \tag{25.59}$$

In this recursion, agent k first combines the iterates from its neighbors and finds an intermediate estimate

$$\psi_{k,n-1} \triangleq \sum_{\ell \in \mathcal{N}_k} a_{\ell k}\, w_{\ell,n-1} \tag{25.60}$$

and subsequently updates $\psi_{k,n-1}$ to $w_{k,n}$:

$$w_{k,n} = \psi_{k,n-1} \; - \; \mu\alpha_k \,\nabla_{w^\mathsf{T}} P_k(w_{k,n-1}) \tag{25.61}$$

After sufficient iterations, each agent k will end up with a local estimate, $w_{k,n}$, for the desired minimizer $w^\star$. The combination coefficients $\{a_{\ell k}\}$ that appear in (25.59) are nonnegative scalars that satisfy the following conditions for each agent $k = 1, 2, \ldots, K$:

$$a_{\ell k} \geq 0, \;\; a_{\ell k} = a_{k\ell}, \;\; \sum_{\ell=1}^{K} a_{\ell k} = 1, \;\; \text{and} \;\; a_{\ell k} = 0 \text{ if } \ell \notin \mathcal{N}_k \tag{25.62}$$

Condition (25.62) implies that the combination matrix $A = [a_{\ell k}]$ is doubly stochastic. The resulting consensus implementation is listed in (25.64) – see Fig. 25.5. Observe that if each agent k were to run an individual gradient-descent iteration on its local risk function, say, an update of the form:

$$w_{k,n} = w_{k,n-1} - \mu \alpha_k \nabla_{w^{\mathsf{T}}} P_k(w_{k,n-1}), \quad n \geq 0 \tag{25.63}$$

then the consensus construction (25.59) could also be motivated by replacing the starting point $w_{k,n-1}$ in (25.63) by the same neighborhood convex combination used in (25.59).

The information that is used by agent k from its neighbors consists of the iterates $\{w_{\ell,n-1}\}$ and these iterates are *already* available for use from the previous iteration $n-1$. As such, there is *no* need to cycle through the agents anymore. At every iteration n, all agents in the network can run their consensus update (25.59) or (25.64) *simultaneously*. Accordingly, there is no need to select a cyclic trajectory or to renumber the agents, as was the case with the incremental strategy. If desired, the step size μ can be agent-dependent as well and replaced by μ_k.

Consensus strategy for solving problem (25.39).

given a connected graph linking K agents with doubly stochastic A;
start from an arbitrary initial condition, $w_{k,-1}$, for each agent.
repeat until convergence over $n \geq 0$:
  repeat over agents $k = 1, 2, \ldots, K$:

$$\begin{aligned} \psi_{k,n-1} &= \sum_{\ell \in \mathcal{N}_k} a_{\ell k}\, w_{\ell,n-1} \\ w_{k,n} &= \psi_{k,n-1} - \mu \alpha_k \nabla_{w^{\mathsf{T}}} P_k\left(w_{k,n-1}\right) \end{aligned} \tag{25.64}$$

  end
end
iterate $w_{k,n}$ at each agent approximates $w^\star$.

Network description

For later reference, it is useful to express the consensus strategy (25.64) in vector form by introducing network quantities that aggregate the variables from across the graph. Specifically, we collect all iterates $w_{k,n}$ from the agents vertically into the network vector:

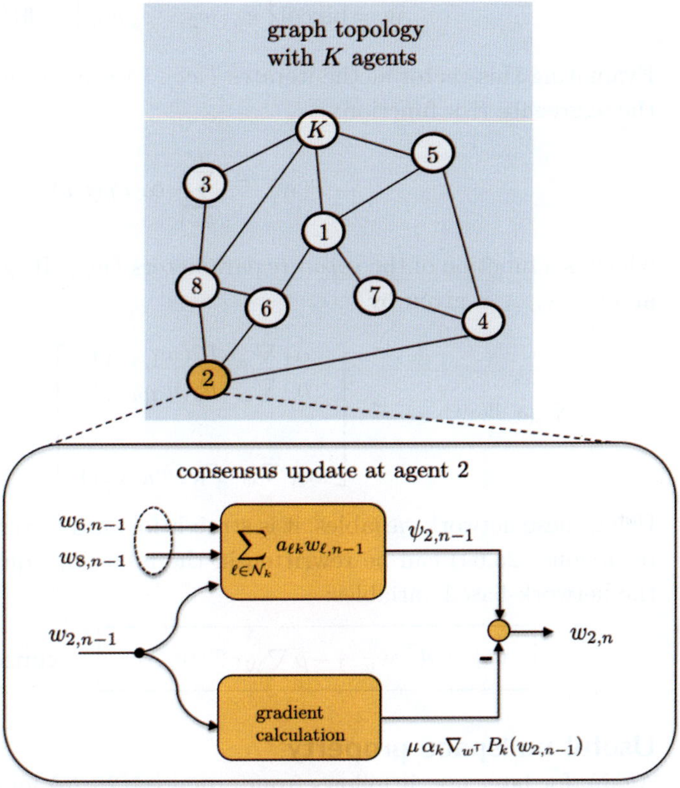

Figure 25.5 The top plot shows the given graph topology involving K agents. The bottom plot illustrates the consensus update that is performed by agent 2.

$$\mathcal{W}_n \triangleq \begin{bmatrix} w_{1,n} \\ w_{2,n} \\ \vdots \\ w_{K,n} \end{bmatrix} \in \mathbb{R}^{KM} \tag{25.65}$$

This vector reflects the state of the network at iteration n. We also define the network combination matrix

$$\mathcal{A} \triangleq A \otimes I_M, \qquad (KM \times KM) \tag{25.66}$$

in terms of the Kronecker product with the identity matrix. This operation replaces every scalar entry $a_{\ell k}$ in A by the diagonal matrix $a_{\ell k} I_M$ in $\mathcal{A}$. In this way, the matrix $\mathcal{A}$ will be a $K \times K$ block matrix with each block having size $M \times M$.

We also associate a parameter vector $w_k \in \mathbb{R}^M$ with each agent k (it serves as a local copy for the parameter w) and collect these parameters into the network vector

$$\mathcal{W} = \text{blkcol}\Big\{w_1, w_2, \ldots, w_K\Big\} \in \mathbb{R}^{KM} \tag{25.67}$$

Evaluating this vector at the iterates $\{w_{k,n}\}$ results in $\mathcal{W}_n$. We further introduce the aggregate risk function:

$$\mathcal{P}(\mathcal{W}) \stackrel{\Delta}{=} \sum_{k=1}^{K} \alpha_k P_k(w_k) \tag{25.68}$$

which is a function of the *separate* parameters $\{w_k\}$. Its gradient vector evaluated at $\mathcal{W} = \mathcal{W}_{n-1}$ is given by

$$\nabla_{\mathcal{W}^\mathsf{T}} \mathcal{P}(\mathcal{W}_{n-1}) \stackrel{\Delta}{=} \begin{bmatrix} \alpha_1 \nabla_{w^\mathsf{T}} P_1(w_{1,n-1}) \\ \alpha_2 \nabla_{w^\mathsf{T}} P_2(w_{2,n-1}) \\ \vdots \\ \alpha_K \nabla_{w^\mathsf{T}} P_K(w_{K,n-1}) \end{bmatrix} \quad (KM \times 1) \tag{25.69}$$

Using these network variables, it is straightforward to verify that the agent-based recursions (25.64) can be rewritten in the following equivalent form in terms of the network-based variables:

$$\boxed{\mathcal{W}_n = \mathcal{A}^\mathsf{T} \mathcal{W}_{n-1} - \mu \nabla_{\mathcal{W}^\mathsf{T}} \mathcal{P}(\mathcal{W}_{n-1})} \quad \textbf{(consensus)} \tag{25.70}$$

Useful nullspace property

Again, for later use, it follows from property (25.25) for doubly stochastic primitive combination matrices A that, in terms of the extended network matrix $\mathcal{A}$ (the proof is given in Appendix 25.B):

$$\boxed{\text{nullspace}(I_{KM} - \mathcal{A}) = \text{span}(\mathbb{1}_K \otimes I_M)} \tag{25.71}$$

One useful conclusion from this property, and which is already reflected in results (25.153) and (25.158) in the appendix, is the following. If $y = \text{col}\{y_1, y_2, \ldots, y_K\} \in \mathbb{R}^{KM}$ is any vector in the nullspace of $(I_{KM} - \mathcal{A})$ with individual subvectors $y_k \in \mathbb{R}^M$, then it will hold that

$$\boxed{y \in \text{nullspace}(I_{KM} - \mathcal{A}) \iff y_1 = y_2 = \ldots = y_K} \tag{25.72}$$

so that the subvectors within y must all agree with each other. If we further recall the notation $V = (I - A)$ from (25.28) and introduce the extended network matrix

$$\mathcal{V} \stackrel{\Delta}{=} V \otimes I_M \tag{25.73}$$

then results (25.71) and (25.72) can be stated equivalently in terms of $\mathcal{V}$:

$$\text{nullspace}(\mathcal{V}) = \text{span}(\mathbb{1}_K \otimes I_M) \tag{25.74a}$$

$$y \in \text{nullspace}(\mathcal{V}) \iff y_1 = y_2 = \ldots = y_K \tag{25.74b}$$

25.4.3 Diffusion Strategy

If we compare the consensus implementation (25.59) with the incremental implementation (25.57), we observe that consensus replaces the two instances of $w_{k-1,n}$ on the right-hand side of (25.57) by two different substitutions, namely, by $\psi_{k,n-1}$ and $w_{k,n-1}$. This *asymmetry* in the consensus construction can be problematic for online implementations. This is because the asymmetry can cause an unstable growth in the state of the network, i.e., some iterates $w_{k,n}$ can grow unbounded. Diffusion strategies resolve the asymmetry problem and have been shown to have superior stability and performance properties. They lead to the class of decentralized stochastic gradient-descent (D-SGD) algorithms.

Combine-then-adapt (CTA) diffusion form

In the CTA formulation of the diffusion strategy, the *same* iterate $\psi_{k,n-1}$ is used to replace the two instances of $w_{k-1,n}$ on the right-hand side of the incremental implementation (25.57), thus leading to description (25.75) where the gradient vector is evaluated at $\psi_{k,n-1}$ as well. The reason for the name "combine-then-adapt" is that the first step in (25.75) involves a combination step, while the second step involves an adaptation (gradient-descent) step. The reason for the qualification "diffusion" is that the use of $\psi_{k,n-1}$ to evaluate the gradient vector allows information to diffuse more thoroughly through the network. This is because information is not only being diffused through the aggregation of the neighborhood iterates, but also through the evaluation of the gradient vector at the aggregate state value.

CTA diffusion strategy for solving problem (25.39).

given a connected graph linking K agents with doubly stochastic A;
start from an arbitrary initial condition, $w_{k,-1}$, for each agent.
repeat until convergence over $n \geq 0$:
  repeat over agents $k = 1, 2, \ldots, K$:

$$\begin{aligned} \psi_{k,n-1} &= \sum_{\ell \in \mathcal{N}_k} a_{\ell k}\, w_{\ell,n-1} \\ w_{k,n} &= \psi_{k,n-1} - \mu \alpha_k \nabla_{w^\mathsf{T}} P_k\left(\psi_{k,n-1}\right) \end{aligned} \tag{25.75}$$

  end
end
iterate $w_{k,n}$ at each agent approximates $w^\star$.

In the diffusion implementation, all agents in the neighborhood $\mathcal{N}_k$ share their current iterates $\{w_{\ell,n-1}\}$ with agent k. These iterates are combined in a convex manner to obtain the intermediate value $\psi_{k,n-1}$, which is then used to perform the gradient-descent update in the second equation taking $\psi_{k,n-1}$ to $w_{k,n}$. All agents perform similar updates at the same time and the process repeats itself continually – see Fig. 25.6 (*left*). After sufficient iterations, each agent k will end up with a local estimate, $w_{k,n}$, for the desired minimizer $w^\star$. Again, if desired the step size μ can be replaced by μ_k and become agent-dependent.

Using the same network-based quantities (25.65)–(25.68), it is straightforward to verify that the diffusion recursions (25.75) can be rewritten in the following equivalent form in terms of the network-based variables (compare with (25.70) for consensus):

$$\boxed{\mathcal{W}_n = \mathcal{A}^\mathsf{T}\,\mathcal{W}_{n-1} - \mu\,\nabla_{\mathcal{W}^\mathsf{T}}\,\mathcal{P}(\mathcal{A}^\mathsf{T}\,\mathcal{W}_{n-1})} \qquad \textbf{(CTA diffusion)} \qquad (25.76)$$

Adapt-then-combine (ATC) diffusion form

A similar implementation can be obtained by switching the order of the combination and adaptation steps in (25.75), as shown in (25.77). The structure of the CTA and ATC strategies are fundamentally identical; the difference lies in which variable we choose to correspond to the updated iterate $w_{k,n}$. In ATC, we choose the result of the *combination* step to be $w_{k,n}$, whereas in CTA we choose the result of the *adaptation* step to be $w_{k,n}$.

ATC diffusion strategy for solving problem (25.39).

---

given a connected graph linking K agents with doubly stochastic A;
start from an arbitrary initial condition, $w_{k,-1}$, for each agent.
repeat until convergence $n \geq 0$:
  repeat over agents $k = 1, 2, \ldots, K$:

$$\begin{aligned} \psi_{k,n} &= w_{k,n-1} - \mu\alpha_k\,\nabla_{w^\mathsf{T}} P_k\,(w_{k,n-1}) \\ w_{k,n} &= \sum_{\ell\in\mathcal{N}_k} a_{\ell k}\,\psi_{\ell,n} \end{aligned} \qquad (25.77)$$

  end
end
iterate $w_{k,n}$ at each agent approximates $w^\star$.

In this implementation, agent k first performs a local gradient-descent update using its risk function and moves from $w_{k,n-1}$ to the intermediate value $\psi_{k,n}$. Subsequently, all agents in the neighborhood $\mathcal{N}_k$ share the intermediate iterates $\{\psi_{\ell,n}\}$ with agent k. These iterates are combined in a convex manner to obtain $w_{k,n}$. All agents perform similar updates at the same time and the process repeats itself continually – see Fig. 25.6 (*right*). After sufficient iterations, each agent k will end up with a local estimate, $w_{k,n}$, for the desired minimizer $w^\star$. Again, if desired the step size μ can be replaced by μ_k and become agent-dependent.

Using the same network-based quantities (25.65)–(25.68), it is straightforward to verify that the diffusion recursions (25.77) can be rewritten in the following

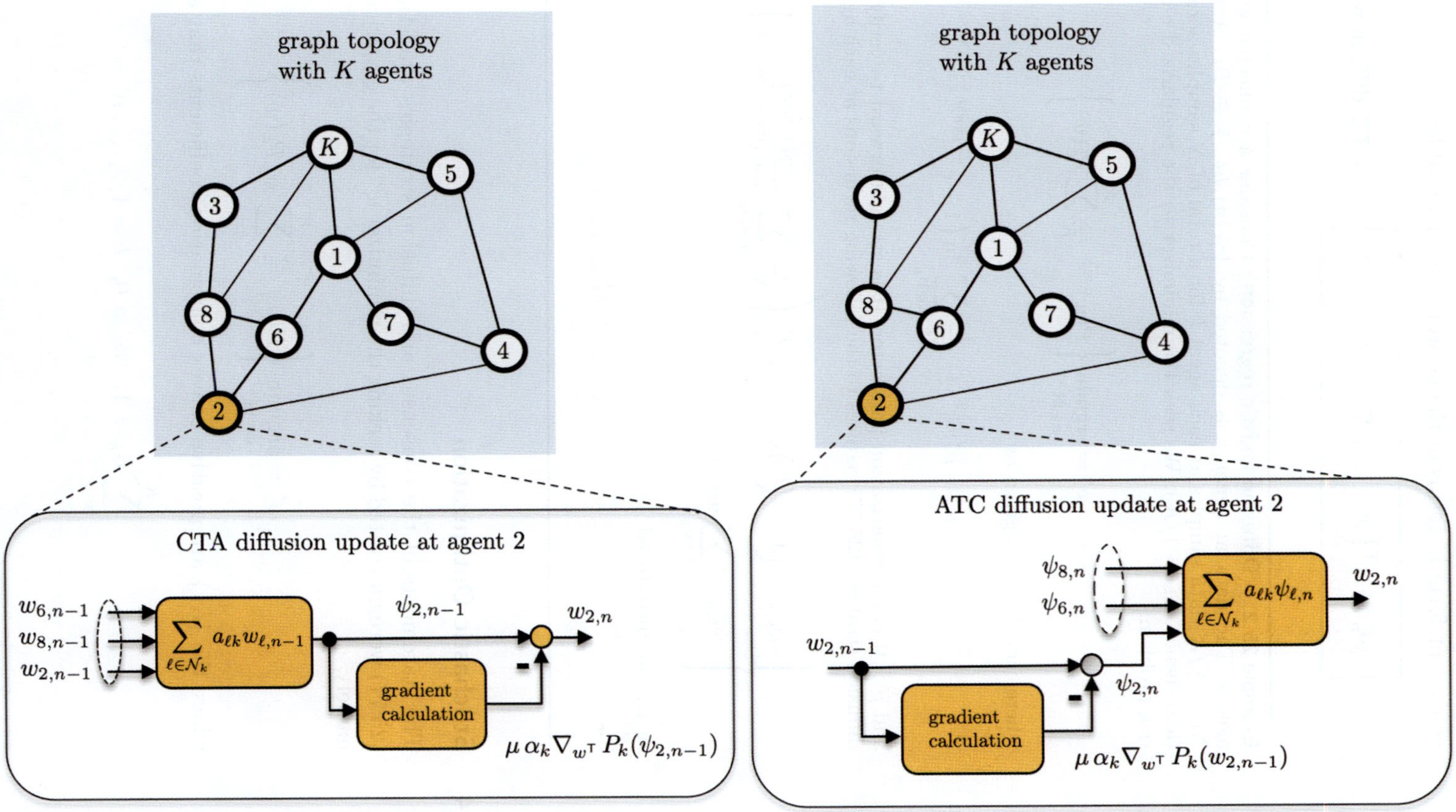

Figure 25.6 The plot on the left illustrates one CTA diffusion step, while the plot on the right illustrates one ATC diffusion step.

equivalent form in terms of the network-based variables:

$$\boxed{\mathcal{W}_n = \mathcal{A}^{\mathsf{T}}\Big(\mathcal{W}_{n-1} - \mu\, \nabla_{\mathcal{W}^{\mathsf{T}}}\, \mathcal{P}(\mathcal{W}_{n-1})\Big)} \qquad \textbf{(ATC diffusion)} \tag{25.78}$$

Example 25.2 **(Diffusion logistic regression)** Consider a connected graph linking K agents. The data at agent k is denoted by $\{\gamma_k(n), h_{k,n}\} \in \mathcal{D}_k$. Assume there are $N_k = N/K$ data samples at each agent k for a total of N samples across the network (and, hence, $\alpha_k = 1/K$). We seek the minimizer to the regularized empirical logistic risk problem:

$$w^\star = \underset{w\in\mathbb{R}^M}{\text{argmin}} \left\{ P(w) \stackrel{\Delta}{=} \frac{1}{K}\sum_{k=1}^{K} P_k(w) \right\} \tag{25.79}$$

where each $P_k(w)$ is given by

$$P_k(w) = \rho\|w\|^2 + \frac{1}{N_k}\sum_{n\in\mathcal{D}_k} \ln\left(1 + e^{-\gamma_k(n)h_{k,n}^{\mathsf{T}}w}\right) \tag{25.80}$$

and $\rho > 0$ is a regularization parameter. It is straightforward to verify that the ATC diffusion strategy (25.77) reduces to the following recursions at each agent k:

$$\left\{\begin{aligned} \psi_{k,n} &= \left(1 - \frac{2\rho\mu}{K}\right) w_{k,n-1} + \frac{\mu}{K}\left(\frac{1}{N_k}\sum_{m\in\mathcal{D}_k} \frac{\gamma_k(m)h_{k,m}}{1 + e^{\gamma_k(m)\widehat{\gamma}_k(m)}}\right) \\ w_{k,n} &= \sum_{\ell\in\mathcal{N}_k} a_{\ell k}\, \psi_{\ell,n} \end{aligned}\right. \tag{25.81}$$

where we introduced

$$\widehat{\gamma}_k(m) \stackrel{\Delta}{=} h_{k,m}^{\mathsf{T}} w_{k,n-1} \tag{25.82}$$

25.4.4 Stochastic Optimization

The statements of the consensus and diffusion strategies (25.64), (25.75), and (25.77) were motivated by seeking the minimizer $w^\star$ for the empirical risk (25.39), namely,

$$w^\star = \underset{w\in\mathbb{R}^M}{\text{argmin}} \left\{ P(w) \stackrel{\Delta}{=} \sum_{k=1}^{K} \alpha_k P_k(w) \right\} \tag{25.83}$$

where the $\{\alpha_k\}$ are positive convex combination coefficients that add up to 1:

$$\sum_{k=1}^{K} \alpha_k = 1, \quad \alpha_k > 0, \quad k = 1, 2, \ldots, K \tag{25.84}$$

and each individual risk function is defined as the sample average of the loss function over the data at agent k:

$$P_k(w) \triangleq \frac{1}{N_k} \sum_{m \in \mathcal{D}_k} Q(w; \gamma_k(m), h_{k,m}) \tag{25.85}$$

The algorithms require the evaluation of the gradients of these local risk functions, which can be demanding, depending on the size N_k of the local datasets. An alternative implementation would be to rely on *stochastic* gradient approximations. At every iteration n, agent k selects *at random* (with or without replacement or according to some sampling policy) a data sample $(\boldsymbol{\gamma}_k(n), \boldsymbol{h}_{k,n})$ from its dataset $\mathcal{D}_k$. The value of the local loss function at the selected sample would then be used to approximate the true gradient vector:

$$\widehat{\nabla_{w^{\mathsf{T}}} P_k}(\boldsymbol{w}_{k,n-1}) = \nabla_{w^{\mathsf{T}}} Q(\boldsymbol{w}_{k,n-1}, \boldsymbol{\gamma}_k(n), \boldsymbol{h}_{k,n}) \tag{25.86}$$

Observe that we are now denoting the iterates and data in boldface to refer to their random nature. This same gradient approximation can be used when the individual risks $P_k(w)$ happen to be *stochastic in nature*, as in (25.42):

$$P_k(w) = \mathbb{E}_k\, Q(w; \boldsymbol{\gamma}, \boldsymbol{h}) \tag{25.87}$$

In that case, independent streaming data $(\boldsymbol{\gamma}_k(n), \boldsymbol{h}_{k,n})$ would be used to evaluate the loss functions in (25.86). Either way, the construction leads to stochastic implementations for the consensus and diffusion strategies as described in the listings below.

Stochastic consensus strategy for solving problem (25.39).

---

given a connected graph linking K agents with doubly stochastic A;
start an from arbitrary initial condition, $\boldsymbol{w}_{k,-1}$, for each agent.
repeat until convergence over $n \geq 0$:
  repeat over agents $k = 1, 2, \ldots, K$:
    select or receive a random data pair $(\boldsymbol{\gamma}_k(n), \boldsymbol{h}_{k,n})$ from $\mathcal{D}_k$ (25.88)

$$\boldsymbol{\psi}_{k,n-1} = \sum_{\ell \in \mathcal{N}_k} a_{\ell k}\, \boldsymbol{w}_{\ell,n-1}$$

$$\boldsymbol{w}_{k,n} = \boldsymbol{\psi}_{k,n-1} - \mu\alpha_k \nabla_{w^{\mathsf{T}}} Q\left(\boldsymbol{w}_{k,n-1}; \boldsymbol{\gamma}_k(n), \boldsymbol{h}_{k,n}\right)$$

  end
end
iterate $w_{k,n}$ at each agent approximates $w^\star$.

Stochastic CTA diffusion strategy for solving problem (25.39).

given a connected graph linking K agents with doubly stochastic A;
start from an arbitrary initial condition, $\boldsymbol{w}_{k,-1}$, for each agent.
repeat until convergence over $n \geq 0$:
  repeat over agents $k = 1, 2, \ldots, K$:
    select or receive a random data pair $(\boldsymbol{\gamma}_k(n), \boldsymbol{h}_{k,n})$ from $\mathcal{D}_k$
$$\begin{aligned} \boldsymbol{\psi}_{k,n-1} &= \sum_{\ell \in \mathcal{N}_k} a_{\ell k}\, \boldsymbol{w}_{\ell,n-1} \\ \boldsymbol{w}_{k,n} &= \boldsymbol{\psi}_{k,n-1} - \mu \alpha_k \nabla_{w^{\mathsf{T}}} Q\left(\boldsymbol{\psi}_{k,n-1}; \boldsymbol{\gamma}_k(n), \boldsymbol{h}_{k,n}\right) \end{aligned} \tag{25.89}$$
  end
end
iterate $w_{k,n}$ at each agent approximates $w^\star$.

Stochastic ATC diffusion strategy for solving problem (25.39).

given a connected graph linking K agents with doubly stochastic A;
start from an arbitrary initial condition, $\boldsymbol{w}_{k,-1}$, for each agent.
repeat until convergence over $n \geq 0$:
  repeat over agents $k = 1, 2, \ldots, K$:
    select or receive a random data pair $(\boldsymbol{\gamma}_k(n), \boldsymbol{h}_{k,n})$ from $\mathcal{D}_k$
$$\begin{aligned} \boldsymbol{\psi}_{k,n} &= \boldsymbol{w}_{k,n-1} - \mu \alpha_k \nabla_{w^{\mathsf{T}}} Q\left(\boldsymbol{w}_{k,n-1}; \boldsymbol{\gamma}_k(n), \boldsymbol{h}_{k,n}\right) \\ \boldsymbol{w}_{k,n} &= \sum_{\ell \in \mathcal{N}_k} a_{\ell k}\, \boldsymbol{\psi}_{\ell,n} \end{aligned} \tag{25.90}$$
  end
end
iterate $w_{k,n}$ at each agent approximates $w^\star$.

These implementations minimize aggregate risks of the form (25.83) where the local risks, $P_k(w)$, are defined either by the empirical expression (25.85) or by the stochastic expression (25.87), namely,

$$(\textbf{stochastic risk minimization})\;\; P_k(w) = \mathbb{E}_k\, Q(w; \boldsymbol{\gamma}, \boldsymbol{h}) \tag{25.91a}$$

$$(\textbf{empirical risk minimization})\;\; P_k(w) = \frac{1}{N_k} \sum_{m \in \mathcal{D}_k} Q(w; \gamma_k(m), h_{k,m}) \tag{25.91b}$$

We will continue to derive *deterministic* decentralized algorithms for solving (25.83), assuming access to the true or full gradient information. It is understood by now that it is always possible to transform the resulting algorithms into stochastic versions by appealing to the instantaneous gradient approximation (25.86) or to mini-batch approximations, as already discussed at great length in previous chapters.

Example 25.3 **(Stochastic diffusion logistic regression)** If we refer to the earlier diffusion implementation (25.81) for the decentralized logistic regression problem, its stochastic version will now take the form (where we are replacing μ/K by the step-size parameter μ' for simplicity)

$$\left\{\begin{array}{l}\text{select or receive a random data pair } (\boldsymbol{\gamma}_k(n), \boldsymbol{h}_{k,n}) \text{ from } \mathcal{D}_k\\ \boldsymbol{\psi}_{k,n} = (1-2\rho\mu')\boldsymbol{w}_{k,n-1} + \mu' \dfrac{\boldsymbol{\gamma}_k(n)\boldsymbol{h}_{k,n}}{1+e^{\boldsymbol{\gamma}_k(n)\widehat{\boldsymbol{\gamma}}_k(n)}}\\ \boldsymbol{w}_{k,n} = \displaystyle\sum_{\ell\in\mathcal{N}_k} a_{\ell k}\,\boldsymbol{\psi}_{\ell,n}\end{array}\right. \tag{25.92}$$

where we introduced

$$\widehat{\boldsymbol{\gamma}}_k(n) \stackrel{\Delta}{=} \boldsymbol{h}_{k,n}^{\mathsf{T}}\boldsymbol{w}_{k,n-1} \tag{25.93}$$

Example 25.4 **(Regression networks)** Consider a connected graph linking K agents. The data at agent k are denoted by $\{\gamma_k(n), h_{k,n}\} \in \mathcal{D}_k$ and they arise from some underlying unknown distribution. We seek the minimizer to the regularized mean-square-error risk problem:

$$w^\star = \underset{w\in\mathbb{R}^M}{\text{argmin}} \left\{\rho\|w\|^2 + \frac{1}{K}\sum_{k=1}^{K}\mathbb{E}_k(\boldsymbol{\gamma} - \boldsymbol{h}^{\mathsf{T}}w)^2\right\} \tag{25.94}$$

where $\mathbb{E}_k$ refers to averaging relative to the data distribution at agent k. Using (25.90) we reach the following diffusion recursions for this MSE or regression network (where we are replacing μ/K by the step-size parameter μ' for simplicity):

$$\left\{\begin{array}{l}\text{select or receive a random data pair } (\boldsymbol{\gamma}_k(n), \boldsymbol{h}_{k,n}) \text{ from } \mathcal{D}_k\\ \boldsymbol{\psi}_{k,n} = (1-2\rho\mu')w_{k,n-1} + 2\mu'\boldsymbol{h}_{k,n}(\boldsymbol{\gamma}_k(n) - \widehat{\boldsymbol{\gamma}}_k(n))\\ \boldsymbol{w}_{k,n} = \displaystyle\sum_{\ell\in\mathcal{N}_k} a_{\ell k}\,\boldsymbol{\psi}_{\ell,n}\end{array}\right. \tag{25.95}$$

where we introduced

$$\widehat{\boldsymbol{\gamma}}_k(n) \stackrel{\Delta}{=} \boldsymbol{h}_{k,n}^{\mathsf{T}}\boldsymbol{w}_{k,n-1} \tag{25.96}$$

Example 25.5 **(Diffusion with proximal steps)** We can also consider situations in which a nonsmooth regularization term is added to the risk function and replace, for example, problem (25.83) by one of the form

$$w^\star = \underset{w\in\mathbb{R}^M}{\text{argmin}} \left\{\kappa\|w\|_1 + \sum_{k=1}^{K}\alpha_k P_k(w)\right\} \tag{25.97}$$

where we are adding an ℓ_1-regularization term, $R(w) = \kappa\|w\|_1$. Situations of this type can be handled by incorporating a proximal step into the diffusion update (and similarly for the other algorithms described in this chapter). For instance, the ATC diffusion strategy (25.90) in the stochastic case would become:

$$\begin{cases} \text{select or receive a random data pair } (\boldsymbol{\gamma}_k(n), \boldsymbol{h}_{k,n}) \text{ from } \mathcal{D}_k \\ \boldsymbol{\psi}_{k,n} \;=\; \boldsymbol{w}_{k,n-1} - \mu\alpha_k \nabla_{w^{\mathsf{T}}} Q\left(\boldsymbol{w}_{k,n-1}; \boldsymbol{\gamma}_k(n), \boldsymbol{h}_{k,n}\right) \\ \boldsymbol{z}_{k,n} \;=\; \sum\limits_{\ell \in \mathcal{N}_k} a_{\ell k}\, \boldsymbol{\psi}_{\ell,n} \\ \boldsymbol{w}_{k,n} \;=\; \mathbb{T}_{\mu\kappa}(\boldsymbol{z}_{k,n}) \end{cases} \tag{25.98}$$

in terms of the soft-thresholding function defined earlier (11.18). More generally, for other choices $R(w)$ for the regularization term instead of $\kappa\|w\|_1$, the last step would become

$$\boldsymbol{w}_{k,n} = \text{prox}_{\mu R}(\boldsymbol{z}_{k,n}) \tag{25.99}$$

in terms of the proximal operation for $\mu R(w)$ – see Prob. 26.6.

Example 25.6 (**Simulating consensus and ATC diffusion**) We illustrate the behavior of the stochastic consensus and ATC diffusion strategies in the context of logistic regression problems. We generate at random a connected graph topology with $K = 20$ agents and employ a metropolis combination matrix, A. The network is shown in Fig. 25.7.

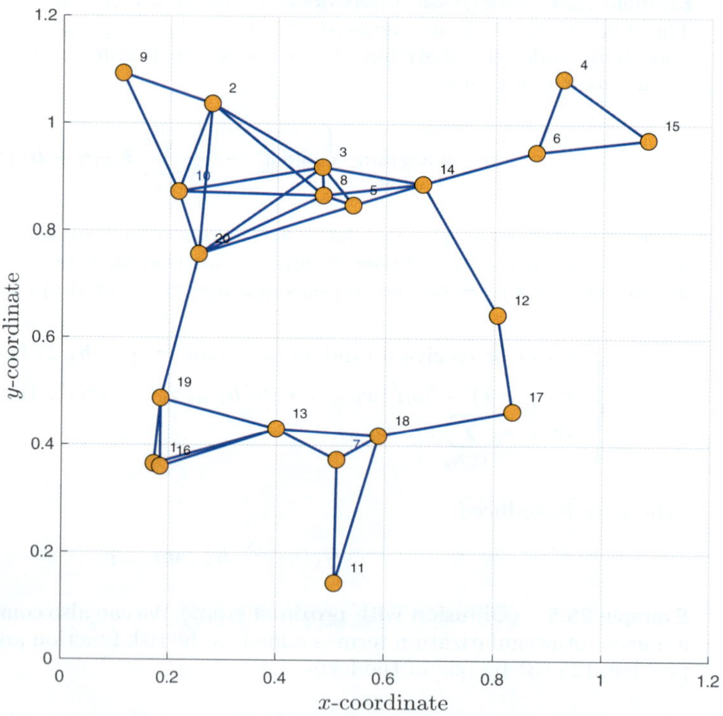

Figure 25.7 A connected graph topology with $K = 20$ agents used in the simulation to illustrate the performance of stochastic diffusion and consensus strategies.

The data at an arbitrary agent k is denoted by $\{\gamma_k(n), h_{k,n}\} \in \mathcal{D}_k$ where $\gamma_k(n) \in \{\pm 1\}$ and $h_{k,n} \in \mathbb{R}^M$. We assume there are $N_k = N/K$ data samples at each agent k for a total of N samples across the network (and, hence, $\alpha_k = 1/K$). The numerical values are

$$M = 10,\ K = 20,\ N_k = 200,\ N = 4000 \tag{25.100}$$

We seek the minimizer to the regularized empirical logistic risk problem:

$$w^\star = \underset{w\in\mathbb{R}^M}{\text{argmin}} \left\{ P(w) \triangleq \frac{1}{K}\sum_{k=1}^{K} P_k(w) \right\} \tag{25.101}$$

where each $P_k(w)$ is given by

$$P_k(w) = \rho\|w\|^2 \ + \ \frac{1}{N_k}\sum_{n\in\mathcal{D}_k} \ln\left(1 + e^{-\gamma_k(n)h_{k,n}^\mathsf{T} w}\right) \tag{25.102}$$

and $\rho = 0.001$ is the regularization parameter. The loss function at a generic data point (γ, h) is given by

$$Q(w;\gamma,h) \ = \ \rho\|w\|^2 \ + \ \ln(1 + e^{-\gamma h^\mathsf{T} w}) \tag{25.103}$$

with gradient vector

$$\nabla_{w^\mathsf{T}} Q(w;\gamma,h) \ = \ 2\rho w \ - \ \frac{\gamma h}{1+e^{\gamma\widehat{\gamma}}}, \quad \widehat{\gamma} = h^\mathsf{T} w \tag{25.104}$$

The stochastic versions of the consensus and ATC diffusion strategies then take the following forms (where $\mu' = \mu/K$ and the step size μ is set at $\mu = 0.01$):

$$\left\{\begin{array}{l} \textbf{(ATC diffusion)} \\ \text{at each agent } k, \text{ select a random data pair } (\boldsymbol{\gamma}_k(n), \boldsymbol{h}_{k,n}) \text{ from } \mathcal{D}_k \\ \widehat{\boldsymbol{\gamma}}_k(n) = \boldsymbol{h}_{k,n}^\mathsf{T} \boldsymbol{w}_{k,n-1} \\ \boldsymbol{\psi}_{k,n} = (1-2\rho\mu')\boldsymbol{w}_{k,n-1} + \mu' \dfrac{\boldsymbol{\gamma}_k(n)\boldsymbol{h}_{k,n}}{1+e^{\boldsymbol{\gamma}_k(n)\widehat{\boldsymbol{\gamma}}_k(n)}} \\ \boldsymbol{w}_{k,n} = \displaystyle\sum_{\ell\in\mathcal{N}_k} a_{\ell k}\, \boldsymbol{\psi}_{\ell,n} \end{array}\right. \tag{25.105}$$

and

$$\left\{\begin{array}{l} \textbf{(consensus)} \\ \text{at each agent } k, \text{ select a random data pair } (\boldsymbol{\gamma}_k(n), \boldsymbol{h}_{k,n}) \text{ from } \mathcal{D}_k \\ \widehat{\boldsymbol{\gamma}}_k(n) = \boldsymbol{h}_{k,n}^\mathsf{T} \boldsymbol{w}_{k,n-1} \\ \boldsymbol{\psi}_{k,n-1} = \displaystyle\sum_{\ell\in\mathcal{N}_k} a_{\ell k}\, \boldsymbol{w}_{\ell,n-1} \\ \boldsymbol{w}_{k,n} = (1-2\rho\mu')\boldsymbol{\psi}_{k,n-1} + \mu' \dfrac{\boldsymbol{\gamma}_k(n)\boldsymbol{h}_{k,n}}{1+e^{\boldsymbol{\gamma}_k(n)\widehat{\boldsymbol{\gamma}}_k(n)}} \end{array}\right. \tag{25.106}$$

The simulation generates $N = 4000$ data points $\{\gamma(m), h_m\}$ according to a logistic model, and distributes them across the $K = 20$ agents (with $N_k = 200$ data pairs at each agent). First, a random parameter model $w^a \in \mathbb{R}^{10}$ is selected, and a random collection of feature vectors $\{h_m\}$ are generated with zero-mean unit-variance Gaussian entries. Then, for each h_m, the label $\gamma(m)$ is set to either $+1$ or -1 according to the following construction:

$$\gamma(m) = +1 \ \text{if} \ \left(\tfrac{1}{1+e^{-h_m^\mathsf{T} w^a}}\right) \geq 0.5; \ \text{otherwise} \ \gamma(m) = -1 \tag{25.107}$$

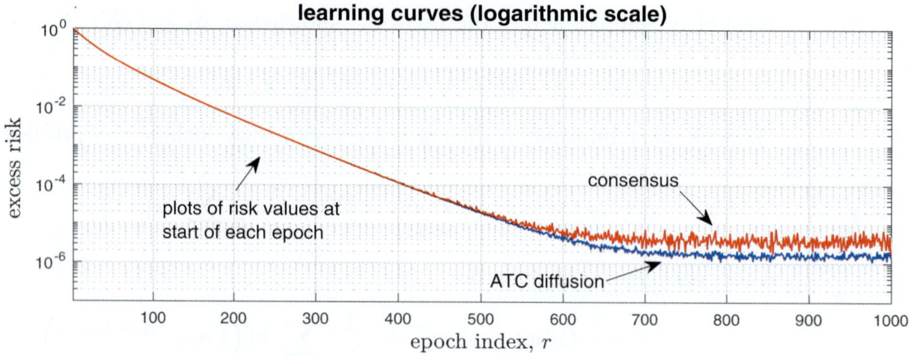

Figure 25.8 Learning curves obtained by averaging over $L = 10$ experiments. The curves show the evolution of the normalized excess risk values at the beginning of each run for consensus and ATC diffusion.

Using the generated data, and applying repeatedly a centralized gradient-descent recursion to the minimization of (25.101), we estimate its optimal parameter $w^\star$ and the corresponding minimum risk value as (rounding to four decimal points):

$$w^\star = \begin{bmatrix} 0.3337 \\ 0.2581 \\ 0.9239 \\ 0.0181 \\ 0.2019 \\ 0.1795 \\ 0.5204 \\ 0.4402 \\ 0.6362 \\ 0.4837 \end{bmatrix}, \quad P(w^\star) \approx 0.5308 \tag{25.108}$$

A total of $R = 1000$ epochs are run over the data. We select one agent at random for illustration purposes (agent 8), and evaluate the evolution of the aggregate risk value $P(w^r_{-1})$ at the beginning of each run r at that agent. We subsequently average these values over $L = 10$ experiments (with each experiment involving $R = 1000$ runs) to smooth out the resulting learning curve. We plot the average excess risk curve $\mathbb{E}\,P(\boldsymbol{w}^r_{-1}) - P(w^\star)$ after normalizing it by its peak value so that the learning curves start from the peak value of 1 for comparison purposes. The result is shown in Fig. 25.8.

Remark 25.1. (Wider stability range) Observe from the consensus iteration (25.88) that the computation of $\boldsymbol{w}_{k,n}$ resembles a stochastic gradient-descent step except that the starting point is $\boldsymbol{\psi}_{k,n-1}$ and the gradient of the loss function is evaluated at the different point $\boldsymbol{w}_{k,n-1}$. We say that this update is *asymmetric*. In comparison, if we examine the same updates in the diffusion implementations (25.90) and (25.92) we find that they are *symmetric*, meaning that their starting points coincide with the points at which the gradients of the loss function are evaluated. For example, for ATC diffusion, the starting point is $\boldsymbol{w}_{k,n-1}$ and the gradient of the loss is evaluated at this same point. It can be shown that the symmetry in the diffusion update leads to a *wider* stability range over consensus, meaning that the diffusion implementations are stable (in both cases of deterministic and stochastic optimization problems) over a wider range of the step-size parameter μ. The symmetry also leads to improved steady-state performance for the diffusion strategies, as is already evident from Fig. 25.8. We provide pertinent references in the comments at the end of the chapter.

■

25.5 FORMAL DERIVATION AS PRIMAL METHODS

In this section we re-derive the consensus and diffusion strategies more formally by presenting them as solutions to a primal optimization problem. Thus, consider again the original optimization problem (25.39), repeated here for ease of reference:

$$w^\star = \underset{w\in\mathbb{R}^M}{\text{argmin}} \left\{ P(w) \overset{\Delta}{=} \sum_{k=1}^{K} \alpha_k P_k(w) \right\} \tag{25.109}$$

where the $\{\alpha_k\}$ are positive convex combination coefficients that add up to 1:

$$\sum_{k=1}^{K} \alpha_k = 1, \quad \alpha_k > 0, \quad k = 1, 2, \ldots, K \tag{25.110}$$

The objective is to derive a decentralized solution run over a collection of K connected agents with the property that the iterates obtained at the individual agents will approach the desired minimizer $w^\star$.

We reformulate the optimization problem as follows. Let $w_k \in \mathbb{R}^M$ be a parameter vector associated with agent k. We collect these parameters into the network vector

$$\mathcal{W} = \text{blkcol}\Big\{ w_1, w_2, \ldots, w_K \Big\} \in \mathbb{R}^{KM} \tag{25.111}$$

and pose the equivalent problem:

$$\mathcal{W}^\star \overset{\Delta}{=} \underset{\mathcal{W}\in\mathbb{R}^{KM}}{\text{argmin}} \left\{ \sum_{k=1}^{K} \alpha_k P_k(w_k) \right\}, \ \text{ subject to } w_1 = w_2 = \ldots = w_K \tag{25.112}$$

Observe that we are assigning an individual argument w_k to each risk $P_k(\cdot)$ and requiring all these arguments to match, which takes us back to the original formulation (25.109). We say that we are requiring the agents to reach *agreement* or *consensus*. There are many ways by which constrained optimization problems of the form (25.112) can be solved. In this section, we pursue a *penalty-based* approach. Specifically, we add a term that penalizes deviations from the constraints and replace problem (25.112) by

$$\mathcal{W}^\dagger \overset{\Delta}{=} \underset{\mathcal{W}\in\mathbb{R}^{KM}}{\text{argmin}} \left\{ \sum_{k=1}^{K} \alpha_k P_k(w_k) + \frac{1}{4\mu} \sum_{k=1}^{K} \sum_{\ell\in\mathcal{N}_k} a_{\ell k} \|w_\ell - w_k\|^2 \right\} \tag{25.113}$$

with a quadratic penalty term defined in terms of nonnegative combination weights $\{a_{\ell k}\}$. Other penalty terms are of course possible and would lead to different algorithms. We continue with the above formulation. Note that the quadratic term penalizes discrepancies among the local parameters $\{w_\ell, w_k\}$ and encourages their values to be close to each other. The smaller the value of the "regularization factor" $\mu > 0$ is, the more significant the contribution of the quadratic term becomes and the closer the local parameters will need to be to each other. The presence of the rightmost regularization term in (25.113) leads to a different optimization problem than (25.112), which is why we are denoting the minimizer by $\mathcal{W}^\dagger$ instead of $\mathcal{W}^\star$. As $\mu \to 0$, we expect both minimizers to approach each other.

It is straightforward to verify that the penalty term can be written in vector form as follows (see Probs. 25.19 and 25.20):

$$\frac{1}{2}\sum_{k=1}^{K}\sum_{\ell\in\mathcal{N}_k} a_{\ell k}\|w_\ell - w_k\|^2 \;=\; \mathcal{W}^\mathsf{T}(I_{KM} - \mathcal{A})\,\mathcal{W} \tag{25.114}$$

in terms of the network vector $\mathcal{W}$ and the extended combination matrix $\mathcal{A} = A \otimes I_M$. In this way, we are reduced to solving the following penalized problem with a quadratic regularization term (recall from property (b) in Lemma 25.4 that $I_{KM} - \mathcal{A}$ is nonnegative-definite):

(primal optimization problem)

$$\mathcal{W}^\dagger \;\stackrel{\Delta}{=}\; \underset{\mathcal{W}\in\mathbb{R}^{KM}}{\operatorname{argmin}} \left\{ \mathcal{P}(\mathcal{W}) + \frac{1}{2\mu}\,\mathcal{W}^\mathsf{T}(I_{KM} - \mathcal{A})\,\mathcal{W} \right\} \tag{25.115}$$

where we are using the aggregate risk $\mathcal{P}(\mathcal{W})$ defined earlier by (25.68):

$$\mathcal{P}(\mathcal{W}) \;\stackrel{\Delta}{=}\; \sum_{k=1}^{K} \alpha_k P_k(w_k) \tag{25.116}$$

We refer to (25.115) as a *primal optimization problem* since the optimization is over the primal variable $\mathcal{W}$.

25.5.1 Consensus Strategy

Writing down a gradient-descent iteration with step size μ for minimizing (25.115) we get (see also Prob. 25.22):

$$\mathcal{W}_n = \mathcal{W}_{n-1} - \mu \left[\nabla_{\mathcal{W}^\mathsf{T}} \mathcal{P}(\mathcal{W}_{n-1}) + \frac{1}{\mu}(I_{KM} - \mathcal{A})^\mathsf{T}\, \mathcal{W}_{n-1} \right] \tag{25.117}$$

which simplifies to

$$\boxed{\mathcal{W}_n = \mathcal{A}^\mathsf{T}\, \mathcal{W}_{n-1} - \mu\, \nabla_{\mathcal{W}^\mathsf{T}}\, \mathcal{P}(\mathcal{W}_{n-1})} \quad \textbf{(consensus)} \tag{25.118}$$

This expression coincides with the consensus strategy derived earlier in (25.70). In other words, the decentralized consensus algorithm corresponds to a gradient-descent implementation for solving the penalized primal problem (25.115). For this reason, we refer to the consensus strategy as an algorithm in the *primal domain.*

25.5.2 Diffusion Strategy

We start from the same gradient-descent iteration (25.117) written as

$$\mathcal{W}_n = \mathcal{W}_{n-1} - \mu \nabla_{\mathcal{W}^\mathsf{T}} \mathcal{P}(\mathcal{W}_{n-1}) - (I_{KM} - \mathcal{A})^\mathsf{T} \mathcal{W}_{n-1} \tag{25.119}$$

and observe that the right-hand side consists of the addition of two terms to $\mathcal{W}_{n-1}$. The updated iterate $\mathcal{W}_n$ can be obtained by adding one term at a time, say, as:

$$\mathcal{O}_{n-1} = \mathcal{W}_{n-1} - (I_{KM} - \mathcal{A})^\mathsf{T} \mathcal{W}_{n-1} \tag{25.120a}$$
$$\mathcal{W}_n = \mathcal{O}_{n-1} - \mu \nabla_{\mathcal{W}^\mathsf{T}} \mathcal{P}(\mathcal{W}_{n-1}) \tag{25.120b}$$

These equations provide an equivalent implementation for (25.119): They first compute the intermediate vector $\mathcal{O}_{n-1}$ and then use it in the second equation to obtain $\mathcal{W}_n$. The first equation can be simplified, leading to

$$\mathcal{O}_{n-1} = \mathcal{A}^\mathsf{T} \mathcal{W}_{n-1} \tag{25.121a}$$
$$\mathcal{W}_n = \mathcal{O}_{n-1} - \mu \nabla_{\mathcal{W}^\mathsf{T}} \mathcal{P}(\mathcal{W}_{n-1}) \tag{25.121b}$$

The diffusion strategy is obtained by transforming these two steps into an *incremental* calculation. Specifically, it is expected that the updated iterate $\mathcal{O}_{n-1}$ is a "better estimate" than $\mathcal{W}_{n-1}$. Therefore, as befits an incremental implementation, this updated intermediate vector can be used to replace $\mathcal{W}_{n-1}$ in the second equation above, leading to:

$$\mathcal{O}_{n-1} = \mathcal{A}^\mathsf{T} \mathcal{W}_{n-1} \tag{25.122a}$$
$$\mathcal{W}_n = \mathcal{O}_{n-1} - \mu \nabla_{\mathcal{W}^\mathsf{T}} \mathcal{P}(\mathcal{O}_{n-1}) \tag{25.122b}$$

where $\mathcal{O}_{n-1}$ appears in both terms on the right-hand side of the second equation. If we now blend both equations together by substituting the first equation into the second we get

$$\boxed{\mathcal{W}_n = \mathcal{A}^\mathsf{T} \mathcal{W}_{n-1} - \mu \nabla_{\mathcal{W}^\mathsf{T}} \mathcal{P}(\mathcal{A}^\mathsf{T} \mathcal{W}_{n-1})} \quad \textbf{(CTA diffusion)} \tag{25.123}$$

which is the CTA form (25.76) of the diffusion strategy.

In a similar vein, we could have performed the two additions from (25.119) in a different order with the gradient update coming first, thus leading to

$$\mathcal{O}_{n-1} = \mathcal{W}_{n-1} - \mu \nabla_{\mathcal{W}^\mathsf{T}} \mathcal{P}(\mathcal{W}_{n-1}) \tag{25.124a}$$
$$\mathcal{W}_n = \mathcal{O}_{n-1} + (I_{KM} - \mathcal{A})^\mathsf{T} \mathcal{W}_{n-1} \tag{25.124b}$$

The ATC diffusion strategy can be obtained by transforming these two steps into an *incremental* calculation:

$$
\begin{aligned}
\mathcal{O}_{n-1} &= \mathcal{W}_{n-1} - \mu \nabla_{\mathcal{W}^\mathsf{T}} \mathcal{P}(\mathcal{W}_{n-1}) && (25.125a) \\
\mathcal{W}_n &= \mathcal{O}_{n-1} + (I_{KM} - \mathcal{A})^\mathsf{T} \mathcal{O}_{n-1} = \mathcal{A}^\mathsf{T} \mathcal{O}_{n-1} && (25.125b)
\end{aligned}
$$

If we now blend both equations together by substituting the first equation into the second we get

$$
\boxed{\mathcal{W}_n = \mathcal{A}^\mathsf{T}\Big(\mathcal{W}_{n-1} - \mu \nabla_{\mathcal{W}^\mathsf{T}} \mathcal{P}(\mathcal{W}_{n-1})\Big)} \quad \textbf{(ATC diffusion)} \qquad (25.126)
$$

which is the ATC form (25.78) of the diffusion strategy. We therefore find that the decentralized diffusion algorithm (in both its forms) corresponds to an *incremental* gradient-descent implementation for solving the penalized primal problem (25.115). For this reason, we refer to the diffusion strategy as an algorithm in the *primal domain* as well. The ATC and CTA diffusion variants are also known as decentralized stochastic gradient descent algorithms.

25.5.3 Convergence Performance

The convergence analysis for the consensus and diffusion algorithms can be carried out by applying similar tools to what we used in the previous chapters while studying the convergence of deterministic and stochastic optimization algorithms (and variations thereof). We carry out this analysis in Appendix 25.C at the end of this chapter. Here, we provide some intuition and explanations for the convergence behavior.

Thus, consider the consensus and diffusion strategies (25.118), (25.123), and (25.126). It is known, by following arguments similar to those leading to (25.292)–(25.293) in the appendix, that for strongly convex and δ-Lipschitz risk functions $P(w)$ in (25.109), there will exist small enough step sizes μ such that these recursions approach their respective fixed points, denoted generically by the notation $\mathcal{W}_\infty$. These points would satisfy the relations

$$
\begin{aligned}
\mathcal{W}_\infty &= \mathcal{A}^\mathsf{T} \mathcal{W}_\infty - \mu \nabla_{\mathcal{W}^\mathsf{T}} \mathcal{P}(\mathcal{W}_\infty) && \text{(consensus)} && (25.127a) \\
\mathcal{W}_\infty &= \mathcal{A}^\mathsf{T} \mathcal{W}_\infty - \mu \nabla_{\mathcal{W}^\mathsf{T}} \mathcal{P}(\mathcal{A}^\mathsf{T} \mathcal{W}_\infty) && \text{(CTA diffusion)} && (25.127b) \\
\mathcal{W}_\infty &= \mathcal{A}^\mathsf{T}\Big(\mathcal{W}_\infty - \mu \nabla_{\mathcal{W}^\mathsf{T}} \mathcal{P}(\mathcal{W}_\infty)\Big) && \text{(ATC diffusion)} && (25.127c)
\end{aligned}
$$

Recall that our objective is to solve (25.112) and determine $\mathcal{W}^\star$, whose block entries are identical:

$$
\mathcal{W}^\star = \mathbb{1}_K \otimes w^\star, \quad \text{for some } w^\star \in \mathbb{R}^M \qquad (25.128)
$$

Relations (25.127a)–(25.127c) already reveal that the consensus and diffusion solutions are biased in that their fixed points will not agree with $\mathcal{W}^\star$. This can be seen by considering any of the above relations. For instance, consider the first equation relating to consensus. It is clear that $\mathcal{w}^\star$ does not satisfy it. Indeed, if we substitute $\mathcal{w}_{n-1} \leftarrow \mathcal{w}^\star$ in (25.118), we get

$$\begin{aligned} \mathcal{w}_n &= \mathcal{A}^\mathsf{T} \mathcal{w}^\star - \mu \nabla_{\mathcal{w}^\mathsf{T}} \mathcal{P}(\mathcal{w}^\star) \\ &= \mathcal{w}^\star - \mu \nabla_{\mathcal{w}^\mathsf{T}} \mathcal{P}(\mathcal{w}^\star), \quad \text{since } \mathcal{A}^\mathsf{T} \mathcal{w}^\star = \mathcal{w}^\star \\ &\neq \mathcal{w}^\star \end{aligned} \tag{25.129}$$

since in general $\nabla_{\mathcal{w}^\mathsf{T}} \mathcal{P}(\mathcal{w}^\star) \neq 0$. This is because it consists of the individual gradients $\alpha_k \nabla_{w^\mathsf{T}} P_k(w^\star)$ and these gradients are not generally annihilated at the same location $w^\star$. It follows that $\mathcal{w}_n \neq \mathcal{w}^\star$ and $\mathcal{w}^\star$ cannot be a fixed point for consensus.

Thus, introduce the bias error term:

$$\widetilde{\mathcal{w}}_n \stackrel{\Delta}{=} \mathcal{w}^\star - \mathcal{w}_n \tag{25.130}$$

If we subtract $\mathcal{w}^\star = \mathcal{A}^\mathsf{T} \mathcal{w}^\star$ from both sides of (25.118) we find that the bias term for consensus evolves according to the recursion:

$$\widetilde{\mathcal{w}}_n = \mathcal{A}^\mathsf{T} \widetilde{\mathcal{w}}_{n-1} + \mu \nabla_{\mathcal{w}^\mathsf{T}} \mathcal{P}(\mathcal{w}_{n-1}) \tag{25.131}$$

Similar equations can be derived for the bias term under diffusion. From these equations, and using future expression (26.117), the size of the bias in the steady state can be shown to satisfy:

(bias in deterministic optimization)

$$\lim_{n\to\infty} \left\{ \frac{1}{K} \sum_{k=1}^{K} \|\widetilde{w}_{k,n}\|^2 \right\} = O(\mu^2 b^2) \tag{25.132}$$

where

$$b^2 \stackrel{\Delta}{=} \sum_{k=1}^{K} \alpha_k \, \|\nabla_{w^\mathsf{T}} P_k(w^\star)\|^2 \tag{25.133}$$

is a measure of how far the optimal solution $w^\star$ is from the minimizers for the individual risk functions. We therefore conclude that for deterministic optimization problems, the squared bias is on the order of the square of the small step-size parameter.

If we examine instead the stochastic versions of consensus and diffusion, where the gradient of $P_k(w)$ is replaced by the gradient of the loss function $Q(w;\cdot)$, then the successive iterates $\mathcal{w}_n$ become random in nature. In this case, it can be verified that the expression for the bias is replaced by the following in terms of the MSE measure (see the discussion around (26.117) in the comments at the end of the chapter):

(bias in stochastic optimization)

$$\limsup_{n\to\infty}\left\{\frac{1}{K}\sum_{k=1}^{K}\mathbb{E}\,\|\widetilde{\boldsymbol{w}}_{k,n}\|^2\right\} = O(\mu\sigma_g^2+\mu^2 b^2) \tag{25.134}$$

where σ_g^2 is a parameter that is related to the "size" of the gradient noise process resulting from the gradient approximation (similar to (18.32)). When the step size is small, the $O(\mu)$ term dominates so that, for stochastic optimization,

$$\limsup_{n\to\infty}\left\{\frac{1}{K}\sum_{k=1}^{K}\mathbb{E}\,\|\widetilde{\boldsymbol{w}}_{k,n}\|^2\right\} = O(\mu\sigma_g^2) \tag{25.135}$$

Example 25.7 **(Left-stochastic combination matrices)** The above convergence arguments assume a doubly stochastic combination matrix A and measure performance relative to the minimizer $\mathcal{W}^\star$ defined by (25.128). The consensus and diffusion strategies can still be applied even when A is left-stochastic. In that case, however, it is known that the algorithms will approach the minimizer, denoted by $w^{\star\star}$, of a modified risk function where the local risks are further weighted by the entries of the Perron vector of A, namely,

$$\mathcal{W}^{\star\star} \overset{\Delta}{=} \underset{\mathcal{W}\in\mathbb{R}^{KM}}{\text{argmin}} \left\{\sum_{k=1}^{K} p_k\,\alpha_k P_k(w_k)\right\}, \text{ subject to } w_1 = w_2 = \ldots = w_K \tag{25.136}$$

where $Ap = p, \mathbb{1}^{\mathsf{T}}p = p,\;\; p_k > 0$. The bias results (25.132) and (25.134) will continue to hold except that the error term (25.130) will need to be measured relative to $\mathcal{W}^{\star\star}$:

$$\widetilde{\mathcal{W}}_n \overset{\Delta}{=} \mathcal{W}^{\star\star} - \mathcal{W}_n \tag{25.137}$$

We comment on this behavior at the end of the chapter and provide relevant references.

25.6 COMMENTARIES AND DISCUSSION

Graph theory. There is a rich and classical body of work on graph theory and discrete mathematics, especially in the computer science field. In this chapter, we focused on concepts that are relevant to the treatment in the text by adjusting the presentation from Sayed (2014a, c); in particular, the argument in the proof of Lemma 25.2 is adapted from Sayed (2014c). For more extensive discussion on graphs and their properties, readers may refer, among many other works, to the texts by Trudeau (1994), Bollobas (1998), Godsil and Royle (2001), Kocay and Kreher (2004), Gross and Yellen (2005), Bondy and Murty (2008), Balakrishnan (2010), Chartrand and Zhang (2012), Bapat (2014), Deo (2016), Nica (2018), and Erciyes (2021).

Decentralized optimization. There is an extensive literature on decentralized algorithms for the minimization of convex and nonconvex risk functions. We described several strategies in the body of the chapter and focused on first-order gradient methods, motivated by the presentation from Sayed (2014a). Second-order information such as

Hessian matrices can also be used in the development of decentralized strategies (e.g., of the Newton type), but we opted to focus on implementations of the gradient-descent type due to their simpler structure.

For the incremental strategy, the reader may refer to Bertsekas (1997, 1999), Nedic and Bertsekas (2001), Rabbat and Nowak (2005), Lopes and Sayed (2007c), Blatt, Hero, and Gauchman (2008), Johansson, Rabi, and Johansson (2009), and Helou and De Pierro (2009). For the primal methods, the consensus strategy is from Tsitsiklis, Bertsekas, and Athans (1986) and Bertsekas and Tsitsiklis (1997), while the diffusion strategy is from Chen and Sayed (2013, 2015a, b), Sayed (2014a, b, c), and Zhao and Sayed (2015a, b, c). We will find in the next chapter that the ATC and CTA protocols, where adaptation and combination steps follow each other, and which were originally introduced for diffusion strategies, have been adopted by several subsequent primal–dual methods and form the basis for decentralized stochastic gradient-descent methods. The references Chen and Sayed (2012b, 2013) clarify the connection to Pareto optimality, as was explained in Example 25.1, following Sayed (2014a). These references also clarify the nature of the limit points for the consensus and diffusion primal methods under left-stochastic combination policies, as was discussed in Example 25.7. For more information on Pareto optimality and the optimal trade-off curve, the reader may refer to Zadeh (1963), Intriligator (1971), and Boyd and Vandenberghe (2004, p. 183).

The proof of convergence for the primal methods involving consensus and diffusion strategies in Appendix 25.C is from Sayed (2014a, chs. 8–9). It turns out that some care is needed for consensus implementations when constant step sizes are used. The main reason is that, as explained in Tu and Sayed (2012), Sayed *et al.* (2013), Sayed (2014a, section 10.6), and Sayed (2014b, section VIII), instability can occur in consensus networks due to an inherent *asymmetry* in the dynamics of the consensus iteration. Observe from (25.59) that the starting point for the recursion is the combination of the iterates over the neighborhood, while the gradient vector is evaluated at the different location $w_{k,n-1}$. In contrast, in the diffusion implementation (25.77), the starting point for the iteration and the location at which the gradient vector is evaluated are the same. The *symmetry* in the update relation also leads to improved performance for diffusion strategies in relation to the consensus strategy, for both cases of constant and decaying step sizes – see Sayed (2014a) and Towfic, Chen, and Sayed (2016).

Attaining consensus over graphs. The references by Sayed (2014a, b) provide overviews of consensus and diffusion algorithms for decentralized optimization. As noted in these references, there has been extensive work on consensus techniques in the literature, starting with the foundational results by DeGroot (1974) and Berger (1981), which were of a different nature and did not deal with the minimization of empirical or stochastic risk functions. The original consensus formulation dealt instead with the problem of computing averages over graphs. This can be explained as follows (see, e.g., Probs. 25.8–25.10 and also Tsitsiklis and Athans (1984) and Tsitsiklis, Bertsekas, and Athans (1986)).

Consider a collection of (scalar or vector) measurements denoted by $\{w_k,\ k = 1, 2, \ldots, K\}$ available at the vertices of a connected graph with K agents. The objective is to devise a decentralized algorithm that enables every agent to determine the *average* value:

$$\overline{w} \triangleq \frac{1}{K}\sum_{k=1}^{K} w_k \tag{25.138}$$

by interacting solely with its neighbors. When this occurs, we say that the agents have reached consensus (or agreement) about $\overline{w}$. We select a $K \times K$ *doubly stochastic* combination matrix $A = [a_{\ell k}]$ and assume its second largest-magnitude eigenvalue satisfies

$$|\lambda_2(A)| < 1 \tag{25.139}$$

Using the combination coefficients $\{a_{\ell k}\}$, each agent k then iterates *repeatedly* on the data of its neighbors:

$$w_{k,n} = \sum_{\ell \in \mathcal{N}_k} a_{\ell k}\, w_{\ell,n-1}, \quad n \geq 0, \quad k = 1, 2, \ldots, K \tag{25.140}$$

starting from the boundary conditions $w_{\ell,-1} = w_\ell$ for all $\ell \in \mathcal{N}_k$. The superscript n denotes the iteration index. Every agent k in the network performs the same calculation, which amounts to combining repeatedly, and in a convex manner, the state values of its neighbors. It can then be shown that (see Probs. 25.8–25.10, DeGroot (1974), Berger (1981), and appendix E in Sayed (2014c)):

$$\lim_{n \to \infty} w_{k,n} = \overline{w}, \qquad k = 1, 2, \ldots, K \tag{25.141}$$

In this way, through the localized iterative process (25.140), the agents are able to converge to the global average value, $\overline{w}$. Some useful extensions of these results exist for nondoubly stochastic combination matrices, as well as for computing the majority opinion by networked agents holding binary values – see, e.g., Bénézit *et al.* (2010) and Bénézit, Thiran, and Vetterli (2011).

Earlier two timescale solutions. It is explained in Sayed (2014a) that, motivated by the elegant result (25.141), several works in the literature have proposed useful extensions of the original consensus construction (25.140) to minimize aggregate costs of the form (25.39) or to solve distributed estimation problems of the least-squares or Kalman filtering type. Among these works we mention Tsitsiklis, Bertsekas, and Athans (1986), Bertsekas and Tsitsiklis (1997), Xiao and Boyd (2004), Olfati-Saber (2005), Alriksson and Rantzer (2006), Speranzon, Fischione, and Johansson (2006), Das and Mesbahi (2006), Carli *et al.* (2008), Johansson *et al.* (2008), Nedic and Ozdaglar (2009), Kar and Moura (2011), and Kar, Moura, and Ramanan (2012). Several of these earlier extensions involved the inconvenient use of *two* separate timescales: one faster timescale for performing multiple consensus iterations similar to (25.140) over the states of the neighbors, and a second slower timescale for performing gradient vector updates or for updating the estimators by using the result of the consensus iterations, as was the case in Xiao and Boyd (2004), Olfati-Saber (2005), Das and Mesbahi (2006), Speranzon, Fischione, and Johansson (2006), Carli *et al.* (2008), and Kar, Moura, and Ramanan (2012). An example of a two timescale implementation would be an algorithm of the following form:

$$\left\{\begin{array}{l} w^{(-1)}_{\ell,n-1} \longleftarrow w_{\ell,n-1}, \;\text{ for all agents } \ell \text{ at iteration } i-1 \\ \textbf{for } j = 0,1,2,\ldots,J-1 \textbf{ iterate:} \\ \quad w^{(j)}_{k,n-1} = \displaystyle\sum_{\ell \in \mathcal{N}_k} a_{\ell k} w^{(j-1)}_{\ell,n-1}, \;\text{ for all } k = 1,2,\ldots,N \\ \textbf{end} \\ w_{k,n} = w^{(J-1)}_{k,n-1} - \mu\, \alpha_k\, \nabla_{w^{\mathsf{T}}} P_k(w_{k,n-1}) \end{array}\right. \tag{25.142}$$

If we compare the last equation in (25.142) with (25.59), we observe that the variable $w^{(J-1)}_{k,n-1}$ that is used in (25.142) to obtain $w_{k,n}$ is the result of J repeated applications of a consensus operation of the form (25.140) on the iterates $\{w_{\ell,n-1}\}$. The purpose of these repeated calculations is to approximate well the average of the iterates in the neighborhood of agent k. These J repeated averaging operations need to be completed before the availability of the gradient information for the last update step in (25.142). In other words, the J averaging operations need to be performed at a faster rate than the last step in (25.142). Such two timescale implementations are a hindrance for online learning algorithms.

Single timescale solutions. Building upon a useful procedure for distributed optimization from eq. (2.1) in Tsitsiklis, Bertsekas, and Athans (1986) and eq. (7.1) in Bertsekas and Tsitsiklis (1997), later works proposed single timescale implementations for consensus strategies by using an implementation similar to (25.59) – see, e.g., eq. (3) in Braca, Marano, and Matta (2008), eq. (3) in Nedic and Ozdaglar (2009), eq. (19) in Dimakis *et al.* (2010), and eq. (9) in Kar and Moura (2011). These references, however, employ decaying step sizes, $\mu_k(n) \to 0$, to ensure that the iterates $\{w_{k,n}\}$ across all agents converge almost-surely to the same value, namely, they employ recursions of the form:

$$w_{k,n} = \sum_{\ell \in \mathcal{N}_k} a_{\ell k}\, w_{\ell,n-1} \; - \; \mu_k(n)\alpha_k \nabla_{w^\mathsf{T}} P_k(w_{k,n-1}) \tag{25.143}$$

or variations thereof. As noted before, when diminishing step sizes are used, adaptation is turned off over time, which is prejudicial for online learning purposes. Setting the step sizes to constant values in (25.59) endows the consensus iteration with continuous adaptation, learning, and tracking abilities (and also enhances the convergence rate).

The fact that separate timescales are unnecessary was the main motivation behind the introduction of the diffusion strategies and their adapt-then-combine (ATC) mechanisms for decentralized stochastic gradient learning, starting in 2006, in various levels of generality by Sayed and co-workers – overviews appear in Sayed (2014a, b). The diffusion strategies (25.75) and (25.77) showed that single time scale distributed learning from *streaming* data is possible, and that this objective can be achieved under *constant* step-size adaptation in a stable manner – see, e.g., Sayed (2014a, chs. 9–11). The CTA diffusion strategy (25.75) was first introduced for MSE estimation problems in Lopes and Sayed (2006, 2007a, b, 2008) and Sayed and Lopes (2007). The ATC diffusion structure (25.77), with adaptation preceding combination, appeared in Cattivelli, Lopes, and Sayed (2007, 2008) and Cattivelli and Sayed (2008, 2010a, b). Extensions to more general loss functions, other than the MSE loss, are studied in Zhao and Sayed (2012, 2015a, b, c), Chen and Sayed (2013, 2015a, b), and Sayed (2014a). Extensions to multi-task scenarios also appear in Nassif *et al.* (2020a, b, c). The CTA structure (25.75) with an iteration-dependent step size that decays to zero, $\mu(n) \to 0$, was subsequently used in the works by Ram, Nedic, and Veeravalli (2010), Srivastava and Nedic (2011), and Lee and Nedic (2013) in the solution of distributed optimization problems that require all agents to reach agreement. The ATC form (25.77), also with an iteration-dependent sequence $\mu(n)$ that decays to zero, was further applied by Bianchi *et al.* (2011) and Stankovic, Stankovic, and Stipanovic (2011) to ensure almost-sure convergence and agreement among agents.

PROBLEMS

25.1 If each of three individual left-stochastic combination matrices $\{A_o, A_1, A_2\}$ is primitive, is the product $P = A_1 A_o A_2$ primitive? Conversely, if P is primitive, does it necessarily follow that each of the matrices $\{A_o, A_1, A_2\}$ is primitive? Prove or give a counter example.

25.2 Let A be a $K \times K$ doubly stochastic matrix. Show that $\text{Tr}(A^\mathsf{T} H A) \leq \text{Tr}(H)$, for any $K \times K$ nonnegative-definite symmetric matrix H.

25.3 Give an example of a four-agent graph that is connected (but not strongly connected) and whose combination matrix A is not primitive.

25.4 We know that if a graph is strongly connected, then its left-stochastic combination matrix A is primitive. Is the converse true? That is, does it hold that if A is a primitive left-stochastic matrix, then the graph is strongly connected?

25.5 Consider a strongly connected graph with K agents. Prove that, for any agent k, there always exists a circular path (i.e., a cycle) with nonzero scaling weights that starts at k and ends at the same location k.

25.6 Establish result (25.23) for the limit of A^n when A is a stochastic matrix.

25.7 Let A be a $K \times K$ primitive and left-stochastic matrix. Let p denote its Perron vector. Let $\mathcal{A} = A \otimes I_M$. Show that

$$\lim_{n\to\infty} (\mathcal{A}^{\mathsf{T}})^n = (\mathbb{1}_K \otimes I_M)(p \otimes I_M)^{\mathsf{T}}$$

25.8 Consider K agents, each with an initial state value $x_k \in \mathbb{R}^M$. Collect the initial state of the network into $\mathcal{X}_0 = \text{col}\{x_1, x_2, \ldots, x_K\}$. Let A be a $K \times K$ primitive and left-stochastic matrix. Assume the agents share their state values in a distributed manner and update them over time according to the construction:

$$x_{k,n} = \sum_{\ell \in \mathcal{N}_k} a_{\ell k} x_{\ell,n-1}, \quad n \geq 0$$

(a) Verify that the recursion can be written in the aggregate form $\mathcal{X}_n = \mathcal{A}^{\mathsf{T}} \mathcal{X}_{n-1}, n \geq 0$, where $\mathcal{A} = A \otimes I_M$.

(b) Use the result of Prob. 25.7 to conclude that all agents converge to the same state value:

$$\lim_{n\to\infty} x_{k,n} = \sum_{k=1}^{K} p_k x_k$$

where the $\{p_k\}$ are the entries of the Perron vector of A.

(c) Conclude that when A is doubly stochastic then the states of all agents converge to $\bar{x} = \frac{1}{K}\sum_{k=1}^{K} x_k$.

25.9 We continue with Prob. 25.8 with A primitive and left-stochastic. Assume the initial states are $x_k = e_k$, in terms of the M-dimensional basis vectors. Show that $x_{k,n}$ converges to the Perron vector p. That is, each agent ends up learning the Perron vector of the graph.

25.10 We continue with Prob. 25.8 with A primitive and doubly stochastic. Introduce $\widetilde{x}_{k,n} = \bar{x} - x_{k,n}$ and collect these errors into the network vector $\widetilde{\mathcal{X}}_n$. Show that

$$\widetilde{\mathcal{X}}_n = -\left[\left(A^{\mathsf{T}} - \frac{1}{K}\mathbb{1}_K\mathbb{1}_K^{\mathsf{T}}\right) \otimes I_M\right]^n \mathcal{X}_0$$

Conclude that the rate of convergence of $x_{k,n}$ to $\bar{x}$ is determined by the second largest eigenvalue of A, i.e., by $|\lambda_2(A)|$.

25.11 We continue with Prob. 25.8. Let A denote a generic $K \times K$ combination matrix with nonnegative entries (it is not necessarily primitive). Show that for any initial state values $\{x_k\}$, the successive iterates $\{x_{k,n}\}$ will converge to the network average $\bar{x}$ as $n \to \infty$ if, and only if, the following two conditions are met: (a) A is doubly stochastic and (b) the spectral radius of the matrix $A^{\mathsf{T}} - \frac{1}{K}\mathbb{1}\mathbb{1}^{\mathsf{T}}$ is strictly less than 1.

25.12 Let $p = \text{col}\{p_\ell\}$ denote the Perron eigenvector of a primitive left-stochastic matrix A. Let $P = \text{diag}\{p_\ell\}$. The matrix A is said to be balanced if it satisfies the condition $AP = PA^{\mathsf{T}}$. Refer to expression (25.16). Determine the Perron eigenvector of the averaging combination matrix and show that the matrix is balanced. Show further that every primitive doubly stochastic matrix is balanced.

25.13 We continue with Prob. 25.12. Assume A is primitive, left-stochastic, and balanced. Show that the matrix $AP - P + I$ is primitive, symmetric, and doubly stochastic.

25.14 We continue with Prob. 25.12. Assume A is primitive, left-stochastic, and balanced. Show that $P - AP$ is symmetric and positive semi-definite. Conclude that the nullspace of $P - AP$ is spanned by the vector $\mathbb{1}_K$, i.e., $\mathcal{N}(P - AP) = \text{span}(\mathbb{1}_K)$.

25.15 Consider a $K \times 1$ vector p with positive entries $\{p_\ell\}$ that add up to 1. Consider further an undirected connected graph with K nodes so that $\ell \in \mathcal{N}_k \iff k \in \mathcal{N}_\ell$. Construct a $K \times K$ combination matrix $A = [a_{k\ell}]$ as follows:

$$a_{\ell k} = \begin{cases} 0, & \ell \notin \mathcal{N}_k \\ p_\ell, & \ell \in \mathcal{N}_k \backslash \{k\} \\ 1 - \sum_{m \in \mathcal{N}_k \backslash \{k\}} a_{mk}, & \ell = k \end{cases}$$

Verify that:
(a) $a_{kk} \geq p_k > 0$.
(b) A is left-stochastic and primitive.
(c) p is the Perron eigenvector for A, namely, $Ap = p$.
(d) A is balanced so that $AP = PA^{\mathsf{T}}$.

25.16 The entries of a $K \times K$ uniform combination matrix are defined in terms of the degrees of the agents as follows:

$$a_{\ell k} = \begin{cases} \dfrac{1}{n_k}, & \ell \in \mathcal{N}_k \\ 0, & \text{otherwise} \end{cases}$$

Verify that A is left-stochastic and that the entries of its Perron vector are given by

$$p_k = \frac{n_k}{\sum_{m=1}^K n_m}$$

25.17 The entries of a $K \times K$ relative-degree combination matrix are defined in terms of the degrees of the agents as follows:

$$a_{\ell k} = \begin{cases} \dfrac{n_\ell}{\sum_{m \in \mathcal{N}_m} n_m}, & \text{if } \ell \in \mathcal{N}_k \\ 0, & \text{otherwise} \end{cases}$$

Verify that A is left-stochastic. Determine its Perron vector. *Remark.* For more details on the relative-degree rule, the reader may refer to Cattivelli and Sayed (2010b), and Sayed (2014a).

25.18 The entries of a $K \times K$ Hastings combination matrix are defined in terms of the degrees of the agents as follows:

$$a_{\ell k} = \begin{cases} \dfrac{\theta_k^2}{\max\{n_k \theta_k^2, n_\ell \theta_\ell^2\}}, & \ell \in \mathcal{N}_k \backslash \{k\} \\ 1 - \sum_{m \in \mathcal{N}_k \backslash \{k\}} a_{mk}, & \ell = k \end{cases}$$

where the $\{\theta_k^2\}$ are nonnegative parameters that reflect a level of confidence in the data at the agents. Verify that A is left-stochastic and that the entries of its Perron vector are given by

$$p_k = \frac{1/\theta_k^2}{\sum_{\ell=1}^K 1/\theta_\ell^2}$$

Remark. For more details on the Hastings rule, the reader may refer to Hastings (1970) and Sayed (2014a).

25.19 Establish the validity of equality (25.114) for the penalty term in the primal optimization problem.

25.20 Assume the combination matrix A is symmetric but not necessarily stochastic. Let $B = \text{diag}(A\mathbb{1}) - A$ and $\mathcal{B} = B \otimes I_M$. Show that in this case the term on the right-hand side of (25.114) becomes $\mathcal{w}^{\mathsf{T}} \mathcal{B} \mathcal{w}$.

25.21 Show that the Laplacian matrix of a fully connected graph with K agents is given by $\mathcal{L} = KI_K - \mathbb{1}_K \mathbb{1}_K^{\mathsf{T}}$. Conclude that the largest eigenvalue of $\mathcal{L}$ is equal to K.

25.22 Assume agents employ agent-dependent step sizes μ_k. How would you modify the argument leading to (25.118) to re-derive the consensus strategy?

25.23 Consider a strongly connected network with K agents with quadratic individual risks $P_k(w) = \sigma_k^2 - 2r_k^\mathsf{T} w + w^\mathsf{T} R_k w$, where $\sigma_k^2 > 0$ is a scalar, $w, r_k \in \mathbb{R}^M$, and $R_k > 0$. The aggregate risk is defined by $P(w) = \frac{1}{K}\sum_{k=1}^{K} P_k(w)$.

(a) Verify that the minimizer for each individual risk is given by $w_k^\star = R_k^{-1} r_k$.

(b) Motivated by expression (25.46) for determining Pareto solutions, consider the weighted aggregate risk $P^{\text{glob},\pi}(w) = \sum_{k=1}^{N} \pi_k P_k(w)$. Verify that its minimizer can be expressed as the combination

$$w^\pi = \sum_{k=1}^{K} B_k w_k^\star, \quad B_k \triangleq \left(\sum_{\ell=1}^{K} \pi_\ell R_\ell\right)^{-1} (\pi_k R_k)$$

where the matrices $\{B_k\}$ satisfy $B_k > 0$ and $\sum_{k=1}^{K} B_K = I_M$.

(c) Specialize the results to the case $K = 2$.

25.24 Pareto optimality is a useful concept in game theory as well. Consider two players, $\mathbb{A}$ and $\mathbb{B}$. Each player has a choice between two strategies: Player $\mathbb{A}$ can choose between strategies A1 or A2, while player $\mathbb{B}$ can choose between strategies B1 or B2. The table below lists the costs associated with the four possible choices by the players. For example, refer to the entry in the table corresponding to player $\mathbb{A}$ selecting A1 and player $\mathbb{B}$ selecting B1. The cell shows the values (6, 4), meaning that 6 is the cost incurred by player $\mathbb{A}$ and 4 is the cost incurred by player $\mathbb{B}$. The players wish to minimize their costs. Can you identify which strategies are Pareto optimal? That is, can you identify those strategies such that there are no other strategies where at least one player sees their cost reduced (i.e., does better) while the other player does not do worse?

	B1	**B2**
A1	(6,4)	(5,5)
A2	(4,6)	(7,5)

25.25 Derive a CTA diffusion algorithm for minimizing the following constrained cost function over a connected graph of K agents:

$$\min_{w \in \mathbb{R}^M} \sum_{k=1}^{K} \mathbb{E}\,(\boldsymbol{\gamma}_k(n) - \boldsymbol{h}_{k,n}^\mathsf{T} w)^2 \quad \text{subject to} \quad c^\mathsf{T} w = \alpha$$

where c is a given column vector and α is a given scalar.

25.26 Refer to the network error recursion (25.192) and assume we approximate $\mathcal{B}_{n-1}$ by the constant matrix $\mathcal{B} = \mathcal{A}_2^\mathsf{T}(\mathcal{A}_0^\mathsf{T} - \mathcal{M}\mathcal{H})\mathcal{A}_1^\mathsf{T}$, where $\mathcal{H} = \text{blkdiag}\{H_k\}$ and each H_k is equal to the Hessian of $P_k(w)$ evaluated at the global minimizer, $H_k = \nabla_w^2 P_k(w^\star)$. It is explained in Sayed (2014a) that the resulting long-term model serves as a good approximation for the evolution of the network for small step sizes.

(a) Verify that the coefficient matrices $\mathcal{B}$ corresponding to CTA and ATC diffusion, as well as to a noncooperative mode of operation where the agents operate independently of each other, are given by:

$$\mathcal{B}^{\text{atc}} = \mathcal{A}^\mathsf{T}(I_{2MN} - \mathcal{M}\mathcal{H}), \quad \mathcal{B}^{\text{cta}} = (I_{2MN} - \mathcal{M}\mathcal{H})\mathcal{A}^\mathsf{T}, \quad \mathcal{B}^{\text{ncop}} = (I_{2MN} - \mathcal{M}\mathcal{H})$$

where $\mathcal{A} = A \otimes I$ and A is the doubly stochastic combination matrix used by the diffusion implementation.

(b) Assume the noncooperative matrix $\mathcal{B}^{\text{ncop}}$ is stable. Show that $\mathcal{B}^{\text{atc}}$ and $\mathcal{B}^{\text{cta}}$ will be stable as well regardless of A.

(c) What about the coefficient matrix for consensus, $\mathcal{B}^{\text{cons}}$?

25.27 Refer to the general strategy (25.161) and observe that each line contains a sum over agents ℓ in a neighborhood $\mathcal{N}_k$; these sums use combination weights from the matrices $\{A_1, A_o, A_2\}$. Do these neighborhoods need to coincide? If we write instead

$$\begin{cases} \boldsymbol{\phi}_{k,n-1} &= \sum_{\ell\in\mathcal{N}_{k,1}} a_{1,\ell k}\, \boldsymbol{w}_{\ell,n-1} \\ \boldsymbol{\psi}_{k,n} &= \sum_{\ell\in\mathcal{N}_{k,o}} a_{o,\ell k}\, \boldsymbol{\phi}_{\ell,n-1} \;-\; \mu\,\alpha_k \widehat{\nabla_{w^\mathsf{T}} P_k}\left(\boldsymbol{\phi}_{k,n-1}\right) \\ \boldsymbol{w}_{k,n} &= \sum_{\ell\in\mathcal{N}_{k,2}} a_{2,\ell k}\, \boldsymbol{\psi}_{\ell,n} \end{cases}$$

where $\{\mathcal{N}_{k,1}, \mathcal{N}_{k,o}, \mathcal{N}_{k,2}\}$ refer to neighborhoods defined by the respective matrices $\{A_1, A_o, A_2\}$, would the result of Lemma 25.6 continue to hold?

25.28 Refer to the general strategy (25.161) and replace it by

$$\begin{cases} \boldsymbol{\phi}_{k,n-1} &= \sum_{\ell\in\mathcal{N}_{k,1}} a_{1,\ell k}\, \boldsymbol{w}_{\ell,n-1} \\ \boldsymbol{\psi}_{k,n} &= \sum_{\ell\in\mathcal{N}_{k,o}} a_{o,\ell k}\, \boldsymbol{\phi}_{\ell,n-1} \;-\; \mu \sum_{\ell\in\mathcal{N}_{k,c}} c_{\ell k}\alpha_\ell \widehat{\nabla_{w^\mathsf{T}} P_\ell}\left(\boldsymbol{\phi}_{k,n-1}\right) \\ \boldsymbol{w}_{k,n} &= \sum_{\ell\in\mathcal{N}_{k,2}} a_{2,\ell k}\, \boldsymbol{\psi}_{\ell,n} \end{cases}$$

where $C = [c_{\ell k}]$ is a right-stochastic matrix (each of its rows adds up to 1), and $\{\mathcal{N}_{k,1}, \mathcal{N}_{k,o}, \mathcal{N}_{k,2}, \mathcal{N}_{k,c}\}$ refer to neighborhoods defined by the matrices $\{A_1, A_o, A_2, C\}$. Extend the result of Lemma 25.6 to this case and derive the corresponding error recursion.

25.A PROOF OF LEMMA 25.1

In this appendix, we establish the statement of Lemma 25.1 by following the argument from Sayed (2014a). Pick two arbitrary agents ℓ and k. Since the graph is connected, this implies that there exists a sequence of agent indices $(\ell, m_1, m_2, \ldots, m_{n_{\ell k}-1}, k)$ of shortest length that forms a path from agent ℓ to agent k, say, with $n_{\ell k}$ nonzero scaling weights $\{a_{\ell m_1}, a_{m_1,m_2}, \ldots, a_{m_{n_{\ell k}-1},k}\}$:

$$\ell \xrightarrow{a_{\ell m_1}} m_1 \xrightarrow{a_{m_1,m_2}} m_2 \longrightarrow \ldots \longrightarrow m_{n_{\ell k}-1} \xrightarrow{a_{m_{n_{\ell k}-1},k}} k \qquad [n_{\ell k} \text{ edges}] \tag{25.144}$$

From the rules of matrix multiplication, the (ℓ, k)th entry of the $n_{\ell k}$th power of A is given by

$$[A^{n_{\ell k}}]_{\ell k} = \sum_{m_1=1}^{K}\sum_{m_2=1}^{K}\cdots\sum_{m_{n_{\ell k}-1}=1}^{K} a_{\ell m_1} a_{m_1 m_2}\cdots a_{m_{n_{\ell k}-1} k} \tag{25.145}$$

We already know that the sum in (25.145) should be nonzero because of the existence of the aforementioned path linking agents ℓ and k with *nonzero* scaling weights. It follows that $[A^{n_{\ell k}}]_{\ell k} > 0$. This means that the matrix A is *irreducible*; a matrix A with nonnegative entries is said to be irreducible if, and only if, for every pair of indices (ℓ, k), there exists a finite integer $n_{\ell k} > 0$ such that $[A^{n_{\ell k}}]_{\ell k} > 0$, which is what we have established so far. We assume that $n_{\ell k}$ is the smallest integer that satisfies this property. Note that under irreducibility, the power $n_{\ell k}$ is allowed to be dependent on the indices (ℓ, k). Therefore, graph connectivity ensures the irreducibility of A. We now go a step further and show that *strong* graph connectivity ensures the primitiveness of A. Recall that for a strongly connected graph there exists at least one agent with a self-loop. We verify that an irreducible matrix A with at least one positive diagonal element is necessarily primitive so that a *uniform* power n_o satisfies (25.1) for all (ℓ, k).

Since the network is strongly connected, this means that there exists at least one agent k_o with $a_{k_o,k_o} > 0$. We know from (25.145) that for any agent ℓ in the network, it holds that $[A^{n_{\ell k_o}}]_{\ell k_o} > 0$. Then,

$$\left[A^{(n_{\ell k_o}+1)}\right]_{\ell k_o} = [A^{n_{\ell k_o}} A]_{\ell k_o} = \sum_{m=1}^{K} [A^{n_{\ell k_o}}]_{\ell m}\, a_{mk_o} \geq [A^{n_{\ell k_o}}]_{\ell k_o}\, a_{k_o,k_o} > 0 \tag{25.146}$$

so that the positivity of the (ℓ, k_o)th entry is maintained at higher powers of A once it is satisfied at power $n_{\ell k_o}$. The integers $\{n_{\ell k_o}\}$ are bounded by K. Let

$$m_o \;\stackrel{\Delta}{=}\; \max_{1\leq \ell \leq K} \{n_{\ell k_o}\} \tag{25.147}$$

Then, the above result implies that

$$[A^{m_o}]_{\ell k_o} > 0, \quad \text{for all } \ell \tag{25.148}$$

so that the entries on the k_oth column of A^{m_o} are all positive. Similarly, repeating the argument (25.145), we can verify that for arbitrary agents (k, ℓ), with the roles of k and ℓ now reversed, there exists a path of length $n_{k\ell}$ such that $[A^{n_{k\ell}}]_{k\ell} > 0$. For the same agent k_o with $a_{k_o,k_o} > 0$ as above, it holds that

$$\left[A^{(n_{k_o\ell}+1)}\right]_{k_o\ell} = [AA^{n_{k_o\ell}}]_{k_o\ell} = \sum_{m=1}^{K} a_{k_o m}\, [A^{n_{k_o\ell}}]_{m\ell} \geq a_{k_o,k_o}\, [A^{n_{k_o\ell}}]_{k_o\ell} > 0 \tag{25.149}$$

so that the positivity of the (k_o, ℓ)th entry is maintained at higher powers of A once it is satisfied at power $n_{k_o\ell}$. Likewise, the integers $\{n_{k_o\ell}\}$ are bounded by K. Let

$$m_o' \;\stackrel{\Delta}{=}\; \max_{1\leq \ell \leq K} \{n_{k_o\ell}\} \tag{25.150}$$

Then, the above result implies that

$$\left[A^{m_o'}\right]_{k_o\ell} > 0, \quad \text{for all } \ell \tag{25.151}$$

so that the entries on the k_oth row of $A^{m_o'}$ are all positive.

Now, let $n_o = m_o + m_o'$ and let us examine the entries of the matrix A^{n_o}. We can write schematically

$$A^{n_o} = A^{m_o} A^{m_o'} = \begin{bmatrix} \times & \times & + & \times \\ \times & \times & + & \times \\ \times & \times & + & \times \\ \times & \times & + & \times \end{bmatrix} \begin{bmatrix} \times & \times & \times & \times \\ \times & \times & \times & \times \\ + & + & + & + \\ \times & \times & \times & \times \end{bmatrix} \tag{25.152}$$

where the plus signs are used to refer to the *positive* entries on the k_oth column and row of A^{m_o} and $A^{m_o'}$, respectively, and the $\times$ signs are used to refer to the remaining entries of A^{m_o} and $A^{m_o'}$, which are necessarily nonnegative. It is clear from the above equality that the resulting entries of A^{n_o} will all be positive, and we conclude that A is primitive.

25.B PROOF OF PROPERTY (25.71)

In this appendix we establish property (25.71). Let y denote any nonzero vector in the span of $(\mathbb{1}_K \otimes I_M)$, i.e.,

$$\begin{aligned} y &= (\mathbb{1}_K \otimes I_M)x, \quad \text{for some nonzero vector } x \in \mathbb{R}^M \\ &= \text{blkcol}\{x, x, \ldots, x\} \\ &= \mathbb{1}_K \otimes x \end{aligned} \tag{25.153}$$

so that y consists of the vector x stacked vertically K times. Then, it holds that

$$\begin{aligned} (I_{KM} - \mathcal{A})y &= \Big((I_K - A) \otimes I_M\Big)(\mathbb{1}_K \otimes x) \\ &\overset{(a)}{=} (I_K - A)\mathbb{1}_K \otimes x \\ &\overset{(b)}{=} (\mathbb{1}_K - \mathbb{1}_K) \otimes x \\ &= 0 \end{aligned} \tag{25.154}$$

where in step (a) we used a Kronecker product property from Table 1.1, and in step (b) we used the fact that $A\mathbb{1} = \mathbb{1}$. Therefore, we established the direction:

$$y \in \text{span}(\mathbb{1}_K \otimes I_M) \Longrightarrow y \in \text{nullspace}(I_{KM} - \mathcal{A}) \tag{25.155}$$

Conversely, let y be any nonzero vector in the nullspace of $(I_{KM} - \mathcal{A})$. Then, it holds that

$$(I_{KM} - \mathcal{A})y = 0 \Longleftrightarrow (I_{KM} - \mathcal{A} + I_{KM})y = y \tag{25.156}$$

which means that y is an eigenvector for $(I_{KM} - \mathcal{A} + I_{KM})$ corresponding to the eigenvalue 1. Now, recall that A is primitive and doubly stochastic by assumption and, therefore, it has a unique eigenvalue at 1. This means that $\mathcal{A}$ is primitive and doubly stochastic as well and its eigenvalue at 1 has multiplicity M. Let e_m denote any of the basis vectors in $\mathbb{R}^M$. Then, we find that

$$\begin{aligned} (I_{KM} - \mathcal{A} + I_{KM})(\mathbb{1}_K \otimes e_m) &= \Big((I_K - A + I_K) \otimes I_M\Big)(\mathbb{1}_K \otimes e_m) \\ &= \Big((I_K - A + I_K)\mathbb{1}_K\Big) \otimes e_m \\ &= \mathbb{1}_K \otimes e_m \end{aligned} \tag{25.157}$$

This means that the M vectors $\{\mathbb{1}_K \otimes e_m\}$ are eigenvectors for $(I_{KM} - \mathcal{A} + I_{KM})$ corresponding to the eigenvalue at 1. Comparing with (25.156) we conclude that

$$\begin{aligned} & y \in \text{span}\,\{\mathbb{1}_K \otimes e_1,\ \mathbb{1}_K \otimes e_2,\ \ldots,\ \mathbb{1}_K \otimes e_M\} \\ \Longleftrightarrow\ & y \in \text{span}(\mathbb{1}_K \otimes I_M) \end{aligned} \tag{25.158}$$

as expected.

25.C CONVERGENCE OF PRIMAL ALGORITHMS

In this appendix we establish the convergence of the consensus and diffusion strategies for both deterministic and stochastic optimization by reproducing the arguments from Sayed (2014a). For generality, we will consider the stochastic case; the results can be specialized to the deterministic scenario by setting the gradient noise to zero; some of the conditions can also be relaxed. The main objective here is to mimic the derivation and arguments used in the single-agent case in Chapter 19. Although we will assume that the combination matrix is doubly stochastic, the arguments in Sayed (2014a) are more general and assume left-stochasticity.

Thus, consider again the optimization problem (25.39), repeated here for ease of reference:

$$\boxed{w^\star = \underset{w\in\mathbb{R}^M}{\text{argmin}} \left\{ P(w) \triangleq \sum_{k=1}^{K} \alpha_k P_k(w) \right\}} \tag{25.159}$$

where the $\{\alpha_k\}$ are positive convex combination coefficients that add up to 1:

$$\sum_{k=1}^{K} \alpha_k = 1, \quad \alpha_k > 0, \quad k = 1, 2, \ldots, K \tag{25.160}$$

We assume that $P(w)$ is ν-strongly convex while the individual $P_k(w)$ are convex and twice-differentiable. We did not require the risk function(s) to be twice-differentiable in the earlier chapters when we examined the convergence behavior of stochastic optimization algorithms; it was sufficient to work with the ν-strong convexity of the risk function and its δ-Lipschitz gradients. We can pursue a similar argument here as well. However, for convenience of the presentation, we will assume twice-differentiability.

It is useful to describe the recursions for the consensus and (ATC, CTA) diffusion strategies from Section 25.4 by means of a single *unifying* set of equations using three combination matrices $\{A_o, A_1, A_2\}$ with entries $\{a_{o,\ell k}, a_{1,\ell k}, a_{2,\ell k}\}$ as follows:

$$\left\{\begin{aligned} \boldsymbol{\phi}_{k,n-1} &= \sum_{\ell\in\mathcal{N}_k} a_{1,\ell k}\, \boldsymbol{w}_{\ell,n-1} \\ \boldsymbol{\psi}_{k,n} &= \sum_{\ell\in\mathcal{N}_k} a_{o,\ell k}\, \boldsymbol{\phi}_{\ell,n-1} \;-\; \mu\alpha_k \widehat{\nabla_{w^\mathsf{T}} P}_k\left(\boldsymbol{\phi}_{k,n-1}\right) \\ \boldsymbol{w}_{k,n} &= \sum_{\ell\in\mathcal{N}_k} a_{2,\ell k}\, \boldsymbol{\psi}_{\ell,n} \end{aligned}\right. \tag{25.161}$$

where $\{\boldsymbol{\phi}_{k,n-1}, \boldsymbol{\psi}_{k,n}\}$ denote $M \times 1$ intermediate variables, and the approximate gradient is generally given by

$$\widehat{\nabla_{w^\mathsf{T}} P}_k(w) \;=\; \nabla_{w^\mathsf{T}}\, Q(w;\cdot) \tag{25.162}$$

in terms of the gradient of a loss function. The $K \times K$ combination matrices are symmetric and doubly stochastic:

$$A_o^\mathsf{T}\mathbb{1} = \mathbb{1}, \quad A_1^\mathsf{T}\mathbb{1} = \mathbb{1}, \quad A_2^\mathsf{T}\mathbb{1} = \mathbb{1}, \quad A_o = A_o^\mathsf{T}, \quad A_1 = A_1^\mathsf{T}, \quad A_2 = A_2^\mathsf{T} \tag{25.163}$$

We also introduce the matrix product (we alert the reader not to confuse the matrix P with the notation for the aggregate risk $P(w)$, which is a function of the argument w):

$$P \triangleq A_1 A_o A_2 \tag{25.164}$$

which is doubly stochastic as well (but not necessarily symmetric). We assume that each of the combination matrices $\{A_o, A_1, A_2\}$ defines an underlying *connected* graph topology so that none of their rows are identically zero. Different choices for $\{A_o, A_1, A_2\}$ correspond to different decentralized strategies:

$$\text{noncooperative: } A_1 = A_o = A_2 = I_K \quad\longrightarrow\quad P = I_K \tag{25.165}$$

$$\text{consensus: } A_o = A, \;\; A_1 = I_K = A_2 \quad\longrightarrow\quad P = A \tag{25.166}$$

$$\text{CTA diffusion: } A_1 = A, \;\; A_2 = I_K = A_o \quad\longrightarrow\quad P = A \tag{25.167}$$

$$\text{ATC diffusion: } A_2 = A, \;\; A_1 = I_K = A_o \quad\longrightarrow\quad P = A \tag{25.168}$$

We assume that P is a *primitive* matrix. For example, this condition is automatically guaranteed if the combination matrix A in the selections (25.166)–(25.168) is primitive, which in turn is guaranteed for strongly connected networks. It then follows from the conclusions derived from the Perron–Frobenius theorem and listed before (25.24) that:

(a) The matrix P has a *single* eigenvalue at 1.

(b) All other eigenvalues of P are strictly inside the unit circle so that $\rho(P) = 1$.

(c) With proper sign scaling, all entries of the right-eigenvector of P corresponding to the single eigenvalue at 1 are *positive*. We let p denote this right-eigenvector, with its entries $\{p_k\}$ normalized to add up to 1, i.e.,

$$Pp = p, \quad \mathbb{1}^\mathsf{T} p = 1, \quad p_k > 0, \;\; k = 1, 2, \ldots, K \tag{25.169}$$

We refer to p as the *Perron eigenvector* of P. Since P is assumed doubly stochastic, we have $p = \mathbb{1}/K$ (uniform entries equal to $1/K$).

We associate with each agent k the three error vectors:

$$\widetilde{\boldsymbol{w}}_{k,n} \stackrel{\Delta}{=} w^\star - \boldsymbol{w}_{k,n} \tag{25.170}$$

$$\widetilde{\boldsymbol{\psi}}_{k,n} \stackrel{\Delta}{=} w^\star - \boldsymbol{\psi}_{k,n} \tag{25.171}$$

$$\widetilde{\boldsymbol{\phi}}_{k,n-1} \stackrel{\Delta}{=} w^\star - \boldsymbol{\phi}_{k,n-1} \tag{25.172}$$

which measure the deviations from the minimizer $w^\star$. Subtracting $w^\star$ from both sides of the equations in (25.161) we get

$$\left\{ \begin{aligned} \widetilde{\boldsymbol{\phi}}_{k,n-1} &= \sum_{\ell \in \mathcal{N}_k} a_{1,\ell k}\, \widetilde{\boldsymbol{w}}_{\ell,n-1} \\ \widetilde{\boldsymbol{\psi}}_{k,i} &= \sum_{\ell \in \mathcal{N}_k} a_{o,\ell k}\, \widetilde{\boldsymbol{\phi}}_{\ell,n-1} + \mu\alpha_k \widehat{\nabla_{w^\mathsf{T}} P_k}\,(\boldsymbol{\phi}_{k,n-1}) \\ \widetilde{\boldsymbol{w}}_{k,n} &= \sum_{\ell \in \mathcal{N}_k} a_{2,\ell k}\, \widetilde{\boldsymbol{\psi}}_{\ell,n} \end{aligned} \right. \tag{25.173}$$

We also associate with each agent k the following gradient noise vector:

$$\boldsymbol{g}_{k,n}(\boldsymbol{\phi}_{k,n-1}) \stackrel{\Delta}{=} \widehat{\nabla_{w^\mathsf{T}} P_k}(\boldsymbol{\phi}_{k,n-1}) \; - \; \nabla_{w^\mathsf{T}} P_k(\boldsymbol{\phi}_{k,n-1}) \tag{25.174}$$

in addition to the mismatch (or bias) vector:

$$b_k \stackrel{\Delta}{=} -\nabla_{w^\mathsf{T}} P_k(w^\star) \tag{25.175}$$

In the special case when all individual risks, $P_k(w)$, are minimized at the same location $w^\star$, then b_k will be identically zero. In general, though, the vector b_k is nonzero since the minimizers for the local risks need not agree with each other and with $w^\star$. Let $\mathfrak{F}_{n-1}$ represent the collection of all random events generated by the processes $\{\boldsymbol{w}_{k,m}\}$ at all agents $k = 1, 2, \ldots, K$ up to time $n-1$:

$$\mathfrak{F}_{n-1} \stackrel{\Delta}{=} \text{filtration}\Big\{\boldsymbol{w}_{k,-1}, \boldsymbol{w}_{k,0}, \boldsymbol{w}_{k,1}, \ldots, \boldsymbol{w}_{k,n-1}, \text{ all } k\Big\} \tag{25.176}$$

Similar to the analysis in prior chapters for single-agent learning, we assume that the gradient noise processes across the agents satisfy the following conditions.

Assumption 25.1. (Conditions on gradient noise) *It is assumed that the first- and second-order conditional moments of the individual gradient noise processes, $\boldsymbol{g}_{k,n}(\boldsymbol{\phi})$, satisfy the following conditions for any iterates $\boldsymbol{\phi} \in \boldsymbol{\mathcal{F}}_{n-1}$ and for all $k, \ell = 1, 2, \ldots, K$:*

$$\mathbb{E}\left[\boldsymbol{g}_{k,n}(\boldsymbol{\phi}_{k,n-1})\,|\,\boldsymbol{\mathcal{F}}_{n-1}\right] = 0 \tag{25.177a}$$

$$\mathbb{E}\left[\|\boldsymbol{g}_{k,n}(\boldsymbol{\phi}_{k,n-1})\|^2\,|\,\boldsymbol{\mathcal{F}}_{n-1}\right] \leq \beta_{g,k}^2\|\widetilde{\boldsymbol{\phi}}_{k,n-1}\|^2 + \sigma_{g,k}^2 \tag{25.177b}$$

$$\mathbb{E}\left[\boldsymbol{g}_{k,n}(\boldsymbol{\phi}_{k,n-1})\boldsymbol{g}_{\ell,n}^{\mathsf{T}}(\boldsymbol{\phi}_{\ell,n-1})|\boldsymbol{\mathcal{F}}_{n-1}\right] = 0, \quad k \neq \ell \tag{25.177c}$$

for some nonnegative scalars $\beta_{g,k}^2$ and $\sigma_{g,k}^2$.

As was done in Chapter 18, one can verify that conditions of the form (25.177a) and (25.177b) are justified under the conditions (**A3**) or (**A3'**) imposed in the body of this chapter on the gradients of the loss function under empirical or stochastic risk minimization. Condition (25.177c), on the other hand, is an assumption requiring the gradient noises across the agents to be conditionally uncorrelated.

In the absence of gradient noise (e.g., in deterministic optimization), we set $\beta_{g,k}^2 = 0 = \sigma_{g,k}^2$. Using the gradient noise process, we can rewrite the error recursions (25.173) in the form (where we are suppressing the argument $\phi_{k,n-1}$ from the gradient noise term for compactness):

$$\begin{cases} \widetilde{\boldsymbol{\phi}}_{k,n-1} = \displaystyle\sum_{\ell \in \mathcal{N}_k} a_{1,\ell k}\, \widetilde{\boldsymbol{w}}_{\ell,n-1} \\ \widetilde{\boldsymbol{\psi}}_{k,i} = \displaystyle\sum_{\ell \in \mathcal{N}_k} a_{o,\ell k}\, \widetilde{\boldsymbol{\phi}}_{\ell,n-1} + \mu\alpha_k \nabla_{w^{\mathsf{T}}} P_k\left(\boldsymbol{\phi}_{k,n-1}\right) + \mu\alpha_k \boldsymbol{g}_{k,n} \\ \widetilde{\boldsymbol{w}}_{k,n} = \displaystyle\sum_{\ell \in \mathcal{N}_k} a_{2,\ell k}\, \widetilde{\boldsymbol{\psi}}_{\ell,n} \end{cases} \tag{25.178}$$

We observe that the gradient vectors in (25.178) are evaluated at the intermediate variable, $\boldsymbol{\phi}_{k,n-1}$, and not at any of the *error* variables. For this reason, equation (25.178) is still not an actual recursion. To transform it into a recursion that only involves error variables, we call upon the mean-value theorem (10.10) to write:

$$\nabla_{w^{\mathsf{T}}} P_k(\boldsymbol{\phi}_{k,n-1}) = \underbrace{\nabla_{w^{\mathsf{T}}} P_k(w^\star)}_{\triangleq\, -b_k} - \underbrace{\left[\int_0^1 \nabla_w^2 P_k\Big(w^\star - t\widetilde{\boldsymbol{\phi}}_{k,n-1}\Big)dt\right]}_{\triangleq\, \boldsymbol{H}_{k,n-1}} \widetilde{\boldsymbol{\phi}}_{k,n-1} \tag{25.179}$$

That is,

$$\nabla_{w^{\mathsf{T}}} P_k(\boldsymbol{\phi}_{k,n-1}) = -b_k - \boldsymbol{H}_{k,n-1}\widetilde{\boldsymbol{\phi}}_{k,n-1} \tag{25.180}$$

in terms of an $M \times M$ random matrix $\boldsymbol{H}_{k,n-1}$ defined in terms of the integral of the $M \times M$ Hessian matrix at agent k:

$$\boldsymbol{H}_{k,n-1} \triangleq \int_0^1 \nabla_w^2 P_k\Big(w^\star - t\widetilde{\boldsymbol{\phi}}_{k,i-1}\Big)dt \tag{25.181}$$

Substituting (25.180) into (25.178) leads to

$$\left\{\begin{array}{rcl}\widetilde{\boldsymbol{\phi}}_{k,n-1} & = & \sum\limits_{\ell\in\mathcal{N}_k} a_{1,\ell k}\,\widetilde{\boldsymbol{w}}_{\ell,n-1}\\ \widetilde{\boldsymbol{\psi}}_{k,n} & = & \sum\limits_{\ell\in\mathcal{N}_k} a_{o,\ell k}\,\widetilde{\boldsymbol{\phi}}_{\ell,n-1} - \mu\alpha_k\Big(\boldsymbol{H}_{k,n-1}\widetilde{\boldsymbol{\phi}}_{k,n-1} + b_k + \boldsymbol{g}_{k,n}\Big)\\ \widetilde{\boldsymbol{w}}_{k,n} & = & \sum\limits_{\ell\in\mathcal{N}_k} a_{2,\ell k}\,\widetilde{\boldsymbol{\psi}}_{\ell,n}\end{array}\right. \tag{25.182}$$

These equations describe the evolution of the error quantities at the individual agents for $k = 1, 2, \ldots, K$.

We collect the error vectors from across all agents into the following $K \times 1$ block error vectors (whose individual entries are of size $M \times 1$ each):

$$\widetilde{\boldsymbol{\mathcal{W}}}_n \triangleq \begin{bmatrix}\widetilde{\boldsymbol{w}}_{1,n}\\ \widetilde{\boldsymbol{w}}_{2,n}\\ \vdots\\ \widetilde{\boldsymbol{w}}_{K,n}\end{bmatrix},\quad \widetilde{\boldsymbol{\phi}}_{n-1} \triangleq \begin{bmatrix}\widetilde{\boldsymbol{\phi}}_{1,n-1}\\ \widetilde{\boldsymbol{\phi}}_{2,n-1}\\ \vdots\\ \widetilde{\boldsymbol{\phi}}_{K,n-1}\end{bmatrix},\quad \widetilde{\boldsymbol{\psi}}_n \triangleq \begin{bmatrix}\widetilde{\boldsymbol{\psi}}_{1,n}\\ \widetilde{\boldsymbol{\psi}}_{2,n}\\ \vdots\\ \widetilde{\boldsymbol{\psi}}_{K,n}\end{bmatrix} \tag{25.183}$$

We also define the following block gradient noise and bias vectors:

$$\boldsymbol{\mathcal{G}}_n \triangleq \begin{bmatrix}\boldsymbol{g}_{1,n}\\ \boldsymbol{g}_{2,n}\\ \vdots\\ \boldsymbol{g}_{K,n}\end{bmatrix},\quad b \triangleq \begin{bmatrix}b_1\\ b_2\\ \vdots\\ b_K\end{bmatrix} \tag{25.184}$$

Recall that each entry $\boldsymbol{g}_{k,n}$ in (25.184) is dependent on $\boldsymbol{\phi}_{k,n-1}$. Recall also from recursions (25.161) for the decentralized algorithm that $\boldsymbol{\phi}_{k,n-1}$ is a combination of various $\{\boldsymbol{w}_{\ell,n-1}\}$. Therefore, the block gradient vector, $\boldsymbol{\mathcal{G}}_n$, defined in (25.184) is dependent on the network vector, $\boldsymbol{\mathcal{W}}_{n-1}$, namely,

$$\boldsymbol{\mathcal{W}}_{n-1} \triangleq \begin{bmatrix}\boldsymbol{w}_{1,n-1}\\ \boldsymbol{w}_{2,n-1}\\ \vdots\\ \boldsymbol{w}_{K,n-1}\end{bmatrix} \tag{25.185}$$

For this reason, we will also sometimes write $\boldsymbol{\mathcal{G}}_n(\boldsymbol{\mathcal{W}}_{n-1})$ rather than simply $\boldsymbol{\mathcal{G}}_n$ when it is desired to highlight the dependency of $\boldsymbol{\mathcal{G}}_n$ on $\boldsymbol{\mathcal{W}}_{n-1}$.

We further introduce the Kronecker products

$$\mathcal{A}_o \triangleq A_o \otimes I_M,\quad \mathcal{A}_1 \triangleq A_1 \otimes I_M,\quad \mathcal{A}_2 \triangleq A_2 \otimes I_M \tag{25.186}$$

The matrix $\mathcal{A}_o$ is a $K \times K$ *block* matrix whose (ℓ, k)th block is equal to $a_{o,\ell k} I_M$. Similarly for $\mathcal{A}_1$ and $\mathcal{A}_2$. Likewise, we introduce the following $K \times K$ *block* diagonal matrices, whose individual entries are of size $M \times M$ each:

$$\mathcal{M} \triangleq \mu\,\text{blkdiag}\{\,\alpha_1 I_M,\ \alpha_2 I_M,\ \ldots,\ \alpha_K I_M\,\} \tag{25.187}$$

$$\boldsymbol{\mathcal{H}}_{n-1} \triangleq \text{blkdiag}\{\,\boldsymbol{H}_{1,n-1},\ \boldsymbol{H}_{2,n-1},\ \ldots,\ \boldsymbol{H}_{K,n-1}\,\} \tag{25.188}$$

We then conclude from (25.182) that the following relations hold for the *network* variables:

$$\begin{cases} \widetilde{\boldsymbol{\phi}}_{n-1} = & \mathcal{A}_1^{\mathsf{T}}\widetilde{\boldsymbol{\mathcal{W}}}_{n-1} \\ \widetilde{\boldsymbol{\psi}}_n = & \left(\mathcal{A}_o^{\mathsf{T}} - \mathcal{M}\boldsymbol{\mathcal{H}}_{n-1}\right)\widetilde{\boldsymbol{\phi}}_{n-1} + \mathcal{M}\,\boldsymbol{\mathcal{G}}_n(\boldsymbol{\mathcal{W}}_{n-1}) - \mathcal{M}b \\ \widetilde{\boldsymbol{\mathcal{W}}}_n = & \mathcal{A}_2^{\mathsf{T}}\widetilde{\boldsymbol{\psi}}_n \end{cases} \tag{25.189}$$

so that the network weight error vector, $\widetilde{\boldsymbol{\mathcal{W}}}_n$, ends up evolving according to the following *stochastic* recursion over $n \geq 0$:

$$\widetilde{\boldsymbol{\mathcal{W}}}_n = \mathcal{A}_2^{\mathsf{T}}\left(\mathcal{A}_o^{\mathsf{T}} - \mathcal{M}\boldsymbol{\mathcal{H}}_{n-1}\right)\mathcal{A}_1^{\mathsf{T}}\widetilde{\boldsymbol{\mathcal{W}}}_{n-1} + \mathcal{A}_2^{\mathsf{T}}\mathcal{M}\,\boldsymbol{\mathcal{G}}_n(\boldsymbol{\mathcal{W}}_{n-1}) - \mathcal{A}_2^{\mathsf{T}}\mathcal{M}b \tag{25.190}$$

For comparison purposes, if each agent operates individually and uses the noncooperative strategy, then the weight error vectors across all K agents would instead evolve according to the following stochastic recursion:

$$\widetilde{\boldsymbol{\mathcal{W}}}_n = \left(I_{KM} - \mathcal{M}\boldsymbol{\mathcal{H}}_{n-1}\right)\widetilde{\boldsymbol{\mathcal{W}}}_{n-1} + \mathcal{M}\,\boldsymbol{\mathcal{G}}_n(\boldsymbol{\mathcal{W}}_{n-1}) - \mathcal{M}b \tag{25.191}$$

where the matrices $\{\mathcal{A}_o, \mathcal{A}_1, \mathcal{A}_2\}$ do not appear since, in this case, $A_o = A_1 = A_2 = I_K$. We summarize the discussion so far in the following statement.

Lemma 25.6. (Network error dynamics) *Consider a network of K interacting agents running the distributed strategy (25.161). The evolution of the error dynamics across the network relative to the reference vector $w^\star$ is described by the following recursion:*

$$\widetilde{\boldsymbol{\mathcal{W}}}_n = \boldsymbol{\mathcal{B}}_{n-1}\widetilde{\boldsymbol{\mathcal{W}}}_{n-1} + \mathcal{A}_2^{\mathsf{T}}\mathcal{M}\,\boldsymbol{\mathcal{G}}_n(\boldsymbol{\mathcal{W}}_{n-1}) - \mathcal{A}_2^{\mathsf{T}}\mathcal{M}b, \; n \geq 0 \tag{25.192}$$

where we introduced

$$\boldsymbol{\mathcal{B}}_{n-1} \stackrel{\Delta}{=} \mathcal{A}_2^{\mathsf{T}}\left(\mathcal{A}_o^{\mathsf{T}} - \mathcal{M}\boldsymbol{\mathcal{H}}_{n-1}\right)\mathcal{A}_1^{\mathsf{T}} \tag{25.193}$$

$$\mathcal{A}_o \stackrel{\Delta}{=} A_o \otimes I_M, \quad \mathcal{A}_1 \stackrel{\Delta}{=} A_1 \otimes I_M, \quad \mathcal{A}_2 \stackrel{\Delta}{=} A_2 \otimes I_M \tag{25.194}$$

$$\mathcal{M} \stackrel{\Delta}{=} \mu\,\text{blkdiag}\{\,\alpha_1 I_M,\ \alpha_2 I_M,\ \ldots,\ \alpha_K I_M\,\} \tag{25.195}$$

$$\boldsymbol{\mathcal{H}}_{n-1} \stackrel{\Delta}{=} \text{blkdiag}\{\,\boldsymbol{H}_{1,n-1},\ \boldsymbol{H}_{2,n-1},\ \ldots,\ \boldsymbol{H}_{K,n-1}\,\} \tag{25.196}$$

$$\boldsymbol{H}_{k,n-1} \stackrel{\Delta}{=} \int_0^1 \nabla_w^2 P_k(w^\star - t\widetilde{\boldsymbol{\phi}}_{k,n-1})dt \tag{25.197}$$

where $\nabla_w^2 P_k(w)$ denotes the $M \times M$ Hessian matrix of $P_k(w)$ relative to w. Moreover, the vectors $\{\widetilde{\boldsymbol{\mathcal{W}}}_n, \boldsymbol{\mathcal{G}}_n, b\}$ are defined by (25.183) and (25.184).

We are now ready to establish the mean-square stability of the network error process and show that its MSE tends asymptotically to a bounded region on the order of $O(\mu)$.

Theorem 25.1. (Network mean-square-error stability) *Consider the same setting of Lemma 25.6. Assume the aggregate risk $P(w)$ is ν-strongly convex with δ-Lipschitz gradient vectors. Assume also each individual risk $P_k(w)$ is twice-differentiable and convex, with at least one of them being ν_d-strongly convex. Assume further that the first- and second-order moments of the gradient noise process satisfy the conditions in Assumption 25.1. Then, the network is mean-square stable for sufficiently small step sizes, namely, it holds that*

$$\limsup_{n\to\infty} \mathbb{E}\,\|\widetilde{\boldsymbol{w}}_{k,n}\|^2 = O(\mu), \quad k = 1, 2, \ldots, K \tag{25.198}$$

for any $\mu < \mu_o$, for some small enough μ_o.

Proof: The derivation is demanding. We follow arguments motivated by the analysis from Chen and Sayed (2015a, b), Zhao and Sayed (2015a, b, c), and Sayed (2014a).

The derivation involves, as an initial step, transforming the error recursion (25.199) shown below into a more convenient form shown later in (25.247).

We start from the network error recursion (25.192):

$$\widetilde{w}_n = \mathcal{B}_{n-1}\widetilde{w}_{n-1} + \mathcal{A}_2^\mathsf{T}\mathcal{M}\,\mathcal{G}_n(w_{n-1}) - \mathcal{A}_2^\mathsf{T}\mathcal{M}b, \quad n \geq 0 \tag{25.199}$$

where

$$\begin{aligned}\mathcal{B}_{n-1} &= \mathcal{A}_2^\mathsf{T}\left(\mathcal{A}_o^\mathsf{T} - \mathcal{M}\mathcal{H}_{n-1}\right)\mathcal{A}_1^\mathsf{T} \\ &= \mathcal{A}_2^\mathsf{T}\mathcal{A}_o^\mathsf{T}\mathcal{A}_1^\mathsf{T} - \mathcal{A}_2^\mathsf{T}\mathcal{M}\mathcal{H}_{n-1}\mathcal{A}_1^\mathsf{T} \\ &\stackrel{\Delta}{=} \mathcal{P}^\mathsf{T} - \mathcal{A}_2^\mathsf{T}\mathcal{M}\mathcal{H}_{n-1}\mathcal{A}_1^\mathsf{T}\end{aligned} \tag{25.200}$$

in terms of the matrix

$$\begin{aligned}\mathcal{P}^\mathsf{T} &\stackrel{\Delta}{=} \mathcal{A}_2^\mathsf{T}\mathcal{A}_o^\mathsf{T}\mathcal{A}_1^\mathsf{T} \\ &= (A_2^\mathsf{T}\otimes I_M)(A_o^\mathsf{T}\otimes I_M)(A_1^\mathsf{T}\otimes I_M) \\ &= (A_2^\mathsf{T}A_o^\mathsf{T}A_1^\mathsf{T}\otimes I_M) \\ &= P^\mathsf{T}\otimes I_M\end{aligned} \tag{25.201}$$

The matrix $P = A_1A_oA_2$ is doubly stochastic and assumed primitive. It follows that it has a single eigenvalue at 1 while all other eigenvalues are strictly inside the unit circle. We let $p = \mathbb{1}/K$ denote its Perron eigenvector, which was defined earlier in (25.169). This vector is also the Perron vector for each of the doubly stochastic matrices $\{A_o, A_1, A_2\}$. Introduce the scalars

$$q_k \stackrel{\Delta}{=} \mu\alpha_k/K, \quad k = 1, 2, \ldots, K \tag{25.202}$$

Obviously, it holds for the extended matrices $\{\mathcal{P}, \mathcal{A}_2\}$ that

$$\mathcal{P}(p\otimes I_{2M}) = (p\otimes I_M) \tag{25.203}$$

$$\mathcal{M}\mathcal{A}_2(p\otimes I_{2M}) = (q\otimes I_M) \tag{25.204}$$

$$(\mathbb{1}^\mathsf{T}\otimes I_M)(p\otimes I_M) = I_M \tag{25.205}$$

Moreover, since A_1 and A_2 are doubly stochastic, it holds that

$$\mathcal{A}_1^\mathsf{T}(\mathbb{1}\otimes I_M) = (\mathbb{1}\otimes I_M) \tag{25.206}$$

$$\mathcal{A}_2^\mathsf{T}(\mathbb{1}\otimes I_M) = (\mathbb{1}\otimes I_M) \tag{25.207}$$

The derivation that follows exploits the eigen-structure of P. We start by noting that the $K \times K$ matrix P admits a Jordan canonical decomposition of the form (see Horn and Johnson (1990, p. 128)):

$$P \stackrel{\Delta}{=} V_\epsilon J V_\epsilon^{-1} \tag{25.208}$$

$$J = \left[\begin{array}{c|c} 1 & 0 \\ \hline 0 & J_\epsilon \end{array}\right] \tag{25.209}$$

$$V_\epsilon = \left[\begin{array}{c|c} p & V_R \end{array}\right] \tag{25.210}$$

$$V_\epsilon^{-1} = \left[\begin{array}{c} \mathbb{1}^\mathsf{T} \\ \hline V_L^\mathsf{T} \end{array}\right] \tag{25.211}$$

where the matrix J_ϵ consists of Jordan blocks, with each one of them having the generic form (say, for a Jordan block of size 4×4):

$$\begin{bmatrix} \lambda & & & \\ \epsilon & \lambda & & \\ & \epsilon & \lambda & \\ & & \epsilon & \lambda \end{bmatrix} \tag{25.212}$$

with $\epsilon > 0$ appearing on the lower[1] diagonal, and where the eigenvalue λ may be complex but has magnitude strictly less than 1. The scalar ϵ is any small positive number that is independent of μ. Obviously, since $V_\epsilon^{-1}V_\epsilon = I_N$, it holds that

$$\mathbb{1}^\mathsf{T} V_R = 0 \tag{25.213}$$

$$V_L^\mathsf{T} p = 0 \tag{25.214}$$

$$V_L^\mathsf{T} V_R = I_{N-1} \tag{25.215}$$

The matrices $\{V_\epsilon, J, V_\epsilon^{-1}\}$ have dimensions $K \times K$ while the matrices $\{V_L, J_\epsilon, V_R\}$ have dimensions $(K-1) \times (K-1)$. The Jordan decomposition of the extended matrix $\mathcal{P} = P \otimes I_M$ is given by

$$\mathcal{P} \;=\; (V_\epsilon \otimes I_M)(J \otimes I_M)(V_\epsilon^{-1} \otimes I_M) \tag{25.216}$$

so that substituting into (25.200) we obtain

$$\mathcal{B}_{n-1} = \Big((V_\epsilon^{-1})^\mathsf{T} \otimes I_M\Big)\Big\{(J^\mathsf{T} \otimes I_M) - \mathcal{D}_{n-1}^\mathsf{T}\Big\}\Big(V_\epsilon^\mathsf{T} \otimes I_M\Big) \tag{25.217}$$

where

$$\begin{aligned} \mathcal{D}_{n-1}^\mathsf{T} &\overset{\Delta}{=} \Big(V_\epsilon^\mathsf{T} \otimes I_M\Big)\mathcal{A}_2^\mathsf{T}\mathcal{M}\mathcal{H}_{n-1}\mathcal{A}_1^\mathsf{T}\Big((V_\epsilon^{-1})^\mathsf{T} \otimes I_M\Big) \\ &\equiv \begin{bmatrix} \boldsymbol{D}_{11,n-1}^\mathsf{T} & \boldsymbol{D}_{21,n-1}^\mathsf{T} \\ \boldsymbol{D}_{12,n-1}^\mathsf{T} & \boldsymbol{D}_{22,n-1}^\mathsf{T} \end{bmatrix} \end{aligned} \tag{25.218}$$

Using the partitioning (25.210)–(25.211) and the fact that

$$\mathcal{A}_1 \;=\; A_1 \otimes I_M, \qquad \mathcal{A}_2 \;=\; A_2 \otimes I_M \tag{25.219}$$

we find that the block entries $\{\boldsymbol{D}_{mn,n-1}\}$ in (25.218) are given by

$$\boldsymbol{D}_{11,n-1} = \sum_{k=1}^{K} q_k \boldsymbol{H}_{k,n-1}^\mathsf{T} \tag{25.220}$$

$$\boldsymbol{D}_{12,n-1} = (\mathbb{1}^\mathsf{T} \otimes I_M)\mathcal{H}_{n-1}^\mathsf{T}\mathcal{M}(A_2 V_R \otimes I_M) \tag{25.221}$$

$$\boldsymbol{D}_{21,n-1} = (V_L^\mathsf{T} A_1 \otimes I_M)\mathcal{H}_{n-1}^\mathsf{T}(q \otimes I_M) \tag{25.222}$$

$$\boldsymbol{D}_{22,n-1} = (V_L^\mathsf{T} A_1 \otimes I_M)\mathcal{H}_{n-1}^\mathsf{T}\mathcal{M}(A_2 V_R \otimes I_M) \tag{25.223}$$

Let us now show that the entries in each of these matrices is on the order of $O(\mu)$, as well as verify that the matrix norm sequences of these matrices are uniformly bounded from above for all n. To begin with, recall from (25.197) that

$$\boldsymbol{H}_{k,n-1} \;\overset{\Delta}{=}\; \int_0^1 \nabla_w^2 P_k(w^\star - t\widetilde{\boldsymbol{\phi}}_{k,n-1})dt \tag{25.224}$$

[1] For any $K \times K$ matrix A, the canonical Jordan decomposition $A = TJ'T^{-1}$ involves Jordan blocks in J' that have 1s on the lower diagonal instead of ϵ – recall Prob. 1.2. However, if we introduce the diagonal matrix $E = \text{diag}\{1, \epsilon, \epsilon^2, \ldots, \epsilon^{K-1}\}$, then $A = TE^{-1}EJ'E^{-1}ET^{-1}$, which we rewrite as $A = V_\epsilon J V_\epsilon^{-1}$ with $V_\epsilon = TE^{-1}$ and $J = EJ'E^{-1}$. The matrix J now has ϵ values instead of 1s on the lower diagonal.

and, moreover, by assumption, all individual risks $P_k(w)$ are convex functions, with at least one of them, say, the cost function of index k_o, being ν_d-strongly convex. This fact implies that, for any w,

$$\nabla^2_w P_{k_o}(w) \geq \nu_d I_M > 0, \quad \nabla^2_w P_k(w) \geq 0, \quad k \neq k_o \tag{25.225}$$

Consequently,

$$\boldsymbol{H}_{k_o,n-1} \geq \nu_d I_M > 0, \quad \boldsymbol{H}_{k,n-1} \geq 0, \quad k \neq k_o \tag{25.226}$$

and, therefore, $\boldsymbol{D}_{11,n-1} > 0$. More specifically, the matrix sequence $\boldsymbol{D}_{11,n-1}$ is uniformly bounded from below as follows:

$$\begin{aligned} \boldsymbol{D}_{11,n-1} \quad &\geq \quad q_{k_o}\nu_d I_M \\ &\overset{(25.202)}{=} (\mu\,\alpha_{k_o}\nu_d/K)I_M \\ &= \quad O(\mu) \end{aligned} \tag{25.227}$$

On the other hand, from the upper bound (25.50) on the sum of the Hessian matrices, and since each individual Hessian matrix is at least nonnegative definite, we get

$$\boldsymbol{H}_{k,n-1} \leq \delta I_M \tag{25.228}$$

so that the matrix sequence $\boldsymbol{D}_{11,n-1}$ is uniformly bounded from above as well:

$$\begin{aligned} \boldsymbol{D}_{11,n-1} \quad &\leq \quad q_{\max} K\delta I_M \\ &\overset{(25.202)}{=} (\mu\,\alpha_{\max}/K)K\delta I_M \\ &= \quad O(\mu) \end{aligned} \tag{25.229}$$

where $\alpha_{\max}$ denotes the largest $\{\alpha_k\}$. Combining results (25.227)–(25.229), we conclude that

$$\boldsymbol{D}_{11,n-1} = O(\mu) \tag{25.230}$$

Actually, since $\boldsymbol{D}_{11,n-1}$ is symmetric positive-definite, we also conclude that its eigenvalues (which are positive and real) are $O(\mu)$. This is because from the relation

$$(\mu\,\alpha_{k_o}\nu_d/K)I_M \leq \boldsymbol{D}_{11,n-1} \leq (\mu\,\alpha_{\max}/K)K\delta I_M \tag{25.231}$$

we can write, more compactly,

$$c_1\mu I_M \leq \boldsymbol{D}_{11,n-1} \leq c_2\mu I_M \tag{25.232}$$

for some positive constants c_1 and c_2 that are independent of μ and n. Accordingly, for the eigenvalues of $\boldsymbol{D}_{11,n-1}$, we can write

$$c_1\mu \leq \lambda(\boldsymbol{D}_{11,n-1}) \leq c_2\mu \tag{25.233}$$

It follows that the eigenvalues of $I_M - \boldsymbol{D}^{\mathsf{T}}_{11,n-1}$ are $1 - O(\mu)$ so that, in terms of the 2-induced norm and for sufficiently small μ:

$$\begin{aligned} \|I_M - \boldsymbol{D}^{\mathsf{T}}_{11,n-1}\| &= \rho(I_M - \boldsymbol{D}^{\mathsf{T}}_{11,n-1}) \\ &\leq 1 - \sigma_{11}\mu \\ &= 1 - O(\mu) \end{aligned} \tag{25.234}$$

for some positive constant σ_{11} that is independent of μ and n.

Similarly, from (25.226) and (25.228), and since each $\boldsymbol{H}_{k,n-1}$ is bounded from below and from above, we can conclude that

$$\boldsymbol{D}_{12,n-1} = O(\mu), \quad \boldsymbol{D}_{21,n-1} = O(\mu), \quad \boldsymbol{D}_{22,n-1} = O(\mu) \tag{25.235}$$

and that the norms of these matrix sequences are also uniformly bounded from above. For example, using the 2-induced norm (i.e., maximum singular value):

$$\begin{aligned} \|\boldsymbol{D}_{21,n-1}\| &\leq \|V_L^{\mathsf{T}} A_1 \otimes I_M\| \, \|q \otimes I_M\| \, \|\mathcal{H}_{n-1}^{\mathsf{T}}\| \\ &\leq \|V_L^{\mathsf{T}} A_1 \otimes I_M\| \, \|q \otimes I_M\| \left(\max_{1 \leq k \leq K} \|\boldsymbol{H}_{k,n-1}\| \right) \\ &\overset{(25.228)}{\leq} \|V_L^{\mathsf{T}} A_1 \otimes I_M\| \, \|q \otimes I_M\| \, \delta \\ &= \|V_L^{\mathsf{T}} A_1 \otimes I_M\| \, \|q\| \, \delta \\ &\leq \|V_L^{\mathsf{T}} A_1 \otimes I_M\| \sqrt{K \, q_{\max}^2} \, \delta \\ &= \|V_L^{\mathsf{T}} A_1 \otimes I_M\| \sqrt{K} \, (\mu \, \alpha_{\max}/K) \, \delta \end{aligned} \tag{25.236}$$

so that

$$\|\boldsymbol{D}_{21,n-1}\| \leq \sigma_{21}\mu = O(\mu) \tag{25.237}$$

for some positive constant σ_{21}. In the above derivation we used the fact that $\|q \otimes I_M\| = \|q\|$ since the singular values of a Kronecker product are given by all possible products of the singular values of the individual matrices – refer to Table 1.1. A similar argument applies to $\boldsymbol{D}_{12,n-1}$ and $\boldsymbol{D}_{22,n-1}$, for which we can verify that

$$\|\boldsymbol{D}_{12,n-1}\| \leq \sigma_{12}\mu = O(\mu), \qquad \|\boldsymbol{D}_{22,n-1}\| \leq \sigma_{22}\mu = O(\mu) \tag{25.238}$$

for some positive constants σ_{21} and σ_{22}. Let

$$\mathcal{V}_\epsilon \overset{\Delta}{=} V_\epsilon \otimes I_M, \qquad \mathcal{J}_\epsilon \overset{\Delta}{=} J_\epsilon \otimes I_M \tag{25.239}$$

Then, using (25.217), we can write

$$\boldsymbol{\mathcal{B}}_{n-1} = \left(\mathcal{V}_\epsilon^{-1}\right)^{\mathsf{T}} \begin{bmatrix} I_M - \boldsymbol{D}_{11,n-1}^{\mathsf{T}} & -\boldsymbol{D}_{21,n-1}^{\mathsf{T}} \\ -\boldsymbol{D}_{12,n-1}^{\mathsf{T}} & \mathcal{J}_\epsilon^{\mathsf{T}} - \boldsymbol{D}_{22,n-1}^{\mathsf{T}} \end{bmatrix} \mathcal{V}_\epsilon^{\mathsf{T}} \tag{25.240}$$

To simplify the notation, we drop the argument $\boldsymbol{\mathcal{W}}_{n-1}$ in (25.199) and write $\mathcal{G}_n$ instead of $\mathcal{G}_n(\boldsymbol{\mathcal{W}}_{n-1})$ from this point onwards. We now multiply both sides of the error recursion (25.199) from the left by $\mathcal{V}_\epsilon^{\mathsf{T}}$:

$$\mathcal{V}_\epsilon^{\mathsf{T}} \widetilde{\boldsymbol{\mathcal{W}}}_n = \mathcal{V}_\epsilon^{\mathsf{T}} \boldsymbol{\mathcal{B}}_{n-1} \left(\mathcal{V}_\epsilon^{-1}\right)^{\mathsf{T}} \mathcal{V}_\epsilon^{\mathsf{T}} \widetilde{\boldsymbol{\mathcal{W}}}_{n-1} + \mathcal{V}_\epsilon^{\mathsf{T}} \mathcal{A}_2^{\mathsf{T}} \mathcal{M} \mathcal{G}_n - \mathcal{V}_\epsilon^{\mathsf{T}} \mathcal{A}_2^{\mathsf{T}} \mathcal{M} b, \quad n \geq 0 \tag{25.241}$$

and let

$$\mathcal{V}_\epsilon^{\mathsf{T}} \widetilde{\boldsymbol{\mathcal{W}}}_n = \begin{bmatrix} (p^{\mathsf{T}} \otimes I_M) \widetilde{\boldsymbol{\mathcal{W}}}_n \\ (V_R^{\mathsf{T}} \otimes I_M) \widetilde{\boldsymbol{\mathcal{W}}}_n \end{bmatrix} \overset{\Delta}{=} \begin{bmatrix} \bar{\boldsymbol{\mathcal{W}}}_n \\ \check{\boldsymbol{\mathcal{W}}}_n \end{bmatrix} \tag{25.242}$$

$$\mathcal{V}_\epsilon^{\mathsf{T}} \mathcal{A}_2^{\mathsf{T}} \mathcal{M} \mathcal{G}_n = \begin{bmatrix} (p^{\mathsf{T}} \otimes I_M) \mathcal{A}_2^{\mathsf{T}} \mathcal{M} \mathcal{G}_n \\ (V_R^{\mathsf{T}} \otimes I_M) \mathcal{A}_2^{\mathsf{T}} \mathcal{M} \mathcal{G}_n \end{bmatrix} \overset{\Delta}{=} \begin{bmatrix} \bar{\mathcal{G}}_n \\ \check{\mathcal{G}}_n \end{bmatrix} \tag{25.243}$$

$$\mathcal{V}_\epsilon^{\mathsf{T}} \mathcal{A}_2^{\mathsf{T}} \mathcal{M} b = \begin{bmatrix} (p^{\mathsf{T}} \otimes I_M) \mathcal{A}_2^{\mathsf{T}} \mathcal{M} b \\ (V_R^{\mathsf{T}} \otimes I_M) \mathcal{A}_2^{\mathsf{T}} \mathcal{M} b \end{bmatrix} \overset{\Delta}{=} \begin{bmatrix} 0 \\ \check{b} \end{bmatrix} \tag{25.244}$$

where the zero entry in the last equality is due to the fact that

$$
\begin{aligned}
(p^{\mathsf{T}} \otimes I_M)\mathcal{A}_2^{\mathsf{T}}\mathcal{M}b &= (q^{\mathsf{T}} \otimes I_M)b \\
&= \sum_{k=1}^{K} q_k b_k \\
&= -\sum_{k=1}^{K} q_k \nabla_{w^{\mathsf{T}}} P_k(w^\star) \\
&= -\frac{\mu}{K}\left(\sum_{k=1}^{K} \alpha_k \nabla_{w^{\mathsf{T}}} P_k(w^\star)\right) \\
&= 0
\end{aligned} \tag{25.245}
$$

Moreover, from the expression for $\check{b}$ in (25.244), we note that it depends on $\mathcal{M}$ and b. Recall from (25.175) and (25.184) that the entries of b are defined in terms of the gradient vector $\nabla_{w^{\mathsf{T}}} P_k(w^\star)$. Since each $P_k(w)$ is twice-differentiable, then the gradient vector of $P_k(w)$ is a differentiable function and therefore bounded. It follows that b has bounded norm and we conclude that

$$\check{b} \;=\; O(\mu) \tag{25.246}$$

Using the just-introduced transformed variables, we can rewrite (25.241) in the form:

$$
\begin{bmatrix} \bar{\boldsymbol{w}}_n \\ \check{\boldsymbol{w}}_n \end{bmatrix} = \begin{bmatrix} I_M - \boldsymbol{D}_{11,n-1}^{\mathsf{T}} & -\boldsymbol{D}_{21,n-1}^{\mathsf{T}} \\ -\boldsymbol{D}_{12,n-1}^{\mathsf{T}} & \mathcal{J}_\epsilon^{\mathsf{T}} - \boldsymbol{D}_{22,n-1}^{\mathsf{T}} \end{bmatrix} \begin{bmatrix} \bar{\boldsymbol{w}}_{n-1} \\ \check{\boldsymbol{w}}_{n-1} \end{bmatrix} + \begin{bmatrix} \bar{\boldsymbol{g}}_n \\ \check{\boldsymbol{g}}_n \end{bmatrix} - \begin{bmatrix} 0 \\ \check{b} \end{bmatrix} \tag{25.247}
$$

or, in expanded form:

$$\bar{\boldsymbol{w}}_n = (I_M - \boldsymbol{D}_{11,n-1}^{\mathsf{T}})\bar{\boldsymbol{w}}_{n-1} \;-\; \boldsymbol{D}_{21,n-1}^{\mathsf{T}}\check{\boldsymbol{w}}_{n-1}^{e} \;+\; \bar{\boldsymbol{g}}_n \tag{25.248}$$

$$\check{\boldsymbol{w}}_n = (\mathcal{J}_\epsilon^{\mathsf{T}} - \boldsymbol{D}_{22,n-1}^{\mathsf{T}})\check{\boldsymbol{w}}_{n-1} \;-\; \boldsymbol{D}_{12,n-1}^{\mathsf{T}}\bar{\boldsymbol{w}}_{n-1} \;+\; \check{\boldsymbol{g}}_n \;-\; \check{b} \tag{25.249}$$

Conditioning both sides on $\boldsymbol{\mathcal{F}}_{n-1}$, computing the conditional second-order moments, and using the conditions from Assumption 25.1 on the gradient noise process, we get

$$\mathbb{E}\left[\|\bar{\boldsymbol{w}}_n\|^2 \,|\, \boldsymbol{\mathcal{F}}_{n-1}\right] = \|(I_M - \boldsymbol{D}_{11,n-1}^{\mathsf{T}})\bar{\boldsymbol{w}}_{n-1} - \boldsymbol{D}_{21,n-1}^{\mathsf{T}}\check{\boldsymbol{w}}_{n-1}\|^2 + \mathbb{E}\left[\|\bar{\boldsymbol{g}}_n\|^2 \,|\, \boldsymbol{\mathcal{F}}_{n-1}\right] \tag{25.250}$$

and

$$
\begin{aligned}
\mathbb{E}\left[\|\check{\boldsymbol{w}}_n\|^2 \,|\, \boldsymbol{\mathcal{F}}_{n-1}\right] = \|(\mathcal{J}_\epsilon^{\mathsf{T}} - \boldsymbol{D}_{22,n-1}^{\mathsf{T}})\check{\boldsymbol{w}}_{n-1}^{e} - \boldsymbol{D}_{12,n-1}^{\mathsf{T}}\bar{\boldsymbol{w}}_{n-1} - \check{b}\|^2 \;+ \\
\mathbb{E}\left[\|\check{\boldsymbol{g}}_n\|^2 \,|\, \boldsymbol{\mathcal{F}}_{n-1}\right]
\end{aligned} \tag{25.251}
$$

Computing the expectations again we conclude that

$$\mathbb{E}\,\|\bar{\boldsymbol{w}}_n\|^2 = \mathbb{E}\,\|(I_M - \boldsymbol{D}_{11,n-1}^{\mathsf{T}})\bar{\boldsymbol{w}}_{n-1} - \boldsymbol{D}_{21,n-1}^{\mathsf{T}}\check{\boldsymbol{w}}_{n-1}\|^2 + \mathbb{E}\,\|\bar{\boldsymbol{g}}_n\|^2 \tag{25.252}$$

and

$$\mathbb{E}\,\|\check{\boldsymbol{w}}_n\|^2 = \mathbb{E}\,\|(\mathcal{J}_\epsilon^{\mathsf{T}} - \boldsymbol{D}_{22,n-1}^{\mathsf{T}})\check{\boldsymbol{w}}_{n-1} - \boldsymbol{D}_{12,n-1}^{\mathsf{T}}\bar{\boldsymbol{w}}_{n-1}^{e} - \check{b}\|^2 + \mathbb{E}\,\|\check{\boldsymbol{g}}_n\|^2 \tag{25.253}$$

Continuing with the first variance (25.252), we can appeal to the Jensen inequality from Section 8.7 and apply it to the function $f(x) = \|x\|^2$ to bound the variance as follows:

$$
\begin{aligned}
&\mathbb{E}\,\|\bar{\mathfrak{w}}_n\|^2 \\
&= \mathbb{E}\left\|(1-t)\frac{1}{1-t}(I_M - \boldsymbol{D}_{11,n-1}^{\mathsf{T}})\bar{\mathfrak{w}}_{n-1} - t\frac{1}{t}\boldsymbol{D}_{21,n-1}^{\mathsf{T}}\check{\mathfrak{w}}_{n-1}\right\|^2 + \mathbb{E}\,\|\bar{\mathfrak{g}}_n\|^2 \\
&\le (1-t)\mathbb{E}\left\|\frac{1}{1-t}(I_M - \boldsymbol{D}_{11,n-1}^{\mathsf{T}})\bar{\mathfrak{w}}_{n-1}\right\|^2 + t\,\mathbb{E}\left\|\frac{1}{t}\boldsymbol{D}_{21,n-1}^{\mathsf{T}}\check{\mathfrak{w}}_{n-1}\right\|^2 + \mathbb{E}\,\|\bar{\mathfrak{g}}_n\|^2 \\
&\le \frac{1}{1-t}\mathbb{E}\left[\|I_M - \boldsymbol{D}_{11,n-1}^{\mathsf{T}}\|^2\,\|\bar{\mathfrak{w}}_{n-1}\|^2\right] + \frac{1}{t}\mathbb{E}\left[\|\boldsymbol{D}_{21,n-1}^{\mathsf{T}}\|^2\,\|\check{\mathfrak{w}}_{n-1}\|^2\right] + \mathbb{E}\,\|\bar{\mathfrak{g}}_n\|^2 \\
&\le \frac{(1-\sigma_{11}\mu)^2}{1-t}\mathbb{E}\,\|\bar{\mathfrak{w}}_{n-1}\|^2 + \frac{\sigma_{21}^2\mu^2}{t}\mathbb{E}\,\|\check{\mathfrak{w}}_{n-1}\|^2 + \mathbb{E}\,\|\bar{\mathfrak{g}}_n\|^2
\end{aligned}
\tag{25.254}
$$

for any arbitrary positive number $t \in (0,1)$. We select

$$
t = \sigma_{11}\mu \tag{25.255}
$$

Then, the last inequality can be written as

$$
\mathbb{E}\,\|\bar{\mathfrak{w}}_n\|^2 \le (1-\sigma_{11}\mu)\mathbb{E}\,\|\bar{\mathfrak{w}}_{n-1}\|^2 + \left(\frac{\sigma_{21}^2\mu}{\sigma_{11}}\right)\mathbb{E}\,\|\check{\mathfrak{w}}_{n-1}\|^2 + \mathbb{E}\,\|\bar{\mathfrak{g}}_n\|^2 \tag{25.256}
$$

We now repeat a similar argument for the second variance relation (25.253). Thus, using the Jensen inequality again we have

$$
\begin{aligned}
&\mathbb{E}\,\|\check{\mathfrak{w}}_n\|^2 \\
&= \mathbb{E}\left\|\mathcal{J}_\epsilon^{\mathsf{T}}\check{\mathfrak{w}}_{n-1} - \left[\boldsymbol{D}_{22,n-1}^{\mathsf{T}}\check{\mathfrak{w}}_{n-1} + \boldsymbol{D}_{12,n-1}^{\mathsf{T}}\bar{\mathfrak{w}}_{n-1} + \check{b}\right]\right\|^2 + \mathbb{E}\,\|\check{\mathfrak{g}}_n\|^2 \\
&= \mathbb{E}\left\|t\frac{1}{t}\mathcal{J}_\epsilon^{\mathsf{T}}\check{\mathfrak{w}}_{n-1} - (1-t)\frac{1}{1-t}\left[\boldsymbol{D}_{22,n-1}^{\mathsf{T}}\check{\mathfrak{w}}_{n-1} + \boldsymbol{D}_{12,n-1}^{\mathsf{T}}\bar{\mathfrak{w}}_{n-1} + \check{b}\right]\right\|^2 + \mathbb{E}\,\|\check{\mathfrak{g}}_n\|^2 \\
&\le \frac{1}{t}\mathbb{E}\left\|\mathcal{J}_\epsilon^{\mathsf{T}}\check{\mathfrak{w}}_{n-1}\right\|^2 + \frac{1}{1-t}\mathbb{E}\left\|\boldsymbol{D}_{22,n-1}^{\mathsf{T}}\check{\mathfrak{w}}_{n-1} + \boldsymbol{D}_{12,n-1}^{\mathsf{T}}\bar{\boldsymbol{w}}_{n-1} + \check{b}\right\|^2 + \mathbb{E}\,\|\check{\mathfrak{g}}_n\|^2
\end{aligned}
\tag{25.257}
$$

for any arbitrary positive number $t \in (0,1)$. Now note that

$$
\begin{aligned}
\left\|\mathcal{J}_\epsilon^{\mathsf{T}}\check{\mathfrak{w}}_{n-1}\right\|^2 &= (\check{\mathfrak{w}}_{n-1}^{e})^{\mathsf{T}}\left(\mathcal{J}_\epsilon^{\mathsf{T}}\right)^{\mathsf{T}}\mathcal{J}_\epsilon^{\mathsf{T}}\check{\mathfrak{w}}_{n-1} \\
&= \left(\check{\mathfrak{w}}_{n-1}\right)^{\mathsf{T}}\mathcal{J}_\epsilon\mathcal{J}_\epsilon^{\mathsf{T}}\check{\mathfrak{w}}_{n-1} \\
&\le \rho\left(\mathcal{J}_\epsilon\mathcal{J}_\epsilon^{\mathsf{T}}\right)\|\check{\mathfrak{w}}_{n-1}\|^2
\end{aligned}
\tag{25.258}
$$

where we called upon the Rayleigh–Ritz characterization (1.16) of the eigenvalues of symmetric matrices, namely,

$$
\lambda_{\min}(C)\,\|x\|^2 \le x^{\mathsf{T}}Cx \le \lambda_{\max}(C)\,\|x\|^2 \tag{25.259}
$$

for any symmetric matrix C. Applying this result to the symmetric and nonnegative-definite matrix $C = \mathcal{J}_\epsilon\mathcal{J}_\epsilon^{\mathsf{T}}$, and noting that $\rho(C) = \rho(C^{\mathsf{T}})$, we obtain (25.258). From definition (25.239) for $\mathcal{J}_\epsilon$ we further get

$$
\begin{aligned}
\rho\left(\mathcal{J}_\epsilon\mathcal{J}_\epsilon^{\mathsf{T}}\right) &= \rho\left[(J_\epsilon \otimes I_M)(J_\epsilon^{\mathsf{T}} \otimes I_M)\right] \\
&= \rho\left[(J_\epsilon J_\epsilon^{\mathsf{T}} \otimes I_M)\right] \\
&= \rho(J_\epsilon J_\epsilon^{\mathsf{T}})
\end{aligned}
\tag{25.260}
$$

The matrix J_ϵ is block diagonal and consists of Jordan blocks. Assume initially that it consists of a single Jordan block, say, of size 4×4, for illustration purposes. Then, we can write:

$$
\begin{aligned}
J_\epsilon J_\epsilon^{\mathsf{T}} &= \begin{bmatrix} \lambda & & & \\ \epsilon & \lambda & & \\ & \epsilon & \lambda & \\ & & \epsilon & \lambda \end{bmatrix} \begin{bmatrix} \lambda & \epsilon & & \\ & \lambda & \epsilon & \\ & & \lambda & \epsilon \\ & & & \lambda \end{bmatrix} \\
&= \begin{bmatrix} |\lambda|^2 & \epsilon\lambda & & \\ \epsilon\lambda & |\lambda|^2+\epsilon^2 & \epsilon\lambda & \\ & \epsilon\lambda & |\lambda|^2+\epsilon^2 & \epsilon\lambda \\ & & \epsilon\lambda & |\lambda|^2+\epsilon^2 \end{bmatrix}
\end{aligned} \tag{25.261}
$$

Using the property that the spectral radius of a matrix is bounded by any of its norms, and using the 1-norm (maximum absolute column sum), we get for the above example

$$
\begin{aligned}
\rho(J_\epsilon J_\epsilon^{\mathsf{T}}) &\le \|J_\epsilon J_\epsilon^{\mathsf{T}}\|_1 \\
&= |\lambda|^2 + \epsilon^2 + 2\epsilon|\lambda| \\
&= (|\lambda| + \epsilon)^2
\end{aligned} \tag{25.262}
$$

If J_ϵ consists of multiple Jordan blocks, say, L of them with eigenvalue λ_ℓ each, then

$$
\rho(J_\epsilon J_\epsilon^{\mathsf{T}}) \le \max_{1\le \ell \le L} \left\{ (|\lambda_\ell| + \epsilon)^2 \right\} = (\rho(J_\epsilon) + \epsilon)^2 \tag{25.263}
$$

where $\rho(J_\epsilon)$ does not depend on ϵ and is equal to the second largest eigenvalue in magnitude in J, which we know is strictly less than 1 in magnitude. Substituting this conclusion into (25.257) gives

$$
\begin{aligned}
\mathbb{E}\,\|\check{w}_n\|^2 \le\; & \frac{1}{t}(\rho(J_\epsilon)+\epsilon)^2\, \mathbb{E}\,\left\|\check{w}_{n-1}\right\|^2 + \\
& \frac{1}{1-t}\mathbb{E}\,\left\|D_{22,n-1}^{\mathsf{T}}\check{w}_{n-1}^{e} + D_{12,n-1}^{\mathsf{T}}\bar{w}_{n-1}^{e} + \check{b}\right\|^2 + \mathbb{E}\,\|\check{g}_n\|^2
\end{aligned} \tag{25.264}
$$

Since we know that $\rho(J_\epsilon) \in (0,1)$, then we can select ϵ small enough to ensure $\rho(J_\epsilon)+\epsilon \in (0,1)$. We also select

$$
t = \rho(J_\epsilon) + \epsilon \tag{25.265}
$$

and rewrite (25.264) as

$$
\begin{aligned}
\mathbb{E}\,\|\check{w}_n\|^2 \le\; & (\rho(J_\epsilon)+\epsilon)\mathbb{E}\,\left\|\check{w}_{n-1}\right\|^2 + \mathbb{E}\,\|\check{g}_n\|^2 + \\
& \left(\frac{1}{1-\rho(J_\epsilon)-\epsilon}\right)\mathbb{E}\,\left\|D_{22,n-1}^{\mathsf{T}}\check{w}_{n-1} + D_{12,n-1}^{\mathsf{T}}\bar{w}_{n-1} + \check{b}\right\|^2
\end{aligned} \tag{25.266}
$$

We can bound the last term on the right-hand side of the above expression as follows:

$$
\begin{aligned}
& \mathbb{E}\,\left\|D_{22,n-1}^{\mathsf{T}}\check{w}_{n-1} + D_{12,n-1}^{\mathsf{T}}\bar{w}_{n-1} + \check{b}\right\|^2 \\
&= \mathbb{E}\,\left\|\frac{1}{3}3D_{22,n-1}^{\mathsf{T}}\check{w}_{n-1} + \frac{1}{3}3D_{12,n-1}^{\mathsf{T}}\bar{w}_{n-1} + \frac{1}{3}3\check{b}\right\|^2 \\
&\le \frac{1}{3}\mathbb{E}\,\left\|3D_{22,n-1}^{\mathsf{T}}\check{w}_{n-1}\right\|^2 + \frac{1}{3}\mathbb{E}\,\left\|3D_{12,n-1}^{\mathsf{T}}\bar{w}_{n-1}\right\|^2 + \frac{1}{3}\|3\check{b}\|^2 \\
&\le 3\mathbb{E}\,\left\|D_{22,n-1}^{\mathsf{T}}\check{w}_{n-1}\right\|^2 + 3\mathbb{E}\,\left\|D_{12,n-1}^{\mathsf{T}}\bar{w}_{n-1}\right\|^2 + 3\|\check{b}\|^2 \\
&\le 3\sigma_{22}^2\mu^2\mathbb{E}\,\left\|\check{w}_{n-1}\right\|^2 + 3\sigma_{12}^2\mu^2\mathbb{E}\,\|\bar{w}_{n-1}^{e}\|^2 + 3\|\check{b}\|^2
\end{aligned} \tag{25.267}
$$

Substituting into (25.266) we obtain

$$\begin{aligned}
\mathbb{E}\,\|\check{\mathcal{w}}_n\|^2 \leq \left(\rho(J_\epsilon)+\epsilon+\frac{3\sigma_{22}^2\mu^2}{1-\rho(J_\epsilon)-\epsilon}\right)\mathbb{E}\,\left\|\check{\mathcal{w}}_{n-1}\right\|^2 + \\
\left(\frac{3\sigma_{12}^2\mu^2}{1-\rho(J_\epsilon)-\epsilon}\right)\mathbb{E}\,\|\bar{\mathcal{w}}_{n-1}\|^2 + \\
\left(\frac{3}{1-\rho(J_\epsilon)-\epsilon}\right)\|\check{b}\|^2+\mathbb{E}\,\|\check{\mathcal{g}}_n\|^2
\end{aligned} \tag{25.268}$$

We now bound the noise terms, $\mathbb{E}\,\|\bar{\mathcal{g}}_n\|^2$ in (25.256) and $\mathbb{E}\,\|\check{\mathcal{g}}_n\|^2$ in (25.268). For that purpose, we first note that

$$\begin{aligned}
\mathbb{E}\,\|\bar{\mathcal{g}}_n\|^2 + \mathbb{E}\,\|\check{\mathcal{g}}_n\|^2 &= \mathbb{E}\left\|\begin{bmatrix}\bar{\mathcal{g}}_n\\ \check{\mathcal{g}}_n\end{bmatrix}\right\|^2 \\
&= \mathbb{E}\left\|\mathcal{V}_\epsilon^{\mathsf{T}}\mathcal{A}_2^{\mathsf{T}}\mathcal{M}\,\mathcal{g}_n\right\|^2 \\
&\leq \left\|\mathcal{V}_\epsilon^{\mathsf{T}}\mathcal{A}_2^{\mathsf{T}}\right\|^2\|\mathcal{M}\|^2\,\mathbb{E}\,\|\,\mathcal{g}_n\,\|^2 \\
&\leq v_1^2\mu^2\mathbb{E}\,\|\,\mathcal{g}_n\,\|^2
\end{aligned} \tag{25.269}$$

where the positive constant v_1 is independent of μ and is equal to the following norm

$$v_1 \;\stackrel{\Delta}{=}\; \left\|\mathcal{V}_\epsilon^{\mathsf{T}}\mathcal{A}_2^{\mathsf{T}}\right\| \tag{25.270}$$

On the other hand, using (25.177c), we have

$$\mathbb{E}\,\|\,\mathcal{g}_n\,\|^2 = \sum_{k=1}^{K}\mathbb{E}\,\|\boldsymbol{g}_{k,n}\|^2 \tag{25.271}$$

in terms of the variances of the individual gradient noise processes. Now, for each term $\boldsymbol{g}_{k,n}$ we have

$$\begin{aligned}
\mathbb{E}\,\|\boldsymbol{g}_{k,n}\|^2 &\stackrel{(25.177b)}{\leq} \beta_{g,k}^2\mathbb{E}\,\|\widetilde{\boldsymbol{\phi}}_{k,n-1}\|^2 + \sigma_{g,k}^2 \\
&= \beta_{g,k}^2\mathbb{E}\left\|\sum_{\ell\in\mathcal{N}_k} a_{1,\ell k}\,\widetilde{\boldsymbol{w}}_{\ell,n-1}\right\|^2 + \sigma_{g,k}^2 \\
&\leq \beta_{g,k}^2\sum_{\ell\in\mathcal{N}_k} a_{1,\ell k}\,\mathbb{E}\,\|\widetilde{\boldsymbol{w}}_{\ell,n-1}\|^2 + \sigma_{g,k}^2 \\
&\leq \beta_{g,k}^2\sum_{\ell=1}^{K}\mathbb{E}\,\|\widetilde{\boldsymbol{w}}_{\ell,n-1}\|^2 + \sigma_{g,k}^2 \\
&= \beta_{g,k}^2\mathbb{E}\,\|\widetilde{\mathcal{w}}_{n-1}\|^2 + \sigma_{g,k}^2 \\
&= \beta_{g,k}^2\mathbb{E}\left\|(\mathcal{V}_\epsilon^{-1})^{\mathsf{T}}\mathcal{V}_\epsilon^{\mathsf{T}}\widetilde{\mathcal{w}}_{n-1}\right\|^2 + \sigma_{g,k}^2 \\
&\leq \beta_{g,k}^2\left\|(\mathcal{V}_\epsilon^{-1})^{\mathsf{T}}\right\|^2\mathbb{E}\left\|\mathcal{V}_\epsilon^{\mathsf{T}}\widetilde{\mathcal{w}}_{n-1}\right\|^2 + \sigma_{g,k}^2 \\
&\stackrel{(25.242)}{=} \beta_{g,k}^2\,v_2^2\left[\mathbb{E}\,\|\bar{\mathcal{w}}_{n-1}\|^2 + \mathbb{E}\,\|\check{\mathcal{w}}_{n-1}\|^2\right] + \sigma_{g,k}^2
\end{aligned} \tag{25.272}$$

where the positive constant v_2 is independent of μ and denotes the norm

$$v_2 \;\stackrel{\Delta}{=}\; \left\|(\mathcal{V}_\epsilon^{-1})^{\mathsf{T}}\right\| \tag{25.273}$$

In this way, we can bound the term $\mathbb{E}\,\|\,\mathcal{S}_n\,\|^2$ as follows:

$$\begin{aligned}\mathbb{E}\,\|\,\mathcal{S}_n\,\|^2 &= \sum_{k=1}^{K}\mathbb{E}\,\|\boldsymbol{s}_{k,n}\|^2 \\ &\leq v_2^2\beta_d^2\Big(\mathbb{E}\,\|\bar{\boldsymbol{w}}_{i-1}^e\|^2 \,+\, \mathbb{E}\,\|\check{\boldsymbol{w}}_{i-1}^e\|^2\Big) \,+\, \sigma_s^2 \end{aligned} \tag{25.274}$$

where we introduced the scalars:

$$\beta_d^2 \triangleq \sum_{k=1}^{K}\beta_{g,k}^2, \qquad \sigma_s^2 \triangleq \sum_{k=1}^{K}\sigma_{g,k}^2 \tag{25.275}$$

Substituting into (25.269) we get

$$\mathbb{E}\,\|\bar{\mathcal{S}}_n\|^2 + \mathbb{E}\,\|\check{\mathcal{S}}_n\|^2 \leq v_1^2v_2^2\beta_d^2\mu^2\left[\mathbb{E}\,\|\bar{\mathcal{W}}_{n-1}\|^2 + \mathbb{E}\,\|\check{\mathcal{W}}_{n-1}\|^2\right] + v_1^2\mu^2\sigma_s^2 \tag{25.276}$$

Using this bound in (25.256) and (25.268), we find that

$$\begin{aligned}\mathbb{E}\,\|\bar{\mathcal{W}}_n\|^2 &\leq \left(1-\sigma_{11}\mu + v_1^2v_2^2\beta_d^2\mu^2\right)\mathbb{E}\,\|\bar{\mathcal{W}}_{n-1}\|^2 \,+ \\ &\quad\left(\frac{\sigma_{21}^2\mu}{\sigma_{11}} + v_1^2v_2^2\beta_d^2\mu^2\right)\mathbb{E}\,\|\check{\mathcal{W}}_{n-1}\|^2 \,+\, v_1^2\mu^2\sigma_s^2\end{aligned} \tag{25.277}$$

and

$$\begin{aligned}\mathbb{E}\,\|\check{\mathcal{W}}_n\|^2 &\leq \left(\rho(J_\epsilon)+\epsilon+\frac{3\sigma_{22}^2\mu^2}{1-\rho(J_\epsilon)-\epsilon} + v_1^2v_2^2\beta_d^2\mu^2\right)\mathbb{E}\,\left\|\check{\mathcal{W}}_{n-1}\right\|^2 + \\ &\quad\left(\frac{3\sigma_{12}^2\mu^2}{1-\rho(J_\epsilon)-\epsilon} + v_1^2v_2^2\beta_d^2\mu^2\right)\mathbb{E}\,\|\bar{\mathcal{W}}_{n-1}\|^2 + \\ &\quad\left(\frac{3}{1-\rho(J_\epsilon)-\epsilon}\right)\|\check{b}\|^2 + v_1^2\mu_{\max}^2\sigma_s^2\end{aligned} \tag{25.278}$$

We introduce the scalar coefficients

$$a = 1-\sigma_{11}\mu_{\max} + v_1^2v_2^2\beta_d^2\mu^2 \;=\; 1-O(\mu) \tag{25.279}$$

$$b = \frac{\sigma_{21}^2\mu}{\sigma_{11}} + v_1^2v_2^2\beta_d^2\mu^2 \;=\; O(\mu) \tag{25.280}$$

$$c = \frac{3\sigma_{12}^2\mu^2}{1-\rho(J_\epsilon)-\epsilon} + v_1^2v_2^2\beta_d^2\mu^2 \;=\; O(\mu^2) \tag{25.281}$$

$$\begin{aligned}d &= \rho(J_\epsilon)+\epsilon+\frac{3\sigma_{22}^2\mu^2}{1-\rho(J_\epsilon)-\epsilon} + v_1^2v_2^2\beta_d^2\mu^2 \\ &= \rho(J_\epsilon)+\epsilon+O(\mu^2)\end{aligned} \tag{25.282}$$

$$e = v_1^2\mu^2\sigma_s^2 \;=\; O(\mu^2) \tag{25.283}$$

$$f = \left(\frac{3}{1-\rho(J_\epsilon)-\epsilon}\right)\|\check{b}\|^2 \;=\; O(\mu^2) \tag{25.284}$$

since $\|\check{b}\| = O(\mu)$. Note that the scalar e will be zero when there is no gradient noise and the actual gradients of the risk functions are used, as happens in deterministic optimization; we will use this fact below.

Using the above parameters, we can combine (25.277) and (25.278) into a single compact inequality recursion as follows:

$$\begin{bmatrix}\mathbb{E}\,\|\bar{\mathcal{W}}_n\|^2\\ \mathbb{E}\,\|\check{\mathcal{W}}_n\|^2\end{bmatrix} \preceq \underbrace{\begin{bmatrix}a & b\\ c & d\end{bmatrix}}_{\Gamma}\begin{bmatrix}\mathbb{E}\,\|\bar{\mathcal{W}}_{n-1}\|^2\\ \mathbb{E}\,\|\check{\mathcal{W}}_{n-1}\|^2\end{bmatrix} + \begin{bmatrix}e\\ e+f\end{bmatrix} \tag{25.285}$$

in terms of the 2×2 coefficient matrix Γ indicated above and whose entries are of the form:

$$\Gamma = \begin{bmatrix} 1 - O(\mu) & O(\mu) \\ O(\mu^2) & \rho(J_\epsilon) + \epsilon + O(\mu^2) \end{bmatrix} \tag{25.286}$$

Now, we invoke again the property that the spectral radius of a matrix is upper-bounded by any of its norms, and use the 1-norm (maximum absolute column sum) to conclude that

$$\rho(\Gamma) \leq \max\Big\{1 - O(\mu) + O(\mu^2),\ \rho(J_\epsilon) + \epsilon + O(\mu) + O(\mu^2)\Big\} \tag{25.287}$$

Since $\rho(J_\epsilon) < 1$ is independent of μ, and since ϵ and μ are small positive numbers that can be chosen arbitrarily small and independently of each other, it is clear that the right-hand side of the above expression can be made strictly smaller than 1 for sufficiently small ϵ and μ. In that case, $\rho(\Gamma) < 1$ so that Γ is stable. Moreover, it holds that

$$\begin{aligned} (I_2 - \Gamma)^{-1} &= \begin{bmatrix} 1-a & -b \\ -c & 1-d \end{bmatrix}^{-1} \\ &= \frac{1}{(1-a)(1-d) - bc} \begin{bmatrix} 1-d & b \\ c & 1-a \end{bmatrix} \\ &= \begin{bmatrix} O(1/\mu) & O(1) \\ O(\mu) & O(1) \end{bmatrix} \end{aligned} \tag{25.288}$$

If we now iterate (25.285), and since Γ is stable, we conclude that

$$\begin{aligned} \limsup_{n\to\infty} \begin{bmatrix} \mathbb{E}\,\|\bar{\boldsymbol{w}}_n\|^2 \\ \mathbb{E}\,\|\check{\boldsymbol{w}}_n\|^2 \end{bmatrix} &\preceq (I_2 - \Gamma)^{-1} \begin{bmatrix} e \\ e+f \end{bmatrix} \\ &= \begin{bmatrix} O(1/\mu) & O(1) \\ O(\mu) & O(1) \end{bmatrix} \begin{bmatrix} O(\mu^2) \\ O(\mu^2) \end{bmatrix} = \begin{bmatrix} O(\mu) \\ O(\mu^2) \end{bmatrix} \end{aligned} \tag{25.289}$$

from which we conclude that

$$\limsup_{n\to\infty} \mathbb{E}\,\|\bar{\boldsymbol{w}}_n\|^2 = O(\mu), \quad \limsup_{n\to\infty} \mathbb{E}\,\|\check{\boldsymbol{w}}_n\|^2 = O(\mu^2) \tag{25.290}$$

and, therefore,

$$\begin{aligned} \limsup_{n\to\infty} \mathbb{E}\,\|\widetilde{\boldsymbol{w}}_n\|^2 &= \limsup_{n\to\infty} \mathbb{E}\, \left\| \left(\mathcal{V}_\epsilon^{-1}\right)^{\mathsf{T}} \begin{bmatrix} \bar{\boldsymbol{w}}_n \\ \check{\boldsymbol{w}}_n \end{bmatrix} \right\|^2 \\ &\leq \limsup_{n\to\infty} v_2^2 \Big(\mathbb{E}\,\|\bar{\boldsymbol{w}}_n\|^2 + \mathbb{E}\,\|\check{\boldsymbol{w}}_n\|^2 \Big) = O(\mu) \end{aligned} \tag{25.291}$$

which leads to the desired result (25.198).

Returning to (25.289) and setting $e = 0$, which happens under deterministic optimization, we would conclude instead that (where we are also removing the expectation operator and the boldface notation):

$$\limsup_{n\to\infty} \begin{bmatrix} \|\bar{w}_n\|^2 \\ \|\check{w}_n\|^2 \end{bmatrix} \preceq (I_2 - \Gamma)^{-1} \begin{bmatrix} 0 \\ f \end{bmatrix} = \begin{bmatrix} O(\mu^2) \\ O(\mu^2) \end{bmatrix} \tag{25.292}$$

so that

$$\limsup_{n\to\infty} \|\widetilde{w}_n\|^2 = O(\mu^2) \tag{25.293}$$

■

REFERENCES

Alriksson, P. and A. Rantzer (2006), "Distributed Kalman filtering using weighted averaging," *Proc. Int. Symp. Mathematical Theory of Networks and Systems* (MTNS), pp. 1–6, Kyoto.

Balakrishnan, V. K. (2010), *Introductory Discrete Mathematics*, Dover Publications.

Bapat, R. B. (2014), *Graphs and Matrices*, 2nd ed., Springer.

Bénézit, F., V. Blondel, P. Thiran, J. Tsitsiklis, and M. Vetterli (2010), "Weighted gossip: Distributed averaging using non-doubly stochastic matrices," *Proc. IEEE Int. Symp. Information Theory* (ISIT), pp. 1753–1757, Austin, TX.

Bénézit, F., P. Thiran, and M. Vetterli (2011), "The distributed multiple voting problem," *IEEE J. Sel. Topics Signal Process.*, vol. 5, no. 4, pp. 791–804.

Berger, R. L. (1981), "A necessary and sufficient condition for reaching a consensus using DeGroot's method," *J. Amer. Statist. Assoc.*, vol. 76, no. 374, pp. 415–418.

Bertsekas, D. P. (1997), "A new class of incremental gradient methods for least squares problems," *SIAM J. Optim.*, vol. 7, no. 4, pp. 913–926.

Bertsekas, D. P. (1999), *Nonlinear Programming*, 2nd ed., Athena Scientific.

Bianchi, P., G. Fort, W. Hachem, and J. Jakubowicz (2011), "Convergence of a distributed parameter estimator for sensor networks with local averaging of the estimates," *Proc. IEEE ICASSP*, pp. 3764–3767, Prague.

Bertsekas, D. P. and J. N. Tsitsiklis (1997), *Parallel and Distributed Computation: Numerical Methods*, Athena Scientific, Singapore.

Blatt, D., A. O. Hero, and H. Gauchman (2008), "A convergent incremental gradient method with a constant step size," *SIAM J. Optim.*, vol. 18, pp. 29–51.

Bollobas, B. (1998), *Modern Graph Theory*, Springer.

Bondy, A. and U. Murty (2008), *Graph Theory*, Springer.

Boyd, S. and L. Vandenberghe (2004), *Convex Optimization*, Cambridge University Press.

Braca, P., S. Marano, and V. Matta (2008), "Running consensus in wireless sensor networks," *Proc. Int. Conf. on Information Fusion*, pp. 1–6, Cologne.

Carli, R., A. Chiuso, L. Schenato, and S. Zampieri (2008), "Distributed Kalman filtering using consensus strategies," *IEEE J. Sel. Areas Commun.*, vol. 26, no. 4, pp. 622–633.

Cattivelli, F. S., C. G. Lopes, and A. H. Sayed (2007), "A diffusion RLS scheme for distributed estimation over adaptive networks," *Proc. IEEE Workshop on Signal Processing Advances for Wireless Communications* (SPAWC), pp. 1–5, Helsinki.

Cattivelli, F. S., C. G. Lopes, and A. H. Sayed (2008), "Diffusion recursive least-squares for distributed estimation over adaptive networks," *IEEE Trans. Signal Process.*, vol. 56, no. 5, pp. 1865–1877.

Cattivelli, F. S. and A. H. Sayed (2008), "Diffusion LMS algorithms with information exchange," *Proc. Asilomar Conf. Signals, Systems, and Computers*, pp. 251–255, Pacific Grove, CA.

Cattivelli, F. S. and A. H. Sayed (2010a), "Diffusion strategies for distributed Kalman filtering and smoothing," *IEEE Trans. Aut. Control,* vol. 55, no. 9, pp. 2069–2084.

Cattivelli, F. S. and A. H. Sayed (2010b), "Diffusion LMS strategies for distributed estimation," *IEEE Trans. Signal Process.*, vol. 58, no. 3, pp. 1035–1048.

Chartrand, G. and P. Zhang (2012), *A First Course in Graph Theory*, Dover Publications.

Chen, J. and A. H. Sayed (2012b), "On the limiting behavior of distributed optimization strategies," *Proc. Ann. Allerton Conf. Communication, Control, and Computing*, pp. 1535–1542, Monticello, IL.

Chen, J. and A. H. Sayed (2013), "Distributed Pareto optimization via diffusion strategies," *IEEE J. Sel. Topics Signal Process.*, vol. 7, no. 2, pp. 205–220.

Chen, J. and A. H. Sayed (2015a), "On the learning behavior of adaptive networks – part I: Transient analysis," *IEEE Trans. Inf. Theory*, vol. 61, no. 6, pp. 3487–3517.

Chen, J. and A. H. Sayed (2015b), "On the learning behavior of adaptive networks – part II: Performance analysis," *IEEE Trans. Inf. Theory*, vol. 61, no. 6, pp. 3518–3548.

Das, A. and M. Mesbahi (2006), "Distributed linear parameter estimation in sensor networks based on Laplacian dynamics consensus algorithm," *Proc. IEEE SECON*, vol. 2, pp. 440–449, Reston, VA.

DeGroot, M. H. (1974), "Reaching a consensus," *J. Amer. Statist. Assoc.*, vol. 69, no. 345, pp. 118–121.

Deo, N. (2016), *Graph Theory with Applications to Engineering and Computer Science*, Dover Publications.

Dimakis, A. G., S. Kar, J. M. F. Moura, M. G. Rabbat, and A. Scaglione (2010), "Gossip algorithms for distributed signal processing," *Proc. IEEE*, vol. 98, no. 11, pp. 1847–1864.

Erciyes, K. (2021), *Discrete Mathematics and Graph Theory*, Springer.

Godsil, C. and G. F. Royle (2001), *Algebraic Graph Theory*, Springer.

Gross, J. L. and J. Yellen (2005), *Graph Theory and Its Applications*, 2nd ed., Chapman & Hall.

Hastings, W. K. (1970), "Monte Carlo sampling methods using Markov chains and their applications," *Biometrika*, vol. 57, pp. 97–109.

Helou, E. S. and A. R. De Pierro (2009), "Incremental subgradients for constrained convex optimization: A unified framework and new methods," *SIAM J. Optim.*, vol. 20, pp. 1547–1572.

Horn, R. A. and C. R. Johnson (1990), *Matrix Analysis*, Cambridge University Press.

Intriligator, M. D. (1971), *Mathematical Optimization and Economic Theory*, Prentice Hall.

Johansson, B., M. Rabi, and M. Johansson (2009), "A randomized incremental subgradient method for distributed optimization in networked systems," *SIAM J. Optim.*, vol. 20, pp. 1157–1170.

Johansson, B., T. Keviczky, M. Johansson, and K. Johansson (2008), "Subgradient methods and consensus algorithms for solving convex optimization problems," *Proc. IEEE Conf. Decision and Control* (CDC), pp. 4185–4190, Cancun.

Kar, S. and J. M. F. Moura (2011), "Convergence rate analysis of distributed gossip (linear parameter) estimation: Fundamental limits and tradeoffs," *IEEE J. Sel. Topics Signal Process.*, vol. 5, no. 4, pp. 674–690.

Kar, S., J. M. F. Moura, and K. Ramanan (2012), "Distributed parameter estimation in sensor networks: Nonlinear observation models and imperfect communication," *IEEE Trans. Inf. Theory*, vol. 58, no. 6, pp. 3575–3605.

Kocay, W. and D. L. Kreher (2004), *Graphs, Algorithms, and Optimization*, Chapman & Hall.

Lee, S. and A. Nedic (2013), "Distributed random projection algorithm for convex optimization," *IEEE J. Sel. Topics Signal Process.*, vol. 7, no. 2, pp. 221–229.

Lopes, C. G. and A. H. Sayed (2006), "Distributed processing over adaptive networks," *Proc. Adaptive Sensor Array Processing Workshop*, pp. 1–5, MIT Lincoln Laboratory, MA.

Lopes, C. G. and A. H. Sayed (2007a), "Diffusion least-mean-squares over adaptive networks," *Proc. IEEE ICASSP*, vol. 3, pp. 917–920, Honolulu, HI.

Lopes, C. G. and A. H. Sayed (2007b), "Steady-state performance of adaptive diffusion least-mean squares," *Proc. IEEE Workshop on Statistical Signal Processing*, pp. 136–140, Madison, WI.

Lopes, C. G. and A. H. Sayed (2007c), "Incremental adaptive strategies over distributed networks," *IEEE Trans. Signal Process.*, vol. 55, no. 8, pp. 4064–4077.

Lopes, C. G. and A. H. Sayed (2008), "Diffusion least-mean squares over adaptive networks: Formulation and performance analysis," *IEEE Trans. Signal Process.*, vol. 56, no. 7, pp. 3122–3136.

Nassif, R., S. Vlaski, C. Richard, J. Chen, and A. H. Sayed, (2020a), "Multitask learning over graphs," *IEEE Signal Process. Mag.*, vol. 37, no. 3, pp. 14–25.

Nassif, R., S. Vlaski, C. Richard, and A. H. Sayed (2020b), "Learning over multitask graphs – part I: Stability analysis," *IEEE Open J. Signal Process.*, vol. 1, pp. 28–45.

Nassif, R., S. Vlaski, C. Richard, and A. H. Sayed (2020c), "Learning over multitask graphs – part II: Performance analysis," *IEEE Open J. Signal Process.*, vol. 1, pp. 46–63.

Nedic, A. and D. P. Bertsekas (2001), "Incremental subgradient methods for nondifferentiable optimization," *SIAM J. Optim.*, vol. 12, no. 1, pp. 109–138.

Nedic, A. and A. Ozdaglar (2009), "Distributed subgradient methods for multiagent optimization,"*IEEE Trans. Aut. Control*, vol. 54, no. 1, pp. 48–61.

Nica, B. (2018), *A Brief Introduction to Spectral Graph Theory*, European Mathematical Society.

Olfati-Saber, R. (2005), "Distributed Kalman filter with embedded consensus filters," *Proc. IEEE Conf. Decision and Control* (CDC), pp. 8179–8184, Seville.

Rabbat, M. G. and R. Nowak (2005), "Quantized incremental algorithms for distributed optimization," *IEEE J. Sel. Areas Commun.*, vol. 23, no. 4, pp. 798–808.

Ram, S. S., A. Nedic, and V. V. Veeravalli (2010), "Distributed stochastic subgradient projection algorithms for convex optimization," *J. Optim. Theory Appl.*, vol. 147, no. 3, pp. 516–545.

Sayed, A. H. (2014a), *Adaptation, Learning, and Optimization over Networks,* Foundations and Trends in Machine Learning, NOW Publishers, vol. 7, no. 4–5, pp. 311–801.

Sayed, A. H. (2014b), "Adaptive networks," *Proc. IEEE*, vol. 102, no. 4, pp. 460–497.

Sayed, A. H. (2014c), "Diffusion adaptation over networks," in *E-Reference Signal Processing*, R. Chellapa and S. Theodoridis, editors, vol. 3, pp. 323–454, Academic Press.

Sayed, A. H. and C. G. Lopes (2007), "Adaptive processing over distributed networks," *IEICE Trans. Fund. Electron., Commun. Comput. Sci.*, vol. E90–A, no. 8, pp. 1504–1510.

Sayed, A. H., S.-Y. Tu, J. Chen, X. Zhao, and Z. Towfic (2013), "Diffusion strategies for adaptation and learning over networks," *IEEE Signal Process. Mag.*, vol. 30, no. 3, pp. 155–171.

Speranzon, A., C. Fischione, and K. H. Johansson (2006), "Distributed and collaborative estimation over wireless sensor networks," *Proc. IEEE Conf. Decision and Control* (CDC), pp. 1025–1030, San Diego, CA.

Srivastava, K. and A. Nedic (2011), "Distributed asynchronous constrained stochastic optimization," *IEEE J. Sel. Topics. Signal Process.*, vol. 5, no. 4, pp. 772–790.

Stankovic, S. S., M. S. Stankovic, and D. S. Stipanovic (2011), "Decentralized parameter estimation by consensus based stochastic approximation," *IEEE Trans. Aut. Control*, vol. 56, no. 3, pp. 531–543.

Towfic, Z. J., J. Chen, and A. H. Sayed (2016), "Excess-risk of distributed stochastic learners," *IEEE Trans. Inf. Theory*, vol. 62, no. 10, pp. 5753–5785.

Trudeau, R. J. (1994), *Introduction to Graph Theory*, Dover Publications.

Tsitsiklis, J. N. and M. Athans (1984), "Convergence and asymptotic agreement in distributed decision problems," *IEEE Trans. Aut. Control*, vol. 29, no. 1, pp. 42–50.

Tsitsiklis, J. N., D. P. Bertsekas, and M. Athans (1986), "Distributed asynchronous deterministic and stochastic gradient optimization algorithms," *IEEE Trans. Aut. Control*, vol. 31, no. 9, pp. 803–812.

Tu, S. Y. and A. H. Sayed (2012), "Diffusion strategies outperform consensus strategies for distributed estimation over adaptive networks," *IEEE Trans. Signal Process.*, vol. 60, no. 12, pp. 6217–6234.

Xiao, L. and S. Boyd (2004), "Fast linear iterations for distributed averaging," *Syst. Control Lett.*, vol. 53, no. 1, pp. 65–78.

Zadeh, L. A. (1963), "Optimality and non-scalar-valued performance criteria," *IEEE Trans. Aut. Control*, vol. 8, pp. 59–60.

Zhao, X. and A. H. Sayed (2012), "Performance limits for distributed estimation over LMS adaptive networks," *IEEE Trans. Signal Process.*, vol. 60, no. 10, pp. 5107–5124.

Zhao, X. and A. H. Sayed (2015a), "Asynchronous adaptation and learning over networks – part I: Modeling and stability analysis," *IEEE Trans. Signal Process.*, vol. 63, no. 4, pp. 811–826.

Zhao, X. and A. H. Sayed (2015b), "Asynchronous adaptation and learning over networks – part II: Performance analysis," *IEEE Trans. Signal Process.*, vol. 63, no. 4, pp. 827–842.

Zhao, X. and A. H. Sayed (2015c), "Asynchronous adaptation and learning over networks – part III: Comparison analysis," *IEEE Trans. Signal Process.*, vol. 63, no. 4, pp. 843–858.

26 Decentralized Optimization II: Primal–Dual Methods

The previous chapter focused on the class of *primal methods* for decentralized optimization, both in the deterministic and stochastic settings. The primal methods resulted from minimizing a regularized risk function where the regularization term penalizes deviations from agreement among the agents – recall expression (25.113). We explained that the penalty term introduces some tolerable bias into the operation of the algorithms on the order of $O(\mu^2)$ for deterministic optimization and $O(\mu)$ for stochastic optimization. In this chapter, we describe the class of *primal–dual methods*, which result from solving constrained optimization problems and lead to exact solutions without bias for deterministic optimization. The methods can also be applied to stochastic optimization; however, their mean-square-error (MSE) performance will continue to be similar to primal methods and on the order of $O(\mu)$. We will also describe dual methods, as well as comment on performance under nonconvex settings.

We continue with the same notation and problem formulation from the previous chapter.

26.1 MOTIVATION

Consider again the original optimization problem (25.39), repeated here for ease of reference:

$$w^\star = \underset{w\in\mathbb{R}^M}{\text{argmin}} \left\{ P(w) \triangleq \sum_{k=1}^{K} \alpha_k P_k(w) \right\} \tag{26.1}$$

where the $\{\alpha_k\}$ are positive convex combination coefficients that add up to 1:

$$\sum_{k=1}^{K} \alpha_k = 1, \quad \alpha_k > 0, \quad k = 1, 2, \ldots, K \tag{26.2}$$

The objective is to derive a decentralized solution run over a collection of K connected agents with the property that the iterates obtained at the individual agents will approach the desired minimizer $w^\star$.

We reformulate the optimization problem as follows. Let $w_k \in \mathbb{R}^M$ be a parameter vector associated with agent k. We collect these parameters into the network vector:

$$\mathcal{W} = \text{blkcol}\Big\{w_1, w_2, \ldots, w_K\Big\} \in \mathbb{R}^{KM} \tag{26.3}$$

and pose the equivalent problem:

$$\mathcal{W}^\star \stackrel{\Delta}{=} \underset{\mathcal{W}\in\mathbb{R}^{KM}}{\text{argmin}} \left\{\sum_{k=1}^{K} \alpha_k P_k(w_k)\right\}, \;\; \text{subject to } w_1 = w_2 = \ldots = w_K \tag{26.4}$$

Observe that we are assigning an individual argument w_k to each risk $P_k(\cdot)$ and requiring all these arguments to match, which takes us back to the original formulation (26.1). We say that we are requiring the agents to reach *agreement* or *consensus*.

There are many ways by which constrained optimization problems of the form (26.4) can be solved. In the previous chapter, we added a penalty factor, in the form of a quadratic regularization term, and then showed that gradient-descent iterations, with and without an incremental step, lead to the consensus and diffusion strategies. However, the addition of the penalty factor biases the solution since the algorithms now approach the minimizer $w^\dagger$ in (25.113) rather than the desired $w^\star$, as was revealed by (25.132). The size of the bias is small since μ is generally small. We will now develop other decentralized algorithms that ensure *exact* convergence of their agents' iterates to $w^\star$ for deterministic optimization. These alternative methods differ by how they incorporate the agreement (or consensus) requirement $w_1 = w_2 = \ldots = w_K$ into their problem formulations.

26.2 EXTRA ALGORITHM

The first decentralized variant that we consider is the EXTRA algorithm. It can be derived as follows. We introduce

$$\mathcal{V} \stackrel{\Delta}{=} \frac{1}{2}(I_{KM} - \mathcal{A}) \tag{26.5}$$

and appeal to the nullspace property (25.74b) to rewrite (26.4) in the equivalent form:

$$\mathcal{W}^\star \stackrel{\Delta}{=} \underset{\mathcal{W}\in\mathbb{R}^{KM}}{\text{argmin}} \left\{\mathcal{P}(\mathcal{W}) + \frac{1}{2\mu}\|\mathcal{V}^{1/2}\,\mathcal{W}\,\|^2\right\}, \;\; \text{subject to } \;\mathcal{V}^{1/2}\,\mathcal{W} = 0 \tag{26.6}$$

where we are using the aggregate risk $\mathcal{P}(\mathcal{W})$ defined earlier by (26.7), namely,

$$\mathcal{P}(\mathcal{W}) \stackrel{\Delta}{=} \sum_{k=1}^{K} \alpha_k P_k(w_k) \tag{26.7}$$

Observe that the consensus requirement is represented by the added constraint $\mathcal{V}^{1/2}\,\mathcal{W} = 0$. Observe also that the quadratic term added to $\mathcal{P}(\mathcal{W})$ does not alter the risk because $\mathcal{V}^{1/2}\,\mathcal{W} = 0$ under the constraint. Other algorithms in the following will incorporate the constraints in different ways. The scalar μ is a small positive parameter and will end up playing the role of the step-size parameter.

To solve (26.6) we introduce the Lagrangian function:

$$\begin{aligned}\mathcal{L}_e(\mathcal{W},\lambda) &\overset{\Delta}{=} \mathcal{P}(\mathcal{W}) + \frac{1}{2\mu}\|\mathcal{V}^{1/2}\,\mathcal{W}\,\|^2 + \frac{1}{\mu}\lambda^{\mathsf{T}}(\mathcal{V}^{1/2}\,\mathcal{W}) \\ &\overset{\Delta}{=} \mathcal{P}(\mathcal{W}) + \frac{1}{4\mu}\,\mathcal{W}^{\mathsf{T}}(I_{KM}-\mathcal{A})\,\mathcal{W} + \frac{1}{\mu}\lambda^{\mathsf{T}}(\mathcal{V}^{1/2}\,\mathcal{W}) \end{aligned} \tag{26.8}$$

where $\lambda \in \mathbb{R}^{KM}$ is a Lagrangian factor (scaled by $1/\mu$). We next consider the saddle point problem:

$$\min_{\mathcal{W}}\max_{\lambda}\left\{\mathcal{P}(\mathcal{W}) + \frac{1}{4\mu}\,\mathcal{W}^{\mathsf{T}}(I_{KM}-\mathcal{A})\,\mathcal{W} + \frac{1}{\mu}\lambda^{\mathsf{T}}(\mathcal{V}^{1/2}\,\mathcal{W})\right\} \tag{26.9}$$

and seek its saddle point by pursuing gradient-descent over $\mathcal{W}$ and gradient-ascent over[1] λ:

$$\mathcal{W}_n = \mathcal{W}_{n-1} - \mu\,\nabla_{\mathcal{W}^{\mathsf{T}}}\,\mathcal{P}(\mathcal{W}_{n-1}) - \frac{1}{2}(I_{KM}-\mathcal{A})\,\mathcal{W}_{n-1} - (\mathcal{V}^{1/2})^{\mathsf{T}}\lambda_{n-1} \tag{26.10a}$$

$$\lambda_n = \lambda_{n-1} + \mathcal{V}^{1/2}\,\mathcal{W}_n \tag{26.10b}$$

where we employ the updated $\mathcal{W}_n$ in the second step (recall the discussion in Probs. 12.24 and 12.25). The first relation simplifies to (recall that $\mathcal{V}^{1/2}$ and $\mathcal{A}$ are symmetric matrices):

$$\mathcal{W}_n = \bar{\mathcal{A}}\,\mathcal{W}_{n-1} - \mu\,\nabla_{\mathcal{W}^{\mathsf{T}}}\,\mathcal{P}(\mathcal{W}_{n-1}) - (\mathcal{V}^{1/2})^{\mathsf{T}}\lambda_{n-1} \tag{26.11}$$

$$\lambda_n = \lambda_{n-1} + \mathcal{V}^{1/2}\,\mathcal{W}_n \tag{26.12}$$

where we introduced

$$\bar{A} \overset{\Delta}{=} \frac{1}{2}(I_K + A), \quad \bar{\mathcal{A}} = \bar{A}\otimes I_M \tag{26.13}$$

which is also doubly stochastic and primitive since A is doubly stochastic and primitive. Recall that A has a single eigenvalue at 1 while all other eigenvalues are strictly inside the unit circle, $-1 < \lambda(A) \le 1$. It follows that all eigenvalues of $\bar{A}$ are positive so that $\bar{A}$ is positive-definite. In other words, the mapping (26.13) transforms A into another doubly stochastic and primitive matrix that is positive-definite.

For initialization we set $\lambda_{-1} = 0$ and get

$$\mathcal{W}_0 = \bar{\mathcal{A}}\,\mathcal{W}_{-1} - \mu\,\nabla_{\mathcal{W}^{\mathsf{T}}}\,\mathcal{P}(\mathcal{W}_{-1}) \tag{26.14}$$

$$\lambda_0 = \mathcal{V}^{1/2}\,\mathcal{W}_0 \tag{26.15}$$

Now, relation (26.11) gives over two successive instants:

$$\begin{aligned}&\mathcal{W}_n - \mathcal{W}_{n-1} \\ &= \bar{\mathcal{A}}(\mathcal{W}_{n-1} - \mathcal{W}_{n-2}) - \mu\Big(\nabla_{\mathcal{W}^{\mathsf{T}}}\,\mathcal{P}(\mathcal{W}_{n-1}) - \nabla_{\mathcal{W}^{\mathsf{T}}}\,\mathcal{P}(\mathcal{W}_{n-2})\Big) - \mathcal{V}\,\mathcal{W}_{n-1}\end{aligned} \tag{26.16}$$

[1] Here we are using the differentiation properties $\nabla_{x^{\mathsf{T}}}(x^{\mathsf{T}}Cy) = Cy$, $\nabla_{x^{\mathsf{T}}}(y^{\mathsf{T}}Cx) = C^{\mathsf{T}}y$, and, when C is symmetric, $\nabla_{x^{\mathsf{T}}}(x^{\mathsf{T}}Cx) = 2Cx$.

and, using $\mathcal{V} = \frac{1}{2}(I_{KM} - \mathcal{A})$, we arrive at the network-based description of the EXTRA algorithm:

$$\boxed{\mathcal{W}_n = \bar{\mathcal{A}}(2\,\mathcal{W}_{n-1} - \mathcal{W}_{n-2}) - \mu\Big(\nabla_{\mathcal{W}^\mathsf{T}}\,\mathcal{P}(\mathcal{W}_{n-1}) - \nabla_{\mathcal{W}^\mathsf{T}}\,\mathcal{P}(\mathcal{W}_{n-2})\Big)} \tag{26.17}$$

This recursion is written in terms of network variables. We can rewrite it at the level of the individual agents, as shown in listing (26.18). Observe that EXTRA has a similar structure to the consensus implementation (25.75) where the gradient is evaluated at $w_{k,n-1}$, while the update for $\psi_{k,n}$ starts from a different point. There is also the last step, which amounts to a correction by the difference $(\psi_{k,n} - \psi_{k,n-1})$.

EXTRA algorithm for solving (26.1).

given a connected graph linking K agents with doubly stochastic A;
let $\bar{A} = \frac{1}{2}(I_K + A) = [\bar{a}_{\ell k}]_{\ell,k=1}^{K}$;
start from an arbitrary initial condition, $w_{k,-1}$, for each agent;
set $w_{k,0} = \sum_{\ell\in\mathcal{N}_k} \bar{a}_{\ell k} w_{\ell,-1} - \mu\alpha_k \nabla_{w^\mathsf{T}} P_k(w_{k,-1})$, $k = 1, 2, \ldots, K$.
repeat until convergence over $n \geq 0$**:**
 repeat over agents $k = 1, 2, \ldots, K$**:**

$$\begin{aligned} \phi_{k,n-1} &= \sum_{\ell\in\mathcal{N}_k} \bar{a}_{\ell k} w_{\ell,n-1} \\ \psi_{k,n} &= \phi_{k,n-1} - \mu\alpha_k \nabla_{w^\mathsf{T}} P_k(w_{k,n-1}) \\ w_{k,n} &= \phi_{k,n-1} + (\psi_{k,n} - \psi_{k,n-1}) \end{aligned} \tag{26.18}$$

 end
end
iterate $w_{k,n}$ at each agent approximates $w^\star$.

26.3 EXACT DIFFUSION ALGORITHM

The second decentralized method we consider is the EXACT diffusion algorithm; its relation to EXTRA is similar to the relation between regular diffusion and consensus: diffusion incorporates an incremental step.

We start again from the gradient-descent and -ascent recursions (26.10a)–(26.10b) and incorporate the terms in the first line one item at a time, say, as:

$$\mathcal{B}_n = \mathcal{W}_{n-1} - \mu \nabla_{\mathcal{W}^\mathsf{T}} \mathcal{P}(\mathcal{W}_{n-1}) \tag{26.19a}$$
$$\mathcal{V}_n = \mathcal{B}_n - \frac{1}{2}(\mathcal{I}_{KM} - \mathcal{A})\, \mathcal{W}_{n-1} \tag{26.19b}$$
$$\mathcal{W}_n = \mathcal{V}_n - (\mathcal{V}^{1/2})^\mathsf{T} \lambda_{n-1} \tag{26.19c}$$
$$\lambda_n = \lambda_{n-1} + \mathcal{V}^{1/2}\, \mathcal{W}_n \tag{26.19d}$$

Next, we employ an incremental construction and substitute the iterate $\mathcal{W}_{n-1}$ in the second line by $\mathcal{B}_n$ from the first line:

$$\mathcal{B}_n = \mathcal{W}_{n-1} - \mu \nabla_{\mathcal{W}^\mathsf{T}} \mathcal{P}(\mathcal{W}_{n-1}) \tag{26.20a}$$
$$\mathcal{V}_n = \mathcal{B}_n - \frac{1}{2}(I_{KM} - \mathcal{A})\, \mathcal{B}_n = \bar{\mathcal{A}}\, \mathcal{B}_n \tag{26.20b}$$
$$\mathcal{W}_n = \mathcal{V}_n - (\mathcal{V}^{1/2})^\mathsf{T} \lambda_{n-1} \tag{26.20c}$$
$$\lambda_n = \lambda_{n-1} + \mathcal{V}^{1/2}\, \mathcal{W}_n \tag{26.20d}$$

Substituting the first two relations into the third, we arrive at the following primal–dual construction that is similar in form to what we obtained in (26.11)–(26.12) for the EXTRA algorithm (recall that $\bar{\mathcal{A}}$ and $\mathcal{V}^{1/2}$ are symmetric):

$$\mathcal{W}_n = \bar{\mathcal{A}}\Big(\mathcal{W}_{n-1} - \mu \nabla_{\mathcal{W}^\mathsf{T}} \mathcal{P}(\mathcal{W}_{n-1})\Big) - (\mathcal{V}^{1/2})^\mathsf{T} \lambda_{n-1} \tag{26.21}$$
$$\lambda_n = \lambda_{n-1} + \mathcal{V}^{1/2}\, \mathcal{W}_n \tag{26.22}$$

For initialization we set $\lambda_{-1} = 0$ and get

$$\mathcal{W}_0 = \bar{\mathcal{A}}\Big(\mathcal{W}_{-1} - \mu \nabla_{\mathcal{W}^\mathsf{T}} \mathcal{P}(\mathcal{W}_{-1})\Big) \tag{26.23}$$
$$\lambda_0 = \mathcal{V}^{1/2}\, \mathcal{W}_0 \tag{26.24}$$

We can proceed in a similar manner from this point and combine the relations for $\{\mathcal{W}_n, \lambda_n\}$. We have for two successive instances:

$$\begin{aligned} &\mathcal{W}_n - \mathcal{W}_{n-1} \\ &= \bar{\mathcal{A}}\Big(\mathcal{W}_{n-1} - \mathcal{W}_{n-2} - \mu\nabla_{\mathcal{W}^\mathsf{T}} \mathcal{P}(\mathcal{W}_{n-1}) - \mu\nabla_{\mathcal{W}^\mathsf{T}} \mathcal{P}(\mathcal{W}_{n-2})\Big) - \mathcal{V}\,\mathcal{W}_{n-1} \end{aligned} \tag{26.25}$$

and, using $\mathcal{V} = \frac{1}{2}(I_{KM} - \mathcal{A})$, we arrive at the network-based description of the EXACT diffusion algorithm:

$$\boxed{\mathcal{W}_n = \bar{\mathcal{A}}\Big(2\,\mathcal{W}_{n-1} - \mathcal{W}_{n-2} - \mu\nabla_{\mathcal{W}^\mathsf{T}} \mathcal{P}(\mathcal{W}_{n-1}) + \mu\nabla_{\mathcal{W}^\mathsf{T}} \mathcal{P}(\mathcal{W}_{n-2})\Big)} \tag{26.26}$$

This recursion is written in terms of network variables. We can rewrite it at the level of the individual agents, as shown in listing (26.27). Comparing with the ATC (adapt then combine) diffusion strategy (25.77), we observe that the main difference is the addition of an intermediate *correction* step that transforms $\psi_{k,n}$ into $\phi_{k,n}$ before the combination that results in $w_{k,n}$.

EXACT diffusion algorithm for solving (26.1).

given a connected graph linking K agents with doubly stochastic A;
let $\bar{A} = \frac{1}{2}(I_K + A) = [\bar{a}_{\ell k}]_{\ell,k=1}^{K}$;
start from an arbitrary initial condition, $w_{k,-1}$, for each agent;
set $w_{k,0} = \sum_{\ell \in \mathcal{N}_k} \bar{a}_{\ell k}\Big(w_{\ell,-1} - \mu\alpha_\ell \nabla_{w^\mathsf{T}} P_\ell(w_{\ell,-1})\Big)$, $k = 1, 2, \ldots, K$.
repeat until convergence over $n \geq 0$**:**
  repeat over agents $k = 1, 2, \ldots, K$**:**

$$\begin{aligned} \psi_{k,n} &= w_{k,n-1} - \mu\alpha_k \nabla_{w^\mathsf{T}} P_k(w_{k,n-1}) \\ \phi_{k,n} &= w_{k,n-1} + (\psi_{k,n} - \psi_{k,n-1}) \\ w_{k,n} &= \sum_{\ell \in \mathcal{N}_k} \bar{a}_{\ell k}\phi_{\ell,n} \end{aligned} \tag{26.27}$$

  end
end
iterate $w_{k,n}$ at each agent approximates $w^\star$.

In a manner similar to Remark 25.1, it can again be verified that EXACT diffusion is stable (i.e., converges) over a wider range of step-size parameters μ than the EXTRA algorithm – see Remark 26.2.

Example 26.1 **(Simulating EXTRA and EXACT diffusion)** We illustrate the behavior of the *stochastic* EXTRA and EXACT diffusion strategies in the context of the same logistic regression problem from Example 25.6. We employ the same connected graph topology with $K = 20$ agents shown in Fig. 25.7 with a metropolis combination matrix, A. The same parameter values are used in this simulation, namely,

$$M = 10,\ K = 20,\ N_k = 200,\ N = 4000,\ \ \rho = 0.001,\ \ \mu = 0.01 \tag{26.28}$$

The stochastic versions of EXTRA and EXACT diffusion take the following forms where $\mu' = \mu/K$ and $\bar{A} = 0.5(I_K + A)$:

$$\left\{\begin{aligned} &\textbf{(EXTRA)} \\ &\text{at each agent } k\text{, select or receive a random pair } (\boldsymbol{\gamma}_k(n), \boldsymbol{h}_{k,n}) \text{ from } \mathcal{D}_k \\ &\widehat{\boldsymbol{\gamma}}_k(n) = \boldsymbol{h}_{k,n}^\mathsf{T} \boldsymbol{w}_{k,n-1} \\ &\boldsymbol{\phi}_{k,n-1} = \sum_{\ell \in \mathcal{N}_k} \bar{a}_{\ell k} \boldsymbol{w}_{\ell,n-1} \\ &\boldsymbol{\psi}_{k,n} = (1 - 2\rho\mu')\boldsymbol{\phi}_{k,n-1} + \mu' \frac{\boldsymbol{\gamma}_k(n)\boldsymbol{h}_{k,n}}{1 + e^{\boldsymbol{\gamma}_k(n)\widehat{\boldsymbol{\gamma}}_k(n)}} \\ &\boldsymbol{w}_{k,n} = \boldsymbol{\phi}_{k,n-1} + (\boldsymbol{\psi}_{k,n} - \boldsymbol{\psi}_{k,n-1}) \end{aligned}\right. \tag{26.29}$$

and

$$
\left\{
\begin{array}{l}
\textbf{(EXACT diffusion)} \\
\text{at each agent } k, \text{ select or receive a random pair } (\boldsymbol{\gamma}_k(n), \boldsymbol{h}_{k,n}) \text{ from } \mathcal{D}_k \\
\widehat{\boldsymbol{\gamma}}_k(n) = \boldsymbol{h}_{k,n}^{\mathsf{T}} \boldsymbol{w}_{k,n-1} \\
\boldsymbol{\psi}_{k,n} = (1 - 2\rho\mu')\boldsymbol{w}_{k,n-1} + \mu' \dfrac{\boldsymbol{\gamma}_k(n)\boldsymbol{h}_{k,n}}{1 + e^{\boldsymbol{\gamma}_k(n)\widehat{\boldsymbol{\gamma}}_k(n)}} \\
\boldsymbol{\phi}_{k,n} = \boldsymbol{w}_{k,n-1} + (\boldsymbol{\psi}_{k,n} - \boldsymbol{\psi}_{k,n-1}) \\
\boldsymbol{w}_{k,n} = \displaystyle\sum_{\ell \in \mathcal{N}_k} \bar{a}_{\ell k}\, \boldsymbol{\phi}_{\ell,n}
\end{array}
\right. \tag{26.30}
$$

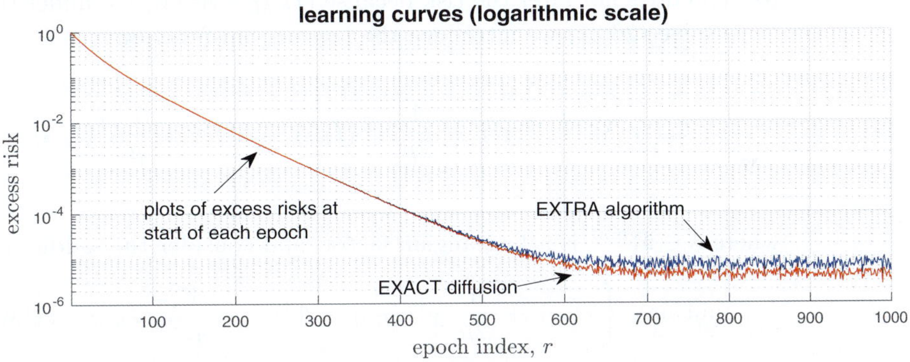

Figure 26.1 Learning curves obtained by averaging over $L = 10$ experiments. The curves show the evolution of the normalized excess risk values at the beginning of each run for EXTRA and EXACT diffusion.

Both algorithms are implemented starting from zero initial conditions for their parameter iterates. The simulation generates $N = 4000$ data points $\{\gamma(m), h_m\}$ according to the same logistic model from Example 25.6. A total of $R = 1000$ epochs are run over the data. We select one agent at random for illustration purposes (agent 8), and evaluate the evolution of the aggregate risk value $P(w_{-1}^r)$ at the beginning of each run r at that agent. We subsequently average these values over $L = 10$ experiments (with each experiment involving $R = 1000$ runs) to smooth out the resulting learning curve. We plot the average excess risk curve $\mathbb{E}\, P(\boldsymbol{w}_{-1}^k) - P(w^\star)$ after normalizing it by its peak value so that the learning curves start from the peak value of 1 for comparison purposes. The result is shown in Fig. 26.1.

26.4 DISTRIBUTED INEXACT GRADIENT ALGORITHM

The third decentralized method that we consider is the distributed inexact gradient (DIGing) algorithm, which performs a form of gradient tracking as explained further ahead. DIGing enforces agreement by incorporating the constraints in an alternative form to what was done earlier in (26.6) where the matrix $\mathcal{V} = \frac{1}{2}(I_K - A)$ and its square-root were used. We now appeal instead to the nullspace property (25.26) and rewrite problem (26.4) in the equivalent form:

$$\begin{aligned} w^\star \stackrel{\Delta}{=} \; & \underset{w \in \mathbb{R}^{KM}}{\text{argmin}} \left\{ \mathcal{P}(w) + \frac{1}{2\mu} \| w \|^2_{(I_{KM} - \mathcal{A}^2)} \right\} \\ & \text{subject to } (I_{KM} - \mathcal{A})\, w = 0 \end{aligned} \tag{26.31}$$

where we are employing the notation $\|x\|_Q^2 = x^\mathsf{T} Q x$ for the weighted squared Euclidean norm. Observe that the consensus requirement is represented by the added constraint $(I_{KM} - \mathcal{A})\, w = 0$. Observe also that the quadratic term added to $\mathcal{P}(w)$ does not alter the risk because $(I_{KM} - \mathcal{A}^2)\, w = 0$ under the constraint.

To solve (26.31) we introduce the Lagrangian function:

$$\mathcal{L}_e(w, \lambda) \stackrel{\Delta}{=} \mathcal{P}(w) + \frac{1}{2\mu} w^\mathsf{T}(I_{KM} - \mathcal{A}^2)\, w + \frac{1}{\mu}\lambda^\mathsf{T}(I_{KM} - \mathcal{A})\, w \tag{26.32}$$

where $\lambda \in \mathbb{R}^{KM}$ is a Lagrangian factor, and consider the saddle point problem:

$$\min_{w} \max_{\lambda} \left\{ \mathcal{P}(w) + \frac{1}{2\mu} w^\mathsf{T}(I_{KM} - \mathcal{A}^2)\, w + \frac{1}{\mu}\lambda^\mathsf{T}(I_{KM} - \mathcal{A})\, w \right\} \tag{26.33}$$

We approach the saddle point by pursuing gradient-descent over w and gradient-ascent over λ:

$$w_n = w_{n-1} - \mu \nabla_{w^\mathsf{T}} \mathcal{P}(w_{n-1}) - (I_{KM} - \mathcal{A}^2)\, w_{n-1} - (I_{KM} - \mathcal{A})^\mathsf{T} \lambda_{n-1} \tag{26.34a}$$

$$\lambda_n = \lambda_{n-1} + (I_{KM} - \mathcal{A})\, w_n \tag{26.34b}$$

where we employ the updated w_n in the second step. The first relation simplifies to (recall that $\mathcal{A}$ is symmetric):

$$w_n = \mathcal{A}^2\, w_{n-1} - \mu \nabla_{w^\mathsf{T}} \mathcal{P}(w_{n-1}) - (I_{KM} - \mathcal{A})\lambda_{n-1} \tag{26.35}$$

$$\lambda_n = \lambda_{n-1} + (I_{KM} - \mathcal{A})\, w_n \tag{26.36}$$

so that over two successive instants:

$$\begin{aligned} & w_n - w_{n-1} = \\ & \mathcal{A}^2(w_{n-1} - w_{n-2}) - \mu\Big(\nabla_{w^\mathsf{T}} \mathcal{P}(w_{n-1}) - \nabla_{w^\mathsf{T}} \mathcal{P}(w_{n-2})\Big) - (I_{KM} - \mathcal{A})^2\, w_{n-1} \end{aligned} \tag{26.37}$$

This leads to the network-based description of the DIGing algorithm:

$$\boxed{w_n = \mathcal{A}(2\, w_{n-1} - \mathcal{A}\, w_{n-2}) - \mu\Big(\nabla_{w^\mathsf{T}} \mathcal{P}(w_{n-1}) - \nabla_{w^\mathsf{T}} \mathcal{P}(w_{n-2})\Big)} \tag{26.38}$$

This recursion is written in terms of network variables. We can rewrite it at the level of the individual agents, as shown in listing (26.43). The above algorithm is also equivalent to the form shown in Prob. 26.1. There we show that the following equations are equivalent to (26.38):

$$\mathcal{W}_n = \mathcal{A}\,\mathcal{W}_{n-1} - \mu\,\mathcal{O}_{n-1} \tag{26.39a}$$

$$\mathcal{O}_n = \mathcal{A}\,\mathcal{O}_{n-1} + \nabla_{\mathcal{W}^\mathsf{T}}\,\mathcal{P}(\mathcal{W}_n) - \nabla_{\mathcal{W}^\mathsf{T}}\,\mathcal{P}(\mathcal{W}_{n-1}) \tag{26.39b}$$

This alternative rewriting helps explain why the algorithm is said to perform "gradient tracking." Equation (26.39b) has the form of a dynamic average algorithm. Specifically, for any doubly stochastic matrix $\mathcal{A}$, let us consider a recursion of the form:

$$\mathcal{X}_n = \mathcal{A}\,\mathcal{X}_{n-1} + \mathcal{V}_n - \mathcal{V}_{n-1} \tag{26.40}$$

where $\mathcal{V}_n$ is some time-varying sequence. Let $\{x_{k,n}, v_{k,n}\}$ denote the $M\times 1$ vector components of the block vectors $\{\mathcal{X}_n, \mathcal{V}_n\}$. If the sequence $\mathcal{V}_n$ converges, it can be verified that:

$$\lim_{n\to\infty} x_{k,n} = \lim_{n\to\infty}\left(\frac{1}{K}\sum_{k'=1}^{K} v_{k',n}\right) \tag{26.41}$$

Applying this conclusion to (26.39b), it would follow that the individual block entries of $\mathcal{O}_n$, denoted by $\{o_{k,n}\}$, would converge to:

$$o_{k,n} \to \frac{1}{K}\sum_{k'=1}^{K} \nabla_{w^\mathsf{T}} P_{k'}(w_{k',n}) \tag{26.42}$$

so that each $o_{k,n}$ can be interpreted as tracking an estimate for the gradient vector of the local risk function.

DIGing algorithm for solving (26.1).

given a connected graph linking K agents with doubly stochastic A;
start from an arbitrary initial condition, $w_{k,-1}$, for each agent;
set $w_{k,0} = \sum_{\ell\in\mathcal{N}_k} 2a_{\ell k} w_{\ell,-1} - \mu\alpha_k \nabla_{w^\mathsf{T}} P_k(w_{k,-1})$, $k = 1, 2, \ldots, K$.
repeat until convergence over $n \geq 0$**:**
  repeat over agents $k = 1, 2, \ldots, K$**:**

$$\psi_{k,n-1} = 2w_{k,n-1} - \sum_{\ell\in\mathcal{N}_k} a_{\ell k} w_{\ell,n-2}$$

$$\phi_{k,n-1} = \sum_{\ell\in\mathcal{N}_k} a_{\ell k}\psi_{\ell,n-1}$$

$$w_{k,n} = \phi_{k,n-1} - \mu\alpha_k\Big(\nabla_{w^\mathsf{T}} P_k(w_{k,n-1}) - \nabla_{w^\mathsf{T}} P_k(w_{k,n-2})\Big)$$

  end
end
iterate $w_{k,n}$ at each agent k approximates $w^\star$. (26.43)

26.5 AUGMENTED DECENTRALIZED GRADIENT METHOD

The fourth decentralized method we consider is also a gradient tracking method; it incorporates an incremental step into the evaluation of (26.34a)–(26.34b) in a manner similar to the derivation of EXACT diffusion.

We start from the same saddle point problem (26.33) and write down gradient-descent and gradient-ascent recursions for the primal and dual variables:

$$\mathcal{W}_n = \mathcal{W}_{n-1} - \mu \nabla_{\mathcal{W}^\mathsf{T}} \mathcal{P}(\mathcal{W}_{n-1}) - (I_{KM} - \mathcal{A}^2)\,\mathcal{W}_{n-1} - (I_{KM} - \mathcal{A})^\mathsf{T} \lambda_{n-1} \quad (26.44a)$$

$$\lambda_n = \lambda_{n-1} + (I_{KM} - \mathcal{A})\,\mathcal{W}_{n-1} \quad (26.44b)$$

where, compared with (26.34b), we are employing the prior iterate $\mathcal{W}_{n-1}$ rather than the updated value $\mathcal{W}_n$ in the second line. We can perform the first iteration in two steps and rewrite the above equations in the equivalent form:

$$\mathcal{B}_n = \mathcal{W}_{n-1} - \mu\,\nabla_{\mathcal{W}^\mathsf{T}}\,\mathcal{P}(\mathcal{W}_{n-1}) - (I_{KM} - \mathcal{A})^\mathsf{T} \lambda_{n-1} \quad (26.45a)$$

$$\mathcal{W}_n = \mathcal{B}_n - (\mathcal{I}_{KM} - \mathcal{A}^2)\,\mathcal{W}_{n-1} \quad (26.45b)$$

$$\lambda_n = \lambda_{n-1} + (I_{KM} - \mathcal{A})\,\mathcal{W}_{n-1} \quad (26.45c)$$

Next, we employ an incremental construction and substitute the iterates $\mathcal{W}_{n-1}$ in the second and third lines by the updated iterate $\mathcal{B}_n$:

$$\mathcal{B}_n = \mathcal{W}_{n-1} - \mu\,\nabla_{\mathcal{W}^\mathsf{T}}\,\mathcal{P}(\mathcal{W}_{n-1}) - (I_{KM} - \mathcal{A})^\mathsf{T} \lambda_{n-1} \quad (26.46a)$$

$$\mathcal{W}_n = \mathcal{B}_n - (\mathcal{I}_{KM} - \mathcal{A}^2)\,\mathcal{B}_n \;=\; \mathcal{A}^2\,\mathcal{B}_n \quad (26.46b)$$

$$\lambda_n = \lambda_{n-1} + (I_{KM} - \mathcal{A})\,\mathcal{B}_n \quad (26.46c)$$

Substituting the first equation into the second, we get

$$\mathcal{W}_n = \mathcal{A}^2\,\mathcal{W}_{n-1} - \mu\,\mathcal{A}^2\,\nabla_{\mathcal{W}^\mathsf{T}}\,\mathcal{P}(\mathcal{W}_{n-1}) - \mathcal{A}^2 (I_{KM} - \mathcal{A})^\mathsf{T} \lambda_{n-1} \quad (26.47)$$

so that over two successive time instants:

$$\begin{aligned}\mathcal{W}_n - \mathcal{W}_{n-1} = {} & \mathcal{A}^2\,\mathcal{W}_{n-1} - \mathcal{A}^2\,\mathcal{W}_{n-2} - \mathcal{A}^2 (I_{KM} - \mathcal{A})^2\,\mathcal{B}_{n-1} - \\ & \mu\,\mathcal{A}^2 \Big(\nabla_{\mathcal{W}^\mathsf{T}}\,\mathcal{P}(\mathcal{W}_{n-1}) - \nabla_{\mathcal{W}^\mathsf{T}}\,\mathcal{P}(\mathcal{W}_{n-2})\Big)\end{aligned} \quad (26.48)$$

Noting from (26.45b) that $\mathcal{W}_{n-1} = \mathcal{A}^2\,\mathcal{B}_{n-1}$, we arrive at the ATC-DIGing (also called augmented distributed gradient method or Aug-DGM):

$$\boxed{\mathcal{W}_n = \mathcal{A}\Big\{2\,\mathcal{W}_{n-1} - \mathcal{A}\,\mathcal{W}_{n-2} - \mu\,\mathcal{A}\,\Big(\nabla_{\mathcal{W}^\mathsf{T}}\,\mathcal{P}(\mathcal{W}_{n-1}) - \nabla_{\mathcal{W}^\mathsf{T}}\,\mathcal{P}(\mathcal{W}_{n-2})\Big)\Big\}} \quad (26.49)$$

The designation ATC, which was originally introduced in the context of the diffusion strategy (25.77), is used here to refer to the fact that the combination policy $\mathcal{A}$ is applied to the aggregate term that appears between brackets in the above expression. The above recursion is written in terms of network variables. We can rewrite it at the level of the individual agents, as shown in listing (26.50) – see also Prob. 26.2.

ATC-DIGing or Aug-DGM algorithm for solving (26.1).

given a connected graph linking K agents with doubly stochastic A;
start from an arbitrary initial condition, $w_{k,-1}$, for each agent;
set $w_{k,0} = \sum_{\ell \in \mathcal{N}_k} a_{\ell k}\Big(2w_{\ell,-1} - \mu\alpha_\ell \nabla_{w^\mathsf{T}} P_\ell(w_{\ell,-1})\Big)$, $k = 1, 2, \ldots, K$.
repeat until convergence over $n \geq 0$:
  repeat over agents $k = 1, 2, \ldots, K$:

$$\begin{aligned}
c_{k,n} &= \sum_{\ell \in \mathcal{N}_k} a_{\ell k}\alpha_\ell \nabla_{w^\mathsf{T}} P_\ell(w_{\ell,n-1}) \\
\psi_{k,n} &= w_{k,n-1} - \mu c_{k,n} \\
b_{k,n} &= w_{k,n-1} - \mu\alpha_k \nabla_{w^\mathsf{T}} P_k(w_{k,n-1}) \\
\phi_{k,n} &= \sum_{\ell \in \mathcal{N}_k} a_{\ell k} b_{\ell,n} \\
v_{k,n} &= w_{k,n-1} + \psi_{k,n} - \phi_{k,n-1} \\
w_{k,n} &= \sum_{\ell \in \mathcal{N}_k} a_{\ell k} v_{\ell,n}
\end{aligned} \tag{26.50}$$

  end
end
iterate $w_{k,n}$ at each agent k approximates $w^\star$.

26.6 ATC TRACKING METHOD

The fifth decentralized method we consider incorporates the constraints by rewriting (26.4) in the equivalent form:

$$\mathcal{W}^\star \stackrel{\Delta}{=} \underset{\mathcal{W} \in \mathbb{R}^{KM}}{\text{argmin}} \left\{ \mathcal{P}(\mathcal{W}) + \frac{1}{2\mu} \| \mathcal{W} \|^2_{(I_{KM} - \mathcal{A})} \right\}, \quad \text{subject to } (I_{KM} - \mathcal{A})\,\mathcal{W} = 0 \tag{26.51}$$

where the weighting matrix is now given by $I_{KM} - \mathcal{A}$ instead of $I_{KM} - \mathcal{A}^2$. To solve (26.51) we introduce the Lagrangian function:

$$\mathcal{L}_e(\mathcal{W}, \lambda) \stackrel{\Delta}{=} \mathcal{P}(\mathcal{W}) + \frac{1}{2\mu}\mathcal{W}^\mathsf{T}(I_{KM} - \mathcal{A})\,\mathcal{W} + \frac{1}{\mu}\lambda^\mathsf{T}(I_{KM} - \mathcal{A})\,\mathcal{W} \tag{26.52}$$

where $\lambda \in \mathbb{R}^{KM}$ is a Lagrangian factor, and consider the saddle point problem:

$$\min_{\mathcal{W}} \max_{\lambda} \left\{ \mathcal{P}(\mathcal{W}) + \frac{1}{2\mu}\mathcal{W}^\mathsf{T}(I_{KM} - \mathcal{A})\,\mathcal{W} + \frac{1}{\mu}\lambda^\mathsf{T}(I_{KM} - \mathcal{A})\,\mathcal{W} \right\} \tag{26.53}$$

We approach the saddle point by pursuing gradient-descent over $\mathcal{w}$ and gradient-ascent over λ:

$$\mathcal{w}_n = \mathcal{w}_{n-1} - \mu \nabla_{\mathcal{w}^\mathsf{T}} \mathcal{P}(\mathcal{w}_{n-1}) - (I_{KM} - \mathcal{A})\,\mathcal{w}_{n-1} - (I_{KM} - \mathcal{A})^\mathsf{T} \lambda_{n-1} \tag{26.54a}$$

$$\lambda_n = \lambda_{n-1} + (I_{KM} - \mathcal{A})\,\mathcal{w}_n \tag{26.54b}$$

where we employ the updated $\mathcal{w}_n$ in the second step. Recall that $\mathcal{A}$ is symmetric. Therefore, these recursions simplify to

$$\mathcal{w}_n = \mathcal{A}\,\mathcal{w}_{n-1} - \mu \nabla_{\mathcal{w}^\mathsf{T}} \mathcal{P}(\mathcal{w}_{n-1}) - (I_{KM} - \mathcal{A})\lambda_{n-1} \tag{26.55a}$$

$$\lambda_n = \lambda_{n-1} + (I_{KM} - \mathcal{A})\,\mathcal{w}_n \tag{26.55b}$$

We add an additional step that involves fusing the iterates from across all agents and combining them using $\mathcal{A}$ as follows:

$$\mathcal{z}_n = \mathcal{A}\,\mathcal{w}_{n-1} - \mu \nabla_{\mathcal{w}^\mathsf{T}} \mathcal{P}(\mathcal{w}_{n-1}) - (I_{KM} - \mathcal{A})\lambda_{n-1} \tag{26.56a}$$

$$\lambda_n = \lambda_{n-1} + (I_{KM} - \mathcal{A})\,\mathcal{z}_n \tag{26.56b}$$

$$\mathcal{w}_n = \mathcal{A}\,\mathcal{z}_n \tag{26.56c}$$

Over two successive time instants we get

$$\mathcal{z}_n - \mathcal{z}_{n-1} = \mathcal{A}(\mathcal{w}_{n-1} - \mathcal{w}_{n-2}) - \mu\Big(\nabla_{\mathcal{w}^\mathsf{T}} \mathcal{P}(\mathcal{w}_{n-1}) - \nabla_{\mathcal{w}^\mathsf{T}} \mathcal{P}(\mathcal{w}_{n-2})\Big) - (I_{KM} - \mathcal{A})^2\, \mathcal{z}_{n-1} \tag{26.57}$$

so that after grouping terms

$$\mathcal{z}_n = (2\mathcal{A} - \mathcal{A}^2)\mathcal{z}_{n-1} + \mathcal{A}(\mathcal{w}_{n-1} - \mathcal{w}_{n-2}) - \mu\Big(\nabla_{\mathcal{w}^\mathsf{T}} \mathcal{P}(\mathcal{w}_{n-1}) - \nabla_{\mathcal{w}^\mathsf{T}} \mathcal{P}(\mathcal{w}_{n-2})\Big) \tag{26.58}$$

Multiplying by $\mathcal{A}$ from the left and using $\mathcal{w}_n = \mathcal{A}\,\mathcal{z}_n$ we arrive at the ATC tracking recursion in terms of the $\mathcal{w}$-variables:

$$\boxed{\mathcal{w}_n = \mathcal{A}\Big\{2\,\mathcal{w}_{n-1} - \mathcal{A}\,\mathcal{w}_{n-2} - \mu\Big(\nabla_{\mathcal{w}^\mathsf{T}} \mathcal{P}(\mathcal{w}_{n-1}) - \nabla_{\mathcal{w}^\mathsf{T}} \mathcal{P}(\mathcal{w}_{n-2})\Big)\Big\}} \tag{26.59}$$

This recursion is written in terms of network variables. We can rewrite it at the level of the individual agents, as shown in listing (26.60). Motivated by the ATC structure of diffusion strategies, it is seen from the update for $w_{k,n}$ that this method is adding a combination step to the original DIGing recursion (26.43).

ATC tracking algorithm for solving (26.1).

given a connected graph linking K agents with doubly stochastic A;
start from an arbitrary initial condition, $w_{k,-1}$, for each agent;
set $w_{k,0} = \sum_{\ell \in \mathcal{N}_k} a_{\ell k}\Big(2w_{\ell,-1} - \mu\alpha_\ell \nabla_{w^\mathsf{T}} P_\ell(w_{\ell,-1})\Big),\ k = 1, 2, \ldots, K.$
repeat until convergence over $n \geq 0$**:**
  repeat over agents $k = 1, 2, \ldots, K$**:**

$$\begin{aligned} \psi_{k,n-1} &= 2w_{k,n-1} - \sum_{\ell \in \mathcal{N}_k} a_{\ell k} w_{\ell,n-2} \\ \phi_{k,n} &= \psi_{k,n-1} - \mu\alpha_k\Big(\nabla_{w^\mathsf{T}} P_k(w_{k,n-1}) - \nabla_{w^\mathsf{T}} P_k(w_{k,n-2})\Big) \\ w_{k,n} &= \sum_{\ell \in \mathcal{N}_k} a_{\ell k}\phi_{\ell,n} \end{aligned} \tag{26.60}$$

  end
end
iterate $w_{k,n}$ at each agent k approximates $w^\star$.

The tracking methods (26.43), (26.50), and (26.60) tend to perform well even over time-varying graphs and also over directed graphs (where A is not necessarily symmetric). However, they require twice the amount of communications per iteration in comparison to the EXTRA and EXACT diffusion methods (26.18) and (26.27), respectively. In particular, note that EXTRA and EXACT diffusion communicate one vector per iteration among the neighbors, while tracking methods require communicating two vectors per iteration.

Example 26.2 **(Simulating DIGing and ATC gradient tracking)** We illustrate the behavior of stochastic versions of DIGing and ATC gradient tracking in the context of the same logistic regression problem from Example 25.6. We employ the same connected graph topology with $K = 20$ agents from Fig. 25.7 with a metropolis combination matrix, A. The same parameter values are used in this simulation, namely,

$$M = 10,\ K = 20,\ N_k = 200,\ N = 4000,\ \ \rho = 0.001,\ \ \mu = 0.01 \tag{26.61}$$

The stochastic versions of DIGing and ATC gradient tracking take the following forms where $\mu' = \mu/K$:

$$\left\{\begin{array}{l} \textbf{(DIGing)} \\ \text{at each agent } k\text{, select or receive a random data pair } (\gamma_k(n), \boldsymbol{h}_{k,n}) \text{ from } \mathcal{D}_k \\ \widehat{\gamma}_k(n) = \boldsymbol{h}_{k,n}^{\mathsf{T}} \boldsymbol{w}_{k,n-1} \\ \boldsymbol{\psi}_{k,n-1} = 2\boldsymbol{w}_{k,n-1} - \displaystyle\sum_{\ell \in \mathcal{N}_k} a_{\ell k} \boldsymbol{w}_{\ell,n-2} \\ \boldsymbol{\phi}_{k,n-1} = \displaystyle\sum_{\ell \in \mathcal{N}_k} a_{\ell k} \boldsymbol{\psi}_{\ell,n-1} \\ \boldsymbol{w}_{k,n} = (1 - 2\rho\mu')\boldsymbol{\phi}_{k,n-1} + \mu' \left(\dfrac{\gamma_k(n)\boldsymbol{h}_{k,n}}{1 + e^{\gamma_k(n)\widehat{\gamma}_k(n)}} - \dfrac{\gamma_k(n-1)\boldsymbol{h}_{k,n-1}}{1 + e^{\gamma_k(n-1)\widehat{\gamma}_k(n-1)}} \right) \end{array}\right. \tag{26.62}$$

and

$$\left\{\begin{array}{l} \textbf{(ATC gradient tracking)} \\ \text{at each agent } k\text{, select or receive a random data pair } (\gamma_k(n), \boldsymbol{h}_{k,n}) \text{ from } \mathcal{D}_k \\ \widehat{\gamma}_k(n) = \boldsymbol{h}_{k,n}^{\mathsf{T}} \boldsymbol{w}_{k,n-1} \\ \boldsymbol{\psi}_{k,n-1} = 2\boldsymbol{w}_{k,n-1} - \displaystyle\sum_{\ell \in \mathcal{N}_k} a_{\ell k} \boldsymbol{w}_{\ell,n-2} \\ \boldsymbol{\phi}_{k,n} = (1 - 2\rho\mu')\boldsymbol{\psi}_{k,n-1} + \mu' \left(\dfrac{\gamma_k(n)\boldsymbol{h}_{k,n}}{1 + e^{\gamma_k(n)\widehat{\gamma}_k(n)}} - \dfrac{\gamma_k(n-1)\boldsymbol{h}_{k,n-1}}{1 + e^{\gamma_k(n-1)\widehat{\gamma}_k(n-1)}} \right) \\ \boldsymbol{w}_{k,n} = \displaystyle\sum_{\ell \in \mathcal{N}_k} a_{\ell k}\, \boldsymbol{\phi}_{\ell,n} \end{array}\right. \tag{26.63}$$

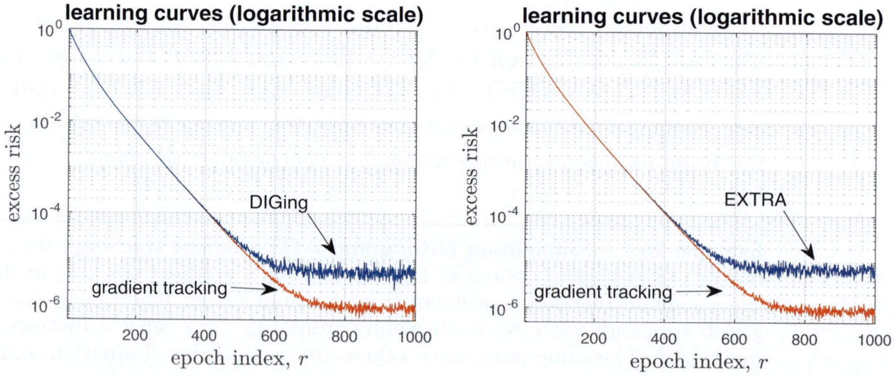

Figure 26.2 Learning curves obtained by averaging over $L = 10$ experiments. The curves show the evolution of the normalized excess risk values at the beginning of each run for EXTRA, DIGing, and ATC gradient tracking.

The simulation generates $N = 4000$ data points $\{\gamma(m), h_m\}$ according to the same logistic model from Example 25.6. A total of $R = 1000$ epochs are run over the data. We select one agent at random for illustration purposes (agent 8), and evaluate the evolution of the aggregate risk value $P(w_{-1}^r)$ at the beginning of each run r at that agent. We subsequently average these values over $L = 10$ experiments (with each experiment involving $R = 1000$ runs) to smooth out the resulting learning curve. We plot the average excess-risk curve $\mathbb{E}\, P(\boldsymbol{w}^{r}_{-1}) - P(w^{\star})$ after normalizing it by its peak value so that the learning curves start from the peak value of 1 for comparison purposes. The result is shown in Fig. 26.2.

26.7 UNIFIED DECENTRALIZED ALGORITHM

In this section, we describe a unifying and generalized framework that includes *all* the decentralized primal–dual methods derived so far as special cases.

Let $\mathcal{B} \in \mathbb{R}^{KM\times KM}$ and $\mathcal{C} \in \mathbb{R}^{KM\times KM}$ denote two general *symmetric* matrices that satisfy the following conditions:

$$\begin{cases} \mathcal{B}\,\mathcal{W} = 0 \Longleftrightarrow w_1 = w_2 \ldots = w_K \\ \mathcal{C}\,\mathcal{W} = 0 \Longleftrightarrow \mathcal{B}\,\mathcal{W} = 0 \text{ or } \mathcal{C} = 0 \\ \mathcal{C} \text{ is positive semi-definite} \end{cases} \tag{26.64}$$

For example, $\mathcal{C} = \mathcal{V} = \frac{1}{2}(I_{KM} - \mathcal{A})$ and $\mathcal{B} = \mathcal{V}^{1/2}$ is one choice, but many other choices are possible, including beyond what we have encountered so far. We will provide more examples in the following. Let also

$$\bar{\mathcal{A}} = \bar{A} \times I_M \tag{26.65}$$

where $\bar{A}$ is some symmetric doubly stochastic matrix. For example, $\bar{A} = \frac{1}{2}(I_K + A)$ is one possibility. Assuming the matrices $\{\bar{\mathcal{A}}, \mathcal{B}, \mathcal{C}\}$ have been chosen, we can then reformulate (26.4) in the equivalent form

$$\mathcal{W}^{\star} \triangleq \underset{\mathcal{W}\in\mathbb{R}^{KM}}{\text{argmin}} \left\{ \mathcal{P}(\mathcal{W}) + \frac{1}{2\mu}\| \mathcal{W} \|^2_{\mathcal{C}} \right\}, \quad \text{subject to } \mathcal{B}\,\mathcal{W} = 0 \tag{26.66}$$

and introduce the corresponding saddle point formulation:

$$\min_{\mathcal{W}} \max_{\lambda} \left\{ \mathcal{L}_e(\mathcal{W},\lambda) \triangleq \mathcal{P}(\mathcal{W}) + \frac{1}{2\mu}\| \mathcal{W} \|^2_{\mathcal{C}} + \frac{1}{\mu}\lambda^{\mathsf{T}}\mathcal{B}\,\mathcal{W} \right\} \tag{26.67}$$

where $\lambda \in \mathbb{R}^{KM}$ is a Lagrangian factor and $\mu > 0$. To solve the above problem, we introduce the following *unified decentralized algorithm* (UDA), which consists of three successive steps (primal-descent, dual-ascent, and combination):

$$\mathcal{Z}_n = (I_{KM} - \mathcal{C})\,\mathcal{W}_{n-1} - \mu\,\nabla_{\mathcal{W}^{\mathsf{T}}}\,\mathcal{P}(\mathcal{W}_{n-1}) - \mathcal{B}\lambda_{n-1} \tag{26.68a}$$

$$\lambda_n = \lambda_{n-1} + \mathcal{B}\,\mathcal{Z}_n \tag{26.68b}$$

$$\mathcal{W}_n = \bar{\mathcal{A}}\,\mathcal{Z}_n \tag{26.68c}$$

The first step is a primal-descent step over $\mathcal{W}$ applied to the Lagrangian function. The result is denoted by the intermediate variable $\mathcal{Z}_n$. The second equation is a dual-ascent step over λ; it uses the updated iterate $\mathcal{Z}_n$ instead of $\mathcal{W}_{n-1}$ as befits an incremental implementation. The last equation represents a combination step.

If desired, we can eliminate the dual variable λ_n from the above equations. Indeed, note that over two successive time instants we get

$$\mathcal{Z}_n - \mathcal{Z}_{n-1} = (I_{KM} - \mathcal{C})(\mathcal{W}_{n-1} - \mathcal{W}_{n-2}) - \mu\Big(\nabla_{\mathcal{W}^\mathsf{T}}\,\mathcal{P}(\mathcal{W}_{n-1}) - \nabla_{\mathcal{W}^\mathsf{T}}\,\mathcal{P}(\mathcal{W}_{n-2})\Big) - \mathcal{B}^2\,\mathcal{Z}_{n-1} \tag{26.69}$$

or, after rearrangements,

$$\mathcal{Z}_n = (I_{KM} - \mathcal{B}^2)\,\mathcal{Z}_{n-1} + (I_{KM} - \mathcal{C})(\mathcal{W}_{n-1} - \mathcal{W}_{n-2}) - \mu\Big(\nabla_{\mathcal{W}^\mathsf{T}}\,\mathcal{P}(\mathcal{W}_{n-1}) - \nabla_{\mathcal{W}^\mathsf{T}}\,\mathcal{P}(\mathcal{W}_{n-2})\Big) \tag{26.70a}$$

$$\mathcal{W}_n = \bar{\mathcal{A}}\,\mathcal{Z}_n \tag{26.70b}$$

with initial condition

$$\mathcal{Z}_0 = (I_{KM} - \mathcal{C})\,\mathcal{W}_{-1} - \mu\,\nabla_{\mathcal{W}^\mathsf{T}}\,\mathcal{P}(\mathcal{W}_{-1}), \qquad \mathcal{W}_0 = \bar{\mathcal{A}}\,\mathcal{Z}_0 \tag{26.70c}$$

for any $\mathcal{W}_{-1}$. Now, it is straightforward to see that recursions (26.70a)–(26.70b) reduce to the various decentralized algorithms presented in the earlier sections for different choices of the triplet $\{\bar{\mathcal{A}}, \mathcal{B}, \mathcal{C}\}$. This is illustrated in Table 26.1, where

$$\mathcal{V} \stackrel{\Delta}{=} \frac{1}{2}(I_{KM} - \mathcal{A}) \tag{26.71}$$

Obviously, other possibilities can be considered. Observe that EXTRA and DIGing employ $\mathcal{A} = I_{KM}$.

Table 26.1 Obtaining several decentralized methods as special cases of the unified decentralized algorithm (UDA) described by (26.70a)–(26.70b). In the expressions in the table we are using $\mathcal{V} = \frac{1}{2}(I_{KM} - \mathcal{A})$.

Primal–dual algorithm	$\bar{\mathcal{A}}$	$\mathcal{B}$	$\mathcal{C}$
EXTRA (26.17)	I_{KM}	$\mathcal{V}^{1/2}$	$\frac{1}{2}(I_{KM} - \mathcal{A})$
EXACT diffusion (26.26)	$\frac{1}{2}(I_{KM} + \mathcal{A})$	$\mathcal{V}^{1/2}$	0
DIGing (26.38)	I_{KM}	$I_{KM} - \mathcal{A}$	$I_{KM} - \mathcal{A}^2$
ATC-DIGing (26.49)	$\mathcal{A}^2$	$I_{KM} - \mathcal{A}$	0
ATC tracking (26.59)	$\mathcal{A}$	$I_{KM} - \mathcal{A}$	$I_{KM} - \mathcal{A}$

REMARK 26.1. **(Consensus and diffusion strategies)** The consensus and diffusion strategies were shown earlier to correspond to primal methods. They do not fit into the unified decentralized formulation of this section, which is specific to primal–dual methods. Nevertheless, it can still be seen from recursions (26.68a)–(26.68c) for the UDA that we can recover the consensus recursion (25.70) by setting $\mathcal{B} = 0$, $\mathcal{C} = I_{KM} - \mathcal{A}^\mathsf{T}$, and $\bar{\mathcal{A}} = I_{KM}$ and the ATC diffusion strategy (25.77) by setting $\mathcal{B} = 0$, $\mathcal{C} = 0$, and $\bar{\mathcal{A}} = \mathcal{A}^\mathsf{T}$. These choices do not satisfy conditions (26.64). ■

26.8 CONVERGENCE PERFORMANCE

We will now comment on the convergence performance of the various primal–dual algorithms. Proofs are given in Appendix 26.A.

Consider the EXTRA, EXACT diffusion, DIGing, ATC DIGing, and ATC tracking methods referenced in Table 26.1. For strongly convex and δ-Lipschitz risk functions $P(w)$ in (26.1), there will exist small enough step sizes μ such that these recursions approach their respective fixed points, denoted generically by $\mathcal{W}_\infty$. From the recursions (26.17), (26.26), (26.38), (26.49), and (26.59) for these primal–dual algorithms we find that their fixed points should satisfy the relations:

$$
\begin{aligned}
\mathcal{W}_\infty &= \bar{\mathcal{A}}\, \mathcal{W}_\infty && \text{(EXTRA)} && \text{(26.72a)}\\
\mathcal{W}_\infty &= \bar{\mathcal{A}}\, \mathcal{W}_\infty && \text{(EXACT diffusion)} && \text{(26.72b)}\\
\mathcal{W}_\infty &= \mathcal{A}(2I_{KM} - \mathcal{A})\, \mathcal{W}_\infty && \text{(DIGing)} && \text{(26.72c)}\\
\mathcal{W}_\infty &= \mathcal{A}(2I_{KM} - \mathcal{A})\, \mathcal{W}_\infty && \text{(ATC DIGing)} && \text{(26.72d)}\\
\mathcal{W}_\infty &= \mathcal{A}(2I_{KM} - \mathcal{A})\, \mathcal{W}_\infty && \text{(ATC tracking)} && \text{(26.72e)}
\end{aligned}
$$

Since the matrices $\bar{\mathcal{A}}$ and $\mathcal{A}(2I - \mathcal{A})$ are symmetric and doubly stochastic, it follows that any of these fixed points $\mathcal{W}_\infty$ has the form:

$$
\mathcal{W}_\infty = \mathbb{1}_K \otimes w_\infty, \quad \text{for some } w_\infty \in \mathbb{R}^M \tag{26.73}
$$

That is, $\mathcal{W}_\infty$ will consist of identical block entries, as desired by the consensus constraint $w_1 = w_2 = \ldots = w_K$. Actually, $\mathcal{W}_\infty$ will coincide with the sought-after unique minimizer $\mathcal{W}^\star$. Let us illustrate this fact for EXTRA and the same argument applies to the other variants listed above. We derived EXTRA by formulating the constrained risk optimization problem (26.6). Using (26.11)–(26.12), the EXTRA recursion converges to a saddle point $(\mathcal{W}_\infty, \lambda_\infty)$ that satisfies

$$
\begin{aligned}
\mu \nabla_{\mathcal{W}^\mathsf{T}} \mathcal{P}(\mathcal{W}_\infty) + \mathcal{V}^{1/2}\lambda_\infty &= 0 && \text{(26.74)}\\
(\mathcal{V}^{1/2})^\mathsf{T}\, \mathcal{W}_\infty &= 0 && \text{(26.75)}
\end{aligned}
$$

where we used the property $\mathcal{W}_\infty = \bar{\mathcal{A}}\, \mathcal{W}_\infty$. From the earlier Lemma 9.1 on convex optimization problems with linear constraints, these are the same conditions for optimality for the constrained problem:

$$
\mathcal{W}^\star \;\stackrel{\Delta}{=}\; \underset{\mathcal{W}\in\mathbb{R}^{KM}}{\text{argmin}} \left\{ \mu\, \mathcal{P}(\mathcal{W}) + \frac{1}{2}\|\mathcal{V}^{1/2}\, \mathcal{W}\,\|^2 \right\}, \quad \text{subject to } \mathcal{V}^{1/2}\, \mathcal{W} = 0 \tag{26.76}
$$

which is the same as problem (26.6) where the risk function is simply scaled by μ (this step does not alter the problem). Therefore, we must have $\mathcal{W}_\infty = \mathcal{W}^\star$ and the bias for EXTRA is zero. The same argument can be applied to the other primal–dual recursions, from which we can conclude that the bias is zero for deterministic optimization problems (see the proof in Appendix 26.A):

(bias in deterministic optimization)

$$\lim_{n\to\infty}\left\{\frac{1}{K}\sum_{k=1}^{K}\|\widetilde{w}_{k,n}\|^2\right\} = 0 \tag{26.77}$$

If we examine now the *stochastic* versions of these algorithms, where the gradient of each $P_k(w)$ is replaced by the gradient of the loss function $Q(w;\cdot)$ evaluated at some randomly selected data point, then the successive iterates $\boldsymbol{w}_n$ become random in nature. In this case, and as explained for EXACT diffusion in (26.116) further ahead, the expression for the bias would be replaced by one of the form:

(bias in stochastic optimization)

$$\limsup_{n\to\infty}\left\{\frac{1}{K}\sum_{k=1}^{K}\mathbb{E}\,\|\widetilde{\boldsymbol{w}}_{k,n}\|^2\right\} = O(\mu\sigma_g^2+\mu^2b^2) \tag{26.78}$$

in terms of the mean square of the error measure $\widetilde{\boldsymbol{w}}_n = w^\star - \boldsymbol{w}_n$ and where σ_g^2 is a parameter that is related to the "size" of the gradient noise process resulting from the gradient approximation (similar to (18.32)). Moreover,

$$b^2 \stackrel{\Delta}{=} \sum_{k=1}^{K}\alpha_k\,\|\nabla_{w^{\mathsf T}}P_k(w^\star)\|^2 \tag{26.79}$$

is a measure of how far the optimal solution $w^\star$ is from the minimizers for the individual risk functions. When the step size is small, the $O(\mu)$ term dominates so that for stochastic optimization:

$$\limsup_{n\to\infty}\left\{\frac{1}{K}\sum_{k=1}^{K}\mathbb{E}\,\|\widetilde{\boldsymbol{w}}_{k,n}\|^2\right\} = O(\mu\sigma_g^2) \tag{26.80}$$

REMARK 26.2. **(Stability ranges)** We examine the stability of the UDA in Theorem 26.5 and establish its linear convergence to the optimal solution for small enough step sizes. The theorem provides two bounds on the step sizes depending on the choices for the matrices $\{\bar{\mathcal{A}},\mathcal{B},\mathcal{C}\}$ shown in (26.123a)–(26.123b). The first choice applies to EXACT diffusion, ATC-DIGing, and ATC tracking (these are schemes with ATC tracking mechanisms), while the second choice applies to EXTRA and DIGing. For the first group under (26.123a), linear convergence is guaranteed for $\mu < (2-\sigma_{\max}(C))/\delta$. In contrast, for the second group under (26.123b), linear convergence is guaranteed for $\mu < (1-\sigma_{\max}(C))/\delta$ in terms of the maximum singular value of C. Observe that the step-size range is smaller for the second group (EXTRA and DIGing) than for the first group (EXACT diffusion and ATC algorithms). In particular, observe that EXACT diffusion has a wider stability range than EXTRA, and ATC-DIGing has a wider stability range than DIGing. ■

26.9 DUAL METHOD

We next describe another approach to decentralized optimization that is referred to as the dual method. We again introduce

$$\mathcal{V} \stackrel{\Delta}{=} \frac{1}{2}(I_{KM} - \mathcal{A}) \tag{26.81}$$

and appeal to the nullspace property (25.74b) to rewrite problem (26.4) in the equivalent form

$$\mathcal{w}^\star \stackrel{\Delta}{=} \underset{\mathcal{w}\in\mathbb{R}^{KM}}{\operatorname{argmin}} \left\{\mathcal{P}(\mathcal{w})\right\}, \quad \text{subject to } \mathcal{V}^{1/2}\,\mathcal{w} = 0 \tag{26.82}$$

where we are using the aggregate risk $\mathcal{P}(\mathcal{w})$ defined earlier by (26.7); recall also that, by construction, $\mathcal{V}^{1/2}$ is a symmetric matrix.

The above formulation is a special case of constrained optimization problems of the form (9.1) involving only equality constraints. We explained in Section 9.1 that such problems can be solved by means of duality arguments. Note first that problem (26.82) is feasible since there exists at least one vector $\mathcal{w}$ that satisfies the constraint, namely, $\mathcal{w} = \mathbb{1}_K \otimes w$. Second, since the equality constraint $\mathcal{V}^{1/2}\,\mathcal{w} = 0$ is affine in $\mathcal{w}$ and $\mathcal{P}(\mathcal{w})$ is assumed differentiable and convex, then we know from the arguments in Section 9.1 that strong duality holds for problem (26.82). We can therefore seek its solution by first determining the optimal dual variable, followed by the optimal primal variable. One convenience of this approach is that the dual variable can be obtained by maximizing the dual function, which will not be a constrained problem.

We start by introducing the Lagrangian function:

$$\mathcal{L}_e(\mathcal{w}, \lambda) \stackrel{\Delta}{=} \mathcal{P}(\mathcal{w}) + \lambda^\mathsf{T}\mathcal{V}^{1/2}\,\mathcal{w} \tag{26.83}$$

where $\lambda \in \mathbb{R}^{KM}$ is the Lagrangian factor, also called the dual variable, while $\mathcal{w}$ is the primal variable. The corresponding dual function is defined by minimizing over $\mathcal{w}$:

$$\begin{aligned}
\mathcal{D}(\lambda) &\stackrel{\Delta}{=} \inf_{\mathcal{w}} \mathcal{L}_e(\mathcal{w}, \lambda) \\
&= \inf_{\mathcal{w}} \left\{\mathcal{P}(\mathcal{w}) + \lambda^\mathsf{T}\mathcal{V}^{1/2}\,\mathcal{w}\right\} \\
&= -\sup_{\mathcal{w}}\left\{-(\mathcal{V}^{1/2}\lambda)^\mathsf{T}\,\mathcal{w} - \mathcal{P}(\mathcal{w})\right\} \\
&\stackrel{(a)}{=} -\mathcal{P}^\star(-\mathcal{V}^{1/2}\lambda) \\
&\stackrel{(b)}{=} -\mathcal{P}^\star(z)
\end{aligned} \tag{26.84}$$

where in step (b) we introduced the variable

$$z \stackrel{\Delta}{=} -\mathcal{V}^{1/2}\lambda \tag{26.85}$$

and in step (a) we used the notation $\mathcal{P}^\star(z)$, with a star in the superscript, to refer to the *conjugate function* of $\mathcal{P}(w)$. Recall that the conjugate function was defined earlier in (8.83) by means of the expression:

$$\mathcal{P}^\star(z) \stackrel{\Delta}{=} \sup_{w} \left\{z^\mathsf{T} w - \mathcal{P}(w)\right\} \tag{26.86}$$

The conjugate function can be computed in closed form in some cases of interest; hence, the arguments in this section would be most valuable when this calculation is possible. Note that if we partition z in the same manner as w and denote its block entries of size $M \times 1$ each by $\{z_k\}$, then using form (26.7) we have

$$\begin{aligned}\mathcal{P}^\star(z) &= \sup_{w}\Big(z^\mathsf{T} w - \mathcal{P}(w)\Big)\\ &= \sup_{w}\sum_{k=1}^{K}\Big(z_k^\mathsf{T} w_k - \alpha_k P_k(w_k)\Big)\\ &\stackrel{(a)}{=} \sum_{k=1}^{K}\sup_{w_k}\Big(z_k^\mathsf{T} w_k - \alpha_k P_k(w_k)\Big)\\ &\stackrel{(b)}{=} \sum_{k=1}^{K}\alpha_k P_k^\star(z_k/\alpha_k)\end{aligned} \tag{26.87}$$

where step (a) is because the terms inside the sum in the second line decouple over the $\{w_k\}$ (recall Prob. 8.50), and $P_k^\star(z_k)$ is the conjugate function of $P_k(w)$:

$$P_k^\star(z_k) \stackrel{\Delta}{=} \sup_{w_k}\Big\{z_k^\mathsf{T} w_k - P_k(w_k)\Big\} \tag{26.88}$$

Moreover, in step (b) we used the property of conjugate functions from Prob. 8.49 that if $h^\star(z)$ is the conjugate function of $h(w)$, then $\alpha h^\star(z/\alpha)$ is the conjugate function of $\alpha h(w)$ for any $\alpha > 0$. We note for later use that it follows from (26.87) that

$$\nabla_{z^\mathsf{T}}\mathcal{P}^\star(z) = \begin{bmatrix}\nabla_{z_1^\mathsf{T}} P_1^\star(z_1/\alpha_1)\\ \nabla_{z_2^\mathsf{T}} P_2^\star(z_2/\alpha_2)\\ \vdots\\ \nabla_{z_K^\mathsf{T}} P_K^\star(z_K/\alpha_K)\end{bmatrix} \tag{26.89}$$

In this way, the computation of the aggregate conjugate function $\mathcal{P}^\star(z)$ can be replaced by the computation of the individual conjugate functions $\{P_k^\star(z_k)\}$. Again, the derivation in this section assumes that these conjugate functions can be evaluated in closed form. This is possible, for example, when $P_k(w_k)$ is an affine function, a strictly convex quadratic function, a log-determinant function, or has the form of a negative logarithm function – recall Table 8.4.

Now, the optimal dual variable can be determined by maximizing the dual function, namely,

$$\lambda^\star \stackrel{\Delta}{=} \arg\max_{\lambda}\Big\{\mathcal{D}(\lambda)\Big\} \stackrel{(26.84)}{=} \arg\min_{\lambda}\Big\{\mathcal{P}^\star(-\mathcal{V}^{1/2}\lambda)\Big\} \tag{26.90}$$

This is an unconstrained optimization problem. The value of $\lambda^\star$ can be learned by applying, say, a gradient-descent update of the form:

$$\lambda_n = \lambda_{n-1} + \mu \mathcal{V}^{1/2}\left(\nabla_{\lambda^\mathsf{T}} \mathcal{P}^\star(-\mathcal{V}^{1/2}\lambda_{n-1})\right) \tag{26.91}$$

Multiplying both sides by $\mathcal{V}^{1/2}$ and using $\mathcal{z} = -\mathcal{V}^{1/2}\lambda$ and $\mathcal{V} = \frac{1}{2}(I_{KM} - \mathcal{A})$, we get the network recursion

$$\boxed{\mathcal{z}_n = \mathcal{z}_{n-1} - \frac{\mu}{2}(I_{KM} - \mathcal{A})\,\nabla_{\mathcal{z}^\mathsf{T}} \mathcal{P}^\star(\mathcal{z}_{n-1})} \tag{26.92}$$

or, in terms of the individual block vectors of $\mathcal{z}$ and using (26.89):

$$z_{k,n} = z_{k,n-1} - \frac{\mu}{2}\left\{\nabla_{z_k^\mathsf{T}} P_k^\star(z_{k,n-1}/\alpha_k) - \sum_{\ell \in \mathcal{N}_k} a_{\ell k}\,\nabla_{z_\ell^\mathsf{T}} P_\ell^\star(z_{\ell,n-1}/\alpha_\ell)\right\} \tag{26.93}$$

Observe that this recursion requires knowledge of the conjugate functions $\{P_k^\star(z_k)\}$. Moreover, since (26.90) is an unconstrained problem, we can apply some of the acceleration methods described earlier in Section 17.5.

Once $\lambda^\star$ or $z^\star$ is estimated, we can focus on determining the primal variable by solving

$$\begin{aligned}
\mathcal{w}^\star &= \arg\min_{\mathcal{w}} \left\{\mathcal{P}(\mathcal{w}) + (\lambda^\star)^\mathsf{T}\mathcal{V}^{1/2}\,\mathcal{w}\right\} \\
&= \arg\min_{\mathcal{w}} \left\{\mathcal{P}(\mathcal{w}) - (\mathcal{z}^\star)^\mathsf{T}\,\mathcal{w}\right\}, \quad \text{since } \mathcal{z}^\star = -\mathcal{V}^{1/2}\lambda^\star \\
&\overset{(25.68)}{=} \arg\min_{\{w_k\}} \sum_{k=1}^{K}\left\{\alpha_k P_k(w_k) - (z_k^\star)^\mathsf{T} w_k\right\}
\end{aligned} \tag{26.94}$$

and, hence, for each entry of $\mathcal{w}$:

$$w_k^\star = \arg\min_{w_k}\left\{\alpha_k P_k(w_k) - (z_k^\star)^\mathsf{T} w_k\right\} \tag{26.95}$$

This minimizer may admit a closed-form expression in some cases or it may be approximated iteratively by a gradient-descent algorithm. Listing (26.96) summarizes the main steps of the dual method. One of the disadvantages of the dual approach is that the conjugate function $P_k^\star(z)$ may not admit a closed-form representation and the computation of its gradient vector need not be straightforward.

Decentralized dual method for solving (26.1).

given a connected graph linking K agents with doubly stochastic A;
let $P_k^\star(z_k) = \sup_{w_k} \left\{ z_k^\mathsf{T} w_k - P_k(w_k) \right\}$, $k = 1, 2, \ldots, K$;
start from an arbitrary initial condition, $z_{k,-1}$, for each agent.
repeat until convergence over $n \geq 0$**:**
  repeat over agents $k = 1, 2, \ldots, K$**:**

$$z_{k,n} = z_{k,n-1} - \frac{\mu}{2}\left\{ \nabla_{z_k^\mathsf{T}} P_k^\star(z_{k,n-1}/\alpha_k) - \sum_{\ell \in \mathcal{N}_k} a_{\ell k} \nabla_{z_\ell^\mathsf{T}} P_\ell^\star(z_{\ell,n-1}/\alpha_\ell) \right\}$$

  end
end
for each agent $k = 1, 2, \ldots, K$**:**
  set $z_k^\star \leftarrow z_{k,n}$
  compute $w_k^\star = \arg\min_{w_k} \left\{ \alpha_k P_k(w_k) - (z_k^\star)^\mathsf{T} w_k \right\}$
end
iterate $w_k^\star$ at each agent k approximates $w^\star$.

(26.96)

26.10 DECENTRALIZED NONCONVEX OPTIMIZATION

We end this chapter by commenting on the performance of decentralized algorithms when the risk function $P(w)$ in (26.1) is nonconvex. We studied nonconvex optimization problems for centralized solutions in Chapter 24. There we focused on the performance of the (centralized) stochastic gradient recursion, which in the current setting takes the form (25.54) with the true gradients replaced by approximate gradients, i.e.,

$$\widehat{\nabla_{w^\mathsf{T}} P}(\boldsymbol{w}_{n-1}) = \sum_{k=1}^{K} \alpha_k \nabla_{w^\mathsf{T}} Q_k(\boldsymbol{w}_{n-1}; \boldsymbol{\gamma}_k(n), \boldsymbol{h}_{k,n}) \tag{26.97a}$$

$$\boldsymbol{w}_n = \boldsymbol{w}_{n-1} - \mu \widehat{\nabla_{w^\mathsf{T}} P}(\boldsymbol{w}_{n-1}), \quad n \geq 0 \tag{26.97b}$$

We established that such implementations are able to escape from saddle points and converge toward desirable second-order stationary points (namely, local minima). The same types of arguments can be extended to decentralized optimization when a collection of agents interact to solve a global nonconvex optimization problem, thus leading to similar conclusions to what we presented in Chapter 24 for the centralized case. We will only summarize the main conclusions here and refer the reader to the references listed in the comments at the end of the chapter for the derivations; these derivations are similar to what we presented earlier in Appendices 24.A–24.D.

26.10.1 Diffusion Strategy

It is sufficient for our purposes to focus on the stochastic ATC diffusion strategy (25.90), namely,

$$\begin{aligned}
&\text{select or receive a random pair } (\boldsymbol{\gamma}_k(n), \boldsymbol{h}_{k,n}) \text{ from } \mathcal{D}_k && \text{(26.98a)}\\
\boldsymbol{\psi}_{k,n} &= \boldsymbol{w}_{k,n-1} - \mu\alpha_k \nabla_{w^\mathsf{T}} Q\left(\boldsymbol{w}_{k,n-1}; \boldsymbol{\gamma}_k(n), \boldsymbol{h}_{k,n}\right) && \text{(26.98b)}\\
\boldsymbol{w}_{k,n} &= \sum_{\ell\in\mathcal{N}_k} a_{\ell k}\, \boldsymbol{\psi}_{\ell,n} && \text{(26.98c)}
\end{aligned}$$

We can rewrite these recursions in extended form as follows. In a manner similar to (26.7) we introduce

$$\mathcal{Q}(\mathcal{W}) \stackrel{\Delta}{=} \sum_{k=1}^{K} \alpha_k Q_k\Big(w_k; \gamma_k(n), h_{k,n}\Big) \tag{26.99}$$

and then write

$$\boldsymbol{\mathcal{W}}_n = \mathcal{A}^\mathsf{T}\Big(\boldsymbol{\mathcal{W}}_{n-1} - \mu\, \nabla_{\mathcal{W}^\mathsf{T}}\, \mathcal{Q}(\boldsymbol{\mathcal{W}}_{n-1})\Big) \tag{26.100}$$

which is the stochastic version of (25.126). We continue to assume the combination matrix A is doubly stochastic (although the results can be extended to left-stochastic matrices as well). The entries of its Perron vector are uniform and given by $p_k = 1/K$. We define the weighted *centroid* vector:

$$\boldsymbol{w}_{c,n} \stackrel{\Delta}{=} \sum_{k=1}^{K} p_k \boldsymbol{w}_{k,n} = (p^\mathsf{T} \otimes I_M)\boldsymbol{\mathcal{W}}_n \tag{26.101}$$

Multiplying (26.100) by $(p^\mathsf{T} \otimes I_M)$ from the left and using $A = A^\mathsf{T}$ and $Ap = p$, we conclude that the centroid vector evolves according to the following dynamics:

$$\boxed{\boldsymbol{w}_{c,n} = \boldsymbol{w}_{c,n-1} - \mu \sum_{k=1}^{K} p_k\alpha_k \nabla_{w^\mathsf{T}} Q_k\Big(\boldsymbol{w}_{k,n-1}; \boldsymbol{\gamma}_k(n), \boldsymbol{h}_{k,n}\Big)} \tag{26.102}$$

Examination of (26.102) shows that $\boldsymbol{w}_{c,n}$ evolves *almost* according to the centralized stochastic gradient recursion (26.97b), with the subtle difference that the gradient approximations are now evaluated at the local iterates $\{\boldsymbol{w}_{k,n-1}\}$ instead of the centroid $\boldsymbol{w}_{c,n-1}$. Nevertheless, as long as the collection of iterates $\{\boldsymbol{w}_{k,n-1}\}$ do not deviate too much from each other, and hence from the centroid $\boldsymbol{w}_{c,n}$, one would expect the evolution of (26.102) to carry similar performance guarantees to the centralized solution (26.97b). This is indeed the case as the statements given further ahead reveal.

26.10.2 Modeling Assumptions

We continue to use the same space decompositions $\{\mathcal{S}, \mathcal{S}^C, \mathcal{D}, \mathcal{U}\}$ defined earlier in Section 24.3, with the only adjustment being to the expression for c_1 in (24.38), which is changed to:

$$c_1 \stackrel{\Delta}{=} \frac{1}{2}(1 - 2\mu\delta) = O(1) \tag{26.103a}$$

$$c_2 \stackrel{\Delta}{=} \frac{\delta\sigma^2}{2} = O(1) \tag{26.103b}$$

Recall that $\mathcal{S}$ corresponds to the set of vectors w with "small" gradient norms, while $\mathcal{U}$ corresponds to the set of undesirable weights within $\mathcal{S}$ (such as local maxima and saddle points), whereas $\mathcal{D}$ is the set of desirable weights within $\mathcal{S}$ (such as local minima or "weak" saddle points). Moreover, $\mathcal{S}^C$ is the complement of $\mathcal{S}$ and consists of all points with "large" gradient norms. We also adopt the following assumptions, which extend the conditions from the centralized case under Assumption 24.1 to the decentralized setting.

ASSUMPTION 26.1. (Modeling assumptions) *The convergence analysis of the stochastic diffusion strategy (26.98b)–(26.98b) for nonconvex risks $P(w)$ assumes the following conditions:*

$$(\mathbb{A}1) \ \text{Each } P_k(w) \text{ is first- and second-order differentiable} \tag{26.104a}$$

$$(\mathbb{A}2) \ \|\nabla_w P_k(w_2) - \nabla_w P_k(w_1)\| \leq \delta\|w_2 - w_1\| \tag{26.104b}$$

$$(\mathbb{A}3) \ \|\nabla_w^2 P_k(w_2) - \nabla_w^2 P_k(w_1)\| \leq \delta_2\|w_2 - w_1\| \tag{26.104c}$$

$$(\mathbb{A}4) \ \|\nabla_w P_k(w) - \nabla_w P_\ell(w)\| \leq G \tag{26.104d}$$

for some nonnegative constants (δ, δ_2, G), and

$$(\mathbb{B}1) \ \mathbb{E}\left(\boldsymbol{g}_{k,n}(\boldsymbol{w}_{k,n-1}) \,|\, \boldsymbol{w}_{k,n-1}\right) = 0 \tag{26.105a}$$

$$(\mathbb{B}2) \ \mathbb{E}\left(\|\boldsymbol{g}_{k,n}(\boldsymbol{w}_{k,n-1})\|^4 \,|\, \boldsymbol{w}_{k,n-1}\right) \leq \sigma^4 \tag{26.105b}$$

$$(\mathbb{B}3) \ \|R_{g,k}(w_2) - R_{g,k}(w_1)\| \leq \beta_R\|w_2 - w_1\|^\eta \tag{26.105c}$$

for some $0 \leq \eta \leq 4$ and $\sigma^4 \geq 0$, and

$$(\mathbb{C}1) \ \lambda_{\min}\left\{\left(V_k^{<0}\right)^{\mathsf{T}} R_{g,k}(w)\, V_k^{<0}\right\} \geq \sigma_\ell^2 > 0, \ \forall w \in \mathcal{U} \tag{26.106}$$

The first set of conditions ($\mathbb{A}$1–$\mathbb{A}$4) deal with the individual risk functions $\{P_k(w)\}$: They are assumed to be first- and second-order differentiable, have δ-Lipschitz gradients, δ_2-Lipschitz Hessians, and bounded gradient disagreements. The second set of conditions ($\mathbb{B}$1–$\mathbb{B}$3) deal with the quality of the local gradient approximations and the resulting gradient noise processes where, for each agent k:

$$\boldsymbol{g}_{k,n}(\boldsymbol{w}_{k,n-1}) \stackrel{\Delta}{=} \nabla_{w^{\mathsf{T}}} Q_k\left(\boldsymbol{w}_{k,n-1}; \boldsymbol{\gamma}_k(n), \boldsymbol{h}_{k,n}\right) - \nabla_{w^{\mathsf{T}}} P_k(\boldsymbol{w}_{k,n-1})$$

$$R_{g,k} \stackrel{\Delta}{=} \mathbb{E}\left(\boldsymbol{g}_{k,n}(\boldsymbol{w}_{k,n-1})(\boldsymbol{g}_{k,n}(\boldsymbol{w}_{k,n-1}))^{\mathsf{T}} \,|\, \boldsymbol{w}_{k,n-1}\right) \tag{26.107a}$$

The last condition ($\mathbb{C}1$) links the behavior of the risk functions and the gradient noise processes; the latter need to have persistent presence along the descent directions in $V_k^{<0}$, which is determined by the Hessian of $P_k(w)$ at any $w \in \mathcal{U}$:

$$\nabla_w^2 P_k(w) = V_k \Lambda_k V_k^{\mathsf{T}} \quad \text{(eigen-decomposition)} \tag{26.108a}$$

$$\Lambda_k = \begin{bmatrix} \Lambda_k^{\geq 0} & 0 \\ 0 & \Lambda_k^{<0} \end{bmatrix} \tag{26.108b}$$

$$V_k = \begin{bmatrix} V_k^{\geq 0} & V_k^{<0} \end{bmatrix} \tag{26.108c}$$

We note that condition ($\mathbb{B}2$) implies

$$(\mathbb{B}2') \;\; \mathbb{E}\left(\|\boldsymbol{g}_{k,n}(\boldsymbol{w}_{k,n-1})\|^2 | \boldsymbol{w}_{k,n-1} \right) \leq \sigma^2 \tag{26.109}$$

26.10.3 First- and Second-Order Guarantees

We are ready to state the performance guarantees for the diffusion strategy in nonconvex environments. The proofs are omitted but appear in the references listed in the comments at the end of the chapter. Theorem 26.1 ascertains that the local iterates $\{\boldsymbol{w}_{k,n}\}$ stay close to the centroid $\boldsymbol{w}_{c,n}$ after sufficient iterations.

THEOREM 26.1. (Network disagreement) *Under the conditions stated in Assumption 26.1, for sufficiently small step sizes, and after sufficient iterations $n \geq n_o$, it holds that the network disagreement for each agent k satisfies:*

$$\mathbb{E}\,\|\boldsymbol{w}_{k,n} - \boldsymbol{w}_{c,n}\|^4 = O(\mu^4), \quad k = 1, 2, \ldots, K \tag{26.110}$$

To develop some intuition about the implication of (26.110), observe that from the Jensen inequality we get $\mathbb{E}\,\|\boldsymbol{w}_{k,n} - \boldsymbol{w}_{c,n}\|^2 = O(\mu^2)$ and, therefore,

$$\frac{1}{K}\sum_{k=1}^{K} \mathbb{E}\,\|\boldsymbol{w}_{k,n} - \boldsymbol{w}_{c,n}\|^2 = O(\mu^2) \tag{26.111}$$

We conclude that (26.111) bounds the average deviation of the local iterates $\boldsymbol{w}_{k,n}$ from the centroid $\boldsymbol{w}_{c,n}$ in the mean-square sense by a term that is on the order of μ^2, which is negligible for sufficiently small μ. This allows us to derive essentially the same performance guarantees for the network centroid $\boldsymbol{w}_{c,n}$ as for the centralized recursion (26.97b), after accounting for the small deviation (26.111).

Starting from a centroid location $\boldsymbol{w}_{c,n-1}$ with a "large" gradient value, i.e., $\boldsymbol{w}_{c,n-1} \in \mathcal{S}^C$, the next result (26.112) shows that the diffusion strategy behaves in a desirable manner and updates $\boldsymbol{w}_{c,n-1}$ to a new location $\boldsymbol{w}_{c,n}$ where the risk value is smaller on average, i.e., $\mathbb{E}\,P(\boldsymbol{w}_{c,n}) < \mathbb{E}\,P(\boldsymbol{w}_{c,n-1})$ since the positive term $\mu^2 c_2/\epsilon$ is subtracted from $\mathbb{E}\,P(\boldsymbol{w}_{c,n-1})$ and the last term depends on μ^3. Moreover, this decrease in the average risk value occurs over a single iteration.

This is a reassuring result because it means that, starting from centroids within the large gradient space $\mathcal{S}^C$, the diffusion algorithm will move the centroid closer to locations with smaller risk values where the gradient is expected to be smaller. That is, the centroid will move away from $\mathcal{S}^C$ and toward $\mathcal{S}$.

THEOREM 26.2. (Descent in the large gradient regime) *Assume a sufficiently small step size μ. When the gradient of the risk function $P(w)$ at the centroid $w = \boldsymbol{w}_{c,n-1}$ happens to be sufficiently large, i.e., $\boldsymbol{w}_{c,n-1} \in \mathcal{S}^C$, the stochastic diffusion strategy (26.98b)–(26.98c) will yield descent of the risk value under expectation in one single iteration, namely, it will hold that:*

$$\mathbb{E}\left(P(\boldsymbol{w}_{c,n}) \,|\, \boldsymbol{w}_{c,n-1} \in \mathcal{S}^C\right) \leq \mathbb{E}\left(P(\boldsymbol{w}_{c,n-1}) \,|\, \boldsymbol{w}_{c,n-1} \in \mathcal{S}^C\right) - \mu^2 \frac{c_2}{\epsilon} + \frac{O(\mu^3)}{\pi^{\mathcal{S}^C}_{n-1}} \tag{26.112}$$

as long as $\pi^{\mathcal{S}^C}_{n-1} = \mathbb{P}(\boldsymbol{w}_{c,n-1} \in \mathcal{S}^C) \neq 0$, and where $0 < \epsilon < 1$ is a small number selected by the designer (see the statement of Theorem 26.4).

Now that we have assured convergence of the centroid toward the set $\mathcal{S}$ of first-order stationary points (i.e., small gradient norms), we can examine more closely how the diffusion algorithm evolves within this set and, in particular, in the vicinity of strict saddle points in the undesirable set $\mathcal{U}$. As the results will reveal, the two sets $\mathcal{S}^C$ and $\mathcal{U}$ are distinguished by the fact that for centroids in $\mathcal{S}^{\mathcal{C}}$, the gradient norm $\|\nabla_w P(w)\|^2$ is large enough for a single (stochastic) gradient step to be sufficient to guarantee descent in expectation (as shown by Theorem 26.2), while points in $\mathcal{U}$ (i.e., strict saddle points) have small gradient norms and a single gradient step will not be sufficient to guarantee descent.

THEOREM 26.3. (Descent away from strict saddle points) *Starting from any strict saddle point $\boldsymbol{w}_{c,n-1} \in \mathcal{U}$ and iterating for $n^s = O(1/\mu\tau)$ iterations after $n-1$ with*

$$n^s = \frac{\log\left(2M\frac{\sigma^2}{\sigma_\ell^2} + 1\right)}{\log(1 + 2\mu\tau)} \leq O\left(1/\mu\tau\right) \tag{26.113}$$

guarantees

$$\begin{aligned} &\mathbb{E}\left(P(\boldsymbol{w}_{c,n+n^s}) \,|\, \boldsymbol{w}_{c,n-1} \in \mathcal{U}\right) \\ &\leq \mathbb{E}\left(P(\boldsymbol{w}_{c,n-1}) \,|\, \boldsymbol{w}_{c,n-1} \in \mathcal{U}\right) - \frac{\mu}{2} M\sigma^2 + o(\mu) + \frac{o(\mu)}{\pi^{\mathcal{U}}_{n-1}} \end{aligned} \tag{26.114}$$

as long as $\pi^{\mathcal{U}}_{n-1} = \mathbb{P}(\boldsymbol{w}_{c,n-1} \in \mathcal{U}) \neq 0$.

Theorem 26.3 ensures that, even when the norm of the gradient at $\boldsymbol{w}_{c,n-1} \in \mathcal{U}$ is too small to carry sufficient information about the descent direction, the persistent gradient noise at the individual iterates $\{\boldsymbol{w}_{k,n-1}\}$ helps ensure descent in n^s iterations, where the escape-time n^s scales favorably with problem parameters. For example, the escape-time scales logarithmically with the feature dimension M, implying that we can expect fast evasion of saddle points even in high dimensions. Moreover, the escape-time scales inversely with $1/\tau$, meaning that the larger the value of τ is, the faster the escape will be.

Having established descent from both the large-gradient regime ($\boldsymbol{w}_{c,n-1} \in \mathcal{S}^C$) and the strict saddle point regime ($\boldsymbol{w}_{c,n-1} \in \mathcal{U}$), we can combine the results to conclude eventual second-order stationarity. Theorem 26.4 states that, starting from an initial centroid $\boldsymbol{w}_{c,-1}$ for the diffusion strategy, it will reach the desirable set $\mathcal{D}$ in at most $O(1/\mu^3\tau)$ iterations.

Theorem 26.4. (Second-order convergence guarantee) *Assume the risk function $P(w)$ is bounded from below, say, $P(w) \geq P^o$ for all w. Let $\epsilon \in (0,1)$ denote a small positive number selected by the designer. Then, there exist sufficiently small step sizes μ such that, starting from an arbitrary initial centroid $\boldsymbol{w}_{c,-1}$ and with high probability $1-\epsilon$, the centroid $\boldsymbol{w}_{c,n^o}$ will belong to the desirable set $\mathcal{D}$, i.e., $\|\nabla_w P(\boldsymbol{w}_{c,n^o})\|^2 \leq O(\mu)$ and $\lambda_{\min}\left(\nabla_w^2 P(\boldsymbol{w}_{c,n^o})\right) > -\tau$ in at most n^o iterations, where*

$$n^o \leq \frac{(P(w_{c,-1}) - P^o)}{\mu^2 c_2 \epsilon} n^s \leq O\left(1/\mu^3\tau\right) \tag{26.115}$$

Comparing Theorems 26.2–26.4 in the decentralized setting to Theorems 24.1–24.3 in the centralized setting, we note that the descent and second-order stationarity guarantees for the network centroid $\boldsymbol{w}_{c,n}$ generated by the diffusion algorithm (26.98b)–(26.98c) are essentially the same as those for the ordinary stochastic gradient descent recursion (24.18) after adjusting the constants to account for the decentralized nature. Theorem 26.1, on the other hand, ensures that all local iterates $\{\boldsymbol{w}_{k,n}\}$ will closely track the network centroid $\boldsymbol{w}_{c,n}$ after sufficient iterations, and hence each agent k in the network will inherit the second-order guarantees of $\boldsymbol{w}_{c,n}$.

26.11 COMMENTARIES AND DISCUSSION

Primal–dual decentralized optimization methods. We derived in the body of the chapter several primal–dual strategies for the solution of decentralized optimization problems, and commented on their convergence behavior. The EXTRA strategy is from Shi *et al.* (2015), the Aug-DGM strategy is from Xu *et al.* (2015), and the gradient tracking methods are from Lorenzo and Scutari (2016) and Sun, Scutari, and Palomar (2016). These latter strategies employ a dynamic consensus strategy of the form (26.40) from Freeman, Yang, and Lynch (2006) to track the difference between gradient vectors – see also Zhu and Martinez (2010) and Ying, Yuan, and Sayed (2019). Likewise, ATC-DIGing is from Nedic *et al.* (2017), the DIGing strategy is from Nedic, Olshevsky, and Shi (2017) and Qu and Li (2018), and EXACT diffusion is from Yuan *et al.* (2017, 2019a,b). EXACT diffusion is referred to as the D^2 (decentralized data) and NIDS (network independent step-size) algorithms in subsequent works by Tang *et al.* (2018) and

Li, Shi, and Yan (2019). One precursor to EXTRA and EXACT diffusion is the primal–dual method derived by Towfic and Sayed (2015) for MSE risks; compare expressions (18a)–(18b) and (19a)–(19b) in this reference to expressions (26.10a)–(26.10b) in the body of the chapter where the matrix $\mathcal{V}^{1/2}$ is replaced by an incidence matrix. The UDA and the convergence proof for primal–dual methods in Appendix 26.A are from Alghunaim *et al.* (2021). Several of these procedures employ the ATC construction originally proposed by Cattivelli and Sayed (2010a, b) in the context of diffusion strategies.

Other references on decentralized algorithms are the works by Mokhtari and Ribeiro (2016), Li and Yan (2017), Xin and Khan (2018), Li, Shi, and Yan (2019), Jakovetic (2019), and Sundararajan, Van Scoy, and Lessard (2019). The last two works also attempted to unify some of the algorithms. However, they exclude gradient tracking methods or focus on methods that require a single round of communication per iteration. Another well-known family of distributed primal–dual methods are those based on the alternating direction method of multipliers (ADMM) – see, e.g., the works by Boyd *et al.* (2010), Mateos, Bazerque, and Giannakis (2010), Mota *et al.* (2013), Shi *et al.* (2014), Ling *et al.* (2015), and Mokhtari *et al.* (2016). However, ADMM methods tend to be more complex and computationally expensive since they require the solution of optimal subproblems at each iteration.

Comparing primal and primal–dual methods. We described in the body of the chapter several bias-correction methods such as EXTRA, EXACT diffusion, and gradient tracking algorithms for the solution of deterministic and stochastic optimization problems. These methods were shown to converge linearly to the *exact* minimizer $w^\star$ under proper conditions for *deterministic* optimization. However, their performance under *stochastic* settings where actual gradient vectors are replaced by stochastic approximations has only been explored to a limited extent in the literature. In these remarks we comment on, and reproduce some, useful conclusions from the work by Yuan *et al.* (2020), which compares the performance of EXACT diffusion (a primal–dual method) against traditional diffusion (a primal method) under stochastic settings in the presence of noisy gradients and constant step sizes (namely, for online learning). It turns out that it is not necessarily the case that bias-correction (primal–dual) methods outperform traditional (consensus and diffusion) methods when gradient noise is present. The following results and comments are extracted from Yuan *et al.* (2020). The work provides conditions under which EXACT diffusion has superior steady-state mean-square deviation (MSD) performance. In particular, it is proven that this superiority is more evident over sparsely connected graph topologies such as lines, cycles, or grids. There are instances when bias-correction may degrade performance.

Specifically, it is shown in the above work that under sufficiently small step sizes, the EXACT diffusion strategy (26.27) will converge exponentially fast, at the rate of $1 - O(\mu\nu)$, to a neighborhood around $w^\star$. Moreover, the size of the neighborhood will be characterized by the MSE (these expressions assume $\alpha_k = 1/K$ in the definition of the aggregate risk $P(w)$ in (26.1)):

$$\limsup_{n\to\infty} \left\{ \frac{1}{K}\sum_{k=1}^{K} \mathbb{E}\,\|\widetilde{\boldsymbol{w}}_{k,n}\|^2_{\text{xd}} \right\} = O\left(\frac{\mu\sigma_g^2}{K\nu} + \frac{\delta^2}{\nu^2} \times \frac{\mu^2\sigma_g^2}{1-\lambda} \right) \tag{26.116}$$

where δ and ν are the Lipschitz and strong-convexity constants, the quantity σ_g^2 is a measure of the variance of the gradient noise, and $\lambda \in (0,1)$ is the second largest magnitude of the eigenvalues of the combination matrix $A = [a_{\ell k}]$ (which reflects the level of network connectivity). The subscript "xd" indicates that $\boldsymbol{w}_{k,n}$ is generated by the EXACT diffusion method. In comparison, the MSD for the diffusion strategy (25.77) converges at a similar rate albeit to the following neighborhood:

$$\limsup_{n\to\infty} \left\{ \frac{1}{K}\sum_{k=1}^{K} \mathbb{E}\,\|\widetilde{\boldsymbol{w}}_{k,n}\|^2_{\text{df}} \right\} = O\left(\frac{\mu\sigma_g^2}{K\nu} + \frac{\delta^2}{\nu^2} \times \frac{\mu^2\lambda^2\sigma_g^2}{1-\lambda} + \frac{\delta^2}{\nu^2} \times \frac{\mu^2\lambda^2 b^2}{(1-\lambda)^2} \right) \tag{26.117}$$

where the subscript "df" indicates that $\boldsymbol{w}_{k,n}$ is now generated by the diffusion method, and

$$b^2 \triangleq \frac{1}{K}\sum_{k=1}^{K} \|\nabla_{w^\mathsf{T}} P_k(w^\star)\|^2 \tag{26.118}$$

is a bias constant independent of the gradient noise. Observe that the expressions on the right-hand side of (26.116) and (26.117) depend on μ and μ^2. The terms that depend on μ^2 in (26.116) and (26.117) help reveal the important insights that arise from using the EXACT (bias-correction) diffusion strategy. As explained in Yuan *et al.* (2020), these expressions have the following three useful implications:

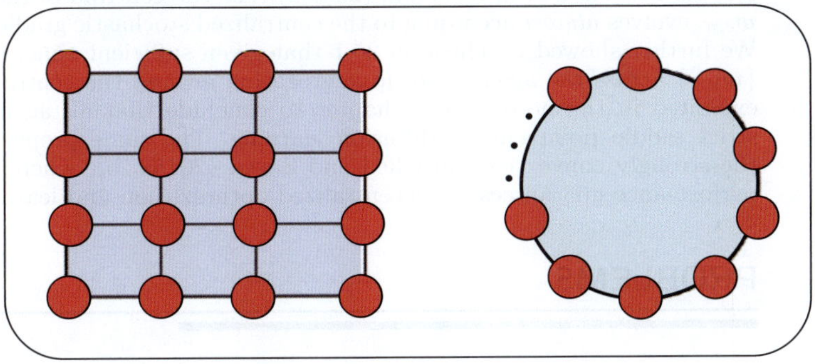

Figure 26.3 Illustration of a grid topology (*left*) and a cyclic topology (*right*).

- **(a)** First, it is obvious that diffusion suffers from the additional bias term given by $\mu^2\lambda^2 b^2/(1-\lambda)^2$, which is independent of the gradient noise σ_g^2, while EXACT diffusion removes it completely. In the deterministic setting, when the gradient noise $\sigma_g^2 = 0$, it is observed from (26.116) and (26.117) that diffusion converges to an $O(\mu^2)$-neighborhood around the global solution $w^\star$ while EXACT diffusion converges exactly to $w^\star$.
- **(b)** Second, the performance of diffusion and EXACT diffusion differ only in the $O(\mu^2)$ terms inside (26.116) and (26.117). When the step size is moderately small so that the $O(\mu^2)$ terms are nonnegligible, then the superiority of EXACT diffusion or diffusion will depend on the graph topology. In particular, when the graph topology is sparsely connected (in which case λ approaches 1), the bias term given by $\mu^2\lambda^2 b^2/(1-\lambda)^2$ will be significantly large and the elimination of this term will greatly improve the steady-state performance. The bias-correction property of EXACT diffusion is particularly critical for large-scale *linear* or *cyclic* graphs where $1-\lambda = O(1/K^2)$ and *grid* graphs where $1-\lambda = O(1/K)$ since the bias term will grow rapidly on these graph topologies as the size K increases – see Fig. 26.3 and also Seaman *et al.* (2017) and Mao *et al.* (2020) for the expressions for λ for cyclic and grid graphs. On the other hand, when the graph is well-connected (in which case λ approaches 0), one can find that the $O(\mu^2)$ term in diffusion (26.117) diminishes while the $O(\mu^2)$ term in EXACT diffusion (26.116) still exists. This implies that for well-connected graphs and moderately small step sizes, diffusion is a better choice than EXACT diffusion. The comparison between (26.116) and (26.117) provides guidelines on the proper choice of diffusion or EXACT diffusion in various application scenarios.
- **(c)** Third, the difference between EXACT diffusion and diffusion will vanish as the step size μ approaches 0. This is because $O(\mu\sigma_g^2/K\nu)$ will dominate the $O(\mu^2)$ terms when μ is sufficiently small.

Nonconvex optimization. There are several useful works on decentralized optimization in nonconvex environments, including by Lorenzo and Scutari (2016), Tatarenko and Touri (2017), Lian *et al.* (2017), Tang *et al.* (2018), Daneshmand, Scutari, and Kungurtsev (2018), Wang, Yin, and Zeng (2019), and Vlaski and Sayed (2020, 2021a, b). The presentation in Section 26.10 follows the approach from Vlaski and Sayed (2020, 2021a, b) and considers stochastic decentralized optimization; the presentation extends the arguments from Chapter 24 to the decentralized setting and establishes the validity of Theorems 26.1–26.4. Although we assumed the combination matrix A to be doubly stochastic in Section 26.10, these references establish the theorems for the more general case of left-stochastic combination matrices A (which are not necessarily symmetric). The conclusions continue to be similar to what was presented in the body of the chapter. We also showed in recursion (26.102) that the centroid of the network, denoted by $\boldsymbol{w}_{c,n}$, evolves *almost* according to the centralized stochastic gradient recursion (26.97b). We further showed in Theorem 26.1 that given sufficient time, the individual iterates $\{\boldsymbol{w}_{k,n}\}$ across the agents end up aggregating around the centroid. This property was exploited in the body of the chapter to conclude that all agents are able to escape strict saddle points under diffusion learning. The same property was established in the strongly convex case in Chen and Sayed (2015a, b), where it was used to derive performance guarantees for decentralized optimization and learning.

PROBLEMS

26.1 The DIGing algorithm is described by the primal–dual recursions:

$$\begin{aligned} \mathcal{W}_n &= \mathcal{A}\, \mathcal{W}_{n-1} - \mu\, \mathcal{O}_{n-1} \\ \mathcal{O}_n &= \mathcal{A}\, \mathcal{O}_{n-1} + \nabla_{\mathcal{W}^\mathsf{T}} \mathcal{P}(\mathcal{W}_n) - \nabla_{\mathcal{W}^\mathsf{T}} \mathcal{P}(\mathcal{W}_{n-1}) \end{aligned}$$

Eliminate $\mathcal{O}_{n-1}$ from the first relation and show that these recursions reduce to the gradient tracking algorithm (26.38). *Remark.* For more discussion on the DIGing algorithm, the reader may refer to Nedic, Olshevsky, and Shi (2017) and Qu and Li (2018).

26.2 The ATC-DIGing algorithm is described by the primal–dual recursions:

$$\begin{aligned} \mathcal{W}_n &= \mathcal{A}(\mathcal{W}_{n-1} - \mu\, \mathcal{O}_{n-1}) \\ \mathcal{O}_n &= \mathcal{A}\Big(\mathcal{O}_{n-1} + \nabla_{\mathcal{W}^\mathsf{T}} \mathcal{P}(\mathcal{W}_n) - \nabla_{\mathcal{W}^\mathsf{T}} \mathcal{P}(\mathcal{W}_{n-1})\Big) \end{aligned}$$

Eliminate $\mathcal{O}_{n-1}$ from the first relation and show that these recursions reduce to algorithm (26.49). *Remark.* For more discussion on the ATC-DIGing algorithm, the reader may refer to Xu *et al.* (2015) and Nedic *et al.* (2017).

26.3 Refer to the primal minimization step (26.95). Use the second property from Prob. 8.49 for conjugate functions to conclude that the minimum value is given by

$$\inf_{w_k} \Big\{ \alpha_k P_k(w_k) - (z_k^\star)^\mathsf{T} w_k \Big\} = -\alpha_k P_k^\star(z_k^\star/\alpha_k)$$

26.4 We add a possibly nonsmooth convex regularization term $R(w)$ to the aggregate risk in problem (26.4) and consider instead:

$$w^\star = \underset{w\in\mathbb{R}^M}{\operatorname{argmin}} \left\{ P(w) + R(w) \overset{\Delta}{=} \sum_{k=1}^{K} \alpha_k \Big(P_k(w) \;+\; R(w)\Big) \right\}$$

subject to $w_1 = w_2 = \ldots = w_K$, where the $\{w_k\}$ are local copies of w at the individual agents. We further introduce, in a manner similar to (25.68), the functions

$$\mathcal{P}(\mathcal{W}) \triangleq \sum_{k=1}^{K} \alpha_k P_k(w_k), \quad \mathcal{R}(\mathcal{W}) \triangleq \sum_{k=1}^{K} \alpha_k R(w_k)$$

Let $\mathcal{B} \in \mathbb{R}^{KM\times KM}$ and $\mathcal{C} \in \mathbb{R}^{KM\times KM}$ denote two general *symmetric matrices* that continue to satisfy the conditions (26.64).

(a) Argue that the above optimization problem is equivalent to solving:

$$\mathcal{W}^\star \triangleq \underset{\mathcal{W}\in\mathbb{R}^{KM}}{\text{argmin}} \left\{\mathcal{P}(\mathcal{W}) + \mathcal{R}(\mathcal{W}) + \frac{1}{2\mu}\| \mathcal{W} \|^2_{\mathcal{C}}\right\}, \quad \text{subject to } \mathcal{B}\,\mathcal{W} = 0$$

(b) Consider the following proximal gradient decentralized procedure for solving the problem of part (a):

$$\begin{aligned}
z_n &= (I_{KM} - \mathcal{C})\,\mathcal{W}_{n-1} - \mu \nabla_{\mathcal{W}^\mathsf{T}}\, \mathcal{P}(\mathcal{W}_{n-1}) - \mathcal{B}\lambda_{n-1}, \quad n \geq 0\\
\lambda_n &= \lambda_{n-1} + \mathcal{B}\, z_n, \quad \lambda_{-1} = 0\\
\mathcal{W}_n &= \text{prox}_{\mu\mathcal{R}}(\bar{\mathcal{A}}\, z_n)
\end{aligned}$$

where $\bar{\mathcal{A}} = \bar{A} \times I_M$ is some symmetric and doubly stochastic matrix. Assume each $P_k(w)$ is first-order differentiable and ν-strongly convex with δ-Lipschitz gradients ($\nu \leq \delta$). Introduce the saddle point formulation:

$$\min_{\mathcal{W}} \max_{\lambda} \left\{\mathcal{P}(\mathcal{W}) + \mathcal{R}(\mathcal{W}) + \frac{1}{2\mu}\| \mathcal{W} \|^2_{\mathcal{C}} + \lambda^\mathsf{T}\mathcal{B}\,\mathcal{W}\right\}$$

Show that a fixed point $(\mathcal{W}^\star, z^\star, \lambda^\star)$ exists for the recursions that satisfies

$$\begin{aligned}
z^\star &= \mathcal{W}^\star - \mu\nabla_{\mathcal{W}^\mathsf{T}}\, \mathcal{P}(\mathcal{W}^\star) - \mathcal{B}\lambda^\star\\
0 &= \mathcal{B}\, z^\star\\
\mathcal{W}^\star &= \text{prox}_{\mu\mathcal{R}}(\bar{\mathcal{A}}\, z^\star)
\end{aligned}$$

(c) Show further that the points $(\mathcal{W}^\star, z^\star)$ from part (b) are unique and $\mathcal{W}^\star = \mathbb{1}_K \otimes w^\star$.

(d) Show that there exists a unique $\lambda_d^\star$ in the range space of $\mathcal{B}$ that satisfies the conditions in part (b).

Remark. For more details on the proximal gradient decentralized algorithm of this problem, the reader may refer to Alghunaim and Sayed (2020) and Alghunaim *et al.* (2021). Observe that z_n is used in the update for λ_n in part (b), as opposed to $\mathcal{W}_n$ as happens with the PG-EXTRA algorithm from Li, Shi, and Yan (2019). This difference enables the algorithm proposed in this problem to attain linear convergence, as shown in Prob. 26.5.

26.5 We continue with the setting of Prob. 26.4 and assume $\{\bar{\mathcal{A}}, \mathcal{B}, \mathcal{C}\}$ satisfy

$$0 < I_{KM} - \mathcal{B}^2, \quad \bar{\mathcal{A}}^2 \leq I_{KM} - \mathcal{B}^2, \quad 0 \leq \mathcal{C} < 2I_{KM}$$

Let $\sigma_{\min}(\mathcal{B})$ denote the smallest nonzero singular value of $\mathcal{B}$. Let also $\widetilde{\mathcal{W}}_n = \mathcal{W}^\star - \mathcal{W}_n$ and $\widetilde{\lambda}_n = \lambda_d^\star - \lambda_n$. Show that under $\mu < (2-\sigma_{\max}(\mathcal{C}))/\delta$, the proximal gradient decentralized algorithm of Prob. 26.4 converges linearly. In particular, verify that under $\lambda_0 = 0$:

$$\|\widetilde{\mathcal{W}}_n\|^2 + \|\widetilde{\lambda}_n\|^2 \;\leq\; \rho^n\left(\|\widetilde{\mathcal{W}}_{n-1}\|^2 + \|\widetilde{\lambda}_{n-1}\|^2\right)$$

where $\rho \triangleq \max\Big\{1 - \mu\nu(2 - \sigma_{\max}(\mathcal{C})),\, 1 - \sigma^2_{\min}(\mathcal{B})\Big\}$.

26.6 We continue with the setting of Prob. 26.4 and select $\bar{A} = \frac{1}{2}(I_K + A)$, $B^2 = \frac{1}{2}(I_K - A)$, and $C = 0$. Introduce $\mathcal{X}_n = \bar{\mathcal{A}}\,\mathcal{Z}_n$. Verify that the proximal gradient decentralized algorithm leads to the recursion

$$\begin{aligned} \mathcal{X}_n &= \bar{\mathcal{A}}\Big(\mathcal{X}_{n-1} + \mathcal{W}_{n-1} - \mathcal{W}_{n-2} - \mu\nabla_{\mathcal{W}^\mathsf{T}}\,\mathcal{P}(\mathcal{W}_{n-1}) - \mu\nabla_{\mathcal{W}^\mathsf{T}}\,\mathcal{P}(\mathcal{W}_{n-2})\Big) \\ \mathcal{W}_n &= \text{prox}_{\mu\mathcal{R}}(\mathcal{X}_n) \end{aligned}$$

Deduce that these global recursions lead to the following proximal EXACT diffusion algorithm at the level of the individual agents:

$$\begin{aligned} \psi_{k,n} &= w_{k,n-1} - \mu\alpha_k\,\nabla_{w^\mathsf{T}}\,P_k(w_{k,n-1}) \\ z_{k,n} &= x_{k,n-1} + \psi_{k,n} - \psi_{k,n-1} \\ x_{k,n} &= \sum_{\ell\in\mathcal{N}_k} \bar{a}_{\ell k} z_{\ell n} \\ w_{k,n} &= \text{prox}_{\mu\alpha_k R}(x_{k,n}) \end{aligned}$$

26.7 Refer to the discussion leading to Theorem 26.2 for decentralized convex optimization. Show that

$$\mathbb{E}\Big(P(\boldsymbol{w}_{c,n})\,|\,\boldsymbol{w}_{c,n-1}\in\mathcal{D}\Big) \leq \mathbb{E}\Big(P(\boldsymbol{w}_{c,n-1})\,|\,\boldsymbol{w}_{c,n-1}\in\mathcal{D}\Big) + \mu^2 c_2 + O(\mu^3/\pi^{\mathcal{D}}_{n-1})$$

as long as $\pi^{\mathcal{D}}_{n-1} = \mathbb{P}(\boldsymbol{w}_{c,n-1}\in\mathcal{D}) \neq 0$.

26.A CONVERGENCE OF PRIMAL–DUAL ALGORITHMS

In this appendix we establish the linear convergence of the UDA introduced in Section 26.7, and which was shown there to subsume as special cases the various primal–dual algorithms described in the section. We reproduce here with adjustments the arguments and unifying approach from Alghunaim *et al.* (2021).

Consider again the following convex optimization problem with linear constraints introduced earlier in (26.66) and repeated here:

$$\mathcal{W}^\star \stackrel{\Delta}{=} \underset{\mathcal{W}\in\mathbb{R}^{KM}}{\text{argmin}} \left\{\mathcal{P}(\mathcal{W}) + \frac{1}{2\mu}\|\,\mathcal{W}\,\|^2_{\mathcal{C}}\right\}, \quad \text{subject to } \mathcal{B}\,\mathcal{W}=0 \tag{26.119}$$

where

$$\mathcal{P}(\mathcal{W}) \stackrel{\Delta}{=} \sum_{k=1}^{K} \alpha_k P_k(w_k) \tag{26.120}$$

and $\mathcal{B}\in\mathbb{R}^{KM\times KM}$ and $\mathcal{C}\in\mathbb{R}^{KM\times KM}$ denote two general *symmetric matrices* that satisfy the conditions:

$$\begin{cases} \mathcal{B}\,\mathcal{W}=0 \Longleftrightarrow w_1 = w_2\ldots, w_K \\ \mathcal{C}\,\mathcal{W}=0 \Longleftrightarrow \mathcal{B}\,\mathcal{W}=0 \text{ or } \mathcal{C}=0 \\ \mathcal{C} \text{ is positive semi-definite} \end{cases} \tag{26.121}$$

Let also

$$\bar{\mathcal{A}} = \bar{A}\times I_M \tag{26.122}$$

where $\bar{A}$ is some symmetric doubly stochastic matrix. The convergence analysis in this section relies on the following assumption on the matrices $\{\bar{\mathcal{A}}, \mathcal{B}, \mathcal{C}\}$.

Assumption 26.2. (Conditions on combination matrices) *It is assumed that the matrices $\{\bar{\mathcal{A}}, \mathcal{B}, \mathcal{C}\}$ defined by (26.121)–(26.122) and used in the construction of the UDA in (26.68a)–(26.68c) satisfy either one of the following conditions:*

$$\left\{0 < \bar{\mathcal{A}} \leq (I_{KM} - \mathcal{B}^2) \text{ and } 0 \leq \mathcal{C} \leq 2I_{KM}\right\} \text{ or} \tag{26.123a}$$

$$\left\{\bar{\mathcal{A}} = I_{KM} \text{ and } 0 \leq \mathcal{B}^2 \leq \mathcal{C} < I_{KM}\right\} \tag{26.123b}$$

If we refer to the algorithms listed in Table 26.1, we find that the choices for EXTRA are $\bar{\mathcal{A}} = I_{KM}$, $\mathcal{B}^2 = \frac{1}{2}(I_{KM} - \mathcal{A}) = \mathcal{C}$, which satisfy the second condition (26.123b). For EXACT diffusion we have $\bar{\mathcal{A}} = \frac{1}{2}(I_{KM} + \mathcal{A})$, $\mathcal{B}^2 = \frac{1}{2}(I_{KM} - \mathcal{A})$, and $\mathcal{C} = 0$. These choices satisfy the first condition (26.123a). For DIGing we have $\bar{\mathcal{A}} = I_{KM}$, $\mathcal{B} = I_{KM} - \mathcal{A}$, and $\mathcal{C} = I_{KM} - \mathcal{A}^2$. Condition (26.123b) is met if the eigenvalues of the combination matrix A lie within the open interval $(0, 1]$ rather than $(-1, 1]$. If the matrix A does not satisfy this condition, we can replace it by $A \leftarrow \frac{1}{2}(I + A)$. For ATC-DIGing we have $\bar{\mathcal{A}} = \mathcal{A}^2$, $\mathcal{B} = I_{KM} - \mathcal{A}$, and $\mathcal{C} = 0$. Condition (26.123a) is met again if the eigenvalues of the combination matrix A lie within the interval $(0, 1]$. For ATC tracking, we have $\bar{\mathcal{A}} = \mathcal{A}$, $\mathcal{B} = I_{KM} - \mathcal{A}$, and $\mathcal{C} = I_{KM} - \mathcal{A}$. In this case, condition (26.123a) requires A to be nonnegative-definite with eigenvalues within the same interval $(0, 1]$. Therefore, for the last three cases, we can simply replace A by $A \leftarrow \frac{1}{2}(I + A)$ when necessary so that we can assume that conditions (26.123a)–(26.123b) are met as needed.

Returning to (26.119), the corresponding saddle point formulation is given by

$$\min_{\mathcal{W}} \max_{\lambda} \left\{ \mathcal{L}_e(\mathcal{W}, \lambda) \stackrel{\Delta}{=} \mathcal{P}(\mathcal{W}) + \frac{1}{2\mu}\|\mathcal{W}\|^2_{\mathcal{C}} + \frac{1}{\mu}\lambda^{\mathsf{T}}\mathcal{B}\,\mathcal{W} \right\} \tag{26.124}$$

where $\lambda \in \mathbb{R}^{KM}$ is a Lagrangian factor and $\mu > 0$. We know from the earlier Lemma 9.1 on convex optimization problems with linear constraints that problem (26.119) has a solution $\mathcal{W}^\star$ if, and only if, there exists a pair $(\mathcal{W}^\star, \lambda^\star)$ such that

$$\mu \nabla_{\mathcal{W}^{\mathsf{T}}} \mathcal{P}(\mathcal{W}^\star) + \mathcal{B}\lambda^\star = 0 \tag{26.125a}$$

$$\mathcal{B}\,\mathcal{W}^\star = 0 \tag{26.125b}$$

Moreover, the unique solution $\mathcal{W}^\star$ to (26.119) has the form

$$\mathcal{W}^\star = \mathbb{1}_K \otimes w^\star, \quad w^\star \in \mathbb{R}^M \tag{26.126}$$

where $w^\star$ is the unique solution to the original problem (26.1). This solution satisfies $\bar{\mathcal{A}}\,\mathcal{W}^\star = \mathcal{W}^\star$. It also follows from (26.121) that $\mathcal{C}\,\mathcal{W}^\star = 0$.

We further know from Lemma 9.2 that there exists a *unique* dual variable $\lambda_d^\star \in \mathcal{R}(\mathcal{B})$ satisfying (26.125a)–(26.125b). We examine convergence of the UDA by measuring deviations relative to this unique dual variable. Thus, we introduce the network error vectors:

$$\widetilde{\mathcal{W}}_n \stackrel{\Delta}{=} \mathcal{W}^\star - \mathcal{W}_n \tag{26.127a}$$

$$\widetilde{\lambda}_n \stackrel{\Delta}{=} \lambda_d^\star - \lambda_n \tag{26.127b}$$

$$\widetilde{\mathcal{Z}}_n \stackrel{\Delta}{=} \mathcal{W}^\star - \mathcal{Z}_n \tag{26.127c}$$

and conclude from the UDA recursions and (26.125a)–(26.125b) that

$$\widetilde{\mathcal{Z}}_n = (I_{KM} - \mathcal{C})\widetilde{\mathcal{W}}_{n-1} + \mu\Big(\nabla_{\mathcal{W}^{\mathsf{T}}} \mathcal{P}(\mathcal{W}_{n-1}) - \nabla_{\mathcal{W}^{\mathsf{T}}} \mathcal{P}(\mathcal{W}^\star)\Big) - \mathcal{B}\widetilde{\lambda}_{n-1} \tag{26.128a}$$

$$\widetilde{\lambda}_n = \widetilde{\lambda}_{n-1} + \mathcal{B}\widetilde{\mathcal{Z}}_n \tag{26.128b}$$

$$\widetilde{\mathcal{W}}_n = \bar{\mathcal{A}}\widetilde{\mathcal{Z}}_n \tag{26.128c}$$

We examine the convergence of these recursions in the sequel. The following are two auxiliary results that will be called upon in the presentation. Lemma 26.1 relates to the aggregate risk and is a variation of proposition 3.6 from Shi *et al.* (2015); the proposition from this reference uses $I_{KM} - \mathcal{A}$ instead of $\mathcal{B}$.

Lemma 26.1. (Property of aggregate cost) *Refer to the aggregate risk $\mathcal{P}(\mathcal{w})$ defined by (26.120) and let $\mathcal{w}^\star$ denote the unique minimizer of (26.119). Under the conditions stated at the beginning of Section 25.3.2 for deterministic optimization problems (strongly convex risk $P(w)$, convex individual risks $P_k(w)$, and δ-Lipschitz gradients), it holds that the augmented risk $\mathcal{P}(\mathcal{w}) + \frac{1}{2\mu}\|\,\mathcal{w}\,\|^2_{\mathcal{B}}$ is strongly convex with respect to $\mathcal{w}^\star$ (in which case it is said to be "restricted" strongly convex), meaning that, for any scalar $\eta > 0$,*

$$(\mathcal{w}-\mathcal{w}^\star)^\mathsf{T}\Big(\nabla_{\mathcal{w}^\mathsf{T}}\,\mathcal{P}(\mathcal{w}) - \nabla_{\mathcal{w}^\mathsf{T}}\,\mathcal{P}(\mathcal{w}^\star)\Big) + \eta\|\,\mathcal{w}-\mathcal{w}^\star\,\|^2_{\mathcal{B}^2} \;\geq\; \nu_\eta\|\,\mathcal{w}-\mathcal{w}^\star\,\|^2 \tag{26.129}$$

for any $\mathcal{w}$ and where

$$\nu_\eta = \min\left\{\nu - 2\delta c,\; \frac{\eta c^2\sigma^2_{\min}(\mathcal{B})}{4(c^2+1)}\right\} > 0, \quad \textit{for any } c \in (0, \nu/2\delta) \tag{26.130}$$

Here, $\sigma_{\min}(\mathcal{B})$ denotes the smallest nonzero singular value of $\mathcal{B}$.

Lemma 26.2. (Useful inequality) *Consider the same setting of Lemma 26.1. If $\mu < (2-\sigma_{\max}(C))/\delta$, then the following inequality holds for all $n \geq 0$ and $\eta > 0$:*

$$\begin{aligned}\|\widetilde{\mathcal{z}}_n\|^2_{(I_{KM}-\mathcal{B}^2)} + \|\widetilde{\lambda}_n\|^2 \leq\; & (1-\mu\nu_\eta(2-\sigma_{\max}(\mathcal{C})) - \mu\delta)\|\widetilde{\mathcal{w}}_{n-1}\|^2 + \\ & (1-\sigma^2_{\min}(\mathcal{B}))\|\widetilde{\lambda}_{n-1}\|^2 + \\ & \mu\eta(2-\mu\delta_\mu)\|\widetilde{\mathcal{w}}_{n-1}\|^2_{\mathcal{B}^2} - (2-\mu\delta_\mu)\|\widetilde{\mathcal{w}}_{n-1}\|^2_{\mathcal{C}}\end{aligned} \tag{26.131}$$

where

$$\delta_\mu \;\stackrel{\Delta}{=}\; \delta + \frac{1}{\mu}\sigma_{\max}(\mathcal{C}) \tag{26.132}$$

Proof: Computing the squared Euclidean norm of both sides of (26.128a) we get

$$\begin{aligned}\|\widetilde{\mathcal{z}}_n\|^2 = \;& \|(I_{KM}-\mathcal{C})\widetilde{\mathcal{w}}_{n-1} + \mu\,(\nabla_{\mathcal{w}^\mathsf{T}}\,\mathcal{P}(\mathcal{w}_{n-1}) - \nabla_{\mathcal{w}^\mathsf{T}}\,\mathcal{P}(\mathcal{w}^\star))\|^2 + \\ & \|\mathcal{B}\widetilde{\lambda}_{n-1}\|^2 - \\ & 2\widetilde{\lambda}^\mathsf{T}_{n-1}\mathcal{B}^\mathsf{T}\Big\{(I_{KM}-\mathcal{C})\widetilde{\mathcal{w}}_{n-1} + \mu\,(\nabla_{\mathcal{w}^\mathsf{T}}\,\mathcal{P}(\mathcal{w}_{n-1}) - \nabla_{\mathcal{w}^\mathsf{T}}\,\mathcal{P}(\mathcal{w}^\star))\Big\}\end{aligned} \tag{26.133}$$

and

$$\begin{aligned}\|\widetilde{\lambda}_n\|^2 &= \|\widetilde{\lambda}_{n-1}\|^2 + \|\mathcal{B}\widetilde{\mathcal{z}}_n\|^2 + 2\widetilde{\lambda}^\mathsf{T}_{n-1}\mathcal{B}\widetilde{\mathcal{z}}_n \\ &= \|\widetilde{\lambda}_{n-1}\|^2 + \|\widetilde{\mathcal{z}}_n\|^2_{\mathcal{B}^2} - 2\|\mathcal{B}\widetilde{\lambda}_{n-1}\|^2 + \\ &\quad\; 2\widetilde{\lambda}^\mathsf{T}_{n-1}\mathcal{B}\Big\{(I_{KM}-\mathcal{C})\widetilde{\mathcal{w}}_{n-1} + \mu\,(\nabla_{\mathcal{w}^\mathsf{T}}\,\mathcal{P}(\mathcal{w}_{n-1}) - \nabla_{\mathcal{w}^\mathsf{T}}\,\mathcal{P}(\mathcal{w}^\star))\Big\}\end{aligned} \tag{26.134}$$

Adding the two equations and rearranging terms gives

$$\begin{aligned}\|\widetilde{\mathcal{z}}_n\|^2_{(I_{KM}-\mathcal{B}^2)} + \|\widetilde{\lambda}_n\|^2 = \;& \|\widetilde{\lambda}_{n-1}\|^2 - \|\mathcal{B}\widetilde{\lambda}_{n-1}\|^2 + \\ & \|(I_{KM}-\mathcal{C})\widetilde{\mathcal{w}}_{n-1} + \mu(\nabla_{\mathcal{w}^\mathsf{T}}\,\mathcal{P}(\mathcal{w}_{n-1}) - \nabla_{\mathcal{w}^\mathsf{T}}\,\mathcal{P}(\mathcal{w}^\star))\|^2\end{aligned} \tag{26.135}$$

where the weighting matrix $(I_{KM} - \mathcal{B}^2)$ is positive-definite from conditions (26.123a) and (26.123b). Now since $\lambda_0 = 0$ and $\lambda_n = \lambda_{n-1} + \mathcal{B}\,\mathcal{z}_n$, we conclude that $\lambda_n \in \mathcal{R}(\mathcal{B})$

for all n (i.e., the dual variable stays in the range space of $\mathcal{B}$). It follows that both λ_n and $\lambda_d^\star$ lie in the range space of $\mathcal{B}$ and, moreover,

$$\|\mathcal{B}\widetilde{\lambda}_{n-1}\|^2 \geq \sigma_{\min}(\mathcal{B}^2)\|\widetilde{\lambda}_{n-1}\|^2 \tag{26.136}$$

Substituting into (26.136) gives

$$\begin{aligned}\|\widetilde{z}_n\|^2_{(I_{KM}-\mathcal{B}^2)} + \|\widetilde{\lambda}_n\|^2 \leq (1-\sigma_{\min}(\mathcal{B}^2))\|\widetilde{\lambda}_{n-1}\|^2 + \\ \left\|\widetilde{w}_{n-1} + \mu\Big(\nabla_{w^\mathsf{T}}\mathcal{P}(w_{n-1}) - \nabla_{w^\mathsf{T}}\mathcal{P}(w^\star) - \frac{1}{\mu}\mathcal{C}\widetilde{w}_{n-1}\Big)\right\|^2\end{aligned} \tag{26.137}$$

Now, consider the positive scalar δ_μ defined by (26.132). It can easily be verified that the augmented function $\mathcal{P}(w) + \frac{1}{2\mu}\|w\|^2_{\mathcal{C}}$ has δ_μ-Lipschitz gradients. It follows from the co-coercivity property established in Prob. 10.4 for δ_μ-smooth convex functions that:

$$\begin{aligned}&\Big(\nabla_{w^\mathsf{T}}\mathcal{P}(w^\star) - \nabla_{w^\mathsf{T}}\mathcal{P}(w_{n-1}) + \frac{1}{\mu}\mathcal{C}\widetilde{w}_{n-1}\Big)^\mathsf{T}\widetilde{w}_{n-1} \\ &\geq \frac{1}{\delta_\mu}\left\|\nabla_{w^\mathsf{T}}\mathcal{P}(w^\star) - \nabla_{w^\mathsf{T}}\mathcal{P}(w_{n-1}) + \frac{1}{\mu}\mathcal{C}\widetilde{w}_{n-1}\right\|^2\end{aligned} \tag{26.138}$$

Moreover, we deduce from (26.129) that

$$-\widetilde{w}_{n-1}^\mathsf{T}\Big(\nabla_{w^\mathsf{T}}\mathcal{P}(w) - \nabla_{w^\mathsf{T}}\mathcal{P}(w^\star)\Big) \geq -\eta\|\widetilde{w}_{n-1}\|^2_{\mathcal{B}^2} + \nu_\mu\|\widetilde{w}_{n-1}\|^2 \tag{26.139}$$

Combining (26.138) and (26.139) we find that

$$\begin{aligned}&\left\|\widetilde{w}_{n-1} + \mu\Big(\nabla_{w^\mathsf{T}}\mathcal{P}(w_{n-1}) - \nabla_{w^\mathsf{T}}\mathcal{P}(w^\star) - \frac{1}{\mu}\mathcal{C}\widetilde{w}_{n-1}\Big)\right\|^2 \\ &= \|\widetilde{w}_{n-1}\|^2 + \mu^2\left\|\nabla_{w^\mathsf{T}}\mathcal{P}(w_{n-1}) - \nabla_{w^\mathsf{T}}\mathcal{P}(w^\star) - \frac{1}{\mu}\mathcal{C}\widetilde{w}_{n-1}\right\|^2 + \\ &\quad 2\mu\Big(\nabla_{w^\mathsf{T}}\mathcal{P}(w_{n-1}) - \nabla_{w^\mathsf{T}}\mathcal{P}(w^\star) - \frac{1}{\mu}\mathcal{C}\widetilde{w}_{n-1}\Big)^\mathsf{T}\widetilde{w}_{n-1} \\ &\overset{(26.138)}{\leq} \|\widetilde{w}_{n-1}\|^2 - \mu^2\delta_\mu\Big(\nabla_{w^\mathsf{T}}\mathcal{P}(w_{n-1}) - \nabla_{w^\mathsf{T}}\mathcal{P}(w^\star) - \frac{1}{\mu}\mathcal{C}\widetilde{w}_{n-1}\Big)^\mathsf{T}\widetilde{w}_{n-1} + \\ &\quad 2\mu\Big(\nabla_{w^\mathsf{T}}\mathcal{P}(w_{n-1}) - \nabla_{w^\mathsf{T}}\mathcal{P}(w^\star) - \frac{1}{\mu}\mathcal{C}\widetilde{w}_{n-1}\Big)^\mathsf{T}\widetilde{w}_{n-1} \\ &= \|\widetilde{w}_{n-1}\|^2 + \mu(2-\mu\delta_\mu)\Big(\nabla_{w^\mathsf{T}}\mathcal{P}(w_{n-1}) - \nabla_{w^\mathsf{T}}\mathcal{P}(w^\star) - \frac{1}{\mu}\mathcal{C}\widetilde{w}_{n-1}\Big)^\mathsf{T}\widetilde{w}_{n-1} \\ &= \|\widetilde{w}_{n-1}\|^2 + \mu(2-\mu\delta_\mu)\Big(\nabla_{w^\mathsf{T}}\mathcal{P}(w_{n-1}) - \nabla_{w^\mathsf{T}}\mathcal{P}(w^\star)\Big)^\mathsf{T}\widetilde{w}_{n-1} - \\ &\quad (2-\mu\delta_\mu)\|\widetilde{w}_{n-1}\|^2_{\mathcal{C}} \\ &\overset{(26.139)}{\leq} (1-\mu\nu_\eta(2-\mu\delta_\mu))\|\widetilde{w}_{n-1}\|^2 + \mu\eta(2-\mu\delta_\mu)\|\widetilde{w}_{n-1}\|^2_{\mathcal{B}^2} - (2-\mu\delta_\mu)\|\widetilde{w}_{n-1}\|^2_{\mathcal{C}}\end{aligned} \tag{26.140}$$

where in the last inequality step we used the fact that $2-\mu\delta_\mu > 0$, which is guaranteed by the condition $\mu < (2-\sigma_{\max}(\mathcal{C}))/\delta$. Substituting (26.140) into (26.138) gives (26.132).

■

Theorem 26.5. (Linear convergence) *Consider the same setting of Lemma 26.1. The UDA converges linearly as follows depending on which condition (26.123a) or (26.123b) is satisfied:*

(a) *Assume condition (26.123a) holds and select any positive step size satisfying*

$$\mu < (2 - \sigma_{\max}(\mathcal{C}))/\delta \tag{26.141a}$$

Then, it holds for all n:

$$\|\widetilde{\mathcal{w}}_n\|^2_{(I_{KM}+\mathcal{B}^2)} + \|\widetilde{\lambda}_n\|^2 \;\le\; \rho^n C_o \tag{26.141b}$$

for some constant $C_o \ge 0$ and where

$$\rho \;\stackrel{\Delta}{=}\; \max\Big\{1 - \mu\nu_\eta(2 - \sigma_{\max}(\mathcal{C}) - \mu\delta), 1 - \sigma^2_{\min}(\mathcal{B}), \mu\eta(2 - \sigma_{\max}(\mathcal{C}) - \mu\delta)\Big\} \tag{26.141c}$$

satisfies $0 \le \rho < 1$.

(b) *Assume condition (26.123b) holds and select any positive step size satisfying*

$$\mu < (1 - \sigma_{\max}(\mathcal{C}))/\delta \tag{26.142a}$$

Then, it holds for all n:

$$\|\widetilde{\mathcal{w}}_n\|^2_{(I_{KM}-\mathcal{B}^2)} + \|\widetilde{\lambda}_n\|^2 \;\le\; \rho^n C_o \tag{26.142b}$$

for some constant $C_o \ge 0$ and where

$$0 \le \rho \;\stackrel{\Delta}{=}\; \max\Big\{1 - \mu\nu_\eta(2 - \sigma_{\max}(\mathcal{C}) - \mu\delta),\; 1 - \sigma^2_{\min}(\mathcal{B})\Big\} \;<\; 1 \tag{26.142c}$$

Proof: Let us examine item (a) first. We know that

$$\big(1 - \mu\nu_\eta(2 - \sigma_{\max}(\mathcal{C}) - \mu\delta)\big) < 1 \iff \mu < (2 - \sigma_{\max}(\mathcal{C}))/\delta \tag{26.143}$$

Let

$$\gamma_1 \;\stackrel{\Delta}{=}\; 1 - \mu\nu_\eta(2 - \sigma_{\max}(\mathcal{C}) - \mu\delta) \tag{26.144a}$$

$$\gamma_2 \;=\; 1 - \sigma^2_{\min}(\mathcal{B}) \tag{26.144b}$$

$$\gamma_3 \;\stackrel{\Delta}{=}\; \mu\eta(2 - \mu\delta_\mu) = \mu\eta(2 - \sigma_{\max}(\mathcal{C}) - \mu\delta) \tag{26.144c}$$

$$\gamma_4 \;\stackrel{\Delta}{=}\; 2 - \mu\delta_\mu \tag{26.144d}$$

Then, inequality (26.132) becomes

$$\begin{aligned}\|\widetilde{\mathcal{z}}_n\|^2_{(I_{KM}-\mathcal{B}^2)} + \|\widetilde{\lambda}_n\|^2 &\le \gamma_1\|\widetilde{\mathcal{w}}_{n-1}\|^2 + \gamma_2\|\widetilde{\lambda}_{n-1}\|^2 + \gamma_3\|\widetilde{\mathcal{w}}_{n-1}\|^2_{\mathcal{B}^2} - \gamma_4\|\widetilde{\mathcal{w}}_{n-1}\|^2_{\mathcal{C}} \\ &\le \gamma_1\|\widetilde{\mathcal{w}}_{n-1}\|^2 + \gamma_2\|\widetilde{\lambda}_{n-1}\|^2 + \gamma_3\|\widetilde{\mathcal{w}}_{n-1}\|^2_{\mathcal{B}^2}\end{aligned} \tag{26.145}$$

We verify next that $\|\widetilde{\mathcal{w}}_n\|^2_{(I+\mathcal{B}^2)} \le \|\widetilde{\mathcal{z}}_n\|^2_{(I_{KM}-\mathcal{B}^2)}$. Recall that $\bar{\mathcal{A}}$ is symmetric doubly stochastic and also positive-definite under condition (26.123a). It further holds that $I_{KM} - \bar{\mathcal{A}} \ge 0$. Let $\bar{\mathcal{A}}^{1/2}$ denote a symmetric square-root of $\bar{\mathcal{A}}$. It follows that

$$0 \le \bar{\mathcal{A}}^{\frac{1}{2}}(I_{KM} - \bar{\mathcal{A}})^2\bar{\mathcal{A}}^{\frac{1}{2}} \iff \bar{\mathcal{A}}^2 - \bar{\mathcal{A}}^3 \le \bar{\mathcal{A}} - \bar{\mathcal{A}}^2 \tag{26.146}$$

and therefore

$$\bar{\mathcal{A}}\mathcal{B}^2\bar{\mathcal{A}} \overset{(26.123a)}{\le} \bar{\mathcal{A}}(I_{KM} - \bar{\mathcal{A}})\bar{\mathcal{A}} = \bar{\mathcal{A}}^2 - \bar{\mathcal{A}}^3 \overset{(26.146)}{\le} \bar{\mathcal{A}} - \bar{\mathcal{A}}^2 \tag{26.147}$$

From the above equation we conclude that

$$\begin{aligned}
\|\widetilde{w}_n\|^2 &\overset{(26.128c)}{=} \|\widetilde{z}_n\|^2_{\bar{\mathcal{A}}^2} \\
&= \|\widetilde{z}_n\|^2_{\bar{\mathcal{A}}^2+\bar{\mathcal{A}}-\bar{\mathcal{A}}} \\
&= \|\widetilde{z}_n\|^2_{\bar{\mathcal{A}}} - \|\widetilde{z}_n\|^2_{\bar{\mathcal{A}}-\bar{\mathcal{A}}^2} \\
&\overset{(26.147)}{\leq} \|\widetilde{z}_n\|^2_{\bar{\mathcal{A}}} - \|\widetilde{z}_n\|^2_{\bar{\mathcal{A}}\mathcal{B}^2\bar{\mathcal{A}}} \\
&\overset{(26.123a)}{\leq} \|\widetilde{z}_n\|^2_{(I_{KM}-\mathcal{B}^2)} - \|\widetilde{z}_n\|^2_{\bar{\mathcal{A}}\mathcal{B}^2\bar{\mathcal{A}}}
\end{aligned} \tag{26.148}$$

Using

$$\|\widetilde{z}_n\|^2_{\bar{\mathcal{A}}\mathcal{B}^2\bar{\mathcal{A}}} = \|\bar{\mathcal{A}}\widetilde{z}_n\|^2_{\mathcal{B}^2} = \|\widetilde{w}_n\|^2_{\mathcal{B}^2} \tag{26.149}$$

in (26.148) gives

$$\|\widetilde{w}_n\|^2_{(I_{KM}+\mathcal{B}^2)} \leq \|\widetilde{z}_n\|^2_{(I_{KM}-\mathcal{B}^2)} \tag{26.150}$$

Using this bound in (26.145) yields:

$$\begin{aligned}
\|\widetilde{w}_n\|^2_{(I_{KM}+\mathcal{B}^2)} + \|\widetilde{\lambda}_n\|^2 &\leq \gamma_1\|\widetilde{w}_{n-1}\|^2 + \gamma_2\|\widetilde{\lambda}_{n-1}\|^2 + \gamma_3\|\widetilde{w}_{n-1}\|^2_{\mathcal{B}^2} \qquad (26.151)\\
&\leq \max\{\gamma_1,\gamma_2,\gamma_3\}\Big(\|\widetilde{w}_{n-1}\|^2_{(I_{KM}+\mathcal{B}^2)} + \|\widetilde{\lambda}_{n-1}\|^2\Big)
\end{aligned}$$

Iterating this inequality we arrive at (26.141b).

Let us now examine item (b) in the statement of the theorem. From condition (26.123b) we know that $\bar{\mathcal{A}} = I$ so that from (26.128c) we have $\widetilde{w}_n = \widetilde{z}_n$. It follows from inequality (26.132) that

$$\|\widetilde{w}_n\|^2_{(I_{KM}-\mathcal{B}^2)} + \|\widetilde{\lambda}_n\|^2 \leq \gamma_1\|\widetilde{w}_{n-1}\|^2 + \gamma_2\|\widetilde{\lambda}_{n-1}\|^2 + \gamma_3\|\widetilde{w}_{n-1}\|^2_{\mathcal{B}^2} - \gamma_4\|\widetilde{w}_{n-1}\|^2_{\mathcal{B}^2} \tag{26.152}$$

where, in the first step, we used $\mathcal{B}^2 \leq \mathcal{C}$ from (26.123b). Adding and subtracting $\gamma_1\|\widetilde{w}_{n-1}\|^2_{\mathcal{B}^2}$ to the previous inequality gives

$$\begin{aligned}
\|\widetilde{w}_n\|^2_{(I_{KM}-\mathcal{B}^2)} + \|\widetilde{\lambda}_n\|^2 \leq \gamma_1\|\widetilde{w}_{n-1}\|^2_{(I_{KM}-\mathcal{B}^2)} + \gamma_2\|\widetilde{\lambda}_{n-1}\|^2 + \\
(\gamma_1+\gamma_3-\gamma_4)\|\widetilde{w}_{n-1}\|^2_{\mathcal{B}^2}
\end{aligned} \tag{26.153}$$

We next select η to ensure that $(\gamma_1+\gamma_3-\gamma_4) \leq 0$, which requires

$$\mu\eta(2-\mu\delta_\mu) \leq 1 - \sigma_{\max}(\mathcal{C}) - \mu\delta + \mu\nu_\eta(2-\mu\delta_\mu) \tag{26.154}$$

The last factor $(2-\mu\delta_\mu)$ is positive for $\mu < (2-\sigma_{\max}(\mathcal{C}))/\delta$. Therefore, by ignoring this last factor (since it is nonnegative), the above requirement is satisfied if we select η to satisfy the stricter condition:

$$\mu\eta(2-\mu\delta_\mu) \leq 1 - \sigma_{\max}(\mathcal{C}) - \mu\delta \tag{26.155}$$

which translates into

$$\eta \leq \frac{1-\sigma_{\max}(\mathcal{C})-\mu\delta}{\mu(2-\mu\delta_\mu)} \tag{26.156}$$

The expression in the denominator is positive in view of the bound on μ, while the expression in the numerator is positive when

$$\mu < \frac{1}{\delta}(1-\sigma_{\max}(\mathcal{C})) \; < \; \frac{1}{\delta}(2-\sigma_{\max}(\mathcal{C})) \tag{26.157}$$

Returning to (26.153), we now have

$$\begin{aligned}\|\widetilde{w}_n\|^2_{(I_{KM}-\mathcal{B}^2)} + \|\widetilde{\lambda}_n\|^2 &\leq \gamma_1\|\widetilde{w}_{n-1}\|^2_{(I_{KM}-\mathcal{B}^2)} + \gamma_2\|\widetilde{\lambda}_{n-1}\|^2 \\ &\leq \underbrace{\max\{\gamma_1,\gamma_2\}}_{\triangleq\,\rho}\left(\|\widetilde{w}_{n-1}\|^2_{(I_{KM}-\mathcal{B}^2)} + \|\widetilde{\lambda}_{n-1}\|^2\right)\end{aligned} \tag{26.158}$$

where we introduced the coefficient ρ. We can find conditions on μ to ensure that $0 \leq \rho < 1$. First, we know from (26.123b) that $0 < \gamma_2 < 1$. Second, the same condition

$$\mu < \frac{1}{\delta}(2 - \sigma_{\max}(\mathcal{C})) \tag{26.159}$$

ensures $\gamma_1 < 1$. On the other hand, it will hold that $\gamma_1 \geq 0$ if

$$1 - 2\mu\nu_\eta + \mu^2\nu_\eta\delta_\mu \geq 0 \tag{26.160}$$

Using the fact that $\delta_\mu \geq \delta \geq \nu \geq \nu_\eta$, the above condition is satisfied if

$$1 - 2\mu\nu_\eta + \mu^2\nu_\eta\nu = (1-\mu\nu_\eta)^2 \geq 0 \tag{26.161}$$

which is always true. Iterating (26.158) we reach the desired result (26.142b). ∎

REFERENCES

Alghunaim, S. A., E. K. Ryu, K. Yuan, and A. H. Sayed (2021), "Decentralized proximal gradient algorithms with linear convergence rates," *IEEE Trans. Aut. Control*, vol. 66, no. 6, pp. 2787–2794.

Alghunaim, S. A. and A. H. Sayed (2020), "Linear convergence of primal–dual gradient methods and their performance in distributed optimization," *Automatica*, vol. 117, article 109003.

Boyd, S., N. Parikh, E. Chu, B. Peleato, and J. Eckstein (2010), "Distributed optimization and statistical learning via the alternating direction method of multipliers," *Foundations and Trends in Machine Learning*, NOW Publishers, vol. 3, no. 1, pp. 1–122.

Cattivelli, F. S. and A. H. Sayed (2010a), "Diffusion strategies for distributed Kalman filtering and smoothing," *IEEE Trans. Aut. Control*, vol. 55, no. 9, pp. 2069–2084.

Cattivelli, F. S. and A. H. Sayed (2010b), "Diffusion LMS strategies for distributed estimation," *IEEE Trans. Signal Process.*, vol. 58, no. 3, pp. 1035–1048.

Chen, J. and A. H. Sayed (2015a), "On the learning behavior of adaptive networks – part I: Transient analysis," *IEEE Trans. Inf. Theory*, vol. 61, no. 6, pp. 3487–3517.

Chen, J. and A. H. Sayed (2015b), "On the learning behavior of adaptive networks – part II: Performance analysis," *IEEE Trans. Inf. Theory*, vol. 61, no. 6, pp. 3518–3548.

Daneshmand, A., G. Scutari and V. Kungurtsev (2018), "Second-order guarantees of distributed gradient algorithms," available at arXiv:1809.08694.

Freeman, R. A., P. Yang, and K. M. Lynch (2006), "Stability and convergence properties of dynamic average consensus estimators," *Proc. IEEE Conf. Decision and Control* (CDC), pp. 338–343, San Diego, CA.

Jakovetic, D. (2019), "A unification and generalization of exact distributed first-order methods," *IEEE Trans. Signal Inf. Process. Netw.*, vol. 5, no. 1, pp. 31–46.

Li, Z., W. Shi, and M. Yan (2019), "A decentralized proximal-gradient method with network independent step-sizes and separated convergence rates," *IEEE Trans. Signal Process.*, vol. 67, no. 17, pp. 4494–4506.

Li, Z. and M. Yan (2017), "A primal–dual algorithm with optimal step sizes and its application in decentralized consensus optimization," available at arXiv:1711.06785.

Lian, X., C. Zhang, H. Zhang, C.-J. Hsieh, W. Zhang, and J. Liu (2017), "Can decentralized algorithms outperform centralized algorithms? A case study for decentralized parallel stochastic gradient descent," *Proc. Advances Neural Information Processing Systems* (NIPS), pp. 5330–5340, Long Beach, CA.

Ling, Q., W. Shi, G. Wu, and A. Ribeiro (2015), "DLM: Decentralized linearized alternating direction method of multipliers," *IEEE Trans. Signal Process.*, vol. 63, no. 15, pp. 4051–4064.

Lorenzo, P. D. and G. Scutari (2016),"Next: In-network nonconvex optimization," *IEEE Trans. Signal Inf. Process. Netw.*, vol. 2, no. 2, pp. 120–136.

Mao, X., K. Yuan, Y. Hu, Y. Gu, A. H. Sayed, and W. Yin (2020), "Walkman: A communication-efficient random-walk algorithm for decentralized optimization,"*IEEE Trans. Signal Process.*, vol. 68, pp. 2513–2528.

Mateos, G., J. A. Bazerque, and G. B. Giannakis (2010), "Distributed sparse linear regression," *IEEE Trans. Signal Process.*, vol. 58, no. 10, pp. 5262–5276.

Mokhtari, A. and A. Ribeiro (2016), "DSA: Decentralized double stochastic averaging gradient algorithm," *J. Mach. Learn. Res.*, vol. 17, no. 61, pp. 1–35, 2016.

Mokhtari, A., W. Shi, Q. Ling, and A. Ribeiro (2016), "DQM: Decentralized quadratically approximated alternating direction method of multipliers," *IEEE Trans. Signal Process.,* vol. 64, no. 19, pp. 5158–5173.

Mota, J., J. Xavier, P. Aguiar, and M. Puschel (2013), "D-ADMM: A communication-efficient distributed algorithm for separable optimization," *IEEE Trans. Signal Process.*, vol. 61, no. 10, pp. 2718–2723.

Nedic, A., A. Olshevsky, and W. Shi (2017), "Achieving geometric convergence for distributed optimization over time-varying graphs," *SIAM J. Optim.*, vol. 27, no. 4, pp. 2597–2633.

Nedic, A., A. Olshevsky, W. Shi, and C. A. Uribe (2017), "Geometrically convergent distributed optimization with uncoordinated step-sizes," *Proc. American Control Conference* (ACC), pp. 3950–3955, Seattle, WA.

Qu, G. and N. Li (2018), "Harnessing smoothness to accelerate distributed optimization," *IEEE Trans. Control Netw. Syst.*, vol. 5, no. 3, pp. 1245–1260.

Seaman, K., F. Bach, S. Bubeck, Y.-T. Lee, and L. Massoulie (2017), "Optimal algorithms for smooth and strongly convex distributed optimization in networks," *Proc. Int. Conf. Machine Learning* (ICML), pp. 3027–3036, Sydney.

Shi, W., Q. Ling, K. Yuan, G. Wu, and W. Yin (2014), "On the linear convergence of the ADMM in decentralized consensus optimization," *IEEE Trans. Signal Process.*, vol. 62, no. 7, pp. 1750–1761.

Shi, W., Q. Ling, G. Wu, and W. Yin (2015), "EXTRA: An exact first-order algorithm for decentralized consensus optimization," *SIAM J. Optim.*, vol. 25, no. 2, pp. 944–966.

Sun, Y., G. Scutari, and D. Palomar (2016), "Distributed nonconvex multiagent optimization over time-varying networks," *Proc. Asilomar Conf. Signals, Systems and Computers,* pp. 788–794, Pacific Grove, CA.

Sundararajan, A., B. Van Scoy, and L. Lessard (2019), "A canonical form for first-order distributed optimization algorithms," *Proc. American Control Conference* (ACC), pp. 4075–4080, Philadelphia, PA.

Tang, H., X. Lian, M. Yan, C. Zhang, and J. Liu (2018), "d2: Decentralized training over decentralized data," *Proc. Int. Conf. Machine Learning* (ICML), vol. 80, pp. 4848–4856, Stockholm.

Tatarenko, T. and B. Touri (2017), "Non-convex distributed optimization," *IEEE Trans. Aut. Control*, vol. 62, no. 8, pp. 3744–3757.

Towfic, Z. J. and A. H. Sayed (2015), "Stability and performance limits of adaptive primal–dual networks," *IEEE Trans. Signal Process.*, vol. 63, no. 11, pp. 2888–2903.

Vlaski, S. and A. H. Sayed (2020), "Second-order guarantees in centralized, federated and decentralized nonconvex optimization," *Commun. Inf. Syst.*, vol. 20, no. 3, pp. 353–388.

Vlaski, S. and A. H. Sayed (2021a), "Distributed learning in non-convex environments – part I: Agreement at a linear rate," *IEEE Trans. Signal Process.*, vol. 69, pp. 1242–1256.

Vlaski, S. and A. H. Sayed (2021b) "Distributed learning in non-convex environments – part II: Polynomial escape from saddle-points," *IEEE Trans. Signal Process.*, vol. 69, pp. 1257–1270.

Wang, Y., W. Yin, and J. Zeng (2019), "Global convergence of ADMM in nonconvex nonsmooth optimization," *J. Sci. Comput.*, vol. 78, no. 1, pp. 29–63.

Xin, R. and U. A. Khan (2018), "A linear algorithm for optimization over directed graphs with geometric convergence," *IEEE Control Syst. Lett.*, vol. 2, no. 3, pp. 315–320.

Xu, J., S. Zhu, Y. C. Soh, and L. Xie (2015), "Augmented distributed gradient methods for multi-agent optimization under uncoordinated constant step-sizes," *Proc. IEEE Conf. Decision and Control* (CDC), pp. 2055–2060, Osaka.

Ying, B., K. Yuan, and A. H. Sayed (2019), "Dynamic average diffusion with randomized coordinate updates," *IEEE Trans. Inf. Signal Process. Netw.*, vol. 5. no. 4, pp. 753–767.

Yuan, K., S. A. Alghunaim, B. Ying, and A. H. Sayed (2020), "On the influence of bias-correction on distributed stochastic optimization," *IEEE Trans. Signal Process.*, vol. 68, pp. 4352–4367.

Yuan, K., B. Ying, X. Zhao, and A. H. Sayed (2017), "Exact diffusion strategy for optimization by networked agents," *Proc. EUSIPCO*, pp. 141–145, Kos Island.

Yuan, K., B. Ying, X. Zhao, and A. H. Sayed (2019a), "Exact diffusion for distributed optimization and learning – part I: Algorithm development," *IEEE Trans. Signal Process.*, vol. 67, no. 3, pp. 708–723.

Yuan, K., B. Ying, X. Zhao, and A. H. Sayed (2019b), "Exact diffusion for distributed optimization and learning – part II: Convergence analysis," *IEEE Trans. Signal Process.*, vol. 67, no. 3, pp. 724–739.

Zhu, M. and S. Martinez (2010), "Discrete-time dynamic average consensus," *Automatica*, vol. 46, no. 2, pp. 322–329.

Author Index

Please note: this index contains terms for all three volumes of the book. Only terms listed below with page numbers 1–1008 are featured in this volume.

Subject Index

Please note: this index contains terms for all three volumes of the book. Only terms listed below with page numbers 1–1008 are featured in this volume.